Proceedings of the Summer Study on High Energy Physics in the 1990s

Proceedings of the Summer Study on High Energy Physics in the 1990s

June 27 – July 15, 1988
Snowmass, Colorado

Editor

Sharon Jensen

DPF

Organized by the Division of Particles and Fields
of the American Physical Society

Sponsored jointly by the Division of Particles and Fields
of the American Physical Society, United States Department of Energy,
National Science Foundation, and Universities Research Association, Inc.

World Scientific
Singapore • New Jersey • London • Hong Kong

Published by

World Scientific Publishing Co. Pte. Ltd.
P O Box 128, Farrer Road, Singapore 9128

USA office: World Scientific Publishing Co., Inc.
687 Hartwell Street, Teaneck, NJ 07666, USA

UK office: World Scientific Publishing Co. Pte. Ltd.
73 Lynton Mead, Totteridge, London N20 8DH, England

HIGH ENERGY PHYSICS IN THE 1990s

ISBN 9971-50-841-9

Printed in Singapore by General Printing Services Pte. Ltd.

Preface

In late 1987 the Division of Particles and Fields (DPF) of the American Physical Society established a Steering Committee (listed below) to hold a Summer Study in 1988. This followed three previous, highly successful, summer studies held at Snowmass, Colorado, in 1982, 1984, and 1986. The first meeting in this series was organized by the DPF to re-evaluate the future course of U.S. high energy physics without reference to a specific laboratory or facility. The practical possibility of achieving a proton-proton collider with an energy per beam in the 20 TeV energy range was first recognized by a large fraction of the community at the 1982 conference. It grew to be endorsed as a national goal for the field by HEPAP in July 1983. The 1984 and 1986 Snowmass meetings then focused on the SSC, its reference design, its physics potential, and the challenge to detector design posed by the high energy and luminosity of this accelerator.

The summer study, with essential financial support provided by the Department of Energy, National Science Foundation, and the Universities Research Association, took place from June 27 to July 15, 1988. In many ways, it was similar to its Snowmass antecedents. However, it had a different scope than the previous two meetings. Broadly stated, the goal of Snowmass 88 was to examine and evaluate the opportunities for pursuing high energy physics in the 1990s; that is, in the proposed period of construction, commissioning, and first operation of the SSC. Such opportunities include, first of all, research at existing and upgraded facilities and at the SSC. Also included were possible opportunities at special facilities, through new accelerator technology, in nonaccelerator experiments, and through innovations in instrumentation and detector technology.

For achieving the overall goal, the Summer Study was organized by physics topics into large Working Groups. These were led by J. Cronin and E. Kolb for *Astrophysics/Particle Physics*; A. Seiden and H. Georgi for *Electroweak Symmetry Breaking*; M. Shapiro and I. Hinchliffe for *Strong Interactions*; C. Baltay and G. Kane for *New Particles*; M. Witherell and R. Peccei for *Weak Decays*; and M. Month for *Accelerator Physics*. These large groups met in the mornings, while subgroups, formed around types of experimental facility, generally met in the afternoons. Five types of experimental facilities were singled out for special study: *K Factories*, *B Factories*, *High Energy Linear Electron-Positron Colliders*, *High Luminosity Hadron Colliders* at "moderate" energy, and the *SSC*. A working group on *Computers* was led by Stewart Loken. A very successful submeeting was organized by M. Gilchriese of the SSC Central Design Group to review the progress in the program of detector research and development for experiments at future *pp* colliders. Needless to say, the leadership of these people, with their energy and taste for good physics, was essential to the scientific content and ultimate success of the meeting.

The very large turnout of 536 participants (more than twice previous Snowmass meetings) allowed there to be a "critical mass" of people interested in many of the subgroups. A strong liaison was established between the accelerator physicists and the experimentalists who were interested in the physics possibilities at particular facilities. In the particular case of *B* physics at electron-positron colliders, the physics subgroup and the accelerator physics subgroup were organized together as effectively one group from the beginning. Previous Snowmass meetings have proven to be unique and invaluable in drawing the high energy physics community together and enabling it to analyze the opportunities open to it. This was true all the more so with the large turnout this time. It also meant that this meeting drew more physicists than previous ones from Europe, the Soviet Union, and the Third World.

In addition to the program outlined above, two Town Meetings were held. The first, held on July 11, was on the subject of "The SSC: Making the Dream a Reality" and had Ed Temple as the principal speaker, followed by a discussion. On July 13, L. Lederman, B. Richter, and C. Rubbia took on the subject of International Collaboration in High Energy Physics in the form of a panel discussion.

Any conference of this magnitude needs a devoted secretariat to make it all work. From the beginning, the invaluable experience and organizational talents of Joanne Day (Argonne) was the key to making Snowmass 88 run so smoothly. Joanne, along with Robin Craven (Wisconsin), Barb Kristen (Fermilab), Sandy Klepec (Argonne), Rene Donaldson (CDG), Colleen Mendel (Argonne) and Kathie Hardy (LBL), spent many, many hours Xeroxing, collating, stuffing packets, registering, selling tickets, taking messages, answering questions, ordering materials (and transporting them), planning social events and just generally being available to do what was needed. The Summer Study would not have been such a success without them.

The enormous task of organizing the computing facility was once again superbly done by Robin Craven. This time however, Lawrence Berkeley Laboratory provided the most essential resource—the computers. John Waters (LBL) helped Robin with the hookup, provided user accounts and played an essential liaison role between the facility and LBL. Over 200 participants were given computer accounts and many were able to access and carry out important computational work due to their efforts.

Most of the photographs you will see in these Proceedings were taken by Charles Bonham (Wisconsin). Charles spent hours roaming the trails of Snowmass and the surrounding area. The mountains, meadows, skys, wild flowers and people are what make Snowmass so memorable and Charles did a wonderful job in bringing them back for us.

Special thanks are extended to the Top of the Village Resort who provided our meeting and computing space as well as a considerable portion of our lodging. As in the previous years, Dave Spence, Stephanie Peterson, Gary Soucie and the rest of the staff of TOV paid special attention to our needs. Not only did they go all out for us, but they did so with enthusiasm and warmth, making our stay there so enjoyable.

Last, but hardly least, the record of what was accomplished at Snowmass is found in the Proceedings which you have in your hands. In all its aspects it depended critically on the editor, Sharon Jensen from SLAC. It has been a pleasure to work with her, and I thank her for the patience and persistence she has shown, aside from the many technical skills it took to put this volume together.

The output of Snowmass 88 indeed comprises a broad survey of the possibilities of doing high energy physics in the 1990s. It contains some significant refining of our estimates of the physics reach of different facilities in the light of our present understanding of the Standard Model and what may lie beyond it. Happy reading!

Fred Gilman
Chairman
Division of Particles and Fields
of the American Physical Society
February, 1989

MEMBERS OF THE STEERING COMMITTEE FOR SNOWMASS 88

Barry Barish
Frederick J. Gilman (Chairman)
Ian Hinchliffe
Melvin Month
Lee Pondrom
Bernard Pope
Jonathan L. Rosner
Michael Witherell

Table of Contents

NEW PARTICLES

STRONG INTERACTIONS

WEAK DECAYS

ACCELERATOR PHYSICS

ACCELERATOR PHYSICS – K FACTORIES

ACCELERATOR PHYSICS – HADRON COLLIDERS

ACCELERATOR PHYSICS – LINEAR COLLIDERS

COMPUTING

FUTURE DIRECTIONS IN DETECTOR R&D

ASTROPHYSICS/ PARTICLE PHYSICS

James W. Cronin and Edward W. Kolb

Working Group Leaders

Particle Physics - Astrophysics Working Group

Organizers: James W. Cronin (Chicago), Edward W. Kolb (Fermilab & Chicago)

Participants

David Berley (Maryland)
George Blumenthal (U.C. Santa Cruz)
C. Y. Chang (Maryland)
James Cronin (Chicago)
Bruce Dawson (Utah)
Thomas Devlin (Rutgers)
Jerry Elbert (Utah)
Jean Ernwein (CEN-Saclay)
Michael Fortner (Northern Illinois)
Thomas Gaisser (Bartol Research)
Gerald Garvey (Los Alamos)
Graciela Gelmini (ICTP)
Jordan Goodman (Maryland)
Dick Gustafson (Michigan)
Francis Halzen (Wisconsin)
Edward Kolb (Fermilab & Chicago)
William Kropp (U.C. Irvine)
Dick Lamb (Iowa State)
Kenneth Lande (Pennsylvania)
Robert Lanou (Brown)
Kwong Lau (Houston)
John Learned (Hawaii)
John Losecco (Notre Dame)
Al Mann (Pennsylvania)
James Matthews (Michigan)
Stuart Mufson (Indiana)
Dietrich Müller (Chicago)
Darragh Nagle (Los Alamos)
Bernard Sadoulet (U.C. Berkeley)
Vic Stenger (Hawaii)
Suresh Tonwar (Tata Institute)
Michael Turner (Chicago)
Michael Witherell (U.C. Santa Barbara)
Gaurang Yodh (U.C. Irvine)

The working group met each afternoon and listened to mini-symposia on a broad range of subjects covering all aspects of particle physics - astrophysics both theoretical and experimental. As a result, a number of papers which follow were "commissioned" to reflect the present status and future prospects of the field.

PARTICLES AND GAMMA RAYS FROM ASTROPHYSICAL POINT SOURCES

T.K. Gaisser

Bartol Research Institute, University of Delaware Newark DE 19716

Abstract

High energy gamma ray astronomy and high energy neutrino astronomy are the subject of much experimental activity at present. In the energy region of one TeV and higher, the motivation is to understand the origin of the very high energy cosmic rays and the role of particle acceleration and interactions inside energetic astrophysical systems. Particle physics is an essential component of the enterprise because of the central role of particle interactions at high energy, both in the sources and at the detectors.

ORIGIN OF COSMIC RAYS

Supernovas have long been identified as a likely source of power for the acceleration of most cosmic rays,[1] because the power available is just what is needed. The mechanism for acceleration to relativistic energies is more difficult to pinpoint, but the most popular candidate at present is first order Fermi acceleration at the shock that expands into the interstellar medium for several thousand years after a supernova explosion.[2] This scenario can account nicely for the bulk of cosmic rays, but there is an upper limit to the energy per particle that can be attained by acceleration at a strong shock during its lifetime.

The maximum acceleration rate with the first order Fermi mechanism at a strong shock (i.e. one with a high Mach number) is of order[3]

$$\dot{E} \sim \beta u e B, \tag{1}$$

where B is the ambient magnetic field and $\beta = u/c$ is the shock velocity. If the shock has a lifetime τ, the maximum energy attainable is

$$E_{\max} \sim \beta u e B \tau. \tag{2}$$

The canonical parameters for supernovas are $\beta \sim 0.015$ and $\tau \sim 1000$ years. The nominal field for the average interstellar medium is $\sim 3\ \mu$Gauss. Such numbers lead to the estimate[3]

$$E_{\max} \sim 100 \text{ TeV} \tag{3}$$

for protons. But cosmic rays with energies up to at least 10^{19} eV are routinely observed with air shower experiments. Where and how are they accelerated to such high energies?

One class of possibilities is that the same mechanism is at work, but on larger scales or with higher fields. Supernovas probably do not go off in the "average" instellar medium, but are likely to be associated with regions of high field associated with plasma winds from the progenitor star. Völk and Biermann[4] have pointed out that this could raise the limit of Eq. 3 by one or two orders of magnitude. It has also been pointed out[5] that in certain field configurations (magnetic field nearly perpendicular to direction of propagation of the shock) the acceleration rate is faster than Eq. 1. Strong shocks may also occur on large scales in other environments, for example, the termination shock of a strong stellar wind.[6] Another possibility which could account even for the very highest cosmic rays (10^{20} eV) is the termination shock of a galactic wind, a high velocity plasma expanding into the halo of the galaxy,[7] perhaps driven by the cumulative effect of many supernovas in the disk.

A quite different scenario is that some portion of the cosmic ray spectrum above 100 TeV comes from acceleration processes in compact regions with very high magnetic fields. These examples involve neutron stars with fields of 10^8–10^{13} Gauss. (The possibilities are reviewed in Ref. 8) The power source may be either rotational energy of the neutron star or accretion from a companion star in a binary system. Acceleration could be by the Fermi mechanism in an accretion shock[9] or in a shock in a pulsar-driven wind[10]. Quite different acceleration mechanisms are also possible. A homopolar generator constructed from differential rotation between an accreting neutron star and its accretion disk has been suggested.[11] Magnetic reconnection associated with plasma turbulence in the accretion flow is another possibility.[12] (This would generate flare-like acceleration, as in the Sun).

The fundamental problem that makes it difficult to distinguish among all these possibilities is that cosmic ray nuclei do not travel in straight lines, but diffuse through the tangled magnetic fields in the galaxy. (The gyroradius of a 10^{15} eV proton in a 3 μGauss field is less than 1 % of the thickness of the disk of the galaxy.) Since the cosmic rays themselves do not point back to their sources it is difficult to identify them.

ASTROPHYSICAL BEAM DUMPS

Interactions of cosmic rays with matter give rise to secondaries including both photons (from π^0 decay) and neutrinos. These neutral particles do point back to their origin and therefore can, in principle, be used as a tracer of the charged particles that produce them. Study of $\sim$ 100 MeV gamma rays has become a standard tool for studying the origin and propagation of the bulk of cosmic rays with energies in the few GeV range.[13] The problem is that the production of secondaries is proportional to the product of the cosmic ray density and the gas density, so mapping these photons reveals only a convolution of interstellar medium with the cosmic ray distribution. Nevertheless, the observations are consistent with supernova blast waves as the site of acceleration for the bulk of the cosmic rays.[14] Diffuse neutrinos are unlikely to be of practical use in this context because their fluxes and interactions rates are so small.[15]

If compact systems accelerate charged particles to high energy they may be visible as point sources of high energy photons and neutrinos, provided they are sufficiently powerful and provided there is a target region nearby. TeV photons could be produced either by electrons (e.g. by inverse Compton scattering on X-rays or other intense photons near the source) or by interactions of protons (or nuclei) in a concentration of gas (e.g. the accretion disk in a binary system). It is generally believed that photons of 100 TeV or higher could only come from hadronic processes because acceleration of electrons is limited by synchrotron radiation in the high magnetic fields needed to make the accelerator work. Neutrinos could only come from hadronic processes.

Detection of excess air showers from the direction of objects such as Cygnus X-3[16,17] and Hercules X-1[18] suggests that we are seeing cosmic accelerators at work producing beams perhaps up to 10^6 TeV.[19] Hercules X-1, and probably also Cygnus X-3, are X-ray binaries; that is, neutron stars radiating thermal X-rays as matter from a companion star falls onto the neutron star surface. These and other X-ray binaries have also been observed in the TeV range with air Cherenkov detectors. (See Ref. 20 for a review.)

The supernova 1987A has been much discussed as a likely particle accelerator and therefore a possible source of high energy photons and neutrinos while the expanding shell is sufficiently dense to act as a beam dump. (See Ref. 10 for a review.) Several upper limits have been placed on $\sim$TeV air Cherenkov signals and on air showers from SN1987A[21–26]. The best limits correspond to power in accelerated particles at the source of around 10^{39} erg/s. Moreover, it is now apparent[27] from the late evolution of the supernova light curve that any power source other than the radioactive decay of $^{56}\mathrm{Ni} \rightarrow {}^{56}\mathrm{Co} \rightarrow {}^{56}\mathrm{Fe}$ must contribute less than 10^{39} erg/s or so to the optical wavelength region. It is therefore clear that this supernova is not nearly strong enough to have been visible in $\sim$TeV neutrinos. (The experimental limit[28–31] of $\sim 10^{-13}$ upward ν-induced muons $\mathrm{cm}^{-2}\mathrm{s}^{-1}$ corresponds to a limit on power in accelerated particles at the source in the region of 10^{42} erg/s.[32]) A possible signal around 3 TeV has been reported[23b] on January 14/15, which would correspond to a power during that time of $\sim 2 \times 10^{39}$ erg/s. Possible implications of this have been discussed in Refs. 10 and 33-35.

LUMINOSITY

The relation between power at the source and signal at Earth has been discussed by several authors.[36–39] Hillas[36] has noted that the average power required to produce the signal reported by the Kiel group for Cygnus X-3 for the period 1976-80 may be more than 10^{39} erg/s for an assumed distance of 10 kiloparsecs ($= 3 \times 10^{22}$ cm). A single source this powerful might be enough to supply all the galactic cosmic rays with energies greater than several hundred TeV.[36,33] On the other hand, 10^{39} erg/s is greater than the Eddington limit for spherical accretion.[40] Stecker et al.[41] raised the question whether an accelerator this powerful in a close binary system would tend to disrupt the system. Even if the energy in photons and charged particles were absorbed and re-radiated near the surface of the companion star, high energy (>TeV) neutrinos would be absorbed deep inside, leading to expansion of the companion star on a relatively short time scale.[42]

This discussion illustrates the kinds of issues involved in the search for astrophysical point sources of air showers. It may be useful to outline a simple calculation of photon and neutrino production in an astrophysical beam dump as a framework for interpretation and design of experiments. A particularly important practical issue is the relation between fluxes of photons and fluxes of neutrinos. Because they interact only weakly, neutrinos can escape from deep inside a system and provide complementary information to the TeV and PeV photons. For the same reason, however, very large effective target volumes are required to detect neutrinos. The question then arises, what should be the design goal for a detector for high energy neutrino astronomy?

Production of Photons and Neutrinos

The flux of secondaries, i ($= \nu$ or γ), produced by a spectrum of proton interactions, $\mathrm{d}N_p/\mathrm{d}E_p$, is given by

$$\frac{\mathrm{d}N_i}{\mathrm{d}E_i} = \int \frac{\mathrm{d}n_i(E_i, E_\pi)}{\mathrm{d}E_i} \left\{ \int \frac{\mathrm{d}n_\pi(E_\pi, E_p)}{\mathrm{d}E_\pi} \frac{\mathrm{d}N_p}{\mathrm{d}E_p} \mathrm{d}E_p \right\} \mathrm{d}E_\pi, \tag{4}$$

where $\mathrm{d}n_\pi(E_\pi, E_p)$ is the number of pions with energy E_π produced in the interaction of a proton of energy E_p, and $\mathrm{d}n_i/\mathrm{d}E_i$ is the distribution for $\pi^\pm \to \nu_i$ or $\pi^0 \to \gamma_i$. In this equation I have shown explicitly only the contribution from pions. Production *via* kaons, etc., is included in a straightforward way. Explicit solutions may be obtained for the case of a power law spectrum of nucleons injected by the accelerator. In the thick target limit the spectrum of nucleon interactions is[43]

$$\frac{\mathrm{d}N_p}{\mathrm{d}E_p} = \frac{\Lambda}{\lambda} \frac{K}{E_p^{\gamma+1}}, \tag{5}$$

where $\Lambda = \lambda/(1 - Z_{NN})$ is the nucleon attenuation length and λ the interaction length of the nucleon. The notation for moments of the inclusive hadronic cross sections is described in Ref. 43,

$$Z_{ij} \equiv \int_0^1 x^{\gamma-1} F_{ij}(x)\,\mathrm{d}x, \tag{6}$$

where $F_{ij}(E_j/E_i) \equiv E_j\,[\mathrm{d}n_j(E_j/E_i)/\mathrm{d}E_j]$. The simple analytic solutions assume scaling in the forward fragmentation region ($x \equiv E_j/E_i > 0$), a good approximation in this application.[44]

Substitution of Eq. 3 gives, for the quantity inside brackets in Eq. **4**,

$$\{\ \} = \frac{\Lambda}{\lambda} \frac{K}{E_\pi^{\gamma+1}} \int_{x_{\min}}^1 x^{\gamma-1} F_{N\pi}(x)\,\mathrm{d}x \le \frac{\Lambda}{\lambda} \frac{K}{E_\pi^{\gamma+1}} Z_{N\pi}, \tag{7}$$

where $x_{\min} = E_\pi/E_0^{\max}$. For steep spectra or when $E_\pi \ll E_0^{\max}$, one may take the limit $x_{\min} \to 0$. ($E_0^{\max}$ is the upper cutoff of the accelerator.) For $\gamma = 1$, the flattest power spectrum normally considered, this limit is valid for $E_\pi < E_0/1000$.

Finally, folding in the kinematics of pion decay, one has

$$\frac{\mathrm{d}N_\gamma}{\mathrm{d}E_\gamma} \sim \frac{\Lambda}{\lambda} \frac{K}{E_\gamma^{\gamma+1}} \frac{2\,Z_{N\pi^0}}{\gamma+1}, \tag{8}$$

and

$$\frac{\mathrm{d}N_{\nu_\mu+\bar{\nu}_\mu}}{\mathrm{d}E_\nu} \sim \frac{\Lambda}{\lambda} \frac{K}{E_\nu^{\gamma+1}} \frac{Z_{N\pi^+} + Z_{N\pi^-}}{\gamma+1} \left(1 - \frac{m_\mu^2}{m_\pi^2}\right)^\gamma. \tag{9}$$

If the beam dump is such that high energy muons decay without appreciable energy loss, then Eq. 9 must be multiplied by a factor[45] 2.2 (2.5) for $\gamma = 1\,(1.7)$. Similarly, there will then also be an appreciable high energy flux of $\nu_e + \bar{\nu}_e$ of 0.54 (0.59) times the *total* flux of $\nu_\mu + \bar{\nu}_\mu$.

Estimates of Fluxes

The fluxes as derived in Eqs. 8 and 9 are actually quite general when interpreted as *production* spectra of photons and neutrinos. For neutrinos, the production spectra will generally coincide with the spectra outside the source. (Only for an obstruction of stellar proportions is neutrino absorption important.) The fluxes for a thin target are obtained by the replacement $\Lambda/\lambda \to \Delta t/\lambda$ in Eqs. 8 and 9, where Δt is the thickness of the target in g/cm^2. If Δt is less than a radiation length, then the photon spectrum outside the source will also be approximated by the production spectrum of the photons (unless absorption on ambient photons, e.g. X-rays, is important).

There is another important scenario for which the production spectrum of photons may be a good approximation to the photon spectrum outside the source. This occurs in a diffuse beam dump in which charged particles are confined but which is optically thin to emission of photons as well as neutrinos. Because of confinement (which could be the result of diffusion in turbulent magnetic fields associated with the accelerator), the beam dump is then a thick target for protons but a thin target for photon emission. In this model, if the geometrical size, L, of the beam dump is less than $\gamma c\tau$ for neutrons, then the fluxes in Eqs. 8-9 must be reduced somewhat because neutrons with energy $E > m_p c^2\, L/(c\tau)$ escape from the system.[46] By analogy with cosmic ray propagation, we might call this scenario the "leaky source" model. It is the most efficient type of photon source because it minimizes the accelerator power required to produce a given photon signal. For the same reason, however, it produces the lowest neutrino/photon ratio.

To illustrate the possibilities, consider an accelerator with an output of L_p erg/s in the form of nucleons accelerated to a differential power law spectrum, $K\,E^{-(\gamma+1)}$, where E is the energy per nucleon. (The accelerator could produce a mixture of protons and heavier nuclei.) Numerically, the normalization constant, K, is given by

$$K = 624\,\frac{\mathrm{GeV}}{\mathrm{erg}} \times L_p \times \left(\int_{E^{\min}}^{E^{\max}} \frac{dE}{E^\gamma}\right)^{-1}. \qquad (10)$$

For numerical examples I take $E^{\min} = 1$ GeV and $E^{\max} = 10^8$ GeV.

The moments of the inclusive distributions, as obtained from accelerator data on nuclear targets, are summarized in Table 1. The table also shows the ν/γ ratio at production, assuming that all muons decay without energy loss. The last line of the table is the ratio of the photon flux at production to the flux of primary nucleons evaluated at the same energy (from Eq. 8).

From the numbers in Table 1, a "leaky source" with $L_p \sim 10^{39}$ erg/s at 10 kpc would give a signal of 10^{-9} photons $\mathrm{cm}^{-2}\mathrm{s}^{-1}$ above 1 TeV. Observation of a periodicity in the TeV and PeV signals that is associated with orbital motion of the binary system, as in the case of Cyg X-3, implies, however, that the configuration of the beam dump is not that of an isotropic "leaky source", but that there is perhaps a target of limited extent that is only periodically aligned to produce a signal at Earth without reabsorption of the photons in the source.[47,37] In this case the flux estimate must be reduced accordingly. If, for example, the duty cycle for photon production is 0.02 (as originally assumed for illustration by Hillas[36]) the flux above 1 TeV would be 2×10^{-11}, in the range of observed values quoted in Ref. 20. If the periodicity is achieved by beaming of the accelerator output (rather than alignment of a limited target region) then the power required for a given signal is reduced accordingly.[8] To estimate the rate of air showers with energies above $\sim$ 300 TeV, it is also necessary[36] to account for absorption of photons by $\gamma\gamma \rightarrow e^+ e^-$ on the microwave background radiation.

Table 1: Parameters for neutrino and photon production

γ	1.0	1.2	1.4	1.6	1.8
Z_{NN}	0.41	0.37	0.34	0.31	0.29
$Z_{N\pi^0}$ (includes 6% η)	0.17	0.092	0.066	0.048	0.036
$Z_{N\pi^+} + Z_{N\pi^-}$ (includes $\mathrm{K}^\pm$)	0.35	0.19	0.14	0.10	0.075
$(\nu_\mu + \bar{\nu}_\mu)/\gamma$ (at production)	0.98	0.86	0.77	0.66	0.58
$\frac{2\,Z_{N\pi^0}}{(\gamma+1)(1-Z_{NN})}$	0.29	0.13	0.082	0.054	0.036

HIGH ENERGY NEUTRINO ASTRONOMY

Most neutrinos do not interact in the Earth, let alone in the vicinity of a detector. Therefore, to estimate the neutrino signal expected from a source it is necessary also to fold in the probability that the neutrinos make a detectable interaction. In this situation, as recognized long ago[48] in connection with the DUMAND proposal, it is advantageous to look for neutrino-induced muons that originate outside the volume of the detector, thus enlarging the effective detector volume. One approach[49] is to calculate the probability, P_ν, that a neutrino on a trajectory toward an underground detector produces a muon with enough energy to emerge from the rock and go through the detector. P_ν is essentially the product of the neutrino cross section and the muon range. For $E_\mu < 500$ GeV the muon range is proportional to its energy, and for $E_\nu < M_W^2/2m_p \sim$3600 GeV the neutrino cross section is proportional to neutrino's energy. Thus P_ν at first increases quadratically with energy, then linearly and finally $\sim$logarithmically after the W-propagator cuts off the linear rise of the neutrino charged current cross section. Taking account of QCD evolution of the structure functions,[50] leads to the values of $P_\nu(E_\mu > 2\ \mathrm{GeV})$ given in Table 2.[51]

With the values given in Table 2 it is straightforward to integrate the flux of muon neutrinos (Eq. 9) and obtain the expected neutrino-induced signal of muons. The results can be approximated by the formula

$$\mathrm{Signal} = 8\times10^{-8}\,\mathrm{cm}^{-2}\mathrm{s}^{-1} \times \frac{L_p(\mathrm{erg/s})}{4\pi d^2}\, e^{-5(\gamma-1)\ln 5}. \quad (11)$$

The numerical values here assume that all muons decay without energy loss. The integration was cut off at $E_\nu = 10^6$ GeV to represent a cutoff of the proton spectrum at 10^8 GeV. L_p is the total power in accelerated particles at the source in erg/s. The signal calculated from Eq. 11 must be further reduced by the fraction of time that the potential source is sufficiently below the horizon to avoid the background of atmospheric muons for the particular detector. For d = 10 kpc and $L_p = 10^{39}$ and assuming a reduction factor of 1/2, one finds a signal of $\sim 4\ \times$

Table 2: Probability for neutrino-induced muon (average of ν_μ and $\bar{\nu}_\mu$)

E_ν	$P_{\nu\to\mu}$ ($E_\mu > 2$ GeV)
3	4×10^{-12}
10	5×10^{-11}
30	8×10^{-10}
100	8×10^{-9}
300	7×10^{-8}
10^3	4.5×10^{-7}
3×10^3	3×10^{-6}
10^4	1.2×10^{-5}
3×10^4	4×10^{-5}
10^5	1.0×10^{-4}
3×10^5	2.6×10^{-4}
10^6	4.5×10^{-4}
3×10^6	10^{-3}

10^{-15} per cm^{-2} per second. This is the type of calculation that led to the estimates[38,52] of one upward, neutrino-induced muon per 1000 m^2 per year. Clearly, there are large uncertainties associated, not with the calculation itself, but with the underlying assumptions. In particular, it may be possible to obtain a photon signal in the range measured in Refs. 16 and 17 with somewhat lower L_p. Alternatively, the source may be variable, as suggested by a recent measurement[53] that gives a significantly lower average flux from Cyg X-3 for the period April 1986-May 1987 than was reported earlier.[16,17] It is also true that for this, or for some other potential source, the power of the accelerator could be much higher, with more attenuation to reduce the flux of photons. The neutrino flux would be much higher, but one then has to worry about how such a powerful source avoids destroying itself.

MUONS IN AIR SHOWERS

Perhaps the most puzzling aspect of the observations of air showers from point sources is that, where muon measurements are available,[16,18] the signal appears to have a muon content comparable to that of the atmospheric background. (I refer here to muons with energies of a GeV or so.) The background consists of air showers produced by interactions of cosmic ray protons and nuclei, which have a relatively high muon content from decay of charged pions and kaons. The signal is presumed to be photon-induced, and should therefore have at least an order of magnitude fewer muons.[54,55] If the signal and the high muon content are both real, the most conservative interpretation is that photon interactions at ultra-high energies (> 100 TeV) have a much higher probability for hadroproduction than had previously been thought.[56]

The primary focus of air shower experiments now under construction or expansion in this country and abroad is to resolve the questions associated with the existence of air showers from point sources. There are two aspects to this effort. The first is to increase collecting power (area) in an effort to increase the signal-to-noise ratio to the point where subtle statistical analysis is no longer necessary to extract a signal. A proposed scheme[57] to coordinate the observing efforts of arrays that can observe the same sources simultaneously is important in this regard in view of the possible variable or sporadic nature of the sources. The second crucial aspect is to improve the sampling of the muon content of showers very significantly.

Acknowledgment This work was supported in part by the U.S. Department of Energy under Contract No. DE-AC02-078ER05007.

References

1) Ginzburg, V.L. and Syrovatskii, S.I., *The Origin of Cosmic Rays*, (Pergamon, 1964).
2) Blandford, R. and Eichler, D., "Particle Acceleration at Astrophysical Shocks: A Theory of Cosmic Ray Origin," Physics Reports **154**, 1-75 (1987).
3) Lagage, P.O. and Cesarsky, C.J., "The Maximum Energy of Cosmic Rays Accelerated by Supernova Shocks," Astron. Astrophys. **125**, 249-257 (1983).
4) Völk, H.J. and Biermann, P.L., "Maximum Energy of Cosmic Ray Particles Accelerated by Supernova Remnant Shocks in Stellar Wind Cavities," preprint (May, 1988).
5) Jokipii, J.R., "Mechanism of Particle Energy Gain at Shocks: The Role of Magnetic Field Geometry," Proc. 6th Int. Solar Wind Conf. (Estes Park, Colorado, August 1977) NCAR TN-306, vol 2, p. 481. See also Ap. J. **313**, 842 (1987) and **255**, 716 (1982).
6) Cesarsky, C.J. and Montmerle, T., "Gamma Rays from Active Regions in the Galaxy: The Possible Contribution of Stellar Winds," Space Sci. Revs. **36**, 173-93 (1983).
7) Jokipii, J.R. and Morfill, G.E., "On the Origin of High Energy Cosmic Rays," Ap.J. **290**, L1-L4 (1985). Also **312**, 170-177 (1987).
8) Nagle, D.E., Gaisser, T.K. and Protheroe, R.J., "Extensive Air Showers Associated with Discrete Astrophysical Sources," Ann Rev. Nucl. Part. Sci., **38**, 609-657 (1988).
9) Kazanas, D. and Ellison, D.C., "Origin of Ultra-high energy γ-rays from Cygnus X-3 and Related Sources," Nature **319**, 380-382 (1986).
10) Gaisser, T.K., Harding, A. and Stanev, T., "Signatures of Particle Acceleration at SN1987A," Bartol preprint BA-88-53. Also Nature **329**, 314-16 (1987).
11) Chanmugam, G. and Brecher, K., "Ultra-high Energy γ-rays and Cosmic Rays from Accreting Degenerate Stars," Nature **313**, 767-68 (1985).
12a) Wang, Y.-M., "Production of Ultra-high-energy γ-rays by Accreting Neutron Stars," Astrophysics and Space Science **121**, 193-200 (1986).
12b) Katz, J.I. and Smith, I.A., "Particle Acceleration in Accreting Magnetospheres," Ap. J. **326**, 733-737 (1988).
13) Bloemen, J.B.G.M., "High-energy Gamma Rays Probing Cosmic-ray Spectral Differences Throughout the Galaxy," Ap. J **317**, L15-L19 (1987).
14) Wolfendale, A.W., "The Origin and Propagation of Cosmic Rays," to appear in the Proceedings of the Erice School on Cosmic Rays, April (1988).
15) Stecker, F.W., "Diffuse Flux of Cosmic High Energy Neutrinos," Ap. J. **228**, 919-927 (1979).
16) Samorski, M and Stamm, W., "Detection of 2×10^{15} to 2×10^{16} eV Gamma-rays from Cygnus X-3," Ap. J. Lett. **268**, L17-21 (1983). See also Proc. 18th Int. Cosmic Ray Conf. (Bangalore) Vol. 11, p. 244 (1983).
17) Lloyd-Evans, J., et al., "Observation of γ Rays > 10^{15} eV from Cygnus X-3," Nature **305**, 784-87 (1983).
18) Dingus, B.L., et al., "Observation of Pulsed Emission from Hercules X-1," Phys. Rev. Letters (to be published, 1988).
19) Cassiday, G.L., et al., "Evidence for 10^{18} eV Neutral Particles from the Direction of Cygnus X-3," Utah preprint UUHEP-88/3.
20) Weekes, T.C., "Very High Energy Gamma-Ray Astronomy," Physics Reports **160**, 1-121 (1988).
21) Ciampa, D., et al., "A Search for Ultra-high energy Gamma-rays from Supernova 1987A," Ap. J. Lett. **326**, L9-11 (1988).
22) Raubenheimer, B.C., et al., "VHE γ-ray Observations of SN1987A during November 1987," Astronomy and Astrophysics (1988).
23a) Bond, I.A., et al., "Search for γ-rays from Supernova 1987A at Energies Greater than 100 TeV,", Phys. Rev. Letters **60**, 1110-13 (1988).
23b) Bond, I.A., et al., "Search for TeV γ-rays from SN1987A during December 1987 and January 1988," ICR-Report-161-88-7, KEK Preprint 88-7.
24) Chadwick, P.M., et al., Ap. J. **333**, L19 (1988).
25) Morello, C., to appear in the Proceedings of the Vulcano Workshop on Frontier Objects in Astrophysics and Particle Physics, 1988.
26) Smith, N.J.T., et al., "An Experiment to Search for Ultra High Energy γ-ray Sources from the South Pole," Bartol/Leeds preprint October, 1988. See also Ref. 35.
27) Pinto, P.A. and Woosley, S.E., "The Theory of Gamma-ray Emergence in Supernova 1987A,", Nature **333**, 534-537 (1988).
28) Oyama, Y. et al., "Search for High-energy Neutrinos from SN1987A: First Six Months," Phys. Rev. Letters **59**, 2604 (1987).
29) van der Velde, J.C., et al., Proc. Moriond Conference, 1988.
30) Berger, Ch., et al., 1988, "Recent Results from the Frejus Experiment," Nuclear Instr. Methods, **A264**, 24 (1988).
31) Baksan collaboration, A.E. Chudakov, private communication.
32) Gaisser, T.K. and Stanev, T., "Energetic (>1 GeV) Neutrinos as a Probe of Acceleration in the New Supernova," Phys. Rev. Letters **58**, 1695-97 (1987) and **59** E844.
33) Gaisser, T.K., "Acceleration to PeV Energies in SN1987A and Cygnus X-3,", to appear in Proceedings of the Vulcano Workshop on Frontier Objects in Astrophysics and Particle Physics, 1988.
34) Berezinsky, V.S. and Stanev, T., "On the Burst of TeV Gamma-Rays from SN1987A," Bartol preprint BA-88-

49.

35) Gaisser, T.K. et al., "Search for >50 TeV Photons from SN1987A in Early 1988," Bartol/Leeds preprint, October, 1988.

36) Hillas, A.M., "Is Cygnus X-3 a Monoenergetic 10^{17} eV Accelerator?" Nature **312**, 50-51 (1984) and Proc. 19th Int. Cosmic Ray Conf. (La Jolla) vol 9, pp. 407-414 (1985).

37) Berezinsky, V.S., Castagnoli, C. and Galeotti, P., "High-energy Neutrino Astronomy with 'Small' Underground Detectors", Nuovo Cimento **8c**, 185-211 (1985).

38) Gaisser, T.K. and Stanev, T., "Calculation of Neutrino Flux from Cygnus X-3," Phys. Rev. Letters **54**, 2265-68 (1985).

39a) Berezinsky, V.S., et al., "High Energy Gamma-astronomy with Large Underground Detectors," Astron. Astrophys. **189**, 306-318 (1988).

39b) Stanev, T., "Are Deep Underground Detectors Good γ-ray Telescopes?" Phys. Rev. **D33**, 2740-2743 (1986).

39c) Halzen, F., Hikasa, K. and Stanev, T., "Particle Physics with Cosmic Accelerators,", Phys. Rev. **D34**, 2061-2070 (1986).

40) Frank, J., King, A.R. and Raine, D.J., *Accretion Power in Astrophysics* (Cambridge, 1985).

41) Stecker, F.W., Harding, A.K. and Barnard, J.J., "Anatomy of a Cosmic Ray Neutrino Source and the Cygnus X-3 System," Nature **316**, 418-20 (1985).

42) Gaisser, T.K., Stecker, F.W., Harding, A.K. and Barnard, J.J., "Effects of High-energy Neutrino Production and Interactions on Stars in Close X-ray Binaries," Ap. J. **309**, 674-681 (1986).

43) Gaisser, T.K. and Yodh, G.B., "Particle Collisions above 10 TeV as seen by Cosmic Rays,", Ann. Rev. Nucl. Part. Sci. **30**, 475-542 (1980).

44) Gaisser, T.K. and Stanev, T., "Mini-jets in Minimum-bias Events," Bartol Preprint BA-88-48.

45) Barr, S., Gaisser, T.K., Tilav, S. and Lipari, P., "Ratio of ν_e/ν_μ in Atmospheric Neutrinos," Physics Letters B, (to be published, 1988).

46a) Berezinsky, V.S. and Prilutsky, O.F., "Pulsars and Cosmic Rays in the Dense Supernova Shells," Astron. Astrophys. **66**, 325-334 (1978).

46b) Gaisser, T.K. Halzen, F. and Stanev, T., "Ultra-high Energy Radiation from Young Supernovae,", Nature **332** 314 (1988).

47) Vestrand, W.T., and Eichler, D., "On the Ultrahigh-energy Gamma Rays from Cygnus X-3," Ap. J. **261**, 251-58 (1982). See also D. Eichler and W.T. Vestrand, Nature **307**, 613-14 (1984).

48) Roberts, A. (Editor) Proceedings of the 1976 DUMAND Summer Workshop (FNAL) 1976. See also M.A. Markov, in Proceedings of the 1960 Int. Conf. on High Energy Physics (Rochester) 578 (1960).

49) Gaisser, T.K. and Stanev, T., "Response of Deep Detectors to Extraterrestrial Neutrinos," Phys. Rev. **D31**, 2770-72 (1985). See also, Phys. Rev. **D30**, 985-990 (1984).

50a) McKay, D.W. and Ralston, J.P., "Small-x QCD and the Ultra-high Energy νN Total Cross Section," Phys. Lett. **167B**, 103 (1986).

50b) Quigg, C. Reno, M.H., and Walker, T.P., "Interactions of Ultrahigh Energy Neutrinos," Phys. Rev. Lett. **57**, 774-77 (1986).

50c) Reno, M.H. and Quigg, C., "On the Detection of Ultrahigh-energy Neutrinos," Phys. Rev. **D37**, 657-664 (1988).

51) Gaisser, T.K. and Grillo, A.F., "Energy Spectra of Neutrino-induced Upward Muons in Underground Experiments," Phys. Rev. **D36**, 2752-56 (1987).

52) Kolb, E.W., Turner, M.S. and Walker, T.P., "Production and Detection of High Energy Neutrinos from Cygnus X-3," Phys. Rev. **D32**, 1145-51 (1985).

53) Dingus, B.L., et al., "Search for Signals from Cygnus X-3 at Energies above 50 TeV," Phys. Rev. Letters **60**, 1785-88 (1988).

54) Stanev, T., Gaisser, T.K. and Halzen, F., "Muons in Gamma Showers from Cygnus X-3?," Phys. Rev. **D32**, 1244-47 (1985).

55) Edwards, P.G., Protheroe, R.J. and Rawinski, R.J., "The Muon Component of Gamma-ray Initiated Air Showers," J.Phys. **G11**, L101-L107 (1985).

56) Drees, M., Halzen, F. and Hikasa, K., "Muons in Gamma Showers," MAD/PH/424, KEK-TH-208 (June, 1988).

57) Lande, K., "A Long Baseline Multi-Air Shower Array in the Western United States to Study Gamma Ray Bursts from Active Stellar Systems," this volume.

Very High Energy Gamma-ray Astronomy

R. C. Lamb
Physics Department, Iowa State University, Ames, Iowa 50011

ABSTRACT

Present and possible future contributions of very high energy (VHE) gamma-ray astronomy to particle physics are summarized. Air showers initiated by 1 TeV photons from the Crab Nebula are in accord with theoretical expectations. The most powerful of the VHE sources yet identified (Cygnus X-3 and Hercules X-1) appear to be cosmic accelerators in which a substantial fraction of the gravitational energy that is released is converted to high energy beam energy. With its high rate capability VHE gamma-ray astronomy can gather vital observations that constrain possible beam acceleration mechnism(s), that provide detailed characterizations of any anomalous shower behavior indicative of new physics, and that help settle long-standing questions about the origin of high energy cosmic rays. Improvements in detector sensitivity of two orders of magnitude are attainable with future detectors and will help this potential to be fully realized.

INTRODUCTION

This review is intended for the high energy particle physicist who may be wondering why such a topic is among those studied at Snowmass '88. Its purpose is to "uncover" the field rather than cover it. For further background material the reader is referred to several excellent comprehensive reviews that have recently appeared, among them Weekes[1)] and Nagle, Gaisser, and Protheroe[2)]. See also the paper on ultra-high energy gamma-ray astronomy by Matthews[3)] in this same volume.

The organization of this paper is as follows: In the first section I will present some of the motivation for developing this field, emphasizing those aspects that are particularly relevant to particle physics. In section II, the prototypical VHE gamma-ray source, the binary gamma-ray pulsar, is discussed with emphasis on the two best established sources: Hercules X-1 and Cygnus X-3. Section III described the atmospheric Cherenkov technique with an illustration from the detection of the Crab Nebula. The last section looks to the future and includes questions that may ultimately be answered with the improved detectors that are proposed.

1. MOTIVATION

VHE gamma-ray astronomy is generally taken to span the photon energy range from 0.1 to 100 TeV. The region from 0.1 to 100 PeV has been designated ultra high energy (UHE), and the region above 10^{17} eV, extremely high energy (EHE). All of these astronomies are conducted on earth because of the very low flux of such photons. For example, the photon flux from the Crab Nebula above 1 TeV would give a detected photon every three years of observation by the high energy experiment of the Gamma-ray Observatory satellite. For typical VHE detectors the rate would be about 1 photon every three minutes.

For these astronomies, the atmosphere plays a crucial role, acting as the shower development medium. Figure 1 shows a schematic of three photonic showers representative of each of these energy regimes. For VHE astronomy, the Cherenkov light from those shower particles exceeding the Cherenkov threshold provides the signal.

What is the particle physics justification for these astronomies in general, and VHE astronomy in particular?

In the first place such high energy quanta indicate the presence, at various astrophysical sites, of charged particles of even higher energies. It is well known that the cosmic ray particle energy spectrum extends to beyond 10^{19} eV. What is not known is where such particles originate. VHE and UHE astronomies are beginning to give an answer to this question, which serves as a fundamental motivation for them.

For energies up to a few times 10^{13} eV, acceleration of protons in supernovae shock waves is a favored explanation for the origin of cosmic rays, however this mechanism fails for higher energies.[4)] (Even for the low

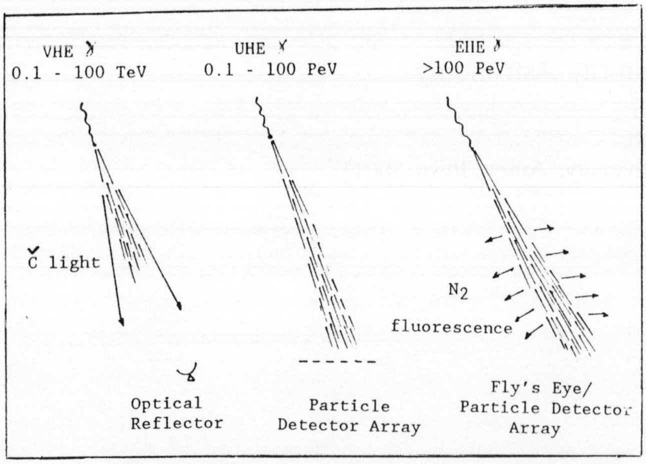

FIG. 1. Schematic views of three air showers created in the atmosphere by VHE, UHE, and EHE photons. In the VHE regiime the cascade particles that survive to the ground are limited in number, so that detection is via the particles' Cherenkov emission. In the UHE and EHE regimes particles are detected. For the highest energies, detection of N_2 fluorescence by a "Fly's Eye" type receiver is also possible.

energy regime, the supernova shock hypothesis may not be correct. The arguments in its favor are indirect.)

It is now clear that some sites of high energy particle production occur in the vicinity of spinning, highly magnetized neutron stars in binary systems. High energy charged particle beams are developed and the gamma rays that are observed are thought to be beam-dump secondaries, in which the companion star provides target material. Simple arguments (see section II) in regard to the power contained in the primary particle beams that are responsible for the radiation indicate that such sources make significant contributions to the high energy cosmic ray flux above 10^{15} eV. Whether their combined contribution is 1% or 100% of such cosmic rays is not known; further progress in the ground-based astronomies should provide decisive answers to many of the questions surrounding the origin(s) of the high energy cosmic rays.

In the context of this source picture and the assumption that at least some of the 10^{19} eV particles are associated with such sources, then it is also clear that particle physicists should be interested in the findings of VHE, UHE, and EHE astronomy. If the 10^{19} eV particles are protons, then their interactions with target material correspond to center-of-mass energies of 140 TeV, well beyond even SSC capabilities.

Because most source energy spectra (like the spectra of secondaries from particle accelerators) fall with increasing energy and because of the large area associated with the Cherenkov light front, VHE detectors are intrinsically high-rate detectors relative to their higher energy cousins. In some sense VHE astronomy is to UHE astronomy as high rate, low energy "factories" are to the highest energy accelerators. Each has its complementary role to play.

VHE astronomy has led the way in the discovery of high energy gamma-ray sources. Eleven of the 13 sources listed by Weekes[1)] have been seen at TeV energies. Four of these sources have also been observed at PeV energies; in three cases the TeV emission was observed first.

Another of the possible spin-offs from the ground-based astronomies for high energy physics comes in the area of understanding the acceleration mechanism or mechanisms that are at work. For the accretion-driven sources it appears that a substantial fraction of the gravitational energy that is released is converted to the energy by the high energy beams. This is somewhat surprising in view of the second law of thermodynamics in which disorganized forms of energy, i.e., thermal radiation would appear to be favored. After the observations were reported a number of models were advanced to account for the binary sources, none of which is fully satisfactory.

2. THE BINARY GAMMA-RAY PULSAR

In our own galaxy there are something like 20 binary pulsars with measured neutron star spin rates that have been observed by X-ray satellites. Three of these are relatively well established VHE gamma-ray sources: Hercules X-1, 4U0115+63, and Vela X-1. Observations in X-rays of these sources clearly establish them to be spinning neutron stars in binary orbits around reasonably ordinary stars. Both Hercules X-1 and 4U0115+63 have line features in their X-ray spectra that correspond to electronic transitions between quantatized cyclotron orbits (Landau levels) that allow their surface magnetic fields to be inferred. Hercules X-1 has a field in the range of 3-5 Teragauss; 4U0115+63 's field is approximately 1 TG. Two other X-ray binary pulsars, Cen X-3 and LMC X-4, have also been reported as UHE gamma-ray emitters. A reasonable conjecture at this time is that as the sensitivity of gamma-ray detectors improves, most, if not all, such X-ray sources will also prove to be gamma-ray sources.

To this list of binary gamma-ray pulsars, I will add Cygnus X-3. Its binary character is not yet proven (as the other sources are) through Doppler-shift measurements of the neutron star's motion around its companion, but it is generally thought to be a binary with the observed 4.8-hour modulation ascribed to its orbital period. A possible 12.6 ms periodicity has been reported for the spin period of the neutron star.[5)]

Table I lists the characteristics of each of these four sources. In all cases the X-ray and VHE gamma-ray luminosities that are given show a substantial range of variation, with the variation greatest for X-rays. The range of variation in the gamma-ray regime is probably just as great (or greater); however, the limited dynamic range of VHE measurements at the present time introduces substantial uncertainty on precisely what is the true range of variation in the gamma-ray fluxes.

What is important to note is that the VHE gamma-ray luminosity is comparable to the X-ray luminosity.

Table I. Comparison of VHE binary sources.

Source	Time scale			Surface Magnetic Field(G)	Luminosity (erg sec^{-1})	
	Spin	Orbit	Precession		X-ray	VHE gamma rays
Hercules X-1	1.24 seconds	1.7 day	35 days	3-5 $\times 10^{12}$	10^{35} to 10^{37}	10^{35} to 10^{37}
4U0115+63	3.61 seconds	24 days		1×10^{12}	10^{33} to 10^{37}	10^{35} to 10^{36}
Vela X-1	283 seconds	9.0 days		$> 10^{13}$?	10^{34} to 10^{37}	10^{34}
Cygnus X-3	12.59 msec?	4.8 hours	19.2 days?		10^{37} to 10^{37}	10^{35} to 10^{37}

When one recognizes that the particle beam luminosity must be even greater (at least a factor of 10 or so), then it is clear that at least for some of the time the dominant mode of energy release for these so-called X-ray binary pulsars is in VHE gamma rays. So that it is appropriate to call them binary gamma-ray pulsars.

(We note that some isolated neutron star pulsars, e.g., the Crab and Vela pulsars, have also been detected in the VHE regime. However, the underlying physics may be quite different; since the Crab and Vela pulsars are isolated neutron stars. The source of the energy radiated in the loss of rotational kinetic energy of the neutron star. The fraction of the power radiated in the VHE region is less than 10^{-4} or 10^{-5} of the total power, in contrast to the binary case where the VHE power is a significant fraction of the total power.)

Figure 2 gives a highly schematic representation of the binary gamma-ray pulsar. The fundamental source of energy is the work done by gravity on material accreted from the companion. The power developed is $GM\dot{M}/R$ where M is the mass of the neutron star, $\dot{M}$ is the mass accretion rate, and R is the radius to which the material falls. If the material falls all the way to the neutron star surface, roughly 10% of the rest mass energy is liberated. For sufficiently high mass accretion rates (10^{18} g/s) the energy released surpasses the limit at which radiation pressure balances gravity (the Eddington limit). This corresponds to a power of 1.4×10^{38} erg/s times the mass of the neutron star in units of the sun's mass.

For the binary sources the compactness of the orbits ensures a variety of target material. The companion star itself, the accreting material, and the accretion disk could all act as targets. In this picture the time of the observed gamma-ray emission relative to the orbital phase provides important geometric constraints on the nature and location of the target material. For example, Hercules X-1 for example, has, in addition to its 1.7 day orbital period, a 35-day cycle that is due to a wobbling of its accretion disk. This motion modulates the intensity of X-ray emission and also appears to govern the time at which VHE gamma-rays are seen. In Figure 3 the times of VHE/UHE outbursts are plotted, as a function of the 35-day cycle (x-axis) and the 1.7-day orbital cycle (y-axis). This type of phase plot provides clues to beam-target geometry and a testing ground for theoretical models.

The magnetic field appears to play a dominant role

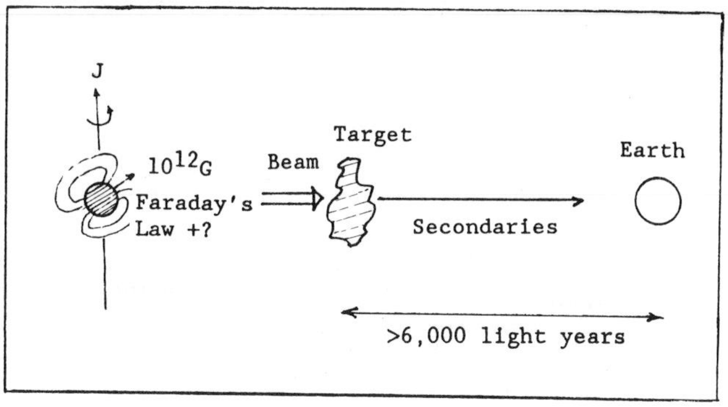

Fig. 2. Cartoon of the binary VHE gamma-ray pulsar; the principal ingredients of which are a highly magnetized, spinning neutron star in a binary system; a beam whose cause in unknown; and a target of accreting material.

here. Because of the difficulty associated with producing modulated radiation in the situation in which the field is aligned with the spin axis, the field axis is usually considered to be oblique to the spin axis.

How the beam is developed is unknown. Many models have been advanced; most to account for Cygnus X-3. There is as yet no theoretical consensus since each model appears to have problems. Hillas[6)] has given a critique of their strengths and weaknesses. In Table 2 we list the theoretical models of which we are aware.

In order to compare these gamma-ray luminosities (and the beam luminosities ultimately responsible for the gamma rays) with the power contained in cosmic rays we need to know not only the cosmic-ray energy density (measured at earth) but also the confinement volume and the lifetime of the cosmic rays. Of these three quantities, only the energy density is well measured. The volume and lifetime of high energy cosmic rays are uncertain to more than an order of magnitude, in my view, so that the power estimate may be uncertain by that amount or more. The integral energy density (above 10^{14} eV) is 2×10^{-15} ergs-cm^{-3}. The integral energy density will scale like $E^{-0.6}$ power up to the cosmic ray knee at 2 PeV; above that it scales as $E^{-1.1}$. If we assume a confinement volume to be the volume of the galactic disk $\sim 2 \times 10^{67}$ cm^2, and a confinement lifetime of 10^5 years, the power developed above 10^{14} ev is 10^{40} erg/s, 3×10^{39} erg/s above 10^{15} eV.

Figure 4 shows the photon energy spectrum for Cygnus X-3. We may convert this integral intensity to a

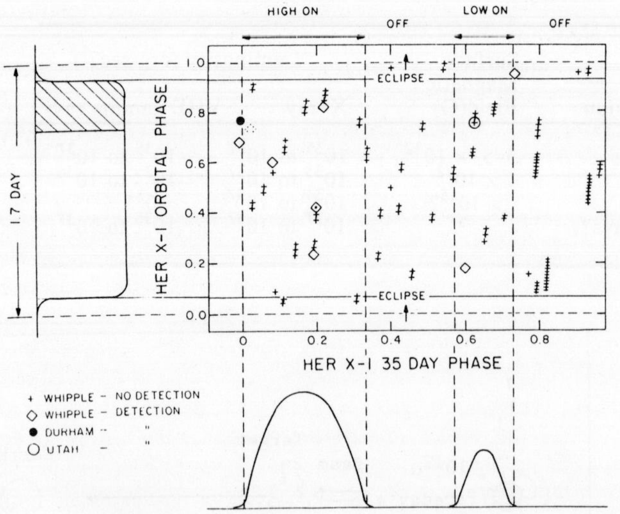

FIG. 3. Distribution in phase (1.7 day and 35 day) of VHE/UHE observations of Hercules X-1 at the Whipple Observatory and other observatories between 1983 and 1986. From ref. 1.

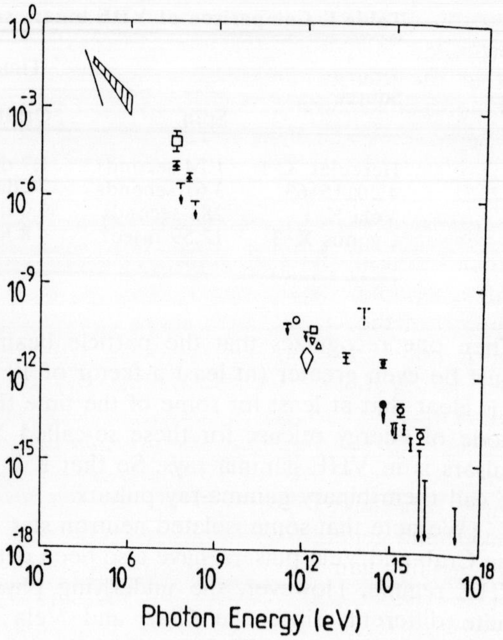

FIG. 4. Energy spectrum of gamma-ray observations from Cygnus X-3 taken at various times between 1972 and 1984. From ref. 1.

luminosity by assuming a source distance (a lower limit is 30,000 light years) and accounting for some attenuation of the photon beam due to absorption by the microwave background. The resulting gamma-ray luminosity above 10^{14} eV is $4x10^{37}$ erg/s. If one takes into account that only for about 10% of each orbital cycle are gamma rays seen (whereas for the entire cycle the cosmic-ray particles that produce the gamma rays are emitted) and that the efficiency for the conversion of cosmic ray energy into gamma-ray energy is less than 10%, then a cosmic ray beam luminosity of $4x10^{39}$ erg/s results. Thus only a few sources like Cygnus X-3 are sufficient to maintain the observed high energy cosmic ray power.

The recently reported[13)] multi-hundred TeV bursts from Hercules X-1 have an estimated luminosity of 2-3 × 10^{37} erg/s. It is clear that much better information regarding the on-time of the gamma-ray binaries is needed to (among other things) establish an accurate accounting of the high energy cosmic-ray energy budget.

3. THE ATMOSPHERIC CHERENKOV TECHNIQUE

Detectors of VHE radiation are, at the present time, exclusively atmospheric Cherenkov detectors. In the final section of this paper we will mention two non-Cherenkov efforts that may be competitive with the Cerenkov technique at energies above 10 TeV. However, in this section we will consider only Cherenkov detectors.

Near sea level the Cherenkov light wave front associated with a given gamma-ray air shower is approximately 300 meters in diameter, so that even the simplest searchlight type receiver can achieve collection areas of 10^9 cm^2. The energy threshold for detection is approximately inversely related to the diameter of the light collecting device. For 10 meter diameter reflectors operated in coincidence a threshold of 0.1 TeV can be

Table 2. Models for VHE binary sources.

Beam	Acceleration Mechanism	Radiation Mechanism	Reference	Limitation
Electron	Pulsar	Bremsstrahlung	7)	Rotation rate limits maximum energy to below what is observed for the slowest rotators.
Electron	Pulsar	Synchrotron radiation	8)	
Proton/ion	Pulsar	π^0 production	9)	
Proton/ion	Shock wave	π^0 production	10)	Magnetic fields may be too strong
Proton/ion/neutron	shock wave	π^0 production	11)	
Proton/ion	Shock wave	π^0 production	12)	

achieved. Because of the virtues of simplicity and the enormous effective area of flux collection, it appears likely that in the energy range 0.1 to 10 TeV the Cherenkov technique will maintain its monopoly.

Because of the directionality of the Cherenkov light (emission at 1.3 degrees to the direction of a $\beta = 1$ particle at sea level, less at altitude) the arrival direction of the initiating photon (or background cosmic-ray particle) is easily determined to 1-2 degrees with a narrow angle acceptance detector. This is in contrast to the effort required in the UHE region where arrival directions are determined by fast timing of shower fronts over baselines determined by the distributed elements of the particle-detector array.

In retrospect the simplicity of the Cherenkov technique has, I believe, delayed its development. With a simple Cherenkov receiver the numerous cosmic ray background generally swamps the feeble signal of gamma rays, making gamma-ray source detection difficult. In the last few years there have been developments on a number of fronts that have begun to complicate the detectors, but that have also begun to bear fruit in terms of substantial background reduction.

In order to gain some appreciation of these efforts we need to look at the details of the shower process at TeV energies as deduced by Monte Carlo simulations. Although the air showers from gamma rays and from cosmic rays are superficially similar, they differ considerably in a number of essential ways. Dramatic differences exist in the Cherenkov light images and the lateral distribution of that light. Present generation detectors are beginning to measure one or more of these parameters, however, no detector yet exists in which all of this information is gathered simultaneously. Table III lists each of these parameters, giving the distinguishing characteristics of the gamma rays, as well as representative observatories at which these features are emphasized.

Table 3. Comparison of Cherenkov light characteristics for γ-ray air showers.

Parameter	γ-ray Shower Characteristic	Observatory Where Tested
Image		
Size	Smaller, less fluctuations[14]	Crimea
Orientation	Images point to true arrival direction[14]	Whipple
Lateral Profile	Smooth distribution, peak at ~140 meters from shower axis. Proton distribution much more variable[15,16]	Potchefstroom Tata Themis (ASGAT)
Spectral content	uv-poorer[17]	Crimea
Time spread	2-3 ns for γ-ray showers 4-5 ns for proton showers[1]	Haleakala Sandia

These differences are based on shower simulation calculations by a number of authors[14,15,16,17] in which distributions of parameter values are obtained. However, one can gain some understanding of how these differences arise by comparing representative showers of each type. Figure 5 shows a side view of the particle content of each shower type at an energy of one TeV. Because the photon's radiation length is only 1/3 of the proton's interaction length, fluctuations are much more important for the nucleon case. The much greater transverse momentum of the nucleon cascade produces wider Cherenkov images and wider, albeit more uneven, lateral distributions of Cherenkov light. Because of greater fluctuations for nucleon showers, the variation in both their image shape and their lateral distribution is substantially larger than for gamma-ray showers. Examples of the lateral distribution of the Cherenkov light and of images of that light (by a perfect imaging detector) are shown in Figures 6 and 7. The fact that that nucleon cascade possesses penetrating muons gives rise to its uv-rich character. These same muons produce the slight increase in the time spread of the Cherenkov pulse.

An example of the kind of improvement in sensitivity that second generation Cherenkov detectors can provide is given by the recent detection of the Crab Nebula by the Whipple Observatory Collaboration.[19] Using a 37-element camera, with a pixel scale of 0.5 degrees, the group selected gamma-ray images based on their predicted properties. More than 98% of the cosmic ray background

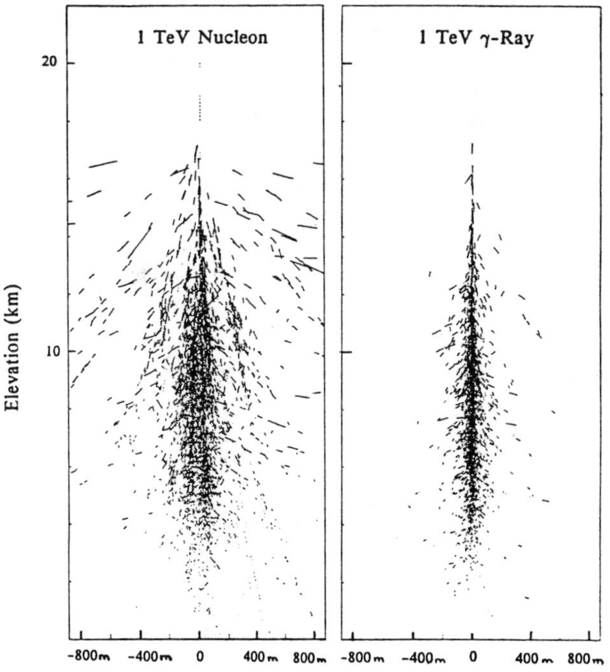

FIG. 5. **Monte Carlo simulations of 1 TeV air showers; nucleon-induced on the left; gamma-ray induced on the right. Note that the horizontal scale is magnified by a factor of 5 relative to the vertical scale.**

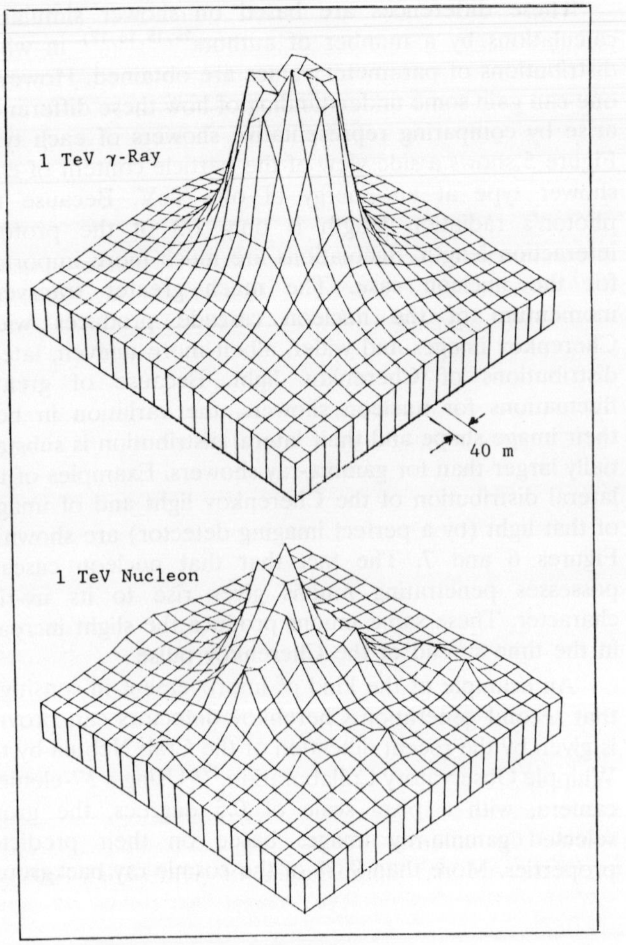

Fig. 6. Monte Carlo simulations[18] of the lateral distributions the Cherenkov light from 1 TeV air showers at an elevation of 2.2 km.

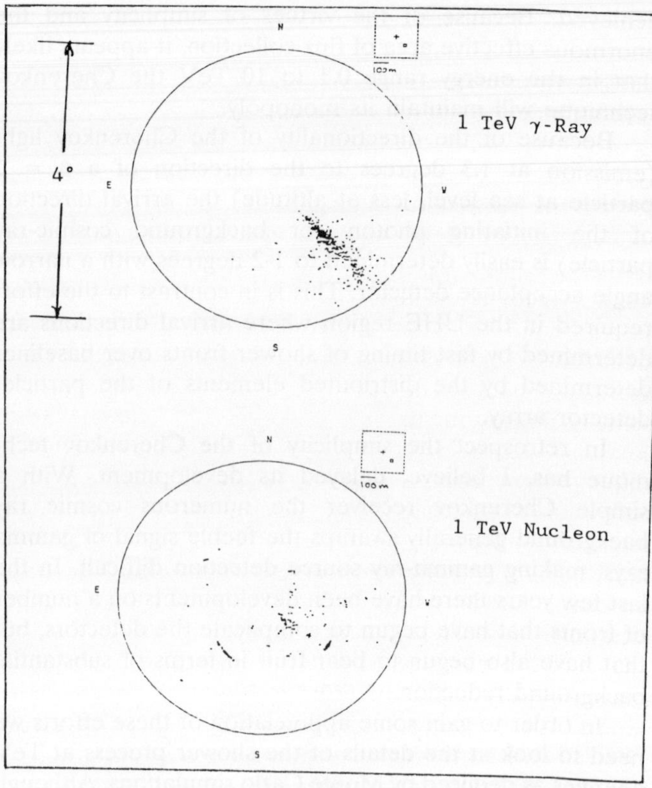

Fig. 7. Monte Carlo simulations of the pattern of the Cherenkov light as viewed by an imaging detector with no aberrations. In each case the source direction is the center of the field-of-view. The detector is pointed in this direction and the detector itself is located 80 meters from the core of the shower as indicated by the small box in which the core's location is shown as a "+" and the detector as a "□".

was rejected, changing a 2 sigma excess (uncut) into a 9.0 sigma detection. The distribution of the parameter used for the cut is shown in Figure 8, for both the on-source and off-source data. The parameter measures the width of individual shower images in the azimuthal direction and combines two distinctive aspects of gamma showers relative to the nucleon showers; namely that they are intrinsically narrower and they tend to point to their true arrival direction in the image plane. The only significant difference between the two distributions occurs for small values of azwidth, as predicted by the simulations, indicating the presence of gamma rays.

For particle physics, the most significant aspect of the Crab detection is that it indicates that photons in the range of 1 TeV have properties in accord with quantum electrodynamics. There does not appear to be any anomalous behavior.

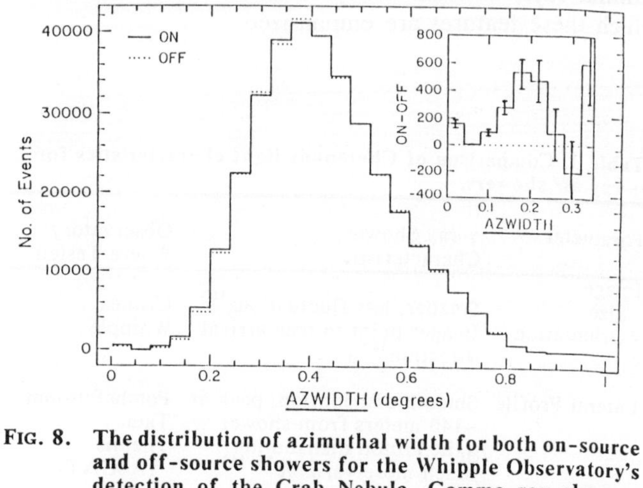

Fig. 8. The distribution of azimuthal width for both on-source and off-source showers for the Whipple Observatory's detection of the Crab Nebula. Gamma-ray showers should, on the average, have substantially smaller values of this parameter than cosmic-ray showers. The only statistically significant difference in the two distributions occurs in the region expected to be populated by gamma-ray showers. That region, 0.0 to 0.3 degrees, is shown as an inset to the figure.

4. THE FUTURE

The long-term development of VHE astronomy and its potential impact on particle physics can be directly tied to improvements in the sensitivity of detectors. Improvements may be made either by increasing the rate[20,21] or by reducing the cosmic ray background via better angular resolution and/or better background discrimination.

Since increased rate buys improved sensitivity only as the square root of the rate, the area of greatest potential improvement is to reduce the background. Here it would appear that a Cherenkov receiver that measures all of the information inherent in the Cherenkov light can achieve sensitivity improvements of a factor of 100 over present second generation detectors. Weekes[22] has long advocated such an approach and others[23,24] have more recently joined in calling for a Cherenkov facility which would incorporate instrumentation to accomplish this. The Whipple Observatory's HERCULES proposal[25] for stereoscopic viewing of Cherenkov shower images is a step toward such a facility.

One of the more promising ideas which has been put forward recently is that of using a lake as a gamma-ray detector. The surface of the lake is made light tight and is instrumented to detect shower particles via their Cherenkov emission in the water. By using several layers of detector one can discriminate between electrons and muons. Since the shower front is sampled densely the usual 100 TeV threshold for array detectors is reduced by a factor of 10 or more. Illustrative of this approach is the GRANDE proposal,[26] in which the detectors are arranged to detect upward going as well as downward moving particles, so that the lake is both a VHE gamma-ray detector and a VHE neutrino detector.

Another proposal, by the Heidelberg-Max Planck collaboration,[27] calls for the use of tracking devices located at mountain elevations. Angular accuracies of 0.1 to 0.3 degrees give rise to a claimed improvement of a factor of 100 in comparison to existing conventional arrays and atmospheric Cherenkov detectors, with good efficiency for showers above 10 TeV for elevations in the 3000-4000 meter range.

The answers that VHE astronomy is beginning to supply in regard to the origins of cosmic rays are scientifically very interesting. However, the greatest impact that this astronomy will have on particle physics will come in two areas, I believe.

One area of impact will be in understanding the particle acceleration mechanism(s). In order to achieve this it will be necessary to explore a wide range of parameter space for the gamma-ray binary pulsars. What combinations of magnetic field, binary orbit, spin rate, and rate of change of spin are most favorable? What are the characteristic time scales of the VHE emission? For Hercules X-1, typical bursts last 15 to 45 minutes. Are there phenomena on even shorter time scales, such as we see from the sun, and, if so, what does that say about the acceleration mechanism? With its high rate capability VHE detectors will play a crucial role in the study of short time-scale phenomena. It will, of course, be necessary to establish the long-term duty factor of each source to see if, in fact, localized sources are a major contributor to the cosmic rays. For long-term source monitoring, wide-angle detectors (such as are used in UHE astronomy) will need to be developed.

Another area of potential impact is in regard to new physics/new particles. Several reports[28,29,13] have already raised the possibility of this, stimulating theoretical conjectures about a change in the character of the photon's interaction at energies above 1 TeV.[30,31] VHE detectors will be able to pin down these possibilities only if they are sufficiently "complicated." If there is new physics to be learned, then detectors that overconstrain the expected may be able to decipher the unexpected.

References

1. Weekes, T. C., "Very High Energy Gamma-Ray Astronomy," Physics Reports **160**, 1 (1988).
2. Nagle, D. E. , Gaisser, T. K., and Protheroe, R. J., "Extensive Air Showers Associated with Discrete Astrophysical Sources," Ann. Rev. Nuclear and Particle Sci. 38, in press (1988).
3. Matthews, J., "Ultra-High Energy Gamma-ray Astronomy," this volume.
4. Hillas, A. M., "The Origin of Ultra-High-Energy Cosmic Rays," Ann. Rev. Astron. Astrophys. **22**, 425 (1984).
5. Chadwick, P. M. et al., " A 12.6 ms Pulsar in Cygnus X-3," Nature 318, 642 (1985).
6. Hillas, A. M., "Source Mechanisms for Accretion Objects," Proc. NATO Advanced Research Workshop on Very High Energy Gamma-ray Astronomy, ed. K. E. Turver (Reidel, Dordrecht) 1986, pp. 71-80.
7. Vestrand, W. T., and Eichler, D., "On the Ultrahigh-Energy Gamma Rays from Cygnus X-3," Astrophys. J. 261, 251 (1982).
8. Cohen, J. M., and Eichler, D., "Gamma-Ray Pulsar Model," Astrophys. J. 319, 930 (1987).
9. Eichler, D. and Vestrand, W. T., "Implications of 10^{16} eV γ Rays from Cygnus X-3," Nature 307, 613 (1984).
10. Eichler, D., and Vestrand, W. T., "Implications of Ultra-High-Energy Emission from Hercules X-1," Nature 318, 345 (1985).
11. Kazanas, D., and Ellison, D. C., "Origin of Ultra-High Energy γ Rays from Cygnus X-3 and Related Sources," Nature 319, 380 (1986).
12. Chanmugam, G. and Brecher, K., "Ultra-High Energy γ Rays and Cosmic Rays from Accreting Degenerate Stars," Nature, 313, 767 (1985).
13. Dingus, B. L. et al., "Ultra High Energy Pulsed Emission from Hercules X-1," submitted to Phys. Rev. Lett. (1988).
14. Hillas, A. M., "Cherenkov Light Images of EAS

Produced by Primary Gamma Rays and by Nuclei," Proc. 19th Int. Cosmic Ray Conf., La Jolla, 3, 445 (1985).

15. Browning, R., and Turver, K. E., "Cherenkov Radiation from Computer Simulations of the γ-Ray Initiated Atmosphere Showers," Nuovo Cimento, 38A, 223 (1977).
16. Hillas, A. M., and Patterson, J. R., "Characteristics of Cherenkov Images produced by Primary Photons and Hadrons and by Local Muons," Proc. NATO Workshop of Very High Energy Gamma-Ray Astronomy, ed. K. E. Turver (Reidel, Dordrecht) 1986, pp. 243-248.
17. Zyskin, Yu. L. et al., "The Investigation of Cherenkov Flashes in Ultraviolet Spectral Region," Proc. 20th Int. Cosmic Ray Conf., Moscow, 2, 342 (1987).
18. Macomb, D. J., private communication 1988.
19. Weekes, T. C. et al., "Observation of TeV Gamma Rays from the Crab Nebula using the Atmospheric Cherenkov Imaging Technique," submitted to Astrophys. J. (1988).
20. Cawley, M. F. "A Distributed Array of Automated Cherenkov Telescopes for TeV Gamma-Ray Astronomy," Nuclear Inst. Methods A264, 64 (1988).
21. Resvanis, et al., "The Haleakala Gamma Observatory," Proc. NATO Workshop on Very High Energy Gamma-Ray Astronomy, ed. K. E. Turver (Reidel, Dordrecht) 1986, pp. 225-230.
22. Weeks, T. C., "The Third Generation VHE Gamma Ray Observatory," Proc. of the Workshop on Cosmic Ray and High Energy Gamma Ray Experiments for the Space Station Era, ed. W. V. Jones and J. P. Wefel, (LSU) 1985, pp. 497-503.
23. Akerlov, C. W., "The Mercedes-Benz Approach to γ-Ray Astronomy," Nuclear Inst. Methods A264, 74 (1988).
24. Lamb, R. C., "What TeV Astronomy Needs," Proc. of Vulcano Workshop 1988—Frontier Objects in Astrophysics and Particle Physics, in press (1988).
25. Weeks, T. C.; Lamb, R. C., and Hillas, A. M., "HERCULES—A New Instrument for TeV Astronomy," Proc. NATO Workshop on Very High Energy Gamma-Ray Astronomy, ed. K. E. Turver (Reidel, Dordrecht) 1986, pp. 235-242.
26. Bratton, C., et al. "Proposal to 'Construct the GRANDE Facility for the Study of Astrophysical Sources and High-Energy Particle Interactions," submitted to the Dept. of Energy (1988).
27. Heintze, J., et al., "Cosmic Ray Tracking—A New Approach to High Energy γ-Astronomy," Nuclear Instruments and Methods in press (1988).
28. Samorski, M., and Stamm, W., "Detection of 2×10^{15} eV to 2×10^{16} eV Gamma Rays from Cygnus X-3," Astrophys. J. 268, L17 (1983).
29. Lewis, D. A. et al., "TeV Observations of the X-ray Binary Pulsar Hercules X-1 by the Whipple Observatory Collaboration," Proc of Vulcano Workshop 1988—Frontier Objects in Astrophysics and Particle Physics, in press (1988).
30. Drees, M., and Halzen, F., "Hadron Structure of High-Energy Photons," Phys. Rev. Lett. 61, 275 (1988).
31. Drees, M., Halzen, F., and Hikasa, K., "Muons in Gamma Showers," University of Wisconsin/KEK preprint MAD/DH/424 KEK-TH-208 (1988).

UHE Gamma Ray and Particle Astronomy

J. Matthews

Harrison M. Randall Laboratory of Physics
University of Michigan, Ann Arbor, MI, 48109

Abstract

Aspects of Ultra High Energy Cosmic Ray Physics are reviewed with emphasis on topics of interest to the High Energy Physics community. The present techniques of the field are summarized and recent results on the identification of point sources of UHE gamma rays are presented. The X-ray binary systems Hercules X-1 and Cygnus X-3 have been reported as powerful gamma ray emitters but the characteristics of the extensive air showers produced by these radiations are not in accord with standard elementary particle theory. The prospects of experimentally resolving these puzzles in the near future are discussed.

Introduction

This paper will survey some of the field of Ultra High Energy (UHE) cosmic ray physics and astronomy, highlighting those parts which are of interest to high energy physicists. It is by no means intended as a comprehensive review of the subject as there are numerous excellent articles for the reader to peruse if she desires[1)–4)]; I apologize in advance to those whose work has been omitted here. The focus will be on experimental results and opportunites.

It is appropriate to consider this field's immediate future at a Workshop convened to assess High Energy Physics in the 1990 s: elementary particle physics (and its practioners) sprang from cosmic ray studies; the questions and methods of these disciplines are in many cases the same. It is an exciting time for both fields, with the shapes of definitive answers to outstanding puzzles appearing on the horizon.

The paper is in four parts: first, a brief outline of the physics questions that UHE cosmic rays address, next a discussion of the techniques of the field as illustrated by some specific experiments, then some recent results and outstanding problems, and finally an outlook for the immediate future.

1. Physics

UHE physics is usually defined to be the physics of elementary particles with energies exceeding 10^{14} eV (100 TeV). The beam is the flux of primary cosmic rays, mostly protons. The goal of these studies involves particle interactions and astrophysics, but the two are necessarily intertwined.

The last ten years have seen much progress in experimental cosmic ray physics, in part because the knowledge and techniques of accelerator-based particle physics experiments have been brought to bear. Still, the outstanding questions of the field await definitive answers: How are protons and heavier nuclei (up to Fe, at least) accelerated to energies in excess of 10^{20} eV? What is the source of such particles? How many such sources are in our own galaxy? (An succinct tour of acceleration theories and the astrophysics of neutron stars can be found in R.C. Lamb's review[3)], this volume). The experiments aimed at answering these puzzles force more immediate questions: What is the composition of the primary beam as a function of energy? Is QCD a workable model of interactions at higher and higher energies? Is there new physics up there?

These are all questions that a high energy physicist, as well as an astrophysicist, would like to know the answers. At these energies the two fields become one. But the data provided so far pose more questions than answers. For example, with very few exceptions, there have been no gamma showers detected above a few hundred GeV which have *any* characteristics expected of a gamma shower. Why?

The working high energy scientists at this conference will need to consider cosmic ray physics and its future as we contemplate the 1990 s. The UHE regime bridges the energy region of today's Tevatron and tomorrow's SSC. It offers the only view to energies in the post-SSC realm exceeding 10^{19} eV. And, apparently, no one turns off the beam.

2. Techniques

The UHE regime is principally explored through detection and measurement of extensive air showers (EAS). Other techniques will directly measure the interactions of UHE neutrinos in a detector or its environs. We consider here the basic parameters of EAS which determine experimental configurations, the data required to establish an astrophysical object as a source of UHE particles, and how specific experiments meet these needs.

The background to a detection of gammas or neutrinos is a primary beam composed of (mostly) protons following an $E^{-1.7}$ integral spectrum, steepening to $E^{-2.0}$ at about 10^{15} eV, with perhaps further structure at higher energies (Figure 1.). The arrival directions appear isotropic at earth presumably due to the bending of the protons' paths in the 3 μGauss interstellar magnetic field. When a primary proton strikes the earth's atmosphere it initiates a hadronic cascade and an electromagnetic shower (principally from secondary π^0 decay). The cascade multiplies and then declines as the energy is distributed among the secondaries; at the ground, the muon and electromagnetic sizes are parameterized by experiment and calculation[5] as

$$N_e \propto E_o^{1.1}$$

$$N_\mu \propto N_e^{.75}$$

with the electron density falling rapidly with distance from the core (see Figures 2,3,4). The particles arrive in a pancake of thickness $\lesssim 3$ m.

The size N_e at the ground begins to become large enough to measure at about 10^{12} eV (1 TeV), but to effectively measure the shower, most ground based particle detectors trigger at about 100 TeV. This electron size is roughly the same for both γ and proton induced EAS of similar primary energy. The main difference expected at ground level between γ showers and proton showers is the muon size. In proton showers muons are copiously produced from $\pi^{\pm}$ and K decay; in a "pure" electromganetic cascade, muons primarily arise from the rare ($\sigma = 100\mu b$) photoproduction events. Monte Carlo simulations universally agree that N_μ in γ showers is only a few percent of that expected in a proton shower of similar primary energy[5),6)]. EAS produced by heavier nuclei will have still more muons and this provides a means to evaluate the composition of the beam (see Figure 4).

There are three independent means of identifying a celestial source of UHE neutral radiation: (1) Pointing: A search is made for an excess number of EAS (above the essentially isotropic background) originating from a particular direction on the celestial sphere; (2) Timing: The arrival times of EAS, perhaps recorded over many years, are searched for the same periodicities measured in specific X-ray or radio emitting objects; (3) Hadron-rejection: EAS are analyzed in detail to eliminate those with characteristics expected from hadron induced interactions, such as muon content or kinematic details.

Experiments are designed to exploit one or more of these approaches. The "Pointing" criteria is enhanced by maximizing the rate at which EAS are recorded and the resolution of their reconstructed directions. Existing EAS ground arrays can be generically described as a group of 20-100 counters (e.g., scintillators) spread over the countryside to measure the electrons in the shower. The spatial extent of the array should be at least about 100 m to fully contain the event but arrays are made as large as possible to increase the EAS target area and hence the rate at which triggers (both signal and background) are recorded. Fast timing of the particle front allows good

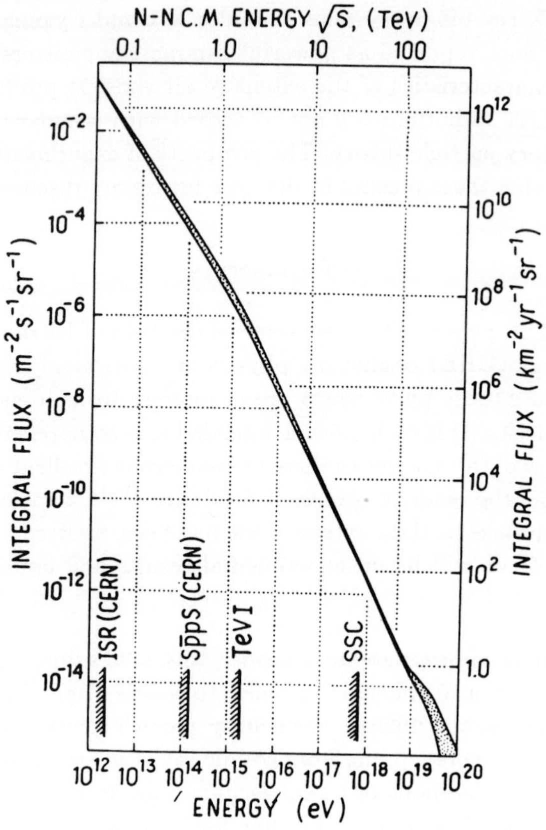

Figure 1: The primary cosmic ray spectrum at the top of the atmosphere (from reference 1). Energies of various high energy accelerators are shown for comparison.

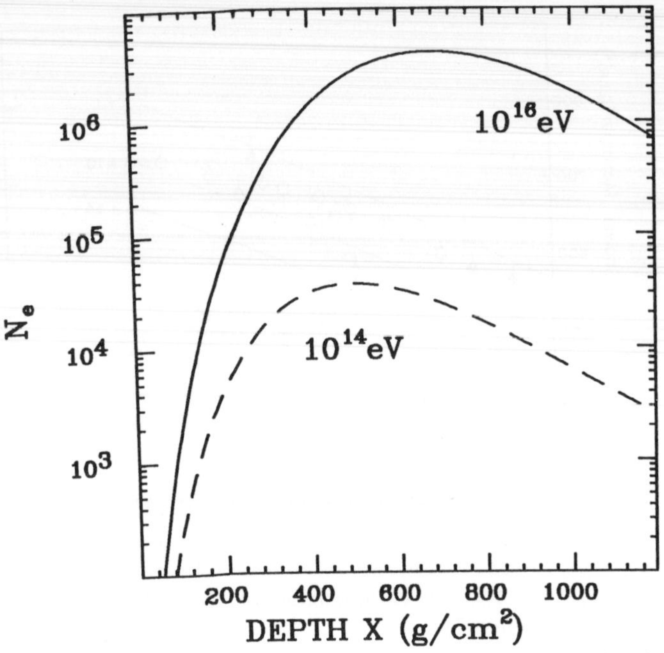

Figure 2: The number of electrons > 5 MeV versus depth in the atmosphere for extensive air showers initiated by primary protons of the indicated energies. Sea level is 1033 g/cm^2. (from simulations in reference 5)

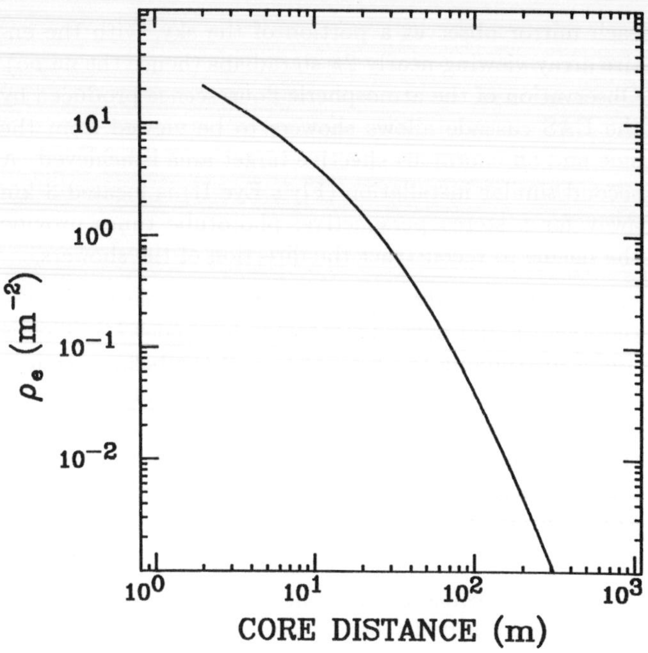

Figure 3: The density of electrons versus distance from the shower axis for showers initiated by 10^{14} eV protons, evaluated at 850 g/cm^2. (from simulations in reference 5)

directional resolution to minimize the background; large arrays also help in reconstruction by providing a long geometric lever arm. Triggering schemes which lower the energy threshold as much as possible will increase the rate, assuming the source γ spectrum is falling something like the cosmic-ray spectrum (although, as previously mentioned, the energy threshold is determined mostly by the physics of the cascade: usually 10-100 TeV).

The "Timing" aspect requires precise matching of the event times to a standard source; this is usually done with an accurate clock synchronized to Universal Time through the N.B.S. radio station WWVB, satellite transmissions, or communications signals.

"Hadron-rejection" is frequently attempted by complementing the surface array with heavily shielded counters to measure the muon component of the EAS; care must be taken to fully eliminate the punch-through of the vast number of electrons in the shower. The coupling of surface EAS arrays and deep-underground detectors addresses both hadron-rejection and pointing. The underground detectors have a very high muon threshold ($\gtrsim$1 TeV), thus sampling the earliest interactions of the cascade. The geometry of such tracks is a complementary reconstruction technique and a measure of p_t in UHE interactions. The muon multiplicity is used for composition studies[7] as well.

The Fly's Eye experiment[8] is unique among UHE projects in its method and capabilities. Sixty-seven mirrors with 1072 photomultipliers are arranged such that

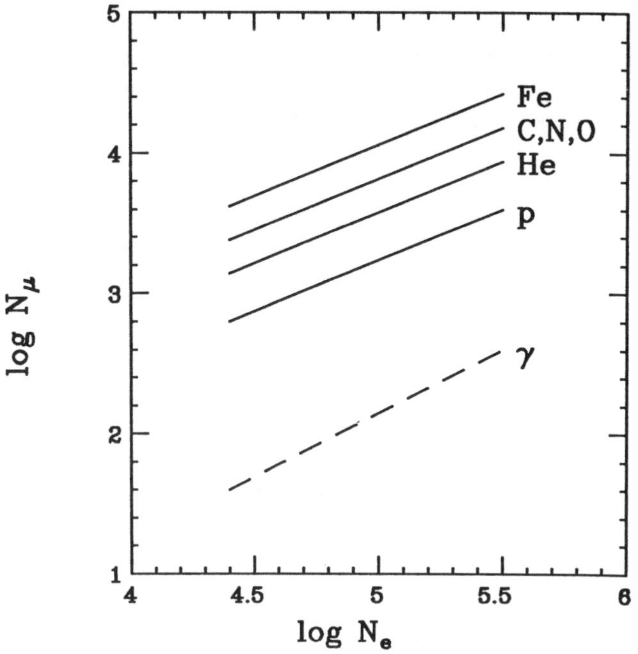

Figure 4: The muon and electron sizes at 850 g/cm^2 for showers initiated by various primaries. (Monte-Carlo simulation; adapted from reference 6)

each mirror observes a portion of the sky, with the entire array viewing nearly 2π steradians (hence the name). Observation of the atmospheric flourescence produced by the EAS cascade allows showers to be viewed from the side and an enormous effective target area is achieved. A second similar installation (Fly's Eye II) is located 3 km away for a stereo perspective; phototube times provide the means to reconstruct the direction of the showers.

This is the only experiment directly able to observe the longitudinal development of individual EAS. This is useful in studying the composition of the beam, particularly in distinguishing gamma from hadron induced EAS: at a given energy, the average location of maximum N_e in the atmosphere is believed to differ among gamma, proton, and heavier primaries (e.g., a radiation length is 37 g/cm^2 in air, while a proton interaction length is about 90 g/cm^2; see Figure 5).

The large collection area allows EAS in excess of 10^{19} eV to be obtained at reasonable rates. A secondary mode of operation detects EAS by the Cerenkov light directed along the shower axis; these cascades must land very close to the mirrors, decreasing the target area, but are available at much lower energies (10^{13} eV) and so arrive at a higher rate.

Other experiments plan to detect UHE neutrinos via their interactions in a large detector mass or through the muons produced by νN interactions in the detector's surroundings. By discriminating upward traveling muons from the background of downward ones, cosmic ray proton showers are rejected. Three underground proton-decay experiments (Baksan, IMB, and Kamiokande) have so far recorded a total of $\sim$1000 neutrinos with energies $\gtrsim$10 GeV in this way. In the near future, the technique is proposed for the UHE regime by three experiments: The MACRO experiment is under construction in the Gran Sasso Tunnel near Rome, the proposed DUMAND and GRANDE experiments would be immersed in the ocean and a large reservoir, respectively; these new projects are described in section 4 below.

3. Recent Results

Most UHE efforts in recent years have focussed on attempting to observe the binary pulsars Her X-1 and Cyg X-3. These objects are attractive candidates as sources of UHE radiation since they have been seen repeatedly at lower energies[3)]. Observations from X-rays through VHE gamma rays follow an E^{-1} spectrum[3)], suggesting a hadronic rather than electromagnetic origin. The inferred energy output in particles from Cyg X-3 is so high that it has been calculated that only a handful of such objects may be needed to account for all the UHE cosmic rays observed at earth.[9)] It is useful to review here the original observations of Cyg X-3 in UHE gamma rays since these results have been the benchmark by which most other experiments are compared.

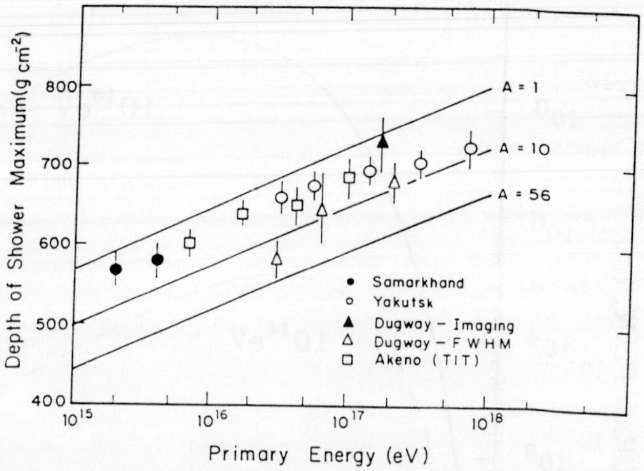

Figure 5: Measurements of the mean depth in the atmosphere of shower maximum versus shower size at the ground. Also shown are predictions based on various assumptions for the composition of the primary beam. (From reference 7)

The Kiel experiment reported the observation of excess EAS from the direction of Cyg X-3 in data taken from March 1976 to January 1980[10)] The reported signal was a 4.4σ excess over measured background (31 EAS observed when 14.4 were expected) from a $4° \times 3°$ bin centered at the celestial location of Cyg X-3. The showers analyzed were selected so as to be reconstructed with confidence: the core locations were within 30 m of the center of the array, the directions within 30° of the zenith. This sample was further cut to enhance the proportion of gamma showers to hadron showers by only accepting those with a relatively flat lateral distribution of electrons (the "age parameter" $s > 1.1$); the data showed no excess from Cyg X-3 without this cut. The signal showed a striking corelation with the 4.8 hour X-ray period: 13 of the excess 16.6 EAS had arrived in the same 0.3 - 0.4 phase bin.

This observation was confirmed by the EAS arrays at Haverah Park[11)]. An excess of EAS with primary energies exceeding 1×10^{15} eV was observed (although only at the 1.7σ level) which showed a strong timing corelation with X-ray period. In both the Kiel and Haverah Park measurements, the inferred flux of γ rays matched well to other observations of TeV and GeV radiation assuming an E^{-1} spectrum. There have been numerous positive reports and upper-limits obtained since these results; by way of example, those results reported[1)] at the 1987 Moscow International Cosmic Ray Conference are assembled in Figure 6.

The most striking feature of the Kiel observations involved the muon content of the showers. The EAS from the direction of Cyg X-3 had essentially the same muon size as the isotropic background EAS. This is impossible to explain using standard QED photon couplings. It remains the most outstanding problem in understanding UHE radiations.

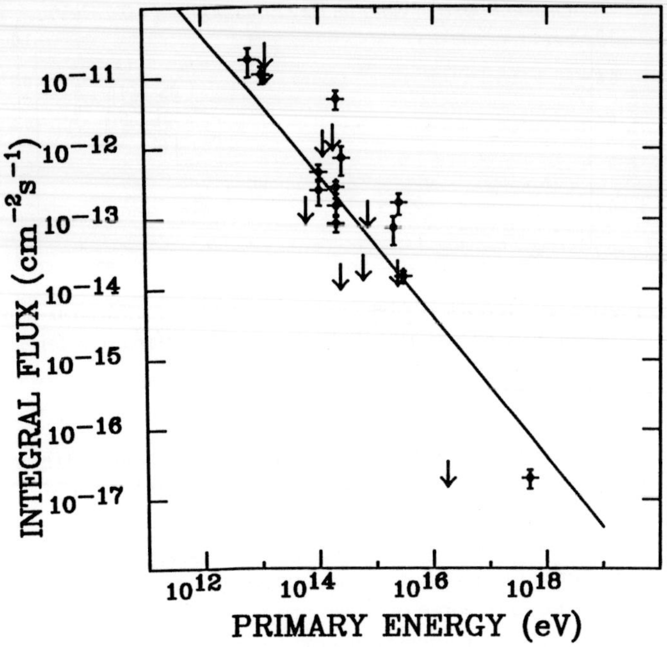

Figure 6: Limits and observations of the integral γ flux from Cygnus X-3 from reports at the 1987 ICRC, Moscow (from reference 1); also Kiel[10)], Haverah Park[11)], and recent Fly's Eye[15)] data. The solid line is an E^{-1} fit to VHE and COSB x-ray data.

This effect has been recently suggested again, in the reports of the CYGNUS experiment at Los Alamos on observations of both Cyg X-3[12)] and Her X-1[13)]. This experiment is an array of 60 scintillation counters surrounding the E225 beam dump experiment at Los Alamos. Muons in the air shower are detected by the E225 central detector in an array of flash tubes and scintillators surrounded by MWPCs. A steel and concrete shield screens out the electromagnetic components of the EAS but allows muons with energies > 2 GeV to be recorded. Overall trigger requirements produce an energy threshold of 50 TeV.

In data taken from April 1986 to May 1987, no enhancement in the counting rate of showers from the direction of Cyg X-3 compared to expected background was observed. Nor was any effect observed when the data was divided into sets with and without recorded muons. There was a mild D.C excess observed (2.1σ)for data taken during a 45 day period starting on 17 April 1986 but, again, the effect could not be enhanced when eliminating showers with recorded muons.

The group has announced a positive observation of Her X-1. On July 24, 1986, the apparatus recorded 2 separate bursts of showers reconstructing from the direction of Her X-1 with each burst lasting about thirty minutes. The overall excess amounted to collecting 17 events when only 3 were expected. Furthermore, the event times indicated a strong periodicity nearly the same as the well known X-ray period of 1.24 seconds. The likelihood of obtaining this excessive number of showers from the direction of Her X-1, with this degree of periodicity, from a chance fluctuation of the normal cosmic ray background is computed to be 2×10^{-5}.

Several features of this observation are remarkable. First, although the strong periodicity does not exactly match the known x-ray period, it does agree well with signals independently reported by two groups at lower energies (1 TeV) within a few months of this detection[3)]. It is not at all unreasonable to expect that the physical region of emission of VHE and UHE gamma rays in the binary system differs from that causing x-rays and hence might have a slightly different period. Second, the duration ($\sim$ 30 min.) of the bursts is also in good agreement with observations at lower energies (indeed down to some infra-red observations). These two facts lend strong support to this detection being a credible one. Lastly, the showers were measured to be somewhat larger in both muon and electron size than what would be anticipated from the background. This sort of N_e spectrum is consistent with an E^{-1} primary γ beam, but the muon numbers are at least an order of magnitude above what could be expected from a gamma-ray induced event.

It is difficult to reconcile the muon observations with the hypothesis of any known neutral primary. Known alternatives to photons are unlikely: due to their short lifetime, neutrons don't survive in suffficent numbers to reach the earth from Cyg X-3 (although see the recent Fly's Eye results described below) and neutrinos interact so weakly with the earth's atmosphere that the flux requirements are prohibitively high for both of these candidates. It has proven impossible to adjust the (well known) couplings of QED and the $\pi^{\pm}$ photoproduction cross section to obtain a muon content in electromagnetic showers of much more than a few percent of that in hadronic cascades. Are we observing new physics at high energies or new particles from space?

Halzen and coworkers have proposed a somewhat more conservative explanation[14)]. They describe the collision of the primary cosmic ray and an air nucleus in terms of a standard QCD where gluon-gluon scattering becomes dominant at high energies. Citing UA1 mini-jet measurements from the CERN $p\bar{p}$ collider, this scenario suggests a rapid increase in the g-g scattering contribution to the inelastic cross section. The proton becomes a "gluon bomb" at an effective threshold $\sqrt{s} \sim 1$ TeV. The photon has a gluon content, too, and so should similarly exhibit a sudden rise in its inelastic (i.e. $\pi^{\pm}$ photoproduction) cross section. Since photons have no interacting valence quarks, this threshold, ($E_\gamma \sim 500$ TeV in the laboratory) is marked by a dramatic increase in the muon content of photon induced showers, perhaps becoming comparable to hadron induced interactions at some higher energy. It has proven difficult to predict these levels and thresholds with great precision.

The Fly's Eye has recently obtained a positive signal from Cyg X-3 at the highest energy yet observed[15]. The group has analyzed 5609 reconstructed EAS with energies $> 5 \times 10^{17}$ eV, taken from November 1981 through May 1988. Although the data exhibits an excess number of showers from a narrow angular bin centered on the direction Cyg X-3, the experimenters have used a technique which accounts for uncertainties in energy and direction of the reconstructed showers and yields a continuous shower density function across the celestial sphere; this is clearly superior to methods which rely on an a priori selection of angular bins. The distribution expected from an isotropic flux of primaries is estimated from the data by randomly exchanging the arrival times of the events and computing the density function over the sky with these "simulation data sets".

The shower density from the direction of Cyg X-3 exceeds the expected background with a significance of about 4σ; the authors estimate the absolute signal size from the source to be 25±6 showers above 5×10^{17} eV. The likelihood of an isotropic background conspiring to fluctuate upward at this level is quoted as 6.5×10^{-4}. Given the exposure of the apparatus, the signal intensity is computed to be $(2.0 \pm 0.6) \times 10^{-17}$ particles cm^{-2} s^{-1}.

This is by far the most energetic signal yet seen from Cyg X-3 (see Figure 6). Indeed, at these energies the survival probability of neutrons becomes significant (10% at $E_o = 5 \times 10^{17}$ eV). It is likely that the hadronic processes which produce photons should also produce neutrons so that any source of UHE gamma rays would also be a source of UHE neutrons.

The detections of UHE radiation by the CYGNUS and Fly's Eye groups seem quite solid. However, many results in the field must be considered marginal at best; the signals are barely above detectability in most cases and no two experiments have observed a burst at the same time. Cuts, such as that made on the age parameter by Kiel, have been examined more carefully by simulation and their value is disputed. Better ephemerides of the orbital elements have been obtained from X-ray and radio astronomy and can weaken some reported UHE observational significance. Still, the multitude of positive observations is compelling and reasonably consistent in the observed spectrum and phase distribution (see Figure 7). The mission is clear: to build more sensitive experiments to confirm or refute these reports.

4. The Future

As UHE cosmic ray physics looks to the near future, definitive measurements are perhaps in sight. Improved experiments, both underway and planned, address the two most serious needs of the field: better signal-to-noise for fluxes at the intensities reported so far and simultaneous, independent measurement of bursts by other groups and techniques. Sensitivities will be improved as EAS arrays

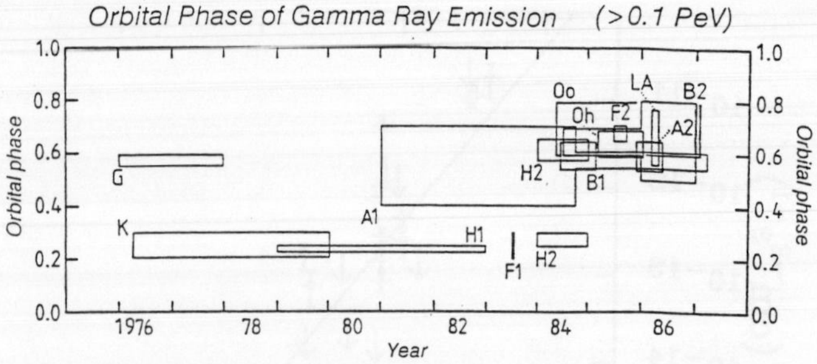

Figure 7: Reported observation of phased emission from Cygnus X-3 above 100 TeV versus time (from reference 1). The orbital period is 4.8 hours.

are increased in scale and size; Joint VHE and UHE experiments at the same facility are presently beginning on a major scale; in the U.S., three UHE experiments in close proximity to one another will provide simultaneous measurements of any short term bursts of γ rays from space. I describe here some of these projects, moving from those in some stage of operation now to those proposed for the early 1990 s.

The Utah-Michigan-Chicago (UMC) experiment[16] is an ambitious project designed to measure gamma ray induced EAS from point sources with a sensitivity far exceeding previous measurements. The approach vastly increases the scale of the basic EAS-array. This endeavor is located at the Fly's Eye site in Utah and will work in cooperation with the existing and planned detectors there.

The envisioned EAS detector is an array of 1033 scintillator stations spread uniformly over a 500m×500m grid. Fast timing will provide angular resolution of $\lesssim 0.5°$; sophisticated triggering schemes will allow EAS above 10^{14} eV to be recorded at 20 Hz. There will be an additional 2500 m^2 of scintillators spread out and buried beneath the surface array under $10'$ (24 radiation lengths) of dirt. This muon coverage far exceeds that of any other operating array and should provide a hadron rejection factor of 30-1000, depending upon energy. If gamma ray induced EAS are anomolously muon rich, the muon content will be measured carefully to test proposed mechanisms for the effect. This entire installation is located around the Fly's Eye II detector. The group has begun operating a moderate surface array (33 units over $\sim 10^4$ m^2) and 4 tracking Cerenkov telescopes (10 TeV threshold). The size, resolution, and hadron rejection abilities of planned UMC array will result in 15 γ-showers day^{-1} from Cyg X-3 (with a signal to noise factor of 33) if the source intensity is like that reported by the Kiel experiment. Put another way, a 5σ excess over background would emerge if the source intensity was only 10^{-2} the Kiel report.

The inner half of the muon array, the Cerenkov Telescopes, and the smaller surface array have been operating in conjunction with the two Fly's Eye's since late 1987; the group has reported limits on γ shower contribution to the ambient cosmic ray EAS background[17]. Fifty units of the large array are in place and scheduled to begin operating by the end of 1988. The full apparatus is planned for completion by 1991.

The Fly's Eye experiment has measured the cosmic ray spectrum up to the 10^{20} eV region and now proposes to technically upgrade the apparatus in order to better measure this energy regime and optimize its relationship with the expanding UMC facilities[18]. Specifically, the proposed High Resolution Eye will: (1) Increase the rate of collecting EAS with primary energies $> 10^{19}$eV to > 200 year^{-1} and (2) Improve the resolution of longitudinal EAS development, particularly the location of shower maximum, to < 30 g cm^{-2} . The new experiment will increase the mirror area and number of phototubes, have better timing and angular coverage of the phototubes, and expand the size of the facility. The HiRes Eye is envisioned as three distinct installations spaced on an equilateral triangle 15 km on a side, each partial eye observing π steradians. The technology of such an apparatus is well understood from the experience of the Fly's Eye I and II and the new experiment is proposed to be operational in 5 years.

When combined with the completed UMC apparatus, the Utah facility will have unparalleled coverage of the cosmic ray spectrum from 10^{13}eV to 10^{19}eV. The overlapping regions of each part of the facility will provide simultaneous cross-calibration and measurements of aspects of EAS never before attempted. To be sure, the direction of many cosmic ray experiments of the early 1990 s will be toward simultaneous observations of different parts of EAS and the presumed bursts of gamma radiation from astrophysical objects.

The Homestake EAS experiment[19] has begun operation with a surface array of 27 detectors, each a 3 m^2 liquid scintillator, located above the long running deep-mine Homestake Solar Neutrino experiment in South Dakota. The array is spread out to obtain an EAS target area of 6300 m^2 and will do UHE γ ray astronomy by looking for muon poor showers, much like the CYGNUS array. The EAS threshold is quoted as 300 TeV.

An array of proportional tubes is installed around the underground device to measure muons exceeding 2.6 TeV at the surface. These very high energy particles are of great interest as they are produced in the first few interactions of the cascade. Bundles and/or anomolously high multiplicity will separate protons from heavier nuclei (i.e. Fe) initiating showers above 10^{15} eV. This technique has been exploited with interesting results through a similar arrangement at the Kolar Gold Fields[7]. The Homestake mine has levels every 850 feet and the experimenters hope to instrument these as well to measure in detail the μ spectrum in the TeV region.

Simultaneous observation of EAS bursts from celestial sources are possible by Homestake, UMC and CYGNUS at Los Alamos (as well as the VHE installation at Mt Hopkins, Arizona) since these arrays are within a few degrees of each other in longitude. This is the first time that several major installations are operating in close geographic proximity with similar thresholds and capabilities. It may also be possible in the near future for these groups to mutually explore very short time scales within a burst of EAS with high accuracy (100 nsec); moderately priced, high precision time bases synchronized to Universal Time are becoming available through the U.S. Government's GPS satellite system. By making tight phase requirements on known neutron star spin times (msec-seconds), the three experiments could share data with reduced thresholds.

The Monopole And Cosmic Ray Observatory (MACRO)[20] is presently under construction in the new Gran Sasso facility near Rome. The project is in one of several experimental halls built by the Italian government in the mountain and will have a vertical overburden of rock equivalent to of 3600 meters of water. The rock shields the device from all but the most energetic muons in an EAS. The proposed device (Figure 8) is a layered arrangement of streamer tubes and liquid scintillator modules. UHE primary neutrinos will interact in the surrounding rock; product muons penetrate the detector and are accurately measured. Keeping only upward going muons eliminates the background EAS. The detector is arranged to optimize the area available for upward going muons (864 m^2). UHE γ showers can be detected, as in the Homestake arrangement, when the experiment operates in coincidence with an air shower array on top of the mountain. The first of the 12 "supermodules" has been successfully installed and operated, measuring muons associated with surface array EAS triggers. It will be the most sensitive UHE neutrino detector for southern hemisphere sources when it is completed in two years.

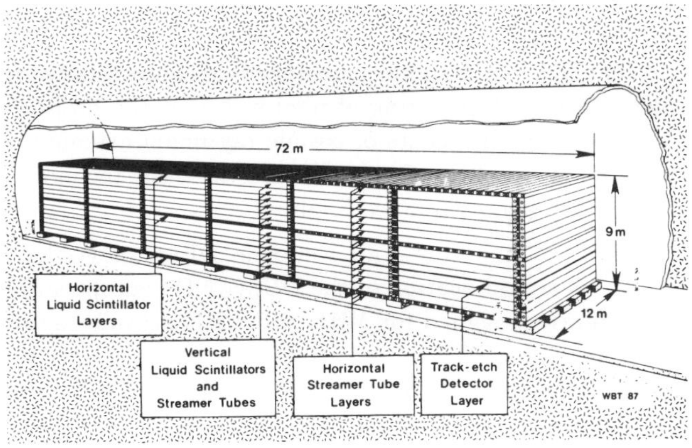

Figure 8: The MACRO detector in the Gran Sasso Laboratory.

The Deep Underwater Muon and Neutrino Detector[21] (DUMAND) is a unique and very ambitious device which proposes to detect UHE neutrinos via the Cerenkov radiation of the muons in the ocean. It would place 9 vertical strings of 24 large-bulb (16″) photomultiplier tubes 4.5 km deep in the clear ocean waters off Keahole Point on the island of Hawaii. The ocean water serves both as a target for the neutrinos as well as the Cerenkov radiator for the product muons. The device would trigger on upward going muons to eliminate the cosmic ray background; the initial stage of the array as proposed has a contained volume of 1.8×10^6 tons of water. The experimenters envision an expansion to a 1 km^3 volume. Fiber optic cables are to carry the signals from the array to a permanent shore station. Fast timing allows the reconstruction of 50 GeV muons with a resolution of $< 1°$ which, at these energies, is also the parent ν_μ direction.

The physics goals of the experiment are many; in addition to measuring and reconstructing both the muon and electron component of EAS, the high energy physicist should note that the experiment will obtain some 3500 νN interactions above 1 TeV each year. The earth is not entirely transparent to neutrinos at high energies and this attenuation with path length should become measurable above 5 TeV. The project will also do UHE γ astronomy by exploiting the kinematic differences between 30-100 TeV muons produced in γ showers from those in proton showers[22].

The DUMAND experimenters have successfully deployed a prototype string of phototubes at the proposed location. Measurements of cosmic ray muon attenuation with water depth were made[21] and the basic technology was proven successful (see Figure 9).

The Gamma Ray And Neutrino DEtector (GRANDE)[23] is a modification of the DUMAND concept and is particularly designed for VHE and UHE gamma ray astronomy and VHE neutrino astronomy. This detector will have 5 layers of photomultipliers (6515 8″ tubes), vertically separated by ~ 20 meters, immersed in purified water in an Arkansas lake. The layers are optically isolated from each other and the device acts as a calorimeter and μ/e separator as it detects the Cerenkov light from the particles in EAS which strike the water (see Figure 10). This project has a very large area (6.25×10^4 m^2), all of which is usable for μ and e detection.

GRANDE would be the largest EAS array with full sensivity over the entire apparatus for all three elements of the shower: electromagnetic, muonic, and hadronic. The expected threshold for γ showers is 10 TeV, between VHE and UHE levels. The predicted hadron-rejection factor, based on muon content, is 200-10,000 and the projected angular resolution is excellent (0.3°). A long term signal from a point source at 1×10^{-15} $cm^{-2}sec^{-1}$ (or a 1 hour burst at 1×10^{-11} $cm^{-2}sec^{-1}$) will appear as a 5σ excess (cf. Figure 6.).

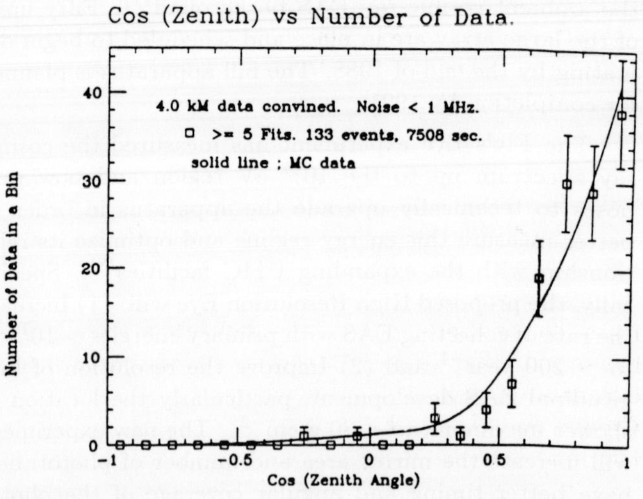

Figure 9: Measurement of muon attenuation in the ocean by the DUMAND collaboration using the prototype string (from reference 21).

Afterword

The field of UHE cosmic ray physics is poised to answer some longstanding problems. Much as in the High Energy Physics community, the efforts are directed to experiments of larger scale and loftier goals. Both fields are under some scrutiny at present: the technological complexity proposed for the next generation of experiments,

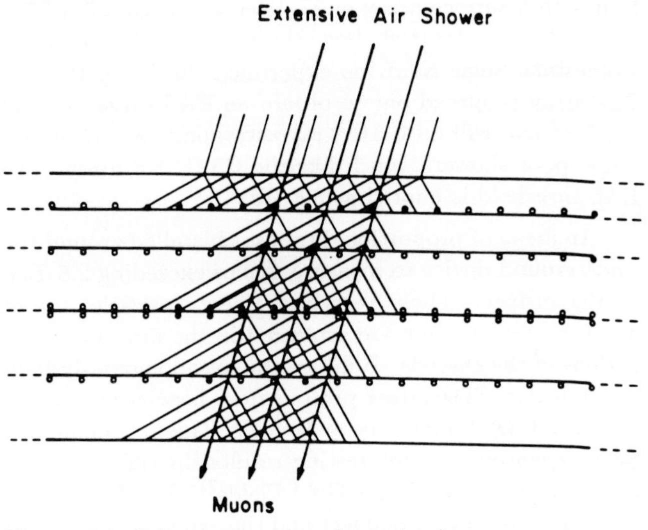

Figure 10: Schematic view of an air shower penetrating the GRANDE detector. The optically isolated layers will separate muons from electrons based on their respective ranges (from reference 23).

both in cosmic rays and in HEP, causes concern within and without our ranks. For those who may be daunted by these ambitions, a final thought-

Ah, but a man's reach should
exceed his grasp;
Or, what's a heaven for?
- Robert Browning

Aknowledgements

The author wishes to thank the organizers of Snowmass '88 for a productive and enjoyable meeting. I am indebted to the various groups who shared details of their experiments with me for this report; special thanks to J. Cronin, R.C. Lamb, K. Lande, M.V.S Rao, and particularly J.C. Van der Velde for enlightening discussions.

References

1) Protheroe, R.J., "Gamma Ray Astronomy at Energies Above 0.3 TeV", rapporteur paper in *Proc. 20th Int. Cosmic Ray Conf., Moscow, 1987*, V**8**, p.21

2) Nagle, D., T.K. Gaisser, and R.J. Protheroe, "Extensive Air Showers Associated with Discrete Astrophysical Sources", in *Annual Reviews of Nuclear and Particle Science* **38**, (Annual Reviews, Palo Alto, 1988, in press)

3) Lamb, R.C., "Very High Energy Gamma Ray Astronomy", this volume.

4) *Physics Through the 1990 s: Gravitation, Cosmology, and Cosmic Ray Physics*, pps. 111-164 (National Academy Press, Washington, 1986)

5) Fenyves, E.J., et al., "Electromagnetic Composition of $10^{14} - 10^{16}$ eV Air Showers", *Phys.Rev.* **37**,649 (1988)

6) Edwards, P.G., R.J. Protheroe, and E. Rawinski, "The Muon Component of Gamma-Ray-Initiated Air Showers", *J.Phys.G: Nucl.Phys.* **11**, L101 (1985)

7) Rao, M.V.S., "Primary Cosmic Ray Composition at High Energy", to appear in *Proc. 5th International Symp. on Very High Energy Cosmic Ray Interactions, Lodz,1988*

8) Cassiday, G.L., "Observatory for Ultra High Energy Processes: the Fly's Eye", in *Annual Reviews of Nuclear and Particle Science* **35**, pps. 321-349 (Annual Reviews, Palo Alto, 1985)

9) Hillas, A.M., "Is Cygnus X-3 a monoenergetic 10^{17}eV cosmic accelerator?", *Nature* **312**, 50 (1984)

10) Samorski, M., and W. Stamm, "Detection of 2×10^{15} to 2×10^{16} eV Gamma Rays from Cygnus X-3", *Ap.J.* **268**, L17-21, (1983); M. Samorski and W. Stamm, "Muon Densities of Extensive Air Showers Induced by $10^{15} - 10^{16}$ eV Gamma Rays", in *Proc. International Conference on Cosmic Rays, Bangalore*,V11, 244 (1983)

11) Lloyd-Evans, J., et al., "Observation of γ Rays $> 10^{15}$ eV From Cygnus X-3", *Nature* **305**, 784 (1983)

12) Dingus, B.L., et al., "Search for Signals from Cygnus X-3 at Energies Above 50 TeV", *Phys. Rcv. Lctt.* **60**, 1785 (1988)

13) Dingus, B.L., et al., "Ultra High Energy Pulsed Emission from Hercules X-1 with Anomolous Air Shower Muon Production", submitted to *Phys. Rev. Lett.* (1988)

14) Drees, M., and F. Halzen, "Hadron Structure of High-Energy Photons", *Phys. Rev. Lett.* **61**, 275 (1988)

15) Cassiday, G., et al.,"Evidence for 10^{18}eV Neutral Particles from the Direction of Cygnus X-3", submitted to *Phys. Rev. Lett.* (1988)

16) Gibbs, K.G., et al.,"The Chicago Air Shower Array (CASA)", *Nucl. Instr. and Meth.***A264**, 67 (1988); J. Chapman, et al., "Proposal to Complement the Fly's Eye with Subsurface Muon Detectors" (unpublished, University of Michigan, 1986)

17) Thornton, G., et al., "Preliminary Results of a Search for $> 10^{14}$ eV Gamma-Ray Sources", *Proc. Meeting Div. of Part. and Fields, Storrs, 1988* (to be published)

18) Baltrusaitis, R., et al., "The Fly's Eye Detector: Present and Future", *Nucl. Instr. and Meth.* **A264**, 87 (1988)

19) Lande, K., presentation at this workshop.

20) Calicchio, M., et al., "The Macro Detector at the Gran Sasso Laboratory", *Nucl. Instr. and Meth.* **A264**, 18 (1988)

21) Bosetti, P., et al., "DUMAND II: Proposal to Construct a Deep Ocean Laboratory for the Study of High Energy Neutrino Astrophysics and Particle Physics" (unpublished, University of Hawaii, 1988)

22) Berezinsky, V.S., et al., "High energy gamma astronomy with large underground detectors. DUMAND-detector of high energy gamma quanta", *Astron. Astrophys.* **189**, 306 (1988)

23) Bratton, C., "Proposal to Construct the GRANDE Facility for the Study of Astrophysical Sources and High-Energy Particle Interactions" (unpublished, University of California- Irvine, 1988)

THE COMPOSITION OF COSMIC RAYS AT HIGH ENERGIES: RECENT MEASUREMENTS AND FUTURE PROSPECTS*

Dietrich Müller

Enrico Fermi Institute and Department of Physics, The University of Chicago, Chicago, Illinois 60637

ABSTRACT

Cosmic ray particles arrive near earth with energies up to about 10^{20} eV, but direct information on their composition is still not available beyond about 10^{13} eV/particle. We briefly describe a new measurement of the elemental composition of cosmic ray nuclei that was recently performed on the Space Shuttle, and we summarize data on electrons and positrons obtained with balloons. We then discuss future possibilities to extend our knowledge of the spectra of electrons, positrons and antiprotons to about 10^{12} eV, and to determine the nuclear composition to energies approaching 10^{15} eV/particle. We argue that a number of highly significant astrophysical questions will be addressed with such measurements.

1. INTRODUCTION

One of the key questions in cosmic ray astrophysics is the identification of the sources and acceleration processes responsible for the enormous range of energies, up to 10^{20} eV with which cosmic rays arrive near earth. To that end, one needs accurate knowledge of the energy spectra and of the detailed composition of these particles *near their sources*. This is not readily available: at low energies, below about 10^{10}eV/particle, rather precise data on the composition *near earth* exist, but the extrapolation to the *sources* is strongly affected by solar modulation of the fluxes and spectra, and by variations with energy of the cross sections for nuclear interactions with the interstellar gas. At higher energies, uncertainties in the composition limit our understanding: Most measurements with balloon borne instruments were restricted to energies well below 10^{12}eV/particle, and the large range of energies beyond about 10^{14}eV/particle relies on air shower observations which do not unambiguously reveal the identity of the primary particle. Thus, a prime experimental goal remains the precise measurement of the composition and of the individual energy spectra of cosmic rays to much higher energies than previously accessible.

The range of particle species identified in the cosmic rays includes the nuclei of virtually all known elements, as well as electrons, positrons, and anti-protons. There is no evidence as yet for more exotic forms of cosmic ray matter, such as heavy anti-nuclei, monopoles, superheavy nuclei, etc. One generally distinguishes between primary cosmic rays that are accelerated at galactic sources (e.g. H, He, C, O, Fe, e^-) and secondary species produced by interaction in interstellar space (e.g. Li, Be, B, and at least part of the e^+ and $\bar{p}$ flux).

The simplest model of propagation in the galaxy assumes equilibrium between the generation of cosmic rays in sources throughout the galaxy, and their loss due to interactions or due to escape into extragalactic space. The average containment time τ is measured to be $\tau \approx 10^7$ years from the abundance of the ^{10}Be isotope at low energies[1,2], below 1 GeV/amu. As energy increases, cosmic rays seem to traverse less and less interstellar matter [3)], therefore generating a decreasing fraction of secondary nuclei through spallation in the interstellar gas. One finds that the propagation pathlength Λ, and probably with it the escape time τ, appear to decrease with energy as $E^{-\alpha}$ with $\alpha \approx 0.5$. As a consequence, the energy spectrum with which cosmic rays are accelerated must be harder than that observed near earth.

The energy requirement for sustaining the galactic cosmic ray flux is substantial, of the order of a few percent of the total energy released by all supernovae. Shock acceleration in conjunction with supernova remnants can fulfill this requirement and leads to power law spectra in rigidity, similar to those observed [4)]. However, supernova shocks cannot lead to particle energies much beyond 10^{15}eV. This may be the reason for the steepening (the "knee") observed in the overall energy spectrum at this energy. To find out how particles at still higher energies

* Work supported by NASA Contract NAS 8-32828 and Grant NSG 7464

are accelerated, remains a formidable challenge in astrophysics.

The source of cosmic rays consists of a mix of elements whose composition is not very different from the composition of solar system material. It is commonly assumed that the source composition does not change with energy, at least not up to the "knee" at 10^{15}eV. However, the experimental support for this assumption is restricted to a small range of energies below $\sim 10^{11}$ eV/particle. Certainly at very high energies, beyond 10^{15} eV/particle, one would expect changes in the measured composition as the gyration radii in the galaxy become so large that the lighter nuclei may preferentially escape.

Electrons are observed with an energy spectrum in the range 10 to 1000 GeV that is steeper (softer) than that of any nuclear cosmic ray species [5,6]. This spectral shape can be well explained as due to radiative energy losses in interstellar space if one assumes that electrons are accelerated at the same sources as nuclei, and with the same source spectrum. There is no observational support for such an assumption.

It is clear that direct measurements of the composition and of the individual energy spectra of all cosmic ray components at much higher energies than presently available are needed. There are well known obstacles to such measurements: the low flux of high energy particles, necessitating large detectors and long exposure times above the atmosphere, and the difficulties in measuring energy or momentum of highly relativistic particles with instrumentation that must be relatively compact and of light weight.

2. RECENT NEW DATA

2.1 The Nucleonic Component of Cosmic Rays

Protons and α-particles appear to be the dominant cosmic ray components. Their differential energy spectra seem to extend well into the TeV region with a power law shape $\sim E^{-2.7}$, but the data beyond $\sim$ 100 GeV/amu still exhibit large uncertainties.

The composition of heavier cosmic ray nuclei has been studied in a number of balloon experiments, in space on the HEAO-3 satellite, and, most recently, on the Space Shuttle. The questions addressed are the energy spectra of the "primary" nuclei (such as $_6$C, $_8$O, $_{14}$Si, $_{26}$Fe) which are accelerated at the galactic sources, and the relative abundances of spallation produced "secondary" cosmic rays (such as $_3$Li, $_4$Be, $_5$B, and $_{21}$Sc, $_{22}$Ti, $_{23}$V).

As the flux of high energy cosmic ray nuclei is small, successful measurements require large area detectors. Calorimeters for energy measurements [7] become excessively heavy, and "non-destructive" techniques must be favored for weight-reasons. For instance, a large instrument which was designed by our group for flight on the Space Shuttle [8] uses gas Cherenkov counters and transition radiation detectors to measure the particle energy.

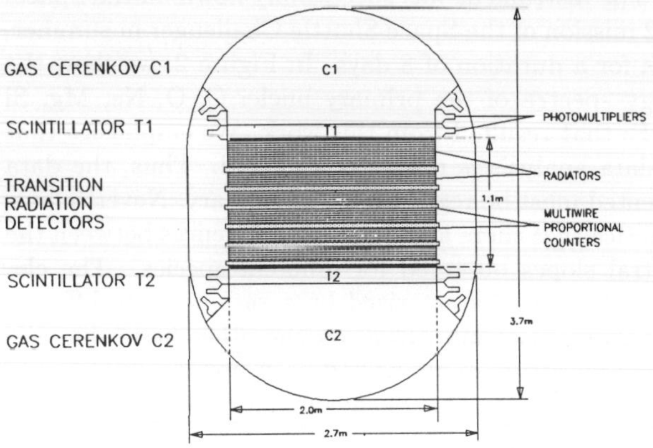

Fig. 1 Schematic cross section of the Cosmic Ray Nuclei detector of the University of Chicago (CRN).

We shall briefly describe this "Cosmic Ray Nuclei Detector" (CRN) and present some of the results from this measurement.

Figure 1 shows a schematic cross section of the instrument. The main elements are as follows:

(i) Two plastic scintillators T1, T2 serve as the main trigger and as charge measuring devices. Each particle must traverse both scintillators in the downward direction (time of flight measurement) and the charge is determined from the measurement of the pulse heights. The charge resolution for oxygen (Z=8) above 50 GeV/amu is about 0.2 charge units.

(ii) Two identical integrating gas Cherenkov counters C1, C2 are formed by the two hemispherical heads of the instrument container, above and below the scintillators. They are filled with a N_2-CO_2 mixture at atmospheric pressure, with a Cherenkov threshold at $\gamma_0=E_0/mc^2=40$ GeV/amu. Analysis of the Cherenkov signals allows energy measurements for traversing particles over the range 40-150 GeV/amu.

(iii) The center part of the instrument, between the two scintillators, is occupied by a six-layer transition radiation detector (TRD). Each layer is composed of a polyolefin fibre radiator [9], and a multiwire proportional chamber (MWPC) filled with a Xe/He/CH_4 mixture. The signal measured in the MWPC's provides a measure for the particle energy. For energies up to 500 Gev/amu, the rather weak relativistic increase in the specific ionization signal is used, and above that energy, transition radiation x-rays superposed upon the ionization signal, provide for a very strong energy response. The MWPC's also serve as hodoscopes, determining the trajectory of particles through the instrument. The energy resolution of this instrument varies with charge and energy, from a few percent to about 50% in the worst case (in the region around 200 GeV/amu where the Cherenkov counter is in saturation, but the TRD is below threshold).

The instrument was successfully flown on the Spacelab-2 mission of the Space Shuttle Challenger in summer, 1985, for a duration of 8 days. In Figure 2, we show the energy spectra of the primary nuclei C, O, Ne, Mg, Si and Fe that resulted from this work. We emphasize that the data analysis is not yet completed. Thus, the data presented must be regarded as preliminary. Nevertheless, we notice that there are apparent differences between the spectral slopes observed for different species. The element silicon is characterized by a rather steep differential energy spectrum. Iron, on the other hand, has the hardest spectrum of all, as had been previously observed at lower energies. This behavior may be contrary to the model described in the introduction, and we shall discuss this in more detail below.

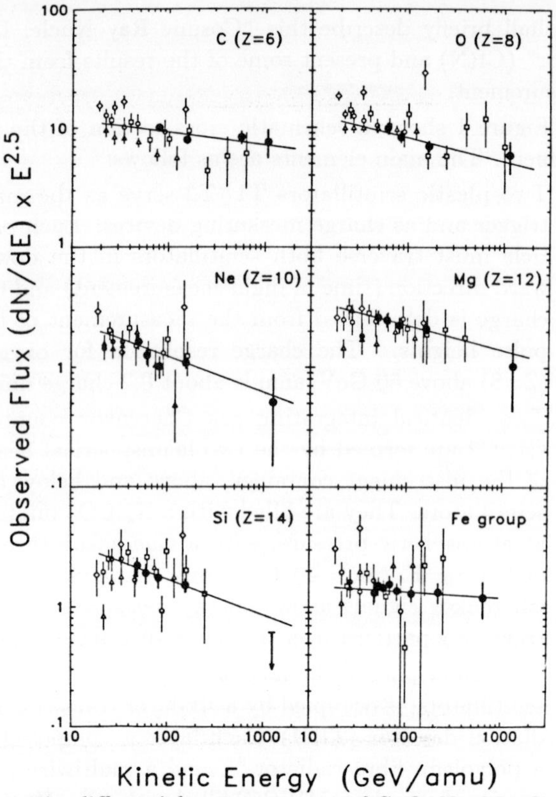

Fig. 2 The differential energy spectra of C, O, Ne, Mg, Si and Fe group (Z = 25,26,27) nuclei. Our results are shown as filled circles, and power-law fits to our data are given as solid lines. The spectra are multiplied by $E^{2.5}$ to emphasize spectral differences. Our spectra are compared, with an arbitrary normalization at 50 GeV amu^{-1}, with the results of previous measurements [*open circles*, ref. (10); *triangles*, ref. (11); *squares*, ref. (7); *diamonds*, ref. (12)].

Our experiment also determined the spectra and relative abundances of some of the nuclei that are entirely or predominantly of interstellar secondary origin, for instance boron (Z=5) and nitrogen (Z=7). These clearly indicate that the interstellar pathlength continues to decrease with energy well into the TeV regime.

With these results, we now have information on the individual spectral shapes of the major primary cosmic ray nuclei well into the TeV/amu region, [i.e., for iron to almost 10^{14} eV/particle]. To investigate whether the differences observed in the individual spectral slopes can be caused by propagation effects in interstellar space, we consider in a simple leaky box model, with equilibrium between the source flux dS/dE and the loss due to either spallation or escape. dS/dE and the observed flux dN/dE are related as

$$\frac{dS}{dE} \sim \frac{dN}{dE}\left(\frac{1}{\Lambda_{sp}} + \frac{1}{\Lambda_{esc}}\right) \quad (1)$$

where Λ_{sp} = pathlength against spallation, and Λ_{esc} = escape pathlength.

If Λ_{esc} depends on energy as $E^{-0.5}$, it will become the dominant contributor to eq. (1) at high energies (when $\Lambda_{sp} >> \Lambda_{esc}$). If dS/dE has the form of a power law $\sim E^{-\Gamma}$, then the observed spectrum dN/dE will become $\sim E^{-(\Gamma+0.5)}$. However, for iron, with a smaller spallation pathlength, this asymptotic spectral form will be reached at higher energies than for the lighter nuclei. One can show quantitatively that this would explain the difference in power law slope between iron and some of the lighter elements as shown in Figure 2. One would then expect that the observed Fe-spectrum at still higher energies, asymptotically steepens to the same slope $\sim E^{-2.7}$ as that of the lighter nuclei. However, this simple propagation model does not well account for the steep silicon spectrum. Additional data at still higher energy are required to clarify this issue.

2.2 The Electron Component

Electrons consist of negatrons and positrons. One commonly assumes that negatrons are available in the same sources that provide the nuclear cosmic rays and that they can be accelerated by the same mechanism. Alternate origins are unique to the electron component and produce both negatrons and positrons: pair production from gamma rays, secondary production from nuclear interactions via the $\pi^\pm \to \mu^\pm \to e^\pm$ decay chain, or annihilations of massive weakly interacting particles that have been proposed to account for the dark matter in the universe. An electron traveling through interstellar space is characteristically affected by electromagnetic interactions: by bremsstrahlung with interstellar matter, by synchrotron radiation with magnetic fields, and by inverse Compton scattering with photons. Through these interactions, electrons are unique probes of the interstellar medium, and the electromagnetic radiations produced are observable over a wide range of wavelengths (in particular, in the radio and in the gamma ray region).

Figure 3 shows a compilation of balloon borne measurements of electrons and positrons combined, extending to energies around 1000 GeV. Even though the uncertain-

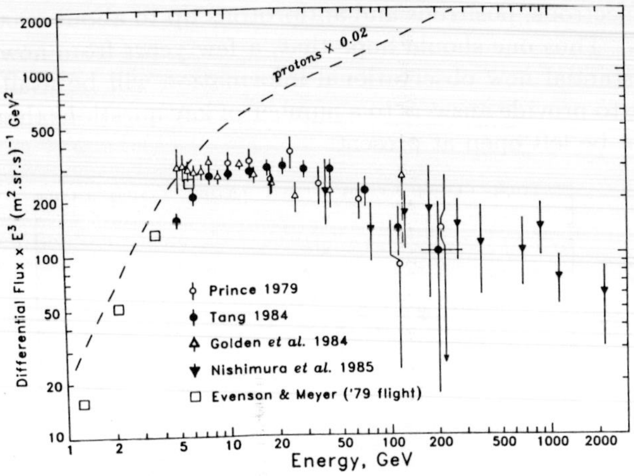

Fig. 3 A summary of recent data on the flux and energy spectrum of electrons [Prince, ref. (13); Tang, ref. (5); Golden, ref. (14); Nishimura, ref. (6); Evenson & Meyer, ref. (15)]. For comparison, the shape of the proton spectrum is indicated by a dashed line. Note that the flux is multiplied by E^3.

ties above 100 GeV are still quite large, the very steep slope distinguishes this spectrum dramatically from the spectra of all nuclear species. Separate measurements of negatrons and positrons are thus far restricted to energies below about 20 GeV (see Figure 4). They indicate a strong negatron enhancement at these energies.

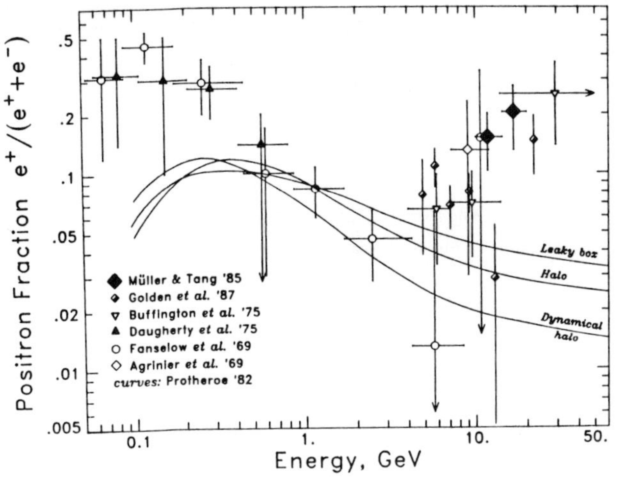

Fig. 4 A summary of measurements of the positron fraction $e^+/(e^++e^-)$, as a function of energy [Müller & Tang, ref. (16); Golden, ref. (17); Buffington, ref. (18); Daugherty, ref. (19); Fanselow, ref. (20); Agrinier, ref. (21); Protheroe, ref. (22)].

The steepness of the combined energy spectrum is commonly explained as being caused by Compton and synchrotron energy losses. The energy loss rate due to these radiative interactions increases with the square of the electron energy. On the basis of the measured spectrum, one derives that the loss rate becomes comparable to that due to escape from the galaxy at energies around 30 GeV, and one then deduces a galactic containment time of about 10^7 years. This value agrees well with that deduced from measurements of the ^{10}Be abundance at much lower energies (below 1 GeV/amu)[1,2]. But the argument rests upon the assumption that negatrons and nuclei have the same energy spectrum at the acceleration site. Further, measurements on the nuclear composition of cosmic rays[3] lead one to expect the containment time to decrease with energy and thus to be lower for high energy electrons than for low energy Be-nuclei.

Several models have attempted to resolve these difficulties. But it has long been recognized that the key to the solution must be found with separate measurements of positrons and negatrons at very high energies. As Figure 4 shows, already the existing positron data are difficult to reconcile with calculations assuming a purely secondary origin of positrons [22]. Most conspicuous seems the apparent increase in the positron fraction above 10 GeV. This must be confirmed through precise measurements to much higher energies. Is the increase to continue with energy such that the e^+/e^++e^- ratio approaches $\sim$ 50% at high energy? This value would be expected if both negatrons and positrons are purely secondary (from pion decay) or are pair-produced. And if this were observed, would it mean that primary negatrons cannot reach us at high energies due to radiative losses that either occur en route or within the acceleration site? [23] Clearly, this observation would provide unique information on the character of and, perhaps, the distance to the source. Or, could the enhanced positron fraction indeed reveal the signature of annihilation of a dark matter candidate? This would be a most profound discovery and would likely have a counterpart in the flux of antiprotons and, possibly, high energy gamma rays.

3. FUTURE PROSPECTS

It is clear that further measurements on the composition of cosmic rays are needed. Such measurements must provide data with higher statistical accuracy than presently available and must extend to higher energies than could be covered until now. This is a severe experimental challenge, but several new opportunities do exist now and shall be discussed below. First, however, we summarize the outstanding scientific questions addressed in such measurements:

(a) Are all primary cosmic ray components accelerated by the same mechanism, or are there spectral differences between different components that would indicate contributions from different sources? For instance, are the relatively steep spectra of the elements Ne, Mg and Si related to a specific source mechanism that also is responsible for the observed unusual isotopic composition of these elements (namely enhanced with neutron rich isotopes) at lower energy? And can the flat slope of the iron spectrum really be explained as due to interstellar propagation effects?

(b) Will the relative abundances of spallation produced nuclei continue to decrease at energies in the TeV/-amu region and beyond? Will then the interstellar pathlength become so small (< 1 g/cm^2) that we see essentially the "bare" source composition?

(c) What is the composition at the "knee" at 10^{15} eV, and beyond? What are the sources of cosmic rays in the 10^{15}-10^{20} eV/particle region?

(d) Are all high energy cosmic rays galactic in origin? Are there indications of more exotic sources, such as cosmic strings?

(e) What are the sources of cosmic electrons? Are all positrons produced via the π-μ-e decay from interstellar interactions, or are there primary sources such as magnetic pair production near pulsars?

(f) Are all antiprotons produced by nuclear interactions in interstellar space, and is their flux consistent with that of other secondary cosmic ray nuclei? Or are there exotic sources of both, antiprotons and positrons, for instance the annihilation of supersymmetric relic particles?

The first step towards answering some of these questions should be a long-duration space flight of the CRN instrument described in section II. The present data are limited entirely by counting statistics corresponding to a flight in space of just a few days duration. The dynamic range and energy response of the detector would permit extending the measurements by at least another decade in energy, beyond 10 TeV/amu (close to 10^{15} eV/particle for iron). In order to obtain statistically meaningful data, a space flight of about one-year duration would be required. Such a flight has to await the availability of a suitable spacecraft. The instrument could easily be flown as a payload attached to the planned Space Station, but a much simpler space platform would suffice as well. Figure 5 illustrates the range of energies that could be covered in such a flight.

While this long duration flight will address many of the questions summarized above, it will not yield data beyond the "knee". This extremely interesting energy regime, where very little is known about possible acceleration processes, must remain the domain of ground based air shower installations, as space-borne equipment would have to be unreasonably large and heavy. Several new and rather sophisticated air shower installations are now in the planning stage or under construction; thus one may hope for significant progress in this area. It is important to realize, however, that the direct measurements of CRN would provide an important calibration at the region of overlap with air shower detectors (10^{14}-10^{15} eV/particle).

Further observations of electrons, positrons, and antiprotons require the use of a magnet spectrometer. The development of such a spectrometer ("Astromag") as an attached payload to the Space Station is presently under consideration at NASA [25]. Among a number of other scientific objectives, Astromag could provide precise data on electrons, positrons and antiprotons up to about 1000 GeV. Thus one should hope that, a few years from now, substantial new observational information will be available to provide answers to a number of key questions that must be left open at present.

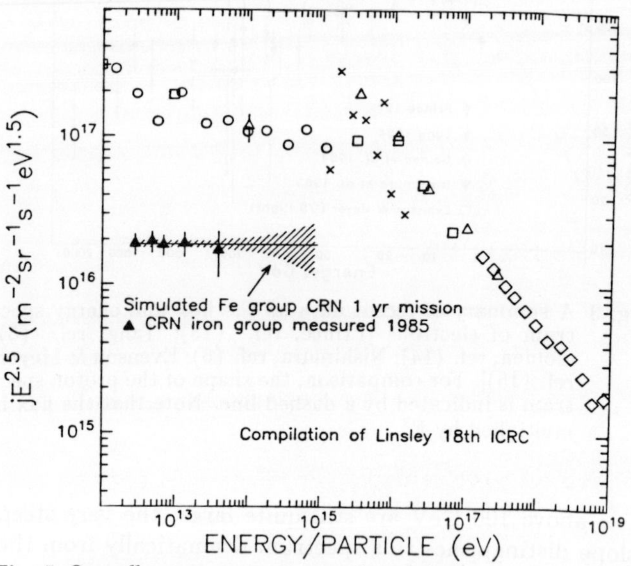

Fig. 5 Overall energy spectrum of cosmic rays, compiled by Linsley, ref. (24). The data points are measured values for iron (from CRN), and the shaded area indicates to which energies the data could be continued if CRN were flown for about one year.

REFERENCES

1. Garcia-Munoz, M., Mason, G.M., Simpson, J.A., Astrophys. J. 217, 859 (1977).
2. Wiedenbeck, M.E., Greiner, D.E., Astrophys. J. L239, 139 (1980).
3. Juliusson, E., Meyer, P., Müller, D., Phys. Rev. Lett. 29, 445 (1972).
4. Axford, W.I., *20th Int. Conf. Cosmic Rays*, 8, 120 (Moscow 1987).
5. Tang, K.K., Astrophys. J. 278, 881 (1984).
6. Nishimura, J., Fujii, M., Taira, T., Aizu, E., Nomura, Y., Kobayashi, T., Niu, K., Nishio, A., Golden, R.L., Koss, T.A., Lord, J.J., Wilkes, R.J., *17th Int. Conf. Cosmic Rays*, 2, 94 (Paris 1981).
7. Simon, M., Spiegelhauer, H., Schmidt, W.K., Spiegelhauer, Siohan, F., Ormes, J.F., Balasubrahmanyan, V.K., Arens, J.F., Astrophys. J. 239, 712 (1980).
8. Grunsfeld, J.M., L'Heureux, J., Meyer, P., Müller, D., Swordy, S.P., Astrophys. J. 327, L31 (1988).
9. Swordy, S.P., L'Heureux, J., Müller, D., Meyer, P. Nucl. Instr. Meth. 193, 591 (1982).
10. Caldwell, J.H., Astrophys. J. 218, 269 (1977).
11. Juliusson, E., Astrophys. J. 191, 331 (1974).
12. Orth, C.D., Buffington, A., Smoot, G.F., Mast, T.S. Astrophys. J. 226, 1147 (1978).
13. Prince, T.A., Astrophys. J. 227, 676 (1979).
14. Golden, R.L., Mauger, S., Badhwar, G.D., Daniel, R.R., Lacy, J.L., Stephens, S.A., Zipse, J.E., Astrophys. J. 287, 622, 1984.
15. Evenson P., and Meyer, P., J. Geophys. Res. 89, 2647 (1984).
16. Müller D., and Tang, K.K., Astrophys. J. 312, 183 (1987).
17. Golden, R.L., Stephens, S.A., Mauger, B.G., Badhwar, G.D., Daniel, R.R., Horan, S., Lacy, J.L., Zipse, J.E., Astron. Astrophys. 188, 145 (1987).
18. Buffington, A., Orth, C.D., Smoot, G.F., Astrophys. J. 199, 699 (1975).
19. Daugherty, J.K., Hartman, R.C., Schmidt, P.J., Astrophys. J. 198, 493 (1975).
20. Fanselow, J.L., Hartman, R.C., Hildebrand, R.H.. Meyer, P., Astrophys. J. 158, 771 (1969).
21. Agrinier B., et al., Lett. Nuovo Cimento, Ser. 1, 1, 53 (1969).
22. Protheroe, R.J., Astrophys. J. 254, 391 (1982).
23. Tang, K.K., Müller, D., *20th Int. Conf. Cosmic Rays*, 2, 88 (Moscow 1987).
24. Linsley, J., 18th Int. Conf. Cosmic Rays 12, 135 (Bangalore 1983).
25. Astromag, Report of the Definition Team, edited by Ormes, J.F., Israel, M., Wiedenbeck, M., Mewaldt R.; Washington 1988 (NASA).

A LONG BASE LINE- MULTI AIR SHOWER ARRAY SYSTEM IN THE WESTERN UNITED STATES TO STUDY GAMMA RAY BURSTS FROM ACTIVE STELLAR SYSTEMS

Kenneth Lande
Department of Physics
University of Pennsylvania
Philadelphia, PA 19104

I) GAMMA RAY LIKE BURSTS FROM HER X-1

Gamma rays serve as a powerful probe of very high energy processes in active stellar systems. The neutrality of gamma rays insures that they are not deflected by Galactic magnetic fields, and their zero mass insures that they preserve the time structure of their production. Very high energy gamma rays, above 10^{12} - 10^{14} eV, permit us to study high energy physics phenomena in active stellar systems in energy domains above those presently available in terrestrial accelerators. A number of such objects, Cygnus X-3, Hercules X-1 and the Crab Nebula, have been studied in recent years. The main detectors used in these studies are air Cerenkov telescopes (10^{12} eV threshold) and extensive air shower arrays (10^{14} eV threshold).

Recent observations of Her X-1 suggest that there are bursts of gamma ray-like particles with durations of 10 - 20 minutes. One such burst was detected by the Haleakala Cerenkov telescope on May 13, 1986 (15 minute duration)[1], another by the Whipple Cerenkov Telescope on June 11, 1986 (25 minutes duration)[2] and two by the CYGNUS air shower array on July 24, 1986 (28 and 21 minutes duration) [3]. The signals in all of these observations exhibited a period of about 1.236 seconds, slightly lower that the period of X-ray emissions, 1.2378 seconds.

Even more impressive than the observation of bursts is the anamalous character of these detections. The Whipple group uses eccentricity of the image of the Cerenkov light emission pattern to distinguish nucleon induced showers from those produced by gamma rays. The gamma ray produced images tend to be long and narrow, reflecting the small transverse momentum involved in these showers. Such patterns are seen in the signals from the Crab Nebula. The larger transverse momenta characteristic of nucleon induced showers produce a wide, random Cerenkov image. The Her X-1 signals exhibit a nucleon-like rather than gamma ray like pattern.[4]

The CYGNUS detector samples the number of low energy muons associated with the air shower. For gamma rays, no muons are expected. The signals from the direction of Her X-1 have an average of about six muons, consistent with nucleon induced showers and at variance with gamma ray showers.

If these observations are confirmed, we are faced with two possibilities, either very high energy gamma rays interact strongly or Her X-1 is emitting a strongly interacting, neutral, low mass particle. Either of these scenarios would be very exciting and a major addition to our picture of elementary particles. It is thus very desirable to verify these observations and to search for other stellar objects with such emission characteristics.

The fact that there were two observations in one year by Cerenkov telescopes (5% duty factor, nighttime observations when there is no moon and a clear sky), indicates that such Her X-1 bursts may occur about 20 times per year for a given detector. The CYGNUS observation, one detection per 200 Her X-1 pulses, indicates that the signal intensity is close to the noise level. It also indicates that the signal is tightly phase bunched with a pulse width less than 1/10 of the pulse period.

II) LONG BASE LINE MULTI-AIR SHOWER ARRAY SYSTEM

The use of multi-air shower array coincidences may permit detection of these gamma bursts at lower signal thresholds and at improved signal to noise ratios The short duration of the bursts, 20 minutes, requires that the air shower arrays in the system be within a few degrees longitude of each other so that they can see the source simultaneously. Fortunately, there are three arrays in the Western United States, Fig. 1, that meet this condition, CYGNUS (35.9° N.Lat., 106.3° W.Long.), Dugway 40.2° N. Lat., 113.1° W.Long.) and Homestake (44.4° N. Lat., 103.7° W. Long.). They are located at the corners of an equilateral triangle with sides of about 1000 km. The longitude of the Whipple Cerenkov Telescope at Mt. Hopkins is also within this grouping. The sources of interest, Her X-1 (dec. +35) and Cyg X-3 (dec. +41) pass right over this triangle.

With a detection probability of only one event per 200 Her X-1 pulses it is highly unlikely that several detectors will see the same pulse. However, the tight phase bunching and precisely determined periodicity should permit us to "assemble" bursts where each of the air shower arrays sees different pulses in the burst. The only requirements are: (1) the three arrays are operating at the same time, (2) their clocks are correlated to within 1 ms. of each other, and (3) they have similar thresholds.

The probability that a second array see a background air shower from within 2° of the source during the phase time of a burst seen by another array is less than 10^{-1}. Requiring that the two other arrays see a total of three showers during a burst detected by the third array thus has a background probability of less than 10^{-3}.

It might even be possible to "assemble" bursts from two pulses observed at each of the three arrays. The probability of "assembling" such a burst from background is about 10^{-5} per day. If Her X-1 actually emits 20 bursts per year while above a given detector site, the signal/background = 5×10^{3} for such "assembled" bursts. This technique can also be used to search for bursts from other sources.

An additional benefit of this multi-array approach is that each of the three arrays has a muon detection capability. Those at CYGNUS and Dugway are sensitive to low energy muons (10 GeV), while that at Homestake has a high energy threshhold (3 TeV). Thus, the combined system can explore the energy distribution of the produced muons and study the production characteristics of these muons.

There are no major additional costs involved in the operation of this multi-array system since two of the detectors (CYGNUS and Homestake) are already operating and the third (Dugway) is already funded and is now under construction.

(1) Resvanis, L. et. al., 1987 in "Very High Energy Gamma Ray Astronomy, ed. E. E. Turver (Dordrecht: Reidel) p.135

(2) Lamb, R. C. et. al., 20th Int'l Cosmic Ray Conf., Moscow, 1, 244 (1987)

(3) Dingus, B.L. et. al. Phys. Rev. Lett. 1988, to be published

(4) Hillas & Lamb, 20th Int'l Cosmic Ray Conf., Moscow, 2, 360 (1987)

WESTERN UNITED STATES LONG BASE LINE MULTI-ARRAY AIR SHOWER DETECTOR SYSTEM.

ON TRACKING OF UHE SHOWER SECONDARIES

D. Berley and C. Y. Chang
Department of Physics and Astronomy
University of Maryland
College Park, MD 20742

ABSTRACT

Using the environment of the CYGNUS experiment at LANL as an example, we have studied the feasibility of measuring with high angular resolution the direction of ultra high energy cosmic rays. The direction of the charged shower secondaries is measured with tracking chambers located in an operating EAS array at high altitude.

Powerful stellar point sources of X-rays and gigantic radio bursts have been clearly and unambiguously identified in the past two decades. Recently, searches for ultra-high energy (UHE) gamma ray signals from these sources have been made with atmospheric Cherenkov techniques and with extensive air shower (EAS) arrays covering the energy range from a few GeV to tens of PeV. Interesting observations of emissions from Hercules X-1, Cygnus X-3, CRAB, and CRAB pulsar for example, correlated with periodic intensity modulations observed in their low energy X-rays emissions, and radio frequency bands, have been reported [1]. However, it is important to note that the high energy observations are episodic, the signals are sporadic and intermittent. No persistent, or so-called "standard candle", has ever been identified for any of those sources. The signals are of weak statistical significance and are often confused by radiation not associated with the sources. Thus, improving the angular resolution of the detector, the ratio of signal to noise for the source identification can be enhanced. Angular resolution is certainly one of the most important facts to be considered in designing a new generation experiment.

Another important issue concerning some of these observations is the muon content of the extensive air showers. To maintain its direction and continue to point to the source after traversing the galactic magnetic fields the observed emissions must be neutral. The only known neutral particle which can survive the distance between the source and the earth without decaying, and can initiate an air shower, is the photon. The observed air showers are anomalous when compared with expectations from conventional gamma ray showers [2,3]. It is therefore desirable to have a powerful muon identifier associated with the EAS array. The muon identifier must be able to identify the muon and count their number with high accuracy and certainty.

It is believed that the angular resolution of the primary cosmic ray direction determined via conventional arrays of scintillation counters, depends on the dispersion in time of the light collection system, the time resolution of the photomultipliers and the fluctuations in the time of arrival of the first shower particle striking the scintillators. The technique is intrinsically limited by the finite size of the EAS and the time spread of the shower front. The best resolution obtained to date with an array of scintillators is about 0.8°. Recently, Poirier, et al. suggested that it is possible to find the primary direction with an improved precision of one-quarter degree by measuring the directions of 100 charged secondaries of the EAS with an array of proportional wire chamber tracking detectors [4]. Systematic effects could be corrected with Monte Carlo simulations of the showers. Poirier, et al. also suggested that the same tracking detectors can be used for muon identification and muon counting when a 2.5 radiation length converter is inserted between two of the chamber layers at the bottom part of the module. Such an idea is certainly attractive. The tracking method experiment can be tested inside a currently operating EAS array at high altitude, for instance the CYGNUS experiment at Los Alamos National Laboratory (LANL) [3], where the properties of an extensive air shower can be determined independently by using information extracted from the array of scintillators independent of the tracking chambers. To be more specific:

1. The primary shower axis, $\vec{A}_{TOA}$, is reconstructed with data from the existing scintillator array. The time-of-arrival method is used. The angular distribution of the observed tracks in the tracking detector is compared with Monte Carlo simulations.

2. The primary shower axix $\vec{A}_{TOA}$ is a guide for the selection of the leading electrons (those close in direction to the shower axis). The leading electrons measured by the tracking chambers are used in reconstructing a better approximation to the shower axis, $\vec{A}$, - as discussed below.

3. Los Alamos is at moderately high altitude and has the advantage of higher event rate than possible lower altitude sites, say at sea level.

A currently operating EAS array, such as the CYGNUS experiment at LANL, can provide an ideal laboratory for testing the Poirier idea. The scintillator technique has been used for many years and the information from it is understood as are its limitations. Comparison of a new technique with an old one would be invaluable. The chambers, after the test, can be installed inside some existing shelter (e.g., the E645 tunnel for the CYGNUS experiment), and used for muon tracking, which provides additional confidence to clarify the very important and fundamental issue concerning the muon content for the UHE emissions observed in cosmic rays.

From our experience of having built 6000 eight cell plastic streamer tube chambers for the OPAL Hadron Calorimeter at LEP, CERN, we feel that streamer tube chambers are reliable and relatively low cost for devices of large volume. Their properties have been extensively studied with cosmic rays and also with test beams. The technique is conventional and well understood. So let us assume that a prototype tracking detector would be built of streamer tubes inside the EAS scintillator array of the CYGNUS experiment at LANL. As shown in Figure 1-a, the detector would consist of four modular detectors, each covering an area of 9 square meters in the horizontal plane. The four chamber modules forming a face centered equilateral triangle of sides ranging from 25 to 50 meters, are to be optimized to the shower sizes of the observed events. As shown in Figure 1-b, each module would consist of six vertically-stacked layers of the standard plastic "Iarocci" chambers separated by 40 cm between adjacent layers. The orientation of the chambers would be arranged in such a way that the anode wires would run along the North-South direction, which allows one to measure the right ascention with the chamber's higher spatial resolution in that direction. The chambers would be operated in the standard limited streamer mode and triggered by the master control signal of the existing scintillator array.

From the tracking information and the time of arrival direction, $\vec{A}_{TOA}$, determined from the scintillator array, a better approximation of the primary shower axis, $\vec{A}$, can be reconstructed. Let $\hat{U}i$ be the average direction of the selected leading electrons in the local reference frame of the ith module, as shown in Figure 1-a. We can form a shower plane containing three vectors: $\vec{A}_{TOA}$, $\hat{U}i$, and $\hat{R}i$, where $\hat{R}i$ is the unit vector along the direction from the center of the ith detector, normal to the measured vector of the

Fig. 1-a

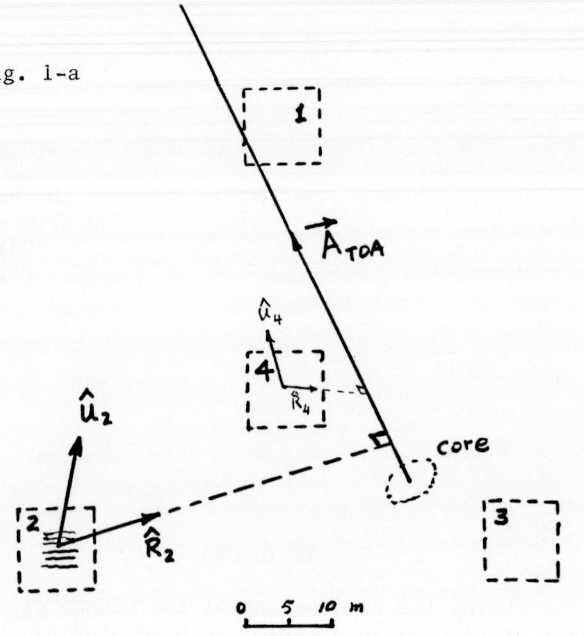

Fig. 1-b

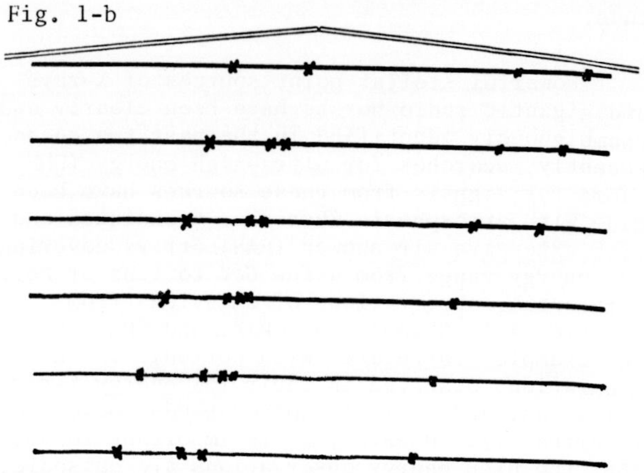

Figure 1-a. The proposed prototype shower-tracking detectors. The four modular detectors form a face centered equilateral triangle. The separation between the modules may vary from 25 to 50 meters, optimized to the shower sizes of the observed events in the current CYGNUS experiment at LANL. A shower plane can be formed from: $\hat{U}i \times \hat{R}i$, where $\hat{U}i$ and $\hat{R}i$ are defined in the text.

Figure 1-b. The Design of the tracking detectors. A Monte Carlo Simulated event seen by one of the chamber module is also shown.

shower core, or the direction of the impact parameter from the center of the local detector to $\vec{A}_{TOA}$. The primary shower axis, $\vec{A}$, can then be reconstructed by finding the common intercept of the 4 shower planes determined by the four modular detectors. The desired shower axis, $\vec{A}_{ij}$, determined by the detector modules i and j is therefore the intercept of the two shower planes whose normals are: $(\hat{U}i \times \hat{R}i)$ and $(\hat{U}j \times \hat{R}j)$ respectively. Therefore $\vec{A}_{ij} = (\hat{U}i \times \hat{R}i) \times (\hat{U}j \times \hat{R}j)$ where $i \neq j$ 1,2,3 and 4. An estimate of the direction of the primary shower axis, $\vec{A}$, can be obtained via the following procedures:

1. Rotating the local coordinate system of the ith module with respect to the master reference frame of the EAS array, such that the new local z-axis is along the direction of $\vec{A}_{TOA}$, and the new x-axis is along the impact parameter vector, $\hat{R}i$.
2. Finding $< \theta_x >$ (also called $< \theta_R >$, meaning along the $\hat{R}i$ direction) and $< \theta_y >$ (also called $< \theta_T >$, meaning transverse to the shower plane) of the selected leading electrons in the new local coordinate system, where θ is the angle between $\hat{U}i$ and the new z-axis. $\theta_{x,y}$ are the corresponding projections of θ on the x-z and y-z planes respectively in the new reference frame of the local system.
3. Minimizing the $< \theta_y >$ by further small rotations of the local coordinate system. This should be a good assumption because the shower axis, $\vec{A}$, is approximately perpendicular to the x-y plane in the new local reference frame. The shower is, on average, symmetric with resepct to its primary axis, $\vec{A}$..
4. Rotating back to the master reference frame of the EAS array.
5. Taking the average of the $\vec{A}_{ij}$ by averaging over all six possible combinations of the four shower planes (weighted by their corresponding number of observed tracks). As shown in Figure 2, the center of the circle formed by the intercepts projected on the horizontal plane will be the "best" core center determined by the selected shower secondaries measured in the four modules.

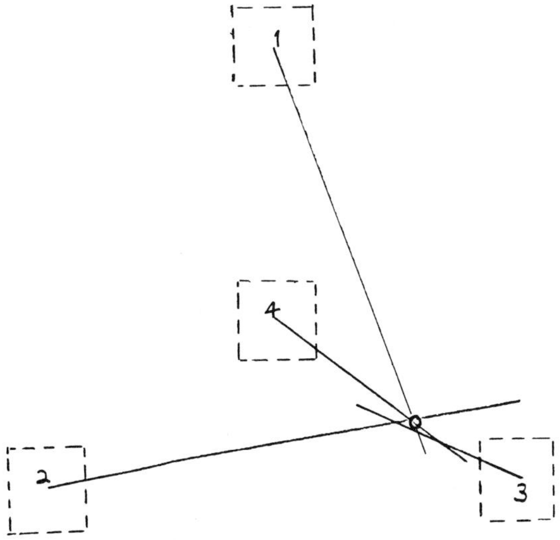

Figure 2. The center of the circle formed by the intercepts projected on the horizontal plane of the CYGNUS EAS array gives the "best" determined center of the EAS core.

More generally an estimate of the direction of the primary can be estimated by minimizing the transverse components of all the tracks simultaneously with respect to the shower plane for that track, viz. $\cos \theta_T = (\hat{u}i \times \hat{r}i) \cdot \vec{A}$. The symbol indicates unit vector. The minimization is obtained by varying the direction of $\vec{A}$ and its intercept coordinates with the ground plane. The use of $\vec{A}_{TOA}$ as the starting point in the minimumization procedure is expected to speed convergence over other a-priori choices.

As a result, the primary shower axis, $\vec{A}$, specified by four parameters: α, β, x_s, and y_s (its direction and core location) are determined. The statistical significance of this fit can be found by the following analysis:

As shown in Figure 3, the frequency of triggered events being registered in the LANL CYGNUS experiment is plotted against the number of minimum ionizing charged particles (MIP) as measured by a typical scintillation counter with an area of 0.83 square meters. The high peak at zero indicates that a typical counter

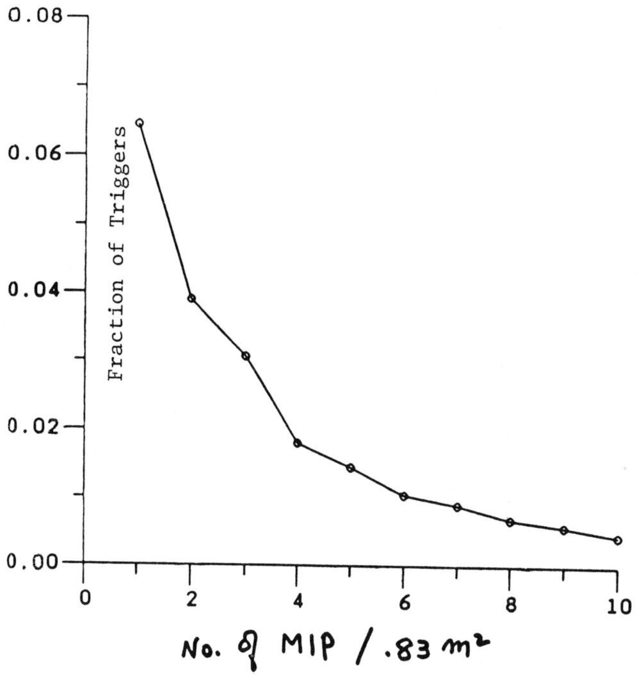

Figure 3. The frequencies of triggered events being registered in the LANL CYGNUS experiment as a function of the number of minimum ionizing charged particles (MIP) measured by a typical scintillation counter with an area of 0.83 square meter.

will be missed by MIPs most of the time. For 6% of the cases, a counter may register one hit per event, and only in 1% of the time, it may be hit by more than 10 MIPs. The integrated spectrum gives 1.2 hits per square meter per event. Figure 4 shows the average number of

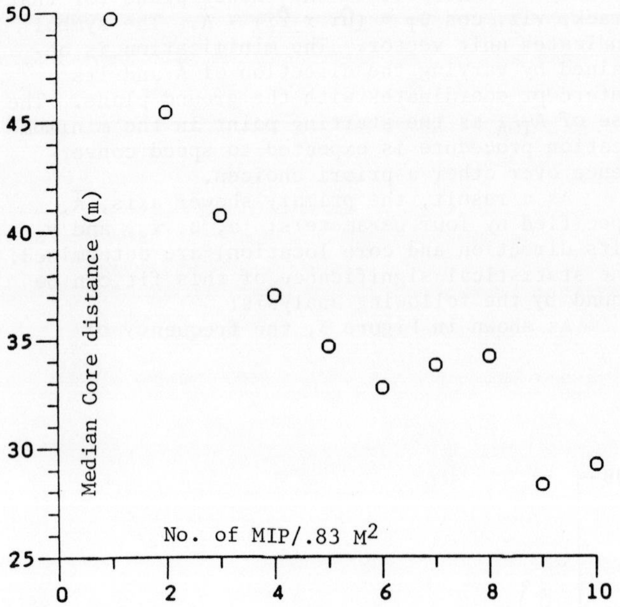

Figure 4. The averaged number of hits per counter as a function of its core distance.

hits per counter as a function of the distance to the shower core. Figure 4 indicates that essentially there are zero hits per counter when the core distance is larger than 50 meters. On the other hand, it would be more than 10 MIPs per square meter if the core distance is less than 25 meters. Thus the optimized separation between modules would be somewhere between 25 to 50 meters. Finally, from the observed muon spectrum of the CYGNUS experiment, we find that the number of muons with energy greater than 2 GeV per meter is less than 0.06. As a result, we conclude:

i) $\langle$ No. of electrons / module $\rangle$ = 11.

ii) $\langle$ Muon / Electron $\rangle$ = 0.05

Based on these numbers, one reaches the following conclusions:

1. One needs a detector of very large area for muon tracking. Therefore burying many small area scintillation counters deep under ground seems good enough for muon identification and numerical counting.
2. For electrons, the four modules covering a total of 36 square meters in the horizontal plane with 45 to 100 leading electrons can certainly provide reasonable statistics for the directional determination of the primary shower axis yet be far from flooding the detectors with too many tracks.
3. Following Poirier, we anticipate that leading electrons can be selected with rms deviation of about two degrees from the shower axis. With a sample of about 50 to 100 electrons per event, we anticipate an uncertainty of about one quarter degree. To achieve this precision requires that the tracking technique provide core location coordinates substantially better than that achieved with scintillator arrays.

It remains to demonstrate whether the technique of tracking can indeed result in superior performance compared to the technique of time of arrival. In particular several questions have occurred to us: whether leading tracks can be cleanly identified, whether events can be selected with high efficiency, whether the anticipated precision can be obtained in the environment of an actual experiment and whether a full scale experiment can be built at a reasonable cost. We believe that a prototype test in the environment of a working detector is the best way to understand the tracking technique of measuring ultra-high energy showers.

We greatly acknowledge informative discussions with our colleagues on the CYGNUS experiment and we thank Celeste Change for discussion on vector algebra.

References

1. Review on high energy gamma ray astrophysics and their earlier references can be found in: Weekes, T. C., Physics Reports 160, 1 and 2 (1988). See also the summary talk of this workshop on UHE emissions.
2. Samorski, M. and Stamm, W., Astrophys. J. V268, L17 (1983), 18th ICRC, Bangalore India, Vol. 11, page 244 (1983).
Bartelt, J., et al., Phys. Rev. D32, 1630 (1985).
Battistoni, G., et al., Phys. Lett. B155, 465 (1985).
Stanev, T., Gaisser, T. K., and Halzen, F., Phys. Rev. D32, 1244 (1985), also: Stanev, T., et al., Proc. 19th ICRC 7, 219 (La Jolla 1985), and Edwards, P. G., Protheroe, R. J. and Turver, K. E., J. Phys. G11, L10 (1985).
3. Dingus, B. L., et al., Phys. Rev. Lett. 60, 1785, 1988, and also the Herc X-1 paper, to be published in Phys. Rev. Lett.
4. Poirier, J., et al., NIM A264, 81 (1988. See also: Heintze, J., et al., Technic Report, Physikalisches Institut der Universitat Heidelberg.

PARTICLE COLD DARK MATTER

Edward W. Kolb and Michael S. Turner
NASA/Fermilab Astrophysics Center
Fermi National Accelerator Laboratory
Batavia, Illinois 60510

Abstract

Two claims are usually made in "Early Universe" papers and lectures: 1) The early Universe was to a good approximation in local thermodynamic equilibrium, and 2) The high temperatures of the early Universe makes it an ideal laboratory to study elementary particle physics. These two claims are obviously contradictory. If the Universe is in LTE, the microphysics is *irrelevant*. The interaction of particle physics and cosmology depends on *departures* from equilibrium. In this paper I discuss one such example; the decoupling and production of cold dark matter.

1. Introduction

For much of its early history, most of the constituents of the Universe were in thermal equilibrium, making an equilibrium description a good approximation. However, there have been a number of very notable departures from thermal equilibrium – neutrino decoupling, decoupling of the background radiation, primordial nucleosynthesis, and on the more spectulative side, inflation, baryogenesis, decoupling of relic WIMPs, etc. If not for such departures from thermal equilibrium, the present state of the Universe would be completely specified by the present temperature, $T_0 = 2.75$ K. The departures from equilibrium have led to important relics – the light elements, the neutrino backgrounds, a net baryon number, relic WIMPs (weakly-interacting massive particles), relic Laboratory Directors, and so on.

As is well known, once a particle species decouples from the plasma (no longer interacts) its evolution is very simple: particle number density decreasing as R^{-3} and particle momenta decreasing as R^{-1}, where $R(t)$ is the cosmic scale factor. The behavior of the particle distribution either in LTE or completely decoupled is simple. It is the evolution of particle distributions around the epoch of decoupling that is challenging. The rough criterion for a particle species to be either coupled or decoupled is the comparison of the interaction rate of a particle, Γ, and the expansion rate of the Universe, H:

$$\Gamma \gtrsim H \qquad \text{(coupled)} \tag{1}$$

$$\Gamma \lesssim H \qquad \text{(decoupled)} \tag{2}$$

where Γ is the interaction rate (per particle) for the reaction(s) which keep the species in thermal equilibrium.

In order to treat properly decoupling one must follow the microscopic evolution of the particle's phase space distribution function $f(p^\mu,\ x^\mu)$. This of course is governed by the Boltzmann equation, which can be written as

$$\hat{\mathbf{L}}[f] = \mathbf{C}[f] \tag{3}$$

where $\mathbf{C}$ is the collision operator and $\hat{\mathbf{L}}$ is the Liouville operator. The familiar non-relativistic Liouville operator for the phase space density $f(\vec{v},\ \vec{x})$ of a species of particles with mass m subject to a force $\vec{F} = d\vec{p}/dt$ is

$$\hat{\mathbf{L}} = \frac{d}{dt} + \frac{d\vec{x}}{dt}\cdot\vec{\nabla}_x + \frac{d\vec{v}}{dt}\cdot\vec{\nabla}_v. = \frac{d}{dt} + \vec{v}\cdot\vec{\nabla}_x + \frac{\vec{F}}{m}\cdot\vec{\nabla}_v \tag{4}$$

The covariant generalization of the Liouville operator is

$$\hat{\mathbf{L}} = p^\alpha\frac{\partial}{\partial x^\alpha} - \Gamma^\alpha_{\beta\gamma}p^\beta p^\gamma\frac{\partial}{\partial p^\alpha}. \tag{5}$$

Note that, as expected, the gravitation force enters the equation through the affine connection. For the FRW model Universe, the phase space density must be (spatially) homogeneous and isotropic and $f = f(|\vec{p}|,\ t)$ (or equivalently $f(E,\ t)$) only. For the Robertson-Walker metric the Liouville operator is

$$\hat{\mathbf{L}}[f(E,t)] = E\frac{\partial f}{\partial t} - \frac{\dot{R}}{R}|\vec{p}|^2\frac{\partial f}{\partial E}. \tag{6}$$

Using the definition of the number density in terms of the phase space density

$$n \equiv \frac{g}{(2\pi)^3}\int d^3p f(E), \tag{7}$$

and upon integration by parts, the Boltzmann equation can be written in the form

$$\frac{dn}{dt} + 3\frac{\dot{R}}{R}n = \frac{1}{(2\pi)^3}\int \mathbf{C}[f]\frac{d^3p}{E}. \tag{8}$$

The collision term for the process $\psi + a + b + \cdots \longrightarrow i + j + \cdots$ is

$$\begin{aligned}\frac{1}{(2\pi)^3}\int \mathbf{C}[f]\frac{d^3p_\psi}{E_\psi} &= -\int d\Pi_\psi d\Pi_a d\Pi_b \ldots d\Pi_i d\Pi_j \ldots \\ &\times(2\pi)^4\delta^4(p_\psi + p_a + p_b \cdots - p_i - p_j \cdots) \\ &\times\left[|\mathcal{M}|^2_{\psi+a+b+\ldots\to i+j+\ldots} f_a f_b..f_\psi(1\pm f_i)(1\pm f_j)..\right. \\ &-|\mathcal{M}|^2_{i+j+\ldots\to\psi+a+b+\ldots} f_i f_j \ldots (1\pm f_a) \\ &\times(1\pm f_b)\ldots(1\pm f_\psi)]\end{aligned} \tag{9}$$

where f_i, f_j, $,f_a$, f_b, ... are phase space densities of species i, j, $,a$, b, ..., f_ψ is the phase space density of ψ (the species whose evolution we are focusing on), (+) applies to bosons, (−) applies to fermions, and

$$d\Pi \equiv g\frac{1}{(2\pi)^3}\frac{d^3p}{2E} \tag{10}$$

(where g counts the internal degrees of freedom). The 4-dimensional delta function enforces energy and momentum conservation, and the matrix element squared,

$$|\mathcal{M}|^2_{i+j+\ldots\to\psi+a+b+\ldots}, \tag{11}$$

for the process $i + j + \ldots \to \psi + a + b + \ldots$, includes the appropriate symmetry factors for identical particles in the initial or final states.[1]

In the most general case, the Boltzmann equations are a coupled set of integral-partial differential equations for the phase space distributions of all the species present! Fortunately, in problems of interest all but one or two species will have equilibrium phase space distribution functions because of their rapid interactions with other species, reducing the problem to a single integral-partial differential equation for the species ψ.

There are two approximations that greatly simplifies (9). The first is to assume T (or CP) invariance, which implies

$$\begin{aligned}|\mathcal{M}|^2_{i+j+\ldots\to\psi+a+b+\ldots} &= |\mathcal{M}|^2_{\psi+a+b+\ldots\to i+j+\ldots} \\ &\equiv |\mathcal{M}|^2.\end{aligned} \tag{12}$$

The second assumption is the use of Maxwell-Boltzmann statistics for all species instead of Fermi-Dirac for fermions and Bose-Einstein for bosons. In that case, the blocking and stimulated emission factors can be ignored, $1 \pm f \simeq 1$, and $f_i(E_i) = \exp[-(E_i - \mu_i)/T]$ for all species in kinetic equilibrium.[2] With these two assumptions the Boltzmann equation may be cast in the familiar form

$$\begin{aligned}&\dot{n}_\psi + 3Hn_\psi = \\ &-\int d\Pi_\psi d\Pi_a d\Pi_b \ldots d\Pi_i d\Pi_j \ldots (2\pi)^4|\mathcal{M}|^2 \\ &\times\delta^4(p_i + p_j \cdots - p_\psi - p_a - p_b \cdots) \\ &\times[f_a f_b \ldots f_\psi - f_i f_j \ldots]\end{aligned} \tag{13}$$

where, as usual, $H \equiv \dot{R}/R$.

The significance of the individual terms is manifest: the $3Hn_\psi$ term accounts for the dilution effect of the expansion of the Universe, and the right hand side of (13) accounts for interactions which change the number of ψ's present. In the absence of interactions the solution to (13) is $n_\psi \propto R^{-3}$.

Finally, it is usually useful to scale out the effect of the expansion of the Universe by considering the evolution of the number of particles per comoving volume. This is done by using the entropy density, s, as a fiducial, and defining as the dependent variable

$$Y \equiv \frac{n_\psi}{s}. \tag{14}$$

Using the conservation of entropy per comoving volume ($sR^3 =$ constant), it follows that

$$\dot{n}_\psi + 3Hn_\psi = s\dot{Y}. \tag{15}$$

Furthermore, since the interaction term will usually depend explicitly upon temperature rather than time, it is useful to introduce as the independent variable

$$x \equiv m/T, \tag{16}$$

where m is any convenient mass scale (usually taken as the mass of the particle of interest). During the radiation-dominated epoch x and t are related by

$$t = 0.301g_*^{-1/2}\frac{m_{PL}}{T^2} = 0.301g_*^{-1/2}\frac{m_{PL}}{m^2}x^2 \tag{17}$$

so that the Boltzmann equation can be rewritten as

$$\begin{aligned}&\frac{dY}{dx} = \\ &-\frac{x}{H(m)s}\int d\Pi_\psi d\Pi_a d\Pi_b \ldots d\Pi_i d\Pi_j \ldots (2\pi^4)|\mathcal{M}|^2 \\ &\times\delta^4(p_i + p_j \cdots - p_\psi - p_a - p_b \cdots) \\ &\times[f_a f_b \cdots f_\psi - f_i f_j \cdots]\end{aligned} \tag{18}$$

where $H(m) = 1.67g_*^{1/2}m^2/m_{PL}$, and $H(x) = H(m)x^{-2}$.

We will now consider some specific applications of the formalism we have developed here to treat non-equilibrium thermodynamics.

2. Freeze Out: Making of a Relic

If a massive particle species remained in thermal equilibrium until the present, its abundance

$$n/s \sim (m/T)^{3/2}\exp(-m/T) \tag{19}$$

[1] For n identical particles of a given species in the initial or final state a factor of $1/n!$.

[2] For any non-relativistic species, in the limit that $(m_i - \mu_i)/T >> 1$ the use of Maxwell-Boltzmann statistics becomes exact.

would be absolutely negligible because of the exponential factor. If the interactions of the species freeze out (i.e., $\Gamma < H$) at a temperature such that m/T is not $\gg 1$, the species can have a significant relic abundance today. I will now calculate that relic abundance.

First, suppose that the species is stable (or very long-lived compared to the age of the Universe when its interactions freeze out). Given that it is stable, only annihilation and inverse annihilation processes, e.g.,

$$\psi\bar{\psi} \longleftrightarrow X\bar{X}, \tag{20}$$

can change the number of ψ's and $\bar{\psi}$'s.[3] Here X stands for some particle with mass less than the mass of the ψ. In addition, I assume that there is no asymmetry between ψ's and $\bar{\psi}$'s.[4]

We will also assume that all the species X, $\bar{X}$ into which ψ, $\bar{\psi}$ annihilate have thermal distributions with zero chemical potential. Because these particles will usually have additional interactions which are 'stronger' than their interactions with ψ's, the assumption of equilibrium for the X's is almost always good. For example, let $\psi,\ \bar{\psi} = \nu,\ \bar{\nu}$ and $X,\ \bar{X} = e^-,\ e^+$; while the neutrinos only have weak interactions, the $e^{\pm}$'s have weak and electromagnetic interactions.

Now consider the factor $[f_\psi f_{\bar{\psi}} - f_X f_{\bar{X}}]$ in the collision term in the Boltzmann equation. Since X, $\bar{X}$ are in thermal equilibrium (and for simplicity we assume they have zero chemical potential)

$$\begin{aligned} f_X &= \exp(-E_X/T), \\ f_{\bar{X}} &= \exp(-E_{\bar{X}}/T). \end{aligned} \tag{21}$$

The energy part of the δ-function enforces $E_\psi + E_{\bar{\psi}} = E_X + E_{\bar{X}}$, so that

$$\begin{aligned} f_X f_{\bar{X}} &= \exp[-(E_X + E_{\bar{X}})/T] \\ &= \exp[-(E_\psi + E_{\bar{\psi}})/T] = f_\psi^{\rm EQ} f_{\bar{\psi}}^{\rm EQ}, \end{aligned} \tag{22}$$

since $f_\psi^{\rm EQ} \equiv \exp(-E_\psi/T)$ and $f_{\bar{\psi}}^{\rm EQ} \equiv \exp(-E_{\bar{\psi}}/T)$. Therefore, it follows that

$$[f_\psi f_{\bar{\psi}} - f_X f_{\bar{X}}] = [f_\psi f_{\bar{\psi}} - f_\psi^{\rm EQ} f_{\bar{\psi}}^{\rm EQ}]. \tag{23}$$

Now the interaction term can be written in terms of n_ψ, the actual number density of ψ's, and $n_\psi^{\rm EQ}$, the equilibrium number density of ψ's, as

$$\frac{dn_\psi}{dt} + 3Hn_\psi = -\langle \sigma_{\psi\bar{\psi}\to X\bar{X}}|v|\rangle [n_\psi^2 - (n_\psi^{\rm EQ})^2], \tag{24}$$

or,

$$\frac{dY}{dx} = \frac{-x\langle \sigma_{\psi\bar{\psi}\to X\bar{X}}|v|\rangle s}{H(m)}(Y^2 - Y_{\rm EQ}^2), \tag{25}$$

where $Y \equiv n_\psi/s = n_{\bar{\psi}}/s$ is the *actual* number of ψ, $\bar{\psi}$'s per comoving volume, $Y_{\rm EQ} \equiv n_\psi^{\rm EQ}/s = n_{\bar{\psi}}^{\rm EQ}/s$ is the *equilibrium* number of ψ, $\bar{\psi}$'s per comoving volume, and I have assumed that the number of ψ and $\bar{\psi}$'s are equal (no particle-antiparticle asymmetry). The thermally-averaged annihilation cross section times velocity is given by[5]

$$\begin{aligned} \langle \sigma|v|\rangle \equiv & \left(n_\psi^{\rm EQ}\right)^{-2} \int d\Pi_\psi d\Pi_{\bar{\psi}} d\Pi_X d\Pi_{\bar{X}} \\ & \times (2\pi)^4 \delta^4(p_\psi + p_{\bar{\psi}} - p_X - p_{\bar{X}}) \\ & \times |\mathcal{M}|^2 \exp(-E_\psi/T)\exp(-E_{\bar{\psi}}/T). \end{aligned} \tag{26}$$

Now if we consider other annihilation channels for $\psi\bar{\psi}$, say $\psi\bar{\psi}$ to some final state F, there is an additional term in $\dot{n}_\psi$, which is simply (24) with $\langle \sigma_{\psi\bar{\psi}\to X\bar{X}}|v|\rangle$ replaced by $\langle \sigma_{\psi\bar{\psi}\to F}|v|\rangle$. Summing over all annihilation channels yields the final result in terms of the *total* annihilation cross section $\langle \sigma_A|v|\rangle$

$$\begin{aligned} \frac{dn_\psi}{dt} + 3Hn_\psi &= -\langle \sigma_A|v|\rangle [n_\psi^2 - (n_\psi^{\rm EQ})^2] \\ \frac{dY}{dx} &= \frac{-x\langle \sigma_A|v|\rangle s}{H(m)}(Y^2 - Y_{\rm EQ}^2). \end{aligned} \tag{27}$$

In the nonrelativistic ($x \gg 3$) and in the extreme relativistic ($x \ll 3$) regimes the equilibrium value of the number of ψ's per comoving volume has the following simple limiting forms,

$$Y_{EQ} = \frac{45\zeta(3)}{2\pi^4}\frac{g_{\rm eff}}{g_{*S}} = 0.278\frac{g_{\rm eff}}{g_{*S}} \quad \text{(ER)} \tag{28}$$

$$\begin{aligned} Y_{EQ} &= \frac{45}{2\pi^4}\left(\frac{\pi}{8}\right)^{1/2}\frac{g}{g_{*S}}x^{3/2}e^{-x} \\ &= 0.145\frac{g}{g_{*S}}x^{3/2}e^{-x} \quad \text{(NR)} \end{aligned} \tag{29}$$

where $g_{\rm eff} = g$ (bosons) and $g_{\rm eff} = 3g/4$ (fermions).

This very convenient form of the Boltzmann equation that I have just derived has all the 'features' that we would have expected: the destruction rate of ψ, $\bar{\psi}$'s per comoving volume is just proportional to the annihilation rate of ψ, $\bar{\psi}$'s, and the net rate of destruction is just balanced by inverse (creation) processes when $n_\psi = n_\psi^{\rm EQ}$. Moreover, creation processes are Boltzmann surpressed for $T \ll m$, because only a small fraction of $X\bar{X}$ pairs have sufficient KE to create a $\psi\bar{\psi}$ pair. In fact, we could have forgone the sleight-of-hand involving the rewriting of f_X, $f_{\bar{X}}$ in terms of $f_\psi^{\rm EQ}$, $f_{\bar{\psi}}^{\rm EQ}$, and just written this form of the Boltzmann equation by invoking the 'principle of detailed balance' –

[3] For simplicity I will first consider $2 \leftrightarrow 2$ annihilation and creation processes; it is straightforward to generalize from here.

[4] It is straightforward to generalize this to include the possiblility of an excess (or deficit) of ψ's over $\bar{\psi}$'s.

[5] In principle, there is a different average involved in the creation and the annihilation terms. The average for the annihilation term is the average for the creation term, (26), with $f_\psi f_{\bar{\psi}}$ replacing the product of exponentials.

however, at the expense of pedagogy. Whenever all species except the one of interest are in thermal equilibrium, the creation processes can be obtained from the destruction processes by substituting $n \to n_{\rm EQ}$ for the species of interest.

Remembering that $H \propto x^{-2}$, so that $H(T) = x^{-2}H(m)$, we see that (27) can be written in a very suggestive form,

$$\begin{aligned} \frac{x}{Y_{\rm EQ}}\frac{dY}{dx} &= -\frac{\Gamma_A}{H}\left[\left(\frac{Y}{Y_{\rm EQ}}\right)^2 - 1\right], \\ \Gamma_A &\equiv n_{\rm EQ}\langle\sigma_A|v|\rangle. \end{aligned} \tag{30}$$

In this form, we see that the change of ψ's per comoving volume is controlled by the 'effectiveness of annihilations', the usual Γ/H factor, times a measure of the deviation from equilibrium.. It is then clear that when Γ/H is less than order unity, the number of ψ's per comoving ceases to change, i.e., annihilations freeze out, and the number of ψ's per comoving 'freezes in.'

The Boltzmann equation for the evolution of the abundance of a species is a particular form of the Riccati equation, for which there are no general, closed-form solutions, however there are approximate solutions. The annihilation rate Γ_A varies as n_{EQ} times the thermally-averaged cross section $\langle\sigma_A|v|\rangle$. In the relativistic regime, $n_{EQ} \sim T^3$, and like other rates, Γ_A will vary as some power of T. In the non-relativistic regime, $n_{EQ} \sim (mT)^{3/2}\exp(-m/T)$, so that Γ_A decreases exponentially. In either regime, Γ_A decreases as T decreases, and so eventually annihilations become impotent, roughly when $\Gamma_A \simeq H$, which for definiteness, say occurs for $x = x_f$ ('freeze out'). Thus, we expect that for $x \lesssim x_f$, $Y \simeq Y_{\rm EQ}$, while for $x \gtrsim x_f$ the abundance 'freezes in': $Y(x \gtrsim x_f) = Y(x_f)$.

• *Hot Relics:* First consider the case of a particle species for which $x_f \lesssim 3$. In this case, freeze out occurs when the species is still relativistic and $Y_{\rm EQ}$ is not changing with time during freeze out. Since $Y_{\rm EQ}$ is constant, the final value of Y is very insensitive to the details of freeze out (i.e., the precise value of x_f), and the asymptotic value of Y, $Y(x \to \infty) \equiv Y_\infty$ is just the equilibrium value at freeze out:

$$Y_\infty = Y_{\rm EQ}(x_f \gtrsim 3) = 0.278 g_{\rm eff}/g_{*S}(x_f). \tag{31}$$

Thus the species freezes out with order unity abundance relative to s (or the number density of photons). Assuming the expansion remains isentropic thereafter (constant entropy per comoving volume), the abundance of ψ's today is (s_0 is the present entropy density)

$$\begin{aligned} n_{\psi 0} &= s_0 Y_\infty = 2970 Y_\infty \text{ cm}^{-3} \\ &= 825[g_{\rm eff}/g_*(x_f)] \text{ cm}^{-3}. \end{aligned} \tag{32}$$

If after freeze out the entropy per comoving volume of the Universe should increase, say by a factor of γ, the present abundance of ψ's per comoving volume would be diminished by γ: $Y_\infty = Y(x_f)/\gamma$.

A species which decouples when it is relativistic is often called a 'hot relic'. The present relic mass density contributed by a 'hot relic' is simple to compute,

$$\begin{aligned} \rho_{\psi 0} &= s_0 Y_\infty m = 2.97 \times 10^3 Y_\infty (m/\text{eV})\text{eVcm}^{-3} \\ \Omega_\psi h^2 &= 7.83 \times 10^{-2}[g_{\rm eff}/g_{*S}(x_f)](m/\text{eV}). \end{aligned} \tag{33}$$

Based upon the present age of the Universe we know that $\Omega_0 h^2 \lesssim 1$; applying this bound to the contribution of the species ψ we obtain a cosmological bound to the mass of the ψ:

$$m \lesssim 12.8\text{eV}[g_{*S}(x_f)/g_{\rm eff}]. \tag{34}$$

Light (mass $\lesssim$ MeV) neutrinos decouple when $T \sim$ few MeV, and $g_{*S} = g_* = 10.75$. For a single neutrino species $g_{\rm eff} = 2\times(3/4) = 1.5$, so that $g_{\rm eff}/g_{*S} = 0.140$. This implies that

$$\Omega_\nu h^2 = \frac{m_\nu}{91.5 \text{ eV}}, \tag{35}$$

$$m_\nu \lesssim 91.5 \text{ eV}. \tag{36}$$

This cosmological bound to the mass of a stable, light neutrino is often referred to as the Cowsik-McClelland bound.[6]

Note that the present density of ψ's depends upon the factor $g_{*S}(x_f)$. If a species decouples very early on, when g_{*S} is large, its present density is smaller. As an example, consider a species with $g_{\rm eff} = 1.5$ which decouples at a temperature $T \gtrsim 300$ GeV, when $g_{*S} \simeq g_* \gtrsim 106.75$. For such a species the present contribution to the energy density is

$$\Omega_\psi h^2 = \frac{m}{910 \text{ eV}}, \tag{37}$$

about a factor of 10 less than that of a conventional neutrino species.

• *Cold Relics:* Now consider the more difficult case where freeze out occurs when the species is non-relativistic ($x_f \gtrsim 3$), and $Y_{\rm EQ}$ is exponentially decreasing with x. In this case the precise details of freeze out are important.

First I will parameterize the temperature dependence of the annihilation cross section. On general theoretical grounds we expect the annihilation cross section to have the velocity dependence $\sigma_A|v| \propto v^p$, where $p = 0$ corresponds to s-wave annihilation, $p = 2$ to p-wave annihilation, etc. Since $\langle v\rangle \sim T^{1/2}$, $\langle\sigma_A|v|\rangle \propto T^n$, $n = 0$ for s-wave annihilation, $n = 1$ for p-wave annihilation, etc. Therefore I parameterize $\langle\sigma_A|v|\rangle$ as

$$\langle\sigma_A|v|\rangle \equiv \sigma_0(T/m)^n = \sigma_0 x^{-n}. \tag{38}$$

With this, the Boltzmann equation for the abundance of ψ's becomes,

$$dY/dx = -\lambda x^{-n-2}(Y^2 - Y_{\rm EQ}^2), \tag{39}$$

[6] In their original paper, Coswik and McClelland consider a 4-component neutrino ($g = 4$), and took $\Omega < 3.8$, $h = 1/2$ and $T_\gamma = T$, which resulted in the bound $m \lesssim 8$ eV.

where

$$\lambda = 0.264(g_{*s}/g_*^{1/2})m_{PL}\, m\, \sigma_0,$$
$$Y_{\rm EQ} = 0.145(g/g_{*s})x^{3/2}e^{-x}. \qquad (40)$$

This differential equation can be solved approximately to very good accuracy (better than 5%):

$$x_f = \ln[0.038(g/g_*^{1/2})m_{PL}\, m\, \sigma_0] - \left(n+\frac{1}{2}\right)\ln\left\{\ln\left[0.038(g/g_*^{1/2})m_{PL}\, m\, \sigma_0\right]\right\} \qquad (41)$$

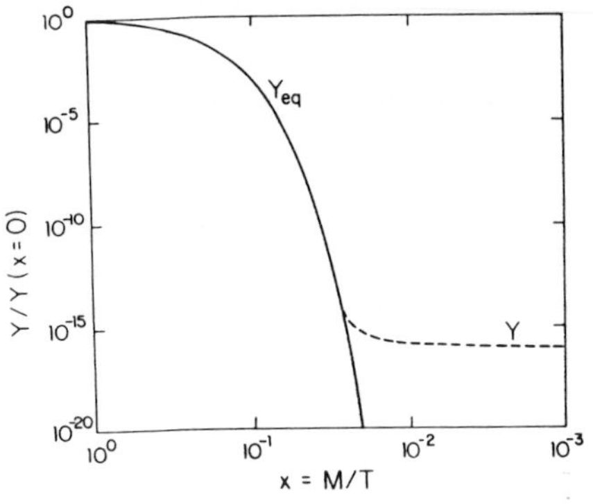

Fig. 1: The freeze out of a massive particle species. The dashed line is the actual abundance, and the solid line is the equilibrium abundance.

$$Y_\infty = \frac{3.79(n+1)x_f^{n+1}}{(g_{*s}/g_*^{1/2})m_{PL}\, m\, \sigma_0}. \qquad (42)$$

The approximate solution, along with the actual numerical integration of the Boltzmann equation, is shown in Fig. 1.

As with a hot relic, the present number density and mass density of relic ψ's is easy to compute,

$$n_{\psi 0} = s_0 Y_\infty = 2970 Y_\infty\ {\rm cm}^{-3}$$
$$= 1.13\times 10^5 \frac{(n+1)x_f^{n+1}}{(g_{*s}/g_*^{1/2})m_{PL}\, m\, \sigma_0}\ {\rm cm}^{-3} \qquad (43)$$
$$\Omega_\psi h^2 = 1.07\times 10^9 \frac{(n+1)x_f^{n+1}\ {\rm GeV}^{-1}}{(g_{*s}/g_*^{1/2})m_{PL}\sigma_0}. \qquad (44)$$

It is very interesting to note that the relic density of ψ's is inversely proportional to the annihilation cross section and mass of the particle

$$Y_\infty = \frac{3.79(n+1)g_*^{1/2}g_{*s}x_f}{m\, m_{PL}\langle\sigma_A|v|\rangle} \qquad (45)$$

The more massive the particle, or the smaller its annihilation cross section, the greater the relic abundance—heavy and weak shall prevail. Moreover, the mass density is proportional to just the annihilation cross section (at freeze out), which for $n=0$ (s-wave annihilation) is independent of temperature (and energy).

• *Two examples of cold relics:* Now let's apply this formalism to two simple examples. First, consider a baryon symmetric Universe (not ours!), that is equal numbers of nucleons and antinucleons. Using the above machinery we can calculate the surviving relic abundance of nucleons and antinucleons. Taking the nucleon–antinucleon annihilation cross section to be $\langle\sigma_A|v|\rangle = c\, m_\pi^{-2}$, where $m_\pi = 135$ MeV is the pion mass and c is a numerical constant of order unity, we find that

$$x_f \simeq 42 + \ln c,$$
$$T_f \simeq 22\ {\rm MeV},$$
$$Y_\infty \simeq 7\times 10^{-20}c^{-1}. \qquad (46)$$

Today, we know the abundance of nucleons is $Y_\infty = n_B/s \simeq \eta/7 \sim (6-10)\times 10^{-11}$, some 9 orders-of-magnitude larger than the relic abundance of nucleons in a baryon-symmetric Universe. As we will discuss later, this calculation provides strong evidence for the necessity of a baryon asymmetry – to prevent the so-called annihilation catastrophe in a baryon symmetric Universe. The presence of an excess of nucleons over antinucleons, of course, precludes this catastrophe, as all the excess nucleons necessarily survive annihilation. While the formalism developed here does not allow for a $\psi - \bar\psi$ asymmetry, it is straightforward to extend the formalism. In the case of the nucleon–antinucleon asymmetric Universe (like ours), the relic abundance of antinucleons can be computed (i.e., those that survive annihilation); it is

$$Y_\infty(\bar N) = 10^{18}\exp(-9\times 10^5). \qquad (47)$$

This very, very small number illustrates an interesting feature of freeze out. In the symmetric case, as time goes on the annihilation rate falls exponentially:

$$\Gamma_A = n_{\rm EQ}\langle\sigma_A|v|\rangle$$
$$\propto (m/T)^{3/2}\exp(-m/T)\langle\sigma_A|v|\rangle, \qquad (48)$$

and annihilations necessarily quench as particles and antiparticles become very rare. In the case of antinucleons in a baryon asymmetic Universe, the annihilation rate of antinucleons is: $\Gamma_A = n_N\langle\sigma_A|v|\rangle$, which does not decrease exponentially, as the nucleon abundance levels off at the value set by the asymmetry, $n_N \simeq n_B$.

Next, consider the relic abundance of a hypothetical heavy, stable neutrino species of mass $m \gg$ MeV. Because of its large mass, it will decouple when it is non-relativistic (though not necessarily at the canonical $T\sim$ few MeV), and the formulae for a cold relic pertain.

Annihilation for such a species proceeds through Z^0 exchange to final states $i\bar i$; where $i = \nu_L, e, \mu, \tau, u, d, s, \ldots$ (ν_L denotes any lighter neutrino species). The annihilation cross section depends upon whether ν is a Dirac or Majorana species; for $T \lesssim m \lesssim M_Z$, the annihilation cross section is

$$\langle\sigma_A|v|\rangle_{\rm Dirac} = \frac{G_F^2 m^2}{2\pi}\sum_i (1-z_i^2)^{1/2}[(C_{V_i}^2+C_{A_i}^2) + \frac{1}{2}z_i^2(C_{V_i}^2+C_{A_i}^2)]$$

$$\langle\sigma_A|v|\rangle_{\rm Majorana} = \frac{G_F^2 m^2}{2\pi}\sum_i (1-z_i^2)^{1/2}[C_{A_i}^2 2z_i^2 + (C_{V_i}^2+C_{A_i}^2)8\beta_i^2/3], \qquad (49)$$

where $z_i = m_i/m$, β is the relative velocity, and C_V and C_A are given in terms of the weak isospin j_3, the electric charge q, and the Weinberg angle θ_W by $C_A = j_3$, $C_V = j_3 - 2q\sin^2\theta_W$.[7]

In the Dirac case, the annihilation proceeds through the s-wave and $\langle\sigma_A|v|\rangle$ is velocity independent:

$$\sigma_0 \simeq c\,G_F^2 m^2/2\pi \qquad (50)$$

where $c \sim 5$. Taking $g = 2$ and $g_* \simeq 60$, from our formulae for x_f and Y_∞ we find

$$x_f \simeq 15 + 3\ln(m/\text{GeV}) + \ln(c/5) \qquad (51)$$

$$Y_\infty \simeq 6\times 10^{-9}\left(\frac{m}{\text{GeV}}\right)^{-3}\left[1+\frac{3\ln(m/\text{GeV})}{15}+\frac{\ln(c/5)}{15}\right] \qquad (52)$$

from which we compute that

$$\Omega_\nu h^2 = 1.6(m/\text{GeV})^{-2}\left[1+\frac{3\ln(m/\text{GeV})}{15}\right]. \qquad (53)$$

Note that freeze out takes place at

$$T_F \simeq m/15 \simeq 70\ \text{MeV}(m/\text{GeV}), \qquad (54)$$

before the interactions of light neutrinos freeze out. This is because as neutrinos annihilate and become rare, the annihilation process quenches. Requiring $\Omega_\nu h^2 \lesssim 1$ we obtain the so-called Lee-Weinberg bound:

$$m \gtrsim 1.3\ \text{GeV}. \qquad (55)$$

Although it is called the Lee-Weinberg bound, it was discovered independently by a number of people.

For the Majorana case, annihilation proceeds through both the s and p-waves; however the formulae which obtain fo x_f, Y_∞ and $\Omega_\nu h^2$ are similar. In Fig. 2 we show the contribution to $\Omega_0 h^2$ for a stable, massive neutrino species. For $m \lesssim$ MeV, $\Omega_\nu h^2 \propto m$ as the relic abundance is constant. For $m \gtrsim$ MeV, $\Omega_\nu h^2 \propto m^{-2}$ as the relic abundance decreases as m^{-3}. The relic mass density achieves its maximum for $m \sim$ MeV; neutrino masses less than about $92h^2$ eV, or more than about ~ 1 GeV (Dirac) or ~ 5 GeV (Majorana) are cosmologically acceptable.

[7] We have assumed that the neutrino is less massive than M_Z. If the neutrino is more massive than the Z^0 the annihilation cross section will be $\sigma_A \sim \alpha^2/M^2$, and the calculation of freeze out will be modified.

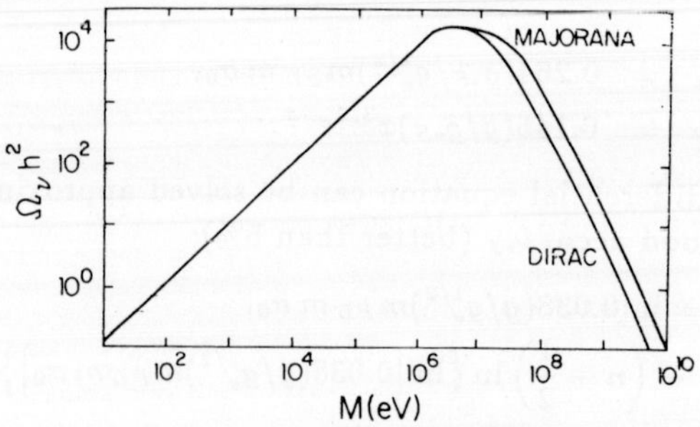

Fig. 2: $\Omega_0 h^2$ for a stable massive neutrino.

3. The Particle Connection

In the first part I considered non-equilibrium processes in the expanding Universe. As I have stressed, departures from equilibrium are responsible for most the cosmological relics—without such departures the past history of the Universe would be irrelevant. The Boltzmann equation is the primary theoretical tool for dealing with non-equilibrium in the expanding Universe.

Of particular interest was the relic abundance of a massive particle species. Most of the mass density in the Universe is dark, and if $\Omega_0 \gtrsim 0.15$, the dark matter is unlikely to be baryonic. The prime suspects in that case are relic particles from the early history of the Universe—e.g., a light neutrino species of mass $92h^2$ eV, or a heavy neutrino species of mass a few GeV would provide $\Omega_0 \simeq 1$. There are almost countless other candidates—axions, photinos, higgsinos, superheavy monopoles, etc., and most of the candidates, light the neutrino, arise as thermal relics.

"Cold relics" have the interesting property that their present abundance is inversely proportional to their mass and annihilation cross section. Consider a hypothethical species X, with mass m, 2 degrees of freedom, and annihilation cross section, $\langle\sigma_A|v|\rangle = \sigma_0(T/m)^n$, where $\sigma_0 = a\times 10^{-34}$ cm^2. Then it follows from the discussion of freeze out that

$$x_f = 24\left[1+\frac{1}{24}\ln(a(n+1)(m/\text{GeV})) - \frac{3.2(n+1/2)}{24}\right] \equiv 24b$$

$$Y_\infty = 9\times 10^{-11}(n+1)(b^{n+1}/24^2)a^{-1}(\text{GeV}/m)$$

$$\Omega_X h^2 = 0.025(n+1)(b^{n+1}/24^2)a^{-1} \qquad (56)$$

where we have taken $g_* = 60$. Note that up to logarithmic factors (i.e., b) $\Omega_X h^2$ is determined by the annihilation cross section, which is parameterized by a. It is also apparent that for $\Omega_X h^2$ to be of order unity, the annihilation cross section must be characteristic of a weak process. It is now clear why a heavy neutrino, and many of the supersymmetric partners, are prime dark matter candidates—all interact (with ordinary matter) with roughly weak strength.[8]

The fact that $\Omega_X h^2 \sim 1$ fixes the annihilation cross section to be order 10^{-34} cm^2 has implications in the contemporary Universe. All interactions of our hypothetical dark matter particles with ordinary matter—annihiation into ordinary matter, pair production, elastic scattering, etc., are set by this cross section (and its scaling with energy or temperature). Thus the production of such particles at accelerators, or the astrophysical direct detection of these particles necessarily involves 'weak processes'. There have been a variety of interesting proposals for the detection of relic dark matter particles—observing γ-rays from their annihilations in the galactic halo, observing high energy neutrinos from the annihilations of relic dark matter particles which are captured by and accumulate in the sun, and detecting in a cryogenic, bolometric detector the tiny (order keV) energy they deposit when they scatter with ordinary matter. All of the relevant cross sections are set by $\Omega_X h^2$—and fortunately for us, a weak cross section appears large enough to make the aforementioned detection schemes feasible.

4. Acknowledgements

This work was supported by the DoE and the NASA at Fermilab and the University of Chicago.

5. References

The freeze-out of a massive particle was first discussed by

1. Ya. B. Zel'dovich, *Zh. Eksp. Teor. Fiz.* **48**, 986 1965; Ya. B. Zel'dovich, L. B. Okun, and S. B. Pikelner, *Usp. Fiz. Nauk*, **84**, 113 (1965); H.-Y. Chiu, *Phys.Rev. Lett.* **17**, 712 (1966).

An early numerical treatment of freeze-out was given by

2. S. Wolfram, *Phys. Lett.* **82B**, 65 (1979).

Convenient approximate analytic formulae for the decoupling of a massive species and its relic abundance are given in

3. J. Bernstein, L. Brown, and G. Feinberg, *Phys. Rev. D* **32**, 3261 (1985); R. Scherrer and M. S. Turner, *Phys. Rev. D* **33**, 1585 (1986); **34**, 3263 (E) (1986).

The neutrino mass bound usually referred to as the Cowsik - McClelland bound is given in

4. R. Cowsik and J. McClelland, *Phys. Rev. Lett.* **29**, 669 (1972). It was first derived by G. Gerstein and Ya. B. Zel'dovich *Zh. Eksp. Teor. Fiz. Pis'ma Red.* **4**, 174 (1966); and it was also discussed by G. Marx and A. Szalay, in *Neutrino '72*, eds. A. Frenkel and G. Marx (OMKDT-Technoinform, Budapest, 1972), p. 123.

The Lee-Weinberg bound is discussed in

5. B. W. Lee and S. Weinberg, *Phys. Rev. Lett.* **39**, 165 (1977), although it was discovered independently by many people: P. Hut, *Phys. Lett.* **69B**, 85 (1977); K. Sato and H. Kobayashi, *Prog. Theor. Phys.* **58**, 1775 (1977); M. I. Vysotskii, A. D. Dolgov, and Ya. B. Zel'dovich, *JETP Lett.* **26**, 188 (1977).

The fact that the annihilation of massive Majorana particles into light fermions is suppressed at low energies because Fermi statistics requires *p*-wave annihilation was emphasized by

6. H. Goldberg, *Phys. Rev. Lett.* **50**, 1419 (1983). The effect on the Lee-Weinberg bound was first noted by L. L. Krauss, *Phys. Lett.* **128B**, 37 (1983). Detailed numerical calculations of the Lee-Weinberg bound for both Dirac and Majorana neutrinos are given in E. W. Kolb and K. A. Olive, *Phys. Rev. D* **33**, 1202 (1986).

Evasion of the Lee-Weinberg bound by decay of the massive neutrino was pointed out by

7. D. A. Dicus, E. W. Kolb, and V. L. Teplitz, *Phys. Rev. Lett.* **39**, 168 (1977).

Limits on supersymmetric particles can be found in

8. H. Pagels and J. Primack, *Phys. Rev. Lett.* **48**, 223 (1983); S. Weinberg, *Phys. Rev. Lett.* **48**, 1303 (1983); J. Ellis, J. S. Hagelin, D. V. Nanopoulos, K. A. Olive, and M. Srednicki, *Nucl. Phys.* **B238**, 453 (1984).

It is likely that for any known, suspected, postulated, hoped-for, or dreaded particle there exist cosmological limits to its properties.

The idea of detection of elementary particle dark matter by the energy deposited in a cryogenic bolometric detector was proposed by

9. M. Goodman and E. Witten, *Phys. Rev. D* **31**, 3059 (1985). Reprinted in *Early Universe – Reprints*.

For references to low-background cryogenic detectors:

10. A. Drukier and L. Stodolsky, *Phys. Rev. D* **30**, 2295 (1984); A. Drukier, K. Freese, and D. Spergel, *Phys. Rev. D* **33**, 3495 (1986); B. Cabrera, L. Krauss, and F. Wilczek, *Phys. Rev. Lett.* **55**, 25 (1985); S. P. Ahlen, et al., Harvard CfA preprint (1986); D. O. Caldwell, in *Neutrino Masses and Neutrino Astrophysics*, eds. V. Barger, F. Halzen, M. Marshak, and K. Olive (World Scientific, Singapore 1987) p. 262.

For detection of elementary particle dark matter via its annihilation products, see

11. M. Srednicki, S. Theisen, and J. Silk, *Phys. Rev. Lett.* **56**, 263 (1985); S. Rudaz, *Phys. Rev. Lett.* **56**, 2128 (1986); S. Rudaz and F. W. Stecker, *Ap. J.*, in press (1988); T. K. Gaisser, G. Steigman, and S. Tilav, *Phys. Rev. D* **34**, 2206 (1986); M. Srednicki, K. A. Olive, and J. Silk, *Nucl. Phys.* **B279**, 804 (1987).

[8] This presumes that the scale of the masses of the supersymmetric partners is, as is currently popular, order the weak scale.

PARTICLE PHYSICS DISCOVERY POTENTIAL USING DARK MATTER DETECTORS

R. E. Lanou[a] and B. Sadoulet[b]
[a] Department of Physics, Brown University, Providence, RI 02912
[b] Department of Physics, University of California, Berkeley, CA 94720

ABSTRACT

Many of the present directions suggested for going "beyond the Standard Model" show promise for tests utilizing astrophysical sources. Some astrophysical puzzles suggest particle physics solutions requiring new, unconventional particles. In order to capitalize on these opportunities for progress in particle physics new detection techniques are required. Present technology has already permitted some important tests but the technology is near its limit. Intense activity is underway particularly in the area of cryogenic detectors. These developments have created an exciting new frontier at the intersection of condensed matter, astrophysics and particle physics. The nature of this frontier and its implications for progress in particle physics and its complementarity to the SSC potential are emphasized. Examples of dark matter and solar neutrino experiments are used to illustrate.

1 NEEDS FOR APPROACHES COMPLEMENTARY TO THE SSC

1.1 Cosmology and Particle Physics

Over the last 10 years cosmology has made spectacular progress in part because of the input of particle physics. It has been realized that some of the most important problems in cosmology could have their solution in particle physics. For instance, the explanation proposed for the extreme homogeneity of the universe has come from particle physics: a phase transition at high energy that induces inflation. The same phenomenon may also account for density fluctuations that lead to galaxy formation.

Astrophysics must be combined with particle physics to address such fundamental questions as: why are we apparently living in a universe made out of matter only; and why does at least 90% of the mass of the universe not radiate electromagnetic radiation, so that it is detected only through its gravitational interactions? Schemes based on CP violation as well as on a "hidden" new family of particles (e.g., through supersymmetry) are proposed for explaining these fundamental properties of our universe.

In this paper we will argue that not only particle physics is essential for cosmology, but that conversely cosmology and more generally astrophysics are important for the progress of particle physics. The goal of particle physics is to understand the fundamental forces of nature. In particular, the unification of forces is a central question. Unfortunately, the unification scale appears to be very high and out of reach of man-made accelerators. The coupling constants' behavior with energy, indicates an unification scale above 10^{15} GeV.[1)] If new families of particles exist, the unification scale will be pushed even higher. The absence of proton decay at the 10^{31} year level also indicates a very high value.

How then can we get experimental information on the physics at the unification scale? Two approaches may be attempted. We can study a system which has gone through the unification scale, namely the early universe; if the Big Bang picture is correct, the universe went through a period of high temperature and may carry the imprints of physics at ultra high energy. Or we can also look at subtle consequences at low energy of the physics at very high energy; searches for proton decays and the measurement of neutrino masses are typical of this route. Such experiments need very high sensitivity and in some cases only the astrophysics environment may provide the mass, density or length scales to achieve the desired resolution.

1.2. The early universe as a probe of the physics at ultra high energy.

The status of the universe today gives hints about the physics at high energy and provides important constraints on theoretical models. Typically, an otherwise attractive theoretical model will have to be rejected because it leads to conflicts with basic cosmological parameters or to unacceptable distortions of the standard cosmological scenario. As a first example, we can note that the spectrum of particles has to be compatible with both the density and the composition of the present universe. The first type of constraint limits the masses of Dirac neutrinos. Masses between approximately 30 eV/c^2 and 2 GeV/c^2 would overclose the universe.[2)] The primordial abundance of helium requires the number of particle species relativistic at the time of nucleosynthesis to be smaller than approximately 13.[3)] This constrains the number of particles with mass smaller than 1MeV. In particular, the number of neutrino families has to be smaller than five.[4)]

Objects which are still unknown may have been created when the temperature in the early universe was higher than the energy available at our accelerators and may have survived. For instance, very massive particles may be present around us, such as monopoles and dark matter particles. Additional cosmological observations can attempt to detect those objects or at least put limits on their abundance. Topological singularities, such as cosmic strings generated at time of phase transitions, may also have observable consequences, creating linear discontinuities of temperature in the 2.7 K background or a linear pattern of gravitational lensing.

Cosmology also fixes important parameters in particle physics models. For instance, baryogenesis models trying to explain the absence of antimatter around us have to reproduce the ratio of the number of protons to the number of photons which is necessary for the standard primordial nucleosynthesis to work. SU(5) was unable to do so and could have been rejected on that ground only, even if protons had happened to decay at the required rates. More complex models can usually be adjusted to meet this demand. Even though the concept of inflation is very attractive and successful, no model without adjustable parameters has been able to reproduce the required amount of inflation and the amplitude of density fluctuations necessary for galaxy formation. Here again, cosmological constraints help to fix the shape of the potential.

In summary, cosmology more and more constrains model building and begins to play the role that the measurement of the (g–2) of the muon had played a few years ago! And the challenge it presents to particle physicists to explain the universe around us has become one of the driving forces of theoretical progress.

1.3 The Cosmos as a particle physics laboratory

In the attempts to see at low energies the consequences of the physics at very high energy, the effects may be so small that only the cosmos can provide the masses, densities, and scales necessary to see them.

An obvious example is the detection of very small neutrino masses. The supernova SN1987A has provided an astounding demonstration limiting the mass of the electron neutrino mass to less than 20 eV;[5)] a value which compares favourably with the present accuracy of laboratory experiments. In the future the detection of neutral current events from a close–by supernova may be the only practical way to measure at an electron–volt level the mass of the mu and tau neutrinos. If the masses are smaller than the electron–volt, it may only be possible to infer the masses from the presence of vacuum or matter oscillations observed via the use of the advantageous solar–terrestial separation distance as well as by virtue of their mass as sources and targets for neutrinos.

Moreover very high density states are available only in the cosmos, for instance in the form of neutron stars. The study of these compact objects may provide important results about the hadronic matter equation of state, the quark–hadron transition or about the powerful acceleration mechanisms which may be at work around them.These points will not be discussed further in this paper.

Taking finally a wider perspective, let us note that in the spirit of the unification of forces, (quantum) gravity will eventually have to come back into the bosom of particle physics. And gravity is observed mainly in the astronomical realm, and only in that environment can we expect to observe gravitational waves and black holes.

2 TWO EXAMPLES: DARK MATTER AND THE SOLAR NEUTRINO PUZZLES

2.1 Dark matter

2.1.1 The Astrophysics. The problem of dark matter is one of the most central problem of science today. It arises from the experimental observation that at least 90% of the mass in the universe does not shine or emit electromagnetic waves of any kind. More technically, the measure of the ratio Ω of the average density to the critical density leads to estimates very much greater than the value obtained with the stars that we see in galaxies (Table 1).

Table 1
Cosmological Density Estimates

	Scale (h^{-1}Mpc)	Ω
Visible parts of galaxies	=0.01	0.002
Halos of Galaxies	=0.1	0.02-0.2
Clusters	=1.	=0.2
Virgo infall	=10	=0.2
Large scale infall	=30	0.2-1
Geometry	3000	0.3-2

If there is a wide consensus on the existence of such an effect, there is a heated debate on the nature of this dark matter. Many astronomers hold the view that it is baryonic (e.g., Ref. 6) since we expect dark ashes from star evolution: white dwarfs which have cooled down, neutron stars, black holes, etc. On the other hand, many cosmologists and particle physicists will argue that dark matter can hardly be baryonic. The strongest argument is based on the nucleosynthesis of primordial elements. Conventional nucleosynthesis is in excellent agreement with observations;[3] for a recent review see Ref. 4, and indicates that at most Ω (baryons) < 0.15 when the values necessary to account for the large scale structure of the universe is definitely larger than that value. Every attempt so far, to modify the conventional model has been unnatural or unsuccessful. It is also very difficult to prevent baryons from shining or absorbing radiation and to stay really dark.[7] It seems moreover difficult to understand galaxy formation without an additional component: a purely baryonic universe which would be in equilibrium with photons at the time of emission of the 2.74K cosmic microwave background, would have been very uniform at that time, as is the relic radiation we receive today; in these conditions, it would not have been able to grow through gravity and form galaxies today. Thus we need an additional component, explosions, strings or non–dissipative dark matter! Finally, in an evolved universe, Ω is not expected to be close to 1 unless it is exactly 1. Any value different from 1 will evolve very rapidly away from 1 and it would be rather surprising to observe a value so close to 1.

None of these arguments are very strong, but they point at the fact that dark matter may be constituted of some kind of non–baryonic, primordial objects. Among the various hypothesis, black holes, nucleites etc..., massive particles produced in the early universe seem to be the most natural candidates. Presently, not only the "zoo" population is quite large with the photino, higgsino, scalar neutrino, heavy Dirac neutrino, the axion etc. but a value of the Ω the order of 1 naturally arises also in a large class of theories.

2.1.2 Particle Physics. This surprising coincidence emerges when it is assumed that the considered particle that we will call δ was once in thermal equilibrium with quarks (q) and leptons (l), presumably through the reactions $\delta\bar{\delta} \leftrightarrow q\bar{q}, l\bar{l}$. This simple hypothesis (which does not apply to the axion) leads to a present δ density in the universe that is a function of their annihilation rate at the time they went out of equilibrium (the "freeze–out" time).[2] The argument is simple: Let us first assume that there is no initial asymmetry and the number of δ's is the same as the number of their antiparticles. If this annihilation rate is much faster than the rate of expansion of the universe, they all disappear and they cannot constitute the present dark matter. If on the other hand, the rate is too small, the expansion quickly dilutes the δ's, which have difficulty finding an antiparticle to annihilate with, and their abundance now will be too large. For δ masses above 300 MeV/c^2, the freeze–out temperature is about a twentieth of the rest mass of the particle, and the annihilation cross section is, for δ mass in the Gev/c^2 region, $\langle\sigma v\rangle = 10^{-26}/(\Omega_\delta h^2) cm^3/s$ where Ω_δ is the current ratio of the δ average density to the critical density and h is the Hubble constant in units of 100 km/s/Mpc. Such a result is interesting because of two facts:

— For $\Omega_\delta \approx 1$ this annihilation rate has the value expected from weak interaction. This may be a numerical coincidence, or a precious hint that something like supersymmetry has some truth!

— The interaction scale is fixed; moreover it is a lower limit. We could thus imagine an initial asymmetry, similar to the one usually assumed for baryons and antibaryons. In this case, the cross section could become large enough for all the pairs δ anti–δ to disappear and the small excess of one component to make up the dark matter. Through field theory, this annihilation cross section at a specific energy can be related to other potentially detectable processes. These considerations result in the important conclusion that this general hypothesis may be testable in a rather model independent way through a combination of experiments. This is shown schematically in Figure 1 and reviewed at length in 8.

Given the form of the effective Lagrangian (e.g., vectorial, axial vectorial) one can extrapolate from the freeze–out energy to very low energy and predict the annihilation rate today in the halo of our galaxy. The annihilation gives rise to a low–energy flux of antiprotons, positrons and gamma–ray lines.

The cross section can also be extrapolated to high energy, where experiments of the type $e^+e^- \rightarrow \gamma +$ (no detected particle) [ASP, MAC, and CELLO] provide limits on the cross section for $e^+e^- \rightarrow \delta\delta$. It can be shown that these accelerator data bound the δ mass from below around a few GeV (although another allowable region exists around 1 eV, that we will not discuss here).

New experiments at the colliders and at SLC (e.g., measurement of the Z^0 width) will provide vital constraints on particle models. If evidence for supersymmetry is discovered, an immediate question will be whether the least massive supersymmetric particle could be responsible for the dark matter.

Given the same matrix element, the annihilation cross section can be related by crossing to the elastic cross section of δ on ordinary quarks. This leads to two consequences: first, that the δ can be trapped in the sun and in the planets, modifying their core temperature and leading to enhanced annihilations which may be detectable as a high energy neutrino flux.[9,10] In the case of the sun, one consequence might be to explain the solar neutrino problem.[11] The second consequence is that the elastic scattering rate of halo δ on ordinary matter in the laboratory may be large enough to be detectable.[12]

2.1.3 The detection problem. The arguments outlined above lead to a rather model–independent estimate of the elastic cross section of δ's on nuclei. For $m_\delta \geq$ 1 GeV/c^2, $\sigma_{el}(\delta N \rightarrow \delta N) \geq 10^{-38} cm^2/(\Omega_\delta h^2) \times$ coherence factor. A coherence factor has to be included because the particles to be detected have the typical virial velocity of the halo of our galaxy, some 300 km/s. Taking into account the known density of the galactic halo, it is then possible to calculate the expected interaction rate. Complications arise from the coherence factor and a possible initial asymmetry, but one can build archetypes which which give an idea of the expected rates.[13)] Two of them are are given in Figure 2. The important number to remember for the present purpose is that the expected interaction rate of dark matter particles concentrated in the halo of our galaxy with a target in the laboratory may be of the order of a few events per kg per day. In general terms, since both vector and axial vector interactions may occur, a variety of target materials are necessary to untangle the coherence effects. In particular, the detection of the photino requires a material with nuclear spin or a spin zero detector of sufficient mass to be sensitive to a scalar interaction.[13,14)]

Since in dark matter applications both the target and projectile tend to be massive and non–relativistic, we have for a recoil energy, $E_R = \frac{m_\delta^2 M_N^2 v^2}{(m_\delta + M_N)^2}(1-\cos\theta^*)$. Because the scattering occurs at low energy in the s–wave, the energy deposition spectrum is essentially independent of the considered model. A calculation for a low mass object shows that the mean energy deposition can be extremely low (500 eV) (Figure 3a), calling for detectors of very low thresholds.

A critical question is that of signatures. Let us assume that a detector with such a low threshold indeed sees a potential signal at low energy. How could we be sure that it is due to dark matter particles from the halo of our galaxy? The most convincing signature would be an annual modulation of $\pm 4.5\%$ in the mean energy deposition and of $\pm 2.5\%$ in the rate (Figure 3b). The halo has not collapsed significantly and is believed to have very small overall angular velocity. On the other hand, as the sun goes around the galaxy and therefore through the halo at 200 km/s, the earth is adding (subtracting) half of its velocity to the sun velocity in the summer (winter). In order to observe such an effect at 5σ, we would need about 4000 events. Therefore, a detector of very large mass ($\sim$10kg) would have to be built!

Other important signatures include the shape of the energy distribution, which is expected to be different from X–ray lines or Compton scattering (which has a flat spectrum). The dependence of the energy deposition on the material used as a target might allow a measurement of the mass of the δ. The use of a mix of materials may permit one to exclude the possibility that the incident particle is a γ or a neutron from the surrounding material.

As far as the radioactive background is concerned, the state of the art is encouraging. The double β–decay experiments of LBL/UCSB and PNL/USC get background rates[15,16)] of the order of 0.5 to 1 event/kg/keV/day at 20 keV. At lower energy, the LBL/UCSB group, which has been able to decrease its threshold to 3 keV equivalent–electron–energy, observes quite prominently tritium (at the level of 25 decays/kilogram/day). Even this very small contamination, presumably due to spallation production by cosmic ray interactions, will eventually have to be eliminated for such experiments to be sensitive to low mass dark matter particles.

2.2 Solar Neutrinos

Interest in utilizing the sun as a source of neutrinos has recently intensified as a result of the convergence of three diverse problems: a) Can the sun be used to find (or set limits on) fundamental properties of the neutrino? b) Is the Standard Solar Model correct? and c) Can it play a role as a "target" for dark matter in the galaxy?

Although knowledge of the neutrino's existence has been with us for more than fifty years and although it has played a central role in much of our present understanding of electroweak interactions, the neutrino is still in some ways the most enigmatic of particles. We do not know if its mass is zero or, if it is, why. Nor is much known about its charge or magnetic structure, if any. Further, the origin of lepton numbers and flavors and to what extent they are mixed or conserved is not known. Discoveries on any of these points could provide important clues to guide us beyond the now Standard Model of $U(1) \times SU(2) \times SU(3)$ of particle interactions.[17)]

2.2.1 Astrophysics. From the astrophysics point of view, the sun as our nearest star offers a valuable opportunity to test solar models. While observations of the sun's surface are as ancient as mankind, ability to probe its interior is a recent development of great importance. The two techniques of solar neutrino detection[18,21b)] and of helioseismology[19)] are the main tools.

The particle interaction processes which power the sun are well understood[20)] and as a consequence the sun must be a copious source of neutrinos (see Table 2 and Figure 4 for fluxes and spectra). Also, because the neutrino interacts so weakly with the matter in the sun they emerge promptly; reaching the earth eight minutes after their creation — a brief time in comparison to the more than 10^6 years taken by photons to "percolate" through the solar opacity. Similarly, helioseismology by its present sensitivity to pressure oscillation modes (and developing sensitivity for gravitational modes and with a growing number of observation stations around the world) provides real–time information on solar structure. This technique is already providing important constraints on solar density profiles and radii. Since the focus of this paper is on particle physics we will not discuss helioseismology further and refer the reader to some excellent reviews.[21)]

Table 2

Source	Flux (10^{10}cm ^{2}sec^{-1})
pp	6.0(1±0.02)
pep	$1.4 \times 10^{-2}(1\pm0.05)$
hep	7.6×10^{-7}
^{7}Be	$4.7 \times 10^{-1}(1\pm0.15)$
^{8}B	$5.8 \times 10^{-4}(1\pm0.37)$
^{13}N	$6.1 \times 10^{-2}(1\pm0.50)$
^{15}O	$5.2 \times 10^{-2}(1\pm0.58)$
^{17}F	$5.2 \times 10^{-4}(1\pm0.46)$

Beyond the inclusion of the basic energy producing processes in the sun, the Standard Solar Model consists of a spherical system in hydrostatic equilibrium between gravitational and thermal pressures. Important input include reaction cross sections, primordial helium abundance and opacities. Detailed discussion of these considerations can be found in Ref. 22. It is in this context that the pioneering experiment of Davis et al.[18] was conceived, run and discovered the now famous "solar neutrino problem". For reasons we discuss below this should possibly be called "the first, solar neutrino problem" and it consists of the observation that averaged over the period from 1971 to 1985 there are a factor 3 less ν_e detected from the most energetic neutrino processes in the sun than predicted by the Standard Solar Model.

2.2.2 Particle physics. From the particle physics point of view the sun provides a unique source for some neutrino property experiments and also as a source of products (high energy neutrinos) from the interactions of certain classes of elementary particle dark matter candidates. Even if the "neutrino problems" have a solution outside the realm of particle physics, the sun remains as perhaps our most favorable (Energy/Length) ratio to test neutrino mass in oscillation experiments provided energy thresholds can be low enough and neutral currents detected. We return to these issues below.

However, among the suggested particle physics explanations of the above Davis results are neutrino mass and mixing, neutrino lifetime, neutrino magnetic moment and solar core cooling due to dark matter trapping.[23] Any one of these solutions, should they occur, would signal important new physics.

All of these are still live issues although analyses[24] of SN1987A neutrino fluxes severely limit the possibility of lifetime as one. While the G-W-S Standard Model predicts a magnet moment for ν_e many orders of magnitude[25] below present methods of experimental technique, new data from the Davis et al. (Homestake) experiment have rekindled interest in the ν_e magnetic moment. These new data,[26] when taken together with data from the previous 18-year period appear to show neutrino flux increases anti-correlated with the solar cycle. Not yet statistically significant, the experiment will be so as the cycle reaches its activity minimum in the next two years. Should these data hold up we would then have a "second solar neutrino problem". Evidence for a ν_e magnetic moment large enough[27] to cause this effect would signal the existence of a finite mass.

The possibility of neutrino mass and flavor mixing which might be detected through neutrino oscillations continues to be a central motivation for most solar neutrino experiments in progress or planned. Although the searches for vacuum oscillations have been conducted at reactors and accelerators[28] for several years, the limits are now sufficiently stringent that terrestrially-based experimental limits are nearly saturated.[29] The realization that the effect of even very small mass differences and mixing angles could be vastly enhanced by neutrino passage through matter (matter oscillations[30]) has been a crucial new factor in the field. More particularly that the matter density profile of the sun is such that the range in mass differences and mixing angle could be extended by several orders of magnitude (M-S-W effect[31]) over terrestial experiments and further, with the right combination of experiments, could establish values for the possible mass differences and mixing angles. This combination of factors has motivated many active groups to push present techniques to the limit, produced many new ideas for detection, and initiated development of radically new techniques. There is a convergence of some of these techniques for solar neutrinos and for dark matter searches — principally in the newer and cryogenic ones. We focus on them as examples. An excellent catalogue of many of the solar neutrino detection ideas is contained in Ref. 19.

Discussion of solar neutrino effects caused by dark matter are deferred to below.

2.2.3 The detection problem. As the energy and fluxes illustrated in Table 2 and Figure 4 show, the usual problems of detecting neutrinos are multiplied in the solar case due to the simultaneous requirement of large target mass, low energies and low background. Not surprisingly, these competing requirements have spawned a hierarchy of techniques whose exploitation and current development form a continuum of "three generations."

In the first generation we have the pioneering Davis et al. (Homestake) experiment based upon the inverse beta decay reaction $\nu_e + ^{37}Cl \rightarrow e^- + ^{37}Ar^*$ and, after chemical manipulation to extract the argon, the counting of its decay. This technique is uniquely sensitive to the ν_e but provides no information on the energy except that it is above a threshold. While the reaction threshold is 0.814 MeV the transitions are dominated by higher analogue states (~ 8 MeV) thus making the technique primarily sensitive to the ^{8}B neutrinos. Further, the technique is not sensitive to ν_μ, ν_τ or a new flavor neutrino. Also in this first generation is the Kamiokande detector. Constructed for a search for proton decay, the experimenters have beautifully exploited the technique by pushing the threshold and noise to levels which have permitted them to see both the supernova and to set limits on solar neutrinos[32] which confirm the Davis et al. results on the disagreement between the Standard Solar Model and the measured ^{8}B neutrino flux. It is very important to note that the confirmation by Kamiokande is by means of a completely different technique. It is a water Cerenkov method in which the neutrinos are detected by observing the Cerenkov light from the recoil electron in the reaction $\nu_e + e^- \rightarrow \nu_e + e^-$. The kinematics of the reaction favors very highly forward electron scatters and thus point back to the source. The observed energy spectrum of the electrons indirectly samples that of the neutrinos. The ν_e reaction has contributions from both charged and neutral current interactions; the ν_μ and ν_τ reactions with electrons, should they be present, have only neutral current interactions and as a result are about seven times smaller at these energies. This suggests a broader use for this reaction in some oscillation searches. While it is expected that some further improvement in thresholds of the present Kamiokande experiment will occur, it will not go beyond the ^{8}B neutrinos.

In terms of event rate, both of these first generation experiments are very low: 1.2 per day for ~500 ton Homestake and ~.4 for the 3000 ton Kamiokande.

To exploit existing technology in order to obtain more detailed information on the ^{8}B reaction and to take a first step into the more copious, but lower energy and less model dependent p–p reaction neutrinos a "second generation" of detectors have exhibited feasibility. One, the use of the reaction $\nu_e + ^{71}Ga \rightarrow e^- + ^{71}Ge^*$, is an extension of the Davis et al. technique to a lower threshold (0.213 MeV) element and thus samples the ν_e flux from the p–p reaction. The other, an extension of the water Cerenkov technique to heavy water, promises to directly measure the ^{8}B and hep ν_e spectrum via $\nu_e + n \rightarrow e^- + p$. A very important new capability of this latter technique will be the direct measurement of the total neutrino flux — an essential quantity for confirmation of any evidence for neutrino oscillations. This would be accomplished by the weak neutral current disintegration of the deuteron and the subsequent detection of the neutron capture gamma ray in chlorine with which the target medium is seeded. Because the weak neutral current is flavor independent ν_μ, ν_τ, and $\nu_?$ as well as ν_e's would be detected.[33)]

There are two versions of the gallium experiment currently under construction — Gallex with 30 tons in the Gran Sasso Laboratory of Italy and another in the Baksan Laboratory in the Soviet Union with 60 tons.[34)] Approximately 0.05 events per ton per day are expected with data taking to begin with partial tonnage in the next 18 months. The heavy water experiment, SNO (Sudbury Neutrino Observatory), in the final proposal and decision stage, would be constructed by a joint Canadian–American team and located in the INCO Sudbury mine. The planned tonnage (1 kiloton) should yield 3600 charged current events and 1000 neutral current events in one year if the Davis–Kamiokande results are correct.

The developments so far are impressive and constitute important advances in neutrino detection technology. They will provide us with results which will be crucial to our understanding of neutrinos from the sun; however, if the outcomes suggest that the source of the "solar neutrino problem(s)" is in the particle sector then new techniques will be needed to provide spectral information and neutral current sensitivity over the entire solar energy range. There are several inconclusive outcomes possible for the above experiments all of which imply further experimentation at low neutrino energies.[35] Should there be an outcome clearly indicating a stellar model origin rather than a particle one, then the sun remains as the best source to search for neutrino vacuum oscillations because the quality factor (Energy/Distance) of the neutrinos and the source are so favorable. Again, however, spectral and neutral current information at the lowest energies are required.

The implication is that "third generation" detectors are required. What are the choices?

3 EXTENSIONS OF EXISTING TECHNOLOGY

3.1 Ionization Detectors

3.1.1 Extension of Ionization Detectors for Dark Matter

The direct detection of dark matter particles through their interaction with a suitable target in the laboratory is presently performed with germanium ionisation detectors.[15,16)] In spite of the fact that ionization detectors suffer a low energy transfer efficiency to ionization leading to a loss of sensitivity below 10 KeV (for a review of experimental measurements see Ref. 36) they certainly have not said their last word.

The LBL/UCSB group is presently assembling silicon detectors with a total mass of 100g to check the presence of cosmions.[36)] The "cosmion" is the generic name for an ad hoc particle whose properties[11)] are such as to cause a deficit of neutrinos from the solar ^{8}B reaction.

It is possible to decrease even further the threshold through the use of electron-hole drift techniques in the semiconductor.[37)] They may allow a decrease of the detector element total capacitance and improve the threshold accordingly — a factor of two or three likely is within reach. At the cost of additional channels, detectors where the electrodes are divided into strips could provide some position resolution, a useful tool to fight the background.

As recently shown by Luke,[38)] it may even be possible to measure ionisation with calorimetric techniques at low temperature (=1K). The amplification is such that at the limit even one electron- hole pair could be detected.

Proportional chambers may also be adapted to the task. Although their effective threshold is poor, their imaging capability , in a time projection configuration, is potentially an asset for background rejection as shown by the experience of double beta experiments using this type of device.[39,40] At low pressure (e.g. =30 torrs) it should be possible to recognize a β–ray of 10 keV from its range (=2 cm).[41] A recoiling nucleus may also leave a "track" and it may be possible to measure its direction (distinguishing the origin from the end). The energy range where this scheme works remains to be experimentally identified, and it remains to be shown that the spontaneous count rate can be decreased in order that large detectors of this type be useful.

3.1.2 For solar neutrinos. Because the energy cost per ion pair is high (~ 20 eV) ionization–based detectors are not the detectors of choice for low energy recoils. However, for some applications, collections of ions can be a good match to signal, background and thresholds if some kind of pattern recognition is used as a discrimination aid. Assuming underground locations, the cosmic ray background is minimized but can still be significant; other important sources of background are capture gammas and neutrons from the surrounding cavity, radio–isotopes intrinsic to the detecting medium, and long–lived spallation products induced in the detecting medium during the time the medium was on the surface of the earth. The former can be adequately controlled by appropriate shielding; however, the latter two are severe problems — not just for ionization detectors but for all detectors. To a certain extent it might be said that one of the principal distinctions among detectors is how they deal with those background problems.

One interesting extension of existing technology for ionization is a scale-up to dimensions of 3000 tons of ion drift in liquid argon[42] with pattern recognition, achieved by wire chamber techniques. For solar neutrinos, the pattern involved the reactions $\nu_e +^{40} Ar \rightarrow e^- +^{40} K^*$ and the subsequent conversion of the decay gamma ray. The detector is under prototyping and construction in the Gran Sasso Laboratory and has other major applications beyond solar neutrinos including proton decay. The solar neutrino process available is that involving ^{8}B and hep. The experiment provides sensitivity to spectra as well as to neutral current.

3.2 Cerenkov Technology

3.2.1 Dark matter. Present-day proton decay technology is being used to search for high energy neutrinos from the sun. These neutrinos would be the presumed annihilation products of dark matter particles or, for example, supersymmetric particles annihilating in the sun. Measurements[43] of high energy neutrino fluxes orginating in the vicinity of the sun as measured by two, water Cerenkov detectors (IMB, Kamiokande) and one ionization–based detector (Frejus) have been interpreted in an attempt to put limits on several candidates. For example, a scalar neutrino is considered to be excluded while uncertainties in the axial vector coupling of the photino preclude setting a limit on it at this time.[44]

Results from this technology have also been interpreted to set weak limits on a heavy, Dirac neutrino as a candidate for galactic dark matter.

3.2.2 Solar neutrinos. The extension of the Cerenkov technology to future work on solar neutrinos is best exemplified by the elegant Sudbury detector which is discussed in Section 2.2.3.

4 THE PUSH TOWARD CRYOGENIC DETECTORS

4.1 Why?

For both dark matter and for solar neutrinos there are strong reasons for going to very low temperatures for the target material. A central issue for both is the question of achieving low thresholds in order to see very low energy recoils. These recoils range from a few eV in solids to several keV in cryogenic liquids. An important but less restrictive requirement is to have good energy resolution. Because the recoil energies are so low the statistical requirements for good resolution imply that the mechanism of energy transfer between the recoil particle and the target medium have a high multiplicity of low energy secondary carriers; thus for example, the production of phonons or the breaking of Cooper–pairs is to be preferred over ionization pairs.

The kinematics of direct detection for dark matter was outlined in Section 2.1.3.

The particular signatures selected for dark matter and for solar neutrinos have somewhat different kinematics and therefore threshold and background treatments may differ in application. For dark matter application, it is obvious from the peaked energy spectrum exemplified in Figure 3 that for low mass δ's, a premium is placed upon a well-behaved noise spectrum near threshold. This favors the use of low energy quanta and therefore cryogenic detectors. For large masses the main problem is the background level and there cryogenic detectors may also be superior to conventional technology, if they allow (as is theoretically possible) the identification of interactions on the nucleus (through the simultaneous measurement of the phonon and electron-hole components) and, perhaps, through the measurement of the direction of the incident projectile. These background rejection tools will be precious if they are practical, since typical rates, based upon the expected properties deduced from astronomical observation, go as $\frac{1}{m_\delta}$ and signal to background[13] (largely Compton electrons) as $\frac{1}{m_\delta^3}$.

Extending the example to solar neutrinos produces somewhat similar conclusions but for different reasons. Because of the expected fluxes, really massive (tons) of detector are required and there is a correlated requirement that the material be extremely pure from any kind of radioactive background. These general constraints immediately restrict the candidates of the future (assuming the full solar spectrum to be covered) detectors to a small set of materials. They should be inexpensive, clean and instrumentable. A particularly interesting one is ^{4}He in its superfluid state.[45] At these temperatures (<100 mK) helium freezes out all other atomic species and provides a high multiplicity of useful carriers (rotons, see below) of deposited energies from the signature event. There are two solar neutrino signatures — a) scattering from electrons with sensitivity to both charged and neutral currents and b) scattering coherently from nuclei via the neutral current. (In helium, all charge current reactions by solar neutrinos on the nucleus are energetically forbidden.) The kinematics dictate quite diverse thresholds for the two reactions. The former, with a threshold of a few keV for the recoil electron, sees most of the p–p neutrino spectrum. While a 100–200 eV threshold would be required for a similar acceptance of ^{8}B neutrino via coherent scattering. The recoil kinetic energies are given by: $E_R = \frac{2m_X E_\nu^2 \cos^2\theta_X}{E_\nu^2 (1-\cos^2\theta_X)+2m_X E_\nu + m_X^2}$, where $m_X = m_e$ or M_N for a) or b) respectively. The signal event rate in a 10 ton ^{4}He detector for reaction a) would be 12 and 8 per day, respectively for p–p and ^{7}Be neutrinos if the threshold where 10 keV; while for reaction b), if the threshold could be 100 eV there would be only 0.5 per day from ^{8}B but 5 additional events per day from ^{7}Be. The backgrounds as in the case of dark matter detection are expected to be principally Compton electrons.

4.2 Current Ideas

Many ideas have been proposed to use the small energy of excitations at low temperature. The hopes they raise and the challenges they face is reviewed in Ref. 46.

4.2.1 Physics of superconductors. Cooper pairs in a superconductor have a binding energy of 2Δ (the energy gap). Since 2Δ is of the order of 3.5 kT_c where T_c is the critical temperature, and ranges from 4×10^{-5} eV (Ir) to $3 \times 10-3$ eV (Nb), the deposition of even a modest amount of energy in a superconductor leads to a large number of broken Cooper pairs (quasiparticles). Detection of these quasiparticles may lead to very sensitive or accurate devices. Two methods are being developed actively (recent results of the various groups can be found in Ref. 47).

If the superconductor elements are small enough, small granules or narrow thin films, the energy deposition may be large enough to drive the element to a normal state. In the scheme proposed originally by a Orsay group,[48] later joined by Waysand and Drukier, the detector is made of many small spheres of a few microns diameter immersed in a magnetic field, and the transition of a single sphere from the superconducting to the normal state can be detected through the variation of magnetic flux in a loop surrounding the spheres generated by the suppression of the Meissner effect. Groups at UBC (around Drukier), Munich M.P.I. (Stodolsky, Pretzl,[49] and co-workers), Paris (Waysand et al.), Annecy (Perret- Gallix and Gonzales-Mestres) are actively involved in solving problems ranging from the uniformity of the granules to the sensitivity of the amplifiers. An interesting variant has been pursued recently at Stanford by Cabrera and co-workers[50] using a thin (300A), narrow (35μ) zigzag of titanium. An electric current is used to detect that part of the film becomes normal.

A potentially more sensitive and linear detector could be provided by tunnel junctions. If they are placed in a magnetic field to suppress the tunneling (Josephson) current of Cooper pairs, the quasi particles could be detected directly as an apparent increase in the leakage current of the junction. Groups at SIN (Zehnder), Munich T.U. (von Feilitzsch et al.), Oxford (N. Booth for the detection of solar neutrinos in Indium)), Stanford (Cabrera) and San Francisco State (Bland and Neuhauser) are working along this track. Results[51] obtained with Sn junctions are encouraging but much more work is needed before we could make large, reproducible and stable devices with low leakage current.

Note that both types of techniques could be used for phonon detection if the superconductor is deposited as a film on the surface of an insulating crystal. Phonons of high enough energy will break Cooper pairs.

4.2.2 Physics of phonons. Phonons, the bulk, quantized excitations in solids and liquids, typically carry a few tens of milli-electron volts or less and are thus attractive agents for energy transfer out of the detecting media[52] when low thresholds and good energy resolution is needed for small excess energy deposition. At any given temperature and for a specific medium there is an equilibrium population of phonons and the energy–momentum constraint on any phonons in the material is governed by an appropriate dispersion relation. An individual phonon can decay, scatter or undergo umklap processes with the equilibrium phonon population. The lifetimes are energy dependent[53] and go as $\tau = \alpha E^{-5}$ hence, should the equilibrium density and impurities be reduced to a negligible amount, the remaining phonons from an excess energy deposition can propagate freely. These are referred to as the ballistic phonons.

Their use is to be preferred over detection of the temperature rise in the target medium due to the equilibration of the excess for several reasons. Among the undesirable effects resulting from target size are: unfavorable (mass) dependence of ΔT and reduced sensitivity due to escape of ballistic phonons.[53] Equally significant perhaps is that by use of timing and geometry of the ballistic phonons themselves — especially in crystals[54]—it is hoped that sensitivity to event position will be achievable. The required operating temperatures are also not as severe using this method.

Superfluid ^{4}He presents special differences from phonon behavior in crystals. Its rather special dispersion curve[55)] results in a class of high energy phonons ($\sim$0.7 meV) which are stable in pure ^{4}He at <100 mK and dominate phase space. These are called rotons and, because of their desorbtive effects at free surfaces, are believed to be efficiently detectable.[45)]

The development and selection of devices to detect ballistic phonons or their effects with good efficiency, linearity and resolution with low noise is of course an important and crucial ingredient to the success of these cryogenic techniques. Among the methods currently under study are the use of thermistors, superconducting thin films and superconducting tunnel junctions.[56)] It is still unclear what is the best method. The Milano group (headed by Fiorini) have recently been able to detect 6 keV X-Rays with a germanium crystal of 10 grams and a thermistor measuring the temperature rise of the crystal. This line of attack is also followed by the British group of Rutherford/Imperial/Cambridge/Sussex. The Munich T.U. group[58)] has achieved the detection of ballistic phonons from deposition by alpha particles and has resolved space separation of deposition sites of 0.75 mm. The Brown group[45)] has recently begun work on roton detection in superfluid helium. The extensive study of one type of thermistors by the Berkeley group seems to show that they can very good detectors of high energy phonons. Stanford has encouraging results with the detection of high energy phonons with a superconducting film.[50)] Interesting potential uses which further study of ballistic phonon techniques might open up are their use for extracting some directional information on the low energy recoils as well as discriminating between Compton recoils and nuclear recoils by virtue of the fact that the two recoils share their deposited energy differently between ionization and bulk excitation of the medium.[57)] However it is clear that a long and challenging development is still in front of us.

4.2.3 Outlook. A substantial community of particle physicists is working on the above problems and with exciting recent success. Progress in many of the above areas requires better understanding of the condensed matter physics underlying them. At these low temperatures all of this physics is not known in detail and constitutes topics of fundamental interest to many condensed matter physicists who are now attracted to the efforts. Invited talks on these new techniques and their potential applications have been featured at meetings of the A.P.S. Division of Condensed Matter, the I.E.E.E. and the "Texas" Meeting of Astrophysics.[59)]

As soon as they can operate detectors of a few tens of grams with low enough thresholds, the groups working with cryogenic detectors will soon set further significant limits on dark matter. The ultimate dark matter detector goals are for devices in the range of tens of kilograms or larger.

An interesting sidelight of the new developments discussed above are the potential applications to other physics problems. For example, neutral current coherent scattering studies at reactors, more sensitive double beta decay, neutrino mass measurements, medical imaging devices, and ν_μ magnetic moment tests.

The next few years (2–3?) will be crucial ones of research and development.

Besides testing as a dark matter detector of modest scale, a pilot demonstration of superfliud helium could be a reactor experiment. While the weak neutral current coherent scattering of $\bar{\nu}_e$ from nuclei is fully predicted by the Standard Model, it has never been experimentally demonstrated because the technique to do so has not existed. With a detector mass of $\sim$200 pounds there would be collected $\sim$700 events/day from the coherent reaction $\bar{\nu}_e + n \to \bar{\nu}_e + n$ and for calibration purposes 30 events/day from the well established $\bar{\nu}_e + e^- \to \bar{\nu}_e + e^-$ which is also directly related to the solar neutrino reactions.

CONCLUSION

It appears to us that at the present time nature has provided us with a rare conjunction of scientific problems which may have a unique explanation in particle physics which may be discovered if we can bring to bear upon them the right combinations of experimental tools. Among the problems are the clear limitations of the Standard Model, how to test the present theories to go beyond — including those whose unifying mass scales are not accessible terrestrially, and the dark matter problem of astrophysics. All of the tools to tackle these problems are not entirely in hand: the SSC which will take us to the next important step of the unification scale is under development and the search for new experimental approaches to utilize it has been an important part of this — and other — Snowmass Workshops. As we have developed above, there are astrophysical sources which can complement the SSC–scale and for which there are new detector ideas under consideration and development. The financial resources necessary to develop these new astro-particle detectors is very modest in comparison. The detectors show promise of other application and there is a body of talented physicists from the disciplines of electrical engineering, material science, condensed matter and particle physics who are actively engaged. Probably they do not yet constitute a critical mass but with modest R&D funding could be expected to very soon constitute one.

The physics areas to be tested and the detector development to do so encompass a set of fundamental and important problems at what might be considered a new frontier at the interface between astrophysics, condensed matter and particle physics. The prospects for excitement and discovery are clearly there.

Acknowledgment: This work was partially supported by the U.S. Department of Energy.

REFERENCES

1) Ellis, John, Lectures presented at the NATO Advanced Study Institute: Quarks, Leptons, and Beyond, (Munich ASI, 451 (1983).
2) Lee, B. W. and Weinberg, S., Phys. Rev. Lett., 39, 165 (1977).
3) Yang, J. et al., Ap. J., 281, 493 (1984).
4) Denegri, D., Sadoulet, B., Spiro, M., Submitted to Rev. of Modern Physics, LBL-88-26014, (1988).
5) Spergel, D. N., Bahcall, J. N., Phys. Lett. 200B, 366 (1988).
6) Bahcall, J. N., Casertano, S., Ap. J., 293, L7 (1985).
7) Hegyi, D., Olive, K. A., Ap. J., 303, 56 (1986).
8) Primack, J. R., Seckel, D. and Sadoulet, B., Ann. Rev. Nucl. Part. Sci., 38, 751 (1988).
9) Silk, J., Olive, K., Srednicki, M., Phys. Rev. Lett., 55, 257 (1985).
10) Krauss, L. M. et al., Ap. J., 299, 1001 (1985).
11) Press, W. H., Spergel, D. N., Ap. J., 296, 679 (1985). Faulkner, J., Gough, D. O., Vahia, M. N., Nature, 321, 226 (1986).
12) Goodman, M. W., Witten, E., Phys. Rev. D., 31, 3059 (1985).
13) Sadoulet, B., in Ulmer, M. L., Proceedings of the 13th Texas Symposium on Relativistic Astrophysics, (World Scientific, Singapore) p. 260 (1987). See also Ref 8.
14) Griest, K., Submitted to Phys. Rev. Lett., Fermilab preprint #FNAL-Pub-88/74A (1988).
15) Ahlen, S. P. et al., Phys. Lett. B, 195, 603 (1987).
16) Caldwell, D. O., Eisberg, R. M., Grumm, D. M. et al., Phys. Rev. Lett., (1988).
17) Weinberg, S., Phys. Rev. Lett. 19, 1264 (1967); Salam, A., *Elementary Particle Theory*, ed. N. Svartholm, Almqvist and Wiksell, Stockholm (1968); Glashow, S. L., Nucl. Phys. 22, 579 (1961).
18) Rowley, J. K., Cleavland, B. T. and Davis, R., in *Solar Neutrinos and Neutrino Astronomy*, ed.: M. L. Cherry, W. A. Fowler and K. Lande (A.I.P. Conf. Proc. #126, N.Y.) p. 1 (1985); Davis, R., Report to the Seventh Workshop on Grand Unification (ICOBAN '86, Toyoma, Japan) p. 237 (1986); Davis, R., Phys. Rev. Lett. 12, 203 (1964).
19) Leighton, R. B. et al., Ap. J. 135, 474 (1962); Ulrich, R. K., Ap. J. 162, 993 (1970);
20) Bethe, H. A. and Critchfield, C. L., Phys. Rev. 54, 248 (1938); Bethe, H. A., Phys. Rev. 55, 434 (1939); Alpher, R. A. and Herman, R. C., Rev. Mod. Phys. 22, 153 (1950).
21) For reviews see: a) Christensen–Dalsgaard, J., et al., Science 229, 923 (1985); b) Bahcall, J. N. and Ulrich, R. K., Rev. Mod. Phys. 60, 297 (1988).
22) Bahcall, J. N. et al., Rev. Mod. Phys. 54, 767 (1982); see also Ref. 5b.
23) Gilliland, R. et al., Ap. J. 306, 703 (1986); Gelmini, G. et al., Nucl. Phys. B281, 726 (1987); Raby, S. and West, G., Phys. Lett. 194B, 557 (1987).
24) Dar, A. and Dado, S., Phys. Rev. Lett. 59, 2368 (1987); Frieman, J. et al., SLAC Pre-print # SLAC-Pub-4261 (1987).
25) Duncan, M. J. et al., Univ. of Pennsylvania Report #UPR-0138-T (1986); Nussinov, S. and Rephaeli, Y., Phys. Rev. D36, 2278 (1987).
26) Private communication, Davis, R., and as reported by Landé, K., in these Proceedings.
27) Okun, L. B., Sov. J. Nuc. Phys. 44, 546 (1986); Voloshin, M. B. et al., 44, 440 (1986); and Ref. 9; Latimer, J. and Cooperstein, J., Phys. Rev. Lett. 61, 23 (1988).
28) Vuilleumier, J.-L., Rep. Prog. Phys. 40, 1293 (1986).
29) Lanou, R. E., Proceedings of Snowmass Workshop, p. 538 (1982).
30) Wolfenstein, L., Phys. Rev., D17, 2369 (1978).
31) Mikheyev, S. P. and Smirnov, A., Sov. J. Nucl. Phys. 42, 1441 (1985).
32) Beier, E. W., Proceedings of 7th Grand Unification Workshop (Toyama, Japan) p. 79 (1986) and Suzuki, A. et al., Univ. of Tokyo Report # UT–ICEPP–86–07 (1986); Hirata, K. et al., ibid #87-04 (1987).
33) Chen, H. H., Phys. Rev. Lett. 55, 1534 (1985); Ewan, G. T. et al., Sudbury Neutrino Observatory Proposal #SNO–87–12 (1987).
34) GALLEX: Kirsten et al., in "Massive Neutrinos in Astrophysics and in Particle Physics", ed O. Fackler and J. Tran Than Van, (Editors Frontieres) p. 119 (1986); BAKSAN: Barabanov, I. R. et al., in "Solar Neutrinos and Neutrino Astronomy" (AIP Conf. Proc. #126, N.Y.), ed. M. Cherry, W. Fowler, K. Lande, p. 175 (1985).
35) See for example: Kuo, T. K. and Pantaleone, J., Phys. Rev. Lett. 57, 1805 (1986); Baltz, A. J. and Weneser, J., Phys. Rev. D35, 528 (1986) and Rev. 5b; Zaglauer, H. and Schevarzer, K., Phys. Lett. 198B, 556 (1987).
36) Sadoulet, B. et al., Astrophys. J., 324, L75 (1988).
37) Rehak, P. et al., Nucl. Instrum. Methods a 238, 459 (1986).
38) Luke, P.K., Submitted to Appl. Phys. LBL preprint (1988).
39) Iqbal, M. A. et al., Nucl. Inst. Methods, A243, 459 (1986).
40) Elliot, S. R. et al., Phys. Rev. Lett., 56, 2582 (1986).
41) Rich, J., Spiro, M., DPhPE, CEN-Saclay preprint, (1987).
42) Bahcall, J., Baldo-Ceollin, M., Cline, D., Rubbia, C., Phys. Lett. B178, 324 (1986).
43) IMB: LoSecco, J. et al., Phys. Lett. B188, 388 (1987); Kamiokande: Totsuka, Y., Univ. of Tokyo Report #UT–ICEPP–87–02 (1987); Frejus: Kuznik, E. et al., in "New and Exotic Phenomena", ed: O. Fackler, and J. Tran Thanh Van (Editions Frontieres) 215 (1987).
44) Ellis, J. et al., Phys. Lett. B198, 393 (1987).
45) Lanou, R., Maris, H. and Seidel, G., Phys. Rev. Lett. 58, 2498 (1987).
46) Sadoulet, B., IEEE Trans. on Nucl. Sci., NS-36, 47 (1988).
47) Proceedings of European Workshop on Low Temperature Devices for the Detection of Low Energy Neutrinos and Dark Matter, Gonzales-Mestrez, L. and Perret-Gallix,D. edit., (In press, (1988).
48) Bernas, H. et al., Phys. Lett., 24A, 721 (1967).
49) Pretzl, K. Preprint Max Planck Institute, Munich, MPI-PAE/Exp. El. 187 (1987).
50) Young,B.A. et al., Applied Superconductivity Conference, San Francisco, Stanford Preprint (1988).
51) Twerenbold, D., Zehnder, A., J. Appl. Phys., 61, 1 (1987); Krauss, H. et al., Europhys. Lett., 1, 161 (1986).
52) Ziman, J., in "Electrons and Phonons", Oxford University Press (1960).
53) Maris, H., in "Fifth Intl. Conf. on Phonon Scattering in Condensed matter", Urbana, Ill. (1986) and Brown Univ. Preprint #C00–3130TC–29 (1986).
54) Newhauser, B. et al., in Proceedings of Applied Superconductivity Conf. (Baltimore) 1986.

55) See for example: Wilks, J., in "Liquid and Solid Helium", Oxford Univ. Press (1967).
56) See for example several papers in "Proc. of Workshop on Low Temp. Detectors for Neutrinos and Dark Matter", ed. K. Pretzel et al., Springer–Verlag (1987).
57) Lanou, R., Maris, H. and Seidel, G., in Proc. of XXIII Rencontre de Moriond (Les Arc, France) in press (1988).
58) Peterreins, T. et al., Phys. Lett. 202B, 161 (1988).
59) For example: Seidel, G. M., at A.P.S. "March" meeting (New Orleans, 1988); Sadoulet, B., at IEEE Nuclear Symposium (San Francisco, 1987); Sadoulet, B., at 13th Texas Symposium on Relativistic Astrophysics (Chicago, 1987).

Fig. 1: Schematic of scenarios leading to detection of elementary particle dark matter.
Fig. 2: Dark matter particle interaction rates vs. particle mass for two particle archetypes.
Fig. 3: Differential energy distribution and asymmetry resulting from annual earth–sun motion.
Fig. 4: Predicted solar neutrino spectrum.

Fig. 1

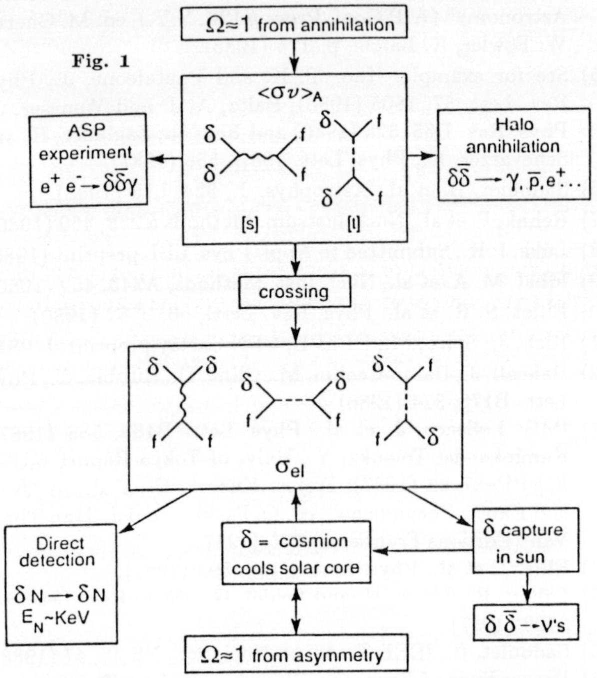

Fig. 4

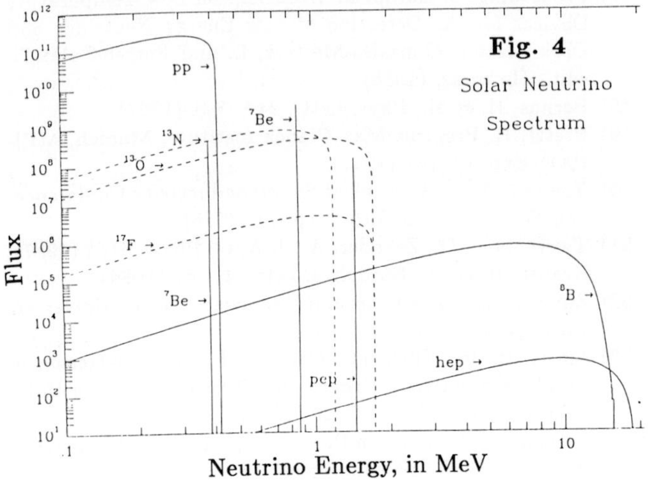

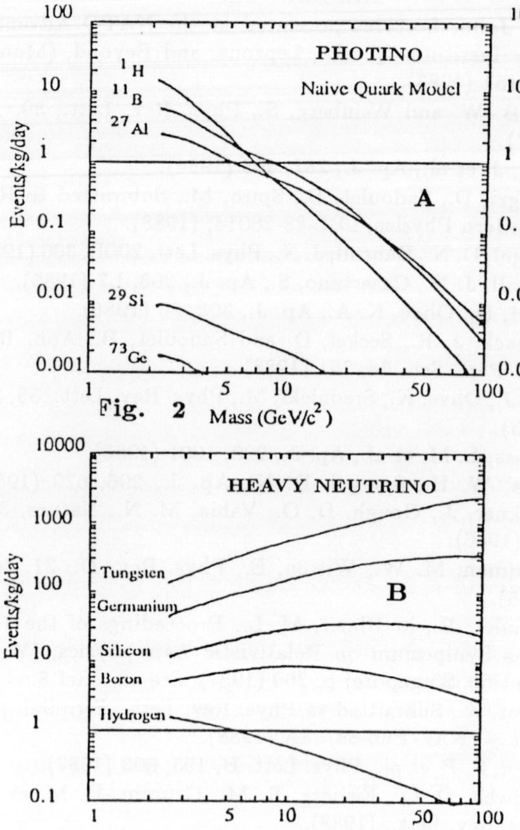

Fig. 2

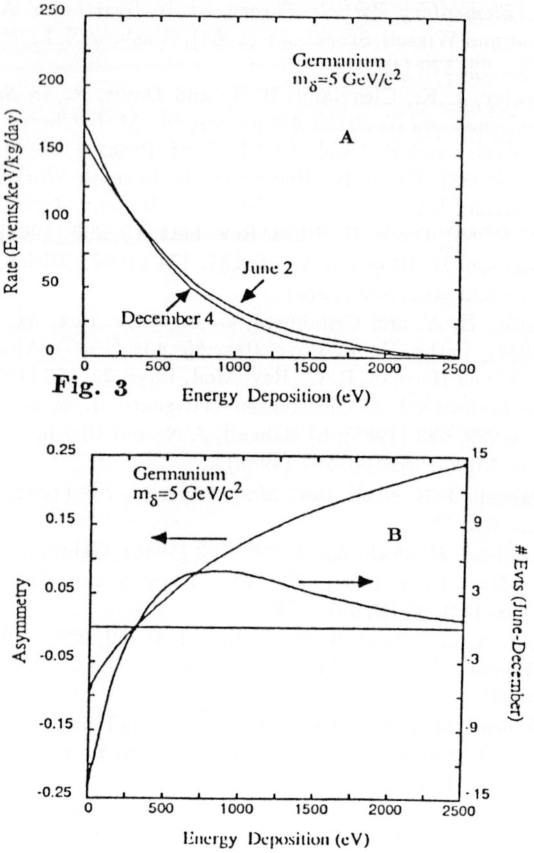

Fig. 3

UNDERGROUND/UNDERWATER DETECTORS FOR MONOPOLES, MUONS, AND HIGH ENERGY NEUTRINO ASTRONOMY

Stuart L. Mufson
Indiana University

ABSTRACT

In this paper we review the generation of experiments which will be the first to push the search for GUT monopoles significantly below the Parker limit: In addition, we discuss the underground/underwater detectors which will play a significant role in improving our understanding of high energy particle astrophysics, neutrino physics, and cosmic ray neutrino physics in the 1990 s.

I. INTRODUCTION

In recent years there has been a great upsurge in interest in the physics and astrophysics which can be studied with detectors placed deep underground and underwater. The continuing questions raised by advances at the intersection of particle physics and astronomy - for example, undiscovered fundamental particles like the monopole which arise naturally in many big bang scenarios and the identity of the nonbaryonic 'dark matter' which is thought to comprise 50-90% of the mass in the universe - have lured many experimentalists into non-accelerator astrophysics. In this field, there is no need yet for the size, sophistication, and expense of present-day accelerator experiments. Underground/underwater detectors are also being designed to search for cosmic sources of high energy neutrinos, the detection of which will open a new window onto the most energetic processes occurring in the universe. Terrestrial particle physics questions too, like neutrino oscillations and weak interacton cross sections at ultrahigh energies, will be studied by these detectors. Finally, the long-standing problems of the origin and composition of the highest energy cosmic rays will be explored intensively by this newest generation of underground/underwater detectors.

II. MAGNETIC MONOPOLES

A. The Physics of Monopole Detection

In 1982 Cabrera cautiously reported[1)] an event which triggered an explosion of effort into the search for magnetic monopoles. The result of this effort was the realization that monopoles are not plentiful in the universe and their detection will require the deployment of very large and sophisticated detectors. Unlike the 'early years', the prevailing attitude among experimentalists today is that the chance of discovering a magnetic monopole is relatively slim, but their detection would ultimately have a revolutionary impact on physics.[2)] "The observation of a monopole would at once place experimental physics at mass scales many orders of magnitude beyond the standard model and at epochs even more remote than the elementary synthesis era."[3)] In addition, there is some hope that magnetic monopoles will provide at least a partial solution to the 'dark matter' problem.

One of the major difficulties associated with the search for magnetic monopoles is that their properties are poorly constrained. About the only property of monopoles on which almost all agree is that their magnetic charge, as first given by Dirac[4)], is $g = ne/2\alpha$, where n is an integer, e is the electron charge and α is the fine structrure constant. Dirac's study further showed that the quantization of the electric charge, a long-standing challenge to theoretical physics, follows naturally if even one monopole exists. For experimentalists it is sufficient to assume that n=1, because n >1 leads to even greater effects. Until recently there was relatively little theoretical guidance concerning the mass of monopoles. It was originally assumed that monopoles would have a mass scale similar to nucleons, and many unsuccessful searches for the significant ionization losses expected from such monopoles were carried out in each new generation of accelerators. In addition, many searches for monopoles at this mass scale which would come to rest in matter (geological strata, lunar rocks) have also proven negative.

Grand Unified Theories (GUTs) have provided a more interesting monopole and more direction for the search. GUT monopoles are massive ($>10^{14}$ GeV/c^2)[5)] and likely to be moving nonrelativistically. Their low velocity is a direct result of their presumed primordial origin in the big bang. By the present epoch, primordial monopoles would have cooled to low energies like the microwave background photons had they no interactions with matter. However, monopoles are accelerated by magnetic forces in their local Galactic environment to speeds typical of the Galaxy's gravitational potential (10^{-3} - 10^{-4} c). This means that all previous monopole searches which were based on the detection of large ionization losses from relativistic monopoles with a nucleonic mass and a Dirac charge were insensitive to GUT monopoles. GUT monopoles may have additional properties that would complicate search stategies. For instance, GUT monopoles may bind a proton or heavier nucleus; they may possess an intrinsic electric charge; and/or they may catalyze proton decay.[6)]

The abundance of monopoles is difficult to predict. In simple big bang-Grand Unified theory scenarios, there would presumably be about one monopole per event horizon produced at the breaking of GUT symmetry. If expansion proceeded in the conventional way until nucleons condensed at 10^{-6} s, then approximately one monopole per baryon is expected. GUT monopoles with a density as great as this would have closed the universe immediately. On the other hand, if inflation occurred between monopole production and baryon production, there might only be only one monopole within the present event horizon. Consequently, the expected range of monopole abundances is uncertain within a factor of 10^{80}! (ref. 3)

At the present time, all searches for GUT monopoles - based on astrophysical arguments (wide ranging, very model dependent limits), superconducting induction experiments ($< 2\times10^{-12}$ cm^{-2} sr^{-1} s^{-1}), track-etch detectors ($< 10^{-17}$ cm^{-2} sr^{-1} s^{-1} at 10^{-3} c), scintillator experiments ($< 1.9\times10^{-15}$ cm^{-2} sr^{-1} s^{-1}), wire proportional counters and streamer tubes ($< 2\times10^{-14}$ cm^{-2} sr^{-1} s^{-1}), and nucleon decay catalysis - have proven to be negative[6)]. But many of these quoted limits are not very restrictive (induction experiments), or are very model dependent (astrophysical), or are dependent upon a specific type of GUT monopole (track-etch).

A useful reference flux for monopole searches is the Parker limit.[7)] Parker and his colleagues have argued that if monopoles exist at fluxes greater than this limit, they will 'short circuit' the magnetic field of our Galaxy in less than 10^{8} yrs, a time scale over which the field has been present. This limit for a 10^{16} GeV/c^{2} monopole (GUT mass scale) at 10^{-3} c is 5×10^{-16} cm^{-2} sr^{-1} s^{-1}. (It is interesting to note that if monopoles were to exist at the GUT mass scale, and they were distributed uniformly throughout the universe at the Parker limit, monopoles would account for a significant fraction of the 'dark matter'.) Except for the track-etch results, no monopole experiment has yet reached the Parker limit. (The assumptions implicit in the track-etch result and its inherent uncertainties have been discussed by Ahlen[8)]).

It is clear that the search for monopoles at and below the Parker limit is going to require detectors larger than any which have yet been built and which operate for many years. Two such detectors, MACRO and LVD, are currently under construction.

Both MACRO[9)] and LVD[10)] are large underground detectors currently being constructed in the Gran Sasso Underground Laboratory[11)] in Italy. Of these two experiments, MACRO is the one which has been designed with a primary objective of searching for GUT monopoles at flux levels significantly below the Parker limit. (It should be noted that the techniques used in GUT monopole searches will also make these experiments sensitive to other ultramassive magnetically and/or electrically charged particles such as fractionally charged superstring particles.) MACRO has been optimized for this monopole search by building it with the largest possible acceptance, by including enough redundancy in it to be convincing on the basis of a small number of events, and by placing it deep underground in a low background environment. Both MACRO and LVD are being built with liquid scintillation counters and limited streamer tubes; MACRO has in addition track-etch detectors.

streamer tubes; MACRO has in addition track-etch detectors.

The acceptance of these detectors to an isotropic flux of penetrating monopoles, and a comparison with other underground detectors searching for GUT monopoles, is given in Table 1. MACRO has been designed to reach the Parker in approximately 1 year of operation and will be an order of magnitude below the Parker limit in 5 years.

In the final column of Table 1 are listed the detection techniques used by these GUT monopole experiments. Both MACRO and LVD will have large volumes (MACRO = 1,000 metric tons; LVD = 1,800 metric tons) of liquid scintillator separated into modules which are viewed by fast photomultiplier tubes. Each of these modules will have slow particle triggers for monopoles. In the liquid scintillation layers, monopoles are detected by their long-duration scintillation pulses and their long times-of-flight through the detector.[12)] Experimental studies using slow protons have demonstrated indirectly that scintillators will respond to monopoles with velocities greater than 3×10^{-4} c [13)].

The tracking systems in these detectors consist of highly segmented arrays of limited streamer tubes similar to devices already in use in the NUSEX experiment in the Mont Blanc[14)]. The tubes have digital readouts in two coordinates and the spatial accuracy of these arrays will be approximately 1 cm. For MACRO the gas flowing through the streamer tubes will contain a significant admixture of He, making the tracking system sensitive to slow monopoles through the Drell-Penning effect.[15)]

Finally, there will be track-etch detector (CR-39) in MACRO. The track-etch detector will be deployed in the "triggered mode"; that is, the track-etch sheet will be removed, etched, and scanned in the module crossed by a candidate track in the active detectors.[16)]

III. NEUTRINO ASTRONOMY/PHYSICS

A. High Energy Neutrino Astronomy

1. Energetics of Cosmic Neutrino Emission

The recent detection of low energy (10-50 MeV) stellar collapse neutrinos from SN 1987A by both the Kamioka and IMB detectors has ushered in the age of extra-solar system neutrino astronomy.[17)] Although the search for astrophysical point sources of high energy neutrinos (>1 GeV) has so far been unsuccessful[18)], there is every reason to expect that such sources exist. The basic motivation driving this search is the desire to open a new window on the cosmos. High energy neutrinos will provide a powerful probe of astrophysical processes where energetic hadronic interactions are occurring, e.g., in the environment of the accretion disks surrounding collapsed objects and near the engine driving luminous extragalactic sources. In many ways neutrino astronomy is complementary to photon astronomy because neutrinos can escape from dense concentrations of matter impenetrable to electromagnetic radiation. In addition, the different cross sections for the interactions of neutrinos and antineutrinos with normal matter make it possible for neutrinos to differentiate between matter and antimatter in cosmic sources. Although not feasible today, future neutrino detectors will probe the universe for baryon/antibaryon asymmetry. The physics of neutrino detection offers yet another advantage - these detectors are sensitive to significant regions of the celestial sphere for 24 hours/day making the detection of serendipitous sources more likely. However, neutrino astronomy requires detectors with very large areas placed deep underground/underwater since the signal is weak and the background is large.

Over the past two decades, there has accumulated abundant evidence that exceedingly energetic phenomena occur widely throughout the universe. This evidence has been interpreted as implying spectacular but poorly understood hadron acceleration occurs commonly. The cosmic rays which have been detected with energies in excess of 10^{20} eV provide direct evidence for such particle acceleration. Observations of UHE (100 TeV - 100 PeV) and VHE (100 GeV - 100 TeV) γ-rays from the Galactic X-ray binaries Cyg X-3, Her X-1, and Vela X-1 provide more indirect evidence. In these sources it is assumed that the γ-rays are produced when a beam of energetic protons, accelerated at the neutron star, strikes the atmosphere or the accretion wake of the companion star . At the points in the neutron star orbit where the proton beam is directed toward the earth through intervening target material, hadronic interactions will produce abundant π^o's. The decay of these neutral pions on their flight to earth will give rise to the observed high energy γ–ray signal. Most investigators consider it unlikely that these UHE and VHE γ-rays come from electromagnetic processes since energy losses at these energies generally proceed too rapidly to allow the required acceleration of electrons. In this scenario, the corresponding $\pi^{\pm}$ decay will result in a ν flux comparable or larger than the γ-ray flux. These neutrino signals, however, will be quite feeble because neutrinos are so weakly interacting. Nevertheless, by adding the signals in phase with the known binary period, the signal/noise can be improved significantly, perhaps over the threshold of detection.

In addition to Galactic sources, there are many extragalactic sources of high energy γ-ray emission. The quasar 3C 273, the double radio source Cen A, and the Seyfert galaxy 3C 120 all emit in the high energy γ-ray band. Again, the high energy photons observed in these sources are thought to be secondary products of hadronic processes occurring deep within the enigmatic 'engine' which drives these objects. However, the detection of extragalactic sources will be more difficult than Galactic binaries since the signals will presumably be weaker and unmodulated. Only the large underwater experiments currently being proposed for the middle to late 1990 s, DUMAND II, GRANDE, and Lake Baikal are large enough to have a reasonable chance of detecting extragalactic sources.

Weakly Interacting Massive Particles (WIMPs) are often proposed as candidates for the dark matter in the universe. If they exist, these particles (photinos, majorana neutrinos, Dirac neutrinos, etc.) would be created early in the universe and then gravitationally concentrated in the sun. Subsequent mutual annihilation deep inside the sun would produce neutrinos which could result in a detectable signal. Detailed calculations of the neutrino spectra and flux expected from this mechanism have already been performed[19)] and one of the more interesting candidates, the photino, appears to be just beyond the reach of present detectors such as IMB, Kamioka, and Frejus.

In summary, there are likely to be many point sources of neutrinos both inside and outside the Galaxy. It is also likely, given the historical development of photon astronomy, that unanticipated objects and phenomena will be discovered once the high energy neutrino window is opened onto the cosmos.

2. Cosmic Neutrinos from Hadronic Interactions

Given that the γ-ray signal from UHE and VHE sources results from the decay of π^o's from hadronic interactions ($\pi^o \rightarrow \gamma\gamma$), then neutrinos will also be naturally produced in the same interactions by the decay of $\pi^{\pm} \rightarrow \mu^{\pm} + \nu_{\mu}$. The connection between the observed γ-ray flux and the accompanying ν flux has been discussed by Kolb *et al.* [20)]. Assuming that the observed γ-rays are due to π^o decay, that the charged pions which are also produced in such hadonic interactions decay in flight before they interact, and that the K and charmed meson contributions can be neglected, then the measured γ-ray spectrum $\frac{dS_\gamma}{dE_\gamma} = c\, E_\gamma^{-\alpha}$ implies a pion spectrum [21] : $\frac{dS_{(\pi^+ + \pi^-)}}{dE_\pi} = 2\,\frac{dS_{\pi^o}}{dE_\pi} = \alpha c E_\pi^{-\alpha}$. The resulting π-decay neutrino spectrum is given by

$$\frac{dS_\nu}{dE_\nu} \approx c\,(1 - \frac{m_\mu^2}{m_\pi^2})^{\alpha-1}\, E^{-\alpha} \approx (.1\text{-}.4)\,\frac{dS_\gamma}{dE_\gamma}, \qquad (1)$$

for typical spectral indices in the range $\alpha = 2\text{-}4$ [22].

In a binary star system, however, there will be an energy dependent ν–γ duty factor, λ(E), which accounts for the fact that neutrino emission lasts through much of the orbit whereas γ-ray emission is significantly decreased if the target thickness is much greater than 200 gm/cm^2. For neutrinos, there will only be absorption by the main body of the companion star. The estimates of λ(E) are very model dependent and fall in the range $1 \leq \lambda(E) \leq 1{,}000$ [20,23]. Values of λ are typically assumed to be 20-50. Further, a fraction of the γ-ray signal, $f_{\gamma\text{-bb}}(E)$, will be attenuated on the microwave background photons ($\gamma_{uhe}\gamma_{bb} \to e^+e^-$)[24]. The emitted π-decay ν spectrum is thus

$$\frac{dS_\nu}{dE} \approx 0.3\,\lambda(E)\, f_{\gamma\text{-bb}}(E)\,\frac{dS_\gamma}{dE}. \qquad (2)$$

Since the λ and $f_{\gamma\text{-bb}}$ factors are poorly constrained, the ν spectrum for a given γ spectrum is quite uncertain.

High energy neutrinos are detected in large underground/underwater detectors when their interactions in the rock/water surrounding the detector produce an energetic muon which passes through or stops in the detector. So when computing event rates, an energy dependent volume must be used, $V(E_\mu) = \varepsilon\, A\, R_\mu$, where A is the geometric area of the detector, ε accounts for the on-source duty factor and the efficiency for detecting a neutrino source at a particular orientation with respect to the detector, and $R_\mu = 3\text{x}10^5 \ln(1 + \frac{2E_\mu}{1\ \text{TeV}})$ cm is the range of the muon. The probability that a neutrino of energy E_ν headed toward the detector interacts in the rock within a distance R_μ of the detector and then produces a muon which then traverses the detector is $P(E_\nu) = \rho R_\mu \sigma_\nu(E_\nu = 2E_\mu)$, where ρ is the particle density of the material, σ_ν is the charged current weak interaction cross section, and it is assumed that a ν of energy E_ν produces a muon of energy $E_\mu = E_\nu/2$. The π-decay, ν-induced muon event rate in a detector of geometric area A is then

$$n_\mu \approx \rho\varepsilon A \int_{E_{th}} dE_\mu R_\mu \left[\frac{dS_\nu}{dE_\nu}\sigma_\nu\right], \qquad (3)$$

where E_{th} = the threshold for muon detection in the detector ≈ 3 GeV for underground detectors and ≈ 100 GeV for underwater detectors; the quantity in brackets is evaluated at $E_\mu = E_\nu/2$; and $\frac{dS_\nu}{dE_\nu}$ is obtained from eq.(2).

Calculations by Gaisser and Stanev [25] have demonstrated that the largest contribution to the muon signal comes from neutrino enrgies near 1 TeV with the peak depending on the spectrum of the source. Since the greatest contribution to the muon signal comes from energies in excess of 100 GeV, the parameter which is most important when comparing neutrino detectors is the geometric area. In Table 1, the geometric area of the various underground/underwater detectors is given.

3. Cosmic Neutrino Detectors

Since the detection of neutrinos is based upon the detection of muons, the technology is well-known. The difficulty is that the weak muon signals from cosmic sources must be distinguished from the background due to muons and neutrinos produced in high energy cosmic ray interactions in the atmosphere. To discriminate upward-going muons due to neutrino interactions from downward-going muons due to cosmic ray interactions, large detector dimensions and fast timing electronics for time-of-flight measurements are required. The expected ratio of downward/upward muon events demands that the discrimination be better than 10^6:1. To differentiate muons produced by atmospheric neutrinos from those produced by neutrinos from cosmic sources requires a detector with excellent angular resolution so that very few of the isotropic atmospheric events fall into a given angular bin. Using accelerator data at 100 GeV and standard theory to extrapolate to higher energies, it been shown that most charged current events have lepton scattering angles $\theta \leq 1.5^o/\sqrt{E_\nu(\text{TeV})}$. [26] Since the largest contribution to the cosmic neutrino signal is from neutrinos with energies of order 1 TeV, cosmic neutrino telescopes need angular resolutions $\leq 1^o$ to be most effective in rejecting this background. For comparison, the muon angluar resolution for the Cerenkov underground proton decay experiments is typically a few degrees.

The large underground detectors currently under construction, MACRO and LVD, have already been described. Their fine-grained streamer tube tracking result in angular resolutions $\Delta\theta \leq 0.2^o$, and their fast timing electronics provide excellent background rejection.

The proposed large underwater detectors, DUMAND [27], GRANDE [28], and Lake Baikal [29] plan to use the ring imaging Cerenkov technique which already has been successfully demonstrated by IMB and Kamioka. These detectors discriminate upward from downward-going muons by the directionality of their Cerenkov light and their time of flight through the detector. As a direct result of their great size, DUMAND II and GRANDE, and presumably Lake Baikal, have angular resolutions better than 1°.

DUMAND II is a deep ocean neutrino laboratory. The site is approximately 25 km west of the Island of

Hawaii and at a depth of 4.8 km . The proposed design consists of 9 vertical strings of 230 m length, each with photomultipliers spaced 10 m apart vertically along the string. The strings are spaced horizontally in an octagonal configuration. The major advantage of this approach is that the deep ocean is one of the few locations suitable for neutrino detectors in the cubic kilometer class. GRANDE is a large detector to be placed in a rock quarry in Arkansas. The proposed design of GRANDE consists of 5 optically isolated layers suspended in a light-tight plastic membrane bag 250 m x 250 m x 70 m filled with purified water. The bag will be suspended in a larger body of unpurified water. The three lowest layers, the neutrino telescope, are instrumented with planes of downward-looking pmt's while two of the upper three layers, the γ-ray telescope, have planes of upward-looking pmt's. The advantage of this approach is that it combines a γ-ray telescope with a neutrino telescope. In addition, the γ–ray telescope serves as an efficient anti-coincidence monitor for the downward-going particles that might otherwise trigger the neutrino telescope. Lake Baikal is a Soviet experiment which looks at muons traversing vertical planes in the 1.4 km deep lake in Siberia.

ICARUS II [30] is a large volume detector which has been proposed for the Gran Sasso Laboratory or some other deep site. This detector relies on a technology in which electrons liberated by ionizing events inside a large volume of ultrapure liquid argon remain untrapped long enough (>1 ms) to drift in an electric field to planes of sensing wires. In principle, these planes provide a high resolution (± 1 mm) three-dimensional image of all ionizing events. In addition, ionization losses along the particle path are accurately recorded. It is claimed that this detector is capable of unambiguous identification of all tracks for both throughgoing and stopping particles. Initially this detector will be 200 tons and dedicated to the study of B^8 solar neutrinos. A later generation detector will be enlarged to several thousand tons and will investigate the full range of problems planned for the other underground/underwater detectors. In this device a slightly worse spatial resolution is accepted in exchange for a much larger detector volume.

On further technology which is on the horizon may eliminate the need for large volume detectors of the DUMAND and GRANDE class. Resistive plate counters [31] are well-suited for large underground experiments where a time-of-flight system can be used to discriminate upward/downward muons. RPC's provide the excellent spatial resolution and low noise operation required of such systems. However, before such systems can be deployed on a large scale, it must be demonstrated that they can operate at atmospheric pressures.

B. Neutrino Physics

1. Neutrino Oscillations

The atmospheric neutrino-induced upward-going muon flux can be used to search for neutrino oscillations in underground/underwater detectors. There are two methods by which $\nu_\mu \rightarrow \nu_e$ oscillations can be studied. The first relies on detection of contained events, i.e., interactions in the detector itself. The contained events in the scintillator or water can be used to infer the direction of motion of the incident neutrinos. The ratio of upward to downward incident atmospheric neutrinos is then an indicator of neutrino oscillations.[32] This is the method employed by IMB[33] to give limits on the mass parameter and the mixing angle in a range of parameter space inaccessible to accelerator experiments; it is also likely to be the method employed by LVD and MACRO.

The second method is more direct. The determination of the angular distribution of the muons induced by ν_μ interactions in the water/rock surrounding the detector is a direct measure of the loss of neutrino flux due to neutrino oscillations. The excellent angular resolution and their large atmospheric neutrino event rate makes this the preferred method to search for neutrino oscillation in the large water detectors.

2. Neutrino Cross Section at High Energies

Above a few hundred GeV the neutrino charged current interaction has not been measured at accelerators. Above 100 TeV calculations [27] suggest that the cross section is enhanced. This enhancement can in principle be checked by the large underwater detectors.

IV. COSMIC RAY MUON PHYSICS

Multiple coincident muon bundles in underground /underwater detectors can be used to measure the elementary composition of primary cosmic rays in an energy range (> 10^{14} eV/nucleon) where the flux is too low for more direct experiments. This investigation relies on understanding how the multiplicity distribution for high energy cosmic ray interactions in the atmoshere is related to the composition of primary cosmic rays.[34] The assumption is that for a given nucleus energy, a heavier nucleus on the average produces many more muons, and consequently its multiple muon/single muon ratio is larger. The existing multimuon data which presently exist are difficult to analyze and the reported conclusions do not agree.[35] These problems are primarily due to limited statistics, detectors which are often small compared with the muon separations within the bundle, and detectors which are at depths too shallow to probe high energies.

Large underground/underwater detectors will be able to probe the the composition of cosmic rays above the 'knee' at 2×10^{15} eV/nucleon. In 5 years, for instance, MACRO will record 10^8 single muon events and 10^6 multimuon events, a large enough sample to compare in detail with theory. The large underwater detectors are expected to be an order of magnitude better than this.

V. CONCLUSIONS

The underground/underwater detectors presently under construction or being proposed for the 1900s will significantly improve our understanding of high energy particle astrophysics, neutrino physics, and cosmic ray muon physics. They will provide the first experiments capable of searching for magnetic monopoles at flux levels significantly below the Parker limit. They will offer the exciting possibility of opening the high energy neutrino window onto energetic hadronic processes occurring in the universe. In addition, they will significantly improve our understanding of the composition of high energy cosmic rays. Further, these experiments will explore new regions in the parameter space for neutrino oscillations and will investigate the high energy behavior of the charged current weak interaction cross section.

I acknowledge I.N.F.N., Frascati, Italy for their hospitality while this paper was prepared. I especially wish to thank Rossana Morani of I.N.F.N.; she was always gracious and more than willing to help in times of need. I also thank Dick Streinberg of Drexel University for reading an earlier version of this paper.

References

1) Cabrera, B., Phys.Rev.Lett. 48, 1378 (1982).

2) Carrigan, R.A., and Trower, W.P., eds., *Magnetic Monopoles* (Plenum, New York, 1983); Stone, J.L., ed, *Monopole '83* (Plenum, New York, 1984).

3) Groom, D.E., "Is There Anyplace Left for a Massive Magnetic Monopole to Hide?", presented at *Proceedings of the 3rd Conference on the Intersections between Particle and Nuclear Physics,* Rockport, Maine, May 14-19, 1988.

4) Dirac, P.A.M., Proc.Roy.Soc.London, Ser. A 133, 60 (1931); Dirac, P.A.M., Phys.Rev., 74, 817 (1948).

5) 't Hooft, G., Nucl. Phys.,B79, 276 (1974); Polyakov, A., JETP Lett., 20, 194 (1974)

6) Groom, D.E., Phys.Reports, 140, 323 (1986).

7) Turner, M.S., Parker, E.N., and Bogdan, T.J., Phys.Rev., D26, 1296 (1982)

8) Ahlen, S.P., in *Magnetic Monopoles*, *op.cit.*, p. 1137.

9) Calicchio, M. *et al.*, Nucl.Inst.&Meth., A264, 18 (1988).

10) Bari, C., *et al.*, Nucl.Inst.&Meth., A264, 5 (1988).

11) Bellotti, E., Nucl.Inst.&Meth., A264, 1 (1988).

12) Musser, J.A., *et al.*, Proc.Int.Cosmic Ray Conf., Moscow, 6, 334 (1987).

13) Musser, J., *et al.,* Proc.Int.Cosmic Ray Conf., Moscow, 6, 338 (1987).

14) Battistoni , G., *et al.*, Nucl.Inst.&Meth., A245, 277 (1986).

15) Drell, S.D., *et al.*, Phys.Rev.Lett., 50, 429 (1983).

16) Giacomelli, G., *et al.*, Proc.Int.Cosmic Ray Conf., Moscow, 6, 513 (1987).

17) Hirata, K., *et al.*, " ", Phys.Rev.Lett., 58, 1490 (1987); Bionta, R.M., *et al.*, " ", Phys.Rev.Lett., 58, 1494 (1987).

18) Svoboda, R., Ap.J., 315, 420 (1987).

19) Gaisser, T.K., Steigman, G., and Tilev, S., Phys.Rev., D34, 2206 (1986).

20) Kolb, E.W., Turner, M.S., and Walker, T.P., Phys.Rev., D32, 1145 (1985).

21) Rossi, B., *High Energy Particles* (Prentice-Hall, New York, 1952); GRANDE Proposal to the US DOE (1988).

22) Weekes, T.C., Phys.Rep., 160, 1 (1988).

23) Gaisser, T.K., and Grillo, A.F., Phys.Rev., D36, 2753 (1987); Gorham, P., and Learned, J.G., Nature, 323, 422 (1986).

24) Wdowczyk, J., Tkaczyk, W., and Wolfendale, A.W., J.Phys.A.: Gen.Phys., 5, 1419 (1972).

25) Gaisser, T.K., and Stanev, T., Phys.Rev., D31, 2770 (1985).

26) Stenger, V.J., DUMAND-80, 1, 169 (1980).

27) "DUMAND II, Proposal to Construct a Deep Ocean Laboratory for the Study of High Energy Neutrino Astrophysics and Particle Physics", proposal to US DOE (1988).

28) "Proposal to Construct the GRANDE Facility for the Study of Astrophysical Sources and High Energy Particle Interactions", proposal to the US DOE (1988).

29) Domogatsky, G.V., *et al.*, Proc.Int.Cosmic Ray Conf., Moscow, 6, 292 (1987).

30) "ICARUS: Searching for New Underground Phenomena with High Resolution Visual Techniques and Magnetic Analysis", proposal for the Gran Sasso Laboratory (1985); "ICARUS I: An Optimized, Real Time Detector of Solar Neutrinos" (1988).

31) Battistoni, G., Campana, P., Denni, U., Gustavino, C., and Iarocci, E., Nucl.Inst.&Meth., A270, 190 (1988).

32) Ayres, D.S., *et al.*, Phys.Rev., D29, 902 (1984).

33) LoSecco, J.M., *et al.*, Phys.Rev.Lett. 2299 (1985).

34) Gaisser, T.K., and Stanev, T., Nucl.Inst.&Meth., A235, 183 (1985).

35) Cherry, M., Deakyne, M., Lande, K., Lee, C.K., and Steinberg, R.I., Phys.Rev., D27, 1444 (1983); Tisserant, S., *et al.*, Proc.Int.Cosmic Ray Conf., Moscow, 6, 219 (1987).

Table 1

Underground/Underwater Detectors

Experiment	Acceptance[1] (m^2 sr)	Area [2] (m^2)	Detection Technique[3]
1. Operational			
Baksan	1,850	250	a
Frejus	1,000	75	b
Homestake	1,200	130	a
IMB	---	400	c
KGF	250	15	b
NUSEX	10	12	b
Soudan 1	100	9	d
2. Under Construction			
LVD	7,000	530	a,b
MACRO	10,000	870	a,b
Soudan 2	1,600	15	d
3. Proposed			
DUMAND	-----	2.3×10^4	c
ICARUS II	$1.\times10^4$	$8.\times10^2$	e
GRANDE	-----	6.3×10^4	c
Lake Baikal	-----	4×10^5 (?)	c

Notes:

1. acceptance to an isotropic flux of nonrelativistic monopoles.
2. surface area for the detection of energetic muons.
3. Key: a = liquid scintillator; b = wire proportional counters/streamer tubes; c = water Cerenkov; d = tracking calorimeter; e = liquid Argon.

INVESTIGATIONS OF THE SOLAR NEUTRINO FLUX

Kenneth Lande
Department of Physics
University of Pennsylvania
Philadelphia, PA 19104

The source of energy in the Sun, a typical star, has long intrigued us. For the past half century it has been assumed that the Sun was powered by nuclear fusion, the fusion of hydrogen into helium. In each fusion reaction two protons are converted into neutrons and thus two neutrinos are emitted. Observation of these neutrinos and a measure of their flux verifies the hypothesis of nuclear fusion in the Sun. A study of the energy spectrum of these neutrinos and a measure of the time dependence of their flux provides information about the details of the fusion reaction chain and also permits an investigation of neutrino interaction and transformation as they pass through the Sun and Earth.

Some general remarks can be made about the expected neutrino flux at the Earth. The Sun emits 4×10^{33} ergs/sec of thermal energy. Since the fusion of four hydrogen nuclei into one helium nucleus liberates 26 MeV, the Sun must undergo 10^{38} fusions/sec. Since each fusion reaction produces two neutrinos, the Sun emits 2×10^{38} neutrinos per second, and thus gives rise to a flux of 6.4×10^{10} neutrinos/sec cm^2 at the earth. In order to compare this prediction with observation, we must know the details of the reaction chains involved in the fusion process and thus the energy spectrum of the emitted neutrinos. In addition, a model of solar neutrino emission must address any possible time variation in the production process and the effect on the emitted neutrinos of the solar and terrestrial matter through which they travel between production and detection.

SOLAR NEUTRINO OBSERVATIONS

In 1965, there began the excavation of a large, deep underground chamber in the Homestake Gold Mine in Lead, South Dakota to house a solar neutrino detector. The detector, a tank containing 615 tons of C_2Cl_4 was put into operation in 1968. The reaction of interest is, $\nu_e(\text{Sun}) + {}^{37}Cl \rightarrow {}^{37}Ar + e^+$. Every two months, the accumulated argon is swept from the perchloroethylene liquid by a stream of helium gas. The extracted argon is purified and put into a small proportional counter in which the decay of each ^{37}Ar back into ^{37}Cl is detected. At the end of each extraction cycle, a small amount of stable argon, either ^{36}Ar or ^{38}Ar, is put into the perchloroethylene and mixed with the liquid. The fraction of this stable "carrier argon" that is recovered during the next extraction, typically 95%, determines the extraction efficiency of the neutrino produced ^{37}Ar[1].

During the past twenty years over 100 extractions of ^{37}Ar have been made from the Homestake detector. The average production rate of ^{37}Ar during these runs has been 0.492 ± 0.037 per day [2]. In a separate experiment, using a KOH detector, the ^{37}Ar production rate due to cosmic ray muons has been measured [3]. After correction for the muon induced ^{37}Ar production, there is left a signal above known backgrounds of 0.438 ± 0.047 ^{37}Ar per day or 2.33 ± 0.25 SNU, see Table 1. These observations should be compared with the latest predictions of the Standard Solar Model of 7.9 ± 2.6 SNU [4] (1 Solar Neutrino Unit = 10^{-36} interactions per target nucleus per second).

TABLE 1. SUMMARY OF ^{37}Ar PRODUCTION RATES
(Atoms/day in 615 tons C_2Cl_4)

Period of Observation	1970.3-88.3	1986.8-88.3
Average ^{37}Ar production rate	0.518±0.036	0.87±0.13
Cosmic ray background	0.08 ±0.03	0.08±0.03
^{37}Ar above known backgrounds	0.438±0.047	0.79±0.13
^{37}Ar Production rate in SNU	2.33 ±0.25	4.2 ±0.7

During the past year the Kamiokande water Cerenkov detector has been observing the neutrino-electron elastic scattering signal due to solar neutrinos. The detector operates with a threshold of 9.5 MeV. The Kamiokande group reports that the ratio of their observed rate to that predicted for ^8B by the Standard Solar Model is 0.5±0.17 [5]. During this same period, late 1986 to early 1988, the Homestake observation is 4.2±0.7 SNU (see Table 1) or 0.53±0.09 of the SSM prediction. The consistency between the Kamiokande and Homestake results for the above time period is most comforting.

SUGGESTIONS FOR REDUCING ν_e FLUX

The source of the difference between observation and prediction and the details of the observations of the past twenty years have stimulated widespread experimental and theoretical interest in the scientific community. These explanations generally fall into two categories: (1) modifications of the Standard Solar Model to change the spectrum of neutrinos produced in the solar core and (2) modifications of the properties of the neutrino that would permit conversion of ν_e into other species as the neutrinos pass through the outer regions of the Sun. A number of experiments are now either in preparation or being planned that would test various aspects of these proposed explanations.

There are four principal reactions in the Sun that produce neutrinos.

	REACTION	FLUX(/cm² sec)
(1)	$P + P \rightarrow D + e^+ + \nu_e$	6.0×10^{10}
(2)	$P + P + e^- \rightarrow D + \nu_e$	1.5×10^8
(3)	$^7Be + e^- \rightarrow {}^7Li + \nu_e$	4.7×10^9
(4)	$^8B \rightarrow {}^8Be + e^+ + \nu_e$	6.0×10^6

The neutrinos from the first reaction are below the chlorine threshold and those from the second and third reactions contribute only about 1/4 of the chlorine reaction rate. The main sensitivity of the chlorine detector is to the last reaction, ^8B decay, which represents about 3/4 of its rate. The production of ^8B in the Sun, however, is inhibited by a strong Coulomb repulsion in the formation reaction, P + ^7Be, and so is very sensitive to the temperature in the solar core. A 2 - 3% change in solar core temperature would be enough to suppress the production of ^8B and thus account for the diminished detection rate. The Standard Solar Model, however, does not allow such a reduction in temperature.

The solar core temperature prediction can be lowered either by changing the opacity of the Sun (a change in the chemical composition of the solar interior) [6] or by providing alternate modes by which the energy produced in the solar core can flow to the solar exterior. Among suggestions for such mechanisms are WIMPS and radial oscillations. Wimps (weakly interacting massive particles) would acquire energy in the solar interior by collision, rely on their weak interactions to move through the Sun without interaction and then deposit this energy in the solar exterior by another collision. The cool wimps would then be pulled back to the solar core by gravitational force and repeat the cycle. Radial matter oscillations are an extension of the surface oscillations that have been observed on Sun for the past decade.

The other suggestion for reducing the detectable ν_e flux at the earth would be to have the neutrinos produced in the solar core convert into a non-detectable form in their transit through the Sun. In one of these approaches, the neutrino would interact via its magnetic moment with the toroidal magnetic fields near the solar surface and convert from a left handed to a right handed particle [7]. The magnetic moment required, 10^{-11} Bohr magnetons, is just below that detectable in terrestrial experiments. Since sunspots, the surface manifestation of solar toroidal fields, vary in number with the phase of the solar cycle and depend on the solar latitude, a time dependence of neutrino magnetic moment flip would occur. The reduction of neutrino flux would be greatest during the period of maximum solar activity. In addition, since the plane of motion of the earth is inclined by 7.5° with respect to the solar equator, there should be a seasonal variation in the detected solar neutrino rate.

A second suggestion for attenuating the ν_e flux in its passage through the Sun is a resonant conversion into another neutrino species. This process, an elaboration by Mikheyev and Smirnov [8] on a idea of Wolfenstein, requires a modification of lepton conservation, in that it permits transitions between different neutrino species. It assumes that, in vacuum, ν_e is the lightest neutrino. In the dense interior of the Sun, the ν_e experiences a deeper potential well than do the other neutrino species and thus has a density dependent effective mass. As the neutrino travels out of the Sun, it reaches a region in which the ν_e mass becomes equal to that of another neutrino species. In this region, there is an almost total conversion of ν_e into the other species. Given the densities available in the Sun, such transitions could occur if the mass differences between different species are in the range 10^{-2} to 10^{-3} eV/c^2.

The neutrino resonant conversion mechanism is so effective that it will almost completely eliminate an energy region of the emitted neutrino spectrum. For one set of parameters, the high energy end of the neutrino spectrum will be converted and only the low energy end of the ν_e spectrum will reach the earth. Such a spectrum distortion could explain the apparent low signal measured by the chlorine experiment. Other sets of parameters will either eliminate the low end of the neutrino spectrum or result in a partial reduction of the entire spectrum.

For one region of mass difference and interaction strength, where the resonant conversion occurs in the low density region of the Sun, it is even possible that a partial reconversion of neutrinos occurs in the Earth. For this case, there will be a low ν_e flux during the day, when the Sun is above the detector and only a few kilometers of earth are between the Sun and the detector. At night, when the solar neutrinos must travel through 10^4 km of earth to reach the detector a partial reconversion to ν_e can occur. This case is particularly interesting since it naturally results in a observed neutrino flux that is about 1/3 of that predicted.

There are clearly more than enough explanations for a reduced ν_e flux. The problem is to acquire sufficient experimental data to establish the nature of the effect. There are three components to the experimental problem: (1) verify the chlorine experiment measurements, (2) measure the solar neutrino energy spectrum, and (3) measure the time variation of the solar neutrino flux.

DOES THE NEUTRINO FLUX VARY WITH TIME?

The Homestake solar neutrino rate from 1986.8 to 1988.3, Table 1 above, is almost twice that of the average over the past eighteen years of operation. This raises the question of whether the solar neutrino flux is time independent [9]. In fig. 1 is shown the time history of the Homestake observations over the past eighteen years. The present high rate is mimiced by a similar high region around 1977. That earlier high region was followed by a very low period in 1979 - 80. The time span between high periods, a decade, is very similar to the period for solar activity, eleven years. The low neutrino flux period occurs during the peak in solar activity as measured by the sunspot number. In fig. 2, is shown the correlation between neutrino flux and two measures of solar activity, sunspot number and solar radius. The crucial test of the reality of this correlation will be made during the next peak in solar activity, around 1990. A repeat of the flux reduction observed in 1980 would confirm this observation, while a high neutrino detection rate would suggest that the 1980 reduction was a statistical fluctuation.

The general question of time variability of the solar neutrino flux is particularly interesting. There are three possible periods that can be considered, daily, seasonal or 11 year cycle. A daily variation must be associated with electron neutrino regeneration in the earth. A seasonal variation can be associated either with the inclination of the earth's orbit with respect to the solar equator or variation of the neutrino paths through the earth [10]. The first correlation would suggest that there is a magnetic moment effect in the Sun. The second relation would support neutrino flavor resonant oscillations in the earth. A solar cycle correlation provides information about both the solar interior and neutrino properties.

The Homestake detector is now being modified to permit separation of daytime and nighttime solar neutrino induced ^{37}Ar. This system should be in operation by mid-1989. THe Kamiokande detector is also

Figure 1. ^{37}Ar Production Rates in the Homestake C_2Cl_4 Solar Neutrino Detector.

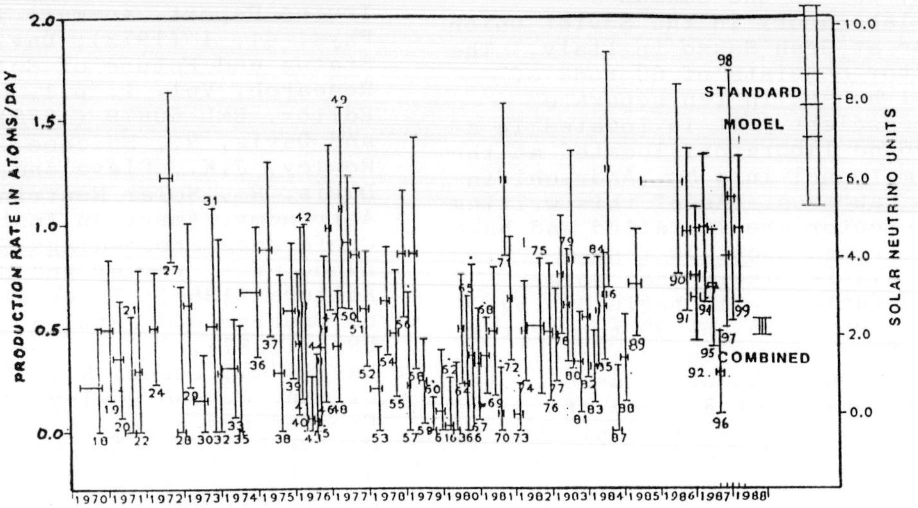

Figure 2. Comparison of Time Variation of 5 Extraction Running Averages of Measured Solar Neutrino Flux with Number of Sunspots and Solar Diameter.

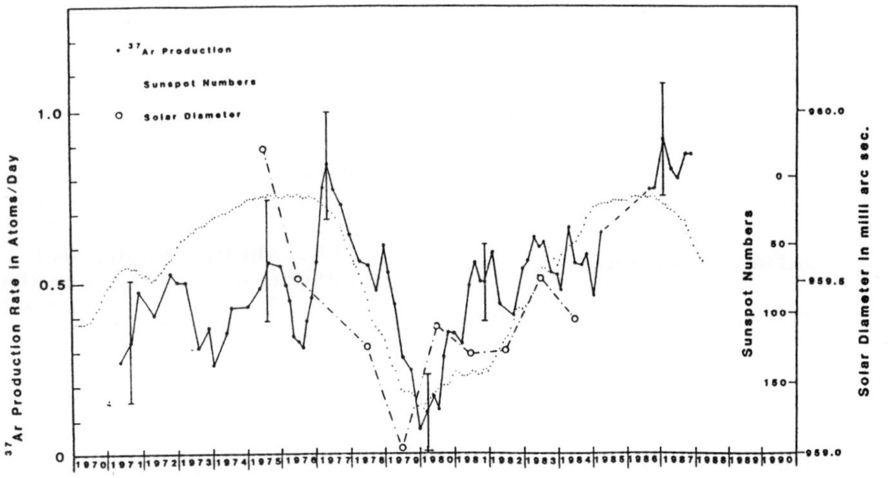

separating its signals into day and night time data. Both of these detectors are limited by detection rate, about one count per day. To carry out a sensitive study of time variations it would be desirable to have a detector with a count rate $\geq$ 10/day and with a signal to background ratio $\geq$ 10/1. Haxton has suggested that ^{127}I has the desired characteristics. It has a neutrino capture rate that is about 20 times that of ^{37}Cl. Iodine is likely to be less sensitive to backgrounds that is chlorine, since the major background, (p,n) reactions, will be suppressed by the higher Coulomb barrier of the iodine nucleus. Thus, a Homestake-like detector using iodine might be quite suitable for pursuing time variations.

The proposed Sudbury detector, a water Cerenkov detector of the Kamiokande type, consisting of 1000 tons of D_2O, should also exhibit a high counting rate, 10 - 20 neutrino captures per day.

OTHER SOLAR NEUTRINO DETECTORS

One model independent prediction of nuclear fusion as the source of solar energy is the flux of very low energy neutrinos ($E \leq 0.44$ MeV) that arise from the P-P fusion reaction. In 1966, Kuzmin pointed out that $\nu_e + {}^{71}Ga \rightarrow {}^{71}Ge + e^-$ has a favorable crosssection for neutrino interaction and a low (0.23 MeV) threshold. The P-P neutrino flux should give rise to a 70 SNU signal in Ga. In addition, the higher energy solar neutrino channels should contribute an additional 63 SNU. The latter contribution, of course, is subject to the same model sensitivities that might effect the ^{37}Cl and Kamiokande observations.

Two gallium detectors are now under construction, one in the Baksan underground laboratory in the Soviet Union and the other at Gran Sasso in Italy. The Baksan detector consists of 60 tons of gallium metal housed in ten separate extraction vessels[11]. It is located in a deep underground laboratory located at the end of 3.5 km tunnel into Mt. Andrychi in the Northern Caucasus. As of this writing 30 tons of detector are installed and have been cycled through about 10 test extractions. It is likely that solar neutrino observations will begin during the spring of 1989 with the 30 ton system while the rest of the 60 ton detector is put into operation. The Gran Sasso detector will consist of 30 tons of gallium in the form of gallium chloride in a HCl solution[12]. It is scheduled to be put into operation in 1990.

We can anticipate three possible gallium results: (1) 130 or more SNU, (2) between 70 and 130 SNU or (3) below 70 SNU. An observation below 70 SNU would indicate that neutrinos are transformed after production or that the Sun is not only powered by fusion. A result between 70 and 130 SNU would permit several solutions; either a reduction of high energy neutrino production or a transformation of the high energy end of the neutrino spectrum. A result above 130 SNU would create yet one more mystery in this already confusing situation.

There are a number of other proposed solar neutrino detectors. There are proposals to use both liquid scintillator and liquid argon to detect neutrino scattered electrons. There are also proposals to use ^{11}B, ^{81}Br, and ^{115}In to look for neutrino capture. Finally, there are several new types of proposed detectors using liquid helium and superconducting or supercooled materials to detect neutrino interactions.

Perhaps, the one picture that emerges is that the investigation and understanding of solar neutrino emission has stimulated and excited both the experimentalist and the theoretician in a fashion that we seldom see. There is an exciting mix of astronomy, particle physics and instrumentation involved in these activities. Hopefully, the results will warrant the attention that has been paid to this problem.

Support for this work has been provided by the U.S. Department of Energy, the National Science Foundation and the University of Pennsylvania Research Foundation.

1. Davis, R., Irvine Solar Neutrino Conference, University California, Irvine Report, summary in Rev. Mod. Phys. 41, 1 (1973); Davis, R., The Status and Future of Solar Neutrino Research, Vol. 1, p.1, G. Friedlander, Editor, BNL 50879 (1978); Bahcall, J.N. and Davis, R., Science 191, 264 (1974); Rowley, J.K., Cleveland, B.T. and Davis, R., Solar Neutrinos and Neutrino Astronomy, American Inst. Physics Conference Proceedings No. 126, p.1, K. Lande, W.A. Fowler and M.L. Cherry, Editors (1985); International Symposium on Underground Physics, Baksan Valley USSR (1987).
2. Davis, R., Lande, K., Cleveland, B.T., Ullman, J. and Rowley, J.K., Proceedings of the Neutrino '88 Conference, to be published.
3. Wolfendale, A.W., Young, E.C.M., and Davis, R., Nature Phys. Science, 238, 1301 (1972), Fireman, E.L., American Institute Physics Conference Proceedings No. 126, K. Lande et al, Editors (1985).
4. Bahcall, J.N. and Ulrich, R.K., Rev. Mod. Phys. 60, 297 (1988).
5. Suzuki, A. et al; Mann, A.K., Proceedings of the Neutrino '88 Conference, to be published.
6. Cahen, S., Doom, C. and Casse, M., 86' Massive Neutrinos Astrophysics and Particle Physics, p.83, O. Fackler and J. Tran Thanh Van, Editors, Frontieres Publisher; Rood, R.T., The Status and Future of Solar Neutrino Research, Vol. 1, p.175, G. Friedlander, Editor (1978) BNL 50879; Newman, M., The Physics of the Sun, Edited by P.A. Sturrock, Kluwer Academic Publishers.
7. Voloshin, M.B., Vysotskii, M.I. and Okun, L.B., Soviet Physics JETP 64, 446 (1986) and references therein; Veselov, A.I., Vysotskii, M.I. and Yurov, U.P., Soviet J. Nucl. Phys. 45, 865 (1987).
8. Mikheev, S.P. and Smirnov, A. Yu., Soviet Physics Usp. 30, 759-790 (1987).
9. Rowley, J.K., Cleveland, B.T. and Davis, R., American Inst. Phys. Conference Proceedings No. 126, p.1 (1985); Davis, R., 7th Workshop on Grand Unification/ICOBAN-86, J. Anafune, Editor; Davis, R., Cleveland, B.T. and Rowley, J.K., Proceedings 20th International Cosmic Ray Conference, Moscow, 1987, Vol. 4, p.328.
10. Baltz, A. and Weneser, J., Phys. Rev. D35, 528 (1987).
11. Gavrin, V. N. et. al. Proceedings of the Neutrino '88 Conference, to be published.
12. Kirsten, T. et. al. Proceedings of the Neutrino '88 Conference, to be published.

ELECTROWEAK SYMMETRY BREAKING

Abraham Seiden and Howard Georgi

Working Group Leaders

ELECTROWEAK SYMMETRY BREAKING THEORY

Howard Georgi

Harvard University
Cambridge, MA 02138

My talk will be brief, so that Abe can get on to the important stuff. But I do have a few theoretical observations. I will begin by reminding you of some truths which are well known but sufficiently important to be repeated over and over again. I will then go on to discuss some theoretical confusions that were cleared up during the study. I will then discuss some of the new, unfinished ideas that we discussed. Finally I will go back and repeat the obvious truths once more for emphasis.

Many theorists contributed to the Electroweak Subgroup, including Sekhar Chivukula, Sally Dawson, Mitch Golden, Howard Haber, Steve King, Marie Machacek, Steve Selipsky, Scott Willenbrock, Eram Yehudai and others who shared time with the New Particles and Strong Interactions Subgroups. In addition to working on the specific projects that I will discuss, many of these people floated from one working group to another doing good works that will be aluded to later in Abe's summary.

OBVIOUS TRUTHS

If you count up the spin and flavor states of the particles and antiparticles of the standard model (not counting the Higgs), you get 117. Of these we understand all but 3 (although we haven't seen all those that we understand - so perhaps I'm jumping the gun in saying that). The three longitudinal polarization states of the W and Z are clearly something new. Our task in the 1990 s is to figure out what they are, as best we can. Now particle physicists have a standard way of dealing with such questions. To see what some new kind of particle is made of, you slam them together and watch what happens. **Enter the SSC.** This is a device for doing W_L-W_L scattering. It is important to remember what it is for, when thinking about either the SSC or other proposed alternative machines. If your machine can't do W_L-W_L scattering, it can't deal with the really crucial question, so you should try harder.

I think that this fact, obvious as it is, is so important that it is worth saying it in a different way. Much of the effort in our working group has been devoted to the question of whether we can see a standard model Higgs particle. To keep things in perspective, it is necessary to remember that there is not even any circumstantial evidence for the existence of such an object. **None at all!**

Early in the study, Chris Quigg introduced the concept of BORING versus INTERESTING accelerators. Theories can also be classified this way, but it is useful to introduce another dimension as well – RADICAL versus CONSERVATIVE.

The standard model with a fundamental Higgs boson is an example of a RADICAL-BORING theory. It is radical because it begins by solving the problem of the longitudinal spin states of the W and Z by postulating the existence of an entirely new kind of object, a fundamental spinless field, unlike anything that we have actually seen in the world. However, after making this absurdly radical assumption, the rest of the analysis of the model is boring. All the couplings are weak and the physics can be analyzed in pertubation theory.

The only alternative to the fundamental Higgs boson that has been explored theoretically, TECHNICOLOR, is an example of a CONSERVATIVE-INTERESTING theory. The starting point is very conservative because the new particles that are introduced are simply more fermions interacting with more gauge bosons, repeating the well-tried pattern of the quarks and leptons. In this theory the longitudinal spin states of the W and Z are states of the new technifermions bound together by strong gauge interactions, just as the pions of QCD are bound states of quarks. But now the strong interactions involved in the binding make the analysis of this theory much more interesting and uncertain. Indeed, if it were not for the fact that the theory is built on an analogy with QCD, so that we can read many of the relevant facts out of the wallet cards, it would be entirely too interesting for comfort.

Neither class of models is much good. Both have problems. I suspect that what we need are models that are both RADICAL and INTERESTING.

TESTING VERSUS LOOKING

Faced with this situation, it is important to maintain a two-pronged attack on the question of electroweak symmetry breaking. On the one hand, we should continue to test the standard model with a single fundamental Higgs boson. It may be a ridiculous model, but at least it is well defined in terms of a managable number of parameters. Who knows, maybe it is a useful first approximation. Once we go beyond the very simplest standard model, however, the space of possible theories quickly becomes so huge that it is ridiculous to even talk about testing models. Instead, you must look directly for the physics of electroweak breaking, whatever it is, in W_L-W_L scattering, in the hopes that this will give the clues needed to develop the right model.

One of the contributions at the study provided a nice example of the difference between "testing" and "looking". Steve Selipsky reported on work that he has done and is doing with Brian Lynn and Mike Peskin on radiative corrections to $e^+e^- \to W^+W^-$ in two kinds of model, one with heavy ($\sim$1 TeV) degenerate doublets of fermions, another with a very heavy Higgs boson (this calculation was not quite finished at the time of the study). They found that both models produce radiative corrections which grow with s and become very important for $s \sim 1$ TeV. My conclusion is that $e^+e^- \to W^+W^-$ scattering is an interesting place to **test** the standard model, because extensions of the standard model are likely to produce important deviations from standard model predictions. However, it is not a good place to **look** for the physics of electroweak symmetry breaking. That is clear because heavy Higgs boson, which is directly related to electroweak breaking, gives radiative corrections of the same size as a heavy family, which has nothing to do with electroweak breaking. The moral, it seems to me, is that it is not good enough just to produce W^+-W^- pairs. You need to pay the price and get to a high enough energy and luminosity to scatter W's as partons, not far off their mass shells, into on shell Ws. At the SSC, maybe you can afford it. At a 1 TeV e^+-e^- collider, you can't, but at a higher energy machine in which the "2 W process", $e^+e^- \to \nu\bar{\nu}W^+W^-$ can produce W pairs with 1 TeV invariant mass, you can really look directly at electroweak symmetry breaking.

CONFUSIONS

Several people in our group were concerned with the search for a very light standard model Higgs. The theoretical discussion that goes with the experimental bounds on the parameters for such and object have been confused for several reasons.

One is the existence of potentially large QCD radiative corrections to the Wilczek process for decay of a heavy quarkonium state into γ + light Higgs. No progress was made here on this issue.

The other is the theoretical controversy over the decay rate for a light Higgs into 2 π. The issue here was the validity of a clever argument due to M. Voloshin which had been called into question in several recent papers. Here we took advantage of the fact that all of the people involved in questioning Voloshin's argument were former students of mine who happened to be here, either at Snowmass or at the Aspen Center for Physics, to try to clear up the confusion. The result, that Voloshin is right, is discussed in the contribution by Cohen, *et al.*

Another confusion that I tried to clear up is that perenial favorite, "What is a Higgs?" The term is thrown around very loosely in the literature, and often causes confusion. In my view, the only reasonable definition of a "Higgs" is the state that is part of an $SU(2)_{Weak}$ doublet with the Goldstone bosons that are eaten by the Higgs mechanism, or equivalently, the state that couples to a pair of longitudinal W's. In models with more than one $SU(2)_{Weak}$ doublets of spinless bosons, there is still only one "Higgs". However, in such models, the interesting thing that happens is that the Higgs may not be a mass eigenstate. The Higgs may be a linear combination of several different mass eigenstates:

$$H^0 = \sum_i a_i h_i^0 \qquad \sum_i |a_i|^2 = 1 .$$

Each of the mass eigenstates, h_i^0, is a "part" of the Higgs, not a Higgs by itself. Jack Gunion emphasized how the normalization condition on the state leads to a sum-rule for the coupling of the different spinless boson states to longitudinal W's. Marc Sher discussed the fact that in CP violating theories, the a_i may be complex.

For the charged spinnless bosons, the situation is even less complicated. In a model with only $SU(2)_{Weak}$ doublets, the term "Charged Higgs" is an oxymoron. The charged spinless bosons that are not eaten by the Higgs mechanism are necessarily which are not in an $SU(2)_{Weak}$ doublet with the real Higgs. They are simply charged spinless fields, parts of complete $SU(2)_{Weak}$ doublets. Their neutral partners may mix with the true Higgs, but they cannot even do that.

NEW IDEAS

Most of the really new theoretical ideas discussed in our group were connected in some way with the idea of "Technicolor", the sexy name for dynamical breaking of the $SU(2)_{Weak}$ symmetry. In thinking about technicolor models, it is important to remember that whatever breaks $SU(2)_{Weak}$ must do two jobs. One is to give mass to the W and Z. This is an automatic consequence of the Higgs mechanism, as soon as $SU(2)_{Weak}$ is broken. If, in addition, the dynamics has a custodial SU(2) symmetry, the standard model relation between M_W, M_Z and $\sin^2\theta$ is automatically satisfied up to radiative corrections. However, the information of $SU(2)_{Weak}$ breaking must also be communicated to the quarks and leptons, in order to produce the quark and lepton masses. Whatever physics does this communication produces a coupling of the longitudinal components of the W and Z to the quarks and leptons, proportional to their masses. I like to call the mechanism that does this communication the "Goldberger-Treiman mechanism" (GT), after the relation in strong interaction physics between the mass of the nucleon, the coupling of the pion to nucleons, and f_π. I think that it is important to celebrate the past contributions of Princeton to theoretical physics below the Planck scale. As in the Higgs mechanism, the gross features of the Goldberger-Treiman mechanism are determined by the symmetry structure, so that the quark and lepton masses by themselves do not yield any information about the mechanism.

In the standard model, the Goldberger-Treiman mechanism is implemented by the Yukawa couplings of the fundamental Higgs $SU(2)_{Weak}$ doublet to the quarks and leptons. This implementation of the Goldberger-Treiman mechanism is only possible with fundamental spinless bosons. One of the primary motivations for technicolor models was the desire to eliminate the need for fundamental spinless bosons, so much effort has been devoted to devising alternative versions of the Goldberger-Treiman mechanism that work without spinless bosons. The first such alternative to be extensively discussed in the literature is "Extended technicolor", or "Sideways". The idea of sideways is that the technifermions should carry flavor quantum numbers that mimic those of the quarks and leptons. If there were such a multiplicity of technifermions, one would expect pseudoGoldstone bosons, light spinless bound states of the technifermions, analogous to the pions in QCD.

One must resist the temptation to identify these pseudoGoldstone bosons with technicolor. They have nothing directly to do with technicolor. They are not needed to give mass to the W and Z, but are associated with a particular idea for implementing the Goldberger-Treiman mechanism. When you look for these pseudoGoldstone bosons, you are not testing the idea that the $SU(2)_{Weak}$ is broken dynamically. Rather, you a testing a class of models (which don't work anyway) of the Goldberger-Treiman mechanism.

Technicolor itself is much simpler. The simplest technicolor model is one in which there is a single doublet of technifermions, for example, to make it look like QCD as much as possible, suppose that the technifermions are N's of an $SU(N)_{TC}$ technicolor group,

$$\begin{pmatrix} p \\ m \end{pmatrix}_L , \quad p_R , \quad \text{and} \quad m_R ,$$

where the left-handed fields form an $SU(2)_{Weak}$ doublet while the right-handed fields are $SU(2)_{Weak}$ singlets. If the charges of the technifermions are $\pm 1/2$, and the N of $SU(N)_{TC}$ is even, the model has no gauge anomalies. The p and m (for "plus" and "minus") technifermions are very much analogous to the u and d quarks of QCD, except that where the u and d have small current-algebra masses, the technifermions can have no mass terms because of the

$SU(2)_{Weak}$ symmetry. Thus the chiral $SU(2)\times SU(2)$ symmetry of the technicolor model is an exact symmetry of the Lagrangian, while the corresponding symmetry in QCD is slightly broken. Presumably, just as in QCD, the chiral $SU(2)\times SU(2)$ symmetry of the technicolor model breaks spontaneously down to an SU(2) (the analog of isospin). The Goldstone bosons that necessarily accompany such a spontaneous breaking are built as bound states out of the technifermions, just as the pions are built as bound states of the quarks. However, because the $SU(2)_{Weak}$ symmetry is gauged, the Higgs mechanism converts these Goldstone bosons into the longitudinal polarization states of the massive gauge bosons, W and Z. The analog in the technicolor model of f_π in QCD is the $SU(2)_{Weak}$ breaking scale, $v \approx 250$GeV. Of course, all of this happens in a completely Lorentz invariant way, so that it is really only in an infinite momentum frame that you can think of the longitudinal states as identified with the Goldstone bosons. In practice, the effects show up as soon as you get to energies far above the W mass. The longitudinal Ws, in a technicolor model, behave as the pions of the strong technicolor interaction for $E \gg M_W$.

The signature of such a model, if it is like QCD, is the appearance of resonances, ρ_{TC}, ω_{TC}, η'_{TC}, etc., in $W_L - W_L$ scattering. There are no pseudoGoldstone bosons. Of course, we have also not specified the Goldberger-Treiman mechanism. I have not told you how quarks and leptons get their masses. We can describe the process just as we used to describe the weak interactions in the days before the W and Z, as nonrenormalizable, four (or more) fermion interactions. We can even say something about the symmetry structure of these interactions that is required if we are to avoid trouble with flavor changing neutral current processes. However, we can't say exactly what they are.

It seems unlikely that we will get much information about technicolor, in the sense that I have just described, before the SSC. Even at the SSC, it is not trivial to do this physics. The prospects were reviewed for our subgroup by Chivukula. The dominant decay modes for the low lying technicolor resonances are shown below:

$$\rho_{TC} \to W_L W_L \, ;$$

$$\omega_{TC} \to W_L W_L W_L \, , W_L \gamma \text{ or } W_L W_L \, ;$$

$$\eta'_{TC} \to W_L W_L W_T \, , W_L W_L \gamma \, , \text{ or } t\bar{t} \, .$$

The technirho can be produced by W_L-W_L fusion. This process has been studied, and it seems tough but perhaps doable at the SSC, at least if N is not too small. One of the new ideas discussed here was the possibility of an enhancement of the technirho production associated with the Goldberger-Treiman mechanism. In particular, Chivukula showed us, if the GT interactions have the flavor symmetries of the standard model, as suggested by the success of the GIM mechanism in suppressing flavor changing neutral currents, then the enhancement can be very large.

Based on a suggestion by Haber, some of us looked at the possibility of producing η'_{TC} through the GT interactions by t-$\bar{t}$ fusion. Unfortunately, this turns out not to be very interesting. It is similar to trying to produce a 1TeV Higgs, but only through t-$\bar{t}$ fusion.

King and Machacek are studying the consequences of the simplest technicolor model for a 1TeV e^+-e^- collider.

WALKING TECHNICOLOR

In the last few years, there has been a development in technicolor theories in the direction of RADICAL and INTERESTING theories. It goes under the name of "walking technicolor".[1)] We were fortunate that Tom Appelquist and Ken Lane had organized a workshop at the Aspen Center for Physics to discuss this and related ideas.

The idea of walking technicolor is very simple. Suppose that the technicolor coupling, instead of being asymptotically free, like the QCD coupling, is approximately constant at large momentum. If so, and if the theory still breaks chiral symmetries, as QCD does, then the behavior of the theory is very different from that of QCD at high energies. At small energies, below the $SU(2)_{Weak}$ breaking scale, the theory looks very much like ordinary technicolor, because the only particles with mass are the gauge bosons and the Goldstone bosons, whose properties are determined just by the chiral symmetries. But above the $SU(2)_{Weak}$ breaking scale, the theory is something brand new. Instead of getting simple, as it does in an asymptotically free theory, the physics stays complicated, strongly interacting, far above the the $SU(2)_{Weak}$ breaking scale.

It is somewhat surprising that we can say anything at all about a theory so radically different from anything in our experience, and certainly, everything that we say is based on some guess-work. Nevertheless, there are good reasons to believe that we can determine one important feature of these models. We can find the scale dependence of the technifermion dynamical mass.

In QCD, you can think of the quarks inside a proton or a ρ meson as having a dynamically generated mass of the order of $m_P/3$. This mass is the glue that the quarks carry around because of the confining color interactions. In a violent collision, at high momentum transfer, the QCD interactions are weaker because of asymptotic freedom, so some of the glue comes unstuck, and the effective dynamical mass of the quark is smaller. One can show that the dynamical mass decreases roughly with the square of the energy scale.

In a walking technicolor model, the confined technifermion inside some technihadron like the ρ_{TC} can also be thought of as having a dynamical mass. And just as in QCD, the effective dynamical mass of the technifermion decreases as the energy scale increases. Now, because the technicolor interaction does not weaken as the energy increases, less techniglue comes unstuck, and the fall-off of the dynamical mass with energy is slower, like one over the energy scale.

This difference, which is the only thing that we can say about walking technicolor with anything approaching confidence, has important consequences. It pushes up the scale at which the Goldberger-Treiman interactions must operate.

The contribution of the Goldberger-Treiman mechanicsm to a quark or lepton mass is proportional to the dynamical mass of the technifermion at the scale at which the mechanism acts. Λ_{GT}. Thus in a conventional technicolor model, the quark and lepton masses to like $1/\Lambda_{GT}^2$. In a walking technicolor model, they go like $1/\Lambda_{GT}$. Thus, in a walking technicolor model, Λ^{TC} can be much larger than in a conventional model. This is important, because there are always flavor changing effects associated with the Goldberger-Treiman interactions, which can cause trouble with FCNC interactions. The larger Λ^{TC}, the smaller the FCNC effects. Specifically, we expect that in an ordinary technicolor model,

$$\Lambda_{GT} \lesssim 2.5\text{TeV} \left(\frac{25\text{GeV}}{m_t}\right)^{1/2}$$

The t quark mass appears because it is the heaviest fermion (that we know of), so it gives the strongest constraint on Λ_{TC}. In a walking technicolor model, on the other hand,

$$\Lambda_{GT} \lesssim 25\text{TeV} \left(\frac{25\text{GeV}}{m_t}\right)$$

This suggests that it may be possible to build walking technicolor models incorporating a Goldberger-Treiman mechanism without running afoul of constraints on FCNC. This hasn't been done yet, perhaps because we know so little about these field theories, but we theorists are working on it.

In fact, the really interesting thing about walking technicolor is that if we can't push the scale of the Goldberger-Treiman mechanism up far enough to supress FCNC effects in these theories, then we just can't do it at all. Appelquist and Chanowitz have shown[2], on the basis of a unitarity argument like that bounds the scale of electroweak breaking, that the scale of the Goldberger-Treiman mechanism cannot be much larger than the upper bound in a walking technicolor model.

Clearly, the model builders have their hands full with these new ideas. The important question for Snowmass is, how do we find out about it experimentally. That, of course, is a hard question. This is not like QCD where we theorists had a large backlog of experimental information to help us to analyze the underlying theory. At this point, we have only guesses. One general feature seems very likely. The resonances, like the ρ_{TC}, in a walking technicolor model will be much broader and less prominent than the corresponding features in QCD. This seems clear, simply because the fact that the gauge coupling walks, rather than running, implies that quantum effects of technimatter loops are much more important than the quark loops in QCD. Zweig's rule, mysterious in QCD, will probably be useless in walking technicolor. It is even possible that the resonances will be so broad that it won't make sense to talk about them at all. Then the W_L-W_L cross section will just rise smoothly to an approximately constant value at high energies, gradually becoming dominated by inelastic processes in which many W_Ls are produced. It is important it have this scenario in mind when thinking about the SSC, because it is clearly one of the hardest to characterize experimentally.

OBVIOUS TRUTHS AGAIN

The good and the bad news is the same. Theorists can tell you where to look, but not what to look for. We must do W_L-W_L scattering at high energies, but we don't know what we will find. The one completely model independent feature of electroweak symmetry breaking is that the growth of the W_L-W_L scattering amplitude is cut off above the scale at which the physics of symmetry breaking appears. Our job, and it is not an easy one, is to be sure that we can find that scale in whatever form it takes, whether it is a Higgs mass, a technirho mass, or just the onset of a new kind of strong interaction. The stakes are high. We have a fantastic opportunity to discover the mechanism of $SU(2)_{Weak}$ breaking. It will not be easy. It will put tremendous constraints on machines, detectors, analysis and on the organizational structure of particle physics itself. Still, we must try. If we take the easy way out, it probably won't be good enough.

ACKNOWLEDGEMENTS

Research supported in part by the National Science Foundation under Grant #PHY-8714654. I am very grateful to the Aspen Center for Physics, where some of this work was performed, and to Tom Appelquist and Ken Lane for their efforts to coordinate their workshop with Snowmass.

REFERENCES

1) B. Holdom, Phys. Lett. B150 (1985) 301; T. Akiba and T. Yanagida, Phys. Lett. 169B (1986) 432; T. Appelquist, D. Karabali and L. C. R. Wijewardhana, Phys. Rev. D35 (1987) 2605; M. Bando, T. Morozumi, H. So and K Yamawaki, Phys Rev. Lett. 59 (1987) 389; See also V. A. Miransky, Nuovo Cimento 90A (1985) 149; W. A. Bardeen, C. N. Leung and S. T. Love, Phys. Rev. Lett. 56 (1986) 1230 and C. N. Leung, S. T. Love and W. A. Bardeen, Nucl. Phys. B273 (1986) 649.

2) T. Appelquist and M. S. Chanowitz, Phys. Rve. Letters **59** (1987) 2405.

ELECTROWEAK SYMMETRY BREAKING: EXPERIMENTAL PHYSICS SUMMARY

ABRAHAM SEIDEN
Santa Cruz Institute for Particle Physics
University of California, Santa Cruz, CA 95064

ABSTRACT

We summarize the work done on the search for the Higgs at the 1988 Summer Study on High Energy Physics in the 1990's, held in Snowmass, Colorado. Topics studied were: limits on a very light Higgs from Υ and B decays, the expected impact in the next few years of the precision measurement of electroweak parameters, and possibilities for Higgs detection for a mass range from 100 GeV to 1 TeV at very high-energy pp or e^+e^- colliders. Specific supersymmetry and technicolor scenarios were looked at as examples of how the Higgs mechanism might be realized. This work represents the efforts of a large number of people. Specific contributions are referenced below. In addition, we had many meetings as a group benefitting from each others' ideas and efforts.

INTRODUCTION

In the next five years we expect to learn a great deal about the electroweak interactions up to an energy scale of nearly 200 GeV.[1)] Precision studies of the Z^0, combined with W mass measurement, will tell us if Higgs doublet models provide an adequate framework for electroweak symmetry breaking. Weak radiative correction measurements will provide constraints on physics at higher mass scales, especially after the top quark mass is determined and neutrino counting at the Z^0 pins down the number of fermion families. The search for the simplest Standard Model Higgs should be extended up to about 80 GeV, in steps, by combining: detailed B decay measurements relevant for a Higgs mass search up to a few GeV, Z^0 decay studies at LEP1 extending the Higgs search from a few GeV up to about 40 GeV, and data from LEP2 extending the search up to about 80 GeV. In this summary we will discuss these topics briefly. The primary focus will, however, be on searches at mass scales beyond the gauge boson scale. These will require a new generation of colliders, which can extend the Higgs search up to about 1 TeV in mass. This is a particularly important range to reach since unitarity requires new dynamics to occur in the WW scattering channel by this mass scale.[2)]

In our discussion below, we will often assume the simplest Standard Model Higgs scenario. This assumes that electroweak symmetry breaking and fermion mass generation, for all types of fermions, originates through couplings to the same minimal number of Higgs fields. There is no compelling evidence for this, and the large number of parameters needed to describe the fermion masses and mixing may perhaps suggest that we should expect a more complex Higgs sector. The minimal model assumptions for the fermion couplings are quite critical for Higgs production for masses below a few hundred GeV, and decay at masses below $2m_W$, since these couplings are involved. We will not discuss this in detail below; the reader should, however, keep these limitations in mind whenever Higgs-fermion couplings occur.

LIMITS FOR LIGHT HIGGS[3)]

A light Standard Model Higgs has been looked for in both Υ and B decay. The decay diagrams involved are shown in Figs. 1 and 2.

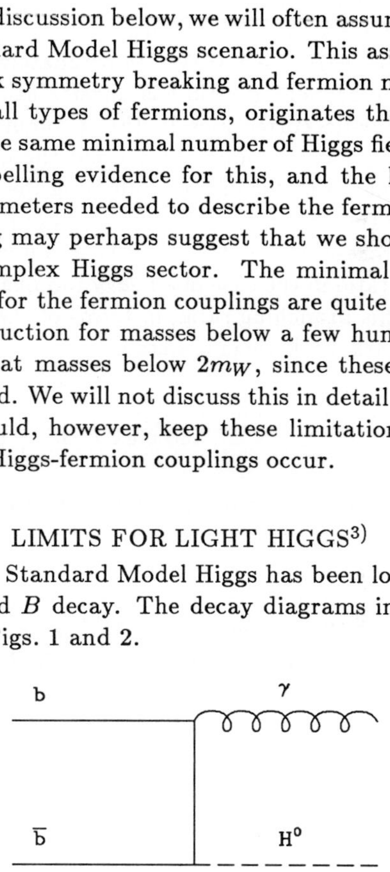

Fig. 1. $\Upsilon \to \gamma + H^0$.

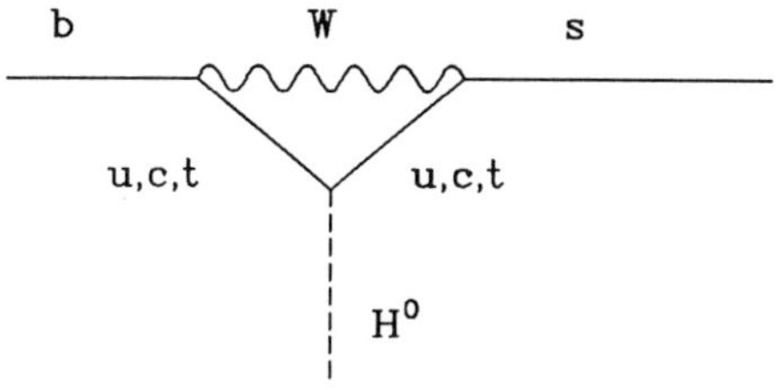

Fig. 2. $b \to s + H^0$.

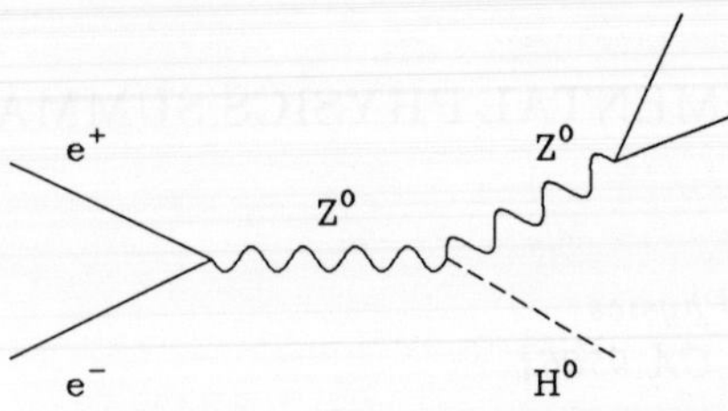

Fig. 3. Higgs Production at LEP

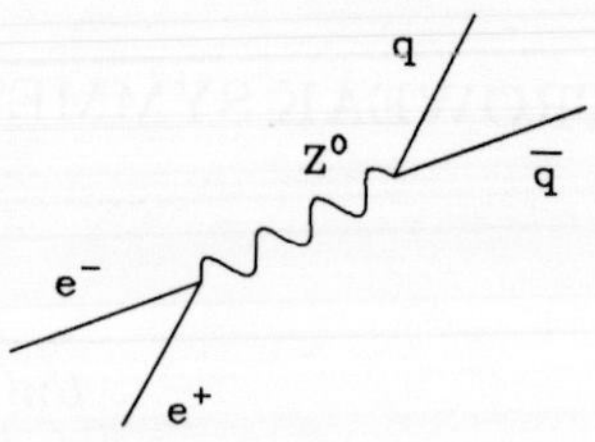

Fig. 4. Z^0 Production and Decay into Fermion Pairs

The process $\Upsilon \to \gamma + H^0$ is particularly attractive because one need only look for a line in the photon spectrum without having to worry about reconstruction of the final state from the H^0 decay. Using this technique, the CUSB experiment has set a 95% confidence level limit ruling out a Higgs with $m_H < 4.8$ GeV. Unfortunately, the calculation of the branching ratio has uncertainties stemming from relativistic corrections in the Υ bound state and QCD radiative corrections. At present the uncertainties are more than a factor of two, nullifying the limit in the case where the suppression factor is moderately large.

The process $B \to H + X$ has a potentially large branching ratio in the case of a heavy top quark. An estimate for this branching ratio, in terms of the top quark, bottom quark, and Higgs masses is

$$Br(B \to H + X) = 4.2\% \left(\frac{m_t}{50\ \text{GeV}}\right)^4 \left[1 - \left(\frac{m_H}{m_b}\right)^2\right]^2 .$$

To utilize this decay one must be able to calculate the fraction of events yielding a specific final state X, or use inclusive measurements for a known H^0 decay. The state X is usually taken to be either K or K^* while the H^0 decay is typically chosen to be $\mu^+\mu^-$, $\pi^+\pi^-$ or K^+K^-. A large number of final states have been looked at by the CLEO and ARGUS experiments resulting in branching ratio limits which are typically between 0.01% and 0.2%. Taken together, these rule out a light H^0 with mass between $2m_\mu$ and about 3.7 GeV.

For the future we can expect an order of magnitude more data from the upgraded CLEO2 detector running at higher luminosity. The combination of excellent tracking combined with an excellent photon detector should allow better measurements and a decisive exclusion of the possibility of a minimal Standard Model Higgs with mass below a few GeV. Note, in particular that it will be important to close the mass gap between about 14 MeV (the nuclear physics limit) and $2m_\mu$ where the H^0 decays mostly into e^+e^- pairs. Over a large part of this range the H^0 lives long enough to either escape a detector or provide a clear secondary vertex. A careful analysis of the results of experiments looking for a light Higgs is given in Ref. 4. In this reference, some of the difficulties in interpreting some of the limits are reviewed.

PHYSICS AT THE GAUGE BOSON SCALE

The SLC, LEP1 and LEP2 will allow high precision measurements of gauge boson masses and couplings. In the Standard Model these are only weakly dependent on the Higgs sector. However, if the Higgs mass is below about 80 GeV, it will be directly detectable[1)] through the diagram in Fig. 3. If its mass is larger, it has a small affect through electroweak loop corrections. However, attributing such effects to the Higgs requires an understanding of the other physics which can contribute to the virtual loops which occur in the theory.

The Standard Model, with Higgs doublets only, has four parameters which characterize physics at the 100 GeV scale. These are

$$m_Z,\ \sin^2\theta_W,\ m_W,\ \Delta r,$$

where Δr is due to loop corrections and is expected to be $\sim$0.07 due to standard physics involving the light fermions.

On the Z^0 one will measure m_Z, with an error $\delta m_Z \simeq$ 40 MeV. Using various asymmetries in the process shown in Fig. 4, the product of couplings, $v_f a_f/(v_f^2 + a_f^2)$ can be measured, where a_f and v_f are the axial and vector couplings of fermion f to the Z^0.[5)] This will be done best using polarized beams which are expected to be available first at the SLC. This should allow the measurement of $\sin^2\theta_W$ with a statistical error $\delta\sin^2\theta_W \simeq 0.0005$ to .001 if all loop corrections were known. The extraction of $\sin^2\theta_W$ from an asymmetry measurement is, however, sensitive to the effect of corrections from physics at higher mass scales, such as a heavy Higgs.

From the values of m_Z and $\sin^2\theta_W$ we can calculate m_W and Δr through[6)] the two Standard Model constraint equations

$$\sin^2\theta_W = 1 - (m_W/m_Z)^2$$

$$m_W^2 = (\pi\alpha/\sqrt{2}G_F)\ (1/\sin^2\theta_W)\ [1/(1-\Delta r)].$$

We note that at fixed m_Z (assumed to be very well measured), changing m_H from 10 GeV to 1 TeV changes $\sin^2\theta_W$ by 0.005. Thus, a very accurate measurement of $\sin^2\theta_W$ does put interesting constraints on the high mass scale physics, which we infer from the value of Δr extracted from the above formulas.

Measurements at the Z^0 alone do not check the constraint equations, which tell us if our Higgs framework is consistent. For this purpose we want as accurate as possible a value of m_W. For example, a 1% measurement of $\sin^2\theta_W$ ($\delta(\sin^2\theta_W) = 0.002$) gives a prediction of m_W, assuming the error on m_Z is negligible, with an uncertainty

$$\delta m_W \simeq 100 \text{ MeV} .$$

It is thus important to achieve this level of accuracy. At LEP2, an accuracy ~100 to 200 MeV should be attainable. For 10 pb^{-1} of luminosity at the Tevatron collider, an error $\simeq$ 250 MeV may be attainable. An order of magnitude more data might produce a measurement with an error ~100 MeV.

H^0 AT THE SSC

INTRODUCTION

The search for the Higgs provides one of the major justifications for building the SSC. The ability to find signatures for the Higgs will provide critical benchmarks in evaluating possible detectors and has provided a major focus for recent studies, most recently the workshop in Berkeley in 1987.[7)]

The most favored method for detection of a Standard Model Higgs varies with its mass. In this study we have looked at the following decay processes:

1) $80 \leq m_H \leq 150$ GeV range:

$$H^0 \to \gamma\gamma$$
$$H^0 \to b\bar{b} \text{ (assumes } m_H < 2m_t)$$

2) $180 \leq m_H \leq 600$ GeV range:

$$H^0 \to ZZ \to 4 \text{ charged leptons.}$$

This is the cleanest process but is rate-limited beyond about 600 GeV.

3) $600 < m_H < 1$ TeV range:

$$H^0 \to WW \to 2 \text{ jets}$$
$$\hookrightarrow l\nu$$
$$H^0 \to ZZ \to 2 \text{ jets.}$$
$$\hookrightarrow l^+l^-$$

Above 1 TeV the Higgs channel is strongly interacting and it is unclear what to expect. In this case, technicolor and extended technicolor theories provide interesting examples of what could happen.

The work presented below, unless specified otherwise, uses a detector model based on work presented in Ref. 7. The detector parameters are:

- Rapidity coverage for E_T measurements, $|Y| < 5.5$.
- Calorimeter towers, each covering .05 × .05 in $\Delta\phi \times \Delta Y$, up to $|Y| < 2.5$.
- Electromagnetic resolution of $15\%/\sqrt{E(\text{GeV})}$
- Hadronic resolution of $50\%/\sqrt{E(\text{GeV})}$.

Rates are quoted for an integrated luminosity $= 10^4 \text{pb}^{-1}$. Monte Carlo studies are usually done using PYTHIA.[8)] More effort is, however, needed to understand the limitations of these Monte Carlo programs.[9)]

HIGGS PRODUCTION AND DECAY

The expected Higgs production cross section at the SSC[10)] as a function of its mass is shown in Fig. 5. Per SSC year, the production rates vary from about 5 $\times 10^6$ for a 100 GeV H^0 to about 10,000 for a 1 TeV H^0. The rates are dependent on the mass of the top quark as shown in Fig. 5.

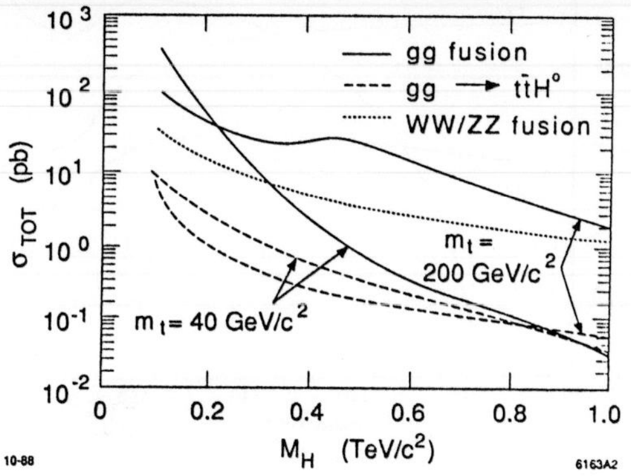

Fig. 5. The Higgs production cross sections for various processes as a function of the Higgs mass for two different t quark masses.

The primary production diagrams of interest are shown in Fig. 6. Diagram (c) is only large at low mass, contributing 25,000 events for $m_H = 100$ GeV, as opposed to millions of events for diagram (a). It is of interest mainly if the presence of a $W^\pm$ in the final state helps in the rejection of backgrounds.

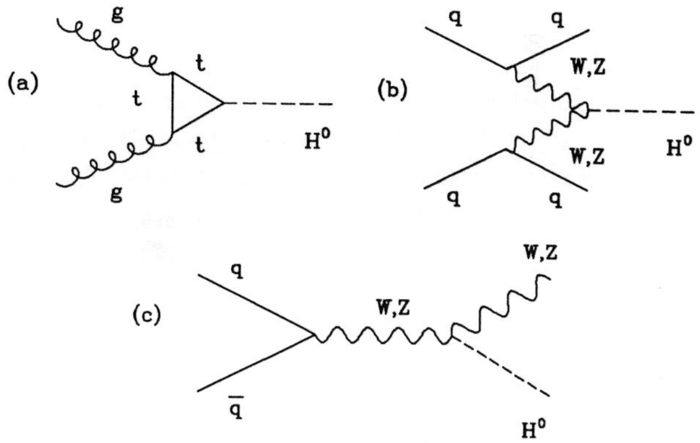

Fig. 6. Diagrams for Higgs Production at the SSC.

The decay modes of the H^0 are dependent on its mass. We will call the Higgs heavy if it has a mass above 180 GeV (*i.e.*, above the threshold for decay into gauge boson pair) and light if its mass is below this value.

Above the gauge boson pair decay threshold the H^0 decays 2/3 into W^+W^- and 1/3 into Z^0Z^0. For a sufficiently light top quark, it will decay almost entirely into $t\bar{t}$ below this threshold. Clearly, the measurement of the t quark mass is critical for predicting the production cross section and determining a search strategy for the Higgs.

For a heavy top quark, the light Higgs would decay mostly into $b\bar{b}$. Fig. 7 shows the expected branching ratios in this case.[11)] The ZZ^* rate is for a final state with one real and one virtual Z. Not shown in the figure is the WW^* branching ratio.

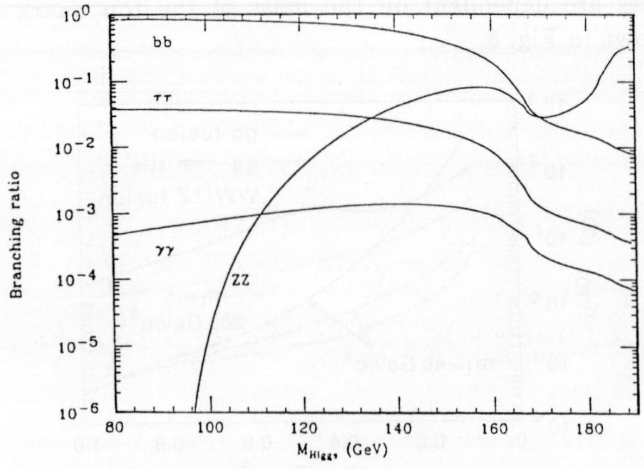

Fig. 7. The branching ratios for various decays of the Higgs boson. It has been assumed that $m_H < 2m_t$.

The decay into $b\bar{b}$ has enormous background from QCD processes, making diagram (c) in Fig. 6 the only useful production process from this decay mode. The decay into $\gamma\gamma$ has a small branching ratio but a very nice signature since it produces isolated photons with an invariant mass of m_H. We discuss these processes in the next section.

STUDY OF $H^0 \to \gamma\gamma$[12)]

If the light Higgs lies above top threshold, its decays to top quarks will dominate to the extent that the reaction $H^0 \to \gamma\gamma$ will be undetectable at the SSC. For a Higgs lying below top threshold, the small branching ratio into $\gamma\gamma$ of about 10^{-3} may be detectable. This process has a clean signature, yielding two isolated photons in the final state, however it sits on a large continuum background. Hence the signal to noise will be directly proportional to the mass resolution achievable with the photon detector. This process has two kinds of backgrounds. The first is irreducible and comes from the processes $q\bar{q} \to \gamma\gamma$ and $gg \to \gamma\gamma$. The second comes from QCD jet–jet events where fluctuations yield jets indistinguishable in the detector from photons. The jet–jet rate is expected to be about 5×10^7 as large as the rate from the Higgs. These potential background events will require a rejection factor of $\gtrsim 10^7$. Monte Carlo analyses indicate that, although difficult, this may be possible. This issue will require further study. Below, we discuss only the irreducible background of real $\gamma\gamma$ events.

The analysis of signal and background is based on the following cuts:

$$|Y_\gamma| < 3, |\cos\theta^*| < 0.8, E_T^\gamma > 10 \text{ GeV},$$

where θ^* is the photon angle in the $\gamma\gamma$ rest frame. For an 80 GeV top quark mass this would give about 500 signal events for a 100 GeV Higgs and about 800 for a 150 GeV Higgs. The background falls by about a factor of five over this mass range, so the signal to noise increases substantially with increasing Higgs mass up to about 160 GeV, where the expected H^0 to $\gamma\gamma$ branching ratio starts falling rapidly (see Fig. 7).

To compare the signal to background, a calorimeter resolution must be assumed. Two models have been looked at:

$$\frac{\sigma_E}{E} = \frac{10\%}{\sqrt{E}} + 1\%, \text{ Excellent Resolution.}$$

$$\frac{\sigma_E}{E} = \frac{3\%}{\sqrt{E}} + 0.5\%, \text{ Extraordinary Resolution.}$$

Fig. 8(a) and (b) show the expected $\gamma\gamma$ mass distribution for the signal from a 100 GeV Higgs superimposed on the irreducible background for the two detector models. Fig. 9(a) and (b) are the analogous distributions for an assumed 150 GeV Higgs. These plots are then fitted to a known background shape plus a signal of width given by the detector resolution. The results are given in the table below.

Higgs Mass	Detector Resolution	Mass Resolution	Statistical Significance
100 GeV	Excellent	1.44 GeV	None
100 GeV	Extraordinary	0.55 GeV	2.8σ
150 GeV	Excellent	1.91 GeV	7.6σ
150 GeV	Extraordinary	0.80 GeV	12.0σ

Unless the Higgs mass is a priori known, a signal of a few σ is not credible since the number of potential mass bins being searched in is large. It is therefore unlikely that the Higgs can be found in this mode alone for masses $\lesssim$ 100 GeV. The region near 150 GeV seems possible provided the calorimeter is of very high quality, the jet–jet background can be rejected, and the top quark mass is larger than 80 GeV. Note, in addition, the detector must have good geometrical acceptance and be capable of triggering on isolated photons with transverse energy $\gtrsim$ 10 GeV, without generating an unacceptably high trigger rate. These are all very stringent requirements and may require a dedicated experiment.

STUDY OF $H^0 \to b\bar{b}$[13)]

The process being searched for in this case is

$$q\bar{q} \to W^\pm H^0 \to b\bar{b}.$$
$$\hookrightarrow l\nu$$

The presence of a $W^\pm$ in the final state is essential for background rejection, since the $H^0 \to b\bar{b}$ by itself would be swamped by QCD processes yielding $b\bar{b}$ pairs. We will

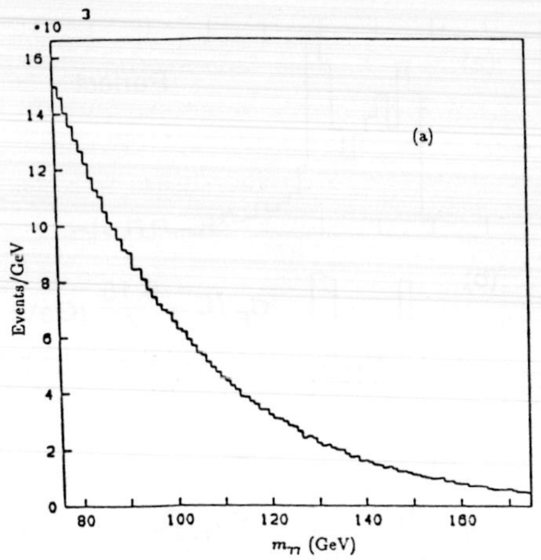

Fig. 8(a). Mass of $\gamma\gamma$ pairs in a detector of excellent resolution for a 100 GeV Higgs plus $\gamma\gamma$ background.

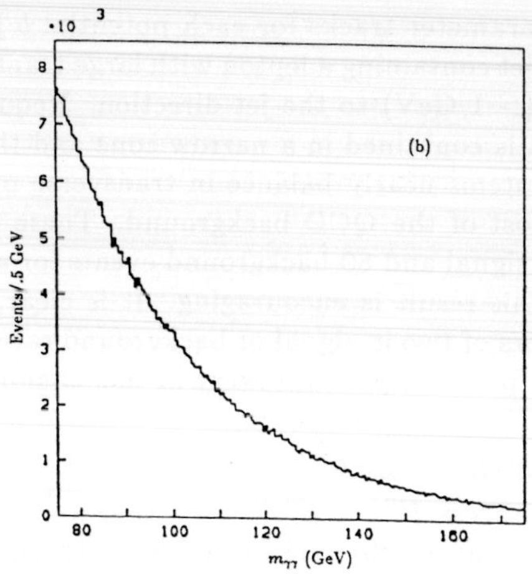

Fig. 8(b). Mass of $\gamma\gamma$ pairs in a detector of extraordinary resolution for a 100 GeV Higgs plus $\gamma\gamma$ background.

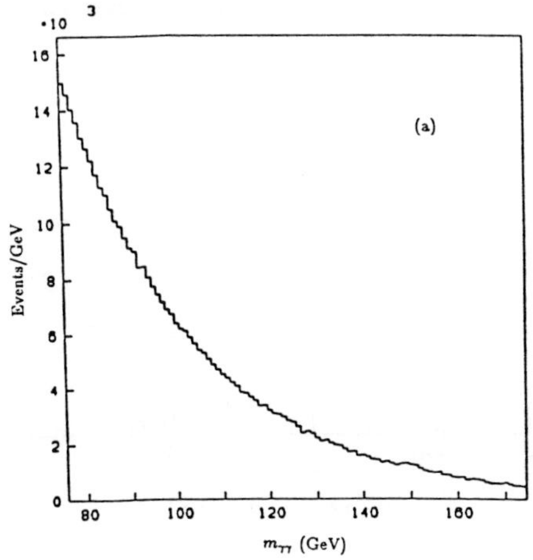

Fig. 9(a). Mass of $\gamma\gamma$ pairs in a detector of excellent resolution for a 150 GeV Higgs plus $\gamma\gamma$ background.

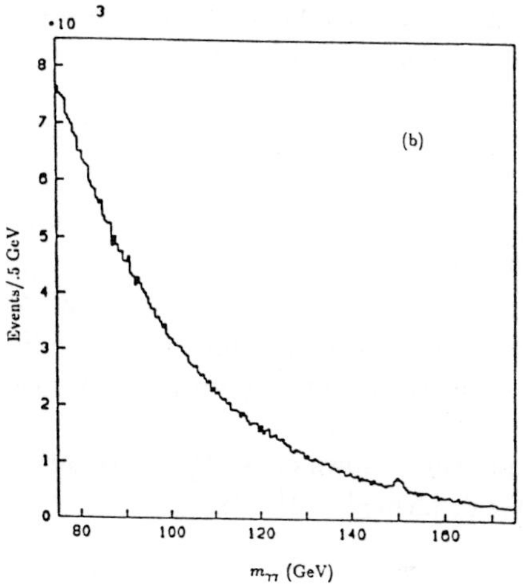

Fig. 9(b). Mass of $\gamma\gamma$ pairs in a detector of extraordinary resolution for a 150 GeV Higgs plus $\gamma\gamma$ background.

assume that the top quark is sufficiently heavy that the dominant decay of the H^0 is into $b\bar{b}$ in the mass range being considered. The potential background processes are

$$qg \to W^{\pm}q$$
$$q\bar{q} \to W^{\pm}g$$

which are both large, and

$$q\bar{q} \to W^{\pm}Z^0$$

which is comparable to the signal and only a significant background for $m_H \simeq m_Z$.

The table below, based on a Monte Carlo study, shows the number of potential signal events for a 100 GeV H^0, as a function of the analysis cuts. The background starts off at a ratio 10^4 as large as the signal. The 20 GeV mass cut around m_H reduces this factor to about 10^3. To reduce the background substantially below this requires tagging the b decays, since the $b\bar{b}$ component of the QCD background is small in this mass range. With b tagging (including rejection of $c\bar{c}$ events) the background can be

Selection Criteria	Signal/Year
$pp \to W^{\pm}H^0$, $P_T > 50$ GeV	24,000
$W \to l\nu$	7,500
$P_T^l > 25$ GeV, $E_T^{\rm miss} > 40$ GeV	2,900
2 Highest E_T jets in $\lvert Y\rvert < 2.5$, mass in a 20 GeV mass bin around m_H	750

reduced by a factor of about 50 relative to the signal. To reach a signal to noise ratio of 1, requires additional background rejection of a factor of about 20. This must come from the event structure. The feasibility of finding the signal is thus very dependent on whether such differences exist and whether further cuts based on these differences leave enough signal to provide a convincing mass peak at m_H in the jet–jet mass plot.

A detailed Monte Carlo analysis has been done implementing b tagging criteria which involve requiring large impact parameter tracks for each potential b jet and at least one jet containing a lepton with large transverse momentum (> 1 GeV) to the jet direction. Requiring that each b jet is contained in a narrow cone and that the W and $b\bar{b}$ systems nearly balance in transverse momentum rejects most of the QCD background. These cuts leave about 40 signal and 80 background events for a 100 GeV Higgs. This result is encouraging. It is clear, however, that factors of two in signal or background rates strongly affect the possibility of discovery in this channel.

STUDY OF $H^0 \to Z^0Z^0 \to 4$ CHARGED LEPTONS[14)]

The process $H^0 \to Z^0Z^0 \to 4$ charged leptons provides the cleanest channel for a Higgs search. In addition, it allows a search over a substantial region in mass, $180 \lesssim m_H \lesssim 600$ GeV. It is, however, rate limited, since the branching ratio for the Higgs into this final state is only 1.4×10^{-3}. Potential backgrounds from Z^0+ jets are large, yielding $Zc\bar{c}$, $Zb\bar{b}$, and $Zt\bar{t}$ final states which produce significant numbers of events with 4 leptons. The real Z^0Z^0 events are distinguishable from background by using the Z^0 mass constraints, isolation criteria to eliminate leptons from light quark jets, and rapidity and transverse momentum cuts.

The processes $q\bar{q} \to Z^0Z^0$ and $gg \to Z^0Z^0$ provide a substantial continuum signal on which the H^0 resonance would sit. The table below indicates the rates expected per year, based on Monte Carlo studies. The detected rates assume the follows cuts: $P_T^l > 10$ GeV, $|Y^l| < 2.5$, $P_T^Z > 50$ GeV, $|\Delta m_Z| < 10$ GeV.

Mass Choices	Raw Rates	Detected Rates, After Cuts
$m_t = 40$ GeV		
$m_H = 200$ GeV	575	70
$m_H = 400$ GeV	144	89
$m_H = 600$ GeV	60	40
$m_t = 200$ GeV		
$m_H = 200$ GeV	687	93
$m_H = 400$ GeV	560	345
$m_H = 600$ GeV	225	146
Continuum Background from $q\bar{q}, gg \to Z^0Z^0$	1500	280

The detector used for this process needs to be adequate for detection of the signal and elimination of background as discussed above. Since the event rates are small, further losses in efficiency from tracking or particle identification must be minimal. The detector has to have adequate mass resolution to avoid spreading the Higgs peak, since it sits on a continuum signal from which it has to be distinguished. Fig. 10 illustrates this point by showing the predicted distribution of Z^0Z^0 reconstructed invariant mass in detectors with varying tracking or calorimetric resolutions. For this Monte Carlo study the Higgs mass has been chosen to be 400 GeV.

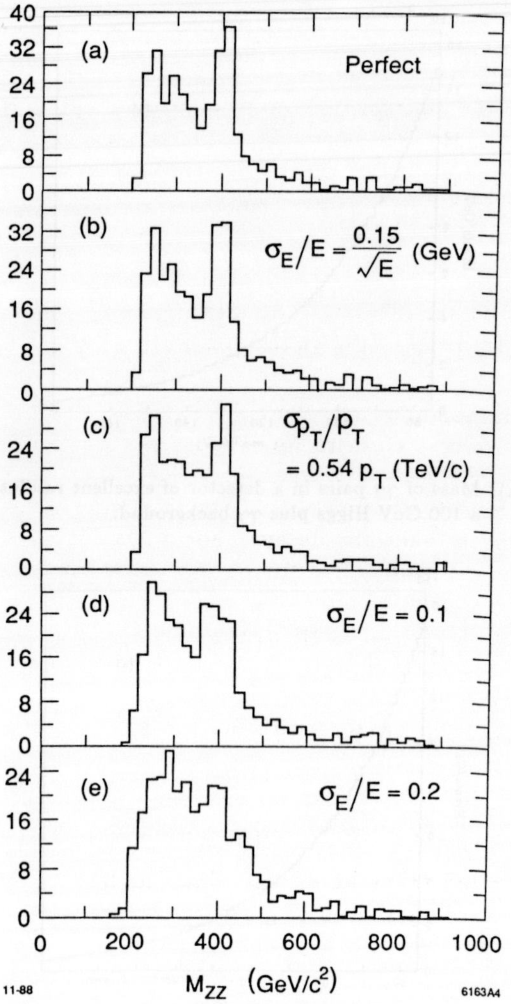

Fig. 10. Z^0Z^0 mass distribution for events reconstructed with detectors of varying resolution. $m_H = 400$ GeV. Continuum background is included.

Further work on the four charged lepton channel is needed to answer several important questions. For the case of a heavy top quark, the decay channel $H^0 \to Z^0Z^{0*} \to 4$ charged leptons, where Z^{0*} is a virtual Z^0, has a branching ratio from the Higgs $\sim 2 \times 10^{-4}$ for $130 \lesssim m_H \lesssim 175$ GeV. This would yield a small number of events. Background rejection for this process is not as good as at higher masses since the Z^0 mass constraint can't be used for the Z^{0*}. Lepton isolation criteria can, however, still be used for background rejection. A detailed analysis of this case, including background estimates, is quite important.

A possible strategy to increase the heavy Higgs event rate would be to run at very high luminosity

($> 10^{34}\text{cm}^{-2}\text{sec}^{-1}$) detecting only muons outside of a substantial absorber. To minimize background, a good magnetic tracking system should be used. Such a strategy would most probably not allow the lepton isolation cuts possible in a solenoidal detector. A detailed analysis, in particular of the background for such a detector configuration, would be of interest.

Detection of Gauge Boson Pairs from Heavy Higgs Decay[15]

For a Higgs mass above about 600 GeV the rate for $H^0 \to Z^0 Z^0 \to 4$ charged leptons is too small to allow unambiguous detection of the Higgs, unless the top quark mass is close to its maximum allowed value (for three quark generations) of 200 GeV. For a smaller top quark mass it will be necessary to detect one gauge boson through its hadronic decays. For a 1 TeV Higgs about 1,000 events per year will be produced in the final state:

$$pp \to H^0 \to WW \to \text{jets}$$
$$\hookrightarrow l\nu.$$

The expected background from the mixed electroweak–QCD process

$$pp \to W + \text{jets}$$
$$\hookrightarrow l\nu$$

is expected to be about 100 times as large after loose selection criteria. Selecting events with jet systems of invariant mass close to the W, for example, $70 \leq m_{\text{jet system}} \leq 90$ GeV still leaves about 10 times as much background as signal. Thus it is necessary to find an additional factor of ten background suppression to allow the extraction of a clear signal. In particular this would then allow the observation of a mass peak at the W mass, indicating that we are indeed detecting WW final states.

The main diagrams for the signal and background are shown below in Fig. 11.

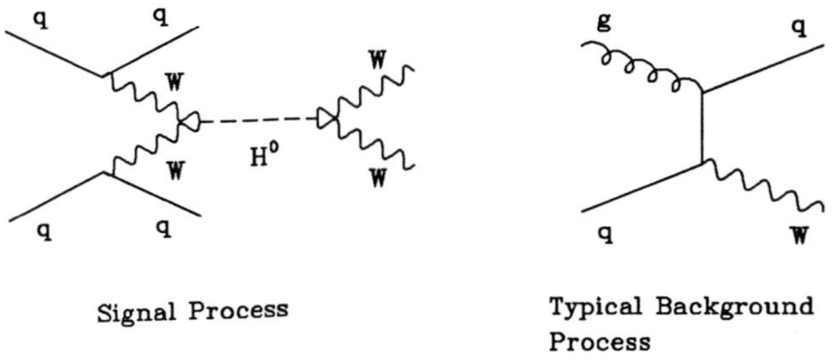

Fig. 11. Primary production and background diagrams for very heavy Higgs.

Note, the signal process involves no exchange of color and a momentum transfer governed by the W mass scale. The background process involves an exchange of color between the protons and a momentum transfer scale given by the Higgs mass. Thus for the signal we are observing a boosted W which yields a multiplicity of hadrons like a W at rest, while the background contains a full 1 TeV scale hadronic jet. Simulating these processes using PYTHIA we find two dramatic differences:

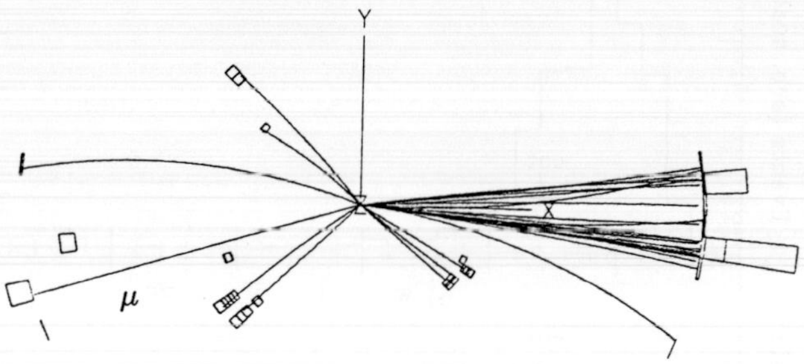

a)

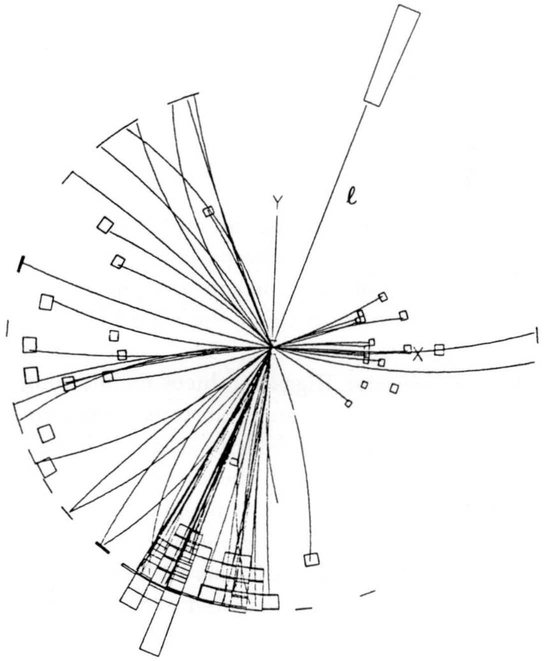

b)

Fig. 12. Event display which shows charged tracks with $p_t \geq 0.5$ GeV and calorimeter cells with $E_t > 0.5$ GeV in $|Y| < 2.5$ together with the isolated lepton from the W decay. *a)* Higgs $\to W^+W^+$ event with $M_{WW} = 1230$ GeV, $N_{ch} = 33$ and $m_j = 73$ GeV. *b)* Mixed EW-QCD event with invariant mass $M_{W\text{jet}} = 1280$ GeV, $N_{ch} = 73$ and $m_j = 82$ GeV.

1) The W jet [Fig. 12(a)] is typically collimated much more tightly than the QCD jet [Fig. 12(b)] which resembles a broad "fan".
2) The event multiplicity is lower by about a factor 2.5 in the Higgs case.

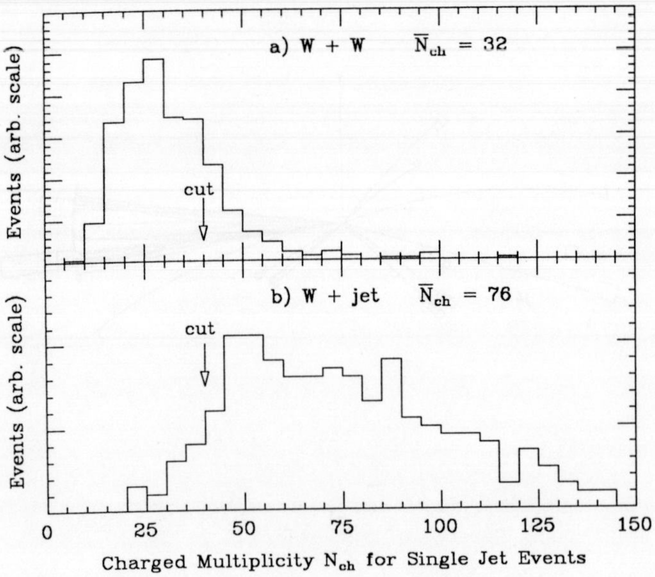

Fig. 13. Charged event multiplicity N_{ch} for tracks with $|Y| < 2.5$ and $p_t \leq 0.5$ GeV: *a)* Higgs $\to WW$, *b)* $W+$ jet.

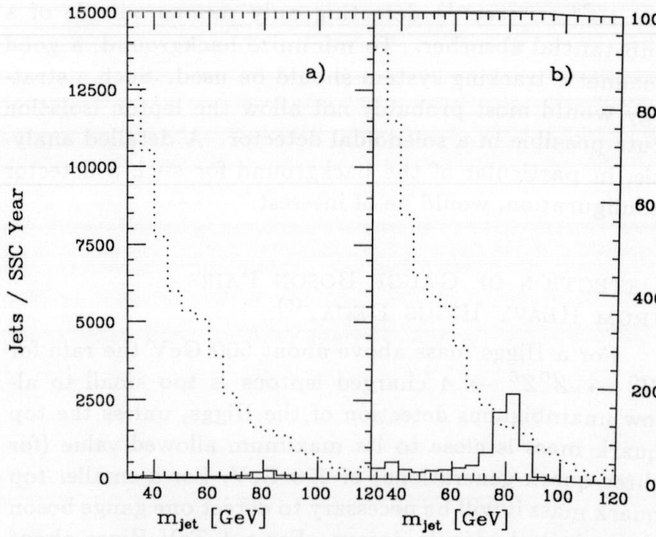

Fig. 14. Mass of jets m_{jet} with $R = 0.5$ and $E_t > 100$ GeV for Higgs (solid line) and $W+$ jet (dotted line) production: *a)* before multiplicity cut, *b)* after multiplicity cut.

The charged particle multiplicity distribution from the simulation is shown in Fig. 13. A cut at 40 charged tracks leaves 78% of the signal and only 7.3% of the background. Thus this cut has the potential to provide the needed extra factor of 10 of background rejection. It also has the advantage that little of the signal is lost and it is independent of the W helicity. The latter is of interest since the very massive Higgs produces mainly longitudinal W's and it would be an important measurement if one could determine the W helicity. Fig. 14 shows the jet mass distribution from a calorimetric detector before and after the multiplicity cut. Note, in the latter case, the W mass peak should be clearly visible over the background. Depending on details of the analysis a 1 TeV Higgs should yield between 200 and 500 detected events in the decay channel discussed for 1000 initially produced.

We note, finally, that a similar analysis can be applied to

$$pp \to H^0 \to Z^0 Z^0 \to \text{ jets.}$$
$$\hookrightarrow l^+ l^-$$

The use of the hadronic decays yields a factor 20 increase in statistics, as compared to the all charged lepton decays. This channel typically results in about a factor 5 fewer events than in the WW channel because of the smaller $H^0 \to Z^0 Z^0$ and $Z^0 \to l^+ l^-$ branching ratios. It would, however, allow a corroboration of the signal seen in the WW channel and in addition allows a more direct measurement of the Higgs mass and Z^0 helicity.

The reconstruction of the invariant mass of hadronic decays of strongly boosted W or Z^0 bosons relies heavily on calorimetric energy measurement. The smearing of the measured energies still allows a reasonably good mass resolution of approximately 5 GeV. The incident energy is, however, rather narrowly confined in angle making shower spreading in the rear of the calorimeter a potentially important contributor to the mass resolution.[16] This issue deserves more attention in deciding on both an optimum algorithm for calculating the invariant mass from the shower measurements and an optimum calorimeter design.

For Higgs masses in the TeV range, the WW scattering channel becomes strongly interacting. A particular model for this is given by technicolor theories.[17] In these models, the Standard Model diagrams shown in Fig. 15 are supplemented by interactions through the bound states of the technicolor force as shown in Fig. 16 We choose for our example $Z^0 W^{\pm}$ scattering, with analogous diagrams for the other boson pairs. In addition, if we assume this interaction also generates the fermion masses (extended technicolor) there must be direct fermion couplings, as shown in Fig. 17. These latter couplings could produce very large rates since they do not require the radiation of gauge bosons from the incoming particles. These models can produce anywhere from small enhancements in rates as compared to the Standard Model, to large increases as much as a factor of 50. In addition, dramatic resonances in the gauge boson pair mass could appear. The technicolor theory presents an exciting possibility for SSC physics.

HIGGS SEARCH USING HIGH ENERGY e^+e^- COLLIDERS[18]

Introduction

High energy e^+e^- colliders potentially provide a clean environment for the study of electroweak symmetry breaking. As in the case of pp colliders, Higgs cross sections are small, necessitating very high luminosity machines.

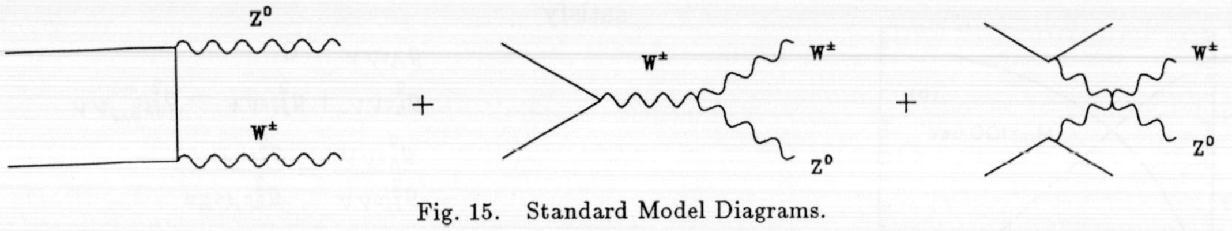

Fig. 15. Standard Model Diagrams.

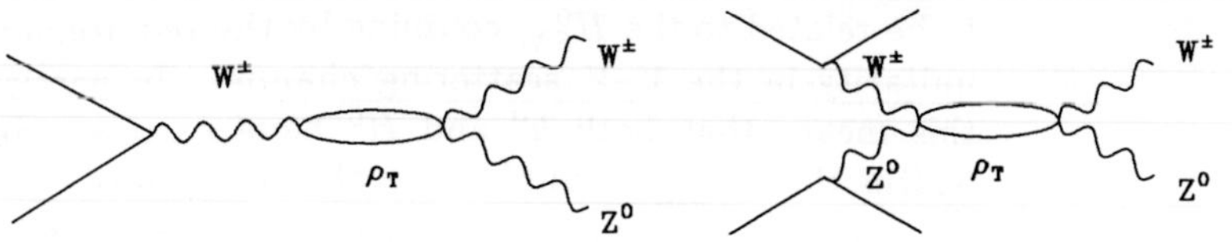

Fig. 16. Technicolor couplings.

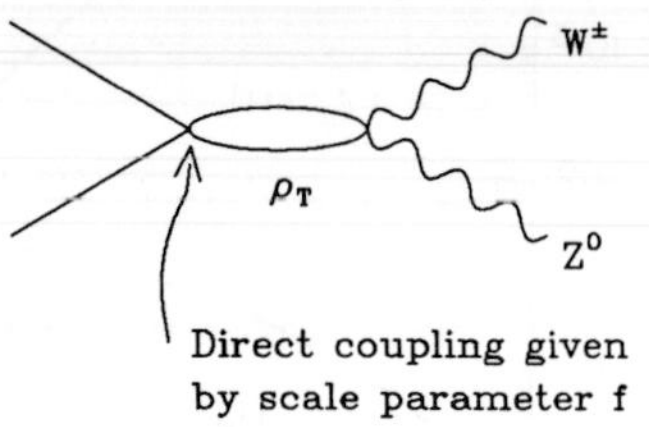

Fig. 17. Extended technicolor coupling to fermions.

These pose severe technological challenges for the machine builder and provide some challenges in doing analysis, mostly related to the fact that the incident beams can lose substantial amounts of energy ("beamstrahlung") prior to interaction. The detector for such a machine seems to be relatively straightforward to design and build.

Since it is likely that such a machine will be constructed after the next large *pp* collider, it is important to understand how well it would do in studying electroweak symmetry breaking for models more complicated than the minimal Higgs model, which will have been investigated at a *pp* collider. In addition, the mass range for a Higgs near 100 GeV, which is difficult to cover with a *pp* collider, might lend itself to study with an e^+e^- machine. These issues will be the focus of the discussion below. As an example of an extended Higgs model, we will choose the simplest supersymmetric scenario.[19] The search for the Standard Model Higgs at a 1 TeV e^+e^- collider has been discussed in Ref. 20. It was shown that for an integrated luminosity of 10^4 pb^{-1} a minimal Higgs boson could be discovered if its mass is not too near the $W^\pm$ or Z^0 ($\gtrsim$150 GeV) and less than about 400 GeV.

Light Minimal Higgs Boson for a 1 TeV Collider

The primary production processes for the Higgs are shown in Fig. 18(a) and (b).

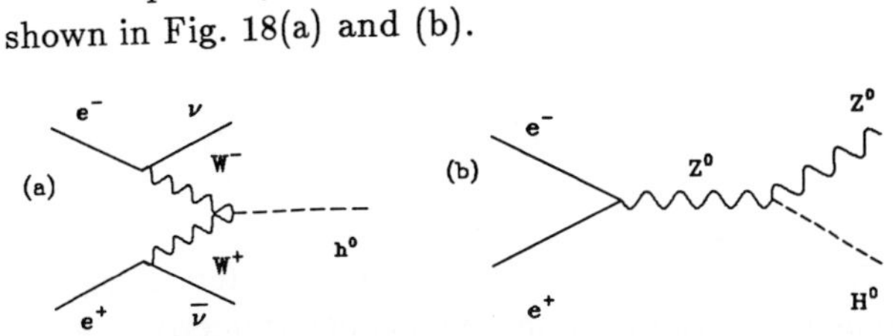

Fig. 18. (a) Primary production process for Higgs at high energy. (b) Higgs production for annihilation diagram.

At 1 TeV, with 10^4pb^{-1} luminosity, for a ~100 GeV Higgs, the first diagram typically will yield thousands of Higgs events, while the second only about a hundred. The analysis for this case is therefore aimed at selecting events which are characteristic of the first process. These are characterized by a substantial amount of missing energy and a large transverse momentum imbalance ($\gtrsim$ 50 GeV, governed by the gauge boson mass scale) due to the undetected neutrinos. These criteria will eliminate most of the Standard Model backgrounds. The largest remaining backgrounds come from the diagram in Fig. 19, where the final state electron is not detected. This has a similar topology to the signal process, but about a factor of 30 higher rate.

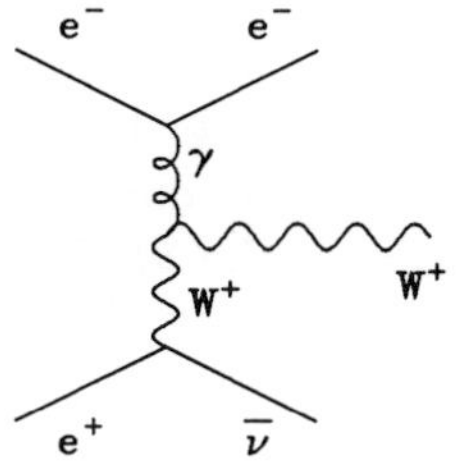

Fig. 19. Primary background process to low mass Higgs search.

To achieve further rejection requires using the fact that the H^0 decays mainly into $b\bar{b}$ or, for a sufficiently light t quark, $t\bar{t}$ final states. The characteristics of these final states can be exploited if the detector has good vertex detection (*e.g.*, a silicon microstrip detector placed a few cm from the interaction region). In this case, requiring four tracks with a large impact parameter improves the signal to background by a factor of about three. In the case of $t\bar{t}$ final states, the events are characterized by having typically $\gtrsim$ 4 final state jets. Making such a cut reduces the background by an additional factor of 5. With these cuts the Higgs can be found for a mass larger than about 100 GeV.

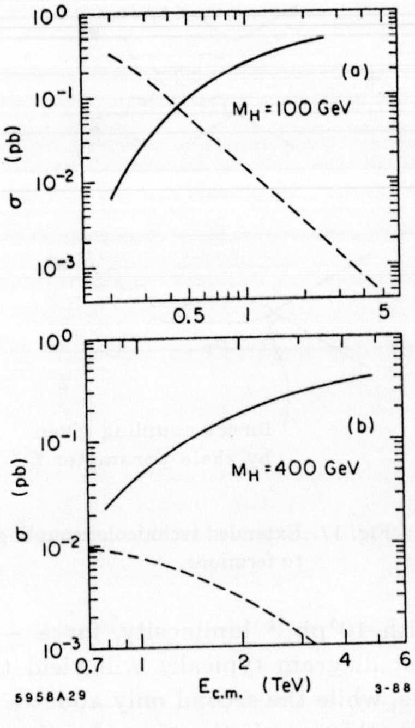

Fig. 20. Cross sections for *a)* 100 GeV and *(b)* 400 GeV Higgs boson production in e^+e^- collisions. The solid curve represents the WW-fusion process, $e^+e^- \to W^+W^-\nu\bar{\nu}$, and the dashed curve represents the annihilation process, $e^+e^- \to ZH$.

MINIMAL HIGGS BOSON FOR A 300 TO 400 GEV COLLIDER[18,21]

To find a Higgs boson with mass between 80 GeV and 180 GeV it is not necessary to build a 1 TeV machine. A machine with a 300 to 400 GeV energy would be sufficient. Fig. 20 shows the cross section as a function of the machine energy for the two Higgs production processes shown earlier in Fig. 18(a) and (b). At 400 GeV the Z^0H^0 final state is dominant and has nearly the same cross section as the $\nu\bar{\nu}H^0$ process at 1 TeV. Thus, an integrated luminosity of 10^4 pb^{-1} would give about one thousand useful events. An analysis aimed at specifically selecting Z^0H^0, would automatically reject the $e\nu W$ background discussed in the previous section. For a H^0 mass near that of the Z^0, the primary background would come from the Z^0Z^0 final state. Again b tagging for $H^0 \to b\bar{b}$ decays and the event structure for $H^0 \to t\bar{t}$ should allow substantial reduction of the background. With these techniques, it should be possible to find the Higgs all the way down to the 80 GeV mass limit expected from LEP2.

TWO-HIGGS DOUBLET MODELS

For models with two Higgs doublets, five physical scalar particles are present after symmetry breaking. Two of these scalars are charged and three are neutral, denoted by $H^\pm$, h^0, H^0, A^0 where h^0 is the lightest neutral scalar, h^0 and H^0 are CP even and A^0 is CP odd. Denoting the gauge vector bosons by V, the Higgs couplings satisfy

$$g_{A^0VV} = 0$$
$$g^2_{h^0VV} + g^2_{H^0VV} = g^2_{H^0_{SM}VV}$$
$$\frac{g^2_{h^0VV}}{g^2_{H^0VV}} = \frac{g^2_{H^0A^0Z^0}}{g^2_{h^0A^0Z^0}}.$$

In the above, $g^2_{H^0_{SM}VV}$ is the Standard Model minimal Higgs coupling. The h^0 and H^0 couplings are constrained to be related to the H^0_{SM} coupling by the requirement of unitarity in the V–V scattering channel. In particular this means that both h^0 and H^0 have narrower decay widths than H^0_{SM} would have at the same mass.

From the relations above, if g_{h^0VV} is close to the Standard Model coupling, then the process $V + V \to H^0$ and $Z^0 \to h^0 + A^0$ are both small, whereas $V + V \to h^0$ and $Z^0 \to H^0 + A^0$ are large. For this case we would expect to produce these particles in the channels shown in Fig. 21(a) and (b).

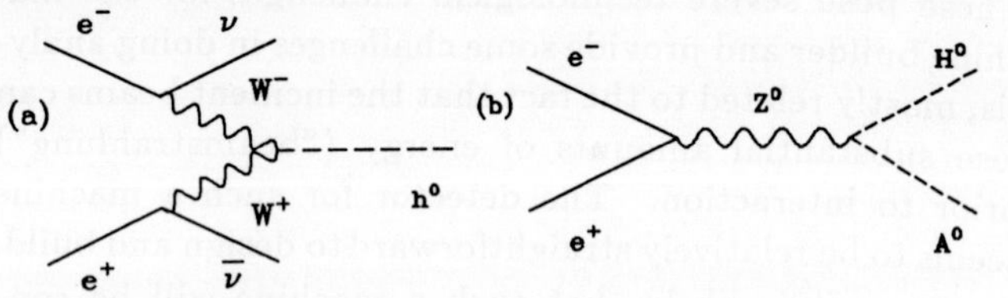

Fig. 21. (a) Diagram for h^0 production. (b) Diagram for H^0A^0 production.

In the general two doublet case there is freedom to separately choose all of the Higgs masses. The simplest supersymmetric model is much more constrained with everything specified in terms of two parameters, which can be taken to be the mass $m_{H^\pm}$ and the ratio of the vacuum expectation values of the two Higgs doublets, expressed as $\tan\beta$.

The masses in this case satisfy

$$m^2_{A^0} = m^2_{H^\pm} - m^2_{W^\pm}$$
$$m^2_{H^0,h^0} = \tfrac{1}{2}\Big(m^2_{A^0} + m^2_{Z^0} \pm \sqrt{(m^2_{A^0} + m^2_{Z^0})^2 - 4m^2_{Z^0}m^2_{Z^0}\cos 2\beta}\Big).$$

These imply

$$m_{h^0} \le m_Z|\cos 2\beta| \le m_Z$$
$$m_{H^0} \ge m_Z$$
$$m_{H^\pm} \ge m_{W^\pm}$$
$$m_{A^0} \le m_{H^\pm}.$$

For $m_{H^\pm} \gg m_{W^\pm}$, H^+, H^0 and A^0 are nearly degenerate in mass. Note the very important prediction that the neutral Higgs, h^0, is lighter than the Z^0.

To look more closely at the prospects for seeing the various particles in the minimal supersymmetric case, a specific example was looked at in detail. In this example the mass of the $H^\pm$ was taken to be 150 GeV, the top quark mass was 60 GeV, and $\tan\beta$ was chosen to be 1.5. This gives a fairly low mass for the A^0, but still higher than the Z^0 which is a potential background. The table below gives the set of masses and branching ratios for this example.

Masses (GeV)	Branching Ratios (%)	
$m_{H^\pm} = 150.0$	$H^0 \to t\bar{t}$	79.6
$m_{H^0} = 153.4$	$H^0 \to b\bar{b}$	6.5
$m_{A^0} = 125.7$	$H^0 \to h^0h^0$	13.6
$m_{h^0} = 29.2$	$A^0 \to t\bar{t}$	95.0
$m_t = 60.0$	$A^0 \to b\bar{b}$	5.0
	$h^0 \to b\bar{b}$	100.0

For this model the h^0 would have been discovered in Z^0 decays and the purpose of the high energy e^+e^- machine would be to find the other particles. We discuss below only the neutral particles, other studies indicate that the $H^\pm$ could also be discovered. The production process for a 1 TeV machine [shown in Fig. 21(b)] has a cross section of about 10^{-2}pb, implying that about 100 events a year would be produced via $e^+e^- \to H^0A^0$. The H^0 and A^0 then decay into final states with many particles and many secondary vertices. An advantage of running at 1 TeV is that the decay products, despite their complexity, separate into two hemispheres corresponding to H^0 and A^0 decay particles, respectively. A lower energy machine, although producing more events, tends to mix together the particles from the A^0 and H^0, making it harder to interpret the result, although still providing a very good signal for the presence of new physics.

The analysis of a Monte Carlo model for the data from a 1 TeV machine relies on the following characteristics for the H^0A^0 process: $\sin^2\theta$ distribution for production, high multiplicity and many large impact parameter tracks in the decay. The table below shows the effect on the signal and primary backgrounds of cuts motivated by the characteristics mentioned above. The initial Monte Carlo sample below would correspond to the data expected for a 10^4pb^{-1} run.

Note, the initial background which is about 1000 times the signal has been reduced by the cuts to be comparable to the signal. The invariant masses of the clusters are then calculated with a beam energy constraint included. This constraint allows an unobserved beamstrahlung photon to accompany the event. Fig. 22(a) and (b) show the higher mass and lower mass clusters, including background. The H^0 and A^0 masses come out correctly near 150 GeV and 125 GeV, respectively. A further verification of the interpretation of the peaks as coming from the production of a pair of new particles could be gotten by a two dimensional plot showing the correlation between the two peaks. Equivalently, we can cut on the higher cluster mass and replot the lower cluster mass, giving the clear signal shown in Fig. 23. Note, finally that the assumed luminosity of 10^4pb^{-1} is close to the lower limit for a successful study of this process.

Cut	Number of Events H^0A^0	$q\bar{q}$	W^+W^-	Z^0Z^0
Generated Sample	112	40,000	35,000	2,000
Total Charged Multiplicity ≥ 40	77	5934	995	53
Multiplicity ≥ 20 within $\lvert\cos\theta\rvert \leq 0.71$	75	5505	853	45
≥ 17 High Impact Parameter Tracks	43	128	65	6
≥ 7 High Impact Parameter Tracks in each Hemisphere	31	71	32	2
Invariant mass for each Hemisphere not in $W^\pm, Z^0$ mass region, that is not within $63 \leq m \leq 112$ GeV	30	66	14	1

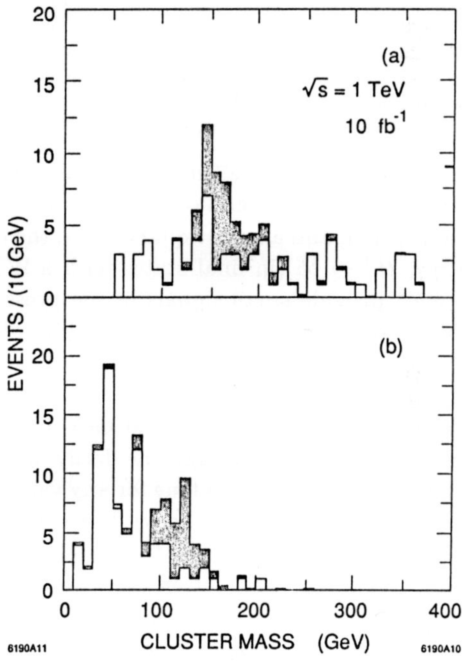

Fig. 22. Reconstructed mass distribution for (a) the larger mass cluster and (b) the smaller mass cluster when all analysis cuts are applied. The shaded region is signal; the open region is background.

CONCLUSIONS ON HIGGS SEARCHES

(1) For the Standard Model Higgs:

(*a*) The mass region $m_W \leq m_H \leq 180$ GeV is a difficult discovery region for both the SSC and e^+e^-

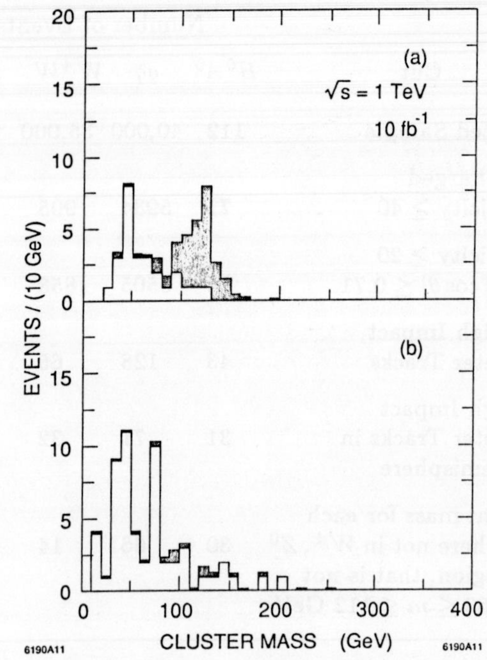

Fig. 23. Reconstructed mass distribution of the smaller mass cluster for events in which (a) the larger mass lies *inside* the range 120–200 GeV, and (b) the larger mass lies *outside* the range 120–200 GeV. The shaded region is signal; the open region is background.

machines. A 400 GeV e^+e^- machine could be useful for searching in the low end of this mass range. For this mass region, and a large t quark mass, the decays $H^0 \to \gamma\gamma$, $H^0 \to b\bar{b}$ (in final states with a W), and $H^0 \to ZZ^*$ provide small signals in the presence of large backgrounds at the SSC. The viability of using these decays will depend in the end on how large the background really is.

(*c*) The mass range $180 \le m_H \le 600$ GeV provides a good signal at the SSC through the decay mode into four charged leptons. This is true for any value of the top quark mass, although a heavy top would allow an upward extension of the useful search region by increasing the Higgs production cross section. The detector for this search needs to have sufficiently good mass resolution so as not to smear the Higgs mass peak. Event rates are small so a high luminosity and good reconstruction efficiency are essential.

(*d*) For $600 \le m_H \le 1$ TeV a search for decays at the SSC of the type $W^\pm W^\mp \to$ jets and $Z^0Z^0 \to$ jets
$\quad\quad\quad\quad\hookrightarrow l^\pm\nu \quad\quad\quad\quad\quad \hookrightarrow l^+l^-$
would yield several hundred events per year. Cuts on jet masses and particle multiplicity produce comparable signal and background rates. The cross sections and background suppression techniques need further study, in particular to understand their sensitivity to QCD models.

(2) In the TeV mass region, Technicolor could produce exciting physics at the SSC. The possibilities include large rates (thousands of reconstructed events per year) in the two gauge boson final state. Distinct resonance peaks are also possible.

(3) Low mass supersymmetry, involving multiple Higgs partners, can be studied at a 1 TeV e^+e^- machine. Signals can most likely be separated from background. A large integrated luminosity $\gtrsim 10^4 \text{pb}^{-1}$ is required. In the simplest case we would expect one of the scalars to be lighter than the Z^0. We therefore might get a hint that this is a correct scenario if a Higgs scalar is discovered in the LEP energy range.

REFERENCES

1. *Physics at LEP*, edited by John Ellis and Roberto Peccei. CERN 86-02 (1986). This two-volume report also contains a large number of references.
2. Chanowitz, M., "Electroweak Symmetry Breaking: Unitarity Dynamics, Experimental Prospects," LBL-24878 (1988).
3. Kass, R., these proceedings.
4. Raby, S., West, G.B. and Hoffman, C.M., Los Alamos Preprint LA-UR-88-2267.
5. *Proceedings of the SLC Workshop on Experimental Use of the SLAC Linear Collider*, March 1981, SLAC-247 (1982). See also Blondel, A., and Grosse-Wiesmann, P., these proceedings.
6. Sirlin, A. *Phys. Rev.* **D22**, 971 (1980); Marciano, W and Sirlin, A., *Phys. Rev.* **D22**, 2695 (1980).
7. Donaldson, R. and Gilchriese, M.G.D., editors, *Proceedings of the Workshop on Experiments, Detectors, and Experimental Areas for the Supercollider*, Berkeley, CA (1987).
8. Bengtsson, H. and Sjostrand, T., *Comput. Commun.* **46**, 43 (1987).
9. Collins, John D., in Ref. 7.
10. Cahn, R.N. *et al.*, in Ref. 7; Gunion *et al.*, U.C. Davis Preprint UCD 86-15 (1986).
11. Hinchliffe, I., in Ref. 7.
12. Barter, C. *et al.*, these proceedings.
13. Brau, J.E., Pitts, K.T. and Price, L.E., these proceedings.
14. Wang, E.M., Hanson, G.G., *et al.*, these proceedings.
15. Sadrozinski, H.W.-W., Seiden, A. and Weinstein, A., these proceedings.
16. Wang, E.M., these proceedings.
17. Chivukula, R.S., Boston University Preprint BUHEP-88-16.
18. Alexander, J., *et al.*, these proceedings.
19. Gunion, J.F., *et al.*, UCD Preprint 88-11.
20. Burchat, P.R., Burke, D.L., and Petersen, A., SLAC-PUB-4638.
21. Grosse-Wiesmann, P., these proceedings.

HIGGS DECAY INTO GOLDSTONE BOSONS

R. Sekhar Chivukula
Department of Physics
Boston University
Boston, MA 02215

Andrew Cohen † and Howard Georgi
Lyman Laboratory of Physics
Harvard University
Cambridge, MA 02138

Benjamin Grinstein
Fermi National Accelerator Laboratory
Batavia, IL 60510

Aneesh V. Manohar
Center for Theoretical Physics
Massachusetts Institute of Technology
Cambridge, MA 02139

Abstract

We comment on the calculation of Voloshin and Zakharov of the rate of decay of a light Higgs Boson into pions.

The Higgs Boson of the standard model may be very light, with a mass of 1 GeV or less, if there is a heavy weak doublet fermion (e.g. an 80 GeV quark or 105 GeV lepton) [1] [2] [3]. If this is the case, the decay of the Higgs into mesons composed of light quarks becomes an important experimental signal. Recently Voloshin and Zakharov [4] [5] have pointed out that the couplings of a light Higgs to pions are calculable to leading order in chiral perturbation theory. The basic idea exploited by Voloshin and Zakharov relies on the quantum violation of scale invariance by the strong interactions.

In this note, we review the Voloshin-Zakharov analysis in slightly different language, which, we hope, will convince any doubters of the signifigance of their result.

The classical Lagrangian of pure QCD with massless quarks possesses a scale invariance, or dilatation symmetry

$$\begin{aligned} A_\mu(x) &\to e^\alpha A_\mu(e^\alpha x), \\ q(x) &\to e^{3\alpha/2} q(e^\alpha x). \end{aligned} \tag{1}$$

The scale current corresponding to this transformation is then conserved

$$\partial_\mu s^\mu = 0. \tag{2}$$

This current can, in principle, be used to derive Ward identities in the standard manner, giving rise to low-energy theorems of a conventional form. Like all symmetries of the Lagrangian, this scale invariance may be realized explicitly on the vacuum of the theory, or it may be realized in the Nambu-Goldstone mode, that is, spontaneously broken. In QCD, once chiral symmetry breaks spontaneously, so does dilatation symmetry; the low energy theorem then implies the existence of a massless scalar particle, the dilaton.

Unfortunately, this simple reasoning is not correct—quantum effects spoil the classical scale invariance by forcing the introduction of a scale-dependent renormalized coupling; the scale current is no longer conserved, and the consequences of low-energy theorems derived on the basis of this symmetry are false. By performing a scale transformation on the effective Lagrangian including quantum corrections, we may compute the divergence of the scale current

$$\partial_\mu s^\mu = \frac{1}{2}\frac{\beta(g)}{g} G^2, \tag{3}$$

where $G^a_{\mu\nu}$ is the gluon field strength tensor and $\beta(g)$ is the usual beta-function for QCD.

Rather than immediately plunge into the morass of QCD, we begin with a simpler model in which we can compute the divergence of the scale current and its matrix element between two goldstone bosons perturbatively. The example we use is the so-called Coleman-Weinberg model [6]. This is a renormalizable theory of scalar electrodynamics with a global SO(N) symmetry. The scalar multiplet $\vec{\phi}$ transforms as the vector of SO(N), and the Lagrangian is taken to be

$$\mathcal{L} = -\frac{1}{4}F^2 + D_\mu\vec{\phi}\cdot D^\mu\vec{\phi}^* - \lambda\left(\vec{\phi}\cdot\vec{\phi}^*\right)^2. \tag{4}$$

With this choice of potential[1] the theory is classically scale invariant. Its spectrum consists of a (massless) photon and N massless charged scalars.

[1] There is another possible term we could include in the potential: $|\vec{\phi}\cdot\vec{\phi}|^2$. However the Lagrangian (4) actually posses an SU(N) symmetry which forbids the appearance of such a term.

Once we go beyond the classical approximation, the structure of the theory changes dramatically. We are free to choose λ and the electromagnetic coupling e to be as small as we wish, and consequently we may compute all Green's functions perturbatively. It is natural to take e and λ to be related in order of magnitude in a way consistent with renormalization. Since λ is renormalized by the graph of Fig. 1, we take $\lambda \sim e^4/(16\pi^2)$. The one loop correction to the scalar potential is then the Coleman-Weinberg potential

$$V_{1-\text{loop}} = \frac{12}{64\pi^2} e^4 \left(\vec{\phi}\cdot\vec{\phi}^*\right)^2 \ln\frac{\vec{\phi}\cdot\vec{\phi}^*}{\mu^2} + 0(\frac{\lambda^2}{16\pi^2}). \qquad (5)$$

Here μ is an arbitrary renormalization point. Since we have chosen $\lambda \sim e^4/(16\pi^2)$ we may ignore the λ^2 term in this effective potential. The full effective potential to leading order is then:

$$U_{eff}(\phi) = \lambda(\mu)\left(\vec{\phi}\cdot\vec{\phi}^*\right)^2 + \frac{3e^4}{16\pi^2}\left(\vec{\phi}\cdot\vec{\phi}^*\right)^2 \ln\frac{\vec{\phi}\cdot\vec{\phi}^*}{\mu^2}. \qquad (6)$$

We have introduced a μ dependence in the coupling λ to make the entire expression μ independent. To this order the μ dependence of λ is

$$\mu\frac{d\lambda}{d\mu} \equiv \beta(\lambda) = \frac{3e^4}{8\pi^2}. \qquad (7)$$

The true minimum of the potential is no longer at $\vec{\phi} = 0$, but at some value $v^2 = \vec{\phi}\cdot\vec{\phi}^*$. It is convenient to choose μ to equal the vacuum expectation value, $\mu^2 = v^2$. Then the minimum of the potential is easily found:

$$\left.\frac{\partial U}{\partial\vec{\phi}}\right|_{\vec{\phi}\cdot\vec{\phi}^*=v^2} = 0 \Rightarrow 2\lambda(v) = -\frac{3e^4}{16\pi^2}. \qquad (8)$$

Dimensional transmutation has occurred—the original dimensionless parameters λ and e may be traded for v and e.

Since the field $\vec{\phi}$ has acquired a vacuum expectation value, the global SO(N) symmetry is spontaneously broken down to SO($N-1$) leaving $N-1$ charged Goldstone Bosons. In addition the U(1) gauge symmetry is broken, and the photon develops a mass. There remains a single massive real scalar. To make these degrees of freedom apparent, we parametrize the field as

$$\vec{\phi} = \left[\left(\frac{\rho}{\sqrt{2}} + v\right)e^{i\theta}, \vec{\phi}_R\right], \qquad (9)$$

where $\vec{\phi}_R$ is a complex $N-1$ vector, and ρ is a real scalar. The angular field θ is then 'eaten' by the photon, the vector $\vec{\phi}_R$ remains massless, and ρ becomes the real massive degree of freedom.

In addition to spontaneously breaking the global SO(N) symmetry the one-loop effective potential also breaks the dilatation symmetry spontaneously. However the presence of the running coupling constant breaks the symmetry explicitly as well. We may compute the one-loop violation of scale invariance by performing a scale transformation on the renormalized Lagrangain, yielding

$$\partial_\mu s^\mu = -\beta(\lambda)\left(\vec{\phi}\cdot\vec{\phi}^*\right)^2, \qquad (10)$$

with $\beta(\lambda)$ given by (7). Using the representation (9) we may compute the expectation value of $\partial_\mu s^\mu$ between the vacuum and two low momentum Goldstone Bosons, $\langle\phi_R^i(p_1)\phi_R^{j*}(p_2)|\partial_\mu s^\mu(q)|0\rangle$, from the graphs of Fig. 2, where the dotted line represents the propagator of the ρ field. Expanding $\vec{\phi}\cdot\vec{\phi}^*$, these diagrams yield

$$-2\beta(\lambda)v^2\delta_{ij} - \beta(\lambda)2\sqrt{2}v^3\frac{i}{q^2-m_\rho^2}(-i)\frac{m_\rho^2}{v\sqrt{2}}\delta_{ij}, \qquad (11)$$

where $q = p_1 + p_2$, $m_\rho^2 = v^2 3e^4/(4\pi^2) = 2\beta(\lambda)v^2$, and we have used (7) in simplifying this expression. Combining terms this becomes

$$\langle\phi_R^i(p_1)\phi_R^{j*}(p_2)|\partial_\mu s^\mu(q)|0\rangle = -q^2\left(\frac{m_\rho^2}{q^2-m_\rho^2}\right)\delta_{ij}. \qquad (12)$$

Note that this is proportional to q^2 as it must be for coupling to two Goldstone Bosons. If we consider the limit $q^2 \ll m_\rho^2$ this gives finally

$$\langle\phi_R^i(p_1)\phi_R^{j*}(p_2)|\partial_\mu s^\mu(q)|0\rangle = q^2\delta_{ij}. \qquad (13)$$

This final result is independent of v and e, the two coupling constants of our theory; the matrix element, including its normalization, is determined uniquely by the derivative coupling of the Goldstone Bosons at zero momentum.

In the case of QCD, we can no longer compute the matrix element perturbatively as in the Coleman-Weinberg example. However the uniqueness of the low energy result allows us to compute the matrix element at low q^2 without the need of perturbation theory. Fortunately this is the domain of interest for a light Higgs. In QCD the symmetries at the classical level include an SU(3) × SU(3) chiral symmetry[2] as well as the dilatation symmetry. At a scale $\Lambda_{\chi SB} \sim 1$ GeV these symmetries break spontaneously to an SU(3) vector symmetry, leaving nine Goldstone Bosons, the meson octet π and the

[2] The $U(1)_A$ symmetry and flavor anomalies are irrelevant for the present discussion.

dilaton σ. The effective Lagrangian for these light degrees of freedom must be invariant under the chiral and dilatation transformations

$$\begin{aligned} \text{Dilatation:} \quad & \sigma(x) \to \sigma(e^{\alpha}x) + f'\alpha, \\ & \Sigma(x) \to \Sigma(e^{\alpha}x), \\ \text{Chiral:} \quad & \sigma(x) \to \sigma(x), \\ & \Sigma(x) \to \mathrm{L}\Sigma(x)\mathrm{R}^{\dagger}, \end{aligned} \tag{14}$$

where L and R are SU(3) matrices, $\Sigma(x) = \exp(2i\pi/f)$, $\pi = \pi^a T^a$, and the T^a are the generators of SU(3). Note that the field Σ does not scale with a power of e^{α}; this implies that π, unlike a fundamental scalar field, has scale dimension zero. This arises from the nature of the Goldstone Bosons as coordinates on a compact manifold; the fixed range of the Goldstone Bosons, (e.g. $0 \leq \pi^3 < 2\pi$) would not be preserved by a scale transformation if the Goldstone Bosons had a non-trivial scale dimension. Alternatively, if we had proposed the more general transformation law

$$\pi(x) \to e^{\alpha d}\pi(e^{\alpha}x), \tag{15}$$

we would have constructed the dilatation current

$$J^{\mu}_{d}(x) = J^{\mu}_{0}(x) + dJ^{\mu a}_{5}(x)\pi^a, \tag{16}$$

where $J^{\mu}_{0}(x)$ is the dilatation current with $d = 0$, and $J^{\mu a}_{5}$ is the chiral current associated with the chiral transformation generated by T^a in (14). The second term on the right hand side of eqn. (14) is not chirally invariant. Hence, in order to have a chirally invariant dilation current, we must choose $d = 0$.

Given these symmetries we now construct the most general effective lagrangian for the light degrees of freedom which is invariant under the chiral and dilatation transformations. As usual we expand the lagrangian in powers of derivatives, keeping only those terms that are most relevant at low momentum. The unique lagrangian up to two derivatives is

$$\mathcal{L} = \frac{f^2}{4} e^{2\sigma/f'} \mathrm{Tr}\partial_{\mu}\Sigma\partial^{\mu}\Sigma^{\dagger} + e^{2\sigma/f'}\frac{\partial_{\mu}\sigma\partial^{\mu}\sigma}{2}. \tag{17}$$

The coefficients of these terms have been chosen to yield conventionally normalized kinetic energies for the Goldstone Bosons. The lagrangian then contains an interaction $\sigma\pi\pi$

$$-\frac{1}{2}i\frac{2}{f'}p_1 \cdot p_2 \delta_{ab} = -\frac{i}{f'}\frac{1}{2}q^2\delta_{ab}. \tag{18}$$

The scale current may be computed and its divergence shown to vanish.

The quantum scale violations of QCD of course destroy this simple picture. The dilaton develops a potential and derivative couplings to the pions that break the scale invariance with a characteristic size of 1 GeV. The divergence of the scale current is no longer zero, but now involves these new dilaton couplings.

The introduction of these symmetry breaking terms will produce a mass for the dilaton of order 1 GeV; since we are interested in the physics of the Goldstone Bosons at scales much less than this, we may integrate out this heavy state which leaves us with an effective Lagrangian of exactly the form (17) with $\sigma = 0$. The divergence of the scale current is then:

$$\partial_{\mu}s^{\mu} = -\frac{f^2}{2}\mathrm{Tr}\partial_{\mu}\Sigma^{\dagger}\partial^{\mu}\Sigma. \tag{19}$$

The matrix element of $\partial_{\mu}s^{\mu}$ between the vacuum and two pions is

$$\langle \pi^a(p_1)\pi^b(p_2)|\partial_{\mu}s^{\mu}(q)|0\rangle = q^2\delta_{ab}. \tag{20}$$

This is precisely the same answer we found in the Coleman-Weinberg example. Indeed this is not a coincidence; this final result could have been computed without introducing the dilaton at all. Since the dilaton becomes heavy due to the violation of scale invariance, we do not need to keep it in our effective theory of pions. The only symmetry we need now impose is chiral symmetry; but this gives us the usual chiral Lagrangian, plus terms that vanish faster than q^2 as $q^2 \to 0$. The divergence of the scale current may be computed by performing a scale transformation; the matrix element of this operator may then be computed between pions yielding the result (20). It is easy to see that a similar result will hold in *any* theory of Goldstone Bosons.

The important aspect of this discussion is that it enables us to compute the matrix element of interest in Higgs decay. The procedure involves recognizing the operators that are responsible for coupling the Higgs to the Goldstone Bosons. The obvious coupling of the Higgs to the pions is through the quark masses. In particular, if we substitute the Higgs field times the Yukawa coupling matrix in place of the quark mass matrix, the effective lagrangian will include the Higgs coupling

$$\mathcal{L}_{\varphi} = \frac{f^2}{4}\Lambda\mathrm{Tr}\left[M(\varphi)\Sigma + \text{h.c.}\right], \qquad M(\varphi) = M_0 + \lambda\varphi, \tag{21}$$

where M_0 is the quark mass matrix and λ is the Yukawa coupling matrix. Expanding to two pions this leads to a coupling

$$-\frac{\varphi}{v}\frac{m_a^2\pi^a\pi^b}{2}\delta_{ab}. \qquad (22)$$

If the mass term were the only coupling, we would hardly need the involved discussion of scale invariance; however the bane of all low energy hadronic calculations, the strong interactions, produces a coupling of the Higgs to the other principal component of the pions, the gluon. In particular the Higgs develops a coupling to two gluons through the loop graph of Fig. 3. To compute the contribution of this operator to the decay of the Higgs requires knowing the matrix element of the operator G^2 between two pions. This task is in fact trivial in the chiral limit, following our discussion of scale invariance—the divergence of the scale current, by the quantum anomaly, involves the operator whose matrix element we wish to know. But as we have just seen, this matrix element is determined by the nature of the pion as a Goldstone Boson

$$\begin{aligned}&\langle\pi^a(p_1)\pi^b(p_2)|\partial_\mu s^\mu(q)|0\rangle \qquad (23)\\ &=\langle\pi^a(p_1)\pi^b(p_2)|\frac{\beta(g)}{g}\frac{1}{2}G^2|0\rangle = q^2\delta_{ab}.\end{aligned}$$

We have only to relate the coefficient of the exact operator whose matrix element we wish to evaluate, to $\beta(g)/g$ in order to complete our calculation. In doing this there is a potential problem—the single loop diagram of Fig. 3 is inadequate in the strong coupling regime where all dressings of the diagram with arbitrary numbers of gluons are equally important.

To deal with this situation, we rely on the construction of an effective field theory by integrating out the quarks that are going to contribute to the diagram one at a time, using the renormalization group to scale the effective theory between the quark thresholds. When a heavy quark of mass m is integrated out, a new operator must be introduced into the effective theory

$$\mathcal{L}_{eff} = \frac{g^2}{48\pi^2}G^2\frac{\varphi}{v} + \cdots. \qquad (24)$$

The ellipses represent terms that are suppressed by further powers of the QCD coupling constant as well as higher dimension operators that are suppresed by powers of the heavy quark mass. At this scale, the QCD coupling is not large, and the terms higher order in g may be ignored. In addition we are working to leading order in $1/m$ and may therefore ignore higher dimension operators. Keeping such terms would produce corrections in the decay amplitude suppressed by powers of m_H/m. We must now scale this operator down to the next quark threshold. This is simple, for small coupling, because (24) is proportional to the gluonic term in the divergence of the scale current, $G^2\beta(g)/g$, which is independent of scale. Thus (24) is independent of scale, up to corrections of order g^4. Therefore, each heavy quark threshold simply adds another term of the form (24), so that just below the charmed quark mass, the induced coupling of the Higgs to gluons is

$$\mathcal{L}_{eff} = n_h\frac{g^2}{48\pi^2}G^2\frac{\varphi}{v} + O(g^4), \qquad (25)$$

where n_h is the number of heavy quarks (3?) to which the Higgs is coupled.

We must now scale this operator down to 1 GeV where chiral symmetry breaking will occur. In this region the coupling is no longer small, and we should not use the perturbative running of this operator. However, the non-perturbative renormalization of the operator G^2 is simple—$G^2\beta(g)/g$ is independent of scale μ [7]. If we rewrite (25) in terms of this operator at the quark threshold we obtain

$$-\frac{n_h}{3b}\frac{\varphi}{v}\left(\frac{\beta(g)}{g}G^2\right)\bigg|_{\mu=m_c} \qquad (26)$$

where $\beta(g) = -bg^3/16\pi^2$ plus higher order terms. Here $b = 11-\frac{2}{3}n_l$, and $n_l = 3$ is the number of light quarks. This form differs from (25) by terms of order $g^2\,G^2\beta(g)/g$ — one may hope that this difference is unimportant since the coupling is assumed small at the charmed quark threshold. The operator in (26) is trivial to scale—it is now written in an explicitly μ independent form. If we were to need the matrix element of G^2 independently of $\beta(g)/g$ we would be in trouble, since the separation of these two components requires knowing the full beta function. Fortunately the only operator we will require is the combination which appears in (26).

The final result at 1 GeV is

$$\mathcal{L}_{eff} = -\frac{n_h}{3b}\frac{\varphi}{v}\left(\frac{\beta(g)}{g}G^2\right)\bigg|_{\mu=1\mathrm{GeV}}. \qquad (27)$$

The Higgs decay into two pions can now be obtained by taking the matrix element of this operator between two pions—this is proportional to the divergence of the scale current, whose matrix element we have already computed (23). The contribution to the decay amplitude is

$$i\mathcal{A}_{G^2}(\varphi\to\pi^a\pi^b) = -\frac{2i}{3v}\frac{n_h}{b}q^2\delta_{ab}. \qquad (28)$$

This contribution depends only on the vev and the leading coefficient of the beta function at heavy quark scales. Adding to this the contribution to the amplitude from the quark masses we obtain the final form for the amplitude[3]

$$\mathcal{A}(\varphi \to \pi^a \pi^b) = -\left[\frac{2}{3v}\frac{n_h}{b} q^2 + \frac{1}{v} m_a^2\right] \delta_{ab}. \tag{29}$$

For the standard six quark model, $n_h = 3$ and $n_h/b = 1/3$.

There are high energy corrections to (29) suppressed by powers of the strong interaction coupling at the heavy quark scales as well as low energy corrections suppressed by powers of $q^2/\Lambda_{\chi SB}^2$, where $\Lambda_{\chi SB} \approx 1$ GeV. The most important high energy corrections occur at the charm quark threshold, and are of order $\alpha_s(m_c)/4\pi \approx 0.03$. We can get a feeling for the size of the low energy corrections by computing the chiral logarithmic [8] corrections to (20). Evaluating the diagrams shown in Fig. 4 we find

$$\langle \pi^a(p_1)\pi^b(p_2)|\partial_\mu s^\mu(q)|0\rangle = q^2\left[1 - \frac{3}{4}\frac{q^2}{\Lambda_{\chi SB}^2}\ln\left(\frac{q^2}{\Lambda_{\chi SB}^2}\right)\right]. \tag{30}$$

From (30) we see that final state pion interactions tend to enhance the decay of a Higgs Boson to two pions. This result is consistent with the dispersion theoretic analysis presented in [9]. For a Higgs mass of order 700 MeV ($q^2 = m_\varphi^2$) we see that the corrections to (20) are large, of order 50%, and the lowest order results are not reliable.

In this note we have discussed the calculation of Voloshin and Zakharov of the rate of decay of a light Higgs Boson into pions. We have illustrated their calculation in the SO(N) Coleman-Weinberg model. We have evaluated the 'chiral logarithmic' corrections to their result and find that these corrections enhance the rate of the decay of a light Higgs to two pions.

Acknowledgements

We would like to thank the Aspen Center for Physics where much of this work was done. This work was supported in part by NSF contracts PHY-82-15249, PHY87-14654, and D.O.E. grants DE-AC02-76ER03069, and DE-AC02-86ER40284. A.M. was supproted in part by a grant from the Alfred P. Sloan foundation.

[3] The mass term contribution is actually $\mathrm{Tr}M\{T^a, T^b\}$ which is not proportional to δ_{ab}. However, once the mass matrix has been diagonalized, this Higgs coupling will be also. Hence the result in the equation is correct with the understanding that the indices a and b run over the mass eigenstates.

References

[1] H. Haber, A. Schwarz, and A. Snyder, Nucl. Phys. B294 (1987) 301.

[2] R. S. Chivukula and A. V. Manohar, Phys. Lett. 207B (1988) 86.

[3] B. Grinstein, L. Hall, and L. Randall, University of California preprint UCB-PTH-88/6, April 1988.

[4] M. Voloshin and V. Zakharov, Phys. Rev. Lett. 45 (1980) 688.

[5] M. Voloshin, Sov. J. Nucl. Phys. 44 (1986) 478.

[6] S. Coleman and E. Weinberg, Phys. Rev. D7 (1973) 1888.

[7] H. Kluberg-Stern and J. B. Zuber, Phys. Rev. D12 (1975) 467, D12 (1975) 482, and D12 (1975) 3159 ; B. Grinstein and L. Randall, Lawrence Berkeley Laboratory preprint.

[8] A. Cohen, in preparation.

[9] S. Raby and G. B. West, Los Alamos preprint LA-UR-88-2491, July 1988.

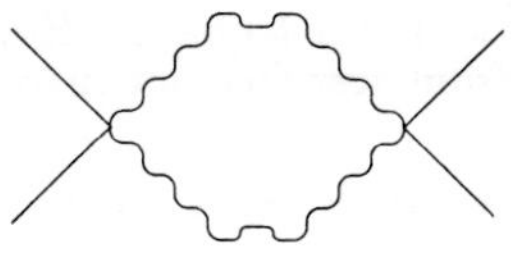

Figure 1: Contribution to the renormalization of λ

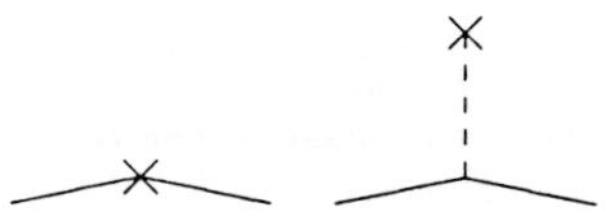

Figure 2: Diagrams contributing to $\langle\phi\phi|\partial_\mu s^\mu|0\rangle$

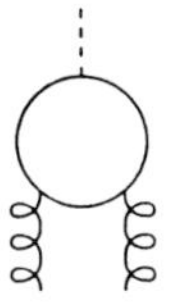

Figure 3

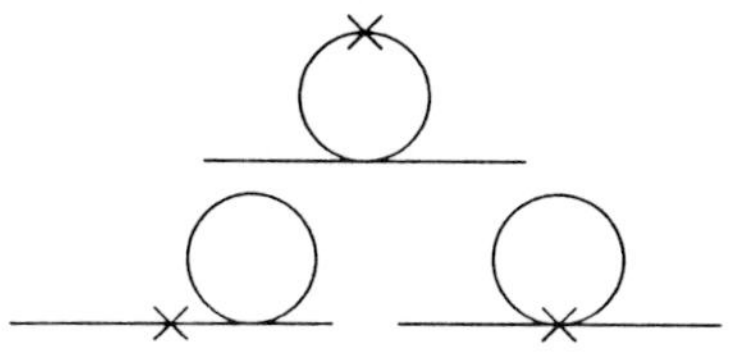

Figure 4: Diagrams contributing logs to $\langle\pi\pi|\partial_\mu s^\mu|0\rangle$

HIGGS LIGHT

Richard Kass
The Ohio State University

ABSTRACT

The production and decay of a light mass Higgs is examined in this note. Particular emphasis is placed on systems containing a b quark.

Introduction

An essential ingredient of the Standard Model[1)] is a scalar particle known to the High Energy Physics community as the Higgs Boson. This particle arises as a by product of the symmetry breaking mechanism invoked to generate the masses of the W and Z bosons. In the minimal version of the Standard Model the coupling of the Higgs to matter is specified, however, the mass of the Higgs is not. Current wisdom suggests that the mass of the Higgs will be large and accessible only to accelerators that operate at or beyond the Z mass. In this note we go against the current dogma and examine several models that allow for a low mass Higgs concentrating on experimental searches using bound states containing a b-quark. However, first we review some results from searches not involving the b-quark.

Reactions not involving a b-quark

The Higgs can be arbitrarily light if there are quarks with mass greater than $\approx$ 80 GeV or leptons with mass greater than $\approx$ 100 GeV.[2)] In the absence of these heavy fermions the Higgs mass must be greater than $\approx$ 7 GeV. However, whether or not this bound is relevant cannot be decided until it is shown that the mass of the t-quark is < 80 GeV. An extremely light Higgs, $m_H < 15$ MeV, can be ruled out from nuclear physics experiments.[3)] An analysis of kaon decay[4)] leads to $m_H > 325$ MeV, while a study of $\eta' \to \eta H, H \to \mu^+\mu^-$ gives $m_H > 409$ MeV.[5)] As discussed in references 6 and 7, the limit from η' decay is suspect as it relies heavily on the knowledge of the branching rate of $H \to \mu^+\mu^-$ vs $H \to \pi^+\pi^-$.

The reaction $\Upsilon \to \gamma H$

Quark-antiquark bound states provide the cleanest environment to search for evidence of the Higgs. This was first pointed out by Wilczek[8)] who calculated the rate for a $q\bar{q}$ bound state to decay into a photon and a neutral scalar boson. As we shall see below, the small branching fraction for the decay $\Upsilon \to \gamma H$ is experimentally accessible with today's data samples. The signature of the radiative decay of the Υ into a Higgs is a monochromatic photon. The mass of the Higgs (m_H) is related to the energy of the monochromatic photon (E_γ) by:

$$E_\gamma = \frac{m_{q\bar{q}}}{2}\left(1 - \left(\frac{m_H}{m_{q\bar{q}}}\right)^2\right) \quad (1)$$

Here $m_{q\bar{q}}$ is the mass of the bound state which for the Υ resonance is 9.4 GeV. This method of search is extremely attractive as it allows for the discovery of the Higgs without relying on any knowledge of its decay modes.

The branching rate (BR) for $\Upsilon \to \gamma H$ can be expressed as:

$$BR_{\Upsilon\to\gamma H} = B_{\mu\mu}\left(\frac{G_F}{4\pi\alpha\sqrt{2}}\right) m_\Upsilon^2\left(1 - \left(\frac{m_H}{m_{q\bar{q}}}\right)^2\right)x^2 R(\alpha_s, m_H, m_\Upsilon) \quad (2)$$

Here $B_{\mu\mu}$ is the branching rate for $\Upsilon \to \mu^+\mu^-$. $R(\alpha_s, m_H, m_\Upsilon)$ accounts for higher order QCD radiative corrections[9)] and is $\approx 1/2$. In the Standard Model with one neutral scalar $x = 1$. However, we can generalize the problem to models with more than one neutral Higgs by setting $x = \langle\phi_1\rangle/\langle\phi_2\rangle$ with $\langle\phi_i\rangle$ the vacuum expectation value of the Higgs fields. There is no bound on x^2, i.e. $0 \le x^2 \le \infty$.

The CUSB group has searched for structure in the photon energy spectrum using a sample of $\approx 10^6$ Υ(1S) and Υ(3S) decays.[10)] The CUSB detector is a highly segmented NaI and BGO calorimeter with energy resolution $\sigma_E/E = 0.018/\sqrt[4]{E_\gamma(GeV)}$. There are several sources of high energy photons in Υ decay which provide a background to a monochromatic photon. These sources include: $\pi^o \to \gamma\gamma$, $\eta^o \to \gamma\gamma$, two gluon decay of the Υ ($\Upsilon \to gg\gamma$), and radiative decays of the Υ(3S) (e.g. $\Upsilon \to \gamma^3$P). After correcting the data for backgrounds and efficiencies the CUSB group finds no evidence for structure in the photon energy spectrum. Figure 1 presents the 95% confidence levels for the Higgs mass as a function of x. For $x = 1$ and radiative corrections included, we have $m_H > 4.8$ GeV.

The reaction $B \to HX$

It has been pointed out by several authors[7,11,12)] that B meson decay might be a copious source of Higgs particles. The proposed mechanism responsible for the decay $B \to HX$ is shown in Figure 2. Assuming that there are only three generations, the inclusive branching ratio for $B \to HX$ normalized to semileptonic B decay can be written as:

$$\frac{BR_{B\to HX}}{BR_{B\to l\nu X}} = \left(\frac{27\sqrt{2}G_F}{64\pi^2}\right)\left|\frac{V_{tb}V_{ts}^*}{V_{cb}}\right|^2 m_b^2\left(\frac{m_t}{m_b}\right)^4\left(1 - \left(\frac{m_H}{m_b}\right)^2\right)^2 R\left(\frac{m_c}{m_b}\right) \quad (3)$$

Figure 1. The 95% confidence level upper limits on the the mass of the Higgs as a function of x.

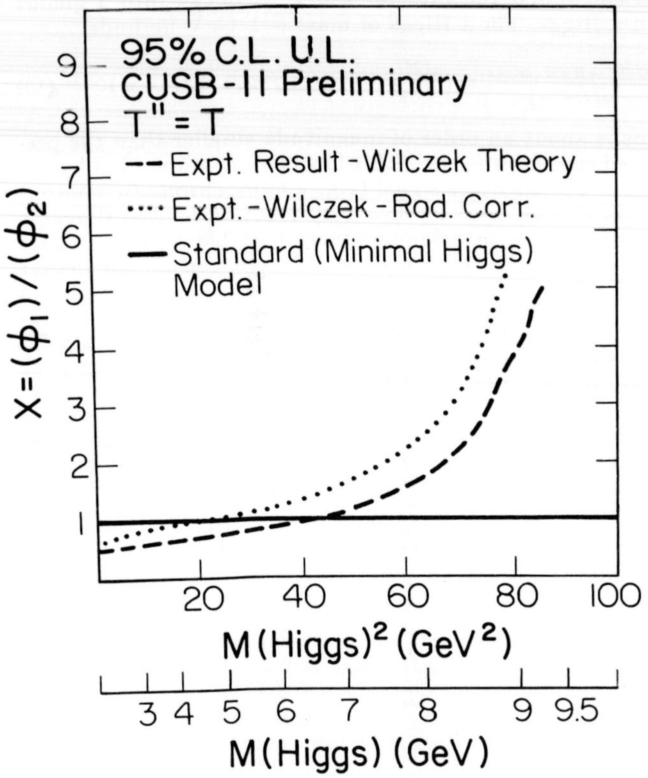

Figure 2. Diagram responsible for $B \to HX$ in the standard model.

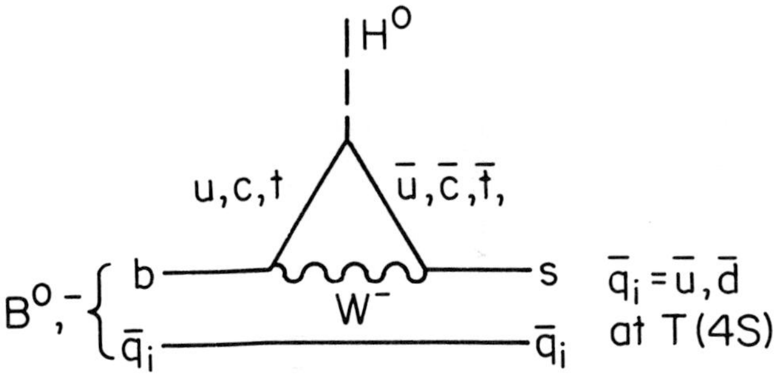

Figure 3. Diagrams considered by Voloshin in Ref. 13.

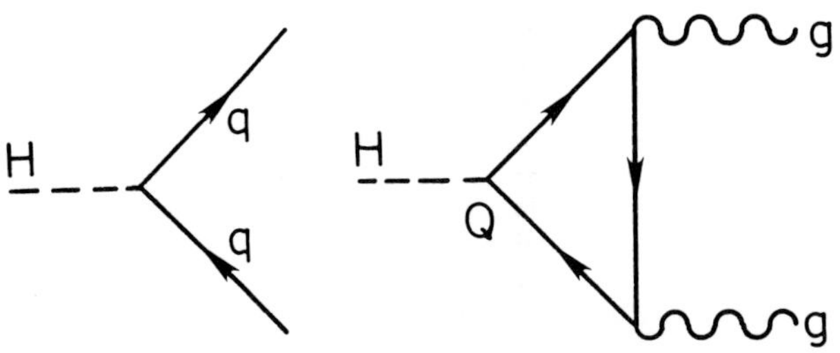

Here m_b, m_t, m_H and m_c is the mass of the b, t, Higgs, and c quark, respectively. The terms V_{tb}, V_{ts}, and V_{cb} are the Kobayashi Maskawa matrix elements. Finally, $R(m_c/m_b)$ is a phase space correction factor. It is important to note that in this model the rate for $B \to HX$ grows like m_t^4. We can estimate the rate $B \to HX$ using the following data; m_b=4.9 GeV, m_c=1.6 GeV, R=2.08, $BR_{B\to l\nu X}$=11%, and $|V_{tb}V_{ts}^*/V_{cb}|^2$=1.Recent data suggests that $m_t \geq 50$ GeV therefore we rewrite Eq. 3 as:

$$BR_{B\to HX} = (0.042)(\frac{m_t}{50})^4(1-(\frac{m_H}{m_b})^2)^2 \qquad (4)$$

If $m_H < m_b$, the rate for $B \to HX$ is about two orders of magnitude larger than Higgs production via the Wilczek mechanism.

An estimate of the two body decay $B \to HK$ has been made by Haber et al. in Ref. 12. They predict the following large rate for this mode:

$$BR_{B\to HK} = (0.07)/[(1-(\frac{m_H}{m_b})^2)(1-(\frac{m_H}{\Lambda})^2)^2] \qquad (5)$$

with $\Lambda = 6.1$ GeV. They also predict that rate for $B \to HK^*$ will be small.

Unlike the decay $\Upsilon \to \gamma H$, detection of the Higgs in B decay must rely on some knowledge of the branching modes of the Higgs. While it is tempting to try to use a K as a tag for Higgs production in B decay the copious production of K's due to non-Higgs sources at the Υ(4S) makes this unworkable. Thus while the rate for $B \to HX$ might be higher than $\Upsilon \to \gamma H$, Higgs detection in B decay is experimentally a much more challenging project.

The branching modes of the Higgs depend strongly on its mass. Assuming that the Higgs couples to mass, there are three regions of interest; $2m_e < m_H < 2m_\mu$, $2m_\mu < m_H < 2m_\pi$, and $m_H > 2m_\pi$. In the first two regions the Higgs decays 100% of the time into e^+e^- or $\mu^+\mu^-$ respectively. There are several model calculations of these decay rates when $m_H > 2m_\pi$. For $H \to \mu\mu$ and $H \to \pi\pi$, Voloshin[13)] calculates, using the diagrams in Fig. 3:

$$\frac{BR_{H\to\pi\pi}}{BR_{H\to\mu\mu}} = \left(\frac{m_H^2}{27m_\mu^2}\right)(1+\frac{11m_\pi^2}{2m_H^2})^2|\frac{p_\pi}{p_\mu}| \qquad (6)$$

In Eq. 6 p_μ and p_π are the momenta of the μ and π in the rest frame of the Higgs. We note that at m_H=1 GeV, Eq. 6 gives ≈ 4.3, several orders of magnitude larger than the estimate of 2.6×10^{-2} of Refs. 3 and 14. Extending this calculation to include $s\bar{s}$ pairs and limiting the the Higgs mass to the range m_H <2 GeV, the lower limit of $H \to \mu^+\mu^-$ is:

$$BR_{H\to\mu\mu} = 81m_\mu^2/(81m_\mu^2 + 8m_H^2) \qquad (7)$$

For m_H < 2 GeV Eq. 7 predicts that the Higgs decay to $\mu^+\mu^-$ should be greater than 10%. The region between 2 and 3 GeV yields no clear prediction of the rate for $H \to \mu^+\mu^-$. Above 3 GeV and below $m_H < 2m_\tau$ Voloshin calculates:

$$BR(H\to s\bar{s}) : BR(H\to gg) : BR(H\to\mu^+\mu^-)$$
$$= 3m_s^2 : 16(\frac{N_h\alpha_s}{12\pi})^2 m_H^2 : m_\mu^2 \qquad 8)$$

Choosing m_H=3 GeV, α_s=0.2 and N_h=3.5 we calculate a rate of 8% for $H \to \mu^+\mu^-$.

Grinstein et. al.[7] find results similar to Voloshin's in the region $m_H > 2$ GeV. Here they predict:

$$BR(H \to s\bar{s}) : BR(H \to gg) : BR(H \to \mu^+\mu^-)$$
$$= 3m_s^2 : (\frac{\alpha_s}{\pi})^2 m_H^2 : m_\mu^2 \qquad (9)$$

However, these authors question the validity of Voloshin's calculations below 2 GeV and illustrate the sensitivity of the results to the details of the strong interaction.

There have been several searches for a low mass Higgs decaying into $\mu^+\mu^-$ at electron-positron storage rings. Both the CLEO[15] and ARGUS[16] experiments have measured the inclusive $\mu^+\mu^-$ cross section at the $\Upsilon(4S)$. Similar measurements using high energy continuum data have been made by the TASSO[17] group. All of these experiments use large, general purpose solenoidal detectors emphasizing good momentum resolution through charged particle tracking. The data from CLEO and ARGUS combined with the calculations of either Voloshin or Grinstein et al. rule out a Higgs in the mass range $2 < m_H < 3.7$ GeV. Depending on the choice of QCD parameters one chooses, the data from CLEO and TASSO can be used to eliminate the mass range $2m_\pi < m_H < 2$ GeV. Above $\tau^+\tau^-$ threshold all of the above experiments suffer from lack of sensitivity, and hence no conclusions on Higgs production can be drawn from the data.

The CLEO collaboration has searched a data sample of 1.8×10^5 B mesons for evidence of two body decays of the form $B \to HK$, HK^* and $H \to X\bar{X}$. In this search all the decay products are measured and the resulting invariant mass is compared to the known B mass. Candidate events with invariant mass within 3.5 MeV of the B mass are kept for further study. An aid in rejecting background is the use of the known beam energy in place of the measured particle energies when calculating the invariant mass of the candidate B. For each of the reactions listed in Table 1 the invariant mass distribution of the $X\bar{X}$ system was searched for structure. The only structure present in any of the reactions was the ψ in the $\mu^+\mu^-$ channel. The 90% confidence level upper limits on the product branching ratio for the various channels is given in Table 1. These limits correspond to the highest upper limit obtained in each mode as a function of the $X\bar{X}$ mass. These data along with the calculations of References 7, 11, and 12 rule out a Higgs in the mass range $0.3 < m_H < 3.0$ GeV and $3.2 < m_H < 3.7$ GeV assuming the mass of the top quark is greater than 43 GeV. The mass region 3-3.2 GeV must be excluded from this analysis as ψ production dominates this region.

$B \to HK^-$ H decay	BR$\times10^{-3}$ < 90% C.L.
$\mu^+\mu^-$	0.4
$\pi^+\pi^-$	0.2
K^+K^-	0.1
$K^\pm K^{*\mp}$	2
$\overline{K}^{*o}K^{*o}$	2
$\overline{K}^oK^{*o} + K^o\overline{K}^{*o}$	2
$B \to HK^{*o}$	
$\mu^+\mu^-$	0.8
$\pi^+\pi^-$	1
K^+K^-	0.9
$K^\pm K^{*\mp}$	4
$\overline{K}^{*o}K^{*o}$	6
$\overline{K}^oK^{*o}+K^o\overline{K}^{*o}$	4

Table 1: The worst case 90% confidence levels for the upper limits on the branching fraction for several modes of $B \to HK, K^*$

Other Candidate Reactions

There are other mechanisms that can produce a low mass Higgs besides the two discussed above. Voloshin[13] has calculated the rate for the Υ to decay into 3 gluons and a Higgs. For a Higgs of mass ~ 1 GeV he finds:

$$\frac{BR_{\Upsilon\to H+Hadrons}}{BR_{\Upsilon\to 3g}} \approx (\frac{\sqrt{2}G_F m_b^2}{\pi^2})\ln(m_b/m_H) \approx 3-4\times10^{-5} \qquad (10)$$

This is about an order of magnitude smaller than the predicted rate for $\Upsilon \to \gamma H$. This low rate, compounded by the lack of a clean Higgs signal (e.g. a monochromatic photon) makes this an unlikely discovery reaction for the Higgs.

Finally, we mention one more possible source of Higgs production. Bergström et al.[18] have considered decays of pseudoscalar quarkonium (η_c,η_b) into two photons and either a scalar (i.e. Higgs) or a pseudoscalar (i.e. axion). Unlike the case of vector quarkonium, the rate for pseudoscalar quarkonium→ Higgs differs from pseudoscalar quarkonium→ axion. At first glance this mode appears experimentally accessible as the two photons provide a clean tag for this channel. However, it is unlikely that the Higgs will be observed here as calculations indicate that experiments will need unattainably large ($> 10^5$) samples of η_c's or η_b's.

Summary and Conclusions

Several interesting methods for discovering a low mass Higgs have been discussed here. Experimental results from Υ and B decay, subject to theoretical interpretation, rule out a Higgs in the mass range $\approx$ 0.3-5 GeV. We can also eliminate the mass range $m_H < 0.3$ GeV by including results from kaon decay and nuclear physics experiments. Higgs search experiments at low energy storage rings can be expected to continue for next several years. Both CESR and DORIS expect to operate at record luminosities providing an order of magnitude increase in the samples of Υ's and B's. Thus until higher energy machines become available physicists will continue to search for the Higgs under lamp posts where the light shines brightest.

Acknowledgements

The author would like to thank all those involved with the organization of the workshop. This work was supported by the O.J.I. program and D.O.E. contract No. DE-AC02-76ER01545.

References

1) S. L. Glashow, Nucl. Phys., 22, 579 (1961); S. Weinberg, Phys. Rev. Lett., 19, 1264 (1967); A. Salam, Proc. 8th Nobel Symp. (Stockholm), ed. N. Svartholm (Almqvist and Wiksells, Stockholm, 1968) P.367.

2) S. Weinberg, Phys. Rev. Lett., 36, 294 (1976); A.D. Linde, JETP Lett, 23, 73 (1976).

3) R. Barbieri and T. D. Ericson, Phys.Lett., 57B, 270 (1975); J. Ellis, M. K. Galliard, D. V. Nanopoulos, Nucl. Phys., B106, 292 (1976); A. I. Vainshtein, et al., Sov. Phys. Usp., 23 429 (1980).

4) R. S. Willey and H. L. Yu, Phys. Rev., D26, 3287 (1982).

5) R. I. Dzhelyadin et al., Phys. Lett., 105B, 239 (1981); V. A. Viktorov et al., Sov. J. Nucl. Phys., 33, 822 (1981).

6) R. Ruskov, Phys. Lett., 187B, 165 (1987).

7) Grinstein, Hall, and Randall, LBL preprint LBL-25095, April 1988.

8) F. Wilczek, Phys. Rev. Lett., 39, 1304 (1977).

9) M. I. Vysotsky, Phys. Lett., 97B, 162 (1980).

10) M. Narain et al., paper submitted to the 1987 Symp. on Lepton and Photon interactions at High Energies, Hamburg, 1987.

11) R. S. Willey and H. L. Yu, Phys. Rev., D26, 3086 (1982); B. Grzadkowski and P. Krawszyk, Z. Phys., C18, 43 (1983); R. M. Godbole, U. Turke, and M. Wirbel, Phys. Lett., 194B, 302 (1987).
12) H. Haber, A. Schwarz, A. Snyder, Nucl. Phys., B294, 301 (1987).
13) M. B. Voloshin, Sov. J. Nucl. Phys., 44, 478 (1986).
14) A. A. Ansel'm, et al., Sov. Phys. Usp., 28 113 (1985).
15) P. Avery, et al., Phys. Rev. Lett., 53, 1309 (1984); P. Avery et al., "Search for the Neutral Higgs in B-Meson Decay ", paper submitted to the 1987 Symp. on Lepton and Photon interactions at High Energies, Hamburg, 1987.
16) H. Albrecht, et al., Phys. Lett., 199B, 451 (1987).
17) M. Althoff, et al., Z. Phys., C22 219 (1984).
18) L. Bergström, P. Poutianinen, and Rubinstein, Bonn University preprint, Bonn-HE-88-05, February 1988.

The importance of the W-Z mass difference measurement and its implications for the Tevatron upgrade program

Rajendran Raja
Fermi National Accelerator Laboratory[1]
Batavia, Illinois, 60510, USA
November 1988

Abstract

We examine the phenomenology of the W-Z mass difference as a function of the top quark mass and the Higgs mass. Scenarios are presented for the integrated luminosities of 10 pb^{-1} and 100 pb^{-1} at the Tevatron. It is shown that the Tevatron is competitive with LEP II for integrated luminosities of 100 pb^{-1} in attainable precisions in this measurement. These considerations argue strongly for the pp option in the Tevatron upgrade program. For a detailed description of this work, please see reference [1]

Phenomenology of radiative corrections in the electroweak theory

In what follows, we adopt the definition of the weak mixing angle due to Sirlin [2].

$$sin^2\theta_W \equiv 1 - \frac{M_W^2}{M_Z^2} \tag{1}$$

This implies

$$M_Z^2 \equiv \frac{M_W^2}{cos^2\theta_W} \tag{2}$$

From low energy phenomenology, one can show [3] that

$$M_W^2 = \frac{A^2}{(1-\Delta r)sin^2\theta_W} \tag{3}$$

where

$$A = \sqrt{\frac{\pi\alpha}{\sqrt(2)G_\mu}} = 37.281 \pm 0.003 GeV \tag{4}$$

with the muon decay constant $G_\mu = 1.16634 \pm 0.00002 \times 10^{-5} GeV^{-2}$ with the inverse fine structure constant $1/\alpha = 137.035963(15)$. Δr is the term due to radiative corrections at the one loop level in the electroweak theory. Δr is a function of the top quark mass and the Higgs mass, the two unknown masses in the theory. Note that in this formalism,

$$\rho \equiv \frac{M_W^2}{M_Z^2 cos^2\theta_W} \equiv 1 \tag{5}$$

by definition, within the standard model. We will only need to consider models with $\rho \neq 1$ if the standard model is found to be inconsistent by measuring the W-Z mass difference . There are thus four unknowns M_W, M_Z, $sin^2\theta_W$, and Δr and two equations 2 , 3. So one can express everything in terms of two variables which we can choose to be M_Z and Δr. Δr depends on the Higgs mass and the top quark mass as stated previously. It also depends on other fermion masses such as the charm quark mass, bottom quark mass etc, but these masses are constrained from low energy phenomenology.

If we assume that $M_{Higgs} = M_Z$, $m_c = 1.5\ GeV/c^2$, $m_b = 4.5\ GeV/c^2$ and $m_t = 36\ GeV/c^2$ [5], we get $\Delta r = 0.0696 \pm 0.0020$. This leads to "predictions" of the W and Z masses of $M_W = 83.0 + 2.9 - 2.7\ GeV/c^2$ and $M_Z = 93.8 + 2.4 - 2.2\ GeV/c^2$. Note that these numbers assume a value for the top quark mass which may not be realistic. However, if $\Delta r = 0$, $M_W = 78.2\ GeV/c^2$ which is well outside the measured values by UA1 and UA2 [6]. This indicates that radiative corrections are important and need to be taken into account while calculating masses in the electroweak theory [3].

In what follows, we treat Δr as a function of m_t and m_H, the top quark and Higgs masses respectively. So one can make tables of M_W and $sin^2\theta_W$ for various values of M_Z, m_t and m_H [7] [8]. LEP will soon measure M_Z accurately. According to Altarelli [9], $\Delta M_Z = \pm 30 MeV/c^2$ should well be within the reach of LEP once QED radiative corrections are taken into account. We assume conservatively that the Z mass is measurable to an accuracy of $\pm 50 MeV/c^2$ by the time the Tevatron measurements of the W-Z mass difference are preformed.

For the rest of this analysis, we will assume that LEP has measured $M_Z = 94.0 \pm 0.050\ GeV/c^2$. Once the true value of M_Z is known, it should be easy to scale up the curves presented here with minor changes in the conclusions.

Figure (1) shows the variation of the W-Z mass difference [7] as a function of the top quark mass. The three curves are for Higgs masses of 10 , 100 and 100 GeV/c^2. The reason the curves are not quite smooth is due to truncation errors in the data which were published as numerical tables. As the top quark mass varies from 50 GeV/c^2 to 225 GeV/c^2 , the W mass changes from $\sim 83.5\ GeV/c^2$ to $\sim 84.3\ GeV/c^2$. The W mass is a weak function of the Higgs mass. From the figure, it can be seen that the present global average of UA1 and UA2 data [6] scaled up to $M_Z = 94\ GeV/c^2$ places an upper limit of ~ 225

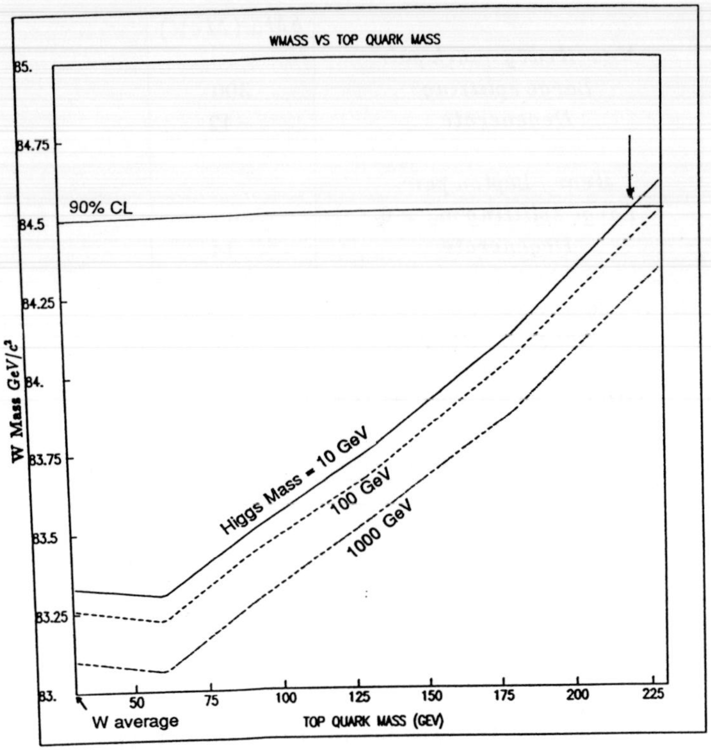

Figure 1: Variation of the W mass as a function of the top quark mass. $M_Z - 94.0\ GeV/c^2$

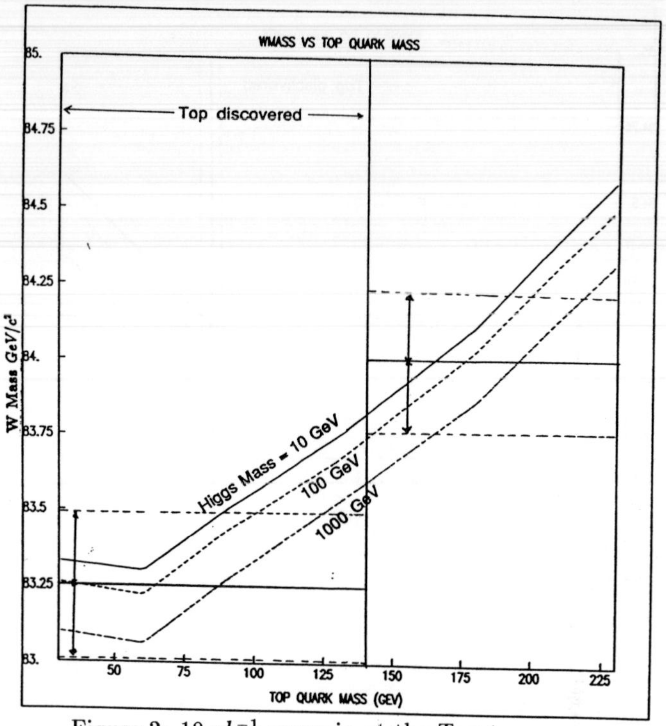

Figure 2: 10 pb^{-1} scenario at the Tevatron

GeV/c^2 for the top quark mass.

$10pb^{-1}$ scenario at the Tevatron

From an analysis of data simulated under expected D0 parameters [1], we conclude that it is possible to measure the W mass to an accuracy of $\pm 0.2\ GeV/c^2$ and the Z mass to an accuracy of $\pm 0.12\ GeV/c^2$ with 10 pb^{-1} of data at the Tevatron. It can also be shown [1],[10] that for 10 pb^{-1} of integrated luminosity, CERN can see a top quark if it is below 80 GeV/c^2 and Fermilab at 1.8 TeV will see it if it is below 140 GeV/c^2. Figure 2 shows two possible scenarios. If the top quark mass is below 140 GeV/c^2, the lower scenario for the W mass will hold with the indicated errors. The W-Z mass difference measurement can be used to set constraints on the top quark mass and help search for it in the data. If however, it is greater than 140 GeV/c^2, the upper scenario for the W mass will hold and the measurement again can be used to predict the top quark mass. The search for the top quark and the W-Z mass difference measurement should go on concomitantly.

100 pb^{-1} scenario at the Tevatron

We now consider the $100pb^{-1}$ scenario with a Tevatron upgrade. With a pp machine at a luminosity of 10^{32}, 11 days of running will produce 100,000 W's and 10,000 Z's. δM_W can be measured to $\pm 0.075\ GeV/c^2$ statistically. We assume that the techniques outlined in reference [1] can be used to control the systametics to the required accuracy. This is comparable to what is obtainable for the W mass at LEP II [11]. The Z transverse mass can now also be fitted using an algorithm similar to the W fitting algorithm, by ignoring one of the decay electrons of the Z. There is enough statistics at these luminosities and this gives a way to control systematic differences between the two fitting techniques used for the W and the Z masses. Figure 3 shows a possible scenario. The top quark discovery limit is now 240 GeV/c^2 and a standard model top quark must be discovered. We assume that such a top quark has been discovered and its mass determined accurately using the Tevatron, LEP II or a new genration of e^+e^- collider. If this is the case, then the precision measurement of the W-Z mass difference will enable one to get an idea of the Higgs mass. It will provide a crucial test of the standard model. If discrepancies are found, a whole host of scenarios can be invoked outside the standard model to explain the data. Table 1 shows the shift in the W mass due to scenarios such as the fourth generation, heavy lepton pairs, SUSY and Technicolor. It can be seen that the

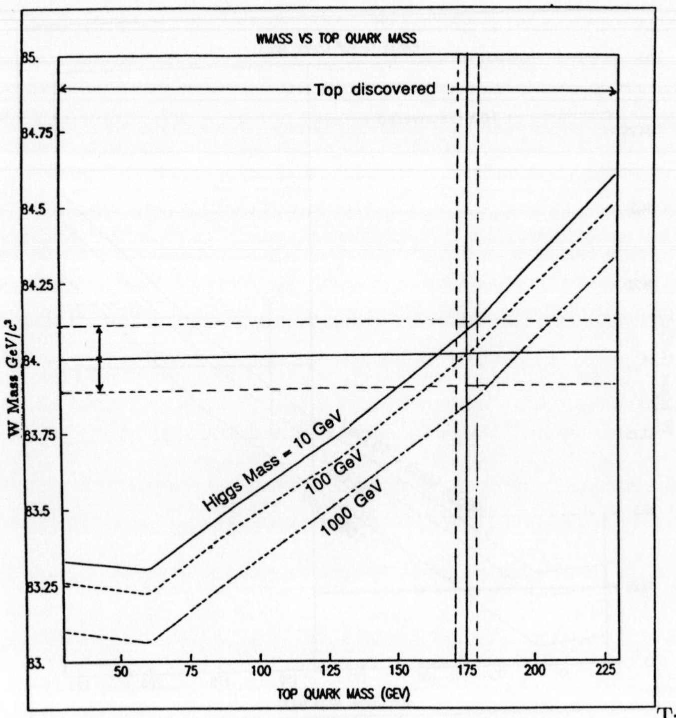

Figure 3: 100 pb^{-1} scenario at the Tevatron

	$\delta M_W(MeV)$
New Heavy quark pair	
Large splitting	300
Degenerate	−42
Heavy Lepton pair	
Large splitting $m_\nu = 0$	300
Degenerate	−14
Heavy squark pair, slepton pair	
Large splitting	300
Degenerate	0
Winos	
$m_{3/2} \ll 100\ GeV/c^2$	100
$m_{3/2} \gg 100\ GeV/c^2$	< 10
Technicolor	
SU_8XSU_8	−500
O_{16}	−500

Table 1: Shift in the W mass due to effects beyond the standard model

limits obtained on the top quark mass earlier assume the completeness of the standard model. It is possible to have a very heavy top quark which will drive the W mass up, and technicolor which will drive the W mass down again to the observed value. In general fermion pairs with large splitting will drive the W mass up whereas, existence of scalars such as in Technicolor will drive the W mass down.

To conclude, the Tevatron collider with the high luminosity *pp* upgrade option, provides a tool for measuring the W-Z mass difference to a precision comparable to what is obtainable elsewhere [11]. The measurement in the 10 pb^{-1} scenario can help predict the top quark mass if it is greater than 140 GeV/c^2 and in the 100 pb^{-1} scenario can help establish the completeness or otherwise of the standard model. In order to control the systematics of the measurement, it is necessary to acquire the 100 pb^{-1} of data rapidly. With a $10^{32}pp$ collider, it is possible to acquire 100 pb^{-1} of data in just 11 days of running. During this short interval, it should be possible to prevent the calibration of the detector from drifting significantly. A *pp* collider alone can provide these luminosities in the foreseeable future. The W-Z mass difference measurement is as fundamental a measurement as one can make with the Tevatron. It should be made in parallel with searches for phenomena which go beyond the standard model.

References

[1] Theoretical Implications of the W-Z mass difference and the capabilities of the D0 detector in measuring it , R.Raja, Proceedings of the 7th Topical Workshop on $\bar{p}p$collisions, Fermilab, June 1988, FNAL Preprint CONF-88/198

[2] A. Sirlin, Phys. Rev D22 ,971 (1980).

[3] W.J.Marciano and A.Sirlin, Phys. Rev. D22, 2695 (1980)
A.Sirlin and W.J.Marciano, Nucl. Phys. B189, 442 (1981)
Testing the standard model by precise determinations of $W^\pm$ and Z masses, W.J.Marciano and A.Sirlin, Phys Rev D29, 945. (1984)

[4] W.J.Marciano, Phys. Rev D20, 274 (1979)
W.J.Marciano and A. Sirlin Phys. Rev. Lett. 46, 163 (1981)
A.Sirlin, Phys. Rev. D29, 89 (1984)

[5] W. Wetzel, Z. Phys. C 11 , 117 (1981)
W. Marciano, Aspen Winter Conference 1985.
M. Consoli, S. Lo Presti and L. Maiani, Nucl. Phys. B 223, 474 (1983)

[6] UA1 data used for this analysis is from the presentations at the Uppsala Conference (1987)
UA2 data come from CERN-EP87-05.

[7] B.W. Lynn, M.E.Peskin and R.G.Stuart, Physics at LEP (J.Ellis, R.Peccei, editors) Vol 1, Cern 86-02, Page 90.

[8] B.W.Lynn and R.G.Stuart, Nucl. Phys. B 253,216 (1985)

[9] G. Altarelli et. al. Physics at LEP (J.Ellis, R.Peccei, editors) Vol 1, Cern 86-02, Page 1.

[10] Estia Eichten, Fermilab-Conf-85/178-T (May 1986)

[11] G. Barbiellini et. al. Physics at LEP (J.Ellis, R.Peccei, editors) Vol 2, Cern 86-02, Page 1.

DETECTION OF $H^0 \to \gamma\gamma$ AT THE SSC

C. Barter and R. Partridge
Brown University, Providence, Rhode Island 02912

A. Bay and A. Spadafora
Lawrence Berkeley Laboratory, Berkeley, California 94720

S. Whitaker
Boston University, Boston, Massachusetts 02215

A. Abashian
University of Virginia, Chalottesville, Virginia 22901

R. Kass
Ohio State University, Columbus, Ohio 43210

ABSTRACT

We explore the prospects for using the SSC to find an intermediate mass Higgs boson through its decay to two photons. Monte Carlo studies of the signal and various backgrounds are performed to determine the required detector parameters.

INTRODUCTION

The essential task facing the SSC is to understand the mechanism for electroweak symmetry breaking. In light of the continuing success of the standard model, it is imperative that experiments at the SSC be capable of testing the predictions of the minimal standard model for the Higgs sector. Since the Higgs mass is essentially undetermined by low energy data, we must be prepared to look for the Higgs in all mass ranges. Furthermore, the task of determining Higgs couplings to fermions and gauge bosons can only be accomplished if a variety of different channels are measured. This paper explores the possibility of observing the Higgs through its decay into two photons.

A light Higgs ($M_H < 85$ GeV) will almost certainly be found at LEP II before the SSC turns on, while a heavy Higgs ($M_H > 180$ GeV) will be studied using Z^0Z^0 and W^+W^- pairs. The intermediate mass Higgs ($85 < M_H < 180$ GeV) proves to be problematical. The preferred decay to a pair of heavy quarks is buried under the much larger QCD two-jet background. The only hope for finding an intermediate mass Higgs at the SSC is through a rare decay with a clean and distinctive signature.

If the Higgs lies above top threshold, decays to top quarks will dominate to the extent that the reaction $H^0 \to \gamma\gamma$ will be undetectable at the SSC.[1] For a Higgs lying below top threshold, the small branching ratio $B(H^0 \to \gamma\gamma) \approx 10^{-3}$ may be detectable.[2] This paper reports on our efforts to reliably simulate signal and backgrounds as they might appear in an SSC detector. Our goal is to establish the required detector performance for observing $H^0 \to \gamma\gamma$.

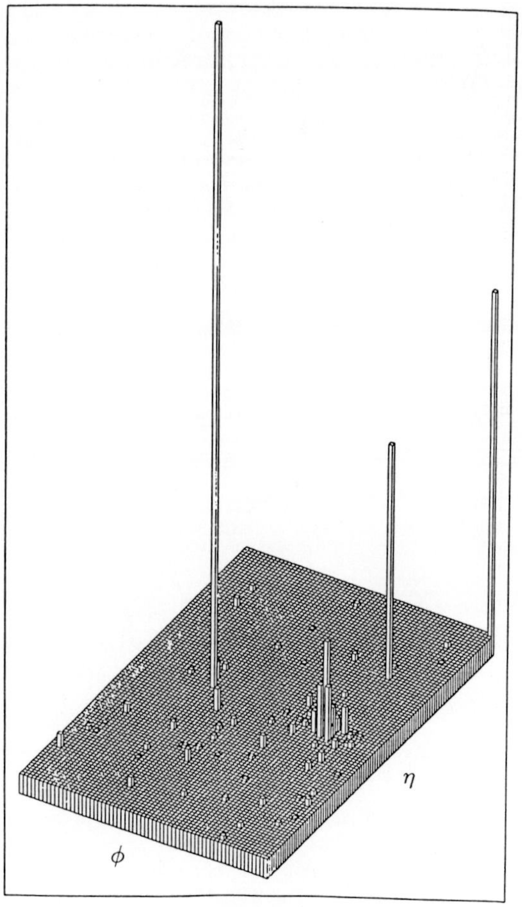

FIG. 1. Lego plot showing E_T distribution for typical $H^0 \to \gamma\gamma$ event. To show the vertical scale, 25 GeV was entered in the right rear corner bin.

The decay $H^0 \to \gamma\gamma$ has an extremely clean and distinctive signature: a pair of isolated high p_T photons, no missing p_T, and a minimal amount of hadronic debris. To simulate this process, we have modified the PYTHIA[3] and ISAJET[4] Monte Carlo programs to include the decay $H^0 \to \gamma\gamma$. A typical event is shown in fig. 1, where the two photons are clearly visible as well as a hadron jet resulting from initial state gluon radiation.

The small number of $H^0 \to \gamma\gamma$ events makes it essential that a detector designed to study this reaction be optimized for high detection efficiency. Since it is unlikely that an SSC detector will provide high quality electromagnetic calorimetry in the forward direction,[5] the geometrical acceptance for Higgs photons is assumed to be limited to $|\eta| < 3$. Figure 2 shows the Higgs rapidity distribution with and without geometrical acceptance cuts on the decay photons. Also shown in fig. 2 is the effect of photon conversions in the beam pipe and tracking system for an effective radiator thickness of 0.05 X_0 and 0.1 X_0 at normal incidence. Both the geometrical acceptance and the increased conversion probability at small angles serve to shrink the acceptance in Higgs rapidity.

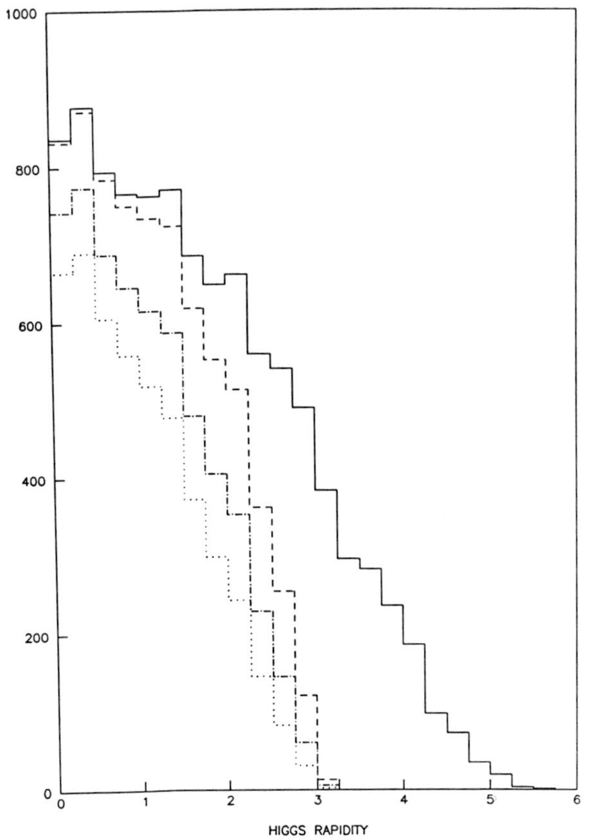

FIG. 2. Higgs rapidity distributions for: all produced Higgs (solid); events with both photons accepted, $5° < \theta_\gamma < 175°$ (dashed); both photons accepted and not converted in 0.05 X_0 radiator (dash-dot); 0.1 X_0 radiator (dotted).

Figure 3 shows the E_T distribution for Higgs photons, which peaks just above 40 GeV and has significant numbers of events extending down to $E_T \approx 10-20$ GeV. Thus, the detector must also have the ability to trigger on relatively low energy electromagnetic showers. Fast charged particle identification may be necessary to reduce trigger rates from low energy electrons. Furthermore, the detector will need to be $> 99\%$ efficient in rejecting electrons to prevent a large punchthrough from $Z^0 \to e^+e^-$. The required detector resolution is essentially a question of signal/background ratio, and is discussed below.

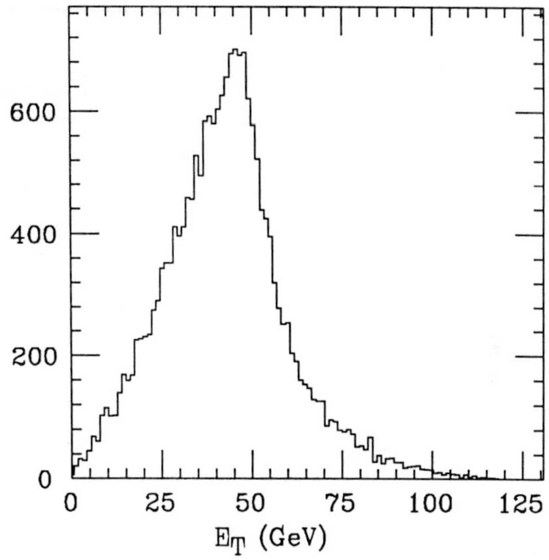

FIG. 3. Transverse energy distribution for Higgs photons.

DIRECT PHOTON BACKGROUNDS

The direct production of photon pairs is the most serious background for detecting $H^0 \to \gamma\gamma$. While various kinematic cuts can be made to minimize the problem, the existence of events with two direct photons forms an irreducible background to the desired signal. The principal sources for direct photons are the reactions $q\bar{q} \to \gamma\gamma$ and $gg \to \gamma\gamma$, where the later reaction proceeds through a virtual quark loop.

To simulate these backgrounds, the PYTHIA Monte Carlo program was used to generate 5000 events originating from $q\bar{q} \to \gamma\gamma$ and $gg \to \gamma\gamma$ processes. Modifications to the PYTHIA program were required to include the process. The $gg \to \gamma\gamma$ cross section we used includes only light quark loops. The effect of neglecting the top quark loop is to slightly overestimate the background near top threshold since this diagram interferes destructively below top threshold.[6] The $gg \to \gamma\gamma$ cross section obtained from PYTHIA is in good agreement with the cross section of ref. 6.

To simulate the $H^0 \to \gamma\gamma$ decay, PYTHIA was modified to incorporate Higgs decays into $\gamma\gamma$ and include the

WW^* and ZZ^* channels in the calculation of the Higgs width (W^* and Z^* are virtual W and Z bosons). The resulting $H^0 \to \gamma\gamma$ cross section is in agreement with the calculations of ref. 1. Samples of $H^0 \to \gamma\gamma$ events corresponding to one SSC year (10^4 pb^{-1}) were generated for Higgs masses of 100 GeV and 150 GeV. The energy and direction of the Higgs photons were smeared using Gaussian resolution functions with the following resolutions:

$$\sigma_\phi = 0.001$$
$$\sigma_\eta = 0.001$$
$$\frac{\sigma_E}{E} = \frac{10\%}{\sqrt{E}} + 1\% \qquad \text{Excellent Resolution}$$
$$\frac{\sigma_E}{E} = \frac{3\%}{\sqrt{E}} + 0.5\% \qquad \text{Extraordinary Resolution}$$

Two different energy resolutions were used in this analysis: excellent resolution and extraordinary resolution. Excellent resolution corresponds to what might be acheived in a well built fine sampling electromagnetic calorimeter. Extraordinary resolution would require unconventional techniques, such as the use of BaF crystals or a liquid Xe detector.

The following cuts were imposed on the photons in both signal and background event samples:

- $|\eta| < 3$
- $E_T > 10$ GeV
- $|\cos\theta^*| < 0.8$

where $\cos\theta^*$ is the photon angle in the $\gamma\gamma$ rest frame. The motivation for the first two cuts is discussed above. The cut on $\cos\theta^*$ was found useful in enhancing the Higgs signal, which is isotropic, relative to the forward peaked direct photon background. The $\gamma\gamma$ mass was then plotted for each event surviving these cuts.

Full simulation of the background distribution would require the generation of $\approx 10^6$ background events. To save computer time, we choose to determine the shape of the background using a small sample (5000) of PYTHIA events; the background distribution is made by randomly throwing events according to the fitted background shape. In this way, a smooth, statistically correct background distribution is generated for a one year SSC run. The four Higgs samples are combined with the background distribution to produce the $\gamma\gamma$ mass plots shown in figs. 4-7.

The solid lines in figs. 5-7 show the results of fitting the $\gamma\gamma$ mass distributions by varying the background parameters and the mean and amplitude of a Gaussian resolution function. The width of the Gaussian is fixed at the detector resolution. The fit results are summarized in the table below.

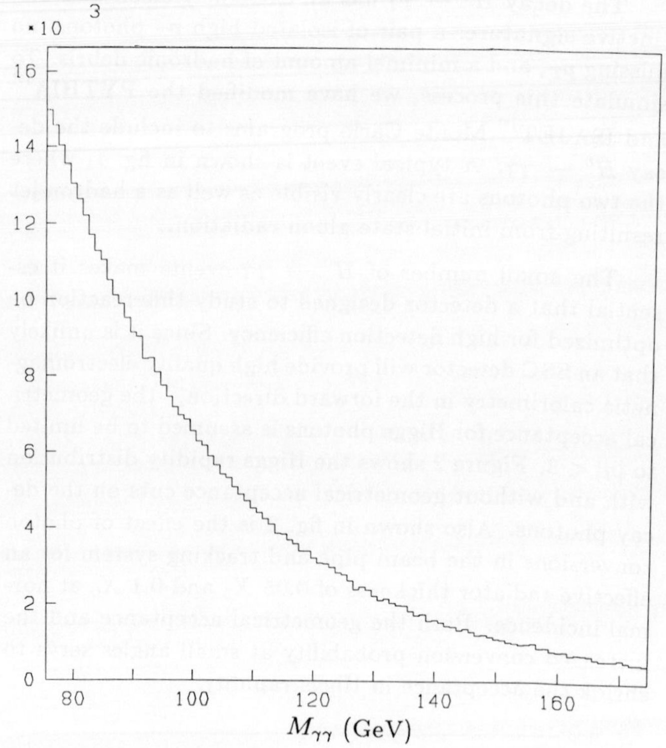

FIG. 4. Simulated mass distribution for 100 GeV Higgs in detector with excellent resolution.

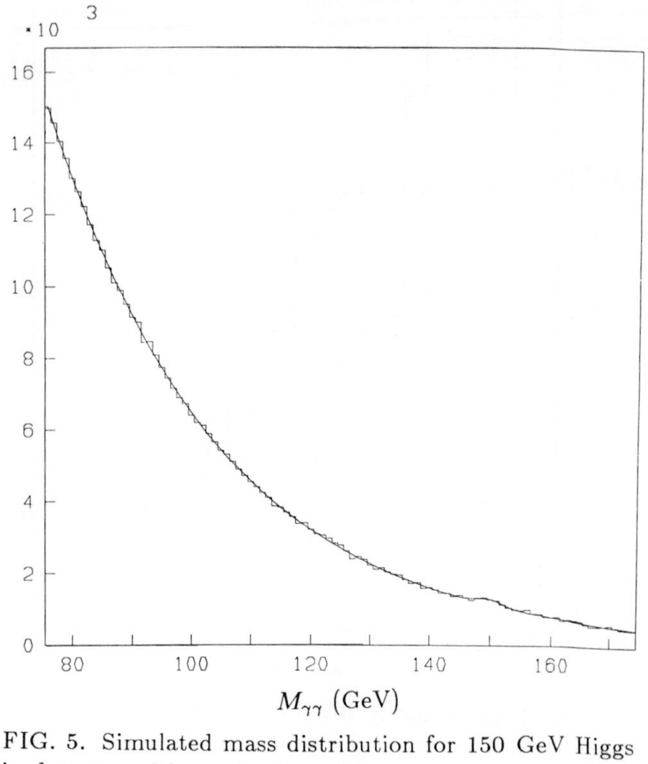

FIG. 5. Simulated mass distribution for 150 GeV Higgs in detector with excellent resolution.

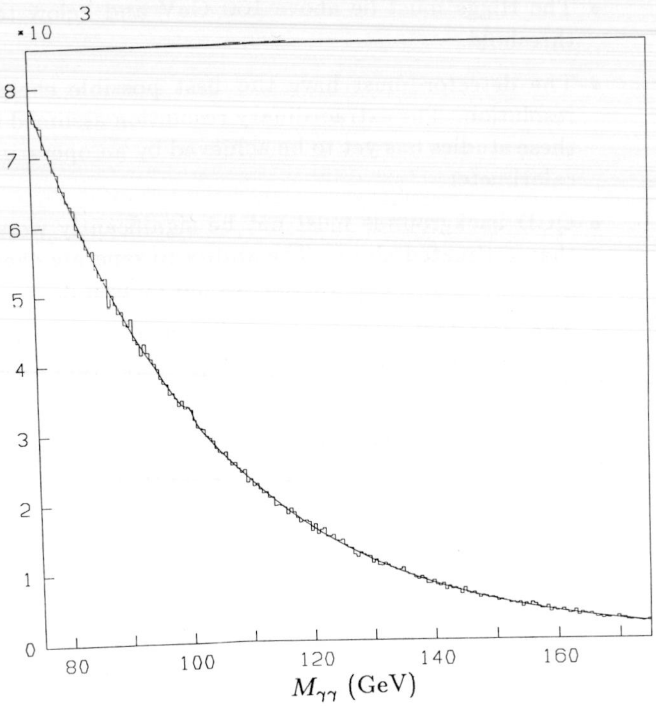

FIG. 6. Simulated mass distribution for 100 GeV Higgs in detector with extraordinary resolution.

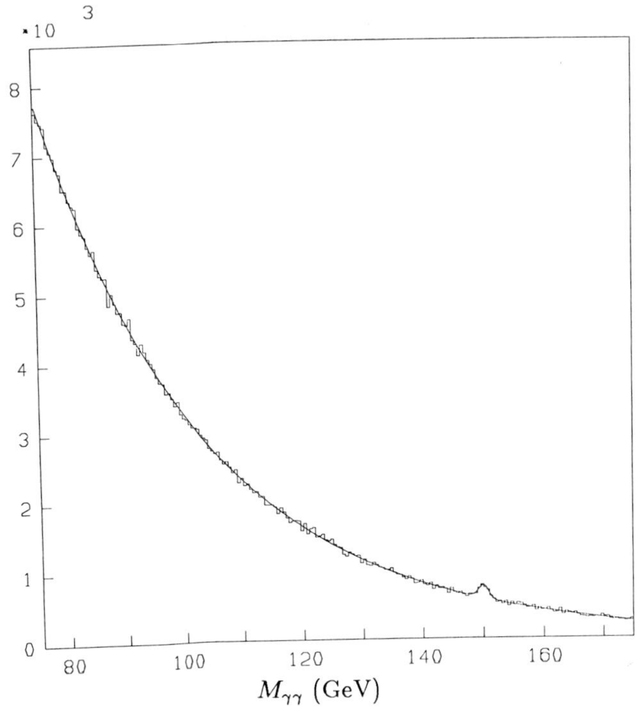

FIG. 7. Simulated mass distribution for 150 GeV Higgs in detector with extraordinary resolution.

Higgs Mass	Detector Resolution	Mass Resolution	Statistical Significance
100 GeV	Excellent	1.44 GeV	None
100 GeV	Extraordinary	0.55 GeV	2.8 σ
150 GeV	Excellent	1.91 GeV	7.6 σ
150 GeV	Extraordinary	0.80 GeV	12 σ

The simulated mass plots are not particularly encouraging. The event samples for a detector with excellent resolution show no indication of a 100 GeV Higgs and a small but significant (7.6 σ) bump for a 150 GeV Higgs. The situation is somewhat better for a detector with extraordinary resolution: a 100 GeV Higgs is statistically marginal (2.8 σ), while the signal for a 150 GeV Higgs is unmistakeable (12 σ).

Figure 6 demonstates one of the perils in searching for signals of marginal statistical significance: a 4.3 σ signal was expected, but a small fluctuation in the background reduced the observed significance to 2.8 σ. Another potential difficulty is due to uncertainties in the background shape. In this example, the background shape is known a priori, allowing a very precise determination of the expected background under the signal. At the SSC, the background shape will be uncertain and not necessarily smooth. For example, a small inefficiency in charged particle tracking will generate a peak at 93 GeV from misidentified $Z^0 \rightarrow e^+e^-$ events. Given these difficulties and the above results, the presence of a large direct photon background makes detection of $H^0 \rightarrow \gamma\gamma$ marginal for a 100 GeV Higgs.

Bremsstrahlung processes are also expected to be a source of direct photons.[7] Since these events are usually accompanied by high p_T jets, it may be possible to eliminate them by rejecting events with significant hadronic activity. The effect of this background needs to be evaluated further.

QCD BACKGROUNDS

The large rate for QCD two-jet events allows the possibility that such events will occasionally fluctuate to fake an event with two isolated photons. For a 100 GeV Higgs, two-jet events are estimated[2] to outnumber $H^0 \rightarrow \gamma\gamma$ events by a factor of $\approx 0.5 \times 10^8$. Reducing the QCD background to a level significantly below the direct photon background will require a rejection factor of $\approx 10^7$. A four-vector analysis of this background[8] based on the PYTHIA Monte Carlo was able to achieve this level of rejection by requiring photon candidates to have $|\eta| < 3$, $E_\gamma > 40$ GeV, and no particles with energy greater than

5 GeV within an $\eta - \phi$ cone of radius 1. We report below the results of second study of this background employing a crude calorimeter model and jet clustering techniques.

The ISAJET Monte Carlo program was used to generate two samples of QCD two-jet events: a wide-band sample of 10,000 events with p_T in the range 50-100 GeV and a narrow-band sample of 10,000 events with p_T in the range 100-120 GeV. A simple calorimeter model employing $\Delta\eta = \Delta\phi = 0.1$ cells is used to represent the detector. The energy of each particle in an event is assigned to the calorimeter cell struck by the particle. Jets are identified by requiring at least 10 GeV of transverse energy within an $\eta - \phi$ cone of radius 0.5. Isolated photon candidates are selected by requiring at least 90% of the transverse energy within the jet cone originate from an electromagnetic core of radius 0.15. Photon candidates are also required to have $|\eta| < 3$. These cuts have an efficiency of 50% for identifying $H^0 \to \gamma\gamma$ events generated using ISAJET.

A total of 40 events, 16 in the wide-band sample and 24 in the narrow-band sample, are classified as having a photon candidate (none of the events have two photon candidates). Since each event contains two jets, these data indicate $\approx 10^{-3}$ of the QCD jets fake isolated photons. Approximately half of these fake photons are soft, having transverse energies in the range 10-20 GeV. Eliminating events with soft photons and assuming identification of jets as fake photons is uncorrelated, a rejection factor of $\approx 4 \times 10^6$ is obtained. The remaining fake photons are not easily distinguished from $H^0 \to \gamma\gamma$ photons. Further isolation cuts failed to significantly improve the signal/background ratio.

The rejection achieved in this analysis is somewhat worse than is obtained by ref. 8 using the PYTHIA Monte Carlo and somewhat different selection criteria. The fake photon rejection factor in both analyses is only marginally adequate. There is no evidence that either Monte Carlo correctly predicts the small probability for producing a fake photon, and neither analysis has sufficient statistics to determine the mass distribution for pairs of fake photons. For these reasons, it may be desirable to supplement the rejection power of the calorimeter. The CDF calorimeter uses a finely segmented wire chamber within the calorimeter to help distinguish direct photons from energetic π^0. A similar attempt to identify photon pairs using a preshower converter or drift measurements in liquid Xe might prove useful at the SSC.

CONCLUSIONS

The decay $H^0 \to \gamma\gamma$ provides one of the few avenues available for finding an intermediate mass Higgs at the SSC. Success in detecting $H^0 \to \gamma\gamma$ will depend on several factors:

- The Higgs must lie above 100 GeV and below top threshold.
- The detector must have the best possible energy resolution. The extraordinary resolution assumed in these studies has yet to be achieved by an operating calorimeter.
- QCD backgrounds must not be significantly worse than estimated above. The ability to seperate energetic π^0 from single photons would be useful.
- The detector must have good geometrical acceptance and be capable of triggering on electromagetic showers with 10-20 GeV of transverse energy.
- The amount of material inside the calorimeter must be minimized to reduce photon conversions at forward angles.
- The tracking system must be capable of rejecting the 2 Hz $Z^0 \to e^+e^-$ rate with high efficiency.

Even if these conditions are met, the prospect for finding a 100 GeV Higgs is marginal. The situation improves substantially with increasing Higgs mass. A Higgs at 150 GeV can probably be seen by a detector with the excellent energy resolution obtainable in well designed sampling calorimeters.

We would like to thank Scott Willenbrock, Howard Baer, John Huth, and Mitch Golden for useful discussions.

REFERENCES

1. J.F. Gunion, P. Kalyniak, M. Soldate, and P. Galison, Phys. Rev. D **34**, 101 (1986).
2. J.F. Gunion, G.L. Kane, and J. Wudka, Nucl. Phys. **B299**, 231 (1988).
3. H.-U. Bengtsson and T. Sjoštrand, *The Lund Monte Carlo for Hadronic Processes - PYTHIA version 4.9.*
4. F.E. Paige and S.D. Protopopescu, *ISAJET 5.34: A Monte Carlo Event Generator for pp and p̄p Reactions.*
5. R. Partridge, *Closing the Gap: Forward Detectors for the SSC*, Proceedings of the Workshop on Experiments, Detectors, and Experimental Areas for the Supercollider, edited by Rene Donaldson and M.G.D. Gilchriese (Berkeley, CA, 1987), p. 657 (hereafter referred to as Berkeley 87).
6. D.A. Dicus and S.S.D. Willenbrock, Phys. Rev. D **37**, 1801 (1988).
7. H. Baer and J.F. Owens, Phys. Lett. **205B**, 377 (1988).
8. D.M. Atwood *et al.*, Berkeley 87, p. 728.

DETECTION OF HIGGS BOSONS DECAYING TO BOTTOM QUARKS AT THE SUPERCONDUCTING SUPERCOLLIDER*

James E. Brau, Kevin T. Pitts
University of Oregon, Eugene, OR 97403

and

Lawrence E. Price
Argonne National Laboratory, Argonne, IL 60439

ABSTRACT

We have studied the feasibility of detecting the intermediate mass Higgs boson through its decay to two bottom quark jets at the SSC. The Monte Carlo program PYTHIA 4.8 provides the signal and background event generation. We follow the earlier suggestion that it be searched for in associated W production in order to suppress continuum background. The remaining large background may be overcome by a choice of cuts designed to optimize the rejection of background and preserve the signal. Measurement of high j_T electrons is an important requirement on the experiment. Close-in tracking permitting impact distance measurements is crucial, but a modest vertex detector performance is acceptable. Also detector hermeticity is demanded by the trigger and the physics cuts. Detection of this process appears extremely difficult, but possible.

INTRODUCTION

The most profound contribution of the next generation of hadron colliders (the SSC and the LHC) to our understanding of the fundamental interactions could be the discovery of the Higgs boson(s). The search for the Higgs, or an alternate explanation of mass, is in fact one of the principal motivations for the construction of the SSC [1] and has driven the design of its energy and luminosity[2].

While theory provides little guidance for the Higgs mass[3] and experiment is still extremely limited in coverage[4], we can anticipate that by the time of turn on for the SSC/LHC we will either have found it to have a mass of less than ≈ 85 GeV at SLC, LEP I, or LEP II, or we will still be searching at higher energies. Even if a Higgs has been found, theory suggests there will be more to discover[5].

Finding the Higgs at the SSC will be most difficult if its mass is in the intermediate mass range (80-170 GeV). If its mass is larger it will be relatively easily extracted from background due to its primary decay to W^+W^- and Z^0Z^0 and their small nonresonant pair production[6]. However, if the Higgs is lighter then it will decay mostly to heavy quarks and be extremely difficult to extract from the ever present background. For example [7] if $m_t = 40\ GeV/c^2$ and $m_H > 2m_t$,

$$\sigma(pp \to [H^0 \to t\bar{t}]X) \approx 10^{-2}[d\sigma(pp \to t\bar{t}X)/dM^2]\Delta M^2$$

where $\Delta M^2 \approx 0.1m_H^2$. Similar problems arise for heavy top with dominant Higgs decays to bottom. But the prospects for finding the Higgs in bottom quark final states is not nearly as unlikely as for the top quark final states. Still one must deal with a substantial continuum background.

In order to suppress this enormous background from continuum production, it has been proposed to look for Higgs production in association with a W^+ or W^-. [7,8] In this case the production of signal events,

$$pp \to H^0W^{\pm}X(H^0 \to Q\overline{Q}) \tag{1}$$

is still orders of magnitude less than the backgrounds which feed this channel,

$$qg \to W^{\pm}Q \tag{2}$$

$$q\overline{q} \to W^{\pm}g \tag{3}$$

$$q\overline{q} \to W^{\pm}Z^0. \tag{4}$$

However, the backgrounds are manageable and a signal may be detectable. This will be demonstrated by the Monte Carlo studies presented in this paper.

The expected Higgs cross sections at the SSC have been calculated [9] as a function of the Higgs mass. There is a very strong top quark mass dependence, but for the Higgs of intermediate mass the cross section should be about a hundred picobarns, about one event every ten seconds at the design luminosity of $10^{33}cm^{-2}sec^{-1}$. As mentioned earlier, this large rate is unusable due to the large continuum background. When events with associated W's are selected the rate drops by about two orders of magnitude. The Higgs branching ratio to bottom quarks is expected to be greater than ninety percent as long as $m_H < 2m_t$ and $m_H \ll 2m_W$. Such a scenario presents a challenge for detection. It is the purpose of this study to investigate its feasibility.

THE MONTE CARLO STUDY

This study has used PYTHIA 4.8 [10] to generate signal and background events. The W produced in the event is required to have a transverse momentum of at least 50 GeV relative to the beam axis. The cross sections for each process depend upon the input constants and for this study we have assumed the top quark mass is safely high ($m_t = 150\ GeV$). Table 1 shows the resulting cross sections. The background events outnumber signal events by roughly four orders of magnitude prior to any selection.

The trigger required for this process has been studied in a previous SSC workshop[11] and found to effectively suppress background. The trigger is based on the leptonic decay of the associated W. A high E_T lepton combined with missing transverse energy will suppress the conventional background significantly. As a first level trigger, cuts of 40 GeV on missing E_T and 25 GeV on the lepton E_T have been found [11] to reduce the total cross section to 7.5 μb. Additional lepton isolation cuts and jet E_T requirements can reduce it to 1.1 nb with an acceptable loss of signal. We have applied this trigger in our study. Figure 1 presents the distributions for these variables from 4000 signal events with leptonic W decays. (All of the figures in this paper are based on this 4000 event sample.)

The measured missing E_T is very sensitive to the hermiticity of the detector. The detector has been assumed to cover down to one degree from the beam axis for trigger purposes. Jets will be reconstructed only for particles contained within ±2.5 of pseudorapidity ($\theta > 9$ degrees). Only leptonic decays of the W have been included in Figure 1 as the study indicated that the quark decay modes only result in a one percent acceptance for the trigger cuts. Thirty-nine percent of the leptonic decays of the W pass the trigger cuts. Of the electron/neutrino and muon/neutrino decays, 50 percent pass, while 15 percent of the tau/neutrino decays do. The background events are slightly suppressed by the trigger as 27 percent of reactions (2) and (3) pass, while 34 percent of reaction (4) pass.

Figure 2 displays the distribution of transverse momentum of the W generated by PYTHIA in reaction (1) when the Higgs is given a mass of 100 GeV. The p_T cut of 50 GeV is evident and the distribution falls off abruptly above one hundred GeV, although two and one-half per cent of the events have $p_T(W)$ in excess of three hundred GeV.

The jets are reconstructed from the tracks which have $p_T > 500\ MeV/c^2$ and are found within the pseudorapidity cuts ($|\eta| < 2.5$). This reconstruction is idealized and a more realistic detector would likely degrade it some.

CROSS SECTIONS		
	$(P_T > 50GeV)$	
$qg \to W^{\pm}q$	27 nb	
$q\bar{q} \to W^{\pm}g$	1.8 nb	
$q\bar{q} \to W^{+}Z^{o}$, $Z^{o} \to b\bar{b}, c\bar{c}$	6.4 pb	
		M_H
$q\bar{q} \to W^{\pm}H$, $H \to b\bar{b}$	3.9 pb	(75 GeV)
	2.3 pb	(100 GeV)
	1.0 pb	(150 GeV)

Table 1. Cross Sections

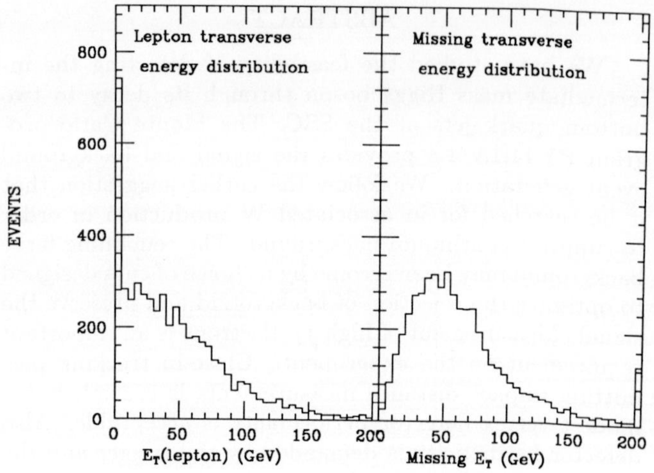

Figure 1 . E_T (lepton) and missing E_T Distributions

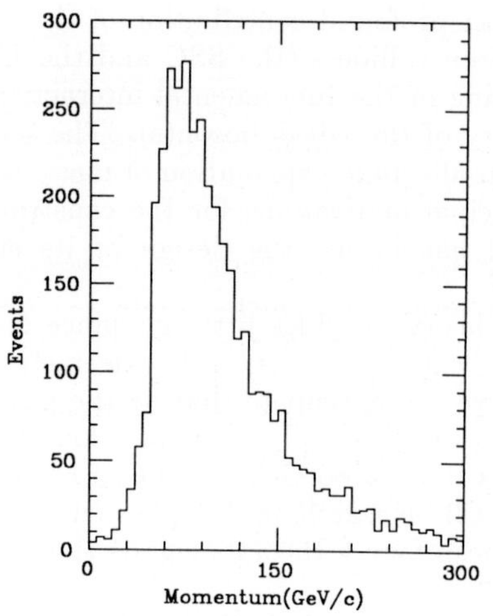

Figure 2 . W Transverse Momentum Distribution

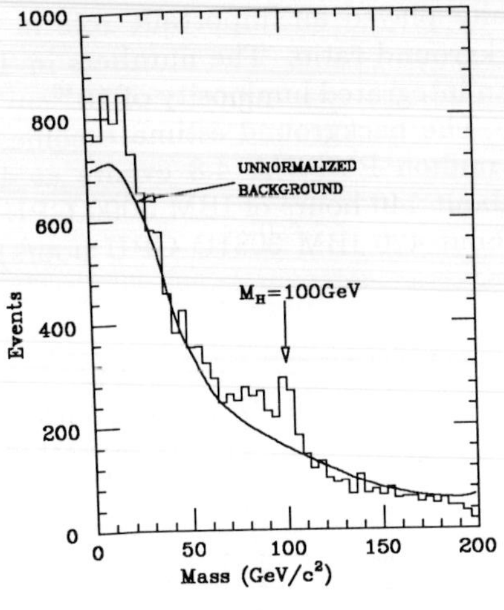

Figure 3 . Jet-Jet Mass Distribution

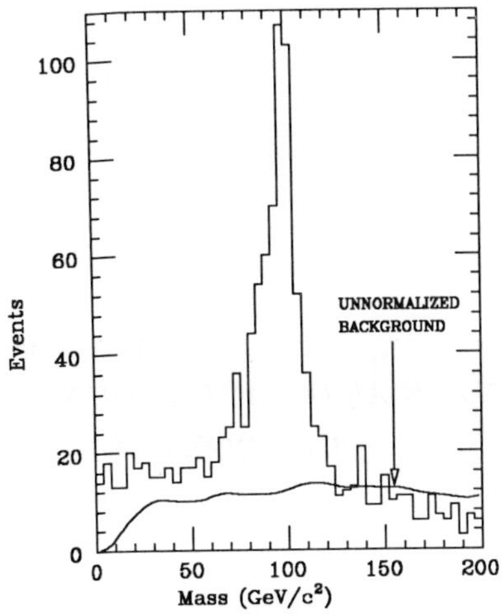

Figure 4 . Jet-Jet Mass Distribution for "Best" Jets

Many jets are found in each event and Figure 3 shows the two-jet mass distribution from the signal events observed for the 100 GeV Higgs case. The excess in the signal events in the 100 GeV mass region is obvious, Also shown is the distribution computed for reactions (2) and (3). The two scales are different, and the background is very large, about three orders of magnitude larger than the signal. At this point the effects of the detector resolution are being ignored, but they are assumed small[8], as is shown later in this study.

Most of the jets selected are low E_T jets originating from the fragmenting valence quarks of the interacting protons. These can be greatly suppressed and the Higgs decay jets selected by choosing the two jets in each event with the largest E_T (with a minimum of 10 GeV). Figure 4 presents the jet-jet mass distribution resulting from this selection. Obviously, this choice is successful, as the signal is very prominent in this figure. The shape of the background distribution is very flat, as shown by the unnormalized curve on Figure 4 .

Further improvement of the signal to background ratio demands additional cuts biased for signal events. In the earlier Monte Carlo study of this channel, Gilman and Price[8] found that with a minimal number of cuts they could achieve a good clear signal. However, that study was limited by Monte Carlo statistics. They warned that the final background sample was one event and was consequently very uncertain. On the other hand, they had not imposed any impact distance requirements, to take advantage of the lifetime of the b quark, and they had not considered a number of other kinematical cuts that can be used to bias against background. This new study has been conducted by starting with their cuts and proceeding with additional cuts to further expose the signal. The following cuts have been used (the first two are nearly those employed by Gilman and Price[8]). The first four are designed to extract b quark jets, and the remainder to bias against random b quark pairs and in favor of pairs originating from W decays.

1. A lepton must be found in the jet with a transverse momentum relative to the jet axis (j_T) of at least 1 GeV. This cut favors heavy quark decay. This cut obviously requires that the lepton of the jet be isolated sufficiently to allow such a measurement, particularly for electrons. We have investigated the degree of isolation of electrons passing this cut and Figure 5 illustrates the result. Here the probability is presented that a selected high j_T electron will find more than a certain fraction (10, 40, or 70 percent) of its energy matched by other particles of the jet within the specified distance in $R = \sqrt{(\Delta\eta)^2 + (\Delta\phi)^2}$ as a function of R. For muons this is not important, but for electron identification it is. Clearly the situation is not ideal and some loss of events can be expected. In the experiment this loss must be dealt with, but we have ignored the problem for the remaining analysis.

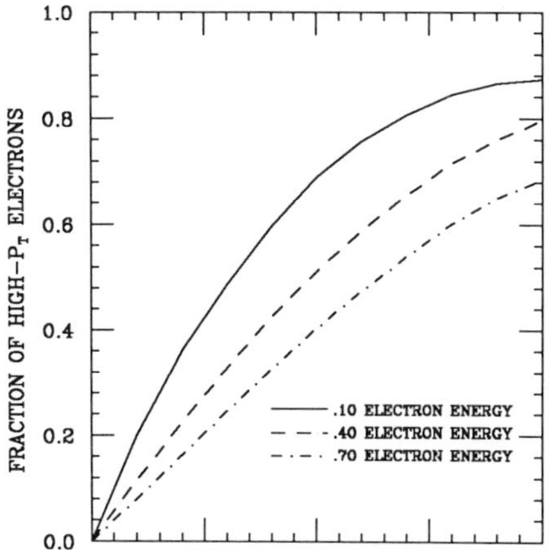

Figure 5 . Electron isolation distributions

2. The tightness of each jet is calculated by computing the distance of each particle in the jet from the jet axis and requiring that 70 percent of the jet energy be contained within a distance of 0.2. The measure of distance is given, as above, by $\sqrt{(\Delta\eta)^2 + (\Delta\phi)^2}$. This cut has been motivated by the work of Reference 12 .

3. The impact distance of one of the particles of the jet containing the high j_T lepton must be at least 50 microns in the transverse plane and five standard deviations from zero. The error on the impact distances is taken as $\sigma = \sqrt{(5\mu m)^2 + (80\mu m/p(GeV))^2}$. The capability to measure with such precision at the SSC is being contemplated[13]. Recognizing that secondary vertex detection will pose a challenge in the SSC environment due to the large number of secondary particles, we note that this is not vertex finding, but rather the finding of a single track with evidence of a secondary decay.

4. The impact distance of one particle from the second jet must also be at least 50 microns in the transverse plane, and five standard deviations from zero. With this cut the background events will be reduced to mostly b jet pairs from the continuum. Further improvement in the signal/background will require cuts favoring the Higgs decay.

5. The W is reconstructed in the plane transverse to the beam axis by adding the missing energy vector in this plane (assumed to be the neutrino) to the lepton vector. (Figure 6 shows the mass so reconstructed in this plane.) Its direction in the transverse plane is required to be opposite to the two jet direction in this plane to within 23 degrees ($|cos\theta| > .92$).

6. The transverse momentum of the reconstructed W is required to balance that of the two-jets to within 5 percent.

7. The cosine of the jets of the Higgs relative to the jet axis is required to be less than 0.7. This removes the many asymmetric background events where one high E_T jet combines with a small E_T jet and will result in a large value of the cosine.

Table 2 shows the expected number of events for signal ($m_H = 100GeV$) and background following each of these cuts. The main problem is that the background events are mostly b jet pairs after the impact distance cuts are imposed. This means that any further suppression is very difficult and must be achieved by understanding the origin of the continuum b-jet pairs. This awareness has led to the cuts on angular distribution of the Higgs decay, balancing of p_T, and the back-to-back requirement with the partially reconstructed W.

The cuts have not been applied in the order outlined above, as some of them were developed after the background sample had been generated and many events rejected. However, it was verified that the "event shape" cuts actually played an important role in the final signal to background ratio. The numbers in Table 2 correspond to an integrated luminosity of $10^{40}cm^{-2}$, a nominal SSC-year. The background estimate comes from generating 1.5 million PYTHIA 4.8 events at 40 TeV. This required about 140 hours of IBM 3090 CPU time (equivalent to about 420 IBM 3081D CPU hours), illustrating

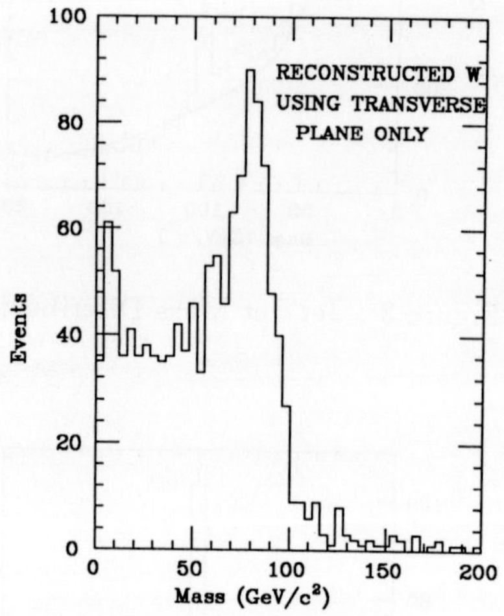

Figure 6 . Reconstructed W Mass Distribution

SIGNALS AND BACKGROUNDS FOR

$\int Ldt = 10^{40}cm^{-2}$

cut	*signal*	$W_g + W_q$ *bckgnd*	$\frac{s}{bckgnd}$	$\frac{s}{\sqrt{bckgnd}}$
1. $j_T^{lepton} > 1GeV$	223			
2. $R_1, R_2 < 0.2$	120	2525	0.05	2.4
5. $\lvert\theta_{W-H} - 180°\rvert < 23°$	107	1275	0.08	3.0
6. $\left\lvert\frac{\Delta P_T}{\sum P_T}\right\rvert < 0.05$	62	433	0.14	3.0
7. $\lvert\cos\theta^*_{b-Higgs}\rvert < 0.7$	58	382	0.15	3.0
3. $d_1 > 50\mu m$ *and* 5σ				
4. $d_2 > 50\mu m$ *and* 5σ	41	84	0.5	4.5

Z^o BACKGROUND

47 *with* $76 < m < 92$ (23 *in* $84 - 100GeV$)
54 *all* m (*contained in* $55 - 100GeV$)

Table 2. Events Passing Cuts in One Year

the problem associated with getting reliable estimates of backgrounds at the SSC. Even after generating so many events and selecting small subsamples it is questionable how reliable the generators are as accurate models of reality. Nevertheless, they are our only tool for estimating backgrounds, so we proceed. In order to maximize the information obtained from each generated event, the b-quarks have been allowed to decay ten times for each event and each decay analyzed for impact distance.

The table shows that the background can finally be suppressed adequately. From the table one sees that following the cuts the signal nearly equals the background, being in a ratio of 0.5. This signal represents 4.5 standard deviations above the Wq + Wg background. The addition of a background from the WZ^0 channel reduces it to a 4.0 standard deviation effect, still reasonably strong. The powerful effect of the impact distance cuts are clear, with signal/background increasing by a factor of 4 when these two cuts are applied. The impact distance cuts could be relaxed some without substantially reducing the sample of events. The average maximum impact distance for the final selected jets is about 350 microns and an increase of the cuts to 100 microns results in only a twenty percent decrease in the signal.

Table 3 summarizes the results of studies done for four different Higgs masses (M_H). As the mass is increased, the reconstructed width increases and a wider cut in mass (ΔM_H) is required to contain the peak. While the lower mass Higges appear detectable, the higher masses suffer for a decreasing Higgs cross section and an increasing background. The background increase comes mostly from the larger spread of the reconstructed Higgs mass.

Cut 1 involving the jet transverse momentum of a lepton was introduced in Reference 8 to select heavy quark decays. Having successfully achieved selection with the impact distance cuts we might conclude that cut 1 is not required and may very dramatically reduce the signal. Furthermore, it selects the semileptonic decays of the b quark, yielding an inherently limited Higgs mass resolution. However, our studies have convinced us that the increases in background events from charm decay are unacceptable without this cut.

We have studied the effect of energy resolution and unequal hadronic and electromagnetic calorimeter response on the resolution of the reconstructed Higgs mass. The reconstructed mass distribution shows an rms of about 14 % due to missing particles and inaccurate jet reconstruction. This is before any calorimeter response has been added. Therefore, reduced calorimeter performance would not be expected to worsen the measurement very much. This has been confirmed by the study in which $\frac{e}{h}$ was varied from 1 to 1.4, the hadronic resolution was varied from perfection to $0.08 + 0.50/\sqrt{E}$, and the electromagnetic resolution was fixed at $0.01 + 0.15/\sqrt{E}$. The mass resolution was found to degrade by about 30 % (i.e. rising to nearly 18 %) over the range of the variation.

SIGNALS AND BACKGROUNDS FOR VARIOUS M_H

($events\ for \int Ldt = 10^{40} cm^{-2}$)

M_H	ΔM_H	*signal*	$W+g$ $W+q$	Z^o
75	16	40	84	0
100	16	41	84	23
125	20	22	105	0
150	28	26	147	0

Table 3. Events Passing All Cuts in One Year

CONCLUSIONS

In summary, we have found that the prospects for finding a Higgs boson of intermediate mass which decays to bottom quark jets at the Supercollider to be dependent upon excellent close-in tracking (i.e. the ability to pick out tracks with impact distances on the order of 50-100 microns) supported by a hermetic detector and good lepton detection and measurement. The transverse segmentation of the electromagnetic calorimeter must be fine enough to separate the high j_T electron. By employing the associated production of the Higgs boson with a W^+ or W^- a good trigger is provided by the leptonic decay of the W and the background is amply suppressed to enable further cuts to bring the signal and background into near equality. Detection of this process appears extremely difficult, but possible.

* Work supported in part by the U.S. Department of Energy

REFERENCES

1. Superconducting SuperCollider, Reference Design Study for the U.S. Department of Energy, May 8, 1984.

2. Conceptual Design of the Superconducting Super-Collider, SSC-SR-2020 (1986).

3. Linde, A.D., Sov. Phys. - JETP Lett. 23, 64 (1976); Weinberg, S., Phys. Rev. Lett. 36, 294 (1976); Veltman, M., Acta Phys. Polon. *B8*, 475 (1977); and Lee, B.W., C. Quigg, and H.B. Thacker, Phys. Rev. *D16*, 1519 (1977).

4. P.M. Tuts et al, Phys. Lett. 186*B*, 233 (1987).

5. J.F. Gunion and H.E. Haber, Nucl. Phys. *B272*, 1 (1986).

6. Alverson, G. et al, in Proceedings of the Summer Study on the Physics of the Superconducting Supercollider, Snowmass 1986, Rene Donaldson and Jay Marx, editors, page 114.

7. Gunion, J.F., P. Kalyniak, M. Soldate, and P. Galison, Phys. Rev. Lett. 54, 1226 (1985); Gunion, J.F., P. Kalyniak, M. Soldate, and P. Galison, Phys. Rev. *D*34, 101 (1986);

8. Gilman, F.J. and L.E. Price, "Detection of Higgs Bosons Decaying to Bottom Quarks," Proceedings of the Summer Study on the Physics of the Superconducting Supercollider, Snowmass 1986, Rene Donaldson and Jay Marx, editors, page 185.

9. Gunion, J.F., G.L. Kane, and J. Wudka, Nucl. Phys. *B*299, 231 (1988).

10. Bengtsson, H. and T. Sjostrand, Comput. Phys. Commun. 46, 43 (1987).

11. Kane, G., F. Paige, L. Price et al, "SSC Physics Signatures and Trigger Requirements," Proceedings of the Workshop on Triggering, Data Acquisition and Computing for High Energy/ High Luminosity Hadron-Hadron Colliders, Fermilab, November, 1985, page 1.

12. Gottschalk, T.D., "Remarks on the Discrimination of Prompt and Shower Bottom Production in Hadronic Jets," Proceedings of the Summer Study on the Physics of the Superconducting Supercollider, Snowmass 1986, Rene Donaldson and Jay Marx, editors, page 67.

13 Sadrozinski, H.F.-W., A. Seiden, A.J. Weinstein, "Tracking at the SSC", SCIPP 87/100 (1987).

HIGGS → FOUR LEPTONS AT THE SSC

E. M. Wang*
Lawrence Berkeley Laboratory
University of California
Berkeley, CA 94720

G. G. Hanson†
Stanford Linear Accelerator Center
Stanford University
Stanford, California 94309

A. Bay*
Lawrence Berkeley Laboratory
M. Chiba
Tokyo Metropolitan University
R. Cirio
University of Torino/INFN
Y. Fukui
KEK/Fermi National Accelerator Laboratory
T. Han
University of Wisconsin

M. Levi*
Lawrence Berkeley Laboratory
H. Ogren
Indiana University
A. P. T. Palounek†
Stanford Linear Accelerator Center
N. Tamura
Kyoto University
H. Yamamoto°
University of California, Los Angeles

Abstract

Detection of an intermediate mass or heavy Higgs boson through its decay into four charged leptons is studied with emphasis on background considerations and detector requirements. The intermediate mass Higgs decay via ZZ^* is expected to be a difficult mode to observe due to low event rates. In addition, cuts which are needed to reduce the backgrounds reduce the signal even further. The rapidity coverage and energy resolution requirements for this mode are more severe than for the heavy Higgs. The heavy Higgs boson decay $H \to Z^0 Z^0 \to 4\ell^\pm$ continues to be observable for a Higgs mass between twice the Z mass and about 600 GeV/c^2 when detector characteristics of generic large SSC detectors are included. A careful study is made of backgrounds from $q\bar{q}, gq \to Z + Jets$ and detector-related issues. It is shown that an isolation cut on the lepton candidates can be expected to reduce the background from this source to low levels.

1. Introduction

One of the primary motivations for the SSC is the understanding of the mechanism for electroweak symmetry breaking. In the minimal standard model this occurs through a single Higgs boson. Although there are more complicated scenarios such as supersymmetric models, we concentrate here on the "standard Higgs boson." Assuming a Higgs boson exists, the experimental limits on and theoretical prediction of its mass are not yet sufficiently constraining and a search over a range of Higgs masses must be performed.

For a Higgs boson in the intermediate mass range, that is, for a Higgs mass (M_H) between half the mass of the Z^0 (M_Z) and $2\,M_Z$, it has been suggested[1],[2] that the decay mode $H \to ZZ^* \to 4\,\ell^\pm$ where one Z is off mass shell may be a viable channel for discovery, at least in the mass range 125 GeV/c^2 $< M_H < 2\,M_Z$. This possibility will be reexamined with attention paid to backgrounds. If the mass of the Higgs boson is greater than twice the mass of the Z^0, it will decay predominantly through the modes Z^0Z^0 and W^+W^-. This Higgs mode will subsequently be referred to as the heavy Higgs mode. The most easily detected mode occurs when both Z^0's decay to e^+e^- or $\mu^+\mu^-$. These are the so-called "gold-plated" events. This decay has been studied extensively in previous workshops and papers[3],[4],[5],[6]. However, the probability for the Higgs boson to decay by this mode is only 1.4×10^{-3}, which limits the discovery range to ~ 600 GeV/c^2 in Higgs mass, depending on the mass of the top quark, for a year's running at the SSC at design luminosity (taken to be 10^4 pb^{-1}). The branching ratio for $H \to Z^0Z^0 \to \ell^+\ell^-\nu\bar{\nu}$ is six times larger, so the discovery range can be extended to higher Higgs mass, provided that this decay can be detected in a convincing manner. The $H \to \ell^+\ell^-\nu\bar{\nu}$ decay has also been **studied**

* This work was supported by the Director, Office of Energy Research, Office of High Energy and Nuclear Physics, Division of High Energy Physics of the U.S. Department of Energy, Contract DE-AC03-76SF00098.

† Work supported by the U.S. Department of Energy, contract DE-AC03-76SF00515.

° Work supported by the U.S. Department of Energy, contract DE-AT03-88ER40384.

in numerous workshops and papers[3], [7], [8], [9]. Another decay mode worth investigating is $H \to Z^0Z^0 \to \ell^+\ell^- q\bar{q}$.

This group has tried to add to previous work mostly in the areas of background considerations and detector requirements for the detection of Higgs decays into four leptons. We will focus primarily on backgrounds to both the intermediate and heavy Higgs decay to four leptons from the processes generically known as $Z + Jets$, that is, from $q\bar{q} \to Zg$, $gq \to Zq$ and $gg \to ZQ\bar{Q}$, where Q represents a heavy quark such as bottom or top.

2. Rates and Backgrounds

2.1 Intermediate Mass Higgs

In the intermediate mass region, the Higgs boson is produced mainly by the gluon-gluon fusion process, shown in Fig. 1. For a top quark mass (M_t) of 55 GeV/c^2, the production cross section is $\sim$ 200 pb and is slowly falling in the Higgs mass range from 100–200 GeV/c^2. Branching ratios for Higgs decay into ZZ^* for various Higgs masses, taken from Ref. 2 , are shown in Table 1. If, in addition, the branching ratios for $Z^0 \to e^+e^-$ or $\mu^+\mu^-$ of 3.4% for each of the Z's are included, the event rate is found to be quite small ($\sim$16 events in one SSC design year for M_H = 140 GeV/c^2), as shown in Table 1.

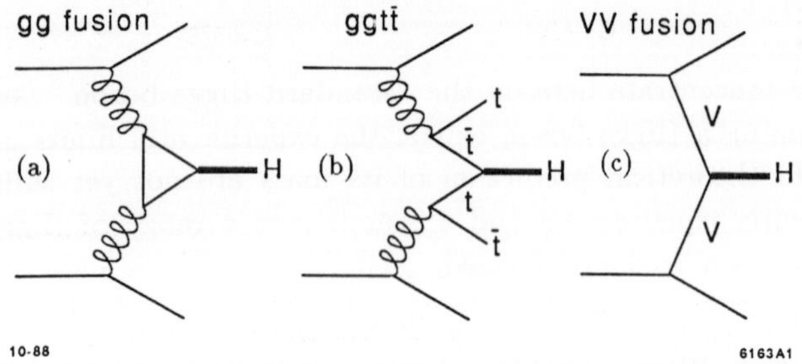

FIG. 1. Higgs production via (a) gluon-gluon fusion, (b) quark-antiquark annihilation, and (c) WW or ZZ fusion.

M_{Higgs} (GeV)	Br. Ratio ($H \to ZZ^*$)	M_{cut} (GeV)	Signal Ev./yr.	$q\bar{q} \to ZZ^*$ Ev./yr.	$gg \to Zb\bar{b}$ Ev./yr.
120	4×10^{-4}	10	3	2	1000 ± 500
140	1×10^{-3}	20	16	3	550 ± 200
160	3×10^{-3}	25	44	2	300 ± 150
180	8×10^{-3}	40	84	8	300 ± 150

Table 1. $H \to ZZ^*$ Branching Ratios and Rates for One SSC Design Year (10^{40} cm^{-2}) for M_t = 55 GeV/c^2.

The intrinsic backgrounds considered in Ref. 2 are $q\bar{q} \to ZZ^*$ (or γ^*) and $gg \to ZZ^*$ (or γ^*). The mode $q\bar{q} \to Z\gamma^*$ was considered to be the largest background but could be eliminated by requiring that the invariant mass of the two leptons which do not come from the decay of the real Z be larger than some cut (M_{cut}) since this background peaks at zero invariant mass. Shown in Table 1 are the numbers of signal and background events that pass such an invariant mass cut and fall within a 2% wide mass bin around the nominal Higgs mass. The Higgs width is typically between 10 MeV to 1 GeV for the Higgs mass range considered. Since the signal is a sharp peak and the background gives a broad distribution, the signal is readily observable above the background. An M_{ZZ*} spectrum was not presented in Ref. 2 with cuts on the invariant mass of the leptons from the off-shell particle. It was concluded that this mode provides sensitivity in the range 125 GeV/c^2 $< M_H <$ $2\,M_Z$. What is not known for this mode are the effects of the background from $Z + Jets$ and in particular the $gg \to ZQ\bar{Q} \to Z\ell^+\ell^- X$ rate. This will be discussed below.

2.2 Heavy Higgs

The heavy Higgs boson can be produced at the SSC through gluon-gluon fusion, quark-antiquark annihilation, and gauge boson fusion, as shown in the diagrams in Fig. 1 . The first two processes depend strongly on the top quark mass. The relevant production cross sections are shown in Fig. 2, taken from Ref. 10. The calculation of the cross section is less reliable for $M_H \gtrsim 600$ GeV/c^2 (see the discussion in Ref. 3). The branching ratio for $H \to Z^0Z^0$ is $\sim$ 30%.

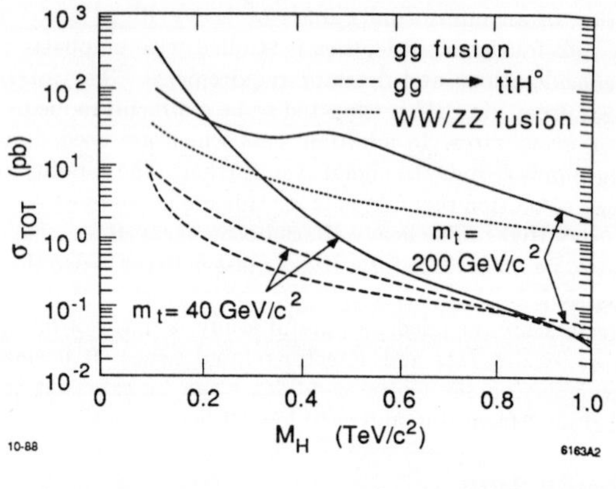

FIG. 2. Heavy Higgs production cross sections from the various processes shown in Fig. 1 as a function of Higgs mass for two different t quark masses (from Ref. 10).

The processes $q\bar{q} \to ZZ$ and $gg \to ZZ$ are serious intrinsic backgrounds to $H \to ZZ$. Calculations of the cross sections for these processes are uncertain because of uncertainties in the structure functions and higher order QCD corrections. The calculations of the ZZ continuum rate are probably reliable to at least a factor of two. Gluon-gluon production of continuum ZZ is approximately 50% of $q\bar{q}$ annihilation for $|y_Z| < 1.5$, where y_Z is the rapidity of the Z. In practice, the measurements of such processes as di-lepton production at the SSC will contribute to the knowledge of the structure functions by the time the **Higgs** search takes place[3] . Comparisons between the theoretical calculations and the Monte Carlo programs ISAJET[11] and PYTHIA[12] were made in Ref. 3.

2.3 Z + Jets Backgrounds

As stated earlier, the QCD processes $q\bar{q} \to Zg$, $gq \to Zq$, and $gg \to ZQ\bar{Q}$ also contribute background to Higgs production. The effects of these backgrounds, however, are sensitive to detector characteristics. For the heavy Higgs case, we are interested in the final state in which two lepton candidates not from the real Z have an invariant mass near M_Z. Most of the $Z + Jets$ background can probably be removed[5] by cuts on lepton isolation and requiring that the invariant mass of pairs of leptons be consistent with the Z mass. For instance, recently it has been found[13] that $gg \to Zt\bar{t}$, shown in Fig. 3 , is the dominant contribution to the $Z + Jets$ background to the heavy Higgs signal.

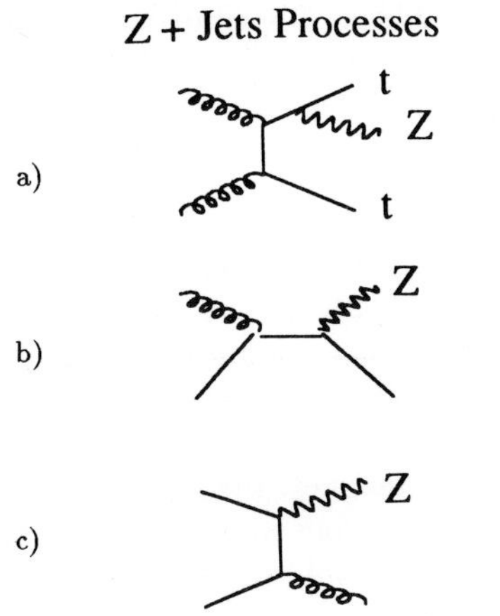

FIG. 3. Z + Jets processes a) $gg \to Zt\bar{t}$,b) $gq \to Zq$, c) $q\bar{q} \to Zg$

A potential background to the four lepton channel from this process arises if the two leptons from the semileptonic decay of the t quarks have the invariant mass of a Z. An explicit parton level calculation shows, however, that if sufficiently high transverse momentum (P_T) is required for the two Z's (i.e., the real Z and the fake Z from leptons from the top decays), the rate is low. Other detector-related backgrounds such as conversion electrons, particle misidentification and particle decays have not been included in that calculation. In this case, even the other $Z + Jets$ subprocesses, $q\bar{q}$ and gq, can be significant. Reliably estimating such backgrounds presents a challenge since fluctuations of one part in 10^4 can become important. Rates for production of a heavy Higgs decaying to four charged leptons and relevant background processes are shown in Table 2. An estimate of the background after cuts due to $q\bar{q}, gq \to Z + Jets$ is given in Section 4.2.

For the intermediate mass Higgs decay to ZZ^*, the power of an M_Z consistency cut is lost for one of the pairs. If one $a)$ generates $gg \to ZQ\bar{Q}$ where the heavy quarks are top or bottom; $b)$ demands that the invariant mass of the decay leptons from the heavy quarks be larger than a value, M_{cut}, as in Ref. 2 ; and $c)$ asks for the number of these events that fall within a 2% mass bin around the nominal Higgs mass, then one finds the rates shown in the last column of Table 1. Fig. 4 shows the invariant mass spectra, M_{ZZ^*}, for signal and $q\bar{q}$ background, adapting numbers from Ref. 2 , and superimposing the $ZQ\bar{Q}$ contribution. Indeed, the $Z + Jets$ background is already becoming significant.

Process	Raw Rate No Cuts	Rate with Cuts*
$H \to ZZ \to \ell^+\ell^-\ell^+\ell^-$		
M_H = 200 GeV/c², M_t = 40 GeV/c²	575	70
M_H = 200 GeV/c², M_t = 200 GeV/c²	687	93
M_H = 400 GeV/c², M_t = 40 GeV/c²	144	89
M_H = 400 GeV/c², M_t = 200 GeV/c²	560	345
M_H = 600 GeV/c², M_t = 40 GeV/c²	60	40
M_H = 600 GeV/c², M_t = 200 GeV/c²	225	146
$q\bar{q} \to ZZ \to \ell^+\ell^-\ell^+\ell^-$	1500	280
$gg \to Zt\bar{t} \to \ell^+\ell^-\ell^+\ell^-$		
M_t = 40 GeV/c²	Large	14
M_t = 100 GeV/c²	Large	~ few
$q\bar{q}, qg \to Z(\to \ell^+\ell^-) + g, q$ (M_{Zjet} > 200 GeV/c²)	10^7	–

Table 2. Rates for Higgs Production and Decay to $\ell^+\ell^-\ell^+\ell^-$ and Background Processes for One SSC Design Year (10^{40} cm^{-2}). * Cuts are $y_\ell < 2.5$, $P_{T,\ell} > 10$ GeV, $P_{T,Z} > 50$ GeV and $\Delta M_Z <$ ± 10 GeV.

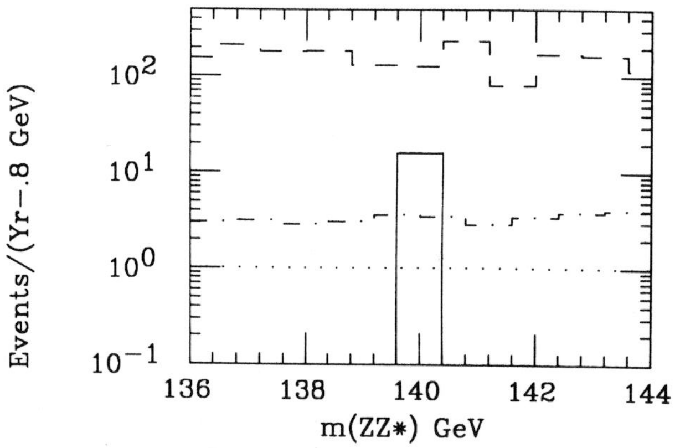

FIG. 4. Invariant mass spectra of ZZ^* for $M_{Z*} > 20$ GeV/c² adapting rates from Ref. 12 and using parton level calculation from program PAPAGENO written by Ian Hinchliffe. Higgs signal shown as solid curve, $q\bar{q}$ as dotted, $gg \to Zt\bar{t}$ as dot-dashes and $gg \to Zb\bar{b}$ as dashes.

3. Analysis Procedures

We now explore some detector requirements for the intermediate mass Higgs decay $H \to ZZ^*$ and for the heavy Higgs decay $H \to Z^0 Z^0$. In both cases the Z's decay to $\ell^+\ell^-$. Part of the analysis procedure is common to the two cases, although the kinematic ranges may be different.

1. The event must have at least four identified electrons or muons. The detector must then be able to identify electrons and muons over the kinematic ranges required by the physics, as described below.

2. The four leptons must be isolated. This cut is needed to reduce the background from $Z + Jets$.

3. A cut on the P_T of the decay leptons is needed to remove leptons which do not come from Higgs decays, for example, leptons from b quark decays from $gg \to Zb\bar{b}$, Dalitz decays of π^0's, and $\pi^\pm$ and $K^\pm$ decays.

4. At least one pair of leptons must have invariant mass consistent with M_Z (one pair for the intermediate mass Higgs, both pairs for the heavy Higgs). This cut is needed to reduce background from QCD processes and from $Z + Jets$ in the case of the heavy Higgs. There should be at least four e's, or four μ's, or two e's and two μ's. If there are more than four leptons which pass all the cuts, combinatorics must be handled. The possibility of no charge determination is discussed below. The degree to which background can be reduced depends on the width of the Z peak, which in turn depends on the energy or momentum resolution and the angular resolution.

5. The final step in the analysis in both cases is examining the plot of the invariant mass of the Z and Z^* candidates or the two Z candidates. There should, of course, be a bump at the mass of the Higgs. The measured width of the bump should not be much larger than the intrinsic width of the Higgs or observability will be degraded. This requirement again places constraints on the energy or momentum resolution and the angular resolution for the four leptons. The rates for background processes can then be evaluated.

3.1 Intermediate Mass Higgs

We now examine in more detail the requirements for the $H \to ZZ^*$ mode. Two questions come to mind. First, is there a cut that can reduce the large $ZQ\bar{Q}$ background while retaining the signal? Second, what happens to the signal if lepton rapidity or P_T cuts are imposed or energy resolution is included? We cannot properly address the first question since we presently cannot fragment the $ZQ\bar{Q}$ system to generate a real experimental environment, but some type of isolation cut involving the decay lepton P_T will probably be necessary to reduce this background. We ignore this for now and study the second question in a simple way. The procedure is as follows:

1. Generate a 140 GeV/c^2 Higgs with PYTHIA (v4.9) selecting the gluon-gluon fusion process.

2. Force the Higgs to decay into a real Z and Z^* where the decay leptons from the Z^* reconstruct to a uniform distribution in invariant mass between 20 and 50 GeV/c^2. (Higgs decays to virtual Z's are not included in PYTHIA.) This is roughly what is expected as described in Ref. 2 .

3. Examine the rapidity and P_T distributions of the leptons.

4. Observe the degradation of the mass spike due to lepton momentum resolution.

3.2 Heavy Higgs

Heavy Higgs production and decay $H \to Z^0 Z^0 \to 4\ell^\pm$ is generated using PYTHIA. Distributions for this process assuming a perfect detector are given in Ref. 3. The analysis procedure includes the points listed above with the following additions specific to the heavy Higgs:

1. The four e's or μ's must have $|\eta| \leq 2.5$, where η is the pseudorapidity. This cut is essentially equivalent (see Ref. 3) to the cut $|y_Z| < 1.5$, which is needed to reduce the background from $Z + Jets$ and from continuum ZZ production, both of which tend to be produced in the forward direction. For a detector with a solenoidal magnet, the leptons are in the region of the best tracking measurement. (Of course, if they weren't, we would have to design a different type of detector.)

2. The leptons are generally required to have $P_T > 10$ GeV/c. Leptons from heavy Higgs decays typically have $P_T > 20$ GeV/c.

3. An additional cut is needed to reduce the background from $gg \to Zt\bar{t}$. We require that the transverse momenta of the Z candidates (P_{TZ}) be greater than 50 GeV/c. Another possibility is a scalar E_T cut.

4. Results

4.1 Intermediate Mass Higgs

Lepton Acceptance We first check basic lepton rapidity and P_T distributions. Figure 5 (a) shows the rapidity distribution of the largest rapidity lepton from either the Z or Z^* decay. If most of the signal events are to be kept, acceptance of leptons must extend to five units of rapidity. Figure 5 (b) shows the P_T distribution of the smallest P_T lepton from either the Z or Z^* decay. Clearly, a small P_T cut on the lepton of $\sim$ 10 GeV/c will significantly reduce the signal. Such a cut will probably be necessary to remove the $gg \to Zb\bar{b}$ background. With the already low rates before cuts, statistics will be severely limited for the 140 GeV/c^2 Higgs.

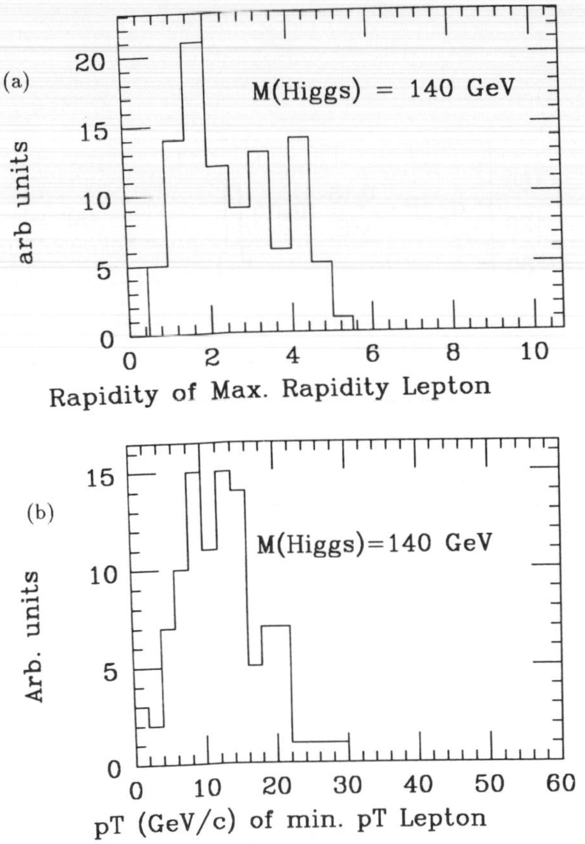

FIG. 5. (a) The rapidity distribution of the largest rapidity lepton from 140 GeV Higgs decay to 4 leptons via the ZZ^* mode with $M_{Z*} > 20$ GeV. (b) The transverse momentum distribution of the lepton with the smallest P_T for the same events just described.

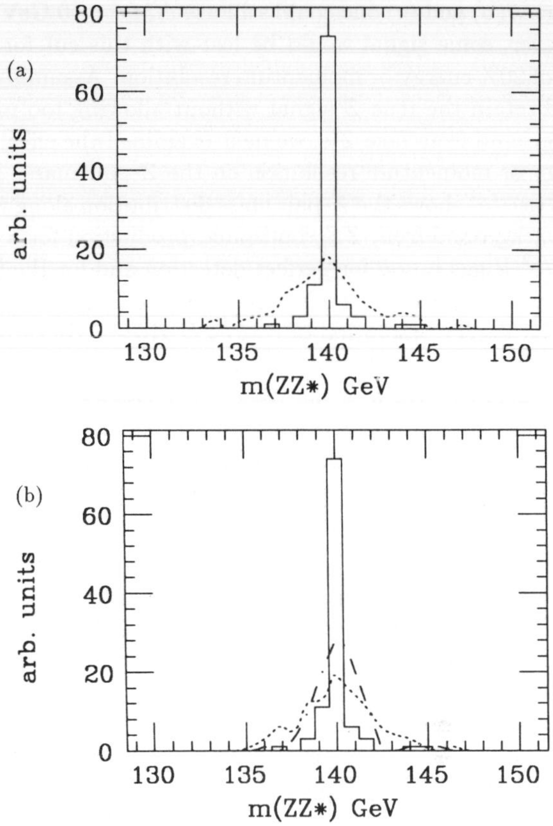

FIG. 6. (a) The Higgs mass peak reconstructed from leptons for a perfect detector (solid histogram) and an EM calorimeter with a resolution of $\sigma_E/E = .01 + .15/\sqrt{E}$ (GeV , dotted curve). (b) The Higgs mass peak reconstructed from leptons for a perfect detector (solid histogram) and a tracker with momentum resolutions of $\sigma_{PT}/P_T = .54\ P_T$ (TeV/c, dot-dashed curve) and $\sigma_{PT}/P_T = .26\ P_T$ (TeV/c, dotted curve). The intrinsic width of the Higgs is included.

Effects of Energy and Momentum Resolution The Higgs peak for various lepton energy resolutions is shown in Fig. 6 . For a perfect detector the peak is very sharp, but with such a small number of events we are particularly sensitive to energy resolution. Figure 6 (a) shows ZZ^* invariant mass distributions for $\sigma_E/E = 0.15/\sqrt{E}$ (E in GeV/c), characteristic of the energy resolution expected for an SSC electromagnetic calorimeter. Figure 6 (b) shows the ZZ^* mass distribution for a momentum resolution of $\sigma_{P_T}/P_T = 0.54 P_T$ (P_T in TeV/c), characteristic of a large magnetic detector[14] , and also for a momentum resolution of $0.26P_T$, which could be obtained by such a magnetic detector if particles are constrained to come from the interaction region. The ZZ^* mass resolution obtained for a typical SSC electromagnetic calorimeter and for a typical tracking system without beam constraint will probably not suffice for observation of a 140 GeV/c^2 Higgs boson. The mass resolution obtained with momentum measurement with a beam constraint is somewhat better, but the ratio of signal to background is still problematic.

4.2 Heavy Higgs

Effects of Energy and Momentum Resolution We have also investigated the effects of energy or momentum resolution on the measured width of the di-lepton invariant mass and on the invariant mass of the two Z candidates for heavy Higgs decays. The width of the di-lepton invariant mass determines how well background from fake Z's can be reduced by requiring that two leptons have invariant mass consistent with the mass of the Z. The width of the invariant mass of the two Z candidates determines how well a Higgs peak can be discriminated from background. Figure 7 shows the di-lepton invariant mass for several energy or momentum resolution functions for a 400 GeV/c^2 Higgs decaying into $ZZ \to \ell^+\ell^-\ell^+\ell^-$: (a) $\sigma_E/E = 0.15/\sqrt{E}$ (E in GeV), (b) $\sigma_{P_T}/P_T = 0.54P_T$ (P_T in TeV/c), (c) $\sigma_E/E = 10\%$, and (d) $\sigma_E/E = 20\%$, characteristic of the momentum resolution in a muon system toroidal detector. The mass resolution in Fig. 7 (a) is quite good – a cut of $\pm$ 5 GeV/c^2 on the mass around M_Z could be used. For

Figs. 7(b) and (c) one could still use $\Delta M_Z < 10$ GeV/c^2. However, some signal would be lost with this cut for the case of 20% energy or momentum resolution. Assuming we could retain the true Z signal without allowing too much background from fake Z's, we next examined the effects of energy or momentum resolution on the Z-pair mass. Figure 8 (a)–(e) shows the Z-pair mass distribution along with the background from ZZ continuum production for a 400 GeV/c^2 Higgs boson for perfect detection and for the four energy or momentum resolutions shown in Fig. 7. We see that an energy resolution of $0.15/\sqrt{E}$ gives a Higgs peak which is indistinguishable from perfect detection because of the intrinsic width of the Higgs. The cases $0.54P_T$ and 10% energy resolution give somewhat broader but still observable peaks at the Higgs mass. However, 20% energy resolution broadens the Higgs peak so much that it is not distinguishable from background.

Our conclusion was that the resolutions under discussion for electromagnetic calorimeters and momentum measurement in a solenoidal detector are acceptable for finding a heavy Higgs signal. A cut of ± 10 GeV/c^2 around M_Z can be used with such detectors. However, an energy or momentum resolution of 20%, such as in a toroidal muon detector, produces an unacceptable degradation of the signal for a 400 GeV/c^2 Higgs.

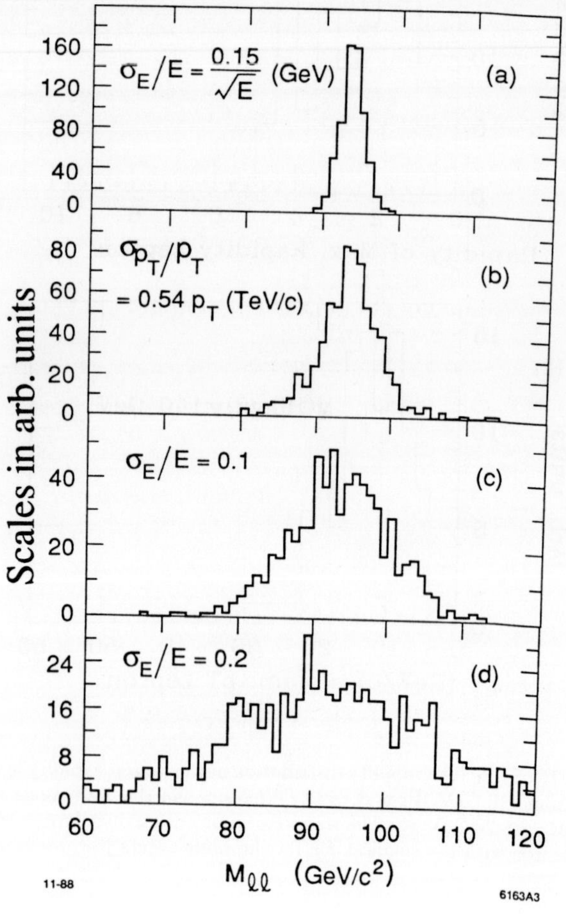

FIG. 7. Di-lepton invariant mass distributions for various energy or momentum resolutions for leptons from Higgs bosons of 400 GeV/c^2 mass decaying into $ZZ \to 4\ell^{\pm}$.

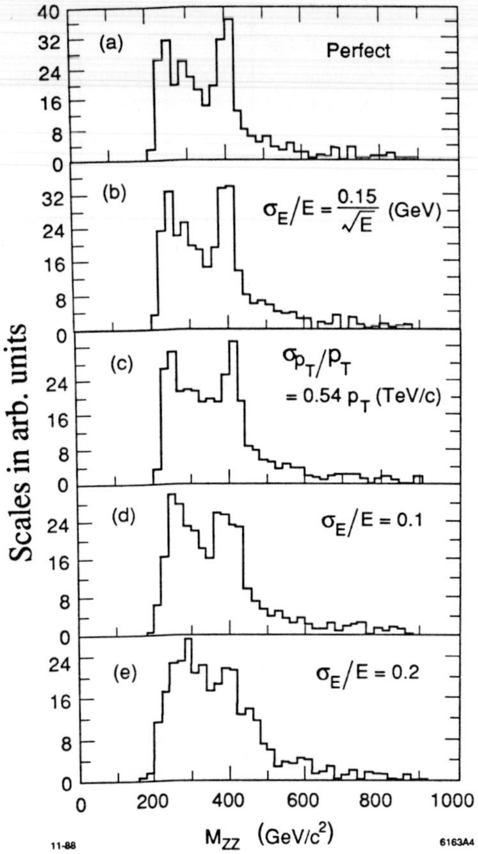

FIG. 8. Z-pair mass distributions for various energy or momentum resolutions for $H \to ZZ \to 4\ell^{\pm}$ for $M_H = 400$ GeV/c^2, along with the background from continuum ZZ production.

Electron Identification and Lepton Isolation A more detailed study was performed of backgrounds from $Z + Jets$ and detector-related issues. The signal consists of two Z's each of which subsequently decays into two charged leptons. For simplicity, we considered the case of four electrons. The main purpose of this study was to find out how many background events pass the Higgs selection cuts without particle identification and then to determine the requirements on electron purity and track separation.

Background from $Z + Jets$ events arises from combinations of a) one real Z and one fake Z or b) two fake Z's. Fake Z's can be reconstructed from a) two real electrons, b) one real electron and another charged or neutral particle which is not an electron, or c) two particles which are not electrons. Cases b) and c) represent complete misidentification for the particles which are not electrons.

We proceeded by generating 100K $q\bar{q}, gq \to Z + Jets$ events using PYTHIA with a minimum $P_T > 50$ GeV/c in the hard scattering frame and imposed the cuts described in Sec. 3.2 on all particles. Roughly speaking, this left particles traversing the central rapidity region of a detector. We addressed the problem of two or more particles that are close together or essentially overlapping appearing as one particle in a detector. To simulate this overlapping, all particles which hit one cell (a cell is defined to be a cone with opening angle equal to the angular resolution and its size heareafter will be labeled by this angle) were treated as a single particle, hereafter referred to as a cluster. If adjacent cells were fired, those clusters were grouped to form a single cluster. Only those clusters which consist of one cell were used to construct Z candidates. In this simulation, the cell size for π^0's and γ's was taken to be 30 mr, and that for charged particles was assumed to be either 0.5 mr or 50 mr.

Tracks in jets tend to have other tracks going along the same direction. To reject those tracks, clusters were required to be isolated. As a measure of isolation, the minimum opening angle between a cluster and all other clusters (θ_{min}) was used. Figure 9 shows θ_{min} distributions for clusters which are electrons from Z decays (referred to as e) and for clusters which are not electrons from Z decays (referred to as h) for two cell sizes. Isolated clusters were taken to be those with $\theta_{min} > 0.5$ radian.

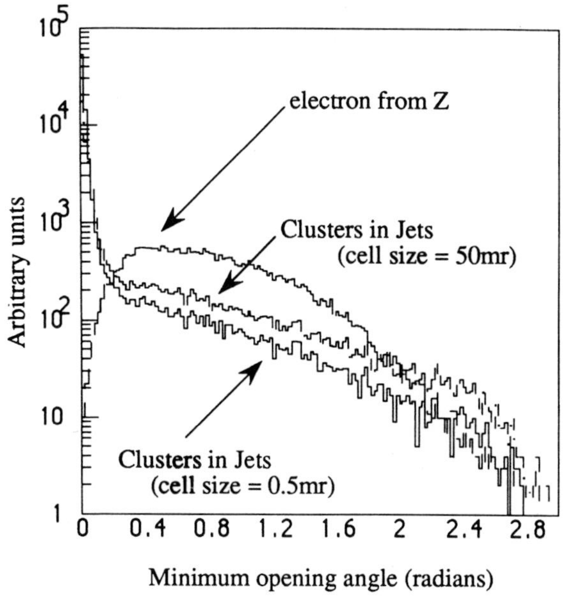

FIG. 9. Minimum opening angle distribution between two clusters for cases described in text.

Two isolated clusters were combined to form a Z candidate if 1) P_T of the pair was larger than 50 GeV/c and 2) the invariant mass of the pair was within 10 GeV/c^2 of M_Z. Table 3 shows the number of Z candidates which passed this selection, where ee (eh, hh) represents a Z candidate which is composed of two type e (e and h, two h) clusters (See Figure 10). We then subjected these clusters to the θ_{min} isolation cut and the Z candidates surviving this were deemed *Good Z's*. As is seen from this Table, many more fake Z candidates passed the isolation cut for the largest cell size. Two Z candidates were combined to form Higgs candidates. Table 4 shows the number of ZZ pairs with invariant mass above 200 GeV/c^2, where, e.g., hh–ee means one Z is hh type and the other is ee type (i.e., a real Z) for all Z candidates and *Good Z* candidates. Figure 11 (a)–(d) shows the invariant mass distributions for ZZ pairs for a cell size of 0.5 mr.

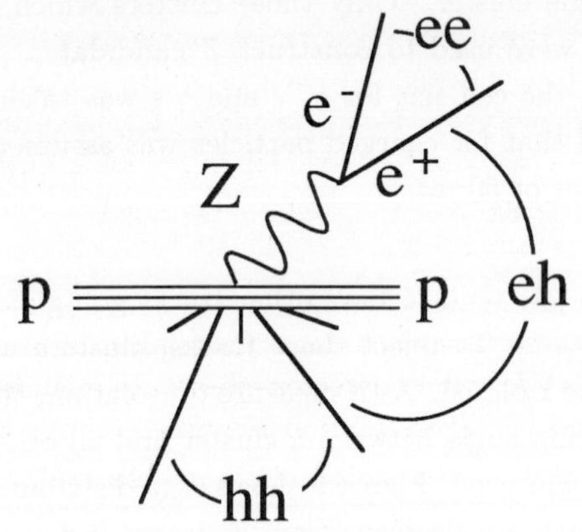

FIG. 10. Cluster pairing types used in constructing Z candidates. Type ee involves two electrons from the Z, type eh involves one electron from Z and another particle and type hh involve particles other than electrons.

Cell Size (mr)	Z Consistency	hh	eh	ee
0.5	none	3400	16300	14900
0.5	Good Z	20	700	9700
50	none	2200	7700	14900
50	Good Z	140	1100	9700

Table 3. Isolation Effects on the Number of Z^0 Candidates from 100K $Z + Jets$ Events Generated: A Z^0 candidate requires two slim clusters with total invariant mass = $M_Z \pm 10$ GeV/c^2 and $P_T > 50$ GeV/c. Good Z consistency requires that each cluster is isolated(as described in the text). For an SSC year run (10^4 pb^{-1}), multiply these numbers by 30 to get proper number of events.

Cell Size (mr)	Higgs Consistency	hh–hh	hh–eh	eh–eh	hh–ee
0.5	none	2000	6100	4300	3000
0.5	Good Higgs	0	1	8	16
50	none	200	1400	1000	2100
50	Good Higgs	0	3	18	104

Table 4. Isolation Effects on the Number of Higgs Candidates from 100K $Z + Jets$ Events Generated: A Higgs candidate requires two Z candidates. Good Higgs consistency means that the two Z's are Good as described in the text and Table 3. For an SSC year run (10^4 pb^{-1}), multiply these numbers by 30 to get proper number of events.

It should be noted that in Fig. 11 statistics are low, but it is clear that most background appears at low mass. The effective rate when scaled to cross section, however, is high. But all of this assumes no lepton/hadron separation. Fairly modest rejection factors of 0.1 to 0.01 will eliminate this form of $Z + Jets$ background. More precise estimates of this background will require experimental verification and more detailed simulation of electron identification.

We note again that other backgrounds arising from pure QCD and $gg \to ZQ\bar{Q}$ have not yet been put through this analysis.

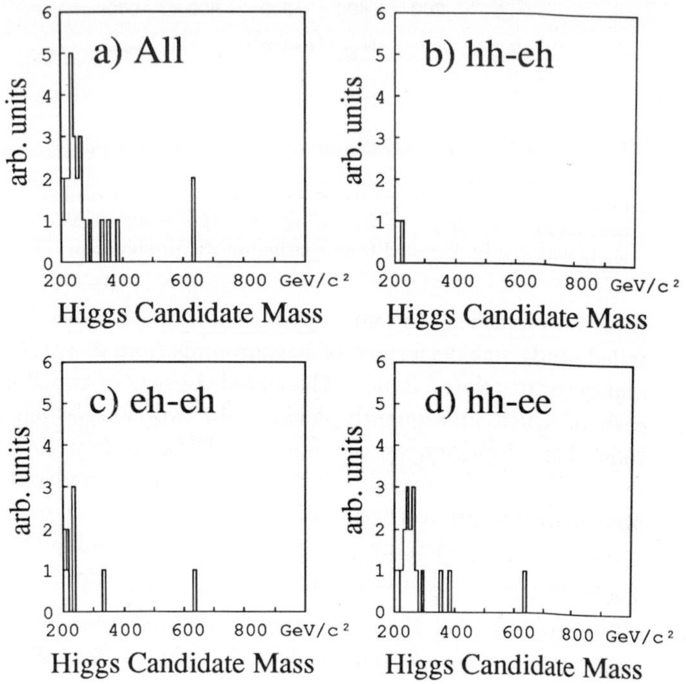

FIG. 11. The invariant mass distributions for Good Higgs candidates for cases (a) All combinations of clusters, (b) hh–eh, (c) eh–eh and (d) hh–ee. The cases are described in the text.

Charge Discrimination We now address the question of charge discrimination. Using the same 100K $q\bar{q}, gq \rightarrow Z + Jets$ events from PYTHIA as described in the previous section, we consider only the charges of the two particles which reconstruct to a Z mass. Again, the particles were required to have $P_T > 10$ GeV/c and $|\eta| < 2.5$. A pair of particles was considered a Z candidate if its $P_T >$ 50 GeV/c and its invariant mass was within 10 GeV/c 2 of the Z mass. No isolation cuts were made. Figure 12 (a)-(c) shows the ZZ invariant mass for Z candidates under various charge requirements. Curve (a) shows M_{ZZ} for all ZZ combinations, regardless of charge. Curve (b) shows it for combinations with no neutral particles contributing to either Z candidate, and (c) shows M_{ZZ} only for combinations in which both Z candidates are made of one positive and one negative particle. Of 4953 original combinations, 1713 (35%) have no neutrals and 514 (10%) have the correct charge combination. This simple cut assumes no lepton/hadron discrimination, yet reduces the background by a factor of 4 above 400 GeV/c^2, and by a factor of 40 at 200 GeV/c^2. It is a powerful but simple cut for background reduction. Again, other backgrounds have not been put through this analysis.

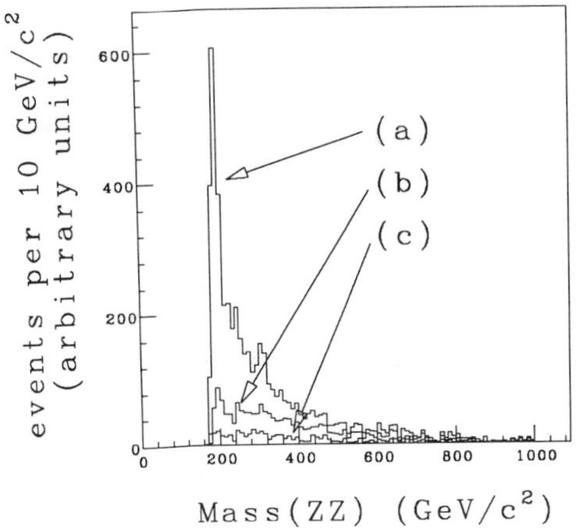

FIG. 12. Diboson mass with varying degrees of charge discrimination. (a) ZZ combinations for all Z candidates, regardless of charge, (b) with discrimination against neutral particles, and (c) with full charge discrimination.

5. Future Work

The $gg \rightarrow ZQ\bar{Q}$ system must be hadronized before its contribution due to particle identification problems can be estimated.

This group looked at some of the backgrounds due to $Z + Jets$ in an idealized way using four-vectors. While this is satisfactory for gaining intuition on relative sizes of various backgrounds, the effects of detector deficiencies have not been studied adequately.

Of course, the spectre of the pure QCD two-jet background remains unstudied but not forgotten. The raw QCD cross section is another factor of 10^4 larger than the $Z + Jets$ background.

Another area which was discussed in this group was the decay $H \rightarrow \ell^+\ell^-\nu\bar{\nu}$. It is clear that more work is needed to establish that this decay of the Higgs can give a believable signal. However, within the short time period of the Workshop we were not able to advance beyond the previous work on this mode. Another higher-rate decay of the Higgs which may prove promising is $H \rightarrow Z^0Z^0 \rightarrow \ell^+\ell^-q\bar{q}$. To our knowledge no detailed study has been made for this mode.

6. Conclusions

The intermediate mass Higgs boson decay into four charged leptons via the ZZ^* mode is expected to be difficult to observe due to low event rates. The case of a 140 GeV/c^2 Higgs was examined in some detail. The detector requirements are stringent in terms of rapidity and minimum P_T acceptance and energy and momentum resolution for the leptons. Cuts needed to reduce backgrounds, for example, from $gg \rightarrow Zb\bar{b}$, also significantly reduce the already very small signal.

The heavy Higgs boson decay $H \rightarrow Z^0Z^0 \rightarrow \ell^+\ell^-\ell^+\ell^-$ still seems to be observable for a Higgs mass of from $2\,M_Z$ to ~ 600 GeV/c^2 when detector effects are included. Momentum and energy resolution requirements for the leptons are within the range proposed for SSC detectors, for example, of the large solenoid type. A careful study of backgrounds from $q\bar{q}, gq \rightarrow Z + Jets$ shows that an isolation cut on lepton candidates can be expected to reduce this contribution to the 400 GeV/c^2 Higgs signal to low levels.

References

1. Atwood, et al., "Intermediate Mass Higgs Boson(s)", in *Proceedings of the Workshop on Experiments, Detectors, and Experimental Areas for the Supercollider, edited by R. Donaldson and M.G.D. Gilchriese, Berkeley, CA*, 1987, p. 728.
2. J.F. Gunion, G.L. Kane,and J. Wudka, Nuc. Phys., **B299**, 239, (1988).
3. R. N. Cahn et al., "Detecting the Heavy Higgs Boson at the SSC", in *Proceedings of the Workshop on Experiments, Detectors, and Experimental Areas for the Supercollider, edited by R. Donaldson and M.G.D. Gilchriese, Berkeley, CA*, 1987, p. 20.
4. R. Thun et al., "Searching for Higgs $\rightarrow Z^0Z^0 \rightarrow \mu^+\mu^-\mu^+\mu^-$ at SSC", *ibid.*, p. 78.
5. F. E. Paige, "ELMUD: an ELectron MUon Detector for Higgs Physics at the SSC", Proceedings of the Workshop "From Colliders to Supercolliders", Madison, WI (1987).
6. M. Chen, E. Nagy, and G. Herten, "High P_t Weak Bosons As Signatures for Higgs-like Heavy Particles", in *Proceedings of the Workshop on Experiments, Detectors, and Experimental Areas for the Supercollider, edited by R. Donaldson and M.G.D. Gilchriese, Berkeley, CA*, 1987, p. 95.
7. A. Savoy-Navarro, "Exploring the Higgs Sector", *ibid.*, p. 68.

8. J. Gunion and M. Soldate, Phys. Rev. D**34**, 826 (1986).
9. V. Barger, T. Han, and R. J. N. Phillips, Phys. Rev. D**36**, 295 (1987); Phys. Rev. D**37**, 2005 (1988).
10. J. F. Gunion et al., University of California at Davis Preprint 86–15 (1986).
11. F. E. Paige and S. D. Protopopescu, "ISAJET 5.30: A Monte Carlo Event Generator for pp and $\bar{p}p$ Interactions," in *Proceedings of the 1986 Summer Study on the Physics of the Superconducting Supercollider, edited by R. Donaldson and J. Marx, Snowmass, CO*, 1986, p. 320. (The current version of ISAJET is 6.12.)
12. H.U. Bengtsson and T. Sjostrand, "PYTHIA: The Lund Monte Carlo for Hadronic Processes," in *Proceedings of the 1986 Summer Study on the Physics of the Superconducting Supercollider, edited by R. Donaldson and J. Marx, Snowmass, CO*, 1986, p. 311.
13. I.Hinchliffe, E. M. Wang, "Effect of Lepton Energy Resolution on Higgs Searches at the SSC",these proceedings.
14. G. G. Hanson, S. Mori, L. G. Pondrom, H. H. Williams et al., "Report of the Large Solenoid Detector Group," in *Proceedings of the Workshop on Experiments, Detectors, and Experimental Areas for the Supercollider, edited by R. Donaldson and M.G.D. Gilchriese, Berkeley, CA*, 1987, p. 340.

Effect of Lepton Energy Resolution on Higgs Searches at the SSC. *

Ian Hinchliffe

Edward M. Wang

Lawrence Berkeley Laboratory
University of California
1 Cyclotron Road
Berkeley, California 94720

Abstract

We discuss the effects of realistic detector resolutions on the processes $H \to ZZ \to e^+e^-e^+e^-$ and $H \to ZZ \to e^+e^-\mu^+\mu^-$ at the SSC. The background from $Zt\bar{t}$ where the $t\bar{t}$ system produces two isolated leptons in its decays is discussed.

In this paper, we analyse the effects of detector resolution on attempts to extract the signal for a Higgs decaying via two Z's into four charged leptons, either electrons or muons at the SSC. Previous studies[1] of this decay mode have shown that the background to it arises from the production of two Z bosons via quark-antiquark[2] or gluon-gluon annihilation.[3] These studies have demonstrated that, given a good determination of the Z pair invariant mass, a convincing signal can be extracted and that Higgs bosons of mass in the range $2M_Z$ to 800 GeV can be found in a canonical SSC year which corresponds to an integrated luminosity of $10^{40}cm^{-2}$. Here we shall investigate in detail what is meant by "a good determination" of the Z momenta and will indicate the effects of expected lepton momentum resolution. We will only consider Higgs bosons of mass 400 GeV. The reader is assumed to be familiar with the basics of Higgs boson production and detection at the SSC. (These issues are reviewed in Ref. 1.)

Higgs bosons are produced at the SSC via gluon-gluon[4] and WW fusion[5]. The rate from the former process depends on the top quark mass. We will assume 40 GeV in our discussion unless otherwise stated. The rate for production of Higgs bosons with masses in the region 300-800 GeV increases as the top quark mass rises above this value. A Higgs boson in this mass range decays to ZZ with a branching ratio of approximately 1/3. The Z bosons can be reconstructed via their decays to $\mu^+\mu^-$ or e^+e^-. The only important source of background Z pairs arises from the processes $q\bar{q} \to ZZ$ and $gg \to ZZ$ (the former dominates in the case of interest[3]). Figure 1 shows the distribution in ZZ invariant mass arising from the production of a 400 GeV Higgs boson. The final state leptons are required to have rapidity $|y| < 2.5$ and are assumed to have their momenta measured with no error. The background shown includes only the process $q\bar{q} \to ZZ$ and we have used the scale $Q = M_{ZZ}$, the invariant mass of the Z pair, in evaluating the structure functions needed in estimating its rate. For a discussion of the scale dependence and the effect of the $gg \to ZZ$ process, see reference 1; a more reasonable scale choice tends to decrease the rate from $q\bar{q} \to ZZ$ by about 30-40%.

It can be seen from Figure 1 that a clear signal is visible as a peak above the background. We will investigate what happens to this peak as the momentum resolution on the leptons is degraded.

*This work was supported by the Director, Office of Energy Research, Office of High Energy and Nuclear Physics, Division of High Energy Physics of the U.S. Department of Energy under Contract DE-AC03-76SF00098.

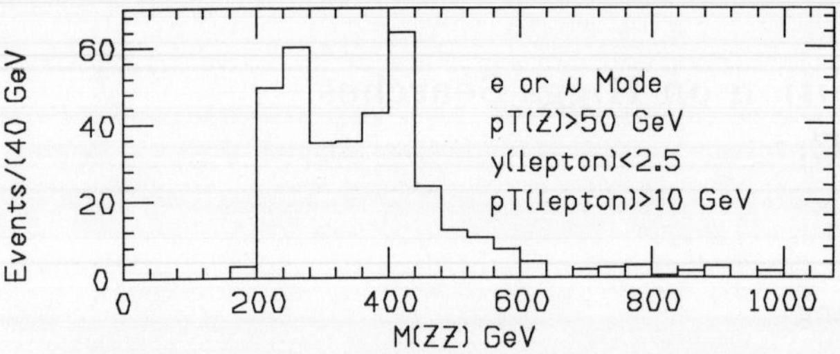

Figure 1. Histogram showing the invariant mass distribution of a pair of Z bosons. The events, which were generated with the PYTHIA event generator, correspond to an integrated luminosity of $10^{40} cm^{-2}$. The Z pairs arise from the production of a Higgs boson of mass 400 GeV and the process $q\overline{q} \to ZZ$. The Z's are reconstructed in the decay modes $Z \to e^+e^-$ and $Z \to \mu^+\mu^-$; all of the leptons All of the leptons are required to satisfy $|y| < 2.5$, $p_t > 10$ GeV. In addition the reconstructed Z's must have $P_t > 50$ GeV.

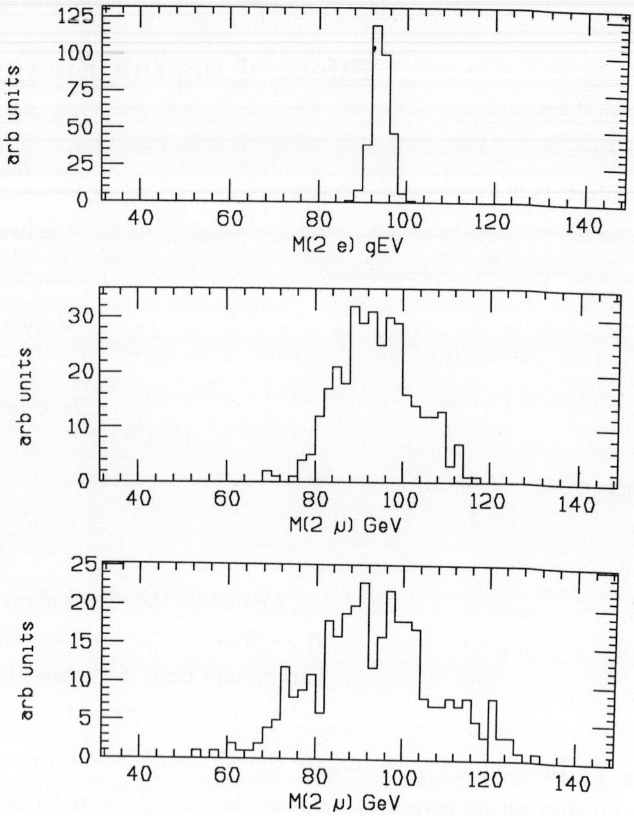

Figure 2 Histogram showing the effect of lepton energy resolution on the reconstructed Z mass. Figures (a), (b) and (c) have $\Delta E/E = .15\sqrt{E} + 0.01$, $\Delta P/P = 0.1$ and $\Delta P/P = 0.2$ respectively.

In the case of electron momentum determination, it is reasonable to expect[6] that one will have an error $\Delta E/E \sim 15\%/\sqrt{E} + 1\%$, where E is expressed in GeV. In the case of muons the resolution is expected to be much worse, $\Delta E/E \sim 10 - 20\%$ if only the muon toroids are used[7]. The muon resolution could be improved by including information on the muon track from close to the interaction point. We show in Figure 2 the effect of resolution on the dilepton invariant mass. This Figure show clearly that the reconstructed Z no longer has a well defined mass when the resolution is poor. This gives rise to the first source of difficulty. If there is a source of dilepton pairs of invariant mass close to that of the Z, some of these pairs will have to be included in the background. These "fake Z's" will then increase the background shown in Figure 1. In the case of the final state with only e^+e^- pairs, the spread of the reconstructed Z mass shown in Figure 2 is small, and this extra background will be irrelevant. The resolution in the $\mu^+\mu^-$ mass may be poor and this extra background must be considered.

Where does the extra background come from? The leptons from the decay of a Z will be isolated and will have large invariant mass. The largest source of leptons at SSC will be from the production and decay of charm, bottom and top quarks since these quarks are produced by strong interaction processes. Of these, the leptons arising from top decay will be more isolated than those from charm and bottom. Consider the final state $Z + Q\overline{Q} + X$ which can arise from the partonic initial states $gluon + gluon$ and $quark + antiquark$. We estimated these rates by using the known matrix elements[8] for $gg \to Zq\overline{q}$ and $q'\overline{q'} \to Zq\overline{q}$, where q and q are massless quarks, together with the correct phase space for massive final state particles. The same approximation applied to the process $gg \to Q\overline{Q}$ is accurate to 30%.

In Table 1, we show the rates from this process. In obtaining the values in this Table, we have required that the Z have transverse momentum greater than 100 GeV. Table 2 shows the rates from $Z + t\overline{t}$ where we have required additionally that there be a lepton pair arising from the $t\bar{t}$ system with invariant mass between 40 and 130 GeV and transverse momentum larger than 100 GeV. The number of similar events from the background process $q\bar{q} \to ZZ$ is also shown for comparison. Notice that the number of events from the $ZQ\overline{Q}$ final state is smaller than that from ZZ. This indicates that this additional source of background which will be present in the $e^+e^-\mu^+\mu^-$ mode is not large. The invariant mass distribution of the $Z + fake\ Z$ system from $ZQ\overline{Q}$ is shown in Figure 3. It has no peaks and contributes a smooth background which falls rapidly with increasing M_{ZZ}. Figure 4 shows the transverse momentum distribution of the real and fake Z's. The real Z was required to have $p_t > 50GeV$. This Figure shows that a cut requiring the fake Z also to have $p_t > 50$ GeV, is very powerful in reducing the background rate (approximately 70% of the events fail this cut). The reduction in the number of signal events due to this cut is approximately 30%. Table 2 shows the effects of various cuts on the processes $H \to ZZ$, $q\bar{q} \to ZZ$ and the $Zt\bar{t}$ final state giving rise to Z and "fake" Z. This Table shows that the extra background is not a problem.

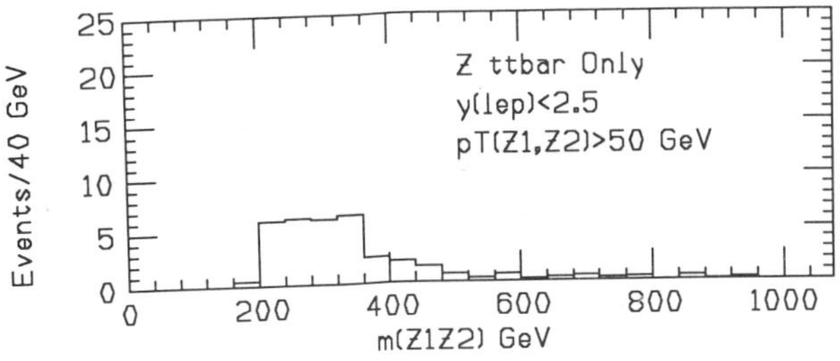

Figure 3. Histogram showing the invariant mass distribution of a Z and " fake" Z. The latter arises as a lepton pair from the final state $Z+t+\bar{t}$; the lepton pair arising from the $t\bar{t}$ is required to have invariant mass between $M_Z - 50$ and $M_Z + 50$ GeV. The events, which were generated with the partonic generator discussed in the text, correspond to an integrated luminosity of $10^{40}cm^{-2}$. All of the leptons All of the leptons are required to satisfy $|y| < 2.5$, $p_t > 10$ GeV. In addition the reconstructed Z's must have $P_t > 50$ GeV.

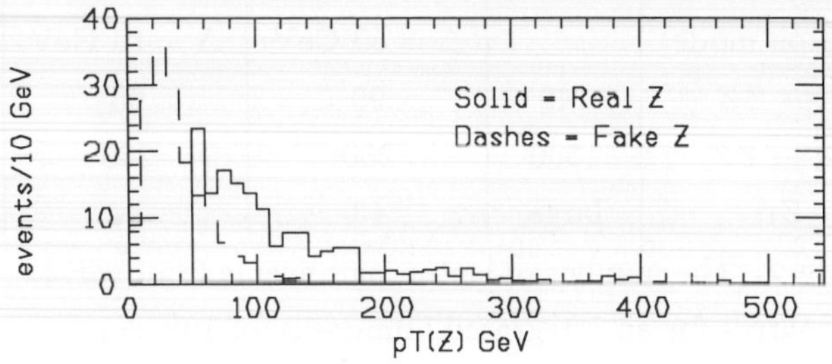

Figure 4. Transverse momentum distribution of the real and "fake" $Z's$ arising from the process $pp \to Zt\bar{t} \to 4l + X$. The physical is required to have $p_t > 50$ GeV, and the "fake" Z is defined to be a lepton pair with invariant mass between 40 and 130 GeV. All of the leptons are required to satisfy $|y| < 2.5$, $p_t > 10$ GeV.

Process	PYTHIA	ISAJET	PARTON (I.H.)
$gq, q\bar{q} \to Z + q, g$ $p_{t_Z} > 100$ GeV	2.9 nb	3.5 nb	2.5nb
$q\bar{q} \to Z + g$ $p_{t_Z} > 100$ GeV	210 pb	295 pb	not run
$gg, q\bar{q} \to Zc\bar{c}$	n.a.	n.a.	205 pb
($q\bar{q}$ only)	15.7 pb	13.0 pb	9 pb
$gg, q\bar{q} \to Zb\bar{b}$	n.a.	n.a.	204 pb
($q\bar{q}$ only)	9.2 pb	14.2 pb	not run
$gg, q\bar{q} \to Zt\bar{t}$ ($m_{top} = 100$ GeV)	n.a.	n.a.	8.2 pb
($q\bar{q}$ only)	0	$< .7$ pb	not run
$gg, q\bar{q} \to Zt\bar{t}$ ($m_{top} = 40$ GeV)	n.a.	n.a.	40 pb
($q\bar{q}$ only)	0	1.8 pb	1.8 pb
$gg, q\bar{q} \to Zt\bar{t}$ ($m_{top} = 5$ GeV)	n.a.	n.a.	156 pb
($q\bar{q}$ only)	0	13.6 pb	6.6 pb

Table 1. Cross sections for various processes as given by PYTHIA[9], ISAJET[10], and the parton calculation outlined in this paper. The Z is required to have $p_t >$ 100 GeV. n.a. means not available.

Process (4lepton mode)	NoCuts	$p_{t_0} = 50$ GeV $\Delta = 50$ GeV	$p_{t_0} = 50$ GeV $\Delta = 10$ GeV
H$\rightarrow ZZ$	140	90	90
$q\bar{q} \rightarrow ZZ$	1500	280	280
$Zt\bar{t}$	large	43	14

Table 2. The number of lepton pair events from various sources for an integrated luminosity of 10^{40} cm^{-2}. The leptons are required to have $|y| < 2.5$. The pairs of leptons are each required to from a "Z", i.e. are required to have invariant mass between $M_Z \pm \Delta$ and the Z's are required to have $p_t > p_{t_0}$.

Isolation cuts on the leptons are not required to reduce this background to an irrelevant level. However such cuts may be required to eliminate the other leptonic backgrounds. For example, cuts requiring that the leptons be isolated from hadronic jets will be particularly effective in reducing the background due to charm and bottom decays. We have not included the fake Z's from the production of charm and bottom followed by their leptonic decay. Such leptons are softer than those from top and are not isolated.

For comparison purposes we have also indicated in Table 1 the event rates arising from the $Z + jet$ processes that are produced by the ISAJET[10] and PYTHIA[9] event generators. These generators produce first produce the final state $Z + parton$, where the parton can either be a quark or a gluon. The parton then showers and the $Q\overline{Q}$ pair can be produced as a component of this shower. An approximation to the $Z + Q\overline{Q}$ rate is then obtained. Unfortunately the approximation is not good as can be seen from the Table. The difficulty is as follows. There is no partonic process of the form $gluon + gluon \rightarrow Z + parton$, hence the event generators are not including any approximation to $gluon + gluon \rightarrow Z + Q + \overline{Q}$.* The process $q\bar{q} \rightarrow Zg$ followed by the "decay" $g \rightarrow Q\overline{Q}$ is present and, as Table 1 shows, provides a good approximation to the $q\bar{q} \rightarrow ZQ\overline{Q}$ process. However the gg initiated process is dominant.

Since the event generators do not accurately reproduce the rates for $ZQ\overline{Q}$ final states they cannot be used in a background study. Nevertheless it is clear from our partonic analysis, that bad resolution in the $\mu\mu$ system does not seriously increase the background in the mode $H \rightarrow ZZ \rightarrow \mu^+\mu^-e^+e^-$.

In the case of the $\mu^+\mu^-\mu^+\mu^-$ final state, there is another additional source of potential background, namely the $Q\overline{Q}\ Q\overline{Q}$, where all of the quarks decay semileptonically and two fake Z's are produced. We have not attempted to estimate this background here.

The second difficulty caused by poor resolution is that the peak in the reconstructed Z pair is smeared out. This effect is shown in Figure 5. It is clear from this Figure that in the 4μ with 20% resolution, the peak has disappeared. However at 10% a peak is still clearly visible. In the $2e2\mu$ final state the signal is still observable. We have tried using the Z mass as a constraint in the reconstruction; the effect of this is shown in Figure 6. By comparison with Figure 5, it can be seen that the improvement is not significant.

In conclusion, we have investigated the effect of detector resolution on the four lepton decay mode of the Higgs boson at the SSC. The bad resolution characteristic of a muon system with resolution of order 20% does not cause an appreciable increase in background due to "fake" Z's in the $4e$ or $2e2\mu$ channels. However it does degrade the resolution in the Z pair invariant mass to such an extent that the peak due to Higgs boson of mass 400 GeV is no longer visible in the 4μ channel and the $2e2\mu$ channel is compromised. A resolution of order 10% or better is required if the peak is to survive. We have only considered backgrounds due to real leptons. It is possible that in a more realistic detector simulation, backgrounds due to misidentified particles or other detector related sources could be significant.

*In principle, the process can be approximated from $gg \rightarrow Q\overline{Q}$ if the Z is then radiated off an outgoing quark line. This is not likely to be an efficient way of simulating the rate.

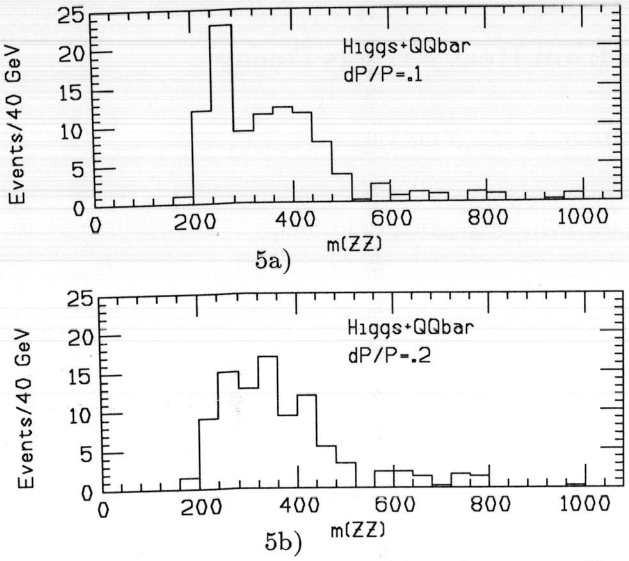

Figure 5. The reconstructed invariant mass distribution for Z pairs from the final states $ZZ \to \mu^+\mu^-\mu^+\mu^-$ The cuts are the same as in Figure 1 In 5a) we have taken $\Delta P/P = 0.1$ for muons and in 5b) $\Delta P/P = 0.2$

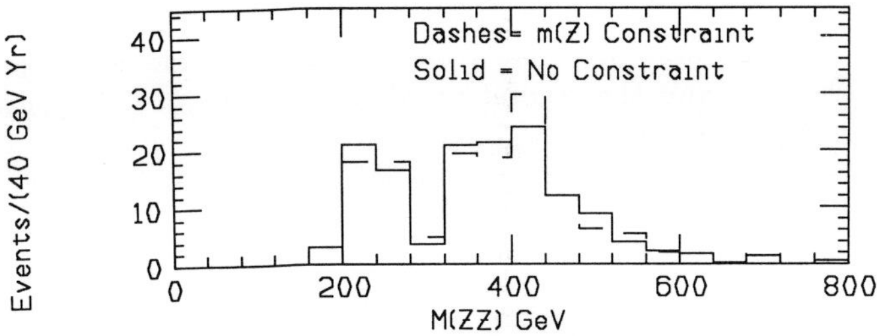

Figure 6. As Figure 5 except for the final state $ZZ \to e^+e^-\mu^+\mu^-$. We have taken $\Delta E/E = .15\sqrt{E} + 0.01$, for electrons and $\Delta P/P = 0.1$ for muons. The dashed histogram shows the effect on resolution if the Z mass is used as a constraint in the fit.

References

1. R.N. Cahn *et al.*, *Experiments, Detectors and Experimental Areas for the Supercollider* Ed. R. Donaldson M.G.D. Gilchriese, World Scientific Publishing, Singapore (1988) and references therein.

2. R.W. Brown and K. Mikaelian, *Phys. Rev.* **D19**, 922 (1979).

3. D.A. Dicus, C. Kao, and W. Repko, *Phys. Rev.* **D36**, 1680 (1987).

4. H. Georgi, *et al.*, *Phys. Rev. Lett.* **40**, 692 (1979).

5. R. N. Cahn and S. Dawson, *Phys. Lett.* **136B**, 196 (1977).

6. C. Baltay *et al.*,*Proceedings of the 1986 Summer Study on Physics of the Superconducting Super Collider* Ed. R. Donaldson, J. Marx, Snowmass (1986).

7. D. Carlsmith *et al.*, *Proceedings of the 1986 Summer Study on Physics of the Superconducting Super Collider* Ed. R. Donaldson, J. Marx, Snowmass (1986) or G. Hanson *et al.*, *Experiments, Detectors and Experimental Areas for the Supercollider* Ed. R. Donaldson, M.G.D. Gilchriese, World Scientific Publishing, Singapore (1988).

8. S. Ellis and J. Stirling, J. Gunion and Z. Kunszt, K. Ellis and R. Gonsalves in *Supercollider Physics* Ed. D. Soper, World Scientific Publishing, Singapore (1986).

9. H.-U. Bengtsson and T. Sjostrand,in *Monte Carlo Simulation and Detector Capabilities* Ed H.-U. Bengstsson, C. Buchanon, T. Gottschalk and A. Soni, World Scientific Publishing, Singapore (1986).

10. F. Paige and S. Protopopescu,in *Monte Carlo Simulation and Detector Capabilities* Ed H.-U. Bengstsson, C. Buchanon, T. Gottschalk and A. Soni, World Scientific Publishing, Singapore (1986).

Detection of Gauge Boson Pairs from Heavy Higgs Decay

H.F.-W. SADROZINSKI, A. SEIDEN, A.J. WEINSTEIN

Santa Cruz Institute for Particle Physics
University of California, Santa Cruz, CA 95064

INTRODUCTION

We present a Monte Carlo study of the detection of a heavy Higgs boson H decaying to vector boson pairs (WW and ZZ). In one SSC year (10^4 pb^{-1}) about 5000 events/year are expected to be produced in the final state

$$pp \to H \to WW \tag{1a}$$

for a Higgs mass $m_H = 1000$ GeV (WW pair mass interval $850 < m_{WW} < 1350$ GeV). A similar number of events is expected in the channel

$$pp \to H \to ZZ \tag{1b}$$

for $m_H = 800$ GeV. In order to accumulate enough events, at these large Higgs masses, one of the vector bosons has to be observed in its two-jet hadronic decay, which constitute the majority of W and Z decays. The other vector boson is seen via its leptonic decay, providing a signature for triggering and event selection:

$$pp \to H \to WW \to \text{jets} \quad \hookrightarrow l\nu \tag{2a}$$

$$pp \to H \to ZZ \to \text{jets}\ . \quad \hookrightarrow l^+l^- \tag{2b}$$

This process has been looked at in several studies before.[1] What we would like to emphasize here is the use of all the components of a multi-purpose detector, i.e. calorimetry—the main thrust of previous studies—and high precision tracking, which contributes new information.

To have convincing evidence for the above decay, two questions have to be addressed:

1) establish a signal in the WW and ZZ channel above the dominant background from mixed electroweak (EW)–QCD processes

$$pp \to W + \text{jets} \quad \hookrightarrow l\nu, \tag{3}$$

with a demonstrable peak at the W or Z mass for the hadronic decay channel.

2) measure the polarization of the W's or Z's to determine the longitudinal polarization of the W's characteristic for the decay of a scalar Higgs.

The first problem deals mainly with the event selection, and the second problem depends on defining a proper jet finding algorithm to have high efficiency in jet reconstruction for all decay angles θ^* in the vector boson C.M. system. The study below is based on Monte Carlo events from PYTHIA 4.8, [2,3] which were smeared in energy only ($\sigma_e/E = 0.15/\sqrt{E}$ for electro-magnetic, $\sigma_e/E = 0.50/\sqrt{E}$ for hadronic showers) in a calorimeter extending from $-2.5 < Y < 2.5$ and over the whole azimuth in 100×100 cells. We do not address the serious problem of transverse shower spread which might worsen the jet mass resolution.[4] The tracking device is assumed to extend over the same solid angle and capable of measuring the charged event multiplicity for all particles with $p_t \geq 0.5$ GeV. We call this multiplicity N_{ch} below. It has been shown [5,6] that a Si μstrip tracker has the required performance. The rates are quoted per year for a nominal SSC integrated luminosity of 10^4 pb^{-1}. There is considerable uncertainty in the cross-section of large mass Higgs and the absolute rates should be taken with caution. On the other hand, the angular acceptance in θ^* is independent of the cross-section and depend only on the way fragmentation is implemented in the Monte Carlo, and on the properties of the detector. We restrict the mass m_H of a 1000 GeV Higgs to $850 \leq m_H \leq 1350$ GeV and assume that 4500 Higgs $\to WW$ events are produced per SCC year. For a 800 GeV Higgs

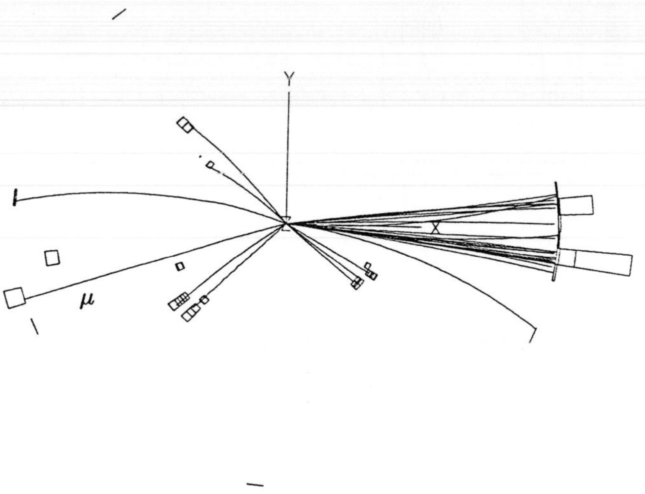

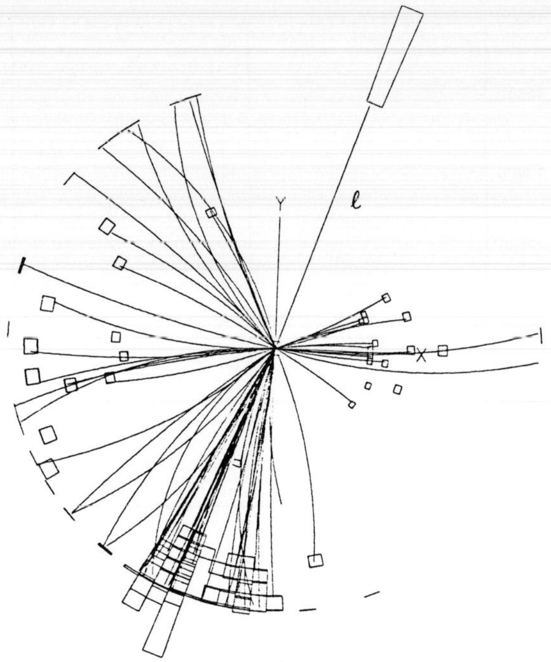

Fig. 1. Event display which shows charged tracks with $p_t \geq 0.5$ GeV and calorimeter cells with $E_t > 0.5$ GeV in $|Y| < 2.5$ together with the isolated lepton from the W decay. *a)* Higgs $\rightarrow W^+W^+$ event with $M_{WW} = 1230$ GeV, $N_{ch} = 33$ and $m_{\text{jet}} = 73$ GeV. *b)* Mixed EW-QCD event with invariant mass $M_{W\text{jet}} = 1280$ GeV, $N_{ch} = 73$ and $m_{\text{jet}} = 82$ GeV.

($600 < m_{ZZ} < 1350$ GeV), we assume that 5000 ZZ events are produced per year.[7)]

The discussion is divided into three parts. First, we describe a new method of discriminating $H \rightarrow WW$ events [eq. (2a)] from mixed EW processes [eq. (3)] by utilizing the large difference in charged event multiplicity between the two processes. Then we discuss a jet algorithm to reconstruct $W \rightarrow 2$ jet events and measure the helicity of the W. Last, the Higgs to ZZ channel [eq. (1b)] is discussed. It allows the determination of the Z polarization in both the jet–jet and the lepton–lepton final state.

RECONSTRUCTION OF THE W IN ONE JET

In order to discriminate Higgs production into WW from the mixed EW jet production, we will compare both the whole event and the jet structure in the two cases. Fig. 1 shows PYTHIA events in a hypothetical SSC detector where charged tracks with $p_t \geq 0.5$ GeV and calorimeter cells with $E_t \geq 0.5$ GeV are shown. Fig. 1a is a Higgs event [eq. (1a)] and Fig. 1b is a mixed EW event [eq. (3)]. In both cases one W decays leptonically.

There are two dramatic differences:

1) The W jet (Fig. 1a) is collimated much more tightly than the QCD jet (Fig. 1b) which resembles a broad "fan".

2) The event multiplicity is lower by about a factor 2.5 in the Higgs event (Fig. 1a).

A theoretical explanation of these observations is given in Ref. 8. Our analysis to establish the WW signal in events with isolated leptons and large missing E_t proceeds as follows: Employing "GETJET"[9)] we reconstruct jets in a limited cone $\left(\Delta R = \sqrt{(\Delta\phi)^2 + (\Delta Y)^2} = 0.5\right)$ with cut off in $E_t^{\text{jet}} \geq 100$ GeV, and then cut on the event multiplicity N_{ch}. Table 1 shows the flow of the analysis.

TABLE 1
NUMBER OF EVENTS/JETS IN 1 SSC YEAR

	WW	$W + j$
EVENTS:		
E_t(lepton) > 20 GeV, $n_{\text{jet}} \geq 1$	1094	99 k
JETS:		
$\Delta R = 0.5$, $E_t^{\text{jet}} > 100$ GeV	1365	186 k
$70 \leq m_{\text{jet}} \leq 90$ GeV	580	6700
$0 \leq N_{ch} \leq 39$	456	490

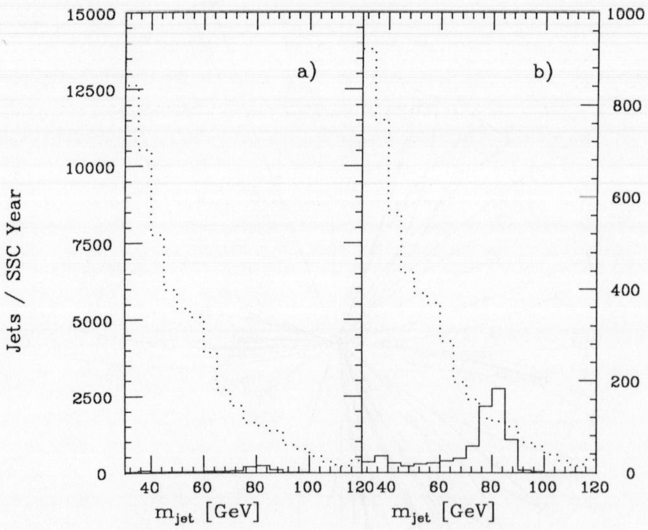

Fig. 2. Mass of jets m_{jet} with $R = 0.5$ and $E_t > 100$ GeV for Higgs (solid line) and $W+$ jet (dotted line) production: *a)* before multiplicity cut, *b)* after multiplicity cut.

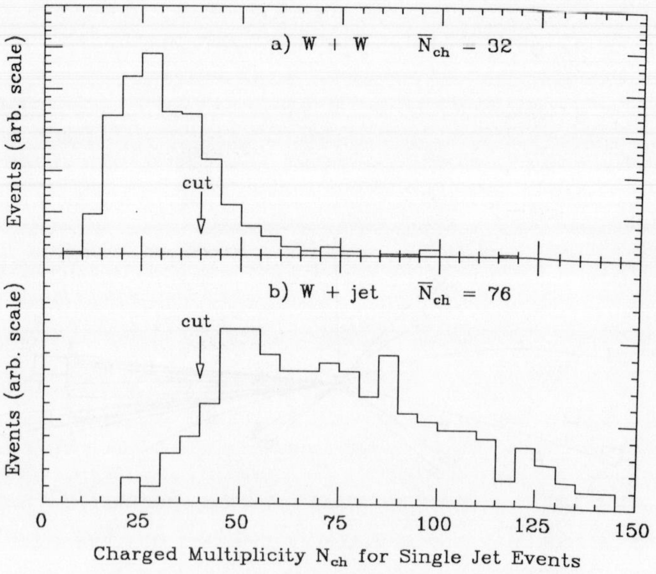

Fig. 3. Charged event multiplicity N_{ch} for tracks with $|Y| < 2.5$ and $p_t \leq 0.5$ GeV: *a)* Higgs $\rightarrow WW$, *b)* $W+$ jet.

One way to follow the analysis is to show the jet mass distribution for Higgs production and the background, respectively, before and after the multiplicity cut (Fig. 2). While before cuts, the signal-to-background ratio r in the jet mass bin 70–90 GeV is $r = 0.086$, we have, after the multiplicity cut, $r = 0.93$, an improvement by a factor 11. In the latter case the W gives a recognizable peak in the mass distribution.

We show in Fig. 3 the charged multiplicity N_{ch} for events with at least one jet in the W mass region $70 < m_{jet} < 90$ GeV, for the signal [eq. (2a)] and the background process [eq. (3)]. The tracks have $|Y| < 2.5$ and $p_t \geq 0.5$ GeV. The two distributions overlap very little and a cut at $N_{ch} < 40$ leaves 78% of the signal and only 7.3% of the background. It is interesting to check how effective the multiplicity cut is as a function of the transverse momentum cut-off of the tracker. For $p_t \geq 0.5$ GeV $r = 0.93$, for $p_t \geq 1$ GeV $r = 0.72$, for $p_t > 2$ GeV $r = 0.68$, showing that a significant difference of the two processes is in the soft component. (For consistency we adjust the multiplicity cut such that 80% of the signal is retained for each case above.)

From this analysis we conclude that the use of the event multiplicity is a powerful tool to distinguish between EW and QCD phenomena.

RECONSTRUCTION OF THE W IN JET PAIRS AND THE POLARIZATION OF THE W

As a consequence of the scalar nature of the Higgs, for large Higgs masses the emerging W and Z are longitudinally polarized, and thus the leptons and the quark jets have the following angular distribution in the W center of mass:

$$dN/d\cos\theta^* \sim \sin^2\theta^* \ . \tag{4}$$

This can be measured in the laboratory by measuring the energies of the two leptons or jets E_1, E_2, since

$$\cos\theta^* = \frac{E_1 - E_2}{E_1 + E_2} \ . \tag{5}$$

In order to map out the angular distribution [eq. (4)], a measurement at large $\cos\theta^*$ is necessary and uniform efficiency is required. This necessitates a different, somewhat more complicated, analysis than in the previous section.

We have analyzed the kinematics of the W decay and find that the angular separation $\Delta\Theta$ between the two jets, and the angular size of each quark jet in the laboratory described by a cone of half-angle δ, vary depending on the CM decay angle θ^*

$$\Delta\Theta \approx 2/\gamma_W \sin\theta^* \tag{6}$$
$$\delta \approx 2p_\perp/\{x \cdot E_W(1+\cos\theta^*)\} \tag{7}$$

where $x = p^{||}/p^{||}_{\rm jet}$ is the Feynman x of a track in the jet, $p_\perp$ the transverse momentum of the track relative to the jet direction, both in the W rest frame. The boost of the W into the lab frame is given by $\gamma_W = E_W/m_W$. For $p_\perp = 500$ MeV, $x \geq 0.05$ and $E_W = 500$ we get

$$\delta = 0.04/(1+\cos\theta^*) \tag{8}$$

which varies from $\delta = 0.02$ to $\delta = 0.80$ for $\cos\theta^*$ varying from 1 to -0.95.

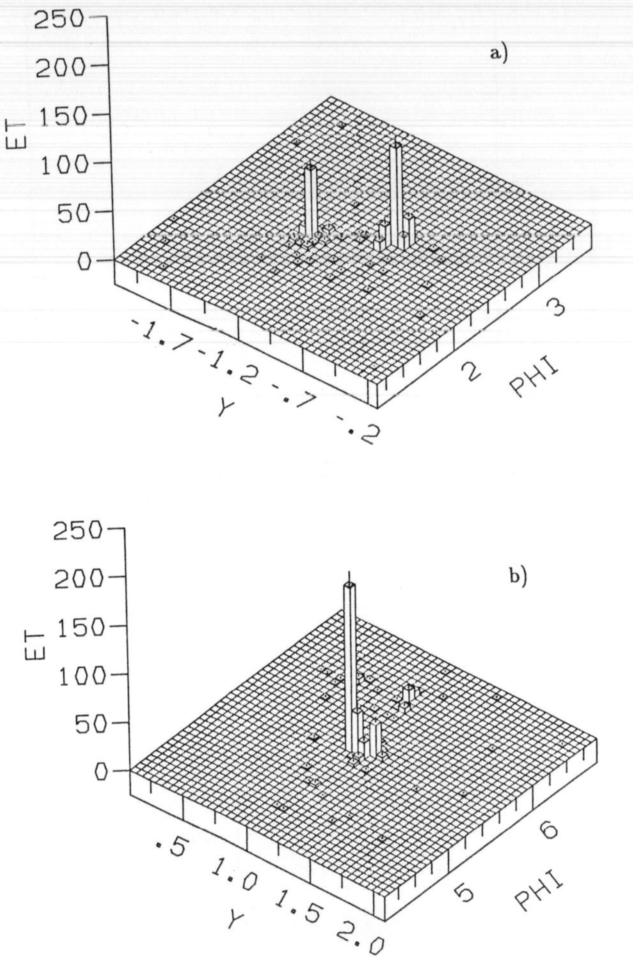

Fig. 4. Lego plot of a hadronic W decay in a calorimeter of $\Delta\phi = 0.05, \Delta Y = 0.063$. The W points into the center of the plots. *a)* $E_W = 500$ GeV, $\cos\theta^* = 0.03$. Note the two almost symmetrical narrow clusters. *b)* $E_W = 539$ GeV, $\cos\theta^* = 0.92$. Note that the cluster in the center of the plot is one jet with $E = 500$ GeV while the scattered cells belong to the jet with $E = 40$ GeV.

We illustrate these facts in Fig. 4 with "LEGO" plots of two W decays of $E_W = 500$ GeV, one for $\cos\theta^* = 0.03$ (Fig. 4a) and one for $\cos\theta^* = 0.92$ (Fig. 4b). The centers of the plots coincide with the generated W direction. While the more energetic jets are well collimated, the low energy jet in Fig. 4b shows up in the scattered cells surrounding the central W direction and the more energetic jet. If one wants to collect all the energy of the jets one has to vary the radius of the jet cone with the jet energy according to [eq. (8)]. In our implementation of these ideas we first note that the angular relationships [eqs. (6) – (8)] hold in phase space rather in $\Delta\phi \cdot \Delta Y$ space of the calorimeter. We thus redefine ΔR as the angular separation in real space. The jet finding algorithm works then as follows: we first search for jets in large cones in the detector ($\Delta R = 1$) with large E_t cut-off ($E_t \geq 100$ GeV). Then we search inside each of these jets for jets i with $E_t^i \geq 40$ GeV within a cone ΔR_i which varies with the energy of the jet: $\Delta R_i = 1.5 \cdot \Delta\phi + (10/E_j^i)$, where $\Delta\phi = 0.06$ is the angular size of a calorimeter cell and E_j^i is the energy of the jet i in GeV. This jet search is started with the four cells containing the largest energy left and is then iterated by adjusting ΔR_i.

TABLE 2. NUMBER OF PAIRS IN 1 SSC YEAR

	WW	$W+$ Jet
2 jets, ΔR_i variable $E_t^i \geq 40$ GeV, within jet with $\Delta R = 1$, $E_t \geq 100$ GeV	2315	312 k
$n_{\text{jet}} = 2, 3$; $\Delta m \leq 14$ GeV	405	13.7 k
$70 \leq m_{\text{jet,jet}} \leq 90$ GeV	269	940
$N_{ch} \leq 34$	223	76

Table 2 shows the analysis flow in the two-jet case. The cut which limits n_{jet}, the number of individual jets inside the large cone to two or three is employed to simplify the analysis. Even without this cut, the final sample contains only jets with $n_{\text{jet}} = 2, 3$. We then require a match of the mass of the large cone jet m_{jet} with the jet pair mass $m_{\text{jet,jet}}$ and limit the difference $\Delta m = m_{\text{jet}} - m_{\text{jet,jet}}$ to ≤ 14 GeV. The distributions of Δm, after all other cuts are applied, are shown for both signal and background in Fig. 5. The pair mass of the W decays reconstructs at a low mass: $70 \leq m_{\text{jet,jet}} \leq 90$ GeV which we use to identify the W. Again the cut

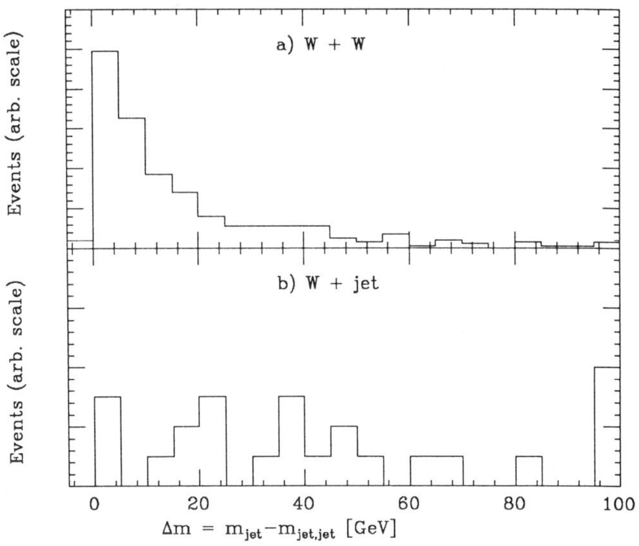

Fig. 5. Mass difference Δm between large jet and jet pair ($70 \leq m_{jet,jet} \leq 90$ GeV): *a)* Higgs, *b)* mixed E.W.

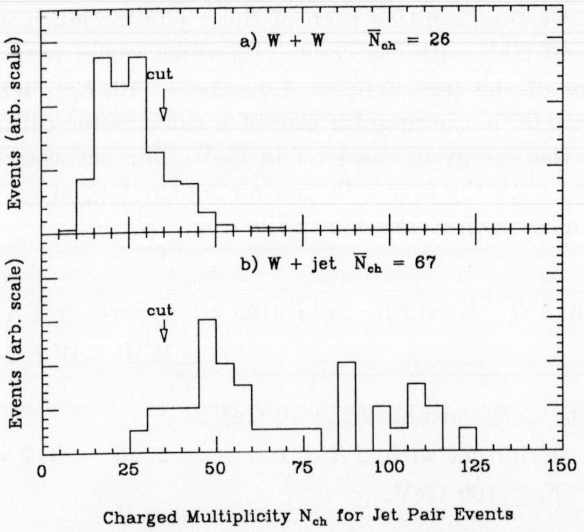

Fig. 6. Charged event multiplicity N_{ch} for events with a jet pair looking like a $W(m_{\text{jet}} \approx m_{\text{jet,jet}} \approx m_W)$: *a)* Higgs, *b)* mixed E.W.

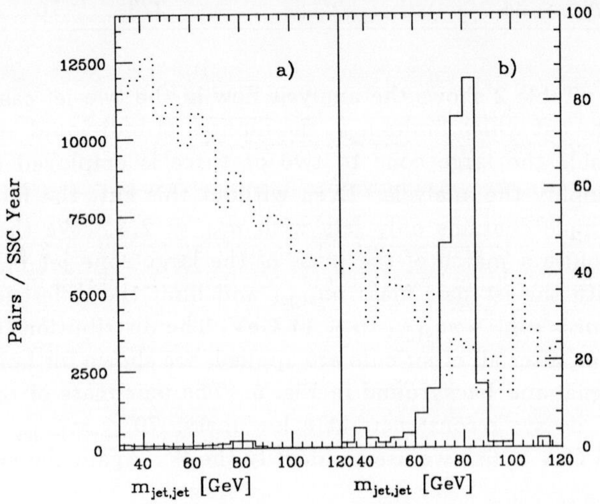

Fig. 7. Jet pair mass $m_{jet,jet}$ for the Higgs $\to WW$ decay (solid line) and mixed E.W. decay (dotted line). *a)* before cuts, *b)* after cuts (Table 2.).

on the charged event multiplicity $N_{ch} < 34$, as shown in Fig. 6, is very effective and leads to improvement of the signal-to-background by a factor 10. The multiplicity distributions in Fig. 6 have lower means than in the one-jet case and we adjusted the cuts to retain 80% of the signal. The effect of all the cuts is shown in Fig. 7 where the jet pair mass is shown for Higgs decays and mixed EW production before the cuts (Fig. 7a) and after the cuts (Fig. 7b). The signal-to-background ratio r improves from $r \approx 0.015$ to about $r = 3$ in the jet pair mass interval corresponding to the W.

The parameters in the cuts are not the result of a great deal of tuning, although we did recognize the sensitivity of the results on the choice of parameters. The cut-off of the transverse energy of the individual jets E_t^i influences both the reconstruction efficiency at large $\cos\theta^*$ and the signal-to-background ratio r: a smaller E_t^i

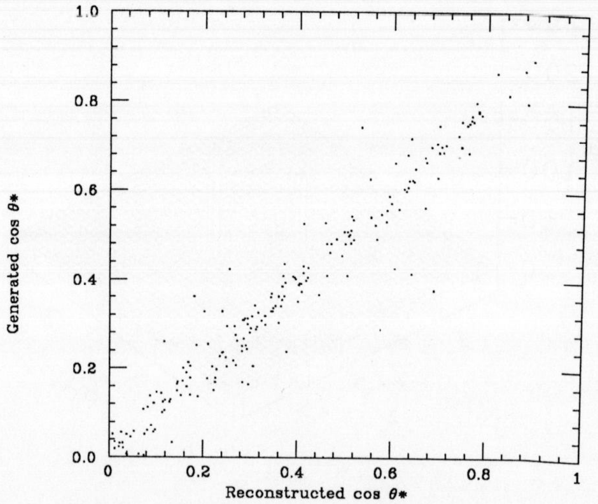

Fig. 8. Generated $\cos\theta^*$ vs. reconstructed $\cos\theta^*$ for jet pairs.

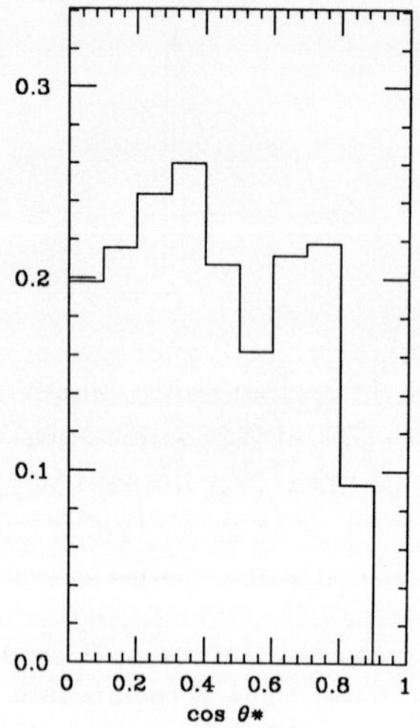

Fig. 9. Efficiency of reconstructing pairs of jets from W decay as function of C.M. angle θ^*.

improves the reconstruction efficiency for larger $\cos\theta^*$, but decreases r. For example taking $E_t^i \leq 20$ decreases r by a factor 0.5, while adding only a few events at $\cos\theta^* \geq 0.8$. Another important parameter is Δm: as shown in Fig. 5 the Δm distribution of the signal is steeply falling while the background is consistent with being flat in Δm. As a consequence, if we assume a flat Δm distribution of the background, choosing $\Delta m \leq 20$ GeV instead of 14 GeV, decreases the signal-to-background ratio r by a factor 0.9.

We can now address the question how well we can reconstruct the decay angle $\cos\theta^*$ in the W rest frame. Fig. 8 shows a scatter plot of the generated $\cos\theta^*$ vs. the reconstructed $\cos\theta^*$ for the final event sample. The correlation is very good with a correlation coefficient of 0.96. The efficiency (Fig. 9) for reconstructing the angle θ^* in the W rest frame shows a dip at $\cos\theta^* = 1$, but is reasonably constant up to $\cos\theta^* = 0.8$.

Due to the limited statistics the background has to be scaled up by a factor 25, leaving only a handful events in the final sample. It is consistent with a flat distribution in $\cos\theta^*$ but with more statistics one might start to see a peaking at large $\cos\theta^*$ as expected for background processes. With a signal-to-background ratio of about 3 as shown in Table 2, the determination of the W polarization looks promising.

RECONSTRUCTION OF THE DECAY $H^0 \to ZZ$

A similar analysis was performed on reaction (2b), in which an 800 GeV Higgs decays to a pair of Z bosons; one Z decays to either e^+e^- or $\mu^+\mu^-$, and the other decays into hadronic jets. For such a high mass Higgs, the events resulting in four leptons from the two Z's are too few in number to produce a significant signal in one SSC year, so it is especially important to allow one Z to decay hadronically (increasing the statistics for detectable events by a factor $\simeq 20$).

Again, the dominant background is expected to be the production of one Z recoiling against a QCD jet; and again, the background rejection depends critically on good mass resolution in the calorimeter for the hadronic jet pair from the Z, and the large difference in event multiplicity. We rely on a precision tracker to measure the multiplicity within $|Y| < 2.3$, and also to supply precise momentum measurements for identifying the electrons or muons from the first Z decay. Those momentum measurements, when coupled with information from the electromagnetic calorimeter or outer muon identifiers, make it possible to positively identify the lepton without relying on strong isolation cuts.[5)]

We require 2 opposite-sign leptons each with $|Y| < 2.3$ and $p_t > 50$ GeV, with invariant mass within 15 GeV of the Z mass. We then search the calorimeter for jets with $E_T > 50$ GeV within a cone $\Delta R = 1$, and require their invariant mass to be within 15 GeV of the Z mass. Finally we cut on the charged multiplicity of the event. The results are given in Table 3, which include the predictions from PYTHIA for background rates from $pp \to Z + \text{jets}$.

TABLE 3
NUMBER OF EVENTS IN 1 SSC YEAR

	ZZ	$Z+j$
2 identified leptons, $77 \leq m_{l^+l^-} \leq 107$ GeV	252	24K
Jet: $\Delta R = 1$, $E_t > 50$ GeV	230	14K
$77 \leq m_{\text{jet}} \leq 107$ GeV	152	2640
$0 \leq N_{ch} \leq 39$	89	192

Again we see that the cut on charged multiplicity provides a substantial improvement in signal to background; the result depends on the minimum p_t of the reconstructed charged tracks in the same way as for the WW case.

The cone containing the hadronic jet identified as a Z is then searched for sub-jets, again within cones according to the incident Z energy [eq. (7)]. Since at these energies the hadronic showers tend to merge, the *directions* of the two highest energy subjets are more precisely measured than the energies themselves. Using the mass of the Z as a constraint, the energies [and therefore the decay helicity angle, eq. (5)] of the two subjets are reconstructed. This technique yields an almost flat efficiency for reconstruction of the helicity angle, from 0 to 0.9. The accuracy of the measurement is also good, giving a correlation coefficient with the generated helicity angle of 0.82.

Of course, the leptons from the first Z also give a measurement of the helicity angle. Thus, although this decay mode suffers in statistics, even small signals can be made to stand out over the background, and can provide good measurements of the vector boson polarization.

We must note here that the background normalization, simulated by PYTHIA as a $2 \to 2$ process $Z + q$ or $Z + g$ is very uncertain. It may be that $2 \to 3$ processes, yielding hadronic jets with invariant masses in the range of the W or Z mass, are more important. We eagerly await a Monte Carlo that simulates these processes, along with a more accurate treatment of the signal event cross section.

CONCLUSIONS

We have presented an analysis of Monte Carlo events aimed at detecting and measuring the cross section for gauge boson pairs at the SSC for invariant masses at the TeV scale. This is crucial for a Higgs search at this mass scale. To have a sufficient number of events one gauge boson must be detected through its hadronic decays. To ensure that we are indeed detecting events with hadronic gauge boson decays, it is essential that the background can be reduced sufficiently that a clear mass peak at the gauge boson mass is visible in the appropriate jet invariant mass distribution. We show that the event multiplicity can provide a powerful discriminator against backgrounds which typically have much larger multiplicities than signal events.

In a massive Higgs decay, the gauge bosons are longitudinally polarized. A measurement of the polarization would provide important information on the source of the gauge boson pairs being detected. We provide an initial analysis aimed at allowing a measurement of the polarization. Substantially more work on the signal and background will be needed to understand how well the polarization can be determined.

REFERENCES

1. Protopopescu, S.D., *Proceedings of the 1986 Summer Study on Physics of the SSC*, Snowmass (1986); Cahn, R.N. *et al.*, *Proceedings of the Workshop on Experiments, Detectors and Experimental Areas for the SSC*, Berkeley (1982) and reference cited there.

2. H. Bengtsson and T. Sjostrand, *Comput. Commun.* **46**, 43 (1987).

3. We have checked that the program ISAJET (F. Paige and S. Protopopescu) gives qualitative, similar results on the effectiveness of the multiplicity cut.

4. Wang, E., Talk presented at this workshop.

5. Sadrozinski, H., Seiden, A. and Weinstein, A., "Tracking at 1 TeV," *Proceedings of the Topical Seminar on Perspectives for Experimental Apparatus*, San Miniato, Italy, 1988, to be published.

6. DeWitt, J. *et al.*, "Tracking Requirements for the SSC and Silicon Strip Development," these proceedings.

7. M. Golden, Vega, R., and Yuan, C.P., private communication.

8. Gunion, J.F. *et al.*, UC Santa Cruz preprint SCIPP 88/26 (1988).

9. This program was written by F. Paige as part of the ISAJET package.

Techniomega Production at the SSC

R. Sekhar Chivukula

Department of Physics
Boston University
Boston, MA 02215

Technicolor [1] is an elegant mechanism for breaking the $SU(2)_W \times U(1)_Y$ electroweak gauge interactions down to the $U(1)_{em}$ of electromagnetism. In these theories chiral symmetry breaking due to a strong technicolor gauge interaction, analogous to chiral symmetry breaking in QCD, is responsible for electroweak symmetry breaking. If technicolor is responsible for electroweak symmetry breaking there will be technihadronic resonances in the TeV region, just as there are hadronic resonances in the GeV region. The observation of these resonances would be a nice signature of technicolor.

We expect the lightest technihadronic resonances to be vector mesons analogous to the isotriplet $\rho(770)$ and isosinglet $\omega(783)$ of QCD. In this talk, I discussed the prospects for production and detection of technivector mesons at the SSC. In this report I will discuss work started at Snowmass on the possibility of the production of techniomega mesons through flavor symmetric extended technicolor interactions and their detection via the decay $\omega_{TC} \to Z_L\gamma$. A more complete description of this work and the talk may be found in [2]. Vector resonances in theories with strongly interacting electroweak symmetry breaking sectors have also recently been considered in [3].

For simplicity, I restrict myself to a technicolor model with only one technidoublet. In this case, the technifermion electric charges are $\pm 1/2$ [1] and the Z gauge boson couplings to technifermions are purely weak isospin triplet. The techniomega is a weak isospin singlet, and does not mix with the isotriplet Z and γ. It therefore cannot be produced through the process $q\bar{q} \to (\gamma^*$ or $Z^*) \to \omega_{TC}$. As discussed below, however, it may be produced through flavor symmetric extended technicolor interactions.

Using large N_{TC} (where the technicolor group is $SU(N_{TC})$) arguments to estimate [4] the mass of the techniomega we find that

$$M_{\omega_{TC}} \approx M_\omega \frac{v}{f_\pi}\left(\frac{3}{N_{TC}}\right)^{\frac{1}{2}} \approx 2.1\,\mathrm{TeV}\left(\frac{3}{N_{TC}}\right)^{\frac{1}{2}}. \tag{1}$$

The $\omega(783)$ decays primarily into pions. Similarly the techniomega will decay into technipions, which are the longitudinal components of the W's and Z's. For the partial width of the ω_{TC} into technipions we find

$$\Gamma^{TC}_{3\pi} \approx \left(\frac{3}{N_{TC}}\right)^2 \Gamma^{QCD}_{3\pi}\frac{M_{\omega_{TC}}}{M_\omega}\frac{1}{\beta^6_\pi} \approx \left(\frac{3}{N_{TC}}\right)^{\frac{5}{2}} 60\,\mathrm{GeV}\,. \tag{2}$$

The $\omega(783)$ also decays into $\pi^0\gamma$ - similarly, the techniomega will decay into $Z_L\gamma$ and, scaling from QCD, we find the partial width

$$\Gamma^{TC}_{Z\gamma} \approx \Gamma^{QCD}_{\pi\gamma}\frac{M_{\omega_{TC}}}{M_\omega}\frac{1}{\beta^3_\pi} \approx \left(\frac{3}{N_{TC}}\right)^{\frac{1}{2}} 3\,\mathrm{GeV}\,. \tag{3}$$

In order to transmit electroweak symmetry breaking to the quarks and leptons, it is necessary to introduce interactions ("extended technicolor" (ETC) interactions) involving both fermions and technifermions. Such interactions may give rise to *direct* couplings of techniomegas to quarks and leptons. Consider, for example, the operator

$$\frac{(\bar{\chi}_L\gamma_\mu\chi_L)(\bar{\psi}_L\gamma^\mu\psi_L)}{4f^2}, \tag{4}$$

where χ_L is the left handed technidoublet, ψ_L is left handed (u,d) quark doublet, and the I is the 2×2 identity matrix. This operator appears naturally in models in which the (ETC) interactions are approximately flavor symmetric and, due to limits on quark and lepton compositeness, f must be greater than 1 TeV [5]. Scaling from QCD and using "vector meson dominance" [6], (4) results in the coupling

$$\frac{M^2_{\omega_{TC}}}{4 f_\omega f^2} \left(\bar{\psi}_L \gamma_\mu \psi_L\right) \omega^\mu_{TC}. \qquad (5)$$

f_ω may be determined from fitting to the decay $\omega \to e^+e^-$ and is found to be

$$f_\omega \approx 5.5 \left(\frac{3}{N_{TC}}\right)^{\frac{1}{2}}. \qquad (6)$$

The direct couplings (5) imply that the techniomega may decay into left handed quarks and leptons - if the couplings are completely flavor symmetric, it will decay into all light fermions with equal probability. Using (5) and (6) we find

$$\Gamma_{f\bar{f}} \approx 27\,\text{GeV} \left(\frac{1\text{TeV}}{f}\right)^4 \left(\frac{3}{N_{TC}}\right)^{\frac{3}{2}} \left(\frac{N_f}{24}\right), \qquad (7)$$

for the partial width into fermions, where N_f is the number of left handed light fermions. The decay widths of the techniomega are summarized in the following table, where I have taken $f = 1.0\,\text{TeV}$ and $N_f = 24$ for the purpose of illustration.

N_{TC}	$\Gamma_{3\pi}$	$\Gamma_{Z\gamma}$	$\Gamma_{f\bar{f}}$	$BR(Z\gamma)$
2	165 GeV	3.7 GeV	48 GeV	$\geq 2\%$
4	30 GeV	2.6 GeV	18 GeV	$\geq 5\%$
6	11 GeV	2.1 GeV	10 GeV	$\geq 9\%$

Using (5) we may calculate the SSC techniomega production cross section

$$\sigma_{\omega_{TC}} \approx 5.1 \times 10^{-6}\,\text{nb} \left(\frac{3}{N_{TC}}\right) \left(\frac{1\,\text{TeV}}{f}\right)^4 \times \left(\frac{d\mathcal{L}_{u\bar{u}}}{d\tau} + \frac{d\mathcal{L}_{d\bar{d}}}{d\tau}\right)\Big|_{\tau = \frac{M^2_\omega}{s}}, \qquad (8)$$

where the $d\mathcal{L}_{q\bar{q}}/d\tau$ are the $q\bar{q}$ differential partonic luminosity functions [7] and $s = (40\,\text{TeV})^2$. Using (8), folding in the $Z\gamma$ branching ratio of the techniomega, and the branching ratio of approximately 6% for $Z \to l^+l^-$ (where $l = \mu$ or e) we can calculate the number of observable techniomega events. This number is listed in the third column of the following table assuming an integrated pp luminosity of $10^{40}\ cm^2$. The number of background continuum $q\bar{q} \to Z\gamma$ events with invariant $Z\gamma$ pair mass within 50 GeV of the techniomega mass is shown in the fourth column (a rapidity cut of 1.5 was placed on the $Z\gamma$ pair).

N_{TC}	M_ω	$\omega_{TC} \to Z\gamma \to l^+l^-\gamma$	$Z\gamma \to l^+l^-\gamma$
2	2.6 TeV	$20\left(\frac{1\,\text{TeV}}{f}\right)^4$	1.0
4	1.8 TeV	$100\left(\frac{1\,\text{TeV}}{f}\right)^4$	0.4
6	1.5 TeV	$215\left(\frac{1\,\text{TeV}}{f}\right)^4$	0.1

The results in this table should not be taken literally: uncertainties in the structure functions, in what the appropriate momentum transfer is for evaluating the structure functions, and in possible QCD corrections make these results only estimates. They should be reasonable estimates, however, and should certainly be correct to within a factor of two. From this table we see that $\omega \to Z\gamma$ is a nice signature for the techniomega meson. If flavor symmetric ETC interactions exist with a scale (f) less than a few TeV, techniomega production should be observed at the SSC. All we need now is the money to build the SSC!

I would like to thank K. Lane and M. Golden for useful discussions. I also thank the Aspen Center for Physics, where much of this work was carried out.

References

[1] E. Farhi and L. Susskind, Phys. Rep. 74 (1981) 277, and references therein.
[2] R. S. Chivukula, Boston University Preprint BUHEP-88-16, to be published in the proceedings of the *12th Johns Hopkins Workshop in Particle Theory*, Baltimore, MD, June 8-10, 1988;
R. S. Chivukula and M. Golden, to be published.
[3] R. Rosenfeld and J. L. Rosner, Phys. Rev. D38 (1988).
[4] S. Dimopoulos, Nucl. Phys. B168 (1968) 69;
S. Dimopoulos, S. Raby, and G. Kane, Nucl. Phys. B182 (1981) 77.
[5] R. S. Chivukula and H. Georgi, Phys. Lett. B188 (1987) 99.
[6] J. J. Sakurai, *Currents and Mesons*, Univ. of Chicago Press, 1969.
[7] E. Eichten, I. Hinchliffe, K. Lane and C. Quigg, Rev. Mod. Phys. 56 (1984)579.

FLAVOR CHANGING FROM LEFT-OVER U(1)'s

Bob Holdom
Department of Physics
University of Toronto
Toronto Ontario
CANADA M5S 1A7

ABSTRACT

We discuss some effects of relatively light U(1) gauge bosons associated with flavor physics, with emphasis on "walking" technicolor (TC) theories.

In extended technicolor theories quark and lepton masses are associated with various new mass scales of flavor physics above the scale of the weak interactions. These scales are inversely related to the associated masses. There need not be many new scales since this is not the only source of mass splitting among quarks and leptons; another source is splittings, enhanced in walking TC, among different technifermion condensates.[1] But we expect a few new mass scales of flavor physics. This new physics will give rise to flavor changing four fermion operators, and in general the largest such operators will involve the most massive quarks and leptons.

This last effect is enhanced in walking TC theories, which for whatever reason have a small or vanishing TC β-function.[2] There the relation between a quark or lepton mass m_ψ and the associated scale of flavor physics Λ_ψ is modified. The coefficient $1/\Lambda_\psi^2$ of a flavor changing four fermion operator is now proportional to m_ψ^ξ where ξ may be as large as 2 (instead of 1). Walking TC suppresses flavor changing, but we now see that the flavor changing among heavy quarks is less suppressed than that among light quarks.

In the breakdown of an extended TC gauge group there are typically diagonal generators which remain unbroken down to the TeV scale. In walking TC the flavor changing effects of these "left-over" U(1)'s are enhanced relative to the more massive off-diagonal generators which may directly link different families. Consider a left-over U(1) which couples to the third family. Due to the existence of KM mixing the quark "flavor" eigenstates to which this U(1) couples are unlikely to be mass eigenstates in both the up and down sectors simultaneously. Thus flavor changing operators are generated, the largest being

$$(\theta/M_{U(1)}{}^2)\overline{\psi}_3\psi_3\overline{\psi}_3\psi_{1,2}$$

where the subscripts are family indices and θ is a mixing angle. This could give rise to enhanced rare decays of heavy quarks, eg.

$$t \Rightarrow \tau^+\tau^- + (c \text{ or } u).$$

But operators of the form

$$(\theta/M_{U(1)})^2\overline{\psi}_3\psi_{1,2}\overline{\psi}_3\psi_{1,2}$$

give rise to $B^0\overline{B}^0$ or $T^0\overline{T}^0$ mixing of order

$$\Delta m/m \approx (f\theta/M_{U(1)})^2,$$

where f denotes f_B or f_T. Experimental results for $B^0\overline{B}^0$ mixing suggest that $M_{U(1)}/\theta$ is greater than a few hundred TeV. This would be discouraging for the rare decays. On the other hand it may be that mass mixing between families occurs mostly in the charge $^2/_3$ sector, implying that the θ appearing in the rare t decay and in $T^0\overline{T}^0$ mixing is much larger than the θ appearing in $B^0\overline{B}^0$ mixing. This situation arises naturally in some models[3] and it may be required to adequately suppress $K^0\overline{K}^0$ mixing in TC theories.

A left-over U(1) may also have slight mixing with the Z. This would lead to flavor changing Z decays involving at least one heavy quark or lepton, eg.

$$Z \Rightarrow t + (\overline{c} \text{ or } \overline{u})$$
$$Z \Rightarrow \tau^+ + (\mu^- \text{ or } e^-).$$

Such mixing is proportional to isospin breaking in

some models.

Another type of left-over U(1) in ETC theories is a massless U(1) coupling only to technifermions. We emphasize that this is also a fairly natural consequence of the breakdown of a large flavor gauge symmetry down to the TC subgroup. Loop effects will induce magnetic moment couplings between this "paraphoton" and ordinary quarks and leptons of the form

$$a F^{\mu\nu} \bar{\psi} \sigma_{\mu\nu} \psi,$$

where

$$a \approx \Lambda_{TC}/\Lambda_{\psi}^2 \approx m_{\psi}/\Lambda_{TC}^2.$$

This produces a coupling which is effectively proportional to m_{ψ}^2. But in walking TC the last equality is altered to produce an effective coupling which grows with mass even faster, $m_{\psi}^{1+\xi}$. Flavor off-diagonal couplings are expected with associated mixing angle factors. It thus appears best to search for flavor changing decays of heavy quarks and leptons, eg.

$$t \Rightarrow u + \gamma^* \text{ and } \tau \Rightarrow e + \gamma^*,$$

where the paraphoton is invisible energy. (The signature for the latter is the 2-body phase space.) The similar decay mode for massive neutrinos may relax the cosmological constraints on neutrino mass.[4)]

A paraphoton is expected to mix slightly with the photon.[5)] This will be induced automatically via mass splittings among technifermions. The result would be a slight shift in the ordinary charge of all paracharged technifermions. The lightest paracharged particle is stable and could be expected to have ordinary charge in the range $10^{-2} < \epsilon < 10^{-3}$. The present limits on the direct production and detection of such a particle are rather weak.[6)]

1) B. Holdom, Phys. Lett. 150B, 301 (1985); Phys. Rev. Lett. 60, 1233 (1988).
2) B. Holdom, Phys. Rev. D24, 1441 (1981); Phys. Lett. 150B, 301 (1985); Phys. Lett. 198B, 535 (1987); T. Akiba and T. Yanagika, Phys. Lett. 169B, 432 (1986); K. Yamawaki, M. Bando and K. Matumoto, Phys. Rev. Lett. 56, 1335 (1986); M. Bando, T. Morozumi, H. So, K. Yamawaki, Phys. Rev. Lett. 59, 389 (1987); T. Appelquist, D. Karabali and L.C.R. Wijewardhana, Phys. Rev. Lett. 57, 957 (1986); T. Appelquist and L.C.R. Wijewardhana, Phys. Rev. D35, 774 (1987); Phys. Rev. D36, 568 (1987); T. Appelquist, D. Carrier, L.C.R. Wijewardhana, W. Zheng, Phys. Rev. Lett. 60, 1114 (1988); A. Cohen and H. Georgi, Harvard preprint, HUTP-88/A007, March 1988.
3) B. Holdom, Phys. Lett. B143, 227 (1984).
4) B. Holdom, Phys. Lett. B160, 303 (1985).
5) B. Holdom, Phys. Lett. B166, 196 (1986).
6) D. Bailey, private communication.

SEARCH STRATEGIES FOR HIGGS BOSONS AT HIGH ENERGY e^+e^- COLLIDERS*

J. ALEXANDER,[a] D. L. BURKE, C. K. JUNG, S. KOMAMIYA
Stanford Linear Accelerator Center, Stanford University, Stanford, California 94309

P. R. BURCHAT
Santa Cruz Institute for Particle Physics, University of California, Santa Cruz, California 95064

ABSTRACT

We have used detailed Monte Carlo simulations to study search strategies for Higgs bosons at high energy e^+e^- colliders. We extend an earlier study of the minimal single–Higgs-doublet model at a center-of-mass energy of 1 TeV to examine the effects of b–quark tagging and jet counting. It is found that these techniques can increase the signal-to-noise ratio substantially in the mass range around the W mass. In addition, we have studied this model at a center-of-mass energy of 400 GeV and found that an e^+e^- collider in this region would be sensitive to a Higgs boson with mass up to twice the Z° mass.

We have also considered a nonminimal two-doublet model for the Higgs sector by extending a study of charged Higgs boson searches to include a mass very close to the mass of the $W^\pm$. We demonstrate that techniques which include b–quark tagging can be utilized to extract a significant signal. In addition, we have examined the prospects for detecting nonminimal neutral Higgs bosons at 1 TeV. We conclude that it would be possible to detect the CP-even and CP-odd neutral Higgs bosons when they are pair-produced in e^+e^- annihilation over a limited mass range. However, in some scenarios of supersymmetry, the charged Higgs boson constitutes a significant background to the CP-odd and the more massive CP-even neutral Higgs boson.

1. INTRODUCTION

We continue two earlier studies of Higgs boson search strategies at a 1 TeV e^+e^- collider: one for the neutral boson H° which remains as a physical particle in the minimal Higgs theory,[1] and another for the charged bosons $H^\pm$ which appear in two-doublet versions of the Higgs sector.[2] These studies included the effects of detector resolution on the observed final states and of beamstrahlung on the spectrum of center-of-mass energies at which e^+e^- collisions take place. The main conclusions of the previous studies were the following:

— It is possible to extract a significant signal from all standard model backgrounds if the mass of the minimal Higgs boson is not near the mass of the $W^\pm$ or Z° and is less than 400 (500) GeV when a data sample of 10 (30) fb^{-1} has been collected at 1 TeV. The most important signal is due to the process $e^+e^- \to \nu_e \bar{\nu}_e H^\circ$. In the region around the mass of the $W^\pm$, the major background to this signal is due to the process $e^+e^- \to e^+ \nu_e W^-$.

— Production of a charged Higgs boson with mass less than about 80% of the beam energy is detectable with an integrated luminosity of about 10 fb^{-1} at 1 TeV through direct pair production and also from the decay of the top quark if the top quark is massive enough. In this study, charged Higgs bosons with mass less than about 100 GeV were not considered.

In this paper, we extend the above studies in the following ways:

— For the minimal Higgs boson at $\sqrt{s} = 1$ TeV, we consider b–quark tagging and jet counting as ways to enhance the signal at M_{H° near $M_{W^\pm}$.

— We study search strategies for the minimal Higgs boson at $\sqrt{s} = 400$ GeV.

— We study the detection of nonminimal neutral Higgs bosons through the pair production of a CP-even and CP-odd neutral Higgs boson at $\sqrt{s} = 1$ TeV.

— For the charged Higgs boson, we consider b–quark tagging, as well as other techniques, to enhance the signal around the mass of the $W^\pm$.

2. SIMULATION OF EVENTS AND DETECTOR RESPONSE

We have used standard Monte Carlo techniques to generate both signal and background events, and to simulate the response of the detector. Detailed accounts of these simulations are given in Refs. (1) and (2), but we briefly review them here.

2.1 Beamstrahlung

The extremely high density to which beam particles must be focused to produce sufficient luminosity to do particle physics at a high-energy e^+e^- collider results in a significant interaction rate between the particles of one

*Work supported by Department of Energy, contracts DE-AC03–76SF00515 and DE-AM03–76SF00010.

[a] Present address: Cornell University, Ithaca, New York 14853.

beam and the collective electromagnetic fields produced by the particles in the opposite beam. A major consequence of this is that the spectrum of center-of-mass energies at which e^+e^- interactions take place is not at all monochromatic. Radiation of photons during the beam-beam collision ("beamstrahlung") will result in a spectrum of particle interactions that depends in detail on the energy and bunch characteristics of each beam. We have used a spectrum that has been calculated[3] for a specific set of beam parameters, but that is typical of most machine designs that have been studied.[4] This calculation includes the effects of multiple radiation and beam-beam disruption for an e^+e^- machine operating at a nominal center-of-mass energy of 1 TeV and a luminosity of $3 \times 10^{33}\ \mathrm{cm^{-2}s^{-1}}$. Approximately 30% of the luminosity lies within 2% of the nominal center-of-mass energy squared for the spectrum that we have chosen.

Included in the Monte Carlo simulations are both the effective reduction in center-of-mass energy and the net longitudinal momentum created by the energy loss of each beam. The mean fractional energy loss that particles undergo during the beam-beam collision is 0.26 for the spectrum that we have used. We compute production cross sections by convoluting the luminosity spectrum with each cross section evaluated at the reduced center-of-mass energy of individual e^+e^- pairs.

2.2 Event Simulation

With the exception of the Higgs particles that we are studying, we have assumed that the standard model with three generations of quarks and leptons is a correct description of nature. The LUND 6.3 model[5] with full showering of quark and gluon partons has been used to generate QCD background events and to fragment into hadrons final-state quarks and gluons in all signal and background processes. The mass of the top quark was assumed to be 40 GeV, unless otherwise specified, and the masses of the $W^\pm$ and Z° gauge bosons have been taken to be 83 GeV and 93 GeV, respectively. The background process $e^+e^- \to e^+\nu_e W^-$ was generated with the formulae of Gabrielli.[6]

2.3 Detector Simulation

The properties of a detector that are necessary to successfully carry out a search for a Higgs boson at a high energy e^+e^- collider are easily achievable with current technology. The emphasis in our analysis is on calorimetry and we generally ignore tracking information for hadrons.

To simulate finite detector resolutions and segmentation we modify the generated particle momenta for signal and background processes in the following manner. We assume the azimuthal and polar angles of muons and electrons are well determined by a tracking chamber. The energy of each muon is smeared by a Gaussian distribution with standard deviation

$$\frac{\sigma_E}{E} = 3 \times 10^{-4}\, E\ \mathrm{GeV}^{-1} \quad .$$

This resolution should be achievable, for example, with a drift chamber with radius 1.8 m, 72 position measurements each with 200 μm resolution, and embedded in a solenoidal magnetic field of 1.0 Tesla. For electrons, we calculate the above smearing for low energies and the calorimetric formula

$$\frac{\sigma_E}{E} = \frac{8\%}{\sqrt{E}}\ \mathrm{GeV}^{1/2} + 2\%$$

for high energies, and apply the smearing formula which gives the best resolution for a particular energy. We assume that the charges of electrons and muons can be assigned unambiguously.

Photons and charged and neutral hadrons are treated indiscriminately as clusters of calorimetric energy. Tracks within 4° of one another are combined; then the direction of the combined track is smeared by a box function of size ± 2°. The energies of these tracks are smeared with

$$\frac{\sigma_E}{E} = \frac{50\%}{\sqrt{E}}\ \mathrm{GeV}^{1/2} + 2\% \quad .$$

Finally, to account for possible obstruction of the acceptance of the detector near the beam line by machine components and the hardware needed to support them, we simply ignore particles within 10° of the beam axis. We assume, however, that the remainder of the detector is completely hermetic, and that the hardware performs well enough to avoid any further loss of sensitivity to the presence of particles in signal or background events.

2.4 Impact Parameter Tagging

Because Higgs bosons are expected to decay preferentially through b–quarks, the detection of the presence of b–quarks in an event is an important tag which can be used to isolate the production of Higgs particles. The relatively long lifetime of the b–quark gives rise to decay vertices displaced from the primary e^+e^- vertex or, equivalently, to tracks with large impact parameters. At a TeV linear collider, where the uncertainty in the position of the primary vertex is expected to be much smaller than the mean decay length of the b–quark, especially in the plane perpendicular to the beam direction,[7] precision vertex detectors may be used to identify such large-impact-parameter tracks which can then be exploited to enhance the signal-to-background ratio in searches for Higgs bosons.

We investigate this possibility by simulating the performance of a combined vertex detector and central tracking chamber in our Monte Carlo and using the impact parameter information in the analysis. The simulation is kept simple and generic in order that the success of the analysis not be dependent on specific details of detector design. Each track in an event is projected into the plane transverse to the beam line and the impact parameter is defined as the distance of closest approach of the track projection to the primary vertex. We compute it for each track using the generated four-vectors, and smear these calculated values with an impact parameter resolution function chosen to be $\sigma = \sqrt{A^2 + (B/p)^2}$. The details of the vertex and tracking system are encompassed in the parameters A and B, which reflect in turn the intrinsic resolution of the tracking system and the effect of multiple scattering of particles as they traverse the beam pipe and detector. We choose values of A and B that are characteristic of combined vertex and tracking systems currently in use, namely $A = 5\,\mu$m and $B = 50\,\mu$m GeV. We assume that the uncertainty in the location of the collision point is small compared to the impact parameter resolution. For the analysis in this paper, a charged track is defined to be a 'high impact parameter track' if it lies within $|\cos\theta| \leq 0.9$ and has a (smeared) impact parameter less than 3 mm and greater than three times the resolution $\sigma = \sqrt{A^2 + (B/p)^2}$. The 3 mm upper limit suppresses contamination from K_S and Λ decays.

3. MINIMAL HIGGS BOSON AT $\sqrt{s} = 1$ TeV

In Ref. (1), two production mechanisms were considered for the standard model Higgs boson: $e^+e^- \to Z^\circ H^\circ$ and $e^+e^- \to \nu_e \bar{\nu}_e H^\circ$. It was found that the process $e^+e^- \to \nu_e \bar{\nu}_e H^\circ$ gave the strongest signal at $\sqrt{s} = 1$ TeV. The signal is well above standard model backgrounds* everywhere except around the mass of the $W^\pm$ (50–150 GeV). In this region, the signal is dominated by the $W^\pm$ mass peak from the process $e^+e^- \to e^+ \nu_e W^-$ when no attempt is made to enhance the signal through b-quark tagging or jet counting. We will now consider the effect of b-quark tagging and jet counting for the intermediate mass Higgs ($M_H < 2M_W$) at $\sqrt{s} = 1$ TeV.

First, we will review the selection criteria discussed in detail in Ref. (1). We assume that a Higgs boson with mass less than twice M_W will decay to either a top or bottom quark pair depending on the masses of the Higgs boson and the top quark. The philosophy of our analysis of the $e^+e^- \to \nu_e \bar{\nu}_e H^\circ$ reaction with $M_H < 2M_W$ is to search for events that contain two acoplanar jets each with mass below that of the W. We select events in which $|\cos\theta_{thr}| < 0.7$ and $|\cos\theta_{miss}| < 0.9$ where θ_{thr} and θ_{miss} are the polar angles of the thrust axis and the missing momentum in the event, respectively. The first cut ensures that the event is well contained in the detector and preferentially rejects processes with differential cross sections which are sharply peaked along the beam axis (for example, $e^+e^- \to W^+W^-$, $e^+e^- \to Z^\circ Z^\circ$, and $e^+e^- \to q\bar{q}$). The second cut reduces backgrounds due to events with large beamstrahlung radiation and/or missing particles. We reject events if the visible energy is less than 100 GeV or greater than 400 GeV. This cut removes backgrounds but is extremely efficient for retaining the signal events.

We divide the event into two clusters using a clustering algorithm and select events in which the acoplanarity of the clusters is greater than 10° and in which each cluster has an invariant mass greater than 1 GeV (to reject leptonic decays of the $W^\pm$) and less than 50 GeV. Finally, events are accepted only if the missing momentum transverse to the beam is greater than 50 GeV and the number of charged particles outside the 10° hole around the beam axis is between 10 and 36. The probability that an event of the type $e^+e^- \to \nu_e \bar{\nu}_e H^\circ$ will pass the above selection criteria varies between about 35% and 50% depending on the mass of the Higgs boson and the decay mode of the Higgs boson. The background is dominated by the process $e^+e^- \to e^+ \nu_e W^-$. When the invariant mass of the event is plotted in 4 GeV bins the peak bin for the background distribution contains about 25 to 50 times as many events as the peak bin for the signal. It should be possible to discover a Higgs boson with mass greater than about 150 GeV using just the above selection criteria[1] but the observability of Higgs bosons with masses closer to M_W depends critically on the resolution of the detector.

To further enhance the signal-to-background ratio, we count the number of high impact parameter (h.i.p.) tracks in each event. The definition of a h.i.p. track was given in Sec. 2.4. Events are accepted if the number of h.i.p. tracks is at least four. We first consider the case in which the H^0 decays to $b\bar{b}$. The efficiency for signal events to pass the cut on h.i.p. tracks, after all other cuts, is about 92% while for the background it is about 30%. The invariant mass distributions for a 120 GeV Higgs boson and a 150 GeV Higgs boson are shown in Figs. 1(a) and (b), respectively. The figures include standard model backgrounds. The 150 GeV Higgs boson signal is clearly separated from the large W peak due to the $e^+e^- \to e^+ \nu_e W^-$ background. However, the 120 GeV Higgs boson signal is just a shoulder on the side of the $W^\pm$ mass peak. The 120 GeV signal stands out clearly, though, if the resolution of the hadronic calorimeter is:

$$\frac{\sigma_E}{E} = \frac{35\%}{\sqrt{E}} \mathrm{GeV}^{1/2} + 2\% \quad ,$$

instead of

$$\frac{\sigma_E}{E} = \frac{50\%}{\sqrt{E}} \mathrm{GeV}^{1/2} + 2\% \quad .$$

After making the cut on h.i.p. tracks, we are left with a background which is dominated by events in which the $W^\pm$ decays to $t\bar{b}$. It should be noted that if the top quark mass is greater than the $W^\pm$ mass, these events will be absent, significantly reducing the background. On the other hand, if the top quark mass is 40 GeV, the signal-to-noise ratio is improved by about 60% by reducing the allowed jet mass range from (1–50) GeV to (1–25) GeV. However, the number of signal events which pass all of the selection criteria is reduced by about 60%.

For Fig. 1, we assumed that the Higgs boson decays to a bottom quark pair. If we assume that the Higgs boson decays to a top quark pair, we can further increase the signal-to-background ratio by accepting events only if they have at least four clusters found by an algorithm which groups particles until further additions to the group would result in a cluster of particles with invariant mass much greater than the bottom quark mass and less than the top quark mass. The technique does not depend critically on the exact cutoff mass. In Fig. 2, we plot the invariant mass of events with at least four high-impact-parameter tracks and at least four clusters found by the clustering algorithm. The distribution includes a 150 GeV Higgs boson and backgrounds. The W peak from the $e^+e^- \to e^+ \nu_e W^-$ background has been further reduced so that now the peak signal is over half of the peak background.

We conclude that impact parameter tagging is a useful tool for enhancing the signal of an intermediate mass Higgs boson relative to standard model backgrounds. In addition, if the Higgs boson decays to $t\bar{t}$, the number of low mass clusters provides a useful discriminator between signal and background.

*In a CERN study[8] of detection of neutral heavy Higgs bosons at CLIC ($\sqrt{s} = 2$ TeV), it was found that a significant background to a heavy Higgs boson signal ($M_H \gtrsim 500$ GeV) is due to the process $e^+e^- \to e^- \bar{\nu}_e Z^\circ W^+$. This background was not considered in Ref. 1; however, this background is not as significant for $M_H \lesssim 500$ GeV (the mass range considered in Ref. 1) because the width of the Higgs scales as M_H^3, leading to a considerably narrower signal at smaller masses. We estimate the contribution from this background[9] at 10% of the signal for $M_H = 300$ GeV and 20% for $M_H = 500$ GeV in the Weizsacker–Williams approximation with beamstrahlung neglected. This increases to $\sim 40\%$ when the beamstrahlung spectrum described in Sec. 2.1 is included, due to γW interactions involving the beamstrahlung photon.

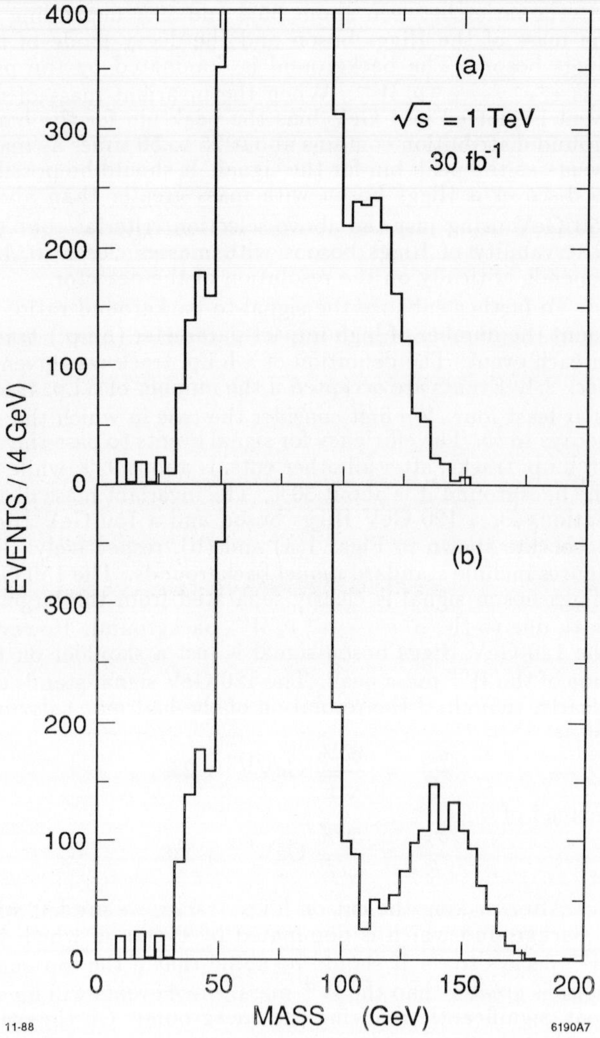

Fig. 1. Reconstructed mass distribution, after selection criteria, for background events plus (a) a 120 GeV Higgs boson and (b) a 150 GeV Higgs boson. The Higgs boson is produced in the process $e^+e^- \rightarrow \nu_e \bar{\nu}_e H^\circ$ at $\sqrt{s} = 1$ TeV and decays to $b\bar{b}$. The selection criteria include a cut on the number of high-impact-parameter tracks. The number of events in the peak bin near the $W^\pm$ mass is about 3000.

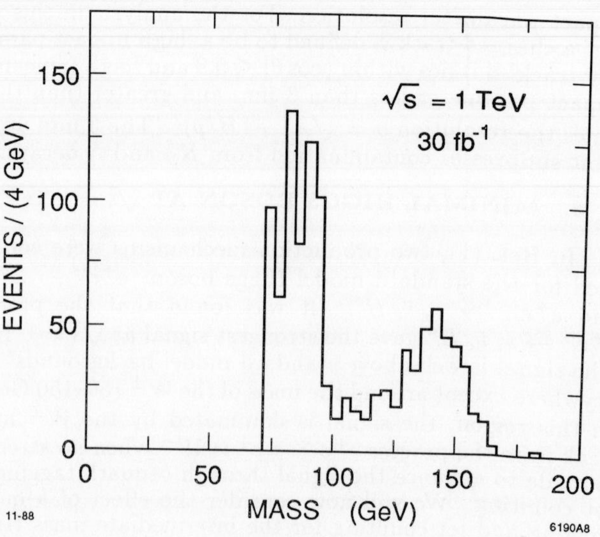

Fig. 2. Reconstructed mass distribution, after selection criteria, for background events plus a 150 GeV Higgs boson. The Higgs boson is produced in the process $e^+e^- \rightarrow \nu_e \bar{\nu}_e H^\circ$ at $\sqrt{s} = 1$ TeV and decays to $t\bar{t}$. The selection criteria include a cut on the number of high-impact-parameter tracks and on the number of clusters found.

4. MINIMAL HIGGS BOSON AT $\sqrt{s} = 400$ GeV

If the mass of the Higgs boson is less than ~ 200 GeV (for example, $M_{H^\circ} < 2M_{Z^\circ}$), the Higgs boson could be discovered and studied at an e^+e^- collider with a center-of-mass energy considerably less than 1 TeV. It is important[10] to gain access to this mass range since it may not be possible to find such particles at any other machine that is currently being contemplated by the particle physics community and because of the difficulty with backgrounds at larger center-of-mass energy discussed in the previous section. We have briefly examined the situation at an e^+e^- collider with $\sqrt{s} = 400$ GeV. The designs of machines[11] in this energy range incorporate very little beamstrahlung (the mean energy loss per beam is typically less than a few percent), so analyses that apply energy and momentum constraints are quite efficient.

The cross sections for the processes $e^+e^- \rightarrow Z^\circ H^\circ$ and $e^+e^- \rightarrow \nu_e \bar{\nu}_e H^\circ$ are shown in Fig. 3 for the case $M_{H^\circ} = 100$ GeV. The process $e^+e^- \rightarrow Z^\circ H^\circ$ has previously been studied[12] in detail at $\sqrt{s} = 200$ GeV in preparation for the LEP II physics program. It has been shown that Higgs particles with masses up to 80 GeV can be found rather easily at that machine but that it will be difficult to reach higher masses. The cross section for $e^+e^- \rightarrow Z^\circ H^\circ$ decreases as $\sqrt{s}$ becomes large compared to the masses of the final state particles. However, at $\sqrt{s} = 400$ GeV it is still greater than the fusion reaction $e^+e^- \rightarrow \nu_e \bar{\nu}_e H^\circ$ for $M_{H^\circ} = 100$ GeV. There would be approximately 1000 $e^+e^- \rightarrow Z^\circ H^\circ$ events produced in a data sample of 10 fb^{-1} at this Higgs boson mass. For the machine under discussion, this reaction is clearly worthy of further study.

The decay signatures for $e^+e^- \rightarrow Z^\circ H^\circ$ are summarized in Fig. 4. The analyses of these final states at $\sqrt{s} = 400$ GeV will be quite similar to those developed for LEP II. The process shown in Fig. 4(a) can be isolated from $q\bar{q}$, W^+W^-, and $Z^\circ Z^\circ$ backgrounds at all masses except $M_{H^\circ} \approx M_{Z^\circ}$ by taking only high-multiplicity events with large missing transverse momentum. To search for the possibility that the Higgs boson mass is close to that of the Z° mass, it is necessary to reduce the $Z^\circ Z^\circ$ background in the sample. This can be done by directly tagging the flavor of the final decay products as discussed in Sec. 2 or by removing events with small sphericity in the center-of-mass system of the observed final state particles as discussed in Ref. (12) Events in which the Z° decays to lepton pairs (e^+e^- or $\mu^+\mu^-$) as shown in Fig. 4(b) can be effectively reduced to the same problem once the $Z^\circ \rightarrow l^+l^-$ decay has been identified. In this case, the use of the measured

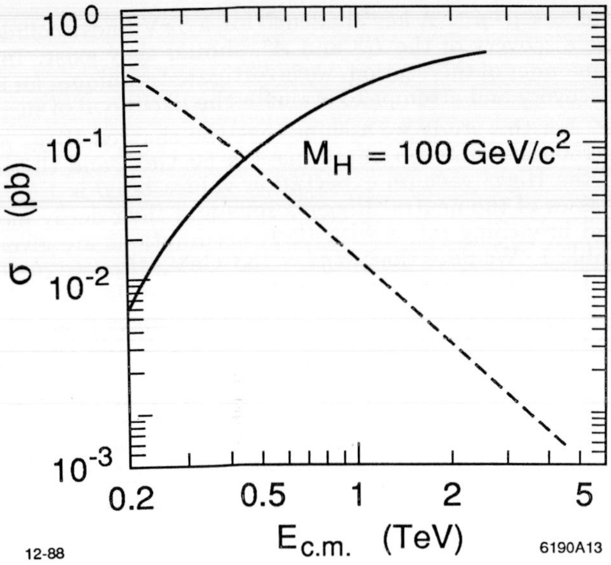

Fig. 3. Cross sections for the processes $e^+e^- \to Z^\circ H^\circ$ (dashed curve) and $e^+e^- \to \nu_e \bar{\nu}_e H^\circ$ (solid curve) for a 100 GeV Higgs boson.

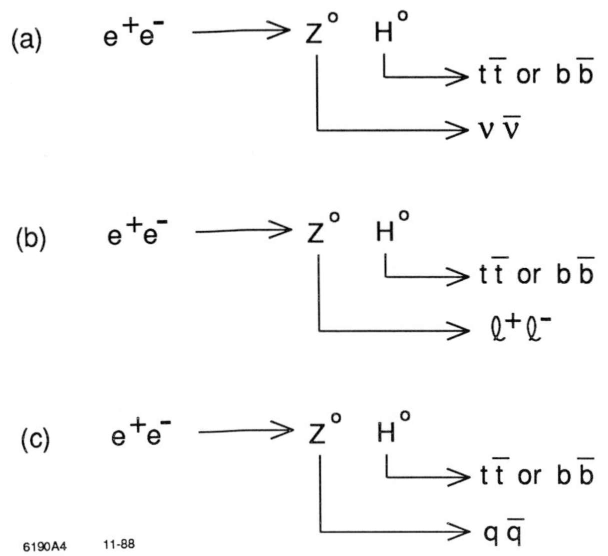

Fig. 4. The decay signatures for the process $e^+e^- \to Z^\circ H^\circ$.

lepton momenta to reconstruct the Higgs signal as a recoil mass is a useful tool that is available at lower energy e^+e^- machines. We have not done detailed studies of these final states at $\sqrt{s} = 400$ GeV, but Higgs particles with masses up to 180–200 GeV should be visible in these channels.[13)]

The hadronic final states shown in Fig. 4(c) are the most numerous, but require the most sophisticated analyses to find and study. The strategy is to reconstruct the individual quark jets. It has been shown[12)] that at LEP II the case $H^0 \to b\bar{b}$ results in distinctive four-jet events that can be kinematically fitted to yield clear signals for M_{H° up to 70 GeV. We consider here the case in which a heavier

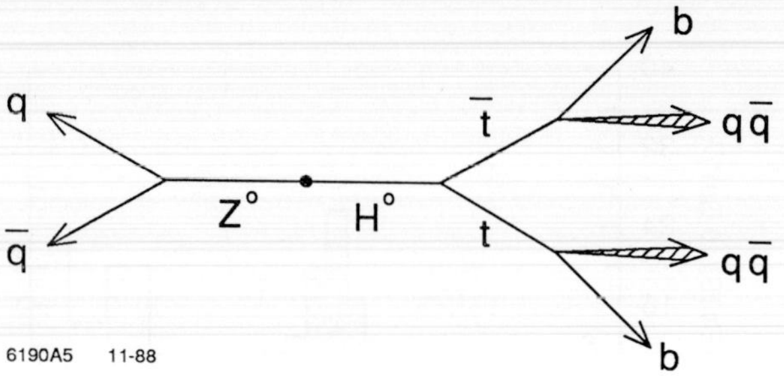

Fig. 5. The jet structure for $e^+e^- \to Z^\circ H^\circ$ when the Z° decays hadronically and the Higgs boson decays to a top quark pair.

Higgs decays to $t\bar{t}$ pairs. For this exercise we assume that the top quark mass is 40 GeV and the parton structure of the final state is as shown in Fig. 5. The signature of this final state is a multijet event tagged by the pair of jets that originate from the decay of the Z^0. Once this pair has been isolated, the remainder of the event (in its own center-of-mass) is a highly spherical decay of two massive quarks. The details of the event-selection procedure are as follows:

(1) Reconstruct the visible mass and thrust axis of each event. Demand greater than 300 GeV of visible mass and $|\cos\theta_{thr}| < 0.8$.

(2) Reconstruct the jet topology of the event. We have used the LUND cluster algorithm modified to find jets with masses only up to 20 GeV. This means that top quarks are split into their decay partons. Require that there be five or more jets in candidate $Z^\circ H^\circ$ events. This technique strongly rejects $Z^\circ Z^\circ$ and W^+W^- events.

(3) Search for jet pairs with invariant masses within 15 GeV of the Z° mass. If more than one pairing results in a Z° candidate, then identify the one with the invariant mass nearest to 93 GeV as the Z°. If no candidate Z° decays are found, then reject the event.

(4) Boost to the center-of-mass of the system that recoils against the Z°, find the thrust axis in this system, and reconstruct the invariant masses of the observed particles in the hemispheres defined by the plane perpendicular to this thrust axis. Demand that both of these hemisphere masses lie between 20 and 50 GeV as would be expected for decays of 40 GeV top quarks.

(5) Fit the measured jet four-vectors in each event to the hypothesis that energy and momentum are conserved and that the Z° mass is 93 GeV. We have assumed that there is no smearing of the overall center-of-mass energy and momentum due to beamstrahlung, but have included lowest order initial state radiation. Reject events if they yield a poor fit ($\chi^2/DOF > 1$).

The result of this analysis is shown in Fig. 6. Backgrounds from all annihilation and two-photon processes are summed and shown as the open region. The signal is shown as the shaded region. This also includes $Z^\circ H^\circ$ events in which the wrong pair of jets is identified as the Z°. The solid histogram gives the signal that would be seen for $M_{H^\circ} = 150$ GeV. An integrated luminosity of 10 fb^{-1} would result in a signal of 43 events over a background of 35 events.

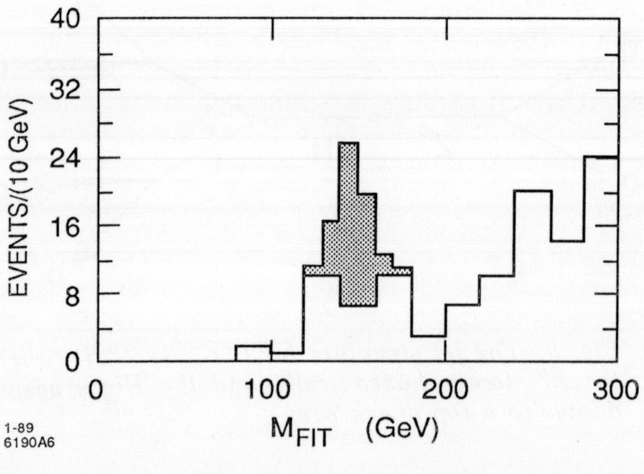

Fig. 6. Invariant mass distribution, after selection criteria, for a 150 GeV Higgs boson produced in the process $e^+e^- \to Z^\circ H^\circ$ at $\sqrt{s} = 400$ GeV. Backgrounds from annihilation and two-photon processes are shown as the open region; the signal is shown as the shaded region.

We conclude that Higgs boson searches at an e^+e^- collider with $\sqrt{s} = 400$ GeV will be successful in the $Z^\circ H^\circ$ channel for M_{H° at least as large as twice the Z° mass. The mass interval near M_{Z° will require special attention, but the $Z^\circ Z^\circ$ cross section is only ≈ 5 times the $Z^\circ H^\circ$ cross section (when $M_{H^\circ} = M_{Z^\circ}$) and the techniques of jet counting and flavor tagging discussed earlier in this paper effectively isolate the heavy quark decays of the Higgs boson. An integrated luminosity of 10 fb^{-1} will suffice to find and conduct reasonable studies of the minimal H°. This sample could be accumulated over a 3–4 year period of operation at a machine with an instantaneous luminosity of $\approx 3 \times 10^{32} \text{cm}^{-2}\text{s}^{-1}$.

5. NONMINIMAL NEUTRAL HIGGS BOSONS AT $\sqrt{s} = 1$ TeV

In the simplest extension of the standard model, two Higgs doublets yield five physical Higgs particles, two charged and three neutral. The neutral Higgs particles consist of two CP-even states (h° and H°) and a CP-odd state (A^o). The masses of the charged and three neutral Higgs bosons, and two mixing angles are free parameters in the most general case. Minimal supersymmetry can be used[14] to impose constraints, reducing the free parameters from six to two. These may be chosen to be the charged Higgs boson mass, $M_{H^\pm}$, and the ratio of the vacuum expectation values of the two Higgs doublets, expressed as $\tan\beta$. Once $M_{H^\pm}$ and $\tan\beta$ are specified, production cross sections, decay branching ratios, and masses of the neutral Higgs particles can be calculated. The results of Ref. (14) may be summarized briefly. The light scalar h° behaves much like the standard model Higgs boson and may be produced by the familiar mechanism $e^+e^- \to Z^* \to Zh^\circ$ with some possibility for discovery at SLC or LEP. By contrast, these production mechanisms are forbidden for the A^o and suppressed for the H°. The only production mechanism available to these heavy Higgs particles that has adequate cross section is $e^+e^- \to Z^* \to H^\circ A^o$. At 1 TeV, the largest predicted cross section is approximately 10 fb,f or about one-tenth the lowest order QED cross section for $e^+e^- \to \mu^+\mu^-$. A key challenge for a TeV linear collider is the discovery of the H° and A^o, should they exist. In the remainder of this section, we investigate techniques for such discovery and attempt to evaluate the likelihood of success.

For this study we assume that the charged Higgs mass is 150 GeV, the top quark mass is 60 GeV, and the ratio of the Higgs vacuum expectation values $\tan\beta$ is 1.5. The masses of the neutral Higgs bosons and their decay modes and branching ratios with these assumptions are given in Table 1. We note that $M_{H^\circ} = 153$ GeV, $M_{A^\circ} = 126$ GeV, and the branching ratios are dominated by $t\bar{t}$ and $b\bar{b}$. The choice of 150 GeV for the charged Higgs mass is motivated by two considerations. If $M_{H^\pm} \leq 135$ GeV then $M_{A^\circ} \leq 100$ GeV and there is little hope of seeing the signal over the W and Z background. As $M_{H^\pm}$ becomes larger than 150 GeV, the signal is further from the W and Z background, and becomes only easier to see. Moreover, for large $M_{H^\pm}$, $M_{A^\circ} = M_{H^\circ}$, giving one additional handles on the signal that may be exploited. The choice of $M_{H^\pm} = 150$ GeV was therefore felt to be the most challenging case with any prospect of success.

Table 1. Parameters used to generate $e^+e^- \to H^\circ A^o$.

$M_{H^\pm} = 150$ GeV	$BR(H^\circ \to t\bar{t}) = 79.6\%$
$M_{H^\circ} = 153.4$	$BR(H^\circ \to b\bar{b}) = 6.5\%$
$M_{A^\circ} = 125.7$	$BR(H^\circ \to h^\circ h^\circ) = 13.6\%$
$M_{h^\circ} = 29.2$	$BR(A^o \to t\bar{t}) = 95\%$
$M_t = 60$	$BR(A^o \to b\bar{b}) = 5\%$
$\tan\beta = 1.5$	$BR(h^\circ \to b\bar{b}) = 100\%$

Events of the type $e^+e^- \to H^\circ A^o$ are characterized by two jets of unknown, and not necessarily equal, mass. To maintain the highest level of generality, the analysis exploits only two features of the events: the fact that the jets should be distributed as $\sin^2\theta$, and the fact that many b–quarks will be produced, leading to events with high multiplicity and tracks with large impact parameters. The dominant standard model backgrounds are $e^+e^- \to q\bar{q}$, $e^+e^- \to W^+W^-$ and $e^+e^- \to Z^\circ Z^\circ$. In contrast to $H^\circ A^o$ events, these backgrounds are characterized by angular distributions peaked in the forward and backward direction, by lower multiplicity, and by few tracks with large impact parameters. The cuts on multiplicity and impact parameter prove to be the most powerful in enhancing the signal-to-background ratio.

In a run of 10 fb^{-1} integrated luminosity, one expects to produce approximately 112 $H^\circ A^o$ events, 40000 $q\bar{q}$ events, 35000 W^+W^- events, and 2000 $Z^\circ Z^\circ$ events. To simulate such a run, we generate the correct numbers of background events, and 1120 $H^\circ A^o$ events which are then weighted by 0.1. We apply the following cuts:

(1) The total number of detected charged tracks in the detector volume ($|\cos\theta| \leq 0.9$) is at least 40. This favors events with b–quarks which tend to have high multiplicities.

(2) The number of detected charged tracks in the central region of the detector ($|\cos\theta| \leq 0.71$) is at least 20. This cut favors events with $\sin^2\theta$ distribution.

(3) The total number of high impact parameter tracks (as defined previously) is at least 16. This strongly favors events with b-quarks.

(4) When the event is divided into two jets by a cluster finding algorithm,[15] then the cluster containing fewer high impact parameter tracks must still contain at least seven such tracks. This tends to reject background events in which one jet has a large number of high impact parameter tracks due to a statistical fluctuation.

(5) The invariant mass of each jet is formed and the event is rejected if *both* jets have a mass in the range 63 GeV to 112 GeV. This helps reject remaining W^+W^- and $Z^\circ Z^\circ$ events. Naturally it also excludes any possibility of finding $H^\circ A^o$ events with masses in this range, but in doing so enhances the prospects for finding such events with masses *near* the W and Z°.

The results of this analysis are not sensitive to the exact number of detected charged tracks required in the first two cuts. The number required would have to be tuned for a particular detector configuration.

Table 2. Selection criteria and number of events which satisfy each criteria.

Cut	$H^\circ A^o$	$q\bar{q}$	W^+W^-	Z°
Generated sample	1120	40000	35000	2000
Total charged multiplicity ≥ 40	774	5934	995	53
Central multiplicity ≥ 20	753	5505	853	45
# High impact parameter tracks ≥ 16	426	128	65	6
# H.i.p. tracks on side with fewer ≥ 7	305	71	32	2
Not both in 63 GeV$< M <$ 112 GeV	302	66	14	1

Table 2 shows the impact of each of these cuts on the signal and background events. The net efficiency for the $H^\circ A^o$ signal is 27% and the rejection power against standard model backgrounds is approximately 1000:1. For a 10 fb^{-1} run we expect 30 signal and 81 background events, giving an overall signal-to-background ratio of about 1:3. A nonstandard-model background which we do not address here is the process $e^+e^- \to \gamma^*, Z^* \to H^+H^-$ which is expected to have a cross section approximately three times larger than the $H^\circ A^o$ cross section. Such events will pass our selection criteria. In the minimal supersymmetric model, it is predicted that $M_{H^\pm} \sim M_{H^\circ} \sim M_{A^\circ}$ as the mass of the charged Higgs particle becomes heavier ($M_{H^\pm} \gtrsim 150$ GeV). For such a case, it would be very difficult to distinguish the Higgs particles from one another unless some new characteristic parametrization other than the invariant mass is found. Outside the context of supersymmetry, however, no relation among the Higgs masses is predicted and the mass peaks need not be close to one another at all.

The invariant masses of the clusters are calculated with a beam energy constraint included. The beam-constrained mass calculation balances the momentum of the two clusters and the (unobserved) beamstrahlung photon in both the beam direction and the plane transverse to the beam direction, and requires in addition that the three energies sum to 1 TeV. By making the approximation that all beamstrahlung and initial state radiation is carried by a single photon, we can write the momentum and energy conservation conditions as three equations in three variables (the two cluster energies and the beamstrahlung photon energy). The invariant masses of the clusters are then computed using the beam-constrained cluster energies. Figure 7 shows the configuration of the two-jet momenta and the beamstrahlung momentum. In general, the three vectors do not lie in a plane as shown in Fig. 7, but a (small) Lorentz boost transverse to the beam direction is applied to bring them into this configuration. We then compute the velocity vector for each jet by summing over particle momenta and energies: $\beta = \Sigma p_i / \Sigma E_i$. In the coordinate system shown in Fig. 7, the momentum and energy constraint equations become

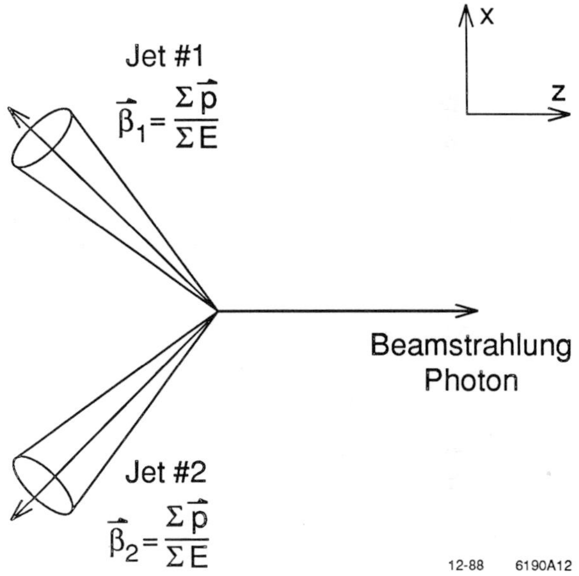

Fig. 7. Momentum vectors for two jets and a beamstrahlung photon. The z-axis is along the beam direction.

$$\beta_{1x}E_1 + \beta_{2x}E_2 = 0 \quad ,$$

$$\beta_{1z}E_1 + \beta_{2z}E_2 + P_\gamma = 0 \quad , \quad \text{and}$$

$$E_1 + E_2 + E_\gamma = \sqrt{s} \quad .$$

In the approximation that only one photon carries away momentum and energy, we have $E_\gamma = |P_\gamma|$. Solving for E_1, E_2, and E_γ, we then derive the jet masses using $M_i = E_i\sqrt{1-\beta_i^2}$. The beamstrahlung photon energy E_γ is a by-product of little direct use, but it is checked against the generated energy to confirm the correctness of the method. Use of the algorithm sharpens the mass distributions and yields mass values quite close to the generated values. It proves to be essential for identification of mass peaks above background. The method is generally useful for analysis of two-jet events with missing momentum and energy along the beam axis.

Figures 8 (a) and (b) show the mass distributions for the higher mass cluster and the lower mass cluster, just prior to applying the final analysis cut (exclusion of events with both masses in the range 63–112 GeV). A substantial background of W and Z events is evident. Figures 9(a) and (b) show the same plots after all analysis cuts are applied. In Fig. 9(a) the higher mass cluster stands out clearly at the correct (*i.e.* generated) mass of $\sim$ 150 GeV while in Fig. 9(b) the lower mass cluster (expected at $\sim$ 125 GeV) is obscured by the presence of $W^{\pm}$ and Z^{o} backgrounds. The lower mass cluster may be enhanced by applying a cut on the higher mass cluster. Figure 10(a) shows that when the higher mass is required to lie inside the range 120–200 GeV, a choice motivated by Fig. 9(a), the lower mass cluster stands out more clearly. By contrast, Fig. 10(b) shows the lower mass cluster distribution when the higher mass is required to lie *outside* the range 120–200 GeV; in this case there is no evidence for a mass peak.

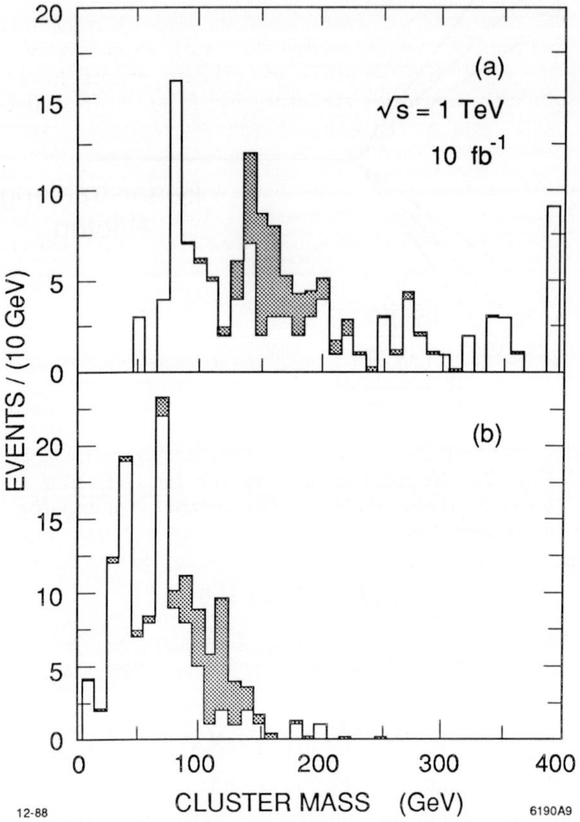

Fig. 8. Reconstructed mass distribution for (a) the larger mass cluster and (b) the smaller mass cluster when the final analysis cut (exclusion of events with both masses in the range 63–112 GeV*) is not applied. The shaded region is signal; the open region is background.*

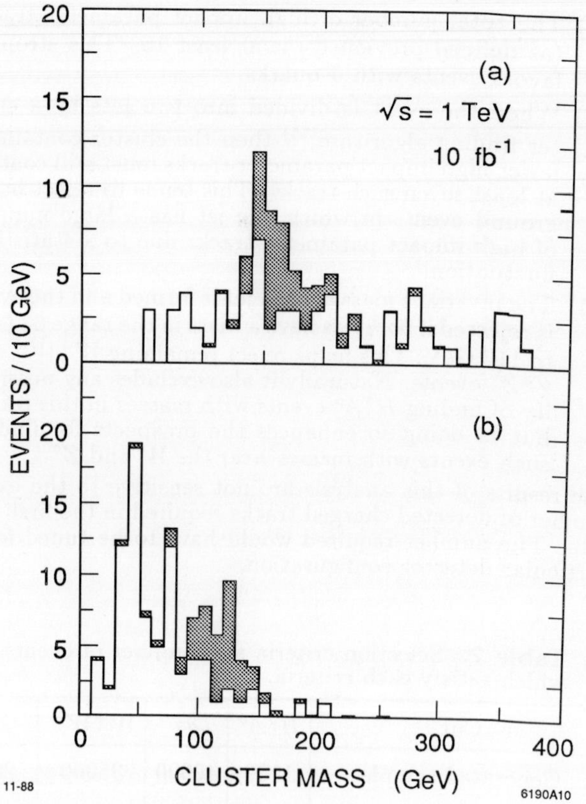

Fig. 9. Reconstructed mass distribution for (a) the larger mass cluster and (b) the smaller mass cluster when all analysis cuts are applied. The shaded region is signal; the open region is background.

We conclude that with modest impact parameter resolution and simple cuts designed to enhance events with b–quarks and central distributions, a sample of 10 fb^{-1} integrated luminosity is sufficient to identify the presence of $H^{o}A^{o}$ events with Higgs boson masses larger than $\sim$ 110 GeV and with decay modes dominated by heavy quarks.

6. CHARGED HIGGS BOSON AT $\sqrt{s} = 1$ TeV

6.1 Introduction

Charged Higgs bosons can be produced in pairs in e^+e^- annihilation via a virtual photon, or via a real or a virtual Z boson. The cross section normalized to the lowest order QED cross section for muon pairs, $R(H^+H^-)$, does not depend strongly on the models except near the Z peak since the photon couples to all charged particles with the same strength. In a previous study,[2)] it was demonstrated that searches for charged Higgs bosons at e^+e^- colliders with $\sqrt{s} \approx 1/2 - 1$ TeV are relatively easy. The main conclusions from the earlier study are:

(1) With an e^+e^- linear collider of $\sqrt{s} \approx 1/2 - 1$ TeV and an integrated luminosity of about 10^{40} cm^{-2}, we can detect production of charged Higgs bosons and determine the mass for $H^{\pm}$ bosons with masses less than 80% of the beam energy and a dominant decay mode $H^+ \to t\bar{b}$. we can detect production of charged Higgs bosons and determine the mass for $H^{\pm}$ bosons with masses less than 80% of the beam energy and a dominant decay mode $H^+ \to t\bar{b}$.

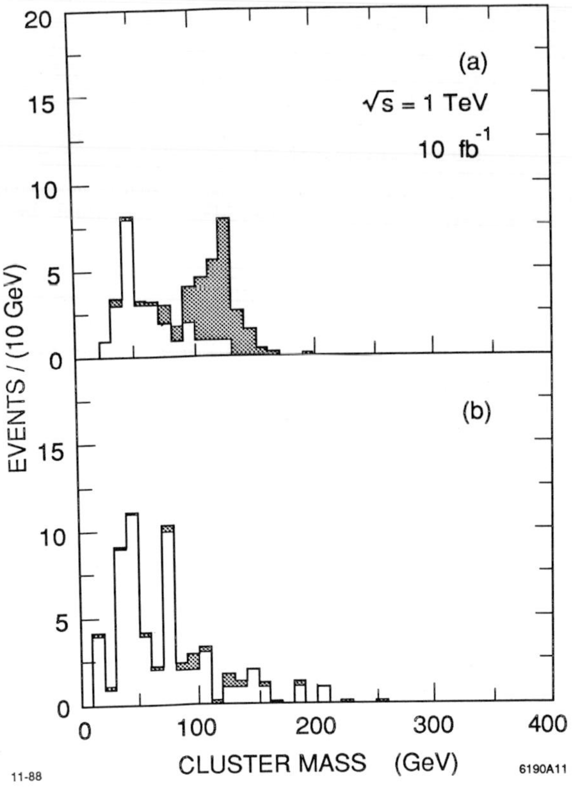

Fig. 10. Reconstructed mass distribution of the smaller mass cluster for events in which (a) the larger mass lies inside the range 120–200 GeV and (b) the larger mass lies outside the range 120–200 GeV. The shaded region is signal; the open region is background.

(2) If the charged Higgs boson is sufficiently lighter than the top quark, the top quark decays to H^+b. We can then detect the signal of the charged Higgs boson both through its direct pair production and in the top quark decay.

(3) If there is a light neutral Higgs boson, a charged Higgs boson may decay into W plus the neutral Higgs boson with a large branching fraction. Even if neutral Higgs bosons cannot be produced via the process $e^+e^- \to Z^\circ H_i^\circ$, or WW– or ZZ–fusion (for example, the CP-odd state), the neutral Higgs boson can be produced and detected in the decay $H^\pm \to W^\pm H_i^\circ$.

(4) It is necessary to understand higher order QCD processes, to improve the QCD shower models, and to test them at lower energies. Also, processes containing weak vector bosons must be experimentally understood.

(5) Beamstrahlung effects must be moderate. We have to compromise between the integrated luminosity and beamstrahlung effects. The moderate beamstrahlung effects assumed in the earlier Monte Carlo studies[2] are perfectly acceptable for studies of charged Higgs boson production.

However, it looks difficult to search for $H^\pm$ if the mass is close to the $W^\pm$ mass since some of the W boson decay modes are the same as the $H^\pm$ decay modes and the cross section for $W^\pm$ production is about 100 times larger than for $H^\pm$ production. Fortunately, since the W^+W^- differential cross section peaks in the forward region and is relatively low in the backward region, we can efficiently reduce the background by rejecting events in the forward region if the charge of the W can be identified. Even if we cannot measure the charge, we can reduce the W pair background by applying a hard cut on the polar angle: for example, $|\cos\theta| < 0.5$. Although W bosons have the same decay modes as charged Higgs bosons, their branching fractions are very different since charged Higgs bosons prefer to couple to heavy fermions ($t\bar{b}$, $c\bar{s}(\bar{b})$ or $\nu_\tau\tau^+$) while W bosons couple to all fermions with the same strength.

We can also search for the charged Higgs boson in top quark decay. We have already studied this case in the previous paper[2] and the analysis does not depend significantly on the $H^\pm$ mass (of course, the $H^\pm$ mass must be lighter than the top quark mass). Therefore, we do not repeat the analysis in this report.

In the next section, we will demonstrate that we can find $H^\pm$ at TeV e^+e^- colliders with an integrated luminosity of 10 fb^{-1} even if the $H^\pm$ mass is very close to the $W^\pm$ mass.

6.2 $M_{H^\pm} > M_t + M_b$

For this case, the production and decay sequence is $e^+e^- \to H^+H^- \to t\bar{b} + b\bar{t}$. These events have approximately a two-jet structure for $M_{H^\pm}$ = 70–100 GeV at $\sqrt{s} = 1$ TeV. To enhance the H^+H^- signal relative to WW and ZZ backgrounds, and ordinary multihadron backgrounds, the following cuts are applied (the cuts are optimized for $M_{H^\pm} \approx M_W$ and $\sqrt{s} = 1$ TeV):

(1) $E_{\rm vis} > 0.5\sqrt{s}$, where $E_{\rm vis}$ is the total visible energy measured by the electromagnetic and hadronic calorimeters (muon momenta are added).

(2) $|\Sigma p_z|/E_{\rm vis} < 0.4$, where Σp_z is the sum of the longitudinal momenta measured in the same way as the visible energy.

Cuts (1) and (2) reject events with large momentum imbalance along the beam direction due to beamstrahlung and initial state radiation.

(3) $|\cos\theta_{thr}| < 0.50$, where θ_{thr} is the polar angle of the thrust axis.

This cut rejects the major part of the W^+W^- background. The event is now divided into two hemispheres by the plane perpendicular to the thrust axis.

(4) Both thrust hemispheres must have large numbers of charged particles: one hemisphere must have at least 16 charged particles and the other must have at least eight.

Since in each event there are four B–hadrons which have a relatively long lifetime of about 1 ps, we select events containing many charged particles with large impact parameter. For this analysis, we assume an impact parameter resolution of

$$\sigma_\delta = \sqrt{(10\ \mu\text{m})^2 + [70\ (\mu\text{m GeV})/p]^2}\ .$$

The following cut is then applied:

(5) The number of charged tracks with large impact parameter δ in each thrust hemisphere is counted. Only events with many large-impact-parameter tracks in both thrust hemispheres are selected: one thrust hemisphere must have at least five such tracks and the other must have at least three. The definition of a large-impact-parameter track for this analysis is: $p > 1$ GeV, $\delta/\sigma_\delta > 3$, and $\delta < 2$ mm. The larger impact parameter cut of 2 mm reduces the contamination of charged particles from K_S or Λ decays.

(6) The difference in the invariant mass of each hemisphere must be less than 15 GeV.

After all the cuts, the reconstructed Higgs mass (the average of the two hemisphere invariant masses in an event) is plotted for $M_{H^\pm} = 70$ GeV, 83 GeV, 93 GeV and 100 GeV in Fig. 11(a)–(d). Contributions from the background processes WW, ZZ and $q\bar{q}$ are added to the plots. As can be seen from the plots, the signal (shaded part in the figure) is larger than the background near $M_{H^\pm}$. The plots are normalized to an effective integrated luminosity of 10 fb^{-1}.

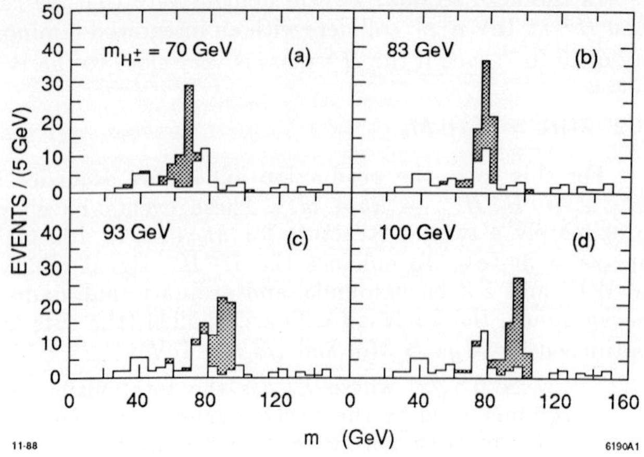

Fig. 11. Invariant mass distribution of reconstructed charged Higgs bosons (the average of the two in each event) after all the cuts for $H^+H^- \to b\bar{t}t\bar{b}$ with (a) $M_{H^\pm} = 70$ GeV, (b) $M_{H^\pm} = 83$ GeV, (c) $M_{H^\pm} = 93$ GeV, (d) $M_{H^\pm} = 100$ GeV. The top quark mass is assumed to be 60 GeV. The shaded region is signal; the open region is background.

6.3 $M_{H^\pm} < M_t + M_b$

If the decay $H^- \to b\bar{t}$ is kinematically forbidden, it is worth studying events of the type $\tau^-\bar{\nu}_\tau+$ hadrons since the branching fraction for the decay mode $H^- \to \tau^-\bar{\nu}_\tau$ can be as large as 30%. The branching fraction depends only on the ratio of the vacuum expectation values for the minimal supersymmetric model, and in general it depends on the Higgs boson-fermion couplings and Higgs boson-gauge boson couplings. Therefore, we will study the production and decay sequence $e^+e^- \to H^+H^- \to \tau^+\nu_\tau + s\bar{c}(b\bar{c})$.

The cuts to select $H^+H^- \to \tau\nu +$ hadrons are the following:

(1) $N_{ch} \geq 3$ where N_{ch} is the number of visible charged particles.

(2) $E_{\rm vis} > 0.5\sqrt{s}$.

(3) $|\Sigma p_z|/E_{\rm vis} < 0.4$.

(4) One of the thrust hemispheres (hemisphere 1) is required to have exactly one charged particle. The momentum of this particle must be greater than 5 GeV. The invariant mass of this hemisphere M_1 is required to be less than 5 GeV. The invariant mass of the other thrust hemisphere M_2 must be larger than 50 GeV and the charged multiplicity N_{ch2} must be larger than 10; i.e., $N_{ch1} = 1$, $N_{ch2} > 10$, $M_1 < 5$ GeV, and $M_2 > 50$ GeV.

(5) Since the WW background is relatively small in the backward region, the events are restricted to this region. The event must satisfy $0.0 < Q_1 \cos\theta_{thr} < 0.7$ where Q_1 is the charge of the single particle in hemisphere 1 and θ_{thr} is the direction of the thrust axis pointing out of hemisphere 1.

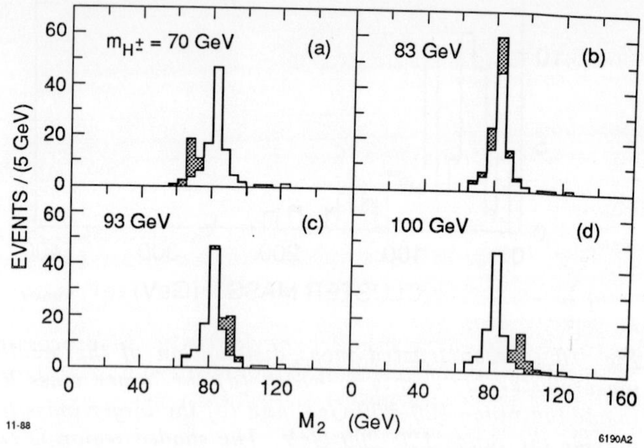

Fig. 12. Distribution of the larger hemisphere mass after cuts (1) to (5) for $H^+H^- \to \tau\nu +$ hadrons (shaded region) in addition to the expected background (open region) for (a) $M_{H^\pm} =$ 70 GeV, (b) $M_{H^\pm} = 83$ GeV, (c) $M_{H^\pm} = 93$ GeV, and (d) $M_{H^\pm} = 100$ GeV.

After applying cuts (1)–(5), the higher hemisphere mass M_2 is plotted for $M_{H^\pm} =70$ GeV, 83 GeV, 93 GeV and 100 GeV in Fig. 12 (a)–(d). Contributions from the background processes WW, ZZ and $q\bar{q}$ are added in the plot. The dominant background due to $e^+e^- \to W^+W^-$ events makes it difficult to see the signal close to the $W^\pm$ mass. In Fig. 13, the same plots are shown for the case when the charged particle in hemisphere 1 is a pion. We assume 100% efficiency for pion identification and no particle misidentification. The signal-to-background ratio is improved because of the large branching fraction for $\tau^- \to \nu_\tau\pi^- + n\pi^0$ $(n \geq 0)$. A branching fraction of 30% for $H^- \to \tau^-\bar{\nu}_\tau$ is assumed in this analysis. The plots are normalized to an integrated luminosity of 10 fb^{-1}.

6.4 Summary

Charged Higgs bosons with a mass of 70–100 GeV can be found at e^+e^- colliders with $\sqrt{s} \sim 1$ TeV with an integrated luminosity of $O(10^{40}\ {\rm cm}^{-2})$, even if the mass is very close to the W mass. The event topology $H^+H^- \to t\bar{b} + b\bar{t}$ can be selected by using the b-tagging method in addition to a hard polar angle cut which rejects WW background.

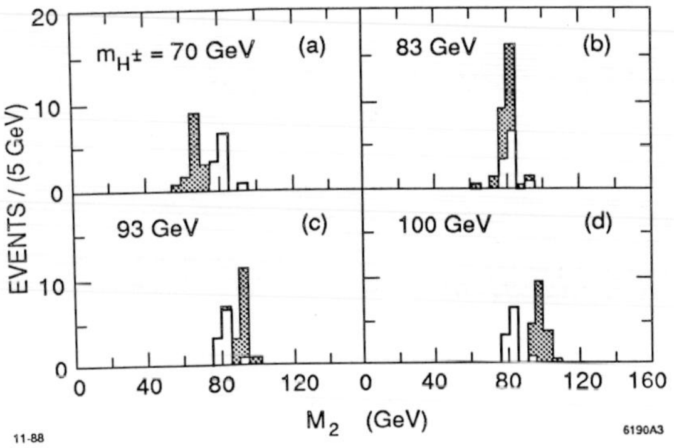

Fig. 13. Distribution of the larger hemisphere mass after requiring the isolated particle to be a charged pion for $H^+H^- \to \tau\nu +$ hadrons *events (shaded region) in addition to the expected background (open region) for (a)* $M_{H^\pm} = 70$ GeV, *(b)* $M_{H^\pm} = 83$ GeV, *(c)* $M_{H^\pm} = 93$ GeV, *and (d)* $M_{H^\pm} = 100$ GeV.

The process $H^+H^- \to \tau\nu + cs(cb)$ can be selected by tagging an isolated pion from the tau decay and by rejecting events in the backward region where WW backgrounds dominate.

These methods can also be used at intermediate energy (350–400 GeV) e^+e^- colliders. At $\sqrt{s} = 400$ GeV, the signal-to-background ratio is worse than at 1 TeV because the relative cross section of H^+H^- to W^+W^- is smaller due to the different threshold behavior of the two processes. On the other hand, the absolute cross section for the signal is larger at lower energies.

At LEP II it may be difficult to detect a charged Higgs boson with mass around the W mass with a luminosity of 500 pb^{-1}. The R value for a 83 GeV $H^\pm$ at $\sqrt{s} = 200$ GeV is only 0.057, so the number of expected events after cuts is very small and the W background is larger. At LEP energies, the event topology is not two back-to-back jets and the particles from H^+ and H^- will be mixed with each other due to the small velocity of the $H^\pm$. Therefore, a more complicated analysis may be required. In any case, the b-tagging and the polar angle cuts discussed in this report are useful at any energy.

7. CONCLUSIONS

High energy e^+e^- colliders will offer nearly ideal opportunities to search for and study the Higgs sector of nature or its equivalent.[16] We have shown that at $\sqrt{s} = 1$ TeV the neutral Higgs boson that appears as a real particle in the simplest doublet model of the Higgs sector can be clearly isolated as long as its mass is less than about 80% of the beam energy and greater than about 120 GeV. Near the mass of the $W^\pm$, the tagging of high-impact-parameter tracks is an important tool for enhancing the signal-to-background ratio. The intermediate mass range ($M_{H^o} \lesssim 2M_{Z^o}$) can be reached at $\sqrt{s} \approx 400$ GeV. We study the case in which the Higgs boson is heavy enough to decay to $t\bar{t}$. In this case, the reconstruction of parton jets is an important tool for rejecting backgrounds.

Models in which the Higgs sector is more complicated contain greater numbers of new particles which can be found and studied. We have determined that charged Higgs particles can be found even when the mass is near the $W^\pm$ mass and that the neutral members of the spectrum of Higgs states can be seen over a limited mass range in some models through the production mechanism $e^+e^- \to H^o A^o$. Again, the tagging of high-impact-parameter tracks is essential for identifying the neutral Higgs bosons.

Integrated luminosities of (10–30) fb^{-1} are needed to complete the investigations that we have discussed. Beamstrahlung is not a major problem for most analyses and at lower energy machines ($\sqrt{s} \lesssim 500$ GeV) the design parameters that have been proposed do not predict significant beamstrahlung in any case. The requirements that are placed on detectors are quite modest and can be satisfied with existing technology.

REFERENCES

1. Burchat, P. R., Burke, D. L., and Petersen, A., Phys. Rev. **D38**, 2735 (1988). After publication an error was found in the simulation of detector resolution. After correcting the error, the W mass peak from $e^+e^- \to e^+\nu_e W^-$ is approximately twice as wide as reported, dominating the Higgs boson signal from $e^+e^- \to \nu_e\bar{\nu}_e H^o$ for a Higgs mass less than about 150 GeV. The error has no significant effect on the heavy Higgs boson analysis. The correct simulation is used for this Snowmass study.
2. Komamiya, S., Phys. Rev. D**38**, 2158 (1988).
3. Yokoya, K., Nucl. Instrum. Methods A**251**, 1 (1986); Chen, P., SLAC–PUB–4293, 1987.
4. Palmer, R. B., SLAC–PUB–4295 (1987); Schnell, W., *Advanced Accelerator Concepts,* Madison, WI, AIP Conf, Proc. **156**, 12 (1987), and SLAC/AP–61 (1987).
5. Sjostrand, T., and Bengtsson, M., Computer Phys. Comm. **43**, 367 (1987).
6. Gabrielli, E., ROME–550–1987.
7. Feldman, G. J., SLAC–PUB–4563, 1988.
8. Richard, F., Proc. of the Workshop on Physics at Future Accelerators, La Thuile (Italy) and Geneva (Switzerland), January 7–13, 1987; Mele, B., ibid.
9. Yehudai, E., private communication.
10. Kane, G. L., and Scanio, J. J. G., CERN–TH.4532/86, 1986.
11. Palmer, R., and Ruth, R., contributions to these proceedings.
12. Boucrot, J. *et al.*, CERN–EP/87–40, 1987.
13. Grosse–Wiesmann, P., these proceedings.
14. Gunion, J. F. *et al.*, Phys. Rev. D**38**, 3444 (1988).
15. Bartel, W. *et al.*, Z. Phys., C**33**, 23 (1986).
16. "New Particle Searches in e^+e^- Collisions," report of study group, these proceedings.

Backgrounds to hadronic Higgs detection in a 1 TeV e^+e^- collider*

ERAN YEHUDAI

Stanford Linear Accelerator Center
Stanford University, Stanford, California, 94305

ABSTRACT

We calculate the cross-section of all standard model processes which constitute a potential background to the hadronic decays of an intermediate-mass neutral Higgs in a 1 TeV e^+e^- collider. These calculations are performed taking into account a realistic photon distribution which is due to beamstrahlung. Imposing appropriate kinematic cuts, and looking only at events with a $t\bar{t}$ combination, the Higgs signal is visible, though marginally, above this background. A more complete treatment of these calculations is given in reference 1.

We concentrate on an intermediate-mass neutral Higgs boson. Assuming $2m_t < m_H < 2m_W$, the dominant decay mode is $H \to t\bar{t}$. We systematically calculate all the processes with an experimental signature of two hadronic jets with invariant mass close to that of the Higgs, which constitute a potential background to its detection.

The total cross-section for Higgs production in a 1 TeV e^+e^- collider is 1-2 units of R, depending on the Higgs mass, where R is the point cross-section ($1R = 4\pi\alpha^2/3s = 86.8$ fb).

Radiated photons play a very significant role in a TeV scale e^+e^- accelerator. Although classical bremsstrahlung remains important, a much larger effect[2)] arises from the synchrotron radiation emitted by electrons in one beam due to the electric field it experiences as it passes through the other beam; this radiation is termed "beamstrahlung". We treat each beam as being composed of both electrons (or positrons) and (real and virtual) photons. We therefore look at processes with both electrons and photons as initial particles. We distinguish between photon-initiated processes and electron-initiated processes containing a virtual photon by insisting that in the later process the photon carries a nontrivial momentum perpendicular to the beam direction.

We parametrize the effect of beamstrahlung in terms of luminosity functions expressing the effective collision luminosity as a function of the fraction of total s involved in the collision. The accelerator parameters used here are $C = 0.3$, $G = 5.00$ and $Y = 1600$, with the terminology as in reference 3. These correspond to a bunch length $\ell_0 \sim 0.4mm$, and a luminosity per pulse $\mathcal{L} \sim 10^{31}cm^{-2}$ (to determine the physics luminosity, multiply $\mathcal{L}$ by the repetition rate, of order 100/sec.). The cross-section of the bunches is assumed to be elliptical with a_x and a_y as the semi-major and semi-minor axes, and $a_x/a_y = 100$.

The processes we calculate are: $e^+e^- \to q\bar{q}$, $\gamma\gamma \to q\bar{q}$, $e^+e^- \to q\bar{q}\gamma$, $e^+e^- \to q\bar{q}Z_0 \to q\bar{q}\nu\bar{\nu}$, $\gamma e^- \to q\bar{q}e^-$ + (c.c.) and $\gamma e^- \to q\bar{q}\nu$ + (c.c.). In order to avoid divergences occurring as either a quark, a photon or an electron goes on shell we impose several cuts. We insist that the momentum of the $q\bar{q}$ system perpendicular to the beam axis — $p_\perp$ be smaller than 50 GeV, and that the minimal perpendicular momentum of either quark — $p'_\perp$ be smaller than 5 GeV.

Further we discard events with a visible particle in the hemisphere opposite the $q\bar{q}$ system. However the third particle is not detectable in all the events of the form $\gamma e \to q\bar{q}\gamma$ or $e^+e^- \to q\bar{q}e$. We must consider the cases in which the third particle, though visible in principle, escapes detection by falling too close to the beam. Here we follow reference 4 in estimating that no detection is possible within 10° of the beam.

Having imposed those cuts we are left with events having the phenomenological signature of two hadronic jets, with large $p_\perp$, and no visible track in the opposite hemisphere. These include $\gamma e \to q\bar{q}\nu$ and $\gamma e \to q\bar{q}Z_0$ (with the Z_0 subsequently decaying to two ν's), but also $\gamma e \to q\bar{q}e$ and $\gamma e \to q\bar{q}\gamma$ with the third particle escaping detection.

Figure 1 compares the combined cross-section for these processes and those in which the visible particle escapes detection with that of the Higgs signal under similar cuts. The Higgs signal is still lower than these backgrounds.

* Work supported by the Department of Energy, contract DE–AC03 – 76SF00515.

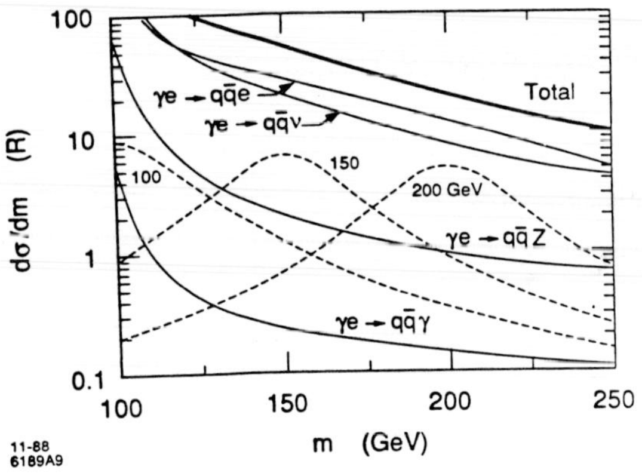

Figure 1. Differential cross-section for processes with no visible extra particle present at more than 10° from the beam, as a function of the invariant mass of the $q\bar{q}$ or Higgs system. Here and henceforth, we show Higgs production cross-sections for Higgs mass of 100, 150 and 200 GeV.

If the Higgs mass is above the two top threshold, virtually all Higgs particles decay into two top quarks. It is reasonable[5] to assume that a top jet can be distinguished from a light jet, either based on the invariant mass of the jet or by looking for a large number of subjets. A further enhancement of the Higgs signal relative to its background can thus be achieved by looking only on events involving a $t\bar{t}$ combination. Figure 2 shows the cross-sections these events, assuming $m_t = 60\,\text{GeV}$. The Higgs signal is now visible above its background.

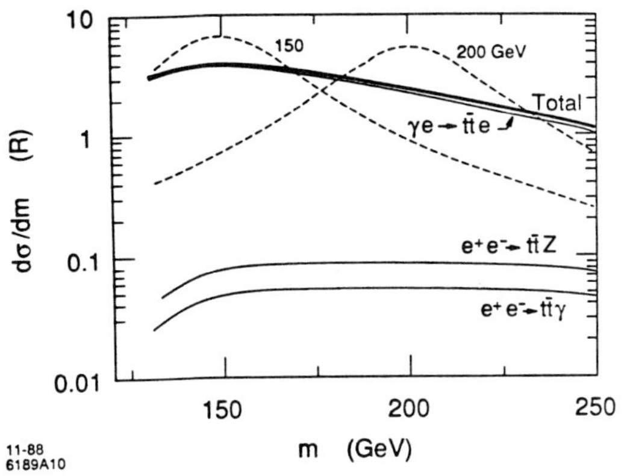

Figure 2. Differential cross-section for $e^+e^- \to t\bar{t}$ + (undetected particles) *vs.* $e^+e^- \to H\nu\bar{\nu}$.

Finally, one can use the measurable angular distribution of the visible electrons in the process $e\gamma \to t\bar{t}e$ to estimate the size of the signal for the same process with the electron escaping detection. Figure 3 shows the angular distribution of electrons from the process $e\gamma \to t\bar{t}e$ compared with the Higgs signal.

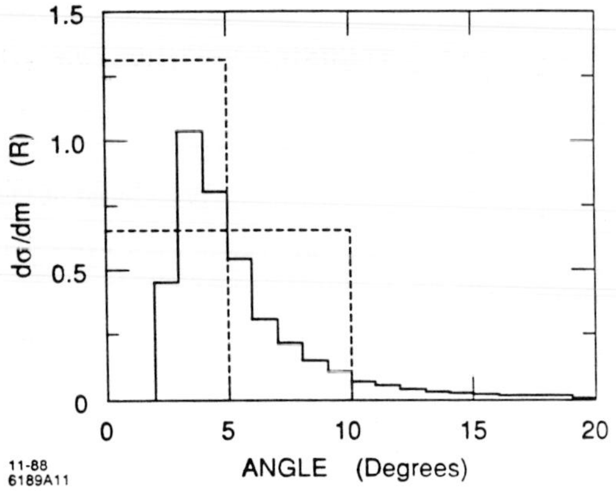

Figure 3. Angular distribution of electrons from $e\gamma \to t\bar{t}e$ with invariant mass of 150 GeV, compared with the $e^+e^- \to H\nu\bar{\nu}$ signal with m_H = 150 GeV. The Higgs signal is spread over 5° and 10°.

Note that though the signal for angles larger than 10° comprises only about 10% of the total cross-section, comparing our predictions with experimental results in that range can confirm our estimate for that part of the background which is experimentally indistinguishable from the Higgs signal.

In conclusion we see that under a reasonable set of cuts: $p_\perp > 50$ GeV, $p'_\perp > 5$ GeV, no visible track in opposite hemisphere, and looking only for t jet events, the Higgs signal can be separated from all conventional backgrounds, with signal to noise ratio of about 2:1.

ACKNOWLEDGEMENTS

I am grateful to M. Peskin for suggesting this topic, for instructive discussions. I also thank R. Blankenbecler for useful discussions, and in particular for supplying the photon spectrum used here.

REFERENCES

1. E. Yehudai, SLAC-PUB-4794(1988).
2. G.J. Feldman, in *Proceedings of the Fifteenth SLAC Summer Institute on Particle Physics*, E.C. Brennan, ed. SLAC Report No. 328 (1987).
3. R. Blankenbecler and S.D. Drell, *Phys. Rev.* **D37** (1988) 3308.
4. C. Ahn *et.al.*, SLAC-Report-329 (1988).
5. P.R. Burchat, D.L. Burke and A. Petersen, SLAC-PUB-4638(1988).

TECHNIRHO PRODUCTION AT ELECTRON-POSITRON SUPERCOLLIDERS

Steve King
Department of Physics, University of Southampton
Southampton, SO9 5NH, England
and
Marie E. Machacek*
Department of Physics, Northeastern University
Boston, Massachusetts, 02115, USA

ABSTRACT

In this note we summarize results for the resonant production of technirho mesons in the process $e^+e^- \to \rho_{tc} \to W^+_L W^-_L$ at very high energy electron-positron colliders for single doublet SU(N) models of technicolor. The details of these and related results are presented elsewhere.[1)]

In the minimal single doublet SU(N) technicolor models electroweak symmetry breaking occurs dynamically when the technicolor interactions become strong. Bound states, analogous to the mesons in QCD, are produced. The lightest of these, the technipions, are the Goldstone bosons that give mass to the W and Z bosons and disappear from the physical particle spectrum. The next lightest technicolor bound states are the technirho (ρ_{tc}) and the techniomega (ω_{tc}), technicolor analogues of the $\rho(770)$ and $\omega(783)$. The masses and widths of these particles are estimated by scaling from the QCD analogues using large N arguments. For the technirho, the mass and width are

$$M_{\rho_{tc}} \approx 2\ \text{TeV}\times\left(\frac{3}{N}\right)^{1/2}, \text{ and}$$

$$\Gamma_{\rho_{tc}} \approx 480\ \text{GeV}\times\left(\frac{3}{N}\right)^{3/2}, \text{ respectively.}$$

Thus in the minimal model, a measurement of the mass or width of the technirho can be used to determine the N of the technicolor group SU(N).

In our calculations we study the consequences of technirho production in the channel $e^+e^- \to \rho_{tc} \to W^+_L W^-_L$. Our assumptions are similar to those used in the study of technirho production at the SSC.[2)] We assume that the equivalence theorem [3)] may be used up to terms $\sim O(\frac{M_W}{\sqrt{s}})$ to replace the longitudinal modes of the W's by their Goldstone degrees of freedom, the technipions π_{tc}. The $\rho_{tc}\pi_{tc}\pi_{tc}$ coupling is scaled from the strong interaction $\rho\pi\pi$ coupling again using large N arguments. In order to determine the production of ρ_{tc} in e^+e^- interactions we assume, as in vector meson dominance in strong interactions,[4)] that the vector current couplings to the γ and Z at these energies are saturated by the lightest technivector mesons.[2)] Technivector meson dominance is expected to hold for energies $\sqrt{s} \leq M_{\rho_{tc}}$.

In many technicolor models extended technicolor interactions are needed to produce masses for the ordinary fermions. These interactions produce four fermion contact interactions between ordinary fermions and technifermions whose strength depends inversely on the square of the extended technicolor scale f. These interactions effectively induce a direct $e^+e^-\rho_{tc}$ coupling that interferes with ρ_{tc} production by a virtual γ or Z. In flavor symmetric models the extended technicolor scale f may be as low as 1 - 2 TeV.[5)] Such a scale would fall well within the discovery reach of e^+e^- supercolliders.

We first consider the idealized case where the initial electron beam is 100% right or left polarized. We also assume that the longitudinally polarized W's are distinguishable from transversely polarized W's. Figure 1 shows the differential cross section for $e^+e^-_L \to W^+_L W^-_L$, in units of R, for minimal SU(N) technicolor, N = 2, 4, 6, 8 at $\cos\theta = -0.5$. The differential cross sections for e^-_R are similar in shape but smaller.[1)] The scattering angle θ is the angle between the outgoing W^- and the electron beam. Both differential cross sections for $W^+_L W^-_L$ in this limit have a simple multiplicative $\sin^2\theta$ dependence in sharp contrast to the sharp forward peaking that occurs in the standard model production of transverse pairs $W^+_T W^-_T$.[6)] Thus by choosing to look away from the forward direction at $\cos\theta = -0.5$, we reduce the background of transverse W pairs relative to the longitudinal W pairs of interest. For N= 4, 6, and 8 the rise in the cross section is dramatic as we cross the resonance. However, a broad resonance such as N= 2 would be difficult to detect. The resonance peaks move to lower energy and decrease in width with increasing N as expected.

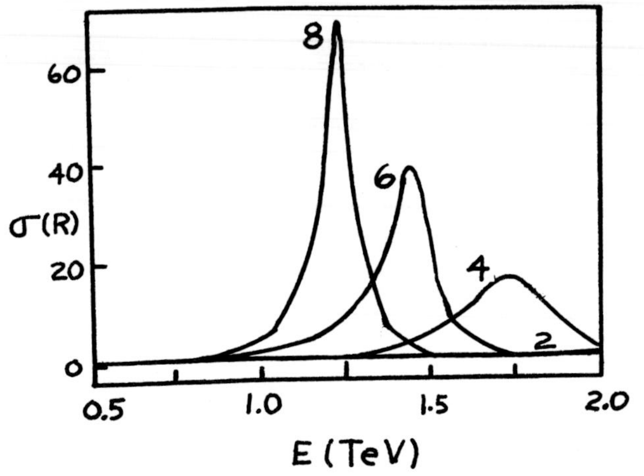

Figure 1. $\frac{d\sigma}{d\cos\theta}$ as a function of $\sqrt{s}$ at $\cos\theta = -0.5$ for $e^+e^-_L \rightarrow W^+_L W^-_L$ in minimal single doublet SU(N) technicolor models for N = 2, 4, 6 and 8. The contact term (f= 30 TeV, $\eta=+1$) is negligible.

We next define a polarization asymmetry A such that

$$A = \frac{\sigma_L - \sigma_R}{\sigma_L + \sigma_R}$$

where L(R) denote left(right) polarized electrons and σ is the differential cross section for $W^+_L W^-_L$ production at $\cos\theta = -0.5$. Since the technirho Breit-Wigner appears as a common multiplicative factor in each of these cross sections, A as defined above is independent of the particular N of the minimal model. In the absence of extended technicolor, or some other source of direct contact terms, A is unchanged from the standard model result of A= 0.68 at these energies. In the presence of the contact terms of extended technicolor, however, A shows a strong f dependence while remaining insensitive to the details of the technirho itself. For example, Figure 2 shows that in a left-left flavor symmetric model with phase $\eta = +1$, A is sensitive to f< 6 TeV at $\sqrt{s}$= 1 TeV and to f< 10 TeV at $\sqrt{s}$= 2 TeV. If $\eta=-1$ we obtain a similar set of curves but with a negative slope.[1)]

The cross section for $W^+_L W^-_L$ production might be difficult to disentangle from $W^+_T W^-_T$ production or the production of W's with mixed polarizations.[6)] In Figure 3 we show the cross section for $e^+e^- \rightarrow W^+W^-$ averaged over initial state electron helicities and summed over final state W polarizations. For $\cos\theta=-0.5$, N=6 the "pure" technicolor ($\eta=0$) cross section still dominates the standard model. We also show the effect on the resonance of a contact term with f=5 TeV. For $\eta=+1(-1)$ the height and width of the resonance are both increased (reduced).

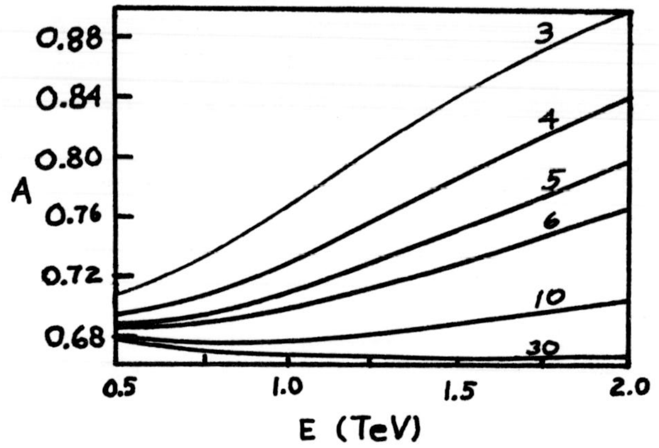

Figure 2. The polarization asymmetry A at $\cos\theta=-0.5$ as a function of $\sqrt{s}$ for the production of longitudinal W pairs with extended technicolor scales f= 3, 4, 5, 6, 10 and 30 TeV. The curve for f= 30 TeV is indistinguishable from the standard model result for $W^+_L W^-_L$.

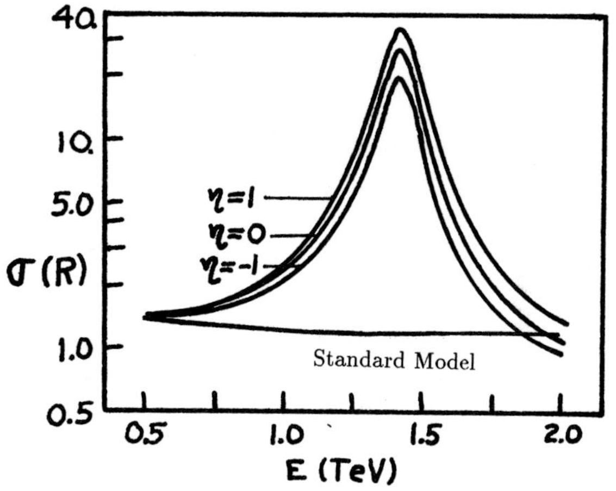

Figure 3. $\frac{d\sigma}{d\cos\theta}$ as a function of $\sqrt{s}$ for $\cos\theta=-0.5$ for $e^+e^- \rightarrow W^+ W^-$ averaged over initial state e^- helicities and summed over final state W polarizations. The resonances correspond to an SU(6) technicolor model, with $\eta=0$ being pure technicolor and $\eta=\pm1$ corresponding to a contact term with f = 5 TeV.

We can also define a polarization asymmetry A similar to that in Figure 2 but where σ_L and σ_R refer to the cross sections for $e^+e^-_{L,R} \rightarrow W^+W^-$ summed over final state W polarizations. In Figure 4 we show this polarization asymmetry for the same parameters used in the set of curves of Figure 3. The standard model result is significantly larger than in Figure 2. This is due to the influence of the $W^+_T W^-_T$ channel whose cross section is larger than for $W^+_L W^-_L$, and for which $\sigma_L \gg \sigma_R$ due to the t-channel ν-exchange graph. For pure technicolor ($\eta=0$) with N=6 the cross section for $W^+{}_L W^-{}_L$ dominates over the cross section for $W^+{}_T W^-{}_T$ across the resonance region causing a healthy dip in A whose trough occurs at the standard model value shown in Figure 2. The effect of a contact term with f=5 TeV and $\eta=+1$ is to cause the dip to become shallower with the trough now occurring at a larger value than shown in Fig. 2. For $\eta=-1$ the effect is reversed leading to a deeper trough. Thus by measuring the depth of the trough, both the sign of η and the value of f can be determined.

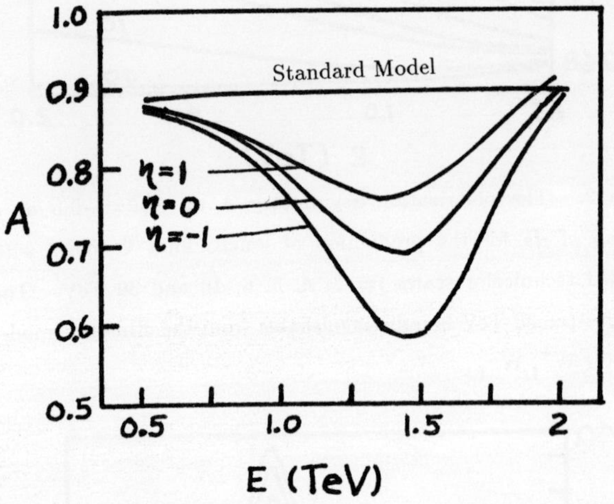

Figure 4. The polarization asymmetry A as a function of $\sqrt{s}$ for $\cos\theta=-0.5$ defined for $e^+e^-{}_{L,R}\rightarrow W^+W^-$ summed over final state W polarizations. The dips correspond to the same technicolor model as Figure 3.

In summary, we have shown that at an e^+e^- supercollider of sufficient energy to reach the resonance region, it will be relatively easy to determine the details of the technicolor model, such as N and f, by studying W^+W^- final states. The existence of technicolor may well be first discovered at large hadron colliders such as the SSC.[2] But if history repeats itself the details of the model will wait to be revealed by an e^+e^- supercollider with $\sqrt{s}\approx M_{\rho_{tc}}$.

* Work supported by the National Science Foundation grant PHY83–06514.

REFERENCES

1) King, S. and Machacek, M., in preparation, 1988.

2) Chivukula, R.S., Boston University preprint BUHEP-88-16, June 1988.

3) Chanowitz, M. S. and Gaillard, M. K. ,Nucl. Phys. B261, 379, 1985.

4) Sakurai, J.J., *Currents and Mesons,* (University of Chicago Press), 1969.

5) Chivukula, R. S. and Georgi, H., Phys. Lett. B188, 99, 1987.

6) Ahn, C., et. al., SLAC report 329, 151, May 1988.

NEW PARTICLES

Charles Baltay and Gordon Kane
Working Group Leaders

NEW PARTICLE PRODUCTION - AN OVERVIEW

C. BALTAY
Yale University

R. HOLLEBEEK
University of Pennsylvania

G. KANE
University of Michigan

U. NAUENBERG
University of Colorado

W. H. SMITH
University of Wisconsin

Abstract

This article is an overview of the considerations of the New Particles Group at the 1988 Snowmass Summer Study. The foundations of this overview are contained in the three following detailed articles on hadron-hadron, e^+e^-, and *ep* colliders.

Introduction

The scope of Snowmass '88 was to examine the goals of particle physics for the 1990s and the facilities required to best achieve these goals. The core of the U.S. program is the SSC, the planned 40 TeV proton-proton collider. However, a healthy program requires other facilities. Examples of other facilities are high energy e^+e^- colliders, *ep* colliders, b factories, K factories, heavy ion colliders and non accelerator facilities.

The New Particles Group, as its name implies, considered the search for new particles at these various possible facilities. An exception was the search for the Higgs boson, which was studied by a separate group on spontaneous symmetry breaking. The work of the New Particles Group was carried out by three subgroups, studying hadron-hadron colliders, e^+e^- colliders, and *ep* colliders, respectively.

General Motivation

Past discoveries of new particles have been stepping stones for our understanding of particle physics. In planning for the future it is sensible to consider what kinds of new particles may exist and what kinds of facilities will be needed not only to discover them but also to carry out detailed studies of their properties. Although the truly exciting discoveries in the 1990s may be those of totally unexpected new particles, the best we can do is to plan facilities well suited for discovering anticipated new particles, under the assumption that such facilities will prove useful at discovering whatever nature has in store for us.

We have many reasons to search for the existence of new particles that would have a fundamental impact on the understanding of our field.

1. Within the standard model, the discovery of the b quark, whose measured quantum numbers ($T_3 = -\frac{1}{2}$) suggest that it is part of a doublet, strongly implies the existence of an additional new quark (with $T_3 = +\frac{1}{2}$) which we have already named the *t* quark. For experimental and theoretical reasons[1] we now believe that the *t* mass is below 200 GeV, well within the reach of several of the new facilities under discussion, unless physics beyond the Standard Model affects *t* quark decays and contributions to virtual processes.
2. The Standard Model has been very successful in predicting or explaining essentially all of the observations made in particle physics so far. It does however have a serious flaw in that before its symmetries are broken it predicts that all of the basic particles have the same zero rest mass, which is not the case experimentally. For this reason it is necessary to postulate one or more new particles, which could take the form of a simple Higgs boson, or could be whole families of new particles as predicted by Supersymmetric or Technicolor models. In order to produce the mass spectrum of the known particles, these new particles should have masses below a TeV or so. The energy of the SSC has been chosen to ensure adequate coverage of this mass range.
3. The origin and the number of families of fundamental constituents (quarks and leptons) is obviously of great importance. The search for new heavy quarks and leptons, perhaps from additional families, is therefore of great interest.
4. Many new theoretical ideas such as superstrings, predict the existence of additional heavy Z' or W' gauge bosons. The existence of one or more of these would have a fundamental impact on our thinking.

In these and other cases we need to have the capability of finding the new particles if they do exist. Furthermore, their systematic exclusion is almost as important if they are not present in a particular mass range. When new particles are found it will be crucial to produce them copiously under experimentally clean conditions to be able to carry out detailed studies of their properties.

Facilities Considered

In the course of the summer study a large variety of facilities were discussed, ranging from the existing to the mythical. These colliders are listed in Table 1 with the energies and range of luminosities considered. These facilities were considered in the spirit of a summer study where the mind can wander freely, unencumbered by reality. It must be kept in mind that there is a large spread in the degree of reality of the highest energy colliders. There is now a detailed design report for the SSC and discussions of funding and management are well underway, while the technical feasibility of the TeV e^+e^- colliders remains to be demonstrated.

Collider Facilities Considered for New Particle Searches

Table I

Hadron-Hadron Colliders	$\sqrt{s}$	$L(cm^{-2}sec^{-1})$
CERN-ACOL	630 GeV	$\sim 10^{30}$
TEVATRON	1.8-2.0 TeV	$\sim 10^{30}$
TEVATRON upgrades	2-3 TeV	$\sim 10^{30} - 10^{32}$
LHC	16 TeV	$\sim 10^{33}$ - few 10^{34}
SSC	40 TeV	$\sim 10^{33}$
e^+e^- Colliders		
SLC-LEP	100 GeV	$\sim 10^{31}$
LEP 2	200 GeV	$\sim 10^{31}$
Super Z^0 Factory	100 GeV	10^{33}
NLC-CLIC	500 GeV-2 TeV	10^{32} - 10^{33}
ep Colliders		
HERA	314 GeV	$\sim 10^{31}$
HERA′	410 GeV	$\sim 10^{31}$
LEP × LHC	1.4 TeV	$\sim 10^{31}$
LEP2 × LHC	1.8 TeV	$\sim 10^{31}$

Comments on Monte Carlo Techniques

The calculations carried out in this summer study (as well as any other similar study) are critically dependent on Monte Carlo programs such as ISAJET or PYTHIA to simulate what the expected signals as well as backgrounds will be like at these very high energies. These programs are based on our current understanding of the electro-weak interactions and of QCD, and the existing (relatively low energy) measurements of the quark and gluon structure functions of the nucleon. They have been calibrated and normalized using data from the existing relatively modest energies. Many of the conclusions of the study, especially those dependent on estimations of backgrounds, depend on subtleties of transverse or missing energy distributions, and quite often are sensitive to details of the tails of steeply falling distributions. Therefore, a certain amount of caution is necessary in the interpretation of these results. By the time the SSC turns on, the HERA *ep* facility will have measured quark and gluon distribution functions down to Feynman-x values of 10^{-5} and Q^2 up to 10^4 GeV^2/c^2. These measurements will greatly reduce the QCD uncertainties that plague calculations today. The topic is discussed in detail in the report of the Structure Functions Group.

Meanwhile, Monte carlo programs such at ISAJET and PYTHIA have been extensively used and proven to be immensely useful in the interpretation of CERN S$\bar{p}$pS and Fermilab TEVATRON data. They have done very well in not only explaining but in many cases predicting the data. We thus have reason for some confidence in these programs. Our community should continue to encourage work on the development of these Monte Carlo programs and be aware of the importance of having a number of experts in this area of theoretical research.

Comment on Previous Studies

There have been many previous studies[2–9], in the U.S. and in Europe, on the questions we are interested in. Many of these studies are difficult and complicated. The first attempts appropriately began with simple examples and assumptions. In the present work, particularly for the supersymmetric partners, we believe that the studies finally reflect the situation more realistically, with cross sections and branching ratios consistent with those that could come from a reasonable theory. A number of additional signatures and techniques have emerged as a result.

Note that simple "discovery limits" based on only a knowledge of cross sections and luminosity are useful for processes for which there is essentially no background, but can be very misleading for estimating for the possibilities of discovering something and the luminosity and detector capabilities required when large backgrounds are present. The prospects for finding a $Z' \to \ell^+\ell^-$ can indeed be deduced from a knowledge of its production cross section and leptonic branching ratio, but the likelihood of detecting a heavy charged lepton is subtle to estimate (see the report of the new particle hadron group). When large backgrounds are present the mass reach decreases, and comparisons between different facilities, different energies, and detectors are subtle.

When studies for future hadron colliders began about eight years ago, no working hadron collider detectors existed and it was difficult to be confident about what would happen. Today the situation is much more clear, and the results are very encouraging. Not only did the discovery of $W^{\pm}$ and Z° go well at UA1 and UA2, but the studies of missing momentum events by UA1[10], UA2[11], and now CDF[12] have led to published limits [i.e. , a signal would have been detected if supersymmetric partners or heavy leptons existed in certain mass regions], the UA1 b quark analysis[13] has given a number of results of importance, and the hard photon cross sections from UA2[14] and then UA1[15] and CDF[16] provide tests of QCD and show that photons can be used as a tool to search for new physics.

Motivation for and General Features of New Particle Searches

Search for the *t*-quark

A theorem for the t-quark can be stated as follows:

Either $m_t \lesssim 200$ GeV and $BR(t \to be\nu) \simeq BR(t \to b\mu\nu) \simeq 1/9$, or there is new physics in the region below a TeV. The upper limit exists[1] because the t quark occurs in loops and a t quark heavier than about 200 GeV would lead to inconsistencies between different experimental measurements of M_z, M_W, and $sin^2\theta$. However, this assumes the Standard Model spectrum; if new particles occur they could also enter the loops and affect the result. Thus, if $m_t > 200$ GeV, additional contributions must exist. Similarly, if the decays $t \to b\ell\nu$ have the above branching ratios, then sizeable non-standard decays do not occur.

Consequently, a major goal for the next decade must be to measure m_t and the t decays. The present TEVATRON can search for the t up to $m_t \sim 100$ GeV/c^2 if it has Standard Model decays. The TEVATRON can extend the search to larger m_t as its luminosity is upgraded; probably of order 10^3 pb^{-1} would be required to search up to $m_t = 200\,GeV$. LEP2 will eventually also cover the region $m_t \lesssim 100$ GeV for many decay modes (not just Standard Model ones). If it is not found at a hadron collider at LEP2, it may be worthwhile to look for the t at a new e^+e^- collider. One with $\sqrt{s} = 500$ GeV is sufficient to find the t quark up to the 200 GeV mass limit, or know that new physics exists. If the semileptonic decays of the t quark are as expected, a heavy t could be detected at the SSC as well as at an e^+e^- collider, but the e^+e^- collider is somewhat better able to deal with anomalous decays.

Search for Higgs Bosons

This is the domain of another group at Snowmass '88, but for completeness we mention here that an intermediate energy e^+e^- collider ($\sqrt{s} \sim 500$ GeV) can cover the very important intermediate mass region $m_h \leq 2M_z$ that is the most difficult region for the SSC (though probably doable there if appropriate detectors exist). e^+e^- colliders are also very effective at finding charged scalars $H^{\pm}$. On the other hand, the heavier neutral scalars of the supersymmetric spectrum are very difficult to produce at e^+e^- colliders. In order to fully study the Higgs sector it is clearly desirable to have an e^+e^- collider to supplement the SSC, thought the SSC is essential because all of the states are produced there.

Supersymmetry Searches

Supersymmetry is a well motivated and fairly well developed idea. However, there is no prediction for the masses of any of the supersymmetric particles. Naively, we expect masses to be of order m_W if supersymmetry is relevant to understanding the weak scale. We would have been very lucky if a superpartner had turned up in the searches so far. The current limits on squarks and gluinos are 85-90 GeV[12], and on uncolored superpartners are 40 GeV or less.

Several new aspects of how to search for squarks and gluinos in hadron colliders were studied at Snowmass and are described in detail in the hadronic section. Because of the problem of tails of distributions, a major background study was carried out. Many backgrounds could contribute a few missing momentum events at a very small fraction of their total rate. Examples include $pp \to W^- +$ jets, with $W \to \ell\nu$ and ℓ not identified, or $t\bar{t}$ production where t and $\bar{t}$ come from a radiated gluon so they tend to go in the same direction and so do the neutrinos from their semileptonic decays, giving substantial missing momentum. Methods to reduce the backgrounds to the necessary level were studied; the goal of the studies is to understand what region of masses can be studied, and how sensitive that is to energy, luminosity, and detector capabilities.

It has been understood for several years that if the gluinos were much heavier than $W^{\pm}, Z$, they would decay via chains where $\widetilde{\gamma}$, $\widetilde{Z}$, $\widetilde{W}, \ldots$ all entered, and final states involving $W^{\pm}$ and Z° would become important. Studies including these effects are reported in the hadron group analyses.

If a signal for possible superpartners is ever detected, before it can be believed there must be several ways to check it. Two additional approaches that could serve either as the first clue to supersymmetry, or a confirmation, were studied here. One uses associated production of a gluino and the lightest supersymmetric particle (LSP). After a heavy gluino decays via a chain, the resulting LSP carries off little momentum, while the prompt LSP from the production carries considerable momentum. Thus the missing momentum signal from such events is larger. In addition, the asymmetry of the event, with considerable activity vs. little activity, helps trigger such events.

The second method cleverly uses the Majorana nature of the gluino. A gluino can decay to either $u\bar{d}\widetilde{\chi}^-$ or $\bar{u}d\widetilde{\chi}^+$, for example. ($\widetilde{\chi}^{\pm}$ are partners of $W^{\pm} or H^{\pm}$, called charginos.) The $\widetilde{\chi}^{\pm}$ will decay part of the time to $\ell^{\pm}$.

Then in the process

$$p^{\pm}p \to \tilde{g}\tilde{g} + \text{anything},$$

1/2 of the time it will happen that decays via charginos give $\ell^+\ell^+$ or $\ell^-\ell^-$ in the final state. As always, like sign dileptons imply new physics. There is a background that must be understood; namely the production of $t\bar{t}$ with $t \to b\ell^+\nu$, and

$$\bar{t} \to \bar{b}\, q\, \bar{q}$$
$$\quad \hookrightarrow \bar{c}l^+\nu$$

thus also giving like-sign dileptons with relatively isolated leptons. Of course, such a signature is only of use with a detector that can measure the sign of the lepton charge in the appropriate momentum region. Systematic studies of all these possibilities are reported in the hadron study following this overview.

One important additional aspect should be stressed here. Any high-luminosity collider (e^+e^- or hadron) has a particularly good way to search for superpartners, via the production of neutralinos or charginos[17]. The lightest two neutralinos ($\tilde{\chi}^o_{1,2}$) are generally much lighter than the scale of *SUSY* breaking, so kinematically the process $e^+e^- \to \tilde{\chi}^o_2 + \tilde{\chi}^o_1$ is allowed. The cross section is 0.1 units of R or larger over most of the phase space, so with $L \geq 10^{33}$ over 100 events are produced at a 500 GeV e^+e^- collider. The signatures are generally good, with $\tilde{\chi}^o_2 \to \ell^+\ell^-\tilde{\chi}^o_1$ or $\tilde{\chi}^o_2 \to h\tilde{\chi}^o_1$; $\tilde{\chi}^o_1$ is the LSP and escapes detection. Similarly, the lightest chargino is usually considerably lighter than the scale of SUSY breaking, so $e^+e^- \to \tilde{\chi}^+_1 + \tilde{\chi}^-_1$ is kinematically allowed even if most superpartners are very heavy. Its cross section is about 1 R, so even with luminosities of only $\sim 10^{32}$ it is a good opportunity.

Superpartners may be discovered in *ep* collisions by the associated production of a slepton and a squark, thru either chargino or neutralino exchange. The mass range of the superpartners is such that the selection will decay into an electron and the LSP (most likely a photino), while the squark decays into its corresponding quark and a photino. Monte Carlo studies detailed in the section on new particles in *ep* collisions show that at HERA the discovery limits on the sum of the selection and squark mass extend up to 200 GeV/c^2. At LEPxLHC the limits extend to 700 GeV/c^2. A unique aspect of studying supersymmetry at HERA is the use of polarization and charge symmetries to enable determination of the different supersymmetric parameters. A particularly useful tool in studying supersymmetry would be elastic associated slepton-gaugino production. At HERA the event is as clean as in the case of e^+e^- collisions, there is practically no standard model background, and the energy of the outgoing proton is measured, thus determining the invariant mass of the slepton-gaugino system.

Another interesting aspect of the search for SUSY particles is their possible relation to dark matter. If nature is supersymmetric the LSP is a good candidate to be the dark matter that makes up much of the universe. If that is so, only a certain range of masses for the dark matter can qualify. The cross section for dark matter annihilation is proportional to $m^2_{\rm LSP}$. If the LSP were too light, it would annihilate too little and the resulting mass density would exceed that allowed by the observed expansion of the universe. If the LSP were too heavy, it would annihilate too effectively and the mass density would get too small to contribute significantly to the dark matter. The relevance to e^+e^- colliders is that the allowed region for LSP masses is completely covered with even a 400 GeV collider, so that if the process $e^+e^- \to \tilde{\chi}^o_2 + \tilde{\chi}^o_1$ is not observed, then dark matter does not originate from supersymmetry. (A significant part of the region is covered at LEP2, but $\sqrt{s} \sim 400$ GeV is necessary to cover the entire region).

Technicolor Searches and Leptoquark States

Technicolor ideas are not well defined at present. Older work has had considerable doubt cast on its relevance because of possible conflicts with experiment, while newer approaches are just being developed and are not ready for phenomenological analysis. Contributions by Chivukula and Holden in the "Higgs" section of these proceedings provide some analysis relevant for new particles, and explanation of the current situation.

Leptoquark states have long been proposed by technicolor ideas, and can also arise from a variety of theories such as superstring theories. They should be taken as serious possibilities for new particle discoveries. As always, their masses are uncertain, but it would be surprising if they were lighter than $O(M_w)$. Hadron colliders are an ideal place to search for leptoquark states, since (a) production cross sections are always large because of the color triplet nature of the leptoquark (apart from spin and multiplicity factors the leptoquark cross sections are the same as heavy quark cross sections), and (b) the signatures are always good, with leptoquark decaying into isolated charged leptons or neutrinos plus isolated jets. Their mass can be reconstructed. Leptoquarks can be studied at e^+e^- colliders as well, with cross sections typically a significant part of a unit of R.

While leptoquarks present in extended technicolor theories are infrequently produced in *ep* collisions (since they couple with strength proportional to lepton masses), the leptoquarks found in superstring-inspired E_6 models are copiously produced with striking signals. These scalar weak isosinglets may be produced directly through an s-channel resonance, yielding distinct peaks in x distributions that lie 1-2 orders of magnitude above the ordinary electroweak background. According to the Monte Carlo study detailed in the section on *ep* physics, discovery up to masses of 250 GeV/c^2 and higher is feasible at HERA. Similarly it shows that the LEPxLHC and LEP2xLHC facilities can explore masses up to 1.0-1.3 TeV/c^2. In addition, leptogluons (color octets with lepton number coupling to a lepton-gluon pair) are also produced as s-channel resonances with enormous rates in ep collisions.

Search for new Quarks and Leptons

Although there is no compelling need from either theory or data to have new quarks or leptons, many approaches suggest new fermions might exist. Interestingly, most approaches suggest they are not too heavy; if the fermions had masses larger than about 250 GeV, much of the theoretical motivation would disappear. At e^+e^- colliders searches for quarks and charged leptons are probably straightforward up to $\frac{1}{2}\sqrt{s}$. Thus a relatively low energy collider, with $\sqrt{s} \sim 500 GeV$, covers much of the region of interest. There is controversy about whether heavy leptons can be detected at hadron colliders, with the European studies[18] reporting that it is possible while one independent analysis[19] concluded it was not possible[20].

One type of lepton for which the hadron collider does very well is a neutral Majorana lepton N° (similar to the gluino case). Then production can occur via real or virtual Z's. If $N^\circ + \overline{N}^\circ$ are produced, one can decay to ℓ^+ and the other first oscillate $N^\circ \to \overline{N}^\circ$ and then decay again to ℓ^+, giving like-sign dileptons. Production can also be to $\ell^\pm + N^\circ$ followed by N° decay, giving like-sign dileptons half the time. Production of massive neutral leptons may be larger at hadron colliders than at other facilities, and the like-sign dilepton signature then favors the hadron facility for detection.

Models of composite quarks or leptons predict excited partners of the known fermions. In ep collisions, the production of excited electrons will give a clean signal due to the large p_T of the decay products. the high detection efficiency for excited electrons, along with the ease of distinguishing them from possible backgrounds leads to a discovery limit at HERA of 200-250 GeV/c^2, depending on their coupling. At LEPxLHC the limit is 1 TeV/c^2.

New Heavy Z' and W' Gauge Bosons

Such states occur in a number of attempts to extend the Standard Model, because of extra $U(1)$ symmetries that arise when larger gauge groups are broken down to $SU(3) \otimes SU(2) \otimes U(1)$. Hadron colliders are very good at looking for new Z' states. Depending on couplings, the SSC can probably find such states if they are lighter than about 7 TeV. (See the following reports for these numbers). The relative extent for the same couplings is about 3.5 TeV for LHC, about 1 TeV for the UNK collider, about 0.7 TeV for the TEVATRON collider with about 10^3 pb^{-1}, about 0.35 TeV for the present TEVATRON. These numbers assume typical extended gauge theory couplings; they would be larger if Standard Model couplings were used. More detailed study is needed to see how well the width can be measured, how many branching ratios of a new Z' can be found (such as $\tau^+\tau^-$, W^+W^-, ZZ, $t\bar{t}$, $b\bar{b}$, decays to new particles), how well forward-backward asymmetries can be determined, etc. These questions are very sensitive to detector resolution.

Cross sections for the production of Z' states in e^+e^- collisions are $10^3 - 10^4$ units of R, providing substantial sources of new physics. The beamstrahlung effect is a help for higher energy e^+e^- colliders, in that it allows automatic scanning of a large energy range for new Z''s, an opportunity that previously was thought to belong only to hadron colliders. Thus higher energy colliders can cover the entire range up to $\sqrt{s}$ for $M_{Z'}$. The rates are large enough that many branching ratios could be seen, including possible new states. Studies indicate backgrounds are quite small.

A number of interesting measurements are possible at e^+e^- colliders. Polarized beams could be used to make a Z', and the polarization of final τ's (using multipion modes) or W's could be measured. The decay $Z' \to \nu\overline{\nu}$ could be determined by using $e^+e^- \to Z' + \gamma$. Searches for Higgs using $Z' \to ZH$ would be clean, and allow the H decays to be studied.

New gauge bosons are best searched for in ep collisions by examining the Q^2-distributions of asymmetries between cross sections with polarized electrons and positrons. Deviations from the standard model predictions yield discovery limits at HERA of 375 GeV/c^2 for a new W_R and between 115 and 450 GeV/c^2 for a Z', depending on the model. The limits at LEPxLHC are 1050 GeV/c^2 for a W_R and between 575 and 1150 GeV/c^2 for a Z'. The details are found in the section on ep physics and the contribution to these proceedings by F. Cornet. While inclusive cross section measurements also provide the opportunity for discovery of new gauge bosons, the most sensitive electron and positron tests are normally obtained when polarized beams are available. In addition, polarized beams provide the ability to distinguish between different models once a signal is observed.

Other Topics of Interest

Several topics of interest, even though they were not new particle searches, were discussed within the work of the new particles group.

Neutrino generation counting

Studies will be done at SLC and LEP to determine the number of light neutrinos, both by measuring Γ_Z and by using the process

$$e^+e^- \to \gamma + Z, \quad Z \to \nu\overline{\nu}$$

whose signature is a single γ, $E_\gamma \sim 1 - 10 GeV$. The cross section is about 15×10^{-35} cm^2, so over 10^3 events can be gathered in a year, leading to a measurement of n_ν to ± 0.2. A collider at $\sqrt{s} \simeq m_z$ with $L \simeq 10^{33}$ could probably reach ± 0.05.

Testing the ZWW and the γWW vertices

The importance of the non-Abelian ZWW and γWW vertices in the structure of the $SU(2) \times U(1)$ gauge theory was understood over 15 years ago, but precise tests cannot be made until LEP2 is available, still over 5 years away. It has been realized more recently that precise tests could be made at hadron colliders, and that considerable extra accuracy could be obtained at intermediate energy electron colliders.

Since W production at HERA will provide a 30%-50% measurement of the γWW coupling, several years before LEP2 makes a 20% measurement in W pair production. However, at LEP2 a linear combination of ZWW and γWW couplings is measured. Therefore, the independent measurement of the γWW coupling from HERA will be important for the analysis at LEP2.

W production in *ep* collisions is strongly peaked at low p_{TW}. Only at larger W transverse momentum do other processes become relevant. The signal searched for at HERA will be a charge lepton, mission p_T from W leptonic decay, and a relatively low p_T jet recoiling against the W. The details of the study are contained in the section on *ep* physics, and the contribution to these proceedings by D. Atwood et al.

At e^+e^- colliders the process studied is $e^+e^- \to W^+ W^-$. Two technical advantages occur at the hadron colliders. First, at e^+e^- the important diagram, with the ZWW vertex, has a neutral current Ze^+e^- vertex at the other end and is thus suppressed; for hadron colliders the neutral current vertex occurs in the diagrams without the Z, thus enhancing the relative contribution of the interesting diagram. Second, at e^+e^- the γ and Z contributions both must occur, so the cross section depends on all the γWW and on all the ZWW parameters, while at the hadron collider the ZWW parameters occur in $u\bar{u} \to WZ$ and the γWW parameters in $u\bar{d} \to W\gamma$, allowing a very useful separation.

These tests are very sensitive to the value of M_{ww} or M_{wz} or $M_{w\gamma}$. Having M_{wz} or M_{ww} at 350 GeV provides an order of magnitude more sensitivity than at 200 GeV. The hadron machines naturally produce WZ and $W\gamma$ over a range. At the TEVATRON, a very sensitive test can be made for $W\gamma$ with 100 pb^{-1} of data, and for WZ with 10^3 pb^{-1}. At LHC and SSC extremely sensitive tests are possible.

The 200 GeV energy of LEP2 has been chosen to be just over the e^+e^- to W^+W^- threshold. However the sensitivity to new physics increases rapidly with $\sqrt{s}$. At an intermediate energy (500 GeV) e^+e^- collider the sensitivity is an order of magnitude higher than at LEP2, making this study quite interesting[21].

Rare Z decays

LEP will eventually have of order 10^7 Z^o's, so it will be able to study any rare decay at the 10^{-5}—10^{-6} level, depending on signature. The Super Z Factories (e^+e^- colliders with $\sqrt{s} = 100 GeV$ and L=$10^{33} cm^{-2} sec^{-1}$) could produce $\sim 10^9 Z^o$ decays. It is very interesting to ask what new windows would appear if of order 10^9 Z^o's were available.

One set of decays, the Z $\beta-$decays, such as $Z \to We\nu$, etc. have branching ratios[22]

$$BR(Z \to W + \text{anything}) \simeq 2 \times 10^{-7}$$
$$BR(Z \to W\ell\nu) \simeq 10^{-8}$$
$$BR(Z \to W\pi) \simeq 10^{-8}$$
$$BR(Z \to \pi^o\gamma) \simeq 3 \times 10^{-11}$$

Given the signature problems, even 10^9 Z^o are probably not enough to detect these, let alone study them.

Another set of interesting decays are the lepton number violating decays such as $Z \to e\mu$, $Z \to e\tau$, $Z \to \mu\tau$. They are as exciting a possibility as $K \to \mu e$, $\mu \to e\gamma, \cdots$. Current theoretical restrictions[23] with input from μ decays and τ decays allow one to conclude that $BR(Z \to \mu e)$ is less than about 10^{-5} [perhaps considerably less since fine tuning was required to reach 10^{-5}], and $BR(Z \to \tau\mu,\ Z \to \tau e)$ are essentially unrestricted, less than about 10^{-2}. Conceivably $Z \to \mu e$ would fall in the window of interest from 10^{-6}—10^{-8}; careful analysis is needed to evaluate the experimental problems given backgrounds such as $Z \to \mu\mu$ with $\mu \to e\nu\bar{\nu}$. Associated with flavor changing decays there could also be universality violations, e.g., $\Gamma(Z \to \mu^+\mu^-) \neq \Gamma(Z \to e^+e^-)$, etc.

One decay is of great interest if the Higgs boson h exists, namely $Z \to h\gamma$. It requires of course that h exist and $m_h < m_z$. It is very interesting because it proceeds via one loop, with contributions such as

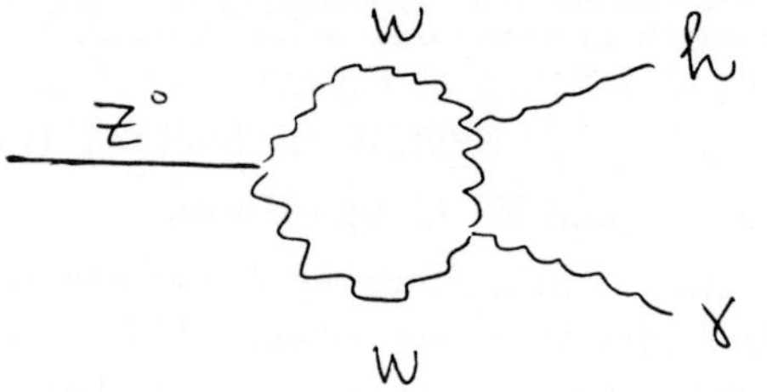

Any particle, known or new, that interacts with γ, Z, and h will contribute to the loop, so the rate for this is a measure of possible new physics. Other similar one-loop processes involving Higgs bosons include $h \to \gamma\gamma$ and $gg \to h$, both of which can be measured at a high luminosity hadron collider. Combining information from all of these can give very valuable sum rules to help measure the Higgs boson couplings to the Standard Model particles and to help untangle physics beyond the Standard Model[24].

Unfortunately, $BR(Z \to h\gamma) \simeq$ few $\times 10^{-7}$ depending on m_h, and presumably not all decays of h are detectable. At the same time, very accurate measurements would be needed to verify the Standard Model prediction at the level needed to confirm the relevant couplings and to distinguish it from new physics contributions. Consequently, well over 10^9 Z° would be required to be confident of getting useful results on $Z \to h\gamma$.

Concluding Remarks

We conclude this overview with some general observations.

1. The SSC is essential to explore the new physics of the TeV mass scale. Our present theoretical understanding of particle physics implies that some new physics must occur below a few TeV to resolve the mass problem of the otherwise spectacularly successful standard model. The 40 TeV center of mass energy of the SSC was chosen to cover the mass range of the new physics with some confidence. Snowmass '88, as well as other recent studies, indicate that detectors can be built and analysis techniques are available to properly exploit the physics potential of the SSC. The recent successes of the CERN SppS and the Fermilab TEVATRON colliders give further confidence that this will be the case.
2. High energy e^+e^- colliders in the range of 0.5 to 2 TeV center of mass energies are an important component of a healthy well-rounded particle physics program; nevertheless, the technical feasibility of these machines needs still be to demonstrated. Such colliders, if they can be built with luminosities of the order of $10^{33}\times$ s(TeV^2 cm^{-2} sec^{-1}), will compliment the physics program of the SSC in very significant ways. In the past, hadron accelerators and e^+e^- colliders have complemented each other in an extremely fruitful U.S. particle physics program that would have been considerably less productive had we concentrated on one or the other exclusively. On the question of the desirable energy of an e^+e^- collider, we conclude that from the physics point of view obviously the higher the better. However, this study indicated that the window for new physics for even a 500 GeV collider is surprisingly large and would complement the capabilities of the SSC in important ways. It is clear however, that to probe the mass scales accessible to the SSC, an e^+e^- collider in the several TeV range is required.
3. Possible TEVATRON upgrades could provide useful physics options through the 1990s, and would provide important opportunities in hadron physics during the SSC construction period. Of the various upgrade options discussed at the study, the one with PP collisions at 2 TeV and 10^{32} cm^{-2} sec^{-1} luminosity offers considerably more physics opportunities than other possibilities.
4. The *ep* physics program planned for HERA will make significant contributions to our understanding of physics in the 1990s. This program and the possibility of colliding the LEP ring with an LHC would also complement the physics program of the SSC and electron colliders.

REFERENCES

1. See, for example, U. Amaldi et al., Phys. Rev. D36 (1987) 1385.
2. Proceedings of the 1982 DPF Summer Study on Elementary Particle Physics and future Facilities, Snowmass. CO, eds. R. Donaldson, R. Gustafson, and F. Paige.
3. Proceedings of the 1984 DPF Summer Study on the Design and Utilization of the SSC, Snowmass, CO, eds. R. Donaldson and J. G. Morfin.
4. Proceedings of the 1986 DPF Summer Study on the Physics of the SSC, eds. R. Donaldson and J. Marx.
5. Proceedings of the Workshop on Experiments, Detectors, and Experimental Areas for the SSC, Berkeley, July 1987, eds. R. Donaldson and M. G. Gilchriese.
6. Proceedings of the Madison Workshop "From Colliders to Supercolliders," Int. J. Mod. Phys. A2 (1987).
7. Proceedings of the La Thuile Workshop on the Physics of Future Accelerators, ed. J. H. Molvey, Jan. 1987.
8. Proceedings of the ECFA-CERN Working Groups on LEP2, ED. J. Mulvey.
9. Eichten, E., I Hinchliffe, K. Lane, and C. Quigg, Rev. Mod. Phys. 56 (1984) 579.
10. Arnison, G. et al., Europhys. Lett. 1 (1986) 327; Albajan, C. et al., Phys. Lett. B19S (1987) 261.
11. Ansari, R. et al., Phys. Lett. B195 (1987) 613.
12. See the talk of J. Freeman at the 7th Topical Workshop on $p - \overline{p}$ Collider Physics, FNAL, July 1988.
13. See the talk of K. Eggert at the 7th Topical Workshop on $\overline{p} - p$ Collider Physics, FNAL, July 1988.
14. Appel. J.A. et al., Phys. Lett. B176 (1986) 239.
15. Albajar, C. et al., CERN-EP/88-45.
16. See the talk of R. Blair at the 7th Topical Workshop in $\overline{p} - p$ Collider Physics, FNAL, July 1988.
17. This has recently been reviewed by H. Haber et al. in the SLAC TLC Studies.
18. Froidevaux, D., ref. 6.
19. Barger, V., T. Han, and J. Ohnemus, Phys. Rev.
20. See the contribution of I. Hinchliffe for a new analysis.

21. A recent paper from which literature can be traced is G. L. Kane, J. Vidal, and C.-P. Yuan, UM-th-88-16.
22. Marciano, W., et al., ref 1.
23. Kane, G. L., S. Swaminathan, and W. Rolnick, to be published.
24. See Weiler, T., and T Yuan, Vanderbilt preprint, for recent calculations and a summary of earlier week.

New Particle Signals at the SSC and at an Upgraded Tevatron Collider*

R. Michael Barnett[a], Robert J. Hollebeek[b], Andrew P. White[c], John Yoh[d], Howard A. Baer[e], Bruce A. Barnett[f], Estia Eichten[d], James E. Freeman[d], Georgio Gamberini[a], J.A. Grifols[g], John F. Gunion[h], Howard E. Haber[i], John Haggerty[j], Clemens A. Heusch[i], JoAnne L. Hewett[k], Ian Hinchliffe[a], T. Kamon[l], S. Kim[l], A. Méndez[g], Jean-Pierre Mendiburu[m], Sergio F. Novaes[a], Frank E. Paige[j], Rajendran Raja[d], Thomas G. Rizzo[n], Xerxes Tata[o,k], Jeffrey Woodside[p]

[a]*Physics Division, Lawrence Berkeley Laboratory, Berkeley, CA 94720*
[b]*Department of Physics, University of Pennsylvania, Philadelphia, PA 19104*
[c]*Department of Physics, University of Florida, Gainesville, FL 32611*
[d]*Fermilab, Batavia, IL 60510*
[e]*Department of Physics, Florida State University, Tallahassee, FL 32306*
[f]*Department of Physics, Johns Hopkins University, Baltimore, MD 21218*
[g]*Grup de Física Teòrica, UAB, 08193 Bellaterra (Barcelona), Spain*
[h]*Department of Physics, University of California, Davis CA 95616*
[i]*SCIPP, University of California, Santa Cruz, CA 95064*
[j]*Physics Department, Brookhaven National Laboratory, Upton, NY 11973*
[k]*Department of Physics, University of Wisconsin, Madison, WI 53706*
[l]*Inst. of Physics, University of Tsukuba, Ibaraki-ken 305, Japan*
[m]*College de France, Paris, France*
[n]*Ames Lab. and Department of Physics, Iowa State University, Ames, IA 50011*
[o]*Department of Physics and Astronomy, University of Hawaii, HI 96822*
[p]*Department of Physics, Oklahoma State University, OK 74078*

ABSTRACT

We have studied the production and detection of several types of new particles at the Superconducting Super Collider (SSC) and at three possible upgrades of the Fermilab Tevatron Collider. We compare the physics potential of the SSC with that of an upgraded collider, and we discuss in depth the relative capabilities of the three Tevatron Collider upgrades. From a physics standpoint, we suggest that one of the proposed upgrades has several advantages.

* This work was supported by the Director, Office of Energy Research, Office of High Energy and Nuclear Physics, Division of High Energy Physics, of the U.S. Department of Energy under Contract Nos. DE-AC03-76SF00098, DE-AC02-76-ER0-3071, DE-AS03-76ER70191, DE-AA03-76-SF00010, DE-AC02-76ER00881 and W-7405-Eng-82, by the National Science Foundation under agreement no. PHY83-18358, and by the University of Wisconsin Research Committee with funds granted by the Wisconsin Alumni Research Foundation.

1. Introduction

The "New Particles at Hadron Colliders" group had several general tasks.

a) Refine the techniques for finding new particles at the SSC and at an upgraded Tevatron Collider. For the most part we concentrated on those areas where previous studies were inadequate or non-existent.

b) Investigate whether there is a gap between the capabilities of the SSC and the Tevatron. If there is a gap, would an upgraded Tevatron Collider be able to make a significant difference ? It has also been suggested that any gap could be filled by running the SSC at lower energies. The SSC Conceptual Design Report states that the SSC can be run at lower energies with high luminosity (at $\sqrt{s} = 10$ TeV, luminosity would be 5×10^{32} $cm^{-2}s^{-1}$; at $\sqrt{s} = 3$ TeV, luminosity would be 10^{32}). However, we did not take this possibility into account.

c) Compare three proposed upgrades for the Tevatron Collider. We examined the upgrades to see which allowed the best search limits for various types of new particles and which upgrades could best study any particles which were discovered. It is important to decide which signals are feasible because the branching ratios for each signal is different and can lead to differing conclusions for each upgrade.

The table shows the parameters which were our starting assumptions. Since these are hypothetical upgrades, the parameters may change in time. By "future runs" we refer to Tevatron Collider runs before the upgrades A, B or C.

The new particles discussed here include new heavy leptons for which we show for the first time that they could be observable at the SSC. For new W' and Z' bosons we have concentrated on the possibilities at an upgraded Tevatron Collider. We made an intensive study of supersymmetric particles because of serious deficiencies in previous studies; this work is not complete but the preliminary indication is that techniques for finding gluinos and squarks do exist. We looked carefully at the requirements for finding supersymmetric particles at Tevatron upgrades. Work on new heavy quarks (beyond the top quark) is continuing and is not reported here. Higgs bosons and technicolor were assigned to the Electroweak Symmetry Breaking group at Snowmass and are therefore not included. No new work was done on compositeness (substructure) and on the particles predicted by superstring theories except as they fall into the above categories.

2. Heavy Leptons

2.1. Heavy Lepton Introduction

Heavy leptons are some of the most difficult new particles to find at a hadron collider. Their production rate is small since they must be produced by electro-weak interactions. Backgrounds are very large, and strong cuts are needed to bring them below the signals. Previous studies of the detection prospects have concentrated on the rate and have not studied the backgrounds[1,2] in detail or have concluded that signals cannot be extracted due to the large backgrounds.[3] As described below, Ian Hinchliffe has devised a set of cuts which do, in principle, allow a signal to be established.

Let us consider a fourth generation lepton doublet $\binom{L}{N}$, with L (N) being the charged (neutral) member. We shall further assume that L is heavier than N. Some constraints on the existence of such a doublet exist already. Measurements of the value of W and Z masses and the weak mixing angle imply that $|m_L - m_N| < 310$ GeV (90% confidence).[4] This bound could be upset by other particles added to the standard model. There is a limit $M_L > 41$ GeV (95% confidence) from the UA1 collaboration[5] working at the $Sp\bar{p}S$ collider. This limit arises from the non-observation of electrons from the process $W \rightarrow LN \rightarrow e\nu N\overline{N}$ and assumes that N is stable and is light.

OPTION		$\sqrt{s}$	$\mathcal{L}$	sec/year	$\int \mathcal{L}\, dt$
A	$p\bar{p}$	2 TeV	2×10^{31}	0.5×10^{7}	100 pb^{-1}
B	pp	2 TeV	10^{32}	10^{7}	1000 pb^{-1}
C	$p\bar{p}$	3 TeV	6×10^{30}	0.5×10^{7}	30 pb^{-1}
This run	$p\bar{p}$	1.8 TeV			$3 - 5$ pb^{-1}
Future runs	$p\bar{p}$	1.8 TeV			$15 - 20$ pb^{-1}?
SSC	pp	40 TeV	10^{33}	10^{7}	$10,000$ pb^{-1}

If the N is sufficiently light, it will contribute to the Z width

$$\Gamma(Z \to N\overline{N}) = 160\sqrt{1 - 4m_L^2/m_Z^2}(1 - m_L^2/m_Z^2) \quad MeV.$$

Therefore experiments which count the number of neutrinos will be sensitive to it provided that it is stable. Currently one cannot exclude such a neutrino.[6] However, experiments at LEP should be able either to detect such a lepton or to set a limit of order 40 GeV on its mass.[7] (A better limit can probably be obtained at LEP2, up to 80 GeV or so, but only if the N is unstable). Charged leptons of masses up to 80 GeV or so will be found at LEP2.[7] In the context of a discussion of searches at hadron colliders only the SSC and LHC are relevant. It should be possible to improve the UA1 limit somewhat at the Tevatron collider, but the limiting factor is rate; only leptons accessible in W decay are produced copiously enough to be found.

2.2. Heavy Lepton Discussion and Results

Apart from W decay, which is not relevant to heavy lepton production in high energy colliders, there are four relevant production mechanisms. First, L^+L^- or LN pairs can be made by quark antiquark annihilation via a virtual Z/γ or W. The rate for this is well predicted, given the L and N masses, since it depends only on the quark and antiquark distribution functions which are well known.[1]

Second, an L^+L^- (or $N\overline{N}$) pair can be produced in a gluon gluon collision via the process shown in Fig. 1.[8]

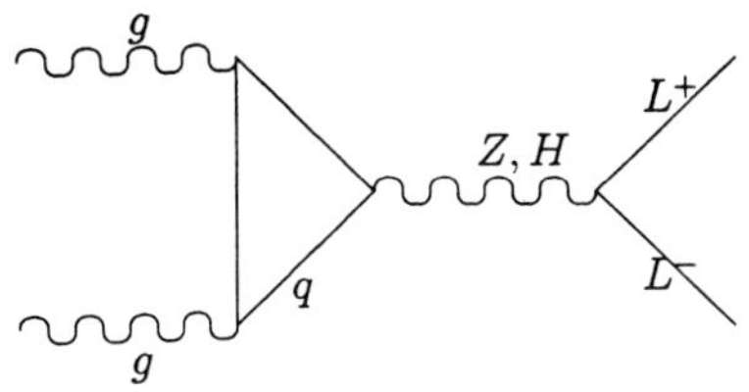

Figure 1. Feynman diagram showing the process $gg \to L^+L^-$. The loop contains a sum over all quark flavors.

The loop contains quarks. The contribution from the virtual Z is zero unless there is a quark doublet the masses of whose members straddle the mass of the virtual Z. The contribution from the Higgs depends on the quark masses, but there are no cancellations between the individual quark contributions. The presence of an additional lepton doublet in the standard model requires the existence of another quark doublet to cancel anomalies. The

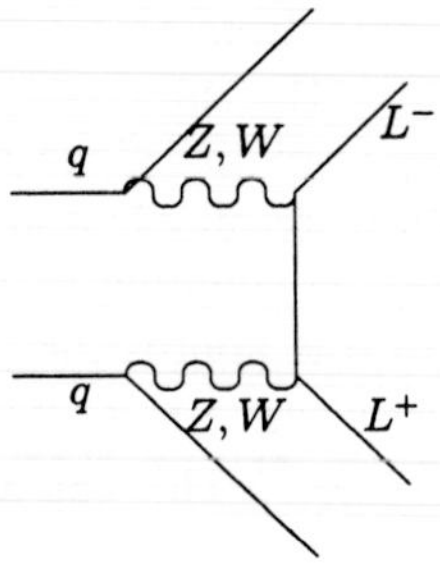

Figure 2. Feynman diagram showing the process $qq \to L^+L^-qq$.

masses of these quarks are very important in determining the production rate from this process.

Third, there is the process $qq \to qqL^+L^-$ via intermediate W's or Z's, see Fig. 2. This process produces a rate which is negligible[9] with respect to the ones from the processes above and is not discussed further.

Fourth, there is the possibility that some other new particle, produced copiously, could decay into the heavy lepton. (Recall the UA1 limit discussed above.) The rate from such a process is very model dependent.

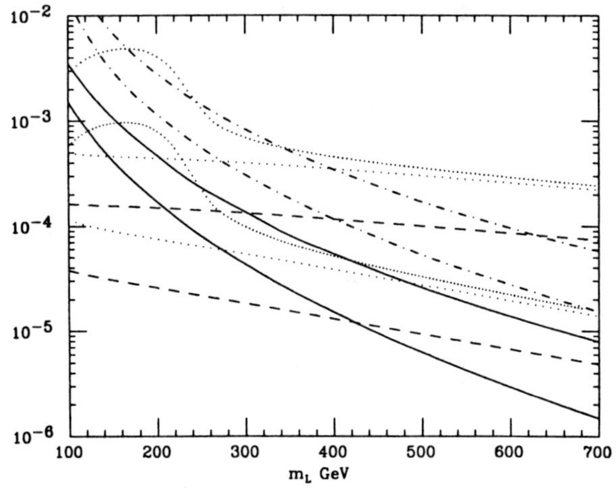

Figure 3. Total cross-sections for heavy lepton production in pp collisions as a function of the charged lepton (L) mass; N is assumed to be massless. A pair of lines is shown for each process; the upper (lower) is for $\sqrt{s} = 40(17)$ GeV. The solid lines are for the process $q\overline{q} \to L^+L^-$ and the dot-dashed for $q\overline{q} \to L^{\pm}N$ (charges summed). The rates from the process $gg \to L^+L^-$ are shown with $m_H = 100$ GeV and no 4th generation of quarks (dashed lines), $m_H = 100$ GeV and the 4th generation quark masses set to m_L and $m_L + 100$ GeV (dotted lines), and $m_H = 500$ GeV and the 4th generation quark masses set to m_L and $m_L + 100$ GeV (close dotted lines).

Figure 3 shows the total cross-section from the first three of these processes; the formulae are given at the end of this section for completeness. In the case of the third process several lines are shown corresponding to different choices of the Higgs and fourth generation quark masses. It can be seen from this figure that the rates are of order 1 pb over the interesting range of masses. Notice that the rate from the gluon-gluon process falls relatively slowly with increasing lepton mass. This occurs because the Higgs-lepton coupling is proportional to the lepton mass. The rate is approximately independent of the Higgs mass (m_H) if $m_L > m_H$. If $m_H > m_L$ the rate increases as m_H increases.[8] This can also be seen clearly from Fig. 3 where rates from Higgs masses of 100 and 500 GeV are shown.

It can be seen from this figure that the rate from the gluon-gluon initial state is dominant over most of the relevant range of masses if we are interested in the L^+L^- final state. In what follows we shall take the fourth generation down and up quark masses to be $m(Q_4) = m_L$ and $m_L + 100$ GeV and shall assume $m_H = 100$ GeV. While the rates so obtained are model dependent, they could be larger if either m_H or the splitting between the charge 2/3 and charge 1/3 quarks is larger. The charged lepton L will decay into "W"$+N$, where the W is real if $m_L - m_N > m_W$ and virtual otherwise. Initially we will assume that N is stable.

Let us consider first the LN final state (which has a larger cross-section than that of the L^+L^- final state if $m_L >> m_N$). This then gives rise to a W and missing transverse momentum, carried off by the N's. If one attempts to detect the W via its *leptonic* decay mode, the signal is now an isolated lepton and missing transverse momentum. This signal is overwhelmed by isolated leptons from the process $q\overline{q} \to l\nu$ via an intermediate W boson.[3] An attempt to detect W decaying *hadronically* faces a background from the final state ZW where the Z decays to neutrinos and the W hadronically. This background is not overwhelming but there is an additional, much more problematic, final state of $Z + jets$ where the Z decays to neutrinos and the jet system is indistinguishable from that arising from a W decay.

Much work[10] has been done on the problem of resolving a W decaying hadronically from a jet system, mostly in the context of searches for the Higgs boson decaying to W^+W^- where one W decays leptonically and the other hadronically. Here the dominant background is $W + jets$. One requires that the invariant mass of the jet system be close to the W. Other methods such as an attempt to resolve the jet system into two jets (due to q and $\overline{q}$ in the W decay case), followed by cuts on the angular distribution of these jets[10], or a cut on the particle multiplicity[11] of the jet system may increase the background rejection somewhat.[12] The multipicity cut is considered because the W decays to quarks whereas the background jet system is mainly gluons; gluon jets are expected to produce larger multiplicity. In the LN case, a cut on the invariant mass of the jet system $m_{jets} = m_W \pm 10$ GeV produces a signal to background ratio of less than 0.1 at $\sqrt{s} = 40$ TeV

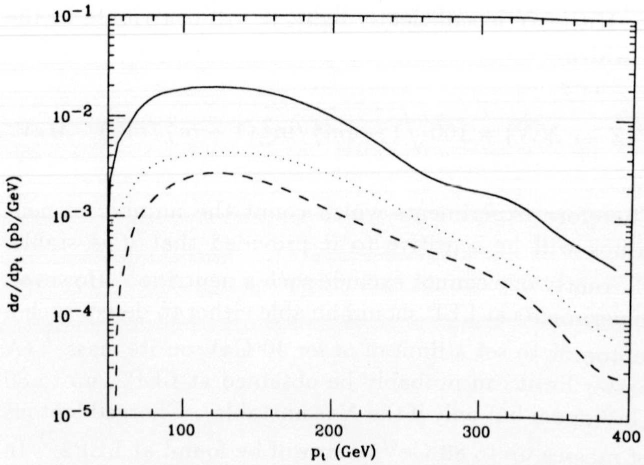

Figure 4. The transverse momentum distribution $d\sigma/dp_t$ for a W boson from the process $pp \to L^\pm N + N \to W^\pm N\overline{N} + X$ at $\sqrt{s} = 40$ TeV. There is missing transverse momentum which balances that of the W. The W is decayed to $q\overline{q}$ and the following cuts on the rapidity (y), transverse momentum (p_t) and azimuthal angle (ϕ) of the quark and anti-quark made: $|y| < 5$, $p_t > 50$ GeV, $\Delta R\sqrt{((\Delta\phi)^2 + (\Delta y)^2)} > 0.5$. The last cut is an attempt to ensure that the two jets from the quark and antiquark are distinct (ΔR is the separation of the partons). The dotted (dashed) line corresponds to $m_L = 100$ (200) GeV; $m_N = 0$. The solid line is the rate $d\sigma/dp_t$ from the process $pp \to Z(\to \nu\overline{\nu}) + 2partons + X$ shown as a function of p_t of the Z. The two parton system is subject to the same cuts as above. In addition the invariant mass of the system must be in the range $M_W \pm 10$ GeV.

energies. This is shown in Fig. 4 where the background is calculated from the final state Z+2 *partons*. The signal to background ratio is slightly worse at $\sqrt{s} = 17$ TeV and the event rate is lower. Some of the other cuts may be able to increase the signal to background ratio somewhat, nevertheless, it is difficult to be optimistic about the observability of this LN mode. This conclusion is independent of the mass of the N.

Turning now to the final state L^+L^-, it clearly will produce, after decay, a state of two W's and missing transverse momentum. If both W's decay leptonically, one will see events with two leptons of opposite charge and missing transverse momentum (carried off by $N\overline{N}$ and the neutrinos from the W decays). This is very similar to the state in which the tau was discovered at SPEAR.[13] In this channel the dominant background arises from W^+W^- events which can also give rise to two charged leptons and missing transverse momentum.

However, we show in Fig. 5 that if the event is projected into the transverse plane and the distribution in angle (ϕ) between the two leptons examined it will be found that there is a range over which the signal exceeds the background. Notice that $\phi = 0$ corresponds to the back-to-back configuration. In this plot we have included the

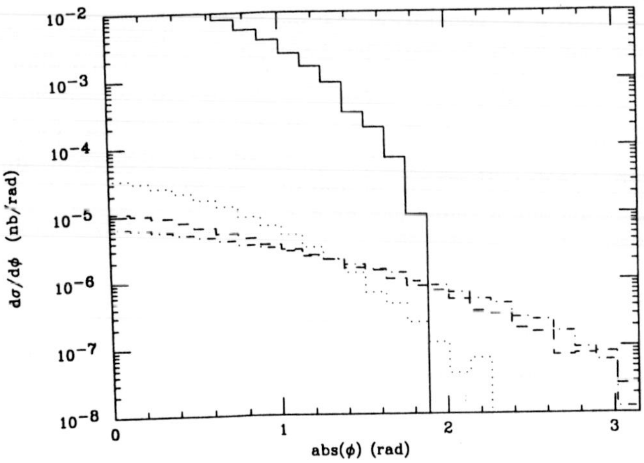

Figure 5. The distribution in the variable ϕ defined as the absolute value of the difference in the azimuthal angle of the two charged leptons resulting from the process $pp \to L^+L^- + X \to l^+l^- N\overline{N}\nu\overline{\nu} + X$ at $\sqrt{s} = 40$ TeV. N is taken to be massless. The charged leptons (l) which can be either e or μ are required to satisfy $p_t > 50$ GeV and $|y| < 3$. The dotted, dashed and dot-dashed histograms correspond to m_L=100, 200 and 400 GeV. The solid histogram is the background arising from $pp \to WW + X \to l^+l^-\nu\overline{\nu} + X$

decay channels $W \to e\nu$ and $W \to \mu\nu$ and have required that the charged leptons have $p_t > 50$ GeV and $|y| < 3$ and have assumed $\sqrt{s} = 40$ TeV. Requiring $|\phi| > 2$ results in approximately 3 signal events and no background in the case of $m_L = 400$ GeV assuming the canonical SSC integrated luminosity of $10^{40} cm^{-2}$. Several comments are in order.

The cut on ϕ becomes more effective at larger values of m_L, since the charged leptons become less tightly correlated than those in the background. Indeed, it is probably not possible to detect a heavy lepton of mass below 120 GeV by this method. There is an additional source of background from the W pairs arising from the production and decay ($b' \to Wt$) of a pair of fourth generation quarks,* and also from the production of top quarks if the top quark is heavier than the W. These events occur at a much larger rate than the production of W pairs (provided $m_{b'} < 600$ GeV)) but have more jets (from the t and $\bar{t}$) in them; they must be rejected by requiring the absence of such jets. There is a possible background arising from the final state of $ZZ^* \to \nu\overline{\nu}e^+e^-$, where Z^* indicates an off-shell Z boson. This background, if relevant, can be eliminated by looking at $e\mu$ final states.

The number of events passing the cuts is small and the detection of a heavy lepton by this method is limited by the available integrated luminosity rather than by the background. If experiments are possible at higher integrated luminosities or the Higgs mass is larger so that the event rate is increased (see Fig. 3) then this mode is exploitable. At $\sqrt{s} = 17$ TeV the rate is lower by a factor of 4 or so and the signal correspondingly more difficult to extract.

The event rate is larger if one of the W's is detected via its hadronic decay modes. In this case the signal consists of an isolated lepton and some hadronic jets whose invariant mass reconstructs to the W mass. One background arises from W pairs as before. But now there is a potentially much larger source – the final state $W + jets$ where the $jets$ system has invariant mass near the W mass and is indistinguishable from the jets from a real W decay.

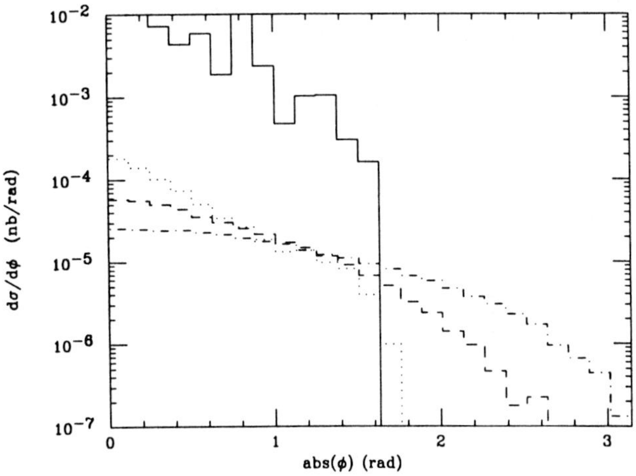

Figure 6. The distribution in the variable ϕ defined as the absolute value of the difference in the azimuthal angle of the charged leptons and the hadronic system from the W resulting from the process $pp \to L^+L^- + X \to l^\pm W^\mp(\to hadrons)N\overline{N}\nu + X$ at $\sqrt{s} = 40$ TeV. N is taken to be massless. The charged lepton (l) which can be either e or μ is required to satisfy $p_t > 20$ GeV and $|y| < 3$. The W is decayed to $q\overline{q}$ and the following cuts on the rapidity (y), transverse momentum and azimuthal angle (ϕ_q) of the quark and antiquark made: $|y| < 5$, $p_t > 50$ GeV, $\sqrt{((\phi_q - \phi_{\overline{q}})^2 + (y_q - y_{\overline{q}})^2)} > 0.5$. The dotted, dashed and dot-dashed histograms correspond to m_L=100,200 and 400 GeV. The solid histogram is the background arising from $pp \to W + jets + X \to l^\pm\overline{\nu} + jets + X$. The two parton system is subject to the same cuts as the quark and antiquark in the signal, In addition the invariant mass of the system must be in the range $M_W \pm 20$ GeV.

Figure 6 shows the distribution in the angle ϕ in the transverse plane between the lepton and the vector formed by the jet system. At the parton level the background consists of 2 jets which are here required to have an invariant mass of $M_W \pm 20$ GeV. The background is severe but again a cut on ϕ can reject it. A cut requiring that $|\phi| > 1.7$ rejects all the background.** For $m_L = 400$

* We are assuming that the charge 1/3 member (b') of the quark doublet is the lighter.

** The background from W pair events is also eliminated by this cut.

GeV, 70 signal events remain at $\sqrt{s} = 40$ TeV for an integrated luminosity of $10^{40} cm^{-2}$. The rate is about a factor of 4 less at $\sqrt{s} = 17$ TeV. Notice that it is not necessary to measure the missing transverse momentum in the event, and an additional cut on the missing transverse momentum is not useful.

If there exists a new W or Z boson then the production rates for a heavy lepton will rise. As an example consider the case of a new W boson of mass 1 TeV (400 GeV) with the same coupling strength to quarks and leptons as the standard model W (it could, for example, be one occurring in a left-right symmetric model). The production cross-section[1] at SSC is of order 0.1 nb (1.5 nb). The branching ratio into LN will be approximately

$$6\%\sqrt{1 + a^2 + b^2 - 2(a + b + ab)(1-a/2-b/2-(a-b)^2/2))}$$

where $a = m_L^2/m_W^2$ and $b = m_N^2/m_W^2$. It is clear therefore that the rate of LN production from new W's of mass 1 TeV is *less* than that from the quark-antiquark annihilation process shown in Fig. 3. In the case of a 400 GeV new W and an L of mass 300 GeV (N is massless), the rate for LN production is *increased* by about a factor of 15. This increase may not be sufficient to overcome the large backgrounds in this channel.

The signals are much more complicated if the N decays inside the detector. The N can decay via its mixing with other neutrinos (separate lepton number conservation for each generation is violated). The possible decays are $N \to l^+ l^- \nu$ where l is any other charged lepton (l^+ and l^- could be τ^+ and μ^- for example). In this case detection will be much easier since the event will have additional charged leptons. The LN final state will produce 4 leptons and a W; a signal which will stand clearly above backgrounds.

2.3. Heavy Lepton Summary

The discussion in this paper has been at the parton level only. A fully convincing demonstration of a signal is only possible with the use of a full Monte-carlo event generator and a detector simulation. Since one of the backgrounds involves jets and W's, such a generator needs to have this final state as one of its options. This final state is only implemented in leading-log approximation in current event generators.[14] This approximation may only be accurate to a factor of three or so. The effectiveness of cuts involving leptons used to establish a signal are very unlikely to be compromised by a full simulation. The cuts on the jet system are more problematic but are not very stringent; the main issue is how well the reconstructed jet direction follows that of the partons which initiate the jet.

In conclusion, it has been demonstrated in this work by I. Hinchliffe that it is possible to devise a set of cuts whereby a 4th generation charged lepton in a doublet with a stable neutrino could be detected in a hadron collider. The detection is limited by event rate and hence the highest energy and integrated luminosity is needed. In the event that the neutrino is unstable and decays within the detector, there are no relevant standard model backgrounds and a signal can be seen.

2.4. Heavy Lepton Appendix

The cross-sections for the relevant production processes are given here. s is the total center of mass energy squared of the initial partonic system. There are two contributions from the gluon-gluon initial state which add incoherently. Firstly from Fig. 2 where the intermediate particle is a Z.

$$\sigma(gg \to L^+L^-) = \frac{\beta\alpha^2\alpha_s^2}{2048\pi \sin^4\theta_W} \frac{m_L^2}{m_W^4}|I|^2$$

where $\beta = \sqrt{1 - 4m_L^2/s}$, α and α_s are the electromagnetic and strong coupling constants and θ_W is the weak mixing angle. Also

$$I = 2\sum_Q (\pm) \int_0^1 dx \int_0^{1-x} dy \frac{xy}{xy - m_Q^2/s}$$

The sum runs over all quarks of mass m_Q. The + (-) sign applies to charge 2/3 (-1/3) quarks. If the intermediate particle is a Higgs boson

$$\sigma(gg \to L^+L^-) = \frac{\beta^3\alpha^2\alpha_s^2}{4608\pi \sin^4\theta_W} \frac{m_L^2}{m_W^4} \times |J|^2 \frac{s^2}{(s - m_H^2)^2 - \Gamma_H^2 m_H^2}$$

Here m_H (Γ_H) is the mass (width) of the Higgs boson and

$$J = 3\sum_Q \int_0^1 dx \int_0^{1-x} dy \frac{1 - 4xy}{1 - xys/m_Q^2}$$

and again the sum runs over all quark flavors.

Secondly the quark-antiquark initial states give

$$\sigma(q_i\overline{q_i} \to L^+L^-) = \frac{8\pi\alpha^2}{9s} \times \left[\frac{\beta B}{2}\left(e_i^2 - \frac{e_i s(s - m_Z^2)(L_q + R_q)(L_e + R_e)}{8\sin^2\theta_W \cos^2\theta_W((s - m_Z^2)^2 + \Gamma_Z^2 m_Z^2)}\right) + \frac{\beta s^2(L_e^2 + R_e^2)[B(L_Q + R_Q)^2 + 2\beta^2(L_Q - R_Q)^2]}{256\sin^4\theta_W \cos^4\theta_W((s - m_Z^2)^2 + \Gamma_Z^2 m_Z^2)}\right]$$

where $B \equiv 3 - \beta^2$, e_i is the charge of the quark of type i, $L_e = 2\sin^2\theta_W - 1$, $R_e = 2\sin^2\theta_W$, $L_q = \tau_3 - 2e_i \sin^2\theta_W$

and $R_q = -2e_i \sin^2\theta_W$ with $\tau_3 = 1(-1)$ if $e_i = 2/3(-1/3)$.

$$\sigma(u\bar{d} \to L^+N) = \frac{\pi\alpha^2|U_{ud}|^2}{48\sin^4\theta_W} \frac{\beta s}{(s-m_W^2)^2 - \Gamma_W^2 m_W^2}$$

$$\times \left(1 + \frac{\beta^2}{3} - \left(\frac{m_L^2 - m_N^2}{s}\right)^2\right)$$

Here U is the appropriate element of the Kobayashi-Maskawa matrix and in this equation

$$\beta = \left(1 - 2\frac{m_L^2 + m_N^2}{s} + (\frac{m_L^2 - m_N^2}{s}\right)^2)^{1/2}$$

3. Heavy W' and Z' Bosons

3.1. Heavy Bosons Introduction

In this chapter we summarize the potential discovery limits for new heavy bosons ($W'^{\pm}$ and Z'^{0}) at the SSC and at an upgraded Tevatron Collider. We consider the production of heavy gauge bosons within classic Left-Right (LR) Symmetric gauge models[15] and within the context of E_6 superstring-inspired models. We also address the question of whether there is sufficient information available at hadron colliders to uniquely determine the couplings of Z' bosons and to distinguish between the various models. In all cases we are considering the production of heavy gauge bosons and their subsequent decays into lepton pairs. More precisely, we focus on the processes

$$\begin{Bmatrix} pp \\ p\bar{p} \end{Bmatrix} \to \begin{Bmatrix} Z'^0 \to e^+e^- \\ W'^{\pm} \to e^{\pm}\nu_R \end{Bmatrix} .$$

or into modes with muons.

The results are clearly model dependent since the couplings and branching ratios are fixed only in the context of various assumptions. The work reported in the next subsection on the classic Left-Right models was done by Grifols, Mendez and Barnett, while the E_6 inspired work (which also includes Left-Right models) was done by Hewett and Rizzo. It is important to note that the two groups used different benchmarks for discovery limits. The first group required 5 events (per year for design luminosity) in the e^+e^- mode while the second group required 5 events in either the e^+e^- or $\mu^+\mu^-$ mode (i.e.- the first group required twice as many events as the second). For SSC the yearly integrated luminosity as taken to be 10^4 pb^{-1} whereas it was 100, 1000 and 30 pb^{-1} for Tevatron upgrades A, B and C (see Chapter I).

3.2. Heavy Bosons from Left-Right Models

To be specific we chose a particular model[15] in which the vertex $Z'ff$ is given by: $g\gamma_\mu(G_L P_L + G_R P_R)$ where $P_L = (1-\gamma_5)/2$, $P_R = (1+\gamma_5)/2$ and where

$$\begin{aligned}
G_L(Z'uu) &= -C\sin^2\theta_W/6 \\
G_R(Z'uu) &= -C(7\sin^2\theta_W - 3)/6 \\
G_L(Z'dd) &= -C\sin^2\theta_W/6 \\
G_R(Z'dd) &= \quad C(5\sin^2\theta_W - 3)/6 \\
G_L(Z'ee) &= \quad C\sin^2\theta_W/2 \\
G_R(Z'ee) &= \quad C(3\sin^2\theta_W - 1)/2 \\
G_L(Z'\nu\nu) &= \quad C\sin^2\theta_W/2 \\
G_R(Z'\nu\nu) &= -C(\sin^2\theta_W - 1)/2
\end{aligned}$$

where $C \equiv 1/(\cos\theta_W(\cos 2\theta_W)^{\frac{1}{2}})$. In this model assuming three light generations including the right-handed neutrino ν_R, the width of the new boson is $\Gamma_{Z'} = \alpha C^2 M_{Z'}(1 - 4\sin^2\theta_W + \frac{14}{3}\sin^4\theta_W)/\sin^2\theta_W$ (using $g^2 = 4\pi\alpha/\sin^2\theta_W$).

To evaluate the branching ratios of the Z' we note that the possible decay channels are $e_i^+e_i^-$, $\nu_i\bar{\nu}_i$, $u_i\bar{u}_i$, $d_i\bar{d}_i$, W^+W^- and ZH, where $i = 1,2,3$ is a generation index and H is the light "standard" Higgs boson of the LR models. In the usual LR model with a scalar sector consisting of one bidoublet and two triplets, the last two channels are always allowed and it can be shown[16] that the mixing $Z_L - Z_R$ (which is fixed by the masses and is very small) leads to $BR(Z' \to WW) \approx BR(Z' \to e\ +e\ -)$. On the contrary, the mixing between W_L and W_R can vary from zero to some upper bound depending on the relative values of the bidoublet vev's κ and κ'. In fact the zero value for the mixing (corresponding to $\kappa' = 0$) is required phenomenologically.[17] For the W' the possible decay channels then are $e_i\bar{\nu}_{Ri}$ and $d_i\bar{u}_i$ (i.e.- we do not consider the channels WZ and WH since they would only be allowed if the $W_L - W_R$ mixing were not zero).

The total width is obtained by summing over the partial widths of all these channels.[16] The branching ratios of interest are then

$$BR(Z'^0 \to e^+e^-) = 1.8\% , \quad BR(W'^{\pm} \to e^{\pm}\nu_R) = 8\% .$$

The exclusion of the WW and ZH channels would increase $BR(Z' \to e^+e^-)$ from 1.8 to 1.9%. If ν_R is very heavy, then the branching ratio would be increased to 2.3%.

The production rates are shown in Figs. 7 and 8 for different upgrade options of the Tevatron. For reference, the corresponding rates for the SSC are shown in Fig. 9. The rapidity cuts are introduced in the initial quark-antiquark system. Figure 10 shows the sensitivity of the cross sections to the rapidity cut for the SSC case.

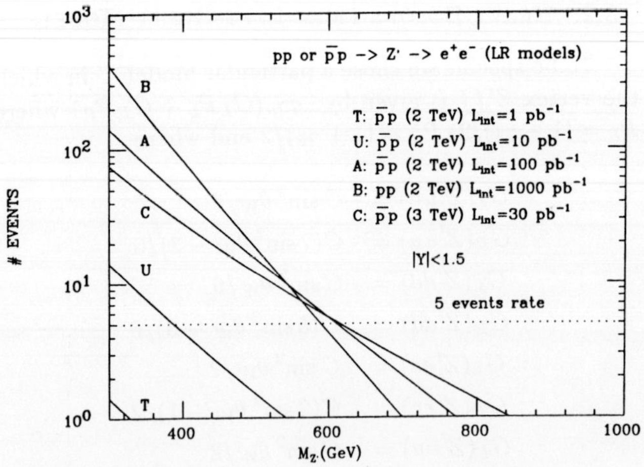

Figure 7. The production rates for Z'^0 from the model of Ref. 15 at the Tevatron Collider.

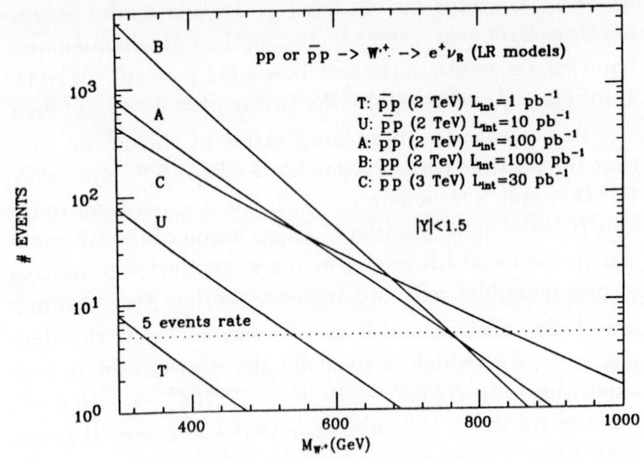

Figure 8. The production rates for W'^+ from the model of Ref. 15at the Tevatron Collider. Note that all rates are the same for W'^- for $p\bar{p}$ colliders, but that the rates for W'^- are a factor of about two lower at pp machines.

3.3. Heavy Bosons from E_6 Inspired Models

Interest in the possibility of observing new neutral gauge bosons at hadron colliders has been renewed due to the advent of E_6 superstring-inspired models.[18] In our work below we will concentrate our discussion on two classes of models originating from E_6:

$$\begin{array}{ll} \text{(A)} & \mathrm{SU}(3)_C \times \mathrm{SU}(2)_L \times \mathrm{U}(1)_Y \times \mathrm{U}(1)_\theta \\ \text{(B)} & \mathrm{SU}(3)_C \times \mathrm{SU}(2)_L \times \mathrm{SU}(2)_R \times \mathrm{U}(1) \end{array} \qquad (1)$$

Models of type (A) are the so-called 'effective rank-5' models (ER5M) which contain a free parameter (θ) describing the amount of mixing between the Z_ψ and Z_χ states (from the decomposition $E_6 \to \mathrm{SO}(10) \times \mathrm{U}(1)_\psi \to \mathrm{SU}(5) \times \mathrm{U}(1)_\chi \times \mathrm{U}(1)_\psi$) within the Z'. If the mixing between the Z' and standard model (SM) Z is ignored, then

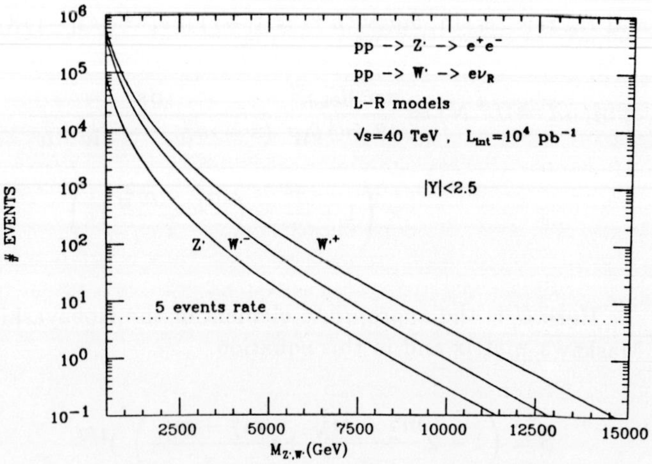

Figure 9. The production rates for Z'^0 and $W'^\pm$ from the model of Ref. 15 at the SSC.

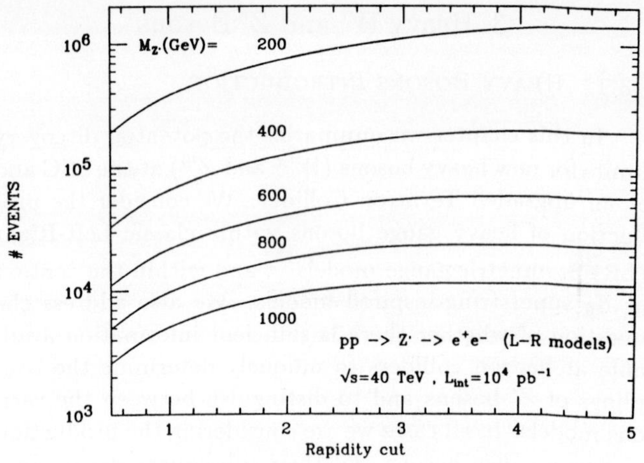

Figure 10. The dependence on the rapidity cut of the production rates for Z'^0 from the model of Ref. 15 at the SSC.

Z' is a mass eigenstate; this is an excellent approximation in general when the Z' mass is $\gtrsim 0.4$ TeV or so, which is the case of interest to us here. Models of type (B) are represented by the usual left-right model[19] (LRM) as well as the alternative left-right model[20] (ALRM). The LRM and ALRM are quite distinct in that their associated Z' bosons have very different couplings and that the right-handed charged gauge boson (W_R) can be light ($\simeq 210$ GeV) in the ALRM whereas it is expected to be heavy ($\gtrsim 2$ TeV) in the LRM. W_R in the ALRM also carries lepton number and negative R-parity so that it cannot be produced by a Drell-Yan-like process at hadron colliders.

In what follows we will address the following questions: (i) what are the search limits for Z' bosons in the various models at hadron colliders and can we distinguish between these models? (ii) Given the unusual nature of the W_R in the ALRM, what search limits can we find for its production? All relevant formulae can be found elsewhere.[18] In setting detailed search limits on the Z' we will assume a minimum of $N = 5$ events in the $\ell^+\ell^-$ channel (where ℓ can be either e or μ). We will later comment on how our results change when a discovery limit of 10 events in this channel is assumed. These search limits will be quite model dependent since the rate for $\overset{(-)}{p}p \to Z'^0 \to \ell^+\ell^-$ depends on the quark couplings to the Z' as well as its leptonic branching fraction (B) and these $Z'f\bar{f}$ couplings are dependent on the parameter θ in the ER5M. Even within a given model, this number may vary depending on, *e.g.*, whether exotic fermions in the E_6 **27**-representation can contribute ($n_g = 3$) or not ($n_g = 0$) to the Z' total width.

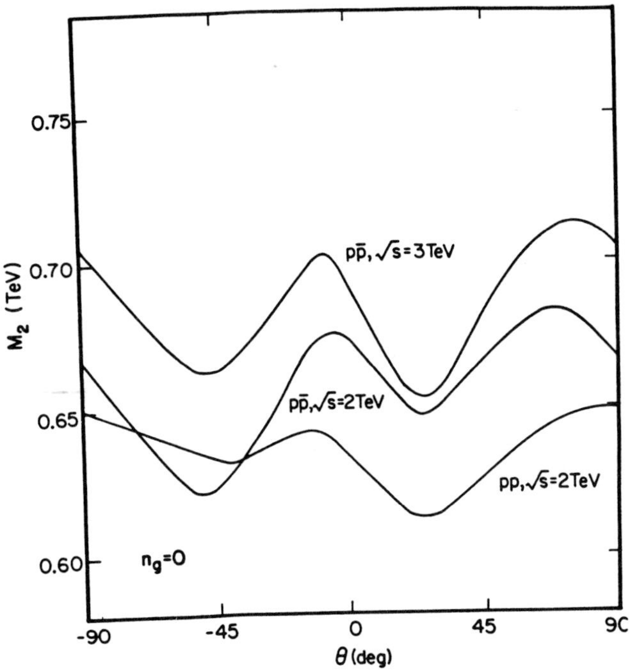

Figure 11. Z' mass limits for the ER5M as a function of θ for the Tevatron uprades assuming $n_g = 0$ and $N = 5$ events.

Figure 11 and 12 show the search limits for the Z' in the ER5M as a function of θ, assuming $N = 5$ and $n_g = 0$, at the Tevatron upgrades and the SSC, respectively. Only the range $-90° \le \theta \le 90°$ is shown since mapping $\theta \to \theta + 180°$ leaves the search limits invariant because it merely flips the sign of all Z' couplings. Thus, without Z-Z' mixing, this ambiguity in θ will always remain. The corresponding limits for the Z' in the LRM and the ALRM can be found in Table 1. The figures and the table show the trade-off between machine energy and luminosity as well as pp *vs.* $p\bar{p}$ at the Tevatron. Note that much stronger limits are obtained for both the LRM and ALRM since the overall strength of the couplings is greater in these two cases than in the ER5M.

How do our Z' search limits change if we give up the assumption of $n_g = 0$ and/or the assumption that $N = 5$ events in the $\ell^+\ell^-$ lepton channel? Table 2 shows the limits (or range of limits in the ER5M) for several other values of (n_g, N) for each of the models under consideration. Note that these limits are significantly weaker for every model in comparison to the $n_g = 0$, $N = 5$ case already discussed. When combined with the $n_g = 0$, $N = 5$ case previously discussed, the results in the table generally show the variation one can expect in Z' limits at various hadron colliders within the E_6 model context.

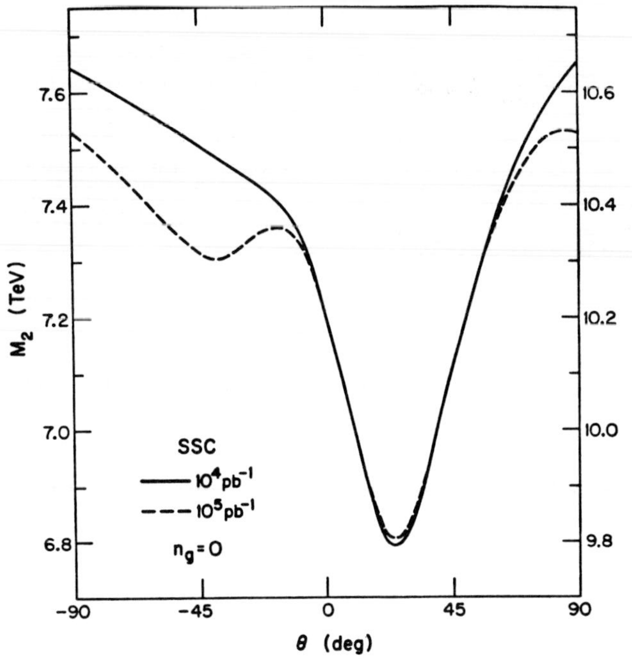

Figure 12. Same as Fig. 1 but for the SSC with $10^4\,\mathrm{pb}^{-1}$ or $10^5\,\mathrm{pb}^{-1}$ integrated luminosity. For 10^4 $(10^5)\,\mathrm{pb}^{-1}$ the left- (right-) hand scale should be used.

Once produced, are these Z' bosons identifiable, *i.e.*, can we distinguish a Z' originating in the LRM or ALRM from one coming from the ER5M? Can we determine the angle θ uniquely? Certainly, in order to make a detailed study of the Z' boson's properties one needs several hundred to a thousand events and, hence, the Z' must be far lighter than the search limits discussed above. If $n_g = 0$, the only Z' properties useful in determining the couplings are the leptonic branching fraction (B_ℓ), the forward-backward asymmetry for leptons, A^ℓ_{FB},[21] and, to a lesser extent, the total Z' width, Γ_0. A measurement of Γ_0 cannot be done very accurately at hadron colliders due to the large energy resolution inherent in such machines. Hadronic Z' decay modes will be invisible due to the large QCD background and will hence be useless in determining the model.

Table 1 and Fig. 13 show the values of A^ℓ_{FB} for the various models under consideration for particular values of the Z' mass. Note that in the scaling limit A^ℓ_{FB} is $M_{Z'}$ and $\sqrt{s}$ *independent* so that A^ℓ_{FB} is not expected to

Table 1

Z' search limits and A^{ℓ}_{FB} values in the LRM and ALRM for the various hadron colliders. In calculating A^{ℓ}_{FB} we take $M_{Z'}(= M_2)$ to be 0.450 TeV (3 TeV) for the Tevatron (SSC). $N = 5$ events and $n_g = 0$ are assumed. Z' masses are in units of TeV.

Machine		$\mathcal{L}$	Z' limit		A^{ℓ}_{FB} limit	
			LRM	ALRM	LRM	ALRM
pp	$\sqrt{s} = 2$ TeV	$10^3\,\text{pb}^{-1}$	0.70	0.76	0.123	−0.230
$p\bar{p}$	$\sqrt{s} = 2$ TeV	$10^2\,\text{pb}^{-1}$	0.75	0.87	0.205	−0.351
$p\bar{p}$	$\sqrt{s} = 3$ TeV	$30\ \text{pb}^{-1}$	0.81	0.97	0.199	−0.344
SSC		$10^4\,\text{pb}^{-1}$	8.30	9.40	0.103	−0.205
SSC′		$10^5\,\text{pb}^{-1}$	11.35	12.55	0.103	−0.205

Table 2

Z' search limits in the ER5M, LRM, and ALRM for $n_g = 3$ with $N = 5$, 10 events and for $n_g = 0$ with $N = 10$ events in the $\ell^+\ell^-$ channel ($\ell = e, \mu$) for various hadron colliders. The Z' masses are in units of TeV.

Machine	Model	$N = 5$, $n_g = 3$	$N = 10$, $n_g = 0$	$N = 10$, $n_g \doteq 3$
$p\bar{p}$ (2 TeV)	ER5M	0.52-0.64	0.58-0.66	0.45-0.57
$p\bar{p}$ (2 TeV)	LRM	0.69	0.68	0.61
$p\bar{p}$ (2 TeV)	ALRM	0.76	0.80	0.69
$p\bar{p}$ (3 TeV)	ER5M	0.48-0.66	0.60-0.68	0.40-0.57
$p\bar{p}$ (3 TeV)	LRM	0.71	0.71	0.62
$p\bar{p}$ (3 TeV)	ALRM	0.81	0.86	0.70
pp (2 TeV)	ER5M	0.53-0.63	0.59-0.63	0.48-0.58
pp (2 TeV)	LRM	0.65	0.65	0.60
pp (2 TeV)	ALRM	0.69	0.71	0.64
SSC	ER5M	5.35-7.20	6.35-7.20	5.05-6.35
SSC	LRM	7.50	7.40	6.65
SSC	ALRM	8.05	8.45	7.15
SSC′	ER5M	8.15-10.05	9.30-9.50	7.25-9.20
SSC′	LRM	10.45	10.40	9.50
SSC′	ALRM	11.15	11.60	10.20

be very sensitive to changes in beam energy or the actual value of $M_{Z'}$. It is clear that a B_ℓ measurement, however

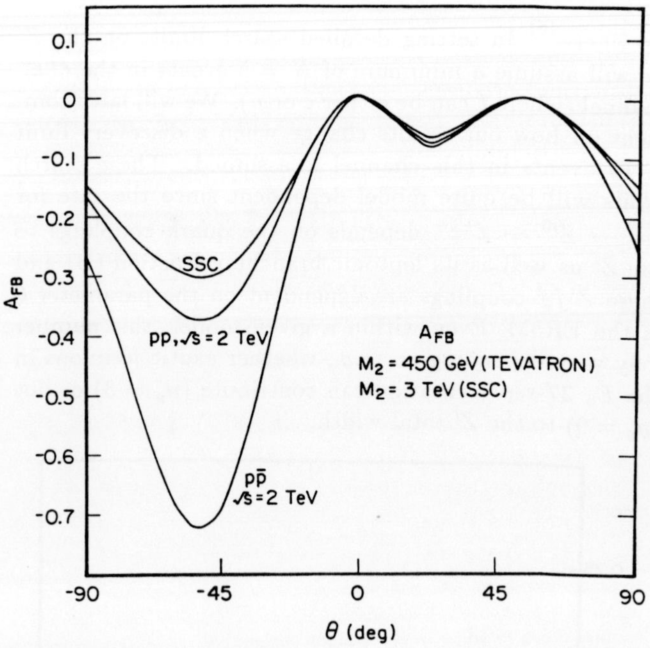

Figure 13. A^{ℓ}_{FB} as a function of θ in the ER5M for the SSC and Tevatron upgrades with $M_Z = 3$ TeV and 450 GeV, respectively.

precise, cannot be used to discriminate between models without an accurate knowledge of Γ_0. The value of A^{ℓ}_{FB} can be used to separate the ALRM and LRM but not the ALRM from the ER5M. The reason for this is that there always exists a value of θ in the ER5M for which $A^{\ell}_{FB}(\text{ER5M}) = A^{\ell}_{FB}(\text{ALRM})$ as can be easily seen from Fig. 13 and Table 1. Thus the sign of A^{ℓ}_{FB} will separate the LRM from the other two models under consideration but will not distinguish the ALRM and the ER5M.

This situation improves somewhat for $n_g = 3$ since there are now additional, unstable, color singlet final states (exotics) accessible in Z' decay for which the QCD backgrounds are quite small. Although the Z' width in this case, Γ_3, would still be only poorly determined, the *relative* branching fractions between these exotic modes and $\ell^+\ell^-$ are Γ_3 independent and could be used to identify the model, at least in principle. If the exotic fermion masses could be determined (by reconstruction) and various exotics separated by their decay final states, then the LRM, ALRM, and ER5M could be easily distinguished and, when combined with A^{ℓ}_{FB} data, can be used to determine θ almost uniquely. The problems with this approach are that the decay modes of these exotics are very dependent on their quantum numbers (baryon and lepton numbers as well as R-parity) assignments, of which there are several possibilities, and that many of the exotics have similar decay patterns. In addition, for many assignments of these quantum numbers, there will be missing energy in the decay final state of these exotics making accurate reconstruction of their masses somewhat difficult. Still, this procedure may hold some promise in determining the couplings of the new Z' boson.

We now turn to W_R search limits in the ALRM. Due to the usual properties of the W_R discussed above it cannot be produced in the usual manner (*i.e.*, by quark-antiquark annihilation). The best mechanism for W_R production has been found to be[20] gluon $+ \overset{(-)}{u} \to W_R^+(W_R^-) + \overset{(-)}{h}$ where h is an exotic fermion with leptoquark quantum numbers and negative R-parity contained in the E_6 **27** representation. Since W_R and h get their masses from the same vev one might expect them to have masses of comparable magnitude. In setting our limits, we have assumed the production of 10 $W_R^\pm \overset{(-)}{h}$ events and the details of the production signal have been discussed by Gunion *et al.* in Ref. 20. Table 3 shows a summary of these results. Note that our limits are roughly dependent on the sum of the W_R mass (M_R) and the exotic h fermion mass (m_h). Note also that the sum of the $W_R^+ h$ and $W_R^- \bar{h}$ cross sections is the same at pp and $p\bar{p}$ machines so that $\sqrt{s}$ and machine luminosity are the only relevant factors (and not the parton luminosities) in determining these search limits. Also, if $M_R \simeq m_h$ then W_R production may be unobservable at the $p\bar{p}$ Tevatron upgrades since $M_R \gtrsim 210$ GeV from already existing data (see Barger *et al.* in Ref. 20). To really explore a reasonable W_R search range it is clear that the SSC is required.

Table 3

Search limits for $W_R^\pm(\bar{h})$ production in the ALRM at various hadron colliders.

Machine		$\mathcal{L}$	$M_R + m_h$ Search limit
$p\bar{p}$	$\sqrt{s} = 2$ TeV	$10^2\,\mathrm{pb}^{-1}$	0.375-0.400 TeV
$p\bar{p}$	$\sqrt{s} = 3$ TeV	$30\,\mathrm{pb}^{-1}$	0.405-0.440 TeV
pp	$\sqrt{s} = 2$ TeV	$10^3\,\mathrm{pb}^{-1}$	0.500-0.525 TeV
	SSC	$10^4\,\mathrm{pb}^{-1}$	4.8-5.6 TeV
	SSC$'$	$10^5\,\mathrm{pb}^{-1}$	6.2-7.4 TeV

It is quite clear that hadron colliders offer a unique opportunity to explore a large mass range in the search for the new gauge bosons expected in E_6 superstring-inspired models.

4. Searches for Supersymmetric Particles

4.1. Supersymmetric Particles Introduction

Supersymmetry is an attractive possible extension of the Standard Model both because of its mathematical elegance and because it provides a natural mechanism for keeping Higgs bosons light even in the presence of a superheavy mass scale such as a grand unification scale or the Planck scale. It has therefore been extensively studied both for existing and for future colliders. The best experimental limits on the masses of gluinos ($\tilde{g}$) and squarks ($\tilde{q}$) come from the UA1 Collaboration[22] at the S$p\bar{p}$S and the CDF Collaboration[23] at the Tevatron. Based on studies of events with jets and substantial missing transverse energy (E_T^{miss}), they find the following limits (independent of the other mass)

$$\begin{aligned} M_{\tilde{g}} &> 53\,\mathrm{GeV}, \quad M_{\tilde{q}} > 45\,\mathrm{GeV} \qquad (\mathrm{UA1}) \\ M_{\tilde{g}} &> 73\,\mathrm{GeV}, \quad M_{\tilde{q}} > 74\,\mathrm{GeV} \qquad (\mathrm{CDF}) \end{aligned} \tag{4.1}$$

The work reported here was done in the context of the minimal supersymmetric model.[24] To give masses to up and down quarks in this model, it is necessary to introduce two Higgs doublets, resulting in three neutral and one charged pair of Higgs bosons after symmetry breaking. The supersymmetric partners of the Higgs bosons mix with the $\tilde{\gamma}$ and $\tilde{Z}$ to make four Majorana states $\tilde{\chi}_i^0$, $i = 1,2,3,4$, which are known collectively as neutralinos and are numbered in order of increasing mass. Similarly the supersymmetric partner of the charged Higgs boson mixes with the $\widetilde{W}$ to make two chargino states $\tilde{\chi}_i^+$, $i = 1,2$. We shall assume that R parity is conserved so that all supersymmetric particles are produced in pairs; each supersymmetric particle then decays into lighter supersymmetric particles, and the cascade continues until the LSP ($\tilde{\chi}_1^0$) is produced which is absolutely stable. This particle escapes from any collider detector.

In the minimal model all of the masses and mixings of the charginos and neutralinos are determined in terms of four parameters. One of these can be eliminated by assuming that the gaugino masses are degenerate at the grand unification or Planck scale. The remaining three parameters can be taken to be the gluino mass $M_{\tilde{g}}$, a supersymmetric Higgs mass μ, and a ratio $\tan\beta = v_2/v_1$ of Higgs vacuum expectation values (where v_2 (v_1) is the vacuum expectation value of the Higgs that gives mass to the up (down) quarks. If supersymmetry is related to the electroweak scale, we expect both $M_{\tilde{g}}$ and μ to be less than about 1 TeV; $\tan\beta$ should be of order one. Once the chargino and neutralino mixing angles (and one Higgs mass) are known, the couplings of the mass eigenstates $\tilde{\chi}_i$ are determined, and all branching ratios can be calculated.

The experimental analyses and most previous Monte Carlo studies[25)] have concentrated on the simplest possible decay modes, namely $\tilde{g} \to \tilde{\chi}_1^0 q\bar{q}$ if the gluino is lighter than the squark, or $\tilde{q} \to \tilde{\chi}_1^0 q$ if the squark is lighter than the gluino. They have, furthermore, always assumed that the branching ratios for these direct decays to the LSP are 100%. Since $\tilde{\chi}_1^0$ escapes from the detector, the signature for these modes is large missing transverse energy ($E_T^{\rm miss}$) plus multiple jets. For masses up to $60 - 70\,{\rm GeV}$, these simple modes are dominant for most but not all values of the parameters, so that the current limits (above) are likely to be valid.

However, this assumption begins to break down[26,27)] for squark and gluino masses $M_{\tilde{g},\tilde{q}} \gtrsim 100\ GeV$. For example,[27)] for $M_{\tilde{g}} = 180\ GeV$ (which would be accessible to an upgraded Tevatron), the branching ratio for the gluino to decay directly to the LSP can be no larger than 0.4 (and may be much smaller depending on the choice of supersymmetric parameters). The suppression is even more dramatic for heavier gluinos ($M_{\tilde{g}} \gtrsim 500\ GeV$), where the branching ratio is no greater than 0.14.

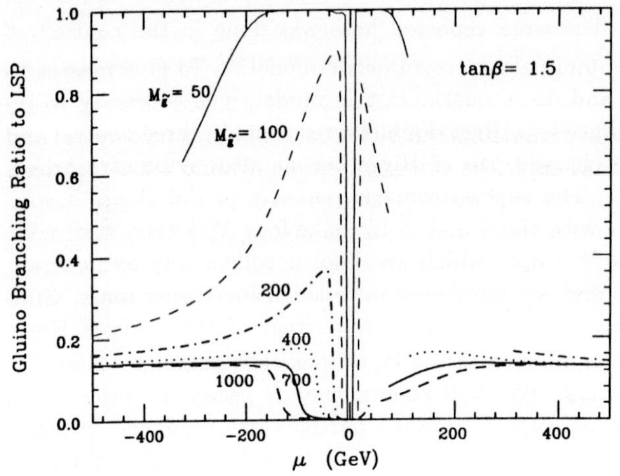

Figure 14. The branching ratio for the direct decay $\tilde{g} \to q\bar{q}LSP$ as a function of the mass of the gluino and the parameter μ.

(See Fig. 14 for the dependence of the branching ratio on various parameters.) This reduction occurs for heavier squarks and gluinos because decays into heavier neutralinos and charginos become possible, and any one mode has a fraction of the total. Even if the details are not correct, this minimal model presumably provides a representative example of supersymmetric models. In more complicated (non-minimal) models such small BR's would also be expected. The heavier neutralinos and chargino states themselves decay, eventually ending up with the LSP.

In general, the decay chains can be rather complicated. In particular, the LSP which eventually gets emitted is much softer and therefore the missing energy signal is degraded. However, the cascade decays can also introduce many interesting new signatures. In particular, it is rather easy to produce leptons in the chain, either via a three body decay of a chargino or neutralino, mediated by the W or Z, or by a direct two-body decay in which the W and Z are produced on-shell and subsequently decay leptonically. We show a particular example of the principal gluino branching ratios at the end of this introductory section.

Thus, it remains an open question whether gluinos can be easily isolated if produced at a future hadron collider. Similar considerations apply to squark production and decay. However the branching ratio suppressions for direct LSP production in squark decay are not as severe, so the "classic" missing energy signature is probably viable. Incidentally, we should emphasize that it is not necessary for *both* gluinos or *both* squarks to decay directly to the LSP in order to obtain large $E_T^{\rm miss}$. In this paper, we concentrate on the more problematic (and more copiously produced) gluinos. Given this, it follows that we should focus on the case for which the squarks are heavier than the gluino (in the opposite case gluinos would have a two-body decay into squarks and it would be more efficient to study the direct squark production). To be definite we often chose $M_{\tilde{q}} = 2M_{\tilde{g}}$ although the precise values of the squark masses have little effect on these results.

The couplings of gluinos to gluons and quarks are completely specified by supersymmetry, and hence the $\tilde{g}\tilde{g}$ production cross section is calculable given the gluino mass (with weaker dependence on squark mass). For $M_{\tilde{g}} = 750\,{\rm GeV}$ at $\sqrt{s} = 40\,{\rm TeV}$ the total $pp \to \tilde{g}\tilde{g}X$ cross section is

$$\sigma_{\tilde{g}\tilde{g}} \simeq 70\,{\rm pb} \tag{4.2}$$

All the decay modes are similarly calculable. For the 750 GeV case the principal gluino branching ratios are (for $m_{H^+} = 500\ GeV$, $\mu = 150\ GeV$ and $\tan\beta = 1.5$)

$$\begin{array}{ll} \tilde{g} \to \tilde{\chi}_1^{\pm}\bar{q}q' & 29\% \\ \tilde{g} \to \tilde{\chi}_2^{\pm}\bar{q}q' & 29\% \\ \tilde{g} \to \tilde{\chi}_1^{0}\bar{q}q' & 12\% \\ \tilde{g} \to \tilde{\chi}_2^{0}\bar{q}q' & 17\% \\ \tilde{g} \to \tilde{\chi}_4^{0}\bar{q}q' & 13\% \end{array} \tag{4.3}$$

The chargino branching ratios are

$$\begin{array}{ll} \tilde{\chi}_1^{+} \to \tilde{\chi}_1^{0}\bar{q}q' & 67\% \\ \tilde{\chi}_1^{+} \to \tilde{\chi}_1^{0}\ell\nu & 33\% \end{array} \tag{4.4}$$

and

$$\begin{array}{ll} \tilde{\chi}_2^{+} \to \tilde{\chi}_1^{+}Z^0 & 41\% \\ \tilde{\chi}_2^{+} \to \tilde{\chi}_1^{+}H_2^0 & 5\% \\ \tilde{\chi}_2^{+} \to \tilde{\chi}_1^{0}W^+ & 11\% \end{array}$$

$$\tilde{\chi}_2^+ \rightarrow \tilde{\chi}_2^0 W^+ \qquad 31\%$$
$$\tilde{\chi}_2^+ \rightarrow \tilde{\chi}_3^0 W^+ \qquad 12\% \qquad (4.5)$$

where H_2^0 is the lightest Higgs scalar. The neutralino branching ratios are

$$\tilde{\chi}_2^0 \rightarrow \tilde{\chi}_1^0 H_2^0 \qquad 100\% \qquad (4.6)$$

and

$$\tilde{\chi}_3^0 \rightarrow \tilde{\chi}_1^0 Z^0 \qquad 19\%$$
$$\tilde{\chi}_3^0 \rightarrow \tilde{\chi}_1^0 H_2^0 \qquad 60\%$$
$$\tilde{\chi}_3^0 \rightarrow \tilde{\chi}_1^{\pm} \bar{q} q' \qquad 14\%$$
$$\tilde{\chi}_3^0 \rightarrow \tilde{\chi}_1^{\pm} \ell \nu \qquad 7\% \qquad (4.7)$$

and

$$\tilde{\chi}_4^0 \rightarrow \tilde{\chi}_1^0 H_2^0 \qquad 0.3\%$$
$$\tilde{\chi}_4^0 \rightarrow \tilde{\chi}_2^0 H_2^0 \qquad 6\%$$
$$\tilde{\chi}_4^0 \rightarrow \tilde{\chi}_3^0 Z^0 \qquad 8\%$$
$$\tilde{\chi}_4^0 \rightarrow \tilde{\chi}_1^{\pm} W^{\mp} \qquad 85\% \qquad (4.8)$$

The work reported in this section was done by many members of the group. This summary is an amalgam from three written contributions submitted by various combinations of H. Baer, R.M. Barnett, J. Freeman, J.F. Gunion, H.E. Haber, R.J. Hollebeek, F.E. Paige, R. Raja, X. Tata, A.P. White, and J. Woodside.

4.2. The Missing-Energy Signature for Gluinos Using a Partonic Monte Carlo with the Full Cascade Decays

In this section (work by Baer, Tata and Woodside) we compute the missing transverse energy signal[28] expected from the process $p\bar{p}$ (or pp) $\rightarrow \tilde{g}\tilde{g}X \rightarrow q\bar{q}\tilde{\chi}_i + q\bar{q}\tilde{\chi}_j + X$ where the $\tilde{\chi}$ collectively denotes the two charginos ($\tilde{\chi}_{1,2}^{\pm}$) and the four neutralinos ($\tilde{\chi}_{1,2,3,4}^0$) of the minimal model. These charginos and neutralinos then decay via the two body modes $\tilde{\chi}_i \rightarrow \tilde{\chi}_j + (V$ or $H)$ (here $V = W^{\pm}$ or Z, and H is any Higgs scalar) if these are kinematically accessible; otherwise, they decay into a fermion pair (quark or lepton) and a lighter $\tilde{\chi}$. The secondary $\tilde{\chi}$ again decays as described above unless it is $\tilde{\chi}_1^0$, which we take to be the lightest supersymmetric particle, LSP. Thus the final state from $\tilde{g}\tilde{g}$ production consists of n quarks + m leptons ($n \geq 4$, $m \geq 0$) + E_T^{miss} (coming from the two $\tilde{\chi}_1^0$'s and any neutrinos). The relevant branching fractions are computed as specified in Ref. 29, see the example shown above.

Shown in Fig. 15 is the number of gluino pairs expected annually at (A) a 2 TeV $p\bar{p}$ collider with $\int Ldt = 100pb^{-1}$, (B) a 2 TeV pp collider with $\int Ldt = 1000pb^{-1}$, and (C) a 3 TeV $p\bar{p}$ collider with $\int Ldt = 30pb^{-1}$ which are representative of the various options for the upgrading

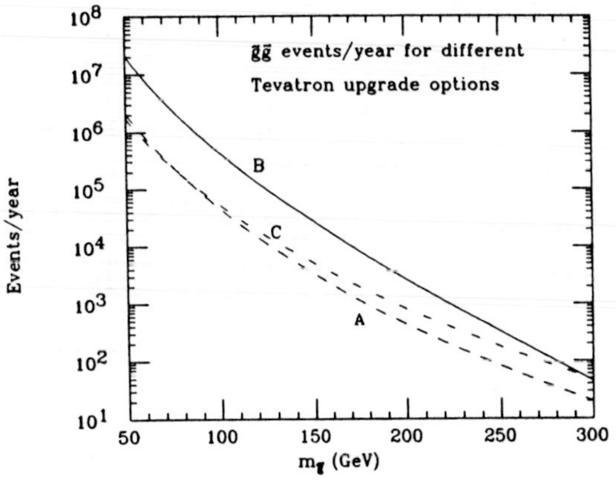

Figure 15. Expected rate of production of $\tilde{g}\tilde{g}$ events versus $M_{\tilde{g}}$ for the three different Tevatron upgrade options discussed in the text. We take $M_{\tilde{q}} = 2\ M_{\tilde{g}}$.

to phase II of the Tevatron in the next decade. Here we have taken $M_{\tilde{q}} = 2\ M_{\tilde{g}}$. We see from Fig. 15 that option B gives the greatest number of gluino pairs unless $M_{\tilde{g}} >$ 300 GeV.

In our computation of the various event topologies, we have incorporated the complete decay cascade of the $\tilde{g}$, $\tilde{\chi}_i^{\pm}$ and $\tilde{\chi}_i^0$ as given by the minimal supergravity model. Although the gluino production cross-sections do not depend on the parameters $\tan\beta$ (the ratio of Higgs field vev's) and μ, that enter the chargino and neutralino mass matrices, the cross-section into each event topology, in principle, does depend on these parameters via the branching fractions. Except for small values of μ, the cross sections for multijet + 0, 1 or 2 leptons are relatively insensitive to these parameters.

Small positive values of μ are already disallowed as discussed in Ref. 30. For definiteness, we have taken $\mu = -150GeV$, $\tan\beta = 1.5$ and $M_{\tilde{q}} = 2\ M_{\tilde{g}}$ as "typical values" in our calculation. In terms of these parameters (and $M_{\tilde{g}}$) all the cross-sections can be calculated.

Our calculations have been performed at the parton level. We have coalesced two partons within

$$\Delta R (\equiv \sqrt{\Delta\eta^2 + \Delta\varphi^2}) < 0.7$$

into a single jet and have required $\mid \eta_{jet} \mid < 2.5$ and $E_T(jet) > 15$ GeV. We require (for zero-lepton topologies) a trigger $E_T^{\text{miss}} > 50$ GeV.

The results of our calculation for $n\ jet + E_T^{\text{miss}}$ events at a 2 TeV $p\bar{p}$ collider (option A) are shown in Fig. 16. Dominant standard model backgrounds are listed in Table 4. For brevity, we have not shown the corresponding cross-sections for options B and C discussed above. These can be approximated by scaling the cross-sections in Fig. 16 by a ratio of the production rates that can be obtained from Fig. 15. We have computed the following Standard Model backgrounds to the signals:

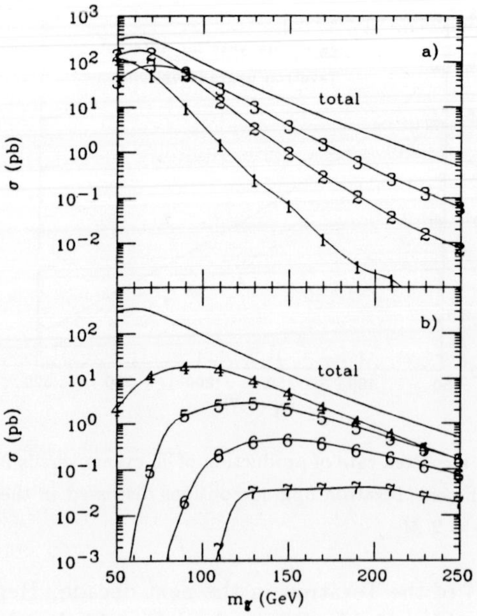

Figure 16. Cross-section for producing n-jet + E_T^{miss} events from gluino pairs versus $M_{\tilde{g}}$ at a 2 TeV $p\overline{p}$ Collider. We take $M_{\tilde{q}} = 2\ M_{\tilde{g}}$, and require $E_T^{\text{miss}} > 50$ GeV. Along with the total rate, we show (a) n = 1-3 and (b) n = 4-7 jet topologies.

(*a*) Z + "jets".
(*b*) $W \to \tau$ + "jets".
(*c*) $t\bar{t}$ production.

Table 4

Background rates from standard model processes for the signals shown in Figure 16. All numbers are in picobarns.

n JETS + E_T^{miss} ($E_T^{\text{miss}} > 50$ GeV)

# Jets	Z	W→ τ	t(60)	t(120)	Tot.(60)	Tot.(120)
1	23	7	9	.3	39	30
2	11	20	13	.9	44	32
3	3	4	10	1.4	17	8
4	.5	.8	3	1	4	2
5	.02	.06	.2	.2	.3	.3

For the background computation, we used a shower Monte-Carlo described in Ref. 31 for W and Z processes and the same experimental criteria as for the signal. We also computed the background from $t\bar{t}$ production for m_t = 60 and 120 GeV.

The following comments are in order: For $M_{\tilde{g}} \lesssim 120$ GeV, dijets and trijets dominate the signal, while four and five jet events dominate for $M_{\tilde{g}} > 120$ GeV. The dijet (trijet) rate exceeds background for $M_{\tilde{g}} < 90$ (115) GeV, whereas four (five) jet events exceed background for $M_{\tilde{g}} < 150$ (200) GeV. The large rate for E_T^{miss} events with five or more jets is a direct consequence of allowing for the complete gluino cascade. Running under option B, we would expect about 600 five jet events/year for $M_{\tilde{g}}$ = 200 GeV. We expect that by tuning the cuts to optimize signal/background, option B should easily probe $M_{\tilde{g}}$ = 200 - 250 GeV.

To summarize, from the theorists perspective option B (a Tevatron upgrade to a pp collider operating at $\sqrt{s}$ = 2 TeV and $\int L\ dt = 1000\ pb^{-1}$) gives the highest event rate for gluinos if $M_{\tilde{g}} < 300$ GeV. Backgrounds should be more manageable because of lower W and Z production rates relative to a $p\overline{p}$ collider.

4.3. The Missing-Energy Signature from the Production of $\widetilde{\gamma}\widetilde{g}$

Here, we focus (following Baer, Tata and Woodside) on the process $pp \to \widetilde{g}\widetilde{\chi}_1^0$, where $\widetilde{g} \to q\overline{q}\widetilde{\chi}_i^0$, or $q\overline{q}\widetilde{\chi}_i^{\pm}$. Production would be signalled by jets in one hemisphere recoiling against a very large E_T^{miss} . As we will see, the E_T^{miss} backgrounds from strong interaction processes within the SM can be eliminated by simple kinematic cuts. This process may, therefore, provide an interesting way to search for relatively light gluinos at the SSC, and, in particular, provide a partial answer to the question, "Is there a window of gluino masses that cannot be explored either at an upgraded Tevatron or the SSC?"

$\widetilde{g}\widetilde{\chi}_1^0$ production at the SSC proceeds via $q\overline{q}$ annihilation mediated by $\widetilde{q}_L$ and $\widetilde{q}_R$ exchanges in the t- and u-channels. The production rate, therefore, depends sensitively on $M_{\tilde{q}}$. It also depends on the parameters μ and $\tan\beta$ that enter the $q\widetilde{q}\widetilde{\chi}_1^0$ couplings via the neutralino mass matrix. We have checked, however, that $\sigma_{\widetilde{g}\widetilde{\chi}_1^0}$ depends only weakly on μ and $\tan\beta$ except for a small region of parameter space where $\widetilde{\chi}_1^0$ is almost a higgsino. For definiteness, we have taken $\mu = 400$ GeV, $\tan\beta = 1.5$ and $M_{\tilde{q}} = 2\ M_{\tilde{g}}$ in our analysis. Except for the "higgsino region" discussed above, $\sigma_{\widetilde{g}\widetilde{\chi}_1^0} = (6.5 \pm 1.0)$ pb for $M_{\tilde{g}}$ = 220 GeV and $\sigma_{\widetilde{g}\widetilde{\chi}_1^0}$= (2 ±0.3pb) for $M_{\tilde{g}} = 300$ GeV. These rates are considerably larger (by a factor of 3-4 for $M_{\tilde{q}} = M_{\tilde{g}}$) for lighter squarks. If we assume $M_{\tilde{q}} \simeq 2\ M_{\tilde{g}}$, ~ few $\times 10^4$ $\widetilde{g}\widetilde{\chi}_1^0$ events are expected annually at the SSC for $M_{\tilde{g}}$ = 200 - 300 GeV, the maximum reach of an upgraded Tevatron.

The E_T^{miss} spectrum from $\widetilde{g}\widetilde{\chi}_1^0$ production is shown in Fig. 17 for $M_{\tilde{g}} = 220$ GeV. Also shown is the E_T^{miss} spectrum from $\widetilde{g}\widetilde{g}$ events and from "Z $\to \nu\overline{\nu}$ + jet(s)" events. In our computation of the supersymmetric processes we have used a parton level Monte Carlo program that incorporates the complete cascade decay of the gluinos and the

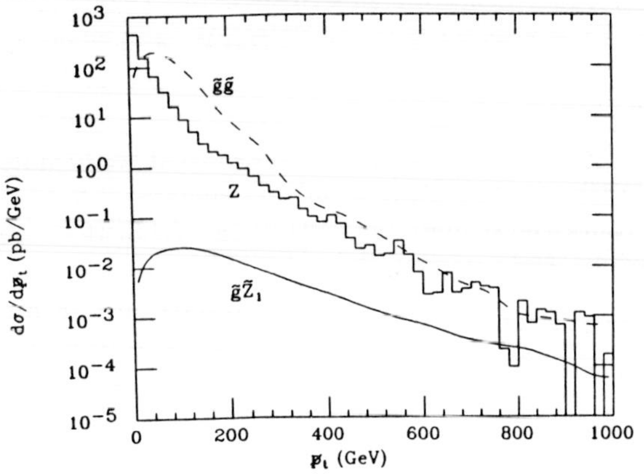

Figure 17. $E_T^{\rm miss}$ spectrum from $\widetilde{g}\widetilde{g}$, $\widetilde{g}\widetilde{\chi}_1^0$, and Z + jets (where $Z \to \nu\nu$) produced in pp collisions at $\sqrt{s} = 40$ TeV. We have taken $M_{\widetilde{g}} = 220$ GeV, $\tan\beta = 1.5$ and $\mu = 400$ GeV.

subsequent $\widetilde{\chi}_i^0$ and $\widetilde{\chi}_i^{\pm}$. The "Z + jet" cross-section has been computed using the shower Monte Carlo program of Ref. 32. We see that $\widetilde{g}\widetilde{\chi}_1^0$ production is completely swamped by the other processes for the whole range of $E_T^{\rm miss}$. In order to isolate the $\widetilde{g}\widetilde{\chi}_1^0$ signal, we have further required that there be no hadronic activity with $E_T >$ 30 GeV outside a 90° cone (in the transverse plane) back to back with the $E_T^{\rm miss}$ vector. We find that this isolation requirement on the $E_T^{\rm miss}$ vector completely cuts out all $\widetilde{g}\widetilde{g}$ events if we also require $E_T^{\rm miss} > 400$ GeV. We expect that this requirement cuts out the $E_T^{\rm miss}$ background from heavy flavour production with even a greater efficiency. The dominant SM background, therefore, comes from $Z \to \nu\overline{\nu}$ + jets.

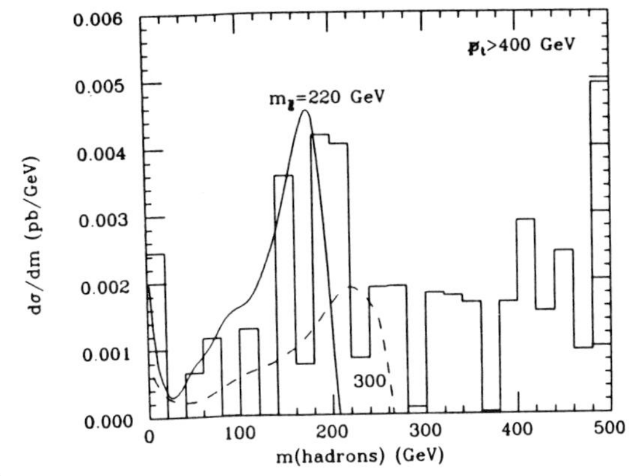

Figure 18. Invariant mass of all hadrons in the transverse hemisphere opposite the $E_T^{\rm miss}$ vector. We require $E_T^{\rm miss} > 400$ GeV and isolation. The histogram is the contribution from Z + jets where $Z \to \nu\overline{\nu}$. The solid line is from $\widetilde{g}\widetilde{\chi}_1^0$ with $M_{\widetilde{g}} = 220$ GeV while the dashed line is for $M_{\widetilde{g}} = 300$ GeV.

The distribution of the invariant mass, m, of the hadronic system for those events with $E_T^{\rm miss} > 400$ GeV and which satisfy the $E_T^{\rm miss}$ isolation requirement is shown in Fig. 18 for $M_{\widetilde{g}} = 220$ and 300 GeV. μ, $\tan\beta$ and $M_{\widetilde{g}}$ are fixed as stated earlier. Also shown is the background distribution from the shower calculation of Z^o + jets, for which we ran 10^6 events, corresponding to 34 CPU hours on a VAX 8700.

About 4000 (2300) $\widetilde{g}\widetilde{\chi}_1^0$ events pass our cuts for $M_{\widetilde{g}}$ = 220 GeV(300 GeV) assuming an integrated luminosity of $10^4 pb^{-1}$. The bulk of the background has $m \geq 300$ GeV so that requiring $m \leq 300$ significantly enhances the signal to background for the case of the relatively light gluinos. We see from Fig. 18 that signal to background $\sim$ unity may be possible, but this should be taken with a grain of salt as there is a considerable uncertainty in our estimate of the background (coming from the fact we are sampling a small tail of the background distribution). We have checked that increasing the $E_T^{\rm miss}$ cut to 600 GeV does not significantly improve the situation.

The conclusions of our analysis are rather pessimistic. The observation of a $\widetilde{g}$ via $\widetilde{g}\widetilde{\chi}_1^0$ production at the SSC is, at best, marginal. If the scalar partners of the fermions are all heavy ($\sim$1 TeV) with only the gluinos light, $\widetilde{g}\widetilde{\chi}_1^0$ production is completely obscured by the Z^0 + jet background. Only in the rather fortuitous case where $M_{\widetilde{q}} \simeq M_{\widetilde{g}}$ can we conclude that the signal is substantially larger than the background. Of course, in this case, one would obtain an additional contribution to the signal from $\widetilde{q}\widetilde{\chi}_1^0$ production.

4.4. The Missing-Energy Signature for Gluinos Using the ISAJET Monte Carlo with the Full Cascade Decays

In this section we report a preliminary study of the $E_T^{\rm miss}$ signature and backgrounds for studying supersymmetry for higher masses at SSC energies (done by Barnett, Freeman, Gunion, Haber, Hollebeek, Paige, Raja and White). In our analysis we have considered only gluino pair production and decay. The main conclusion of our preliminary analysis is that, even with the more complicated decays predicted by the minimal supersymmetric model, the $E_T^{\rm miss}$ signature remains observable. We will refine this result and study additional signatures in a future publication.

To simulate the signatures it is necessary to choose specific parameters. The values of μ and $\tan\beta$ are not very crucial because the branching ratio of the gluino to the LSP depends only weakly on them, although other branching ratios vary somewhat more. We have chosen three gluino masses and three sets of values for the other parameters in the minimal supersymmetric model:

$$i)\ M = 150\,{\rm GeV},\ \mu = -150\,{\rm GeV},\ \tan\beta = 1.5;$$
$$ii)\ M = 220\,{\rm GeV},\ \mu = +400\,{\rm GeV},\ \tan\beta = 1.5;$$

iii) $M = 750\,\text{GeV},\ \mu = +150\,\text{GeV},\ \tan\beta = 1.5.$(4.9)

The first mass may be observable at the present Tevatron; the second might be observed either at an upgraded Tevatron or at the SSC; the third is typical of the SSC mass range. In this section we will discuss only the 750 GeV case in detail. As discussed earlier, we have chosen (for definiteness) $M_{\tilde{q}} = 2M_{\tilde{g}}$ although the precise values of the squark masses have little effect on these results. The relevant branching ratios appear above in Eqs. (4.3)-(4.8).

As discussed in the introduction to this chapter, long chains of cascade decays are possible, giving rise to relatively small E_T^{miss} from the final $\tilde{\chi}_1^0$. But these decays also can produce $W^\pm$'s, Z^0's, jets, and leptons, leading to many possible signatures. Here we focus on the missing-energy signature using the ISAJET Monte Carlo program to generate $\tilde{g}\tilde{g}$ events including all the above decay modes. In addition to missing energy we use several additional cuts designed to reflect the fact that the gluinos are produced with $p_T(\tilde{g}) \sim M_{\tilde{g}}$ and $y(\tilde{g}) = \mathcal{O}(1)$, so that the decay products typically have large opening angles in the laboratory frame. We should emphasize that it is not necessary for *both* gluinos to decay directly to the LSP in order to obtain large E_T^{miss}.

The Standard Model background to events involving jets, leptons, and E_T^{miss} comes from the QCD production and semileptonic decay of heavy quarks $Q = c$, b, and t and from the production of $W^\pm$ and Z^0 at large p_T. We have also used ISAJET to simulate all of these backgrounds. The largest background for gluinos as for most other processes is the production of heavy quark pairs at high transverse momentum from higher-order processes like

$$g + g \to Q + \overline{Q} + g \qquad (4.10)$$

Such higher-order processes can not be directly included in ISAJET without severe double counting, since they are already produced by the QCD evolution, which approximates all higher-order QCD cross sections by a classical Markov branching process. To generate Eq. (4.10), therefore, it is necessary first to generate a primary hard scattering $gg \to gg$ and then to select events with the branching $g \to \overline{Q}Q$. Since the $gg \to gg$ cross section is 100 times larger than the $gg \to \overline{Q}Q$ one, this is rather slow.

To speed up the generation of the QCD background, we have used a trick developed by M. Della Negra of the UA1 Collaboration. Each primary hard scattering is evolved 10 times, and events not containing heavy quarks are immediately discarded. The rest are each hadronized 10 times, and only events containing the desired leptons or E_T^{miss} from neutrinos are saved. Finally the surviving events were weighted appropriately. We have generated 22,500 accepted QCD background events with $p_T >$ 250 GeV (equivalent to 2.25×10^6 events), 15,000 $W^\pm$ and 15,000 Z^0 events with $p_T > 250\,\text{GeV}$, and 2000 signal events.

The most characteristic signature of supersymmetry is missing transverse energy in association with multiple jets. We have concentrated on this signature, leaving to future work the study of signatures involving $W^\pm \to \ell^\pm\nu$ and $Z^0 \to \ell^+\ell^-$ produced in the gluino cascade decay. To analyse the simulated events, we have used a highly idealized calorimeter with uniform segmentation $\Delta y = \Delta\phi = 0.05$ for $|y| < 6$ and Gaussian energy resolutions

$$\left(\frac{\Delta E}{E}\right)_{\text{e.m.}} = \frac{0.15}{\sqrt{E}}, \qquad \left(\frac{\Delta E}{E}\right)_{\text{had.}} = \frac{0.70}{\sqrt{E}}. \qquad (4.11)$$

In future work we will replace this with the more realistic QFL simulation developed for the CDF Collaboration. Then we found all jets with $p_T > 50\,\text{GeV}$ using a simplified version of the UA1 jet algorithm with a clustering radius $\Delta R = 0.7$.

To select the $M = 750\,\text{GeV}$ gluino signal from the QCD and electroweak backgrounds we impose the following set of cuts:

- The highest jet has $p_T > 300\,\text{GeV}$;
- The missing transverse energy is $E_T^{\text{miss}} > 200\,\text{GeV}$;
- The sphericity S calculated in the transverse plane is $S > 0.2$;

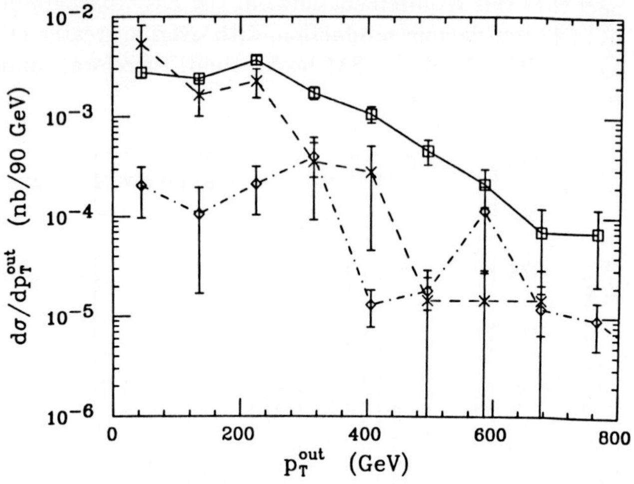

Figure 19. Cross section $d\sigma/dp_T^{\text{out}}$ vs. p_T^{out} after the cuts given in Sec. 4.4. Here p_T^{out} is the missing p_T out of the plane defined by the beam and the highest p_T jet. □ symbol: $\tilde{g}\tilde{g}$ signal. × symbol: QCD background. ◇ symbol: $W^\pm$ and Z^0 background. Curves are drawn to guide the eye.

- The number of jets with $p_T > 50\,\text{GeV}$ is $n_{\text{jet}} \geq 4$.

After these cuts we plot the missing transverse momentum p_T^{out} out of the plane defined by the beam and the highest p_T jet. The resulting distribution is shown in

Fig. 19. Making a further cut at $p_T^{out} > 300\,\text{GeV}$ gives the following cross sections:

$$\sigma_{\text{signal}} = 2.4\,\text{pb}$$
$$\sigma_{QCD} = 0.62\,\text{pb}$$
$$\sigma_{W^\pm,\,Z^0} = 0.25\,\text{pb} \qquad (4.12).$$

The signal has been substantially reduced from Eq. (4.2), but it still represents 24,000 events for the nominal integrated luminosity of $10^4\ pb^{-1}$. The signal-to-background ratio seems quite adequate provided that realistic detector effects do not degrade the resolution too much. Presumably the shape of the p_T^{out} distribution could be used to determine the mass, but we have not yet investigated the mass resolution. A detailed study of other gluino signatures will be presented elsewhere.

4.5. Signals from Gluino Decay to Leptons Using a Partonic Monte Carlo with the Full Cascade Decays

We look in this section (following the work of Baer, Tata and Woodside) at the full cascade decays of gluinos and searched for events with energetic, isolated electrons with modest missing energy. As described in a previous subsection (above), we compute the signals[28)] expected from the process $p\overline{p}$ (or pp) $\to \widetilde{g}\widetilde{g}X \to q\overline{q}\widetilde{\chi}_i + q\overline{q}\widetilde{\chi}_j + X$ where the $\widetilde{\chi}$ collectively denotes the two charginos ($\widetilde{\chi}^\pm_{1,2}$) and the four neutralinos ($\widetilde{\chi}^0_{1,2,3,4}$) of the minimal model. These charginos and neutralinos then decay via the two body modes $\widetilde{\chi}_i \to \widetilde{\chi}_j$ + (V or H) (here V=$W^\pm$ or Z, and H is any Higgs scalar) if these are kinematically accessible; otherwise, they decay into a fermion pair (quark or lepton) and a lighter $\widetilde{\chi}$. The secondary $\widetilde{\chi}$ again decays as described above unless it is $\widetilde{\chi}^0_1$. Thus the final state from $\widetilde{g}\widetilde{g}$ production consists of n quarks + m leptons (n $\geq$ 4, m $\geq$ 0) + E_T^{miss} (coming from the two $\widetilde{\chi}^0_1$'s and any neutrinos). The relevant branching fractions may be found in Ref. 29. In our computation of the various event topologies, we have incorporated the complete decay cascade of the $\widetilde{g}, \widetilde{\chi}^\pm_i$ and $\widetilde{\chi}^0_i$ as given by the minimal supersymmetry model. For definiteness, we have taken $\mu = -150GeV$, $\tan\beta = 1.5$ and $M_{\widetilde{q}} = 2\ M_{\widetilde{g}}$ as "typical values" in our calculation. In terms of these parameters (and $M_{\widetilde{g}}$) all the cross-sections can be calculated.

Our calculations have been performed at the parton level. We have coalesced two partons within

$$\Delta R (\equiv \sqrt{\Delta\eta^2 + \Delta\varphi^2}) < 0.7$$

into a single jet and have required $|\ \eta_{jet}\ | < 2.5$ and $E_T(jet) > 15$ GeV. For events containing one or more hard ($p_T^\ell > 20$ GeV) and central ($|\ \eta_e\ | < 3$ and $|\ \eta_\mu\ | <$ 0.76) charged leptons, we have used the cut $E_T^{miss} > 20$ GeV. We further require that the leptons be isolated (no partons with $\sum E_T > 3$ GeV in a cone of $\Delta R = 0.4$ about each lepton).

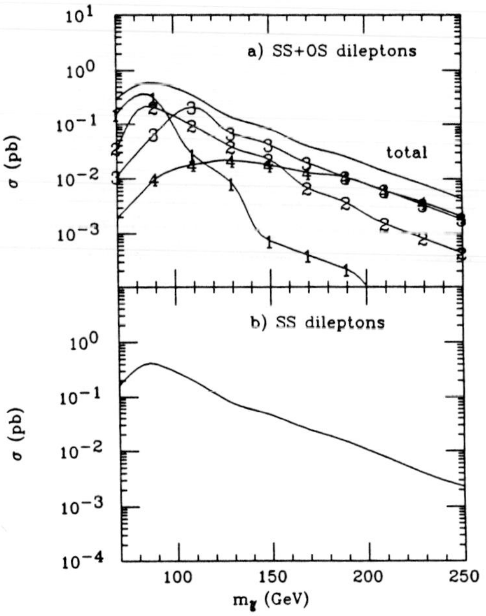

Figure 20. Cross-section for producing isolated dilepton plus n-jet events in $p\overline{p}$ collisions at $\sqrt{s} = 2$ TeV. Cuts are as described in the text. We show (a) all dileptons (summed over $e, \overline{e}, \mu, \overline{\mu}$) and (b) same-sign dilepton rates.

The results of our calculation for *n jet + 2 lepton* events at a 2 TeV $p\overline{p}$ collider (option A) are shown in Fig. 20. Dominant standard model backgrounds are listed in Table 5. For brevity, we have not shown the corresponding cross-sections for options B and C discussed above. These can be approximated by scaling the cross-sections in Fig. 20 by a ratio of the production rates that can be obtained from Fig. 15. We have computed the following Standard Model backgrounds to the signals:

(*a*) 1 - lepton events: W $\to$ e or μ + "jets" and $t\overline{t}$ production

(*b*) 2 - lepton events: $t\overline{t}$ production

For the background computation, we used a shower Monte-Carlo described in Ref. 31 for W and Z processes and the same experimental criteria as for the signal. We also computed the background from $t\overline{t}$ production for m_t = 60 and 120 GeV. The following comments are in order:

1. The backgrounds for 1 *lepton + n jets +* E_T^{miss} events appear to exceed signal for all choices of $M_{\widetilde{g}}$, hence we have not illustrated this case. However, in the present analysis we have made no attempt at optimizing signal/background. Cutting on quantities such as M_T(e,E_T^{miss}) or jet p_t may help in this matter. This is under study.
2. Signals for *two isolated leptons + n jets +* E_T^{miss}

Table 5

Background rates from standard model processes for the signals shown in Figure 20. All numbers are in picobarns.

Isolated Dileptons + n JETS + $E_T^{\rm miss}$ ($E_T^{\rm miss} > 20$ GeV)

# JETS	t(60)	t(120)
1	7	.3
2	4	.5
3	.7	.2
4	–	.002

events[33]) exceed background only for $n \geq 4$. The cross-sections for two isolated leptons + 4 jets are typically in the range of ~0.01 pb so that at most a handful of these events may be expected. The cross-section for isolated same sign dileptons can exceed 0.1 pb and is as large as 0.01 pb even for $M_{\tilde{g}} = 200$ GeV. We were unable to find any background from $t\bar{t}$ for this class of events. A separate study, described in the next subsection, involved a more detailed analysis with different cuts, and also found optimistic results. Although the rate for these events is too small to envision that supersymmetry would be discovered in the dilepton channel, a handful of these clean events would provide a striking confirmation of a signal discovered in another channel.

To summarize, option B (a Tevatron upgrade to a pp collider operating at $\sqrt{s} = 2$ TeV and $\int$ Ldt = 1000 pb^{-1}) would give the highest event rate for gluinos if $M_{\tilde{g}} < 300$ GeV. Backgrounds should be more manageable because of lower W and Z production rates relative to a $p\bar{p}$ collider. Incorporating complete gluino cascade sequences in our programs yields new signatures for gluinos. The most promising are (1) isolated same sign dileptons + $E_T^{\rm miss}$ and (2) isolated dileptons + four or more jets + $E_T^{\rm miss}$. Each of these signals (along with the $E_T^{\rm miss}$ signal) should be present in a data sample if discovery of gluinos is to be claimed. More detailed studies of these signals are in progress.

4.6. Like-Sign Dileptons as a Signal for Gluino Production

In this section (following the work of Barnett, Gunion and Haber), we are interested in new gluino search techniques which could complement (or perhaps even substitute for) the standard missing-energy signature. One problem with the more complicated signatures is that, if seen, one must find means for identifying such a signal as being that of a gluino. Our approach will be to make use of a special feature of the gluino: it is a Majorana fermion. (Here, we use a broader definition of Majorana to include neutral particles which transform under real representations of the underlying Standard Model gauge group.) A very distinctive feature of a Majorana particle is that it can decay with equal probability into fermions and antifermions. Thus, we focus on a particular class of gluino decay chains such that $\tilde{g}\tilde{g}$ production will result in *like-sign* dileptons in the final state. Due to the Majorana nature of the gluino, the probability for the production of like-sign and opposite-sign leptons is equal, and the lepton decay characteristics of the two classes of final states are identical. This is quite a distinctive result, and if seen would be extremely helpful in identifying the origin of the events.

Consider as an example the decay $\tilde{g} \to q\bar{q}'\tilde{\chi}_1^{\pm}$, where $\tilde{\chi}_1^{\pm}$ is the lightest chargino state. As shown in Ref. 27, this decay mode is the dominant gluino decay (with branching ratio near 60%) over a large range of supersymmetric parameter space. The key point here is that the Majorana nature of the gluino implies that it is equally likely to decay into a chargino of either sign. The chargino decays dominantly into the LSP plus a real (or virtual) $W^{\pm}$, which then decays to electrons or muons 22% of the time Thus the branching ratio for the decay chain $\tilde{g} \to q\bar{q}\ell^{\pm}\nu\tilde{\chi}_1^0$ is typically around 13%, with equal probability to produce a lepton of either sign. For simplicity, the τ–lepton will be neglected from our considèrations. Since gluinos are produced in pairs (we assume that squarks are heavier so that $\tilde{q}\tilde{g}$ production can be ignored), this implies that the number of dilepton final states resulting from the decay of the two gluinos may be as large as 2% of all $\tilde{g}\tilde{g}$ events, of which half have like-sign leptons. Given the large cross-sections for gluino pair production at future colliders, there would exist a potentially large and interesting sample of events. This sample consists of events with hadronic jets (two from each gluino), missing energy due to the LSP and neutrinos in the final state, and a dilepton pair which can come in one of the following like-sign combinations: $e^{\pm}e^{\pm}$, $e^{\pm}\mu^{\pm}$, $\mu^{\pm}\mu^{\pm}$, and the corresponding opposite-sign combinations. Because the background is clearly larger for detecting the opposite-sign pairs, we shall focus here on the isolation of like-sign dilepton events originating from gluino pair production.

We have used a parton Monte Carlo in order to determine the viability of gluino detection using the like-sign dilepton signal. We have investigated the signal at the present Tevatron and at two future hadron colliders under consideration: an upgraded Tevatron and the SSC. Three Tevatron options are being considered, but we find that only the choice of a pp collider with $\sqrt{s} = 2\ TeV$ and an integrated luminosity of $1000 pb^{-1}$ gives a significant improvement in discovery limits over the existing Tevatron. In addition to studying the characteristics of the signal, we have tried to identify the major source of Standard Model background. In the case of the like-sign dilepton signal, the dominant background is expected to be $t\bar{t}$ production, where one t decays to $b\ell\nu$ and the other t decays to $b\bar{q}q'$ with the secondary b decaying into $c\ell\nu$. Such events will also produced like-sign dileptons (both muons and electrons), hadronic jets, and missing energy in the form of neutrinos. The background estimate will depend crucially on the unknown t-quark mass. We have studied two extreme cases in this regard: the "light" top ($m_t = 60\ GeV$) and the "heavy" top ($m_t = 160\ GeV$). We have identified a number of characteristics of the background that allow us to distinguish between it and the signal.

First, there is the lack of isolation of the lepton emitted from the b-quark. (We make use of the theoretical expectation that there is negligible mixing in neutral T-mesons which could cause primary like-sign dileptons in $t\bar{t}$ production from the t-quark decays.) This is to be compared with the case of dileptons from gluinos where both leptons resulting from the decay of a chargino (presumed to be much heavier than the b-quark) would be quite isolated. Thus we define $\Delta R^{min}_{l,j}$ to be the minimum $\Delta R \equiv \sqrt{(\Delta\eta)^2 + (\Delta\phi)^2}$ obtained by considering all lepton-jet combinations. Two other variables that allow us to discriminate against the background due to a top quark that is significantly lighter than the gluino are the total reconstructed mass of the event,

$$M^2_{tot} = \left(\sum_{i=j_1,j_2,j_3,j_4,l_1,l_2} E_i + E_T^{\rm miss} \right)^2 - \left(\sum_{i=j_1,j_2,j_3,j_4,l_1,l_2} \vec{p_i} + \vec{p_T}^{miss} \right)^2 \tag{4.13}$$

designed to approximate the total mass of the produced $\tilde{g}\tilde{g}$ system, and the invariant mass ($M_{l_2,2j}$) of the slow lepton and its two nearest (in ΔR) jets, which provides a measure of the invariant mass of the gluino. In Eq. (4.13) jets and leptons are labelled in order of decreasing transverse energy: $E_T^{j_1} > E_T^{j_2} > E_T^{j_3} > E_T^{j_4}$ and $E_T^{l_1} > E_T^{l_2}$.

Before presenting any details of the analysis we preview our conclusions. At the upgraded Tevatron option with an integrated luminosity of 1000 pb^{-1}, the like-sign dilepton signature is found to be viable for gluino masses up to 180 GeV. Of course, it is necessary to have a magnetic detector in order to be able to perform the type of analysis we describe here. Since it is clearly of interest to make sure that the SSC can probe down to this mass (so that there is no gap in accessible gluino masses for the combination of the two machines) we present the bulk of our results (both for an upgraded Tevatron and the SSC), assuming $M_{\tilde{g}} = 180\ GeV$.

For simplicity, we employ the minimal supersymmetric extension of the Standard Model. Having specified $M_{\tilde{g}}$, the remaining free parameters are μ, $\tan\beta$ and the mass of one of the Higgs bosons of the model (see Ref. 27 for notation). A particular choice of parameters can affect the chargino and neutralino mass spectrum and the gluino leptonic branching ratio. We have surveyed the supersymmetric parameter space and find that the gluino leptonic branching ratio is fairly stable over a considerable range of the parameters. We present results here for a typical choice: $\tan\beta = 1.5$, $m_{H^+} = 500\ GeV$, and $\mu = -150\ GeV$. For these values, we find $BR(\tilde{g} \to \bar{q}q'\tilde{\chi}_1^+) = 0.52$ and $BR(\tilde{\chi}_1^+ \to \ell^\pm + \text{anything}) = 0.22$, where ℓ is either an electron or muon. Since in this case this is the only decay chain giving a single lepton, we find that like-sign dilepton events are 0.6% of all $\tilde{g}\tilde{g}$ events.

Let us first focus on the Tevatron upgrade analysis. First, we define a jet such that all parton jets within a cone of $\Delta R \leq 0.6$ around the jet axis are counted as one jet. We then employ the following cuts: (i) $N_{jet} \geq 4$, (ii) $E_T^{j_{1,2,3,4}} > 10\ GeV$, (iii) $N_{lepton} = 2$, and (iv) $E_T^{l_{1,2}} > 5\ GeV$. In addition, we impose a CDF-like trigger by requiring that at least one of the following three conditions be satisfied: (i) $E_T^{l_1} > 20\ GeV$, (ii) $E_T^{j_1} > 60\ GeV$, or (iii) $E_T^{j_{1,2,3}} > 20\ GeV$. Distributions in $\Delta R^{min}_{l,j}$, M_{tot}, $M_{l_2,2j}$ and $E_T^{l_2}$ are presented in Fig. 21. For the latter three we have plotted the $\tilde{g}\tilde{g}$ signal results before any $\Delta R^{min}_{l,j}$ cut (a $\Delta R^{min}_{l,j} > 0.48$ cut results in only a small decrease in the signal curve), in comparison to background results for $m_t = 60\ GeV$ with the $\Delta R^{min}_{l,j} > 0.48$ cut imposed and $m_t = 160\ GeV$ without any $\Delta R^{min}_{l,j}$ cut. All Tevatron curves are normalized to an integrated luminosity of 1000 pb^{-1}. The number of events in the $m_t = 60\ GeV$, $\Delta R^{min}_{l,j} > 0.48$ curves is 20, the number in the $m_t = 160\ GeV$, no $\Delta R^{min}_{l,j}$ cut curves is 91, while there are 18 and 14 events in the signal curve before and after the $\Delta R^{min}_{l,j}$ cut, respectively. For $m_t = 60\ GeV$ the background event rate is very high to begin with, but the $\Delta R^{min}_{l,j}$ cut is extremely effective in reducing it to the level of the signal. The mass distributions then allow one to easily distinguish between signal and background. At $m_t = 160\ GeV$ the background is not that much larger than signal and the $\Delta R^{min}_{l,j}$ distributions are sufficiently different to allow separation. Having made such a cut, the $E_T^{l_2}$ distributions of signal and background are quite different, as seen in Fig. 21b. On the other hand the mass distributions (Fig. 21c-d) are quite similar when $m_t \sim M_{\tilde{g}}$. Note that in both cases, the gluino mass can be determined (within an accuracy of order 10%) from the M_{tot} distribution. These results from our upgraded Tevatron simulation show that the dilepton signature is marginal and requires large luminosity. The other proposed upgrades which do not plan for such a high integrated luminosity would not allow use of the dilepton

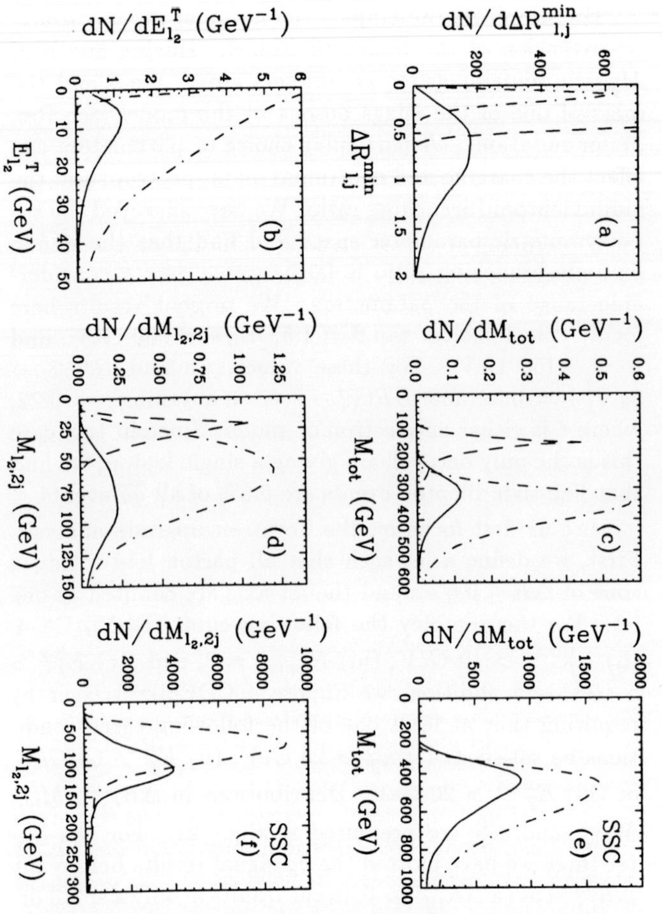

Figure 21. We present normalized distributions for the Tevatron upgrade, (a)-(d), and the SSC, (e) and (f), for $M_{\tilde{g}} = 180\ GeV$. All include the cuts and triggering conditions described in the text. The $\tilde{g}\tilde{g}$ signal curves are solid lines and have no $\Delta R^{min}_{l,j}$ cut; in (b)-(f) a $\Delta R^{min}_{l,j} > 0.48$ cut results in only a small decrease in the solid curve. Background curves for $m_t = 60\ GeV$ are dashed, while those for $m_t = 160\ GeV$ are dotdashed. The distributions are: (a) $\Delta R^{min}_{l,j}$ for the $\tilde{g}\tilde{g}$ signal multiplied by a factor of 10 compared to $m_t = 60\ GeV$ and $m_t = 160\ GeV$; (b) $E^{l_2}_T$ for $m_t = 160\ GeV$ before and after a $\Delta R^{min}_{l,j} > 0.48$ cut, compared to the $\tilde{g}\tilde{g}$ signal; (c) M_{tot} for $m_t = 60\ GeV$, $\Delta R^{min}_{l,j} > 0.48$, and $m_t = 160\ GeV$, no $\Delta R^{min}_{l,j}$ cut, compared to signal; (d) as in (c) but for $M_{l_2,2j}$; (e) M_{tot} for $m_t = 160\ GeV$, no $\Delta R^{min}_{l,j}$ cut, compared to signal; and (f) as in (e) but for $M_{l_2,2j}$.

signal to explore gluino masses as large as 180 GeV — the dilepton signal is essentially event rate limited.

We have also run our Monte Carlo under SSC conditions. We assume an integrated luminosity of $10^4\ pb^{-1}$. Since the raw $\tilde{g}\tilde{g}$ production rate is enormous for $M_{\tilde{g}} = 180\ GeV$ (yielding some 2.48×10^6 like-sign dilepton events before cuts), it is trivial to impose strong cuts on the $\Delta R^{min}_{l,j}$ and mass variables that completely eliminate any $m_t = 60\ GeV$ background. Thus, we focus on $m_t = 160\ GeV$. We employed a new set of cuts: (i) $N_{jet} \geq 4$, (ii) $E^{j_{1,2,3}}_T > 40\ GeV$, $E^{j_4}_T > 10\ GeV$, (iii) $N_{lepton} = 2$, and (iv) $E^{l_{1,2}}_T > 10\ GeV$. (Making a stronger cut on $E^{l_2}_T$ would completely eliminate the background. We have not done so in order to have some background to compare with our signal.) No trigger conditions were imposed. After the above cuts, we are left with 6×10^5 background events and 2.6×10^5 signal events. The $\Delta R^{min}_{l,j}$ distribution of the background is even more sharply peaked towards small $\Delta R^{min}_{l,j}$ in the SSC simulation than at the Tevatron, and would clearly allow isolation of the signal. The M_{tot} and $M_{l_2,2j}$ distributions for signal and background are shown in Fig. 21, with no $\Delta R^{min}_{l,j}$ cut imposed. We see that $M_{l_2,2j}$ also allows some discrimination. This SSC study suggests that we should even be able to isolate a gluino signal for gluinos lighter than 180 GeV.

We have also explored the maximum gluino mass that could be probed using the like-sign dilepton signature at the SSC.[34] For heavy gluinos the effective branching ratio for $\tilde{g}\tilde{g}$ final states containing like-sign electrons or muons and at least four energetic jets is typically of order 1% for a large range of parameter space. The large difference in $\Delta R^{min}_{l,j}$ distributions between signal and top backgrounds remains, and now the M_{tot} and $M_{l_2,2j}$ distributions for the $\tilde{g}\tilde{g}$ signal peak at such high masses that there is essentially no overlap with any $m_t < 200\ GeV$ background. Thus isolation of the signal is purely event rate limited. For $M_{\tilde{g}} = 2\ TeV$ and integrated luminosity of $10^4\ pb^{-1}$ there are 25 events in the dilepton signal, which would almost certainly allow discovery. Higher masses could be probed for larger integrated luminosities.

In conclusion, at an upgraded Tevatron with at least 1000 pb^{-1}, the like-sign dilepton signal may be visible for gluino masses up to 180 GeV, depending on the values of the other supersymmetric parameters. (Such a signal may even be observable at a Tevatron operating under present conditions for gluino masses near 90 GeV for special parameter choices.) At the SSC, the dilepton signal should be easy to detect for the entire range of gluino masses from $180 - 2000\ GeV$, very probably extending even lower. Such a copious like-sign dilepton signal would be strong evidence for phenomena related to a Majorana fermion, and the large event rates predicted here would almost surely allow experimenters to claim the discovery of the gluino, if such like-sign dilepton events were discovered. Clearly, the dilepton signal at the SSC is a powerful tool for uncovering evidence for supersymmetry.

5. Summary and Conclusions

Our study of the new particle discovery capabilities of the SSC indicates that it has a powerful potential to unravel many current mysteries of high energy physics. The range of masses to which it has access is enormous. Many types of signatures can be clearly separated from backgrounds. When new physics is discovered, it should be possible to establish not only its existence but also much about its nature. While we have not examined every possible type of new particle, the range of signals we studied was large, and probably includes signals expected from other particles.

The identification of new physics may require isolation of several signatures. This makes the achievement of design luminosity essential, because the branching ratios to given modes are often small. Among the many reasons why a *magnetic* detector is critical for finding new particles is the need to know the charge of leptons in multi-lepton events. We have shown that charge identification can be important both for eliminating backgrounds and for identifying the nature of the new physics. It should be emphasized that for many kinds of new physics the leptons are not fast (often $E_T^\ell \approx 20-100$ GeV) since they come from cascade decays. However, they are quite isolated compared with leptons from b decay. Leptons from new W' and Z' bosons might have p_T in the multi-TeV range, and sign determination for these may be unimportant. Because of the importance of decays with leptons and the small branching ratios to any given mode, it is essential to detect both electrons and muons in reasonable rapidity ranges. We have not directly addressed this issue in this particular study, but lepton coverage in $|\eta| < 3$ should be adequate. Note that in the supersymmetry section we found that the backgrounds for events with single leptons were too high so that the useful signals would be multi-lepton events; for these events the need for adequate rapidity coverage of electrons and muons is compounded. An issue which requires more attention is that of lepton misidentification; in particular, the mistaken labelling of isolated hadrons as leptons is a potential problem. We cannot say, for example, whether misidentification at the level of 10^{-3} is acceptable. The lepton and jet resolutions commonly discussed for SSC detectors appear to be sufficient for discovering most new particles. A role for vertex detection has not been suggested, but it may be useful in eliminating backgrounds. Further study is needed as to whether there will be problems triggering on some of the signatures proposed here and elsewhere.

We have shown here for the first time that new heavy leptons can be discovered at the SSC. New heavy bosons ($W'^{\pm}$ and Z'^{0}) are among the easiest particles to find at the SSC, and a high statistics study of asymmetries may yield useful information about their nature. For the first time we have performed detailed studies of the production of gluinos (the supersymmetric partner of the gluon) in which the small branching ratios were incorporated and in which full cascade decays were followed. We have emphasized the critical role of decays including leptons. The preliminary studies are very encouraging, leading us to believe that much can be learned about supersymmetry. High statistics ISAJET studies are proceeding.

We turn now to the issues concerning the proposed upgrades for the Tevatron Collider. There are motivations for upgrading the Tevatron Collider or for choosing one option over another option (accelerator technology, detector capabilities, etc.) beyond the purely physics issues raised here, but we have not addressed those questions at all. While the range of new particle masses available to an upgraded Tevatron Collider appears to be fully covered by existing facilities (including the existing Tevatron) plus the SSC, there may be valuable physics which could be done before the advent of the SSC. The excluded mass range for new particles such as new W' and Z' bosons, squarks and gluinos could be extended by 100-200 GeV. This can be compared with the SSC which will eventually extend these limits by 2000-7000 GeV. Of course it is also possible that these new particles have masses within the discovery reach of an upgraded Tevatron Collider. Some particles cannot be found at the Tevatron Collider because the luminosities proposed are inadequate for the combination of small production rates and very small branching ratios to any given mode. In fact the search range for most new physics is limited by luminosity.

The three upgrades options we considered were:

- Option A - A 2 TeV $p\bar{p}$ collider of luminosity 100 pb^{-1}.
- Option B - A 2 TeV pp collider of luminosity 1000 pb^{-1}.
- Option C - A 3 TeV $p\bar{p}$ collider of luminosity 30 pb^{-1}.

Which upgrade performs the best in new particle search and discovery depends largely on the mass of the particle. Furthermore, this question is dependent on whether just detection is sufficient or whether larger event rates are desired so that the properties of new particle can be studied. We found no circumstances in which Option A was preferable. The higher energy of Option C can lead to the highest search limits if and *only if* the resultant rate is adequate.

New heavy bosons (W' and Z') provide the one of the best possible examples of a favorable circumstance for Option C. Their production requires quark-antiquark annihilation thereby favoring a $p\bar{p}$ machine since it has a larger supply of antiquarks. However, the branching ratio to leptons is small, so even here Option C does not necessarily provide the best limits (based on 5 events), because of its limited luminosity. Option B is clearly preferred if 50 or more events are desired (in this latter case the accessible masses are somewhat lower). Which option does best depends on the model and on model-dependent parameters. Option C tends to have the best discovery limits for new gauge bosons, although the differences are only 50 GeV or less. Looking at the forward-backward asymmetry, we see that for some choices of model parameters, it is considerably larger for $p\bar{p}$ machines, but this must be matched against the smaller event rates of Option C compared with Option B (at masses low enough to achieve event rates great enough to measure an asymmetry). If we compare the discovery limits of the upgrades to the limits from the current Tevatron Collider run (which we assume to have an integrated luminosity of 10 pb^{-1}), we find that the limits would be raised from about 550 GeV for W' bosons to about 800 GeV while the limits for Z' bosons would go from 400 to 600 GeV. For fixed boson masses of say 400 GeV, the upgrades yield about ten times as many events as the current run. Thus the potential for investigating new physics in the 1990s with Tevatron upgrades could be significant.

The production of other types of particles such as squarks and gluinos (the supersymmetric partners of the quarks and gluon) does not require quark-antiquark annihilation and in fact is dominated by production via gluons. In this case Option A does not surpass Option B for $\widetilde{g}\widetilde{g}$ production until the gluino mass is 380 GeV, and at that mass there is a fraction of an event per year for any given signal after branching ratios. For masses well below the discovery threshold, Option B has a tremendous advantage over the other options. For example, for $M_{\widetilde{g}} = 150$ GeV Option B gives 26,000 gluino pairs per year (before any branching ratios) compared with 3200 and 5100 for Options A and C, respectively and with only 190 for the current Tevatron run with 10 pb^{-1}. At $M_{\widetilde{g}} = 220$ GeV (which may be the discovery limit) the ratio of gluino pair production for Option B compared to Option C is 974 events to 381 events. The crossover point at which Option C begins to dominate Option B is about $M_{\widetilde{g}} = 290$ GeV where about 50 total events per year would be produced; after branching ratios this rate is not adequate for discovery. Another measure of the difference for gluino production among the three options is a comparison of discovery reach. If Option B can achieve $M_{\widetilde{g}} = 220$ GeV, then Option A can achieve 180 GeV, Option C 200 GeV and the current 10 pb^{-1} run 150 GeV. Incidentally for $M_{\widetilde{g}} = 220$ GeV, the SSC would produce 2×10^8 events per year.

In conclusion, the potential for finding new physics at an upgraded Tevatron Collider is clearly far less than at the SSC, but several important discoveries might be within reach. In general we felt that Option B (with the highest luminosity) was a superior upgrade in the search for new particles. The differences in ultimate search limits were never great, but when larger event rates were required, Option B enjoyed a clear advantage. Option A seemed the least desirable. Finally, our group found that the SSC is essential in the search for the new particles which can solve some of the great puzzles of physics.

6. Acknowledgements

We would like to express our appreciation to the administrative, secretarial and computing staff of the Snowmass workshop for making the workshop a productive and enjoyable occasion. We also appreciate the effort of Sharon Jensen in helping the New Particles at Hadron Colliders group prepare this report and several other contributions. This work was supported by the Director, Office of Energy Research, Office of High Energy and Nuclear Physics, Division of High Energy Physics, of the U.S. Department of Energy under Contract Nos. DE-AC03-76SF00098, DE-AC02-76-ER0-3071, DE-AS03-76ER70191, DE-AA03-76-SF00010, DE-AC02-76ER00881 and W-7405-Eng-82, by the National Science Foundation under agreement no. PHY83-18358, and by the University of Wisconsin Research Committee with funds granted by the Wisconsin Alumni Research Foundation.

REFERENCES

1. E. Eichten *et al.*, *Rev. Mod. Phys.* **56**, 579 (1984).
2. D. Froidervaux in Proc. of the 1987 La Thuile Workshop on Physics at Future Accelerators, CERN 87/07.
3. V. Barger, Y. Han and J. Ohnemus, *Phys. Rev.* **D37**, 1134 (1988).
4. M.S. Chanowitz, M. Furman and I. Hinchliffe, *Nucl. Phys.* **B153**, 402 (1979), U. Amaldi, *et al.*, *Phys. Rev.* **D36**, 1385 (1987).
5. C. Albajar, *et al.*, *Phys. Lett.* **185B**, 241 (1987), *Phys. Lett.* **191B**, 463 (1987) (Erratum *Phys. Lett.* **197B**, 565 (1987)).
6. C. Hearty, *et al.*, *Phys. Rev. Lett.* **58**, 1711 (1987), C. Albajar, *et al.*, *Phys. Lett.* **198B**, 271 (1987).
7. Physics at LEP, ed. J. Ellis and R. Peccei, CERN 86-02.
8. S. Willenbrock and D. Dicus, *Phys. Lett.* **156B**, 429 (1987).
9. S. Dawson and S. Willenbrock, *Nucl. Phys.* **B284**, 449 (1987).
10. G. Alverson *et al.*, in Proc. of the 1986 DPF Summer Study on the Physics of the Superconducting Supercollider, Ed. J. Marx and R. Donaldson, FNAL (1987), J.F. Gunion and M. Soldate, *Phys. Rev.* **D34**, 826 (1986).
11. J.F. Gunion, G.L. Kane, H.F.W. Sadrozinski, A. Seiden, A.J. Weinstein, and C.P. Yuan, UCD-88-38, Nov 1988.

12. G. Alverson *et al.*, Ref. 10, H.-U. Bengtsson and A. Savoy-Navarro, *Phys. Rev.* **D37**, 1787 (1988)
13. M. Perl, *et al.*, *Phys. Lett.* **63B**, 466 (1976).
14. G. Marchesini and B. Webber, Cavendish preprint HEP 87/9 (1987), F. Paige and S.D. Protopopescu in Proc of 1986 in Proc. of the 1986 DPF Summer Study on the Physics of the Superconducting Supercollider, Ed. J. Marx and R. Donaldson, FNAL (1987), H.-U. Bengtsson and T. Sjostrand *Comp. Phys. Comm.* **46**,43 (1987).
15. J.C. Pati and A. Salam, *Phys. Rev.* **D10**, 275 (1974); R.N. Mohapatra and J.C.Pati, *Phys. Rev.* **D11** 566, 2558 (1975); G. Senjanovic and R.N. Mohapatra, *Phys. Rev.* **D12**, 1502 (1975). For a recent review, see R.N. Mohapatra in *Quarks, Leptons and Beyond*, ed. H. Fritzsch *et al.*, Plenum, New York (1985), p. 217.
16. N.G. Deshpande, J.A. Grifols and A. Méndez, *Phys.Lett.* **208B**, 141 (1988).
17. J.F. Gunion, J.A. Grifols, A.Mendez, B. Kayser and F. Olness; UCD preprint in preparation.
18. J. L. Hewett and T. G. Rizzo, Univ. Wisconsin report MAD/PH/446, submitted to Physics Reports (1988).
19. For a general survey of the left-right Symmetric Model, its basis in SO(10) and original references see R. N. Mohapatra, *Unification and Supersymmetry*, Springer- Verlag, New York, 1986.
20. E. Ma, Phys. Rev. **D36**, 274 1987 ; Mod. Phys. Lett. **A3**, 319 (1988); K. S. Babu, X.-G. He, and E. Ma, Phys. Rev. **D36**, 878 (1987); J. F. Gunion *et al.*, *Proceedings of the Workshop From Colliders to Supercolliders*, Madison, WI, May (1987) and Int. J. Mod. Phys. **A2**, 119 (1987); V. Barger and K. Whisnant, *ibid.* **A3**, 879 (1988).
21. The problems associated with a measurement of A^{ℓ}_{FB} for new Z' bosons has been discussed earlier by H. Haber, New W and Z Group Summary, *Proceedings of the 1984 Summer Study on the Design and Utilization of the SSC*, Snowmass, CO, 1984, edited by R. Donaldson and J. Morfin.
22. C. Albajar et al., Phys. Lett. **B198**, 261 (1987)
23. J. Freeman and F. Abe et al., Proceedings of the 7th International Conference on $\overline{p}p$ Collisions, Fermilab, June, 1988 (preliminary).
24. See H.E. Haber and G.L. Kane, Phys. Rep. **117**, 75 (1985) and references therein.
25. R. Batley, in *Proc. of the 1987 La Thuile Workshop on Physics at Future Accelerators*, Vol. 2, ed. J.H. Mulvey, CERN 87-07 (1987), p.109; R.M. Barnett, H.E. Haber and G.L. Kane, Phys. Rev. Lett. **54**, 1983 (1985) and Nucl. Phys. **B267**, 625 (1986); H. Baer, J. Ellis, G. Gelmini, D. Nanopoulos and X. Tata, Phys. Lett. **161B**, 175 (1985); G. Gamberini, Z Phys. **C30**, 605 (1986); H. Baer, V. Barger, D. Karatas and X. Tata, Phys. Rev. **D36**, 96 (1987).
26. G. Gamberini, Z. Phys. **C30**, 605 (1986); H. Baer and E.L. Berger, Phys. Rev. **D34**, 1361 (1986); H. Baer, V. Barger, D. Karatas and X. Tata, Phys. Rev. **D36**, 96 (1987); R. Barbieri, G. Gamberini, G.F. Giudice and G. Ridolfi, Nucl. Phys. **B301**, 15 (1988).
27. R.M. Barnett, J.F. Gunion and H.E. Haber, Phys. Rev. Lett. **60**, 401 (1988); Phys. Rev. **D37**, 1892 (1988).
28. H. Baer, J. Ellis, G. Gelmini, D. Nanopoulos and X. Tata, Phys. Lett. **161B**, 175 (1985); R.M. Barnett, H.E. Haber and G.L. Kane, Phys. Rev. Lett. **54**, 1983 (1985) and Nucl. Phys. **B267**, 625 (1986); G. Gamberini, Z Phys. **C30**, 605 (1986); H. Baer, V. Barger, D. Karatas and X. Tata, Phys. Rev. **D36**, 96 (1987); R.M. Barnett, J.F. Gunion and H.E. Haber, Phys. Rev. **D37**, 1892 (1988).
29. J. Gunion, et.al., in *From Colliders to Supercolliders*, pg. 255 edited by V. Barger and F. Halzen, published by World Scientific; J. Gunion and H. Haber, Phys. Rev. **D37**, 2515 (1988).
30. H. Baer, K. Hagiwara and X. Tata, Phys. Rev. **D38**, 1485 (1988).
31. H. Baer, V. Barger, H. Goldberg and R. Phillips, Phys. Rev. **D37**, 3152 (1988).
32. V. Barger, T. Gottschalk, J. Ohnemus and R. Phillips, Phys. Rev. **D32**, 2950 (1985).
33. V. Barger, W.Y. Keung and R. Phillips, Phys. Rev. Lett. **55**, 166 (1985); see R.M. Barnett, J.F. Gunion and H.E. Haber, these proceedings.
34. For a somewhat more extensive survey of the high mass region at future hadron colliders, see J.F. Gunion, "INFN Eloisatron Project Working Group Report", Erice-Trapani (1988).

NEW PARTICLE SEARCHES IN e^+e^- COLLISIONS*

C. BALTAY
Yale University, New Haven, Conneticut 06511

D. BURKE AND R. SCHINDLER
Stanford Linear Accelerator Center, Stanford University, Stanford, California 94309

M. CHEN
Massachusetts Institute of Technology, Cambridge, Massachusetts 02139

G. KANE
University of Michigan, Ann Arbor, Michigan 48109

U. NAUENBERG
University of Colorado, Boulder, Colorado 80309

1. INTRODUCTION

This report summarizes the results of the Study Group on New Particle Searches at e^+e^- Colliders. The work of this Group was organized by U. Nauenberg. The Group was divided into two subgroups: Low Energy Colliders ($E_{cm} \leq 200$ GeV) led by R. Schindler and M. Chen, and High Energy Colliders ($E_{cm} \geq 200$ GeV) led by D. Burke.

A variety of e^+e^- colliders, from 100 GeV to 2 TeV in the center-of-mass, were considered. These facilities, with the assumed luminosities for each, are listed in Table I. Some of these facilities (SLC and LEP) are now under construction and will start operating in the next year or so, while the high energy colliders are quite hypothetical with no credible detailed designs existing at this time.

Table I. e^+e^- colliders considered in the study.

E_{cm}	Facilities	Assumed Luminosity ($\text{cm}^{-2}\text{sec}^{-1}$)
100 GeV	SLC/LEP	$(0.6 \text{ to } 2) \cdot 10^{31}$
200 GeV	LEP II	$(1\text{–}2) \cdot 10^{31}$
100 GeV	Super Z^0 Factory	10^{33}
1/2 to 2 TeV	ILC/TLC/CLIC	$(0.2 \text{ to } 4) \cdot 10^{33}$

This report, as well as the deliberations of the Study Group, have drawn heavily on previous studies. Particularly helpful were the Proceedings of the La Thuile Workshop (CERN Report 87–07) 1987 and the SLAC Study Group Report (SLAC–PUB–329) 1988.

The general motivation for studying new particle searches, and in particular e^+e^- collisions, in the context of Particle Physics for the 1990's, is given in the preceding article, New Particle Production — An Overview. As mentioned in the Overview, investigations of the Higgs sector were made and reported in the Gauge Symmetry Breaking working group section of the Workshop.

Members of the study group were: Low Energy Colliders: D. Atwood, H. Baer, G. Bélanger, A. Blondel, M. Chen, G. Couture, J. Grifols, P. Grosse-Wiesemann, C. Heusch, B. Irwin, D. London, P. McBride, A. Mendez, U. Nauenberg, J. Robinson, R. Schindler, S. Sugimoto, X. Tata, R. Zhu; High Energy Colliders: J. Alexander, H. Baer, C. Baltay, T. Barklow, P. Burchat, D. Burke, C. Dib, K. Gan, J. Gomez-Cadenas, J. Grifolo, H. Haber, J. Hewett, S. Komamiya, W. Kozanecki, P. Rankin, T. Rizzo, S. Selipsky, A. Sill, C. Simopoulos, R. Stroynowski, W. Tuung, R. Van Kooten, B. Ward, E. Yehudai, D. Zeppenfeld.

2. LUMINOSITY REQUIREMENT OF e^+e^- COLLIDERS

The e^+e^- cross sections at high energies are rather small; thus high luminosities are needed to achieve useful event rates at high energy e^+e^- colliders. It is customary to discuss cross sections in units of R, where one unit of R is the cross section for the process:

$$\begin{aligned}\sigma(R=1) &= \sigma(e^+e^- \to \gamma^* \to \mu^+\mu^-) \\ &= 87 \times 10^{-39} cm^{-2}/s(\text{TeV}^2) \quad ,\end{aligned}$$

where s is the square of the total center-of-mass energy, $s = E_{cm}^2$.

Cross sections in the vicinity of one unit of R are typical of many new particle production processes. A minimum useful event rate seems to be 1000 events/year/unit of R. For example at 1 TeV, using 10^7 second/year (one "Snowmass year") we need $\mathcal{L} = 10^{33}$ cm^2 sec^{-1}. Because of the s^{-1} dependence of $\sigma(R=1)$, the minimum useful luminosities for the high-energy machines that we consider are:

* Work supported by the Department of Energy, contract DE–AC03–76SF00515.

E_{cm}	$\mathcal{L}_{min}$ (cm^{-2} sec^{-1})
400 GeV	2×10^{32}
1 TeV	10^{33}
2 TeV	4×10^{33}

It should also be noted that these minimum luminosities might be just enough to discover possible new effects. Higher luminosities (by an order of magnitude?) could be needed to carry out detailed studies of such effects.

3. GENERAL FEATURES OF e^+e^- COLLISIONS

In the past it has generally been true that electron accelerators and colliders had lower center-of-mass energies than the corresponding hadron facilities in existence at any given time. This has limited the energy reach of the electron facilities. However, in reality, the effective physics reach of the various colliders have been rather similar, since in e^+e^- collisions the entire energy goes into the point interactions of interest, while in hadron collisions the larger total energy is divided among many constituent quarks and only a small fraction of it goes into any particular quark-quark collision of interest. Similar comments can be made about cross sections. Total cross sections are much larger at hadron colliders, but the cross sections for the subprocesses of interest are not so different from the corresponding cross sections at e^+e^- colliders. In fact, the absence of the large background cross sections, and the simplicity of the e^+e^- collisions, have made the results of e^+e^- colliders much clearer and easier to interpret. Another nice feature of e^+e^- collisions is that because of the well known electromagnetic couplings, negative results can be more easily converted into reliable limits on the relevant physics parameters.

The general characteristics outlined above are expected to be similar in the next generation of TeV Colliders. However, the nature of e^+e^- collisions will change in many ways, which we discuss in a little more detail in the remainder of this section.

3.1 Features of Low Energy e^+e^- Collisions (i.e., $E_{cm} \leq 200$ GeV or so)

(a) The particles of interest decay into light quarks and leptons. In fact most of the important discoveries were made by observing leptonic decays such as $\psi \rightarrow e^+e^-$ or $\tau \rightarrow \mu\nu\bar{\nu}$. For this reason, good lepton detection and measurement was emphasized in the design of detectors.

(b) The backgrounds to the processes of interest were very small. The cross sections for light quark pair production are similar to those of the interesting lepton pair or heavy quark pair production processes. The only known annihilation background was the two-gamma process which could easily be separated by a visible energy cut. In the colliders at the Z^0 mass, the collisions are dominated by the signal, Z^0 production, with a cross section of ~ 5000 units of R, while the known Z^0 background has a cross section of the order of 10 units of R.

(c) In the low-energy colliders the energies of the incident $e^{\pm}$ beams at the collision point were known to a precision of typically 10^{-3} or 10^{-4}. This provided two very powerful constraints, *i.e.*, precise knowledge of the total center-of-mass energy, $E_{cm} = 2E_{beam}$, and the total momentum, $\bar{P} = 0$, *i.e.*, the center-of-mass of the collision was stationary in the lab. These constraints were very important in understanding the details of the processes of interest.

3.2 Features of High Energy e^+e^- Collisions ($E_{cm} \geq 200$ GeV or so)

(a) At the high energy colliders we anticipate that the new particles we are interested in will have prominent decay modes into Z's and W's, such as the Higgs boson, $H^0 \rightarrow W^+W^-$, or new heavy quarks $Q \rightarrow q+W$. Detailed studies (see following sections) indicate that in e^+e^- collisions the backgrounds will be sufficiently small so that the dominant hadronic decays $W, Z \rightarrow q + \bar{q} \rightarrow 2$ jets will be useful signatures in addition to the leptonic W and Z decay modes. This possibility has a number of significant advantages:

(1) The branching fractions for the hadronic decays is of the order of 70%, compared to a few percent for any particular leptonic decay. Since often there are two or more W's or Z's involved in a signature for a new process, the feasibility of using the hadronic modes can enhance the detection efficiencies by orders of magnitude.

(2) In the leptonic decay $W^{\pm} \rightarrow \ell^{\pm}\nu$ there is a neutrino which is undetected, so that generally the mass of the new particle in question cannot be reconstructed in a straightforward way. With the hadronic decay modes most of the energy is measurable and the masses of the new particle can be reconstructed. (See Fig. 1.) This allows the measurement of the mass, as well as a much more convincing signature for the discovery of some new particle, *i.e.*, a peak in some effective mass distribution (see the detailed discussions in the following sections).

(3) The precision of hadron calorimetry improves as the energies increase. Thus the energies and angles of the jets in W, $Z \rightarrow q\bar{q} \rightarrow 2$ jets can be measured well enough that the W or Z mass can be reconstructed using this technique.

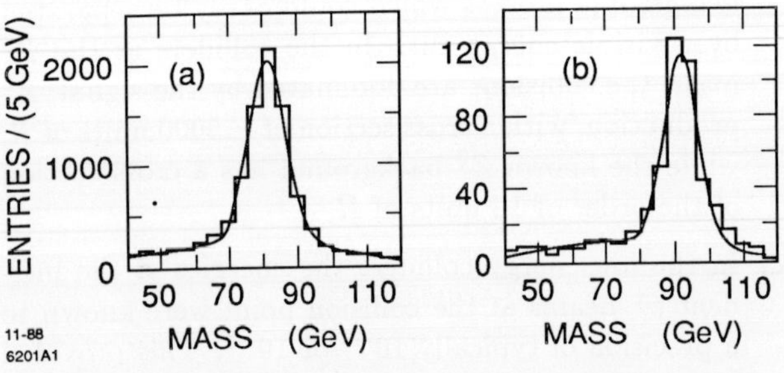

Fig. 1. Reconstructed W and Z mass distributions for machine and detector parameters used in this report.

Table II. Cross sections and event rates at a 1 TeV e^+e^- collider.

Process	σ (units of R)	Events/year†
1. Annihilation		
$e^+e^- \to \nu_\mu\bar{\nu}_\mu$	0.25	250
$\mu^+\mu^-$	1.3	1300
$d\bar{d}$	1.1	1100
$u\bar{u}$	2.1	2100
total $\ell^+\ell^-$	4	4000
total $q\bar{q}$	9	9000
$e^+e^- \to W^+W^-$	27	27000
Z^0Z^0	1.5	1500
$\gamma\gamma$	10	10000
γZ^0	31	31000
$W^+W^-Z^0$	0.4	400
$Z^0Z^0Z^0$	0.03	30
2. Two-γ, $\gamma - W$, $\gamma - Z$ fusion		
$e^+e^- \to e^+e^-q\bar{q}$	$\sim$1	1000‡
$e^+e^-W^+W^-$	9.3	9300
$e\nu W^\pm Z^0$	3.4	3400
3. $e^+e^- \to e\nu W^\pm$	140	140000
$e^+e^-Z^0$	70	70000

†With $\int \mathcal{L}dt = 10^{40}$ cm^{-2} = 10 fb^{-1}, neglecting beamstrahlung.

$^\ddagger m\,(q\bar{q}) \geq 200$ GeV.

(b) New background processes, such as W^+W^- pair production and $\gamma\gamma$, $\gamma - W$, and $\gamma - Z$ fusion processes, become important, with relatively large cross sections at high energies. The cross sections and event rates for the relevant processes of 1 TeV are summarized in Table II. While some of these reactions, like $e^+e^- \to W^+W^-$, are of great interest on their own,

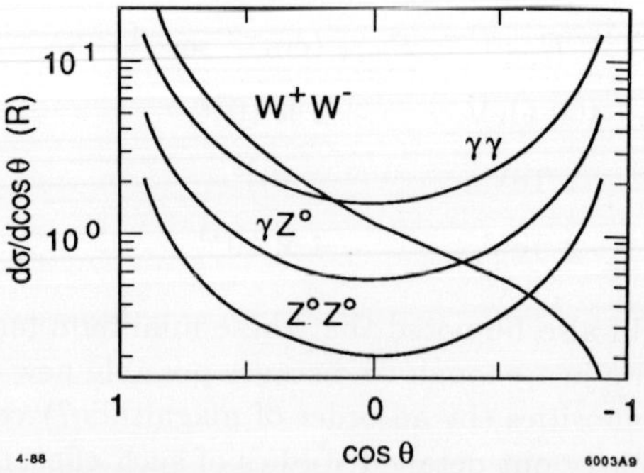

Fig. 2. Differential cross sections for the production of gauge bosons by e^+e^- annihilation at $\sqrt{s}$ = 1 TeV.

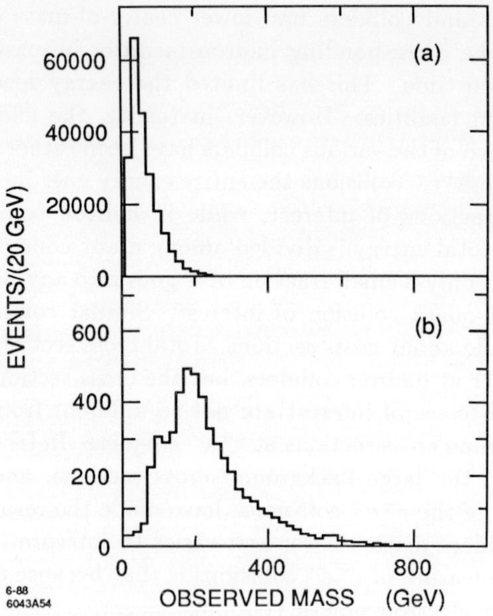

Fig. 3. Distribution of observed masses for two-photon process, (a) $\gamma\gamma \to q\bar{q}$ and (b) $\gamma\gamma \to W^+W^-$. The distributions are normalized to e^+e^- integrated luminosity of 10 fb^{-1}.

they nevertheless represent a background in searches for new particle production. Detailed studies, however, indicate that these processes have some very distinctive features that allow various selection criteria to eliminate them as serious backgrounds to new particle production. Examples of these are an angle cut to eliminate the bulk of the $e^+e^- \to W^+W^-$ process (see Fig. 2) and a total visible energy cut to eliminate the $\gamma\gamma$, γW, or γZ fusion processes (see Fig. 3). As will become apparent from the detailed discussion of various particle searches, these backgrounds will have to be worried about but should turn out to be manageable.

(c) The very high energy e^+e^- colliders will most likely be colliding linear accelerators. To obtain the required high luminosities the beams will have to be focused very hard to produce sub-micron size spots at the collision point. As the $e^{\pm}$ beams pass through each other at the intersection region they can radiate a non-negligible fraction of their energy (called beamstrahlung) before the e^+e^- collision takes place. Thus, the total center-of-mass energy and the total momentum of the collision (*i.e.*, the motion of the center-of-mass in the lab) are no longer precisely defined, and the constraints $E_{cm} = 2\,E_{beam}$ and $\bar{P}_{cm} = 0$, which were so useful at low energy colliders, might be lost. There is some flexibility in the design of the colliders to minimize these effects, and it is therefore not clear at this time how seriously these constraints will be compromised. The size of this effect can be characterized by the parameter δ, which is the mean energy loss of one beam as it passes through the other beam. The spread in E_{cm} for the actual e^+e^- collisions are shown for various existing preliminary collider designs in Fig. 4 (TLC design at SLAC with $\delta = 0.26$) and in Fig. 5 (CLIC design at CERN with $\delta \approx 0.1$).

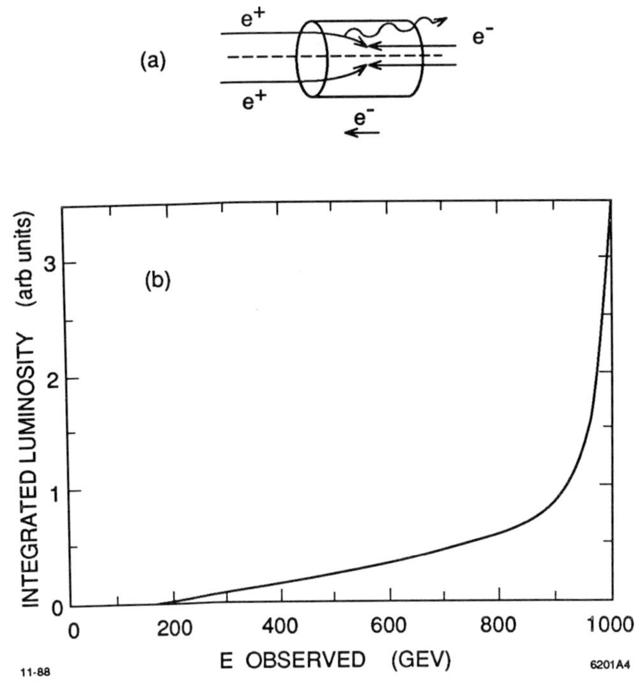

Fig. 4. (a) Beamstrahlung induced by the deflection of particles in one bunch by the collective fields of the opposing bunch. (b) A typical spectrum of center of mass energy in e^+e^- collisions in the presence of beamstrahlung. The curve is computed for one particular design of colliding beams each with initial energy of 500 GeV.

The deterioration of the E_{cm} and P_{cm} constraints were taken into account in the detailed studies of new particle searches presented in the remainder of this report. These studies indicate that the deterioration of these constraints makes the analyses more difficult, but the problems are manageable and all of the searches are feasible.

A useful feature of the smearing of the collision energy is that the high energy colliders become self-scanning, *i.e.*, a collider sitting at its peak energy can see resonances at lower masses without doing an energy scan. This sensitivity is illustrated in Fig. 6 for the TLC design with $\delta = 0.26$.

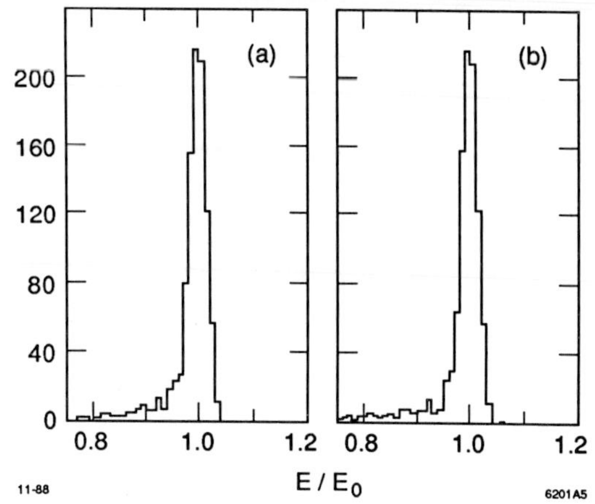

Fig. 5. Spectra similar to that shown in Fig. 4(b) for alternative machine designs at (a) 1 TeV, *and (b) 2* TeV.

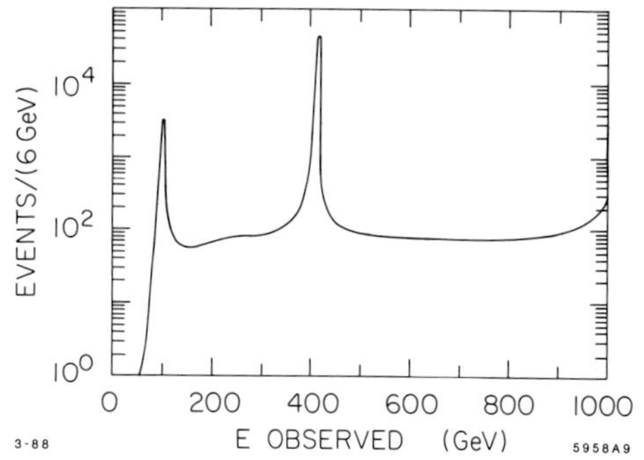

Fig. 6. Distribution of observed hadronic invariant masses resulting from e^+e^- annihilation to hadrons. The effect of a new Z' boson of mass 400 GeV *is also shown. The distribution is computed with the beamstrahlung spectrum shown in Fig. 4(b).*

4. COMMENT ON DETECTORS

There was not much discussion of detailed detector design at this Summer Study. However, the SLAC study (SLAC–PUB–329) did consider the detector parameters needed to do the physics at a TeV collider. The following detector parameters were adopted for that study:

4.1 Tracking

A momentum resolution of $\delta p/p = 3 \times 10^{-4}$ p(GeV) was assumed (*i.e.*, $\sim 15\%$ at 500 GeV). This can be achieved in a relatively straightforward way with a drift chamber of 1.8 m radius in a magnetic field of 0.6 Tesla with 80 layers of wires with a resolution of 120 μm on each wire. Precision vertex detectors of the kind being built for the SLC and LEP detectors will be desirable.

4.2 Calorimetry

The following parameters were assumed:

Energy Resolution:

Electromagnetic: $\frac{\delta E}{E} = \frac{8\%}{\sqrt{E}} + 2\%$

Hadronic: $\delta E = \frac{50\%}{\sqrt{E}} + 2\%$

Segmentation: — $4° \times 4°$ Projective Towers

The segmentation requirement was arrived at by studying the mass resolution and detection efficiency of reconstructing a $W \to q\bar{q} \to 2$ jets as a function of segmentation. As can be seen from Fig. 7, $4° \times 4°$ towers seem quite adequate.

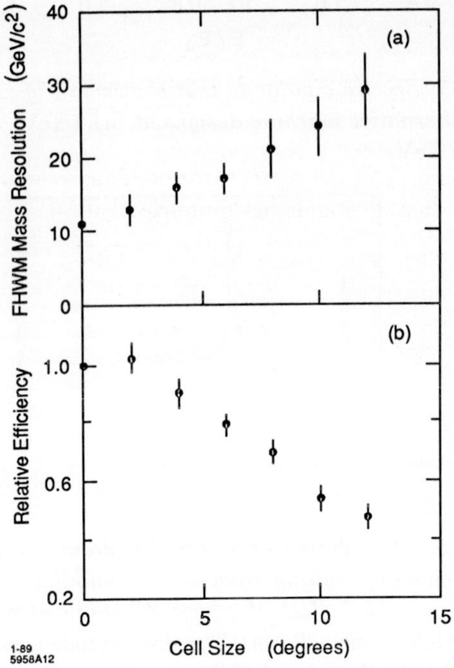

Fig. 7. (a) Mass resolution for gauge bosons obtained from $e^+e^- \to W^+W^-$ events reconstructed with calorimeters with differing granularity. (b) Acceptance for $e^+e^- \to W^+W^-$ events as a function of calorimeter cell size after demanding each W be properly reconstructed.

The event rates at a 1 TeV collider with a luminosity of 10^{33} cm^{-2} sec^{-1} are less than the event rates expected at SLC and LEP, thus no new problems associated with triggering or radiation damage are expected.

The conclusion thus seems to be that the parameter of the detectors required for a TeV e^+e^- collider are not far from those at SLC and LEP. Therefore, the detectors for the new high energy e^+e^- colliders are a straightforward extrapolation of the detectors now under construction at SLC and LEP, and no real problems are expected.

5. DISCUSSION OF SELECTED SEARCH TOPICS

5.1 Search For New Heavy Quarks

The possibility of a fourth generation of quarks was considered:

$$\begin{pmatrix} d \\ u \end{pmatrix} \begin{pmatrix} s \\ c \end{pmatrix} \begin{pmatrix} b \\ t \end{pmatrix} \begin{pmatrix} b' \\ t' \end{pmatrix} ,$$

where b', t' are new heavy quarks with charges of $-1/3$ and $+2/3$, respectively. The dominant production process in e^+e^- collisions is expected to be via a γ or Z^0 exchange in the s channel:

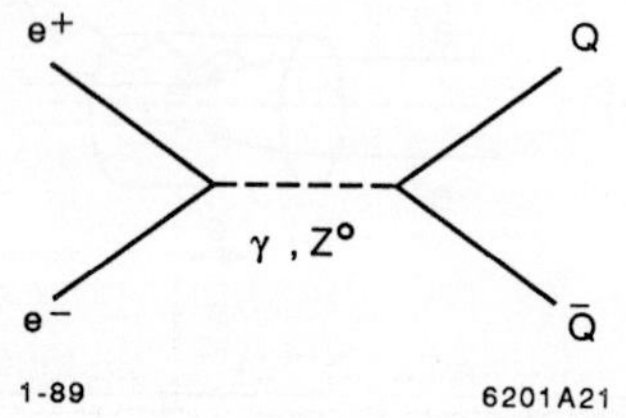

The cross sections for these processes are essentially independent of M_Q, the quark mass, almost up to the kinematic limits:

$$\sigma(e^+e^- \to b'\bar{b}') \sim 1 \text{ unit of } R$$
$$\sigma(e^+e^- \to t'\bar{t}') \sim 2 \text{ units of } R$$

The expected numbers of events produced at a 2 TeV collider in a sample of 1000 events per unit of R (an integrated luminosity of $\sim 4 \times 10^{40}$ cm^{-2}) are shown in the first column of Table III for various b' and t' masses.

The backgrounds are expected to be sufficiently small so that the dominant hadronic decay modes can be used in the search:

$$b' \to W^- + q \quad , \quad W^- \to q + \bar{q} \quad ;$$
$$t' \to W^+ + b' \quad , \quad b' \to W^- + q \quad , \quad W^\pm \to q + \bar{q} \quad .$$

The signature, for example for a $b'\bar{b}'$ pair, are events with six jets, divided into two hemispheres with three jets each. The main backgrounds are annihilation processes like $e^+e^- \to q\bar{q}$ or W^+W^-, and the $\gamma\gamma$ or γW fusion processes $e^+e^- \to e^+e^-q\bar{q}$, $e^+e^-W^+W^-$, or $e\nu WZ$ with the e^+e^- going down the beam pipe. The total numbers of these events are also shown in the first column of Table III.

Table III. Heavy quark event rates in 2 TeV e^+e^- collisions.

Use sample of 1000 events/unit of R		
Process	Events Produced	Events after Selection
• $e^+e^- \rightarrow b'\bar{b}'$		
$m_{b'} = 300$ GeV	1000	330
500	900	250
700	700	150
900	450	100
• $e^+e^- \rightarrow t'\bar{t}'$		
$m_{t'} = 300$ GeV	2000	850
500	1800	700
700	1400	450
900	900	300
• Backgrounds		
$e^+e^- \rightarrow q\bar{q}$	9000	200
W^+W^-	36000	900
$e^+e^-W^+W^-$	250000	300

Some fraction of these events will have a 6-jet topology due to jets from hard gluon Bremsstrahlung, and thus be a background to heavy quark production.

Several detailed Monte Carlo calculations were carried out to simulate the effects of realisitic detector parameters and the necessary selection criteria on the signal and the backgrounds. Figure 8 shows the results of an analysis by P. Igo–Kemenes from the La Thuile study.[1)] The LUND Monte Carlo generation with JETSET 6.3 and LUCLUS were used. A 2 TeV Collider with an integrated luminosity of 1000 events/unit of R were assumed. The following selection criteria were applied.

1. Select multijet events with at least five or more jets.
2. Total visible energy larger than 50% of the nominal collider E_{cm}, *i.e.*, $Evis \geq 1$ TeV.
3. The event was separated into two hemispheres along the thrust axis. The effective mass of each hemisphere was calculated, and the two masses were required to be equal to within 30%.
4. The sphericity of the events were required to be larger than 0.04.

The number of signal and background events passing these cuts are shown in the last column of Table III. The efficiency for the signal events is quite

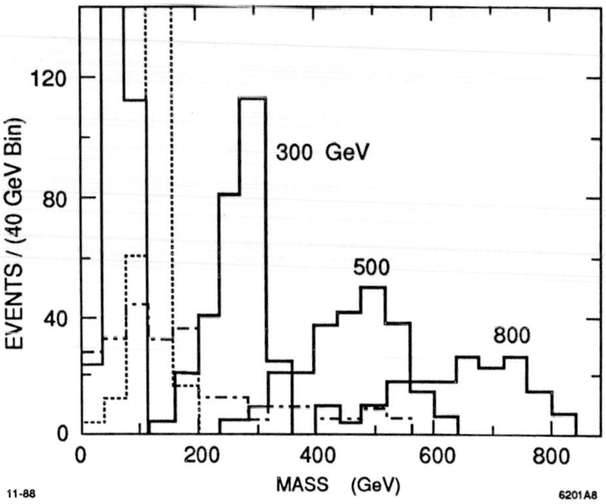

Fig. 8. Reconstructed invariant mass per hemisphere for $e^+e^- \rightarrow b'\bar{b}'$ events with various b' masses. Backgrounds are also shown and described in the text.

good, while the backgrounds have been reduced by large factors to a manageable level. The distribution of the effective mass of each hemisphere for the events surviving the cuts is shown in Fig. 8. It can be seen that the remaining background peaks at low masses, and the signal corresponding to heavy quark production stands out quite clearly.

This analysis indicates that heavy quarks can be found cleanly up to almost the kinematic limits with high energy e^+e^- colliders.

5.2 Search for the *t* Quark

The expected sixth quark, the t quark, has so far eluded detection. The current limits on the mass of the t quark are:

$$41 \leq m_t \leq 200 \text{ MeV} \quad ;$$

the lower limit comes from the UA1 experiment at CERN. The upper limit comes from theoretical considerations of loop corrections of the general kind:

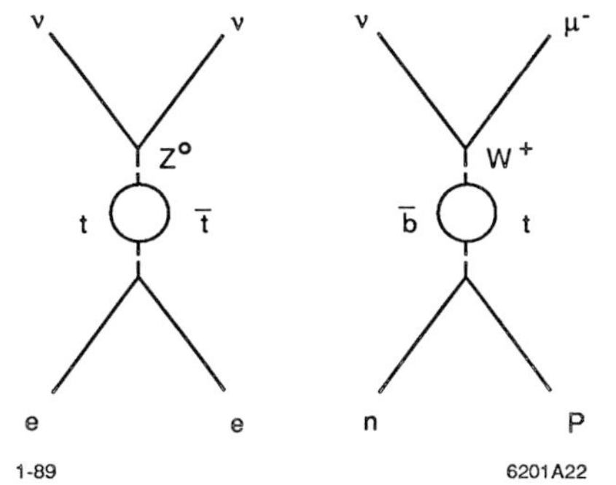

The size of such corrections typically depends on m_t^2; $m_t \geq$ 200 GeV would lead to corrections which would be large and different for neutral current and charged current processes, in direct disagreement with the experimentally measured consistency of the W and Z masses and $\sin\theta_W$.

The SLC, LEP, and the Tevatron can probe some, but probably not all, of the allowed range of m_t in the near future. If the t is not found at these facilities in the near future an e^+e^- collider with an E_{cm} of 400 to 500 GeV would be sufficient to probe the whole mass range to either find the t or to show that something unexpected is going on.

A detailed Monte Carlo analysis of the feasibility of a t quark search at a 400 GeV e^+e^- collide was carried out at this Summer Study by R. Van Kooten. The dominant production process is via γ and Z^0 exchange in the s channel.

The cross sections for various values of m_t and the numbers of events produced in a 10 fb^{-1} sample are given in Table IV. The backgrounds are similar to those discussed in the previous section and listed in Table III.

Table IV. Cross sections and event rates for top quark production at a 400 GeV e^+e^- collider.

m_t (GeV)	σ (units of R)	Events in 10 fb^{-1}
50	2.1	11000
100	1.8	9000
150	1.4	7600
170	1.1	6000
190	0.7	3800

The decay modes used in the analysis were:

$$e^+e^- \to t + t'$$
$$t \to b + W^+, W^+ \to e^+ \text{or } \mu^+ + \nu$$
$$\bar{t} \to b + W^-, W^- \to q\bar{q} \to 2 \text{ jets} \quad ;$$

thus the signature looked for is a jet and an isolated fast-charged lepton with two or three jets in the opposite hemisphere. Signal and background events were generated using the LUND 6.3 generator with LULEPT. The following selection criteria were applied:

1. Select multijet events with one energetic ($p \geq$ 2 GeV/c) e or μ that is isolated from the rest of the particles in the event (less than 2 GeV of energy within a 20° cone around the e or μ).
2. Find the thrust axis, and require $\cos\theta_{thrust} \leq 0.8$.
3. Separate the event into two hemispheres along the thrust axis and calculate the effective mass of each hemisphere. Require that both masses are larger than 104 GeV.

The distribution of the mass in the hemisphere opposite to the isolated e or μ is shown in Fig. 9. A top mass of 150 GeV was used to generate this distribution. A clear peak near the t mass is visible. The backgrounds, shaded in Fig. 9, are quite small.

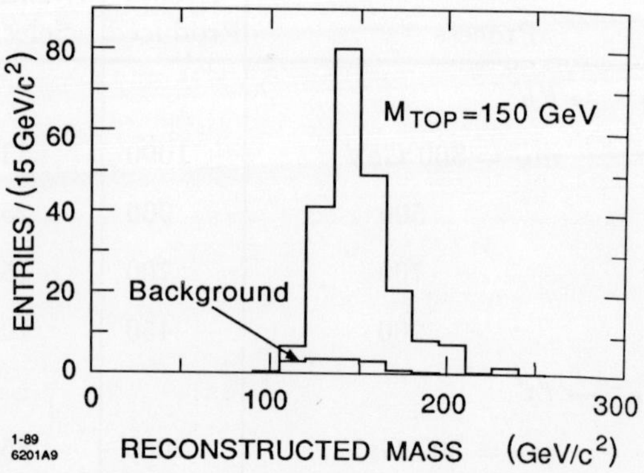

Fig. 9. Observed signal from a top quark search with integrated luminosity of 10 fb^{-1} at an e^+e^- collider with $\sqrt{s}$ = 400 GeV if the top mass is 150 GeV. The shaded region shows the backgrounds from all known processes.

5.3 Leptoquark Search

Leptoquarks are hypothetical new particles with both lepton (L=$\pm$1) and quark ($B = \pm$ 1/3 or 2/3) quantum numbers. They are predicted in all technicolor and extended technicolor models. Recent interest has been added by predictions of leptoquarks by superstring-inspired E6 models. As an example we considered the scalar lepto quarks with charge $-1/3$ and the following expected decay modes:

$$L_0^e \to u + e-$$
$$L_0^\mu \to c + \mu^- \quad ,$$

where u and c are the up and charmed quarks, respectively. The main production processes in e^+e^- collisions are expected to be via the diagrams.

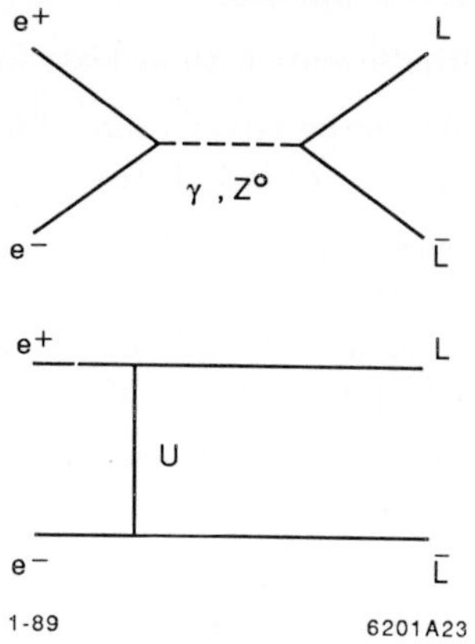

The cross section is predicted to be:

$$\sigma(e^+e^- \to L_0 + \bar{L}_0) \cong 0.4B^3 \text{ in units of } R \quad .$$

The cross sections for various values of the leptoquark mass are given in the first column of Table V. The expected numbers of events produced by a 2 TeV collider with an integrated luminosity of 1000 events/units of R are given in the second column of Table IV.

Table V. Cross sections and event rates for leptoquark search in 2 TeV e^+e^- collisions.

Use sample of 1000 events/units of R Process	Cross Section Units of R	Events Produced	Events after Selection
1. $e^+e^- \to L_0\bar{L}_0$			
$m_L = 200$ GeV	0.4	400	375
400	0.3	300	280
600	0.2	200	150
800	0.08	80	65
900	0.03	30	20
2. Backgrounds			
$e^+e^- \to q\bar{q}$	9	9000	45
W^+W^-	36	36000	4
e^+e^-WW	250	250000	$\leq$ 1

These leptoquark events should have a very simple signature: an isolated energetic e^- or μ^- with a jet on one side, with an e^+ or μ^+ and a jet on the other. The potential backgrounds to this process are also shown in Table V.

A detailed Monte Carlo calculation for a leptoquark ($L \to u + e^-$) search was carried out by Shaile and Zerwas at the La Thuile Study.[2)] They used the latest version of the LUND Monte Carlo generator with JETSET 6.3 and LUCLUS to simulate both the signal and the background events at a 2 TeV Collider. They applied the following selection criteria to the events.

1. Select events with an energetic e^+ and an energetic $e^-(P_{e^\pm} \geq 20$ GeV) and two or more hadronic jets.
2. Total visible energy $E_{vis} \geq 1.6$ TeV.
3. $\sum P_Z \leq 0.15 E_{vis}$.
4. $\sum P_\perp \leq 150$ GeV/c.

The numbers of events passing these cuts are shown in the third column of Table V. For the events passing these cuts the effective masses for the $e^\pm$–jet combinations were calculated. Figure 10 shows the distribution in the e–jet mass for the events passing the cuts. The solid curve is for the $L \to u + e$ signal for a L mass of 600 GeV. The shaded curve is the background passing the cuts, mostly due to $e^+e^- \to q\bar{q}$ with an energetic $e^\pm$ in each jet. The signal is quite clearly distinguished from the background.

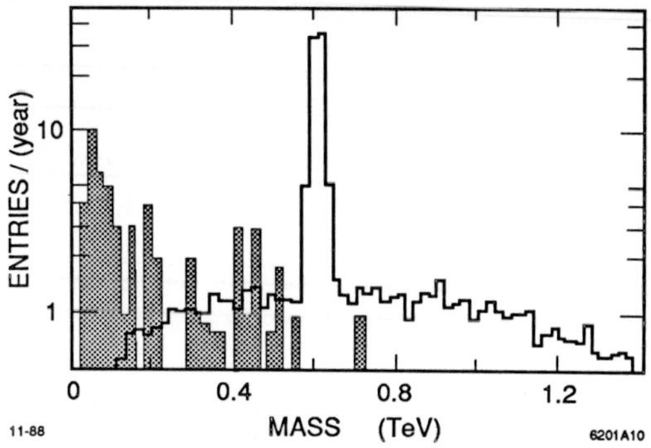

Fig. 10. Reconstructed signal from a search for leptoquarks with a data sample of 10 fb^{-1} at $\sqrt{s}$ = 2 TeV. *The solid line is the invariant mass of all e-jet combinations in accepted signal events and the shaded region gives the background.*

5.4 New Neutral Gauge Boson Z′

There are a variety of theoretical reasons for expecting additional heavy Z^0 type gauge bosons. We will call these Z' in this report. Left-right symmetric models predict additional Z' as well as right-handed charged bosons $W_R^\pm$. Superstring-inspired E_6 Models, which are very fashionable with theorists this year, predict several new neutral bosons. In this model the E_6 group breaks down into several U(1) groups [the new U(1) groups have new Z bosons associated with them];

$$E_6 \to SO(10) + U(1)_\psi$$
$$SO(10) \to SU(5) + U(1)_\chi \quad ,$$

thus predicting the existence of two new Z' bosons Z_ψ and Z_χ. In general, these can mix to produce mass eigenstates:

$$Z'_\eta \;=\; \cos\theta Z_\psi + \sin\theta Z_\chi \quad .$$

The value of this E_6 mixing angle θ is of some interest theoretically. One value of particular interest is $\cos\theta = -\sqrt{5/8}$ which yields:

$$Z'_\eta \;=\; -\sqrt{5/8}Z_\psi + \sqrt{3/8}Z_\chi \quad .$$

The search for such Z' bosons and their detailed study, shedding light on their origin and mixing angles like θ, is obviously of great interest.

5.4.1. Search for Z′ bosons in e$^+$e$^-$ collisions

(a) If a Z' boson exists in the kinematic range of an e$^+$e$^-$ collider, *i.e.*, with mass less than the total center-of-mass energy, $m_{Z'} \leq E_{cm}$, the Z' will manifest itself as a resonance in the total cross section at an energy equal to the mass of the Z'. The total cross section at resonance is expected to be in the range of 1,000 to 10,000 units of R. Compared to the expected cross sections without a Z' (see Table II), a collider with any appreciable luminosity cannot fail to find such a Z' within its kinematic range.

(b) An e^+e^- collider is sensitive to a new Z' even if the mass of the Z' is several times larger than E_{cm} of the collider. This is because the tail of a Z' will affect the cross sections, the forward-backward asymmetries, A_{FB}, and the polarization asymmetries A_{LR} of various processes such as $e^+e^- \rightarrow \mu^+\mu^-$ and $e^+e^- \rightarrow q\bar{q}$. These effects can be quite large, but are generally dependent on the details of the model predicting the Z'. At this study, J. Hewett and T. Rizzo carried out extensive calculations of such effects in the context of a superstring-inspired E_6 model. In these models the effects, like the deviation of the polarization asymmetry A_{LR}, depend sensitively on the E_6 mixing angle θ. Figure 11 shows their result of the sensitivity to the mass of a new Z' (M_2 in the Figure) for a 1 standard deviation change in A_{LR} for various channels as a function of the mixing angle θ for a 400 GeV collider. The curves indicate a sensitivity up to 1.2 to 2.0 TeV. Figure 12 shows the same thing for a 1 TeV collider, again showing sensitivity to a Z' mass up to several times the center-of-mass energy of the collider.

5.4.2 Detailed study of a new Z′ boson.

It is quite possible that if a new Z' boson exists it will be discovered at a hadron collider like the Tevatron or the SSC. The detailed study of such a new Z' will be best carried out at an e^+e^- collider with a center-of-mass energy at or just above the mass of the Z'. The situation would be quite analogous to the standard model Z^0, which was discovered at a hadron collider but whose detailed study has to wait for the e^+e^- colliders SLC and LEP.

The total e^+e^- cross section on a Z' resonance is expected to be in the range of 1,000 to 10,000 units of R. Thus a collider with the minimum luminosity of 1,000 events/year/unit of R discussed in Section II above should produce 10^6 to 10^7 Z' bosons/year. The non-Z' background should be of the order of 1% or less of the Z' production (see Table II). Thus clean, detailed studies should be possible. The analogy with the Z^0 at SLC and LEP is quite close. The physics potential of SLC and LEP should be an indication of the richness of an e^+e^- collider should a new Z' boson exist.

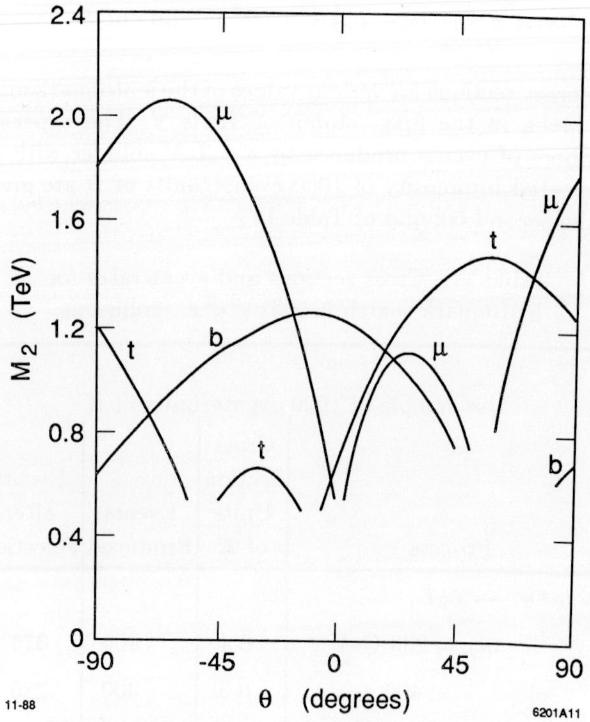

Fig. 11. Mass limits on new gauge bosons that would generate one standard deviation changes in polarization asymmetry measurements in various channels at an e$^+$e$^-$ *collider with* $\sqrt{s}$ = *400* GeV. *The angle* θ *is a mixing angle needed to define the gauge structure in the theory.*

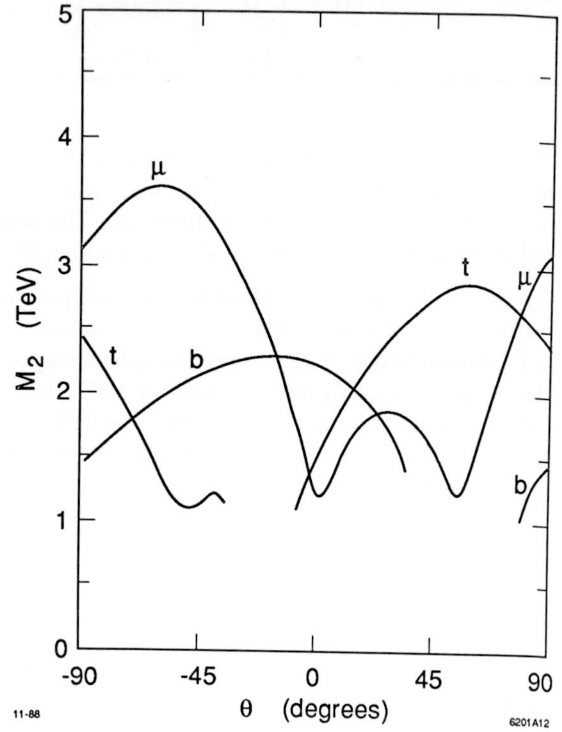

Fig. 12. Limits similar to those shown in Fig. 11 for a collider with $\sqrt{s}$ = *1* TeV.

5.5 Neutrino Counting

Our present view of particle physics is based on three generations of quarks and leptons. It is obviously of fundamental importance to determine experimentally whether there are additional generations of quarks and leptons, and if so, how many. One way to do this is to look for new heavy quarks and leptons; these could be quite heavy and beyond the reach of our facilities. However, the neutrinos associated with the existing three families are very light, probably massless. Assuming this trend to continue, counting the number of neutrino types that occur in Z^0 decay would give a good indication of the number of generations.

The total width of the Z^0 is sensitive to the number of neutrino types. With the three known generations the total width of the Z^0 should be around 2.7 GeV. Each additional neutrino type will increase this by 170 MeV. Thus, a fairly accurate measurement of the width is required. Radiative corrections and other systematic errors will make this a difficult measurement.

A better way to measure the number of neutrinos is to run the collider 5 to 10 GeV above the Z^0 mass and look at the reaction $e^+e^- \to Z^0 + \gamma$. The reaction is identified by a monochromatic γ. One then measures the branching ratio of the Z^0 into ν's, normalized to say the branching ratio into $\mu^+\mu^-$ — *i.e.*,

$$r = \frac{e^+e^- \to \gamma + Z^0 \to \gamma + \nu\bar{\nu}}{e^+e^- \to \gamma + Z^0 \to \gamma + \mu^+ + \mu^-}$$

This ratio is expected to be:

$$r = 2 \cdot (\text{number of } \nu \text{ types}) = \begin{array}{l} 6 \text{ for } 3\ \nu \text{ types} \\ 8 \text{ for } 4\ \nu \text{ types} \\ 10 \text{ for } 5\ \nu \text{ types} \end{array}$$

The factor of two comes from the relative neutral current couplings of neutrinos and muons to the Z^0. Thus the difference between three or four neutrinos is 33%. However, it is important to measure this branching ratio as well as possible since it is sensitive to the existence of any other weakly interacting neutrals that couple to the Z^0. For example, a supersymmetric neutrino would change the ratio from six to seven.

Neutrino counting using this technique was looked at in this study to answer the question of how much better than SLC or LEP could a super Z^0 factory (100 GeV e^+e^- collider with 10^{33} luminosity) do? The cross section for $e^+e^- \to Z^0 + \gamma$ around $E_{cm} = 97$ GeV is around 2×10^{-35} cm^2. Figure 13 shows the expected precision that can be obtained for a measurement of r. At integrated luminosities of 10^{37} cm^{-2} or less the precision is statistics limited. At higher integrated luminosities systematic uncertainties become important, which eventually limit the precision in r to about 1% or so.

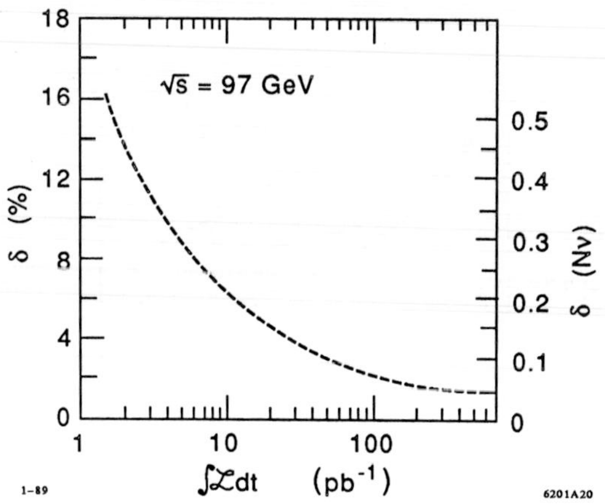

Fig. 13. Limits obtainable on the coupling of the Z^0 to neutral weakly-interacting particles by a dedicated experiment as a function of intergrated luminosity.

From Fig. 13 we see that a long dedicated run at SLC or LEP could get a precision in r of around 2%, which is about 1/10 of a neutrino family. A super Z^0 factory with two orders of magnitude more luminosity can improve the measurement precision only by a factor of two, to 1% in r, since at that level of precision systematic effects dominate.

5.6 Study of the Reaction $e^+e^- \to W^+W^-$.

This reaction proceeds by the following three diagrams:

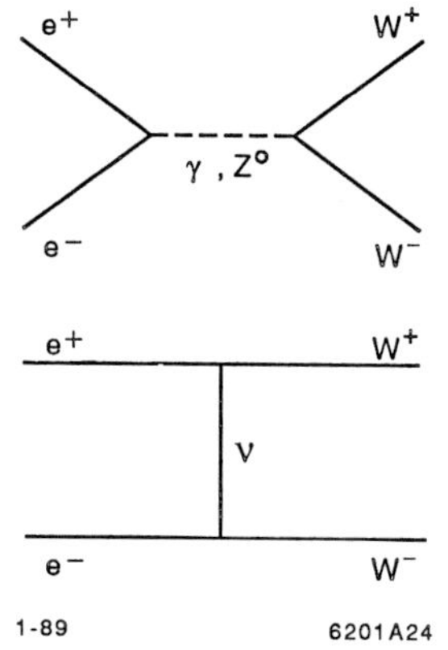

A study of this reaction probes the γWW and the Z^0WW couplings, and is very sensitive to new physics beyond the Standard Model. Some examples are:

1. Sensitivity to an anomalous magnetic moment of the W (*i.e.*, evidence for internal structure of the W). An anomalous magnetic moment would change the differential cross section for this reaction from the Standard Model prediction. The size of the effect expected is shown in Fig. 14 for a 1 TeV collider.

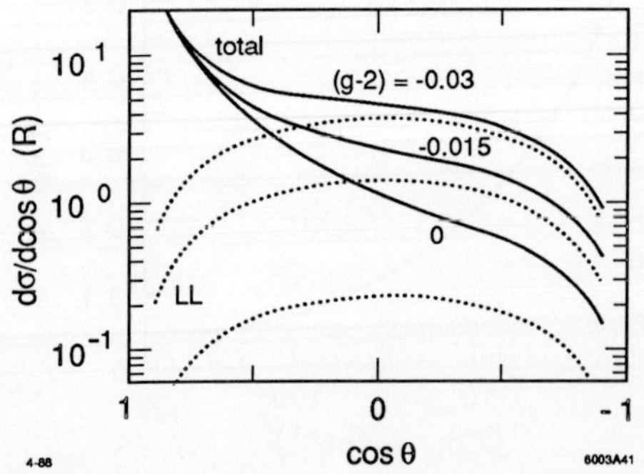

Fig. 14. Sensitivity of the $e^+e^- \to W^+W^-$ differential cross section to an anomalous magnetic moment of the W. The solid lines give the total cross section and the dashed curves show the cross section for production of longitudinal W's only.

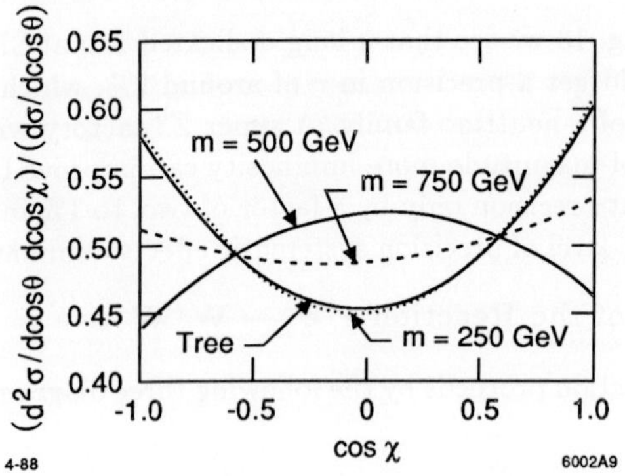

Fig. 15. Effect of a new generation of heavy fermions on the distribution of lepton angles χ from W decay, at $\sqrt{s}$ = 1 TeV *and $\cos\theta$ = 0. The heavy fermions are assumed to be degenerate.*

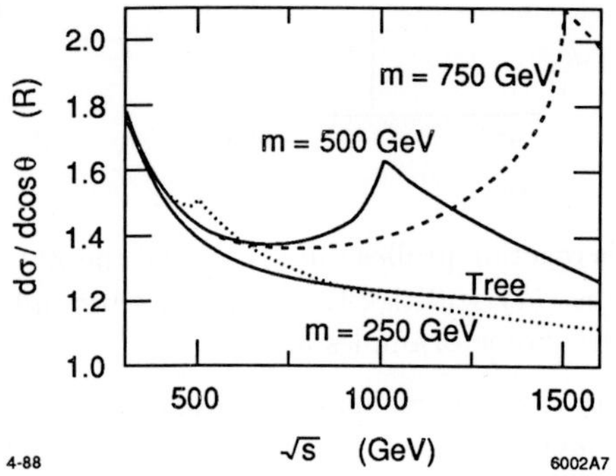

Fig. 16. Effect on the differential cross section for $e^+e^- \to W^+W^-$, at $\cos\theta = 0$, of an additional generation of heavy quarks and leptons.

2. Sensitivity to a 4th generation of quarks too heavy to be pair-produced (*i.e.*, $m_Q \geq 1/2E_{cm}$) comes from quark loop corrections to the above diagrams. The measurable effects are a distortion of the expected distribution in the lepton decay angles χ at a fixed energy, as shown for a 1 TeV collider in Fig. 15, or a change in the cross section at a fixed angle as a function of E_{cm}, as shown in Fig. 16.

3. Sensitivity to the existence of a heavy techni-ρ, ρ_T, beyond the mass reach of the collider. Such a particle would manifest itself as a very broad resonance in the $e^+e^- \to W^+W^-$ cross section, as shown in Fig. 17. The resonance is so broad that a measurable deviation occurs even if the mass of the resonance is considerably larger than the E_{cm} of the collider. Thus, a 1 TeV collider is sensitive to a ρ_T of mass up to 1–1/2 or 2 TeV.

The main motivation behind LEP II is to study this important reaction. The energy of 200 GeV was chosen to be just above threshold. However, the sensitivity to all of the new effects discussed above increases as $\beta \cdot (s/m_W^2)$ as the energy of the collider is increased ($s = E_{cm}^2$). This rapidly increasing sensitivity is shown in Fig. 18. An intermediate energy collider with 500 GeV in center-of-mass energy has 10 times the sensitivity of LEP II to deviations from the Standard Model.

Much of the discussion of this reaction both at this study as well as the SLAC study were concerned with the question of how well an actual detector would be able to reconstruct this reaction and how well the angular distribution could be measured. Backgrounds are not a big issue since this process is the dominant annihilation process with a large cross section. The $\gamma\gamma$ fusion processes can easily be separated. Since measurement of the angular distributions and asymmetries is important one must know the sign of the W's. The analysis therefore should look at one W decaying leptonically, $W \to e$ or $\mu + \nu$, with the other W decaying hadronically, $W \to q\bar{q} \to$ jets. A detailed Monte Carlo study of this process at 1 TeV was carried out in the SLAC study.[3)] The following selection criteria were applied:

1. Select events with an isolated lepton (less than 2 GeV of energy in a 30° cone around an $e^\pm$ or $\mu^\pm$) opposite two hadronic jets.

2. Require that the effective mass of the two jets agree with the W mass to ± 10 GeV.

3. Reconstruct the missing ν using the constraints $m(\ell, \nu) = m(W)$ and $P_x = P_y = 0$ for the event. Resolve the quadratic ambiguity by choosing the smaller ν momentum.

4. Do a Lorentz transformation to $P_z = 0$ frame (where the Z axis is along the beam direction).

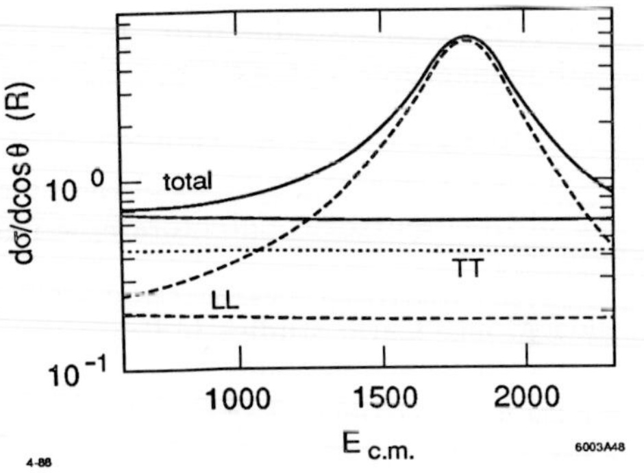

Fig. 17. Effect of a techni-rho resonance, in the minimal scheme of technicolor, on the differential W-pair cross sections at cosθ = −0.5. The dashed lines indicate the contributions of longitudinally polarized W bosons alone.

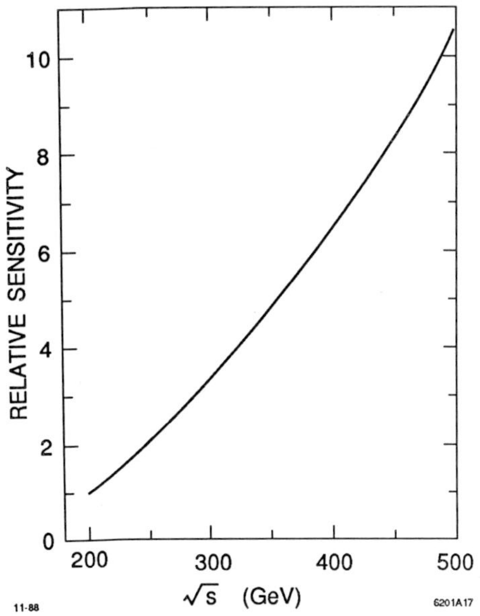

Fig. 18. Relative sensitivity to anomalies in $e^+e^- \to W^+W^-$ *cross sections as a function of the center-of-mass energy of the machine.*

These criteria yield a clean sample of $e^+e^- \to W^+W^-$ events. The effective mass distribution of the two hadron jets is shown in Fig. 19. The peak of the W mass is prominent with very little background. The angular distribution for the events passing these selection criteria are shown in Fig. 20. The Monte Carlo events were generated with the angular distribution given by the Standard Model, which is the solid curve on Fig. 20. The distribution of the reconstructed events, shown by the points with error bars, agrees well with the solid curve, indicating that the selection criteria and the reconstruction of the events do not significantly distort the angular distribution. The sensitivity to anomalous values of $g-2$ are also shown on the figure.

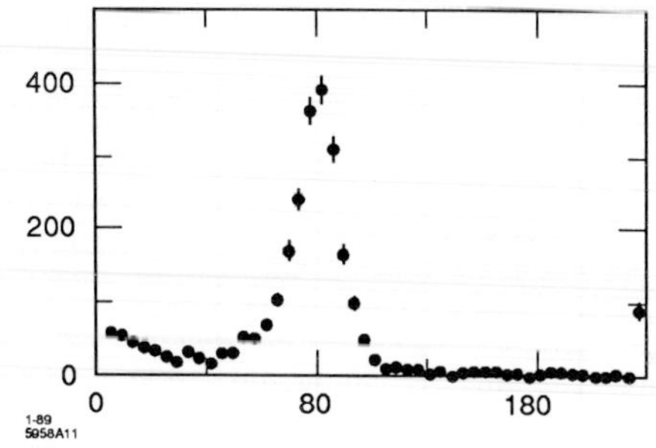

Fig. 19. Reconstructed W boson mass in $e^+e^- \to W^+W^-$ *events at* $\sqrt{s}$ = *1* TeV.

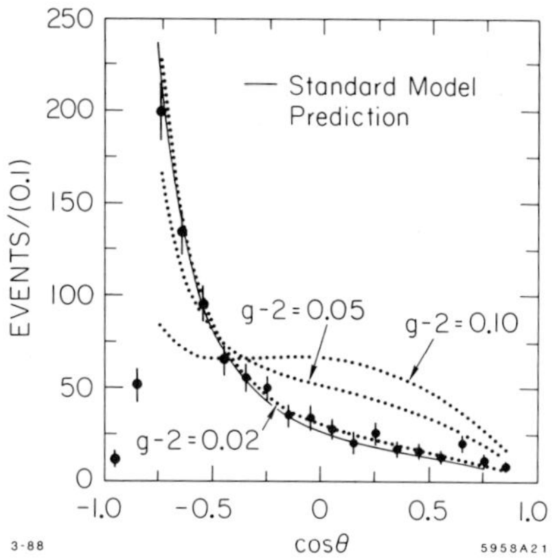

Fig. 20. Precision with which the production angle distribution can be measured for the W-pair process with integrated luminosity of 30 fb^{-1} *at* $\sqrt{s}$ = *1* TeV. *The effect of an anomalous magnetic moment of the W is also shown.*

6. SUMMARY AND CONCLUSIONS

The physics reach of the various energy e^+e^- colliders considered are summarized in Table VI. The luminosities for the high energy colliders have been assumed to be the minimum useful luminosities discussed in Section II. The numbers in Table VI are the masses up to which new particles of the various categories could be discovered if they existed, or the mass limits that could be set if they did not exist. It can be seen from this table that most of the new particles discussed can be discovered up to masses close to the kinematic limits of the colliders considered.

The study has led to the following general conclusions about new particle production at e^+e^- colliders:

1. The minimum desirable luminosity for high energy e^+e^- colliders is:

$$\mathcal{L} \cong 10^{33} \times E_{\rm cm}^2 \ {\rm TeV}^2 \ {\rm cm}^{-2} \ {\rm sec}^{-1}$$

2. Low energy e^+e^- colliders have given clear and unambiguous results on new particle searches. Life

Table VI. Physics reach of e^+e^- colliders.

E_{cm}(GeV) New $\mathcal{L}$ Particle	100 $\sim 10^{31}$ SLC, LEP	200 $\sim 10^{31}$ LEP II	100 10^{33}	500 2×10^{32} NEXT LINEAR COLLIDER	1 TeV 10^{33}	2 TeV 4×10^{33}
New Quarks	45	90	45	225	450	900
New Leptons	45	90	45	200	400	800
$Z^{0\prime}$	*	200	*	500^a	1000^a	2000^a
$W^{\pm\prime}$				250	500	1000
H^0	40	80	50	250	500	1000
$H^\pm$	45	90	45	200	400	800
Supersymmetry						
$\tilde{q}, \tilde{\ell}$ *	45	80	45	200	400	800
$\tilde{W}, \tilde{Z}(\tilde{\chi}^\pm, \tilde{\chi}^0)$	45	80	45	200	400	800
Technicolor						
$P_8^\pm$	45	80	45	200	400	800
Leptoquarks	45	80	45	200	400	800
ρ_T				500	1500	3000
Composite e^*, μ^*, q^*	100	200	100	500	1000	2000
Number of ν Generations	± 0.1		± 0.05			

(a) These limits refer only to direct observation of the excitation of a new resonance. See text.

gets harder at high energies due to new background processes and loss of precise knowledge of the total center-of-mass energy because of beamstrahlung. However, the detailed studies indicate that these problems are manageable.

3. The detectors for TeV e^+e^- colliders are a reasonable and buildable extrapolation of the present generation of SLC and LEP detectors now under construction.

4. Detailed Monte Carlo studies indicate that we can get clean signals, and in many cases do detailed studies, for the new particles that have been considered, up to masses close to the kinematic limits of the colliders.

5. From the point of view of new particle searches, a collider with $E_{cm} \sim 400$ to 500 GeV seems like a sensible and productive first step toward the eventual multi TeV e^+e^- collider.

6. Proof of feasibility and credible estimates of costs and time scales do not exist at this time for the TeV e^+e^- colliders. Therefore they are of a different level of reality from hadron colliders like the SSC, for which a detailed design with cost estimates and schedules exist.

REFERENCES

1. P. Igo–Kemenes, *Proceedings of the La Thuile Study,* Volume II, p. 69.
2. D. Shaile and P. M. Zerwas, *Proceedings of the La Thuile Study,* Volume II, p. 251.
3. C. Ahn *et al.,* SLAC–PUB–329, 1988, p. 138.

NEW PARTICLES IN *ep* COLLISIONS

W. H. Smith,[a] D. Atwood,[b] A. Caldwell,[c] F. Cornet[d] G. Couture,[e] M. Drees,[a]
W. Frisken,[f] C. Heusch,[g] J. Hewett,[a] B. Irwin,[b] U. Mallik,[h] F. Olness,[i]
D. Reeder,[a] T. Rizzo,[j] J. Robinson,[b] K. Sugano,[k] J. Woodside,[l] D. Zeppenfeld[a]

University of Wisconsin,[a] McGill University,[b] Columbia University,[c]
Max Planck Institute,[d] University of Guelph,[e] York University,[f]
University of California at Santa Cruz,[g] University of Iowa,[h]
Illinois Institute of Technology,[i] Iowa State University,[j]
Argonne National Lab,[k] Oklahoma State University[l]

ABSTRACT

The prospects for discovery and analysis of new particles at *ep* colliders are examined. Colliders considered include the HERA facility at DESY, an upgraded HERA, LEP I × LHC at CERN, and LEP II × LHC. Physics topics studied include measurement of QCD parameters, substructure, the gluon distribution, heavy flavor production, the hadronic substructure of the photon, new gauge bosons, the γWW coupling, excited electrons, supersymmetry, elastic supersymmetry, diquarks, leptoquarks and leptogluons.

I. INTRODUCTION

The decade of the 1990 s will see the commissioning of at least one facility providing *ep* collisions. The first to turn on will be HERA, a new storage ring that will provide collisions between 820 GeV protons and 30 GeV electrons by 1990. This facility will measure nucleon structure functions in neutral and charged current reactions at Q^2 up to 30,000 GeV^2/c^2. This is more than two orders-of-magnitude higher than previously obtained. This will enable a study of electron and quark substructure, the properties of the electroweak current, and QCD interactions over distances smaller by more than a factor of ten over present deep inelastic scattering experiments. If these measurements agree with the Standard Model, then one can measure the couplings between fundamental constituents. If these measurements do not agree, then this may reveal new forms of interactions due to internal structure of quarks or leptons, new higher-mass W and Z particles, or unexpected properties of the weak, electromagnetic, or strong forces. The facility will search for new gauge bosons, excited electrons, supersymmetric particles, diquarks, leptoquarks, and leptogluons among others. The mass range for these searches is about a factor of two higher than the range of the experiments at LEP and SLC, but not necessarily as clean. The mass range of the HERA experiment is lower than the experiments at the $Sp\bar{p}S$ and Tevatron, but is significantly cleaner. What is more important is that the collisions of leptons with quarks or gluons are unique and different from the collisions of quarks and gluons with quarks and gluons, or the collisions of leptons with leptons. Each advance in the energy scale of lepton-nucleon scattering has brought exciting discoveries, such as those of quarks and scaling violations.

The $\sqrt{s}$ of the HERA collider is 314 GeV. The projected luminosity is 1.5 $\times 10^{31}$ cm^{-1} s^{-1}, yielding 200 pb^{-1}/year. In order to estimate the prospects for new particle discovery, we have required discovery after one year of running at HERA. The electron and positron beams at HERA will eventually be 80% polarized. For the purposes of estimating the results from polarized running we assumed a two-year program involving 500 pb^{-1} total luminosity evenly divided among e_L^-, e_R^-, e_L^+, and e_R^+. We also explored the physics capability of an upgraded HERA collider, dubbed HERA′, with 1200 GeV protons on 35 GeV $e^\pm$, giving a $\sqrt{s}$ of 410 GeV while having the same luminosity and polarization capability as the HERA machine. Another facility examined was the CERN LEP I ring, with 60 GeV $e^\pm$, colliding with the LHC ring containing 8 TeV p. This machine would have $\sqrt{s} = 1.4$ TeV and $\mathcal{L} = 10^{32} cm^{-1} s^{-1}$, with 1000 pb^{-1} expected per year. As with the HERA machine, the electron and positron beams are expected to be 80% polarized, but with twice the integrated luminosity available for polarized running. Finally, we considered the physics of colliding the LEP II ring with 100 GeV $e^\pm$ against the LHC. This machine would have $\sqrt{s} = 1.8$ TeV and $\mathcal{L} = 10^{31} cm^{-1} s^{-1}$, with 100 pb^{-1} expected per year and no polarized beams.

II. *ep* COLLIDERS AND QCD

Currently, QCD-related uncertainties are often the dominant uncertainties of theoretical calculations; thus, the accuracy of theoretical predictions are limited by our incomplete understanding of QCD. Such uncertainties can thwart the search for "new physics." If we are searching for "new physics" that has a small signal (as most will), we must insure that the uncertainties arising from QCD are not so great as to mask the new signal. When HERA turns on in the early 1990 s, experimentalists can observe large numbers of electron-proton collisions in completely unexplored kinematic ranges of large Q^2 and small Feynman x. We can use these events to perform exacting tests of QCD, thereby improving our understanding and reducing the inherent uncertainty which has previously limited our ability to make accurate predictions. We see that HERA will provide us with a better understanding of QCD, which is fundamental to most any attempt to discover new particles and/or "new physics."

At HERA we will be capable of observing events with Feynman-x values $x \gtrsim 10^{-5}$ and momentum transfers with $Q^2 \lesssim 200^2\ GeV^2$. This is to be contrasted with the kinematic range of current Deep Inelastic Scattering (DIS) experiments which limit us to Feynman-x values larger than $x \gtrsim 5 \times 10^{-2}$. Specifically, there will be in excess of 1000 events with $Q^2 \geq 10^4\ GeV^2$ for *both* the charged current process ($e^- + p \to \nu + X$) and the neutral current process ($e^- + p \to e^- + X$) in a nominal HERA year ($\int \mathcal{L}\ dt = 200\ pb^{-1}$).[1]

Clearly, these events can be used to perform stringent tests of QCD in an x and Q^2 region that is unique to an electron-proton collider. These tests are described in detail in the report of the Structure Function Working Group in these proceedings. In the following, we will briefly point out some of the more interesting features of QCD that HERA will illuminate.

Although HERA is designed to run with 30 GeV electrons on 820 GeV protons ($\sqrt{s} = 314$ GeV), there is the additional option of performing a low-energy run. One case of interest is colliding 10 GeV electrons on 300 GeV protons ($\sqrt{s} = 110$ GeV). Such an option would allow HERA to further probe the unknown kinematic range, thus enhancing our ability to perform accurate tests of QCD.

For example, the high-statistics measurements that HERA could obtain over a broad range in Q^2 would allow a more accurate measure of the scale breaking of the structure functions. We can compare the experimentally measured scale breaking of the structure functions with that predicted by QCD:

$$F(x) \longrightarrow \frac{F(x)}{1 + b\ \ln(Q^2/\Lambda^2_{QCD})} \quad . \qquad (2.1)$$

A deviation of the structure function from the QCD predictions might be interpreted as a signal of various new phenomena such as compositeness.

Additionally, such information would allow us to determine Λ_{QCD}. A separate study on HERA[2] physics concludes that 100 pb^{-1} of data at the nominal energy ($\sqrt{s} = 314\ GeV$) will constrain $\delta\Lambda_{QCD}$ to $\pm \sim 100$ MeV. If we combine these data with a low-energy run, the combination of measurements would allow a determination of $\delta\Lambda_{QCD}$ to $\pm \sim 40$ MeV. If one could constrain the allowed gluon distributions, particularly in the small x range below $x \sim 10^{-2}$, it would be possible to measure $\delta\Lambda_{QCD}$ to $\pm \sim 20$ MeV. We will discuss the small x uncertainties of the gluon distribution below.

Sensitive tests of QCD can also alert us to the presence of "new physics." For example, if the Q^2 behavior of the structure functions deviates from the QCD prediction, we might interpret this as a sign of quark and electron substructure. One possibility is to characterize this as a point-like four-fermion interaction with an effective Lagrangian of the form:

$$\mathcal{L} \sim \frac{g^2}{\Lambda^2_C} \left(\bar{e}_\alpha \gamma^\mu e_\alpha\right) \left(\bar{q}_\beta \gamma_\mu q_\beta\right) \quad . \qquad (2.2)$$

The parameter Λ_C then sets the compositeness scale.

If we compute the ratio of the structure function F_2, with the contact interaction of the above equation, to the straight QCD prediction without the contact term, then any deviation from unity (outside the errors of the calculation) will signal the existence of compositeness. We find that for $\Lambda_C \sim 1$ TeV (where we assume for simplicity that $g^2/4\pi = 1$), this ratio deviates from unity by factors of ~ 5 to 10 in the accessible range of Q^2. It is estimated that two years of data at the nominal luminosity will allow HERA to probe Λ_C up to ~ 7 TeV.[3]

The large kinematic range of HERA will enable us to measure the Parton Distribution Functions (PDF's) in the presently unexplored kinematic ranges of small x and large Q^2. As we have mentioned, the current DIS data limit us to $x \gtrsim 10^{-2}$, so HERA will give us a wealth of new information about the PDF's. Extraction of the PDF's at a hadron-hadron collider is complicated by the fact that the calculation of any observed process involves two PDF's (one for each hadron), so that the extraction of a single PDF is convoluted. This is to be contrasted with an electron-proton collider which involves only one PDF for the proton.

The gluon distribution, $G(x, Q^2)$, has been particularly elusive; this distribution has only been measured indirectly, e.g., through a momentum sum rule. Furthermore, the small x behavicr of both the gluon and quark PDF's is unconstrained due to the lack of any DIS data below $x \lesssim 10^{-2}$. This uncertainty in the small x region can be disastrous when one attempts to extrapolate QCD predictions to the high-energy regime such as that found at the SSC, where the small x region plays an increasingly important role. For example, the b-quark pair production cross section can vary by an order of magnitude depending on the small x form of the gluon distribution.[4]

Present deep inelastic scattering data are insufficient to constrain the parton distributions in the small x region. (For the purpose of our discussion, we shall define the small x region as $x \lesssim 10^{-2}$.) Lacking any experimental guidance in this region, it has been conventional to assume a canonical $1/x$ power law behavior at an (arbitrarily chosen) initial Q_0^2 point for the small x behavior of $G(x)$. This behavior has been inferred from the lowest order perturbation formulas for bremsstrahlung, and from related results on the first order Altarelli-Parisi kernel.[5,6]

Recent work indicates that a $1/x$ assumption for the gluon distribution $G(x)$ at small x is inaccurate. There are indications that a much better choice is to generalize the above form to $G(x) \sim 1/x^J$, and take the parameter J to be ~ 1.5.[7] The motivation for this choice comes from two sources: first, from Regge theory, and secondly from the numerical stability of the evolution.

The details of the theoretical form of the PDF's is discussed in the report by the Structure Function Working Group. Here, we will focus on one particular process, in which HERA is well suited to determine $G(x)$.

The dominant process for J/ψ production at HERA is photon-gluon fusion.[8,9] We extract a photon from the electron using an Weizsäcker-Williams effective photon approximation, and a gluon from the proton with a distribution function $G(x, Q^2)$. The effective photon, "γ," and the gluon fuse to produce a $c\bar{c}$-pair, which then forms a J/ψ (and a soft gluon to balance color). This process enables us to probe the gluon distribution directly in the important small x region, at a Q^2 scale of $\sim M_\psi^2$. The approximate dependence of the J/ψ cross section on the gluon distribution is given by[8]

$$\sigma(\gamma p \to \psi X) \propto \bar{x} G(\bar{x}, M_\psi^2) \quad , \tag{2.3}$$

where $\bar{x}$ represents a characteristic x value, which we expect to be of order $(M_\psi^2/s_{\gamma p})$; $\sqrt{s_{\gamma p}}$ is the photon-proton center-of-mass energy. Substituting the general form for the gluon distribution, $G(x) \sim 1/x^J$, we find:

$$\sigma(\gamma p \to \psi X) \propto \left(\frac{M_\psi^2}{s_{\gamma p}}\right)^{1-J} \simeq \text{a few } nb \quad . \tag{2.4}$$

Actual calculations of the inelastic J/ψ production cross section at HERA energies[8] indicate that this cross section can vary by up to 300% as the J-parameter varies over the conservative range of 1.0 to 1.5. The actual cross section for J/ψ photo-production is on the order of a few nanobarns, depending, of course on the choice of the J parameter.[10]

Experiments at HERA will produce sufficient statistics to allow this technique to be of use in constraining the gluon distribution in the $Q^2 \simeq M_\psi^2$ region. One method for determining $G(x, Q^2)$ in the large Q^2 region, is to measure the ratio $R = \sigma_L/\sigma_T$ of the longitudinal to transverse cross-sections in electron-proton Deep Inelastic Scattering (DIS) experiments.[8,11] At larger Q^2 values (where higher twist contributions can be ignored) and at small x, heavy quark production in DIS is expected to be a sensitive measure of the gluon distribution.

We have demonstrated how HERA will be an important tool which will further our understanding of the QCD interactions. This knowledge can be immediately applied toward our search for "new physics" by reducing the ambiguities arising from the QCD calculations, thereby enhancing our ability to separate signals from the backgrounds.

III. HEAVY QUARK PRODUCTION

We will examine two particular aspects of heavy quark production at electron-proton colliders: 1) We will examine the productions rates for charm and bottom quarks, and mention some uncertainties which enter into the calculation. 2) We will consider the possibility of discovering the top quark at both HERA and future *ep* colliders, and identify the window each has for this discovery.

There are three primary processes by which HERA can produce heavy quarks. We denote these as

a) charged current processes,

b) neutral current processes, and

c) hadronic photon processes.

In the charged current processes (Figure 3.1a), the proton emits a light quark q, and the resulting partonic sub-process is $e + q \to \nu + Q$ with a W-boson exchanged in the t-channel.

In the neutral current process (Figure 3.1b), the electron emits a neutral boson (γ, or Z^0), the proton emits a gluon, and the partonic sub-process is $\gamma/Z^0 + g \to Q + \bar{Q}$ via the t- or u-channel.

In the hadronic photon processes (Figure 3.1c), the photon actually has a finite (and sizeable) probability for emitting a gluon. Taking a second gluon from the proton, we then have the partonic sub-process: $g + g \to Q + \bar{Q}$, in the s-, t-, and u-channel.

An estimate of the contribution from each of these three processes has been computed for charm and bottom production at HERA, and the results are listed in Table 3.1.[12,13,14]

There are two important results of note in this table. First, the contribution from the hadronic photon is comparable to the dominant neutral current process. This contribution has often been neglected, yet we see that it can generate over 30% of the charm and bottom events at HERA.

Secondly, HERA is a charm and bottom factory capa-

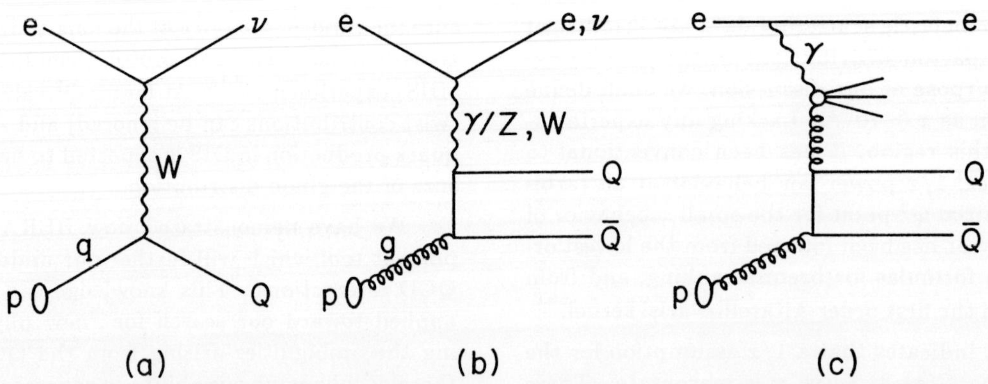

Fig. 3.1. Feynman diagrams for the three principle heavy quark production processes.

ble of generating sufficient events to allow for high statistics tests of heavy quark production mechanisms.

Such high statistics tests would be of particular value in reducing the inherent uncertainties which arise in the calculation of heavy quark production.

For example, the above calculations for the charm and bottom quark production rates are subject to the following uncertainties:

- Choice of parameters, *e.g.*, quark masses, $\alpha_s(Q^2)$.
- Gluon and photon structure functions.
- Higher order QCD corrections.[15]

As mentioned in Sec. II, these uncertainties are precisely the ones which HERA can help to constrain.

Charm and Bottom Quark Production				
	Charged Current	Neutral Current	Hadronic Photon	Total
Charm	720.	2.0×10^8	1.2×10^8	3.2×10^8
Bottom	2.4	1.7×10^6	9.8×10^5	2.7×10^6

Table 3.1. Numbers of charm and bottom quarks produced in a HERA year using the nominal parameters, $\int \mathcal{L} = 200 pb^{-1}$ and $\sqrt{s} = 314$ GeV.

The charged and neutral current results are from Ingelman and Schuler,[12] and the results involving the hadronic content of the photon are from Drees and Godbole.[13] (Note that, the charged current process does not include any t-quark contributions.)

We now address the possibility of discovering the top-quark at HERA. For a light top mass, we expect the the neutral current process to give the dominant contribution, as in the case of charm and bottom production. However, for a heavy top quark, we expect the charged current process will win out since the neutral process requires the top quarks to be pair-produced. Calculations show this expectation is correct, and that the charged current process overtakes the neutral process for top quarks in excess of ~ 55 *GeV*. The inclusive top production cross section in this region is ~ 1 *pb*.

The key to discovering the top quark at HERA certainly will be extracting the top signal from the background; much of this background comes from the numerous charm and bottom events. Although the heavier quarks are expected to have larger transverse momentum on the average, the tail of the charm and bottom $p_\perp$ distribution could be sufficient to swamp the top signal. Ingleman and Schuler[12] have performed a preliminary Monte Carlo analysis which suggests that extraction of the top signal from the charm and bottom background is possible with the proper choice of cuts.

They impose a cut on the total transverse energy ($\sum E_\perp \geq 100$ *GeV*) and the transverse sphericity ("circularity," $C \geq 0.35$), to reduce the charm plus bottom background to approximately half of the top signal. The effect of these cuts is shown in Fig. 3.2.

The heaviest top quark mass that HERA can attain, as estimated in the literature,[12,14] is in the range 70 ~ 85 GeV, (Table 3.2). We can perform similar estimates for the proposed future electron-proton colliders, and these are also listed in Table 3.2.

Top Quark Discovery Limits		
Facility	Energy $\sqrt{s}$ *GeV*	Top Mass (Maximum)
HERA	314	$70 \sim 85$ *GeV*
HERA′	410	110 *GeV*
LHC ⊗ LEP-I	1400	~ 400 *GeV*

Table 3.2. The top-quark discovery reach of various electron-proton colliders.

We have seen that HERA will be an abundant source of charm and bottom quarks, and this capability will allow high precision measurements of the heavy quark production parameters. Additionally, there is a window where

HERA can discover the top quark ($m_t \lesssim 85$ GeV), and an upgraded HERA, (HERA$'$, $\sqrt{s} \simeq 410$ GeV), could discover a top quark as heavy as ~ 110 GeV.

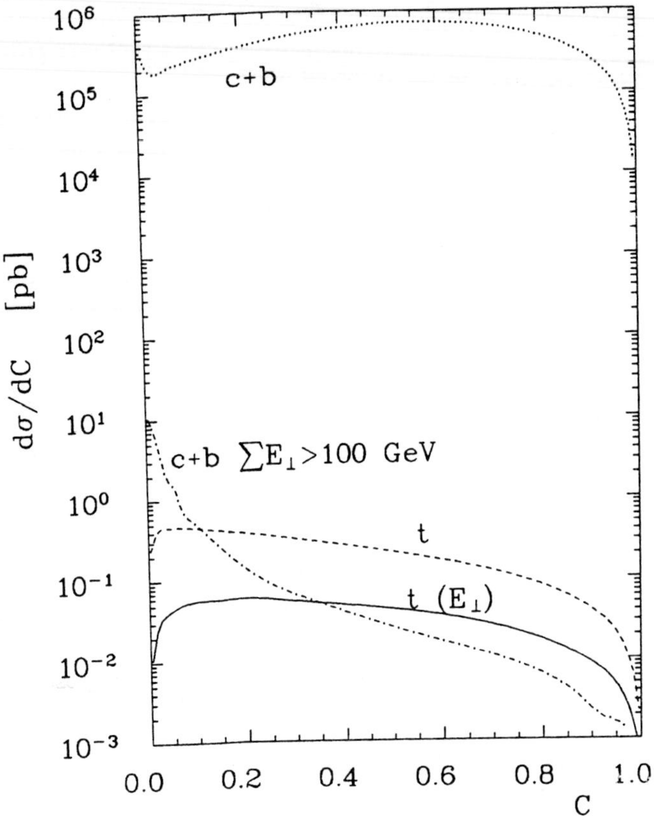

Fig. 3.2. Distribution of circularity (*i.e.*, sphericity in the transverse plane),[12] $C = 1$ corresponds to a circularly symmetric final state, and $C = 0$ to a back-to-back pencil-like two-jet system. Here, the top mass is taken to be 60 GeV.

IV. PHOTON STRUCTURE AND DI-JETS

The photoproduction of two or more jets with high transverse momentum p_T is interesting for two reasons. It might teach us something about QCD, and its understanding is a prerequisite for the extraction of many signals for new physics.

Two fundamentally different types of processes[16] contribute to the continuum photoproduction of two jets with large transverse momentum p_T (see Fig. 4.1). In the first case (Fig. 4.1a), the almost real photon, which is radiated off the electron, couples directly to the quarks and gluons in the proton via $\gamma q \rightarrow gq$ and $\gamma g \rightarrow q\bar{q}$; these constitute the "direct" contributions.[17] In the second class (Fig. 4.1b) the photon itself is resolved into quarks and gluons, which in turn interact with the quarks and gluons inside the proton; these "resolved photon" contributions include the same 8 subprocesses[18] that are responsible for the production of high p_T jets in pp or $\bar{p}p$ collisions. These resolved photon contributions are interesting since they might allow[19] for the measurement of the unknown gluon content of the photon. This measurement in turn would not only test different models[20] for the photon structure function, but would also help to sharpen predictions[21] for the total γp cross section at very high energies ($\sqrt{s}(\gamma p) \gtrsim 0.5$ TeV) which is crucial[22] for the understanding of the μ content of γ initiated cosmic air showers.[23] On the other hand, the resolved contributions increase the two-jet cross section by as much as a factor of ten (at $p_T \simeq 5$ GeV, $\sqrt{s} = 314$ GeV) and thus lead to a much more hadronic, "messy" environment than one would naively expect.

Obviously both the detailed study of the photon structure and the reduction of the continuum two-jet background to new physics signals is much easier if one can clearly separate direct from resolved photon contributions.

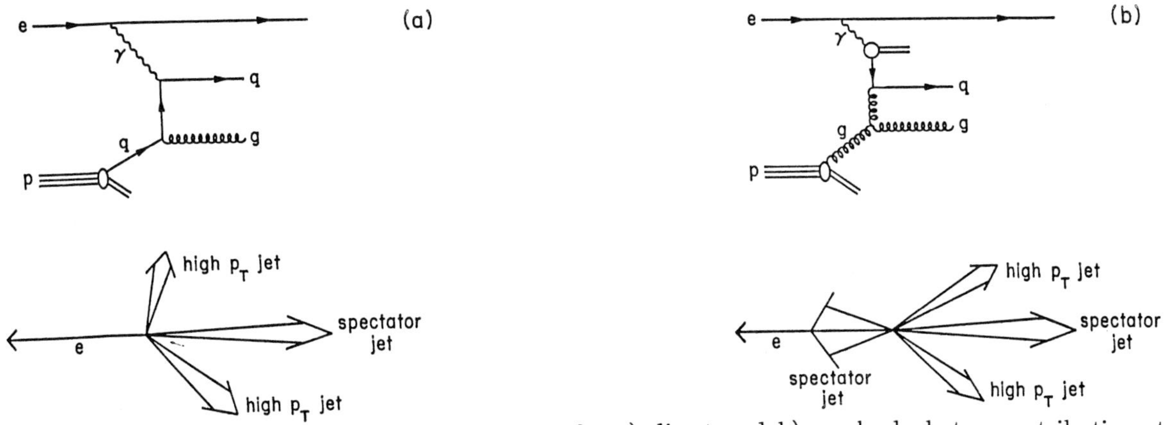

Fig. 4.1. Typical Feynman diagram and event topology for a) direct and b) resolved photon contributions to the production of two high-p_T jets in ep collisions. The two hard jets have equal and opposite transverse momenta, but their rapidities are in general unrelated, although they tend to be larger for the resolved contributions. The second spectator jet, unique for the resolved photon contributions, is rather broad.

In principle, the resolved photon contributions are uniquely characterized by the existence of a second spectator jet originating from the photon; Fig. 4.1b. However, since the incoming photons are collinear to the electrons and the spectator jet in turn goes in the direction of the photon, the axis of this jet coincides with the electron beam direction; it might therefore be impossible to measure its energy. On the other hand, the average energy of this jet in the lab frame is predicted[19] to be rather low ($\langle E_{\text{spec. jet}} \rangle \sim 10$ GeV at HERA), which presumably leads to a rather broad jet, with an estimated opening angle of almost 30°; if this estimate is correct, some parts of this jet should be visible in the HERA detectors ZEUS and H1, which have[24] gaps around the electron direction of about ±5°.

In order to demonstrate the benefits of the possibility to discriminate between direct and resolved contributions for the reduction of the continuum two-jet background we show in Figs. 4.2 the invariant mass distribution $d\sigma(jj)/dM_{jj}$ at HERA ($\sqrt{s} = 314$ GeV) and LEPI × LHC ($\sqrt{s} = 1.4$ TeV), where only events with $p_T(\text{jet}) \geq 20$ GeV are included. The short-dashed and long-dashed curves show the direct and resolved photon contributions, respectively, while the solid curves show the sum of the two.

Obviously the importance of the resolved contributions increases with increasing $\sqrt{s}$ (or, equivalently,[17] with decreasing $p_T(\text{jet})$). The reason is that in this class of contributions only a part of the photon energy is available for the production of the two high-p_T jets; the cross section is thus proportional to the convolution of three distribution functions (the photon density in the electron, and the quark and gluon densities inside the photon and proton), which makes it more sensitive to changes in the available phase space volume than the direct contributions, which are proportional to the convolution of only two distribution functions.

From this argument one would also expect the resolved photon contributions to become relatively less important for increasing M_{jj}, whereas Fig. 4.2 shows just the opposite behaviour. Note, however, that some of the hard $2 \to 2$ subprocesses entering the resolved photon contributions proceed via gluon-exchange in the t- or u-channel, whereas the subprocesses relevant for the direct contributions only have quarks in t- or u-channel propagators. As a result, the subprocess cross sections grow like $(M_{jj}^2/4p_T^2)^2$ in the resolved photon case, but only like $M_{jj}^2/4p_T^2$ in the direct case. Furthermore, at the values of p_T considered here, the resolved photon contributions are dominated[19] by subprocesses which have a quark from the photon in the initial state; unlike the gluon density inside the photon, the quark distribution inside the photon is very hard,[25] which makes this subclass of resolved contributions relatively insensitive to changes in the available phase space volume.

The production of new particles can resemble both direct or resolved photon processes; an example for the latter is single production of scalar diquarks,[26] while the photoproduction of a $t\bar{t}$ pair or charged current single top production fall into the former category. From Fig. 4.2a it is clear that in either case the continuum background can be reduced by roughly a factor of two if it is possible to concentrate onto either direct or resolved photon contributions.

Of course, many interesting signal events would not only contain two high p_T-jets, but also additional jets, leptons and/or missing p_T; however, as long as the total energy scale of the relevant hard scattering process is similar to that of Fig. 4.2a, one can expect direct and resolved photon contributions to the corresponding final states to be of similar magnitude.

As a first, simple example we have studied the question whether hadronically decaying W-bosons are detectable at HERA. Here we concentrate on the resolved photon contribution to W-production, which constitutes[27] the biggest part of the total W cross section;[28] at $\sqrt{s} = 314$ GeV, it is about 0.65 pb, which has to be multiplied with a branching ratio into light $q\bar{q}'$ final states of about

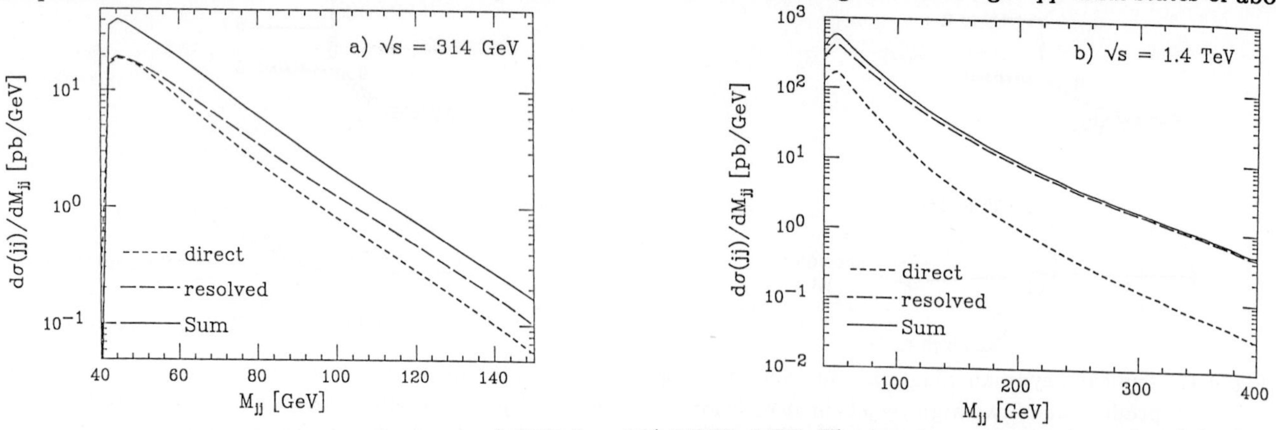

Fig. 4.2. Di-jet invariant mass distribution at a) HERA and b) LEPI×LHC. The transverse momentum of each jet has been required to be bigger than 20 GeV.

2/3, leading to a signal cross section of less than 0.5 pb before cuts. A glance at Fig. 4.2a shows immediately that some cuts will be necessary to isolate this signal.

Note that the background cross section falls very rapidly[17,19] with increasing p_T, whereas the signal has a Jacobian peak at $p_T = 0.5\,M_W \simeq 41$ GeV. However, due to the transverse polarization of the produced W's, the signal cross section for $p_T > p_{T,\min}$ is given by

$$\sigma_W(p_T > p_{T,\min}) \simeq 0.43\,\text{pb} \cdot \left(1 - \frac{4\,p_{T,\min}^2}{M_W^2}\right)^{3/2} \tag{1}$$

(for an unpolarized W, the bracket would come with power 1/2), so that already at $p_{T,\min} = 30$ GeV the signal is reduced by a factor 0.46. Nevertheless, the signal-to-background ratio turns out to be maximal for $p_{T,\min} \simeq$ 35 GeV. Unfortunately, Fig. 4.3 shows that even after this cut the signal is almost invisible, even if one assumes 100% efficiency for the rejection of direct contributions and a very optimistic two-jet invariant mass resolution of 2.5 GeV (which is about equal to the decay width of the W). One might try to improve the signal-to-background ratio by looking at W's that have a finite p_T, or by implementing cuts on the charged particles multiplicities, but our simple estimate does certainly give reason to be pessimistic.

In summary, one can say that resolved photon processes are both a curse and a blessing for ep colliders. On the one hand, they lead to a very large ($O(1\,\mu\text{b})$ at HERA) "mini-jet" cross section and thus a messy environment, and also make substantial contributions to backgrounds for all hadronically decaying new particles. On the other hand, they should allow us to learn more about the hadronic structure of the photon. This understanding may aid the interpretation of high energy photons from Cygnus X-3 and other cosmic sources.

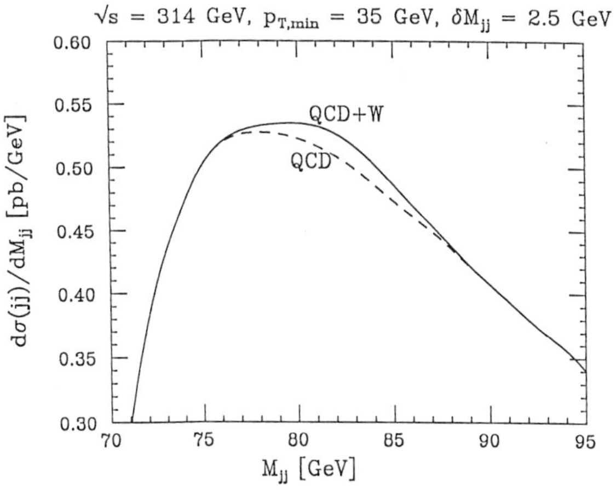

Fig. 4.3. Di-jet invariant mass distribution around $M_{jj} = M_W$. Note that the curve labelled "QCD" only includes the resolved photon contributions.

V. NEW GAUGE BOSONS

The prediction of the existence of additional gauge bosons is a common feature to almost all of the proposed extensions of the Standard Model, such as Grand Unification Theories or Left-Right Symmetric Models. Although in many cases the new gauge bosons are expected to be very massive, in some scenarios they can become light enough to induce observable effects at the energies of the next accelerator generation. Experimentally, on the other hand, masses as low as $m_{Z'} \sim 110\ GeV$ and $m_{W_R} \sim 300\ GeV$ are still allowed for some couplings of a new Z' and for a W' with couplings to right-handed fermions, respectively.[29,30] We have studied the prospects for the detection of new gauge bosons at the future ep colliders. A more detailed description of our method and results can be found in Ref. 31.

The Neutral Current lagrangian in the presence of an extra neutral gauge boson,

$$\mathcal{L}_{NC} = \mathcal{L}_{NC}^{SM} + g' J_{Z'}^{\mu} Z'_{\mu}, \tag{5.1}$$

contains the Standard Model Neutral Current lagrangian plus a term, proportional to the coupling constant g', describing the interaction between the Z' and the fermions. The current $J_{Z'}^{\mu}$ is given by

$$J_{Z'}^{\mu} = \sum_f \bar{f}\gamma^{\mu}(v'_f - a'_f\gamma_5)f, \tag{5.2}$$

where v'_f and a'_f are the vector and axial couplings of the Z'.

The detection limits that can be set for a Z' depend on the model considered, i.e. on the assumed values for g', v'_f and a'_f. It is, thus, important to study different models in order to estimate the sensitivity of the future ep accelerators. We have taken four models. Three of them are based on the group $SU(3)_C \times SU(2)_L \times U(1)_Y \times U(1)_{Y'}$ resulting from the breaking of an E_6 Grand Unification Group. There are an infinite number of possibilities for Y'_f that can be conveniently parametrized in terms of

$$Y'_f = Y_f^{\psi} \cos\alpha + Y_f^{\chi} \sin\alpha, \tag{5.3}$$

where Y_f^{ψ} and Y_f^{χ} correspond to the generators of the $U(1)$ group appearing in the decomposition $E_6 \to SO(10) \times U(1)_{\psi} \to SU(5) \times U(1)_{\chi} \times U(1)_{\psi}$. The models we have studied correspond to $\cos\alpha = \sqrt{5/8}$ (Model A), $\cos\alpha = 0$ (Model B) and $\cos\alpha = -\sqrt{15/16}$ (Model C). g' is taken to be $g' = \sqrt{5/3}e\cos\theta_W$, where e is the electromagnetic coupling constant and θ_W is the Weinberg angle. The fourth model is a left-right symmetric model based on the group $SU(3)_C \times SU(2)_L \times SU(2)_R \times U(1)_{B-L}$ with the $U(1)$ generator being proportional to the difference between the baryon (B) and lepton (L) numbers. The fermion quantum numbers assignment is:

$$
\begin{aligned}
l_L &= (\frac{1}{2}\,,\,0\,,\,-1), \qquad q_L = (\frac{1}{2}\,,\,0\,,\,\frac{1}{3}) \\
l_R &= (0\,,\,\frac{1}{2}\,,\,-1), \qquad q_R = (0\,,\,\frac{1}{2}\,,\,\frac{1}{3})
\end{aligned}
\tag{5.4}
$$

and the coupling constant: $g' = e/\sin\theta_W$.

We have studied the deviations induced by the new gauge boson in inclusive quantities, such as total cross-sections and asymmetries between one of the cross sections for e_R^+, e_L^+, e_R^-, e_L^- and another. In general asymmetries are more sensitive than cross-sections (except in model B). Moreover, systematic errors, are expected to be smaller in asymmetry measurements than in cross-section measurements. As an example we show in Fig. 5.1 the Q^2-distribution for the most sensitive asymmetry in Model A

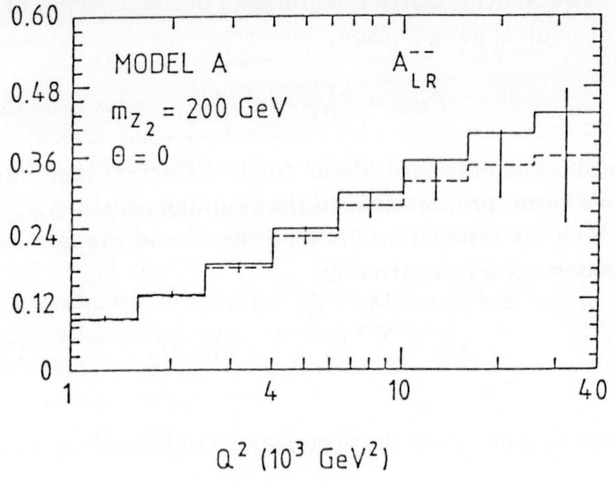

Fig. 5.1. Polarization asymmetry A_{LR}^{--} integrated over x for model A (full line) compared to the Standard Model prediction (dashed line) at HERA. In this figure, the polarization degree is assumed to be 100%.

compared to the Standard Model prediction (dashed line) at a center of mass energy $\sqrt{s} = 314\ GeV$. In the evaluation of the Standard Model we have taken $m_Z = 93.3\ GeV$ and $\sin^2\theta_W = 0.22$. The deviations with respect to the standard model predictions are better described in terms of

$$
\chi^2(m_2) = \sum_{bins\ in\ Q^2} \left(\frac{F(m_2) - F_{SM}}{\delta F}\right), \tag{5.5}
$$

where $F(m_2)$ and F_{SM} are the predictions for a particular quantity (cross-section or asymmetry) in the extended and the standard models, respectively, and δF is the statistical error expected in the Standard Model. In Table 5.1 are listed the bounds on m_{Z_2} (in GeV) at 95% C.L. that are expected for the four ep colliders under consideration. Also the observables from which these bounds are derived are indicated.

$\sqrt{s}$ GeV	314	400	1400	1800	Present bounds
$A(A_{LR}^{--})$	175	200	400	325	110
$B(e^+)$	250	275	575	450	270
$C(A_{LL}^{-+})$	115	125	225	175	—
$LR(A_{RR}^{-+})$	450	500	1150	1050	320

Table 5.1. Bounds at the 95% c.l. for $m_{Z'}$ in GeV.

For the charged gauge boson we have considered only the case of the Left-Right Symmetric Model. In this case the extended charged current lagrangian takes the form

$$
\mathcal{L}_{CC} = \frac{g}{2\sqrt{2}} J_W^\mu W_\mu + \frac{g'}{2\sqrt{2}} J_{W'}^\mu W'_\mu, \tag{5.6}
$$

where g and g' are the coupling constants of the $SU(2)_L$ and $SU(2)_R$ groups, respectively, J_W^μ is the standard model charged current and $J_{W'}^\mu$ is given by

$$
J_{W'}^\mu = \sum_{i,j} \left(\overline{\nu}_i \gamma^\mu (1-\gamma_5)\delta_{ij} e_j + \overline{u}_i \gamma^\mu (1-\gamma_5) V'_{ij} d_j\right). \tag{5.7}
$$

Here we have neglected the mixing among lepton species and introduced the matrix V'_{ij} to describe the mixing in the right-handed quark sector.

A W' boson coupled to right-handed currents gives rise to the process $ep \to \nu_R X$. The nature (Dirac or Majorana fermion) and mass ($m_{\nu_R} <<$ or $\sim m_{W'}$) of the right-handed neutrino, if it exists, is not known. In the case where ν_R is a light Dirac particle, the signature is the same as the one for the standard charged current process and one has to rely on indirect searches as in the case of a Z'. The observables we have investigated are the polarized and unpolarized cross-sections as well as the ratios $R^- = \tilde{\sigma}(e_R^-)/\tilde{\sigma}(e_L^-)$ and $R^+ = \tilde{\sigma}(e_L^+)/\tilde{\sigma}(e_R^+)$, where $\tilde{\sigma} \equiv d\sigma/dQ^2$. The ratios R^- and R^+ have the advantage that they are independent of the uncertainties in the proton structure functions. The most sensitive quantities turn out to be $d\sigma(e_R^-)/dQ^2$ and R^- which produce the bounds at the 95% c.l. shown in Table 5.2.

$\sqrt{s}$ GeV	314	400	1400	1800	Present bounds
$W_R\ (e_R^-, R^-)$	375	425	1050	1050	300

Table 5.2: Bounds at the 95% c.l. for m_{W_R} in GeV.

As is shown above, the most sensitive electron and positron tests are normally obtained when polarized beams are available. In addition, polarized beams provide the possibility to discriminate among different models once a signal is observed.

An intriguing extension of the strong interaction sector of the Standard Model. suggests that above the weak scale, the strong interactions exhibit a chiral $SU(3)_L \times SU(3)_R$ symmetry.[32] In this model, the "low energy" $SU(3)_C$ group corresponds to the diagonal subgroup whicl remains unbroken after spontaneous symmetry breaking. A universal prediction of such theories is a massive octet of strongly interacting gauge bosons which couple axially to quarks. If, as predicted, these "axigluons" have a mass of ~ 100 GeV[32] then HERA will be the optimum planned facility to discover these new gauge bosons.[33] At HERA axigluons can be produced in the reaction $ep \to eAX \to 3jets + (e)$. The p_T distribution for axigluon production is given by

$$\frac{d\sigma}{dp_T} = \tag{5.8}$$
$$2p_T \sum \int\int dx_A dx_B f_{\gamma/e}(x_A) f_{q/p}(x_B) \left(\frac{\hat{s}}{|\hat{t}-\hat{u}|}\right) \frac{d\hat{\sigma}}{d\hat{t}},$$

where the subprocess differential cross section is

$$\frac{d\hat{\sigma}}{d\hat{t}}(\gamma q_i \to \ Aq_i) = \frac{16\pi\alpha_s\alpha Q_i^2}{3\hat{s}^2}\left[1 - \frac{\hat{t}^2 \ + M_A^4}{2\hat{s}\hat{u}}\right]. \tag{5.9}$$

For the photon distribution ($f_{\gamma/e}$) we use the Weizsacker-Williams approximation and for the quark structure function ($f_{q/p}$) we choose Duke-Owens set I. As shown in Ref. 33the signal to background ratio improves for growing P_T. For an axigluon mass around 80 GeV the signal to background ratio is roughly 1-2 above $P_T = 40$ GeV, with an integrated signal cross section of roughly 0.5 pb, yielding 100 events per year. In the dijet invariant mass distribution one would thus expect a ~ 35 peak around $m_{jj} \approx 80$ GeV. Since axigluons decay fairly isotronically, the 2 decay jets will in general have large scalar E_T which should be usefull for improving the signal versus background. As shown in Ref. 33a, the axigluon signal is further enhanced over the QCD background by reconstructing its $b\bar{b}$ decay modes. Axigluons with mass in the range 70 GeV$\leq M_A \leq 125$ GeV appear to be inaccessible to planned e^+e^- and hadronic colliders.[33] HERA has a unique opportunity to examine axigluons with mass in this potentially important window.

VI. SINGLE W PRODUCTION

One of the less frequent standard model (SM) processes in ep collisions is single production of W's or Z's. Inclusive rates have been estimated recently by Gabrielli[34] with the result that the largest cross section is expected for $ep \to eWX$ ($\sigma \approx 0.8$ pb at HERA) while the cross sections for $ep \to \nu WX$ or $ep \to eZX$ are smaller by more than one order of magnitude. This and the fact that cross sections are quite sensitive to details of the $W\gamma$ interaction has led us to reconsider the process $ep \to eWX$ at HERA and beyond.[35]

W production in ep collisions is strongly peaked at low p_{TW} where it can be viewed as the subprocess $q\bar{q} \to W$, with one of the quarks being treated as a hadronic constituent of the photon.[36] Only at larger W transverse momenta ($p_{TW} \gtrsim 5$ GeV at HERA) do other subprocesses become relevant, among them $\gamma q \to Wq'$ via t-channel W exchange. These make the cross section sensitive to the values of the magnetic dipole moment μ_W and the electric quadrupole moment Q_W of the W^+:

$$\begin{aligned} \mu_W &= \frac{e}{2M_W}(1+\kappa+\lambda)\,, \\ Q_W &= \frac{e}{M_W^2}(\lambda-\kappa)\,. \end{aligned} \tag{6.1}$$

The SM predicts $\kappa = 1$ and $\lambda = 0$.

For measurement of these multipole moments, the interesting signal at HERA will be a charged lepton ($\ell = e^\pm, \mu^\pm$) and missing p_T from W leptonic decay and a relatively low p_T jet recoiling against the W. Requiring the charged lepton and the accompanying jet to be observable inside the detector ($p_T > 5$ GeV, $|y| < 3.5$ for both) the SM predicts ~ 20 leptonic W events/year at HERA. Hadronic W decays are overwhelmed by a large QCD background and are expected to yield a significant signal only after several years of running with the nominal luminosity of $200\,\text{pb}^{-1}$/year.[37]

Even though this rate is small, it is quite sensitive to the value of κ and, to a lesser extent, λ in Eq. (6.1). After cuts, the dependence of $\sigma(ep \to eWj)\cdot\text{Br}(W \to e\nu, \mu\nu)$ on deviations from the SM prediction for κ and λ are shown in Fig. 6.1 for HERA.

Deviations from the SM value of $\Delta\kappa \approx \pm 0.5$ or $\Delta\lambda \approx \pm 2$ would yield a $1\,\sigma$ effect on the number of events observed with $200\,\text{pb}^{-1}$ of data. These sensitivities can be substantially increased by running at an upgraded HERA machine ($\sqrt{s} = 410$ GeV) and/or by exploiting the increased integrated luminosities expected after a few years of running.

Single W production at HERA will yield the most precise direct measurement of the $WW\gamma$ vertex before W pair production can be studied at LEP II. Even at LEP II an accuracy of $\Delta\kappa, \Delta\lambda \approx 0.2$ only is expected,[38] and in W^+W^- production it is actually a linear combination of $WW\gamma$ and WWZ couplings which will be measured.

At present only indirect and very model dependent bounds on the low energy values of κ and λ exist. From loop contributions to $g-2$ of the muon[39] one estimates $|5(\kappa-1)+\lambda| \lesssim 4.4$, if W's and Z's show novel strong interactions at a scale $\Lambda \approx 1$ TeV. Bounds derived from the W/Z mass ratio[40] are more stringent, (by about one order of magnitude), but also more controversial, because loop corrections are particularly ill-defined in this case. Unitarity of high energy electroweak boson production[41] sets limits of roughly $|\kappa - 1|$, $|\lambda| \lesssim (1 \text{ TeV}/\Lambda)^2$ for a cutoff scale Λ in the formfactors which are associated with the W electric and magnetic multipole moments.

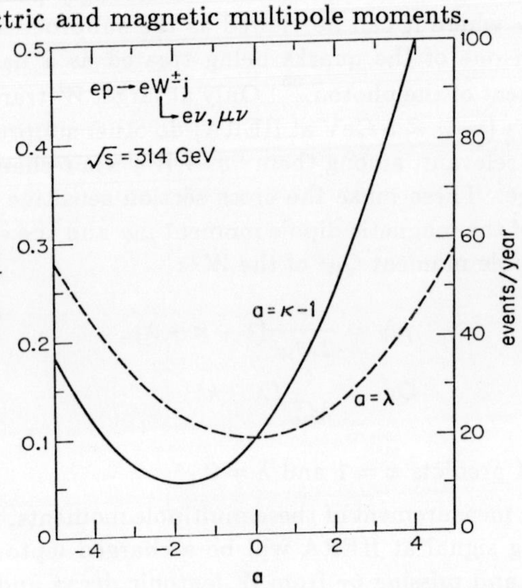

Fig. 6.1. Dependence of the leptonic W signal at HERA on deviations of either κ or λ from the SM values $\kappa = 1$, $\lambda = 0$.

In view of our poor present knowledge of the values of κ and λ, the direct measurement of the $WW\gamma$ couplings at HERA will clearly constitute major progress. It will be very helpful in disentangling $WW\gamma$ and WWZ anomalous coupling in W^+W^- production at LEP II in the mid 90's.

VII. EXCITED ELECTRON PRODUCTION

A firm prediction of models of composite quarks or leptons is the existence of excited partners of the known fermions. At high energy ep colliders the production of excited electrons will presumably give the cleanest signature for compositeness and hence we have concentrated on this possibility. The excited fermions considered here are, by definition, massive spin 1/2 particles, which decay radiatively into their light partners. Specifically, we assume the excited electron e^* and an excited neutrino ν^* form a weak doublet $F = \binom{\nu^*}{e^*}$, which couples to the Standard Model electron doublet $\ell = \binom{\nu_e}{e}_L$ by the interaction Lagrangian[42]

$$\mathcal{L} = \frac{gf}{\Lambda}\,\bar{F}\sigma^{\mu\nu}\frac{\vec{\tau}}{2}\ell\,\partial_\mu \vec{W}_\nu + \frac{g'f'}{\Lambda}\,\bar{F}\sigma^{\mu\nu}Y\ell\,\partial_\mu B_\nu + \text{h.c.}. \quad (7.1)$$

Here g and g' are the SM SU(2) and U(1) coupling constants, $\vec{\tau}$ denotes the three Pauli matrices, and $Y = -\frac{1}{2}$ is the hypercharge. The overall strength of the radiative transition is parametrized by f/Λ and f'/Λ, where Λ is the compositeness scale of the electron ($\Lambda > O(1$ TeV), from the search for contact terms in Bhabha scattering,[43] while f and f' are arbitrary normalization factors, presumably of $O(1)$). The effective Lagrangian (7.1) naturally avoids constraints on f or f' from $(g-2)$ measurements, because it only couples left-handed electrons to their excited partner.[44] In addition, it should be noted that the restriction to spin 1/2 and isospin 1/2 excited electrons is only for illustrative purposes. Other quantum numbers are discussed in Ref. 45

According to the effective Lagrangian (7.1) excited electrons may be produced in ep collisions by the fusion of the electron with a virtual photon or Z, which is radiated off the proton. While the Z contribution can be reliably calculated in terms of the parton subprocess $eq \to e^*q$, t-channel photon exchange has a large probability to leave the proton intact. Therefore elastic and quasielastic scattering have to be considered as well as deep inelastic scattering. All three contributions were considered in Ref. 46 for HERA energies and the result is depicted in Fig. 7.1: deep inelastic and elastic scattering contribute about equally to e^* production, with the elastic contribution even dominating at large e^* masses. The equivalent curves for the LEPI/LHC combination are shown in Fig. 7.2. In both cases the cut between the deep inelastic and the quasielastic regions is made at $Q_0^2 = 5$ GeV^2.

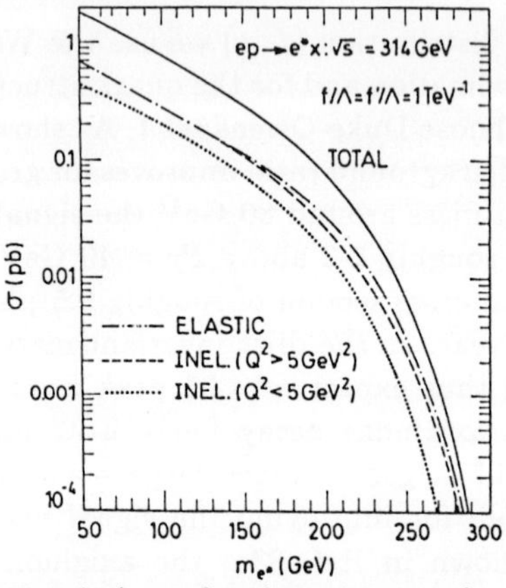

Fig. 7.1. Total e^* production cross section versus e^* mass in ep collisions at $\sqrt{s} = 314$ GeV with magnetic transition couplings $f/\Lambda = f'/\Lambda = 1$ TeV^{-1}. Elastic (dash-dotted line), quasielastic (dotted line), and deep inelastic contributions (dashed line) are shown separately.

According to the Lagrangian (7.1) excited electrons, once they are produced, will rapidly decay via $e^* \to e\gamma$, $e^* \to eZ$, and $e^* \to \nu W$. For $m_{e^*} < m_W$ the latter two modes involve virtual W's and Z's and are highly suppressed. For the larger e^* masses which can already be probed at HERA, the branching ratios depend critically on the ratio of f'/f. For a particular, but representative, value $m_{e^*} = 200$ GeV the three branching ratios[45] are plotted in Fig. 7.3. It should be noted that the decay products of the e^* will typically have very large p_T (of order $m_{e^*}/2$)

and should thus be easily distinguishable from possible backgrounds. Even the $e^* \to \nu W$ decay channel will give a striking signal: very large missing p_T balanced by a dijet system with invariant mass equal to m_W, and half of these events will have no beam jet but rather a single proton which may be identified in the forward spectrometer. We thus expect a very high detection efficiency for the e^*. Requiring ten events to be seen, according to Fig. 7.1, excited electrons may be discovered at HERA if they are lighter than 200-250 GeV (depending on the actual values of f/Λ and f'/Λ) while the combination LEPI/LHC is sensitive to e^*'s up to masses around 1 TeV.

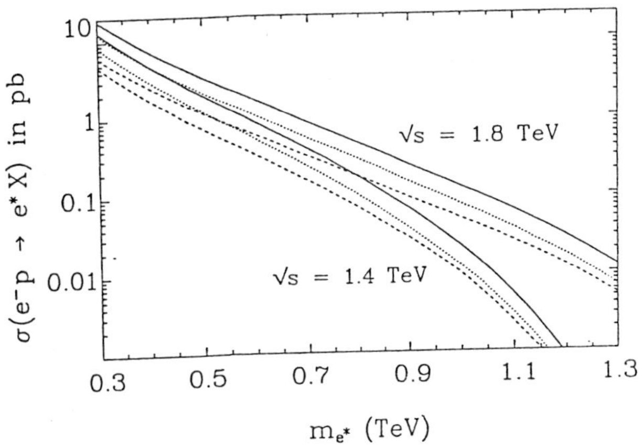

Fig. 7.2. Excited electron production in ep collision at LEP/LHC. Shown are the cross sections for elastic production $ep \to e^*p$ (dashed), inelastic production (dotted), and the sum of both (solid) for $f/\Lambda = f'/\Lambda = 1/m_{e^*}$. From Ref. 47.

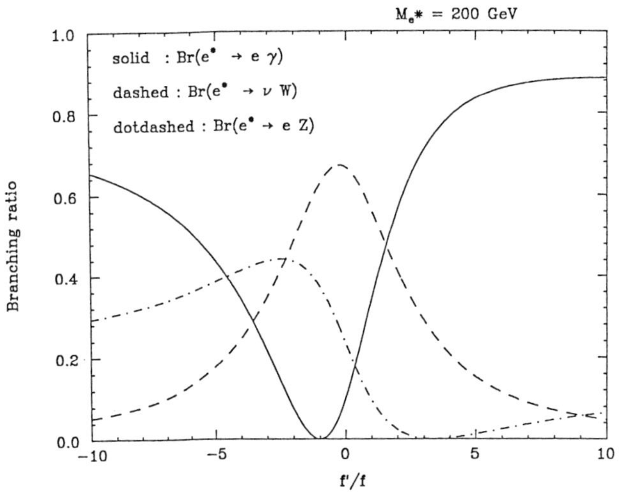

Fig. 7.3. Branching ratios of the three e^* decay modes as a function of the $U(1)_Y/SU(2)_L$ coupling ratio f'/f.

VIII. SUPERSYMMETRY

Supersymmetry is a symmetry relating the fermions and the bosons and could explain the mass hierarchy between the Planck scale and the electroweak symmetry breaking scale.[48] The minimal supersymmetric extension of the standard model contains two scalar Higgs doublets, and the superpartners, namely, the spin 0 sleptons and squarks associated with the left and right handed leptons and quarks, and the spin 1/2 gluinos ($\tilde{g}$), electroweak gauginos (photinos($\tilde{\gamma}$), zinos($\tilde{Z}$) and winos($\tilde{W}^{\pm}$)) associated with the gluons, and the photon, Z and $W^{\pm}$ bosons. Similarly, the spin 1/2 higgsinos ($\tilde{H}_1^0, \tilde{H}_1^-, \tilde{H}_2^+, \tilde{H}_2^0$) are associated with the Higgs bosons. All the couplings of these new fields are related by supersymmetry to the gauge and the Yukawa couplings of the standard model. It is a broken symmetry with an expected scale of masses within 0.5-1 TeV above those of the ordinary particles (if it is to explain the weak mass scale). The photino, Zino, and neutral Higgsinos mix, and the resulting states are called neutralinos; similarly, the Wino and charged Higgsinos mix and are called charginos.

Among numerous SUSY models the most popular ones have dynamic schemes for symmetry breaking suggested by supergravity models. The details of these models can be found elsewhere.[49,50]

Here we are interested in the discovery of the SUSY particles at HERA. In particular, we consider the associated production of a slepton and a squark, through either a chargino exchange or a neutralino exchange. This process represents the largest cross section of all the SUSY production processes. Figures 8.1 a and b show the chargino exchange and the neutralino exchange diagrams, respectively. At HERA energy, the mass range of the supersymmetric particles is such that their direct two body decay modes are still the primary decay modes, and one need not be concerned with the cascading modes. Hence, the selectron is presumed to decay into an electron and the lightest Supersymmetric Particle (LSP), which in most models is the photino. Similarly, the squark decays into the corresponding quark and a photino. The sneutrino is hard to detect other than in very special kinematic configurations and we will only consider the process shown in Fig. 8.1b.

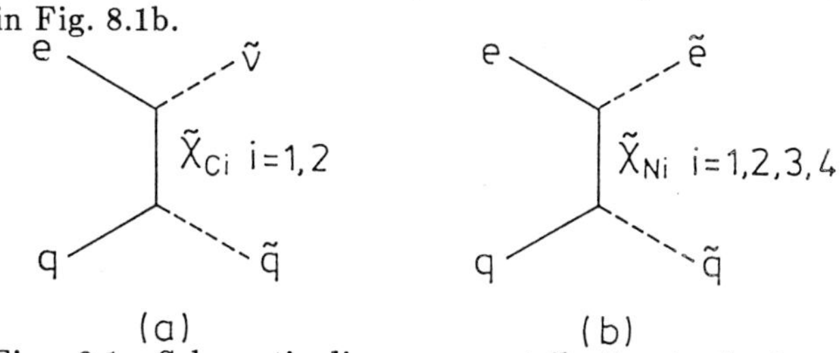

Fig. 8.1. Schematic diagrams contributing to the inclusive processes: (a) $ep \to \tilde{\nu}\tilde{q}x$ and (b) $ep \to \tilde{e}\tilde{q}x$ from ref. 49.

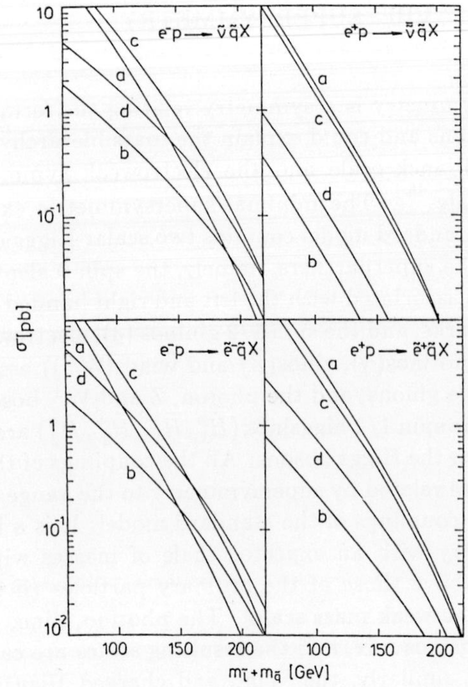

Fig. 8.2. Slepton-squark production cross sections at root $s = 314GeV$ versus $m_{\tilde{l}} + m_{\tilde{q}}$ and four different mixing scenarios labelled by a,b,c, and d from ref. 49

Figure 8.2 shows the cross sections for the associated production in e^-p and e^+p collisions for four different models, with parameters representing reasonably different regions of the parameter space.[49] Table 8.1 describes the parameters of these models,[49] namely, three of the neutralino masses (the fourth one decouples and is not shown), the two chargino masses and their corresponding mixing matrices.

For a discovery limit of 20 events per year (corresponding to $\sigma = 0.1pb$), the sensitive mass range for the associated production of the squark and the selectron will be between 160 and 180 GeV for a cm energy of 314 GeV, as shown in Fig. 8.2 for the above four models. At a cm energy of 410 GeV, this range is extended between 220 and 235 GeV. The 20 event discovery limit is conservative. In fact, recent realistic Monte Carlo studies show[51] ten events should be sufficient to establish a clear supersymmetry signal. This allows the discovery limits to extend up to 200(250) GeV for a cm energy of 314(410) GeV.

Current limits on the masses of the scalar quarks and gluinos are $M(\tilde{q}) \geq 80$ GeV, and $M(\tilde{g}) \geq 80$ GeV, for $M(\tilde{\gamma}) \sim 0$. The best limit on the scalar electron mass from combining the results of different experiments is $M(\tilde{e}) \geq 84$ GeV in the case of a massless photino.[52] Direct searches only rule out selection masses below 25 GeV. HERA and the Fermilab collider will have the same sensitivity for the squark mass in the light selectron scenario.

Details of the Monte Carlo study can be found elsewhere.[51] A short summary is given below. A supersymmetric event of the kind shown in Fig. 8.1b, where the selectron decays into an electron and a photino will have a similar topology to that of a neutral current (NC) event because of the undetected photino. The NC events, unlike the SUSY events, should conserve p_T. However, the effect of the beam pipe, finite detector resolution and

Neutralinos			Charginos		SUSY Breaking	Higgsino Mass
m_{N1} $(U_{N11}, U_{N21}, U_{N31})$	m_{N2} $(U_{N12}, U_{N22}, U_{N32})$	m_{N3} $(U_{N13}, U_{N23}, U_{N33})$	m_{C1} (U_{L11}, U_{L21})	m_{C2}	M_2	μ
0 (1.0,0.0,0.0)	38 (0.0,-0.927i,0.375)	230 (0.0,0.375,-0.927i)	30 (0.939i,-0.345)	222	0.0	192.0
20 (0.103,-0.277,-0.955i)	221 (0.895,-0.393,0.210i)	441 (0.434,0.877,-0.207i)	30 (0.203,0.979i)	441	423.7	46.9
20 (0.972,-0.231,0.050i)	52 (0.237,0.952,-0.192i)	429 (0.003i,-0.198i,0.980)	50 (0.985,-0.175i)	424	35.5	-410.0
100 (0.919,-0.375,0.123i)	206 (0.030i,-0.244i,0.969)	208 (0.393,0.894,-0.213i)	200 (0.205i,-0.979)	207	189.7	-182.9

Table 8.1. Parameters of the SUSY Models, Masses and Eigenstates of Neutralinos and Charginos. The matrix U_N' contains the information of the neutralino eigenstates ($\tilde{\chi}_{Ni}$,i=1,4) $\tilde{\chi}_{Ci}$,i=1,2) as well as describing the $\tilde{\gamma}, \tilde{Z}, \tilde{H}$ and $\tilde{H}'$ admixtures. Similarly, the matrix U_L, contains the information of the chargino eigenstates ($\tilde{\chi}_{Ci}$,i=1,2) and describe the admixtures of the $\tilde{W}\pm$ and $\tilde{H}\pm$.

missing mass (e.g. neutrinos in the hadronization process) can give rise to imbalanced p_T in the NC events. The study showed that most of the NC events are eliminated by a cut which demands a small amount of p_T imbalance whereas the SUSY events would be evenly distributed in the p_T imbalance. This is shown in Fig. 8.3a and b. In addition, assuming a photino with negligible mass, the kinematics of the SUSY events could be reconstructed by assuming the three momentum of the missing photino to be known. This was varied over a grid of 200 (GeV/c)[51] with a spacing of 10 GeV/c. Attempts at reconstructing the event with this assumed photino momentum only ended in realistic, i.e. positive definite, mass and momenta of the selectron and the squark if this were close to the real photino momentum. Therefore, the selectron and and the squark masses can be obtained from the "good" reconstructions. Figure 8.4 shows the distribution of the selectron and the squark masses for the successful attempts. The systematics were checked and controlled by the Monte Carlo studies.

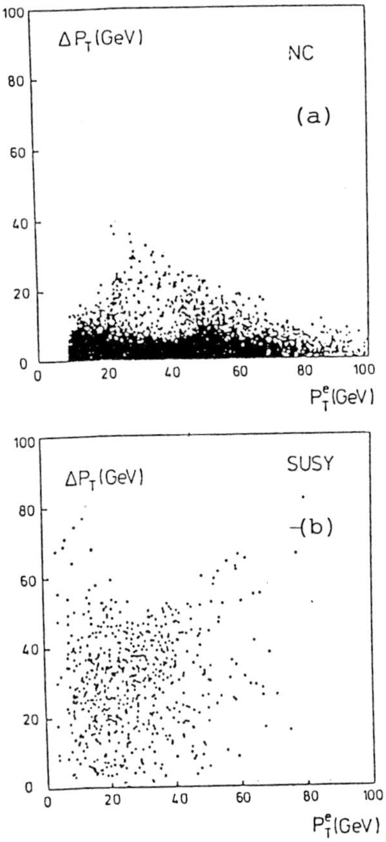

Fig. 8.3. ΔP_T vs. P_T for (a) NC events and (b) SUSY events. Note the cut $x > 0.1$ corresponds to $P_T^{electron} > 8$ Gev/c from ref. 51

The strength of HERA, unlike other machines, is not simply in discovering the supersymmetric event signature, but in its ability to pursue it for a deeper understanding with the help of the study of the polarization and the charge asymmetries. As mentioned earlier, the parameter space for the SUSY models, even within supergravity, is vast. Using a polarized electron beam, only the L or the R type SUSY particles (partners of the left and right handed particles) can be produced. Figure 8.5 shows the left-right asymmetry as a function of the mass difference of the left and right handed SUSY partners and the combined mass of the selectron and the squark.[50] The other parameters, such as the SUSY breaking mass can also be checked by this measurement.

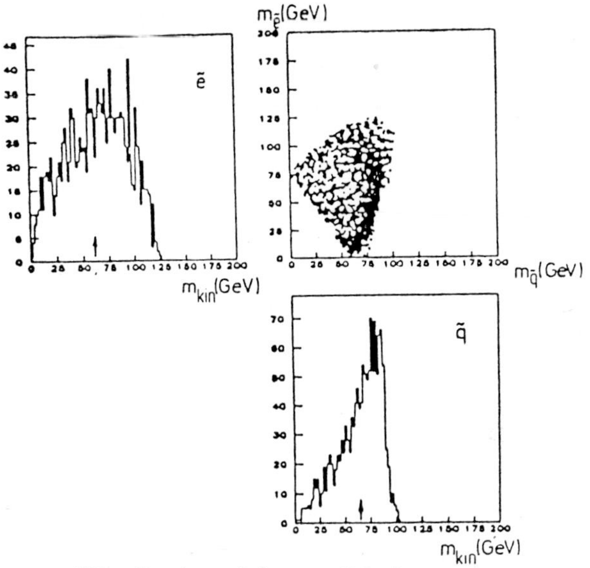

Fig. 8.4. Distribution of the possible "reconstructed" masses for one single SUSY event. The arrows indicate the averages of the distribution, here named kinematical "masses" from ref. 51

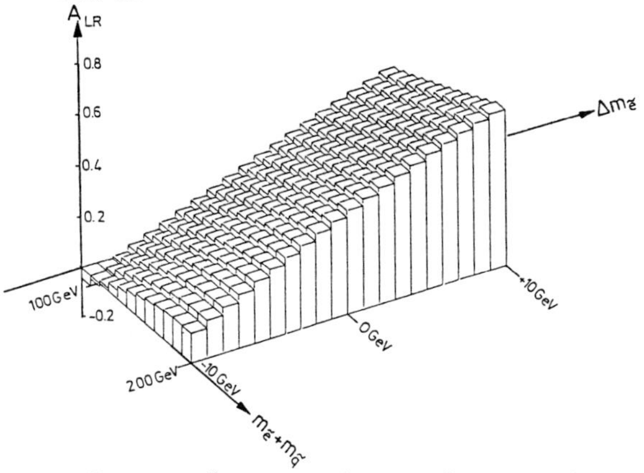

Fig. 8.5. Integrated cross section as a function of $m_{\tilde{e}}+m_{\tilde{q}}$. Values of the parameters are $M_2 = 0, \tilde{M}_{cl} = -50GeV, v_1/v_2 = 1$ from ref. 50

Table 8.1 displayed parameters for the four situations shown in Fig. 8.2. The charge asymmetry in the production cross section depends on these parameters as shown in Fig. 8.6.[49] The difference in the cross sections for electron-proton and positron-proton is evident from Fig. 8.2. Figure 8.6 displays it for a number of them. This measurement, clearly, will help determine the neutralino and the chargino eigenstates and the mixing matrix U, helping to determine the exact parameters of the specific model.

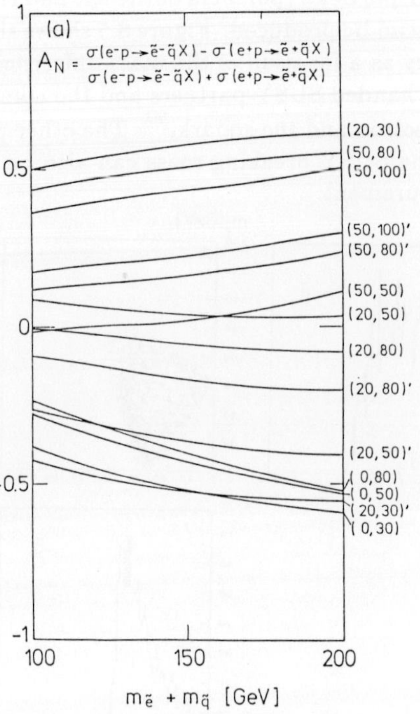

Fig. 8.6. Asymmetries of $\tilde{e}\ \tilde{q}$ production in $e^{\pm}p$ collisions at $\sqrt{s} = 314$ GeV versus $m_{\tilde{l}} + m_{\tilde{q}}$ for equal scalar masses and a number of mixing scenarios with $M_{N1} < 100$ GeV. The numbers in brackets give the values of M_{N1} and M_{C1} in GeV from ref. 49

The higher energy *ep* option of LEP1 on LHC with a cm energy 1.4 TeV can reach the sum of selectron and the squark masses up to 700 GeV.[53] However, at these masses the direct two body decay branching ratio of the selectron(squark) into an electron(quark) and a photino is not very high. Instead, the intermediate cascading decays become the prevalent decay modes. These require further investigation.[54]

In summary, HERA is an excellent machine to search for supersymmetric particles. Not only will it be competitive with any contemporary machine in covering the accessible kinematic range, but it s real strength lies in its ability to understand and learn about supersymmetry, rather than the mere discovery of it.

IX. ELASTIC SUPERSYMMETRY

Our group also looked[55] at associated slepton- gaugino production via the elastic processes $ep \to \tilde{e}\tilde{Z}_i p$ and $ep \to \tilde{\nu}\tilde{W}_- p$, where $\tilde{Z}_i$ and $\tilde{W}_-$ denote a generic neutralino (there are at least four of them) and the lightest chargino, respectively. The study of these reactions at HERA has four advantages over most other SUSY processes at machines of the same generation:

- the potential mass reach is higher than at LEP or for the production of weakly interacting sparticles at the Tevatron; already after 1 year of running, HERA will be sensitive[55] up to $m_{\tilde{e}} \simeq m_{\tilde{W}} \simeq 50$ GeV;
- the event is as clean as at an e^+e^- collider;
- the energy of the outgoing proton, and thus the invariant mass of the produced sfermion-gaugino system, can be measured in a forward proton spectrometer. In the simplest case of $\tilde{e}\tilde{Z}_1$ production and subsequent $\tilde{e} \to e\tilde{Z}_1$ decay this measurement together with the measurement of the p_T spectrum of the outgoing electron is sufficient[55] to pin down both $m_{\tilde{e}}$ and $m_{\tilde{Z}_1}$;
- most of the emerging signals, especially the one electron plus missing p_T signal from $\tilde{e}\tilde{Z}_1$ production or $\tilde{\nu}\tilde{W}_-$ production with subsequent $\tilde{W}_- \to \tilde{\nu}e$ decay, and the three leptons plus missing p_T signal from $\tilde{e}\tilde{Z}_2$ production with subsequent $\tilde{e} \to e\tilde{Z}_1$, $\tilde{Z}_2 \to \ell^+\ell^-\tilde{Z}_1$ decays, have practically no standard model backgrounds.

Here we try to further quantify the discovery limits resulting from these processes. As in Ref. 55 we will assume the gaugino-sector to be described by minimal supergravity models, where all neutralino and chargino masses and mixings are determined by the values of three free parameters: the supersymmetric higgsino mass μ, the gluino mass $m_{\tilde{g}}$, and the ratio ω of the vacuum expectation values of the two neutral Higgs fields.

In Ref. 55 it was shown that the cross section for the production of any one neutralino varies strongly with variations of the parameters, due to the existence of level-crossings in the neutralino sector, these cross sections need not even be continuous functions of the parameters. However, variations in the *sum* over all four neutralino cross sections are much smaller; more specifically this sum is a smooth function of the three parameters mentioned above.

In Fig. 9.1 we therefore show lines of constant $\sum_{i=1}^{4} \sigma$ $(ep \to \tilde{e}\tilde{Z}_i p)$ at $\sqrt{s} = 314$ GeV for unpolarized electrons and degenerate left- and right-handed selectrons of mass 30 GeV. We see that for this value of $m_{\tilde{e}}$, HERA is sensitive to gluino masses as large as 400 GeV if $\text{sign}(m_{\tilde{g}}) = -\text{sign}(\mu)$, and more than 500 GeV if $\text{sign}(m_{\tilde{g}}) = \text{sign}(\mu)$. (The physical gluino mass is given by the absolute value of $m_{\tilde{g}}$.) We can also see that the total neutralino cross section depends only weakly on μ.

In Fig. 9.1 we choose $\omega = 1.5$, but the results are not very sensitive to this choice. They do, however, depend somewhat on the polarization of the incident electrons.

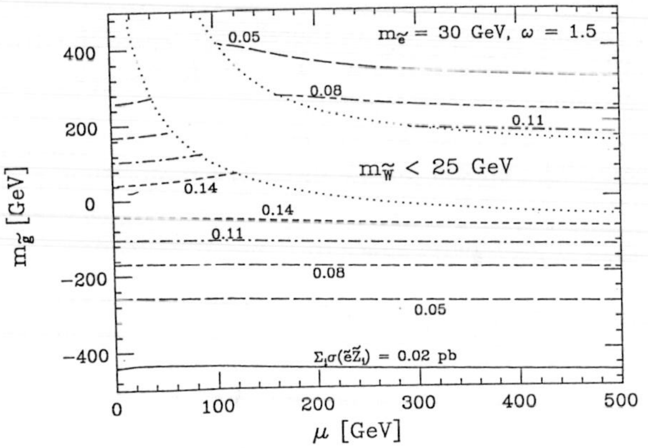

Fig. 9.1 Lines of constant cross section for total elastic selectron-neutralino production, for $\sqrt{s}$ = 314 GeV and unpolarized incident electrons. Here, as well as in Figs. 9.2–9.4, the modified effective photon approximation of Ref. 55 has been used.

If they are 100% right-handed, the total neutralino cross section becomes smaller in the region of small $|m_{\tilde{g}}|$, but bigger for large $|m_{\tilde{g}}|$. The reason is that in the former region $\tilde{Z}_2$-production can make sizeable contributions; since in this region the $\tilde{Z}_2$ is more or less the superpartner of the Z, it couples only weakly to right-handed electrons. If, on the other hand, $|m_{\tilde{g}}|$ is large, $\tilde{Z}_2$ production is negligible, and the $\tilde{Z}_1$ is dominantly a hypercharge-gaugino, which couples four times stronger to $\tilde{e}_R$ than to $\tilde{e}_L$. The situation for left-handed incident electrons is, of course, just the opposite of the right-handed case.

In Fig. 9.2 we show lines of constant $\sigma(ep \to \tilde{\nu}\tilde{W}_- p)$ for unpolarized electrons, $m_{\tilde{\nu}}$ = 30 GeV and $\sqrt{s}$ = 314 GeV. These lines almost coincide with lines of constant chargino mass. This is not surprising, since our cross sections always depend more strongly on the mass of the charged sparticle than on that of the neutral one, due to the u-channel Feynman diagram which contains the charged sparticle propagator. Obviously the cross sec-

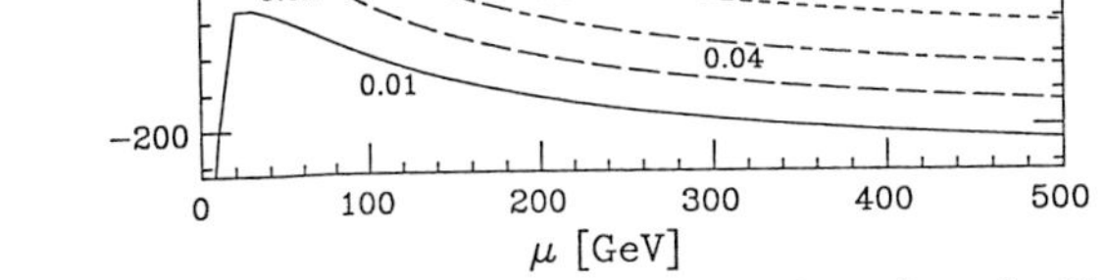

Fig. 9.2 Lines of constant cross section for elastic chargino-sneutrino production, for $\sqrt{s}$ = 314 GeV and unpolarized electrons.

tions of Fig. 9.2 would be twice as large for a 100% left-handed polarized electron beam, and vanish for right-handed electrons.

From the comparison of Figs. 9.1 and 9.2 one might conclude that selectron-neutralino production should be sizeable whenever $\tilde{\nu}\tilde{W}^-$ production is detectable. This is, however, only true if sleptons are relatively light. If we want to discuss a more general case quantitatively, we have to make some assumptions about the scalar spectrum.

We choose to work within the framework of minimal supergravity, where one assumes that all squarks and sleptons have the same mass m_0 at the Planck scale M_p, whereas all gauginos have the common mass M at M_p. These relations are, however, altered at lower energies due to radiative corrections; at the physically relevant weak scale one finds the relations[56]

$$m_{\tilde{g}} = 3\,M\,; \tag{9.1}$$
$$m^2_{\tilde{e}_R} = m_0^2 + 0.15\,M^2 - 0.28\,D\,; \tag{9.2}$$
$$m^2_{\tilde{e}_L} = m_0^2 + 0.53\,M^2 - 0.22\,D\,; \tag{9.3}$$
$$m^2_{\tilde{\nu}_L} = m_0^2 + 0.53\,M^2 + 0.5\,D\,; \tag{9.4}$$
$$m^2_{\tilde{q}} \simeq m_0^2 + 7\,M^2\,; \tag{9.5}$$

where

$$D = M_Z^2\frac{1-\omega^2}{1+\omega^2} \le 0 \quad \text{for} \quad \omega \ge 1\,. \tag{9.6}$$

The squark mass (9.5) is an average over left- and right-handed u- and d-squarks. Note that these models always predict[56] $\omega \ge 1$.

The model is thus described by four parameters: m_0, M, μ, and ω. In Figures 9.3 and 9.4 we have fixed $\omega = 1$ and $\mu = -500$ GeV, well in the asymptotic region where all relevant cross sections are almost independent of μ. The two figures then show discovery limits in the plane of the two remaining parameters, m_0 and M, for $\sqrt{s}$ = 314 GeV and 1.4 TeV, respectively.

The solid lines in Fig. 9.3 are lines of constant total elastic SUSY cross section = 0.025 pb, leading to five events in a "typical" HERA year, for different polarizations; note that these curves contain both $\tilde{e}\tilde{Z}_i$ and $\tilde{\nu}\tilde{W}_-$ production. For comparison we also show the discovery limit from the best of the "classic" SUSY-processes at HERA, $ep \to \tilde{e}_R\tilde{q} + X$ (the limits from $\tilde{e}_L\tilde{q}$ and $\tilde{\nu}\tilde{q}$ production are worse); here we have assumed that HERA is sensitive up to $m_{\tilde{e}_R} + m_{\tilde{q}}$ = 160 GeV, which is the most optimistic value found in Ref. 57. Obviously the elastic processes are detectable over a much larger region of parameter space, especially if experiments with left- and right-handed electron beams are combined. (For our purpose, a left-handed positron would do as well as a right-handed electron.) Experiments with right-handed electrons are especially sensitive in the region of small m_0

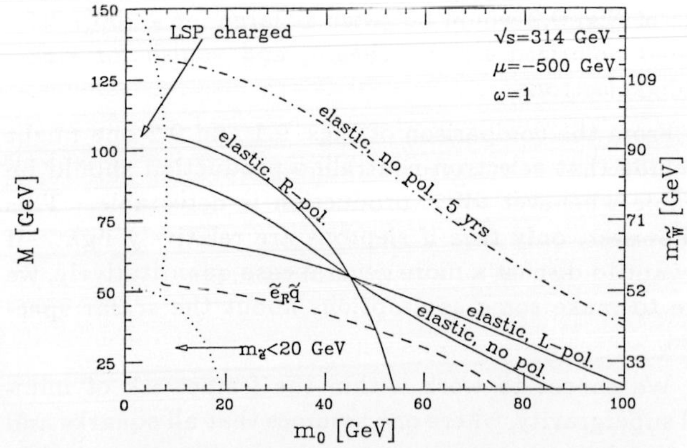

Fig. 9.3 SUSY discovery limits at HERA. The solid lines correspond to a total elastic SUSY cross section of 25 fb, corresponding to 5 ev. /yr. The dot-dashed line corresponds to five elastic SUSY events in 5 years. The dashed line shows the discovery limit from $ep \rightarrow \tilde{e}_R \tilde{q} X$, given by[57] $m_{\tilde{e}_R} + m_{\tilde{q}} = 160$ GeV. The regions to the left of the dotted curves are excluded since there the lightest supersymmetric particle, which is stable, would be charged, or the $\tilde{e}_R$ would be lighter than 20 GeV.

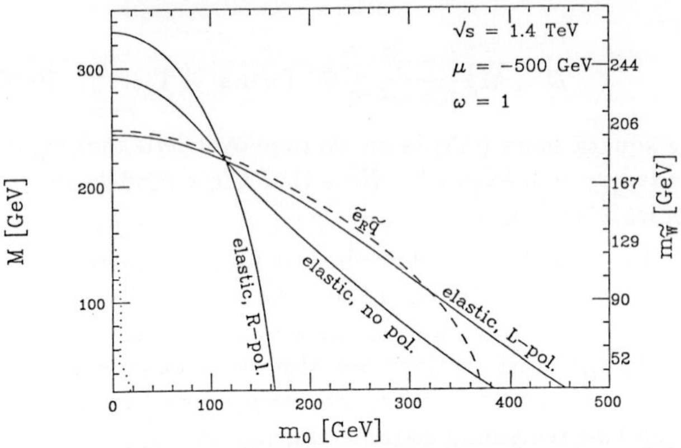

Fig. 9.4 Discovery limits at LEP I×LHC. The solid curves correspond to a total elastic SUSY cross section of 5 fb (5 ev./yr.), while the dashed curve corresponds to $m_{\tilde{e}_R} + m_{\tilde{q}} =$ 750 GeV, which is[57] the region of sensitivity of $\tilde{e}_R \tilde{q}$ production.

and large M, where $m_{\tilde{e}_R}$ is still fairly small (see Eq. (9.2)); the discovery limit here comes mainly from $\tilde{e}_R \tilde{Z}_1$ production. The region of large m_0 and rather small M can best be probed with left-handed electron beams; in this region all sleptons are fairly heavy, but charginos are light (see the scale at the right side of the frame), and the limit is mainly determined by $\tilde{\nu}\tilde{W}_-$ production.

One remarkable property of the elastic SUSY cross sections is that their dependence on sparticle masses is relatively mild, since the only distribution function (the photon-density in the proton) that enters the cross section is fairly hard, going essentially like $1/x$. An increase in the integrated luminosity can therefore increase the discovery limits quite dramatically. This is demonstrated by the dot-dashed curve in Fig. 9.3, which corresponds to a total elastic SUSY cross section of 5 fb, or five events in five years. The effects on the discovery limits from $\tilde{e}_R \tilde{q}$ production would be much smaller, increasing the accessible region of parameter space to[57] $m_{\tilde{e}_R} + m_{\tilde{q}} \simeq$ 200-210 GeV.

What region could other colliders of the same generation cover? The Tevatron should be able to find gluinos and squarks if they lie below 200 GeV, *i.e.* will cover the region below $M = 65$ GeV, and thus the whole region where the "classic" HERA mechanism for SUSY production can be observed. As soon as LEP I or SLC start to produce large numbers of Z bosons, the region $m_{\tilde{e}_R}, m_{\tilde{W}_-} \lesssim M_Z/2$ will be probed; that covers the region that can be probed by one year of elastic SUSY production at HERA with right-handed electrons, but still leaves a window for the left-handed mode. This window will be closed once LEP I reaches its maximal energy of $\sqrt{s} = 110$ GeV; it seems likely, however, that by then HERA has accumulated more luminosity. A good part of the region which can be probed by 5 HERA years can otherwise only be investigated by LEP II.

The situation looks somewhat less favorable for the planned LEP I×LHC collider with $\sqrt{s} = 1.4$ TeV, Fig. 9.4. The sum of the elastic processes can still cover a somewhat larger region of parameter space than the "classic" $\tilde{e}_R \tilde{q}$ production; however, the regions between the solid curves and the dashed curve will all have been covered by LEP II before LEP I×LHC goes into operation. Furthermore, the whole shown region of parameter space can be covered by squark and gluino searches at the SSC and LHC. On the other hand, if one of these other machines does find a signal, SUSY-searches at LEP I×LHC, whether they are successful or not, will help to determine the parameters of the model. Furthermore, it is of course possible that nature is supersymmetric but not minimally so; if *e.g.* sleptons are lighter than given by Eqs. (9.2)-(9.4), which might well be the case in $E(6)$ models.[58] In this case, LEP I×LHC can make unique contributions to SUSY-searches.

X. DIQUARKS

The production of diquarks, both singly and in pairs, in ep collisions has been examined using the effective photon approximation. Diquarks are elementary particles with integral spin, electric charge $Q = \pm\frac{1}{3}, \pm\frac{2}{3}$, or $\pm\frac{4}{3}$, baryon number $\pm\frac{2}{3}$, lepton number 0, and transform as $3(\bar{3})$ or $6(\bar{6})$ under $SU(3)_C$, and can occur in composite models and superstring-inspired E_6 theories. For our numerical calculation we take diquarks to be color triplets only. They couple to quark-quark pairs with an unknown strength which we parameterize as $\kappa\alpha_{em}$.

Pair production of diquarks at *ep* colliders proceeds through photon-gluon fusion and is hence independent of the unknown coupling strength. The cross section for diquark pairs depends only on the diquark mass and charge, and is of order 1 pb at $\sqrt{s} = 314$ GeV for $M_D \simeq 42$, 34, 28 GeV with $|Q| = \frac{4}{3}, \frac{2}{3}, \frac{1}{3}$, respectively. The $\sqrt{s} = 1.8$ TeV LHC $\times$ LEP II option yields larger cross sections, with values of 1 pb for $M_D \simeq 140$ GeV with $|Q| = \frac{4}{3}$. The corresponding signal is 4 jets with double peaking in the jet-jet invariant mass distribution and is relatively background free.

Single diquark production occurs via photon-quark fusion. The cross section is directly proportional to the unknown quark-quark-diquark couplings (κ) and thus is quite sensitive to its value. Here, cross sections of order 1 pb are obtained at HERA for $M_D \simeq 70, 65, 45$ GeV with $|Q| = \frac{4}{3}, \frac{1}{3}, \frac{2}{3}$, respectively, and at LHC $\times$ LEP II ($\sqrt{s} =$ 1.8 TeV) for $M_D \simeq 150$ GeV with $|Q| = \frac{4}{3}$, assuming $\kappa = 1$ in all cases. Hence, for $\kappa = 1$ the single production cross sections are larger than those for pair production, but the single production rate decreases substantially for smaller values of κ. The signal for single production is 3 (or 2) jets, with the jet from the associated antiquark preferring to be at small angles. The background results from QCD processes and is potentially larger than that for diquark pairs since the associated antiquark jet produced along with the single diquark comes out at a small angle relative to the beam. Although p_t cuts may decrease the QCD background, they will also substantially reduce the signal.

Search limits for diquarks (single and pair production) have been estimated for various values of κ and the diquark charge for each of the *ep* machines under consideration, assuming 10 events constitutes a discovery (limits with 100 events have also been considered for the case of single production). These discovery limits and the details of diquark production and detection are contained in a separate contribution to these proceedings by Rizzo.

XI. LEPTOQUARKS AND LEPTOGLUONS

Leptoquarks are color triplet particles which couple to a lepton-quark pair (q, ℓ) and are naturally present in many theories beyond the standard model (SM). They appear in technicolor theories, superstring-inspired E_6 models, the strong-coupling version of the SM, SO(10) unified gauge theories with SU(4) color symmetry, and in some horizontal symmetry schemes. The various quantum numbers that leptoquarks can have in these schemes include: spin 0, 1, or 2, electric charge Q = -4/3, -1/3, 2/3, or 5/3, weak iso-spin 0, 1/2, or 1, baryon number $\pm 1/3$, and lepton number ± 1. The leptoquarks present in extended technicolor theories,[59] P_3, have couplings which are proportional to the mass of the quark-lepton pair, and hence the coupling ueP_3 will be tiny. For definiteness, we will review leptoquark production at *ep* machines for the case of the superstring-inspired E_6 leptoquarks.[60] These leptoquarks, which we will denote as S, are scalar, charge $-1/3$, baryon number $= +1/3$, lepton number $= +1$, weak iso-singlets, and are the supersymmetric partner of the corresponding fermion which appears in the 27-representation of E_6.

The $q\ell S$ couplings are given in the E_6 superpotential, where the terms involving S are:

$$\lambda_6 S u^c e^c + \lambda_7 L S^c Q + \lambda_8 \nu^c S d^c \,, \tag{11.1}$$

where L and Q denote the left-handed lepton and quark doublets and the superscript c denotes the charge conjugate states. The Yukawa couplings are unknown and for calculational purposes we set $\lambda_6 = \lambda_7 = \lambda_8 = \lambda$ and parameterize λ by

$$\frac{\lambda^2}{4\pi} = F\alpha_{\rm em} \,. \tag{11.2}$$

The left- and right-handed decay widths are then given by (assuming no right-handed neutrinos)

$$\Gamma_L = \frac{2\lambda^2 M_S}{16\pi} \,, \qquad \Gamma_R = \frac{\lambda^2 M_S}{16\pi} \,. \tag{11.3}$$

Flavor changing neutral currents and the process $e^+e^- \to q\bar{q}$ place limits (which are summarized in Ref. 60) on the possible values of λ and M_S. For masses in the HERA range, this constrains $F \leq 0.1$ (1.0) for $M_S \leq 100$ (300) GeV.

ep colliders are especially well-suited for leptoquark production and offer quite striking production signals. Direct production occurs through an s-channel resonance by $e + u \to S$ with a subsequent decay to eu or νd with equal rates. The differential cross section for this process is given by

$$\frac{d\sigma}{dx\,dy} = \frac{u(x,Q^2)}{32\pi} \frac{\lambda^2 \hat{s}}{(\hat{s} - M_S^2)^2 + M_S^2 \Gamma^2} \,, \tag{11.4}$$

where $u(x, Q^2)$ is the u-quark distribution function. The total event rate at HERA is displayed in Fig. 11.1 from the work of Harnew.[61] It is clear from the figure that large event rates are obtainable almost up to the kinematic limit but that these rates are quite dependent on the strength of the unknown coupling. The s-channel resonance, $d\sigma/dx \sim \delta(x - M_S^2/s)$, yields distinctive peaks in the x-distribution which have been shown[62] to lie 1-2 orders of magnitude above the ordinary electroweak background $eq \to eq$.

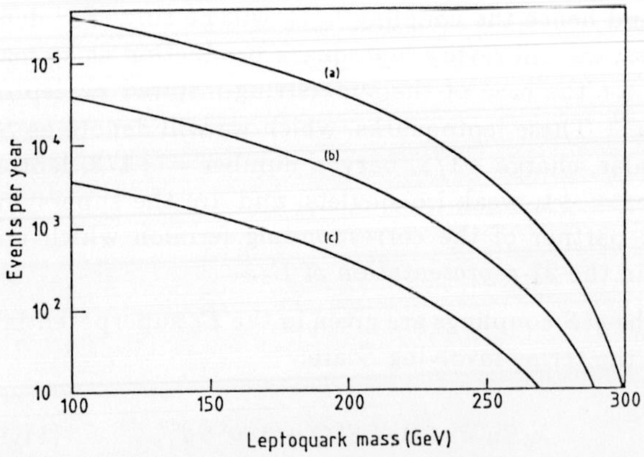

Fig. 11.1. Event rates per year at HERA for s-channel leptoquark production taking a) $F = 1.0$, b) $F = 0.1$, and c) $F = 0.01$.

Appropriate cuts in the y-distributions can also enhance the signal to background ratio. This is due to the fact that S is isotropic in y since it is a scalar, whereas the electroweak background is forward peaked. Figure 11.2, from the analysis by Harnew,[61] shows the signal to background rate for a 200 GeV leptoquark with cuts of $y > 0.01$ and $y > 0.5$. This analysis was performed for the ZEUS detector, assuming a $15\%/\sqrt{E}$ resolution for the e^- and $35\%/\sqrt{E}$ for jets, and taking $F = 0.01$. A y cut of $y > 0.5$ seems to be optimal for enhancing the S production signal over both the deep inelastic scattering neutral current and charged current backgrounds.

Discovery limits for both decay channels at HERA are presented[61] in the leptoquark coupling strength-mass plane in Fig. 11.3. Here, the discovery criteria are a

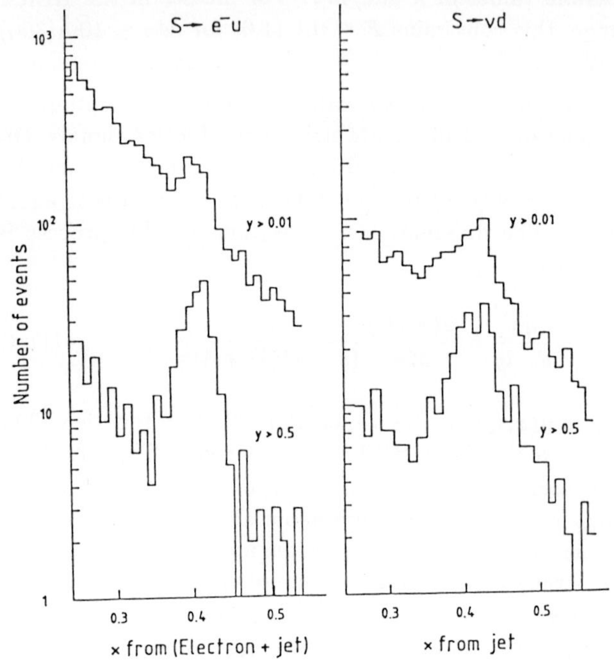

Fig. 11.2. x distributions for both decay channels of a 200 GeV leptoquark (with $F = 0.01$) in a uranium calorimeter with cuts in y as indicated.

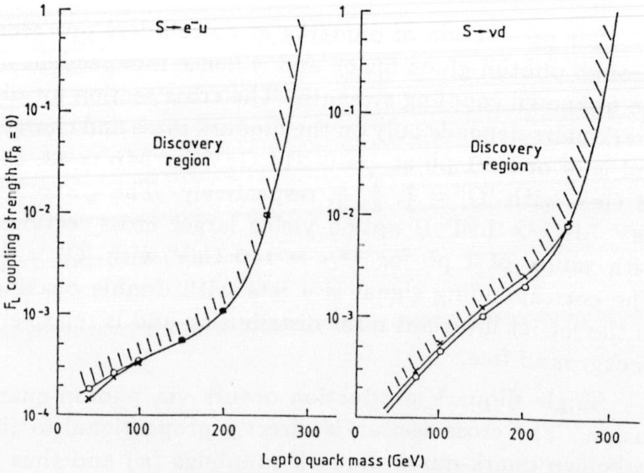

Fig. 11.3. Leptoquark discovery limits in the coupling strength-mass plane using the criteria discussed in the text, for one year's run time in a uranium (circles) or liquid argon (crosses) calorimeter.

five sigma signal with $N_{\text{signal}} > 25$ events after a cut of $y > 0.5$ has been imposed. It is seen that even for $F = 0.01$, leptoquarks with mass up to ~ 250 GeV can be detected in both decay channels, and for larger values of F the discovery region approaches the kinematic limit of the machine. Also note that the discovery regions are comparable for the two HERA detectors, ZEUS and H1. The LHC×LEP I and II options dramatically increase the detectable region as is demonstrated in Fig. 11.4 from Ref. 63, where the same criterion discussed above is imposed. For $F = 0.01$, leptoquarks up to 1 (0.8) TeV can be detected in the eu channel at $\sqrt{s} = 1.4$ (1.8) TeV. For larger values of the coupling, the 1.8 TeV option has the largest reach as the bigger value of $\sqrt{s}$ eventually wins over luminosity.

Leptoquarks may also be produced via t-channel interactions and by associated production ($eu \to Sg$, $eg \to Su^c$). However, the event rates for these processes are small compared to those from the s-channel resonance and we will not discuss them further. We also mention that the detection of leptoquarks which appear in technicolor theories where τt is the dominant decay mode has also been studied,[61] with the conclusion that discovery up to masses of 150 GeV should be possible.

If S is too massive to be produced directly at HERA, perhaps it could be indirectly detected through its effects on the neutral current process $eq \to eq$. If polarized e^- beams are available then various asymmetries can be measured. These asymmetries have been previously discussed[64] where it was found that the deviation from the SM prediction at HERA due to the presence of scalar leptoquarks is small, even for a mass of only $M_S = 400$ GeV (with $F = 0.3$). Hence, it is doubtful that leptoquarks of mass above the kinematic limit can be indirectly detected. However, if their mass is such

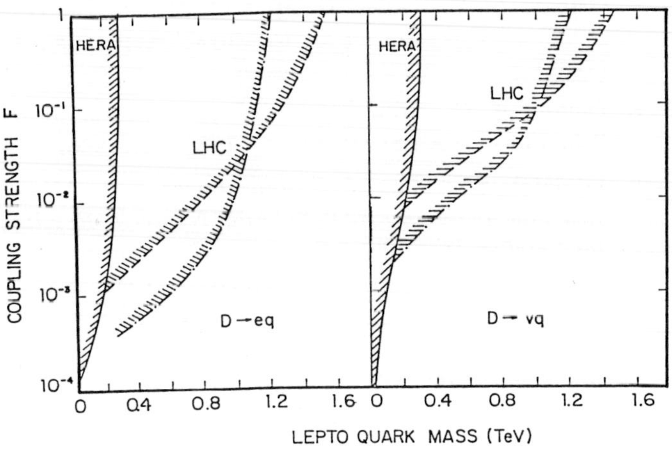

Fig. 11.4. Comparison of leptoquark search limits for HERA and the two LHC×LEP options.

that they have been directly produced, then the polarized beams and the various asymmetries which can then be formed can be valuable in probing the nature of the leptoquark couplings (*e.g.*, left- *vs.* right-handedness).

Leptogluons, which are color octets with lepton number and couple to a lepton-gluon pair and appear in some composite models, can also be produced via s-channel resonance in ep colliders. The $egLG$ coupling has the following form

$$\frac{g_s}{\Lambda}\,\ell^{8a}\sigma^{\mu\nu}e_L G^a_{\mu\nu}\,, \tag{11.5}$$

where ℓ^{8a} represents the leptogluon fields and Λ is the unknown coupling strength; the width is then

$$\Gamma = \frac{\alpha_s m_8^3}{4\,\Lambda^2} \tag{11.6}$$

with m_8 being the leptogluon mass. The resulting contribution to deep inelastic scattering due to leptogluon-induced $eg \to eg$ subprocess is:[65]

$$\frac{d\sigma}{dx\,dQ^2} = \frac{16\alpha_s(-\hat{s}\hat{u})}{s^2x^2\Lambda^4}g(x,Q^2) \times\left[\frac{\hat{s}^2}{(m_8^2-\hat{s}^2)^2+m_8^2\Gamma^2}+\frac{\hat{u}^2}{(m_8^2-\hat{u}^2)^2}\right], \tag{11.7}$$

where $g(x,Q^2)$ is the gluon distribution function. The event rate at HERA is shown[61] in Fig. 11.5 for $\Lambda = m_8$ and $\Lambda = 10\,m_8$ and is absolutely enormous. Demanding a five sigma signal with a y cut of $y > 0.5$ and using the same detector resolutions as before, Harnew[61] finds that for $\Lambda = m_8$ leptogluons can be discovered for masses up to 300 GeV at HERA.

In summary, we see that leptoquarks and leptogluons have very distinctive signatures at ep colliders. They can be produced by a direct s-channel resonance and hence have enormous event rates and a characteristic peak in the x-distribution. This together with their iso-tropic

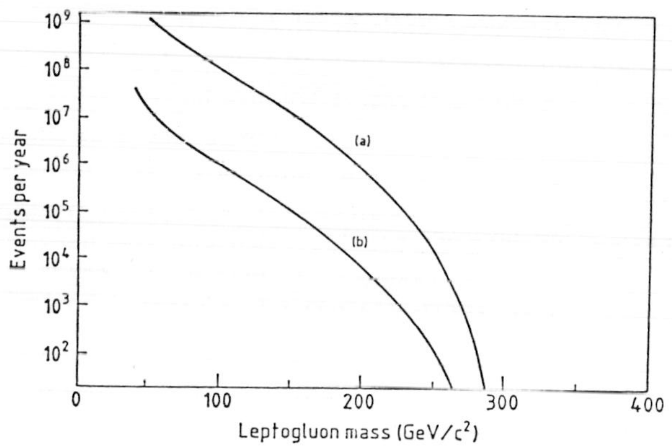

Fig. 11.5. Event rates per year at HERA for leptogluon production assuming a) $\Lambda = m_8$ or b) $\Lambda = 10\,m_8$.

y-distributions (compared to the forward peaked background) makes it easy to identify leptoquark and leptogluon production above the SM background. The search limits for these particles at HERA approaches the kinematic reach of the machine, and gives HERA the best leptoquark/leptogluon discovery reach of all colliders in the early to mid-1990 s. The LHC×LEP I and II options can explore masses up to 1-1.3 TeV (taking $F = 0.1$) and hence can compete with the SSC.

XII. CONCLUSION

We have seen how ep physics in the 1990 s is an important tool. It will further our understanding of QCD. It will provide a unique access to the small-x region of the gluon distribution. Resolved photon processes should teach more about the hadronic structure of the photon. New gauge bosons may be discovered, and if they are, use of polarization will distinguish among different models. Axigluons in the mass window 70 GeV $\le M_A \le$ 125 GeV should be open to discovery at HERA and inaccessible to other colliders. Single W production will yield a precise direct measurement of the $WW\gamma$ vertex. The production of excited electrons would give a very clean signature for compositeness. The ep colliders are not only able to discover supersymmetric particles, but through the use of polarization, can delineate which point of the vast SUSY parameter space is realized in nature. In addition, the elastic production of SUSY particles will offer a signature as clean as found in e^+e^- machines with a mass discovery range characteristic of hadron colliders. Reasonable rates and distinct signals exist for diquark production in ep collisions. Leptoquarks are produced at ep colliders as an s-channel resonance, yielding distinctive signals and higher discovery limits than their contemporary electron or hadron colliders. Leptogluons would be produced at an enormous rate.

The discovery limits for these particles and parameters are summarized in Table 12.1. The phenomena where ep colliders offer a discovery window beyond other colliders are listed in boldface type. Those phenomena where *ep* collisions provide additional information complementary to that found at other colliders are underlined. We expect that *ep* collisions will provide a rich source of physics.

E cm	314 GeV	410 GeV	1.4 TeV	1.8 TeV
New pb^{-1}/yr	200	200	1000	100
Particle $\mathcal{L}$	1.5×10^{31}	1.5×10^{31}	10^{32}	10^{31}
(or parameter)	HERA	HERA	LEPI × LHC	LEPII × LHC
Top Quark	80	110	500	
Composite e^*	200	250	1000	1000
$\underline{Z', Z_R}$	425	470	1200	
$W_R^{\pm}$	360	410	1000	
Axigluon	120			
$H^{\pm}$	40			
$\underline{\tilde{e} + \tilde{q}}$	180	235	700	
$\underline{\tilde{Z}}$	70	90	300	110
Diquark	80	90	290	180
Leptoquark	300	400	1100	1300
Leptogluon	280	380	1000	1200
$WW\gamma : \Delta\kappa$	.5	.3		
Contact: Λ_{eq}	7000		14000	

Table 12.1 Physics reach of *ep* colliders

ACKNOWLEGEMENT

The authors wish to thank Ms. Jean Johnson and Ms. Linda Eckstrom of the University of Wisconsin for editing and preparing this document.

REFERENCES

1. Wolf, G., *HERA: Physics, Machine and Experiments*, published in Lake Louise Winter Institute 1986, 144 (1986).

2. Ingelman, G. *et al.*, *Deep Inelastic Physics and Simulation, Proceedings of the DESY Workshop on Physics at HERA*, DESY 88-104.

3. Martyn, H. -U., *Contact terms and Substructure at HERA*, preprint PITHA 87-40, December 1987.

4. Olness, F. and Tung, W. -K., *Small-x Physics at the SSC and the Tevatron*, Int. J. Mod. Phys. A2, 1413 (1987); *ibid, From Colliders to Supercolliders, ed.*, V. Barger and F. Halzen, Madison Workshop 1987, 523 (1987).

5. Gribov, L. V., Levin, E. M. and Ryskin, M. G., *Semihard Processes in QCD*, Phys. Rep. 100, 1 (1983).

6. Collins, J. C., *Mini-jets and Regge Theory*, published in UCLA SSC Workshop 1986, 15 (1986).

7. Lomatch, S., Collins, J. C. and Olness, F. I., *Mini-jets and B-Pairs: Analytic and Numeric Results for Small-x Dominated Processes*, IIT preprint IIT-TH/23-88, May 1988. To appear in Nucl. Phys. B.

8. Martin, A. D., Roberts, R. G. and Stirling, W. J., Phys. Rev. D37, 1161 (1988).

9. Kunszt, Z., *J/Psi and Upsilon Production in High-Energy Electron Proton Collisions*, NSF-ITP-88-23, February 1988; Ellis R. K. and Kunszt, Z., Nucl. Phys. B303, 653 (1988).

10. Although there are numerical errors in the MRS (Ref. 8) second order distributions functions, the qualitative features of their analysis for J/ψ production are not expected to be significantly affected. Private communication, J. Collins and W.-K. Tung.

11. Cooper-Sarkar, A. M., Ingelman, G., Long, K. R., Roberts, R. G., Saxon, D. H., *Measurement of the Longitudinal Structure Function and the Small-x Gluon Density of the Proton*, preprint RAL-87-112; Barbagli G., and D'Agostini, G., *Tagging of $\gamma g \to q\bar{q}$ events and direct measurement of the gluon structure function at HERA, Proceedings of the DESY Workshop on Physics at HERA*, DESY 88-104; Ingelman G. and Rückl, R., Phys. Lett. 201B, 369 (1988).

12. Ingelman, G. and Schuler, G. A., *Characteristics of Heavy Flavor Production in ep Collisions*, preprint DESY 88-020, February 1988.

13. Drees, M. and Godbole, R. M., Phys. Rev. Lett. 61, 682 (1988).

14. Schuler, G. A., Nucl. Phys. B299, 21 (1988); Baur, U. and van der Bij, J. J., Nucl. Phys. B304, 451 (1988); Eichler R. A. and Kunszt, Z., *Charm, Bottom and Top Production in High-Energy Electron Proton Collisions*, preprint ETHZ-IMP P/88-1, February 1988.

15. Nason, P, Dawson, S. and Ellis, R. K., Nucl. Phys. B303, 607 (1988).

16. Owens, J. F., Phys. Rev. D21, 54 (1980).

17. Drees, M. and Godbole, R. M., Phys. Rev. Lett. 61, 682 (1988).

18. Combridge, B. L., Kripfganz, J. and Ranft, J., Phys. Lett. 70B, 234 (1977).

19. Drees M. and Godbole, R. M., Univ. Wisconsin preprint MAD/PH/419, to appear in Phys. Rev. D.

20. See e.g., Glück, M., Grassie, K. and Reya, E., Phys. Rev. D30, 1447 (1984); Antoniadis A. and Grunberg, G., Nuc. Phys. B213, 445 (1983).

21. Drees M. and Halzen, F., Phys. Rev. Lett. 61, 275 (1988).

22. Drees, M., Halzen, F. and Hikasa, K., Univ. Wisconsin preprint MAD/PH/424 (1988).

23. For a review, see F. Halzen in *Proceedings of the 14th SLAC Summer Institute on Particle Physics*, Stanford, 1986.

24. For a review, see G. Wolf, DESY preprint 86-089 (1986).

25. For parametrizations of the parton densities inside the photon, see Duke D. W. and Owens, J. F., Phys. Rev. D26, 1600 (1982); Drees M. and Grassie, K., Z. Physik C28, 451 (1985).

26. T. Rizzo, these proceedings.

27. Drees, M., Mod. Phys. Lett. A2, 573 (1987).

28. Gabrielli, E., Mod. Phys. Lett. A1, 465 (1986).

29. Amaldi, U. *et al.*, Phys. Rev. D36, 1385 (1987).

30. Langacker, P. *Proceedings of the 1985 Int. Symp. on Lepton and Photon Interactions at High Energies*, Eds. M. Konuma and K. Takahashi, Kyoto Univ., Kyoto 1986, p.186.

31. Cornet, F. these proceedings, and Cornet, F. and Rückl, R. *Proceedings of the DESY Workshop on Physics at HERA*, DESY 88-104.

32. Framptom, P. H. and Glashow, S. L. Phys. Lett. 190B, 157 (1987); Phys. Rev. Lett. 58, 2168 (1987).

33. *a.* Robinson, R. W. and Rizzo, T. C., Penn State Univ Preprint PSU/TH/49 (1988), to be published in Phys. Lett. B.
b. Atwood, D. M. and Robinson, J. A., these proceedings. Robinson, R. W. and Rizzo, T. C., Penn State Univ Preprint PSU/TH/49 (1988), to be published in Phys. Lett. B.

34. Gabrielli, E., Mod. Phys. Lett. A1, 465 (1986).

35. Atwood, D., Baur, U., Couture, G., and Zeppenfeld, D., these proceedings.

36. Drees, M., these proceedings.

37. Baer, H., Ohnemus, J., and Zeppenfeld, D., in preparation.

38. Treille, D. *et al.*, in *Proceedings of the ECFA Workshop on LEP 200*, Aachen (1986), report CERN 87-08, vol. 2, p. 414; Zeppenfeld, D., Phys. Lett., 183B, 380 (1987).

39. Herzog, F., Phys. Lett. 148B, 355 (1984); Wallet, J. C., Phys. Rev. D32, 813 (1985); Grau, A,. Grifols, J. A., Phys. Lett. 197B, 437 (1987).

40. Suzuki, M., Phys. Lett. 153B, 289 (1985); van der Bij, J. J., Phys. Rev. D35, 1088 (1987).

41. Suzuki, M., Phys. Lett. 153B, 289 (1985); Baur, U. and Zeppenfeld, D., Phys. Lett. 201B, 383 (1988); Bilchak, C., Kuroda, M., and Schildknecht, D., CERN-TH. 4283/87.

42. Renard, F. M., Phys. Lett. 126B, 59 (1983); 139B, 449 (1984); Cabibbo, N., Maiani, L. and Srivastava,Y., Phys. Lett. 139B, 459 (1984); De Rūjula, A., Maiani, L. and Petronzio, R., Phys. Lett. 140B, 253 (1984); Kühn J., and Zerwas, P., Phys. Lett. 147B, 189 (1984).

43. Komamiya, S., *Proceedings of the 1985 International Symposium on Lepton and Photon Interactions at High Energies*, Kyoto, August 1985 (1985).

44. Renard, F. M., Phys. Lett. 116B, 264 (1982); del Aguila, F., Méndez, A. and Pascual, R., Phys. Lett. 140B, 431 (1984); Suzuki, M., Phys. Lett. 143B, 237 (1984).

45. Kühn J. and Zerwas, P., Phys. Lett. 147B, 189 (1984).

46. Hagiwara, K., Komamiya, S. and Zeppenfeld, D., Z. Physik C29, 115 (1985).

47. Cornet F. and Rückl, R., in *Proceedings of the Workshop on Physics at Future Accelerators*, La Thuile, Italy, 1987; CERN 87-07, Vol. 2, p. 287.

48. Haber, H. E. and Kane, G. L. Phys. Rep. 117, 75 (1985); and references therein.

49. Komatsu, H. and Rückl, R. DESY Rep 87-088, August 1987.

50. Bartels, J. and Holik, W. DESY Rep 87-174, December 1987.

51. Stanco, L., *Proceedings of the DESY Workshop on Physics at HERA*, DESY 88-104.

52. Davier, M. rapporteur talk at the XXIII International Conference on High Energy Physics, Berkeley, 16-23 July 1986.

53. Komatsu, H. and Rückl, R. DESY Rep. 87-089, August 1987.

54. The cross section for other processes e.g., $ep \to e\tilde{q}\tilde{g}X, ep \to e\tilde{q}\bar{\tilde{s}}qX$ become appreciable at these higher energies, though not quite compatible with the associated production process.

55. M. Drees and D. Zeppenfeld, these proceedings.

56. See *e.g.* L. E. Ibáñez, C. Lopez, and C. Muñoz,Nucl. Physics B256 218 1985 .

57. H. Komatsu and R. Rückl, Nucl. Physics B299 401 1988 .

58. See *e.g.* M. Drees, Nucl. Physics B298 333 1988 .

59. Chivukula, R. S. and Georgi, H., Phys. Lett. 188B 99 1987; Phys. Rev. D36 2102 1987 ; King, S. F., Phys. Lett. 195B 66 1987 .

60. For a review of superstring-inspired E_6 theories see, Hewett, J. L. and Rizzo, T. G., Univ. of Wisconsin Report MAD/PH/446 (1988).

61. Harnew, N., *Proceedings of the DESY Workshop on Physics at HERA*, DESY 88-104.

62. Gunion, J. F. and Ma, E., Phys. Lett. 195B 257 1987; Rückl, R., *Proceedings of the DESY Workshop on Physics at HERA*, DESY 88-104 and DESY Report DESY-87-109 (1987).

63. Ellis, J. and Pauss, F. *et al.*, *Proceedings of the Workshop on Physics at Future Accelerators*, La Thuile, Italy, 1987, CERN 87-07.

64. Buchmüller, W., Rückl, R., and Wyler, D., Phys. Lett. 191B, 442 (1987); Grifols, J. A. and Peris, S., Phys. Lett. 201B 287 1988 ; and see Rückl, R. in Ref. 62.

65. Bijnens, J., *Proceedings of the DESY Workshop on Physics at HERA*, DESY 88-104.

HADRON - HADRON COLLIDERS

SEARCH LIMITS FOR NEW GAUGE BOSONS AT HADRON COLLIDERS

JoAnne L. Hewett
Department of Physics, University of Wisconsin, Madison, WI 53706

and

Thomas G. Rizzo
Ames Laboratory and Department of Physics, Iowa State University, Ames, IA 50011

SUMMARY

We summarize the search limits for new neutral (Z') and charged (W') gauge bosons at the SSC and various Tevatron upgrades within the context of E_6 superstring-inspired models. We address the question of whether there is sufficient information available at hadron colliders to uniquely determine the couplings of Z' bosons and to distinguish between the various models.

Interest in the possibility of observing new neutral gauge bosons at hadron colliders has been renewed due to the advent of E_6 superstring-inspired models.[1] In our work below we will concentrate our discussion on two classes of models originating from E_6:

$$\begin{array}{ll}(A) & SU(3)_C \times SU(2)_L \times U(1)_Y \times U(1)_\theta\,, \\ (B) & SU(3)_C \times SU(2)_L \times SU(2)_R \times U(1)\,. \end{array} \qquad (1)$$

Models of type (A) are the so-called 'effective rank-5' models (ER5M) which contain a free parameter (θ) describing the amount of mixing between the Z_ψ and Z_χ states (from the decomposition $E_6 \to SO(10) \times U(1)_\psi \to SU(5) \times U(1)_\chi \times U(1)_\psi$) within the Z'. If the mixing between the Z' and standard model (SM) Z is ignored, then Z' is a mass eigenstate; this is an excellent approximation in general when the Z' mass is $\gtrsim 0.4$ TeV or so, which is the case of interest to us here. Models of type (B) are represented by the usual left-right model[2] (LRM) as well as the alternative left-right model[3] (ALRM). The LRM and ALRM are quite distinct in that their associated Z' bosons have very different couplings and that the right-handed charged gauge boson (W_R) can be light ($\simeq 210$ GeV) in the ALRM whereas it is expected to be heavy ($\gtrsim 2$ TeV) in the LRM. W_R in the ALRM also carries lepton number and negative R-parity so that it cannot be produced by a Drell-Yan-like process at hadron colliders.

In what follows we will address the following questions: (*i*) what are the search limits for Z' bosons in the various models at hadron colliders and can we distinguish between these models? (*ii*) Given the unusual nature of the W_R in the ALRM, what search limits can we find for its production? All relevant formulae can be found elsewhere.[1] In setting detailed search limits on the Z' we will assume a minimum of $N = 5$ events in both the e^+e^- and $\mu^+\mu^-$ channels. We will later comment on how our results change when a discovery limit of 10 events in each channel is assumed. These search limits will be quite model dependent since the rate for $p\overset{(-)}{p} \to Z' \to \ell^+\ell^-$ depends on the quark couplings to the Z' as well as its leptonic branching fraction (B) and these $Z'f\bar{f}$ couplings are dependent on the parameter θ in the ER5M. Even within a given model, this number may vary depending on, *e.g.*, whether exotic fermions in the E_6 **27**-representation can contribute ($n_g = 3$) or not ($n_g = 0$) to the Z' total width.

Figures 1 and 2 show the search limits for the Z' in the ER5M as a function of θ, assuming $N = 5$ and $n_g = 0$, at the Tevatron upgrades and the SSC, respectively. Only the range $-90° \le \theta \le 90°$ is shown since mapping $\theta \to \theta + 180°$ leaves the search limits invariant because it merely flips the sign of all Z' couplings. Thus, without Z-Z' mixing, this ambiguity in θ will always remain. The corresponding limits for the Z' in the LRM and the ALRM can be found in Table 1. The figures and the table show the trade-off between machine energy and luminosity as well as pp *vs.* $p\bar{p}$ at the Tevatron. Note that much stronger limits are obtained for both the LRM and ALRM since the overall strength of the couplings is greater in these two cases than in the ER5M.

How do our Z' search limits change if we give up the assumption of $n_g = 0$ and/or the assumption that $N = 5$ events in each lepton channel? Table 2 shows the limits (or range of limits in the ER5M) for several other values of (n_g, N) for each of the models under consideration. Note that these limits are significantly weaker for every model in comparison to the $n_g = 0$, $N = 5$ case already discussed. When combined with the $n_g = 0$, $N = 5$ case previously discussed, the results in the table generally show the variation one can expect in Z' limits at various hadron colliders within the E_6 model context.

Once produced, are these Z' bosons identifiable, *i.e.*, can we distinguish a Z' originating in the LRM or ALRM from one coming from the ER5M? Can we determine the angle θ uniquely? Certainly, in order to make a detailed study of the Z' boson's properties one needs several hundred to a thousand events and, hence, the Z' must be far lighter than the search

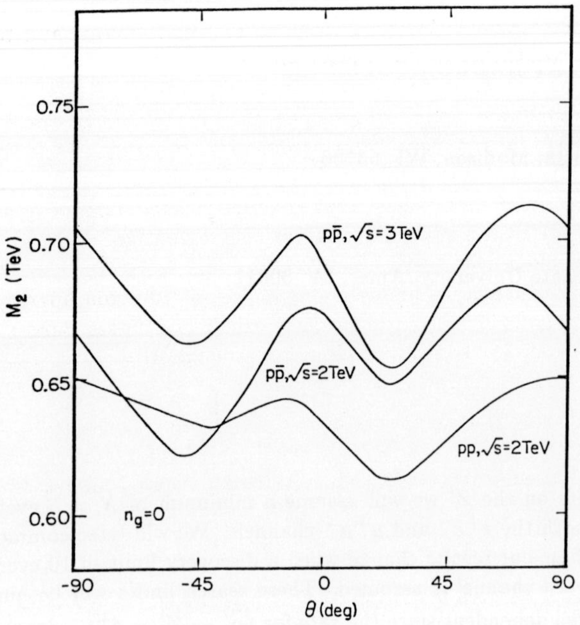

Fig. 1. Z' mass limits for the ER5M as a function of θ for the Tevatron uprades assuming $n_g = 0$ and $N = 5$ events.

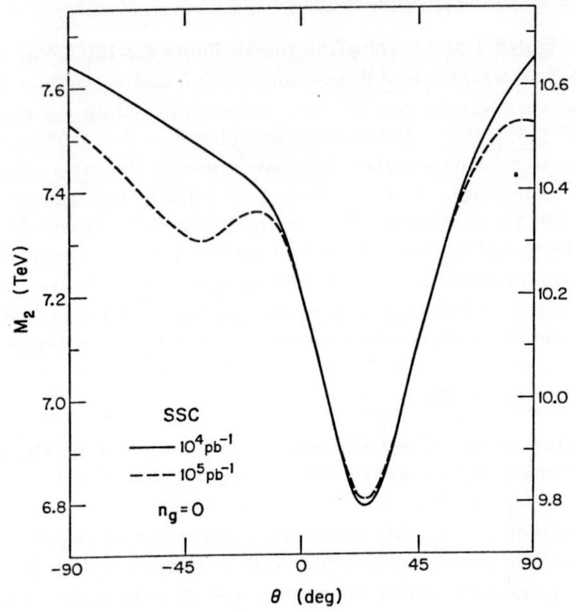

Fig. 2. Same as Fig. 1 but for the SSC with $10^4\,\text{pb}^{-1}$ or $10^5\,\text{pb}^{-1}$ integrated luminosity. For 10^4 (10^5) pb^{-1} the left- (right-) hand scale should be used.

limits discussed above. If $n_g = 0$, the only Z' properties useful in determining the couplings are the leptonic branching fraction (B_ℓ), the forward-backward asymmetry for leptons, A^ℓ_{FB},[4] and, to a lesser extent, the total Z' width, Γ_0. A measurement of Γ_0 cannot be done very accurately at hadron colliders due to the large energy resolution inherent in such machines. Hadronic Z' decay modes will be invisible due to the large QCD background and will hence be useless in determining the model. Table 1 and Fig. 3 show the values of A^ℓ_{FB} for the various models under consideration for particular values of the Z' mass. Note that in the scaling limit A^ℓ_{FB} is $M_{Z'}$ and $\sqrt{s}$ *independent* so that A^ℓ_{FB} is not expected to be very sensitive to changes in beam energy or the actual value of $M_{Z'}$. It is clear that a B_ℓ measurement, however precise, cannot be used to discriminate between models without an accurate knowledge of Γ_0. The value of A^ℓ_{FB} can be used to separate the ALRM and LRM but not the ALRM from the ER5M. The reason for this is that there always exists a value of θ in the ER5M for which $A^\ell_{FB}(\text{ER5M}) = A^\ell_{FB}(\text{ALRM})$ as can be easily seen from Fig. 3 and Table 1. Thus the sign of A^ℓ_{FB} will separate the LRM from the other two models under consideration but will not distinguish the ALRM and the ER5M.

Table 1. Z' search limits and A^ℓ_{FB} values in the LRM and ALRM for the various hadron colliders. In calculating A^ℓ_{FB} we take $M_{Z'}(= M_2)$ to be 450 GeV (3 TeV) for the Tevatron (SSC). $N = 5$ events and $n_g = 0$ are assumed. Z' masses are in units of GeV (TeV) for the Tevatron (SSC).

Machine		$\mathcal{L}$	Z' limit		A^ℓ_{FB} limit	
			LRM	ALRM	LRM	ALRM
pp	$\sqrt{s} = 2$ TeV	$10^3\,\text{pb}^{-1}$	695	760	0.123	−0.230
$p\bar{p}$	$\sqrt{s} = 2$ TeV	$10^2\,\text{pb}^{-1}$	750	865	0.205	−0.351
$p\bar{p}$	$\sqrt{s} = 3$ TeV	$30\,\text{pb}^{-1}$	810	970	0.199	−0.344
SSC		$10^4\,\text{pb}^{-1}$	8.30	9.40	0.103	−0.205
SSC′		$10^5\,\text{pb}^{-1}$	11.35	12.55	0.103	−0.205

Table 2. Z' search limits in the ER5M, LRM, and ALRM for $n_g = 3$ with $N = 5$, 10 events and for $n_g = 0$ with $N = 10$ events in each lepton channel for various hadron colliders. The Z' masses are in units of GeV (TeV) for the Tevatron (SSC).

Machine	Model	$N = 5$, $n_g = 3$	$N = 10$, $n_g = 0$	$N = 10$, $n_g = 3$
$p\bar{p}$ (2 TeV)	ER5M	515-640	575-655	445-565
$p\bar{p}$ (2 TeV)	LRM	685	680	610
$p\bar{p}$ (2 TeV)	ALRM	760	795	685
$p\bar{p}$ (3 TeV)	ER5M	480-655	595-675	400-565
$p\bar{p}$ (3 TeV)	LRM	710	705	615
$p\bar{p}$ (3 TeV)	ALRM	810	860	700
pp (2 TeV)	ER5M	525-625	585-630	475-575
pp (2 TeV)	LRM	645	645	595
pp (2 TeV)	ALRM	685	710	635
SSC	ER5M	5.35-7.20	6.35-7.20	5.05-6.35
SSC	LRM	7.50	7.40	6.65
SSC	ALRM	8.05	8.45	7.15
SSC′	ER5M	8.15-10.05	9.30-9.50	7.25-9.20
SSC′	LRM	10.45	10.40	9.50
SSC′	ALRM	11.15	11.60	10.20

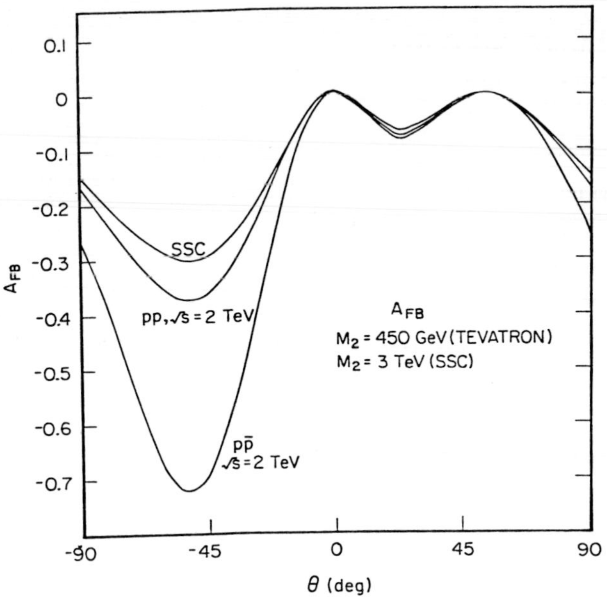

Fig. 3. A^{ℓ}_{FB} as a function of θ in the ER5M for the SSC and Tevatron upgrades with M_Z = 3 TeV and 450 GeV, respectively.

This situation improves somewhat for $n_g = 3$ since there are now additional, unstable, color singlet final states (exotics) accessible in Z' decay for which the QCD backgrounds are quite small. Although the Z' width in this case, Γ_3, would still be only poorly determined, the *relative* branching fractions between these exotic modes and $\ell^+\ell^-$ are Γ_3 independent and could be used to identify the model, at least in principle. If the exotic fermion masses could be determined (by reconstruction) and various exotics separated by their decay final states, then the LRM, ALRM, and ER5M could be easily distinguished and, when combined with A^{ℓ}_{FB} data, can be used to determine θ almost uniquely. The problems with this approach are that the decay modes of these exotics are very dependent on their quantum numbers (baryon and lepton numbers as well as R-parity) assignments, of which there are several possibilities, and that many of the exotics have similar decay patterns. In addition, for many assignments of these quantum numbers, there will be missing energy in the decay final state of these exotics making accurate reconstruction of their masses somewhat difficult. Still, this procedure may hold some promise in determining the couplings of the new Z' boson.

We now turn to W_R search limits in the ALRM. Due to the usual properties of the W_R discussed above it cannot be produced in the usual manner (*i.e.*, by quark-antiquark annihilation). The best mechanism for W_R production has been found to be[3] gluon $+ \overset{(-)}{u} \rightarrow W^{+}_{R}(W^{-}_{R}) + \overset{(-)}{h}$ where h is an exotic fermion with leptoquark quantum numbers and negative R-parity contained in the E_6 **27** representation. Since W_R and h get their masses from the same VEV one might expect them to have masses of comparable magnitude. In setting our limits, we have assumed the production of 10 $W^{\pm}_{R}\overset{(-)}{h}$ events and the details of the production signal have been discussed by Gunion *et al.* in Ref. 3. Table 3 shows a summary of these results. Note that our limits are roughly dependent on the sum of the W_R mass (M_R) and the exotic h fermion mass (m_h). Note also that the sum of the $W^{+}_{R}h$ and $W^{-}_{R}\bar{h}$ cross sections is the same at pp and $p\bar{p}$ machines so that $\sqrt{s}$ and machine luminosity are the only relevant factors (and not the parton luminosities) in determining these search limits. Also, if $M_R \simeq m_h$ then W_R production may be unobservable at the $p\bar{p}$ Tevatron upgrades since $M_R \gtrsim 210$ GeV from already existing data (see Barger *et al.* in Ref. 3). To really explore a reasonable W_R search range it is clear that the SSC is required.

It is quite clear that hadron colliders offer a unique opportunity to explore a large mass range in the search for the new gauge bosons expected in E_6 superstring-inspired models.

Table 3. Search limits for $W^{\pm}_{R}(\bar{h})$ production in the ALRM at various hadron colliders.

Machine		$\mathcal{L}$	$M_R + m_h$ Search limit
$p\bar{p}$	$\sqrt{s} = 2$ TeV	10^2 pb^{-1}	375-400 GeV
$p\bar{p}$	$\sqrt{s} = 3$ TeV	30 pb^{-1}	405-440 GeV
pp	$\sqrt{s} = 2$ TeV	10^3 pb^{-1}	500-525 GeV
	SSC	10^4 pb^{-1}	4.8-5.6 TeV
	SSC$'$	10^5 pb^{-1}	6.2-7.4 TeV

ACKNOWLEDGEMENTS

This research was supported in part by the University of Wisconsin Research Committee with funds granted by the Wisconsin Alumni Research Foundation, and in part by the U. S. Department of Energy under contracts DE-AC02-76ER00881 and W-7405-Eng-82. T.R. would like to thank the phenomenology group at the Univ. of Wisconsin-Madison for their hospitality.

REFERENCES

1. Hewett J. L. and Rizzo, T. G., Univ. Wisconsin report MAD/PH/446, submitted to Physics Reports (1988).
2. For a general survey of the left-right Symmetric Model, its basis in SO(10) and original references see Mohapatra, R. N., *Unification and Supersymmetry*, Springer-Verlaz, New York, 1986.
3. Ma, E., Phys. Rev. D36, 274 (1987); Mod. Phys. Lett. A3, 319 (1988); Babu, K. S., He, X.-G., and Ma, E., Phys. Rev. D36, 878 (1987); Gunion, J. F. *et al.*, *Proceedings of the Workshop From Colliders to Supercolliders*, Madison, WI, May (1987) and Int. J. Mod. Phys. A2, 119 (1987); Barger, V., and Whisnant, K., *ibid.* A3, 879 (1988).
4. The problems associated with a measurement of A^{ℓ}_{FB} for new Z' bosons has been discussed earlier by H. Haber, New W and Z Group Summary, *Proceedings of the 1984 Summer Study on the Design and Utilization of the SSC*, Snowmass, CO, 1984, edited by R. Donaldson and J. Morfin.

Searching for 100-300 GeV Gluinos at the Tevatron and SSC *

HOWARD BAER

Department of Physics, Florida State University, Tallahassee, FL 32306

XERXES TATA

Department of Physics and Astronomy, University of Hawaii, Honolulu, HI 96822

and Department of Physics, University of Wisconsin, Madison, WI 53706

JEFFREY WOODSIDE

Physics Department, Oklahoma State University, Stillwater, OK 74078

Summary

We examine event topologies from gluino pair production at various Tevatron upgrades, including the complete cascade decay of each gluino as given by the minimal supergravity model. We give expected rates for signal and background for 0 and 2 lepton(s) + n jet(s) + $\not{p}_t$ events. We also examine whether the process of $pp \to \tilde{g}$+ LSP yields an observable $\not{p}_t$ signature at the SSC, and conclude that it does only when $m_{\tilde{g}} \simeq m_{\tilde{q}}$ and $m_{\tilde{g}} \lesssim 300$ GeV.

I. Searching For Gluinos at a High Luminosity Tevatron

In this section we compute the signals [1] expected from the process $p\bar{p}$ (or pp) $\to \tilde{g}\tilde{g}$ X $\to q\bar{q}\tilde{\chi}_i + q\bar{q}\tilde{\chi}_j$+ X where the χ collectively denotes the two charginos ($\widetilde{W}_{l,h}$) and the four neutralinos ($\widetilde{Z}_{1,2,3,4}$) of the minimal model. These charginos and neutralinos then decay via the two body modes $\tilde{\chi}_i \to \tilde{\chi}_j$ + (V or H) (here V=$W^{\pm}$ or Z and H any Higgs scalar) if these are kinematically accessible; otherwise, they decay into a fermion pair (quark or lepton) and a lighter χ. The secondary χ again decays as described above unless it is the lightest neutralino, $\widetilde{Z}_1$, which we assume to live long enough to escape detection (and which we take to be the lightest SUSY particle, LSP). Thus the final state from $\tilde{g}\tilde{g}$ production consists of n quarks + m leptons in the central region ($n \geq 4$, $m \geq 0$) + $\not{p}_t$ (coming from the two $\widetilde{Z}_1$'s and any neutrinos that escape detection.)

The relevant branching fractions may be found in Ref. 2. For $m_{\tilde{g}} \gtrsim 100$ GeV, the branching fraction into the $q\bar{q}\widetilde{Z}_1$ mode is typically 0.2 - 0.6 unless the supersymmetric higgsino mass ($2m_1$) is small ($\lesssim M_W$) so that the LSP contains a substantial Higgsino component. We thus expect that analyses based on the assumption that the $\tilde{g}$ always decays via $\tilde{g} \to q\bar{q}$ + LSP are not valid for the range of gluino masses accessible at an upgraded (or even the present) Tevatron Collider or the SSC.

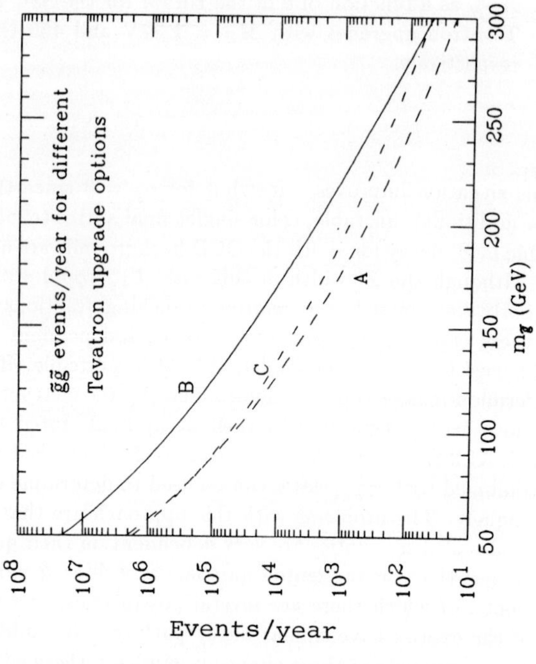

Figure 1. Expected rate of production of $\tilde{g}\tilde{g}$ events versus $m_{\tilde{g}}$ for the three different Tevatron upgrade options disucssed in the text. We take $m_{\tilde{q}} = 2\ m_{\tilde{g}}$.

Shown in Fig. 1 is the number of gluino pairs expected annually at (A) a 2 TeV $p\bar{p}$ collider with $\int Ldt = 100pb^{-1}$, (B) a 2 TeV pp collider with $\int Ldt = 1000pb^{-1}$, and (C) a 3 TeV $p\bar{p}$ collider with $\int Ldt = 30pb^{-1}$ which are representative of the various options for the upgrading to phase II of the Tevatron in the next decade. Here we have taken $m_{\tilde{q}} = 2\ m_{\tilde{g}}$. We see from Fig. 1 that option B gives the greatest number of gluino pairs unless $m_{\tilde{g}} > 300$ GeV.

* Research supported in part by the U.S. Department of Energy

In our computation of the various event topologies, we have incorporated the complete decay cascade of the $\widetilde{g}, \widetilde{W}_i$ and $\widetilde{Z}_i$ as given by the minimal supergravity model. Although the gluino production cross-sections do not depend on the parameters v'/v (the ratio of Higgs field vevs) and $2m_1$, that enter the chargino and neutralino mass matrices, the cross-section into each event topology, in principle, does depend on these parameters via the branching fractions. Except for small values of $2m_1$, the cross sections for multijet + 0,1 or 2 leptons are relatively insensitive to these parameters.

Small negative values of $2m_1$ are already disallowed as discussed in Ref. 3. For definiteness, we have taken $2m_1 = 150$ GeV, $v'/v = 2/3$ and $m_{\widetilde{q}} = 2\, m_{\widetilde{g}}$ as "typical values" in our calculation. In terms of these parameters (and $m_{\widetilde{g}}$) all the cross-sections can be calculated.

Our calculations have been performed at the parton level. We have coalasced two partons within Δ r ($\equiv \sqrt{\Delta\eta^2 + \Delta\varphi^2}$) < 0.7 into a single jet and have required $|\ \eta_{jet}\ | < 2.5$. For zero lepton topologies, we require a trigger $\not{p}_t > 50$ GeV. For events containing one or more hard ($p_{lt} > 20$ GeV) and central ($|\ \eta_e\ | < 3$, $|\ \eta_\mu\ | < 0.76$) charged leptons, we have relaxed the $\not{p}_t$ cut down to 20 GeV. We further require that the leptons be isolated (no partons with $\sum E_T > 3$ GeV in a cone of Δ r = 0.4 about each lepton).

The results of our calculation for n jet + m lepton events at a 2 TeV $p\overline{p}$ collider are shown in Figs. 2 and 3 for m=0 and 2, respectively. Dominant standard model backgrounds are listed in Table 1.

TABLE 1

Background rates from standard model processes for the signals shown in Figures 2 and 3. All numbers are in picobarns.

# Jets	Z	W→ τ	t(60)	t(120)	Tot.(60)	Tot.(120)
1	23	7	9	0.3	39	30
2	11	20	13	0.9	44	32
3	3	4	10	1.4	17	8
4	0.5	0.8	3	1	4	2
5	.02	0.06	0.2	0.2	0.3	0.3

ISOLATED DILEPTONS + n JETS + $\not{p}_t$ ($\not{p}_t > 20$ GeV)

# JETS	t(60)	t(120)
1	7	0.3
2	4	0.5
3	0.7	0.2
4	–	0.002

For brevity, we have not shown the corresponding cross-sections for options B and C discussed above. These can be approximated by scaling the cross-sections in Figs. 2 and 3 by a ratio of the production rates that can be obtained from Fig. 1. We have computed the following Standard Model backgrounds to the signals:

(*a*) Zero lepton events: Z + "jets", W → τ + "jets" and $t\bar{t}$ production.

(*b*) 1 - lepton events: W → e or μ + "jets" and $t\bar{t}$ production

(*c*) 2 - lepton events: $t\bar{t}$ production

For the background computation, we used a shower Monte-Carlo described in Ref. 4 for W and Z processes and the same experimental criteria as for the signal. We also computed the background from $t\bar{t}$ production for $m_{\widetilde{t}}$ = 60 and 120 GeV. The following comments are in order:

1. For $m_{\widetilde{g}} \lesssim 120$ GeV, dijets and trijets dominate the signal, while four and five jet events dominate for $m_{\widetilde{g}} > 120$ GeV. The dijet (trijet) rate exceeds background for $m_{\widetilde{g}} < 90$ (115) GeV, whereas four (five) jet events exceed background for $m_{\widetilde{g}} < 150$ (200) GeV. The large rate for $\not{p}_t$ events with five or more jets is a direct consequence of allowing for the complete gluino cascade. Running under option B, we would expect about 600 five jet events/year for $m_{\widetilde{g}} = 200$ GeV. We expect that by tuning the cuts to optimize signal/background, option B should easily probe

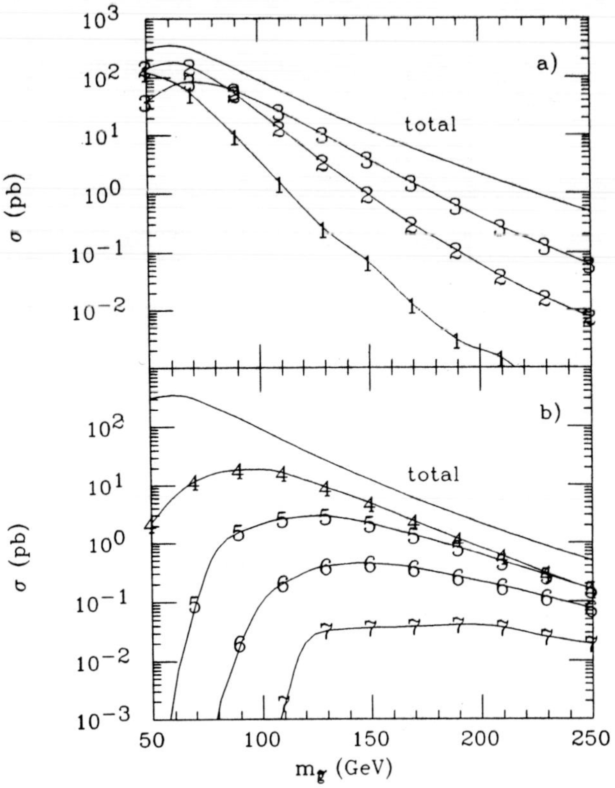

Figure 2. Cross-section for producing n-jet + $\not{p}_t$ events from gluino pairs versus $m_{\widetilde{g}}$ at a 2 TeV $p\overline{p}$ Collider. We take $m_{\widetilde{q}} = 2\ m_{\widetilde{g}}$, and require $\not{p}_t > 50$ GeV. Along with the total rate, we show (a) n = 1-3 and (b) n = 4-7 jet topologies.

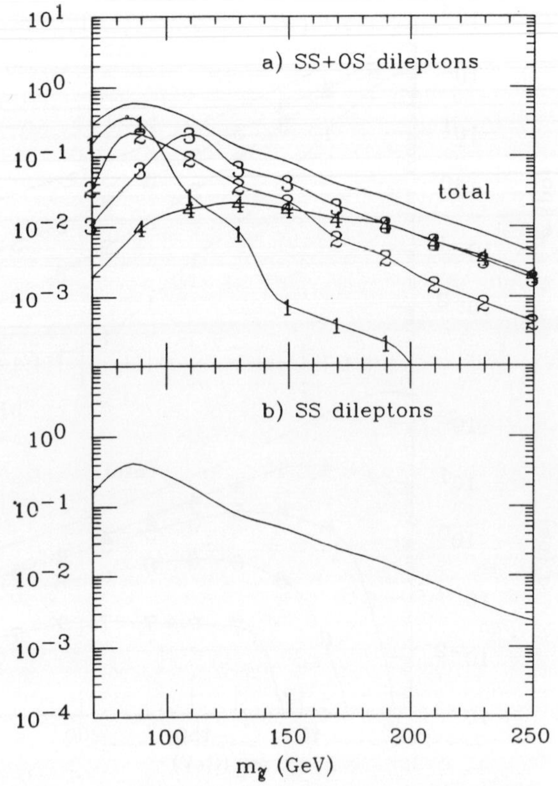

Figure 3. Cross-section for producing isolated dilepton plus n-jet events in $p\overline{p}$ collisions at $\sqrt{s}$ = 2 TeV. Cuts are as described in the text. We show (a) all dileptons (summed over $e, \overline{e}, \mu, \overline{\mu}$) and (b) same-sign dilepton rates.

$m_{\tilde{g}}$ = 200 - 250 GeV.

2. The backgrounds for 1 lepton plus n jets + $\not{p}_t$ events appear to exceed signal for all choices of $m_{\tilde{g}}$, hence we have not illustrated this case. However, in the present analysis we have made no attempt at optimizing signal/background. Cutting on quantities such as $M_T(e,\not{p}_t)$ or jet p_t may help in this matter. This is under study.

3. Signals for two isolated leptons[5] + n jets + $\not{p}_t$ exceed background only for n $\geq$ 4. The cross-sections for two isolated leptons + 4 jets are typically in the range of ~0/400.01 pb so that at most a handful of these events may be expected. The cross-section for isolated same sign dileptons can exceed 0.1 pb and is as large as 0.01 pb even for $m_{\tilde{g}}$ = 200 GeV. We were unable to find any background from $t\overline{t}$ for this class of events. Barnett et al. (Ref. 5) in these proceedings perform a more detailed analysis with different cuts, and also find optimistic results. Although the rate for these events is too small to envision that SUSY would be discovered in the dilepton channel, a handful of these clean events could lead to a striking confirmation of a signal discovered in another channel.

To summarize, from the theorists perspective option B (a Tevatron upgrade to a pp collider operating at $\sqrt{s}$ = 2 TeV and $\int$ Ldt = 1000 pb^{-1}) gives the highest event rate for gluinos if $m_{\tilde{g}} < 300$ GeV. Backgrounds should be more manageable because of lower W and Z production rates relative to a $p\overline{p}$ collider. Incorporating complete gluino cascade sequences in our programs yields new signatures for gluinos. The most promising are (1) multi-jet + $\not{p}_t >$ 50 GeV events (in particular, five jet + $\not{p}_t$ events allow one to probe $m_{\tilde{g}} > 200$ GeV (2) isolated same sign dileptons + $\not{p}_t$ and (3) isolated dileptons + four or more jets + $\not{p}_t$. Each of these signals should be present in a data sample if discovery of gluinos is to be claimed. More detailed studies of these signals are in progress.

II. Gluino Production in Association with the LSP at the SSC

Here, we focus on the process $pp \to \tilde{g}\tilde{Z}_1$, where $\tilde{g} \to q\overline{q}\tilde{Z}_i$, or $q\overline{q}\widetilde{W}_i^{\pm}$. Production would be signalled by jets in one hemisphere recoiling against a very large $\not{p}_t$. As we will see, the $\not{p}_t$ backgrounds from strong interaction processes within the SM can be eliminated by simple kinematic cuts. This process may, therefore, provide an interesting way to search for relatively light gluinos at the SSC, and, in particular, provide a partial answer to the question, "Is there a window of gluino masses that cannot be explored either at an upgraded Tevatron or the SSC?"

$\tilde{g}\tilde{Z}_1$ production at the SSC proceeds via $q\overline{q}$ annihilation mediated by $\tilde{q}_L$ and $\tilde{q}_R$ exchanges in the t- and u-channels. The production rate, therefore, depends sensitively on $m_{\tilde{q}}$. It also depends on the parameter $2m_1$ and v'/v that enter the $q\tilde{q}\tilde{Z}_1$ couplings via the neutralino mass matrix. We have checked, however, that $\sigma_{\tilde{g}\tilde{Z}_1}$ depends only weakly on $2m_1$ and v'/v except for a small region of parameter space where $\tilde{Z}_1$ is almost a higgsino. For definiteness, we have taken $2m_1 = -400$ GeV, v'/v = 2/3 and $m_{\tilde{q}} = 2\ m_{\tilde{g}}$ in our analysis. Except for the "higgsino region" discussed above, $\sigma_{\tilde{g}\tilde{Z}_1} = (6.5 \pm 1.0)$ pb for $m_{\tilde{g}}$ = 220 GeV and $\sigma_{\tilde{g}\tilde{Z}_1}$= (2 $\pm$0.3pb) for $m_{\tilde{g}}$ = 300 GeV. These rates are considerably larger (by a factor of 3-4 for $m_{\tilde{q}} = m_{\tilde{g}}$) for lighter squarks. If we assume $m_{\tilde{q}} \simeq 2\ m_{\tilde{g}}$, ~ few x 10^4 $\tilde{g}\tilde{Z}_1$ events are expected annually at the SSC for $m_{\tilde{g}}$ = 200 - 300 GeV, the maximum reach of an upgraded Tevatron.

The $\not{p}_t$ spectrum from $\tilde{g}\tilde{Z}_1$ production is shown in Fig. 4 for $m_{\tilde{g}}$ = 220 GeV. Also shown is the $\not{p}_t$ spectrum from $\tilde{g}\tilde{g}$ events and from "Z $\to \nu\overline{\nu}$ + jet(s)" events. In our computation of the SUSY processes we have used a parton level Monte Carlo program that incorporates the complete cascade decay of the gluinos and the subsequent $\tilde{Z}_i$ and $\widetilde{W}_i^{\pm}$. The "Z + jet" cross-section has been computed using the shower Monte Carlo program of Ref. 6. We see that $\tilde{g}\tilde{Z}_1$ production is completely swamped by the other processes for the whole range of $\not{p}_t$. In order to isolate the $\tilde{g}\tilde{Z}_1$ signal, we have further required that there be no hadronic activity with $E_T > 30$ GeV outside a 90° cone (in the transverse plane) back to back with the $\not{p}_t$ vector.

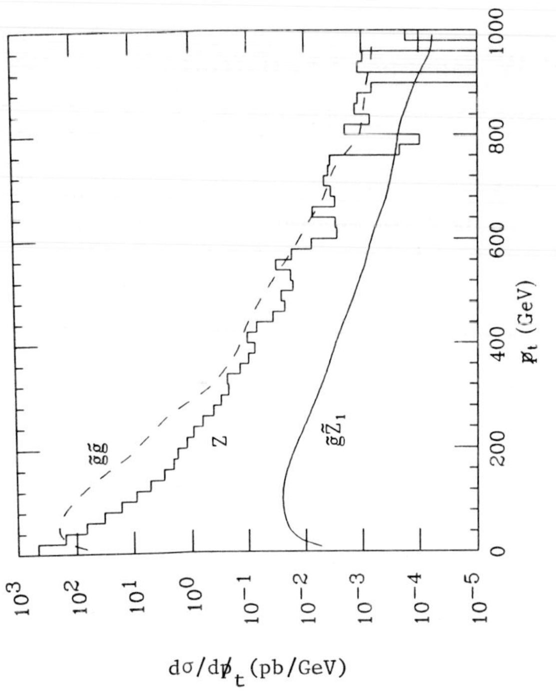

Figure 4. $\not{p}_t$ spectrum from $\tilde{g}\tilde{g}$, $\tilde{g}\tilde{Z}_1$, and Z + jets (where Z $\to \nu\nu$) produced in pp collisions at $\sqrt{s}$ = 40 TeV. We have taken $m_{\tilde{g}}$ = 220 GeV, v'/v = 2/3 and $2m_1$ = -400 GeV.

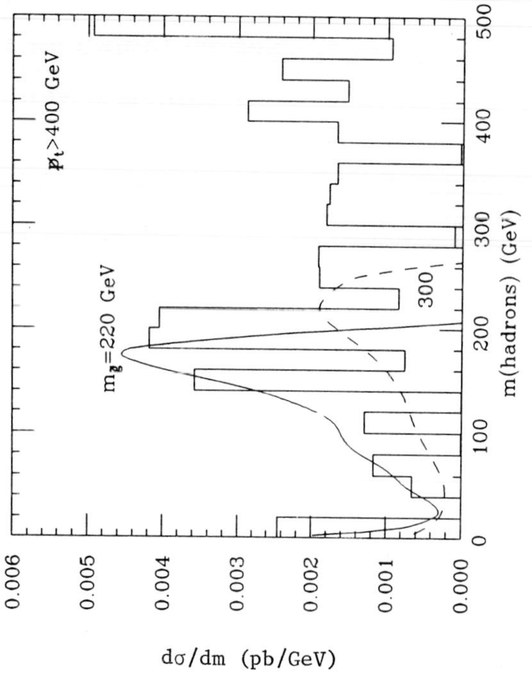

Figure 5. Invariant mass of all hadrons in the transverse hemisphere opposite the $\not{p}_t$ vector. We require $\not{p}_t$ > 400 GeV and isolation. The histogram is the contribution from Z + jets where Z $\to \nu\bar{\nu}$. The solid line is from $\tilde{g}\tilde{Z}_1$ with $m_{\tilde{g}}$ = 220 GeV while the dashed line is for $m_{\tilde{g}}$ = 300 GeV.

We find that this isolation requirement on the $\not{p}_t$ vector completely cuts out all $\tilde{g}\tilde{g}$ events if we also require $\not{p}_t >$ 400 GeV. We expect that this requirement cuts out the $\not{p}_t$ background from heavy flavour production with even a greater efficiency. The dominant SM background, therefore, comes from $Z \to \nu\bar{\nu}$ + jets.

The distribution of the invariant mass of the hadronic system for those events with $\not{p}_t$ > 400 GeV and which satisfy the $\not{p}_t$ isolation requirement is shown in Fig. 5 for $m_{\tilde{g}}$ = 220 and 300 GeV. $2m_1$, v'/v and $m_{\tilde{g}}$ are fixed as stated earlier. Also shown is the background distribution from the shower calculation of Z^o + jets, for which we ran 10^6 events, corresponding to 34 CPU hours on a VAX 8700.

About 4000 (2300) $\tilde{g}\tilde{Z}_1$ events pass our cuts for $m_{\tilde{g}}$ = 220 GeV(300 GeV) assuming an integrated luminosity of $10^4 pb^{-1}$. The bulk of the background has $m \geq 300$ GeV so that requiring $m \leq 300$ significantly enhances the signal to background for the case of the relatively light gluinos. We see from Fig. 5 that signal to background $\sim$ unity may be possible, but this should be taken with a grain of salt as there is a considerable uncertainty in our estimate of the background (coming from the fact we are sampling a small tail of the background distribution). We have checked that increasing the $\not{p}_t$ cut to 600 GeV does not significantly improve the situation.

The conclusions of our analysis are rather pessimistic. The observation of a $\tilde{g}$ via $\tilde{g}\tilde{Z}_1$ production at the SSC is, at best, marginal. If the sfermions are all heavy ($\sim$1TeV) with only the gluinos light, $\tilde{g}\tilde{Z}_1$ production is completely obscured by the Z^o + jet background. Only in the rather fortuitous case where $m_{\tilde{q}} \simeq m_{\tilde{g}}$ can we conclude that the signal is substantially larger than the background. Of course, in this case, one would obtain an additional contribution to the signal from $\tilde{q}\tilde{Z}_1$ production.

References

1. H. Baer, J. Ellis, G. Gelmini, D. Nanopoulos and X. Tata, Phys. Lett. **161B**, 175 (1985); G. Gamberini, Z Phys. **C30**, 605 (1986); H. Baer, V. Barger, D. Karatas and X. Tata, Phys. Rev. **D36**, 96 (1987); R. Barnett, J. Gunion and H. Haber, Phys. Rev. **D37**, 1892 (1988).
2. J. Gunion, et.al., in From Colliders to Supercolliders, pg. 255 edited by V. Barger and F. Halzen, published by World Scientific; J. Gunion and H. Haber, Phys. Rev. **D37**, 2515 (1988).
3. H. Baer, K. Hagiwara and X. Tata, Phys. Rev. **D38**, 1485 (1988).
4. H. Baer, V. Barger, H. Goldberg and R. Phillips, Phys. Rev. **D37**, 3152 (1988).
5. V. Barger, W.Y. Keung and R. Phillips, Phys. Rev. Lett. **55**, 166 (1985); see R. Barnett, J. Gunion and H. Haber, these proceedings.
6. V. Barger, T. Gottschalk, J. Ohnemus and R. Phillips, Phys. Rev. **D32**, 2950 (1985).

Search Limits for Heavy Z′ and W′ Bosons of Left-Right Symmetric Models for Different Tevatron Upgrades

J.A. GRIFOLS[a], A. MÉNDEZ[a] AND R. MICHAEL BARNETT[b*]

[a]*Grup de Física Teòrica, UAB, 08193 Bellaterra (Barcelona), Spain*

[b]*Physics Division, Lawrence Berkeley Laboratory, Berkeley, CA 94720*

In this paper we summarize the potential discovery limits for new heavy bosons ($W'^{\pm}$ and Z'^{0}) at the SSC and at an upgraded Tevatron Collider. We consider the production of heavy gauge bosons of Left-Right (LR) Symmetric gauge models[1)] in pp and $p\bar{p}$ collisions with their subsequent decays into lepton pairs. More precisely, we focus on the processes

$$\begin{Bmatrix} pp \\ p\bar{p} \end{Bmatrix} \to \begin{Bmatrix} Z'^0 \to e^+e^- \\ W'^{\pm} \to e^{\pm}\nu_R \end{Bmatrix} .$$

To be specific we chose a particular model[2)] in which the vertex $Z'ff$ is given by: $g\gamma_\mu(G_L P_L + G_R P_R)$ where $P_L = (1-\gamma_5)/2$, $P_R = (1+\gamma_5)/2$ and where

$$\begin{aligned}
G_L(Z'uu) &= -C\sin^2\theta_W/6\\
G_R(Z'uu) &= -C(7\sin^2\theta_W - 3)/6\\
G_L(Z'dd) &= -C\sin^2\theta_W/6\\
G_R(Z'dd) &= C(5\sin^2\theta_W - 3)/6\\
G_L(Z'ee) &= C\sin^2\theta_W/2\\
G_R(Z'ee) &= C(3\sin^2\theta_W - 1)/2\\
G_L(Z'\nu\nu) &= C\sin^2\theta_W/2\\
G_R(Z'\nu\nu) &= -C(\sin^2\theta_W - 1)/2
\end{aligned}$$

where $C \equiv 1/(\cos\theta_W(\cos 2\theta_W)^{\frac{1}{2}})$. In this model assuming three light generations including the right-handed neutrino ν_R, the width of the new boson is $\Gamma_{Z'} = \alpha C^2 M_{Z'}(1 - 4\sin^2\theta_W + \frac{14}{3}\sin^4\theta_W)/\sin^2\theta_W$ (using $g^2 = 4\pi\alpha/\sin^2\theta_W$).

To evaluate the branching ratios of the Z' we note that the possible decay channels are $e_i^+e_i^-$, $\nu_i\bar{\nu}_i$, $u_i\bar{u}_i$, $d_i\bar{d}_i$, W^+W^- and ZH, where $i = 1,2,3$ is a generation index and H is the light "standard" Higgs boson of the LR models. In the usual LR model with a scalar sector consisting of one bidoublet and two triplets, the last two channels are always allowed and it can be shown[2)] that the mixing $Z_L - Z_R$ (which is fixed by the masses and is very small) leads to $BR(Z' \to WW) \approx BR(Z' \to e\ + e\ -)$. On the contrary, the mixing between W_L and W_R can vary from zero to some upper bound depending on the relative values of the bidoublet vevs κ and κ'. In fact the zero value for the mixing (corresponding to $\kappa' = 0$) is required phenomenologically[3)] . For the W' the possible decay channels then are $e_i\bar{\nu}_{Ri}$ and $d_i\bar{u}_i$ (*i.e.*- we do not consider the channels WZ and WH since they would only be allowed if the $W_L - W_R$ mixing were not zero).

The total width is obtained summing over the partial widths of all these channels[2)] . The branching ratios of interest are then

$$BR(Z'^0 \to e^+e^-) = 1.8\% \ , \quad BR(W'^{\pm} \to e^{\pm}\nu_R) = 8\% \ .$$

The exclusion of the WW and ZH channels would increase $BR(Z' \to e^+e^-)$ from 1.8 to 1.9% . If ν_R is very heavy, then the branching ratio would be increased to 2.3% .

The production rates are shown in figs. 1,2,3 for different upgrade options of the Tevatron. For reference, the corresponding rates for the SSC are shown in fig. 4. The rapidity cuts are introduced in the initial quark-antiquark system. Figure 5 shows the sensitivity of the cross sections to the rapidity cut for the SSC case.

We would like to thank Howard Haber for valuable conversations and for checking our calculations.

* This work was supported by the Director, Office of High Energy and Nuclear Physics, Division of High Energy Physics, of the U.S. Department of Energy under Contract No. DE-AC03-76SF00098, and by the National Science Foundation under agreement no. PHY83-18358.

REFERENCES

1. J.C. Pati and A. Salam, *Phys. Rev.* **D10**, 275 (1974); R.N. Mohapatra and J.C.Pati, *Phys. Rev.* **D11** 566, 2558 (1975); G. Senjanovic and R.N. Mohapatra, *Phys. Rev.* **D12**, 1502 (1975); R.N. Mohapatra and G. Senjanovic, *Phys. Rev. Lett.* **44**, 912 (1980) and *Phys. Rev.* **D23**, 165 (1981). For a recent review, see R.N. Mohapatra in *"Quarks, Leptons and Beyond"*, ed. H. Fritzsch *et al.*, Plenum, New York (1985), p. 217.
2. N.G. Deshpande, J.A. Grifols and A. Méndez, *Phys.Lett.* **208B**, 141 (1988).
3. J.F. Gunion, J.A. Grifols, A.Mendez, B. Kayser and F. Olness; UCD preprint in preparation.

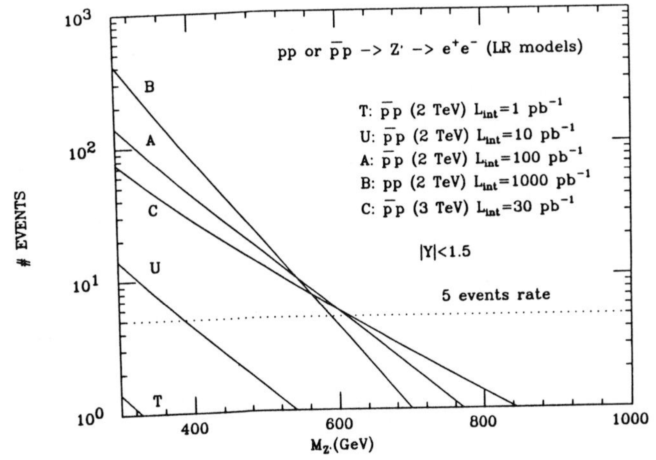

Figure 1. The production rates for Z'^0 from the model of Ref. 2 at the Tevatron Collider.

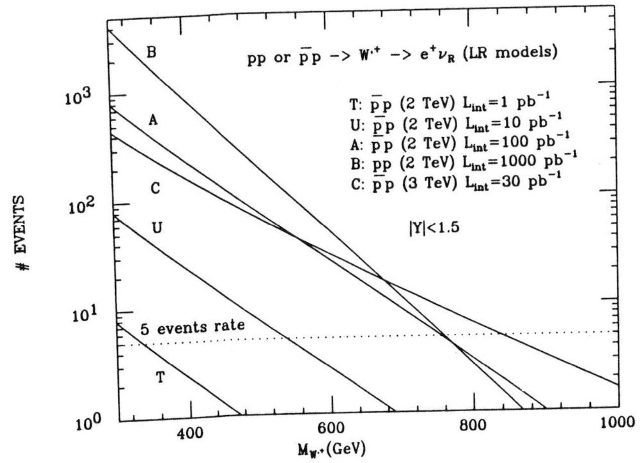

Figure 2. The production rates for W'^+ from the model of Ref. 2 at the Tevatron Collider.

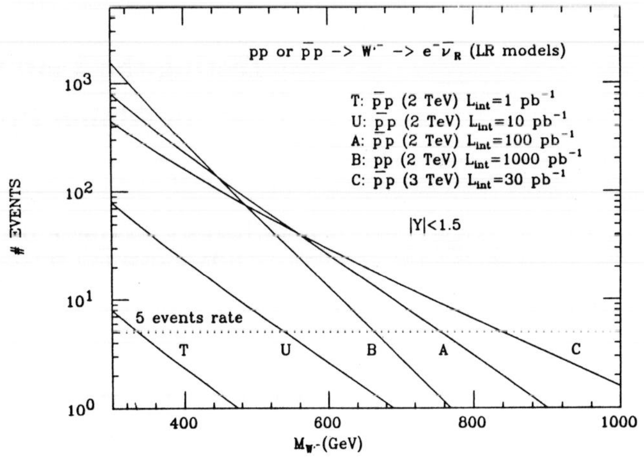

Figure 3. The production rates for W'^- from the model of Ref. 2 at the Tevatron Collider. Note that all rates are the same for W'^- as for W'^+ for $p\bar{p}$ colliders, but that the rates for W'^- are a factor of about two lower at pp machines.

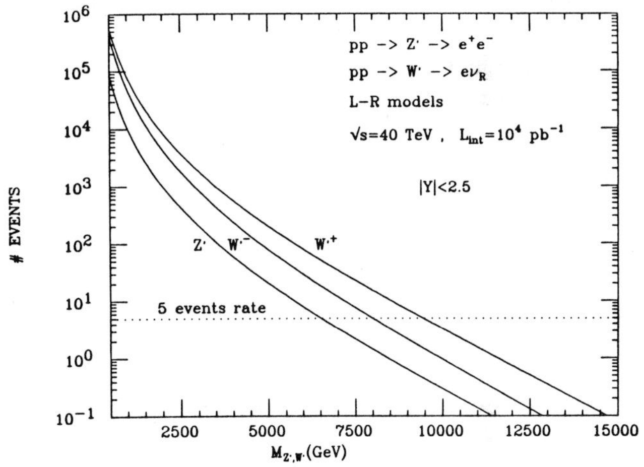

Figure 4. The production rates for Z'^0 and $W'^\pm$ from the model of Ref. 2 at the SSC.

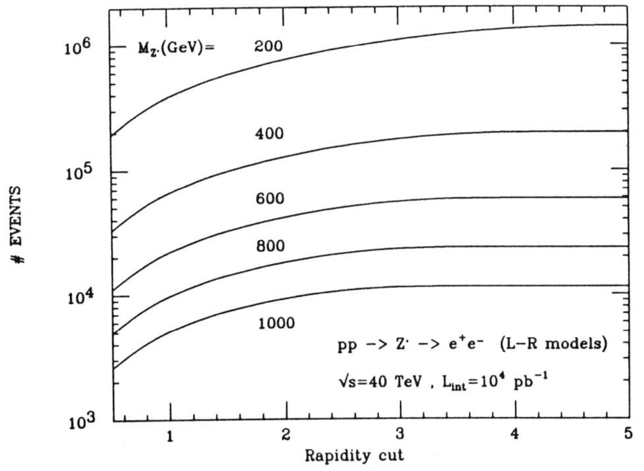

Figure 5. The dependence on the rapidity cut of the production rates for Z'^0 from the model of Ref. 2 at the SSC.

Testing the Viability of the E_T^{miss} Signature in Gluino Production at the SSC

R.M. Barnett, Lawrence Berkeley Laboratory, Berkeley, CA 94720
J. Freeman and R. Raja, Fermilab, Batavia, IL 60510
J.F. Gunion, University of California, Davis, CA 95616
H.E. Haber, University of California, Santa Cruz, CA 95064
R. Hollebeek, University of Pennsylvania, Philadelphia, PA 19104
F.E. Paige, Brookhaven National Laboratory, Upton, NY 11973
A.P. White, University of Florida, Gainesville, FL 32611

Abstract

We report on a preliminary study at SSC energies of signatures and backgrounds for heavy gluinos in the minimal supersymmetric extension of the Standard Model. Unlike previous studies we account for the small branching ratio of gluinos to the lightest supersymmetric particle. We find that an E_T^{miss} signature is observable over the Standard Model background.

Introduction

Supersymmetry is an attractive possible extension of the Standard Model both because of its mathematical elegance and because it provides a natural mechanism for keeping Higgs bosons light even in the presence of a superheavy mass scale such as a grand unification scale or the Planck scale. It has therefore been extensively studied both for existing and for future colliders. The best experimental limits on the masses of gluinos ($\widetilde{g}$) and squarks ($\widetilde{q}$) come from the UA1 Collaboration[1] at the S$\bar{p}$pS and the CDF Collaboration[2] at the Tevatron. Based on studies of events with jets and substantial E_T^{miss}, they find the following limits (independent of the other mass)

$$\begin{aligned} &M_{\widetilde{g}} > 53\,\text{GeV}, \quad M_{\widetilde{q}} > 45\,\text{GeV} \qquad (\text{UA1}) \\ &M_{\widetilde{g}} > 73\,\text{GeV}, \quad M_{\widetilde{q}} > 74\,\text{GeV} \qquad (\text{CDF}) \end{aligned} \tag{1}$$

In this paper we report a preliminary study of the E_T^{miss} signature and backgrounds for studying supersymmetry for higher masses at SSC energies.

The experimental analyses and most previous Monte Carlo studies[3] have concentrated on the simplest possible decay modes, namely $\widetilde{g} \to \widetilde{\gamma} q\bar{q}$ if the gluino is lighter than the squark, or $\widetilde{q} \to \widetilde{\gamma} q$ if the squark is lighter than the gluino. The photino $\widetilde{\gamma}$ has a neutrino-like cross section and escapes from the detector, so the signature for these modes is E_T^{miss} plus multiple jets. These simple modes are usually dominant for masses up to $60 - 70\,\text{GeV}$, but other modes become dominant for the heavier masses of interest at the SSC.[4,5] The reason is that any supersymmetric extension of the Standard Model contains not just a single photino but at least four neutral color singlet fermions $\widetilde{\chi}_i^0$ and two charged ones $\widetilde{\chi}_i^+$. If the gluino mass is heavy, then decays into some or all of these are possible, so the branching ratio for a direct decay into any one is small. Heavier $\widetilde{\chi}$ particles eventually decay into the lightest supersymmetric particle (LSP) $\widetilde{\chi}_1^0$, which like a photino escapes from the detector. The cascade decays reduce the E_T^{miss}, but they also introduce many interesting new signatures. We should emphasize that it is not necessary for *both* gluinos to decay directly to the LSP in order to obtain large E_T^{miss}.

In our analysis we have considered only gluino pair production and decay, since squark decays are less affected by the existence of the new modes. The main result of our preliminary analysis is that even with the more complicated decays predicted by the minimal supersymmetric model, the E_T^{miss} signature remains observable. We will refine this result and study additional signatures in a future publication.

Minimal Supersymmetric Model

To give masses to up and down quarks in the minimal supersymmetric model,[6] it is necessary to introduce two Higgs doublets, resulting in three neutral and one charged pair of Higgs bosons after symmetry breaking. The supersymmetric partners of the Higgs bosons mix with the $\widetilde{\gamma}$ and $\widetilde{Z}$ to make four states $\widetilde{\chi}_i^0$, $i = 1,2,3,4$, which are known collectively as neutralinos and are numbered in increasing order of mass. Similarly the supersymmetric partner of the charged Higgs boson mixes with the $\widetilde{W}$ to make two chargino states $\widetilde{\chi}_i^+$, $i = 1,2$. We shall assume that R parity is conserved so that all supersymmetric particles are produced in pairs; each supersymmetric particle then decays into lighter supersymmetric particles, and the cascade continues until the LSP ($\widetilde{\chi}_1^0$) is produced which is absolutely stable. As already remarked, this particle escapes from any collider detector.

In the minimal model all of the masses and mixings of the charginos and neutralinos are determined in terms of four parameters. One of these can be eliminated by assuming that the gaugino masses are degenerate at the grand unification or Planck scale. The remaining three parameters can be taken to be the gluino mass $M_{\widetilde{g}}$, a supersymmetric Higgs mass μ, and a ratio $\tan\beta = v_2/v_1$ of Higgs vacuum expectation values. If supersymmetry is related to the electroweak scale, we expect both $M_{\widetilde{g}}$ and μ to be less than about 1 TeV; $\tan\beta$ should be of order one. We shall assume that the squarks are heavier than the gluino; to be definite we choose $M_{\widetilde{q}} = 2M_{\widetilde{g}}$ although the precise values of the squark masses have little effect on these results. Once the chargino and neutralino mixing angles (and one Higgs mass) are known, the couplings of the mass eigenstates $\widetilde{\chi}_i$ are determined, and all branching ratios can be calculated. The resulting branching ratio for $\widetilde{g} \to \widetilde{\chi}_1^0 q\bar{q}$ is shown in Fig. 1 for a typical value of $\tan\beta$. This branching ratio is usually large for light gluinos but is always less than about 20% for gluino masses above 300 GeV (see ref. 5). Even if the details are not correct, this minimal model presumably provides a representative example of supersymmetric models. In more complicated (non-minimal) models such small BR's would also be expected.

Supersymmetric Signatures and Backgrounds

To simulate the signatures it is necessary to choose specific parameters. The values of μ and $\tan\beta$ are not very crucial because the branching ratio of the gluino to the LSP depends only weakly on them, although other branching ratios vary somewhat more. We have chosen three gluino masses and three sets of values for the other parameters in the minimal supersymmetric model:

$$\begin{aligned} &i)\ M = 150\,\mathrm{GeV},\ \mu = -150\,\mathrm{GeV},\ \tan\beta = 1.5;\\ &ii)\ M = 220\,\mathrm{GeV},\ \mu = +400\,\mathrm{GeV},\ \tan\beta = 1.5;\\ &iii)\ M = 750\,\mathrm{GeV},\ \mu = +150\,\mathrm{GeV},\ \tan\beta = 1.5. \end{aligned} \tag{2}$$

The first mass may be observable at the present Tevatron; the second might be observed either at an upgraded Tevatron or at the SSC; the third is typical of the SSC mass range. In this paper we will discuss only the 750 GeV case in detail.

The couplings of gluinos to gluons and quarks are completely specified by supersymmetry, and hence the $\widetilde{g}\widetilde{g}$ production cross section is calculable given the gluino mass (with weaker dependence on squark mass). For $M_{\widetilde{g}} = 750\,\mathrm{GeV}$ at $\sqrt{s} = 40\,\mathrm{TeV}$ the total $pp \to \widetilde{g}\widetilde{g}X$ cross section is

$$\sigma_{\widetilde{g}\widetilde{g}} \simeq 70\,\mathrm{pb} \tag{3}$$

All the decay modes are similarly calculable. For the 750 GeV case the principal gluino branching ratios are

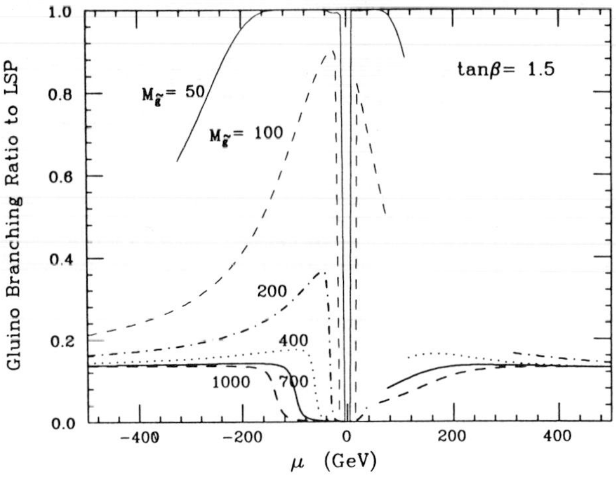

Figure 1. The branching ratio for $\widetilde{g} \to q\bar{q}\widetilde{\chi}_1^0$ as a function of μ for a series of $M_{\widetilde{g}}$ values (in GeV units), where $\widetilde{\chi}_1^0$ is the lightest supersymmetric particle (LSP). Sections of the curve that are not plotted correspond to parameter choices which yield $M(\widetilde{\chi}_1^+) < 30$ GeV.

(for $m_{H^+} = 500\ GeV$)

$$\begin{aligned} &\widetilde{g} \to \widetilde{\chi}_1^{\pm}\bar{q}q' && 29\%\\ &\widetilde{g} \to \widetilde{\chi}_2^{\pm}\bar{q}q' && 29\%\\ &\widetilde{g} \to \widetilde{\chi}_1^{0}\bar{q}q' && 12\%\\ &\widetilde{g} \to \widetilde{\chi}_2^{0}\bar{q}q' && 17\%\\ &\widetilde{g} \to \widetilde{\chi}_4^{0}\bar{q}q' && 13\% \end{aligned} \tag{4}$$

The chargino branching ratios are

$$\begin{aligned} &\widetilde{\chi}_1^+ \to \widetilde{\chi}_1^0\bar{q}q' && 67\%\\ &\widetilde{\chi}_1^+ \to \widetilde{\chi}_1^0\ell\nu && 33\% \end{aligned} \tag{5}$$

and

$$\begin{aligned} &\widetilde{\chi}_2^+ \to \widetilde{\chi}_1^+Z^0 && 41\%\\ &\widetilde{\chi}_2^+ \to \widetilde{\chi}_1^+H_2^0 && 5\%\\ &\widetilde{\chi}_2^+ \to \widetilde{\chi}_1^0W^+ && 11\%\\ &\widetilde{\chi}_2^+ \to \widetilde{\chi}_2^0W^+ && 31\%\\ &\widetilde{\chi}_2^+ \to \widetilde{\chi}_3^0W^+ && 12\% \end{aligned} \tag{6}$$

where H_2^0 is the lightest Higgs scalar. The neutralino branching ratios are

$$\widetilde{\chi}_2^0 \to \widetilde{\chi}_1^0H_2^0 \qquad 100\% \tag{7}$$

and

$$\begin{aligned} &\widetilde{\chi}_3^0 \to \widetilde{\chi}_1^0Z^0 && 19\%\\ &\widetilde{\chi}_3^0 \to \widetilde{\chi}_1^0H_2^0 && 60\%\\ &\widetilde{\chi}_3^0 \to \widetilde{\chi}_1^{\pm}\bar{q}q' && 14\%\\ &\widetilde{\chi}_3^0 \to \widetilde{\chi}_1^{\pm}\ell\nu && 7\% \end{aligned} \tag{8}$$

and

$$\begin{aligned} \tilde{\chi}_4^0 &\to \tilde{\chi}_1^0 H_2^0 && 0.3\% \\ \tilde{\chi}_4^0 &\to \tilde{\chi}_2^0 H_2^0 && 6\% \\ \tilde{\chi}_4^0 &\to \tilde{\chi}_3^0 Z^0 && 8\% \\ \tilde{\chi}_4^0 &\to \tilde{\chi}_1^{\pm} W^{\mp} && 85\% \end{aligned} \quad (9)$$

Evidently long chains of cascade decays are possible, giving rise to relatively small $E_T^{\rm miss}$ from the final $\tilde{\chi}_1^0$. But these decays also can produce $W^{\pm}$'s, Z^0's, jets, and leptons, leading to many possible signatures. Some of these signatures are examined in other contributions to these proceedings. Here we focus on the missing-energy signature using the ISAJET Monte Carlo program to generate $\tilde{g}\tilde{g}$ events including all the above decay modes. In addition to missing energy we use several additional cuts designed to reflect the fact that the gluinos are produced with $p_T(\tilde{g}) \sim M_{\tilde{g}}$ and $y(\tilde{g}) = \mathcal{O}(1)$, so that the decay products typically have large opening angles in the laboratory frame.

The Standard Model background to events involving jets, leptons, and $E_T^{\rm miss}$ comes from the QCD production and semileptonic decay of heavy quarks $Q = c$, b, and t and from the production of $W^{\pm}$ and Z^0 at large p_T. We have also used ISAJET to simulate all of these backgrounds. The largest background for gluinos as for most other processes is the production of heavy quark pairs at high transverse momentum from higher-order processes like

$$g + g \to Q + \overline{Q} + g \quad (10)$$

Such higher-order processes can not be directly included in ISAJET without severe double counting, since they are already produced by the QCD evolution, which approximates all higher-order QCD cross sections by a classical Markov branching process. To generate Eq. (10), therefore, it is necessary first to generate a primary hard scattering $gg \to gg$ and then to select events with the branching $g \to \overline{Q}Q$. Since the $gg \to gg$ cross section is 100 times larger than the $gg \to \overline{Q}Q$ one, this is rather slow.

To speed up the generation of the QCD background, we have used a trick developed by M. Della Negra of the UA1 Collaboration. Each primary hard scattering is evolved 10 times, and events not containing heavy quarks are immediately discarded. The rest are each hadronized 10 times, and only events containing the desired leptons or $E_T^{\rm miss}$ from neutrinos are saved. Finally the surviving events were weighted appropriately. We have generated 22,500 accepted QCD background events with $p_T >$ 250 GeV (equivalent to 2.25×10^6 events), 15,000 $W^{\pm}$ and 15,000 Z^0 events with $p_T > 250\,\text{GeV}$, and 2000 signal events.

Results and Conclusions

The most characteristic signature of supersymmetry is missing transverse energy in association with multiple jets. We have concentrated on this signature, leaving to future work the study of signatures involving $W^{\pm} \to \ell^{\pm}\nu$ and $Z^0 \to \ell^+\ell^-$ produced in the gluino cascade decay. To analyse the simulated events, we have used a highly idealized calorimeter with uniform segmentation $\Delta y = \Delta\phi = 0.05$ for $|y| < 6$ and Gaussian energy resolutions

$$\left(\frac{\Delta E}{E}\right)_{\rm e.m.} = \frac{0.15}{\sqrt{E}}, \quad \left(\frac{\Delta E}{E}\right)_{\rm had.} = \frac{0.70}{\sqrt{E}}. \quad (11)$$

In future work we will replace this with the more realistic QFL simulation developed for the CDF Collaboration. Then we found all jets with $p_T > 50\,\text{GeV}$ using a simplified version of the UA1 jet algorithm with a clustering radius $\Delta R = 0.7$.

To select the $M = 750\,\text{GeV}$ gluino signal from the QCD and electroweak backgrounds we impose the following set of cuts:

- The highest jet has $p_T > 300\,\text{GeV}$;
- The missing transverse energy is $E_T^{\rm miss} > 200\,\text{GeV}$;
- The sphericity S calculated in the transverse plane is $S > 0.2$;
- The number of jets with $p_T > 50\,\text{GeV}$ is $n_{\rm jet} \geq 4$.

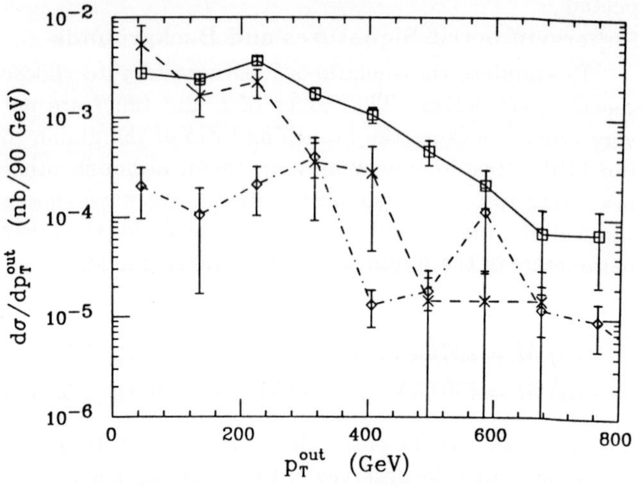

Figure 2. Cross section $d\sigma/dp_T^{\rm out}$ vs. $p_T^{\rm out}$ after the cuts given in the text. Here $p_T^{\rm out}$ is the missing p_T out of the plane defined by the beam and the highest p_T jet. □ symbol: $\tilde{g}\tilde{g}$ signal. × symbol: QCD background. ◇ symbol: $W^{\pm}$ and Z^0 background. Curves are drawn to guide the eye.

After these cuts we plot the missing transverse momentum $p_T^{\rm out}$ out of the plane defined by the beam and the highest p_T jet. The resulting distribution is shown in Fig. 2. Making a further cut at $p_T^{\rm out} > 300\,\text{GeV}$ gives the following cross sections:

$$\sigma_{\text{signal}} = 2.4\,\text{pb}$$
$$\sigma_{QCD} = 0.62\,\text{pb}$$
$$\sigma_{W^\pm,\ Z^0} = 0.25\,\text{pb}\ . \qquad (12)$$

The signal has been substantially reduced from Eq. (3), but it still represents 24000 events for the nominal integrated luminosity of $10^4 pb^{-1}$. The signal-to-background ratio seems quite adequate provided that realistic detector effects do not degrade the resolution too much. Presumably the shape of the p_T^{out} distribution could be used to determine the mass, but we have not yet investigated the mass resolution. A detailed study of other gluino signatures will be presented elsewhere.

REFERENCES

1. C. Albajar et al., Phys. Lett. **B198**, 261 (1987)
2. J. Freeman and F. Abe et al., Proceedings of the 7th International Conference on $\overline{p}p$ Collisions, Fermilab, June, 1988 (preliminary).
3. R. Batley, in *Proc. of the 1987 La Thuile Workshop on Physics at Future Accelerators*, Vol. 2, ed. J.H. Mulvey, CERN 87-07 (1987), p. 109
4. G. Gamberini, Z. Phys. **C30**, 605 (1986); H. Baer and E.L. Berger, Phys. Rev. **D34**, 1361 (1986). H. Baer, V. Barger, D. Karatas and X. Tata, Phys. Rev. **D36**, 96 (1987); R. Barbieri, G. Gamberini, G.F. Giudice and G. Ridolfi, Nucl. Phys. **B301**, 15 (1988).
5. R.M. Barnett, J.F. Gunion and H.E. Haber, Phys. Rev. Lett. **60**, 401 (1988) and Phys. Rev. **D37**, 1892 (1988)
6. See H.E. Haber and G.L. Kane, Phys. Rep. **117**, 75 (1985) and references therein.

Like-Sign Dileptons as a Signal for Gluino Production

R. Michael Barnett
Lawrence Berkeley Laboratory, Berkeley, CA 94720
and
John F. Gunion
Department of Physics, University of California, Davis, CA 95616
and
Howard E. Haber
Santa Cruz Institute for Particle Physics
University of California, Santa Cruz, CA 95064

Abstract

We discuss the viability of a like-sign dilepton signal for the discovery or confirmation of gluino production at an upgraded Tevatron and at the SSC.

Many collider studies in the past have investigated the missing-energy signature in gluino production under the assumption that the gluino decays directly into $q\bar{q}\widetilde{\chi}_1^0$ with 100% branching ratio.[1] Here, $\widetilde{\chi}_1^0$ is the lightest neutralino, which we assume is the lightest supersymmetric particle (LSP) and escapes the detector. Indeed, for the gluino masses which were accessible to the CERN $p\bar{p}$ Collider ($M_{\widetilde{g}} \lesssim 80$ GeV), this large branching ratio is likely, although not guaranteed. However, this branching ratio becomes much smaller for larger gluino mass, $M_{\widetilde{g}} \gtrsim 100$ GeV. These issues have been studied in detail for the minimal supersymmetric extension of the Standard Model in refs. 2 and 3. For example, for $M_{\widetilde{g}} = 180$ GeV (which we will find is accessible to an upgraded Tevatron), the branching ratio for the gluino to decay directly to the LSP can be no larger than 0.4 (and may be much smaller depending on the choice of supersymmetric parameters). The suppression is even more dramatic for heavier gluinos ($M_{\widetilde{g}} \gtrsim 500$ GeV), where the branching ratio is no greater than 0.14. The remainder of the branching ratio is taken up by decays into heavier neutralinos and chargino states which themselves decay, eventually ending up with the LSP. However, the LSP which eventually gets emitted is much softer and therefore the missing-energy signal is degraded. On the other hand, it is rather easy to produce leptons in the decay chain, either via a three-body decay of a chargino or neutralino, mediated by the W or Z, or by a direct two-body decay in which the W and Z are produced on-shell and subsequently decay leptonically.

In this paper, we are interested in new gluino search techniques which could complement (or perhaps even substitute for) the standard missing-energy signature. One problem with the more complicated signatures is that, if seen, one must find means for identifying such a signal as being that of a gluino. Our approach will be to make use of a special feature of the gluino: it is a Majorana fermion. (Here, we use a broader definition of Majorana to include neutral particles which transform under real representations of the underlying Standard Model gauge group.) A very distinctive feature of a Majorana particle is that it can decay with equal probability into fermions and antifermions. Thus, we focus on a particular class of gluino decay chains such that $\widetilde{g}\widetilde{g}$ production will result in *like-sign* dileptons in the final state. Due to the Majorana nature of the gluino, the probability for the production of like-sign and opposite-sign leptons is equal, and the lepton decay characteristics of the two classes of final states are identical. This is quite a distinctive result, and if seen would be extremely helpful in identifying the origin of the events.

Consider as an example the decay $\widetilde{g} \rightarrow q\bar{q}'\widetilde{\chi}_1^\pm$, where $\widetilde{\chi}_1^\pm$ is the lightest chargino state. As shown in ref. 3, this decay mode is the dominant gluino decay (with branching ratio near 60%) over a large range of supersymmetric parameter space. The key point here is that the Majorana nature of the gluino implies that it is equally likely to decay into a chargino of either sign. The chargino decays dominantly into the LSP plus a real (or virtual) $W^\pm$, which then decays to electrons or muons 22% of the time. Thus the branching ratio for the decay chain $\widetilde{g} \rightarrow q\bar{q}\ell^\pm\nu\widetilde{\chi}_1^0$ is typically around 13%, with equal probability to produce a lepton of either sign. For simplicity, the τ-lepton will be neglected from our considerations. Since gluinos are produced in pairs (we assume that squarks are heavier so that $\widetilde{q}\widetilde{g}$ production can be ignored), this implies that the number of dilepton final states resulting from the decay of the two gluinos may be as large as 2% of all $\widetilde{g}\widetilde{g}$ events, of which half have like-sign leptons. Given the large cross-sections for gluino pair production at future colliders, there would exist a potentially large and interesting sample of events. This sample consists of events with hadronic jets (two from each gluino), missing energy due to the LSP and neutrinos in the final state, and a dilepton pair which can come in one of the following like-sign combinations: $e^\pm e^\pm$, $e^\pm\mu^\pm$, $\mu^\pm\mu^\pm$, and the corresponding opposite-sign combinations. Because the background is clearly larger for detecting the opposite-sign pairs, we shall focus here on the isolation of like-sign dilepton events originating from gluino pair production.

We have used a parton Monte Carlo in order to determine the viability of gluino detection using the like-sign dilepton signal. We have investigated the signal at the present Tevatron and at two future hadron colliders under consideration: an upgraded Tevatron and the SSC. Three Tevatron options are being considered, but we find that only the choice of a pp collider with $\sqrt{s} = 2\ TeV$ and an integrated luminosity of $1000 pb^{-1}$ gives a significant improvement in discovery limits over the existing Tevatron. In addition to studying the characteristics of the signal, we have tried to identify the major source of Standard Model background. In the case of the like-sign dilepton signal, the dominant background is expected to be $t\bar{t}$ production, where one t decays to $b\ell\nu$ and the other t decays to $b\bar{q}q'$ with the secondary b decaying into $c\ell\nu$. Such events will also produced like-sign dileptons (both muons and electrons), hadronic jets, and missing energy in the form of neutrinos. The background estimate will depend crucially on the unknown t-quark mass. We have studied two extreme cases in this regard: the "light" top ($m_t = 60\ GeV$) and the "heavy" top ($m_t = 160\ GeV$). We have identified a number of characteristics of the background that allow us to distinguish between it and the signal.

First, there is the lack of isolation of the lepton emitted from the b-quark. (We make use of the theoretical expectation that there is negligible mixing in neutral T-mesons which could cause primary like-sign dileptons in $t\bar{t}$ production from the t-quark decays.) This is to be compared with the case of dileptons from gluinos where both leptons resulting from the decay of a chargino (presumed to be much heavier than the b-quark) would be quite isolated. Thus we define $\Delta R^{min}_{l,j}$ to be the minimum $\Delta R \equiv \sqrt{(\Delta\eta)^2 + (\Delta\phi)^2}$ obtained by considering all lepton-jet combinations. Two other variables that allow us to discriminate against the background due to a top quark that is significantly lighter than the gluino are the total reconstructed mass of the event,

$$M^2_{tot} = \left(\sum_i E_i + E_T^{miss}\right)^2 - \left(\sum_i \vec{p}_i + \vec{p}_T^{\,miss}\right)^2 \quad (1)$$

designed to approximate the total mass of the produced $\tilde{g}\tilde{g}$ system, and the invariant mass ($M_{l_2,2j}$) of the slow lepton and its two nearest (in ΔR) jets, which provides a measure of the invariant mass of the gluino. In eq. (1) we sum over jets and leptons, labelled in order of decreasing transverse energy: $E_T^{j_1} > E_T^{j_2} > E_T^{j_3} > E_T^{j_4}$ and $E_T^{l_1} > E_T^{l_2}$.

Before presenting any details of the analysis we preview our conclusions. At the upgraded Tevatron option with an integrated luminosity of $1000\ pb^{-1}$, the like-sign dilepton signature is found to be viable for gluino masses up to $180\ GeV$. Of course, it is necessary to have a magnetic detector in order to be able to perform the type of analysis we describe here. Since it is clearly of interest to make sure that the SSC can probe down to this mass (so that there is no gap in accessible gluino masses for the combination of the two machines) we present the bulk of our results (both for an upgraded Tevatron and the SSC), assuming $M_{\tilde{g}} = 180\ GeV$. For simplicity, we employ the minimal supersymmetric extension of the Standard Model. Having specified $M_{\tilde{g}}$, the remaining free parameters are μ, $\tan\beta$ and the mass of one of the Higgs bosons of the model (see ref. 3 for notation). A particular choice of parameters can affect the chargino and neutralino mass spectrum and the gluino leptonic branching ratio. We have surveyed the supersymmetric parameter space and find that the gluino leptonic branching ratio is fairly stable over a considerable range of the parameters. We present results here for a typical choice: $\tan\beta = 1.5$, $m_{H^+} = 500\ GeV$, and $\mu = -150\ GeV$. For these values, we find $BR(\tilde{g} \to \bar{q}q'\tilde{\chi}_1^+) = 0.52$ and $BR(\tilde{\chi}_1^+ \to \ell^\pm + \text{anything}) = 0.22$, where ℓ is either an electron or muon. Since in this case this is the only decay chain giving a single lepton, we find that like-sign dilepton events are 0.6% of all $\tilde{g}\tilde{g}$ events.

Let us first focus on the Tevatron upgrade analysis. First, we define a jet such that all parton jets within a cone of $\Delta R \leq 0.6$ around the jet axis are counted as one jet. We then employ the following cuts: (i) $N_{jet} \geq 4$, (ii) $E_T^{j_{1,2,3,4}} > 10\ GeV$, (iii) $N_{lepton} = 2$, and (iv) $E_T^{l_{1,2}} > 5\ GeV$. In addition, we impose a CDF-like trigger by requiring that at least one of the following three conditions be satisfied: (i) $E_T^{l_1} > 20\ GeV$, (ii) $E_T^{j_1} > 60\ GeV$, or (iii) $E_T^{j_{1,2,3}} > 20\ GeV$. Distributions in $\Delta R^{min}_{l,j}$, M_{tot}, $M_{l_2,2j}$ and $E_T^{l_2}$ are presented in fig. 1. For the latter three we have plotted the $\tilde{g}\tilde{g}$ signal results before any $\Delta R^{min}_{l,j}$ cut (a $\Delta R^{min}_{l,j} > 0.48$ cut results in only a small decrease in the signal curve), in comparison to background results for $m_t = 60\ GeV$ with the $\Delta R^{min}_{l,j} > 0.48$ cut imposed and $m_t = 160\ GeV$ without any $\Delta R^{min}_{l,j}$ cut. All Tevatron curves are normalized to an integrated luminosity of $1000\ pb^{-1}$. The number of events in the $m_t = 60\ GeV$, $\Delta R^{min}_{l,j} > 0.48$ curves is 20, the number in the $m_t = 160\ GeV$, no $\Delta R^{min}_{l,j}$ cut curves is 91, while there are 18 and 14 events in the signal curve before and after the $\Delta R^{min}_{l,j}$ cut, respectively. For $m_t = 60\ GeV$ the background event rate is very high to begin with, but the $\Delta R^{min}_{l,j}$ cut is extremely effective in reducing it to the level of the signal. The mass distributions then allow one to easily distinguish between signal and background. At $m_t = 160\ GeV$ the background is not that much larger than signal and the $\Delta R^{min}_{l,j}$ distributions are sufficiently different to allow separation. Having made such a cut, the $E_T^{l_2}$ distributions of signal and background are quite different, as seen in fig. 1b. On the other hand the mass distributions (fig. 1c-d) are quite similar when $m_t \sim M_{\tilde{g}}$. Note that in both cases, the gluino mass can be determined (within an accuracy of order 10%) from the M_{tot} distribution. These results from our upgraded Tevatron

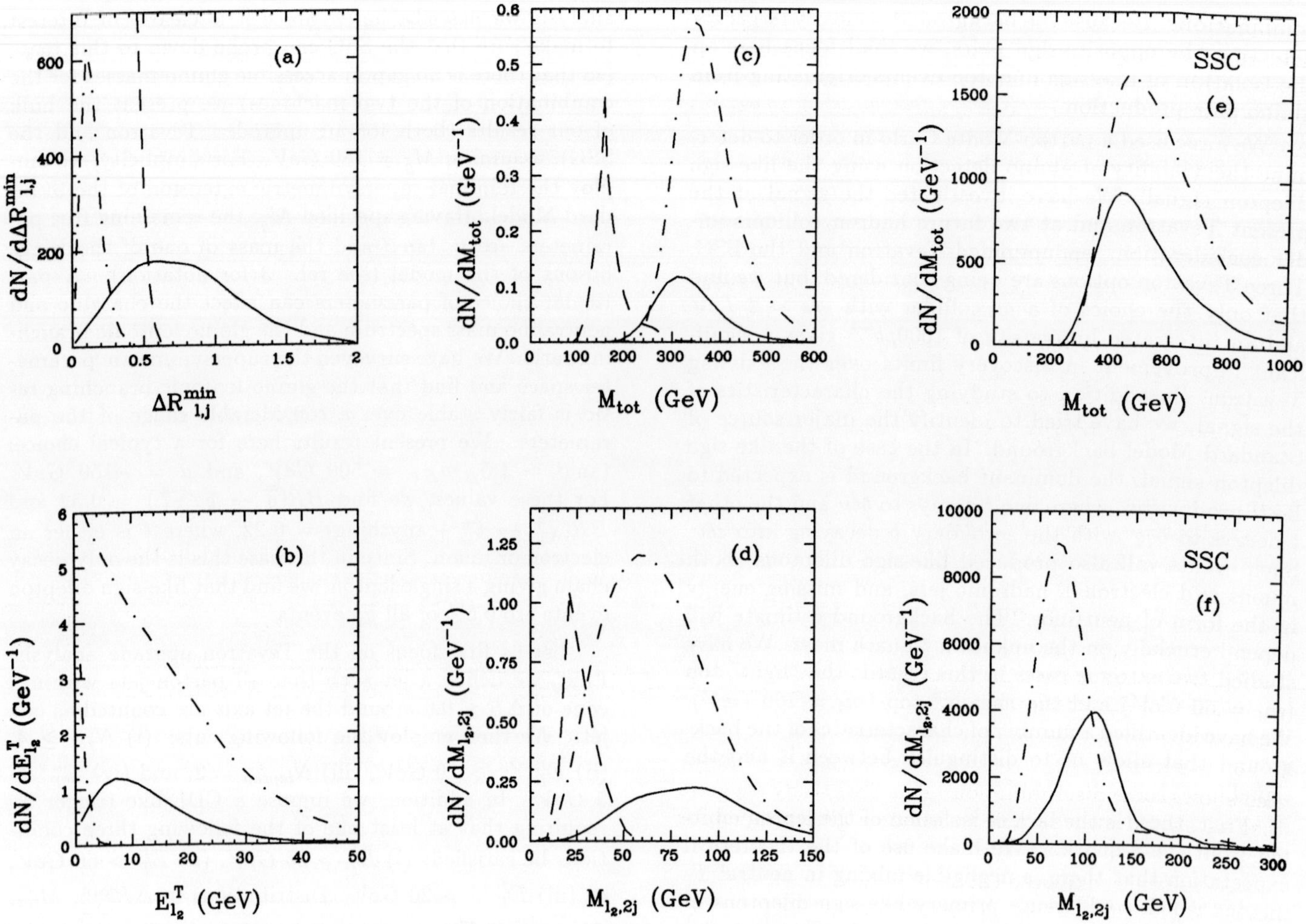

Figure 1. We present normalized distributions for the Tevatron upgrade, (a)–(d), and the SSC, (e) and (f), for $M_{\widetilde{g}} = 180$ GeV. All include the cuts and triggering conditions described in the text. The $\widetilde{g}\widetilde{g}$ signal curves are solid lines and have no $\Delta R^{min}_{l,j}$ cut; in (b)–(f) a $\Delta R^{min}_{l,j} > 0.48$ cut results in only a small decrease in the solid curve. Background curves for $m_t = 60$ GeV are dashed, while those for $m_t = 160$ GeV are dot-dashed. The distributions are:

(a) $\Delta R^{min}_{l,j}$ for the $\widetilde{g}\widetilde{g}$ signal multiplied by a factor of 10 compared to $m_t = 60$ GeV and $m_t = 160$ GeV

(b) $E_T^{l_2}$ for $m_t = 160$ GeV before and after a $\Delta R^{min}_{l,j} > 0.48$ cut, compared to the $\widetilde{g}\widetilde{g}$ signal

(c) M_{tot} for $m_t = 60$ GeV, $\Delta R^{min}_{l,j} > 0.48$, and $m_t = 160$ GeV, no $\Delta R^{min}_{l,j}$ cut, compared to signal

(d) Same as in (c) but for $M_{l_2,2j}$

(e) M_{tot} for $m_t = 160$ GeV, no $\Delta R^{min}_{l,j}$ cut, compared to the signal

(f) Same as in (e) but for $M_{l_2,2j}$.

simulation show that the dilepton signature is marginal and requires large luminosity. The other proposed upgrades which do not plan for such a high integrated luminosity would not allow use of the dilepton signal to explore gluino masses as large as 180 GeV — the dilepton signal is essentially event rate limited.

We have also run our Monte Carlo under SSC conditions. We assume an integrated luminosity of 10^4 pb^{-1}. Since the raw $\tilde{g}\tilde{g}$ production rate is enormous for $M_{\tilde{g}} =$ 180 GeV (yielding some 2.48×10^6 like-sign dilepton events before cuts), it is trivial to impose strong cuts on the $\Delta R^{min}_{l,j}$ and mass variables that completely eliminate any $m_t = 60$ GeV background. Thus, we focus on $m_t =$ 160 GeV. We employed a new set of cuts: (i) $N_{jet} \geq 4$, (ii) $E_T^{j_{1,2,3}} > 40\ GeV$, $E_T^{j_4} > 10\ GeV$, (iii) $N_{lepton} = 2$, and (iv) $E_T^{l_{1,2}} > 10\ GeV$. (Making a stronger cut on $E_T^{l_2}$ would completely eliminate the background. We have not done so in order to have some background to compare with our signal.) No trigger conditions were imposed. After the above cuts, we are left with 6×10^5 background events and 2.6×10^5 signal events. The $\Delta R^{min}_{l,j}$ distribution of the background is even more sharply peaked towards small $\Delta R^{min}_{l,j}$ in the SSC simulation than at the Tevatron, and would clearly allow isolation of the signal. The M_{tot} and $M_{l_2,2j}$ distributions for signal and background are shown in fig. 1, with no $\Delta R^{min}_{l,j}$ cut imposed. We see that $M_{l_2,2j}$ also allows some discrimination. This SSC study suggests that we should even be able to isolate a gluino signal for gluinos lighter than 180 GeV.

We have also explored the maximum gluino mass that could be probed using the like-sign dilepton signature at the SSC.[4] For heavy gluinos the effective branching ratio for $\tilde{g}\tilde{g}$ final states containing like-sign electrons or muons and at least four energetic jets is typically of order 1% for a large range of parameter space. The large difference in $\Delta R^{min}_{l,j}$ distributions between signal and top backgrounds remains, and now the M_{tot} and $M_{l_2,2j}$ distributions for the $\tilde{g}\tilde{g}$ signal peak at such high masses that there is essentially no overlap with any $m_t < 200$ GeV background. Thus isolation of the signal is purely event rate limited. For $M_{\tilde{g}} = 2$ TeV and integrated luminosity of 10^4 pb^{-1} there are 25 events in the dilepton signal, which would almost certainly allow discovery. Higher masses could be probed for larger integrated luminosities.

In conclusion, at an upgraded Tevatron with at least 1000 pb^{-1}, the like-sign dilepton signal may be visible for gluino masses up to 180 GeV, depending on the values of the other supersymmetric parameters. (Such a signal may even be observable at a Tevatron operating under present conditions for gluino masses near 90 GeV for special parameter choices.) At the SSC, the dilepton signal should be easy to detect for the entire range of gluino masses from 180 – 2000 GeV, very probably extending even lower. Such a copious like-sign dilepton signal would be strong evidence for phenomena related to a Majorana fermion, and the large event rates predicted here would almost surely allow experimenters to claim the discovery of the gluino, if such like-sign dilepton events were discovered. Clearly, the dilepton signal at the SSC is a powerful tool for uncovering evidence for supersymmetry.

Acknowledgements

This work was supported by the Director, Office of High Energy and Nuclear Physics, Division of High Energy Physics, of the U.S. Department of Energy under Contract Nos. DE-AC03-76SF00098, DE-AS03-76ER70191 and DE-AA03-76-SF00010, and by the National Science Foundation under agreement no. PHY83-18358.

REFERENCES

1. For a recent analysis and references to past work, see R. Batley, *Proceedings of the "Workshop on Physics at Future Accelerators"*, LaThuile, Jan. 7-13 (1987), Vol. II, p. 109.
2. G. Gamberini, *Z. Phys.* **C30** (1986) 605; H. Baer, V. Barger, D. Karatas, and X. Tata, *Phys. Rev.* **D36** (1987) 96.
3. R.M. Barnett, J.F. Gunion and H.E. Haber, *Phys. Rev.* **D37** (1988) 1892.
4. For a somewhat more extensive survey of the high mass region at future hadron colliders, see J.F. Gunion, "INFN Eloisatron Project Working Group Report", Erice-Trapani (1988).

ELECTRON - POSITRON COLLIDERS

NEW GAUGE BOSONS AT HIGH ENERGY e^+e^- COLLIDERS

JoAnne L. Hewett
Department of Physics, University of Wisconsin, Madison, WI 53706

and

Thomas G. Rizzo
Ames Laboratory and Department of Physics, Iowa State University, Ames, IA 50011

SUMMARY

We extend the SLAC 'Red-Book' study of new gauge boson production at high energy e^+e^- colliders within the context of E_6 superstring-inspired models. In particular, search limits for new Z' bosons in such models are set by making use of cross section and asymmetry deviations from the Standard Model at $\sqrt{s} = 0.4$ and 1 TeV e^+e^- colliders. The processes $e^+e^- \to e^+e^-Z'$, $2Z'$, and $e^\pm W_R^\mp S$ are briefly considered.

1. INTRODUCTION

Interest in the possibility of observing new gauge bosons at high energy e^+e^- colliders has been renewed due to the advent of E_6 superstring-inspired models.[1] Within this context, the purpose of this work is to address, at least partially, the following questions:

(*i*) Once a Z' is discovered and its properties measured can we uniquely determine its origin, even within the limited E_6 context?

(*ii*) What limits can be placed on the mass of a possible new Z' lying above the top e^+e^- center-of-mass energy of the machine? How model dependent is this result and how sensitive is it to $\sqrt{s}$, beam polarization, luminosity, *etc.*?

(*iii*) If a W' exists and is too heavy to be pair produced at a given $\sqrt{s}$, is single production, $e^+e^- \to W'^\pm + X$, viable as a search mode? This is particularly important for very heavy W' bosons.

Although these and similar questions have been addressed by several studies[2,3] in the literature, in particular the SLAC 'Red-Book' Study,[3] this work is still incomplete and should be extended to a wider class of models—even if one remains within the E_6 context.

To be concrete, we will examine the following two broad classes of models based on E_6:

$$\begin{array}{ll} \text{(A)} & SU(3)_C \times SU(2)_L \times U(1)_Y \times U(1)_\theta\,, \\ \text{(B)} & SU(3)_C \times SU(2)_L \times SU(2)_R \times U(1)\,. \end{array} \qquad (1)$$

Models of type (A), the 'effective rank-5' models (ER5M), have a free parameter (θ) which measures the amount of mixing between the Z_ψ and Z_χ gauge fields in the new Z' due to the decomposition $E_6 \to SO(10) \times U(1)_\psi \to SU(5) \times U(1)_\chi \times U(1)_\psi$, where $Z' = Z_\psi \cos\theta - Z_\chi \sin\theta$. Z' is a mass eigenstate if its mixing with the standard model (SM) Z is ignored. Models of type (B) include the usual Left-Right Symmetric Model[4] (LRM) as well as the variant due to Ma and co-workers[5] which takes advantage of the fermion assignment ambiguities in the E_6 **27** representation—the alternate Left-Right Symmetric Model (ALRM). Note that the Z' couplings in the LRM and ALRM are quite *distinct* and the W_R in the ALRM can be quite light ($\simeq 210$ GeV) as opposed to the LRM wherein the W_R must be very heavy ($\gtrsim 2$ TeV). The detailed nature of these models has been widely discussed in the literature.[1]

2. ON-RESONANCE PROPERTIES

What Z' properties are accurately measurable and can be used to distinguish among various models at e^+e^- colliders? In addition to width and mass, branching fractions for various modes should be measurable, as well as the forward-backward asymmetries for (at least) the conventional final states: $\mu^+\mu^-$ (A_{FB}^μ), $c\bar{c}/t\bar{t}$ (A_{FB}^u), and $b\bar{b}$ (A_{FB}^d). Longitudinal beam polarization will also allow for the measurement of the left-right asymmetry, A_{LR}, which to lowest order in QCD is the same for all massless final state fermions on resonance. A measurement of the final state τ polarization also yields the value A_{LR}. Will this amount of data, even if precisely known, allow us to completely determine the model parameters (*e.g.*, θ) and will it allow us to distinguish among the various E_6 models?

Table 1. A comparison of Z' properties in the LRM and ALRM including asymmetries, branching fractions in per cent for $n_g = 3$ (0), and the scaled total widths.

Asymmetry/Mode	LRM	ALRM
A_{LR}	0.290	0.504
A^{μ}_{FB}	0.063	0.191
A^{u}_{FB}	0.206	−0.358
A^{d}_{FB}	0.211	−0.227
B		
e	1.27 (2.46)	3.29 (9.04)
u	5.62 (10.91)	5.62 (15.43)
d	9.84 (19.20)	0.75 (2.06)
ν_L	0.45 (0.87)	2.48 (6.80)
ν_R	5.04	0
h	1.20	10.29
E	4.96	2.93
N_L	2.48	0.45
S_R	0	5.04
N_R	2.48	2.48

First, it is relatively easy using partial widths and asymmetries to distinguish among the ER5M, LRM, and ALRM since for all θ values, apart from QCD corrections, one finds the following predictions in the ER5M:

$$A^{u}_{FB} = 0\,, \quad A^{\mu}_{FB} = -A^{d}_{FB}\,, \quad \Gamma(Z' \to d\bar{d}) = 3\Gamma(Z' \to e^+e^-)\,. \tag{2}$$

If exotic fermions[1] also participate in Z' decay, then in the limit that E and h are degenerate one also has

$$\Gamma(Z' \to h\bar{h}) = 3\,\Gamma(Z' \to E^+E^-)\,. \tag{2'}$$

Table 1 shows that *none* of the conditions (2) or (2′) are satisfied for the LRM or the ALRM. Table 1 also shows quite explicitly that the branching fractions and asymmetries for these two models (LRM and ALRM) are very different and clearly allows one to distinguish between these two. Thus (2) and (2′) can be used to test all ER5M and to separate them from other possibilities such as the LRM or ALRM. These results are essentially unaffected by small ($|\phi| \lesssim 0.01$) Z-Z' mixing and finite values of $m_t/M_{Z'}$.

If Eqs. (2) and (2′) *are* satisfied, can we then determine θ uniquely? Figure 1 shows the branching fraction (B) for massless e, u, d, and ν_L as a function of θ when exotic fermions do not participate ($n_g = 0$) in Z' decay. Note that only the range $-90° \leq \theta \leq 90°$ is plotted because $\theta \to \theta + 180°$ leaves B's *and* asymmetries unaltered since it flips the signs of both v and a for all fermions. Therefore there already exists a two-fold ambiguity in the value of θ which cannot be removed, even with off resonance measurements, unless Z-Z' mixing is non-negligible. The figure assumes only three massless generations of ordinary fermions (*i.e.*, $m_t^2 \ll M_{Z'}^2$), neglects QCD corrections, and $\sin^2\theta_w = 0.230$.

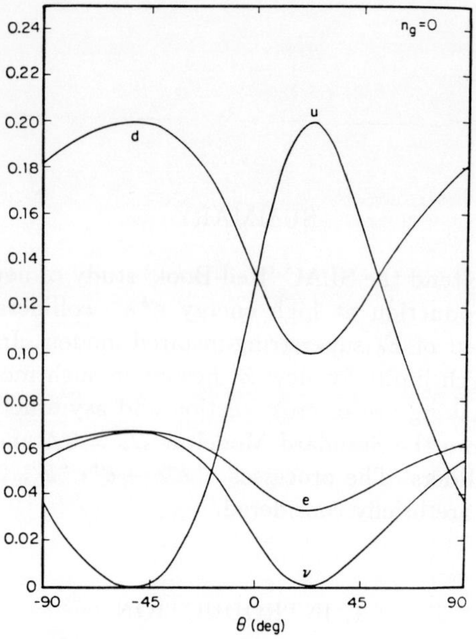

Figure 1. Branching fractions for Z' in the ER5M for three ordinary massless generations of fermions assuming $n_g = 0$ as a function of θ.

Returning to Fig. 1 we see that even in the $-90° \leq \theta \leq 90°$ range, θ cannot be uniquely determined even with the help of the asymmetries shown in Fig. 2. Note that B's (and both A_{LR} and A_{FB}) are symmetric around $\theta_0 = 1/2\cos^{-1}(\sqrt{3/8}) \approx 26.12°$ resulting in a further two-fold ambiguity. In this case, however, this ambiguity may be removable by considering the total width (Γ_0) shown in Fig. 3 since it does not show this symmetry. However, to use Γ_0 to fix θ we must *assume* that other final states (W^+W^-, SUSY partners, *etc.*) are *not* accessible in Z' decay—a very strong assumption to make. Unless one can experimentally determine that other final states are not participating *or* determine their masses and branching fractions accurately, it may be that a Γ_0 measurement will not be able to resolve this ambiguity.

This situation improves drastically if exotic fermion final states are also accessible in Z' decay. Figure 4 assumes that three generations ($n_g = 3$) of massless exotics as well as the usual fermions are present. As expected, the

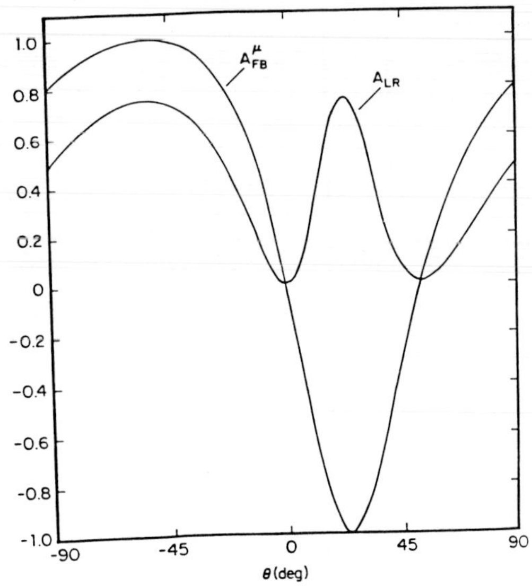

Figure 2. Asymmetries on the Z' peak in the ER5M as a function of θ. Remember $A^{u}_{\rm FB} = 0$ and $A^{d}_{\rm FB} = -A^{\mu}_{\rm FB}$ on resonance in this model.

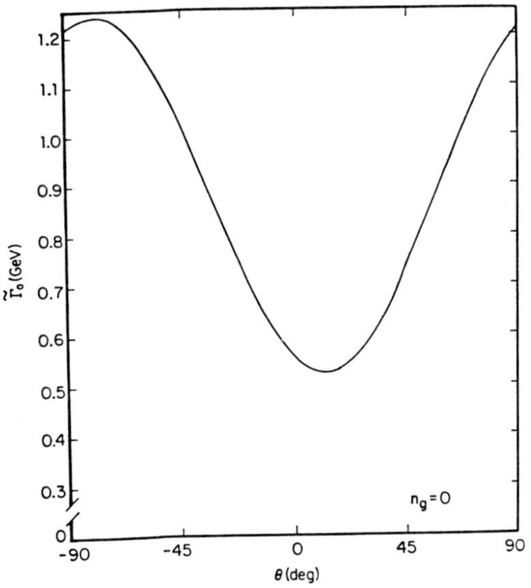

Figure 3. Scaled width, $\tilde{\Gamma}_0 = \Gamma M_Z/M_{Z'}$, for Z' in the ER5M with $n_g = 0$ as a function of θ.

e, u, d, and ν_L branching fractions show the same θ_0 symmetry as in the $n_g = 0$ case but the exotics show no such symmetry. Thus, *e.g.*, a measurement of B for e^+e^- and E^+E^- can determine θ uniquely (up to the $\theta \to \theta + 180°$ ambiguity) once the E's mass is taken into account. [In reality, this may not be so easy since E^+E^- final states may not be easy to reconstruct even in a relatively clean e^+e^- collider environment. It will surely depend on the quantum number assignment for these exotics.] This procedure assumes that the actual effects of phase space for the exotic modes can be properly taken into account along with QCD radiative corrections.

A last comment is in order. When measuring $A^{t}_{\rm FB}$

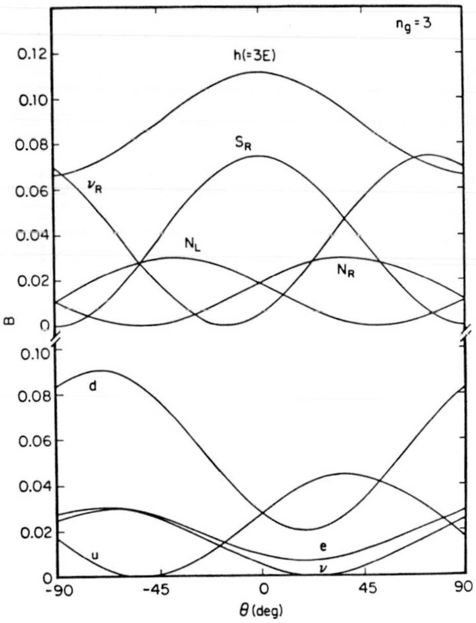

Figure 4. Branching fractions for Z' in the ER5M for three ordinary plus three exotic ($n_g = 3$) massless generations of fermions as a function of θ.

(assuming it is non-zero!), one must remember that the finite size of $m_t/M_{Z'}$ can modify the predicted value of this quantity substantially in almost all models. For example, in the ALRM with $M_{Z'} = 450$ GeV, $A^{t}_{\rm FB}$ ($m_t = 50$ GeV), and $A^{t}_{\rm FB}$ ($m_t = 200$ GeV) differ by $\simeq 30\%$ with a similar result holding in the LRM.

We conclude that on-resonance data can be used to separate the ER5M, LRM, and ALRM scenarios and may allow the determination of the parameter θ up to an overall $\theta \to \theta + 180°$ ambiguity. These results are essentially unaffected by small Z-Z' mixing and a finite t-quark mass.

3. Z' SEARCH LIMITS

If $M_{Z'} > \sqrt{s}$ we can still see its effects indirectly in the cross sections and asymmetries for various fermions. For a fixed $\sqrt{s}$ what limits can we place on $M_{Z'}$ by the measurement of these quantities, at least within the limited E_6 framework? Which quantity gives the best limit? In what follows we assume $\phi = 0$, neglect QCD corrections, and assume 100% beam polarization. For fixed $\sqrt{s}$ we ask what value of $M_{Z'}$ will produce a shift in the asymmetries ($\delta A_{\rm LR}$, $\delta A_{\rm FB}$) or cross section ($\delta\sigma/\sigma$) away from their SM values at the level of 1σ. The results of this analysis can be seen in Table 2 for the LRM and ALRM and in Figs. 5 and 6 for the ER5M as a function of θ. Note that as θ and $\sqrt{s}$ change the final state fermion producing the strongest limit on $M_{Z'}$ also changes. These results have been obtained by using the luminosities of the SLAC 'Red-Book' study.[3] We will assume 1000 (6250) events of each flavor at $\sqrt{s} = 1.0$ (0.4) TeV.

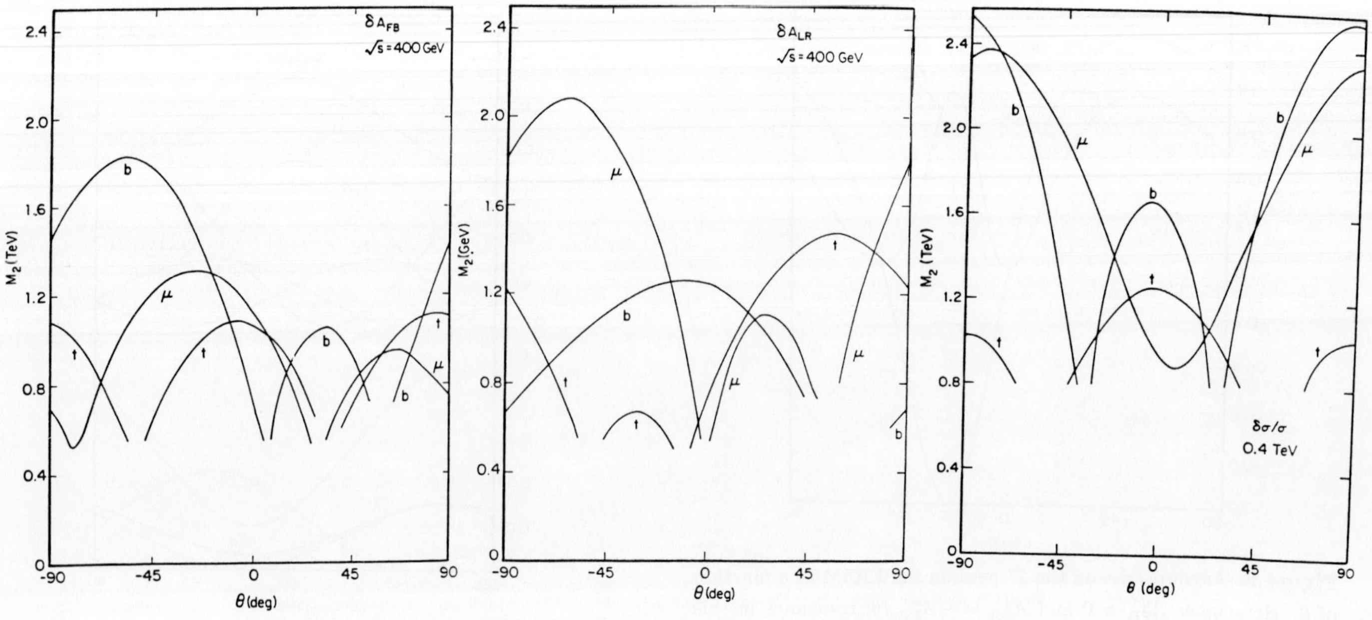

Figure 5. (a) $\delta A_{\rm FB}$, (b) $\delta A_{\rm LR}$, and (c) $\delta\sigma/\sigma$ 1σ limits on the Z' mass (M_2) as a function of θ in the ER5M for $\sqrt{s} = 0.4$ TeV.

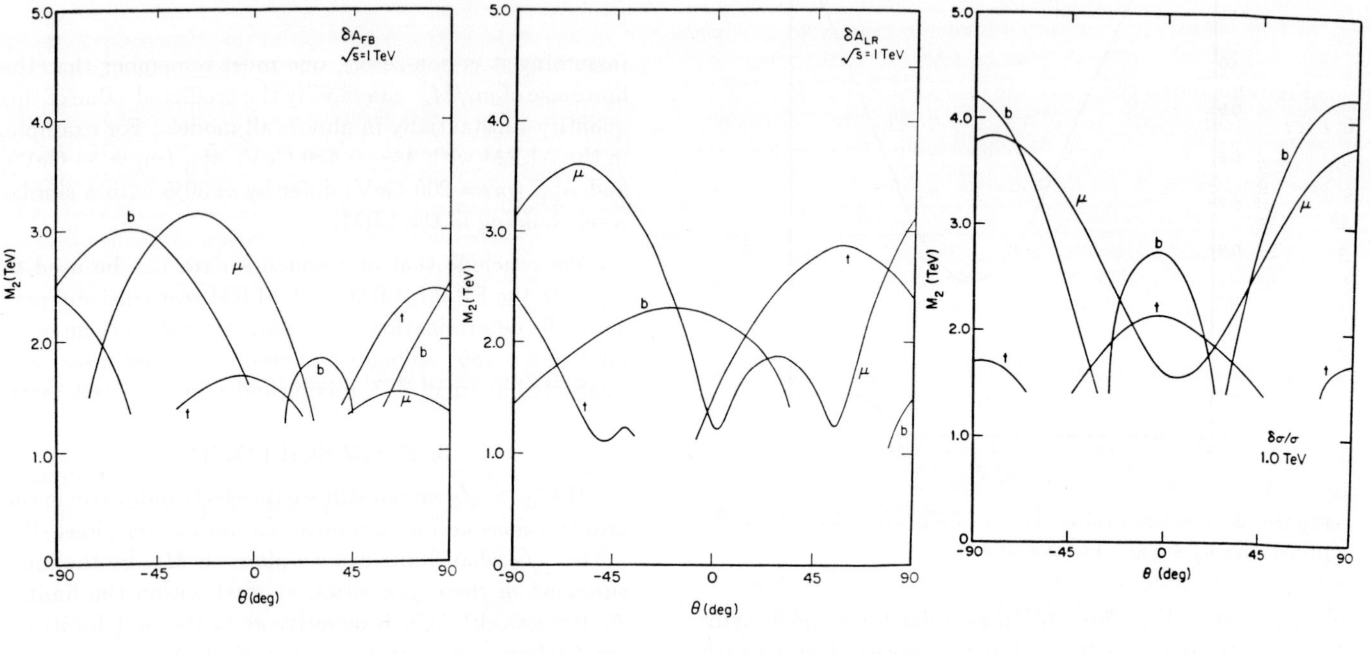

Figure 6. Same as Fig. 5 but for $\sqrt{s} = 1$ TeV.

Table 2. Indirect limits on the Z' mass (in TeV) in the LRM and ALRM using 1σ deviations in the asymmetries and total cross section for various fermions assuming $\sqrt{s}=0.4$ or 1 TeV at an e^+e^- collider.

$\sqrt{s}=0.4$ TeV

	ALRM			LRM		
	$\delta A_{\rm FB}$	$\delta A_{\rm LR}$	$\delta\sigma/\sigma$	$\delta A_{\rm FB}$	$\delta A_{\rm LR}$	$\delta\sigma/\sigma$
μ	1.15	1.60	2.80	1.35	0.88	1.15
b	1.55	1.40	0.67	1.50	2.28	0.63
c	1.05	1.10	1.85	1.35	1.90	1.40

$\sqrt{s}=1.0.$ TeV

μ	2.05	2.75	4.50	2.35	1.60	1.90
b	2.10	2.40	1.45	2.60	3.75	1.50
c	2.60	2.70	3.00	2.40	3.10	2.30

Figure 5 shows the 1σ limits on $M_{Z'}$ $(=M_2)$ at $\sqrt{s}=0.4$ TeV coming from (a) $\delta A_{\rm FB}$, (b) $\delta A_{\rm LR}$, and (c) $\delta\sigma/\sigma$ using massless μ's, b's, and t's (which would be the case for c's or when $m_t^2 \ll M_{Z'}^2$) for the ER5M. For each quantity, as θ varies, a different flavor provides the strongest limit. Also note that as θ changes the quantity ($\delta A_{\rm FB}$, δA_{LR}, $\delta\sigma/\sigma$) providing the best limit changes as well and that *for a wide range of θ the best limits on $M_{Z'}$ come from* $\delta\sigma/\sigma$ for μ's and b's. Table 2 shows the corresponding limits in the LRM and ALRM.

Figure 6 shows the 1σ limits on $M_{Z'}$ $(=M_2)$ at $\sqrt{s}=1$ TeV coming from (a) $\delta A_{\rm FB}$, (b) $\delta A_{\rm LR}$, and (c) $\delta\sigma/\sigma$ in the ER5M. Clearly, Fig. 6 shows that the $M_{Z'}$ limits one can obtain from a $\sqrt{s}=1$ TeV machine are not as good (*relative to* $\sqrt{s}$) as for a $\sqrt{s}=0.4$ TeV machine given the same luminosity due to the fall-off in the number of events and decreased statistics. We again see that, *for most values of θ, $\delta\sigma/\sigma$ from μ's and b's provides the strongest constraints* on $M_{Z'}$. For the LRM and ALRM at $\sqrt{s}=1$ TeV, Table 2 shows a similar pattern.

It should be noted that whereas δA_{LR} yields stronger constraints than $\delta A_{\rm FB}$ for both $\sqrt{s}=0.4$ and 1 TeV, we have assumed 100% beam polarization in calculating $\delta A_{\rm LR}$. With decreased polarization and fixed luminosity the $\delta A_{\rm LR}$ limits will quickly deteriorate below those from $\delta A_{\rm FB}$ which do not require polarization. With less than 100% polarization the dominance of $\delta\sigma/\sigma$ in setting limits becomes even greater than that implied by the figures and Table 2.

4. $e^+e^-\to e^+e^-Z'$, $e^+e^-\to 2Z'$, AND $e^+e^-\to W_R^\pm e^\mp S$

These processes which we will discuss in this section generally have small cross section either due to phase space suppression or small couplings but do have significant rates at e^+e^- colliders with luminosities in the 10-100 fb^{-1} range and probe some interesting physics. Except for the case of $W_R^\pm$ production we will limit ourselves to the ER5M in what follows. The effective photon approximation will be used to calculate the $e^+e^-\to e^+e^-Z'$ and $W_R^\pm e^\mp S$ production cross sections. S is an SO(10) singlet in the E_6 **27** representation which we will take to be massless.[5]

For $e^+e^-\to e^+e^-Z'$, the subprocess ($\gamma e^\pm\to e^\pm Z'$) cross section is given by[1]

$$\frac{d\sigma}{d\hat{t}}=\frac{\pi\alpha^2}{32\,x_w(1-x_w)\hat{s}^2}\left(\frac{-\hat{u}}{\hat{s}}-\frac{\hat{s}}{\hat{u}}-\frac{2M_{Z'}^2\hat{t}}{\hat{s}\hat{u}}\right)(v_e^2+a_e^2)\,. \tag{3}$$

The production cross section is sensitive to θ via the overall $v_e^2+a_e^2$ factor ($x_w=\sin^2\theta_w$)

$$v_e^2+a_e^2=\left(\frac{5x_w}{3}\right)\left\{\frac{2}{5}\sin^2\theta+\left(\frac{\cos\theta}{\sqrt{6}}-\frac{\sin\theta}{\sqrt{10}}\right)^2\right\}. \tag{4}$$

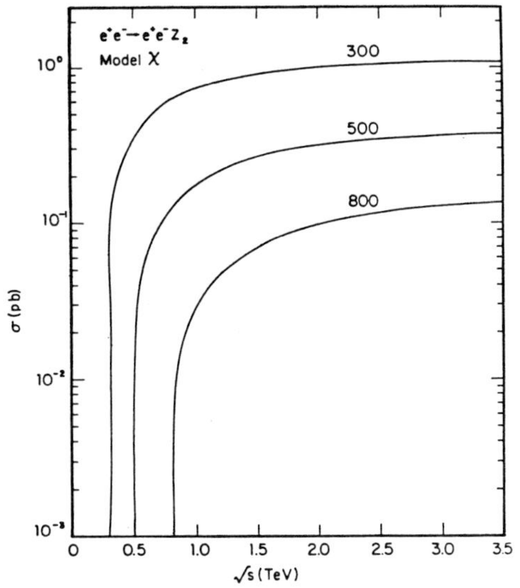

Figure 7. Cross section for $e^+e^-\to e^+e^-Z'(Z_2)$ as a function of $\sqrt{s}$ for different values of M_2 in GeV within the context of model χ using the effective photon approximation.

To be specific, since there are so many parameters, let us first consider model $\chi(\theta=-90°)$. Figure 7 shows σ for $e^+e^-\to e^+e^-Z'$ as a function of $\sqrt{s}$ for different values of $M_{Z'}$ for this model. Note that once the threshold region is passed, σ stays relatively constant in the 10^2-10^3 fb region for the range of masses considered. How do we obtain the corresponding results for other values of θ? Figure 8 shows the θ dependence for $e^+e^-\to e^+e^-Z'$ with $\sqrt{s}=1$ TeV and $M_2=M_{Z'}=300$ GeV and can

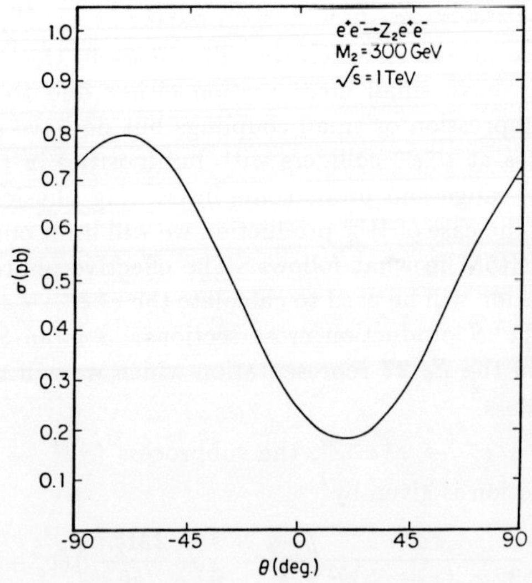

Figure 8. Cross section for $e^+e^- \to e^+e^-Z_2$ at $\sqrt{s} = 1$ TeV for $M_2 = 300$ GeV as a function of the parameter θ.

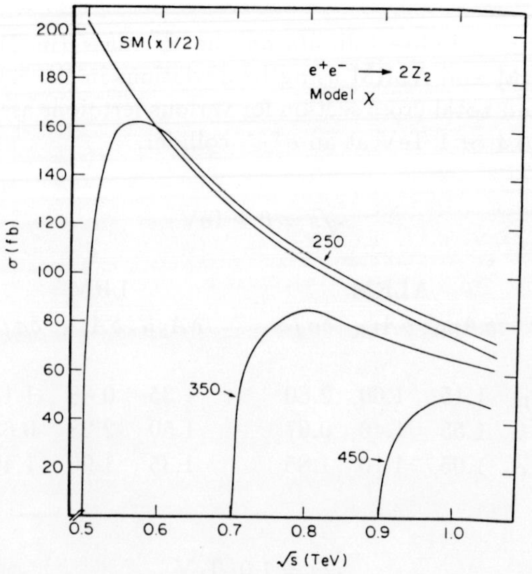

Figure 9. Cross section for $e^+e^- \to 2Z_2$ as a function of $\sqrt{s}$ for different M_2 in model χ. The SM cross section for $e^+e^- \to 2Z$ is shown for comparison.

be used for any fixed $M_{Z'}$ and $\sqrt{s}$ to convert between different models. As θ is varied, σ is found to vary by approximately a factor of 4. The Z' should be observable in either the $e^+e^- \to e^+e^-\ell^+\ell^-$ or $e^+e^- + 2$ jet final state with the peak size dependent on the relevant Z' branching fractions. The relevant backgrounds come from the $\gamma\gamma$ processes $\gamma\gamma \to \ell^+\ell^-$ and $\gamma\gamma \to q\bar{q}$ and are peaked at low p_T. Integrating over the Z' peak we can compare signal to background; for model χ with $n_g = 0$, $M_{Z'} = 300$ GeV, and $\sqrt{s} = 1$ TeV, we find $\sigma \cdot B \simeq 44$ (10) fb with a background of 2.8 (1.0) fb for the $\ell^+\ell^-$ mode. For the hadronic mode, we find $\sigma \cdot B \simeq 484$ (110) fb with a background of 4.2 (1.5) fb. It should be remembered that if we take $n_g = 3$ and $\theta \approx 25°$ then these optimistic results will need to be rescaled downwards by a factor of order $\simeq 0.2$. Nevertheless, it seems plausible that $e^+e^- \to e^+e^-Z'$ should be observable even for luminosities in the 10 fb^{-1} range and allow for a probe of the e^+e^-Z' couplings.

The $2Z'$ final state[1] is even more sensitive to the value of θ for which we find

$$\frac{d\sigma}{dt} = \frac{\pi\alpha^2}{16x_w^2(1-x_w)^2s^2}\left\{(v_e^2+a_e^2)^2 + 4v_e^2a_e^2\right\} \times \left[\frac{t}{u} + \frac{u}{t} + \frac{4M_{Z'}^2 s}{ut} - M_{Z'}^4\left(\frac{1}{t^2}+\frac{1}{u^2}\right)\right] \tag{5}$$

which leads to the total cross sections shown in Fig. 9 as a function of $\sqrt{s}$ for model χ. It is clear that the reason this process will be sensitive to θ is due to the fourth power of the couplings in the above expression. Figure 10 shows σ as a function of θ for $M_2 = M_{Z'} = 250$ GeV at the cross-section maximum; note σ changes by as much as a factor of 25 as θ is varied. Figure 10 can be used to convert the model χ cross-sections over to those for any value of θ and shows that σ displays no symmetry around θ_0. For

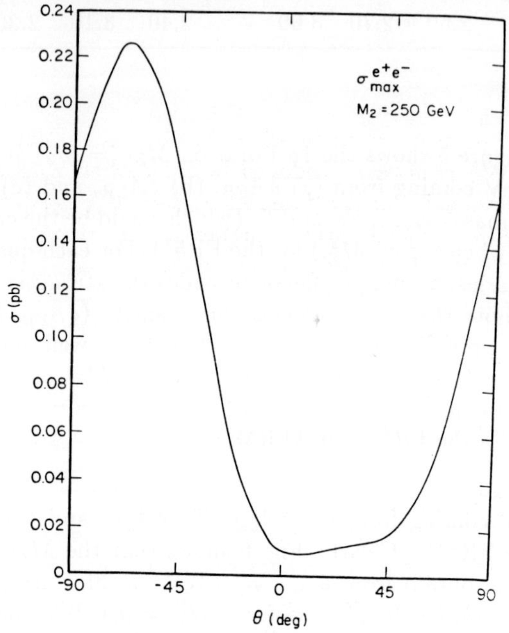

Figure 10. θ dependence of the $e^+e^- \to 2Z_2$ cross section at the cross section maximum for $M_2 = 250$ GeV.

$M_2 = 250$ GeV at $\sqrt{s} = 1$ TeV, $\sigma \simeq 100$ fb and looking at the $\ell^+\ell^- + 2$ jet final state we expect $\simeq 500$ ev/yr at a luminosity of 30 pb^{-1} for model χ if $n_g = 0$. This rate is smaller for most other values of θ especially if $n_g = 3$. If, *e.g.*, $\theta = 0$ and $n_g = 3$ we get only $\simeq 8$ ev/yr for the same luminosity. The $\gamma\gamma \to q\bar{q}$ background to this process can be almost completely avoided if there are sufficient statistics to see the double peaking in the $\ell^+\ell^-$ and 2-jet invariant mass distributions. Although the event rate for this process is small for some θ values it can provide a very sensitive probe to the value of θ in the ER5M scenario.

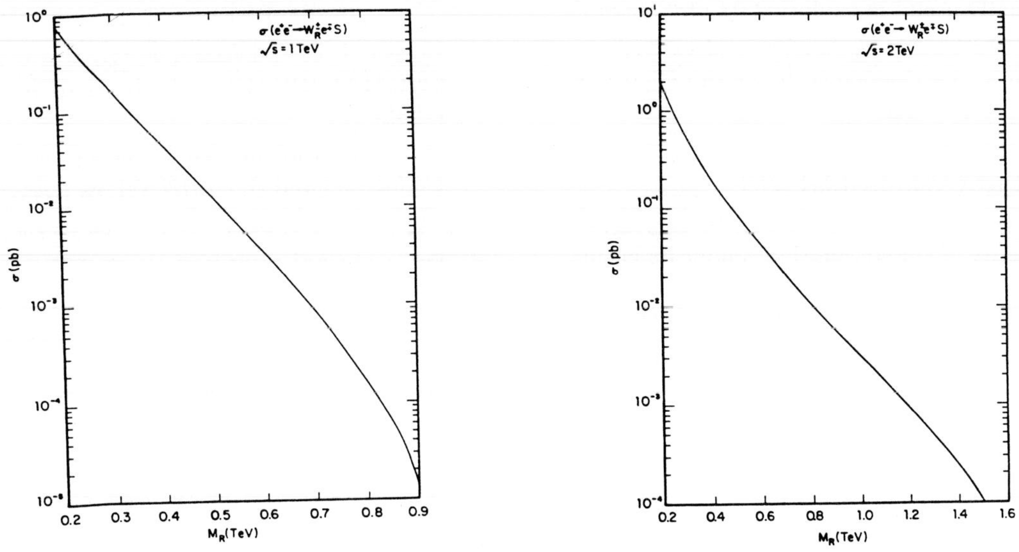

Figure 11. Cross section for single $W_R^{\pm}$ production as a function of M_R with (a) $\sqrt{s} = 1$ TeV or (b) $\sqrt{s} = 2$ TeV.

Single $W_R^{\pm}$ production[1] may provide a window into the properties of the W_R below the pair-production threshold. W_R is always produced in association with a light SO(10) singlet exotic fermion S since W_R carries negative R-parity and lepton number in the ALRM. For the $\gamma e^{\pm} \to W_R^{\pm} S$ subprocess, the cross section is given by (neglecting the S mass)

$$\frac{d\sigma}{d\hat{t}} = \frac{\pi\alpha^2}{2x_w\hat{s}^2} \times \left[\frac{2M_R^2 - \hat{u}}{\hat{s}} + \frac{2(M_R^4 - \hat{s}\hat{u})}{(\hat{s}+\hat{u})^2} + \frac{2(\hat{u}-M_R^2)}{\hat{s}+\hat{u}} - \frac{2M_R^4}{\hat{s}(\hat{s}+\hat{u})}\right] \quad (6)$$

which leads to the total cross section shown in Fig. 11 for $\sqrt{s} = 1$ TeV. In this case single W_R production is only important if $M_R > 0.5$ TeV since for lighter masses W_R pair production will dominate. Figure 11 shows that for $M_R = 0.6$ TeV and a luminosity of 30 pb^{-1} we obtain only $\simeq 120$ events before folding in branching ratios. However, it is expected that the mode $W_R \to eS$ dominates due to phase space suppression of exotic modes so that marginal studies of W_R may be possible using this process with this kind of luminosity.

5. CONCLUSIONS

We have analyzed the capability of new e^+e^- colliders to find and determine the couplings of new neutral gauge bosons within the context of E_6 superstring-inspired models. We have found that the three classes of E_6 models (ER5M, LRM, and ALRM) we have examined are easily distinguished while the parameter θ in ER5M can be determined apart from two two-fold ambiguities. In the absence of Z-Z' mixing, at least one of these ambiguities will always remain while the second may be resolvable if exotics contribute to the Z' width or if all of the Z' final states can be determined.

We have examined the ability of $\sqrt{s} = 0.4$ and 1 TeV e^+e^- colliders to search for Z' bosons with masses in excess of $\sqrt{s}$ by measurement of $\delta A_{\rm FB}$, $\delta A_{\rm LR}$, and $\delta\sigma/\sigma$ for various fermion flavors. In general, *$\delta\sigma/\sigma$ for μ's and b's in the final state provides the most stringent limits on the existence of a new Z'.* This result is surprising in that it is usually expected that asymmetry measurements are generally more powerful in probing for new Z's than simple cross-section measurements. This result is due to the larger tail on the Z' resonance peak for the heavier Z' mass range that these higher energy colliders can explore.

We have studied the process $e^+e^- \to e^+e^-Z'$ and $e^+e^- \to 2Z'$ as possible probes of the θ parameter in the ER5M. While the first process is reasonably sensitive to θ, Z' pair production is extremely sensitive to θ with the cross-section varying by as much as a factor of 40. The parallel process for the production of single right-handed W bosons ($e^+e^- \to e^{\pm}W_R^{\mp}S$) below the W_R-pair threshold can be used to search for the W_R but is found to be only marginally useful in exploring its properties.

Clearly, new e^+e^- colliders can be used as tools to probe for and study the new gauge bosons predicted in E_6 superstring-inspired models.

ACKNOWLEDGEMENTS

TGR would like to thank the Phenomenology Institute at the University of Wisconsin-Madison for its hospitality. This research was supported in part by the University of Wisconsin Research Committee with funds granted by the Wisconsin Alumni Research Foundation, and in part by the U. S. Department of Energy under contracts DE-AC02-76ER00881 and W-7405-Eng-82.

REFERENCES

1. Hewett J. L. and Rizzo, T. G., Univ. Wisconsin report MAD/PH/446, submitted to Physics Reports (1988).
2. *Proceedings of the Workshop on Physics at Future Colliders*, I and II, CERN Yellow Report 87007 (1987).
3. Ahn, C. *et al.*, SLAC-Report-329 (1988).
4. For a general review of the Left-Right Symmetric Model, its basis in SO(10), and original references see Mohapatra, R. N., *Unification and Supergravity*, Springer-Verlag, New York, 1986.
5. Ma, E., Phys. Rev. D36, 274 (1987); Mod. Phys. Lett. A3, 319 (1988); Babu, K. S., He, X.-G., and Ma, E., Phys. Rev. D36, 878 (1987); Gunion, J. F. *et al.*, *Proceedings of the Workshop: From Colliders to Supercolliders*, Madison, WI, May (1987) and Int. J. Mod. Phys. A2, 1199 (1987); Barger, V. and Whisnant, K., *ibid.* A3, 879 (1988).

$e^+e^- \to W^+W^-$: A HIGH ENERGY MAGNIFYING GLASS*

STEPHEN B. SELIPSKY

Department of Physics
Stanford University, Stanford, California 94305

ABSTRACT

There is an amplification in the radiative corrections to $e^+e^- \to W^+W^-$. The tree level matrix element exhibits a cancellation (to first order in squared center-of-mass energy s) which keeps the cross section unitary. Radiative corrections disturb the cancellation, effectively being enhanced by a factor s/m_W^2; this can be a large factor. The cancellations are re-established only after the center-of-mass energy exceeds the virtual-particle masses. A unified framework in which to calculate all such effects clarifies their physical origins, separating ordinary coupling-constant renormalization and boson self-energies from new effects at the non-Abelian vertex.

Many of the most interesting questions in physics, particularly in current weak interaction physics, concern energy scales which we cannot attain directly. Radiative corrections can be indirect probes of this high energy world; heavy states appear as quantum fluctuations even when they cannot appear on-shell. This kind of effect, already suppressed by coupling constants, is often reduced further by Appelquist-Carazzone decoupling.[1] Useful probes require us to evade such suppressions or to find compensating enhancements.

The Appelquist-Carazzone theorem states that high energy degrees of freedom can be integrated out from any low energy physics that is protected by global symmetries. Of course any parameters connected with breaking such symmetries will evade the suppression. Thus we can expect "nondecoupling" effects from *e.g.* fermion axial couplings breaking chiral symmetry, or from large mass splittings within electroweak multiplets breaking the custodial $SU(2)$ isospin symmetry. The latter effect applies also to a large Higgs mass, via the W-Z mass splitting.

* Work supported in part by a National Science Foundation Graduate Fellowship.

The reaction $e^+e^- \to W^+W^-$ is an ideal setting in which to search for such "nondecoupling". The tree-level diagrams involve a neutrino exchange in the t-channel and photon and Z exchanges in the s-channel. Taken alone, either channel grossly violates the unitarity limit on the cross section, but together they cancel the leading dependence on s. Radiative corrections have no such cancellations (at energies below threshold for production of the virtual states), and relative to the tree level enjoy an s/m_W^2 enhancement which overcomes the α/π coupling constant suppression. In addition, longitudinal W production carries a kinematic factor s/m_W^2 arising from the polarization vectors, indicating that experiments seeking the effect should isolate that polarization.

The radiative corrections are best organized with an effective Lagrangian. Summing vacuum polarization and boson self energies simply exchanges the bare theory's parameters for renormalized, running ones. Effective vertices are then modified by the remaining loop corrections;[2] the non-Abelian three-boson vertices here receive contributions independent of those in four-fermion processes. The s and t channel matrix elements are arranged to contain identical running couplings. Thus the fully corrected matrix element has just the same form as the tree level one, exhibiting the *same* cancellations, except for the correction terms separated out from the effective vertices (and a minor contribution from the Z propagator).[3] These terms are responsible for the enhanced effects. Note that the running of the coupling constants is controlled only by logarithms and decoupled physics suppressed by large masses. The corrections have been calculated for heavy fermions and scalars;[3] they can give rise to observable corrections, exemplified in figs. 1 and 2.

A heavy Higgs boson, or any equivalent electroweak symmetry-breaking field, must be treated differently because of mixing with the gauge bosons. In such cases a

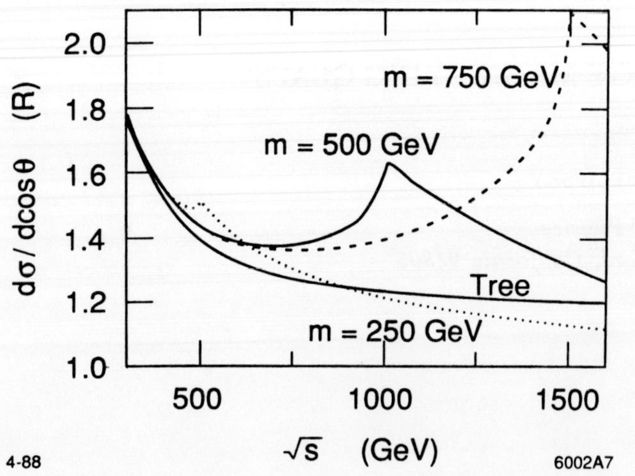

Figure 1. The differential cross section at $\cos\theta = 0$ for $e^+e^- \to W^+W^-$, with a generation of fermions at various masses.

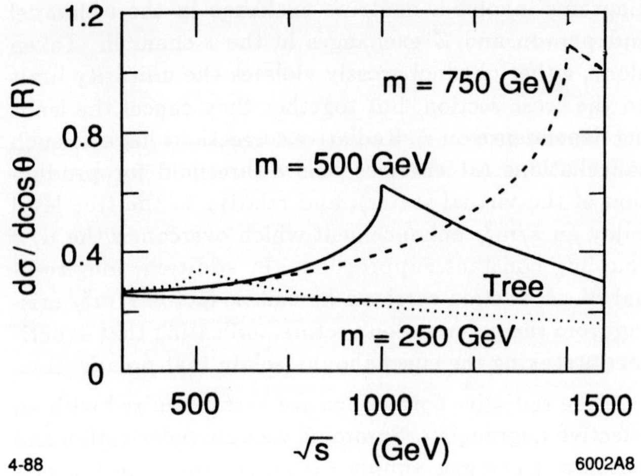

Figure 2. The contribution to fig. 1 from longitudinal W's only.

nonlinear sigma model with divergences cut off by the Higgs mass (or an equivalent new physics scale) gives an effective Lagrangian[4] whose terms can be inserted directly into the W-production matrix element as above. Details of that calculation will be given elsewhere. Technicolor models can be treated similarly; in *any* model of physics above the weak scale, the effective matrix element has the above form and behavior, requiring only insertion of the specific loop functions.

REFERENCES

1. T. Applequist and J. Carazzone, *Phys. Rev.* **D11** (1975) 2856.
2. D.C. Kennedy and B.W. Lynn, SLAC-PUB-4039.
3. C. Ahn, M.E. Peskin, B.W. Lynn and S.B. Selipsky, *Nucl. Phys.* **B**, to be published.
4. O. Cheyette and M.K Gaillard, *Phys. Lett.* **197B** (1987) 205.

LR Asymmetry in $e^+e^- \to \mu^+\mu^-$ off the Z^0 Pole: Discovery Limits of Extra Z''s in E_6

David London

Deutsches Elektronen Synchrotron (DESY), Hamburg, FRG

and

G. Bélanger

Université de Montréal, Montréal, Canada

Abstract

We examine the discovery limits for extra E_6 Z''s at a Z factory of luminosity 10^{34}, running at 100 GeV, assuming no Z^0-Z' mixing. By looking at the LR asymmetry in $e^+e^- \to \mu^+\mu^-$, we find that, for most values of θ_{E_6}, a Z factory is sensitive to Z' masses well beyond the limits of the TeVatron or LEP II. We stress that, because of systematic errors, such a measurement is better than a measurement of $\sigma(e^+e^- \to \mu^+\mu^-)$, and therefore a Z factory will require polarization.

Many theories beyond the standard model predict the existence of extra Z's (which we will generically refer to as Z'). By running high luminosity e^+e^- colliders at M_{Z^0} (Z factories), and looking for deviations from standard model predictions for the Z^0 mass, width, etc, it is possible to measure (or put limits on) Z^0-Z' mixing to high accuracy. Combined with theoretical expressions, these measurements can then be related to very large Z' masses.

However, it is possible (in principle) to have a "light" Z' and no Z^0-Z' mixing. In this case, precision measurements at the Z^0 pole will see no effects from new Z''s. We therefore suggest going to 100 GeV and looking at the LR asymmetry in $e^+e^- \to \mu^+\mu^-$. This will probe the existence of an extra Z', even if the Z^0-Z' mixing angle is zero.

As an example of the sensitivity of such a measurement, we consider the group E_6. In E_6, there are two extra Z's. The simplest way to classify them is to consider the following breakdown of E_6:

$$E_6 \to SO(10) \times U(1)_\psi \\ \to SU(5) \times U(1)_\chi \times U(1)_\psi \ .$$

If there is just one low-energy extra Z, it must be a linear combination of Z_ψ and Z_χ:[1)]

$$Z' = Z_\chi \cos\theta_{E_6} + Z_\psi \sin\theta_{E_6} \ .$$

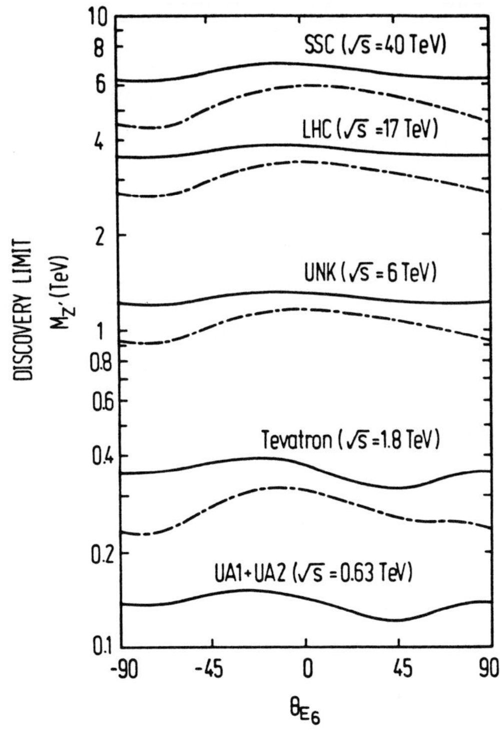

Figure 1. Discovery limits of Z''s at hadron colliders. Taken from Ref. 2.

First of all, as a benchmark, in Fig. 1 we show the discovery limits of Z''s at hadron colliders (taken from Ref. 2), in which the criteria for discovery is at least 5 $Z' \to \mu^+\mu^-$ events. As can be seen, the TeVatron can detect the existence of Z''s in the approximate mass range 300-400 GeV, depending on the value of θ_{E_6}. In Fig. 2, we show the discovery limits of Z''s at e^+e^- colliders (also taken from Ref. 2), based on a 95% c.l. deviation from the standard model. For some values of θ_{E_6}, LEP II can do better than the TeVatron, reaching up to $M_{Z'} = 500$ GeV in some cases (the third graph, which shows the LR asymmetry, assumes 100% polarization for the e^- beam, and an unpolarized e^+ beam).

We now consider the LR asymmetry in the process $e^+e^- \to \mu^+\mu^-$ at 100 GeV, assuming a polarization of 0.8. This corresponds to a 50% polarization of each of the beams.[3] We take the integrated luminosity of the machine to be 4×10^4 pb^{-1} (which corresponds to $\mathcal{L} = 10^{34}$). Using A_{LR}, the discovery limits of an E_6 Z' are then shown in Fig. 3 for both 90% c.l. and 95% c.l. For most values of θ_{E_6}, it is possible with such a Z factory to see the effects of a Z' (at 95% c.l.) well beyond the discovery limits of the TeVatron or LEP II. In fact, such a method can probe Z' mass scales up to 10 times the machine energy.

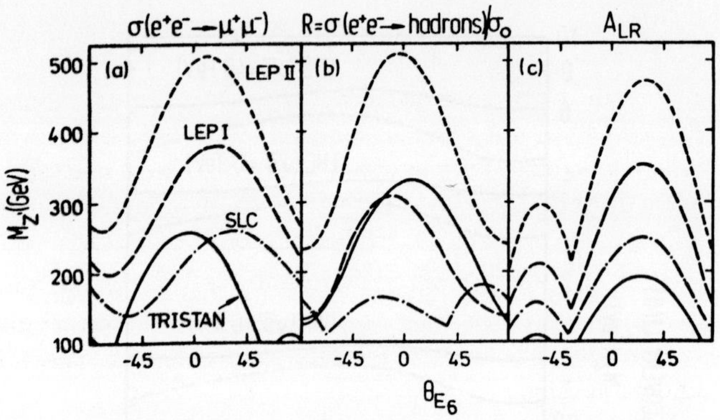

Figure 2. Discovery limits of Z''s at e^+e^- colliders. Taken from Ref. 2.

There is one further important point to be made concerning Z factories. In Fig. 2, the values of $M_{Z'}$ which can be seen using the measurement of $\sigma(e^+e^- \to \mu^+\mu^-)$ are somewhat above those which are probed using A_{LR}. However, the error in $\sigma(e^+e^- \to \mu^+\mu^-)$ is systematics limited, and hence not as good as the measurement of A_{LR} at a Z factory. Therefore, in order to avoid such systematic errors, a Z factory will require polarization.

References

1. D. London and J.L. Rosner, Phys. Rev. **D34**, 1530 (1986).
2. S. Capstick and S. Godfrey, Phys. Rev. **D37**, 2466 (1988).
3. P. Grosse-Wiesmann, private communication.

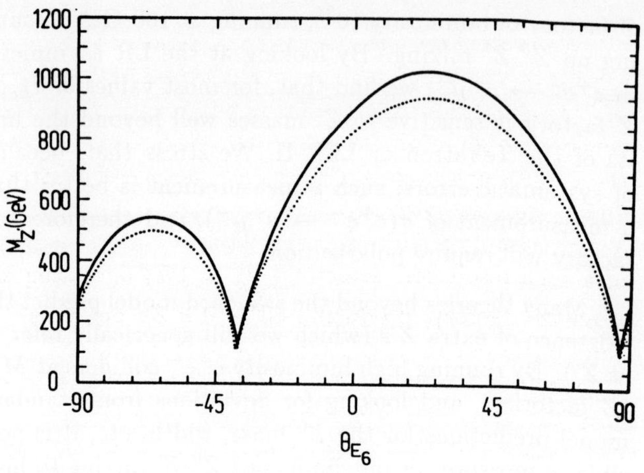

Figure 3. Discovery limits of Z''s at a Z factory at 95% c.l. (dotted line) and 90% c.l. (solid line).

Lepton Flavor Violating Processes at Z^0 Factories*

J. J. GOMEZ-CADENAS† AND C. A. HEUSCH

Santa Cruz Institute for Particle Physics
University of California, Santa Cruz, California, 95064

ABSTRACT

We present our results of a Monte Carlo study on lepton flavor violating processes of the type $e^+e^- \to (Z^0) \to l_i\overline{l_j}$ ($l_i \neq l_j$) which can be investigated in experiments achieving high luminosity at the Z^0 pole. First, we study the results expected in present Z^0 factories, using a realistic simulation of the Mark-II detector at SLC, with a data sample of 10^6 Z^0. Next, we study the possibilities of a high-resolution detector ($\frac{\sigma_E}{\sqrt{E}} \approx 1\%$) operating at a super-$Z^0$ factory, producing $10^8 - 10^9$ Z^0 per year. We summarize our results in terms of the achievable limit on a "new interaction mass scale" Λ. We find that limits on Λ of the order of 1 TeV can already be achieved by the experiments running at present Z^0 machines in the very near future. Experiments running at super-Z^0 factories would be able to improve this limit up to a factor of five, thus achieving a limit of the order of $\Lambda \sim 5\,TeV$.

1. Introduction

The search for new effects that cannot be described in the context of the Standard Model will be a fundamental aim of the new e^+e^- colliders, SLC and LEP, during the next few years. These machines will be producing typically 10^6 Z^0 per year, enough to permit searches for rare effects which can be a signal of "new physics". Among possible new accelerators to be built during the next decade, there are the so-called super-Z^0 factories, which could produce $10^8 - 10^9$ Z^0 per year. In this paper, we aim to discuss the possibility of studying the conservation of lepton flavor in experiments running at both Z^0 and super-Z^0 factories. The conservation of lepton flavor can be a very sensitive tool in the search of new physics. Although it has been observed to hold to an astonishing degree of accuracy for all cases investigated to date, it is not protected by any gauge principle. As a matter of fact, several extensions of the Standard Model can lead to observable violations of lepton flavor at or near the Z^0 pole.[1,2] It is therefore a natural question to ask how much projected running at the Z^0 pole can contribute to our knowledge of new interactions beyond our present understanding.

Following Ref. 3, let us characterize such a modification to the standard interaction scheme by means of an effective Lagrangean

$$L_{eff} = L_0 + \frac{1}{\Lambda}L_1 + \frac{1}{\Lambda^2}L_2 + \ldots$$

The ansatz is made in terms of a "new interaction mass scale" Λ. L_{eff} contains the effective low-mass theory in the L_0 term. The lowest-dimension operator that will conserve baryon and lepton number is L_2.[3] The amplitude for a new interaction will therefore be of the form:

$$A \approx \frac{1}{\Lambda^2}(f_1{}^+\Gamma f_2)(f_3{}^+\Gamma f_4)$$

where $f_i, f_i{}^+$ are fermion field operators. Note that this amplitude leads to a characteristic rate dependence on Λ

$$\sigma_{new} \approx \frac{1}{\Lambda^4}.$$

In terms of energy or momentum transfer parameters,

$$\sigma_{new} \approx s \approx Q^2.$$

Thus, from the expression above, limits on experimentally observed rare process are readily translated into lower limits for a new-interaction energy scale. A recent calculation of such limits was given in Ref. 4. Most of the data used to set them come from rare decay experiments, and are therefore characterized by low Q^2 values. There are no tight experimental limits for processes including the heaviest known lepton, τ. This suggests that searches of

* Work supported in part by the U.S. Department of Energy, contract DE-AT03-86ER40249

† Fulbright Fellow on leave of absence from Instituto de Fisica Corpuscular, Valencia, Spain

the type

$$e^+e^- \to (Z^0) \to e\tau$$
$$e^+e^- \to (Z^0) \to \mu\tau$$

may give new insights: they are characterized by much higher Q^2 values than rare decay experiments, and they involve the heavy lepton τ. At the Z^0 pole, we can write:

$$\frac{\Gamma_{new}}{\Gamma_{SM}(Z^0)} = \frac{(\frac{4\pi}{\Lambda})^2}{|(\frac{ig^2}{\Gamma_Z \cdot m_Z})^2|} \qquad (\sqrt{s} \sim m_z) \ .$$

Then, an upper limit in the branching fraction

$$B(e^+e^- \to (Z^0) \to l_i\overline{l_j}) \qquad i \neq j$$

leads immediately to a lower limit for Λ, assuming that this new interaction manifests itself as an admixture to the physical Z^0. In principle, there are three possible flavor violation processes to consider in the charged lepton sector; they are:

$$e^+e^- \to (Z^0) \to \mu\tau \ ,$$
$$e^+e^- \to (Z^0) \to \mu e \ ,$$
$$e^+e^- \to (Z^0) \to e\tau \ .$$

The tight limits on Λ derived from rare μ decay experiments[5] ($B(\mu^- \to e^-\gamma) < 1.7 \cdot 10^{-10}$, $B(\mu^- \to e^-e^+e^-) < 2.4 \cdot 10^{-12}$), may appear to make the channel $e^+e^- \to (Z^0) \to \mu e$ not very promising. On the other hand, the best limits obtained for rare τ decays[6] are five to seven orders of magnitude less stringent than the ones involving μ decay, and do not translate into tight limits on Λ. Thus, we will emphasize the lepton flavor violating processes involving the τ lepton in this paper. In order to gain realistic insight into the first results that can be obtained, we begin our study with a simulation of the Mark-II/upgrade detector at SLC, and consider, in Section 2, a sample of 10^6 Z^0; this may be seen as realistic for SLC and LEP. In Section 3, we study the results that can be achieved by a high resolution detector (HRD) operating at a super-Z^0 factory. Finally, in Section 4, we present our conclusions.

2. Study of the Process $e^+e^- \to Z^0 \to l\tau(l = e, \mu)$ at Mark II/SLC

2.1. The Mark II detector at SLC

The Mark II detector at SLC has been described elsewhere.[7,8] Here we will only summarize its main features. The detector is shown in Figure 1. The primary components are:

(1) A conventional solenoid, providing a field of at least 5 kG.

(2) A cylindrical drift chamber based on a multi-sense-wire cell structure . With vertex constraint, a precision of $\frac{\sigma_p}{p^2} \approx 0.2\% \, GeV^{-1}$ is achieved over 65% of 4π solid angle.

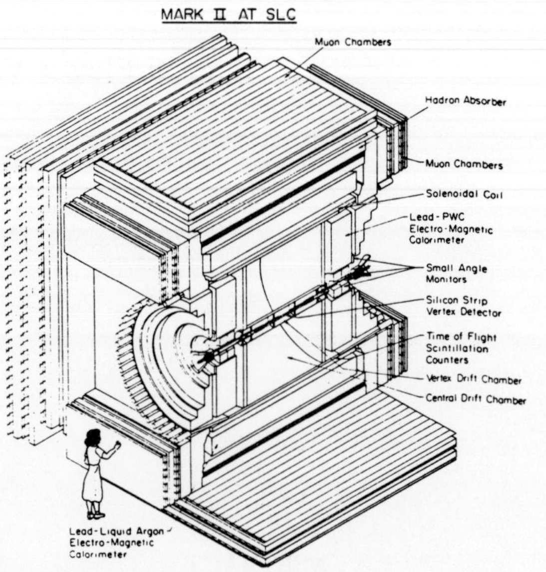

Figure 1. The Mark II detector at SLC.

(3) Electromagnetic calorimetry is provided over the angular range $\cos\theta < 0.7$ by an octagonal array of finely segmented liquid argon modules providing an energy resolution of $\frac{\sigma_E}{E} \approx \frac{20\%}{\sqrt{E}}$. The forward region is covered by sandwich calorimeters alternating Lead and Proportional tube chamber layers, providing a similar resolution.

(4) A time of flight system, with a resolution of 200 ps.

(5) A muon identification system provided by a system of steel absorbers followed by layers of proportional tubes, the system covers 55% of 4π.

2.2. Monte Carlo simulation

We use a realistic simulation of the Mark II detector. All the different parts described above are accurately described, including inefficiencies due to dead zones, cracks, etc.[9] To study the lepton flavor violating processes (LFV), we have developed a Monte Carlo generator for the processes $e^+e^- \to Z^0 \to l\tau(l = e, \mu)$. Since the Feynman diagrams leading to two different lepton flavors in the final state are model dependent, we have not attempted to generate the events according to any of them. Instead, we simply assume couplings of both leptons to the intermediate Z^0. Using code developed from the LULEPT[10] generator, we produce events for the process:

$$e^+e^- \to l^+l^-,$$

with $l = \mu, \tau$, or a heavier sequential lepton χ. The lepton subsequently decays into lighter fermions. In the case of the decay $l \to \nu\overline{q_1}q_2$, the LUND code performs the subsequent quark fragmentation to the final state particles. To simulate LFV processes, we produce τ pairs, then scan the LUND banks containing the final stable particles, and remove those corresponding to the decay of one of the τ, inserting instead one stable lepton of different flavor. To study the different backgrounds to our processes, we have used different Monte Carlo generators.[11] In all the cases, the final four-vectors produced by the generators were passed through the detector simulation.

2.3. Overall cuts and particle identification criteria

In order to guarantee a good measurement of the charged-track energies and momenta, we accept only events up to $|\cos\theta| < 0.7$. In the case of the process $e^+e^- \to (Z^0) \to e\tau$, where the calorimeters are used to measure the energy of the electron, the endcap calorimeters provide a coverage up to $|\cos\theta| < 0.95$, with good resolution. Nevertheless, in the forward region, $|\cos\theta| > 0.7$, the Bhabha cross section, one of the most important backgrounds to this process, increases very rapidly. In the case of the process $e^+e^- \to (Z^0) \to \mu\tau$, muon identification is guaranteed only for $|\cos\theta| < 0.7$. Also, the momentum measured by the drift chamber worsens at forward angles.

Next, we make use of the low multiplicity characteristic of our process. Since more than 99% of the τ's decay to either one or three charged prongs, we only accept events with 2 or 4 good charged tracks. A charged track is defined as *good* if:

i) The momentum measured by the drift chamber (DC) is less than $P_{max} = \frac{\sqrt{s}}{2}(1+3\sigma)$;

ii) The track projects into the fiducial volume of the calorimeters and/or the muon identification system;

iii) The track projects back to the origin: $|z_{track}| < 6$ cm, $R_{track} < 4$ cm;

iv) $P_{\perp(track)} > 100$ MeV, where $P_{\perp(track)}$ is the track transverse momentum.

Particle identification is achieved in Mark II through a combination of algorithms using the momentum and $\frac{dE}{dx}$ as measured by the drift chamber, the energy as measured by the calorimeter, and the muon identification and time of flight information for each particle. In Figure 2, we illustrate the efficiency of these algorithms. A detailed description can be found in Refs. 12 and 13.

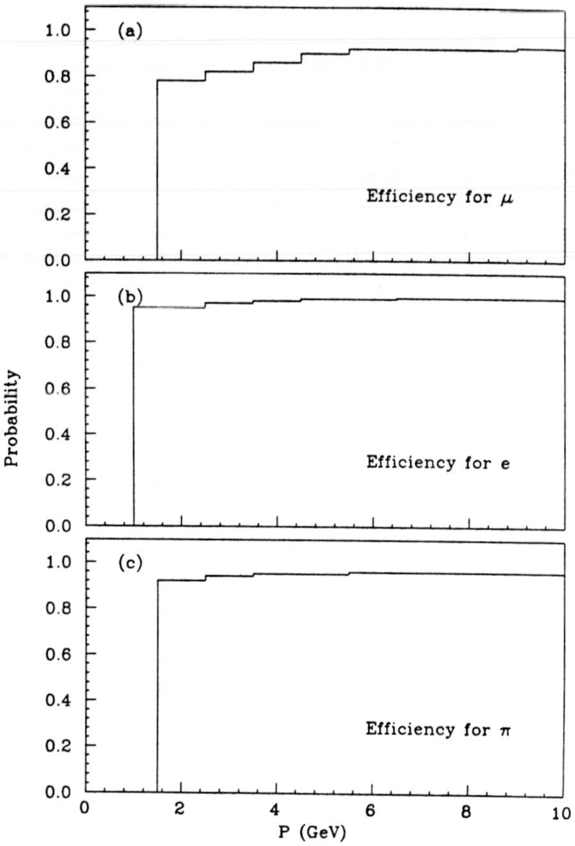

Figure 2. Efficiency for (a) muon identification, (b) electron identification, (c) pion identification.

2.4. Study of the process $e^+e^- \to (Z^0) \to e\tau$

<u>Event signature and backgrounds.</u>

The event signature for the process $e^+e^- \to (Z^0) \to e\tau$ is very distinctive. The event is characterized by a very energetic electron, carrying the beam energy, recoiling against a τ candidate. The tau decays most of the time to either one or three charged prongs (plus any number of neutrals). Thus, we demand:

i) Low charged multiplicity: two or four charged tracks with zero net charge.

ii) A "boosted event" (high thrust) with topology 1 versus 1 or 1 versus 3.

iii) Missing energy and momentum, due to the emission of neutrinos.

iv) An electron candidate with $E \approx E_{beam}$.

v) A τ candidate recoiling against the electron candidate. In Table 1, we show the most relevant τ decays along with their signature.

As we discuss below, the channel $\tau^- \to e^-\overline{\nu_e}\nu_\tau$ is not considered, so as to avoid the radiative Bhabha background. The channels $\tau^- \to \mu^-\overline{\nu_\mu}\nu_\tau$ and $\tau^- \to \pi^-\nu_\tau$ present a very clean signature, with their e versus μ or e versus π topology and large missing energy and momentum. The decays $\tau^- \to (\rho^-\nu_\tau) \to \pi^-\pi^0\nu_\tau$ and $\tau^- \to (A_1\nu_\tau) \to (3\pi)^-\nu_\tau$ are characterized by a high invariant mass, as opposed to the QED background events, which are characterized by an invariant mass close to zero. In Figure 3 we show typical events with the four accepted signatures, while in Figures 4, 5, and 6, we show the relevant distributions characterizing an event of the type $e^+e^- \to (Z^0) \to e\tau$.

The two most significant backgrounds to this process are, first, τ pair production, where one τ subsequently decays into an electron and two neutrinos, $\tau^- \to e^-\overline{\nu_e}\nu_\tau$,

Table 1

τ decays	Br. Fraction	Signature	Assum' Resonance	Inv' Mass
$\tau^- \to e^- \overline{\nu_e} \nu_\tau$	17.5	e candidate + miss' E and P	none	0
$\tau^- \to \mu^- \overline{\nu_\mu} \nu_\tau$	17.5	μ candidate + miss' E and P	none	0
$\tau^- \to \pi^- \nu_\tau$	10	π candidate + miss' E and P	none	0
$\tau^- \to (\rho^- \nu_\tau) \to \pi^- \pi^0 \nu_\tau$	21	π + neutrals + miss' E and P	ρ	0.7 GeV
$\tau^- \to (A_1 \nu_\tau) \to (3\pi)^- \nu_\tau$	13.5	3π + miss' E and P	A_1	1.1 GeV

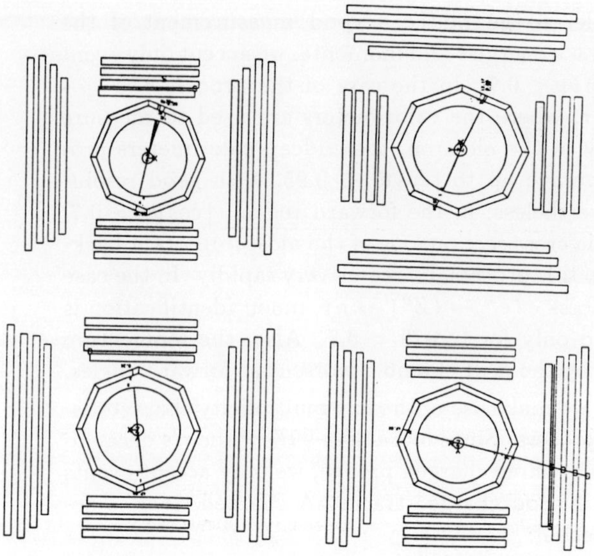

Figure 3. Monte Carlo generated $\tau^+ e^-$ pairs, with the τ subsequently decaying via (a) $\tau^- \to \mu^- \overline{\nu_\mu} \nu_\tau$, (b) $\tau^- \to \pi^- \nu_\tau$, (c) $\tau^- \to (\rho^- \nu_\tau) \to \pi^- \pi^0 \nu_\tau$, and (d) $\tau^- \to (A_1 \nu_\tau) \to (3\pi)^- \nu_\tau$.

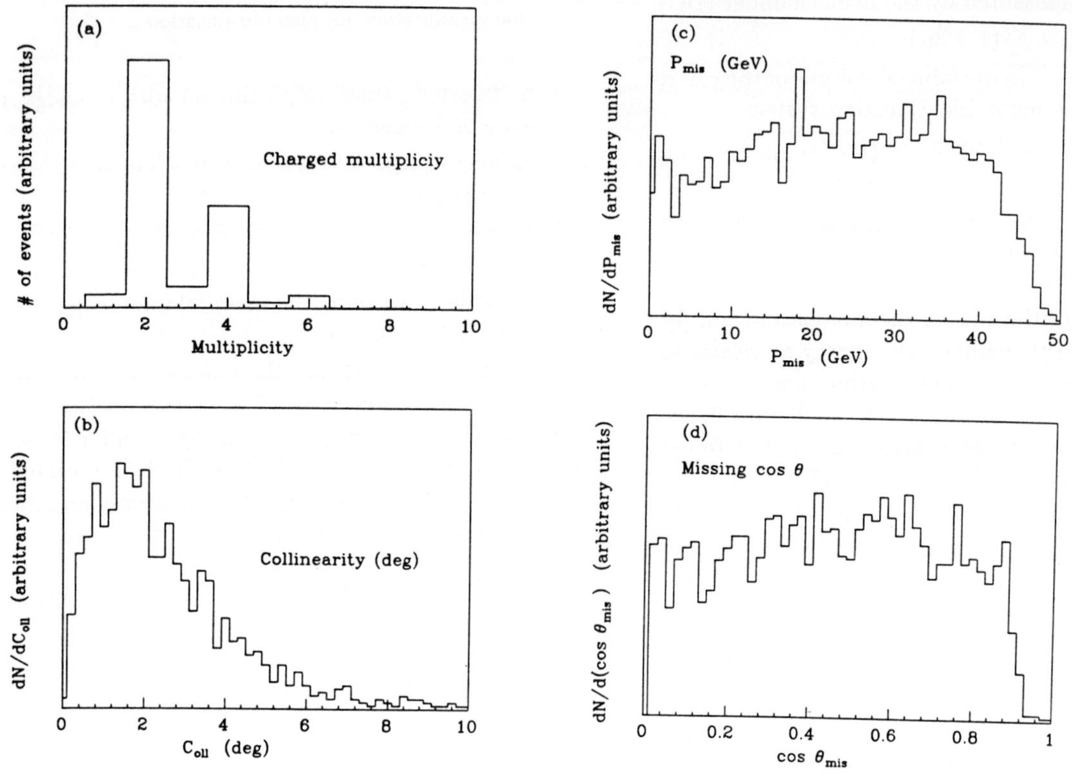

Figure 4. Overall event signature: (a) Charged multiplicity, (b) collinearity in degrees (1 versus 1), (c) Missing momentum, (d) $|\cos\theta_{miss}|$.

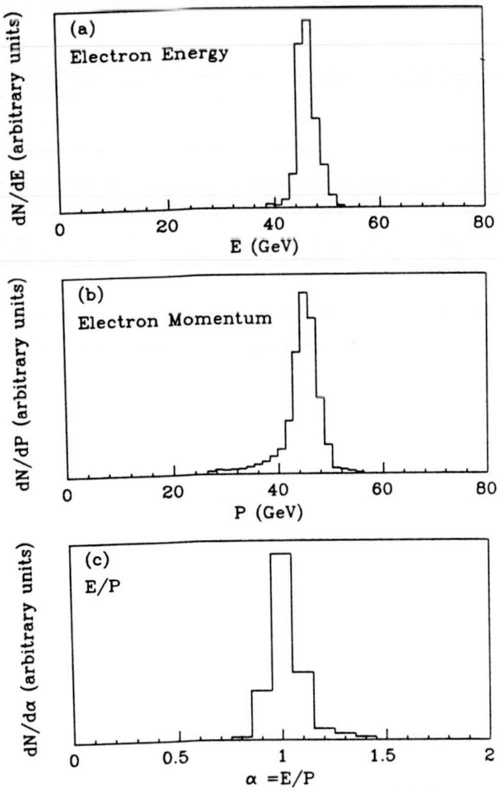

Figure 5. Electron candidate signature: (a) Energy measured in the calorimeter, E. (b) Momentum measured in the drift chamber P. (c) $\frac{E}{P}$.

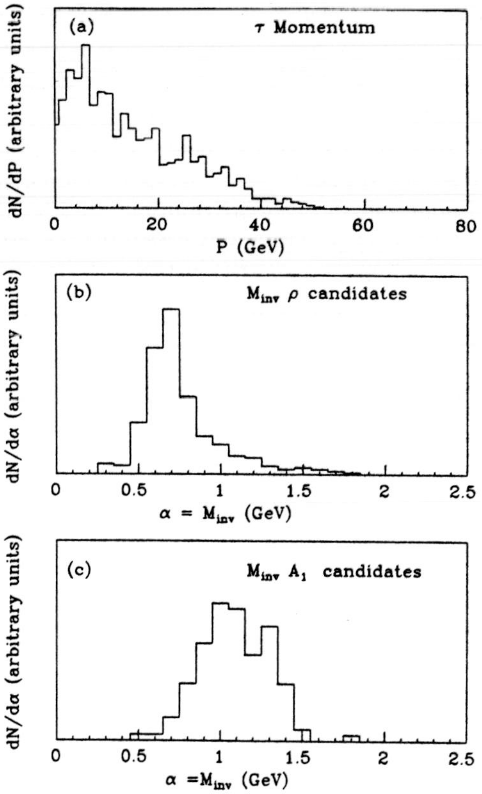

Figure 6. τ candidate signature: (a) Inclusive momentum distribution, (b) invariant mass distribution for $\tau^- \to (\rho^- \nu_\tau) \to \pi^- \pi^0 \nu_\tau$ candidates, (c) invariant mass distribution for $\tau^- \to (A_1 \nu_\tau) \to (3\pi)^- \nu_\tau$ candidates.

and the electron is at the end-point of its energy distribution; second, radiative Bhabha scattering events, where one of the electrons radiates such as to simulate a τ. Additionally, there is a small background coming from events of the type $e^+e^- \to \gamma\gamma \to e^+e^-\mu^+\mu^-$. We have not found any other significant source of background.

Radiative Bhabha event suppression.

The most important QED background to our process is radiative Bhabha scattering. It can simulate the LFV signal in two different ways:

A. Topology 1 versus 1.

Consider an event of the type $e^+e^- \to Z^0(\gamma) \to e^+e^-(\gamma)$, where the final-state photon is misidentified or unseen by the detector (i.e., it goes through a crack). The event can then be incorrectly interpreted as a LFV candidate, with the τ decaying into an electron and two neutrinos. Since the Bhabha cross-section is very large, that situation will occur frequently; this leads to an additional cut, namely, not to accept events where *two* electrons are identified. As we discuss above, this cut implies the veto of the $\tau^- \to e^-\overline{\nu_e}\nu_\tau$ channel, thus suppressing 17.5% of the signal, but also suppresses part of the τ background. This cut alone rejects most of the radiative Bhabha events. Nevertheless, there is still some contamination, since the electron identification algorithms have non-ideal efficiency. Further suppression is achieved by taking advantage of the fact that the total missing energy and momentum of a radiative Bhabha event are close to zero. Thus, additional cuts on these variables are imposed.

B. Topology 1 versus 3.

The radiative Bhabha events can fake this topology when a photon converts into an electron pair. This can be easily eliminated by using a pair-finding algorithm.[12] Also, the event is characterized by low invariant mass, as opposed to the $\tau^- \to (A_1\nu_\tau) \to (3\pi)^-\nu_\tau$ channel. We impose a cut in the invariant mass of the system, and an electron-rejection cut for all 3 tracks.

$e^+e^- \to \gamma\gamma \to e^+e^-\mu^+\mu^-$ events suppression.

Another possible QED background is due to $e^+e^- \to \gamma\gamma \to e^+e^-\mu^+\mu^-$ events, where one e and one μ make it into the detector and the other pair escapes in the forward direction. However, these events are characterized by low average energy, and missing momentum pointing along the beam direction. Also, the charge combination is wrong 50% of the time ($|\sum Q| > 0$). Furthermore, the cross-section for this type of processes is very small. The suppression of this background is achieved through the cut on the energy of the electron candidate, along with the requirements of zero net charge, and a cut on the direction of the missing momentum vector: $|\cos\theta_{miss}| < 0.9$.

τ pair suppression.

Tau pair production may fake the LFV signal, when one of the τ decays via $\tau^- \to e^-\overline{\nu_e}\nu_\tau$, and the electron is at the end-point of its energy distribution (see Fig. 6(a)). To suppress this background, we have to impose a cut as high as possible in the energy of the electron candidate. The efficiency of this cut is limited by the detector resolution. The electron energy distribution, in the case of a LFV process, is approximately a Gaussian with mean value $<E> = E_{beam}$, and dispersion $E_{disp} = \sigma$, where σ is the detector resolution. On the other hand, the τ background energy distribution is linear near the end-point. Thus, we have to impose a cut $E_{cut} = E_{beam} - r\sigma$, where r is selected to maximize the τ background rejection while keeping a reasonable efficiency for the LFV signal. The optimum value of r turns out to be $r = 1\sigma$. Given our electromagnetic calorimeter resolution, $\sigma \approx 1.5\,GeV$ at $\sqrt{s} = 93\,GeV$.

Summary of the cuts.

To summarize, we impose the following cuts:

i) Two or four good charged tracks, with zero net charge.

ii) "Boosted" event with topology 1 versus 1 or 1 versus 3.

iii) Collinearity angle between tracks (1 versus 1 topology), $\theta_{coll} > 1\,\text{deg}$.

iv) Missing energy and momentum on the event, $E_{miss} >$ 5 GeV, $P_{miss} >$ 5 GeV.

v) $|\cos\theta_{miss}| < 0.9$.

vi) One and only one electron candidate with energy $E_e > E_{beam} - 1\sigma = 45.0$ GeV.

vii) One τ candidate decaying to $\tau^- \to \mu^-\overline{\nu_\mu}\nu_\tau$, $\tau^- \to \pi^-\nu_\tau$, $\tau^- \to (\rho^-\nu_\tau) \to \pi^-\pi^0\nu_\tau$, or $\tau^- \to (A_1\nu_\tau) \to (3\pi)^-\nu_\tau$. In the cases in which the τ decays to one or three charged tracks plus neutrals, a cut on the invariant mass of the system is imposed, $M_{inv} > 0.5\,GeV$.

This set of cuts leaves 30% of the signal, while rejecting the τ background by a factor of $3 \cdot 10^{-4}$. The QED background is suppressed by a factor of better than 10^{-5}.

τ pair/$e\tau$ separation. Maximum-likelihood fit to the $e\tau$ distributions

Once the final sample has been selected, we perform a maximum-likelihood fit to the resulting distribution. Since the only relevant background passing the cuts is the τ background, we can obtain further rejection due to the fact that we know the exact shape of both the signal and the background distributions. For the signal, we have, in the absence of radiative corrections, a monochromatic electron carrying the beam energy, while in the case of the τ background, the electron energy distribution can be expressed as:[14]

$$\frac{d\Gamma}{dx_2} = A_2(x) - A_3(x) + \tfrac{2}{3}\rho\left(\tfrac{4}{3}A_3(x) - A_2(x)\right).$$

In this expression, for $\beta \approx 1$ we have $A_n(x) = \frac{1}{n}(1 - x^n)$, $x = \frac{E_{electron}}{E_{beam}}$; ρ is an effective Michel parameter which can be tuned to take into account the radiative corrections.

Let us call $S'(x')$ and $B'(x')$ the theoretical functions describing the signal and the background, respectively. Both are functions of the "real" or unsmeared normalized energy x'. To take into account the influence of the experimental resolution, we define the functions $S(x)$ and $B(x)$, where x is now the observed normalized energy

$$S(x) = \int_{x'} S'(x')\cdot G(x',x)\cdot dx',$$

$$B(x) = \int_{x'} B'(x')\cdot G(x',x)\cdot dx';$$

here, $G(x',x)$ is the resolution function describing the effect of the detector. We approximate $G(x',x)$ as a Gaussian:

$$G(x',x) = \frac{1}{\sqrt{2\pi}\sigma}e^{\frac{-(x-x')^2}{2\sigma^2}},$$

with σ denoting the detector resolution. In Fig. 7, we show the resulting fits to the signal and background distributions.

We now define normalized functions in the region we are considering,

$$U_s = \frac{S(x)}{N_s},$$
$$U_b = \frac{B(x)}{N_b}.$$

The normalizing denominators are:

$$N_s = \int_{x1}^{x2} S(x)dx,$$

$$N_b = \int_{x1}^{x2} B(x)dx,$$

with $x1 = \frac{(E_{beam}-1.5\sigma)}{E_{beam}} = 0.97, x2 = \frac{(E_{beam}+4.5\sigma)}{E_{beam}} = 1.13$, so that $\int_{x1}^{x2} U_s dx = \int_{x1}^{x2} U_b dx = 1$. Next, we form the likelihood function $L(x)$. Assuming that we have n events in the region $x_1 - x_2$,

$$L(\alpha) = \prod_{i=1}^{n} f(x_i,\alpha) = \prod_{i=1}^{n}(1-\alpha)U_b(x_i) + \alpha U_s(x_i).$$

The parameter α gives the mixing between signal and background. From the determination of α we obtain an upper limit to the fraction

$$\Gamma = \frac{\Gamma(e^+e^- \to (Z^0) \to e\tau)}{\Gamma(e^+e^- \to Z^0 \to \tau^+\tau^-)} = \frac{\Gamma_\tau}{\Gamma_{LFV}}\alpha$$

where Γ_τ and Γ_{LFV} are, respectively, the fraction of τ background and signal events passing the cuts. The value

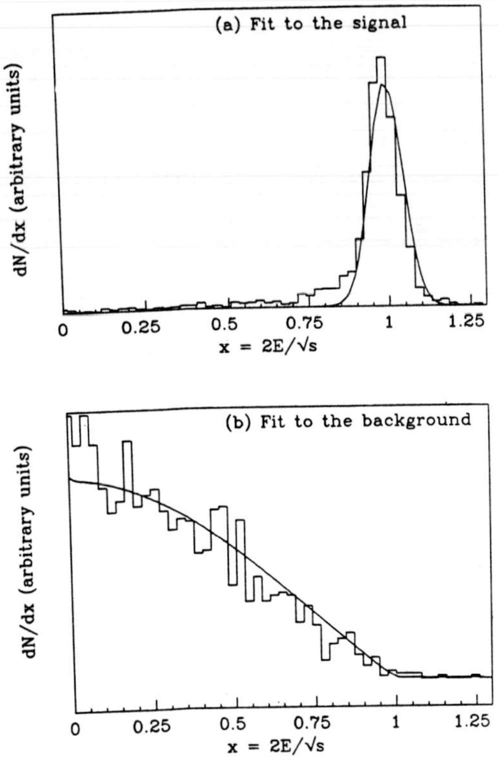

Figure 7. Fits to the (a)signal and (b) background distributions with the functions $S(x)$ and $B(x)$.

of Γ translates directly into an upper limit for the branching fraction $B(e^+e^- \to (Z^0) \to e\tau)$ and the new interaction scale Λ. To obtain α, we maximize $L(\alpha)$, or, alternatively minimize $-\log(L(\alpha))$

$$-\log(L(\alpha)) = -\sum_{i=1}^{n} \log[(1-\alpha)U_b(x_i) + \alpha U_s(x_i)] \ .$$

In Fig. 8, we show the minimization of the function $-\log(L(\alpha))$ for the τ background events passing the cuts. We obtain $\alpha = 0.1$ at 95% C.L. This translates into a limit on the Branching fraction for our process,

$B(e^+e^- \to (Z^0) \to e\tau) < 10^{-4}$, with 10^6 Z^0.

Under the assumptions made in section one, this value translates into a limit for the new interaction scale, of $\Lambda > 1\,TeV$.

2.5. The process $e^+e^- \to (Z^0) \to \mu\tau$

The study of this process is very similar to the study of the process that we have discussed in detail, $e^+e^- \to (Z^0) \to e\tau$. The only difference is that, now, we have a μ instead of an electron, carrying the beam energy. In this case, the relevant backgrounds are radiative μ pair production, and τ pair production, with one tau decaying via $\tau^- \to \mu^- \overline{\nu}_\mu \nu_\tau$ and with the muon at the end-point of the distribution. The event selection is basically the same as in the previous case, except that, obviously, we ask for

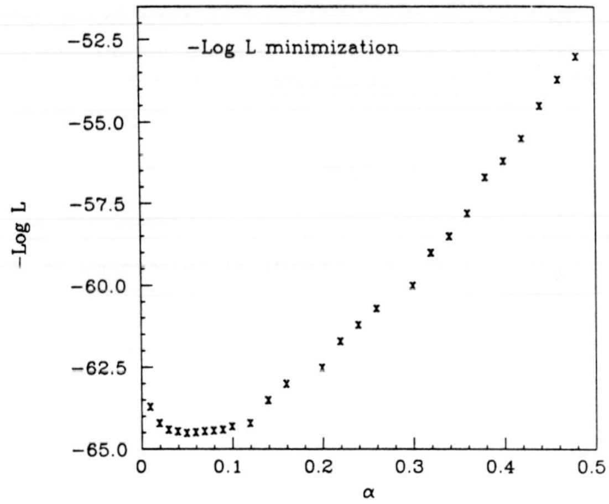

Figure 8. Minimization of the function $-\log(L(\alpha))$.

one and only one μ in the event. In our study of the Mark II/SLC detector, we obtain a worse limit for $B(e^+e^- \to (Z^0) \to \mu\tau)$ than for $B(e^+e^- \to (Z^0) \to e\tau)$ due to the limited resolution for measuring the μ energy, and the small solid angle subtended by the muon detector. This need not be the general case, since, in the future, a high resolution detector running at Z^0 factories will achieve comparable resolutions for electron and muon energies. Then, the study of this channel is very useful: not only does a cross-check for the channel $e^+e^- \to (Z^0) \to e\tau$, due to the fact that the backgrounds and the systematic errors associated with particle identification are different, but there is the obvious possibility that generation mixing between the second and third families is quite different for the first and third families case.

3. Study of the Process $e^+e^- \to Z^0 \to l\tau (l = e, \mu)$ at HRD/SZF

We now consider the case of a high resolution detector (HRD) operating at a super-Z^0 factory (SZF). We have made a crude simulation of this detector, assuming a resolution for measuring lepton (electron or muon) energies of $\frac{\sigma_E}{E} = 1\%$. Due to the better particle identification of a HRD, we obtain a better signal efficiency, of the order of 45%, while improving the background rejection. As we discussed in the previous section, the most serious background to consider is τ pair production.

We have considered three different data samples: 10^7, 10^8, and 10^9 Z^0. After performing the analysis that we described in the previous section, we obtain the limits:

$B(e^+e^- \to (Z^0) \to e\tau) < 7 \cdot 10^{-6}$ for 10^7 Z^0,

$B(e^+e^- \to (Z^0) \to e\tau) < 7 \cdot 10^{-7}$ for 10^8 Z^0.

We do not obtain much better results with 10^9 Z^0. To improve the limit beyond $7 \cdot 10^{-7}$ one needs not only 10^9 Z^0 but better detector resolution. This is probably out of the reach of even the next generation of experiments . Thus, we conclude that the best limit on Λ that we can project for experimentation at the center-of-mass energy defined by the Z^0 pole is of the order of $\Lambda \sim 5\,TeV$.

4. Running Off The Z^0 pole

The above considerations pertain to the case where $e^+ e^-$ colliders are run specifically at the energy of the Z^0 pole. This leads to excellent event rates, but is by no means ideal for lepton flavor violation searches: well off the pole, the relation of new-interaction width to standard model width is approximately given by[3)]

$$\frac{\Gamma_{new}}{\Gamma_{SM}(1\gamma)} = \frac{(\frac{4\pi}{\Lambda})^2}{(\frac{e^2}{Q^2})^2}.$$

Assuming we can experiment at a center-of-mass of 100 GeV, and reach a total sample of some 10^3, 10^4, annihilation events with τ leptons, the new-interaction scale parameter Λ could be probed out to some 6 to 10 TeV. This regime is the subject of a separate study.

5. Conclusions

We have studied in detail the limits that can be achieved in the search for lepton flavor violating processes at the Z^0 pole. We find that a limit on the branching fraction for the LFV process $B(e^+e^- \to (Z^0) \to e\tau) < 10^{-4}$ is within the range of the experiments that will start data taking within the next two years. We project a limit of $B(e^+e^- \to (Z^0) \to e\tau) < 7 \cdot 10^{-7}$ for a possible next generation of machines and experiments at the Z^0 pole. Assuming standard (minimal) Q^2 dependencies for new interactions, this search leads to sensitivities of their energy scale at a level of $\sim 1\,TeV$ (Mark II/SLC) to $\sim 5\,TeV$ (HRD/SZF).

Acknowledgements: We are indebted to the Mark II Collaboration, which has developed most of the software tools we used. We profited from discussions with P. Burchat, J. Dorfan, G. Gidal, L. Labarga, B. Lockman D. Stoker. G. Trilling, A. Weinstein and C. von Zanthier. One of the authors (JJG) wants to acknowledge a Fulbright Grant which has made possible his stay at SCIPP possible.

REFERENCES

1. J. Bernabeu *et al.*, Phys. Lett. 187B (1987) 303. J. Bernabeu and A. Santamaria; Phys. Lett. 197B (1987) 418.
2. T. G. Rizzo; Phys. Rev. D38 (1988) 71, G. Eilam and T. G. Rizzo; Phys. Lett. 188B (1987) 91.
3. C.A. Heusch, in: Progress in Electroweak Interactions J. Tran Thanh Van ed.; Mark II/SLC Physics Working Group Note 2-21. SLAC Report 306 (1987).
4. I. Bigi *et al.*, Phys. Lett. 166B (1986) 238.
5. M. Aguilar-Benitez *et al.*, Phys. Lett. 170B (1986).
6. H. Albrecht *et al.*, Phys. Lett. 185B (1987) 228.
7. Proposal for the Mark II at SLC, SLAC-PUB-3561.
8. The Mark II Detector for the SLC, SLAC-PUB-4558.
9. We have used the Mark II Monte Carlo package, HOWL, which includes a very realistic simulation of the detector; it has been extensively tested with experimental data at PEP.
10. D. Stoker, Mark II/SLC Note 161 (1987).
11. We have used the following codes: LULEPT for the study of the τ pair background; LUND68 for the hadronic background; HOWLEG and HOWLEX for the radiative Bhabha events; MUMUG for the radiative μ pair backgrounds; and GGDEPA for the 2 γ backgrounds. All these programs are part of the standard Monte Carlo package of the Mark II detector.
12. P. Burchat, Ph.D. Thesis, SLAC-292 UC-34D.
13. P. Burchat and M. Nelson. Mark II/SLC note #114.
14. A. Ali and Z. Z. Aydin, DESY 77/04, January 1977.

SEARCHING FOR AN EXOTIC LEPTON AND GAUGE BOSON AT HIGH ENERGY e^+e^- COLLIDERS*

K. K. GAN

Stanford Linear Accelerator Center, Stanford University, Stanford, California 94309

ABSTRACT

The possibility of searching in high energy e^+e^- collisions for a singly produced heavy lepton with an ordinary light lepton, $e^+e^- \to L^+\ell^-$, has been investigated. The search is motivated mainly by its distinctive experimental signatures. In a data sample corresponding to an integrated luminosity of 30 fb^{-1} collected at a 1 TeV e^+e^- collider, the search is sensitive to a lepton of mass up to about 1 TeV/c^2 and a production cross section as low as 10% of the μ–pair point cross section. For the case where the process is mediated by an exotic gauge boson, sensitivity to a gauge boson of mass up to about 2 TeV/c^2 can be achieved.

1. INTRODUCTION

The search for a process that is beyond the standard model has long been the passion of high energy physics. One process that has not been explored in e^+e^- collisions is the single production of a heavy lepton together with an ordinary light lepton,[1] $e^+e^- \to L^+\ell^-$, where the heavy lepton has a mass greater than the beam energy and is not an excited state of the light lepton. For the case of an excited lepton there has been a profusion of searches, starting with the pioneering work at the fixed target experiments, and it will not be discussed here.[2] The search is motivated mainly by the distinctive experimental signatures. For example, the momentum spectrum of the light lepton exhibits a monochromatic peak. In the spirit of the Fermi's parameterization of the weak interaction, the production mechanism is assumed to be a four-fermion contact interaction, as shown in Fig. 1(a). Examples of the production mechanism are an exotic gauge boson Z' exchange [Fig. 1(b)] or a constituent exchange as in the composite model [Fig. 1(c)].

The unpolarized differential cross section for the process is

$$\frac{d\sigma}{d\Omega} = \frac{\alpha^2}{4s}\chi^2(1+\cos\theta)\left(\frac{P_L}{P_b}\right)\left[\frac{1}{2}\left(\frac{m_L}{P_b}\right)^2 + (1+\cos\theta)\left(\frac{P_L}{P_b}\right)\right] , \quad (1)$$

*Work supported by the Department of Energy, contract DE–AC03–76SF00515.

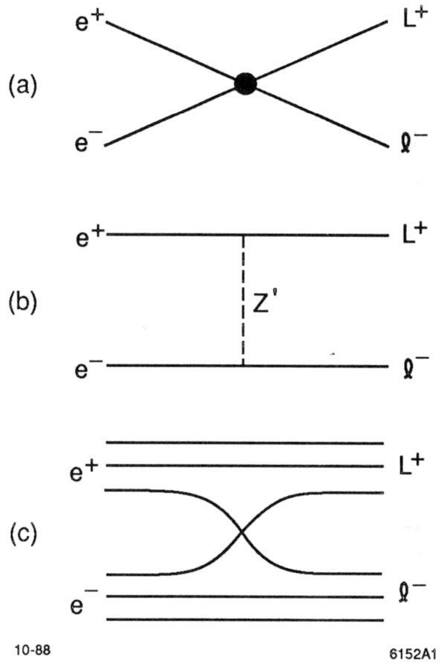

Fig. 1. The production of a heavy-light lepton pair through a four-fermion contact interaction (a). The interaction can be mediated by an exotic gauge boson exchange (b) or a constituent exchange (c).

where

$$\chi = \frac{\sqrt{2}\,Gs}{4\pi\alpha} , \quad (2)$$

G is the interaction coupling strength, m_L is the mass of the heavy lepton, θ is the polar angle of the light lepton, and P_b and P_L are the momentum of the beam and the heavy lepton, respectively. Integrating over the entire solid angle gives,

$$\sigma = \sigma^0 \cdot \chi^2\left(\frac{P_L}{P_b}\right)\left[\frac{3}{8}\left(\frac{m_L}{P_b}\right)^2 + \left(\frac{P_L}{P_b}\right)\right] , \quad (3)$$

where σ^0 is the μ–pair point cross section through the one-photon exchange. For the case of a Z' with mass $m_{Z'}$ being exchanged,

$$\chi = \frac{s}{4\pi\alpha} \cdot \frac{g^2}{4m_{Z'}^2} \ , \tag{4}$$

where g is the coupling strength of Z'.

The sensitivity of the search is studied using a Monte Carlo technique. The study assumes an e^+e^- linear collider operating at $\sqrt{s} = 1$ TeV with a luminosity of 3×10^{33} cm^{-2} s^{-1}. The effects of beamstrahlung[3] and initial state radiation[4] are included. A simple simulation of a generic detector is used.[5] In the Monte Carlo program, the heavy-light lepton pair is produced with an angular distribution according to Eq. (1). The heavy lepton is assumed to be a sequential lepton and decays into a weak vector boson W and a light neutrino ν_L. The W then decays according to the standard model. The width of the heavy lepton is given by

$$\Gamma = \frac{\alpha m_L}{16 \sin^2\theta_W} \left(\frac{m_L}{m_W}\right)^2 \left(1 + 2\frac{m_W^2}{m_L^2}\right) \left(1 - \frac{m_W^2}{m_L^2}\right)^2 \ , \tag{5}$$

where m_W is the W mass and θ_W is the Weinberg angle. The width is quite large as shown in Table I. Also shown is the cross section for the case of a Z' exchange that has a standard weak coupling strength and a mass of 1 TeV/c^2. The finite width dilutes some of the characteristics of the event signatures; however, it increases the cross section for the case of a very heavy lepton. The effects of beamstrahlung and initial state radiation decrease the cross section as expected since the cross section is proportional to s.

In searching for a heavy-light lepton pair, the light lepton candidate must first be identified. The identification depends not only on whether the light (primary) lepton is e, μ, or τ, but also on whether the W decays hadronically or leptonically. If the search is for an e or μ primary lepton with the W decaying hadronically, then the primary lepton is defined as the lepton with the highest momentum. Figure 2 shows the momentum spectrum of all the leptons in the events. (All kinematic plots shown here have passed all selection criteria described in this paper.) The primary lepton is clearly separated from the leptons from the W fragmentation. In the case that the W decays leptonically, no attempt is made to separate the primary lepton from the lepton from the W decay because the latter could be more energetic if the heavy lepton is very heavy.

Table I. Width of the heavy lepton and cross sections for the case of a Z' exchange with $m_{Z'} = 1$ TeV/c^2. Here σ is the lowest order cross section with zero width and σ^γ is the cross section with the standard model width, beamstrahlung, and initial state radiation.

m_L (GeV/c^2)	500	700	800	900
Γ_L (GeV/c^2)	38	103	154	219
$\sigma\,(fb)$	189	142	106	60
$\sigma^\gamma\,(fb)$	136	91	67	59

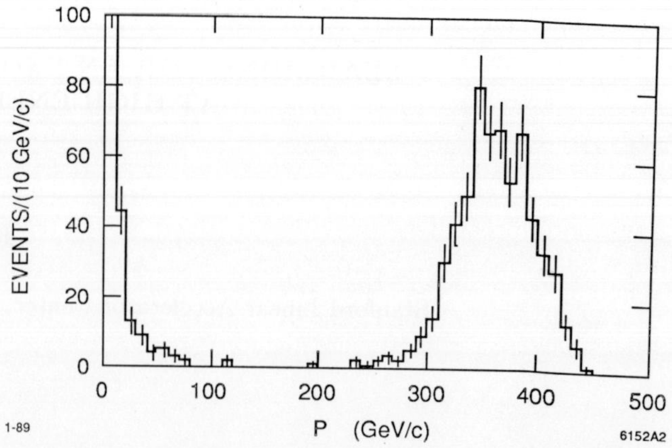

Fig. 2. The momentum spectrum of all the leptons in the heavy-light lepton candidates, with the light lepton being e or μ and the W decaying hadronically. The mass of the heavy lepton is 500 GeV/c^2.

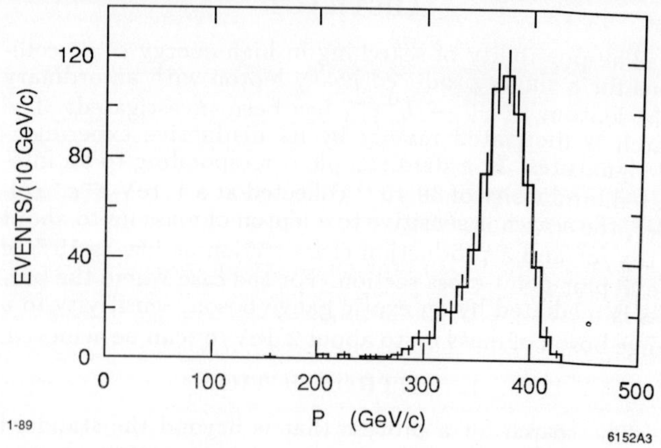

Fig. 3. The corrected momentum spectrum of the τ candidates in the heavy-light lepton events, with the light lepton being τ and the W decaying hadronically. The mass of the heavy lepton is 500 GeV/c^2.

However, the momentum spectrum still exhibits a peak (not shown). If the primary lepton is τ and the τ decays hadronically, all the decay products are combined into a single particle due to the limited calorimeter spatial resolution assumed. For the case that the W decays hadronically, the τ decay product is defined to be the most isolated particle. To reject soft particles ejected backward from the W fragmentation, an isolation criteria weighted by the particle momentum P is used:

$$\text{Weight} = P(1 - \cos\theta) \ ,$$

where θ is the opening angle between the two particles of interest. The particle with the largest weight is assumed to be from the τ decay. Since the τ is very energetic, its direction is well represented by that of the decay product. It is therefore possible to fully reconstruct the event by assuming energy and momentum conservation. Figure 3 shows the reconstructed τ momentum spectrum. The spectrum shows an asymmetric peak, a reflection of beamstrahlung. When the W decays leptonically, no attempt is made to identify the primary τ lepton candidate as in the case when the primary lepton is e or μ.

The selection criteria for heavy-light lepton candidates are designed to suppress the dominant background $e^+e^- \rightarrow$

W^+W^-. The candidate is required to have a minimum total visible energy of 250 GeV and transverse momentum of 180 GeV/c. The acollinearity and acoplanarity angles between the primary lepton and the W, defined as the hadronic jet or the lepton, are required to be greater than 25°. Both the primary lepton and the W are restricted to be in the angular region $|\cos\theta| < 0.9$, where θ is the polar angle of the particle. In order to reject $e^+e^- \to Z^0Z^0$ when one Z^0 decays into two neutrinos, the event mass is required to be greater than 100 GeV/c^2.

Additional criteria are applied when the W decays hadronically to further reduce the background from $e^+e^- \to W^+W^-$. The mass of the hadronic jet is required to be between 70 and 90 GeV/c^2. This eliminates the background from $e^+e^- \to q\bar{q}$ and $e^+e^- \to e^+e^-q\bar{q}$, although it has little effect on $e^+e^- \to W^+W^-$. The background from $e^+e^- \to W^+W^-$ can be reduced by reconstructing the neutrino momentum using momentum conservation. The reconstructed neutrino is combined with the primary lepton candidate and if the mass of the combined system is less than 150 GeV/c^2, the event is rejected. In the case that the primary lepton is e or μ, the background is further suppressed by using the fact that the primary lepton momentum is related to the mass of the heavy lepton. The mass calculated from the momentum is compared with the mass calculated from the hadronic jet (W) and neutrino combination and if the mass difference is greater than 100 GeV/c^2, the event is discarded. This powerful cut is also effective in eliminating the remaining background from $e^+e^- \to e^+e^-q\bar{q}$ and $e^+e^- \to e^+\nu_eW^-$, with e^+e^- or $e^+\nu_e$ going down the beam pipe and a lepton from the fragmentation identified as the primary lepton. Unfortunately, this cut cannot be implemented when the primary lepton is τ because it is necessary to use both energy and momentum conservation in order to reconstruct the neutrino. Three new cuts are introduced to reduce the backgrounds in this case. The first cut eliminates the remaining background from $e^+e^- \to e^+e^-q\bar{q}$ and $e^+e^- \to e^+\nu_eW^-$ by requiring the opening angle between the primary lepton candidate and the hadronic jet to be greater than 10°. The second cut exploits the fact that, in the background from

$$e^+e^- \to W^+W^-, \quad W^+ \to q\bar{q}', \quad W^- \to e^+\nu_e \text{ or } \mu^+\nu_\mu ,$$

the reconstructed "τ" momentum discussed earlier in this paper is close to the "primary" lepton momentum, because there is only one missing neutrino. Therefore, if the reconstructed momentum is within 20% of the measured momentum the event is rejected. Finally, the lepton mass calculated from the reconstructed τ momentum is required to be greater than 400 GeV/c^2.

The characteristic event signatures depend on the mass of the heavy lepton. The peak in the lepton momentum spectrum shown in Figs. 2 and 3 is for leptons of mass $\sim$ 500 GeV/c^2. The spectrum for a very heavy lepton is very broad. However, in the case when the primary lepton is e or μ and the W decays hadronically, the total visible energy spectrum also exhibits an asymmetric peak, as shown in Fig. 4. The peak is due to the very limited boost given by the heavy lepton to the monochromatic W.

The detection efficiency is quite good, as shown in Fig. 5. It could be as high as 20% when the primary lepton is e or μ and the W decays hadronically.

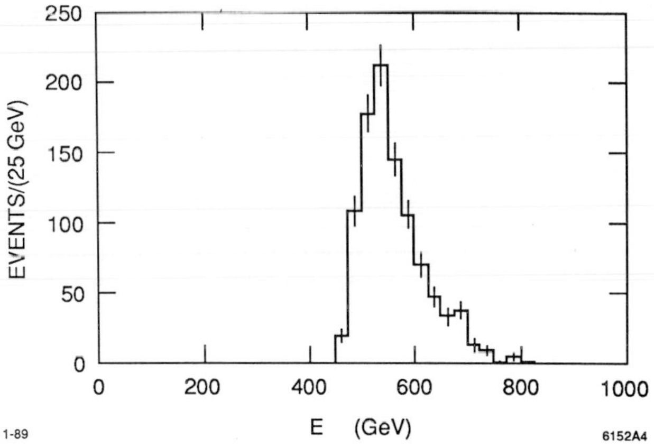

Fig. 4. The total visible energy spectrum of the heavy-light lepton candidates, with the light lepton being e or μ and the W decaying hadronically. The mass of the heavy lepton is 900 GeV/c^2.

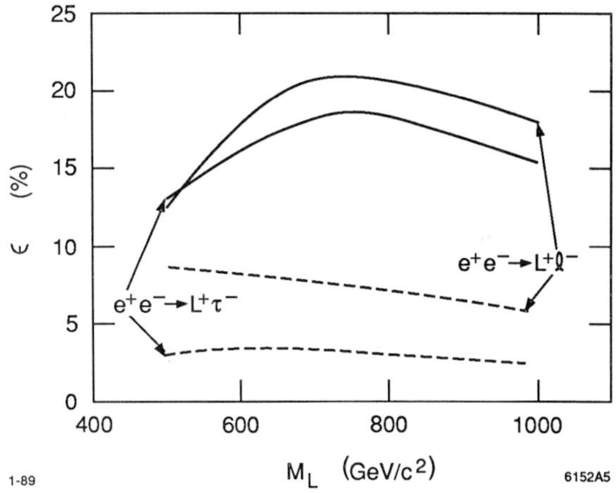

Fig. 5. The detection efficiency for the heavy-light lepton candidates. The solid lines are for the case when the W decays hadronically and the dash lines are for the leptonic decay. Here ℓ^- is either e^- or μ^-.

The sensitivity of the search is defined as the number of signal events required to see an excess of five standard deviations over the background. Since this is a Monte Carlo study and no attempt is made to estimate any systematic error, the standard deviation is calculated in the most conservative fashion: square root of the sum of the signal and background. The signal required for the search, together with the background, are summarized in Table II, for an integrated luminosity of 30 fb^{-1}, which may be collected over a period of three years. The sensitivity, in units of the μ-pair point cross section through one-photon exchange, is shown in Fig. 6. The hadronic decay of the W has better sensitivity to the heavy lepton than the leptonic decay. No attempt is made to calculate the combined sensitivity since the two decays may have different systematic errors in the real world. The search is sensitive to a lepton of mass up

Table II. The signal required to see a 5σ excess above the background for various search scenarios.

Search Scenario	$L^+ \to q\bar{q}'\bar{\nu}_L$ $\ell^- = e^-$ or μ^-	$L^+ \to q\bar{q}'\bar{\nu}_L$ $\ell^- = \tau^-$	$L^+ \to \ell^+\nu_\ell\bar{\nu}_L$ $\ell^- = e^-$, μ^- or τ^-
Signal	55	68	64
Background			
W^+W^-	64	97	90
Z^0Z^0	0	3	8
$e^+e^-W^+W^-$	3	16	3
Total	67	116	101

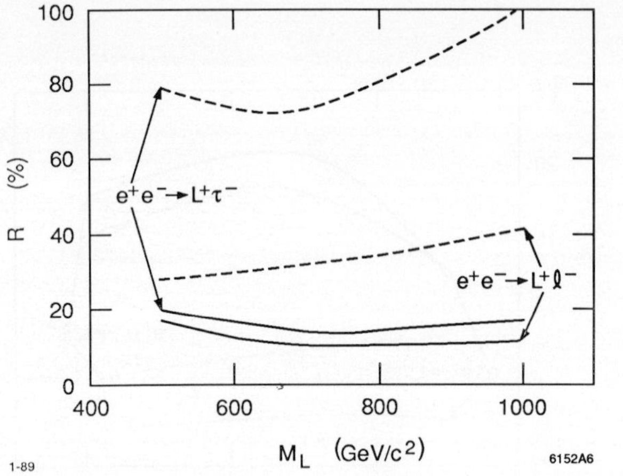

Fig. 6. The sensitivity of the search in units of the point cross section. The solid lines are for the case that the W decays hadronically and the dash lines are for the leptonic decay. Here ℓ^- is either e^- or μ^-.

to about 1 TeV/c^2 and a production rate as low as 10% of the point cross section. The sensitivity to an exotic Z' is shown in Fig. 7, for the case that the Z' couples to the leptons with the standard weak strength. The search is sensitive to a Z' with mass up to about 2 TeV/c^2.

The sensitivity for the case where the primary lepton is τ can be improved by the further use of the properties of the τ and the tracking information. The finite decay length of the τ would be a powerful discrimination against the background. For example, a τ with a momentum of 300 GeV/c has a decay length of 15 mm. A new search category can be created with the primary lepton candidate contains only three tracks and the three-track invariant mass is small. The background contamination in this category will be small. The only contamination from $e^+e^- \to W^+W^-$ is when one of the W's decays into τ which in turn decays into three charged particles. However, the search is limited by statistics because the three-charged-particle branching ratio of the τ is only $\sim 13\%$.

The search technique is not limited to the heavy lepton. The search is also sensitive to a massive particle that decays into a W and a light neutral particle. There is no need to

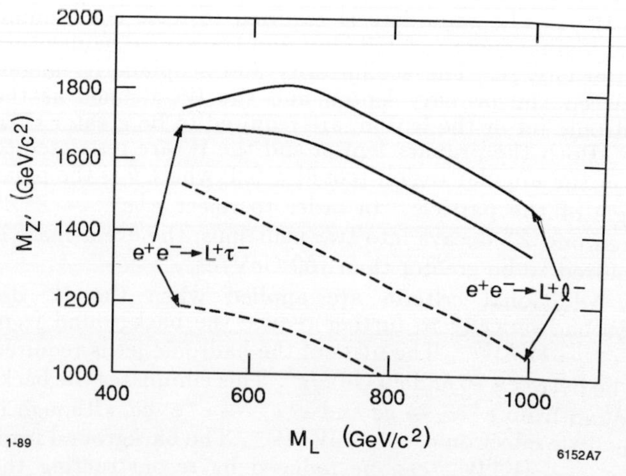

Fig. 7. The sensitivity of the search for an exotic Z'. The solid lines are for the case that the W decays hadronically and the dash lines are for the leptonic decay. Here ℓ^- is either e^- or μ^-.

restrict the search to the case where the hadronic jet is reconstructed to be W; a scan through all jet masses is highly desirable. In fact, the sensitivity could be much higher outside the W mass region because of the large cross section for $e^+e^- \to W^+W^-$. There is also no need to restrict the light "lepton" to be e, μ, or τ. The search with the light lepton being τ demonstrates that it can be a light particle that decays into one light charged particle and one light neutral particle. In fact, this is a general search technique for the production of a light-heavy particle pair! The search also should be conducted at PEP, PETRA, and TRISTAN colliders.

In conclusion, there is good sensitivity to the production of a heavy-light lepton pair in e^+e^- colliders. The event has distinctive signatures. At a 1 TeV e^+e^- collider, the search is sensitive to a lepton of mass up to about 1 TeV/c^2 and a production cross section as low as 10% of the μ-pair cross section. If the production is mediated by an exotic gauge boson, sensitivity to the gauge boson mass of up to about 2 TeV/c^2 can be achieved.

This work benefits greatly from the work of the SLAC group that studies the physic opportunities at a high-energy e^+e^- collider, in particular, the work of P. Rankin on $e^+e^- \to W^+W^-$. The author wishes to thank P. Burchat and D. Stoker for useful discussions.

REFERENCES

1. Reference to particles is taken to include their antiparticles, when appropriate.
2. For a recent review see, for example, K. K. Gan and M. L. Perl, Int. J. Mod. Phys. **A3**, 531 (1988).
3. K. Yokoya, Nucl. Instrum. Methods **A251**, 1 (1986); P. Chen, SLAC–PUB–4293 (1987).
4. F. A. Berends, R. Kleiss, and S. Jadach, Nucl. Phys. **B202**, 63 (1982); Computer Phys. Comm. **29**, 185 (1983).
5. The generic detector assumed is identical to that used by C. Ahn *et al.*, SLAC–Report–329, 1988, to study the physics opportunities at a high energy e^+e^- linear collider.

ELECTRON - PROTRON COLLIDERS

NEW GAUGE BOSONS AT *ep* COLLIDERS

F. Cornet
Max Planck Institute für Physik und Astrophysik,
Werner Heisember Institut für physik, P.O. Box 40 12 12, D-8000 Munich 40, FGR.

The study of the electroweak gauge sector is one of the most important tasks of the future accelerators. This includes precise measurements of the parameters of the known $W^{\pm}$, Z as well as the search for new gauge bosons. We have studied the sensitivity of the proposed *ep* colliders to the presence of extra gauge bosons. In particular, the accelerators considered are:

1. HERA, with a center of mass energy $\sqrt{s} = 314$ GeV and an integrated luminosity per polarized beam of $L = 125\ pb^{-1}$ for e_R^-, e_R^+, e_L^- and e_L^- each.

2. An upgraded version of HERA with $\sqrt{s} = 400$ GeV and the same luminosity as HERA,

3. LEP I × LHC, with $\sqrt{s} = 1.4$ TeV and $L = 250\ pb^{-1}$/year per polarized beam and

4. LEP II × LHC, with $\sqrt{s} = 1.8$ TeV and the same luminosity as HERA.

Previous studies show that the most sensitive tests can be performed running with polarized electron and positron beams.[1] However, since it is not clear at present whether polarized beams will be available in all the colliders we have also estimated the sensitivity limits obtained using unpolarized beams. In this case, the integrated luminosity is taken to be, $L = 200\ pb^{-1}/year$ for HERA and LEP II × LHC and $L = 500\ pb^{-1}/year$ for LEP I × LHC.

The values of the masses of the new gauge bosons that can be probed depend strongly on the assumed couplings of the new particles to the fermions. Therefore, it is important to study various models in order to estimate the sensitivity of the future *ep* colliders. We have chosen three models based on the group $SU(3)_C \times SU(2)_L \times U(1)_Y \times U(1)_{Y'}$, corresponding to three choices for the charge Y', and a Left-Right Symmetric Model based on the group $SU(3)_C \times SU(2)_L \times SU(2)_R \times U(1)_{B-L}$, where the $U(1)$ generator is associated with the difference between the baryon (B) and lepton (L) numbers. All these models contain a Z', while the Left-Right Symmetric Model contains, additional, new, charged gauge bosons.

We first consider the case of an extra, neutral gauge boson. The effective Neutral Current (N.C.) lagrangian is given by

$$\mathcal{L}_{NC} = \mathcal{L}_{NC}^{SM} + g' J_{Z'}^{\mu} Z'_{\mu}, \tag{1}$$

where $\mathcal{L}_{NC}^{SM}$ is the Standard Model N.C. lagrangian and the second term describes the coupling of the new gauge boson Z' to the current

$$J_{Z'}^{\mu} = \sum_f \overline{f} \gamma^{\mu} (v'_f - a'_f \gamma_5) f \tag{2}$$

with a coupling constant g'. v'_f and a'_f are the vector and axial couplings of the Z' to the fermions.

As in the Standard Model, spontaneous symmetry breaking can generate masses for the weak bosons. The mass eigenstates are, in general, mixtures of Z and Z' with mixing angle θ:

$$\begin{pmatrix} Z_1 \\ Z_2 \end{pmatrix} = \begin{pmatrix} \cos\theta & \sin\theta \\ -\sin\theta & \cos\theta \end{pmatrix} \begin{pmatrix} Z \\ Z' \end{pmatrix} . \tag{3}$$

If we don't make any assumption on the vacuum expectation values of the Higgs bosons, the mixing angle is a free parameter of the model. The lower mass eigenstate, Z_1, is to be identified with the observed neutral boson. The properties of Z_1, however, differ from the ones of the standard Z in three ways. First, m_{Z_1} is smaller than m_Z as can be seen from the inequality

$$m_Z^2 = m_{Z_1}^2 \cos^2\theta + m_{Z_2}^2 \sin^2\theta \geq m_{Z_1}^2 \tag{4}$$

implied by Eq. (3). Second, the Weinberg angle defined by the gauge couplings or, equivalently, by

$$\rho = \frac{m_W^2}{m_Z^2 \cos^2\theta_W} = 1, \tag{5}$$

where m_W is the mass of the standard, charged $W^{\pm}$ boson, differs from the angle $\overline{\theta}_W$ defined from the W and Z_1 mass ratio

$$\sin^2\overline{\theta}_W \equiv 1 - \frac{m_W^2}{m_{Z_1}^2} \leq 1 - \frac{m_W^2}{m_Z^2} \equiv \sin^2\theta_W . \tag{6}$$

The relation (5) is known to hold if the Higgs fields responsible for the symmetry breaking are $SU(2)_L$ doublets or singlets, which is going to be our only assumption on the Higgs sector. Finally, the Z_1 vector and axial couplings receive a contribution from v'_f and a'_f given by:

$$\begin{pmatrix} v_{1f} \\ v_{2f} \end{pmatrix} = \begin{pmatrix} \cos\theta & \sin\theta \\ -\sin\theta & \cos\theta \end{pmatrix} \begin{pmatrix} v_f \\ \frac{g'\cos\theta_W}{g_L} v'_f \end{pmatrix} \quad . \tag{7}$$

$$\begin{pmatrix} a_{1f} \\ a_{2f} \end{pmatrix} = \begin{pmatrix} \cos\theta & \sin\theta \\ -\sin\theta & \cos\theta \end{pmatrix} \begin{pmatrix} a_f \\ \frac{g'\cos\theta_W}{g_L} a'_f \end{pmatrix},$$

where g_L is the $SU(2)_L$ coupling constant.

The models are defined specifying the values of v'_f, a'_f and the coupling constant g'. We first assume that the low energy gauge group is $SU(3)_C \times SU(2)_L \times U(1)_Y \times U(1)_{Y'}$, resulting from the breaking of an E_6 grand unification group. In this case the new charges can be conveniently parametrized in terms of the charges of the $U(1)_\psi$ and $U(1)_\chi$ groups appearing in the decomposition $E_6 \to SO(10) \times U(1)_\psi \to SU(5) \times U(1)_\chi \times U(1)_\psi$[3]

$$Y' = Y_\psi \cos\alpha + Y_\chi \sin\alpha. \tag{8}$$

Each family of fermions is assigned to the 27-dimensional fundamental representation of E_6. The quantum numbers of the known fermions of the first family obtained in this way are given in Table 1. The models we have studied correspond to $\cos\alpha = \sqrt{5/8}$ (Model A), $\cos\alpha = 0$ (Model B) and $\cos\alpha = -\sqrt{15/16}$ (Model C). In the three cases g' is fixed to be:

$$g' = \sqrt{\frac{5e}{\cos\theta_W}}, \tag{9}$$

where e is the electromagnetic coupling constant, arising from the requirement $g' = g_L$ in the E_6 limit. We have also considered a Left-Right Symmetric Model,[4] based on the group $SU(3)_C \times SU(2)_L \times SU(2)_R \times U(1)_{B-L}$ with the following fermion quantum number assignments:

$$l_L = (\frac{1}{2}\,,\,0\,,\,-1), \qquad q_L = (\frac{1}{2}\,,\,0\,,\,\frac{1}{3})$$
$$l_R = (0\,,\,\frac{1}{2}\,,\,-1), \qquad q_R = (0\,,\,\frac{1}{2}\,,\,\frac{1}{3}) \tag{10}$$

that lead to

$$v'_f = \frac{1}{\cos\theta_W\sqrt{\cos 2\theta_W}}(\frac{1}{2}T_{3_f} - Q_f \sin^2\theta_W)$$
$$(11)$$
$$a'_f = -\frac{\sqrt{\cos 2\theta_W}}{\cos\theta_W}(\frac{1}{2}T_{3_f}),$$

where $g' \equiv g_R = g_L$ has been assumed.

We have studied the devations with respect to the Standard model predictions induced in inclusive quantities, such as cross-sections, $e^\mp p \to e^\mp X$, and asymmetries constructed with polarized cross-sections. There are a total of six asymmetries that can be classified in three groups:

1. polarization asymmetries:

$$A_{LR}^{--} = \frac{d\sigma(e_L^-) - d\sigma(e_R^-)}{d\sigma(e_L^-) + d\sigma(e_R^-)}, \qquad A_{LR}^{++} = \frac{d\sigma(e_L^+) - d\sigma(e_R^+)}{d\sigma(e_L^+) + d\sigma(e_R^+)}, \tag{12}$$

2. charge asymmetries:

$$A_{LL}^{-+} = \frac{d\sigma(e_L^-) - d\sigma(e_L^+)}{d\sigma(e_L^-) + d\sigma(e_L^+)}, \qquad A_{RR}^{-+} = \frac{d\sigma(e_R^-) - d\sigma(e_R^+)}{d\sigma(e_R^-) + d\sigma(e_R^+)} \tag{13}$$

3. mixed asymmetries:

$$A_{LR}^{-+} = \frac{d\sigma(e_L^-) - d\sigma(e_R^+)}{d\sigma(e_L^-) + d\sigma(e_R^+)}, \qquad A_{RL}^{-+} = \frac{d\sigma(e_R^-) - d\sigma(e_L^+)}{d\sigma(e_R^-) + d\sigma(e_L^+)}, \tag{14}$$

although only four of them are independent.

L-field	$SO(10)$	$SU(5)$	$SU(3)_C$	Q	T_3	Y_ψ	Y_χ
e^{-c}	16	10	1	1	0	$1/2\sqrt{6}$	$1/2\sqrt{10}$
d	16	10	3	$-1/3$	$-1/2$	$1/2\sqrt{6}$	$1/2\sqrt{10}$
u	16	10	3	$2/3$	$1/2$	$1/2\sqrt{6}$	$1/2\sqrt{10}$
u^c	16	10	3^*	$-2/3$	0	$1/2\sqrt{6}$	$1/2\sqrt{10}$
d^c	16	5^*	3^*	$1/3$	0	$1/2\sqrt{6}$	$-3/2\sqrt{10}$
e^-	16	5^*	1	-1	$-1/2$	$1/2\sqrt{6}$	$-3/2\sqrt{10}$
ν_e	16	5^*	1	0	$1/2$	$1/2\sqrt{6}$	$-3/2\sqrt{10}$

Table 1. Quantum numbers of the left-handed fields of the standard fermions.

The Z' models specified so far still involve four free parameters: the vector boson masses, the $Z - Z'$ mixing angle and the Weinberg angle θ_W. The lowest mass, m_{Z_1}, is assumed to be very precisely known from other experiments. In all the numerical calculations it has been taken to be $m_{Z_1} = 93.3\ GeV$. In addition, Eqs. (4), (5) and (6) fix the Weinberg angle, θ_W as a function of m_{Z_2} and θ. The numerical results have been obtained using the expressions given in Ref. 2. Furthermore, for the quark distribution functions we have used the set I from Duke and Owens[5] and for the Standard Model parameters we have taken: $\alpha = 1/137$, $m_Z = 93.3 GeV$, $m_W = 82.4 GeV$ and $\sin^2\theta_W = 0.22$. In Fig. 1 we show the contours in the plane (θ, m_{Z_2}) corresponding to a deviation with respect to the Standard Model $\delta A = 0.04$ for the most sensitive of the polarization (A_{LR}^{--}, solid line), charge (A_{RR}^{-+}, dashed line) and mixed (A_{RL}^{-+}, dash-dotted line) asymmetries for Model A at $\sqrt{s} = 1.4\ TeV$ at the kinematical point $x = 0.05$ and $Q^2 = 7 \times 10^4\ GeV^2$. Two regions can be distinguish in this plot. For large m_{Z_2} the effects are due to the $Z - Z'$ mixing and one is actually probing the dependence of the asymmetries on the $\sin^2\theta_W$. More interesting is the region of low m_{Z_2} and small θ. Here the effects are mainly due to the exchange of the second gauge boson and are, thus, much more sensitive to the model considered. In Fig. 1 we have also included the present bounds obtained in Ref. 8 (thick line) extrapolated to larger values of m_{Z_2} according to the asymptotic behavior $1/m_{Z_2}$ (dotted line). Since the comparison between these bounds and the $\delta A = 0.04$ contours shows

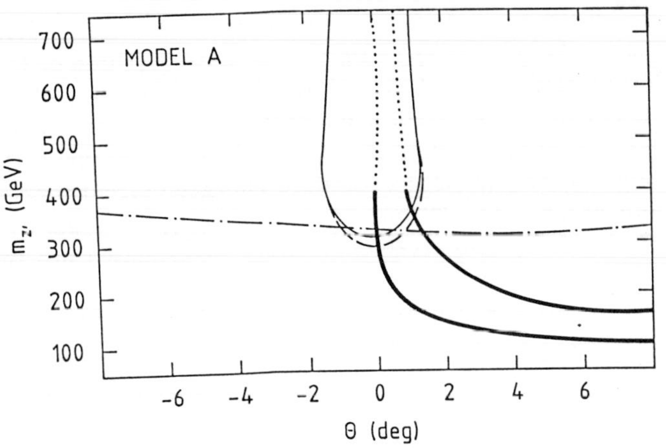

Fig. 1. Contours in the (θ, m_{Z_2}) plane for which the asymmetries A_{LR}^{--} (full line), A_{RR}^{-+} (dashed line) and A_{RL}^{-+} (dash-dotted line) in Model A deviate from the Standard Model predictions $\delta A = 0.04$ at LEP I×LHC. The present bounds are indicated by the thick lines and extrapolated to larger m_{Z_2} masses (dotted line).

that only in the region of low m_{Z_2} one can expect to improve the present bounds, we will calculate the sensitivity limits for $\theta = 0$ as a representative value.

The drawback of the asymmetry measurement is the necessity to share luminosity among runs with $e^{\mp}_{L,R}$ beams. **Fig. 2** illustrates the statistical significance expected at LEP I×LHC (Fig. 2a) and LEP II × LHC. In both cases, the most sensitive asymmetry for Model A with $m_{Z_2} = 200$ *GeV* (solid line) is compared with the Standard Model predictions (dashed line) in bins of Q^2 such that $\Delta log Q^2 = 0.2$. One can infer from this figure that, due to its larger luminosity, LEP I × LHC option is more sensitive to the presence of a new gauge boson than the higher energy, but lower luminosity collider LEP II × LHC. In order to numerically estimate the sensitivity limits, we define

$$\chi^2(m_2) = \sum_{bins\ in\ Q^2} \left(\frac{F(m_2) - F_{SM}}{\delta F} \right), \qquad (15)$$

where $F(m_2)$ and F_{SM} are the predictions for a particular quantity (cross-section or asymmetry) in the extended and the standard models, respectively, and δF is the statistical error expected in the Standard Model. We have plotted in Fig. 3 the χ^2 values as a function of m_{Z_2} for the most sensitive quantities in the four models under consideration, assuming a more realistic 80% beam polarization. We find that, except for Model B, asymmetries provide more sensitive tests to the presence of Z_2 than unpolarized cross-sections, even though the integrated luminosity is a factor of two larger for the latter. In Table 2 we summarize the sensitivity limits for each accelerator, including the possibility that polarized beams are not available.

For the charged gauge boson we have considered only the case of the Left-Right Symmetric Model. In this case the extended charged current lagrangian takes the form

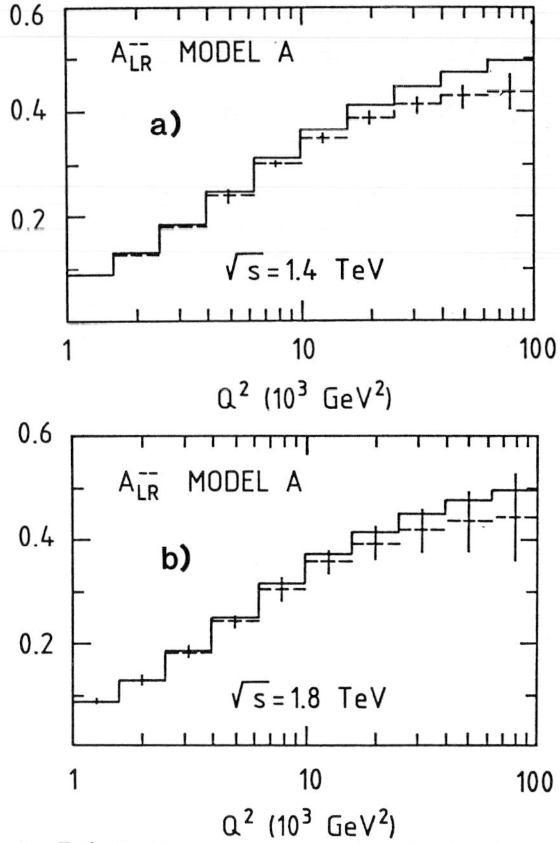

Fig. 2. Polarization asymmetry A_{LR}^{--} for Model A (full line) with $m_{Z_2} = 200$ *GeV* and $\theta = 0$ compared with the Standard Model predictions (dashed line) at LEP I × LHC (a) and LEP II × LHC (b). The error bars indicate the expected statistical uncertainties.

$\sqrt{s}$ GeV	314	400	1400	1800	Present bounds[8]
$A(A_{LR}^{--}, e^-)$	175, 150	200, 150	400, 375	325, 300	110
$B(e^+)$	250	275	575	450	270
$C(A_{LL}^{-+}, e^+)$	115, –	125, –	225, 135	175, –	—
$LR(A_{RR}^{-+}, e^-)$	450, 300	500, 350	1150, 850	1050, 750	320

Table 2 Bounds at the 95% c.l. for $m_{Z'}$ in GeV.

$$\mathcal{L}_{CC} = \frac{g}{2\sqrt{2}} J^{\mu}_{W} W_{\mu} + \frac{g'}{2\sqrt{2}} J^{\mu}_{W'} W'_{\mu}, \qquad (16)$$

where g and g' are the coupling constants of the $SU(2)_L$ and $SU(2)_R$ groups, respectively, J^{μ}_{W} is the standard model charged current and $J^{\mu}_{W'}$ is given by

$$J^{\mu}_{W'} = \sum_{i,j} \left(\overline{\nu}_i \gamma^{\mu}(1-\gamma_5)\delta_{ij} e_j + \overline{u}_i \gamma^{\mu}(1-\gamma_5) V'_{ij} d_j \right). \qquad (17)$$

Here we have neglected the mixing among lepton species and introduced the matrix V'_{ij} to describe the mixing in the right-handed quark sector.

A W' boson coupled to right-handed currents gives rise to the process $ep \to \nu_R X$. The nature (Dirac or Majorana fermion) and mass ($m_{\nu_R} \ll$ or $\sim m_{W'}$) of the right-handed neutrino, if it exists, is not known. In the case where ν_R is a light Dirac particle, the signature is the same as the one for the standard charged current process and one has to rely on indirect searches as in the case of a Z'.[6] The observables we have investigated are the polarized and unpolarized cross-sections as well as the ratios $R^- = \tilde{\sigma}(e_R^-)/\tilde{\sigma}(e_L^-)$ and $R^+ = \tilde{\sigma}(e_L^+)/\tilde{\sigma}(e_R^+)$, where $\tilde{\sigma} \equiv d\sigma/dQ^2$. The ratios R^- and R^+ have the advantage that they are independent of the uncertainties in the proton structure functions.[7] The most sensitive quantities turn out to be R^- and $d\sigma(e_R^-)/dQ^2$, (the χ^2 distribution for $d\sigma(e_R^-)/dQ^2$ is shown in Fig. 4) which produce the the bounds at the 95% c.l. shown in Table 3.

$\sqrt{s}$ GeV	314	400	1400	1800	Present bounds[9]
$W_R(e_R^-, R^-)$	375	425	1050	1050	300
$W_R(e^-)$	250	275	700	675	

Table 3 Bounds at the 95% c.l. for m_{W_R} in GeV.

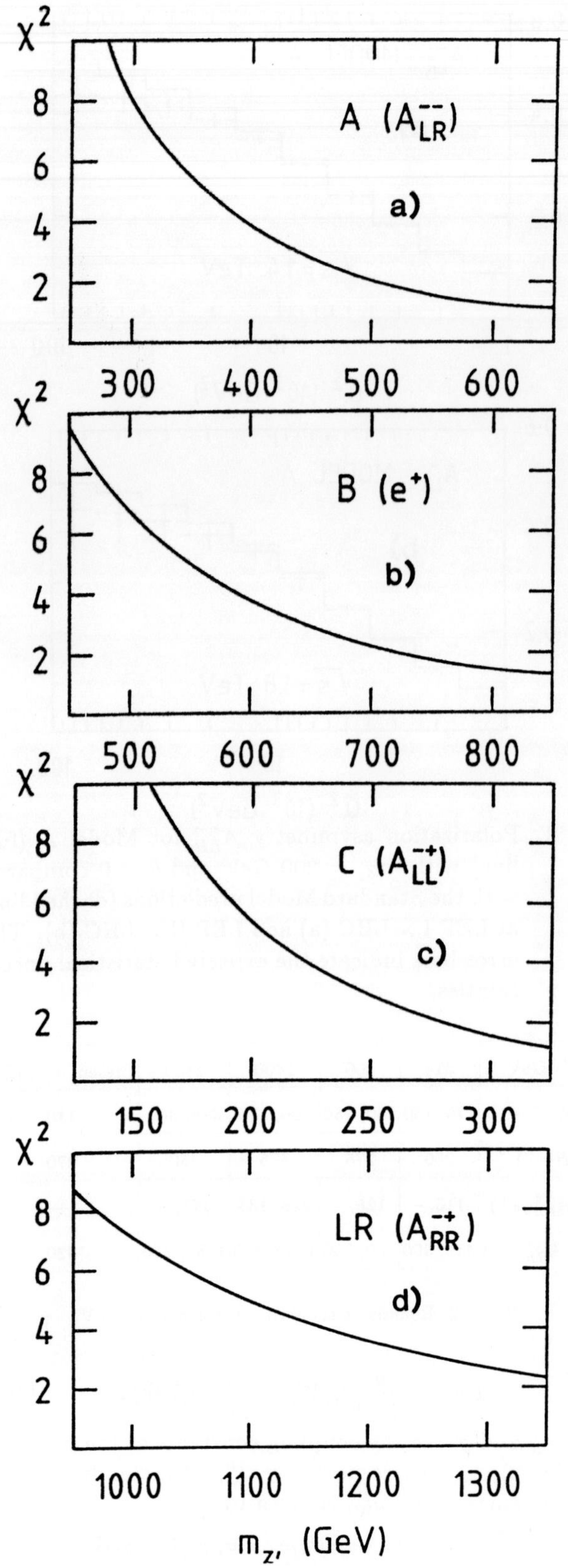

Fig. 3. χ^2 distributions expected at LEP1 $\times$ LHC as a function of m_{Z_2} for the most sensitive quantities in (a) Model A, (b) Model B, (c) Model C and (d) Left-Right Symmetric Model.

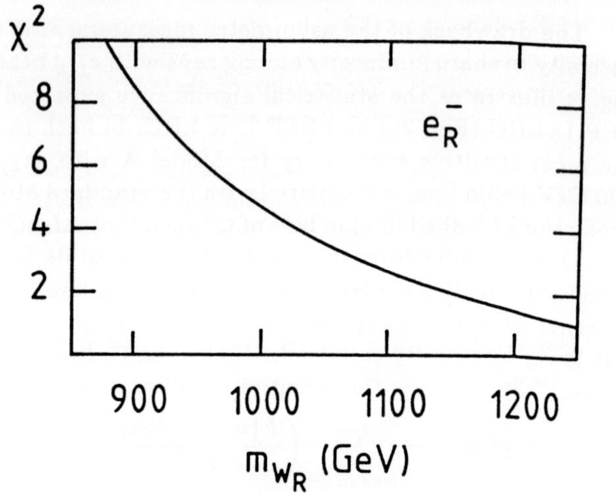

Fig. 4. χ^2 distribution expected at LEPI $\times$ LHC as a function of m_{W_2} for the a charged gauge boson coupled to right-handed currents.

In conclusion, we have studied the sensitivity of searches for new gauge bosons in the future ep colliders. The main effects of the new gauge bosons are in small deviations of the cross-sections with respect to the Standard Model predictions at large Q^2. Since the event rate drops rapidly with Q^2 it is very important to achieve a high luminosity. We have estimated the sensitivity limits that are summarized in Tables 2 and 3. In almost all the cases the use of polarized beams provides a higher sensitivity to the presence of new gauge bosons. Furthermore, polarization also helps to distinguish among different models.

REFERENCES

1. Cashmore, R. J. *et al.*, Phys. Rep. 122, 275 (1985); Angelopoulos, V. D. *et al.*, Phys. Lett. 176B, 203 (1986).
2. Cornet, F. and Rückl, R., Phys. Lett. 184B, 263 (1987); Cornet, F. and Rückl, R., *Proceedings of The Workshop at Future Accelerators*, La Thuile, 1987, Ed. J.H. Mulvey, CERN 87-07, Vol. II, p.190; Cornet, F. and Rückl, R., *Proceedings of the DESY Workshop on Physics at HERA*, DESY 88-104.
3. London, D. and Rosner, J. L., Phys. Rev. D34, 1530 (1986).
4. For a review see Mohapatra, R. N., *Quarks and Leptons and Beyond*, Eds. H. Fritzsch, R.D. Peccei, H. Saller and F. Wagner, Plenum Press, New York (1985) p.219.
5. Duke, D. W. and Owens, J. F., Phys. Rev. D30, 49 (1984).
6. For the case of a heavy right-handed neutrino see Altarelli, G., Mele, B., Rückl, R., *Proceedings of the ECFA-CERN Workshop on a Large Hadron Collider in the LEP Tunnel, Lausanne and Geneva*, 1984, Ed. M. Jacob, CERN 84-10, Vol. II, p. 549.
7. Longo, E., *Proceedings of The Workshop Experimentation at HERA, NIKHEF Amsterdam*, 1983, DESY-HERA 83/20, p.285.
8. Amaldi U., *et al.*, Phys. Rev. D36, 1385 (1987).
9. Langacker, P., *Proceedings of the 1985 Int. Symp. on Lepton and Photon Interactions at High Energies*, Eds. M. Konuma and K. Takahashi, Kyoto Univ., Kyoto 1986, p.186.

W PRODUCTION IN ep COLLISIONS

D. Atwood,[1] U. Baur,[2] G. Couture,[3] and D. Zeppenfeld[4]

[1] Physics Department, McGill University, Montreal, Canada
[2] TH-Division, CERN, CH-1211 Geneva 23, Switzerland
[3] Physics Department, University of Guelph, Guelph, Ontario, Canada
and
Brookhaven National Laboratory, Upton, L.I., NY 11973
[4] Physics Department, University of Wisconsin, Madison, WI 53706

ABSTRACT

We discuss single W production in the process $ep \to eWX$ at HERA ($\sqrt{s} = 314$ GeV), a possible HERA upgrade ($\sqrt{s} = 410$ GeV) and an ep collider in the LEP tunnel ($\sqrt{s} = 1.4$ TeV), focusing attention on the measurement of the $WW\gamma$ vertex. Even though production rates at HERA are small, cross sections at large values of W transverse momentum are quite sensitive to details of the $WW\gamma$ interaction and allow a measurement of the vertex with 50% accuracy during the first year of running at HERA.

In addition to virtual W's contributing to deep inelastic scattering at high Q^2, real W's will be produced as well in ep collisions at HERA,[1] and perhaps at proposed ep colliders like the LEP/LHC combination at CERN.[2] This electroproduction process has been considered previously by several authors,[3-6] who have pointed out also that the cross sections for the processes $ep \to eWX$ or $ep \to \nu WX$ are quite sensitive to the actual value of the anomalous magnetic moment of the W. Unfortunately, total W production cross sections at ep colliders are relatively small. In Ref. 5 Gabrielli estimates $\sigma(ep \to eW^{\pm}X) \approx 0.8$ pb and $\sigma(ep \to \nu WX) \approx 0.015$ pb at $\sqrt{s} = 300$ GeV (*i.e.* roughly the HERA energy), values which increase to 15 pb and 1 pb, respectively at $\sqrt{s} = 1.5$ TeV which is in the LEP/LHC range.[2] With expected integrated luminosities per year of 200 pb^{-1} for HERA and 1000 pb^{-1} for LEP/LHC and considering the leptonic decays $W \to e\nu$, $W \to \mu\nu$ only, this roughly corresponds to 30 leptonic W decays per year at HERA and 3000 events at LEP/LHC, rates which may be compared with 30 clean $p\bar{p} \to W\gamma X$ and 140 radiative W decay events at the Tevatron[7] (for an integrated luminosity of 10 pb^{-1}) or more than 10^4 $pp \to WZX$ and $pp \to W\gamma X$ events per year at the LHC or SSC[8] (with subsequent leptonic W and Z decays).

The above inclusive rates contain little information, of course, on how much interesting physics can be done with them, *e.g.* on how well experiments at the various machines can determine the $WW\gamma$ or WWZ couplings. In particular possible backgrounds and the finite acceptance of realistic detectors have to be considered. Such studies have previously been performed for hadron colliders.[7,8] The purpose of this contribution is to investigate these questions for ep colliders, in particular

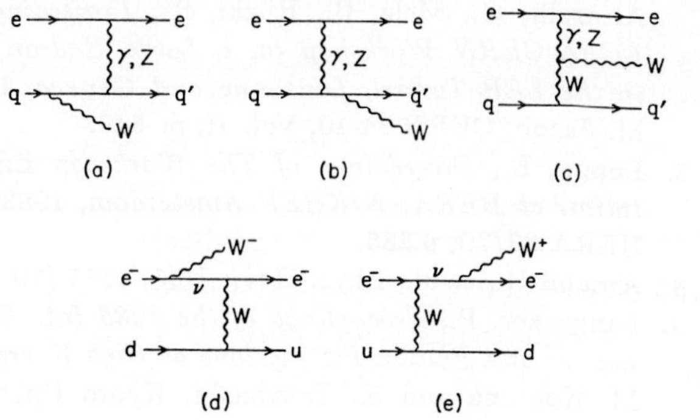

Fig. 1. Feynman graphs for the parton level processes contributing to $ep \to eWX$.

HERA at DESY (30 GeV electrons on 820 GeV protons), a hypothetical HERA upgrade (HERA′) with $E_e = 35$ GeV and $E_p = 1200$ GeV, and for an ep collider in the LEP tunnel, LEP/LHC, with $E_e = 60$ GeV, $E_p = 8$ TeV.

Because of the small cross section for $ep \to \nu WX$[5] we concentrate on the process $e^-p \to e^-W^{\pm}X$. At the parton level the process proceeds via the Feynman graphs depicted in Fig. 1. The dominant contributions arise from the γ-exchange graphs in Fig. 1a-1c due to the t-channel pole induced by the photon propagator. At HERA the other contributions can safely be neglected. The $WW\gamma$ vertex that we are interested in enters via graph 1c. Within the standard model (SM) this vertex is uniquely determined by $SU(2)_L \times U(1)$ gauge invariance. The most general Lorentz and CP invariant vertex compatible with electromagnetic gauge invariance can be parametrized as follows[9,10]

$$\mathcal{L}_{\text{eff}} = -ie\left(W^{\dagger}_{\mu\nu}W^{\mu}A^{\nu} - W^{\dagger}_{\mu}A_{\nu}W^{\mu\nu}\right) - ie\kappa W^{\dagger}_{\mu}W_{\nu}A^{\mu\nu} - ie\frac{\lambda}{M_W^2}W^{\dagger}_{\lambda\mu}W^{\mu}{}_{\nu}A^{\nu\lambda} \tag{1}$$

where W^{μ} is the W^- field, and $W^{\mu\nu}$ and $A^{\mu\nu}$ denote the W and photon field strength tensors. The free parameters κ and λ are related to the anomalous magnetic dipole moment μ_W

and the electric quadrupole moment Q_W of the W^+ by

$$\mu_W = \frac{e}{2M_W}(1+\kappa+\lambda)\,, \qquad Q_W = \frac{e}{M_W^2}(\lambda-\kappa)\,. \qquad (2)$$

The SM, at tree level, predicts $\kappa = 1$, $\lambda = 0$.

Our calculations of the signal are based on a parton level Monte Carlo program which includes all the Feynman graphs of Fig. 1, for arbitrary values of κ and λ, and the decay of the W into a lepton pair. The code was adopted from the formulas given in Ref. 10 and extended to include graphs (1d) and (1e) by using the helicity amplitude techniques described in Ref. 11. For the parton structure functions we used set 1 of Ref. 12. The Monte Carlo results were found to agree well with a calculation of the photon exchange contributions (Fig. 1a-1c) in the Weizsäcker-Williams approximation and with the results presented in Ref. 6, within the cuts specified there.

From Fig. 1 it is clear that the signal consists of a high transverse momentum (p_T) electron or muon plus large missing transverse momentum ($\not{p}_T$) due to the associated neutrino in W leptonic decay, a jet produced by the quark struck inside the proton and, possibly, the original beam electron. Since the exchanged photon will usually be almost real, the scattered electron will in general be lost down the beam pipe. (Requiring an electron scattering angle $\theta_e > 4°$, appropriate for the ZEUS detector at HERA,[13] will reduce the signal by a factor 3 at HERA energy ($\sqrt{s} = 314$ GeV).) Hence the signal we are considering is

$$e^- p \to \ell^\pm + \text{jet} + \not{p}_T \qquad (3)$$

with $\ell^\pm = e^\pm$ or $\mu^\pm$.

The dominant contribution to $ep \to eWX$ arises from the graph of Fig. 1a, which at the level of the $\gamma q \to Wq'$ subprocess is u-channel quark exchange. The corresponding u-channel pole leads to a strong peaking of the cross section at small jet transverse momenta (p_{Tj}). This peaking is shown in Fig. 2a for $W^\pm$ production at HERA. The signal at low p_{Tj} is insensitive to the $WW\gamma$ vertex and no sensitivity is lost by requiring the jet to be observed in the detector. (We require $p_{Tj} > 5$ GeV in the following.)

Due to the small event rates expected at HERA, the only meaningful observables will be some inclusive event rates within certain cuts. In order to translate an inclusive rate into a measurement of the vertex, any normalization errors in the W production cross section have to be eliminated. For this purpose W events, where the accompanying jet cannot be resolved, will still be useful, since their rate is determined by the well-measured vector boson-fermion couplings alone. However, in the low p_{Tj} region, close to the u-channel pole, QCD corrections are expected to be large and W production can be thought of as $q\bar{q} \to W$ with one of the quarks being described as a constituent of the photon. Thus contact is made with parton structure functions inside the photon[14] and a careful merging of the two pictures is required. Modulo this caveat the W events with $p_{Tj} < 5$ GeV will provide an important consistency check on the normalization of the SM expectation for the W production rate.

Apart from the problem of observing the low p_T jet accompanying W production, detectors at ep colliders should have a high efficiency in identifying the W events, because both the charged lepton and the missing neutrino from W decay will typically have large transverse momentum and the charged lepton and the accompanying jet, even though peaked in the proton beam direction, will be observable in detectors hermetic down to a few degrees.

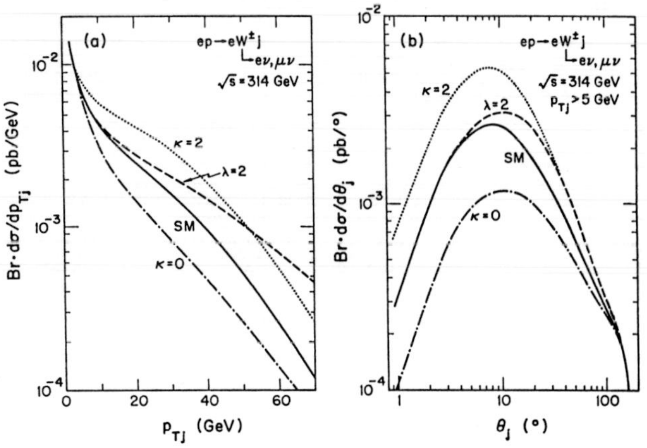

Fig. 2. Transverse momentum spectrum (a) and scattering angle distribution in the lab (b) for the jet arising from the subprocess $eq \to eWq$ at HERA for four combinations of $WW\gamma$ couplings. The solid lines represent the SM prediction.

This is demonstrated in Figs. 2 and 3 where the distributions at HERA for the transverse momentum and the scattering angle (measured with respect to the proton beam) for both the jet and the charged lepton are shown. $p_{Tj} > 5$ GeV is required in Fig. 2b and Figs. 3a and 3b.

The charged lepton p_T spectrum in Fig. 3a shows the familiar Jacobian peak at $p_T \approx \frac{1}{2}M_W$, smeared however by the W's transverse momentum $p_{TW} \approx p_{Tj}$. Since we have summed over W^+ and W^- decay spectra the $p_{T\ell^\pm}$ and $\not{p}_T$ distributions are indistinguishable, and it is clear from Fig. 2a that leptonic W decays will essentially always give rise to a significant $\not{p}_T$ signal.

In order to roughly simulate detector response we will impose the following cuts on the lab frame scattering angles and the transverse momenta of the charged lepton and the jet. We require the $\ell^\pm$ and quark pseudorapidities to lie in the interval $-3.5 < \eta < 3.5$ which corresponds to scattering angles between about 4° and 176°, and we impose $p_{T\ell^\pm}, p_{Tq} > 5$ GeV at HERA and HERA′ and $p_{T\ell^\pm}, p_{Tq} > 10$ GeV at LEP/LHC. In addition we require the charged lepton and the jet to be well separated in the pseudorapidity-azimuthal angle plane:

$$R = \sqrt{(\Delta\eta)^2 + (\Delta\phi)^2} > 0.5\,. \qquad (4)$$

This isolation requirement for the charged lepton is almost always fulfilled for W events and thus backgrounds from heavy quark semileptonic decays are expected to pose no problem. Within these cuts the standard model values for $\sigma(e^-p \to e^-W^\pm X) \cdot \mathrm{Br}\,(W \to e\nu, \mu\nu)$ are 0.099 pb for HERA, 0.225 pb for HERA′, and 1.72 pb for LEP/LHC.

The jet $+\,\ell + \not{p}_T$ signal of Eq. (3) should be virtually background free for $\ell = \mu^\pm$ or e^+ (in the case of e^-p scattering). For $\ell = e^-$ neutral current (NC) deep inelastic scattering (DIS) events are a potential source of background, when allowing for mismeasurements leading to missing p_T. In order to estimate this background we have analyzed $e^-p \to (e^-)W^-$ jet, $W^- \to e^-\bar{\nu}_e$ under the DIS hypothesis. Q^2 and y for a corresponding deep inelastic scattering events were derived inde-

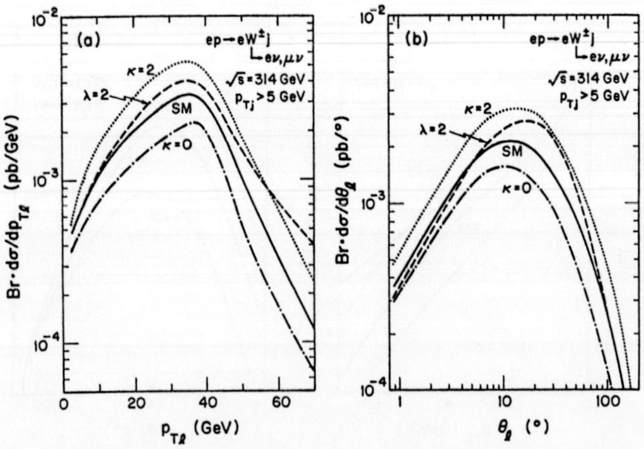

Fig. 3. Same as Fig. 2 but for the charged lepton arising from $W^{\pm}$ decay. $\theta = 0$ is the proton beam direction.

pendently from the W decay electron and from the jet. While jet derived values fall into $y < 0.4$ and cluster at small Q^2 values, almost all electron derived points lie in the region $y > 0.6$, where a ~ 3000 times larger background from NC induced DIS events is located.[1] Thus a reduction factor of a few times 10^4 is certainly sufficient to eliminate any DIS background. This factor should be easily achievable by requiring *i*) e^- and jet derived values of y and Q^2 not to match within errors, *ii*) a significant amount of missing p_T in the event, *iii*) no back-to-back e^- and jet in the transverse plane, or any combination of *i*)-*iii*). Since the DIS background affects only 25% of the signal we have not pursued this analysis in detail, and we neglect any effect of ensuing cuts in the following.

From Fig. 2 it is clear that the event rate observed above $p_{Tj} = 5$ GeV is an excellent measure of anomalous couplings. $\sigma(e^-p \to e^-W^{\pm}\text{jet}) \cdot \text{Br}(W \to e\nu, \mu\nu)$ is of course quadratic in the deviation $a = \kappa - 1$ or λ from the SM

$$\sigma \cdot \text{Br} = \sigma_0 + a\sigma_1 + a^2\sigma_2 \,. \tag{5}$$

In Table 1 the values of the coefficients σ_0, σ_1 and σ_2 are given for anomalous values of κ or λ at all three machines. The large value of σ_1 indicates that anomalous values of κ lead to a strong interference between SM and anomalous contributions to the amplitudes, in contrast to variations of λ. Consequently κ can be measured much more precisely than λ. Considering statistical errors of the cross section measurement only, event rates expected with $200\,\text{pb}^{-1}$ at HERA and HERA′ and with $1\,\text{fb}^{-1}$ at LEP/LHC can immediately be translated into the $1\,\sigma$ errors on κ and λ listed in the last column of Table 1. In order to realize these sensitivities the normalization of the SM rate (within cuts) has to be known within $\sim 15\%$ at HERA and to better than 2% at LEP/LHC. The higher rates at LEP/LHC would allow to exploit asymmetries or changes in distributions due to anomalous couplings also and thus normalization uncertainties may pose no serious problem. We have not investigated this question during the workshop, however.

No direct measurement of the $WW\gamma$ vertex exists to date. $W\gamma$ production and W radiative decays at the Tevatron are expected to determine κ to within[7] $-3 \lesssim \kappa \lesssim 5$. W pair production in e^+e^- annihilation at LEP II is expected to yield[15] $|\kappa - 1| < 0.2$, $|\lambda| < 0.2$. In W pair production, however, it is a combination of $WW\gamma$ and WWZ couplings which is measured,

Table 1. Coefficients of the cross section expansion in powers of anomalous couplings (see Eq. (5)) within cuts as described in the text. The sensitivities represent $1\,\sigma$ statistical errors after accumulating $200\,\text{pb}^{-1}$ of data ($1000\,\text{pb}^{-1}$ at LEP/LHC).

	a	σ_0 [pb]	σ_1 [pb]	σ_2 [pb]	sensitivity
HERA	$\kappa - 1$	0.099	0.048	0.013	+0.4, −0.5
	λ	0.099	0.0003	0.007	±1.8
HERA′	$\kappa - 1$	0.225	0.118	0.034	±0.3
	λ	0.225	0.001	0.026	±1.1
LEP/LHC	$\kappa - 1$	1.72	1.19	0.52	±0.04
	λ	1.72	0.05	1.89	±0.16

and hence the independent measurement of $WW\gamma$ couplings at HERA will be valuable input for the analysis at LEP II. It certainly needs to be stressed that HERA can perform a 30-50% measurement of the $WW\gamma$ vertex several years before LEP reaches sufficiently high energies to produce W pairs.

ACKNOWLEDGEMENTS

We would like to thank the members of the *ep* working group for useful discussions. This research was supported in part by the University of Wisconsin Research Committee with funds granted by the Wisconsin Alumni Research Foundation, in part by the U. S. Department of Energy under contract DE-AC02-76ER00881, and in part by the NSERC of Canada under grant number UO573.

REFERENCES

1. See *e.g.* Wolf, G.: DESY 86-089 and references therein.
2. *Proceedings of the Workshop on Physics at Future Accelerators*, La Thuile (1987), CERN 87-07.
3. Llewellyn-Smith, C. H., Wiik, B. H, DESY 77/36 (1977); Salati, P., Wallet, J. C., Z. Phys. C16, 155 (1982); Kamal, A. N., *et al.* Phys. Rev. D24, 2482 (1984); Neufeld, H., Z. Phys. C17, 145 (1983).
4. Altarelli, G., *et al.* Nucl. Phys. B262, 204 (1985).
5. Gabrielli, E., Mod. Phys. Lett. A1, 465 (1986).
6. Böhm, M. , Rosado, A., Z. Phys. C39, 275 (1988).
7. Cortez, J., Hagiwara, K., Herzog, F., Nucl. Phys. B278, 26 (1986).
8. Zeppenfeld, D., Willenbrock, S., Phys. Rev. D37, 1775 (1988); Baur, U., Zeppenfeld, D., Nucl. Phys. B308, 127 (1988).
9. Gaemers, K., Gounaris, G., Z. Phys. C1, 259 (1979).
10. Hagiwara, K., *et al.* Nucl. Phys. B282, 253 (1987).
11. Hagiwara, K., Zeppenfeld, D., Nucl. Phys. B274, 1 (1986).
12. Duke, D., Owens, J., Phys. Rev. D30, 49 (1984).
13. ZEUS collaboration, ZEUS detector technical specifications, DESY preprint.
14. Drees, M., these proceedings.
15. Treille, D. *et al.*, in *Proceedings of the ECFA Workshop on LEP 200*, Aachen (1986), CERN 87-08, vol. 2, p. 414; Zeppenfeld, D., Phys. Lett. 183B, 380 (1987).

ELASTIC PRODUCTION OF SUPERSYMMETRIC PARTICLES AT *ep* COLLIDERS

M. Drees[1,2] and D. Zeppenfeld[2]

[1] Theory Division, CERN, CH-1211 Genève 23, Switzerland

[2] Physics Department, University of Wisconsin, Madison, WI 53706, U.S.A.

ABSTRACT

We discuss supersymmetric processes at ep colliders where the proton is left intact, *i.e.* $ep \to \tilde{e}\tilde{Z}_i p$ and $ep \to \tilde{\nu}_e \tilde{W}_- p$ where $\tilde{Z}_i$ and $\tilde{W}_-$ are a generic neutralino and the lighter chargino, respectively. We find the cross sections to be similar to the corresponding inelastic ones, whereas the signals are much cleaner in the elastic case due to the absence of any serious background. We also discuss the possibility to use these processes to pin down the masses of the produced sparticles.

Reactions at ep colliders can be grouped into three general classes: Deep inelastic reactions, where a highly virtual gauge boson scatters off the quarks and gluons in the proton; photoproduction processes, where an almost real photon interacts with the proton to produce a more complicated final state; and elastic processes, where the proton remains intact. This last class of processes is usually forgotten, probably because in the standard model (SM) the possible reactions either look boring (like $ep \to e\gamma p$) or have very small cross sections (like $ep \to \nu W p$). Outside the SM, however, elastic processes can both be very interesting and have measurable cross sections. In this contribution we discuss the elastic production of sleptons and gauginos via the reactions $ep \to \tilde{e}\tilde{Z}_i p$ and $ep \to \tilde{\nu}\tilde{W}_- p$, where $\tilde{Z}_i$ and $\tilde{W}_-$ denote a generic neutralino and the lighter chargino, respectively. Our main focus will be on the HERA collider ($\sqrt{s} = 314$ GeV), which should start operations in 1990, but we will also briefly mention more futuristic machines.

The main idea of this contribution is to use the protons as a source of almost real photons, which in turn interact with the electron to produce a slepton-gaugino pair. The corresponding processes for e^+e^- colliders have been discussed in Ref. 1 ($e^+e^- \to e\tilde{e}\tilde{Z}_i$) and Ref. 2 ($e^+e^- \to e\tilde{W}_-\tilde{\nu}_e$), but for ep colliders only the deep inelastic analogues have been discussed (Ref. 3). This is somewhat surprising, since elastic processes at ep colliders have two obvious advantages: the available center-of-mass (cms) energy is quite high and the energy of the outgoing proton can be measured in a forward proton spectrometer even if it has no transverse momentum, which simplifies the reconstruction of the final state. (Synchrotron radiation makes it impossible to use a forward spectrometer at an e^+e^- ring.)

Our analysis of elastic supersymmetry production is based on complete tree level formulas for the processes[4] $ep \to \tilde{e}\tilde{Z}_i p$ and $ep \to \tilde{\nu}_e \tilde{W}_- p$ which proceed via t-channel photon exchange. The electromagnetic coupling of the photon is parametrized in terms of the electric and magnetic form factors of the proton, for which the usual dipole form is assumed. In addition to these "exact" formulas, we have developed[4] a modified Weizsäcker-Williams approximation (WWA), which takes into account that the exchange of photons with $Q^2 \gtrsim 1$ GeV^2 is strongly damped due to the finite size of the proton.

In Fig. 1 we compare the exact result for $\sigma(ep \to p\tilde{e}\tilde{Z}_1)$ with this modified WWA, where we have assumed the neutralino to be a pure photino. We see that the approximation deviates by only about 20% at the highest accessible sparticle masses and works even better for smaller values of $m_{\tilde{e}}$ and $m_{\tilde{Z}_1}$. The approximation is thus good enough to perform scans of parameter space for which the exact formulas are too slow numerically. The error introduced by the approximation corresponds to a shift in $m_{\tilde{e}}$ of less than 5 GeV.

Unfortunately Fig. 1 also shows that the cross section is rather small, 0.12 pb or less for realistic selectron masses ($m_{\tilde{e}} \gtrsim$ 30 GeV from the nonobservation[5] of acoplanar e^+e^- pairs

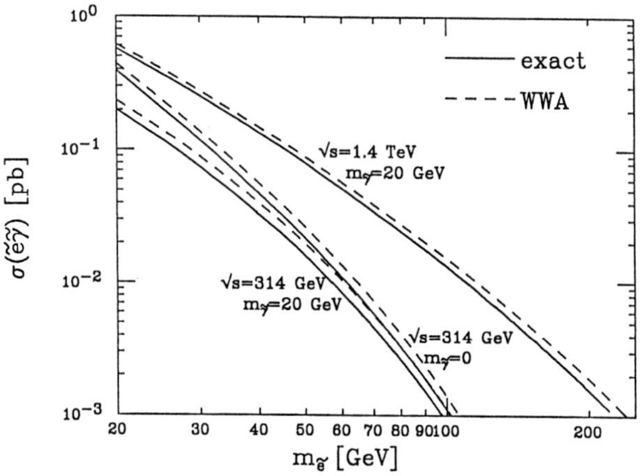

Fig. 1. Total cross sections for $ep \to p\tilde{e}\tilde{\gamma}$. The solid and dashed curves show the exact cross section and results using the modified Weizsäcker-Williams approximation.

at the CERN $Sp\bar{p}S$); the reason is, of course, that the cross section is of $O(\alpha^3)$. On the other hand, comparison with the results of Ref. 3 show that our elastic cross section is very similar to the corresponding inelastic cross section ($Q^2 \gtrsim 1$ GeV^2), where the proton is broken up; the possibility to measure the outgoing proton make the elastic events easier to recognize and reconstruct (see later).

Note that the results of Fig. 1 are optimistic, since usually a massive neutralino has[6] a weaker coupling to electrons than a pure photino state. As an example we show in Fig. 2 the four cross sections for $ep \to p\tilde{e}\tilde{Z}_i$, $i = 1, \ldots, 4$, of the minimal supersymmetric model.[7] In this model the masses and couplings of the four physical neutralinos and two charginos are completely determined by three parameters: The supersymmetric higgsino mass μ, the gluino mass $m_{\tilde{g}}$, and the ratio of the vacuum expectation values $\omega \equiv v_2/v_1$ (where v_2 gives rise to up-quark masses). We choose the sign convention where μ and ω are always positive; in this case $m_{\tilde{g}}$ can have either sign (the physical gluino mass is given by $|m_{\tilde{g}}|$). In Fig. 2 we fixed $\mu = 50$ GeV and $\omega = 1.5$ and varied $m_{\tilde{g}}$ between -500 GeV and $+500$ GeV; this corresponds to $0 \leq m_{\tilde{Z}_1} \leq 48$ GeV, 34 GeV $\leq m_{\tilde{Z}_2} \leq 70$ GeV, 56 GeV $\leq m_{\tilde{Z}_3} \leq 107$ GeV, and 107 GeV $\leq m_{\tilde{Z}_4} \leq 187$ GeV. The jumps in the curves correspond to level crossings; at $m_{\tilde{g}} \simeq -200$ GeV, the $\tilde{Z}_3$ and $\tilde{Z}_4$ eigenstates change identity, and the same thing happens for $\tilde{Z}_1$ and $\tilde{Z}_2$ at $m_{\tilde{g}} \simeq 200$ GeV.

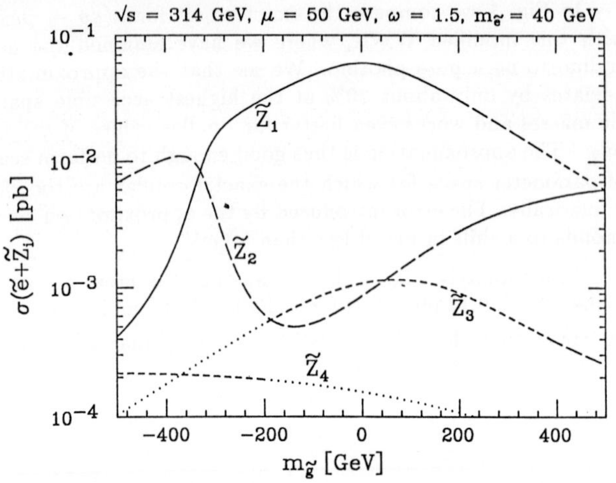

Fig. 2. Total cross sections for $ep \to p\tilde{e}\tilde{Z}_i$, $i = 1, \ldots, 4$, within the minimal SUSY model. The changes in textures of the curves correspond to level-crossings of the eigenvalues of the neutralino mass matrix. The modified WWA has been used.

From this figure and similar results for other combinations of parameters we conclude that the cross section for the production of the next to lightest neutralino can be sizeable; in these regions of parameter space the $\tilde{Z}_1$ cross section is rather small, since the $\tilde{Z}_1$ is mostly a higgsino. (It is possible to find[6] regions of parameter space where the $\tilde{Z}_1$ is mostly a gauge fermion and yet does not couple to left-handed electrons; it does, however, couple to right-handed electrons. Note that in Fig. 2 we have assumed unpolarized electrons.) The cross section for the production of the heavier neutralinos $\tilde{Z}_3$ and $\tilde{Z}_4$ at HERA is, however, always very small, less than 0.02 pb for $m_{\tilde{e}} \geq 30$ GeV. This value can go up to 0.1 pb at $\sqrt{s} = 1.4$ TeV

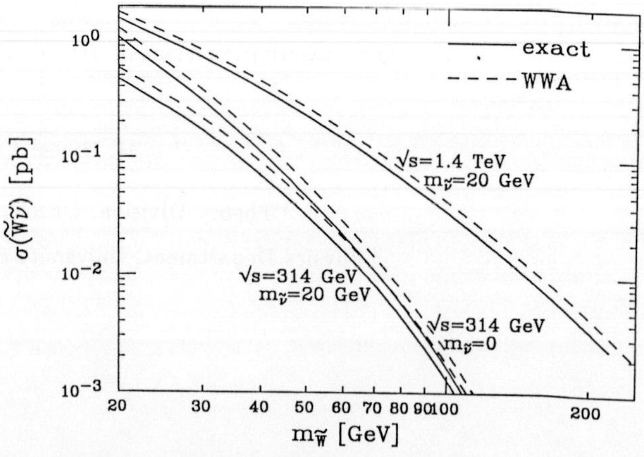

Fig. 3. Total cross sections for $ep \to p\tilde{W}_-\tilde{\nu}_e$ where we have assumed the $\tilde{W}_-$ to be a pure W-ino. The meaning of the full and dashed curves is as in Fig. 1.

if the selectron mass is indeed around 30 GeV; however, if selectrons are too heavy to be seen at LEP II, *i.e.*[8] $m_{\tilde{e}} \gtrsim 80$ GeV, the $\tilde{Z}_{3,4}$ cross sections would again fall below 0.02 pb.

In Fig. 3 we show results for the total cross section for $ep \to p\tilde{W}\tilde{\nu}_e$, where we have assumed the chargino to be a pure W-ino, which leads to the largest cross section. Again we show exact results (solid lines) and the predictions of the modified WWA (dashed) separately; the deviations are somewhat bigger than for selectron-photino production (see Fig. 1), but are still smaller than 30% for all interesting cases. Since now the lepton-slepton-gaugino coupling strength is given by the SU(2) gauge coupling rather than the QED coupling, the cross section is about a factor of two bigger than in the $\tilde{e}\tilde{\gamma}$ case if we set $m_{\tilde{\gamma}} = m_{\tilde{\nu}}$, $m_{\tilde{W}} = m_{\tilde{e}}$. However, existing lower bounds on the mass of a W-ino like chargino are somewhat higher than on $m_{\tilde{e}}$; in Refs. 9 it has been concluded that one can exclude hadronically decaying W-inos with mass below 40 GeV from UA1 monojet data. (Note, however, that these are not "official" experimental bounds; they also do not hold[10] for a general chargino state.) The bound is even stronger,[5] around 45 GeV, if sneutrinos are light ($m_{\tilde{\nu}} \lesssim 30$ GeV) so that $\tilde{W}_- \to \ell\tilde{\nu}_e$ decays are possible. If this is true, the elastic $\tilde{W}\tilde{\nu}$ cross section cannot exceed 0.1 pb at HERA, which is very similar to the maximal realistic $\tilde{e}\tilde{Z}_1$ cross section. This lower bound on $m_{\tilde{W}}$ also implies that $\tilde{W}\tilde{\nu}$ production is only interesting at HERA if $m_{\tilde{\nu}} \lesssim m_{\tilde{W}}$.

We now turn to a discussion of the signals for our processes. We start with the simplest and perhaps most interesting reaction, $ep \to p\tilde{e}\tilde{Z}_1$ with the subsequent decay $\tilde{e} \to e\tilde{Z}_1$. The $\tilde{Z}_1$ may be stable, if it is the lightest supersymmetric particle (LSP), or may decay into $\tilde{\nu} + \nu$ if the $\tilde{\nu}$ is the LSP, but it is always invisible. The signal thus just consists of one hard electron in the main detector, plus a proton that goes down the beam pipe and may or may not be detected by the forward proton spectrometer. Simulations of the ZEUS detector predict[11] a maximal detection efficiency of about 50% (if the proton loses between 5 and 20% of its energy), but it may well be higher for certain spectacular types of events where milder cuts on the proton detection may be sufficient; an event with just one hard electron in the main detector certainly qualifies as spectacular.

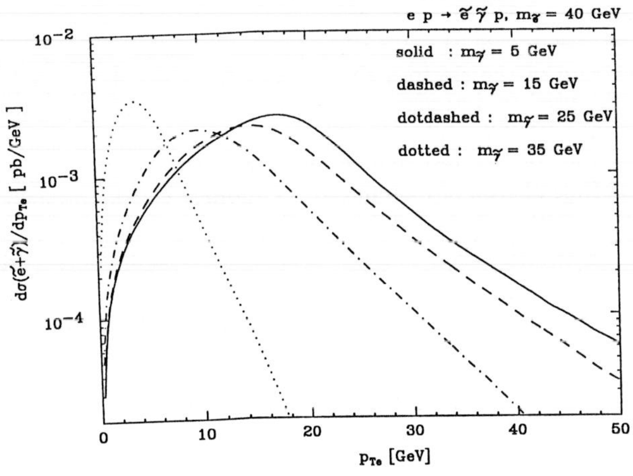

Fig. 4. The transverse momentum spectrum of the electrons that originate from selectron-decay in $ep \to p\tilde{e}\tilde{\gamma}$ events.

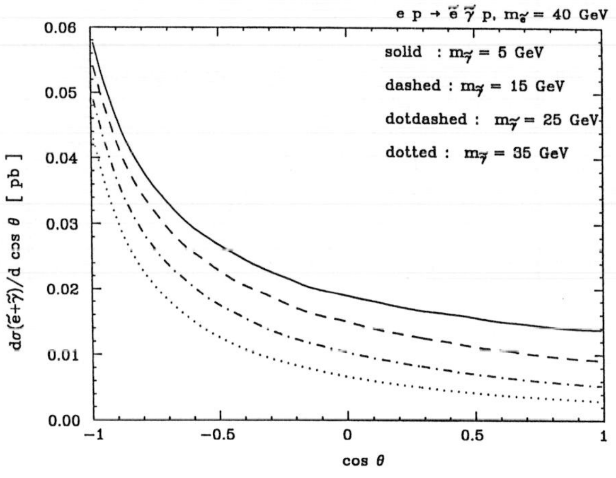

Fig. 5. The angular distribution of the decay electrons in $\tilde{e}\tilde{\gamma}$ events in the lab frame.

In Fig. 4 we show the transverse momentum spectrum of the electron for the case $\tilde{Z}_1 = \tilde{\gamma}$, which is almost identical to the missing p_T spectrum, since the proton practically never carries more than 1 GeV of transverse momentum. We see that all curves have a well-defined peak near the value of the electron momentum in the selectron rest frame,

$$p_e^* = \frac{m_{\tilde{e}}^2 - m_{\tilde{\gamma}}^2}{2\, m_{\tilde{e}}} \,. \tag{1}$$

At least 50% of the events have an electron-p_T within 5 GeV of the value (1). Unfortunately the shape of the curves depends only weakly on $m_{\tilde{\gamma}}$ as long as $m_{\tilde{e}}^2 \gg m_{\tilde{\gamma}}^2$. On the other hand, if $m_{\tilde{e}} - m_{\tilde{\gamma}}$ becomes too small, the electron becomes too soft to pass cuts against beam-gas and radiative QED events. From Fig. 4 we see that a mass difference of 5 GeV might already be marginal.

The energy spectrum of the electron looks very similar to the p_T-spectrum and thus does not provide additional information.

In Fig. 5 we show the polar angle distribution of the electron with respect to the electron beam direction in the lab frame. Since the electron beam energy is fixed whereas the photons have a rather soft spectrum the selectron-photino system tends to be boosted in the electron direction for small $m_{\tilde{e}} + m_{\tilde{\gamma}}$, and in the proton direction for large sparticle masses. For the chosen value of $m_{\tilde{e}} = 40$ GeV this results in a peak at $\cos\theta_e = -1$ (the proton direction) in the distribution of Fig. 5 which becomes more pronounced as the photino mass increases. However, even for the largest accessible values of $m_{\tilde{e}}$ and $m_{\tilde{\gamma}}$, the angular distribution is sufficiently flat not to loose too many events from acceptance cuts. On the other hand, the change of the shapes of the curves of Fig. 5 should be big enough to make it possible to get some estimate of $m_{\tilde{e}} + m_{\tilde{\gamma}}$ from the angular distribution of the electron.

A much more direct measurement of this sum is possible if one can measure the energy of the outgoing proton, which gives the energy of the photon emitted by it. In Fig. 6 we show the distribution $d\sigma/dz$, where z is the photon energy in units of the incident proton energy. This distribution has a very pronounced threshold at

$$z_{\min} = \frac{(m_{\tilde{e}} + m_{\tilde{\gamma}})^2}{s} \,. \tag{2}$$

If one measures this threshold together with the maximum of the p_T-spectrum, Eq. (1), one can in principle determine both $m_{\tilde{e}}$ and $m_{\tilde{\gamma}}$! On the other hand Fig. 6 shows that the z-distribution outside the threshold region is almost independent of the final state masses.

For reasons of space we can only briefly describe the signals for the other two interesting reactions, $ep \to p\tilde{Z}_2\tilde{e}$ and $ep \to p\tilde{W}_-\tilde{\nu}_e$. They obviously depend on the decay pattern of the produced gauginos, which in turn depends not only on the three parameters $m_{\tilde{g}}$, μ and ω that determine the masses and mixings of the gauginos, but also on the parameters of the scalar sector.

In the $\tilde{Z}_2\tilde{e}$ events the decay of the $\tilde{e}$ will again give a hard electron and missing p_T. The $\tilde{Z}_2$ will decay either into[12] $\tilde{Z}_1 + H_L^0$, where H_L^0 is the light neutral scalar Higgs boson[13] of minimal supersymmetry, or[9] into $\tilde{Z}_1\ell^+\ell^-$, $(\ell = e, \mu, \tau)$ or invisibly into $\tilde{Z}_1\nu\bar{\nu}$. Since the cross section at HERA is only interesting if $m_{\tilde{e}} \lesssim 50$ GeV, whereas from CDF[14] we now know that squarks are heavier than about 80 GeV, we expect[9] the 3-body decay $\tilde{Z}_2 \to \tilde{Z}_1 q\bar{q}$ to be unimportant. If it is the only kinematically allowed two-body decay mode, the first of these decays will be the dominant one, leading to a final state consisting of a $b\bar{b}$ (or $c\bar{c}$ or $\tau^+\tau^-$ if $m_{H_L^0} \lesssim 2\, m_b$) pair in addition to the hard electron; note that at least within the minimal SUSY model this decay chain is not unlikely, since H_L^0 becomes massless if either $\omega \to 1$ or the scalar mass scale goes to zero and we have assumed at least the selectrons to be rather light. However, in the region of parameter space interesting for HERA, one often has $m_{\tilde{e}} < m_{\tilde{Z}_2}$, so that $\tilde{Z}_2 \to \tilde{e}e$ decays are also possible.

We have already seen that in view of existing[8–10] lower bounds on the chargino mass, $\tilde{\nu}_e\tilde{W}_-$ production at HERA will only be interesting if $m_{\tilde{\nu}_e} < m_{\tilde{W}_-}$, which implies that the $\tilde{W}_-$ can decay into an e-sneutrino and an electron, leading to the same signature as in the $\tilde{e}\tilde{Z}_1$ case since the sneutrino is either

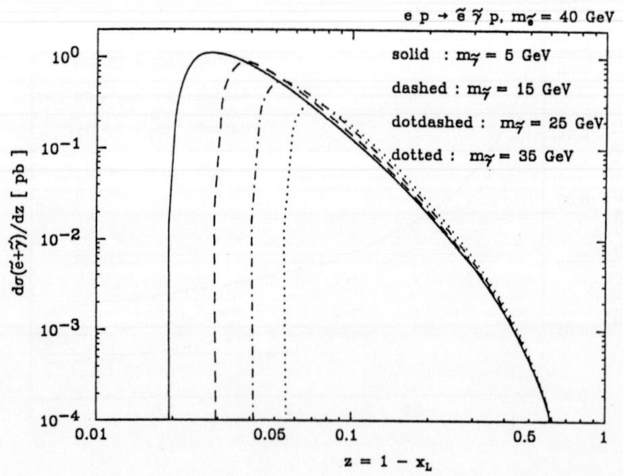

Fig. 6. The distribution of the energy of the exchanged photon (in units of the proton beam energy = 820 GeV) in the lab frame for $\tilde{e}\tilde{\gamma}$ events; x_L is the ratio of incoming to outgoing proton energies.

stable or decays invisibly. However, in minimal SUSY models[7] one expects e-, μ- and τ-sneutrinos to be (almost) degenerate in mass, so that $\tilde{W}_-$-production should lead to almost equal numbers of e, μ, and τ events; and even if $m_{\tilde{\nu}_{\mu,\tau}} \gg m_{\tilde{\nu}_e}$ so that the $\tilde{W}_-$ always decays into an electron, $\tilde{W}_- \tilde{\nu}_e$ production could easily be distinguished from $\tilde{e}\tilde{Z}_1$ production since the former cross section vanishes for right-handed polarized electrons, whereas the latter cross section should change by less than a factor two (it remains constant if the $\tilde{Z}_1$ is a photino).

We would finally like to make some remarks concerning backgrounds. The only SM process leading to a final state consisting of a single charged lepton ℓ and missing p_T in the main detector are $ep \to p\nu W^-$ with the subsequent decay $W^- \to \ell\bar{\nu}_e$, and $ep \to epZ^0$ with subsequent $Z^0 \to \nu\bar{\nu}$ decay. The corresponding cross sections are tiny ($\lesssim 0.01$ pb) at HERA-energies, but become sizeable (0.1 pb) at $\sqrt{s} = 1.4$ TeV; however, since the properties of W- and Z-bosons are well known, one should be able to subtract this background. We have seen in the last section that $\tilde{e}\tilde{Z}_2$ production can lead to events containing 3 charged leptons and missing p_T; we do not know of any standard model physics background for this topology. In these two types of events the detection of the outgoing proton will therefore not have to be part of the event recognition.

The only class of events where the proton detection might be necessary to get rid of SM backgrounds are those where the $\tilde{Z}_2$ decays into a Higgs boson, since the Higgs decay will lead to some hadronic activity in the main detector. In principle, cuts on missing p_T, the invariant mass of the hadronic system, and the angle between the hadronic system and the hard electron (which should be in opposite hemispheres in deep inelastic background events) should greatly reduce or even eliminate that background; however, in practice, the small number of signal events may make this approach impossible.

In summary we have shown that the elastic production of a selectron plus the lightest or second-lightest neutralino, as well as elastic sneutrino-chargino production, should lead to observable signals at HERA if $m_{\tilde{e}}, m_{\tilde{W}} \lesssim 50$ GeV (for small $m_{\tilde{Z}_1}, m_{\tilde{\nu}_e}$). For the case of $\tilde{e}\tilde{Z}_1$ production the measurement of the energy of the outgoing proton together with the p_T-spectrum of the electron that results from the decay of the $\tilde{e}$ can be used to determine both $m_{\tilde{e}}$ and $m_{\tilde{Z}_1}$. We also pointed out that HERA might be able to discover simultaneously the selectron, the lightest and second lightest neutralino, and the light Higgs boson of minimal SUSY models via the production of $\tilde{e}+\tilde{Z}_2$ with subsequent $\tilde{e} \to e+\tilde{Z}_1$ and $\tilde{Z}_2 \to H_L^0+\tilde{Z}_1$ decays. All backgrounds appear to be negligible or at least tractable. For a discussion of the discovery limits of HERA and future ep colliders within the framework of minimal SUSY we refer to the summary article of the ep working group.[15]

ACKNOWLEDGEMENTS

We thank X. Tata, D. Reeder, and W. Smith for useful discussions. This research was supported in part by the University of Wisconsin Research Committee with funds granted by the Wisconsin Alumni Research Foundation, and in part by the U. S. Department of Energy under contract DE-AC02-76ER00881.

REFERENCES

1. Gaillard, M.K., Hall, L., Hinchlffe, I., Phys. Lett. 116B, 279 (1982); Kobayashi, T., Kuroda, M., Phys. Lett. 134B, 271 (1984).
2. Eilam, G. and Reya, E., Phys. Lett. 145B, 425 (1984); erratum, Phys. Lett. 148B, 502 (1984); Grifols, J. A, Pascual, R., Phys. Lett. 135B, 319 (1984).
3. Altarelli, G. *et al.*, Nucl. Phys. B262, 204 (1985).
4. Drees, M., Zeppenfeld, D., MAD/PH/443 (1988).
5. UA2 collaboration, Phys. Lett. 95B, 613 (1987).
6. Drees, M., Kim, C. S., Tata, X., Phys. Rev. D37, 784 (1987).
7. For a review, see Haber, H. E., Kane, G. L., Phys. Rep. 117 75 (1985).
8. Chen, M. *et al.*, Phys. Rep. 159, 201 (1988).
9. Baer, H., Hagiwara, K., Tata, X., Phys. Rev. Lett. 57, 294 (1986) and Phys. Rev. D35, 1598 (1987); Chamseddine, A., Nath, P., Arnowitt, R., Phys. Lett. 174B, 399 (1986).
10. Baer, H., Hagiwara, K., Tata, X., Univ. Wisconsin preprint MAD/PH/409 (1988).
11. ZEUS collaboration, *Zeus Detector Technical Specification*, DESY preprint.
12. Barbieri, R. *et al.*, Nucl. Phys. B296, 75 (1988).
13. Inoue, K. *et al.*, Prog. Theor. Phys. 67, 189 (1982); Gunion, J. F., Haber, H. E., Nucl. Phys. B272, 1 (1986).
14. Freeman, J., Talk presented at the *7th Topical Workshop on Proton-Antiproton Collider Physics*, Fermilab, June 1988.
15. Smith, W., these proceedings.

DIQUARK PRODUCTION IN *ep* COLLISIONS

Thomas G. Rizzo
Ames Laboratory and Department of Physics
Iowa State University, Ames, Iowa 50011

SUMMARY

We examine the production of single diquarks (D) and diquark pairs ($D\bar{D}$) in *ep* collisions using the effective photon approximation. Cross-sections for both mechanisms are found to be quite sensitive to the assumed charge of the diquark but can lead to significant production rates in all cases at HERA and future *ep* colliders. Backgrounds from QCD jet production are found to be potentially significant in the case of single D production whereas the production of $D\bar{D}$ pairs is relatively background free.

INTRODUCTION

A diquark (D) can be most generally defined to be an elementary integer-spin particle with baryon number $|B| = 2/3$ and lepton number $L = 0$ which couples to a pair of quarks. If we restrict ourselves to a class of models with only color triplet ($\underline{3}$) quarks with electric charges $Q = 2/3$ or $-1/3$ as in the Standard Model (SM) then the corresponding diquarks will transform as $\underline{3}$ or $\underline{\bar{6}}$ ($\underline{\bar{3}}$ or $\underline{6}$) under $SU(3)_C$ and have electric charges $|Q_D| = 1/3$, $2/3$, or $4/3$. In such models if R-parity is a conserved quantity, then diquarks necessarily will have positive R-parity.

Diquarks can occur in many scenarios which involve new physics beyond the SM, e.g., composite models and E_6 superstring-inspired models (SSIM).[1] In particular, in SSIM each generation of fermions lies in a $\underline{27}$ representation of E_6 which demands the existence of new fermions (beyond those which are already known) in order to complete the $\underline{27}$. These fermions are vector-like under $SU(2)_L \times U(1)_Y$ and although the color and electric charges of these exotic states are fixed, their values of B and L are not. Several assignments of B and L for these exotics are possible by the imposition of discrete and/or continuous symmetries on the superpotential and the demand of tree-level proton stability. One of these exotics, h, which is a color-triplet with $Q = -1/3$ can be assigned $B = -2/3$, $L = 0$ and negative R-parity. Thus its superpartner, $\tilde{h}$, has all the properties expected of a $Q = -1/3$, color-triplet, weak-isosinglet diquark.

The production of diquarks in e^+e^- annihilation[2] and in hadronic collisions[3] has been discussed elsewhere so that we will consider diquark production in *ep* colliders in our discussion below. We will also assume diquarks are color-triplets for color counting purposes when presenting our numerical results.

PRODUCTION MECHANISMS

In *ep* collisions, diquarks can be either produced singly or in pairs and will examine each of these mechanisms separately.

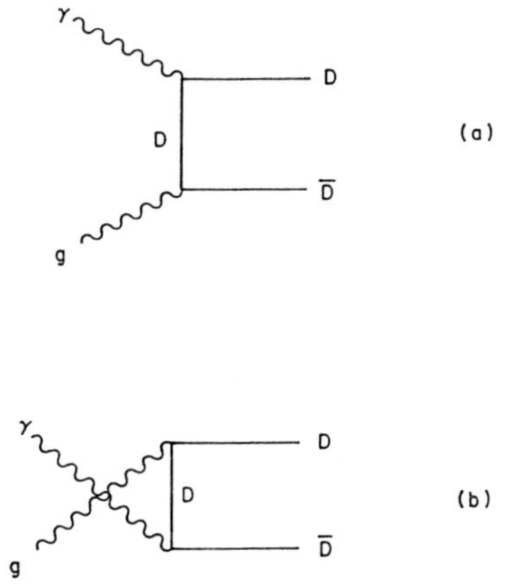

Fig. 1. $D\bar{D}$ pair production via photon-gluon fusion.

(*i*) Pair Production. Figure 1 shows the production of diquark pairs ($D\bar{D}$) via photon-gluon fusion In the effective photon approximation,[4] the production cross-section for this process can be written as ($\hat{s} = \tau s$)

$$\sigma_{D\bar{D}} \simeq \int_{4m_d^2/s}^{1} d\tau \int_{\tau}^{1} \frac{dx}{x} f_{\gamma/e}(x, \sqrt{s}/2) g(\tau/x, Q^2) \hat{\sigma}_{D\bar{D}} \tag{1}$$

with the subprocess cross-section for $\gamma g \to D\bar{D}$ given by

$$\hat{\sigma}_{D\bar{D}} = \frac{\pi \alpha \alpha_s(Q^2)}{8\hat{s}} Q_D^2 \left\{ (1+\beta^2)^2 \times \ell n \left(\frac{1+\beta}{1-\beta} \right) + 2\beta(1-\beta^2) \right\} \tag{2}$$

where $\beta \equiv (1 - 4M_D^2/\hat{s})^{1/2}$, and $g(f)$ is the gluon (photon) distribution function for the proton (electron). Note that $\sigma_{D\bar{D}}$ depends only on the charge (Q_D) and mass (M_D) of the diquark. Figures 2a and 2b show $\sigma_{D\bar{D}}$ as a function of M_D for $Q_D = 1/3, 2/3,$ and $4/3$ for $\sqrt{s} = 314$ GeV and $\sqrt{s} = 1.8$ TeV respectively. Figure 2b also shows $\sigma_{D\bar{D}}$ for $Q_D = 1/3$ and $\sqrt{s} = 1.4$ TeV for purposes of comparison.

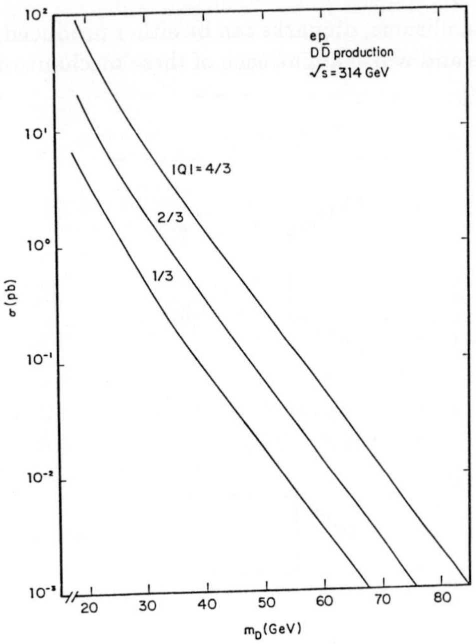

(a)

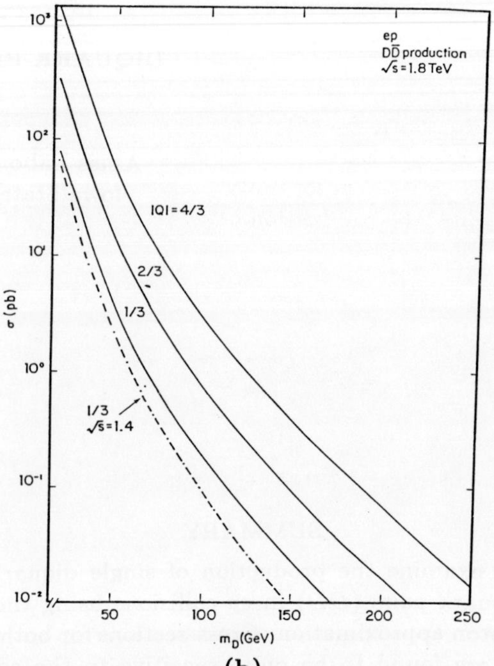

(b)

Fig. 2. Production cross-section for $D\bar{D}$ pairs as a function of M_D for $Q_D = 4/3$ $2/3$ and $1/3$ with (a) $\sqrt{s} = 314$ GeV; (b) $\sqrt{s} = 1.8$ TeV. Also shown in (b) is the cross-section for $D\bar{D}$ production for $Q_D = 1/3$ and $\sqrt{s} = 1.4$ TeV for comparison.

(*ii*) Single Production. Figure 3 shows the production of single diquarks via photon-quark fusion which involves the *a priori* unknown $D_{qq'}$ vertex which can be parameterized as $\sim a + b\gamma_5$.

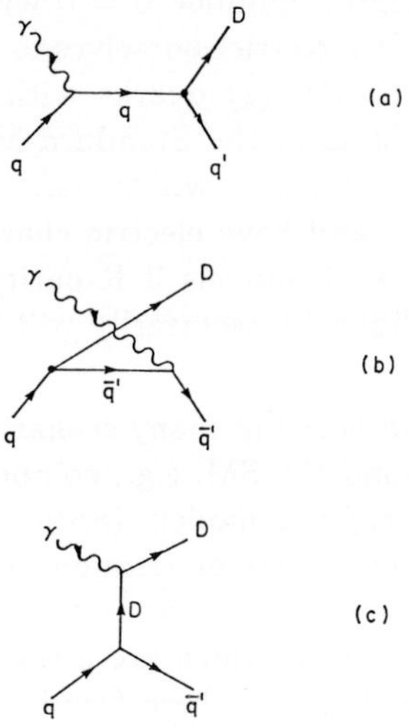

Fig. 3. Single D production via photon-quark fusion.

We will assume the parameters a and b are independent of the choice of q and q' and define a new parameter, κ, which measures the overall strength of the coupling relative to that of electromagnetism:

$$\frac{a^2+b^2}{4\pi} = \kappa\alpha \quad (3)$$

where α is the usual fine-structure constant. Henceforth, κ will be treated as a free parameter but it will be generally assumed to be of order unity. In the effective photon approximation, the production cross-section for this process can be written as

$$\sigma_D \simeq \int_{M_D^2 s}^{1} d\tau \int_{\tau}^{1} \frac{dx}{x} f_{\gamma/e}(x, \sqrt{s}/2) \sum_q q(\tau/x, Q^2) \int dz \left(\frac{d\hat{\sigma}_D}{dz}\right) \quad (4)$$

with the subprocess differential cross-section given by

$$\frac{d\hat{\sigma}_D}{dz} = \kappa\frac{\pi\alpha^2}{2\hat{s}}\beta'\Big\{ -\hat{s}\hat{u}(Q^2/\hat{s}^2 + Q'^2/\hat{u}^2) - 2\frac{QQ'}{\hat{u}\hat{s}}(\hat{u}-M_D^2)\times (\hat{s}-M_D^2) - 2M_D^2(\hat{u}+\hat{s}-M_D^2)(Q-Q')\times (\hat{s}^2Q' - \hat{u}^2Q)/[\hat{u}\hat{s}(\hat{u}+\hat{s})^2]\Big\} \quad (5)$$

where $\beta' \equiv 1 - M_D^2/\hat{s}$, $\hat{u} \equiv -1/2\hat{s}\beta'(1+\cos\theta)$, and $Q(Q')$ is the charge of the initial (final) $q(\bar{q}')$. The sum in Eq. (4) extends over all quarks and antiquarks in the proton. Figures 4a and 4b show the production cross-sections for single diquarks (D or $\bar{D}$) at $\sqrt{s} = 314$ GeV and $\sqrt{s} = 1.8$ TeV respectively as functions of M_D for different values of Q_D.

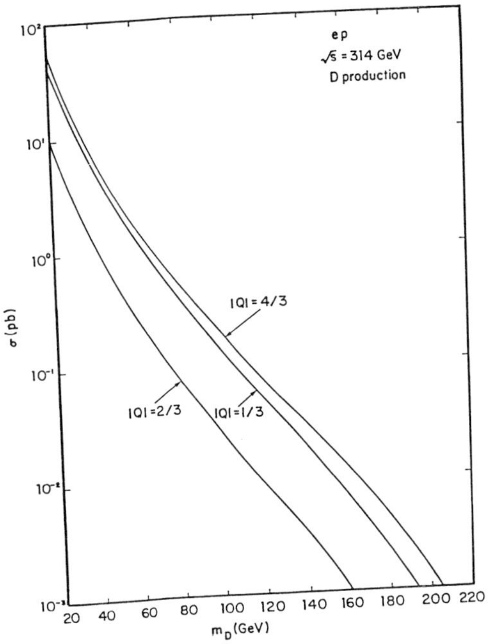

(a)

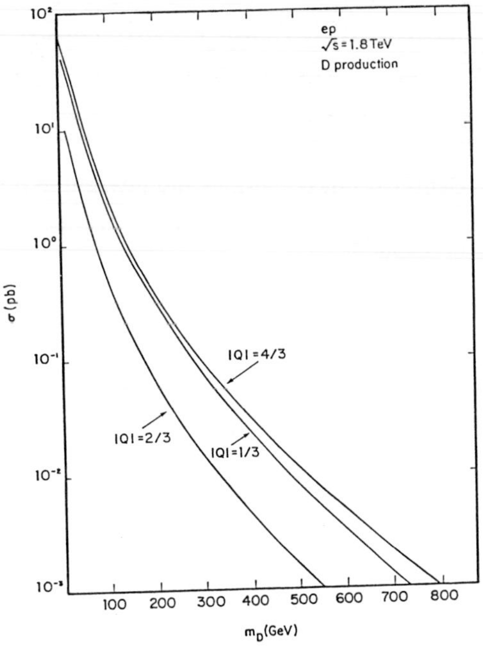

(b)

Fig. 4. Same as Fig. 2 but for single D production with $\kappa = 1$ with (a) $\sqrt{s} = 314$ GeV and (b) $\sqrt{s} = 1.8$ TeV.

$\kappa = 1$ is summed in both figures so that σ_D for any other value of κ is easily obtainable by a similar rescaling of the results presented in Fig. 4a and 4b.

SIGNALS, BACKGROUNDS AND DISCOVERY LIMITS

The decay of diquarks into a pair of quarks provides the signal for D production, i.e., 2 jets with a peaking in the two-jet invariant mass distribution ad M_D. Thus in the case of $D\bar{D}$ production, the resulting four jet final state (from $D\bar{D}$ decay) shows a double peaking in the jet-jet invariant mass with a common value (M_D) if the jets are correctly paired. the QCD background is potentially more severe in the case of single D production[5,6] since the associated $\bar{q}'$ in the final state prefers to be produced at small angles (i.e., down or near the beam pipe) while the decay of the D results in the usual two jets. By demanding that the p_T of the $\bar{q}'$-initiated jet is above some minimum value (p_T^{min}) helps in reducing this background but also results in a significant reduction in the cross-section. For example, if we take $p_T^{min} = 5(10$ GeV) the single D production cross-section is reduced by roughly a factor of 3-6 (5-10) with the exact value being somewhat sensitive to both Q_D and M_D. This case clearly deserves a detailed Monte Carlo study in order to fully examine the size and shape of the QCD background relative to the single D production signal.

In Table 2 we display the search limits for diquarks from pair production as well as single production (without any p_T cuts) for the four ep colliders listed in Table 1:[7]

	HERA	HERA′	LEPI × LHC	LEPII × LHC
$\sqrt{s}$ (GeV)	314	410	1400	1800
$L(pb^{-1})$	200	200	1000	100

Table 1. Center-of-mass energies and integrated luminosities for the ep colliders discussed in the text and in Table 2.

Since $D\bar{D}$ production is relatively background free, these limits are obtained assuming the production of 10 events using the calculated cross-sections and the machine integrated luminosities shown in Table 2.

Pairs	$\sqrt{s} = 314$ GeV	$\sqrt{s} = 410$ GeV	$\sqrt{s} = 1.4$ TeV	$\sqrt{s} = 1.8$ TeV
$Q_D =$				
4/3	61	74	218	188
2/3	52	63	181	147
1/3	43	52	148	111
Single				
$\kappa = 1$				
$Q_D =$				
4/3	128 (81)	143 (93)	475 (290)	300 (180)
1/3	120 (75)	135 (88)	450 (265)	275 (160)
2/3	88 (50)	97 (60)	300 (165)	175 (90)
$\kappa = 1/3$				
$Q_D =$				
4/3	106 (82)	121 (78)	380 (205)	215 (115)
1/3	95 (58)	113 (72)	350 (190)	200 (110)
2/3	48 (35)	80 (46)	210 (115)	125 (60)
$\kappa = 3$				
$Q_D =$				
4/3	152 (106)	170 (118)	600 (390)	400 (220)
1/3	141 (97)	153 (110)	550 (350)	360 (200)
2/3	108 (65)	122 (80)	390 (230)	240 (130)

Table 2. Discovery limits in GeV for diquarks at various ep colliders for different values of the diquark charge. For pair production, the discovery limit is set by 10 events whereas for single diquark production 10(100) events are used since the QCD backgrounds are larger in this case. No p_T cuts or backgrounds are included.

In the case of single D production, we quote limits for both 10 and 100 events (since the backgrounds from QCD are expected to be larger) for different values of the parameter κ. It is clear from Table 1 that the high luminosity, $\sqrt{s} = 1.4$ TeV, LEPI × LHC can probe the largest diquark mass range for all values of κ and the diquark charge. The search limits in the single D case are lowered appreciably when the p_T cuts discussed in the last paragraph are included.

CONCLUSION

We have examined the production of diquarks, both singly and in pairs, at various ep colliders. The cross-section are found to be sufficiently large so that new mass scales can be probed at such machines. Clearly, full Monte Carlo studies of diquark production in ep collisions are warranted.

The author would like to thank M. Drees, J. L. Hewett and W. Smith for detailed discussions related to this work.

This work was supported by the U. S. Department of Energy, Contract No. W-7405-Eng-82, Office of Energy Research (KA-01-01), Division of High Energy and Nuclear Physics

REFERENCES

1. For a review of the phenomenology of E_6 superstring-inspired models see Hewett,J. L. and Rizzo, T. G., Ames Lab reports IS-J 3005 (1988), submitted to Physics Reports.
2. Schaile, D. and Zerwas, P. M. in *Proceedings of the Workshop on Physics at Future Accelerators*, CERN yellow report 87-07, Vol. II, p. 251 (1987).
3. Argyes, E. N., Papadopoulos, C. G. and Vlassopulos, S. D. P., Phys. Lett. 202B, 425 (1988); V. D. Angelopoulos *et al.*, Nucl. Phys. B292, 59 (1987).
4. For a general discussion of the effective gauge boson approximation see Johnson, P. W., Olness F. and Tung, W. -K. in *Proceedings of the 1986 Summer Study on the Physics of the Superconducting Supercollider*, edited by R. Donaldson and J. Marx (1986).
5. The author would like to thank Manuel Drees for extensive discussions related to the importance of the QCD background in the case of single D production coming from the hadronic structure of the photon.
6. Drees, M. and Godbole, R. M., Phys. Rev. Lett. 61, 682 (1988).
7. Smith, W., summary talk given at the *1988 DPF Summer Study on High Energy Physics in the 1990's*, Snowmass, CO, June 26 - July 17, 1988.

AXIGLUON PRODUCTION AT HERA

David M. Atwood and Jerry A. Robinson
Physics Department, McGill University
3600 University St., Montréal, PQ
Canada, H3A 2T8

ABSTRACT

The recently proposed chiral color model predicts the existence of strongly coupled axigluons having a mass the order of the weak bosons. In this paper we consider the production of axigluons in ep collisions at HERA. We find that an axigluon with a mass in the experimentally allowed range of 25 GeV $\leq M_A \leq$ 150 GeV should be observable at HERA for 25 GeV $\leq M_A \leq$ 100 GeV. In particular, the mass window 70 GeV $\leq M_A \leq$ 100 GeV, which appears to be inaccessible to e^+e^- and $p^{(}\overline{p}^{)}$ machines, is open to HERA within the first year of running.

INTRODUCTION

At present energies all known phenomena are successfully described by the standard $SU(3)_C \times SU(2)_L \times U(1)_Y$ theory. Extensions of this theory include enlargement of the weak group (e.g. $SU(2)_L \times SU(2)_R \times U(1)_{B-L}$) or unification (e.g. SU(5), SO(10),...). However, these alternatives invariably leave the strong interaction sector unchanged. Recently it has been suggested that above the weak scale the strong interactions exhibit a chiral $SU(3)_L \times SU(3)_R$ symmetry.[1] In this model, the "low energy" $SU(3)_C$ group corresponds to the diagonal subgroup which remains unbroken after spontaneous symmetry breaking. The quintessential prediction of theories of this type is a massive octet of strongly interacting gauge bosons. These "axigluons" must have a mass $\sim$ 100 GeV and couple axially to quarks.[1]

The production of axigluons at hadronic colliders has been thoroughly analyzed by Bagger, Schmidt, and King.[2] Using existing CERN data, these authors conclude that axigluons with a mass between 125 and 275 GeV can be ruled out. This agrees reasonably well with the limit recently announced by the UA1 collaboration:[3] exclusion of masses in the range 150-315 GeV at 95% confidence level. A lower bound of $M_A \geq$ 25 GeV can be obtained from Upsilon decays.[4] Rizzo[5] has found that axigluons with mass $\leq$ 70 GeV should be observable, via bremsstrahlung off of quarks in Z-decays, at LEP and SLC. Also, Z-decays to gluon plus axigluon (via a triangle diagram) also provide an observable signal for $M_A \leq$ 70 GeV.[6] In this paper we study the production of axigluons at HERA via the process $ep \to eAX \to 3jets + (e)$. We find that an axigluon having a mass in the range 40 to 100 GeV should be observable at HERA.

CALCULATIONS

The Feynman diagrams for the subprocess $\gamma q \to Aq$ are shown in Figure 1(a). To obtain the matrix element for this process consider the graphs for $q\overline{q} \to Z\gamma$ shown in Figure 1(b). The differential cross section for this reaction is[7]

$$\frac{d\hat{\sigma}}{d\hat{t}}(q_i\overline{q}_i \to Z\gamma) = \frac{1}{16\pi\hat{s}^2}\sum_{spin}|\mathcal{M}|^2 = \frac{\pi\alpha^2 Q_i^2(v^2+a^2)}{12x_w(1-x_w)\hat{s}^2}\left[\frac{\hat{s}^2+M_Z^4}{2\hat{t}\hat{u}} - 1\right] \tag{1}$$

where

$$v = \tau_3 - 4Q_i x_w \qquad a = -2Q_i x_w \tag{2}$$

$x_w = sin^2\theta_w = 0.23$, Q_i is the quark charge, and $\hat{s}$, $\hat{t}$, and $\hat{u}$ are the usual kinematic invariants: $\hat{s} = (p_1+p_2)^2$, $\hat{t} = (p_1-p_3)^2$, and $\hat{u} = (p_1-p_4)^2$. By crossing symmetry we have

$$|\mathcal{M}|^2(\gamma q_i \to Zq_i) = -|\mathcal{M}|^2(q_i\overline{q}_i \to Z\gamma, s \leftrightarrow t) = \frac{48\pi^2\alpha^2Q_i^2(v^2+a^2)}{x_w(1-x_w)}\left[1 - \frac{\hat{t}^2+M_Z^4}{2\hat{s}\hat{u}}\right]. \tag{3}$$

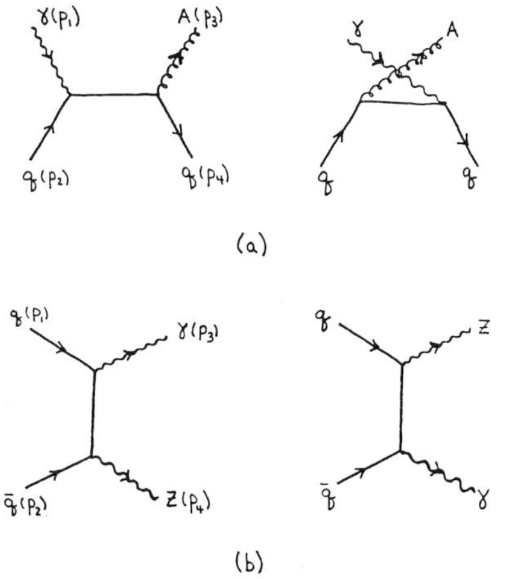

Fig. 1. Feynman diagrams for the subprocess (a) $\gamma q \to \gamma Z$ and (b) $q\overline{q} \to \gamma Z$.

Thus averaging (summing) over initial (final) state spin and color yields

$$\frac{d\hat{\sigma}}{d\hat{t}}(\gamma q_i \rightarrow Z q_i) = \frac{\pi\alpha^2\, Q_i^2(v^2+a^2)}{4x_w(1-x_w)\hat{s}^2}\left[1-\frac{\hat{t}^2+M_Z^4}{2\hat{s}\hat{u}}\right]. \tag{4}$$

Now, the coupling of a Z to a quark is given by $g/4c_w\gamma_\mu(v-a\gamma_5)$, $(c_w = cos\theta_w)$, and an axigluon couples as $g_A\gamma_\mu\gamma_5$. Then $d\hat{\sigma}/d\hat{t}(\gamma q_i \rightarrow Aq_i)$ is obtained by setting the vector coupling of the Z equal to zero, substituting g_A for $-ga/4c_w$, and multiplying by the color factor attributed to the axigluon. Technically,

$$\frac{d\hat{\sigma}}{d\hat{t}}(\gamma q_i \rightarrow Aq_i) =$$
$$\frac{d\hat{\sigma}}{d\hat{t}}(\gamma q_i \rightarrow Z q_i; v=0, \frac{\alpha a^2}{x_w(1-x_w)} = \alpha_A) \times C_F \tag{5}$$

with $C_F = 4/3$ and $\alpha_A = \frac{g_A^2}{4\pi}$. Hence, we finally obtain the result

$$\frac{d\hat{\sigma}}{d\hat{t}}(\gamma q_i \rightarrow Aq_i) = \frac{16\pi\alpha_s\alpha Q_i^2}{3\hat{s}^2}\left[1-\frac{\hat{t}^2+M_A^4}{2\hat{s}\hat{u}}\right]. \tag{6}$$

It should be noted that we have set the axigluon coupling to that of the strong interactions, $\alpha_A = \alpha_s$, in Eq. (6).

To obtain the physical cross section we need to fold $d\hat{\sigma}/d\hat{t}$ with the photon wave function of the electron and the quark distribution in the proton. For the probability distribution of emitting a photon from an electron $(f_{\gamma/e})$ we use the Weizsacker-Williäms approximation and for the quark structure function $(f_{q/p})$ we use Duke-Owens set I.[8] Thus, the p_T distribution for axigluon production is given by

$$\frac{d\sigma}{dp_T} =$$
$$2p_T\sum\int\int dx_A dx_B f_{\gamma/e}(x_A) f_{q/p}(x_B)\left(\frac{\hat{s}}{|\hat{t}-\hat{u}|}\right)\frac{d\hat{\sigma}}{d\hat{t}}. \tag{7}$$

The results of this numerical integration are displayed in Figure 2. Here we have plotted (solid line) the p_T distribution of the three jet final state resulting from the two jet decay of the axigluon plus the recoil jet. As representative values we choose axigluons of mass 40, 80, and 120 GeV. The three jet QCD background (dashed line) has been calculated using the expressions from Ellis and Kunszt[9], however, it should be noted that the uncertainty in this estimate may be as large as $\pm 50\%$. To obtain the p_T distribution of the background we select two of the three jets (in all possible two-jet combinations) and require that the invariant mass of the two jet system (M_{2j}) be given by $M_{2j} = M_A \pm 0.1 M_A$. This $\pm 10\%$ uncertainty approximately simulates the detector inefficiency in mea-

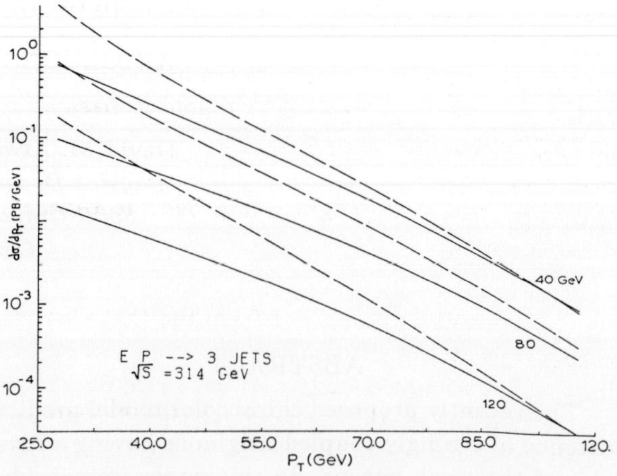

Fig. 2. Transverse momentum distribution of the 2-jet system defined by $M_{2j} \simeq M_A \pm 10\%(M_A)$. Shown are sample values of $M_A = 40, 80$, and 120 GeV. The dashed lines represent the 3-jet QCD background as calculated in Ref. 9.

suring jet energy as well as the energy uncertainty arising from the width of the axigluon. We see that for large p_T, axigluon events are nearly as large as the 3-jet background. For $p_T \leq 40$ GeV a measurement of $d\sigma/dp_T$ at roughly $\sim 15\%$ for $M_A = 120$ GeV, or $\sim 30\%$ for $M_A = 40$ GeV, will deviate sufficiently from the QCD result.

We now turn our attention to the plot of total cross section versus axigluon mass shown in Figure 3. To integrate $d\sigma/dp_T$ we have included all events with $p_T \geq 40$ GeV (upper two curves) and $p_T \geq 50$ GeV (lower curves). For an axigluon of mass ~ 80 GeV we see that the 3-jet QCD background is roughly a factor of two larger than the 3-jet signal. Since σ ($M_A \simeq 80$ GeV) with $p_T \geq 40$ GeV ($p_T \geq 50$ GeV) is roughly $0.5pb$ ($0.15pb$) we would expect $\simeq 100$ ($\simeq 30$) events of this kind per year at the HERA luminosity of $200pb^{-1}$ year. So we see that for axigluons of this mass the signal produces roughly a 3σ

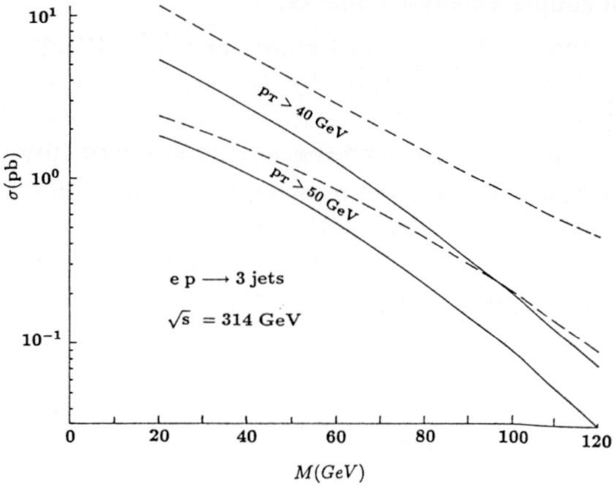

Fig. 3. Total cross section versus axigluon mass for events with $p_T \geq 40$, 50 GeV. The dashed line is the 3-jet background, as in Fig. 2.

(nearly 4σ for $p_T \geq 50$ GeV) deviation from the expected QCD result, however, in order to have a reasonable event rate the p_T cut of 40 GeV is preferable in this mass range. With a p_T cut of 40 GeV an axigluon of mass as large as 100 GeV should be observable yielding 40 events/year (a 3σ effect). For lighter mass axigluons the cut $p_T \geq 50$ GeV is preferable to slightly improve the ratio of signal to background.

In Figure 4 we plot $d\sigma/dM_{jj}$ versus $M_j j$. Recall that the QCD result was obtained from the three jet events by considering each possible di-jet combination (3 combinations) and testing that the recoil jet have the specified p_T. We see that the axigluon mass peak significantly enhances the distribution for both 40 and 80 GeV axigluons. The low event rate at the 80 GeV peak shows that the $p_T \geq 50$ GeV cut is to restrictive and hence for heavy axigluons the p_T cut of 40 GeV is preferable.

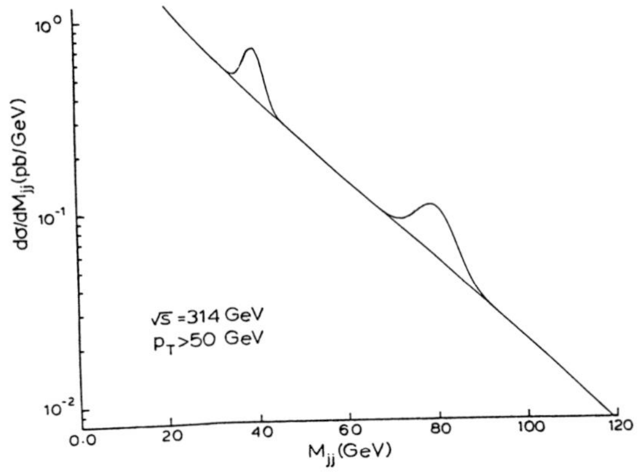

Fig. 4. Di-jet invariant mass distribution for the case $p_T \geq 50 GeV$. Superimposed on the QCD background are the mass peaks for M_A =40 and 80 GeV.

As argued in the introduction (and in Ref. [2-6]) the axigluons of mass in the range 70 GeV$\leq M_A \leq$ 125 GeV appear to be inaccesible to planned e^+e^- and hadronic machines. We have seen that axigluons with mass in the range 25 GeV$\leq M_A \leq$ 100 GeV should be accessible at HERA.

Thus, HERA provides a unique opportunity to examine axigluons with mass in this potentially important window.

Note: Upon Completion of this work we learned of a work by Robinett and Rizzo [10] which reaches conclusions similar to ours.

ACKNOWLEDGEMENTS

We would like to thank the Aspen Center for Physics for the use of their facilities during the Snowmass workshop. Also, we would like to thank the organizers of Snowmass '88 for a well organized and stimulating workshop. Finally, we thank the members of the *ep* working group for comments and in particular B. Irwin and W. Smith for their interest in this work.

This work was supported by NSERC and in part by the Quebec Department of Education, FCAR.

REFERENCES

1. P.H. Frampton and S.L. Glashow, Phys. Lett. 190B, 157 (1987); Phys. Rev. Lett. 58, 2168 (1987).
2. J. Bagger, C. Schmidt, and S. King, Phys. Rev. D37, 1188 (1988).
3. C. Albajar *et al.*, (UA1 Collaboration), Phys. Lett. 209B, 127 (1988).
4. M.A. Doncheski, H. Grotch, and R.W. Robinett, Phys. Rev. D38, 412 (1988).
5. T.G. Rizzo, Phys. Lett. 197B, 273 (1987).
6. E.D. Carlson, S.L. Glashow, and E.E. Jenkins, Phys. Lett. 202B, 281 (1988).
7. E. Eichten, I. Hinchliffe, K. Lane, and C. Quigg, Rev. Mod. Phys. 56, 579 (1984).
8. D. Duke and J. Owens, Phys. Rev. D30, 49 (1984).
9. R.K. Ellis and Z. Kunszt, Nucl. Phys. B303, 653 (1988).
10. R. W. Robinett and T. G. Rizzo, Penn. State Univ. preprint PSU/TH/49 (1988).

STRONG INTERACTIONS

Marjorie Shapiro and Ian Hinchcliffe

Working Group Leaders

Report of the QCD Working Group. *

Ian Hinchliffe
Lawrence Berkeley Laboratory
University of California
Berkeley, California 94720
and
Marjorie D. Shapiro
Physics Dept.
Harvard University
Cambridge, Massachusetts 02138

Working Group Members.

D. Baden, E.L. Berger, S. Bhadra, B. Blumenfeld, D. Brown, A. Caldwell, R. Carey, S. Dawson, C. Day, L. Galtieri, P. Ghez, J. Huston, J. Huth, P.K. Malhotra, J. Owens, B. Pope, P. Sheldon, E.M. Wang, B.R. Webber, J. White, S. Willenbrock, H. Yamamoto, M. Zielinski.

Abstract

We discuss some current problems associated with the applications of QCD to event rates in high energy collisions. Emphasis is given to the current ambiguities and uncertainties that exist in estimates of signals and backgrounds. The production of jets and isolated photons at hadron colliders is discussed in some detail. The problems of jet definition are addressed. Some features of the events underlying the hard scattering process are discussed.

1 Introduction.

The main issue confronting QCD in the next generation of experiments in e^+e^-, ep and hadron–hadron collisions will be the computation of and measurement of backgrounds to new physics processes. While "tests" of QCD will continue, these will come to be interpreted as limits on new physics rather than as confirmation of the validity of QCD. In this report we will emphasize the measurements and calculations that need to be done in order to understand such backgrounds. Most of our discussion will be given in the context of hadron colliders, although many of the issues discussed are relevant to other interactions.

We shall begin with a discussion of the uncertainties in predicted rates for QCD processes. This will involve a discussion of the the methods used to calculate these rates and of the status of distribution functions and of higher order QCD corrections. We shall use the production of heavy quarks and large transverse momentum photons in hadron–hadron collisions to illustrate the issues. We shall then discuss some issues related to the definition (experimental and theoretical) of the final state jets. Finally, we shall address some aspects of the structure of the whole event where the initial state radiation and underlying event play a vital role.

2 Estimating QCD Rates.

The rate for some hard scattering process is given by the following parton model formula applicable to hadron–hadron interactions. [1]

$$\sigma \sim \sum_{ij} \int dx_1 dx_2 \hat{\sigma}_{ij} f_i(x_1, M^2) f_j(x_2, M^2). \qquad (1)$$

Here i and j label the types of incoming partons (gluons and the various flavors of quarks and antiquarks). $f_i(x, M^2)$ is the parton distribution function *i.e.* the probability that a parton of type i carries momentum x times that of its parent hadron. The invariant mass of the parton-parton system ($\sqrt{\hat{s}}$) is related to the hadron-hadron center of mass energy ($\sqrt{s}$) by $\hat{s} = x_1 x_2 s$. The parton cross section σ_{ij} can be expressed as an expansion in $\alpha_s(\mu)$

$$\hat{\sigma}_{ij} = A\alpha_s^n(\mu)(1 + B\alpha_s(\mu)\ldots). \qquad (2)$$

The scales M and μ are of the same order as the momentum transfers in the parton scattering process.

It is often convenient to think of a five stage process when describing the hadronic event. The distribution functions evolve from some reference scale M_0 to M by emission of quarks and gluons that are almost collinear to the incoming beam. This evolution is described by the Altarelli-Parisi [2] equations and results in a pair of partons each with invariant mass squared of order $-M^2$ (they are spacelike) in addition to the collinear shower that can be thought of, by analogy with QED, as initial state radiation. The two partons now interact in a manner described by $\hat{\sigma}$. The outgoing partons which have invariant mass squared of order M^2 (they are timelike) now emit, if they are colored, a shower of quarks and gluons most of which are soft or collinear with the initial outgoing partons. During this showering the invariant mass is reduced to some reference value of order a few GeV, where perturbative QCD ceases to be applicable. The quarks and gluons are then combined to form hadrons. The final part of the event is that resulting from the remnants of the original beam particles (underlying event). The separation between these processes is not clear cut, but provides a convenient paradigm.

Before Equation 1 can be used we must; *(a)* calculate $\hat{\sigma}$; *(b)* have a set of distribution functions ($f_i(x, M^2)$); and *(c)*, choose the scales M and μ. *(a)* is straightforward although it can be extremely tedious. For most processes only the term A in Equation 2 is known. The term B has been computed for the following processes in hadron–hadron collisions; the total cross section integrated over transverse momentum ($p_\perp$) for the the production of a W/Z or lepton pair of large invariant mass (Drell-Yan) [3]; the total cross-section for the production of a heavy quark-antiquark pair [4]; the production of a photon at large transverse momentum [5]; and the inclusive single jet rate at large $p_\perp$ [6]. In the case of the latter two processes the rate depends upon the definition of an "isolated" photon and a jet; these issues will be discussed further below. For lepton–hadron scattering, the higher order corrections to the total cross-section for the production of a heavy quark-antiquark pair [7] and to the structure functions [8] are known.

A discussion of the determination of the distribution functions and an estimate of the errors in them has been given in an accompanying article [9]. The existing parameterizations arise from fits to deep inelastic scattering data (with occasional input from Drell-Yan and photon production in hadron collisions). One of the major difficulties with such fits is the systematic disagreement between different data sets. The EMC [10] and BCDMS [11] collaborations have both measured the structure function $F_2(x, Q^2)$ on using muon scattering from hydrogen targets in the same range of x and Q^2. The results of these two measurements show systematic differences that are larger than the quoted errors [12]. When using these data to extract distribution functions, a choice must be made between them.

There are many sets of distribution functions coming from fits to the data using lowest order QCD. The most frequently used of these are the two sets of Duke and Owens [13] (DO1 and DO2) which were based on data from EMC [14], SLAC [15] and CDHS [16] [17] (the latter were renormalized in an attempt to deal with the systematic differences in the data sets, see above), and Eichten *et al.* [18] (EHLQ1 and EHLQ2) based primarily on the CDHS data. These pairs correspond to different shapes for the gluon distribution and consequently different values of α_s (or Λ). As usual, the gluon distribution with more support at large x (harder distribution) corresponds to the larger value of α_s (EHLQ2 and DO2). Parameterizations of these distribution functions are given in the papers and can easily be applied to a variety of other processes.

Recently, fits using next to leading order QCD have emerged. Diemoz, Ferroni, Longo and Martinelli (DFLM) [19] used neutrino data from BEBC [20] , CCFRR [21], CHARM [22] and CDHS [16] [17]. They also provide different fits corresponding to different values of α_s. They give sets of distribution functions corresponding to a range of $\Lambda^\dagger$ viz $\Lambda = 160,\ 260,\ 360$ MeV. These fits are used to estimate the uncertainties in top quark rates at the Tevatron and $Sp\bar{p}S$ colliders [24].

Martin, Roberts and Stirling (MRS) [25] have used EMC data together with that from CDHSW and CCFRR to which they apply a renormalization of order 10%, to remove the systematic disagreement with EMC. They present three fits that differ in the form of $xg(x, M^2 = 4GeV^2)$.

$$xg(x, M^2 = 4GeV^2) \qquad \sim (1-x)^5 \qquad (set\ 1)$$

---

†Here we are quoting a Λ that corresponds to 4 flavors, in the range $m_{charm} < Q < m_{bottom}$ the formula for α_s is $\alpha_s(Q^2) = \frac{12\pi}{25\log(Q^2/\Lambda^2)}[1 - \frac{462}{625}\frac{\log\log(Q^2/\Lambda^2)}{\log(Q^2/\Lambda^2)}]$. See reference [23] for a summary of the behavior of this formula as a threshold is crossed.

$$\sim (1-x)^4(1+9x) \qquad (set\ 2)$$
$$\sim x^{-1/2}(1-x)^4(1+9x) \quad (set\ 3)(3)$$

They then use data from J/ψ production [26] and photon production [27] at large transverse momentum, processes that are sensitive to the shape of the gluon distribution (see below), in an attempt to distinguish between the sets. They conclude that the soft gluon distribution of *set 1* is preferred.

Recently, [28] they have refitted *set 1* using the BCDMS [11] [29] data instead of EMC [10] [14]. Here they find that the neutrino data and BCDMS are compatible and that a renormalization of the former is not needed. We will refer to these new fits as MRS1 (EMC) and MRS2 (BCDMS) respectively when they are used to calculate rates. These authors have compared the predictions from these two sets of distributions with the data on Drell–Yan production at the ISR [30]. The BCDMS fit is preferred, but the order α_s QCD corrections to the Drell–Yan rate are quite large [3] and the α_s^2 terms are not known so any definite conclusion seems premature.[‡].

Existing deep-inelastic scattering data do not extend below $x \sim 0.01$ and cover a very small range of Q^2 at small x. This is a potential problem since for some applications it is necessary to know the parton distributions in this region. It is traditional to assume that the gluon distribution obeys

$$\lim_{x\to 0} xg(x, Q_0^2) = const. \tag{4}$$

for some scale Q_0 of order a few GeV. However this form is unstable. When evolved to higher M^2, it develops rapidly into a steeper form (see Figure 1). At very small x and large M^2, it is possible to solve the Altarelli–Parisi equations analytically. At order α_s we have

$$\lim_{x\to 0, M\to\infty} xg(x, M^2) \sim e^{\sqrt{\frac{48}{(11-2n_f/3)}\log(1/x)\log\log(M^2)}} \tag{5}$$

This form is singular as $x \to 0$. It is also possible to sum to all orders in α_s the most singular terms at small x and large M^2. This gives

$$\lim_{x\to 0, M\to\infty} xg(x, M^2) \sim x^{-\delta} \tag{6}$$

where, $\delta = 12\alpha_s \log(2)/\pi$, which is an even more singular form [32]. It has been suggested [32] that one should use a form for $xg(x, Q_0^2)$ that is more like the asymptotic form.

$$xg(x, Q_0^2) \sim 1/\sqrt{x} \tag{7}$$

is most commonly used. It is not clear that this form is a better assumption than the traditional one, or below what value of x this form should hold. Notice that the momentum sum rule provides almost no constraint since the amount of momentum carried by gluons in the region $x < 0.01$ is small, whichever form is used there. Figure 1 compares the resulting gluon distributions at higher M^2 that evolve from different forms at M_0^2. The two starting forms are equal for $x > 0.02$ ($M_0^2 = 5GeV^2$) and have the forms of Equations 4 and 7 at smaller x; the first of these is the EHLQ set 2 (see above). We will refer to the other as EHLQ2′ and will use it below to illustrate rates from such an extreme choice. As can be seen from Figure 1, the differences become less important at large M^2. The uncertainties in predicted rates due to the small x problem are therefore serious only for processes sensitive to small x *and* small M^2.

In what follows we will give rates from a representative set of distribution functions. In view of the systematic differences between the data sets on which the distribution functions are based, it is very difficult to assign an error to a prediction due to the distribution functions. However, the uncertainties are typically less than a factor of two unless the process involves very small values of x at relatively low M^2, for example, the total cross section for the production of bottom quarks at the SSC.

The form of the term B in Equation 2 will depend upon the scale M appearing in $f_i(x, M^2)$ since

$$f_i(x, M^2) = f_i(x, M_1^2) + \frac{\alpha_s(M_1^2)}{2\pi} log(M^2/M_1^2) \sum_j \int_x^1 \frac{dy}{y} f_j(y, M_1^2) P_{ij}\left(\frac{x}{y}\right), \tag{8}$$

where $P_{ij}(z)$ is the appropriate Altarelli-Parisi splitting function.[§] B will also depend upon the definition of the distribution functions. Since the distribution functions are extracted from measurements of one process for use in another, the reference process must be computed to the same order as the process where the functions are used. In the case of quark distributions, this reference process is deep inelastic scattering and the quark distributions can be de-

[‡]Recently Tung [31] has discovered that there is a problem with the evolution program of MRS in the small x region that generates a value for $xg(x, M^2)$ that is too large for $x < 0.01$.

[§]This expression is exact only in the limit $M \to M_1$

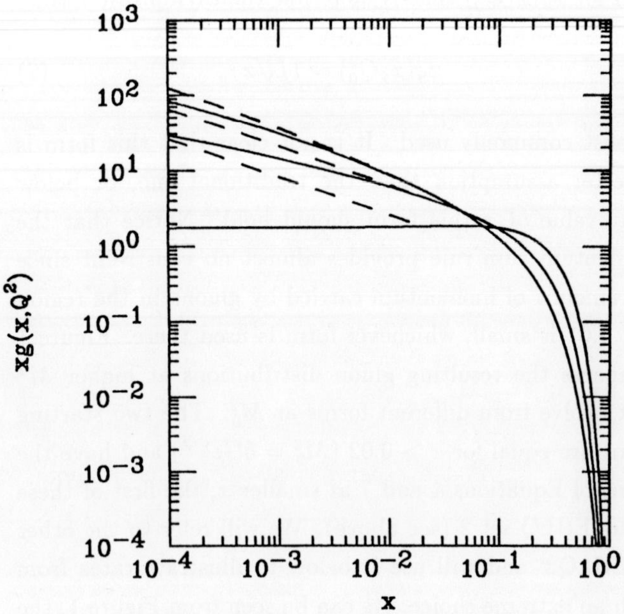

Figure 1: A comparison of the gluon distributions for fixed M^2 as a function of x. The solid lines are EHLQ set 2 and the dashed are EHLQ2′ (see text). The higher (lower) curve at small x corresponds to $M^2 = 50(5)$ GeV^2.

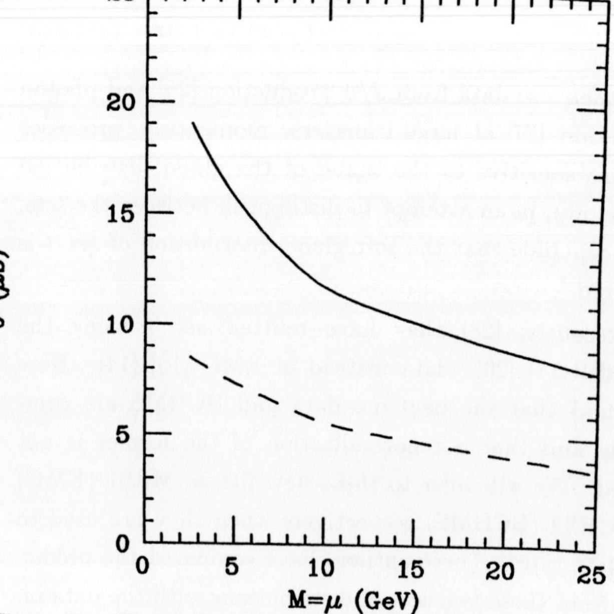

Figure 3: The dependence of the bb production rate upon $\mu = M$ for $m_{bottom} = 4.9$ GeV at $\sqrt{s} = .63$ TeV in $p\bar{p}$ collisions in leading (dotted line) and next to leading order (solid line) using the DFLM ($\Lambda =$ 260 MeV) distribution functions.

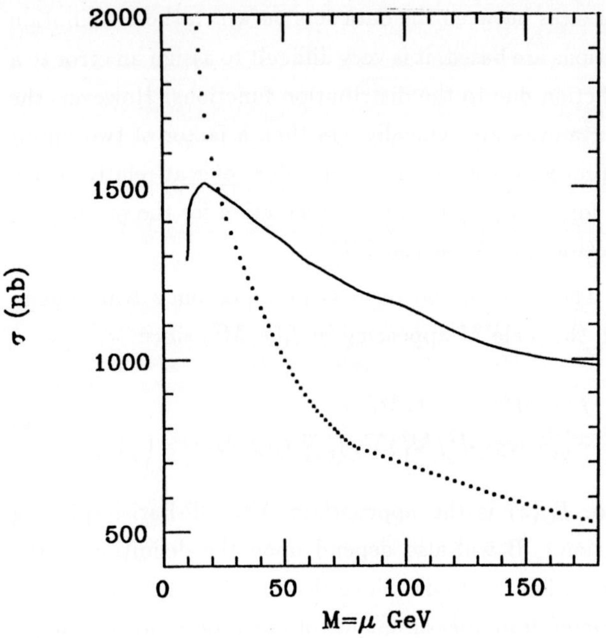

Figure 2: The dependence of the $t\bar{t}$ production rate upon $\mu = M$ for $m_{top} = 60$ GeV at $\sqrt{s} = 1.8$ TeV in $p\bar{p}$ collisions in leading (dotted line) and next to leading order (solid line) using the DFLM ($\Lambda =$ 260 MeV) distribution functions.

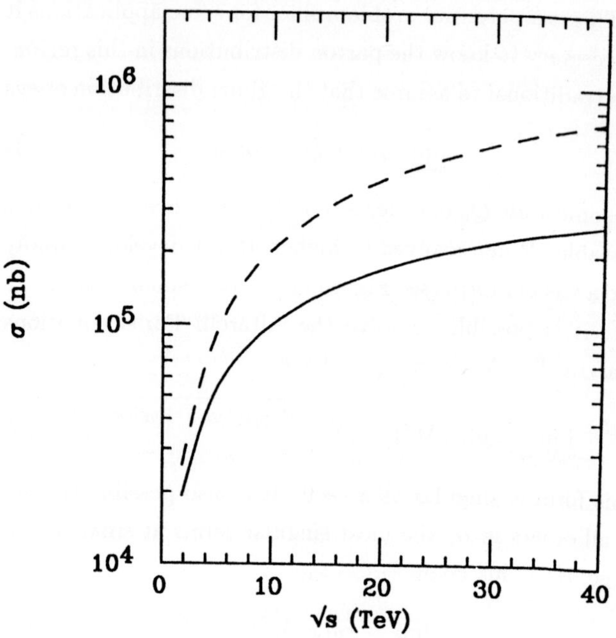

Figure 4: The production rate for $b\bar{b}$ pairs in pp collisions as a function of $\sqrt{s}$. The distribution functions of EHLQ2 (solid line) and EHLQ2′ of (dashed line) are used.

fined so that

$$F_2(x,M^2) = x \sum_{i=quarks} e_i^2 f_i(x,M^2) \tag{9}$$

with no $O(\alpha_s)$ correction. Here e_i is the electric charge of parton type i. A different definition will result in an order α_s correction to this expression and to a corresponding change in B in Equation 2. DFLM use this scheme to define their quark distribution functions; MRS do not. Hence their results are not comparable. One must compute a physical quantity such as F_2 to facilitate a comparison.

The gluon distribution function cannot be probed directly in deep inelastic scattering, except from a measurement of $R = \sigma_L/\sigma_T$ or, equivalently, the longitudinal structure function F_L, where the gluon distribution enters at the same order in α_s as the quark distribution

$$\begin{aligned} F_L(x,M^2) &= F_2(x,M^2) - 2xF_1(x,M^2) \\ &= \frac{\alpha_s(M^2)}{2\pi}\int_x^1 \frac{dy}{y^3}\left[F_2(y,M^2) + (1-x/y)yg(y,M^2)\right]. \end{aligned} \tag{10}$$

At present, data are too poor for this to be a practical method of determining $g(x,M^2)$. Data from HERA may enable this to be used in the future [33] [34]. Consequently, the current $g(x,M^2)$ are inferred from the measured growth in the antiquark distributions with M^2. The resulting $g(x,M^2)$ is correlated with the value of α_s which also controls the growth.

The form of B in Equation 2 is also dependent upon the choice of μ, the scale appearing in $\alpha_s(\mu)$ if $n \neq 0$ since

$$\alpha_s(\mu^2) = \alpha_s(\mu_1^2)(1 - \frac{33-2n_f}{12\pi}\alpha_s(\mu_1^2)log(\mu_1^2/\mu^2)). \tag{11}$$

Here n_f is the number of quark flavors with mass less than μ.

If the terms in Equation 2 were computed to all orders in α_s, the result of Equation 1 would be independent of the choice of μ and M. A change would merely adjust the relative size of the different terms in the α_s expansion; the sum would be unchanged. The dependence upon M and μ is greatest if only the lowest order in Equation 2 is known. Computation of the higher order terms helps to reduce the uncertainty in the predicted rates due to the choice of M and μ if the perturbation series is convergent. The concept of a K factor defined as the ratio of the result of Equation 1 evaluated using next to leading order QCD to the result using leading order is meaningless since the factor will depend upon the choice of M and μ. There is no requirement that $M = \mu$, but such a choice enables one to simplify the discussion so it will be made in the following. There are several approaches to the M/μ dependence.

One can vary M over a reasonable range and look at the resulting uncertainty in the cross-section. It is clear that M should be a scale characteristic of the momentum transfers in the partonic process. In the case of the production of jets or photons or transverse momentum $p_\perp$ a scale of order $p_\perp$ is appropriate. Varying M from $2p_\perp$ to $p_\perp/2$ will give some estimate of the uncertainty. If the terms B in Equation 2 are known, other strategies are possible. One can choose M so that B is very small, in the hope that the (unknown) still higher order terms will also be rendered small (fastest apparent convergence or FAC). Alternatively one can choose M so that $d\sigma/dM = 0$ [35], a property that is known to be true if σ is calculated to all orders (minimal sensitivity or PMS). Neither of these two choices is acceptable if it gives rise to unreasonable values of M.

3 Heavy Quark Production.

We will illustrate the M dependence by discussing top quark production rates at the Tevatron Collider. At lowest order in QCD, there are two relevant processes, $gg \to t\bar{t}$ and $q\bar{q} \to t\bar{t}$. (Here q represents any light quark.) The term B in Equation 2 receives contributions from one-loop corrections to these processes as well as from $gg \to gt\bar{t}$, $q\bar{q} \to gt\bar{t}$ and $gq \to qt\bar{t}$. These latter processes contain divergences when the light quark or gluon in the final state has $p_\perp = 0$. These singularities are the same ones that give rise to the growth with M^2 of the distribution functions. They are reabsorbed into the distribution functions. Of course, the remaining finite piece depends upon the definition of the distribution functions. Figure 2 shows the production cross–section for a top quark pair as a function of the scale M [4]. The DFLM ($\Lambda = 260MeV$) distribution functions are used. It can be seen that when the next to leading order calculation is used there is an uncertainty of order $\pm 10\%$ from this scale ambiguity if we vary M from $m_{top}/2$ to $2m_{top}$. The uncertainty in the lowest order prediction is much larger. Notice that the FAC (where the higher order corrections is zero) and PMS (the stationary point on the solid curve) choices are similar and correspond to $M \sim m_{top}/2$. Table 1 shows the spread in predicted cross–sections from the different sets of distribution functions. This spread is larger than $\pm10\%$, hence the primary uncertainty in top production rates, and therefore on a limit on the top quark mass if no events are seen, is due to distribution functions. Even so, it is fair to conclude that we understand these rates to better than a factor of two or so.

The situation with regard to bottom quark production is not so clear. Firstly the M dependence of the cross–section is quite large [4] (see Figure 3) and indeed is stronger if the next to leading order terms are included. This is an indiction that the predictions are unreliable. Notice that the results in next to leading order are approximately twice those in leading order for all values of M and μ. Since the higher order corrections increase the rate whatever the choice of M and μ, an uncertainty in the cross-section at $\sqrt{s} = 1.8$ TeV of at least a factor of two exists due to the choice and to the unknown next order corrections exists. In addition, the values of x_1 and x_2 (of order $m_{bottom}/\sqrt{s}$) and relatively low M^2 are in a region where the uncertainties due to the low x behavior of the distribution functions are large (*c.f.* Figure 2). In Table 1, we have indicated the spread in predicted bottom production rates in $p\overline{p}$ collisions at $\sqrt{s} = 1.8$ TeV due to the use of different distribution functions. We have shown values corresponding to two choices for the bottom quark mass. Lowest order QCD has been used in deriving the quantities in this table. If we take account of the higher order corrections any value between 15 μb and 70μb seems to be allowed by theory at present.

Figure 4 shows the total $b\overline{b}$ production rate in pp collisions as a function of $\sqrt{s}$ with $M = m_{bottom}$; the two curves correspond to EHLQ2 and EHLQ2′ distributions. One can see from this that the uncertainties get worse as $\sqrt{s}$ rises since M^2 is fixed and the relevant values of x_1 and x_2 decrease. An uncertainty of at least a factor of 5 exists in the rates predicted at the SSC. In order to reduce this, we need information on the gluons at small x since the $gg \to b\overline{b}$ process is dominant.

4 Photon Production.

Photons produced at large $p_\perp$ in hadron collisions can be used as both a test of QCD and as a probe of the gluon distribution. At lowest order in QCD perturbation theory, they are produced via the processes $q\overline{q} \to \gamma g$ and $qg \to \gamma q$. In pp collisions the latter process is always dominant. In $p\overline{p}$ collisions the latter process also dominates over the accessible range of $p_\perp$ as can be seen from Figure 5 which shows the production cross-section as a function of $p_\perp$ at $\sqrt{s}$ or 1.8 TeV and .63 TeV. Here the EHLQ2 distribution functions are used with $M = \mu = p_\perp$.

In view of the dominance of the gluon-quark initial state, this process could be useful in reducing some of the uncertainties in the gluon distribution function, since it probes values of x and M^2 not currently accessible in deep inelastic scattering experiments. Before attempting this, one must estimate the experimental error on the rate and the theoretical uncertainties arising from variations of M in order to determine whether these are larger than the uncertainties due to distribution functions. The terms contributing to B in Equation 2 have been calculated for this process by Aurenche and collaborators [5].

There are contributions from one loop corrections to the lowest order processes as well as from the processes $gq \to gq\gamma$, $qq \to qq\gamma$, $gg \to q\overline{q}\gamma$ *etc.*. There are singularities associated with quarks/gluons being parallel to the beam direction but these are reabsorbed into the distribution functions. There are also singularities associated with two final state collinear gluons (for example in $q\overline{q} \to gg\gamma$) or a collinear gluon and quark but these are removed when one calculates the inclusive photon spectrum. If the double differential cross-section $d\sigma/d^3p_\gamma/d^3p_{jet}$ were calculated then a "jet" would have to be defined and this last singularity so removed (see below). But processes such as $qq \to qq\gamma$ contain a divergence when the photon is collinear with a final state quark (bremsstrahlung); this divergence remains in the final result. It can be reabsorbed into the fragmentation function describing the probability of finding a photon in the jet resulting from a quark hadronization. If the quark had mass m, this fragmentation function would take the form

$$d_{q\gamma}(z, M^2) = \frac{e_q^2\alpha_{em}}{2\pi}\frac{1+(1-z)^2}{z}log(M^2/m^2) \qquad (12)$$

with $z = E_\gamma/E_{quark}$. In principle, this fragmentation function must be measured (for example from the production of a photon in $e^+e^- \to hadrons + \gamma$). In practice the above formula can be used as an approximation provided that m is replaced by Λ.

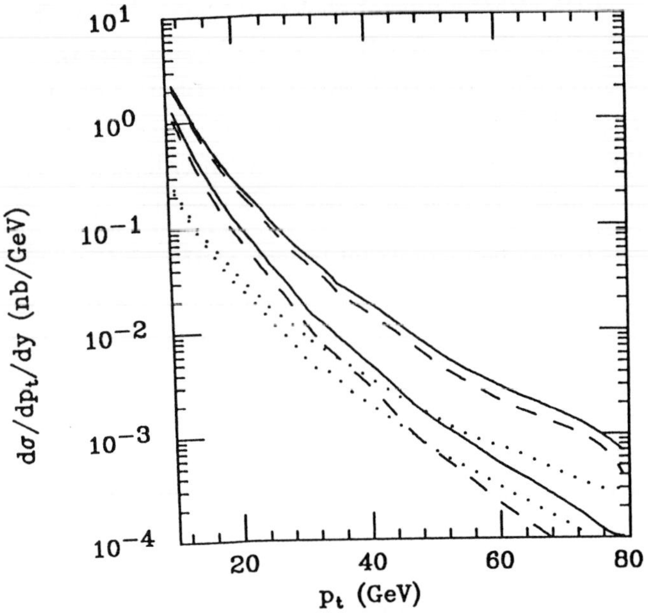

Figure 5: The cross section $d\sigma/dp_t dy$ for the production of a photon at $y = 0$ in $p\bar{p}$ collisions at $\sqrt{s} =$ 1.8 TeV (upper lines) and .63 TeV (lower lines). The solid lines represent the total rate while the dashed (dotted) is the contribution from gluon quark (quark antiquark) initial states. EHLQ2 distribution functions are used with $M = \mu = p_\perp$.

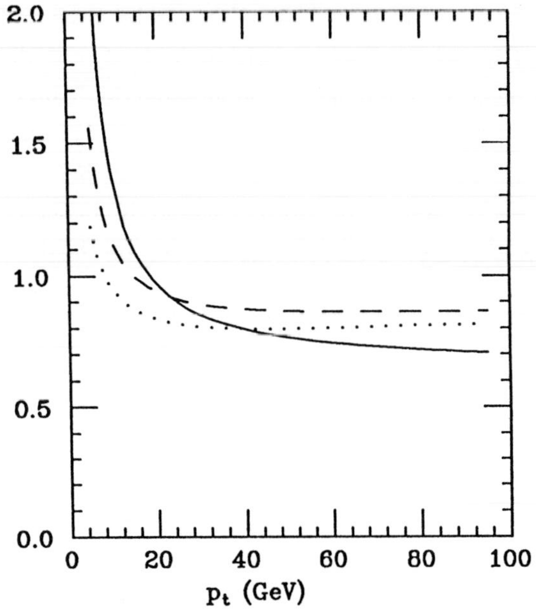

Figure 7: The ratio of the cross sections $d\sigma/dp_t dy$ for $M = \mu = p_\perp/2$ to that for $M = \mu = 2p_\perp$. The solid, dotted and dashed lines correspond to leading and next to leading order full rate and next to leading order isolated rate (same cone size at Figure 6)

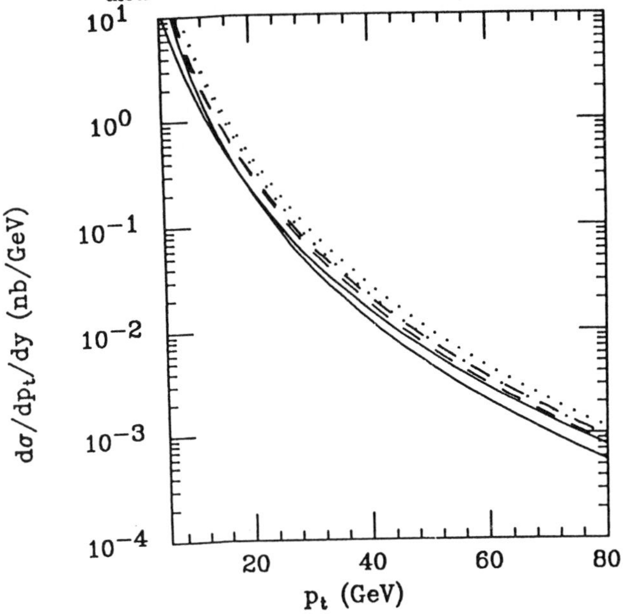

Figure 6: The cross section $d\sigma/dp_\perp dy$ for the production of a photon at $y = 0$ in $p\bar{p}$ collisions at $\sqrt{s} =$ 1.8 TeV. The bands between the pairs of solid, dotted and dashed lines correspond to the range of cross section as $M = \mu$ is varied from $p_\perp/2$ to $2p_\perp$ for lowest order, the full next to leading order rate and the next to leading order rate when the photon is required to be isolated in a cone of half-angle 40^0. Duke and Owens set 1 of distribution functions are used.

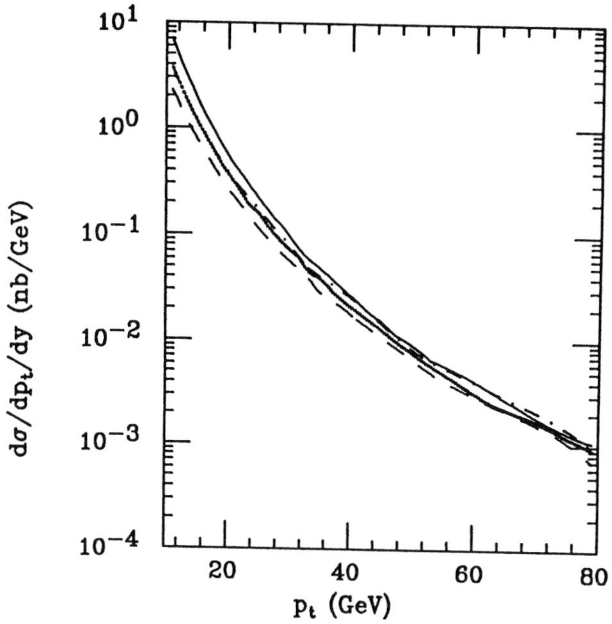

Figure 8: The cross section $d\sigma/dp_t dy$ for the production of a photon at $y = 0$ in $p\bar{p}$ collisions at $\sqrt{s} = 1.8$ TeV for $M = \mu = p_\perp$. The solid, dashed, dotted, and dot-dashed lines correspond to MRS2, EHLQ2 and DFLM ($\Lambda = 260$ MeV) and EHLQ2′. distribution functions respectively.

In an experiment, the photon must be "isolated" in order to be seen. Photons within jets are not distinguished from π^0's and hence the cross-section cannot be measured. The definition of isolation will vary between experiments. For example one can require that there be less than some amount E_0 of hadronic energy in a cone of radius $\Delta R = \sqrt{(\Delta y)^2 + (\Delta\phi)^2}$. ($\Delta y$ and $\Delta\phi$ are the distance in rapidity and azimuth from the photon.) In their calculation, Aurenche *et al.* define isolation by requiring that there be no parton in cone of half angle δ around the photon direction. This definition excludes the bremsstrahlung singularity and enables a well defined prediction to be made. However, their result cannot be easily compared to data. In an experiment, E_0 cannot be set to zero since there will always be some energy from the underlying event inside the cone; this underlying event is not present in the parton calculation.

These higher order corrections can be used to estimate the uncertainties in the rates due to the choice of M and μ. In Figure 6 we show range of predicted cross sections as $M = \mu$ is varied from $p_\perp/2$ to $2p_\perp$. In the case of the next to leading order calculations, two results are shown corresponding to an isolated and non–isolated cross section. The range of uncertainty can be seen more clearly in Figure 7 which shows the ratio of the cross-sections at the upper and lower edges of the bands in Figure 6. It can be seen that the higher order corrections reduce the theoretical uncertainties. If the PMS scheme is used to determine M and μ, a value somewhat less that $p_\perp/2$ is obtained and consequently the predicted rates are above the bands shown in Figure 6. It is not clear that the value obtained in this way is sensible. It is also clear from Figure 6 that the bremsstrahlung contribution is large, since the full rate (dotted lines) and the "isolated" rate (dashed lines) are substantially different. In order to use this process to make a detailed measurement of the gluon distribution or test of QCD, it will be necessary to incorporate the higher order corrections into a complete event generator, which includes the underlying event, so that an isolation cut can be applied to the theoretical calculation which is the same as that applied to the data.

Despite the uncertainties due to M and μ and due to the definition of an isolated photon, the process will be useful in constraining the gluon distribution. In Figure 8 we show the rates for prompt photon production from different sets of distribution functions. It can be seen that the range of uncertainty associated with the distribution functions is larger than that associated with the choice of M and μ. Note that the differences due to the distribution functions are smallest at large $p_\perp$ where M is largest.

It may be more useful to look at the cross section as a function of y rather than $p_\perp$. Figures 9 and 10 show this distribution for EHLQ2 distributions with $M = \mu = p_\perp$ and $M = \mu = \sqrt{x_1 x_2 s}$ as well as EHLQ2$'$ with $M = \mu = p_\perp$. Since the gq initial state dominates the rate is symmetric under $y \leftrightarrow -y$ and it is shown as a function of $|y|$. It can be seen that the more singular form of the gluon distribution at small x in the EHLQ2$'$ set gives rise to a different shape of the rapidity distribution. Changes in the choice of M and μ are not able to simulate this effect. Again the differences are most apparent in Figure 9 which is integrated over $p_\perp$ between 10 and 15 GeV. In Figure 10 which is integrated from 30 to 50 GeV, the scale M and the relevant ranges of x_1 and x_2 are larger and the differences in distribution functions less (recall Figure 1).

Figure 11 shows the direct photon cross-section at fixed target energies appropriate to FNAL. ($\sqrt{s} = 38$ GeV). At these energies the appropriate values of x_1 and x_2 are larger while M and μ are smaller than at collider energies. As in the collider case, the inclusion of the higher order terms reduces the uncertainty in the predicted rates. Notice in this case that the bremsstrahlung contribution is small; the dotted and dashed lines corresponding to the isolated and full rates are almost equal. Figure 12 shows the dependence of the rate upon the choice of distribution functions. A comparison of this Figure with figure 11 demonstrates that the uncertainties due to the choice of scale are larger than those associated with the distribution functions.

Production of a pair of photons in hadron-hadron collisions proceeds via the processes $q\bar{q} \to \gamma\gamma$ and $gg \to \gamma\gamma$. The latter process [36], which proceeds via a quark loop and is higher order in α_s, dominates since the gluon distribution is larger in the relevant regions of x_1 and x_2. The cross section is shown in Figure 13. The rates are small except at small values of $p_\perp$. If this process can be measured it will provide further insight into the behaviour of the gluon distribution function at small x and M^2.

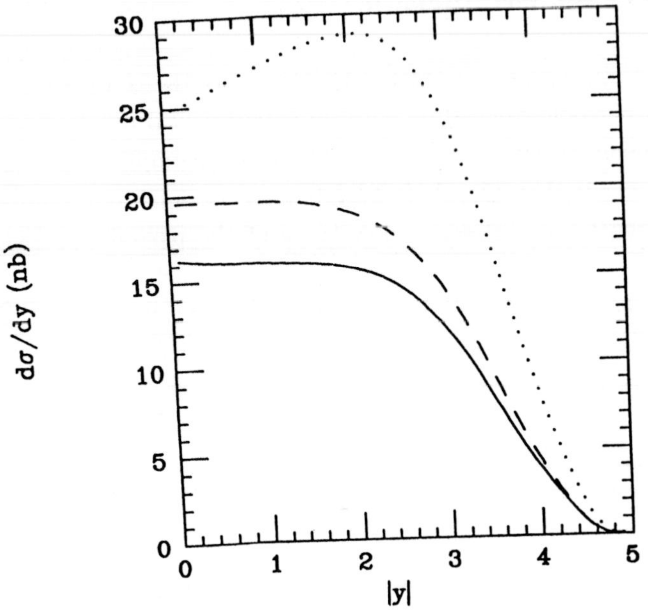

Figure 9: The cross section $\int_{10}^{15} \frac{d\sigma}{dp_\perp dy} dp_\perp$ for the production of a photon as a function of $|y|$ in $p\bar{p}$ collisions at $\sqrt{s} = 1.8$ TeV for $M = \mu = p_\perp$. The solid and dotted, lines correspond EHLQ2 and EHLQ2′ distribution functions. The dashed curve has $M = \mu = \sqrt{x_1 x_2 s}$ and EHLQ2 distribution functions.

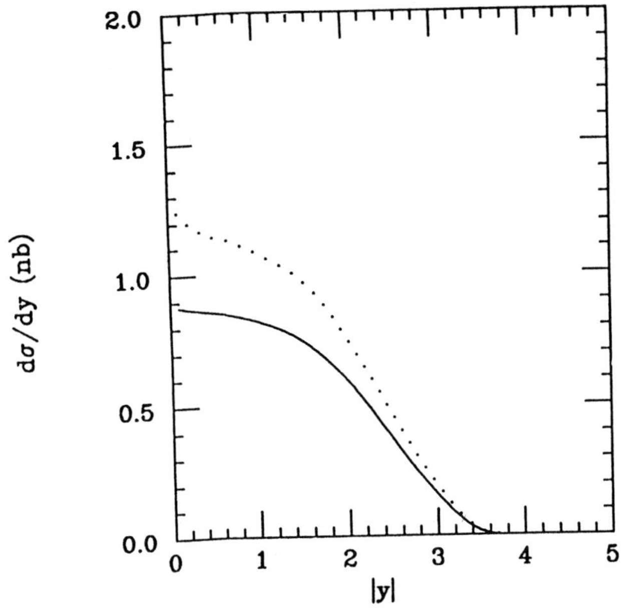

Figure 10: The cross section $\int_{30}^{50} \frac{d\sigma}{dp_\perp dy} dp_\perp$ for the production of a photon as a function of $|y|$ in $p\bar{p}$ collisions at $\sqrt{s} = 1.8$ TeV for $M = \mu = p_\perp$. The solid and dotted, lines correspond EHLQ2 and EHLQ2′ distribution functions.

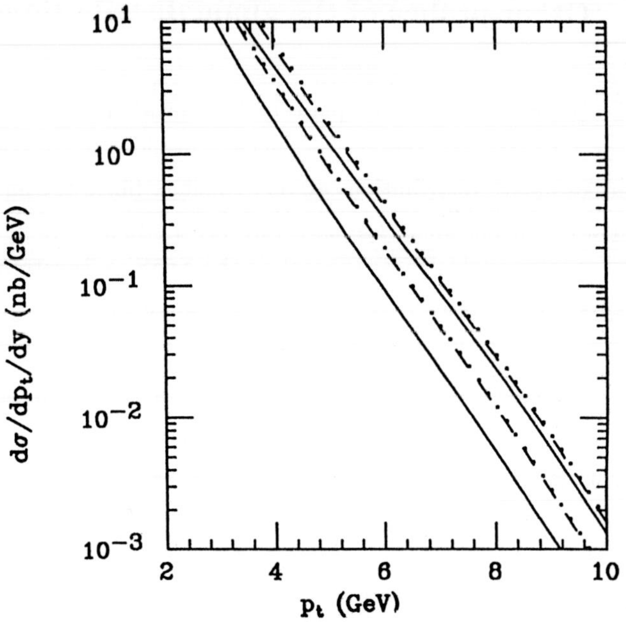

Figure 11: The cross section $d\sigma/dp_\perp dy$ for the production of a photon at $y = 0$ in pp collisions at $\sqrt{s} = 38$ GeV The bands between the pairs of solid, dotted and dashed lines correspond to the range of cross section as $M = \mu$ is varied from $p_\perp/2$ to $2p_\perp$ for lowest order, the full next to leading order rate and the next to leading order rate when the photon is required to be isolated in a cone of half-angle 40^0. Duke and Owens set 1 of distribution functions are used.

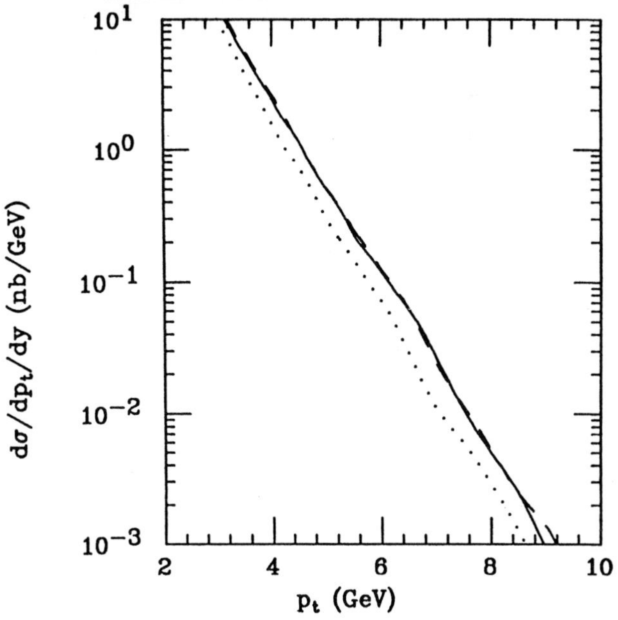

Figure 12: The cross section $d\sigma/dp_t dy$ for the production of a photon at $y = 0$ in pp collisions at $\sqrt{s} = 38$ GeV for $M = \mu = p_\perp$. The solid, dashed, and dotted, lines correspond to EHLQ1, MRS2 and DO2 distribution functions respectively.

5 Other probes of the gluon distribution.

The production of $\mu^+\mu^-$ pairs of low invariant mass in hadron hadron colliders can be used as a probe of the quark and antiquark distributions at very small x. Since the antiquark distribution is strongly correlated with the gluon distribution such a measurement at the $Sp\overline{p}S$ or Tevatron colliders could provide additional information on the gluon distribution in the small x region [37].

The production of charm and bottom quarks in γp (or ep) collisions proceeds via $\gamma g \to c\overline{c}$ and hence a measurement [38] of this process can, in principle, yield a measurement of the gluon distribution. The QCD corrections to this process have been computed [7]Calculations including these corrections indicate that there are large theoretical uncertainties in the charm cross section and that, consequently, the charm rate does not provide a useful handle on the gluon distribution. The bottom cross-section is more reliably predicted and consequently, experiments at HERA should be able to use the process to constrain the gluon distribution.

The production of J/ψ in proton–proton collisions [26] has been advocated [28] as a probe of the gluon distributions since theoretical models [39] for the production include the processes $q\overline{q} \to J/\psi$ and $gg \to J/\psi$. In the latter case one or more gluons must be emitted in order to conserve color. Absolute predictions for the rate are model dependent and unreliable. Nevertheless, if one assumes that the rate has two components

$$\sigma \sim g(x, M^2)g(x_2, M^2)\sigma_1 + q_i(x_1, M^2)\overline{q}_i(x_2, M^2)\sigma_2 \quad (13)$$

it is possible by studying the rapidity dependence of the production rate to get some information on the shape of the gluon distribution ($y_\psi = 0.5\log(x_1/x_2)$). Here an assumption is made that there is not additional rapidity dependence in σ_1 and σ_2. One is really assuming that the extra emitted gluons are all soft. Martin *et al.* [25] have used the this to infer that the softer gluon distribution of their *set 1* is preferred. The model for J/ψ production at large $p_\perp$ is in better shape theoretically; here the dominant process is $gg \to gJ/\psi$ which can be calculated [40].

6 Jet Production.

Most quantitative tests of QCD involve the comparison of a partonic calculation with an experimental measurement involving jet production. The basic assumption of these tests is that the observed jet cross sections and angular distributions closely follow those of the partonic process. This assumption relies on our ability to define a prescription for finding jets that is both experimentally well defined and well matched to the theoretical calculation of interest, and upon the belief that hadronization effects that cause the quarks and gluons to turn into hadrons do not affect the gross features of the final state such as the total rate or the energy flow. It is well known from the analysis of e^+e^- data that the details of jet fragmentation and of the experimental jet finding algorithm can significantly effect any detailed interpretations of jet measurements, and, in particular, of attempts to use such analyses to extract the value of α_s [41].

Here we discuss the issues associated with defining jets, using $p\overline{p}$ collisions at Tevatron energies as an example. We begin with a description of methods used to compare jet data to QCD predictions. We then describe one commonly used jet finding algorithm and discuss the effect of jet fragmentation on this algorithm. We then discuss an algorithm that merges partons (with no hadronization) and apply this parton merging algorithm to the ISAJET [42] Monte Carlo generator and compare the results to an exact QCD calculation.

6.1 Calculations of Jet Rates.

We will begin with a discussion of the partonic processes leading to jet final states. The inclusive jet cross section in shown in Figure 14. At lowest order in QCD perturbation theory (α_s^2) this rate is given by the $2 \to 2$ partonic scattering processes such as $gg \to gg$, and a jet is interpreted as being a single parton. The dependence of the rate at this order upon $M = \mu$ is shown in Figure 14. At order α_s^3 (the term B in Equation 2) there are contributions from loop corrections to the $2 \to 2$ processes as well as from the $2 \to 3$ processes such as $gg \to ggg$. At this order it is necessary to define a jet. If one outgoing parton has a given $p_\perp$, as will be that case if we are computing an inclusive rate, the $2 \to 3$ processes have configurations in which the scattering amplitude is divergent. Firstly, one of the other outgoing partons could have very small transverse momentum less than some p_0. It will not appear as a jet in the detector since, for sufficiently small p_0, after hadronization, the hadrons will not be separated from those in the underlying event. Secondly, if the other two outgoing partons are close together,

$$\Delta R = \sqrt{(y_1 - y_2)^2 + (\phi_1 - \phi_2)^2} < R_0, \tag{14}$$

after hadronization the two jets will be so overlapped that they cannot be separated experimentally. Therefore, parts of the $2 \to 3$ process give rise to 2–jet, as opposed to 3–jet, final states. Hence, at order α_s^3, the dependence on M and μ is decreased but a dependence on R_0 and p_0 is introduced. The contribution from 3–jet final states becomes very large as p_0 and R_0 become small. Qualitatively, the ratio of the 3-jet rate to 2–jet rate or the emission probability for an additional parton is proportional to $\alpha_s log(1/R_0)$ or $\alpha_s log(\sqrt{\hat{s}}/p_0)$. The calculation of the inclusive rate¶ has so far been carried through only in a "toy QCD" with no quarks [43]. In this calculation, the uncertainty due to M and μ is reduced to about 1/2 of that shown in Figure 14 which corresponds to the lowest order case.

It is possible to use a partonic calculation to compare jet data with QCD or to estimate background rates. In this case, the theoretical prediction is taken from a partonic calculation done to some fixed order in α_s. It is important to realise that such a calculation depends not only on α_s but also on the cut-off parameters p_0 and R_0 that go into the definition of a jet. A fully correct treatment of this is, in fact, only possible in the context of a complete higher order calculation (*c.f.* previous paragraph). If one needs, for example, the four jet final state that occurs at order α_s^4, one must calculate the two loop corrections to the 2–jet final state and the one loop corrections to the 3–jet final state. In practice, the tree level results can be used (α_s^2 for 2–jet, α_s^3 for 3–jet *etc.*) together with the cutoffs. Recently there has been much progress in calculating these tree level rates. The exact partonic matrix elements are known for all the processes contributing to 3–parton [46] and 4–parton [47] final states. An algorithm has also been developed [48] that enables the n–parton matrix elements to be computed recursively. The exact matrix elements are very complicated and slow to evaluate for more than three jets. Nevertheless, approximations [49] have been developed that are accurate to 10% or better for the 4–jet and 3–jet final states and can be extended with some confidence to the final states with five or more jets. These fixed order calculations should be reliable provided that all of the jets are of approximately the same $p_\perp$. If $\alpha_s(p_\perp^{max})log(p_\perp^{max}/p_\perp^{min}) \sim 1$ or $\alpha_s log(2\pi y_{max}/\Delta R) \sim 1$, then the parton calculation ceases to be reliable. Here $p_\perp^{max}$ ($p_\perp^{min}$) is the transverse momentum of the stiffest (softest) parton, y_{max} is the range of rapidity covered by the detector and ΔR is the separation in rapidity–phi space of the closest two partons. The latter criterion is always irrelevant given the segmentation present in current detectors.

Such parton calculations are also available for many other final states and are much faster than those of the full event generators discussed below. They can be useful in estimating rates and devising cuts to eliminate backgrounds. Two such parton generators are widely available [50], [51].

A more realistic parton calculation will have to include the showers of partons almost collinear with the beam (small $p_\perp$) and almost collinear with the outgoing partons from the hard scattering (recall the five stage paradigm discussed above and that the emission probability for such partons is large). These showers of partons produce a broadening of the QCD jet. The uncertainties in jet definition due to jet hadronization do not disappear as the parton energies increase since, with increasing M^2 or $\hat{s}$, there is an increased probability for the partons involved in the hard scattering process to radiate. The consistent handling of this radiation is an essential aspect of the calculation. Unfortunately it is extremely difficult to carry out this procedure. These showers can be added in the leading log approximation (as is done in the Monte–Carlo generators discussed below). We are then faced with a double counting problem since some part of the order α_s^n parton calculation is given when a shower is added to the order α_s^p calculation ($n > p$).

If such a partonic calculation is to be used to compare with data, either the experimental data must be corrected back to "partonic energies", or the results of the calculation must be fed into a Monte Carlo event generator that frag-

¶The inclusive single particle rate at large $p_\perp$ has also been computed [44]. In this case the parameters p_0 and R_0 are not needed, but a fragmentation function has to be introduced and defined. This is usually done by reference [45] to one particle inclusive rates in e^+e^- annihilation.

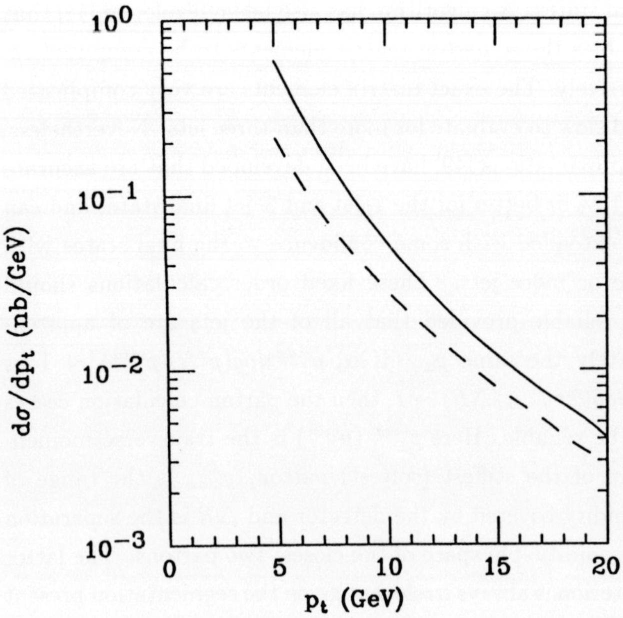

Figure 13: The cross section $d\sigma/dp_t$ for the production of a pair of photons in $p\bar{p}$ collisions at $\sqrt{s} = 1.8$ TeV for $M = \mu = p_\perp$ and the EHLQ2 distribution functions. The solid line is the total inclusive rate, the dashed is that from the quark antiquark initial state.

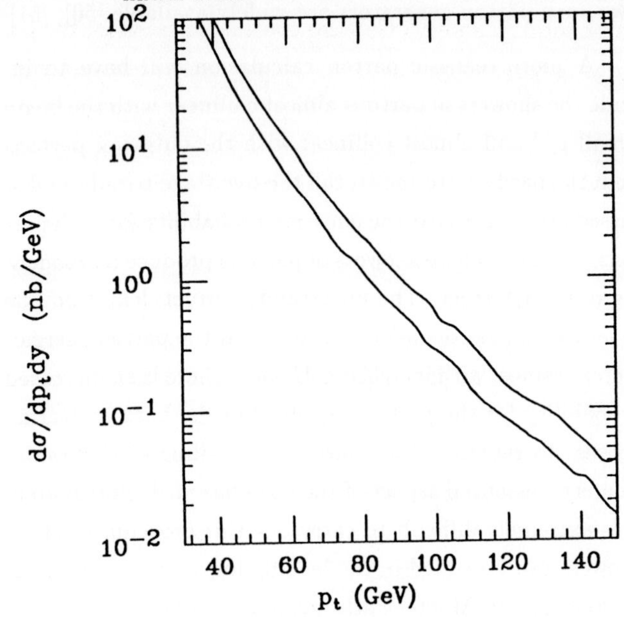

Figure 14: The cross section $d\sigma/dp_\perp dy$ for the production of a jet at $y = 0$ in $p\bar{p}$ collisions at $\sqrt{s} = 1.8$ TeV. The curves correspond to the EHLQ2 set distribution functions with $\mu = M = p_t/2$ (upper curve) and $\mu = M = 2p_t$ (lower curve).

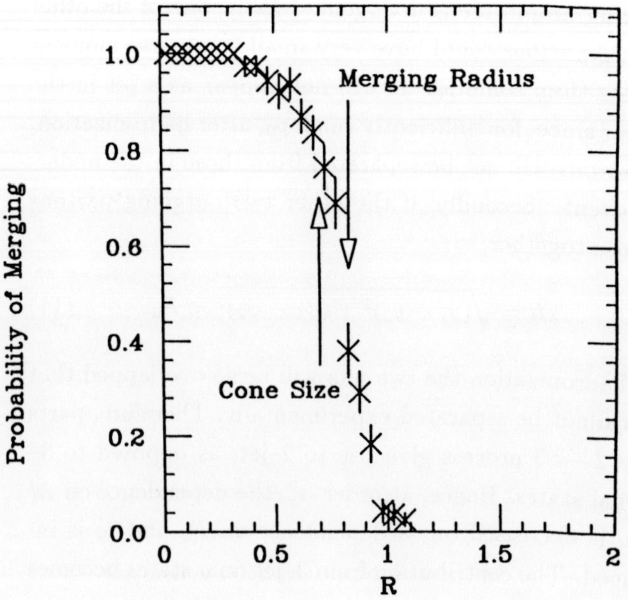

Figure 15: Figure showing the probability that two jets separated by $R = \sqrt{(\eta_1 - \eta_2)^2 + (\phi_1 - \phi_2)^2}$ in $\eta - \phi$ will be merged by the CDF jet algorithm. This probability is well behaved; it has a value of 1 when the jets are close together followed by a narrow range of R in which the probability falls rapidly to zero. The jets have E_t between 25 and 35 GeV.

	$m_t = 50$	$m_t = 80$	$m_b = 5$	$m_b = 4.5$
$EHLQ1$	2.3	0.31	12000	14000
$EHLQ2$	2.0	0.26	12000	13000
$DO1$	2.0	0.28	7600	9200
$DO2$	1.3	0.16	31000	41000
$DFLM(160)$	2.0	0.27	14000	19000
$DFLM(260)$	1.8	0.25	17000	25000
$DFLM(360)$	1.6	0.23	22000	31000
$MRS(EMC)$	2.0	0.27	13000	17000
$MRS(BCDMS)$	2.9	0.29	18000	25000

Table 1.

The total cross-section in nanobarns for the production of a $t\bar{t}$ pair in $p\bar{p}$ collisions at $\sqrt{s} = 1.8\ TeV$ for $m_t = 50, 80\ GeV$. The values given by various sets of distribution functions are shown. Also shown is the total cross-section for $b\bar{b}$ production for two values of the bottom quark mass. The scales M and μ are set to $m_q/2$ and lowest order QCD is used.

ments the final state quarks and gluons into the hadrons seen in the detector. The advantage of this technique is that the true QCD matrix element is used. The disadvantage is that the calculation does not include the effects of additional gluon radiation and hence of "jet broadening". There is another difficulty in that an n–jet final state is attributed to a 2 →n–parton calculation. After such a state is hadronized and passed through a jet finding algorithm, it may appear as an $(n-1)$–jet final state. Since such states are supposed to be produced by the $2 \rightarrow (n-1)$–parton scattering, there is a double counting problem.

Before a calculation can be meaningfully compared to data, we must determine that the jet definitions used in the calculation are not at odds with the experimental measurement. For example, if the angular resolving power of the detector is better than the angular cutoff (R_0) of the calculation, we cannot expect the calculation to properly represent the data.

An alternative method of comparison involves using a QCD inspired Monte Carlo generator (ISAJET [42], PYTHIA [52] or HERWIG [53] for example). Such generators usually start with the lowest order 2 → 2 calculation and follow the paradigm discussed above. They use a classical branching process to simulate the initial and final state radiation. This generates a multiparton final state in the so-called leading log approximation. This approximation is good when $log(\sqrt{\hat{s}}/p_\perp)$ or $log(\Delta R)$ is large, where ΔR is the separation of one of the radiated partons from one of the two original outgoing partons. The parton system can then be hadronized. The advantages of this approach are that it can reproduce many jet final states and that it will automatically include any jet broadening effects caused by gluon radiation. It also has no inherent problem in normalizing the rates for different numbers of jets. The hadron (or parton) can be passed through a jet algorithm and the number of jets determined. The disadvantage of the method is that the leading log approximation does not reproduce the exact calculation for wide angle radiation (typical errors can be as large as a factor of 2 or 3 in rate) and so may not provide a good basis for comparing to multi–jet data. It is also very difficult to include higher order QCD corrections in a fully correct manner.

6.2 Experimental Jet Definition.

In order to compare the output of a Monte Carlo generator with data, an experimental jet definition must be introduced and discussed. At hadron colliders, jets are usually measured using calorimetric techniques. The detectors typically have projective geometry with uniform segmentation in the polar angle ϕ and pseudorapidity (η). Such segmentation into "towers" is ideal for jet finding. It can be shown that if particles are produced with limited transverse momentum relative to some jet axis, then the loci of particles with a given p_t with respect to that axis will form a circle in η - ϕ space (where η is measured with respect to the beamline) with a radius of that is proportional to p_t. Thus, jets in hadron–hadron collisions look like circles in η - ϕ space. This fact means that a *fixed cone* algorithm that gathers all transverse energy (E_t) with a circle of fixed size in $R_h = \sqrt{\eta^2 + \phi^2}$ is a sensible way to find jets.

The CDF jet finding algorithm [54] is one example of such a fixed cone method. The algorithm proceeds in three steps. First, "preclusters" are formed by combining all contiguous towers with $E_t > 1$ GeV. All preclusters with a total E_t above 3 GeV are considered seeds for the jet finder. This preclustering with a high threshold is designed to select against clusters formed by fluctuations in the energy distribution of the soft underlying collision. Next, a circle of fixed radius in η - ϕ space is drawn around each precluster. The cone size is a parameter of the algorithm and can be varied depending upon the physics requirements of a specific analysis. The standard cone size (R_h) used by CDF is 0.7. All towers with $E_t > 0.1$ GeV are now included in the cluster and the E_t weighted centroid of the cluster is calculated. A new cone is then drawn about the recalculated centroid and any additional towers are included in the cluster. This process is iterated until the number of towers included is stable. Finally, clusters are merged together if they share more than 75% of their towers in common. When clusters are not merged, the energy in a given tower is associated with that cluster that whose centroid is closest in η - ϕ space. All resulting clusters must have transverse momentum above some value (p_h).

In order to test the quality of this algorithm on real data, the CDF [54] group has analyzed events where an extra cluster (extracted from another event) has been inserted into the event. The result of this study is shown in Figure 15; the clusters have 25 GeV $< E_t <$ 30 *GeV*. This plot shows the fraction of events where the extra cluster has been merged with another jet as a function of the separation in η - ϕ space between the clusters. The study shows that this merging probability is well behaved (*i.e.* has a sharp boundary in separation distance) but that the merging radius (R_m), defined as distance where this fraction falls to 0.5, is *larger* than the specified cone size (R_h). This increased merging radius results from the fact that jets have finite size in η - ϕ space. The exact relationship between the merging radius and the cone size depends upon the the energy profile of the jets in η–ϕ space, and hence upon the details of the parton hadronization, and upon the segmentation in the detector. For those cases where the two clusters are not merged, the energy and position of the clusters don't change significantly due to the addition of the extra jet. For initial cluster energies in the range of 25 to 35 GeV, the change in cluster energy was less than 1 GeV in 99% of the events.

If a leading log parton shower calculation is used to calculate jet rates, the resulting partons (an indefinite number) must be merged into jets and a cluster algorithm with a cone size R_p and minimum cluster transverse momentum p_p must be introduced. These quantities are not necessarily the same as R_0 and p_0, the partonic separation and momentum cut off in a fixed order QCD calculation, or R_h and p_h, the values used after hadronization of the parton shower and detector simulation.

In order to make any detailed precision tests of QCD using jet data, the relations between the various cone sizes and momentum cut offs must be investigated and understood. A similar situation arose in the attempt to measure α_s in e^+e^- annihilation [41]. Such a detailed comparison is beyond the scope of this report. However we have attempted to compare the results of leading log parton showers and fixed order QCD.

In the section that follows, partons from the leading log shower calculation are merged to form jets using the following algorithm. First, all partons are ordered in energy. Starting with the highest energy parton and working down to the lowest, the distance in $\eta - \phi$ space between the parton and all other partons in the event is calculated. If the distance between the two is less than the specified cone size (ΔR_p), the partons are merged by adding together

#jets	*ISAJET*	*partons*
2	22000	23000
3	1700	540
4	58	22

Table 2.

The cross section for jet events in nanobarns for $p\overline{p}$ collisions at $\sqrt{s} = 1.8$ TeV as given by ISAJET [42] and the parton calculations using fixed order QCD [51]. The ISAJET jets are found by passing the partons from ISAJET events through a cluster algorithm with cones of size $R_p = 0.8$. (The jets are not hadronized.) The resulting clusters are required to have $p_\perp > 15$ GeV. In the parton calculations, the partons are required to have $p_\perp > 15$ GeV and be separated by $R_0 > 0.8$. EHLQ set 1 of structure functions are used. The scale of the hard scattering is set to $p_\perp$ of the initial hard scatter for ISAJET and to the average $p_\perp$ of the partons for the parton case (hence the same scales are used in the 2–jet case.). The invariant mass of the multijet system is required to be in the range 60–120 GeV.

#jets	*ISAJET*	*partons*
2	6800	7000
3	2200	1500
4	300	280

Table 3

As Table 2 except that the invariant mass of the jet system is required to be in the range 120–300 GeV.

their four-vectors. The procedure is then iterated until the number of merged partons is stable.

Tables 2 and 3 show a comparison of ISAJET, with partons merged using the technique described above, and an exact leading order QCD calculation for the two, three and four jet cross sections (*i.e.* the $2 \to 2$, $2 \to 3$ and $2 \to 4$ partonic processes). Recall that the ISAJET final state does not have a definite number of partons. The number increases as the $p_\perp$ of the hard scattering rises and as the invariant mass cut–off in the showering is lowered. In both calculations, the parton cone radius (R_p and R_0) was set to 0.8. Other relevant parameters are:

- Each cluster (ISAJET) or parton (parton calculation) has $p_\perp > 15$ GeV (p_0 and p_p).

- The structure functions used are EHLQ Set 1.

- The scale $M = \mu$ is p_t of the hard scatter. In ISAJET this corresponds to the $p_\perp$ of either of the two outgoing partons in the initial $2 \rightarrow 2$ scattering. In the exact calculation, this the average $p_\perp$ of the jets in the event.
- The value of α_s is evaluated to 1 loop assuming 4 light quark flavors, with $\Lambda = 200$ MeV.
- The total center-of-mass energy ($\sqrt{s}$) is 1.8 TeV .

It can be seen from these tables that the ISAJET and parton rates agree within a factor of 2 or so. Recall from the above discussion that uncertainties of this order are to be expected from variations in the choice of distribution functions and the choice of M and μ. Notice also that the ratios of ISAJET and parton results depend upon the number of jets and upon the invariant mass of the jet system. Changes in M and μ will tend to produce effects independent of these variables. That the ISAJET and parton rates shown in Tables 2 and 3 are in reasonable agreement with each other and that the agreement is better in Table 3 where $p_0\sqrt{\hat{s}}$ is smaller and leading log should be a better approximation. In order to observe the differences in more detail we must look at the jet angular and energy distributions.

For the two jet events, the cross section can be described as a function of the following variables:

1. Three boost variables that determine the velocity of the hard scattering center-of-mass frame with respect to the $p\bar{p}$ center-of-mass frame. In the case of the parton calculation, two of these are trivial since the hard scattering frame moves parallel to the proton direction. This is not true in the ISAJET case because of initial state radiation.
2. The invariant mass of the dijet system.
3. $\cos(\theta^*)$: the scattering angle of one of the jets with respect to the beamline in the center of mass of the dijet system.
4. ϕ: the azimuthal angle of one jet with respect to the beamline. (This last variable is trivial).

For the 3–jet events, the variables are:

1. Three boost variables that are the velocity of the hard scattering (*i.e.* 3–jet) system's center-of-mass.
2. The invariant mass of the 3–jet system $\sqrt{\hat{s}}$.
3. E_1 and E_2, the energies of the leading and sub-leading jets. (The leading jet is defined to be the jet with the largest ENERGY in the hard scattering center-of-mass)
4. $\cos(\theta^*)$: The scattering angle of the leading jet with respect to the beamline in the center of mass of the hard scattering system.
5. Φ: the azimuthal angle of the leading jet.
6. Ψ^*: the angle of rotation of the plane of the 3-jet event about the axis of the leading jet. Again this angle is defined in the hard scattering center of mass. $\Psi^* = 0°$ is the case where the event plane and the plane defined by the beam and leading jet are parallel and the 3rd (least energetic) jet is closest to the beam line. The two planes are again parallel for $\Psi^* = 180°$, but now the 2nd jet is closer to the beam.

In the case of four jet events, we will use the same variables as in the 3–jet case with the additional variables defined:

1. E_4: the energy of the least energetic jet in the hard scattering center-of-mass.
2. ϕ_4 and θ_4, the azimuthal and polar angles of the least energetic jet with respect to the leading jet.

We have chosen not to display distributions in the boost variables or invariant mass, since they depend only on the choice of structure functions, or the ϕ variable since it has no dynamical significance. In all cases, we define our fiducial volume to be $|\eta| \leq 3.5$. When histogramming the above variables, we look in two separate regions of invariant mass: 60-120 GeV and 120-300 GeV and we shall only show the 3–jet distributions due to limited ISAJET statistics in the 4–jet case.

In Figures 16 and 17 we show the distribution in energy of the leading and subleading jets in the three jet events when the three jet system has invariant mass in the range 60–120 GeV. The curves are normalized to the same area. The absolute normalization can be obtained from Table 2. Notice that the parton calculation gives a slightly flatter distribution. Figures 18 and 19 show the distribution in the variables $\cos(\theta^*)$ and ψ^* respectively.

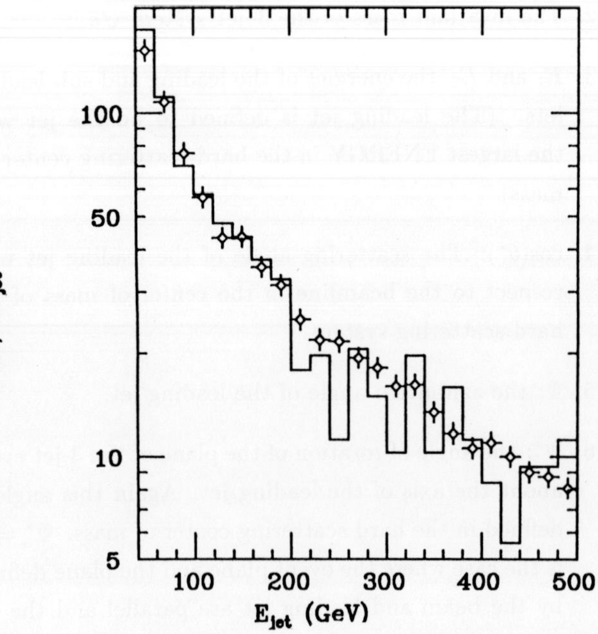

Figure 16: The distribution in *energy* of the most energetic jet in 3–jet events. The energy is evaluated in the center of mass frame of the 3–jet system. The histogram is the result from ISAJET, the points are an order α_s^3 QCD calculation. The curves are normalized to the same area. The 3–jet system is required to have invariant mass in the range 60–120 GeV. Each parton cluster (ISAJET) or parton (QCD) is required to have $p_\perp > 15$ GeV, $|\eta| < 3.5$; the clusters are merged with the algorithm described in the text and a cone size R_p =0.8; the partons are required to be separated by $R_0 > 0.8$.

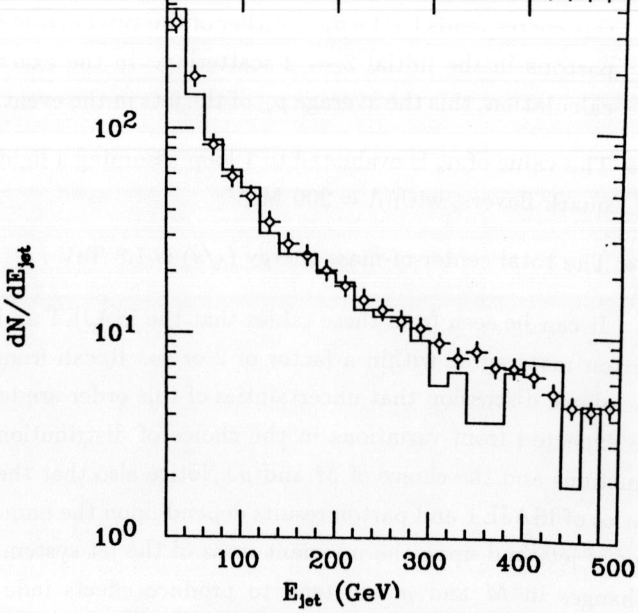

Figure 17: The distribution in *energy* of the second most energetic jet in 3–jet events. The histogram is the result from ISAJET, the points are an order α_s^3 QCD calculation. The curves are normalized to the same area. The 3–jet system is required to have invariant mass in the range 60–120 GeV. Each parton cluster (ISAJET) or parton (QCD) is required to have $p_\perp > 15$ GeV, $|\eta| < 3.5$; the clusters are merged with the algorithm described in the text and a cone size R_p =0.8; the partons are required to be separated by $R_0 > 0.8$.

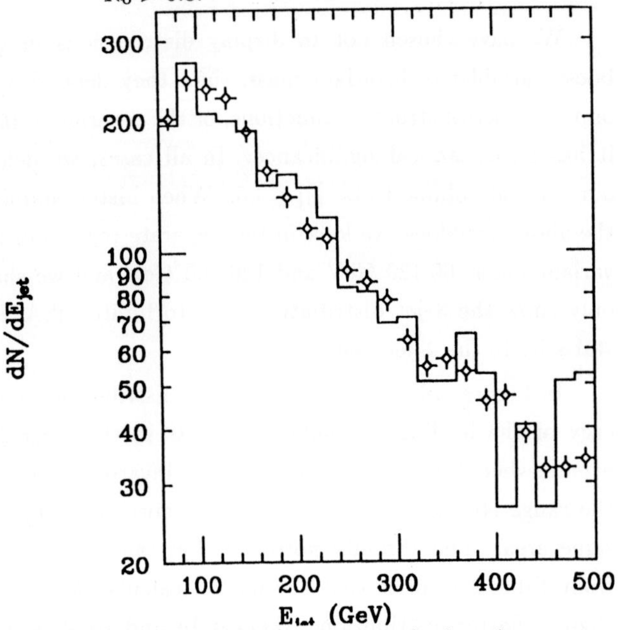

Figure 20: As Figure 16 except that the invariant mass of the 3–jet system is between 120 GeV and 300 GeV.

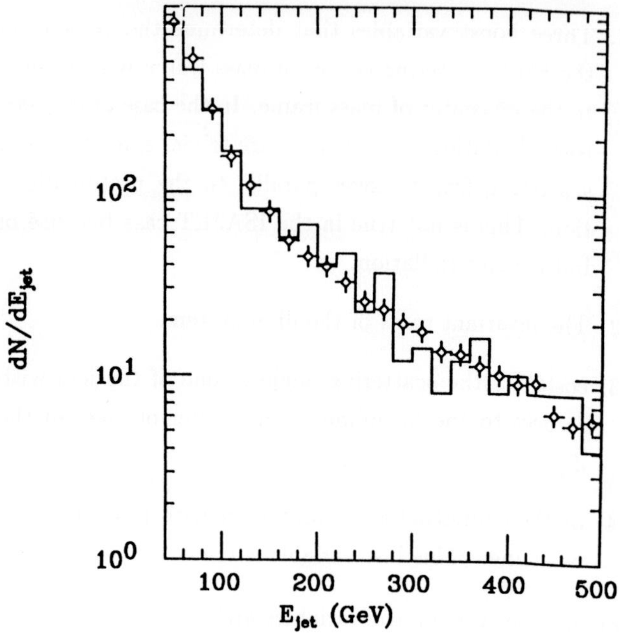

Figure 21: As Figure 17 except that the invariant mass of the 3–jet system is between 120 GeV and 300 GeV.

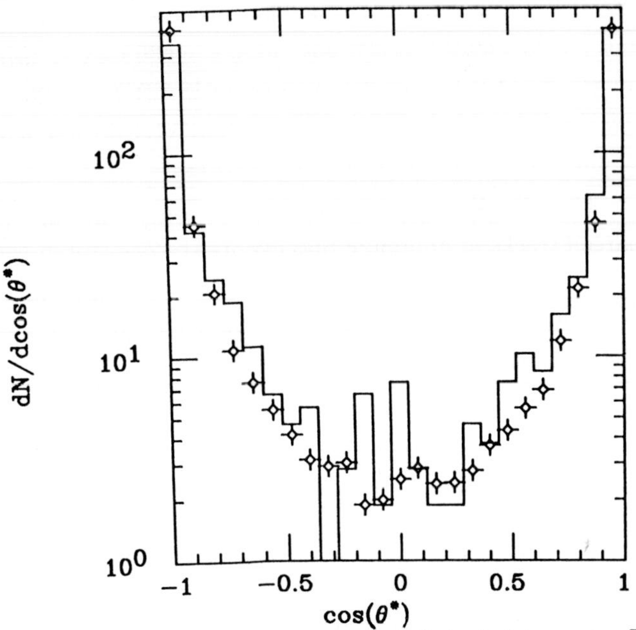

Figure 18: The distribution is $\cos(\theta^*)$ for 3–jet events. The histogram is the result from ISAJET, the points are an order α_s^3 QCD calculation. The curves are normalized to the same area. The 3-jet system is required to have invariant mass in the range 60-120 GeV. Each parton cluster (ISAJET) or parton (QCD) is required to have $p_\perp > 15$ GeV, $|\eta| < 3.5$; the clusters are merged with the algorithm described in the text and a cone size R_p =0.8; the partons are required to be separated by $R_0 > 0.8$.

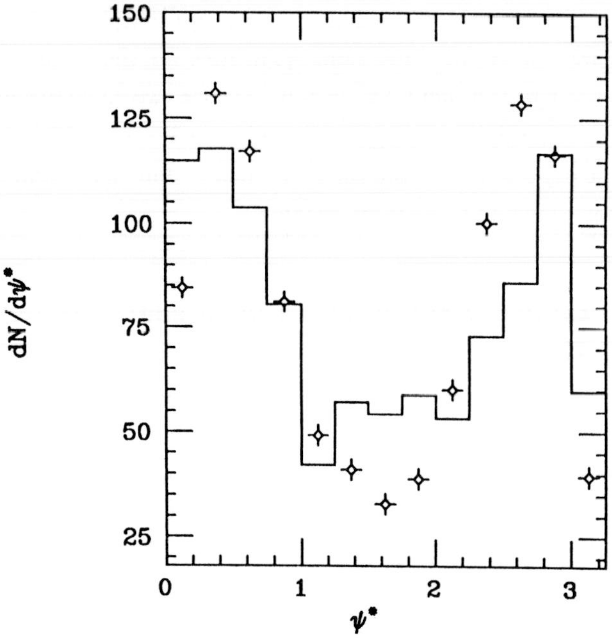

Figure 19: The distribution is Ψ^* for 3–jet events. The histogram is the result from ISAJET, the points are an order α_s^3 QCD calculation. The curves are normalized to the same area. The 3-jet system is required to have invariant mass in the range 60–120 GeV. Each parton cluster (ISAJET) or parton (QCD) is required to have $p_\perp > 15$ GeV, $|\eta| < 3.5$; the clusters are merged with the algorithm described in the text and a cone size R_p =0.8; the partons are required to be separated by $R_0 > 0.8$.

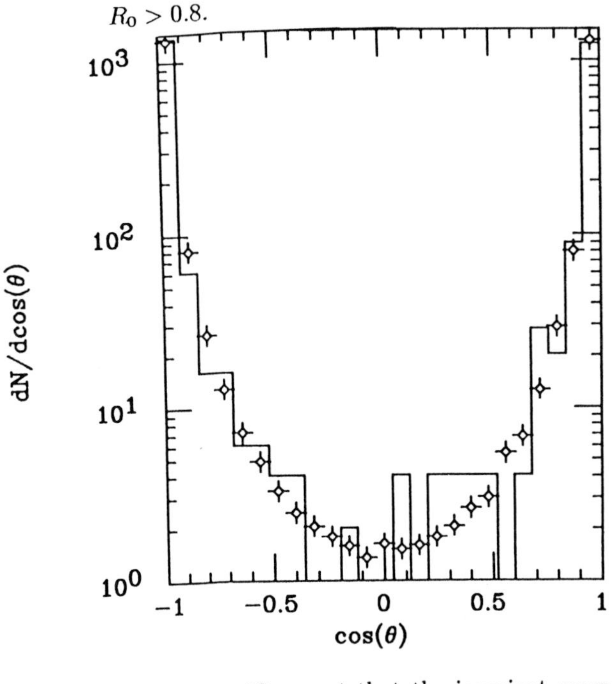

Figure 22: As Figure 18 except that the invariant mass of the 3-jet system is between 120 GeV and 300 GeV.

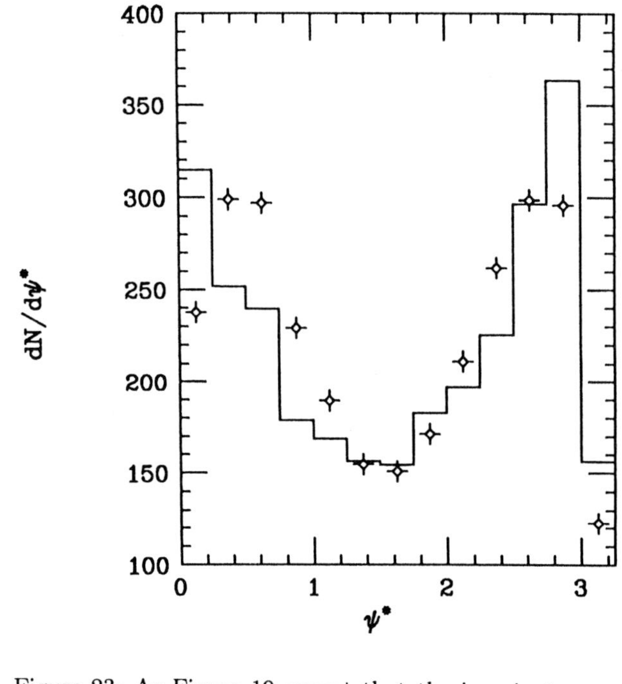

Figure 23: As Figure 19 except that the invariant mass of the 3-jet system is between 120 GeV and 300 GeV.

Here the exact parton calculation is more sharply peaked. Figures 20-23 show the same quantities for the invariant mass range 120–300 GeV. In this case the two calculations are in slightly better agreement with each other. This is to be expected since the $p_0/\sqrt{\hat{s}}$ is smaller and thus leading log is expected to be more reliable.

7 Global Properties of hard scattering events.

In a hadron–hadron collision, events that do not contain a hard scattering make up the dominant part of the cross-section at currently available energies. These events ("minimum bias") consist of hadrons of small transverse momentum distributed uniformly in azimuth and approximately uniformly in rapidity. Since the properties of these events are not calculable in QCD, the various Monte–Carlo generators use models to simulate them. ISAJET [42] uses a Regge model [55]; PYTHIA [52] builds up the event from a large number of parton–parton scatterings each of which produces an outgoing parton of very small $p_\perp$; HERWIG [53] uses a phenomenological model based on the UA5 data [56]. All of these models contain parameters which are adjusted so that they correctly describe the data at the $Sp\bar{p}S$ collider.

When these generators are used to predict the minimum bias structure at higher energies, it is not guaranteed that they will agree, either with each other or with the data. Figure 24 shows the pseudo–rapidity distribution predicted for $p\bar{p}$ collisions at $\sqrt{s}$=1.8 TeV. Two sets of values for the PYTHIA generator are shown; "standard" corresponds to the default set of parameters, "tuned"‖ to the retunings suggested [57] to improve agreement with the UA5 data. It can be seen from this figure that the Monte-Carlo generators do not agree with each other and that PYTHIA provides the best agreement with the CDF data [58]. HERWIG is in reasonable agreement with the data, while ISAJET is somewhat low. However, in the PYTHIA case we have not included the contribution from the "double–diffractive" process. Including this process will lower the multiplicity slightly. It is needed at $\sqrt{s} = 630$ GeV to bring the generated values closer to those of ISAJET, HERWIG and the UA5 data. ISAJET and HERWIG do not have "double–diffractive" as a separate process.

The distribution in transverse momentum of the charged tracks in minimum bias events at $\sqrt{s} = 1.8$ TeV is shown in Figure 25. The plot also shows the parameterization of this distribution given by the CDF [59] and UA1 collaborations [60], *viz.*

$$\frac{dN}{dp_\perp} \sim \frac{p_\perp}{(p_\perp + p_o)^a} \qquad (15)$$

At $\sqrt{s} = 540$ GeV, $a = 9.14$ and $p_o = 1.3$ GeV [60], while at $\sqrt{s} = 1800$ GeV, $a = 8.28$ and $p_o = 1.3$ GeV [59]. It can be seen from this figure that the ISAJET $p_\perp$ spectrum peaks at the largest $p_\perp$ and that none of the generators agree too well with the fit to the data. Indeed agreement is better with the lower energy UA1 fit, indicating that the generators do not adequately reproduce the change in shape of this distribution with $\sqrt{s}$.

Figure 26 shows the pseudorapidity distribution in pp collisions at $\sqrt{s}$= 40 TeV. It is important to note that the differences between the generators get larger as $\sqrt{s}$ rises and that PYTHIA gives that largest and ISAJET the lowest number. The retunings in PYTHIA produce differences that are smaller than those between PYTHIA and the other generators.

In a hadron–hadron hard scattering event, such as the production of jets or W bosons, the initial state partons in the hard scattering have evolved off shell by an amount of order the momentum transfer in the hard scattering. This evolution occurs by the emission of quarks and gluons all of which have limited transverse momentum with respect to the beam direction. These quarks and gluons then turn into hadrons of limited $p_\perp$ distributed approximately uniformly in rapidity. One therefore expects that the multiplicity of particles in the underlying event (*i.e.* that part of the event that is separated in $\phi - \eta$ space from the products of the hard scattering) should be larger in events which contain a hard scattering than in events which do not. This qualitative feature is seen in the data [61]. A comparison of this effect in the different Monte–Carlo generators is shown in Figure 27 which also shows data from CDF. [58]. A comparison of this figure with Figure 24 shows that there are

‖The tunings are as follows:- In SUBROUTINE PYRAND the variable RQQBAR is set to RQQBAR = PYPAR(30)*(1.-(QT2/(QT2+(PYPAR(31)*PYPAR(32))**2))**2) The following parameters
have also been changed from the default:- IPY(12)=4, IPY(3)=12, IPY(17)=3, PYPAR(30)=0.7, PYPAR(31)=0.5, PYPAR(32)=1.96

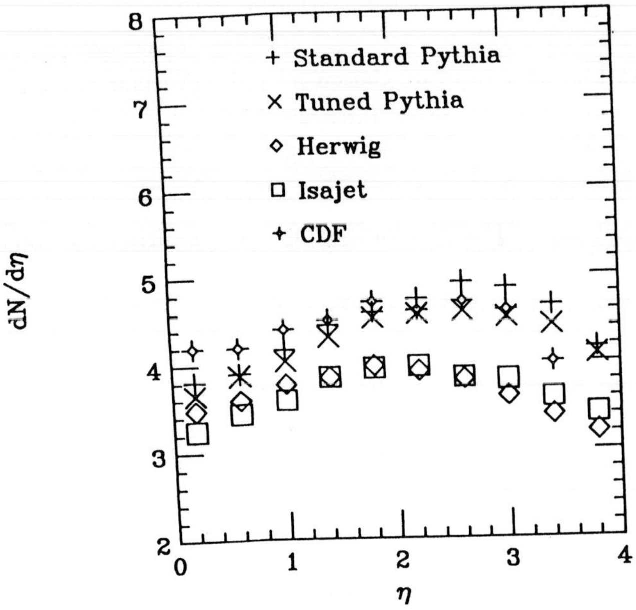

Figure 24: The distribution of multiplicity with respect to pseudo-rapidity, $dN/d\eta$, for events with no hard scattering (minimum bias) in proton–antiproton collisions at $\sqrt{s}$=1.8 TeV. The predictions of Monte-Carlo generators are compared with the CDF data [60].

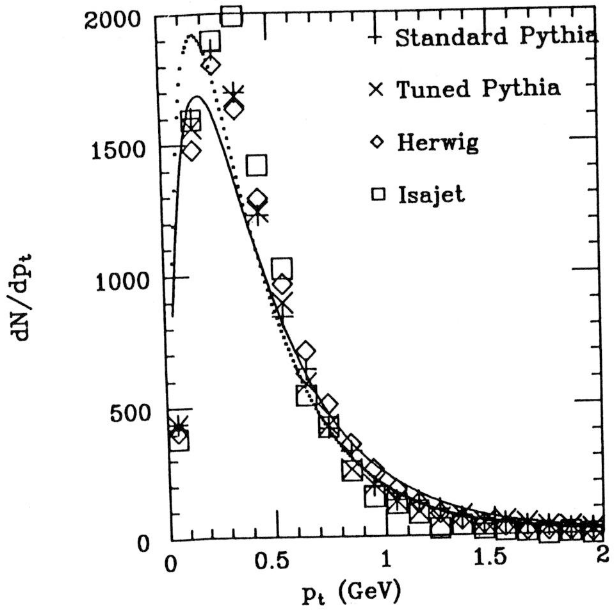

Figure 25: The distribution $\frac{dN}{dp_\perp}$ for minimum bias events at $\sqrt{s}$ = 1.8 TeV for the various event generators. The solid (dotted) line corresponds to the fit to the CDF (UΛ1) parameterization. (see text). The curves are all normalized to the same area.

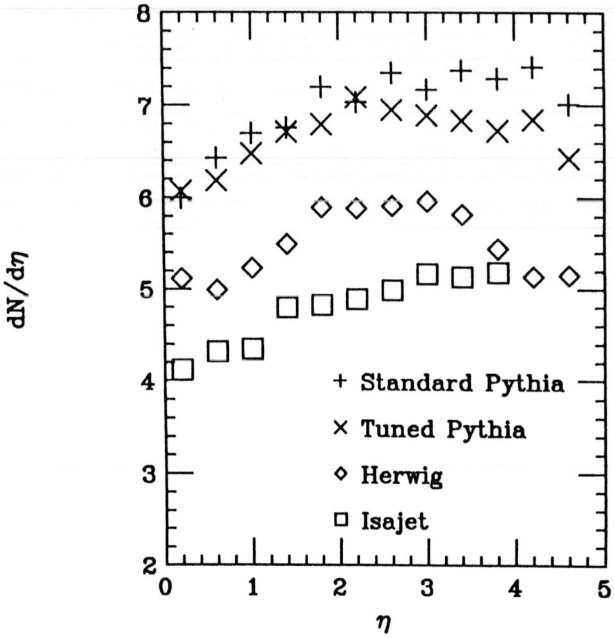

Figure 26: The pseudo–rapidity distribution $dN/d\eta$ for events with no hard scattering (minimum bias) at in proton–proton $\sqrt{s}$=40 TeV as predicted by Monte-Carlo generators.

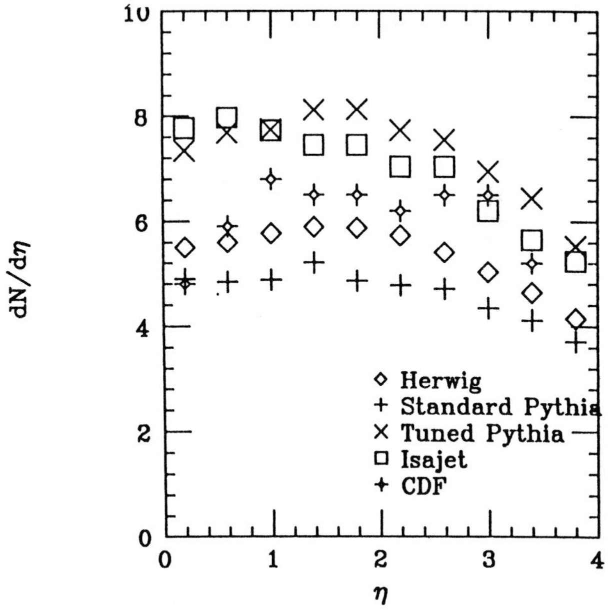

Figure 27: The pseudo-rapidity distribution $dN/d\eta$ for events in which there is a W (that decays to $e\nu$) in proton–antiproton collisions at $\sqrt{s}$=1.8 TeV. The predictions of Monte–Carlo generators are compared with the CDF data [58]. The electron from the W decay is not included.

indeed more particles in the underlying event when a W is produced but that the distribution remains of approximately the same shape in pseudo–rapidity. In the case of W production, the disagreement between PYTHIA [52] and the CDF data [58] is quite striking. Indeed the differences between the tuned and standard PYTHIA are unexpectedly large. Since these two bracket the data, it is clear that another retuning will give excellent agreement. HERWIG agrees very well with this data, and ISAJET is acceptable.

If the outgoing particles from the partonic process have color then they will also radiate and give rise to increased particle multiplicity. In Figures 28 and 30, we show the multiplicity distribution $dN/d\phi$ for jet events. Events are generated and passed through the CDF cluster algorithm. The cluster with the largest E_t defines $\phi = 0$. In figure 28 (30) we have required that the invariant mass of the two clusters with the largest $p_\perp$ is between 50 and 70 GeV (150 and 200 GeV). The figures show a peak at $\phi = 0$ and at $\phi = \pi$, since most of the events are two jet events. The value of $dN/d\phi$ near $\phi = \pi/2$ is higher than the value of $dN/d\phi$ in minimum bias events and reflects the presence of the initial and final state radiation. The level is also higher in Figure 30 than in Figure 28; this reflects the fact that in the former case the trigger jet has larger $p_\perp$ and hence the value of M in the hard scattering is larger and there is more initial and final state radiation. The level in the peaks at $\phi = 0$ and $\phi = \pi$ reflects the multiplicity of particles within a jet. It too rises as the jet $p_\perp$ in increased. Notice that the rise is greatest in the ISAJET generator. We shall discuss this further below. In Figure 29 we show the same distribution at in Figure 28 except that there is no cut on the transverse momentum of individual particles. Notice that the fractional increase in the multiplicity when the cut is removed is larger in the peaks ($\phi = 0, \pi$) than in the valley ($\phi = \pi/2$).

Figure 31 shows the underlying event multiplicity distribution $dN/d\eta$ for events with a pair of jets with invariant mass in the range $M_W \pm 10$ GeV. The multiplicity is computed by taking that in the range $3\pi/8 < \phi < 5\pi/8$ and multiplying by 4. This procedure is necessary to eliminate the particles in the trigger jet (and in the jet which balances it) and to facilitate a comparison with Figure 27. The dominant fundamental parton process contributing to jet production is $gg \to gg$ and the initial partons are gluons, in contrast with W production where the initial partons are quarks. Since have we chosen jet pair invariant masses to be around the W mass, the scale M for the two processes should be comparable. Since the gluon has a larger color charge than the quark and can radiate more copiously, one might expect that the underlying event multiplicity would be larger in the jet case than in the W case. A comparison of Figures 31 and 27 shows this to be the case. However, the shape of the pseudo–rapidity distribution is radically different between each of the generators and from the much flatter distribution in the W case. The shape of the distribution is also dependent upon the location of the trigger jet.

Some of the features expected in the properties of jets and the underlying event have been suggested as possible tools in the extraction of new physics signals at the SSC. The most analyzed case is that of the search for a Higgs boson [62] decaying to WW followed by one W decaying hadronically and the other leptonically. The dominant background arises from the final state $W + jets$ where the W decays leptonically and the jet system is indistinguishable from $W \to q\bar{q}$. Since the Higgs (if heavy) is produced dominantly by gauge boson fusion [63], the scale M is of order M_W. In the background process the scale M is of order of the transverse momentum of the final state W and so can be substantially larger. One would therefore expect lower multiplicity away from the jets in the signal events than in the background events. The particle multiplicity of the jets making up the W in the signal should be the same as that of a W produced at rest, since the W is a color–neutral object and cannot radiate gluons. The jets in the background which are faking the W should have a larger particle multiplicity if $p_\perp > M_W$ since in this case the outgoing partons radiate and the scale M is of order $p_\perp$. While, as we have seen, all these qualitative features are correctly incorporated in the event generators, it should be clear from our discussion that any attempt at a quantitative analysis is premature. Indeed we expect that such an analysis will still not be possible even after the generators have been tuned up to agree with current data. The extrapolations from 630 GeV to 1800 GeV and the disagreements between the generators that emerge illustrate the problems that can occur. The retuning of PYTHIA to fix a problem with the minimum bias caused an unexpectedly large ef-

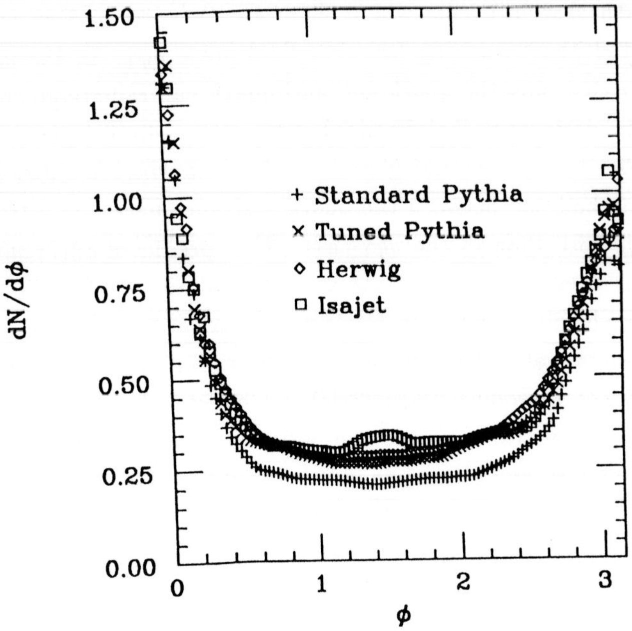

Figure 28: A plot showing the particle multiplicity $dN/d\phi$ in jet events. Events are generated in the various Monte–Carlo generators at $\sqrt{s} = 1.8$ TeV. The fundamental partonic scattering process is taken to be $2 \to 2$ parton scatterings that then give rise to jet final states. The events are passed through the CDF detector simulation and cluster algorithm and the jet with the largest $p_\perp$ is defined as the "trigger" jet and is arbitrarily assigned to have $\phi = 0$. The particle multiplicity of the full event is plotted as a function of the azimuthal distance (ϕ) from the trigger jet. Only particles with $|\eta| < 5.0$ and $p_\perp > 0.5$ GeV are included. The invariant mass of the the two clusters with the largest $p_\perp$ is between 50 and 70 GeV.

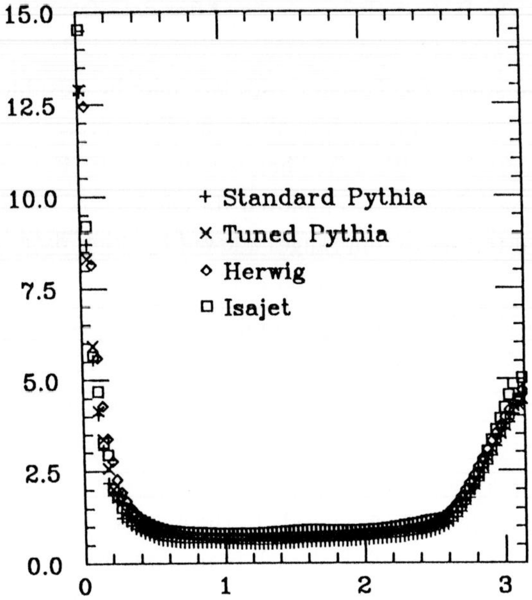

Figure 29: As Figure 28 except that there is no requirement on the $p_\perp$ of the particles.

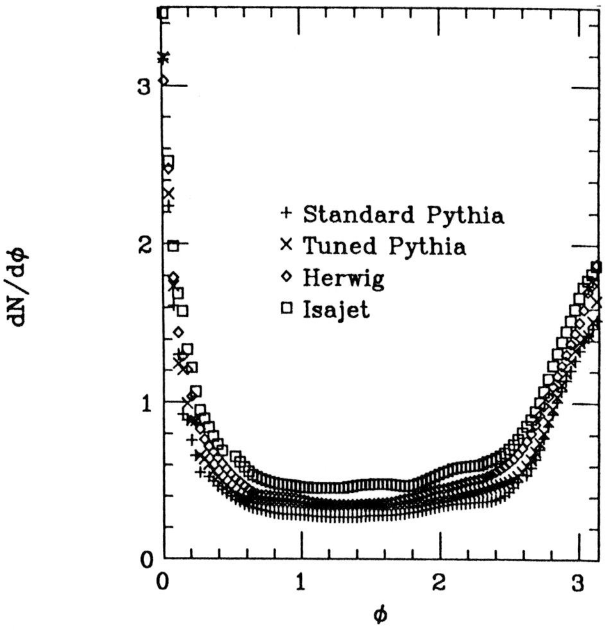

Figure 30: As Figure 28 except that the invariant mass of the two highest clusters is between 150 and 200 GeV.

fect in the W underlying events. After data from the SSC become available and a further retuning is made, a quantitative *signal/background* estimate may be possible. At the very least, any such analyses made at present should be undertaken with more than one generator. Even when only one generator is employed its parameters should be allowed to vary within ranges allowed by other data.

There are two contributions to the multiplicity of hadrons within a jet. A jet is initiated by a quark or gluon which is produced in a hard scattering. This parton is off mass-shell by an amount (Q) of order the momentum transfer involved in the hard scattering. This parton can produce a shower of other partons via a branching process controlled by perturbative QCD; the invariant mass of each parton being degraded at each branching. Eventually the invariant mass becomes of order 1 GeV and perturbative QCD ceases to be reliable. The final hadronization takes place via a non-perturbative process. The multiplicity of the partonic shower can be calculated in perturbative QCD.

Since the hadronization is characterised by a mass scale (p) of order 1 GeV which does increase with the jet energy, the growth of multiplicity is controlled by the growth of the parton shower and is therefore calculable [64]. This calculation yields

$$\langle n(Q)\rangle \sim \frac{1}{2}exp(\sqrt{\frac{72}{(33-2n_f)}}(\sqrt{log(Q^2/\Lambda^2)}-\sqrt{log(p^2/\Lambda^2)})) \tag{16}$$

The second component to the hadronic multiplicity is that due to the hadronization of the parton shower into the observed hadrons. Since the scale associated with the hadronization (p) does not rise with the jet energy, the growth in the hadronic multiplicity should be described by Equation 16.

The result of Equation 16 is not reproduced by the leading log approximation branching used in the Monte–Carlo event generators [42], [52]. This is because the result of Equation 16 includes quantum interference effects that are destructive and that lower the multiplicity.

These interference effects have been observed in e^+e^- annihilation where three jet events occur due to the final state $q\bar{q}g$. Additional soft gluon emission is possible from either the quark or gluon lines. This radiation gives rise to soft hadrons distributed away from the jet axes. This distribution of soft hadrons can be measured and compared with that in events where there are two jets and an isolated photon arising from the final state $q\bar{q}\gamma$. In the latter case only the quark and anti-quark can radiate gluons. The destructive interference between the radiation patterns from the quarks and gluons results in a *lower* number of soft hadrons between the quark and antiquark jets in the $q\bar{q}g$ case than in the $q\bar{q}\gamma$ case. This predicted effect [65] has been observed in data at PEP [66] and PETRA [67]. The leading log Monte–Carlo generators are based on algorithms that allow each parton leg to radiate independently do not reproduce the interference effects. In the LUND e^+e^- generator [68], the non-perturbative hadronization of the quark and gluon jets is handled differently. In the energy ranges of the PEP and PETRA data, LUND is able to reproduce interference effects seen in the data, despite not handling the radiation correctly.

The leading log algorithm can be modified so that they can correctly reproduce the interference effects [69]. These corrections can be incorporated into the classical branchings used in Monte-Carlos [53] which should then reproduce the multiplicity growth of Equation 16. The incoherent branching used by ISAJET and PYTHIA gives rise to a much larger multiplicity than that given by Equation 16. Figure 32 shows the multiplicity predicted by Equation 16 as well as that given by incoherent branching as a function of the jet energy. The multiplicity has been normalized so that it is equal to 6 for a 20 GeV jet. It can be clearly seen from this figure that the incoherent branching will give rise to much too large a multiplicity for very energetic jets such as those that will be visible at the SSC or a TeV e^+e^- collider. The growth of multiplicity with jet energy can be seen by comparing the peaks at $\phi = 0$ in Figures 28 and 30. Note that the HERWIG generator which includes coherence effects and should follow Equation 16 gives a lower multiplicity (1.3 and 3.0 at $\phi = 0$ in Figures 28 and 30, respectively) than ISAJET (1.4 and 3.5). At the higher jet energy PYTHIA (1.3 and 3.2 for both tuned and standard) has risen above HERWIG.

A quantity that can be calculated in perturbative QCD is the multiplicity of heavy quark (mass m_q) pairs within a QCD jet [70]. Provided that the heavy quark mass (m_q) is larger than the QCD scale Λ, the multiplicity (n_q) rises with the initial off-shellness (Q) of the parton initiating the

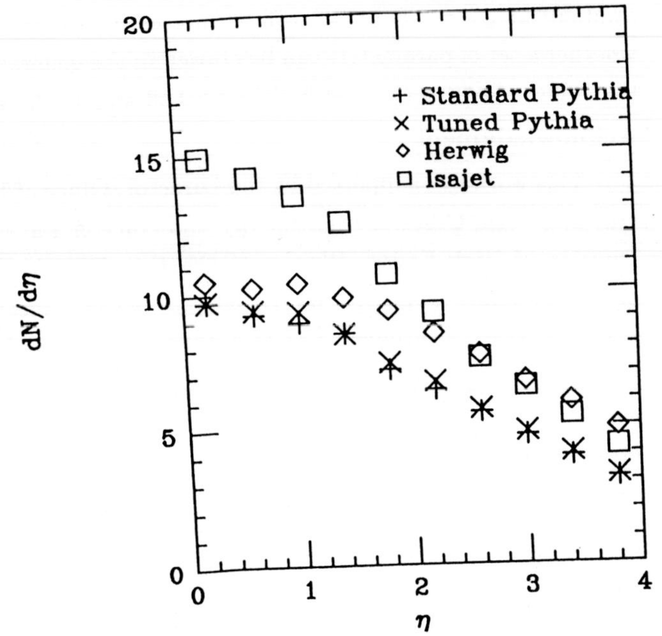

Figure 31: The pseudo–rapidity distribution $dN/d\eta$ for events in which there is a jet in proton–antiproton collisions at $\sqrt{s}$=1.8 TeV. The predictions of Monte–Carlo generators shown. The Monte–Carlo events are passed through the CDF detector simulation and cluster algorithm and the jet with the largest $p_\perp$ is defined as the "trigger" jet and is arbitrarily assigned to have $\phi = 0$. The invariant mass of the two clusters with largest $p_\perp$ is required to be in the range $M_W \pm 10$ GeV. The multiplicity is computed by taking that in the range $3\pi/8 < \phi < 5\pi/8$ and multiplying by 4. This procedure is necessary to eliminate the particles in the jet (and in the jet which balances it) and to facilitate a comparison with Figure 27.

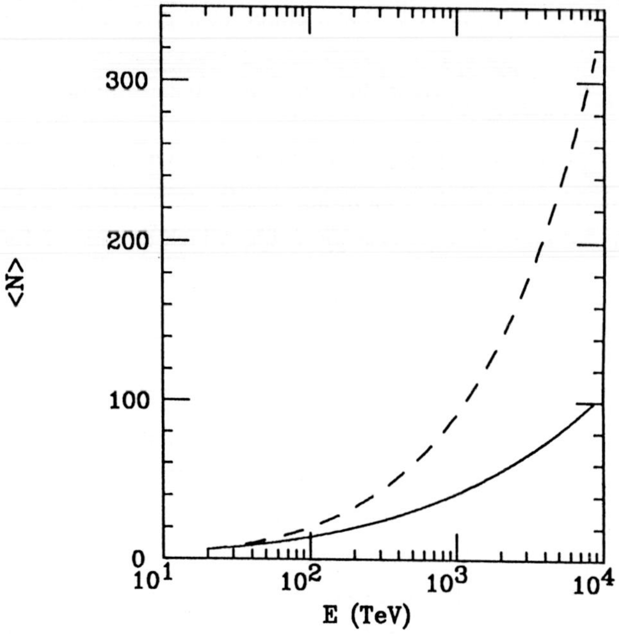

Figure 32: The multiplicity of particles within a jet as predicted by perturbative QCD (solid line) and the incoherent leading log approximation (dashed line) as a function of the jet energy (E). The curves are normalized to a multiplicity of 6 at $E = 20$ GeV.

jet. The multiplicity within a gluon jet is given by

$$n_q = \frac{1}{3\pi}\int_{4m_q^2}^{Q^2} \frac{dk^2}{k^2}\alpha_s(k^2)(1+2m_q^2/k^2)\sqrt{0.25 - m_q^2/k^2} \quad (17)$$

The production of charm and bottom quarks at large $p_\perp$ is dominated by the generation within gluon jets given by this formula. The process $gg \to Q\overline{Q}$ which dominates the total $Q\overline{Q}$ production rate is unimportant at if $p_\perp >> m_Q$.

8 Conclusions.

This work has concentrated on the predictions of QCD which can be tested in the near future. We have emphasized the uncertainties in predicted rates due to distribution functions and the choice of the scales M and μ at which the distribution functions and α_s are evaluated when computing a QCD rate. In the case of photon production at $\sqrt{s} = 1.8$ TeV, the importance of the definition of an isolated photon has been shown. While no precise QCD test is possible until the theoretical calculations use a definition compatible with the experiments, useful information can be obtained about distribution functions. At fixed target energies, there is no problem with the definition of isolation but it is unlikely that measurements will add significantly to our knowledge of distribution functions.

We have discussed the various ways in which jet measurements can be compared to QCD calculations. Any detailed comparison will require that the calculations use cut off parameters (R_0 and p_0) that can be subsumed into an experimental jet definition after hadronization. At present all comparisons require a Monte–Carlo generator which includes a model of hadronization. As was the case in e^+e^- annihilation, any detailed QCD tests (or measurements of of α_s) are likely to depend on this model. We have seen that the leading log Monte–Carlos (in particular ISAJET) give a reasonable agreement with exact parton calculations of two, three and four jet events at $\sqrt{s} = 1.8$ TeV.

We have shown the differences between the predictions for particle multiplicities in the various event generators. In some cases we have been able to compare with data. While all the generators agree qualitatively with QCD predictions, large quantitative differences can appear as the energy changes and it is likely that substantial retunings will be required before these features of the generators can be exploited at higher energies. Once more data becomes available at $\sqrt{s} = 1.8$ TeV, it will be interesting to discover whether a set of parameters can be chosen in the generators that allow them to agree with data at 0.63 and 1.8 TeV.

Acknowledgment

This work was supported by the Director, Office of Energy Research, Office of High Energy and Nuclear Physics, Division of High Energy Physics of the U.S. Department of Energy under Contracts DE-AC03-76SF00098. Accordingly, the U.S. Government retains a nonexclusive, royalty-free license to publish or reproduce the published form of this contribution, or allow others to do so, for U.S. Government purposes.

References

[1] For a review of the QCD parton a model see G. Altarelli *Phys. Rev.* **81C**, 1 (1982).

[2] G. Altarelli and G. Parisi, *Nucl. Phys.* **B126**, 298 (1977).

[3] G. Altarelli, *et al.*, *Nucl. Phys.* **B143**, 521 (1978), Erratum *Nucl. Phys.* **B146**, 544 (1978).

[4] S. Dawson, R. K. Ellis and P. Nason, *Nucl. Phys.* **B303**, 607 (1988).

[5] P. Aurenche, R. Baier, A. Douiri and D. Schiff, *Phys. Lett.* **169B**, 441 (1986), *Phys. Lett.* **140B**, 87 (1984).

[6] R.K. Ellis and J.C. Sexton, *Nucl. Phys.* **B269**, 445 (1986).

[7] R. K. Ellis and P. Nason, FNAL 88/54–T (1988).

[8] E.G. Floratos, D.A. Ross and C.T. Sachrajda, *Nucl. Phys.* **B129**, 66 (1977), Erratum *Nucl. Phys.* **B139**, 545 (1978); C. Curci, W. Furmanski, and R. Petronzio, *Nucl. Phys.* **B175**, 27 (1980); W. Furmanski and R. Petronzio, *Phys. Lett.* **97B**, 437 (1980), *Zeit. fur Physik* **C11**, 293 (1982).

[9] J. Morfin and W.–K. Tung, these proceedings.

[10] V. Papavassiliou (EMC collaboration) Proc. of Int. Europhysics Conference on High Energy Physics, Upsala (1987).

[11] A. Benvenuti *et al.*, *Phys. Lett.* **195B**, 91 (1987).

[12] See, for example, F.J. Sculli, Talk at 1988 Arles Conference, June 1988, and R. Voss in Proc. of 1987 International Symposium on Lepton and Photon Interactions at High Energies, Hamburg 1987.

[13] D. Duke and J. Owens *Phys. Rev.* **D30**, 49 (1984).

[14] J.T. Aubert *et al., Nucl. Phys.* **B293**, 740 (1987); **B272**, 158 (1986), **B259**, 189 (1985); or M. Arneodo *et al., Nucl. Phys.* **B264**, 739 (1986).

[15] A. Bodek *et al., Phys. Rev.* **D20**, 1471 (1979).

[16] H. Abramowicz, *et al., Zeit. fur Physik* **C17**, 283 (1983).

[17] H. Abramowicz, *et al., Zeit. fur Physik* **C35**, 443 (1987); B. Vallage, Thesis submitted to Université de Paris Sud (1984).

[18] E. Eichten *et al., Rev. Mod. Phys.* **56**, 579 (1984).

[19] M. Diemoz, F. Ferroni, E. Longo, and G. Martinelli, CERN–TH–4757 (1987).

[20] D. Allasia *et al., Phys. Lett.* **135B**, 231 (1986); *Zeit. fur Physik* **C28**, 321 (1985).

[21] D. B. Macfarlane et al., *Zeit. fur Physik* **C26**, 1 (1984).

[22] J.V. Allaby, *et al., Phys. Lett.* **197B**, 281 (1987), CERN EP/87/225. F. Bergsma *et al., Phys. Lett.* **123B**, 269 (1983), **153B**, 111 (1985).

[23] R.M. Barnett, I. Hinchliffe and J. Stirling in Review of Particle Properties *Phys. Lett.* **204B**, 1 (1988).

[24] G. Altarelli, *et al.,* CERN–TH–4978 (1988).

[25] A.D. Martin, R.G. Roberts and J. Stirling, *Phys. Rev.* **D37**, 1161 (1988).

[26] J.A. Branson *et al., Phys. Rev. Lett.* **38**, 1331 (1977); K.J. Anderson *et al., Phys. Rev. Lett.* **42**, 944 (1979); E.J. Siskind *et al., Phys. Rev.* **D21**, 628 (1980).

[27] T. Akesson, *et al., Zeit. fur Physik* **C34**, 293 (1987).

[28] A.D. Martin, R.G. Roberts and J. Stirling, *Phys. Lett.* **207B**, 205 (1988).

[29] A. Benvenuti *et al.,* JINR–E1–87–689, 699 (1987).

[30] D. Antreasyan *et al., Phys. Rev. Lett.* **48**, 302 (1982).

[31] W-K. Tung, private communication and talk at 1988 DPF Summer Study on High Energy Physics in the 1990's, Snowmass, Colorado, July 1988.

[32] J. Collins in "Supercollider Physics" Ed. D. Soper, World Scientific publishing (1987).

[33] R. Brinkman "HERA" DESY–HERA–88–03 (1988).

[34] A.M. Cooper–Sarkar, *et al.,* RAL–87–112.

[35] P.M. Stevenson, *Nucl. Phys.* **B231**, 65 (1984).

[36] B.L. Combridge, *Nucl. Phys.* **B174**, 243 (1980), C. Carimalo *et al., Phys. Lett.* **98B**, 105 (1981).

[37] F. Olness and W-K. Tung, Int. J. Mod. Phys. **A2**, 1413 (1987).

[38] For a review see, J.A. Appel, FNAL–CONF–88/22 (1988).

[39] M. Gluck, J. Owens and E. Reya, *Phys. Rev.* **D17**, 2324 (1978).

[40] R. Baier and R. Ruckl, *Phys. Lett.* **102B**, 364 (1981).

[41] See, for example, S.L. Wu in *Proc. of International Symposium on Lepton–Photon interactions*, Hamburg (1987) and references therein.

[42] F. Paige and S. Protopopescu, in Proc. of 1986 Snowmass Summer Study on the Physics of the SSC, Ed. R. Donaldson FNAL (1987).

[43] S. Ellis, Z. Kunszt and D. Soper.

[44] P. Chiapetta, Talk at 1988 Workshop on $p\bar{p}$ physics, FNAL, June 1988.

[45] R.K. Ellis *et al., Nucl. Phys.* **B173**, 397 (1980).

[46] F. A. Berends, *et al., Phys. Lett.* **103B**, 124 (1982).

[47] S.J. Parke and T.R. Taylor, *Nucl. Phys.* **B269**, 410 (1986); Z. Kunszt, *Nucl. Phys.* **B271**, 333 (1986); J.F. Gunion and J. Kalinowski, *Phys. Rev.* **D34**, 2119 (1986).

[48] F.A. Berends and W.T. Giele, *Nucl. Phys.* **B306**, 759 (1988).

[49] C.J. Maxwell, *Phys. Lett.* **192B**, 190 (1987), Durham preprint DTP-88/32 (1988); Z. Kunszt and W.J. Stirling, *Phys. Rev.* **D37**, 2439 (1988); S.J. Parke and M. Mangano FNAL–Pub–88/92–T (1988).

[50] EUROJET, B. van Eijk, in Proc. of the 5th International Workshop on $p\bar{p}$ Interactions, St. Vincent, Italy (1985).

[51] PAPAGENO, I. Hinchliffe, in preparation.

[52] H-U. Bengtsson and T. Sjostrand, *Comp. Phys. Comm.* **46** 43, (1987).

[53] G. Marchesini and B.R. Webber, CAVENDISH preprint 88/7 (1988).

[54] CDF collaboration presented by A. Garfinkel at the 7th Topical Workshop on $p\bar{p}$ Interactions, Fermilab, June 20–24, 1988 *Phys. ReV. Lett* (to appear).

[55] V.A. Abramovskii, O.V. Kanchelli and V.N. Gribov in *Proc.* 16th International Conference on High Energy Physics, FNAL (1983), Vol. 1, p.363.

[56] C.G. Alner, *et al.*, *Phys. Rep.* **154**, 247 (1987).

[57] T. Sjostrand, private communication.

[58] CDF collaboration presented by J. Proudfoot at the 7th Topical Workshop on $p\bar{p}$ Interactions, Fermilab, June 20–24, 1988, ANL–HEP–CP–88-61.

[59] F. Abe *et al.*, ANL–HEP–PR–88–32, *Phys. Rev. Lett.* , (1989) (to appear).

[60] G. Arnison *et al.*, *Phys. Lett.* **118B**, 167 (1982).

[61] C. Albajar, *et al.*, CERN–EP/88-29.

[62] For a recent review see R.N. Cahn *et al.*, in Proc of the 1987 Berkeley Workshop on 'Experiments, Detectors, and Experimental Areas for the Supercollider", Ed. M. Gilchriese and R. Donaldson, World Scientific Publishing, Singapore, 1988.

[63] S. Dawson and R.N. Cahn, *Phys. Lett.* **136B**, 196 (1984); S. Petcov and D.R.T. Jones, *Phys. Lett.* **84B**, 440 (1979).

[64] A. H. Mueller, *Phys. Lett.* **104B**, 161 (1981); Yu. Dokshitzer, *et al.*, *Phys. Lett.* **115B**, 242 (1982), *Nucl. Phys.* **B312**, 85 (1983), *Nucl. Phys.* **228**, 351 (1983); A. Bassetto, *et al.*, Phys. Rep., 100:201 (1983).

[65] Ya. I. Azimov, *et al.*, Leningrad preprint 1051 (1985), *Phys. Lett.* **165B**, 147 (1985).

[66] H. Aihara, *et al.*, *Zeit. fur Physik* **C28**, 31 (1985).

[67] W. Bartel, *et al.*, *Phys. Lett.* **134B**, 275 (1984).

[68] B. Andersson, *et al.*, *Phys. Rep.* **97**, 31 (1983).

[69] G. Marchesini and B. Webber, *Nucl. Phys.* **B238**, 501 (1984).

[70] A. Mueller and P. Nason, *Phys. Lett.* **157B**, 226 (1985), *Nucl. Phys.* **B266**, 265 (1986).

STRUCTURE FUNCTIONS AND PARTON DISTRIBUTIONS

Wu-Ki Tung[a,b,c], J. G. Morfín[c], H. Schellman[c], S. Kunori[d], A. Caldwell[e], F. Olness[f]

Other members participating in the work-group at Snowmass were: R. Blair, F. Borcherding, J. Huston, G. Snow, and M. Zielinski

I. Introduction

The QCD Parton Model provides a comprehensive framework for describing general high energy processes in current and planned accelerators and colliders. In this framework, a typical hadron-hadron collision process A + B → C + X, where C represents a final state of physical interest, can be expressed in terms of more basic processes involving the elementary constituents (partons) of the fundamental theory as depicted in Fig.1.

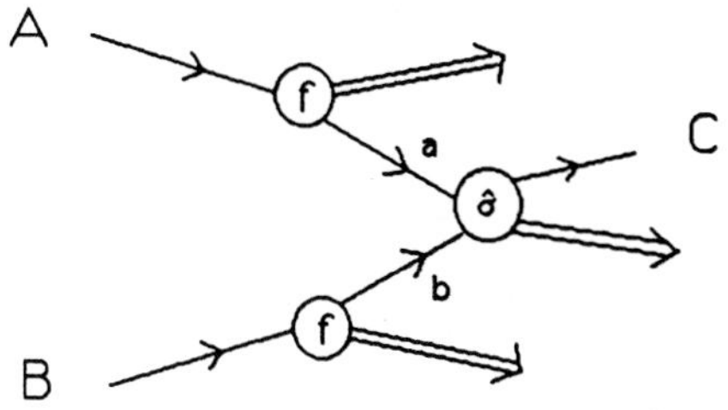

Fig.I-1: The QCD Parton Model

The physically measured cross-section for such a process is written in terms of a convolution of *Parton Distribution functions* $f_A^a(x,Q)$ and *Hard cross-sections* $\hat{\sigma}_{ab}$ which represent the fundamental physical processes initiated by the partons:

$$\sigma_{AB}(s\,..) = f_A^a(x_1\,..) \otimes \hat{\sigma}_{ab}(\hat{s},\ Q\ ..) \otimes f_B^b(x_2\,..) \qquad (I\text{-}1)$$

Here s is the squared energy of the overall processes, $\hat{s}$ is the corresponding quantity for the parton-parton subprocess, Q is an appropriate energy scale characteristic of the hard process (usually of the order of $\sqrt{\hat{s}}$) and $x_{1,2}$ are the frac-

[a] Argonne National Laboratory, Argonne, IL.
[b] Illinois Institute of Technology, Chicago, IL.
[c] Fermi National Accelerator Laboratory, Batavia, IL.
[d] University of Maryland, Collge Park, MD.
[e] Nevis Laboratory, Irvington-on Hudson, NY.
[f] University of Oregon, Eugene, OR.

tional longitudinal momenta carried by the partons relative to the parent hadrons.

It is obvious from Fig.I-1 and Eq.(I-1) that the **parton distributions** $f_A{}^a(x,Q)$ play a key role in any serious study of high energy process. *All quantitative relations between measured cross-sections and fundamental interactions of the theory require precise knowledge of these (universal) distribution functions.* The main source of information on the parton distributions come from **deep inelastic scattering Structure Functions**. The Group on Structure Functions and Parton Distributions at this Workshop was set up to review the current status of and to identify outstanding issues confronting the determination of the parton distributions for the study of physics of the 1990's, and to address these issues as much as possible.

The QCD Parton Model picture suggests that the kinematic variables (x,Q) are important in considering the physics potential of current and planned high energy facilities. In Fig.I-2, we show a Road Map of HEP for the 1990's which indicates the regions in the x-Q plane of the physics reach of current and planned accelerators and colliders.

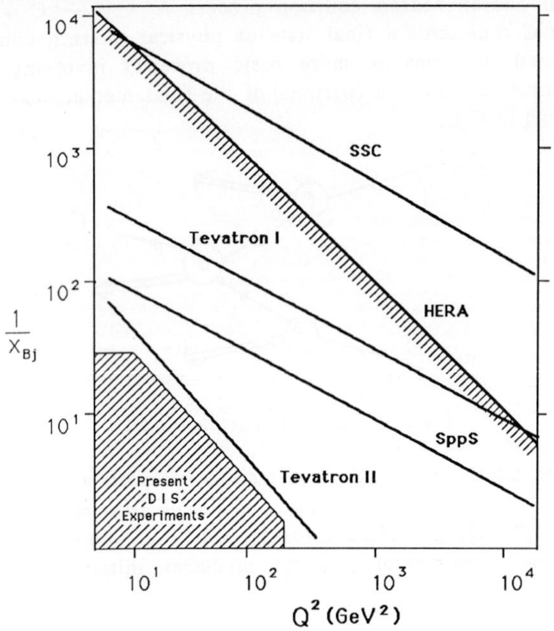

Fig.I-2: Road Map of High Energy Physics in the 1990's.

The shaded region in the lower left corner represents that covered by current fixed-target deep inelastic scattering experiments where extensive information on structure functions are, in principle, available. It is clear from this map that extrapolations from this kinematic range to those of physical processes at the Tevatron and SSC are enormous in both Q and x variables. The theoretical tool for effecting the extrapolation to large Q (horizontal direction in Fig.I-2) is the **QCD evolution equation** which describes well the Q-dependence of structure functions within the range of available energies. On the other hand, much less is known about the extrapolation to small x (vertical direction in Fig.I-2). Available theoretical guides are qualitative and are subject to change when further progress are made. Hence, it is particularly important to acquire *experimental* information on parton distribution functions in the small x region. To this end, *increasing reliance must be placed on studying hadron collider processes which are sensitive to parton distributions at small x* - at least before the e-p collider at HERA comes on line. (For practical purposes, "small x" means here x values below currently measured range, say less than 0.05, cf. Fig.I-2.)

Recent developments in high energy jet physics, heavy-flavor production, lepton-pair production, direction-photon production, ... all point to the importance of including **higher order effects of the QCD parton model**, Eq.(I-1), if quantitative QCD analyses are to be made.[1] This requires not only the calculation of higher order hard cross-sections, as usually emphasized, but also the use of parton distribution functions derived with evolution kernels to the corresponding order in the QCD coupling parameter α_s. For this reason, any serious study of parton distributions must spell out the precise definitions of these distributions, their relations to the measured structure functions, and their proper use in the general formula Eq.(I-1) which are *valid beyond the leading order* in QCD. With this in mind, the summary of the basic formalism given in Sec.II of this report serves as a guide to contemporary practicers of the QCD Parton Model who are not necessarily experts of perturbative QCD.

II. Basic Definitions and Formalism

For modern applications of the QCD Parton Model, it is essential to clearly distinguish the *universal* **Parton Distributions** from the **Structure Functions** measured in the *specific* Deep Inelastic Scattering experiments. The relation between the two is intimate, but by no means trivial beyond the leading order. Fig.II-1 illustrate this relationship:

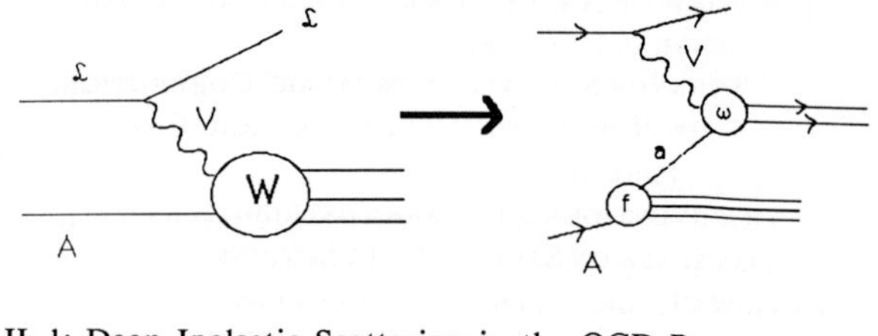

Fig.II-1: Deep Inelastic Scattering in the QCD Parton Model.

part (a) represents the *physically measurable* structure functions (to be denoted by $W_A{}^i, F_A{}^i$); part (b) represents the corresponding QCD Parton Model expression for these, in terms of a convolution of a parton distribution function ($f_A{}^a$) (*soft* wave function) and a Parton structure function ($\omega_a{}^i$) (*hard* scattering of the vector boson probe V off the parton).

In the **naive parton model**, $\omega_a{}^i$ is represented by simple constant point-couplings of the partons, as shown in Fig.II-2a. In this approximation, the gluon does not contribute to the scattering in any direct way (all its couplings vanish); and the quarks do not couple to the longitudinal structure functions (the gauge coupling of the quarks are pure transverse). **Beyond the leading order**, the perturbatively calculable $\omega_a{}^i$ become non-trivial (Fig.II-2b,c); the gluon will contribute directly (Fig.II-2c); and the longitudinal structure function is predicted to be non-vanishing. In fact, this leads to the most direct way to measure the gluon distribution, cf. Sec.II.C.

(a) Leading Order; (b) 1st Order Quark; (c) 1st Order Gluon.

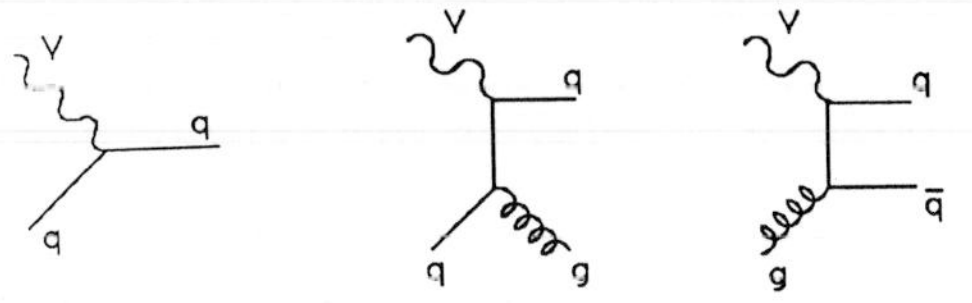

Fig.II-2 Perturbative series for the (hard) *Parton Structure Function*

Figs.II-1,2 make it clear that the very definition of the parton distribution functions $f_A^{\,a}$ depends on the approximation adopted for $\omega_a^{\,i}$ — as the product must remain the same as the *physical* structure function. Beyond the leading order, $\{\omega_a^{\,i}\}$ are necessarily renormalization-scheme dependent; hence, so are the parton distributions $\{f_A^{\,a}\}$. The following subsections are devoted to the definition of the physical structure functions, the leading order parton formulas and the next-to-leading order QCD parton formulas in turn.

A. Structure Functions in Deep Inelastic Scattering

The generic deep inelastic scattering process depicted in Fig.II-1 are characterized by the well-known kinematic variables:

$$x = \frac{Q^2}{2p\cdot q} \qquad ; \qquad Q^2 = q^2 = (\ell - \ell')^2$$
$$\nu = -q\cdot p/M = E - E'; \qquad W^2 = -p'^2 = M^2 - Q^2 + 2M\nu \qquad (II\text{-}1)$$
$$y = q\cdot p \,/\, \ell\cdot p = (E - E')\,/\,E$$

where E and E' are the laboratory energies of the leptons ℓ and ℓ' respectively, and the Lorentz metric is (-1,1,1,1).

Taking into account the known lepton vertex in Fig.II-1, the cross-section for the overall process can be written in terms of **Structure Functions** (W_i, $i = 1, 2, 3$) as follows:

$$\frac{d\sigma^{\ell A}}{dxdy} = N_V^{\,\ell}\,[2W_1^{VA}(x,Q)\sin^2(\theta/2) + W_2^{VA}(x,Q)\cos^2(\theta/2) \pm W_3^{VA}(x,Q)\,\frac{E+E'}{M}\,\sin^2(\theta/2)] \qquad (II\text{-}2)$$

where (ℓ,V,A) are labels for the (beam) lepton, the (exchanged) vector-boson and the (target) hadron respectively; the (±) signs are associated with $(V = W^{\pm})$, i.e. ν- and anti ν-scattering; and θ is the laboratory scattering angle. The process-dependent normalization constant N_V in $(II\text{-}2)$ is given by:

$$N_\gamma^{\ell\pm} = 8\pi\alpha^2 yEE'/q^4 \qquad (II\text{-}3)$$

for electron and muon scattering, and

$$N_W^{\nu,\bar{\nu}} = \pi\alpha^2 yEE'/2\,\sin^4\theta_w (q^2 + M_W^2)^2 \approx G_F^2 yEE'/\pi \qquad (II\text{-}4)$$

for charged current neutrino- and antineutrino-scattering. When neutral-current weak interaction becomes important, the Z-exchange amplitude contributes by itself as well as through an interference term with the γ-exchange amplitude. The formulas for $N_Z^{\,\nu}$ and $N_Z^{\,\ell\pm}$ are not given here explicitly.

The **structure functions** W_i^{VA} *are fundamental physical quantities* for lepton-hadron scattering. They are proportional to the **total cross-section** of the relevant vector-boson - hadron scattering, hence (by the Optical Theorem) represent the imaginary part of the **forward Compton scattering amplitude** of V on A. The fundamental significance of the structure functions makes them the natural meeting ground between experiment and theory. In principle, *the measurement of these structure functions in lepton-hadron scattering is independent of further theoretical assumptions*, including the QCD parton model.

An alternate set of structure functions, commonly called F^i *(or* F^λ*)*, is motivated by the discovery of (approximate) scaling and the subsequent successes of the parton-model interpretation of deep inelastic scattering. The new structure functions are related to the *W's* by the following relations.

$$F^1 \equiv 2\,W_1 \equiv F^L + F^R$$
$$F^2 \equiv 2\left(\frac{\nu^2}{q^2} + 1\right) W_2 \equiv F^L + F^R + 2F^{Long} \qquad (II\text{-}5)$$
$$F^3 \equiv \frac{\sqrt{\nu^2 + q^2}}{M}\,W_3 \equiv F^L - F^R$$

where the left-hand sides define F^i [i = *1,2,3*] in the *tensor* basis while the right-hand sides define F^λ *[*λ *= L (left-handed), R (right-handed)*, and *Long (longitudinal)]* in the *helicity* basis. The boson and hadron indices as well as the arguments of the functions have been suppressed. It is seen that F^1 *represents the* **transverse amplitude**, F^2 *the* **sum of helicity amplitudes**, *and* F^3 *the* **parity-violating amplitude**. For reference, the inverses of the above relations are:

$$F^{L,R} = \frac{F^1 \pm F^3}{2} \qquad ; \qquad F^{Long} = \frac{F^2 - F^1}{2} \qquad (II\text{-}6)$$

Since neutrinos carry definite helicities, the cross-section formula Eq.*(II-2)* takes on a simple form in terms of the helicity structure functions in the deep inelastic scattering limit $M^2/q^2 << 1$:

$$\frac{d\sigma^{\nu A}}{dxdy} = \frac{G^2}{\pi}\,MEx\,\left[\begin{matrix}F^L_+\\F^R_-\end{matrix} + 2(1-y)\begin{matrix}F^{Long}_+\\F^{Long}_-\end{matrix} + (1-y)^2\begin{matrix}F^R_+\\F^L_-\end{matrix}\right] \qquad (II\text{-}7a)$$

where the top line (with structure function boson sub-index +) applies to **neutrino-scattering**, and the bottom line (with structure function boson sub-index -) applies to **anti-neutrino scattering**. This is to be compared with the more often seen formula in terms of the tensor structure functions,

$$\frac{d\sigma^{\nu A}}{dxdy} = \frac{G^2}{\pi}MEx\left[\frac{y^2}{2}F^1_\pm + (1-y)F^2_\pm \pm \left(y - \frac{y^2}{2}\right)F^3_\pm\right] \qquad (II\text{-}7b)$$

where ± refers to neutrino and anti-neutrino scattering respectively.

It is to be noted that the above definitions of F^i with the tensor index *(i)* are slightly different from the original definitions of the scaling structure functions found in the earlier literature and in experimental papers:

$$F^1 = 2 \cdot F_1^{org}; \quad F^2 = F_2^{org}/x; \quad F^3 = F_3^{org} \qquad (II\text{-}8)$$

More and more recent theoretical papers use the new set defined by Eq.*(II-5)* because of the simple relations to the helicity amplitudes and to the parton model expressions for these quantities (cf. next section). It should be emphasized, however, that *all sets of structure functions are mere redefinitions of* W_i; and, as such, *they are just as fundamental and free from dynamical assumptions as the latter.* This fact is, unfortunately, not observed and respected in some experimental "measurements" where model-dependent (usually parton-model inspired) "corrections" were applied to the *structure functions* analyses. (Cf. Sec. III.B)

B. Parton Distributions in the Naive Parton Model

In general, quarks of different generations couple identically to the electro-weak gauge bosons and the gluon does not couple directly to them. To leading order in the QCD coupling, the parton structure functions ω_a^i are just (squares) of these electro-weak coupling constants. Denoting by $U(x)$ and $D(x)$ the generic up-like (u,c,t) and down-like (d,s,b) quark probability distributions inside the target hadron (A), respectively, the **electromagnetic structure functions** are given by:

$$\begin{aligned} &F_{\gamma A}^{3,Long}(x) = 0 \\ &F_{\gamma A}^{1,2}(x) = \frac{4}{9}\,\Sigma\,[U_A(x)+\bar{U}_A(x)] + \frac{1}{9}\,\Sigma\,[D_A(x)+\bar{D}_A(x)] \end{aligned} \qquad (II\text{-}9)$$

where the first sum (over the *generations*) includes all up quarks and anti-quarks, and the second sum includes all down and anti-down quarks. The **charged-current weak structure functions** are simplest when written in the *helicity basis* because of the basic left-handed weak coupling:

$$\begin{aligned} &F_{+A}^{L}(x) = 2\,\Sigma\,D_A(x)\,, \qquad F_{+A}^{R}(x) = 2\,\Sigma\,\bar{U}_A(x) \\ &F_{-A}^{L}(x) = 2\,\Sigma\,U_A(x)\,, \qquad F_{-A}^{R}(x) = 2\,\Sigma\,\bar{D}_A(x) \end{aligned} \qquad (II\text{-}10)$$

Thus, these measured structure functions neatly sort out the up-quarks, down-quarks and the corresponding anti-quarks. We have, of course, again the result that the gluon distribution does not contribute to the structure functions; and that the longitudinal structure vanishes in this approximation:

$$F_{VA}^{Long}(x) = 0 \qquad \text{for all } V \qquad (II\text{-}11)$$

For **isoscalar targets** (normalized to per nucleon cross-section and denoted by $A{=}N$), we define

$$F_{VN}^{i}(x) = \frac{1}{2}\,[\,F_{Vp}^{i}(x) + F_{Vn}^{i}(x)\,] \qquad (II\text{-}12)$$

Invoking **strong iso-spin invariance**, and singling out the first generation up and down quarks, we use the abbreviations:

$$\begin{aligned} &f_p^{u,d}(x) = f_n^{d,u}(x) \equiv u(x),\ d(x) \\ &f_p^{s,c,b}(x) = f_n^{s,c,b}(x) \equiv s(x),\ c(x),\ b(x) \end{aligned} \qquad (II\text{-}13)$$

Eq.(II-9) becomes

$$\begin{aligned} F_{\gamma N}^{1,2}(x) &= \frac{5}{18}\,(u+\bar{u}+d+\bar{d}) + \frac{4}{9}\,\Sigma'(U+\bar{U}) + \frac{1}{9}\,\Sigma'(D+\bar{D}) \\ &= \frac{5}{18}\,\Sigma\,(q+\bar{q}) + \frac{1}{6}\,\Sigma'(U+\bar{U}-D-\bar{D}) \\ &\approx \frac{5}{18}\,\Sigma\,(q+\bar{q}) + \frac{1}{3}\,\Sigma'(U-D) \end{aligned} \qquad (II\text{-}14)$$

Where Σq indicates the sum over all quark flavors, and Σ' denotes the sum over all generations beyond the first. In the third line, one makes the additional assumption that at all values of x,

$$U(x) = \bar{U}(x) \quad ; \quad D(x) = \bar{D}(x) \qquad (II\text{-}15)$$

for all higher generations.

For **neutrino scattering on isoscalar targets**, we obtain:

$$\begin{aligned} &F_{+N}^{L}(x) = d(x) + u(x) + 2s(x) + 2b(x) \\ &F_{+N}^{R}(x) = \bar{u}(x) + \bar{d}(x) + 2\bar{c}(x) \end{aligned} \qquad (II\text{-}16)$$

and for **anti-neutrino scattering**:

$$\begin{aligned} &F_{-N}^{L}(x) = u(x) + d(x) + 2c(x) \\ &F_{-N}^{R}(x) = \bar{d}(x) + \bar{u}(x) + 2\bar{s}(x) + 2\bar{b}(x) \end{aligned} \qquad (II\text{-}17)$$

Note that the corresponding neutrino and anti-neutrino structure functions for iso-scalar targets differ only in the higher generation parton contents which are in general small at moderate values of x. *The left-handed and right-handed structure functions separately measure the quark distributions and the anti-quark distributions.*

Now, consider the **tensor structure functions for neutrino scattering**:

$$\begin{aligned} F_{+N}^{1,2}(x) &= u + \bar{u} + d + \bar{d} + 2s + 2b + 2\bar{c} \approx \Sigma(q+\bar{q}) \\ F_{+N}^{3}(x) &= u - \bar{u} + d - \bar{d} + 2s + 2b - 2\bar{c} \\ &\approx \Sigma\,(q-\bar{q}) + 2(s+b-c) = \Sigma\,q_v + 2(s+b-c) \end{aligned} \qquad (II\text{-}18)$$

where $\Sigma q(x)$ $[\Sigma\bar{q}(x)]$ represents the sum over all quark [anti-quark] flavors and $q_v(x)$ is the *valence* quark distribution. For **anti-neutrino scattering**, we obtain:

$$F^{1,2}_{-N}(x) = u + \bar{u} + d + \bar{d} + 2\bar{s} + 2\bar{b} + 2c \approx \Sigma(q + \bar{q})$$

$$F^{3}_{-N}(x) = u - \bar{u} + d - \bar{d} - 2\bar{s} - 2\bar{b} + 2c$$

$$\approx \Sigma\, q_V - 2\,(s + b - c) \qquad (II\text{-}19)$$

Note that $F_{\pm N}^3$ are dominated by their first term which is basically $(u_v + d_v)$ where u_v and d_v denote the "valence" up and down quark distributions. The residue pieces in $F_{\pm N}^3$ does not have definite flavor quantum numbers. Hence $F_{\pm N}^3$ is, strictly speaking, not "non-singlet". That distinction applies to the combined neutrino and anti-neutrino structure function.

Neutrino and anti-neutrino structure functions are, *a priori*, different quantities. In the parton model they are closely related as shown above. It is common practice to form sums and differences of neutrino and anti-neutrino cross-sections within one experiment and extract the resulting structure functions. We begin with the case of **iso-scalar target** then consider the single hadron results. In the parton model, from the **sum of the cross-sections**, one obtains:

$$F^{2+}_{N}(x) = \frac{F^2_{+N} + F^2_{-N}}{2} = \Sigma\,[\bar{q}(x) + q(x)]\,,$$

$$F^{3+}_{N}(x) = \frac{F^3_{+N} - F^3_{-N}}{2} \approx 2\,[s - c + b] \qquad (II\text{-}20)$$

Thus the sum of cross-sections is dominated by the F^2 (F^1) structure function.

From the **difference of the two cross-sections**, one obtains:

$$F^{2-}_{N}(x) = \frac{F^2_{+N} - F^2_{-N}}{2} = s - \bar{s} - c + \bar{c} + b - \bar{b} \approx 0$$

$$F^{3-}_{N}(x) = \frac{F^3_{+N} + F^3_{-N}}{2} \approx u_V + d_V \qquad (II\text{-}21)$$

Hence *the difference of cross-sections is dominated by the F^3 structure function which measures the valence quark content of the (iso-scalar) hadron target.*

For the general case of **non-isoscalar target** (e.g. the proton), we obtain, for the **sum of cross-sections**:

$$F^{2+}_{A}(x) = \Sigma\,[D(x) + \bar{D}(x) + U(x) + \bar{U}(x)]$$

$$F^{3+}_{A}(x) = \Sigma\,[D(x) + \bar{D}(x) - U(x) - \bar{U}(x)] \qquad (II\text{-}22)$$

where the sum is over the flavor generations, *and* (U, D) denote the up and down quark distributions of any generation as before. For the **difference of cross-sections**, one gets:

$$F^{2-}_{A}(x) = \Sigma\,[D_V(x) - U_V(x)]$$

$$F^{3-}_{A}(x) = \Sigma\,[D_V(x) + U_V(x)] \qquad (II\text{-}23)$$

where $D_V = D - \bar{D}$, and similarly for $U_V(x)$. *We see that neither F_A^{3+} nor F_A^{1-} are necessarily small in general, being* in each case related to *the difference in the up and down quark distributions.*

C. QCD Parton Model Formulas

To **leading order in the QCD running coupling**, all naive parton model formulas given above remain valid in form with the proviso that all parton distribution functions acquire a Q-dependence dictated by the first-order QCD evolution equation (the Altarelli-Parisi Equation): i.e. $f_A{}^a(x) \longrightarrow f_A{}^a(x,Q)$.

Going **beyond the leading order**, as is increasingly necessary in current QCD parton model analyses, one must use *the perturbatively calculated parton structure functions* $\omega_a{}^i$ (cf. Fig.II-1) and the *corresponding parton distribution functions* $f_A{}^a(x,Q)$ generated with evolution kernels of the next order in QCD coupling. Thus in the next-to-leading order, the formula corresponding to Fig.II-1 reads:

$$F^i_{VA}(x,\, Q) = \omega^i_{Va}(x/\xi, Q, \mu) \otimes f^a_A(\xi,\, \mu) \qquad (II\text{-}24a)$$

where ⊗ indicates a convolution integral over the parton fractional momentum variable ξ ($x < \xi < 1$), i.e.

$$\omega \otimes f \equiv \int_x^1 \frac{d\xi}{\xi}\, \omega\left(\frac{x}{\xi}\right) f(\xi) \qquad (II\text{-}24b)$$

The *parton structure functions* $\omega_a{}^i$ in Eq.(II-24) are no longer simple constant coefficients, but are calculated in perturbative QCD to a given order in the effective coupling $\alpha_s(\mu)$:

$$\omega^i(z, Q, \mu) = \omega^i_0 \delta(1-z) + \alpha_s(\mu)\, \omega^i_1(z, Q/\mu) \qquad (II\text{-}25)$$

In Eq.(II-24), μ is a renormalization scale parameter which necessarily arises in perturbation calculations beyond the tree approximation. In practice, μ is usually *chosen* to be the physical variable Q, which then simplifies the expressions for $\omega_a{}^i$ to a perturbation series in which the coefficient function in each order depends only on one variable (cf. Eq.(II-27) below).

The functions $\omega_a{}^i$ have been calculated to one-loop order in the MS-bar subtraction scheme[2]. A well-known result is that the longitudinal structure function is no longer zero: both quarks and gluons give non-vanishing contributions to the sum over parton label (a) in Eq.(II-24) to this order. We have:

$$F^{Long}_{VA}(x, Q) = \frac{\alpha(Q)}{2\pi}\left[\omega_g\left(\frac{x}{\xi}\right) \otimes f^g_A(\xi, Q) + \omega_q\left(\frac{x}{\xi}\right) \otimes f^q_A(\xi, Q)\right] \qquad (II\text{-}26)$$

where

$$\omega_g(z) = 2\, N_f \cdot T_R \cdot 4z(1-z) \qquad \text{(gluon)}$$

$$\omega_q(z) = C_F \cdot 2z \qquad \text{(quarks)} \qquad (II\text{-}27)$$

Here C_F and T_R are the usual QCD group factors and N_f is the number of quark flavors. Since the quark distributions can, to a large extent, be determined from the leading-order analysis of F^2 and F^3, *the first-order QCD relation between F^{long} and the parton distributions, Eq.(II-25), can therefore be used to precisely define, and to determine the gluon distribution.* This is extremely important because knowledge on the gluon distribution is crucial in understanding the quantitative features of most hadronic processes at high energies, and that direct handles on the gluon distribution are very hard to come by. We shall come back to the above equations in a later section on HERA.

In addition to their dependence on the renormalization scale μ as indicated above, *the perturbative expressions for* ω_a^i are also *renormalization scheme dependent.* Although they are usually calculated in the **MS-bar scheme**, other schemes have been advocated in various applications as being more practical. Since *the physical structure functions* (the left-hand sides of Fig.II-1 and Eq.(II-24)) *must be independent of renormalization schemes*, the scheme-dependence of ω necessarily implies that *the parton distributions* $f_A^{\ a}(x,Q)$ *are also renormalization-scheme-dependent constructs beyond the leading order in QCD.* This fact is becoming increasingly relevant because most recent "QCD analyses" by both theorists and experimental groups use the next-to-leading order formalism. Although some of these analyses explicitly specify the choice of renormalization scheme, others implicitly use a scheme which may be different from the commonly accepted ones. (Cf. Sec.III.B) Hence, it is desirable to describe here explicitly the definitions of these schemes.

The **MS-bar scheme** does not require further elaboration; one simply apply Eqs.(II-24,25) with the published MS-bar 1-loop parton structure functions ω (Wilson coefficients).[2] In this scheme, all structure functions $F^i(x,Q)$ *acquire 1st order corrections of the form given in Eq.(II-24).* The alternative schemes (to be called **DIS schemes** from now on) take advantage of the freedom to shift finite terms between the two factors in Eq.(II-24), and *choose to define parton distributions such that one of the deep inelastic scattering structure functions retains its simple leading order relationship to these distributions.* Because $F^2(x,Q)$ is the best experimentally measured structure function, it is common in theoretical papers which elect to use this approach to choose this structure function for simplification. Explicitly, we have, for neutrino scattering off iso-scalar targets, as an example,

$$\begin{aligned} & F^2(x,Q) \equiv \Sigma\,(q^{(2)} + \bar{q}^{(2)}) \\ \text{and}\quad & F^1(x,Q) = F^2(x,Q) - 2\,F^{Long}(x,Q) \end{aligned} \qquad \text{(II-28)}$$

to 1-loop order, where $F^{Long}(x,Q)$ is given by Eq.(II-26). The superscript in parentheses makes explicit the scheme dependence of this definition of the parton distributions and identifies $F^{(2)}$ as the defining structure function.

There is an obvious alternative choice which picks the pure transverse structure function $F^1(x,Q)$ (Cf. Eq.(II-5) to retain its simple relationship to parton distributions. In that case, Eq.(II-28) becomes:

$$\begin{aligned} & F^1(x,Q) \equiv \Sigma\,(q^{(1)} + \bar{q}^{(1)}) \\ \text{and}\quad & F^2(x,Q) = F^1(x,Q) + 2\,F^{Long}(x,Q) \end{aligned} \qquad \text{(II-29)}$$

We shall call these two schemes DIS-2 and DIS-1 respectively. Note that *these differing schemes result in different definitions of parton distributions* (distinguished by the superscript in parentheses); *the physical structure functions must remain the same.*

In both of the above schemes, $F^3(x,Q)$ acquires a non-trivial 1-loop correction (similar to the expression for $F^{Long}(x,Q)$) which we shall not reproduce here. Since the helicity structure functions $F^{R,L}(x,Q)$ are linear combinations of $F^{1,3}(x,Q)$ (cf. Eq.(II-6)), they both have non-trivial corrections in these schemes. It is however possible to define two more distinct schemes (say, DIS-R and DIS-L) such that either $F^R(x,Q)$ or $F^L(x,Q)$ is given simply as sums of anti-quark or quark distributions respectively. This is an important point in the experimental determination of anti-quark distributions in neutrino scattering because the "leading" term and the 1st order "correction" can be numerically comparable, as we shall see in Sec.III.B.

It is worth noting also that, *in contrast to the situation with the MS-bar scheme, Eqs.(II-28,29) and variations thereof cannot be used to define "DIS schemes" beyond the 1-loop order. Both* $F^1(x,Q)$ and $F^2(x,Q)$ must acquire higher order corrections beyond 1st order. In Sec.III.B we shall review the conventions used by the major experimental groups, and in a later section on parton distributions (Sec.IV.C) we shall discuss the numerical differences between the MS-bar-scheme distributions and their counterparts in the DIS-2 scheme.

Because of the inter-dependence of $f_A^{\ a}(x,Q)$ and the parton cross-section $\hat{\sigma}$, Eq.(I-1), and structure functions $\omega_a^{\ i}$, Eq.(II-24), any meaningful application of the QCD Parton Model to high energy physical processes **beyond the leading order** must use parton distribution functions *defined in the same scheme and calculated to the corresponding order in perturbation theory* as the latter. In particular, to match *next-to-leading* hard cross-sections in Eqs.(I-1) and (II-24), one must use Q-dependent parton distributions derived with *two-loop evolution kernels.* Otherwise, the formulas will not be consistent. New results on next-to-leading order parton distributions will be reviewed in Sec.IV.C.

III. Review of Measurements of Structure Functions

Major strides have been made in the last few years in obtaining very high statistics data on deep inelastic scattering structure functions from electron-, muon- and neutrino-experiments on a variety of targets. In spite of this impressive progress, however, some noticeable differences remain which have caused a great deal of attention and concern. This section briefly reviews the status of the experimental measurements of structure functions and highlights any significant disagreements. In section V, we will comment on the implication of these differences on the extraction of parton distributions.

A. Review of Recent Experimental Results

Tables III-1 and III-2 summarize some key features of the more recent high statistics muon and (charged-current) neutrino experiments, including their kinematic ranges and important corrections applied. An understanding of the latter is extremely important if results from these experiments are to be combined in any quantitative model analyses, as there exist considerable differences in definitions and practices among the experiments. The second part of this section addresses some of these practices.

MUON EXPERIMENTS			
	BCDMS	BFP	EMC
Target	C and H_2	Fe	H_2 D_2 Fe
Energy	100 - 280	93, 215	120 - 280
x-range	.06 - .80	.08 - .65	.03 - .65
Q^2-range	25 - 280	5 - 220	3 - 200
# events	C: 680K	690K	Fe: 1080K
$R(x,Q^2)$	Expt.	0.0	0.0

Table III-1: Major recent Muon Experiments.

NEUTRINO EXPERIMENTS				
	BEBC	CCFRR	CDHSW	CHARM
Target	Ne H	Fe	Fe	Marble
Energy	10 - 200	30 - 250	30 - 300	10 - 200
x-range	.025 - .80	.02 - .65	.02 - .65	.02 - .55
Q^2-range	2 - 70	1 - 200	0.2 - 200	0.2 - 100
$R(x, Q^2)$	R(QCD)	R(QCD)	R(QCD)	0.1
# Events	25K	170K	940K	160K
SU(3) symmetry	$\bar{s} = 0.25\,(\bar{u} + \bar{d})$; $c = \bar{c} = 0$		$\bar{s} = 0.2\,(\bar{u} + \bar{d})$; $c = \bar{c} = 0$	
Charm	slow rescale: m = 1.5		No correction	

Table III-2: Major recent charged-current Neutrino Experiments.

It is obvious from the above tables that these experiments are very extensive in their coverage. The cross-section measurements furnish, in principle, detailed information on the structure functions $F_{\gamma,\nu}^{2,3}(x,Q)$ and $F_{\gamma,\nu}^{long}(x,Q)$ (cf. Sec.II.A). The best determined structure functions *are* $F_{\gamma,\nu}^{2}$ which correspond to combinations of the sum of quark and anti-quark distributions in the parton model. Unfortunately, because F_{ν}^{3} and $F_{\gamma,\nu}^{long}$ are determined by differences of cross-sections (Eqs.(II-5,6), hence the errors are more substantial. Similar limitations have so far discouraged efforts to separate the helicity structure functions which contain much more specific information on the parton content of the hadron. (Cf. Eq.(II-10))

Before comparing these measurements, it should be noted that *differences are expected* due to systematic effects, which are experiment dependent, and due to the different kinematical regimes covered by experiments. The impact of this last point is shown in Fig.III-1. Note that at the same value of x the average value of Q^2 can differ by as much as an order of magnitude between various experiments. Care must be taken to remove this "natural" difference before comparing measurements.

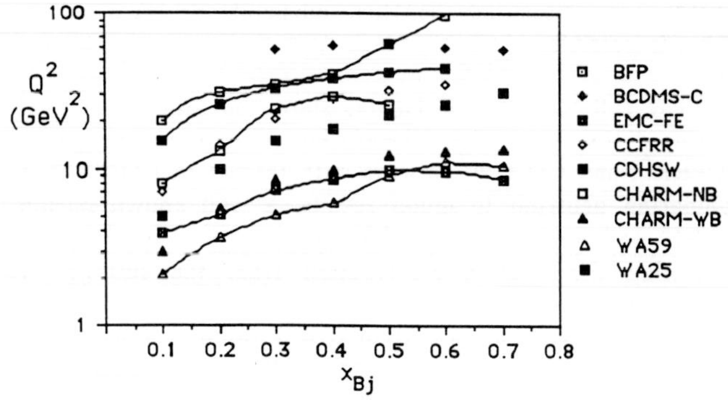

Fig.III-1: The dependence of $<Q^2>$ on x for the various experiments.

$F^2(x,Q)$: Heavy Targets Experiments

Because of the relatively small neutrino cross-section, most of the high statistics experiments have used heavy targets (i.e. iron) detectors. Muon experiments, on the other hand, can get sufficient statistics even with hydrogen or deuterium targets. We will discuss the relation between heavy and light target results — the "EMC-effect" — shortly, but first we examine the ratio of the structure function $F^2(x,Q)$ as measured by the heavy target experiments. The black points on Fig.III-2 represent this ratio between the two high statistics muon experiments, EMC on iron[3] and BCDMS on carbon[4].

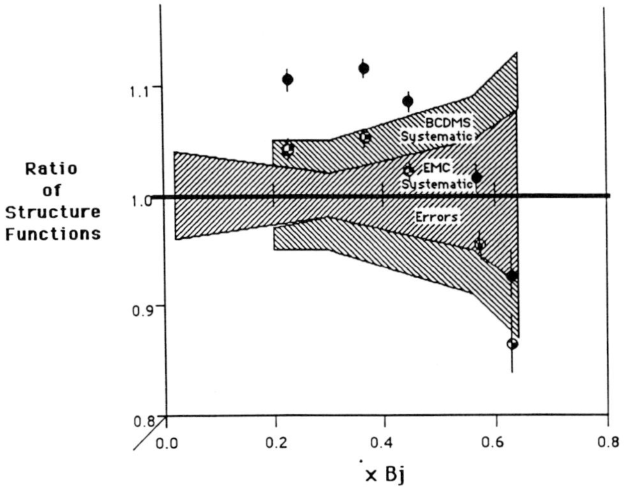

Fig.III-2: The Ratio of $F^2(x,Q)$ measured by BCDMS and EMC on heavy targets (see text).

The error bars on the individual points are statistical and the systematic error from each experiment is shown in the cross-hatched area. The black/white points on the plot show the effect of a 5% change in relative normalization between the two experiments. There are no low x points since the high minimum Q of the BCDMS carbon experiment translates to a minimum x of about 0.2. The first thing one notices is the x-dependent trend of the ratio. However, the sum of systematic errors also depends on x. As illustrated by the black/white points, a 5% normalization change brings the "disagreement" within the band of systematic error. We note that, even though both experiments were done in the same beam at

CERN, each measured the flux independently so a relative offset is possible.

The following set of figures show the ratio of $F^2(x,Q)$ measured by the other heavy target experiments BFP[5], CCFRR[6], and CDHSW[7], always with respect to EMC. When comparing neutrino to muon results, a well-known parton-model motivated constant 5/18 has been applied, hence neglecting the small-x contribution from the strange and charm sea quarks to this factor (cf. Eqs.(II-14,18)). The systematic errors for the CCFRR and BFP points are not shown (they should be comparable to the EMC error shown in the plots). In all cases, a shift in scale of a few percent statistically eliminates any discrepancy.

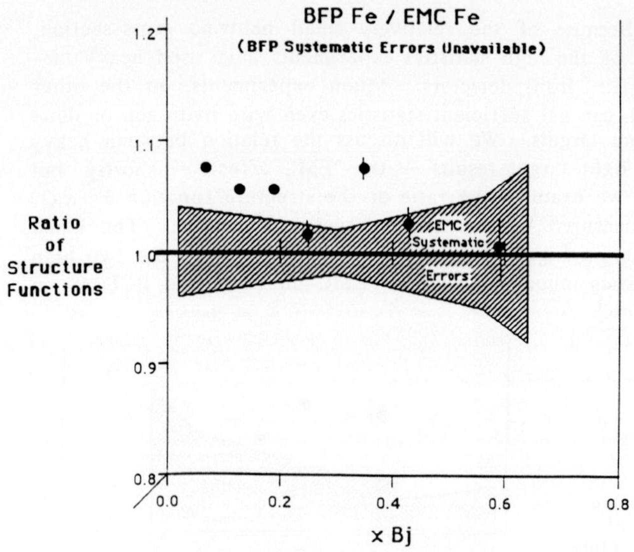

Fig.III-3: Ratio of BFP and EMC structure functions.

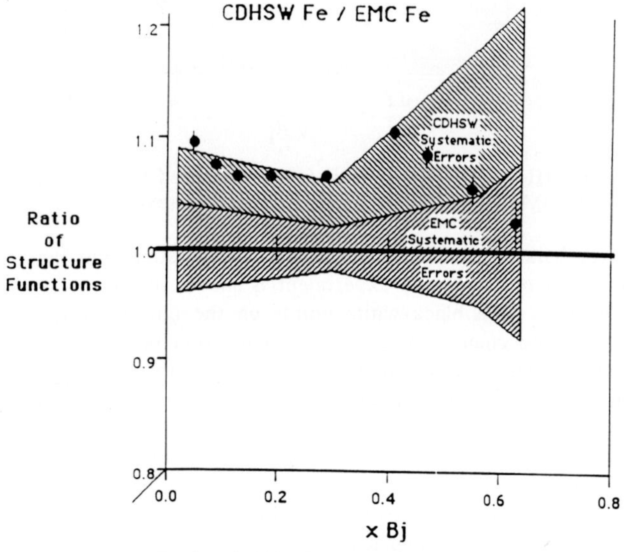

Fig.III-4: Ratio of F2(x,Q) CDHSW and EMC experiments.

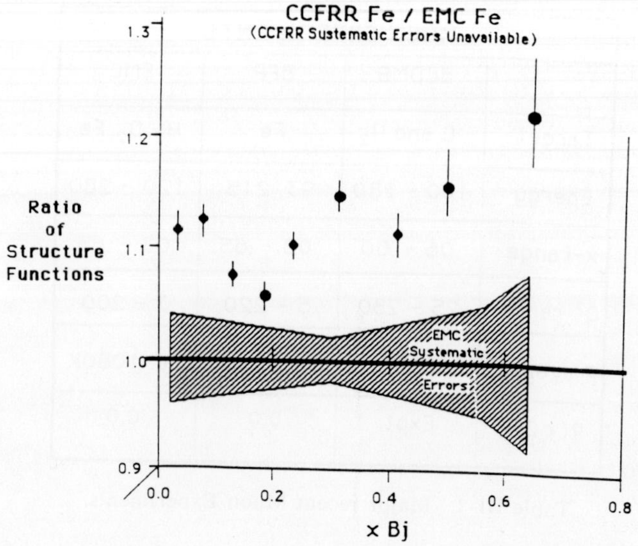

Fig.III-5: Ratio of $F^2(x,Q)$ CCFRR and EMC experiments.

This report will concentrate on charged-current structure functions where detailed measurements exist. We shall only mention here that there is one new result from the Fermilab-MIT-MSU neutrino collaboration (E594)[8] which measures the neutral-current structure functions and compares the resulting quark distributions with those found in charged-current experiments. The comparison bears out the consistency of the quark parton model.

$F^2(x,Q)$: Hydrogen Targets Data

Fig.III-6 shows the ratio of $F^2(x,Q)$ as measured by the two muon experiments BCDMS[9] and EMC[10] using a hydrogen target in each case. There is a pronounced x-dependent trend in this ratio similar to the heavy target case shown earlier.

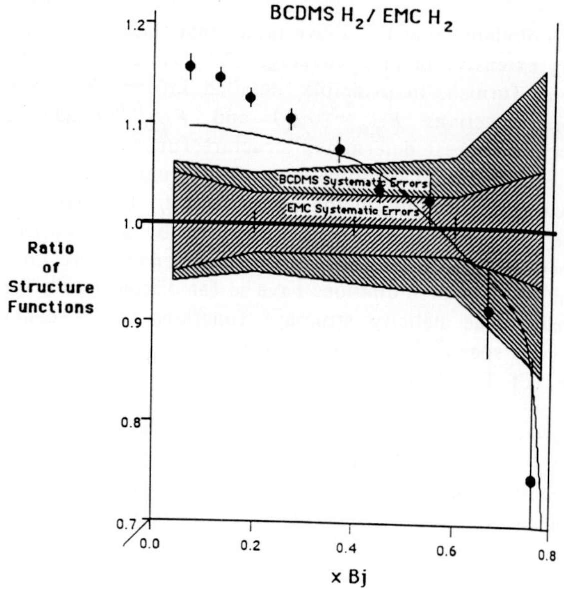

Fig.III-6: The Ratio of $F^2(x,Q)$ measured by BCDMS and EMC on hydrogen target (see text).

However, in this case, no shift in relative normalization can eliminate the differences; there is clearly a statistically significant difference between these two high statistics experiments. The curve drawn represents an attempt by R. Mount[11] to simulate the ratio by assuming a 10% relative normalization mismatch *and* an assumed error in the BCDMS muon energy measurement of 0.5 GeV + 0.6% E_μ. We see that even these extreme assumptions cannot force an agreement between the two sets of data.

Status of the "EMC-Effect"

New data on non-trivial x-dependence in the ratio of $F^2(x,Q)$ measured on heavy targets and that on light target (deuterium)[12] come from SLAC E140[13] and EMC'[14]. These new results are plotted along with earlier ones in Fig.III-7. They confirm the general feature that the ratio is below 1 at very low x, rises above 1 toward a maximum around x ≈ 0.15, and then steadily decreases until x ≈ 0.7 before rising again.

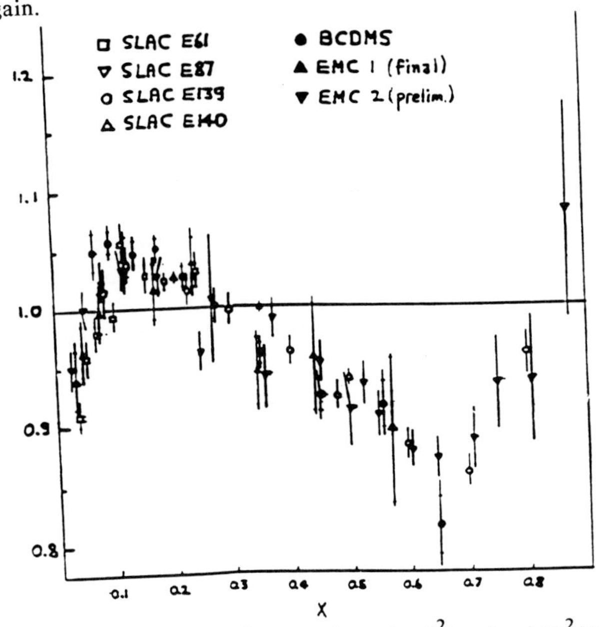

Fig.III-7: Combined data on the ratio $F^2(\text{nucleus})/F^2(D_2)$

There have been many attempts to explain this effect. It is generally agreed that shadowing effect is important below x = 0.1; and nuclear effects are responsible for the observed effect above 0.1. For detailed model studies and earlier references, we refer the reader to two recent papers on this subject.[15,16] From a phenomenological point of view, it is obvious that this experimental ratio must be taken into account in any serious QCD parton model analysis which makes use of heavy target data on structure functions.

In contrast to the generally agreed x-dependence, experimental evidence on possible Q-dependence is unclear. F. Taylor[17] recently tried to fit current data to the ansatz:

$$\frac{d\,(F_2^{Fe}\,/\,F_2^{D})}{d\,(Ln\,Q^2)} = (.077\pm.023\pm.047) - (.25\pm.09\pm.14)\,x \qquad \text{(III-1)}$$

(± statistical ± systematic). The data and fit are shown in Fig.III-8.

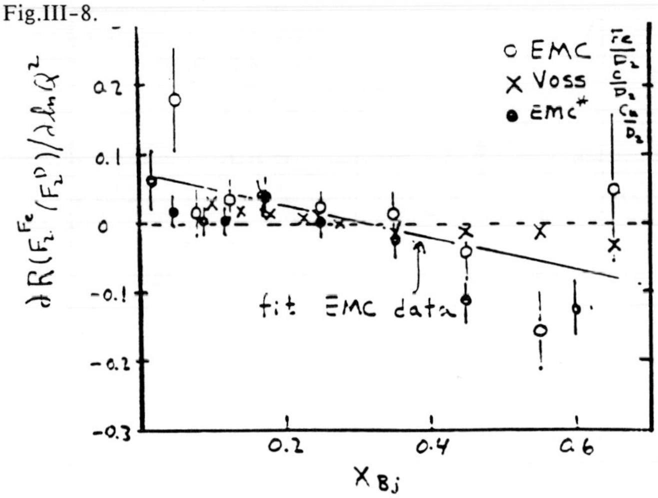

Fig.III-8: Possible Q-dependence of the EMC ratio.

The Longitudinal Structure Function

There have been numerous experiments attempting to measure the ratio $R(x,Q) = F^{long}(x,Q)/F^2(x,Q)$. An indication of the accuracy of the current measurements is shown in Fig.III-9.

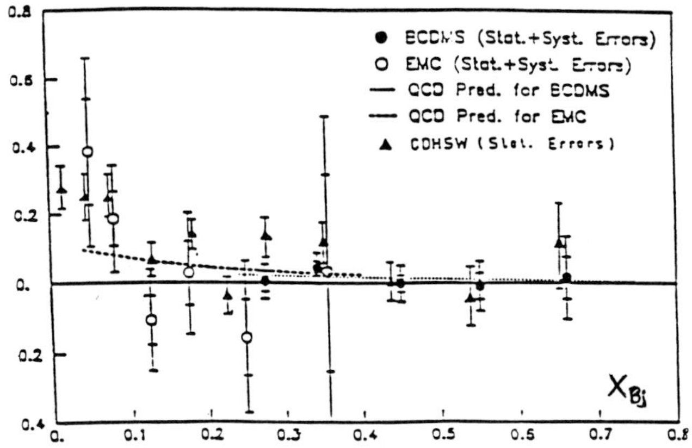

Fig.III-9: R(x) as measured by CDHSW, EMC and BCDMS.

A new impressive effort by the SLAC E140 experiment obtained very good data in the low-Q region. The measured ratio is much larger than that given by the QCD formula Eq.(II-26), as is known for a long time. Any understanding of these data must go beyond the traditional twist-2 perturbative QCD, as mass threshold and other "higher-twist" effects still play an important role. Unfortunately, there is no real theory on these effects. An *ad hoc* prescription for including the so-called target mass effect seems to fit the data quite well, as seen in Fig.III-10.[18]

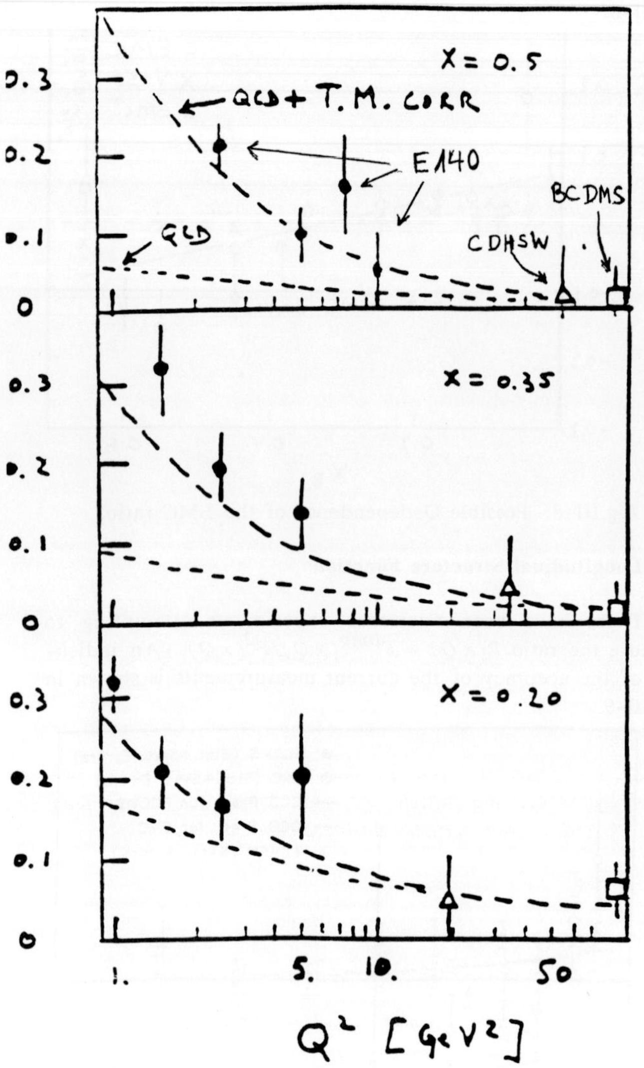

Fig.III-10: Q-dependence of the Ratio R.

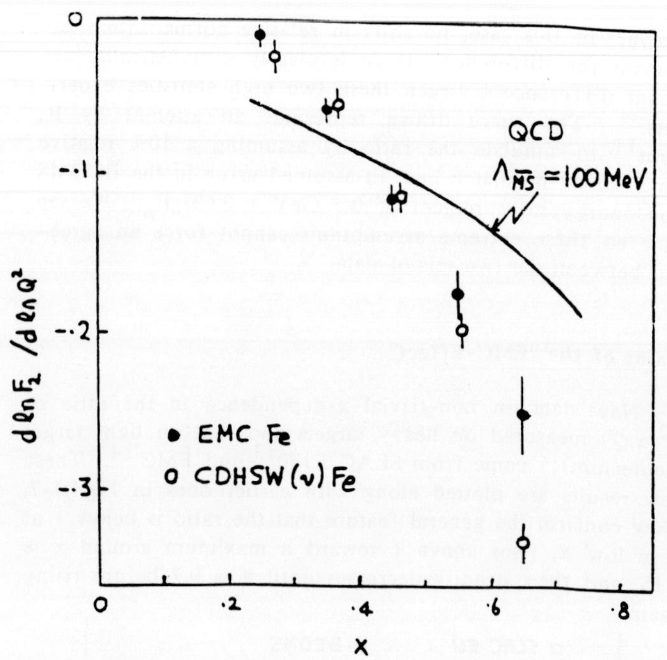

Fig.III-11: Slope of Q-dependence: EMC & CDHSW vs. QCD.

In contrast to the above, the recent BCDMS carbon and hydrogen data yield slope functions perfectly consistent with QCD. The result is reproduced in Fig.III-12. The BCDMS group obtained good fits to QCD formulas (cf. Sec.II.B,C) with QCD coupling parameter Λ_{QCD} slightly over 200 MeV in the MS-bar scheme. Consistent values were obtained by fitting to data above x = 0.25 using the non-singlet formula without the gluon and by fitting to the data over the entire x range including gluon effects.[21]

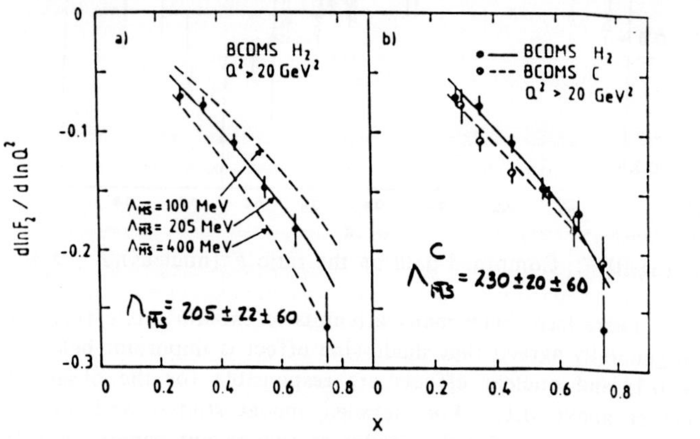

Fig.III-12: Slope of Q-dependence: BCDMS vs. QCD.

Q-dependence of $F^2(x,Q)$ and QCD

The Q-dependence of the structure functions is usually characterized by their logarithmic derivative, e.g. $d(Ln\ F^2(x,Q))/d(Ln\ Q2)$. It is well known that this slope function has a definite x-dependence which is often taken as a good test of QCD predictions on the Q-evolution of structure functions. In the last two years, a lot of attention has been drawn to the fact that this measured slope function from the EMC and CDHSW experiments do not agree well with QCD.[19] Fig.III-11 shows the comparison between the experimental values and "the best fit" to next-to-leading order QCD. It has been pointed out[20] that as the minimum Q of the data is raised, the quality and the stability of the fit improve dramatically.

A possible way to reconcile the QCD slope predictions with the EMC and CDHSW data was suggested by F. Taylor: if one applies a Q-dependent heavy-target correction to the structure function, as mentioned in the section on the "EMC-effect", one can write,

$$\frac{\partial\ Ln\ F_2^{Fe}}{\partial\ Ln\ Q^2} = \frac{\partial\ Ln\ F_2^{D}}{\partial\ Ln\ Q^2} + \frac{\partial\ Ln\ R^{Fe/D}}{\partial\ Ln\ Q^2} \qquad (III\text{-}2)$$

Then the QCD fit to the EMC data is improved as shown in Fig.III-13. A similar improvement was found for the BFP fit.

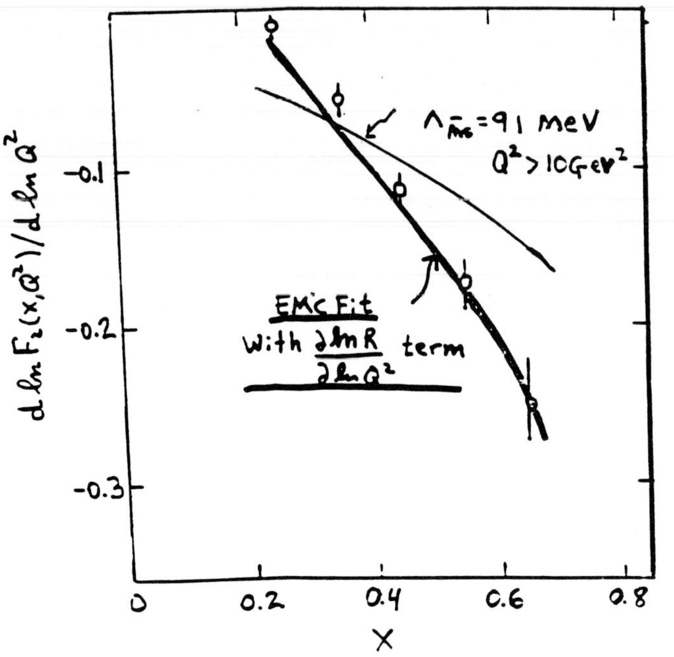

Fig.III-13: QCD fits to the EMC Fe data with and without the Q-dependent heavy-target correction.

B. Review of Analysis Assumptions and Corrections

General Remarks

As seen in the previous section, the statistical power of the world's data on deep inelastic scattering structure functions has increased enormously. Because of the diminishing statistical error, correlations, systematic errors, and small differences in analysis assumptions become the dominant factors in the comparison of experimental data.

The major sources of systematic error — beam flux, energy calibration and detector acceptance — are strongly correlated from (x, Q) bin to bin. Thus, strictly speaking, they cannot be simply added to the statistical errors as they are not independent. For most of the recent experiments (but not all), information on separate systematic errors, including the sign of the effects, are available. In neutrino experiments, the statistical errors on $F^2(x,Q)$ and $F^3(x,Q)$ are also correlated for each bin. Thus, the correlation coefficient is needed in a comprehensive global fit.

Each major experiment adopts a unique procedure to extract structure functions from measured cross-sections. In order to make meaningful use of the published results, close attention must be given to the analysis assumptions used in the various experiments as they are often not the same. We briefly review procedures and pratices used by the major experiments which are most relevant to QCD parton model analyses, and critically examine some of them in this section.

When individual experiments perform a "QCD analysis" of their measured structure functions, most of the experimental considerations mentioned above are, of course, taken into account. There is, however, the question of definitions which are not standardized. In particular, the precise meaning of "Λ_{QCD}" and the definition of anti-quark distributions vary in the published papers when "2nd order QCD analyses" are involved (Cf. Sec.II.C). The experimental considerations become a real problem when data sets are combined to make global fits by third parties. Since a lot of the relevant information are not even available in published form, the common practice is, more often than not, to ignore them. This clearly can lead to problematic results with uncertain errors.

In principle, measurement of the deep inelastic scattering structure functions should be independent of any theoretical model assumptions, as emphasized in Sec.II.A. For instance, six independent structure functions could be separately measured from the y-distributions of neutrino and anti-neutrino scattering for fixed (x,Q). In practice, because experiments have limited coverage and/or efficiency over the kinematic range, only two combinations of cross-sections (roughly speaking, the sum and differences of neutrino and anti-neutrino cross-sections) are normally used to extract the published structure functions $F^2(x,Q)$ and $F^3(x,Q)$ respectively. When doing this, parton-model based expressions for the other four structure functions are introduced as corrections. Since the structure functions are, in turn, used as the main source of information to determine parton distributions, this is at best an iterative procedure. For serious QCD analyses, it clearly raises the concern of being a circular process. An open question is: *which of these corrections are really necessary, and which are used because of "accepted practice" and of over-reliance on the parton model.*

"Strange Sea" and "Isoscalar" Corrections

Specifically, the structure functions usually quoted in the literature correspond to, in the notation of Sec.II, $F^{2+}=(F^2{}_+ + F^2{}_-)/2$ (Eqs.(II-20,22)) and $F^{3-}=(F^3{}_+ - F^3{}_-)/2$ (Eqs.(II-21,23)). In order to extract these functions, the following procedure is usually adopted to treat the other combinations: (i) $F^{Long}(x,Q)$ is estimated from independent experimental measurements (y-distribution) or from the QCD formula Eq.(II-26) using assumed $F^2(x,Q)$ and $G(x,Q)$ functions, or is set equal to some small constant; (ii) since F^{2-} is given by $(d_v - u_v)$ in the parton model (Eqs.(II-21,23)) it is expected to vanish for iso-scalar targets. Hence for targets close to being iso-scalars, it is assumed to be small; and an estimated **iso-scalar correction** is used to account for its contribution to the cross-section; (iii) the function F^{3+} is related in the parton model to the difference between all down-like quarks and up-like quark distributions (Eq.(II-22)), hence is proportional to the difference between the strange and charm quark distributions for iso-scalar targets (Eq.(II-20)). Since this is again expected to be small, an estimated **strange-sea correction** is applied to the cross-section to remove its effect. Using these (model-dependent) assumptions, no information on the y-distribution is needed for the determination of the two main structure functions.

"Slow Rescaling" Correction and Charm Threshold

The QCD parton model provides an unambiguous formalism only when Q is much larger than the mass scales of the charm threshold, which leads to complications in the parton-model interpretation of the structure functions for Q values of a few GeV. It is important to emphasize that *this complication only arises in the parton picture, it is not intrinsic to the determination of structure functions as basic physical quantities* as underlined in Sec.II.A. Hence there is no compelling physical reason to make a correction for this effect when structure function are measured.

In neutrino scattering, charm production is 5-10% of the total cross-section. There are two ways to confront the charm threshold problem when one makes a QCD parton analyses of charged-current structure functions. Either one imposes a Q^2-cut sufficiently above the charm quark mass so that the charm quark behaves as a light quark, threshold effects (this is one form of "higher-twist" effect) can be neglected, and the QCD formalism applies unambiguously; or one can apply a charm-mass correction to the structure functions with the purpose of preserving the parton language even at low Q but at the risk of working with highly model-dependent parton distributions. The choice between these two alternatives is clearly subjective. The analyses to be presented in later parts of this report use Q^2-cuts in the selection of data, and closely monitor the dependence of the physical output on the choice of the cut.

CCFRR and BEBC[22] choose to "correct" their measured structure functions for the charm-mass effect. The most often used method is termed the **slow rescaling correction.** As this directly affects the comparison of published experimental results, we give a brief summary of the commonly used procedure to do this.[23] First, in order to ensure that the final-state (charm) quark in the parton structure function ω_a^i (Fig.II-2a) be an "on-shell" parton, the momentum fraction of the initial parton (*s* or *d*) becomes

$$\xi = x\ (1 + M_c^2/Q^2)$$

rather than the familiar Bjorken x. Parton distribution functions are written as functions of (ξ,Q) instead of (x,Q). Secondly, to take into account the existence of a minimum energy for charm production, a threshold multiplicative factor of $\Theta(1-xyM_c^2/\xi Q^2)$ is introduced for those terms leading to charm production in the parton formulas for the structure functions. For example, the normal parton expression $x\ s(x,Q)$ is replaced by

$$\cos^2\theta_c\ \xi\ s(\xi,Q)\ \Theta(1-\frac{xyM_c^2}{\xi Q^2}) + \sin^2\theta_c\ x\ s(x,Q)$$

and similarly for the other flavors.

CCFRR and BEBC *correct* for charm-mass effects to arrive at structure functions "$F^2(\xi,Q)$" and "$F^3(\xi,Q)$" which, according to this procedure, are interpreted as the equivalent zero-charm-mass structure functions (which have the usual simple relations to the "parton distributions" $f(\xi,Q)$ in this scheme). CDHSW, on the other hand, defines their structure functions in the standard model-independent way as formulated in Sec.II. Vallage[30] made an explicit comparison of these two different definitions of structure functions. His result is shown in Fig.III-14. The differences at low Q^2 are obviously significant!

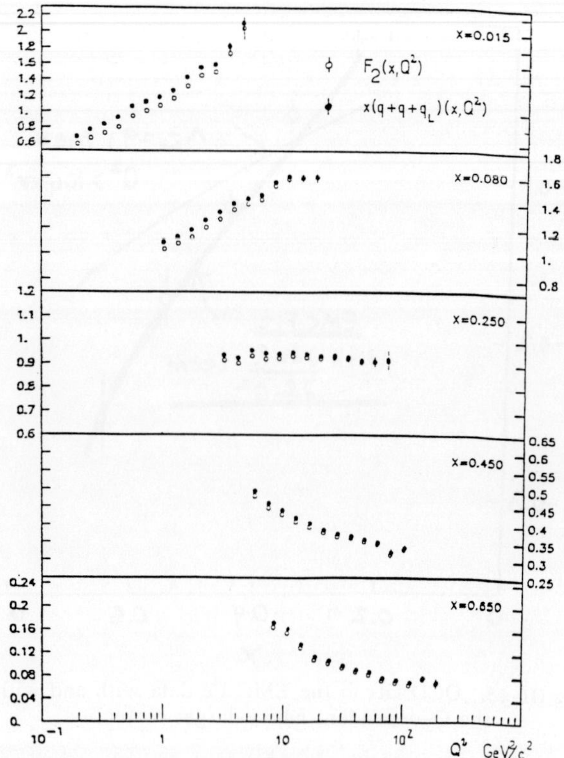

Fig.III-14: Comparison of $F^2(x,Q)$ and "slow-rescaling corrected" $F^2(x,Q)$.

Because there is no reliable quantitative theory for threshold effects (these are fundamentally non-perturbative), and because the significance of structure function transcends the parton model, it is clearly desirable to retain the model-independent definition. Without standardization, published data on "structure functions" from different experiments actually correspond to different quantities, hence cannot be used in a meaningful way by anyone except the original experimental group. Thus, the potential usefulness of the laboriously obtained experimental data can be severely limited by model-dependent corrections. We strongly recommend that all experiments publish the standard structure functions *without the slow-rescaling* correction.

It is, needless to say, *interesting and tempting to explore parton-model interpretations of the measured structure functions even to low Q values, but that exercise should be conceptually distinguished from structure function measurements*. In this case, the charm threshold problem must be faced and some form of slow-rescaling correct has to be applied. Since this correction usually requires information on the beam spectrum and the weighting of events which go beyond the structure functions, *it would be extremely helpful if the published experimental papers also include tables of the measured cross-sections*. This will maximize the usefulness of the experimental results, clarify the parton model analysis by the experimental groups (and others using their results), and minimize the possibility of unnecessary confusion and controversy.

Conventions on Quark Distributions

Most major experiments perform some form of QCD analysis and determination of parton distributions with their data. There are many subtleties and potential sources of confusion both due to intrinsic physics considerations and due to the different procedures and definitions used by different groups.

One problem concerns the treatment of the longitudinal structure function. Most experiments either measure $F^{Long}(x,Q)$ independently or use an estimate for it in their determination of the other structure functions. Thus, experimentally $F^1(x,Q)$ and $F^2(x,Q)$ are not equal: they differ by $2 \cdot F^{Long}(x,Q)$. In leading order QCD, on the other hand, F^{Long} vanishes and F^2 *and* F^1 are supposed to be equal. Thus *parton model analyses at this level has an inherent ambiguity in their determination of the parton distributions*, depending on which of these experimental structure functions is used in the calculation. This results in real differences in that the muon experiments (BCDMS, BFP, and EMC) and CHARM[24] identify the sum of quark and anti-quark distributions with $F^2(x,Q)$ whereas CCFR and CDHSW identify this combination of parton distributions with $F^1(x,Q)$.

To avoid this ambiguity, one must go beyond the leading order in QCD. Then the relationship between the structure functions and the parton distributions are no longer simple, and careful attention must be given to the explicit definition and the consistent use of a renormalization scheme, as explained in Sec.II.C. Since this level of the QCD parton model has not yet become a part of the folklore, confusion frequently arise when experimental results and naive parton model parlance are freely used in some detailed QCD analyses.

A specific case in point is the determination of the "anti-quark distribution" in neutrino experiments. The measured **physical quantity** is the coefficient of the $(1-y)^2$ term in the neutrino cross-section which, according to Eq.(II-7), is just the **right-handed helicity structure function** $F^R(x,Q)$. Experimentally, this quantity is relatively small compared to $F^2(x,Q)$; it is *numerically comparable to* $F^{Long}(x,Q)$ at low values of Q. However, in the QCD parton model formalism, $F^R(x,Q)$ is formally of *leading order, in contrast* to $F^{Long}(x,Q)$. To this order, it is simply proportional to the sum of anti-quark distributions as given by Eqs.(II-10,16). The ambiquity described in the previous paragraph now comes into full play when one tries to determine the quark distributions by combining $F^R(x,Q)$ with $F^2(x,Q)$ or $F^1(x,Q)$ -- *the ambiguous term* (i.e. $2 \cdot F^{Long}(x,Q)$) *is numerically comparable* to $F^R(x,Q)$, supposedly the "anti-quark distribution" one is trying to subtract.

As mentioned earlier, the obvious way out of this dilemma is to go to the next order in QCD where this ambiguity is absent. In that case, however, one must accept the fact that the direct relationship between $F^R(x,Q)$ and the anti-quark distributions is drastically modified in general — the 1st order "correction term" to the naive parton formula which links the two may be comparable to the "leading term" because the latter starts numerically small (cf. Eq.(II-28,29) and the two paragraphs following these equations). One can, as explained in Sec.II.C, choose a convention (i.e. the DIS-R scheme) in which the anti-quark distributions are still simply related to $F^R(x,Q)$. In doing so, however, all other structure functions acquire non-trivial corrections.

For next-to-leading order QCD analyses, all muon experimental groups use the DIS-2 scheme definition of parton distributions (in the terminology of Sec.II.C) for which $F^2(x,Q)$ retains its simple interpretation as the sum of quark and anti-quark distributions, but all other structure functions acquire non-trivial corrections to 1-loop order. The CHARM collaboration also explicitly use this scheme in their QCD analyis. The other neutrino experiments have not yet presented full next-to-leading order QCD analyses. To the extent that they usually identify $F^1(x,Q)$ with the sum of quark and anti-quark distributions, they are effectively using the DIS-1 scheme defined in Sec.II.C. As pointed out in the previous paragraph, the *relationship between the anti-quark distribution and the measured* $F^R(x,Q)$ *is not simple in either of these schemes.* Users of the quark and anti-quark distributions extracted by the experimental groups should be very careful in ensuring that the convention used in the experimental analysis is consistent with the applications.

We close this section on review of experimental practices with one more observation. In view of the observed target-dependence of the structure functions (i.e. the EMC-effect, cf. Sec.III.A), the CHARM collaboration makes a *correction* to their measured structure functions based on the measured A-dependence from the muon experiments.[12] None of the other experiments adopt this practice. This appears to be wise, for the same reasons as those described in the section on charm mass corrections. Such corrections are best applied at the time of specific applications.

C. Database on Structure Functions

One of the main purposes of this study group is to compile a consistent database of measured structure functions which can be used in further quantitative QCD analyses. The work on extracting parton distributions,to be reported in Sec.V, is an example of such a analysis. Our goal is to produce computer readable files with structure functions and error estimates for all recent experiments in a common format for easy access and to have available other relevant information on analysis methods for easy comparison.

We would like to include in the data base: the structure function values, the statistical errors, any statistical correlation with other values (e.g. $F^2(x,Q)$ and $F^3(x,Q)$), and separate error estimates for each of the major sources of systematic error for each data point. Each file will be accompanied by a list of the analysis assumptions used (such as those discussed in previous subsections) in the extraction of the structure functions.

As noted above, in structure function measurements, the major sources of systematic error are often calibrations, acceptance, or normalizations which are highly correlated from point to point. With separated systematic errors, one can better assess the effect of experimental uncertainties (say, a 1% shift in beamline energy calibration) on the determination of physical quantities (such as Λ_{QCD}).

We currently have essentially complete files for some experiments (BCDMS, CDHSW, and EMC). Information on the other experiments are not complete because they are not published and are not yet accessible to us. We strongly urge experimental groups to publish their data and to include as much complete information on systematic effects and cross-sections as possible.

IV. Review of Parton Distribution Functions

Deep inelastic scattering Structure Functions are the primary source of our knowledge on the Parton Distribution Functions inside a hadron. The extraction of the latter from the structure functions is based on the QCD formula (II-24) (cf. Fig.II-1). The extensively used parton distribution functions of Gluck-Hoffman-Reya,[25] Duke-Owens,[26] and Eichten-Hinchliffe-Lane-Quigg (EHLQ) [27] were deduced from experimental data available up to the early 1980's; they were based on the leading order approximation to the QCD formula plus specific assumptions about the form of the initial distributions at a given Q_0. Recently, two new sets of distributions, based on next-to-leading order QCD calculations and on comparison to newer experimental data, have been made available to the high energy physics community.[28,29]

In order to assess the reliability of these distributions as input to quantitative QCD Parton Model applications and as tools for extrapolation from the measured fixed target energy range (the shaded area in the x-Q plot of Fig.I-2) to the kinematic region of current and future collider processes (cf. Fig.I-2), we examine: (i) how well do the available parton parametrizations reproduce the measured deep inelastic scattering structure functions (as reviewed in Sec.III); (ii) what are the important factors which must be considered in a meaningful fitting program to extract the parton distributions from DIS experiments, and whether they have been taken into account in existing fits; and (iii) whether the existing next-to-leading order calculations are consistent and reliable, especially in the small-x region where the evolution kernel has a singular behavior. These points have not been systematically studied in the published literature so far. All of them have significant bearing on the central question of concern: how much does current data constrain the behavior of the parton distributions in the extrapolation to small-x and large Q? The small-x question is especially important because of the dearth of firm theoretical knowledge on this issue, and because all interesting high energy cross-sections depend sensitively to this extrapolation.

A. Comparison of Known Parton Distributions with Data

We have systematically compared recent data on structure functions with parton model calculations based on the standard parton distribution functions mentioned above. Space limitation only permits the presentation of typical results from representative data sets and parton parametrizations. In Fig.IV-1 we show the high statistics BCDMS-H_2, CDHSW-Fe,[30] and EMC-H_2 structure functions $F_2(x,Q)$ plotted against Q for various different x-bins. For each experiment, data are divided into two plots: those with x < 0.3 and with x > 0.3. To avoid overlapping data points and for clarity of display, the small x plots correspond to $F^2(x,Q)/x$ in log scale, whereas the large x plots correspond to $xF^2(x,Q)$ in linear scale. The 4 curves shown on each plot are obtained from the parton model formulas, Eqs.(II-14) and (II-18), using the parametrizations of EHLQ-1, Duke-Owens-1, MRS-E, and DFLM-FXAVR respectively. We see significant departure of some of these curves from the experimental data, for small-x in general, and for moderate x as well in the case of the BCDMS data. Not shown here are the corresponding results for $F_3(x,Q)$. The agreement between data and existing parametrizations there is much better, although the error bars on the experimental data points are considerably bigger.

Since the curves shown above are obtained the same way most "predictions" of the QCD-parton model are calculated, these graphs give a reasonable over-all picture of the quality of existing parton distributions *in the currently accessible energy range.* (Extrapolation is not involved here!) It is not possible to discuss here in detail the sources of the observed disagreement. The main factors are: (i) most standard parametrizations were obtained by fitting only one, or few, set(s) of data; (ii) in some kinematic regions, the new data sets are different from those of the early 1980's on which the first-generation parametrizations were obtained; and (iii) most authors allow a relative normalization scale between data points from different experimental sets in the fitting procedure, which cannot be included in the above comparison or any practical application. A notable example of how the discrepency arises is given by the widely used EHLQ distributions which were derived primarily from the 1983 CDHS data. The latter were obtained using values for the total cross-section which later mearsurements have determined to be around 10% too low.[31] Thus the EHLQ distributions lead to structure functions conspicuously away from the new data points as shown in Fig.IV-1.

The above comparison makes the need to determine more up-to-date parton distributions obvious. Also apparent from the plots is the fallacy of the often used statement to the effect that *parton model predictions based on different existing parton parametrizations should bracket the true answer and give an estimate of the theoretical uncertainties.*

B. Experimental Issues on the Extraction of Parton Distributions

In order to extract reliable parton distributions (inside the proton) from the measured structure functions in a quantitative analysis, it is, in principle, necessary to take into account: (i) systematic errors of the experimental data used, (ii) corrections due to differences of structure functions on heavy targets from those on light targets (the "EMC effect"), (iii) significant correlations of structure functions measured in the same experiment, and (iv) the variety of "corrections" applied by the experimental groups in extraction of the structure functions (which are often different from experiment to experiment). (Cf. Sec.III) In practice, a review of the original papers reveal that previous efforts to extract parton distributions, with few exceptions, mostly ignore these

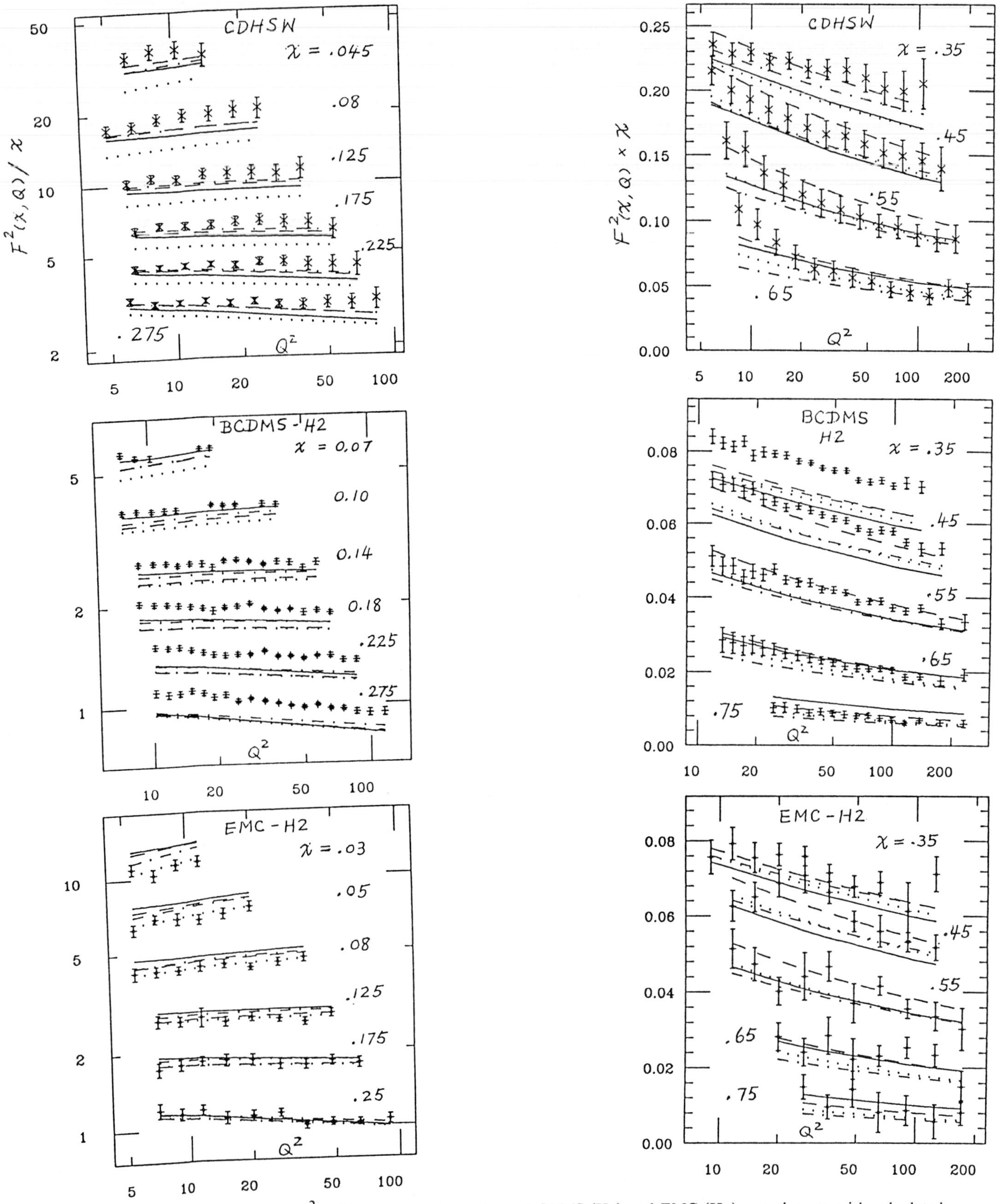

Fig.IV-1: Comparison of $F^2(x,Q)$ measured in CDHSW, BCDMS (H_2) and EMC (H_2) experiments with calculated values from representative parton distribution sets: Duke-Owens-1 (dashed curve), ELHQ-1 (dotted), DFLM-FXAVR (solid), and MRS-E (dashe-dotted).

considerations. A necessary condition for incorporating all relevant factors is to have a comprehensive data base which include these details. Our effort to compile such a database is described in Sec.III.C.

C. Next-to-Leading Order Evolution Calculations

Nominally, the inclusion of the second-order evolution kernel in the calculation of the scale-dependent parton distribution functions yields order α_s/π (about 10%) corrections to the leading order results. However, this is not necessarily true over all kinematic regions. Just as in the case of hard scattering cross-sections,[1] there are circumstances under which the higher order kernel can be significant, or even dominant, compared to the leading one. This consideration is especially relevant for small-x because, in many respects, the region of small x is exceptional.[32] (For instance, the evolution kernel is singular at x = 0. Certain elements of the second-order kernel, in fact, is more singular then the leading-order one; hence they can become quite important.) It is extremely important to determine how the rapid growth of the parton distribution functions with the energy scale in the small-x region (on which projected large cross-sections for physical processes in future colliders are based) is affected by the inclusion of second order effects.

The implications of the second order kernel on the predicted behavior of parton distributions, especially in the small-x region has been recently studied in a systematic manner.[33] This investigation confirmed and quantified the significance of the 2nd order effect. Fig.IV-2 illustrates the difference between 1st and 2nd order evolution of the parton distributions. Shown are the momentum distribution of the u-quark and the gluon at Q = 15 GeV, obtained from 1st and 2nd order calculations which start with the same set of distributions at Q_0 = 3 GeV and which use the same effective coupling strength. We see significant differences in the small-x region but only small variations above x = 0.05. The 2nd-order u-quark distribution is 35% *higher* than the 1st-order one at x = 10^{-4} because of the dominance of the 2nd-order quark-quark evolution function at small-x (which comes about because the 1st-order kernel happens to be unusually regular at x=0). The gluon distribution in the 2nd order calculation, on the other hand, is *lower* than that in 1st order at small x by about 15% at the same x, due to the fact that the 2-loop evolution kernel is large and negative at small x. Details of this study can be found in Ref.33

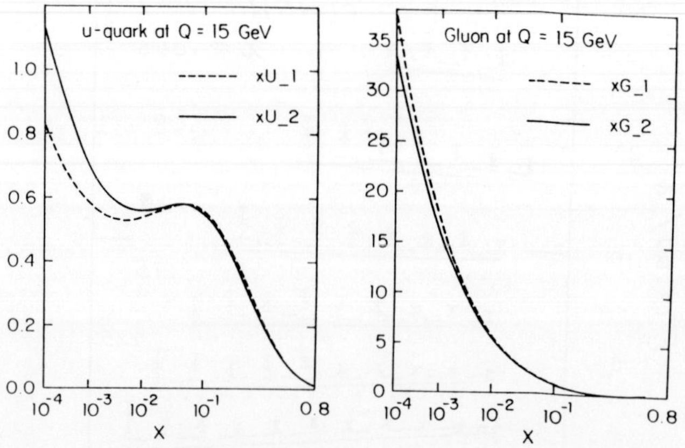

Fig.IV-2: Comparison of 1st and 2nd order evolved u-quark and gluon distributions at Q = 15 GeV.

Most recent parton distribution calculations include 2-loop evolution effects. Ref.33 made direct comparisons with the two widely distributed sets in order to check the consistency of these (relatively complicated) calculations. Agreement was found with one published set (DFLM)26 but substantial disagreement emerged in comparison to the other (MRS)27 in the small-x region. Fig.IV-3 shows the comparison, along with the corresponding 1st order results.

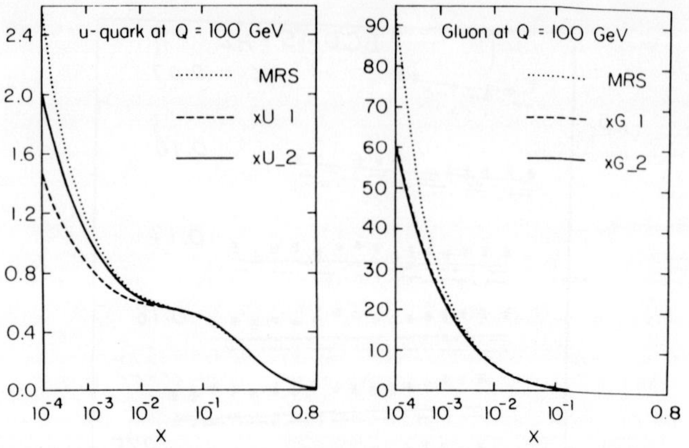

Fig.IV-3: Comparison of 2nd order calculations with MRS distributions.

The latter has since been found to contain a programming error[34] and agreement is obtained after correction of the error. Previous calculations of cross-sections based on the these distributions below x < 0.01 must be revised.

For quantitative applications of the QCD parton model incorporating next-to-leading effects, one must also, in principle, take into account the **scheme dependence** of the parton distributions (cf. Sec.II.C). Ref.33 also compared parton distribution functions in the MS-bar scheme to those in the DIS scheme. The numerical difference is of the order of 10% at Q = 4 GeV. It decreases steadily as Q increases. Since the Wilson coefficient connecting the two sets are regular, this effect is not strongly x-dependent (in contrast to that of the 2nd order evolution kernel).

V. Program to Extract Parton Distribution Functions

We have started a program to utilize the data base described above to extract parton distribution functions suitable for quantitative applications of the QCD parton model and for studying the range of possibilities involved in extrapolation to the small-x region beyond directly measured range. For this purpose, we would like to include a wide range of data sets, and to take into consideration as many as possible the relevant experimental factors described in Sec. IV.B. This work is still underway. Here we describe our particular approach and give a progress report on preliminary results.

A. Description of Methods

The first phase of this analysis will be concerned mainly with the latest major experimental data on structure functions measured in high-energy muon- and charged current neutrino-hadron scattering processes for which we have obtained detailed information on -- EMC, CDHSW, and BCDMS. Because of the unresolved discrepancy between some of these experiments, especially in the BCDMS and EMC hydrogen data, it is important to analyze the dependence of the parton distribution parameters on the choice of data. We have systematically evaluated the quality of the fit for various combinations of experimental data sets and compared the parton distributions extracted from these combinations. We will try to answer the question of general concern: Is the well-publicized BCDMS-EMC discrepancy a major hindrance in the determination of a set of good parton distributions? At a later stage, data from other charged current (BEBC, BPF, CCFRR, CHARM, ...) and neutral current experiments (FMM, ...) will be taken into account as well.

In practical applications of the QCD Parton Model, both leading order and next-to-leading order calculations are widely used. Accordingly, our program is set up to perform fits to each of these orders and to compare the results. Discussions given in Sec.IV.C indicate that these fits will result in different extrapolations to the small-x region even if they are fitted to the same data in the currently available energy range. Thus, confronting predictions on hadron collider processes (which typically involve partons with smaller fractional momentum x than fixed-target ones) based on these distributions may provide good tests of the next-to-leading order QCD Parton Model. Although one expect to be able to fit data with either the 1st or the 2nd order formulas, an interesting question is: whether the 2nd order fits will lead to more stable (hence unique) parton distribution parameters - as they should if the QCD perturbation series indeed makes sense? We shall try to answer this question.

In addition to effects due to experimental systematic errors, heavy-target corrections (the EMC effect), and correlations among measured structure functions, experience shows that the quality and the outcome of a QCD analysis of the parton distributions can also depend on the choice of cuts in Q and W in data selection. If the model is applicable in the region where this analysis is made (hence the results are to be trusted), this cutoff dependence must be negligible. Our study includes investigations on this dependence.

One of the most important motivation for undertaking this investigation is to study the range of possibilities of the small-x behavior of parton distributions. Traditional assumptions of the functional form for the initial distributions (at some arbitrarily chosen Q_0) over-restrict this behavior. It is important to recognize, for instance, that QCD evolution itself alters the effective functional form assumed for the initial distribution at small-x as soon as the evolution is turned on.[35] Since no particular value of Q_0 is *a priori* special, *it is not natural to assume a specific form for the small-x behavior at a given Q*. Consequently, *in our study this functional form is left open, to be determined by the fit to data*. We will try to delineate the range of possibilities in the extrapolation of parton distributions to small-x, given all the current data.

An interesting idea about the choice of Q_0 and the associated initial distributions is that of "dynamically generated parton distributions": one assumes that there is some Q_0, presumably of fairly low value, at which the only effective parton content of the hadron consists of the valence quarks and allow the parton distributions at physically relevant values of Q be generated solely by QCD evolution.[36] In practice, the value of Q_0 in this scheme has to be so low (certainly below 1 GeV) that the perturbative formalism is not applicable at that scale. Therefore, this parametrization of the initial condition can, at best, be regarded as an effective one chosen for its simplicity. We use this method along with the conventional one for comparison. Our implementation of this approach is somewhat different from the conventional one: we include an initial gluon distribution along with the valence quarks; and, instead of arbitrarily fixing Q_0, we allow its value to be determined by the fit. We can get reasonable fit to data, comparable to other methods, in our preliminary studies. Because this approach starts the evolution at very low Q_0 (our fits typically yield values of about 0.5-0.6 GeV), and because the QCD evolution rapidly accumulates partons at small x, this approach tend to yield parton distributions at a physical value of Q which rise more rapidly toward small-x than the conventional approach. (Cf. Ref.36)

With the above considerations in mind, our study is organized in a way best described as a multi-dimensional grid. In the first direction, we have the various factors described above which need to be incorporated in any given fit: (i) heavy target correction (EMC-effect), (ii) experimental systematic errors, (iii) correlations in the measured structure functions ... etc. In another direction, we have the possible choices of data sets which include different combinations of experiments as well as different kinematic cuts on the data sample. In the third direction, we have the choices of theoretical schemes which encompass 1st/2nd order QCD formalisms and various possible functional schemes for the parametrization of the initial parton distributions. We have being performing fits on this grid and systematically recording the results for detailed study. In an ideal world, one should obtain good fits yielding parton distributions stable against the various choices represented in the second and third directions if all the relevant factors listed in the first direction are properly taken into account. The fitting is done with the full MINUIT program and the QCD calculations are carried out with a fast evolution program by WKT[37] (the same as the one used in the investigation of small-x behavior of 2nd order QCD parton distributions described in Sec.IV.C).

B. Preliminary Results

To date, about one hundred different fits to data, exploring the dependence of extracted parton distributions on most of the factors exhibited in the matrix of Fig.V-1 have been carried out. The most extensive studies were done with conventional parametrizations of the initial parton distributions where we allow more freedom in the behavior of these distributions at small-x and in the flavor-dependence of the sea-distributions then allowed in the existing distributions. In order to restrict ourselves to a manageable number of parameters while doing this, we adopt the following procedure: we

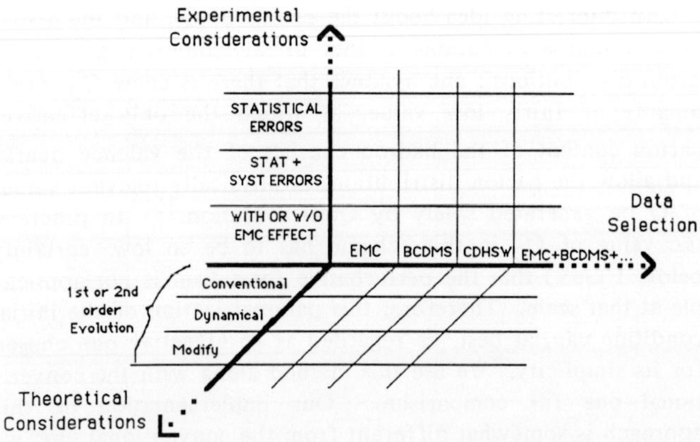

Fig.V-1: Organizational scheme for PDF-study.

use as initial distributions at a chosen Q^2 those from an established set (mostly MRS-E, for the results presented below) multiplied by appropriate factors of the form $C * x^A * (1-x)^B$ to render the desired flexibility and then fit data to extract these (A,B,C) (as well as Λ_{QCD}) as the fitting parameters. One experimental factor — correlation of errors — have been left out of these preliminary studies because time constraint does not allow its proper treatment. We now give a brief discussion of some preliminary results and tentative conclusions. Details will be described in a forthcoming publication.[38]

Data Set Selection: So far the only extensive data sets for which we have detailed information on are: EMC (iron, hydrogen & deuterium), CDHSW (iron) and BCDMS (hydrogen & carbon). We have carried out fits to these data sets individually and in various combinations in order to check the discriminating power of the data sets, the consistency between them, and to determine the stability of the extracted parton distributions and the QCD coupling scale parameter (Λ_{QCD}) against the choice of data sets. We obtain very good fits to the combined CDHSW & EMC and for the combined CDHSW & BCDMS data sets, typically $\chi 2$ per dof of 1 or less (for details, see dependence on Q2-cut below); and we also obtain reasonable fits to the combined data sets of all three experiments with slightly higher $\chi 2$ (1.0 to 1.3 per dof depending on Q2-cut). Fig.V-2 gives a global view of the quality of a typical fit to the full combined set. The format of this rather busy figure follows that of Fig.IV-1 in Sec.IV. We shall come back to discuss the some features of this fit after examining the Q2-cut dependence.

Systematic Errors: For most of our fits, the systematic error is added in quadrature to the statistical error point-by-point. Correlation of errors have not yet been taken into account, as explained earlier. Separate fits have been done with systematic errors ignored (as done by most previous studies) in order to assess their effect on the results. We found the qualify of fit depends critically on whether the experimental systematic errors are included, in particular when multi-sets of data from different experiments are used. The χ^2 per degree of freedom typically jumps from around 1 to 3 - 5 (depending on the Q^2-cut used) when systematic errors are left out. Also, the shapes of the extracted distributions are modified, and the value of QCD-lamda becomes rather unstable against changes in other variables such as Q^2-cut.

Heavy Target (EMC-effect) Corrections: Since the objective is to extract parton distributions inside the *proton*, we apply a heavy target correction to the iron and carbon data before combining these in a fit with hydrogen and deuterium data. The x-dependent correction factor is taken from an empirical formula derived from the measured ratio of F_2(iron)/F_2(deuterium) (EMC) and F_2(carbon)/F_2(deuterium) (BCDMS). No Q-dependence of this factor is included as there is no clear experimental evidence for it. Leaving out this correction (as some existing parton distribution analyses have done) typically does not affect the quality of the fit (as measured by the chi-square per dof) by much. However, the shapes of the extracted parton distributions are clearly changed.

1st vs. 2nd Order QCD Evolution: We have tried to extract parton distributions from the same set of data using both the 1st order and the 2nd order QCD parton model formulas (cf. Sec.II.B,C. We find satisfactory fits in each case. This is expected, as the discussion in Sec.IV.C indicated that the 2nd-order evolution differs little from the 1st order one for $x > 0.01$, provided the value of QCD-Λ is appropriately adjusted. The shape parameters of the parton distributions extracted from these analyses are close to each other, as they should be. The parameters obtained with the 2nd order fits are somewhat more stable against changes in other variable (cf. the paragraph on Q^2-cut below).

Q^2-cut Dependence: If we are in the kinematic region where perturbative (twist-2) QCD is a good approximation, the results of our fits should not depend on the choice of Q^2-cut in our selection of data. In practice, the question of Q^2-cut is important for at least two reasons: (i) twist-4 as well as non-perturbative effects come in at low values of Q^2. We do not know a priori when do these effects set in; (ii) the discrepancies between existing data sets, especially that between EMC and BCDMS are x- and Q-dependent. For fits to the three different combinations of data sets described in a previous paragraph and for calculations performed with both 1st and 2nd order QCD formulas, we have systematically varied the Q^2-cut in the range 6.0 to 25 GeV^2 to observe its effect.

The stability of the shape parameters and Λ_{QCD} against the Q^2-cut can depend on the choice of the parametrization scheme. The following summary only applies to the relatively economical parametrization scheme described earlier in this section. We found the value of Λ_{QCD} to be relatively stable against variations of Q^2-cut in a series of fits for a given choice of data set and calculational scheme. Some typical results are presented in Fig.V-3

Since the meaning of Λ is ambiguous, depending on the number of active flavors and on the order of perturbation used, we have converted all Λ values to their 1st order 4-flavor equivalent (so that the QCD couplings are the same) for this comparison. For the 2nd order fits, the true 2nd order 4-flavor Λ value, $\Lambda(2,4)$, is related to the equivalent 1st order value, $\Lambda(1,4)$, given above by a very accurate formula $\mathrm{Log}_{10}\Lambda(2,4) = 0.156 + 0.81 \cdot \mathrm{Log}_{10}\Lambda(1,4)$.) We will comment

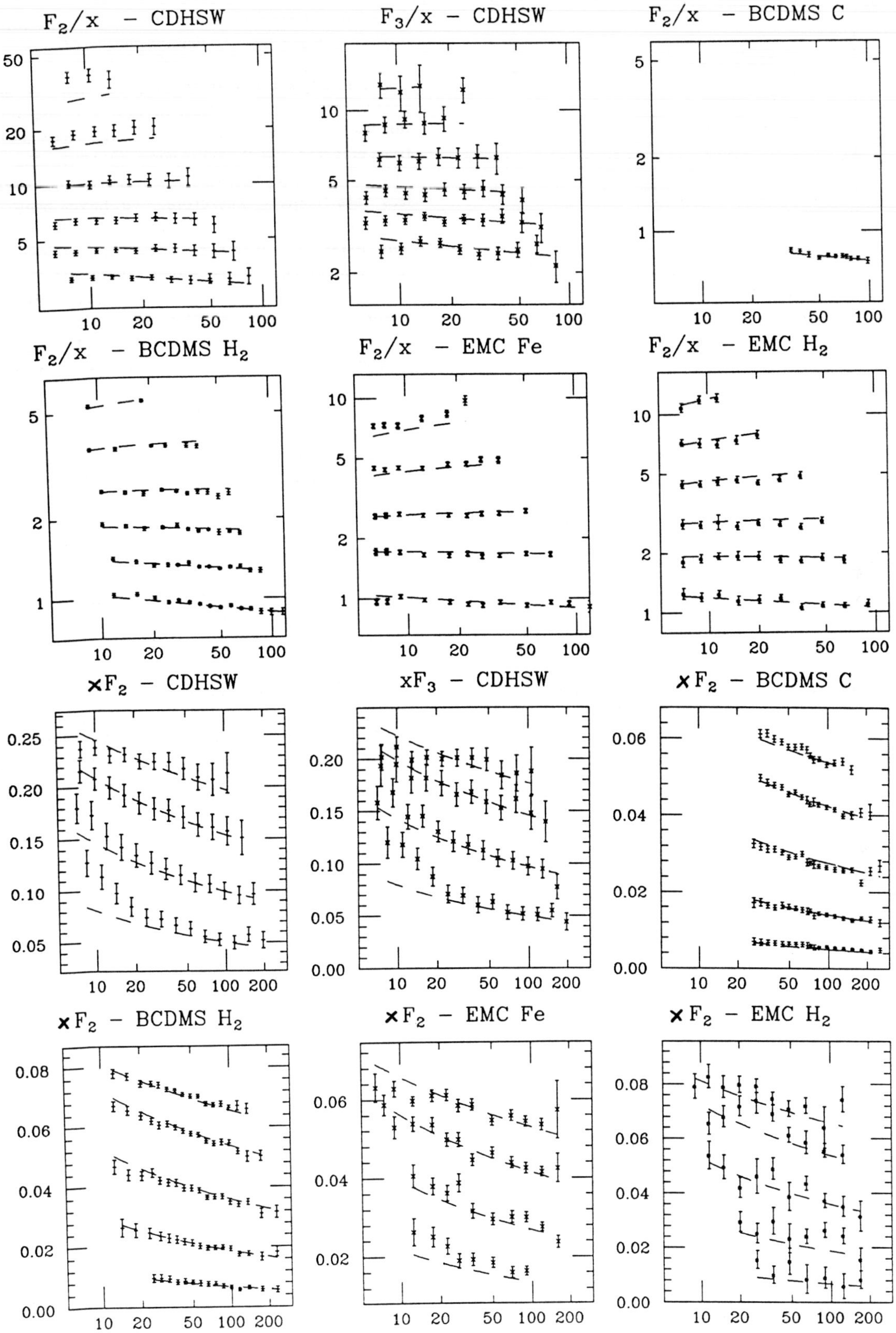

Fig.V-2: Sample fit to combined data sets of CDHSW, BCDMS, and EMC. The Q**2-cut is 6.25 GeV**2. The dashed lines represent the best fit. The top six plots are for x < 0.3. The bottom six plots are for x > 0.3. The specific values of x for the data points are the same as those shown in Fig.IV-1.

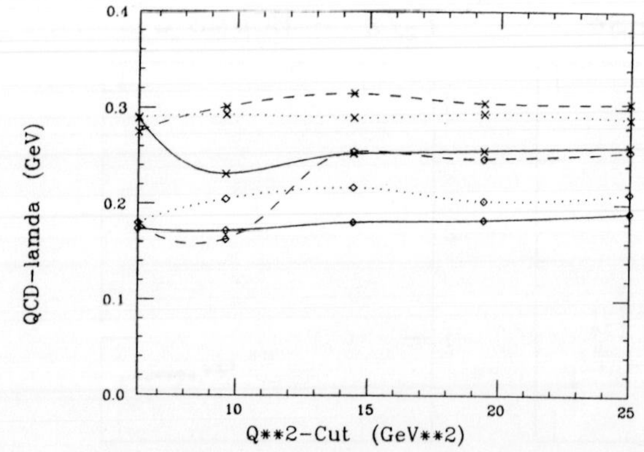

Fig. V-3: Value of Λ_{QCD} vs. Q^2. Points connected by the top three curves are from 1st order fits to BCE, CE, and BC data sets in descending order; those connected by the bottom three curves are from the corresponding 2nd order fits. (BC = Bcdms + Cdhsw; CE = Cdhsw + Emc; BCE = all three).

more on these results in the discussion section.

The parameters which characterize the initial parton distributions at a given Q^2 (say $Q_0{}^2$=10 GeV^2) are also found to be generally stable against variations in Q^2-cut, although some more so than others. A detailed description will necessitate a full discussion of our parametrization formulas which space does not allow here. We give as an example the most Q^2-cut-dependent parameter we found which is the exponent B to the (1-x) factor in the gluon distribution.

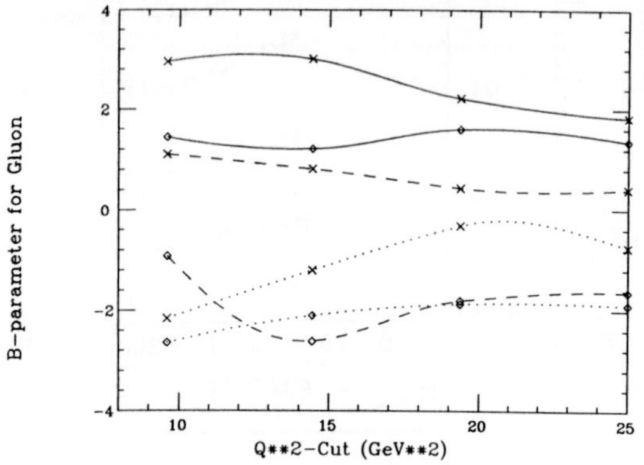

Fig. V-4: Gluon B-parameter vs. Q^2-cut from various fits. The B parameter is normalized to 0 with respect to the MRS-E distributions at Q^2 = 10 GeV^2. The labelling of the points is the same as in Fig.V-3.

We see that this parameter depends rather sensitively on the choice of data set. This is one of the manifestations of the well-known fact that the shape of the gluon distribution is closed coupled to the value of Λ_{QCD} which, as show above, also differs in the various fits. Within a given data set, the value obtained for B_G is rather stable, especially for 2nd-order fits and for Q^2-cut above 15 GeV^2.

C. Tentative Observations and Conclusions

Do disagreements between existing experiments preclude a meaningful parton distribution analysis?

The answer is not an obvious yes. A meaningful QCD parton model analysis must invoke a reasonable Q^2-cut in the selection of data to minimize higher-twist and non-perturbative effects. With increasing Q^2-cut, the well-publicized disagreements between BCDMS and EMC experiments become less significant. Therefore, we have been able to obtain reasonable QCD fits to the combined BCDMS-CDHSW-EMC data above Q^2 10-15 GeV^2. Perhaps somewhat unexpectedly, in a typical fit, points which contribute the most to the χ^2 turn out to come from the two lowest x bins and the highest x bin of the new CDHSW experiment and the EMC-Fe experiment rather than from the BCDMS and EMC hydrogen data points, as illustrated in Fig.V-2. This is due to the statistical weight of the hydrogen data (with very small errors), and due to the fact that an overall normalization factor for each experiment is allowed as a fitting parameter. We found that these fits favor a normalization factor of ≈ 0.95 for the BCDMS data points, and a factor of ≈ 1.03 for the EMC points, both with respect to those of CDHSW.

We do get better fits with combined data from only two out of these three experiments -- BCDMS-CDHSW or CDHSW-EMC -- as compared to the all-inclusive fits discussed above (χ^2 per dof 0.7-1.0 compared to 1.0-1.3). And the fitting parameters from the these fits also seem to be a little more stable than from the grand combined fits, as one may see in Figs.V-3 and V-4. However, the difference does not seem to be very significant. We emphasize that it is possible to reach this conclusion *only if the systematic errors of the experiments are fully taken into account and if a healthy Q2-cut is imposed*. (See Fig. V-5)

What do existing data tell us about the small-x behavior of parton distributions?

The answer is: not very much. The lowest values of x included in our analyses are in the range 0.03 - 0.05, depending on Q^2-cut. The small-x extrapolation of the fitted parton distributions is mainly controlled by the parameter A which is the exponent in the standard factor x^A in the distribution function (at the initial Q^2, usually chosen to be 10 GeV^2 in our calculations). The values of A for (x times) the gluon and the sea-quark distributions we obtain are very stable within a given set of fits, but differ from set to set. The results are summarized in the following table.

	BCDMS CDHSW	CDHSW EMC	All Combined
1st Order	0.303	-0.05	0.305
2nd Order	0.340	-0.01	0.085

Table V-1: Exponent parameter A obtained in various fits (normalized to 0 with respect to the MRS-E distributions at Q^2 = 10 GeV^2).

The different values will give rise to very different small-x extrapolations to small-x values below 10^{-3}. However, we

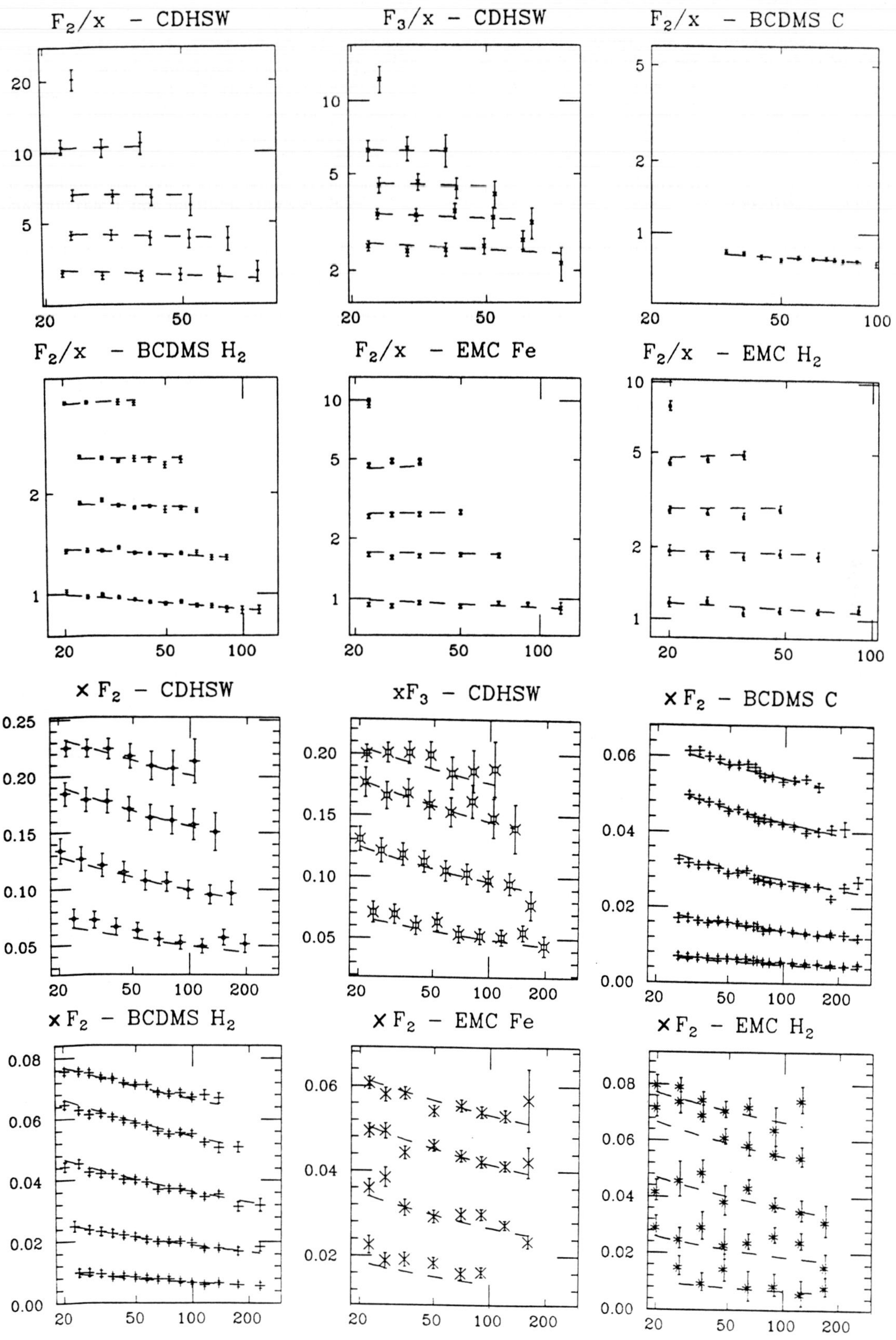

Fig. V-5: Fits to the combined data sets of CDHSW, BCDMS, and EMC with a Q^2 cut of 20 GeV^2. The format is the same as that of Fig. V-2.

can also get acceptable fits to each set of data by fixing this parameter at any value close to the range above. For instance, if the BCDMS-CDHSW set is fit with fixed A = -0.2, the χ^2 per dof only increases from 1.06 to 1.17. Of course, when this is done, some of the other parameters (e.g. Λ_{QCD}) also change from their best-fit values.

How well can one determine Λ_{QCD} and the initial parton distributions from the DIS data?

We confirm the known answer: the u- and d-quark distributions are fairly well determined, but the other parameters are correlated and not so well constrained. We have seen from Figs.V-3 and V-4 and the last paragraph that the extracted values of Λ_{QCD} and parton shape parameters do vary from one set of fits to another. In this regard, the "stability" of the fitted values for some parameter within a given set of fits can be misleading. The range of variation spanned by the various sets of fits is a better indication of the uncertainties involved. A reliable way to determine the uniqueness of the extracted parameters is to perform a full error analysis in the fitting process (e.g. by invoking MINOS inside MINUIT). We found that MINUIT will not yield meaningful error quotes on Λ_{QCD} and the gluon distribution parameters unless most other parameters are artificially fixed.

To conclude: many uncertainties on the parton distributions in QCD fits to current deep inelastic scattering data are intrinsic to the problem because the number of combinations of these distributions measured by the structure functions are limited. This is especially true for the gluon since it does not couple directly to the electroweak currents, and also for the anti-quarks since they contribute only a small part to the measured cross-section. These uncertainties can make a fitting program ambiguous. The only way to remedy this problem is to supplement deep inelastic scattering with complementary processes which involve gluons and anti-quarks in a direct way. Well-know examples are Drell-Yan processes (W-, Z-production, as well as continuum lepton pairs), direct-photon production, heavy-flavor production, ... etc. Any serious study of the parton distributions must take these processes into account. We certainly plan to incorporate these processes in our analysis at a later stage.

VI. Role of Current and Future Experiments in the Determination of Parton Distributions

In the continuing effort to determine the parton distributions in the QCD Parton Model beyond the currently available energy range, we must have more and better experimental data in a variety of processes. As illustrated on the map of kinematic variables, Fig.I.2, we must use available hadron collider processes (at the S$\bar{p}$pS and the Tevatron) to extend the coverage to smaller x, in addition to supply information on the gluon and anti-quarks. Eventually, HERA will expand the kinematic range by two orders of magnitude when it becomes operative. This section summarizes the opportunities offered by the various forthcoming experiments and facilities.

A. Fixed Target Lepton-Hadron Scattering

Although extensive data on deep inelastic scattering structure functions from fixed-target experiments already exist, the glaring discrepancy between major high-statistics experiments, especially for hydrogen target, cries for a resolution by further accurate experiments. In the years ahead, we can look forward to:

* The CCFR Collaboration at Fermilab have a sample of 250,000 anti-neutrino and 1,500,000 neutrino events for energies in the range 0-600 GeV from their 1985 and 1987 runs. Structure functions are expected in early 1989.

* The NMC collaboration (an upgraded version of EMC) currently underway at CERN;

* The Fermilab Muon collaboration (Expt. E665). This Tevatron Muon collabation at FNAL have several million deep inelastic scattering events recorded during their 1987 run at 500 GeV. Their data extend to x of 0.001. However, the experiment is not designed to optimize the measurement of structure functions.

In addition, some entirely new experiments are being considered at Fermilab and UNK.

B. HERA

HERA, projected to come into operation in 1990, will extend measurement of the structure functions to about 0.0001 in x and about 1,000 GeV^2 in Q^2, as shown in Fig.I-2. Since the event rate is very high for small x, the structure functions can be measured with good accuracy in this important region. In view of the lack of reliable experimental data below x ≈ 0.05, and the importance of this information for the study of physics in future colliders, the importance of the HERA measurements cannot be over-emphasized. The capabilities of HERA have been extensively documented in various HERA study reports.[39] For the purpose of this review, we shall only highlight its unique potential to measure the gluon distribution.

The definitive method for determining the gluon distribution is to measure the longitudinal structure function (the difference between F^2 and F^1, Eq.(II-6)). The basic formula in the QCD parton model which defines the gluon distribution inside the hadron in terms of F^{Long} is given by Eqs.(II-26). (Cf. also Figs.II-1b and II-2c.) This method requires the measurement of the cross-section at several collision energies. FLong(x,Q) is determined from the slope of a straight-line plot of the cross-section versus a suitably chosen variable (commonly called ϵ) for fixed (x,Q) -- the classic "Rosenbluth Plot".[40]

In order to gain some feeling on what can be achieved, we give a brief description of some relevant experimental details. For definiteness, we consider here a specific detector, namely ZEUS.[41] The overall kinematically accessible x and $Q2$ range is shown in Figure VI-1, bounded by the curve labelled "before cut". Several cuts must be applied to simulate detector acceptance:

- a 200 mrad cut is applied on the angle of the scattered electron with respect to the beam to ensure that it will enter the rear detector.

- the electron is required to have a transverse momentum of at least $3GeV/c$ for it to be well separated from the fragmentation jet;

- a cut of $y < 0.83$ is applied to stay away from regions where radiative corrections are changing rapidly.[42]

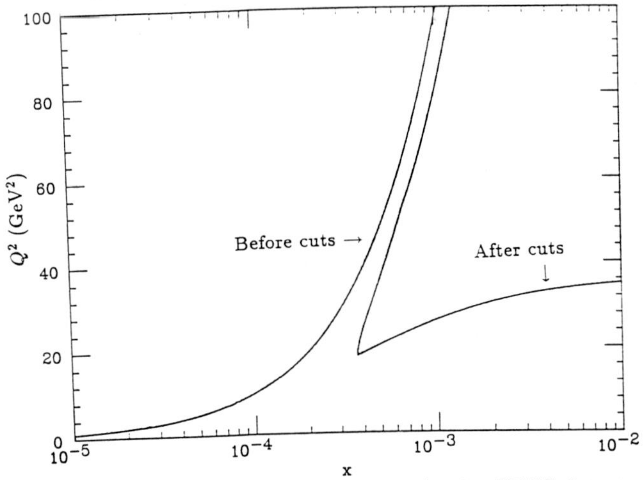

Fig.VI-1: Available low x and Q range in the ZEUS detector.

The resultant x and Q^2 range available after the cuts is bounded by the curve labelled "after cut" in Fig.VI-1. The P_T cut places the largest restriction on x while the angle cut has the largest effect on the minimum Q^2.

Fig.VI-2 shows the anticipated ability to resolve between two representative gluon distributions[40] at $Q^2 = 50\ GeV^2$ for initial gluon distributions of the form: A) $xG(x, Q^2) = 0.676x^{-1/2}(1 - x)^5$ and B) $xG(x, Q^2) = 5(1 - x)^9$. The error bars indicate the size of the statistical error expected for $100pb^{-1}$ of luminosity, and the shaded band shows the expected size of systematic errors, which include a 5% luminosity error and systematic errors on N (the number of events in an x, Q^2 bin) and Γ due to resolution effects. Note that the relative luminosity errors for different ϵ can be removed by using high x data. According to this study, these distributions will be easily distinguishable.

An alternative method to measure the gluon distribution at HERA has been proposed.[43] This utilizes the cross-section for J/ψ production and assumes that it is dominated by the photon-gluon fusion mechanism.[44] Details can be found in ref.43. Whereas this certainly provides an interesting window to test the gluon distribution, it cannot provide a true measurement for the latter because of the uncertain theoretical basis for the underlying interaction mechanism. There is no unambiguous order-by-order perturbative QCD formula for this process.

With results now available from the detailed parton distribution analyses described in Sec.V, we plan to investigate the resolving power of the HERA experiments concerning the major remaining uncertainties on the parton distributions: the shape of the gluon distribution and on the extrapolation to small-x.

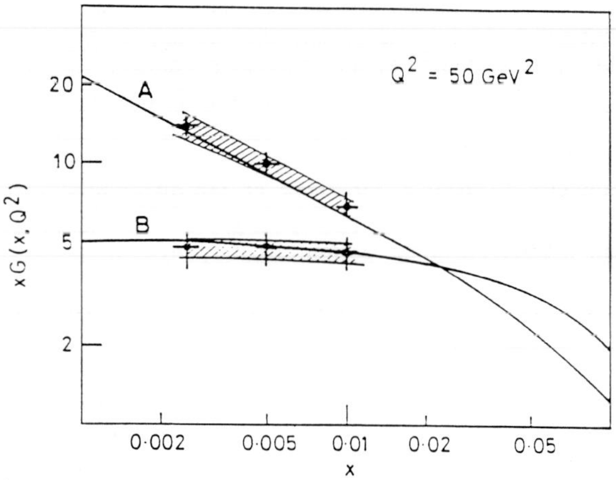

Fig.VI-2: The anticipated resolving power on the small-x behavior of the gluon distribution at HERA.

C. Drell-Yan Process

Next to deep inelastic scattering, the production of lepton-pairs (via a virtual γ), and W-, and Z-production by the Drell-Yan mechanism (and its higher order generalizations) are the cleanest in terms of its QCD parton model foundation. The basic processes involved are shown in Fig.VI-3:

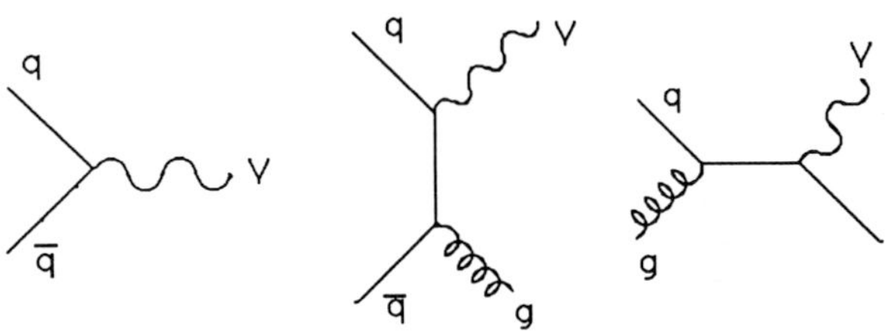

Fig.VI-3: Leading and Next-to-leading Drell-Yan processes.

It is very important to measure the cross-section for these processes at the current and future colliders for two major reasons: (i) as is apparent from Fig.VI-3, these processes receive substantial contribution from the gluon through the quark-gluon "Compton scattering" mechanism, hence provide a much needed handle on the gluon distribution; and (ii) the continuum lepton-pairs probe an x-range much lower than that given by fixed target DIS experiments, hence furnish us with valuable information on the trend of small-x extrapolation of the parton distributions. Specifically, since $x_1x_2 = Q^2/s$, one can reach below $x = 10^{-3}$ at the Tevatron for lepton-pairs produced away from the central region. This is to be compared to $x > 0.05$ in existing deep inelastic scattering.

Drell-Yan cross-section from fixed-target experiments have been used in the extraction of parton distributions in some of the sets reviewed in Sec.IV. The sensitivity of the projected cross-section, both total and differential, to different assumptions about the small-x behavior of the parton distributions have been systematically studied during the 1987 Madison Workshop[45] in the context of a general survey of all promising "small-x processes". It was found, by a leading order Monte Carlo study, that in addition to the obvious dependence of the total cross-section on the assumed parton luminocity at small-x, the shape of the differential y-distribution of the Drell-Yan process is extremely sensitive to the small-x behavior of the parton distributions.

In view of recent awareness of the importance of higher order calculations, in particular for small-x, we have carried out a next-to-leading order calculation of the Drell-Yan cross-section during this Workshop.[46] The results confirm the conclusions of the previous study. In Fig.VI-4a we show the next-to-leading order results on this y-distribution at the Tevatron energy obtained by using parton distributions with assumed x^{-1} behavior (solid curve) and with assumed $x^{-1.5}$ behavior (dashed curve) at Q = 3.0 GeV and evolve with 2-loop evolution kernel to Q = 15.0 GeV. In Fig.VI-4b, we show the corresponding results at the SSC energy. We see that the shape of these curves provide a good handle on the assumed small-x behavior. Even without measuring the full y-range, the ratio between the cross-sections at the central region and some forward region can already yield valuable information on the behavior of parton distributions at small-x. More detailed study of the discriminating power of this type of measurement starting from the results of Sec.V is clearly worthwhile.

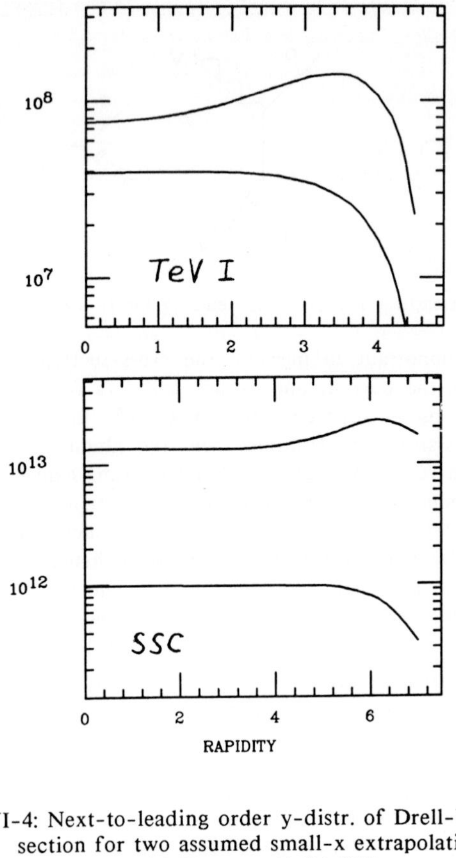

Fig.VI-4: Next-to-leading order y-distr. of Drell-Yan Cross-section for two assumed small-x extrapolations: (a) Tevatron I; (b) SSC

D. Direct-Photon Production

Details on direct photon production is covered in another contribution to these proceedings.[47] It is worthwhile mentioning, however, that this process can have an important role in the overall scheme of determining the parton distributions. The basic underlying mechanism is very similar to the case of Drell-Yan processes discussed above. It is sensitive to the gluon distribution. And 2nd order QCD formulas for the cross-section do exist.

Summary

In this report we reviewed the current status of deep inelastic scattering structure functions and parton distributions in the QCD parton model. We spelled out explicitly the necessary theoretical framework for QCD analyses in the leading and the next-to-leading order. Likewise, we discussed important experimental analysis issues in the measurement of structure functions and in the subsequent process of extracting parton distribution functions.

In order to make maximum use of recent experiments with vastly increased statistics, it is important to have a data base with complete information on all existing data in easily accessible form and in a common format. We have been working to assemble such a complete database. An important part of this effort consists of incorporating detailed information on systematic errors and analysis assumptions which will allow better estimates of the errors on physical quantities extracted from these data.

We also presented a review of widely used parton distributions and showed the need for a more up-to-date analysis using current data and incorporating significant experimental and theoretical considerations relevant for contemporary applications of the QCD parton model, particularly attention to the small-x behavior and next-to-leading order effects. We reported on our own program to carry out such an analysis and presented preliminary results and conclusions from this extensive work.

We underlined the important roles that current collider processes and upcoming HERA measurements can play in complementing the structure function analysis for the complete determination of the parton distributions. We highlighted the Drell-Yan process and direct photon production at the colliders as the most crucial in providing information on the gluon distribution in general and all parton distributions in the small-x region in particular. This will be important for its own sake as well as for providing a vital constraint in the extrapolation to even smaller x needed for the study of most SSC physics processes.

Acknownledgement

We would like to thank our colleagues at the Snowmass Workshop, especially members of the QCD group, to which this subgroup belonged, for stimulating discussions. WKT would also like to thand John Collins for useful discussions and suggestions.

-REFERENCES

1 For a recent review, see K. Ellis, Review talk given at the 1988 Munich Conference on High Energy Physics, Fermilab Preprint, 1988.

2 For comprehensive reviews and complete lists of references, see G. Altarelli, Phys. Rep. **81C**, 1 (1982); W. Furmanski and R. Petronzio, Z. Phys. **C11**, 293 (1982).

3 EMC Fe: Aubert et al., Nucl. Phys. **B159**, 189 (1985).

4 BCDMS C: Benvenuti et al., Phys. Lett. **B195**, 91 (1987)

5 BFP Fe: P. Meyers et al., Phys. Rev D34, 1265 (1986).

6 CCFRR Fe: MacFarlane et al., Z. Phys. **C26**, 1-12 (1984); MacFarlane, Nevis preprint 1297 (1984), Purohit, Nevis preprint 1298 (1984).

7 CDHS: Abramowicz et al., Z. Phys. **C17**, 283 (1984); Z. Phys. **C25**, 29 (1984); Z. Phys. **C35**, 443 (1984); CDHSW: B. Vallage, DPhPE (Saclay) CEA-N-2513 (in French)

8 FMM: T. Mattison el.al. Phys. Rev. Lett. **55**, 574 (1985); and MIT Ph.D. Thesis (1987)

9 BCDMS H: Benvenuti et. al. Preprint JINR-E1-87-689.

10 EMC H: Aubert et al., Nuclear Phys. **B259**, 189 (1985).

11 R. Mount, talk given at 1988 Munich International Conference on High Energy Physics.

12 "EMC-effect": Aubert et al., Nucl. Phys. **B272**, 158 (1986)

13 SLAC E140: S. Dasu, et. al., Phys. Rev. Lett. **60**, 2591 (1988).

14 EMC':

15 K. Lassila et. al., Phys. Lett. **209B**, 343 (1988).

16 E. Berger & J. Qiu, Preprint ANL-HEP-88-42 (1988).

17 F. E. Taylor, Proc. 13th Int. Conf. on Neutrino Physics and Astrophysic, Medford, MA (1988).

18 S. Dasu, et. al., Phys. Rev. Lett. **61**, 1061 (1988)

19 R. Voss, in Proc. of 1987 International Symposuim on Lepton and Photon Interactions at High Energies, Hamburg, 1987.

20 I.A. Savin, (private communication)

21 Benvenuti et al., Phys. Lett. 195B, 97 (1987); and JINR preprint JINR-E1-87-699 (1987).

22 BEBC: D. Allasia el.al., Z. Phys. **C28**, 321 (1985); Phys. Lett. **135B**, 231 (1986); Varvell et al. CERN/EP 87-46 (1987).

23 R.N. Barnett, Phys. Rev. Lett. **36**, 1163 (1976); B.J. Edwards, T.D. Gottschalk, Nucl. Phys. **B186**, 309 (1981); B.J. Kaplan, F. Martin, Nucl. Phys. **B115**, 333 (1976); C.H. Lai, Phys. Rev. **D18**, 1422 (1978); H. Georgi, H.D. Politzer, Phys. Rev. **D14**, 1829 (1976)

24 CHARM: Bergsma et al., Phys. Lett. **123B**, 269 (1983); ibid. **153B**, 111 (1985); Allaby et al., Phys. Lett. **B197**, 281 (1987).

25 Gluck et al, Z. Phys. **C13**, 119 (1982)

26 D. Duke and J. Owens, Phys. Rev. D30, 49(1984).

27 E. Eichten et. al., Rev. Mod. Phys. 56, 579 (1984), and Erratum 58, 1065 (1986)

28 M.Diemoz, F.Ferroni, E.Longo & G.Martinelli, Z. Phys. C39, 21 (1988).

29 A.D.Martin, R.G.Roberts & W.J.Stirling, Phys. Rev. D37, 1161 (1988).

30 B. Vallage, Thesis submitted to Universite de Paris Sud.

31 P. Berge et al., Z. Phys. **C35**, 443 (1987); Allaby et al., Z. Phys. **C38**, 403 (1988).

32 Proceedings of the 1984 Summer Study on the Design and Utilization of the SSC, Ed. R. Donaldson and J. Morfin, (1985); and Proceedings of the 1986 Summer Study on the Physics of the SSC, Ed. R. Donaldson and J. Marx, (1987).

33 Wu-Ki Tung, "Small-x Behavior of Parton Distribution Functions in the Next-to-leading Order QCD Parton Model", Nucl. Phys. (To be published), Fermilab-PUB/88-135-T.

34 R. Roberts, private communication.

35 J.C. Collins, S. Lomach, and F. Olness, IIT preprint (1988).

36 M. Gluck, R.M. Godbole, & E. Reya, (Dortmund U.), DO-TH-88/9 and DO-TH-88/18 (1988).

37 Wu-Ki Tung (to be published).

38 J.G. Morfin and Wu-Ki Tung (to be published).

39 See for example, R. Brinkman "HERA" DESY-HERA-88-03 (1988).

40 For details, see Cooper-Sarkar et al., Z. Phys. **C39**, no. 2, 1988.

41 ZEUS Collaboration, Technical Proposal (March 1986); ZEUS Collaboration, Status Report (September 1987).

42 C. Kiesling, Proc. DESY Theory Workshop, 1987.

43 S.M. Tkaczyk, W.J. Stirling, D.H. Saxon, RAL-88-041. G.A. Schuler, DESY Preprint, DESY 87-114.

44 E.L. Berger, D. Jones, Phys. Rev. D23 (1981) 1521; and A.D. Martin, C.K. Ng and W.J. Stirling; Phys. Letters B191 (1987) 200.

45 F. Olness and W.K. Tung, Int. Journal of Mod. Phys. A2, 1413 (1987); also in Proceedings of International Conference on From Colliders to Super-colliders, Madison, World Scientific (1987).

46 J.C. Collins, F. Olness, and Wu-Ki Tung, (to be published).

47 I. Hinchliffe and M. Shapiro, these proceedings.

NUCLEAR AND HADRONIC STRUCTURE FUNCTIONS AT AN ADVANCED HADRON FACILITY

Gerald T. GARVEY

Los Alamos National Laboratory
Los Alamos, New Mexico 87545*

Deep inelastic scattering of charged leptons from nuclei shows the observed structure function per nucleon to be modified in the nuclear medium. The first[1)] experimental report of such an effect was made several years ago by the European Muon Collaboration (EMC); hence, the effect is popularly known as the EMC effect. Considerable time and effort[2)] have now gone into carrying out accurate, relative structure-function measurements. A rather consistent picture of nuclear effects has emerged that appears to be quite systematic as a function of the nucleon number A, as shown in Figure 1.

Unfortunately, the information obtained from deep-inelastic charged-lepton scattering is insensitive to quark flavor as the contribution of each quark flavor is weighted by the square of its electric charge. In standard notation

$$\frac{d\sigma}{dQ^2dx} = \frac{4\pi\alpha^2}{Q^2}$$
$$\left([1+(1-y)^2]F_1(x) + (1-y)\left[\frac{F_2(x)}{x} - 2F_1\right] - \frac{M}{2E}yF_2\right)$$

where

$$2xF_1(x) = F_2(x)$$
$$= x\left(\frac{4}{9}[u(x)+\overline{u}(x)] + \frac{1}{9}[d(x)+\overline{d}(x)] + \dots\right) .$$

More particular information[3)] can be obtained by using the Drell-Yan process where x_1 is associated with the quark (antiquark) in the incident hadron and x_2 is associated with the antiquark (quark) in the target hadron. The quark–antiquark annihilation generates a massive photon ($M_{\mu\mu}$) which in turn creates a dilepton pair. The Drell-Yan cross section is written

$$\frac{d\sigma}{dM_{\mu\mu}dx_F} =$$
$$\frac{8\pi\alpha^2}{9M^3\mu\mu}\sum_i \frac{e_i^2 x_1 x_2}{x_1+x_2}\left[q_i(x_1)\overline{q}_i(x_2) + \overline{q}_i(x_1)q_i(x_2)\right] .$$

The summation index i runs over all quark flavors and $x_F = x_1 - x_2$.

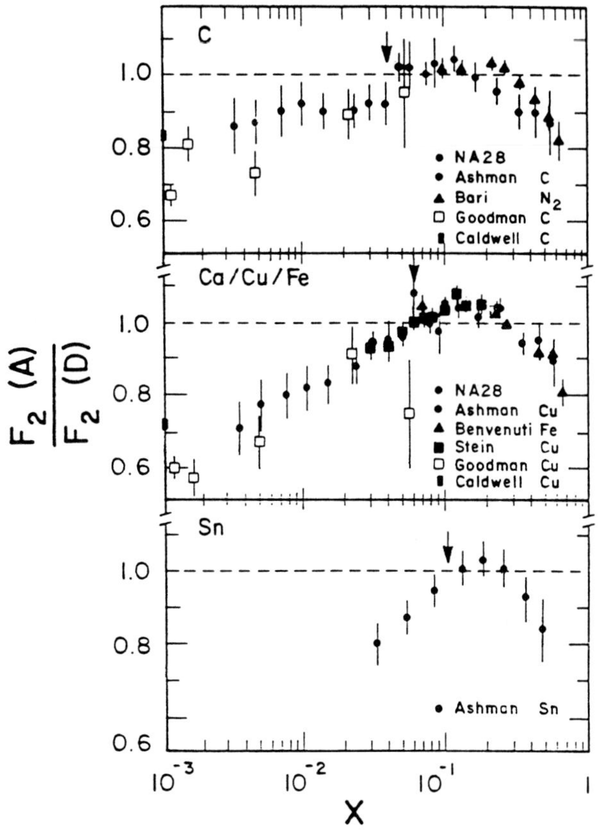

Figure 1. Ratio of the quark structure function $F_2(x)$ per nucleon as measured in nucleus (A) relative to that of deuterium ($A = 2$). The measurements are obtained from the deep inelastic scattering of charged leptons. The figures a), b), and c) show the ratio as a function of $x = \frac{Q^2}{2M\nu}$ and correspond to: a) Carbon (A = 12); b) Ca/Cu/Fe (A ~ 50); and, c) Tin (A ~ 120).

The antiquark distribution target can be measured by using incident high-energy protons and requiring that $x_1 \gg x_2$. An experiment (E772) at Fermi National Accelerator Laboratory (FNAL) has been carried out using 800-GeV protons and is currently being analyzed[4)]. The Drell-Yan yield for ^{12}C, ^{40}Ca, Fe, Sn, and W have been measured relative to D over the interval of $0.05 \leq x_2 \leq 0.3$ with systematic errors believed to be the order of 1%. A total of 5×10^5 good events have been obtained and are being analyzed on a special purpose 48-node ADP form. Analysis should be complete by the end of this year. Figure 2 shows very preliminary analysis[4)] for the ratio of Fe to D under the constraints shown in the caption.

* Work performed under the auspices of the U.S. Department of Energy.

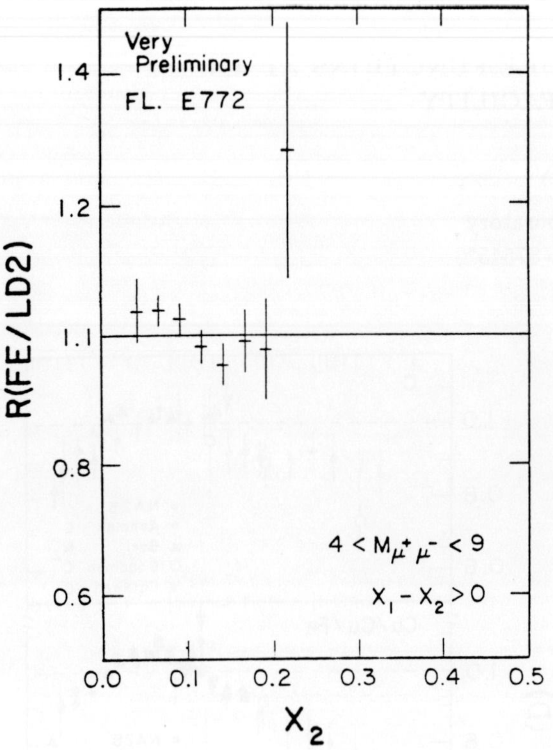

Figure 2. Ratio of the observed Drell-Yan yield per nucleon from Fe relative to deuterium. The ratio is shown as a function of the target x (x_2). The mass of the dilepton pairs has been chosen to lie between 4 and 9 GeV/c^2 and further $x_1 - x_2 > 0$.

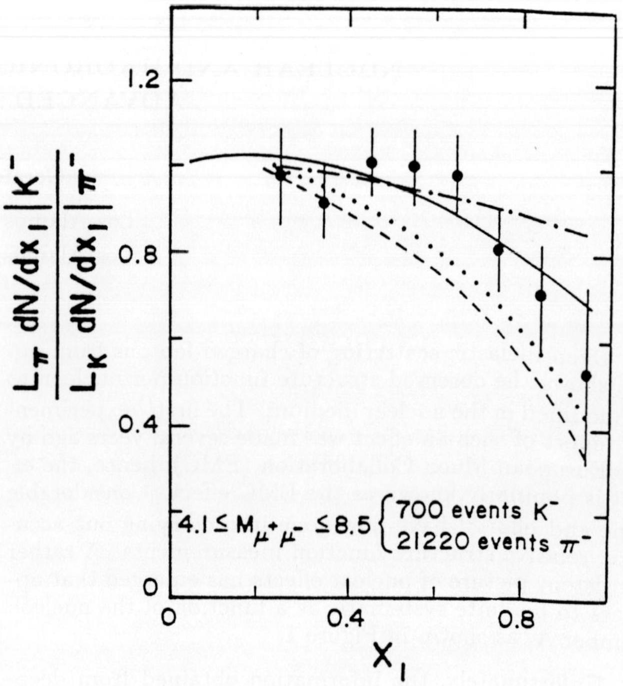

Figure 3. Ratio of the $\overline{u}$ structure function for K^- to π^- as measured via the Drell-Yan process (Ref. 5).

To measure $\overline{q}(x_2)$ at higher x_2, say $0.3 \leq x_2 \sim\leq 0.6$, it is better to employ a lower-energy proton beam [of course, $x_1 > x_2$ to insure $\overline{q}(x_2)$]. A 60-GeV proton beam such as would be generated by the Advanced Hadron Facility (AHF) proposed by Los Alamos National Laboratory would greatly facilitate such a measurement. The advantage of lower energy arises because the cross section is proportional to $M_{\mu\mu}^{-2}$; thus, using 60 GeV rather than 800 GeV yields, for fixed values of x_1 and x_2, an increase of a factor of 13 in cross section while still allowing the mass of the dilepton pair to be well above that of the J'/ψ' vector mesons.

The backgrounds at 60 GeV are not certain, but are not believed to present a serious problem. Estimates of the $\overline{q}(x)$ at $x = 0.6$ are uncertain, but are the order of $\overline{q}(0.6) \sim 6 \times 10^{-4}$. The resulting cross section is therefore small, and requires luminosities of the order of 10^{35} cm^{-2} sec^{-1} to obtain a few events per day. This represents a challenge, but appears doable with conventional techniques. However, it may be interesting and profitable to investigate the use of open-detector geometries that pick out electron pairs rather than adsorbing hadrons to detect the muon pairs. The requisite sensitivity to electrons and insensitivity to other particles is achieved by using He Čerenkov detection and detection of synchrotron radiation arising from electron deflection in magnetic fields.

The proposed Los Alamos Advanced Hadron Facility will also permit much better measurement of the ratio of the relative $\overline{u}(x)$ structure function in the K^- to that of the π^-. Previous measurements[5)] were based on a total of only 700 kaon events, and showed sizable differences for x greater than 0.6 as shown in Figure 3. The observed differences were ascribed to the constituent masses of the s and d quarks respectively. A higher flux of separated kaons ($K/\pi \sim 1$) appears readily achievable at AHF and will permit this measurement to be greatly improved with an increase of the order of 100 times the number of K^--induced Drell-Yan events.

References:

1. J. J. Aubert, *et al.*, *Physics Letters B* **123** (1983) 275.
2. G. Bari, *et al.*, *Physics Letters B* **163** (1985) 282. A. C. Benvenuti, et al., *Physics Letters B* **189** (1987) 483. EMC (NA28), private communication. NMC, private communication.
3. E. L. Berger, *Nuclear Physics B* **267** (1986) 231. R. P. Bickenstaff, M. C. Birse, and G. A. Miller, *Physical Review D* **33** (1986) 3228.
4. FNAL (E772), private communication.
5. J. Badier, *et al.*, *Physics Letters B* **93** (1988) 354.

SHAPE DISTRIBUTIONS AND FACTORIZATION OF FRAGMENTATION FUNCTIONS IN SOFT COLLISIONS†

Sergio P. Ratti
Dipartimento di Fisica Nucleare e Teorica
Universitá di Pavia and
Instituto Nazionale di Fisica Nucleare–Sezione di Pavia

1. INTRODUCTION

Concerning large cross section phenomena and soft collisions, P. V. Landshoff[1] in the 1987 SSC Workshop at Berkeley, wrote, in spite of whether or not large cross section physics has to be regarded as dirty; ... *one should give these particular emphasis, because they have the best choice of being interpreted and understood. The aim must always be to try and understand* **dynamics** *rather than just making multi-parameter fits to the data.*

No doubt the dynamics is the one of the long range part of the strong interactions at high energies; no doubt there is little understanding of it at the fundamental level. No doubt one has to provide at least a good phenomenological description of the studied processes in order to help the formulation of a hopefully more theoretically based understanding .

In the region of the large cross section, low p_T processes, non-perturbative QCD is needed and all the interesting results of the only available calculational tool of the long range forces (perturbative QCD) are close to being completely useless. The QCD coupling at long distances is not small.

Nonetheless a number of experimental empirical observations are important and might be useful insofar as they are model independent in nature.

In large cross section physics, the step next to total cross sections, elastic scattering, diffraction dissociation and multiplicity distributions, is the study of the semi-inclusive shape parameters of multiparticle production.

Variables such as thrust,[2] sphericity[3] or others,[4] if not to be very useful to search for jets or minijets, may turn out of great help in describing the background undoubtedly existing under three relatively rare processes.

In the following Sections I shall describe some interesting experimental observations on the behavior of semi-inclusive sphericity distribution in hadron-hadron collisions at low energy and discuss their possible implications at SSC energies.

‡ Presented at the 1988 Summer Study on High Energy Physics in the 1990's, Snowmass, Colorado, July 1988.

2. FACTORIZATION OF QUARK FRAGMENTATION FUNCTIONS

It is well known[5] that, when writing the fully differential n-particle invariant cross section,

$$E_1 \ldots E_n \frac{d^{2n}\sigma_n}{dp_{L_1}dp^2_{T_1}\ldots dp_{L_n}dp^2_{T_n}} = f_n(p_{L_1},p_{T_1},\ldots,p_{L_n},p_{T_n},s)\,, \qquad (1)$$

[where E_i and p_{Li} (p_{Ti}) indicate the energy and the longitudinal (transverse) momenta of the i-th particle respectively, s being the square of the total energy available to the multiparticle state], most of the time it is useful to use "reduced variables" such as:

$$\mathbf{x}_i = p_{Li}/\Sigma p_i \qquad ; \qquad \mathbf{u}_i = (p_{Ti})^2/\Sigma(p_i)^2\,, \qquad (2)$$

where $p_i^2 = p_{Li}^2 + p_{Ti}^2$. Formula (1) is then rewritten as:

$$d^{2n}\sigma_n/d\mathbf{x}_1 d\mathbf{u}_1 \ldots d\mathbf{x}_n d\mathbf{u}_n = F_n(\mathbf{x}_1,\mathbf{u}_1,\ldots,\mathbf{x}_n,\mathbf{u}_n)\,. \quad (3)$$

The reduced variables $\mathbf{u}_i(i = 1,2,\ldots,n)$ introduced in formula (2), do not change the features of the p_T distribution. Integrating (3) over the longitudinal variables in order to evidentiate the transverse part of the cross section, one obtains:

$$d^n\sigma_n/d\mathbf{u}_1\ldots d\mathbf{u}_n = \int d\mathbf{x}_1\ldots d\mathbf{x}_n F_n = G_n(\mathbf{u}_1,\ldots,\mathbf{u}_n)\,. \qquad (4)$$

Now, most of the quark models, independently of whether or not they are based on QCD, postulate a Gaussian distribution for the transverse momenta (q_{Ti}) of the i–th quark pairs, hence including that the distribution of the transverse momenta (p_{Ti}) of the i-th meson (where $p_{Ti} = q_{Ti} - q_{Ti-1}$), is given by the same functional form as the Gaussian distribution.[6] Since most often it is assumed that there is no correlation between non-adjacent mesons in momentum space, it is natural to factorize the function G_n as:

$$d^n\sigma_n/d\mathbf{u}_1\ldots d\mathbf{u}_n = G_n(\mathbf{u}_1,\ldots,\mathbf{u}_n) = \Pi\, g_i(\mathbf{u}_i)\,, \quad (5)$$

and assume a Gaussian distribution for the transverse variables $\mathbf{u}_i$; i.e.,

$$g_i(\mathbf{u}_i) = d\sigma_n/d\mathbf{u}_i = K\exp(-k\mathbf{u}_i)\,, \qquad (6)$$

where K and k are parameters given by the normalization condition and the average value of $\mathbf{u}_i$. Formulae (5) and

(6) depend only on the factorization hypothesis and the weak assumption of a Gaussian shape of the transverse variables. From then, one can then write:

$$G_n(\mathbf{u}_1,\ldots,\mathbf{u}_n) \;=\; \left\{K\ \exp\left(-\,k\,\frac{\sum_i \mathbf{u}_i}{n}\right)\right\}^n . \qquad (7)$$

From (7) and (6) it is easy to derive the shape for the distribution in "average $\mathbf{u}_i$"; i.e.,

$$d\sigma_n/d\left(\sum_i \mathbf{u}_i/n\right) \;=\; K\ \exp\left(-\,k\sum_i \mathbf{u}_i/n\right) . \qquad (8)$$

And, in turns, it is easy to find in (8) the signature of the sphericity variable, provided one rotates the reference frame to measure the transverse momentum components just with respect to that axis which minimizes the sphericity variable

$$S \;=\; 3/2\,[\Sigma \mathbf{u}_i] \;=\; 3/2\,\{\Sigma(p_{Ti})^2/\Sigma(p_i)^2\} . \qquad (9)$$

Therefore, as a consequence of factorization of the fragmentation function $G_n(\mathbf{u}_1,\mathbf{u}_2,\ldots,\mathbf{u}_n)$ in (5), assumed to be Gaussian in (6), instead of writing formula (8), it is easy to write directly the normalized semi-inclusive differential cross section in sphericity at fixed final multiplicity (eventually charged multiplicity) n as:

$$[1/\sigma_n]d\sigma_n/dS \;=\; A\ \exp[(-\alpha/n)S] . \qquad (10)$$

The normalization condition gives:

$$A^{-1} \;=\; \int_0^1 dS\ \exp\left(-\,\frac{\alpha}{n}\,S\right) \;=\; \frac{1}{\alpha}\left\{1\;-\;\exp\left(-\,\frac{\alpha}{n}\right)\right\} . \qquad (11)$$

Distribution (10) would depend upon a parameter which gives the "spread" of the Gaussian fragmentation function through which the initial state generates qq states (you can call them clans, firestrings or whatever...) giving rise to the real final mesons. Its importance is that of giving a strict connection between the sphericity distributions for different multiplicity, depended on the "dynamics" of the strong soft process since factorization is a dynamical assumption. If an experiment measures sphericity distributions for say 4, 6, 8, 10, ..., charged final particles, the sphericity distributions must go as:

$$\exp(-\alpha/4)S;\ \exp(-\alpha/6)S;$$
$$\exp(-\alpha/8)S;\ \exp(-\alpha/10)S ,$$

with a unique value for the parameter α.

3. LOW ENERGY DATA

Several experimental features are known from low energy hadron-hadron, lepton-lepton, lepton-hadron collisions:

1. Multi-hadron final states in low p_T interactions show very similar structures independent on the nature of the initial colliding particles.

   Experimental distributions of thrust, spherocity, and sphericity, observed in hadron-hadron collisions initiated by different particles,[7] have been compared to the same distributions initiated by leptons, within the jets description.
2. It is well known, from e^+e^- annihilation data at different energies[8,9] that the transverse momentum distributions of jets show a typical two component structure, the slope of the high p_T component being energy dependent.
3. The average values[10] of sphericity (thrust) increase (decrease) linearly with increasing charged multiplicity till phase space limitations start showing their effect.
4. Particle multiplicities scale, rather than according to the old Koba, Nielsen, Olesen scaling law, according to a negative binomial distribution.[11]
5. At high energies, the average multiplicity of the produced particles seems to vary[12] as:

$$\langle n_{ch}\rangle \;=\; 7.6\,s^{0.125} - 7.4 . \qquad (12)$$

Now, it is easy to obtain from formulae (10) and (11) the multiplicity dependence of the average sphericity. One gets:

$$\langle S\rangle_n \;=\; n/\alpha - 1/[\exp(\alpha/n) - 1] . \qquad (13')$$

In the region $n \ll \alpha$ (13") leads to the obvious linear relation:

$$\langle S\rangle_n \;\approx\; n/\alpha , \qquad (13'')$$

which provides a very simple and naive explanation of the experimental observation 4.

Ratti, Nakamura and Introzzi (Ref. 5) have interpreted the features of the sphericity distributions at fixed-charged multiplicity in terms of the simple uncorrelated quark cascade assumption made in Section 2, with the help of a two-component mechanism [points 1 and 2 above].

Generalizing formula (10) of Section 2, the normalized inclusive sphericity distribution is given by:

$$\frac{1}{\langle n\rangle\sigma}\,\frac{d\sigma}{dS} \;=\; \sum_n \frac{n}{\langle n\rangle}\,P_n\,A\ \exp\left(-\,\frac{\alpha}{n}\,S\right), \qquad (14)$$

where $\langle n\rangle$ stands for the averaged charged multiplicity and $P_n = \sigma_n/\Sigma_n\sigma_n = \sigma_n/\sigma$.

As mentioned at point 2, at the beginning of this Section, the transverse momentum distribution of jets indicates a two component structure. Such structure will reflect on the sphericity distribution as demonstrated in Section 2.

Therefore one can apply a parameterization which makes use of a two component distribution function as follows:

$$\frac{1}{n\sigma_n}\frac{d\sigma_n}{dS} = A_1 \exp\left(-\frac{\alpha_1}{n}S\right) + A_2 \exp\left(-\frac{\alpha_2^2}{n}S\right), \tag{15}$$

and

$$\frac{1}{\langle n\rangle\sigma}\frac{d\sigma}{dS} = \sum_n \frac{n}{\langle n\rangle} P_n \left\{A_1 \exp\left(-\frac{\alpha_1}{n}S\right) + A_2 \exp\left(-\frac{\alpha_2}{n}S\right)\right\}. \tag{16}$$

Using data available to the author, formulae (15) and (16) have been applied to determine the parameter α introduced in Section 2.

The data samples used are:

3746	four	prong events
3996	six	prong events
3434	eight	prong events
4332	10-to-18	prong events

obtained in an experiment performed exposing the 30′ rapid cycling bubble chamber of the International Hybrid Spectrometer Consortium to tagged $\pi^\pm$, K^+, p beams of momentum 150 GeV/c from the Fermilab Tevatron.

In order to confront comparable data samples, all multiplicities with $8 < n < 20$ have been combined to represent the high multiplicity distributions which scarcely show any two component feature at least at an energy as low as $\sqrt{s} \sim 15$.

Hence, in the actual fitting procedure, instead of (15) and (16) the following parameterization has been adopted:

$$\frac{1}{n\sigma_n}\frac{d\sigma_n}{dS} = A\left\{R_n \exp\left(-\frac{\alpha_1}{n}S\right) + (1-R_n)\exp(-\beta S)\right\}, \tag{17}$$

for the semi-inclusive distributions and

$$\frac{1}{\langle n\rangle\sigma}\frac{d\sigma}{dS} = \sum_n \frac{n}{\langle n\rangle} P_n A\left\{R_n \exp\left(-\frac{\alpha_1}{n}S\right) + (1-R_n)\exp(-\beta S)\right\}, \tag{18}$$

for the inclusive distribution.

For a direct comparison with the experimental distributions of events, use is made of the formula:

$$\frac{dN_n}{dS} = A_n\left\{R_n \exp\left(-\frac{\alpha_1}{n}S\right) + (1-R_n)\exp(-\beta S)\right\}, \tag{17'}$$

for the event number distributions at fixed number of charged prongs and

$$\frac{dN}{dS} = \sum_n A_n\left\{R_n \exp\left(-\frac{\alpha_1}{n}S\right) + (1-R_n)\exp(-\beta S)\right\}, \tag{18'}$$

for the total distribution. The ratio R_n of the two components is multiplicity dependent; i.e., R_n approaches 1 as n approaches 4, whereas $R_n = 0.0$ for the high multiplicity events having $n > 8$. The experimental distributions Ref. 7a are collected in Figs. 1 to 4.

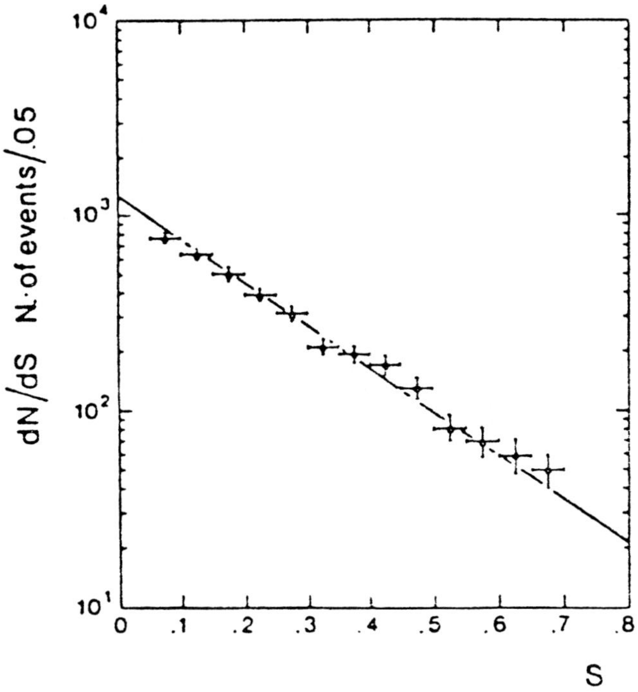

Fig. 1. Sphericity distributions for high multiplicity events ($8 < n < 20$) in the range $0.06 < S < 0.70$.

Figure 1 shows the "high multiplicity" sphericity distribution for $n > 8$. Figure 2 shows the sphericity distributions for 4 prong events; Fig. 3 the distribution for 6 prong events and Fig. 4 the distribution for 8 prong events.[13] The full lines in Figs. 1 to 4 represent the predictions of formula (17′), fitted contemporarily to the four independent distributions, using the best fitted values reported in Table 1.

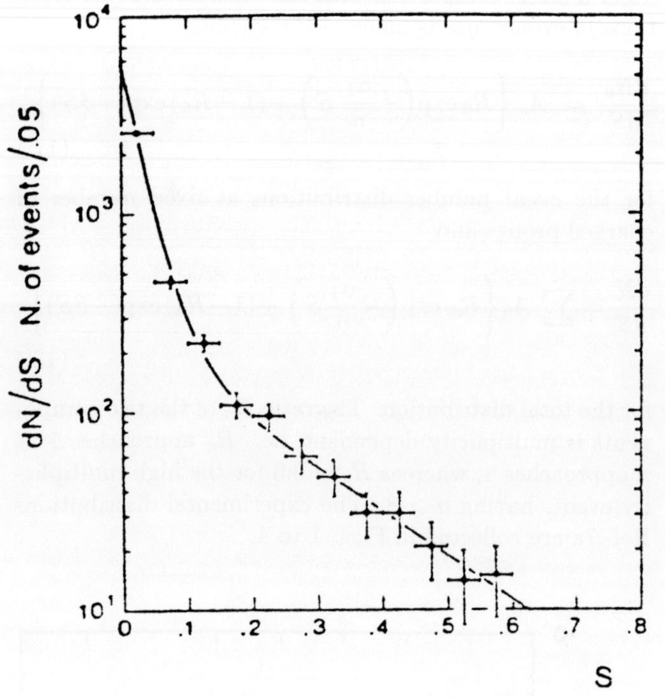

Fig. 2. Sphericity distribution for 4 prong events.

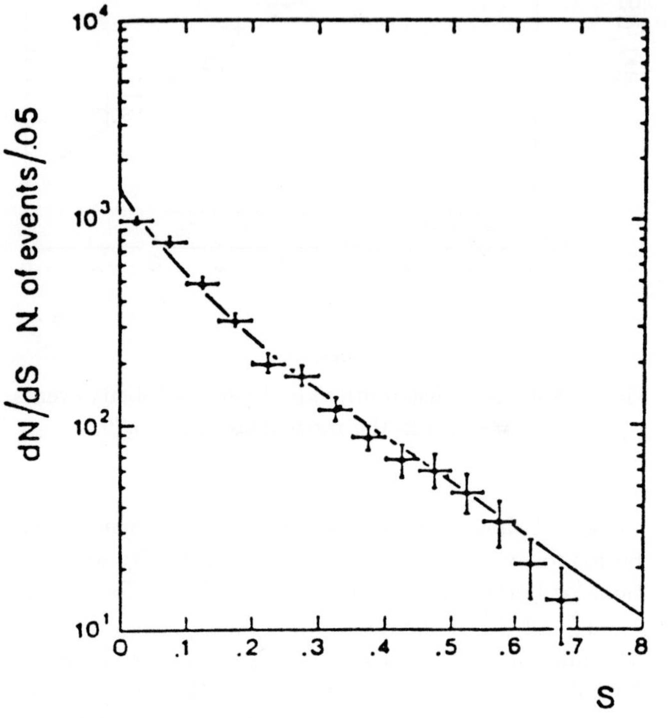

Fig. 3. Sphericity distribution for 6 prong events.

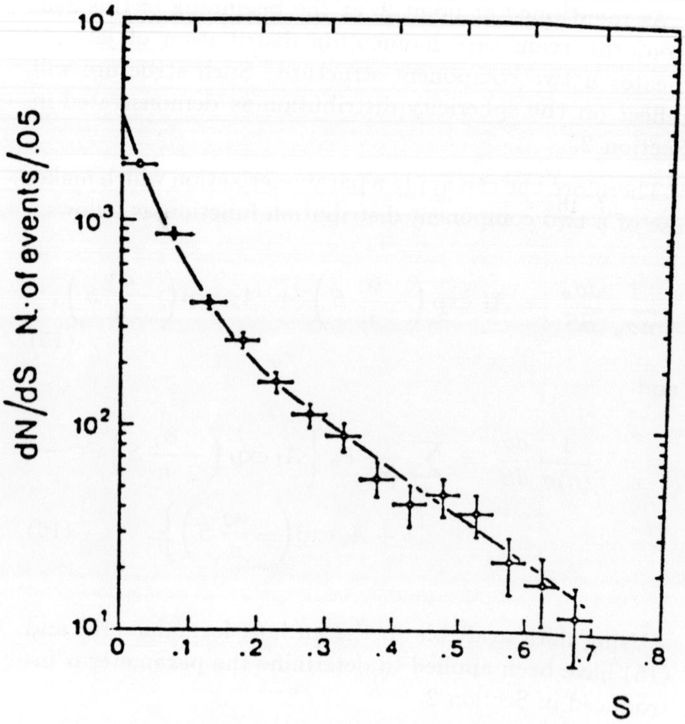

Fig. 4. Sphericity distribution for 8 prong events.

Table 1

α	141 ± 6	β	5.11 ± 0.12
A_4	5846 ± 319	A_6	3190 ± 135
A_8	1447 ± 72	A_{10-18}	1254 ± 42
R_4	0.958 ± 0.004	R_6	0.850 ± 0.012
R_8	0.531 ± 0.043	$R_{10-18} =$	0.0

(The χ^2 of the fit is 56.4 for 55 degrees of freedom.)

The inclusive distribution shown in Fig. 5 is very well reproduced by formula (18), with an excellent confidence level.

4. AVERAGE SPHERICITY VS. CHARGE MULTIPLICITY

As a consequence of the calculations in Sections 2 and 3, one can easily understand Fig. 6 reproducing Ref. 7a the experimental data on the variation of $\langle S_n \rangle$ vs. n_{CH}.

In the comparatively small n region available to a statistically significant analysis at low energy, and in view of the fact that $\alpha \sim 140$, the linear behavior of $\langle S_n \rangle$ comes from the first component. $\langle S_n \rangle$ becomes then roughly constant in the region $n > 8$ because of the second component, the decreasing effect of R_n and the limitations due to phase space.

The comparison between the experimental values taken from Ref. 7 and those calculated by $(17')$ is done in Table 2. It shows an excellent agreement apart from the region dominated by phase space limitations.

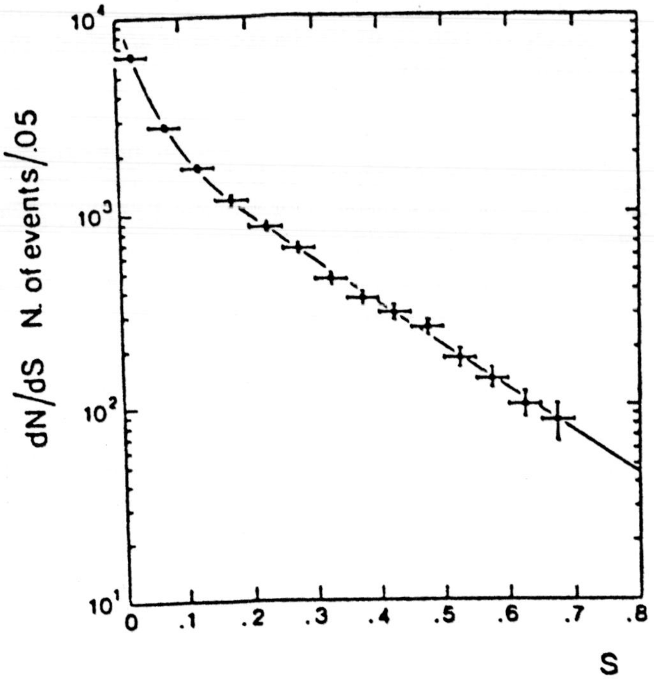

Fig. 5. Inclusive sphericity distribution for all events with $n < 2$.

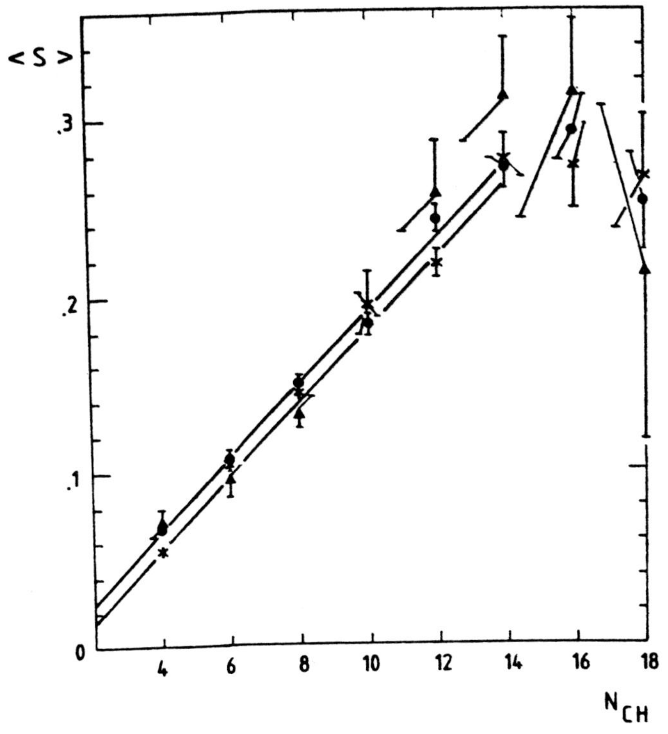

Fig. 6. Average sphericity vs. charged multiplicity $\times$–π; ▲–k; ●–p.

Table 2

	Experimental values $\langle S_n \rangle$	Theoretical values (17′)
n		
4	0.067 ± 0.008	0.067
6	0.102 ± 0.008	0.106
8	0.142 ± 0.009	0.156
$8 < n < 20$	0.218 ± 0.057	0.195

Now, formula (12) is only approximately linear and show a flattening effect when n approaches the value α (Fig. 7). Obviously (12) doesn't contain any of the phase space constraints at the high multiplicity boundary.

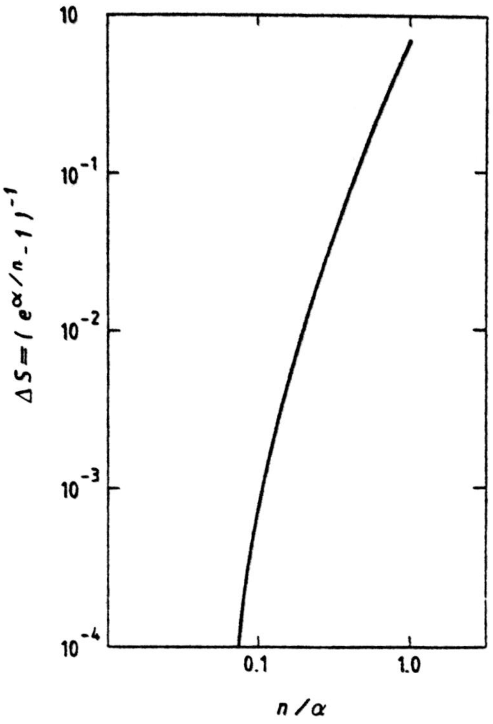

Fig. 7. Correction to the linear behavior as given by the second term in (13′).

5. THE SSC REGION

When going to the SSC energy region, the picture of long range strong interactions might change considerably, however there will be no way to kill the very soft hadron collisions. If the fragmentation of quarks into soft pions is a fundamental process, the peculiarities observed at low energies should not disappear completely.

The dominant factor $\exp(-\alpha/n)S$, consequence of the factorization and the assumption of a Gaussian shape in the transverse variables, should survive if the two assumptions are valid.

Should all this be correct, the calculation would provide an easy and reasonable tool to estimate the background on top of which minijets and other harder phenomena would be detected.

As far as the linear dependence of $\langle S_n \rangle$ vs. n_{CH}, if the indication (12) of the collider data is confirmed, the average charged multiplicity at the SSC would be $\langle n_{CH} \rangle \sim 100$.

Being $\alpha \sim 140$ and being α presumably slowly dependent on energy, due to the nature of the fragmentation phenomenon, the linearity should be observable within few percents up to multiplicities of about $n_{CH} \sim 30 - 40$. In fact, for a value of $n/\alpha = 0.3$ the linearity would be off by only about 3%.

REFERENCES

1. P. V. Landshoff in: "Experiments, Detectors and Experimental Areas for the Supercollider." Edited by R. Donaldson and M.G.D. Gilchriese. World Scientific (1987), p. 811.
2. E. Fahri: Phys. Rev. Lett. 39, 1587 (1977); b) J. B. Babcock, R. E. Kutkosky: Zeit. Phys. C. 15, 133 (1982).
3. J. D. Bjorken, S. J. Brodsky, Phys. Rev. D1, 1416 (1970).
4. See for instance a) A. Giovannini, G. C. Mantovini, S. P. Ratti: Rivista del Nuovo Cimento 2, 1, (1979); b) S. P. Ratti, invited talk to the "Europhysics Study Conference on Jet Structures from Quark and Lepton Interactions;" c) G. C. Fox, S. Wolfram: Nucl. Phys. B149, 413 (1979); d) R. Goettgens *et al.*, Nucl. Phys. B206, 349 (1982).
5. S. P. Ratti, E. R. Nakamura, G. Introzzi: Zeit. Phys. C. 24, 171 (1984); b) S. P. Ratti, G. Introzzi, E. R. Nakamura in: Multiparticle Dynamics 1985, Eds. G. Gustafson, C. Peterson. World Scientific (1985) p. 36.
6. B. Anderson in: Multiparticle Dynamics 1981. Eds. W. D. Shephard and V. P. Kenney. World Scientific (1981), p. 285; b) R.D. Field, R.P. Feynmann: Nucl. Phys. B136, 1 (1978).
7. I.H.S.C., D. Brick *et al.*, paper #959 to the XXth Int. H.E.P. Conference; b) S.P. Ratti in: High Energy Physics 1980. Eds., L. Durand, L.G. Pondrom. A.I.P. Conf. Proc. N. 68, Particles and Fields subseries N.22 A.I.P. (1981), p. 84; c) D. Brick *et al.*, Zeit. Phys. C. 15, 1 (1982).
8. TASSO collaboration: R. Brandelik *et al.*, Phys. Lett. 88*B*, 243 (1979).
9. D.P. Barber *et al.*, Phys. Lett. 108*B*, 63 (1982).
10. a) D. Brick *et al.*, Zeit. Phys. C. 11, 335 (1982); b) R. Goettgens *et al.*, Zeit. Phys. C. 9, 21 (1981); c) M. Barth *et al.*, Nucl. Phys. *B*192, 289 (1981).
11. A. Giovannini and L. Van Hove: Zeit. Phys. *C*30, 391 (1986).
12. G.J. Alner *et al.*, Phys. Lett. *B*138, 304 (1984); *B*160, 193 (1985).
13. In order to make sensible chi square fits, we limited the distributions in Figs. 2 to 4, to those bins containing at least 10 events ($S < 0.725$) and thus 3 four-prong events, 12 six-prong events and 14 eight-prong events were omitted in the fit. As for the data of Fig. 1, the sphericity interval used is $0.075 < S < 0.825$.

WEAK DECAYS

Michael Witherell and Roberto Peccei

Working Group Leaders

Weak Decays: Theoretical Summary

G. Bélanger [1], C. Dib [2], C. Geng [3], F. Gilman [2], J. Hewett [4†], B. Kayser [5‡], C.S. Kim [6△], M. Kuroda [7], D. London [8], S. Matsuda [9], J.N. Ng [3], R.D. Peccei [8*], A.I. Sanda [10], T. Walsh [11]

[1] Univ. of Montreal, [2] SLAC, [3] TRIUMF, [4] Ames Lab. and Iowa State U. [5] LBL, [6] Univ. of Wisconsin, [7] Meiji-Gakuin University, [8] DESY, [9] Kyoto University [10] Rockefeller University, [11] Univ. of Minnesota

Abstract

The present status of the Cabibbo Kobayashi Maskawa matrix is reviewed and the prospects for improving our knowledge of the various elements by K and B decay experiments in the 1990 s are critically examined. Windows provided by these investigations for physics beyond the standard model are also discussed. Particular attention is devoted to the expected patterns of CP violation in the B system.

1 The Cabibbo Kobayashi Maskawa Paradigm

The $SU(2) \times U(1)$ gauge theory of Glashow Salam and Weinberg [1] is the present paradigm for the electroweak interactions. Within the standard model, hadronic flavor changing weak transitions have a somewhat less secure role, since they require some knowledge of the mass generation mechanism in the quark sector. The simplest possibility to give mass to the fermions in the theory makes use of Yukawa interactions involving the doublet Higgs field. These interactions give rise to, what might be called, the Cabibbo Kobayashi Maskawa (CKM) paradigm: Quarks of different flavor are mixed in the charged weak currents by means of an unitary matrix V. However, both the electromagnetic current and the weak neutral current remain flavor diagonal. Departures from this pattern such as, for example, finding evidence for a flavor changing neutral current transition like $K \to \mu e$, would belie this simple hypothesis and be a sign for new physics.

Second order weak processes are on even less secure ground, since they can be affected by both beyond the standard model virtual contributions, as well as direct new physics contributions. Thus they provide very stringent tests of the CKM paradigm. For example, both the $\Delta S = 2$ transition responsible for K-$\overline{K}$ mixing, and the CP violating parameter ϵ in neutral kaon decays are governed by the size of the CKM matrix elements entering in Fig. 1. However, if low mass superpartners existed, or if not too massive diquarks existed, additional contributions, like those shown in Fig. 2, would vitiate these predictions. Hence, accurate comparisons between experiment and the predictions of the CKM paradigm can give indications either for, or against, new physics. It is clear that one of the main tasks in the 1990 s will be to either establish – at a higher level of accuracy – the CKM paradigm, or to disprove it.

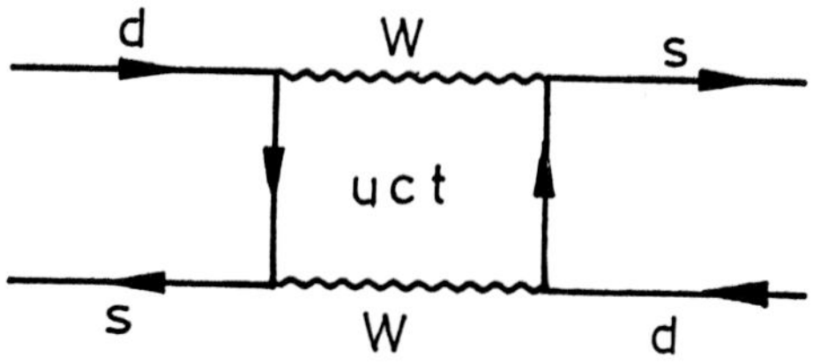

Figure 1: W exchange graph for the $\Delta S = 2$ process, which depends on V_{is} and V_{id} $(i = u, c, t)$.

The task of testing the CKM paradigm is much simpler if there exist **only** 3 generations. In this case, as is well known, the CKM matrix depends on three real angles and only one phase, since the other phases entering in the 3×3

* Convener
Present address: † University of Wisconsin; ‡ NSF; △ Durham University

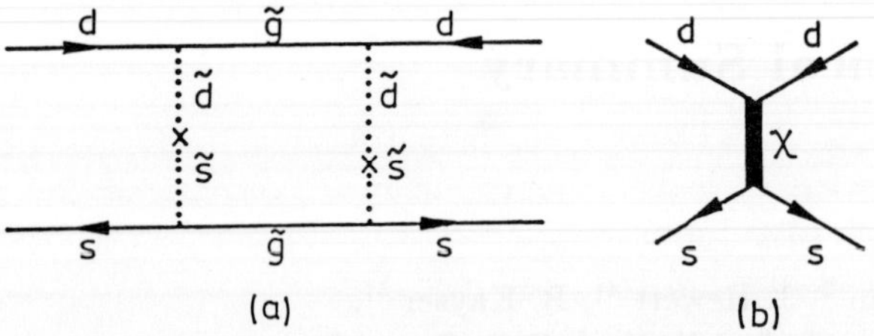

Figure 2: Possible supersymmetric contribution (a) and diquark contribution (b) to the $\Delta S = 2$ process.

unitary matrix V can be absorbed by a redefinition of the quark fields. For 4 generations, on the other hand, one needs 6 angles and 3 phases to determine V. In this study, as a benchmark, we have focused on the 3 generation case. There are a number of very good practical reasons to do so:

1. We shall soon know from SLC and/or LEP whether the number of neutrino species – and thus, presumably, the number of generations – is 3.
2. Unitarity $[V^\dagger V = 1]$ strongly restricts V_{ij}, even without any explicit evidence for a t quark.
3. In the case of 3 generations, experiment indicates a nice hierarchical pattern in V, as a function of the Cabibbo angle: $\sin\theta_c = \lambda$

$$V \sim \left| \begin{array}{ccc} 1 & \lambda & \lambda^3 \\ \lambda & 1 & \lambda^2 \\ \lambda^3 & \lambda^2 & 1 \end{array} \right| \tag{1}$$

A convenient form of the CKM matrix, which displays this hierarchy in a simple way, is[1]

$$V = \left| \begin{array}{ccc} c_1c_3 & s_1c_3 & s_3e^{-i\delta} \\ -s_1c_2 - c_1s_2s_3e^{i\delta} & c_1c_2 - s_1s_2s_3e^{i\delta} & s_2c_3 \\ s_1s_2 - c_1c_2s_3e^{i\delta} & -c_1s_2 - s_1c_2s_3e^{i\delta} & c_2c_3 \end{array} \right| \tag{2}$$

In the above, s_i, c_i are shorthand for $\sin\theta_i$ and $\cos\theta_i$. The experimental hierarchy of Eq. (1) is realized by writing

$$s_1 = \lambda; \quad s_2 = A\lambda^2; \quad s_3 = A\rho\lambda^3 \ , \tag{3}$$

where A and ρ are real numbers. Since $\lambda \simeq 0.22$ one can approximate [3]

$$V = \left| \begin{array}{ccc} 1 - \frac{\lambda^2}{2} & \lambda & A\rho\lambda^3e^{-i\delta} \\ -\lambda(1 + A^2\rho\lambda^4e^{i\delta}) & 1 - \frac{\lambda^2}{2} & A\lambda^2 \\ A\lambda^3(1 - \rho e^{i\delta}) & -A\lambda^2 & 1 \end{array} \right| \tag{4}$$

[1]This is now the form of the CKM matrix "officially" adopted by the Particle Data Group [2].

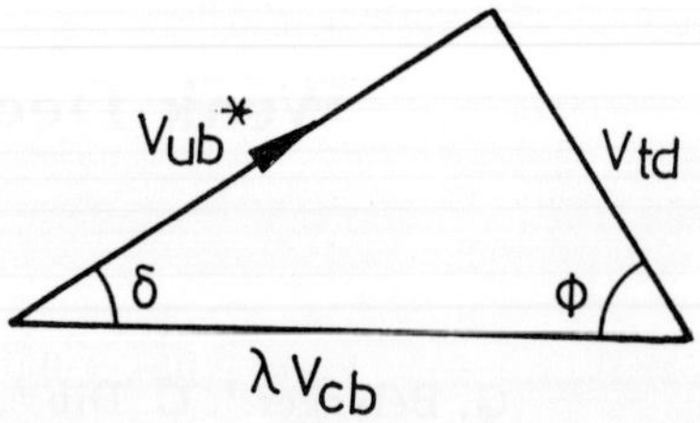

Figure 3: The unitarity triangle

Note that the only elements which have a potentially large phase in the CKM matrix V, in this parametrization, are the ub entry, $|V_{ub}|e^{-i\delta}$, and the td entry, $|V_{td}|e^{-i\phi}$, where the phase ϕ is related to δ in an obvious way. The unitarity of the CKM matrix, along with the hierarchy, implies a simple interrelation among the matrix elements

$$V_{ub}^* + V_{td} = \lambda V_{cb} \ . \tag{5}$$

That is, the elements V_{ub}^*, V_{td} and λV_{cb} form a triangle in the complex plane [4]. The phases δ and ϕ are two of the angles of this triangle, as shown in Fig. 3. As we shall see, CP violating asymmetries in B decays can directly measure the various angles of the unitarity triangle.

1.1 Status of the CKM Matrix

Apart from the Cabibbo angle, which is well determined, the remaining parameters of the CKM matrix today are still quite uncertain. Roughly speaking, λ is determined to 1%, A to 10%, while the uncertainty in ρ and in the phase δ is of the order of 100%! The value of λ, obtained by Leutwyler and Roos [5] from an analysis of K_{l3} decays and Hyperon β-decay, is

$$V_{us} = \lambda = 0.221 \pm 0.002 \ . \tag{6}$$

This value, plus unitarity, implies a value for V_{ud} which is in excellent agreement (to the per mil level) with the direct determination of this matrix element, from a comparison of μ-decay with 0^+-0^+ nuclear transitions [6].

The parameter A in the CKM matrix, being related to V_{cb}, is fixed by the B lifetime and by the semileptonic B decay spectrum. In principle, since $|V_{ub}|^2 \ll |V_{cb}|^2$, the B lifetime measurement gives directly A. However, since in a simple spectator model $\tau_B \sim m_b^{-5}$, the uncertainty in the value of the b quark mass is far too large to allow for a good determination of A. To do better, one must eliminate m_b in favor of m_B – the B meson mass. This requires constructing some (simple) bound state model, whose parameters can be fitted from the lepton spectrum in B semileptonic decays [7]. A recent analysis along these lines, by Altarelli and Franzini [8], gives

$$A = 1.05 \pm 0.17 \quad \leftrightarrow \quad V_{cb} = 0.051 \pm 0.008 \ . \quad (7)$$

One of us (CSK) has reanalyzed this question independently, by using CRYSTAL BALL data [9] and the bound state model of Ref. [7], obtaining similar results to those given in Eq. (7). In particular, using the new value for the B semileptonic branching ratio [10,11] $B(B \to l\nu_l X) = 0.10 \pm 0.01$, this reanalysis gives $A = 1.03 \pm 0.14 \leftrightarrow V_{cb} = 0.050 \pm 0.007$, in excellent agreement with the results of [8]. Unfortunately, at present, it appears very difficult to reduce the error on A below the 10% level.

The parameter ρ is related to V_{ub}. Since no evidence exists for charmless semileptonic B decays, one can obtain an upper bound on ρ from the study of the lepton spectra in these processes. This is a, somewhat, model dependent bound since the detailed predictions for the lepton spectrum shape depend on the bound state model one assumes for the B meson. Adopting again the simple model of [7] and analyzing the CRYSTAL BALL data [9] – using the parameter set that provides the best fit – yields[2]

$$|\frac{V_{ub}}{V_{cb}}| < 0.18 \ . \quad (8)$$

We note that the recent analyses of the ARGUS and CLEO collaborations [10,11] each give a very similar limit $|\frac{V_{ub}}{V_{cb}}| < 0.16$. In terms of ρ, Eq. (8) implies

$$\rho < 0.8 \ . \quad (9)$$

In principle, the observation by the ARGUS collaboration [12] of the charmless B decay $B \to p\bar{p}\pi(\pi)$ provides a lower bound for ρ. However, the ARGUS observation has not been confirmed by CLEO [13]. Furthermore, extracting a reliable lower bound for ρ from the ARGUS value is difficult theoretically. Nevertheless, if one had to hazard a guess from present day B decay evidence, one would put $\rho \geq 0.2$-0.3.

The phase δ and ρ are constrained by the observation of CP violation in the Kaon system and by B-$\overline{B}$ oscillations. Since there would be **no** CP violation if δ or ρ vanished[3], the observed value of ϵ in Kaon decays provides bounds for these parameters. The allowed region for δ and ρ is correlated. The standard analysis of Buras et al. [14] gives for ϵ a formula which is essentially proportional to the imaginary part of the box graph in Fig. 1. The final answer depends on the value assumed for m_t, even though the t quark graph is Cabibbo suppressed by a factor of λ^4 with respect to the other contributions, since it is proportional to m_t^2. There is, furthermore, another uncertainty connected with the value of the K-$\overline{K}$ matrix element of the effective Hamiltonian, which is not precisely known theoretically[4], with the allowed region in the ρ-δ plane shrinking or expanding depending on what one assumes for this matrix element.

The B_d-$\overline{B_d}$ mixing parameter $x_d = \frac{(\Delta m)_{B_d}}{\Gamma_{B_d}}$ is given by the real part of a box graph similar to that of Fig. 1, with $s \leftrightarrow b$. For x_d, the t quark contribution in the box graph really dominates and one has that

$$x_d \sim |V_{td}|^2 \sim (1 + \rho^2 - 2\rho\cos\delta) \ . \quad (10)$$

Thus B_d-$\overline{B_d}$ mixing can be used to restrict the range for ρ and δ. Unfortunately, again there are corresponding uncertainties; x_d also depends on m_t (essentially quadratically) and on the value of the hadronic matrix element[5].

We show in Fig. 4, as an illustration, the allowed region in the ρ-δ plane from the combined fit of ϵ and x_d. In this Figure, taken from [15], we let m_t range over 40 GeV $< m_t <$ 180 GeV, fix $B_K = \frac{2}{3}$ and assume 100 MeV $< f_{B_d} B_{B_d}^{\frac{1}{2}} <$ 200 MeV. In this analysis, furthermore, ρ was restricted to be $\rho < 0.9$ [compare with Eq. (9)]. A knowledge of m_t would considerably restrict the allowed region in the ρ-δ plane. Fig. 5 illustrates this, even in the case when one assumes, besides the previously considered uncertainty for the hadronic matrix element in x_d, also a rather large uncertainty in the estimate of the Kaon matrix element [$1/3 < B_K < 1$].

To really test the CKM paradigm it will be necessary, in the future, to considerably restrict the rather large region now allowed in the ρ-δ plane. In principle, this can be achieved in a number of ways:

- More accurate theoretical calculations of the various hadronic matrix elements would be very helpful. It is possible – although by no means certain – that the large effort now underway in lattice calculations of weak matrix elements [16] may bear fruit in the 1990's and provide us with predictions with an accuracy of 10%. Without this kind of precision, even beautiful new data, like the NA31 result on ϵ' [17], cannot provide further constraints on the CKM model [18].

- Confirmation of the ARGUS direct signal for charmless B decay may provide a lower bound on ρ. However, again, here one is faced with difficulties in

[2]Conventionally one quotes a bound for $R = \frac{\Gamma(b\to u)}{\Gamma(b\to c)}$, which is related to the Kobayashi Maskawa matrix element ratio (8) up to a phase space factor. For the best fit parameter set, this phase space factor is 2.10 ± 0.15.

[3]$\delta = \pi$ is also a CP conserving point.

[4]Conventionally, one normalizes one's ignorance by quoting a value for the, so called, B_K parameter – the ratio of the matrix element $\langle K^0|[\bar{d}\gamma_\mu(1-\gamma_5)s]^2|\overline{K^0}\rangle$ to that obtained by vacuum saturation.

[5]The conventional normalization of one's ignorance here is done via the parameter combination $f_{B_d}^2 B_{B_d}$, because in this case the B_d decay constant is also unknown.

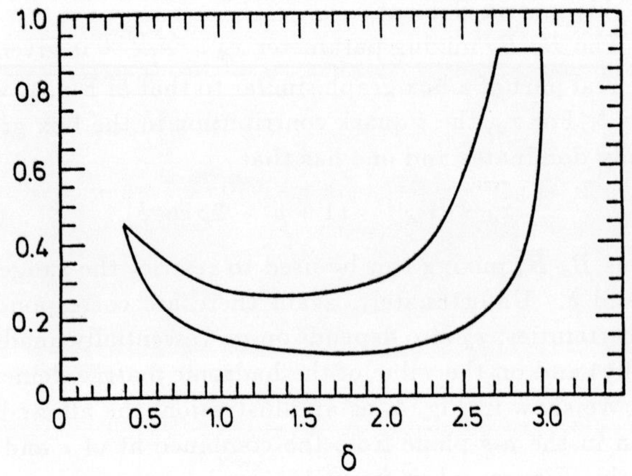

Figure 4: Allowed region in the ρ-δ plane, from [15], for the range of parameters given in the text.

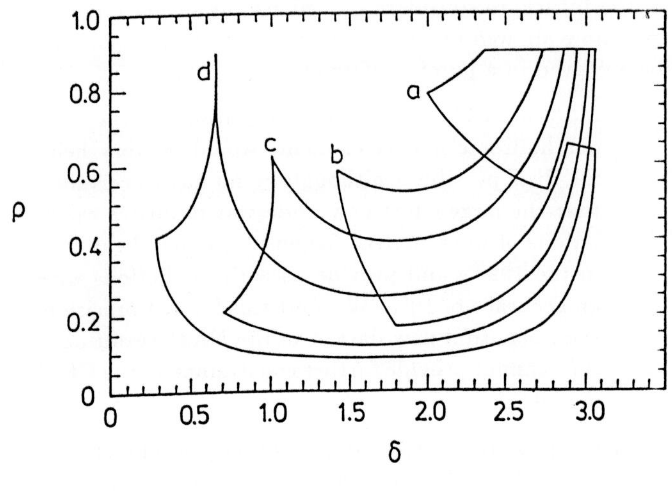

Figure 5: Allowed regions in the ρ-δ plane for various values of m_t: (a) m_t = 60 GeV; (b) m_t = 90 GeV; (c) m_t = 120 GeV; (d) m_t = 180 GeV. From [15].

trying to reliably estimate some two-body hadronic matrix element. A measurement of the process $B \to \rho l \nu_l$ may, in this respect, be more helpful to pin down $|V_{ub}|$.

- Measurement of the B_s-$\overline{B_s}$ mixing parameter x_s will provide, in conjunction with x_d and an assumption about (or a calculation of) the ratio $f_{B_s} B_{B_s}^{\frac{1}{2}} / f_{B_d} B_{B_d}^{\frac{1}{2}}$, a further constraint on ρ and δ.

- The decay $K^- \to \pi^- \nu \bar{\nu}$ measures a combination of CKM matrix elements weighted differently than that appearing in ϵ and x_d. Thus, this decay can provide additional constraints, particularly since the estimate of the hadronic matrix elements here is much more under control. A measurement of the very rare process $K_L^0 \to \pi^0 \nu \bar{\nu}$ would also provide new information.

- The measurement of different CP violating hadronic asymmetries in B decays directly tests the phase structure of the CKM matrix. Hence, these experiments could provide the most direct test of the CKM paradigm.

In the subsequent sections, we shall address some of the theoretical issues connected with the points raised above.

2 A Few Crucial K Decays

The NA31 [17] positive evidence for **direct** $\Delta S = 1$ CP violation is a very important result, since it kills the still extant superweak explanation [19] for CP violation in the Kaon system. Nevertheless, it is not possible to obtain an accurate theoretical estimate, within the CKM paradigm, to compare with the NA31 result [17]:

$$\mathrm{Re}\frac{\epsilon'}{\epsilon} = (3.3 \pm 1.1) \times 10^{-3} . \qquad (11)$$

In the CKM model ϵ' arises through the Penguin graph of Fig. 6. Unfortunately, the matrix element of the associated Penguin operator $O_p = (\bar{s}\gamma^\mu(1-\gamma_5)\lambda_a d)(\bar{q}\gamma_\mu\lambda_a q)$ between a Kaon and two pions is theoretically uncertain. The theoretical explanation for the ratio (11) is, furthermore, also affected by our inability to calculate ϵ precisely. This notwithstanding, the central value measured by NA31 is in excellent agreement with the central value of the theoretical expectation in the CKM model [18], but a theoretical excursion of a factor 2-3 is permitted.

The above status of affairs suggest the following three recommendations:

1. It is important to pursue an independent measurement of $\frac{\epsilon'}{\epsilon}$ to check that $\epsilon' \neq 0$. This is presently being done at Fermilab and will soon be attempted

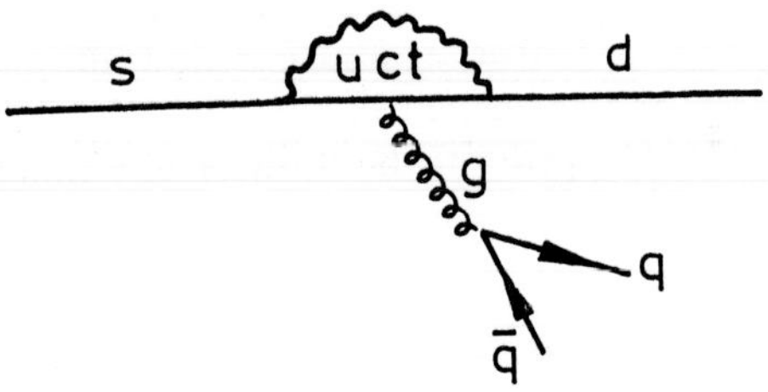

Figure 6: Penguin graphs contributing to ϵ'.

at LEAR[6]. However, given the theoretical uncertainties in precisely predicting a value for ϵ', the knowledge of a very accurate experimental value for ϵ' seems superfluous.

2. One should explore other K decays which are sensitive to direct CP violation. Among these, the decay channel $K_L^0 \to \pi^0 e^+ e^-$ may provide a particularly interesting example, as we shall discuss in more detail below.

3. With the benefit of hindsight, at least in the CKM model, the Kaon system appears to be the **wrong place** to look for CP violation. The dominant, 2nd order weak transitions between an s and a d quark (shown in Fig. 7) are approximately real. For the corresponding transitions from b to d, the real and imaginary parts are of the same order in the Cabibbo angle. Thus, if one believes in the CKM paradigm, the place to look for CP violation is in the B system.

2.1 The decay $K_L^0 \to \pi^0 e^+ e^-$

The weak decay $K_L^0 \to \pi^0 e^+ e^-$ is particularly interesting dynamically since there are numerous physical mechanisms at play simultaneously. The decay amplitude $K_L^0 \to \pi^0 e^+ e^-$ can be related to that of the process $K_L^0 \to \pi^0 V^*$, where V^* is an effective $J = 1$ CP even state. This latter amplitude violates CP, since $(\pi^0 V^*)_{J=0}$ has CP $= +1$. Through this channel, therefore, $K_L^0 \to \pi^0 V^*$ proceeds either by **indirect** CP violation, because K_L^0 has a small CP even admixture $[K_L^0 = \epsilon K_1]$, or through the **direct** CP violating decay of the CP odd component of K_L^0:

[6]The CP LEAR experiment does not quite measure $\frac{\epsilon'}{\epsilon}$, but a hadronic asymmetry in the Kaon system - quite analogous to those we will discuss in the B system - which is proportional to ϵ'.

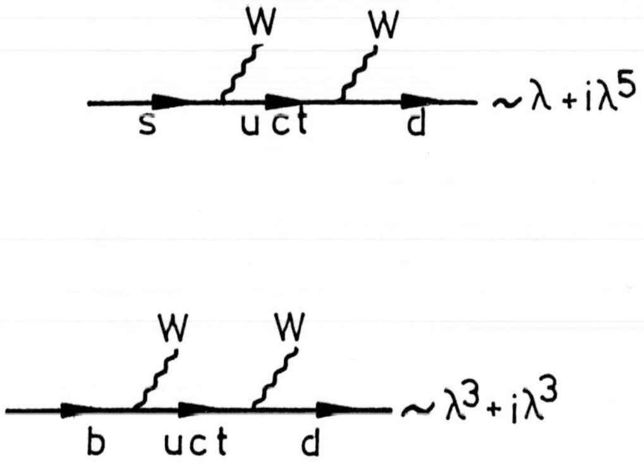

Figure 7: (a) 2nd order weak transition from s to d; (b) 2nd order weak transition from b to d.

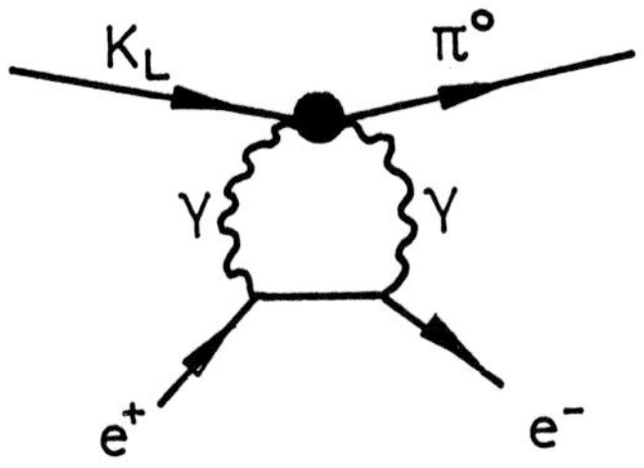

Figure 8: CP conserving contribution to $K_L^0 \to \pi^0 e^+ e^-$, via a two photon intermediate state.

$K_2 \to \pi^0 V^*$. In addition, however, there is an effective two photon contribution to the process, as shown schematically in Fig. 8. This contribution is CP conserving, but is of higher order in α.

One can estimate (CD,FG) [20] the associated branching ratio due to the indirect CP violating contribution as follows. If one assumes a pure $\Delta I = \frac{1}{2}$ transition, then

$$\begin{aligned} &B(K_L^0 \to \pi^0 e^+ e^-)_{\text{ind CP viol}} \\ &\quad = B(K^+ \to \pi^+ e^+ e^-) \cdot \frac{\Gamma(K^+ \to \text{all})}{\Gamma(K_L^0 \to \text{all})} \cdot \\ &\quad \frac{\Gamma(K_1 \to \pi^0 e^+ e^-)}{\Gamma(K^+ \to \pi^+ e^+ e^-)} \cdot \frac{\Gamma(K_L^0 \to \pi^0 e^+ e^-)_{\text{ind CP viol}}}{\Gamma(K_1 \to \pi^0 e^+ e^-)} \\ &\quad = 2.7 \times 10^{-7} \cdot 4.2 \cdot 1 \cdot |\epsilon|^2 \\ &\quad \simeq 6 \times 10^{-12} \,. \end{aligned} \qquad (12)$$

This number is in agreement with the estimates of [21,22] and is typically of the same order or less than that ob-

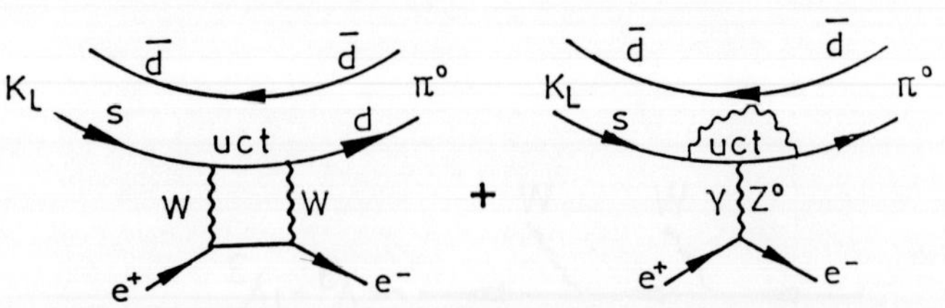

Figure 9: Graphs contributing to the direct CP violating decay of $K_L^0 \to \pi^0 e^+ e^-$.

tained from a calculation of the relevant graphs, shown in Fig. 9, causing the direct CP violating transition. The direct CP violating amplitude, being mostly a short distance electroweak contribution, is calculable [23]. However, the actual answer one obtains depends on the unknowns of the CKM paradigm: the values of m_t and the range in the ρ-δ plane. A recent analysis done by two of us (CD,FG) with I. Dunietz [20], including QCD corrections, indicates that

$$B(K_L^0 \to \pi^0 e^+ e^-)_{\text{dir CP viol}} \simeq (2 - 20) \times 10^{-12} , \quad (13)$$

where the larger branching ratios correspond to bigger values for m_t.

The CP conserving 2γ contribution to the decay $K_L^0 \to \pi^0 e^+ e^-$ was estimated earlier [22] to be negligible. If this were the case, this decay could provide a useful test of the CKM paradigm, since the direct CP violating contribution is certainly non-neglible and may in fact dominate[7]. However, some recent papers [24,25] have questioned this point. In particular, using a vector dominance model for the subprocess $K_L^0 \to \pi^0\gamma\gamma$, these authors arrive at a CP conserving branching ratio[8]

$$B(K_L^0 \to \pi^0 e^+ e^-)_{\text{CP conserving}} \simeq (8 - 16) \times 10^{-12} , \quad (14)$$

Several comments are in order:

1. The vector dominance estimate for the CP conserving amplitude is not in disagreement with the formal results of chiral perturbation theory [26]. The tiny estimate for this amplitude obtained by Donoghue et al [22] concerns terms which are of $O(p^4)$ (where p is the pion 4-momentum), but which are strongly suppressed by their helicity structure, giving a rate proportional to m_e^2. The results of [24,25] rely on terms which are of $O(p^6)$ in chiral perturbation theory, but which are not chirally suppressed because an additional tensor structure is now allowed. Nevertheless, the coefficient of the p^6 term which leads to Eq. (14) appears to be numerically much larger than what one would expect simply by counting factors of f_π [21].

2. One may be able to resolve the issue of how large the CP conserving contribution for the $K_L^0 \to \pi^0 e^+ e^-$ decay is by measuring directly the process $K_L^0 \to \pi^0\gamma\gamma$. A sizable $O(p^6)$ contribution would distort the spectrum of the two photons, predicted by chiral perturbation theory [26], and lend support to the vector dominance calculation.

3. Sehgal [24] has made the amusing observation that, in principle, there is a way to disentangle the CP conserving from the CP violating contribution in the decay $K_L^0 \to \pi^0 e^+ e^-$, by looking at the energy distribution of the electrons and positrons. The presence of the CP violating amplitudes leads to an asymmetry in this distribution, between e^+ and e^-. However, although the asymmetry can be large [24], the number of events one is likely to collect in the 1990 s would be far too few for any meaningful test.

2.2 The decays $K^\pm \to \pi^\pm \nu\bar{\nu}$ and $K_L^0 \to \pi^0 \nu\bar{\nu}$

The weak decay $K^\pm \to \pi^\pm \nu\bar{\nu}$ does not directly test CP violation. However, in the CKM paradigm, for the case of three generations, it is sensitive to V_{td} and, therefore, indirectly to the phase δ. The relevant graphs, at the quark level, for this decay are the analogues to those in Fig. 9, but with $e \leftrightarrow \nu_e$ and $\bar{d} \leftrightarrow \bar{u}$. These graphs give rise to an effective Lagrangian at short distances for this process, of the form

$$\mathcal{L}_{\text{eff}} = \frac{G_F}{\sqrt{2}} \sum_l A_l \bar{s}\gamma^\mu(1-\gamma_5)d\bar{\nu}_l\gamma_\mu(1-\gamma_5)\nu_l , \quad (15)$$

with A_l a calculable function of the quark and lepton masses and the CKM angles [23]. Using (15), the matrix element for $K^- \to \pi^-\nu\bar{\nu}$ is directly related to that for $K^- \to \pi^0 e^- \bar{\nu}_e$ and the number of neutrino species:

$$B(K^- \to \pi^-\nu\bar{\nu}) = \frac{2\sum_l |A_l|^2}{|V_{us}|^2} B(K^- \to \pi^0 e^- \bar{\nu}_e) . \quad (16)$$

This result ignores possible long distance contributions to this decay, of the type shown in Fig. 10. However, a simple estimate [27] shows these contributions to be negligibly small, leading to a branching ratio about 2 to 3 orders of magnitude below that arising from Eq. (16).

[7] In the superweak model, where only the $\Delta S = 2$ parameter ϵ is nonvanishing, there would be no direct CP violating contribution and one would expect a branching ratio like that given in Eq. (12).

[8] The value obtained by [25] is larger than this, but we believe this is due to a trivial error in their phase space estimate.

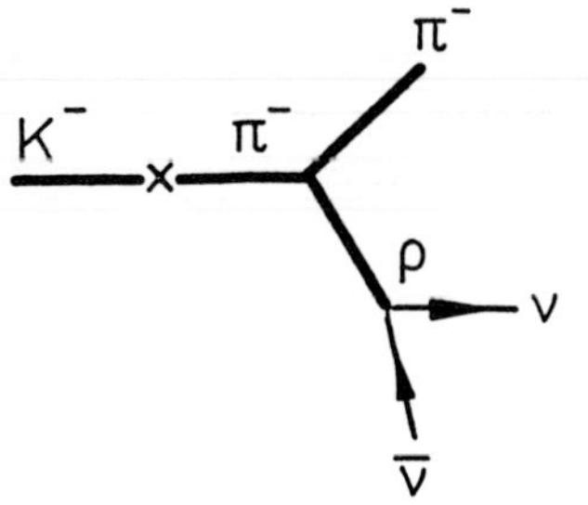

Figure 10: Possible long distance contribution to the decay $K^- \to \pi^- \nu\bar{\nu}$.

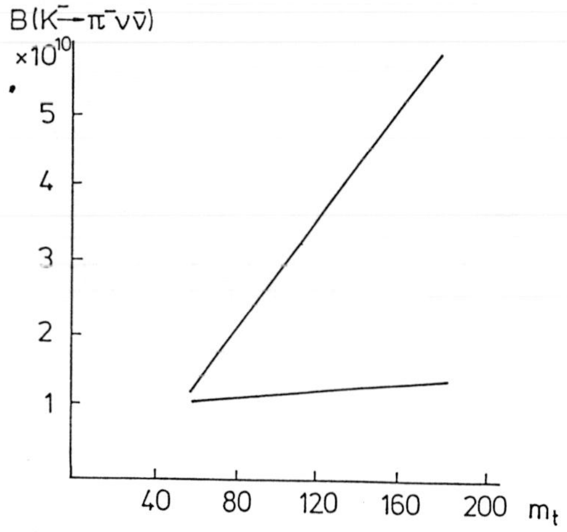

Figure 11: Expected value of $B(K^- \to \pi^- \nu\bar{\nu})$ as a function of m_t.

Using the results of Inami and Lim [23] for the functions A_l (and ignoring a numerically irrelevant contribution which arises from keeping $m_\tau \neq 0$) and the experimental value for the K_{e3} branching ratio, one finds

$$B(K^- \to \pi^- \nu\bar{\nu}) = 6 \times 10^{-7} N_\nu \cdot |F(y_c) + A^2\lambda^4(1-\rho e^{i\delta})F(y_t)|^2 \eta \ . \quad (17)$$

Here N_ν is the number of neutrino species [$N_\nu = 3$ for the three generation case we are considering], while the function $F(y_i)$, with $y_i = m_i^2/M_W^2$, calulated in [23], is given by

$$F(y) = \frac{y}{4}\left[\frac{3(y-2)}{(y-1)^2}\ln y + \frac{y+2}{y-1}\right] . \quad (18)$$

The parameter η in Eq. (17) is a QCD correction to the short distance calculation of Eq. (15), which has been estimated to be close to unity [28] and so will be ignored in what follows[9], given the other uncertainties.

The rate (17) has an irreducible uncertainty connected to the value of the A parameter and that of the charm mass to be used. For instance, for $m_c = 1.5 \pm 0.2$ GeV, and A as given in Eq. (7), typically the branching ratio for $B(K^- \to \pi^- \nu\bar{\nu})$ is only determined to a 50% accuracy. The overall rate is quite sensitive to m_t, ρ, and δ, since the c quark and t quark contributions in (17) are quite comparable. Given the present constraints on these parameters – and using the central values for m_c and A – a straightforward analysis (DL, RDP) gives the wedge shaped prediction for the branching ratio $B(K^- \to \pi^- \nu\bar{\nu})$ shown in Fig. 11, plotted as a function of m_t. The upper values for this branching ratio are obtained for large values of ρ and δ, while the lower line in Fig. 11 corresponds to small values for ρ and δ. Given a value for m_t, a measurement of this branching ratio can provide useful information on the CKM parameters ρ and δ. This requires, however, that one should measure a rate which is, typically, a few times 10^{-10} that of the main K^- decay mode – a tough experimental challenge!

Although, as Fig. 11 shows, the predictions of the three generation model for $B(K \to \pi\nu\bar{\nu})$ lie around 10^{-10}, these expectations can be considerably changed if there are more generations, or if there are additional interactions – like those caused by a charged Higgs boson.

The existence of a fourth generation can dramatically alter the three generation standard model predictions for the branching ratios of rare decays. Relaxed unitarity constraints on the 3×3 CKM matrix yield wider ranges of allowed values for some of the 4×4 CKM matrix elements, V_{ij}, which need no longer scale according to the "natural" hierarchy of Eq. (1). This, combined with the large span of possible fourth family fermion masses (as well as the uncertainty in the t-quark mass, m_t), is responsible for the extension of the SM branching ratios.

One of us (JH) has examined the rare process $K^\pm \to \pi^\pm \nu\bar{\nu}$ (along with $b \to s\gamma$ and $b \to se^+e^-$) in the four generation model. Apart from Penguin-type diagrams, the exchange of the fourth family up-type quark, t', in the process $K \to \pi\nu\bar{\nu}$ also receives a contribution from the fourth generation charged lepton, L, in the box diagrams. In the calculation, the unknown fermion masses were taken in the ranges

$$\begin{aligned} 45 < \ & m_t < 180 \text{ GeV} \\ 100 < \ & m_{t'} < 400 \text{ GeV} \\ 50 < \ & m_L < 300 \text{ GeV}, \end{aligned} \quad (19)$$

subject to $m_{t'} > m_t$. In the spirit of Ref. [29], an extensive numerical search was performed for given pairs of t and t' masses over the six mixing angles and three

[9] In effect, it appears worthwhile to repeat the calculation of [28], since this was done only in certain kinematical limits concerning m_t.

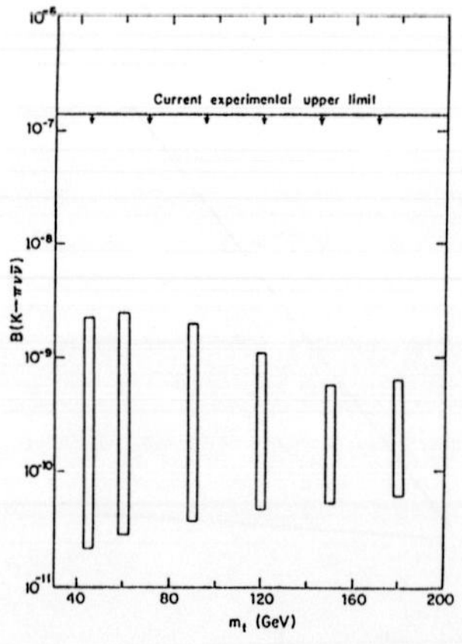

Figure 12: $B(K^- \to \pi^- \nu\bar{\nu})$ predictions with four generations (hollow boxes) as a function of m_t.

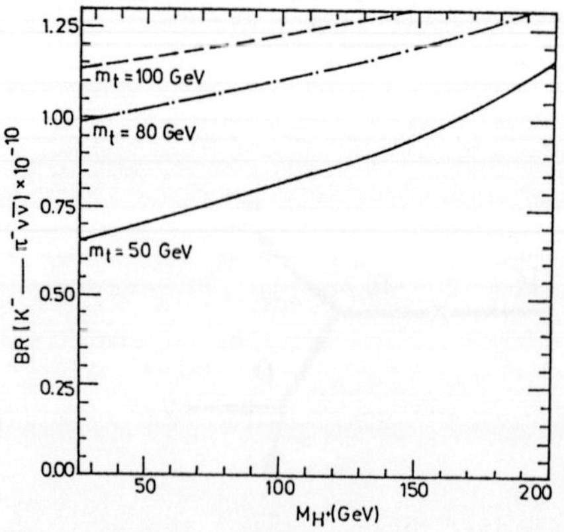

Figure 13: $B(K^- \to \pi^- \nu\bar{\nu})$ predictions in a model with charged Higgses as a function of m_{H^+}, and various values of m_t, with $V_{td} = .01$.

phases in the 4×4 CKM matrix and constrained to fit the existing experimental data discussed in Sec. 1. Using the set of angles which pass these constraints, the ranges of the branching ratios for each process, including QCD corrections, were calculated.

The branching ratio $B(K^- \to \pi^- \nu\bar{\nu})$ within the four generation model is displayed in Fig. 12. The hollow rectangles correspond to the allowed range of branching ratios in the four family case. It is clear from the figure that the addition of a fourth family can greatly extend the branching ratio range, and can either enhance or suppress the branching ratio compared to the three generation standard model value.

This same comment applies if one extends the Higgs sector of the standard model to two doublets, so that one has also physical charged Higgs bosons in the theory. For **fixed** CKM matrix elements V_{ij}, the presence of charged Higgses always **adds** to the standard model rate for $K^- \to \pi^- \nu\bar{\nu}$. However, having charged Higgses in the theory changes the constraints on the V_{ij}. In particular, one now has a larger allowed range for V_{td} and this can change the $K^- \to \pi^- \nu\bar{\nu}$ branching ratio, and may in fact allow values below those predicted by the standard model. An example is provided by Fig. 13, obtained by two of us (CG, JNN), which shows this branching ratio for various values of m_t, as a function of the charged Higgs mass, for fixed[10] $V_{td} = .01$. Obviously, going beyond the standard model renders any analysis much more uncertain, since it involves many extra unknown parameters!

The last Kaon decay which is interesting to consider, even though it is very difficult to attempt to measure, is the process $K_L^0 \to \pi^0 \nu\bar{\nu}$. This decay rate is closely related to that for $K_L^0 \to \pi^0 e^+ e^-$. However, in contrast to this case – as has been realized recently by Littenberg [30] – the direct CP violating amplitude really dominates over the indirect amplitude. Furthermore, there is no two photon CP conserving contribution. Hence, this process provides a very clean (albeit rare!) test for the CKM paradigm.

That the indirect CP violating amplitude is very small can be gathered by an estimate completely analogous to that in Eq. (12). Instead of having the branching ratio $B(K^- \to \pi^- e^+ e^-) \sim 3 \times 10^{-7}$, the estimate of the indirect CP violating amplitude here involves $B(K^- \to \pi^- \nu\bar{\nu}) \sim 3 \times 10^{-10}$, which is three orders of magnitude smaller. The direct CP violating amplitude, on the other hand, involves the imaginary part of the amplitude entering in Eq. (17). Typically this imaginary part is not very much smaller than the real part. Thus in this case, indeed, the direct CP violating amplitude dominates.

The branching ratio for the decay $K_L^0 \to \pi^0 \nu\bar{\nu}$, assuming three generations and neglecting QCD corrections, is given by

$$\begin{aligned} B(K_L^0 \to \pi^0 \nu\bar{\nu}) &= 1.8 \times 10^{-6} F^2(y_t) A^4 \lambda^8 \rho^2 \sin^2 \delta \\ &\simeq 1.2 \times 10^{-11} F^2(y_t) \rho^2 \sin^2 \delta \ , \quad (20) \end{aligned}$$

where the second line uses the central value for A. Since $F(y_t) \sim O(1)$, and ρ and $\sin\delta$ are also permitted to be of order unity (Fig. 4), this branching ratio could be as large as 10^{-11}. However, taking the whole range of allowed values for ρ and δ, along with the correlated values for m_t, considerably smaller branching ratios can also ensue. We find (DL,RDP)

$$4 \times 10^{-13} < B(K_L^0 \to \pi^0 \nu\bar{\nu}) < 2 \times 10^{-11} \ , \quad (21)$$

where the lower limit is essentially m_t independent.

[10]This matrix element can remain fixed at this value, even as m_t varies, since x_d has an additional charged Higgs contribution.

3 Some (non CP violating) B Physics

If one wants to look for deviations in **rate** from the CKM paradigm, one should select processes in which, if at all possible:

1. The standard model prediction for the decay rate is large,
2. Exotic (new physics) contributions can give sizable effects,
3. The standard model process, per se, is already interesting.

Clearly, it is helpful experimentally to be able to study channels in which the decay rate is not hopelessly small. Furthermore, it is imperative to select processes in which new physics effects could have a substantial impact. However, if these effects turn out not to be there – a normal occurrence! – it is nice if the process itself is interesting, even within the standard model. The decay $b \to s\gamma$, discussed below, fulfills all three of these criteria very nicely.

3.1 $b \to s\gamma$

The radiative decay of a b quark to an s quark proceeds through the electromagnetic penguin contributions shown in Fig. 14. In the three generation CKM model, there is not much uncertainty in the rate prediction connected with the electroweak sector, except for the value of m_t[11]. However, QCD corrections – of which a typical diagram is shown in Fig. 14b – substantially modify the rate [31]. Basically what happens is that, including gluonic effects, the GIM suppression factor of $(m_t^2 - m_c^2)/M_W^2$, coming from the lowest order diagram, becomes only logarithmic: $\alpha_s \ln(m_t^2/m_c^2)$. This produces quite a sizable difference in the rate, as can be appreciated from Fig. 15. So the decay $b \to s\gamma$ is theoretically interesting as a QCD laboratory.

The predicted branching fraction for this decay is quite large [$B(b \to s\gamma) \sim$ (2-4) $\times 10^{-4}$, depending on m_t]. Unfortunately, one cannot directly access this inclusive process experimentally, but must look for some exclusive signal. This obviously reduces the rate and, more damagingly, introduces considerable theoretical uncertainty. Typically, one looks for $B \to K^*\gamma$ and various estimates [31] suggest[12]

$$\Gamma(B \to K^*\gamma) \simeq (4.5 - 7) \times 10^{-2}\Gamma(B \to X_s\gamma)\ . \qquad (22)$$

[11]The c and t quark loops dominate in Fig. 14 and V_{cb} and V_{ts} are reasonably well known, since they depend only on A and not ρ and δ.

[12]However, there exist in the literature [33] other estimates in which $B \to K^*\gamma$ makes up as much as 30% of all $b \to s\gamma$ decays!

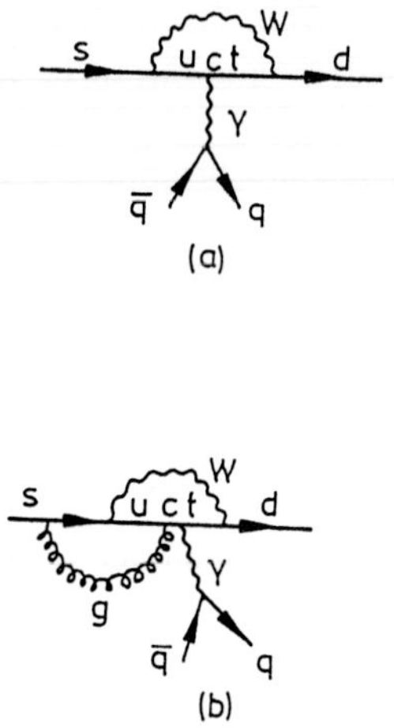

Figure 14: (a) Lowest order electromagnetic Penguin contribution to $b \to s\gamma$; (b) Gluonic correction to the $b \to s\gamma$ process.

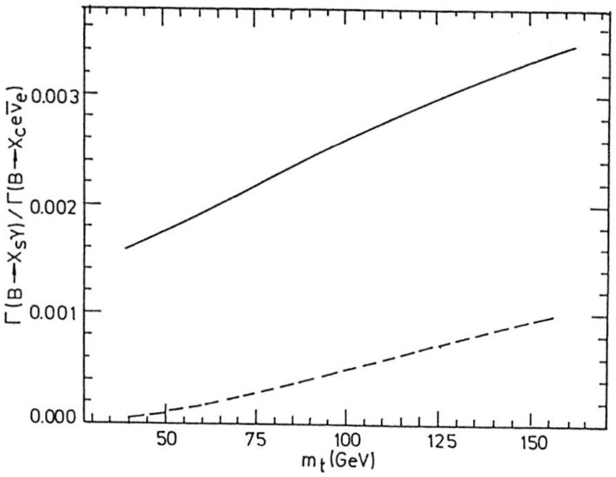

Figure 15: Ratio of the decay rates of $b \to s\gamma$ to $b \to ce\overline{\nu}_e$ with (solid line) and without (dotted line) QCD corrections, plotted as a function of m_t. From [32].

Taking the above at face value, the present CLEO limit [34]

$$B(B \to K^*\gamma) < 1.7 \times 10^{-4} \tag{23}$$

places a bound on the rate $b \to s\gamma$ which is about an order of magnitude away from the prediction of the standard model. This is already beginning to be interesting, since the expectations for $b \to s\gamma$ can be substantially modified, if there exists physics beyond the CKM paradigm. We shall discuss three different possibilities which were examined during the Snowmass Summer Study, involving charged Higgses (CG,JNN), fourth generation (JH), and supersymmetry (GB,AIS).

3.1.1 Charged Higgs

The minimal charged Higgs model arises when one introduces two doublet fields Φ_1 and Φ_2, coupled to right-handed charge $\frac{2}{3}$ and $-\frac{1}{3}$ quarks, respectively, rather than just Φ and its charge conjugate $\tilde{\Phi}$. The model has two additional parameters: the ratio of the two Higgs vacuum expectation values $\xi = v_2/v_1$, and the charged Higgs boson mass. Some restrictions on these parameters are imposed by B_d-$\overline{B_d}$ mixing, since now diagrams involving charged Higgses also contribute. Typically [35] ξ must be greater than unity and it grows as M_{H^+} grows, but decreases as m_t increases. However, the detailed value for ξ depends on assumptions one makes about V_{td}, and is influenced by the uncertainties on the hadronic matrix elements. The value of V_{td} used in our analysis ($V_{td} = .01$) is, quite naturally, smaller than that allowed by a global fit using the standard model only, as charged Higgses provide additional contributions to x_d.

The decay rate for $b \to s\gamma$ is always increased by the presence of charged Higgses. This is easy to understand since this rate is not influenced by V_{td}. Furthermore, corresponding to each graph in Fig. 14, there is now also one in which $W \leftrightarrow H$. Also for the charged Higgs diagrams QCD corrections, which have been computed by Grinstein and Wise [36], turn out to be important and they have been included in our analysis. In Fig. 16 we plot the branching ratio for the process $b \to s\gamma$, as a function of the charged Higgs mass, for various values of m_t, taking into account the restriction on ξ imposed by B_d-$\overline{B_d}$ mixing. Clearly a large enhancement over the standard model is obtained for small values of m_t and M_{H^+}. Indeed, recalling that the present CLEO limit [34] on $B \to K^*\gamma$ corresponds to, roughly, $B(b \to s\gamma) \simeq$ (2-4) $\times 10^{-3}$, a modest improvement in the experimental limit could be important for ruling out low mass charged Higgses.

3.1.2 4th Generation

The situation is not as clear if one imagines that there are four and not three generations. In this case, since unitarity does not restrict V_{ts} to be approximately equal

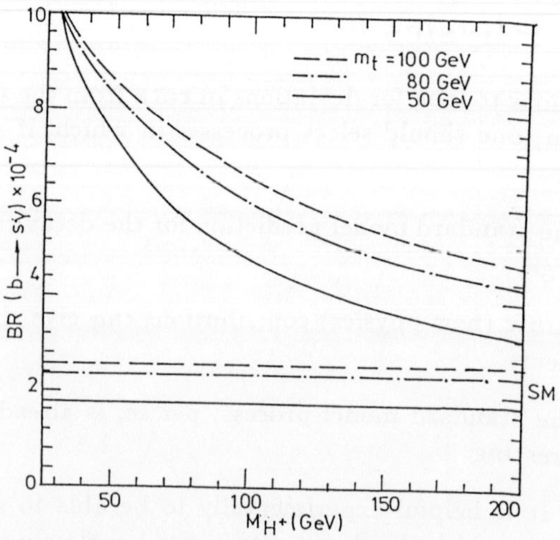

Figure 16: Branching ratio for the process $b \to s\gamma$ in a model with charged Higgses, plotted as a function of M_{H^+}, for various values of m_t. The standard model predictions are also included.

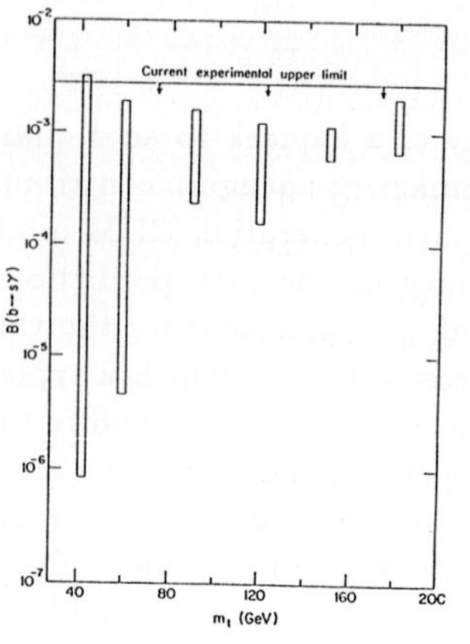

Figure 17: Predictions for $b \to s\gamma$ in a 4 generation CKM model.

to V_{cb}, the predictions for $b \to s\gamma$ can vary considerably from those of the standard model [37]. Performing the same type of analysis as that discussed in Sec. 2.2, one finds a rather large range allowed for the branching ratio $b \to s\gamma$, particularly for small values of m_t. The results are displayed in Fig. 17 and one notes that, in contrast to the charged Higgs case, here it is also possible to have a branching ratio below that predicted by the standard model. The even rarer decay $b \to se^+e^-$, which has a standard model branching ratio of the order of (4-10) $\times 10^{-6}$, depending on m_t, is perhaps an even better place to look for 4th generation effects [38]. In fact, typically, for the whole range of m_t, we find that with 4 generations this branching ratio can be as large as (2-4) $\times 10^{-4}$.

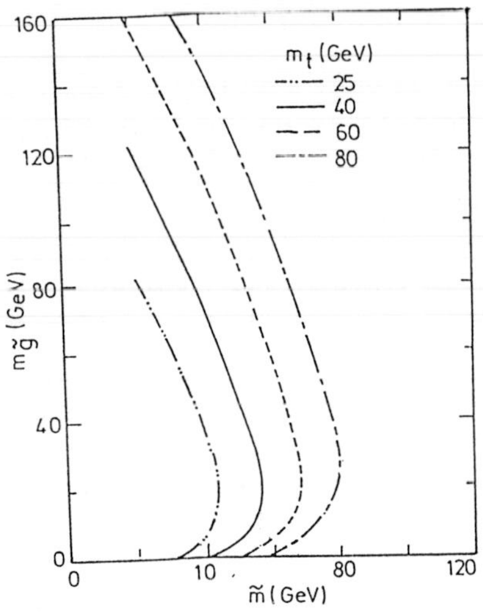

Figure 18: Excluded region (to the left) of values for $\tilde{m}$ and $m_{\tilde{g}}$, for different m_t, assuming $B(b \to s\gamma) = 2 \times 10^{-3}$ and $A = 3$, $C = 0.5$. From [40].

3.1.3 SUSY

A supersymmetric extension of the standard model produces a situation which is akin to that of the case of 4 generations. Here also there are many unknown parameters and, in certain circumstances, one can expect a sizable enhancement over the standard model rate. A general discussion, which is valid for all parameter space, is difficult, and one is forced to make simplifying assumptions. In supersymmetric models the flavor changing $b \to s$ transition can occur through gluino exchange, since the quark-squark-gluino coupling is not flavor diagonal [39]. In the simplest scenario [40], this mixing is characterized by a parameter, A, detailing the amount of squark chiral mixing. Furthermore, the supersymmetric contributions to the rate for $b \to s\gamma$ vanish in the limit of degenerate squark masses. Hence the rate is sensitive to $\Delta\tilde{m}^2$. In a minimal model, one can approximate $\Delta\tilde{m}^2 \simeq Cm_t^2$, with C being a parameter of O(1). In addition to A, C, and m_t, the rate depends on the actual values one takes for the gluino mass and the central mass of the squarks.

Given this plethora of parameters, it is difficult to make strong assertions. Nevertheless, as Bertolini et al [40] have emphasized, light SUSY particles ($\tilde{m}$, $m_{\tilde{g}} \sim$ 40-60 GeV) can give rather substantial effects, particularly if m_t is large[13]. This is illustrated in Fig. 18, which shows the excluded region in $\tilde{m}$-$m_{\tilde{g}}$ space for $B(b \to s\gamma) = 2 \times 10^{-3}$, for various values of m_t. If a signal for $b \to s\gamma$ is seen around this level, which is an order of magnitude above the standard model predictions, a supersymmetric explanation may well be in order. However, the reverse assertion is probably not warranted. Given the many unknowns, one should not use present experimental limits on $b \to s\gamma$ to argue for excluding certain regions in the squark-gluino mass parameter space.

[13]This is perhaps not surprising, since $\Gamma_{SUSY}(b \to s\gamma) \sim (\Delta\tilde{m}^2)^2$, and so grows like m_t^4.

3.2 B_s-$\overline{B_s}$ Mixing

A second topic in B physics which was studied at Snowmass was the prospects for determining the mixing parameter $x_s = \frac{(\Delta m)_{B_s}}{\Gamma_{B_s}}$ for the B_s system. An accurate determination of x_s is of considerable interest, as it would severely constrain the allowed values for ρ and δ. The mixing parameters x_s and x_d arise theoretically from identical box graphs, save for the $s \leftrightarrow d$ interchange. Hence one has, simply

$$x_s = x_d \left|\frac{V_{td}}{V_{ts}}\right|^2 \left[\frac{\tau_{B_s} f_{B_s}^2 B_{B_s}}{\tau_{B_d} f_{B_d}^2 B_{B_d}}\right]. \tag{24}$$

The factor in the square brackets in (24) is expected to be close to one, since $\tau_{B_s} \simeq \tau_{B_d}$[14] and the ratio of hadronic matrix elements can differ from unity only through $SU(3)$ breaking effects. Actually these latter effects are difficult to compute and some caution is in order, since some preliminary attempts to calculate this ratio on the lattice [41] indicate corrections as large as 50%. At any rate, we shall assume in what follows that the square bracket in Eq. (24) is unity and hope that, by the time x_s is measured, one will have a more reliable theoretical estimate for this factor. Thus

$$x_s \simeq \frac{x_d}{\lambda^2|1 - \rho e^{i\delta}|^2} \simeq \frac{14.4}{|1 - \rho e^{i\delta}|^2}, \tag{25}$$

where for the numerical value above we have used $x_d = 0.7$, which is the central value of the combined ARGUS [42] and CLEO [43] measurements.

It is clear from Eq. (25) and from the allowed range for ρ and δ given in Fig. 4, that one expects rather large values for x_s in the CKM model ($x_s \geq 4$). This, unfortunately, makes life more difficult rather than easier! In a time integrated measurement, like that performed for B_d at the $\Upsilon(4S)$, what is measured is not the mixing parameter but the ratio r_s of the rates for the B_s mesons to decay to leptons of opposite charge. Neglecting the contribution due to a possible lifetime difference between the two neutral B_s eigenstates, one has that

$$r_s = \frac{\Gamma(B_s \to l^- X)}{\Gamma(B_s \to l^+ X)} \simeq \frac{x_s^2}{2 + x_s^2}. \tag{26}$$

Thus if indeed $x_s \geq 4$, then $r_s \geq 0.89$ and one needs a measurement which is extremely accurate to extract x_s. Furthermore, at the $\Upsilon(5S)$ the situation is much more complicated, since the produced leptons can arise not only from B_s (or $\overline{B_s}$) decays, but also from produced B_d's, B_d^*'s

[14]For inclusive decays of B mesons the spectator model, which implies the equality of the B_s and B_d lifetimes, is supposed to work very well.

and charged B's. For an accurate measurement therefore, one needs to be able to tag, somehow, the B_s meson sample.

Two ideas have been studied for measuring x_s. In the first method, one works at the $\Upsilon(5S)$ using a time integrated method, but enriches the B_s signal by tagging on D_s mesons. In the second approach, one tries to take advantage of the fast time oscillations of the B_s signal to extract x_s. Both these techniques are examined in detail in [44] and they have also been amply discussed by some of our experimental colleagues [45,46]. Thus we shall concentrate here only on some highlights of both methods.

3.2.1 Time Integrated Techniques

The idea here is to work at the $\Upsilon(5S)$ and hope that the expectations of potential models [47] that the $B_s\overline{B_s}$ pairs are dominantly produced in a given angular momentum state are correct[15]. This is necessary since, if both angular momentum states are produced, what is measured is not r_s but the incoherent ratio [44]

$$R_s^{\rm inc} \simeq \frac{2r_s}{1+r_s^2} \overset{x_s \text{ large}}{\simeq} 1 - \frac{2}{x_s^4}, \tag{27}$$

which is even more insensitive to x_s. To proceed at all the B_s sample needs to be enriched with respect to the B_d sample. This is most efficiently done by requiring an additional D_s to be present (arising from the decay $B_s \to D_s l X_s$) so that the signal one is looking for is $\Upsilon(5S) \to D_s l^+ l^- X$. The small contamination of B_d remaining, arising from the decay $B_d \to D_s l X_d$, can be further purified by an invariant mass cut [48]. The state X_d necessarily must involve a Kaon and a neutrino, while X_s is dominantly just a neutrino. Thus requiring that $m_X^2 = 0$ is a useful cut to make.

Typically [47] the ratio of B_s to B_d production at the $\Upsilon(5S)$ is around 10-30%. According to the analysis of [44], requiring a D_s in the final state increases this ratio by a factor of about 35, while the neutrino mass cut provides another factor of 5 enhancement. Even with this purified sample one can reduce the systematic uncertainty produced in r_s by misidentified B_d's only to about 6%[16]. A matching statistical error, so that $\delta r_s \sim 8\%$, calls for 800 $\Upsilon(5S) \to l^+ l^- D_s X$ decays. This in turn, according to the estimates of Krawczyk et al [44], requires at least 2×10^8 $b\bar{b}$ pairs to be produced. Thus this method, although interesting, does not look too promising, unless one can somehow or other reduce the statistical onus.

[15] It may be necessary to operate a little below the $\Upsilon(5S)$ in energy, if one wants to eliminate kinematically $B_s^*\overline{B_s^*}$ production

[16] Some of the factors entering in this estimate also take into account uncertainties in the expected initial ratio of B_s to B_d and the production of charged B's.

3.2.2 Time Dependent Methods

The situation looks quite a bit better if one can study the time development of the produced leptons from B_s decays. The number of l^+'s produced by a decaying sample of B_s at rest is a modulated exponential

$$N_s^+(t) = N^+(0)e^{-\Gamma t}\left[1 + \cos\frac{(\Delta m)_s t}{2}\right]. \tag{28}$$

For large x_s, the oscillating factor above will vary numerous times in an e-fold. Since $x_d \simeq 0.7$, on the other hand, a decaying sample of B_d's will have a much milder undulation. Hence, as far as the time development goes, the presence of some B_d contamination is not so damaging, since it will only change the magnitude of the oscillating B_s signal.

The time development of B_s's can be studied in a hadronic collider or a Z factory. However, the necessity of tagging to determine whether one is dealing with an initial B_s or $\overline{B_s}$, combined with the needed precision in energy and position measurements, require again more than 10^8 $b\bar{b}$ pairs. However, with this method values of x_s up to 15 are accessible [44]. A much more promising avenue is afforded by having an asymmetric e^+e^- collider operating at the $\Upsilon(5S)$. In this case, one can measure the distribution in the proper time difference between the leptons produced by the decay of the two B mesons by a distance measurement. These distributions, assuming again that the $B_s\overline{B_s}$ mesons are produced in an even angular momentum state $[B_s\overline{B_s^*}]$, depend directly on x_s. One has [44]

$$N(l^+l^-;\Delta t\ ) = \frac{N_0}{2(1+x_s^2)}e^{-\frac{\Delta t}{\tau}} \cdot \left\{2\cos^2(\frac{x_s\Delta t}{2\tau}) + x_s^2 - x_s\sin(\frac{x_s\Delta t}{\tau})\right\}, \tag{29}$$

$$N(l^+l^-;\Delta t\ ) = \frac{N_0}{2(1+x_s^2)}e^{-\frac{\Delta t}{\tau}} \cdot \left\{2\sin^2(\frac{x_s\Delta t}{2\tau}) + x_s^2 + x_s\sin(\frac{x_s\Delta t}{\tau})\right\}. \tag{30}$$

Although both the same sign and opposite sign dilepton distributions have the same type of oscillatory behaviour, experimentally the same sign distribution suffers much less from backgrounds produced by $B^\pm$ and B_d decays. Fig. 19 from [44] shows the expected distribution of same sign dileptons as a function of the proper time difference for different values of x_s. In the figure the ratio of B_s to B_d produced at the $\Upsilon(5S)$ was that of [47]. Furthermore a mistagging probability of 5% and a vertex resolution of 25μm [corresponding to $\Delta t/\tau = 0.1$] were assumed. As can be seen the different x_s patterns are quite visible. This is to be contrasted with the opposite sign pattern shown in Fig. 20, computed with the same parameters. Krawczyk et al [44] conclude that a measurement of x_s up

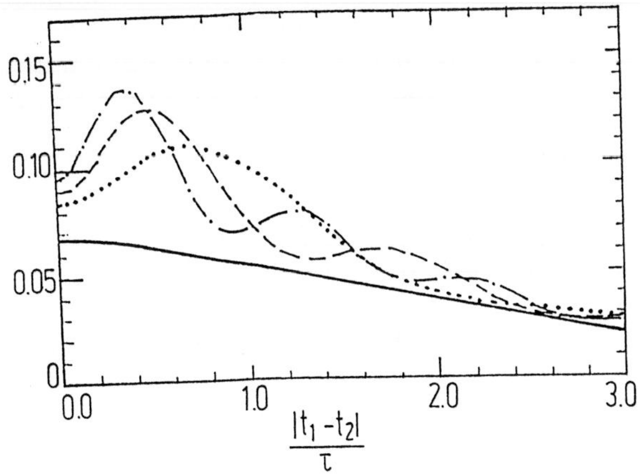

Figure 19: Expected distribution of same sign dileptons as a function of the proper time difference, under the conditions discussed in the text, for $x_s = 0$ (solid); 3 (dotted); 5 (dashed) and 7 (dot-dashed).

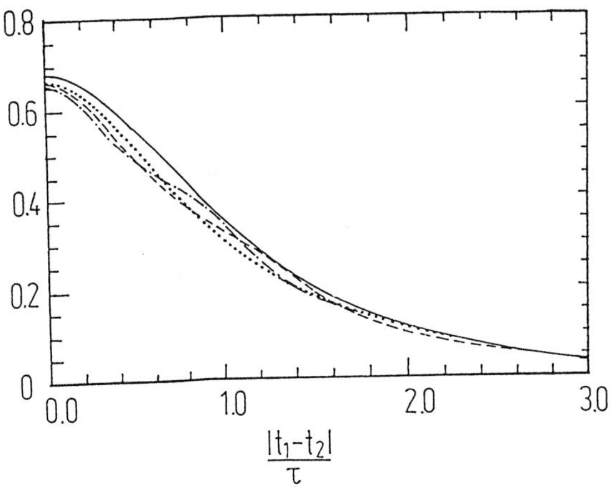

Figure 20: As in Fig. 19, except for opposite sign dileptons.

to about 10 is feasible. under the above assumed conditions. Since the detection of a D_s meson is not necessary, this measurement can be done with considerably fewer $b\bar{b}$ pairs, approximately 5×10^6. However, one should emphasize that the range of x_s accessible is quite dependent on the assumed production rates for B_s versus B_d. Furthermore, a factor of two worsening of the vertex resolution washes out the signal. Nevertheless, this appears the most promising way to try to measure x_s.

4 CP Violation in the B System

Observing CP violation in B decays has become the holy grail of the subject. This is understandable since, within the CKM paradigm, there exist quite definite predictions which should – hopefully in the 1990's – either be proven or falsified. Unfortunately, this is a nontrivial endeavour, since it will require studying of the order of 10^8 $b\bar{b}$ pairs. This is a subject which has also been discussed rather extensively in the past [49]. Thus our discussion here, rather than being encyclopedic, will focus on a few points of interest which arose during the summer study.

Roughly speaking, there exist three categories of CP violating processes in the B system.

- Processes where, in the standard model, one expects rather small CP violating effects;
- Processes where the CP violating effects are intrinsically intertwined with dynamics;
- Processes where there may well be rather large, and substantially model independent, CP violating signals.

Each of these classes of CP violation is interesting and worth commenting upon, although, obviously, the last class is the one which holds the most promise.

An example of the first kind of CP violation is provided by the lepton charge asymmetry [50], which measures the difference in the number of same sign dileptons, arising from the production of $B\overline{B}$ pairs:

$$a = \frac{N(l^+l^+) - N(l^-l^-)}{N(l^+l^+) + N(l^-l^-)} = \frac{r - \bar{r}}{r + \bar{r}} . \tag{31}$$

In the standard model r and $\bar{r}$, and therefore the number of same sign dileptons, can be large (e.g. $r_s \simeq 1$). However, unfortunately, the difference between r and $\bar{r}$ is very small, leading to predictions for a_d and a_s in the range 10^{-3}-10^{-4} [51]. Obviously, such an asymmetry requires an enormous number of $B\overline{B}$ pairs (of order 10^9-10^{10}), for a statistically significant measurement. Thus these asymmetries, at least for the 1990 s, are experimentally completely out of reach. Nevertheless, they should be looked for, since the observation of any signal in the foreseeable future will mean that $a \gg 10^{-3}$. Although this would constitute a prima facie case for new physics, the overwhelming probability is that nothing will be found at all!

Processes where the observable CP violation might be large, but where the signal depends in detail on the hadronic dynamics, are less useful for testing the CKM paradigm. Nevertheless, these processes may very well be interesting to study for their own sake, because of the intricate interplay between the strong and weak dynamics. An example is provided by the decays of charged B's into CP conjugate states [52]. The ratio $\Gamma(B^- \to f)$ and $\Gamma(B^+ \to \overline{f})$ can differ from each other if there is an interference between two amplitudes which have different CP violating phases and are subject to different strong final state interactions. Then

$$\begin{aligned}\Gamma(B^- \to f) \quad - \quad \Gamma(B^+ \to \overline{f}) \\ \sim \quad \sin(\delta_1 - \delta_2)_{\text{weak}} \sin(\delta_1 - \delta_2)_{\text{strong}} \, .\end{aligned} \tag{32}$$

In Snowmass some of us (BK,MK,RDP,AIS) studied a different example of this sort, involving the measurement of some final state triple correlation. Strictly speaking, the effect studied is really a T violating effect, rather than a CP violating effect, but these are equivalent through the CPT theorem.

For B decay processes, where the weak decay rate is well approximated by its lowest order expression, it is easy to show [53] using a combination of CP and T invariance that

$$\begin{aligned}\Gamma[B \to a_1(\vec{p}_1, \vec{s}_1) + \ldots + a_n(\vec{p}_n, \vec{s}_n)] \\ = \Gamma[B \to a_1(-\vec{p}_1, -\vec{s}_1) + \ldots + a_n(-\vec{p}_n, -\vec{s}_n)] \, ,\end{aligned} \tag{33}$$

where $a_i(\vec{p}_i, \vec{s}_i)$ denotes a daughter particle of given momentum $\vec{p}_i$ and spin $\vec{s}_i$. From this form, we see that the observation in a given B decay, of any triple correlation of the type $\vec{a} \cdot \vec{b} \times \vec{c}$, where $\vec{a}$, $\vec{b}$ and $\vec{c}$ are each the momentum or spin of some daughter particle, would be a signal of CP violation. We should caution, however, that this conclusion ceases to be true in the presence of strong interaction rescattering phases. So in practice, it is necessary to be able to estimate the strong interaction contribution, so that it can be subtracted from an observed signal to reveal the interesting CP violating part.

There are many candidates for decay modes which involve kinematically nontrivial CP violation in the B system – including examples in which one compares particle-antiparticle rates, like $B^\pm \to K^{*\pm}\Psi$, in different helicity states [53]. Here we want to illustrate some of the challenges involved in analyzing these effects by discussing two examples briefly: $\Lambda_b \to \Lambda D^0, \Lambda \overline{D^0}$, with $D^0, \overline{D^0} \to K_s \pi^0$, and $\Lambda_b \to \Lambda \pi^0$. Clearly, in both cases the presence of a term like $\vec{s}_{\Lambda_b} \cdot \vec{s}_\Lambda \times \vec{p}_\Lambda$ in the rate would be a possible signal for CP violation effects. Naively, in each case, it appears that this may be possible theoretically, since two, relatively complex, amplitudes contribute. [For the first process $b \to c(\overline{u}s) \sim \lambda^3$ and $b \to u(\overline{c}s) \sim \lambda^3 e^{-i\delta}$; for the second process $b \to u(\overline{u}s) \sim \lambda^4 \epsilon^{-i\delta}$ and the Penguin contribution $b \to s(\overline{u}u) \sim \lambda^2 \alpha_s$.] Nevertheless, only in the second example does a nontrivial triple correlation appear. In the first case, at least in the simple dynamical approximation considered, although one has two terms with different phases, they contribute equally to the parity conserving and parity violating amplitudes. Thus, the amplitudes for the two subprocesses are kinematically identical and no **kinematically nontrivial** effect, such as the triple correlation, can arise from this interference. It appears that it is always necessary (but not always sufficient) to have two kinematically distinct amplitudes at the quark level for a triple correlation to be nonvanishing.

4.1 Hadronic Asymmetries and CP Violation

The observation of CP violating processes requires that there should be interference between two amplitudes with different phases. This can come about in the B system through B-$\overline{B}$ mixing. Because of this mixing, a state which at $t = 0$ was a pure B meson can evolve in time into a mixture of B and $\overline{B}$ mesons. The physical state at time t is easily seen to be [54]

$$|B_{\text{phys}}(t)\rangle = f_+(t)|B\rangle + f_-(t)\eta_M e^{i\phi_M}|\overline{B}\rangle \, , \tag{34}$$

where

$$\begin{aligned} f_+(t) &= e^{-imt} e^{-\frac{\Gamma}{2}t} \cos \frac{\Delta m t}{2} \\ f_-(t) &= i e^{-imt} e^{-\frac{\Gamma}{2}t} \sin \frac{\Delta m t}{2} \, . \end{aligned} \tag{35}$$

In the above $\eta_M e^{i\phi_M}$ is related to the ratio of the off diagonal mass and decay width matrix elements of the B-$\overline{B}$ system

$$\eta_M e^{i\phi_M} = \left[\frac{M_{12}^* - i\Gamma_{12}^*/2}{M_{12} - i\Gamma_{12}/2}\right]^{1/2} \, . \tag{36}$$

Since $\Gamma_{12} \ll M_{12}$, the magnitude η_M is essentially unity. The phase ϕ_M is convention dependent. However, in the quark phase convention we shall adopt, it is just minus twice the phase of the $\Delta B = 2$ mass matrix, calculated in terms of the usual quark box graphs. Since these graphs are dominated by the t-quark contribution [55] one has simply

$$\phi_M \simeq \begin{cases} -2\phi & B = B_d, \\ 0 & B = B_s. \end{cases} \tag{37}$$

In the above $-\phi$ is the phase of V_{td} [see Fig. 3], and we have used the fact that V_{ts} is approximately real.

A CP violating asymmetry can be measured in decays of B and $\overline{B}$ mesons to a common hadronic final state f. If CP is not conserved, the decay probability of a state B_{phys}, which at $t = 0$ was a pure B meson, into f need not agree with the decay probability of $\overline{B}_{\text{phys}}$, a state which at $t = 0$ was a pure $\overline{B}$ meson, into $\overline{f}$, the CP conjugate state of f. The interesting asymmetry

$$A_f(t) = \frac{\Gamma(B_{\rm phys}(t) \to f) - \Gamma(\overline{B}_{\rm phys}(t) \to \overline{f})}{\Gamma(B_{\rm phys}(t) \to f) + \Gamma(\overline{B}_{\rm phys}(t) \to \overline{f})} \qquad (38)$$

depends both on $\eta_M e^{i\phi_M} \simeq e^{i\phi_M}$ and on the dynamics contained in the ratio

$$\rho_f = \frac{A(\overline{B} \to f)}{A(B \to f)} \ . \qquad (39)$$

A very interesting circumstance occurs [54] if f is a **CP eigenstate** [$f = \pm\overline{f}$] and the decays $B \to f$, $\overline{B} \to f$ are dominated by **one** amplitude. In this case $|\rho_f| \simeq 1$ and all the remaining dynamical information is contained in its phase[17]

$$\rho_f \simeq e^{i\phi_D} \ . \qquad (40)$$

Since $V_{ub} = |V_{ub}|e^{-i\delta}$ and V_{cb} is real, neglecting Penguin effects to which we shall return later, one sees that, in these circumstances,

$$\phi_D \simeq \begin{cases} 0 & \text{Cabibbo allowed decays } (b \to c), \\ -2\delta & \text{Cabbibo suppressed decays } (b \to u). \end{cases} \qquad (41)$$

The rate for B decay to charge conjugate states, which obey the above mentioned dynamical circumstances, read simply [54]:

$$\begin{aligned} \Gamma(B_{\rm phys}(t) \to f) &= e^{-\Gamma t}\left[1 + \alpha_f \sin \Delta m t\right] \\ \Gamma(\overline{B}_{\rm phys}(t) \to f) &= e^{-\Gamma t}\left[1 - \alpha_f \sin \Delta m t\right] \ , \end{aligned} \qquad (42)$$

where

$$\alpha_f = -{\rm Im}\rho_f \eta_M e^{i\phi_M} \simeq -\sin(\phi_M + \phi_D) \ . \qquad (43)$$

Using Eqs (37) and (41), we see that there are four classes of decays [15,56] which measure different combinations of the phases entering in the CKM matrix and in the unitarity triangle of Fig. 3:

$$\begin{aligned} &\text{Cabibbo allowed } B_d \text{ decays [e.g. } B_d \to \Psi K_s] : \\ &\qquad \alpha_f \simeq \sin 2\phi \\ &\text{Cabibbo allowed } B_s \text{ decays [e.g. } B_s \to \Psi\phi] : \\ &\qquad \alpha_f \simeq 0 \\ &\text{Cabibbo suppressed } B_d \text{ decays [e.g. } B_d \to \pi^+\pi^-] : \\ &\qquad \alpha_f \simeq \sin(2\phi + 2\delta) \\ &\text{Cabibbo suppressed } B_s \text{ decays [e.g. } B_s \to K_s\pi^0] : \\ &\qquad \alpha_f \simeq \sin 2\delta \ . \end{aligned} \qquad (44)$$

The measurement of the various types of CP violating asymmetries of Eq. (38) would provide the most stringent test of the CKM paradigm. Ideally, one could check

[17] There is in general a $\pm$ sign, associated with the CP properties of f, which we have dropped in Eq. (40).

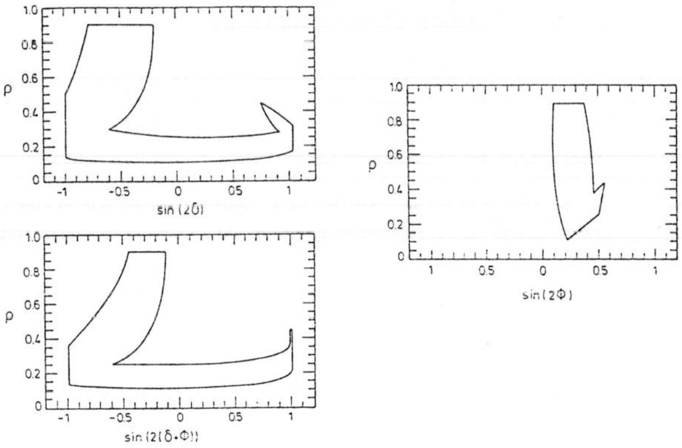

Figure 21: Allowed region for $\sin 2\phi$, $\sin 2(\phi + \delta)$ and $\sin 2\delta$, as a function of ρ, corresponding to the allowed region in ρ-δ space given in Fig. 4. From [15].

the unitarity of the CKM matrix experimentally by proving that the sum of the angles in the unitarity triange – related to the different α_f's – indeed add up to 180°! Lacking such experimental input at the moment, one can at least see what values for the different α_f's are allowed by our present knowledge of ρ and δ. Fig. 21, taken from [15], shows the different allowed regions for $\sin 2\phi$, $\sin 2\delta$ and $\sin(2\phi + 2\delta)$, as a function of ρ, corresponding to the allowed region in ρ-δ space of Fig. 4. One sees that for $\rho > 0.3$, the values permitted for the various α_f's are largely ρ independent and can be quite large in magnitude.

Although the hadronic asymmetries in B decay can be substantial, one needs to check theoretically that the assumption of having a dominant weak amplitude, and thus a well defined weak phase for ρ_f, really holds for the decays of interest. Furthermore, it is important to try to devise some sensible experimental strategy for measuring these asymmetries. Concerning the first point, two of us (DL,RDP) have looked at the possible influences that Penguin diagrams can have on the predictions of the hadronic asymmetries, Eq. (44). Roughly speaking, the situation can be summarized as follows [57]. Since the momentum transfer in B decays is typically greater than, or of order of, the charm mass, effectively, after making use of the CKM unitarity condition, only the t-quark loop contributes to the Penguin amplitude. Hence the $b \to s$ Penguin is approximately real, while the $b \to d$ Penguin has the phase of V_{td}^*. These simple observations lead to a number of consequences:

1. The predictions for the decay process $B_d \to \Psi K_s$, which is particularly interesting experimentally, are quite safe. The $b \to s$ Penguin contribution is real, as is the main $b \to c\bar{c}s$ quark decay amplitude. Thus

the asymmetry associated with this decay should measure purely the mixing phase ϕ_M and $\alpha_f = \sin 2\phi$, without any corrections.

2. In decays like $B_d \to \pi^+\pi^-$, or $B_s \to K_s\pi^0$, where the $b \to d$ Penguins may be important, the predictions for the corresponding α_f's do change. However, the corrections to Eq. (44) are proportional to the relative strength of the Penguin amplitude to the pure quark decay amplitude, which is typically of $O(\frac{\alpha_s}{12\pi}\ln\frac{m_t^2}{m_c^2})$, and may well be small. In decays like $B_d \to K_s\pi^0$ or $B_s \to K^+K^-$, which are very Cabibbo suppressed at the quark level, the Penguin effects, which are Cabibbo enhanced [58], may change the answer for α_f totally.

3. There are pure Penguin dominated decays, in which one could also expect interesting hadronic asymmetries. An interesting example is provided by $B_d \to \phi K_s$, which proceeds through the $b \to s$ Penguin. Since this graph is approximately real, one expects that also $\alpha_{\phi K_s} = \sin 2\phi$, just as was the case for the decay $B_d \to \Psi K_s$.

We shall only make two general remarks on the experimental strategy to measure CP violation in the B system, as this subject is discussed in much more detail in the reports of our experimental colleagues in these Proceedings [46,59]. One of these comments concerns the question of time integrated versus time dependent asymmetry measurements. Time integrated measurements are only sensible for B_d decays, since one is not punished unduly by the mixing parameter x. Integrating Eq. (38) over time, using (42), gives for the time integrated asymmetry

$$\langle A_f \rangle = \frac{x}{1+x^2}\alpha_f \, . \qquad (45)$$

For the B_d system, the mixing parameter x_d is such that the mixing factor in front of α_f above is essentially maximal, $\frac{x_d}{1+x_d^2} \simeq 0.47$. For the B_s system, on the other hand, since x_s is large in the CKM model, the relevant time integrated asymmetry can be much reduced. In this case, it is probably better to try to follow the time development of the decay amplitude. Indeed, the considerations needed for measuring the CP asymmetry are quite similar to those we discussed in the case of B_s-$\overline{B}_s$ mixing.

Our second comment concerns the question of tagging. To measure the hadronic asymmetry (38) one needs to know how a decaying B meson was born. That is, was the state at $t = 0$ of this meson a B^0 or a $\overline{B^0}$? There is no universal answer on how to tag. Rather, this is a machine dependent question. For instance, a leptonic tag – with a concomitant signal dilution – may be sensible at a hadronic collider. In the case of an asymmetric e^+e^- machine, operating at the $\Upsilon(4S)$, because of the p-wave wave function of the produced B_d's one really needs no tag [60]. Here the first decay, because of Bose statistics, essentially serves to tag the second. Finally, for a linearly polarized collider operating at the Z^0, one can tag the initial B by a forward-backward tag [61], since particles originating from a b quark will preferentially go along the direction of the incoming electron.

Some of us in the summer study (BK,SM,AIS) have investigated some aspects of this last proposed method for tagging. In particular, there was some concern on whether hadronization corrections could affect the quark level formulas for the asymmetry and whether the formula for the asymmetry in [61]

$$A^f_{\rm meas} = \frac{N(f,\cos\theta) - N(\overline{f},-\cos\theta)}{N(f,\cos\theta) + N(\overline{f},-\cos\theta)} = A_f A_{FB} \qquad (46)$$

itself had any corrections. Here $N(f,\cos\theta)$ is the number of decays into the final state f in the forward direction, while $N(\overline{f},-\cos\theta)$ is the number of decays to the charge conjugate state, $\overline{f}$, in the backward direction. Finally, A_{FB} is the quark level forward-backward asymmetry. Essentially, the answer to the first question is no, even though in the case of some two body decays – like $Z^0 \to B\overline{B}$ – there is no memory of the quark parent. The point is that the decays of $Z^0 \to$ jet + jet predominate, and for these decays the correlation with the quark parent exists, as the successful measurement of the forward-backward asymmetry for b-quarks at PETRA [62] has demonstrated. As far as the second question goes, it turns out that there are corrections to Eq. (46). However, these corrections vanish if $|\rho_f| = 1$. Hence for the cases of most interest, these corrections are also not important.

5 Concluding Remarks

Our conclusions are quite simple. In weak decays, the task of the 1990's is to go and test the CKM paradigm. Here we will need the help of our experimental friends. However, there is plenty of work to do for theorists also, who should continue to refine the phenomenology and work hard to reduce the uncertainties associated with hadronic matrix elements.

It may be that, on the way to test the CKM model, we shall come upon new physics in the form of violent disagreements of experiment with standard model predictions and/or the discovery of flavor changing neutral currents processes. Although everybody should keep their eyes open for this kind of opportunity – and indeed it is useful to quantify the possible signals and predictions of theories beyond the standard model – we should admit that, in this realm, present theories are incapable of trustworthy predictions. Our hope is also that, in the 1990 s, we shall be granted some illumination in this area!

Acknowledgments

BK thanks the Lawrence Berkeley Laboratory for its warm hospitality during 1987-1988. CSK thanks G. Altarelli for useful discussions. This work was supported in part by DOE contract numbers DE-AC03-76-SF-00515 (CD,FG), DE-AC02-87-ER-40325-TASKB (AIS), DE-AC03-76-SF-00098 (BK), DE-AC02-76-ER-00881 (CSK,JH), W-7405-Eng-82 (JH), by the University of Wisconsin Research Committee with funds granted by the Wisconsin Alumni Research Foundation (JH), by NSF INT-8613131 (AIS), by the US-Japan Collaborative Research program jointly supported by the NSF and JSPS (NSF INT-8613131) (MK), and by the National Sciences and Engineering Research Council (Canada) (GB, CG,JNN).

References

[1] S.L. Glashow, Nucl. Phys. 22 (1961) 579; A. Salam in **Elementary Particle Theory**, ed. N. Svartholm (Almqvist and Wiksell, Stockholm, 1968); S. Weinberg, Phys. Rev. Lett. 19 (1967) 1264

[2] Particle Data Book, Phys. Lett., to be published.

[3] L. Wolfenstein, Phys. Rev. Lett. **51** (1983) 1945.

[4] This triangle has been emphasized by a number of authors. See, for example, J. L. Rosner, A. I. Sanda and M. P. Schmidt, *Proceedings of the Workshop on High Sensitivity Beauty Physics at Fermilab*, ed. A. J. Slaughter, N. Lockyer, and M. P. Schmidt (Fermi National Accelerator Laboratory, Batavia, 1987) p. 165, and references therein.

[5] H. Leutwyler and M. Roos, Z. Phys. **C25** (1984) 91.

[6] A. Sirlin, Phys. Rev. **D35** (1987) 3423.

[7] G. Altarelli, N. Cabibbo, G. Corbo, L. Maiani and G. Martinelli, Nucl. Phys. **B208** (1982) 365.

[8] G. Altarelli and P. Franzini, Z. Phys. **C37** (1988) 271.

[9] K. Wachs, DESY preprint DESY-F31-88-01 (1988), Ph.D. Thesis.

[10] ARGUS Collaboration: H. Schröder, *Proceedings of the XXIV International Conference on High Energy Physics*, Munich, W. Germany (1988).

[11] CLEO collaboration: R. Fulton et al, preprint CBX-88-13 (1988).

[12] ARGUS Collaboration: H. Albrecht et al, Phys. Lett. **209B** (1988) 119.

[13] CLEO Collaboration: C. Bebek et al. Cornell preprint CLNS-88/862 (1988).

[14] A. J. Buras, W. Slominski and H. Steger, Nucl. Phys. **B238** (1984) 529; J. Hagelin, Nucl. Phys. **B193** (1981) 123.

[15] P. Krawczyk, D. London, R.D. Peccei and H. Steger, Nucl. Phys. **B307** (1988) 19. Note that, unfortunately, in this paper δ is defined with the wrong sign! Nevertheless, using δ and ϕ as defined here (cf Eq. (4)), all figures in Krawczyk et al are correct.

[16] C. Bernard, A. El-Khadra and A. Soni, UCLA report UCLA/88/TEP/31, to be published in the Proceedings of the Ringberg Workshop on Hadronic Matrix Elements and Weak Decay; European Lattice Collaboration: M.B. Gavela et al, CERN preprint CERN-TH.5152/88 (1988), to be published in the Proceedings of the Ringberg Workshop on Hadronic Matrix Elements and Weak Decay.

[17] NA31 Collaboration: H. Burkhardt et al, Phys. Lett. **206B** (1988) 169.

[18] G. Altarelli and P. Franzini, CERN preprint CERN-TH.4914/87 (1987); J. Ellis, J. Hagelin, S. Rudaz and D. D. Wu, Nucl. Phys. **B304** (1988) 205.

[19] L. Wolfenstein, Phys. Rev. Lett. **13** (1964) 562.

[20] C. O. Dib, I. Dunietz and F. J. Gilman, SLAC preprints SLAC-PUB-4762 (1988) and SLAC-PUB-4818 (1988).

[21] G. Ecker, A. Pich and E. de Rafael, Nucl. Phys. **B291** (1987) 692; **B303** (1988) 665.

[22] J. F. Donoghue, B. R. Holstein and G. Valencia, Phys. Rev. **D35** (1987) 2769.

[23] T. Inami and C.S. Lim, Prog. Theor. Phys. **65** (1981) 297.

[24] L. M. Sehgal, Phys. Rev. **D38** (1988) 808.

[25] T. Morozumi and H. Iwasaki, KEK preprint KEK-TH-206 (1988).

[26] A. Pich, CERN preprint CERN-TH.5194/88 (1988).

[27] J. Flynn and L. Randall, Berkeley preprint UCB-PTH-88-21 (1988).

[28] J. Ellis and J. Hagelin, Nucl. Phys. **B217** (1983) 189.

[29] G. Eilam, J. Hewett and T. Rizzo, Phys. Lett. **193B** (1987) 533; Phys. Rev. **D34** (1986) 2773; J.L. Hewett, Phys. Lett. **193B** (1987) 327;

[30] L. Littenberg, private communication.

[31] S. Bertolini, F. Borzumati and A. Masiero, Phys. Rev. Lett. **59** (1987) 180; N.G. Deshpande, P. Lo, J. Trampetic, G. Eilam and P. Singer, Phys. Rev. Lett. **59** (1987) 183; N.G. Deshpande, P. Lo and J. Trampetic, Z. Phys. **C40** (1988) 369.

[32] B. Grinstein, R. Springer and M. B. Wise, Phys. Lett. **202B** (1988) 138.

[33] C.A. Dominguez, N. Paver and Riazuddin, Phys. Lett. **214B** (1988) 459.

[34] D. Cassel, private communication; see also ARGUS Collaboration, H. Albrecht et al, Phys. Lett. **210B** (1988) 258.

[35] W.-S. Hou and R. S. Willey, Phys. Lett. **202B** (1988) 591; T. G. Rizzo, Phys. Rev. **D38** (1988) 820; C. Q. Geng and J. N. Ng, Phys. Rev. **D38** (1988) 2857.

[36] B. Grinstein and M. B. Wise, Phys. Lett. **201B** (1988) 274.

[37] W.-S. Hou, A. Soni and H. Steger, Phys. Lett. **192B** (1987) 441; J.L. Hewett, Phys. Lett. **193B** (1987) 327.

[38] W.-S. Hou, R. S. Willey and A. Soni, Phys. Rev. Lett. **58** (1987) 1608; erratum: ibid **60** (1988) 2337; G. Eilam, J. Hewett and T. Rizzo, Phys. Rev. **D34** (1986) 2773.

[39] J.-M. Gerard, W. Grimus, A. Masiero, D.V. Nanopoulos and A. Raychaudhuri, Nucl. Phys. **B253** (1985) 93.

[40] S. Bertolini, F. Borzumati and A. Masiero, Phys. Lett. **192B** (1987) 437; A. Masiero and G. Ridolfi, Phys. Lett. **212B** (1988) 171; **213B** (1988) 562.

[41] C. Bernard, T. Draper, G. Hockney and A. Soni, UCLA report UCLA/87/TEP/28 (1988).

[42] ARGUS Collaboration: H. Albrecht et al, Phys. Lett. **192B** (1987) 245.

[43] CLEO Collaboration: A. Jawahery, *Proceedings of the XXIV International Conference on High Energy Physics*, Munich, W. Germany (1988).

[44] P. Krawczyk, D. London and H. Steger, DESY preprint DESY 88-163 (1988).

[45] T. Nakada, private communication.

[46] G. Feldman, these Proceedings.

[47] N. Byers and D. S. Hwang, *Proceedings of the Workshop on High Sensitivity Beauty Physics at Fermilab*, ed. A. J. Slaughter, N. Lockyer, and M. P. Schmidt (Fermi National Accelerator Laboratory, Batavia, 1987) p. 199.

[48] This was suggested by T. Nakada at Snowmass '88, following a technique used by ARGUS: H. Albrecht et al, Phys. Lett. **197B** (1987) 452.

[49] See, for example, *Proceedings of the Workshop on High Sensitivity Beauty Physics at Fermilab*, ed. A. J. Slaughter, N. Lockyer, and M. P. Schmidt (Fermi National Accelerator Laboratory, Batavia, 1987).

[50] L. Okun, V. Zakharov, and B. Pontecorvo, Lett. Nuovo. Cim **13**, (1975) 218; A. Pais and S. B. Treiman, Phys. Rev. **D12**, (1975) 2744.

[51] P. J. Franzini, CERN preprint CERN-TH.4846/87 (1987).

[52] M. Bander, D. Silverman, and A. Soni, Phys. Rev. Lett. **43** (1974) 242; J. Bernabeu and C. Jarlskog, Z. Phys. **C8** (1981) 233; L. L. Chau and H. Y. Cheng, Phys. Rev. Lett. **53** (1984) 1037; **59** (1987) 958; I. I. Bigi and A. I. Sanda, Nucl. Phys. **B281** (1987) 41.

[53] B. Kayser, M. Kuroda, R.D. Peccei and A. I. Sanda, in preparation; For a discussion of kinematical CP violation in the context of the specific decay $B \to \eta\pi^+\pi^-$, see S. Barshay and J. Geris, Phys. Lett. **84B** (1979) 319.

[54] A. Carter and A. I. Sanda, Phys. Rev. Lett. **45** (1980) 952; Phys. Rev. **D23** (1981) 1567; I. I. Bigi and A. I. Sanda, Nucl. Phys. **B193** (1981) 85; **B281** (1987) 41; I. Dunietz and J. Rosner, Phys. Rev. **D34** (1986) 1404; Y. Azimov, V. Khoze, and M. Uraltsev, Yad. Fiz. **45** (1987) 1412; D. Du, I. Dunietz, and D. Wu, Phys. Rev. **D34** (1986) 3414.

[55] A. J. Buras, W. Slominski and H. Steger, Nucl. Phys. **B245** (1984) 369.

[56] K. J. Foley et al, SLAC preprint SLAC-PUB-4426 (1987), to appear in the Proceedings of the Workshop on Experiments, Detectors, and Experimental Areas for the Supercollider.

[57] D. London and R.D. Peccei, in preparation.

[58] B. Guberina, R.D. Peccei and R. Rückl, Phys. Lett. **90B** (1980) 169.

[59] M. P. Schmidt, these Proceedings.

[60] R. Aleksan, J. E. Bartelt, P. Burchat, and A. Seiden, SLAC preprint SLAC-PUB-4673 (1988).

[61] W. B. Atwood, I. Dunietz, and P. Grosse-Wiesmann, SLAC preprint SLAC-PUB-4544 (1988).

[62] JADE Collaboration: W. Bartel et al, Phys. Lett. **146B** (1984) 437.

KAON PHYSICS IN THE 1990s: RARE DECAYS AND CP VIOLATION

M. Atiya, E. Blackmore, G. Bock, D. Bryman, M. Cooper, H. Gordon, L. Littenberg, W. Louis, H. Lubatti, K. McFarlane, K. Nishikawa, J. Ritchie, L. Roberts, T. Shinkawa, M. Sivertz, A. J. S. Smith, G. Thomson, R. Tschirhart, T. Yamanaka

Boston University, Brookhaven National Laboratory, Fermi National Accelerator Laboratory, KEK
Los Alamos National Laboratory, Princeton University, Rutgers University, Stanford University, Temple University, TRIUMF, University of Tokyo, University of Victoria, University of Washington

ABSTRACT

The objective of this group was to assess the opportunities for experiments on the K system over the next few years. This necessitated evaluating the impact of recent experimental and theoretical activity on the motivation for this work, adducing the technical lessons of the experiments, projecting the experimental techniques toward future efforts, and reviewing the facilities at which future experiments might be carried out. By and large we tried to indicate the general directions in which we feel future efforts will be most productive. More detailed attention is given to certain cases which seem to us particularly promising.

MOTIVATION

Although studies of the K system have probably contributed as much as any to the Standard Model, the primary motivation to the latest round of experiments has been the prospect of going beyond the Model. There are a number of experimentally tractable K decays whose observation would unambiguously signal the existence of new physics. There are also several less exotic modes whose Standard Model contributions are orders of magnitude below the current experimental limits. If these were seen at level clearly higher than the Standard Model prediction, they too would decisively signal new physics. What is more, almost all of the theoretical attempts to extend or replace the Standard Model predict or enhance these decays at some level. For example, one approach that was quite popular in the period during which the current round of experiments was proposed was that of extended Technicolor which both removes the need for spontaneous symmetry breaking and explains the fermion mass-scale. Technicolor models predicted that lepton flavor-violating decays such as $K^0_L \to \mu e$ would occur at levels $\geq 10^{-9}$. A crude idea of the scale being probed by experiments sensitive to this particular decay is given by the relationship $BR/10^{-8} \sim (20\ TeV/M_{heavy})^{1/4}$ (assuming that all couplings are equal to g). Since BR sensitivities down to 10^{-11} are possible from the current round of experiments, mass scales of > 100 TeV can be probed.

In the case of CP-violation, the motivation was somewhat different. It was realized in the late 70's that the Standard Model with three generations implied CP-violating phenomena that are not consistent with the previously all-explaining but intellectually unsatisfying superweak model. Although experiments with the potential to demonstrate these effects looked very difficult, the prospect of finally making progress on the long-standing mystery of CP-violation tempted a number of groups into action. Most of the efforts focussed on $K \to 2\pi$ decay.

Recently the motivation in both areas has undergone significant evolution. The lepton flavor-violation experiments have soaked up the "easy" orders of magnitude or sensitivity and shot down the original Technicolor target. Theorists, never at a loss, have released Technicolor from its shackle to the experiments. At the moment there does not seem to be a conspicuous theoretical target for lepton-flavor violation, although most new models continue to predict it in a generic fashion. Since the mass-scale probed goes only as the 1/4 power of the branching ratio sensitivity, considerable experimental advances are required if significant progress is to be made. Recent experimental experience suggests that beyond a BR sensitivity of 10^{-12}such advances will require new techniques. This may make such experiments relatively less attractive than some of the other possibilities discussed here. Of course, if theorists come up with a well-motivated, specific target not too far beyond the current limit this all changes. Moreover, it should always be kept in mind that a positive result from the current round of experiments is quite possible and would drastically change the context for this sort of work.

The motivation for the suppressed decay modes has also undergone significant evolution since the current round of experiments was proposed. For example, the discovery of the long b-quark lifetime and of large b-b mixing have permitted predictions of $K^+ \to \pi^+ \nu\nu$ with less uncertainty than was previously possible. This allows a cleaner interpretation of any anomalously large result of this branching ratio. In addition the list of exotic candidates for the X in $K^+ \to \pi^+ X$ continues to grow as theorists exercise their imagination. The case of $K^0{}_L \to \pi^0 ee$ which attracted such attention in our subgroup is more complicated. Once again, significant improvements were made in our knowledge of the short distance contribution to this decay, but if anything, situations regarding the long distance contribution has recently become more confusing than ever. We can still be confident that a branching ratio much larger than 10^{-10} betokens new physics, so that a window of at least a factor 100 still exists As the various windows are closed by current and near-future experiments, the interest in these decays evolves from new physics jackpot to Standard Model study. One can still search for new physics but now one has to do it by making detailed comparisons with Standard Model prediction, rather than by sudden dramatic discovery. The particular nature of the interest in these decays will be strongly influenced by the outcome of the present round of CP-violation experiments. If the NA31 result is not confirmed, and the Standard Model explanation of CP-violation remains problematical, modes such as $K^0{}_L \to \pi^0 ee$ wherein large direct CP-violating effects are predicted will be of forefront interest. If the NA31 results is confirmed then $K^0{}_L \to \pi^0 ee$ and its sister modes will be viewed more symmetrically with $K^+ \to \pi^+ \nu\nu$ as arenas for determining Standard Model parameters such as KM angles. These decays have the virtue with respect to $K \to 2\pi$, that there is no unknown hadronic matrix element to be calculated before the KM parameters can be meaningfully constrained.

LEPTON FLAVOR VIOLATION SEARCH

There have been four recent/on-going LFV searches, three searching for $K^0{}_L \to \mu e$ (AGS E780 and E791, and KEK E137), and one seeking $K^+ \to \pi^+ \mu^+ e^-$ (AGS E777). The most sensitive of each category now has data on tape sufficient to reach sensitivities of ~ $2\text{x } 10^{-10}$. These decays are kinematically very well-constrained and so far the experiments do not appear to be background limited. In fact the experimenters are reasonably confident that this will remain true to at least the 10^{-11} level. Difficulties are more serious in the area of trigger rates and reconstruction efficiency. These experiments have had to deal with chamber singles in the region from several MHz to several tens of MHz. Counter plane singles over 10 MHz were also encountered. E780 and E777 ran at or near rate limits. E137 and especially E791 could have run somewhat faster. In E777 eliminating subsidiary triggers and other minor improvements would allow a factor of 2 or so in rate-taking capability, and that an improved beam could reduce the rates/incident K^+ by a factor of ten or twenty. This would allow a measurement at the 10^{-11} level to be made. The rates observed in E791, although high in an absolute sense, are within a factor of 2 of being totally attributable to K decays. The experimenters are proposing to run in conditions which will increase the instantaneous flux of useful $K^0{}_L$, and therefore, most probably the instantaneous rates in general, by a factor of ~6. With improvements in running efficiency, and increased running time, this would allow them to get to <~$3\text{x } 10^{-11}$ sensitivity in the next running period. The flux available to go beyond 10^{-11} will be available at BNL when the AGS Booster comes on line (~1991). At this point, horoscope plane rates will likely be in the neighborhood of 200 MHz. If such rates could be handled, and modest improvements in background rejection made, a Stretcher + increased running/year would allow the sensitivity to be further improved by a factor ~5 to the near 10^{-12} level. For couplings equal to the electroweak, this B.R. corresponds to a mass scale of ~200 TeV, comfortably above what can be directly accessed by the SSC. Progress beyond this point may require a new approach. For example, rejecting the random background due to coincident K_{e3} and $K_{\mu 3}$ decay will require vetoing in an environment approaching the Ghz level. Further discussion of the problems of extrapolating this experiment is given in the accompanying paper by McFarlane and Cooper and in the Los Alamos and TRIUMF K Factory proposals.

$K^+ \to \pi^+ \nu\nu$

We next consider $K^+ \to \pi^+ \nu\nu$. This mode, or rather this topology (since the u and u can be replaced by any undetectable light objects) is a lightening rod for new physics of various sorts.[1] The present BR upper limit[2] of $1.4\text{x}10^{-7}$ is at least two orders of magnitude above the Standard Model prediction for this process (currently[3] 1-8 x 10^{-10}). This window for new physics is expected to be completely exploited by the one currently running experiment of this type, AGS E787[4] In the absence of new non-Standard physics, this process can be calculated with relatively little uncertainty in terms of the KM angles and top quark mass (estimates by the theorists present at Snowmass of this uncertainty ranges from 10-35%). Thus, if nothing new is found, the motivation for future experiments of this type will depend on our state of knowledge of the Standard Model parameters. The constraints on the KM parameters which such a measurement can yield then compare favorably with those obtainable from sources such as BB mixing, exclusive B decay, etc. To be useful in this context, depending on the actual value of the BR, one would need sensitivity level.

This is no mean task, however, since compared to the other decay modes we are considering, $K^+ \to \pi^+ \nu\nu$ has a very poor signature. There is no kinematic constraint, so that one is thrown back on particle identification and vetoing. Fortunately, known sources of π^+'s emerging from K^+ decay are limited to momenta below 205 MeV/c (the recoil momentum from $K^+ \to \pi^+\pi^0$). The signature is then such a π^+ unaccompanied by any other particle. The necessity for excellent vetoing and particle i.d. push one towards a stopping K^+ geometry. At low momenta relatively intense and pure K^+ beams are obtainable, and the stopping geometry facilitates both high vetoing efficiency and high geometrical acceptance. Fig.1 shows the E787 apparatus. This detector features nearly hermetic gamma and charged particle vetoing and ~50% geometrical efficiency.

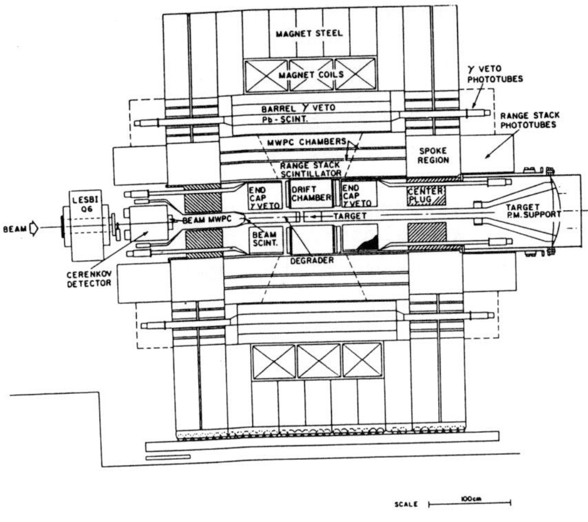

Figure 1: The E787 Spectrometer

Since E787 is somewhat behind the other BNL experiments, having just ended its first real data run in May, extrapolation from its experiences is correspondingly less certain. E787 took the equivalent of about ten days worth of data, nominally enough to reach a few x 10^{-8} sensitivity if there are no problems with background. Raising the beam rate to the maximum allowed (~7 x 10^{12} pot), reducing various inefficiencies, and making slight improvements in triggering, this detector should be able to reach a sensitivity below 10^{-9} in a year or two. Since the yield of usable stopping K's has proved to be about half of what was originally anticipated (~ 150K/5 x 10^{12} pot instead of 300K), a beam upgrade will be needed to reach the experimental goal of 2×10^{-10} sensitivity. If the acceptance of the LESB-1 (the beam used by E787) were increased to equal that of the LESB-2, the K^+ flux would be expected to triple. If some of this increase could be traded in for improved beam purity (there are currently $2\pi+\mu/K$) one could hope to find an optimum wherein the K^+ is doubled, but the total beam rate only rises by 50%. Since the singles rates in E787 seem to be proportional to the number of p's and K's incident on the degrader (a BeO cylinder used to slow the K^+'s from ~800 MeV/c to ~300 MeV/c so that they can be stopped in a scintillator target), the corresponding increase in single rates could probably be accommodated with minor improvements in the electronics.

How could the sensitivity be improved beyond ~10^{-10} in the future at BNL or another facility? As in almost every case, the simplest improvement is just to run more week/year. At present the BNL AGS is underutilized in this respect by at least a factor 2. Second there is a potential improvement in the duty factor by at least a factor 2.5 (say at BNL via a stretcher) which like the utilization factor would not put additional demands on the instantaneous rate capability of the detector. One could again be in the same position vis a vis instantaneous rate, with a total improvement in sensitivity of a factor ~5. This would bring the statistical sensitivity to a few x 10^{-11} level, whereas the background rejection of E787 was designed to be sufficient for 2×10^{-10}. It is not yet clear whether much improved background rejection could be attained without a major upgrade to the detector. The observed π^0 rejection capability has proved to be ~5×10^{-6} as opposed to the 10^{-5} assumed in the proposal background estimates. The effectiveness of other background rejection modalities (mainly muon rejection through comparisons of range, momentum, and energy, and through following the $\pi\mu e$ decay chain with transient digitizers[5]) is still under study.

To get beyond the few x 10^{-11} level, one is forced to take an increase in instantaneous rates and a detector superior to that of E787 would certainly be necessary. At a Boosted and Stretched AGS, the K flux required to get to ~10^{-11} would be available, at instantaneous rate ~ six times higher than those which are now faced. Fluxes sufficient to get to ~10^{-12} would be available at K factories such as proposed by TRIUMF and LAMPF.

Reaching the level of 10^{-12} would entail the challenge of dealing with instantaneous rates 50 times higher than in the current experiment. This is very serious for an experiment which needs to veto on 1 MeV visible photon energy. At the moment (say for 150 Kstops/pulse), this entails ~10% dead time for a $\pm$ 10 nsec veto gate. Clearly, one cannot turn up the wick beyond a factor 10 without improving the timing. This is not as easy as it might seem since one needs ~< 10^{-3} inefficiency in rejection of high energy photons. To avoid compromising this rejection, the veto gate must be > $\pm$ 3.5σ in extent, and if there are non-Gaussian tails on the time distributions it must be longer still! Putting transient digitizers on all veto channels could help here.

High rates also imperil the ability of the detector to identify pions through their decay chains since a muon can be "promoted" to a pion via an accidental which deposits ~3 MeV within a few τ_{π^+} of the muon's stop. This method of muon rejection must not only be maintained in the face of increased rates but actually must be upgraded in order to keep pace with the improved sensitivity being sought. The problem can be addressed via increased segmentation. E787 has ~200 pion stopping counters in the region of primary sensitivity. This could be increased by an order of magnitude in a future experiment, if funds were available. At present the effective segmentation of the detector is increased by a factor ~5 since the range stack elements are read out on both ends, allowing end to end comparisons of timing and PH to identify non-local accidentals. This factor could be enlarged if light collection, pmts, cables, and electronics were upgraded. The second pm discrimination technique, comparison of p-T-R (momentum-kenetic energy-range), would also be improved by greater segmentation of the range counters (better determination of R) as well as by enhanced light collection (better determination of T). This technique could be further improved by increasing the magnetic field imposed on the drift chamber from the 1 T of E787 to the perhaps 2.5 T a superconducting magnet could provide. The momentum resolution, currently ~2.4% at 200 MeV/c, could be improved to <- 1%, to great benefit. The pion trajectories, currently mainly radial, would become largely azimuthal, allowing a thinner range array and a smaller apparatus overall. The π^0 rejection could be improved, perhaps as much as an order of magnitude by replacing the current lead-scintillator sandwich shower counters by a fast scintillating crystal such as BaF_2 or pure CsI. Finally, many potential backgrounds arise because the stopping K^+ and the outgoing π^+ must pass through the material of the stopping target. The imaging target of E787 is constructed of ~400 bundles of 6 2mm scintillating fibers epoxied into triangular clusters. Given sufficient funds, a demultiplexed version of this target with 2mm (or even finer) granularity could readily be constructed. Future progress in scintillating fiber fabrication techniques should allow targets with negligible dead material (the E787 target has 25% dead material). In sum, perhaps more than in the case of any of the other popular rare decays, techniques for improving the detection of $K^+\to\pi^+\nu\nu$ lie within the reach current or near-future technology.

CP VIOLATION

The study of CP violation can still only be done within the K system. Perhaps the next decade of experiments with B mesons will yield new information on this phenomenon but for much of the next decade we will have to make whatever advances we can through more precision experiments within the Kaon system. Here we summarize the exciting situation which exists today in the two pion decay and the possible directions experiments will take over the next several years. Studies making use of initially pure K^0 and $\bar{K}^0$ will soon be underway. Also CP violation in the three pion decay will also likely be seen in the next decade. A discussion of the (rare) and CP violating decay $K_L\text{->}\ \pi^0 e^+ e^-$ is given in the following section.

CP Violation in Two Pion Decay

The CERN and Fermilab experiments study CP violation through observations of the decays of neutral K mesons to two pions. More specifically these experiments address the issue of whether or not CP violation is confined to $\Delta S = 2$ interactions, as the Superweak Model predicts.

Recall that the physical neutral kaon states ($K_{L,S}$) are almost, but not quite, the CP eigenstates ($K_1[CP = +1]$, $K_2[CP = -1]$) . Specifically the K_L is: $K_L \sim K_2 + \varepsilon K_1$, where $\varepsilon \approx 0.002$. The CP mixture of the K^0 states results from a small T violation in the $\Delta S = 2$ process $K^0 \text{<-->} \bar{K}^0$ allowed by second order weak interactions. Until quite recently all observations of CP violation could be accounted for by the single parameter ε describing this imbalance. This is the prediction of the superweak model. There is also a parameter describing $\Delta S =1$, or "milliweak" CP violation, ε', which is predicted to be small but non-zero in many models of the electroweak interactions. The usual definitions :

$$\eta_{+-} = \mathrm{Amp}(K_L\text{->}\ \pi^+\pi^-)\ /\ \mathrm{Amp}(K_S\text{->}\ \pi^+\pi^-),$$

$$\eta_{oo} = \mathrm{Amp}(\ K_L\text{->}\ \pi^0\pi^0)\ /\ \mathrm{Amp}(K_S\text{->}\ \pi^0\pi^0),$$

expressed in terms of ε and ε' are:

$$\eta_{+-} = \varepsilon + \varepsilon', \quad \text{and} \quad \eta_{oo} = \varepsilon - 2\varepsilon'$$

Since the phases of ε and ε' are nearly equal, a test of the equality of $|\eta_{+-}| = |\eta_{00}| = \varepsilon$ is a test for $\Delta S = 1$ CP violation and we can write:

$$\varepsilon'/\varepsilon = 1/6\,(1 - |\eta_{00}/\eta_{+-}|^2)$$

This is the essential strategy of both the CERN and Fermilab efforts.

A number of calculations predict $\varepsilon'/\varepsilon >$ +0.002, so one is looking for deviations from unity of the quantity $|\eta_{00}/\eta_{+-}|^2$ of less than about one percent--statistics and systematics are crucial.

The most tantalizing result in over 20 years of CP violation studies comes from NA31 with the first report of direct CP violation at the three standard deviation level and comfortably within standard model ranges. The group's result is ε'/ε = 0.0033 +/-0.0011, where the statistical and sytematic contribution to the error are nearly equal.[6] This result awaits further confirmation from within that group and from the Fermilab experiment E731. NA31 has recently finished a run and has plans to continue. Their result is based on approximately 100K observed $K_L \to 2\pi^0$ events. E731 has collected 300K of these events and they are under analysis at this time. An early result from that experiment gave ε'/ε = 0.0032 +/-0.0028(statistical) and +/-0.0012 (sytematic).[7]

Three things can happen: If ε'/ε is nearly 0.003 that will be confirmed to roughly 5 standard deviations with the data in hand at E731 and NA31 and the case is closed--at least, that is, until the theoreticians are able to calculate the hadronic matrix elements needed to describe the process. On the other hand if the results converge at a value less than about 0.002 or if NA31 and E731 disagree we are likely to see new initiatives.

The experiments use quite different techniques and therefore must deal with different sytematics, as well as some commom ones. Both use double ratios to minimize systematic effects, but briefly, the differences are (for more details see Reference 6 and 7):

NA31 uses a single beam detects charged and neutral decays simultaneously and K_S and K_L in separate runs. The decay region is 50 meters long; K_S are produced at different locations within this region by a moveable target train, approximating a "flat " distributuion in decay vertex postion to lessen acceptance effects. The heart of the detector is a liquid argon calorimeter used for both charged and neutral events and PWC's and a TRD fot the charged mode. The resolution in the neutral mode is quite good , in the charged rather poor (there is no magnet) requiring stringent offline cuts. Because the K_S and K_L are detected at different times the detector must be stable and well monitored.

E731 uses two incident K_L beams, one with a regenerator to make K_S and detects both K_S and K_L decays simultaneously. The detector uses a drift chamber magnetic spectrometer and a lead glass calorimeter. A subset of the E731 data was taken with all four modes--charged, neutral, long and short collected simultaneously. Rate effects (accidentals) and spectrometer gain drifts with time are closely monitored but largely cancel because of the simulataneous triggering on K_S and K_L decays. Scattering of kaons in the regenerator from one beam to another and backgrounds (already below 0.5%) in the $K_L \to 2\pi^0$ will probably limit the ultimate sensitivity of this spectrometer to the statistics currently in hand.

If ε'/ε is larger than 0.002 we will know that within 2 years. To push our understanding below say .001 will require new spectrometers and probably new beamlines. A next generation E731 detector would require a higher resolution electromagnetic detector to further improve background levels and allow an analysis of the data in individual momentum and proper time bins further reducing the systematic effects of acceptance corrections.

Both the NA31 and the E731 (E773) groups have taken or will take data to measure the phase difference $\Delta\Phi = \mathrm{Arg}(\eta_{+-}) - \mathrm{Arg}(\eta_{00})$ which currently is inconsistent with CPT by two standard deviations. NA31 is analyzing their data now and E773 is scheduled for running in 1990. We make no comments on the effects of a confirmed violation of CPT.[8]

CP Violation in Three Pion Decay

Fermilab experiment E621 is analyzing data from a measurement of the paramater η_{+-0} which in analogy to the two pion case is defined as:

$$\eta_{+-0} = \mathrm{Amp}(K_S \to \pi^+\pi^-\pi^0)\ /\ \mathrm{Amp}(K_L \to \pi^+\pi^-\pi^0).$$

By collecting three pion decays in a V spectrometer close to a kaon production target, the experimenters look for a small interference between the K_L and K_S decays as a signal of CP violation in the $\pi^+\pi^-\pi^0$ decay mode. Acceptance corrections are crucial for the expected value of $\eta_{+-0} \cong \eta_{+-} \cong 0.002$, and the acceptance for K_S decays is monitored by simultaneously collecting the more copious CP conserving $K_L \to \pi^+\pi^-\pi^0$ decays from a second target located upstream of the K_S target. Based on approximately 100,000 observed three pion decays from a run in 1984 the group reports preliminary values of $|\eta_{+-0}| = 0.022 \pm 0.026$ and $\Phi_{+-0} = -39^{o} \pm 80^{o}$.[9] This is approximately one order of magnitude of improved sensitivity over previous published results. It is also one order of magnitude away from the expected signal. In hand and under analysis are approximately 3 million events from the 1985 run which should give a statistical improvement of one order of magnitude. The proponents have submitted a letter of intent to build a new beam and spectrometer to use a similar technique with factors of improvement of 2-3 in each of several areas: solid angle, decay volume increases, incident proton beam intensity, and spectrometer optimization, to push the experiment to a precision of of 0.0005 or roughly 20% of the expected value of $|\eta_{+-0}|$.

CP Violation Studies from Pure K^0 States

Starting in the Spring of 1989 the CP LEAR[10] collaboration at CERN will begin a study of what is hoped to ultimately be 1E13 proton anti proton annhilations at rest yielding initial states of 10^{10} K^0 and K^0 respectively. Two and three pion decays as well as semileptonic decays will be collected in a 4 pi detector with full particle identification for complete reconstruction of primary and secondary vertices. CP violation appears as a rate difference between the process $K^0 \to f$ and $K^0 \to (CP)f$. Expected precisions on the parameter ε'/ε, neasurements is comparable to those of E731 and NA31 but with yet different systematics--a help if the situation remains cloudy for a few years. Additionally the measurement programs include η_{+-0} and η_{000} with precisions of 10^{-3} or less on the magnitudes and about 1^0 on the phase differences. Detection of low energy photons from K and pion decays decays will be particularly challenging for at rest K decays.

Muon Polarization in $K\mu3$ and $K\to 2\mu$ Decay

While not discussed in detail during this workshop, no discussion of CP violation studies would be complete without mention of searches for T or CP violation which could be revealed in measurements of the muon polarization in either $K\mu3$ or $K\to 2\mu$ decay. In the case of searches for the transverse polarization of the muon from $K\mu3$ decay the most sensitive search to date probed the few X 10^{-3} level. Further reductions in both statistical and systematic errors could be pursued at either BNL or a K Factory. The observation of longitudinal polarization of the muons in $K\to 2\mu$ would likewise signal T violation. Polarizaations of order 10^{-4} to 10^{-3} are predicted in the standard model and some extensions to it. Further improvements in the $K\mu3$ search seem quite possible and polarization in $K\to 2\mu$ may be in the provence of upgraded BNL experiments or K Factories.[11]

$K^0{}_L \Rightarrow \pi^0\, \ell\bar{\ell}$

INTRODUCTION

Much attention at this workshop has been focussed on the process $K^0{}_L \Rightarrow \pi^0 e^+e^-$ [12] Like the LFV processes discussed above, this process is graced with an excellent signature. At the levels for which it has been searched so far (~ few 10^{-8}) it has proved to be relatively background free. As in the case of $K^+ \Rightarrow \pi^+\nu\nu$ there is a Standard Model prediction [13] two or three orders of magnitude below the current upper limit on the branching ratio [14]. There are many possible varieties of new physics that could fall into this window [15]. One of particular interest is the production of a new scalar, S, in $K^0{}_L \Rightarrow \pi^0 + S$, followed by $S \Rightarrow e^+e^-$ [16,17]

Since $K^0{}_L \Rightarrow \pi^0 e^+e^-$ is CP-violating in lowest order, if the Standard Model region can be reached, interesting CP-physics will become accessible. This is particularly true for two reasons. First the direct CP-violating amplitude is predicted to be of about the same size as the indirect CP-violating piece. This is to be contrasted to the case of $K^0 \Rightarrow 2\pi$, wherein this ratio is predicted to be more like 1/300. Second, unlike the 2π case the prediction does not depend on a hard-to-calculate hadronic matrix element. Unfortunately, as discussed by the theorists at Snowmass, there is also a CP-conserving amplitude mediated by two photons. The size of this contribution is presently controversial, but it may be comparable to the CP-violating pieces [18-22]. In what follows, we sketch out a method of extracting the CP-violating information from this decay even in the case where the CP-conserving amplitude is relatively large. Of course, if this amplitude is very small, as some theorists maintain, it will be much easier to pull out the CP information.

Disentangling the amplitudes

At fixed target proton accelerators, p-nucleus collisions create a $\bar{K}$ beam which is mainly K^0 with an admixture of K^0. For the purposes of exposing the essential physics we will ignore this complication and assume we are dealing with a pure K^0 beam. Including the $\bar{K}^0$ component is straightforward but tedious and does not change the results in a qualitative manner.

We take

$$|K^0\rangle \approx \frac{1}{\sqrt{2}} \left(e^{-iM_L t}|K_L\rangle + e^{-iM_S t}|K_S\rangle\right)$$

$$\approx \frac{1}{\sqrt{2}} \left(e^{-iM_L t}\left[|K_2\rangle + \varepsilon|K_1\rangle\right]\right.$$

$$\left. + e^{-iM_S t}\left[|K_1\rangle + \varepsilon|K_2\rangle\right]\right)$$

and

$$\langle\pi^0 ee|T|K^0\rangle = \frac{1}{\sqrt{2}}\,[e^{-iM_L t}(A_2 + \varepsilon A_1)$$

$$+ e^{-iM_S t}(A_1 + \varepsilon A_2)]$$

where $A_{1,2} \equiv \langle\pi^0 ee|T|K_{1,2}\rangle$.

Noting that A_2 (CP-violating or higher order CP-conserving) is small compared with A_1 (CP-conserving in lowest order) and that $|\varepsilon| \sim .0023$, we discard the εA_2 term.

Then,

$$|\langle\pi^0 ee|T|K^0\rangle|^2 = \frac{1}{2}\,[\,|A_2 + \varepsilon A_1|^2 e^{-\Gamma_L t}$$

$$+ |A_1|^2\, e^{-\Gamma_S t}$$

$$+ 2\mathrm{Re}\,\{(A_1^* A_2 + \varepsilon|A_1|^2)\, e^{-i(M_L - M_S)t}\}]$$

Before proceeding, we need some more definitions and simplifying assumptions. First we take A_1 to be real, which is in general not quite true. A_1 and A_2 are functions of kinematic variables $W \equiv 2M_K^2 + 2M_e^2 - 2M_K(E_{e^-} + E_{e^+})$ and $\Delta \equiv 2M_K(E_{e^-} - E_{e^+})$. For future reference, note that $A_1{}^2$ is an even function of W and Δ [18]. We define the real constant η such that the direct CP-violating amplitude for $K_2{}^0 \to \pi^0 ee$ is given by $i\eta|\varepsilon|A_1$. The CP-conserving 2γ-mediated amplitude for $K_2{}^0 \Rightarrow \pi^0 ee|_{CP\ odd}$, we call $A_{2\gamma}$. We then have

$$A_2 = A_{2\gamma} + i\eta|\varepsilon|A_1.$$

Noting that $\varepsilon \approx |\varepsilon|e^{i\pi/4}$, and setting $\delta \equiv m_L - m_S$ we get

$$R(\Delta,W,t) \equiv |\langle\pi^0 ee|T|K^0\rangle|^2$$

$$= \frac{1}{2}\,\{[\,|A_{2\gamma}|^2 + |\varepsilon|^2 A_1(1 + \eta^2 + \sqrt{2}\eta)$$

$$+ 2|\varepsilon|\ \mathrm{Re}[(i\eta + e^{i\pi/4})\ A_{2\gamma}^* A_1)]e^{-\Gamma_L t}$$

$$+ A_1^2\, e^{-\Gamma_S t}$$

$$+ 2\mathrm{Re}\,[[A_1 A_{2\gamma} + (i\eta|\varepsilon| + \varepsilon)\, A_1^2]\, e^{-i\delta t}]\, e^{-\bar{\Gamma} t}\}.$$

We now examine the three time regions separately, starting with the K_S region in which

$$R_S(\Delta,W,t) \approx \frac{1}{2}\, A_1^2\, e^{-\Gamma_S t}.$$

If this formula is integrated over the Dalitz plot, we get simply $1/2\ \Gamma(K_1{}^0 \to \pi^0 ee)\ e^{-\Gamma_S t}$. Thus a measurement at early times determines $\Gamma(K_1{}^0 \to \pi^0 e^+ e^-)$.

At times $\gg \tau_S$, one gets

$$R(\Delta,W,t) = \frac{1}{2}\,\{|A_{2\gamma}|^2$$

$$+ |\varepsilon|^2\,(1 + \eta^2 + \sqrt{2}\eta)\, A_1^2 + 2|\varepsilon|\eta A_{abs} A_1$$

$$+ \sqrt{2}\,|\varepsilon|\,(A_{abs} + A_{disp})\, A_1\}\, e^{-\Gamma_L t}$$

where A_{abs} and A_{disp} are respectively the absorptive and dispersive parts of $A_{2\gamma}$, i.e. $A_{2\gamma} = A_{disp} + iA_{abs}$.

Note that A_1 is proportional to a function of W times $\bar{u}(k)pv(k')$. $A_{2\gamma}$ also has a term proportional to $\bar{u}(k)p(k')$ as well as one proportional to $\bar{u}(k)v(k')$. The latter term is known to be very small as it is helicity suppressed. The former term may also be small, in which case the analysis of this decay is much simplified, but at the moment the expected size of this term is controversial so that we are not free to ignore it. In the lowest order [20]

$$A_{abs} \propto \Delta\ \bar{u}(k)pv(k')$$

$$A_{disp} \propto A_{abs} \times \text{a function of W}.$$

The above implies that $A_{abs}{}^2$ and $A_{disp}{}^2$ are each proportional to $\Delta^2 A_1{}^2$ times a function of W (different in each case). We define a_{abs}, a_{disp}, $f_a(W)$ and $f_d(W)$ such that

$$A_{abs}^2 = |\varepsilon|^2\ |f_a(W)\ a_{abs}|^2\ A_1^2\ \Delta^2$$

$$A_{disp}^2 = |\varepsilon|^2\ |f_d(W)\ a_{disp}|^2\ A_1^2\ \Delta^2$$

and we find, that in the K_L region,

$$R(\Delta,W,t) = \frac{1}{2}\,\{|A_{2\gamma}|^2 + |\varepsilon|^2$$

$$A_1^2\ [(1 + \eta^2 + \sqrt{2}\eta) + 2\eta\ a_{abs}\ \Delta\ f_a(W)$$

$$+ \sqrt{2}\,(a_{disp}\ f_d(W) + a_{abs}\ f_a(W))\ \Delta]\}\ e^{-\Gamma_L t}$$

Note that the terms odd in Δ integrate to 0 over the Dalitz plot so that integrating:

$$R_L(t) = \frac{1}{2}\left[\Gamma_{2\gamma} + |\varepsilon|^2\, \Gamma(K_S^0 \to \pi^0 ee)\right.$$

$$\left.(1 + \eta^2 + \sqrt{2}\eta)\right] e^{-\Gamma_L t}$$

Thus if $\Gamma_{2\gamma}$ is non-negligible, without other measurements η cannot be extracted, i.e., one cannot just make a measurement in the large t region and learn anything about the direct CP-violation.

Finally the interference term is given by

$$R_I(\Delta,W,t) = |\varepsilon|\, A_1^2 \left\{\left(\frac{1}{\sqrt{2}} + a_{disp}\ \Delta\, f_d(W)\right)\cos\delta t\right.$$

$$\left. + \left(\eta + \frac{1}{\sqrt{2}} + a_{abs}\ \Delta\, f_a(W)\right)\sin\delta t\right\} e^{-\bar{\Gamma}t}$$

Once again, the terms odd in Δ integrate to 0 over the Dalitz plot:

$$R_I(t) = |\varepsilon|\ \Gamma(K^0 \to \pi^0 ee)$$

$$\left\{\frac{1}{\sqrt{2}}\cos\delta t + \left(\eta + \frac{1}{\sqrt{2}}\right)\sin\delta t\right\} e^{-\bar{\Gamma}t}$$

Note that η can be extracted from the integrated formulae if all three time regions are measured. Assuming that $\Gamma(K_S^0 \to \pi^0 e^+e^-) \approx \Gamma(K^+ \to \pi^+ e^+ e^-)$, and that neither the direct CP-violation nor the 2γ term is orders of magnitude larger than the indirect (ε) term, the K_S and K_L contributions cross at $\approx 10\ \tau_S$. One must be sensitive to times as small as ~ $7\tau_S$ in order to get any sort of measurement of $\Gamma(K_S^0 \to \pi^0 ee)$. The large t region must be measured well enough to subtract the contribution of the $e^{-\Gamma_L t}$ term from that of the interference region. Figure 2 shows the time distribution under various assumptions for the parameters. It's clear that extracting the parameters from this distribution will require reasonably high statistics, not easily achieved for a process with branching ratio ~ 10^{-11}.

To improve the statistical power of the experiment, it is very beneficial to fit the distribution $R(\Delta,W,t)$ without integrating over the Dalitz plot. Sehgal [20] has pointed out the possible large Dalitz plot interference between the CP-violating and CP-conserving contributions to $K_L^0 \to \pi^0 e^+e^-$. Analyzing this interference as a function of time should constitute an efficient method of extracting the critical parameters. For example, in the K_S-K_L interference region, one can subtract symmetric points on the Dalitz plot:

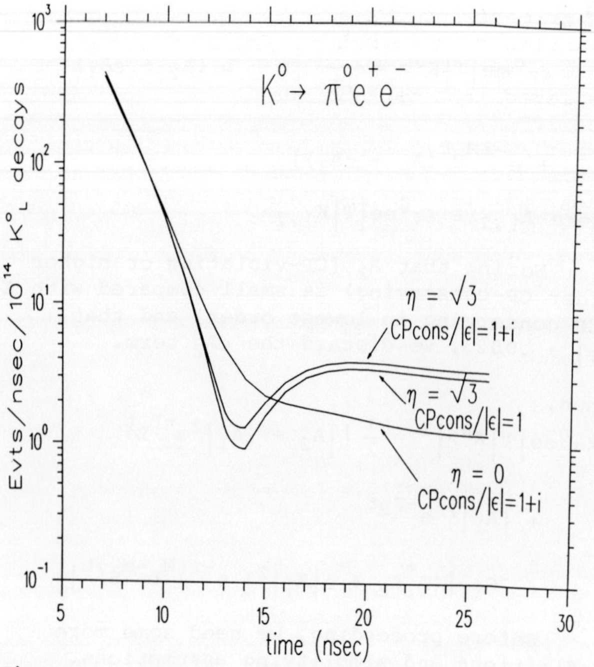

Figure

Time distribution of $\pi^0e^+e^-$ for a beam of 10^{15} produced 16 GeV K^0's, detected in an apparatus with 10% acceptance between 6 and 26 nsec.

$$R_I(\Delta,W,t) - R_I(-\Delta,W,t)$$

$$= 2|\varepsilon|\, A_1^2\, [a_{disp}\ f_d(W)\ \cos\delta t$$

$$+ a_{abs}\ f_a(W)\ \sin\delta t)]\ \Delta\ e^{-\bar{\Gamma}t}$$

This allows the separate extraction of the absorption and dispersive part of the 2γ amplitude. Knowing these then allows η to be extracted.

Experimental Considerations

Detailed studies of CP-violation in $K^0_L \to \pi^0e^+e^-$ could benefit greatly from sensitivities ($< 10^{-13}$) not available at any operating facility. However more study is needed before the feasibility of such an experiment event at a new facility can be assumed. There are two areas where reality may intrude: high detector rates and backgrounds. Measurement of this decay will require both calorimetry and tracking. Naive scaling of rates from existing experiments suggests singles rates of order 10 to 100 MHz per channel for a 10^{-13} experiment. The required improvements are large and may necessitate detector breakthroughs. Background studies have been performed for $K_L \to \pi^0e^+e^-$ in preparation for experiments at BNL and KEK (and a proposed experiment at Fermilab) with expected sensitivities between 10^{-10} and 10^{-11}. It is believed that backgrounds can be suppressed to the 10^{-11}level or below, although an experimental demonstration of this is 2 to 3 years away. As with rate problems, it is not clear without further study how difficult it will be to suppress the backgrounds to the level required for a kaon factory experiment.

Several sources of $K_L \to \pi^0 e^+e^-$ backgrounds can be listed:

1. Either of the decays $K_L \to \pi^0\pi^0\pi^0$ or $K_L \to \pi^0\pi^0$ followed by π^0 Dalitz decay or γ conversion, along with the appropriate number of missing γ's to fake the required topology,

2. $K_L \to \pi e\nu$, with the π misidentified as an electron, along with 2 γ's from another K_L decay (usually $K_L \to \pi^0\pi^0\pi^0$),

3. $K_L \to \pi e\nu\gamma$, with the π misidentified as an electron, along with a γ from another K_L decay,

4. $K_L \to \pi^0\pi e\nu$, with the charged pion misidentified as an electron,

5. $K_L \to \pi^+\pi^-\pi^0$, with double pion misidentification,

6. $K_L \to e^+e^-\gamma$, along with an additional γ from another K_L decay, and

7. $K_L \to \pi^0\pi^0$ followed by $\pi^0 \to e^+e^-$.

Consideration of item (1), those sources involving π^0 Dalitz decay or γ conversion, and of course item (7), lead quickly to the conclusion that events with $m_{ee} < m_{\pi^0}$, where m_{ee} is the invariant mass of the ee pair, must be rejected. The exact cut will depend on experimental details, but in general the low ee mass region presents intractible problems. The cut on the π^0 mass causes roughly 25% loss in acceptance.

A detector capable of rejecting the remaining backgrounds must have the following attributes: (1) large acceptance for observing photons, (2) good energy and position resolution for photons, (3) extremely good discrimination between pions and electrons, and (4) good timing of both charged particles and photons. A tracking system will be required, both to locate the decay vertex and determine electron sign, but the tracking resolution need not be extraordinary, since the errors in reconstruction will generally be dominated by the electromagnetic calorimetry at kaon factory energies (i.e., photon energies of a few GeV).

We have not specified a detector, nor performed calculations of acceptance, sensitivity or background rejection. Meaningful calculations of backgrounds at the 10^{-13} level are not simple. The best strategy for assessing the feasibility of a $K_L \to \pi^0 e^+e^-$ experiment at a kaon factory appears to be to wait until the presently planned experiments at BNL and KEK accumulate some experience.

Related measurements - $K^0{}_L \to \pi^0\mu^+\mu^-$

The closely related process $K^0{}_L \to \pi^0\mu^+\mu^-$ affords an even larger window for new physics than does $K^0{}_L \Rightarrow \pi^0 e^+e^-$. The Standard Model expectation is smaller than that of the latter process because of the smaller Q-value. The current upper limit is also significantly higher[23]. Note also, that once the mass of the Higgs allows it, the decay $H \Rightarrow \mu^+\mu^-$ will completely dominate the e^+e^- channel. The mass dependence of the coupling which is responsible for this may not obtain in the case of non-standard Higgs, but there is some prejudice that the $\mu^+\mu^-$ channel will dominate the e^+e^-.

If the Standard Model level can be reached, the CP physics mentioned above will also be accessible. However the formalism will be more complicated because of the persistence in the muon case of terms proportional to powers of m_ℓ. This may make the decay somewhat less promising than $K^0 \Rightarrow \pi^0 e^+e^-$ for detailed CP-violation studies, but if the theoretical controversy regarding the size of the two-gamma contribution to $K^0{}_L \to \pi^0 l^+l^-$ has not been resolved, even a crude measurement of $BR(K^0{}_L \to \pi^0\mu^+\mu^-)$ could be quite valuable.

The experimental prospects for this decay seem somewhat worse than those for $K^0{}_L \Rightarrow \pi^0 e^+e^-$, at least at BNL energies. In the electron case there are many more options for particle i.d. In addition, the copious $K^0{}_L \Rightarrow \pi^0\pi^+\pi^-$ decay is kinematically much closer to $K^0{}_L \Rightarrow \pi^0\mu^+\mu^-$ than to $K^0{}_L \to \pi^0 e^+e^-$. A study of the relative problems of the muon and electron modes at higher energies would be very welcome.

$K^0{}_L \Rightarrow \pi^0\gamma\gamma$

It was emphasized by the theorists at this workshop that a measurement of this branching ratio would help decide among the contending estimates of the CP-conserving contribution $K^0{}_L \Rightarrow \pi^0 l^+l^-$. It would be necessary to distinguish between predictions which vary by about a factor of two, centered near $BR(K^0{}_L \Rightarrow \pi^0\gamma\gamma) = 10^{-6}$. A rough measurement of the two-gamma mass spectrum would also be very useful. FNAL E731 has the statistical power to make these measurements and is in the process of determining whether its background rejection is adequate. If so, an answer is to be expected in the next few months. If not, it would most likely be possible to make these measurements in a future FNAL experiment.

A new possibility

One of us has pointed out [24] hat the decay $K^0{}_L \Rightarrow \pi^0\nu\bar{\nu}$ affords a very new interesting opportunity in K decay. This decay has never been actively sought, and the Standard Model prediction is of the order of a few × 10^{-12}, leaving a huge window for new physics. Possibilities include new CP-violating currents and $K^0{}_L \Rightarrow \pi^0 S$ where S is a new scaler which either escapes undetected or decays into daughters which do so. If the Standard Model level could actually be reached this decay would be of the utmost importance to the study of CP-violation. It is easy to show [24] that the indirect CP-violation in this decay is expected to be thousands of times smaller than the direct CP-violation. This is to be contrasted with the case of $K^0{}_L \Rightarrow \pi^0 l^+ l^-$ where these contributions are about equal. In addition there is no large long-distance contribution as in the charged lepton case. As in the charged lepton analog, there is no problematic hadronic matrix element to calculate so that a measurement of this branching ratio would give an unambiguous determination of $s_2 s_3 s_\delta$.

We have not seriously taken up the experimental problems of measuring this poor-signature process, but can offer a few remarks. The $K^0{}_L$ flux necessary for a measurement at the 10^{-12} level would probably be available at the upgraded facilities discussed below. A guess at the current state of the art in background rejection for such a process is more like 10^{-8}. It is likely that this measurement would benefit from higher energy than is available at BNL or at K factories as now proposed. Although this measurement is extremely difficult, its ratio of interest/-difficulty seems to us to compare favorably to that of making detailed CP-violation studies in B-decay.

FACILITIES FOR KAON EXPERIMENTS

We consider the following types of facilities for future experiments in the Kaon system: the presently operating cadre, the present facilities with upgrades,and completely new facilities. The presently operating facilities are the AGS at BNL, TEVII at FNAL, the KEK proton synchrotron, LEAR and the SPS at CERN. In some cases (e.g., the AGS Booster) upgrades relevant to K physics are already in progress, in others there has been serious discussion of such upgrades. New facilities [25] at which K physics can be pursued have been proposed or at least discussed by LANL (AHF), TRIUMP (KAON), BNL (AGS II), A European consortium (HEP), KEK (JHF), and INR Moscow.

Presently Operating Facilities

The first question is whether current facilities have been exploited to the fullest possible extent for these experiments. We ignore, for the moment, the question of running time as a fraction of the year. At the BNL AGS, E791 has used a 60 usr beam at 2.75 degrees to improve the sensitivity to $K^0{}_L \rightarrow \mu e$ to the 3×10^{-10} level. They have proposed a series of improvements which would allow them to improve their sensitivity by a factor of 10 or more. These include running the AGS at the full 28.5 GeV and cycle time of 2.5 sec, improving the targeting efficiency by about 10%, reducing the production angle to 1 degree, and most importantly, increasing the average number of protons on the production target from 3.1 to 8 X 10^{12} protons. At this point, the beam available will be close to the most intense $K^0{}_L$ beam deliverable by the AGS that is practical to use for rare decays. This beam will contain about 2 x 10^8 K's/burst and perhaps 10^{10} neutrons. In principle there are a number of improvements which could bring the number of K's/burst even higher, but these tend to be unacceptable for political or practical reasons. One could deliver about 50% more protons to the production target, but this would eliminate the rest of the AGS program. One could run at 0 degrees, but this would make the $n/K^0{}_L$ ratio, already marginal at 1 degree, more than twice as bad. One could increase the beam angle, but this would probably make it impossible to eliminate the effect of neutron interactions on the experiment. AGS E777 presently uses a 6-GeV/c charged beam containing ~ 2×10^7 K^+/pulse. Less than 5% of the AGS intensity is needed to produce this beam. The proponents of E777 envision a beam consuming an order of magnitude more protons to produce > 5 times more K^+ in a cleaner environment. THis would allow an experiment at the 10^{-11} level, and probably represents the maximum practical K^+ flux usable in a rare decay experiment at the AGS. AGS E787 sits at the end of the LESB-1 which produces about 200K stopping K^+/7Tp on target. This is not the most intense low energy separated beam at BNL, however. The LESB-2 produces about 3 times more K^+, albeit at a slightly lower energy. Thus one could probably get ~ 500K K^+ stops/pulse without beggaring the rest of the AGS program. One can summarize the situation at the AGS by saying that the exploitation of the machine thus far has been between 1/3 and 1/10 of maximum, and it is probably only a couple of years from being fully exploited (note however, that the maximum protons on target cannot be supplied to all the experiments at once).

At present the KEK 12 GeV PS accelerates up to 4×10^{12} protons per pulse at a repetition rate of 2.5 secs/pulse including a 0.5 sec flat-top. About 3/4 of the accelerated protons are available at the secondary beam lines after splitting the extracted primary beam. E137, a search for $K^0_L \to \mu e$ decay, is currently using 0.8-1.3 x 10^{12} protons on the target. Simultaneously one or two groups are using other secondary beam lines. Next year the duration of the flat-top will be extended to 1.0 sec at a repetition rate of 3 secs/pulse. In this condition E162, which is a new experiment to search for $K^0_L \to \pi^0 ee$ plans to use $\geq 2 \times 10^{12}$ protons/pulse on their production target. This experiment will start in the winter of 1989. These experiments, which hope to get to the few x 10^{-11} sensitivity level are probably reasonably close to the limit of what can be accomplished in rare K^0_L decay at KEK, although there has been discussion of a new beam to get even further. This summer construction of a new experimental hall began. In this hall at least two high intensity, low energy charged K beam lines will be constructed, scheduled to come online in the spring of 1990. Although designed for hypernuclear experiments, they would probably allow full utilization of KEK's intensity by rare K^+ decay experiments, if they could used for this purpose.

Fermilab is likely at the 10^{-8} level for $K_L \to \pi^0 ee$ already, having a recent result of BR< 4.2×10^{-8} based on an anlysis of 20% of their data. A new proposal from the group aims at sensitivities of 10^{-10} in the first stage of an experiment with the Tevatron K_L beam with an ultimate goal of 10^{-11}.

LEAR can product 4×10^{-3} (K and K) per p for 10^{13} (pp total) or 4×10^{11} kaons. This will allow experiments which compliment the current generation of CP-violation experiments at other facilities. For rare decay experiments, this flux will limit experiments to sensitivity ~ 10^{-10}. Since no K decay experiments have yet been carried out there, it is hard to guess whether the flux can be fully exploited before experiments elsewhere get beyond the 10^{-10} level.

Upgraded Facilities

One upgrade already under way is the Brookhaven AGS Booster project. This will increase the available proton intensity by a factor 4. Some of this factor will be absorbed in mitigating the present AGS proton famine, but individual experiments at significantly higher instantaneous rates are possible. Since, as discussed above, the experiments now experiencing the highest rates are still about one order of magnitude under the maximum which can be achieved at the AGS, full exploitation of the Boosted AGS would imply an increase of ~40 in instantaneous rates. This is a formidably factor and will likely require significant advanced in technology. To go beyond the sensitivity of experiments at the Booster, it seems most sensible to exploit the available factors which do not raise the instantaneous rates. In recent years the AGS has run only 15-20 weeks/year in the slow extracted mode suitable for K experiments. This could be doubled without interfering with maintenance or of other uses of the AGS such as heavy ion acceleration. The increase in operating cost of such additional running would be ~70M/yr. The other large available factor is that given by the duty cycle which at the AGS is ~ .35. This could be brought to ~1 by the addition of a stretcher ring, at a cost of ~$50M, according to AGS staff estimates. These seem the most economical and certain steps toward increasing the reach of K experiments in the near future.

Fermilab is likely to propose a new main injector[26)] which would operate at 120 GeV for slow spill with perhaps 2×10^{13} protons per 1 sec spill every 2 sec. This would bring FNAL within a factor of 2 of the boosted AGS in low energy Kaons/sec, and so allow comparable opportunities. For some of the experiments we are considering, the higher available K energies might constitute a considerable advantage. In the case of FNAL too, it should be kept in mind that maximal running time is the surest path to improved sensitivity.

New Facilities

We must temper our discussion of the influence of possible future K experiments on "kaon factories" with the observation that these facilities can study many other physics topics, and that in fact the proposers of some of these facilities have emphasized that is the requirements of non-K experiments that drive the parameters of their proposals.

The proposed facilities are generally in the BNL AGS energy range but promise a large increment in intensity over the present AGS, as much as a factor of 100 for slow beam running. With the advent of the AGS Booster, this advantage will be reduced to a factor 25. The most valuable piece of this factor is the 2.5 gained by implementing ~100% duty factor since detectors do not have to increase their instantaneous rate capability to realize the gain directly. As discussed above, this great advantage to physics could be realized by the relatively low cost construction of a stretcher at BNL. The advantage of the K Factories would then be a factor of 10 on roughly the same time scale but at considerably larger cost. Trading intensity for increased cleanliness does not seem desirable for the rare decay experiments since the typical cost of a factor of 100 would result in beams comparable to the present ones.

Regarding the specific proposals for future facilities, a number of points need to be made:

1. It is essential that the smoothest possible micro-duty cycle be available. It has to be possible to smooth out the rf structure for very high rate experiments which are bound to be limited by accidentals.

2. Efforts should be made to insure that the number of beamlines and size of the experimental areas are adequate to meet the physics demands. These issues do not seem fully addressed in the present proposals. A high energy (low production angle) neutral beam is a must, particularly if the machine energy is 30 GeV.

3. For the purposes of detecting in-flight K decays, it is not clear that 30 GeV is the perfect energy. The acceptance of the recent rare K^0_L decay experiments at BNL and KEK is peaked at energies considerably high than is the K^0_L spectrum, suggesting that higher beam energy could be dealt with by current techniques and would be an advantage. In addition certain experiments such as those involving multiple photon detection or those which deal with K^0_S, favor higher energy (so that the most recent examples of these have been run at FNAL and CERN, even though higher fluxes were available elsewhere). The whole question of the optimum energy for a new facility should be reopened

4. To assure that K factories offer clear advantages over the modestly enhanced AGS outlined above, efforts should be made to increase the K intensity proposed by up to a further factor 10. This would allow the tradeoffs of raw intensity versus beam purity mentioned above to pay off in increased attainable sensitivity.

5. K Factories are very costly so that it is essential that they be fully utilized. Therefore, there must be a commitment to running as much of the year as possible, preferably at least nine months.

THE FUTURE OF K PHYSICS

In recent years the center-pieces of kaon physics have been the lepton flavor-violation, $K^+ \to \pi\nu\nu$, and $K^0 \to 2\pi$ experiments. For the latter, the future may already be here. For $K^+ \to \pi\nu\nu$ there is a Standard Model target which should be reached in the next few years unless new physics intervenes. The same is true of $K^0_L \to \pi^0 e^+ e^-$. The related process $K^0_L \to \pi^0 \nu\nu$ also has a SM target, which would be extremely interesting, albeit extremely difficult, to reach. It seems practical to pursue the LFV experiments to the 10^{-11} level without undue expense (i.e., without completely new apparatus') but much beyond this the going will become heavy and progress will be at the rate of $M - BR^{.25}$ toward no particular theoretical target. To push all categories beyond the above-mentioned sensitivities, the most certain and most economical approach is to increase the fraction of the year this kind of physics is run, and where possible increase the duty factor from -1/3 to 1. This would allow several more years of pursuing K physics with the very high value of money ratio such work has enjoyed recently.

The optimum course for the longer term depends on what is found in the near future in K physics and elsewhere. To take a concrete example, imagine that $K_L \to \mu e$ is observed at 5×10^{-11}. Many theorists will try to take credit for predicting this. It would then become extremely important, in order to sort out the new interaction, to press the search for $K^+ \to \pi^+ \mu e$ as far as possible and at least to the 10^{-12} level. One would also push on other LFV reactions, including those outside the K system (e.g., $\mu \to e\gamma$ et al.). In the scenario, building a "classic" K factory would become quite attractive. Dramatic discoveries outside LFV might elicit quite different responses. If there are no such discoveries, then pursuing say $K^0_L \to \pi^0 ll$ may compete with B decays in exploring the KM parameter space. In this case higher energy and much higher intensity would be boons. The demands on detector technology would be challenging, but probably no more so than in the attempt to study CP violation in B decay.

ACKNOWLEDGEMENTS

We would like to thank our theoretical colleagues C. Dib , H. Georgi, F. Gilman, J. Hagelin, J. Hewett and R. Peccei for their interesting discussions.

REFERENCES

1. G. L. Kane and R.E. Schrock, "Rare K and mu Decays: Theoretical and Experimental Status and Prospects", in "Intense Medium Energy Sources of Stangeness", A.I.P. Conf. Proc. 102 ed. by T. Goldman, H.E. Haber, and F.-W. Sadrozinski, 123 (1983).

2. Y. Asano, et al., Phys. Lett. 107B, 159 (1981).

3. J. Ellis, J.S. Hagelin, and S. Rudaz, Phys. Lett. B192, 201 (1987).

4. AGS E787, BNL, Princeton, TRIUMF.

5. M. Atiya, et al., BNL-41788, submitted to the Intl. Conf. on Advanced Technology in Particle Physics, Como, Italy (June 1988).

6. M.Burkhardt et. al., Phys. Lett. 206B, 169 (1988) and J. Steinberger, Cern preprint EP88-66.

7. M. Woods, et al., Phys. Rev. Lett. 60, 1695 (1988) and M. Woods, Ph. D. Thesis, University of Chicago (1988).

8. Y. Wah, et. al. FNAL Proposal E773, (unpublished), and P. Clarke, et. al. Addendum to Cern Proposal P174.

9. G. Thomson, et al. FNAL E621 and recent Letter of Intent P796.

10. R. Landua, CERN Preprint 86-3635 (1986).

11. D. Bryman, IJMPA 4 , (Jan. 1989).

12. Weak decay theory at Snowmass.

13. F.J. Gilman, M.B. Wise, Phys. Rev. D21, 3150 (1980).

14. L.K.Gibbons et. al., Phys. Rev. Lett 61, 2661 (1988). and E.Jastrzembski et. al., Phys. Rev. Lett. 61, 2300 (1988). and G.D.Barr et. al., Phys. Lett. 214B, 303 (1988).

15. F. Del Aguila and J. Cortes, Phys. Lett. 156B, 243 (1985). L.J. Hall and L.J. Randall, Nucl. Phys. 274B, 157 (1986). X-G He and B.H.J. McKellar, Univ. of Melborne preprint UMP-88/40 (May 1988).

16. R.S. Wiley and H.L. Yu, Phys. Rev. D26, 3287 (1982); J. Ellis, M.K. Gaillard, and D.V. Nanopoulos, Nucl. Phys. B106, 292 (1976); T.N. Pham, and D.G. Sutherland, Phys. Rev. 151B, 444 (1983); S. Roby, G.B. West, and C.M. Hoffman, LA-UR-88-2667 (1988); S. Dawson, BNL preprint (1988).

17. H.E. Haber, G.L. Kane, and T. Sterling, Nucl. Phys. B161, 493 (1979).

18. J.F. Donohue, B.R. Holstewin, and G. Valencia, Phys. Rev. D35, 2769 (1987).

19. G. Ecker, A. Pich, and E. deRafael, Phys. Lett. 189B, 363 (1987); G. Ecker, A. Pich, and E. deRafael, Nucl. Phys. B291, 692 (1987); G. Ecker, A. Pich, and E. deRafael, Nucl. Phys. B303, 665 (1988).

20. L.M. Sehgal, Phys. Rev. D38, 808 (1988).

21. T. Morozumi and I. Iwasaki, KEK preprint KEK-TH-206 (Apr. 1988).

22. C.O. Dib, I. Dunietz, and F.J. Gilman, SLAC Pub. 4672 (Nov. 1988).

23. A. Carroll, et al., Phys. Rev. Lett. 44, 525 (1980).

24. L. Littenberg, "The CP-violating decay $K^0{}_L \rightarrow \pi^0 \nu\nu$, submitted to Phys. Rev. D.

25. Los Alamos National Laboratory publication, "Physics and Plan for a 45 GeV Facility", Volume LA-10720-MS., and TRIUMF publication, "KAON Factory Proposal", September, 1985, and "Report of the AGSII Task Force (Unpublished), 1984, and B. Winstein, et. al., Univ. of Chicago Preprint, EFI 89-01

26. L. Lederman, "The Fermilab Upgrade", these proceedings.

SEARCHING FOR $K_L^0 \to \mu e$ AT THE 10^{-13} LEVEL [1]

W.K. McFarlane
Physics Department, Temple University, Philadelphia, PA 19122, U.S.A.

and

M. D. Cooper
Los Alamos National Laboratory, Los Alamos, NM 87545, U.S.A.

Abstract

A few of the experimental problems of searching for the lepton non-conserving decay $K_L^0 \to \mu e$ are discussed in the light of the experience of BNL E791. This experiment was designed to have high rate capability and excellent background rejection; as such it provides information useful in evaluating designs for similar experiments at high-intensity facilities. The conclusion is that a level of a few times 10^{-13} may be reachable, but requires careful design and makes significant demands on the accelerator, particularly as regards beam time structure. Development of precise position-sensitive detectors with shorter resolving times is also necessary.

1 Introduction

At improved kaon production facilities, from the upgrade of the Brookhaven National Laboratory AGS to proposed new machines, some exciting prospects include the probing of very rare processes such as $K_L^0 \to ee$, the search for possibly forbidden decays like $K_L^0 \to \mu e$, and the measurement of K_L^0 and K_S^0 decays to a new degree of accuracy.

It is generally accepted that to undertake such experiments an improvement of a least a factor of ten is required to open a reasonable window for discovery. By examining existing experiments, it is possible to evaluate whether such improvements are foreseeable. The most sensitive search for $K_L^0 \to \mu e$ currently under way is BNL E791,[1)] which proposed to reach a branching ratio limit of 10^{-12}, but is currently expected to be limited to about 10^{-11} by a variety of factors.

The factors which appear most likely to prevent a simple duplication of E791, at a Kaon Factory with 100 times the beam intensity, from reaching a limit of 10^{-13} can be put into two categories:

- rate capability of detectors, electronic logic, and data-acquisition systems
- physics backgrounds from K_L^0 decays

After reviewing E791, methods of possibly removing the above limitations will be described.

2 BNL E791

E791 is being carried out at the Brookhaven AGS by collaborating teams from UCLA, UC-Irvine, LANL, Stanford, Temple, Texas, and William and Mary. The main goal is to test the idea of separate lepton-number conservation in $K_L^0 \to \mu e$ to a new level of sensitivity. Secondary goals are to search for the decay $K_L^0 \to ee$, the CP-violating process $K_L^0 \to \pi^0 ee$, and the branching fraction and CP-violating polarization for the decay $K_L^0 \to \mu\mu$. Detailed discussions of these goals can be found in the proposal.[1)] A sensitivity of 10^{-12} for $K_L^0 \to \mu e$ requires high statistics and good background rejection; the spectrometer design is matched to these needs.

The projected sensitivity depends on the primary beam intensity and the total running time, for which the proposed values have yet to be achieved. Extrapolations into the future are based on the experience from the 1988 run, from which E791 is expected to reach a 90% confidence limit of about $2-3\times10^{-10}$ for $K_L^0 \to \mu e$ with good background rejection and to have competitive sensitivities for other rare modes.

[1] Work supported by funds from the U.S. Dept. of Energy, including under grant DE-FG02-88ER-40389.

2.1 The E791 Beam and Spectrometer

For the 1988 run, a new high-intensity neutral beam at the BNL AGS was produced by focusing a proton beam on a 190-mm-long copper target. The design was for 28.5-GeV protons, but to date only 24.0 GeV has been available. The neutral beam was defined by collimators at 2.75° to the incident proton direction that gave it a solid angle of 4 mrad × 15 mrad. The proton beam can be steered in the vertical plane to vary mean production angles from 1° to 2.75°. Charged particles were swept out of the beam by two dipole magnets and photons were suppressed by lead foils between the target and the sweeping magnets. The beam-defining elements were followed by an 8.4-m long evacuated decay region beginning 9.5 m from the target.

The decay region was followed by a two-armed spectrometer with drift chambers, two analyzing magnets, and detectors for muon and electron identification (figure 1). The neutral beam passed through a He bag between the two arms to minimize counting rates from beam interactions in the detectors.

The spectrometer magnets imparted opposite momentum impulses of approximately 300 MeV/c, so that charged particles emerged parallel to their original directions. This technique has three advantages: it increases the accuracy of the momentum determination (by allowing the use of a large bend angle and by measuring twice); it improves rejection of events with a pion decay in the spectrometer; and it allows for on-line event selection based on the initial particle directions inferred from the finely-divided scintillator hodoscopes behind the chambers. The geometric acceptance of the spectrometer for two-body decays was about 5% for decays in the decay region, for kaons between 4 and 20 GeV/c. There is no acceptance below 4 GeV/c, although about half the kaons are below that momentum.

The trigger hodoscopes were followed by Čerenkov counters to identify electrons and a hadron filter and scintillator hodoscope to identify muons. The Čerenkov counters were filled with a nitrogen-helium mixture at atmospheric pressure and had an efficiency for electrons of 93% and for pions below 9 GeV/c of 1%. Muons with momentum above 1.5 GeV/c penetrated the lead-glass and iron filter while electrons and pions were strongly attenuated. The efficiency for detecting muons was 98%; about 10% of pions decay in flight.

Backgrounds from single kaons decaying into two particles are rejected by selecting on the reconstructed value of the two-particle invariant mass m, and the colinearity angle Θ_c between the vector from the target to the decay vertex and the measured total momentum vector of the reconstructed decay. The on-line event selection was performed with programmable logic that triggered the data-acquisition electronics. The architecture of the data acquisition system gave the very low deadtime of $\sim$ 300 ns/event. The digitized information was transferred to a fully parallel buffer and then into on-line computers that reconstructed events from drift chamber data. The on-line program used the first three drift chamber modules to find m and Θ_c; ambiguities were dealt with by assigning the highest possible mass. For reconstructed $K_L^0 \to \pi^+\pi^-$ decays the RMS resolution for m was ~ 1.5 MeV/c^2, and for Θ_c it was ~ 0.3 mrad.

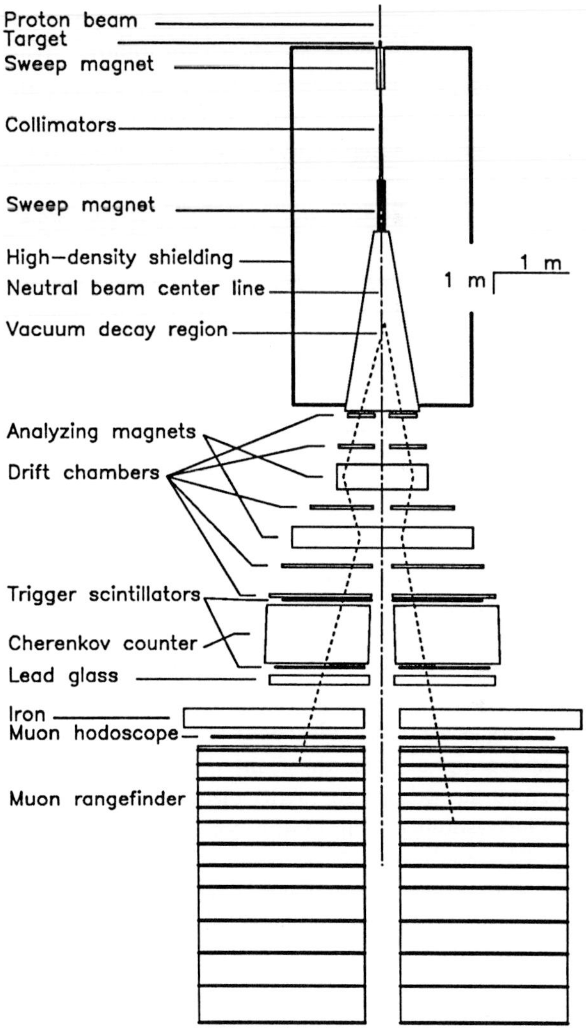

Figure 1: Plan view of E791 apparatus, with sample event

2.2 Results and Method of Analysis

During the run in 1988, the beam had an average intensity of 3×10^{12} protons per spill (the spill length was about 1.2 s), with test runs at 8×10^{12} p/spill (1.1 μA instantaneous). For comparison, a typical Kaon Factory design intensity is 100 μA.

Each event was required to pass (off-line) four successive criteria: that tracks could be found using the hits in the drift chambers with an algorithm different from the on-line one; track momenta, a two-track vertex, and Θ_c fell in some broad range; particle identification information allowed the assignment of a decay mode and the computation of m; and finally m and Θ_c satisfied background cuts. Candidates were retained as rare-decay events with m within 3 standard deviations of the kaon mass and Θ_c less than 1.5 mrad. The acceptance and the particle identification efficiency depend on the masses of the daughter particles.

Preliminary results show no evidence for observation of the process $K_L^0 \rightarrow \mu e$ and a number of $K_L^0 \rightarrow \mu\mu$ that is consistent with previous experiments.

3 Trigger Rates

A brief description of the trigger system follows. There were three trigger levels as proposed. Level 1 began with a coincidence among the four scintillator hodoscopes (two on each side) at the 'back' of the spectrometer (T), with a resolving time of 10 ns. Signals from the particle identification devices – muon hodoscopes and Čerenkov counters – and a signal from the front drift chambers (D) then formed the trigger for digitization of detector signals. The resolving time of the coincidence with the particle ID signals was 15 ns, while that with the drift chamber signal was 56 ns. Level 2 required acceptable patterns in the hodoscopes. The third level consisted of requirements imposed by software in on-line processors.

For an ideal beam, detector rates should be consistent with coming from K_L^0 decays, since other particles in the beam are stable or few in number. Whether the E791 rates are consistent with rates only from K_L^0 decays is an unresolved question. Using Skubic et al.'s data,[2)] and extrapolating down from 300 GeV to 24.1 GeV, one obtains, for the Cu target and beam aperture of E791, 1.4×10^7 K_L^0 produced per 10^{12} protons (Tp) over all momenta. The scaling from Be to Cu used $A^{0.67}$. This calculation gives 1.5×10^6 decays/Tp in the region 9.5 m to 18 m. The yield data of Eichten et al.[3)] give a factor of two more. The detector rates depend on the entire spectrum of K_L^0 's; about half or more of the K_L^0 decays come from kaons of momentum less than 4 GeV/c. However, neither of the above experiments measured the spectrum in this region. The E791 spectrometer has an acceptance from 4 GeV/c up, so that directly-measured data on the spectrum below this momentum are not available.

Current estimates are that singles rates in the detectors may be twice that estimated using Ref. 2, but the uncertainty in the kaon spectrum, and in the resulting acceptance for daughter particles, may amount to a factor of two. At a beam intensity of 8×10^{12} p/spill, for which an estimate based on Ref. 2 would give about 1.2×10^7 K_L^0 decays in the decay region, the front drift chamber had 5×10^6 hits per plane; about 30% of K decays are calculated to produce a hit in a front chamber. Thus the rate may range from being consistent with K decays, to 2-3 times higher. The major source of accidental coincidences in the Level 1 trigger was coincidences between the 'front' and the 'back' $(T \cdot D)$. The rate in the trigger scintillators at 8 Tp/spill was 12×10^6 counts per plane. Thus whether the rate comes solely from K decays is a crucial question; new data on particle production are needed to optimize Kaon Factory designs.

Approaching the working limits of the following detector properties hindered the rate capability of the experiment: singles rates in the elements, detector resolving times, accidental coincidences between 'front' and 'back', available fast logic, and computing power.

4 Backgrounds

4.1 Backgrounds from Single K_L^0 Decays

There are two principal prompt backgrounds to $K_L^0 \rightarrow \mu e$. The first is $K_L^0 \rightarrow \pi e \nu$, where the pion decays to a muon or is misidentified as a muon. With perfect resolution, this background will have a maximum m of 489 MeV/c^2, and non-zero Θ_c, and can thus be rejected with good resolution in these parameters. Special emphasis was placed on minimizing the non-Gaussian tails of the resolution functions by minimizing the mass of the drift chambers, by mapping the magnet carefully, and by designing the drift chambers to have a linear space-time relationship.

The other background also arises from $K_L^0 \rightarrow \pi e \nu$ but requires a double misidentification. If the π is identified as an e, and the e as a μ, *and* the e has a momentum more than 1.25 times the momentum of the π, then a value of $m \sim M_K$ can be reconstructed. Requiring the ratio of the higher track momentum to the lower to be less than some value reduces the number of such events, but a cut at 1.25 would reject real events also, so 2.5 is a better compromise, in conjunction with additional particle identification. Each track momentum was required to be less than 9.0 GeV/c, the Čerenkov threshold for pions, to reduce events with misidentified particles.

4.2 Backgrounds from Accidental Coincidences

Accidental coincidences fall in two categories: two K_L^0 decays or one K_L^0 decay in coincidence with some other interaction. An electron arising from $K_L^0 \rightarrow \pi e \nu$ and a muon coming from $K_L^0 \rightarrow \pi\mu\nu$ that appear to have the same time and form a vertex can give a high invariant mass. In the E791 proposal this background was estimated to be of order 10^{-13} at a beam intensity of 10×10^{12} p/spill giving 10^7 K decays per second in the decay volume, with resolving times of 2 ns and the expected spatial resolution for reconstructing tracks. To achieve an improved level of background rejection, it is necessary either to improve the resolutions or to veto on additional in-time hits and tracks.

Other interactions are harder to estimate because the E791 rates may be consistent with K_L^0 decays only, or may include substantial contributions (half or more) from other processes such as neutron or γ interactions. In either case, reduction of detector rates by more than a factor of two by better beam design and construction seems unlikely.

To reject the accidental backgrounds, it is critical to have a uniform time structure for the beam; any time structure would seriously compromise the measurements inversely as the duty factor. Specifically, a microstructure in which the beam is pulsed for 1 ns every 16 ns would be intolerable.

5 A New Design?

A radical new design might solve all of the problems of measuring $K_L^0 \rightarrow \mu e$ at once. However, such a study is beyond the scope of this report. The goal of this document is to ask whether a modified version of an existing experiment has a hope making significant improvements in the branching ratio sensitivity. If so, even better results might come from a new design.

5.1 Rates

For a new experiment at a Kaon Factory , the 'front'-'back' coincidence rate might be reduced by modifying the E791 apparatus to include a very thin, possibly fiber, scintillator hodoscope at the front, and using a short resolving time between this hodoscope and the downstream hodoscopes in Level 1. This upstream hodoscope could also be used in the Level 2 trigger. To put it in the E791 spectrometer without other changes would, however, degrade the vertex and colinearity resolution unacceptably using existing scintillator technology. To overcome this, the spectrometer would be divided into three regions, measuring the following: initial track directions, track momenta, and particle type. The additional trigger scintillators would be placed between the direction and momentum regions. The upstream scintillators would be thin enough so that multiple scattering would not be confused with pion decays; e.g., a 4-mm-thick conventional scintillator introduces an RMS deflection of 1 mrad, while an average pion decay bends the track by 15 mrad. The extra required wire chambers would provide redundancy for pattern recognition at higher rates.

Greater sophistication in the fast logic over LeCroy ECLline for Level 1 could lead to greater rate capability. The current electronics has basically 50 MHz capability and minimum resolving times of 10 ns, while 5 ns timing has been attained by the Crystal Box[4] for a system with a large number of counters. Logic units need a large number of inputs and good fan-out capability. Custom made modules could provide such specifications as well as allowing for more sophisticated pattern selection amongst the hodoscope elements to further handle rate problems.

Drift chamber improvements will be required to give shorter drift times while keeping the precision and low multiple scattering. New TDC's will also be needed that are economical, compatible with at least 10 Mbyte/s readout capability, and capable of 100 ns encoding, 0.5 ns least count, and 9-bit operation.

Data acquisition system architectures capable of dealing with 200 kHz trigger rates have been achieved. The computing needs will be large: E791 uses the equivalent of one IBM 3090 (about 30 VAX11/780) and a new experiment will need 100 times the processor power to become two orders of magnitude more sensitive. Besides the simple increase in the number of events to deal with, each event will be more complex, with more tracks to consider in reconstructing events; this must be done on-line to a great extent. Fortunately, experts predict that available computer power will double each year at affordable prices so that the required power should not be a difficulty; the need should be kept in mind.

Though rates will remain a concern, it is likely that trigger rates at higher intensities can be managed. The greater problem is the individual rate capabilities of the detectors, which will undoubtedly drive the designer to greater modularity and hence to greater complexity and expense.

5.2 Backgrounds

Suppression of backgrounds for a more sensitive experiment must be accomplished through improved resolutions. The prompt backgrounds are mainly suppressed through the requirements on invariant mass, colinearity, and particle identification. It currently appears hard to make much progress in tracking resolution without introducing detector mass that would give dominant multiple scattering contributions. However, at significant expense it is possible to introduce new magnets with bigger bends to improve the momentum resolution and more sophisticated lepton detectors to improve the electron and muon identification. An interesting possibility for the latter is a CsI electromagnetic calorimeter using the few nanosecond light component.

Accidental coincidences become relevant at higher rates and at sensitivities below 10^{-12}. A good match between the mean K_L^0 beam momentum and the detector acceptance can assure that as many decays as possible are potential candidates for good events and not sources of accidental background. Tuning the proton energy could be a useful tool to achieve this optimum. Time separation of two K_L^0 decays is, of course, critical to the suppression of accidental coincidences. To maximize the average separation, the proton beam must be de-bunched. A bunched beam with one pulse every 16 ns would cost an intolerable factor of 16 higher rates than a de-bunched beam. A Kaon Factory with a bunched beam would not be an improvement over the current AGS.

5.3 Statistics

Duty cycle is the fraction of clock time that the beam is on. Low duty factor can have contributions from three time scales: the micro-structure of the beam bunch, from the macro-structure of the beam spills, or from the machine schedule. The duty cycle should be good on an annual scale; 30 weeks running at 80% stable operation are needed to collect the required statistics. If the instantaneous beam intensity is roughly 10 times that currently at BNL (1 μA) and the macro-structure is three times better, there will be 30 times the average rate per hour, and with double the hours per year, there will be 60 times the kaon decays per year. Such a mode of operation will get the most kaons at the lowest instantaneous rate.

6 Summary

BNL E791 has been a major step in our knowledge of the limit for the rare decay $K_L^0 \to \mu e$, and it has demonstrated the limitations of the technique with the technology of 1988. With a number of changes, it is conceivable that a limit of a few times 10^{-13} could be reached with a similar apparatus and without a radical new design. The biggest factors would come from higher intensity, better duty cycle (which implies no micro-structure), and longer running at upgraded or new accelerators. To take full advantage of the new machines, developments in detector rate capability and resolution will be required. The needed improvements are not implausible in the next decade.

References

1) Cousins, R.D. *et al.*, "Study of very rare K_L^0 decays," BNL proposal no. E791 (Jan. 1984). See also R.D. Cousins *et al.*, "Search for the decays $K_L^0 \to \mu e$ and $K_L^0 \to ee$," Phys. Rev. D38, 2914 (1988).

2) Skubic, P. *et al.*, "Neutral-strange-particle production by 300-GeV protons," Phys. Rev. D18, 3115 (1978).

3) Eichten, M. *et al.*, "Particle production in proton interactions in nuclei at 24 GeV/c," Nucl. Phys. B44, 333 (1972).

4) Bolton, R.D. *et al.*, "Search for rare muon decays with the Crystal Box detector," Phys. Rev. D38, 2077 (1988).

TAU PHYSICS IN THE 1990s*

R. STROYNOWSKI
California Institute of Technology, Pasadena, CA 91125

All present knowledge of the τ lepton comes from e^+e^- annihilation experiments.[1] Groups at SPEAR, DORIS, PEP and PETRA typically have a few thousands of events of the process $e^+e^- \to \tau^+\tau^-$. The largest sample is presently available in the ARGUS experiment, which has over 100,000 events. The data collected so far allow for the determination of the branching fractions of the τ at the level of $0.5 \div 1.0\%$ of the total width and of the lifetime at the level of 3% of its value. The mass determination $m_\tau = 1784.2 \pm 3.2$ MeV comes from cross section measurements near the production threshold. Limits for the unexpected decay modes, e.g., lepton number violating decays are at the 10^{-4} level. There is little hope that new knowledge of τ physics will come from the existing hadron colliders; thus this study concentrates on the expectations from e^+e^- machines.

The total cross section for $\tau^+\tau^-$ pair production in the e^+e^- collisions is shown in Fig. 1. It consists of the point cross section

$$\sigma = \frac{4\pi\alpha^2}{3s} \cdot F(\beta)\,,$$

where $F(\beta) = \beta(3-\beta^2)/2$ and $\beta = \sqrt{1-4m_\tau^2/s}$, and the contributions of Υ and Z^0 decays into $\tau^+\tau^-$ pairs. Under the assumption that projected luminosities of various e^+e^- machines are realistic and that the efficiency of their running can be extrapolated from past experiences, one can expect that the following data will be collected during the next few years:

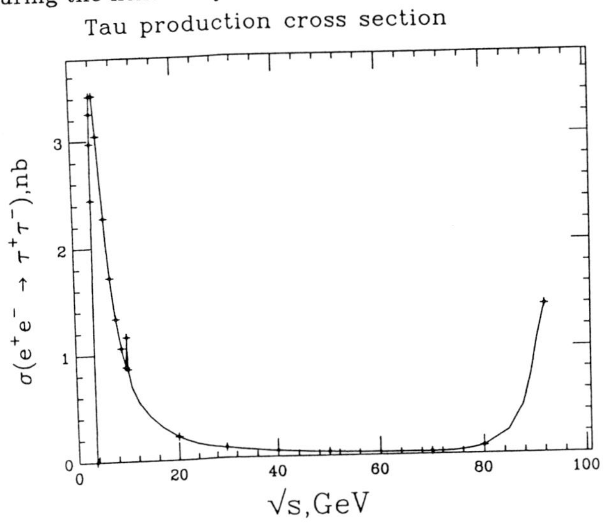

1) SLC and LEP are expected to collect of the order of 10^5 events by 1992.

2) TRISTAN will continue running near its highest available energy where the cross section is small. Thus only a small number of tau pairs can be expected in the VENUS, TOPAZ and AMY experiments.

3) ARGUS and later CLEO II are expected to collect data on the $\Upsilon(4S)$ where $\sigma(\tau^+\tau^-) = 0.77$nb. ARGUS will double its present data sample to a total of $\sim 2 \cdot 10^5$ events by 1991, while CLEO II is expected to collect $\sim 10^6$ tau pairs by 1993.

4) The MARK III detector at SPEAR is collecting data in the energy range with the largest τ pair cross section. During the next two years it is expected to double its data sample to about 120,000 events. Similar data will be obtained during the next few years in the Beijing experiment.

Thus, by the early 1990s the knowledge of the tau branching fractions and of its lifetime will probably reach the 10^{-4} level. The outstanding problem of today — "the missing one-prong branching fraction puzzle", i.e., the lack of understanding of the composition of the final states contributing to the one-prong decay of the τ, will presumably be solved by that time. One can also expect that the sensitivity to unusual decay modes of the tau may reach $10^{-5} \div 10^{-6}$ of the total width. Above the charm threshold, this type of measurement is limited by systematic errors due to leptonic and semileptonic decays of the charm quark.

Mass of the tau neutrino

One of the more intriguing aspects of τ physics is the possibility of studying the mass of the tau neutrino. The present upper limit of $m_{\nu_\tau} < 35$ MeV[2] is obtained by the ARGUS study of the invariant mass distribution at the phase space limit in the decay $\tau \to \nu_\tau 5\pi$. Its value is limited by the systematic errors of the measurements and understanding of background processes. Thus an increased data sample will not improve it in a major way. Better experimental resolution and background rejection in the CLEO II experiment may lead to an upper limit

* Work supported in part by U. S. Department of Energy Contract No. DE-AC03-81-ER40050.

of about 20 MeV using such a method. On the other hand, the measurements of the end point spectrum in the decay $\tau \rightarrow e\nu_e\nu_\tau$ are limited mainly by statistics. One can estimate, following Ref. (3), that large data samples collected at the $\Upsilon(4S)$ could result in an upper limit of $m_{\nu_\tau} \lesssim 12$ MeV. Further improvements, reaching a limit of about 2 MeV, could be achieved at a high luminosity tau factory[4)] close to the tau production threshold.

Hadronic decays

The tau lepton decays hadronically through the intermediate charged vector boson W. The hadronic final state has definite quantum numbers and is free from the backgrounds plaguing the study of resonances in hadronic production processes. The 10^6 data sample projected for the CLEO II experiment will contain about 50,000 decays $\tau \rightarrow \nu_\tau 3\pi$ fallowing for the proper partial wave analysis of the 3π final state. It should clarify the properties of the a_1 meson and answer the theoretical questions posed by present analyses.[5)] However, the potentially more interesting Cabibbo-suppressed channel $\tau \rightarrow \nu_\tau K\pi\pi$ will have only about 2000 events. Similarly, small data samples can be expected for the channels requiring π^0 reconstruction. The study of hadronic spectroscopy in τ decays is a very important subject. It will require statistics an order of magnitude larger than that expected from CLEO II.

Rare decays and sensitivity to new physics

Lepton number violation would be a clear indication of new physics. The limits for such decays in the muon sector are very strict, e.g., $B(\mu \rightarrow e\gamma) < 1.7 \cdot 10^{10}$, $B(\mu \rightarrow eee) < 2.4 \cdot 10^{-12}$. The analogous limits for the tau are smaller than 10^{-4}. Future experiments could reach the 10^{-6} level. The interpretation of such limits is very model-dependent. Theories in which lepton number violation occurs due to its coupling to charged Higgs particles[6)] predict a dependence of rate on the masses involved of

$$\frac{B(\tau \rightarrow 3\mu)}{B(\mu \rightarrow 3e)} \sim \frac{m_\tau m_\mu^3}{m_\mu m_e^3} \sim 10^8 .$$

This is one of the most favorable scenarios. In other models, e.g., based on an idea of technicolor, such enhancement is proportional to $(m_\tau/m_\mu)^2 \sim 400$, while supersymmetry has no mass dependence of the lepton number violating process. It is worth noting that precision measurements of allowed decays and lifetime also provide restrictions on specific models. For example, in the case of supersymmetric models, if the scalar partners of neutrinos are sufficiently light, then the decay $\tau \rightarrow \tilde{\nu}_\tau e \tilde{\nu}_e$ may occur through the exchange of $\widetilde{W}$ — the supersymmetric partner of $W^\pm$ and sneutrinos $\tilde{\nu}_e$, $\tilde{\nu}_\tau$ would escape detection. Such a process affects the partial decay rate and electron decay spectrum. The calculated partial rate[7)] is

$$\frac{d\Gamma^{\text{observed}}}{dx} = \frac{G_F^2 m_\tau^5}{4\pi^3} \left[\left(1 - \frac{4}{3}x\right) x^2 + \left(\frac{m_W}{m_{\widetilde{W}}}\right)^4 f(x) \right]$$

where $f(x)$ is a function of the masses involved. Neglecting the masses of neutrinos and sneutrinos, one gets

$$\Gamma = \frac{G_F^2 m_\tau^5}{192\pi^3} \left[1 + \frac{3}{2} \left(\frac{m_W}{m_{\widetilde{W}}}\right)^4 \right] .$$

Present values of the branching ratios and lifetime allow a limit to be placed on the mass of the wino of $m_{\widetilde{W}} > 142$ GeV at 90% C.L. Error analysis indicates that increased statistics available in, e.g., CLEO II will allow for improvement of such a limit by an order of magnitude.

Summary and Outlook

We may expect many new important results on the physics of the τ lepton to come from experiments running now or in preparation. However, some aspects will remain unexplored by the present generation of experiments. There are a number proposals to construct new e^+e^- machines in the 1990s. It is interesting to speculate what impact they would have on tau physics. Among the colliders under discussion are: the charm-tau factory, the B-$\overline{\text{B}}$ factory, the Z^0 factory and the TeV collider. All of them aim at a luminosity at least an order of magnitude greater than that of CESR and, except for the TeV collider, are expected to operate in the energy ranges where the tau production cross sections are equal within a factor of $2 \div 3$. With $\sim 10^8$ tau pairs produced in such machines, one may expect that most of the problems not addressed by CLEO II can be well studied in all of them. In addition, a low energy charm-tau factory could significantly improve the measurement of the tau mass, which in the early 1990s will contribute a dominant error to the tests of lepton universality. It also may provide the best laboratory for a study of the mass of ν_τ. Higher energy machines with polarized beams will allow the study of polarization parameters in tau decays and thus will provide sensitive tests of the lepton sector of the standard model. Finally, the large cross section for τ production at B-$\overline{\text{B}}$ factories will allow studies of hadronic amplitudes and searches for rare tau decays.

References

1. For recent reviews see: B. Barish and R. Stroynowski, *Phys. Rep.* 157 (1988) 1; K. K. Gan and M. L. Perl, *Int. J. Mod. Phys.* A3 (1988) 531.
2. H. Albrecht *et al., Phys. Lett.* 202B (1988) 149.
3. R. Stroynowski, CALT-68-1431 (1988).
4. J. Kirkby, CERN-EP/87-210 (1987).
5. N. Isgur, C. Morningstar and C. Reader, University of Toronto preprint 88-0442 (1988).
6. J. D. Bjorken and S. W. Weinberg, *Phys. Rev. Lett.* 38 (1977) 622.
7. G. L. Kane and W. B. Rolnick, *Nucl. Phys.* B232 (1984) 21.

B PHYSICS AT HADRON COLLIDERS

M. P. Schmidt
Yale University

Participating members and contributors to the subgroup on B physics at hadron colliders were:

E. Berger	R. J. Morrison
S. Childress	K. Nishikawa
B. Cox	N. W. Reay
K. Foley	L. A. Roberts
M. Fortner	M. P. Schmidt
S. Frederiksen	W. Selove
D. E. Groom	M. Sheaff
R. Imlay	R. A. Sidwell
G. Kalbfleisch	P. Skubic
P. E. Karchin	A. J. Slaughter
C. S. Kim	N. Stanton
H. Kobrak	R. Stefanski
J. Kuzminski	G. Trilling
P. Lebrun	D. E. Wagoner
R. Lipton	M. S. Witherell
N. S. Lockyer	E. Wolin
K. T. McDonald	P. Yager
J. G. Morfin	T. Ypsilantis

INTRODUCTION

A brief review is presented of the activities of the Snowmass subgroup for B physics at hadron colliders. The work here was in large part a continuation of efforts at previous workshops to design a detector capable of detecting sufficiently large samples of B-meson decays to allow for the study of CP violating effects expected in the context of the Standard Model. The emphasis of the working group was on a conceptual design for a spectrometer for the SSC. To this end a working model was adopted, largely based on the work of the Fermilab B Collider Study Group. Many of the studies carried out at Snowmass were continued beyond the course of the summer study and are described in detail in a letter of intent to Fermilab (P784) for a Bottom Collider Detector at the Tevatron.

B Physics at Hadron Colliders

In the past few years the importance of studying the decays of bottom quarks has become increasingly clear [1]. The bottom quark, like the strange quark, is forced to decay in a (generalized) Cabibbo suppressed manner making the study of B meson decays, like the study of K mesons, sensitive to rare processes. In principle the study of B meson decays could provide overconstrained tests of the Kobayashi-Maskawa six quark model of CP invariance violation [2]. This prospect is made more favorable as a result of the large mixing observed in the B_d system. Large CP violating asymmetries ($\sim 10\%$) are expected for many decay modes, in sharp contrast to the situation with K mesons. However in keeping with the suppressed nature of CP violation, these large asymmetries occur in general only for decay modes having small ($< 10^{-4}$) branching fractions [3].

To reach the goal of studying CP violation in decays of B mesons requires a large sample of B's. One particular mode that is often discussed to the set the scale of the effort required is $B_d^0 \longrightarrow \psi K_S$. In this case the final state (f) is a CP eigenstate, and the amplitudes for direct decay and decay after $B-\overline{B}$ mixing can interfere. The observable CP violating asymmetry is:

$$A = \frac{\Gamma(B_d \longrightarrow f) - \Gamma(\overline{B}_d \longrightarrow f)}{\Gamma(B_d \longrightarrow f) + \Gamma(\overline{B}_d \longrightarrow f)}.$$

The value of A is determined directly by the elements of the Kobayashi-Maskawa matrix, and the calculation is free of many of the uncertainties due to the strong interactions that affect the interpretation of results from K decays [5]. Measurement of the CP violating asymmetry A requires knowledge of the particle (b) or antiparticle ($\bar{b}$) content of the B meson decaying to the final state. This is accomplished by 'tagging' the event through (partial) reconstruction of the other B produced in the event. Here the analysis depends in detail whether the tagging B is charged or neutral because of mixing effects for the neutral mesons. It is useful to consider a specific decay sequence for detection: $B_d \longrightarrow \psi K_S$ followed by $\psi \longrightarrow l^+l^-$ and $K_S \longrightarrow \pi^+\pi^-$ accompanied by $\overline{B}_d \longrightarrow D^{(*)}l\nu$ and $D^{(*)} \longrightarrow K + charged$ with the event tagged by the charge of the lepton and kaon [5]. This results in an effective branching fraction ($BR(f)$) to the final state of $\sim 5 \cdot 10^{-5}$ and a tagging efficiency (ϵ_{tag}) on the order of 3%.

Ignoring dilutions due to incorrect tagging and otherwise assuming a perfect detector, the number of $b\bar{b}$ quarks required for a 3σ determination of the asymmetry A is:

$$N_{b\bar{b}} = \left(\frac{3}{A}\right)^2 \cdot \frac{1}{BR(f) \cdot \epsilon_{tag} \cdot f_d}$$

where f_d is the fraction ($\sim$ 35%) of B_d mesons expected from the hadronization of a b quark. For the specific example under consideration the expected (time integrated) asymmetry is in the range of 5 to 30% implying that $2 \cdot 10^8$ to $7 \cdot 10^9$ bb pairs are required. It is noted that the range of the prediction is a result of uncertainties in the Kobayshi-Maskawa matrix elements; improvements on the allowed range for V_{ub}/V_{cb} will be particularly helpful in sharpening the predictions. Similar results are obtained for other decay modes of interest [5].

The large statistical sample required motivates the consideration of hadron colliders as B factories. Table 1 presents the cross section for production of $b\bar{b}$ pairs, the ratio of this cross section to the total cross section, and the number of bb pairs produced in a 10^7 second run for an assumed average luminosity at several (future) hadron colliders. The hadronic cross section for $b\bar{b}$ pair production includes the important α_S^3 corrections recently determined [6]. For comparison appropriate values are also given for e^+e^- colliders at the $\Upsilon(4S)$ and at the Z^0. While the signal to noise is intrinsically superior for e^+e^- colliders, the statistical sample required will only be achieved through new innovations in e^+e^- accelerator technology. For hadron colliders the challenge lies primarily in the detector technology.

It is noted that $\sigma_{b\bar{b}}/\sigma_{tot}$ at the SSC is about the same as $\sigma_{c\bar{c}}/\sigma_{tot}$ for the Tagged Photon experiment at FermiLab (E691). The results from this experiment on charm decays have now equaled and even surpassed that obtained at the e^+e^- machines. These results have been made possible through the use of micro-vertex detectors to extract a sample of charm decays by virtue of the long charm lifetime (0.3 - 0.9 psec). In principle this technique can be exploited as well for studying B decays ($\tau_B \sim 1.2$ psec).

If the resolving power of the micro-vertex detector technique can be employed in a hadron collider detector, then a rich program of B physics can be opened up beyond that available at the e^+e^- machines. And there is a great deal of physics to accomplish along the path to the goal of studying CP violation. Measurements of the production cross sections and rapidity distributions for B_d, B_s and Λ_b will be useful for testing QCD predictions and determining gluon structure functions at low x. The lifetimes for B_d, B_s and Λ_b could be determined as could the mixing in the B_d and B_s systems. Branching fractions for modes with all-charged final states could be probed, with charmless decay modes being of special interest. The branching fraction for modes with definite CP could be determined, sharpening the strategy for the study of CP violation. In addition the search could continue for rare or forbidden decays such as $B \longrightarrow Kl^+l^-$ or $B \longrightarrow \mu e$ [4].

DESIGN CONSIDERATIONS FOR A HADRON COLLIDER DETECTOR FOR B MESONS

It is a formidable challenge to design a B detector capable of collecting and fully reconstructing pairs of B mesons with high efficiency in a hadron collider environment. Central issues are the overall architecture, the vertex detector, pattern reconstruction and tracking, particle identification, triggering and data acquisition.

The architecture for a B physics collider detector is largely determined by the characterisitics of the production and the decay of the B mesons. The dominant process for B meson production at a hadron collider is $b\bar{b}$ production via gluon-gluon fusion [6]. From this mechanism $B's$ are expected to be produced with $< p_t >\sim m_b \sim 5GeV$ and broadly distributed in pseudo-rapidity (η): roughly $\pm$ 3 units at RHIC, $\pm$ 4 units at TEV I and $\pm$ 6 units at the SSC. The bb pairs are produced with important correlations in pseudo-rapidity, and therefore direction. At TEV I the $\bar{b}$ will be produced within $\pm$ 2.5 units of rapidity of a b produced at $\eta \sim 0$ and within $\pm$ 1.5 units for a b produced at $\eta \sim 4$ [7]. The decay of the B meson into low multiplicity modes will result in events with $< p_t^{track} >\sim 1GeV/c$. The decay products of B mesons are then soft in p_t and yet distributed over a large range in polar angle and longitundinal momentum with respect to the colliding beam axis. Neither a forward ('fixed target' like) nor a central (barrel) detector is the optimal solution for high acceptance of $b\bar{b}$ pairs, although the need to cover more forward decays (higher rapidity) increases as the energy of the collision is increased.

At Snowmass a conceptual detector design for B physics at the SSC was considered and an overview is shown in Figure 1. The colliding beam axis is along the long axis of the detector and the collisions take place in a central detector located at the origin. The central part of the detector, discussed further below, is built in and about a large $((4m)^3)$ aperature 1 Tesla dipole magnet oriented with the magnetic field perpendicular to the colliding beams. The magnetic field is weak enough to pose no difficulty for the operation of the accelerator. The use of a dipole field geometry, conceived of at the Fermilab Beauty Workshop [8], optimizes the tracking acceptance, allowing for good momentum resolution for forward and central tracks. The design of the detector in the central ($\eta < 1.4$; $500mr$) and intermediate region ($1.4 < \eta < 3$) would closely follow that envisioned for a detector at the Fermilab collider [9][10]. The vertex detector and tracking chambers are located with the magnet volume.

The detector is extended on one side to increase the acceptance in the forward region beyond $\eta = 3$ ($100\ mr$) to $\eta = 5$ ($13.5\ mr$). Another large dipole magnet would be used for momentum analysis of the large rapidity tracks. Here the design is largely derived from previous work on forward B spectrometers [11]. In the intermediate and forward regions muon identification would be accomplished by instrumenting absorber steel; strategies for triggers and selection criteria for muons could be generalized from methods to be employed in a search for $B \longrightarrow J/\psi + X; J\psi \longrightarrow \mu + \mu-$ at Fermilab [12].

It is envisioned that electron identification would be accomplished with TRDs (transition radiation detectors) and electromagnetic calorimeters in all regions (central, intermediate, and forward). In addition ring imaging Čerenkov counters would be employed in all regions for π, K and p identification. In the central region time of flight counters would augment the RICH detectors. Hadronic calorimeters are not required.

VERTEX DETECTION

The most important element of the detector is the vertex detector shown in Figure 2. The vertex detector is comprised of silicon microstrip planes oriented to cover annular regions (disks) centered on the colliding beam axis and a set of barrels around the central region. Beyond the central region the disks are spaced in a logarithmic fashion (i.e.rapidity spaced). The silicon detectors would have 50μm pitch, 200μm thickness, and double-sided readout.

The basic geometry of the vertex detector is a result of the angular distribution of desired tracks (from B's) and the finite extent ($\pm$7cm) of the collision region. The particular geometry arrived at during Snowmass is based on the criteria that each track of interest must strike at least three planes with an angle of incidence of less than 45 degrees. Since the signal from a silicon detector depends on the track length, tracks at a large angle of incidence will give smaller signals and hence have poorer signal to noise.

To successfully exploit the power of the micro-vertex detector in resolving secondary vertices due to B and D decays, the detector must be close to the primary collision point with the minimum of material intervening. This is due to the effects of multiple Coulomb scattering, even in the silicon microstrip detector itself, which are proportional to the radial distance of the detector from the beam and inversely proportional to the transverse momentum of the track (recall $< p_t^{track} >\sim 1GeV/c$). The entire vertex detector is then to be mounted outside of a 400μm thick Be beam pipe with a radius of 1.3 cm. For operation at a luminosity of $10^{32}cm^{-2}sec^{-1}$ the dose rate for the silicon strips closest to the beam (1.5 cm) would be less than 2 Mrad/year [13], not intolerable for the silicon microstrip detectors themselves [14]. Apart from avoiding accidents involving major losses of the colliding beam in detector, the issue of radiation hardness of the silicon strip readout electronics will require further study.

There is considerable savings in price and complexity for operating at a collider with the minimum collision region length along the beam. The number of detector elements for the design under discussion scales with the length, L(cm), of the collision region: $2 \cdot 10^5 + 1.5 \cdot 10^4 \cdot L$. The combined barrel and disk geometry also presents a serious engineering challenge. A stable mechanical construction free of excessive dead regions is required. The design must also provide penetrations for signal readout and cooling of preamplifier electronics (2 mwatt per element). At Snowmass an initial model based on a hexagonal geometry was discussed and work along these lines continues [15].

Monte Carlo (PYTHIA-GEANT) calculations carried out at Snowmass by L. A. Roberts *et al.* [16] at Snowmass for $B \longrightarrow \pi^+\pi^-$ show that the vertex detector design has a high acceptance ($\sim$ 90%) including the criteria on angle of incidence for the tracks. Including the effects of multiple scattering and detector resolution (but not pattern recognition) these studies also found that about 60% of the events will have a measured B decay vertex separated from the primary vertex by more than five times the expected error in the measurement.

While the working model continues to be based on the double sided silicon microstrip detectors, there is considerable interest [10] in pursuing designs based on newer alternative pixel detector technologies. Pixel detectors provide space points which greatly aid the pattern recognition problem, and might also offer the possibility of incorporating vertex detector information in a high level trigger. Of particular interest are the silicon drift chamber [17] of E. Gatti and P. Rehak, and true pixel detectors under development by Nygren *et al.*, Parker *et al.*, and Shapiro *et al.*

CENTRAL TRACKING AND PATTERN RECOGNITION

While precise information from the vertex detector will be used in the final event reconstruction, the central tracking system must provide accurate three dimensional pattern recognition and reconstruction of tracks. The tracking system must have high spatial resolution and low mass to measure charged particle momenta to $\pm$1% over the full range of acceptance of the 1 Tesla spectrometer magnet.

Figure 3 shows a section of a plan view for one quadrant of the central tracking system. In the view shown the magnetic field due to the spectrometer dipole is perpendicular to the plane of the page. Figure 4 shows a section of the central tracking system as viewed along the colliding beams. Straw-tube chamber technology has been chosen as well matched to the particular requirements for the detector. A design under study by Stanton *et al.* [10] follows the proposal of DeSalvo [18] in using 3 mm diameter tubes with 30 μm thickness pressurized to about 4 atm. About $2 \cdot 10^5$ straw tubes are arranged in superlayers, each having 8 layers of straw tubes, with the super layers distributed throughout the tracking volume in order that tracks of interest intersect with 8 superlayers.

The proposed structure, which requires research on the viability of such thin walled straw tubes, is very attractive from the standpoint that the nearly 4π acceptance is not spoilt by massive support structures. As shown in Figure 4 the unusable acceptance region along the magnetic field direction on either side of the beam collision region is used for electronics cables and other access. The radation tolerance of straw-tubes limits how close to the beam they can be placed, and additional silicon planes would then be required for tracking coverage at small angles to the collidng beams.

Studies on pattern recognition algorithms and possibilities for deriving trigger information from the tracking system are in progress by the proponents of a Fermilab B collider detector. In the not too distant future we can expect to see results of realistic simulations of full events in the vertex and central tracking detector.

PARTICLE IDENTIFICATION

While the identification of electrons by TRDs and electromagnetic calorimeters is comparitively straightforward, the ambitious desire to implement a general single electron trigger greatly increases the complexity. The detailed design will then depend on the design of the vertex and tracking detectors. An efficient trigger for semileptonic decays of B's would require a trigger cut as low as $p_t^e > 1GeV/c$. Work by McDonald [9], Reay and Fortner has shown for operation at $10^{32}cm^{-2}sec^{-1}$ at the Tevatron that the rate of trigger electrons form B's is about 1 kHz. This rate is potentially overwhelmed by contributions from charm decays, e^+e^- decays of vector mesons, Dalitz decays of pseudoscalar mesons, converted photons, and overlaps of π^0 photons with a high rate ($\sim$20MHz) of charged pions. It seems inevitable that vertex information will be required to bring the trigger rate down to a managable level [19].

Identification of hadrons π, K, and p for B physics has been a subject of previous studies [20], and a presentation by Ypsilantis [21] at the summer study was especially useful. For the SSC B spectrometer considered at Snowmass there are RICH detector solutions for each of the three acceptance regions. In the central region ($p \leq 6$ GeV/c) a liquid radiator (C_6F_{14}) would suffice, whereas in the intermediate ($p \leq 30$ GeV/c) and forward regions ($p \leq 225$ GeV/c) liquid and gas (C_5F_{12}) radiators would be required. The preference is to use TEA as the UV detector, especially for its lower optical dispersion, compared to TMAE. Complete coverage over the momentum range will require 1mm position resolution for the Čerenkov photons leading to an estimated 10^7 readout pads!

CONCLUDING COMMENTS

The opportunities for B physics at a hadron collider are very high, but an ambitious program must be pursued. There was much progress at Snowmass on issues concerning designs and simulations for vertex detection, central tracking and particle identification. These aspects are in themselves a challenge although the detectors required are not very far from the state of the art. It will be very interesting to see the results of realistically simulated events in the context of a complete vertex and tracking design. The solution for a powerful enough trigger remains elusive although this interacts very strongly with the data acquisition system design. Here important steps have been taken by the Fermilab study group [10].

There are other concerns to be addressed associated with extrapolations to SSC energies. For example the diluting effects due to multiple B production are undoubtedly largest at the SSC [22], although there are uncertainties with the size of the effect. Finally it is fair to say, though not necessarily universally accepted, that it has not been established which hadron collider is best suited for the pursuit of B physics. For example there are many advantages to be considered for a barrel-like detector operating at high luminosity at RHIC (pp) with a short collison length [23]. Future studies will undoubdtely shed light on these and many other outstanding issues.

References

1) MacFarlane, D. B., "B Physics at e^+e^- Machines" in *Proceedings of the High Sensitivity Beauty Physics Workshop* Fermilab, 1987. Editors A. J. Slaughter, N. S. Lockyer, M. P. Schmidt.

2) Rosner, J. L., Sanda, A. I. and Schmidt, M. P., "Physics Group Summary," and Hamzaoui, C., Rosner, J. L. and Sanda, A. I, "B Meson Decay Asymmetry and $B\overline{B}$ Mixing" in *Proceedings of the High Sensitivity Beauty Physics Workshop* Fermilab, 1987. Editors A. J. Slaughter, N. S. Lockyer, M. P. Schmidt.

3) Bigi, I. I. and Sanda, A. I., "CP Violation in Heavy Flavor Decays," (Nuclear Physics, B281, 1987) pp.41-71.

4) Soni, A and Hou, G., "Loop Induced Rare B Decays," in *Proceedings of the U.C.L.A. Workshop, Linear Collider $B\overline{B}$ Factory Conceptual Design* , Editor Donald Stork.

5) Foley, K. J., *et al.*, "Bottom and Top Physics," in *Proceedings of the Workshop on Experiments, Detectors and Experimental Areas for the SSC* , Berkeley, 1987. Editors R. Donaldson and G. Gilchriese.

6) Berger, E., "Heavy Flavor Production," ANL-HEP-PR 88-26. Nason, P., Dawson, S., Ellis, R. K., "The Total Cross Section for the Production of Heavy Quarks in Hadronic Collisions," Fermilab 87-222-T.

7) Berger, E., "Benchmark Cross Sections for Bottom Quark Production," in *Proceedings of the High Sensitivity Beauty Physics Workshop* , Fermilab, 1987. Editors A. J. Slaughter, N. S. Lockyer, M. P. Schmidt.

8) Reay., N. W. *et al.*, "Summary of the Collider Architecture Working Group," in *Proceedings of the High Sensitivity Beauty Physics Workshop* , Fermilab, Nov. 1987. Editors A. J. Slaughter, N. S. Lockyer, M. P. Schmidt.

9) McDonald, K. T. *et al.*, "Status Report of the Fermilab B Collider Study Group" presented at the Summer Study, Princeton Univ. preprint DOE/ER/3072-45, June 1988.

10) Castro, H. *et al.*, "Letter of Intent for the BCD, A Bottom Collider Detector for the Fermilab Tevatron," Fermilab Proposal 784, October 1988.

11) Foley, K. *et al.*, "A Beauty Spectrometer for the SSC," in *Proceedings of the Workshop on Experiments, Detectors and Experimental Areas for the Supercollider,* Berkeley, 1987. Editors R. Donaldson and G. Gilchriese.

12) Selove, W., "Steps Toward High-Rate Fixed-Target B Physics," presented at this Summer Study, University of Pennsylvania preprint UPR-161E, November 1988.

13) Groom, D., "Radiation Levels in SSC Experiments," presented at Future Directions

in Detector Research and Development for Experiments at pp Colliders, Snowmass, Colorado, 1988.

14) Kondo, T., "Radiation Damage of Silicon Devices," presented at Future Directions in Detector Research and Development for Experiments at pp Colliders, Snowmass, Colorado, 1988.

15) Kalbfleisch, G. R. and Skubic, P., "Considerations Regarding Double-Sided Solid State Detectors," Univ. of Oklahoma preprint, Feb. 1988; Kalbfleisch,G. R. Skubic, P. L. and Lambrecht, M. A., "Results on Signal Correlations from a 'Micron' Doubled-Sided Mini-Strip Detector," Univ. of Oklahoma preprint August 1988.

16) Roberts, L. A., "Monte Carlo Simulation of Silicon Vertex Detector for Bottom Collider Detector", presented at this Summer Study, Fermilab Note 488, June 1988.

17) McDonald, K. T. and Purohit, M. V., *Proposal for Generic Detector Development,*" Princeton Univ. preprint DOE/ER/3072-46, August 1988.

18) DeSalvo, R., *Proposal for Generic Detector Development,*" Cornell Univ. preprint CLNS87/52, 1987.

19) Morrison, R.J., McHugh, S., and Purohit, M. V., "A Semileptonic Electron Trigger for Recording B^0B^0," in *Proceedings of the Workshop on Experiments, Detectors and Experimental Areas for the Supercollider,* Berkeley, 1987. Editors R. Donaldson and G. Gilchriese.

20) Ludlam, T. "Particle Identification for Beauty Physics," in *Proceedings of the High Sensitivity Beauty Physics Workshop* , Fermilab, Nov. 1987. Editors A. J. Slaughter, N. S. Lockyer, M. P. Schmidt.

21) Ypsilantis, T. "RICH/CRID for Future Hadron Collider Experiments," presented at this summer study. See also Brandt, A. *et al.*, SPSC Letter of Intent for a "Study of Beauty and Charm Physics at the SPS collider," August 1988.

22) Kim, C. S. "Production of Multiple $B\overline{B}$ Pairs From Higher Order QCD Processes," presented at this summer study and Univ. of Wisconsin preprint MAD/PH/407, January 1988.

23) Lockyer, N. S., "*B* Factory at RHIC," in the proceedings of the RHIC Workshop, July 1988.

Table 1. Cross sections and rates for $b\bar{b}$ pair production at colliders.

Machine	$\sqrt{s}$ (TeV)	$\sigma_{b\bar{b}}$ (μb)	$\sigma_{b\bar{b}}/\sigma_{tot}$	$<L>$ ($cm^{-2}sec^{-1}$)	$N_{b\bar{b}}$
RHIC	0.5	10	1/4000	10^{32}	10^{10}
TEVI	1.8	40	1/1000	10^{31}	$4 \cdot 10^{9}$
SSC	40	500	1/200	10^{32}	$5 \cdot 10^{11}$
CESR	0.01	10^{-3}	1	10^{32}	10^{6}
LEP/SLC	0.09	$5 \cdot 10^{-3}$	0.13	10^{32}	$5 \cdot 10^{6}$

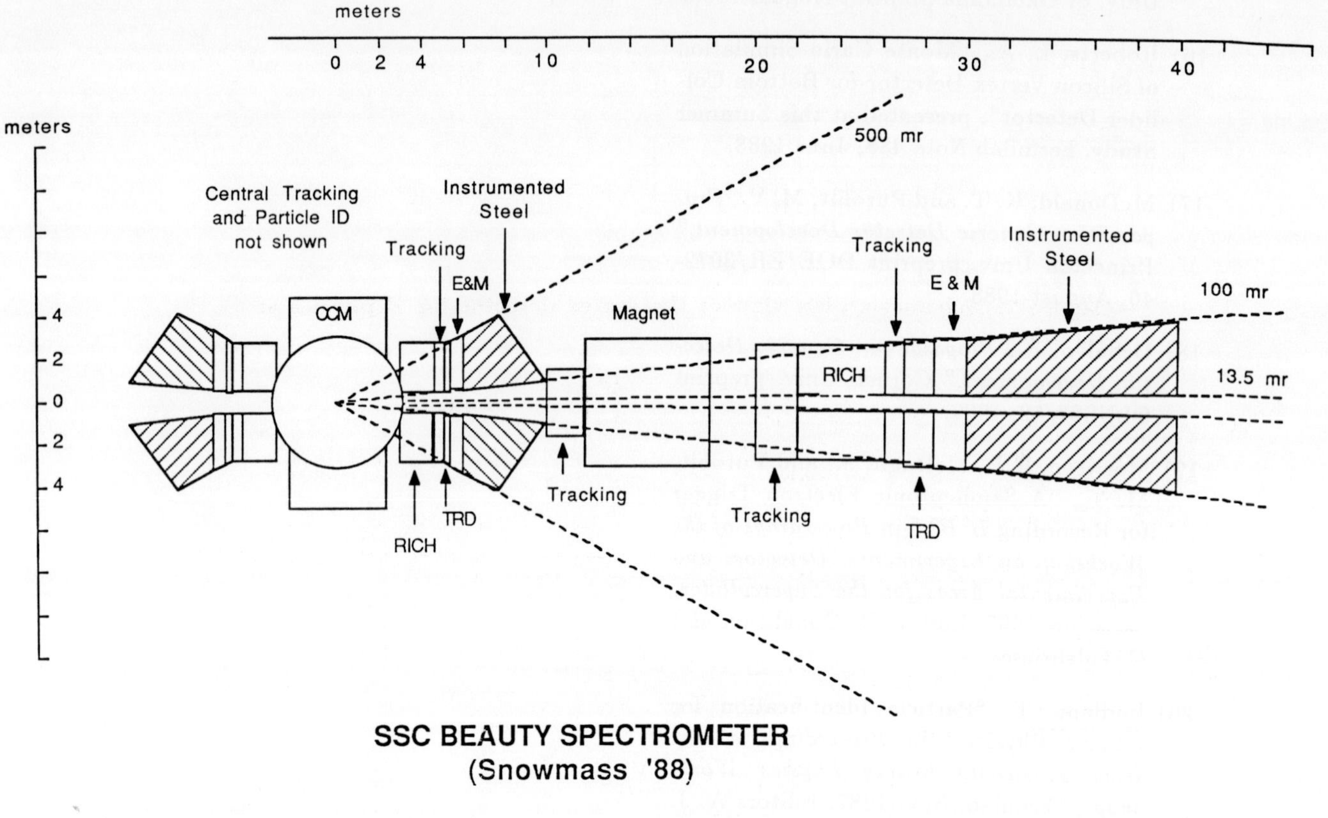

Figure 1. Schematic overview of a Spectrometer for B physics at the SSC.

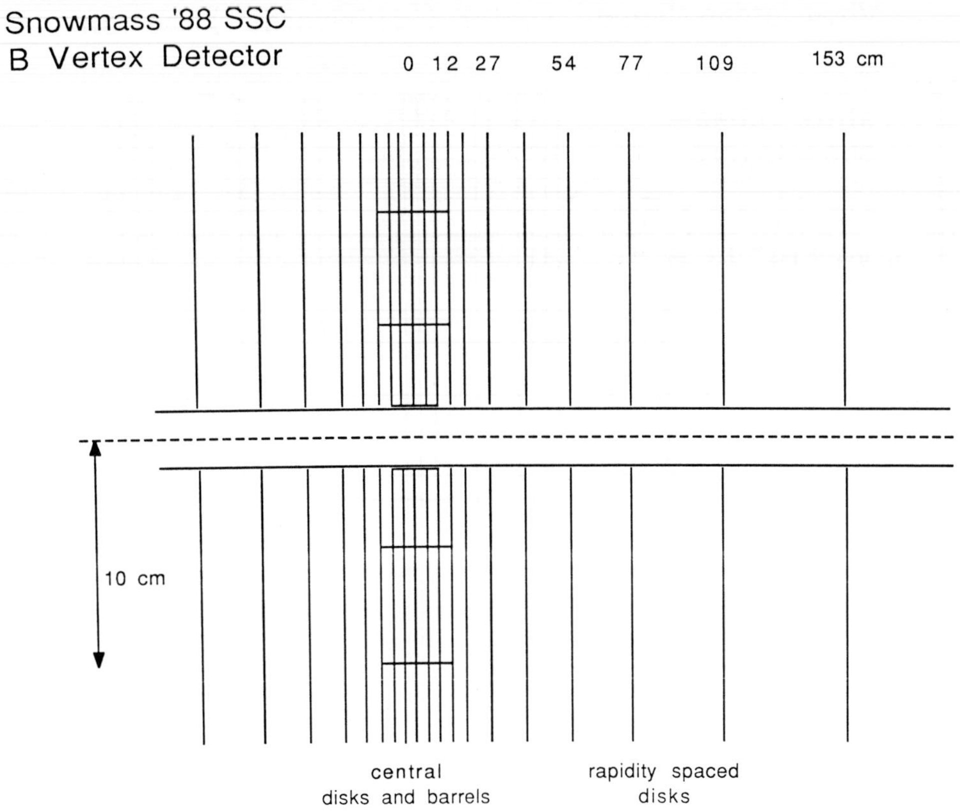

Figure 2. Schematic plan view of the vertex detector for B physics at the SSC.

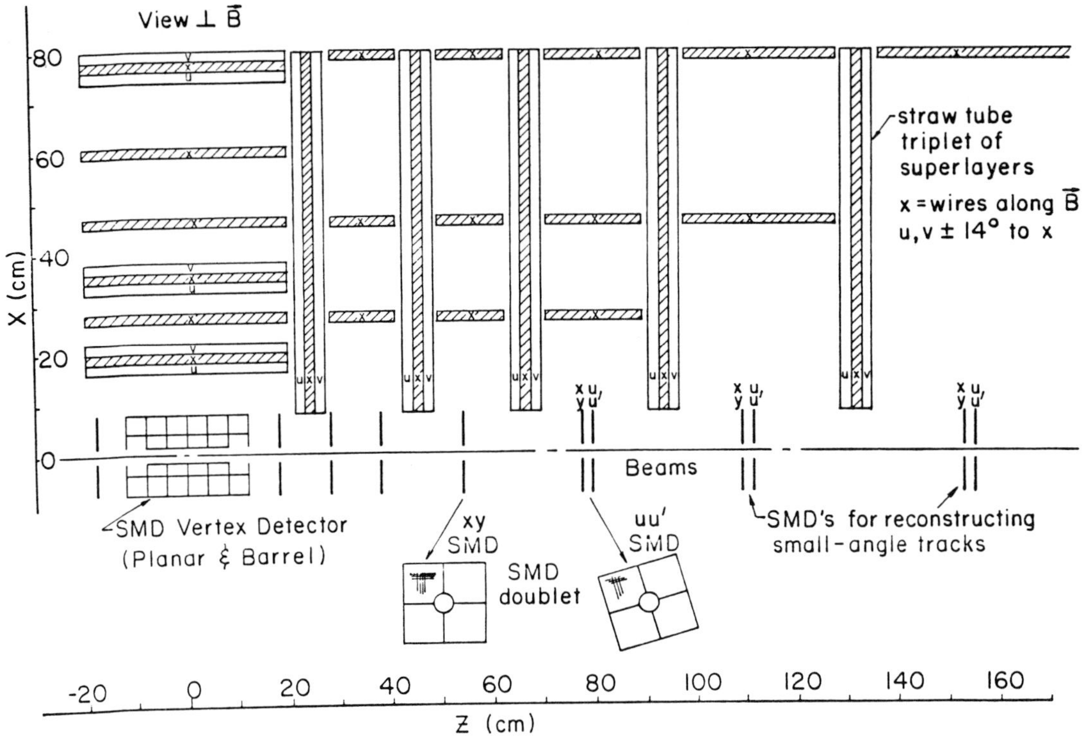

Figure 3. Plan view of the tracking chamber system.

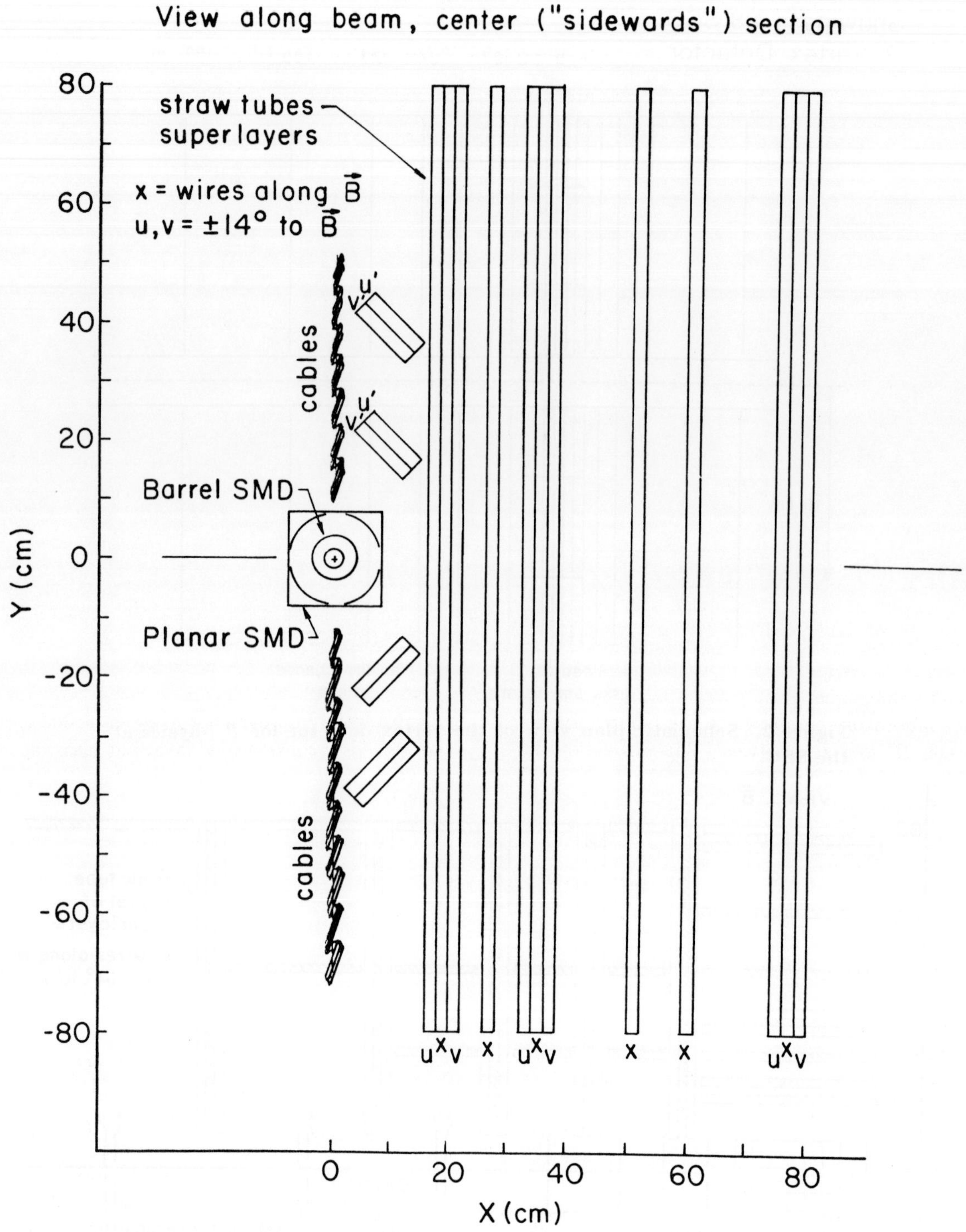

Figure 4. Beam's eye view of the tracking chamber system.

OPTIMAL SCENARIOS FOR MEASURING CP VIOLATION IN B^0 DECAY AT THE $\Upsilon(4S)$

D. Hitlin
California Institute of Technology

and

T. Nakada
Paul Scherrer Institute

and

A. Sanda
Rockefeller University

ABSTRACT

By integrating tagged $B^o\bar{B}^o$ events produced at the $\Upsilon(4S)$ resonance in an asymmetrical e^+e^- collision and excluding events in which the proper decay time difference for the two decays is small, the analyzing power for measuring a CP asymmetry of a given size is enhanced. The degree of enhancement depends on the vertex resolution in the laboratory compared to the laboratory distance between B^o and $\bar{B}^o$ vertices. The expression derived is evaluated by a Monte Carlo calculation in order to ascertain the optimum energy difference between the e^+ and e^- beams. It is shown that more nearly symmetrical beam energies than have previously been considered are satisfactory, thus making the problem of constructing a detector less difficult, as well as making it easier to realize an asymmetrical $B\bar{B}$ factory using either high luminosity storage rings or linear colliders.

The advantages of studying the weak decays of B mesons at the $\Upsilon(4S)$ resonance in e^+e^- annihilation have been clearly demonstrated by the experimental work of ARGUS and CLEO. Since the $\Upsilon(4S)$ decays exclusively to $B\bar{B}$ pairs, it is also an ideal place to search for CP-violating effects. It is well known, however, that CP-violating asymmetries generated by the interference between mixing and the decay amplitude vanish if symmetrically time-integrated quantities are measured when the $B^o\bar{B}^o$ pair is produced in an odd angular momentum state, as at the $\Upsilon(4S)$. This observation has stimulated interest in accelerators with unequal e^- and e^+ energies, which can produce the $\Upsilon(4S)$ resonance in motion in the laboratory. In this way, the time order of the B meson decays can be obtained from the order of the measured decay vertices along the boost direction. This allows the integration over decay lengths to be performed asymmetrically in time, thereby preserving the ability to measure a CP-violating effect. The comparison of different techniques for producing large numbers of $B^o\bar{B}^o$ pairs in order to measure CP violation done at this year's Snowmass Workshop concluded that the leading scenario involved an asymmetrical collider at the $\Upsilon(4S)$. This provides the advantage of improved mass resolution obtained by exploiting the accurate knowledge of the center-of-mass energy, resulting in a substantial improvement in signal-to-background ratio compared to conventional invariant mass reconstruction, as well as advantages in the total number of $B^o\bar{B}^o$ pairs which can be produced per unit time.

Dunietz and Nakada[1)] have analyzed the situation in which the integral over the asymmetry is performed from $t = 0$ for one B meson and for a time greater than some optimized minimum value (which depends on the decay vertex resolution of the detector) for the other, when the $\Upsilon(4S)$ is produced at rest. They concluded that a measurable asymmetry could indeed be preserved, were there to exist a detector of adequate spatial resolution. No such detector, especially one compatible with a beam pipe typical of storage rings, presently exists. The purpose of this note is to extend this result to the case of a moving $\Upsilon(4S)$. We derive analytic expressions which explicitly exhibit the measurable asymmetry and the analyzing power as a function of the minimum decay length and decay length difference which are excluded from the integration. This technique results in improved analyzing power for the asymmetry, thereby reducing the number of $B^o\bar{B}^o$ events needed to measure a CP-violating asymmetry of a given size. The resolution of current vertex detectors is adequate to perform the measurement.

We will consider the now classic example of the measurement of a CP-violating asymmetry in decay into a CP eigenstate, $B^o, \bar{B}^o \to \psi K_S$, where the flavor of the other B meson is determined through the sign of the lepton in semileptonic decay. For ease of comparison with the results produced by the Weak Decays Group at the Snowmass Workshop, we make the same assumptions, *i.e.* we use the leptonic decay modes of the ψ and the optimistic assumption that both the $\pi^+\pi^-$ and $\pi^o\pi^o$ decay modes of the K_s can be utilized.

If the $\Upsilon(4S)$ is produced in e^+e^- annihilation with asymmetrical beams, the mean decay length of the B in the laboratory frame is $L_{B,\bar{B}} = \tau_B \beta_{B,\bar{B}} \gamma_{B,\bar{B}}$,

where $\quad \beta_{B,\bar{B}} = \left(\gamma^2(\pm p\cos\theta + \beta E)^2 + p^2\sin^2\theta\right)^{1/2} / E_{B,\bar{B}}$

and $\quad E_{B,\bar{B}} = \left(m_B^2 + \gamma^2(\pm p\cos\theta + \beta E)^2 + p^2\sin^2\theta\right)^{1/2}$. Here β and γ describe the motion of the $\Upsilon(4S)$ in the laboratory. For the much-discussed case of 12 GeV $\times$ 2.3 GeV collisions, $L_{B,\bar{B}} = (243 \pm 28\cos\theta)\ \mu m$, where θ is the polar angle of the B in the rest frame of the $\Upsilon(4S)$. To illustrate our point, we will present our results in analytic form for integrals over the B decay lengths. An integral over *all* decay lengths at the $\Upsilon(4S)$ clearly yields a null result. We define:

$$\Delta = \frac{l_B - l_{\bar{B}}}{2} \ , \ \Sigma = \frac{l_B + l_{\bar{B}}}{2} \ ,$$

where $l_{B,\bar{B}}$ is the B decay length in the laboratory frame. Note that the polar angles of the B and $\bar{B}$ meson are θ and $-\theta$, respectively. Here, l_B refers to the decay length of that particle produced as a B^o meson at $t = 0$. The maximum amount of information is retained if a maximum likelihood fit is performed on all detected decay pairs, with the sign of Δ preserved. An integrated measurement can be effective, as we will demonstrate.

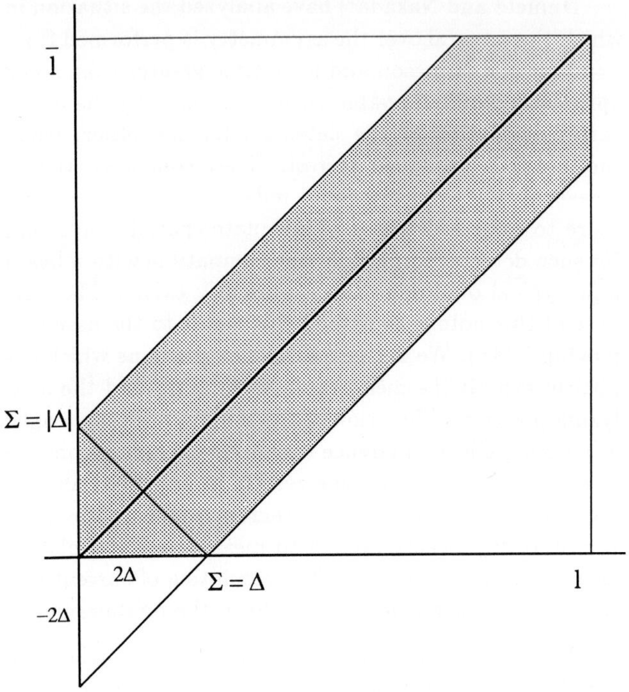

Figure 1. The region of integration in the laboratory frame.

As we know from the tag whether the B^o decay preceded the $\bar{B}^o$ decay or *vice versa*, we will integrate all events above some minimum value , Δ_{min} , preserving the sign of Δ. We will split the integral into regions with $\Delta > 0$ and $\Delta < 0$, as illustrated in Figure 1.

We define

$$\kappa \equiv \tan^{-1}\left(\frac{L_B - L_{\bar{B}}}{L_B + L_{\bar{B}}}\, x\right),$$

where $x \equiv \Delta m/\Gamma$ and $\phi_x = \tan^{-1} x \approx 36^o$, and $A_a \equiv A(B \to f_a)$, *etc.*[2)]

The integral over the B^o and $\bar{B}^o$ decay lengths is replaced by an integral over Σ and Δ . The result of the integral over Σ, for $\Delta > 0$ (+ sign) and < 0 ($-$ sign) separately is:

$$\int_{|\Delta|}^{\infty} \Gamma d\Sigma \sim e^{-2|\Delta|/L} \times$$

$$\Bigg\{ |A_a\bar{A}_b - A_b\bar{A}_a|^2 + |\frac{p}{q}A_aA_b - \frac{q}{p}\bar{A}_a\bar{A}_b|^2$$
$$+ \cos\kappa \cos\left(\kappa \pm \frac{2x|\Delta|}{L}\right)$$
$$\times \left[|A_a\bar{A}_b - A_b\bar{A}_a|^2 - |\frac{p}{q}A_aA_b - \frac{q}{p}\bar{A}_a\bar{A}_b|^2 \right]$$
$$- 2\cos\kappa \sin\left(\kappa \pm \frac{2x|\Delta|}{L}\right)$$
$$\times Im\left[(\frac{p}{q}A_aA_b - \frac{q}{p}\bar{A}_a\bar{A}_b)(A_a\bar{A}_b - A_b\bar{A}_a)^* \right] \Bigg\}.$$

We must now perform the integral over Δ, for path length differences greater than Δ_{min} , separating the cases in which the B^o and $\bar{B}^o$ decay first:

$$\Gamma(\Delta_{min})_{\pm} = \int_{\Delta_{min}}^{\infty} d|\Delta| \int_{|\Delta|}^{\infty} d\Sigma\ \Gamma \sim e^{-2\Delta_{min}/L}$$

$$\times \Bigg\{ |A_a\bar{A}_b - A_b\bar{A}_a|^2 + |\frac{p}{q}A_aA_b - \frac{q}{p}\bar{A}_a\bar{A}_b|^2$$
$$+ \cos\phi_x \cos\kappa \cos\left(\kappa \pm \frac{2x\,\Delta_{min}}{L} \pm \phi_x\right)$$
$$\times \left[|A_a\bar{A}_b - A_b\bar{A}_a|^2 - |\frac{p}{q}A_aA_b - \frac{q}{p}\bar{A}_a\bar{A}_b|^2 \right]$$
$$- 2\cos\phi_x \cos\kappa \sin\left(\kappa \pm \frac{2x\,\Delta_{min}}{L} \pm \phi_x\right)$$
$$\times Im\left[(\frac{p}{q}A_aA_b - \frac{q}{p}\bar{A}_a\bar{A}_b)(A_a\bar{A}_b - A_b\bar{A}_a)^* \right] \Bigg\},$$

where $\Gamma(\Delta_{min})_+$ is the result for $\Delta > 0$, and $\Gamma(\Delta_{min})_-$ is for $\Delta < 0$. Here $L = L_B\ (L_{\bar{B}})$ if $\Delta > 0\ (< 0)$.

For the case of tagging the state a with a positive lepton ($\bar{A}_a = 0$) where the final state b is a CP eigenstate, we have:

$$\Gamma(L_{B,\bar{B}}, \Delta_{min})_\pm = e^{-2\Delta_{min}/L_{B,\bar{B}}} |A_a|^2 \Big\{ 1 + |\frac{p}{q}\rho_b|^2$$
$$+ \Big(\cos\phi_x \cos\kappa \, \cos(\kappa \pm \frac{2x\Delta_{min}}{L_{B,\bar{B}}} \pm \phi_x)\Big)(1 - |\frac{p}{q}\rho_b|^2)$$
$$- \Big(2\cos\phi_x \cos\kappa \, \sin(\kappa \pm \frac{2x\Delta_{min}}{L_{B,\bar{B}}} \pm \phi_x)\Big) Im(\frac{p}{q}\rho_b)\Big\}$$
$$\approx e^{-2\Delta_{min}/L_{B,\bar{B}}} \times$$
$$\Big\{1 - \cos\phi_x \, \cos\kappa \, \sin(\kappa \pm \frac{2x\,\Delta_{min}}{L_{B,\bar{B}}} \pm \phi_x) Im(\frac{p}{q}\,\rho_b)\Big\} ,$$

where we have defined $\rho_b = A_b/\bar{A}_b$ and used the fact that $|\frac{p}{q}\rho_b| \approx 1$.

In the Appendix we show that this expression is valid even for the case when the momentum of states a and b is not measured. We can write the decay rate in terms of l_a and l_b, the distances in the laboratory at which final states a and b are created. Observation of a positive lepton together with a ψK_s does not ensure that the B meson which decayed into the positive lepton was produced as a B meson at $t = 0$.

For $l_b > l_a$, there are two possibilities:

$B(t=0) \to b, \; \bar{B}(t=0) \to a :$

$$\Gamma(l_b - l_a > \Delta_{min}) \sim e^{-2\Delta_{min}/L_b}$$
$$\times \, (1 - \cos\phi_x \, \cos\kappa \, \sin(\kappa + \frac{2x\Delta_{min}}{L_b} + \phi_x) Im(\frac{p}{q}\,\rho_b)) ,$$

$\bar{B}(t=0) \to b, \; B(t=0) \to a :$

$$\Gamma(l_b - l_a > \Delta_{min}) \sim e^{-2\Delta_{min}/L_b}$$
$$\times \, (1 + \cos\phi_x \, \cos\kappa \, \sin(\kappa - \frac{2x\Delta_{min}}{L_b} - \phi_x) Im(\frac{p}{q}\,\rho_b)) .$$

In these expressions, L_b corresponds to $L_B(L_{\bar{B}})$ for $B \to b \; (\bar{B} \to b)$.

Adding these two indistinguishable contributions, we have:

$$\Gamma(l_b - l_a > \Delta_{min}) \sim e^{-2\Delta_{min}/L_b}$$
$$\times \, (1 - \cos\phi_x \, \cos^2\kappa \, \sin(\frac{2x\Delta_{min}}{L_b} + \phi_x) Im(\frac{p}{q}\,\rho_b)) .$$

For $l_a > l_b$, we must again add the two indistinguishable contributions:

$$\Gamma(l_a - l_b > \Delta_{min}) \sim e^{-2\Delta_{min}/L_a}$$
$$\times \, (1 + \cos\phi_x \, \cos^2\kappa \, \sin(\frac{2x\Delta_{min}}{L_a} + \phi_x) Im(\frac{p}{q}\,\rho_b)) .$$

The measured asymmetry is then:

$$\delta = \frac{\Gamma(l_b - l_a > \Delta_{min}) - \Gamma(l_a - l_b > \Delta_{min})}{\Gamma(l_b - l_a > \Delta_{min}) + \Gamma(l_a - l_b > \Delta_{min})} =$$
$$- \cos\phi_x \, \cos^2\kappa \, sin(\frac{2x\Delta_{min}}{L} + \phi_x) Im(\frac{p}{q}\,\rho_b) ,$$

where $L = \tau_B\beta\gamma$, and the $\beta\gamma$ is that for the B or $\bar{B}$ which lives longer.

The measurement of an asymmetry δ for N events is thus optimized when

$$\sigma_\delta/\delta \;=\; [(\sqrt{N}\delta)_+ \;-\; (\sqrt{N}\delta)_-]^{-1}$$
$$\simeq e^{\Delta_{min}/L}\sqrt{\frac{1}{\delta^2} - 1}$$

is minimized. The optimal value of Δ_{min} is (weakly) dependent on $Im(\frac{q}{p}\,\rho_b)$

We have used this result to explore the optimal energy asymmetry at which to produce the $\Upsilon(4S)$ for purposes of measuring a CP asymmetry. A Monte Carlo calculation was performed to optimize Δ_{min} while properly taking into account the $\Upsilon(4S)$polar angle decay distribution and the z-vertex resolution.

The best value of Δ_{min} depends, for a given value of $Im(\frac{p}{q}\rho_b)$, on the beam energy asymmetry and on the z-vertex resolution. For example, Figure 2 shows the value of σ_δ/δ, with $Im(\frac{p}{q}\rho_b) = 0.2$ for $\sigma_z = 20\,\mu m$ with beam energies of 10 and 2.8 GeV. The minimum is obtained at just over 0.5 B lifetimes, corresponding to a cut of $2\Delta_{min} \sim 120\,\mu m$ in the laboratory. For $Im(\frac{p}{q}\rho_b) = 0.6$, the optimum occurs at 0.7 lifetimes.

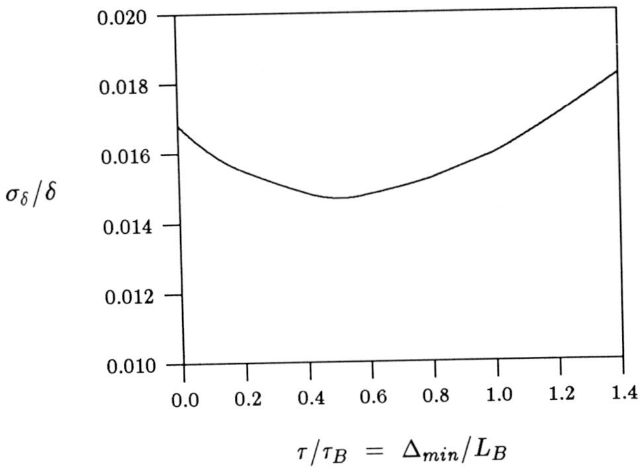

Figure 2. The relative error in the measurement of a CP-violating asymmetry as a function of the minimum vertex separation in units of the B meson lifetime for the case of 10 GeV on 2.8 GeV. A vertex resolution of 20 μm is assumed.

Figure 3 shows the relative error on the asymmetry obtained using 400,000 reconstructed Monte Carlo events and three values of the vertex resolution, for different energies of the high energy beam in an asymmetric collider. A value of $Im(\frac{p}{q}\rho_b) = 0.2$ is assumed. It is clear that energy asymmetries as large as 12 on 2.3 GeV [3] are not necessary to measure a CP-violating asymmetry with good efficiency: very little efficiency is lost when energies

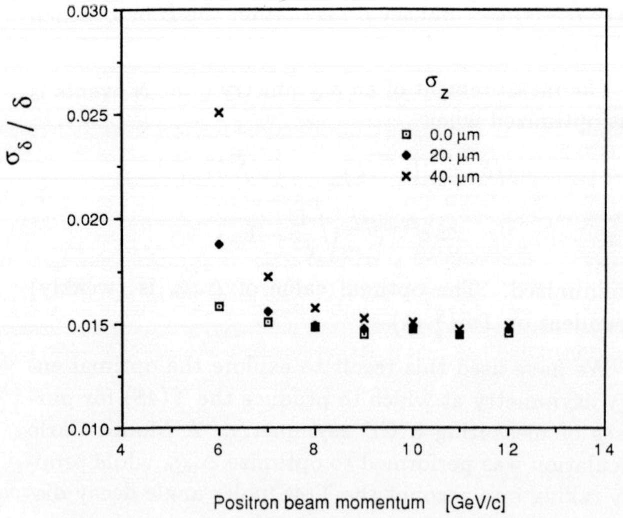

Figure 3. The relative error in the measurement of a CP-violating asymmetry as a function of the energy of the more energetic beam in an asymmetric collider at the $\Upsilon(4S)$, for three values of the vertex resolution.

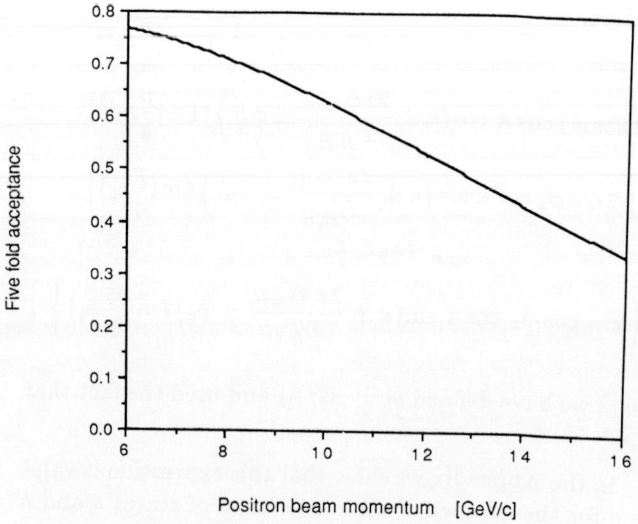

Figure 4. The acceptance of a detector covering $|\cos\theta_{lab}| < 0.95$ for five uncorrelated particles as a function of the energy of the more energetic beam in an asymmetric collider at the $\Upsilon(4S)$.

of 8 GeV on 3.5 GeV or 7 GeV on 4 GeV are used. Such a configuration has many advantages over the 12 Gev on 2.3 GeV example considered at the Summer Study. There are no existing examples of asymmetric storage rings, although several designs have been discussed. It is certain that a smaller energy asymmetry will lead to a more conservative design. There is also a considerable advantage for the detector. With an 8 GeV on 3.5 GeV configuration, the detector would be essentially identical to the 4π detectors used at existing symmetric colliders. This is not the case for the 12 Gev on 2.3 GeV example, which suffers from the compression of laboratory solid angle in the direction of the energetic beam. This causes several problems:

1. The quadrupoles closest to the intersection region tend to restrict effective coverage of the detector. As the minimum number of particles detected in the lepton-tagged ψK_S case is five, we have also shown in Figure 4 the acceptance of a detector which covers polar angles from 18 to 162 degrees in the laboratory ($|\cos\theta_{lab}| < 0.95$) for 5 (uncorrelated) particles as a function of the energy asymmetry. The advantage of a smaller beam energy asymmetry is clear, even in this approximate calculation.
2. The multiple scattering in the beam pipe becomes a significant problem, as many particles traverse the beam pipe at small angles
3. The segmentation of the photon detector in the forward direction must be a factor of four smaller in area to have comparable numbers of overlapping showers in an event. This is quite important if a high-quality photon detector is used, as the excellent resolution of a CsI or BaF_2 detector is realized only when showers do not overlap.

In conclusion, we have shown that by excluding regions of integration with small lifetime differences between the B^o and $\bar{B}^o$ mesons and by manipulation of the order of integration, that it is possible to retain greater sensitivity to CP violating asymmetries than in the case of naive integration. We have used this result as a benchmark to investigate the optimal energy difference for a collider which produces the $\Upsilon(4S)$in motion in the laboratory and have shown that energy differences smaller than had previously been considered offer several advantages both for the accelerator and detector.

APPENDIX

In the derivation given above, we have assumed that we can identify the measured decay time (more precisely, decay length) of the B^o or $\bar{B}^o$ in the laboratory. This assumption is in practice valid, as we shall now demonstrate. We can, in fact, identify t_B and $t_{\bar{B}}$, the decay times of the objects which started out as a B^o and $\bar{B}^o$ by measuring the time at which the lepton or ψK_S^o decay is seen.

A $B^o\bar{B}^o$ state produced at $t = 0$ can be written as

$$|B_k, \bar{B}_{-k}> = |B(\tau, k)\, \bar{B}(\tau', -k) \pm \bar{B}(\tau, k)\, B(\tau', k)>,$$

where the $\pm$ denotes production in an even or odd angular momentum state. This state is proportional to

$$|B_k, \bar{B}_{-k}>\sim$$
$$-\,i \sin\frac{\Delta m}{2}(\tau \pm \tau')\times$$
$$[\frac{p}{q}|B^o, k>|B^o, -k> \pm \frac{q}{p}|\bar{B}^o, k>|\bar{B}^o, -k>]$$
$$+\ \cos\frac{\Delta m}{2}(\tau \pm \tau')\times$$
$$[|\bar{B}^o, k>|B^o, -k> \pm |B^o, k>|\bar{B}^o, -k>]\ .$$

The matrix element governing the decay of this state to final states f_1 and f_2 is

$$< f_{1k}, f_{2-k}|H|B_k, \bar{B}_{-k}>\sim$$
$$-\,i \sin\frac{\Delta m}{2}(\tau_1 \pm \tau_2')[\frac{p}{q}A_{1k}A_{2-k} \pm \frac{q}{p}\bar{A}_{1k}\bar{A}_{2-k}]$$
$$+\ \cos\frac{\Delta m}{2}(\tau_1 \pm \tau_2')[\bar{A}_{1k}A_{2-k} \pm A_{1k}\bar{A}_{2-k}]\ ,$$

where τ_1 and τ_2' label the decay times of B_k and $\bar{B}_{-k}$, respectively. To make further use of the information on τ_1 and τ_2' we must determine that f_1 carries momentum k and f_2 carries $-k$, *i.e.*, we must measure the momenta of final states f_1 and f_2.

As tagging depends on an inclusive measurement, *i.e.*, a lepton from semileptonic decay, this is not feasible. We can avoid the problem by defining τ_1 and τ_2' so that they represent the time at which f_1 and f_2 are *created.* To see how this is easily possible, consider the matrix element which produces final states f_{1-k}, f_{2k}:

$$< f_{1-k}, f_{2k}|H|B_k, \bar{B}_{-k}>\sim$$
$$-\,i \sin\frac{\Delta m}{2}(\tau_2 \pm \tau_1')[\frac{p}{q}A_{2k}A_{1-k} \pm \frac{q}{p}\bar{A}_{2k}\bar{A}_{1-k}]$$
$$+\ \cos\frac{\Delta m}{2}(\tau_2 \pm \tau_1')[\bar{A}_{2k}A_{1-k} \pm A_{2k}\bar{A}_{1-k}]\ .$$

If we do not measure the momentum of states f_1 and f_2, the experiment is actually determining

$$|< f_1, f_2|H|B_k, \bar{B}_{-k}>|^2 = |< f_{1-k}, f_{2k}|H|B_k, \bar{B}_{-k}>|^2 + |< f_{1k}, f_{2-k}|H|B_k, \bar{B}_{-k}>|^2\ .$$

Thus, since the two matrix elements are equal within a $\pm$ sign when $\tau_i = \tau_i'$, the asymmetry does not vanish when we symmetrize in k and $-k$. We are therefore able to identify $\tau_{1,2}$ with the time at which the states $f_{1,2}$ are created.

REFERENCES

1) Dunietz, I. and Nakada, T., Z. Phys.**36**, 503 (1987).

2) Aleksan, R., Bartelt, J., Burchat, P. and Seiden, A., SLAC-PUB-4673 (1988), submitted to Phys. Rev.

3) Bigi, I., and Sanda, A., Nucl. Phys.**B281**, 41 (1987).

B MESON PHYSICS WITH POLARIZED ELECTRON BEAMS AT LINEAR COLLIDERS RUNNING AT THE Z°*

W. B. Atwood

Stanford Linear Accelerator Center, Stanford University, Stanford, California 94309

INTRODUCTION

The expected large cross section for $e^+e^- \to Z°$ and subsequent decay to $b\bar{b}$ quarks makes the $Z°$ an attractive place to pursue B meson physics. The cross section for b–quark production at the $Z°$ is compared to resonance production at the Υ_{4s} and Υ_{5s} in Fig. 1. In addition the big electroweak asymmetries, thought to exist in $Z°$ decays

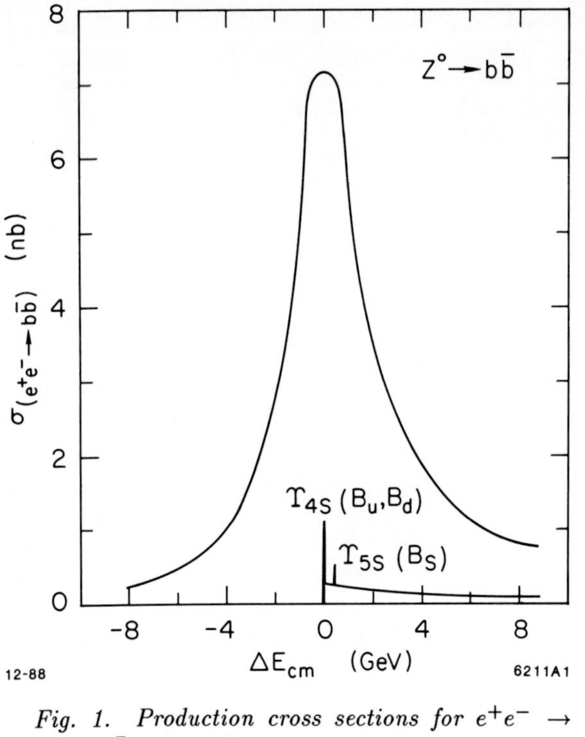

Fig. 1. Production cross sections for $e^+e^- \to Z° \to b\bar{b}$ with 45% left-handed polarized electrons and $e^+e^- \to b\bar{b}$ with unpolarized beams at Υ_{4s} and Υ_{5s}. Cross sections are plotted with relative offsets to these resonances.

*Work supported by the Department of Energy, contract DE–AC03–76SF00515.

to $b\bar{b}$ quarks with polarized electron beams,[1] provide an outstanding handle for observation of such effects as $B° - \bar{B}°$ mixing. In this paper, the feasibility of such measurements is investigated and, with relatively small samples of $Z°$'s (a few hundred thousand), both B_d and B_s meson mixing are shown to be measurable. The subject of CP violation in neutral B mesons is discussed last, but presently such measurements seem to be out of reach.

With polarized electron beams, the experimenter can control the polarization of the $Z°$. Subsequently, these polarized $Z°$'s decay with large forward-backward asymmetries into particle-antiparticle pairs. Table I gives the forward backward asymmetries A_{FB} for b quarks integrated over $-.9 < \cos\theta < .9$. The present goal of the Polarization Group working at the SLC is to deliver 45% polarized electrons[2] and, hence, should achieve a 38% A_{FB} for b quarks.

Table I. Electron beam polarization and the $\cos\theta$ integrated forward-backward asymmetry A_{FB} for b quarks.

Polarization	A_{FB} (B quarks)
0	.11
45%	.38
90%	.62

Another critical ingredient in doing B meson physics is precision vertex tracking. The machine characteristics of linear colliders such as the SLC make possible the use of a relatively slow readout technology, and very small distances between the beams and the detectors. The SLD detector group is building a CCD chip vertex detector to be located at a radius of about 15 mm from the beam line. This device[3] provides unparallelled precision in tracking particles in all three spacial dimensions. Using this detector, it becomes possible to inclusively reconstruct B mesons with high efficiency and with a good signal to noise. The reconstruction technique is to find and identify B mesons which produce resolvable secondary vertices and decay into a charm particle which produces a resolvable tertiary vertex (see Fig. 2). Simple properties of these cascade topologies allow for the separation of B_u, B_d and B_s mesons.

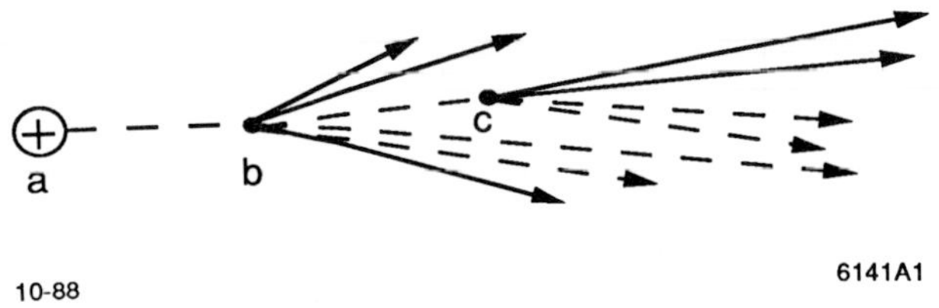

Fig. 2. A cascade B meson decay topology: (a) the primary vertex, (b) the secondary B meson decay vertex (neutral prongs are shown as dashed lines) and (c) the tertiary charmed meson decay vertex.

Reconstruction of the B meson decay vertex is critical to studying B° mesons, because the amount of particle-antiparticle mixing depends on the lifetime of the meson. When the particle mixes into its antiparticle before decaying, the sign of A_{FB} is reversed. As such, when A_{FB} is plotted as a function of the B meson lifetime, it should oscillate at a frequency proportional to the mixing rate $\Delta m/\Gamma$. Data from ARGUS[4)] and CLEO running at the Υ_{4s} have evidence that B_d mesons have a time integrated mixing probability of about 17%. Based on this information, B_s mesons are thought to be almost completely mixed (i.e., a mixing probability of approximately 50%). If the B_s oscillation rate is large, then decay time information will be necessary to measure it.

MONTE CARLO FOR STUDYING B MESON MIXING

To provide "events" for studying B° mixing, the LUND 6.3 Monte Carlo program[5)] was modified to include finite particle lifetimes and B° mixing. In particular, B°s were not allowed to decay. Subsequently, a lifetime was calculated for each such particle. From the generated lifetime a mixing probability was computed. Testing this probability against a uniformly distributed random number determined if mixing had occurred. In the event mixing had happened, a new particle equal to the antiparticle of the originating B° meson was added to the particle list. B° meson decays were then activated and LUND called again to perform these decays.

In Fig. 3, the B_d and B_s mixing used in the Monte Carlo are shown. In Fig. 3(a) it is seen that B_d mixing occurs on a time scale which is slow compared to the B_d meson lifetime ($\Delta m/\Gamma$ = .68). As such, most of the B_d's decay before they mix, and the time integrated probability of mixing (χ) is only 17%. In Fig. 3(b), the B_s mixing rate has been set to be five times faster than the B_d rate. Here, the time integrated rate is approaching saturation (i.e., $\chi \approx 50\%$).

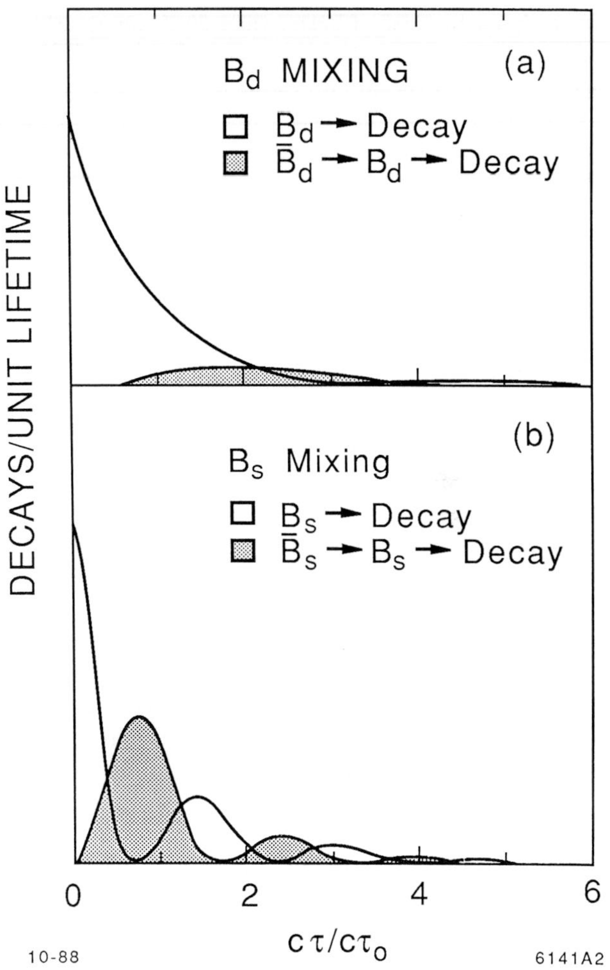

Fig. 3. Mixing probabilities versus meson lifetime for (a) B_d's and (b) B_s's.

The Z° decay events are processed by a parametric simulation of the SLD detector. Tracks are discarded which lie outside the acceptance, track parameters are distorted appropriately according to measurement resolution and multiple scattering and, finally, a detailed simulation of particle identification in the Čerenkov Ring Image Detector is used to identify the long lived particles.

From these processed tracks, vertex topologies are identified using simple (in concept) algorithms. The principle requirements are closeness of tracks to each other (the vertex) and the vertex's separation from the point where the beams collided. Representative cuts for closeness and separation are 45 μm and 500 μm, respectively.

After the initial pass to find vertices from tracks, a second pass is made where the first "track" is one of the previously found vertices. In this way, cascade topologies containing secondary and tertiary vertices are found (see Fig. 2).

The requirement that at least one cascade topology be found in an event separates $b\bar{b}$ pairs from the other events with a signal-to-noise of better than 20:1 (top quarks were not included in the simulation). As such, it wasn't deemed necessary to run the Monte Carlo with the generic mix of quarks, and only $b\bar{b}$ final states were generated. A sample of 48 K $b\bar{b}$ pairs from Z° decays with 45% polarized left-handed electrons was produced and forms the basis for the B meson mixing study.

Vertices from this process are characterized by total momentum (P_{TOT}), observed mass (M_{OBS}), charge [$Q_{b,c}$ for the secondary (b) and tertiary vertex (c)], flight path ($\gamma\beta c\tau$), and particle content (from the aforementioned particle identification system). In order to combine different B mesons with respect to lifetime, the boost factor ($\gamma\beta = P_B/M_B$) must be estimated and the flight paths corrected. A fair estimate of $\gamma\beta$ factor is found to be P_{TOT}/M_{OBS} for values of $\gamma\beta \leq 8$. Missing neutral prongs and, to a lesser extent, measurement error can cause P_{TOT}/M_{OBS} to become larger than eight and, when this occurs, the boost factor is truncated to eight.

The resulting lifetime resolution in $c\tau$ for all B mesons is shown in Fig. 4. The largest contribution to the error in $c\tau$ is from incorrect estimates of the $\gamma\beta$ boost factor. Still, the resolution predicted from this simulation is about one-tenth of a B° lifetime and will make the observation of B_s mixing, in principle, possible up to oscillation rates of $\Delta m/\Gamma \lesssim 15$.

$\mathbf{B_u}$, $\mathbf{B_d}$ AND $\mathbf{B_s}$ MESON PHYSICS

In Table II, events are separated according the charge of the secondary vertex (Q_b) and the charge of the tertiary vertex (Q_c). For each of the nine possible charge combinations, the events are broken down as to their origin: B_u, B_d, B_s and their antiparticles. Separating B_u's from neutral B's is easy: select $Q_b = \pm 1$. Hence, B° and $\bar{B}^\circ$ are separated according to the sign of the tertiary vertex charge (Q_c). No effective way of using the $Q_b = Q_c = 0$ sample has yet been found.

The analysis of the B_u data is straightforward. In Fig. 5(a), the distribution of events plotted against $Q_b \cos\theta$ is shown. The large expected A_{FB} is obvious. The signal-to-noise ratio for this selection is about 3.3:1 in a sample of $\sim$ 2.5 K B_u's (a 7.4% reconstruction efficiency!). The raw $A_{FB} = .34 \pm .02$ which, upon correction for the background by a factor of $[1 + (Signal/Noise)^{-1}]$, becomes

$$A_{FB}(corr) = .44 \pm .03 \ .$$

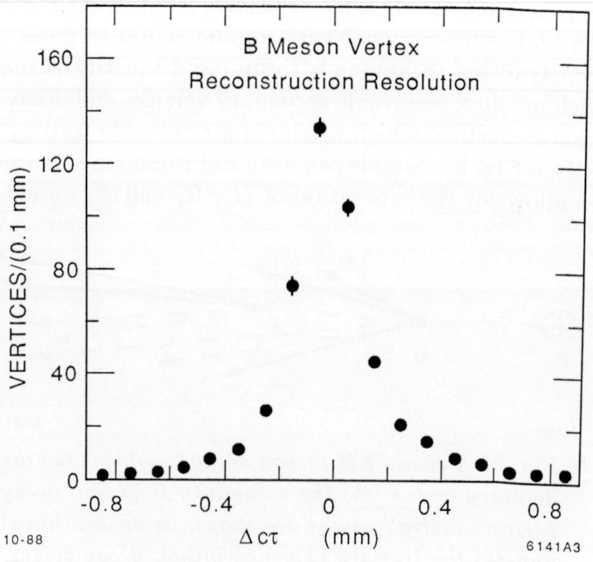

Fig. 4. B meson lifetime resolution in proper time ($c\tau$). The FWHM is 70 μm ($\sigma \sim$ 30 μm) and comes mainly from uncertainty in the relativistic boost factor $\gamma\beta$.

This is to be compared with the input to the Monte Carlo of .38.

Table II. Breakdown of cascade topology events by secondary and tertiary vertex changes, Q_b and Q_c. For each charge combination, the sample is resolved into particles and antiparticles, as well as the meson flavors B_u, B_d, and B_s (see upper left entries under $Q_b = Q_c = -1$ for key).

Q_c Vertex		Q_b Vertex −1: B	−1: $\bar{B}$	0	0	+1	+1
-1	u	43	2	14	23	1	22
	d	8	6	42	446	2	3
	s	2	0	22	159	0	2
0		890	21	73	82	22	899
		93	53	364	394	58	93
		23	22	141	164	27	22
+1		21	0	35	21	1	56
		3	2	492	57	7	3
		3	0	141	31	2	0

10-88
6141A7

In Fig. 5(b), the data from Fig. 5(a) has been binned in units of $c\tau$ for the secondary, b vertex. No apparent dependence on lifetime is present.

Separating neutral B mesons into B_d's and B_s's is harder, and information from the particle identification

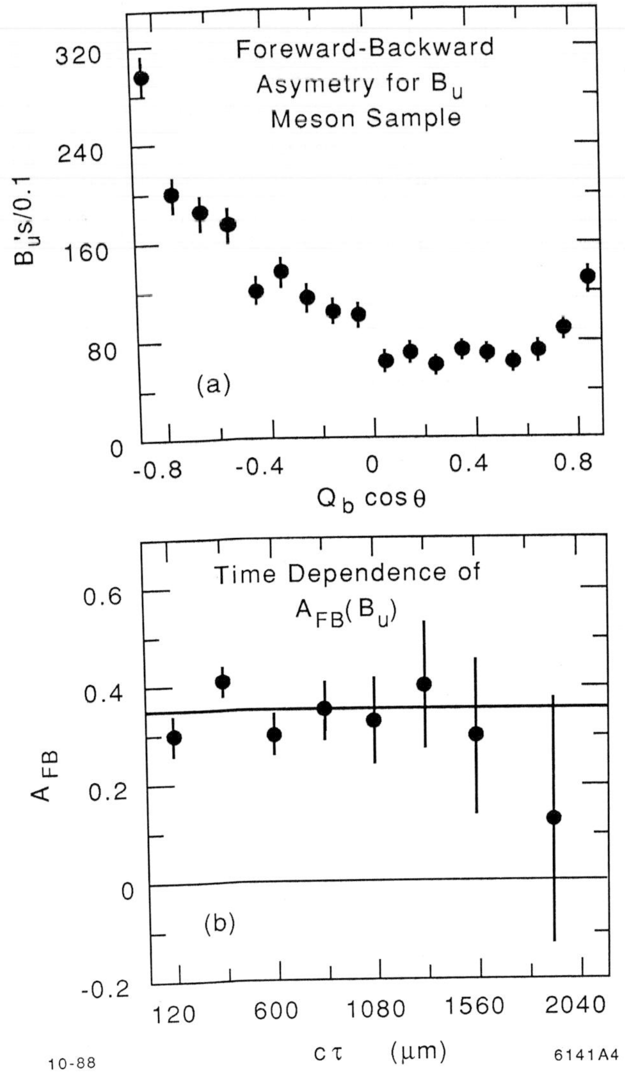

Fig. 5. A_{FB} for the B_u meson sample: (a) the time integrated sample plotted against $Q_b \cos\theta$ and (b) the data from (a) analyzed for A_{FB} and plotted versus $c\tau$ derived from the B_u meson vertex. The solid line indicates the time integrated average for the sample.

systems plays a crucial role. It is found that the kaon content of the tertiary vertex allows for some separation. In the Monte Carlo, b quarks fragment into B_d's 2.3 times more often than into B_s's. By requiring exactly one charged kaon to be associated with the tertiary vertex, D_s meson decays are suppressed and a signal-to-noise ratio of 4.9:1 is achieved for the B_d sample. The 877 reconstructed B_d's represent a reconstruction efficiency of about 2.6%.

The B_d events are plotted against $Q_c \cos\theta$ in Fig. 6(a) and a smaller, yet clearly visible, A_{FB} is seen. The raw A_{FB} for the B_d sample is $.12 \pm .03$. Correcting for the finite signal to noise results in $A_{FB} = .14 \pm .03$. This reduction

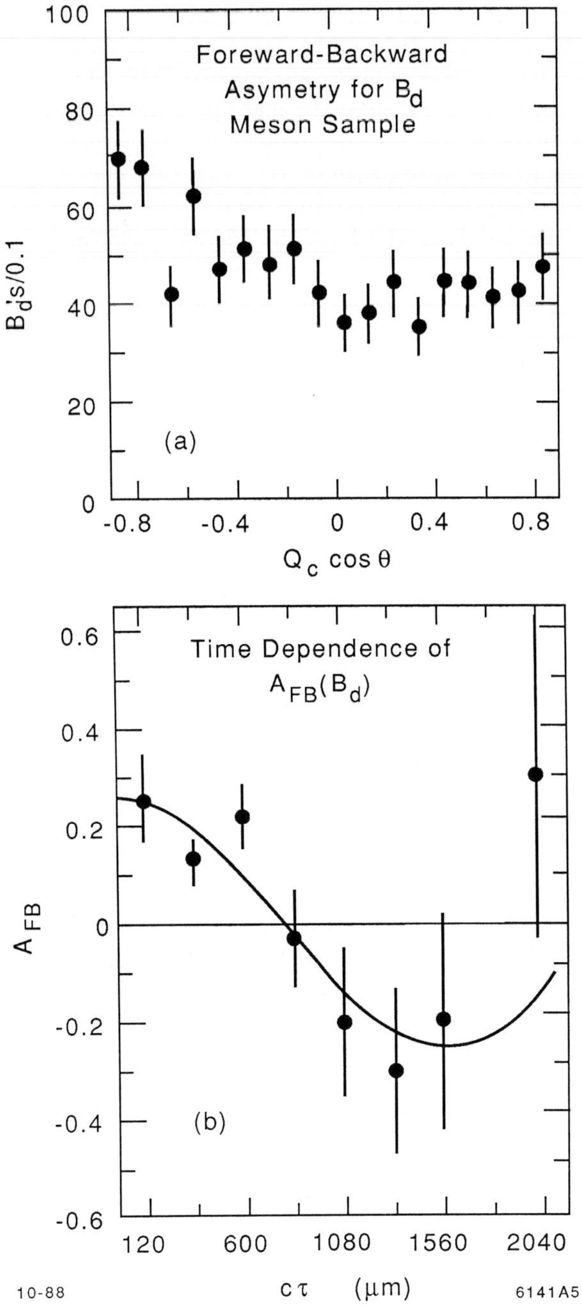

Fig. 6. A_{FB} for the B_d meson sample: (a) the time integrated sample plotted against $Q_c \cos\theta$ and (b) the data from (a) analyzed for A_{FB} and plotted versus $c\tau$ derived from the B_d meson vertex. The solid line is a fit to this data (see text).

in A_{FB} for B_d mesons compared to B_u mesons can be interpreted as due to B_d meson mixing and is described by[6] $A_{FB}(obs) = A_{FB}(phys)\,(1-2\chi)$. The extent to which

$$A_{FB}(obs)/A_{FB}(phys) < 1$$

is an indication that $\chi > 0$. For the numbers produced by this Monte Carlo exercise, $\chi = .34 \pm .04$. Since the mixing phenomenon requires time to develop, the cut made in the vertex finding of flight paths greater than .5 mm introduces a substantial distortion in the measurement of χ. Simply put, by requiring a minimum flight path for the B mesons, the direct $B°$ decays are preferentially discarded with respect to the mixed case of $B° \rightarrow \bar{B}° \rightarrow$ decay. A correction for this finite lifetime requirement on vertex finding can be estimated from the Monte Carlo, and for B_d mesons is found to be .53. Applying the correction gives $\chi_{meas} = .18 \pm .02$, to be compared with the Monte Carlo input of .17.

Proceeding now as for the B_u case, $A_{FB}(B_d)$ is binned in units of $c\tau$ and is shown in Fig. 6(b). We see for the first time the $B_d \rightarrow \bar{B}_d$ oscillation, directly. A fit to this data of the form

$$A_{FB} = A_{B_d} \cos(\Delta m_{B_d} t)$$

gives a $\chi^2/d.f. = 6.4/6d.f.$ The significance of the fit $(A_{B_d}/\sigma A_{B_d})$ is 4.9 and the oscillation parameter is $\Delta m_{B_d}/\Gamma_{B_d} = .65 \pm .08$. The expected number for $\Delta m_{B_d}/\Gamma_{B_d}$ is .72. The merits of this technique over the time integrated method is that systematic errors will be very small: errors in the measurement of $c\tau$ are cancelled to first order in the ratio of $\Delta m/\Gamma$ and errors in the amplitude of the effect have little bearing on the oscillation rate!

A sample of B_s mesons is extracted from the neutral B mesons by requiring the association of a pair of oppositely charged K mesons with the tertiary (charm) vertex. About half of these K pairs came from ϕ mesons and the rest are randomly distributed at higher masses. No cut is made on the K–pair mass. A sample of 116 B_s's results with a signal-to-noise ratio of 1.9:1. In Fig. 7(a),the events are shown versus $Q_c \cos\theta$, similar to that plotted for the B_d sample. The apparent A_{FB} is 0. $\pm$ .09 which, after background correction, becomes .0 $\pm$ .14. The mixing probability deduced from this A_{FB} is $\chi_{B_s} = .50 \pm .18$. Again, the finite vertex flight path introduces a correction which is now smaller, due to the more rapid B_s oscillation rate, and is estimated to be .80 from the Monte Carlo. Hence, $\chi_{B_s}(corr) = .41 \pm .15$ to be compared to .46, the input to the Monte Carlo.

Figure 7(b) shows $A_{FB}(B_s)$ plotted versus $c\tau$. The limited number of events clearly impairs the observation of the B_s oscillation. Since the major contamination in the B_s sample are B_d's, a fit of the form

$$A_{FB} = A_{B_d} \cos(\Delta m_{B_d} t) + A_{B_s} \cos(\Delta m_{B_s} t)$$

is natural to try. The oscillation parameter Δm_{B_d} can be fixed from the prior fit to the B_d data, leaving three parameters and six data points. The $\chi^2/d.f.$ of the fit is $3.5/3d.f.$ and the fit significance $(A_{B_s}/\sigma A_{B_s})$ is 3.3. The oscillation parameter is $\Delta m_{B_s}/\Gamma_{B_s} = 3.81 \pm .28$. The error

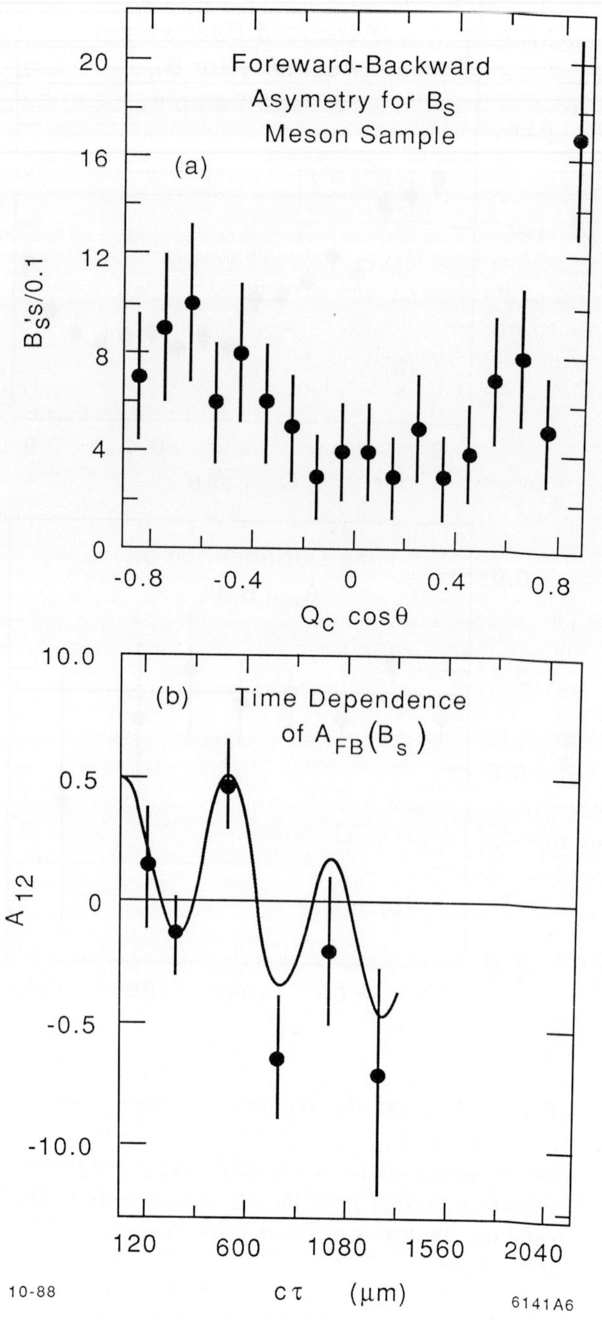

Fig. 7. A_{FB} for the B_s meson sample: (a) the time integrated sample plotted against $Q_c \cos\theta$ and (b) the data from (a) analyzed for A_{FB} and plotted versus $c\tau$ derived from the B_s meson vertex. The solid line is a fit to this data (see text).

is relatively smaller here than in the B_d case because the B_s data ranges over $\sim 2\frac{1}{2}$ periods where, in the B_d case, only half a period was fit. The ratio of $\Delta m_{B_s}/\Delta m_{B_d} = 5.9 \pm .9$ to be compared to the Monte Carlo input of 5.0 (note: $\Gamma_{B_d} = \Gamma_{B_s}$ in the Monte Carlo). Again, the systematic errors are thought to be small.

CP VIOLATION EFFECTS IN B MESONS

This section is a short summary of a previous publication.[7] CP violation is an indication of an asymmetry between the particle world and the antiparticle world. In the case of neutral B meson decay rates,

$$A_{CP} = \frac{\Gamma(B^\circ \to f) - \Gamma(\bar{B}^\circ \to \bar{f})}{\Gamma(B^\circ \to f) + \Gamma(\bar{B}^\circ \to \bar{f})} \quad .$$

It is obvious that the particle-antiparticle composition of the B° at its decay point is required. In Z° decays with polarized electron beams, this information is provided by knowing (statistically) the initial B° meson state (from A_{FB}) and measuring how long it lives before decaying (to estimate the amount of mixing). In the case of B_d mesons, the lifetime information is helpful, but not crucial, as the time integrated mixing is only $\sim$ 17%. But for B_s mesons, where the mixing could well be $\sim$ 50%, the measurement of CP violating asymmetries will not be possible without lifetime information.

The ability of a particular technique to discern the particle-antiparticle nature of the B° at its decay may be characterized by a "separation" asymmetry.

$$A_{sep} = \frac{N_{correct} - N_{wrong}}{N_{correct} + N_{wrong}} \quad ,$$

where $N_{correct(wrong)}$ are the number of decays properly (improperly) labelled. The measured CP violation asymmetry is

$$A_{CP}(meas) = A_{CP} \cdot A_{sep} \quad .$$

In the case of time integrated B_d meson decays from Z°'s, $A_{sep} = A_{FB}(B_d)$.

The number of events $N_{b\bar{b}}$ required to observe N_σ standard deviations can be expressed as

$$N_{b\bar{b}} = \frac{{N_\sigma}^2}{{A_{CP}}^2 A_{sep}^2} \frac{1}{2B_r(B^\circ \to f)} \frac{1}{\epsilon_f} \frac{1}{\epsilon_{tag}} \frac{1}{\sigma_{B^\circ}} \quad ,$$

where σ_{B° is the probability of a b quark fragmenting to a B°, ϵ_f is the final state reconstruction efficiency, ϵ_{tag} is the particle-antiparticle tagging efficiency, and $Br(B^\circ \to f)$ is the branching fraction to the final state f. Different methods for tagging the particle-antiparticle state can be compared by defining

$$Q_{sep} = {A_{sep}}^2 \epsilon_{tag} \quad .$$

An often referred to tagging technique is to observe the charge of the lepton from b semileptonic decays. As the b quarks are produced in pairs, if a prompt lepton is observed, then the other b quark state may be inferred. Of course, confusion with leptons emanating from the charm vertex, B_d and B_s mixing, and mis-tagging (e.g., electrons from γ's, muons from π/K decay, pions faking electrons in the calorimeter, etc.) all lead to values of A_{sep} and ϵ_{tag} less than unity.

In Table III, different tagging methods are presented. The "Gedanken Experiment" illustrates perfection. For the real world of experiments, however, Q_{sep} is found to be considerably less than one. This Table shows that the polarization technique compares very favorably with direct lepton tagging.

Table III. Comparison of various methods to separate particle and antiparticles for B meson CP violation studies.

Method	A_{sep}	ϵ_{tag}	Q_{sep}	$N_{b\bar{b}}^{equiv}$
Gedanken Experiment	1.0	1.0	1.0	1
Lepton tag	0.50	0.12	0.03	33
FB–tag (no polarization)	0.13	0.63	0.01	100
FB–tag (45% polarization)	0.46	0.63	0.13	8
FB–tag (90% polarization)	0.75	0.63	0.35	3

The bad news is that the branching ratio estimates for prime CP violating B_d decay final states are extremely small. A general rule seems to be that the larger the expected A_{CP}, the smaller the branching ratio to that final state. Detailed estimates of required $b\bar{b}$ samples range from 3.6 x 10^7 to 3.0 x 10^8 (multiply by $\sim$ 4.5 for the number of Z° decays). This places these measurements well beyond the luminosity reach of present Z° factories.

An alternative approach to observing CP violation in $b\bar{b}$ event is to count the same sign double semileptonic decay events,

$$A_{CP}(ll) = \frac{N^{--} - N^{++}}{N^{--} + N^{++}} \ .$$

Here, one gains tremendously from the branching ratio, but the gain is more than wiped out from the expect size of A_{CP}. This double lepton signature can be compared to the case where only one lepton is observed and A_{FB} is used in place of the other lepton:

$$A_{CP}(A_{FB}) = \frac{N(l^-, backward) - N(l^+, forward)}{N(l^-, backward) + N(l^+, forward)} \ .$$

The $A_{CP}(A_{FB})$ is favored by a factor of two in the necessary event samples, due to only needing a single semileptonic decay. Still, the number of $b\bar{b}$ events will have to exceed 10^8 before the experiment is within a order of magnitude of the expected value of A_{CP} for this inclusive lepton signal!

CONCLUSIONS

The feasibility of measuring particle-antiparticle mixing for both neutral B mesons has been demonstrated. Along with the mixing measurements, excellent determinations of B_u, B_d and B_s lifetimes will easily be made. Furthermore, the "control" sample for the B_d and B_s mixing study (i.e., the B_u sample) provides a clean measurement of A_{FB} (b quarks). The event samples necessary to pursue these studies are well within the design luminosity reach of present Z° factories.

On the other hand, studies of CP violation in B meson decay is shown to require event samples well out of reach at present. However, the argument is repeated from Ref. 7 that polarized Z°'s provide a very competitive laboratory for these studies, compared to other techniques.

ACKNOWLEDGMENTS

The author expresses thanks to P. Grosse–Wiesmann, I. Dunietz and L. Rochester for many stimulating and helpful discussions. G. Bower deserves thanks for installing B° mixing in the LUND Monte Carlo. Thanks also are due to the SLC Polarization Group and to the SLD Collaboration.

REFERENCES

1. The importance of polarized electron beams was stressed by many people. See, e.g., G. Manning, PEP Summer Study 1975, p. 19; SLAC–190; LBL–4800 and references therein, C. Y. Prescott, SLAC–PUB–2649 (1980); M. Böhm and W. Hollik, Nucl. Phys. B**204**, 45 (1982); B. W. Lynn, M. E. Peskin and R. G. Stuart in "Physics at LEP," CERN 86–02, Vol. 1, Feb 1986, p. 90, J. Ellis and R. Peccei, eds.
2. D. Blockus *et al., Proposal for Polarization at SLC,* SLAC Report SLAC–PROP–1 (1986).
3. SLD Design Report, SLAC–273 (1984 and updates).
4. H. Albrecht *et al.,* ARGUS Collaboration, Phys. Lett. **192B** (1987) 245; see also Nobu Katayama (for the CLEO Collaboration), *$B^\circ \bar{B}^\circ$ Mixing and Charmless B Meson Decay into Baryons from CLEO,* 1988 SLAC Summer Institute.
5. Torbjorn Sjostrand, Muts Bengtsson (Lund Univ.), Comput. Phys. Commun. **43** (1987) 367.
6. The effect of B mixing on A_{FB} has been analyzed by I. I. Y. Bigi and M. Cvetic, Phys. Rev. **D34** (1986) 1651.
7. W. B. Atwood, I. Dunietz and P. Grosse–Wiesmann, SLAC–PUB–4544 (1988).

ACCELERATOR PHYSICS

Melvin Month

Working Group Leader

1988 SUMMER STUDY ON HIGH ENERGY PHYSICS IN THE 1990S

Accelerator Physics and Facilities
Working Group
General Introduction

M. Month
Brookhaven National Laboratory,
Upton, New York 11973

PURPOSE

The purpose of the Summer Study as a whole was "to examine and evaluate the opportunities for pursuing high energy physics in 1990s". One of the six large working groups was the one on Accelerator Physics. As we saw it, our purpose was to develop a picture of potential accelerator facilities that could be developed in the 1990s and to evaluate them on the basis of their readiness and their implications for the facilities we saw emerging for the 21st century.

ORGANIZATION

To accomplish our ambitious set of objectives, we formed four working subgroups to study four types of accelerator facilities that might be developed in the 1990s. These were K-Factories, Hadron Colliders, B-Factories and Linear Colliders. In addition, we took as given the emergence of the SSC as the dominant accelerator facility in the late 1990s. Since there had been a great deal of design work and review of the SSC over the past few years, we made no attempt to review or assess the design parameters or performance projections.

Our four subgroups each had an accelerator physics group leader. In addition, to ensure that the design and performance parameters of the accelerator facilities were consistent with the particle physics that was desired, each subgroup also had a high energy physics coordinator. In fact one of the sub-groups, the B-Factory Group, took an integrated approach where both particle and accelerator physics were studied together by the same group of people and representatives for both areas served as co-leaders of the group. The four subgroups and the group leaders are as follows:

Subgroup	Accelerator Physics Group Leader	Particle Physics Coordinator
K-Factory	M. Craddock	S. Smith
Hadron Collider	S. Holmes/ M. Cornacchia	J. Yoh
B-Factory	R. Siemann	G. Feldman/ D. Cassel
Linear Collider	R. Ruth	T. Himel

The reviews and assessments that have come out of the studies of the four subgroups form the essence of this report of the Accelerator Physics and Facilities Working Group and follow this brief introduction. Each of the reports from the subgroups contains a review by the group leader and contributions from the participants of the summer study, many of which originated in the seminars and other presentations given over the three week period of the Summer Study.

DETECTORS

In the past, considering detectors separately from the accelerator facilities was justified because the accelerators dominated the total budget and took much longer to develop and build. In the era of the large colliders, detectors have also become costly and take many years to build. Furthermore, the preparatory R&D for a facility is not determined only by the accelerator R&D required, but may be strongly influenced by the necessary detector R&D. For example, future detectors for the large hadron colliders will have to be designed for increasingly higher event rates probably requiring substantial R&D. We therefore believe that there should be an integrated approach to detectors and accelerators in designing and developing High Energy Physics facilities, as well as in organizing summer studies such as this one.

THE BIG PICTURE

The major growth area in High Energy Physics in the past couple of decades has been the large colliders in the United States and Western Europe. This continues to be the case and will probably climax in the United States with the construction and operation of the SSC as we approach the 21st Century. At present there are two fully operational large colliders, the proton-antiproton colliders at CERN [Energy ~ 0.3×0.3 TeV] and Fermilab [Energy ~ 0.9×0.9 TeV], both operating at luminosity > 10^{30} cm^{-2} sec^{-1}. There are three other large collider projects at various stages of completion: the "modified e^+e^- linear collider" SLC at SLAC [50×50 GeV], the storage ring e^+e^- collider LEP at CERN [50-100 GeV per beam] and the electron-proton collider HERA at DESY [30 GeV e^- × 0.8 TeV p]. The SLC and LEP hope to achieve their design luminosities of $6{\times}10^{30}$ and $1.5{\times}$ 10^{31} cm^{-2} sec^{-1} respectively in the early 1990s with the HERA luminosity above 10^{31} projected a few years later. For the future, there are two proton-proton colliders being planned for the two regions for "the year 2000", the Large Hadron Collider LHC at CERN [8×8 TeV] and the Superconducting Super Collider SSC in the United States [20×20 TeV], both with design luminosities above 10^{33} cm^{-2} sec^{-1}. The SSC is more ambitious in mass reach than the LHC (by a factor of about 2 1/2 corresponding to the beam energy ratio), while both have the same design goal of reaching the 1 TeV mass region, a study of which, on the basis of very general principles, will open new doors to our understanding of the nature of the elementary particles.

Given the current program and the program in the far future, we then interpreted "Physics in the 1990s" as being what might be termed an "Interim Program", a program of physics in the period between the current collider program and the program of the 21st Century which will have the SSC as its centerpiece. In the United States, this program would be an extension and enhancement of the existing HEP program, which presently includes fixed target physics at the BNL AGS and the Fermilab Tevatron, and e^+e^- collider physics at Cornell and SLAC. Although we didn't particularly consider the SSC Center as part of the Interim Program, it can participate, for example, with a high luminosity (> $2{\times}10^{32}$ cm^{-2} s^{-1}) pp collider (energy 1-2 TeV per beam) which would serve both as the SSC injector and as an SSC testbed.

THE SSC

Since we did not have a separate group looking at the SSC, let me insert a few comments here. The current design of the SSC has gone through many reviews all agreeing on the basic soundness of the accelerator design. During the course of the summer study, there were seminars and discussions connected with the SSC design – in particular on the magnet aperture and the magnet R&D program. There seemed to be general agreement that the magnet R&D program would be accomplished over the next year or so. This pre-construction R&D would for the most part try to find the minimum magnet cost consistent with the projected performance. Although certain changes in the magnet might result from this R&D, for example reduced length, reduced field, increased aperture, or even a change in the coil configuration, we believe this to be part of the normal engineering design of the magnet.

THE INTERIM PROGRAM

Overall for an Interim Program we should try to incorporate the following objectives, which is not meant to be an exhaustive list:

- To do precision studies of discovered processes seeking underlying mechanisms;
- To do searches for signs of new physics with lower energy collisions;
- To develop the capability of very high energy, high luminosity e^+e^- linear colliders to serve as a complementary probe of new physics; and
- To develop study programs for accelerators and detectors to optimize exploitation of the SSC by shortening discovery time.

Our four accelerator physics and facilities groups were organized and designed with these purposes in mind.

To optimally exploit the physics potential in the energy range of the current program, beam intensity and collider luminosity should be substantially increased. This is needed to study rare processes in areas such as K-physics or B-physics, for example. To accomplish this "Interim Program" requires significant improvement in existing facilities, perhaps extensions or upgrades or even the addition of new facilities. Thus, we have considered K-Factories utilizing proton beams capable of delivering nearly 10^{15} protons per second on target; we have investigated B-Factories capable of producing e^+e^- collisions in storage rings at $10^{33\text{-}34}$ cm^{-2} sec^{-1}; and we have explored the possibility of e^+e^- linear colliders and pp colliders of modest energies (e.g. > 100 GeV per beam e^+e^- and 1-2 TeV per beam pp) and high luminosity (~ $10^{32\text{-}33}$ cm^{-2} sec^{-1}) for a variety of pre-SSC physics. There are or will evolve other opportunities for physics in the 1990s; for example, one recently emerging possibility is a τ-charm factory. Thus, our 1988 Snowmass Study was incomplete; but we hope it is a step in developing an integrated HEP program for the SSC era.

Each of the Accelerator Physics subgroups in our Working Group developed its own Interim Program. In each case, we tried to describe technically feasible projects, to include projects that could be completed in the

1990s, and to indicate the R&D associated with each project needed to ensure its success. Because of overall budget and manpower constraints, not all of the many worthy projects we considered will come to fruition in any single geographic region. Since we in the United States can't do them all, then we must have a programmatic analysis which will allow us to choose the most promising options. It would probably be best if a comprehensive program could be developed with full international coordination. But this type of question is beyond the scope of our study.

READINESS COMPARISON

What we have been able to do is compare the different types of accelerator facilities in terms of their project readiness, a project being ready when a specific proposal for construction can be made. We consider as a measure of readiness a class related to the R&D years needed to reach this decision point. The physics goals which require increasingly high luminosity at higher energies and high intensity to study rare processes are included in our assessments.

A class 0 readiness assessment means that no R&D is required to reach a decision on the direction a project is to take. A class 1 assessment means that some uncertainty in the accelerator physics or technology exists and it will take about 1 year of R&D to reach a decision on how to proceed. For class 2, probably significant extension of existing capabilities is needed; while for class 3, probably an invention of some type is needed. A class 2 assessment means that about 3 years of R&D will be needed to reach a decision point, while a class 3 assessment indicates that significantly more time could be required. We cannot overemphasize that a readiness assessment of this type has a great deal of uncertainty associated with it and should be done annually as the R&D proceeds. In addition, once a decision point is reached and a project is started, there is still the pre-construction R&D to be done, which includes, among other things, cost optimization and detailed engineering design.

Estimates of the readiness for the facilities in our four subgroups as well as for the SSC are given in the following Table.

Estimates of Accelerator Readiness

Accelerator Facility	Readiness Class
SSC	0
K-Factory	0
Hadron Collider	0
B-Factory	1
Linear Collider (< 500 GeV per beam)	2
Linear Collider (>1 TeV per beam)	3

CONCLUSIONS

We believe that an Interim Program based on existing and potential facilities would be of great benefit to the worldwide High Energy Physics program. It is clear that many important and exciting accelerator opportunities could be made available in the pre-SSC years of the 1990s. In addition, by providing guideposts in developing the SSC operating program and by providing glimpses of the physics that will be studied when we reach the 1 TeV mass scale, an Interim Program of the type we have described would both complement and enhance the SSC physics program in the 21st Century.

ACCELERATOR PHYSICS – K - FACTORIES

ACCELERATOR PHYSICS - K FACTORIES

K FACTORIES - OVERVIEW

M.K. Craddock*
TRIUMF, 4004 Wesbrook Mall, Vancouver, B.C., Canada V6T 2A3

ABSTRACT

Proposals for high intensity proton synchrotrons (typically providing 100 μA (6×10^{14} p/s) at 30 GeV) are being made in the U.S.A., Canada, Europe, Japan and the USSR. These beams would be roughly 100 times more intense than those available now and would yield equivalent increases in the fluxes of secondary particles (kaons, pions, muons, antiprotons, hyperons and neutrinos) - or cleaner beams for a smaller increase in flux. The ability to investigate rare processes on the "precision frontier" opens new avenues to fundamental questions in both particle and nuclear physics, complementary to traditional approaches via the "energy frontier". The demand for higher currents has led to novel features in many of the accelerator designs: asymmetric magnet cycles, separate collector and stretcher rings, beam painting at injection, bucket-to-bucket beam transfer, avoidance of crossing transition, perpendicular biassing of ferrite, the use of Siberian Snakes to preserve polarization and of a pre-septum to make slow extraction >99.8% efficient. Other characteristic features include rapid cycling rates, booster stages, H^- injection, low impedance enclosures, powerful feedback systems for control of beam instabilities and beam loading, and local collimation and fast abort systems for handling beam loss. This paper reviews the general features of K factory accelerator design, the major conclusions from discussions at Snowmass '88, and the status of the various proposals.

1. INTRODUCTION

Strong initiatives are being pursued in a number of countries for the construction of "kaon factory" synchrotrons capable of producing 100 times more intense proton beams than those available now from machines such as the Brookhaven AGS and CERN PS. Such machines would yield equivalent increases in the fluxes of secondary particles (kaons, pions, muons, antiprotons, hyperons and neutrinos) - or cleaner beams for a smaller increase in flux - opening new avenues to various fundamental questions in both particle and nuclear physics. Major areas of investigation would be

- rare decay modes of kaons and hyperons
- CP violation
- meson and baryon spectroscopy
- meson and baryon interactions
- neutrino scattering and oscillations
- quark structure of nuclei
- hypernuclear properties
- K^+ and $\bar{p}$ scattering from nuclei.

Experience with the pion factories has already shown how high beam intensities make it possible to explore the "precision frontier" with results complementary to those achievable at the "energy frontier". One example is the setting of a lower limit of 380 GeV on the mass of any right-handed W-boson by a muon decay measurement[1] at TRIUMF in 1982. Figure 1 illustrates the improved branching ratios attainable for selected K decay channels. A comprehensive justification of the physics case for K factories may be found in the three proposals so far published[2-4] and in the proceedings of the recent international conferences at Mainz[5], Lake Louise[6] and Rockport[7]. The possibilities for future K-decay and CP violation measurements were discussed in depth by the Snowmass '88 Weak K Decay Group and are reported in full in that section.[8] Stewart Smith provided liaison with them and has contributed a report below on the implications for kaon physics facilities.

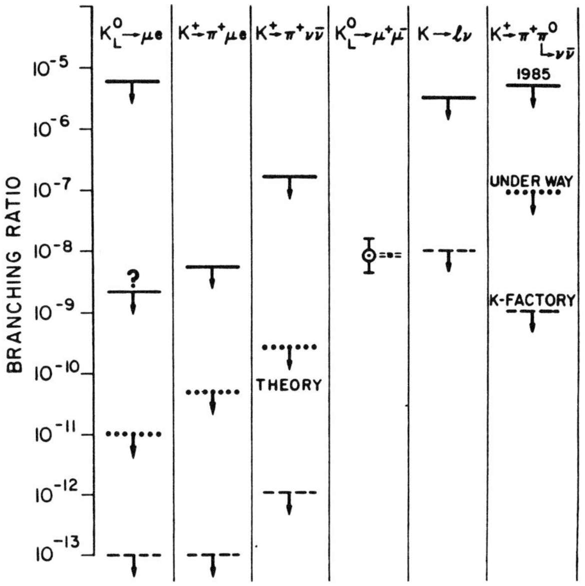

Fig. 1. Branching ratios for several rare kaon decays.

The K Factory Group began by hearing status reports on the various proposals and then broke into 8 working groups to discuss special topics. The 40 participants are listed in the reports of the working groups to which they contributed (chairman's name in parenthesis):

*On leave from Physics Dept., University of British Columbia, Vancouver, B.C., V6T 2A6.

Experimental Facilities (Ewart Blackmore)
Architecture (Michael Craddock)
Efficient Slow Extraction (Ulrich Wienands)
Polarized Beams (Peter Blüm)
Beam Stability (Richard Baartman)
RF Systems (James Griffin)
Injection and Painting (Grahame Rees)
Linac Injectors (with SSC) (Arch Thiessen)

The present paper introduces the general features of kaon factory accelerator design. It is followed by brief descriptions of individual projects and status reports on existing high-intensity proton machines. Then follow the reports of the working groups and finally the texts of invited and contributed papers – some presented at Snowmass and some arising out of the discussions there.

2. K FACTORY ACCELERATORS

The proposed kaon factories aim at energies in the 30–60 GeV range with proton currents of 25–100 μA. Proposals have come from the three existing pion factories at LAMPF, PSI and TRIUMF (these laboratories being unique in already possessing operating machines with adequate energy and current to act as injectors) and from that nearing completion near Moscow, and also from a European consortium and from an INS-KEK collaboration in Japan. The energy and intensity parameters for the proposed machines are listed in Table I along with those for some existing proton synchrotrons. More details can be found in the proposals themselves[2–4] and in the proceedings of the recent Santa Fe "Workshop on Hadron Facility Technology"[9] and Los Alamos "Accelerator Design Workshop".[10]

In considering the maximum intensity which can be accelerated in a synchrotron, two parameters are of particular importance, the number of particles per pulse N and the circulating current $\overline{I}$. N is critical because it determines the incoherent space charge tune shift (and spread) $\Delta\nu$, given by[11]

$$\Delta\nu = -\frac{r_p}{\pi}\frac{N}{\epsilon^*}\frac{FGH}{B_f}\frac{1}{\beta\gamma^2}\ . \qquad (1)$$

Here $r_p = 1.5347 \times 10^{-18}$ m, the classical radius of the proton, ϵ^* is the normalized beam emittance, B_f is the bunching factor, F, G and H are form factors describing the effects of image forces, transverse density distribution and aspect ratio respectively, and β and γ are the usual Lorentz speed and energy factors. In order to avoid coming too close to lower-order resonances $\Delta\nu$ should be kept below 0.2. The $\beta\gamma^2$ factor makes this condition most critical[12] near injection. The circulating current $\overline{I}$ is important through its involvement in longitudinal space charge effects and beam stability. It determines the effect of space charge on bucket height, and appears in the

Table I. High-intensity proton synchrotrons.

	Energy (GeV)	Average Current (μA)	Rep. rate (Hz)	Protons/ pulse N ($\times 10^{13}$)	Circ. current $\overline{I}$ (A)
Slow Cycling[a]					
KEK PS	12	0.32	0.6	0.4	0.6
CERN PS	26	1.2	0.38	2	1.5
Brookhaven AGS	28.5	0.9	0.38	1.6	0.9
– with Booster		(4)	0.36	(7)	(4)
– with Stretcher		(8)	0.77	(14)	(8)
Fast Cycling[b]					
Argonne IPNS	0.5	14	30	0.3	4.0
Rutherford ISIS	0.75	130(200)	50	(2.5)	(6.1)
Fermilab Booster	8	7	15	0.3	0.3
AGS Booster	(1.5)	(20)	(7.5)	(1.6)	(4)
Kaon Factories[a]					
LAMPF AHF	60	25	12	1.3	0.5
TRIUMF	30	100	10	6	2.8
European HF	30	100	12.5	5	2.5
Japanese HF	30?	100?			
Moscow INS	30	100	6.25	10	3.2
Boosters[b]					
LAMPF AHF	6		24	0.7	
	10	25	12	1.3	0.5
	10		6	2.6	
TRIUMF	3	100	50	1.2	2.7
European HF	9	100	25	2.5	2.5
JHF Booster	>2	200	50	2.5	6.7
Moscow INS	5	100	50	1.2	3.2

[a]Slow extraction [b]Fast extraction

Keil-Schnell-Boussard criterion[12] for microwave stability:

$$\left|\frac{Z_{\|}}{n}\right| < \frac{V_p}{\overline{I}}\left(\frac{\Delta p}{p}\right)^2 B_f \mid \eta \mid \beta^2\gamma. \qquad (2)$$

Here $Z_{\|}$ represents the effective impedance of the beam pipe, n is the mode number, $V_p = m_pc^2/e = 938\text{MV}, \Delta p/p$ is the fractional momentum spread in the beam, and the parameter $\eta = \gamma_t^{-2} - \gamma^{-2}$ (γ_t being the transition energy). The energy dependence of this expression is somewhat complicated because every factor is energy dependent except V_p.

The existing higher-energy machines achieve average beam currents of $\approx 1\ \mu$A. These currents are limited both by slow cycling rate (< 1 Hz) and by low injection energies (≤ 200 MeV) into their first synchrotron stages, restricting N to $\approx 2 \times 10^{13}$. The circulating current $\overline{I} \approx 1A$. The booster synchrotron at present under construction at Brookhaven[13] will raise the injection energy into the AGS to 1.5 GeV and allow N to be raised a factor 4. For this and other reasons explained below all the K factory proposals include booster synchrotrons in the range 3–10 GeV.

The highest intensities have been achieved in machines using faster cycling rates (10–50 Hz). The record current (for a synchrotron) is 130 μA at the Rutherford ISIS spallation neutron source,[14] and this will be raised to 200 μA when commissioning is completed. The number of protons per pulse N will be 2.5×10^{13}, only a little more than in the slow-cycling machines, but the circulating current $\overline{I}$ will rise to 6 A. The K factory synchrotrons are all designed to operate at ~10 Hz, a little slower than the Fermilab booster, but 25 times faster than existing 30 GeV machines.

3. COMMON FEATURES OF THE NEW PROPOSALS

The rather similar performance specifications for the various kaon factories have led to a number of common design features – many of them novel and aimed at completely avoiding processes producing beam loss.

Rapid Cycling

This restricts the charge required per pulse Ne (and hence the tune spread $\Delta\nu$) and the circulating current I (and hence beam loading and instability problems) as described above. Frequencies much above 1 Hz imply the use of resonant rather than ramped magnet power supplies. The rapid acceleration also requires a relatively high rf voltage gain per turn (~ 1 MV), implying multiple cavities and a relatively high synchrotron tune ($\nu_s > 0.01$). The latter in turn implies that low order (low $k+\ell+m$) synchro-betatron resonances of the form $\ell\nu_x + m\nu_y + k\nu_s = n$ impinge on the working area in tune space. To avoid exciting these the rf cavities must either be placed in a dispersionless region, or else with a symmetry which does not produce dangerous nth harmonic fields.[15)]

Booster Synchrotrons

Raising the injection energy into the main ring has a triple purpose. In the first place it raises the tune shift limit on the charge per pulse for a reasonable normalized emittance ϵ^* to $N \sim 6 \times 10^{13}$ ppp. This enables the desired current of 100μA to be achieved with only moderately fast cycling rates ~ 10 Hz. Secondly, for a given ϵ^*, it significantly reduces the magnet aperture required for the main ring, and hence its cost. Thirdly it simplifies the design of the rf acceleration system. In a fast-cycling machine the much more rapid acceleration requires a much higher rf voltage than has been conventional at slower cycling-rates – about 2 MV for a 10 Hz 30 GeV machine. At the same time a large frequency swing (20-30%) is required when starting from pion factory energies of 500-800 MeV. The use of a booster enables these demands to be handled separately. Almost the entire frequency swing can be provided in the booster at relatively low rf voltage, while the main ring provides the 2 MV with only a few percent frequency swing. The booster is usually smaller and therefore must cycle faster (12-60 Hz) in order to fill the circumference of the main ring. The charge per pulse would be $N \sim 10^{13}$, comparable to existing machines injected in the same energy range. The circulating current in both booster and main ring would be $\overline{I} < 3$ A, a level which is not expected to present serious problems.

Asymmetric Magnet Cycle

All designs propose the use of a slow rise, typically 3 times longer than the fall; this reduces the rf voltage needed, and hence the number of cavities, by 1/3. Full-scale power supplies providing such a cycle have been developed at Argonne by Praeg,[16)] with the encouragement of Los Alamos.

H^- injection

Injection into the first ring by stripping an H^- beam enables Liouville's theorem to be bypassed and many turns to be injected into the same area of phase space.

Painting Phase Space

In fact it is not necessary or desirable to inject every turn into exactly the same area; the small emittance beam from the injector must be painted[17)] over the much larger three-dimensional acceptance of the first ring to limit the space charge tune shift. Painting also enables the optimum density profile to be obtained and the number of passages through the stripping foil to be limited.

Bucket-to-bucket Transfer

Bunches are transferred from bucket to bucket leaving the injector and between rings. This avoids the losses inherent in capturing coasting beams. The buckets are made large enough that no more than 60% of their area is occupied.

Empty Buckets

A group of adjacent buckets, ~100 ns long, is left empty to allow time for the extraction and injection kickers to turn on and stabilize.

Transition Crossing Avoided

The magnet lattices are designed to place transition outside the acceleration range, avoiding the emittance mismatch and beam losses associated with crossing it in fast-cycling machines, and the difficulties anticipated in making a phase jump under high beam loading. In conventional a.g. proton synchrotrons with regular dipole lattices, the momentum dispersion $\eta_x \equiv \Delta x/(\Delta p/p) \approx R/\nu_x^2$ (constant) and hence the transition energy $\gamma_t \simeq \sqrt{R/<\eta_x>} \simeq \nu_x$, typically in mid-acceleration range. The use of a highish energy booster (say $\geq$6 GeV) makes it possible to avoid crossing transition simply by choosing a high tune for the booster and a low one for the main ring. With a lower energy booster, special lattices may be used to alter $<\eta_x>$ and drive γ_t out of range.

Siberian Snakes

The periodicity of the magnet lattices also drives depolarizing resonances. Pulsed quadrupoles can be used to jump the lower energy resonances in the boosters, as at the AGS. At higher energies, however, it becomes impractical to build quadrupoles fast enough and strong enough.

Instead it is proposed to use "Siberian Snakes".

Control of Beam Instabilities[18)]

In spite of the large circulating currents, the rapid cycling times give instabilities little time to grow to dangerous levels. Coupled-bunch modes, driven by parasitic resonances in the rf cavities and by the resistive wall effect, are damped using the standard techniques (Landau damping by octapoles, bunch-to-bunch population spread and active damping by electronic feedback). The longitudinal microwave instability, which tends to have a rapid growth rate, is avoided by making the longitudinal emittance sufficiently large and by minimizing the high frequency impedance of the vacuum chamber as seen by the beam. To allow for beam blow-up the magnet apertures are designed to accommodate some growth in the beam emittance.

Control of Beam Loading

The high beam powers involved (several MW) imply high loading if the rf system is not to be excessively expensive. To provide stability under these conditions powerful rf control systems have been devised,[19)] based on experience at CERN, including fast feedback around the power amplifiers, and phase, amplitude, tuning, radial and synchronization feedback loops. One-turn-delay feedforward is used to control transient loading effects.

Control of Beam Loss

Successful operation of a high-intensity accelerator depends crucially on minimizing beam losses and the activity they produce. Where some loss is expected, near injection and extraction elements, collimators and absorbers will be provided and equipment designed for remote handling. Computer simulations[20)] suggest that addition of a pre-septum to a conventional slow extraction system could reduce the beam loss from typically 1% to $< 0.02\%$. (This might take the form of a very thin (50 μm) foil or of a massless magnetic septum.) The beam current and any spills will be carefully monitored, and in case of the beam becoming unstable at any time through component failure or power excursions, fast-abort systems will dump the entire beam safely within one turn.

4. MAJOR DISCUSSION TOPICS

Discussions in the working groups naturally centred more on differences in approach than on similarities. In most areas these were matters of specialized interest and are left for the reports of the individual working groups, but in others they remain more substantial:

Energy[21)]

The Los Alamos group favours a 25 μA beam at 60 GeV (rather than the 100 μA at 30 GeV common to the other projects), citing the availability of higher momentum secondary beams for glueball searches and Drell-Yan studies of the nucleus. The rare decay experimentalists inclined to the view that higher energies could bring as many problems as advantages, and that with the AGS beam expected to rise to 4 μA (and possibly 10 μA with a stretcher), accelerator designers should be concentrating on increasing beam intensity (even above 100 μA).

Architecture

Los Alamos is considering 4 booster alternatives: full-size (10 GeV), half-size (10 GeV) and quarter-size (6 GeV) synchrotrons, and a 2.2 GeV booster linac. The first and last choices eliminate the need for a collector ring, but probably increase the overall cost. The quarter-size booster gives the smallest aperture main-ring magnets, but requires a collector. In order to minimize the number of rings, Los Alamos propose flat-topping the main ring field rather than building a separate stretcher. The other groups prefer to separate the special requirements for slow extraction (high field stability, collimation and radiation handling) from those of a fast-cycling accelerator.

Debunched Slow Extraction

The rare decay experimentalists pointed out that it was vital to remove even the rf microstructure in the slow-extracted beam for many of their experiments. Baartman and Wienands[22)] have examined debunching the beam in the TRIUMF Extender by both drift alone and by rf manipulation. Both can be performed in ~1 ms and the limiting factor for both is the onset of microwave instability as the local momentum spread in a bunch is decreased. The latter technique, involving rotation of the bunch through 90° by imposing a mismatched low bucket, eases the impedance tolerance from ~1 Ω to ~10 Ω.

5. STATUS OF THE PROPOSALS

The Brookhaven 1.5 GeV booster is due for completion in late 1990. Upgrading of the AGS and the experimental areas is going on in parallel and should allow 4 μA proton beams to be accelerated and used for secondary particle production. A proposal for a stretcher ring, probably in a separate tunnel, is being considered; the AGS could then be cycled faster, raising the proton current to 8–10 μA.

The Los Alamos proposal[2)] is currently on hold behind two projects given higher priority by the US nuclear physics community, CEBAF and RHIC. Construction funding for the Los Alamos AHF could begin in 1993. Meanwhile an intensive R&D program is underway.

The European Hadron Facility proposal was completed in 1987. So far no country or laboratory has committed itself as host, although Italy remains a possibility. The incorporation of a mini-EHF at CERN (with separate funding) has also been studied.

The Japanese Hadron Facility is being pursued in a number of stages. So far only the first stage has been designed and proposed – a 1 GeV, 400 μA H^- linac and compressor-stretcher ring for μSR, pion and spallation neutron studies. It would also provide a higher energy injector for the existing 12 GeV PS at KEK.

The Moscow Institute for Nuclear Research has already received approval for its K factory, to be located at Troitsk, where its 600 MeV proton linac meson factory, now nearing completion, would act as injector. Funding is planned to begin in 1993.

TRIUMF has just received approval for a one-year $11-million preconstruction study for its KAON Factory. This will fund prototyping of components, design of conventional facilities, updating cost estimates, environmental and economic impact studies, and formal exploration of international contributions on the HERA model (which initial contacts suggest could exceed $100M). The British Columbia government has already committed $87M (Cdn) to the civil works and there appear to be good prospects for full approval in 1990.

ACKNOWLEDGEMENTS

It is a pleasure to acknowledge the efforts of those who agreed to participate in the K factory working groups, and especially those who have provided written contributions. The author is particularly indebted to Jana Thomson for the rapidity and accuracy with which she put this paper together.

REFERENCES

1) Carr, J. et al., Phys. Rev. Lett. 51, 627 (1983).

2) A proposal to extend the intensity frontier of nuclear and particle physics to 45 GeV (LAMPF II), LA-UR-84-3982 (1984).
The physics and a plan for a 45 GeV facility that extends the high-intensity capability in nuclear and particle physics, LA-10720-MS (1986).

3) KAON Factory Proposal, TRIUMF (1985).

4) Proposal for a European Hadron Facility, EHF 87-18 (1987).

5) Proc. Intl. Conf. on a European Hadron Facility, Mainz, ed. Th. Walcher (North Holland, 1987).

6) Proc. 2nd Conf. on Intersections between Particle and Nuclear Physics, Lake Louise, May 1985, ed. D.F. Geesaman, AIP Conf. Proc. 150, (1986).

7) Proc. 3rd Conf. on Intersections between Particle and Nuclear Physics, Rockport, May 1988, AIP Conf. Proc. (in press).

8) Bock, G. and Littenberg, L., these proceedings.

9) Proc. Intl. Workshop on Hadron Facility Technology, Santa Fe, February 1987, ed. H.A. Thiessen, Los Alamos Report LA-11130-C.

10) Proc. AHF Accelerator Design Workshop, Los Alamos, February 1988, ed. H.A. Thiessen (in press).

11) Reich, K.H., Schindl, K. and Schönauer, H., Proc. 12th Int. Conf. on High Energy Accelerators, ed. F.T. Cole, R. Donaldson, Fermilab, p. 438 (1983).

12) Keil, E., Schnell, W., CERN-ISR-TH-RF/69-48 (1969);
Boussard, D., CERN-Lab:II/RF/Int. 75-2 (1975).

13) Weng, W.T., these proceedings.

14) Planner, C.W., these proceedings.

15) Suzuki, T., Part. Accel. 18, 115 (1985).

16) Praeg, W., IEEE Trans. Nucl. Sci., NS-30, 2873 (1983).

17) Kempson, V.C., Planner, C.W. and Pugh, V.T., IEEE Trans. Nucl. Sci., NS-28, 3085 (1981).

18) Baartman, R., these proceedings and in Ref. 9.

19) Pedersen, F., IEEE Trans Nucl. Sci. NS-32, 2138 (1985); TRI-DN-85-15, (1985).

20) Wienands, U., these proceedings.

21) Blackmore, E.W., these proceedings.

22) Baartman, R. and Wienands, U., these proceedings.

TOWARDS AN ADVANCED HADRON FACILITY AT LOS ALAMOS

Henry A. Thiessen
Los Alamos National Laboratory

In our view, the needs of the Nuclear Physics Community and the Materials Science Community are best served by an accelerator which produces high intensity proton beams at two energies, namely: 1) a 1.6-GeV proton beam of 1200 microamperes divided between a spallation source and a stopped-pion neutrino source; and 2) a 60-GeV proton beam of 25 microamperes shared between users of the primary beam and production of secondary beams of pions, kaons, antiprotons, and muons. Simplicity in the design is of paramount importance as obtaining a minimum operating cost and a maximum beam availability depend strongly on minimizing the count of components in the complete facility.

A version of an advanced hadron facility which meets the above goals is shown in Fig. 1. The 800-MeV H^- beam of LAMPF is accelerated to 1600 MeV in a linac. Most of the beam is delivered to a compressor ring which delivers the beam to the neutron and neutrino (LCD) experimental areas. The compressor ring is discussed in an accompanying paper by Colton.

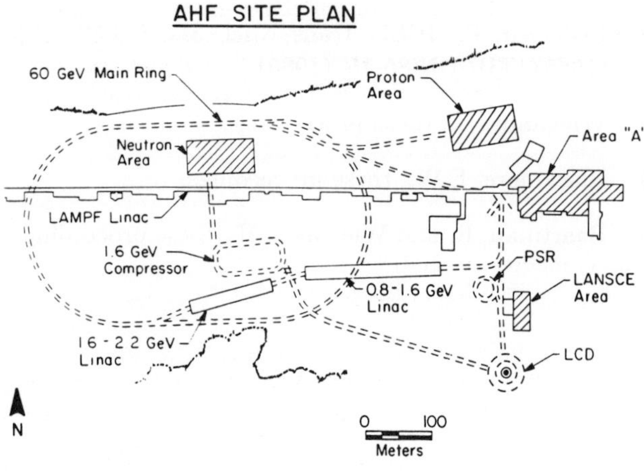

Figure 1. AHF site layout.

A portion (25 microamperes) of the H^- beam is accelerated to 2.2 GeV by a second linac. The beam is then injected into a large synchrotron and accelerated to 60 GeV. The long straight sections of this racetrack-shaped lattice provide space for injection, slow extraction, collimation, and rf.

Some of the parameters of the 60-GeV synchrotron are shown in Table I.

Table I. 60-GeV Synchrotron Parameters.

Repetition Rate	6 Hz
Protons/pulse	2.6×10^{13}
Circumference	1300 meters
Transition Gamma	22 i
rf Frequency	50.3–52.7 MHz
rf Volts/Turn	7.6 MV
Dipole Magnet Field	0.1–2.2 Tesla

The annual yield of kaons from AHF is compared with that of several other proposed hadron facilities in Fig. 2. In preparing Fig. 2, we used the production cross section for kaons at the production target. Because the AHF provides a neutrino beam without interference with the users of the 60-GeV beam, we have increased the yield of AHF by a factor of 2 to account for the fraction of the year that the beam of other hadron facilities will be devoted to other uses than production of secondary beams.

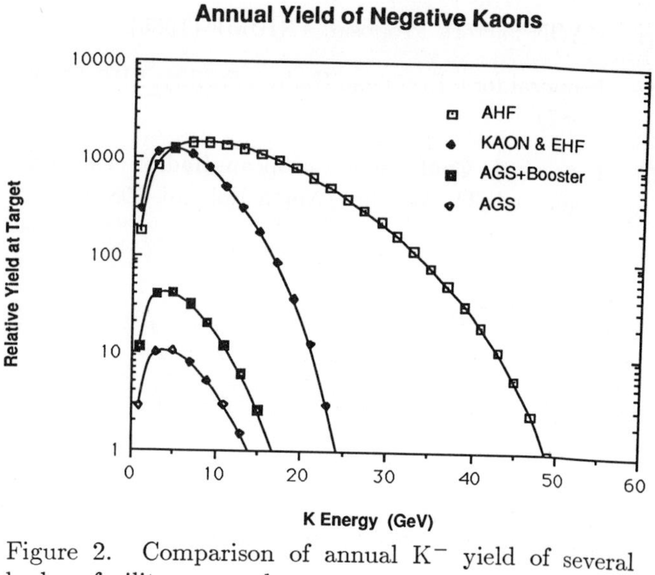

Figure 2. Comparison of annual K^- yield of several hadron facility proposals.

THE TRIUMF KAON FACTORY PROPOSAL

M.K. Craddock*

TRIUMF, 4004 Wesbrook Mall, Vancouver, B.C., Canada V6T 2A3

ABSTRACT

The KAON Factory is designed to accelerate 100 μA proton beams from the 520 MeV H^- TRIUMF cyclotron to 30 GeV. This paper gives a brief description of the five-ring accelerator complex and its present status – at the start of a one-year $11 million pre-construction study.

ACCELERATORS

The TRIUMF Kaon-Antiproton-Otherhadron-Neutrino Factory is fully described in the original proposal[1] and outlined in various papers.[2,3] The primary aim is to provide a 100 μA beam of protons at 30 GeV – about a hundred times more than available at present accelerators. The TRIUMF H^- cyclotron, which routinely delivers 150 μA beams at 500 MeV, would provide a ready-made and reliable injector. It would be followed by two fast-cycling synchrotrons, interleaved with 3 storage rings, as follows:

A	Accumulator:	accumulates cw 450 MeV beam from the cyclotron over 20 ms periods
B	Booster:	50 Hz synchrotron; accelerates beam to 3 GeV; circumference 214 m
C	Collector:	collects 5 Booster pulses and manipulates longitudinal emittance
D	Driver:	10 Hz synchrotron; accelerates beam to 30 GeV; circumference 1072 m
E	Extender:	30 GeV stretcher ring for slow extraction for coincidence experiments

It can be seen from the energy-time plot (Fig. 1) that this arrangement allows the cyclotron output to be accepted without a break, and the B and D rings to run continuous acceleration cycles without wasting time on flat bottoms or flat tops; either fast or slow extracted beams would be available. Beam would be scheduled ~9 months/year, power costs being very low.

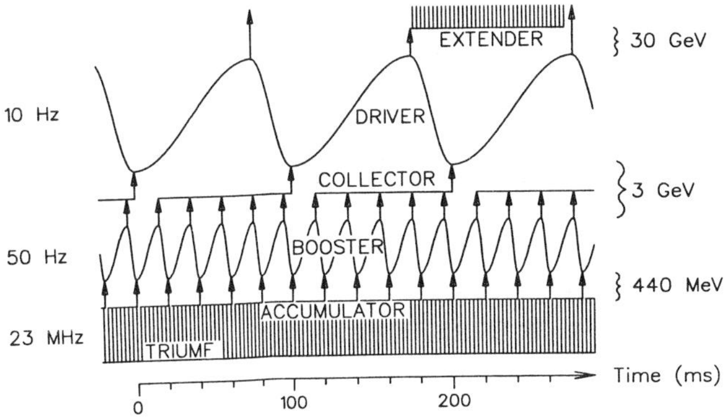

Fig. 1. Energy-time plot showing the progress of the beam through the five TRIUMF KAON Factory rings.

*On leave from Physics Dept., University of British Columbia, Vancouver, B.C., Canada V6T 2A6.

Figure 2 shows a possible site layout together with the arrangement of the A and B rings in the Booster tunnel and of the C, D and E rings in the main tunnel. Identical lattices and tunes are used for the rings in each tunnel. This is a natural choice providing structural simplicity, similar magnet apertures and straightforward matching for beam transfer.

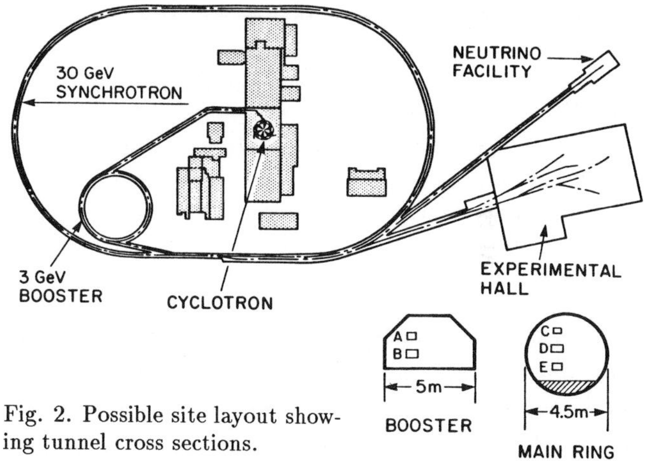

Fig. 2. Possible site layout showing tunnel cross sections.

For the C,D and E rings a racetrack lattice has now been adopted in place of the originally proposed superperiodicity 12 lattice. This provides the long straights required for a super-efficient (>99.8%) slow extraction and collimation system in the E ring, while keeping transition above top energy. Separated function FODO lattices are used. Racetrack lattices are also under consideration for the Booster.

Injection into the Accumulator is achieved by stripping the H^- beam from the cyclotron. The small emittance H^- beam is painted over the much larger three-dimensional acceptance of the Accumulator to limit the space charge tune shift and the number of passages through the foil.

A 15-month $11-million pre-construction study is just beginning. This contains funds for prototyping magnets, rf cavities, kickers, ceramic beam pipe and production targets; for designing the tunnels and buildings; for studying the environmental and economic impacts; for bringing the cost estimates up to date; and for investigating possible international contributions on the HERA model (for which initial explorations have shown a potential >$100 M). The provincial government of British Columbia has already committed $87 M to the civil works and there appear to be good prospects for full approval early in 1990.

1) KAON Factory Proposal, TRIUMF, September 1985.

2) Craddock, M.K., Baartman, R. *et al.* IEEE Trans. Nucl. Sci. NS-32, 1707 (1985).

3) Craddock, M.K. *et al.*, Proc. 1st European Particle Accelerator Conf. (Rome, 1988) (in press).

THE EUROPEAN HADRON FACILITY

P. Blüm
University of Karlsruhe, Institut für Experimentelle Kernphysik,
Federal Republic of Germany

The European Hadron Facility (EHF) is planned as a new research facility for nuclear and particle physics in a medium energy range. A carefully designed accelerator complex will provide a high intensity (100 μA) proton beam of 30 GeV kinetic energy which can be used to produce a wide variety of high-quality secondary beams of neutrinos, muons, pions, kaons, and antiprotons. As a further distinctive feature this facility has the capability to accelerate polarized protons up to the final energy. These many-fold experimental capabilities make a rich and many-faceted research programme conceivable.

The design of the accelerator is based on the consideration of minimizing beam losses to a level that hands-on maintenance becomes possible except at a few "hot" spots where a remote handling is required (e.g., primary target). The optimization of all beam transfers, in particular the multiturn injection and the slow resonant extraction, leads to loss estimations of the order of 1% over the entire acceleration cycle.

The proposed accelerator complex is sketched in Fig. 1. The main components are:

- a high energy linac accelerating a H^- beam to 1.2 GeV
- two fast cycling synchrotrons – a 1.2 to 9 GeV Booster and a 9 to 30 GeV Main Ring with circumferences and repetition rates in the ratios 1:2 and 2:1, respectively.
- two DC rings – a 9 GeV Holding Ring and a 30 GeV Stretcher

The Holding Ring has to match the different repetition rates of the Booster and the Main Ring. It stores one Booster pulse during the next Booster cycle and feeds then the Main Ring with the second Booster pulse appended immediately. Since Holding Ring and Booster have the same size they may share the same tunnel. The Stretcher, on the other hand, has to provide slow extraction of the 30 GeV beam with almost 100% duty cycle. As some beam losses are expected during slow extraction we foresee separate tunnels for the Main Ring and the Stretcher, whose lattice could thus be optimized for minimum beam losses. The long straight sections allow the design of a high β insertion yielding a highly efficient slow, resonant extraction, and the implementation of a collimation system localizing inevitable beam losses. Two experimental halls complete the facility complex.

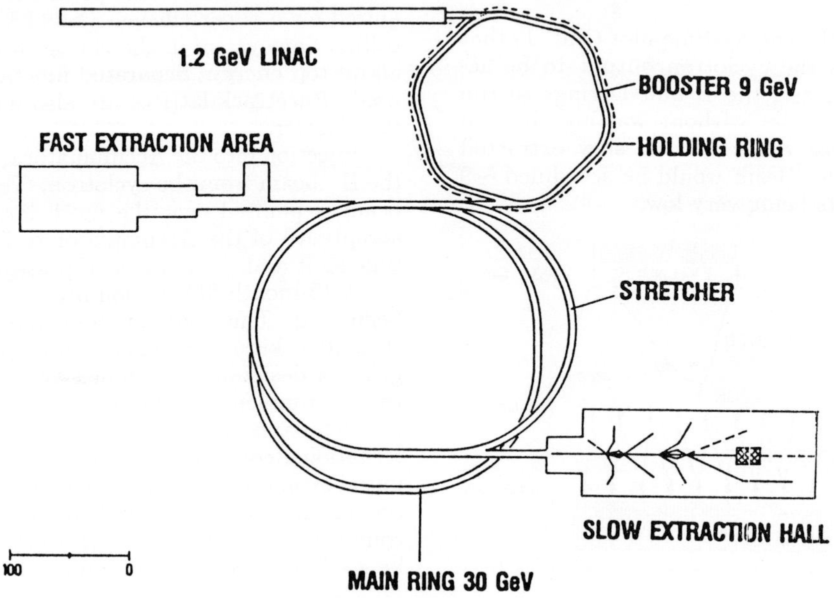

Fig. 1. The European Hadron Facility.

THE JAPANESE HADRON PROJECT

K. Nishikawa
Institute for Nuclear Study, University of Tokyo, Tanashi, Tokyo 188, Japan

We have decided to construct the high-intensity proton facility in two stages. The design of the first stage of JHP (Japanese Hadron Project) is based on the following fundamental considerations:

(1) The facility must meet the requirements regarding pulsed neutron and pulsed muon experiments as well as those involving nuclear physics experiments.

(2) The nuclear physics group requires a high-intensity pion facility.

(3) "Exotic nuclei" (including unstable nuclei) must be accelerated.

(4) Present KEK-PS must be upgraded at reasonable size and cost in order to continue nuclear physics experiments using kaons and high-energy pions in the present KEK-PS.

The scheme of the JHP accelerator complex is shown in Fig. 1. It consists of a 1 GeV proton linac, a heavy-ion linac of 6.5 MeV/μ energy and 1 GeV compressor/stretcher ring (shown in Fig. 2) which produces various time structures of the beam (20 ns to near dc). Rough specifications are shown in Table I. The average current of the ring is 200 μA. The repetition rate is taken as 50 Hz, so that the number of protons per pulse is 2.5×10^{13} and the circulating current becomes about 6.7 A.

A low-frequency rf system is needed for pulsed neutron and pulsed muon sources. The number of bunches in the ring is planned to be 2, which requires that the rf frequency be 3.35 MHz. One of the bunch is first extracted in the single-turn extraction to neutron production target, giving about 200 ns time structure. The other bunch is compressed or stretched further in the ring, giving about 20 ns to near dc (about 70% factor).

The 1 GeV proton linac also delivers part of their beam to ISOL (isotope separator online). Nuclear reaction products are injected into the heavy-ion linac for acceleration up to 6.5 MeV/μ.

This first stage will offer, among other things, a unique pulsed beam of muons and good competition with ISIS in the field of pulsed neutron research as well as a unique beam of unstable nuclei.

Table I. Specifications.

Proton Linac		Heavy Ion Linac		Compressor/Stretcher Ring	
Energy	1 GeV	Energy		Energy	1 GeV
Average beam intensity	400 μA	Beam intensity		Average beam intensity	200 μA
Repetition rate	50 Hz	Unstable nuclei	10^4–10^{12}pps	Repetition rate	50 Hz
Peak intensity	20 mA	Stable nuclei	$<10^{15}$pps	Pulse width	
Pulse duration	400 μs	Charge/mass	>1/60	Standard	200 ns
Total length	500 m	Repetition rate	30 Hz	Ultra-short	20 ns
		Total length	120 m	Cont. beam duty factor	$\geq$70%
				Beam emittance	30π mm-mrad

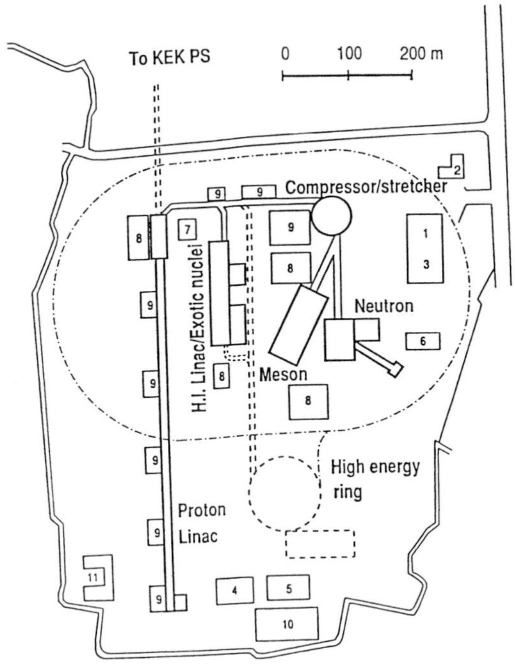

Fig. 1. Layout of JHP accelerator complex.

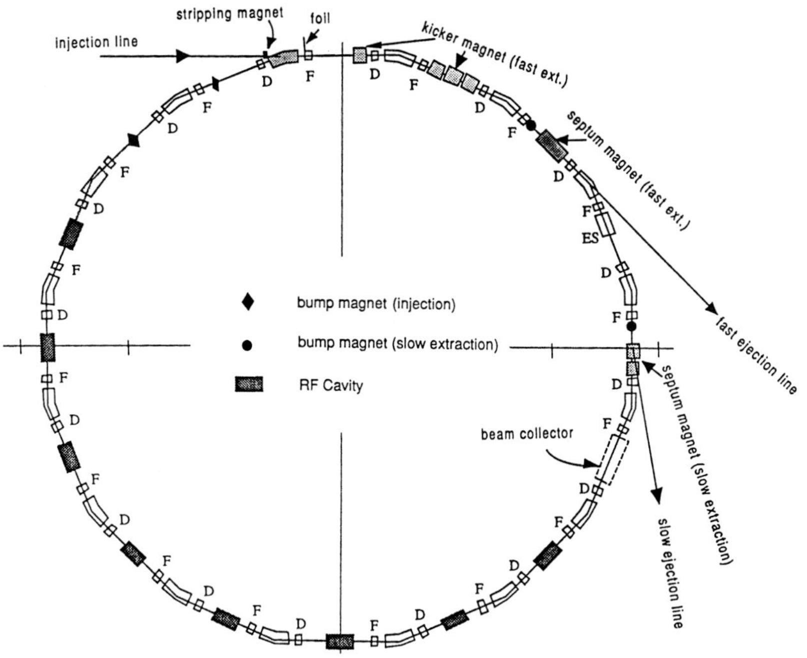

Fig. 2. Structure of compressor/stretcher ring.

AGS II*

Y.Y. Lee and D.I. Lowenstein
AGS Department, Brookhaven National Laboratory
Associated Universities, Inc., Upton, NY 11973

At present, the AGS slow extracted beam current is 1 μA. With the completion of the Booster in 1990 and the associated AGS modifications, the current will rise to 4–5 μA. With the subsequent addition of the Stretcher, which is under design, the current will rise to 8–10 μA and approximately 100% duty factor. We are now examining the possibility of a further enhancement to a current level of 40–50 μA cw.

Let us first examine the capabilities of each of the present AGS accelerators. The linac is capable of running ten pulses a second of 30 mA H^- ions with a 500 μs pulse length. The linac output current exceeds the input capabilities of the Booster. The Booster is capable of pulsing ten cycles a second. Because of the large power swing both in real and reactive power, the Booster is limited presently to operate at 7.5 Hz and an energy of 1.5 GeV. One way to overcome this limitation would be to pulse, out of phase, an equivalent electrical device (another matched accelerator) as an analog to a flywheel so as to smooth the power swing.

At present, the AGS is capable of cycling every 1.2 s. The pulse rate is limited mainly by two factors. One is the limitations of the main magnet power supply and the second is the peak voltage of the present radio-frequency acceleration system. Both of these can be improved. The highest current that can be achieved is when one matches all the accelerators to the repetition rate of the linac. Our scheme assumes that one does not replace either the linac or the AGS.

The first stage in the intensity increase program is the introduction of the Stretcher ring. This machine, which is now under design, allows protons to debunch and then slow extract with $\approx$ 100% duty factor. Since the AGS would not operate with a flattop cycle, the repetition rate and thus the delivered proton intensity would be increased by a factor of two. The implementation of just the Stretcher would provide 10 μA cw current. The cw feature is the necessary aspect of any high intensity facility.

We next propose to increase the Booster energy to 2.8 GeV and the repetition rate to 10 Hz. This is below the Booster transition energy and well within the capabilities of this machine. The Booster is already designed to operate at the increased dB/dt rate. The increased Booster energy is motivated by the energy swing solution described below.

We further propose to introduce a Post-Booster accelerator. This machine would operate at 10 Hz and accelerate protons to an energy above the AGS transition energy. Post-Booster swing could be made to complement that of the Booster and thus overcome the Booster repetition rate limitations. The AGS main ring power supply cycling limitations would also be eased due to the reduced AGS beam energy swing. The Post-Booster would be designed to have the same magnetic energy difference swing as the Booster. These two machines would operate at the same repetition rate but 180° out of phase with each other. To reduce the construction costs by minimizing the number of tunnels, it would be desirable to install the Post-Booster in the same tunnel as the Collector ring. The Collector ring requires a minimum circumference of three times that of the Booster ring. The Post-Booster would thus have a circumference three times that of the the Booster, 75% of the AGS.

The Collector would be a short-term (0.4 s) intermediate storage ring. This machine would reside in the Post-Booster tunnel. The function of this ring is to temporarily store three Post-Booster pulses prior to injection into the AGS. The AGS would accept the three Post-Booster pulses (9 bunches) stored in the Collector and one additional pulse (3 bunches) directly from the Post-Booster for a total of 12 bunches. The Post Booster and the Collector would inject into the AGS every 400 ms. We show in Table I the estimated proton currents at various implementation stages of the above-mentioned proposal. The delivered currents are for slow extracted beam operation.

Table I

Option	Duty Factor (%)	Delivered Current (μA)
AGS	35	1.0
+Booster	35	4.0 – 5.0
+Stretcher	100	8.0 – 10.0
+Post-Booster & Collector	100	40.0 – 50.0

*Work performed under the auspices of the U.S. Department of energy.

BNL High Energy Physics Plans for the Future

Robert K. Adair*
Associate Director for High Energy and Nuclear Physics
Brookhaven National Laboratory

High Energy Physics at 30 GeV

The thrust of that part of particle physics that is still closely connected to observations of nature[1], is now largely directed to tests of the *Standard Model* and extensions of that model. To most of us, that effort has the highest priority – indeed my own research has been so directed. However, it may be the nature of our community to over-focus; with no immunity to *hubris*, we may ofttimes be too certain that we know the proper questions. Hence, especially in times of constrained resources, our priorities may tend too much to the absolute and we may risk advancing on too narrow a front in physics. Those areas that are not awarded the highest priority tend to be dropped absolutely. To most of us (and to me), the SSC holds that highest priority, but we must be most reluctant to accept the SSC only at the cost of other unique resources.

There are many interesting questions in particle physics that cannot be answered by the SSC and there are interesting questions of such a character that the answers *probably* will not have an impact on the Standard Model. But we must not be certain that we understand Nature so well, that we can *completely* disregard those avenues that do not now seem quite so promising. In simpler times there was always a subset of physicists who moved along those paths that were off of the main-stream. But if those less-traveled roads require a class of accelerator that is no longer supported, the roads are closed – probably forever.

The world must not be in a position so that no facilities are available in broad areas. Brookhaven is now almost unique in providing relatively high intensity proton beams of moderate energy and in addressing the substantial areas of physics that are uniquely served by such beams. There are now consequential areas of physics that can be conducted effectively only at the BNL AGS. Some of these fields of research bear directly on possible extensions of the Standard Model; the rare K-decay searches explore the high mass scale desert through their sensitivity to heavy virtual particles; the neutrino beams are well situated in flux and energy for certain kinds of neutrino oscillation searches; and very accurate measurements of the muon g-2 define a very tight wicket through which variations of the Standard Model must pass. Some of the other inquiries address problems that are less certain to have an impact on the bases of the standard model. For example, at BNL, physicists are studying the effects of polarization in the interactions of hadrons and the physics of light quarks, glueballs, and the effects of gluons; they are investigating exotic hadron states of 6 quarks or two pairs of quarks and antiquarks; and there are programs concerned with hypernuclei and the modifications of nuclear matter induced by strange hadrons.

By and large, these are the areas that have been taken as the province of the 'Kaon Factory' propositions and the proponents of those facilities have argued (to me, quite convincingly) that the physics that might be conducted at such facilities is not just important, but essential.

The BNL AGS as a Mini-Kaon Factory

The Kaon factories under serious discussion are – properly – ambitious and hence, expensive and adventurous. The conceptual designs (and only conceptual designs exist at this time) of these machines place those machines near the edge of the accelerator art (hopefully not beyond!). In the nomenclature that Chris Quigg brought to life in his previous talk, the Kaon factories promise not to be "dull" machines where Chris cast a taint on the label 'dull'. They will probably be very "interesting" machines where I interpret 'interesting' is in the sense of the apocryphal Chinese curse, "May you live in interesting times."

With relatively modest further capital investments (of about an order of magnitude smaller than that required for the full Kaon factories) and relatively modest operating budgets (of about one-half order of magnitude smaller than those projected for the Kaon factories), BNL will be in a position to increase its critical beam intensities by a factor of ten through quite conventional (very dull!) accelerator developments that are certain to work and to conduct an effective program using those beams. With such improved facilities, BNL will be able to extend the quality and quantity of its present programs to a degree

[1] We leave superstrings to the superwizards.

that is competitive with those which are expected to be addressed by the proposed Kaon factories. The technical problems that must be considered in the course of the proposed BNL developments are straightforward and the magnitude of that funding required by BNL is well within the practical.

Though the beam intensities achievable from this modest investment at BNL will not be as great (by about a factor of ten) as those the Kaon factories hope to obtain, and therefore not all of the BNL programs will be able to reach the levels of precision promised by the Kaon factory proposals, the improvement and support of the present BNL facilities should be considered as an essential first step in any journey to the ambitious dedicated Kaon factories.

I will now summarize the BNL plans and then describe those plans in more detail. The BNL AGS facility now produces 1 μa of 30 GeV protons with a 30% duty factor and 2 μa for fast ejection (e.g. for neutrino physics). The $31,200,000 BOOSTER, which is now well under way and on schedule for completion at the end of 1990, will give the AGS an increase in beam by a factor of four; by the end of 1991 the AGS will produce 4 μa of slow beam and 8 μa of fast beam.

BNL proposes to begin building the STRETCHER in 1991 using the cadre that is now occupied with the BOOSTER. (The RHIC construction group is quite separate.) The STRETCHER is a small storage ring designed to accept the maximum cycle rate of fast-pulses from the AGS and release the beam on target continuously. With the STRETCHER – and some other small improvements – BNL would be able to provide a beam of 10 μa with a 100% duty factor. The STRETCHER project has been costed at $42,000,000 (1986 dollars) and with inflation and extra shielding requirements, we estimate a total cost of about $50,000,000 today. The STRETCHER can be built in two years and finished in 1993 if the funding level is sufficient (the BOOSTER was delayed for at least a year as a consequence of a less-than-optimum funding profile).

It now seems quite likely that the RHIC construction will be funded in the FY90 DOE budget enabling RHIC to be finished in 1995 and running in 1996. At that time, the core AGS operating budget should be provided through RHIC funding. A recent HEPAP analysis conducted by a subcommittee under Paul Reardon concluded that the present base operation of the AGS cost about $20,000,000 (1988 dollars). With the longer running schedule required for RHIC, this might increase to about $25,000,000. But the AGS would be occupied filling the RHIC storage rings for only about two hours a day. The rest of the day could be devoted to high intensity hadron physics. With the operation 'nut' covered by RHIC funding – required whether or not the AGS were used for other physics – a marginal operations budget of about $25,000,000 would support a hadron physics program that would run for 40 weeks a year (nearly twice the present annual running).

I emphasize that, on the scale of the technical problems posed by the dedicated Kaon factories (and no such machine has been designed!), the BOOSTER and the STRETCHER (and RHIC) are quite straightforward.

Even as the BNL costs will be relatively modest, the projected beam intensities are not extremely high. For an order-of-magnitude larger capital investment and a four-fold increase in operating costs the Kaon factories *hope* to produce ten times the current projected by BNL. How valuable would such an increment be? I think that it is trivially easy to conclude, (a) that one will always be able to do a better experiment with more intensity, and (b) that the improvement in the experimental results will not be proportional to the increase in intensity. It is interesting to consider rare K-decays as paradigmatic experiments. At the levels we are implicitly discussing, the number of events will vary as I^N where I is the intensity and $0 < N < 1$. As an educated guess I would consider it unlikely that N would be greater than 0.5 for rare event searches. The branching ratio limits for rare K-decays (e.g.) will then go as $I^{1/2}$ for experiments where there is no significant background but only as $I^{1/4}$ if there are backgrounds. Hence, a Kaon factory capable of producing a 100 μa beam could expect to reach limits on rare K-decays that are only better by factors of two or three than might be achieved at the proposed 10 μa AGS-BOOSTER-STRETCHER complex.

Although I am primarily here this afternoon to extoll a BNL mini-Kaon factory, the RHIC facility, although labeled *Nuclear Physics*, should also be of great interest to particle physicists. Aside from interest in the character of the nuclear matter at very high temperatures that will be produced by $Au-Au$ collisions that are the main thrust of the project, RHIC can generate high luminosity proton-proton collisions (and with polarized protons!). With a very small additional investment for an external beam dump, RHIC can be expected to produce about 10^{10} B-mesons per year. RHIC will then be an almost free (in impact on the high energy physics budget) B-factory!

BNL also has on the drawing boards an extension of the AGS (+BOOSTER-STRETCHER) labled AGS II designed to increase the 30 GeV proton intensity to 40 μa. We presume that such a project would go ahead only if the proposed Kaon factories – regretably – do not materialize and if it seems evident that the factor of four in intensity to be gained by such an extension will be quite important. We estimate that AGS II will cost about $125,000,000 and that program could be finished by 1999.

Figure 1 shows the geography of the present and proposed accelerators on the BNL site and Table 1 lists the capabilities in a convenient form.

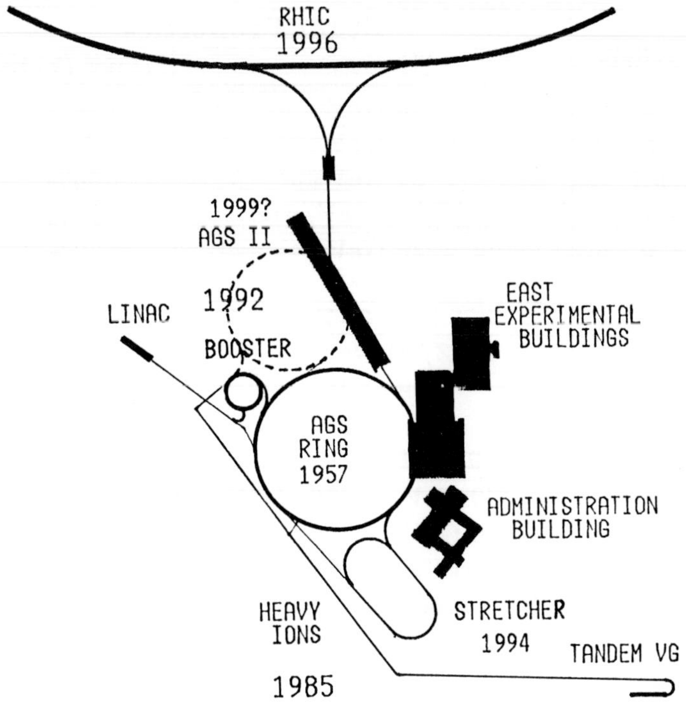

Figure 1: The once and future BNL accelerator complex.

AGS (1957-1988)

Presently, the AGS regularly accelerates $1.6 \cdot 10^{13}$ protons to energies of 30 GeV every 2.5 seconds for slow beam with a duty factor of about 30%, and every 1.2 seconds for fast extracted beams. This corresponds to a current of about 1 μa for slow beam and 2 μa for fast beam. Since the AGS energy is well above the K-meson production Feynman scaling threshhold – and, hence produces about as many K-mesons per interacting proton as accelerators of higher energy, with its high current, the AGS is *now* probably the world's premier K-factory.

The program emphasis is placed on those inquiries that are likely to shed light on the character of extensions of the Standard Model. At this time, rare K-decay searches constitute a main part of the AGS program. Four major experiments are now taking data which has already extended limits on transition rates to nominally forbidden modes in important categories more than two orders of magnitude.

The intense muon beams produced at moderate energy will allow a definitive measurement of the muon g-2 ratio placing severe constraints on "beyond the standard model" concepts. A large Japanese-German-US collaboration is engaged in the early stages of R&D required for this measurement.

The AGS also plays a singularly dominant role in low mass quark spectroscopy – with especial emphasis on gluon effects and glueballs. Two experiments are now engaged in experiments directed towards such studies. Three experiments are directed towards studies of the production and interactions of particles through the strong interactions and with weak and electromagnetic decays. And eight experiments are directed towards studies of exotic matter – '6-quark particles' and the physics of hypernuclei.

The AGS is also a "sole source" of substantial fluxes of intermediate energy polarized protons. The machine accelerates polarized protons to energies above 20 GeV with intensities of about 10^{10} per second. There are three experimental programs that make measurements at polarized proton runs scheduled every two years.

Also, in the last few years the AGS has accelerated beams of heavy ions (up to Silicon) with energies of 14.5 GeV per nucleon. The interaction of such heavy nuclei with lead and uranium targets generate very high nuclear densities at moderate temperatures – a mileau that is complementary to that generated by higher energy heavy ion interactions which would seem to produce lower densities but higher temperature matter. Three major experiments (and several less demanding experiments) are now in progress.

Altogether, 94 institutions and 585 physicists are now engaged in AGS experiments now on our active list (approved but not yet finished). Of the physicists, 28% are foreign, 15% are from BNL, 26% are from other eastern institutions and 31% are from the rest of the United

Accelerator Date, $	Beam Properties			
	Particle	Energy	Duty Factor	Current
AGS since 1957	p	29 GeV	30%	1 μa
	p	29 GeV	*fast*	2 μa
	$p\uparrow$	21 GeV	30%	10^{10} /sec
	^{14}Si	29(Z/A) GeV/nuc	30%	10^8 /sec
AGS+BOOSTER 1992 $32 M	p	29 GeV	30%	4 μa
	p	29 GeV	*fast*	8 μa
	$p\uparrow$	21+ GeV	30%	10^{11} /sec
	^{79}Au	29(Z/A) GeV/nuc	30%	10^9 /sec
AGS+BOOSTER + STRETCHER 1994 $50 M	p	29 GeV	100%	10 μa
	$p\uparrow$	21+ GeV	100%	10^{11} /sec
	^{79}Au	29(Z/A) GeV/nuc	100%	10^9 /sec
RHIC 1996 $377 M	$p \times p$	300 GeV × 300 GeV		$L = 10^{32}$
	$p\uparrow \times p\uparrow$	300 GeV × 300 GeV		$L = 10^{32}$
	$Au \times Au$	125 GeV × 125 GeV /nuc		$L = 5 \cdot 10^{26}$
		B-Production	10^{10} per year	
AGS-II 1999? $125 M?	p	29 GeV	100%	40 μa

Table 1: Brookhaven National Laboratory, Present and Future.

States; about 124 graduate students from 42 schools work on AGS experiments.

The BOOSTER (1991)

The foreseen need for higher intensities – especially to allow more sensitive rare K-decay searches – has lead to the construction of a higher energy injector for the AGS – the BOOSTER. The BOOSTER is now under construction with a budget of $32,000,000 and is to be completed in 1990. The BOOSTER will inject ions into the AGS proper at energies of about 1.5 GeV allowing four-fold increases in the AGS total beam. With the BOOSTER and concurrant upgrades of the AGS RF, the AGS can be expected to provide slow beam currents of about 4 μa and fast beams of 8 μa accelerated beam by 1992.

Since the BOOSTER can speed atoms as heavy as gold to velocities greater than that of the K-shell electrons, post-BOOSTER stripping of these heavy ions can produce fully charged gold ions to be accelerated in the AGS. Hence, heavy ions up to gold will be produced with energies of 29(Z/A) GeV per nucleon with intensities projected to be about 10^9 per second.

The STRETCHER (1994)

Noting the increased need for the intense beams of protons, mesons, and hyperons, addressed by proposals to construct "Kaon Factories", BNL will propose to procede further in the direction of producing intense hadron beams by the addition of an AGS short-time storage Ring. The STRETCHER would be a storage ring able to accept injection of 30 GeV protons from the AGS at the rate of one pulse per 1.2 seconds while ejecting the protons continuously – with a duty factor of nearly 100%. The improvement over the AGS *cum* BOOSTER for slow beam operation should be a factor of two in intensity and a factor of about three in duty factor. This intensity, expected to reach 10 μa of 30 GeV external protons at a duty cycle of 100%, is comparable to (typically, a factor of ten less than) the maximum intensities anticipated from the proposed dedicated Kaon Factories. (We note that there is no great uncertainty in the anticipated intensity of 8 μa since this is an increase of only a factor of four over the 2 μa the AGS now produces for fast ejection.)

Long straight sections in the STRETCHER and other improvements would be implemented to optimize the slow extraction efficiency and reduce the beam loss at the higher intensity operation. Since the aperture would be much smaller than that of the AGS, power would be saved during the spill and it will be possible to increase the cycle rate further (and reach 10 μa beam intensities).

The major element in the proposal, the STRETCHER, has been costed at $42,000,000. We expect to require $15,000,000 more for the construction and shielding of beam lines able to handle the high intensities. These beams have not yet been designed and costed.

We believe that a large portion (though not all) of the programs projected for Kaon factories can be handled expeditiously by an AGS-BOOSTER-STRETCHER mini-Kaon factory that could be constructed for about \$50,000,000 additional moneys.

RHIC (1996)

RHIC (the **R**elativistic **H**eavy **I**on **C**ollider) is expected to be in the DOE contruction budget in FY 1990. The primary goal of this machine is to accelerate two beams of heavy ions (up to Au) injected from the AGS-BOOSTER up to energies of about 125 GeV per nucleon and arrange their collision. Luminosities are projected of $5 \cdot 10^{26}$ for gold-gold collisions. Such collisions are expected to produce nuclear matter at temperatures sufficiently high to "melt" the baryons in a phase change to a quark-gluon plasma.

Protons – and polarized protons – are among the ions available to RHIC. According to the approved design specifications, RHIC should produce proton-proton collisions with a luminosity of $1.5 \cdot 10^{31}$. Hopefully, this may be increased eventually to levels near $2 \cdot 10^{32}$. The luminosity would be the same for polarized protons.

Aside from studies of the interactions of high energy nucleons, particle physicists may find RHIC a superior B-factory. Though the cross section for the production of B's at RHIC is expected to be about one-third as large as at Fermilab, the small interaction length of about 30 cm may prove advantageous. One can expect to produce 10^{10} B's per year at RHIC.

AGS II (1999?)

Discussions and conceptual designs have been carried out concerning the possiblity of further increases in proton intensities in the 30 GeV range. It seems possible to increase the available intensity from the BOOSTER-AGS-STRETCHER system about a factor of four – to near 40 μa – through the construction of a further preinjection ring and susbstantial upgrades of AGS and BOOSTER RF power. A preliminary cost estimate suggests a bare-bones cost of about \$100,000,000 for such a modification. The cost of extra shielding and high-intensity beam transfer lines, might be expected to increase this number by by 25% to \$125,000,000.

PROGRESS WITH THE AGS BOOSTER*

W.T. Weng
Brookhaven National Laboratory
Upton, New York 11973

1. INTRODUCTION

Rare K-decay, neutrino and heavy ion physics demands that a rapid-cycling high vacuum and high intensity Booster be built for the AGS at Brookhaven.

The Booster ring is one-quarter the circumference of the AGS and is located in the angle between the linac tunnel and the AGS ring. The modes of operation for various particles are illustrated in Fig. 1, where the time scale covers the 2.8 s of a single slow extracted pulse. For unpolarized protons four Booster pulses would be injected at a 7.5 Hz repetition rate within a 400 ms flat bottom, enabling the present 1.5×10^{13} ppp to be increased to 6×10^{13} ppp. Initially protons would be accelerated to 1.5 GeV although the bending capability provided for heavy ions would eventually allow protons to be accelerated to 2.5 GeV. For heavy ions a slower acceleration time is required in the Booster, and only one pulse would be injected into the main ring. For polarized protons there is the option of stacking 20 or so pulses in the Booster ring before injecting them into the AGS.

For each mode of operation there are corresponding accelerator physics and design issues needing special attention. Problems pertinent to any single mode of operation have been encountered and solved before, but putting high intensity proton requirements and high vacuum heavy ion requirements into one machine demands careful design considerations and decisions. In Sec. 2, the lattice design and magnet characteristics will be briefly reviewed. In Sec. 3, major design issues will be discussed and design choices explained. Finally, in Sec. 4, the construction status and schedule will be presented.

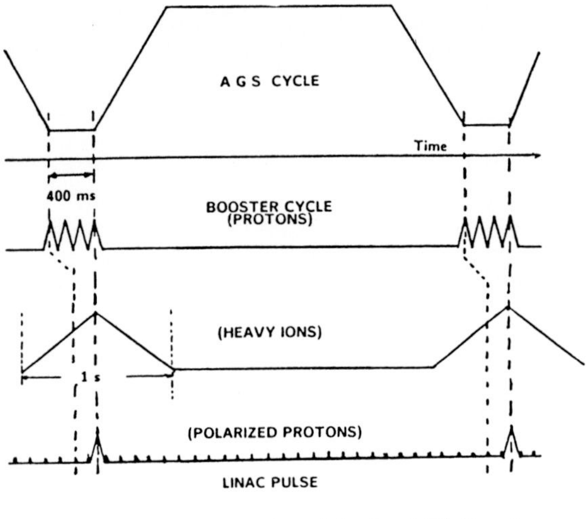

Fig. 1. Operation cycles with the AGS Booster.

*Work performed under the auspices of the U.S. Department of Energy.

2. LATTICE AND MAGNET DESIGN

The Booster ring consists of 24 FODO cells. Phase advance per cell is 72° which gives tune of the accelerator about 4.82 in each plane. Twelve straight sections are created by missing twelve dipoles from inter-quadrupole space resulting in 36 dipoles bending 10° each. The final superperiodicity of the machine is 6. Detailed description of the lattice design is given in Ref. 1. Major lattice and magnet parameters are listed in Table I. Detailed description of Booster magnet design is given in Ref. 2. The choice of optical functions and magnet aperture is predicated on the requirement of accepting high intensity protons without excessive space charge tune shift at injection.

Table I. Booster Parameters

Circumference	201.78 m (1/4 AGS)
$\nu x/\nu_y$	4.82/4.83
$\beta_{\max}/\beta_{\min}$	13.5 m/3.5 m
$x_{p,\max}/x_{p,\min}$	3.1 m/0.7 m
No. of Dipole	36
Length of Dipole	2.4 m
Physical Aperture	75/152 mm (V/H)
$B_{\max}$	12.75 kG
No. of Quadrupole	48
Length of Quadrupole	0.5 m
Physical Aperture	152 mm (circular)
Gradient (max)	1.2 T/m

3. MAJOR DESIGN ISSUES

Since the Booster is designed to meet three operational goals, the accelerator physics and component design requirements are more than typical accelerator having single purpose in mind. In the following, some of those issues as well as the proposed solutions will be discussed.

3.1 Space Charge Tune Shift

At the design intensity of 1.5×10^{13} ppp and injection energy of 200 MeV, the expected maximum vertical tune shift is about 0.4 units as shown in Fig. 2.

The expected shift at injection is larger than what conventionally thought to be a reasonable design goal, but is substantially lower than what has been achieved both at the AGS and PSB at CERN which is 0.7 unit. Experience showed that the space charge tune shift does cause problems for achieving high intensity goal, but is not a phenomenon with threshold behavior to constitute a hard

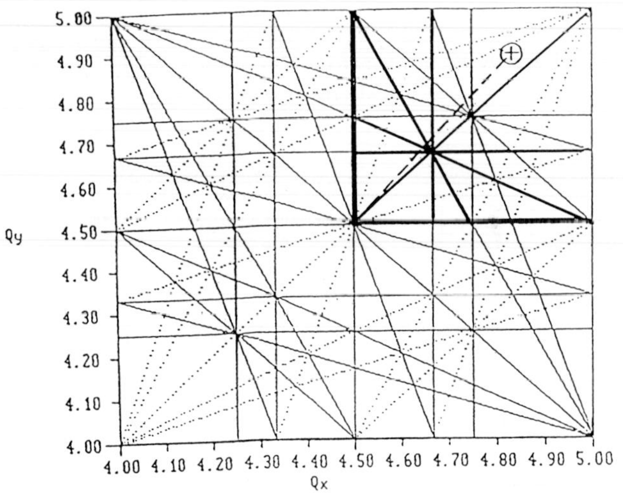

Fig. 2. Booster space charge tune shift.

limit. There are numerous ways to counteract the large tune shift at injection:

a) Stopband correction for 1/2 and 1/3 integer resonances must be provided to minimize the amplitude blow up due to resonance crossing, b) RF voltage should be large enough to reduce the bunching factor and to facilitate optional capture and bunching operation, c) provide beam loading compensation to ensure proper voltage amplitude and phase function is maintained, d) sufficient physical aperture to keep beam for both betatron oscillation and large momentum spread during capturing, e) shape the charge distribution by phase space painting and 2nd harmonic cavity.

3.2 Eddy Current Effects

For proton operation, the Booster has to be ramped at 7.5 Hz to meet the injection requirements to AGS. This amounts to a maximum $\dot{B}$ to be 8 T/s which induces eddy current on the vacuum chamber wall causing both field distortions and wall heating. The typical solution of using a ceramic chamber is not acceptable in UHV environment.

The induced fields have strong dipole and sextupole components. The dipole components can be compensated by the adjustment of main power supply. However, an eddy current correction winding on the vacuum chamber is adopted to compensate sextupole and higher-order multipoles.[3)] This correction has three major advantages: 1) eliminate the systematic sextupole component right at the source and make the chromaticity control much easier, 2) because the correction coil is attached to the vacuum chamber, the positioning of the vacuum chamber is not too critical, 3) because the correction is due to a transformer effect and requires no separate power supply, its correction is rate independent. The prototype correction winding is shown in Fig. 3.

3.3 Orbit and Stopband Corrections

24 BPMS and 24 horizontal correctors will be placed next to focusing quadrupoles for horizontal orbit

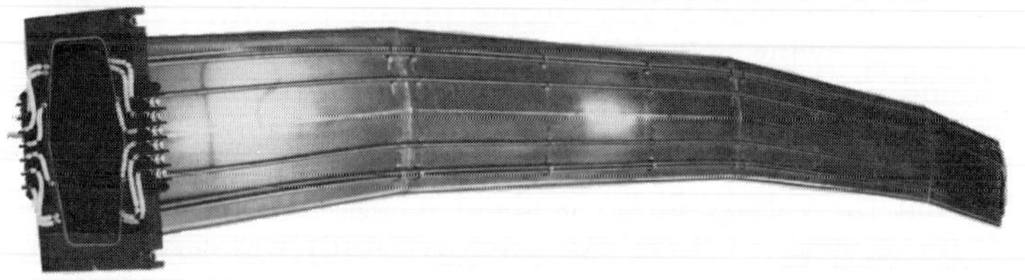

Fig. 3. Dipole vacuum chamber and eddy current correction winding.

correction. (Corresponding arrangement is also provided for vertical plane.) Assuming rms random field error of $\Delta B/B = 2 \times 10^{-4}$, quadrupole misalignment of 0.2 mm, the maximum uncorrected orbit errors are about 10 mm in both planes which can be corrected to be about 0.5 mm using either three bumps method or MICADO algorithm. The maximum corrector strength required is about 15 G-m to correct up to proton final energy.

Since focusing and defocusing quads are independently powered, the control of horizontal and vertical tune will be done through adjusting the QF and QD supplies. Every quadrupole will have an independent trim coil up to 1% of injection strength for 1/2 integer stopband correction. 24 skew quadrupoles will be provided for correction of coupling resonance $Qx - Qy = 0$ and $Qx + Qy = 9$.

To maintain flexibility of operation, we decided to provide 48 sextupoles for chromaticity correction. Every sextupole will have an independent trim coil up to 1% of injection strength for 1/3 integer stopband correction. If it is proven to be necessary, some of the sextupoles will be rotated to provide skew sextupoles for coupling resonance correction.

3.4 RF Capture and Beam Loading

After H^- injection into the Booster, a 2.2 MHz RF system has to be employed to capture 200 MeV linac beam with 200 MHz structure. A modified FNAL ESME[4)] program has been used to study the capture process. We found that multiturn injection process is important with space charge effect, also capture at larger initial voltage is favorable instead of adiabatic capture. Optimal capture can be achieved with chopped beam and off-frequency capture.[5)]

When the beam starts to bunch itself under RF voltage, it will in turn induce voltage across the gap, hence the beam loading effect. A feedforward system will be applied to cancel the beam loading to keep external RF in proper amplitude and phase. When the first pulse of Booster beam enters AGS a transient beam loading will occur which could cause up to 35% amplitude modulation and 20° phase shift. The proposed solution for the AGS RF system is to place the power amplifier next to the cavity and implement fast feedback at the final stage of the amplifier to eliminate the transient beam loading and reduce apparent impedance for the beam.

3.5 Power Grid Perturbation

The Booster will be energized from a feeder with the coupling node at the LILCO Brookhaven substation with a SCC of 2310 MVA. During proton operation the full energy swing is 22 MW which can produce both voltage and phase flicker on the power grid and could also introduce dynamic system oscillation or torsional oscillation of the generator. Both calculation and measurement showed that the voltage flicker is about 0.25% and that of phase flicker is about 0.7%. An extensive study was carried out by GE[6] to ascertain the effect on the last two concerns. The results show that certain frequency is to be avoided as shown in the response function in Fig. 4.

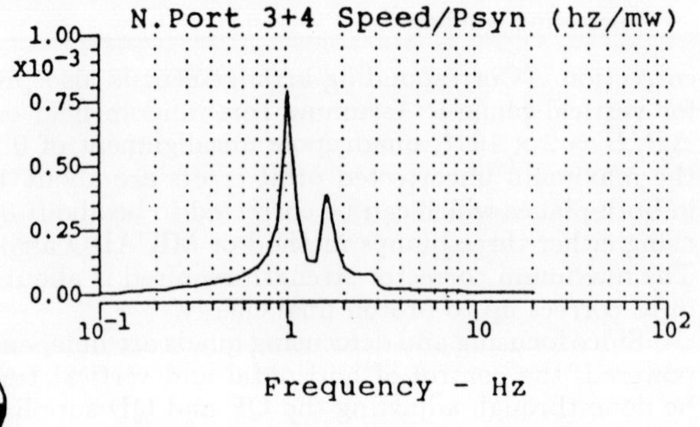

Fig. 4. Frequency response of LILCO power grid.

LILCO agreed to power the Booster provided: 1) a contract is signed to delineate operation conditions, such as power and frequency range, 2) a full power test before routine operation, 3) interlock and monitor to prevent operation in forbidden frequency range.

3.6 Coherent Instabilities

Although the number of particles per bunch is a factor of 4 higher than that in the AGS, the Booster longitudinal phase space area is designed to be 1.5 times larger than that of the AGS. This is more than enough to stay below the instability thresholds for both longitudinal and transverse single bunch instabilities. However, at Booster design intensity, the growth time results from resistive wall coupled bunch instability is around 3 ms which is fast enough to warrant a damping system.

The dominant contribution to the coupling impedance is from the space charge effect. The total longitudinal impedance, including space charge, vacuum chambers, bellows, pick up electrodes, is shown in Fig. 5. Detailed analysis and measurement program is in progress to control impedances introduced by beam components, such as BPM's, kickers, septum magnets, vacuum ports, RF cavities, etc.

4. STATUS AND SCHEDULE

Although the Booster project was approved in FY 1986, major funding is concentrated in FY 1988, 1989 and 1990. Major construction-related works accomplished so far include the following. The conventional construction is 40% complete, full completion is scheduled to be July of 1989. Three prototype dipole magnets, two prototype quadrupoles and one prototype sextupole have been constructed and measured. An order for complete dipole lamination was awarded last August. One production dipole has been constructed with production lamination and tooling. Assembly of 36 dipoles has to await the complete delivery of lamination, since we perform shuffling of lamination by mixing them into 324 blocks.

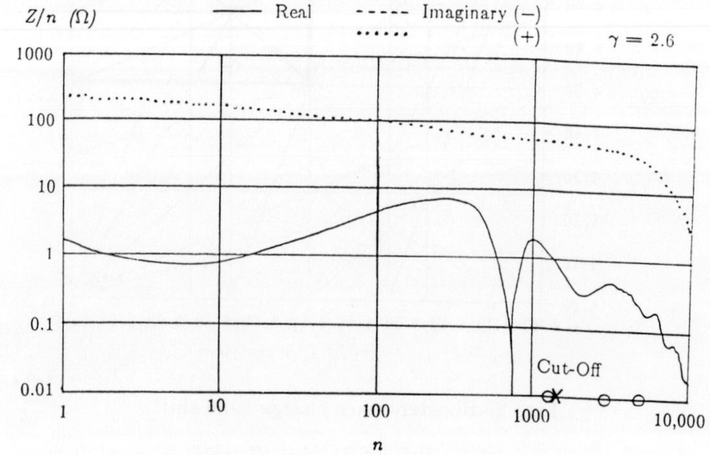

Fig. 5. Z/n versus n for $\gamma = 2.6$.

A 1/2 cell prototype assembly consisting of support base, dipole, quadrupole. sextupole, vacuum chamber and pump was assembled inside a section of mock-up tunnel. This exercise helped us to check space inside the tunnel and in between magnets, assembly procedure and tooling, support and adjustment mechanism and the vacuum performance. A picture of the whole 1/2 cell assembly is shown in Fig. 6. A vacuum test on the half cell vacuum assembly succeeded in producing vacuum in the 10^{-11} Torr range.

120 rings of Phillips 4M2 ferrite have been received. The first RF cavity and power amplifier has been completed and is being measured and tested. A prototype beam position monitor (BPM) system has been constructed and is undergoing performance test. Procurement of vacuum chamber has started and the expected delivery is January of 1989. Major expenditure on power supply, control and instrumentation will take place in FY 1989.

Fig. 6. Booster 1/2 cell assembly.

Installation is scheduled to be completed in October 1990 and commissioning will immediately follow. Direct injection of protons from linac to the AGS will continue for research program to allow for independent commissioning of the Booster until the link through the Booster is successfully established. Useful beam for proton physics can be expected in the Spring of 1991. Useful beam for heavy ion physics will be available in the Fall of 1991.

REFERENCES

1) Lee, Y.Y., *et al.*, "The AGS Booster Lattice", IEEE PAC Proc., March 1987, p. 865.

2) Danby, G., *et al.*, "AGS Booster Prototype Magnets", IEEE PAC Proc., March 1987, p. 1422.

3) Danby, G., Jackson, J., and Damm, R., "Vacuum Chamber Eddy Current Correction Coil for the AGS Booster", contributed paper to EPAC, June 7-11, 1988, Rome, Italy.

4) MacLachlon, J.A., "Particle Tracking in E-ϕ Space as a Beam Tool for Cyclic Accelerators", IEEE PAC Proc., March, 1987, p. 1007.

5) Luccio, A., Khiari, F., and Weng, W.T., "Simulation Studies of Longitudinal Beam Dynamics in the AGS Booster", contributed paper to EPAC, June 7-11, 1988 , Rome, Italy.

6) Rostamkolai, N., and Piwko, D., "LILCO-Brookhaven Synchrotron Study", GE System Development and Engineering, Schenectady, N.Y., November 1987.

PROGRESS TO OPERATING AT 100 μA ON ISIS

C.W. Planner
Rutherford Appleton Laboratory, Chilton, Didcot, Oxon., England OX11 0QX

INTRODUCTION

Over the past 16 months, from March 1987 to July 1988, considerable progress has been made in increasing the intensity of ISIS and developing the use of the facility.

ISIS is a spallation neutron source which uses a 50 Hz proton synchrotron to accelerate protons which are extracted to strike a ^{238}U target. The target is surrounded by moderators to produce thermal and epithermal neutrons. In addition a large detector (KARMEN) has been built to study neutrinos produced in the new target. Also an intense pulsed source of 28 MeV/c muons of both signs has recently been commissioned and is now operational.

The synchrotron is designed to accelerate 2.3×10^{13} protons per pulse (184 μA mean) to 800 MeV.

COMMISSIONING PROGRESS

In March 1987 the proton energy in the synchrotron was raised from 550 MeV to 750 MeV. Having installed the remaining 2 of the 6 rf accelerating cavities, the beam current was increased to 50 μA mean by the addition of feedforward on the rf system. The reduction in the beam-loading of the cavity rf voltage, in the early part of the accelerating cycle when the volts are relatively low, is shown in Fig. 1.

About two months after reaching 50 μA at 750 MeV, the beam current decreased suddenly over a very short period. An aperture restriction in the horizontal plane was subsequently diagnosed. This was due to the buckling of a stainless steel side plate of an rf shield in one of the dipoles. The plate on the outer radius was hit by the beam, probably due to mis-steering, and bowed radially inward, restricting the aperture at the input to the dipole. Accelerating beam after this damage occurred melted two holes in the 2 mm thick plate. It took approximately two weeks to replace the rf shield with a new one. The damage to the shield was confined to the side plate over a length of about a third of a metre. All the fastenings of the plate to its insulating support frames were intact and the capacitor ring, very close to the damaged area, was undamaged. After replacing the shield, the sensitivity of the beam loss radiation monitors near to the dipoles was increased to allow for the radiation shielding produced by the dipole magnet yokes.

At this time it was discovered that the injection dipoles, which produce a localized bump in the synchrotron closed orbit at injection, were also producing a fairly large closed orbit distortion around the ring. This was caused by the dipoles overheating when running at 50 Hz. This problem was alleviated temporarily by reducing the pulse length and eventually cured in early 1988 by fitting additional cooling to the ends of the injection dipole coils.

In July 1987 a muon target (2.5 mm of graphite) and beam line, which had been installed 22 m upstream of the uranium target, were successfully commissioned and brought into routine operation. The optics of the extracted beam line were changed to give a small, 20 × 30 mm, source size at the muon target but retaining the original beam diameter at the uranium target.

August 1987 saw 2.6×10^{13} protons injected and circulated. The injected current was observed to rise linearly with time and showed no sign of saturation. However, fewer than half this number could be captured by the rf, but the accelerated intensity was increased to 78 μA and the machine set up to operate routinely at 60 μA.

A short time after this the first target failure occurred. Failure is thought to result from fatigue of the Zircalloy cladding on the uranium target plates due to a large number of thermal cycles, principally as a result of beam trips. The target took four weeks to change and after about three months running at the 80 μA level this second target failed, but this time the change was made in three weeks. Besides trying to improve the overall reliability of the machine to reduce the number of beam trips, modifications were made to the beam loss interlock system and to the machine operation to reduce the thermal cycling effects. Originally beam loss and intensity monitors were set to trip the beam if a single beam pulse

a)

b)

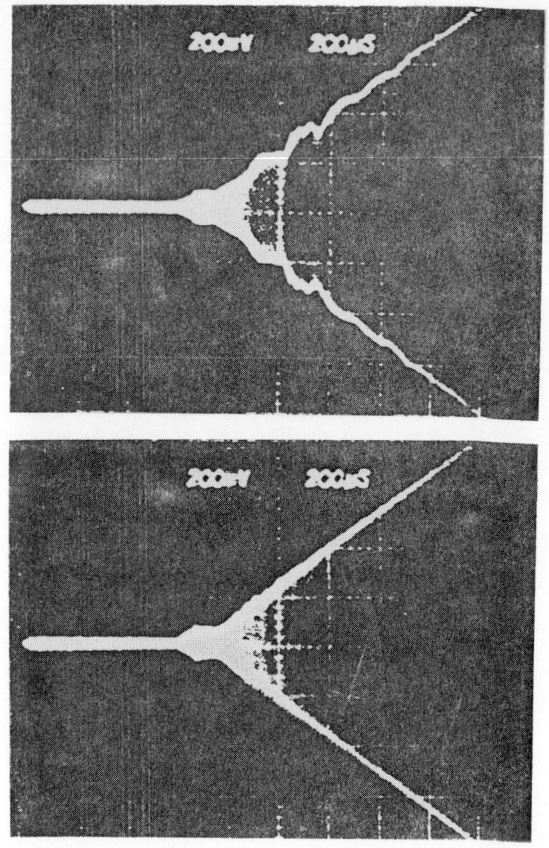

Fig. 1. (a) Beam loading of cavity voltage, (b) Compensation by feedforward system.

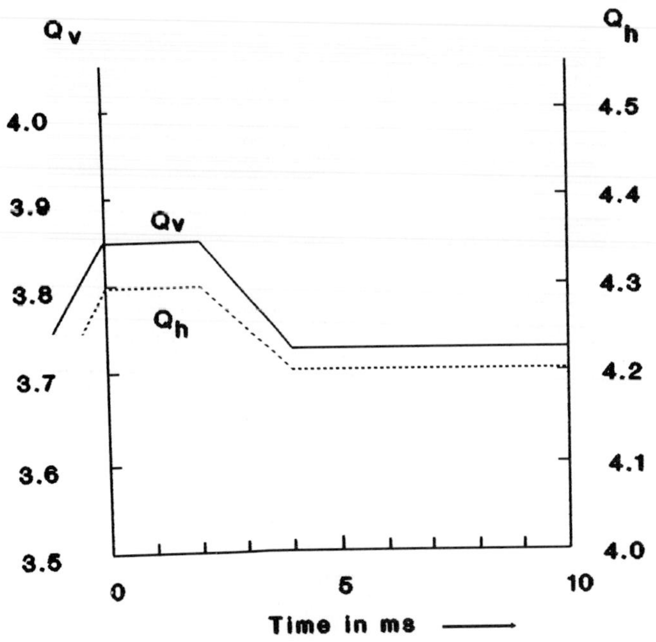

Fig. 2. Betatron tune vs time.

showed abnormal loss. This was modified to trip on the basis of a running average beam loss over a predetermined time interval. As a result the number of trips was reduced by an order of magnitude without increasing the average beam loss. Also after each trip the beam intensity is restored gradually to its previous working level.

To run at the 80 μA level required the tunes to be changed during the acceleration cycle as shown in Fig. 2. Both QH and QV are set high at the start of injection to minimise space-charge effects. They are then reduced around 2 ms into acceleration to avoid a vertical transverse instability and then kept low for the remainder of the cycle to avoid a horizontal vertical coupling resonance. Also a vertical sweeper magnet was used at injection to "paint" a large vertical emittance. The use of the sweeper magnet improved the stability of the beam as can be seen in Figs. 4 and 6.

At this time a considerable increase in the lifetime of the aluminium oxide stripping foil was made by slitting along the lower edge of the foil. The foils, 50 μg/cm^2 thick with an area 122 $\times$ 40 mm, initially had just one 122 mm edge unsupported. Lifetimes were typically 1500 μAh injected current. By slitting along the lower 40 mm edge lifetimes have increased to 10 mAh. These lifetimes are for injected current: the actual irradiation is approximately 25 times these figures due to circulation of the protons through the foil.

Longitudinal microwave and vertical transverse resistive wall instabilities have been observed, both in normal operation and in a 70 MeV storage ring mode of operation. Above 1 $\times$ 10^{12} ppp, the coasting beam at injection develops longitudinal density perturbations at 202.5 and 405 MHz, which is interpreted as "microwave" growth of the residue of the linac bunch structure, see Fig. 3.

As shown in Fig. 4, above 4 $\times$ 10^{12} ppp, the growth saturates and the damping is enhanced with the further injection of beam. The presence of a very low rf voltage on the accelerating cavities also damps the growth of the instability.

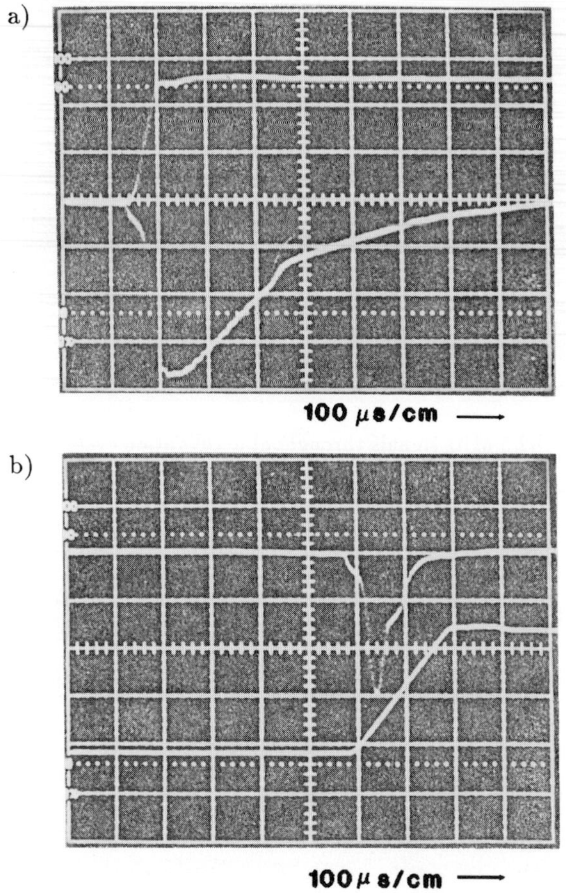

Fig. 3. (a) Upper trace 3.3×10^{12} injected, lower trace 202.5 MHz signal. (b) Upper trace 202.5 MHz signal, lower trace 13.3×10^{12} injected.

Microwave signals are also observed throughout acceleration, as shown in Fig. 5, but these are due to the non-stationary bunch distribution that arises from the fast acceleration of the beam.

When used as a storage ring, the vertical resistive wall instability appears at the lowest coasting beam mode, $(4-QV)\cdot f$, where f is the particle revolution frequency, with enhanced growth as QV approaches 4, see Fig. 6.

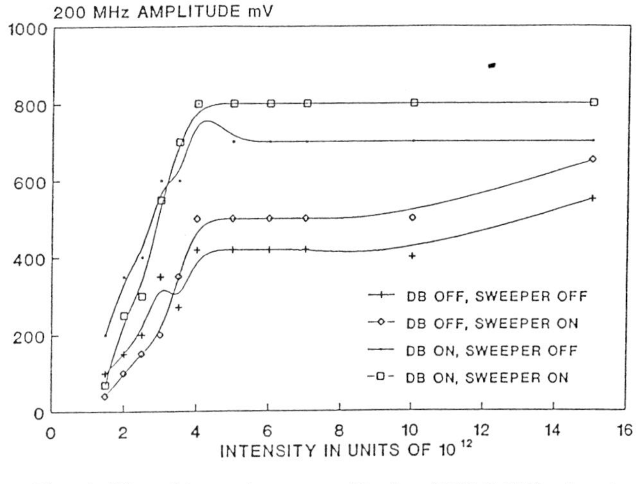

Fig. 4. Plot of intensity vs amplitude of 202.5 MHz signal.

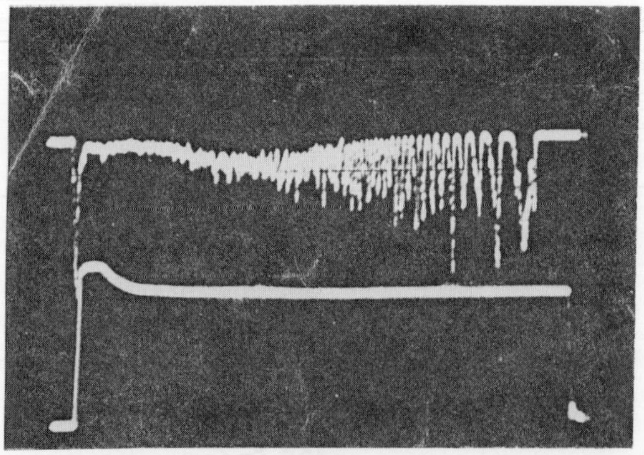

Fig. 5. 202.5 MHz signals throughout acceleration cycle. Upper trace 202.5 MHz signal, lower trace beam intensity.

a)

b)

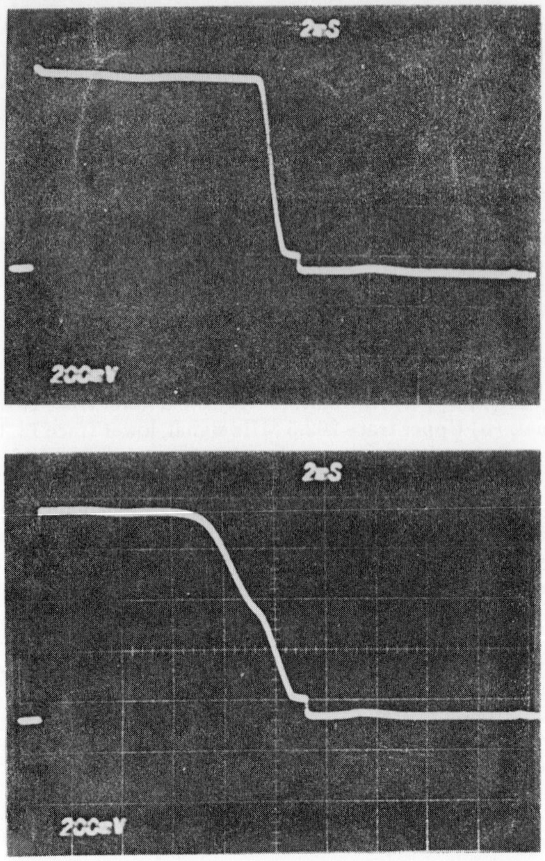

Fig. 6. Vertical transverse instability in storage ring mode. (a) Vertical sweeper off, (b) vertical sweeper on.

In synchrotron operation, the instability develops too slowly to be observed during injection, but the (4–QV) frequency is observed as a resistive wall head-tail mode in the 2 to 4 ms interval of acceleration as shown in Fig. 7. It appears with a node at the bunch centre, characteristic of head-tail mode m=1, see Fig. 8.

In July 1988, by carefully adjusting the tunes, the feed-forward system and the momentum spread at injection by means of the linac debuncher, the machine was set up to

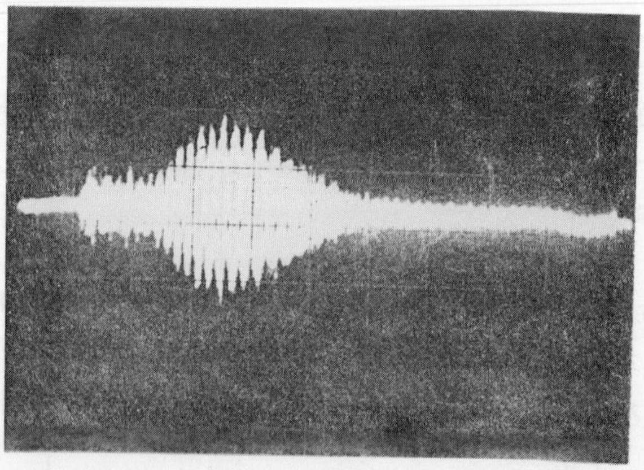

Fig. 7. Development of vertical head tail instability.

Fig. 8. Vertical head tail mode $m = 1$.

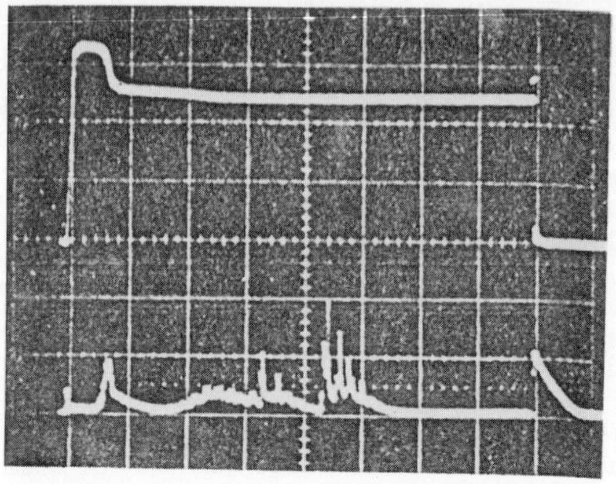

Fig. 9. Beam intensity and beam loss acceleration of 1.3×10^{13}.

operate at 100 μA. Adjustments had to be made to the position of the beam scrapers to ensure adequate machine protection. The beam loss during injection and acceleration, for 1.3×10^{13} accelerated, is shown in Fig. 9.

ACKNOWLEDGEMENTS

This talk represents a digest of the work contributed by the staff working on ISIS.

ANALYSIS OF BEAM LOSSES AT PSR*

R. J. Macek
Los Alamos National Laboratory
Mail Stop H848
Los Alamos, NM 87545 USA

Abstract

Beam losses and the resulting component activation at the Los Alamos Proton Storage Ring (PSR) have limited operating currents to about 35 μA average at a repetition rate of 20 Hz. Loss rates were found to be approximately proportional to the circulating current and can be understood by a detailed accounting of emittance growth in the two step injection process along with Coulomb scattering of the stored beam during multiple traversals of the injection foil. Calculations and simulations of the losses are in reasonable agreement with measurements.

Introduction

The Proton Storage Ring (PSR) at Los Alamos functions as a high-current accumulator to provide intense pulses of 800 MeV protons for the Los Alamos Neutron Scattering Center (LANSCE) spallation neutron source. An 800 MeV H^- beam from the LAMPF linac is converted to H^o in a high-field stripper magnet; the H^o beam then enters the lattice through a dipole and is stripped to H^+ beam with 92% efficiency in a 200 μg/ cm^2 carbon foil. Up to 2800 turns are injected and accumulated in PSR during a single macropulse. Beam is normally extracted in a single turn shortly after the end of injection.

In the present operation at 35 μA, slow losses of $\sim$0.5 μA during accumulation have caused activation at the limit acceptable for hands-on-maintenance. These losses occur principally in the injection and extraction regions which contain the known limiting apertures. Further information on the design and initial performance are published elsewhere.[1]

General Characteristics of the Accumulation Losses

Slow losses are measured to $\sim$30% accuracy with a series of scintillator-based radiation detectors located on the outside wall of the tunnel at beam height opposite each ring dipole. Detector gains are all identical; signals from each as well as a sum signal of all detectors are used for loss measurements. The sum signal is calibrated by allowing a measured quantity of beam to be completely lost. Fast analog current signals, obtained directly from the phototubes, are available in the control room for detailed analysis of the time structure.

The sum current signal is a measure of the beam loss rate, $\dot{L}(t)$. A trace from normal operation is shown in Figure 1 along with a signal from a current monitor that senses the circulating beam current, I(t). The ring current is a ramp because beam is continually injected during an injection

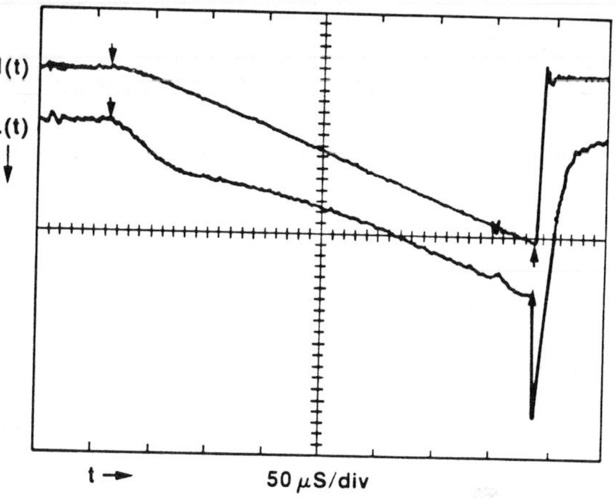

Figure 1. Loss Rate During Normal Operation

period of 375 μs. Figure 1 shows that $\dot{L}$ is nearly proportional to the stored beam intensity. Total losses, $L = \int \dot{L}\, dt$, will then be quadratic in time. To increase the average current to 100 μA we need to inject for $\sim$1000 μs, but the losses under these conditions are an order of magnitude higher than for the present operation at 375 μs which already produces the maximum acceptable activation of ring components.

Losses for an extended period of accumulation can arise from those occurring at the time of injection as well as from continual losses of the stored beam. The two components can be separated in an experiment where beam is accumulated for a short time ($\sim$100 μs) and stored for a much longer period ($\sim$1000 μs) before extraction. Loss rates and circulating current signals from one such experiment for a coasting beam (RF buncher off) are shown in Figure 2. The discontinuity at the end of injection is caused by the cessation of "first turn" or injection losses and amounts to $\sim 2 \times 10^{-3}$

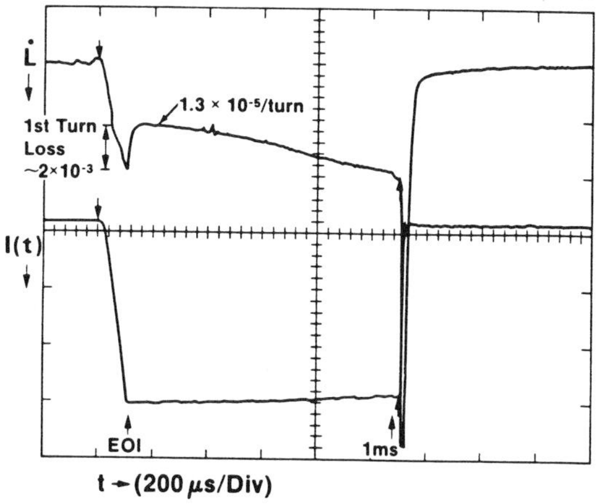

Figure 2. Loss Rate for 1 ms Storage

---

* Work performed under the auspices of the United States Department of Energy.

of the injected beam current. The loss rate during storage is a slowly increasing function of storage time with a fractional loss rate of $\sim 1.3 \times 10^{-5}$ per proton per turn at the end of the 100 μs injection period.

The significant loss observed during injection and the continual losses thereafter suggest that the injected beam somewhat more than fills the acceptance of PSR. This is corroborated by halo plate scans of the horizontal beam profile in the ring in which a thick plate is scanned across the ring aperture and the fraction of the beam intercepted by the plate is obtained by measuring the scattered beam intensity in the sum of several loss monitors. The signal is normalized to unity when all of the beam is intercepted. This technique is especially useful for measuring the beam distribution in the extremities of the beam. Data from one scan are shown in Figure 3 for the situation where beam is extracted shortly (10 μs) after the end of 100 μs of accumulation. The scan provides a good measure of the beam distribution just after capture in the ring and before foil scattering can cause appreciable emittance growth. In Figure 3 it is readily apparent that the beam distribution extends to about 38 mm which corresponds to the value of the limiting aperture defined by the extraction septum. Note that the beta functions at the septum and at the halo plate are nearly equal.

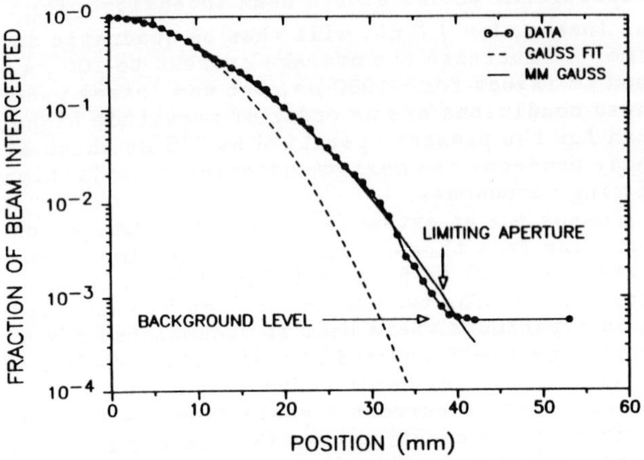

Figure 3. Horizontal Haloplate Scan

A gausian fits the core of the beam but fails to fit the extremities. The mismatched gaussian, described in the next section, fits the distribution of Figure 3 very well.

Emittance Growth and Losses During Injection

It may seem surprising that the beam fills the horizontal acceptance at injection since the acceptance of PSR, $\sim 130\pi$ mm-mrad, is so much larger than the rms emittance, $\sim 0.5\ \pi$ mm-mrad, of the H^- beam from LAMPF. The momentum spread (rms) of the H^- beam is also small with $\Delta p/p \simeq 5 \times 10^{-4}$. Two main factors contribute to emittance growth in the injection process: 1) an increase in horizontal divergence in the stripper magnet when H^- is converted to H^o and 2) a large horizontal optics mismatch of the H^o beam to the PSR acceptance.

Stripping of the H^- in a high magnetic field is a stochastic process which leads to random fluctuations in the point of conversion and thus an increase in H^o beam divergence. Calculations of the angular distribution for a pencil beam of H^- are shown in Figure 4 for two different vertical entrance positions (y = 0 at midplane and y = -4 mm). The calculation used the measured field map of the stripper magnet and a parameterization of the H^- lifetime from earlier Los Alamos work.[2] Emittance growth in the stripper magnet is minimized by use of a small gap magnet with a high field gradient at the entrance and by optics which produce a very small spot in both x and y at the stripper magnet. Even with this optimization the horizontal emittance of the H^o is three (3) times larger than that of the incoming H^- beam.

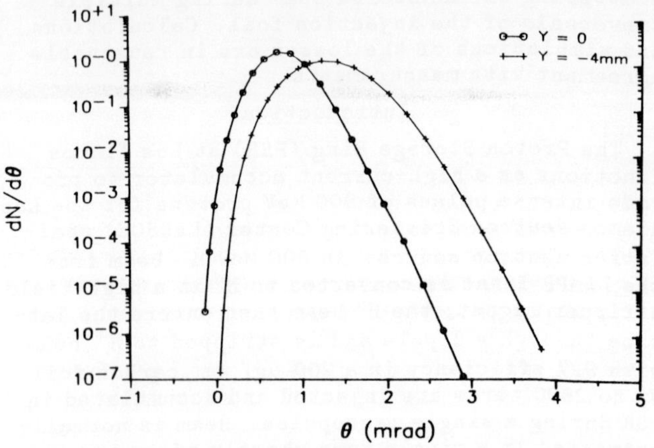

Figure 4. H^o Distribution After Stripper Magnet

An optics mismatch at injection, shown in Figure 5, is an additional consequence of magnetic stripping. The H^o is constrained to diverge from a small spot at the stripper magnet and cannot be matched simultaneously in the (X,X′) and (Y,Y′) planes at the standard location of the foil stripper where the H^o is reasonably well matched in (Y,Y′) but badly mismatched in the (X,X′) plane. The mismatch factor, $C = \{\beta_o\gamma_R + \beta_R\gamma_o - 2\alpha_o\alpha_R\}/2$, (subscript o refers to H^o and R to the ring) has a value of 3.8 indicating a further increase of 3.8 in the rms emittance of the stored beam. The mismatch also changes the beam distribution; for a Gaussian beam injected on axis, one can easily obtain the following closed form for the distribution of the invariant betatron amplitude, y:

$$P(y)dy = \frac{1}{\epsilon_o}\, e^{\frac{-Cy^2}{2\epsilon_o}}\, I_o\left(\frac{y^2}{2\epsilon_o}\sqrt{C^2-1}\right)\, y\, dy \qquad (1)$$

where ϵ_o is the rms emittance of the incoming H^o beam and I_o is a modified Bessel Function.[3] The rms emittance for this distribution is $C\epsilon_o$. This distribution has longer "tails" than a Gaussian with the same rms emittance thereby increasing still further the size of the beam near the limiting apertures. The non-gaussian tails are readily apparent at PSR in the halo plate scan shown in Figure 3 and in wire scanner profiles of the

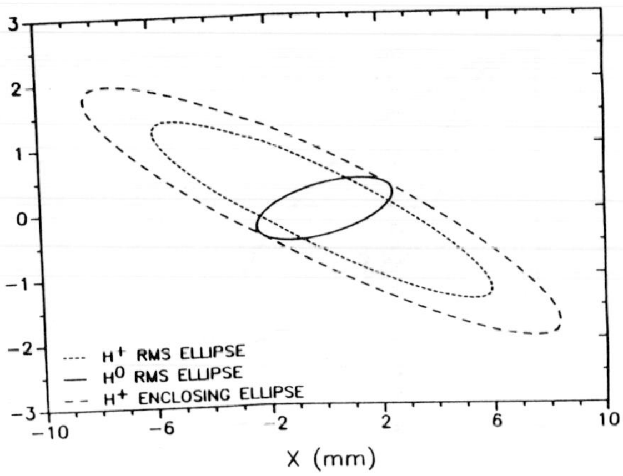

Figure 5. Injection Mismatch in Horizontal Plane

extracted beam taken at the end of short (100 μs) accumulation as shown in Figure 6. The "mismatched" Gaussian distribution derived from equation (1) fits the data very well whereas a Gaussian with the same rms width fits poorly; especially at large X.

Integration of the distribution described by equation (1) for the region outside of the limiting aperture provides an estimate of injection or "first turn" losses. A value of 2 to 4 x 10^{-3} is obtained which is in good agreement with the observed value of $\sim$ 2 x 10^{-3} considering the high sensitivity of the result to the size of the limiting aperture.

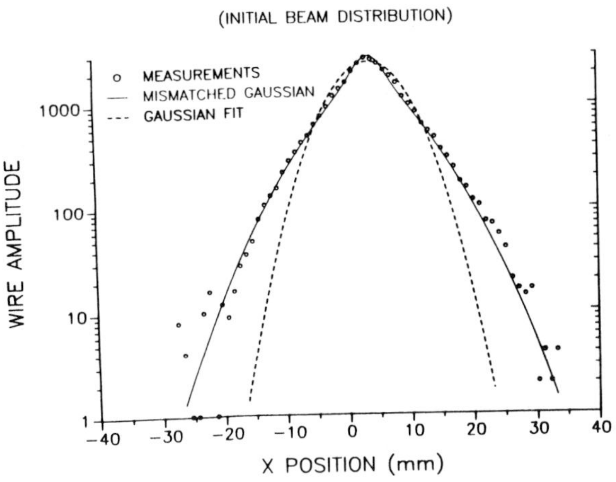

Figure 6. Extraction Wire Scanner Profile

Emittance Growth and Losses During Storage

A typical proton in PSR traverses the injection foil during about half of its revolutions. Multiple Coulomb scattering in the foil will cause emittance growth (rms) given by $\epsilon(t) = \epsilon_o + \beta_f \theta_1^2 f N(t)/2$ where ϵ_o is the initial rms emittance, β_f the beta function at the foil, θ_1 the rms scattering angle from a single foil traversal and fN(t) the number of foil traversals in N turns up to time t. The rms emittance is defined as $\sqrt{<X^2><\theta^2> - <X\theta>^2}$ where $<>$ indicates expectation value and $\theta=X'$. Measurements of beam sizes as a function of storage time are well fit by the above equation for $\epsilon(t)$ and show nearly a factor of 3 increase of emittance during 1 ms of storage.

To accurately estimate losses from multiple Coulomb scattering requires more than knowledge of the rms emittance growth; one must calculate the evolution of the distribution function with time. We have made estimates using two different methods which produce similar results in reasonable agreement with measurements. The first method used a Fokker-Planck equation (2)

$$\frac{l}{K}\frac{\partial P}{\partial t}(t,y) = \frac{\partial}{\partial y}\left[\, y\, \frac{\partial}{\partial y}\left(\frac{P(y)}{y}\right)\right] \qquad (2)$$

which was derived for estimating beam lifetimes in the presence of Coulomb scattering by residual gases.[4] Here y is the invariant betatron amplitude and K a constant. Solutions for the distribution function, P, are obtained as a Fourier-Bessel series and integrated to obtain the losses.

Similar results are obtained with the Monte Carlo tracking code, ARCHSIM, developed by one of the authors for modeling circular machines. The tracking code simulates emittance growth in the stripper magnet and treats scattering in the stripper foil as a combination of multiple Coulomb scattering with single Coulomb and nuclear tails. Results from the simulation and from solutions of the Fokker-Planck equation made using measured values of parameters in the models are compared in Figure 7 with measured data on loss rates. The loss rates obtained using solutions of equation (2) were augmented with a constant term which takes account of nuclear scattering and large-angle single Coulomb scattering. Agreement between calculations and measurements are within the calibration uncertainties of the losses and the errors on parameters in the calculations. Losses are sensitive to several parameters including the mismatch factor, H^o distribution, the probability for traversing the foil, and to the size of the limiting aperture. With small adjustments of parameters, within errors, both calculations can be made to agree completely with the loss rate data.

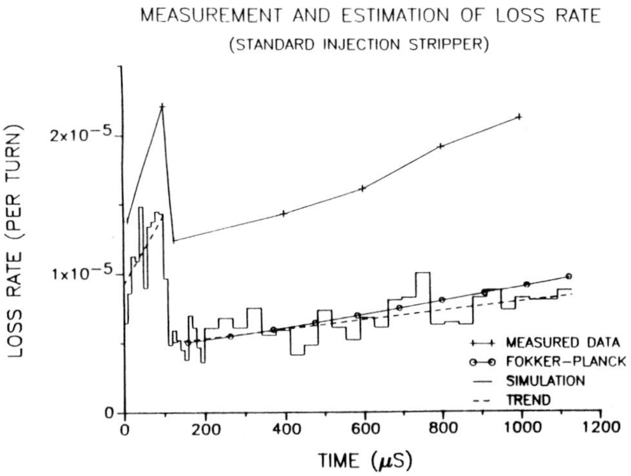

Figure 7. Loss Rate Calculations and Measurements

Effect of the RF Buncher

An RF buncher is used to maintain an empty gap to accommodate the extraction kicker rise time. Synchrotron motion induced by the buncher increases the momentum spread to ~0.3%; because of dispersion the beam size increases by several mm and also contributes to beam losses. For long storage, this shows up as a striking modulation of the loss rate with a frequency twice that of the synchrotron oscillations as shown in Figure 8. Losses are increased by about 45% when the RF is on at typical operating set points.

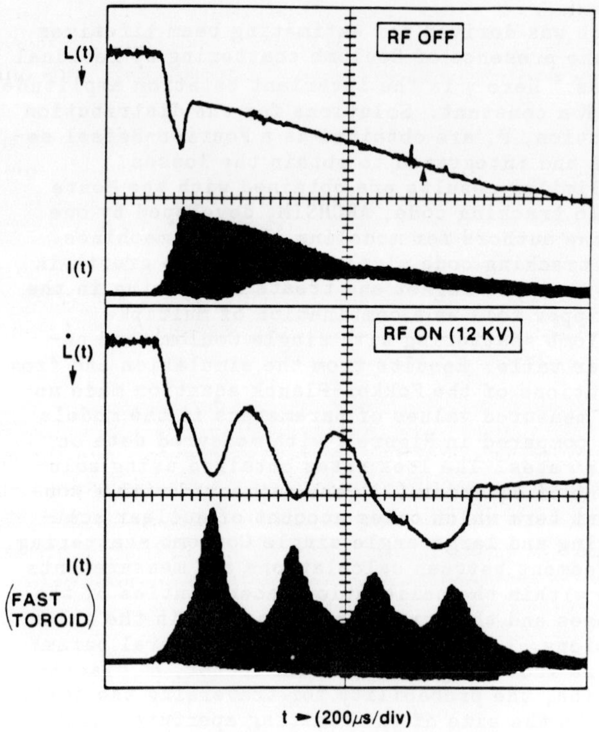

Figure 8. Effect of RF Buncher on Loss Rate

Other Contributions

Losses from nuclear scattering are readily estimated from the total cross section as 3.3×10^{-6} per foil traversal. The contribution from large-angle single Coulomb scattering is often overlooked. Because the cross section falls off only as a power law, rather than as an exponential, it contributes a long tail to any beam distribution. This was seen at PSR when the injection foil location was changed to provide a better match in the (X,X') plane. The halo plate scan of the beam distribution (after a short accumulation) plotted in Figure 9 shows the expected reduction in size of the core of the beam. Also seen is a tail extending to the limiting aperture. The size and shape of the tail agrees with analytical and Monte Carlo calculations of the contribution from single Coulomb scattering. At PSR large-angle Coulomb scattering contributes a loss rate of $\sim 2 \times 10^{-6}$ per foil traversal for a δ function initial distribution and more for a beam of finite emittance.

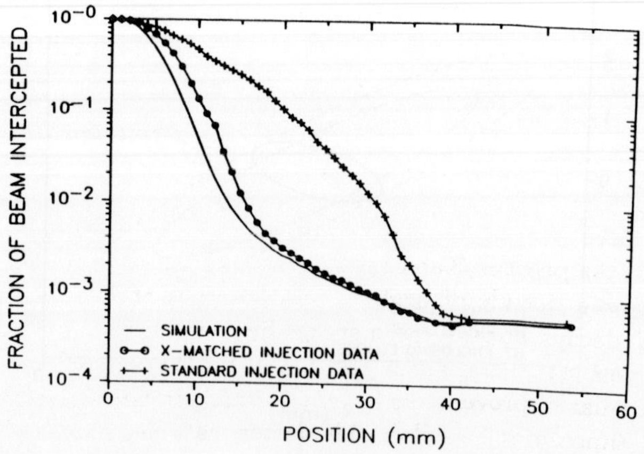

Figure 9. Evidence for Single Coulomb Scattering

Summary of Losses During Accumulation

The composition of accumulation losses during standard operation of PSR (375 μs injection period) can be determined from the data and analysis presented here. Results are listed below:

"First Turn"	0.20%
Nuclear and Large-Angle Coulomb Scattering	0.15%
Emittance Growth in Absence of RF	0.70%
Effect of RF	0.50%
Total	1.55%

Conclusions

Beam losses are the single most fundamental limitation on the average beam intensity from PSR. The origin and causes of the slow beam losses during accumulation can be understood in terms of emittance growth during injection and scattering of the stored beam during multiple traversals of the stripping foil. A clear path to reducing these losses is an injection scheme that keeps the stored beam off the foil for a larger fraction of the time. Constraints on the injection optics resulting from the use of a magnet to strip H^- to H^o greatly reduce the flexibility one has to optimize the injection process.

References

[1] G. Lawrence, "The Performance of the Los Alamos Proton Storage Ring," Proc 1987 IEEE Particle Accelerator Conference, pp. 825-829.

[2] A. Jason, D. Hudgins, and O. van Dyck, "Neutralization of H− Beams by Magnetic Stripping," IEEE Trans. Nucl. Sci., NS-28, pp. 2704-2706 (1981).

[3] R. Macek, "PSR Beam Distributions II: Mismatched Gaussian," Los Alamos National Laboratory PSR Technical Note #153, October 1987.

[4] H. Bruck, "Circular Particle Accelerators," LA-TR-72-10 Rev., Chap XIV (1972). Translation from French original "Accélérateur Circulaires de Particules" (1966).

FACILITIES ISSUES IN KAON PHYSICS

A.J.S. Smith
Princeton University, Princeton, NJ 08544

At present Kaon physics is enjoying a boom period: after a decadc of neglect, new searches at BNL for rare K decays appear to be succeeding in their goal to reach levels $\approx 10^3$ times more sensitive than their predecessors. Similar improvements have been made in CP violation experiments at Fermilab. The latter, as well as some of the rare decay experiments, will soon confront the predictions of the standard model. The advent of the AGS booster and stretcher will extend by yet another order of magnitude the window accessible to the present experiments; further down the road, the proposed "Factories" hold out the prospect of yet another order of magnitude or two. Because of all this excitement K physics has accreted a large and vibrant community. By contrast, the attendance at the K-decay session at the "Steamboat Springs" Conference in 1984 consisted of less than ten people.

A central issue discussed at Snowmass this year was the physics capability of the proposed "Factories": accessible physics window, optimal parameters (e.g., energy, current), scope of the experimental areas, detector improvements and associated R&D, perils lurking as the experiments require higher fluxes and better background rejection. This work is described in the report of the K-decay group.

It is my opinion that, although the experiments will be extremely difficult, they are sufficiently important to be vigorously pursued. Accept that a new experiment should promise an increase in sensitivity of two orders of magnitude for it to be worth while *prima facie*, and that for a smaller gain, a more explicit motivation is required. Then, comparing the capabilities of the "Boosted and Stretched" BNL AGS with the proposed "Factories", one finds only one order of magnitude increase in *current*. Furthermore, part of this flux increase may have to be sacrificed on the altar of better beam quality. On the other hand, unless the present underutilization of the AGS is redressed by increases in funding, it is quite conceivable that a two-order window will remain for the factories. Nevertheless, the above comparison should goad their proponents to do even better in increasing machine capabilities and sharpening physics motivations.

It became clear at the beginning of the conference that there was a compelling need for more interaction between the physicists designing the new accelerators and the experimenters wanting to use them. Such discussions ensued, with the following important conclusions: first, the beams at future facilities must be free of rf structure lest the instantaneous rates be prohibitive. Second, possibilities for eventual upgrades in intensity and energy should not be foreclosed by the initial design of the new accelerators .

KAON FACTORY EXPERIMENTAL FACILITIES

E.W. Blackmore
TRIUMF, 4004 Wesbrook Mall, Vancouver, B.C., Canada V6T 2A3

Working Group Members

M. Atiya	Brookhaven	T. Nagae	KEK
B. Bock	Fermilab	K. Nishikawa	INS Tokyo
D.A. Bryman	TRIUMF	J. Ritchie	SLAC
M. Cooper	Los Alamos	L. Roberts	Boston Univ.
H. Gordon	Brookhaven	T. Shinkawa	KEK
L. Littenberg	Brookhaven	A.J.S. Smith	Princeton
W. Louis	Los Alamos	H.A. Thiessen	Los Alamos
H. Lubatti	Univ. Washington	H. White	Los Alamos
K. McFarlane	Temple		

INTRODUCTION

The beam lines and experimental facilities required to support the experimental program at a kaon factory were discussed at meetings of the working group on K decays and CP violation. This group represents only part of the physics program envisaged at a kaon factory so the requirements for neutrino, antiproton and polarized beams and facilities for hypernuclear and baryon or meson spectroscopy were not considered. However, the study of rare kaon decays and CP violation is likely to put the most stringent demands on the kaon beam parameters.

The topics which were covered fall into five general categories:

- optimization of the proton energy
- requirements on the beam time structure
- design specifications: for $K^{\pm}$ and K^0 beam lines
- layout of the experimental area
- detectors and support facilities

Each of these topics will be discussed briefly and some conclusions drawn.

PROTON ENERGY

An analysis of the accelerator costs for various scenarios of proton energy and current was made at TRIUMF[1] and is shown in Fig. 1. A reasonable assumption is that the total cost of the facility, including shielding and experimental areas, will scale with energy and current in the same way. About 70% of the cost of the facility is associated with energy and 30% with current. The additional cost of a 60 GeV, 50 μA facility over a 30 GeV, 100 μA facility is estimated to be about \$180M (30%).

A proton energy of 30 GeV optimizes the cost/particle for kaon energies of 4–6 GeV while an energy of 60 GeV gives about a factor 3 more flux at 10 GeV for the same beam power (see Fig. 2).[2] The question to be answered is how essential are high fluxes of kaons or antiprotons above 10 GeV. The experience with in-flight K decay experiments at Brookhaven such as BNL E791 is that acceptances tend to peak around 8–10 GeV which is higher than the peak in the kaon production. There are tradeoffs in the detector design as a function of particle energy or momentum which are very experiment dependent. Particle identification, range, energy and momentum resolution all have to be considered. A true measure of the advantages of 60 GeV versus 30 GeV production can only come

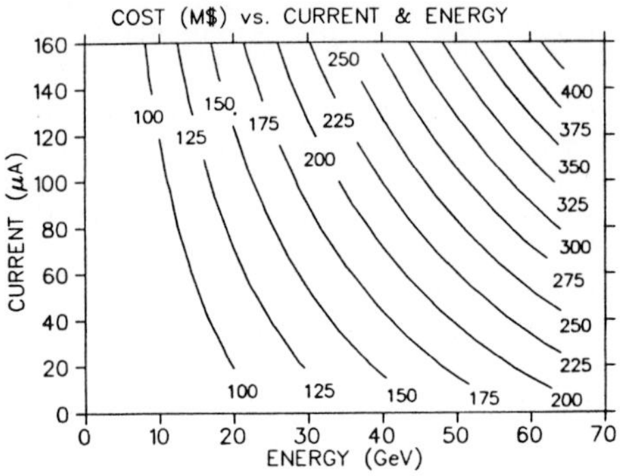

Fig. 1. Cost (M\$) vs. current and energy.

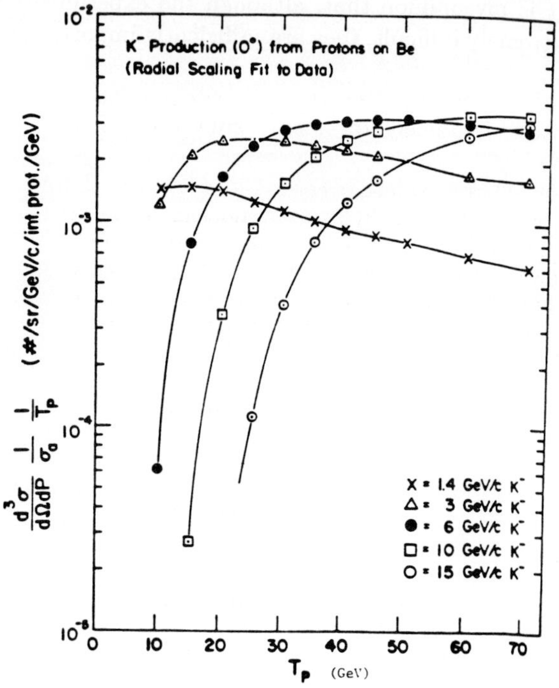

Fig. 2. K^- production from a beryllium target as a function of incident proton energy. Note that the yield is given per unit power in the primary proton beam.

Table I. Time structure considerations. Slow extracted beam parameters: 1–2 ns pulse every 16 ns or 90 ms spill every 100 ms at 60–80% duty factor.

Channel		Particle time of flight			
		e	π	K	p
700 MeV/c L=18 m	β ΔT (ns)	1 60	0.981 61.2	0.817 73.4	0.598 100.3
1.5 GeV/c L=30 m	β ΔT (ns)	1 100	0.996 100.4	0.950 105.3	0.848 117.9
2.5 GeV/c L=54 m	β ΔT (ns)	1 180	0.998 180.3	0.981 183.5	0.936 192.2

from a direct comparison of two experiments each optimized at the appropriate energy. This has not been done as yet. Higher proton energies obviously provide more flexibility for the physics program but the choice of energy is most likely to be made on the basis of the total cost of the facility and the probability of acceptance by the funding agencies.

BEAM TIME STRUCTURE

In the TRIUMF proposal the slow extracted proton beam from the Extender (stretcher) ring is available either with an rf microstructure of 1 ns every 16 ns or fully debunched with a duty cycle of 60–80%. The fully debunched beam would have a gap of 10 ms every 100 ms and an intensity modulation during the spill of less than 20%. Most kaon decay experiments would want to use the fully debunched beam to reduce random background. However, the rf microstructure has three potential applications:

- time-of-flight particle identification of π, K, p to energies up to 1.5 GeV or so (see Table I)
- permits use of an rf separator operating at the synchrotron frequency to give a high purity, low energy K beam as has been used at TRIUMF to obtain a pure muon beam[3)]
- permits use of an rf deflector operating at the synchrotron frequency to cleanly split the primary proton beam

These advantages are perhaps more relevant to non-K decay experiments such as hypernuclear spectroscopy. The conclusion is that both beam structures should be made available. This can be accomplished by changing the transition energy of the stretcher ring together with the rf programming and does not represent significant additional cost.

$K^{\pm}$ AND K^0 BEAM LINES

Table II compares the proposed charged kaon beam line parameters with those existing today at Brookhaven and KEK. Optimizing the design of the secondary channels presents the greatest dilemma in the design of the experimental areas for a kaon factory. Due to the high beam power, shielding and remote handling, the target cells will be extremely complex and difficult to modify once installed and activated. This would indicate that the channels should be made as general purpose as possible.

Table II. Kaon beam lines.

	Beam line P_{max} GeV/c	$\Delta p/p$ %	Ω msr	Length m
Proposed				
TRIUMF	K0.7	5	6	18 (dc)
	K1.5	3.8	1.6	30 (dc)
	K2.5	4	0.5	54 (dc)
	K6	3	0.16	115 (rf)
	K20	8	0.16	46 (μs)
LAMPF				
AHF	K0.8	5	5	18
	K1.5	3	2	25
	K2.5	5	1.0	35
	K6	5	0.07	75
	K35	1	0.05	130
EHF	S0.8	5	5	18
	S1.5	3	2	25
	S3	5	1	35
	S6	5	0.2	60
	S20	5	0.1	100
Existing				
Brookhaven	LESB2(0.8)	6	13	15
	LESB1(1.1)	4	2.6	15.3
	New(2.0)	6	1.3	31
	MESB(6.0)	6	0.3	81
KEK	K3(1.0)	6	7.2	14.5
	K2(2.0)	6	1.0	27.9

However, experience with the rare kaon experiments is that considerable gains can be achieved by optimizing the secondary channels or K^0 beam lines for a particular experiment. For example BNL E777 ($K^+ \to \pi^+\mu^+e^-$) is already operating above $10^7 K^+$/s at 6 GeV/c, about the same proton flux listed in the KAON factory proposal. The difference is that E777 uses a non-separated beam 5 m long while the 6 GeV/c line described in the KAON factory proposal is a separated beam line 115 m long. Although the E777 experimenters agree that their beam line is too short, it is clear that for some experiments a short unseparated beam line is desirable.

Neutral kaon experiments are especially dependent on beam line design. Flux and momentum increase with decreasing production angle but neutron backgrounds increase. At least one proton beam line dedicated to neutral K experiments is required.

LAYOUT OF THE EXPERIMENTAL AREA

On the basis of the previous considerations it seems that it will be necessary to have several slow extracted proton lines at a kaon factory to provide sufficient secondary beam lines and flexibility to meet the demands of the experimental program. Concern was expressed that the experimental hall presented in the TRIUMF KAON factory proposal was too cramped and provided too few beam lines for a facility supporting an international program in kaon physics. Experimental areas could extend up to 100 m for some in-flight decay experiments. Space must be provided for counting rooms as close to the detectors as possible but also in a low radiation environment, a considerable challenge. To allow for future flexibility the experimental hall floor should be as close to grade level as shielding permits.

The relative merits of sequential targets on a single proton beam line versus a single target feeding several secondary channels per proton beam line were discussed. Sequential targets optimize the use of protons and are cost effective but have limited flexibility. Experience at Brookhaven with sequential targets on the C-line has not been good duc to the interaction of the high momentum secondary channel on the first target with the proton transport to the second target. However, it is felt that this arrangement could be improved with better beam line design.

The conclusion is that a lot of effort will have to go into optimizing the layout of the beam lines in the experimental hall. While the selection of beam lines should be made to satisfy the demands of the initial experimental program enough flexibility will have to be built in to provide for future experiments. The complexity and radiation levels in the target areas will make them very expensive to replace or reconfigure.

DETECTOR AND SUPPORT FACILITIES

Many of the detector requirements for kaon factory experiments are common except perhaps in scale with those being developed for the SSC detectors. These include high rate capability, finer segmentation, radiation hardness and fast trigger processing. As for the SSC the time required for detector development and construction is expected to be similar to that required for the accelerator. Therefore, it will be necessary to start a program of detector R&D for kaon factory experiments in parallel with the accelerator design. Computing both on-line and off-line is a major limitation in present kaon decay experiments and will be even more of a problem with the rate increase of two orders of magnitude. Fortunately developments such as the Fermilab ACP are in progress and factors of 100 or more in processing capability are expected on the timescale of the kaon factory. However, the requirements for computing should not be overlooked or minimized in setting up the kaon factory support facilities.

REFERENCES

1) Wienands, U. and Craddock, M.K, TRIUMF internal report TRI-DN-86-7 (1986).

2) Thiessen, H.A., in Intersections between Particle and Nuclear Physics, Steamboat Springs (AIP, New York, 1984), p. 176.

3) Blackmore, E.W. *et al.*, Nucl. Instrum. Methods A234, 235 (1985).

K-FACTORY ARCHITECTURE

M.K. Craddock*
TRIUMF, 4004 Wesbrook Mall, Vancouver, B.C., Canada V6T 2A3

Working Group Members

C. Ankenbrandt	Fermilab	E.G. Raka	Brookhaven
R. Baartman	TRIUMF	G.H. Rees	Rutherford
P. Blüm	Karlsruhe	T. Suzuki	KEK
H.A. Butler	Los Alamos	L.C. Teng	Fermilab
E.G. Colton	Los Alamos	H.A. Thiessen	Los Alamos
Y.Y. Lee	Brookhaven	W. Van Asselt	Brookhaven
D. Neuffer	Los Alamos	W.T. Weng	Brookhaven
C.W. Planner	Rutherford	U. Wienands	TRIUMF

ABSTRACT

The group considered three major topics; the optimum size for the Los Alamos booster, the pros and cons of a separate stretcher ring, and the design of suitable magnet lattices.

1. DISCUSSION

1.1 Booster size

The AHF is considering a "menu" of four alternative booster options. Three are based on synchrotrons – 10 GeV the same size as the main ring, 10 GeV half-size, and 6 GeV quarter-size. The fourth option is based on a 2.2 GeV booster linac. (All options assume a new 0.8–1.6 GeV linac; in the last case the linac would be extended a further 0.6 GeV.) Table I lists the major parameters for each of the options.

Table I. AHF Booster Menu

	Main Only	Full-size Booster		1/2-size Booster		1/4-size Booster	
		Boost	MR	Boost	MR	Boost	MR
Rep Rate (Hz)	6	6	6	12	6	24	6
np (x10**13)	2.6	2.6	2.6	1.3	2.6	0.65	2.6
Inj E (GeV)	2.2	1.6		1.6		1.6	
Energy (GeV)	60	10	60	10	60	6	60
nux	13.2	13.3	12.4	14.2	12.4	7.2	12.4
nuy	12.7	12.7	14.2	12.2	14.2	6.2	14.
Gamma T	22i	22i	8	∞	8	∞	22i
εn (π mm-mrad)	10.5	18	18	9	9	2.7	2.7
long (eV-s)	.14	.10	.10	.10	.10	.10	.10
Volts/turn (MV)	7.7	2.8	6.9	1.3	6.9	0.7	7.1
rf power (MW)	6.9	1.3	7.9	0.5	6.0	.26	6.4
df/f (%)	4.8	7.2	0.4	7.2	0.4	6.6	0.9
Sync Tune	0.06	0.07	0.009	0.04	0.009	0.02	0.01
Z/n (max)(Ohm)	71	98	-10	103	-10	154	-10
Collector	None	None		1/2 - Size		Full Size	

Injecting into the main ring straight from a 2.2 GeV linac avoids building and maintaining any subsidiary rings. It also eliminates the need for a group of empty buckets for kicker rise or fall and the danger of beam instabilities driven by them. On the other hand the magnet aperture required in the main ring becomes relatively large and the frequency swing there rises to 4.8%. The full-size booster avoids the cost of a separate booster tunnel and offers opportunities for energy sharing between the main and booster magnet systems. However, it requires the largest magnet apertures and the largest booster rf voltage. The half- and quarter-size boosters make possible smaller magnet apertures in all rings and themselves require lower rf voltages, but necessitate the use of dc collector rings. The quarter-size booster probably offers the lowest construction cost, but Los Alamos anticipates that this would be outweighed in the long run by the additional operating costs for the extra ring. A careful cost analysis is required for a full comparison of the options.

1.2 Stretcher Ring or Flat-top?

A separate stretcher ring makes it possible to separate the design requirements for ultra-efficient slow extraction from those of a fast-cycling accelerator. Thus clean extraction requires magnets stable to 10^{-5} rather than cycling over a 6:1 range within ~100 ms. Moreover, there is the possibility of physically separating the extraction region and shielding the accelerator from it. On the other hand an extra ring introduces extra components and extra capital and maintenance costs. The arguments were put as cogently as usual by both proponents and opponents of stretchers, but resulted in few, if any, conversions. More to the point, perhaps, the LAMPF group is planning to build a fast-cycling flat-topped power supply to determine what stability can be achieved.

1.3 Lattice Design

Both TRIUMF and LAMPF have now adopted racetrack lattices of the type proposed by Servranckx for their main rings. These have at least four identical sections in their arcs, which if tuned to an integer give dispersion-free straights. Then offer considerable flexibility, permitting tune changes by ±1 in each plane, and either high or low γ_t. Initial tracking studies show that the emittance growth produced by field errors is similar to that for circular lattices. Chromaticity correction schemes need to be studied. For polarized beam they can be tuned to give spin-transparent straights, and thus the appearance of a highly periodic lattice with fewer strong depolarizing resonances. Tracking tolerances for the separate families of quadrupoles in the arcs and straights need to be investigated. Preliminary work by Servranckx and Wienands on racetrack booster lattices is reported in a contributed paper.

*On leave from Physics Dept., University of British Columbia, Vancouver, B.C., Canada V6T 2A6.

EFFICIENT SLOW EXTRACTION

U. Wienands
TRIUMF, 4004 Wesbrook Mall, Vancouver, B.C., Canada V6T 2A3

Working Group Members

R. Baartman	TRIUMF	Y.Y. Lee	Brookhaven
P. Blüm	KfK	A. Pisent	Los Alamos
H. Butler	Los Alamos	T. Suzuki	KEK
M. Craddock	TRIUMF	H. Thiessen	Los Alamos

The group discussed and reviewed recent work done on the slow extraction systems for AHF, EHF, and KAON. Common features for all systems are use of a racetrack lattice with straight sections >130 m long, moderate to high beta function (100–500 m), and a tune variation of 0.01–0.02 for the extraction cycle.

For AHF an investigation demonstrated that the relatively high sextupole content of the 2.2 T magnets (35×10^{-4}) does not imperil the extraction. A Simpson-rule correction scheme with discrete sextupoles can reduce distortions by an order of magnitude. Half-integer extraction is planned for the main accelerator in order to reduce operating costs, limiting the duty factor to below 50%.

An EHF study indicated an extraction efficiency of 99.3% using one electrostatic septum of 4.8 m length and 100 μ thickness and a magnetic septum. A dedicated stretcher in a separate tunnel allows to separate the high-radiation area from the rest of the accelerator complex. Half-integer extraction is used, which is of advantage for loss collimation.

An investigation of half-integer extraction for KAON revealed a possible difficulty to achieve the low extracted emittance necessary for losses at the 0.1% level. Using an electrostatic pre-septum of 0.5 m length and 10 μ thickness in addition to electrostatic and magnetic septa, 0.2% losses are obtained in simulation runs using third-integer extraction.

A report from BNL indicated that stability of the machine is an important issue in the performance of the slow extraction system. With a third-integer system, about 1% septum losses are achieved, but several per cent of the beam are not extracted because of the vanishing stop-band width and have to be dumped on an internal target.

Apart from the reviews the group discussed the following topics:

Half- vs. Third-Integer Resonant Extraction

Half-integer extraction is preferred from the point of view of loss-collimation and being able to empty the ring completely. More simulation work is needed to determine extracted emittance and extraction losses. Chromatic extraction is difficult at best for half-integer extraction.

Pre-septum

EHF and possibly AHF propose a system without pre-septum giving losses on the order of 1%. In this case collimation has to reduce distributed losses by 10^{-3}–10^{-4} in order to keep those losses at or below 0.1 nA/m. High radiation doses for the extraction septa is the drawback of such a system.

Space Charge

Local incoherent space charge tune shift on the order of 10^{-3} at 30 GeV (10% of tune range for extraction) can lead to some emittance blowup and thus increase in losses due to synchrotron oscillations. A simulation for KAON indicates the effect to be of the same order of magnitude as other sources of emittance increase.

Debunched Extraction

This item was discussed at length with the group being augmented by several members of the rf and the instabilities working group. A bunch rotation scheme is preferred to debunching since in the latter case microwave instability can develop even after 10 debunching times due to the very narrow local momentum spread in the beam filaments unless the longitudinal impedance is assumed to be unrealistically small. For a rotated bunch, a momentum spread of 0.05% results in an impedance limit from the Keil-Schnell criterion of <10 Ω with γ_t being 10 and a longitudinal emittance of 0.2 eVs (values for KAON E-ring). Thus an extracted beam with >50% micro duty factor can be delivered. Since bunched chromatic extraction requires natural chromaticity of the machine, γ_t should be imaginary in this case for stability. This makes variability of γ_t a very desirable feature in the stretcher lattice.

POLARIZED BEAMS

P. Blüm
Institut für Experimentelle Kernphysik, Universität Karlsruhe

Working Group Members

H. Butler	Los Alamos	A. Pisent	Los Alamos
M. Craddock	TRIUMF	L. Teng	Fermilab
Y.Y. Lee	Brookhaven	U. Wienands	TRIUMF

In this session the plans of proposed polarized beam facilities within the different Kaon factories have been reviewed, and particular items like small aperture Siberian snakes or partially energized snakes have been discussed. The agenda and the highlights as perceived by the chairman are given below. It is a pleasure for me to acknowledge the contributions of all members of the working group.

Agenda

P. Blüm	"Plans for polarized beams at EHF"
U. Wienands	"Strategy to preserve polarization at KAON"
A. Pisent	"Siberian snakes with small orbit excursions at low energy"
Y.Y. Lee	"Remarks on polarization at AGS"

Round table discussions on particular problems

Summary

A modern hadron facility like EHF, KAON, and AHF should have the capability to provide polarized beams up to the maximum energy. As a matter of fact each of the three facilities aims for a polarized beam operation mode. As these accelerators will operate at a high repetition rate they have the inherent advantage of a much faster passage through a resonance condition so that weaker depolarizing effects develop. In addition the design from scratch on permits an optimum layout for polarized beam operation. The depolarizing effects can be analyzed, and well adapted correction schemes can be selected. It is possible to use as a global cure of depolarizing resonances a Siberian snake scheme, which makes the spin tune independent of energy. Thus no resonance condition will be encountered.

While each of the three facilities proposes to use conventional correction methods in case of the Booster, i.e.

- fast tune jump or spin flip for the intrinsic resonances; $\gamma G = kP \pm Q_y$
- harmonic correction of the driving perturbing field components for the imperfection resonances; $\gamma G = k$

(G is the gyromagnetic anomaly of the particle, k is an integer, and P is the periodicity of the lattice), different strategies are followed in the Main rings.

The EHF has been optimized for the use of a pair of Siberian snakes with orthogonal precession axis of the spin. Then the vertical spin component is preserved, and is independent of energy along the circumference. In order to keep the required orbit excursions inside the snake in reasonable limits, a transfer energy of 9 GeV from the Booster to the Main Ring has been chosen. The dispersion-free straight sections provide an element-free space of 16.2 m long enough to accommodate the Siberian snakes. A variety of snakes have been studied,[1] and can be used in the scheme.

In case of the TRIUMF Driver an interesting strategy is proposed.[2] The long dispersion-free straight section of the race-track lattice, mainly motivated to enable a highly efficient slow extraction, may be tuned to be transparent, i.e. the tune of a straight section is an integer multiple of 2π. Thus the basic lattice periodicity is given by the periodicity of the arcs, and is much larger than two. Therefore only few resonances have to be cured by means of a fast Q-jump or an adiabatic spin flip. The imperfection resonances, however, have to be cured by a harmonic matching procedure.

From the different proposed scenarios of the AHF the one ring scenario has been discussed. The injection energy of 2.2 GeV into the circular accelerator leads to the expectations of large orbit excursions particularly at an energy of 2.2 GeV. A. Pisent described the possible use of a helical, multi-twist snake, which requires only small apertures even at injection energy to accommodate the orbit excursions.[3] This snake can be approximated by a sequence of lumped dipoles or a stack of rotated iron laminations. Thus orbit excursions of about 7 cm at 2.2 GeV and 2 T field are achievable. The total length is of the order of 15.3 m.

A further reduced magnet aperture requirement is possible by the use of a partial spin rotator, i.e. the spin precession angle is $a\pi$ with $a<1$, at low energy. As long as the strength is weak ($|\epsilon| << a/2$) a spin rotator operating at reduced strength is sufficient to prevent a resonance condition. As the orbit excursion scales as "a" it will not be excessively large. In order to cope with stronger resonances at higher energies the rotator magnets will be ramped up to provide a spin precession angle of π. This method has still to be carefully checked with the help of spin tracking codes.

1) Blüm, P. and Pisent, A., Proc. of European Particle Accelerators, Rome, 1988.

2) Wienands, U., Proc. of Accelerator Design Workshop, Los Alamos, 1988

3) Pisent, A., AHF Technical Note 88-015.

RF TECHNOLOGY

J.E. Griffin
Fermi National Accelerator Laboratory, Batavia, IL 60510

Working Group Members

R. Baartman	TRIUMF	F. Pedersen	CERN
R. Burge	TRIUMF	R. Poirier	TRIUMF
M. Craddock	TRIUMF	E. Raka	Brookhaven
T. Enegren	TRIUMF	R. Smythe	Univ. of Colorado
C. Friedrichs	Los Alamos	G. Swain	Los Alamos
V. Hart	Los Alamos	H. Thiessen	Los Alamos
S. Koscielniak	TRIUMF		

INTRODUCTION

Discussions centered on three main topics: the status of orthogonal biassed microwave ferrite in cavity tuning systems, requirements and techniques for damping spurious resonances in cavities, and the mechanical structural design of the proposed Los Alamos "Main Ring" rf cavity. In addition, several status report papers were presented.

ORTHOGONAL BIASSED MICROWAVE FERRITE

Because of the substantially larger "quality factor" of orthogonally biassed microwave ferrite it appears attractive to use this material for tuning high gradient accelerating cavities in KAON factory rings, where substantial tuning is required. The salient point is that the substantially increased accelerating gradient (over rf structures tuned with low Q Ni-Zn ferrite) allows a reduction in total number of cavities and consequently the total ring impedance introduced by the rf system. Results of high power tests of the Los Alamos "Booster" cavity were discussed. Several fractions of ferrite have been observed. In at least one case the damage has been traced to excessive heating of nearby conducting metal rather than to heating resulting from ferrite losses.

In addition to the improvement in Q, the use of orthogonally biassed ferrite creates the potential for improved tuning system bandwidth. The conducting material containing the ferrite may be slotted in such a direction as to allow bias field flux into the ferrite without creating eddy currents and with minimal effect on the internal rf currents. Unfortunately, the substantial bias fields required present difficult problems for the biassing system. The potentially available bandwidth improvement may be lost due to large inductance of the bias windings necessary to reduce current demands on the bias power supply to a reasonable value. RF cavities built at Atomic Energy of Canada Laboratory, Chalk River, for accelerating protons in the PETRA ring at DESY employ orthogonal bias ferrite. In that case the tuning rate is in fact extremely slow due to the tuner inductance.

SPURIOUS MODE DAMPING

A goal of less than 10 kΩ was set for the total real component of impedance introduced by all rf cavities by any and all spurious resonances. The possibility of tuning individual cavity spurious resonances so that they are not superimposed was introduced. Concern was raised that excessive damping of spurious resonances may raise the real component of ring impedance to an unacceptable level at frequencies far removed from the resonances because of the "resistance integral theorem" which states that the integral of the real component of impedance of a (lumped) resonant circuit over the entire spectrum is constant. RF accelerating cavities are almost always distributed resonators and it is not clear that the resistance integral theorem applies over the entire frequency spectrum. Certainly a lumped approximation is valid near any resonance but reduction in the amplitude of the real component of impedance at several resonances may not result in a cumulative increase in the real component at frequencies far from any resonance. This question requires further investigation.

Specific techniques for the introduction of mode damping probes into the Los Alamos cavities were discussed. It was proposed that a scale model prototype of the proposed cavity design be built to investigate spurious modes.

LOS ALAMOS MAIN RING CAVITY STRUCTURES

It appears reasonable to build the large outer cavity structure from cast aluminum. Details of introduction of the inner conductor and the location of rf joints and vacuum seals was extensively discussed.

SHORT REPORTS PRESENTED

- G. Swain, Progress Report on the Los Alamos Main Ring Cavity
- T. Enegren, Calculation of EM Fields in Booster Cavity Ferrites
- R. Smythe, Calculations and Measurements of the Los Alamos Booster Cavity Q
- R. Burge, Phase and Amplitude Control Loops, Los Alamos Booster RF System
- S. Koscielniak, Longitudinal Phase Space Simulations of the Robinson Instability
- C. Friedrichs, Status Report on Ferrite Heating on the Los Alamos Booster Cavity
- F. Pedersen, Cautionary Note on Stability of Various Particle Distributions Resulting from Painting Injection in KAON Factories

BEAM STABILITY

R. Baartman
TRIUMF, 4004 Wesbrook Mall, Vancouver, B.C., Canada V6T 2A3

Working Group Members

J. Griffin	Fermilab	T. Suzuki	KEK
S. Koscielniak	TRIUMF	T.-S. Wang	Los Alamos
F. Pedersen	CERN		

Among the impedance sources in kaon factory synchrotrons, the major concern is the parasitic modes of the rf cavities. If undamped, these impedances will be a much larger problem in kaon factory synchrotrons than in the existing 50 MHz machines. There are two reasons for this: (a) larger circulating currents (3 A in the TRIUMF proposal) and (b) more structure in the circulating beam. The structure arises because ~ 5 rf buckets are left empty to make possible clean injection and extraction. The result is that the parasitic modes, if they are to be safe, must have impedances $1\frac{1}{2}$ to 2 orders of magnitude smaller than the impedance of the fundamental mode. Since there are on the order of 20 modes between the fundamental mode and the cut-off frequency of the beam pipe, the best approach is to damp all by exploiting some feature of the fundamental. For example, one can envision a pickup, capacitively coupled to the cavity gap and connected to a damping resistor through a high pass filter. However, damping of parasitics is not the only possibility: it may be possible to jump resonances by changing either the beam frequency or the parasitic frequency by half the revolution frequency on a time scale which is short compared with the instability growth time.

A few uncertainties about the existing instability theory underline the need for experimental work in this area. The naïve approach of multiplying growth rate at resonance ($Im(Z) = 0, Re(Z) = R_s$) by the time during which the instability mode is within the bandwidth of the parasitic may be insufficient. Also, it is not clear what will happen when both the beam's Fourier component $\omega = n\omega_0$ (which, due to the kicker gaps, contains up to a few hundred mA) and the instability mode $\omega = n\omega_0 - \omega_s$ are together inside the bandwidth of the parasitic mode. It is the recommendation of this group that these questions be studied in the PSR during the 50 MHz cavity tests. Resonance passage can be mimicked by sweeping the parasitic resonance frequency rather than the beam frequency.

It has become clear that certain classes of experiments require debunched beam. The implication of this requirement in terms of the beam pipe impedance in the final ring have been investigated and are treated in a separate paper (Debunched Slow Extraction from the KAON Factory Extender Ring – U. Weinands and R. Baartman, these proceedings). It was found that an additional blow-up of $\delta p/p$ can be avoided if γ_t can be made small enough. A value of 10 for γ_t is allowable for the TRIUMF stretcher; in the case of the AHF design where there is no separate stretcher and γ_t =20i, blow-up is unavoidable unless $| Z_{\|}/n | \lesssim 1\Omega$ can be achieved.

Simulations show that beams which are hollow in longitudinal phase space are non-stationary and perhaps unstable. (See "Stability of Hollow Phase Space Distribution" – S. Koscielniak, these proceedings.) In the simplest injection schemes, such a beam occurs at some point in the accumulation cycle. Instability occurs if either there is sufficient longitudinal space charge force or if there exists an unsophisticated phase feedback loop to lock the beam to the rf cavity. Analytical work is under way to explain these effects.

INJECTION AND PAINTING

G.H. Rees
Rutherford Appleton Laboratory, Chilton, Didcot, Oxon., England OX11 OQX

Working Group Members

R. Baartman	TRIUMF	D. Neuffer	Los Alamos
E. Colton	Los Alamos	C. Planner	Rutherford
M. Craddock	TRIUMF	A. Thiessen	Los Alamos
S. Koscielniak	TRIUMF	W. van Asselt	Brookhaven
M. Masullo	Univ. di Napoli (EHF)	W. Weng	Brookhaven
		U. Wienands	TRIUMF

1. INTRODUCTION

The working group met for two afternoon sessions on 12 and 13 July with the first session devoted to review talks and the second to discussions. Review talks on general Kaon Factory topics had been given earlier on the afternoon of 11 July.

Review talks included:

R. Macek	'PSR Developments and Plans'
E. Colton	'Injection for an AHF 1.2 GeV Racetrack Compressor Ring'
M. Masullo	'Transverse Painting for the EHF'
C.W. Planner	'RAL and TRIUMF Injection Schemes'
S. Koscielniak	'Tracking for Hollow Longitudinal Distributions together with Beam Phase Control'

Discussions concentrated on the following topics:

Injected beam distributions
Optimised injection to minimise stripping foil traversals
The use of corner foils or foils with two free edges
A review of H^- injection schemes
Injection for racetrack lattices (E. Colton and U. Wienands)

A summary of relevant H^- injection schemes is given in Table I and points raised in the discussions are given in Secs. 2 to 5.

2. INJECTED BEAM DISTRIBUTIONS

The aim is to obtain approximately stationary distributions at the end of injection without an enhanced transverse central density. Since there is a constant transverse density in a stationary K-V distribution, its characteristics are of interest. Each particle in this distribution has a constant total horizontal and vertical betatron energy. Thus, large horizontal and small vertical betatron oscillations are correlated and *vice versa*, as are intermediate amplitudes in both planes. Such correlations lead naturally to an elliptical cross-section beam and are the basis of some H^- injection schemes, see Table I.

At RAL, the correlation is obtained by the use of a programmed vertical steering magnet in the injection line and a fixed horizontal orbit bump in the ring. Injection occurs while the guide field in the ring is decreasing towards its minimum value. Initially, there are small horizontal and large vertical betatron amplitudes and finally, the reverse. An alternative to the vertical injection steering is a programmed vertical orbit bump in the ring. The AGS booster plans to use this arrangement.

The H-V correlation may also be achieved by skew-quadrupole coupling. This is the basis of the H^- injection schemes in the EHF, AHF and PSR2 proposals and for one of the PSR1 improvements. Vertical filling occurs via the coupling and there is no need for a vertical orbit bump. A fixed horizontal orbit bump gives the condition of constant

Table I. H^- injection schemes.

	Foil Type Frame	Long. Ramp	Missing of Foil	H-V Coupling	H^- orbit Bump	Vertical Filling	Transv. Dist.[a]	H-L Dist. Correla-tion	Disper-sion
TRIUMF	3-sided	ΔE	H,L	no	ramped	ramp & steering	~CTE	yes	finite
	2-sided	ΔE	H,V,L	no	ramped	(ramp & steering)	~CTE	yes	finite
EHF	3-sided	ΔE or $\Delta\phi$	H	yes	fixed	coupling	~CTE	no	finite
AHF	3-sided	ΔE or $\Delta\phi$	H	yes	ramped	coupling	~CTE	no	small
PSR2	3-sided	ΔE or $\Delta\phi$	H	yes	ramped	coupling	<>VTE	no	small
PSR1	3-sided	no	H	no	none	mismatch	VTE	no	finite
	corner or massless	no	H,V	yes	fixed	coupling	?VTE	no	finite
AGS B	3-sided	no	H	no	ramped	ramp	~CTE	no	finite
RAL	3-sided	no	H	no	fixed or ramped	prog. steering	~CTE	no	finite

[a] ~CTE = constant transverse energy, VTE = variable transverse energy

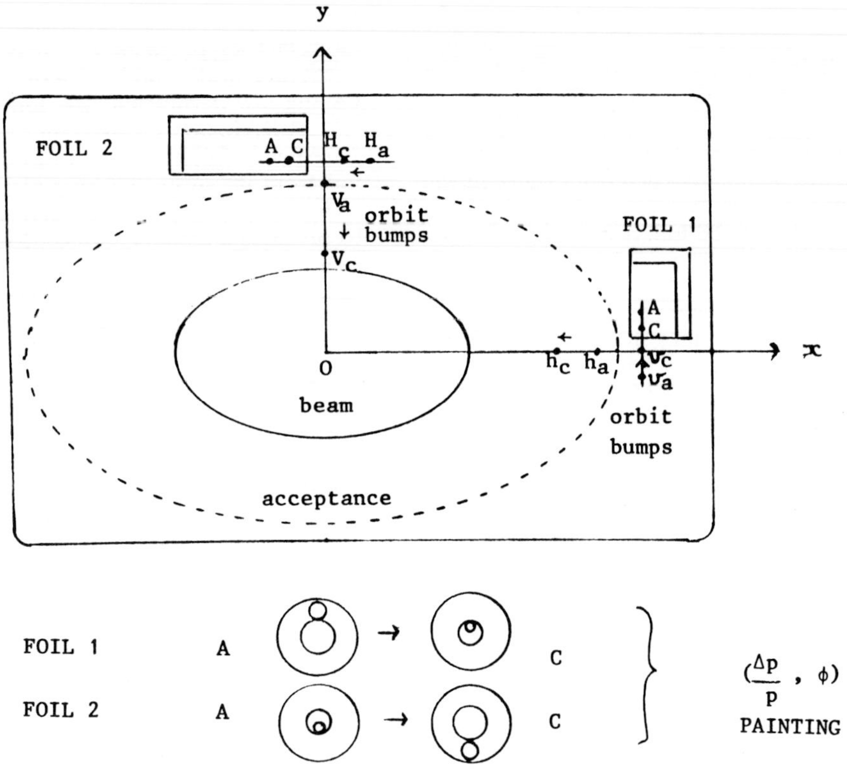

Fig. 1. H^- injection using a foil with 2 free-edges.

transverse energy (CTE) while a ramped bump gives a variable transverse energy (VTE).

All schemes have injection from a H^- linac into a synchrotron apart from at TRIUMF where injection is from a H^- cyclotron into an accumulator ring, with a large number of injected turns. All, apart from RAL, propose to paint in longitudinal as well as in transverse phase space. The rings with low-frequency rf systems, AGSB, PSR1 and PSR2, use chopped injected beams for longitudinal painting via the subsequent filamentation. The rings with high-frequency rf systems, TRIUMF A, EHF and AHF, propose energy or phase ramping of the injected microbunches (either 1 or 2 per bucket). The desired longitudinal distribution is the Hofmann-Pedersen local elliptical energy distribution of approximately parabolic line density.

An important feature of the TRIUMF A scheme is the use of the longitudinal motion of the injected protons to help miss the stripping foil on subsequent rotations. This requires dispersion at the foil and an appropriate ramping of the energy of the injected beam. The ramping allows some correlation to be introduced between the longitudinal and transverse motion of particles. Thus, for horizontal injection, with programmed horizontal and vertical orbit bumps, injection may start with small horizontal and large vertical betatron motion and end with the reverse. If the momentum of the injected beam is then ramped from a large offset to a small one, large synchrotron oscillations are correlated with small horizontal betatron motion and *vice versa*. Initially, there results a hollow longitudinal $(\Delta p/p, \phi)$ distribution. As injection proceeds, the centre of the distribution is gradually filled, but it is only at the end of the filling that there may result a stationary longitudinal distribution.

An important question is what happens to a hollow non-stationary distribution. Tracking studies for a symmetrical annulus in longitudinal phase space indicate that the distribution remains as an annulus for many revolution periods but then develops an asymmetry which increases progressively. It has not been established whether this is the progressive evolution of a perturbed non-stationary distribution or a space charge induced instability.

Tracking studies for the case where a beam phase control loop is in operation and there is an initial hollow longitudinal distribution indicate a more rapid evolution of asymmetries. It has not been established if the enhanced effect persists if a symmetrical limiting detector is used in the phase loop.

3. OPTIMISED H^- INJECTION TO MINIMISE STRIPPING FOIL TRAVERSALS

The scheme as envisaged for TRIUMF A appears to be the optimum. The lattice parameters and steering conditions to be satisfied at the stripping foil are:

$$\alpha_h x + \beta_h x' = 0 \quad \alpha_v y + \beta_v y' = 0 \text{ and } \alpha_h \alpha_p + \beta_h \alpha_p' = 0$$

for all x and y where x, x' and y, y' are co-ordinates of the input beam centre relative to the horizontal and vertical orbit bumps, respectively.

Beam loss due to large angle scattering on some proton foil transversals is best reduced by choosing relatively low values for β_h and β_y at the foil. If the lattice has a double waist at this position, $\alpha_h = \alpha_v = 0$ and the third condition on the lattice parameters reduces to $\alpha_p' = 0$.

The remaining parameter, α_p, is best chosen to be greater than the square root of β_h as this assists the protons in missing the foil during their subsequent synchrotron motion. The units for both α_p and β_h are m.

Proton foil traversals are also reduced by careful choice of the injected beam parameters. The betatron and dispersion functions should be mismatched from those in the lattice, with the input dispersion $\alpha_{pi} = a'_{pi} = 0$. For a lattice with a double waist at the foil, it is best to choose:

$$\beta_{hi} < \beta_h,\ \alpha_{hi} = 0 \text{ and } \beta_{vi} < \beta_v,\ \alpha_{vi} = 0\ .$$

There are additional advantages in using special foil configurations such as the 'massless' foil proposed for the PSR and the '2 free-edge' foil proposed for the JHP (Japan) by RAL. Foil scatterings are reduced because the protons may miss foil material in both transverse dimensions. This may be seen from Fig. 1, where two alternative locations are shown for a '2 free edge' foil.

With the foil at position 1, the transverse and longitudinal paintings are as described previously but with the added advantage that the vertical orbit bump aids considerably in reducing the number of foil transits.

With the foil at position 2, a further advantage may be obtained. The momentum ramp is modified from previously, increasing from a small to a large momentum offset. Filling begins with small synchrotron, large horizontal betatron and small vertical oscillations. Thus, there is never a hollow longitudinal beam distribution. All the correlations of particle amplitudes are the reverse of those for position 1, with a hollow betatron distribution initially in the horizontal instead of the vertical plane. A hollow initial distribution in one transverse plane is considered acceptable, where a hollow longitudinal one is an unknown quantity experimentally. Space charge coupling is expected to modify somewhat the transverse distributions.

4. THE USE OF CORNER FOILS OF 'MASSLESS' FOILS OR FOILS WITH '2 FREE-EDGES'

The advantages of 'massless' foils or foils with '2 free-edges' have been outlined in Sec. 3. Neither has yet been produced, but both are now considered feasible. The original idea for the 'massless' PSR foil was to use an RAL-type foil with 1 free edge, thickness <0.25 μ, and deposit carbon over a restricted area of the foil to a thickness of 1 to 1.5 μ. This thickness is the typical requirement for the kaon factory projects, and it is advantageous that the foil areas are quite small. There are thus two advantages in having low β-values at the foil locations.

At RAL, an improved lifetime of the large, 0.25 μ thick, 1 free-edge foil has been obtained by separating the Al_2O_3 foil from its support frame along its entire lower edge, in order to reduce the stress forces. An attempt is to be made to produce a smaller, thicker foil with 2 free-edges, just by removing the lower limb of the frame. The thickness of this foil is to be 0.5 μ so that two such foils in sequence would serve for a 1 μ foil.

Also suggested for the PSR1 is the use of a conventional foil but located at the corner of the transverse acceptance area. The support edges may then be clear of the circulating beam. Such a system does not allow, however, the desired correlations for the different oscillation amplitudes.

5. INJECTION FOR RACETRACK LATTICES

Racetrack lattices have been designed, both for PSR2 and for the TRIUMF B ring. The lattices have some attractive features for injection, but both of the above lattices require modifications to create the required dispersion of the foil for optimised H^- injection. The use of a chicane in one long insertion is adopted for PSR2 and is under consideration for TRIUMF A.

Racetrack lattices also appear attractive for collimation systems.

6. CONCLUSIONS

1) There appear to be as many H^- schemes as there are designers.
2) Painting the correct transverse and longitudinal distributions is considered to be highly important.
3) It is advisable to avoid hollow longitudinal distributions during the injection filling period.
4) The feasibility of 'massless' foils or foils with '2 free-edges' needs to be established, for they allow improved injection schemes.
5) Racetrack lattices look attractive for further study.
6) Collimators still require detailed evaluation.

SUMMARY OF JOINT SSC / KAON FACTORY LINAC INJECTOR WORKING GROUP

Henry A. Thiessen
Los Alamos National Laboratory, Los Alamos, New Mexico 87545

INTRODUCTION

The SSC linac and the linac injectors for the Kaon factories have many things in common. This is no accident, since the participants in many of the workshops leading up to the designs had a large overlap in the participants. The work done to date has benefited all the projects. I shall try in this summary to report on the common features of many of the designs.

BUNCH-TO-BUCKET TRANSFER

All of the designs rely on synchronous capture of the linac bunches in pre-existing rf buckets in the next stage machine. This is at variance with earlier designs, which usually relied on adiabatic capture in the next stage. The kaon factories chose synchronous capture in order to avoid beam losses. In the SSC, this approach makes possible variable bunch spacing in the collider at the cost of only a chopper in the front-end of the linac.

FRONT-END BEAM PREPARATION

A radio-frequency quadrupole (RFQ) has been selected as the front end accelerator of choice for all the linacs. Typically, the RFQ accelerates the beam to 2-2.5 MeV. The natural choice for the frequency of the RFQ is that of the drift-tube linac immediately downstream. This gives a minimum output phase space and minimizes problems with space charge and peak currents. However, the next stage synchrotrons require a bunch frequency on the order of 50 MHz, while the drift-tube linacs operate between 200 MHz and 450 MHz. Three approaches to this problem are under consideration:

1) Chopping to 50 MHz.
2) Use 50 MHz RFQ.
3) Dual-Frequency RFQ.

The chopping wastes beam, but is the method of choice for AHF because only a small fraction of the beam pulses require 50 MHz. The 50 MHz RFQ is the simplest system which transmits 100% of the beam. This solution has been favored by the Los Alamos group for the SSC injector, although the output phase space (longitudinal) exceeds the specifications of the SSC Conceptual Design Report. The Los Alamos Group believes that the longitudinal phase space specification is unduly restrictive, hence this design will work. The EHF group needs a higher peak current than SSC, so is looking for a solution which loads two adjacent buckets of the drift tube linac. A design for a coupled 50 MHz and 200 MHz rfq has been presented which fills two adjacent buckets. All of these designs are discussed in contributions to this workshop.

An additional chopper, located after the RFQ, is needed in all designs. The Kaon Factories use this chopper to provide a "kicker gap". SSC needs this chopper to provide variable bunch spacing in the collider. Maintaining the longitudinal and transverse phase space in the presence of space charge forces requires numerous matching devices in the chopper region. Designs for the matching system have been presented by the EHF group and by the Los Alamos Group.

LINAC STRUCTURE

A drift-tube linac is used after the output of the chopper in all designs. For new designs, 400 or 450 MHz has been the frequency of choice, with 200 MHz an acceptable option for LAMPF, which already has an operational 201.25-MHz drift-tube linac. The drift-tube linac is used up to a final energy of approximately 100 MeV. After this point, a transition is made to a more efficient structure at a higher frequency, 600-1200 MHz. The 1200-MHz design is favored by most groups because of the structure efficiency and the availability of rf sources. The EHF group is looking at a 600-MHz design because of the larger bucket area. The availability of 600-MHz klystrons is a problem for this project.

SWITCHYARD DESIGN

There are several functions which should be incorporated in the beam line between the linac and the first ring. Both the energy and energy spread of the output beam from the linac should be monitored with resolution of a few parts in 10000. The only presently available technique for this monitoring with the required precision is a magnetic spectrometer. If this spectrometer is located in the transfer line, then continuous monitoring of the beam delivered to the first ring is possible. A design of this type is in use at LAMPF and has been proposed by the EHF design group. The SSC Conceptual Design includes the required real estate and the cost estimate for such a spectrometer, but no detailed design has been presented.

It should be noted that the H^- beam of the linac is easily stripped by a magnetic field. For 800 MeV (1463 MeV/c) at LAMPF, 0.35 Tesla has proved to provide negligible stripping and a suitable margin for magnet adjustment (beam line tuning). This field decreases as approximately the first power of the momentum of the beam, hence the switchyard radius of curvature increases as the square of the beam momentum. Thus such spectrometers naturally require a lot of real estate, on the order of 100 meters. This is no disadvantage, since it is convenient to include a bunch-rotating cavity in the switchyard in order to match the energy spread of the linac to the requirements of the first circular accelerator. It turns out that the 100-meter length of the switchyard is needed for a drift space to make this matching possible.

PAINTING

In all cases, the phase space of the linac is much smaller than that required for stability in the circular machines. The process of increasing this phase space in a controlled manner has been dubbed "painting". There is as yet no universally adopted painting scheme. The techniques proposed by the various groups are reported elsewhere in these proceedings.

FUTURE UPGRADES

The history of the existing circular accelerators shows that the injectors have always been upgraded in energy to reduce the space-charge limit in the circular machine. It is prudent in any new design to incorporate provision for such an upgrade. The SSC conceptual design incorporates provision for an upgrade from 600 MeV to 1133 MeV. At 600 Mev, the injection frequency is 1/9 that of the drift-tube linac. At 1133 MeV, the proton velocity is 9/8 of the velocity at 600 MeV. Thus it is possible to inject the SSC at 1133 MeV if only the frequency of the RFQ and the chopper is changed. All downstream machines can operate without modification. This energy increase would reduce the space-charge tune shift in the Low Energy Booster by a factor of two. To provide for this upgrade, space has been left in the linac tunnel for an additional 533 MeV of coupled-cavity linac and the switchyard radius of curvature has been set large enough to avoid stripping 1133 MeV protons (0.28 Tesla). The injection girder of the LEB should be lengthened to accommodate the higher injection energy.

SUMMARY

The H^- injector systems of the SSC and of the Kaon Factories share many common features. The reasons for the commonality have been discussed. Details are documented in the various contributions to these proceedings. The commonality of the design work to date has made better injectors for all these projects. Hopefully, we will also be able to reap the reward of lower project cost for all projects as they go to the detailed design and construction phase.

DEBUNCHED SLOW EXTRACTION FROM THE KAON FACTORY EXTENDER RING

U. Wienands and R.A. Baartman
TRIUMF, 4004 Wesbrook Mall, Vancouver, B.C., Canada V6T 2A3

The availability of a continuous beam from the Extender ring of the KAON Factory is an essential feature for certain classes of experiments like coincidence experiments and rare-decay experiments. In many of these experiments, even the intensity modulation due to the rf structure of the beam (62 MHz) is undesirable. We therefore investigated debunched slow extraction using either bunch rotation or debunching of a coasting beam.

Assuming debunching of a coasting beam (rf off and shorted out), the time needed for the head of a bunch to reach the tail of the previous bunch is given by the formula[1)]

$$t_{db} = (\pi - \Delta\phi)\left(2\pi f_a \mid \eta \mid \frac{dp}{p}\right)^{-1} ,$$

where $\Delta\phi$ is the phase half-width, f_a is the accelerating frequency, dp/p is the momentum half-width of the beam, and η is $(1/\gamma^2 - 1/\gamma_t^2)$.

The lattices of the KAON Factory are all high γ_t lattices in order to avoid crossing transition in the accelerators. However, the new racetrack lattices give us the flexibility to operate at different values of γ_t. In the Extender ring we can therefore choose γ_t to be 10, about the natural value without modulating the lattice functions since a high value of γ_t is not required. For our case $dp/p = \pm 0.0016$ and the phase width is $\pm 27°$; thus we get a value of 0.5 ms. Therefore, it appears that the beam will debunch within 1 ms, a time short compared with the cycle time of 100 ms.

Debunching the beam this way will not give any reduction in momentum spread of the beam. An alternative scheme is to rotate the bunches by injecting into a mismatched bucket and turn off the rf (and detune or short the cavity) once the phase width of the beam is maximal, corresponding to slightly more than 90° of rotation in phase space. With $\gamma_t = 10$ the optimum rf voltage for this scheme is 300 kV and the small-amplitude synchrotron tune is 0.0016. The 90° rotation is therefore achieved in 140 turns. The advantage of bunch rotation manifests itself in a reduced momentum spread in the extracted beam: $(dp/p)_{\mathrm{FWHM}} = 0.00043$.

Extraction of a coasting beam was simulated with our slow-extraction code SLEX. Debunching took place very rapidly and the extracted beam was almost free of any rf structure. The code simulates longitudinal motion correctly; however, the beam pipe impedance has not been taken into account. The low-frequency part of this impedance will be inductive, $\mathrm{Im}(Z_{\|}/n) \sim 5\ \Omega$, and will dominate over the capacitive space charge impedance, $\mathrm{Im}(Z_{\|}/n) = -0.7\ \Omega$. It is therefore important to operate above transition else the negative mass instability will tend to maintain some rf structure in the beam.

To avoid the microwave instability, the Keil-Schnell stability criterion requires $\mid \gamma_t \mid \lesssim 10$ for $\mid Z_{\|}/n \mid = 10\ \Omega$ and for the nominal longitudinal emittance of 0.2 eV-s. However, even after bunches overlap, local values of dp/p in the filaments continue to decrease and the instability can still occur. Boussard[2)] has found that for the case where the rf is abruptly switched off and the beam allowed to debunch, the Keil-Schnell criterion should be satisfied for the individual filaments in E-ϕ space up to as late as $t = 10\ t_{db}$. For our case, with $\gamma_t = 10$, this would require $\mid Z_{\|}/n \mid < 1\,\Omega$. However, if the bunches are rotated and the rf switch-off is accurately timed, the microwave instability should cause an increase in the base width of dp/p of no more than a factor of 2 even with $\mid Z_{\|}/n \mid = 10\ \Omega$. Naturally, a smaller longitudinal impedance will minimize this growth.

For chromatic bunched extraction, it is also desirable to minimize the beam pipe impedance. One can show that the minimum attainable bunch length is proportional to the cube root of the longitudinal microwave impedance. In particular, $\mid Z_{\|}/n \mid = 10\ \Omega$ gives a circulating bunch length $\Delta t = 2.7$ ns; $\mid Z_{\|}/n \mid = 3\ \Omega \Rightarrow \Delta t = 1.8$ ns, and $\mid Z_{\|}/n \mid = 1\ \Omega \Rightarrow \Delta t = 1.25$ ns. (The chromatic slow-extraction process will reduce these values by a factor of 2 or 3.)

In summary, debunched slow extraction from the Extender ring appears to be feasible without major problems. The flexibility of the racetrack lattice for the Extender ring to be able to run in different modes having different values of γ_t is of great advantage. For chromatic bunched extraction we will have to operate below γ_t in order to maintain stability for negative values of the chromaticity while a stable, high-quality debunched beam is best obtained with a γ_t near the natural value.

REFERENCES

1) Bovet, C. *et al.*, CERN/MPS–SI/Int. DL/70/4.

2) Boussard, D. *et al.*, IEEE Trans. NS-26, 3568 (1979).

BOOSTER RACETRACK LATTICES

U. Wienands, R.V. Servranckx and M.K. Craddock*
TRIUMF, 4004 Wesbrook Mall, Vancouver, B.C., Canada V6T 2A3

For the small rings (Accumulator and Booster) of the proposed TRIUMF KAON Factory, the possibility of using racetrack lattices rather than lattices of the current circular design is being investigated. The design goal is to design a lattice with dispersion-free straight sections for injection/extraction and rf, as well as a moderately high γ_t of about 10 in order to avoid transition and maintain stability against microwave instability with the natural negative chromaticity of the machine. In addition, one of the straight sections in the Accumulator ring has to be dispersive for the phase-space-painting injection process.

Due to the high γ_t, no special dispersion suppressors are foreseen in the lattices; rather, the dispersion will be matched to zero in the straight sections by tuning the arcs to an integer multiple of 2π phase advance. In the regular arcs we can then introduce a periodic modulation of the horizontal β function which in turn will modulate the dispersion function η, thus decreasing the average value of η and raising γ_t. Alternatively, the arcs can be tuned to a sufficiently large phase advance ($4 \times 2\pi$ per arc) that γ_t of the full machine reaches 10 or higher. The same lattice structure can be used for the Accumulator by retuning the machine to create dispersion in one straight section, while maintaining the second straight section dispersion-free. Finally, a regular structure of the arcs helps in reducing the number of depolarizing resonances.

Another issue for the lattices is the length of the straight sections, which should not be less than 30 m per section. With a maximum field of 1.05 T, given by the repetition rate of 50 Hz, and a minimum number of cells of 12 per arc this is not easy to achieve. A doublet focusing structure in the arcs can save some space by allowing the quadrupole pairs to be mounted on to a common piece of beam pipe and reducing the number of joints needed and is therefore more attractive than a FODO structure.

*On leave from Physics Dept., University of British Columbia, Vancouver, B.C., V6T 2A6.

An even more attractive alternative in this respect is a cell with combined-function defocusing bending magnets and separate focusing quadrupoles, reducing the number of quadrupoles by a factor of 2. A lattice with 16 such cells and a phase advance of $4 \times 2\pi$ per arc has a very small beam size in the arcs and leaves about 35 m for the straight section. For the whole machine, γ_t is 11.3 without modulating the arcs (Fig. 1).

Due to its lower average dispersion, chromaticity correction is difficult in this lattice. This is to a certain extent the case for all high-γ_{tr} lattices, but more so for small rings than for large machines. Reducing the tune and modulating the lattice functions does not significantly help the situation since at the position of the sextupoles dispersion will be either small or of opposite sign for sextupoles 180° apart from each other, due to the low average value of η. Since the machine should be most stable at negative chromaticity, a careful assessment of the need for chromaticity correction at full intensity will be done.

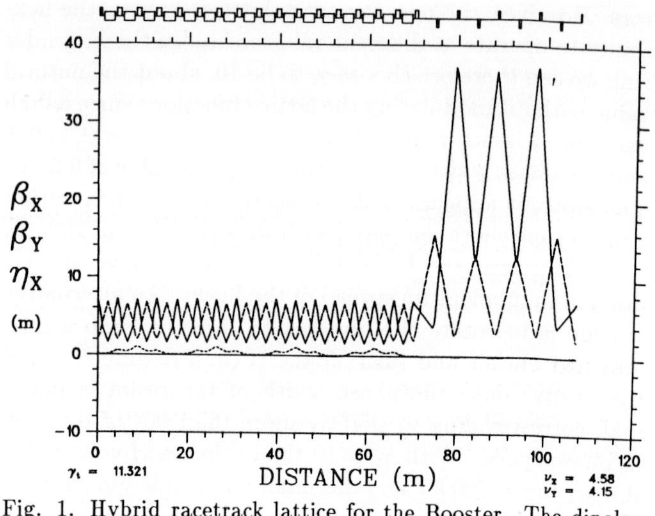

Fig. 1. Hybrid racetrack lattice for the Booster. The dipoles are defocusing.

NEW COMPRESSOR AT LOS ALAMOS

Eugene Colton
Los Alamos National Laboratory, Los Alamos, NM 87545

ABSTRACT

A conceptual design is presented for a second-generation proton-storage-ring complex at LAMPF. The facility would consist of two stacked racetrack-shaped machines. These machines would deliver a 1.2-mA beam of 1.6 GeV protons at 48 Hz. The pulse length would be 1.75 μsec, which represents a time compression of 570.

DISCUSSION

It has been proposed to use nearly the full potential of an upgraded LAMPF in order to produce a world-class proton storage ring. Time-compressed bunches of 1.6-GeV protons would be used to produce intense fluxes of spallation neutrons and neutrinos. The flux requirement is 1.5 x 10^{14} ppp at a 48-Hz repetition rate. Half the beam would go to a neutron production target and the other half would be used to produce neutrinos. A preliminary machine design has been completed. The design uses two stacked racetracks. They would be filled sequentially using 1-msec LAMPF macropulses to a flux of 7.5 x 10^{13} circulating protons in each ring. Figure 1 shows a plan view of the facility. Table I lists the basic machine parameters. The final kinetic energy of 1.6 GeV is obtained from an 800-MeV add-on linac to LAMPF; the H^- ion-source output has to be doubled.

Referring to Fig. 1, it is observed that protons are accumulated using the direct stripping process $H^- \rightarrow H^+$ in a carbon foil. The circulating proton beam is displaced outward using a combination of a fixed four-dipole chicane and two fast orbit bumpers. This choice allows inclusion of a dump for partially stripped H^0 or unstripped H^- particles. The collimators C_1 and C_2 absorb large-angle nuclear (> 2 mr) or single-Coulomb scatters occurring in the stripping foil from repeat proton traversals. The end sections include two 90° two-bend achromats and an extraction system. Sextupole magnets positioned within the achromats effect chromaticity modification. Protons are fast-extracted from each end section at a 24-Hz rate. A fast ferrite kicker and dc-septum combination are used to perform the extraction. The rf cavity is used to maintain a clean extraction gap of about 150 nsec. The beam pulse from each machine is then about 800 nsec in length. Protons from both machines are sent together in box-car fashion to the experimental targets -- the pulse length is 800+150+800 = 1750 nsec.

Injection simulations have been performed assuming 1050 turns of injection. It is found that the final horizontal rms geometric emittance, given by

$$\varepsilon_x = 4[< x^2 > < (x')^2 > - < xx'>^2]^{1/2},$$

is 33 mm-mr. Similarly the vertical ε_y = 43 mm-mr. The rf voltage is made up of the fundamental at 1.05 MHz plus its 2nd, 3rd, 4th, and 5th harmonics; the peak voltage is about 17 kV. This waveform should keep the peak currents below ~20 amperes in each ring. The final dp/p distribution will have a fwhm of ~0.5%, which is enough for longitudinal stability. The space-charge tune shift $|dQ_y|$ will not exceed 0.1. For transverse stability the machines need to have their vertical chromaticity Q_y' enhanced slightly (from -6.76 to ~ -10). Further studies are underway.

Table 1. Machine Parameters

Circumference	265.77m
Number of Rings/Superperiods	2/2
Circulating Current/Ring	12.6 A
Revolution Frequency at 1.6 GeV	1.048 MHz
Betatron Tunes Q_x,Q_y	5.23, 4.23
Chromaticity Q_x',Q_y'	-7.32, -6.76
Transition Gamma γ_t	8.19

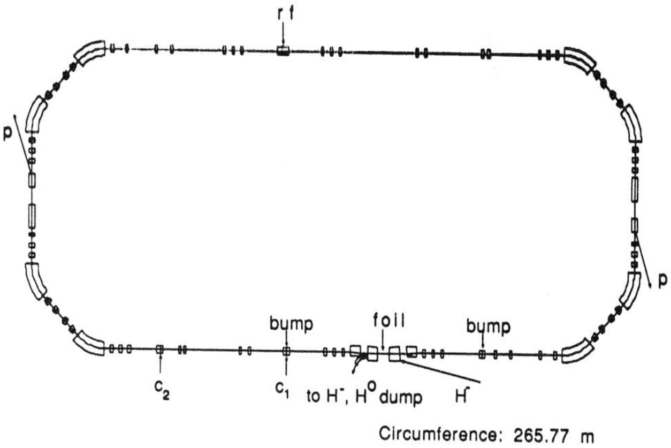

Fig. 1. Layout of 1.6 GeV Compressor

SYNCHRO-BETATRON RESONANCES

Toshio Suzuki*
TRIUMF, 4004 Wesbrook Mall, Vancouver, B.C., Canada V6T 2A3

ABSTRACT

Theories of synchro-betatron resonances are reviewed. The emphasis is on recent works with which I have been more or less concerned. Only single-beam effects are treated.

1. INTRODUCTION

In an accelerator theory for a bunched beam, synchrotron oscillations play an important role. If betatron motions are coupled by some mechanism to synchrotron motion, the resonances caused by the coupling and the synchrotron oscillations become important. These are the so-called synchro-betatron resonances.

For example, when chromaticity is present in particle accelerators, the betatron tune is modulated by synchrotron oscillations. This modulation causes the appearance of sidebands or satellites. A relevant mathematical formula of the modulation theory is a Fourier series expansion of expressions like $\cos(a \cos \nu_s \theta)$. The coefficients are given by Bessel functions and a typical term has a θ-dependence of $\cos m\nu_s\theta$, with m called the order of satellites or sidebands. Though we took a sinusoidal function (linear betatron oscillation), the sidebands can be defined for any function, e.g. a wake function to be considered later. The synchrotron oscillations can also be nonlinear; the only requirement is that they are periodic in θ. In these cases, the coefficients are not given by Bessel functions, but should be evaluated numerically. These satellites were actually observed[1)] in particle accelerators. These sidebands are expected to modify the betatron resonance conditions. This was actually the first process analysed[2)] in the theories of synchro-betatron resonances.

Another example is Sacherer's theory[3)] of a bunched beam instability. In Sacherer's integral equation, we observe Bessel functions. These Bessel functions have the same mathematical origin as described before: the Fourier series expansion of $\exp(ikr \cos \nu_s \theta)$. In this case, we call the index m of each term of the series a mode rather than the satellite or sideband order.

We give brief explanations of various synchro-betatron coupling mechanisms, but the discussion is restricted to single-beam effects. The coupling in beam-beam interactions and also in spin motion are other important subjects.

2. TRANSVERSE SPACE CHARGE EFFECT

The betatron tune is modulated not only by chromaticity, but also by the transverse space charge force. The latter is usually more important in proton machines. The detuning by the space charge force depends on longitudinal position when the bunch is not uniform longitudinally and the longitudinal position is modulated by synchrotron oscillations. For a parabolic bunch, the modulation occurs at twice the synchrotron frequency.

A relevant theory was given by Bruck,[4)] and some of his formulae were corrected by Suzuki.[5)] These explain the emittance growth and beam loss for a bunched beam in proton synchrotrons and storage rings. For a coasting beam, on the other hand, various theories of emittance growth are available.[6)]

3. DISPERSION EFFECT

The first actual observation of synchro-betatron resonances was made in NINA[7)] and it was associated with dispersion at the rf cavities. The particle position and angle are not changed by acceleration, but the equilibrium orbit with respect to which betatron oscillations are measured changes suddenly. Thus betatron oscillations are excited. The mechanism is similar to quantum excitation in electron storage rings. In rf acceleration, the amount of acceleration depends on the rf phase (longitudinal position) and this is modulated by synchrotron oscillations, causing the appearance of sidebands. This effect is also caused by the deceleration induced by the longitudinal wake field. Since the wake field is more nonlinear than the rf acceleration field, this effect is more important for higher satellites. By simple arguments, it can be shown to be a satellite resonance of an integer betatron resonance. The NINA synchrotron had five symmetrically placed rf cavity sections and the horizontal tune was nearly 5.25. This was why this resonance was excited.

This process was further analysed by Piwinski and Wrulich.[8)] They showed that not only is betatron motion affected by synchrotron motion through dispersion, but also synchrotron motion is excited by betatron motion through a path-lengthening effect. These give a description of the coupling mechanism which is both complete and symplectic. This symplecticity is more clearly manifested by the Hamiltonian approach of Chao and Morton,[9)] Suzuki,[10)] and Corsten and Hagedoorn.[11)] They derived the Hamiltonian of synchro-betatron coupled motion from the well-known Hamiltonian for particle motion under the Lorentz force. Suzuki[12)] developed a canonical perturbation theory based on the Hamiltonian formalism of Ref. 11) for the proposed TRIUMF KAON Factory where the dispersion at the rf cavities was rather large. This large dispersion was produced in the process of pushing the transition energy above the operating energies.

*On leave from KEK, Japan.

4. SUNDELIN EFFECT

Sundelin[13] considered a synchro-betatron coupling produced by the closed orbit distortions and the transverse wake field. The wake field strength depends on the longitudinal position in the bunch, and this position is modulated by synchrotron oscillations. Suzuki[14] developed a theory using a Hamiltonian formalism and gave some numerical estimates for TRISTAN.

5. COHERENT SYNCHRO-BETATRON RESONANCES

Brandt and Zotter[15] found by computer simulation that the threshold current of the transverse mode coupling instability is lowered for a localized wake around the betatron tunes satisfying the synchro-betatron resonance conditions. This process was analysed by the Vlasov equation approach[16,17] and by a two-particle model.[18]

By studying the process in detail by considering an arbitrary arrangement of cavities and/or a finite length of the cavity, Suzuki[19] found that these resonances can be interpreted as half-integral satellite resonances. The transverse wake field depends on the position of the leading particle just as the field strength depends linearly on the transverse position in a quadrupole magnet, and a half-integral resonance is excited. This conclusion can also be shown by a two-particle model.

6. WORK OF HAGEL AND SUZUKI[20]

These authors started with the differential equations of betatron and synchrotron motion including wake fields. By separating the transverse coordinates into the closed orbit distortion, the dispersion and the coordinate of betatron motion, various effects such as the dispersion effect, Sundelin's effect and coherent synchro-betatron resonances become manifest in a unified way. If the coupling of betatron motion to synchrotron motion (path-lengthening effect) is neglected as a first-order approximation, the equation can be solved analytically in a closed form. Corrections are made iteratively. This method can describe non-resonant behaviour as well, whereas canonical perturbation theory usually describes the behaviour near resonance.

7. CONCLUSIONS

Theories of synchro-betatron resonances are reviewed. The topics which I am more or less familiar with are chosen. Of course there are many other coupling mechanisms and there are still some other mechanisms that have not been fully treated so far. I conclude that the concepts of modulation and sideband are very important in theories for a bunched beam in which some parameters are modulated by synchrotron oscillations.

REFERENCES

1) Hammer, C.L., Pidd, R.W., and Terwilliger, K.M., Rev. Sci. Instrum. **26**, 555 (1955).

2) Orlov, Yu.F., Sov. Phys. – JETP **5**, 45 (1957).

3) Sacherer, F., CERN/SI-BR/72-5 (1972).

4) Bruck, H., Part. Accel. **11**, 37 (1980); Part. Accel. **11**, 181 (1981).

5) Suzuki, T., TRIUMF internal reports TRI-DN-86-26 (1986) and TRI-DN-88-K2 (1988).

6) e.g. Guignard, G., CERN 78-11 (1978).

7) Crowley-Milling, M.C. and Rabinowitz, I.I., IEEE Trans. Nucl. Sci. **NS–18**, 1052 (1971).

8) Piwinski, A. and Wrulich, A., DESY 76/07 (1976).

9) Chao, A.W. and Morton, P.L., private communication.

10) Suzuki, T., Part. Accel. **12**, 237 (1982).

11) Corsten, C.J.A. and Hagedoorn, H.L., Nucl. Instrum. Methods **212**, 371 (1983).

12) Suzuki, T., Part. Accel. **18**, 115 (1985).

13) Sundelin, R.M., IEEE Trans. Nucl. Sci. **NS–26**, 3604 (1979).

14) Suzuki, T., Nucl. Instrum. Methods **A241**, 89 (1985).

15) Brandt, D. and Zotter, B., Proc. 12th Int. Conf. on High Energy Accelerators, Fermilab (1983), p. 309.

16) Ruggiero, F., Part. Accel. **20**, 45 (1986).

17) Chin, Y.H., CERN SPS/85-33 (DI–MST) (1985).

18) e.g. Zotter, B., IEEE Trans. Nucl. Sci. **NS–32**, 2191 (1985).

19) Suzuki, T., CERN/LEP–TH/87–55 (1987).

20) Hagel, J. and Suzuki, T., in preparation.

ANOTHER LOOK AT COHERENT LONGITUDINAL INSTABILITIES OF BUNCHED BEAMS

G.H. Rees
Rutherford Appleton Laboratory, Chilton, Didcot, Oxon., England

ABSTRACT

Existing theories predict that the higher the mode number is of a longitudinal bunched beam instability, the higher is likely to be the frequency (or range of frequencies) of the instability excitation fields. In this paper a different result is found with all modes having a similar upper frequency limit for the fields but with the odd- and even-order modes having a dissimilar range for the possible frequencies. The relevance for uncoupled bunch, coupled bunch and mode coupling instabilities is discussed.

INTRODUCTION

A bunch train with a stationary distribution of particles in the bunches has a spectrum that extends up to a frequency approximately equal to the inverse of the bunch duration. The bunch harmonics are the only carrier frequencies available to sustain a longitudinal bunched beam instability. Yet the general theory, as developed by Sacherer[1)] and others, predicts that the instability excitation frequencies may be very much higher than the bunch frequencies. Form factors are derived for the different instability modes which indicate that the higher the mode number, the higher is likely to be the excitation frequency. If this prediction of the general theory is correct, how is the beam able to convey information at excitation frequencies higher than the bunch frequencies?

The general theory is developed from the one-dimensional Vlasov equation, with the use of polar co-ordinates (r,θ) in the longitudinal phase space $(\phi,\dot{\phi})$. A two-dimensional density in this phase space, $\psi(r,\theta)$, is assumed to be of the form:

$$\psi_0(r) + \Sigma_m\, R_m(r)\, e^{im\theta}\, e^{ip(m)t} ,$$

where m is the instability mode number.

The simplest case to consider is the dipole mode of instability (m=1) for uncoupled bunch motion (n=0). Then, it is possible to rearrange the Vlasov equation and solve directly for the radial function, $R_1(r)$, leaving a phase equation for coherent dipole motion to be solved. This is the equation originally solved by K.W. Robinson[2)] for a beam interacting with the fundamental resonance of the accelerating cavities, in the absence of any feedback systems; the solution is a fourth-order equation which may have unstable roots in two different frequency regions. It is of interest, for this specific case, to make a detailed comparison between the Robinson and the general theory, as the latter leads to a first- and not a fourth-order equation.

For comparison, it is convenient to consider three separate expressions, the coherent phase equation, the cavity transfer function for modulations of drive current and an expression for the modulation components of beam current. In each case, Robinson uses a different relationship than that developed in the general theory. Firstly, for the coherent phase equation, there is one perturbation due to the phase modulation and one due to the amplitude modulation of the accelerating voltage; Robinson uses both, whereas the general theory neglects the phase modulation term. This is an important omission, for coherent dipole motion involves the both terms (while the odd and even higher-order modes involve only the former and latter, respectively). Secondly, for the beam cavity interaction, Robinson correctly uses the equivalent of a complex transfer function, whereas the general theory evaluates the products of the complex cavity impedance and beam current component at each beam frequency, with a correction term for the cavity bandwidth. Thirdly, for the beam current modulations, a major difference emerges between the two theories. Robinson considers dipole motion as phase modulation of the fundamental beam current component. The general theory uses a Fourier expansion of the beam current modulation in terms of the longitudinal phase space parameter ϕ. Such an expansion is invalid, however, for though the unperturbed line charge density is periodic in ϕ, the modulation is not. Thus, the form factors of the general theory are developed from an incorrect premise.

A modified theory is now developed, first extending the Robinson theory to include parasitic mode excitation of the dipole mode, and then to the case of coupled bunch dipole motion. Also considered is the special case of rf feedback for the main accelerating system. Higher mode motion is analysed from the Vlasov equation, separating the radial mode functions $R_m(r)$ by analogy with the method for the dipole mode. The theory is simplified by the use of a dual complex number representation for modulated signals, with one complex number to represent the carrier and a second for the modulation components.

Revised form factors are developed and their relevance for mode coupling instabilities is discussed.

UNCOUPLED DIPOLE MODE MOTION

The Sacherer form of the Vlasov equation for the dipole mode may be rearranged and approximated to the following form:

$$(s^2+\Omega^2)\, R_1(r)\, \overline{R}\left(\frac{e^{i\theta}}{2i}\right)\, e^{ipt} = -\sin\theta \frac{d\psi_0}{dr}(s^2+\Omega^2)\phi ,$$

where (r,θ) are polar co-ordinates in $(\phi,\dot{\phi}/\Omega)$ space, ψ_0 is the 2-D stationary bunch density, $R_1(r)$ is the radial function for the m=1 mode, Ω is $2\pi\times$ synchrotron oscillation frequency, $\overline{R}$ is the real part of the argument $(\exp i\theta)/2i$, ϕ is the coherent phase motion ($\hat{\phi}\exp(ipt)$), and $s\ (\equiv d/dt) = ip$, with p the complex frequency.

Hence, $R_1(r) = -2\hat{\phi}\, d\psi_0/dr$, and it remains to solve for p from the coherent phase Eq. (1), the complex cavity transfer function (2) and the beam current modulation (3):

$$(s^2+\Omega^2)\phi = s^2(\Delta\phi_v) - \Omega^2 \sin\phi_s(v/V) \quad (1)$$
$$v/V+j\Delta\phi_v = G(jw,ip)\Delta I_b/V \quad (2)$$
$$\Delta I_b = I_b(\Delta\phi_\nu-\phi)\, e^{-j\phi_s} . \quad (3)$$

Here, V is the cavity voltage, v is the amplitude modulation of V, $\Delta\phi_v$ is the phase modulation of V, ΔI_b is the beam current modulation, and $G(jw, ip)$ is the dual complex number representation of the cavity transfer function.

For the phase modulation of the beam current, a linear approximation is assumed for the usual Bessel function expansion. The reference phase is that of V, and I_b is the amplitude of the relevant harmonic of the unperturbed beam current. Input modulation is represented by $\exp(jwt)\exp(ipt)$, with w and p the angular frequencies of the carrier and modulation, respectively. In this notation, a transfer function response is represented by:

$$\overline{R}_j(\overline{R}_i\ (G(jw,ip)\ e^{ipt})\ e^{jwt})/(\cos wt \cos pt) ,$$

where $\overline{R}_i$ is the real part of the complex argument with j assumed constant and $\overline{R}_j$ is the real part of the subsequent argument in which j is assumed complex. Typical $G(jw, ip)$ functions for $\Delta V/\Delta I$ are:

$(jw+ip)L$ for an inductance L ,
$((jw+ip)C)^{-1}$ for a capacitance C .

Such transfer functions are the composite response for both the upper and lower sidebands of the modulation. Alternatively, the response may be found for individual sidebands; for example, the responses for a cavity of loaded shunt resistance R, loaded quality factor Q, and angular resonant frequency w_0 are:

$$R(1\mp ij)/2(1+(2Q/w_0)(j\Delta w+ip)) ,$$

where $\Delta w = w - w_0$ and the $-$ and $+$ signs refer to the upper and lower sidebands, respectively. The terms in (ij) are the quadrature terms associated with single sideband modulation. There is cancellation of the (ij) terms for the composite response.

Sometimes it is necessary to consider the single-pass transient cavity response in addition to the multi-turn response. This is the case for very large rings where the revolution time is of the same order as the cavity time constant. To obtain such a single-pass response, it is necessary to know the detailed form of the beam cavity interaction or to assume an appropriate model for the cavity.

In the special case where rf feedback is used in the final stages of the cavity amplifier systems, there is a modified form of $G(jw, ip)$ for the composite cavity response:

$$R(1 + A\exp(-(j\Delta w + ip)T) + (2Q/w_0)(j\Delta w + ip))^{-1} .$$

Here A and T $(= 2\pi\, k/w_0)$ are the open loop feedback gain and delay, respectively, with k an integer. This form of $G(jw, ip)$ is valid only if the single-pass response may be neglected; then Eqs. (1), (2) and (3) may be combined to give a fourth-order equation:

$$s^4 + b_3 s^3 + b_2 s^2 + b_1 s + b_0 = 0$$

with

$$b_0 = \Omega^2(C^2 + \Delta w^2 + CD\Delta w/((1+A)\cos\phi_s)$$
$$b_1 = \Omega^2 b_3 = 2\Omega^2 C$$
$$b_2 = (C^2 + \Delta w^2 + \Omega^2)$$
$$C = (1+A)/(2Q/w_0 - AT)$$
$$D = RI_b/V .$$

For the case of no rf feedback, $A = T = 0$ and the fourth-order equation reduces to that of Robinson. For the case of rf feedback, there are three possibilities for instability. Two conditions are as for Robinson, *viz.* instability when the cavity is detuned in the direction opposite to that required for beam loading compensation, and instability under reactive compensation above certain ratios of beam-to-cavity power. For no rf feedback, the ratio is unity; for rf feedback, the ratio may be much higher. The third condition for instability occurs if the feedback gain A is increased above the value $Q/k\pi$. The equation becomes of higher order when low-frequency loops are added for control of cavity field amplitude, tuning, beam radial position and coherent phase motion. Then, there is coupling between the loops due to the beam loading but it is minimised by the use of the rf feedback.

The coherent phase motion is further modified if there is dipole mode excitation of both the main and parasitic cavity resonances. From Eq. (3), it may be seen that the beam current modulation is proportional to the amplitude of the associated beam current harmonic. Parasitic mode excitation may thus occur at any frequency within the unperturbed bunch spectrum, which extends up to the inverse of the bunch duration. This is in contrast to the prediction of the general theory, where the form factor for the dipole mode is shown to be finite up to frequencies twice as high. The parasitic resonance must be close to a multiple of the main rf frequency to influence the uncoupled dipole mode motion. The more general case is when the parasitic resonance is close to a harmonic frequency of the ring but not to be multiple of the main rf frequency. In this case, coupled bunch dipole or higher mode motion may be excited.

COUPLED BUNCH DIPOLE MODE MOTION

If there are M circulating bunches in a ring ($M \leq h$), there are M coupled bunch modes of oscillation for each value of m (m=1 for dipole, m=2 for quadrupole ... and h is the rf harmonic number).

Coupled bunch motion may be excited via a parasitic or the main cavity resonance. Parasitic excitation of coupled or uncoupled dipole mode motion may be analysed as follows. Consider the case of h equally populated bunches undergoing dipole motion at coupled bunch number n $(= 0, 1, \ldots (h-1))$, with n=0 corresponding to the uncoupled case. If only one value of n is excited at the complex coherent frequency ($\cong \Omega$), it may be shown that the additional angular frequencies that appear in the beam spectrum, for a revolution frequency $w_r/2\pi$, are:

$$\ell h w_r \pm p_1 \text{ or } p_2 \quad \text{with } \ell = 0, 1 \ldots$$
$$p_1 = p + nw_r , \quad p_2 = p - nw_r .$$

The frequencies p_1 and p_2 converge for n=0. The parasitic resonance (at w_{op}) is excited most strongly by the sidebands linked to one particular value of ℓ. The excitation may be obtained, as previously, by a composite response $G(jw, ip)$, but with $\Delta w = \ell h w - w_{op}$ and the modulation frequencies at p_1 or p_2. For the former (p_1) and the latter (p_2), the sideband pairs are, respectively:

$$(\ell h + n)w_r + p \text{ and } (\ell h - n)w_r - p\ ,$$
$$(\ell h + n)w_r - p \text{ and } (\ell h - n)w_r + p\ .$$

The coherent phase motion is found from modified forms of Eqs. (1), (2 and (3). In (1), $\sin\phi_s$ is approximated by $\sin\ell\phi_s$; in (2), the new Δw is used in $G(jw, ip)$ and p is replaced by p_1 or p_2; and in (3) the following form of ΔI_b is used:

$$\Delta I_b = 2j\ I_\ell\ e^{j\ell(3\pi/2-\phi_s)}\ J_1(\ell(\Delta\phi_v - \phi)$$

where I_ℓ is the amplitude of the ℓth bunch frequency harmonic in the unperturbed beam current. It may be written in terms of an integral involving $J_0(\ell r)$ and $\psi_0(r)$. Combining (1), (2) and (3) gives a fourth-order equation in s, but now with complex coefficients, involving terms in i. It may be reduced to two simultaneous fourth-order equations, with variables the real and complex parts of the coherent frequency.

The motion becomes more complex when the bunches are not equally populated or when there are missing bunches. In this case, each bunch motion is described by a fourth-order equation in terms of the motion of each of the other bunches. Then, for M bunches, there are M coupled fourth-order equations to be solved.

Coupled bunch feedback systems sometimes use just single sideband modulation for the feedback. Such feedback may be analysed by use of the single sideband transfer functions given in the previous section. Coupled bunch frequencies appear in the spectrum in pairs at nearly the same frequencies, e.g. $(\ell h + n \pm p)w_r$. Since the transfer functions are different for the two frequencies, feedback to damp one mode may antidamp the other and *vice versa.* Then it is necessary to transform appropriately the individual frequency components.

ODD-ORDER HIGHER MODES

The appropriate Vlasov equation is as before but with Ω (on LHS) replaced by $m\Omega$, $R_1(r)$ by $R_m(r)$ and the argument $(\exp i\theta)/2i$ by $(\exp im\theta)/(m+1)i$, where $m = 3$, 5, 7 ... .

Excitation currents and fields are of the form $\exp(jwt)$ $\exp(ipt)$ with $wt = \phi$ and $p = p_1$ or p_2. Then the amplitude modulations of field correspond to $\sin(\ell r\cos\theta)$ terms and the phase modulations to $\cos(\ell r\cos\theta)$ terms. These may be expanded in terms of Bessel functions:

$$\sin(\ell r\cos\theta) = 2\sum_1^\infty (-1)^{n+1} J_{2n-1}(\ell r)\ \cos(2n-1)\theta$$
$$\cos(\ell r\cos\theta) = 2\sum_1^\infty (-1)^n J_n(\ell r)\ \cos 2n\theta + J_0(\ell r)\ .$$

From the form of the Vlasov equation and the Bessel function expansions, it is seen that the odd-order higher modes may only be excited by the phase modulation field components. Mode m=3 excitation is related to the sum $J_2(\ell r) + J_4(\ell r)$ and hence to $J_3(\ell r)$:

$$J_{m+1}(\ell r) + J_{m-1}(\ell r) = 2mJ_m(\ell r)/\ell r\ .$$

Similarly, mode m=5 is related to $J_5(\ell r)$ and the order of the mode turns out to be the same as the order of the associated Bessel function. By analogy with the dipole mode, the radial function may be separated:

$$R_m(r) = (-1)^{(m+1)/2}\,(2m(m+1)J_m(\ell r)/\ell r)\frac{d\psi_0}{dr}\,\hat{\phi}\ .$$

The coherent motion for the odd-order modes is characterised by an oscillation in phase and energy of the centre of gravity of the bunches. This corresponds to phase modulation of beam current with an amplitude given by the ΔI_b formula of the last section and a complex mode frequency, $\cong m\Omega$. Since ΔI_b is proportional to I_ℓ and is independent of m, the possible range of excitation frequencies corresponds to the unperturbed bunch spectrum and is the same for all the odd-order modes.

The mode equations may now be developed (using small angle approximations) in terms of the phase motions of the bunch centres:

$$(s^2 + m^2\Omega^2)((1+(2Q/w_0)ip_{1,2})^2 + ((2Q/w_0)\Delta w)^2)\phi =$$
$$\Omega^2 H\phi((1+(2Q/w_0)ip_{1,2})\cos\alpha + (2Q/w_0)\Delta w\sin\alpha)$$

with $H = m^2\, D\ell\, I_\ell/I_b$ and $\alpha = \ell(3\pi/2 - \phi_s)$.

EVEN-ORDER HIGHER MODES

The Vlasov equation is unchanged but now the excitation is due to the amplitude modulation of the fields. Again the order of the mode is the same as the associated Bessel function excitation term.

Even-order modes are characterised by no coherent motion of the bunch centre but by an oscillation of the bunch phase extent ϕ_b. This corresponds to amplitude modulation of the bunch current harmonics at the complex mode frequency, with ampitudes:

$$\Delta I_\ell = (\partial I_\ell/\partial\phi_b)\ e^{j\ell(3\pi/2-\phi_s)}\Delta\phi_b\ .$$

The derivative of I_ℓ with respect to ϕ_b now determines the excitation spectrum. The upper frequency limit to the spectrum is as for I_ℓ but the lower limit is different; e.g. for short bunches, the derivative is zero for the lower harmonics. All the even-order modes may be excited over this reduced frequency range, and it is adequate for both even and odd modes to damp parasitic cavity resonances over the frequency range of the I_ℓ spectrum.

The radial functions for the even-order modes are:

$$R_m = (-1)^{m/2}\,(2m(m+1)\,J_m(\ell r)/\ell r)\frac{d\psi_0}{dr}\ \Delta\phi_b/\phi_b$$

and the mode equations are identical to those for the odd-order modes, but with ϕ replaced by $\Delta\phi_b$ and:

$$H = -D(\partial I_\ell/\partial\phi_b)/I_b\ \cos\phi_s\ .$$

MODE COUPLING

The principle of mode coupling is as described in Ref. 1, but the detailed mechanism is different here due to the higher order of the equations and the different form factors. The form factors H are proportional to I_ℓ for the odd- and to $(\partial I_\ell/\partial\phi_b)$ for the even-order modes.

Coupling occurs when mode frequencies approach one another. Phase modulation of beam current excites both phase and amplitude modulation of V, the former driving odd- and the latter even-order modes. Amplitude modulation of I_ℓ then complements the coupling.

Mode frequencies may cross because of the different range of excitation spectra for the odd and even modes. The extended spectrum for the odd modes increases the inductive contributions to the frequency shifts. Certain combinations of impedances may lead to mode coupling both below and above transition energy, in contrast to earlier findings.

REFERENCES

1) Sacherer, F.J., "Bunch Lengthening and Microwave Instability", CERN/PS/BR 77-6 (1977).
2) Robinson, K.W., "Stability of Beam in Radio-Frequency System", CEAL-1010, 27 February 1964.

STABILITY OF PHASE AND RADIAL LOOPS

S. Koscielniak
TRIUMF, 4004 Wesbrook Mall, Vancouver, B.C., Canada V6T 2A3

ABSTRACT

The properties of phase and radial loops, including the effects of signal delay, are described in relation to the damping of longitudinal bunched beam oscillations in a proton synchrotron. The influence of beam loading is not considered. This note is a much simplified and abridged version of an unpublished report by the author.[1)]

BEAM FREQUENCY SPECTRA

If the beam time-structure is decomposed into harmonic components, then the possible frequencies are :

$$(n+pM)f_0 + mf_s \qquad -\infty < p < +\infty$$

n is the coupled bunch index, M the number of bunches, m is the mode number, f_0 the revolution frequency, f_s the synchrotron frequency, and p an integer. Our commission is to remove the $n=0, m=1$ mode. The position of a bunch performing dipole oscillations can be written

$$\theta(t) = \omega_s t - \Delta\theta.\sin(\Omega_s t)$$

where θ is azimuthal angle about the ring. So synchrotron oscillations are intrinsically FM. Here $|\Delta\theta| < (\pi - \phi_s)/h$, $\omega_s = 2\pi f_0$, and $\Omega_s = 2\pi f_s$ and ϕ_s is the synchronous phase.

If we damp the single-bunch dipole mode oscillation by any means at all, then we obliterate the synchrotron side-bands for all p. We can do this because the beam signal is periodic. When the curative procedure is FM, then we require a reference frequency and phase, which are easiest obtained from the cavity rf. We shall assume small-amplitude phase-modulations of the form

$$\phi(t) = \omega_c t + \Delta\phi\cos(\omega_m t)$$

where $\Delta\phi = \Delta\omega_c/\omega_m$. Here ω_c is a carrier frequency and ω_m is the modulation frequency. If $|\Delta\phi/2| << 1$ then the signal is narrow-band FM, and the dominant side-bands are at $\omega_c \pm \omega_m$. Any device which is to pass the signal without severe distortion must have a band-width $\geq 2\omega_m$ about the carrier frequency. Notably, this implies the cavity quality factor must satisfy $Q \leq hf_0/2f_s$, where h is harmonic number.

STABILITY ANALYSIS OF SYSTEMS WITH DELAY

A stability analysis of a linear system often proceeds by Laplace transforming the components to give a problem in algebraic analysis. Finally, the roots of a characteristic equation are investigated to determine stability.

By delay we mean there is at least one 'component' in the system whose reaction to a stimulus is not instantaneous but starts a finite time after the stimulus has been applied. Under certain conditions, the Laplace transform of a signal with delay is: $\hat{L}[\phi(t+T)] = e^{+sT}\phi(s)$. So if $\phi(t+T) = \theta(t)$, then the stimulus to response transfer function is: $\phi/\theta = e^{-sT}$.

Delays will lead to a characteristic equation containing complex exponential terms. This is equivalent to a polynomial of infinite order and the stability analysis is not treatable by the Hurwitz method. Demonstration of stability requires us to show that every member of the infinite set of roots has zero positive real part. A set of sufficient stability criteria were provided by L. Pontriagin. A readable account of the method and applications will be found in Ref 1. We shall now state the transfer functions of the system components. It will be the task of later sections to combine these elements.

PHASE LOCK LOOP

The basic property of a phase lock loop (PLL) is that the output frequency attempts to follow a time varying input frequency. A voltage control oscillator is driven by the difference of its output phase and a control signal, as shown in Fig. 1. The VCO "gain" (k) is sometimes called the effective band-width. There are cabling delays in the forward (T_f) and backward (T_b) paths. The following notations are employed: $\omega_{a,c} = \Delta\omega_a \equiv \omega_a - \omega_c$ measures the deviation of frequency a from some centre frequency c. The closed loop response function is:

$$\frac{\Delta\omega_a}{\Delta\omega_b} = \frac{ke^{-sT_f}}{s + ke^{-s(T_f+T_b)}}$$

The stability condition depends upon the product of gain (k) and loop delay ($T_L = T_f + T_b$) : $kT_L < \pi/2$.

THE CAVITY TO BEAM TRANSFER FUNCTION

We can compare the phase of the bunch with that of the accelerating rf waveform, $\Phi_{b,rf}$. Likewise we may compare the synchronous particle with the rf, $\Phi_{s,rf}$. The following identity holds $\Phi_{b,rf} + \Phi_{rf,s} + \Phi_{s,b} = 0$, where $\Phi_{b,s}$ is the difference of beam and synchronous phases. In general $\Phi_{rf,s} = \phi_s + \phi_{rf}$. The term ϕ_{rf} accrues from the frequency variations caused by the damping loop. For the beam $\Phi_{b,s} = \int \Delta\omega_b(t)dt$, where $\Delta\omega_b = h(\dot{\theta} - \omega_s)$. The rate of change of beam phase is:

$$\dot{\Phi}_{b,s} = -\frac{h\omega_s|\eta|}{\beta_s^2 E_s} \times E_{b,s}$$

where $E_{b,s} = E_b - E_s$ is the beam energy error. The energy of the bunch, compared with the synchronous particle, changes at the rate:

$$\frac{d}{dt}\left(\frac{E_{b,s}}{\omega_s}\right) = \frac{e}{2\pi}V[\sin(\Phi_{b,rf}) - \sin\phi_s] .$$

Combining and linearising we find

$$\ddot{\Phi}_{b,s} + \Omega_s^2.\Phi_{b,s} = \Omega_s^2\phi_{rf} . \qquad (1)$$

All quantities appearing in this equation are measured at the same location in the ring (e.g. pick-up). We shall have to transform up/down-stream to rewrite in terms of rf-phases appearing at cavities and pick-ups. The (closed loop) transfer function from the phases at the rf-cavities to the beam-phase at the pick-up is:

$$\phi_b^{pu} = \Omega_s^2 e^{+sT_1}/(s^2 + \Omega_s^2)$$

where T_1 is the forward delay along the beam-path. The characteristic equation is $s^2 + \Omega_s^2 = 0$. If the modulation frequency $s = i\omega_m = \pm i\Omega_s$ then there can be output even when there is no input – that is the bunch oscillates. So damping terms must be added to the equation of motion.

PHASE LOOP

I shall use the simple model of one cavity, one PU and one bunch. Consider Fig. 2, where the PLL is controlled by the difference $\Phi_{rf,s} - \Phi_{b,s}$. The characteristic equation is:

$$s[s^2+\Omega_s^2+sk_pe^{-s(T_f+T_b)}]+k_pe^{-sT_f}\Omega_s^2[e^{-sT_b}-e^{-s(\tau_1+T_p-T_1)}] = 0 .$$

A sufficient condition for stability is: $T_b = (\tau_1 + T_p) - T_1$. This is an important result, so let us interpret it. The time for a control signal to propagate from the pick-up to cavity does not have to be equal to the time for the beam to travel from PU to cavity; and the fact that the beam (almost) always beats the control signal does not inhibit damping. Below, we show that the constraint on T_b implies a constraint on the phase loop gain gain k_p.

We can measure a forward delay from PU to cavity along the beam, or along the cabling and electronics. The backward path delay of the PLL (T_b) must be equal to the difference of forward paths from PU-to-cavity, excluding the forward delay through the PLL. Then each cavity damps the same "piece of beam" as was sensed at the PU. In this case, one extraneous root ($s = 0$) factors out, leaving the characteristic equation:

$$s^2 + sk_pe^{-s(T_f+T_b)} + \Omega_s^2 = 0 . \qquad (2)$$

Applying the theorems of Pontriagin, we conclude that the system is stable with respect to all modulation frequencies $s = \omega_m$ provided:

$$k_p(T_f+T_b) < \frac{\pi}{2} - \frac{2}{\pi}\Omega_s^2(T_f+T_b)^2 \qquad \text{when } \Omega_s(T_f+T_b) < \frac{\pi}{2} .$$

This provides an upper limit on the damping-term k_p and hence on damping rate; and, by implication, an upper limit on the synchrotron frequency.

RADIAL LOOP

The beam revolution frequency is related to momentum deviation; and the beam radial position depends on momentum, through lattice dispersion. Consequently, we can make the cavity rf depend on the beam frequency ($\Delta\omega_{b,s}$) directly, by sensing position. The physical layout is shown in Fig. 3. There is a cable delay of T_r seconds from PU to master oscillator, and delay from the VCO to cavity of τ_1. The relation between rf and beam frequency is is $\omega_{rf} = h\omega_s - k_r \times \Delta\omega_{b,s}^{pu} e^{-s(T_r+\tau_k)}$. Using $\Phi_{rf,s} = \phi_{rf} + \phi_s$, we substitute into the time derivative of Eq. (1) and cancel the beam-frequency throughout to give:

$$(s^2 + \Omega_s^2) + k_r.\Omega_s^2 e^{-s(\tau_1+T_r-T_1)} = 0 .$$

Ideally, $T_1 = \tau_1 + T_r$, with the beam delay equal to the cabling delay. This would imply that we damp, at the cavity, the same piece of beam that was sensed (earlier) at the pick-up. It is a simple task to investigate the stability when the net-delay is non-zero. The characteristic equation is:

$$s^2 + \Omega_s^2(1 + k_r e^{-sL}) = 0 .$$

From the Pontriagin criteria, we conclude

$$0 < \Omega_s L\sqrt{1 + |k_r|} < \pi \quad \text{and} \quad -1 < k_r < 0$$

and note that the synchrotron frequency must be depressed.

PHASE AND RADIAL LOOPS COMBINED

Thus far, phase and radial loops have each been examined in isolation; now it is time to combine them. The phase- error and radial-error signals are summed and used to control the output frequency of a VCO, as shown in Fig. 4. Neglecting delays, the transfer function is:

$$\omega_{rf,s} = k_p(\phi_{b,s} - \phi_{rf,s}) + k_r\omega_{b,s} .$$

The system characteristic, with delays, is:

$$s(s^2 + \Omega_s^2) + s^2 k_p e^{-s(T_f+T_b)} + \Omega_s^2 e^{-s(T_f+\tau_1-T_1)}$$
$$\times \left[k_p(e^{-s(T_b+T_1-\tau_1)} - e^{-s(T_p-T_2)}) + sk_r e^{-sT_r}\right]$$

To avoid a cubic polynomial in s we must have: $T_b = T_p + \tau_1 - (T_1 + T_2)$ that is the PLL return-path delay must be equal to the difference of forward paths for beam and cable-signals (excluding T_f). As above, we do not require the net delay to be zero, and the beam is allowed to beat the control signal. If the condition is satisfied, then $s = 0$ factors out leaving:

$$s^2 + sk_p e^{-s(T_f+T_b)} + \Omega_s^2[1 + k_r e^{-s(T_f+T_r+\tau_1-T_1)}] = 0 \, .$$

Evidently, it is preferable that the loop delay from radial PU along cables to cavity and back (counterstream) along the beam to PU is zero: $T_f + T_r + \tau_1 = T_1$. This leaves a system with a damping constant determined by the phase-loop and a resonant frequency determined by the radial-loop:

$$s^2 + sk_p e^{-s(T_f+T_b)} + \Omega_s^2(1 + k_r) = 0$$

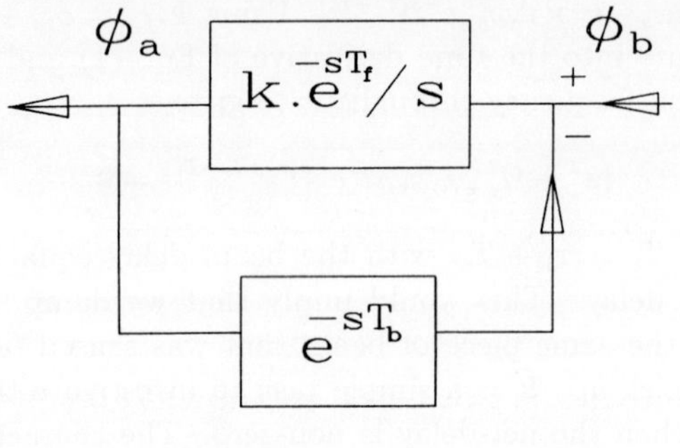

Fig. 1. Schematic of phase lock loop.

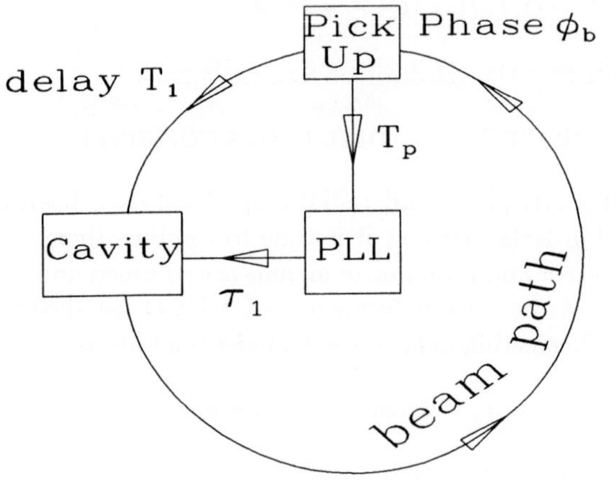

Fig. 2. Phase loop alone.

We compare to Eq. (2) and state the stability constraint on gain:

$$k_p.(T_f + T_b) < \frac{\pi}{2} - \frac{2}{\pi}\Omega_s^2(1 + k_r)(T_f + T_b)^2$$

where $\Omega_s(T_f + T_b)\sqrt{1 + k_r} < \frac{\pi}{2}$.

REFERENCES

1) Koscielniak, S., A tutorial on the stability of phase and radial feedback loops, TRI-DN-88-16. (1988).

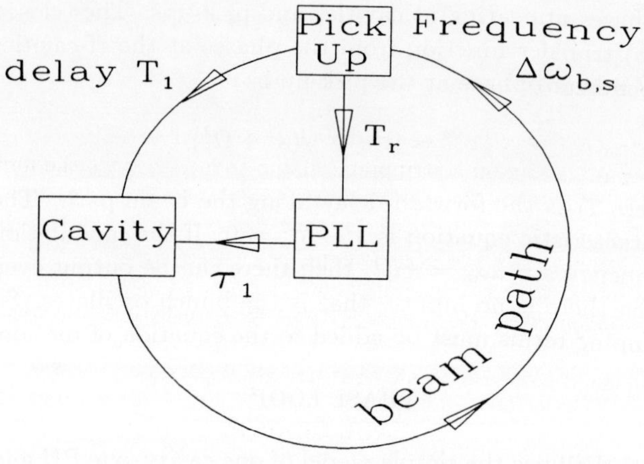

Fig. 3. Radial loop alone.

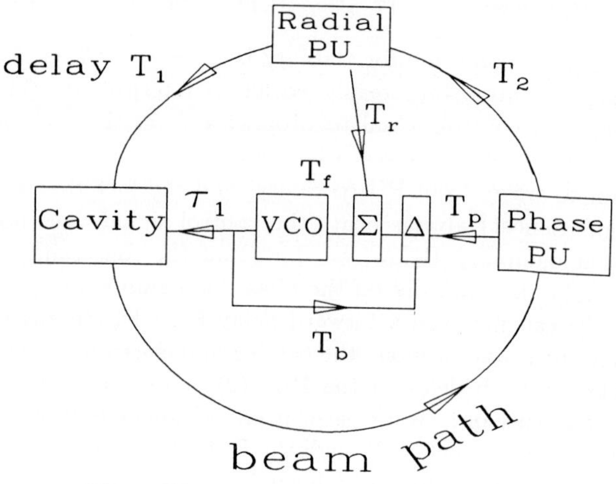

Fig. 4. Phase and radial loops combined.

STABILITY OF HOLLOW PHASE SPACE DISTRIBUTION

S. Koscielniak

TRIUMF, 4004 Wesbrook Mall, Vancouver, B.C., Canada V6T 2A3

ABSTRACT

The purpose of this note is to state, definitively, that a hollow distribution in longitudinal phase-space is non-stationary under the actions of either space-charge or a phase-synchronization loop; and to vehemently dissuade machine designers from adopting such a distribution, despite any apparent advantages it may offer.

INTRODUCTION

Hollow phase-space distributions have been advocated for the TRIUMF Accumulator and EHF Booster. In the Accumulator, the annular distribution allows synchrotron oscillations to "pull" beam off a stripping foil and thereby reduce the number of foil traversals and so minimize the increase in transverse emittance caused by scattering. In the Booster, the hollow distribution allows the bunch-shape to be tailored; giving a long flat bunch with very low bunching factor to minimize the transverse incoherent space-charge tune spread. In the absence of any published instability theory, both these applications were presumed to be (at least) quasi-stationary.

SPACE-CHARGE

Computer simulations with the LONG1D[1] macro-particle code in the summer of 1987 suggested that an annular distribution breaks up. The disruption time was found to depend on the beam current and the width of the annulus. The result was received with scepticism, and so the space-charge modelling was thoroughly reviewed. The earlier space-charge algorithm was replaced by a multi-step second-order symplectic integrator, but the general finding remained unaltered: the hollow distribution is not stationary under space-charge. Independently, Maidment, Planner and Prior of the Rutherford Appleton Laboratory have confirmed the disruption of the annulus; however, their results are not identical to Koscielniak – indicating that space-charge simulations have still not reached complete maturity.

With hindsight, it is easy to formulate a mechanism for the redistribution of particles that occurs, and to produce arguments against the possibility of a stationary hollow distribution.

The annular distribution produces a double-humped bunch shape. The space-charge force is proportional to the derivative of this shape. An initial left-right asymmetry in particle density per unit length causes the synchrotron oscillation frequency to differ from one side of the bunch to the other. This allows a net transferral of particles, enhancing the initial asymmetry in the bunch-shape, a mechanism for disruption.

PHASE LOOP

At the Snowmass meeting, Pedersen reported that experiments at the CERN PS with hollow emittance beams showed them to be non-stationary under the action of the phase synchronization loop. Nominally, the loop acts to damp dipole mode oscillations, using the difference of beam and cavity phase as control signal. The LONG1D code is equipped with a phase (and radial) loop model, and has been used to confirm that the bunch is non-stationary. The beam sensor detects any left-right asymmetry between the two peaks of the double-humped bunch, and acts to shake any clustering of particles toward the centre of the rf-bucket. Figure 1 shows a typical time sequence. From operational experience with ISIS, Rees (of RAL) has pointed out that a signal limiter which clips off the two bunch-peaks restores the correct functioning of the phase-loop; and this is also verified by simulation.

CONCLUSION

Hollow distributions have been tried in the dock and found guilty of gross misdemeanour. It remains for machine designers to pass sentence.

Note, the prosecution has shied away from using the term "instability", preferring instead "non-stationary". Work is now in progress which attempts to produce theoretical estimates for the dipole instability growth rate as a function of beam current and annulus width. Time will tell which of the above two terms is more appropriate.

REFERENCES

1) Koscielniak, S., LONG1D User Guide, TRI-DN-88-20 (1988).

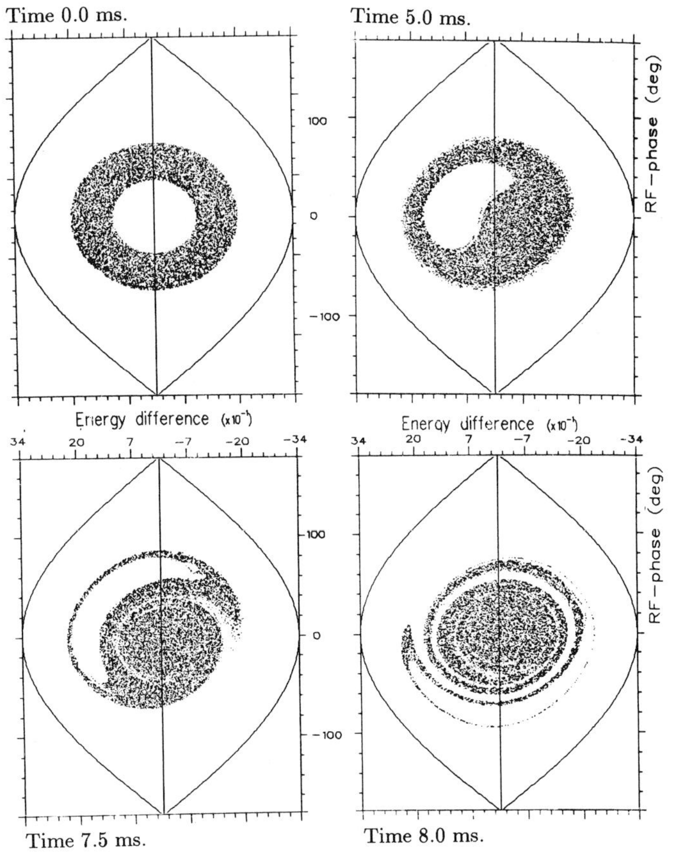

Fig. 1 : Longitudinal phase-space scatter plots for hollow bunch with phase loop control.

LONGITUDINAL EMITTANCE DILUTION STUDIES FOR THE COLLECTOR RING

S. Koscielniak
TRIUMF, 4004 Wesbrook Mall, Vancouver, B.C., Canada V6T 2A3

ABSTRACT

We report the results of computer simulation of forced emittance growth due to a high frequency cavity placed in the collector ring of the proposed TRIUMF KAON Factory. The emittance dilution is studied as a function of perturbation amplitude ρ and modulation frequency ω_d. Generally, the growth rates are disappointingly small. Nevertheless, the combination $\rho = 0.3$, $\omega_d = 4\Omega_s$ can produce the required 4-fold increase.

INTRODUCTION

The microwave instability occurs if the longitudinal emittance (ε_L) of a beam of given intensity is too small, for a given longitudinal coupling impedance $Z_{||}$. Stability is determined from the Keil-Schnell criterion.[1] In the 3 GeV Booster ring, the emittance must be kept small (nominally 0.048 eV.s) to avoid excessive momentum spread ($\Delta p/p$). However, there is no problem with instability provided $(Z_{||}/n) < 10\ \Omega$. The relation between revolution frequency and momentum is

$$\eta \equiv \partial \ln(\omega)/\partial \ln(p) = \gamma_t^{-2} - \gamma^{-2} .$$

In the 30 GeV Driver ring, the Keil-Schnell criterion becomes increasingly stringent due to the dependence on η. Because the transition energy (γ_t) is imaginary, η decreases monotonically. Thus at extraction, $\varepsilon_L = 0.048$ eV s requires the ratio of impedance to mode number be $Z/n < 1\ \Omega$. This condition cannot be met, and so the emittance must be increased by a factor 4 to 0.192 eVs.

Fortunately, the Collector intervenes between B and D rings and provides the opportunity to artificially dilute the Booster emittance. The C ring receives a batch of bunches from the B ring once every 20 ms. After 5 batches are accepted they are transferred to the Driver. There is a wait time of 20 ms before the transfer during which all 5 batches circulate concurrently, and this time can be used for controlled emittance dilution.

EMITTANCE DILUTION PROCEDURE

The basic technique is to perturb the total rf voltage by adding an auxiliary high frequency cavity (HFC) so as to mimic phase and amplitude noise. The method has been in use at the CERN PS[2] for several years. Emittance growth due to pure/real noise would be too slow. Let ϕ be the individual particle phase, and Ω_s be the synchrotron frequency. Including the effect of the HFC, the basic equation of motion is

$$\ddot{\phi} = \Omega_s^2 \big(\sin\phi + \rho \sin[N\phi - \delta(t)]\big) .$$

The free variables which we can be used to optimise the growth rate are: ρ, the perturbation amplitude; N, the multiple of the main rf; and $\delta(t)$, the phase modulation function.

ρ should be small, to reduce the power requirements. Keeping ρ small also minimises the uncontrolled emittance growth which arises when the HFC is *suddenly* turned on. N should be large (~ 20) so that the macroscopic structure imposed on the bunch is of short wavelength. Making N a non-integer has two advantages. Firstly, the beam induced voltage on the HFC is minimised. Secondly, each bunch is treated slightly differently which provides decoupling. Following Kats[3] the modulation function is: $\delta(t) = \alpha \sin(\omega_d t)$ with free parameters α, the phase modulation amplitude; and ω_d the phase modulation frequency. We take $\alpha = \pi$, the maximum possible value.

RESULTS OF THIS STUDY

Using the simulation programme LONG1D[4] and an ensemble of 8000 macro-particles, we set out to make a systematic study of the growth rate as a function of ρ and ω_d. We selected $\rho = 0.1, 0.2, 0.3$; and ω_d equal to integer and half integer multiples of the synchrotron frequency spanning the range Ω_s to $6\,\Omega_s$.

It quickly became evident that rate was an inadequate yard-stick with which to judge one case against another. There is no obvious general scaling law which relates ρ, ω_d to rate. The emittance grows for, apparently, two reasons. There is linear, driven growth. This is preceded by much faster growth due to mismatching when the HFC is turned on. The contribution to the total emittance increase due to mismatching varies considerably. The growth rates at the core and periphery of the occupied phase-space differ and leads to the production of an halo. The degree of halo formation also varies widely among the cases. Consequently, we are forced to tabulate a variety of statistics to indicate the performance of each (ρ, ω_d) combination:

(i) the time for the rms emittance to quadruple
(ii) the linear part of the $4\pi \times \varepsilon_{\rm rms}$ growth rate
(iii) the linear part of the 100% ε_L growth rate
(iv) $4\pi\times$(rms emittance) after 15 ms
(v) 100% emittance ($\varepsilon_{\rm Tot}$) after 15 ms
(vi) Ratio $4\pi\varepsilon_{\rm rms}/\varepsilon_{\rm Tot}$.

When item (i) is greater than 20 ms, then an estimate is given. The rms emittance is indicative of the core, and the 100% emittance suggests the degree of halo. The initial phase-space density was uniform, and so (ideally) $4\pi\varepsilon_{\rm rms}/\varepsilon_{\rm Tot} = 1$. Thus the ratio (vi) is an indicator of halo increase. The initial emittance for our simulations was 0.0395 eV.s and the rf-bucket area was 0.308 eV.s. The main rf was 61 MHz and the synchrotron frequency ($\Omega_s/2\pi$) was 0.05273 MHz.

Table 1 : Emittance dilution results for $\rho = 0.1$

ω_d/Ω_s	Time ($msec$)	$\dot{\varepsilon}_{\rm rms}$ (eV)	$\dot{\varepsilon}_{\rm Tot}$ (eV)	$\varepsilon_{\rm rms}$ (eV.s)	$\varepsilon_{\rm Tot}$ (eV.s)	Ratio
1.0	5e+3	.023	.11	.040	.043	1.05
2.0	7e+3	.015	.018	.039	.040	1.04
2.5	2e+3	.06	.20	.044	.051	1.15
3.0	3.6e+3	.031	.15	.040	.045	1.14
3.5	2.5e+3	.046	.089	.041	.047	1.14
4.0	85	.13	.13	.061	.091	1.5
4.5	3.5e+3	.032	.080	.041	.047	1.14
5.0	7.6e+3	.015	.064	.039	.043	1.07
6.0	7.5e+3	.015	.015	.040	.042	1.07

There is definite evidence in table 1 for some resonance phenomena at $\omega_d = 4\Omega_s$. However, the growth times (several seconds) are very long and wholly inadequate for our purposes.

Table 2 : Emittance dilution results for $\rho = 0.2$

ω_d/Ω_s	Time ($msec$)	$\dot{\varepsilon}_{\rm rms}$ (eV)	$\dot{\varepsilon}_{\rm Tot}$ (eV)	$\varepsilon_{\rm rms}$ (eV.s)	$\varepsilon_{\rm Tot}$ (eV.s)	Ratio
1.0	4e+3	.033	.13	.042	.043	1.04
2.0	700	.16	.23	.045	.049	1.09
2.5	66	1.5	2.1	.078	.105	1.35
3.0	84	.14	.33	.060	.098	1.63
3.5	78	.14	.32	.064	.105	1.64
4.0	40	2.1	2.5	.106	.139	1.31
4.5	43	1.9	2.6	.101	.141	1.40
5.0	55	1.9	2.3	.078	.108	1.40
6.0	34	2.37	3.8	.109	.163	1.49

Though less marked, there again appears to be a resonance phenomenon occurring when $\omega_d = 4\,\Omega_s$. If smaller growth could be tolerated, this could be used to advantage; allowing a smaller voltage on the HFC.

Table 3 : Emittance dilution results for $\rho = 0.3$

ω_d/Ω_s	Time ($msec$)	$\dot{\varepsilon}_{\rm rms}$ (eV)	$\dot{\varepsilon}_{\rm Tot}$ (eV)	$\varepsilon_{\rm rms}$ (eV.s)	$\varepsilon_{\rm Tot}$ (eV.s)	Ratio
1.0	1e+4	.01	.06	.040	.042	1.06
2.0	150	.68	1.4	.060	.077	1.28
2.5	37	2.2	.89	.105	.123	1.17
3.0	26.5	2.6	5.7	.124	.180	1.45
3.5	16.7	4.9	6.4	.146	.227	1.55
4.0	16.5	4.3	8.8	.145	.229	1.58
4.5	15	4.91	11.3	.154	.272	1.76
5.0	13.75	5.1	15.	.160	.305	1.9

Evidently, only $\rho = 0.3$ can produce the 4-fold increase of ε_L in 20 ms. The phase-space scatter plots show a (very) small fraction of particles to have diffused across the separatrix when $\omega_d = 5\,\Omega_s$. Consequently, to minimise the creation of halo, the modulation frequency should be in the range $3.5\,\Omega_s \leq \omega_d \leq 4.5\,\Omega_s$. This pretty much sets the HFC specification.

CAVITY SPECIFICATION

Note the HFC resonant frequency is 1.2 GHz, thus design and fabrication present non-trivial problems. Note also that $\rho = 0.3$ implies one third of the main rf voltage, and so probably more than one HFC is needed. Further, note that $\alpha = \pi$ implies large-amplitude phase-modulation. There are significant side-bands up to $\pm 6 \times \omega_d$ from the carrier frequency. Assuming $\omega_d = 4\,\Omega_s$ we find a bandwith of 1.3 MHz. This implies a low quality factor (~ 770) for the cavity, and low shunt resistance. If α could be reduced to $\pi/3$ the bandwidth could be halved, allowing the shunt resistance to double and the generator current to halve.

CONCLUSION

We have demonstrated that the necessary 4-fold emittance dilution can be achieved with an HFC. However, the performance is unsatisfactory (with high power demands) and other methods, such as modulation of the main rf system, should be investigated before the HFC scheme is adopted.

We have not considered the influence of α, nor the effect of time dependent $\rho(t)$. Both of these should be investigated at a later date. In order to isolate the effect of the HFC, the simulations have ignored space-charge and beam-loading. These beam-effects should also be researched.

REFERENCES

1) Boussard, D., CERN LAB II/RF/Int/75-2.

2) Boussard, D., The PS 200 MHz RF system, CERN-SPS/ARF/78-6.

3) Kats, J., Particle diffusion produced by HFC, in Proc. 1987 Particle Accelerator Conf. (IEEE, New York, 1987), p. 1281.

4) Koscielniak, S., LONG1D User Guide, TRI-DN-88-20.

MAIN RING RF-CAVITY

George R. Swain
Los Alamos National Laboratory, Los Alamos, New Mexico 87545

ABSTRACT

A prototype rf cavity for a hadron facility main ring is being designed as a part of a collaborative effort between LANL and TRIUMF. The cavity uses an orthogonally biased ferrite tuner to provide a 49.3 to 50.7 MHz (3%) range. The design provides for accelerating gap voltages up to 200 kV, with an expected shunt impedance over Q of 36 ohms. Progress on the design and fabrication of the cavity is discussed.

PROGRESS ON DESIGN AND FABRICATION

The prototype version of a main ring cavity presented here is a modified quarter-wave coaxial resonator in which the shorted end has been divided into two legs, with the tuner and rf drive on the leg which is off the beam line,[1] as shown in Fig. 1. Two ferrite

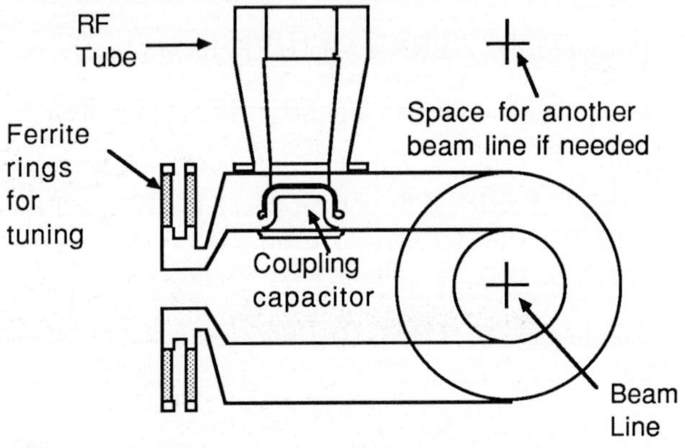

Fig. 1. Cavity section looking along beam line.

rings from a previous LANL booster cavity prototype are to be used in the tuner, which is designed to give the same percentage frequency range as is needed for the D ring of the KAON proposal at TRIUMF.

For design purposes, bias field data from the POISSON code was fed to a code using a transmission line model for the complete cavity, and this data in turn was fed to a code to average the results over an accelerator cycle. Parts of the cavity were modeled using SUPERFISH in order to optimize the transmission line model. For example, the field in the end of the tuner leg is shown in Fig. 2. The transmission line model could be made to agree with SUPERFISH to better than 1.6 MHz for the modes from 50 to 160 MHz.

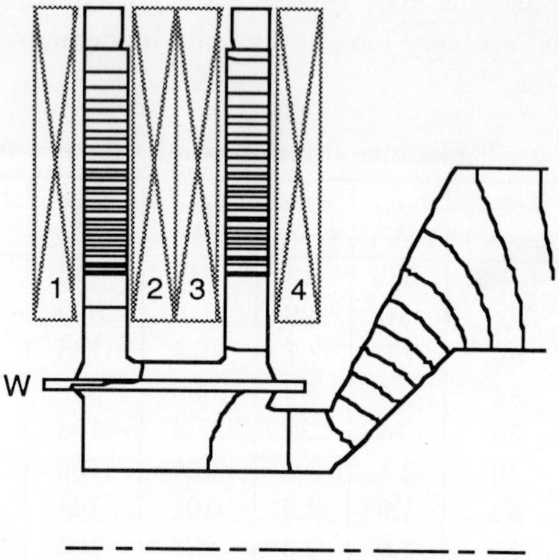

Fig. 2. Section through the end of the tuner leg, as used in a SUPERFISH model. Bias coil positions are also shown.

In Fig. 2, W indicates a ceramic window which isolates the part of the cavity with ferrite from the vacuum. The copper wall of the cavity between bias coils 2 and 3 and the ferrite is slotted to increase the bandwidth of the tuning control loop.

The four bias coils are being fabricated now, and the remainder of the cavity is to be built later this year. It is planned to begin bench tests of the cavity and its rf drive system in early 1989, and then to test the cavity with coasting beam in the Proton Storage Ring at Los Alamos.

REFERENCES

1) Swain, G. R., "Design of a Main Ring Cavity," *Proceedings of the Accelerator Design Workshop*, Los Alamos, NM, 22-27 Feb. 1988, LA-11432-C (to be published).

ACCELERATOR PHYSICS – HADRON COLLIDERS

TEVATRON IN THE 1990s*

LEON M. LEDERMAN
Fermi National Accelerator Laboratory
Batavia, Illinois 60510

I. INTRODUCTION

In 1978, Fermilab set out a goal of building a superconducting accelerator (Energy Saver) which would raise the proton energy to close to 1000 GeV for operation in two modes. Tevatron I would provide proton-antiproton collisions at a total CM energy of near 2.0 TeV to study the particle mass domain beyond 100 GeV. Tevatron II would provide extensive facilities for the programmatic study of Standard Model physics in an upgraded fixed-target program. There was of course the realization that with the right mixture of precision and imagination, the collider could add significantly to Standard Model physics (e.g. W and Z physics, W, Z pairs, B-physics) and that the fixed-target program could explore beyond the Standard Model (e.g., rare K-decays, CP violation). In 1988, we are engaged in setting out the future program of the Laboratory based upon the success of the Energy Saver, TeV I and TeV II construction programs. This future program assures that operation of the TEVATRON facility for physics is the overriding priority between now and perhaps 1993 and it also assumes that the Superconducting Super Collider (SSC) will be funded for construction in 1990 and will begin producing physics by 1999. A "brief history" of upgrades is presented in section XI.

II. HISTORY

The notion of going to higher luminosity in the Collider and more intensity and quality for the fixed-target program has been around since the start of TeV I and TeV II. The simply stated goal in collider physics is to increase the mass range which can be searched for new phenomena and in the fixed-target program to enhance the precision and the detail of our Standard Model base. In Laboratory presentations we have proposed a Superbooster (1980), Dedicated 4 TeV Collider (1983), Brightness Enhancer (January 1984), and Source Brightener (September 1984). Upgrade plans and funding profiles were presented in the 1986, 1987 and 1988 institutional plans. Responses from HEPAP have been positive going back to 1982.[†] Experience with the first engineering run of TeV I in 1985 and the 1986 construction year led to a thorough review of the entire accelerator complex. A Collider upgrade plan was submitted (short form 44) with a TPC of $267M in January 1986.

⋆ This is based upon the work of many people over a long period of time. In particular Steve Holmes and Estia Eichten have been most helpful.

† January 1982 Subpanel on Long Range Planning—Excerpts (p. 29) "The achievement of a luminosity greater than $10^{30}\,\mathrm{cm}^{-2}\,\mathrm{sec}^{-1}$ will, in our judgment, take some years of operational experience. On the other hand, a number of improvements seem possible. Thus, an ultimate goal of $L = 10^{31}$ appears reasonable to us."
"The TEVATRON projects will be the focus of a major part of the U.S. program ... they will open up entirely new areas of physics and accelerator development and will be essentially unique in the world."
July 1983 Subpanel on New Facilities (p. 51) "The viability of the [TEVATRON] facility after about 1992 will depend on the physics interest and the availability of other facilities. If the level of research activity remains high, then an upgrade of the facility and its detectors may be warranted, with a consequent extension of the useful life of the machine for perhaps another five years."
September 1985 Report of the 1985 HEP Study (p. 27) "Because new phenomena may not conform to our current expectations, it is natural to expect the configuration of these detectors [CDF-D0] to evolve in response to our growing understanding ... A program of detector upgrades and accelerator improvements will be an essential part of the hadron collider physics program."

In fixed-target experiments ... experiments can be grouped in terms of the physics questions ...
(1) CP violation in Kaon Decays
(2) Rare Kaon Decays
(3) Heavy Quark Physics
(4) Hadron Dynamics Other than Perturbative QCD
(5) Neutrino Oscillation Experiments
(6) Particle Searches with Beam Dump

As the first phase, the Linac Upgrade was submitted in January 1987 and resubmitted in February 1988. The plan has emerged into two stages: an adiabatic series of improvements which will bring the peak luminosity of the $\bar{p}p$ collider to about 5×10^{30} cm^{-2} sec^{-1}. This should also make over 3×10^{13} *ppp* available to the fixed target, an improvement of almost a factor of two. The Collider energy would be 1.0 TeV and the fixed-target energy near 900 GeV. Given reasonable R&D funds and the Linac line item, all of this should be available for a D0 and CDF run of $\int L dt > 10 pb^{-1}$ in 1992.

In the period until 1993, there would be no planned shutdown in excess of several months for installation of upgraded components. This period would also see modest upgrades to Collider Detector at Fermilab (CDF) and some decisions on major new detectors and upgrades for the fixed-target program. In 1993, one can contemplate a 6–10 month shutdown for the second phase of the upgrade. This would be designed to deliver in excess of 100 pb^{-1} per run to the collider detectors and in excess of 4×10^{13} *ppp* for the fixed-target program. Given enough protons, it will pay to improve the fixed target duty cycle even more—perhaps from 30% to 60%.

There are now several competing elements for the second phase of the upgrade. The purpose of this note is to review these which, at this writing, are evolving out of extensive high-energy physics (HEP) community discussions.

III. REVIEW OF UPGRADE MOTIVATION

The Fermilab collider is the highest energy machine in the world. Until SSC or LHC or the Soviet 3 GeV × 3 GeV collider turn on and begin to produce physics data, this will remain so. We believe we have a time window that will go to 1999 or so since it will take several years for any of the above machines to go from commissioning to real physics. The window is not only an opportunistic window, it is essential that there be continuity in the production of physics results. Whereas, if SSC is proceeding towards, say, a completion date of 1997–98, a fairly large community will be occupied there by 1992, but one cannot put graduate students, new postdocs and pre-tenure professors on many of the SSC detectors until they are much closer to physics. This is borne out by CDF and D0 experience. The Fermilab Collider physics in the period 1994–1999 will also be invaluable as a guide to SSC both from the point of view of collider and detector technology but also from the physics knowledge base. Since a year of SSC is worth $250M (1988), it is terribly cost effective to be as well prepared for the SSC era as one can possibly be. Finally we note that there *may* well be niches of physics for which TEVATRON energy is well enough above threshold; a vast increase in energy may then only increase backgrounds.

The knowledge base will come from both the Fermilab Collider *and* the fixed-target program, especially those experiments which illuminate high-rate technology and those which use precision and detail to test and extend the Standard Model.

To present a glimpse of the relative merits of the various upgrade options we present a series of graphs calculated by E. Eichten. See Figs. 1, 2, and 3. We stress that whereas the optimum plan is not yet clear, what is perfectly clear is *that the design goals are such as to double the discovery limits*, i.e., equivalent to doubling the effective machine energy. Furthermore, it makes possible the collection of huge amounts of data for particles in the W, TOP, e.g., $\leq$ 125–GeV mass range.

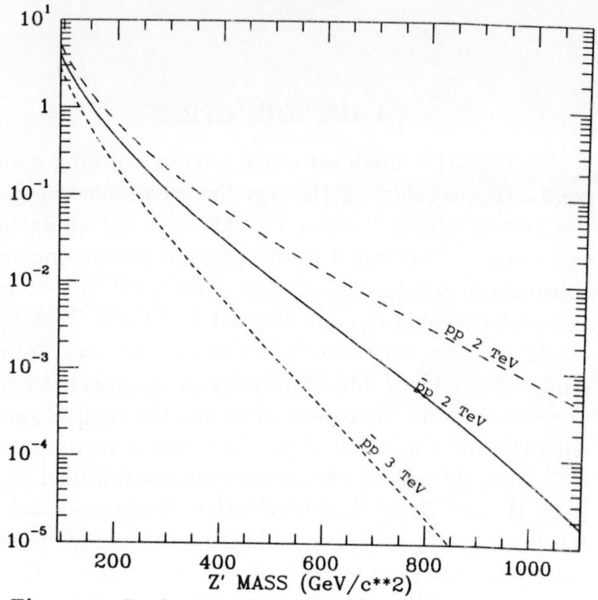

Figure 1. Production of a heavy Z^0. The mass reach is determined by the integrated luminosity and the discovery level required, e.g. 100 produced events. See Chart I for results.

A doubling of the mass reach could be compared to building a 400–GeV e^+e^- machine with sufficient luminosity to double the mass reach of LEP II. Another comparison scale is the current attention to B-physics and proposals for electron-positron B-factories. An upgraded TEVATRON has impressive capabilities here although the issue is complicated by backgrounds.

The potential for discovery of new physics by our upgrade or for the clarification of discoveries which may be made in the early stage of TEVATRON are very significant. We also stress the important support this kind of data gives to SSC where the parameter $M/\sqrt{s}$ will very rarely reach the Upgrade goal of ~ 0.4.

Chart I

Numbers of Produced Events in a 5 Mo (3×10^{6} sec) Year Based on Assumed $\mathscr{L}p$

$\mathscr{L}p\rightarrow$	No Upgrade	Phase I	Phase II		
	3×10^{29}	4×10^{30}	4×10^{31} A.	2×10^{32} B.	1×10^{31} C.
CM Energy	2 TeV	2 TeV	2 TeV	2 TeV	≥ 3 TeV
Int $\mathscr{L}$	$\bar{p}p$ $1pb^{-1}$	$\bar{p}p$ $10pb^{-1}$	$\bar{p}p$ $100pb^{-1}$	pp $500pb^{-1}$	$\bar{p}p$ $30pb^{-1}$

Discovery Limits (•) — ''Factory'' Regime

Z^o

Mass	No Upgrade	Phase I	Phase II A	Phase II B	Phase II C
200	500•	5000	50 K	100 K	20 K
400	30	400•	4 K	3.5 K	2.4 K
600	--	40	400	200	300
800	--	7	70 •	15	75
1000	--	--	10	--	30

TOP

Mass	No Upgrade	Phase I	Phase II A	Phase II B	Phase II C
75	300	3000	30 K	150 K	30 K
100	70•	700	7 K	25 K	6 K
125	20	200	2 K	5 K	1.8 K
150	8	80•	900	2 K	900
175	3	30	300	500	380
200	1	10	100 •	250	150
250	--	--	--	25	--

Gluino

Mass	No Upgrade	Phase I	Phase II A	Phase II B	Phase II C
100	300	3000	30 K	150 K	30 K
150	20•	200	2000	10 K	3 K
200	2	20•	200	800	500
250	--	3	30 •	100	90
300	--	--	5	10	30

W-Pairs

	No Upgrade	Phase I	Phase II A	Phase II B	Phase II C
W^+W^-	5	50	500	1000	240
Z^oZ^o	--	7	70	150	40
W^+Z^o	0.7	7	70	200	30

Technipions

Mass	No Upgrade	Phase I	Phase II A	Phase II B	Phase II C
50	800	8000	80 K		62 K
100	200	2000	20 K		18 K
150	70•	700	7 K		6.5 K
200	30	300	3 K		3 K
250	12	120 •	1.2K		1.4 K

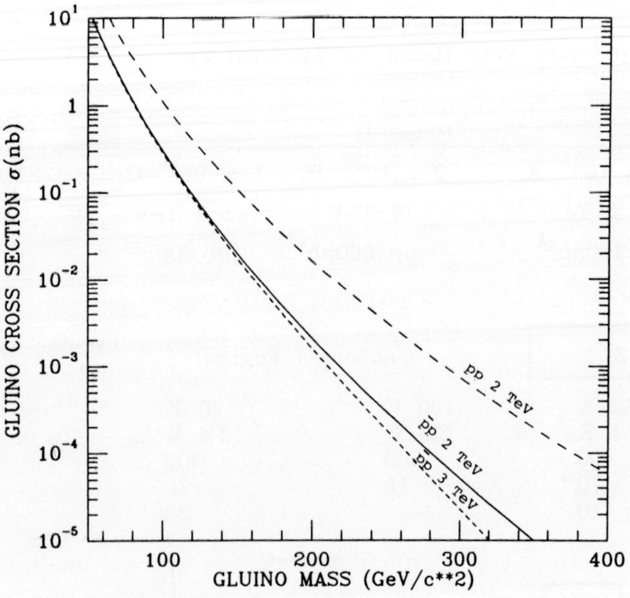

Figure 2. Production of Gluinos

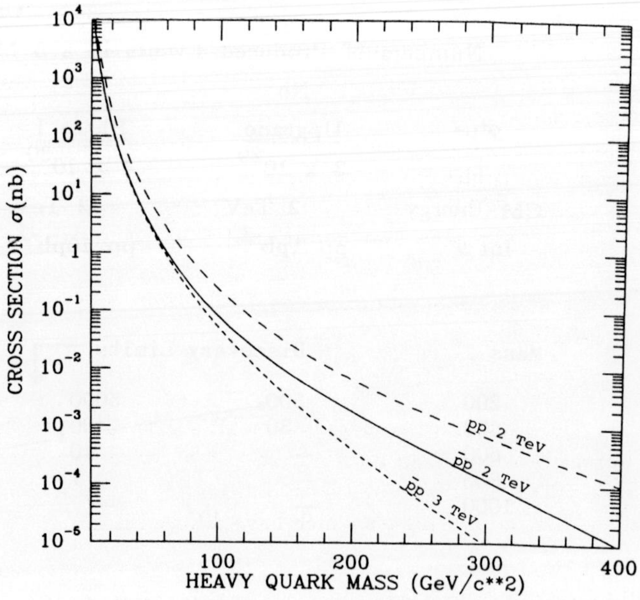

Figure 3. Production of Heavy Quarks

Advancing fixed-target physics will be critically dependent upon advancing the art of detectors. Exploiting higher luminosity in the collider also requires confidence that the detectors are up to resolving signal and background in the high rate environment.

IV. UPGRADE: PHASE I

Goal 5×10^{30} cm^{-2} sec^{-1} and 3×10^{13} *ppp* 1988–1993

The first phase involves a series of steps:

1. Replace Cockcroft-Walton by RFQ, new first tank on linac.
2. Replace last 5 linac tanks by side-coupled cavity type of tank at 800 Mhz (instead of 200 Mhz). This will raise the energy of protons injected into the Booster to 400 MeV. The transverse emittance should go to 12π mm.mr or even as low as 6π at $1 \times 10^{10} p$.
3. Strong low-β quadrupoles for D0, CDF; Goal $\beta^* = 0.25m$.
4. Possible shorter bunches.
5. $\overline{p}$ Source and cooling improvements.
6. Dipole magnet development for separator space; goal is 6.6T dipoles.
7. Cryogenic developments to achieve about $3.9°K$ TEVATRON for 900-GeV fixed-target and 1000–GeV Collider operations.
8. Electrostatic separators - helical orbits. 50 $kV/$cm for 2.5cm gap.

These steps, carried out by AIP, R&D and Linac Line Item funding can and should be complete in time for a 1992 run of CDF and D0 with a goal of $\geq 10pb^{-1}$. Included here are already scheduled improvements in the CDF detector, completion of the D0 detector and new starts on a major fixed-target spectrometer, given PAC approval. Other fixed target activity involves continued upgrades of major existing detector facilities.

V. UPGRADE PHASE: II

Goal 5×10^{31} cm^{-2} sec^{-1} and 4×10^{13} *ppp* at $> 50\%$ Duty Factor 1994–1999

Introduction

We have looked in some detail at several approaches to this next factor of ten. The luminosity *goal* is designed to keep the CDF and D0 detectors from melting, but this luminosity will require substantial upgrades to both detectors. These involve replacement of front-end electronics, perhaps central tracking and perhaps some calorimetry improvements. Consideration is also being given to a possible third collider detector, which would be specifically designed to do B-physics. What is also open is whether this gets its own collision region or goes in to alternate with CDF, say. Finally, considerable weight is given to the fixed-target program and how it is benefitted from the various options. Whereas the TEVATRON Collider mode may be supplanted by SSC, the fixed-target program will probably extend well into the SSC era, taking advantage of SSC detector R&D, the almost certain need for more precision and detail, and the continuous need for test beams. We now list the options as currently understood and later indicate some variations and phasing possibilities.

A. $\overline{p}p$ with Superboosters

Here, in order to supply two IR's with 5×10^{31} luminosity we need an improved source fed by an improved Main Ring (MR) and a place to store 3×10^{12} $\overline{p}$'s. Some means of recovering $\overline{p}$'s which have diffused is also useful.

The major devices here are two 20–GeV rings; one, the proton superbooster, injects into the Main Ring at 20 GeV yielding high transmission, small emittance ($\leq 12\pi$), good lifetime and high proton intensity for proton production. The second ring, a $\overline{p}$ ring, is an antiproton depository. This would also involve 8–16 GHz cooling in the $\overline{p}$ accumulator and depository. The total cost including R&D, pre-op is $124M. The technical problems of actually achieving 5×10^{31} are formidable. A more conservative goal is to have a 5-month run (repeated annually) to yield an integrated luminosity of 100 pb^{-1}.

B. pp Option

This suggests a pp option, where more than 5×10^{31} is assured and an overall efficiency of twice the $\overline{p}p$ option seems reasonable. We then assume that we can collect 500 pb^{-1} in a collider portion of one year run. Higher luminosity, i.e., 2×10^{32}, can be achieved for special purposes *not including* the normal operation of the CDF and D0 detectors. Another virtue of pp is the small interaction diamond which benefits all short lifetime experiments, e.g., B-physics. The pp option, as an accelerator project, is not particularly challenging. However, it requires removing the MR from the tunnel (it becomes the Main Injector) and providing a 120– to 150–GeV tunnel into which MR components would go. All overpasses and other TeV-MR hindrances would disappear. The new injector could also be a $\overline{p}$ producer and be organized to provide 150–GeV test beams during Collider operation. MR removal would allow space for a second superconducting magnet string. Longer straight sections would be needed in order to bring beams into collision. This could be done with small displacements of the CDF and D0 detectors. The total cost estimate here is about $240M.

C. $\overline{p}p$ High Energy Option

A third option removes the MR and/or the TEVATRON from the old tunnel and replaces it with a ring of 6.6 to 8T superconducting magnets of the SSC/HERA style. This would permit $\overline{p}p$ operation at 3–3.5 TeV in the CM. Since there would be no superboosters, the luminosity would be only slightly better than 5×10^{30} that was achieved in Phase I. Both CDF and D0 detectors would work well here with much less extensive upgrades than option B. The mass reach of such a 3.5 TeV $\overline{p}p$ collider is about that of a 2.0 TeV collider at $> 5 \times 10^{31}$. The fixed-target program can gain substantially from a higher energy extracted beam and/or the higher intensity of secondary and tertiary beams. The improvement factors come from the benefits of a redesigned Main Ring (Main Injector) and the luminosity gain due to the higher energy (1.5–1.8 TeV). We take 10^{31} as the design luminosity and therefore an integrated luminosity of 30 pb^{-1} per year. The energy increase could be a significant help in many fields, e.g., in heavy quark studies, in hyperon research and in structure function data. Open questions here have to do with the removal of the MR and the cost of higher field magnets. This will probably come to about the cost of the pp option.

VI. SELECTION CRITERIA

Which of the three options (or none of them) to choose will depend on a number of criteria:

(1) Physics reach in the collider mass domain "beyond the W, Z"
(2) Implication for advancing fixed target physics
(3) Cost
(4) Time and downtime to implement
(5) Detector Implications
(6) Technology experience relevant to SSC
(7) B-physics
(8) SSC at Fermilab? If so, it may favor one option more than others.

Many of these criteria are not simple. Physics reach with high luminosity is clouded by backgrounds, pile-up, etc. It may be useful to assume the following about detectors:

1) CDF requires new electronics at $\geq 5 \times 10^{30}$ at $10M

2) D0 requires new electronics at 5×10^{31}

3) CDF will require new tracking, vertex, etc. at $\sim 10^{31}$

4) Both detectors will require much more major upgrades at $> 5 \times 10^{31}$. However, even these upgrades, at an estimated total cost of $25M each, are much less in time, money and people than starting over.

5) With SSC demands, it makes no sense to contemplate a brand new "standard" 4π detector. New ideas, however ...

VII. PHASED OPTIONS

As phase II in the upgrade one can consider building a new main ring of 120–150 GeV in its own tunnel with new magnets but, initially, only minimal power supplies. This could be constructed and commissioned without interference with the on-going program. Its objectives: i) excellent injector into TEVATRON; ii) excellent $\overline{p}$ producer e.g. 2 sec/cycle; iii) provides 150–GeV beams to fixed-target program during collider runs, saving 2 months of calibration, timing, commissioning of fixed-target experiments; iv) may provide very intense neutrino and k-beams for special experiments. This would also free CDF and D0 from Main Ring backgrounds and provide a space for another interaction hall at E–ZERO. Also, it frees space in the existing tunnel for another superconducting ring. When completed and commissioned, there would be a shutdown for tunnel connections, moving of MR power supplies, etc., and perhaps removing MR magnets. This may be $\sim$ 3 months. Options B or C would follow as phase III.

Other phases, as demanded by physics and allowed by resources would be to upgrade from the modest luminosity of the 3 TeV option to perhaps 3–5 $\times 10^{31}$ using option A devices. Alternatively, if the original TEVATRON ring is still in the tunnel, pp collisions (1.5 TeV $\times$ 1.0 TeV) can be contemplated, especially for the B-detector but perhaps for additionally upgraded CDF/D0.

VIII. SUMMARY: PHYSICS

Our options as of July 1988 are now recapitulated. We assume a 5 month collider run, 5 month fixed target run and 2 months of changeover, studies, etc.

$$A.\quad p\overline{p}\sqrt{s} = 2\ TeV \qquad \mathcal{L}dt = 100\,pb^{-1}/\text{year}$$
$$B.\quad pp\sqrt{s} = 2\ TeV \qquad \mathcal{L}dt = 500\,pb^{-1}/\text{year}$$
$$C.\quad \overline{p}p\sqrt{s} \geq 3\ TeV \qquad \mathcal{L}dt = 30\,pb^{-1}/\text{year}$$

The physics graphs and Table take into account the different quark content of pp and $\overline{p}p$.

From the graphs and from the Table, it is clear that the TEVATRON upgrade has two physics benefits. Any of the options extends the discovery potential for a characteristic subset of theoretical speculations by a factor of two in mass: it permits a thorough exploration of the interesting 200–400 GeV mass domain—"the foothills of the TeV summit." Recall that in new Technicolor theories, the crucial parameter is $F_\pi = 246$ GeV.

Equally significant, for masses near the lower end, it provides "factory" potential. TOP is an excellent illustration. If, as some theorists intimate, the TOP mass is under 125 GeV, then the upgrade makes tens of thousands of TOP quarks per year and thus defines a TOP factory. This applies to many of the potential discoveries—one will be able to exploit the discovery of a GLUINO or TECHNIPION in some detail if the masses are not too high. Perhaps all the theories are wrong—still the exercise indicates that whatever nature has in the 50–400 GeV mass domain, the TEVATRON upgrade will be a powerful tool to guide particle physics on the correct road from the Standard Model toward the ultimate unification.

We have not yet listed some of the obvious "goodies" that have been widely discussed elsewhere:

<u>b-quarks</u>: The upgrade will result in of the order of 10^{10} $B\overline{B}$ per year pairs with option B giving 10^{11} $B\overline{B}$'s. Fermilab proposal P-784 has under design a detector which can carry this to the observation of CP violation.

<u>$W + Z$'s</u>: The 100 pb^{-1} luminosity yields 10^6 W's per year and 2×10^5 Z^0's. With precise Z^0 masses derived from e^+e^- machines and a highly precise mass ratio of W to Z, one can derive unique values for important radiative corrections which involve the Higgs mass.

Compositeness, Drell–Yan, Fourth Generation and many other processes and issues will also be addressed.

<u>Fixed Target</u>: Although we have stressed the benefits to the Collider, the gains to the fixed target are also important with option C probably having the largest influence. Here even a modest increase in energy gives a very large increase in, for example, photoproduced B's (factor $\sim$ 20). Secondary beams gain in energy and intensity, hyperon beams also gain from the increase in laboratory lifetime.

IX. FUNDING SCENARIOS

In our firm, unalterable 15–year plans we have presented funding profiles which have not noticeably produced cardiac arrest among DOE readers. Table X (Profile I) out of the 1988 Institutional Plan is typical. Below this is an alternative plan which assumes less civil construction and more R&D in the realization of the upgrade program. It assumes we do something between the costs of $\overline{p}p$ and pp or $\overline{p}p$ at high energy. The difference is $\pm$ \$10M/year. It includes funds for detector upgrades *and* fixed-target initiatives.

X. CONSTRAINTS

In guiding this discussion we have in fact made a number of constraining assumptions:

1. The non-SSC funding level of \$560M will not be increased during SSC construction.

2. SSC physics will be in full swing with first physics publications by ~ 1999.

3. The upgrade over the period 1989–1994 should require increments to the Fermilab budget of less than $50M/year.

4. No new 4π detector can be contemplated. CDF and D0 may be upgraded but not replaced. A special purpose new detector for B-physics *is* conceivable if its cost is modest compared to original CDF/D0 costs.

5. The upgrade should begin to produce physics by 1994-1995.

6. Until 1993 we plan no shutdowns in excessive of 6 weeks.

7. CDF and D-Zero must have at least 10 pb^{-1} of good data before a long (6–10 month) shutdown.

XI. A HISTORY OF UPGRADES

A. **Cornell**

300 MeV	1949
1 GeV	1954
2 GeV	1964
10 GeV	1968 (SLAC 20 GeV Linac)
8× 8 GeV	1979
8× 8 GeV Upgrade	1988

B. **BNL**

AGS 30 GeV Upgrade (linac)	1970 (Fermilab 200 GeV)
AGS Upgrade (booster etc.)	1988

C. **SLAC**

Linac	1967
Spear	1973
PEP	1979
SLC	1988
400 GeV e^+e^-	? 1992

D. **CERN**

Cyclotron	1958
PS	1960
ISR	1971
SPS	1976
S$\bar{p}p$S	1981
S$\bar{p}p$S → ACOL	1988 (TeV I Going)
LEP I	1989
LEP II	1992
LHC	?

E. **DESY**

DORIS	1974
PETRA	1977
DÓRIS Upgrade	1985
HERA	1990

F. **FERMILAB**

400 GeV	1972
TeV I	1987
TeV II	1984
UPGRADE	→ 1993 proposed

XII. RESUMÉ OF UPGRADE VIRTUES

1. Physics is first rate with very large discovery potential and strong programmatic power.

2. This is the highest energy machine in the world. It deserves the full exploitation compatible with realistic costs, time scale and manpower needs. It represents an investment of $500M in R&D, Equipment, construction and AIP funds. The history of upgrades also speaks eloquently to this.

3. HEP must maintain its excitement and its vitality, especially during the long construction schedule for the SSC. Discoveries, press releases, etc., will serve to keep the flow of new students and will insure the attention which is needed to secure a decent SSC funding profile.

4. The learning curve of new physics and of handling collider subtleties alone will pay the upgrade costs. These can modulate SSC detector design and will be relevant up to turn-on and beyond. CDF and D-Zero *must* learn to cope with subtle signatures at the level of 10^{-10} of the total cross-section. No amount of simulation substitutes for *learning by doing*. This acquired skill becomes the experience base of the SSC and is terribly cost effective at SSC annual costs of $250M/year. CDF and D-Zero at $> 10^{31}$ luminosity are *unique* sources of this learning curve.

Table X

Profile I

Laboratory Funding Summary

($ in Millions)

	Fiscal Years							
	FY 87	FY 88	FY 89	FY 90	FY 91	FY 93	FY 93	FY 94
DOE Effort	135.4	145.0	151.3	169.9	171.01	174.0	180.0	181.0
Work for Others	3.4	5.7	1.1	0.6	0.6	0.6	0.6	0.6
TOTAL OPERATING	138.8	150.7	152.4	170.5	171.7	174.6	180.6	181.6
Capital Equipment	27.5	25.3	30.8	37.0	34.0	32.0	40.0	42.0
Program Construction	8.2	11.2	4.6	43.4	63.4	30.3	0.0	0.0
AIP/GPP	8.4	9.1	9.5	17.0	14.0	14.0	14.0	14.0
General Purpose Equip	0.0	0.0	0.0	0.0	0.0	0.0	4.0	0.0
TOTAL LABORATORY FUNDING	182.9	196.3	197.3	267.9	283.1	250.9	238.6	237.6
PROPOSED CONSTRUCTION	0.0	0.0	0.0	0.0	0.0	0.0	0.0	0.0
TOTAL PROJECT FUNDING	182.9	196.3	197.3	267.9	283.1	250.9	283.6	237.6
*Outyear Escalation Rates (Base Year - FY 89)				3.6%	3.3%	2.8%	2.3%	2.3%

REPORT OF THE MULTI-TEV HADRON COLLIDER GROUP

M. Cornacchia and S.D. Holmes
Lawrence Berkeley Laboratory*, Berkeley, Ca. 94720
Fermi National Accelerator Laboratory+, Batavia, Il. 60510

ABSTRACT

This report summarizes the work of the Multi-TeV Hadron Collider group at the Snowmass Workshop. The group undertook to assess the feasibility of providing a hadron collider facility which would double the mass reach of the existing Fermilab Collider in the mid-1990 s. Such a facility could be based upon enhancement of the luminosity of the Fermilab Proton-Antiproton Collider, construction of a proton-proton collider utilizing the Fermilab Tevatron, or construction of a higher energy (1.5-2.0 TeV/beam) proton-antiproton collider. Performance parameters for these three possible options are discussed.

INTRODUCTION AND SUMMARY

The Multi-TeV Hadron Collider Group[1] attempted to assess the feasibility of and provide realistic performance parameters for a Hadron Collider capable of providing a factor of two extension in mass reach relative to the expected performance of the existing Fermilab Collider extrapolated into the early 1990 s. Such a collider would provide an opportunity for physics research in the mid 1990 s prior to the advent of the SSC, and would likely also afford physicists the opportunity for completion of certain specialized experiments during the SSC era. The Multi-TeV Hadron Collider group interacted with members of the physics study groups in order to define consistent parameters for use by those groups.

The group recognized that enhancements to the Fermilab Collider could come either in the form of increased luminosity or increased energy (or both). The group evaluated specific designs of hadron colliders within the context of integration into the Fermilab accelerator complex without building new large (>6200 meters circumference) tunnels. The group came to several conclusions:

1. A **proton-antiproton** collider operating at **1 TeV/beam** and a peak luminosity of $\mathbf{2x10^{31}\ cm^{-2}s^{-1}}$;
   A **proton-proton** collider operating at **1 TeV/beam** and a peak luminosity of $\mathbf{2x10^{32}\ cm^{-2}s^{-1}}$; and
   A **proton-antiproton** collider operating at **1.5 TeV/beam** and a peak luminosity of $\mathbf{6x10^{30}\ cm^{-2}s^{-1}}$

   all look feasible with a high level of confidence on approximately the same time scale (1994). The associated integrated luminosities per calendar year are estimated to be 100 pb^{-1}, 2000 pb^{-1}, and 30 pb^{-1} respectively.

2. 1.8 TeV/beam appears to be the highest energy which can be contemplated within the existing Fermilab tunnel. Achievement of 1.8 TeV is substantially harder than 1.5 TeV and would require a longer (1-2 years) development time.

3. Slow extracted beam can be provided from a 1.5 TeV superconducting accelerator in the Fermilab tunnel.

4. A new Fermilab Main Ring (or at least improved performance) is required in all three cases.

5. Existing detectors need to be modified to varying extents in all cases.

SCOPE OF STUDY

The primary goal of the group was to define achievable performance parameters versus energy and particle type for a Multi-TeV Hadron Collider. It was assumed that the scale of the collider should be such as to insure a factor of two increase in mass reach relative to the ultimate capabilities of the existing Fermilab Collider. In particular we limited ourselves to consideration of colliders small in scope compared to the SSC.

It is expected that with upgrades to the Fermilab Antiproton Source, the upgrading of the Fermilab Linac from 200 to 400 MeV, and the implementation of separators in the existing Tevatron, an ultimate luminosity of $5x10^{30}$ $cm^{-2}sec^{-1}$ with integrated luminosities of 10 pb^{-1} per (6 month) run will be achievable in the early 1990 s. Enhancement of the physics reach of the facility by a factor of two can be accomplished by raising the luminosity, raising the operating energy, or both. (As a rule of thumb a factor of two in energy is worth a factor of ten in luminosity). Fermilab has already completed design reports[2] for two separate facilities, either of which would fulfill the desired physics need: 1)a high luminosity ($5x10^{31}$ $cm^{-2}sec^{-1}$) proton-antiproton collider operating at 2 TeV in the center-of-mass; and 2)a high luminosity ($2x10^{32}$ $cm^{-2}sec^{-1}$) proton-proton collider also operating at 2 TeV. The first facility achieves increased luminosity by enhancing the antiproton storage capability of the existing Antiproton Source by a factor of ten through the addition of a new antiproton post-accumulator ring, while the second relies on the construction of a new superconducting accelerator within the Fermilab tunnel.

The Multi-TeV Hadron Collider group decided to accept the designs of the two high luminosity/2 TeV upgrades as given and to limit effort on these to a realistic evaluation of the stated performance parameters. The bulk of the effort of

---

*Work supported by the Director, Office of Energy Research, Office of Basic Energy Sciences, Material Science Division of the U.S. Department of Energy under Cotnract No. DE-AC03-76SF00098.

+Operated by Universities Research Association under contract to the U.S. Department of Energy.

the group was directed at identifying the maximum energy which could be reliably achieved in the existing Fermilab tunnel, and defining the associated parameters of a proton-antiproton collider based on such a ring and utilizing the full capabilities of the Antiproton Source as envisioned in the early 1990 s.

FERMILAB SPECIFIC GEOMETRIC CONSTRAINTS

The Fermilab Tevatron is designed to achieve 1.0 TeV beam energy with fields in the dipole magnets of 4.4 Tesla. However, **within the framework of the Tevatron tunnel geometry** certain constraints prohibit one from blindly scaling the operating energy with the magnetic field achievable in the superconducting dipoles. Such constraints arise from the use of fixed field magnets, such as kickers and Lambertson septa, to perform abort and extraction functions. At fixed field a higher energy can require an increased length for such magnets and, in general, longer straight sections. Such extensions of the straight sections can only be accomplished through the use of high field magnets near the ends of the arcs, the result being that all dipoles in the ring do not necessarily run at the same field.

A study was carried out to define the relationship between achievable energy and maximum required dipole field in the Fermilab tunnel. Requirements examined included aborts, low-beta regions, and, in the case of fixed target running, slow extraction of the beam. The following conclusions were reached:

1. A 1.5 TeV beam can be accommodated within the existing ring geometry without lengthening any straight sections. This means that **6.6 Tesla dipoles are required for 1.5 TeV/beam.** It was specifically concluded that 1.5 TeV beam could be slow extracted within this configuration.

2. A 1.8 TeV beam would require the lengthening of existing straight sections. The primary needs are related to the aborts and slow extraction. The requirement for dipole magnets operating at **1.8 TeV/beam** then becomes **8.0 Tesla (many)** and **9.0 Tesla (a few)**. There could possibly be some relief provided that, as indicated[3], internal beam aborts for collider operation could be implemented.

This means that a 1.5 TeV accelerator could be built with a geometry and lattice functions identical to the existing Tevatron. The lattice for such a machine would incorporate 6.6 T dipoles (described in the following section) and 1.4 T/cm quadrupoles (currently being built at Fermilab for the DO low-beta). The lattice for a 1.8 accelerator would be based on 8.0 and 9.0 T dipoles, and 1.5 T/cm quadrupoles. An example of such a lattice can be found elsewhere in these proceedings[4].

APERTURE REQUIREMENTS

The aperture requirements for high field magnets to be used in the higher energy proton-antiproton option have been estimated both at the injection energy (150 GeV) and a peak energy of 1.5 TeV. In the former case the aperture is dominated by the beam size and the separation between protons and antiprotons that is required to reduce the effect of the beam-beam interaction. At high energy the slow extraction process dictates the aperture size. Since AC losses preclude fast ramping at a field greater than 6.6 T (thus precluding also slow extraction at energies greater than 1.5 TeV) the aperture requirement at 1.8 TeV is not estimated. We have assumed that the dipole magnets will be bent to follow the beam trajectory so that the sagitta of the beam will not increase the aperture requirements, and have ignored the lattice outside the regular lattice cells.

Requirement at 150 GeV

The aperture calculation at 150 GeV is based on the following assumptions: 1)The maximum value of β_H and β_V in the arcs is 100 meters and the (95%) normalized transverse emittance at 150 GeV is 20π mm-mr. The corresponding beam size is 3.5 mm in both planes. 2)The maximum dispersion is 5.0 meters and the (95%) longitudinal emittance is 2.0 eV-seconds. This number is based on actual beam coalescing experience in the Fermilab Collider. With the existing Tevatron RF this gives a bunch length of 2.4 meters with a momentum spread of 1×10^{-3}. The total contribution to the horizontal beam size coming from momentum spread is then 5.0 mm. We add the transverse and longitudinal contributions in quadrature and allow 3σ to get a total horizontal beam size of 7.3 mm and a vertical beam size of 4.2 mm. 3)Injection errors require 1.0 mm of aperture. 4)Closed orbit errors require 1.0 mm of aperture. 5)Separated orbits must provide a beam-beam separation of five times the rms beam size--12.4 mm for horizontal separation. This separation is consistent with current operation of the CERN Collider.

Thus, the total required half aperture at 150 GeV becomes,

Horizontal beam size	7.3	mm
Injection errors	1.0	mm
Closed orbit errors	1.0	mm
Beam separation	6.2	mm
Total half aperture @ 150 GeV	**15.5**	**mm**

Adding a safety factor of 25% results in a good field aperture of $\pm$20 mm at 150 GeV.

Requirement at 1.5 TeV

For 1.5 TeV operation we again assume an invariant transverse emittance of 20π mm-mr with β=100 meters in each plane. The associated beam sizes (95%) are 1.1 mm. We assume that the longitudinal emittance is blown up to 4 eV-sec by intrabeam scattering. For a bunch length of 2.4 meters the momentum spread is then 2×10^{-4} and the associated beam size is 1.0 mm. Thus the maximum beam size at 1.5 TeV (3σ) is 1.8 mm. Closed orbit errors are assumed to be 1.0 mm as at 150 GeV.

For running protons and antiprotons on orbits separated by 5σ, an additional 3.0 mm is required.

However, this is small compared to the aperture requirement generated by slow extraction. The slow extraction requirement is based on the need for 99% efficient extraction within the superconducting accelerator. With a septum thickness of 0.1 mm this means a step size of 10 mm at the septum. For a half-integer resonant extraction system this step size can be obtained with a septum located about 10 mm from the unperturbed orbit resulting in a total aperture requirement of 20 mm. If we assume that the septum is at a β=200 meter point then the beam size within the arcs of the machine is 14.1 mm.

The required half aperture at 1.5 TeV is thus,

Horizontal beam size	1.8	mm
Closed orbit errors	1.0	mm
Beam separation	(1.5	mm)
Slow extraction	14.1	mm
Total half aperture @ 1.5 TeV	**16.9**	**mm**

Adding a safety factor of 25% results in a good field aperture requirement of $\pm$ 21 mm at 1.5 TeV. This is essentially the same as the required aperture at 150 GeV.

Thus the required **good field aperture** in the accelerator is about **$\pm$21 mm**. The requirement arises both from injection (with separated orbits) and slow extraction requirements. We guess that the total **coil diameter** in the dipole magnets would have to be **between 6 and 7 cm** to provide good field over 42 mm. This requires further study.

HIGH FIELD (>6 T) MAGNETS

A higher energy (>1 GeV) accelerator requires the contruction of dipole magnets with fields in excess of 4.4 T if the accelerator is to be situated within the Fermilab tunnel. The relationship between energy and field has been described previously. The Multi-Tev Hadron Collider Group examined possible 6.6 T and 9.0 T superconducting magnet designs to be associated with energies of 1.5 TeV and 1.8 TeV[5,6].

The basic magnet parameters are given in Table 1, and cross sections of the magnets are displayed to the right. Both magnets utilize a cold iron design with an aperture of 7.5 cm. This aperture is the same as that of the existing Tevatron and is large enough to allow for slow extraction and/or the eventual implementation of separators. (See above for discussion of aperture requirements.) The combination of cold iron and an SSC style support keep the static heat leak in these magnets low. In fact the magnets are designed (along with associated quadrupoles and spool pieces) so that a proton-proton collider based on either of these magnets could be operated without need of expanding the refrigeration capability of the Fermilab site. In addition a 1.5 TeV ring (6.6 T) could be run in a fixed target mode (90 second ramp) without need for refrigeration capacity in excess of that already planned for Fermilab in the early 1990 s.

The 6.6 T dipole design is regarded by the group as straight forward with minimal development work required. However, the 9.0 T dipole is felt to be substantially more difficult and to have significant operational disadvantages relative tothe 6.6 T dipole. Potential problems are related to mechanical stresses a factor of two more than anything ever attempted in a collared coil magnet type (e.g. a preload of 20 KPSI is required on the coil). As a result a substantial (1-2 years) R&D period is anticipated for such a magnet. The low operating temperature will result in long (weeks) cooldown and warmup periods, making the operating efficiency of a collider based on such a magnet much more dependent upon eliminating magnet failures than in the case of a 6.6 T magnet based collider. Finally, AC losses in this magnet completely preclude fast ramping of this magnet for use in a fixed target program.

Table 1: High Field Magnet Parameters

	6.6 T Dipole	9.0 T Dipole	
Length	8.5	8.5	meters
Coil Aperture	7.5	7.5	cm
Operating Temperature	4.5	1.8	^{o}K
Static Heat Leak @ 4.5^{o}K	2.6	2.3	W
Static Heat Leak @ 1.8^{o}K	---	0.1	W

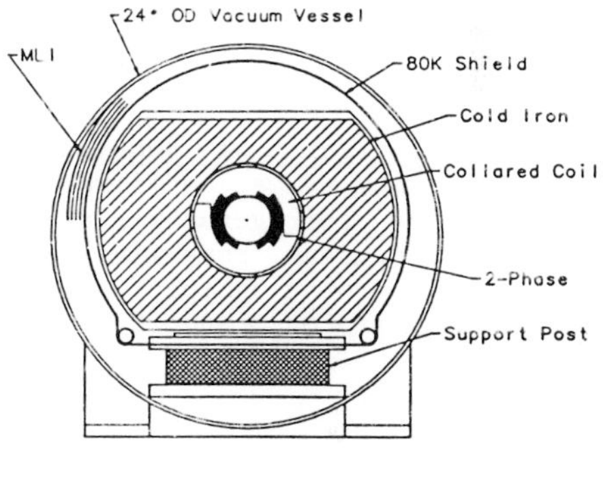

6.6 Tesla Dipole Conceptual Design Cross Section

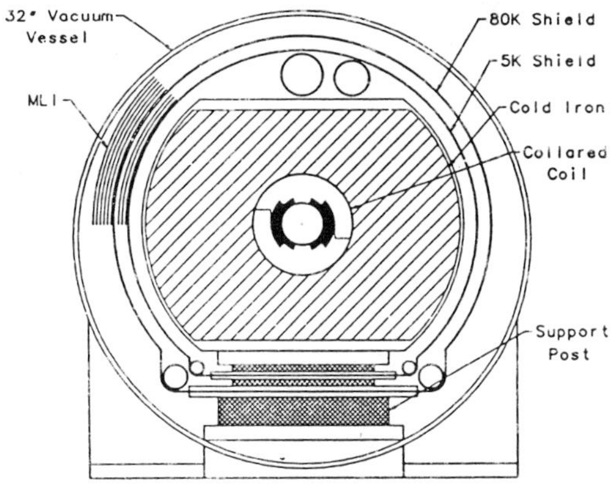

8.8 Tesla Dipole Conceptual Design Cross Section

LOW TEMPERATURE (2°K) CRYOGENICS

The existing Tevatron Cryogenic System is comprised of the Central Helium Liquifier (CHL), a 7 km LHe/LN$_2$ transfer line, and 24 distributed satellite refrigerators. A second CHL is being built and will become operational in early 1990. The existing CHL has the capacity to produce 4000 liters/hour of Helium at 3 atmospheres and 4.5^{o}K;

the new CHL will have a capacity of 5400 l/hr. The satellite refrigerators have a capacity of 966 Watts each at 4.5°K.

Refrigeration requirements for each of the potential upgrade scenarios are given in Table 2. Option A (high luminosity proton-antiproton) is based on supplying more antiprotons to the existing Tevatron and so entries in the table simply represent current operating demand. Option B (high luminosity proton-proton) is implemented through the construction of a new superconducting ring operating at 1.0 TeV. The new superconducting ring has a significantly reduced static heat leak compared to the existing Tevatron (150 W vs. 560 W) but a larger AC load (400 W vs. 180 W) due to the cold iron. As indicated in the table the total heat load for operating both superconducting rings simultaneously in collider mode is approximately the same as for fixed target running with the existing Tevatron. As also seen in the table the total heat load running in fixed target mode is less for the new Tevatron than for the existing machine although the distribution in load between static and AC losses is different. Table 2 also shows that the refrigeration load for a 1.5 TeV/beam superconducting accelerator based on the 6.6 T magnets described above is much reduced over the existing facility when operated as a collider and is comparable when used to provide slow extracted beam with a 90 second cycle.

Table 2: Refrigeration Requirements. Heat loads, liquid Helium usage, and fixed target cycle time **per satellite refrigerator**. Units are Watts, liters/hour, and seconds repectively. Option A is high luminosity proton-antiproton, B is high luminosity proton-proton, and C is high energy proton-antiproton.

		Heat Leak (Watts) 1.8°(Static)	4.5°(Static)	4.5°(AC)	LHe	Cycle Time
Option A	Collider	---	560	0	106	---
	Fixed Target	---	560	180	134	60
Option B	Collider	---	710	0	150	---
	Fixed Target	---	150	400	105	60
Option C (6.6 T)	Collider	---	195	0	68	---
	Fixed Target	---	195	500	153	90
Option C (9.0 T)	Collider	15	255	0	88	---

Because refrigeration requirements for the first three options listed in Table 2 could be satisfied by existing facilities at Fermilab effort was directed at examining the problems of providing a 1.8°K capability as required for the operation of 9.0 T magnets. A description of the issues and an example refrigeration system configuration may be found in a contribution to these proceedings[7]. In reference 7 it is suggested using a magnetic refrigerator-based system with a forced flow of the 3 atmosphere superfluid provided by circulating pumps. The magnetic refrigerator and pumps are an add on to the existing system rejecting heat at 4.5°K with each Watt at 1.8° worth about 6 Watts at 4.5°. The total heat load at 4.5° is still well within the capacity of the existing system.

PERFORMANCE ASSESMENT

The Multi-TeV Hadron Collider group attempted to evaluate the peak and integrated luminosities likely to be realized in the three colliders considered. Table 3 lists the performance parameters of the three options. The 'peak luminosity'is meant to reflect performance if all parameters given in the table are realized simultaneously, while the 'high confidence luminosity' listed includes a downgrading of the peak luminosity to reflect the same (high) level of confidence for all cases. Downgrading of the luminosity was a subjective exercise undertaken by the group in an attempt to incorporate some assesment of the inherent technical risk into the luminosity associated with each option. The basis for this evaluation is given below.

Table 3: Upgrade Performance Parameters. Option A is high luminosity proton-antiproton, B is high luminosity proton-proton, and C is high energy proton-antiproton. Peak luminosities are taken from Reference 2 for options A and B, and are calculated at Snowmass for option C. Interactions/crossing assumes a total cross section of 100 mb.

	Option A	Option B	Option C	
Center of Mass Energy	2.0	2.0	3.0	TeV
Protons per Bunch	5×10^{10}	3×10^{10}	6×10^{10}	
Antiprotons per Bunch	3×10^{10}	---	6×10^{10}	
Normalized Emittance (95%)	12π	12π	20π	mm-mr
Number of Bunches	80	1000	6	
Antiproton Stack	3.6×10^{12}	---	5.4×10^{11}	
Stacking Rate	3.0×10^{11}	---	4.5×10^{10}	hr^{-1}
β_*	0.5	0.5	0.5	meters
σ	30	30	32	μm
σ_z	25	25	25	cm
Crossing Angle	0	0.3	0	
Luminosity (Peak)	5×10^{31}	2×10^{32}	8×10^{30}	$cm^{-3}s^{-1}$
High Confidence Luminosity	2×10^{31}	2×10^{32}	6×10^{30}	$cm^{-2}s^{-1}$
Interactions/Crossing	1.3	0.6	2.8	
Separators	Yes	No	No	
$\Delta\nu$/Crossing	.0031	.0018	.0022	
$\Delta\nu$ Total	.0093	.0054	.0264	
Bunch Spacing	170	19	3500	nsec

Several comments are in order regarding the choice of parameters in Table 3. First the emittances used for options A and B are chosen to be those expected to be delivered from the Fermilab 8 GeV Booster following the 400 MeV Linac Upgrade. In the case of option C the emittance is limited by the total beam-beam tune shift seen by the proton and antiproton beams. The size of the antiproton stack required is derived from the total number of antiprotons in the collider assuming that 67% of the stack can be delivered to the collider and brought into collision with the protons. During present operations this fraction is typically 40%. The stacking rate listed in the table assumes that 80% of the antiprotons actually removed from the stack can be brought into collision with the protons (an efficiency consistent with present operations) and that the antiprotons will be replenished over a ten hour time period. Note finally that in all cases we have assumed a β at the interaction point of 50 cm. While designs exist for β^* of 25 cm at the Tevatron it is our belief that intrabeam scattering will result in bunch lengths lying in the range

$$\sigma_z \geq 25 \text{ cm}$$

thus limiting the utility of low beta designs much less than 50 cm.

Each of the three options listed presents very different beam characteristics to the experimenters and is ultimately limited in performance by different factors. As represented in Table 3 the luminosity achievable in Option A is ultimately limited by antiproton availability--if more antiprotons were available the luminosity would be higher. The size of the antiproton stack required in this option is seven times the design value (and current performance) of the existing Antiproton Source. This is why implementation of Option A is based on the construction of an antiproton post-accumulator with a storage capability of 4×10^{12}. The stacking rate required to support this option is a factor of twenty higher than the current Source performance and a factor of seven higher than the design value. Separators are needed in this scheme because of the need to keep the number of interactions per crossing and the total beam-beam tune shift low in the presence of high luminosity. The number of bunches could be decreased to forty (doubling the number of antiprotons per bunch to 6×10^{10}) if the experimenters if the experimenters were willing to tolerate 2.6 interactions per crossing. Even at forty bunches, the probable practical lower limit, one is forced into a helical orbit separation for the protons and antiprotons. The 'high confidence luminosity' listed in Table 3 for **Option A is degraded by 60%** from the peak value listed due to **concerns about the viability of running helical separated orbits in the Fermilab Tevatron and about the achievement of an antiproton stacking rate in excess of 1×10^{11}/hour.**

We do not know of any fundamental limitation to the luminosity for Option B as represented in Table 3. The emittance listed in the table is the lowest that we expect to achieve for the particle intensities given. However, it is expected that the emittances will rise only linearly with the number of particles per bunch, the result being that the beam-beam tune shift will remain unchanged and the luminosity will increase linearly with the number of protons. It is likely that the limit to the achievable luminosity (in the absence of melting detectors) will be be a total beam intensity limit resulting from coupled bunch instabilities. The total beam intensity, 3×10^{13} protons/beam, is 50% higher than currently achieved in the Fermilab Tevatron. However, present Tevatron operation is not limited by Tevatron so much as by Main Ring performance. This performance is addressed in Option B by the construction of a new Main Ring. **For these reasons we did not feel it necessary to degrade the peak luminosity number for Option B.**

With the exception of the beam energy Option C is based on parameters identical to those expected to yield a luminosity of 5×10^{30} cm^{2}sec^{-1} at the Tevatron Collider in the early 1990 s. The luminosity of 8×10^{30} listed in Table 3 simply reflects the fact that luminosity is proportional to energy in a hadron collider if all other parameters remain fixed. It is also worth noting that the beam parameters are nearly identical to those being used in the current Fermilab Collider run with the exception that antiproton bunches are shown as being populated to the same level as the proton bunches. The luminosity achieved in Option C as shown in the table is limited both by antiproton availability and by the beam-beam interaction. This option is based on utilizing the existing Antiproton Source with the stacking rate improved by a factor of three as a result of ongoing improvements to the Source and a tripling of the number of protons which can be delivered to the antiproton production target by a new Main Ring. The required stack size has already been realized in the Fermilab Source. The option as described does not use separators. As a result each proton (or antiproton) bunch encounters twelve antiproton (or proton) bunches per revolution. The total tune shift of .026 is 25% larger than what the Fermilab Collider is currently achieving. With the implementation of separators it is likely that in excess of 1×10^{31} cm^{-2}sec^{-1} could be achieved by lowering the beam emittances to 12π. Because of **uncertainty in the beam-beam tune shift** which can be achieved we have **degraded** the 'high confidence luminosity' in **Option C by 20% to 6×10^{30} cm^{-2}sec^{-1}.**

In collaboration with the experimentalists from the physics groups we also attempted to make a gross estimate of the amount of integrated luminosity which would end up on an experimenter's tape per calendar year of running at the three colliders studied. The estimates were produced primarily to standardize numbers used by the physics study groups and are only meant to have a relative accuracy of 25% or so. We assumed that in the case of a proton-proton collider about one third of a year (1×10^7 sec of peak luminosity) per calendar year (per interaction region) could be recorded on tape. This fraction is meant to include losses due to accelerator and detector downtime, detector deadtime, and finite luminosity lifetimes. It is generally consistent with experience at e^+e^- colliders and at the ISR. In the case of the proton-antiproton colliders we deflated this number to 5×10^6 seconds of peak luminosity per year. This number is consistent with the first two collider runs at Fermilab and is believed to be consistent with the CERN experience. It is lower than in the proton-proton case primarily because of the lack of an inexhaustable source of antiprotons.

Multiplying by the 'high confidence luminosities' in Table 3 leads to conservative predictions of the integrated luminosity recorded on tape per calendar year per interaction region for the three options:

Option A: **100 pb^{-1}**
Option B: **2000 pb^{-1}**
Option C: **30 pb^{-1}**

CONCLUSIONS

The Multi-TeV Hadron Collider Group reached conclusions regarding the feasibility and expected performance parameters of three posssible paths by which one might provide a facility with at least twice the mass reach of the Fermilab Collider. The conclusions are:

1. Proton-antiproton collisions at 2 TeV (center-of-mass) within the Fermilab Tevatron at a luminosity of $2x10^{31}$ $cm^{-2}sec^{-1}$ appear feasible in the mid 1990 s. Such an upgrade would be based upon the construction of a new antiproton accumulator ring and enhanced Main Ring performance either through raising the injection energy or by recontruction with a smaller circumference.

2. Proton-proton collisions between the existing Fermilab Tevatron and a new superconducting ring at 2 TeV and with a luminosity of $2x10^{32}$ $cm^{-2}sec^{-1}$ appear feasible in the mid 1990 s. Such an upgrade is based on the construction of a new (1 TeV) superconducting accelerator in the existing Fermilab tunnel and relocation/reconstruction of the Fermilab Main Ring.

3. Proton-antiproton collisions at 3 TeV within a new superconducting ring at a luminosity of $6x10^{30}$ $cm^{-2}sec^{-1}$ appear feasible in the mid 1990 s. At Fermilab such an upgrade would involve construction of a new (1.5 TeV) superconducting accelerator within the existing tunnel and relocation/reconstruction of the Main Ring.

4. The 3 TeV proton-antiproton collider and 2 TeV proton-proton collider have more potential for extension of the luminosity beyond the stated values than does the 2 TeV proton-antiproton option.

5. A 3.6 TeV collider probably represents the highest energy collider which can be contemplated within the existing Fermilab tunnel. Magnet development is clearly more difficult than in the 3.0 TeV case and would likely delay construction of such a collider by 1-2 years.

6. Slow extracted beam at 1.5 TeV could be provided based on a new superconducting ring within the Fermilab tunnel.

7. A new Main Ring, or at least improved performance, is required for realization of either of the three options at Fermilab.

We refer the reader to summary reports from the physics groups for discussion of the physics potential of the three options as well as for implications with regard to detector capabilities.

REFERENCES

1. Participants in the Multi-TeV Hadron Collider Group included: C. Ankenbrandt, B. Brown, J. Cary, S. Childress, J. Claus, T. Collins, M. Cornacchia, G.F. Dell, L. Evans, A. Garren, M. Harrison, J. Herrera, S. Holmes, E. Malamud, P. Mantsch, J. McCarthy, L. Michelotti, N. Mokhov, K-Y. Ng, T. Nicol, T. Peterson, Y. Pishalnikov, A. Riddiford, L. Teng, J. Theilacker, A. Vorobiev, and J. Yoh.

2. Fermilab Design Reports, Antiproton-Proton Collider Upgrade: 20 GeV Rings and Proton-Proton Collider Upgrade: (Main Injector, New Tevatron), both dated May, 1988.

3. N. Mokhov and M. Harrison, Contribution to these proceedings.

4. A. Garren, Contribution to these proceedings.

5. P. Mantch, Contribution to these proceedings.

6. T. Nicol and A. Riddiford, Contribution to these proceedings.

7. T. Peterson and J. Theilacker, Contribution to these proceedings.

ACCELERATOR PHYSICS IN RHIC

J. Claus
Brookhaven National Laboratory*
Upton, New York 11973

INTRODUCTION

RHIC (for Relativistic Heavy Ion Collider)[1] is a colliding beam facility to be built at BNL in the tunnel system that was constructed for the defunct ISABELLE/CBA project. It is intended for the study of collisions between fully stripped ions of the same or different species with magnetic rigidities of up to at least 839.5 Tm, corresponding with an energy of 100 GeV/amu (amu for atomic mass unit) for particles with A/Z = 2.5, and 251 GeV for protons. There are six potential crossing regions, each with, initially, time average luminosities of up to a few times 10^{26} cm^{-2} sec^{-1} for Au, with luminosity lifetimes of 10 hours for Au, and longer for the lighter ions. The initial proton-proton luminosity will be about 10^{31} cm^{-2} sec^{-1}. The rms length of the interaction diamond is 0. 2 m when the beams collide colinearly, it is determined by the lengths of the colliding bunches. The diamond length can be reduced by having the beams cross at an angle, but this reduces the luminosity. The magnitude of the crossing angle is restricted geometrically in dependence of the energies and species of the interacting ions and possibly also by dynamic effects in the circulating beams.[2] The distance between each crossing point and the nearest machine component is 9 m on each side, this space is available for experimental equipment. The performance quoted assumes that each ring contains 57 bunches, each bunch with 10^9 Au ions, resp 10^{11} protons and with invariant emittances of 10π, resp 20π mm-mrad, and $\beta_x^* = \beta_y^* = 3$ m. The beam-beam tune-shift per crossing point at these intensities is 0.0025, resp 0.0037.

GENERAL ARRANGEMENT, TRANSVERSE MOTION

RHIC consists of two identical intersecting rings with a common, horizontal, median plane. The circumference of each ring's reference orbit is 3833.845 m, beam guidance and focussing are done with super conducting magnets. A general layout and some definitions are given in fig. 1, while a layout of a half insertion is shown in fig. 2, and fig. 3 shows the geometry of the closed orbits in the crossing regions under various operating conditions. The arcs are composed of identical dipoles and quadrupoles, 24 of each per arc. The radius of orbit curvature in the dipoles is 243.5 m, the dipole axes are bent with about the same radius to avoid aperture loss due to sagitta, each dipole is 9.45 m long and the beamtubes have an inner radius of about 36 mm. The 90 cm difference in radius between the outer and inner arcs is obtained by small differences in the driftspace lengths. The quadrupoles and dipoles are organized in 12 symmetric cells per arc with horizontal and vertical betatron phase advances of about $\pi/2$ rad/cell and $\hat{\beta} \approx 50$ m, $\hat{x}_p \approx 1.5$ m. The quadrupole polarity sequence in a given ring is antisymmetric relative to the crossing points: all outer arcs begin and end with horizontally defocussing (half) quadrupoles and all inner arcs terminate in focussing ones. Each insertion can in principle be tuned to provide $0.5 \text{ m} \le \beta^* \le 10$ m, $x_p^* = 0$ m without loss of the first order betatron match to the adjoining arcs and without change of the betatron phase advances across it; in practice this flexibility is constrained by the capabilities of the power supplies and by aperture considerations: the admittance drops below 6π mm-mrad for $\beta^* < 2$ m while it is 17π mm-mrad for $\beta^* = 6$ m. The preservation of matches and phase advances makes it possible to reset the insertions while the beams are circulating. The betatron phase advance across each insertion is 3.72π rad, the betatron wavenumber for a whole ring about 28.82 both horizontally and vertically, $\gamma_{tr} \approx 25.4$. The chromatic properties of the machine are controled by sextupoles, one next to each arc quadrupole. They are used not only to control the horizontal and vertical chromaticities (change of betatron wavenumbers with momentum deviation) but also to minimize the changes of the β^* s with momentum deviations; they are organized in six separate families for this purpose.

INTRABEAM SCATTERING

RHIC's design is strongly affected by the intra-beam scattering phenomenon.[3] This process tends to redistribute the charge distribution in each bunch until it is gaussian with rms dimensions that match the inverse roots of the focussing strengths radially, vertically and azimuthally, i.e., until equipartition has been reached. It is complicated by the coupling between the various components of the motion, particularly that between the radial and azimuthal ones due to the dispersive characteristics of the ring, and by the aperture restrictions imposed by vacuumchamber cross-section and rf system; these prevent the distribution from becoming gaussian. The negative mass effect makes the process instable at energies above the transition energy. Both mechanisms cause the bunch dimensions to grow continuously with e-folding times from tens of minutes to hours, depending on particle species, energy and operating conditions. The rate of increase depends on the energy, the density in 6D phase space and the cross-section for intrabeam scattering and is

*Work performed under the auspices of United States Department of Energy.

relatively small for the usual beams of light particles (protons, electrons). It is not small for the heavy ion beams proposed for RHIC, where, e.g., the betatron emittances of Au beams increase by a factor 3 or more, depending on the synchronous energy, in 10 hours, as shown in fig. 4. This occurs because the scattering cross-section behaves as $Z^2(Z/A)^2$, with Z the charge number and A the massnumber of the nucleus under consideration. The luminosity decreases, though with a decreasing rate, with time, as does the time average luminosity. The luminosity lifetime for interesting luminosities is limited, particularly for the heaviest nuclei. The 6D admittance of the ring must be large enough to accommodate the 6D emittance of the bunch at the end of the luminosity lifetime. RHIC's beam tube radius was set at 36 mm in order to keep the magnet cost acceptable while the rms bunchlength must be kept below 0.3 m in order to meet users' limit of 0.2 m on the rms length of the interaction diamond. The betatron wavenumbers and rf parameters were chosen to satisfy these requirements.

FILLING AND TRANSITION CROSSING

The rings are filled from the AGS, which itself is to be filled from a booster synchrotron (under construction) which again is filled with ions by BNL's tandem van de Graaff or with protons from its 200 MeV linac. The AGS accelerates the particles always to a magnetic rigidity of 98.2 Tm, about the highest practical rigidity that can be reached in that machine. From this, one calculates that RHIC is injected with $\gamma_{inj} = 31.4$ for protons and $\gamma_{inj} \approx 12.59$ for particles with A/Z = 2.5, since the radius of orbit curvature in the AGS is 85.344 m. Originally it was assumed that all ions would be fully stripped when entering the AGS, but recent studies suggest that there is an advantage in stripping the heaviest ions, i.e., Au and heavier, only down to the K-shell, accelerating them partially stripped in the AGS and removing the last 2 electrons in the transferline between AGS and RHIC. It may be seen that in either case all ions but protons must be accelerated through the transition energy, $\gamma_{tr} \approx 25.4$. This presents certain problems because the rate of acceleration is restricted to $\dot{\gamma} \approx 1.45\ \text{sec}^{-1}$ by the quench protection system and the insulation level of the s.c. magnets. The particles spend a relatively long time in the vicinity of γ_{tr}, where Landau damping of instabilities is weak since the phaseslip factor $\eta \left(= \gamma_{tr}^{-2} - \gamma^{-2}\right)$ is small. Another problem, particularly noticeable if the momentum spread is large, stems from the fact that the differential angular velocity of the particles is a nonlinear function of their momentum deviation due to the horizontal chromaticity. Therefore the longitudinal motions of particles with different momentum deviations do not become unstable simultaneously. No matter when the phase of the rf voltage is switched, the motion of some fraction of the particles will be unstable for a while. Simulation of the process showed that a considerable fraction of the beam can be lost in this manner. One solution is the well known γ_{tr} jump. Here one adjusts the focussing system of the machine to decrease γ_{tr} with time (without changing the betatron wavenumbers), decreasing the length of the interval during which η is very small. The capability to do this will be available. Two other possibilities have been considered. The first exploits the facts that a) the rf voltage necessary to cope with intra-beam scattering is much larger than that necessary for acceleration, b) the aperture available is larger than the aperture required when passing through transition because the blowup of the beam dimensions due to intra-beam scattering has not yet taken place. It is therefore possible to increase the acceleration rate temporarily, even though B is kept constant. The closed orbit, which had previously been moved to the low energy side of the aperture by closed orbit control, expands under the influence of the excess acceleration and moves to its high energy side. Then the acceleration rate is returned to the rate that matches the value of $\dot{B}$ and the closed orbit is moved back to its normal location. The displacement can be fast to the extent that excess accelerating voltage is available, the change in synchronous energy must be somewhat larger than the total energy spread in the bunch. The process has been simulated on a computer and appears to be feasible within the constraints of RHIC's parameters and performance specification. Another method, called AG-rf-focussing is a generalization of the CERN triple switch. The phase of the accelerating voltage relative to the bunch is alternated between operation at the stable and unstable fixed points. This reduces the longitudinal focussing strength greatly but renders it independent of the sign of η, at least in linear approximation. The latter two methods do not require any adaptation of the basic rings (as the γ_{tr} jump does) but depend directly on the specification of the accelerating system. Since that specification is still being studied firm choices cannot yet be made.

R.F.

The longitudinal motion is controled by the rf system and subject to intrabeam scattering. Very strong longitudinal focussing appears to be necessary to prevent the rms bunchlength from growing beyond the limit of 0.3 m after 10 hours for Au imposed by the users. This requirement was stressed during a recent workshop at BNL,[4] where it became clear that the longitudinal focussing strength must be rather stronger than can be provided by the rf system described in ref. 1. That system had a rf frequency of 26.7 MHz and a maximum net rf voltage amplitude of 1.2 MV; the rf frequency corresponds with a harmonic number h = 342. Computer simulation shows that the initial rms bunch length of about 50 cm would increase to about 150 cm after 10 hours for Au at $\gamma = 100$ due to intrabeam scattering. This performance is now considered unacceptable. During the workshop it was proposed to accelerate the beam with 26.7 MHz and a relatively modest voltage (0.3 MV) to the desired operating energy and to use a rf system with a much higher frequency, e.g., ~ 200 MHz, and much higher voltage (numbers up to 36 MV were mentioned), during storage. Subsequent attempts to minimize cost suggested a frequency of 160 MHZ, h = 2052. The 26.7 MHz system remains necessary since the AGS cannot provide bunches that are sufficiently short to fit within the phase acceptance of the high frequency system without serious modification of its rf system. Several schemes for performing the harmonic jump were discussed. In all of them the 26.7 MHz rf voltage must be manipulated at injection to match its buckets to the longitudinal emittance of the bunches arriving from the AGS (buncharea ~ 0.3 eVsec/amu, $|\Delta\gamma/\gamma| \approx 10^{-3}$, bunchlength ~ 17 nsec) and to the acceptance of the high frequency system when the jump is to be made; it has to be rather high in some schemes. One scheme, proposed at the workshop, avoids high voltages by exploiting the fact that the bunches are short in the immediate vicinity of the transition energy and performs the jump there. Simulation of this process shows that this can be done with an efficiency of better than 90%, using a 26.7 MHz rf voltage in the range of 0.3–0.4 MV. It does not work for protons which do not cross transition. However, at $\gamma = 31.4$ the length of a proton bunch of 0.3 eVsec is 4.66 nsec if the rf voltage is 0.4 MV at 26.7 MHz, short enough to fit inside the 6.23 nsec long buckets of the 160 MHz system.

UTILITIES

At some time during the storage cycle the circulating beams will have to be renewed, one expects that in general the machine cycle will be tuned for maximum time averaged luminosity. The circulating beams are disposed of and the rings are refilled with fresh beam. The details of this operation are too complex and too specific to be described here. One important issue is whether to dump the spent beams on internal or on external dumps. An internal dump is cheaper and should be adequate for beams with up to 300 kJ, as specified for the initial operation of RHIC. External dumping will become necessary if the charge per bunch and/or the number of bunches is increased beyond the design specifications. The mechanical stress induced by the energy density in such beams is sufficient to cause material damage in all internal dumps considered so far. The rings are filled, one bunch at a time, from the AGS. The filling pattern is restricted by the rise time of the injection kicker: filling every n^{th} of the 342 available buckets requires $\tau_{rise} \leq 37n-17$ nsec, filling all buckets would require $\tau_{rise} \leq 20$ nsec. In all cases it is necessary to leave some 30 consecutive buckets (~1 μsec) empty in order to accommodate the fall time of the injection kicker and the rise time of the abort kicker. This implies, of course, that the specification for initial operation calls actually for 52 bunches rather than for the nominal 57 (n = 6).

FUTURE DEVELOPMENTS

A number of developments can be foreseen that will improve RHIC's performance perhaps by an order of magnitude in the time averaged luminosity. Improvements in the source chain, principally in the ion source, booster and booster operation, suggest that the charge per bunch can be more than doubled without serious loss of brightness. The acceptable charge per bunch will then be limited by the beam-beam effect in RHIC. The tolerable beam-beam tuneshift can be determined only during the commissioning of the machine and is perhaps not much larger than the calculated $\Delta\nu_{bb} \approx 0.0025$/crossing point given above if all 6 crossing points are in use. Assuming that it is the total tuneshift rather than the tune-shift per crossing point that is the critical factor, one may trade the luminosity per crossing point for number of crossing points and gain in overal luminosity by minimizing their number. Crossing points can be made active or inactive at choice by local vertical separation of the beams. Total tuneshifts of 0.02–0.04 have been observed at CERN's Sp$\bar{p}$S, the lifetimes achieved under such conditions appear to depend critically on the presence of high order betatron resonances and the magnitude of the spread in betatron wavenumbers in the beam.[5)] It seems likely that the charge per bunch can at least be doubled, yielding at least a quadrupling of the luminosity per active crossing point. The luminosity is proportional with the number of bunches, which is restricted by the rise time of the injection kicker. One might therefore expect a luminosity increase with a factor 6 if one fills 'all' buckets, using a kicker with $\tau_{rise} \leq 20$ nsec. This estimate would be slightly optimistic, because the machine must then be operated at a small crossing angle in order to prevent the creation of spurious crossing points. Such crossing points would create background and additional beam-beam forces. Colinear operation requires that the distance between successive bunches be larger than the 20 m long crossing region, and allows only a bunch in every second bucket. An obvious way to increase the luminosity, without changing (to first order) the beam-beam tuneshift is to decrease β^*. The design minimum value for β^* in any of the insertions is 2 m. Lower values, down to 0.5 m, are certainly possible for protons, but decrease the dynamic aperture of the machine and require revision of the quadrupole power supplies in the affected insertions. There is also an impact on the chromaticity correction system. Another improvement may be expected from the application of stochastic cooling. Without cooling the beam brightness and the luminosity decrease with time due to intrabeam scattering; the machine operates at the beam-beam limit only at the beginning of the cycle. With sufficient cooling it can be held at that limit during all of the cycle, this should increase the time average luminosity a factor 2–3 and reduce the formation of halos. Because the beam cross section remains small the maximum energy could be somewhat higher than that of the original design. Longitudinal cooling will reduce the requirement of very high rf voltages, reduce the energy spread and prevent the loss of particles out of the buckets. Another benefit of cooling would be that the luminosity lifetime is no longer restricted by intrabeam scattering. Stochastic cooling in RHIC was studied during the workshop mentioned earlier, and judged feasible and beneficial, though at the cutting edge of the technology in view of the maximum frequency, 16 GHz, required.

CONCLUSION

Taking all this together one concludes that RHIC has certainly the potential for reaching luminosities for heavy ions well in the 10^{-28} $cm^{-2}sec^{-2}$ range and close to 10^{-33} $cm^{-2}sec^{-2}$ for protons.

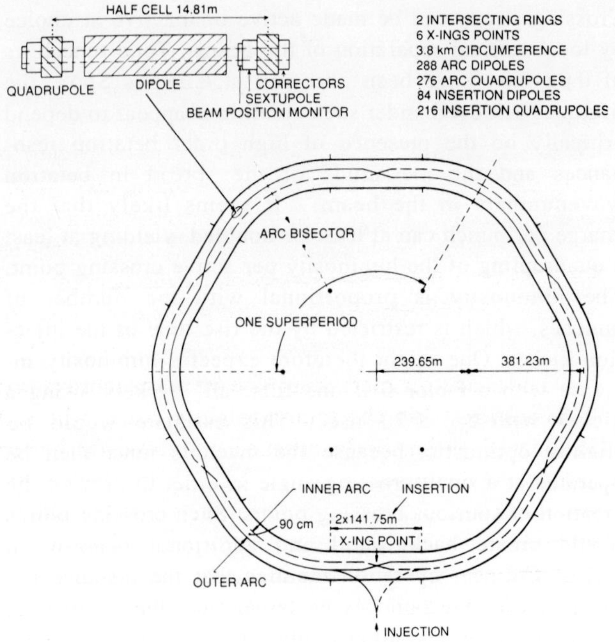

Fig. 1. Layout of the RHIC rings.

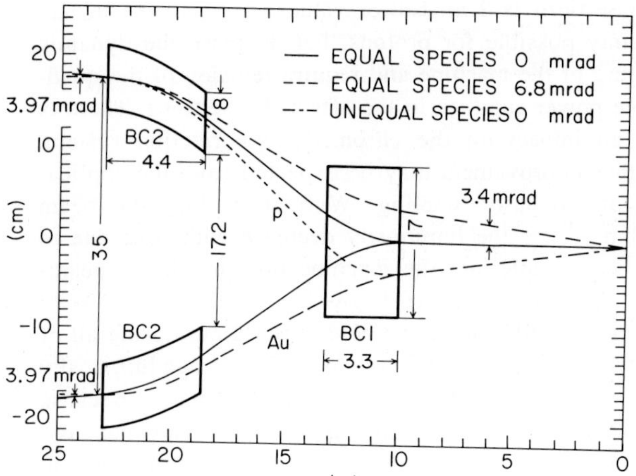

Fig. 3. Beam path at crossing point for Au-Au and Au-p collisions.

REFERENCES

1) Conceptual Design of the Relativistic Heavy Ion Collider RHIC, May 1986, BNL 51932 (A new, updated edition of this report is being prepared and to be expected Jan '89).
2) A. Piwinski, Synchrotron-Betatron Resonances, CERN 87-03, p. 194.
3) G. Parzen, Nucl. Instr. Meth., A251, 220 (1986), A256, 231 (1987) and references therein.
4) W. Willis and T. Ludlam, Proc. Workshop on RHIC Performance, March '88, BNL 41604.
5) L. Evans in [4].

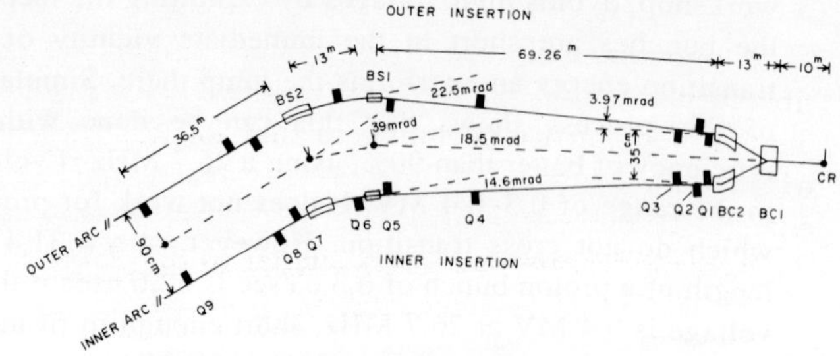

Fig. 2. Insertion magnet layout.

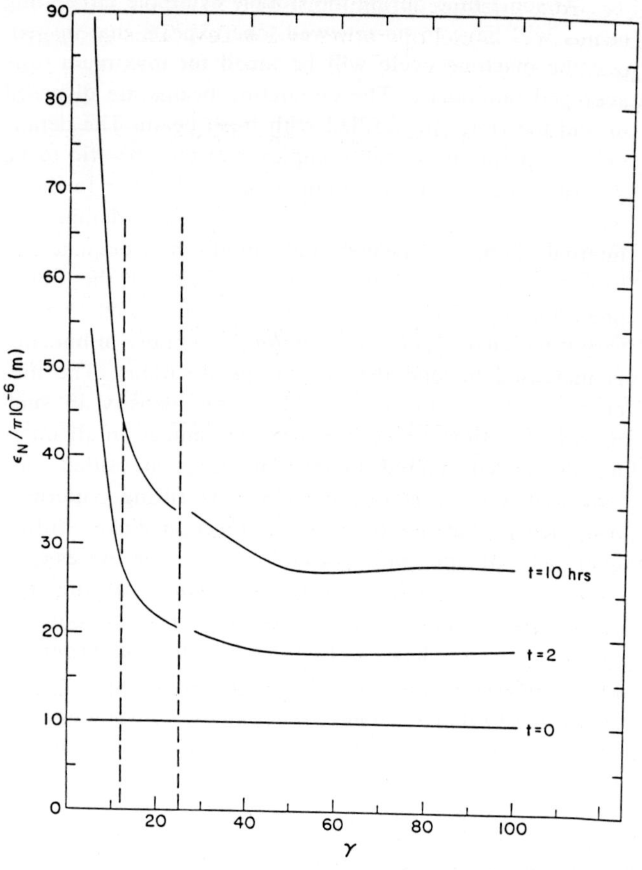

Fig. 4. Au beam emittance growth due to intrabeam scattering.

A 1.8 TeV $\overline{P}P$ UPGRADE LATTICE FOR THE TEVATRON

A. A. Garren
SSC Central Design Group *
$^c/_o$ Lawrence Berkeley Laboratory
Berkeley, CA 94720

and

M. J. Syphers
Fermi National Accelerator Laboratory *
Batavia, IL 60510

ABSTRACT

A preliminary lattice design has been made of a possible future 1.8 TeV $\overline{p}p$ Tevatron upgrade using 8–9 tesla dipoles. The lattice is similar to that for the Tevatron, with modifications to allow for beam extraction at higher energy. The new ring will fit in the present tunnel, and its linear optical properties are similar to the Tevatron's.

INTRODUCTION

The purpose of this study is to determine the feasibility, from a lattice standpoint, of a new $\overline{p}p$ collider in the Tevatron tunnel operating at 1.8 TeV. This is one of a number of options being considered for upgrading the Tevatron.[1]

It is assumed that most of the dipoles will operate at 8 T, the remainder at 9 T, and the quadrupoles at 150 T/m. Since the new machine should also operate in a fixed-target mode, modifications must be made to permit beam extraction at the higher energy. The Upgrade lattice follows that of the Tevatron as closely as possible, with only such modifications as are needed to meet the new conditions. Possibly, further design improvements can be made with additional changes from the Tevatron structure. For example, one might consider use of longer dipoles or of a different cell length.

As with the Tevatron, the Upgrade ring has a six-fold geometrical structure, which is defined by the dipoles. The optical periodicity of the Tevatron is broken by differences in the quadrupole structure in the A, B, and D long straight sections compared to that of the identical 'normal' straight sections C, E, and F. However, for this study it suffices to design a ring made up of six identical superperiods, each having a lengthened long straight section.

BEAM EXTRACTION REQUIREMENTS

If all of the components of the new machine were superconducting there would be no lattice problem; the dipoles and quadrupoles could be replaced by new ones with unchanged length and with strength scaled with energy. However, the kickers and septum magnets of the beam extraction system do not scale. To obtain the same separation between the circulating and extracted beams as with the Tevatron, a longer string of septum magnets is required, so the long straight section must be lengthened.

The kicker magnets must also be made more effective. This requires that the mini-straight section, where they are placed, either be lengthened or moved farther away from the long straight section to increase the phase advance between the kicker and septum magnets. The latter solution has been adopted; sufficient separation at the entrance to the septum magnets can be obtained by moving the mini-straight section 40 m farther away from the long straight section.

How much must the long straight section be lengthened? Consider a string of septum magnets of length l and field B_s at the beginning of a drift space of length L. The transverse separation at the end of the drift L will be given approximately by

$$\Delta x = L\alpha p(1 - p/2)\,,$$

where $p = l/L$ and $\alpha = B_s L/(B\rho)$. As in the Tevatron, we assume that the central drift space L is 80% filled with 1.2 T septum magnets. Using the formula above, one finds for the Tevatron with L =50 m, that Δx =43 cm, while the Upgrade machine with L =70 m will give Δx =48 cm. Therefore the long straight section should be lengthened by about 20 m.

*Operated by the Universities Research Association, Inc. for the U. S. Department of Energy.

PHYSICAL LATTICE CHANGES

The problem is to make the lattice changes required for extraction with minimum transverse displacements of the new ring compared to the present beam line. Figure 1 shows a normal Tevatron superperiod; Figure 2 shows the region surrounding the long straight section and indicates the changes that result in the proposed Upgrade lattice.

Moving the Mini-Straight Section

This is equivalent to removing one dipole from one position and placing it in a drift space 40 m closer to the long straight section. The same operation is to be repeated in each of the six superperiods. As a result, the new beam line will follow a sawtooth pattern relative to the old one, see Figure 3. It will be displaced outward by d at the new dipole locations, inwards by d at the new missing dipole locations, and these points will be connected by 1 km radius circular arcs. The maximum displacement is given by $d = l\theta/2$, where l is the distance the dipole is moved and $\theta = 2\pi/774$ is the bend angle of each dipole. For $l = 40$ m, one obtains $d = 16$ cm, which is probably acceptable, especially as the outward displacement will be counteracted by the inward displacement caused by lengthening the long straight section.

Lengthening the Long Straight Section

To lengthen the long straight section by ΔL, one removes dipoles of length $\Delta L/2$ on each side, moves the inner quadrupole doublets outward by the same amount, and replaces sets of dipoles beyond the doublets with stronger dipoles to preserve the total bending, see Figure 4. As a result, n standard strength dipoles on each side will be replaced by a smaller number of stronger dipoles. For an Upgrade lattice of energy E and field $\widehat{B}$ in the stronger dipoles, the length increase in meters of the long straight sections is approximately

$$\Delta L = 14.2(1 - 4.44E/\widehat{B})n,$$

and the inward displacement D in cm is

$$D = 2.88(1 - 4.44E/\widehat{B})n^2.$$

Taking $E = 1.8$ TeV, $\widehat{B} = 9$ T, and $n = 13$, we obtain $\Delta L = 21$ m, and $D = 55$ cm. The 13 normal strength dipoles are replaced by 11-1/2 strong dipoles that produce the same bending angle. The normal dipoles are scaled from the Tevatron dipoles with energy, from 4.4 T to 8 T to keep the curvature of the regular cells unchanged. To reduce the number of dipoles, the inner 2-1/2 strong dipoles are replaced with two strong 1-1/4-length dipoles.

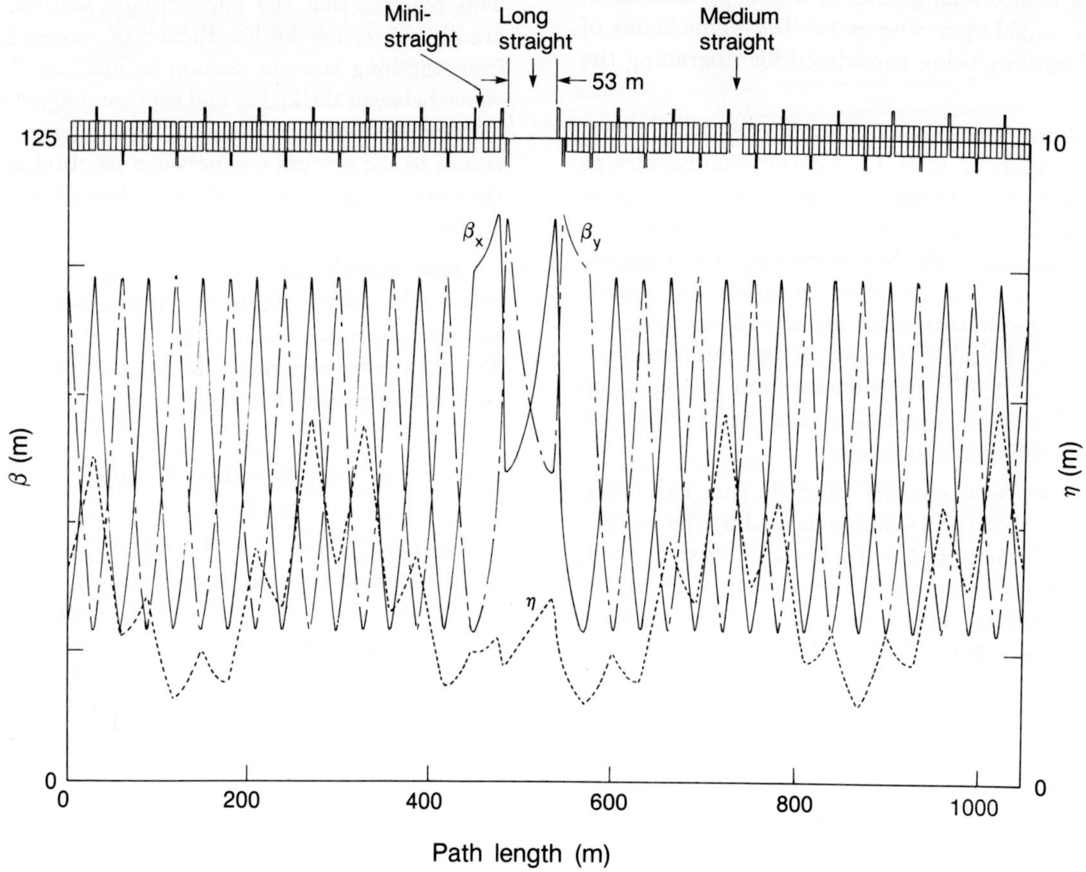

Fig. 1. Tevatron normal superperiod (C,E,F).

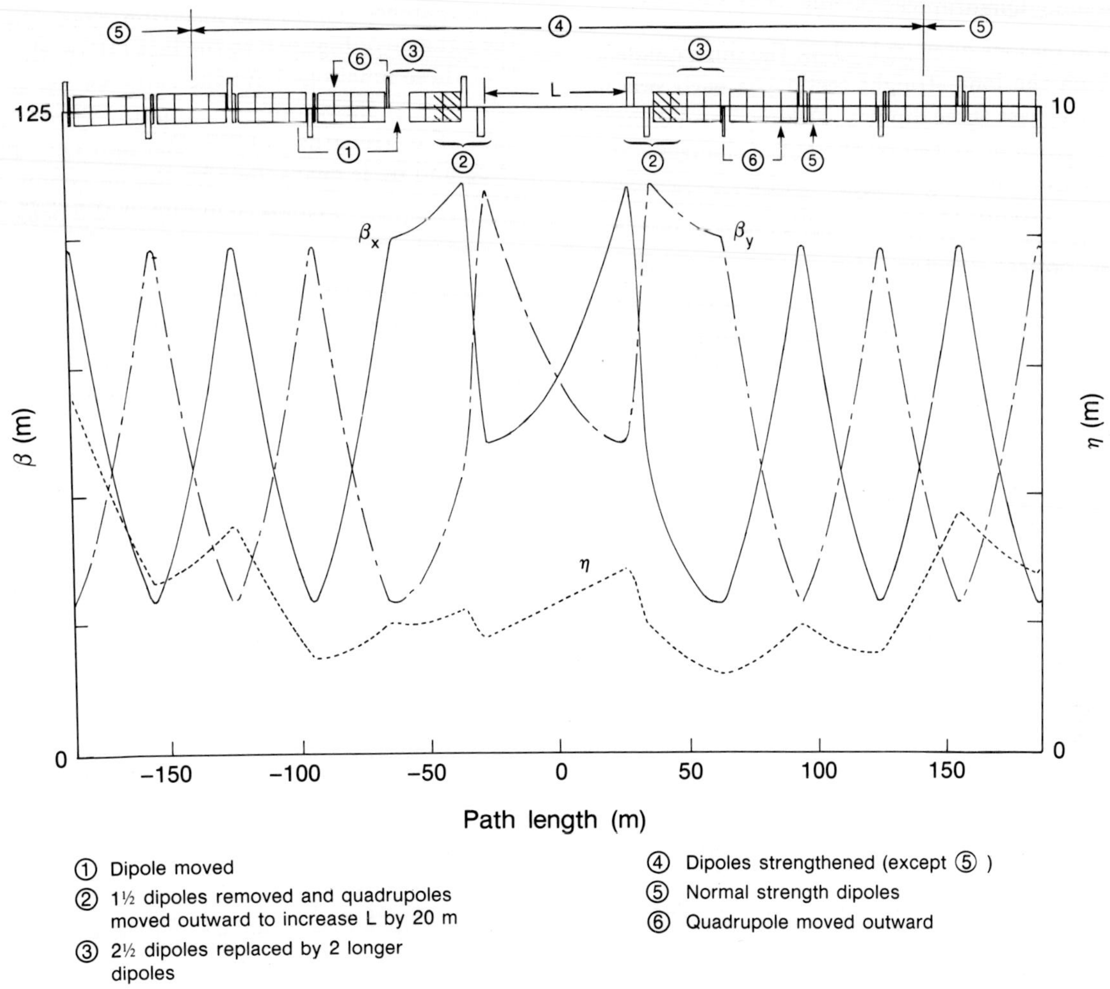

Fig. 2. Vicinity of Tevatron long straight section showing changes required to produce the Upgrade lattice.

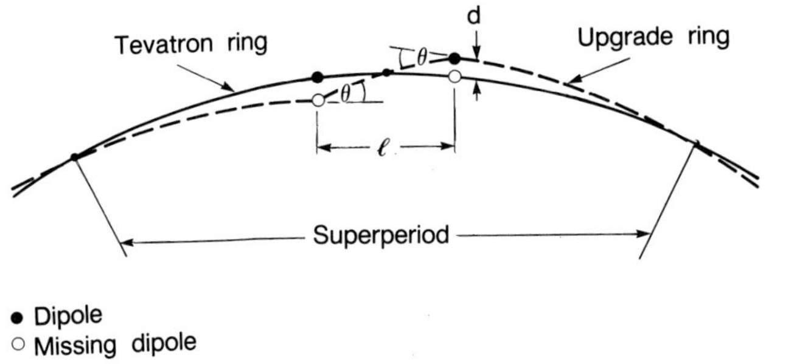

Fig. 3. Change of Upgrade beam line resulting from moving the mini-straight section.

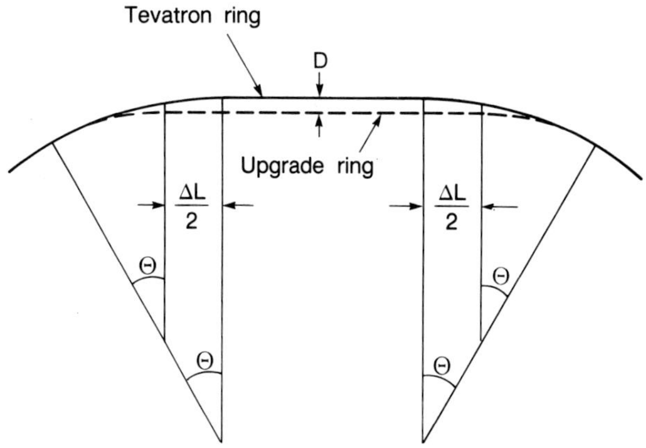

Fig. 4. Inward shift of Upgrade beam line due to lengthening the long straight section.

Moving Matching Quadrupoles

For optical reasons discussed below, the third quadrupole away from the long straight section on each side is moved outwards past three dipoles. Figures 5 and 6 show the Upgrade lattice produced by the above changes.

OPTICAL MATCHING

The basic optical structure of the proposed Upgrade lattice is unchanged from that of the Tevatron. Each superperiod has 15-1/2 cells and an insertion that contains the long straight section. There are three straight sections in each superperiod: mini, medium, and long, and four quadrupole families: (QF,QD), (QFS,QDS), (Q1,−Q1), and (Q2,−Q2), where the members of each pair have equal gradients of opposite sign. Thus the focussing structure remains antisymmetric with respect to the center of the long straight section.

The matching procedure was more or less as follows. The regular cells in the arc were adjusted to a given phase advance. This determines the strength of QF and QD as well as the β-functions of these cells. β_x and α_x were then matched across the insertion, from the first QF on the left side of Figure 6 to the last QF on the right side, so that these functions have the same values as do the periodic cells at the appropriate points. Three variables were used: the strengths of QFS, Q1, and Q2. The matching of β_y and α_y is guaranteed by the antisymmetry, and the third variable was used to minimize the peak value of β_x .

The above steps were repeated with different locations for QFS and QDS, different cell phase advances, and different separations of the Q1, Q2 doublets. The solutions were compared by computing the periodic β-functions and the dispersion η for the complete superperiod. The best arrangement found is shown in Figures 5 and 6, with QFS, QDS moved outwards past three dipoles.

Comparison of the figures and inspection of Table I shows that this Upgrade lattice is nearly the same as the Tevatron with respect to the β-functions and somewhat superior with respect to the dispersion. Further dispersion reductions might be obtainable through the use of additional quadrupole families. The beam-optics calculations were made using the SYNCH program.[2]

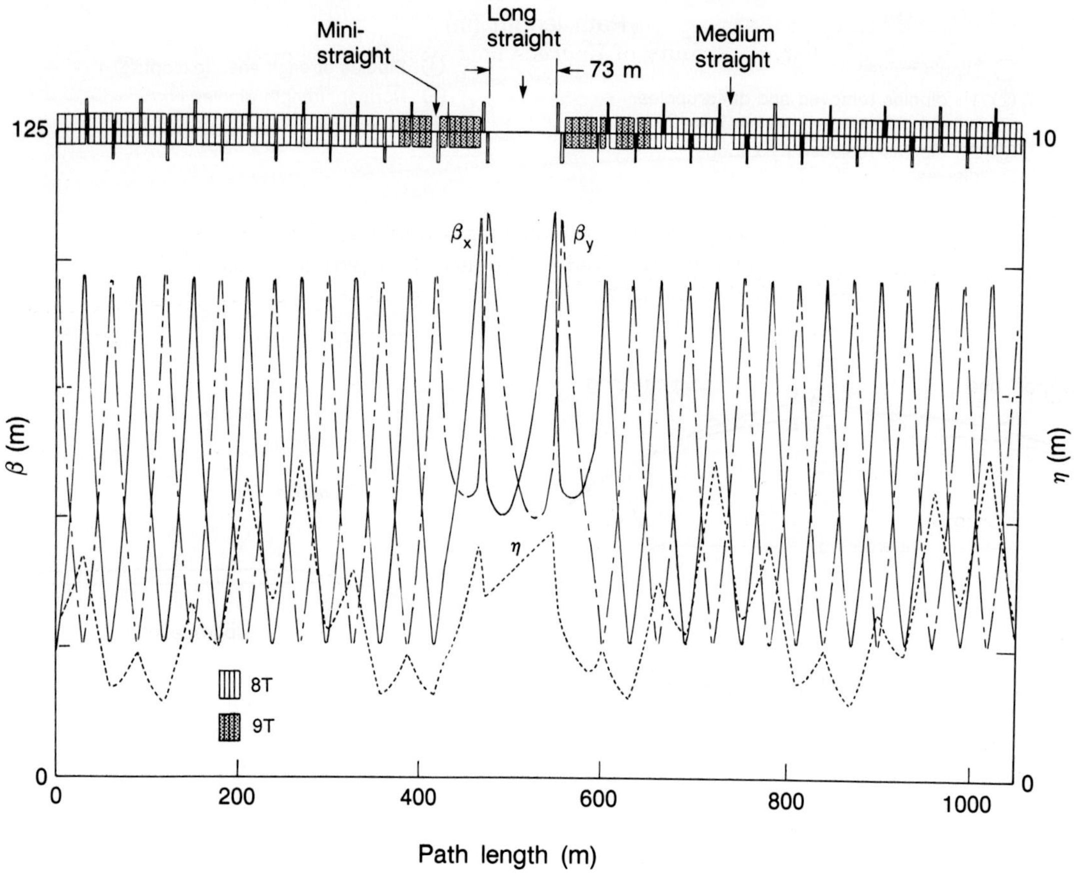

Fig. 5. Upgrade superperiod.

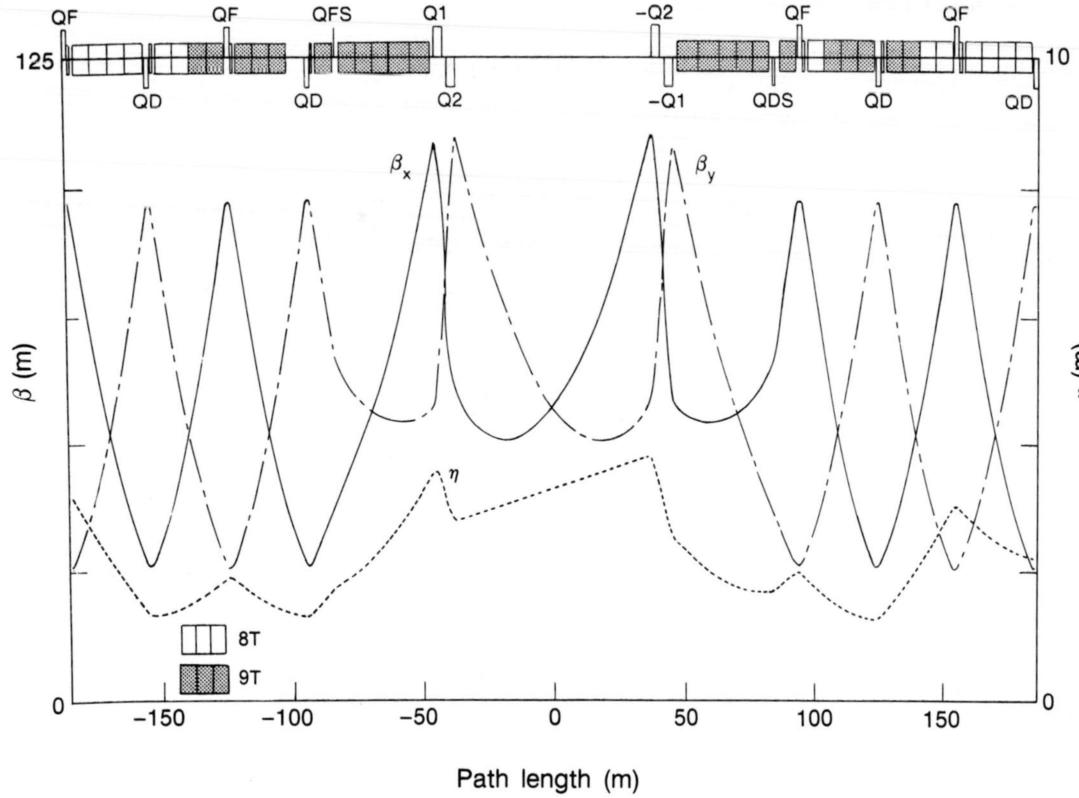

Fig. 6. Vicinity of Upgrade long straight section.

Table I Lattice Parameters

	Tevatron	Upgrade
Energy	1.0 TeV	1.8 TeV
Circumference	6283 m	6283 m
Length of long straight section	53 m	73 m
Cell phase advance	67.8°	73.5°
Betatron tune ν	19.4	20.7
Chromaticity	−22	−24
Momentum compaction	0.00285	0.00245
β_{max} in arc	98 m	97 m
β_{max} globally	110 m	110 m
η_{max} globally	5.9 m	5.0 m
Magnetic field	4.43 T	7.97 T, 9 T
Gradient	76 T/m	147 T/m
Dipole length	6.1214 m	6.1214 m
Cell quadrupole length	1.6789 m	1.6789 m
Cell half-length	29.7434 m	29.7434 m

ACKNOWLEDGMENT

The authors are happy to thank Mike Harrison for very helpful discussions and suggestions.

REFERENCES

1. Leon Lederman, "The Fermilab Upgrade," these proceedings and Fermilab report TM-1536, (September 1988).
2. A. Garren, A. Kenney, E. Courant and M. Syphers, "A Users Guide to SYNCH," Fermilab report FN-420, (June 1985).

PROSPECTS FOR 6- TO 10-TESLA MAGNETS FOR A TEVATRON UPGRADE

Paul M. Mantsch

Fermi National Accelerator Laboratory*
P. O. Box 500
Batavia, Illinois 60510

The first SSC physics is at least 10 years away. An upgrade of the Fermilab Tevatron will ensure the continuity of a vigorous high-energy physics program until the SSC turns on. Three basic proposals are under consideration:

1) $\bar{p}p$ at 3×10^{31} - Increase luminosity by improvements to the $\bar{p}$ source.

2) pp at 1 TeV and 2×10^{32} - Move the main ring to a new tunnel, build a second Tevatron ring.

3) $\bar{p}p > 1.5$ TeV and 7×10^{30} - Replace the Tevatron with a higher energy ring.

The last two options require new higher-field magnets. The second option requires about a hundred 6.6-tesla dipoles in addition to a ring of Tevatron-strength (4.4-T) magnets. These higher-field magnets are necessary in both rings to lengthen the straight sections in order to realize the collision optics. The third option requires a ring of magnets of 6.6 T or slightly higher to replace the present Tevatron plus a number of special 8- to 9-tesla magnets.

The viability of the high-energy option then depends on the practicality of sizable numbers of reliable 8- to 9-tesla dipoles as well as 800 6.6-tesla dipoles. The following discussion develops a specification for an 8.8-T dipole, examines the design considerations and reviews the current state of high-field magnet development.

SPECIFICATIONS FOR 6.6- AND 8.8-TESLA DIPOLES

Coil designs under consideration here are of the traditional Tevatron style with two shells wound of Rutherford cable around a circular aperture. The coils are supported by laminated collars. The design differs from the Tevatron in that a cold-iron yoke is used. In fact the cold iron becomes a vital support for the collared coil in high-field magnets.

A conceptual design has been made for a 6.6-tesla magnet as part of the study for a pp collider.[1] The aperture is the same as that of the Tevatron (3.0 inches) and the conductor is that developed for the DØ low-beta quadrupole system. The specification for this magnet is shown in Table I, a cross section in Fig. 1 and the operating load line in Fig. 2. The cryostat is of the style developed at Fermilab for the SSC. The 6.6-T magnet, if conservatively designed, can successfully be built. Although designed for 5 tesla the similar HERA magnets operate easily up to 6.9 tesla. This is a proof of principle that such magnets can be built.

Table I. Proposed specification for a 6.6-tesla dipole for the Tevatron upgrade.

Aperture	3 in.
Operating field	6.6 tesla
Operating current	4366 amp
Maximum field	7 tesla
Maximum ramp rate	0.2 T/sec
Operating temperature	4.6 K
Amp factor	100.855%
Sextupole	4.4×10^{-4}
Inner coil radius	1.50 in.
Outer coil radius	2.29 in.
Yoke inner radius	3.54 in.
Yoke outer dimensions	12.6 x 18.1 sq in.

The design of an 8.8-tesla magnet, on the other hand, plunges into largely uncharted territory. The high-current densities required assume 1.8-K operation. At 8.8 T the forces which must be accommodated are a factor of four higher than in

*Operated by Universities Research Association, Inc., under contract with the United States Department of Energy.

the Tevatron coil. The two most important magnet design considerations are: to achieve high-current density, and to ensure properly loaded coils for stability. The requirement of low temperature, of course, assumes that a practical large-scale refrigeration system for magnet operation in superfluid helium can be built.

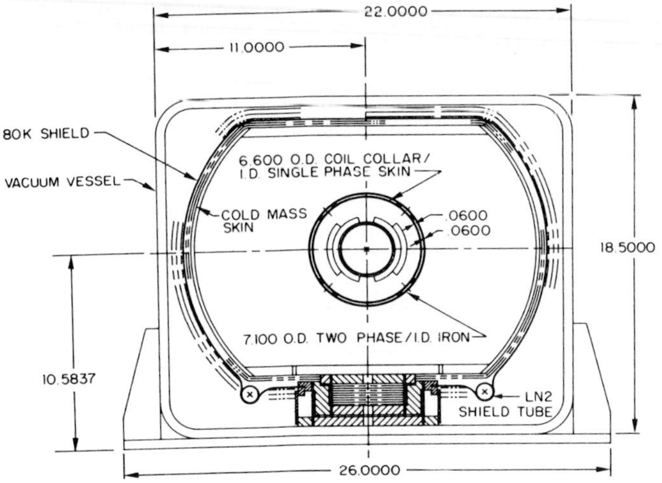

Figure 1. 6.6-tesla dipole cross section.

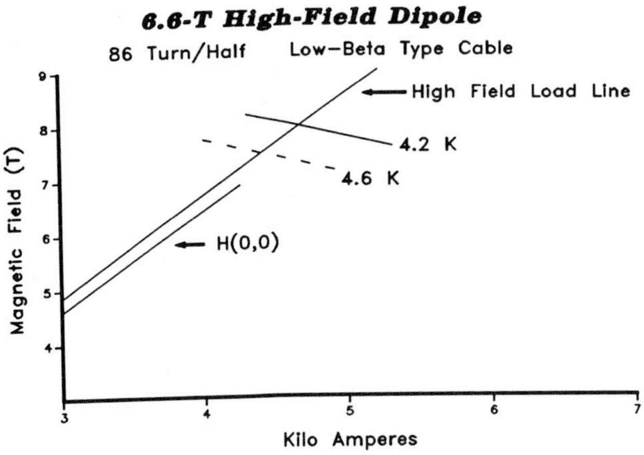

Figure 2. 6.6-tesla dipole load line.

Table II shows a proposed specification for an 8.8-tesla magnet. Since this machine is to be used to supply beams for fixed-target operation, the design must be capable of ramping. The aperture is the same as that of the Tevatron (3 inches) to permit extraction.

Table II. Proposed specification for an 8.8-T dipole for the Tevatron upgrade.

Aperture	3 inches (7.62 cm)
Operating field	8.8 tesla
Maximum field	10.0 tesla
Maximum ramp rate	0.2 T/sec
Operating temperature	1.8 K
Saturation sextupole	<5 units @ 8.8 T
Amp factor	<1.05 @ 8.8 T

DESIGN CONSIDERATIONS FOR AN 8.8-T DIPOLE

The development of the SSC dipole over the past four years and more recently the high-gradient quadrupoles for the low-beta program at Fermilab have lead to important improvements in magnet design and fabrication. These improvements provide a basis for advancing to higher fields. Features critical to a successful design will be considered.

Conductor

High field is achieved by high-current density. Motivated largely by the needs of the SSC, the advance in current density in NbTi conductor has been dramatic in the last few years. Figure 3 shows the increase in current density in the superconductor achieved over the past few years.[2] The current density specification for the Tevatron was 1800 A/mm^2 at 5 T, 4.2 K in cable form. The most recent cable received for the low-beta quad program has a current density in the superconductor of 2914 A/mm^2 - a 60% improvement! This increase has resulted from intensive study of the microscopic structure of superconductor by David Larbalestier and co-workers at the University of Wisconsin. A highly homogenous alloy melt, together with a carefully developed schedule of cold work and heat treatments, is responsible for these improvements.

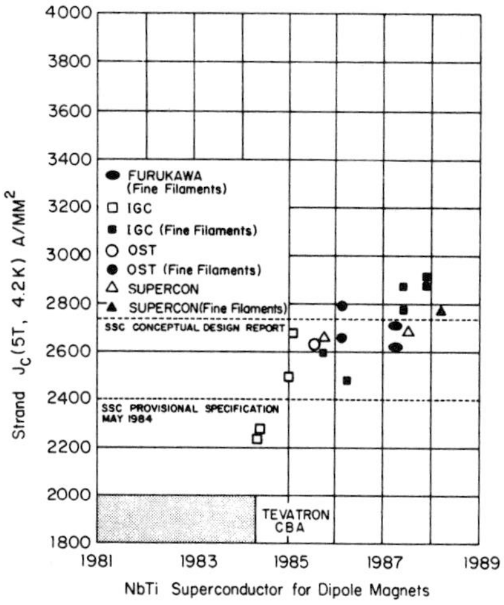

Figure 3. Current density improvements.

At low temperatures and high fields, NbTiTa alloy conductor has offered a current density improvement over NbTi. Although more development needs to be done, existing NbTiTa alloy could add at least a half tesla to the field of a magnet at 9 tesla and 2 K.

The characteristic (critical current vs. applied field) for the low-beta quadrupole conductor is shown in Fig. 2. The performance of a magnet is determined by drawing a "load line" from the origin through a point representing the field at the high-field point on the coil and the current at that field. The point where the load line intersects the critical current characteristic of the conductor is the short sample limit of the magnet. The aperture field is 6 to 9% lower than the high-field point line at the same current. It is clear that the field goes up more slowly than the critical current density as the field must follow the load line.

A secondary goal is high-current density in the cable as a whole. Current density in the cable therefore depends on the copper-to-superconductor ratio. There is not a clear understanding of the role of the copper content (or the properties of the copper) in the stability of a finished magnet, however. Stability is measured (rather poorly) by the number of quenches necessary to reach the conductor short sample limit. Over 30 models of SSC magnets have been built with copper-to-superconductor ratios varying from 1.28:1 to 1.6:1. The test data do not do much to illuminate this issue. We have excellent performers with 1.30:1 and with 1.6:1. Differences in performance as measured by quench history are apparently obscured by other effects.

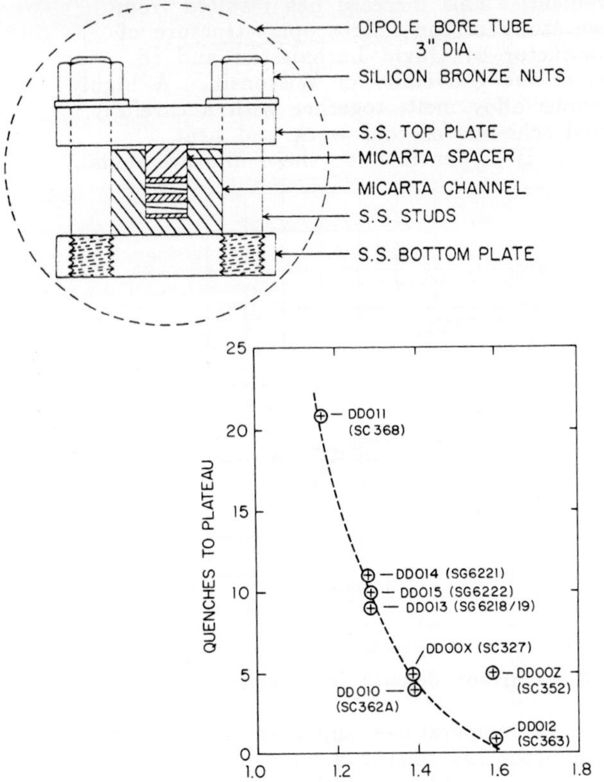

Figure 4. a) BNL fixture for a cable short sample measurement. b) Short sample measurements of the SSC cables. Quench sensitivity to copper-to-superconductor ratio.

Sampson has measured large numbers of cable samples at Brookhaven. The sample is supported rigidly in a fixture that is inserted into the bore of a high-field dipole. A measurement is then made of the characteristic of the cable. Many of the cables quench several times before reaching a plateau current. Sampson finds the number of cable quenches to be a strong function of the copper-to-superconductor ratio. A sketch of the cable support fixture and a plot of quench number vs. current for a number of SSC cables is shown in Fig. 4.[3] In fact, when magnets are made of these cables the quench performance is not nearly as bad as predicted by Sampson's plot. For example a cable with a ratio of 1.3:1 requires about 10 quenches to reach a plateau in Sampson's measurement. A 1-meter model wound with this cable, D-15-A3, reaches that short sample limit on the second quench. The quench history of D-15-A3 is shown in Fig. 5.

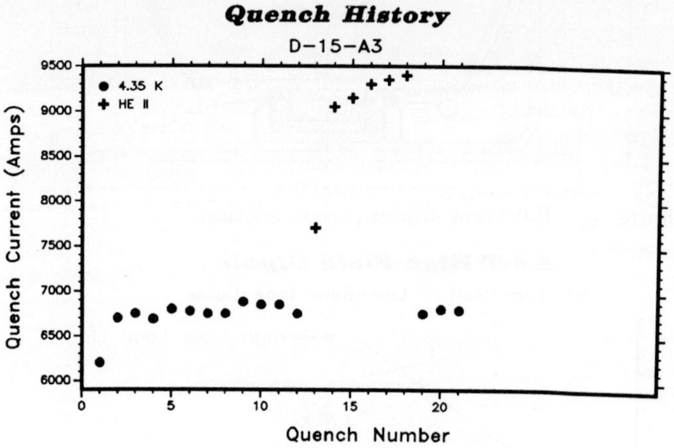

Figure 5. Quench history of magnet D-15-A3.

The evidence is certainly not conclusive on the minimum copper-to-superconductor ratio. On the basis of Sampson's work, however, a prudent choice would be 1.5:1 or higher.

Within the coil package as a whole the current density is affected by the thickness of the insulation. The Tevatron used a 7/12-lap 1-mil Kapton wrap plus a spiral wrap of 7-mil B-stage/fiberglass tape for a total of about 9 mils of insulation. A new thinner insulation system was developed for the Fermilab low-beta quad program. It consists of a 2-mil Kapton butt lap followed by a 1-mil Kapton butt lap. The outer layer has a 0.1-mil coating of B-stage epoxy (i.e. there is no fiberglass tape). If a cable of Tevatron dimensions is insulated with these two schemes, a coil current density increase of 9% is realized. It should be noted that a certain uneasiness exists about leaving out the glass tape because of the resulting loss of helium flow channels and helium volume within the coil. It is not understood how sensitive the coil stability (under ramping or beam loss) is to this displaced helium. Tests of the low-beta quad models are now under way which use this insulation scheme.

Coil Loading

Once every advantage of current density has been taken, the next task is to stably control the large magnetic forces within the coil structure. The coil must be fully supported to resist motion to the highest current. Since the forces increase as the square of the field this becomes rapidly more difficult as the field requirement goes up.

The coil is subjected to a preload at the time of collaring. This preload is successively reduced by creep, by the cooldown and by powering. The preload loss during cooldown results from the slightly faster shrinkage of the coil relative to the stainless steel support collars. Once the coil is powered the turns near the pole are unloaded by the Lorentz force pulling them toward the parting plane. Finally, the radial magnetic forces which increase toward the parting plane react against the support collars resulting in coil distortion. This distortion enlarges the coil cavity further reducing the preload.

It is necessary to examine each of these sources of coil load relaxation to determine if a design can be made that will ensure that the preload to 8.8 tesla can be sustained.

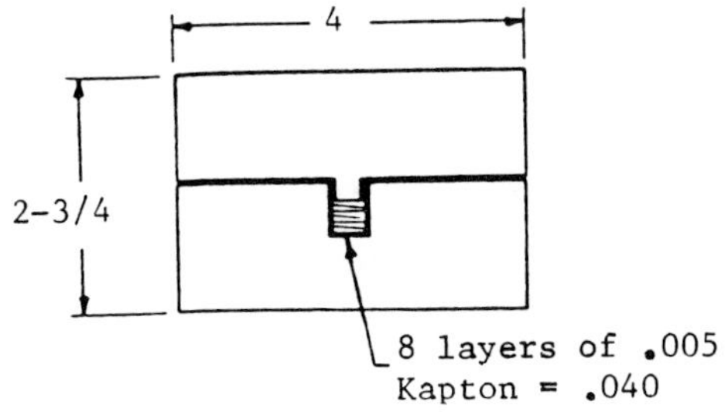

	Insulation Scheme	Breakdown
1.	One layer of 1-mil Kapton, 1/2 lap and one layer 4-mil glass tape 90%	None at 49 K psi (limit of the press)
2.	One layer of 2-mil Kapton butt lap and layer of 1-mil Kapton butt lap	Turn to turn short at 35 K psi after 1 hour
3.	One layer of 1-mil Kapton 3/4 lap	Turn-to-turn short at 25 K psi

Figure 6. Insulation breakdown test (data taken 2/17/88).

It is important to achieve the highest possible preload on the coils at collaring time in anticipation of the subsequent losses. The main limitation to preload is turn-to-turn shorts in the coil as the insulation breaks down under pressure. This limiting pressure is in fact quite high. Tevatron coils have been successfully loaded during collaring to 25 K psi. Recently tests were made by Carson at Fermilab in which cables with various insulating schemes were pressed in a fixture. Between each pressure step hipot measurements were made between turns to determine the onset of insulation breakdown. At least four hours were allowed at each pressure step for the insulation to flow. The results shown in Fig. 6 suggest that 20 K psi preload can be safely used with Kapton only.

It is important to note that the pressure must be uniform across the face of the cable. During the collaring process the collars may be distorted so that stress concentrations may occur at the inner or outer edge of the cable far in excess of the average pressure. The collaring process must therefore be carefully modeled and designed to insure that such stress concentrations do not occur.

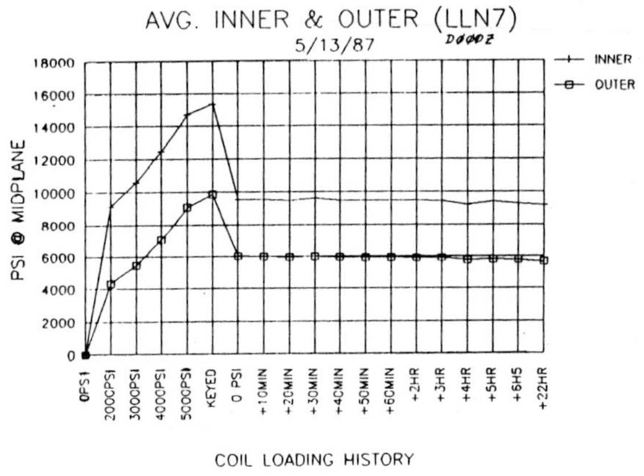

Figure 7. Prestress loss on an early SSC magnet.

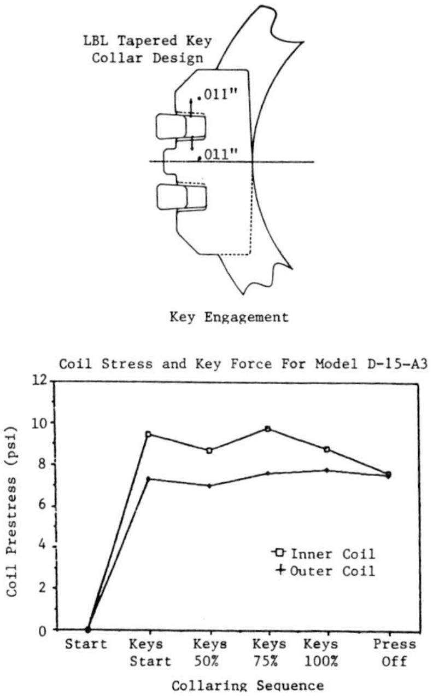

Figure 8. Collaring prestress loss using tapered keys.

Current practice calls for the collars to be locked together with keys, a technique originally used on the Tevatron quadrupoles. The coil is over-compressed, the keys are inserted and the press pressure is removed. Because of the slight gap that remains between the keys and collar the coil relaxes to a preload substantially below the collaring pressure. This is illustrated in Fig. 7 in an early BNL SSC magnet.[4] A new scheme developed by Taylor at LBL uses tapered keys that are engaged by a separate set of hydraulic cylinders just as the collaring press nears its maximum pressure. The tapered keys pull the collars together as they are forced in and thereby preserve the preload (see Fig. 8).[5]

Once the coil is collared and is under pressure the plastic materials within the coil will begin to creep. Strain gauges have been used to monitor the coil preload through the assembly process. Considerable prestress loss due to creep has been observed on SSC models. Performance loss due to creep, however, has not been observed in Tevatron magnets, even those stored for years at ambient temperature. It is very important to cure the coils at sufficiently high pressure so that most of the soft materials flow away from contact surfaces and the remaining material has taken up most of its creep.

The next major loss of preload results from shrinkage of the coil relative to stainless steel collars. In principle, aluminum collars shrink more than the coil and should load the coil during cooldown:

Shrinkage to 4 K

Coil ~ 3.3 mils/inch
Stainless steel ~ 3 mils/inch
Aluminum ~ 4 mils/inch

In practice this additional preload is not observed. The preload at room temperature, however, seems to be preserved. It is, therefore, still attractive to use aluminum collars. Aluminum collars have the additional feature that they have much better thermal conductivity than those of stainless steel.

The remainder of the preload loss occurs during powering. It results from the azimuthal component of the Lorentz force pulling the coil away from the pole and from the enlargement of the coil cavity because of collar distortion. If the pressure on the coil at the pole goes to zero the coil will become unstable and will not sustain high current.

The azimuthal Lorentz force acts on the individual turns relaxing the pressure on those near the pole and compressing those near the parting plane. This azimuthal pressure increases nearly linearly from zero at the parting plane to a maximum at the pole turn given by the product of the field at that point and the current. Figure 9 shows the azimuthal pressure as a function of turn number for the SSC coil.

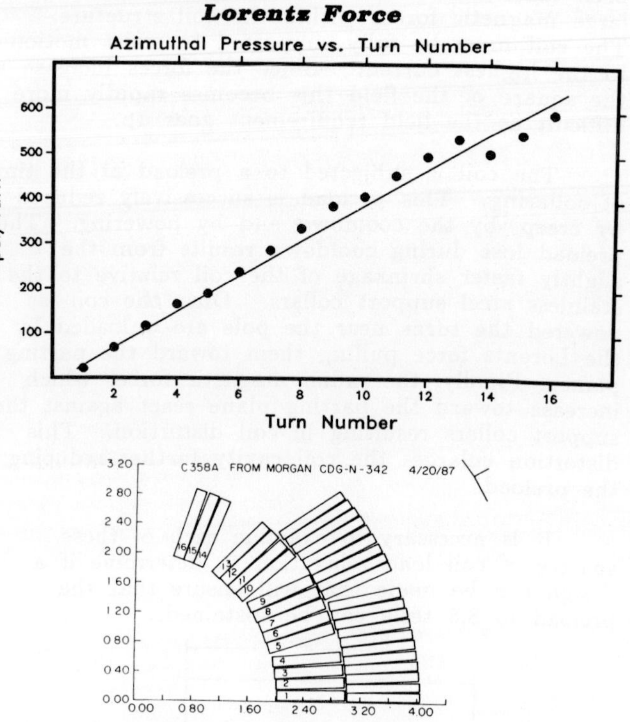

Figure 9. Lorentz force coil loading.

Consider a simple spring model for the coil in which the total azimuthal Lorentz force is concentrated at the center of the opening spring.

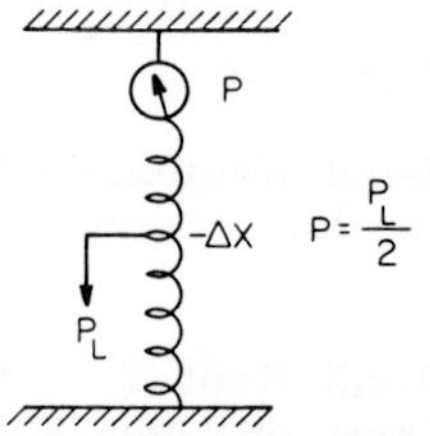

In this model the coil pressure reduction at the pole is half of the total Lorentz force. In a more complex model due to Tollestrup the actual force on each turn is considered. In this model the fraction of the total azimuthal Lorentz force felt at the pole is about two-thirds. Even this model may not adequately describe the pressure distribution in the coil because it ignores friction between the coils and the collars.

The other Lorentz force driven preload relaxation results from the enlargement of the coil cavity due to distortion of the collars. Since collars, and aluminum collars in particular, will flex to some degree it will be important to utilize the iron yoke for additional strength. The steel shrinks less than the aluminum during cooldown so there must be some mechanism to keep the collars in intimate contact with the yoke and to make the yoke itself resistant to flexing. One approach is to use an aluminum or stainless steel outer skin to pull the yoke halves down against the collars at cooldown.

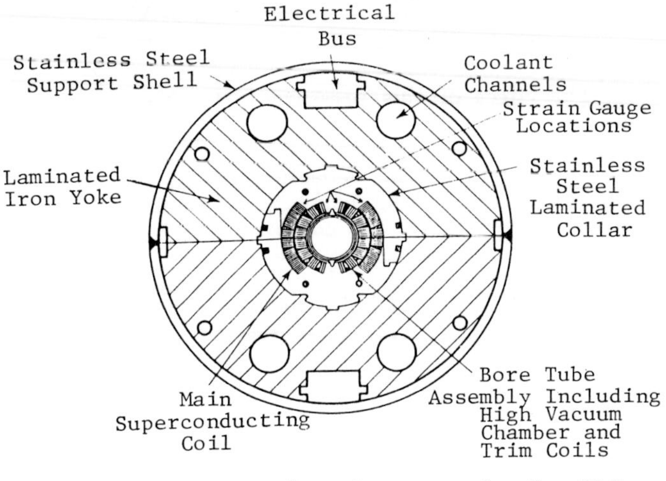

Figure 10. Position of strain gauges for the SSC coil.

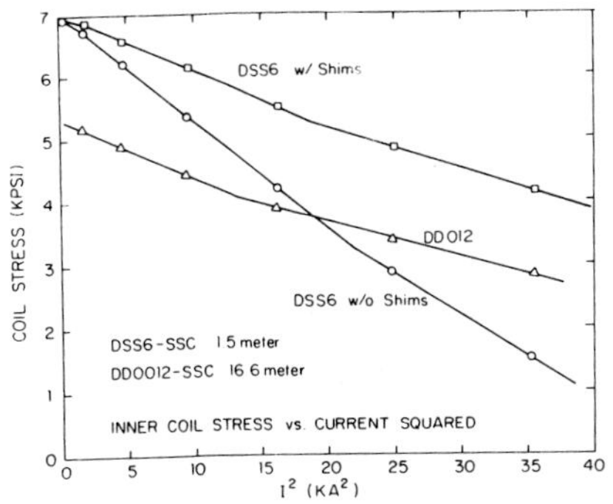

Figure 11. Inner coil stress vs. current squared for SSC magnet model DSS6 and DD0012.

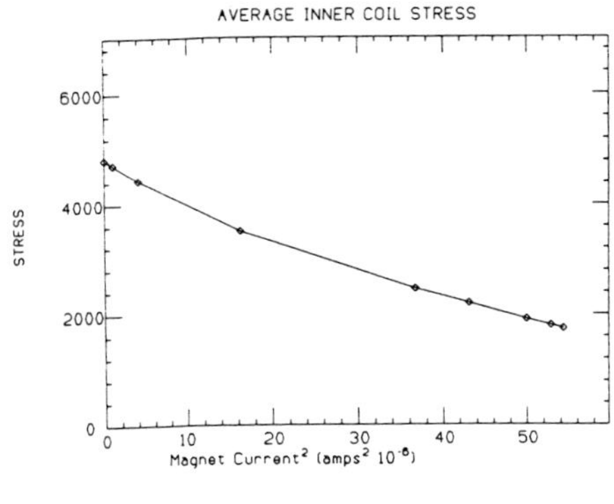

Figure 12. Inner coil stress on SSC magnet DD0012.

Since the beginning of the SSC dipole program, substantial effort has gone into instruments to study the performance of these magnets under power. In particular, strain gauge techniques have been improved to the point where they give believable results. Strain gauges are attached to the poles of the coils as shown in Fig. 10. Figures 11 and 12 show some typical curves of stress at the pole as a function of the square of the coil current. The stress is almost, but not exactly, proportional to the current squared.

Figure 11 is particularly interesting because it involves a BNL SSC 1.5-meter model (DSS6) that was first tested in the normal configuration where the collars alone support the coil. The model was then modified by putting shims between the collared coil and the yoke and leaving a 10-mil gap between the yoke halves at the parting plane (Fig. 13).[6] The shims firmly press the collars against the yoke which results in less collar flexing and the magnetic attraction between the yoke halves actually increases the coil load with increasing current. The net effect is that the coil in the shimmed magnet unloads much more slowly with current. The same shim arrangement was used in SSC magnet DD0012 whose stress curve is shown in Fig. 12. Note that high coil preload is maintained out to the highest current (about 7600 amps). Magnet DD0012 has the best performance of any SSC magnet so far. Figure 14 shows the quench history of DD0012 compared to the next best magnet, D000X.

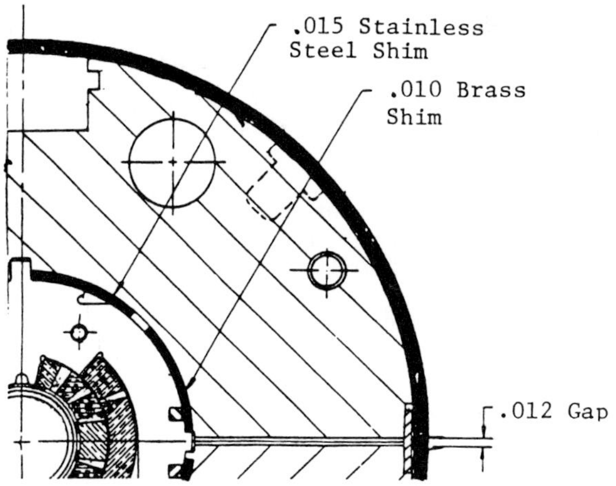

Figure 13. Shim location in magnets DSS6 and DD0012.

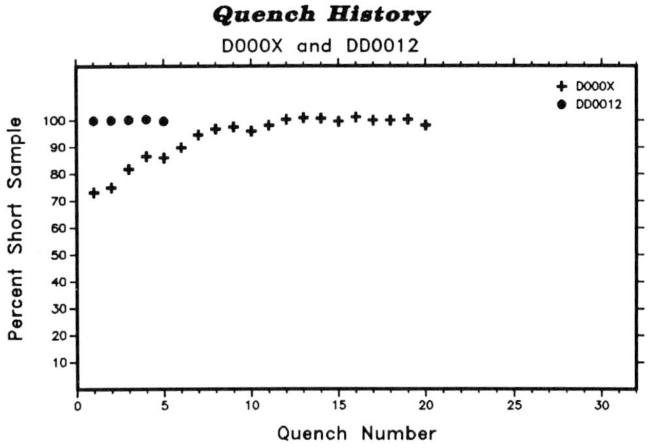

Figure 14. Quench history of D000X and DD0012.

It is possible to gain some understanding of coil unloading by comparing the curves in Figs. 11 and 12 with the spring models described earlier. Figure 15 shows the contribution to the pressure at the pole of the inner coil from the coil azimuthal Lorentz force, the yoke gap and the collar defection. The stress gauge reading at the pole starts with the prestress, P_0 at zero current and is increased or decreased by the different contributions of the magnetic force. Each of the pressures should be proportional to the square of the current. The strain gauge reading should therefore be:

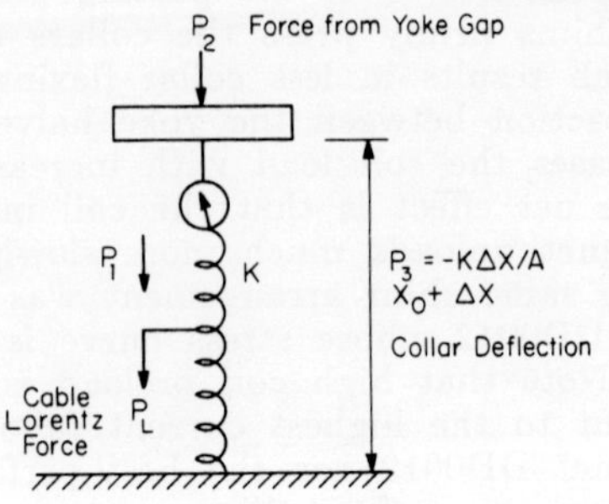

Figure 15. Coil loading model.

$$P_{SG} = P_0 - P_1 + P_2 - P_3$$

$$= P_0 + \alpha \bullet I^2 + \beta I^2 + \gamma I^2$$

$$= P_0 + (\alpha + \beta + \gamma) I^2$$

where

P_{SG} - strain gauge pressure

P_0 - azimuthal prestress

P_1 - pressure due to coil azimuthal Lorentz force

P_2 - pressure on the coil due to the yoke gap

P_3 - pressure due to collar distortion

The slope of the stress curves is the sum of α, β and γ.

Consider the three pressures one at a time.

The total inner coil azimuthal Lorentz force has been calculated by Morgan for the SSC cross section at operating current (6.6 K amp) to be 5820 psi. If the stress at the pole is two-thirds of the total (Tollestrup model) the contribution to the slope from the azimuthal Lorentz force can be predicted:

$$\alpha = \frac{2/3 * 5820}{(6.6)^2} = -89 \text{ psi/KA}^2$$

The pressure on the coil resulting from the presence of the gap between the yoke halves has been calculated by Morgan for this particular magnet to be 1750 psi at 6.6 KA.

$$\beta = \frac{1750}{(6.6)^2} = +40.2 \frac{\text{psi}}{\text{KA}^2}$$

The effect of this pressure is to decrease the slope.

Calculation of the deflection of the collars is a complicated undertaking using finite element analysis. The results are compromised by an incomplete understanding of the properties of the coil. A calculation at BNL for this magnet shows that the collars flex radially by about 2 mils at the parting plane. This gives rise to a small increase in the azimuthal size of the coil and a resulting loss of 1000 psi in coil loading.

$$\gamma = \frac{1000}{(6.6)^2} = -23.0 \text{ psi/KA}^2$$

These numbers taken together should give a prediction for the two slopes shown in Fig. 11 for short model magnet DSS6.

DSS6 - without shims

α = -89 psi/KA^2
β = 0 (there is no yoke gap)
γ = -23.0 psi/KA^2

Slope = -89 -23 = -111 psi/KA^2

DSS6 - with shims

α = -89 psi/KA^2
β = +40.2
γ = 0 (the coil is backed up by the yoke and should not deflect)

Slope = -48.8 psi/KA^2

In comparing these results to the data, note that the curves do not have a constant slope. A careful look at the DD0012 curve shows that there are two fairly distinct slopes, one for low currents and a shallower one at high currents. The results can then be tabulated as follows:

	Low-Current Slope (psi/KA^2)	High-Current Slope (psi/KA^2)	Predicted Slope (psi/KA^2)
DSS6 (no shims)	-166.0	-118.0	-111.0
DSS6 (shims)	-87.6	-59.1	-48.8
DD0012 (shims)	-79.6	-52.9	-48.8

Within the limits of uncertainty of the model the agreement at high current is quite reasonable. The steeper slope in the case of the shimmed magnets is most likely the result of the collars not yet being up against the yoke. At high current the collars are firmly against the yoke. The change in slope of the unshimmed magnet is difficult to explain.

This exercise then provides the basis for estimating the stress loss under power. The magnets above have the yoke gap which is probably not practical in a real magnet. With the collaring structure rigidly constrained and no yoke gap, the inner coil stress loss due to the azimuthal Lorentz force can be estimated for similar cross sections as follows:

Azimuthal Lorentz pressure at pole turn =

$$f\left(\frac{B_{max} \times I_o}{2}\right)\frac{n}{w}$$

where:

f is the fraction of the total azimuthal pressure seen at the pole turn (2/3 is assumed)

B_{max} is the high-field point at the inner pole turn

I_o is the current

n is the number of turns in the quadrant

w is the width of the conductor

If one uses the various schemes mentioned earlier to preclude the loss of preload this formula can be used to estimate the amount of preload required for a magnet of given strength.

Since the preload on high-field magnets is critically important, the coils must be built with sufficient precision that the preload is uniform along the length. The dimensional tolerance on the conductor must be very tight. Carson at Fermilab has developed a measuring machine that measures cable, online at operating stress, to better than ±0.0001 inch. With this machine supplied to the vendor it is possible to set a cable thickness specification at ±0.0002 inch and get cable within those limits. The tooling must also have good precision. Recent coils at Fermilab have been made with variations in the thickness of less than ±.001 inch. This compares to ±.002 inch for the Tevatron.

Ends have traditionally been a weak part of superconducting magnets. The complex geometry of the saddle end with spacers to adjust the end field is hard to support adequately. The transition from the straight section to the end turns is also a source of high-stress points that can lead to turn-to-turn shorts. Study is under way at Fermilab to design an end that is free of high-stress points and in which the end turns are properly supported. In the constant perimeter end the conductors take nearly their natural shape. Analytical work has been done on this problem a number of years ago at Rutherford Lab and more recently at Argonne Lab. Using this work and the CAD system greatly improved ends can be made. Once the design is complete the complex support pieces can be automatically machined from the CAD information. Figure 16 shows some of the drawings made in the CAD during the design process.

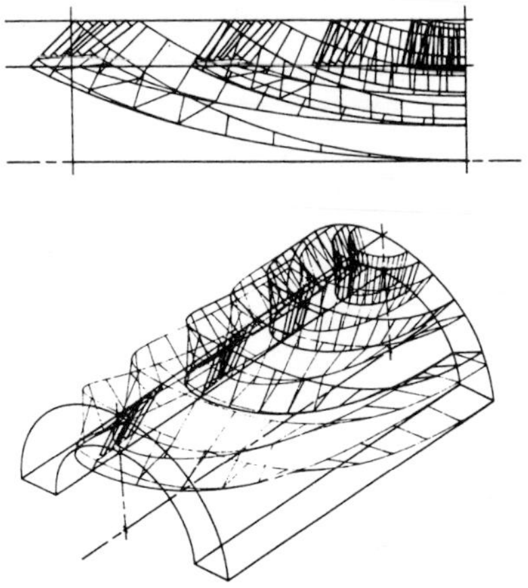

Figure 16. CAD drawings of end turn design.

An 8.8-Tesla Dipole Design

First attempts have been made by Riddiford to design a coil cross section to meet the specification shown in Table I. The magnet is shown in Fig. 17. It makes use of the conductor developed for the low-beta quad system. The load line and cable characteristics are shown in Fig. 18. It is clear that the coil size needs to be increased to achieve a reasonable operating margin.

It is necessary now to estimate the preload needed to ensure adequate loading through the operating range. Including operating margin the magnet should go to 10 tesla. If it is assumed that no preload is lost except during powering, we can calculate the loss in azimuthal preload on the inner coil:

$$\Sigma P_\theta = \frac{B_{max} \cdot I_o}{2}\frac{n}{w} = 28.0 \text{ K psi}$$

If the pole turn sees two-thirds of this, the expected preload loss in the inner coil would be 18.7 K psi. In order to account for other possible losses, a preload of at least 20 K psi is necessary. This high load requires that considerable attention be paid to the insulation system. If this preload can be obtained and the integrity of the insulation scheme can be demonstrated, an 8.8-tesla magnet can probably be successfully made.

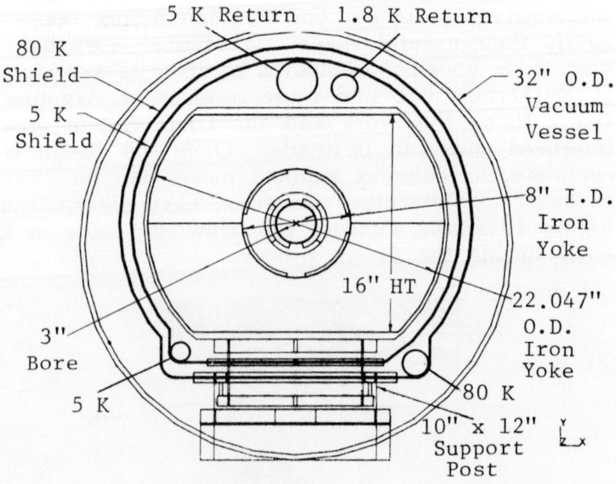

Figure 17. 1.8-K high-field (8.8-T) dipole cross section.

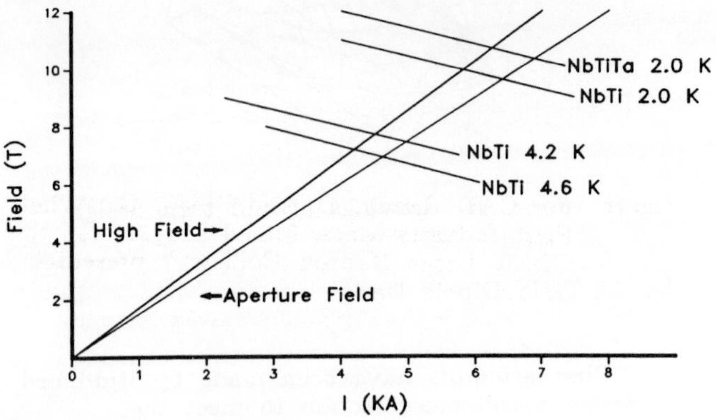

Figure 18. 8.8-T dipole load line.

State Of The Art In High-Field Accelerator Magnets

It is worthwhile to look at the performance of recent high-field magnets and compare design features with those of the proposed 8.8-tesla magnet. Five magnets will be compared: Tevatron, SSC, HERA, LHC Model and the Tevatron 8.8 T.

A 1-meter LHC model has recently been completed and tested by CERN and Ansaldo. Its specification is shown in Fig. 19. The magnet is said to have trained to 9.1 tesla in about four quenches.[7)]

HERA magnets also demonstrate excellent performance. Figure 20 shows that for among three full-size (9-meter) magnets there was only one training quench. The magnets performed well beyond their nominal operating current.

As shown in Fig. 5, SSC 1-meter models have operated at 9.4 tesla.

1)	Central field B_0	8.0 T
2)	Peak field at B_0	8.5 T
3)	Current at B_0	9140 amps
4)	Inner coil diameter	50 mm
5)	Stored energy	160 KJ
6)	Cable geometry	1.3, 1.67 x 12.6 mm^2
7)	Cu/SC ratio	1.8:1

Figure 19. Specification of CERN/Ansaldo LHC 1.5-m model.

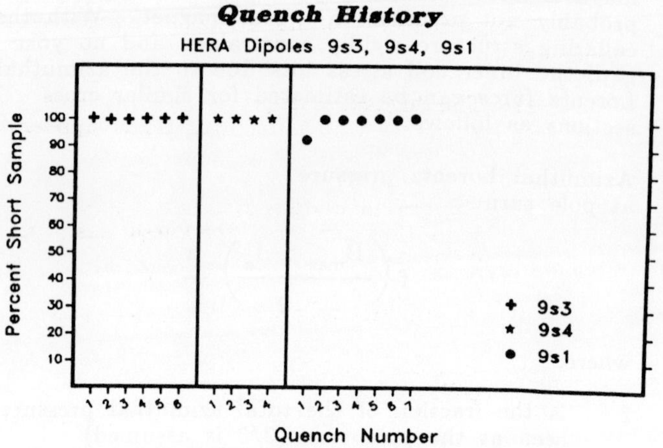

Figure 20. Quench performance of three HERA dipoles.

Table III compares these magnets at operating current and maximum current (short sample). The total inner coil azimuthal Lorentz force pressure is estimated in each case to show the level of stress in the coil.

Table III. Comparison of fields and inner coil Lorentz pressures for recent high-field magnets.

Magnet	Aperture (inches)	At Operating Field I(A)	Bo(T)	B_{max}(T)	ΣPo(inner) K psi	At Highest Field Bo (T)	ΣPo (K psi)
Tevatron	3.0	4526	4.5	4.8	7.1	5.2	9.5
SSC	1.57	6600	6.69	7.2	5.8	9.4	10.7
HERA (9 m)	3.0	4992	4.65	5.0	5.8	6.9	12.5
LHC (1 m)	1.97	9140	8.0	8.5	9.4	9.3	12.7
Tevatron 6.6 T	3.0	4366	6.6	7.0	11.3	7.0	12.7
Tevatron 8.8 T	3.0	6100	8.8	9.6	21.7	10.0	28.0

In terms of coil stress level the HERA magnet and the LHC model operate at about the same stress because of the larger aperture of the HERA magnet. The coil stress of the proposed 8.8-T magnet is nearly twice that of the LHC and HERA!

SUMMARY

In order to realize a viable 8.8-tesla dipole design, the following are essential:

1) High current density.

2) A very rigid mechanical support for the coil.

3) Coils of high precision and uniformity.

4) High coil prestress sustained through collaring, cooldown and powering.

5) An insulation scheme able to maintain its integrity at these high stresses.

Design features of an 8.8-tesla magnet would include

1) The highest current density material available which for high fields appears to be NbTiTa.

2) Kapton insulation with no fiberglass to keep the insulation thickness less than 0.003 inch.

3) Aluminum collars with tapered keys.

4) A cold-iron yoke designed to support the collared coil assembly.

CONCLUSIONS

Design considerations for high-field dipoles have been presented.

Magnets of 6.6 T and 3-inch aperture can be built. Magnets with similar field and stress levels have been successfully built.

Magnets of 8.8 T and 3-inch aperture will stretch the current technology well beyond current experience. In particular we are approaching the limits in accommodating magnetic forces in the present geometry. This implies a substantial R&D program.

Proposed program:

a) Initiate the design and fabrication of a conservative 6.6-T dipole according to the principles outlined.

b) Refine the collared coil assembly above to push performance as close as possible to the 10-T goal at low temperature.

c) Explore alternate geometries.

REFERENCES

1) Conceptual Design Report, Proton - Proton Collider Upgrade (Main Injector, New Tevatron), Fermi National Accelerator Laboratory, May 1988.

2) Scanlan, R., presented at the NbTi Workshop, Asilomar, CA, March 1988 (unpublished).

3) Sampson, W., presented at the SSC Magnet System Integration Meeting, Brookhaven National Laboratory, June 17-19, 1988 (unpublished).

4) Goodzeit, C., presented at the SSC Magnet System Integration Meeting, Berkeley, CA, June 2-3, 1987 (unpublished).

5) Gilbert, W. and Barale, P., "Prompt Report on LBL-SSC Model Magnet D15A3," SSC-MAG-169, November 1, 1987.

6) Goodzeit, C. and Wanderer, P., "DSS6 Retest with Shims Between Collars and Yoke," SSC-N-510, April 1988.

7) Perin, R., Leroy, D. and Spigo, G. "The First Industry-Made Model Magnet for the CERN Large Hadron Collider," presented at the 1988 Applied Superconductivity Conference, San Francisco, CA, August 21-25, 1988.

Magnetic Designs of High Field Dipoles for a Tevatron Upgrade

A. Riddiford
Fermi National Accelerator Laboratory
P.O. Box 500, Batavia, Illinois 60510

Objective

The objectives of the calculations reported in ths note were 1) to estimate how much iron was needed to keep the amp factor below 1% when the axial field was 8.8 Tesla, and 2) to find conductor placements which kept the harmonic coefficients small.

This note reviews two kinds of calculations; 1) analytic, assuming infinite μ iron; 2) finite elements in 2-D. The calculations in 1) guided the placement of conductors in the first quadrant and estimated the harmonic coefficients and high field point more accurately than 2). The focus of the calculations in 2) was to see how much iron was needed to keep the amp factor below 1%, the sextupole coefficient below about 5 units, and the fringe fields below ·100 Gauss at the outside edges of the iron.

Table I lists the results of POISSON calculations. The Tevatron (SAVER) dipole is included for reference. Of the 6.6 Tesla runs,

Table I 6.6 & 8.8 Tesla Dipole Iron Shape Investigation
Assume 50 inner turns & 36 outer turns

Iron ID	Strength Tesla	Height x Dia	Amps/ turn	Amp Factor %	6-pole 10^{-4} units	10-pole 10^{-4} units	Max Field kGauss	Top Fringe Gauss	Side Fringe Gauss	Design Label
7.500"	4.40	10"x15"*	4440	0.1455	0.12±1.2	-0.10±0.45	15.7	98	33	(Saver)
6.000"	6.54	13"x15.748	4100	3.76	9.0	0.19	32.29	58	577	D501
	6.58	13"x16.535	4100	3.20	15.0	0.75	32.46	73	275	D502
	6.59	13"x17.323	4100	2.94	17.7	0.97	32.54	82	145	D503
	6.61	13"x19.685	4100	2.68	20.4	1.22	32.60	91	20	D504
	6.61	13"x20.472	4100	2.66	20.6	1.27	32.65	91	8.5	D505
6.693"	6.51	13"x15.748	4250	2.3	-0.37	-0.51	28.07	59	686	D511
	6.55	13"x16.535	4250	1.7	4.90	-0.21	28.26	74	312	D512
	6.57	13"x17.323	4250	1.4	7.30	-0.05	28.36	83	160	D513
	6.583	13"x18.898	4250	1.16	9.24	2.42	28.4	90	45	D514
	6.586	13"x19.685	4250	1.12	9.64	0.12	28.4	91	20	D515
	6.582	12.6"x19.685	4250	1.17	9.54	0.12	28.4	138	20	D516
7.087"	6.62	13"x18.110	4366	0.806	4.37	-0.10	26.54	93	95	D521
	6.61	12.6"x18.110	4366	0.855	4.40	-0.075	26.52	143	94	D522
	6.62	12.6"x18.898	4366	0.772	5.06	-0.04	26.54	146	48	D523
7.480"	6.63	13"x18.110	4460	0.53	2.39	0.14	24.76	95	102	D531
	6.62	12.6"x18.110	4460	0.58	1.62	-0.07	24.26	147	104	D532
	6.63	12.6"x18.898	4460	0.50	1.35	-0.27	24.77	150	53	D533
7.480"	8.78	14"x20.5	6000	2.0	6.78		31.55	363	182	D541
	8.81	15"x22.1	6000	1.70	9.50		31.73	152	66	D542
	8.82	16"x22.1	6000	1.60	9.50		31.77	64	69	D543
8.000"	8.80	16"x22.1	6100	0.93	5.50		29.0	60	65	D551

D522 (see Figure 1) was thought to be a solution of the most interest. Of the 8.8 Tesla runs, only the 8" iron ID run resulted in an amp factor <1% (see Figure 2). Figure 3 plots the field along the conductor in the inner layer next to the pole. The maximum field was 9% greater than the field at the center of the magnet.

The rest of the figures show conductor positions. The Tevatron dipole (Figure 4) is included for reference. Figures 5, 6 and 7 use the conductor developed for the low beta quads in the Tevatron ring. Figure 8 is a first attempt with 20% larger strands to find conductor placements which will result in small harmonic coefficients.

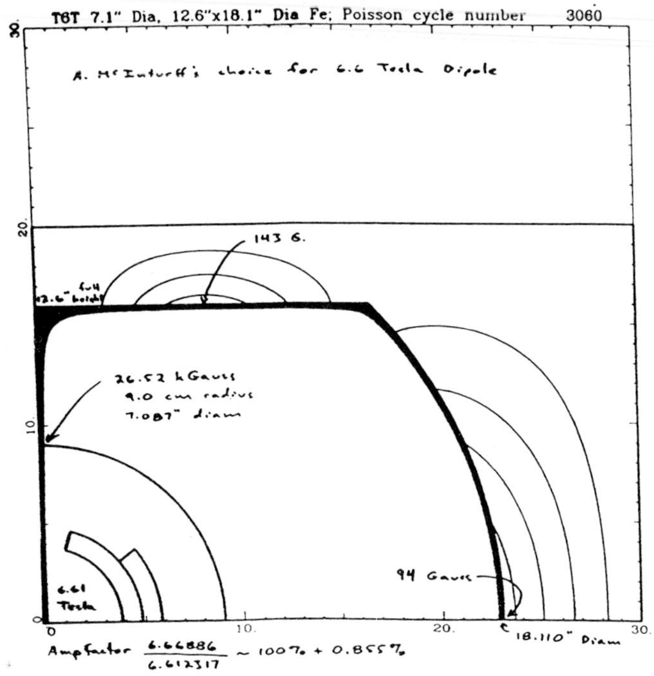

Figure 1. Preliminary iron shape for 6.6T.

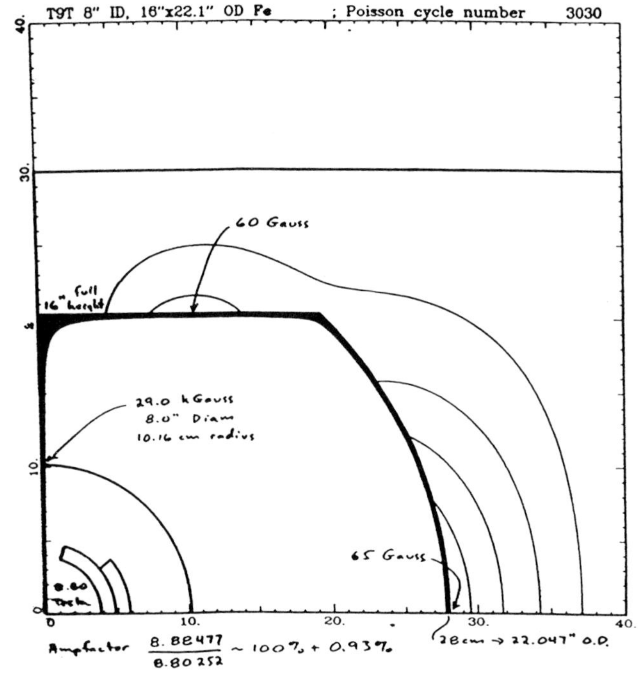

Figure 2. Preliminary iron for 8.8T

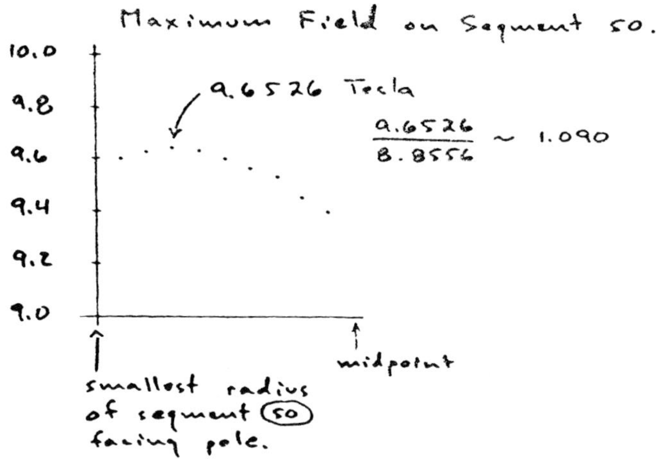

Figure 3. Maximum field for Figure 6

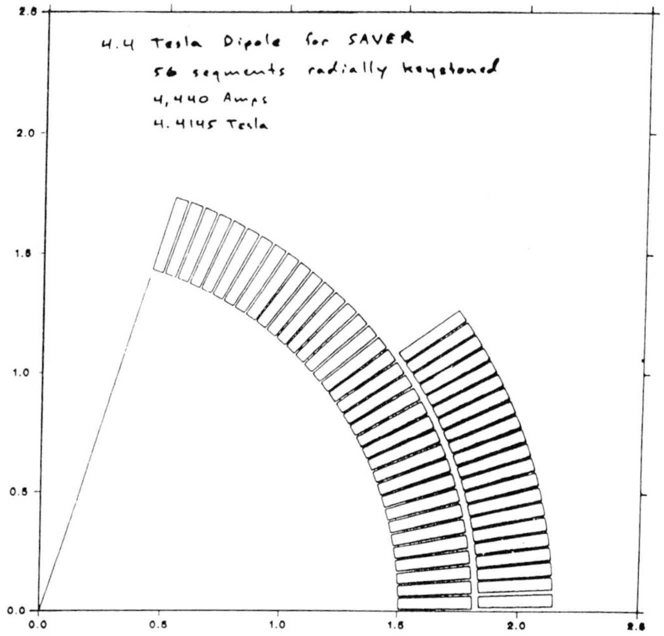

Figure 4. Tevatron dipole conductor placement

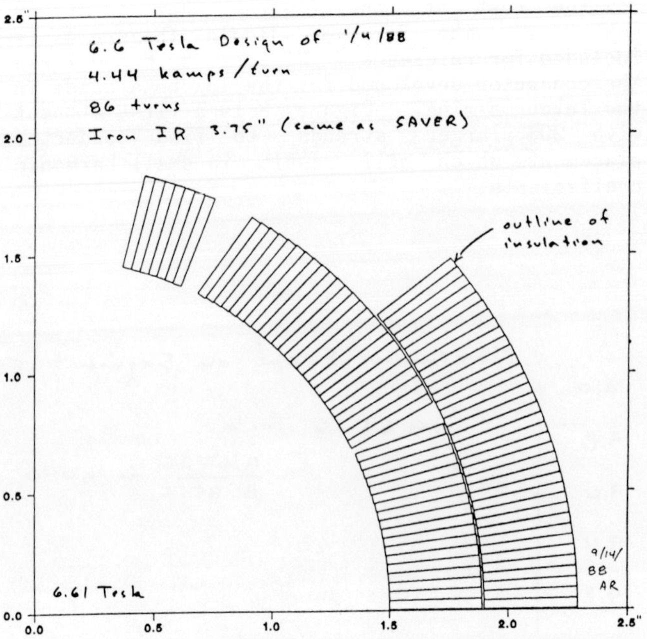

Figure 5. Possible 6.6 Tesla dipole condutor placement

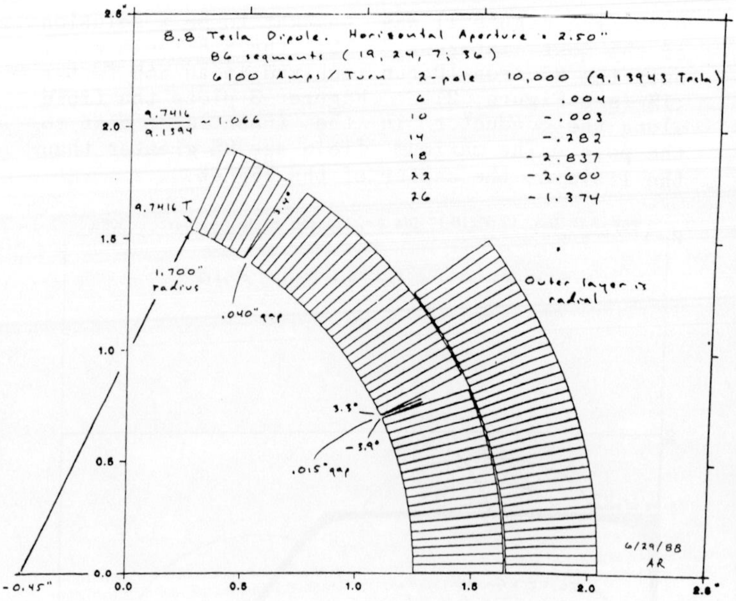

Figure 7. Possible 8.8 Tesla dipole conductor placement with an offset inner coil radius

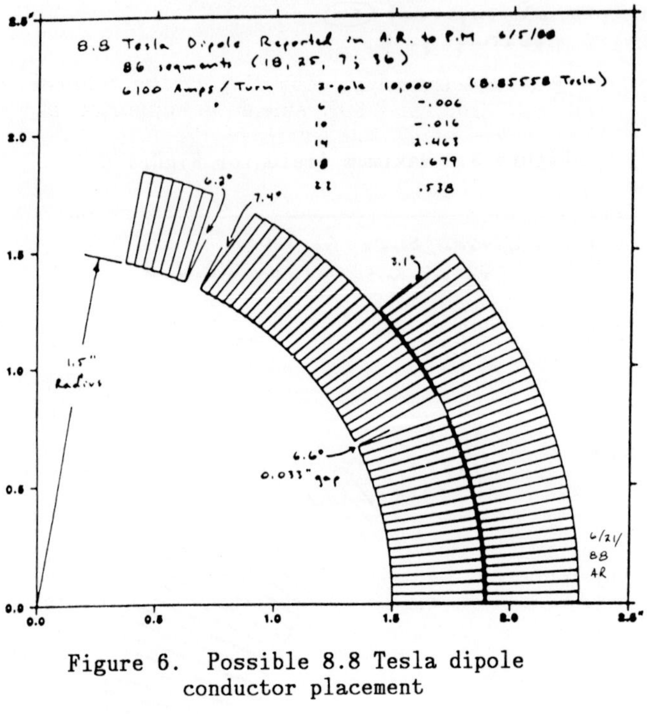

Figure 6. Possible 8.8 Tesla dipole conductor placement

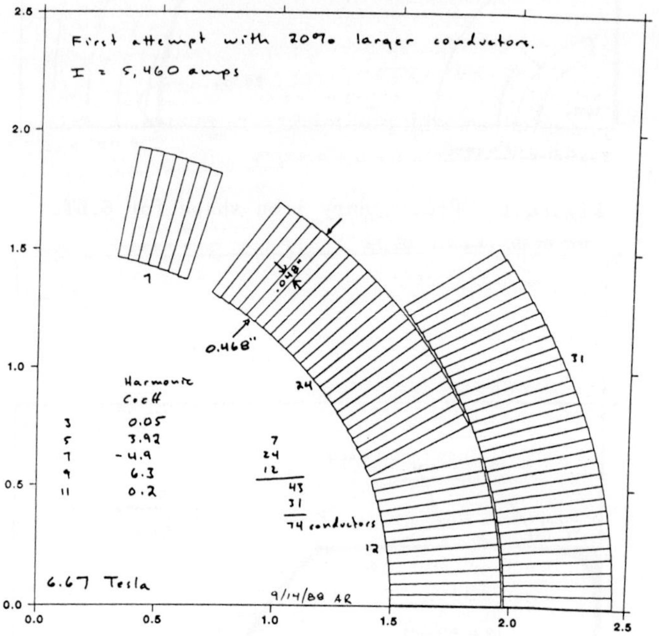

Figure 8. Possible 8.8 Tesla dipole conductor placement using 20% larger strands

Figure 6 shows a solution for 3.000" inner coil diameter centered on the axis.

Figure 7 shows a solution where the horizontal coil aperture has been reduced to 2.5" and where the inner coil radius is 1.7" to a center offset 0.45" from the center of the beampipe.

Conclusion

In order to keep the amp factor below 1% for an axial field of 8.8 Tesla, the outer diameter of the iron must be approximately 22 inches (8 inch ID). The preliminary placement of conductors is given in Figure 6 where the conductors are the same as used in the low beta quads at Fermilab.

HIGH FIELD DIPOLE MAGNET DESIGN CONCEPTS

Thomas H. Nicol
Fermi National Accelerator Laboratory*
P.O. Box 500
Batavia, Illinois 60510

ABSTRACT

High field dipole magnets will play a crucial role in the development of future accelerators whether at Fermilab or elsewhere. This paper presents conceptual designs for two such dipoles; 6.6 and 8.8 Tesla, with special focus on their suitability for upgrades to the Fermilab Tevatron. Descriptions and cross-sectional views will be presented as will preliminary estimates of heat loads and costs

INTRODUCTION

Superconductor performance has steadily improved over the course of the past several years. These improvements, coupled with a better understanding of the mechanics of building superconducting coils, better analysis tools for studying the interactions of individual coil components, and improved refrigeration systems, make it possible to consider dipole magnets with fields as high as 10 Tesla.[1] This paper presents conceptual designs for two such magnets. Both may be considered for general accelerator applications, but are treated here as proposals for upgrades to the existing Tevatron at Fermilab.

6.6 TESLA DIPOLE

Figure 1 illustrates a conceptual view of a 6.6 Tesla dipole suitable for use in the 1.5 Tev pbar-p Tevatron upgrade proposal. This cross section is much like that of an SSC dipole, but with cryogenic piping compatible with existing Fermilab refrigerators. It consists of a 16.76 cm diameter collared coil assembly with a 7.62 cm aperture surrounded by cold iron. The cold iron assembly is 45.97 cm in diameter, truncated at the top and bottom to an overall height of 32.00 cm. An annular space of 0.64 cm between the outside of the collared coil and the inside of the cold iron serves as a flow passage for two-phase helium in counterflow heat exchange with the collared coil. The collared coil assembly operates at 4.5K. An 80K thermal radiation shield serves to intercept heat radiating from 300K. The shield is covered by 2-3 cm of multilayer insulation (MLI).

The cold assembly, consisting of the collared coil, cold iron, and 80K shield, is supported at three places along the length of the magnet by reentrant support post assemblies. The support tubes are filament wound glass in an epoxy matrix. The support posts are derived from work on SSC dipoles and constitute a structurally sound, thermally efficient support system.[2]

The entire cryostat internal assembly, consisting of the cold assembly and supports, is contained inside a 60.96 cm diameter vacuum vessel. The proposed length of the assembly is 8.5 m, but is not restricted to that length. The proposed length allows three 6.6 Tesla dipoles to replace four Tevatron dipoles, thus preserving the Tevatron lattice.

Estimated heat loads and magnet costs are summarized in Tables 1 and 2 respectively. Magnet costs are listed separately and may be considered generic estimates. Project costs are applicable only to the referenced upgrade proposals at Fermilab.

8.8 TESLA DIPOLE

Figure 2 illustrates a conceptual view of an 8.8 Tesla dipole suitable for use in the 1.8 Tev pbar-p Tevatron upgrade proposal. The collared coil assembly is identical to that in the 6.6 Tesla configuration described above. To achieve 8.8 Tesla, this coil must operate at 1.8K. Again, this is a cold iron assembly, but with added iron to prevent saturation at the higher field. The cold

---
*Operated by Universities Research Association, Inc., under contract with the United States Department of Energy.

iron assembly is 56.00 cm in diameter truncated at top and bottom to a height of 40.64 cm. Because it operates at 1.8K, the 8.8 Tesla option does not use two-phase for heat exchange. Rather, two thermal radiation shield are used; operating at 5K and 80K respectively. Both shields are covered with multilayer insulation.

Support posts similar to those described for the 6.6 Tesla option serve as the suspension system, securing the entire cold assembly to the vacuum vessel. Due to the increased iron size and additional radiation shield, the vacuum vessel is 81.28 cm in diameter. Again, the proposed length is 8.5 m, but is not restricted to that length.

Although suitable for use in a 1.8 Tev upgrade operating in collider mode, the 8.8 T dipole described above is not suitable for used in fixed target experiments. Losses due to hysteresis and magnetization during ramping require more refrigeration at 1.8K than can be met by the current or planned liquifier upgrades at Fermilab.[3)]

Estimated heat loads and magnet costs are summarized in Tables 1 and 2 respectively. Magnet costs are listed separately and may be considered generic estimates. Project costs are applicable only to the referenced upgrade proposals at Fermilab.

SUMMARY

Both of the high field dipoles designs described above and illustrated in the accompanying figures are built upon existing work, e.g. the Fermilab Tevatron and the SSC. They are far from the only solutions to the design problem. In some instances; for example the problem the 8.8 T design has in meeting fixed target needs, they are not optimal candidates. New technologies and new materials may well play a role in future design ideas. It does seem clear, however, that we can build dipole magnets with fields approaching 10 Tesla using existing technology and design practices. What is required is a concentrated effort to that end.

REFERENCES

1) Mantsch, P.M., "Prospects for 6- and 10-Tesla Magnets For a Tevatron Upgrade," Hadron Collider Working Group Presentation, Snowmass 1988.

2) Nicol, T.H., Niemann, R.C. and Gonczy, J.D., "SSC Magnet Cryostat Suspension System Design," Advances in Cryogenic Engineering, Volume 33, 1987, pp. 227-234.

3) Theilacker, J. and Peterson, T., Hadron Collider Working Group Presentation, Snowmass 1988.

4) Conceptual Design Report, Proton-Proton Collider Upgrade (Main Injector, New Tevatron), Fermi National Accelerator Laboratory, May 1988.

Table 1. Heat Load Estimates (Per Magnet)[4)]

1.5 Tev pbar-p	1.8K	4.5K	80K
6.6 T Dipole	na	2.6W	20.5W
Quad	na	1.3W	5.7W
Spool	na	7.0W	18.9W
1.8 Tev pbar-p			
8.8 T Dipole	0.1W	2.3W	31.2W
Quad	0.1W	1.3W	5.7W
Spool	0.1W	7.0W	18.9W

Quads and spools are assumed to be roughly equivalent in structural size for all options.

Table 2. Magnet Cost Summary
For 1.5 and 1.8 Tev Upgrade Proposals[4)]

Dipole Cost Summary

	6.6 T 8.5 m	8.8 T 8.5 m
Coil Assembly Mat'l	71.8K	71.8K
Coil Assembly Labor	9.2K	9.2K
He Containment Mat'l	22.2K	30.0K
He Containment Labor	4.3K	4.3K
Cryostat Mat'l	7.5K	12.5K
Cryostat Labor	3.2K	3.2K
Total Per Magnet	118.2K	131.0K

Miscellaneous Component Costs

Standard Quad	38.5K
Special Quad (Avg)	41.3K
Spool (Avg)	65.7K

Project Costs (Magnets Only)

	1.5 Tev	1.8 Tev
Standard Dipoles	75.3M	83.5M
Standard Quads	6.8M	6.8M
Special Quads	1.0M	1.0M
Spools	13.4M	13.4M
Totals	96.5M	104.7M

Quads and spools are assumed to be roughly equivalent for all options.

All costs are in 1988 dollars.

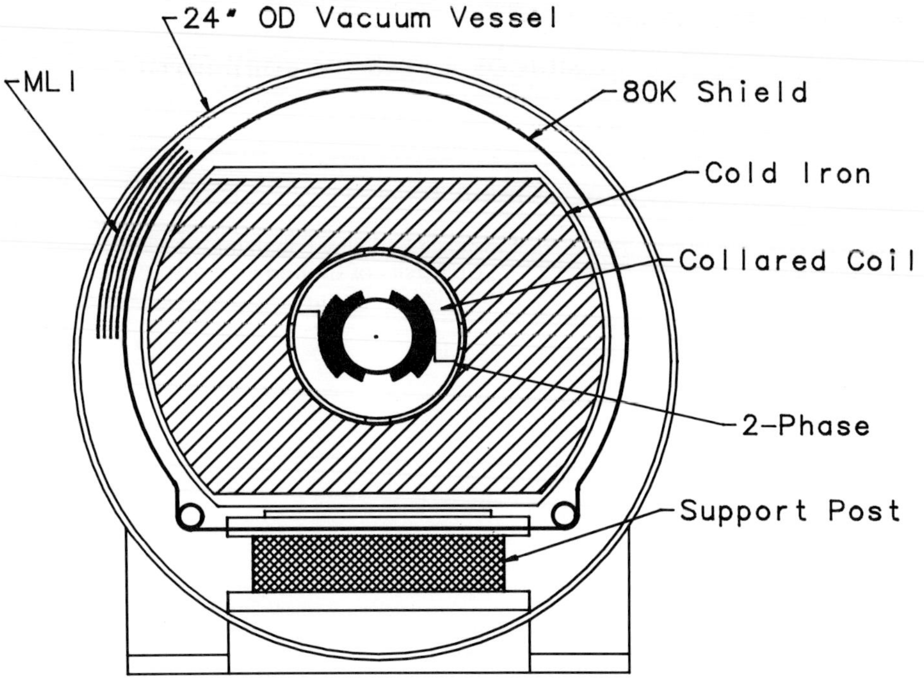

Figure 1. 6.6 Tesla Dipole Conceptual Design Cross Section

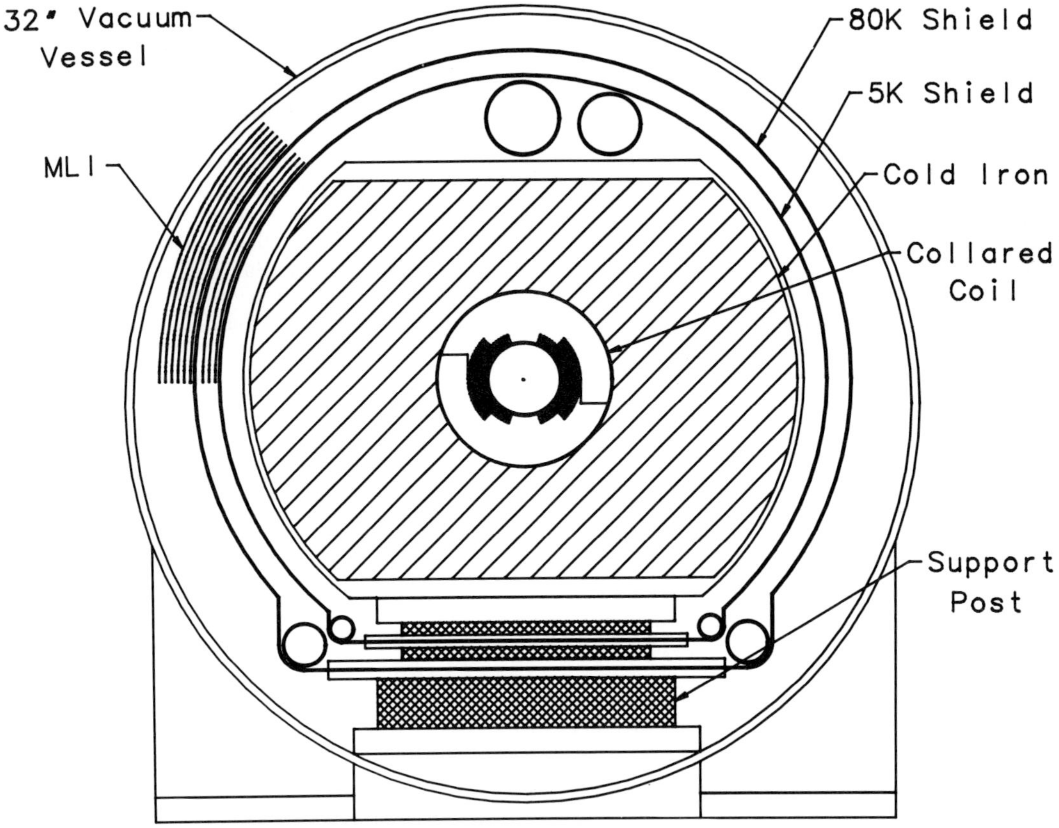

Figure 2. 8.8 Tesla Dipole Conceptual Design Cross Section

DESIGN OF A 9T 5 CM BORE DIPOLE*

Clyde E. Taylor and Shlomo Caspi

Lawrence Berkeley Laboratory
University of California
Berkeley, California 94720

INTRODUCTION

A conceptual design is presented for an accelerator dipole magnet intended for a collider application with central field of 9T and a coil bore diameter of 5 cm. Figure 1 shows a schematic cross-section of the collared coil in a circular cold iron yoke. Emphasis is on identifying a superconducting cable and a coil design that requires little development and which will result in a reasonable operating "margin". Conventional NbTi superconducting cable is used, operating at 2 K to increase the critical current density. The cable and strand proposed are similar to designs now being manufactured for the SSC and the Tevatron. Figure 2 shows a comparison with the 6.6T, 4.35 K SSC dipole.

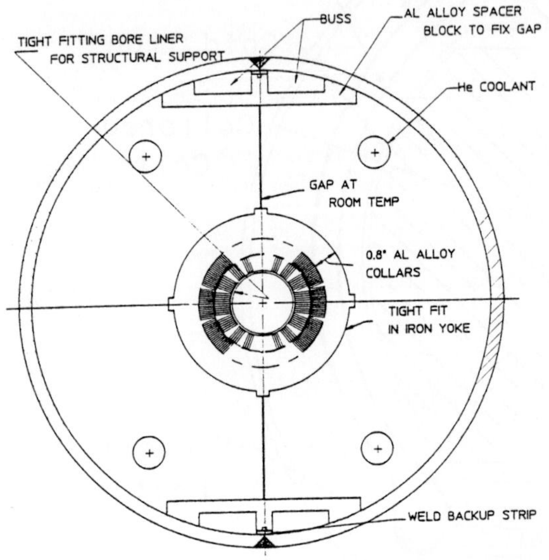

Fig. 1 Schematic of 9T 5 cm bore cold-iron dipole magnet design.

*This work was supported by the Director, Office of Energy Research, Office of High Energy and Nuclear Physics, High Energy Physics Division, U.S. Dept. of Energy, under Contract No. DE-AC03-76SF00098.

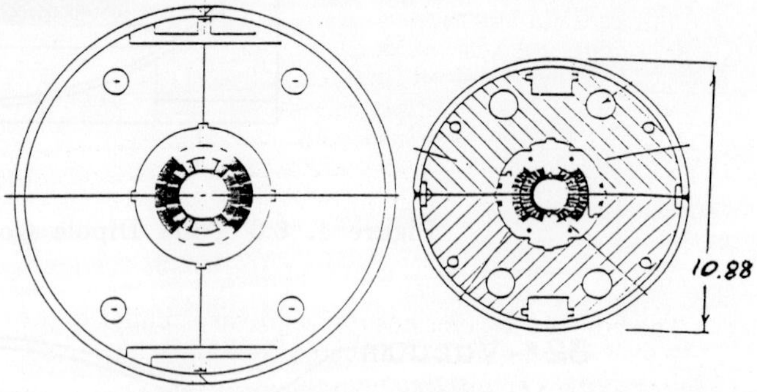

Fig. 2 Comparison of this design with the SSC design - both are shown to the same scale.

CABLE DESIGN

Recent R&D aimed at the Superconducting Super Collider (SSC) has resulted in reliable production of long cable lengths with high J_c. A superconducting strand design is selected that is similar to strands now being manufactured in quantity. Table I gives the strand and cable parameters used in this design for the SSC program.

TABLE I

INNER	
Strand Diameter	0.0318 in.
Cu/SC ratio	1.3
Filament diameter	6 μm
Twist	1-2 per in.
Number strands	30
Keystone angle	1.23 deg.
Thickness	0.0522 in., 0.0625 in.
Width	0.477 in.
OUTER	
Strand Diameter	0.0268 in.
Cu/SC ratio	1.8
Filament diameter	6 μm
Twist	1-2 per in.
Number strands	36
Keystone angle	1.01 deg.
Thickness	0.0438 in., 0.0522 in.
Width	0.477 in.

Note the inner strand is identical to the SSC R&D strand. The outer strand is identical to that now being used for the Tevatron. However, compared to SSC inner and outer cables, these are wider and have a 1.3 times greater cross-sectional area. The inner has 30 strands (rather than 23); 30 strand cable is now being made for the SSC outer layer. The outer strand diameter is 0.0268 in. (the Tevatron strand size) and 36 strands are used; 36 strand cable is now being made at LBL for the Tevatron low-b quadrupoles. Both cables can be made in quantity with minimal development.

GENERAL DESCRIPTION OF THE MAGNET (COLD MASS)

Structural support of the 2-layer, graded, "cos q" coils, is provided by aluminum alloy collars that provide part of the required compression of the coils during assembly; a split cold-iron yoke fits closely around the collars supporting the magnetic forces and providing additional coil compression as shown in Fig. 1. The yoke is split at the vertical plane and a small gap is deliberately maintained at assembly by an aluminum alloy spacer bar. During cool-down, the gap closes as the collared coil shrinks more than the iron yoke. The gap thickness at assembly is precisely determined by the spacer bar. The yoke is enclosed in a stainless steel tube which maintains the required force on both the collared coil and spacer bar during assembly, cooldown, and operation. Figure 1 shows a yoke with a circular outer boundary; this shape as shown has excess iron cross section in the pole region; optimized cross sections with less iron could be elliptical in shape with the same midplane width.

COIL CROSS-SECTION

The coil is a 2-layer cos q design very similar to the one proposed for the SSC. The cable is slightly keystoned and copper wedges are used to provide the arch structure that is self-supporting on the inside; the wedges also provide proper conductor placement for uniform central field. The outer cable results in a higher current density because it experiences a lower field. This design is similar to the SSC design and gives a very uniform field (i.e., small "systematic" or built-in field distortion); Fig. 3 shows the 4-wedge cross section with 18 inner turns per coil and 27 outer turns per coil. The inner-layer wedges are between turns 3 and 4; turns 7 and 8, and turns 13 and 14. The outer layer edge is between turns 19 and 20.

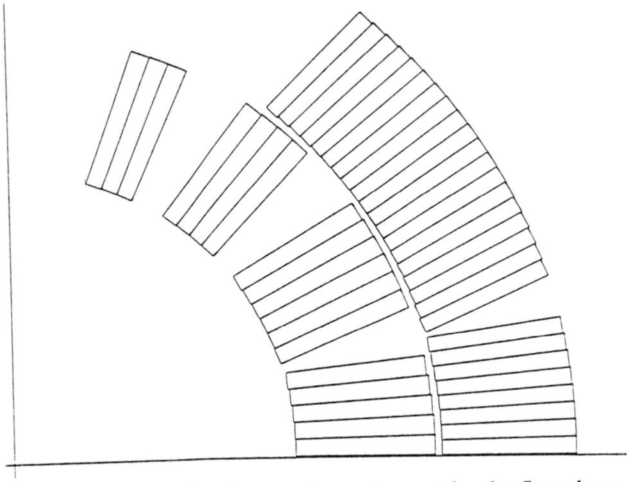

Fig. 3. Quadrant showing conductor layout for the 5 cm bore diameter cross-section, showing the inner layer with 18 turns and 3 wedges and the outer layer with 27 turns and one wedge.

COLLARS AND YOKE

The aluminum alloy collars are assembled around the coils using tapered keys (as in the SSC) producing an azimuthal compression of about 10 kpsi. The split yoke is assembled tightly around the collars with a small gap (~0.5 mm) in the vertical plane; during this assembly some additional compression may be applied to the coils. Upon cooldown, as the coil contracts more than the yoke, the gap closes so as to increase the compressive stress to a predetermined value because the spacer bar, which determines the gap size, also contracts proportionally. We estimate that about 12 kpsi total prestress is needed. The thermal contraction of the stainless steel shell is sufficient to provide the required forces. The yoke gap can as well be placed at the horizontal mid-plane; such a horizontal gap reduces the relative azimuthal sliding between collars and yoke as the gap closes; however, in this case, the aluminum spacer bars, which would span the gap on the sides of the yoke, would require the horizontal yoke dimension to be somewhat greater to provide sufficient iron for the return flux path. The yoke as shown is wide enough to achieve a minimum leakage field with a central field of 9T.

SUPERCONDUCTOR

For this design, an operating temperature of 2 K is chosen because of the very significant increase in J_c as temperature is reduced. Pressurized helium would be pumped through passages in the yoke and recooled at intervals around the ring. A design has not been made for the refrigeration system; however, large-scale 2 K refrigeration plants have been built and the design problems appear straightforward. The magnet test facility at Fermilab has recently been modified to permit circulation of 1.8 K pressurized helium.

Figure 4 from reference 1 shows J_c vs. B for recently produced Nb46.5wt% Ti (the "standard alloy") at 4.2 K and 1.8 K; also shown for comparison is data for a Nb45wt% Ti15wt% Ta ternary alloy.

For the NbTi sample, J_c at 4.2 K and 5T is 2800 A/mm^2 (the present SSC specification is 2750 A/mm^2 minimum); note that at 1.8 K and 8T, J_c is the same as at 4.2 K and 5T. The data of Fig. 4 is redrawn in Fig. 5(a) and 5(b), in which 2 K performance is shown. The 2 K J_c prediction is obtained by a simple linear (and conservative) interpolation between 4.2 K and 1.8 K.

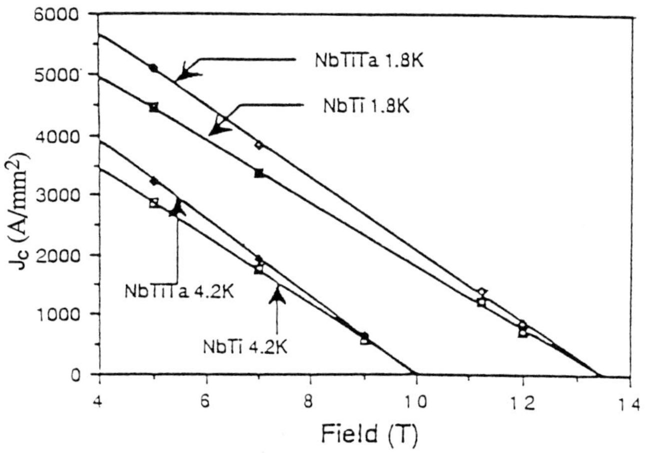

Fig. 4 J_c vs. field at 4.2 K and 1.8 K for test billets of NbTi and NbTiTa (Ref. 1).

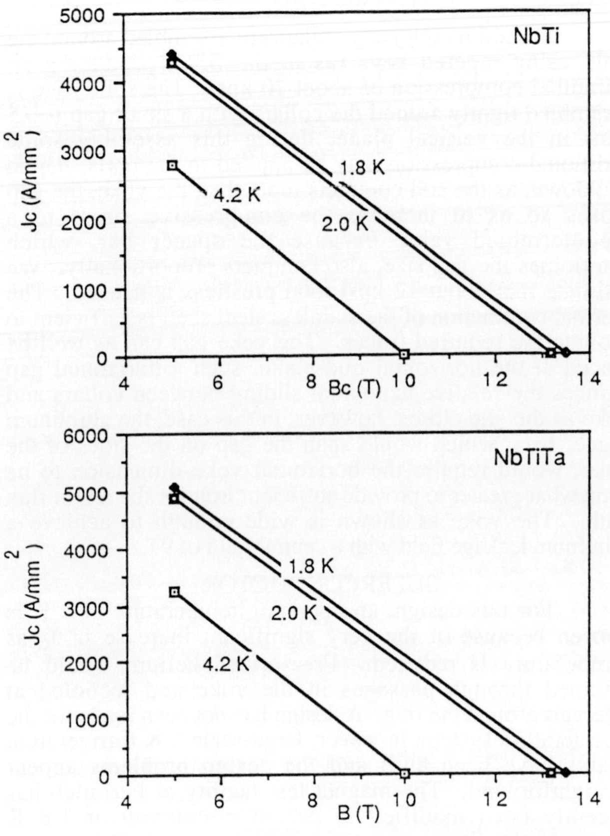

Figs. 5(a) and 5(b) J_c vs. B derived from Fig. 4

Figure 6 shows magnet current I vs. central field B_o, and maximum field B_m at the inner coil. Also shown on the same graph are critical current vs. field for the inner cable for both the NbTi and the NbTiTa alloys. The intersection of the magnet load line, I vs. B_m, with the critical current line determines the critical current (or "short sample" current). The outer cable is designed to quench at a slightly higher current than the inner cable. For this design, Fig. 6 indicates that at 2 K for NbTi, the critical current is 10.4 kA when B_m is 10.2T and B_o = 9.65T. For B_o = 9.0T, I = 9.53 kA while the cable critical current is 12.8 kA. This should be an adequate operating margin at B_o = 9.0T to allow for the myriad of unavoidable departures from "ideal" conditions.

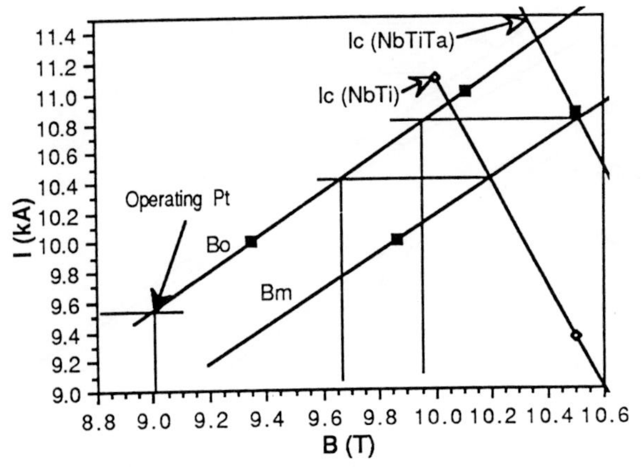

Fig. 6 Load line and critical curve for 9T, 2 K dipole.

The ternary alloy (Nb45wt%Ni15wt%Ta) has been studied for high field applications, but, to date, has been made in small quantities. Addition of Ta results in a higher cost than for the NbTi alloy, but, as seen in Fig. 4, higher J_c is observed. This material is now being studied to determine the optimum properties that can be obtained from various compositions. It appears to be ductile and can be easily drawn and cabled as can NbTi.

Figure 6 shows that the critical field of this design is 9.95T with the NbTiTa alloy at 2 K.

MAGNETIZATION AND FIELD QUALITY

The built-in systematic field distortion is very small; the calculated $\frac{\Delta B}{B}$ for this design is 2 x 10^{-5}. Since the magnet is similar to the SSC design, the expected random multipoles at 1.25 cm (half the coil inner radius) should be about the same as for the SSC at 1.0 cm.[2]

At fields above about 5 T, saturation results in field distortion that is characterized by the normal sextupole component of the field; Fig. 7 shows the sextupole harmonic, b_2, vs. the central dipole field b_o as calculated by the computer code POISSON; $b_2 = B_2/B_o \times 10^{-4}$ where B_2 is the intensity of the sextupole field component. All higher order multipoles have saturation effects less than 10^{-5}. A field line plot is calculated by POISSON shown in Fig. 8.

For filaments of about 6 μm diameter, we expect the field distortion magnetization at very low (injection) field to be similar to that of SSC models with the same filament size. However, the increased J_c at 2 K should result in increased magnetization at constant field. Figure 9 shows sextupole harmonics measured at LBL in a 1-m SSC model[3] at 4.3 K and at 1.8 K; the additional magnetization effect at lower temperature can be seen.

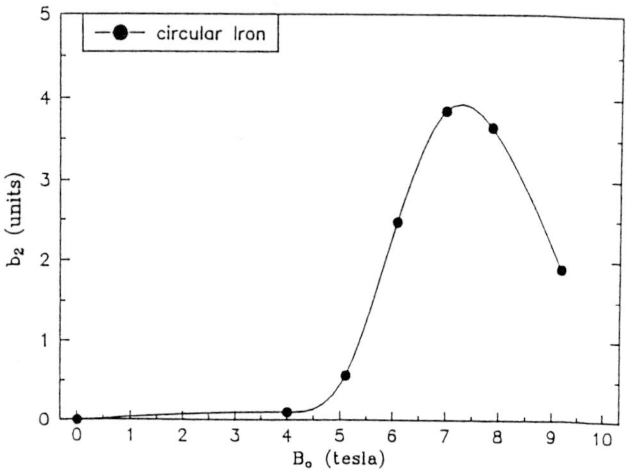

Fig. 7. The Effect of Iron Saturation Effect on b_2.

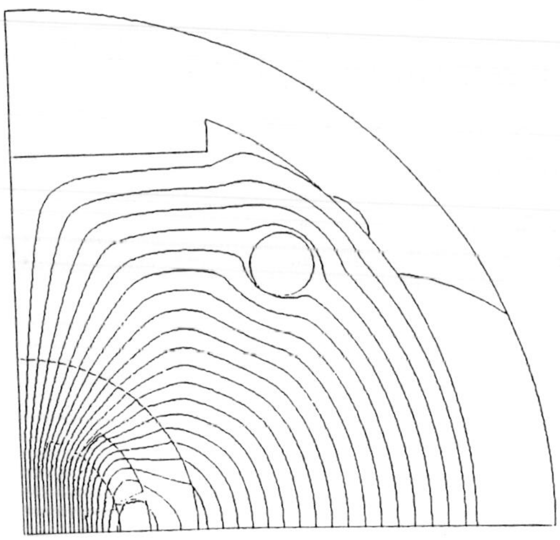

Fig. 8 Field lines calculated with POISSON.

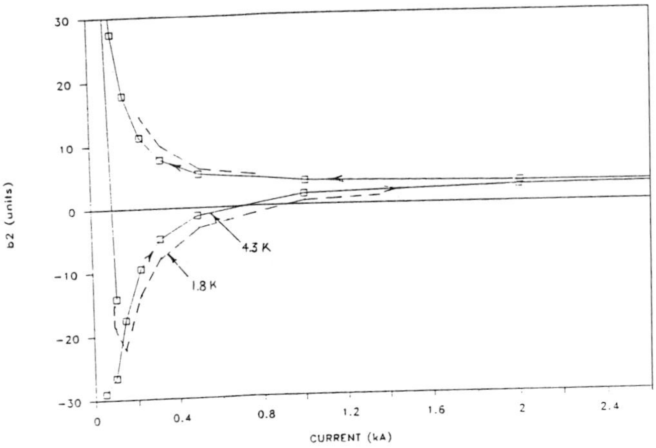

Fig. 9 Sextupole vs. current measured in 1-m SSC Model D15A-5.

CONCLUSIONS

We note that SSC dipole models, although designed for 6.6 T operating field at 4.35 K, have achieved central fields greater than 9T at 1.8 K[4]; for this design, with wider cable, additional iron, and proper mechanical support, a 9T operating field should be possible with reasonable margin as shown by the above results.

REFERENCES

1. E. Gregory, et al., "Preliminary Properties of Ductile Superconducting Alloys at 10T and Above," presented at the International Conference on Cryogenic Engineering, ICEC-12, Southampton, U.K.; July 12, 1988.

2. SSC Central Design Group, "Conceptual Design of the Superconducting Super Collider," SSC-SR-2020, March, 1986.

3. W. Gilbert and P. Barale, "Prompt Report on LBL-SSC Model Magnet D-15A-5," SSC-MAG-196, LBID-1400, April 25, 1988.

4. W. Gilbert and P. Barale, "Prompt Report on LBL-SSC Model Magnet D15A-3, SSC-MAG-170, LBID-1333, November 1, 1987 (for example).

CRYOGENICS FOR A TEVATRON ENERGY UPGRADE

Thomas J. Peterson and Jay C. Theilacker
Fermi National Accelerator Laboratory
P.O. Box 500
Batavia, Illinois 60510

INTRODUCTION

Cryogenic requirements for three potential Tevatron upgrades have been reviewed. The three upgrades include: the addition of a second 1 TeV accelerator for P-P collisions, a new 1.5 TeV PBar-P accelerator, and a 1.8 TeV Pbar-P accelerator at 1.8K. In order to determine whether an upgrade of Fermilab's Tevatron cryogenic system to 1.8K is even feasible, a first-pass design of such an upgrade was made. Choices among innumerable options were made which one would want to review before proceeding to a detailed design. However, our conclusion is that a 1.8K upgrade of our cryogenic system is feasible.

UPGRADE OPTIONS

The cryogenic system for Fermilab's Tevatron has been described elsewhere.[1] Table 1 shows the impact each of the upgrade options has on the satellite refrigerators and the Central Helium Liquefier (CHL). All options utilize the existing cryogenic system capacity (or planned CHL upgrade). One option, the 1.8 TeV/1.8K accelerator, includes a 1.8K refrigeration "stage" added to the existing satellite refrigerator.

Each of these options utilize cold iron magnets[2]. This added thermal mass results in large warm up and cooldown times for magnet changes, as shown in Table 2. Warmup times consider electrical warmup heaters in the iron and are limited by available power distribution in the tunnel. Cooldown is first accomplished using a large external LN_2 heat exchanger and is limited by pressure drop in the magnets.

1.8K SYSTEM DESCRIPTION

An example system was chosen with the following characteristics (see Figure 1):

1. Pressurized superfluid at 3 atmospheres so that the 4.5K shield flow, which is at the same pressure, is supercritical and two-phase flow is avoided.

2. Forced flow of superfluid, as opposed to static superfluid.

3. Temperature rise of 0.1K in superfluid flowing through a magnet string, from 1.8K to 1.9K.

4. Flow forced by a circulating pump as opposed to compressor flow.

5. No superfluid recooling in the tunnel since there is no room for a cold low-pressure transfer line in the tunnel.

6. Pumping on the low pressure helium bath by room-temperature vacuum pumps. (Cold compressors for 1.8K refrigeration have been developed,[3,4] but only for systems of 100 Watts or more at 1.8K. Our heat load from magnets, circulating pump, and miscellaneous sources would total about 40 Watts per house. This is about the size of the 1.8K system at our Magnet Test Facility, which has successfully used warm vacuum pumps).

7. A heat exchanger to warm the low pressure vapor before it enters the pumps, precooling some high pressure helium.

8. Subcoolers at 4.5K and 1.8K with liquid inventory to buffer against upsets.

9. Use of the existing satellite refrigerator cold box and wet expander.

10. Operation of the system in "satellite mode", using 4.5K helium from CHL as is presently done.

MAGNETIC REFRIGERATOR OPTION

Although the system which is diagrammed in Figure 1 utilizes vacuum pumps to obtain a low pressure, low temperature bath which cools the pressurized superfluid, another possible technique is to operate an adiabatic demagnetization refrigerator ("magnetic refrigerator") between 1.8K and 4.5K. The advantage is that 1.8K is achieved without any subatmospheric helium circuits. The primary disadvantage is that magnetic refrigerators which operate in this temperature range have been only as large as a few Watts in capacity[5]; we are looking for a factor of ten larger.

1.8K HARDWARE DEVELOPMENT REQUIRED

1. Magnetic refrigerators. We have a contract with Astronautics Corporation of America for a preliminary design study. Present state-of-the-art is around a few Watts capacity. We want an order of magnitude larger.

2. Circulating pump. There are papers in the literature describing successful tests of centrifugal pumps operating in superfluid.[6,7] Since the pump work is a major heat load in this system, we would want as high an efficiency as we could obtain. More testing and development would be required.

3. Other hardware for superfluid. With either vacuum pumping or magnetic refrigeration special hardware is required for handling superfluid. Every place there is a transition from superfluid to normal fluid, such as at a magnet relief valve or a cooldown line, requires a special check valve or some other restriction to reduce heat in-leak via the superfluid. Power lead feedthroughs and bayonets also require special designs. CEBAF, Tore Supra, and others may provide valuable information.

4. Refrigeration cycle simulation. Steady-state, quench recovery, cooldown, failures which may result in reduced capacity, turn-down capabilities, control system design all need to be understood during design with the help of a computer simulation of this refrigeration system.

5. Optimize coil-cooling passage size. We want magnet reliefs only at "spool pieces" (hence large passage size), but want quick cooldown from 4.5K to 1.8K (hence minimal superfluid volume).

COST CONSIDERATIONS

Table 3 includes the system costs associated with the following components of each upgrade option.

- Non-magnetic lattice components
- Quench relief system
- Additional refrigerator hardware
- Instrumentation and controls

The cost of the 1 and 1.5 TeV options are the same since the cost associated with a larger diameter magnet is compensated by the elimination of the need to cool two accelerators. The need to go to 1.8K in the 1.8 TeV option costs approximately 10 M$ independent of the cooling scheme chosen. Higher hardware costs for the magnetic refrigeration option are compensated by the need for purifier systems and added floor space in the vacuum pumping scheme.

CONCLUSION

About 6 to 10 Watts of 4.5K capacity is lost for each Watt of 1.8K heat lead. Therefore, 40 Watts at 1.8K results in a reduction of 4.5K capacity of 240 to 400 Watts. Our present capacity is about 760 Watts, so this would leave about 360 to 520 Watts of 4.5K capacity. Magnet and spool shields are estimated to total about 120 Watts per house, so it appears that our system would have sufficient steady-state capacity with a comfortable margin. The capacity requirements resulting from constraints on cooldown and quench recovery times have not yet been studied.

REFERENCES

1) Theilacker, Jay, et.al., "Commissioning of the Tevatron Satellite Refrigeration System", Advances in Cryogenic Engineering, Vol. 29, 1984, pp. 429-436.

2) Nicol, T., "High Field Dipole Magnet Design Concepts", Contribution to these procedings.

3) Strehrenberg, W., "Cold Compressor Development", present at the 12th International Cryogenic Engineering Conference,Southampton, England,July 1988.

4) Bon Mardion, G., et.al. "Initial Operation of the 1.8K Tore Supra Cryogenic System", presented at the 12th International Cryogenic Engineering Conference,Southampton, England, July, 1988.

5) Barclay, John A., "Magnetic Refrigeration: A review of a Developing Technology", Advances in Cryogenic Engineering, Vol. 33, 1988, pp. 719-731.

6) Weisand, J.G. and VanSciver, S.W., "Characterization of a Centrifugal Pump in He II", in Advances in Cryogenic Engineering, Vol. 33, 1988 pp. 507-514.

7) Ludtke, P.R., et.al., "Performance of a Small Centrifugal Pump in He I and He II", in Advances in Cryogenic Engineering, Vol. 33, 1988, pp. 515-524.

Table 1. Upgrade Comparison-Heat Load per Refrigerator

Energy TeV	Physics Mode	1.8K Static Watts	4.5K Static Watts	4.5K AC Watts	LHe Usage l/hr	Cycle Time Sec.
1	COLLIDER[a]		710	0	150[b]	
	FIXED TARGET[c][d]		150	400	105	60
1.5	COLLIDER		195	0	68	
	FIXED TARGET[d]		195	500	135	90
1.8	COLLIDER	15	255	0	88	
	FIXED TARGET[e][d] (1.5 TEV)	0	255	505	146[b]	140[f]
EXISTING TEVATRON OPERATION						
0.8	FIXED TARGET		560	200	130	57

NOTES

a. Requires load sharing between the two accelerators at each satellite refrigerator.

b. Requires 2nd CHL with higher capacity (currently under construction).

c. Requires one of the accelerators to be at 80K or warmer (loads shown are for the "new" Tevatron).

d. Low static heat load and relatively high AC heat load will require smart refrigerator and inventory controls.

e. Requires accelerator to be operating at 4.5K due to AC heat load.

f. Cycle time limited by satellite refrigerator capacity.

Table 2. Magnet Warmup and Cooldown Parameters(4 cells)

Upgrade Option	Stored Thermal Energy	Warmup Time	Cooldown Time
1 TeV	11 GJ	24 Hrs	72 Hrs
1.5 TeV	19 GJ	42 Hrs	124 Hrs
1.8 teV	30 GJ	65 Hrs	196 Hrs

Table 3. Cryogenic Costs

Upgrade Option		Cost (M$)
1 TeV	base	7.3
1.5 TeV	base	7.3
1.8 Tev (vacuum pump)		
	base	7.3
	base inflation	1.2
	1.8K hardware	6.2
	Purifiers	1.0
	Civil	2.4
	TOTAL	18.1
1.8 TeV (magnetic refrigeration)		
	base	7.3
	base inflation	1.2
	1.8K hardware	7.2
	Purifiers	0
	Civil	1.2
	TOTAL	16.9

Figure 1
SATELLITE REFRIGERATOR CONFIGURATION FOR 1.8K ACCELERATOR OPTION
JCT/TJP 6-29-89

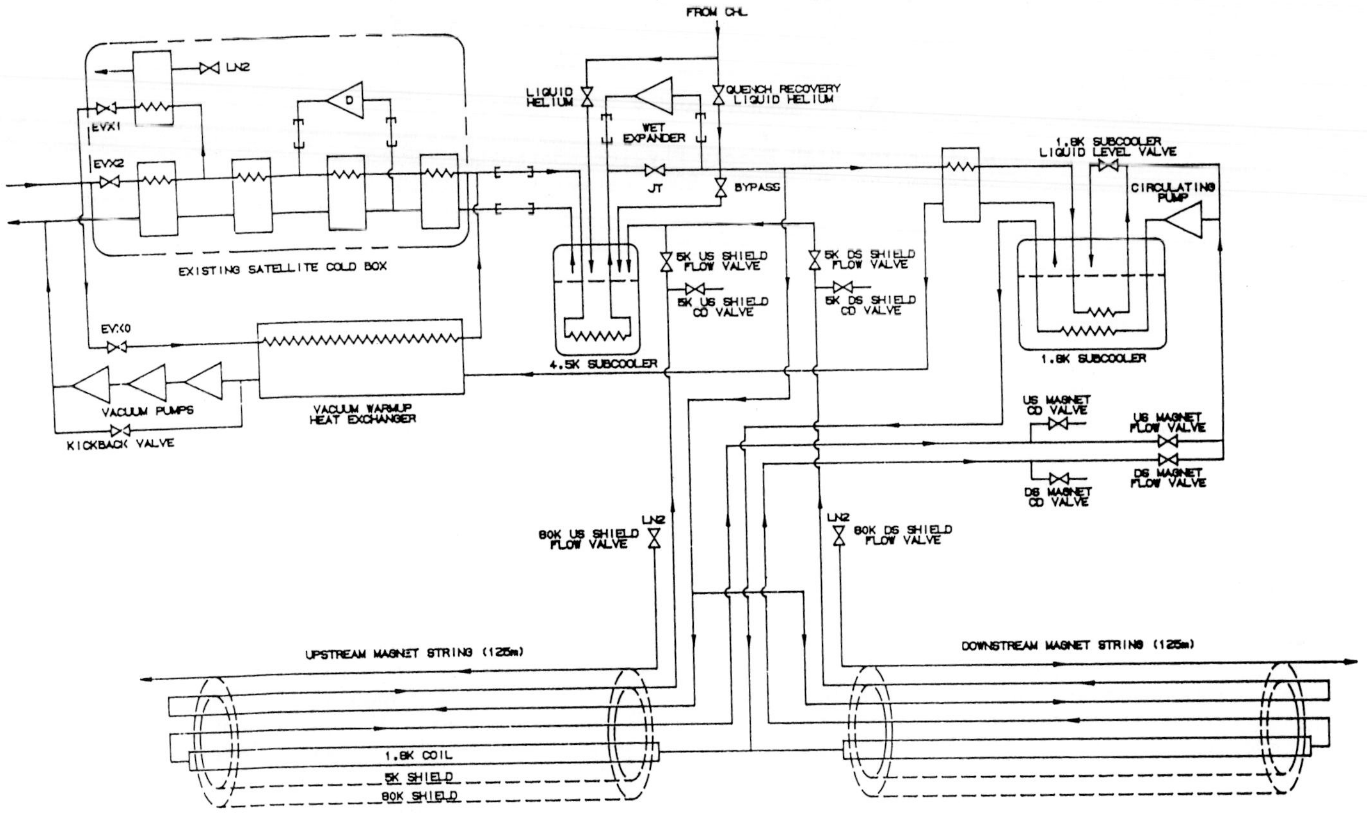

INTERNAL BEAM ABORT SYSTEM FOR THE TEVATRON UPGRADE

N. V. Mokhov* and M. Harrison+

*Institute for High Energy Physics, Serpukhov, U.S.S.R.
+Fermilab, P. O. Box 500, Batavia, IL 60510

INTRODUCTION

In this note we shall examine the properties of an internal beam dump system for the Tevatron running in the pbar-p collider mode. We assume that the beam energy can be as high as 1.8 TeV. The motivation behind this report comes from the fact that the present proton abort system[1,2] is a single-turn fast-extraction system, which becomes progressively more difficult to perform as the beam energy is raised without lengthening the straight section. We examine three different designs (Fig. 1). The first is a system comprised of two beam dumps at each end of the existing straight section, the second dump acting as an absorber for the secondary particles produced in the primary dump as well as functioning as the primary dump for the particles of the opposite sign. The kicker magnets for this scheme are assumed to be outside the straight section in locations similar to the present system. The second layout again consists of beam dumps at either end of the straight section but with the kicker magnets located in the centre of the straight section. In this arrangement both beams are deflected vertically by the same kicker magnets. The advantage of this arrangement is the compact nature of the design with all the components lying within the straight section. The third scenario is similar to the second one with the relative positions of the dumps and kicker magnets reversed. With the dump located in the centre of the free space, the flux of secondary particles hitting the superconducting elements at the end of the straight sections is reduced. The limitations of these schemes will be discussed.

BEAM PARAMETERS

For a given circulating beam current, the more intense the transverse phase-space density the greater the instanteous temperature rise inside the dump. Since the dump lies inside the machine lattice, there is no possibility of blowing up the beam spot at the dump as is customary when using external abort lines. The machine lattice parameters immediately downstream of the straight section quadrupoles have beta values of 60 m and 105 m (H and V) and a horizontal dispersion of 1.8 m. With a normalized transverse beam emittance of 12 π (95%) and a longitudinal energy spread (σ p/p) of 8×10^{-4} (rms) at 150 GeV, this results in a beam spot of 1.68 mm x 1.2 mm (H and V) rms at 150 GeV. The beam spot sizes at higher energies scale in usual fashion from these results.

The presently operating abort system relies on a vertical beam displacement of 24 mm between the circulating and aborted beam trajectories to enter the extraction channel. Under this setup the circulating beam is displaced by 10 mm from the septum magnet. We will assume a similar geometric relation for the scheme with kicker magnets in the arcs, which means that the beam strikes the dump up to 14 mm from the edge. The beam deflection at the dump for schemes 2 and 3 is somewhat less due to the restricted drift length between the kickers and the dump. In both of these cases the beam displacement is up to 10 mm from the edge.

In order to estimate aborted beam intensities we have somewhat arbitrarily assumed a single high-energy full-intensity abort of 2×10^{12} (30 bunches at 6×10^{10} ppb) can occur once per hour, and low energy injection aborts of this intensity can occur every 120 s for a period of four hours per day. Power dissipation and residual radioactivity can be scaled to different operating scenarios from these numbers.

DESIGN CRITERIA

In this study, taking into account the behaviour of the dump materials and optimizing the overall scheme, we required the following criteria to be fulfilled:

1. For a single abort the maximum energy deposition in any region of the dump system and corresponding temperature rise have to be less than the melting points and the shock wave limits

+Operated by Universities Research Association, Inc. under contract with the U.S.Department of Energy

for the given materials.[3)]

2. The cooling system should provide the necessary heat transfer from the core to prepare the dump to the next beam abort.

3. The energy deposition in the superconducting quadrupoles immediately downstream of the straight section must be well below the quench limit.[4)]

4. The induced radioactivity levels near the dumps should be within the acceptable limits.[5,6)]

5. Ground water activation around the abort straight section by hadron fluxes escaping the dumps has to be prevented.

6. Muon fluxes downstream of the abort straight section must be below the tolerable levels.[5,7)]

7. The lifetime of the beam dumps should exceed a few years, at least.

8. The dumps should be as compact as possible.

CALCULATIONS

We have carried out the series of the hadronic and electromagnetic cascade calculations in the Tevatron straight section with the present version of the Monte Carlo program MARS10.[8)]

The only appropriate material for the core of the considered beam dumps is graphite similar to the existing external abort dump.[2] The core consists of the graphite slabs (thickness ~2 cm, density is 1.71 g/cm^3) to reduce the shock wave creation.

We have examined two beam energies for both cases with corresponding beam spot sizes:

<u>Scheme 1</u> 150 GeV (σ_x = 1.68 mm, σ_y = 1.2 mm) and 1.8 TeV (σ_x = 0.48 mm, σ_y = 0.34 mm).

<u>Scheme 2 & 3</u> 1 TeV (σ_x = 0.77 mm, σ_y = 0.32 mm) and 1.5 TeV (σ_x = 0.63 mm, σ_y = 0.25 mm).

Most results are similar for three schemes; therefore, we will describe in detail only the first one. The main and principle exception (quenching) will be given at the end of this report.

SCHEME 1

The proposed layout of the abort dumps in the straight section is shown schematically in Fig. 2. The system is bi-directional with the downstream abort dump acting as an absorber for the secondary particles produced in the upstream dump. We have found that for a 1.8 TeV beam of 2-5 x 10^{12} protons, the minimum length of such an absorber is 480 cm followed by the endcap of 50 cm aluminum and 50 cm steel.

Energy Deposition and Temperature Rise

Figure 3 shows the two-dimensional energy deposition density distribution in the graphite absorber. The results demonstrate the familiar hadronic cascade properties: the very sharp radial fall off and the relatively slow longitudinal dependence. The instantaneous temperature rise in the absorber can be determined from the data of Fig. 3 and from an enthalpy reserve. Using the data from[3,9)] we have calculated the temperature field in the overall beam dump. The results with the core at the initial temperature To = 27°C for the abort of 1.8 TeV 5 x 10^{12} proton beam is shown in Fig. 4. The maximum temperature of ~800°C is reached on the beam axis at a longitudinal distance of 140 cm. Experience with the existing external beam dump system indicates that if the graphite slabs are contained in the inertial (argon) atmosphere their long-life exploitation at such temperatures is possible. Note that the fracture temperature of graphite is ~2200-2300°C.

Figure 4 shows that temperatures at radii >3.5 cm are less than 50°C (one needs to add the temperature rise to the initial temperature 27°C). The maximum temperature rise in the steel endcap is ~100°C. At smaller beam intensities all these numbers are correspondingly less. The maximum instantaneous temperature rise in the graphite core is only 380°C for the 2 x 10^{12} abort. Moreover, the maximum energy deposition and crudely the maximum temperature rise do scale almost linearly with the beam energy, say to 1.5 TeV or 150 GeV. Therefore, one can consider the Fig. 4 data as an extreme case.

On the basis of these calculations the two proposed cross section of the upgraded Tevatron internal beam dump are shown in Figs. 5 and 6.

A. The 13.5 x 9 x 2 cm graphite slabs with 1 mm aluminum beam pipe of 60 x 40 mm aperture. The slabs in an argon atmosphere are contained in an aluminum box with a closed loop cooling system. This box is surrounded by a steel shield. The total length of the dump is 580 cm. Such a dump should work reliably at energies up to 1.8 TeV and aborts up to 6x10^{12} protons.

B. The 6x3x2.5 cm graphite inserts in the aluminum box (Fig. 6). Requirements for beam stability and beam intensity (<3-4 x 10^{12}) are harder for this case.

Both abort dumps in the straight section are identical.

Cooling System

To find the cooling system parameters we use the abort scenarios of Section 2 and results on the total absorbed energy in each dump. Table 1 below gives the energy (in kiloJoules), deposited in the various parts of the dump for the single 2 x 10^{12} protons abort.

Table 1					
Beam Energy TeV	Al Beam Pipe	Graphite	Al Container	Steel Shield	Total
1800	5.4	245.4	57.6	160.0	468.4
150	0.03	13.4	6.7	22.4	42.8

Then, for these scenarios the power of the closed loop cooling system of each internal beam abort dump should be

$$P = 468.4 + 42.8 \times 30 = 1752.4 \text{ kJ/hr or} \sim 0.5 \text{ kW}.$$

At higher intensity the power should be greater. Say, at 5×10^{12} multiply these numbers by 2.5.

Quenching

To determine the superconducting units heating by the stray radiation escaping the dumps we performed the full scale Monte Carlo simulation for the whole straight section shown in Fig. 2. Two factors are favorable:

1. Because the beam displacement in the dump is large enough, there are practically no high energy protons scattered from the edge of the absorbers, which are the most serious component in the long distance irradiation, for example, as in fast resonant extraction case[4]);

2. The second beam dump which is placed at the other end of the abort straight section, just upstream the superconducting quad, serves as a good collimator absorbing particles created in the first dump.

Calculations show that the maximum energy deposition density in the first quadrupole superconducting coils is 3×10^{-7} GeV per gram per incident proton at the reference beam offset 14 mm. For the 2×10^{12} beam abort it gives 0.1 mJ per gram, which is a factor of 5 to 10 below the instantaneous quench limit.[4])

Figure 7 gives the dependence of energy density in the superconducting coil on the beam displacement in the dump. Also shown is maximum energy deposition in the horizontal part of the inner aluminum tube of the dumps. One can conclude that for 1.8 TeV 2×10^{12} abort the minimum beam offset is about 5-7 mm. The corresponding maximum temperature in the aluminum is ~100°C.

Shielding

The averaged over a year tolerable flux of hadrons with E >10 MeV at the outer surface of a shield is about 10^7 hadrons/cm^2 sec.[5-8] This value gives ~100 mrad/hr of contact dose of induced radioactivity and is acceptable from ground water activation point of view (flux 10^6 to the water). For the considered abort scenarios and 50% "collider year" we have

1.8 TeV:

$24 \times 365 \times 0.5 = 4380$ aborts per year

Averaged over year abort intensity is

$2 \times 10^{12} * 4380/3.15 \times 10^7 = 2.8 \times 10^8$ p/sec

Tolerable flux$=10^7/2.8 \times 10^8 = 3.6 \times 10^{-2}$ per cm^2 per proton

150 GeV:

$30 \times 4 \times 365 \times 0.5 = 2.19 \times 10^4$ aborts per year

Averaged abort intensity $= 1.4 \times 10^9$ p/sec

Tolerable flux $= 7.1 \times 10^{-3}$ per cm^2 per proton.

Comparing these numbers to the results of cascade calculations we find that aluminum containers must be surrounded with the steel shield (density 7.86 g/cm^3) of the outer radius 25 cm as shown in Fig. 3.

Muons

Calculations[7]) show that for the design parameters one needs to have 1.5 km of wet soil shield in the direction of aborts to provide on the surface the muon annual dose of 10 mrem.[5] The maximum thickness of the soil above the aborted beam axis is ~4 m, and may be higher (5 m) at the first 100 meters near the CØ straight section dependently on the specific abort design. The Fermilab site is fulfilled to this condition and the only requirement is to kick the aborted beam down.

Lifetime

The integrated over a year hadron flux in the "hottest" point of the dump for given scenarios is in the range of $1.5\text{-}4 \times 10^{18}$ cm^{-2}. The tolerable flux is about 10^{20} cm^{-2}. The present abort dump with the similar maximum hadron flux is exploited for seven years. The necessary requirement is the argon atmosphere for graphite. The beam offset could be different at 150 GeV and 1800 GeV say, 10-14 mm at 1.8 TeV and 7-10 mm at 150 GeV.

SCHEME 2

We have studied two dump designs for this scheme, which are similar to Figs. 5 and 6, but both with inserts and in the last case with aperture 50 x 20 mm.

Considering the maximum beam parameters as 1.5 TeV and 2×10^{12} ppp, we have found that the core insert should consist of 360 cm graphite followed by 40 cm steel and 140 cm tungsten. Maximum temperature in all parts (C, Fe, W) is about 300°C. Shielding and other requirements are the same.

The energy deposition in superconducting coil is extremely high however and even with the aperture of the dump restricted, the quench levels in the superconducting magnets are exceeded by at least an order of magnitude (Table 2). Results of this table show that neither of these arrangements result in a viable design option.

SCHEME 3

The third alternative is similar to scheme 2 in that all the abort elements are confined within the straight section, but is slightly less elegant in the fact that it uses two sets of kicker magnets at each end of the long straight and a double ended dump in the centre, thus losing the operationally desirable feature of a single set of kicker magnets aborting both beams. Each system of kickers consists of five modules similar in design to the existing Tevatron magnets. This results in a beam deflection of 1.3 mrad which corresponds to a beam displacement of 20 mm at the face of the dump. The dump is constructed in three sections. The cross-section is the same as that outlined in scheme 1.

Table 2 Structure of the Central Absorber in Scheme 3

Section No.	1	2	3
Extent, meters	0-4.2	4.2-4.8	4.8-9
Core 6 x 2.5 cm	Graphite	Steel	Graphite
Container r<6 cm	Aluminum	Steel	Aluminum
Shield 6<r<25 cm	Steel	Steel	Steel

The energy deposited in the various sections of this dump is shown in Table 3. These data can be used for determination of closed loop cooling system.

Table 3 Energy (kJoules) Deposited per a Single 2×10^{12} Abort

Element		Eo=1500 GeV	Eo=150 GeV
Core	Sec 1	115.2	5.17
	Sec 2	25.2	0.32
	Sec 3	1.3	0.05
	Total	141.7	5.54
Container	Sec 1	111.0	11.24
	Sec 2	21.8	0.90
	Sec 3	9.0	0.40
	Total	141.8	12.54
Steel shield	Sec 1	160.7	25.47
	Sec 2	8.3	0.50
	Sec 3	7.5	0.45
	Total	176.5	26.42
TOTAL		460.0	44.50

Total power: $P = 460 + 44.5 \times 30 = 1795$ kJ/hr. For antiproton abort numbers for Sections 1 and 3 are exchanged.

The quenching behavior of this design is shown in Table 4. Placing the dump in the centre of the straight section results in a dramatic decrease in the energy deposition in the superconducting elements. There are two reasons for this: the superconducting magnets are further away from the source of the radiation and hence subtend a smaller solid angle, but more importantly the kicker magnets themselves, with their reduced aperture, make excellent absorbers resulting in a greatly reduced flux into the superconducting elements.

CONCLUSIONS

Two of the three internal abort designs considered result in satisfactory behaviour and fulfill the criteria outlined in Section 3. A beam dump positioned directly in front of the superconducting magnets cannot absorb a sufficient number of the secondary particles outscattering from the face of the dump to avoid quenching these magnets at the kind of intensities likely to be encountered during the collider upgrade. Other features of the abort design appear to be within adequate operational tolerances.

REFERENCES

1) M.Harrison. Tevatron Abort System, Fermilab UPC-151, November 1981.

2) J.Kidd, N.V.Mokhov, C.T.Murphy et al. Proc. Part. Accel.Conf. IEEE, Vol.2, p.2274 (1981).

3) N.V.Mokhov. Fermilab FN-328 (1980).

4) Superconducting Accelerator, Design Report, Fermilab (1979). M.A.Maslov and N.V.Mokhov. Part.Accel., Vol.11, p.91 (1980). A.I.Drozhdin, M.Harrison and N.V.Mokhov. Fermilab FN-418 (1985).

5) Radiation Guide, Fermilab (1983).

6) I.S.Baishev, S.L.Kuchinin and N.V.Mokhov. Preprint IHEP 86-76, Serpukhov (1986).

7) A.Van Ginneken, P.Yurista and C.Yamaguchi. Fermilab FN-447 (1987). N.V.Mokhov, G.I.Semenova and A.V.Uzunian. NIM, Vol.180, p.469 (1981).

8) N.V.Mokhov and J.D.Cossairt. NIM, Vol.A244, p.349 (1986). N.V.Mokhov. Sov. J. Part. Nucl. vol.18(5), p.408 (1987).

9) Y.S.Toulokian. Thermophysical Properties of Matter. Vol.1, Plenum, New York - Washington (1970).

Table 4 Quenching Summary for 2×10^{12} Abort

Scheme	1		2			3
Eo, TeV	1.8		1.5			1.5
Dump aperture, mm	60×40		60×40		50×20	60×30
Beam offset, mm	10	14	10	14	14	10
Max. energy density in SC coil, mJ/g	0.18	0.1	100	50	13	< 0.01
Energy deposited in the first meter of Quad, Joules	6	3.2	3962	1958	330	< 0.005
Number of hadrons* in Quad aperture at 1 meter						
with E > 10 MeV	1.E9	6.E8	3.2E12	1.7E12	2.2E11	9.E7
with E > .75 Eo	100	96	255	241	249	168

*Mean hadron energy = 8-35 GeV

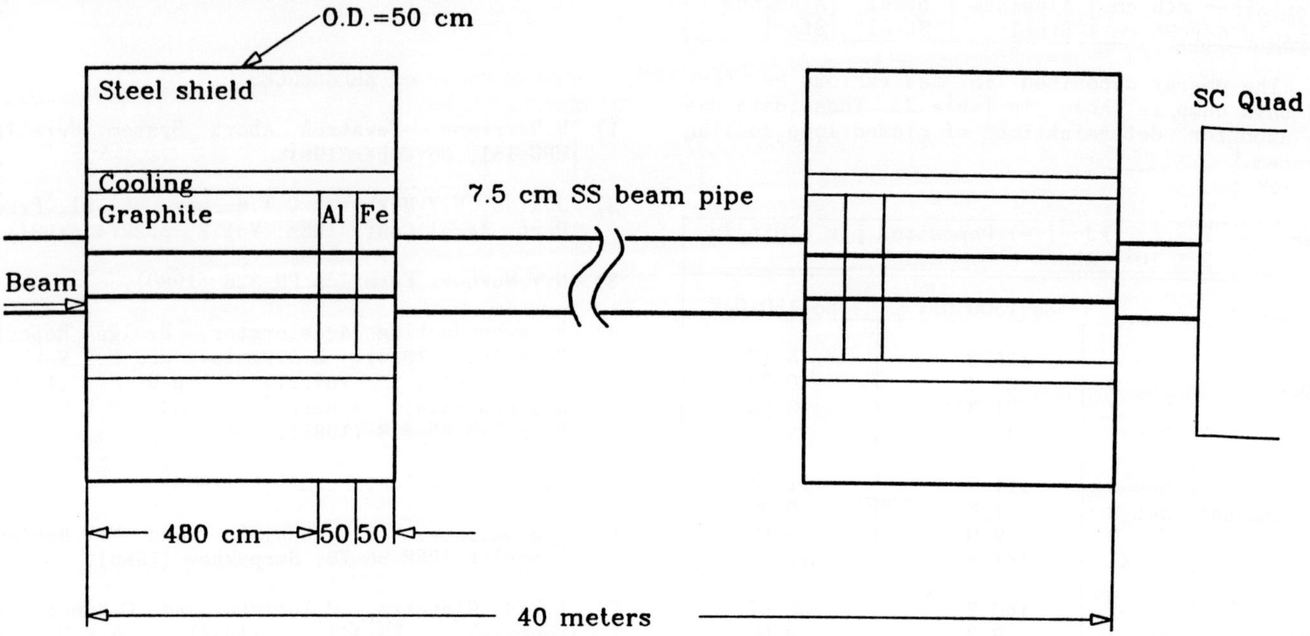

Figure 2 Schematic view of two beam abort dumps placed in the CO straight section.

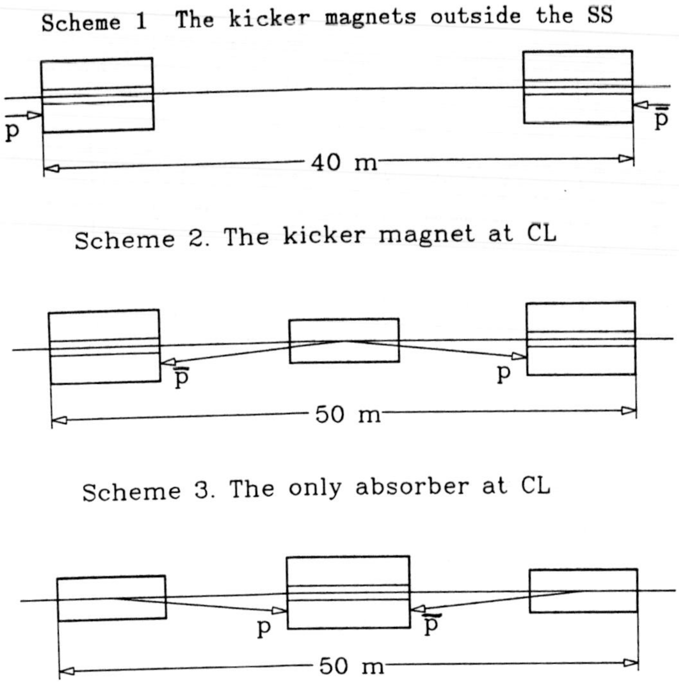

Figure 1 Three examined designs for the Tevtron internal beam dump system.

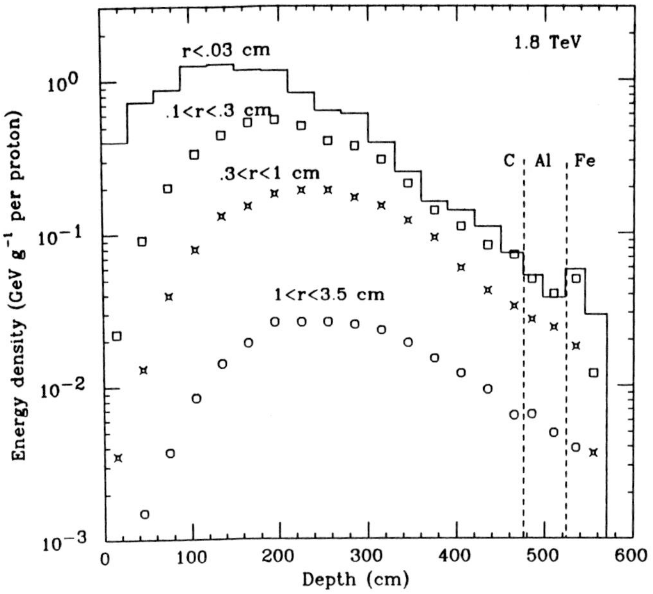

Figure 3 Longitudinal distributions of energy deposition density in the various radial bins of the core of the internal beam dump at the 1.8 TeV proton abort with a beam spot of 0.48*0.34 mm (H*V) rms.

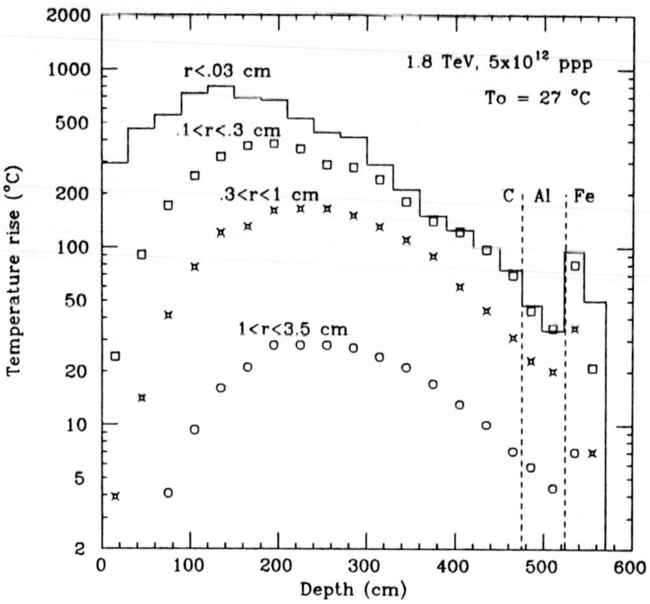

Figure 4 Instantaneous temperature rise distribution corresponding to Fig. 3 for the beam abort of 5×10^{12} protons.

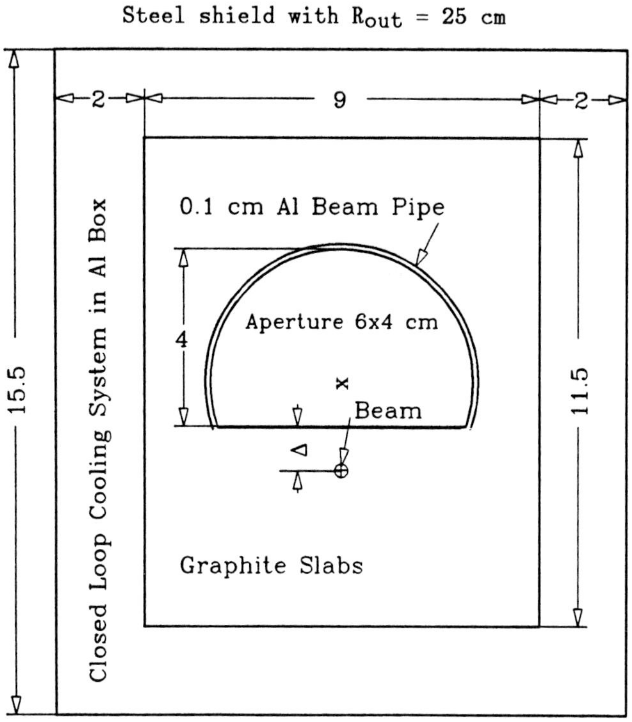

Figure 5 Core of the internal beam abort dump, Scheme 1(A). All dimensions are in cm.

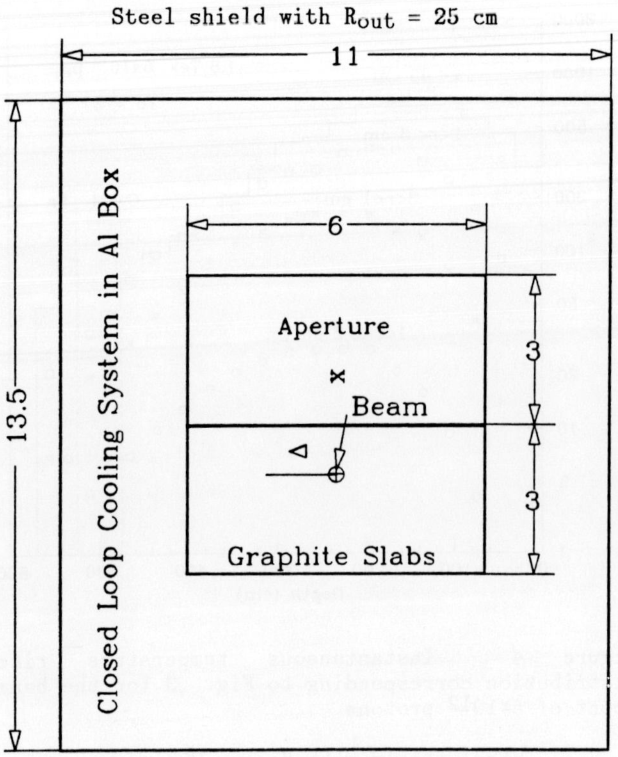

Figure 6 Core of the internal beam abort dump, Scheme 1(B). All dimensions are in cm.

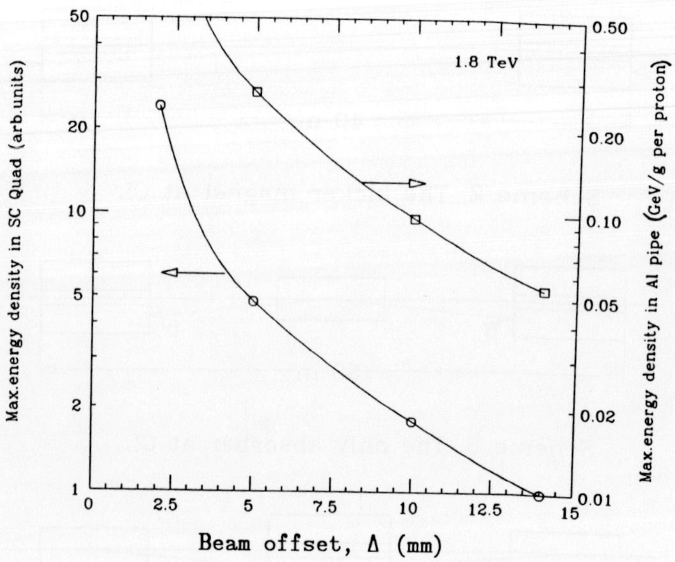

Figure 7 Maximum energy deposition density in the first downstream quadrupole superconducting coils and in the aluminum beam pipe inside the internal abort dump versus beam displacement in the dump.

SEPARATED ORBITS IN THE TEVATRON

Ernest Malamud
Fermi National Accelerator Laboratory,*
Batavia, Illinois 60510

Abstract

An arrangement of electrostatic separators in the Tevatron creates helical beams. Proton and antiproton bunches are kept apart except at the two major detectors at BØ and DØ in order to minimize $\Delta\nu_{bb}$ and maximize lifetime. The same set of separators is used at injection and during acceleration, squeeze to low beta, and to make collisions at both intersection regions. The separators fit in existing warm spaces; thus existence of 6 T dipoles is not required. An option in which separation is only in the vertical plane is also described.

Helical Orbits and Aperture Limitations

A sequence is defined to be a series of steps from injection to collisions and data taking. The parameters defining each step are the energy, the step number in the low beta squeezes (they can be different at the two experiments), and a set of separator voltages. At each step the beam-beam separation as well as the helix orientation are determined. The beam properties assumed in these calculations are $\epsilon_H = \epsilon_V = 15\pi$, and a Δp = that varies linearly from 75 MeV at 150 GeV to 112 MeV at 1000 GeV. With these properties the total separation in σ's between proton and anti-proton beams is calculated. The transverse normalized emittances are 95% figures, and the momentum spread is one σ. All distributions are assumed to be Gaussian. At injection, before the beams have had time to dilute due to intrabeam scattering, more optimistic values could have been used. It may be incorrect to assume that a given separation (in σ's) in momentum space is equivalent in its beam-beam effects to the same separation in betatron space, and combine them in quadrature.

Determining the sequence is a delicate balancing act between having sufficient separation so the long range beam-beam forces do not degrade the beam quality, while at the same time staying within the accelerator aperture. The design goals were minimum separation of 5 σ's and maximum separation of 15 mm (beam-beam center to center). Electric fields are based on a "standard" module with a physical length of 305 cm and an effective length of 257 cm (1). If it is operated at 25 kV/cm ($\pm$ 50 kV on a 4 cm gap or $\pm$ 62.5 kV on a 5 cm gap) it will deflect a 1 TeV beam 6.4 μrad, assuming a point kick. A distributed kick is slightly larger for the same electric field. In the proposed sequences the goal was to hold the separator electric field below 40 kV/cm during periods lasting the order of minutes (i.e. injection, cogging, acceleration), and below 25 kV/cm during periods lasting hours, i.e. during collisions and data taking runs. However, the solution falls short of these goals in some instances. It is important to emphasize that many of the factors affecting the design and proposed sequence need to be measured in accelerator experiments. It is important to determine spark rates vs. electric field with beams passing through the separators.

Incorporation into the Lattice

The lattice for these calculations uses Collins' insertions, matched in betatron and momentum space, at both BØ and DØ (2,3). The tune is set at: $\nu_x = 20.58$, $\nu_y = 20.59$. The beam is squeezed with $\beta_x = \beta_y$ starting from injection values of $\beta* = 170$ cm (step 1) to $\beta* = 25$ cm (step 17).

Injection requires cogging. Furthermore, if injection is done at step 1 in the squeeze then the "natural" helix (defined below) is not well enough developed to allow collisions to begin. So the sequence proceeds through a series of steps during which the beams are separated in the entire ring. Then at some energy, and some step number in the squeeze the beams are brought into collision. Precisely where this transition from no collisions to collisions occurs will depend on

---

*Operated under contract with the Universities Research Association and the U.S. Department of Energy.

the aperture and the separator voltages. There is uncertainty about what happens to the beams during this transition period lasting for some thousands of revolutions (depends on separator time constants) when the beams are "grazing", i.e. $< 5\ \sigma$ apart. Accelerator experiments are needed to determine how rapidly the transition between these two states must be made.

Once collisions begin the oppositely charged beams are kept apart except at the BØ and DØ experiments with local electrostatic "three bumps" in each plane. One pair of bumps creates helical orbits from B11 to C49 (1/3 ring, ≃ 6.5 betatron oscillations). The other pair of bumps keeps the beams apart from D11 to A49 (2/3 ring, ≃ 13 betatron oscillations). The high-β regions closest to the interaction region are used to obtain efficient separation, while getting the beams separated as soon as possible to minimize the number of head on crossings. Since the number of betatron oscillations in these bumps is only approximately an integer or half-integer, the local bumps require 3 elements. After some trial and error the locations for the middle element of the 3-bumps in the vertical plane were chosen to be C17 and F17. The vertical 3-bump from DØ to BØ works well until $\beta*$ goes below 50 cm; then the F17 location is no longer effective in the vertical plane and an electrostatic kick needs to be turned on somewhere in the FØ straight section. There is also room in the F17 medium straight section to accommodate the middle element of the horizontal DØ to BØ 3-bump. For best separation, B48 is chosen for the horizontal plane "middle" element of the BØ - DØ 3-bump. The abort system must be reconfigured during collider runs to free up the 6 meter space at B48 and the 12 meter space at C17.

This choice of separator locations and strenghts results in almost round helices when $\beta* \sim 50$ cm (step 12). The reason is that with separators placed at the "11" and "49" locations, there is almost a 90° phase shift ($\Delta\Phi$) between the vertical and horizontal betatron oscillations in the arcs. This "natural" $\Delta\Phi$ decreases as $\beta*$ increases. One approach for solving this problem would be to use additional correction elements to maintain $\Delta\Phi$ as the beam is squeezed. This idea will not be pursued further in this report.

The Helical Separation Solution

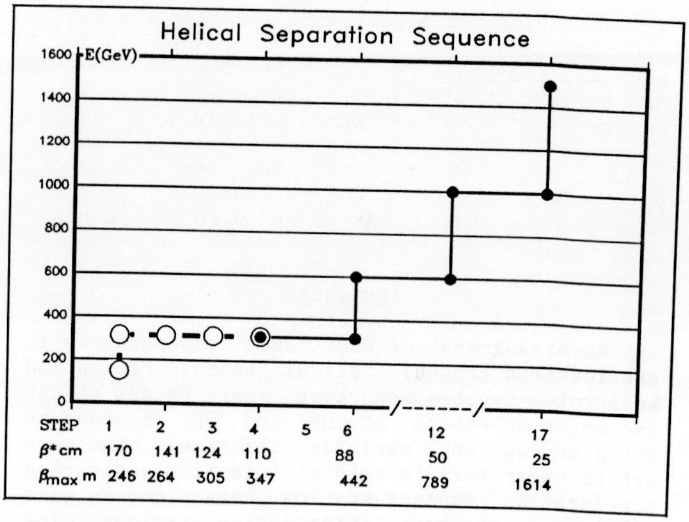

Fig 1 shows the sequence in Tevatron energy - $\beta*$ space. $\beta*$ is correlated with step number in the squeeze and β_{max}. It is desirable for experiments to run at energies as low as 315 GeV for comparison with CERN and also to have a range of $\beta*$ over which collisions can be observed. It is also desirable, and in principle possible with the matched low-β insertions, to have as an operations parameter the ratio of luminosities at the two intersection points.

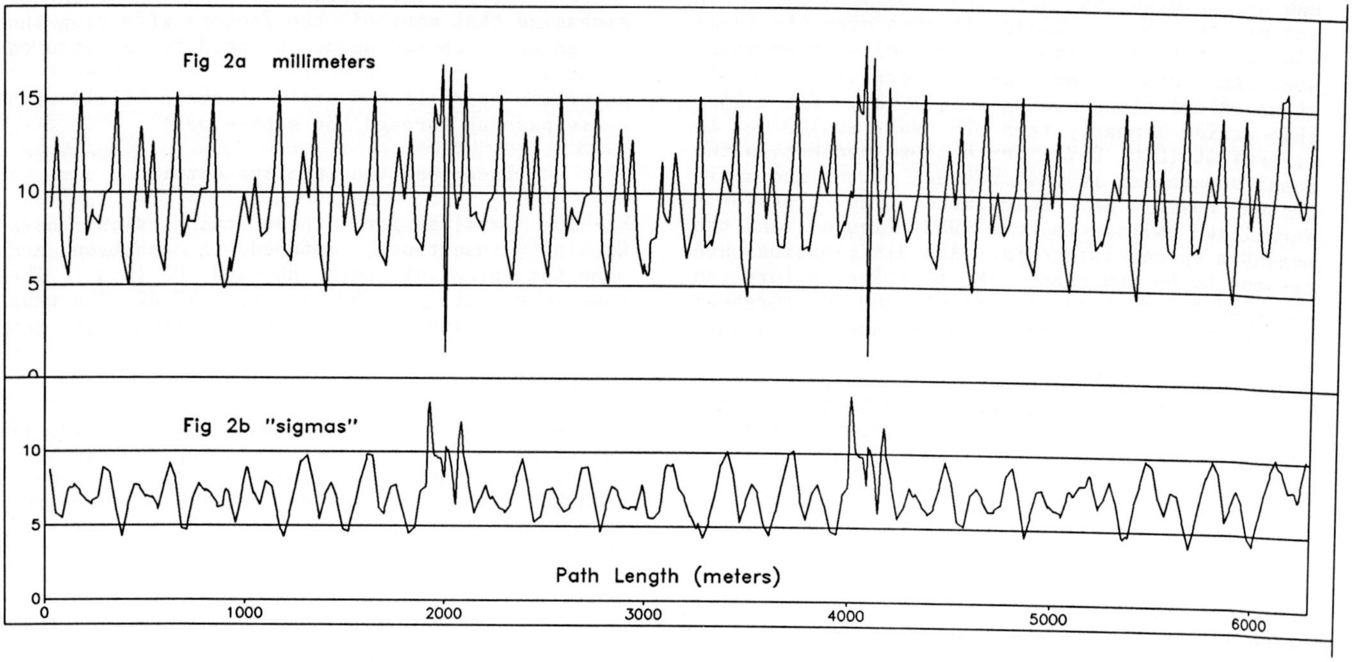

The path chosen depends on several considerations. The beams are injected at 150 GeV, separated at all locations around the ring allowing cogging. Fig 2a shows the total beam-beam (center-center) separations in mm. The total aperture requirement would be these numbers plus beam size. Fig 2b shows the separation in σ's. The largest aperture requirements of over $\pm$ 7 mm are in the horizontal plane and occur in the new low beta quadrupoles.

Aperture limitations will prevent low energy operation at a high step number in the low β squeeze. The beams are accelerated to 315 GeV, and through step 4 ($\beta* = 110$ cm). Raising the energy lessens the aperture problem at BØ and DØ but as the low beta squeeze is applied the problem arises again, reflecting the increase in β_{max}. Only the B48 and C17 separators are on through this step. At step 4 in the squeeze, β_{max} is about 350 m, and $\Delta\Phi$ is 45°. It should be noted that $\sigma = 2.3$ mm at 150 GeV and $\beta_{max} = 350$ m. This can be compared with a beam-beam separation that varies from 5 to 15 mm. Thus, if the Tevatron aperture can accommodate this separation, it should be possible to inject at step 4 or more in the squeeze.

At step 4 (315 GeV) the helix has developed sufficiently to allow the beams be brought into collision. With the beams colliding the aperture problems at BØ and DØ have disappeared but there is a new aperture problem at AØ because of the existing high-β insertion there. This problem diminishes as the energy is raised or at a higher step number and might be a reason to wait until step 6 before bringing the beams into collision. However, during collider runs AØ can be configured to a normal-β insertion by quadrupole polarity reversal in which case the required aperture in this region will be reduced.

With low β at step 6 the machine is ramped to 600 GeV. One cannot go to 1000 GeV and then squeeze due to do separator voltage limits. At 600 GeV the low beta can squeeze the beams to step 12 ($\beta* = 50$ cm). From here it is straightforward to accelerate to 1000 GeV and squeeze (if desired) to step 17 ($\beta* = 25$ cm). Fig 3 shows the beam-beam separations. Bunches spaced a minimum of 21 buckets apart are filled in a pattern having 3-fold symmetry around BØ-DØ-FØ. The black diamonds show the crossing points. An expanded view would show the difficulties that arise with 7 bucket (20 m) or 9 bucket (25 m) bunch spacing scenarios. At 1500 GeV and $\beta*$ at both experiments 25 cm (step 17) the separation in σ's is the same with 20% more separator voltage. The calculations assume the present Tevatron lattice. By the time this energy upgrade is contemplated experimental information will be in hand on spark rate vs. separator voltage and minimum beam separation requirements.

The total number of modules required in the helical separation scheme is 21. To avoid the F17 horizontal and vertical reversing switches, the number could be increased raising the total to 23. The separator locations are listed below:

Location, if $\beta* \geq 50$ cm			allowing $\beta*$ to reach 25 cm	
	Horizontal	Vertical	Horizontal	Vertical
A49	1	2	1	2
B11	2	1	2	2
C49	1	2	1	2
D11	2	1	2	1
C17		2		4
B48	2		2	
F17	1	1	1	1
F0				1
Total	9	9	9	12
Grand Total	18		21	

4. Bunch spacing scenarios.

It is desirable to retain the present 53 MHz, n=1113 harmonic number, rf systems. If B = 53, there are 106 crossing points, spaced 59.3 meters apart. With one collision point at BØ and one at DØ, at the nearest other crossings, the beams have reached almost their full separation of about 5 σ's. Minimum time between bunches is 396 ns.

More ambitious scenarios are for either 7 or 9 bucket spacing. (The latter requires the existence of the 20 GeV rings as injectors.) These give three collisions each at BØ and DØ, plus two more places at each straight section where there are "near" collisions. In the 7 bucket scheme these have <2 σ separation. These extra "4" crossings can be increased to almost 3 σ with the use of cold, i.e. superfluid

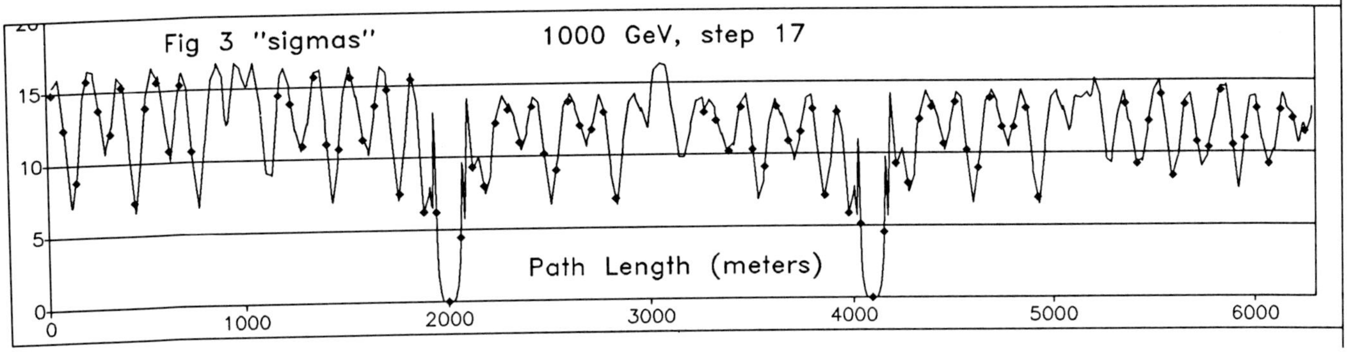

(1.8° K) triplets which are shorter and therefore allow moving the separators closer to the interaction point. (4). In the 9 bucket scheme these three head on collisions just fit in the 50 m stretch where the beams are brought completely together, and the extra four crossings occur with of the order of 3 σ, which may, however, not be enough. It is believed, although accelerator experiments need to be carried out, and further calculations performed, that these small separations will cause unacceptably large tune spreads.

Only some of the subharmonic buckets available will be used and the filled ones can be distributed in a pattern to maximize the number of collisions at both BØ and DØ, and still leave gaps for kickers to rise and fall. If the number of filled buckets is small, e.g. in the 6-bunch vertical plane option discussed in the next section, additional freedom is available in the pattern to reduce difficulties caused by cogging.

5. Vertical separation option

Given the complexity of the solution presented above and the uncertainties of separating in the horizontal plane due to momentum spread, it is interesting to consider separating only in the vertical plane.

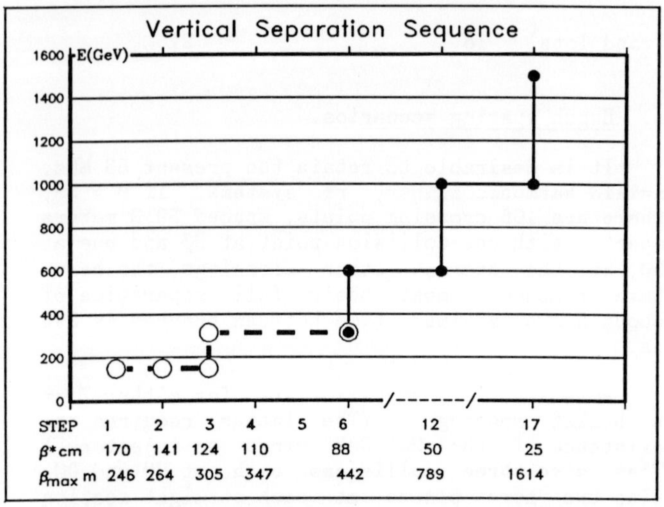

Fig 4 shows the sequence in E - $\beta*$ space. As with the helical solution the hardest part of the sequence occurs at injection. A cusp in one plane at one point in the closed orbit generates roughly 20 betatron oscillations around the circumference. With two beams "cusped" in opposite directions, the beams come together every $\lambda/2$ or in roughly 40 places. The present 6-bunch cogging operation sweeps the azimuthal position through 56 buckets or about 1/20 of the circumference. Therefore the beams can start out not colliding, and end up not colliding, but the cogging operation sweeps them through azimuthal positions where they collide (and graze which is worse). Therefore some horizontal separation is required. This is accomplished with one horizontal separator module at B48 operated at 29.3 kV/cm until cogging is over. Fig 5 shows beam-beam separations; the open triangles are crossing points before cogging and the solid diamonds are crossing points after cogging is complete. Even with this "squashed helix", the beams will sweep through 2σ separation and until accelerator experiments are performed it will not be know what the effects of this will be. It is difficult to raise the separator voltages further because of aperture limitations.

After cogging is finished separation is only in the vertical plane. Fewer separators are required than in the helical case but not by as much as one might think since the requirement of $> 5\sigma$ is still a design goal. At 315 GeV there is an aperture problem at BØ and DØ. This goes away when the beams are colliding but then there is a severe aperture problem at AØ. As remarked on in the discussion on the helical solution, this may be solved by reversing AØ quadrupole polarities during collider runs.

A total of 17 vertical modules are required: 12 at the collision points (A49, B11, C49 and D11), 4 at C17 and 1 at F17. Together with the horizontal module at B48 this makes a total of 18. The voltages on the B11 group are modest so one could probably have 2 modules there instead of 3, reducing the total to 17. However at F17 instead of having 1 module, it would be better to have 3 modules, one operating at positive polarity and 2 operating at negative polarity. This reduces operating voltages, eliminates a reversing switch, and raises the total to 19.

Fig 6 shows step 17 at 1000 GeV with the BØ - DØ local bump turned off so there are 5 head-on collisions. The separators at B48 and C17 are still present and used at injection. If one drops to the 5 head-on collision case then the total module count is reduced by 6 although it might be possible to use the F17 separators during injection instead of those at C17, further reducing the total module count.

Comments on Operations

In contrast to current operation where the beam is first accelerated to maximum energy and then the low beta insertions are run through their program from step 1 to step 17, it now becomes necessary to change $\beta*$ as a function of

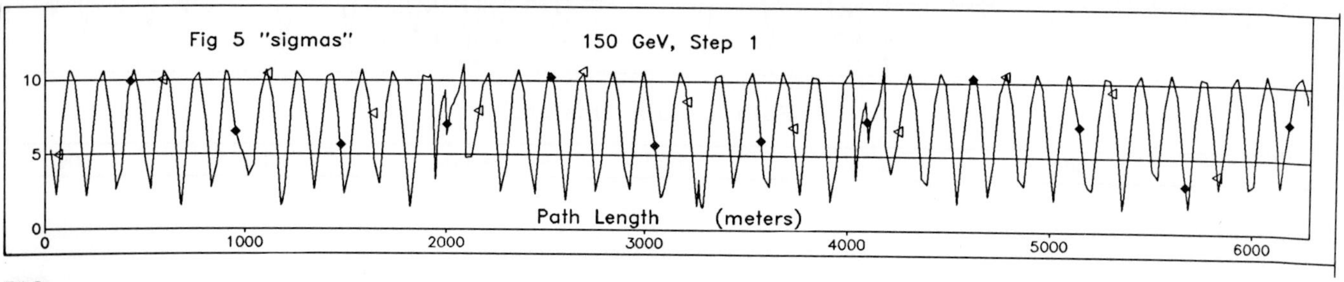

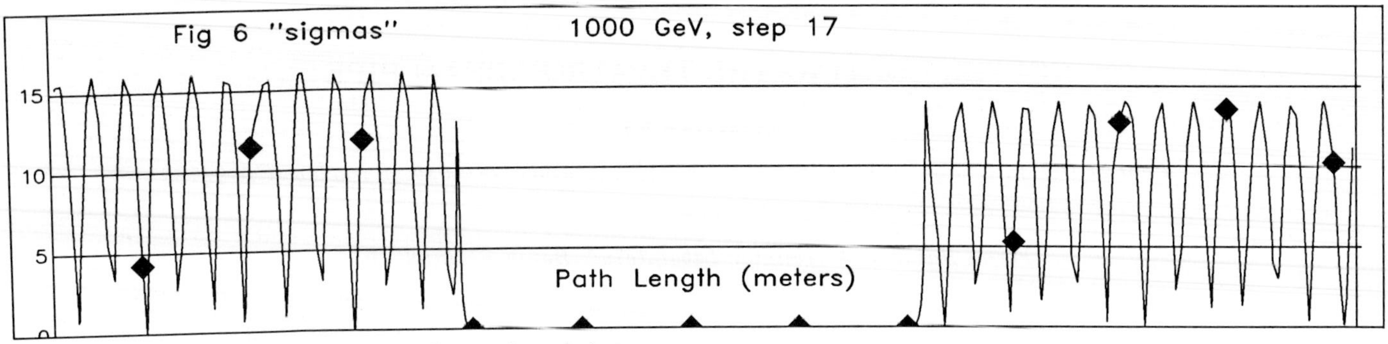

energy. The implications of figs 1 and 4 are that the Tevatron will need a "front-porch" at 315 GeV and another one at 600 GeV. This may be difficult due to persistent currents. If so the sequences shown will have to be divided up into finer steps and the sequence run adiabatically so the low beta quads and separators can track each other as well as the accelerator energy with adequate precision to hold the tune constant and the closed orbits approximate those shown in the figures.

The minimum number of reversing switches required is 2 in both the helical and vertical separation options: on the B48 module set and on the C17 module set. This is necessary to set up the pbar closed orbit using reverse injection of protons.

Acknowledgements

I thank Dr. Helen Edwards for several key suggestions which led to the present solution.

References

1. S. Childress, these proceedings.

2. T. Collins, "On the design of LOW-BETAS for BØ and DØ", Apr.,'87.

3. Karl Koepke: "BØ/DØ Low Beta Lattice", July 1, 1987. "BØ/DØ Low Beta Lattice--Revision 1", Sept. 24, 1987.

4. E. Malamud, "Comparison of Superfluid and Regular Triplets for $\beta*$ = 50 cm", November 28, 1987.

COPPER COATING THE TEVATRON BEAM PIPE

King-Yuen Ng

SSC Central Design Group, c/o Lawrence Berkeley Laboratory 90/4040, Berkeley, CA 94720*

and

Fermi National Accelerator Laboratory, Batavia, IL 60510*

(September 1988)

ABSTRACT

We study the effects of copper coating the Tevatron beam pipe. In order that the beam pipe would not be cracked by the eddy currents during a quench, the copper thickness has to be $\lesssim 2.5\,\mu$m. This thin coating will still reduce the parasitic power loss by 12.3 times. But, it will raise the growth time of transverse coupled-bunch instability by a factor of 3.9 only.

I. INTRODUCTION

There have been several plans to upgrade the Tevatron. One of which involves filling nearly all the rf buckets with proton bunches to collide with bunches from a new similarly built Tevatron ring.[1] With the introduction of so many bunches, one needs to worry the collective coupled motions of the bunches and also the parasitic power lost to the resistivity of the vacuum chamber walls. In this paper, we address the possibility of copper coating the interior of the vacuum chamber walls. Since the conductivity of copper at cryogenic temperature is $\sim$1000 higher than that of stainless steel, collective coupled bunch motions as well as parasitic heating should be greatly reduced. These effects are discussed in Sections II and III.

However, the copper layer can bring about mechanical problems to the beam pipe. During a quench, the Lorentz force generated by the eddy currents in the highly conductive copper layer can lead to deformation of the beam pipe. The partially vaporized helium outside, if not allowed to escape fast enough, can easily push the beam pipe towards stress failure. Also if the copper coating is not even, this Lorentz force generated during a quench will produce a torque, which if strong enough can cause failure to the position-fixing keys attached to the beam pipe. These problems are studied in Sections IV and V.

For the Tevatron ring, we take the mean radius as $R = 1.0$ km, the beam pipe as approximately cylindrical with an internal radius $b = 3.49$ cm and a wall thickness of $t = 1.0$ mm. At cryogenic temperatures, the conductivity of bulk copper is taken as $\sigma_c = 2.3 \times 10^9$ $(\Omega\text{-m})^{-1}$ corresponding to a residual resistance ratio (RRR) of 200. The conductivity of stainless steel does not vary very much with temperature and is taken as $\sigma_s = 2.0 \times 10^6$ $(\Omega\text{-m})^{-1}$.

II. TRANSVERSE COUPLED-BUNCH GROWTHS

The rf harmonic of the Tevatron is $h = 1113$. But only $M = 996$ buckets will be filled with bunches each containing $N = 2.0 \times 10^{10}$ protons.[1] For simplicity let us assume that all the buckets are filled, or $M = h$. There will be M modes of transverse coupled-bunch motion. Here, internal bunch motions have not been considered. The growth rate of the μth mode driven by the transverse coupling impedance $Z_\perp$ is[2]

$$\frac{1}{\tau_\mu} = -\frac{MI_b c}{4\pi\nu_\beta E/e} \sum_{k=-\infty}^{\infty} \mathcal{R}e\, Z_\perp[(kM - \mu + \nu_\beta)\omega_0] \,, \quad (2.1)$$

where E is the energy per particle in the bunch, I_b is the average current per bunch, c is the velocity of light, ν_β is the betatron tune, and $\omega_0/2\pi$ is the mean revolution frequency. Since $\mathcal{R}e\, Z_\perp(\omega)$ is odd in ω, broad-band impedance with a width larger than $\sim \nu_\beta\omega_0$ will not contribute. The wall resistivity behaves like a resonance at low frequencies because it goes to infinity when ω=0. As a result, its contribution to Eq. (2.1) is significant. At harmonic n, the skin depth into the vacuum chamber wall is given by

$$\delta = \frac{1}{\sqrt{n}} \left(\frac{2R}{\beta Z_0 \sigma}\right)^{1/2} \,, \quad (2.2)$$

where βc is the average longitudinal velocity of the beam particle, $Z_0 = 377$ ohms is the free-space impedance, and σ is the electric conductivity of the chamber wall. We get for stainless steel $\delta = 1.63 n^{-1/2}$ mm. Therefore for frequency less than a harmonic, the return image currents fill the whole wall thickness of $t = 1$ mm. The transverse resistive wall impedance of the whole ring is given by

$$Z_\perp(\omega) = \frac{2Rc}{\sigma_s t \omega b^3} \,. \quad (2.3)$$

*Operated by the Universities Research Association, Inc., under contracts with the U.S. Department of Energy.

The ring has a designed betatron tune of $\nu_\beta = 20.6$. Therefore the coupled-bunch mode that has the fastest growth is $\mu = 21$, corresponding to the spectral line with $k=0$ or at $\omega = -0.4\omega_0$ where $Z_\perp = -58.8$ MΩ/m for a stainless steel pipe wall. The next adjacent spectral lines corresponding to $k = \pm 1$ are $M = 1113$ units of ω_0 away and give negligible contributions (Fig. 1). The designed average bunch current is $I_b = eN\beta c/2\pi R \approx 0.153$ mA. The worst growth rate occurs at injection energy of $E = 150$ GeV and is given by

$$\frac{1}{\tau_{21}} = 77.4 \text{ sec}^{-1} , \tag{2.4}$$

or a growth time of $\tau_{21} = 12.9$ ms.

For the Superconducting Super Collider (SSC), any transverse coupled-bunch growth times $\tau > 8.5$ ms can be stabilized by the designed transverse damper. Using this as a criterion, the present Tevatron stainless steel beam pipe should be controllable against transverse coupled-bunch instabilities and no copper coating is needed. However, our result depends very critically on the residual betatron tune. If the tune is changed to $\nu_\beta = 20.9$, τ_{21} becomes $12.9 \times (21 - 20.9)/(21 - 20.6) = 3.2$ ms. Also if the number of protons per bunch is increased to 1.0×10^{11}, the growth time will be decreased 5 times to only $\tau_{21} = 2.6$ ms. Then, the instability will become an important issue. On the other hand, with a copper layer of thickness $t_c = 1$ mil (25.4 micron) coated inside the beam pipe everywhere, the factor $\sigma_s t$ in Eq. (2.3) should be replaced by $\sigma_s t + \sigma_c t_c$. Thus, the transverse wall impedance is reduced by $(\sigma_s t + \sigma_c t_c)/\sigma_s t = 30.2$ times, and the growth time is increased by the same ratio.

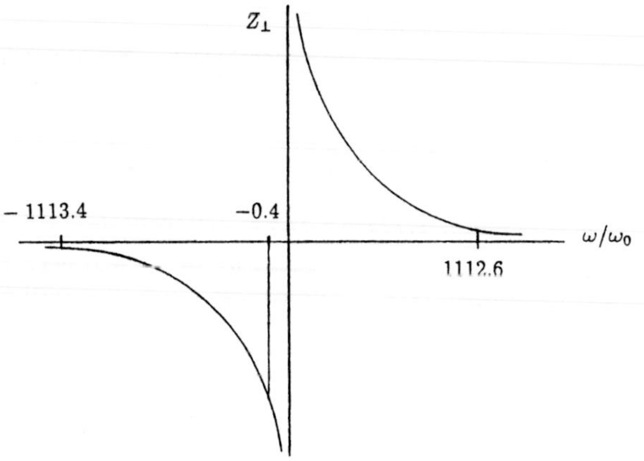

Fig. 1. Plot of the transverse wall impedance $Z_\perp$ and the spectral lines of a bunch at low frequencies.

III. PARASITIC HEATING

The power lost to wall resistivity can be obtained by integrating the longitudinal coupled wall impedance over the bunch power spectrum to get[3]

$$P = \Gamma(3/4) M (I_b R)^2 \left(\frac{Z_0}{2\sigma}\right)^{1/2} \left(\frac{1}{b\sigma_z^{3/2}}\right) , \tag{3.1}$$

where $\Gamma(3/4) = 1.23$ is the gamma function, $Z_0 = 377$ Ω, σ_z is the rms bunch length, and σ is the conductivity of the pipe wall. Here we have neglected all contributions due to higher multipoles and assumed that the wall currents flow in a skin depth of the pipe wall only.

The Tevatron bunch has a typical rms length of $\sigma_z = 20$ cm. With $M = 996$ bunches, the parasitic heating amounts to $P = 2247$ watts for stainless steel pipe wall. The load for each Tevatron refrigerator is listed in Table I. There are 24 refrigerators with a total load of 8.28 kW DC and 21.8 kW AC. Thus, the parasitic heating of the pipe wall contributes to only ~10% of present load and is not significant at all. However, with the number per bunch increases by 5 times to $N = 1.0 \times 10^{11}$, the heating power will be increases by 25 times to $P = 56.2$ kW. Then, the heat generated by the wall resistivity will become a dominant contribution to the load of the refrigerators.

	DC (watt)	AC (watt)
34 dipoles	238.0	238.0
34 dipoles AC		442.0
11 quads	77.0	77.0
11 quads		121.0
1 pair 5000 A lead	10.0	10.0
Set end boxes	20.0	20.0
Total	345.0	908.0

Table I: DC and AC loads of one Tevatron refrigerator.

If the pipe wall is coated with a copper layer of thickness t_c everywhere, the power lost to the wall is still given by Eq. (3.1) with the substitution

$$\frac{I_b^2}{\sigma^{1/2}} \rightarrow \frac{I_c^2}{\sigma_c^{1/2}} + \frac{I_s^2}{\sigma_s^{1/2}} , \tag{3.2}$$

where I_c and I_s are the image currents flowing in the copper layer and the stainless steel pipe respectively. The copper layer and the stainless steel pipe can be viewed as two resistances in parallel. Thus,

$$\frac{I_s}{I_{cu}} = \frac{\sigma_s t}{\sigma_c t_c} , \tag{3.3}$$

or

$$\frac{I_c}{I_b} = \frac{1}{1 + \sigma_s t/\sigma_c t_c} \quad \text{and} \quad \frac{I_s}{I_b} = \frac{\sigma_s t/\sigma_c t_c}{1 + \sigma_s t/\sigma_c t_c} . \tag{3.4}$$

Therefore,

$$\frac{I_c^2}{\sigma_c^{1/2}} + \frac{I_s^2}{\sigma_s^{1/2}} = \frac{I_b^2}{\sigma_c^{1/2}}\xi \ , \tag{3.5}$$

$$\xi = \frac{1}{(1+\sigma_s t/\sigma_c t_c)^2}\left[1+\left(\frac{\sigma_c}{\sigma_s}\right)^{1/2}\left(\frac{\sigma_c t_c}{\sigma_s t}\right)^2\right] \ . \tag{3.6}$$

With $t_c = 1$ mil, $I_s/I_c = \sigma_s t/\sigma_c t_c = 0.0342$ and the factor $\xi = 0.972$, showing that most of the power loss comes from the copper layer. The parasitic heating will be reduced by a factor $\sqrt{\sigma_c/\sigma_s/\xi} \sim 35$ times. Therefore, for a copper layer that is not too thin, the reduction is almost independent of the layer thickness.

To safeguard the beam pipe against stress failure, however, it is found below that only a copper layer of thickness 0.1 mil is allowed. Then, $I_s/I_c = \sigma_s t/\sigma_c t_c = 0.342$ and the factor ξ in Eq. (3.6) takes the value 2.76. Thus, the parasitic heating is reduced by only 12.3 times. The drop in the reduction factor is due to the fact that only about 75% of the image current flows in the copper layer.

IV. STRESS FAILURE DURING A QUENCH

When a quench occurs, the helium outside the beam pipe will be partially vaporized and the external helium pressure will rise. If the beam pipe of average radius $\bar{b}$ and thickness t is perfectly cylindrical, it will be buckled when the radially symmetric helium pressure p reaches the critical buckling pressure

$$p_c = \frac{Yt^3}{4(1-\mu^2)\bar{b}^3} \ , \tag{4.1}$$

where Y is the modulus of tensile elasticity and μ the Poisson's ratio for lateral contraction. The buckled pipe can reach the elastic limit easily when the helium pressure exceeds p_c slightly. Note that p_c is very sensitive to the radius to thickness ratio $m = \bar{b}/t$.

For the analysis in this section and the next, we shall approximate the proportional limit, below which Hooke's law holds, by the yield limit. For many materials, the yield limit comes just slightly beyond the proportional limit.

However, aside from the radially symmetric exterior helium pressure, there is another horizontal force if the beam pipe is copper coated. These forces are illustrated in Fig. 2. During a quench, the sudden drop in vertical magnetic dipole flux density induces eddy currents in the pipe wall which flows mainly in the highly conductive copper layer. Thus, a horizontal outward Lorentz pressure

$$P_L(\varphi) = P_m \sin\varphi \ , \tag{4.2}$$

is generated, where φ is the polar angle. This pressure has a maximum at the equator, $\varphi = \pi/2$, given by

$$P_m = (\sigma_c t_c + \sigma_s t) b |B\dot{B}| \ . \tag{4.3}$$

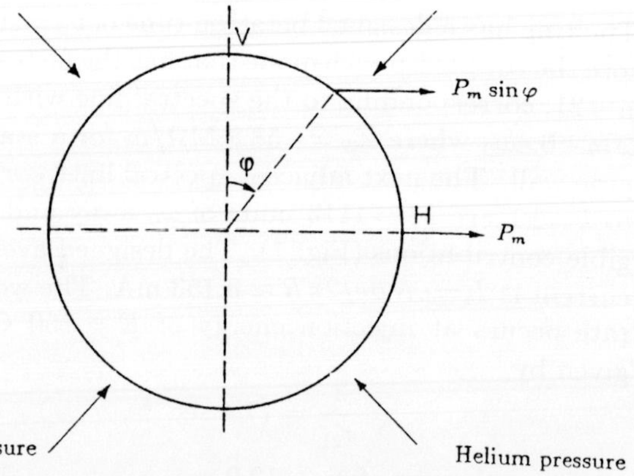

Fig. 2. The pressure produced by the horizontal Lorentz forces acting on the eddy currents in the thin copper layers varies sinusoidally with the polar angle φ, the angle with respect to the direction of the dipole magnetic flux density. Also shown is the radially inward exterior helium pressure.

Here, the conductivity of copper σ_c, the dipole magnetic flux density B, and its time derivative $\dot{B}$ are all time dependent during the quench, but the maximum is implied in Eq. (4.3). Note that this Lorentz pressure is directly proportional to the thickness of the copper layer if the layer thickness is not too small. Equation (4.3) is in mks units. If B is in tesla, $\dot{B}$ in tesla/sec, b and t_c in m, and σ_c in $(\Omega\text{-m})^{-1}$, then the pressure P_m will be in newton/m^2. To convert to psi or atmospheres, the factors 1 newton/m$^2 = 1.45 \times 10^{-4}$ psi and 1 atm $= 14.68$ psi should be used.

With such a Lorentz pressure, the beam pipe will be flattened. Then the helium pressure required to further deform the pipe to the elastic limit will be less than the critical buckling pressure p_c. For a given equatorial Lorentz pressure P_m, the yield helium pressure p is given by[4]

$$\sigma_{yp} = \begin{cases} p\dfrac{\bar{b}}{t} + \dfrac{3P_m}{2}\dfrac{\bar{b}^2}{t^2}\dfrac{p_c}{p_c - p} & \text{when } P_m < 2p \\[2ex] (P_m - p)\dfrac{\bar{b}}{t} + \dfrac{3P_m}{2}\dfrac{\bar{b}^2}{t^2}\dfrac{p_c}{p_c - p} & \text{when } P_m > 2p \ , \end{cases} \tag{4.4}$$

where σ_{yp} is the tensile yield stress of the pipe material. For the first relation of Eq. (4.4), the pipe yields at the equatorial points $\varphi = \pm\pi/2$, and for the second relation, the pipe yields at the polar points $\varphi = 0$ and π.

Taking for stainless steel at 20 K, the modulus of tensile elasticity $Y = 30 \times 10^6$ psi, the Poisson's ratio $\mu = 0.278$, and the yield-point stress $\sigma_{yp} = 115 \pm 25$ kpsi, the limiting helium and Lorentz pressures are computed and are plotted in Fig. 3 with the range of uncertainty indicated by dotted curves. Here, we take the average pipe radius as $\bar{b} = 3.49 + 0.05 = 3.54$ cm and the pipe thickness as $t = 1.0$ mm. As a safety factor, the working yield stress is usually taken as about one half of the actual one. The limiting pressures corresponding to the working yield stress have also been included in the figure.

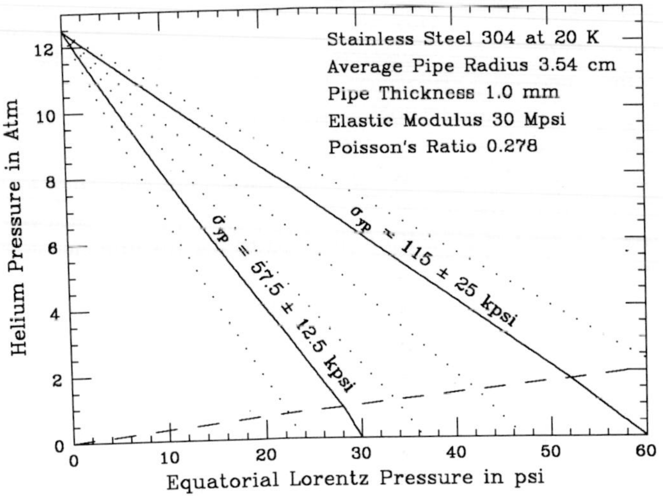

Fig. 3. The limiting helium and Lorentz pressures corresponding to the expected stainless steel yield-point stress of $\sigma_{yp} = 115$ kpsi (upper curves). The dotted curves correspond to the uncertainty of ± 25 kpsi in σ_{yp}. The lower curves result if a safety factor of two in σ_{yp} is applied.

The maximum equatorial Lorentz pressure P_m can be estimated using a model to compute or performing an experiment to measure the maximum of $|B\dot{B}|$ during a quench. For the SSC, with a pipe radius of $b = 1.66$ cm and copper thickness of 2 mils, the estimated P_m is about 90 psi. Here for the Tevatron with a 1 mil copper coating, the product $t_c b$ is roughly the same. The dipole flux density is 6.6 T, the same as the SSC dipole. Thus, we expect P_m to be about 90 psi also. In operation, the helium is flowing at a pressure of 3 atm, and we certainly expect it to be higher during a quench. From Fig. 3, it is clear that the Tevatron beam pipe with a 1 mil copper coating will definitely crack during a quench. If there is no copper layer, the situation is completely different. The eddy current flowing along the stainless steel beam pipe will be much smaller due to its high resistivity. According to Eq. (4.2), the maximum Lorentz pressure at the equator will be reduced by a factor of $(\sigma_s t + \sigma_c t_c)/\sigma_s t = 30.2$ times to $P_m \sim 3.0$ psi. The beam pipe will be safe even when the helium pressure rises to more than 10 atm.

A way to avoid stress failure is to reduce the thickness of the copper coating. A thickness of 0.1 mil or 2.5 μm reduces the maximum Lorentz pressure to $P_m = 11.7$ psi. According to the curve corresponding to the working yield stress, Fig. 3 shows that the beam pipe is still safe up to an external helium pressure of ~ 7 atm. However, such a thickness of copper will only increase the growth time for transverse coupled bunch instability by 3.9 times.

We see from Fig. 3 that most of the time failure will occur at the equatorial points corresponding to the first criterion of Eq. (4.3), which can be rewritten as

$$P_m = \frac{2}{3m^2}\left(1 - \frac{p}{p_c}\right)(\sigma_{yp} - mp) \ , \qquad (4.5)$$

where $m = \bar{b}/t$ is the beam pipe radius to thickness ratio. Together with the definition of the critical buckling pressure p_c in Eq. (4.1), it is easy to observe that the limiting pressure curve is every sensitive to m. If, for example, the pipe radius is increased by a factor of two or the pipe thickness is decreased by a factor of two so that m is doubled, the working limiting curve will starts off from a new critical pressure of ~ 1.63 atm, a factor of 8 smaller and ends at Lorentz pressure $P_m \sim 8$ psi, a factor of 4 smaller.

The Tevatron beam pipe is actually slightly flattened into a four-sided cross section to allow for better helium flow between the beam pipe and the coils. It is not clear whether such flattening will alter the above estimation considerably.

V. BEAM-PIPE TORQUE LIMITATION

In the above section, it was shown that if a copper coating is deposited on the Tevatron beam pipe, it has to be extremely thin and may be of the order ~ 2.5 μm, so that the beam pipe will not be cracked by the external helium pressure during a quench. Such a thin coating is undoubtedly difficult and there will be big unevenness. If, for example, the copper thickness varies with the polar angle φ as

$$t(\varphi) = t_c(1 + \epsilon \sin 2\varphi) \ , \qquad (5.1)$$

the torque per unit length per radius squared twisting the beam pipe about its axis during a quench is given by[5]

$$\frac{\tau}{\ell \bar{b}^2} = \frac{1}{2}\pi\epsilon P_m \ . \qquad (5.2)$$

This torque is resisted by the keys attached to the beam pipe. If the unevenness is big enough, these keys can suffer a stress failure. The detail depends on the actual design and material of the keys. This picture is illustrated in Fig. 4.

If a tensile stress failure occurs, the limiting torque per unit length and radius squared is given roughly by[4]

$$\frac{\tau}{\ell \bar{b}^2} = \frac{w^2 g}{3\ell b h}\sigma_{yp} \ , \qquad (5.3)$$

where each key is assumed to have a width w, length g, height h and located at interval ℓ along the length of the beam pipe. Taking a key similar to that in the SSC, $w \sim 0.2$ in., $g \sim 3$ in., $h \sim 3$ mm, $\ell \sim 18$ in., and $\sigma_{yp} \sim 5000$ psi for the key material, we obtain $\tau/\ell\bar{b}^2 \sim 69$ psi. If as a safety factor, the working yield stress is again taken as one half the actual one, then with the equatorial Lorentz pressure of $P_m \sim 90$ psi per mil of copper layer thickness as estimated in Section VI, Eqs. (5.2) and (5.3) lead to

$$t_c\epsilon \approx 0.24 \text{ mil} \ . \qquad (5.4)$$

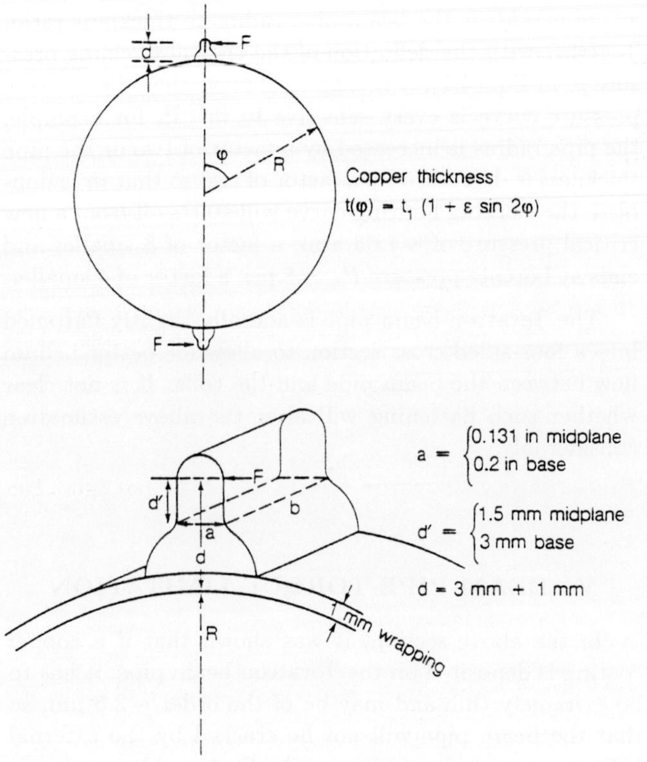

Fig. 4. Geometry of the keys that fix the angular position of the beam pipe with respect to the magnet collars and resist the Lorentz torque that could be produced by non-uniform copper thickness.

Since the copper thickness is of the order 0.1 mil, it takes an unevenness of 240% to break the position-fixing keys. Such unevenness is very unlikely and so is the stress failure of the keys. Such a result in fact is expected. To attain a torque strong enough to break the keys, a certain copper thickness at polar angles $+\pi/4$ and $-5\pi/4$ must exceed that at $-\pi/4$ and $+5\pi/4$ by $t_c\epsilon$. Thus, the unevenness has to be large for a thin coating and vice versa.

VI. CONCLUSIONS

From the above analysis, we conclude that:

(1) If the number of particles per bunch is $N = 2.0 \times 10^{10}$ and betatron tune $\nu_\beta = 20.6$, the worst growth time for transverse coupled-bunch instability is $\tau = 12.9$ ms, which can be controlled by a transverse damper. The parasitic heating due to wall resistivity is $P = 2.2$ kW, which amounts to only 10% of the load of the refrigerators. Therefore, no internal copper coating of the beam pipe is necessary. However, if the number of particles per bunch is increased to $N = 1.0 \times 10^{11}$, the growth time will be reduced to $\tau = 2.6$ ms and the power loss will increase to $P = 56.2$ kW. Both of these numbers then become intolerable. If the betatron tune is 20.9 instead, the worst growth time will be reduced to 3.2 ms and become intolerable.

With a copper layer, the coupled-bunch growth times will be increased by a factor of $\sim \sigma_c t_c/\sigma_s t$ or 29.2 per mil of coating. The parasitic heating of the beam pipe will be reduced by a factor of $\sqrt{\sigma_c/\sigma_s} \sim 34$, roughly independent of the thickness of the coating unless the coating thickness is thinner than $\sim 10\ \mu$m (0.4 mil).

(2) Because of the large beam-pipe-radius-to-thickness ratio, in order that stress failure of the beam pipe will not occur with the combined action of the exterior helium pressure and the horizontal Lorentz pressure arising from eddy currents in the copper layer during a quench, the thickness of the copper layer has to be $\lesssim 2.5\ \mu$m. With such a thin layer, there should not be any problem about breaking the position-fixing keys due to the unevenness of the coating. This thin coating will still reduce the parasitic power loss by 12.3 times. However, it will not be very effective in controlling transverse coupled-bunch instability. It raises the growth time by a factor of 3.9 only.

REFERENCES

1. *Conceptual Design Report: Proton-Proton Collider Upgrade (Main Injector and New Tevatron),* May 1988, Fermilab.

2. see for example, Sacherer F., IEEE Trans. Nucl. Sci. **NS-24**, 1393 (1977).

3. Furman M., SSC Central Design Group Internal Report No. SSC-N-142 (1986).

4. Ng K.Y. and Peterson J., SSC Central Design Group Report No. SSC-168 (1988).

5. Chao A., SSC Central Design Group Internal Report No. SSC-N-434 (1987).

DYNAMIC APERTURE OF A 1.8 TeV p-pbar COLLIDER FOR THE FNAL TUNNEL*

G.F. Dell
Brookhaven National Laboratory
Upton, NY 11973, USA

INTRODUCTION

Tracking is performed on a preliminary lattice for a 1.8 TeV p-pbar collider that can fit into the existing FNAL tunnel[1]. The lattice is 6 fold periodic and has a β^* 38m at all six intersections. The lattice is based on 8.8 T dipoles -- coil i.d.'s of 6 cm and 7.62 cm are used to assess the limitations imposed on the dynamic aperture by the magnet multipoles.

Systematic multipoles were generated for dipoles having a 3 inch coil i.d. by A. Riddiford[2] of FNAL. Random SSC multipoles[3] are scaled to the larger coil diameters. The range of random multipoles σb_n is $2 \le n \le 10$. Only multipoles with order ≤ 16 are used in the tracking study.

Multipoles are scaled from a reference radius r_1, to a reference radius r_2 using the relation: $b_n'(r_2) = b_n'(r_1)*(r_1/r_2)^n$. When scaling to the 6 cm coil i.d., r_1=1.905 cm and r_2=1.50 cm. When scaling SSC random multipoles, r_1=1.0 cm, and r_2 is 1.5 cm for the 6 cm dipoles and 1.905 cm for the 7.62 cm dipoles. All multipoles are given in units of 10^{-4} at 2.54 cm -- as SSC multipoles are expressed in units of 10^{-4} at 1 cm; they have been multiplied by 2.54^n to express them in the same units as the systematic multipoles. All multipoles are listed in Table 1.

Random multipoles for each dipole are generated according to a gaussian distribution truncated at $\pm 3\sigma$. Thin lens kicks are used to represent the influence of multipole fields on the particle motion. The integrated strength of a multipole is divided equally between both ends of each dipole.

The natural chromaticity is $\xi_x=\xi_y=-24.5$; b_2 multipoles contribute an additional $\Delta\xi_x=+256$ (+159) and $\Delta\xi_y=-240$ (-149) for dipoles having 6 cm (7.62 cm) coil i.d.'s. The chromaticity is corrected to zero with two families of sextupoles having integrated strengths of: SF=0.0679 (0.0367) and SD = 0.1456 (0.1007). The tunes are ν_x=20.625, ν_y=20.619, β_{max} = 100 m in the arcs and 140 m in the insertions -- β^* 38 m.

Stability of betatron motion is determined for one random number seed by particle tracking with PATRICIA[4]. The initial amplitudes are adjusted until the limit of stability is found -- ie. the particle with maximum amplitude does not survive 400 turns, while three particles of successively lower amplitudes do survive 400 turns. Stability limits are established over a momentum range of $-0.25\% \le \Delta P/P \le +0.25\%$.

	6 cm Coil i.d.		7.62 cm coil i.d.	
n	b_n	σb_n	b_n	σb_n
2	8.87	6.14	5.5	3.81
3		1.70		0.83
4	-0.078	4.85	-0.03	1.86
5		0.82		0.25
6	1.045	1.84	2.49	0.44
7		0.64		0.12
8	4.128	1.35	0.61	0.20
9		0.34		0.04
10	6.20	1.36	0.57	0.124
12	-19.0		-1.08	
14	7.89		0.278	
16	-8.84		-0.193	

Table 1 Dipole multipoles, random and systematic, used for the present study. Units are 10^{-4} at 2.54 cm.

RESULTS

Results are shown in Figure 1. The maximum initial x amplitude, measured at a horizontally focusing quadrupole in the arc, is 6 mm and 10 mm for the 6 cm and 7.62 cm coil i.d.'s, respectively. Maximum and minimun values of the particle amplitude during every run indicate there is complete transfer of emittance between the x and y planes. The curves of Figure 1 indicate the x amplitude at the start of the tracking run for the initial conditions: $\varepsilon_{o(x)}=\varepsilon_{o(y)}=\varepsilon_o$, $x'=0$, and $y'=0$. Since the maximum emittance in the x plane is $2\varepsilon_o$, it is justified to multiply the results of Figure 1 by $\sqrt{2}$ -- i.e. when $y_o=0$, x_o can be 8.5 and 14 mm, respectively, for the dipoles with 6 cm and 7.62 cm coil i.d.'s.

REFERENCES

1). A. Garren, Current proceedings.
2). A. Riddiford, FNAL, memo to P. Mantsch, June 5, 1988.
3). SSC Conceptual Design Report, SSS-SR-2020, March, 1986, Table 4.4-4, pg. 154.
4). PATRICIA, H. Wiedemann, SLAC (modified 1980 version).

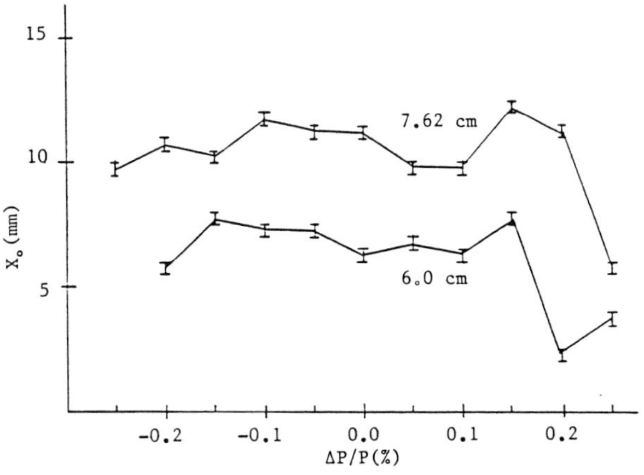

Figure 1 Dynamic aperture of the lattice for dipoles having 6.0 and 7.62 cm coil i.d.'s.

*Work performed under the auspices of the U.S. Department of Energy.

FIXED TARGET ISSUES FOR THE TEVATRON UPGRADE

RAY STEFANSKI

*Fermi National Accelerator Laboratory**

P. O. Box 500, Batavia, Illinois 60510

Abstract

The Tevatron Upgrade poses some interesting prospects for the Fixed-Target program if an option to extract the high energy proton beam is preserved. This paper presents a summary of the advantages of increased energy for fixed target experiments, and evaluates some of the more challenging technical issues. In particular, Bottom production, muon and neutrino interactions, and polarized $\overline{p}$ experiments would benefit substantially from a higher energy primary beam. The new Main Injector will also be important for fixed target experiments as a source for test beams and intense kaon and neutrino beams.

Introduction

In order to extend the physics reach of collider detectors at Fermilab, a machine upgrade will be required after 1993.[1–3] The upgrade may take any one of several forms, including an increase in energy to about two TeV or a conversion to a p-p collider configuration. This article will concentrate on the issues that will pertain to a fixed target program if the accelerator upgrade includes the option of an increase in energy to two TeV. It should be noted first that the highest energy beam that could be extracted to the fixed target areas will be limited by length of straight sections available for extraction, maximum fields attainable in extraction magnets, and available Switchyard geometry, and is likely to be no more than 1.5 TeV.[#1] To be conservative, in this article we will calculate yields for primary beam energies up to 1.4 TeV.

An energy upgrade for the fixed target areas must be based on new physics that would be achieved with higher energy. The last fixed target energy upgrade, Tevatron II, cost nearly $50 million and established four new major experimental halls. That upgrade also made 800 GeV primary beam available to all experiments in the fixed target areas. The physics results from Tevatron II are not yet complete, but already we have impressive inroads established in charm physics, the resolution of discrepancies in neutrino interactions, new measurements in sigma beta decay, new measurements in CP violation, and much more. What could we expect from an additional increase in primary energy?

The new physics that might be achieved in an energy upgrade would depend on the scope of the upgrade. We could imagine construction of new facilities and the upgrade of beam energy for existing experiments. However, the complexity of the present system of extracted beamlines does not easily permit the addition of new facilities except where they would replace an existing facility. The addition of new beamlines would merely reduce the operating efficiency of the program. In this regard a new energy upgrade in the fixed target program would differ markedly from Tevatron II, which included a distinct expansion of facilities.

Test Beams

We must also consider the likely future demand of the fixed target areas for test beams for collider detectors. This demand may increase in the 1990s until beam is available at the SSC site. These test beam facilities would be capable of delivering higher energy to the user if an energy upgrade were to take place, and might more nearly approximate the facilities planned for the SSC where energy of up to one TeV will be available for test beams. However, of greater importance for test beams would be to achieve reliable and stable operation of accelerator and beamlines. Also of greater significance is the need for test beams to be available throughout the calender year. Very often no test beam facilities are available anywhere when the Tevatron operates in a collider mode. For these purposes it would be a great benefit to the entire experimental community if primary beam of 100 GeV or more

---

* Fermilab is operated by Universities Research Assoc., Inc. under contract with the U. S. Department of Energy

#1 Communication from Sam Childress.

were available for test beams somewhere at Fermilab at all times. It would be of even greater benefit if such beams could be available to the existing fixed target areas at all times.

The fixed target program would use a low energy primary beam not only to provide test beams but to test and calibrate experiments in preparation for the start of a run. It is estimated that the availability of low energy beams before the start of a fixed target run could save two months of setup time for many experiments. Given the fact that fixed target experiments receive an average of six months of beam every two years, the increment in efficient use of running time would be very significant for the fixed target program.

Experiments

Aside from the role the fixed target program must play in providing test beams, an increase in primary energy implies an extension of physics for many of the experiments. A list of advantages of 1.4 TeV can be viewed as follows:

For beams that operate at x_F [#2] greater than 0.6 we will be able to produce much more intense beams if higher energy primaries are available. For example, the NM muon beam flux could be increased by nearly an order of magnitude at 550 GeV if 1.4 TeV protons were available.[#3] See Table 1 for comparable rates for experiments in the high energy pion beams.

The physics potential of experiments that look at Bottom production by machine energy protons will also be improved by an energy increase. The production of Bottom might increase by as much as a factor of 3.5 for E-771, because of the increase in production cross-section, and the "folding-forward" of the produced B's to better match the spectrometer acceptance.[#4] Similar increase in rates would be achieved in E-690, P-789 and for the anticipated proton beam run for E-706/672. For comparison, the increase in Bottom production for these experiments in raising the primary beam energy from 800 to 900 GeV would be about 40%.

The increase in production of Bottom in photon beams may be even more dramatic, because the production cross-section rises very rapidly at current energies, and the photon flux is also a strong function of primary beam energy. E-687 should see an increase of factor of seven in Bottom production at 1400 GeV.[#5] Similarly, Bottom production by pion beams, as might be carried out by E-706, would increase by a factor of 6.5 due to higher beam flux and increased production cross section. See Table 2.

#2 x_F is used to indicate the ratio between secondary and primary energy.
#3 Estimate by Jorge Morfin.
#4 Estimate by Brad Cox.
#5 Estimate by Peter Garbincius

For experiments that rely on the identification of jet structure in nucleon interactions, such as E-706 and E-683, the produced jets would be more energetic and therefore easier to identify. This qualitative feature could be important for experiments like E-706 and E-683.

A proposal for a Tagged Neutrino Facility is being evaluated for the fixed target program at the present time. This facility would replace the old neutrino beam, and would be built to study neutrino oscillations and to extend current measurements in neutrino interactions. The capability of this type of beam would benefit considerably from an energy upgrade: There would be twice as many neutrino interactions for each tagged kaon. (Although the tag might be more difficult because of the higher energies involved.) The conventional neutrino physics would also benefit because the neutrino interaction cross-section increases linearly with energy. It should be mentioned, however, that a certain class of neutrino interactions would benefit most from an intense neutrino source as might become available when the new Main Injector is built. For example, the search for neutrino oscillations in the mode $\nu_\tau \to \nu_\mu$ could be carried out at a new experimental area built to accept beam from the 100 GeV injector. The proposed extraction frequency of 0.5 Herz at $> 10^{13}$ protons per cycle for 150GeV[#6] could supply beam for a powerful new neutrino facility.

The potential for physics of an intense kaon source is being evaluated for Fermilab. Neutral kaon beams do not benefit greatly from an increase in available primary energy. As can be seen from Table 1, one can anticipate a 70% increase in K^0 flux[#7] at 1.4 TeV. Much more promising for a kaon factory would be a facility that could extract 100 GeV primary beam from a new Main Injector to a new targeting area. Such a facility could improve the potential statistics of experiments such as E-731 and E-773 by up to three orders of magnitude.

The effects of higher primary energy on hyperon experiments are complex and not always advantageous. For experiments dealing with polarized hyperons, the increased production at say 350 GeV by 1.4 TeV protons is greatly offset by the fact that hyperons are not polarized when produced at low x_F.[#8] Furthermore, pion backgrounds tend to be enhanced by high energy production. Polarized hyperon experiments would have to be carried out at higher x_F, or a secondary energy of 700 GeV or more. This would require a completely reconfigured apparatus, particle identification at higher energy, and higher spatial resolution to measure smaller decay angles. The availability of high energy protons may be important for Ω^- production in which the relative yield at 350 GeV might be a factor of four higher for 1.4 TeV production.

#6 Communication with Steve Holmes
#7 Estimate by G. Bock.
#8 Based on a discussion with G. Rameika.

Related to hyperon production is the polarized proton experiment E-704. Proton polarization is determined by the decay asymmetry of the parent Λ^0, and as such is not dependant on the production mechanism of the parent beam. A polarized proton beam at 200 GeV could therefore be produced by 1.4 TeV primaries, but the gain in yield for this low energy beam would be on the order of 10% over production at 800 GeV. A much more crucial factor for the polarized proton experiments is the gain in $\bar{p}$ production which is predicted to be a factor of 2.2 increase over 800 GeV production.[#9] (See Table 1.)

Table 1: Particle Yields vs Primary Energy[10]

Exp. Number	Secondary Beam Energy GeV	Relative Particle Yields 900 vs 800 GeV	1400 vs 800 GeV
E-773	100 K^0	1.1	1.7
E-704	200 $\bar{p}\uparrow$	1.4	2.2
E-687	350 e^-	1.6	4.7
E-791	500 π^-	2.0	9
E-706	530 π^-	2.3	12
E-665	550 μ^+	1.6	8

Table 2: Relative Bottom Production Rates

Exp. Number	Secondary Beam Energy[11]	Relative Rates 900 vs 800 GeV	1400 vs 800 GeV
E-706	530, 610, 925 π^-	1.6	6.5
E-687	225, 265, 395 γ	1.8	7

This discussion of particle production and yield for the variety of experiments currently operating or planned for the fixed-target program does not consider the ability of the experiments to respond to the increase in flux. As an example, the estimate that the muon beam could yield a factor of eight increase in flux with 1.4 TeV on target, implies that the detector (E-665) will take data eight times faster to take advantage of this increase in flux. This implies that an upgrade in the detector and data acquisition system might be required to take full advantage of the energy upgrade for this experiment. Similarly, other detectors might require improvements to take advantage of the anticipated increase in data collection rates.

#9 Estimates made by Dave Carey

Technical Issues

So it is that one can see substantial benefits for many fixed target experiments in an energy upgrade. However, there are some technical issues that must be considered before an energy upgrade could be fully implemented. The problems associated with primary beam transport in the Switchyard have been evaluated by Childress.[#12] Similar issues must also be addressed in the transport system from the Switchyard to the experimental area targets. Namely, all of the cryogenic bend strings would have to be replaced with the high field magnets, all of the conventional bends would have to be replaced with cryogenic magnets, and the targeting quads would require that additional magnets be added. This implies an increased reliance on cryogenic personnel to operate the beamlines. Furthermore, some civil construction would be needed to maintain the splitting stations in the NEast and PEast beams.

Aside from these issues of beam optics, the second technical issue would be the target piles. In the target piles the primary beam in separated from the secondary beam by passing through a string of bending magnets. The primary beam then enters a dump, while the secondary beam passes along side of the dump. Because this is a high radiation area, the bend magnets in the pile cannot be replaced by cryogenic magnets. To provide for a dumping scheme at energies above one TeV, the target piles would be lengthened. This involves civil construction, and modification of the primary and secondary beam optics. There are eight target piles in the fixed target areas, and they are MWest, MCenter, MP, NMuon, PWest, PCenter, PEast, and PB. Notable exceptions are those beams that transport machine energy protons directly to the experiment: Namely, MEast, NEast, PWest, and MWest when running in the primary beam mode.

The target pile problem may not be serious for most applications, because it is critical only for positive secondary beams that must target at zero degrees. Any secondary beam can operate as a negative beam with 1.4 TeV targeting and a positive beam can operate with targeting angles greater than one milliradian with no target pile modifications.[#13] Roughly speaking, this implies that positive beams would be limited to operating at 50% of

#10 Based on Ref. 4), and estimates by Bock, Garbincius and Rameika.

#11 Secondary beam energy for 800, 900, and 1400 GeV primary beam respectively.

#12 Presentation at Snowmass 1988.

#13 From a discussion with R. Tokarek. For example, in the MWest target pile we could target 1.2 TeV at zero degrees with no modifications. The installed magnet strength already has this capability.

their peak intensity at 300 GeV, and 25% of peak intensity at 600 GeV. The NM muon beam, for example, would gain a factor of two in μ^+ yield at the experiment if the target pile were not modified, whereas operating at zero degrees, which would require a target pile modification, would increase the interaction rate by 8 at 550 GeV. (See Table 1.) The NM beam is an interesting example because the FODO channel could be easily modified to operate at 1.2 TeV. The increase in beam energy is important for muon experiments to measure structure functions at high q^2 and the wee x region. The inclination in the muon beam would therefore be toward a full scale energy upgrade with modifications to the target pile to permit zero degree targeting, if only to take advantage of the existing capability of the FODO channel.

A third technical issue that must be addressed concerns muon shielding.[#14] To account for the increase in radiation due to muons created by dumping 1.4 TeV protons would require the addition of earth berms of about one kilometer in length or the addition of magnetic shielding. In the NM muon beam, for example, the intensity of the muon beam that reaches the site boundary is controlled by a vertical bend located at the downstream end of the Muon Lab. To accommodate the higher primary energy, this bend string would be doubled in length, a relatively minor modification. For beamlines that are located underground, such as those in the Proton Area, the increase muon energy should not a problem, since the muons will simply range out in the earth. However, care is required to locate depressions and valleys and to account for muons punching through due to the earth's curvature. The Neutrino beam already has a substantial berm that is adequate for personnel shielding, and as such the NWest beam that operates parasitically off the Neutrino primary beam could operate with no modifications. New neutrino experiments that might try to operate in the higher energy beam may well require modification to the berm, perhaps by adding steel shielding. The muon shielding problem is most serious in the NEast beam and in all of the Meson Area beams; That is, in beamlines that are located above ground and are thinly shielded. The modifications for these beams might require halo spoilers along the beam and vertical sweeping magnets at the end of the beamline. The design of adequate muon shielding for these beamlines could be a difficult and expensive problem.

A fourth technical issue involves the problem of upgrading secondary beam energy and will be limited by the fact that the target position and the experiment must remain fixed. These beam energies can be increased by the addition of magnets, and in some cases, with some civil construction. Schemes for energy upgrades already exist for PWest, PEast, and NEast. Secondary beam upgrades will be more difficult for those beams that require the construction of a new target pile.

#14 Based on a discussion with Don Cossairt.

Conclusion

Summing it all up, an energy upgrade in the fixed target areas could have substantial benefits for many experiments. Bottom physics, muon interactions, neutrino physics, and polarized $\overline{p}$ experiments could see substantial gains in physics opportunities. A great deal could also be achieved if the new Main Injector could provide protons of 100 GeV or so for test beams, to calibrate fixed target detectors, and for intense new kaon and neutrino beams. The cost of an upgrade could be substantial, especially if the scope includes raising primary and secondary beam energies for all existing facilities. In particular, the increased dependance on cryogenic magnets would require a substantial increase in support personnel to implement and maintain the beamlines.

Acknowledgments

The author wishes to thank Greg Bock, Dave Carey, Sam Childress, Don Cossairt, Peter Garbincius, Jorge Morfin, Gina Rameika, Ken Stanfield, and Roger Tokarek for much information freely provided for this article.

References

1. Conceptual Design Report. *Antiproton - Proton Collider Upgrade 20 GeV Rings.* Fermi National Accelerator Laboratory. (May 1988)
2. Conceptual Design Report. *Proton - Proton Collider Upgrade (Main Injector, New Tevatron).* Fermi National Accelerator Laboratory. (May 1988)
3. Johnson R. P., *A New Collider Upgrade Proposal; Increasing the Pbar - P Collider Energy.* Fermi National Accelerator Laboratory,TM-1525, (May 1988)
4. Malensek A. J., *Emperical Formula for Thick Target Particle Production*, Fermi National Accelerator Laboratory, TM-341-A, (July 1982)

HIGH-ORDER DESCRIPTION OF ACCELERATORS USING DIFFERENTIAL ALGEBRA AND FIRST APPLICATIONS TO THE SSC

M. Berz and E. Forest

SSC Central Design Group
Universities Research Association
c/o Lawrence Berkeley Laboratory, Berkeley, CA 94720

Abstract

The method of Differential Algebra for the treatment of beam dynamics in accelerators is discussed. It allows a very straightforward computation of high order transfer maps from an arbitrary tracking code. For this purpose, all real number quantities depending on the coordinates of the particle to be tracked have to be replaced by differential algebraic quantities. In practice this is facilitated considerably using a FORTRAN precompiler which allows the use of a new differential algebraic data type.

Having changed the tracking code, transfer maps of arbitrary order can be obtained easily. In practice, orders between ten and fifteen can be achieved within reasonable computer time. Contrary to numerical differentiation techniques, the accuracy of the expansion coefficients is very high and only limited by machine roundoff.

The transfer maps so obtained can be used to compute quantities of interest like tune shifts, chromaticities and invariants. Another use of the maps is fast short term tracking. Examples for the use of the method in the simulation of the SSC are given.

Introduction

There are several approaches to the complex of the theoretical analysis of particle accelerators. The simplest but nevertheless often very robust of these approaches is probably the tracking of particles of interest through the accelerator for many turns. This technique allows some phenomenological answers to question of stability and allows an estimation of the dynamical aperture. It also can be used for a numerical computation of nonlinear tuneshifts and smear.

Besides the tracking approach, there is quite a variety of techniques based on Hamiltonian perturbation theory which often allow some analytical computation of quantities of interest. Since these techniques are often used analytically (by hand or with formula manipulators), they are usually limited by the complexity of the system under consideration and frequently only allow the study of simple systems which just describe the major characteristics of the real system.

Finally, there are the perturbative techniques on maps that describe the system by a power series of up to a certain order - an approach originating in optics, where the nonlinearities of the map, the image aberrations, very directly determine the usefulness of the instrument under consideration. In the accelerator community these techniques have not been able to attract much attention in the past, except of course for the study of the linear system, which from the earliest work of Courant and Snyder[1] is described partly in terms of 2 by 2 matrices.

In principle, the map is a concept ideally suited for the description of dynamical systems. It does not exhibit any divergences, which are quite intimately associated with certain forms of Hamiltonian perturbation theory and the Eikonal or generating function approaches. Furthermore, as long as the Hamiltonian of the system under consideration can be expanded in a power series, so can the map. Contrary to mere tracking and similarly to the situation with the perturbative or generating function approaches, it is possible to associate certain properties of the system with particular nonlinearities, thus providing efficient clues as to how to correct undesired properties.

Finally, and perhaps most importantly, many quantities of interest like tune shifts and chromaticities, and to some extent invariants, are concepts that can be expressed directly in terms of power series and hence in the same terminology as the Taylor series map.

The major obstacle with the map techniques, and besides historical aspects the main reason why map techniques have not blossomed in the accelerator field to the same extent as in the optical disciplines, is the high order to which maps have to be available to describe repetitive systems accurately enough. In fact, most existing codes for the power series description of the map[2,3] only allow the computation of third order aberrations, and only recently has it been possible to generate fifth order maps from a kick code based on THINTRACK; this could only be achieved because a fifth order concatenator was available[3] and the physical model was simple enough.[4] Using a special formula manipulator, it was recently possible to generate a thick element library for commonly used beam line elements that allows the computation through fifth order.[5,6,7]

A very recent development, the differential algebraic techniques[8,9,10] have proven very efficient for the computation of nonlinearities of the transfer map to much higher orders than with previous codes. Besides the computation of the maps, these techniques also allow a whole wealth of useful manipulations of the map. As was shown in ref. 11, the whole concept of normal form theory which is a powerful application of Lie algebraic techniques, can be cast very easily into differential algebraic terminology. Within this framework, the exact computation of tuneshifts is very simple. Also, pseudo invariants of the system can be computed readily. As a byproduct, it is possible to obtain the Lie algebraic representation[12] of the map as well as a pseudo Hamiltonian describing the motion which might prove helpful in Hamiltonian perturbation theory.

The map can also be used to compute generating functions describing the system to arbitrary order.[9] Generating functions play a vital role in symplectic tracking techniques, the underlying idea behind which is that if we cannot track the real system, than at least we want to preserve the major symmetry there is in Hamiltonian systems, namely symplecticity.

Much of the theory about differential algebras and their use in accelerator theory has been described in detail in papers cited above. Here, for the sake of completeness, we give a short introduction to differential algebras, but mainly restrict ourselves to the presentation of some first practical examples, using the SSC as our object of study. In a way these examples provide a justification of the work described in the previous papers.

While the authors believe that differential algebraic techniques are very helpful and a step forward, they certainly do not provide the answer to the ultimate question in accelerator design, namely the question of very long term stability. The quite demanding mathematical work in KAM theory only providing a first framework of rigorous statements not yet applicable in practice, it is the authors' belief that we are still far from a solution to this problem.

An Introduction to Differential Algebra

In this section we want to give a short overview of differential algebraic methods. These techniques will allow a very efficient computation of partial derivatives and thus expansion coefficients of the transfer map. The first studies of structures like the ones we discuss here goes back more than a century.[13,14] Then the field lay dormant for a long time until new interest arose around 1960 with a related concept, the field of non-standard[15,16] analysis. A certain connection between the two fields that we hope will allow powerful applications in the future is non-archimedean[17] analysis.

In order to give a feeling for the matter, we will here repeat the treatment of the simplest differential algebraic structure. For a more complete discussion, the reader is referred to 8,9.

Consider the vector space R^2 of ordered pairs (u_0, u_1), $u_0, u_1 \in R$ in which an addition and a scalar multiplication are defined in the usual way:

$$(u_0, u_1) + (v_0, v_1) = (u_0 + v_0, u_1 + v_1) \tag{1}$$

$$t \cdot (u_0, u_1) = (t \cdot u_0, t \cdot u_1) \tag{2}$$

for $u_0, u_1, v_0, v_1, t \in R$. Besides the above addition and scalar multiplication a multiplication between vectors is introduced in the following way:

$$(u_0, u_1) \cdot (v_0, v_1) = (u_0 \cdot v_0, u_0 \cdot v_1 + u_1 \cdot v_0) \tag{3}$$

for $u_0, u_1, v_0, v_1 \in R$. With this definition of a vector multiplication the set of ordered pairs becomes an algebra, denoted by ${}_1D_1$.

In the same way as in the case of complex numbers, one can identify $(u_0, 0)$ as the real number u_0.

On the algebra ${}_1D_1$ one can introduce an ordering that is compatible with the arithmetic on the algebra. We define the ordering as follows. We say

$$(a, b) < (c, d) \text{ if } a < c \text{ or } (a = c \text{ and } b < d)$$

and similarly we introduce ">". With this definition, the algebra becomes totally ordered, i.e. either $(a,b) = (c,d)$ or $(a,b) > (c,d)$ or $(a,b) < (c,d)$, and $(a,b) > (c,d) \Rightarrow (a,b) + (e,f) > (c,d) + (e,f)$ and $(a,b) > (c,d), (e,f) > (0,0) \Rightarrow (a,b) \cdot (e,f) > (c,d) \cdot (e,f)$.

Where in the complex numbers, $(0,1)$ was a root of -1, here it has another interesting property. From the ordering relations we infer that

$$(0,0) < (0,1) < (r,0) \text{ for any positive } r$$

which means that $(0,1)$ lies between 0 and any positive real number, i.e., $(0,1)$ is infinitely small!

It is easy to verify that $(1,0)$ is a neutral element of multiplication, because according to equation (3)

$$(1,0)\cdot(u_0,u_1)=(u_0,u_1)\cdot(1,0)=(u_0,u_1)$$

It turns out that (u_0,u_1) has a multiplicative inverse if and only if u_0 is nonzero; so ${}_1D_1$ is not a field. In case $u_0 \neq 0$ the inverse is

$$(u_0,u_1)^{-1}=(\frac{1}{u_0},-\frac{u_1}{u_0^2}) \qquad (4)$$

Using the above equations it is easy to check that in fact $(u_0,u_1)^{-1}\cdot(u_0,u_1)=(1,0)$.

Similar to the existence of the inverse, one can find a "root" to (u_0,u_1) if and only if u_0 is nonzero. In this case the root is given by

$$\sqrt{(u_0,u_1)}=(\sqrt{u_0},\frac{u_1}{2\sqrt{u_0}})$$

which can be easily verified by squaring the right hand side.

Besides the inverse and root, all functions based on power series like the exponential, logarithm and trig functions can be generalized to differential algebra in a straightforward way. This is shown in detail in refs. 8, 9, and 17, and is based on defining convergence on the new structure.

Instead of going into too many mathematical details, we illustrate the usefulness of this structure for the computation of derivatives with the following example function:

$$f(x)=\frac{1+\sqrt{x}}{x}$$

The derivative of the function is:

$$f'(x)=-\frac{1+\frac{1}{2}\sqrt{x}}{x^2}\,. \qquad (5)$$

Suppose we are interested in the value and the derivative at x = 2. We obtain

$$f(2)=\frac{1+\sqrt{2}}{2},\ f'(2)=-\frac{1+\frac{1}{2}\sqrt{2}}{4}$$

Now take the definition of the function f in equation (5), replace all operations occurring in it by the corresponding ones in differential algebra, and evaluate it at $2+d=(2,1)$. One obtains:

$$\begin{aligned} f((2,1)) &= \frac{1+\sqrt{(2,1)}}{(2,1)} \\ &= \frac{1+(\sqrt{2},\frac{1}{2\sqrt{2}})}{(2,1)} \\ &= (1+\sqrt{2},\frac{1}{2\sqrt{2}})\cdot(\frac{1}{2},-\frac{1}{4}) \\ &= (\frac{1+\sqrt{2}}{2},-\frac{1+\frac{1}{2}\sqrt{2}}{4}) \end{aligned} \qquad (6)$$

As we can see, after the evaluation of the function the real part of the result is just the value of the function at $x=2$, whereas the differential part is the derivative of the function at $x=2$. This is not accidental; as shown in refs. 8 and 9, one can compute derivatives of arbitrary functions using this technique.

Even though the authors find the mathematical concepts described here very appealing, we also want to present a very down to earth approach to understand how the differential algebra discussed here can be used for the computation of derivatives. Suppose there are two functions f and g, and both their values and derivatives at a certain point x are given. We arrange these values into two ordered pairs:

$$(f(x),f'(x)) \text{ and } (g(x),g'(x))$$

Now suppose we are interested in the value and derivative of the sum function $(f+g)$. Then, obviously, the ordered pair describing the value and derivative of the sum function $(f(x)+g(x),f'(x)+g(x))$ is given as the vector sum of $(f(x),f'(x))$ and $(g(x),g'(x))$. On the other hand, suppose we are interested in the ordered pair describing the values and derivatives of the product function. Then, according to the product rule, this ordered pair is given by $(f(x)\cdot g(x),f'(x)\cdot g(x)+g'(x)\cdot g(x))$. But this is precisely what is obtained when multiplying the vectors $(f(x),f'(x))$ and $(g(x),g'(x))$ using the differential algebraic multiplication (3).

In this light, the neutral element in the differential algebraic multiplication, $(1,0)$, is just the ordered pair representing the function that is identical to one and hence has zero derivative. The multiplicative inverse then is the ordered pair describing the value and derivative of the inverse function; an interesting side aspect is that the inverse can be found purely algebraically using the definition of the differential algebraic multiplication and no explicit use of calculus rules is made. In a similar way one can understand the square root.

In this view, the example showing the computation of the derivative of a function (6) can be understood as follows. One begins with the ordered pair describing value and derivative of the identity function at 2, which in our case is the pair $(2, 1)$. Then, the root of this number is computed, giving the value and derivative of the root function at 2. To this, $(1, 0)$ is added, giving value and derivative of $(1 + \sqrt{x})$ at 2. Finally, this is multiplied by the inverse of x, giving value and derivative of the total function.

Thus the differential algebraic operations act as a particularly efficient and simple bookkeeping device to keep track of values and derivatives of certain sub-functions which occur in the process of computing the final function.

Now it becomes quite apparent how the differential algebraic multiplication has to look for higher derivatives and several variables. One first fixes an ordering describing which derivative is stored at what coordinate of the vector. Then for each of these partial derivatives, one determines how it can be written in terms of the values and partial derivatives of f and g.

We note that using the right strategy it is quite straightforward to generalize inverses, roots and all other functions of interest to the new differential algebraic structure. For details, we refer to refs. 8 and 9.

Technical Aspects

From the discussion of the last section, we learned that all that is required to compute the partial derivatives of a function is to break this function down into elementary operations and then perform all the operations in differential algebra. Since FORTRAN, which is the language most of the relevant accelerator simulation programs are written in, does not allow the use of a new data type, the differential algebraic operations can only be implemented as calls to subroutines.

To automate the breaking down of a certain function into the subroutine calls, an interpreter was written.[18] This interpreter understands full FORTRAN formula syntax, including all relevant functions, and is an extension to regular FORTRAN allowing a new differential algebraic data type. The interpreter takes extended FORTRAN and transforms it to regular FORTRAN.

The second major difficulty that had to be overcome is the efficient computation of the product of differential algebraic vectors. While we saw that in principle it is trivial to figure out the product for any order and number of variables, in practice this requires sophisticated (and slow) logic that is not very well suited for a computer environment. However, using a tricky addressing scheme, it was possible to overcome these difficulties. In practical cases, the logic overhead now is down to an insignificant amount of the time required for the arithmetic. For details we refer to ref. 8.

To conclude this section, we want to discuss how to use these techniques for the computation of transfer maps in practice. To be somewhat more specific, suppose we are given a certain tracking code and want to modify it for map extraction. We note that the code (besides sometimes quite involved logistic management sections) represents a functional dependence between the final coordinates and the initial coordinates, even though this dependence typically is very complex and often involves hundreds if not thousands of lines of code. Depending on the model used, this dependence could involve kicks and drifts, or a numerical integrator. In order to compute the derivatives of transfer maps, now simply replace each and every one of these operations by the corresponding on in differential algebra. The code will then automatically compute all the requested partial derivatives of the dependence of the final coordinates on the initial coordinates.

Examples and Results

In this section we will give some examples of the use of the differential algebraic method in practice. Among other codes we enhanced the kick code THINTRACK[4] to map extraction using the techniques described in the last section.

We investigated an SSC lattice without random errors, i.e. only identical cells with realistic corrected errors in the six bends per half cell. We computed the transfer map of one cell of the lattice through eleventh order; higher orders should be feasible if necessary,

Comparing the accuracy of the Taylor expanded map with the result obtained by tracking individual particles through the cell, we found that for particles up to about 75 % of the dynamical aperture the relative accuracy was about 10^{-9}. This means that the accuracy of the mathematical approximation of using the eleventh order Taylor series instead of the real map is well below the level of design errors and thus tolerable.

This number suggests that the map can be used to perform tracking through the system. And indeed, as Figure 1 shows, there is hardly any detectable difference between the tracking pictures using direct tracking versus (the faster) tracking through the power series map. In fact, for the case of particles well inside the dynamical aperture, the agreement is better than printer resolution. Figure 1 shows the $y - p_y$ projection of the motion of a particle launched with $x = p_x = 0$. Because of the good decoupling of the x and y motions, the occupied phase space area is almost one dimensional.

The very high accuracy in the description of the map by its power series is a consequence of the high order used. To demonstrate this, Figure 2 shows the same tracking simulation as in Figure 1, but with Taylor maps of order three and five, respectively. As one can see, the accuracy of the third order map is totally inadequate, the particle even gets lost after a few turns. This picture shows a growth of phase space volume, which is probably due to the fact that the third order map is not fully symplec-

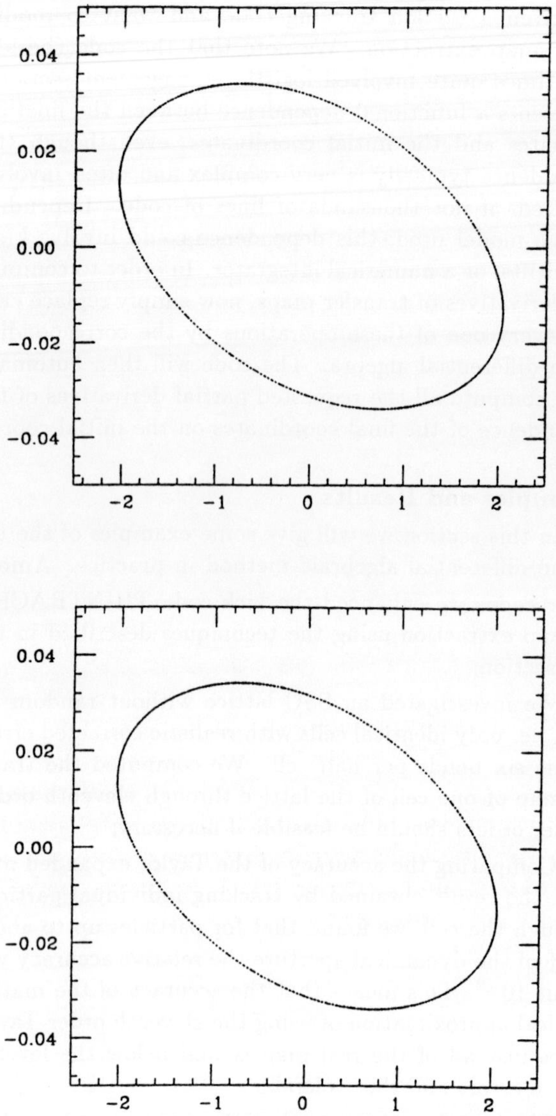

Figure 1. The simulation of a particle tracked through one cell of the SSC for 400 turns. The upper figure shows the result obtained with direct tracking, and the lower figure shows the result using the power series map. Note that the two figures agree to printer resolution.

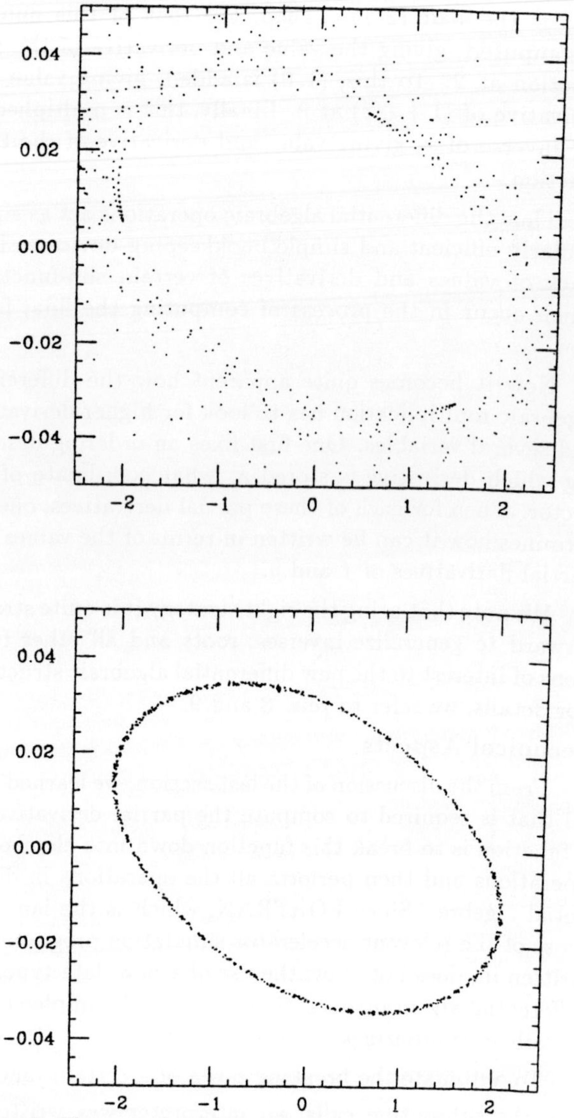

Figure 2. The same simulation as in Figure 1, but now using a map through third order (upper figure) and fifth order (lower figure). The degradation of accuracy due to the lower simulation order is clearly visible.

tic. It is likely that the picture would be more favorable by carefully choosing a symplectification procedure based on generating functions.[9] But even though with symplectification the picture might look more realistic, strictly speaking the accuracy will not improve significantly. In fact, symplectification procedures should be taken with a grain of salt and used carefully, because they have a tendency to make even inaccurate simulations look accurate.

As is to be expected, the accuracy of the same simulation done to fifth order, which is the maximum order obtainable so far by techniques other than differential algebra, is higher than that of third order, but still some inaccuracies are clearly visible.

In Figure 3 we present the $x - p_x$ phase space plot of the simulation shown in Figures 1 and 2. The picture shows the $x - p_x$ motion induced by the remaining nonlinear cross coupling. Again direct tracking and tracking through a map are compared, and no differences between the two pictures are visible, which indicates that the (faster) tracking through the power series map yields the same result to printer resolution.

Altogether we have shown some examples in which the tracking using power series yields results indistinguishable from direct tracking, indicating that the power series map in these cases contains all the significant information about the system. It can be expected that similar results can be obtained for other lattices, where possibly the order of the map has to be adjusted to the degree of

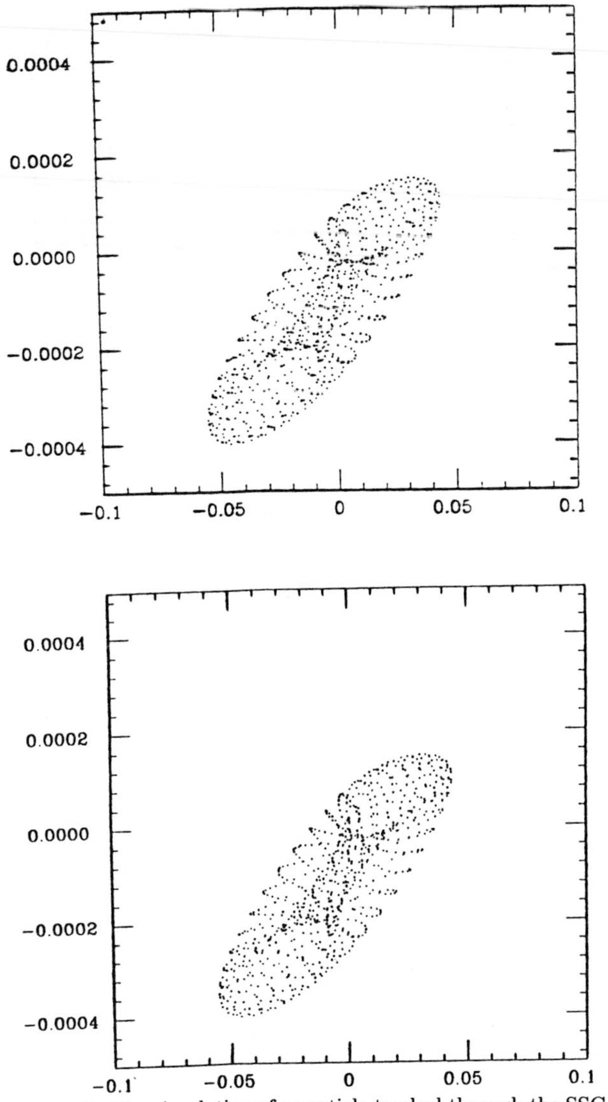

Figure 3. The simulation of a particle tracked through the SSC for 400 turns. The upper figure shows the result obtained with direct tracking, and the lower figure shows the result using the power series map. Note that again the two figures agree to printer resolution.

nonlinearity of the lattice.

We should note, however, that whatever high order we use, if we only track long enough, a difference between the direct tracking and the tracking using a map will quite likely become apparent. This is particularly likely in the case of chaotic motion, in which case particles initially close together in phase space eventually move further and further apart, and in which case this growth is often exponential.

In this respect, the (small) inaccuracies in the map lead to similar effects than the (usually larger) machine errors or the inaccuracies of the physical model of the accelerator or even the computer accuracy. All these effects sooner or later entail discrepancies between the real and the computed particle coordinates, a problem inherent to all tracking techniques. One can hope, however, that while obviously individual coordinates cannot be predicted for large numbers of iterations, at least global aspects of the phase space areas populated by the same particle are reproduced properly. Experience shows that this is often the case.

In order to put the mathematical inaccuracies of the truncation of the power series into perspective with the inaccuracies of the kick model, we show in Figure 4 the same simulation as in Figure 3, now using two kicks per dipole instead of one kick per dipole as in Figure 2. As we can see, the differences are quite noticeable in that the phase space points are not predicted in the same way. Still the total phase space area occupied in the two pictures is very similar, which is the reason why the predictions of the kick model are not completely useless.

However, it is quite apparent that the accuracy of the mathematical approximation (truncate power series) is much better than the accuracy of the kick model. Indeed, the authors believe that the wide spread trust in the kick model is somewhat unjustified. Of course it automatically produces symplectic maps and thus preserves the only universal symmetry of Hamiltonian systems. This, however, can also be achieved with other means, in particular with all power series maps using symplectic tracking as outlined above. The argument that on top of symplecticity the kick model actually describes a "real" physical system is somewhat misleading; indeed, this model requires infinitely strong fields and infinitely short magnets, and thus in this model obviously the SSC would easily fit on the reader's desk.

One of the main reasons why the kick model is so popular is its ease of implementation, and, as a result, the increase in speed which brings long term tracking in the realm of current computers. More exact techniques, like numerical integration through realistic fields, are orders of magnitude slower and thus make tracking quite inefficient. In this respect the power series maps might have an advantage for real long term tracking (with all the problems associated with it); in the case of maps it does not matter whether the map was computed using a kick model or a more realistic model. The tracking through the map is equally fast in both cases. So it might be well worth to invest a lot of computer time in a one turn map that is as accurate as possible and then enjoy more accurate and very fast long term tracking.

As we have argued before, the mathematical accuracy of the power series truncation is not infinite. We illustrate this with an example which shows a similar SSC simulation than in the previous figure, except that in this case the amplitude has been increased into a very nonlinear region in which even the onset of chaos seems to be detectable. Figure 5 shows the coordinates of the particle, predicted by direct tracking and by tracking through the

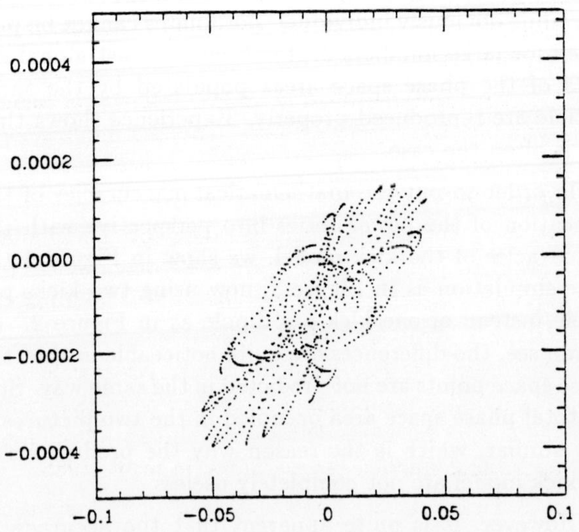

Figure 4. The same simulation as done in Figure 3, except that in the kick model the dipoles were represented by two kicks instead of one. While the "topology" of the picture is similar to that in Figure 3, the position of the particle after the same number of turns is very different. This shows a limitation of the physical model, which here reduces accuracy much more severely than the mathematical approximation of truncating the power series map.

power series map. The agreement is still very good, but does not quite reach printer resolution any more. A careful comparison of the two pictures reveals that now the particle coordinates are slightly off. However, the discrepancy are still far below the model accuracy which can be estimated from Figure 4.

It has been shown elsewhere[11)] that it is quite easy to implement powerful normal form algorithms on the (differential algebraic) map with the differential algebraic techniques. With normal form procedures the exact computation of tunes and tune shifts to arbitrary order becomes quite simple and efficient, and we have used such tune calculations on several occasions.

Here we show another application of normal form theory, namely the determination of invariants and pseudo invariants. In essence, the normal form algorithm which we are using here, turns an arbitrary map of a Hamiltonian system into an integrable one, i.e. one that has a number of invariants equal to the number of position-momentum pairs. Obviously not all Hamiltonian maps have this property, so the invariants that are obtained in this way may indeed prove to be only approximate (or pseudo) invariants.

In Figure 6 we show the motion of the particle shown in Figure 5 in the new coordinates obtained from the normal form algorithm. As we can see, the complicated and perhaps chaotic motion has been turned into a most regular motion: that on a circle. The invariant (or pseudo invariant) is the radius of the circle. We note that the coordinates show a very small spread around a perfect circle.

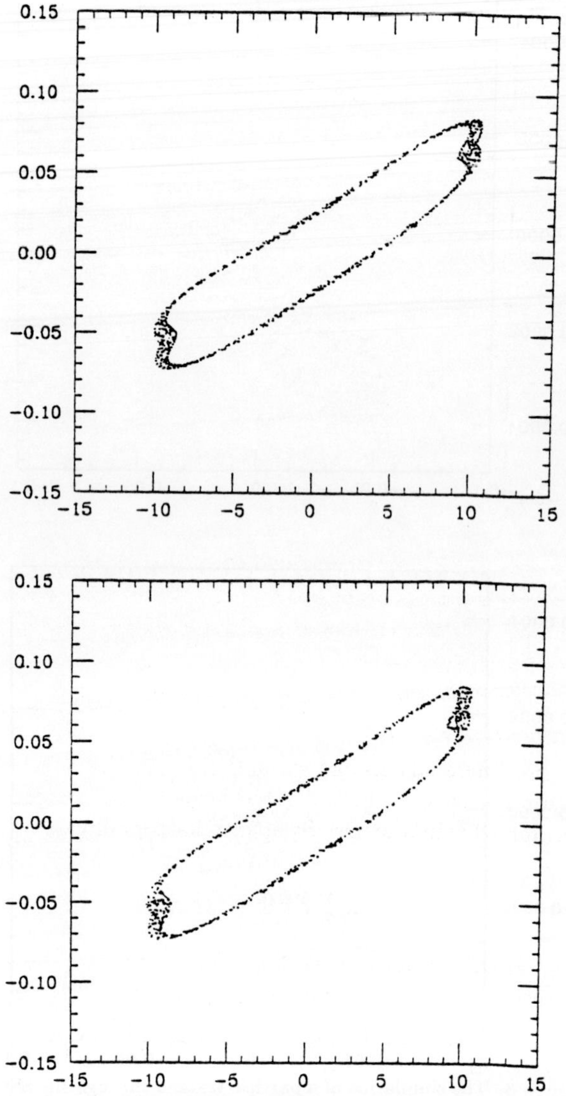

Figure 5. The simulation of a particle tracked through one cell of the SSC for 400 turns. Compared to Figure 1, the amplitude of the particle is now approximately 75 % of the dynamical aperture. The upper figure shows the result obtained with direct tracking, and the lower figure shows the result using the power series map. Note that the two figures agree very well, but due to the increased amplitude, very fine differences due to the less accurate power series representation are noticeable

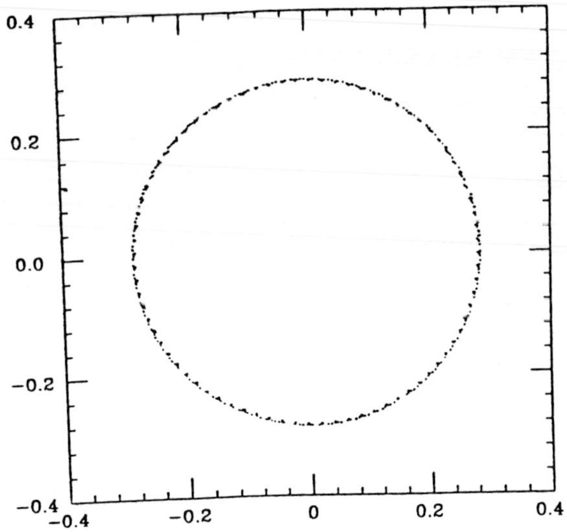

Figure 6. The same motion as in Figure 5, now displayed in normal form coordinates. As can be seen from the picture, the radius is very well preserved, showing that $x^2 + p^2$ is indeed very close to being an invariant of the system.

Acknowledgements

For financial support we would like to thank Dr. Alex Chao. For several fruitful discussions we are indebted to Dr. John Irwin and Dr. Swapan Chattopadhyay.

REFERENCES

1. E.D.Courant and H.S.Snyder. Theory of the alternating gradient synchrotron. *Annals of Physics*, 3:1, 1958.
2. K. L. Brown. *The Ion Optical Program TRANSPORT.* Technical Report 91, SLAC, 1979.
3. A. J. Dragt, L. M. Healy, F. Neri, and R. Ryne. MARYLIE 3.0 - a program for nonlinear analysis of accelerators and beamlines. *IEEE Transactions on Nuclear Science*, Ns-3,5:2311, 1985.
4. B. T. Leeman and E. Forest. *Brief Description of the tracking codes THINTRACK and FASTRACK.* Technical Report SSC-133, SSC Central Design Group, Berkeley, Ca, 1988.
5. M. Berz and H. Wollnik. Analytical solution of the equations of motion to high orders: the formula manipulator HAMILTON and the fifth order code COSY. *AIP Conference Proceedings*, in print, 1988.
6. M. Berz, H. C. Hofmann, and H. Wollnik. COSY 5.0, the fifth order code for corpuscular optical systems. *Nuclear Instruments and Methods*, A258:402, 1987.
7. H. Wollnik, M. Berz, and B. Hartmann. Principles behind GIOS and COSY. *AIP Conference Proceedings*, in print, 1988.
8. M. Berz. Differential Algebraic description of beam dynamics to very high orders. *Particle Accelerators*, in print, 1988.
9. M. Berz. Differential Algebraic treatment of beam dynamics to very high orders including applications to spacecharge. *AIP Conference Proceedings*, in print, 1988.
10. M. Berz. Differential Algebraic description and analysis of trajectories in vacuum electronic devices including spacecharge effects. *IEEE Transactions on Electron Devices*, in print, 1988.
11. E. Forest, M. Berz, and J. Irwin. Normal form methods for complicated periodic systems: a complete solution using Differential Algebra and Lie operators. *Particle Accelerators*, in print, 1988.
12. A. J. Dragt and J. M. Finn. *Journal of Mathematical Physics*, 17:2215, 1976.
13. P. Du Bois-Reymond. *Mathematische Annalen*, 11:, 1877.
14. O. Stolz. *Mathematische Annalen*, 39:, 1891.
15. D. Laugwitz. Ein Weg zur Nonstandard-Analysis. *Jahresberichte der Deutschen Mathematischen Vereinigung*, 75:66, 1973.
16. A. Robinson. Non-standard analysis. In *Proceedings Royal Academy Amsterdam, Series A*, page 432, 1961.
17. M. Berz. *Analysis auf einer nichtarchimedischen Erweiterung der reellen Zahlen.* Report (in German), Universität Gießen, 1988.
18. M. Berz. *The Differential Algebra FORTRAN precompiler DAFOR.* Technical Report AT-3:TN-87-32, Los Alamos National Laboratory, 1987.

THE POSSIBILITY OF PARASITIC 20 TEV BEAMS EXTRACTED FROM THE SSC USING BENT CRYSTALS

B. Cox, J. Lach, M. Maslov, N. Mokhov

C. T. Murphy, R. J. Stefanski

Introduction

The need for an SSC test beam of energy greater than one TeV was discussed in Snowmass '86 (see for example Ref. 1); such a beam could be created from the neutral debris from an interaction region as a parasitic operation. In the same reference, the possibility of extracting a 20 TeV beam with a bent crystal was mentioned. This year it is more apparent that a low intensity, 20 TeV extracted beam would be highly desirable. Experiments which study the properties of short-lived particles, for which the average decay path is long in the lab frame in a fixed target experiment, can be done in such a beam. In particular, a fixed target Beauty detector running at Fermilab would require about 10^8 protons per second. A Beauty experiment running at the SSC in a comparable beam of extracted protons would have twenty times more Beauty production rate and would be further enhanced by a more favorable proper time for the decaying B's.

In this note we will explore in some detail the possible yield of a bent crystal beam and its impact on collider operations.

A Sketch of the Technique

The bending and transmission properties of bent crystals have been studied for many years.[2–4] The basic idea is that a crystal of silicon or tungsten of a few centimeters length, curved with an arc of a few milliradians, will channel between 10^{-4} to 0.8 of the incident beam along the curvature of the crystal, the exact fraction depending on conditions of both the beam and the crystal. To extract a low intensity beam from an accelerator, the crystal is placed in the halo of the circulating beam (perhaps in part of the beam destined to be intercepted by scrapers), and a small fraction of the halo is bent towards the extraction channel. Since the extracted beam is very parallel, refocusing is not needed for a long distance, or not at all.

Experimental work on crystal channeling of high energy particles continues. A secondary goal of Fermilab E-761 (scheduled to run in the late 1989) will use a bent crystal to deflect a polarized hyperon beam. They hope to not only detect the change of the Σ^+ beam direction but also the precession of its magnetic moment.

Exploration of the Technique for UNK

The technique has been explored at Serpukhov for possible application to UNK for the test beams in the 70 - 3000 GeV energy range. The bent angle is about 10 mrad. Multi-turn Monte Carlo studies have indicated that one could have about 2×10^9 inelastic nuclear interactions per second of the 3 TeV protons with a crystal without quenching any magnets. This number coincides to first order with the mean beam intensity at the crystal face position. As a result, an extracted beam of a few times 10^7 per second was felt to be easy to achieve. The remaining difficulty is the positioning of the crystal relative to the beam with an error less than $\pm 0.5/\sqrt{p}$ mrad, for a tungsten crystal (large beta-functions and small beam dispersion at that point). Some experimental tests at 70 GeV have been done.[5]

Application to the SSC

The crystal should be placed at a dispersion free point and at a large beta. Several extraction schemes using crystal septa have been proposed in the past, including those by Sun, Carrigan and Toohig and Vodopyanov et al. An insertion at the Injection Utility Insert is possible. One such insertion[5] mimics the abort inserts. A dog-leg is formed in the circulating beam, with Lambertson magnets forming the middle bend, and the crystal deflects the extracted beam into the field-free region of the Lambertsons. A bend angle of 95 μrad is needed from the crystal.

A feeling for the intensities that could be extracted from the SSC with the bent crystal technique can be obtained from the following order-of-magnitude calculation.

The crystal could be placed at a point where $\beta_x = \beta_y = 500m$. For a normalized emittance of π $mm \times mrad$, the angular divergence would be $\pm0.3\mu rad$, or $\pm0.76\mu rad$ would contain 95% of the beam. This is less than the silicon crystal critical angle of $1.1\mu rad$ (see Ref. 2), so that there is no channeling loss due to excessive angular divergence. According to the formulae developed by Carrigan[6], channeling efficiencies of about 75% of the incident beam should be achievable with a crystal of order 2 to 3 cm long. This represents about 10% of a collision length (elastic scattering included) for particles not channeled.

The design intensity of the SSC is 1.24×10^{14}, or $4.35 \times 10^{17}/sec$. According to Dave Johnson, "scraping" is allowed up to an amount creating a beam lifetime of 24 hours or longer, which would permit $1.4 \times 10^9/sec$ incident on the crystal (assuming as an upper limit that all of it is lost from the machine in one way or another). Of this, about 10^9 would be extracted for the ideal crystal described above.

Of course, such ideal conditions would not be achieved, but the above upper limit on extraction is an order of magnitude more than what is needed for certain interesting fixed-target experiments, such as described below.

A Beauty Experiment

The possibility of using the crystal channeling system described above to produce a continuously extracted 20 TeV proton beam of "low intensity" is an intriguing possibility. The extracted proton beam need be only intense enough to produce 10^7 interactions per second in a silicon tracker. A 5% interaction length tracker would need a beam of 2×10^8/sec. Given the high charged particle multiplicities (150) and relatively large solid angles which must be covered even at the SSC to insure good collection efficiency for B decay products, such a "fixed target configuration" should be examined as a possible alternative.

The cross section for the production of B's using 20 TeV extracted protons ($\sqrt{s} = 194$ GeV) is approximately 50 $\mu barns$ (assuming an intermediate A target such as silicon and a linear A dependance), and the multiplicity of the interactions (assuming that it grows like $\log(s)$ between 40 GeV and 194 GeV) will be approximately 20 into the solid angle of the detector. There are several technical advantages of the extracted proton beam experiment over the collider configuration, among which are:

1. The lack of necessity for a beam pipe.
2. The high momentum of the B's and their decay products makes triggering easier and multiple scattering smaller.
3. The long decay lengths.
4. The normal incidence angles of the decay products with respect to the planar silicon detector configuration is more natural to the Lorentz boosted events.
5. The more modest detector size required to cover the smaller solid angle.

Using the 50 $\mu barn$ cross section mentioned above and assuming a total cross section of $100\mu barns \times A^{0.75}$, we obtain for silicon target material that the ratio of pbar production to total cross section is 10^{-4}, assuming $A^{1.}$ for B hadroproduction. So in an experiment operating for 10^7 seconds at 10^7 interactions per second, 10^{10} B - Bbar pairs can be produced in the relatively clean fixed target environment mentioned above. Trigger strategies which have been developed for fixed target experiments such as E-653 and E-771 at Fermilab are directly applicable to such a configuration without the uncertainties involved with the more complicated environments of the collider and the more expensive experiments necessary in the collider.

Conclusion

The potential for new physics conducted in a modest primary beam of 10^8 protons per second is considerable. Although the problems associated with the use of a crystal septum to extract the beam must continue to be studied, the effort will be well worthwhile. Furthermore, the provision for a bypass in the SSC would provide for a natural area to deliver the extracted beam for a fixed target experiment. We have given the example of some rates for a Beauty experiment, because these are most obviously sufficient for the interesting new physics. This facility could also be used to provide for experiments that use hyperons, K_S^0, low rate neutrinos, muons, or for a test beam facility. The promise is sufficiently great that further work is warranted.

References

1. M. Harrison and T. Toohig. *Physics of the Superconducting Supercollider, Snowmass.* page 534 (1986)
2. Chih-Rae Sun and D. Neuffer. *Design and Utilization of the SSC, Snowmass* page 483 (1984)
3. S. I. Baker et al. *Deflection of an 800 GeV Particle Beam Using Channeling*, NIM, A248,301 (1986)
4. A. S. Vodopyanov et al. *Slow Extraction of 20 TeV Accelerated beam out of Superconducting Supercollider Using a Bent Crystal.* Preprint of the Joint Institute for Nuclear Research, Dubna. JINR P1-88-300 (1988)
5. A. G. Aphonin et al. Preprint IHEP 87-121, Serpukhov (1987)
6. R. A. Carrigan. *Channeling and Dechanneling at High Energy*, Fermilab-Conf-87/149 (1987)

SSC LINAC INJECTOR*

T. S. Bhatia, F. W. Guy, G. H. Neuschaefer, M. Pabst,† S. O. Schriber,
J. E. Stovall, T. P. Wangler, M. T. Wilson, and G. T. Worth
Los Alamos National Laboratory
Los Alamos, NM 87545

ABSTRACT

The parameters for the proposed Superconducting Super Collider (SSC) linac injector system are obtained from the established requirements of the low-energy booster (LEB). The first element of this injector system is a radio-frequency quadrupole (RFQ) that bunches the H^- ions and accelerates these ion bunches to 2.5 MeV. With a suitable matching section, this beam is injected into a drift-tube linac (DTL), which takes the ions to 120 MeV. The final element is a coupled-cavity linac (CCL) designed to accelerate the H^- ions to 600 MeV for injection into the LEB. The conceptual beam dynamics design for the various elements of this linac injector system is described.

INTRODUCTION

The SSC Conceptual Design Report (CDR) was published in March 1986.[1] It included linac information based on preliminary accelerator design parameters. Since then, in order to improve the overall SSC design and also to consider the possibilities for higher beam luminosity, new designs for the LEB have been discussed. The SSC linac injector parameters are determined by the requirements of the LEB. Any changes in the LEB lattice require corresponding changes in the linac injector. Furthermore, in the last few years, there have been some advances in linac design procedures in as far as the control of emittance growth is concerned. It is quite appropriate at this time to launch a new iteration for the SSC linac injector design that provides an H^- ion beam suitable for the LEB painting scheme being considered.

SSC LINAC INJECTOR PARAMETERS

The SSC linac injector parameters must be established from the requirements of the LEB lattice and the painting scheme. One such painting scheme based on the new LEB lattice design,[2] has been described by Colton and Thiessen.[3] In this scheme, 600-MeV H^- ions with a 50-MHz microbunch structure undergo $H^- \rightarrow H^+$ conversion in a 225-μg/cm^2 carbon stripping foil before injection into the LEB, which has a circumference of 342.71 m. It takes 72 microbunches to fill the LEB ring. Each microbunch contains 3.8×10^8 H^- ions. The painting scheme consists of a 26-turn synchronous injection into the LEB. This dictates a linac macropulse length of 37.44 µs, and with the required repetition rate of 10 Hz, the linac-beam duty factor is $\sim$0.04%. The transverse and longitudinal normalized rms emittances, ε^t(n, rms) and ε^ℓ(n, rms), at 600 MeV are required to be less than 0.45 π·mm·mrad and 1.7×10^{-5} π·eV·s, respectively. This painting scheme enables the LEB to accumulate 10^{10} protons per bunch into each of 72 bunches leading to 7.2×10^{11} circulating protons per macropulse. Table I summarizes the required SSC linac injector parameters based on the painting scheme described in Ref. 3.

SSC LINAC INJECTOR

Simulations with the RFQ design code PARMTEQ have shown that for a 50-MHz, 2.5-MeV, 3-mA H^- RFQ, the expected transverse and longitudinal emittances at 2.5 MeV are ε^t(n, rms) $\sim$0.18 π·mm·mrad and ε^ℓ(n, rms) $\sim$8.2 π·mm·mrad. Thus, an SSC linac injector system that starts off with a 50-MHz RFQ could easily satisfy the transverse

---

*Work supported by Los Alamos National Laboratory Institutional Supporting Research, under the auspices of the United States Department of Energy.

†Permanent address: ISI, KFA-Jülich, W. Germany.

TABLE I

Required SSC Linac Injector Parameters

Energy	600 MeV
Beam microbunch structure	50 MHz
Protons per microbunch	3.8×10^{8}
Linac pulse length	37.44 μs
Linac repetition rate	10 Hz
ε^{t}(n, rms)	$\leq$0.45 π·mm·mrad
ε^{ℓ}(n, rms)	$\leq1.7\times10^{-5}$ π eV·s (5.4 π·mm·mrad)

emittance requirement, but fails to meet the longitudinal emittance goal by about a factor of 1.5 at the output of the RFQ.

Higher frequency RFQs, such as 100 to 200 MHz, lead to beams with lower transverse and longitudinal emittances. Such an RFQ must be followed by a chopper designed to produce a 50-MHz microbunch beam structure. Thus, one must have a 9-mA average beam current (during a macropulse) from a 150-MHz RFQ, because the chopping system will allow only one out of three microbunches to be injected into the drift-tube linac.

We have looked at four nominal RFQ design examples with RFQ frequencies ranging from 50 to 200 MHz. The output emittances for the four cases, based on an input rms normalized emittance of 0.10 π·mm·rad are listed in Table II.

TABLE II

2.5-MeV H⁻ RFQ Design Examples

Frequency	Current	ε^{t} (n, rms)	ε^{ℓ} (n, rms)	
(MHz)	(mA)	(π mm·mrad)	(10^{-5} π eV·s)	(π· mm·mrad)
50	3	0.18	2.60	8.26
100	6	0.15	1.00	3.18
150	9	0.14	0.45	1.43
200	12	0.12	0.30	0.95

A 150-MHz RFQ leads to a beam with longitudinal emittance decreased by almost a factor of 6 over that for a 50-MHz RFQ.

The SSC linac injector system designed to satisfy the requirements noted in the previous section is shown in Fig. 1. A 10-mA H⁻ beam from the ion source is properly matched and delivered to a 150-MHz RFQ where the beam is bunched and accelerated to an energy of 2.5 MeV. Simulations show that the beam transmission through the RFQ is ~0.96. Before injection into the 450-MHz DTL, the beam goes through a 50-MHz rf deflector chopper that is designed to retain one out of three microbunches. The 450-MHz DTL, designed for 27-mA (average current during macropulse if all bunches were filled),

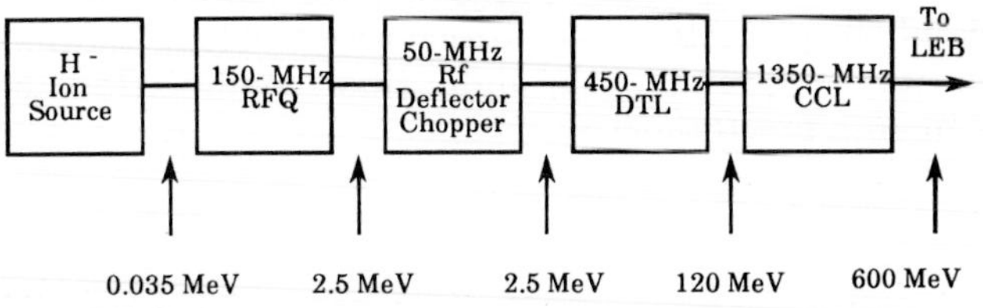

Figure 1. The suggested SSC linac injector system.

accelerates the 2.5-MeV beam to 120 MeV. The final accelerating element is a 1350-MHz CCL designed for 81-mA operation that takes the energy to 600 MeV.

We shall now describe the individual components of this suggested SSC linac injector system.

1. RFQ

Table III shows the PARMTEQ design and simulation results for the 150-MHz, 2.5-MeV H⁻ RFQ. This is a conservative design with maximum peak surface electric field (MPSEF) of ~1.8 Kilpatrick. The RFQ length is 3.45 m. This RFQ could be a 4-rod or a 4-vane structure.

TABLE III

PARMTEQ Simulation Results for the 150-MHz H⁻ RFQ

	Input	Output
Energy (MeV)	0.030	2.5
Current (mA)	10	9.6
ε^{t}(n, rms) (π·mm·mrad)	0.10	0.14
ε^{ℓ}(n, rms) (π·mm·mrad)	0.0	1.43

RFQ length	3.45 m
MPSEF	1.8 E_K (23.7 MV/m)
Total Power	190 KW
Copper Power	166 KW
Beam Power	24 KW
Duty Factor	0.04%

2. RF Deflector Chopper

The 150-MHz RFQ is followed by a 50-MHz rf deflector. One of three beam microbunches needed to be injected into the DTL suffers a deflection to the left, whereas the subsequent two bunches are deflected only slightly to the right where they are collected in a beam stop (see Fig. 2). With this chopping scheme, a very clean 50 MHz microbunch structure should be achieved, and the unused beam is neatly collected in a beam stop. The rf deflector would be combined with a matching section consisting of two or three permanent magnet quadrupoles and an rf buncher. Although numerical simulations for this part of the accelerator system have not yet been done, we believe that it should be possible to keep transverse and longitudinal emittance growth, through this combination of elements, to ~10% or less.

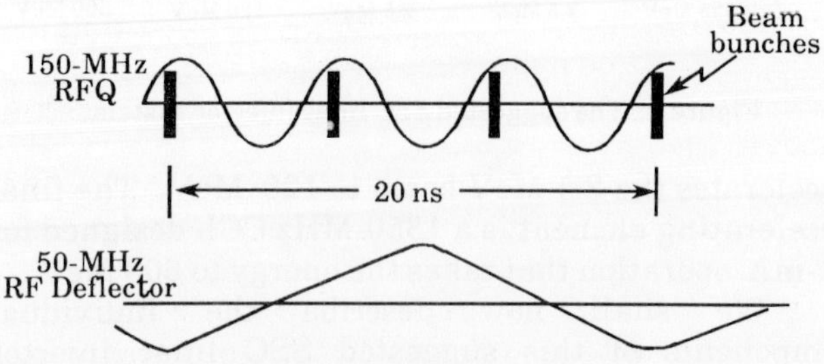

Figure 2. The rf deflector scheme.

3. The 450-MHz DTL

Following the rf deflector system, 3-mA, 50-MHz beam microbunches are injected into the 450-MHz DTL. Thus, one out of nine buckets are filled. This 450-MHz DTL is therefore designed for 27-mA operation.

The DTL parameters obtained from PARMILA simulations are given in Table IV. The transverse and longitudinal emittance growths in this 2.5 to 120-MeV linac are less than ~10% (see Figs. 3 and 4).

TABLE IV

PARMILA Simulations for the 450-MHz 2.5 to 120-MeV DTL for the SSC Linac Injector

	Input	Output
Energy (MeV)	2.5	120
Current (mA)	27	27
ε^x(n, rms) (π·mm·mrad)	0.140	0.145
ε^y(n, rms) (π·mm·mrad)	0.140	0.150
ε^ℓ(n, rms) (π·mm·mrad)	1.43	1.44

DTL length	41.66 m
Number of cells	218
E_0	4.8 MV/m
DTL lattice	FODO
Synchronous phase	-40°
Total rf power including beam power	14.4MW
Duty factor	0.04%

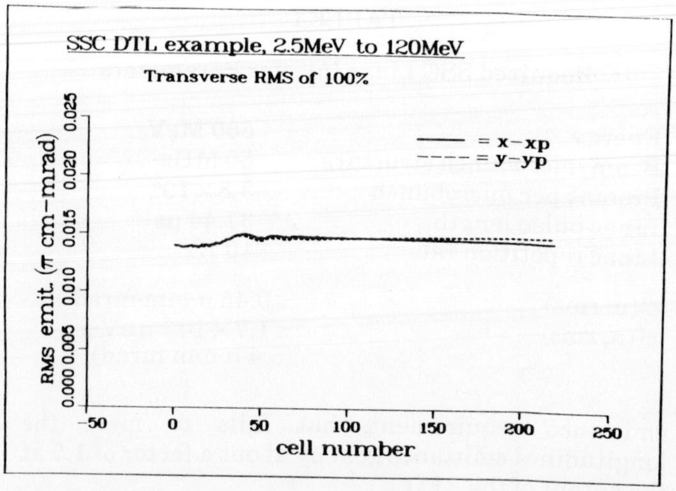

Figure 3. Transverse emittance vs cell number for the 450-MHz DTL

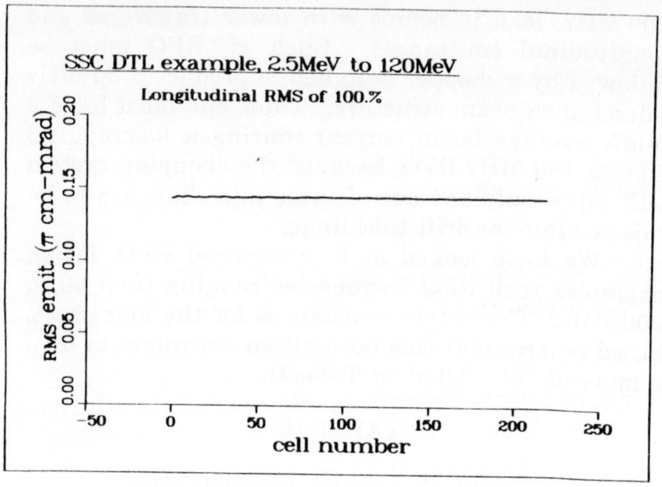

Figure 4. Longitudinal emittance vs cell number for the 450-MHz DTL.

4. The 1350-MHz CCL

The matching section between the 450-MHz DTL and the 1350-MHz CCL is 1.7 m long and consists of one buncher tank and four quadrupoles. The coupled-cavity linac takes the energy from 120 to 600 MeV in a total length of 94.1 m including spacings between tanks. This linac design is for 81-mA operation. The CCL lattice chosen for this design example is also FODO and uses an accelerating gradient $(E_0T)\cdot\cos\ \phi_s$ = (6.5 MV/m)·cos (-30°). The copper power for this CCL design example is estimated at ~64 MW. Simulations show no beam loss and no emittance growth. The 600-MeV beam phase-space plots are shown in Fig. 5. Side-coupled-linac and an on-axis-coupled-linac are two options for this CCL. More work is needed to be able to pick one of these two options. CCL parameters obtained from particle simulations are given in Table V.

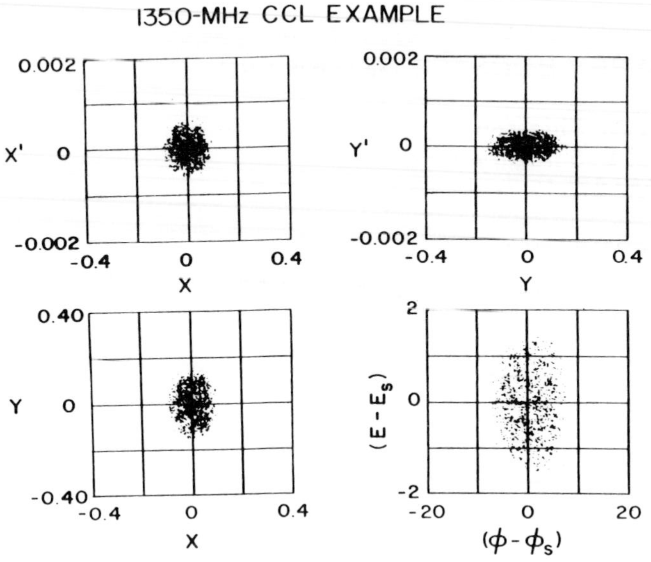

Figure 5. Phase-space plots at 600 MeV.

TABLE V

Simulations for the 1350-MHz
120 to 600-MeV CCL Design Example for the SSC
Linac Injector

	Input	Output
Energy (MeV)	120	600
Current (mA)	81	81
ε^x(n, rms) (π·mm·mrad)	0.15	0.15
ε^y(n, rms) (π·mm·mrad)	0.15	0.15
ε^ℓ(n, rms) (π·mm·mrad)	1.44	1.44

CCL length	94 m
Number of tanks	117 with 10 cells per tank.
E_0T	6.5 MV/m
CCL lattice	FODO
Synchronous phase	-30°
Total rf power including beam power	65.4 MW
Duty factor	0.04%

SUMMARY

Beam dynamics designs for the components of the proposed SSC linac injector system show that the LEB requirements can be satisfied. The suggested system is intended only as an example. The designs for the individual components have not been optimized.

The complete SSC linac injector system is ~145 m long with the total rf power requirements approaching 80 MW. The transverse and longitudinal emittances at 600 MeV are ε^t(n, rms) ~ 0.15 π·mm·mrad and ε^ℓ(n, rms) ~ 1.45 π·mm·mrad, respectively, which are substantially lower than the LEB requirements stated above.

R. Diebold has suggested[4] that a future SSC upgrade should aim at a luminosity increase of 10. This would call for an increase of $\sqrt{10}$ in the beam current delivered by the linac injector system to the LEB. PARMTEQ and PARMILA simulations show that the three accelerating elements of this linac injector system, i.e., the 150-MHz RFQ, the 450-MHz DTL, and the 1350-MHz CCL, can carry beam currents approximately three times higher. Simulations show that whereas transverse emittance at 600 MeV for the three times higher beam would increase by about a factor of two, the longitudinal emittance would remain roughly the same. The increased transverse emittance is still well within the LEB requirements.

ACKNOWLEDGMENTS

Discussions with P. W. Allison, E. P. Colton, J. W. Hurd, R. R. Stevens, and H. A. Thiessen are gratefully acknowledged.

REFERENCES

1) Conceptual Design of the Superconducting Super Collider, SSC Central Design Group, SSC-SR-2020, March 1986.

2) L. K. Chen and M. A. Furman, "A Possible New Design of the SSC Boosters," SSC-164.

3) E. P. Colton and H. A. Thiessen, "H⁻ Injection into the Low Energy Booster of the SSC," LA-UR-88-2532.

4) R. Diebold, "Extending the Reach of the SSC," proceedings of the 1986 Summer Study on the Physics of the Superconducting Supercollider, June 23-June 11, 1986, Snowmass, Colorado, p. 585.

H^- INJECTION INTO THE LOW-ENERGY BOOSTER OF THE SSC

E. P. Colton and H. A. Thiessen
Los Alamos National Laboratory
Los Alamos, NM 87545

ABSTRACT

Protons are accumulated into the low-energy booster of the SSC by utilizing $H^- \rightarrow H^+$ conversion in a 225-μg/cm^2 carbon stripping foil. Synchronous injection is performed for 26 turns into stationary rf buckets, thereby allowing operation with variable bunch spacing. By injecting the beam offset in x and y we obtain the required rms normalized transverse emittance area of 0.75 π mm-mr. Similarly the required rms longitudinal emittance area of $1.75\pi \times 10^{-3}$ eVs is obtained by injecting single linac micropulses, centered at $\phi = 0$, and $dp/p = +0.12\%$, into each 49.9-MHz rf bucket formed with an rf voltage of 350 kV. The transverse space-charge tune shift is -0.17 for 10^{10} protons/bunch accumulated at 600 MeV.

INTRODUCTION

The function of the low-energy booster (LEB) is to accumulate 10^{10} protons per bunch into as many as 72 bunches, and accelerate them up to a final momentum of 8.45 GeV/c for transfer into the medium energy booster (MEB). This process takes place at a repetition rate of 10 Hz. For each bunch the required beam quality is for a rms normalized transverse emittance ε_t of 0.75 mm-mr and rms longitudinal phase space area A_z of $1.75\pi \times 10^{-3}$ eVs. The input linac beam is planned to be of order 3.8×10^8 H^- per microbunch with $\varepsilon_t = 0.45$ mm-mr and $A_z = 1.7\pi \times 10^{-5}$ eVs (actually these may be smaller). Therefore some scheme must be incorporated to increase the linac emittances to the required final values during the injection. In this paper we describe how the required beam quality is obtained during the injection process. Multiparticle simulations are carried out to demonstrate the efficiency of the process.

THE MACHINE

Figure 1 shows a plan view of the latest version of the ring.[1] It is in the shape of a racetrack with two 180° arcs with dispersion suppressors, and two long straight sections, each of length 30.83 m. The straight sections are used for beam transfers and rf cavities. Table I lists the basic ring parameters.

THE INJECTION STRAIGHT SECTION

The injection will take place in the 6-m drift in the center of the long straight section. Figure 2 shows an elevation view of the H^- injection and circulating proton beam

Fig. 1. Plan view of the Low Energy Booster. Circumference is 342.71 m.

Table 1. Parameters of the Low-Energy Booster.

Circumference (m)	342.711
Momentum Range (GeV/c)	1.22–8.45
Repetition Rate (Hz)	10.0
Number of Protons/Pulse	7.2×10^{11}
Betatron Tunes Q_x, Q_y	11.84, 11.78
Chromaticity Q'_x, Q'_y	−15.3, −15.6
Transition Gamma	10.32
RF Harmonic	72

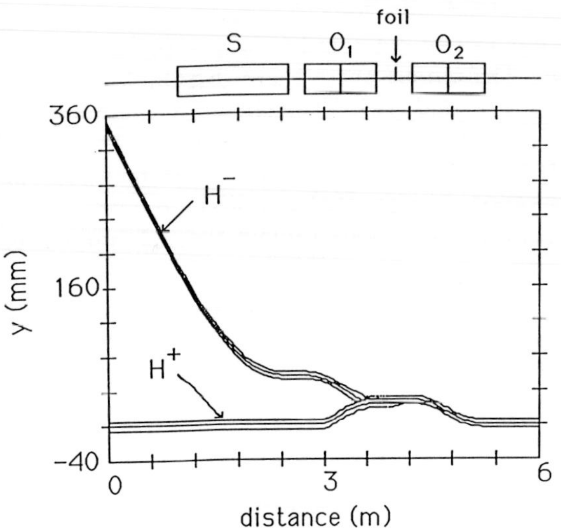

Fig. 2. Elevation view of beam envelopes in the 6.0-m injection straight section.

envelopes in that section. During the injection period the circulating proton beam is displaced vertically by the orbit bumps 0_1 and 0_2 and traverses the stripping foil at a vertical position 3 1/3 cm above beam centerline. The incoming H^- are sent downward through a current septum S and into 0_1. The initial point is at $y = 347$ mm and $y' = -166.67$ mr. The magnetic fields in S, 0_1 , and 0_2 are held at 0.45 T. The magnetic lengths are 1.5 m, 1.0 m, and 1.0 m, respectively. The H^- strip to H^+ in the foil at about 3.5 cm above beam centerline. Greater than 95% stripping efficiency is obtained with a carbon foil with thickness $\rho\ell = 225\ \mu g/cm^2$. After the injection, the bumps are rapidly reduced to zero so the protons cease to traverse the foil during the remainder of the cycle. The lower edge of the foil can be at, e.g., $y = 2.5$ cm.

INJECTION INTO STATIONARY BUCKETS

The SSC users reserve the option for variable bunch spacing. These spacings would be the nominal 4.75 m, as well as 9.5 m and 19.0 m. In the LEB these three cases correspond to all buckets filled, every other bucket filled, or every fourth bucket filled. Furthermore the "empty buckets" should really be free of beam. These requirements preclude use of the standard fill and adiabatic capture technique. Clearly the most efficient filling mechanism is injection of a single microbunch into an existing 49.9-MHz bucket.

It was decided to use 350 kV/turn rf voltage during the injection. A multiparticle simulation was used to study the injection process. Figure 3(a) shows the position and extent of the injected micropulses into the existing buckets. The center of the beam is placed at $[\phi_c, (dp/p)_c] =$

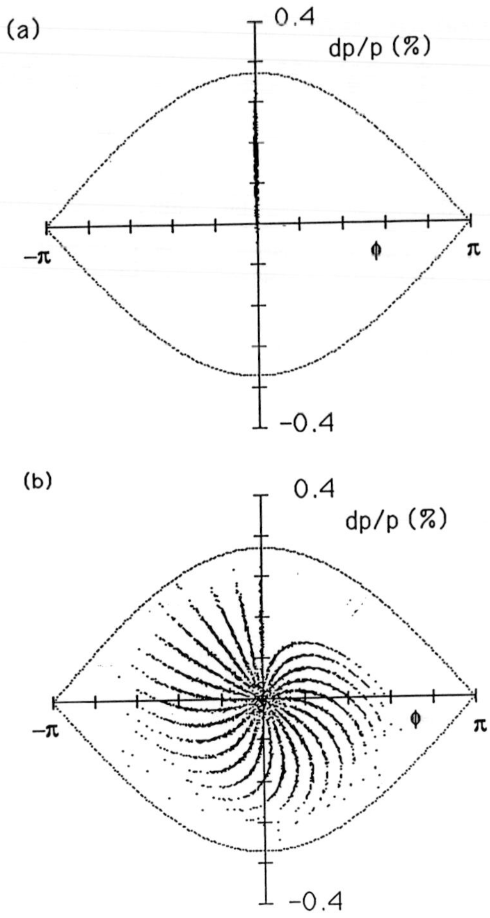

Fig. 3. (a) Injected microbunch within an existing 49.9-MHz rf bucket. (b) Longitudinal phase space after 26 turns of injection. Rms single bunch area $A_z = 1.82\pi \times 10^{-3}$ eVs.

(0,+0.12%), i.e., the beam is injected higher in momentum by 1.2 MeV/c. The rms area of the microbunch was taken to be the linac value $A_z = 1.7\pi \times 10^{-5}$ eVs. The width and height of the microbunch were optimized to $\sigma_\phi = 0.0107$ (relative to the 49.9-MHz rf) and $\sigma_p/p = 0.05\%$. To fill longitudinal phase space we inject for 26 turns (roughly one synchrotron oscillation). Figure 3(b) shows the longitudinal phase space as predicted by the simulation program. The rms bunch area is $1.82\pi \times 10^{-3}$ eVs, just slightly larger than the desired value $A_z = 1.75\pi \times 10^{-3}$ eVs. The ϕ and dp/p projections of Fig. 3(b) are shown in Figs. 4(a) and 4(b),

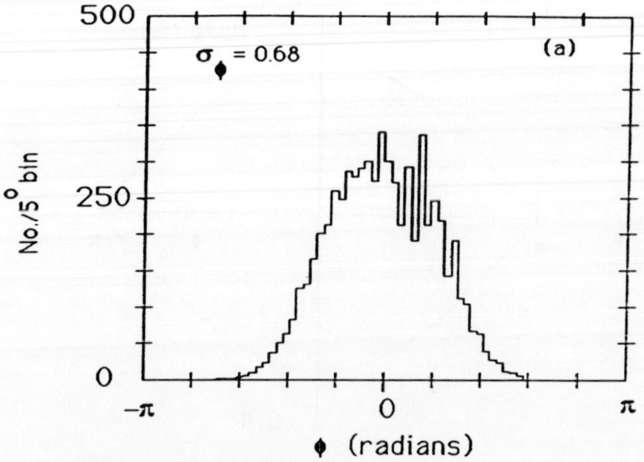

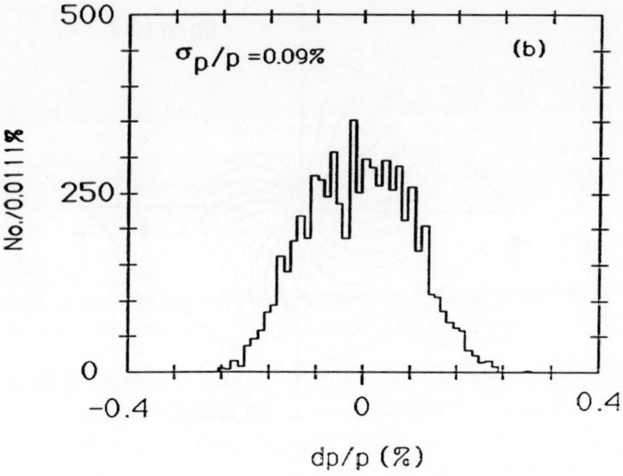

Fig. 4. Projections of Fig. 3(b): (a) $dN/d\phi$; (b) $dN/dp/p$.

respectively. The bunching factor is of order 33% and the rms relative momentum spread σ_p/p is 0.09%.

THE TRANSVERSE MATCH

The output linac rms normalized emittance (ε_t = 0.45 mm-mr) is only slightly less than the desired ring emittance (ε_t = 0.75 mm-mr). In practice injection errors and mismatches will bring them nearly into agreement; nevertheless, we outline an idealized scenario and predict the parameters using the multiparticle simulation program. All quantities are evaluated at the foil location. This point is 1 m downstream of center of the long straight section. For the simulation it was assumed the linac beam was matched to that of the machine at the foil location. The microbunches were taken to be bi-gaussian in x-x' and y-y' phase spaces with rms normalized emittances ε_t = 0.45 mm-mr. In x-y space the microbunches were injected at average x, y values of 1.3 mm and 35.0 mm, respectively; the bottom foil edge was at y = +25 mm. The vertical bump was fixed at 33.3 mm so relative to beam center the injected H^- was injected 1.7 mm high. The injected central divergences x', y' were both assumed to be zero. The injection was carried out over 26 turns with the simulation program. The output phase space distributions are presented in Figs. 5(a) and 5(b). The normalized rms emittances were calculated to be in the range of 0.76–0.78 mm-mr which meets the required specification of 0.75 mm-mr. The x and y projections of Figs. 5(a) and 5(b) are both approximately Gaussian with rms value 1.6 mm and 2.2 mm, respectively.

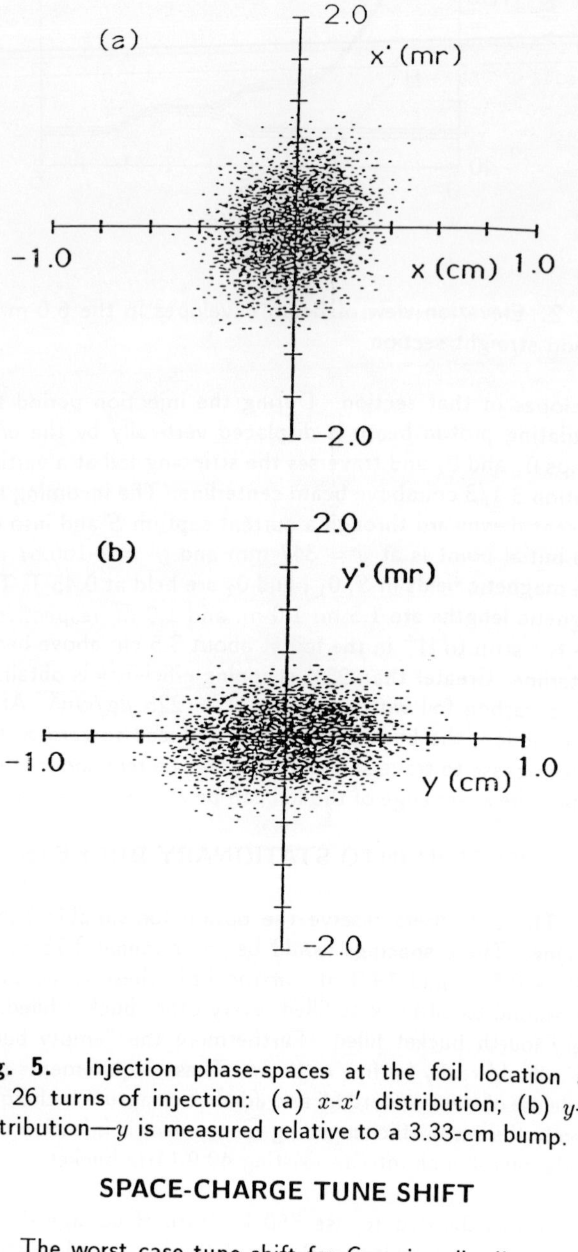

Fig. 5. Injection phase-spaces at the foil location after 26 turns of injection: (a) x-x' distribution; (b) y-y' distribution—y is measured relative to a 3.33-cm bump.

SPACE-CHARGE TUNE SHIFT

The worst case tune shift for Gaussian distributions is given by

$$\Delta Q_y = -\frac{r_p N_b N_p}{4\pi\beta\gamma^2 B\varepsilon_t}$$

where r_p = classical proton radius (1.54×10^{-18} m), N_b = number of bunches (72), N_p = number of protons/bunch (10^{10}), B = bunching factor (0.33), and ε_t = rms normalized transverse emittance (0.75 mm-mr). We obtain $\Delta Q_y = -0.17$ which is certainly within specifications. It is less, in fact, than the quoted CDR value.

COMMENTS

The injection scheme put forth here should easily meet the requirements for 600 MeV H^- injection. If necessary, the present design could work up to 800 MeV H^- injection. If the linac energy is upgraded even further (in order to reduce the transverse emittances), then the injection straight probably has to be lengthened.

REFERENCE

1. L. K. Chen and M. A. Furman, "A Possible New Design of the SSC Boosters," SSC-164.

A NOVEL METHOD FOR CORRECTING THE SSC MULTIPOLE PROBLEM *

David Neuffer

AT-6, Los Alamos National Laboratory, Los Alamos, NM 87545

ABSTRACT

A new method of correcting dynamic nonlinearities due to the multipole content of a large synchrotron such as the Superconducting Super Collider is discussed. The method uses lumped multipole elements placed at the center (C) of the accelerator half-cells as well as elements near the focusing (F) and defocusing (D) quadrupoles at the ends. In a first approximation, the corrector strengths follow Simpson's Rule. Correction of second-order sextupole nonlinearities may also be obtained with the F, C, and D octupoles. Correction of nonlinearities by approximately three orders of magnitude is obtained, and a solution to a fundamental problem in large synchrotrons is demonstrated.

A large synchrotron such as the Superconducting Super Collider (SSC)[1] requires adequate linearity for beam stability and reliable operation. Linear motion is required over a working region sufficient to include the beam size and momentum spread with closed orbit deviations and injection errors. This linear aperture is severely limited by the SSC multipole content, and the nonlinearities are greatly magnified by the large circumference.

Nonlinearity can be measured by the amplitude and momentum-dependent tune shifts per turn $\Delta\nu_x$, $\Delta\nu_y$. An SSC linear aperture tolerance (SLAT) may be set by requiring $\Delta\nu_x$, $\Delta\nu_y \leq \pm 0.005$ for orbits with amplitudes A_x, A_y, up to 0.5 cm in the SSC arcs and with momentum offsets $\delta \equiv \frac{\Delta p}{p} \leq \pm 0.001$.[1] Except for short utility and interaction regions (IR), the SSC circumference is composed of ~320 alternating-gradient cells consisting of long dipoles, short focusing (F) and defocusing (D) quadrupoles, and short corrector magnets (see Fig. 1). The nonlinear magnetic fields in the dipoles dominate the nonlinear motion and they may be represented by

$$B_y + iB_x = B_o\{1 + \sum [b_n(s) + ia_n(s)](x+iy)^n\} \ ,$$

where B_o is the bending field and $b_n(s)$ and $a_n(s)$ are the normal and skew multipole components. The transverse motion may be described by a Hamiltonian,

$$H = \frac{I_x}{\beta_x(s)} + \frac{I_y}{\beta_y(s)} + \Re \sum_n \frac{B_o}{B\rho} \frac{[b_n(s) + ia_n(s)](x+iy)^{n+1}}{n+1} \ ,$$

where $\beta_x(s)$ and $\beta_y(s)$ are the betatron functions[2] of linear motion. The coordinates x and y of particle

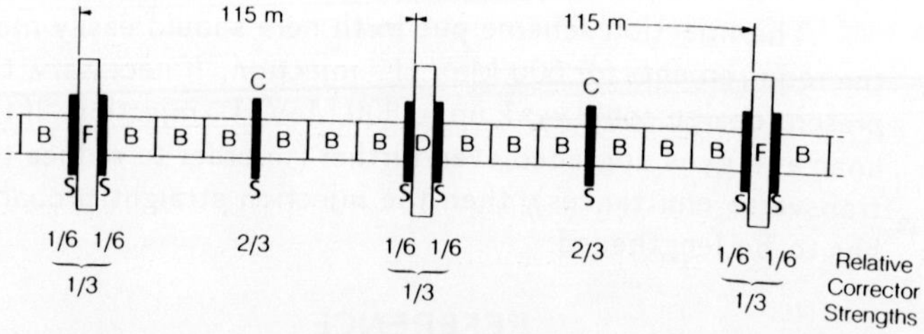

Fig. 1. A symmetrical SSC cell. The element labels are: B - dipoles, F, D - quadrupoles, S - slots for correctors, C - center corrector slot. The correctors on opposite sides of the F and D quads may be combined on either side and exact symmetry is not necessary.

motion are represented in action-angle variables (I,ϕ) by $x = \sqrt{2\beta_x I_x}\cos(\phi_x) + \eta\delta, y = \sqrt{2\beta_y I_y}\cos(\phi_y)$; the off-momentum orbit at δ determined by the dispersion $\eta(s)$ is included. The terms $A_x = \sqrt{2\beta_x I_x}$ and $A_y = \sqrt{2\beta_y I_y}$ are the amplitudes. The tune shifts are obtained by averaging the phase advance caused by the field perturbation

$$\Delta\nu_{x,y} = \frac{1}{2\pi}\int \frac{d\phi_{x,y}}{ds} ds = \Re\langle \frac{dH}{dI_{x,y}} \rangle \ . \quad (1)$$

In first order in the coefficients b_n and a_n only systematic normal multipoles ($\overline{b_n}$) contribute; the resulting expressions for the tune shifts as a function of I_x, I_y, and δ due to sextupole (b_2), octupole (b_3) and decupole (b_4) components are

$$\begin{aligned}
\Delta\nu_x = {} & \langle b_2\beta_x\eta\delta\rangle + \langle \frac{3}{4}b_3\beta_x^2 I_x - \frac{3}{2}b_3\beta_x\beta_y I_y + \frac{3}{2}b_3\beta_x\eta^2\delta^2\rangle \\
& + \langle 3b_4\beta_x^2\eta I_x\delta - 6b_4\beta_x\beta_y\eta I_y\delta + 2b_4\beta_x\eta^3\delta^3\rangle \\
\Delta\nu_y = {} & -\langle b_2\beta_y\eta\delta\rangle + \langle \frac{3}{4}b_3\beta_y^2 I_y - \frac{3}{2}b_3\beta_x\beta_y I_x - \frac{3}{2}b_3\beta_y\eta^2\delta^2\rangle \quad (2) \\
& + \langle 3b_4\beta_y^2\eta I_y\delta - 6b_4\beta_x\beta_y\eta I_x\delta - 2b_4\beta_y\eta^3\delta^3\rangle \ .
\end{aligned}$$

The SLAT may be applied to Eqs. (1) and (2) to obtain tolerance limits[3] on uncorrected $|\overline{b}_n|$; see Table I.

The SSC dipoles are expected to have significant multipole content, particularly in the $\overline{b}_n$ with n even; those multipoles are allowed in dipole symmetry. Estimates[5] of the systematic and rms random multipole strengths as extrapolated from the similar Tevatron dipoles [6] or calculated from persistent current and saturation effects[1] have been collected in Table I where they are compared with SLAT values. Serious deficiencies in $\overline{b}_2, \overline{b}_3$, and $\overline{b}_4$ are observed. Initial SSC design included sets of multipole

* Work supported by the U. S. Department of Energy, High Energy and Nuclear Physics.

TABLE I. Tolerances and estimated strengths of systematic multipole content in the SSC dipoles. All multipole strengths are in units of 10^{-4} cm^{-n}. The tolerances are obtained from the SLAT. Estimated strengths are extrapolated from Tevatron data or calculated from the magnet properties.

Multipole	Tolerance in SSC Lattice (230 m, 90° Cells)	Estimated Random Error (Tevatron)	Systematic Strength (Tevatron)	Persistent Current Multipole Strength	Saturation Multipole Strength
b_2	0.0097	2.0	0.45	-4.7	1.2
b_3	0.017	0.35	-0.14	-	-
b_4	0.031	0.60	-0.33	0.30	-0.05
b_5	0.054	0.06	-0.024	-	-
b_6	0.096	0.08	1.57[a]	0.07	-0.01
b_7	0.17	0.16	0.009	-	-
b_8	0.29	0.02	-2.1[a]	<0.02	0.02

[a] The higher allowed multipoles (b_6, b_8) were not minimized in the Tevatron design; the SSC conductor placement should reduce these within tolerances.[4]

trim coils within every dipole for correction of $\bar{b}_2, \bar{b}_3$, and $\bar{b}_4$, but they greatly complicate the dipoles and may be impractical.

The correction is simplified if it is implemented using short correctors separate from the dipoles. Initial attempts used correctors placed near F and D quads, where chromatic (momentum-dependent) correction sextupoles are placed. Because there are only two first-order b_2 terms, F and D sextupoles can completely correct them; however, second-order sextupole effects are important (see below). For b_3, b_4, and higher multipoles, there are five or more terms and they cannot be completely corrected. The F and D correction is quite inadequate and can reduce the b_3 and b_4 nonlinearities only by a factor of ~2 (Table II).[7,8]

A great improvement is obtained by adding a corrector to the center (C) of each half-cell (Fig. 1). An optimization converges very close to a characteristic solution. For example, a particular corrected b_3 tune-shift term from Eq. (2) may be written as

$$\Delta\nu_x = \frac{3I_x}{4L}\left\{\int_0^L b_3(s)\beta_x^2(s)ds + \left[\frac{S_{3,F}}{B_o}\beta_x^2(0) + \frac{S_{3,C}}{B_o}\beta_x^2(L/2) + \frac{S_{3,D}}{B_o}\beta_x^2(L)\right]\right\} \quad (3)$$

All first-order terms are of similar form. The corrector strengths $S_{n,i}$ are defined by $S_{n,i} \equiv B_{n,i}l_i = -f_{n,i}B_o\bar{b}_nL$, where $B_{n,i}$ and l_i are the corrector lengths and strengths, and L is the half-cell length. The Simpson's Rule[9] solution is $f_F = f_D = 1/6$ and $f_C = 4/6$ per half-cell; it corrects all b_3 and b_4 nonlinearities by two orders of magnitude. Optimization about that solution permits another order of magnitude reduction (see Table II).

The F, C, D correction [Eq. (3)] is equivalent to approximating integrals of powers of betatron functions by a sum over discrete points, and Simpson's Rule is a third-order integration that is very accurate for smoothly varying functions. Figure 2a shows β_x, β_y, and η over a full cell; they are smoothly varying over half-cells with a derivative discontinuity at the quadrupoles. All the terms that appear in b_2, b_3, and higher-order tune shifts

TABLE II. First-order correction of b_3, b_4 and second-order b_2 correction. The correction factor is the ratio of uncorrected to corrected $\Delta\nu$ in the SLAT aperture. The tolerance is the maximum corrected b_n permitted under the SLAT.

Correction Condition	Correction Factor	Tolerance (10^{-4}cm^{-n})
b_3 (Octupole) Correction		
No correction	1.0	0.018
F, D chromatic correction[7] (f_F = 0.28, f4D = 0.70)	1.9	0.033
F, C, D Simpson's Rule (f_F, f_C, f_D) = (1/6, 4/6, 1/6)	93	1.6
F, C, D correction (0.165, 0.66, 0.165)	370	6.7
b_4 (Decupole) Correction		
No correction	1.0	0.029
F, D chromatic correction[7] (f_F, = 0.24, f_D = 0.93)	1.4	0.04
F, C, D Simpson's Rule (f_F, f_C, f_D) = (1/6, 4/6, 1/6)	31	0.9
F, C, D correction (0.158, 0.663, 0.168)	800	24
Second Order b_2 (Sextupole) Correction		
No correction	1.0	1.2
F, D chromatic b_2 correction	5.1	2.7
F, C, D chromatic b_2 correction, equal weights (f_C = 0.5)	24	5.0
F, C, D chromatic correction, Simpson's Rule (f_C = 0.667)	23	5.7
F, D first-order b_2 correction and ($f_{C,2}$ = 0), F, C, D octupoles	120	13
F, C, D first-order b_2 correction ($f_{C,2}$ = 0.5 to 0.67), and F, C, D octupoles	700	32

are smoothly varying on the half-cell level (Figs. 2a and 2b). The F, C, and D correctors provide three-point, Simpson's Rule type cancellation, which is well suited to the correction of all first-order terms.[8] The first-order correction is insensitive to machine perturbations, because it requires only smooth variation on the half-cell level. Second-order effects are also reduced by first-order correction. The method is easily extended to include other multipoles (b_5, b_6, etc.), if necessary. Random multipole effects can be reduced by varying the correctors to follow

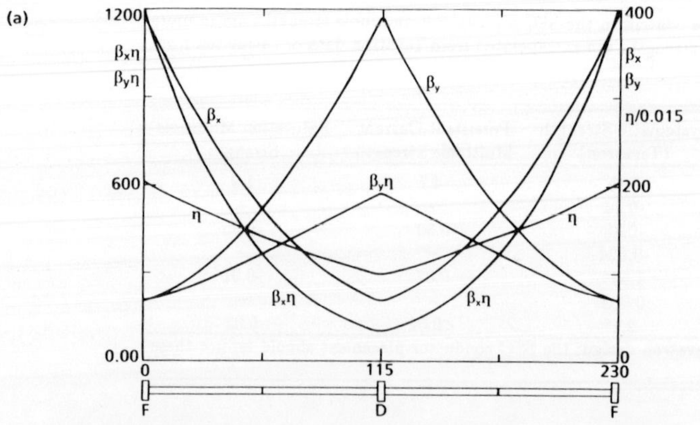

Fig. 2.a) Betatron functions (β_x, β_y, η) for a full SSC cell. The functions that appear in the sextupole tune shifts $(\beta_x\eta, \beta_y\eta)$ are also shown. Note the derivative discontinuity and the reflection symmetry at the D quadrupole.

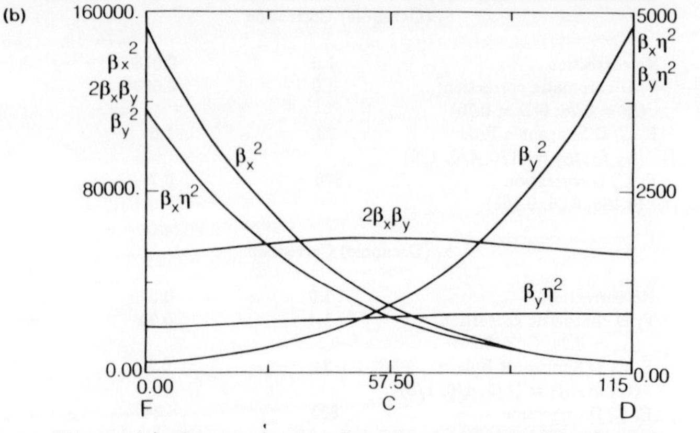

Fig. 2.b) Octupole $\Delta\nu$ functions on a half-cell. The $\beta_x{}^2$ and $\beta_x\eta^2$ terms are derived from x^4 in the Hamiltonian; $2\beta_x\beta_y$ and $\beta_y\eta^2$ from ($-x^2y^2$) and $\beta_y{}^2$ from y^4.

the local multipole content.[10] The method can be extended to correct quadrupole multipole content, particularly in the long IR quads.

Superconducting dipoles have a very large b_2 content, and second-order terms are important. From perturbation theory,[11–13] second-order sextupole tune-shift terms are of the same form as the first-order octupole terms:

$$\Delta\nu_x = aI_x + bI_y + c\delta^2 \text{ and } \Delta\nu_y = dI_y + bI_x + e\delta^2 \quad . \quad (4)$$

The coefficients a through e scale as b_2^2 and are positive after first-order correction. The expression for a in a simplified lattice is

$$a = \frac{C}{32\pi L}\int_0^{2L} ds\beta_x(s)^{3/2}B_2(s)\int_s^{s+2L} ds'\beta_x(s')^{3/2}B_2(s') \times \left\{3\frac{\cos[\psi_x(s')-\psi_x(s)-\pi\mu_x]}{\sin(\pi\mu_x)} + \frac{\cos 3[\psi_x(s')-\psi_x(s)-\pi\mu_x]}{\sin(3\pi\mu_x)}\right\} ,$$

where C is the ring circumference, $B_2(s)$ is the normalized sextupole strength, $\psi_x(s)$ is the betatron phase, μ_x is the cell tune, and $\Delta\nu_x$ is the full-ring tune shift. These terms are double integrals with phase factors and a discontinuity at $s = s'$ and are not closely fitted to Simpson's Rule integration. The first-order b_2 correction with only F and D elements reduces second-order terms by a factor of ~5. Addition of a C sextupole corrector reduces these terms by another factor of ~5, increasing b_2 tolerance to an acceptable level, but without the Simpson's Rule improvement seen in first order (Table II).

Second-order b_2 correction can be greatly improved by using the F, C, D octupoles. Unlike b_2^2 terms, first-order octupole $\Delta\nu$ terms have opposing signs [Eq. (2)], and there are only three correctors for five terms. However, the octupole Hamiltonian has three terms; the $\Delta\nu$ terms that derive from the same Hamiltonian term have similar dependences (Fig. 2b). The correction strategy is to use the C octupoles to correct the b and e terms in Eq. (4) and the F and D octupoles to correct the others; the ratios of F, C, and D strengths per half-cell are $\sim (1 : -2.7 : 1)$. The correctors can correct completely either amplitude or chromatic $\Delta\nu$ with the remanent terms reduced by $\geq 10\times$. The optimum reduction is by a factor of ≥ 30, increasing $\overline{b}_2$ tolerances to $\geq 30 \times 10^{-4}\text{cm}^{-2}$ (see Table II). Other nonlinear effects such as orbit distortion remain small, provided cell resonances are avoided. Because the octupole tune shifts are linear, there is no interference between their b_2^2 and b_3 correction roles. The use of F, C, and D octupoles to correct second-order sextupole nonlinearities adds an extra operational dimension, conceptually similar to the use of F and D sextupoles to correct quadrupole chromaticity. Their use may be extended to control nonlinearities from other elements; for example, all A_i^2 tune shifts can be cancelled to zero, regardless of their sources.

Simple solutions have thus been found for formidable problems of SSC nonlinearities. First-order nonlinearities due to multipoles are correctable by a system of F, C, D elements, with initial strengths determined by the generally valid Simpson's Rule. Second-order sextupole effects can be cancelled with F, C, D octupoles. Tolerances on SSC fields can be increased from the impractical 10^{-6} level to 10^{-3} cm^{-n}. The same corrections can be applied to any synchrotron, with similar improvements.

Acknowledgment

I acknowledge valuable assistance from P. Channell, A. Chao, E. Forest, J. Peterson, P. Limon, A. Dragt, and R. Talman.

References

1) Superconducting Super Collider Conceptual Design, SSC-SR-2020, March 1986.
2) E. D. Courant and H. S. Snyder, *Ann. Phys.* **3**, 1 (1958).

3) D. Neuffer and J. M. Peterson, Proc. 1987 IEEE Particle Accelerator Conf., E. R. Lindstrom and L. S. Taylor, Eds., IEEE Catalog No. 87CH2387-9, p. 1502 (1987).
4) S. Caspi, W. Gilbert, M. Helms, L. J. Laslett, and C. Taylor, *IEEE Trans. Magn.* **MAG-23**, 1219 (1987).
5) Superconducting Super Collider report SSC-7, April 1985, J. M. Peterson, Ed.
6) R. Hanft, B. C. Brown, W. E. Cooper, D. A. Gross, L. Michelotti, E. E. Schmidt, and F. Turkot, *IEEE Trans Nucl. Sci.* **30**, p. 3381 (1983).
7) A. Jackson, Superconducting Super Collider report SSC-107, February 1987.
8) D. Neuffer, "Lumped Correction of Systematic Multipoles in Large Synchrotrons," to be published in *Particle Accelerators* (1988).
9) M. Abramowitz and I. A. Stegun, *Handbook of Mathematical Functions,* National Bureau of Standards, Washington, DC (1972).
10) J. M. Peterson and E. Forest, Superconducting Super Collider report SSC-N-383 (1987), to be presented at the European Particle Accelerator Conference, Rome, 1988.
11) S. Ohnuma, in *Intersections between Particle and Nuclear Physics,* R. Mischke, Ed., *AIP Conf. Proc.* **123** (American Institute of Physics, New York, 1985), p. 415.
12) R. D. Ruth, in *Physics of Particle Accelerators,* M. Month and M. Dienes, Eds., *AIP Conf. Proc.* **153** (American Institute of Physics, New York, 1987), p.152.
13) L. Michelotti, in *Physics of Particle Accelerators,* p. 260, (1987).

STUDIES OF ONE-CAMPUS AND TWO-LEVEL CONFIGURATIONS FOR THE SSC

L. C. Teng
Fermi National Accelerator Laboratory*, Batavia, Ill. 60510

ABSTRACT

The geometries and the magnet lattices for an SSC where the desired eventual six interaction regions are provided on two bypass branches in the same campus is studied. The single campus approach has many obvious advantages. In addition, it would be natural to build the campus and the remainder of the ring which is simply a big beam transport, on different levels each optimized for its own functional requirements. Such a two level configuration is also investigated. Injection into such rings is discussed and their incremental costs are estimated.

In this report we investigate in some detail the geometry and the cost of two modifications in the configuration of the SSC.

1. One-campus configuration — the possibility and the advantages of this configuration were first pointed out by L. Lederman[1)] and later further elaborated by L. Lederman and L. Teng.[2)] Assuming, as in the Conceptual Design, that the fully developed and exploited SSC has 6 interaction regions (IR's) equipped with detectors and that for about half of the time a detector needs to be serviced and/or modified, one should then have a two-branch bypass with 3 IR's on each branch. And to minimize the campus length one should inject into 2 utility straight sections located on the trunk line just beyond the ends of the bypass, denoted as Case C in Ref. 2. The odd number of vertical crossings in each branch brings opposite rings to the top-side for identical injection geometry from above (or below) and yields good overall symmetry between the two rings. This implies, of course, that there must be another crossing (presumably without collision) in the long straight section diametrically opposite from the Campus.

2. Two-level configuration — With the one-campus configuration the circumference of the collider is clearly divided into two sectors with totally different functions. The Campus sector serves all the active functions such as injection, acceleration, beam abort, beam collision and the performance of physics experiments with massive detectors. The remaining sector and by far the greater part of the circumference is just a big beam transport and serves only the inert function of channeling the beam leaving the Campus back to the Campus. Because of the difference in function the optimal physical arrangements for the active "Campus" and the inert "Transport" sectors may be quite different. Specifically, the optimal elevation for these two sectors are different: the active Campus should be easily accessible and hence closer to the grade level, and the inert Transport should be deeper underground to minimize interference with the use and habitation of the ground above.

*Operated by Universities Research Association, Inc. under contract with the U.S. Department of Energy.

For the basic SSC we assume the following parameters[3)] (each ring):

Phase advance per normal cell	90°
Half cell length, l	114.25 m
No. of dipoles per half cell	6
Bend angle per dipole, θ	1.64 mrad
Length of Interaction Region (IR) or Utility Region (UR)	$11l$
Length of Curved Region (CR) between IR's	$9l$
Bend angle per CR	50θ (82 mrad)
Length of Arc Region (AR)	$295l$
Bend angle per AR	1766θ (2896 mrad)
Total number of dipoles	3832
Circumference	$732l$ (83631 m)

We also need an all purpose matching region (MR). Such an MR is shown in Fig. 1. The section is composed of four 90° cells with 6 half-cells filled with bending dipoles. This section matches zero dispersion ($D = D' = 0$) to zero dispersion and has unity transfer matrices in both planes independent of the length l_M and the bend angle θ_M which can be adjusted to match the geometry. We shall identify a specific matching region by MR(l_M , θ_M).

ONE-CAMPUS CONFIGURATION (ONE-LEVEL)

Symmetric Bypass

The principal function of the CR in the basic SSC lattice is to eliminate interference from muons produced in neighboring IR's. But as we shall see later (Appendix A) the CR as specified is longer than necessary. For the symmetric bypass shown in Fig. 2 we choose to use MR ($6l$, 24θ)'s to connect IR's. The IR length is held fixed so that the beam combining geometry and the low-β and the medium-β quadrupole arrangements are unchanged. The same is true with the UR. The parameters of the MR are chosen so that the end Interaction Points (IP's) are separated by a convenient distance. With these choices the parameters of the symmetric bypass are (each ring):

Minimum length of Campus to include all IP's

$$34l = 3.9 \text{ km}$$

Separation of the end IP's = s = 76.4 m

Increment of total tunnel length over the basic SSC

$\Delta L = 123l = 14.1$ km

Increment of total bend angle over the basic SSC

$\Delta\theta = 192\theta = 315$ mrad

RMS μ-beam radius at neighboring IP $= r_\mu = 4.6$ m

(Beam/IP separation $= 38$ m $> 8r_\mu$)

Unipolar Bypass

In this case a larger bend angle between the end UR's and IR's helps to create the separation between the branches. Therefore we will keep the basic CR's with bend angle $\theta_C = 50\theta$ at these locations as shown in Fig. 3. The separation between the branches is then created by inserting the two staggered straight sections of lengths l_I (inner branch) and l_O (outer branch). The most straightforward way to match across l_I and l_O is to use the straight matching sections MR (l_I,0) and MR (l_O,0). We shall assume the same separation s = 76.4 m for easy comparison with the symmetric case. This value is also close to the minimum required to accommodate the large detector halls on both IP's. The parameters so obtained are:

Outer branch straight length

$$l_O = \frac{\cos\theta_C}{\sin\theta_C}s = 930.0 \text{ m} = 8.14l$$

Inner branch straight length

$$l_I = \frac{\cos 2\theta_C}{\sin\theta_C}s = 920.6 \text{ m} = 8.06l$$

Minimum length of Campus to include all IP's

$34l = 3.9$ km

Separation of the IP's = s = 76.4 m

Increment of total tunnel length over the basic SSC

$\Delta L = 139.6l = 16.0$ km

Increment of total bend angle over the basic SSC

$\Delta\theta = 148\theta = 243$ mrad

RMS μ-beam radius at:

Nearest neighboring IP $= r_{\mu 1} = 4.6$ m

(Beam/IP separation $= 38$ m $> 8r_{\mu 1}$)

Second neighboring IP $= r_{\mu 2} = 7.2m$

(Beam/IP separation $= 76$ m $> 10r_{\mu 2}$)

TWO-LEVEL CONFIGURATION

We assume that the active "Campus" of the ring, the section that includes the 6 IR's, is horizontal and at an elevation d_V above that of the inert "Transport" which is also horizontal. We need a vertical dogleg at each end of the Campus to produce the elevation difference.

Symmetric Bypass

Referring to Fig. 4 we see that the end section ABCD is made into a symmetric dogleg in a plane tilted at an angle α from the horizontal. Such a dogleg will produce a vertical displacement. The matching region CE should be on a plane tilted at angle $\pi - \alpha$ from horizontal. This reverses the vertical component of the field and leaves the horizontal component unchanged. Thus the horizontal projections of CD and CE will bend in opposite directions and the vertical projection of ABCE will be identical to that of ABCD.

We will keep the axes of the quadrupoles untilted (no rotation about the beam axis) so that any coupling of beam optics in the two transverse planes is introduced only by the skewed bending of the dipoles. It should be easy to compensate for this weak coupling by the trim skew quadrupoles which are already included in the lattice.

The Tilted Regions AB, CD and CE are denoted by TR and are formed by the very versatile all purpose matching regions with lattice given in Fig. 1. The length and bend angle will be denoted by l_T and θ_T. The projected horizontal (θ_H) and vertical (θ_V) bend angles are given by (see Appendix B).

$$\tan\theta_H = \tan\theta_T\cos\alpha$$

$$\tan\theta_V = \tan\theta_T\sin\alpha$$

The total horizontal and vertical excursion d_H and d_V from A to D of the dogleg are given by

$$d_H = \left(2l_T\frac{1-\cos\theta_T}{\theta_T} + l_U\sin\theta_T\right)\cos\alpha$$

$$d_V = \left(2l_T\frac{1-\cos\theta_T}{\theta_T} + l_U\sin\theta_T\right)\sin\alpha$$

For all our cases θ_T, θ_H and θ_V are small enough so that these formulas can be approximated as

$$\theta_H = \theta_T\cos\alpha \qquad d_H = (l_T + l_U)\theta_T\cos\alpha$$

$$\theta_V = \theta_T\sin\alpha \qquad d_V = (l_T + l_U)\theta_T\sin\alpha$$

If the same SSC dipoles are used for the MR, only 36 dipoles can be accommodated in a length $8l$. Hence we have the additional condition

$$l_T = \frac{8}{36}\frac{\theta_T}{\theta}l$$

For a numerical example we use the following values.

1. $\theta_H = 24\theta$. We assume that the projected horizontal bend angle remains the same as in the one-level configuration.
2. $U = 11l$. The length of the end utility straights is unchanged so that the matching and the injection configurations are the same as in the one-level case.

3. $d_V = l$. For some of the proposed sites the bedrock is ≈ 400 ft below grade. If one assumes that the Transport section is in the rock (best for tunneling) and the Campus section is near the grade level $d_V = l = 114.25$ m is about right.

From the above equations we have

$$d_V = (l_T + l_U)\theta_T \sin\alpha = \left(11 + \frac{8}{36}\frac{\theta_H}{\theta_T}\right) l\theta_T \sqrt{1 - \left(\frac{\theta_T}{\theta}\right)^2}$$

or

$$\frac{d_V}{l} = \left(11 + \frac{8}{36}\frac{\theta_T}{\theta}\right)\sqrt{\theta_T^2 - \theta_H^2}$$

$$= \theta\left(11 + \frac{8}{36}\frac{\theta_T}{\theta}\right)\sqrt{\left(\frac{\theta_T}{\theta}\right)^2 - 24^2}$$

With $d_V/l = 1$ this gives the following parameters for the TR:

Bend angle $= \theta_T = 39.17\theta = 64.22$ mrad

Tilt angle $= \alpha = 0.9113 = 52.2°$ (and $127.8°$)

Projected horizontal bend angle $= \theta_H = 24\theta = 39.35$ mrad

Projected vertical bend angle $= \theta_V = 30.95\theta = 50.75$ mrad

Length $= l_T = 8.70l = 994$ m

and the following parameters for the ring:

Vertical excursion of dogleg = level difference between

Campus and Transport $= l = 114.25$ m $= 375$ ft

Increment of total tunnel length over the one-level

configuration $= \delta L = 35l = 4.0$ km

Increment of total bend angle over the one-level

configuration $= \delta\theta = 91\theta = 149$ mrad

Unipolar Bypass

Referring to Fig. 5 we see that the outer branch end section ABCDE is topologically the same as the mirrored dogleg of the symmetric bypass denoted as ABCE in Fig. 4, except in this case the central straight length l_U is replaced by $l_U + l_O$ and we want the projected horizontal bend angle θ_H of the tilted regions AB and DE to be 50θ, the same as that of the one-level configuration. We also keep l_O and l_I unchanged so as to get the same separation between the two branches.

The inner branch end section ABCFG is a little more complicated. The tilted region CF should be a helix with fixed pitch angle θ_V and projected horizontal bend angle $\theta_H = \theta_C = 50\theta$. The bend angle θ_S along the helix is given by (see Appendix B)

$$\sin\theta_S = \sin\theta_H \cos\theta_V$$

But for values considered here it is a good approximation to simply set

$$\theta_S = \theta_H = 50\theta$$

To bend the orbit back to horizontal we need a vertical MR (l_I , θ_V). It is easy to see that this will bring the 3 inner branch IR's in a horizontal plane roughly on the same elevation as the 3 outer branch IR's. The TR parameters are:

Bend angle $= \theta_T = 53.71\theta = 88.07$ mrad

Tilt angle $= \pi - \alpha = 2.7676 = 158.6°$

Projected horizontal bend angle $= \theta_H = 50\theta = 81.98$ mrad

Projected vertical bend angle $= \theta_V = 19.62\theta = 32.18$ mrad

Length $= l_T = 11.93l = 1363$ m

and the parameters of the two-level ring are

Vertical excursion of dogleg = level difference between

Campus and Transport $= l = 114.25$ m $= 375$ ft

Increment of total tunnel length over the one-level

configuration $= \delta$L $= 21.4l = 2.44$ km

Increment of total bend angle over the one-level

configuration $= \delta\theta = 7.42\theta = 12.2$ mrad

COSTS, REMARKS AND CONCLUSIONS

Several remarks are in order.

1. We have not worked out the transport across the long straight section diametrically opposite the Campus.
   The design should, however, be straightforward.
2. We have not adjusted the circumferences to give integer rf harmonic numbers. But with the very large circumferences this adjustment should be trivially small.
3. In the end UR of the top ring the injection and the switching between branches can both be accommodated in the central ± 330 m long free drift space. One possibility is shown in Fig. 6. The ring orbit is switched between the branches by a switching magnet W_D located near the right end of the drift space which deflects the orbit by $\pm 1/2$ mrad. The magnet W_D is an integral part of the ring and has to ramp up with all the ring magnets.

   The injection beam at 1 TeV enters from the left at a downward angle of, say, 3 mrad + 60 μrad and is switched to follow the active branch, say branch 1, by the upstream injection switching magnet W_U and the tilted ramped septum S_1. W_U deflects the beam by $\pm 1/2$ mrad in the beam plane and S_1, deflects the beam horizontally 1 mrad and vertically 3 mrad (total deflection $\sqrt{10}$ mrad tilted at angle $\tan^{-1}$ (1/3) from vertical). After S_1 the beam approaches the branch 1 orbit vertically downward at a 60 μrad angle and crosses the orbit just beyond W_D.

There it is deflected vertically 60 μrad onto the ring orbit by a vertical fast kicker K_V. The strengths of these magnet elements are given in Fig. 6 and are all quite modest. The dispersions generated by these small deflections although small, should be included in the detailed matching design.

The end UR of the lower ring has an identical orbit switching system but no injection system.

4. For injection into the two end UR's the most convenient location for the ≈ 1 km radius High Energy Booster (HEB) is in the middle of the Campus straddling the two branch lines as shown in Fig. 7. The major advantage of this arrangement is that bipolar operation of the HEB is not needed. Beams extracted from diametrically opposite straight sections of HEB are channeled by long but relatively straight transport lines to the end UR's. These transport lines and the tunnels housing them could be quite simple and inexpensive.

   If the HEB is capable of bipolar operation one can extract oppositely circulating beams from the same straight section and inject them into neighboring utility straights of the SSC rings as in Case A of Ref. 2. This arrangement, in addition to requiring the hard bipolar operation of a superconducting HEB, would need a longer Campus.

5. With the injection geometry shown in Fig. 7, for all 4 configurations — symmetric and unipolar bypasses, one- and two-levels — the minimum size of the Campus to enclose the injector and all 6 IR's is about 5 km long by 2.5 km wide. To this one must add the abort/external beams area, the arc areas, the service and access areas, and the necessary easements.

6. To make some rough estimates of the cost differentials we shall use the 1987 unit-costs given by the Central Design Group, namely

   Tunnel cost (CT): \$4M/km

   Magnet cost (CM): \$0.15/dipole + 20% for quadrupoles

   EDIA = 15% (CT + CM)

   Contingency = 25% (CT + CM + EDIA)

The incremental costs for the one-campus and the two-level configurations are given in the table below (two rings).

	One-Campus			Two-Level		
Symmetric bypass	ΔL	=	14 km	δL	=	4 km
	ΔM	=	2 × 192	δM	=	2 × 91
	CT	=	\$56 M	CT	=	\$16 M
	CM	=	\$69 M	CM	=	\$33 M
	EDIA	=	\$19 M	EDIA	=	\$ 7 M
	Cont.	=	\$36 M	Cont.	=	\$14 M
			\$180 M			\$70 M
Unipolar bypass	ΔL	=	16 km	δL	=	2 km
	ΔM	=	2 × 148	δM	=	2 × 54
	CT	=	\$64 M	CT	=	\$ 8 M
	CM	=	\$53 M	CM	=	\$19 M
	EDIA	=	\$18 M	EDIA	=	\$4 M
	Cont.	=	\$34 M	Cont.	=	\$8 M
			\$169 M			\$39 M

This table shows that the incremental cost of the Unipolar Bypass configuration is slightly less. On the other hand, the Symmetric Bypass configuration is simpler and neater and has many construction, alignment and operation advantages which are well worth the slightly higher cost.

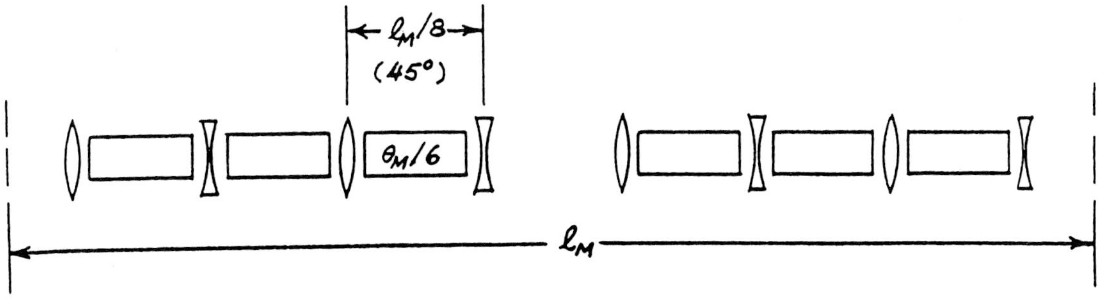

Fig. 1. An all purpose matching region denoted as MR(l_M, θ_M) which matches zero dispersion to zero dispersion and has unity transfer matrices in both transverse planes and adjustable length l_M and bend angle θ_M.

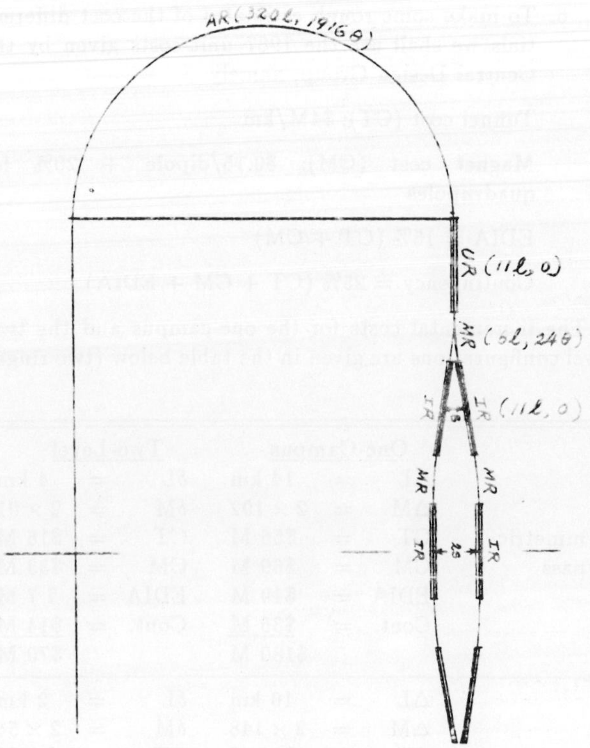

Fig. 2. One-campus one-level configuration with symmetric bypass (not to scale).

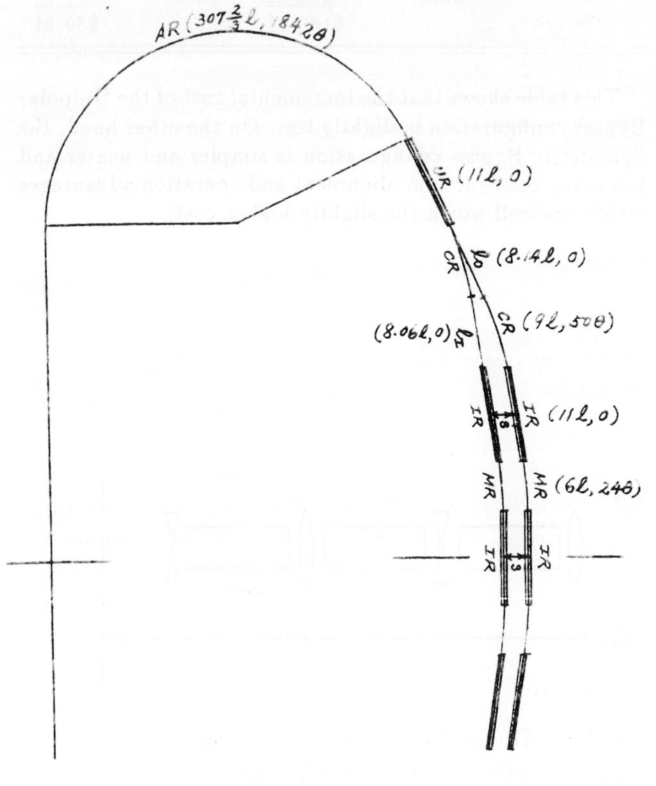

Fig. 3. One-campus one-level configuration with unipolar bypass (not to scale).

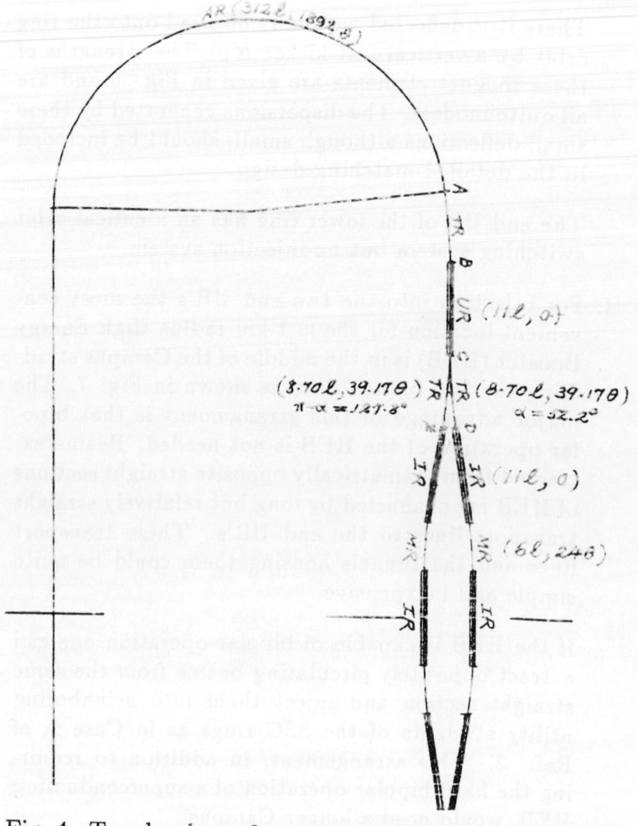

Fig. 4. Two-level configuration with symmetric bypass (not to scale).

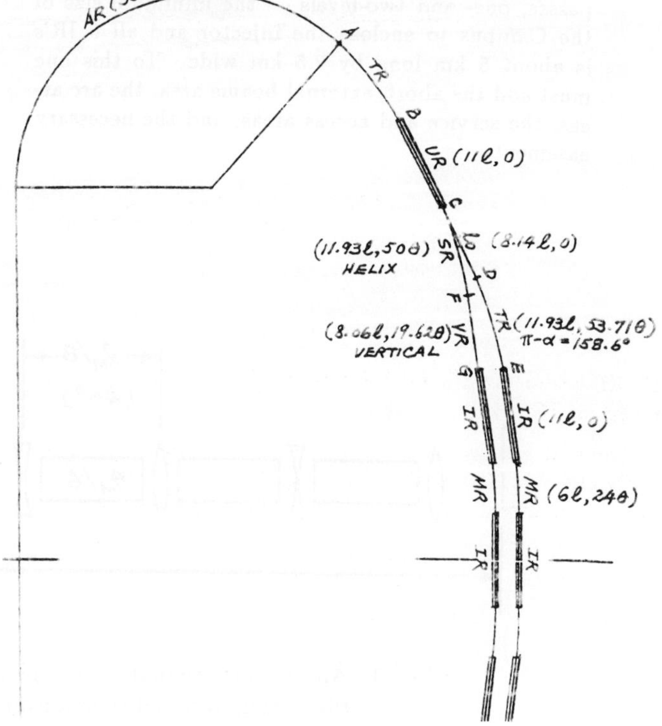

Fig. 5. Two-level configuration with unipolar bypass (not to scale).

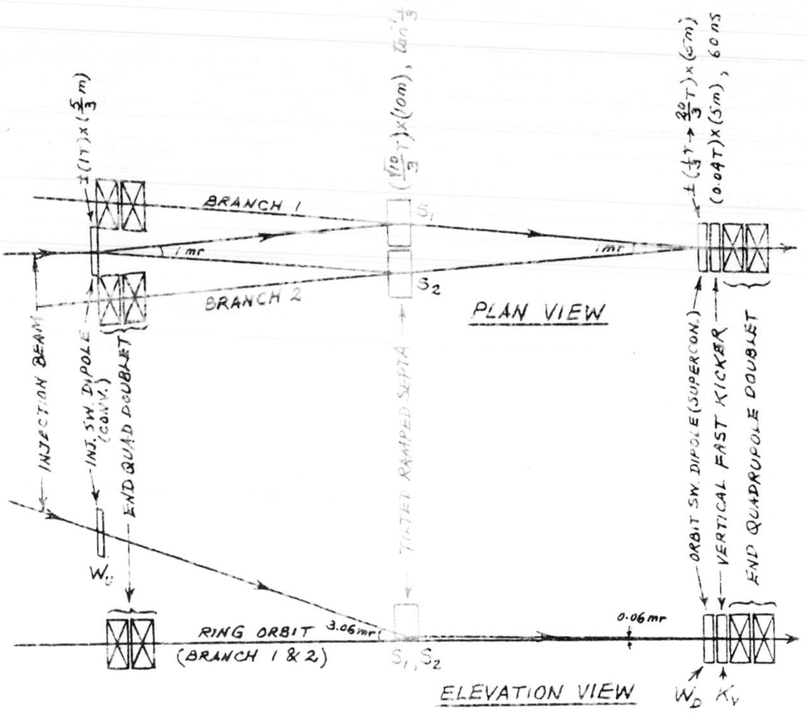

Fig. 6. Injection and branch-switching systems in the central drift space of an end utility region (not to scale).

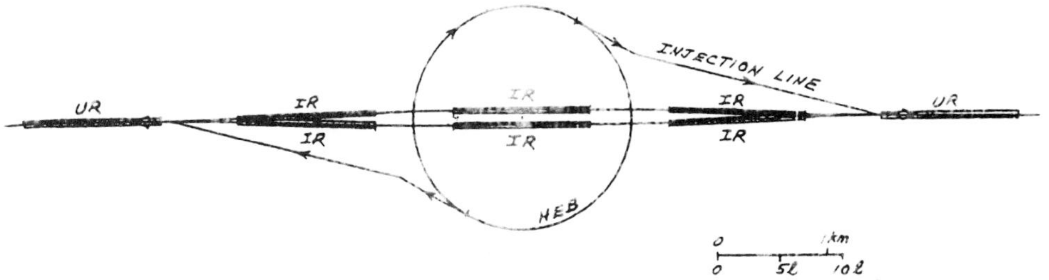

Fig. 7. Optimal positioning of a unipolar high energy booster injector and the injection transport lines.

APPENDIX A

RMS Cone Angle and Radius of the Muon Beam from an IP

The high energy muons from an IP are produced essentially at 0° (or 180°). When they go through a thickness l_μ of earth shielding all muons with parallel momentum

$$p_\parallel < \left(0.4 \frac{\text{TeV}}{c} / \text{km}\right) l_\mu$$

are ranged out. Here we have taken

$$\frac{dE}{dx} \sim 2 \text{ MeV/g/cm}^2, \quad \text{and}$$
$$\text{Density of earth} \sim 2 \text{ g/cm}^3$$

The muons that penetrated the earth shielding will have gained an rms perpendicular momentum through multiple Coulomb scattering of the amount

$$p_\perp = (1.5\text{GeV}/c)\sqrt{l_\mu(\text{km})}$$

Where we have taken the radiation length of earth to be ~ 25 g/cm^2. Thus, the rms cone angle of the muon beam is

$$\theta_\mu = \frac{p_\perp}{p_\parallel} = \frac{4 \text{ mrad}}{\sqrt{l_\mu(\text{km})}}$$

and the rms radius of the μ-beam after going through l_μ length of earth is

$$\sigma_\mu = \theta_\mu l_\mu = (4\text{m}) \sqrt{l_\mu(\text{km})}$$

All IP's should be located several σ_μ away from one another's muon beams.

APPENDIX B

Geometry of Tilted Dogleg Orbit and Helix Orbit

1. One half of the dogleg is shown in Fig. B1. The orbit is the curve ABC, A being the start and C being the midpoint of the dogleg. The angles involved are

Tilt angle = $\alpha = \angle AOD$

Bend angle = $\theta_T = \angle AOB = \angle BPQ$

Projected horizontal bend angle = $\theta_H = \angle EPQ$

Projected vertical bend angle = $\theta_V = \angle GPQ$

The relationship between the angles are

$$\tan\theta_H = \frac{EQ}{PQ} = \frac{BQ}{PQ}\cos\alpha = \tan\theta_T\cos\alpha$$

$$\tan\theta_V = \frac{GQ}{PQ} = \frac{BQ}{PQ}\sin\alpha = \tan\theta_T\sin\alpha$$

The lateral excursions are

$$d_H = 2\ DU = 2\ AW\cos\alpha$$
$$= 2\left(AB\frac{1-\cos\theta_T}{\theta_T} + BC\sin\theta_T\right)\cos\alpha$$
$$= 2\left(l_T\frac{1-\cos\theta_T}{\theta_T} + \frac{l_U}{2}\sin\theta_T\right)\cos\alpha$$

$$d_V = 2\ FV = 2AW\sin\alpha$$
$$= 2\left(l_T\frac{1-\cos\theta_T}{\theta_T} + \frac{l_U}{2}\sin\theta_T\right)\sin\alpha$$

2. A section of a helix orbit is shown in Fig. B2 as AE. Plane ABCD is tangent to the orbit cylinder and AC is tangent to the orbit at A. Similarly EFGH is the tangent plane and EG is tangent to the orbit at E. We have also made AC=EG. Plane ABCD is parallel-translated to EIKH. The angles are

Orbit bend angle from A to E = $\theta_S = \angle KEG$

Horizontal projection of bend angle = $\theta_H = \angle IEF$

Vertical pitch angle of helix = $\theta_V = \angle KEI = \angle GEF$

The relation between these angles is

$$\frac{\sin\theta_S}{\sin\theta_H} = \frac{GK/EK}{FI/EI} = \frac{EI}{EK} = \cos\theta_V$$

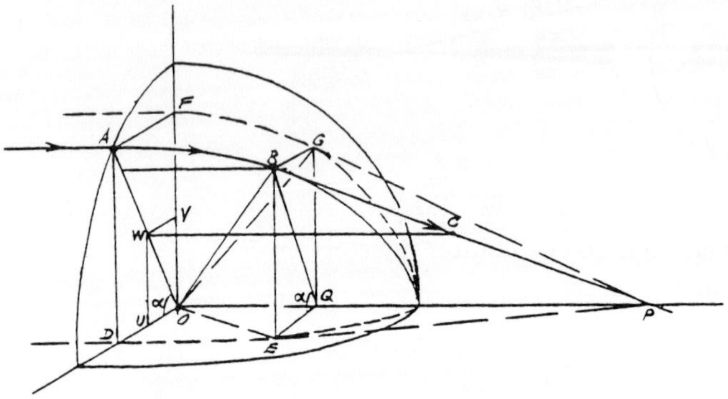

Fig. B1. Geometry of a tilted dogleg orbit.

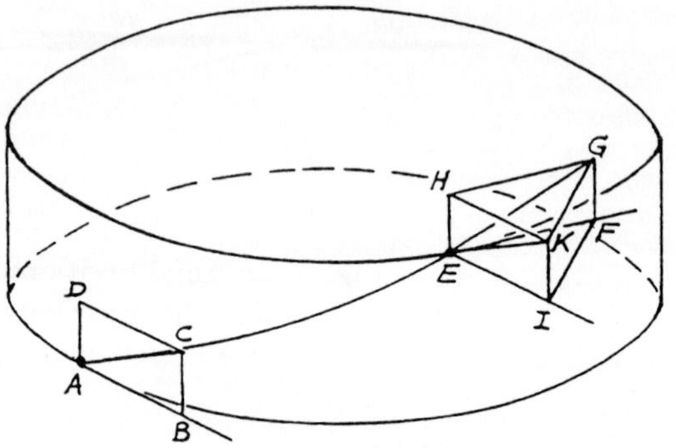

Fig. B2. Geometry of a helix orbit.

REFERENCES

1. L. M. Lederman "A One Lab SSC Configuration" *Proceedings of the 1986 Summer Study on the Physics of the SSC*, p. 518

2. L. M. Lederman and L. C. Teng "A One-Campus SSC" Fermilab report **TM–1452** (April 1987)

3. A. A. Garren and D. E. Johnson "The 90° (September 1987) SSC Lattice" **SSC–146**.

STUDY OF LONG TERM STABILITY OF BETATRON MOTION IN THE SSC*

G.F. Dell
Brookhaven National Laboratory
Upton, NY 11973, USA

Abstract - The long term stability of betatron motion in the presence of random and systematic multipoles in the dipoles and random multipoles in the quadrupoles is reported for the collision lattice used for the SSC CDR (Conceptual Design Report). Long tracking runs of 20K and 28K show no instability.

INTRODUCTION

Informal reports on tracking by Collins[1] at the 1988 Snowmass Summer Study indicated that high order systematic multipoles in the tevatron dipoles can lead to large amplitude motion after long periods of time. A simple model consisting of a transfer matrix plus multipole kicks was used. Although the particle motion seemed stable and well behaved for thousands of turns, the presence of high order multipoles (b_n with $n \cong 20$) sometimes produced unstable motion after tens of thousands of turns. This behavior is consistent with effects seen in the Tevatron -- it was suggested that similar studies should be made for the SSC.

PROCEDURE

Two long term runs have been made using the SSC collision lattice described in the CDR[2]. The first run is an extension of a study made for the 1986 Snowmass Summer Study for which complete emittance transfer between the x and y planes was observed. The betatron amplitude is near the dynamic aperture -- x_o = 3.6 mm, y_o = 2.1 mm, and $\Delta P/P=0$. The second study is more realistic for the collision optics; the betatron amplitude is x_o = 0.25 mm, y_o = 0.14 mm and $\Delta P/P$ = -0.1%.

Particle tracking is performed with PATRICIA[3]. The coordinates of the closed orbit for tracking and subsequent analysis are determined internally by tracking a small amplitude particle for 400 turns. The averages of x, x′, y, and y′ are computed for each of the last ten turns, and the average of the maximum and minimum of these ten averages is used to represent the closed orbit.

The coordinates x, x′, y, and y′ of the particle are stored in memory and are then processed at the end of the run. The betatron amplitudes, Δx, $\Delta x'$, Δy, and $\Delta y'$, relative to the closed orbit, are determined for each turn, and the corresponding emittance

$$\varepsilon_z = (\Delta z^2 + (\alpha_z * \Delta z + \beta_z * \Delta z')^2)/\beta_z$$

is evaluated for z = x and y. The maximum of $\sqrt{\varepsilon_z}$ during a run is used to monitor the growth of the motion. At the end of a run, the final coordinates x, x′, y, and y′ are written to an external binary file. The run can be continued by using these coordinates as the initial coordinates for the next run.

*Work performed under the auspices of the U.S. Department of Energy.

FIRST STUDY

Long term runs have been made using the SSC collision lattice described in the CDR. Previous studies on this lattice showed complete transfer of emittance back and forth between the horizontal and vertical planes during a tracking run of 5,000 turns. This motion is near the dynamic aperture; the initial conditions were purposely selected to make the test more sensitive to instabilities.

Random multipoles (σa_n, σb_n) are present in all dipoles and quadrupoles of the lattice; n =2 to 5 for quadrupoles, and n =2 to 10 for dipoles. The random multipoles in the insertion triplets are assumed to be 90% corrected -- as was assumed for the CDR. The allowed systematic multipoles, b_{2n} with $1 \le n \le 5$, are "tolerance" values at injection from the CDR[4] -- they are multipoles that produce a tune shift $\delta\nu \le 0.005$ when A_β=5 mm and $\Delta P/P \le 0.1\%$. The multipoles are listed in Table 1.

The dipoles used to bring the beams into collision are of the FNAL type, and appropriate multipoles are used. Since these dipoles are rotated by 90° to provide vertical deflection, the multipoles have to be reassigned. The rule used is that normal magnets never have poles in the bending plane, and skew magnets always have poles in that plane. The result of rotating by ±90° a dipole having a downward directed field is shown in Table 2; the pattern repeats with n, modulo 4.

The signs in Table 2 have been ignored when generating random multipoles; they have been observed for systematic multipoles, Sign changes of systematic multipoles for up and down bends are automatically adjusted, since the sign of the bending radius changes.

With the tune set at ν_x=78.276, ν_y=78.273, a series of seven runs, each containing 4000 turns, has been made using the restart option mentioned previously. The dependence of $\sqrt{(\varepsilon_y/\varepsilon_{y(o)})}$ on

	Arc Dipoles			Arc Quads		Vertical Bends	
n	b_n^{sys}	σb_n	σa_n	σb_n	σa_n	σb_n	σa_n
2	.0076	2.14	.63	2.9	2.6	.186	.48
3		.35	.69	.5	1.2	.048	.095
4	.018	.59	.14	.3	.3	.012	.031
5		.059	.16	.56	.14	.003	.006
6	.04	.078	.034				.002
7		.016	.030				
8	.089	.021	.0064				2E-4
9		.003	.0056				
10	.19	.007	.0012				3E-5

Table 1 Magnet multipoles (10^{-4} at 1 cm) used in the first study. Dipole multipoles refer to Bo = 6.6 T; quadrupole multipoles refer to the field at 1 cm (2.12 T). Multipoles of the IR triplets are 0.1 x multipoles of arc quads.

n	bn -90	bn +90	an -90	an +90
1	-b1	-b1	-a1	-a1
2	+a2	-a2	-b2	+b2
3	+b3	+b3	+a3	+a3
4	-a4	+a4	+b4	-b4

Table 2 Multipoles resulting from rotation of bn and an by ±90°. (Positive rotation is counter-clockwise).

$\sqrt{(\varepsilon_x/\varepsilon_{x(o)})}$ during the first 4000 turns (smear plot) is shown in Figure 1. The dotted horizontal and vertical lines at $\sqrt{(\varepsilon_z/\varepsilon_{z(o)})} = 1$, for z = x and y, indicate the initial emittance in each plane. The maximum value of $\sqrt{(\varepsilon_z/\varepsilon_{z(o)})}$ during each run is used to monitor the growth of the betatron motion. $\Delta\sqrt{(\varepsilon_z/\varepsilon_{z(o)})}$ denotes the change in $\sqrt{(\varepsilon_z/\varepsilon_{z(o)})}$ compared to its maximum value during the first 400 turns of the study. This difference is plotted on Figure 3(a) for each of the 4000 turn runs. There is no indication of growth during the series of runs.

SECOND STUDY

A second study has been made using the limits for the collision lattice -- A_β =0.25 mm and $\Delta P/P$=-0.1%. More recent multipole information[5] is included with n = 2 to 9 in all quadrupoles. The IR triplets are assumed to have $\sigma b_n' = \sigma a_n' = 0.05$ (units of 10^{-4} at 1 cm) for n=2 to 5 and no correction for multipoles with n>5. In all elements $\sigma a_1=\sigma b_1=0$. The random b_2 in the dipoles has been reduced by a factor of five to the value currently used by the CDG[6]. All vertical bending magnets used to bring the beams into collision have 7.5 cm coil i.d.'s. The multipoles of these dipoles are scaled[5] from the 4 cm dipoles using the relation $b_n(7.5)=b_n(4)*(4/7.5)^{n+0.5}$ and the assignments of Table 2. The multipoles are listed in Table 3.

As before, a series of 4000 turn runs was made. The "smear" plot from the first 4000 turn run is shown in Figure 2. As before, the maximum of $\sqrt{(\varepsilon_z/\varepsilon_{z(o)})}$ is used to detect any change relative to its maximum value during the first 400 turns of the study. $\Delta(\sqrt{(\varepsilon_z/\varepsilon_{z(o)})})$ is determined and is plotted on Figure 3(b). Again there is no indication of growth throughout the particle history of 20K turns.

	Arc Dipoles			Arc Quads		Vertical Bends	
n	b_n^{sys}	σbn	σan	σbn	σan	σbn	σan
2	.0076	.40	.61	3.35	3.1	.07	.14
3		.35	.69	.58	1.2	.075	.018
4	.018	.59	.14	.32	.34	.005	.013
5		.059	.16	.29	.13	.004	.002
6	.04	.075	.034	.056	.058	.001	.001
7		.016	.030			.000	.000
8	.089	.021	.0064	.025	.021		.000
9		.003	.0056	.021	.019		
10	.19	.007	.0012				.000

Table 3 Magnet multipoles (10^{-4} at 1 cm) used in the second study. Dipole multipoles refer to Bo = 6.6 T; quadrupole multipoles refer to the field at 1 cm (2.12 T). Multipoles of IR the triplets are set to 0.05 primed units for n = 2 to 5.

RESULTS

The present studies are made to test the suggestion that high order systematic multipoles may cause instabilities after a large number of turns (information on realistic multipoles wasn't available -- only "tolerance" values for low and intermediate order multipoles are listed)[4]. The initial conditions used for the first study correspond to betatron amplitudes near the dynamic aperture -- this is an extreme case for the collision optics. The study is a first attempt of locating possible instabilities and is intended to be exploratory. The second study uses a betatron amplitude and a displaced closed orbit consistent with collision optics. Again there is no indication of instability in either of the runs, but the condition of having systematic multipoles of very high order, as suggested by Collins, has not been satisfied. Hence the question of possible instability induced over a "long" period of time by high order systematic multipoles is not resolved.

REFERENCES

1). T.L. Collins, FNAL, unpublished.
2). SSC Conceptual Design Report, SSC-SR-2020, MARCH, 1986.
3). PATRICIA, H. Wiedemann, SLAC (modified 1980 version).
4). SSC CDR. op.cit. Table 4.4-3, pg. 151.
5). Beat T. Leemann, SSC Central Design Group, Report SSC-109, Feb. 1987.
6). ibid. also: R.E. Meller, SSC-N-237.

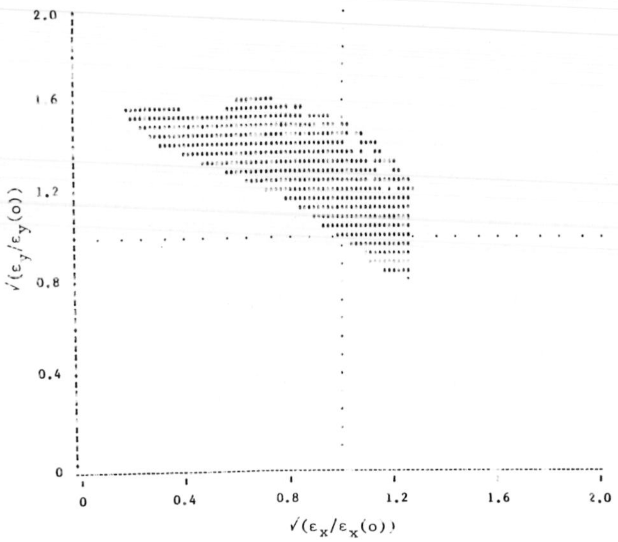

Figure 1. Smear plot ($\sqrt{(\varepsilon y/\varepsilon y(o))}$ vs $\sqrt{(\varepsilon x/\varepsilon x(o))}$) during the first study with random and systematic multipoles.

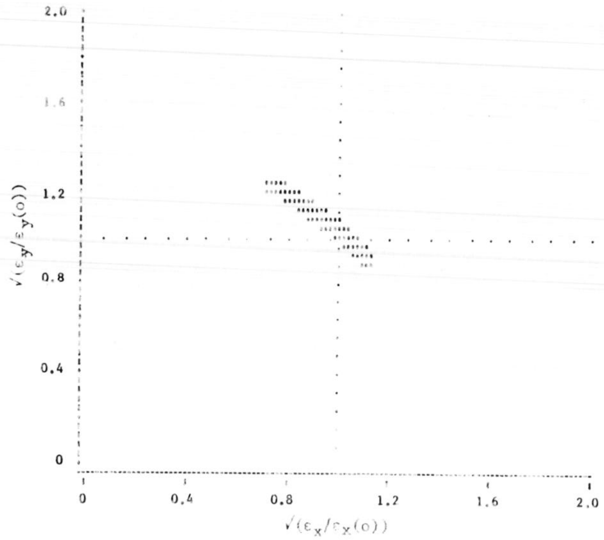

Figure 2. Smear plot ($\sqrt{(\varepsilon y/\varepsilon y(o))}$ vs $\sqrt{(\varepsilon x/\varepsilon x(o))}$) during the second study with random and systematic multipoles.

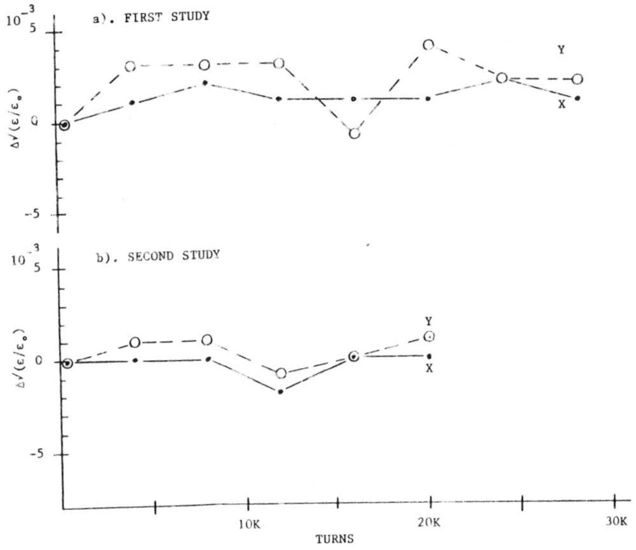

Figure 3. Test for emittance growth: $\Delta\sqrt{(\varepsilon/\varepsilon_o)} = (\sqrt{(\varepsilon z/\varepsilon z(o))})_{max} - (\sqrt{(\varepsilon z/\varepsilon z(o))})_o$ with z = x or y. $(\sqrt{(\varepsilon z/\varepsilon z(o))})_{max}$ is the maximum value during each of the 4000 turn runs, and $(\sqrt{(\varepsilon z/\varepsilon z(o))})_o$ is the maximum value observed during the first 400 turns of the study.

ACCELERATOR PHYSICS – B - FACTORIES

REPORT OF THE B-FACTORY GROUP: I. PHYSICS AND TECHNIQUES*

G. J. Feldman,[(1)†] D. G. Cassel,[(2)†] R. H. Siemann,[(2)†] M. S. Alam,[(3)] R. Aleksan,[(1)] W. B. Atwood,[(1)] J. Bartelt,[(1)] J. J. Bisognano,[(4)] J. R. Boyce,[(4)] D. B. Cline,[(5)] G. Coignet,[(6)] P. S. Drell,[(2)] A. M. Eisner,[(7)] A. Fridman,[(1)] A. A. Garren,[(8)] P. Grosse-Wiesmann,[(1)] D. G. Hitlin,[(9)] S.-O. Holmgren,[(10)] H. Kagen,[(11)] N. Katayama,[(3)] H. Koiso,[(12)] N. B. Mistry,[(2)] T. Nakada,[(13)] P. J. Oddone,[(10)] R. B. Palmer,[(1)] R. S. Panvini,[(14)] D. E. Pellett,[(15)] F. C. Porter,[(9)] J. R. Rees,[(1)] D. L. Rubin,[(2)] E. I. Shibata,[(16)] G. H. Trilling,[(10)] J. J. Welch,[(2)] W. A. Wenzel,[(10)] and M. S. Witherell,[(17)]

[(1)] *Stanford Linear Accelerator Center, Stanford University, Stanford, California 94309*
[(2)] *Cornell University, Ithaca, New York 14853*
[(3)] *State University of New York at Albany, Albany, New York 12222*
[(4)] *Continuous Electron Beam Accelerator Facility, Newport News, Virginia 23606*
[(5)] *University of California, Los Angeles, California 90024*
[(6)] *Laboratoire d'Annecy-le-Vieux de Physique des Particules, F-74019 Annecy-le-Vieux, France*
[(7)] *University of California Institute for Research at Particle Accelerators, Stanford, California 94305*
[(8)] *SSC Central Design Group and Lawence Berkeley Laboratory, Berkeley, California 94720*
[(9)] *California Institute of Technology, Pasadena, California 91125*
[(10)] *Lawrence Berkeley Laboratory and Department of Physics, University of California, Berkeley, California 94720*
[(11)] *Ohio State University, Columbus, Ohio 43210*
[(12)] *National Laboratory for High Energy Physics, Ibaraki 305, Japan*
[(13)] *Paul Scherrer Institut, CH-5234 Villigen, Switzerland*
[(14)] *Vanderbilt University, Nashville, Tennessee 37235*
[(15)] *University of California, Davis, California 95616*
[(16)] *Purdue University, Lafayette, Indiana 47907*
[(17)] *University of California, Santa Barbara, California 93106*

1. INTRODUCTION

The study of B meson decay appears to offer a unique opportunity to measure basic parameters of the Standard Model, probe for interactions mediated by higher mass particles, and investigate the origin of CP violation. These opportunities have been enhanced by the results of two measurements. The first is the measurement of a long B meson lifetime.[1)] In addition to allowing a simpler identification of B mesons and a measurement of the time of their decay, this observation implies that normal decays are suppressed, making rare decays more prevalent. The second measurement is that neutral B mesons are strongly mixed.[2)] This enhances the possibilities for studying CP violation in the B system.

The CESR storage ring is likely to dominate the study of B physics in e^+e^- annihilations for about the next five years. First, CESR has already reached a luminosity of 10^{32} cm^{-1} sec^{-1} and has plans for improvements which may increase the luminosity by a factor of about five. Second, a second-generation detector, CLEO II, will start running in 1989. It has been designed especially to study B meson physics in the 10 GeV region.[3)]

*Work supported in part by the National Science Foundation and the Department of Energy contract number DE-AC03-76SF00515.
† Group co-leader.

Given this background, the main focus of this working group was to ask what is needed for the mid- to late-1990 s. Many laboratories are thinking about new facilities involving a variety of techniques. To help clarify the choices, we focused on one example of CP violation and estimated the luminosity required to measure it using different techniques. This will be the subject of the next chapter. In Chap. 3 we will briefly describe the requirements for detectors matched to these techniques. In particular, we will give a conceptual design of a possible detector for asymmetric collisions (*i.e.*, beams of unequal energy) at the Υ(4S) resonance, one of the attractive techniques which will emerge from this study. A discussion of accelerator technology issues for using these techniques forms the second half of the B–factory Group report, and it follows in these proceedings.[4)]

2. TECHNIQUES FOR STUDYING CP VIOLATION

2.1 Introduction

Any high luminosity B–factory will permit the study of a large variety of issues in B physics including the properties of exclusive B decays, the measurement of $B\bar{B}$ mixing, the determination of Kobayashi–Maskawa (K–M) matrix elements, and the search for rare B decays. However, in

Table 1. Estimation of the number of $B\bar{B}$ events required for CP violation studies at the SSC*

Class	Modes	Branching Ratio	Asymmetry	# $B\bar{B}$ Events Required
I. Charge Asymmetry in	$B_d \bar{B}_d \to l^{\pm} l^{\pm} + X$	0.01	10^{-3}	6×10^{9}
Same Sign Dileptons	$B_s \bar{B}_s \to l^{\pm} l^{\pm} + X$	0.02	10^{-4}	2×10^{12}
II. Mixing with Decay	$B \to \psi K_s$	5×10^{-4}	0.05–0.3	$(1\text{-}34) \times 10^{8}$
to a CP Eigenstate	$B \to \psi K_s X$	2×10^{-3}	0.05–0.3	$(2\text{-}85) \times 10^{7}$
	$B \to D\bar{D}K_s$	5×10^{-3}	0.05–0.3	$(3\text{-}100) \times 10^{7}$
	$B \to \pi^+\pi^-$	5×10^{-5}	0.05–0.5	$(0.3\text{-}32) \times 10^{8}$
	$B \to D^{*+} D^-, D^+D^-, D^* D^*$	3×10^{-3}	0.05–0.3	$(0.7\text{-}26) \times 10^{8}$
III. Mixing with Decay	$B_d \to D^+\pi^-$	6×10^{-3}	0.001	3×10^{11}
to a CP Non-Eigenstate	$B_d \to D^0 K_s$	6×10^{-5}	0.01	7×10^{11}
	$B_s \to D_s^+ K^-$	3×10^{-4}	0.5?	5×10^{7}
IV. Cascade Decays to the	$B^- \to D^0 K^- + X$	10^{-5}	0.1?	9×10^{8}
Same Final State	$\quad \hookrightarrow K_s + Y$			
V. Interference of Spectator	$B^- \to D^{*0} \bar{D}$	3×10^{-3}	0.01	2×10^{9}
and Annihilation Graphs				
VI. Interference of Spectator	$B^- \to K^- \rho^0$	$\sim 10^{-5}$	0.1	1×10^{8}
and Penguin Graphs	$\bar{B}_d \to K^- \pi^+$	$\sim 10^{-5}$	0.1	1×10^{8}

*The specific channels considered here for each class of asymmetry are illustrative and not exhaustive. At this time we need to keep an open mind as to which channels will be best suited for CP violation studies. In this spirit we have included among the modes illustrating Class II asymmetries $\psi K_s X$ and $D\bar{D}K_s$, which are not necessarily CP eigenstates. There is some danger of a cancellation between the asymmetries produced by sub-channels with opposite CP quantum numbers, but a total cancellation is unlikely.

this study, we focus on CP violation because it is the one important issue which may not be adequately addressed by present or near future accelerators.

CP violation in the B system can manifest itself in a number of different ways. Table 1, taken from the 1987 Berkeley SSC Workshop,[5] shows six classes of CP violation along with branching ratios, estimated asymmetries, and the estimated number of $B\bar{B}$ events needed to observe the effect *at the SSC*. We chose to study a decay mode from Class II, "Mixing with Decay to a CP Eigenstate," because this class offers the following particular advantage. In the standard model, the CP-violating asymmetries for such decays depend only on values for elements of the K–M matrix, which, in principle, can be determined from measurements that do not involve CP violation.[6] For this case, the decay of a B or $\bar{B}$ can proceed either directly or following mixing. The CP-violating amplitude arises from interference between the two spectator-diagram processes; all of the hadronic physics drops out of the asymmetry because only a single amplitude is involved.[7]

Thus, from our present measurements of the K–M matrix, we can set limits on the magnitude of this form of CP violation if it is generated from the K–M phase within the three-generation standard model.[8] In the same way, once measured, the result can be easily interpreted. If the result falls outside the bounds given by the three-generation standard model, CP violation must, in part, come from a different source. This is not the only form of CP violation that we wish to investigate, but it is sufficiently basic that any new effort aimed at CP violation in B decays should include the capability of studying it.

We used the decay mode $B^0 \to \psi K_s$ in large part because there was a careful study of this mode submitted to the workshop by Aleksan et al., for the case of an asymmetric $\Upsilon(4S)$ collider.[9] Other Class II decays would probably yield similar results.

From present limits on K–M parameters, the predicted limits on the CP–violation asymmetry parameter, $\sin\phi(\psi\ K_s)$, are

$$0.1 \le \sin\phi(\psi K_s) \le 0.6 \ , \quad (1)$$

where the upper part of the range is preferred for relatively low top-quark masses (60 to 80 GeV/c^2).[5] (The parameter $\sin\phi$ is twice the time-averaged asymmetry listed in Table 1.)

2.2 General Technique

In the presence of CP violation, the B and $\bar{B}$ rates to a CP eigenstate, ψK_s in our case, will evolve differently with time:[7]

$$R(B^0 \rightarrow \psi K_s) \propto e^{-\Gamma t}\ [1 + \sin\phi \sin(x\Gamma t)]$$

$$R(\bar{B}^0 \rightarrow \psi K_s) \propto e^{-\Gamma t}\ [1 - \sin\phi \sin(x\Gamma t)] \ , \quad (2)$$

where R is the rate of the decay process, $x = \Delta m/\Gamma$, and Δm is the mass difference between the two neutral B meson mass eigenstates. Note that the B and $\bar{B}$ in Eq. (2) refer to the identities of the particles at the time they are created, rather than at the time they decay.

Since the b and $\bar{b}$ quarks are produced in pairs, the general technique is to observe a neutral B decaying to ψK_s and determine whether it was originally a B^0 or a $\bar{B}^0$ by tagging it, *i.e.*, by detecting the decay of the other B in a flavor-specific mode. In the following sections, we will explore how this general technique is modified by special situations involving the symmetries of the initial state and special tagging techniques. In particular, we will indicate how the intrinsic asymmetry $\sin\phi$ is diluted to $d \cdot \sin\phi$ $(d < 1)$ for each technique.

2.3 Case 0: Symmetric Collisions at the Υ(4S)

We start with the case which applies to present symmetric (*i.e.*, beams of equal energy) storage rings, such as CESR, running on the Υ(4S) resonance. This has been listed as Case 0 because we will see that the above technique does not work in this situation.

The advantages of studying B's at the Υ(4S) are well known:

1. The resonant cross section is about four times higher than the continuum cross section for $b\bar{b}$ production slightly above the resonance.
2. The final state is clean and well understood. It consists solely of $B^0\bar{B}^0$ and B^+B^- pairs.
3. Due to the simple final state, the well-determined beam energy can be used as a constraint leading to very good mass resolution and low backgrounds.

The Υ(4S) is a $C = -1$ resonance decaying exclusively into $B\bar{B}$. The symmetry of the wave function changes the general formula because the two B's develop as a coherent state. The time development for this case is:

$$R(B^0\bar{B}^0 \rightarrow B^0\psi K_s) \propto e^{-\Gamma(t'+t)}\{1 + \sin\phi \sin[x\Gamma(t'-t)]\}$$

$$R(B^0\bar{B}^0 \rightarrow \bar{B}^0\psi K_s) \propto e^{-\Gamma(t'+t)}\{1 - \sin\phi \sin[x\Gamma(t'-t)]\}, \quad (3)$$

where t and t' are the times at which the ψK_s and the flavor-specific decay occur, respectively.[7]

The time development is in the difference of the two decay times. This can be understood in a simple way: Both the B and $\bar{B}$ begin mixing into the charge conjugate states after they are produced, but with a relative phase which remains constant, so that the two states are always orthogonal. Thus, when the first decay occurs, the wave function collapses with the second B the charge conjugate of the first. The second B then begins to mix into its charge conjugate state. This process is often described as "nothing happens until the first decay takes place."

Since the asymmetry is odd in $(t' - t)$, it integrates to zero if the time difference of the decays is not measured. At a symmetric Υ(4S) collider, the mean decay length of B's is about 30 μm, a distance which is insufficient for the measurement of $(t' - t)$.

We will see there is still a way to measure this class of CP violation at a symmetric 10 GeV collider in Case 2.

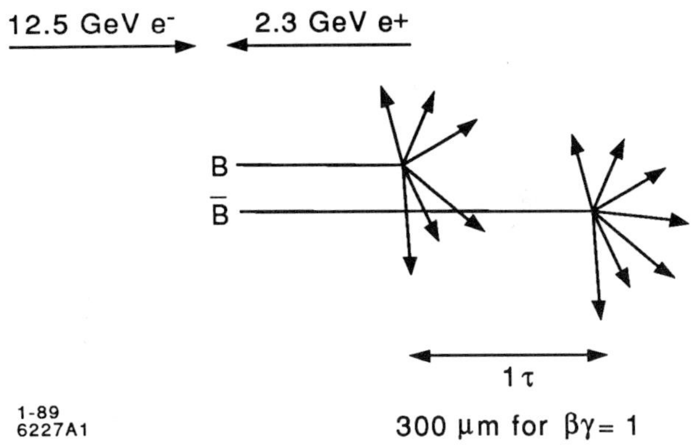

Fig. 1. Schematic of a $B\bar{B}$ decay in an asymmetric collider at the Υ(4S).

2.4 Case 1: Asymmetric Collisions at the Υ(4S)

Consider a collider running at the Υ(4S) resonance, but with unequal beam energies,[10] as illustrated in Fig. 1. The two B's from the Υ decay are almost at rest in the center-of-mass, but are moving in the laboratory. For the example shown in Fig. 1, a 12.5 GeV electron beam against a 2.3 GeV positron beam, the center-of-mass moves with $\beta\gamma = 1$. The two B decays will be separated, on average, by one lifetime, or 300 μm in this example.[11]

Assuming one can measure $\Delta t \equiv (t' - t)$, but not $(t' + t)$,[12] Eqs. (3) imply

$$R(\Delta t) \propto e^{-\Gamma|\Delta t|}[1 \pm \sin\phi \sin(x\Gamma\Delta t)] \; . \qquad (4)$$

By fitting to the time dependence of Eq. (4), one should be able to obtain an asymmetry of $0.61 \sin\phi$ (*i.e.*, a dilution factor $d = 0.61$).[13] This calculation allows for the uncertainties in Δt, due both to the fact that the distance between the B decay points is smeared slightly by the center-of-mass motion of the B's and to experimental resolution. It does not include the effect of incorrect tags, which we will account for separately in Sec. 2.9.

This method thus combines all the advantages of the symmetric $\Upsilon(4S)$ case with the ability to measure the time development of the final state. This technique also yields an advantage in suppressing combinatoric backgrounds since the origin of most charged particles can be determined. However, both the detector and the collider are somewhat more difficult to build. We will return to the former in Chap. 3 and the latter in our report on accelerator technology.[4]

2.5 Case 2: Symmetric Collisions Just Above the $\Upsilon(4S)$

One can measure Class II CP violation with a symmetric 10 GeV collider by running just above the $\Upsilon(4S)$ resonance.[7,14] Above $B\bar{B}^*$ threshold, but below $B^*\bar{B}^*$ threshold, $B\bar{B}^*$ production should be the dominant $b\bar{b}$ final state. This yields a C = +1 $B\bar{B}$ state since the B^* decays into $B\gamma$.

For this case, the time development changes from Cases 0 and 1, in that $(t' - t)$ becomes $(t' + t)$:

$$R(B^0\bar{B}^0 \to B^0\psi K_s) \propto e^{-\Gamma(t'+t)}\{1 - \sin\phi \sin[x\Gamma(t' + t)]\}$$

$$R(B^0\bar{B}^0 \to \bar{B}^0\psi K_s) \propto e^{-\Gamma(t'+t)}\{1 + \sin\phi \sin[x\Gamma(t' + t)]\} \; . \qquad (5)$$

As before, both the B and $\bar{B}$ begin mixing into the charge conjugate states after they are produced, but with the opposite rather than the same phase. Thus, when the first decay occurs, the second B is already mixed and continues to mix until it decays.

Since $(t' + t)$ is even, integration of the asymmetry over both times leads to a finite value. The dilution factor resulting from this integration[15] is

$$d = \frac{2x}{(1 + x^2)^2} \; . \qquad (6)$$

This value assumes that any C = −1 contributions, *i.e.*, $e^+e^- \to B\bar{B}$ reactions, are separable. CP violation can be measured, but at a cost of not having the resonant enhancement of the $\Upsilon(4S)$.

2.6 Case 3: Symmetric Collisions in the Continuum

A (presumably) symmetric collider in the continuum, say 8 GeV per beam, would allow the time development of the B decays to be measured. However, one gives up all the advantages of the $\Upsilon(4S)$ resonance and suffers a cross section reduction as E^{-2}. The B decays develop incoherently in this region, so the general formulae, Eqs. (2), apply directly. Their application requires knowledge of the primary interaction point (unlike Cases 1 and 2).

However, the asymmetry is diluted by mixing of the tagging particle. If f_0, f_-, f_s, and f_Λ are the fractions of b quarks which hadronize as $\bar{B}^0$, B^-, $\bar{B}_s$, and B-baryons, respectively, then

$$R(t) \propto e^{-\Gamma t}\left[1 \pm \left(f_- + f_\Lambda + \frac{f_0}{1 + x^2}\right) \sin\phi \sin(x\Gamma t)\right] \; , \qquad (7)$$

where we have assumed that B_s's are fully mixed. Assuming that fitting this time spectrum leads to similar dilution as for Case 1, then the overall dilution factor is

$$d = 0.61\left(f_- + f_\Lambda + \frac{f_0}{1 + x^2}\right) \simeq 0.61\left(0.50 + \frac{0.35}{1 + x^2}\right) \qquad (8)$$

for a reasonable model.

2.7 Case 4: Z Decay without Polarization

If we consider a collider at the Z mass without longitudinal polarization, then this case is the same as the previous case except that there is a considerable resonant enhancement from the Z:

$$\frac{\sigma(e^+e^- \to b\bar{b})_{\sqrt{s}=m_Z}}{\sigma(e^+e^- \to b\bar{b})_{\sqrt{s}=16 \text{ GeV}}} \approx 60 \; . \qquad (9)$$

2.8 Case 5: Z Decay with Polarization

The situation at the Z mass becomes even more attractive if longitudinal electron polarization is available.[16] This is relatively easy at a linear collider[17] and is under consideration for LEP.[18]

With a polarized electron beam on the Z resonance, there is a large forward-backward asymmetry in the b versus $\bar{b}$ direction (A^b_{FB}):

$$A^b_{FB} \approx \frac{3}{4}\mathcal{P}\frac{2v_b a_b}{v_b^2 + a_b^2} \approx \frac{3}{4}\mathcal{P}(0.94) \quad , \qquad (10)$$

where $\mathcal{P}$ is the magnitude of the polarization. The asymmetry is almost maximal and would be an extremely effective B tagging mechanism for complete polarization.

Table 2. Comparison of B-Factory Techniques

Factor/Case	Asymmetric $\Upsilon(4S)$	Symmetric $\Upsilon(4S)+$	$\sqrt{s} = 16$ GeV	Z $\mathcal{P} = 0$	Z $\mathcal{P} = 0.9$ ($\mathcal{P} = 0.45$)
$b\bar{b}$ cross section, σ (nb)	1.2	0.3	0.11	6.3	6.3
Fraction of B^0, f_0	0.43	0.34	0.35	0.35	0.35
ψK_s reconstruction efficiency, ϵ_r	0.61	0.61	0.61	0.46	0.46
Tag efficiency, ϵ_t (and method)	0.48 (ℓ, K)	0.48 (ℓ, K)	0.30 (ℓ, D)	0.18 (ℓ, D)	0.61 (A_{FB})
Wrong tag fraction, w	0.08	0.08	0.08	0.08	0.125 (0.27)
Asymmetry dilution, d	0.61	0.63	0.45	0.45	0.61
$\int \mathcal{L}dt$ needed for 3σ effect (10^{40} cm^{-2}) *	0.45–16	2.1–77	18–640	0.68–25	0.14–5.0 (0.37–13)
Relative $\int \mathcal{L}dt$ needed	1.0	4.7	40	1.5	0.3 (0.8)

*peak luminosity needed in units of 10^{33} cm^{-2} sec^{-1} for 10^7 seconds of fully efficient running at peak luminosity.

Also note that the additional dilution due to mixing, included for Cases 3 and 4, does not apply here. Present techniques for producing longitudinally polarized beams in linear accelerators limit the polarization to 50% due to a spin degeneracy. There are active programs to find ways of producing polarizations approaching 100%.[19)]

2.9 Evaluation of the Different Techniques

The needed integrated luminosity to measure CP violation to a given degree of accuracy can now be calculated for each of the above techniques by using the following formula:

$$\int \mathcal{L}dt = \left\{\sigma(e^+e^- \to b\bar{b}) f_0 2B\epsilon_r\epsilon_t \left[(1-2w)d\delta(\sin\phi)\right]^2\right\}^{-1}, \quad (11)$$

where:

- f_0 is the fraction of B^0's in the b quark fragmentation (for Case 2, only C = +1 $B\bar{B}$ states are counted);
- B is the product of branching fractions to the desired mode (assumed to be 5×10^{-4} for the ψ K_s mode times 0.14 for the leptonic decay modes of the ψ);
- ϵ_r is the ψ K_s reconstruction efficiency;
- ϵ_t is the tagging efficiency, *i.e.*, the fraction of events in which the flavor of the B decaying to a flavor-specific state can be measured;
- w is the fraction of incorrect tags;
- d is a dilution factor which takes into account the loss in asymmetry due to fitting, integrating over time, and/or the mixing of the tagged decay (taking $x = 0.7$); and
- $\delta(\sin\phi)$ is the required accuracy on the CP asymmetry parameter $\sin\phi$, taken to be 1/3 of $\sin\phi$ in this case in order to give a 3σ effect.

Table 2 gives the factors that we put into Eq. (11) for each of the five cases and the resulting requirement on integrated luminosity. The numbers for the first case, an asymmetric $\Upsilon(4S)$ collider, were taken from Aleksan et al.,[9)] and the other cases were scaled from this work by making reasonable assumptions. Aleksan et al., assume a very good detector (see Chap. 3) in their simulation. Thus, the required luminosities will be larger for less-than optimum detectors.

The ψK_s reconstruction efficiency, ϵ_r, is assumed to be smaller on the Z than at lower energies because it was assumed that the neutral decays of the K_s could not be reconstructed at the Z mass. In a similar way, the tagging efficiency, ϵ_t, decreases with increasing energy because the techniques change and become less efficient. At the $\Upsilon(4S)$,

Aleksan et al.,[9] showed that kaons could be used for tagging if events which had extra kaons were discarded. It was assumed that this would not work in the continuum because too many extra kaons would be produced. However, tagging with reconstructed D's would be possible there. At the higher energy of the Z mass, it was assumed that both the D and lepton reconstruction efficiency would drop. For Case 5, Z decay with polarization, $(1-2w)$ is replaced with A^b_{FB} from Eq. (10) and the tagging efficiency is taken from Ref. 16.

The penultimate row of Table 2 gives the range of integrated luminosities needed to measure CP violation to three standard deviations. There is a factor of 36 in each of these ranges since there is a factor of six in the allowable range of $\sin\phi$ from our present knowledge of the K–M matrix [see Eq. (1)]. The final row of the table gives the relative amount of integrated luminosity needed for each technique normalized to unity for the asymmetric $\Upsilon(4S)$ collider. The asymmetric $\Upsilon(4S)$ collider and Z factories, with or without longitudinal polarization, require the least integrated luminosity. In the second part of this working group's report,[4] we will explore the prospects for obtaining colliders with luminosities in the required range.

3. EXPERIMENTS FOR STUDYING CP VIOLATION

3.1 Introduction

In this chapter we consider the particular features of experiments which will be important for the measurement of CP violation. We will continue to use the $B \to \psi K_s$ mode as an example. Our main concern is to identify the important issues that affect collider and detector design. We will begin with some general comments about detectors that are generic to the different techniques outlined in the previous chapter. After a short summary of some considerations that influence collider design, we will discuss issues that are specific to each case. The principal detector resources required for a measurement of CP violation in $B \to \psi K_s$ are:

- Vertex Detection.

  All techniques except Case 2 require excellent vertex resolution in order to measure the difference between the decay times of the two B mesons.

- $B \to \psi K_s$ Reconstruction.

  The product of the branching ratios for $B \to \psi K_s$ and the leptonic modes of the ψ is small ($\approx 7 \times 10^{-5}$). B mesons that decay this way must be reconstructed with little background.

- $B - \bar{B}$ Tagging.

  The measurement of the CP violation asymmetry depends on efficiently and accurately determining if the other B meson decayed as a B or a $\bar{B}$. This requires excellent lepton and charged kaon identification.

Detectors that can efficiently exploit these techniques can be designed with some confidence, based on the accumulated experience of designing and operating detectors at e^+e^- colliders. The most revolutionary detector appears to be the one required for the asymmetric collider option, because it must cover the forward region with sufficient precision. We discuss some of our thoughts on this subject and present a conceptual design of a detector. For the other cases, we only include an outline of the features required for observing CP violation in $B \to \psi K_s$

3.2 Vertex Detection

Silicon vertex detectors appear to be the only viable solution to the problem of reconstructing decay vertices with sufficient accuracy to measure the difference in decay times between the two B mesons. Silicon detectors can have the required resolution (in the 10 μm range in two dimensions) with sufficient segmentation (approximately 10,000 channels per layer) so that track overlap is not a problem. The alternative technology, drift chambers, lacks the inherent resolution for this problem, particularly in the measurement of the position of the track along the beam. Due to the constraints of space around a 1 cm beam pipe, it appears to be impossible to build a wire chamber with sufficient segmentation along the beam to avoid inefficiencies due to track overlap.

A number of fixed target experiments have successfully used silicon detectors, and SLC and LEP will soon provide experience in using them at e^+e^- colliders.[20] General features of a silicon vertex detector for a B factory operating at the $\Upsilon(4S)$ have been studied,[21] and precision vertex detection has been included in the design of a proposed detector.[22]

The particles resulting from the decay of the $\Upsilon(4S)$ have quite low momenta, even with the boost of an asymmetric collider. Multiple Coulomb scattering in the beam pipe and the first (few) layer(s) of the vertex detector can dominate the vertex resolution, even at high energy colliders. This places severe constraints on the beam pipe radius which, in turn, becomes a significant problem of the collider design, as discussed in the accelerator part of our report.[4] The small beam pipe can also become a detector problem by introducing high backgrounds in the detector due to scattered particles.[23]

3.3 $B \to \psi K_s$ Reconstruction

The $B \to \psi K_s$ mode studied here has a particularly simple event topology. In particular, the tracks from these decays will be well separated from the tracks from the other B meson, if vertex resolution is adequate to separate the vertices of the two B mesons. To avoid mixing these events with $\psi K_s \pi^0$ events, which have the opposite value of CP, the energy and mass resolution of the detector must be adequate to discriminate against slow π^0's.

The interaction between momentum resolution in the detector and the energy spreads of the e^+e^- collider beams in determining the resolution in total energy and reconstructed B meson mass is discussed in a contribution to these proceedings.[24] For experiments at the $\Upsilon(4S)$, the resonance width acts as a monochrometer, sharply limiting the influence of the collider beam energy spread on the total energy or mass resolution of a reconstructed B meson. (Of course, a small collider beam energy spread is still important in concentrating the luminosity in the useful energy region.) Thus, the reconstruction resolution requirements have little influence on collider design, but greatly influence detector design.

Some combination of dE/dx, time of flight, electromagnetic shower counters and iron absorbers are standard equipment for particle identification in e^+e^- collider experiments. The authors of the PSI proposal have recognized[22] that this may not be adequate for reconstructing low multiplicity B decays with high-momentum hadrons, even for experiments at the $\Upsilon(4S)$. Their response has been to include ring-imaging Čerenkov (RICH) counters in their design.

Although we focussed on $B \to \psi K_s$ for this study, it is essential that a detector to search for CP violation be capable of reconstructing more complicated decay modes in order to maximize the opportunity of discovering CP violation and developing an understanding of it after it has been found. Reconstruction of these more complicated modes will generally require better momentum resolution and particle identification than that required for $B \to \psi K_s$.

3.4 B–$\bar{\text{B}}$ Tagging

Tagging the other B meson as having decayed as either a B or a $\bar{B}$ is accomplished by identifying the sign of a lepton or charged kaon. Without clear secondary vertex separation, only high momentum leptons (above about 1.4 GeV in the $\Upsilon(4S)$ frame) are useful for the lepton tag, since low momentum leptons can come from semileptonic decays of D's from $B \to DX$ decays. Measurements of $B\bar{B}$ mixing by UA1,[25] ARGUS,[26] and CLEO[27] demonstrate that this tagging is possible at both low and high energy colliders. Aleksan et al.,[9] showed that muons with momenta between 0.8 and 1.4 GeV could also be used if they could be established as coming from the primary vertex.

B–$\bar{B}$ tagging via charged kaons is limited by Cabibbo-suppressed D decays and the reliability of K/π separation. The Cabibbo-suppressed D decays background can be minimized by rejecting events with excess kaons.[9] The reliability of the K/π separation is an essential detector design consideration; in addition to the usual dE/dx in a central drift chamber, a time of flight and/or a RICH system may also be required.

The B–$\bar{B}$ tagging requirement has little direct influence on collider design but it has a major impact on detector design, except for Case 5.

3.5 Influences on Collider Design

The requirements for an experiment to observe CP violation have the following general influences on collider design:

- High Luminosity.

  The high luminosities required appear to be the major challenge for collider design.

- Low Background Rates.

  High luminosities imply large beam currents which intrinsically lead to high background rates in the detector. To avoid background, the vacuum near an interaction region must be excellent and the region and nearby lattice must be carefully designed to avoid illuminating the beam pipe with direct or scattered synchrotron radiation or stray particles from the beams.

- Beam Pipe Radius.

  The beam pipe radius must be small for all techniques except Case 2, and even in this case a small beam pipe is useful in separating secondary D decays from the primary B decays. This requirement substantially increases the difficulty of achieving low background rates.

- Beam Pipe Thickness.

  If decay times must be measured, the beam pipe must be thin in order to reduce multiple scattering, On the other hand, it must be thick enough to shield sensitive vertex detectors from RF radiation from the beam and be coated to absorb soft synchrotron radiation.

- Beam Energy Spread.

  A small beam energy spread is important for colliders operating at or near the $\Upsilon(4S)$ in order to take advantage of the resonance width. In the other cases this spread is much less important.

3.6 Case 1: Asymmetric Collisions at the Υ(4S)

The two novel issues for a detector at an asymmetric $\Upsilon(4S)$ collider are (1) the problems caused by folding the solid angle forward and (2) the need to detect separate B decay vertices along the beam direction. To understand these issues, we sketched such a detector at the Summer Study; it is shown in Fig. 2. We took the beam energies to be 12.5 GeV versus 2.3 GeV for this study ($\beta\gamma = 1$) and used the interaction region design specified by Garren.[28]

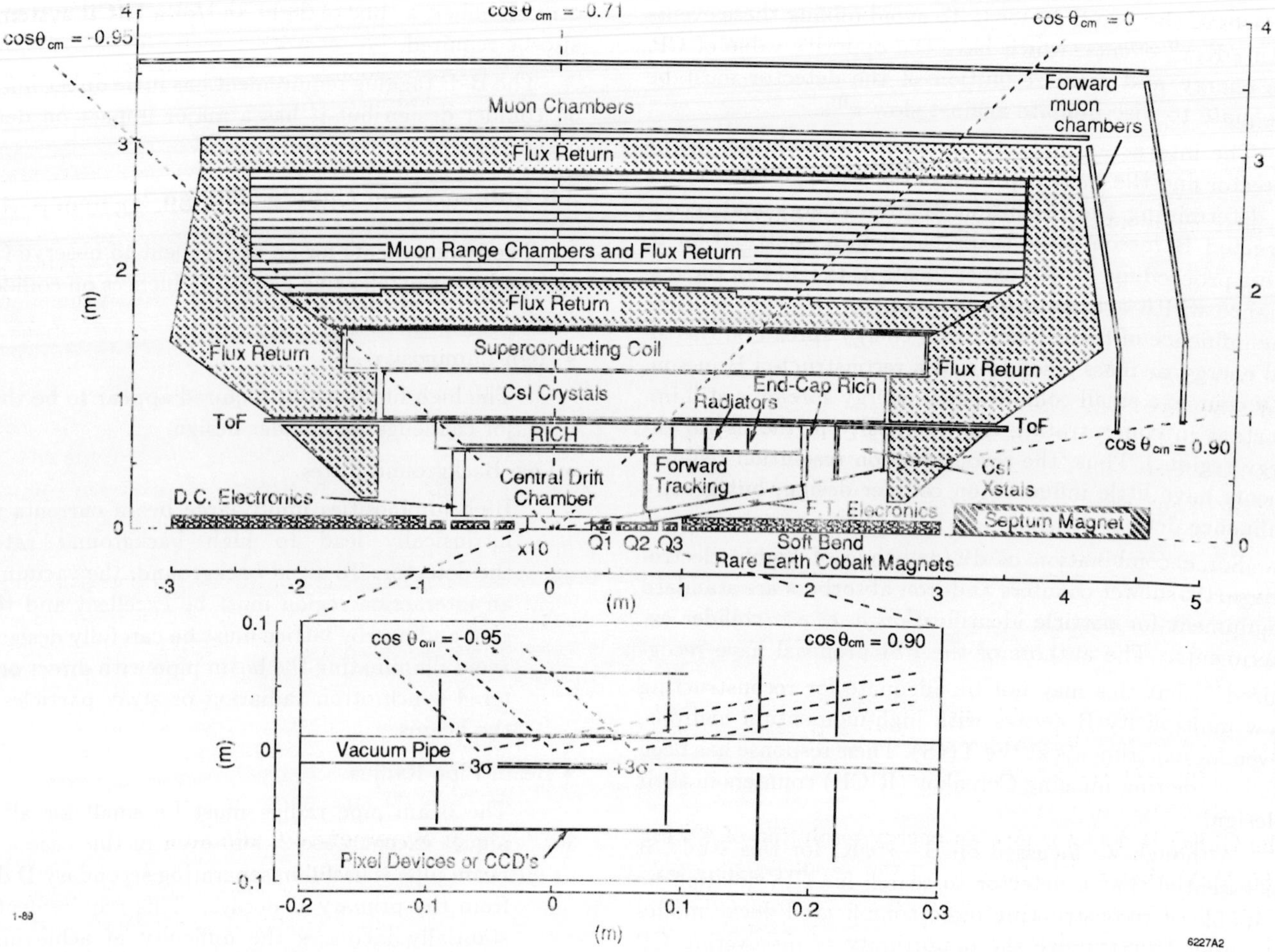

Fig. 2. A detector for an asymmetric collider at the $\Upsilon(4S)$.

For less asymmetric detectors, the solid angle issues are less severe, but the B's separate a smaller amount.[11)]

Our goal was to design a detector which would approximate the near ideal detector assumed in the Aleksan et al., study.[9)] In designing it, we borrowed freely from ideas which have been used in the building the CLEO II detector[3)] and in the detector designed for the PSI proposal.[22)]

Solid Angle Considerations. To obtain high efficiency for reconstruction and tagging, it is essential to cover greater than 90% of the center-of-mass solid angle. The detector in Fig. 2 has full vertex detection, tracking, particle identification, and calorimetry over 92.5% of the solid angle, $-0.95 < \cos\theta_{c.m.} < 0.90$. In the laboratory, this requires coverage from 11° to 132°. The forward angle is limited by the rare earth cobalt quadrupole magnets, which start 30 cm from the interaction point in the Garren design.[28)]

These considerations lead to the need for a forward detector. Both it and the barrel detector individually are quite conventional; indeed, the barrel design is essentially a modest extrapolation of the PSI proposal. However, the interface between the barrel and the forward detectors will require some ingenuity in order to reduce multiple scattering to an acceptable level.

Vertex Detection. There are several issues concerning the vertex detection. First, in order to achieve sufficient resolution in the vertex position, the beam pipe radius must be small (between about 1 to 1.5 cm).[9)] This radius is much smaller than any that has been used so far, so considerable effort will be required to design collider optics and masking that will yield acceptable background rates. A suitable synchrotron radiation masking scheme for the Garren design has not been studied yet.

Second, the bunch length is typically about 2 cm,[28)] which is longer than the radius at which detectors are placed. This makes it difficult to design a detector in which the particles always cross the detector at close to normal incidence. The interaction region enlargement in Fig. 2 shows an attempt to have a reasonable placement of detectors. The dotted lines show solid angle envelope for particles emitted from the interaction point and from ±3 standard deviations in the beam direction.

Third, the requirement for three-dimensional vertex reconstruction argues for silicon devices with two-dimensional readouts, such as CCD[29] or silicon diode arrays.[30]

Charged Particle Tracking. In a 1.5 T field, the required (measurement dominated) momentum resolution[9] of

$$\frac{\sigma_p}{p} = 0.004p \quad (p \text{ in GeV}/c) \qquad (12)$$

can be obtained by the combination of the vertex detector and a 50 layer drift chamber in the radial region from 10 to 60 cm, if each measurement has an error of 100 μm. The SLD Collaboration is building a large chamber with a measurement error in this range.[31]

Particle Identification. Since π/K separation is only needed to about 3 GeV/c at the $\Upsilon(4S)$, a thin RICH counter consisting of a 1 cm NaF crystal radiator and a 13 cm gap is adequate. See Ref. 22 for a detailed discussion.

Photon Detection. Good photon detection and reconstruction is essential for reconstructing B mesons. The best choice for the 10 GeV center-of-mass energy region appears to be CsI crystals.[3,22] Beam tests on crystals for the CLEO II detector give an energy resolution of 4% for 180 MeV photons.[32] Because of the small tracking volume used in this design, even with the forward detector, the total volume of CsI crystals is smaller here than in the CLEO II detector.

Muon Detection. Detection of muons with momenta as low as 800 MeV/c is useful for B–flavor tagging and ψ reconstruction. However, conventional muon detection with thick absorbers and chambers is difficult in this region due to hadron punch-through. The solution used in this design is to take advantage of the difference between pion and muon range by using a muon range detector to cover the 0.8 to 1.2 GeV/c momentum region.[33] Such a detector could be built with 0.5 to 1 cm iron plates separated by thin chambers. Above 1.2 GeV/c, the difference in range approaches the straggling error, and the method loses usefulness. More detailed calculations are needed to establish the utility of this technique.

3.7 Case 2: Symmetric Collisions Just Above the $\Upsilon(4S)$

The detector requirements for symmetric colliders just above the $\Upsilon(4S)$ are well-understood, because this is the energy range in which the CLEO and ARGUS experiments have been successful in reconstructing B mesons. A detector to search for CP violation would likely be constructed along the lines of the CLEO II detector or the PSI proposal. Very precise vertex detection is not crucial in this technique, since separation of B vertices is not required. However, good vertex detection would be a very useful tool for reducing backgrounds by separating D decay products from B decay products.

The mass resolution of the detector must be very good for this technique to succeed because it must be possible to separate reconstructed B mesons coming from $B^0\bar{B}^{*0} \to B^0\bar{B}^0\gamma$ events, where the γ is not detected, from those coming from the direct production of $B^0\bar{B}^0$. The latter could be a serious background. In the first case, if the energy of the B meson candidate is assumed to be half of the total energy, the reconstructed B mass will be about 26 MeV ($\approx E_\gamma/2$) above the true B mass. In the latter case, the B will reconstruct to the correct B mass. The separation of these two types of events depends on the mass resolution of the experiment which, in turn, depends on the energy spread of the collider beams, the momentum resolution of the detector, and the Doppler shift of the B's resulting from $\bar{B}^{*0}$ decay. The resolution of the CLEO detector would be quite adequate[24] to separate these two types of events. The experiment would be very difficult if the mass resolution were noticeably worse.[34]

An alternative would be to detect the photon from $\bar{B}^{*0} \to \bar{B}^0\gamma$, but it appears very difficult to accomplish this with high efficiency.

3.8 Case 3: Symmetric Collisions in the Continuum

At an e^+e^- collider operating near 16 GeV, B mesons are typically produced with $\beta\gamma$ near 1. This matches the asymmetric collider in our study, so in order to measure the decay times of the B mesons with sufficient accuracy, a beam pipe with a radius not much larger than 1 cm will again be required. The rest of the detector can be quite conventional; CLEO II or the PSI proposal can serve as a model. Compared to experiments operating at the $\Upsilon(4S)$, the total energy constraint is lost, and the mass resolution is much worse because the beam energy cannot be used. This implies that an experiment must rely much more heavily on vertex detection for eliminating combinatorial background in B reconstruction.

3.9 Case 4: Z Decay without Polarization

The large B meson rate at the Z has encouraged a large amount of thought about how to detect them efficiently. B mesons resulting from Z decay typically have $\beta\gamma$ in the neighborhood of five to seven. As far as reconstruction of B vertices is concerned, a larger, more conventional, beam pipe radius could be adequate. However, eliminating the large background from charm decays will require excellent vertex detection. Again, there is no total energy constraint to reduce combinatorial background and the mass resolution will be poor, so reconstruction of B decays must rely very heavily on vertex detection.

Compared to Cases 1 to 3, the higher momentum of the tracks implies that it is harder to achieve comparable momentum resolution, and the exclusion of π^0 background from the reconstructed B mesons will be more difficult. Furthermore, in this case tagging the decay of the other B is likely to be harder than at lower energy due to backgrounds.

On the other hand, this case has a number of attractive features: the event rate is high; the large data sample would have enormous potential for physics outside of B meson physics; and the primary vertex will usually be tagged by extra particles from the Z decay, so the decay time can be directly measured. Experience soon to be gained at SLC and LEP should show how easy or difficult it will be to overcome the experimental obstacles.

3.10 Case 5: Z Decay with Polarization

For this case the experiment and detector are basically the same as for Case 4, except that the effort that must be expended on lepton identification for tagging can be substantially reduced, since the tagging is primarily accomplished by the polarization. However, lepton identification is needed in any case to reduce background in reconstructing the ψ, and tagging with leptons will be a useful check on the polarization tag.

4. CONCLUSIONS

It appears that the measurement of CP violation in the B system with e^+e^- colliders is difficult, but not impossible. There are several attractive options, each with its own advantages and drawbacks. Asymmetric colliders at the $\Upsilon(4S)$ and colliders on the Z resonance, with or without polarization, require the least luminosity. Symmetric colliders running just above the $\Upsilon(4S)$ require slightly more. In any case, high luminosities (minimums of a few times 10^{33} to 10^{34} cm^{-1} sec^{-1}) are required to get well into the range in which CP violation can occur. The possibilities of obtaining these luminosities will be the subject of the following paper. The detectors required to measure CP violation are sophisticated, but within the limits of present technology.

FOOTNOTES AND REFERENCES

1. For a summary of recent results see S. L. Wu, *Proceedings of the 1987 International Symposium on Lepton and Photon Interactions at High Energies,* ed. by W. Bartel and R. Ruckl (North Holland, Amsterdam, 1988), p. 39.
2. H. Albrecht *et al.*, *Phys. Lett.* **192B**, 245 (1987); N. Katayama, *Proceedings of the SLAC Summer Institute on Particle Physics* (July 1988), ed. by E. C. Brennan (to be published).
3. CLEO Collaboration, Cornell report CLNS–85/634 (1985).
4. R. H. Siemann *et al.*, following paper in these proceedings.
5. K. J. Foley *et al.*, *Proceedings of the Workshop on Experiments, Detectors, and Experimental Areas for the Supercollider,* July 7–17, 1987, Berkeley, California, ed. by R. Donaldson and M. G. D. Gilchriese (World Scientific, Singapore, 1988), pp. 701–727.
6. R. D. Peccei *et al.*, these proceedings.
7. I. I. Bigi and A. I. Sanda, *Nucl. Phys. B* **193**, 85 (1981); *Nucl. Phys. B* **281**, 41 (1987); I. I. Bigi, V. A. Khoze, N. G. Uraltsev and A. I. Sanda in *CP Violations* (World Scientific Series on Directions in High Energy Physics, v. 3), ed. by C. Jarlskog (World Scientific, Singapore, 1988), also available as SLAC–PUB–4476 (1987); see Ref. 5 for additional references.
8. M. Kobayashi and T. Maskawa, *Prog. Theor. Phys.* **49**, 652 (1973).
9. R. Aleksan, J. Bartelt, P. R. Burchat and A. Seiden, SLAC–PUB–4673 (1988), to be published in *Phys. Rev. D.*
10. P. Oddone, *Proceedings of the UCLA Workshop: Linear Collider $B\bar{B}$ Factory Conceptual Design,* ed. by D. Stork, p. 243 (1987).
11. See D. Hitlin, T. Nakada and A. Sanda, these proceedings, in which it is argued that 8 GeV on 3.5 GeV is an optimum asymmetry.
12. The quantity $(t'+t)$ cannot be measured in an asymmetric storage ring running on the $\Upsilon(4S)$ because the bunch lengths are much longer than the decay lengths.
13. This number is smaller than the value of $d = 0.71$, which was used at the Summer Study, and is based on more precise subsequent work. Correspondingly, the values in Table 2 have changed somewhat from those shown in the oral report.
14. S. Stone, *Modern Physics Lett. A* **3**, 541 (1988) has studied the luminosity required to observe CP violation with this technique and has arrived at somewhat more optimistic conclusions than those in this study.
15. For an asymmetric collider operating in this region, one might want to integrate only over $(t' + t)$. The equation analogous to Eq. (4) is then

$$R(\Delta t) \propto e^{-\Gamma|\Delta t|}\left[1 \mp \frac{\sin\phi}{\sqrt{1+x^2}} \sin(x\Gamma|\Delta t| + \tan^{-1}x)\right].$$

16. W. B. Atwood, I. Dunietz and P. Grosse–Wiesmann, SLAC–PUB–4544 (1988), submitted to *Phys. Lett. B.*
17. D. Blockus *et al.*, SLAC–PROP–1 (1986).
18. G. Alexander *et al.*, CERN–88–06 (1988).
19. C. K. Sinclair, *J. Phys. (Paris)* **46**, C2–669 (1985).

20. SLD Collaboration, SLAC–273 (1984); A. Litke *et al.*, *Nucl. Instrum. Methods A* **265**, 93 (1988); G. Batignani *et al.*, INFN PI/AE 88/4 (1988); and M. Burns *et al.*, CERN–EP/88–82 (1988).
21. V. Lüth, SLAC–PUB–4553 (1988).
22. K. Wacker *et al.*, PSI report number PR–88–09 (1988).
23. N. B. Mistry, P. S. Drell and N. Katayama, these proceedings.
24. D. G. Cassel, N. B. Mistry, J. A. Mueller and N. Katayama, these proceedings.
25. C. Albajar *et al.*, *Phys. Lett.* **186B**, 247 (1987).
26. H. Albrecht *et al.*, *Phys. Lett.* **192B**, 245 (1987).
27. R. Fulton *et al.*, submitted to the XXIV International Conference on High Energy Physics, Munich, August 4–10, 1988.
28. A. A. Garren, "APIARY I Design," SLAC/AP–071 (1988). See Ref. 4 for a discussion.
29. C. Damerell, Rutherford Appleton Laboratory report RAL 86–077 (1986).
30. S. Parker, University of Hawaii report UH–511–633–88 (1988); S. L. Shapiro *et al.*, SLAC–PUB–4701 [Rev.] (1988).
31. W. B. Atwood *et al.*, *Nucl. Instrum. Methods A* **252**, 295 (1986).
32 E. Blucher *et al.*, *Nucl. Instrum. Methods A* **249**, 201 (1986).
33. At momenta of 0.8 and 1.2 GeV/c, the difference between muon and pion range is 24.6 and 18.2 g/cm^2, respectively. See W. H. Barkas and M. J. Berger, NASA SP–3013 (1964).
34. T. Nakada reached similar conclusions based on a Monte Carlo simulation of $B \to \psi K_s$ reconstruction from $B\bar{B}^*$ events, which is described in PSI internal report, PSI/TN–1/88, unpublished.

REPORT OF THE B–FACTORY GROUP: II. ACCELERATOR TECHNOLOGY*

R. H. Siemann,(1)† D. G. Cassel,(1)† G. J. Feldman,(2)† M. S. Alam,(3) R. Aleksan,(2) W. B. Atwood,(2) J. Bartelt,(2) J. J. Bisognano,(4) J. R. Boyce,(4) D. B. Cline,(5) G. Coignet,(6) P. S. Drell,(1) A. M. Eisner,(7) A. Fridman,(2) A. A. Garren,(8) P. Grosse-Wiesmann,(2) D. G. Hitlin,(9) S.-O. Holmgren,(10) H. Kagen,(11) N. Katayama,(3) H. Koiso,(12) N. B. Mistry,(1) T. Nakada,(13) P. J. Oddone,(10) R. B. Palmer,(2) R. S. Panvini,(14) D. E. Pellett,(15) F. C. Porter,(9) J. R. Rees,(2) D. L. Rubin,(1) E. I. Shibata,(16) G. H. Trilling,(10) J. J. Welch,(1) W. A. Wenzel,(10) and M. S. Witherell,(17)

(1) *Cornell University, Ithaca, New York 14853*
(2) *Stanford Linear Accelerator Center, Stanford University, Stanford, California 94309*
(3) *State University of New York at Albany, Albany, New York 12222*
(4) *Continuous Electron Beam Accelerator Facility, Newport News, Virginia 23606*
(5) *University of California, Los Angeles, California 90024*
(6) *Laboratoire d'Annecy-le-Vieux de Physique des Particules, F-74019 Annecy-le-Vieux, France*
(7) *University of California Institute for Research at Particle Accelerators, Stanford, California 94305*
(8) *SSC Central Design Group and Lawence Berkeley Laboratory, Berkeley, California 94720*
(9) *California Institute of Technology, Pasadena, California 91125*
(10) *Lawrence Berkeley Laboratory and Department of Physics, University of California, Berkeley, California 94720*
(11) *Ohio State University, Columbus, Ohio 43210*
(12) *National Laboratory for High Energy Physics, Ibaraki 305, Japan*
(13) *Paul Scherrer Institut, CH-5234 Villigen, Switzerland*
(14) *Vanderbilt University, Nashville, Tennessee 37235*
(15) *University of California, Davis, California 95616*
(16) *Purdue University, Lafayette, Indiana 47907*
(17) *University of California, Santa Barbara, California 93106*

1. INTRODUCTION

The accelerator requirements for observing CP violation in the decay $B^0 \to \Psi K_s$ with roughly one year of running are summarized in Table 1. The experimental technique appropriate for each collider leads to the luminosity range and the other collider requirements. The latter are important considerations that affect the design approach and performance.

The B–factory luminosities are well above that of any operating machine, but there are ideas for accelerators capable of meeting the performance in Table 1. These fall into one of two categories:

i. Those with well-defined accelerator physics questions needing positive answers before detailed design work could begin. This class includes symmetric storage rings (Sec. 2.3), asymmetric storage rings (Sec. 2.4), and LEP (Sec. 4.1).

ii. Concepts with several major accelerator R&D questions that must be addressed successfully before a design could begin. The issues are interconnected, and the results of the R&D could affect the design and performance significantly. This class includes the linear colliders for the $\Upsilon(4S)$ and continuum (Sec. 3) and the Z–factory linear collider (Sec. 4.2). Some of these have potential that far exceeds that of the colliders in the class above, and for those the R&D is well worth pursuing.

This distinction is important, but it should not cloud the overall conclusion that through advances in accelerator physics the bottom end of the luminosity range in Table 1 is within reach.

2. STORAGE RINGS FOR THE $\Upsilon(4S)$ AND CONTINUUM

2.1 Introduction

The accelerator physics issues can be understood by writing the luminosity in terms of quantities that limit performance. For a storage ring these are the beam-beam interaction, single bunch currents, and the total beam current. For clarity of presentation the beams are assumed to have the same properties: equal energies, number of particles, etc. This assumption is not valid for one of the important cases, an asymmetric collider; more general formulae are presented in Sec. (2.4).

*Work supported in part by the National Science Foundation and the Department of Energy contract number DE-AC03-76SF00515.
† Group co-leader.

Table 1: Comparison of B-Factories

Case	Description	W (GeV)	$\mathcal{L}^*$ (10^{33} cm^{-2} s^{-1})	Collider Requirements
1	Asymmetric collider at the $\Upsilon(4S)$	10.6	0.45–16.	Beam energy ratio 2 to 10 1–1.5 cm radius IR beam pipe σ_δ<0.001
2	Symmetric collider above BB* threshold	10.6	2.1–77.	σ_δ<0.001
3	Collider in the continuum	16.	18.–640.	No requirement on σ_δ
4	Collider at the Z, no polarization	93.	0.68–25.	
5	Collider at the Z, with polarization	93.	0.14–5.0	90% polarization

*The luminosity needed to observe a three standard deviation effect in the CP violating asymmetry for $B^0 \to \Psi K_s$ in 10^7 sec of running at peak luminosity.

Table 2: Symbols*

Beam energy (in units of mc^2)	γ
Center-of-mass energy	W, $\gamma_{cm} = 2\sqrt{\gamma_1\gamma_2}$
Luminosity	$\mathcal{L}$
Beam sizes (horiz., vert., ratio)	$\sigma_h, \sigma_v, R_\sigma = \sigma_v/\sigma_h$
Collision and revolution frequencies	f_c, f_0
Particles per bunch	N
Amplitude (β) functions at IR	β_v, β_h
Dispersion at the IR (horiz.)	η
Natural emittance	ϵ
Bunch length & energy spread	σ_L, σ_δ
Momentum compaction	α
Effective longitud. impedance	$\langle Z_L/n \rangle$
Loss factor	k
Beam-beam tune shift	ξ
Disruption parameter	D
Enhancement parameter	H
Beamstrahlung parameters	δ_{cl}, Υ
Normalized emittance	ϵ_n
Classical electron radius	$r_e = 2.82 \times 10^{-15}$ m
Fine structure constant	$\alpha = 1/137$
Impedance of free space	$Z_o = 377\ \Omega$
Electronic charge	$e = 1.6 \times 10^{-19}$ C

*subscripts h and v refer to horizontal and vertical, + and − to e^+ and e^-, and 1 and 2 to the two beams. If the energies are unequal, $\gamma_1 \geq \gamma_2$.

The luminosity is given by

$$\mathcal{L} = \frac{N^2 f_c}{4\pi\sigma_h\sigma_v} . \qquad (1)$$

Symbols are defined in Table 2.

Usually a phenomenological approach to the beam-beam interaction is used in storage ring design. The strength of the beam-beam interaction is parametrized by the beam-beam tune shifts

$$\xi_j = \frac{r_e}{2\pi} \frac{N\beta_j}{\gamma\sigma_j(\sigma_h + \sigma_v)}; \ (j = v, h) . \qquad (2)$$

A maximum tune shift, a "tune shift limit," is chosen based on experience,[1] and $\mathcal{L}$, written in terms of ξ, is maximized. The beam-beam limit is a dynamical effect where details matter, and the effective tune shift limit could depend on the the way $\mathcal{L}$ is maximized. The phenomenological approach assumes this is not the case.

If there are no intensity limits, the horizontal and vertical tune shift limits are equal, and making $\beta_v / \beta_h = R_\sigma$ so that the ring operates simultaneously at both limits

$$\mathcal{L} = \frac{\pi\gamma_{cm}^2 f_c}{4r_e^2} \xi^2 (1 + R_\sigma)^2 \frac{\sigma_h^2}{\beta_h\beta_v} \qquad (3)$$

The vertical β function, β_v , should be minimized consistent with the limit[2,3]

$$\beta_v \gtrsim \sigma_L . \qquad (4)$$

The horizontal beam size and β_h enter in the ratio

$$\frac{\sigma_h^2}{\beta_h} = \varepsilon_h + \frac{\eta^2\sigma_\delta^2}{\beta_h} , \qquad (5)$$

which should be as large as possible. Large emittance and/or dispersion is needed. Within the limits of this analysis these are equivalent, but the same mechanisms leading to Eq. (4), synchrotron modulation of the beam-beam kick, could play a role here.[4] If so, the dispersion should be zero and the emittance must be large. Either the machine aperture or a breakdown of the assumption of no intensity limits will determine the maximum.

It is more likely that there would be a limit from the beam-beam interaction and a limit on the single bunch intensity. The microwave instability is expected to be the dominant single bunch effect;[5] the threshold is

$$N < \frac{\sqrt{\pi}\alpha\gamma\sigma_\delta^2\sigma_L Z_o}{\sqrt{2}r_e\langle Z_L/n\rangle} . \qquad (6)$$

$\langle Z_L/n \rangle$ is the effective impedance of the accelerator. Crudely, it measures the "smoothness" of the beam enclosure—the number and types of discontinuities. In detail $\langle Z_L/n \rangle$ is not a simple concept; the frequency content of the bunch, the cut-off frequency for electromagnetic wave propagation in the vacuum chamber, and unresolved physics such as coherent synchrotron radiation are incorporated in that one quantity. The value of Eq. (6) is that it gives the dependences on accelerator parameters and provides a rule-of-thumb based on experience. Modern storage rings have $\langle Z_L/n \rangle \sim 1\Omega$, and a value as small as 0.1 to 0.2 Ω might be possible with a combination of a large aperture, smooth vacuum chamber and an RF system designed to reduce impedance.

When there is a single bunch intensity limit,

$$\mathcal{L} = \frac{\gamma_{\rm cm} f_c}{4 r_e} \xi (1 + R_\sigma) \frac{N}{\beta_v} , \qquad (7)$$

and σ_h^2/β_h is not free; rather it is set by the requirement of reaching the tune shift limit when N is at its limit. Combining Eqs. (4), (6) and (7), the upper limit to the luminosity is

$$\mathcal{L} < \frac{\sqrt{\pi} Z_o}{8\sqrt{2} r_e^2} (\gamma_{\rm cm}\sigma_\delta)^2 \frac{\alpha f_c (1 + R_\sigma)\xi}{\langle Z_L/n \rangle} . \qquad (8)$$

The middle factor depends on particle physics. For Cases 1 and 2 in Table 1, $\gamma_{\rm cm}$ is fixed and there is an energy spread requirement for kinematic reconstruction to be useful. For continuum running, $\gamma_{\rm cm}$ is roughly determined, but there is no constraint on σ_δ. Therefore, this has a higher luminosity potential, but probably it is not the most cost effective way to study CP violation.

A large RF system would be part of any storage ring B–factory. This has consequences for instabilities (discussed above), "higher mode losses," and RF power; any of these could cause an intensity limit. An example of higher mode losses is the energy lost by the beam when it excites the resonant modes of an RF cavity. These losses are

$$P_{HOM} = k\, N^2 e^2 f_c , \qquad (9)$$

where k is the loss factor (with units V/C) that is related to the impedance. This energy loss depends on N^2 and can be comparable to the synchrotron radiation power. Coupled bunch instabilities are caused primarily by high–Q resonant modes of the RF cavities. All designs assume that multibunch feedback[6] and cavity mode damping are used and that these are sufficient to control the coupled bunch instabilities. Overall, the RF system must be designed to minimize $\langle Z_L/n \rangle$ and k, effectively damp high–Q modes, and have a reasonable power demand.

Table 3: Present and Upgraded CESR Parameters*

Revolution frequency		390 kHz	
RF frequency		500 MHz	
Beam energy range		4.5–6.0 GeV	
Fractional energy spread		6.2×10^{-4}	
Energy loss/turn		1.04 MeV	
Bending radius		89 m	
		Present	Upgraded
Number of interaction regions		2	1
Collision frequency (MHz)		2.7	5.5
Particles per bunch (10^{11})		1.8	3.3
Horizontal emittance (mm-mrad)		0.16	0.16
Crossing parameters (cm)	β_v	1.5	1.5
	β_h	100	100
	η	55	66
Tune shifts	ξ_v**	0.021	0.030
	ξ_h	0.026	0.036
Bunch length (cm)		1.7	1.7
Momentum compaction (10^{-2})		1.5	1.0
$\langle Z_L/n \rangle$ limit (Ω, Eq. 6)		0.89	0.32
Synch. radiation power (kW)		85	320
Higher order mode power (kW)		39	300
Peak luminosity (10^{32} cm^{-2}s^{-1})		1.0	5.0

*Energy dependent parameters are calculated at 5.3 GeV.
**ξ_v is determined for the effective β.

Specific storage rings are now discussed. They are separated into upgrades and near-term prospects (Sec. 2.2) and possibilities for the further future (Secs. 2.3 and 2.4).

2.2 Upgrades and Near Term Prospects

2.2.1 Cornell electron storage ring (CESR). CESR is the highest luminosity e^+e^- collider in the world. It operates with seven bunches per beam in a single ring; electrostatically produced orbit distortions separate the beams at parasitic crossings.[7] The luminosity records are a peak of 1×10^{32}cm^{-2}s^{-1} and integrated luminosities of 4.7 pb^{-1}/day and 27 pb^{-1}/week. CESR parameters and present performance are given in Table 3.

An upgrade is planned that could produce a factor of five increase in luminosity. The two major elements of the upgrade are eliminating one of the interaction regions (IR's) and doubling the number of bunches. The plan also includes other, less significant changes and injection and efficiency improvements; the latter will increase the ratio of integrated to maximum luminosity.

With the installation of the CLEO II detector, there is reduced interest in running the CUSB detector, and the CUSB IR will be eliminated.[8] This will increase the synchrotron radiation emitted between collisions at the remaining IR. Synchrotron radiation leads to damping and randomization of oscillation phases. In both simple models[9,10] and data based on storage ring performance[11,12] this increases the tune shift.

The models predict the tune shift limit is proportional to the square root of the fractional energy loss between collisions; the parametric dependence of the data is not conclusive. The effects of synchrotron radiation have been studied experimentally in CESR. With one (two) collision(s) per revolution the tune shift limit was 0.021 (0.017).

The planned increase of ξ from more synchrotron radiation between collisions is $\sqrt{2}$. Based on the experiment at least one-half of that should be realized. There are also possibilities of increasing σ_h^2/β_h by raising η, but because of the precautions following Eq. (5), this is not anticipated to be a large factor. With no single bunch intensity limit, $\mathcal{L} \propto \xi^2$, and the increase in the tune shift limit would give a factor of two in luminosity.

Doubling the number of bunches requires new electrostatic separators close to the IR for less distance between bunches and a higher horizontal tune for more parasitic crossing points. The separators are being designed, and a lattice for 14 bunches per beam has been developed. The chromatic and nonlinear aspects of the lattice are acceptable; this is another consequence of eliminating one of the IR's. The higher tune decreases the momentum compaction, which is approximately inversely proportional to the square of the horizontal tune, and the microwave instability threshold [Eq. (6)]. The consequences are considered below.

The uncertainties in doubling the number of bunches are the effects of additional parasitic crossings and intensity limits. There is a luminosity degradation of 10 to 20% in collisions of single bunches when the electrostatic orbit distortions are applied. This is attributed to magnet errors and nonlinear elements that affect the separated beams differently. At low currents there is no further degradation as more bunches are added to the beam; the effects of parasitic crossings appear near the beam-beam limit. There, increasing the number of bunches requires greater beam separation to preserve tune shift and beam lifetime. By doubling the number of parasitic crossings the horizontal aperture could limit bunch charge.

With an upgraded CESR operating at the beam-beam limit, the synchrotron radiation and higher mode powers would be 320 kW and 300 kW. The number of RF cavity cells would have to be raised from 28 to 48 to support this beam while remaining conservative with factors such as RF window design. Feedback and cavity resonant mode damping are proposed to control coupled bunch instabilities, and when this is done the dominant effect of the cavities should be an impedance increase.

The impedance limit is 0.32 Ω. This is a factor of three below the present limit due to the increased number of particles per bunch and reduced momentum compaction. It is unlikely that the impedance is this low but there are no relevant data,[13)] and the needed impedance and stability calculations have not been performed. These calculations and/or experience may show that a new RF system with reduced impedance is needed. Reducing the impedance and solving the problems from the increased number of parasitic crossings are the major obstacles to reaching the upgrade luminosity goal of 5×10^{32} cm^{-2} s^{-1}.

Table 4: TRISTAN Accumulation Ring

General parameters	
Maximum beam energy	6.5 GeV
Revolution frequency (f_0)	795 kHz
Maximum number of bunches	10
Fractional energy spread (5.3 GeV)	9.3×10^{-4}
Energy loss per turn (5.3 GeV)	2.94 MeV
Achieved single beam performance (May & June 1988)	
Beam energy	5.0 GeV
Collision β functions (horiz., vert.)	0.40 m, 0.02 m
Horizontal emittance (ϵ_h)	1.7×10^{-7} m-rad
Particles per bunch (N)	2.4×10^{11}
Projected Luminosity Performance	
Peak luminosity	2×10^{32}cm^{-2}s^{-1}
Beam lifetime	3 hours
Experimental time per fill	1 hour
Filling time	15 min
Lost time due to Main Ring transfers	20 min
Running time/year	3600 hours
Integrated luminosity per day	8 pb^{-1}
Integrated luminosity per year	1200 pb^{-1}

2.2.2 TRISTAN accumulation ring at KEK. The primary function of the Accumulation Ring is positron collection for the TRISTAN Main Ring. However, the ring has the size and energy range of a B–factory, and this use has been studied.

The collider would be a single ring with (electrostatically) separated electron and positron orbits to avoid unwanted collisions. Up to ten bunches per beam are possible, but the ring has a high energy loss per turn and total RF power will limit the average current. Optics including low β interaction regions have been designed, and single beam performance was tested earlier this year. The results are summarized in Table 4; two such beams and a tune shift of $\xi = 0.04$ would have given $\mathcal{L} = 6.8 \times 10^{31}$ cm^{-2} s^{-1}. It is planned to test single bunch colliding beam performance in November 1988.

Projected performance is given in Table 4. With that running time and peak luminosity the TRISTAN Accumulation Ring would be competitive with the best CESR performance to date.

2.2.3 Paul Scherrer Institute. The Paul Scherrer Institute (PSI) has prepared a proposal for a B–factory that has been submitted to the Swiss government.[14)] This machine would be optimized for symmetric collisions at the $\Upsilon(4S)$

Table 5: Parameters* of the PSI Collider Operating at the $\Upsilon(4S)$

Revolution frequency (f_0)	463 kHz
Maximum number of bunches	20
Collision β functions (horiz., vert.)	1.00 m, 0.03 m
Beam-beam tune shift limit (ξ)	0.03 to 0.04
Horizontal emittance (ϵ_h)	5.5×10^{-7} m-rad
Particles per bunch (N)	6.6×10^{11}
Bunch length (σ_L)	0.02 m
Fractional energy spread (σ_δ)	7.0×10^{-4}
Momentum compaction (α)	0.025
Energy loss per turn (U_0)	1.75 MeV
Synch. radia. power (10 bunches)	860 kW
RF frequency	500 MHz

*These parameters are for one of the two rings.

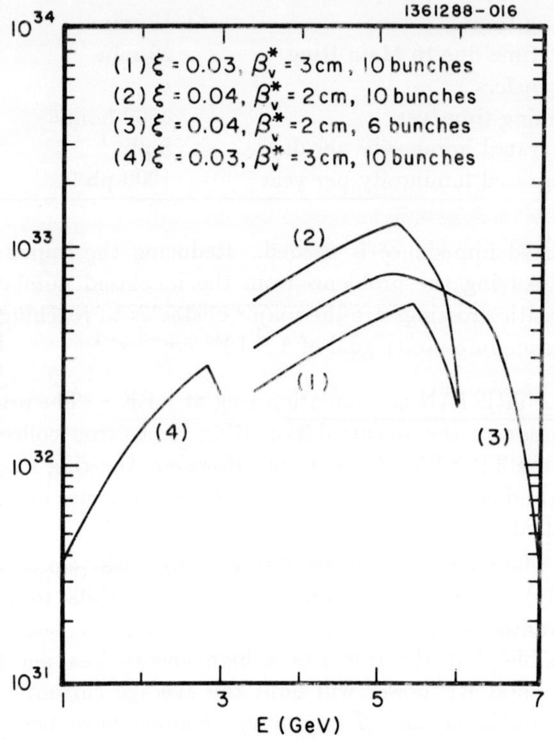

Fig. 1. Luminosity projected for the PSI B–factory. Parameters for the curves are given in the figure; (1) and (4) have different amounts of RF power.

energy, but an extended center-of-mass energy range, 2 to 14 GeV, and asymmetric collisions are part of the design. Principal parameters are given in Table 5, and Fig. 1 shows the performance estimates. The peak luminosity would be two to four times below the minimum for a symmetric collider in Table 1.

The design is that of a double storage ring with two interaction regions. The maximum number of bunches per beam is twenty; it is determined by the requirements of having head-on collisions and avoiding all parasitic collisions. The other parameters and performance result from assuming a beam-beam tune shift limit of 0.03 to 0.04, in the middle of the range for past colliders, and following Eqs. (3) and (4).

Little consideration has been given to factors that could limit the beam current. These could have substantial influence as the design develops, but they do not present fundamental problems. For the parameters in Table 5, the upper limit on $\langle Z_L/n\rangle$ to prevent bunch lengthening is 0.65 Ω, which should be achievable. The proposal calls for an RF cavity optimized for damping of higher modes and multibunch feedback to damp coupled bunch instabilities. Calculations showing the adequacy of this solution remain to be done. For a single RF cavity cell with the approximate geometry proposed, $k = 1.3\times10^{11}$ V/C for $\sigma_L = 2$ cm. This gives 330 kW higher mode loss in the RF system for 10 bunches, and more RF power than called for in the proposal would be needed.

The PSI B–factory should perform close to the design goals in Fig. 1. For the higher luminosities needed to study CP violation either the collision frequency, single bunch intensity and/or tune shift must be increased.

2.3 Future Symmetric Storage Rings

Two concepts have been proposed to reach these higher luminosities. Both require performance beyond present experience with storage rings, and, therefore, require accelerator physics R&D. The accelerator physics issues are highlighted by the contrast between the parameters in Table 6.

2.3.1 NOVOSIBIRSK. A. N. Dubrovin et al. have presented a conceptual design of a double storage ring collider with a luminosity of 10^{34} cm^{-2} s^{-1} at a beam energy of 5.3 GeV.[15] The features that lead to the high luminosity are the short bunch length and low β_v at the interaction point.

The bunch length, emittance and momentum compaction of a storage ring are not completely independent, and α is small as it must be for a short bunch. The short bunch and low α lead to a stringent impedance limit, $\langle Z_L/n\rangle < 0.06$ Ω. In addition, small α leads to a small ϵ_h, and dispersion must be used to produce horizontal beam size needed for high luminosity and moderate tune shift. The resultant synchrotron modulation of the vertical beam-beam kick raises the question of the feasibility of ξ_v=.05. This together with the impedance limit are the crucial issues for this design.

2.3.2 Round beams. Dynamical effects that lead to the beam-beam limit can be controlled by the profile of the beam at the collision point; Eq. (4) is a well-known example of this. A round beam, defined as a beam with $\epsilon_h = \epsilon_v$ and $\beta_v = \beta_h$, has a high tune shift limit,[16] and the

Table 6: **Parameters of B–factory Storage Rings Designed to Operate Near the $\Upsilon(4S)$**

		Dubrovin et al.*	Round Beam
Collision frequency (MHz)		4.0–[11.3]	10
Particles per bunch (10^{11})		4–[10]	6
Energy spread (10^{-3})		[1]–1.5	1
Bunch length (cm)		0.8	1.5
Momentum compaction (10^{-2})		0.1–[0.4]	1
Synchrotron tune		0.009–0.013	0.07
Emittances (m-rad)	ϵ_v	0.3–[3] $\times 10^{-9}$	1×10^{-7}
	ϵ_h	0.1–[3] $\times 10^{-7}$	1×10^{-7}
Crossing parameters (m)	β_v	0.01	0.03
	β_h	0.56	0.03
	η	1.28	0.00
Tune shifts	ξ_v	0.05	0.10
	ξ_h	0.0025–0.01	0.10
Calculated Quantities:			
Beam sizes	σ_v	5.5 μm	55 μm
	σ_h	1.3 mm	55 μm
	R_σ	0.004	1
$\langle Z_L/n \rangle$ limit (Ω, Eq. 6)		0.06	0.44
Luminosity limit (Eq. 8) (10^{34} cm^{-2} s^{-1})		1.4	1.5

*Dubrovin et al. give the range of parameters in the first part of the table. The ones in square brackets were selected for the calculations in the second part.

"Round Beam" collider in Table 6 is based on this.[5] The crucial question is whether high tune shifts can be reached for the more realistic situation of approximately round beams, beams that are nominally round but with some difference between horizontal and vertical.

Initial results from simulations are encouraging; Fig. 2 is an example. In this figure the betatron tunes differ by 0.01, but the tune shift is linear with current to $\xi \sim 0.1$. Simulations are continuing with the major thrust being to include synchrotron oscillations. Phenomena new to storage rings are expected. The disruption parameter which characterizes single pass effects and the tune shift are related as:

$$D = \frac{2r_e \, N \, \sigma_L}{\gamma \sigma_h^2 \, R_\sigma (1 + R_\sigma)} = 4\pi \, \xi \, \frac{\sigma_L}{\beta_v} \quad . \qquad (10)$$

For $\xi \sim 0.1$ and $\beta_v \sim \sigma_L$, $D \sim 1.2$, and single pass collision effects become important, and the consequences are unclear. Luminosity enhancement (good!) and emittance dilution (bad!) are both possibilities. If simulation results remain encouraging, experimental studies are next, and the detailed design of a Round Beam collider could begin.

2.4 Asymmetric Storage Rings

The interest in asymmetric colliders has been stimulated by the luminosity advantage in Table 1 and the possibility of low cost if an existing facility could be used as

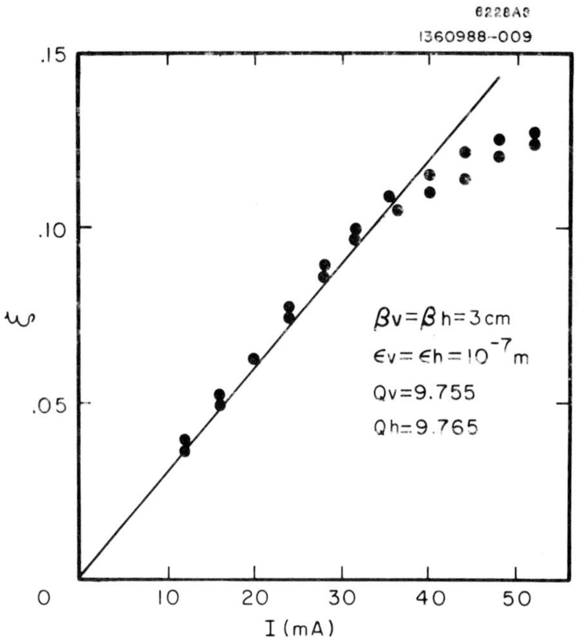

Fig. 2. Beam-beam tune shift vs. single bunch current for a machine with CESR-like parameters except for a round beam collision geometry.

the high energy ring. The expressions in Sec. 2.1 can be generalized for beams with different properties; that generalization, keeping the restriction that the beams have equal sizes, is in Table 7.[17] The comments in Sec. 2.1 regarding the beam-beam interaction, instabilities, higher

Table 7: Generalizations of Formulae in Sec. 2.1*

Eq. No.	Generalization	Revised Eq. No.
(1)	$\mathcal{L} = \frac{N_1 N_2 f_c}{4\pi\sigma_h\sigma_v}$	(1′)
There are four tune shift equations; eg., Eq. (2′):		
(2)	$\xi_{v1} = \frac{r_e}{2\pi}\frac{N_2\beta_{v1}}{\gamma_1\sigma_v(\sigma_h+\sigma_v)}$	(2′)
(3)	$\mathcal{L} = \frac{\pi\gamma_{cm}^2}{4r_e^2} f_c\xi^2(1+R_\sigma)^2 \times \sigma_h^2\left(\frac{1}{\beta_{v1}\beta_{v2}\beta_{h1}\beta_{h2}}\right)^{1/2}$	(3′)
(4)	$\beta_{v1} \geq \sigma_{L2},\ \beta_{v2} \geq \sigma_{L1}$	(4′)
(5)	$\frac{\sigma_h}{\sqrt{\beta_{hi}}} = \left(\epsilon_{hi} + \frac{\eta_i^2\sigma_{\delta i}^2}{\beta_{hi}}\right)^{1/2} \quad (i = 1,\ 2)$	(5′)
(7)	$\mathcal{L} = \frac{\gamma_{cm}}{4r_e}\,\xi f_c\,(1+R_\sigma)\left(\frac{N_1}{\beta_{v1}}\frac{N_2}{\beta_{v2}}\right)^{1/2}$	(7′)

*Equal horizontal and equal vertical sizes are assumed: $\sigma_{h1} = \sigma_{h2}$, $\sigma_{v1} = \sigma_{v2}$.

mode losses, and RF system design are applicable to an asymmetric ring also. The interaction region design is unique because of the different beam energies and particle physics requirements.

There are two conceptual designs, one based on PEP[18] and the other on PETRA;[19] both are constrained to some degree by the existing machine. The parameters are in Table 8.

The beam-beam tune shifts are assumed equal for the two beams, $\xi_1 = \xi_2$; the values chosen are 0.05 and 0.03 for the PEP and PETRA based B–factories, respectively. There is no experience with collisions of unequal energy beams, and data on the energy dependence of the beam-beam interaction has large uncertainties. Different parametrizations of the dependence of ξ on synchrotron radiation energy loss[12] lead to conclusions ranging from $\xi_2 \sim \xi_1$ to $\xi_2 \sim \xi_1/3$. An investigation of the beam-beam interaction with unequal beam energies is needed; computer simulations are likely to be a major component of that study.

Beam current limits in the high energy ring have been discussed for the PETRA design. The single bunch intensity limit is known from experiment to be above the values in Table 8,[20,21] and the single bunch currents are determined by the PETRA aperture. Coupled bunch instabilities are more serious. Eighty-eight 4-cell superconducting cavities make up the synchrotron radiation loss. Reduction of the Q's of higher modes and a newly developed feedback technique[6] are required for stability.

There are enough details for the PEP based collider to look at intensity limit for the low energy ring. The upper limit on $\langle Z_L/n\rangle$ for the low energy ring is 1 Ω; this is a reasonable design goal.

At the interaction region (IR) of an asymmetric collider two very different energy beams must be focused and separated in a short distance. The solution adopted in the PEP design is shown in Fig. 3; the PETRA one is similar. Quadrupoles close to the IR focus the low energy beam. They are followed by dipole magnets acting as beam separators and focusing quadrupoles for the high energy beam.

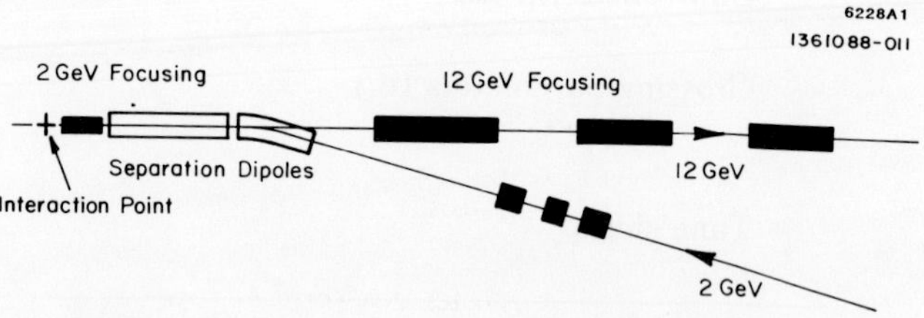

Fig. 3. IR geometry for the PEP B–factory.

This particular design has a 0.16 T dipole beginning 1 m from the collision point. The critical energy of the synchrotron radiation from the 12 GeV beam is 15 keV, and, without masking, the fan of synchrotron radiation is 16 mm wide at the interaction point. Vertex detection is an essential feature of an asymmetric collider, and a beam pipe with 10 to 15 mm radius is required for this. Synchrotron radiation masking must be a central feature of the interaction region. The masks can act as sources for secondary high energy particles,[22] and the mask design must include this as a consideration. The combination of focusing different energies, beam separation, a small beam pipe, and synchrotron radiation masking make the IR a major accelerator physics problem of asymmetric colliders.

2.5 Conclusions

There are plans at Cornell, KEK and PSI for upgrades, conversions, or new storage rings that will advance knowledge of B physics but would not have enough luminosity to observe Standard Model CP violation. There are some uncertainties in these ideas, but they should perform near proposed levels.

There are concepts for symmetric storage rings with $\mathcal{L} \sim 10^{34}\,\text{cm}^{-2}\,\text{s}^{-1}$ and asymmetric storage rings with $\mathcal{L} \sim 10^{33}\text{cm}^{-2}\text{s}^{-1}$. They all require performance beyond our experience, but the accelerator physics issues are clearly defined and could be addressed on the time scale of a year. If resolved successfully, a detailed design of a collider with luminosity in the range needed to see CP violation could begin.

3. LINEAR COLLIDERS FOR THE Υ(4S) AND CONTINUUM

3.1 Introduction

Compared to a storage ring, a linear collider has a small number of particles per bunch and a low collision

Table 8: Parameters of Asymmetric B–Factories

		Based on PEP	
Collision frequency		11.6 MHz	
Luminosity		0.5×10^{33} cm^{-2} s^{-1}	
		Large Ring	Small Ring
Beam energy (GeV)		12	2
Revolution frequency		136 kHz	1.9 MHz
Particles per bunch (10^{11})		1.4	2.9
Emittances (mm-mrad)	ϵ_v	0.01	0.03
	ϵ_h	0.1	0.3
Crossing parameters (cm)	β_v	7.6	2.5
	β_h	76	25
Tune shifts ($\xi_v = \xi_h$)		0.05	0.05
Energy spread (10^{-3})		0.80	0.56
Bunch length (cm)		1.9	2.4
Momentum compaction $\times$ (10^{-3})		3.0	63
$\langle Z_L/n \rangle$ limit (Ω, Eq. 6)		1.0	1.1
Fractional energy loss/turn (10^{-4})		9.3	0.63
Bending radius (m)		166	10.8
Synchrotron radiation power (MW)		2.9	0.067
		Based on PETRA	
Collision frequency		2.6 – 31.2 MHz	
Luminosity		0.09–2.2 $\times 10^{33}$ cm^{-2} s^{-1}	
		Large Ring	Small Ring
Beam energy (GeV)		14	2
Revolution frequency		130 kHz	2.6 MHz
Particles per bunch (10^{11})		1.4	4.3
Emittances (mm-mrad)	ϵ_v	0.03	0.07
	ϵ_h	0.2	0.47
	β_v	7.0–3.5	3.0–1.5
(cm)	β_h	47–24	20–10
Tune Shifts ($\xi_v = \xi_h$)		0.03	0.03
Energy spread (10^{-3})		0.85	1.1
Fractional energy loss/turn (10^{-4})		12.7	1.5
Bending radius (m)		192	4.65
Synchrotron radiation power (MW)		1.1–13.	0.05–0.6

frequency; the luminosity comes from focusing the beam to a small spot. The beam-beam interaction is stronger, and this gives additional focusing that leads to a luminosity enhancement. The luminosity is given by Eq. (1) with an additional enhancement factor H that is a function of the disruption parameter D [Eq. (10)] and σ_L/β_v[23)]

$$\mathcal{L} = \frac{N^2 f_c}{4\pi\sigma_h\sigma_v} H \,. \tag{11}$$

The principal accelerator physics issues are the beam-beam interaction (disruption and beamstrahlung), positron production and damping, and the appropriate RF system for acceleration. The issues are the same for TeV energy colliders, but B–factory parameters are so different that each is a unique problem. The discussion at Snowmass was (and this paper is) restricted to symmetric linear colliders, but the linear collider idea could be applied equally well to an asymmetric machine.

At small values of D the beam-beam interaction acts like a lens with focal length σ_L/D. At larger values particles oscillate with approximately $(D/10)^{1/2}$ oscillations during the beam passage. The luminosity enhancement from focusing during the collision has been calculated using simulations.[23,24)] Recent results for H are shown in Fig. 4. This particular calculation assumes a head-on collision and beams with equal properties (number of particles, transverse dimensions, ...) and has a strong restriction on the electromagnetic fields.[25)] As a result it gives the largest possible enhancement and no information on the effects

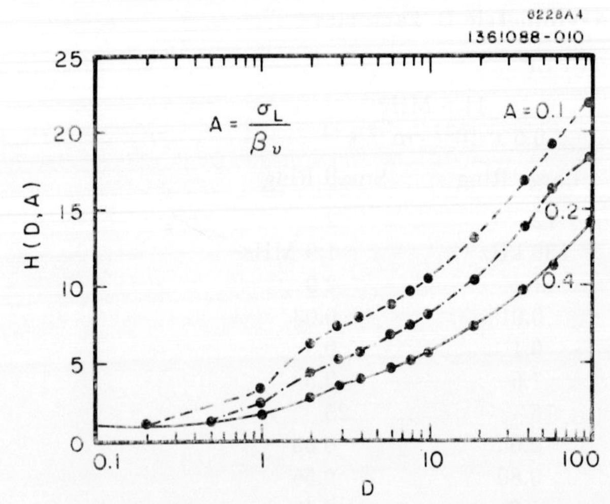

Fig. 4. Luminosity enhancement calculated by Chen and Yokoya.

of errors. This work is being generalized and information about errors should be available in the future.[26)]

The maximum value of disruption is likely to be determined by tolerance to errors, and knowing that value is central to any linear collider design. The maximum value of D is analogous to the tune shift limit of storage rings. Assume that N is limited, e.g., by wakefields, then

$$\mathcal{L} = \frac{\gamma_{cm}}{16\pi r_e} H \; f_c(1+R_\sigma)\; \frac{ND}{\sigma_L} \; . \qquad (12)$$

This equation is to be interpreted in the same way as Eq. (7); the transverse dimensions of the bunch are chosen to reach the disruption limit at the intensity limit. For a fixed $\mathcal{L}$, the collision frequency, positron production rate (Nf_c), bunch length, etc., depend on the intensity and disruption limits. Changing D would affect all of these. Linear collider B–factory parameter lists have values of D ranging from 9 to 28. Without the appropriate calculations it is difficult to know whether these values are practical.

Beamstrahlung contributes to the center-of-mass energy spread. This gives a strong constraint on the luminosity for colliders that have an energy spread specification. The beamstrahlung parameter is

$$\Upsilon = \frac{5r_e^2}{6\alpha} \frac{N\gamma}{(1+R_\sigma)\sigma_h\sigma_L} \; . \qquad (13)$$

(α is the fine structure constant in this equation.) With its low energy a B–factory would be in the classical beamstrahlung regime, $\Upsilon \ll 1$. The center-of-mass energy spread is[27)]

$$\frac{\sigma_W}{W} = \left[\frac{\sigma_\delta^2}{2} + 0.10\;\delta_{cl}^2\left(1+3.9\frac{\gamma r_e}{\sigma_L\Upsilon}\right)\right]^{1/2} \; . \qquad (14)$$

The first term comes from the energy spread of the beams and the second from beamstrahlung; it includes contributions from variation of the deflecting fields within the bunch and fluctuations in the number of photons emitted. The quantity δ_{cl} is the mean fractional energy loss from beamstrahlung

$$\delta_{cl} = \frac{5N^2r_e^3\;\gamma}{6(1+R_\sigma)^2\sigma_h^2\sigma_L} \quad ; \qquad (15)$$

roughly, δ_{cl} is proportional to $\mathcal{L}/f_c\sigma_L$. The ideal would be a high collision frequency and a beam with a low energy spread and a long bunch. This has implications for the accelerator choice and the positron source. When their performances are considered, requiring a low energy spread limits the luminosity to about $10^{33}\,\mathrm{cm}^{-2}\,\mathrm{s}^{-1}$. A specific example is given for ARES in Sec. (3.2).

It is likely that an electron beam with appropriate longitudinal and transverse emittances could be generated with a photocathode gun. Positrons have to be produced by an electron beam striking a converter followed by a damping ring to reduce emittances. Positron production and damping have been major R&D areas for the Stanford Linear Collider (SLC), and a B–factory has still harder demands. A comparison of typical B–factory parameters with the SLC shows that: (i) the number of particles per bunch and (ii) the instantaneous power incident on the converter are comparable; (iii) the collision frequency and (iv) the average converter power are two orders of magnitude higher; and (v) the normalized emittance, ϵ_n, is a factor of ten smaller.

The average and instantaneous powers are about 1 MW and 1–10 TW, respectively. There is a conceptual design of a converter for these power levels that has identified the major problems.[28)] These are thermal shock, removal of heat, high radiation doses, and high levels of residual radioactivity. This design could serve as the starting point for the R&D program needed in positron production.

The damping ring must produce e^+ bunches with an invariant emittance of about 10^{-6} m at a rate of roughly 10 kHz. Because of this combination, the appropriate ring would have a large circumference, many closely spaced bunches, and high field wigglers. One such ring has been studied at the Courmayeur Workshop;[29)] it has a 670 m circumference (1.7 T wigglers make up 2/3's of this) and bunches spaced at 7 m. Wideband multibunch feedback and ultrafast extraction kickers are central features of the ring.

The study also considered the suitability of the ring for colliders with a low energy spread requirement. The energy spread and bunch length at the collision point are related to their values in the damping ring by

$$\gamma\sigma_L\sigma_\delta\big|_{\text{collision point}} \geq \gamma\sigma_L\sigma_\delta\big|_{\text{damping ring}} \; . \qquad (16)$$

The right-hand-side determines an upper limit on the damping ring impedance through Eq. (6). At typical linear collider intensities that limit is $\langle Z_L/n\rangle \leq 1\,\Omega$, which is reasonable.

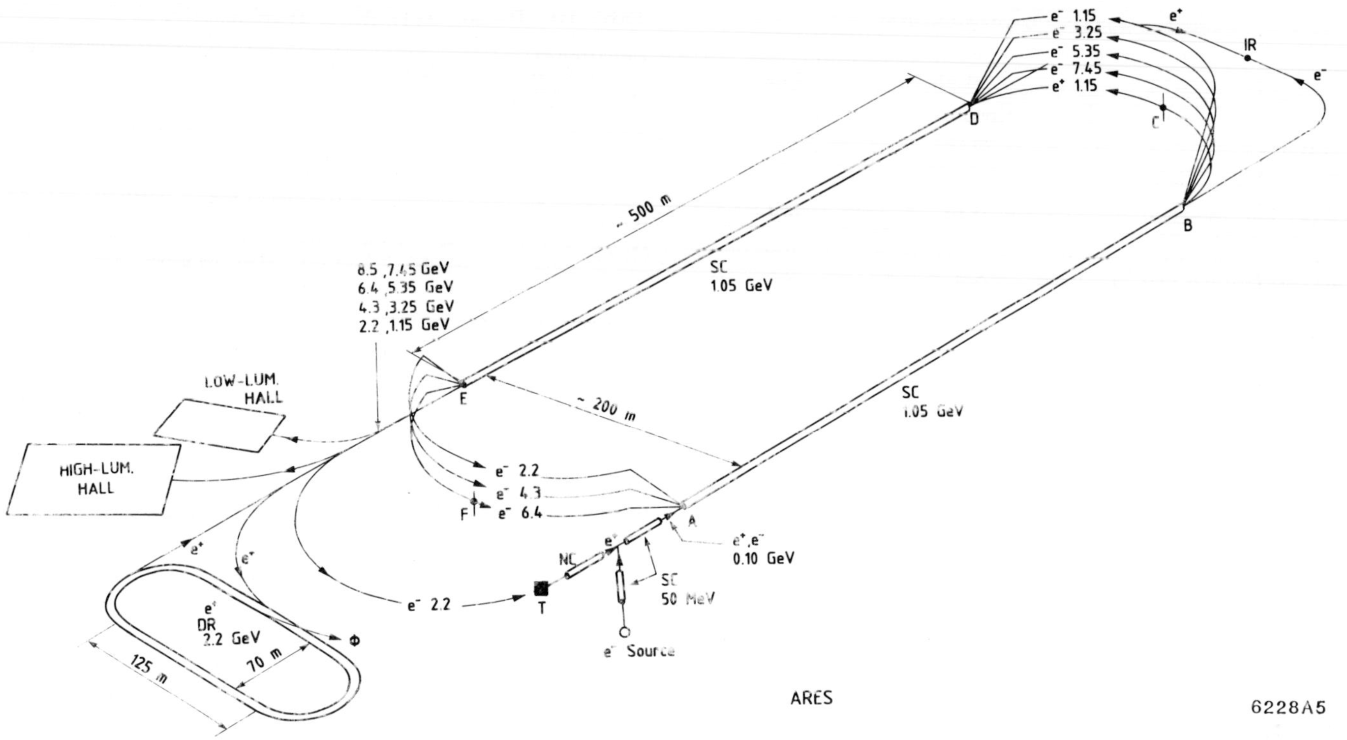

Fig. 5. Schematic of the ARES collider.

The accelerator choice is between a low frequency, superconducting RF linac and a high frequency, room temperature one. From almost all points-of-view superconducting RF seems preferable:

1. A room temperature structure could have a higher acceleration gradient, but the present day gradients and Q's of superconducting RF are adequate.
2. A power source would have to be developed for a high frequency linac but not for a superconducting one.
3. The wall-plug power of a superconducting accelerator would be substantially lower.
4. Transverse emittance increase from wakefields is less serious for low RF frequencies.[30]
5. A low frequency would permit a longer bunch without introducing energy spread. This could be significant for colliders with an energy spread requirement (Cases 1 and 2 in Table 1).

However, a high frequency linac might serve as a prototype for a TeV energy collider. This is discussed in Sec. 3.3. With this introduction specific colliders are now considered.

3.2 ARES

The concept of a linear collider B-factory based on superconducting RF which originated with the work of Amaldi and Coignet[31] has evolved into part of the ARES (Acceleratore Ricircolato per Elettroni, Superconduttore) R&D project at LNF, the Frascati National Laboratory. In December 1987 a workshop was held at Courmayeur, and the proceedings has details about the collider and the associated physics program.[32]

The facility is illustrated in Fig. 5. It is a recirculating superconducting linac with 500 m long accelerating sections with a 5 MeV/m gradient. The accelerator design is based on present day technology; superconducting cavities with 5 MeV/m are available commercially, and the power source is a CW klystron such as the ones used at DESY, KEK, and Cornell. The positron source is a converter, labeled T in the figure, and a 2.2 GeV damping ring. There is no electron damping ring; it is assumed that after some R&D a photocathode gun with appropriate intensity and emittances would be developed. Features that are not part of the B–factory are the experimental halls for nuclear physics and the 0.13 GeV linac for collisions with positrons in the damping ring for producing ϕ's.

Three modes of operation with different values of center-of-mass energy spread, σ_W/W, have been anticipated: high resolution for the $\Upsilon(4S)$, medium resolution for a wide resonance such as the $\Upsilon(5S)$, and low resolution for the continuum. Parameters for the first and third of these are in Table 9.

These parameters reflect the discussion in Sec. 3.1. The narrow energy spread of the high resolution mode is achieved by reducing the luminosity by about an order of magnitude. The design has fewer positrons than electrons

Table 9: ARES Parameters[33)]

Parameter:		High Resolution $\Upsilon(4S)$	Low Resolution Continuum
Beam energy (GeV)		5.29	7.5
$\mathcal{L}$ (10^{34}cm^{-2}s^{-1})		0.13	1.1
$\sigma_W/W(10^{-3})$		0.9	5.8
Particles/bunch (10^{10})	e^-	8.0	8.0
	e^+	2.5	5.0
RF frequncy (MHz)		500	
Gradient (MV/m)		5	
Collision frequency (kHz)		10	
Invariant emittance (10^{-6} m)		2.0	
Spot aspect ratio (R_σ)		1.0	
Beta function ($\beta_v = \beta_h$, mm)		5.0	2.0
Spot radius (μm)		1.0	0.5
Bunch length (mm)	e^-	1.0	0.5
	e^+	3.0	0.7
Disrupt. parameter, D	e^-	22	28
	e^+	21	25
Luminosity enhance., H		7.7	9.5
σ_W/W from beamstrahlung		0.0004	0.005
σ_W/W from σ_δ		0.0009	0.003
Average e^+ converter power (MW)		0.9	1.8
AC power (MW)		18	20

per bunch; this was done to reduce the converter power. The disruptions of the beams are made equal by having a longer positron bunch. The disruption parameter is in the range $D \sim 20$–30. These large values remain to be justified; both the effects of errors and the intentional differences between the beams need study.

Raising the luminosity would require increasing the disruption and/or the collision frequency. The collision frequency is already high, and increasing it would require a multi-megawatt converter and a damping ring with more bunches or a shorter damping time. Substantial R&D would be needed to show this was feasible. The conclusion is that the luminosities in Table 9 are close to the maximum that could be expected.

Superconducting RF development for use in free electron lasers (FEL), nuclear physics accelerators, and future linear colliders (including a B–factory and a TeV energy machine) is the focus of a 70 GLit proposal that the National Institute for Nuclear Physics (INFN) has submitted to the Italian Government. Present plans are that there will be emphasis on the FEL and nuclear physics applications because of the difficulty of the collider.[34)]

Table 10: Parameters for a B–factory Linear Collider

Parameter:	
Beam energy	10.0 GeV
Luminosity	1.0×10^{34}cm^{-2}s^{-1}
σ_W/W	0.005
Particles per bunch	2.2×10^{10}
RF frequency	10 GHz
Gradient	100 MV/m
RF repetition rate	11.1 kHz
Collision frequency	44.4 kHz
Beam bunches/RF pulse	4
Invariant emittance	3.0×10^{-6}m
Spot aspect ratio (R_σ)	1.0
Beta function ($\beta_v = \beta_h$)	0.7 mm
Spot radius	0.32 μm
Bunch length	0.3 mm
Disruption parameter, D	9.0
Luminosity enhancement, H	6.0
σ_W/W from beamstrahlung	4.8×10^{-3}
σ_W/W from σ_δ	1.4×10^{-3}
Average e^+ converter power	1.9 MW
AC power	100 MW

3.3 Linear Colliders with High Frequency RF

Scaling laws for a high frequency, room temperature linear collider B–factory have been developed by P. Wilson;[35)] these account approximately for beamstrahlung and disruption, energy efficiency, wakefields, and final focus design. Starting with values for the acceleration gradient, RF frequency, and AC mains power that are typical of the TeV collider work at SLAC these scaling laws lead to the parameters in Table 10. There are four beam pulses per RF pulse for adequate efficiency, and the RF repetition rate must be about 10 kHz for $\mathcal{L} \sim 10^{34}\,\mathrm{cm}^{-2}\,\mathrm{s}^{-1}$.

Could a B–factory serve as a prototype of a TeV energy collider? Yes, if the principal R&D issues are the same; no, if they are not.[36)] Both colliders must accelerate multiple bunches per RF pulse. This has implications for the accelerator structure, and many aspects of the structure development are the same. The RF power sources and positron sources are significantly different because the B–factory RF repetition rate and collision frequency are over an order of magnitude higher. The beam-beam interaction limits are different; a B–factory would have large disruption and low beamstrahlung compared to the TLC.

R. Palmer has written a computer program that estimates linear collider performance based on the properties and performance of accelerator subsystems. As part of the Snowmass study he used this to estimate the performance of "pure TLC prototypes," linear colliders using TLC technology. For the TLC gradient (186 MeV/m), RF frequency

Table 11: Linac-Storage Ring B–factory

Luminosity	1.6×10^{34} cm^{-2} s^{-1}
Collision frequency	5 MHz
Spot radius ($R_\sigma = 1$) (μm)	2.0

Parameter	Electron Beam	Positron Beam
Beam energy (GeV)	2.5	10.0
Circumference (km)	–	2.0
Particles per bunch (10^{10})	0.2	50
Average current (mA)	1.6	400
Bunch length (μm)	–	5.0
Invariant emittance (μm)	1.0	7.8
Emittance (nm)	0.20	0.40
$\beta_h = \beta_v$ (cm)	2.0	1.0
Beam-beam effects	$D = 360$	$\xi = 0.057$

(18 GHz), and RF repetition rate (400 Hz) the luminosity depends on center-of-mass energy as[37)]

$$\mathcal{L} \sim 10^{34}\ \mathrm{cm}^{-2}\ \mathrm{s}^{-1} \left(\frac{W}{1\ \mathrm{TeV}}\right)^{1.6} . \tag{17}$$

A TLC prototype with W $\sim$ 10–20 GeV would have a luminosity of about 10^{31} cm^{-2} s^{-1}. The repetition rate would have to be raised significantly to get a more interesting luminosity.

The conclusion is that a B–factory is not a prototype for a TeV energy machine. It would require its own R&D program with comparable time scale and cost.

3.4 Linac-Storage Ring Collider

Positron production and damping is one of the major limiting factors of a linear collider B–factory. Therefore, it is natural to consider concepts with positron recovery. There are a number of approaches: (i) recovery, deceleration, and damping at a low energy,[38)] (ii) recovery and damping without changing energy,[39,40)] and (iii) collisions between a linac beam and a beam stored in a storage ring.[41,42)] The latter idea has recently been revived by P. Grosse–Wiesmann as the basis of a B–factory[41)] that was discussed at Snowmass.

The paper by Grosse–Wiesmann has parameters for colliders with luminosities of 0.3, 1.6 and 7.0 $\times$ 10^{34} cm^{-2} s^{-1}; the intermediate one is in Table 11. The electron beam is accelerated in a superconducting linac with a high average current, and the positron beam is stored in a low emittance ring. The accelerator physics issues are the beam-beam interaction, instabilities in the storage ring, and the properties of the linac.

The beam-beam interaction is in a regime where there is no experimental or computer simulation information at the present time. The electron beam is highly disrupted, and the electrons will perform many oscillations during the collision. This could lead to a channeling effect with luminosity enhancement, or emittance blow-up due to nonlinearities. The implications for the positron beam are unknown also.

The storage ring emittance is about a factor of ten below the record minimum of 6.4 nm,[43)] and a combination of a high betatron tune and damping wigglers would be needed to reach it. A high tune alone leads to a small momentum compaction and an unacceptably low impedance limit, $\langle Z_L/n\rangle \sim 0.02\ \Omega$. Damping wigglers increase the energy spread and RF power demand by a few MW but can raise the impedance limit to $\langle Z_L/n\rangle \sim 0.2\ \Omega$ by allowing a lower tune.[44)]

The average linac current is about a factor of ten higher than the CEBAF design. Implications of this have been looked at in a recent paper with the conclusion that, from the point-of-view of wakefields, 10^9 electrons/bunch, an invariant emittance of 10^{-6}m, and a fractional energy spread of 10^{-3} is reasonable.[45)]

The linac-storage ring collider is in an early stage of development. The potential is high, but there are several serious issues that need study before it will be clear whether that potential could be realized.

3.5 Conclusions

Linear colliders and linac-storage ring colliders operating at the $\Upsilon(4S)$ and in the continuum could reach the luminosities needed to study CP violation. These concepts are at an early stage of development with interlocking accelerator physics issues, and developing these ideas will require substantial R&D programs. The results of that R&D could affect the design and performance potential significantly.

The principal issues for a linear collider are the same as for a TeV energy machine: disruption and beamstrahlung, positron production and damping, the accelerator structure, and the RF power source. However, the B–factory parameters are sufficiently different that it cannot serve as a prototype for a higher energy machine. It would need its own R&D program with time and cost scales comparable to those for a higher energy machine.

4. STORAGE RINGS AND LINEAR COLLIDERS FOR THE Z

There are two approaches for studying B physics at the Z. In one of them the beam is unpolarized, and the experimental techniques are similar to those used in the continuum. The advantage of working at the Z is the large cross section for $B\bar{B}$ production. The required luminosity range is 0.68 to 25. $\times$ 10^{33} cm^{-2} s^{-1}. It is natural to compare the projected performance of LEP to this goal; this is done in Sec. 4.1.

A collider operating at $W = m_z$ with a longitudinally polarized electron beam would need significantly less luminosity; the factor depends on the degree of polarization. Polarization in a storage ring is uncertain,[46] and this approach is most likely in the domain of linear colliders. This is discussed in Sec. 4.2.

4.1 LEP Performance

Construction of the LEP storage ring at CERN is nearing completion. Commissioning is scheduled to begin in July, 1989, and it is anticipated that Z's will be observed before the end of that year. This first phase of LEP, LEP I, is designed for running in an energy range near the Z; the beam energy at maximum luminosity is 55 GeV.

The best estimates are that LEP I will be limited to a single bunch current of 0.75 mA (4.16×10^{11} particles) by instabilities at the injection energy of 20 GeV. Using this, an assumed tune shift of $\xi = 0.03$, and four bunches per beam the luminosity at 55 GeV would be $1.6 \times 10^{31} \mathrm{cm}^{-2}\mathrm{s}^{-1}$.[47] The installed RF power in LEP I is 16 MW, and above 55 GeV the luminosity falls because of RF power limits. If the instability threshold could be raised with feedback[48] such that the luminosity at $W = m_z$ was limited only by the beam-beam interaction and the available RF power, that luminosity[49] would be 3.2×10^{31} cm^{-2} s^{-1}. This would require doubling the number of particles per bunch.

For the second phase of LEP, LEP II, superconducting RF will replace the room temperature RF of LEP I, and the energy will reach up to perhaps 100 GeV per beam. That upper energy will be determined by the configuration and gradient of the RF system that is installed.

In addition to raising the energy, superconducting RF would allow a significantly larger current at $W = m_z$. In LEP I most of the RF power goes to producing the accelerating voltage; roughly 10% is radiated by the beam as synchrotron radiation. The LEP II superconducting RF reduces the power needed to produce the voltage, and the stored current can be increased. Some of that increase could be in single bunch current since changing the RF reduces the impedance, but to fully utilize the RF power the number of bunches would have to be raised.

Electrostatic orbit distortions would have to be used to separate the beams at unwanted collision points. It has been suggested that up to 60 mA/beam is possible;[50] the number of bunches would depend on the single bunch current. The luminosity would be approximately 3×10^{32} cm^{-2} s^{-1} if a factor equal to the increased number of bunches were realized. Based on experience with CESR (Sec. 2.2.1) this is an optimistic assumption. This suggestion is under study at CERN. Separation schemes and their implications for LEP are being considered. If a successful detailed plan emerged from that study, LEP would have substantial potential as a B-factory.

Table 12. Z–Factory Parameters

Parameter:		
Beam energy		50 GeV
Luminosity		5.7×10^{33} cm^{-2} s^{-1}
Particles per bunch		1.2×10^{11}
RF frequency		5.9 GHz
Gradient		93 MeV/m
Peak power		270 MW/m
RF repetition rate		872 Hz
Collision frequency		4.36 kHz
Beam bunches/RF pulse		5
Invariant emittance	horiz.	2.69×10^{-5} m
	vert.	2.75×10^{-7} m
Spot aspect ratio (R_σ)		5.7×10^{-3}
Beta function	β_h	7.34 cm
	β_v	0.233 mm
Spot dimensions		4.49 μm × 25.7 nm
Bunch length		0.20 mm
Disruption parameter	D_h	0.05
	D_v	11.7
Luminosity enhance., H		1.7
σ_W/W		0.6%
Average e^+ converter power		1.4 MW
AC power		200 MW

4.2 Z–Factory Linear Collider

The degree of polarization in a linear collider, P, is determined by the electron source. With the presently achievable $P = 0.45$ the luminosity to observe CP violation would be in the range 0.37 to 13. $\times 10^{33}$ cm^{-2} s^{-1}. There are R&D programs to raise the polarization to $P = 0.90$;[51] if successful, the luminosity requirements would be reduced to 0.14 to 5.0 $\times 10^{33}$ cm^{-2} s^{-1}. This is well above the SLC design, and a new linear collider designed without constraints from an existing accelerator would be called for.

The parameters for a Z–factory, a linear collider optimized for running at $W = m_z$ are in Table 12.[37] The design approach is the same as that developed for TeV energy machines. The total AC power is fixed at 200 MW, and the luminosity is maximized while remaining consistent with the performance limitations of collider subsystems. The resulting Z–factory would see CP violation in a year or less of running with a 90% polarized beam! The accelerator physics issues are the same as those discussed in Sec. 3.1.

Constraints and uncertainties from the beam-beam interaction are reduced compared to a linear collider at the Υ(4S). A narrow energy spread is not needed, so there is no limitation from beamstrahlung or the energy spread of the beam. The beam profile at the collision point is flat, $R_\sigma = 0.0057$. Beam steering errors have been simulated for flat beams, and for the Z–factory disruptions in Table 12 the beams tend to self-align and correct these errors.[52]

Positron production and damping need substantial R&D. The converter power and damping rate are comparable to those for ARES, and the discussion in

Sec. 3.2 applies. The design luminosity and the constraint from positron damping rate leads to a high single bunch intensity, 1.2×10^{11} particles per bunch, and an optimum RF frequency of 5.9 GHz. Emittance blow-up would be severe at a higher frequency. An RF power source at that frequency with a peak power capability of 270 MW/m would need to be developed.

A Z–factory R&D program would be a substantial one. Some of the issues are the same as for a TeV energy collider, but the machine could not serve as a "pure TLC prototype" without a substantial reduction of luminosity. A restriction to use TLC technology would reduce the luminosity to 2.5×10^{32} cm^{-2} s^{-1} [Eq. (17)].

4.3 Conclusions

There are opportunities to measure CP violation with machines running at the Z. LEP, upgraded to operate with a large number of bunches, would have a luminosity at the bottom end of the required range and substantial potential as a B–factory. A successful R&D program aimed at a linear collider Z–factory would lead to a machine that exceeds the upper limit given in Table 1. Building such a machine would be a major national commitment.

5. CONCLUSIONS

The conclusions from the three major sections are:

Section 2:

There are plans at Cornell, KEK and PSI for upgrades, conversions, or new storage rings that will advance knowledge of B physics but would not have enough luminosity to observe Standard Model CP violation. There are some uncertainties in these ideas, but they should perform near proposed levels.

There are concepts for symmetric storage rings with $\mathcal{L} \sim 10^{34}$ cm^{-2} s^{-1} and asymmetric storage rings with $\mathcal{L} \sim 10^{33}$ cm^{-2} s^{-1}. They all require performance beyond our experience, but the accelerator physics issues are clearly defined and could be addressed on the time scale of a year. If resolved successfully, a detailed design of a collider with luminosity in the range needed to see CP violation could begin.

Section 3:

Linear colliders and linac-storage ring colliders operating at the $\Upsilon(4S)$ and in the continuum could reach the luminosities needed to study CP violation. These concepts are at an early stage of development with interlocking accelerator physics issues, and developing these ideas will require substantial R&D programs. The results of that R&D could affect the design and performance potential significantly.

The principal issues for a linear collider are the same as for a TeV energy machine: disruption and beamstrahlung, positron production and damping, the accelerator structure, and the RF power source. However, the B–factory parameters are sufficiently different that it cannot serve as a prototype for a higher energy machine. It would need its own R&D program with time and cost scales comparable to those for a higher energy machine.

Section 4:

There are opportunities to measure CP violation with machines running at the Z. LEP, upgraded to operate with a large number of bunches, would have a luminosity at the bottom end of the required range and substantial potential as a B–factory. A successful R&D program aimed at a linear collider Z–factory would lead to a machine that exceeds the upper limit given in Table 1. Building such a machine would be a major national commitment.

Taken as a whole, the accelerator performance needed to see CP violation in B–decay is within reach.

FOOTNOTES AND REFERENCES

1. Storage rings have achieved tune shifts between 0.02 and 0.07. J. Seeman in *Nonlinear Dynamics Aspects of Particle Accelerators,* edited by J. Jowett, M. Month and S. Turner (Springer–Verlag, 1986).
2. S. Myers in *Nonlinear Dynamics Aspects of Particle Accelerators,* edited by J. Jowett, M. Month and S. Turner (Springer–Verlag, 1986).
3. D. Rice, CON 87–3 (Cornell, 1987).
4. S. Peggs and R. Talman, *Phys. Rev.* **D24**, 2379 (1981).
5. R. H. Siemann, CLNS 88/865 (Cornell 1988).
6. R. D. Kohaupt, *Z. Phys.* **C37**, 159 (1987).
7. R. Littauer, *Proc. 12th Int. Conf. on High–Energy Accelerators*, 161 (Fermilab, 1983).
8. Electrostatic separators will separate the beams at that location.
9. A. W. Chao, *AIP Conf. Proc.* **127**, 201 (1985).
10. D. L. Rubin, CON 87–12 (Cornell, 1987).
11. E. Keil and R. Talman, *Part. Accel.* **14**, 109 (1983).
12. J. Seeman in *Nonlinear Dynamics Aspects of Particle Accelerators*, edited by J. Jowett, M. Month and S. Turner (Springer–Verlag, 1986).
13. $N = 5.6 \times 10^{11}$ has been run in a configuration with $\alpha = 0.0063$ and approximately one-half the impedance, but the bunch length was not measured.
14. PSI, "Proposal for an Electron Positron Collider for Heavy Flavour Particle Physics and Synchrotron Radiation," PR–88–09 (July 1988).

15. A. N. Dubrovin, A. N. Skrinsky, G. M. Tumaikin, and A. A. Zholents "Conceptual Design of a Ring Beauty Factory" (Inst. of Nuclear Physics, 1988).
16. J. F. Schonfeld, *AIP Conf. Proc.* **87**, 314 (1982).
17. That restriction could be removed, but there is no strong motivation for doing so (J. Rees, SLAC/AP–67 (SLAC, 1988)), and observations in the S$\bar{p}$pS argue against it (L. Evans, private communication).
18. A. A. Garren, "APIARY I design" (LBL, 1988).
19. H. Nesemann, W. Schmidt–Parzefall, F. Willeke, "The Use of PETRA as a B–factory" (DESY, 1988).
20. R. D. Kohaupt, Habilitationsschrift, Universitat Hamburg (1982).
21. The same is true for PEP. L. Z. Rivkin, in the *Proc. of the Workshop on PEP as a Synchrotron Radiation Source* (SSRL, 1987).
22. N. Mistry *et al.*, these proceedings.
23. P. Chen and K. Yokoya, SLAC–PUB–4339 (SLAC, 1987).
24. R. Hollebeek, *Nucl. Inst. Methods* **184**, 333 (1981).
25. In general the fields can be functions of three spatial coordinates (h, v, z) and time. They are restricted to depend on only $h^2 + v^2$, z, and time.
26. K. Yokoya, 1988 Linac Conference (Williamsburg, VA).
27. R. J. Noble, *Nucl. Inst. Methods* **A256**, 427 (1987).
28. P. Sievers, *Proceedings of the Workshop on Heavy–Quark Factory and Nuclear–Physics Facility with Superconducting Linacs*, edited by E. DeScantis, M. Greco, M. Piccolo and S. Tazzari (Italian Physical Society, 1987), 85.
29. M. Bassetti *et al.*, *Proceedings of the Workshop on Heavy–Quark Factory and Nuclear–Physics Facility with Superconducting Linacs*, edited by E. DeScantis, M. Greco, M. Piccolo and S. Tazzari (Italian Physical Society, 1987), 167.
30. R. H. Siemann, *Ann. Rev. of Nucl. and Part Sci.* **37**, 243 (1987)
31. U. Amaldi and G. Coignet, *Nucl. Inst. Methods* **A260**, 7 (1987).
32. *Proceedings of the Workshop on Heavy–Quark Factory and Nuclear–Physics Facility with Superconducting Linacs*, edited by E. DeScantis, M. Greco, M. Piccolo and S. Tazzari (Italian Physical Society, 1987).
33. U. Amaldi and G. Coignet, *Proceedings of the Workshop on Heavy–Quark Factory and Nuclear–Physics Facility with Superconducting Linacs*, edited by E. DeScantis, M. Greco, M. Piccolo and S. Tazzari (Italian Physical Society, 1987), 59.
34. S. Tazzari, 1988 Linac Conference (Williamsburg, VA).
35. P. B. Wilson, *Proceedings of the UCLA Workshop Linear–Collider BB factory Conceptual Design*, 373 (1987).
36. Comparisons are with respect to the TLC; see R. Ruth *et al.*, these proceedings.
37. R. B. Palmer, these proceedings.
38. H. Gerke and K. Steffen, DESY PET–79/04 (1979).
39. P. L. Csonka and J. Rees, *Nucl. Inst. Methods* **96**, 149 (1971).
40. Work subsequent to the summer study is in D. Cline and C. Pellegrini, these proceedings.
41. P. Grosse–Wiesmann, SLAC–PUB–4545 (1988).
42. R. Sundelin, private communication.
43. H. Winick *et al.*, *Bull. Am. Phy. Soc.* **33**, 1028 (1988).
44. P. Grosse–Wiesmann, these proceedings.
45. J. J. Bisognano *et al.*, 1988 Linac Conf. (Williamsburg, VA).
46. The prospect for polarization in LEP is encouraging enough to install polarimeters to search for it ("LEP Design Report," CERN–LEP/84–01). If found, spin rotators would be needed to change the polarization from transverse to longitudinal for experiments. Compatibility with high luminosity operation (Sec. 4.1) has not been studied.
47. *LEP Design Report*, CERN–LEP/84–01 (CERN, 1984).
48. S. Myers, *Proc. of the 1987 IEEE Part. Accel. Conf.*, 503 (1987).
49. E. Keil, private communication.
50. C. Rubbia, talk at the European Part. Accel. Conf. (1988).
51. C. K. Sinclair, *J. Phys. (Paris)* **46**, C2–669 (1985).
52. P. Chen, these proceedings.

Scaling Relations for Intersecting e^+/e^- Storage Rings of Unequal Energies

D. RUBIN, R. SIEMANN, J. WELCH
Newman Laboratory of Nuclear Studies
Cornell University, Ithaca NY 14853

J. REES
Stanford Linear Accelerator Center
Stanford University, Stanford CA 94395

H. KOISO
KEK National Lab for HEP
Oho-machi, Tskuba-gun, Ibariki-ken,305 Japan

A. GARREN
Fermilab
Wilson Rd. P.O. Box 500, Batavia IL 60510

Introduction

An analysis of the theoretical detection efficiency of CP violation for the specific decay $B_0 \rightarrow \psi K_s$ has been made for asymmetric collisions involving unequal energy beams of 12 GeV and 2 GeV. The efficiency is about six times that of detecting CP violation using symmetric collisions for the same decay mode, operating just above the $\Upsilon(4S)$ resonance.[1][2]

Naturally the question occurs, is there a penalty or advantage in luminosity of a machine with unequal beam energies instead of equal energy beams, assuming the same center of mass energy? The answer to this question depends entirely on the technology and beam dynamics that limit the luminosity; or equivalently, what constraints these limits imply among the variables used to define the luminosity. One analysis of this question has been published and found that under a particular set of constraints, there was no explicit relation between the luminosity and the energy ratio: if all explicit parameters could be held fixed while varying the energy ratio, the luminosity would be unchanged.[3] As we shall see, other sets of constraints do imply an explicit dependence of the luminosity on the energy ratio. In any case, the required luminosity is considerably more than ever attained with conventional e^+/e^- storage rings. Neither the technology nor the appropriate beam dynamics is well determined, so it is far from certain which set of constraints would limit the luminosity in practice. This uncertainty makes it impossible to determine if there is any inherent luminosity advantage or disadvantage in using equal or unequal energy beams. Further research, development and design will be needed before this issue can be satisfactorily decided.

For any given set of constraints, there are however a number implications which can be determined. In this paper we give several general expressions for the luminosity allowing for differences between the electron and positron beams, in energy, beam sizes, tune shift limits, and β^*'s. If in a particular formula for luminosity, the variables which appear explicitly are chosen to represent the constraints, ie. 'fixed', then the dependence of the luminosity on energy ratio, if any, can be extracted. Thus the various formulae are to be used to evaluate the effect of the energy ratio under different sets of constraints. Some of the effects of these relations can be seen in a comparison of designs of unequal energy machines based on PEP and PETRA, and an a equal energy machine, CESR.

Generalized Equations for Luminosity

We begin with two intersecting rings 1,2 with particle energies $\gamma_1 mc^2$, $\gamma_2 mc^2$ measured in the laboratory frame. The energies are chosen so that in the center of mass frame the sum of the energies is just sufficient to make a $B\bar{B}$ pair, assumed to be at rest.* We define $\bar{\gamma} = \sqrt{\gamma_1\gamma_2}$, so that $\bar{\gamma}mc^2$ is the particle energy when both rings have the same energy. Throughout this article we make the assumption that the product $\gamma_1\gamma_2$ is fixed, while the ratio is free to vary. We also assume that the bunch length is always less than the smallest beta function at the interaction point. If this were not the case, the beam beam force can cause a strong synchrobetatron coupling which is believed to be quite detrimental to the luminosity[4]. Furthermore, we assume the bunch charge distribution is gaussian in x, y and z, where x,y are the transverse coordinates and z is the longitudinal coordinate. No assumption is made regarding the aspect ratio or the relative sizes of the two beams.

* In the laboratory frame the bound $B\bar{B}$ pair will be moving with the Lorentz factor $\gamma_{B\bar{B}} = \frac{1}{2}\frac{(\gamma_1+\gamma_2)}{\sqrt{\gamma_1\gamma_2}}$. On the average, the B or $\bar{B}$ will travel $\beta_{B\bar{B}}\gamma_{B\bar{B}}c\tau$, where τ is the lifetime in the rest system.

CONSTRAINTS - BUNCH CHARGES, BEAM SIZES, AND FREQUENCY

Perhaps the simplest expression for the luminosity is,

$$L = \frac{f_c}{4\pi} \frac{N_{B1} N_{B2}}{\langle\sigma_x\rangle\langle\sigma_y\rangle} \quad (1)$$

where $\langle\sigma_x\rangle = \sqrt{\frac{1}{2}(\sigma_{x1}^2 + \sigma_{x2}^2)}$, f_c is the collision frequency, N_{B1} is the number of charges per bunch in ring 1, and σ's are the beam sizes at the interaction point in the horizontal and vertical directions for rings 1 and 2. Similar definitions hold with $x \leftrightarrow y$ and/or $1 \leftrightarrow 2$. In this expression, the luminosity does not depend explicitly on the energy of either ring. If the variables on the right hand side of equation (1) are pushed as far as possible, given the limits on technology and beam dynamics, then the maximum luminosity would be the same for a equal energy machine as an unequal energy machine - provided those limits do not depend on energy.

TUNE SHIFT, β^*, BEAM SIZE, FREQUENCY

The usual approach to scaling laws for colliding rings of equal energies is through the assumption that for a given machine the highest luminosity will be reached if the beam beam tune shift parameters are at their maximum attainable values. This criteria tends to minimize the total rf power. It is suspected on theoretical grounds, that when scaling the maximum tune shift parameter, the relevant variable is the damping decrement.[5] However, a recent compilation of the maximum tune shift parameters for essentially all electron positron storage rings shows no correlation between the maximum tune shift parameters and the damping decrement[6] (see figures 1 and 2). Even the tune shift data used in ref. 5, which according to the theory, should fall on a universal curve, fail to do so to better than a factor of four or so. Historically, the maximum beam beam tune shift parameter for e^+/e^- storage rings has been scattered between 0.015 and 0.07. Apparently, other factors have influenced the observed tune shift limits at least as much as the damping decrement.

In the absence of any satisfactory rule to describe the scaling of tune shift parameters limits with energy, we assume, for the purpose of estimating the tune shift limits of new machines, the maximum obtainable values of the tune shift parameters are independent of all machine parameters under discussion. This assumption is almost certainly false. In particular, there is evidence that *for a given machine*, the limiting tune shift parameter increases with increasing energy.[7] Also, there are simulation results which indicate that with equal emittance round beams, substantially higher tune parameters can be obtained.[8] As a result, there is considerable uncertainty in the validity of scaling laws for luminosity which hold the maximum tune shift parameters constant.

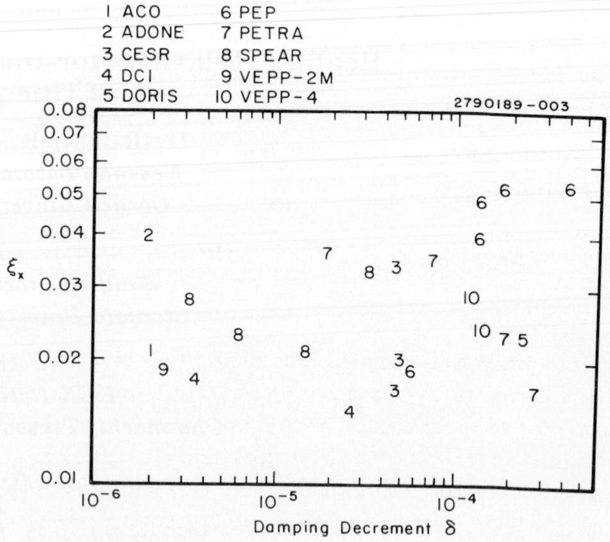

1. This figure is taken from reference 6 showing the maximum horizontal tune shift parameters obtained at various machines operating in the tune shift limit.

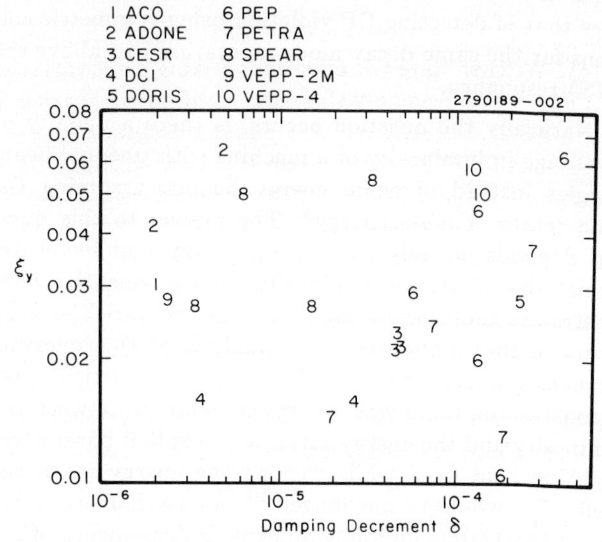

2. The maximum vertical tune shift parameters obtained at various machines operating in the tune shift limit - also taken from ref. 6.

The standard formulae for the tune shift parameter ξ can be modified to take into account two beams of unequal energy and different optical characteristics,

$$\xi_{y1} = \frac{r_e}{2\pi\gamma_1} \frac{N_{B2}\beta_{y1}^*}{\sigma_{y2}(\sigma_{x2} + \sigma_{y2})} \quad (2)$$

Definitions may also be obtained with $x \leftrightarrow y$ and/or $1 \leftrightarrow 2$. Here r_e is the classical electron radius and β^*_{y1} is the vertical beta function at the interaction point for the beam in ring 1.

We can solve for N_{B2} and N_{B1} in (2) and substitute the results into (1) to obtain,

$$L = \frac{\pi f_c}{r_e^2}\gamma_1\gamma_2\xi_{y1}\xi_{y2}\frac{\sigma_{y2}(\sigma_{x2}+\sigma_{y2})\sigma_{y1}(\sigma_{x1}+\sigma_{y1})}{\beta^*_{y1}\beta^*_{y2}\langle\sigma_y\rangle\langle\sigma_x\rangle} \tag{3}$$

Again equally valid expressions may be obtained by $x \leftrightarrow y$. Since $\gamma_1\gamma_2 = \bar{\gamma}^2$, more luminosity would be obtained if the center of mass energy were raised, whether or not the rings had equal energies. By expressing the luminosity as a function of product of a tune shift parameter from each ring, we have eliminated the explicit dependence of the luminosity on the energy ratio.

Relative beam sizes If we exchange x and y in (2) and take ratios we have,

$$\frac{\beta^*_{y1}}{\beta^*_{x1}} = \frac{\xi_{y1}}{\xi_{x1}}\frac{\sigma_{y2}}{\sigma_{x2}} \tag{4}$$

Another equation may be written with 1 and 2 exchanged in (4). We then have two equations linking eight variables, (4 σ's and 4 β's), so only six of the eight may be varied independently. In principle, it is possible to maintain equal beam sizes for the two machines and vary all four β^*'s independently to get the maximum luminosity, but then the aspect ratio and absolute size of the beams could not be fixed. Considering the experiments at the $Sp\bar{p}S$, where a 'pencil beam' was driven through a normal sized beam resulting in a shorter luminosity lifetime,[9] one might want to constrain the rings to have equal beam sizes.* On the other hand, one might wish to use unequal beam sizes if the absolute beam size and aspect ratio were constrained, for example by the beam pipe radius and the desire to use round beams to enhance the tune shift limit.

If the four σ's are varied independently, then two of the β's must be dependent. The beta functions will also be constrained by the bunch length, and optics constraints due to the beam pipe radius and chromaticity added by the high energy final quadrupoles. To constrain them further by independently setting the beam sigmas and the tune shift parameters, would likely imply a less than optimum solution given these other constraints.

CONSTRAINTS - TUNE SHIFT, CURRENT, β^*, BEAM SIZES

Another expression for the luminosity may be constructed from (2) and (1) where the luminosity is expressed as a function of ξ_{y2} and I_2, the average current in ring 2,

$$L = \frac{1}{2er_e}\frac{\gamma_1\xi_{y2}I_2\sigma_{y1}(\sigma_{x1}+\sigma_{y1})}{\beta^*_{y2}\langle\sigma_y\rangle\langle\sigma_x\rangle} \tag{5}$$

Interchanging 1 and 2, solving for currents and taking ratios yields

$$\frac{I_2}{I_1} = \frac{\gamma_1}{\gamma_2}\frac{\beta^*_{x2}}{\beta^*_{x1}}\frac{\sigma_{x2}(\sigma_{x2}+\sigma_{y2})}{\sigma_{x1}(\sigma_{x1}+\sigma_{y1})}\frac{\xi_{x1}}{\xi_{x2}} \tag{6}$$

Here we can see a direct implication of the differing energies of the rings. To compensate for the energy ratio, while maintaining the tune shift parameters fixed, the beam size and focusing functions must be considerably different for the two rings, or the current in the low energy ring will be become larger than the current in the high energy ring. Additional constraints on the focusing functions to separate the beams and focus the beams must be imposed, so the optimal conditions could involve unequal currents. This consideration helps to set the current in the low energy ring in the design using PEP as the high energy ring.[10] In that design, where the energy ratio is about 6:1, and equal beam sizes and equal tune shift parameters assumed, the low energy ring current is held to only twice the high energy ring by making the β_x of the low energy ring one third of β_x of the high energy ring. (see table)

CONSTRAINTS - TUNE SHIFT, β^*, CURRENT

For a very high luminosity machine it is quite natural to specify the tune shift parameters, the average currents (perhaps set by rf power limits), and considering the complications of focusing and possible constraints imposed by the small diameter beam pipe, the beta functions at the interaction point. Equation (2) can be used to express the beam sizes as a function of the other variables. Substituting the result into (3) and writing the energy dependence as a dependence on the ratio of energies and $\bar{\gamma}$, we have,

$$\sigma_{y2} = \sqrt{\frac{r_e N_{B2}}{2\pi\gamma_1}\frac{\xi_{x1}}{\xi_{y1}}\frac{(\beta^*_{y1})^2}{\beta_{x1}\xi_{y1}+\beta_{y1}\xi_{x1}}}$$

The luminosity $L =$

$$\frac{\frac{\bar{\gamma}}{er_e}I_1 I_2}{\sqrt{\left[\left(\frac{\gamma_2}{\gamma_1}\right)^{1/2}(\beta^*_{y1})^2\left(\frac{\xi_{x1}}{\xi_{y1}}\right)\frac{I_2}{\beta^*_{x1}\xi_{y1}+\beta^*_{y1}\xi_{x1}}+1\leftrightarrow 2\right][x\leftrightarrow y]}} \tag{7}$$

The $1 \leftrightarrow 2$ symbol indicates the term immediately to

* The nonlinearities associated with the pencil beam are thought to be more effective as they disrupt the core of the normal beam.

the left should be inserted with the 1 and 2 indices interchanged. The $x \leftrightarrow y$ symbol is defined analogously. While this expression is complicated, its utility derives from the choice of explicit variables. The familiar form of this expression for storage rings with equal energies, equal beam currents and flat beams is,

$$L = \frac{\bar{\gamma}}{2er_e} \frac{I\xi}{\beta_y^*} \tag{8}$$

Equation (7) indicates that the luminosity is proportional to the current *scale*, the tuneshift *scale* and inversely proportional to the *scale* of β^*'s. It also shows that the luminosity depends explicitly on the energy ratio, but not in an obvious way.

Optics constraints at the interaction point

As mentioned above, the difficulty of simultaneously separating and focusing beams of unequal energy and the requirement of a small beam pipe radius around the interaction point may force additional constraints among the variables that determine the luminosity. However, it is conceivable that energy difference could be exploited to help separate the beams over a shorter distance than is possible with electrostatic separators normally used in equal energy machines.

Considering the energy ratio of current designs (6:1, 7:1), the high energy beam is relatively little affected by the optics designed to focus the low energy beam. Conversely, the low energy beam would be drastically overfocused if it were allowed to go through final quadrupoles suitable for focusing the high energy beam. Consequently the beams must be separated before the low energy beam reaches the first high energy quadrupole. This is done by inserting a dipole magnet as shown in figure 3.

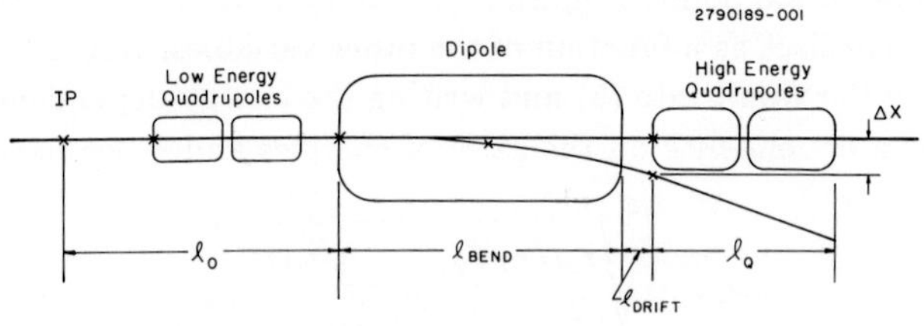

3. The general interaction point optics design used in references 10 and 11.

Physical aperture due to beam pipe radius

In order to get an enhancement in detection efficiency, it necessary to have a small beam pipe radius, $\lesssim 1$ cm, around the interaction point, to achieve sufficiently high vertex resolution that the time evolution of the separate B's may be measured. The technique for measuring CP violation with symmetric collisions using equal energy rings, does not require a particularly small beam pipe radius as the vertex of the B decays is not measured. The vertex resolution is set both by multiple scattering in the walls, and the radius of beam pipe which acts as a 'lever arm'. If the radius is not small enough, the detection efficiency enhancement is lost. For sufficiently small radius, the beam pipe becomes the physical aperture of the machine. When this occurs, proper optimization should include the tradeoffs between resolution and luminosity, and would be dependent on the energy ratio of the machines.

A rule of thumb used in conventional storage ring design is that the lifetime suffers if the aperture is made less than about 20 σ.* That is, if R is the value of the pipe radius at which it becomes the machine aperture, and $\sigma(l/2)$ is the largest beam size at the end the pipe of length l, $\frac{\sigma(l/2)}{R} = 1/20$. The beams spread out from the interaction point according to, $\sigma(s) = \sigma(0)\sqrt{1 + (s/\beta^*)^2}$, where β is the beta function that goes with the largest σ, and s is the distance from the interaction point. At the end of the pipe the beam size is

$$\sigma(l/2) = \sigma(0)\sqrt{1 + \frac{1}{4}\left(\frac{l}{\beta^*}\right)^2} = \frac{R}{20} \tag{9}$$

Given a beam pipe of radius R and length l, equation (9) is an equation of constraint linking β^* and σ. It can be interpreted to give the maximum emittance possible for a given R, l and β^*. Values of R from the PEP and PETRA unequal energy designs, and for CESR are given in the table. Note that in several cases the minimum allowable beam pipe radius is $\geq$ 1 cm.

Beam separation

Referring to figure 3, the first quadrupoles are placed as close as possible to the interaction point, where the limitations are due to the length of beam pipe needed, the quadrupole strength, and the solid angle requirement of the detector. The solid angle requirement depends on the energy ratio. One end of a bend magnet, with uniform field strength B, is placed as close as possible to the quadrupoles, staring a distance l_0 from the interaction point. The bend magnet of length l_{bend} is followed by the high energy quadrupoles located as close as possible to the end of the bend magnet. The separation of the beams at the end of the bend magnet is $\Delta x \approx (l_{bend})^2/2\rho$, where

$$\frac{1}{\rho} = \left|\frac{1}{\gamma_1} - \frac{1}{\gamma_2}\right| \frac{B}{mc^2}$$

* The 20 sigma rule is usually used for design purposes and does not represent a fixed limit. Beams with good lifetime have been sent through 10 sigma apertures, though not in the colliding beam mode.

The beams may then drift apart a distance l_{drift} for a total separation at the beginning of the high energy quadrupole of

$$\Delta x \approx \left[\frac{(l_{bend})^2}{2} + (l_{bend})l_{drift}\right]\left|\frac{1}{\gamma_1} - \frac{1}{\gamma_2}\right|\frac{B}{mc^2} \tag{10}$$

As the distance l_{bend} is made larger the separation gets bigger. Unfortunately, so does the high energy beam size. The rule of thumb mentioned above would imply the ratio of the beam size to the separation distance is the relevant parameter. As the high energy beam very nearly expands linearly with l_0+l_{bend} and the separation distance expands quadratically in l_{bend} we can always find a long enough bend magnet to get whatever $\Delta x/\sigma$ we desire. We may also obtain any desired $\Delta x/\sigma$ by increasing B. The former will cause a reduction in the dynamic aperture (see below), while the latter may increase the background by raising the critical energy of synchrotron radiation from the bend magnet.

CRITICAL ENERGY u_c

Though the beam pipe may be shielded from direct synchrotron radiation from the bending magnet, there will be luminescence from the radiation striking the shielding which will reach the beam pipe and cause background. Luminescence is produced more efficiently by higher energy synchrotron radiation. The critical energy of the synchrotron radiation $u_c = \frac{3\gamma^2\hbar B}{2mc}$ is then a parameter which may limit the bend radius. Here γ is the Lorentz factor for the high energy ring. If we solve for B and put it into equation (10), dropping higher order terms in β^*/s, we get,

$$\begin{aligned}\frac{\Delta x(s)}{\sigma(s)} = u_c\left(\frac{\beta_x}{s}\right)&\left[\frac{(s-l_0)^2}{2} + (s-l_0)l_{drift}\right]\\ &\times\left|\frac{1}{\gamma_1} - \frac{1}{\gamma_2}\right|\frac{2}{3e\hbar c\bar{\gamma}^2\sigma(0)}\end{aligned} \tag{11}$$

where σ_x and β_x are the beam size and beta function of the high energy ring.

The separation ratio is proportional to u_c. To make the ratio larger, keeping u_c fixed at some reasonable value, the proportionality constant must increase. That can be done by making s or β_x^* larger, but at the expense of increasing the chromaticity. Notice the separation distance goes to zero if $\gamma_1 = \gamma_2$. In this case the beams are separated by imposing an electrostatic separators, located outside the final quadrupoles.

DYNAMIC APERTURE

If the low energy quadrupole has no significant effect on the high energy beam, then

$$\beta = \beta^*\left[1 + \left(\frac{s}{\beta^*}\right)^2\right] \tag{12}$$

gives the approximate value of the beta function at the high energy quadrupole located a distance s from the interaction point. The chromaticity added by the high energy quadrupole is,

$$E\frac{\partial\nu}{\partial E} \approx \frac{1}{4\pi}\beta(s)kl_Q \tag{13}$$

where ν is the tune in either the x or y direction, k is the strength of the quadrupole and l_Q is its length. The quadrupole strength must be roughly equal to a lens that has a focal length s, ie. $kl_Q \sim 1/s$. Using equations (12) and (13) we have,

$$E\frac{\partial\nu}{\partial E} \approx \frac{1}{4\pi}\left(\frac{\beta^*}{s} + \frac{s}{\beta^*}\right) \tag{14}$$

So if $s \gg \beta^*$, the chromaticity grows linearly with s. As the chromaticity increases, the sextupole strength in the rest of the ring must also increase in order to keep the total chromaticity near zero required by beam stability. But, stronger sextupole strength implies stronger nonlinearities, and smaller dynamic aperture.

COMPARISON OF OPTICS CONSTRAINTS IN ACTUAL DESIGNS

In a design based on the utilization of PEP as the high energy ring, constrained variables are the tune shift parameters, (set at 0.05), and the average current in the high energy ring (as estimated from the rf power design limits for PEP). There are also constraints placed on the β^*'s by the assumptions that the high and low energy beam sizes should be equal, the emittance of the high energy ring should be very nearly what is obtained at PEP, the low energy ring average current should not get too large, and by the scheme to separate the beams. These latter constraints imply a dependence of the maximum luminosity on the energy ratio, such that an equal energy machine built with the same current limits, same maximum tune shift parameters, but 'conventional' ie. CESR β^*'s (see table), would have almost three times the luminosity.

Another design, which employs PETRA instead of PEP as the high energy ring[11] also assumes that the tune shift parameters are fixed, the average current in the high energy ring is fixed, and beam sizes at the interaction point are the same. They further assume that the luminosity is constrained by the ratio of the emittances in the horizontal and vertical directions, and β_y^*. In the PETRA design, the minimum value of β_y^* is set by the generation of excessive chromaticity. If one again makes the comparison with an equal energy machine running at the same tune shift parameters and currents but CESR β^*'s, the luminosity for the conventional machine would be 2.3 times as much, for the low current case, but almost the same in the high current case. Selected parameters for both designs are given in the table.

	PETRA		CESR	PEP	
	14 GeV	2 GeV	5.3 GeV	12 GeV	2 GeV
β_y^* $[m]$	.07, .035	.03, .015	.015	.0762	.0254
β_x^*	.47, .235	.20, .10	1.00	.762	.254
ξ_y	.03	.03	.017	.05	.05
ξ_x	.03	.03	.025	.05	.05
σ_x $[\mu m]$	306, 217	306, 217	400	276	276
σ_y	46, 32	46, 32	14	27.6	27.6
σ_z $[cm]$	3, 1.5	3, 1.5	1.7	1.09	2.43
I $[A]$	.06, .72	.18, 2.2	.15	.267	.535
N $[10^{11}]$	1.4, 1.4	4.3	1.6	1.44	2.88
f $[MHz]$	2.6, 31	2.6, 31	2.7	11.6	11.6
u_c $[keV]$	43	.9	1.5	15	0.4
R_x $[cm]$	.89, 1.0	1.6, 2.2	1.12	0.66	.675
R_y	.66, .92	.25, .33	.93	0.37	1.09
L $[\frac{10^{32}}{cm^2 s^1}]$	0.88, 21		1.0	5	

1. The table includes selected parameters from the PEP and PETRA designs for unequal energy machines, and from CESR operating at its highest luminosity. The pair of entries for each column of the PETRA design refer to two different designs. The values for ξ_y, σ_x, and σ_y, for CESR are the nominal values. Due to the variation of the transverse beam size during the collision, neither the tune shift parameters nor β_y^* may be exactly appropriate for comparison. R_x and R_y represents a lower bound on the beam pipe dimension which avoids reducing the physical aperture. They are calculated assuming the total beam length is 1 m and 20 sigma clearance is needed. For asymmetric collisions to yield an enhancement in detection efficiency, the beam pipe radius must be $\lesssim 1$ cm.

These comparisons are not meant to be 'fair' in any sense; they only illustrate the way in which the energy ratio can affect the maximum luminosity through differences in β^*'s. There is certainly no reason to believe that the current limits are independent of the energy of the machines. It may also be argued that the tune shift limit for CESR is lower than those used for the unequal beam energy designs, in part because β_y^* is too small. If this is the case, then it may be unphysical to simultaneously assume both a tune shift of .05 or .03 and a β_y^* of 1.5 cm as we have done here.

The differences in scaled luminosity assuming equal currents and tune shift parameters, are in part due to the fact that in unequal energy machines, the two beams go through different final quadrupoles and there is a bend magnet between them. The differences would become more substantial if it were necessary to limit u_c, due to the separation bend magnet, to near CESR values. The values of u_c for the two unequal energy machine designs, are an order of magnitude higher than at CESR.

Conclusion

In most of the general equations for luminosity we discussed, there is no explicit dependence of the luminosity on the energy ratio. The implicit dependence of the maximum luminosity on the energy ratio, comes about through the dependence on energy of the limiting values of the parameters defining the luminosity. We have examined those limits which are unique to unequal energy machines; namely, the small beam pipe radius, the critical energy due to a separation bend magnet, and the chromaticity penalty caused by separation optics. The luminosity advantage or disadvantage of unequal energy machines depends in detail on how far these limits can be extended. Ultimately it depends on, the capabilities of vertex detectors, our understanding of the tune shift limit, and the ability to shield against high energy synchrotron radiation. A thorough understanding of the dependence of these limits on energy is not available, and is necessary to determine the relative maximum luminosity between unequal and equal energy machines.

REFERENCES

1) R.S. Siemann, B-FACTORIES: A Perspective of B-Physics and Possible Accelerator Design Approaches, CLNS 88/865. Proceedings of 1988 Linac Conference

2) Report of the Physics Subgroup, Snowmass 1988

3) J. Rees, SLAC/AP-67, Feb 1988, The author warns that it is questionable whether or not the tune shift parameters can be held constant while varying the energy ratio.

4) D. Rice CON 87-3, Cornell Univ.

5) E.Keil and R.Talman, Particle Accelerators, Vol 14. pp. 109-118, (1983)

6) J.T. Seeman, Proc. Nonlinear Dynamics Aspects of Particle Accelerators, Sardinia 1985, 247, J.M.Jowett, M. Month, S.Turner eds.

7) Examples are given in, J.F. Schonfeld, The Effects of Beam-Beam Collisions on Storage-Ring Performance – a Pedagogical Review, AIP conf. proc. No. 105 p 524, 1983, ed. M. Month

8) S. Krishnagopal, R. Siemann, CBN 88-1,88-2,88-3 Cornell Univ. 1988

9) L. Evans, talk at 1988 Snowmass Workshop

10) A. Garren et.al. APIARY I, LBL (1988)

11) The Use of PETRA as a B-factory, H. Nesemann, W. Schmidt-Parzefall and F. Willke, DESY, June 7 1988

EFFECTS OF BEAM ENERGY SPREAD ON B FACTORY PERFORMANCE

David G. Cassel, Nari B. Mistry, and James A. Mueller
Laboratory of Nuclear Studies, Cornell University, Ithaca, NY 14853

Nobuhiko Katayama
State University of New York at Albany, Albany, NY 12222

ABSTRACT

The energy spread of the stored beams in an e^+e^- storage ring operating at the Υ(4S) influences the yield of $\mathrm{B\overline{B}}$ events and the separation of B mesons from background. These effects are discussed quantitatively.

INTRODUCTION

Designing an e^+e^- collider with enough luminosity to observe CP violation in B meson decay is very challenging and requires many compromises in choosing storage ring parameters.[1] Although the energy spread of the stored beams may be overlooked, it is an important parameter for B factories operating at the Υ(4S) where experiments can take advantage of the resonant cross section and width, the fact that Υ(4S) only decays to $\mathrm{B\overline{B}}$, and the kinematics of these decays. Somewhat different considerations are important for running slightly above the $\mathrm{B^0\overline{B}^{*0}}$ threshhold to search for CP violation in $\mathrm{B^0\overline{B}^0}$ events resulting from $\mathrm{B^0\overline{B}^{*0}} \to \mathrm{B^0\overline{B}^0}\gamma$. At the Υ(4S) energy, the r.m.s. spread,[2] δ_W, in the total center of mass energy (W) is $\delta_{WC} = 4.6$ MeV at CESR[3] and $\delta_{WD} = 9.1$ MeV at DORIS.[4] Substantially larger[1] values of δ_W appear in some B factory proposals. δ_W can influence:

- The B meson yield.
- The total energy constraint for B candidates.
- The mass resolution of B candidates.

The latter two items clearly depend on the relationship between δ_W and the momentum resolution of the detector. In this study, we have used the resolution of the CLEO detector during its most recent data run, when the charged particle momentum resolution was,

$$(\delta_{pt}/p_t)^2 = (0.23\%p_t)^2 + (0.7\%)^2, \tag{1}$$

where p_t is the transverse momentum measured in GeV.

THE Υ(4S) YIELD

The mass and width of the Υ(4S) are $M = 10.58$ GeV and $\Gamma = 24$ MeV respectively.[5] At CESR[6] the Υ(4S) yield corresponds to a cross section of $\sigma_C = 1.15$ nb. The actual shape of the Υ(4S) is not known; for simplicity, we use a Gaussian with FWHM equal to Γ. The standard deviation of this Gaussian is then $\gamma = \Gamma/2.35 = 10$ MeV. If the mean center of mass energy of the storage ring is W_0 and σ_Υ is the peak Υ(4S) cross section, the energy dependent yield (per unit luminosity) will simply be the product of the Gaussian,

$$\sigma(W) = \sigma_\Upsilon \exp\frac{-(W-M)^2}{2\gamma^2}, \tag{2}$$

representing the cross section with the Gaussian,

$$\ell(W) = \frac{1}{\sqrt{2\pi}\delta_W}\exp\frac{-(W-W_0)^2}{2{\delta_W}^2}, \tag{3}$$

representing the W dependence of the luminosity. The product of these two functions can be rearranged to give,

$$y(W) = \frac{\sigma_\Upsilon}{\sqrt{2\pi}\delta_W}\exp\frac{-(W-\overline{W})^2}{2\overline{\delta}^2}\exp\frac{-(W_0-M)^2}{2{\delta_Y}^2}, \tag{4}$$

where,

$$\delta_Y = ({\delta_W}^2+\gamma^2)^{1/2}, \tag{5}$$

$$\overline{\delta} = \frac{\gamma\delta_W}{\delta_Y}, \tag{6}$$

and $\overline{W} = (M{\delta_W}^2 + W_0\gamma^2)/{\delta_Y}^2$. If $y(W)$ is integrated over W, the yield, $Y(W_0)$, is the familiar convolution of the resonance shape with the beam energy spread, which is observed in a graph of the cross section versus energy. For a given value of W_0, the energy width of the yield is $\overline{\delta}$ (*not* δ_W or δ_Y). Of course, the maximum yield,

would make this experiment impossible, unless the $B^0\overline{B}^0$ rate is much smaller than the $B^0\overline{B}^{*0}$ rate.

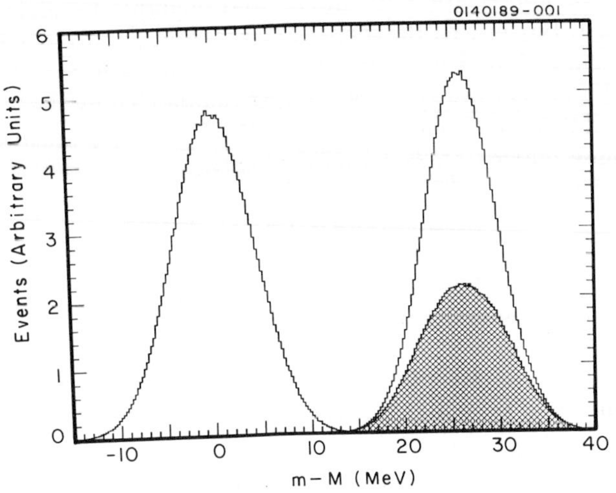

Fig. 4 Reconstructed B meson mass distributions expected for equal $B^0\overline{B}^0$ and $B^0\overline{B}^{*0}$ rates. The hatched events are $\overline{B}^0$'s from $\overline{B}^{*0}$ decays.

Since this structure is substantially wider than the $\Upsilon(4S)$, the quality factors Q_Y, $Q_{\Delta E}$, and Q_m will behave differently with increasing δ_W: Q_Y will decrease more slowly with increasing δ_W, $Q_{\Delta E}$ will deviate noticeably from one for large δ_W, and Q_m falls more rapidly and becomes level at a lower level.

ASYMMETRIC COLLIDERS

Another suggestion for observing CP violation in neutral B meson decay involves the construction of an asymmetric e^+e^- collider (*i.e.*, one with unequal energies) to operate at the $\Upsilon(4S)$. The resulting center of mass motion would boost B meson decay lengths enough so that measurements of the B decay vertices may be possible. This would permit direct observation of CP violation in $B^0\overline{B}^0$ pairs from $\Upsilon(4S)$ decay and would be a welcome alternative to studying $B^0\overline{B}^{*0}$ events above the $\Upsilon(4S)$. A particular scenario for detecting CP violation using an asymmetric collider has been studied[15] in a Monte Carlo calculation.

The results of our work can be applied to B candidates in an asymmetric collider experiment, if the energies and momenta of the tracks in the B candidates are transformed back to the $\Upsilon(4S)$ frame. Quantitative comparisons between options with different values of δ_W but the same detector and value of $\beta\gamma$ (of the $\Upsilon(4S)$) are easy – it is more difficult to compare experiments with equivalent detectors but different values of $\beta\gamma$. As $\beta\gamma$ changes, the energy and momentum resolutions in the $\Upsilon(4S)$ frame will change, so a Monte Carlo calculation is required for each value of $\beta\gamma$. In particular, there is no simple way to scale total energy or mass resolutions from a symmetric to an asymmetric collider. However, some conclusions are possible without a detailed Monte Carlo calculation:

The Maximum Yield

$Y(M)$, given in Eq. (7), depends only on δ_W and γ, and δ_W can easily be calculated from the energy spreads of the two beams. Hence, Q_Y in Fig. 1 is relevant for asymmetric as well as symmetric colliders.

The Total Energy Constraint

In order to utilize the total energy constraint, the laboratory energies of all particles must be transformed to the $\Upsilon(4S)$ frame. The uncertainty $\delta_{\Delta E}$, given in Eq. (8), still determines the width of this cut, with δ_T measured in the $\Upsilon(4S)$ frame. Unfortunately, there is no simple relationship between the energy resolution in the laboratory frame and the $\Upsilon(4S)$ frame. In a solenoidal detector, the resolution in transverse momentum is the same in both frames. However, the resolution in longitudinal momentum will generally be worse for an asymmetric collider, because the Lorentz boost between the $\Upsilon(4S)$ frame and the laboratory insures that the longitudinal momenta of particles in the laboratory will almost always increase with $\beta\gamma$. Hence, momentum and energy resolution in the laboratory at an asymmetric collider will generally be worse than they would be at a symmetric collider, and the boost back to the $\Upsilon(4S)$ frame will usually not compensate for this loss of resolution.

For purposes of comparison, we have assumed that the total energy resolution for a symmetric collider is a lower limit for the resolution of an asymmetric collider with the same δ_W.

The B Mass Resolution

In order to make use of the beam energy to increase the precision of the B mass calculation, the momenta of the particles in the B candidate must also be transformed into the $\Upsilon(4S)$ frame. δ_m can then still be calculated from Eq. (10) since the energy and momentum of the B candidate are measured independently (Eq. (10) cannot be used if E and p are the measured total energy and momentum because they are not independent). In the $\Upsilon(4S)$ frame, the total momentum of the particles will be the same as it was for a symmetric collider, so the p_T/m factor will again reduce the importance of δ_{pT}. However, following the discussion of the total energy constraint, the resolution in total momentum will generally be worse for a given detector operating at an asymmetric collider than at a symmetric collider. For purposes of comparison, we have assumed that the mass resolution of a symmetric collider is a lower limit for the mass resolution of an asymmetric collider with the same value of δ_W.

MODES INVOLVING NEUTRALS

So far we have discussed the reconstruction of B meson decay modes involving only charged particles in the final state. For these modes, excellent vertex detection that is highly efficient in separating B vertices from charm vertices could largely compensate for a poor signal to background ratio due to a large beam energy spread. However, it may be difficult to achieve the required vertex resolution. For decay modes with neutral pions, it will be necessary to fall back on kinematic constraints to decide if a given π^0 is associated with a particular B candidate. For π^0's resulting from daughter decays, the masses of the daughter particles furnish useful kinematic constraints. For π^0's originating in the primary B meson decay, the only kinematic constraints are the measurement of the total energy of all of the tracks in the candidate and the reconstructed B mass.

CONCLUSIONS

We have described how the spread of the beam energy in an e^+e^- collider affects the $\Upsilon(4S)$ yield and reduces the efficiency of the total energy and B mass constraints in rejecting backgrounds. We described these effects by quality factors which give the degradation in the signal to noise ratio relative to the CLEO detector operating at a collider with zero spread in beam energy. Table I gives these quantities and their products, $Q = Q_\Upsilon Q_{\Delta E} Q_m$, which describe the overall signal to noise ratio. Values are included for the existing CESR and DORIS colliders and for several proposed colliders. For asymmetric colliders, a lower limit on δ_m is given, assuming that the detector resolution in the $\Upsilon(4S)$ frame is equal to that of a detector at a symmetric collider. For a detector comparable to CLEO, the decrease in yield and increase in the width of the B mass peak with increasing δ_W are far more important than changes in the width of the total energy cut.

Storage Ring	δ_W [MeV]	δ_m [MeV]	Q_Υ	$Q_{\Delta E}$	Q_m	Q
CESR	4.6	2.7	0.91	1.00	0.63	0.57
DORIS	9.1	3.8	0.75	0.99	0.45	0.33
PSI	5.0	2.8	0.90	1.00	0.60	0.54
TRISTAN	6.9	3.3	0.83	0.99	0.51	0.42
NOVOSIBIRSK	11	4.1	0.68	0.99	0.41	0.28
ARES-I	10	4.0	0.71	0.99	0.43	0.30
ARES-II	62	5.3	0.16	0.99	0.32	0.05
Wilson	53	5.3	0.19	0.99	0.32	0.06
SLAC 2 × 12	5.2	> 2.9	0.89	1.00	< 0.59	< 0.52
DESY 2 × 12	7.4	> 3.4	0.81	0.99	< 0.49	< 0.40

Table I. Uncertainties in W and m, and the quality factors, Q_Υ, $Q_{\Delta E}$, Q_m, and Q for existing and proposed e^+e^- colliders,[1] described elsewhere in these proceedings.

ACKNOWLEDGEMENTS

We appreciate stimulating and helpful discussions with J. Bartelt, K. Berkelman, P. Franzini, D. Rice, R.H. Siemann, P.M. Tuts, and our other colleagues in the 1988 DPF Summer Study. This work was supported by the National Science Foundation and by the U.S. Department of Energy under Contract ER78S025001.

REFERENCES AND FOOTNOTES

1) R.H. Siemann *et al.,* in these proceedings
2) δ_W was actually measured at the lower (narrow) Υ resonances. The value used here has been extrapolated to the $\Upsilon(4S)$ using the fact that $\delta_W/W \propto W$. See M. Sands, SLAC report SLAC-121, UC-28, (ACC), p. 124
3) CLEO Collaboration, D. Andrews *et al.,* Phys. Rev. Lett. **45**, 219 (1980), CUSB Collaboration, G. Finocchiaro *et al.,* Phys. Rev. Lett. **45**, 222 (1980), and D. Rice and R.H. Siemann, private communications
4) D.P. Barber *et al.,* Phys. Rev. Lett. **135B**, 498 (1984)
5) Particle Data Group, G.P. Yost *et al.,* Phys. Lett. **204B**, 1 (1988)
6) CLEO Collaboration, F. Morrow private communication
7) See for example, I.I. Bigi and A.I. Sanda, Nucl. Phys. **B281**, 41 (1987) and references contained therein.
8) I.I. Bigi, V.A. Khoze, N.G. Uraltsev, and A.I. Sanda, SLAC report, SLAC-PUB-4476 (1987), to be published in the *Review Book of CP Violation,* ed. C. Jarlskog, World Scientific, Singapore
9) A particularly elegant discussion of $B^0\overline{B}^0$ mixing and CP violation in terms of "B-spin", an analogue of isospin, can be found in H.J. Lipkin, Argonne National Laboratory report, ANL-HEP-PR-88, to be published.
10) S. Stone, Modern Physics Letters A **3**, 541 (1988)
11) G. Feldman *et al.,* in these proceedings.
12) The discussion applies equally to the charge conjugate mode.
13) CUSB Collaboration, J.M.J. Lovelock *et al.,* Phys. Rev. Lett. **54**, 377 (1985) and CLEO Collaboration, D. Besson *et al.,* Phys. Rev. Lett. **54**, 381 (1985)
14) This angular distribution is due to the fact that parity implies that the helicity of the B^{*0} is ±1, but not 0. The absence of the helicity 0 state is discussed by J.D. Richman, Caltech Thesis, CALT-68-1231 (1985) p.265 (unpublished)
15) R. Aleksan *et al.,* SLAC report SLAC-PUB-4673 (1988), submitted to Phys. Rev. D

A High Luminosity $\overline{\mathrm{B}}\mathrm{B}$ Factory: A Research and Development Program

D. Cline

Physics Department, University of California, Los Angeles
405 Hilgard Avenue, Los Angeles, California 90024-1547

C. Pellegrini

Brookhaven National Laboratory, 75 Brookhaven Avenue
Building 725, Upton, Long Island, New York 11973

1. Introduction

In this paper we discuss a proposal for the construction of a high luminosity, $L \approx 10^{34}\mathrm{cm}^{-2}\mathrm{s}^{-1}$, electron-positron collider, operating in the energy range of 10 to 15 GeV total center of mass energy. The motivation for such a $\overline{\mathrm{B}}$-B system, in particular the rare decay modes and the CP violation.[1)]

Our design is based on the following points:

1. unequal beam energies, to produce the B in motion in the laboratory frame;
2. positron recovery and cooling;
3. possibility of positioning the vertex detectors very near to the interaction point (IP);
4. possibility of changing the center of mass energy maintaining a high luminosity, to be able to operate effectively at the Y(4S) resonance or above.
5. a small beam energy spread, less than 0.1%, to be able to utilize effectively the Y(4S) resonance.

To satisfy these conditions we propose a scheme utilizing:

a. a 3 to 6 GeV high gradient electron linac, operating at about 16 GHz, with a 5 KHz repetition rate, or an Inverse Free Electron Laser;

b. a positron cooling and recovery ring (CRR), provided with a bypass, where the IP is located;

c. a positron converter, low energy positron accumulator, booster synchrotron, to provide the positrons.

This scheme is illustrated in Figure 1.

In this paper we will give only a preliminiary estimate of the main parameters of this system, with the purpose of establishing its feasibility. The high luminosity required to study the B physics makes any collider extremely difficult, and pushes the beam characteristics to a region not yet explored. What we propose is no exception and

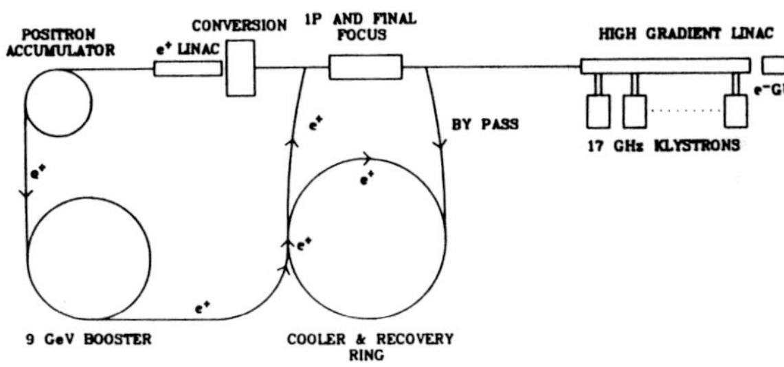

will require a large amount of research and development of beam physics and technology before a more detailed proposal can be made.

2. General Considerations

We consider a linear collider with two beams of energies E_1, E_2, (or γ_1, γ_2). For simplicity we consider only the case of beams with cylindrical symmetry, and rms transverse radius $\sigma_{t,i}$, and length $\sigma_{l,i}$ with i = 1 or 2. We also assume to have N_1 and N_2 particles per bunch.

The luminosity is then given by:

$$L = f\frac{N_1 N_2 H}{2\pi(\sigma_{t,1}^2 + \sigma_{t,2}^2)} \tag{1}$$

where f is the number of bunch collisions per second. To evaluate the luminosity we must use the constraint introduced by the disruption, using the parameter

$$D_i = \frac{r_e H N_k \sigma_{l,k}}{\gamma_i \sigma_{t,k}^2}, \quad \ldots \; i,k = 1,2 \text{ with } i \neq k \tag{2}$$

the other additional condition is given by the beamstrahlung parameter, describing the average energy loss

$$\delta_i = 8\frac{r_e^3 N_k^2 \gamma_i H}{21\pi^{\frac{1}{2}} \sigma_{l,k} \sigma_{t,k}^2} \tag{3}$$

Using eq. (2) we can rewrite the luminosity as

$$L = \frac{P_1 D_1}{2\pi r_e mc^2 \sigma_{l,2}\left(1+\frac{\sigma_{t,1}^2}{\sigma_{t,2}^2}\right)} \tag{4}$$

where

$$P_1 = fN_1\gamma_1 mc^2 \tag{5}$$

is the beam 1 power.

In practical units (5) can be rewritten as

$$L = 7\times 10^{24}\frac{P_1\,(\text{Watt})\,D_1}{\sigma_{l,2}(\text{cm})\left(\frac{1+\sigma_{t,1}^2}{\sigma_{t,2}^2}\right)}\ . \tag{6}$$

Let us now assume that beam 1 is the positron beam, and 2 the electron beam. From this equation, and assuming $\sigma_{t,1}$ on the same order as $\sigma_{t,2}$, one can see immediately that to have a large luminosity one can increase the beam power and/or the disruption of the positron beam, and decrease the electron bunch length. This last option leads to an increase in beamstrahlung, according to (3), and to a larger energy spread for the positrons. If we increase the disruption parameter, we can decrease the positron beam power, but we can hardly recover the positron beam.

As an example, let us assume that the two beams have an energy $E_1 = 9$ GeV, $E_2 = 3$ GeV, $L = 10^{34}\ cm^{-2}s^{-1}$, $\sigma_{t,1} < \sigma_{t,2}$, and $\sigma_{l,2} = 0.1$ cm. We then have $P_1D_1 = 143$ MWatt, corresponding to $\frac{10^{17}}{D_1}$ positrons/second at 25 GeV $\frac{e^-}{e^+}$. This requires 400 MW of e^- beam on target. If we assume a large value of D_1, for instance 10 or larger, to reduce the positron beam power, we eliminate the possibility of positron recovery. If we use a small value, like 1, we can do recovery, but we have a large beam power, which indicates that we need to recover not only the positrons but also their energy. This can be done in the scheme that we are proposing. The alternative of using a much smaller value of the electron bunch length leads to a value of the beamstrahlung parameter, which is incompatible with running at the Y(4S) resonance.

All of these points will be discussed in more detail in the following sections.

To conclude this section, we give in Table 1 a list of the possible parameters for a system. Notice that to increase the luminosity we have chosen a very small value of the electron bunch length, 0.01 cm. This is much shorter than the positron bunch length, 0.5 cm. The two bunches only start to interact at a distance from the IP equal to a quarter of the positron bunch length, and we have chosen the beta function at the IP equal to this distance.

In all these calculations we have also assumed the enhancement factor, H, to be one. We believe this to be a reasonable assumption, because of the small value of the positron disruption parameter. On the contrary, the electron disruption parameter is very large. In this situation the positron bunch will act as a channel in which electrons propagate and are focused. If we use, for simplicity, the model of a uniform positron beam to calculate the force on the electron, we can calculate the characteristics of the electron motion.[2] The oscillation frequency is given by the beam plasma frequency, and is related to the disruption parameter by

$$\omega_p = 2D\left(\frac{c}{\sigma_{l,1}}\right) \tag{7}$$

The number of oscillations during the crossing is given by

$$N_{osc} = \frac{\sqrt{2D}}{2\pi} \tag{8}$$

With the numbers of Table 1, we have $N_{osc} \approx 9$, and $\omega_p \approx 3.4\times 19^{12} s^{-1}$.

Table 1

COLLIDER PARAMETERS	
Energy, Positrons, GeV	9
Energy, Electrons, GeV	3
Positrons per Bunch	1.5×10^{11}
Electrons per Bunch	1.6×10^{10}
Normalized Transverse Emittance, Positrons, cm rad	3×10^{-4}
Normalized Transverse Emittance, Electrons, cm rad	1×10^{-4}
Beta at IP, cm	0.125
Bunch Radius, Positrons, rms, cm	4.6×10^{-5}
Bunch Radius, Electrons, rms, cm	4.6×10^{-5}
Bunch Length, Positrons, cm	0.5
Bunch Length, Electrons, cm	0.01
Repetition Frequency, s^{-1}	5×10^4
Beam Power, Positrons, MW	11.5
Beam Power, Electrons, MW	0.4
Luminosity, $cm^{-2}s^{-1}$	5×10^{33}
Disruption Parameter, Positrons	1.2
Disruption Parameter, Electrons	1800
Beamstrahlung Parameter, Positrons	3.9×10^{-4}
Beamstrahlung Parameter, Electrons	6×10^{-4}
Center of Mass Energy Spread	8×10^{-4}

3. The Cooling – Recovering Ring

The Cooler-Recovery Ring (CRR) must provide low emittance bunches of positrons at a high repetition rate. The bunches are extracted and sent in the bypass where they collide with the electrons at the IP. After the collision the positrons are sent back into the CRR where radiation damping will bring them back to the initial condition after a few damping time. We make the assumption that the effects of the interaction on the bunch, like disruption and beamstrahlung, are such that only a small fraction of the positrons, say less than 1%, are lost during the cycle described above.

The CRR is designed to provide:

a. fast cooling (short damping time);

b. short bunches, to reduce the electron beam disruption;

c. High peak current (for large luminosity);

d. small emittance.

To design the ring we have assumed a FODO type magnetic structure with or without the addition of wigglers. We have also considered two options:

1. a ring energy of 9 GeV, producing a beam which can be used directly for the colider;
2. a lower energy ring, say 2 to 3 GeV, with acceleration to the final energy taking place in the bypass, followed by the interaction with the other beam, and then by deceleration to the ring energy.

In both cases we base the ring design on the work of Bassetti, *et al.*[3] The ring is designed as a series of FODO achromat cells, made of four 90 degree phase advance cells, terminated at each end by a dispersion suppressor. In the zero dispersion straight sections one can insert wigglers, RF cavities and other components. For such a lattice the emittance can be written as

$$\epsilon = p\epsilon_0 + (1-p)\epsilon_w \tag{9}$$

where ϵ_{zero} and ϵ_w are the contributions to the emittance from the arcs and the wiggler magnets, and

$$p = \frac{\rho_w}{\rho_w + \eta\rho} \,. \tag{10}$$

In this formula ρ_w and ρ are the bending radii in the arcs and wigglers, and

$$\eta = \frac{l_w}{2\pi\rho_w} \tag{11}$$

with l_w being the total wiggler length.

The emittance due to the wiggler can be written as

$$\epsilon_w = 3.06\times10^{-8}E^2\beta_w\left(\frac{\lambda_w^2}{\rho_w^3}\right) \quad \text{with E in GeV} \tag{12}$$

and that due to the arcs as

$$\epsilon_0 = 3.83\times10^{-13}\gamma^2\left(\frac{p_i}{N_p}\right)^3 f_{latt}\,. \tag{13}$$

In (12) and (13), λ_w is the wigggler period, β_w is the beta function in the wiggler magnet, assumed roughly constant, and N_p is the number of achromat blocks in the ring, and f_{latt} is a lattice form factor which in our case can be assumed to be $f_{latt} = 0.115$. The radiated energy and the momentum compaction also have a contribution from the arcs and from the wigglers.

To determine the beam characteristics we must also consider and estimate the main collective effects which can limit the beam current and its emittance. For a small emittance ring the main collective effects is the microwave instability, limiting the peak current and energy spread; for a longitudinal coupling impedance $\frac{Z}{n}$, this can be written as

$$I_p = \frac{2\pi\sigma E\sigma_E^2}{\frac{Z}{n}} \tag{14}$$

To estimate the coupling impedance we use the Spear scaling

$$\frac{Z}{n} = \left[\frac{Z}{n}\right]_{co}\left(\frac{\sigma_l}{b}\right)^{1.68} \tag{15}$$

where $\left[\frac{Z}{n}\right]_{co}$ is the impedance at the vacuum pipe cutoff frequency.

At very high frequencies this impedance becomes very small, and one has to take into account another contribution, the so-called vacuum impedance

$$\left[\frac{Z}{n}\right]_{FS} = 300\frac{b}{R}(1+\eta) \tag{16}$$

peaking at a frequency near $\omega_\nu = c\left(\frac{R}{b^3}\right)^{\frac{1}{2}}$. For frequencies larger than ω_1 the impedance decreases like $\omega^{-\frac{2}{3}}$.

In addition to the microwave instability, we must also consider the limit introduced by the fast head-tail effect

$$I_p < \frac{4\pi\nu_\beta\nu_s E}{Z_T} \tag{17}$$

where Z_T is the transverse coupling impedance. To evaluate this we use the approximate formula $Z_T = 2\left(\frac{R}{b}\right)^2\left[\frac{Z}{n}\right]$.

The transverse emittance can be influenced by the Intrabeam scattering effect. For the two rings that we are considering this effect is, however, small and to a first approximation can be neglected.

The main characteristics of the rings are given in Table 2, for the 9 GeV ring, and Table 3 for a 2.2 GeV ring. This last ring is the same one described by Bassetti *et al.*, for the Frascati 5 to 10 GeV linear collider.

4. R&D Program

The collider proposed here offers the highest luminosity and reduced energy spread required to fully exploit the $\Upsilon(4S)$ resonance to produce greater than 10^8 $B\overline{B}$ events per year. Thus it would really be a $B\overline{B}$ Factory as specified in this workshop.[2]

Table 2

Cooler-Recovery Ring Characteristics	
Energy, GeV	9
Magnetic Structure	FODO
Bending Radius, m	33.6
Bending Field, T	0.89
Number of Dipoles	704
Dipole Length, m	0.3
Average Radius, m	67.2
Circumference, m	422
Phase Advance Per Cell, rad	1.99
Bending Angel Per Cell, rad	0.0178
Betatron Tune	112
Normalized Emittance, mm·mrad	3
Momentum Compaction	8×10^{-5}
Energy Loss Per Turn, MeV	18.3
Betatron Damping Time, ms	0.7
RF Frequency, GHz	3
Harmonic Number	4220
RF Voltage, MV	270
Synchrotron Tune	0.066
Energy Spread	1.3×10^{-3}
Bunch Length, mm	0.3
Positrons Per Bunch	8.8×10^{10}

In order to realize this machine, there are several crucial R&D questions which are:

(1) Can the high gradient linac operate at 5 KHz?

(2) Can the disruption of the e^+ beam be kept to the required level?

(3) Can the damping ring complex be simplified to reduce the cost?

(4) Can a lower energy booster be used? These and other issues can only be addressed by a careful design study and R&D work on the accelerator components. We propose that such an R&D effort be launched in the near future.

Table 3

Cooler-Recovery Ring Characteristics	
Energy, GeV	2.2
Magnetic Structure	FODO
Bending Radius, m	22.9
Bending field, T	0.32
Number of Dipoles	96
Dipole Length, m	1.5
Average Radius, m	114
Circumference, m	717
Phase Advance Per Cell, rad	1.57
Bending Angle Per Cell, rad	0.131
Betatron Tune	11.6
Normalized Emittance, mm·rad	3
Momentum Compaction	2.8×10^{-3}
Energy Loss Per Turn, MeV	6
Betatron Damping Time, ms	1.7
RF Frequency, GHz	3
Harmonic Number	7200
RF Voltage, MV	20
Synchrotron Tune	0.17
Energy Spread, at zero current	9×10^{-4}
Bunch Length, at zero current, mm	1.6
Positrons Per Bunch	1.6×10^{11}
Coupling Impedance, OHM	0.2
Energy Spread, at full current	2×10^{-3}
Bunch Length, at full current, mm	3.6

References

1) See the Proceedings of the UCLA $\overline{\text{B}}$B Factory Workshop, 1987.

2) See the Proceedings of this workshop.

3) Pellegrini, C. and Ruggiero, F., Radiation Reaction Effects in High Energy Electron-Positron Colliders, CERN LEP-Th/86-03.

BACKGROUND RATES WITH SMALL DIAMETER BEAM PIPES

Nari B. Mistry and Persis S. Drell

Laboratory of Nuclear Studies , Cornell University , Ithaca, NY 14853

Nobu Katayama

State University of New York at Albany , Albany, NY 12222

5-Dec-1988

INTRODUCTION

Precision vertex detection is a goal for many major detectors. A crucial requirement is to get the first track measurement as close to the collision point as possible, which requires a small diameter, thin walled beam pipe.

A small diameter beryllium beam pipe was installed in the e^+e^- storage ring CESR at the CLEO interaction point in April-May 1988 as a prototype test for the eventual beam pipe to be installed in CLEO-II in 1989. The central section of the beam pipe ($\pm$ 7.5 cm long) had an inner radius of 2.3 cm. The rest of the pipe, up to $\pm$ 30 cm long, was made of 1 mm thick beryllium with the same inner radius. A new tapered copper mask with an inner radius of 1.9 cm at $\pm$ 33 cm from the IP, shadowed the beam pipe from any direct synchrotron radiation(SR). The thin central section of the beryllium was coated on the inside first with aluminum ($\sim 15\mu m$ thick) and then with silver ($\sim 15\mu m$ thick) to absorb SR incident at glancing angles. This test configuration is called the "μVD test" configuration in the following. Four weeks of data were accumulated with this small beam pipe inside the CLEO dètector, using the CLEO drift chambers as diagnostic tools. The results of the tests can be summarized as follows:

1. There was no evidence for SR backgrounds.
2. Tracking chamber currents and trigger rates during the test were four to six times higher than in normal colliding beam operation at the $\Upsilon(5S)$ with a 5.5 cm radius beryllium beam pipe. The background rates scaled roughly as the square of the total beam current.
3. The high trigger rates were due to tracks originating from the beam pipe and the new masks at $\pm$ 33 cm.
4. The high background rates appeared to arise from beam gas interactions, since the rates decreased gradually as the pressure around the ring improved slowly after the μVD installation.
5. Subsequent investigation showed that the dominant source region for the lost electrons was limited to about 4 m to 12 m just upstream of the final horizontally focusing quadrupole before the IP. This was demonstrated by turning off vacuum pumps in discrete sectors near the IP and observing changes in the chamber currents and trigger rates. A local pressure increase in the separator and soft bend region, 4 m to 12 m from the IP, of roughly a factor of 2, produced an increase in trigger rates ranging from 2.57 to 3.5. By contrast, a pressure increase of a factor of 5 in the entire "hard bend" region, 15.4 m to 34 m from the IP, resulted in a increase of only 14% to 20% in the various counting rates. (Note: The estimated pressure increases are very approximate, as pressure monitoring is severely affected by beam induced RF effects.)

SIMULATIONS

The background due to beam gas interactions was studied using Monte-Carlo simulations of Coulomb scattering, ignoring energy losses, and of bremsstrahlung, ignoring deflections. Primary beam gas interactions were generated uniformly along the length of source elements in the CESR lattice, with each primary interaction originating only on the central orbits. The actual $\Upsilon(5S)$ lattice was used in the simulation, as it was in the μVD test. The scattered electrons were then allowed to propagate through CESR until they hit an aperture. Hits in the interaction region (IR) were tallied in order to identify dominant sources of beam gas induced backgrounds.

Coulomb Scattering

The effective source region for the beam gas scattering background is quite limited, extending from about 2 m to 14 m from the IP. Fig. 1 shows the rate of hits on the beam pipe and the luminosity monitors for 1000 scatters per meter in the source region, as a function of the position of the source.

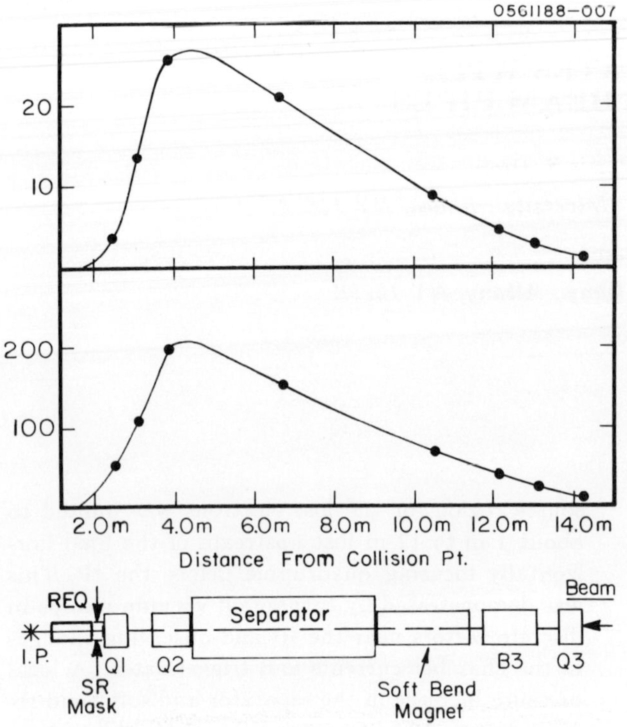

Fig. 1 The effective source region for beam gas scattering into (a) the IR pipe and (b) the outgoing luminosity monitor/SR mask and REQ. The number of hits in each of these regions is plotted for 1000 scatters per meter in the source region.

The rate of hits on the outgoing (downstream) SR mask (these are the original SR masks as distinguished from the new masks at ±33 cm which shadowed the μVD pipe) can be used as the expected rate of hits on the luminosity monitor which is mounted right next to it. The simulations for the actual μVD test conditions predict ~35 kHz per 43 mA beam, for a pressure of 10^{-8} Torr of N_2 or an equivalent 3×10^{-9} Torr of Freon-113. (It is probable that the actual residual gas in the separator was Freon-113 due to small leaks in the cooling system.) The observed rates on the luminosity monitor were ~23 kHz per 43 mA beam, for a measured pressure of 3×10^{-9} Torr. The expected rate for hits in the central IR beam pipe is 5.5 kHz, which can be compared to the observed Level-1 trigger rate of 6.1 kHz. Fig. 2 shows the dependence of Coulomb background on the IP pipe radius with the existing masking.

The best method to reduce the rate is to reduce the gas pressure in the source region.

Bremsstrahlung

For beam pipes larger than 1.5 cm radius, no direct hits from electrons lost by bremsstrahlung (up to $k_\gamma \cong 0.25E_0$) are observed in the simulation. Electrons crashing into the walls of the vacuum chamber Q1 to Q3 (see Fig. 1) can, however, produce backgrounds in the IR

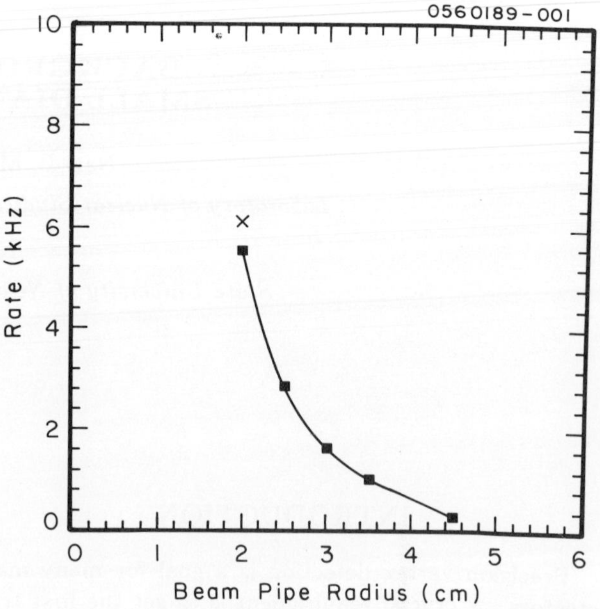

Fig. 2 The dependence of the Coulomb scattered electron background on IR pipe radius. The data point (x) shows the Level-1 trigger rate measured during the μVD single beam test.

pipe through secondary electrons. The effective source region for beam gas bremsstrahlung is 15.4 m to 51 m from the IP. Beyond that the contribution is negligible. The calculated rate of beam pipe hits is 550 Hz per 10^{-8} Torr N_2 pressure in the source region, compared to 5.5 kHz of electron hits expected from Coulomb scattering. During normal operation, the pressure in the source region is $\leq 10^{-9}$ Torr and we conclude that the contribution to the background from this source is negligible.

CONCLUSIONS

The background rates at pipe diameters smaller than ~4 cm radius are at present dominated by Coulomb scattering on residual gas in the separator and soft bend region. It appears feasible to reduce the pressure in this region to well below 1×10^{-9} Torr by proper design. This would reduce the Coulomb scattering background by a factor of 10 from that measured in the μVD test, bringing the background below the $\Upsilon(5S)$ rate with a 5.5 cm pipe. As a conservative first step, CLEO-II will install a 3.5 cm radius pipe this spring.

The lesson for other storage rings is that detailed simulations of the interaction region may be necessary before installing small diameter beam pipes. Pathological situations can exist, particularly when secondary showers are produced very close to the IR. The critical region for Coulomb scattering from residual gas lies quite close the the IP and could dominate the particle background problem, even though bremsstrahlung from residual gas around the whole ring is the main source of particle loss and finite beam lifetime in high emittance rings.

ACCELERATOR PHYSICS – LINEAR COLLIDERS

SUMMARY OF THE LINEAR COLLIDER WORKING GROUP*

RONALD D. RUTH

INTRODUCTION

The focus of the Linear Collider Working Group was on a next generation linear collider with the general parameters shown in Table 1. The energy range is dictated by physics with a mass reach well beyond LEP, although somewhat short of SSC. The luminosity is that required to obtain $10^3 - 10^4$ units of R_0 per year. The length is consistent with a site on Stanford land with collisions occurring on the SLAC site. The power was determined by economic considerations. Finally, the technology was limited by the desire to have a next generation linear collider before the next century.

Table 1. General parameters.

Energy	0.5 – 1.0 TeV in center-of-mass.
Luminosity	$10^{33} - 10^{34}$ cm^{-2} sec^{-1}.
Length	Each Linac $\lesssim$ 3 Km.
Power	$\lesssim$ 100 MW per Linac.
Technology	Must be realizable by 1990–92.

The basic configuration of such a linear collider is shown in Fig. 1. The beam is accelerated by an injector linac and then injected into a damping ring which damps the emittance of the beam and provides the beam with appropriate intensity and repetition rate. After extraction, the bunch must be compressed in length twice in order to achieve the short bunches suitable for the linac and final focus. The linac is used to accelerate the beams to high energy while maintaining the emittance. Finally, the final focus is used to focus the beams to a small spot for collision. This must yield a luminosity with tolerable beam-beam effects (disruption and beamstrahlung) and must also provide a reasonably background-free environment for the detector.

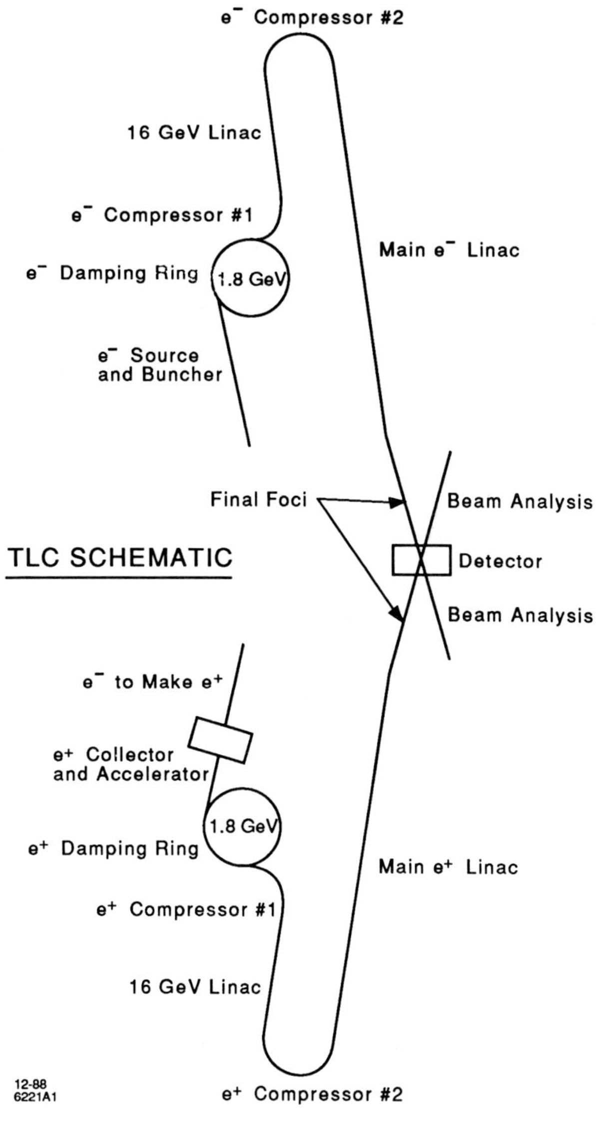

Fig. 1. Schematic layout of the TLC. The angles shown are exaggerated.

⋆ Work supported by the Department of Energy, contract DE–AC03–76SF00515.

Before launching into a rather detailed discussion of the contributions to the Linear Collider section of the Snowmass Proceedings, it is useful to discuss generally the overall results of the workshop and the months following. Perhaps one of the most important developments is the increased interest in an Intermediate Linear Collider (ILC) with an energy of 0.5 TeV in the center-of-mass. This is a factor of two below the TeV Linear Collider (TLC) and thus would require a factor of four less peak power provided that the machines were the same length. One can imagine designing an ILC which would be upgradable in energy by the addition of RF power and minor modifications to the final focus system.

If we begin the discussion of an ILC or TLC at the lower energy end, the damping ring and bunch compressor designs seem relatively straightforward with, however, somewhat tighter tolerances than usual. The main linac will probably have a structure similar to SLAC, except at 4–6 times the frequency. The irises will have slots coupled to radial waveguides to damp the transverse and longitudinal higher order modes. This makes possible the use of multiple bunches per RF fill, which increases the luminosity by a factor of 10 for "free."

There is no definite power source as yet. The recent demonstration of binary pulse compression at SLAC has focused attention on more conventional approaches to long-pulse power production. Low power, low loss tests of RF pulse compression are continuing at SLAC and initial results look very promising. There are plans to build a high power klystron at SLAC to feed the RF pulse compressor, and there are many new ideas for power sources which would drive RF pulse compressors. The relativistic klystron results have been somewhat discouraging, but much as been learned about the problems associated with these high current, high energy beams.

Once the power source problem is solved, we are still left with the luminosity problem. These two aspects are only partially decoupled due to the use of many bunches (a batch) per RF fill. To obtain the luminosity, we must preserve the emittance of the beam throughout the linac. This means tighter tolerances on vertical magnet alignment than are presently achieved. The final focus demagnifies the beam to obtain a very flat beam at the final focus. The chromatic correction for this is quite delicate, and tolerances are tight. Finally, we must measure the beam size at the interaction point in order to tune the final focus. Many of these problems can be addressed via a model final focus at a lower energy. Towards this end, there is presently work ongoing at SLAC to create a Final Focus Test Beam in order to test flat beam final focus optics, measurement techniques, alignment techniques, etc. This would use the 50 GeV SLC beam straight ahead into the old C-line at SLAC.

During the SLAC Workshop in December 1988, following Snowmass, there was one important discovery which should be mentioned here. Beamstrahlung photons create e^+e^- pairs upon interacting with the opposing bunch. One particle of the pair is deflected strongly by the field of the bunch. This, in turn, can cause serious background problems. P. Chen and B. Palmer discuss this in their papers.

Table 2. Linear collider working group participants.

B. Ash—SLAC	R. Miller—SLAC
R. Blankenbecler—SLAC	A. Odian—SLAC
F. Bulos—SLAC	K. Oide—KEK/SLAC
D. Burke—SLAC	R. Palmer—BNL/SLAC
P. Chen—SLAC	E. Paterson—SLAC
Y. Chin—LBL	T. Raubenheimer—SLAC
B. Gabella—U. of Colo.	R. Ruth—SLAC
T. Himel—SLAC	K. Thompson—SLAC
S. Kheifets—SLAC	W. Vernon—UCSD
W. Kozanecki—SLAC	J. Wang—SLAC
N. Kroll—UCSD	B. Warnock—SLAC
G. Loew—SLAC	P. Wilson—SLAC
P. McIntyre—Texas A&M	

The list of working group participants is given in Table 2, while the Table of Contents gives a list of the contributions to the Proceedings. The purpose of the remainder of this paper is to provide guidance to the reader for the papers shown in the Table of Contents. To do this, I first discuss parameters briefly and then discuss damping rings. After discussing the basic concepts of bunch compression, I move to the contributions on linac structures and linac power sources. Since there is no specific paper on beam dynamics issues, in the next section I discuss emittance preservation in the linac in some detail. Next, the final focus papers are discussed and last, but not least, I introduce some of the issues for multibunch effects.

PARAMETERS

In this section, R. Palmer contributed a paper on linear collider energy scaling. Linear colliders are being considered for accelerators ranging from B-factories to Z-factories, up to a TeV Linear Collider. The purpose of this paper is to explore the change in the design of linear colliders as a function of energy given that one is always trying to maximize luminosity, but always respecting the limit on wall plug power shown in Table 1. A very wide range of energies is considered, and this leads to widely differing designs. In particular, one sees that the optimized RF frequency tends to decrease at lower energy while the repetition rate increases.

Table 3. Parameters for ILC and TLC.

		Low grad ILC	High grad ILC	TLC
General				
CM energy	TeV	.5	.5	1
luminosity 10^{33}	cm^{-2} sec^{-1}	1.5	2.9	6.2
RF wavelength	cm	1.75	1.75	1.75
repetition rate	kHz	.36	.36	.36
accel gradient	MV/m	93	186	186
number bunches		10	10	10
particles/bunch	10^{10}	.7	1.4	1.4
wall power	MW	52	103	210
length	Km	7.3	3.7	7.3
Damping				
emittance ϵ_x/ϵ_y		100	100	100
emittance $\gamma\epsilon_x$	μm	3.5	6.0	7.0
emittance $\gamma\epsilon_z$	m	.04	.04	.04
bunch spacing	m	.2	.2	.2
damping time	msec	2.1	2.3	2.3
RF				
pulse length	ns	60	60	60
peak power/length	MW/m	146	580	580
total RF energy	KJ	51	103	210
Linac				
loading η	%	2.5	2.5	2.5
iris radius a	mm	3.5	3.5	3.5
section length	m	1.6	1.6	1.6
Linac tolerances				
alignment	μm	20	35	30
vibration	μm	.009	.017	.012
Final focus				
β_y^*	mm	.1	.12	.11
crossing angle	mrad	4.2	6.1	3.8
disruption angle	mrad	.23	.31	.25
free length	m	.36	.43	.7
Intersection				
σ_y	nm	2.7	3.9	2.8
Oide min σ_y	nm	1.3	1.9	1.9
σ_x/σ_y		132	132	132
σ_z	μm	70	70	70
disruption D		5	5	5
lum enhance H		1.6	1.6	1.6
beamstrahlung δ	%	2	4	11
$\Delta p/p$ physics	%	.7	1.1	3.2

In addition, due to the interest in the ILC, Palmer considers two possible options for an ILC, both of which would be upgradable to a TLC with additional length/power sources. Perhaps the most attractive option is the low gradient ILC which has a physical layout identical to TLC but has one-half the acceleration gradient. The parameters for ILC and TLC are compared in Table 3.

There is also an addendum to Palmer's paper which discusses the problem of e^+e^- pair creation at the interaction point by beamstrahlung photons interacting with the oncoming bunch. He finds that by using his idea of 'crab crossing' it is possible to collide beams with a very large crossing angle. In this way, with the help of solenoidal guide fields, the deflected e^+ or e^- can exit through a large aperture hole adjacent to the incoming quadrupole. This means that the parameter sets which have been presented will have to be modified to include various changes, but the basic parameters still will be rather similar to those given in Table 2.

DAMPING RINGS

The paper by T. Raubenheimer, L. Rivkin and R. Ruth discusses many of the basic design considerations for the damping ring. The basic parameters of the TLC damping ring are shown in Table 4 where they are compared to those of the SLC. The key differences are the decrease of the horizontal emittance by an order of magnitude, the increase of the repetition rate and the requirement of $\epsilon_x/\epsilon_y = 100$. Although asymetrical emittances have been measured in the SLC damping ring, they are not required for SLC operation.

The desired repetition rate is obtained by having many batches of bunches in the ring. Each batch of 10 bunches is extracted on one kicker pulse and accelerated on one RF fill in the linac. The remaining batches are left in the ring to continue damping while an additional batch is injected to replace the extracted one. The threshold current refers to the threshold for the "microwave instability" or "turbulent bunch lengthening."

The basic layout of a possible damping ring is shown in Fig. 2. Notice that there are several insertions which contain wigglers. In order to obtain the high repetition rate, it is necessary to decrease the damping time by the addition of wigglers in straight sections.

Table 4. Basic parameters of the SLC and TLC damping rings.

	TLC	SLC
Energy	$1 \sim 2$ GeV	1.15 Gev
Emittance, $\gamma\epsilon_x$	3.0 μmrad	36 μmrad
Emittance, $\gamma\epsilon_y$	30 nmrad	500 nmrad
Repetition rate	360 Hz	180 Hz
Bunch length	4 mm	5 mm[1)]
Threshold Current	batches of 10 bunches of 2×10^{10}	1.5×10^{10}[1)]

In Tables 5 and 6, you see the basic parameters for the ring. The lattice is combined function which allows the

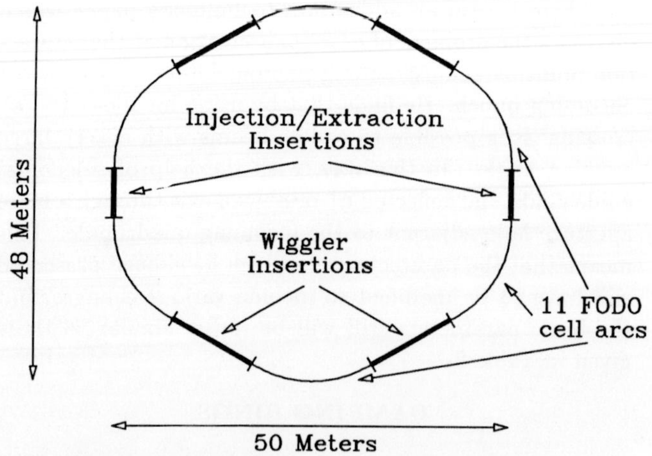

Fig. 2. Schematic of the TLC damping ring.

partition of the damping times to trade horizontal damping time for longitudinal. The RF frequency for this example is necessarily 1.4 GHz since the bunch spacing in this example is about 20 cm. The threshold impedance $(Z/n)_t$ is that for the microwave instability. It is quite small due to the small momentum compaction factor, but is only about a factor of three below that obtained in the SLC damping rings.

Table 5. TLC damping ring parameters.

Energy	$E_0 = 1.8$ GeV
Length	$L = 155.1$ meters
Momentum compaction	$\alpha = 0.00120$
Tunes	$\nu_x = 24.37, \quad \nu_y = 11.27$
RF frequency	$f_{RF} = 1.4$ GHz
Current	10 batches of 10 bunches of $2 \times 10^{10} e^+/e^-$

Table 6. TLC damping ring parameters.

	Wigglers Off	Wigglers On
Natural $\gamma\epsilon_x$	2.46 μmrad	2.00 μmrad
$\gamma\epsilon_x$ w/ intrabeam	3.33 μmrad	2.74 μmrad
Damping, τ_x	3.88 ms	2.50 ms
Damping, τ_y	9.19 ms	3.98 ms
Rep. rate, f_{rep}	155 Hz	360 Hz
Damp. partition, J_x	2.37	1.59
Energy spread, σ_ϵ	0.00128	0.00104
Radiation/turn, U_0	203 KeV	468 KeV
Bunch length, σ_z	5.6 mm	5.2 mm
Synch. tune, ν_s	0.0068	0.0058
$(Z/n)_t$	$\mathcal{F} \times 0.32\Omega$	$\mathcal{F} \times 0.20\Omega$
Natural chrom., ξ_x	-28.35	-28.07
Natural chrom., ξ_y	-25.10	-22.27

Another key aspect of the TLC design is the small vertical emittance. The design calls for an emittance ratio $\epsilon_x/\epsilon_y = 100$. This size emittance ratio is quite common in $e^{\pm}$ storage rings. However, the tolerances for obtaining such a small vertical beam size are proportional to this size. In the damping ring paper, those tolerances which are related to maintaining the emittance ratio are calculated. The tolerances presented in Sec. 5 of the paper are in the 100 μm range and could be improved by adding correction skew quadrupoles in the ring.

BUNCH COMPRESSION AND PRE-ACCELERATION

Designs for bunch compression are presented in a paper by S. Kheifets *et al.* In order to obtain the very short bunches necessary for the linac, it is necessary to perform at least two bunch compressions. A bunch length of about 50 μm in the linac puts a tight constraint on the longitudinal emittance of the damping ring. In addition, during the bunch compressions, it is necessary to keep the energy spread small to avoid the dilution of the transverse emittance. If we assume that we can transport 1% energy spread without diluting either transverse emittance, then at least two bunch compressions are needed. For example, if we consider a 1.8 GeV damping ring with energy spread $\Delta E/E = 10^{-3}$ and a bunch length of 5 mm, the two compressions are shown in Table 7. The first one decreases the bunch length by an order of magnitude. This is followed by a pre-acceleration section to decrease the relative energy spread in the beam by an order of magnitude. One must avoid an increase of energy spread due to the cosine of the RF wave (and also due to beam loading). If this pre-acceleration is done at the present SLAC frequency and if the bunch current is as shown in Table 3, then the additional energy spread induced is about 5×10^{-4}. Neglecting this small increase, the next bunch compression happens at 18 GeV and serves to reduce the bunch length to about 50 μm. This is suitable for injection into the high frequency, high gradient structure.

Table 7. Bunch compression.

E	$\Delta E/E$	σ_z	Compress →	$\Delta E/E$	σ_z
1.8 GeV	10^{-3}	5 mm	Compress →	10^{-2}	0.5 mm
[pre-acceleration at long wavelength, $\lambda = 10.5$ cm]					
18 GeV	10^{-3}	0.5 mm	Compress →	10^{-2}	50 μm

The two designs shown in the paper by S. Kheifets *et al.* are for bunch compressors which have small bending angles. However, 180° bends which do the same job have also been designed.

LINAC

The linac is envisioned to be similar to the SLAC disk-loaded structure with a frequency at least four times the present SLAC frequency. The example shown in Table 3 is for six times the present SLAC frequency. The irises in the design are relatively larger to reduce transverse wakefields. The structure may have other modifications to damp long-range transverse wakefields. This would be driven by a power source capable of about 600 MW/m for the TLC or about 150 MW/m in the case of the ILC. In the case of the low gradient ILC, one can imagine an upgrade consisting of the addition of power sources.

Structures

The first paper on structures by G. Loew and J. Wang treats the question of RF breakdown. There have been many experiments done at various frequencies. If the scaling laws thus obtained are extrapolated to 11.4 and 17.1 GHz, the breakdown limited surface fields obtained are 660 and 807 MV/m, respectively. To convert this to effective accelerating gradient, a reduction factor of 2.5 is typically used.

In both cases, the accelerating gradient is above the 200 MeV/m used for the TLC design in Table 3. However, the measurements also indicated significant "dark currents" generated by captured field-emitted electrons. The question of the effects of dark current on loading and beam dynamics is not yet resolved and needs further study.

The next paper in the structures section is written by R. Palmer on Damped Accelerator Structures. As mentioned in the Introduction, in order to make efficient use of the RF power and to achieve high luminosity, it seems essential to accelerate a train of bunches with each fill of the RF structure. This leads to two problems: (1) the energy of the bunches in the train must be controlled and (2) the transverse stability of the bunch train must be ensured. Both of these problems are helped greatly by damping higher modes (both transverse and longitudinal) in the RF structure. This paper describes a technique of using slotted irises coupled to radial waveguides to damp these modes: Q's as low as 10–20 have been measured in model structures. This encouraging evidence has led to a development program at SLAC to do more detailed studies of slotted structures. The beam dynamics consequences of damping the higher modes is explored in the section on multibunch effects.

RF Power Sources

There are several papers contributed on the subject of RF power sources. The first and second paper are on the relativistic klystron and RF pulse compression, respectively. It is useful to contrast these approaches.

RF Pulse Compression

In Fig. 3(a), you see illustrated the basic principle of RF pulse compression. A long modulator pulse is converted by a high power, 'semi-conventional' klystron or some other power source into RF power with the same pulse width. This RF pulse is then compressed by cleverly slicing the pulse using phase shifts and 3 db hybrids and re-routing the portions through delay lines so that they add up at the end to a high peak power but for a small pulse width. This scheme was suggested by D. Farkas at SLAC and is presently under experimental investigation.[2] With a factor of 16 in pulse compression, the TLC would require a 50 MW klystron with a 1 μsec pulse length for each meter of the accelerator while the ILC would require a 50 MW klystron for each 4 meters of structure.

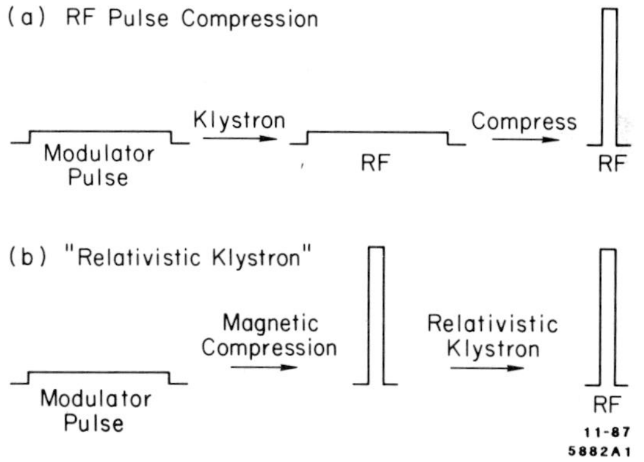

Fig. 3a. Illustration of RF pulse compression. 3b. Illustration of the relativistic klystron with magnetic compression.

The Relativistic Klystron

In Fig. 3(b), you see the principle of the relativistic klystron illustrated. In this case, the pulse compression happens *before* the creation of RF. This technique makes use of the pulsed power work done at LLNL in which magnetic compressors are used to drive induction linacs to produce multi-MeV e^- beams with kiloampere currents for pulses of about 50 nsec. These e^- beams contain gigawatts of power. The object, then, is to bunch the beam at the RF frequency to extract a significant fraction of this power. This can be done either by velocity modulation or by dispersive magnetic "chicanes." After bunching, the beam is passed by an RF extraction cavity which extracts RF power from the beam.

In the relativistic klystron paper by M. Allen *et al.*, four experiments are described. These are the result of a SLAC–LLNL–LBL collaboration which makes use of the

ARC facility (e^- beams 1.2 MeV and $\lesssim$ 1 KA) at LLNL. Thus far the record peak power for any of the devices tested is 200 MW; however, in this case, the RF envelope was noticeably shortened. The highest power obtained with a wide RF pulse was about 80 MW. The most serious problem encountered in the experiment is the pulse shortening phenomenon; however, recent experiments suggest that this is caused by loading due to anomalous charged particle currents. A second serious problem is poor beam transmission. Finally, this RF power has been used to drive a 26-cm travelling wave structure at 11.4 GHz. The peak power of 200 MW corresponds to a local acceleration gradient of 140 MV/m. Work is continuing on this experiment.

In the second paper on RF pulse compression and alternative RF sources, P. Wilson describes RF pulse compression in some detail including estimates of efficiencies. There is an experimental test ongoing at SLAC which seeks to test a low loss, low power system. Initial results of this test have been very encouraging. A 100 MW, 11.4 GHz, "semi-conventional" klystron is presently being constructed at SLAC to perform high power tests of pulse compression. In the second section, the paper discusses other alternative RF sources for input to RF pulse compression.

The next paper on RF power, by W. Vernon, discusses another RF compression technique which uses RF energy storage combined with a high power switch to obtain very large compression of RF. This idea has been tested experimentally but needs much more work on high power switches to be feasible.

The next contribution on the cluster klystron, by R. Palmer and R. Miller, describes a multiple beam array of "klystrinos" which when coupled together can give impressive results. By dividing a single beam into many beams shielded from each other, the problems of space charge are effectively eliminated. This source could be used as a driver for RF pulse compression. Alternatively, with the addition of a grid and an oil-filled transmission line for energy storage, the device could directly produce short RF pulses. Thus far, there has been no experimentation; but calculations and cost estimates are encouraging.

The final paper in this section on RF power is on the Gigatron by H. Bizek *et al.* This device makes use of the lasertron concept to produce a bunched beam directly at the cathode. Field emitting arrays are used for the cathode while a ribbon beam geometry is envisioned to control space charge effects. This device is another candidate for RF pulse compression and has an impressive efficiency on paper. Experimental tests are presently being prepared.

Emittance Preservation

There was no specific contribution to the Snowmass Proceedings on emittance preservation in the linac. The purpose of this section is to fill in that gap. The effects discussed are treated in more detail in Ref. 3.

Chromatic Effects

The filamentation of the central trajectory in a linac can cause dilution of the effective emittance of the beam. If we first consider a coherent betatron oscillation down the linac, then to be absolutely safe, we must require that it be small compared to the beam size. If the spread in betatron phase advance is not too large, then this tolerance is increased to perhaps twice the beam size for the cases shown in Table 3.

The chromatic effect of a corrected trajectory is rather different. In this case, it is the distance between an error and a corrector which matters, and the effects partially cancel yielding a growth $\propto \sqrt{N_{quad}}$. This yields a tolerance on magnet misalignment the order of 20 to 30 times the beam size in the linac (about 30 μm) for the cases shown in Table 3. This is also the tolerance on BPM measurements. If the phase advance of the linac or some subsection is not too large, then this yields a linear correlation of position with momentum (dispersion) which can, in principle, be corrected since it does not vary in time. Therefore, it may be possible to have looser tolerances if such correction is provided.

Transverse Wakefields and BNS Damping

The wakefield left by the head of a bunch of particles, if it is offset in the structure, deflects the tail. If the transverse oscillations of the head and tail have the same wave number, the tail is driven on resonance. This leads to growth of the tail of the bunch.[4] This effect can be controlled by a technique called BNS damping.[5] The bunch is given a head-to-tail energy correlation so that the tail is at lower energy. The offset of the head by an amount $\hat{x}$ induces a deflecting force on the tail away from the axis. The tail, however, feels an additional force $\Delta K\hat{x}$, where ΔK is the difference in focusing strength. These two forces can be arranged to cancel, thereby keeping the coherence of the bunch as a whole. For the designs shown in Table 3, the spread in energy for BNS damping is $\sim \pm.3\%$. This correlation can be accomplished by moving the bunch slightly on the RF wave to obtain a linear variation across the bunch.

Recently, BNS damping has been tested at the SLAC linac with great success. It is now part of normal operating procedure.

Jitter

In order to maintain collisions at the interaction point, the bunch must not move very much from pulse to pulse. Since the optics of the final focus also demagnify this jitter, the tolerance is always set by the local beam divergence compared to the variation of some angular kick. The jitter tolerance on the damping ring kicker is thus related to the divergence of the beam at that point. This is discussed in the paper on damping rings. At the injection point to the linac, the offset caused by this jitter must be small compared to the local beam size.

If all the quadrupoles in the linac are offset by random amounts, the effects accumulate down the linac and the orbit offset grows $\propto \sqrt{N_{quad}}$. This sets the tolerance on the random motion of quadrupoles to be much smaller than the beam size. In the examples in Table 3, the random jitter tolerances are $\simeq 0.01\ \mu$m. On the other hand, tolerances for correlated effects are much less severe. In either case, this size motion from pulse-to-pulse is unlikely due to the large repetition rate of the collider. More gradual motion, which is larger, can be corrected with feedback.

Jitter in RF kicks can cause similar effects. These effects can be reduced by reducing the DC component of the RF kick. This is done by eliminating asymmetries in couplers and by careful alignment of structures.

Coupling

Finally, we discuss coupling of the horizontal and vertical emittance. The beam size ratio in the linac is 10:1. The tolerance on random rotations is given by

$$\Theta_{rms} << \frac{\sigma_y}{\sigma_x} \frac{1}{\sqrt{2N_q}} \quad .$$

For the examples shown in Table 3, the right-hand side is about 3 mrad. This seems quite straightforward. If the errors are not random, larger rotations can indeed result; however, because the beam size is so small, the effects are very linear. This means that skew quadrupoles can be used effectively as correction elements. Certainly in the final focus, skew quads will be an integral part of the tuning procedure to obtain flat beams.

FINAL FOCUS

The final focus, as described in the parameters in Table 3, is a flat beam final focus with a crossing angle. The purpose of the flat beam is to increase the luminosity while controlling beamstrahlung and disruption. The crossing angle is to allow different size apertures for the incoming and outgoing beam. Another invention, "crab-wise crossing," discussed in R. Palmer's paper, allows a much larger crossing angle than the diagonal angle of the bunch. As discussed in Palmer's contribution and in P. Chen's paper in this section, this type of geometry may now be essential due to the production of e^+e^- pairs by beamstrahlung photons in the field of the bunches.

The first paper in this section by K. Oide treats the design of a flat-beam final focus. The vertical size is limited by the synchrotron radiation in the final quadrupole doublet coupled with the chromatic effect of a quad. He also shows how the system can be made much shorter and how the aperture of the final doublet can be increased. Finally, Oide discusses the pulse-to-pulse jitter tolerances on elements in the final focus. The most restrictive requirement is on the final doublet which must be stable pulse-to-pulse to about 1 nm.

Since vibration of the final doublet is the most serious problem, it is considered in some detail in the next paper by W. Ash. In this paper, it is shown that passive vibration isolation seems to be more than adequate to handle the vibrations above ~ 10 Hz at the high frequency end. For low frequencies, Ash suggests an interferometric feedback system to control motion to about 1 μm. Beam steering feedback can then be used to control slow variations in the 1 nm to 1 μm region.

A summary of beam-beam effects is presented by P. Chen. He begins by covering beamstrahlung and disruption for flat beams. The next subject is the kink instability induced if the beams are slightly offset vertically from each other. This effect actually causes the luminosity to be less sensitive to offsets because the beams attract each other and collide anyway. There is also a multibunch kink instability which is more serious since it can cause the trailing bunches to miss each other entirely. The effect of this is to place restrictions on the product of the vertical and horizontal disruption per bunch. This effect is shielded somewhat if the quadrupoles are close to the interaction point. Chen concludes the main body of the paper with a discussion of energy spectrum and maximum disruption; issues which are important for final quad design and backgrounds.

The final section of Chen's paper is an addendum added after the SLAC Workshop in December 1988. As mentioned earlier in the Introduction, it was discovered that the beamstrahlung photons pair-produce in the coherent field of the bunch. The corresponding incoherent process has been known for some time, but its importance has only just been realized. The problem is that low energy e^+e^- pairs are produced in an extremely strong field which then deflects the charge of the appropriate sign while confining the other. This leads to large angular kicks, as mentioned earlier in Palmer's paper on parameters. These stray particles can lead to more background problems, which must be addressed by further interaction point design. Some initial ideas were presented in Palmer's paper, but the problem needs much more study.

The discussion of beam-beam effects continues in a paper by R. Blankenbecler and S. Drell. They show how beamstrahlung photons can be used to create a photon-photon collider. They show an increase in luminosity for real photons *vs* virtual photons. At the time of Snowmass, the problem of pair creation had not been realized. This will probably not effect the luminosity results but would cause difficult background problems.

There is one final paper in the final focus section by T. Himel which deals with some of the background issues. He discusses background from synchrotron radiation in the final quad and also discusses the maximum disruption angle of an electron which has radiated most of its energy.

Clearly, with all this new information on the swarm of e^+e^- pairs produced, there needs to be much more study of the background problems, as well as interaction point geometry.

The measurement of the final spot size was not studied at the Snowmass workshop. This is an extremely important, but as yet unsolved, problem. From SLC experience, it is probably possible to use beam-beam effects to minimize spot sizes. However, for the initial tune-up of the final focus, a single-beam method is almost essential. There was some initial work done at the workshop in June 1988 in Capri, Italy which was also reported at the SLAC workshop.[6] In addition, preliminary results were presented at the SLAC workshop on the use of beamstrahlung from an ionized gas jet.[7] Although this looks promising, there is still much work to be done.

MULTIBUNCH EFFECTS

As mentioned earlier, in order to efficiently extract energy from the RF to obtain high luminosity, it is essential to have many bunches per RF fill. This, however, leads to transverse beam breakup. K. Thompson and R. Ruth have written a paper for this section which traces the problem of multibunches all the way through the linear collider, subsystem by subsystem. Damped accelerating cavities are required for the main linac and the damping rings. Other systems can get by with very strong focusing. Thus, from the transverse point of view, stability seems possible.

In addition, it is necessary to control the energy spread from bunch to bunch very precisely ($\Delta E/E \lesssim 10^{-3}$). This can be accomplished by injecting the bunches before the RF structure is full to match the extraction of energy by the bunches to the incoming energy as the structure fills. This leads to tight tolerances on phase and amplitude of the RF, as well as tight control of the pulse-to-pulse number of particles in a batch of bunches.[8] However, the benefits of multibunching seem to far outweigh any difficulties they impose due to the order of magnitude increase in luminosity.

OUTLOOK

A key result of the Snowmass Workshop is the increased interest in an ILC, that is, a linear collider with 0.5 TeV in the CM which might be upgradable to 1.0 TeV with additional power sources. Since there is a factor of four difference in peak power for the ILC and TLC (for a fixed length), the power source looks much easier to do. Designs of damping rings, bunch compressors and final focus systems are continuing. Perhaps the most important point was realized only after the Snowmass Workshop — beamstrahlung pair creation at the interaction point. The control of backgrounds in the face of pair creation needs much more work. However, with a sufficiently intensive R&D program towards power sources, final focus systems, etc. during the next few years, we may see a proposal for a next generation Linear Collider in the early 1990 s.

ACKNOWLEDGEMENTS

I would like to thank Fred Gilman, Mel Month, and the entire organizing committee for a stimulating and well-organized workshop. Also, I would like to thank all the members of the working group for their stimulating discussions and for their excellent contributions to the Proceedings.

REFERENCES

1. L. Z. Rivkin *et al.,* "Bunch Lengthening in the SLC Damping Ring," SLAC–PUB–4645 (1988).
2. Z. D. Farkas, *IEEE Transcripts on Microwave Theory and Techniques,* **MTT–34**, No. 10 (1986) 1036, and also SLAC/AP–59.
3. R. D. Ruth, *Emittance Preservation in Linear Colliders,* in Lecture Notes in Physics, **296**, Frontiers of Particle Beams, Eds. M. Month, S. Turner, South Padre Island, Texas, 1986, and in SLAC–PUB–4436.
4. A. Chao, B. Richter and C. Yao, *Nucl. Instr. Meth.* **178**, 1 (1980).
5. V. Balakin, A. Novokhatsky and V. Smirnov, *Proceedings of the 12th International Conference on High Energy Accelerators,* Fermilab, (1983) p. 119.
6. J. Norem, presented at the *International Workshop on Next Generation Linear Colliders,* in preparation.
7. D. Burke, P. Chen, M. Hildreth and R. Ruth, work in progress.
8. R. D. Ruth, *Multi-Bunch Energy Compensation,* in proceedings of workshop on "Physics of Linear Colliders," Capri, Italy, 1988; and in SLAC–PUB–4541.

ENERGY SCALING, CRAB CROSSING AND THE PAIR PROBLEM*

R. B. Palmer

Stanford Linear Accelerator Center, Stanford University, Stanford, California 94309
and
Brookhaven National Laboratory, Upton, New York 11973

ABSTRACT

Making reasonable assumptions, the luminosities of linear colliders are calculated for center-of-mass energies of 10 GeV, 100 GeV and 1 TeV. A calculation is also mode for a 1/2 TeV collider that could be upgraded to 1 TeV later. The improvements possible using "crab-like" crossing are also given.

1. INTRODUCTION

The design of linear colliders is a complex problem because of the interdependence of the critical parameters.[1] Changing the number of particles per bunch effects the damping ring design and thus the emittance; it effects the wakefields in the linac and thus the momentum spread; the momentum spread effects the final focus design and thus the final β^*; but the emittance change also effects the final focus design; and all these come together to determine the luminosity, disruption and beamstrahlung at the intersection. Changing the bunch length, or almost any other parameter, has a similar chain reaction. Dealing with this problem by simple scaling laws is very difficult because one does not know which parameter is going to be critical and thus which should be held constant. In the face of this problem, I have written a PC program that—given various assumptions—simultaneously calculates:

- Emittances, damping rate and impedance requirements in the damping ring.
- RF properties of the accelerating structure.
- Longitudinal wakes in the linac.
- Transverse wakes in the linac.
- Focusing requirements for BNS damping and tolerances in the linac.
- Required final focus chromatic correction.
- Required pole tip fields in the final focus quadrupoles.
- Disruption and beamstrahlung at the intersection.
- Multibunch instabilities at the crossing.

Most of the calculations are done using analytic expressions. Inevitably, many approximations have been used. It is not the intent of the program to actually design a machine. It is the object to explore the parameter space and get a feeling for what is and is not possible.

*Work supported by DOE contract numbers DE-AC02-76CH00016 and DE-AC03-76SF00515.

In an earlier paper,[1] I used the program to select reasonable parameters for a .5 on .5 TeV Collider. In this paper, the emphasis is on discovering the scaling to lower energies. There are, however, a number of changes in the assumptions and calculations since Ref. 1:

1. It is assumed that damped acceleration cavities[2] can be used, and thus multiple bunches can be accelerated without buildup of transverse wakes.
2. The longitudinal wakes are correctly integrated, assuming a Gaussian bunch profile (a four-bunch approximation had been used in Ref. 1).
3. Dilutions are introduced: (a) between damping ring and linac, (b) in the linac and (c) in the final focus. Dilutions are specified for emittances in all three directions and for β.
4. The Oide limit on spot size from synchrotron radiation in the final quads is calculated and applied.

2. ASSUMPTIONS

2.1 Ratio of Horizontal to Vertical Emittances

Assume the ratio of horizontal to vertical emittances in the damping ring to be less than or equal to 100:1. An asymmetric emittance is natural in a damping ring and comes with essentially no price. It allows the generation of a flat beam profile. The original motive for using such a flat beam was to minimize the beamstrahlung without loss of luminosity. It also allows the use of a finite angle crossing, so that the disrupted beam does not have to pass through the opposite quadrupole. I find that the best luminosity always requires the maximum ratio of emittances.

2.2 Ratio of Horizontal to Vertical Betas

With conventional crossing, I assume the undiluted ratio of horizontal to vertical betas at the intersection to be 3.24 times the ratio of emittances. When this condition is satisfied, the beam size in the final quads is approximately the same horizontally and vertically, and maximum luminosity is achieved. With 100:1 emittance ratio and this assumption, the beam will have an aspect ratio of 180:1.

With crab crossing (see Sec. 3.3), greater luminosity is obtained with smaller ratios of the betas; but the beamstrahlung rises, and has been limited to a value of $\delta \leq .3$

in this study. The maximum luminosity is reached for a β ratio of 3.2.

2.3 Crossing Angle

In order to avoid excessive luminosity loss, a finite crossing angle is selected that is 0.7 times the bunch diagonal angle (*i.e.*, $0.7 \times \sigma x/\sigma z$). The bunch length is then selected to give a crossing angle sufficiently large, compared with the maximum disruption angle, to avoid collisions of the disrupted beam with the poles of the quadrupole. A larger factor is required at the higher energies because of assumed quantum fluctuations in the disrupted beam (this is yet to be calculated). The required ratios were: crossing/disruption = 24 (1 TeV), = 19 (.1 TeV) and = 14 (10 GeV).

2.4 Damping Ring

A wiggler damping ring is assumed. The energy is chosen to make the contributions from intrabeam scattering and quantum fluctuations the same. The ring diameter is then chosen to give a longitudinal impedance requirement of Z/n = 0.5 Ω. The wiggler fields are 2 T, the quadrupole apertures 12 mm radius and pole tip fields 1.4 T, the partition functions were normal, $\beta y/\beta x = 4$, and phase advance per cell 65°.

2.5 Quadrupole Doublet Final Focus

A conventional, chromatically-corrected quadrupole-doublet final focus is assumed. The ratio of the assumed corrected β to a calculated uncorrected value is taken to be $S = .04 \times dp/p$ (scaling law from K. Brown). The maximum pole tip field is assumed to be 1.4 T. The aperture is taken to be ten times the rms beam size.

2.6 Accelerating Structure

A conventional iris loaded accelerating structure is assumed. The iris radius is taken to be 0.2 times the wavelength. This gives a relatively high group velocity (0.08) and lower wakefields than for a SLAC–like structure (radius 0.1 times the wavelength). The fill time for the structure is usually taken to be 0.3 times the attenuation time. But in the case when the wavelength is 17 mm (1 TeV example), then the fill time is taken to be 0.6 times the attenuation length to avoid using pulses of less than 45 ns.

2.7 BNS Damping

Transverse wakes are assumed to be controlled by BNS damping. Focusing in the linac is taken to give a β that rises with the square root of the energy. Five percent of the linac length is assumed taken up with quadrupoles whose apertures are 1.26 times the structure irises and whose pole tip fields are 1.4 T.

2.8 Number of Bunches

It is assumed that transversely damped accelerating structures are used and that multiple bunches can be used without beam breakup. A limit is set on the number of bunches such that not more than 25% of the total stored energy is extracted. This limit is consistent with considerations of energy constancy.

2.9 Repetition Rate

The repetition rate is set to keep the wall power to 200 MW. A klystron efficiency of 36% is assumed.

2.10 Dilution

The following dilutions are assumed:

Emittance z in buncher:	1.4
Emittance x from kicker:	1.4
Particle transmission through buncher:	1/1.2
Emittance y in linac:	1.4
Particle transmission through final focus:	1/1.2
Emittance x y in final focus:	1.2
β^* x y in final focus:	1.2

2.11 Longitudinal Emittance

At high energies the contribution of the longitudinal emittance in the damping ring to the momentum spread at the intersection is negligible. In this case the longitudinal emittance is constrained by considerations of the bunch length in the damping ring to be less than 0.04 m $\times dp/p$ at $\gamma = 1$.

At low energies the longitudinal emittance is constrained only by considerations of momentum spread in the linac and for physics.

2.12 Bunch Length

The bunch length is selected to maximize the final luminosity. In general, a longer bunch length will allow a larger longitudinal emittance and thus, from the damping ring, smaller transverse emittances and higher luminosity. The primary constraint on the bunch length is the loss of luminosity that arises when the diagonal angle becomes less than the finite crossing angle. Another constraint arises when luminosity is lost because the bunch length has become much longer than the β^* at the intersection.

2.13 Accelerating Fields

From the machine physics point of view there seems no disadvantage in high accelerating fields. The optimized luminosity is little effected, the tolerances are easier and of course the length is less. The field used should thus be the highest possible consistent with breakdown and dark current considerations. I have assumed that these limits are wavelength dependent and used: G = 46.5 MV/m for f = 2 GHz, G = 93 MV/m for f = 6 GHz and G = 186 MV/m for f = 17 GHz.

2.14 Wavelength

The wavelength, like the acceleration gradient, has little effect on the optimized luminosity (for the same loading and average power consumption, the larger wavelength collider will have more luminosity per cycle, but a lower repetition rate). At higher energies a lower wavelength is preferred because it will have the lower stored RF energy, lower peak power requirements and thus a cheaper power source; and lower wavelength solutions are also found to have lower beamstrahlung. However, as the wavelength falls the power source becomes harder to build and the linac tolerances become tighter (both alignment and jitter). I selected 17 GHz as the maximum RF frequency.

At lower energies, if the wavelength is reduced, the repetition rate (for fixed wall power and luminosity) becomes excessive, and it would be very hard to design a damping ring (or rings) to provide the low emittance positrons. But, at lower energies beamstrahlung is less of a consideration; also, the accelerator is shorter so the cost of the power supply is less of a consideration. It is then reasonable to use a larger wavelength. Rather arbitrarily I used for:

10 GeV c.m.: f(rep) = 1.7 kHz at P(wall) = 200 MW,

giving f(rf) = 2 GHz ;

100 GeV (rep) = 1.0 kHz at P(wall) = 200 MW,

giving f(rf) = 6 GHz ;

1 TeV f(rep) = 0.4 kHz at P(wall) = 200 MW,

giving f(rf) = 17 GHz .

2.15 Loading (η)

At high energies, when the longitudinal emittance is held to a constant maximum practical value, the loading has a negligible effect on the momentum spread at the final focus. In this case, the luminosity per bunch is found to be nearly independent of the loading. (A higher loading, and thus higher N, would naturally increase the luminosity, but this is offset by the higher emittance from the damping ring and by the higher β^* because of the greater difficulty in correcting the higher momentum spread from higher wakes.) With multibunching, however, the number of bunches is inversely proportional to the loading, and thus the total luminosity turns out to be inverse with the loading. There are various limits to this: when the momentum spread gets too low, the chromatic correction scaling will break down, or the tolerances will get too tight. I limit the chromatic correction factor to 30, which results in the luminosity reaching its maximum for a loading of about 2.5%.

At lower energies the longitudinal emittance dominates the momentum spread at the final focus, and in this case the luminosity per bunch increases with loading. With multibunching the luminosity now becomes independent of the loading. The choice of loading is now rather arbitrary, providing it is not so large as to give an excessive momentum spread. When in doubt, I have used 2.5%.

3. RESULTS

3.1 Conventional Finite Angle Crossing

With these assumptions I have generated the parameter lists for machines at center-of-mass energies of 10 GeV, 100 GeV and 1 TeV [see Table 1 and Fig. 1 (dots)]. For the 10 GeV case, the momentum spread was restricted to less than 0.2% in order to be not larger than the width of the $\Upsilon(4s)$ state.

Note that far more thought has been given to the 1 TeV case than the lower energy examples, so it is not clear that these lower energy examples are fully optimized. In addition, some of the parameters for the B factory case are unreasonable or unphysical (they are indicated by ***). The damping ring design is unphysical and unreasonably small, but the performance indicated is not unreasonable and can, almost certainly, be obtained in a real design. The accelerator section lengths are also too long, but these could be lowered by adding some magnetic coupling between cells.

The exact performances given should not be taken too seriously. Nevertheless we note that the luminosity seems to fall from a value close to 10^{34} at 1 TeV, to only a little above 10^{33} at 10 GeV: falling about as the root of the energy. This conclusion is reached assuming constant wall power (200 MW). In practice it is hard to believe one would consider a collider at 10 GeV using so much power, so the realistic luminosity of such a B–factory would be even lower.

Table 1. Parameters of optimized colliders at three energies.

		B factory	*Z* factory	TLC
General				
c.m. energy	TeV	.01	.1	1
RF wavelength	cm	15	5	1.7
repetition rate	kHz	1.7	1.0	.36
accel gradient	MV/m	46	93	186
number bunches		10	10	10
particles/bunch	10^{10}	37	6	1.4
wall power	MW	200	200	200
Damping				
emittance x/y		100	100	100
emittance y (ϵ_n)	μm	4.2	.4	.06
emittance z	m dp/p γ	.005	.014	.04
bunch spacing	m	2.4	.5	.2
ring Energy	GeV	.6	.8	1
ring radius	m	.7***	4	15
(avg B)/(peak B)		3***	.7	.24
damping time	msec	3.8	2.8	2.3
RF				
pulse length	ns	680	150	60
peak power/length	MW/m	260	380	580
total RF energy	KJ	40	67	210
Linac				
loading η	%	2.5	2.5	2.5
iris radius a	mm	44	10	3.5
section length	m	57***	4	1.6
dp/p for BNS	%	.04	.17	.4
dp/p from wakes	%	.09	.11	.12
dp/p incl emit	%	.18	.16	.08
RF phase for BNS	deg	20	22	15
Linac tolerances				
alignment	μm	—	300	30
vibration	μm	1.9	.12	.012
Final focus				
β_y^*	mm	.11	.10	.08
crossing angle	mrad	59	14	3.8
disruption angle	mrad	6.7	.72	.25
chrom corr factor		22	24	28
quad radius	mm	19	2.1	.32
free length	m	.53	.56	.68
Intersection				
σ_y	nm	220	21	2.2
Oide min σ_y	nm	43	8	1.7
σ_x/σ_y		180	170	177
σ_z	μm	460	180	70
disruption D		12	8	6.5
lum enhance H		1.3	1.4	1.6
beamstrahlung δ	%	.004	.27	11
dp/p physics	%	.18	.18	3.2
luminosity 10^{33}	cm^{-2} sec^{-1}	1.0	3.1	7.8

*** See text.

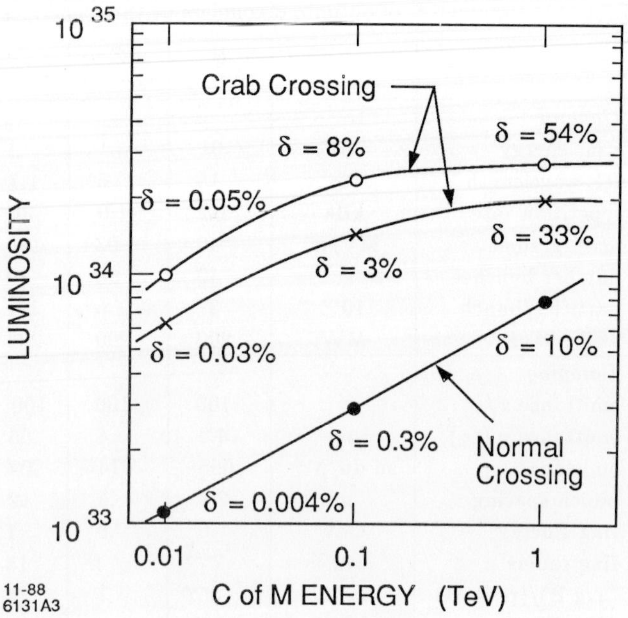

Fig. 1. Luminosity of linear collider designs as a function of energy.

3.2 An Intermediate Linear Collider

Instead of completely redesigning the machine as the energy is lowered, we can consider the performance of a collider with wavelength and repetition rate chosen for the 1 TeV case, but operating at a reduced energy, e.g., at one half energy. Two options are considered: (a) keep the gradient the same and half the length and (b) keep the length the same and half the gradient. In both cases, the final focus and damping rings have been reoptimized. The resulting parameters are shown in Table 2. The higher gradient version gives twice the luminosity and has a less severe vibration tolerance, both of which would be preferred. However, the dark currents and breakdown problems will be less in the low gradient case, and the stored energy is lower by a factor of two, so the power supply will be easier and cheaper.

3.3 Crab-Wise Finite Angle Crossing

The situation can be improved if "crab-wise" crossings are allowed. Such crossings would employ an RF deflector near the end of each linac to deflect the front of the bunches one way and the back the other way (so they now move somewhat crab-wise). If the angle of tilt of each bunch is correctly chosen, then the bunches pass through each other head on—even when the beams have a finite crossing angle (see Fig. 2). This technique removes the constraint on the width of the bunches that is normally present with finite angle crossing and allows a smaller aspect ratio.

Using this crab-wise crossing, machines were redesigned at the three energies [see Fig. 1 (crosses) and Table 3]. For the 10 GeV case, the total momentum spread is again kept below 0.2%; for the 100 GeV case, it is kept to 1% for Z production, and for 1 TeV it is kept to $\delta \leq .3$; *i.e.*, an rms momentum spread of 10%.

Figure 1 (circles) also shows designs with no restriction on momentum spread.

Table 2. Two designs for an intermediate energy collider.

		Low grad ILC	High grad ILC	TLC
General				
c.m. energy	TeV	.5	.5	1
RF wavelength	cm	1.75	1.75	1.75
repetition rate	kHz	.36	.36	.36
accel gradient	MV/m	93	186	186
number bunches		10	10	10
particles/bunch	10^{10}	.7	1.4	1.4
wall power	MW	52	103	210
Damping				
emittance x/y		100	100	100
emittance y (ϵ_n)	μm	.035	.06	.07
emittance z	m dp/p γ	.04	.04	.04
bunch spacing	m	.2	.2	.2
ring energy	GeV	1.1	1.0	1.0
ring radius	m	23	15	15
(avg B)/(peak B)		.17	.24	.24
damping time	msec	2.1	2.3	2.3
RF				
pulse length	ns	60	60	60
peak power/length	MW/m	146	580	580
total RF energy	KJ	51	103	210
Linac				
loading η	%	2.5	2.5	2.5
iris radius a	mm	3.5	3.5	3.5
section length	m	1.6	1.6	1.6
dp/p for BNS	%	.2	.4	.4
dp/p from wakes	%	.11	.11	.11
dp/p incl emit	%	.14	.14	.14
RF phase for BNS	deg	19	15	15
Linac tolerances				
alignment	μm	20	35	30
vibration	μm	.009	.017	.012
Final focus				
β_y^*	mm	.1	.12	.11
crossing angle	mrad	4.2	6.1	3.8
disruption angle	mrad	.23	.31	.25
chrom corr factor		29	29	29
quad radius	mm	.17	.25	.32
free length	m	.36	.43	.7
Intersection				
σ_y	nm	2.7	3.9	2.8
Oide max σ_y	nm	1.3	1.9	1.9
σ_x/σ_y		132	132	132
σ_z	μm	70	70	70
disruption D		5	5	5
lum enhance H		1.6	1.6	1.6
beamstrahlung δ	%	2	4	11
dp/p physics	%	.7	1.1	3.2
luminosity 10^{33}	cm^{-2} sec^{-1}	1.5	2.9	6.2

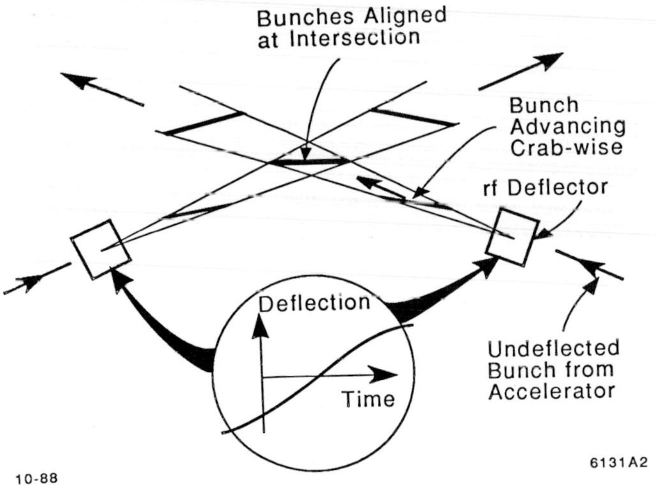

Fig. 2. Crab-wise crossing.

As in the case in Section 3.1, less effort has been spent on the low energy examples than on the TeV case, and there are the same problems in the *B* factory case. Nevertheless, the general trend is probably correct: a significant gain in luminosity is obtained in all cases, with the larger gain being realized in the lower energy examples.

4. POSSIBLE IMPROVEMENTS

1. Some improvements in the damping ring performance are probably possible. In the examples considered I used a wiggler ring but did not consider a combined function lattice that could alter the partition functions. If a combined function wiggler lattice is possible, we might gain a factor of two in the final luminosity.

2. I have assumed that not more than 25% of the energy stored in the cavities can be extracted with multiple bunches. This may be true for a traveling wave structure, but need not be so in a standing wave structure. A factor of two increase in efficiency, and thus luminosity, might be possible.

3. I assumed only 36% efficiency for the power source. Clusters of low perveance mini-klystrons might have much better efficiency. Another factor of two?

4. One can always hope that the dilutions that have been assumed are overly pessemistic. A maximum gain of about three is possible.

5. One can consider asymmetric arrangements in which a higher electron than positron current is employed. If electrons are available from a high brightness gun whose brightness is higher than that available from a damping ring then some gain in luminosity might be possible. For the low energy machines, one could also consider asymmetric energies with the same possible advantage.

At best there might be another order of magnitude to be won, but it will not be easy. Maybe there are other ideas!

Table 3. Parameters for colliders with crab-wise crossing.

		B factory	*Z* factory	TLC
General				
c.m. energy	TeV	.01	.1	1.0
RF wavelength	cm	15	5	1.75
repetition rate	kHz	1.8	1.0	.36
accel gradient	MV/m	46	93	186
number bunches		20	10	10
particles/bunch	10^{10}	19	6	1.4
wall power	MW	200	200	200
damping				
emittance x/y		100	100	100
emittance y	μm	2.7	.36	.05
emittance z (ϵ_n)	m dp/p γ	.004	.016	.04
bunch spacing	m	1.2	.5	.2
ring Energy	GeV	.64	.8	1.0
ring radius	m	1 ***	4	15
(avg B)/(peak B)		2 ***	.7	,24
damping time	msec	3.5	2.8	2.3
RF				
pulse length	ns	680	150	60
peak power/length	MW/m	260	380	580
total RF energy	KJ	40	70	210
Linac				
loading η	%	1.25	2.5	2.5
iris radius a	mm	44	10	3.5
section length	m	57 ***	4	1.6
dp/p for BNS	%	.02	.15	.4
dp/p from wakes	%	.07	.07	.12
dp/p incl emit	%	.19	.15	.08
RF phase for BNS	deg	14	28	15
Linac tolerances				
alignment	μm	8000	340	30
vibration	μm	3.8	.16	.012
Final focus				
β_y^*	mm	.34	.27	.2
crossing angle	mrad	150	45	9.2
Δz tolerance	μm	50	24	18
disruption angle	mrad	17	2.3	.58
chrom corr factor		26	23	28
quad rad	mm	200	18	1.8
free length	m	1.7	1.6	1.6
Intersection				
σ_y	nm	300	32	3.3
Oide min σ_y	nm	28	7	1.7
σ_x/σ_y (R)		24	35	49
σ_z	μm	340	150	70
disruption	D	15	14	10
lum enhance H		1.7	1.9	2
beamstrahlung δ	%	.03	2.7	33
dp/p physics	%	.2	.8	10
luminosity 10^{33}	cm^{-2} sec^{-1}	6	14	20

*** See text.

5. ADDENDUM–THE PAIR PROBLEM

Since the Snowmass meeting and since writing the above, we have become aware of the serious nature of the backgrounds due to electron positron pairs generated by beamstrahlung photons. These photons can be converted by at least two mechanisms. They can make pairs incoherently on the individual electrons in the oncoming beam.[2] They can also be converted in the coherent electromagnetic field of the entire oncoming beam.[3] There has not been time to give a thorough study of these backgrounds, and thus what follows is preliminary, and concerns only the 1 TeV cases discussed above.

5.1 Pair Production Cross Section

The cross section for the incoherent pair production[3] is relatively energy independent and has a value of the order of 6×10^{-26} cm^2. For the TLC parameters given above, approximately 10^6 electron positron pairs are generated per multibunch crossing.

The coherent production is negligible provided the beamstrahlung remains in the classical regime. Defining Υ as 2/3 times the critical energy divided by the beam energy, then

$$\Upsilon = \frac{.86 R_e \lambda\!\!\!^{-}_e \gamma N}{\sigma_z \sigma_x} \tag{1}$$

where R_e (2.82×10^{-15}) is the classical electron radius, $\lambda\!\!\!^{-}_e$ is the classical electron wavelength (3.86×10^{-13}), γ is of the beam, N is the particles per bunch, and σ_z and σ_x are the rms length and width of the bunch.

The coherent production remains negligible compared to the incoherent effect provided Υ is less than 0.6.[3] If Υ is above that value the production rises very rapidly until the number of pairs is equal or even larger than the initial number of electrons. In the normal crossing TLC case given above the Υ is only 0.35 and the coherent production would not be a problem. But in the higher luminosity "crab crossing" case Υ is 1.1 and we would have massive coherent pair production.

5.2 Energy and Angular Distribution

Reflecting the bremstahlung spectrum of the photons, the number spectra of the pair electrons falls as the 2/3rd power of the energy. Their angles of production are of order $1/\gamma$, and therefore very much forward, but the magnetic fields at the bunch crossing deflect them to larger angles.

The maximum transverse momentum given to the pair electrons is, in our TLC examples, of the order of 160 MeV, but that is only given to relatively high energy electrons (approximately 50 GeV), and these emerge at relatively small angles (of the order of 2 mrad). With the finite angle crossings, discussed above, such high energy pair electrons would pass by the opposite quadrupole without causing trouble. For lower energy pair electrons, however, the situation is more serious and now depends on their charge.

In the case of the pair electron with the same charge as the beam from which it came, then the fields deflect it in towards the axis, and if its energy is low, trap it along the axis. When such a trapped electron leaves the bunch it emerges in a relatively forward direction, as do the beam electrons themselves when they have lost a large fraction of their energy.[4] The angles in our case are of the order of 1 mrad vertically, but 10 mrad horizontally, which would already be a problem. But the situation for the other sign is much worse.

For the pair electron with the opposite charge to its beam, then, the deflection is away from the axis and the final angle can be much larger. For low energy electrons (in our case less than approximately 50 GeV) the maximum angle can be expressed for flat beams (but I think only approximately) by

$$\Theta = 1.5 \left(\frac{R_e N}{\gamma \sigma_z} \right)^{1/2} \tag{2}$$

where R_e is the classical electron radius, N is the number of electrons in the oncoming bunch, γ is the γ of the pair electron and σ_z is the rms bunch length. In the TLC cases above, for a pair electron energy of 250 MeV the angle is 70 mrad. And there are approximately 100,000 electrons with energy less than 250 MeV!

5.3 Proposed Solution

I will assume that we should avoid operating in the quantum regime where massive coherent pair production is present. I will also assume that a number of electrons of the order of 100,000, of any energy, colliding with the face of an opposite quadrupole is unacceptable. They will either melt the magnet or at least give an albido that will be intolerable to the detector.

I then propose :

1. *Use crab crossing* to allow a crossing angle of the order of 80 mrad. At 1.6 m this corresponds to a distance of 13 cm. I will assume that we can make a quadrupole with a radius not larger than 4 cm, surrounded by a bucking coil (to shield the quadrupole from the experiments solenoidal field) of thickness not more than 3 cm. In this way only particles at angles greater than 40 mrad can hit the quad or bucking coil.
2. *Reduce the pair electron deflection angles* (Θ) by increasing σ_z and decreasing N. σ_z is increased until the momentum spread from the curvature of the sin wave of the rf is significant. I halve N and, at the same time, double the number of bunches, so as to keep the same total efficiency. The increase in transverse wakes caused by the bunch lengthening is compensated by the lower number of particles per bunch, so the alignment tolerances are the same. Despite the energy spread caused by the long bunch, the lower number of particles per bunch produces a smaller energy spread, and thus the same β can be obtained at the final focus.
3. *Reduce the bunch aspect ratio* $R = \sigma_x/\sigma_y$, and thus increase the luminosity, until the Υ reaches a still safe value of 0.32. The aspect ratio is then 60. The amplitude tolerance on the crab crossing rf separators

$$\frac{dA}{A} = \frac{\sigma_x}{\sigma_z \Theta_c} = 1.6\% \tag{3}$$

where Θ_c is the crossing angle. The position tolerance on the separators is

$$dz = 2\frac{\sigma_x}{\Theta_c} = 4\ \mu \quad . \tag{4}$$

Note that this position error is equivalent to a horizontal beam shift of 0.16 μ, and that the beams horizontal position will in any case have to be servoed. The 4 μ tolerance is a jitter tolerance and is not severe.

With these new parameters the luminosity is 10^{34} cm^{-2} sec^{-1}, as required, and other parameters quite reasonable. The maximum angle of pair electrons at 250 MeV is now 40 mrad, and we have chosen a crossing angle and magnet dimensions so that these will not hit the quad or bucking coil. But we must still do something to stop the lower momentum tracks from hitting these magnets.

4. *Use a kinked solenoidal field* to confine the low momentum tracks. If a solenoidal field can be obtained that is aligned along the outgoing beam axes, then low energy electrons will be guided along those axes. Electrons with 250 MeV or less will have transverse momenta of only 10 MeV or less. If the solenoidal field were 3 T then all such electrons would be constrained within a helix of 1 cm radius, *i.e.*, well away from the quadrupole and bucking coils. This maximum extent of the trapped electrons rises linearly with energy, while the maximum extent of untrapped electrons falls as the root of energy from Eq. (2). The maximum distance (Fig. 3) that any electron reaches from the outgoing axis is now 4 cm (at 500 MeV), which is over 2 cm from the bucking coil.

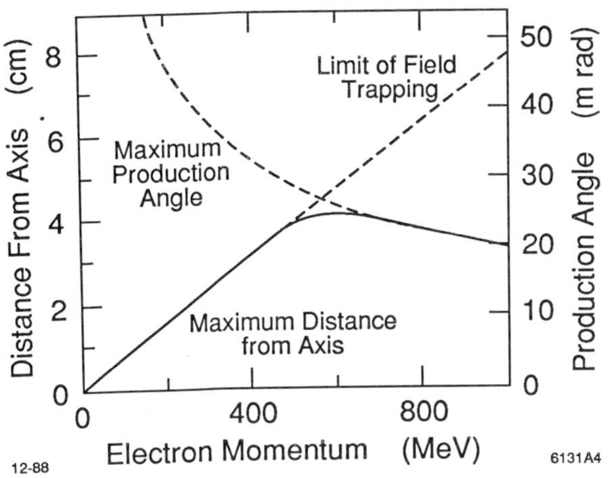

Fig. 3. Maximum radial extent of electrons from their axis at the entrace to the first quadrupole, plotted against their energy. At low energy the extent is limited by field trapping, at high energy by the production angles.

The problem remains as to how to make a solenoidal field that points along both outgoing beam directions; *i.e.*, how to make a solenoidal field with a kink of 80 mrad at the intersection. Such a field would be obtained with a current sheet in the mid plane of the detector, but that would be undesirable. Luckily it seems that the field that inevitably results when bucking coils are placed about the final quadrupoles has, at least to some approximation, the required kink [see Fig. 4(a)]. As shown in the figure, the bucking coils have been extended inward towards the intersection so as to improve the field shape. Further improvements could be made by the addition of horizontal low field (0.12 T) coils between the end of the bucking coils and the intersection. In the absence of bucking coils (e.g., if pure permanent magnet quads were used) such horizontal low field coils, but now much longer, could be used [Fig. 4(b)].

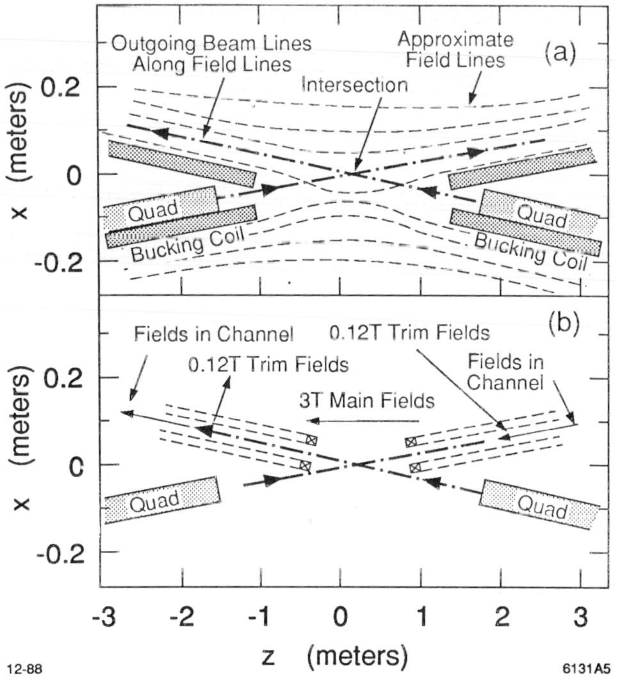

Fig. 4. Possible intersection region geometries in a large experimental solenoid; (a) with bucking coils to shield conventional quadrupoles; and (b) with permenant magnet quadrupoles and transverse trim coils. It is seen that in both cases the resulting fields have the qualitatively correct shape to guide low energy pair electrons away from the quadrupoles.

5.4 Conclusion

It appears that the pair production problem can be overcome, at least at 1 TeV, by a combination of modifications to the previously defined TLC. Crab crossing seems essential to allow a crossing angle of the order of 80 mrad. The high momentum pair electrons now pass well clear of the quadrupoles. The production angles of low momentum pair electrons can be reduced by halving the number of electrons per bunch and doubling the bunch length. Then, with the help of a solenoidal field, these too can be kept clear of the magnets.

A bunch aspect ratio of 60 keeps the beamstrahlung Υ to about 0.3, and the coherent pair production is supressed. Using these new parameters, and by increasing the number of bunches to 20, the luminosity can be maintained at 10^{34} cm^{-2} sec^{-1}.

REFERENCES

1. R. B. Palmer, "The Interdependence of Parameters for TeV Linear Colliders," SLAC–PUB–4295; Proceedings of Workshop on New Developments in Particle Accelerator Techniques, Orsay, France, June 1987.
2. V. Telnov, reported at SLAC Collider Workshop, November 1988.
3. P. Chen, reported at SLAC Collider Workshop, November 1988, and discussed in these Snowmass Proceedings.
4. K. Yokoya, Proceedings of Linear Accelerator Conference, Williamsburg, Virginia, 1988.

Damping Ring Designs for a TeV Linear Collider*

T. O. RAUBENHEIMER, L. Z. RIVKIN, R. D. RUTH
Stanford Linear Accelerator Center, Stanford University, Stanford, California 94309

ABSTRACT

In this paper we present a damping ring design for the TLC (TeV Linear Collider). The ring operates at 1.8 GeV. It has normalized emittances of $\gamma\epsilon_x = 2.8\,\mu$mrad and $\gamma\epsilon_y = 25.4$ nmrad. The damping times are $\tau_x = 2.5$ ms and $\tau_y = 4.0$ ms. To achieve these extremely low emittances and fast damping times, the ring contains 22 m of wigglers.

1. INTRODUCTION

In this paper we discuss a damping ring for the TLC, a TeV linear collider.[1)] The basic design goals of the TLC damping ring are compared with those of the Stanford Linear Collider (SLC) damping rings in Table 1. The normalized horizontal emittance of the TLC ring is an order of magnitude smaller than that of the SLC ring, and the desired repetition rate has increased by a factor of two. Furthermore, the TLC ring needs to achieve an emittance ratio of 100:1. Thus the vertical emittance must be damped to a value three orders of magnitude smaller than the SLC emittance. This implies that TLC ring will either have much faster damping times than the SLC ring or be much larger, thereby damping more bunches at once.

The present design of the TLC operates in a multi-bunch mode. The linac accelerates batches of bunches, where the bunches within a batch are separated by roughly 20 cm and each bunch contains 2×10^{10} particles. To prevent multi-bunch instabilities we need to use a specially designed RF system; a discussion of multi-bunch instabilities is found in Ref. 2. In addition, we would like to operate the ring below the longitudinal microwave instability threshold. Thus, the threshold current must be 30% larger than that in the SLC ring. To achieve this without increasing the longitudinal emittance significantly, the TLC ring must have a very low impedance and a large momentum compaction.

Table 1. Basic parameters of the SLC and TLC damping rings.

	TLC	SLC
Energy	$1 \sim 2$ GeV	1.15 Gev
Emittance, $\gamma\epsilon_x$	3.0 μmrad	36. μmrad [3)]
Emittance, $\gamma\epsilon_y$	30. nmrad	–
Repetition rate	360 Hz	180 Hz
Bunch length	4 mm	5 mm [4)]
Threshold Current	batches of 10 bunches of 2×10^{10}	1.5×10^{10} [4)]

In the next section we discuss the design goals. Then using simple scaling laws, we show the dependance of the various design parameters such as the lattice, main bending field, bending angle per bend, etc. This is applied to illustrate problems with the wiggler damping ring,[5)] an option that seemed promising for low emittances and fast damping times. We then discuss the effect of damping wigglers and methods of changing the damping partitions. Because the parametric dependances become complex, we use a computer program to search for a lattice which satisfies the design criteria and uses a minimal length of wigglers.

In Sec. 3 we present a design which meets all of the specified requirements. We discuss the various portions of the ring, the arcs, the insertion regions, and the wigglers. We then detail the chromaticity correction scheme and the resulting dynamic aperture. Finally, in Sec. 5 we discuss tolerances and methods of loosening the tolerances on alignment and on the extraction kickers.

⋆ Work supported by the Department of Energy, contract, DE – AC03 – 76SF00515

2. DESIGN CONSIDERATIONS

2.1. Basic Parameters

There are two main parameters we need to consider when designing a damping ring: the ring's emittance and the damping times. The emittance of an extracted beam is

$$\epsilon = \epsilon_i e^{-2t/\tau} + (1 - e^{-2t/\tau})\epsilon_{\text{ring}} , \qquad (2.1)$$

where ϵ_i is the emittance of the injected beam and ϵ_{ring} is the ring emittance including the effects of intrabeam scattering. Here, τ is the horizontal or vertical damping time, and t is the time the bunch is in the ring. The present TLC design requires that the extracted beam have normalized emittances of $\gamma\epsilon_x \leq 3 \times 10^{-6}$ and $\gamma\epsilon_y \leq 3 \times 10^{-8}$. We assume an injected beam emittance of $\gamma\epsilon_i = 3 \times 10^{-3}$, which is realistic for a positron beam and an order of magnitude too large for an electron beam. Thus the vertical emittance needs to be decreased by five orders of magnitude. Damping the bunch for seven vertical damping times will reduce the first term of Eq. (2.1) by six orders of magnitude. The limit on the vertical emittance of the ring is then

$$\gamma\epsilon_{y\,\text{ring}} \leq 2.7 \times 10^{-8}\,\text{mrad} . \qquad (2.2)$$

In a storage ring built in the horizontal plane the vertical emittance is mainly determined by the coupling between the horizontal and vertical planes. Intrabeam scattering, which increases the horizontal emittance, has a very small effect on the vertical.[6,7] We will discuss the tolerances necessary to achieve the limit — Eq. (2.2) — in Sec. 5.

The required damping times are determined from the desired repetition rate (360 Hz) the number of damping times per bunch (7) and the number of batches stored in the ring at once (N_b)

$$\tau_x, \tau_y \leq \frac{1}{f_{\text{rep}}} \frac{N_b}{\#\text{ of damping times}} = N_b 0.397\,\text{ms} . \qquad (2.3)$$

The maximum number of batches stored in the ring is limited by the kickers needed for injection/extraction. We assume that the time for the kickers to turn on, extract/inject a batch, and turn off is less than 100 ns.[8] Thus the batches must be separated by at least 50 ns. Since the number of batches is roughly proportional to the size of the ring we can define an effective damping time as

$$\tau_{\text{eff}} \equiv \tau \frac{50\,\text{ns}}{T_0} \leq 0.397\,\text{ms} , \qquad (2.4)$$

where T_0 is the revolution time of the ring.

For reasons we will discuss later, it is desirable that the horizontal damping time be less than or equal to the vertical; thus only the vertical damping time is limited by Eq. (2.3). Furthermore, assuming that $\tau_x \lesssim \tau_y$, the horizontal emittance of the extracted beam is very nearly equal to the horizontal emittance of the ring. Thus

$$\gamma\epsilon_{x\,\text{ring}} \leq 3 \times 10^{-6}\,\text{mrad} . \qquad (2.5)$$

Equations (2.4) and (2.5) determine the basic parameters. Initially, to study these parameters, we make the assumption that all the bending magnets are the same and we ignore the effect of intrabeam scattering. Now we can write simple expressions for $\gamma\epsilon_{x0}$ and $\tau_{y\,\text{eff}}$,[9] the two quantities we want to minimize:

$$\tau_{y\,\text{eff}} = 8.47 \times 10^6 \frac{\rho_B f_w}{J_y \gamma^3} \qquad (2.6)$$

$$\gamma\epsilon_{x0} = 3.84 \times 10^{-13} \frac{\gamma^3}{J_x} \frac{\langle\mathcal{H}\rangle_{\text{mag}}}{\rho_B} . \qquad (2.7)$$

Here, ρ_B is the *local* bending radius of the bend magnets and $\langle\mathcal{H}\rangle_{\text{mag}}$ is the Courant–Snyder dispersion invariant which equals the average of $\mathcal{H} \equiv \gamma\eta_x^2 + 2\alpha\eta_x\eta_x' + \beta\eta_x'^2$ over the bending magnets. Also, $J_{x,y}$ is the horizontal (vertical) damping partition number, and f_w is equal to ρ_B divided by the average bending radius in the bending magnets, ρ_0. Note that with normal bends $f_w = 1$, but in a wiggler ring[5] where the bending magnets bend in both directions, $f_w < 1$.

The emittance of a ring can be reduced by reducing the dispersion in the bend magnets, reducing the strength of the bends, or decreasing the energy of the ring. Unfortunately, the damping times are *increased* by reducing the bend magnet strength or decreasing the energy of the ring. This implies that the dispersion in the bends is the only free parameter. Unfortunately, it is constrained by the longitudinal microwave instability which increases the longitudinal emittance.

At this point it is worth discussing three additional parameters which constrain the design: (1) the longitudinal microwave instability which was just mentioned, (2) the dynamic aperture, and (3) the energy of the ring. The microwave instability, also called turbulent bunch lengthening, occurs at a given current when the longitudinal impedance is larger than a threshold, denoted $(Z/n)_t$. We want to keep this threshold as large as possible to avoid bunch lengthening and longitudinal instabilities. It can be estimated as

$$(Z/n)_t \approx \frac{(2\pi)^{3/2} E\sigma_\epsilon^2 \sigma_z \alpha}{Ne^2 c} \mathcal{F} , \qquad (2.8)$$

where E is the energy, σ_ϵ is the relative energy spread, and σ_z is the bunch length. In addition, α is the momentum compaction, N is the number of e^+/e^- per bunch, and e and c are the electron charge and the speed of light. Finally, $\mathcal{F}$ is a form factor which is greater than 1.[10] At the 1987

ICFA workshop on Low Emittance Beams,[11] an impedance of

$$Z/n \geq 0.2\,\Omega \tag{2.9}$$

was determined to be the minimum reasonable, physically attainable value.

Next we should consider the dynamic aperture. The dynamic aperture of the ring is a function of the sextupoles needed to correct the chromaticity. To prevent particle losses the dynamic aperture should be many times the *injected* beam size. Unfortunately, rings with small emittances tend to have high tunes and large uncorrected chromaticities. This makes the desired dynamic aperture difficult to achieve. Thus we would like to choose a lattice which naturally has a large dynamic aperture.

Finally, the other parameter we mentioned is the ring energy which we would prefer to have low. There are three primary reasons for this: (1) it makes the magnets cheaper, (2) it keeps the longitudinal emittance small, and (3) it makes bunch compression easier. The TLC requires that the damping ring bunch be compressed longitudinally by, roughly, a factor of 100. Since one does not want an uncorrelated energy spread much greater than 1% in the linac, we need to perform at least a portion of the bunch compression at an energy 10 times that of the damping ring.[12] Unfortunately, at higher energies it becomes more difficult to perform the compression without degrading the beam emittances.

2.2. SCALING

Now we can combine these expressions to determine the dependencies of the parameters. Using the definition of and maximum value for $\tau_{y\,\mathrm{eff}}$ — Eqs. (2.6) and (2.4) — we find an expression for the energy of the ring as a function of the bending field and f_w:

$$B_0(KG)\gamma^2 = f_w \frac{1.44 \times 10^5}{\tau_{y\,\mathrm{eff}}} \; . \tag{2.10}$$

In this equation, J_y is assumed to be equal to 1, since it cannot be changed without introducing vertical dispersion which would degrade the vertical emittance. Note that for our parameters, a normal ring, with saturated bending magnets (20 KG), must operate at 2.2 GeV to meet the damping time requirements.

In a similar manner, using Eqs. (2.7) and (2.10), we find an equation for the emittance as a function of f_w, $\langle\mathcal{H}\rangle_{\mathrm{mag}}$, and J_x:

$$\gamma\epsilon_{x0} = \frac{3.25 \times 10^{-6}}{\tau_{y\,\mathrm{eff}}} \frac{\langle\mathcal{H}\rangle_{\mathrm{mag}} f_w}{J_x} \; . \tag{2.11}$$

Next we use Eq. (2.10), along with an equation for the relative energy spread in the ring:[9]

$$\sigma_\epsilon^2 = \frac{C_q\gamma^2}{J_\epsilon \rho_B} = 2.25 \times 10^{-11} \frac{\gamma B(KG)}{J_\epsilon} \; , \tag{2.12}$$

to re-write the expression for $(Z/n)_t$ (Eq. (2.8))

$$(Z/n)_t \approx \mathcal{F} \frac{5.43 \times 10^{11}}{N} \frac{\sigma_z}{\tau_{y\,\mathrm{eff}}} \frac{\alpha f_w}{J_\epsilon} \; . \tag{2.13}$$

Notice that the energy no longer appears in Eqs. (2.11) and (2.13). It is determined by the main bending field and the desired damping time, $\tau_{y\,\mathrm{eff}}$ (Eq. (2.10)).

In Eq. (2.13) we have ignored the constraints on the energy spread. A large energy spread reduces the lifetime and increases the longitudinal emittance. In practice we are limited to a relative energy spread of a couple tenths of a percent and thus there is a lower bound on α. Using $(Z/n)_t = 0.2\,\Omega$, this bound is:

$$\alpha \geq 1.19 \times 10^{-18} \frac{N}{\mathcal{F}\gamma\sigma_\epsilon^2\sigma_z} \; . \tag{2.14}$$

Notice that the bound is inversely proportional to the energy, implying that a higher energy is desirable. In contrast, the energy does not appear in Eq. (2.13) where the energy spread is a free parameter. In most cases, we will find that we cannot increase the energy spread sufficiently to gain from the $1/\gamma$ dependance of Eq. (2.14).

If we assume that $\sigma_z \approx 4\,\mathrm{mm}$, we see that given an effective damping time, we only have six quantities: J_x, J_ϵ, f_w, B, $\langle\mathcal{H}\rangle_{\mathrm{mag}}$, and α, which can be varied to fit the requirements on the energy, the emittance, and the longitudinal impedance threshold. Unfortunately, the system is more tightly constrained since the six variable parameters are not independent.

The bending field, B_0, is determined by the energy and f_w. We are then left with fitting the emittance and the impedance threshold, one of which we would like small and the other large. The parameter f_w can be removed by considering the ratio $(Z/n)_t/\gamma\epsilon_{x0}$. To optimize this ratio, *i.e.* make it large, we can increase J_x at the expense of J_ϵ and/or increase the ratio $\alpha/\langle\mathcal{H}\rangle_{\mathrm{mag}}$. The easier option is that of changing the damping partition numbers. They can be changed by using combined function bending magnets or a Robinson wiggler.[13,14] Unfortunately, we do not wish to increase J_x much beyond 2. Thus, we also have to consider maximizing the ratio of $\alpha/\langle\mathcal{H}\rangle_{\mathrm{mag}}$.

For a given lattice design, we can write down some approximate scaling laws. In most lattices, assuming that the phase advance per cell is held constant, $\langle\mathcal{H}\rangle_{\mathrm{bend}} \propto \Theta^3\rho_{\mathrm{ave}} \propto \Theta^2 L_{cell}$ and $\alpha \propto \Theta^2$. Thus the ratio of $\alpha/\langle\mathcal{H}\rangle_{\mathrm{bend}}$ is inversely proportional to the length of the cell. Notice that unlike normal scaling laws for damping rings, the bend angle is not at our disposal for minimizing the emittance; it is determined by $(Z/n)_t$.

2.3. LATTICES

In choosing a lattice we want one that naturally has a large dynamic aperture and a large ratio of $\alpha/\langle\mathcal{H}\rangle_{\text{mag}}$. The three lattices that seem best suited to our requirements are the FODO lattice, the triple bending achromat (TBA) lattice proposed by Vignola,[15] and the TME lattice described by Steenbergen.[16] Wiedemann has compared the double focusing achromat (DFA) and triplet achromat lattices with the FODO lattice.[17] He found that the FODO lattice had significantly better chromatic properties than the other two.

The TBA lattice has been studied by Bisognano[18] with application to a 750 MeV storage ring for FEL's and by Jackson[19] for a 1–2 GeV synchrotron light source. It has the advantage of probably being less expensive to construct than a FODO structure since fewer cells are needed to achieve the same emittance. In addition, the lattice, potentially, has a larger ratio of $\alpha/\langle\mathcal{H}\rangle_{\text{mag}}$. Unfortunately, the lattices proposed to date have $\alpha/\langle\mathcal{H}\rangle_{\text{mag}}$ ratios that are an order of magnitude too small.

The TME lattice would also be cheaper to build than a FODO lattice and it should be investigated further. In addition, another lattice worth examining would be an extreme version of the FODO lattice where combined function bends completely replace the defocusing quadrupoles. This would be the most compact and could offer the best ratio of $\alpha/\langle\mathcal{H}\rangle_{\text{mag}}$. Unfortunately, it may not be feasible to achieve the necessary gradients in the bends.

For our initial design we chose to use a FODO lattice. This choice was based mostly upon the superior dynamic aperture characteristics of the lattice. Another advantage of the FODO lattice is that it allows for local chromatic correction, *i.e.* placing the correcting sextupoles next to the quadrupoles which generate the chromaticity. This results in looser tolerances on the vertical orbit; see the discussion in Sec. 5.

While we can create a simple FODO design on paper that would satisfy the requirements, the necessary magnetic fields make it technically unfeasible. The problems arise because extremely strong magnetic fields are needed in the quadrupoles to keep the cell length short. This leaves us the option of: (1) using combined function bends to minimize the cell length, (2) varying J_x and J_ϵ, and (3) possibly using wigglers to decrease the required main bending field and increase the energy spread. Before discussing the use of separate wigglers, we feel it is useful to quickly discuss the wiggler ring option.

2.4. WIGGLER RING

The wiggler ring design has excellent damping times at low energies. It achieves this, as wigglers do, by bending the particle a lot while generating very little dispersion. Since the damping times can be much shorter than a conventional ring, the wiggler ring can operate at lower energies and thus achieve the necessary low emittances. Unfortunately, the $(Z/n)_t$ of a wiggler lattice is lower than that of a comparable conventional ring. To understand this we have to compare $\langle\mathcal{H}\rangle_{\text{bend}}$ and α in the two designs.

$\mathcal{H}$ can be written as the sum of two squares: $\mathcal{H} = \eta_x^2/\beta + (\eta_x\alpha + \eta_x'\beta)^2/\beta$. Since the second term depends upon η_x' which does not contribute to α, we would like to keep it small. The wiggler magnet bends in both the positive and negative x direction, and thus η_x and η_x' will oscillate. This will force the second term in $\langle\mathcal{H}\rangle_{\text{bend}}$ to be larger. One could minimize the effects of the oscillations by making the wiggler period shorter, but one runs into technical limitations quickly. Another problem is that the length of the cells in the wiggler ring will be longer than in a normal ring. Using the scaling laws discussed at the end of Sec. 2.2, we can see that this increases $\langle\mathcal{H}\rangle_{\text{mag}}$ without increasing α.

Before concluding this section, we note that many wiggler ring designs were presented at the 1987 ICFA workshop.[11] Unfortunately, these all had very long bunch lengths ($\sigma_z > 15\,\text{mm}$). The long bunch lengths are necessary to meet the impedance limitations and to reduce the effect of intrabeam scattering. However, if the $(Z/n)_t$ is not an issue or if the bunch length is allowed to be longer, the wiggler ring is an attractive option.

2.5. DAMPING WIGGLERS

As was mentioned in Sec. 2.2, it is very hard to build a conventional or wiggler ring which meets all of the design criteria. This leads us to consider the effects of including separate damping wigglers in regions of zero dispersion. By locating the wigglers in regions of zero dispersion, $\langle\mathcal{H}\rangle_{\text{mag}}$ in the wigglers will be small, and thus one can significantly reduce both the damping times and the emittance. In addition, damping wigglers increase the energy spread in the ring. Thus, despite the slight decrease in α resulting from the increased length of the ring, damping wigglers can significantly improve the ratio of $(Z/n)_t/\gamma\epsilon_{x0}$.

The scaling formulas Eq. (2.10) thru (2.13) can be modified to include the effects of damping wigglers:

$$B_0(KG)\gamma^2(1+F_w) = \frac{1.44\times10^5}{\tau_{y\,\text{eff}}} \tag{2.15}$$

$$\gamma\epsilon_{x0} = \frac{3.25\times10^{-6}}{\tau_{y\,\text{eff}}J_x}\frac{\left(\langle\mathcal{H}\rangle_{\text{bend}} + 6\overline{\beta_x}F_w\rho_0/5\pi k_w^2\rho_w^3\right)}{(1+F_w)^2} \tag{2.16}$$

$$(Z/n)_t \approx \mathcal{F}\frac{5.43\times10^{11}}{N\tau_{y\,\text{eff}}}\frac{\sigma_z\alpha}{J_\epsilon}\frac{\left(1+8F_w\rho_0/3\pi\rho_w\right)}{(1+F_w)^2}\,, \tag{2.17}$$

where the parameter F_w is a measure of the effectiveness of the wigglers,

$$F_w \equiv L_w\rho_0/4\pi\rho_w^2\,. \tag{2.18}$$

Here, L_w and B_w are the length and peak field of the wiggler; $\overline{\beta_x}$ is the average beta function in the wiggler, and k_w

is the wiggler wave number, $k_w = 2\pi/\lambda_w$ where λ_w is the wiggler period. Also, ρ_0 and ρ_w are the bending radii of the main bends and the wiggler. Both bending radii are proportional to the energy over the respective magnetic fields. We have assumed sinusoidal wigglers, and thus the integral of $B_w^2(s)$ over the wiggler is equal to $1/2B_w^2$. In addition, we assume that $\overline{\beta_x} \gg \lambda_w/2\pi$ so that the $\eta_x'^2\beta_x$ term dominates in $\langle\mathcal{H}\rangle_{\text{wig}}$. We do not consider decreasing $\overline{\beta_x}$ significantly since that would increase the chromaticity of the ring and might decrease the dynamic aperture.

Looking at Eqs. (2.15) thru (2.17), we can see that the largest effect occurs when $\rho_0 \gg \rho_w$ and F_w is large, *i.e.* when $B_0 \ll B_w$. Unfortunately, we are limited when increasing B_w since the η_x' created by the wigglers will blowup the emittance. This constrains the period of the wiggler and can be expressed

$$B_w^3\overline{\beta_x}\lambda_w^2\frac{F_w}{(1+F_w)} \ll 1.33\times 10^9\gamma\epsilon_{x0}J_x\ . \qquad (2.19)$$

We consider two cases, relatively long Nd-Fe-B hybrid wigglers and short superconducting wigglers. In the Nd-Fe-B hybrid wigglers the maximum field can be specified as a function of the gap and the period[20]

$$B_w \le 3.44\exp\left[-\frac{g}{\lambda_w}\left(5.08 - 1.54\frac{g}{\lambda_w}\right)\right]\ , \qquad (2.20)$$

for $0.08 < g/\lambda_w < 0.7$. To achieve a field of $B_w = 22$ KG with a gap of 2 cm, λ_w must be at least 20 cm. This is much less than the limit imposed by Eq. (2.19); with $\overline{\beta_x} \approx 4$ m, Eq. (2.19) limits $\lambda_w \ll 35$ cm. In the superconducting case we use a 40 KG field and a 20 cm period. This is (arbitrarily) scaled from the 50 KG superconducting wiggler in the DCI ring at LURE[21] with a 65 mm gap and 26 cm period. Assuming that $\overline{\beta_x} \approx 1$ m since the wigglers are shorter, the 20 cm period is less than the limit specified by Eq. (2.19): $\lambda_w \ll 29$ cm.

If we assume that the condition Eq. (2.19) is met, we can use the simple scaling for $\langle\mathcal{H}\rangle_{\text{bend}}$ and α, discussed earlier in Sec. 2.2, to express $\gamma\epsilon_{x0}$ and $(Z/n)_t$:

$$\begin{aligned}\gamma\epsilon_{x0} &\propto \frac{\Theta^2 L_{cell}}{J_x(1+F_w)^2}\\ (Z/n)_t &\propto \frac{\Theta^2}{J_\epsilon(1+F_w)^2}\frac{(1+8F_w\rho_0/3\pi\rho_w)}{1+L_{\text{wig}}/L_{\text{ring}}}\ ,\end{aligned} \qquad (2.21)$$

where L_{wig} is the length of the wigglers and the associated insertion regions and L_{ring} is the total length of the ring. We would like to keep the wiggler lengths short, and thus it is desirable to maximize the wiggler fields and minimize the bending field. While decreasing the main bending field increases the cell length, it still decreases the emittance without decreasing the impedance threshold. Unfortunately, decreasing the main bending field and the length of the wigglers forces us to higher energies.

Because the additional parameters make the analysis complex, a simple computer program was written that calculates $\gamma\epsilon_{x0}$ and $(Z/n)_t$. It uses Eqs. (2.15) thru (2.17) and the relations for $\langle\mathcal{H}\rangle_{\text{bend}}$ and α in FODO cells[22] for the calculation. We chose phase advances per cell of $\nu_x = 0.3$ and $\nu_y = 0.1$; while these phase advances do not minimize the emittance, they are chosen to be regions where the chromaticity does not increase rapidly. The cell lengths are calculated by adding the length of two bending magnets to 0.8 m, length for the drifts and quadrupoles. We then calculate the ring length from the number of cells required plus three times the length of wigglers required, allowing space for dispersion suppression, quadrupole doublets, etc. in the insertion. We use the wiggler parameters: $B_w = 22$ KG, $\lambda_w = 20$ cm, $\overline{\beta_x} = 4$ m and $B_w = 40$ KG, $\lambda_w = 20$ cm, $\overline{\beta_x} = 1$ m, and solve for solutions with $\gamma\epsilon_{x0} = 2.0\times 10^{-6}$ and $(Z/n)_t = 0.2\,\Omega$; the value of $\gamma\epsilon_{x0}$ was chosen to allow for the increase in emittance due to intrabeam scattering. Finally, since wigglers are expensive, we attempt to minimize the length included. Tables 2 and 3 list our solutions with the shortest wigglers for three energies.

Notice that the solutions for the superconducting case in Table 3 have a higher main bending field and smaller bending angles than those in Table 2. This occurs because the stronger wigglers cause a larger energy spread, lowering the momentum compaction required to meet the impedance threshold $(Z/n)_t = 0.2\,\Omega$. Thus one can decrease the bending angle per bend, increase the main bending field and decrease the length of the wigglers, see Eq. (2.21). Since we feel that the wigglers calculated are too expensive, and we do not wish to go much higher in energy, we need to also consider changing the damping partitions. This will further increase the energy spread and lower the emittance, allowing us to increase the main bending field and thereby decrease the length of the wigglers.

Table 2. Solutions for a FODO lattice with 22 KG wigglers: $\gamma\epsilon_{x0} = 2.0\times 10^{-6}, (Z/n)_t = 0.2\,\Omega$.

Energy	# of cells	B_0	L_w	L_{ring}
1.75 GeV	58	5.0 KG	39 m	210 m
2.0 GeV	70	4.5 KG	33 m	225 m
2.25 GeV	80	4.0 KG	28 m	250 m

Table 3. Solutions for a FODO lattice with 40 KG wigglers: $\gamma\epsilon_{x0} = 2.0\times 10^{-6}, (Z/n)_t = 0.2\,\Omega$.

Energy	# of cells	B_0	L_w	L_{ring}
1.75 GeV	90	8 KG	11 m	142 m
2.0 GeV	102	6.5 KG	9 m	170 m
2.25 GeV	112	6.0 KG	7.5 m	185 m

2.6. Damping Partitions

The damping partitions describe the relative rates of the horizontal, vertical and longitudinal damping. Since the injected longitudinal emittance is usually much closer to the damped value than is the horizontal or vertical, it is reasonable to increase the horizontal and/or vertical partitions at the expense of the longitudinal. As was mentioned earlier the vertical partition is effectively fixed, but we can increase the horizontal damping partition and thereby lower the horizontal emittance. In addition, decreasing the longitudinal damping partition increases the energy spread, and thereby $(Z/n)_t$.

The damping partitions in a ring can be changed by using combined function bending magnets or Robinson wigglers.[13)] The partitions can be written:[9)]

$$J_x = 1 + \frac{\oint \eta_x/\rho(1/\rho^2 + 2K_1)ds}{\oint ds/\rho^2} \qquad J_\epsilon = 3 - J_x \ , \quad (2.22)$$

where K_1 is negative for a horizontally focusing quadrupole field.

In the case of a Robinson wiggler, Eq. (2.22) becomes

$$J_x = 1 + \frac{\overline{\eta_x} L_{\text{Rob}} K_1}{2\pi(1+F_w)} \frac{\rho_0}{\rho_{\text{Rob}}} \ , \quad (2.23)$$

where L_{Rob} and ρ_{Rob} are the length and bending radius of the Robinson wiggler. Thus we see that we would like to place it in a region of high dispersion. Unfortunately, this causes the emittance to increase. The emittance blowup is

$$\Delta\gamma\epsilon_x = \frac{3.25\times10^{-6}}{\tau_{y\,\text{eff}}} \frac{1}{J_x(1+F_w)^2} \frac{L_{\text{Rob}}}{4\pi^2} \frac{\overline{\eta_x}^2}{\overline{\beta_x}} \frac{\rho_0^2}{\rho_{\text{Rob}}^3} \ . \quad (2.24)$$

To keep the emittance increase small, the wiggler should be built so both ρ_{Rob} and $|K_1|$ are large and it should be placed in a region where β_x is large also. For example with the parameters: $\eta_x = 20$ cm, $\beta_x = 10$ m, $\rho_0 = 5$ m, and $K_1 = 17\,\text{m}^{-2}$, a wiggler length of 4 m will increase J_x by 1.1 and increase the emittance by less than 5%. The problems with this are: (1) extra length is required for the wiggler insertion and (2) to achieve the strong quadrupole fields the wiggler would have to be built as large aperture quadrupoles with the beam passing off-center; technically this might present a problem.

The other option, that of using combined function bends, seems to be easier. In this case Eq. (2.22) is

$$J_x = 1 + \frac{\alpha K_1 \rho_0 L_{\text{ring}}}{\pi(1+F_w)} \ . \quad (2.25)$$

Here the emittance is not increased by the combined function bends; in fact it should decrease slightly since the defocusing quadrupoles could be made smaller, increasing the filling factor. Furthermore, an additional insertion, which would decrease α, is not required.

Since this is the more promising route, we calculate $\gamma\epsilon_{x0}$ and $(Z/n)_t$ as we do in Tables 2 and 3, except we include the effect of a defocusing field on J_x. Obviously, it is desirable to have the largest defocusing field possible, consistent with limits on the energy spread. We calculate K_1 by choosing the larger of two expressions:

$$K_1 = \frac{14\,\text{KG} - B_0}{B\rho\ 0.0125} \qquad K_1 = \frac{20\,\text{KG} - B_0}{B\rho\ 0.025} \ . \quad (2.26)$$

The first expression estimates the field achievable in an off-center quadrupole, assuming 14 KG pole tip fields and a beam pipe radius of 1.25 cm. The radius of the quadrupole is then the beam pipe radius plus the offset. The second expression estimates the field in a bending magnet with modified poles. We assume 20 KG maximum field with poles that extend to twice the beam pipe radius of 1.25 cm.

Table 4. Solutions for a combined function FODO lattice with 22 KG wigglers. $\gamma\epsilon_{x0} = 2.0\times10^{-6}, (Z/n)_t = 0.2\,\Omega$.

Energy	# of cells	B_0	K_1	L_w	L_{ring}
1.75 GeV	73	13.0 KG	$4.8\,\text{m}^{-2}$	27 m	145 m
2.0 GeV	90	12.5 KG	$4.2\,\text{m}^{-2}$	18 m	160 m
2.25 GeV	108	12.5 KG	$3.7\,\text{m}^{-2}$	11 m	160 m

Table 5. Solutions for a combined function FODO lattice with 40 KG wigglers: $\gamma\epsilon_{x0} = 2.0\times10^{-6}, (Z/n)_t = 0.2\,\Omega$.

Energy	# of cells	B_0	K_1	L_w	L_{ring}
1.75 GeV	96	15.0 KG	$3.4\,\text{m}^{-2}$	7.25 m	113 m
2.0 GeV	112	14.5 KG	$3.3\,\text{m}^{-2}$	4.75 m	125 m
2.25 GeV	120	14.0 KG	$3.2\,\text{m}^{-2}$	2.75 m	140 m

The results for the conventional and superconducting wigglers are listed in Tables 4 and 5. Notice that when compared to Tables 2 and 3, the number of cells has increased and the main bending field has increased. Because the main bends are stronger, the required length of wiggler has decreased by roughly a factor of two. Obviously this is desirable. Unfortunately, increasing the number of cells and increasing the bend strength means that the bends becomes very short, 13 cm for the 1.75 GeV case in Table 5. It is difficult to achieve the necessary field quality in such short bends since the end fields become significant.

We use the results from Table 4 to pick a starting point for our ring. We start with the lowest energy, conventional wiggler, design. This was chosen for the reasons mentioned previously in Sec. 2.1; namely, low energy makes the subsequent bunch compression easier and the magnets are easier to build. In addition, although the low energy design requires more wigglers, it is smaller and therefore (hopefully) cheaper; we have not attempted to optimize costs in a rigorous manner.

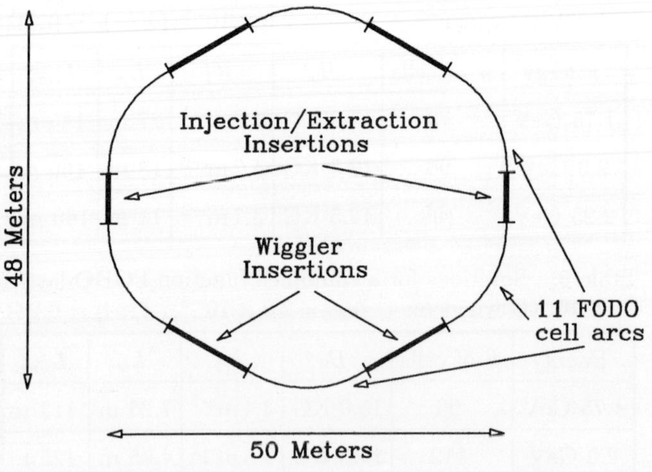

Fig. 1. Schematic of the TLC damping ring.

Table 6. TLC damping ring parameters.

Energy	$E_0 = 1.8$ GeV
Length	$L = 155.1$ m
Momentum compaction	$\alpha = 0.00120$
Tunes	$\nu_x = 24.37, \quad \nu_y = 11.27$
RF frequency	$f_{RF} = 1.4$ GHz
Current	10 batches of 10 bunches of $2 \times 10^{10} e^+/e^-$

3. BASIC LATTICE

A design for the TLC damping ring is illustrated in Fig. 1. The ring has a circumference of 155 m and operates at an energy of 1.8 GeV. There are six insertion regions, two for injection/extraction and four for wigglers. The arcs between each insertion region are composed of 11 combined function FODO cells. The ring has a superperiodicity of two. The parameters are listed in Tables 6 and 7. Table 6 lists general parameters while Table 7 lists parameters for the ring with the wigglers on and off. The optical functions β_x and β_y and the dispersion function η_x for half of the ring are plotted in Fig. 2.

The ring would operate with 10 batches of 10 bunches of 2×10^{10} electrons/positrons. The bunches in a batch are separated by 1 RF period, approximately 20 cm. Each batch is then separated by 50 ns, leaving 100 ns for the kicker pulse to rise and fall. When the wigglers are on, the normalized horizontal emittance, including the intrabeam scattering effects calculated by ZAP,[23] is 2.75×10^{-6}mrad. Because the ring operates at a relatively high energy, necessary to achieve the required damping rates, the intrabeam scattering contribution to the emittance is fairly small — about 27% of the ring emittance. The damping times are $\tau_x = 2.50$ ms and $\tau_y = 3.98$ ms. This allows each batch to remain in the ring for seven vertical damping times (vdt) when operating at a repetition rate of 360 Hz. Ignoring coupling of the horizontal and vertical.planes, 7 vdt will damp a positron beam with an initial normalized emittance of 3×10^{-3} to an emittance of 2.7×10^{-9}mrad. The alignment tolerances required to achieve the 100:1 emittance ratio are discussed later in Sec. 5.

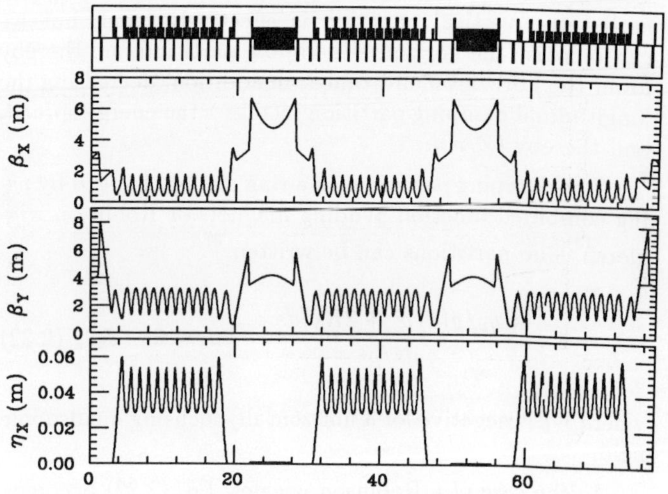

Fig. 2. Optical functions for half of the ring.

Table 7. TLC damping ring parameters.

	Wigglers Off	Wigglers On
Natural $\gamma\epsilon_x$	2.46 μmrad	2.00 μmrad
$\gamma\epsilon_x$ w/ intrabeam	3.33 μmrad	2.74 μmrad
Damping, τ_x	3.88 ms	2.50 ms
Damping, τ_y	9.19 ms	3.98 ms
Rep. rate, f_{rep}	155 Hz	360 Hz
Damp. partition, J_x	2.37	1.59
Energy spread, σ_ϵ	0.00128	0.00104
Radiation/turn, U_0	203 KeV	468 KeV
Bunch length, σ_z	5.6 mm	5.2 mm
Synch. tune, ν_s	0.0068	0.0058
$(Z/n)_t$	$\mathcal{F} \times 0.32\,\Omega$	$\mathcal{F} \times 0.20\,\Omega$
Natural chrom., ξ_x	-28.35	-28.07
Natural chrom., ξ_y	-25.10	-22.27

We wish to operate the ring below or close to the turbulent bunch lengthening threshold. The threshold is $(Z/n)_t = \mathcal{F} \times .2\,\Omega$ where $\mathcal{F}$ is a form factor greater than one.[10] While this impedance is very low (by conventional standards) we plan to achieve such a value by using a constant size beam pipe throughout the ring. The planned pipe has a 1 cm inner radius which is slightly smaller than the SLC damping rings and is approximately seven times the beam size of an injected positron beam.

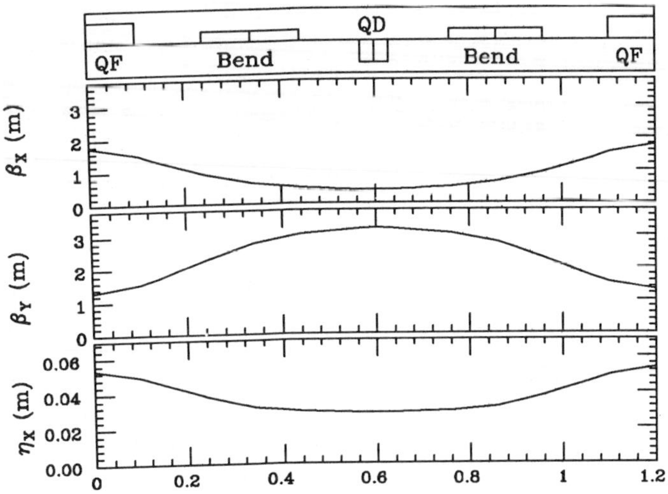

Fig. 3. Arc cell optics and magnets.

3.1. Arcs

The six arcs are constructed of 11 combined function FODO cells. The bends have an additional defocusing quadrupole gradient which re-partitions the damping. The optical functions and the magnet positions are plotted for a single cell in Fig. 3. The bend magnets bend an angle of 2.5° and have normalized defocusing gradients of $K_1 = 5.0\,\mathrm{m}^{-2}$, *i.e.* a gradient of 300 KG/meter. The QF's, the focusing quadrupoles, which are the strongest quadrupoles in the ring, have normalized gradients of $-15.7\,\mathrm{m}^{-2}$. Using a magnetic radius of $r = 1.2\,\mathrm{cm}$, 2 mm greater than the beam pipe, the QF's have pole tip fields of 11.3 KG.

This design has a little more space between magnets than the SLC damping rings do, but it is still tightly packed. There is 14 cm between the QF's and the bends and 13 cm between the bends and the QD's. Note that since the bends have large defocusing fields the QD's are very short. Since space is tight and very strong sextupoles will be needed to correct the chromaticity, we plan to use permanent magnet sextupoles similar to those successfully used in the SLC damping rings.[24)]

3.2. Injection/Extraction Insertion

The injection/extraction insertion regions have a 2 m drift space for septum magnets. On either side of the insertion drift there are additional 1.6 m drift spaces. The horizontal phase advance from the middle of the insertion drift to the center of these side drifts is approximately $\pi/2$, thus making them useful for the placement of the kicker magnets needed for injection/extraction. The optical functions for half an insertion are plotted in Fig. 4.

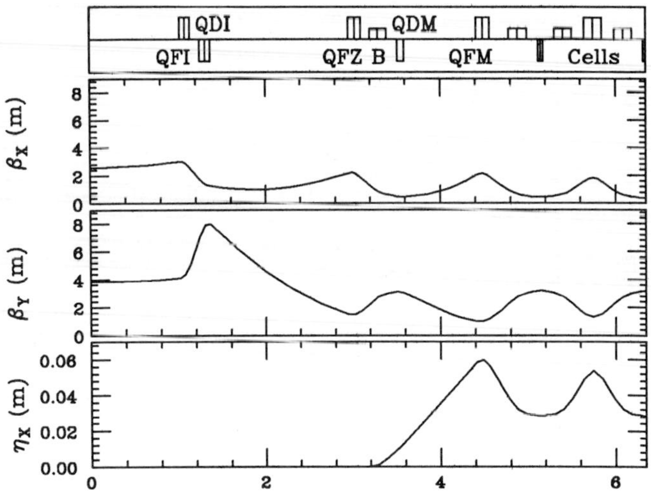

Fig. 4. Half of the injection/extraction insertion.

The dispersion is set to zero in all six of the insertion regions. This is done in the injection/extraction regions to make it easier to match the extraction transport line to the ring. Experience on the SLC has shown that properly matching the various components of the collider is crucial for obtaining small spots at the collision point. In addition, zero dispersion should make injection of the large positron beams easier.

3.3. Wiggler Insertions

The wiggler insertions are very similar to the injection/extraction insertions. Both insertions use the same dispersion suppression arrangement. The main drift in the wiggler insertion is 6 m rather than 2 m. There is also an additional quadrupole so that the wigglers can be continuously varied from full on to off while keeping the phase advance across the region constant. The lattice functions for 1/2 the insertion with the wigglers off are plotted in Fig. 5 and the same with the wigglers on is plotted in Fig. 6. Notice that the vertical focusing due to the wigglers allows β_y to have negative curvature across the insertion.

3.4. Wigglers

Wigglers are required to reduce the damping times by a factor of approximately 2.5. Thus, high peak fields are needed. Unfortunately, to prevent the wigglers from blowing up the emittance, short wiggler periods are also necessary. We chose a wiggler with a 24 KG peak field and a period of 20 cm. With this short period the wigglers actually lower the emittance. The wiggler contribution to the emittance is dominated by the η_x' generated by in the wigglers. Since β_x is approximately 4 m in the wiggler, the dispersion in the wiggler would have to be roughly 15 cm before its contribution to the emittance is comparable. We could have lowered β_x in the wiggler, but this would increase the chromaticity and likely decrease the dynamic aperture because stronger sextupoles would be needed for chromatic correction.

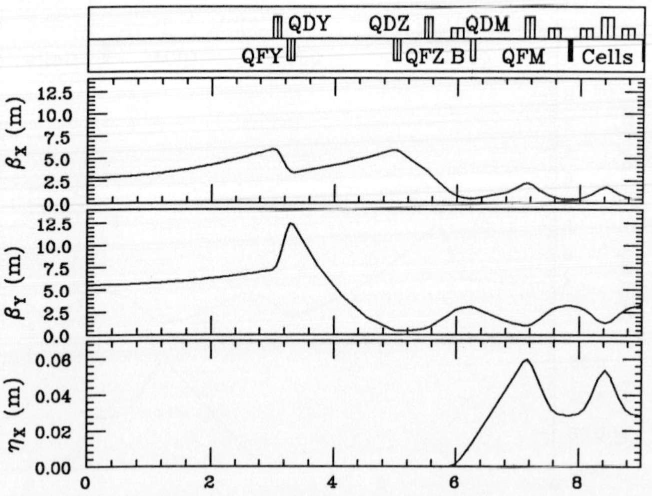

Fig. 5. Half of the wiggler insertion with wigglers off.

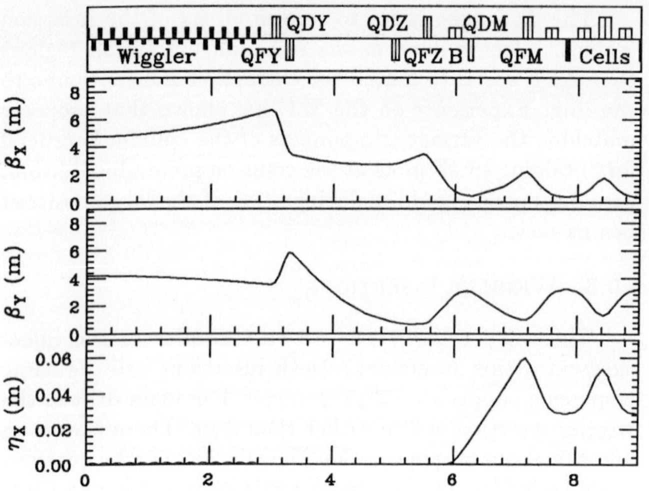

Fig. 6. Half of the wiggler insertion with wigglers on.

The period and the peak field were chosen to be within 15%[25] of the limits for Nd-Fe-B hybrid wigglers as specified in Ref. 20. If such a high peak field is not feasible, we can increase the energy of the design to 1.9 GeV and decrease the peak wiggler fields to 21 KG, the maximum field specified in this reference. The vertical damping time remains 3.98 ms and the natural horizontal normalized emittance increases to 2.29×10^{-6}mrad; intrabeam scattering further increases this to 2.85×10^{-6}mrad. The small increase in emittance occurs because the horizontal damping partition J_x increases as the wiggler gets weaker, see Eq. (2.25).

The wigglers are 5.6 m long, leaving 20 cm between the wiggler and the insertion quadrupoles. We modelled them in the lattice using hard-edged rectangular bends with the same peak field, 24 KG. The poles are 1/4 of a period long and the drifts separating the poles are also a 1/4 period long. Thus the model wigglers generate the same amount of synchrotron radiation as sinusoidal wigglers would, but the vertical focusing contribution is incorrect. With this model we should be over-estimating the vertical focusing and the four quadrupoles in the wiggler insertions should be able to easily compensate the weaker focusing.

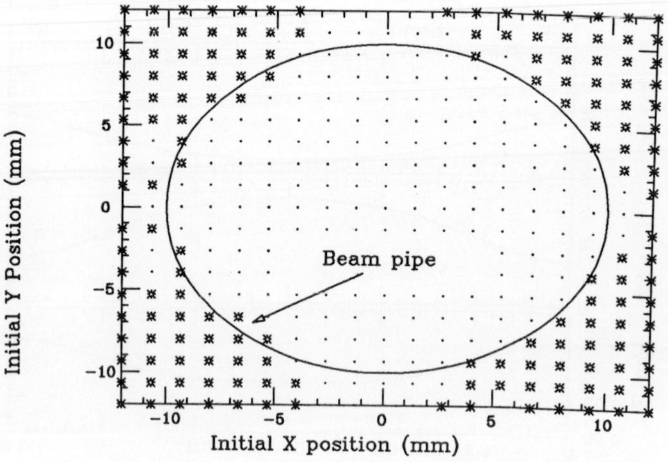

Fig. 7. Dynamic aperture for the chromatically corrected ring.

4. CHROMATIC CORRECTION

Before attempting to optimize the ring, we need to demonstrate that the chromatically corrected ring can have reasonable dynamic aperture. We currently correct the chromaticity with only two families of sextupoles located in the arcs. The integrated sextupole strengths are: $K_{2\,\mathrm{SF}} = -52.0\,\mathrm{m}^{-2}$ and $K_{2\,\mathrm{SD}} = 68.7\,\mathrm{m}^{-2}$.

The phase advance of the cells is adjusted to cancel most of the first order geometrics over an arc, 11 cells. The amplitudes of the first order geometric perturbations are proportional to:[26]

$$\int_{\mathrm{arc}} ds\beta_x^{3/2} K_2 (e^{i3\psi_x} + e^{i\psi_x} + CC)$$
$$\int_{\mathrm{arc}} ds\beta_x^{1/2}\beta_y K_2 (e^{i\psi_x + 2\psi_y} + e^{i\psi_x} + e^{i\psi_x - 2\psi_y} + CC) \tag{4.1}$$

where CC represents the complex conjugate. Since the cells are periodic, these integrals can be minimized by making the arc phase advances integral multiples of 2π. We chose cell phase advances of $\nu_{x\,\mathrm{c}} = .270$ and $\nu_{y\,\mathrm{c}} = .086$. Thus

$$\begin{array}{ll} 11\,\nu_{x\,\mathrm{c}} \approx 3 & 11(3\nu_{x\,\mathrm{c}}) \approx 9 \\ 11(\nu_{x\,\mathrm{c}} + 2\nu_{y\,\mathrm{c}}) \approx 5 & 11(\nu_{x\,\mathrm{c}} - 2\nu_{y\,\mathrm{c}}) \approx 1\ . \end{array} \tag{4.2}$$

While this choice of phase advance cancels the first order geometrics over an arc, it drives the octupole difference resonance, a second order geometric effect of sextupoles. Although we have not tried, we should be able to minimize the higher order geometric and higher order chromatic effects with additional sextupole families.

We have not attempted any optimization other than adjusting the cell phase advances as described above. The dynamic aperture is, for the most part, larger than the physical aperture. The results of tracking 1000 turns are plotted in Fig. 7. Note that the plot is distorted; the beam pipe is be round. The dynamic aperture will decrease when errors and realistic wiggler induced non-linear fields are included, but with more sophisticated chromatic correction schemes it should not present a problem.

5. TOLERANCES

5.1. ALIGNMENT

We need to calculate the effects of misalignments on both the dynamic aperture and the emittances. Errors in the machine decrease the dynamic aperture by causing tune shifts and higher order multipole fields on the closed orbit. In addition, errors will couple the horizontal and vertical planes and generate vertical dispersion, causing an increase in the vertical emittance. As mentioned in the previous section we have not yet attempted to optimize the chromatic correction scheme and the dynamic aperture; thus we only discuss the tolerances necessary to achieve the desired emittance ratio.

In an ideal uncoupled ring there is no vertical dispersion or linear coupling. Thus the synchrotron radiation opening angle, which is very small, determines the vertical emittance. In practice, this is not the case. First, vertical dipole errors and a non-zero vertical orbit in the quadrupole magnets will directly introduce some vertical dispersion. Second, a non-zero vertical orbit through the sextupole magnets, vertical sextupole misalignments, or rotational misalignments of the quadrupoles couple the horizontal and vertical planes. This coupling has two effects both of which increase the vertical emittance. It couples the horizontal dispersion to the vertical, causing an increase in the vertical, and it couples the x and y betatron motion so that energy is transferred between the two.

Assuming uncorrelated misalignments, the vertical emittance due to the dispersion generated by quadrupole rotations and sextupole misalignments is:

$$\begin{aligned}\gamma\epsilon_y &\approx \frac{2C_q\gamma^3}{J_y}\frac{\langle\eta_y^2\rangle}{\beta_y}\frac{\oint ds/\rho(s)^3}{\oint ds/\rho(s)^2}\\ \frac{\langle\eta_y^2\rangle}{\beta_y} &\approx \frac{1}{2\sin^2\pi\nu_y}\sum_{\{\text{quads}\}}(K_1L)^2\beta_y\Theta^2\\ \frac{\langle\eta_y^2\rangle}{\beta_y} &\approx \frac{1}{8\sin^2\pi\nu_y}\sum_{\{\text{sext}\}}(K_2L)^2\beta_y y_m^2\ ,\end{aligned}\tag{5.1}$$

where $C_q = 3.84\times10^{-13}$ m, Θ is the rotation angle, and y_m is the vertical sextupole misalignment. K_1L and K_2L are the integrated normalized quadrupole and sextupole strengths. Likewise, the increase in the vertical emittance due to linear coupling is:

$$\begin{aligned}\epsilon_y &\approx \epsilon_x\overline{\alpha}\frac{1-\cos\Psi_x\cos\Psi_y}{(\cos\Psi_x-\cos\Psi_y)^2}\sum_{\{\text{quads}\}}(K_1L)^2\beta_x\beta_y\Theta^2\\ \epsilon_y &\approx \epsilon_x\overline{\alpha}\frac{1-\cos\Psi_x\cos\Psi_y}{4(\cos\Psi_x-\cos\Psi_y)^2}\sum_{\{\text{sext}\}}(K_2L)^2\beta_x\beta_y y_m^2\ ,\end{aligned}\tag{5.2}$$

where $\Psi = 2\pi\nu$, and $\overline{\alpha} = (\alpha_x+\alpha_y)/\alpha_y$; $\alpha_{x,y}$ is the horizontal (vertical) damping rate. Note that both the formula for the dispersion and linear coupling have resonant denominators. The vertical dispersion grows as ν_y approaches an integer and the linear coupling contribution increases as ν_y approaches the linear sum or difference resonance. To minimize both these we chose the operating tunes $\nu_x = .37$ and $\nu_y = .27$.

It is harder to estimate the effects of the dipole errors since, in addition to generating vertical dispersion, dipole errors cause a vertical orbit which is correlated from point to point. We consider the limit of an orbit fully corrected with dipole correctors. We then approximate the orbit by assuming that it is correlated between correctors, but uncorrelated on either side of a corrector. In this case the vertical dispersion can be approximated as

$$\begin{aligned}\frac{\langle\eta_y^2\rangle}{\beta_y} \approx \frac{\langle y^2\rangle_{\text{res}}}{\beta_y}\Bigg[1+\frac{1}{16\sin^2\pi\nu_y}\times\\ \Big(N_{\text{cell}}(4\pi\xi_{y\,\text{cell}})^2+N_{\text{ins}}(4\pi\xi_{y\,\text{ins}})^2\Big)\Bigg]\ ,\end{aligned}\tag{5.3}$$

where $\langle y^2\rangle_{\text{res}}/\beta_y$ is the residual orbit after correction and $\xi_y = -1/4\pi\int\beta_y(K_1+K_2\eta_x)ds$ is the local vertical chromaticity of one cell or insertion region. Also, N_{corr} is the number of correctors being used and N_{ins} is the number of insertion regions. Note that this formula illustrates the advantage of local chromatic correction. The vertical dispersion due to a vertical closed orbit is a quadratic function of the *local* vertical chromaticity between correctors. This is one advantage of the FODO and TME[16] lattices over the Chasman-Green type structures. In the FODO or TME lattices, sextupoles can surround each quadrupole, whereas, in the Chasman-Green lattice all of chromatic correction is done at the center of each cell structure where the dispersion is non-zero. In our current design we have not yet attempted to minimize the chromaticity locally by correcting the arcs and insertions separately; all chromatic correction is done in the arcs.

Using the same approximation, that the orbit is uncorrelated across the correctors, the emittance due to the linear coupling is:

$$\begin{aligned}\epsilon_y = \frac{\epsilon_x\overline{\alpha}}{64}\frac{\langle y^2\rangle_{\text{res}}}{\beta_y}\sum_{\Delta\nu,\Delta\phi}\frac{N_{\text{corr}}}{\sin^2\pi\Delta\nu\sin^2\pi\Delta\phi_{\text{cell}}}\times\\ \Big(\sum_{i,j\in\text{cell}}(K_2L)_i(K_2L)_j\beta_{y\,i}\beta_{y_j}\sqrt{\beta_{x\,i}\beta_{x\,j}}\cos\Delta\phi_{i,j}\Big)\ ,\end{aligned}\tag{5.4}$$

where the first sum is over the two values of $\Delta\nu$, $\Delta\nu =$

$\nu_x \pm \nu_y$, and two values of $\Delta\phi$ for each value of $\Delta\nu$:

$$\Delta\phi_{i,j} = \begin{cases} \phi_{x\,i,j},\ \phi_{x\,i,j} + 2\phi_{y\,i,j}; & \Delta\nu = \nu_x + \nu_y \\ \phi_{x\,i,j},\ \phi_{x\,i,j} - 2\phi_{y\,i,j}; & \Delta\nu = \nu_x - \nu_y\ . \end{cases} \tag{5.5}$$

Here, $\phi_{i,j}$ is the phase advance between sextupole i and j in the cell and ϕ_{cell} is the cell phase advance. Notice that the expression becomes large when the tunes are close to the linear coupling resonances and when the cell phase advances are close to the sextupole coupling resonances.

Summing all of these contributions, we have

$$\gamma\epsilon_y \approx \Theta^2 8.9\times10^{-8} + y_m^2 10.5\times10^{-6} + \frac{\langle y^2\rangle_{\text{res}}}{\beta_y} 1.8\times10^{-6}\ , \tag{5.6}$$

where Θ is the quadrupole rotational misalignment in milliradians and y_m and y_{res} are the sextupole misalignment and residual orbit in millimeters. Thus the vertical emittance depends strongly upon y_m and the residual orbit. We can specify tolerances by dividing most of the contribution between y_m and the closed orbit:

$$\Theta \le 0.25\,\text{mrad},\quad y_m \le 30\,\mu\text{m},\quad y_{\text{res}} \le 100\,\mu\text{m}\ , \tag{5.7}$$

where $\overline{\beta}_y$ was assumed to be 2 m. Notice that random sextupole misalignments y_m are far more damaging than the residual orbit. In addition, with a vertical corrector in each cell, the residual orbit is strongly dependant on the BPM tolerances and only weakly dependant upon the magnitude of the dipole errors. Thus the BPM's must be aligned to a tolerance of $100\,\mu$m *minus* the BPM measurement sensitivity.

These results were verified with a computer program which, like PETROS, calculates the fully coupled emittances given a distribution of errors in the ring. Using 10 distributions of errors, the program consistently corrects the orbit due to $200\,\mu$m quadrupole misalignments to less than the $100\,\mu$m orbit tolerance. Furthermore, the average value of $\gamma\epsilon_y$ is very close to the results of these analytic formulas. Unfortunately, the rms of $\gamma\epsilon_y$ is quite large, about 70% of the average value, implying that we would need to tighten the tolerances to have confidence in the final emittance ratio.

Fortunately, we can correct much of the coupling. Linear coupling can be corrected with four orthogonal skew quadrupoles. Of course to correct the linear coupling, one has to measure the amount of coupling, and then separate the linear coupling from the effects of vertical dispersion. One option would be to use a number of profile monitors around the ring. The effects of the vertical dispersion could be subtracted off using the fact that the dispersion should oscillate at a harmonic next to the tune. Alternatively, we should be able to find the amplitude and phase of the linear coupling by measuring the phase and amplitude of the $\nu_x \pm \nu_y$ tune lines relative to the ν_x and ν_y lines.[27] We hope to test this technique on the SLC in the future.

We can also correct the vertical dispersion. Skew quads in regions of horizontal dispersion act much like dipole correctors when correcting a closed orbit. Just as when correcting a closed orbit, the first few correctors, skew quads in this case, do most of the correction. Obviously to correct both the linear coupling and the dispersion, we have to perform the corrections in an orthogonal manner.

5.2. Kickers

It is extremely important that the extraction kicker have very small jitter. Assuming linear optics, a one sigma jitter at the extraction septum becomes one sigma jitter at the IP. Actually, without BNS damping[28,29] the jitter would be magnified many times by transverse wakefields in the linac. But with weak BNS damping, the linear approximation is fairly accurate.

We would like to achieve a jitter tolerance of one tenth of the beam size $\sigma_{x,y}$ at the IP. This specifies a tolerance on the kicker. Assuming that the septum plate is 90° in phase downstream of the kicker,

$$\frac{\Delta\Theta}{\Theta} \le \frac{\sigma_{\text{jit}}}{x_{\text{kick}}} \quad \text{where} \quad \sigma_{\text{jit}} \le \frac{1}{10}\sigma_x\ , \tag{5.8}$$

where x_{kick} is the transverse displacement of the kicked beam at the septum. It is a sum of the distance between the closed orbit and the septum plate plus the septum thickness and some extra space for the extracted beam.

This tolerance can be written in a more transparent form:

$$\frac{\Delta\Theta}{\Theta} \le \frac{1}{10}\frac{\sqrt{\gamma\epsilon_{\text{ext}}\beta_{x\,\text{sept}}}}{x_{\text{sept plate}} + 1\,\text{mm} + N_s\sqrt{\gamma\epsilon_{\text{inj}}\beta_{x\,\text{sept}}}}\ , \tag{5.9}$$

where $\gamma\epsilon_{\text{ext}}$ and $\gamma\epsilon_{\text{inj}}$ are injected and extracted beam emittances. N_s is the distance between the closed orbit and the septum plate in units of the injected beam size. To prevent particle loses, the septum plate must be many sigma from the stored beam just after *injection* when the beam is the largest; we assumed $N_s = 7$. The only other quantity in the tolerance is the septum plate thickness. Because a pulsed septum will introduce more jitter problems, we need to use a DC septum. The septum plate in a DC septum with strong fields cannot be made much thinner than 3 mm.[30] Assuming $\gamma\epsilon_{\text{inj}} = 3\times10^{-3}$ mrad and $\beta_x \approx 3$ m, the jitter tolerance on the kicker is

$$\frac{\Delta\Theta}{\Theta} \le 3\times10^{-4}\ . \tag{5.10}$$

Notice that this tolerance is mainly determined by the ratio of the extracted beam emittance to the injected emittance. The only way to minimize the tolerance is to reduce N_s, increasing the particle loses due to scraping, or to reduce the injected emittance.

Such a tolerance will be difficult to achieve. One solution would be to use two kickers, separated by a phase advance of π, to cancel the jitter. One kicker would be placed before the septum and one located in the extraction line. Unfortunately, this approach requires that the two kickers track each other over the time needed to extract a batch, $\sim 7\,\text{ns}$. Since the tolerance for this bunch to bunch tracking is equal to the jitter tolerance, getting the two kickers to track maybe as difficult a problem as having a single kicker to achieve the jitter tolerance. An alternate approach would be to use a current source with feedback instead of a thyratron. Unfortunately, with current technology, it is not apparent that such a system could provide the necessary current.[8] To conclude, we have not yet found a viable solution to the kicker problem and it requires further research.

6. SUMMARY

In this paper we present a design for a damping ring for a TeV linear collider. The ring is based upon a FODO lattice where the bending magnets contain a gradient to re-partition the damping. In addition, damping wigglers are used to lower the operating energy and to increase the energy spread. If the length of wiggler required is too much, we have solutions at higher energies that need less.

We demonstrate, using a very simple chromatic correction scheme, that the chromatically corrected ring has reasonable dynamic aperture. We also calculate the tolerances on the alignment and the vertical orbit necessary to achieve a 100:1 emittance ratio and then discuss methods of correcting the coupling. Finally, we calculate tolerances on the extraction kickers and find that they are very tight. Unfortunately this tolerance depends mainly upon the ratio of the injected beam emittance to the extracted emittance and it is not easy to decrease.

Finally, we note that we have more work in optimizing the injection/extraction insertion regions to loosen the requirements on the extraction kickers. We need to improve the chromatic correction scheme and we should calculate the effect of errors on the dynamic aperture. We also should also take detailed looks at designs based upon the TME lattice and a version of a FODO lattice without separate defocusing quadrupoles.

ACKNOWLEDGEMENTS

We would like to thank Bill Gabella for help with the preliminary design, Phil Morton and Martin Lee for many useful suggestions, and Martin Donald for discussions on chromatic correction.

REFERENCES

1. See references in these proceedings.
2. K. Thompson and R. D. Ruth, "Multibunch instabilities in Subsystems of 0.5 and 1.0 TeV Linear Colliders", *in these proceedings* and SLAC–PUB–4800 (1988).
3. L. Z. Rivkin, "Damping Ring for the SLC", Ph.D. Thesis, California Inst. Tech. (1986).
4. L. Z. Rivkin, *et. al.*, "Bunch Lengthening in the SLC Damping Ring", SLAC–PUB–4645 (1988).
5. R.B. Palmer, "Exotic Damping Ring Lattices", SLAC–PUB–4329 (1987).
6. J.D. Bjorken and S.K. Mtingwa, "Intrabeam Scattering", *Particle Accelerators*, **13**, 115 (1983).
7. Using the formalism of Bjorken and Mtingwa, Pisin Chen estimated the effect of intrabeam scattering on the vertical emittance and found it negligible.
8. F. Bulos, private communication.
9. M. Sands, "The Physics of Electron Storage Rings", SLAC–121 (1970).
10. J. M. Wang and C. Pellegrini, *Proceedings of the 11th Int. Conf. on High Energy Acc.*, Geneva, Switzerland (1980), p. 554.
11. *Proceedings of the ICFA Workshop on Low Emittance e^--e^+ Beams*, BNL, March 20-25 1987.
12. S. Kheifets, "Bunch Compression for the TLC", *in these proceedings* and SLAC–PUB–4802 (1988).
13. K.W. Robinson, *Phys Rev.*, **111**, 373 (1958).
14. A. Hofmann, "Emittance Reduction in PEP with a Robinson Wiggler", SSRL-ADC-NOTE-40 (1986).
15. G. Vignola, "The Use of Gradient Magnets in Low Emittance Storage Rings", *Nucl. Instr. and Meth.*, **A246**, 12 (1986).
16. A. van Steenbergen, "Collider Damping Ring, Theoretical Minimum Emittance Structures", *Proceedings of the ICFA Workshop on Low Emittance e^--e^+ Beams*, BNL (1987), p. 60.
17. H. Wiedemann, "Design of Low Emittance Storage Rings", *Nucl. Instr. and Meth.*, **A246**, 4 (1986).
18. J. Bisognano *et. al.*, "Feasibility Study of a Storage Ring for a High-Power XUV Free-Electron Laser", *Particle Accelerators*, **18**, 223 (1986).
19. A. Jackson, "A Comparison of the Chasman-Green and Triple Bend Achromat Lattices", LBL–21279 (1986).
20. K. Halbach, *J. Appl. Phys.*, **57**, 3605 (1985).
21. C. Bazin, *et. al.*, "A Five-Pole Superconducting Wiggler for the DCI Positron Ring at LURE", *Nucl. Instr. and Meth.*, **A266**, 132 (1988).
22. H. Wiedemann, "Emittance in a FODO Lattice", PEP-303 (1979).
23. M. Zisman *et al.*, "ZAP User's Manual", LBL–21270 (1986).
24. J.E. Spencer, "Permanent Magnet Sextupole Protocol and Tolerances for the Damping Rings", SLAC–CN–309 (1985).
25. Klaus Halbach, private communication.
26. T. O. Raubenheimer and R. D. Ruth, "Superconvergent Tracking", *Particle Accelerators*, **23**, 197 (1988).
27. J. Bengtsson, "Non-Linear Transverse Dynamics for Storage Rings", CERN 88-05 (1988).
28. V. Balakin, A. Novokhatsky, V. Smirnov, *Proceedings of the 12th Int. Conf. on High Energy Accelerators*, Fermilab (1983), p. 119.
29. K. Bane, *IEEE Trans. Nucl. Sci.*, **NS-32**, 2389 (1985).
30. G. Fisher, private communication.

BUNCH COMPRESSION FOR THE TLC. PRELIMINARY DESIGN*

S. A. KHEIFETS, R. D. RUTH, J. J. MURRAY AND T. H. FIEGUTH

Stanford Linear Accelerator Center, Stanford University, Stanford, California 94309

ABSTRACT

A preliminary design of a TLC bunch compressor as a two-stage device is described. The main parameters of the compressor as well as results of some simulations are presented. They show that the ideal system (no imperfections) does the job of transmitting transverse emittances without distortions (at least up to the second-order terms) producing at the same time the desired bunch length of 50 μm.

1. INTRODUCTION

There are several reasons to choose a short TLC bunch length :

(a) short bunch length is required for controlling the adverse effects caused by the transverse wake field.

(b) the high luminosity of the linear collider is achieved by decreasing the value of the beta-function at the interaction point. Such a decrease is effective only when the value of the beta-function is greater than the effective bunch length.

(c) the cost of the RF power supplies, which is one of the largest cost ingredients of the linear collider, tends to decrease for higher frequencies of the accelerating field; to keep the particle energy spread tolerable for a small wave length one needs a small bunch length.

On the other hand, the adverse effects of the longitudinal wake fields, which become more pronounced when the bunch length decreases, limit the length of the bunch from below.

As a compromise, a bunch length of the order of 70 μm has been chosen for the present design of the TLC.[1] A bunch of this length is not readily obtainable from a damping ring and, hence, a compressor will be needed to transform a long bunch with small momentum spread $\Delta p/p$ into a short bunch with a larger momentum spread.

But there is a limit on how large the resultant momentum spread can be, which is determined by the momentum acceptance of the linac. For the bunch extracted from the TLC damping ring the momentum spread will be typically $\Delta p/p \approx 10^{-3}$ with a bunch length of $l \approx$ 5mm. Compression of this bunch length to 50 μm would then require increasing the momentum spread to 10%. That is clearly unacceptable when chromatic aberrations and practical aperture limitations are considered. Therefore, a two-stage compressor with an intermediate acceleration is assumed in the present design. The relevant input parameters of the beam as well as the corresponding desired output parameters for each stage are collected in Table 1.

Table 1.
Compressor input and output parameters.

N	Parameter	1st Stage	2nd Stage
1	Energy (GeV)	1.8	16.2
2	Δl_{in} (mm)	5.0	0.46
3	$\Delta p/p_{\text{in}}$ (%)	0.1	0.12
4	Δl_{out} (mm)	0.46	0.037
5	$\Delta p/p_{\text{out}}$ (%)	1.1	1.6
6	$\gamma\epsilon_x$ (Mrad)	2.10^{-6}	2.10^{-6}
7	$\gamma\epsilon_y$ (Mrad)	2.10^{-8}	2.10^{-8}
8	$\sqrt{\epsilon_x\beta_x}$ (μm)	54.9	24.4
9	$\sqrt{\epsilon_y\beta_y}$ (μm)	3.11	1.20
10	$\sqrt{\epsilon_x/\beta_x}$ (μrad)	10.35	2.33
11	$\sqrt{\epsilon_y/\beta_y}$ (μrad)	1.83	0.475

* Work supported by the Department of Energy, contract DE–AC03–76SF00515.

2. THE CONCEPTUAL COMPRESSOR DESIGN

The design of each compressor stage is conceptually similar to the design of the SLC compressor.[2] A bunch extracted from the damping ring first passes through a matching section, where the beta- and eta-functions of the ring are matched to the corresponding periodic functions of the compressor.

The compressor itself consists of two stages. Each stage starts with an acceleration section where a momentum shift $\Delta p/p$ is introduced for each particle dependent upon its longitudinal position Δl within the bunch. The bunch then passes through a transport line where the path length depends upon particle momentum. The linear part of the path length dependance upon momentum is adjusted in such a way as to obtain a minimal bunch length on its exit. As in Ref. 3, the design of the transport line uses the concept of a second-order achromat[3] to prevent the emittance growth due to the second-order aberrations and chromatic effects. There is an intermediate accelerator between the stages 1 and 2 where particles are accelerated from 1.8 GeV to 16.2 GeV. The momentum spread with which the bunch enters the second stage is thus reduced to $\approx 0.1\%$.

The achromat in each stage is built out of four identical cells with 90° betatron phase advance in each plane. That makes the overall first-order transverse transfer matrix of the line equal to unity. Two families of the sextupole magnets compensate the natural first-order chromaticities eliminating at the same time all the second-order chromatic aberrations.[3,9] Cells are chosen to be symmetric, since that makes the phase ellipses of the line admittance on the entrance to the achromat upright and facilitates matching of the phase ellipses of the particle distribution and the amplitude functions.

Figures 1 and 2 present the horizontal and the vertical amplitude functions for each stage, respectively. Figures 3 and 4 give the horizontal dispersion functions and the bunch length along both achromats. Tables 2 and 3 contain the list of magnetic elements and the parameters of each of them in the TRANSPORT[5] notation. Two-dimensional TURTLE[6] scatter plots of a thousand randomly chosen particles in the $(\Delta l, \Delta p/p)$ plane at the exit of the each stage are presented in Figs. 5 and 6.

Table 4 contains the main parameters of both compressor stages.

3. EMITTANCE CHANGE DUE TO RADIATION AND NON-LINEAR EFFECTS

Of special concern is the emittance increase due to the quantum character of the radiation in a magnetic field. The design should provide that such a growth is small

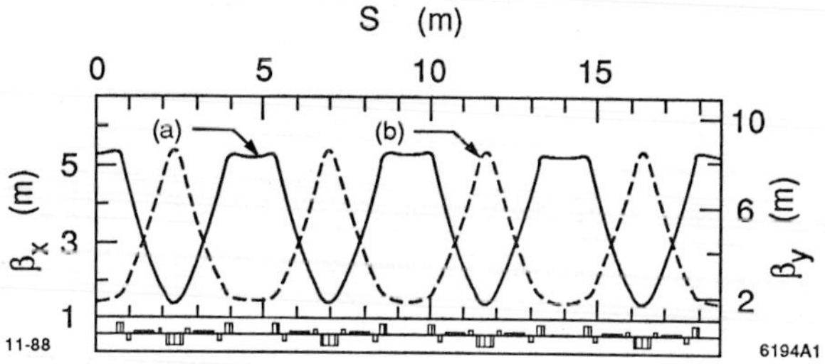

Fig. 1. Horizontal (a) and vertical (b) β-functions for the first compressor stage.

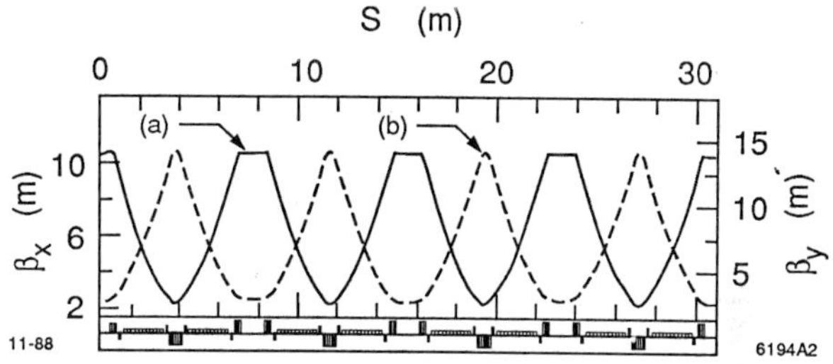

Fig. 2. Horizontal (a) and vertical (b) β-functions for the second compressor stage.

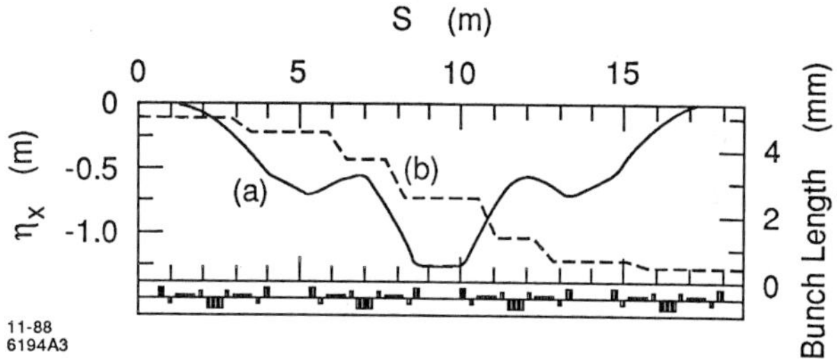

Fig. 3. Horizontal dispersion function (a) and the bunch length (b) along the first compressor stage.

compared to the transverse emittance of the bunch. The effect is evaluated by Sands.[7] The absolute and relative emittance increases due to this effect are to be found in Table 4.

The simulation results, in particular Figs. 5 and 6, show that the ideal designed system (without imperfections) does the job of transmitting transverse emittances without distortions (at least up to the second-order terms) producing at the same time the desired bunch length of 50 μm.

In the last row of Table 4 the main non-linear term of the accelerating field is evaluated and is seen not to

Table 3. Lattice of Stage 2.

9.		4.		;
Element	Label	Length (m)	Field (kG)	Pole Radius (mm)
3.0	"BGND"	0.60000;		
5.00	"Q1A"	0.30000	3.95726	10.0000;
3.0		0.12000;		
8.00	"SEXA"	0.10000	-0.36189	10.0000;
3.0		0.12000;		
4.000	"BH1"	2.00000	-6.41640	0.0 ;
3.0	"DRVA"	0.12000;		
18.00	"SEXB"	0.10000	0.58043	10.0000;
3.0		0.12000;		
5.00	"Q1B"	0.30000	-3.71447	10.0000;
3.0	"SYM"	0.00000;		
5.00	"Q1B"	0.30000	-3.71447	10.0000;
3.0		0.12000;		
18.00	"SEXB"	0.10000	0.58043	10.0000;
3.0	"DRVA"	0.12000;		
4.000	"BH1"	2.00000	-6.41640	0.0;
.0		0.12000;		
18.00	"SEXA"	0.10000	-0.36189	10.0000;
3.0		0.12000;		
5.00	"Q1A"	0.30000	3.95726	10.0000;
3.0	"BGND"	0.60000;		
9.		0.		;

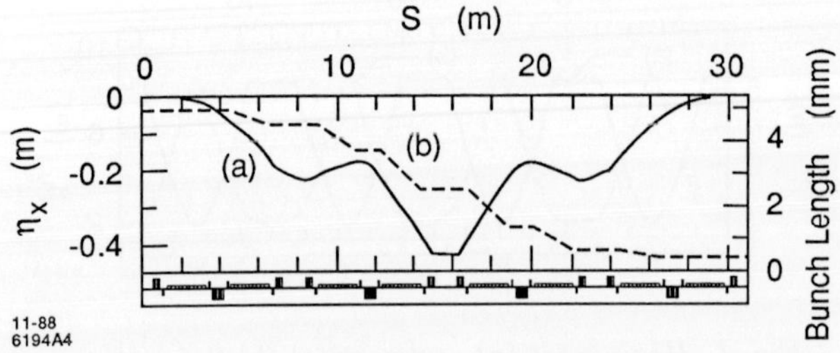

Fig. 4. Horizontal dispersion function (a) and the bunch length (b) along the second compressor stage.

Table 2. Lattice of Stage 1.

9.		4.		;
Element	Label	Length (m)	Field (kG)	Pole Radius (mm)
3.0	"BGND"	0.61351;		
5.00	"Q1A"	0.19070	1.24505	10.0000;
3.0		0.12000;		
18.00	"SEXA"	0.10000	-0.03078	10.0000;
3.0		0.12000;		
4.000	"BH1"	0.60000	-12.58428	0.0;
3.0		0.12000;		
18.00	"SEXB"	0.10000	0.043344	10.0000;
3.0		0.12000;		
5.00	"Q1B"	0.25000	-0.89500	10.0000;
3.0	"SYM"	0.00000;		
5.00	"Q1B"	0.25000	-0.89500	10.0000;
3.0		0.12000;		
18.00	"SEXB"	0.10000	0.043344	10.0000;
3.0		0.12000;		
4.000	"BH1"	0.60000	-12.58428	0.0;
3.0		0.12000;		
18.00	"SEXA"	0.10000	-0.03078	10.0000;
3.0		0.12000;		
5.00	"Q1A"	0.19070	1.24505	10.0000;
3.0	"BGND"	0.61351;		
9.		0.		;

produce any significant changes in the momentum spread for both compressor stages. The same is true for the non-linearity due to second-order terms in the transformation matrix (T_{556} terms in TRANSPORT notation). Since the bunch length is about 10 times longer in the compressor RF section than in the linac, the wake fields are about 10 times stronger. However, since the section is short this would lead to only a small effect. In addition, since the correlation induced by the RF yields lower momentum of particles at the tail of the bunch, the bunch is stabilized by BNS damping[8] throughout part of the structure.

To conclude it is important to note that these two designs are simply examples and are not unique, e.g., the final compressor might be included in a 180° bend in order to keep the site shorter. Several examples of 180° arcs

Table 4. Main compressor parameters.

N	Parameter	1st Stage	2nd Stage
1	Length (m)	18.67	31.04
2	Bend Angle/magnet (^{o})	7.205	1.361
3	Number of Magnets	8	8
4	Radius of curvature (m)	4.771	84.218
5	Total Angle Bend (^{o})	57.64	10.89
6	Correlation $(\Delta p/p)/l$ (%/mm)	0.21517	3.4012
7	RF (GHz)	3.0	17.0
8	λ (cm)	10.0	1.76
9	$\theta = 2\pi l/\lambda$	0.314	0.164
10	V_o (MV)	57.3	1058.5
11	$\{H/\rho\}$	$2.5 \cdot 10^{-2}$	$1.6 \cdot 10^{-5}$
12	$\Delta\epsilon$(Mrad)	$2.03 \cdot 10^{-12}$	$4.45 \cdot 10^{-12}$
13	$\Delta\epsilon/\epsilon$ (%)	0.36	7.8
14	$\Delta(\Delta p/p) = \theta^2/6$	0.016	0.0045

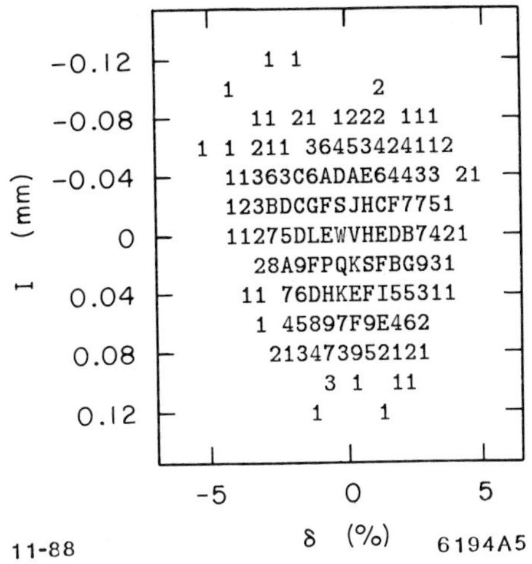

Fig. 5. Scatter plot in the horizontal phase space $(\Delta\ell, \Delta p/p)$ of 1,000 rays at the exit of the first compressor stage. Rms half widths are $(\Delta p/p)_{rms} =$ 1.088%, $(\Delta\ell)_{rms} = 0.487$ mm.

which accomplish this have been calculated with favorable results.[9]

ACKNOWLEDGEMENT

The design of the second stage has been fulfilled during the Workshop. We are grateful to the Workshop organizers for a stimulating working environment.

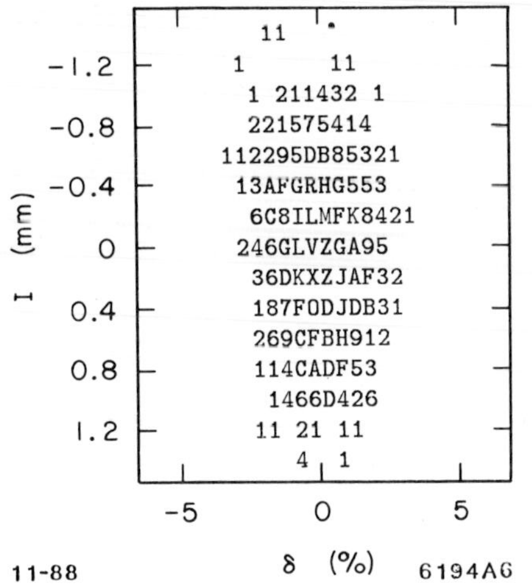

Fig. 6. Scatter plot in the longitudinal phase space $(\Delta\ell, \Delta p/p)$ of 1,000 rays at the exit of the second compressor stage. Rms half widths are $(\Delta p/p)_{rms} =$ 1.591%, $(\Delta\ell)_{rms} = 0.037$ mm.

REFERENCES

1. Palmer, R., SLAC–PUB–4707 (1988) and in these proceedings (1988).
2. Fieguth, T. H. and Murray, J. J., *Proceedings of the 12th International Conference on High Energy Acceleration,* Fermilab, Batavia IL, **401** (1983).
3. Brown, K. L., SLAC–PUB–2257, (February 1979); IEEE Trans. Nucl. Sci. NS–26 (3) p. 3490 (1979).
4. Kheifets, S. A., Fieguth, T. H. and Ruth, R. D., SLAC–PUB–4569, (March 1988).
5. Brown, K. L. et al., CERN Report CERN 80–04, Geneva (1980).
6. Carrey, D. G. et al., SLAC Report–246, UC–28, (March 1982).
7. Sands, M., SLAC Report–121, UC–28, (1970); SLAC Report SLAC/AP–47 (December 1985).
8. Balakin, V., Novokhatskii, A. and Smirnov, V., *Proceedings of the 12th International Conference on High Energy Accelerators*, Fermilab (1983).
9. Kheifets, S. A., Ruth, R. D. , Murray, J. J. and Fieguth, T. H., in preparation.

RF BREAKDOWN AND FIELD EMISSION

G. A. LOEW AND J. W. WANG

*Stanford Linear Accelerator Center**
Stanford University, Stanford, California 94309

One of the crucial parameters in the design of TeV $e^{\pm}$ linear colliders is the maximum electric field gradient that can be attained and used safely in the accelerating structure. In the absence of definitive knowledge on the availability and cost of RF power sources, the optimum machine length for a given energy cannot be determined at this time. However, the designs considered in this report could require gradients as high as 200 MV/m at RF frequencies of 11.424 GHz or 17.136 GHz, corresponding to peak powers of up to 1 GW/m. On the basis of research done at SLAC and at Varian, it now appears that such gradients are reachable, at least in short standing-wave (SW) structures.[1] Considerably more work, however, is needed to determine if the required gradients are achievable in long and complicated structures, and to verify that the accompanying field emitted currents which can absorb power, cause parasitic wakefields and spurious x-rays along the accelerator, are tolerable.

All experiments reported in Ref. 1, except for one X-band test started in collaboration with LLNL but not yet completed, were performed on SW structures. The equivalent traveling-wave accelerating fields were then calculated using the SUPERFISH computer program. S-band experiments were done on a seven-cavity disk-loaded ($2\pi/3$-mode) structure and on a two-cavity nose-cone-shaped (π-mode) structure, powered by a klystron operated up to 47 MW. C-band and X-band tests, done in collaboration with Varian, used nose-cone-shaped half-cavity structures powered by ~1 MW magnetrons. The overall results are shown in Table 1. For the pulse lengths used in the measurements (~1.5-4 μs), the breakdown-limited copper surface electric fields scale roughly as

$$E_s \sim 195[f(\text{GHz})]^{1/2} \quad ,$$

(about seven times the levels predicted by the Kilpatrick limit). If this empirical relationship is valid at higher frequencies, then breakdown-limited surface fields of 660 and 807 MV/m could be reached at 11.424 GHz and 17.136 GHz respectively (corresponding to 264 and 323 MV/m accelerating fields, assuming $E_s/\bar{E}_{acc} = 2.5$). This extrapolation assumes no improvements due to the much shorter pulse lengths required at these higher frequencies (~20 to 100 ns), and conversely no worsening due to the probable need for more complex damped structures (see next section) with slots in the disks.

The structures used in the above measurements were equipped with RF couplers, temperature sensors to measure disk temperature, internal probes to measure field emission (FE), glass and copper windows, external magnets, a spectrometer and Faraday cup to measure the intensity and energy of extracted currents, an x-ray pinhole camera, radiation monitors, a TV camera with video recorder to look at breakdown sparks, pumps and a residual gas analyzer (RGA). Intensive studies were carried out of both pre-breakdown and breakdown phenomena, during and after RF processing. Upon completion of RF processing, Fowler-Nordheim plots of FE currents versus $1/E_s$ yielded β-field enhancement factors of ~60 at 2.856 GHz and 38 at 4.998 GHz. (For more details, see below.) RF-focused beams with peak currents as high as 25 mA and energies up to 13 MeV were extracted from the seven-cavity structure, and energy spectra in agreement with theoretical calculations were analyzed.

The pre-breakdown currents produced intense x-ray fields around the seven-cavity structure, in excess of 0.75 Mrad/h. When breakdown took place at a given RF power input, the resulting reflected RF pulse was invariably accompanied by a momentary current jump by a factor of 20 to 40 (as measured by internal transverse probes), the appearance of a spark, generally on the edge

* Work supported by the Department of Energy, contract DE–AC03–76SF00515.

Table 1. Experimentally obtained breakdown-limited gradients.

	S-band Disk-loaded (2π/3-mode)	S-band With nose cone (π-mode)	C-band Half-cavity	X-band Half-cavity	X-band Disk-loaded (2π/3-mode)
Frequency, f (MHz)	2856	2858	4998	9303	11424
Total length (cm)	24.5	10.5	1.507	0.806	26.25
Filling time* (μs)	0.77	1.0	0.172	0.082	0.028
Pulse length (μs)	1.5-2.5	1.5-2.5	3.5	3.8	0.025†
Peak power input (MW)	47	10.8	0.8	1.2	200†
Peak surface field, E_s (MV/m)	313	340	445	572	303†
Corresponding traveling-wave accelerating field‡	161	175	229	295	133†

* For critical coupling in the case of standing-wave structures.

† Preliminary results, limited by available RF power and not by breakdown.

‡ Assuming SLAC structure, working in the traveling-wave mode, in which $E_s/\overline{E}_{acc} = 1.94$, except for X-band disk-loaded TW structure which was built with $E_s/\overline{E}_{acc} = 2.28$.

of the disk, and a marked increase in CH_4, CO and CO_2 gas at the RGA. As the gas burst was pumped away and RF processing was continued, the FE current gradually subsided to its pre-breakdown level and could eventually be decreased to a lower level for a given RF power input. Further RF breakdown events could then be triggered by increasing the RF power until the quasi-asymptotic levels shown in Table 1 were reached.

All SW structures listed in Table 1 were RF-processed for periods of 3 to 20 hours. Afterwards, they all showed considerable damage in the form of numerous pits and craters, mostly around the high-field points on the disk edges and/or nose cones. It is interesting to note, however, that this damage did not prevent the structures from continuing to operate at the highest electric field levels.

Our conjecture of what happens during RF processing can be understood by considering the expression for the Fowler-Nordheim average FE current density in an RF cavity:

$$\bar{j}_{FE} = \frac{K}{\phi^{1.75}} (\beta E_s)^{2.5} \exp\left(-\frac{6.53 \times 10^9 \times \phi^{1.5}}{\beta E_s}\right)$$

where ϕ is the metal work function in eV and β, the surface field enhancement factor, is a composite equal to $\beta_1\beta_2$: here β_1 is due to surface microprotrusions and β_2 is due to dielectric surface impurities. When RF processing starts, β_1 is relatively low (~2) except perhaps for a few whiskers which burn up early in the process, and β_2 is relatively high (~60-120) because of the thickness of the dielectric layer. As RF processing proceeds, the dielectric surface impurities are gradually desorbed and β_2 decreases; however, as a breakdown "event" takes place through explosive electron emission[1] (EEE), a crater is formed and β_1 grows. When RF processing reaches the asymptotic state (in our case, $\beta_1\beta_2$ ~60 at S-band), β_2 may be reduced to ~10 but β_1 is raised to ~6 because of the craters. The idea that comes to mind is that it may be possible to prevent β_1 from growing by processing the structure more gently, possibly with argon. There is evidence already that this procedure can at least reduce the "clean-up" time. If, in addition, it could reduce the number of craters, then the ultimate β_1 could perhaps be reduced by a factor of two. By operating at gradients at least 10 to 20% below breakdown, the so-called "dark current" due to captured FE current could be reduced considerably. Furthermore, to get the "dark current" per unit length, the above expression for the current density must be multiplied by the area available for FE per disk (which varies as f^{-2}) and the number of disks (which varies as f), yielding a net scaling as f^{-1}, thus favoring the higher frequencies. It is hoped that the experiments at X-band with a traveling-wave section will soon be performed, thus confirming some of these conjectures and predictions.

References

1) Loew, G. A., and Wang, J. W., "RF Breakdown Studies in Room Temperature Electron Linac Structures," SLAC–PUB–4647, May 1988 .

DAMPED ACCELERATOR CAVITIES*)

R. B. Palmer

Stanford Linear Accelerator Center, Stanford University, Stanford, California 94309
and
Brookhaven National Laboratory, Upton, New York 11973

ABSTRACT

Structures with slots to strongly damp higher-order longitudinal and transverse modes should allow the use of multiple bunches in linear colliders, and thus attain luminosities of over 10^{34} cm^{-2} sec^{-1}. Preliminary measurements on model structures suggest that such damping can be achieved.

1. INTRODUCTION

The luminosity of a linear collider[1] is severely limited by the total beam power and thus, since wall power is limited, by the efficiency with which RF power is transferred to the beam. There are, of course, many other applications that would benefit from high efficiency, so this is also a general problem.

The primary bound on the loading (and thus on the efficiency) comes from the fields—known as wakefields[2]—generated by the bunches, themselves, in the structure. Acting back on the bunches, these fields cause unacceptable distortion, momentum spread or deflection. Those that cause momentum variations are referred to as longitudinal wakes and generate momentum variations. Those that cause deflections are referred to as transverse wakes. In this case, they are only generated if the bunch is itself off-center, but cause subsequent bunches to be deflected further off-center, thus leading to the possibility of an instability, known as beam breakup. If the beam is displaced, quadrupole and higher wakes will also be excited, but these will have little effect unless the displacement is large and the bunch broad.

In a linac loaded with many bunches, each bunch sees fields generated by itself and by a sum of all previous bunches. The effects from the bunch itself, at least for short bunches, are dominated by the iris hole diameter. Besides using the largest possible iris, there is nothing much one can do about the fields themselves, although BNS damping[3] can be used to reduce their effects. The maximum fraction of RF energy extracted by one bunch is then bounded to a few percent, the exact amount depending on detailed assumptions. If more efficiency is required, it is imperative to use multiple bunches.

When there are many bunches, the fields seen by a particular bunch are a sum of those from all previous bunches. These wakefields have many different frequencies that are not in general multiples of the bunch spacing and will not add coherently. Thus, for a large number of very small bunches, the contributions of the wakefields are relatively less and higher efficiencies can be obtained.

For colliders, unfortunately, we need both relatively high individual bunch loading plus multiple bunches for high efficiency, and—in this case—both momentum spread and beam breakup become unacceptable unless something is done. In this paper, we study the possibility of damping the wakefields generated by each bunch before later bunches are introduced. Clearly, this damping must be very rapid; i.e., very low Q's are required for all unwanted wakefields.

2. RF MODES

For these studies, I have considered a traveling wave structure similar to that in the SLAC accelerator, except that I have chosen a relatively larger iris to wavelength ratio (.2 instead of .1) and—as a result—a larger group velocity (.07 c instead of .007 c). This choice was made[4,5] to reduce the single bunch wakefields. A multicell model of this structure was constructed at one-half SLAC scale. Figure 1 shows this structure. Figure 2 shows the measured mode frequencies as a function of phase advance.

2.1 Wakefield Requirements

Current parameter lists of TeV linear colliders are limited to luminosities of the order of 10^{33}cm^{-2} sec^{-1} for a wall power consumption of the order of 100 MW. Such designs employ only single bunches and use only from one[4] to a few[6] percent of the stored RF energy in the cavity. If multiple bunches can be employed, then as much as 25% of the energy can be extracted.[7] However, it has been shown[8] that with normal cavities and the required loading, beam breakup is unacceptable with more than one or two bunches. In addition, higher modes of longitudinal wakes will cause unacceptable momentum variations.[9]

*) Work supported by the Department of Energy, contracts DE–AC03–76SF00515 (SLAC) and DE–AC02–76C0016 (BNL).

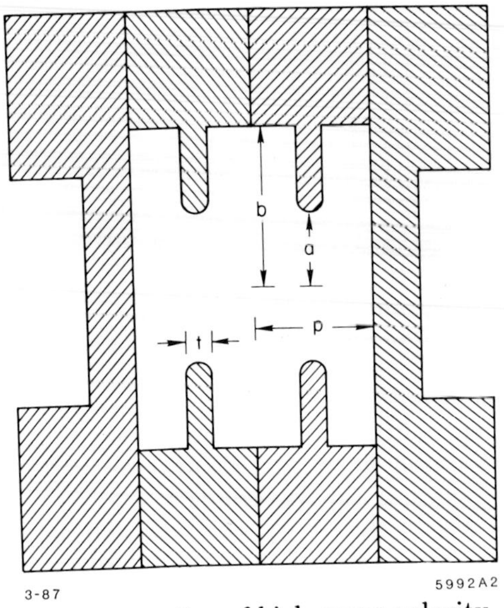

Fig. 1. Cross section of high group velocity accelerating structure.

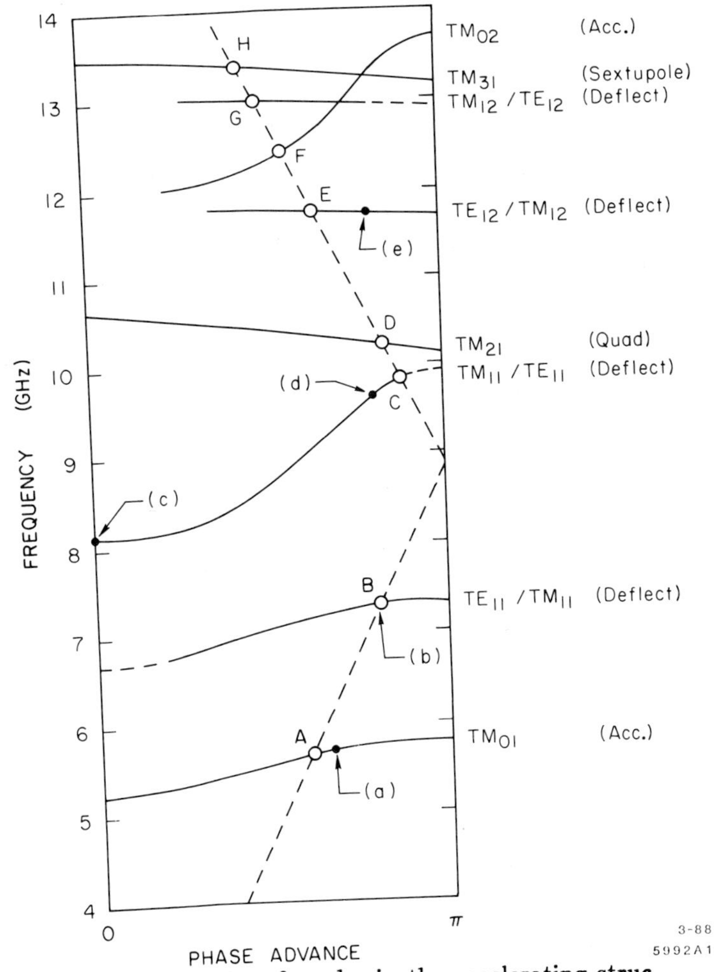

Fig. 2. Frequencies of modes in the accelerating structure as a function of the phase advance per cell.

If, however, the transverse modes can be strongly damped, then the beam breakup will be controlled. If the higher longitudinal modes can also be damped, then the momentum variations will also be controlled. It should then be possible to extract of the order of 25% of the energy and obtain working luminosities of over 10^{34} cm^{-2} sec^{-1} with the order of 200 MW.

In principle, all transverse modes and longitudinal modes other than the first—i.e., all higher modes—should be damped to a negligible value between bunches. This requirement can be eased if we arrange that the bunch spacing is equal to an integer times a half wavelength of the most serious modes. If the bunch spacing is n cycles of the fundamental, f_0 is the frequency of the fundamental and f_i is that of the mode; then this requires that $f_i/f_0 = n/m$, where m and n are integers. With the cavity being considered, $f_i/f_0 \approx 1.3$ so, with some slight adjustment of dimensions, $f_i/f_0 = n/m$ will be satisfied for the first deflecting mode (with $n = 4$ and $m = 3$); i.e., with one bunch every three cycles.

A study of the transverse wake problem[10)] has shown that even with this condition, the wakefields for many bunches are still unacceptable. However, the study has shown that if the modes are damped, the effects can be controlled, provided condition two is satisfied with some reasonable tolerance.

The longitudinal wake problem is less severe than that in the transverse direction, because no instability is involved. The wakes from each bunch simply add up and, since they have random phases, they add incoherently. Nevertheless, with the parameters being considered, they add to give an unacceptable momentum spread between bunches. A study of the problem[9)] shows that the problem is controlled if all higher modes are damped with Q's less than about 80, and this will be taken as a requirement in this study.

3. STRUCTURES WITH HIGHER MODE DAMPING

Higher modes could, in principle, be damped by the introduction of resistive materials into the structure, but it would be hard in this way to damp one and not another mode. Instead, we have studied damping by coupling the unwanted modes into a waveguide and damping the fields externally. Such coupling is accomplished by cutting into the basic cavity with slots leading to waveguides that can be chosen to have a cutoff higher than the fundamental. Thus, such slots will not damp this accelerating mode, but will allow all unwanted higher modes to propagate out of the structure.

In placing the cuts to couple fields to a mode in the waveguide, we can place the slot so as to interrupt currents that would normally be flowing in the cavity wall, thereby inducing such currents to enter the waveguide and induce a traveling wave in the guide. Figure 3(a) shows these currents for the fundamental accelerating mode, and Fig. 3(b) for the first transverse mode. The dotted lines indicate two possible cuts that would interrupt the transverse mode currents. The circumferential cut is seen also to interrupt the currents of the accelerating mode, but the "radial cut" does not interrupt such currents, and would therefore not damp the accelerating mode. Because of their relative simplicity, we first considered circumferential cuts, but these were unable to provide sufficient damping without adversely affecting the accelerating mode, so then—because of their lack of effort on the acceleration—we considered radial cuts.

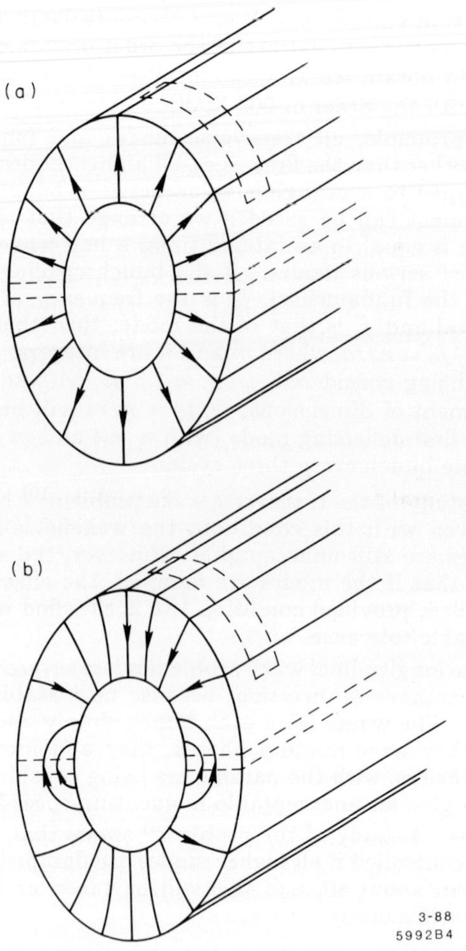

Fig. 3. Sketch showing approximate directions of current flow in the iris and wall of a structure: a) for the first accelerating mode and b) for the first transverse mode.

In all cases, the measurements were made on a single aluminum cell and only π modes were examined. Future studies will be reported using multiple cavities. Also in all cases, the fields in the waveguides were damped by inserting a cone of resistive plastic foam into the exits of the waveguides. When this was not done, the Q's of damped modes were, in general, a factor of two higher. This was because of the approximately 50% reflection of power at an open-ended waveguide.

3.1 Longitudinal Slot

Since the circumferential slots described above were unable to obtain sufficient damping of the transverse mode, we also tried a longitudinal slot. Initially, a slot was tried without a cut through the iris but, as Fig. 3 suggests, this was not effective for the first transverse mode (although it did damp some of the higher modes).

The longitudinal slot with cut iris is shown in Fig. 4. The length of the slot was chosen to have a cutoff about 7% above the accelerating mode and was 15% below the first transverse mode. Its length was then 2.4 cm and thus longer than the individual cell length (of 1.75 cm).

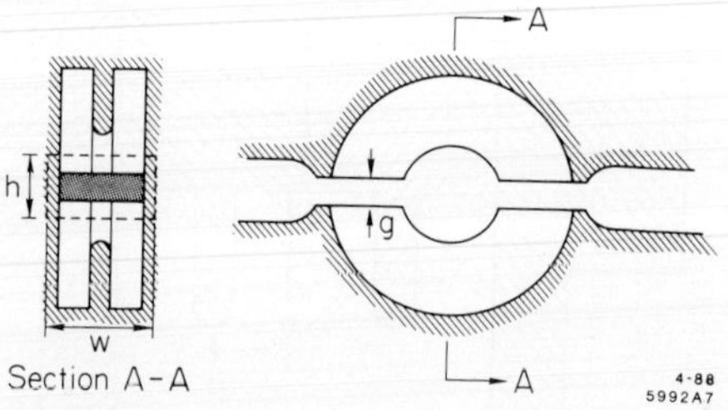

Fig. 4. Diagram of test cell with slot cut longitudinally in the cell wall and extending through the iris.

There is nothing obviously wrong with this, but it does mean that multiple cells cannot all have such slots at the same azimuthal angle (see the next section).

The width of the waveguide was 1 cm, but just prior to its entry to the cavity, this was reduced to a slot of only .5 cm—this being done to minimize the effect on the accelerating mode. The cut in the iris was only .27 cm. The object here was again to minimize the effect on the accelerating mode and, in particular, to minimize the increase in peak electric fields where the iris was cut.

We initially observed:

(*a*) a negligible effect on the accelerating mode, in frequency or Q;

(*b*) a negligible effect on the first higher longitudinal mode;

(*c*) the apparent disappearance of the first transverse mode.

A study was then undertaken to observe the first transverse mode as the gap width was gradually varied from closed to its full width of 2.7 mm. This revealed that as the gap was opened, the transverse mode both widens and moves, and that its direction of motion is towards the undamped zero phase advance transverse mode at 8.1 GHz. Estimations of the final Q were then made both by observing the width and by tracking the amplitude of the resonance—extrapolating, in the case of the width observation, to the fully open case. Unfortunately, a fully consistent picture did not emerge. Both methods were complicated by the presence of the 8.1 GHz signal. We could only conclude that the Q of the mode was in the range between 10 and 20, with the width suggesting the higher value and the amplitude the lower. In either case, the requirement of a Q less than or equal to 20 seems to have been achieved. Further measurements by other methods are needed.

3.2 Combined Slot for Transverse/Longitudinal Modes

A basically longitudinal slot was tried whose length and cut iris was as described in Sec. 3.2, but the width of the slot was opened up to 1.4 cm (see Fig. 5). This width was chosen to have a cutoff in the lateral direction of 10.7 GHz, which is below the first longitudinal mode to be damped.

As expected, the first transverse mode was again damped to a Q of between 10 and 20. The first longitudinal mode was observed to have a Q of 80. Note that in a linac, one will wish to damp both horizontal and vertical transverse modes, and that there will thus have to be

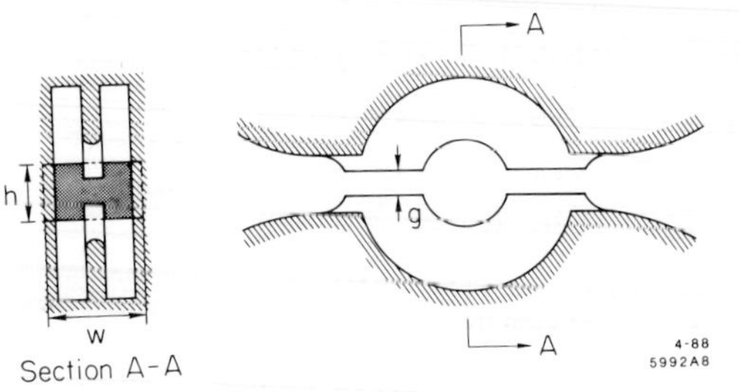

Fig. 5. Diagram of test cell with rectangular waveguide entering the cell wall and with a cut extending through the iris.

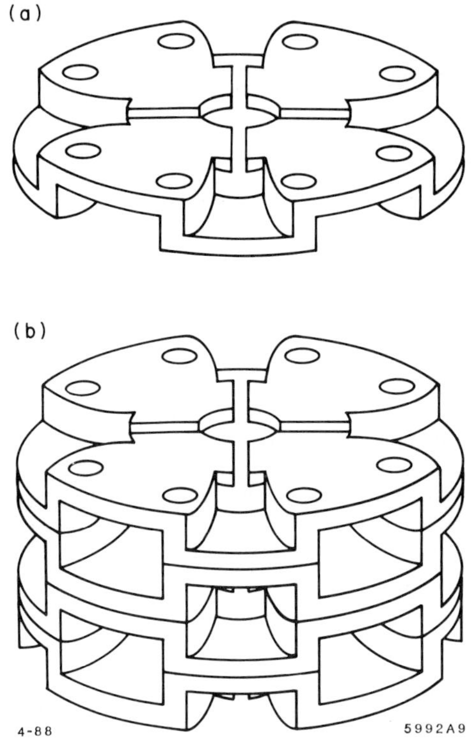

Fig. 6. Conceptual design of damped structure with a) a single cell and b) a stack of four such cells. Each cell is rotated 45° with respect to its neighbors.

twice as many slots as in these model tests. As a result, the longitudinal damping will be twice as strong; i.e., a Q of 40 would be expected where we require 80, which is certainly okay.

The Q of the acceleration mode was lowered by 20% which, with double the number of slots, would become 40% but because of the bolted nature of the test cavity, this is presumably an overestimate. A preliminary study with the three-dimensional computer code "MAFIA" indicated a loss of only 2%.

We appear to have a solution that has required damping of both the first transverse and longitudinal modes. A brief look for higher modes suggested that these will also meet our requirements, although a more careful study is needed.

4. A PRACTICAL REALIZATION

The design of a multicell linac employing such damped accelerating cavities is not simple. Four slots per cell are required, and succeeding slots cannot be at the same azimuthal angles. Each slot should open out before exiting the structure, so as to minimize reflections at their outer ends. The entire structure would then be supported inside a larger vacuum chamber, which may have to contain RF damping resistors or at least be made of a relatively high-resistance material. Since the structure is inside the vacuum container, water cooling would have to be integral to the structure. It should be noted that despite its complication, such a design will have the advantage of providing very good vacuum pumping to the inside of the cavities; something that is probably essential, and something that would be impossible to obtain without slots or holes of some kind. Figure 6 illustrates what a one cell (a) and a multicell structure (b) might look like, but it should be clear from this preliminary paper that much work remains to be done.

REFERENCES

1) Wilson, P. B., "Linear Accelerators for TeV Colliders," *Proceedings of the 2nd Workshop on Laser Acceleration,* Malibu, CA (AIP Conf. Proc. No. 130, 1985); SLAC–PUB–3674 (1985).
2) Bane, K. L. F., Wilson, P. B. and Weiland, T., "Wakefields and Wakefield Acceleration," in *Physics of High Energy Particle Accelerators,* Month, M., Dahl, P. F. and Dienes, M. eds., (AIP Conf. Proc. No. 127, New York, 1985), pp. 875–928.
3) Balakin, V., Novokhatskii, A. and Smirnov, V., *Proceedings of the 12th International Conference on High Energy Accelerators,* Fermilab (1983).
4) Palmer, R. B., "The Interdependence of Parameters for TeV Linear Colliders," *Proceedings of the Workshop on New Developments in Particle Accelerator Techniques,* Orsay, France, June 1987, CERN 87–11, p. 80; SLAC–PUB–4295.
5) Farkas, Z. D. and Wilson, B., "Comparison of High Group Velocity Accelerating Structures," SLAC–PUB–4088 (1987).
6) Schnell, W., CERN–LEP–RF/86–27 (1986).
7) Ruth, R. D., SLAC–PUB–4541 (1988).
8) Yokoya, K., DESY 86–084 (August 1976).
9) Palmer, R. B., to be published.
10) Thompson, K. A. and Ruth, R. D., "Controlling Multibunch Beam Breakup in TeV Colliders," SLAC–PUB–4537 (1988).

RELATIVISTIC KLYSTRON RESEARCH FOR LINEAR COLLIDERS†

M. A. ALLEN, R. S. CALLIN, H. DERUYTER, K. R. EPPLEY, K. S. FANT, W. R. FOWKES,
W. B. HERRMANNSFELDT, T. HIGO,* H. A. HOAG, R. F. KOONTZ, T. L. LAVINE,
T. G. LEE, G. A. LOEW, R. H. MILLER, P. L. MORTON, R. B. PALMER,
J. M. PATERSON, R. D. RUTH, H. D. SCHWARZ, Y. TAKEUCHI,*
A. E. VLIEKS, J. W. WANG, AND P. B. WILSON

Stanford Linear Accelerator Center, Stanford University, Stanford, California 94309

D. B. HOPKINS AND A. M. SESSLER

University of California, Lawrence Berkeley Laboratory, Berkeley, California 94720

W. A. BARLETTA, D. L. BIRX, J. K. BOYD, T. HOUCK, D. S. PRONO,
R. D. RYNE, G. A. WESTENSKOW, AND S. S. YU

University of California, Lawrence Livermore National Laboratory, Livermore, California 94550

ABSTRACT

Relativistic klystrons are being developed as a power source for high gradient accelerator applications which include large linear electron-positron colliders, compact accelerators, and FEL sources. We have attained 200 MW peak power at 11.4 GHz from a relativistic klystron, and 140 MV/m longitudinal gradient in a short 11.4 GHz accelerator section. We report here on the design of our relativistic klystrons, the results of our experiments so far, and some of our plans for the near future.

1. INTRODUCTION

Large linear electron-positron colliders, compact accelerators, and FEL sources require a new generation of high gradient accelerators. Conceptual designs for large linear electron colliders for research at the frontier of particle physics call for beam energies of 250–1000 GeV and luminosities of 10^{33}–10^{34} cm^{-2}sec^{-1}. Accelerating gradients of 100–200 MV/m are desired in order to keep the accelerator length within acceptable limits. Frequencies of 10–30 GHz are desired in order to keep peak and average power requirements and beam loading reasonably small. The peak power necessary to drive a traveling wave structure in the desired frequency range with the desired gradient is of order 1 GW/m with a pulse length of 50–100 ns.

Pulsed beams of such high peak power can be obtained using the technologies of magnetic pulse compression and induction acceleration.[1] Beam pulses of 1 kA current and 50–100 nsec duration are routinely accelerated to several MeV at Lawrence Livermore National Laboratory (LLNL). These beams contain several gigawatts of peak power.

† Work supported by the U.S. Department of Energy under contracts DE-AC03-76SF00515 (SLAC), DE-AC03-76SF00098 (LBL), and W-7405-ENG-48 (LLNL), and by the U.S.-Japan Collaboration on High Energy Physics.

* Visitor from KEK, National Laboratory for High Energy Physics, Oho, Tskuba, Ibaraki 305, Japan.

The first demonstration of RF power extraction from such a beam yielded an impressive 1 GW at 35 GHz, using a free electron laser.[2] A. M. Sessler and S. S. Yu, following a suggestion by W. K. H. Panofsky, proposed a more direct method for energy extraction by bunching a relativistic beam and passing it through extraction cavities.[3] Sessler and Yu suggested that if only part of the beam energy were extracted, the beam could be reaccelerated and energy could be extracted again. Repeated reacceleration and extraction was the concept they called a "relativistic klystron two-beam accelerator."[3] The idea of a relativistic klystron, however, is not limited to the two-beam accelerator concept. Relativistic klystrons can be imagined which span the range from a 1 GW device powering 1 m of accelerator, to a 10 GW device powering 10 m, to a two-beam device extending several kilometers. These ideas have led to a collaboration between Stanford Linear Accelerator Center (SLAC), Lawrence Berkeley Laboratory (LBL), and LLNL to study the combination of the klystron concept with induction accelerator and magnetic pulse compression technology. The first experiments have been done at the Accelerator Research Center (ARC) at LLNL using as a gun an induction accelerator designed to produce 1 kA currents with 1.2 MeV kinetic energy for up to 75 nsec duration. Four klystrons have been tested with this injector. They are, in chronological order as tested,

(1) SL3, a multicavity klystron with a conventional gun designed to operate at 8.6 GHz (three times the frequency of the SLAC linac),

(2) SHARK, a two cavity subharmonic drive relativistic klystron with 5.7 GHz drive and 11.4 GHz output,

(3) SL4, a high gain relativistic klystron at 11.4 GHz (four times SLAC frequency) designed specifically for the high power pulsed beam,

(4) SHARK-2, a three-cavity version of SHARK.

In this paper we discuss the design of these klystrons, report on the results of our experiments so far, and discuss some of our plans for the near future.

2. KLYSTRON SCALING

To motivate the increase in energy of the beam in an otherwise conventional klystron, it is useful to discuss the physics of the klystron interaction. In a two cavity klystron, the beam is velocity modulated by an RF drive cavity and allowed to drift until the velocity modulation bunches the beam. The bunched beam then is passed through another cavity which may be used to extract RF power. In practice, such a two cavity device has low gain. In most klystrons, there are several intermediate "idler" cavities. The first cavity bunches the beam. The bunched beam drives the second cavity to an RF voltage an order of magnitude greater than the first, which in turn bunches the beam more strongly. This process continues until the final idler cavity of the "linear gain region" of the klystron. The bunching is determined primarily by the voltage on the final idler cavity. After this cavity the bunches are allowed to drift until the RF current is maximum. At this point the beam is passed through two more cavities: a highly detuned "penultimate" cavity which sweeps still unbunched electrons into the bunch, and an output cavity which extracts energy by decelerating the beam. The output cavity could be replaced by a series of cavities or by a traveling wave structure.

An important parameter in klystron scaling is the beam plasma wavelength. Velocity modulation bunches a DC beam. However, space charge repulsion (modified by the drift tube) causes the beam to debunch. In the linear region, this process produces oscillations. The distances between cavities in a klystron are chosen to be approximately one-quarter of a plasma wavelength for optimal bunching. For a long relativistic beam of current I and radius a in a narrow tube of radius b the plasma wavelength on axis is

$$\lambda_p \simeq \lambda_{RF}\sqrt{\frac{17\,\mathrm{kA}}{I}\frac{(\beta\gamma)^5}{1+2\ln(b/a)}}$$

where λ_{RF} is the free-space RF wavelength, $\beta = v/c$, and $\gamma = 1/\sqrt{1-\beta^2}$. Increasing the beam energy ameliorates longitudinal space charge effects but increases the bunching distance. Increasing the frequency reduces the bunching distance. Our choice of 2.6 cm RF wavelength makes possible a multicavity klystron design that can bunch a 1 MV, 1 kA beam efficiently and extract power from it in a total distance of 1 m. For higher energy beams, bending magnets can be used to create differences in path length for particles of different energies. This permits bunching of higher energy beams even though their velocity is nearly independent of energy.

Another important parameter in klystron scaling is the magnetic field necessary to focus the beam. For a beam of uniform charge density and normalized edge emittance ϵ_n, the solenoid field B necessary to confine the beam current I to radius a is

$$B = \frac{2m_ec^2}{ea}\sqrt{\frac{2I}{17\,\mathrm{kA}}\frac{1}{\beta\gamma} + \frac{{\epsilon_n}^2}{a^2}}.$$

The constant, $2m_ec^2/e$, is 3.4 kG-cm.

In the relativistic klystrons discussed here both γ and I are greater than in conventional klystrons. At shorter wavelengths higher magnetic fields are needed to focus the beam since the radius of the drift tube scales with the wavelength. An estimate of the required field must include the effects of beam bunching. The peak current in the bunched beam typically is about four times the initial DC current. Thus, for a space charge dominated beam, the magnetic field required is typically twice that calculated for focusing a DC beam.

3. EXPERIMENTAL APPARATUS

3.1 Induction Accelerator

Most of the experimental studies described here were performed using the Snowtron injector at the ARC facility at LLNL. Snowtron is a linear induction injector composed of ten 150 kV induction cells driven by pulsed magnetics.[1] For klystron experiments, a triode electrode configuration was used with a cathode of 12.5 cm diameter and 35.6 cm spherical radius. The inner diameter of the anode drift tube was 8.8 cm. The cathode was placed 35 cm from the downstream end of the injector. Accelerating voltages up to 1.2 MV, beam currents up to 1.4 kA, and pulse widths up to 75 nsec FWHM have been obtained for the klystron experiments. The greatest stress on the injector is 260 kV/cm on the cathode shroud at peak voltage. The DPC computer code, which was used to design Snowtron, predicts peak currents of 2.3 kA at 1.2 MV.[4] However, the operating pressure of the injector led to cathode contamination which precluded uniform space charge limited emission.

3.2 Beam Transport

The distance from cathode to klystron was 4 m for the SL3 test and is 1.6 m for the SHARK and SL4 tests. Just downstream from the injector is a 30 cm taper where the beam pipe narrows from 8.8 to 1.9 cm diameter. The pipe diameter is narrowed further to 9.2 mm in the SHARK and SL4 klystrons. Nine 2.5 kG solenoid coils powered by five separate power supplies focus the beam between the cathode and the klystron. Three independently controlled 5 kG solenoids focus the beam in the relativistic klystron. Four sets of dipole magnets for horizontal and vertical steering are used to correct for beamline misalignments.

Beam transport calculations with the ST code have been used to estimate the required strengths of the focusing fields for 100% transmission of current through the klystron.[4] The result of such a calculation is shown in Fig. 1.

3.3 Klystrons

Parameters of the three relativistic klystrons tested are summarized in Table 1. Further descriptions are given below.

SL3 is a conventional high gain klystron designed to operate at 8.6 GHz with a conventional gun. With its design gun replaced by an induction accelerator, it served as an expedient first demonstration of a relativistic klystron. SL3 was driven by a 1 kW X-band TWT amplifier.

SHARK is a two cavity sub-harmonic drive relativistic klystron. The input cavity is driven by an RF source of several MW at 5.7 GHz which modulates the beam velocity. After drifting, the beam current has large Fourier

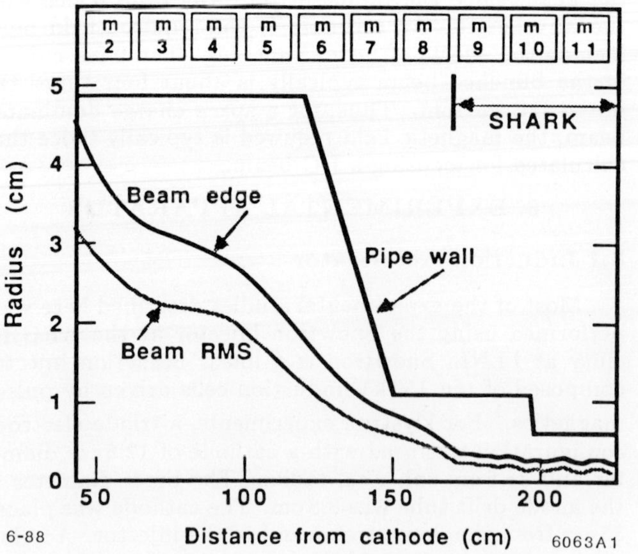

Fig. 1. Beam size calculated through SHARK.

Table 1. Parameters of relativistic klystrons tested.

	SL3	SHARK	SL4	SHARK-2
Output freq. (GHz)	8.57	11.4	11.4	11.4
Drive freq. (GHz)	8.57	5.7	11.4	5.7
Output power (MW)				
Peak (max)	75	47	200	117
Flat pulse (max)	75	47	68	80
Design gain (dB)	54	20	65	25
Efficiency (%)				
Design	60	20	40	40
Operation (max)	55	25	50	32
Beam Voltage (kV)				
Design	330	1200	1200	1200
Operation (max)	1000	1200	1000	1200
Beam Current (A)				
Design	300	1000	1000	1000
Operation (max)	350	750	750	800
Number of cavities	5	2	6	3
Total length (cm)	31	25	98	33
Beam-off loaded Q				
Input cavity	250	725	280	742
Idler cavities	4000	—	120	—
Penultimate cavity	4000	—	3800	3800
Output cavity	44	40	20	20
Drift tube diam (mm)	11	19, 9.2	14, 9.2	19, 9.2

components at 5.7, 11.4, and 17.2 GHz. Resonant cavities tuned to the higher harmonics can be used to extract power and measure breakdown fields at the higher frequencies. The 11.4 GHz output cavity is positioned after a 25 cm drift for maximum RF current at that harmonic. With only two high Q resonant structures in this klystron, problems with beam breakup instabilities are minimized. However, the gain of a two cavity tube is low. Therefore, in order to achieve beam-to-RF power conversion comparable to that in multicavity tubes, a conventional 5 MW, 5.7 GHz klystron is used to drive SHARK. The RF fields in the input and output cavities are comparable for 2 MW of drive and 50 MW of output because of the different Q's.

SHARK was designed to serve as a test bed for cavity design to be used in relativistic klystron research. Its design allows study of a wide range of beam parameters and minimizes difficulties with beam propagation. The drift pipe and output cavity are easy to replace, making it possible to use SHARK to study different output cavities at several frequencies.

SL4 is a high gain, six cavity, relativistic klystron at 11.4 GHz which was designed specifically for the 50 nsec pulsed 1.2 MV, 1 kA Snowtron beam. Therefore, unlike most klystrons, it was designed without an integral gun assembly. In order to make the RF filling time of the SL4 cavities much shorter than the 50 nsec beam pulses, three of the gain cavities are coupled by irises and waveguides to absorptive ceramic wedges. This results in loaded Q's of 120, and filling times of 2–3 nsec for these cavities.

To reduce the difficulty of maintaining a well-focused electron beam over a 1 m drift length, the SL4 drift tube was tapered. The drift tube diameter in the first four cavities is rather wide, 14 mm. The tube then is tapered to 9.2 mm just upstream of the penultimate cavity. Tapering permits the use of solenoid magnets with axial field of 2.7 kG for most of the length of the klystron. A 5 kG solenoid surrounds the region of the penultimate and output cavities.

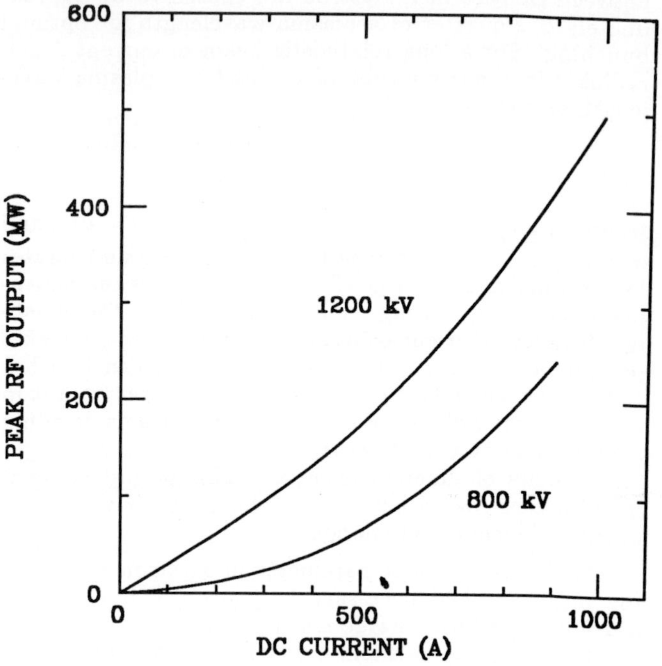

Fig. 2. MASK simulations of SL4 output power.

The design gain and efficiency for SL4, 65 dB and 40%, respectively, are obtained using the MASK computer code.[5] MASK simulations were used to optimize the SL4 design parameters and to predict the efficiency and gain at several different beam currents and voltages.

Some simulation results are shown in Fig. 2. The saturation RF drive power is approximately 200 W, which is supplied by a 1 kW X-band TWT amplifier.

Because of the high peak electric fields in the penultimate and output cavities good vacuum is necessary to prevent cavity breakdown. Consequently, a 500 liter/sec cryopump evacuates the klystron collector section and two 8 liter/sec vac-ion pumps evacuate the output waveguide. In this configuration waveguide and collector pressures can be maintained at 10^{-8} Torr.

SHARK-2 is a modification of SHARK in which a third cavity resonant at 11.8 GHz was inserted between the input and output cavities to increase klystron gain and efficiency. In addition, the outer cylindrical wall of the cavity was fabricated from low carbon steel to shield the cavity interior from the axial magnetic field for reasons discussed below in Section 4.3.

3.4 Diagnostics

The pulsed DC beam current is monitored in three places: at the injector, upstream from the input cavity, and downstream from the output cavity. The DC current monitors measure image currents in the beam pipe wall. An RF current monitor is placed downstream from the SHARK output cavity. The RF current diagnostic for SHARK was a pickup loop, recessed azimuthally in the beam pipe wall, which measured $\dot{B}_\phi$. The RF current diagnostic for SHARK-2 was a pair of probes, recessed radially in the beam pipe wall, which measured $\dot{E}_r$.

Forward and reflected RF drive power signals are sampled using 20 dB broad band waveguide directional couplers. Relativistic klystron output power and, in the SL4 experiment, the RF reflected back from the high gradient accelerator test section, were sampled using 56 dB waveguide directional couplers. The sampled RF signals are transported on individually calibrated, 25 m long, high quality Heliax cables from the couplers to the control room where they are measured with calibrated HP 8470B crystal diode detectors. Calorimetric measurement methods so far have been precluded by the low pulse repetition rate (less than 10 pulses/sec) necessitated by inadequate shielding.

3.5 High Gradient Accelerator

To complement the SL4 experiment, a 26 cm long section of 11.4 GHz accelerator structure operating in the $2\pi/3$ traveling wave mode has been built. The constant impedance structure consists of 30 cells and has $r/Q = 14\,\mathrm{k\Omega/m}$. The attenuation parameter is 0.14 nepers. The group velocity is $0.031c$, giving a filling time of 28 nsec. The iris diameter was chosen to be 7.5 mm. Parameters were calculated by the SUPERFISH code and confirmed by cold test measurements. Coupler dimensions were approximated by extrapolation from S-band data, and finalized by cold test. The accelerator was fabricated from machined "cups" which were stacked and brazed. A special tool permitted each cavity to be tuned up or down in frequency in order to obtain the correct phase advance per cell.

4. EXPERIMENTAL RESULTS

4.1 Beam Transmission

The design goal of 100% beam transmission through the klystrons has not been achieved experimentally, even after focusing adjustments were performed by empirical optimization using a diagnostic such as transmitted current or output RF. The maximum current transported through SHARK is 750 A, only 65% of the maximum current entering the klystron. Up to 80% transmission has been achieved at 400 A. Transmission achieved through SL4 (which is four times longer than SHARK) is 55% at 800 A, and is 65% at 500 A. Transmission is observed to be independent of RF drive for SHARK. However, for SL4 and SHARK-2, a slight decrease in transmission was noted at high RF output levels. (There was no downstream current monitor in the SL3 tests; transmission through SL3 was not measured.)

4.2 SL3 Demonstration

The SL3 test was designed to be a demonstration of the effects of putting a conventional klystron tube (stripped of its gun) in a high power pulsed beam. No unusual or unexpected phenomena were observed. No evidence of multipactor, breakdown, parasitic oscillations, nor other instabilities was observed. RF pulse risetimes were 5–10 nsec. RF pulses reproduced the shape of the beam current pulses quite well. SL3 performance at beam energies much greater than design is illustrated in Fig. 3. Peak power of 75 MW was attained with an 800 kV, approximately 250 A beam. Observed output power agreed well with the predictions of the MASK simulation code.

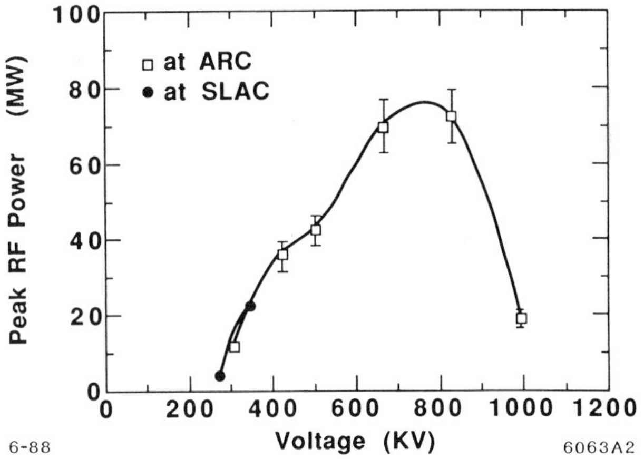

Fig. 3. SL3 performance.

4.3 SHARK and SL4

Peak output power of 200 MW at 11.4 GHz was attained with the SL4 relativistic klystron design using a 930 kV, 420 A beam. SL4 has not yet operated at its 1000 A design current. However, agreement is excellent between output power measured at lower currents and the MASK predictions (Fig. 2) for operation at these currents. The 200 MW peak power delivered by SL4 to the

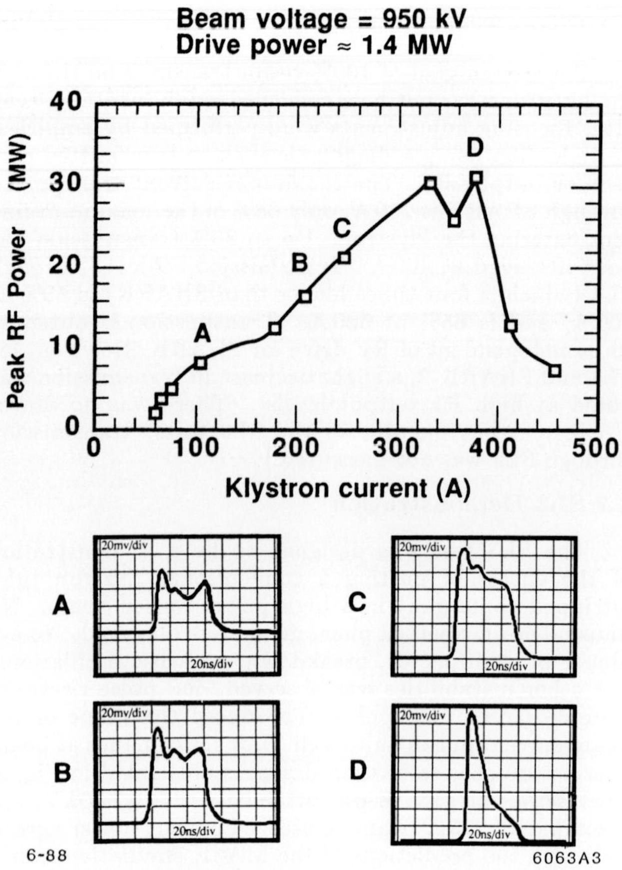

Fig. 4. RF pulse shortening observed in SHARK tests.

11.4 GHz accelerator corresponds to a longitudinal accelerating gradient of 140 MV/m. Early indications are that there is appreciable dark current in the accelerator when the accelerating gradient exceeds 90 MV/m.

In our tests of both SHARK and SL4, we observe that as the beam current through the klystron is increased up to a certain level, the output power pulses remain relatively flat. However, if the beam current is increased beyond this level, the trailing edges of the output power pulses diminish in amplitude, while the leading edges continue to grow with the beam current. This behavior in SHARK tests is illustrated in Fig. 4. We have demonstrated that our ability to obtain flat output power pulses is affected by beam current, RF drive level, and focusing magnetic field strength. The practical importance of these observations is that even though 200 MW of RF was produced with SL4, the maximum reasonably flat RF pulse achieved in our initial tests was only 68 MW. Low and high peak power SL4 pulses are illustrated in Fig. 5. The pulse shortening phenomenon is a serious impediment to making flat high power RF pulses. It is not beam breakup because the transmitted DC beam current pulse does not shorten with the RF pulse.

To understand the mechanism for output pulse shortening in SHARK, an experiment was performed in which simultaneous data on reflected power from the input cavity and on output power were recorded at a critical point for the onset of the shortening phenomenon. With no external changes in the beam condition and/or input power, the output alternated from pulse to pulse between the rectangular and triangular pulse shapes. When the beam turns on, there is a significant dip in the drive power reflected from the SHARK input cavity. Two distinctly different states have been observed in the reflected drive, as shown in Fig. 6, one having a much greater reflection during the beam-on time. Furthermore, the state with large reflection is correlated repeatedly with the narrow output pulse.

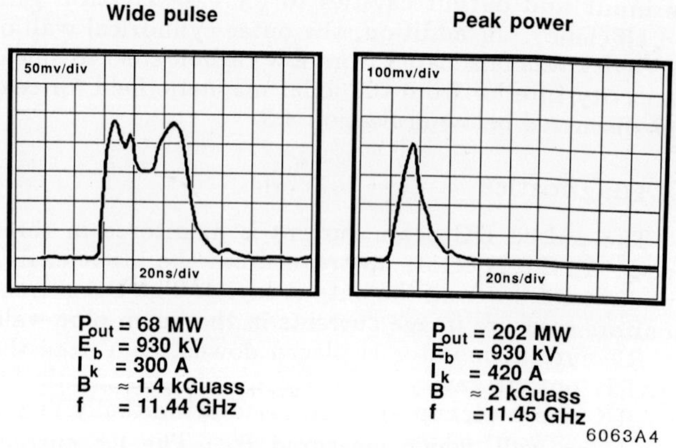

Fig. 5. Low and high peak power pulses in SL4 tests.

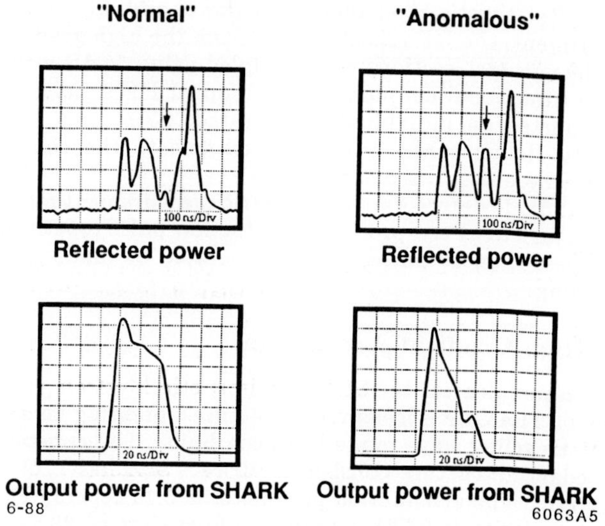

Fig. 6. SHARK reflected drive and output power. Arrows indicate the 50 nsec beam time in the reflected power.

This observation can be reproduced with our computer code for transient analysis. In the code, we use a circuit model to compute the time varying voltage across the input gap. The reflected power is calculated from the time-varying voltage by power balance. Results of the transient calculation are shown in Fig. 7 where the relatively flat output pulse with the low beam-on reflected power was obtained by using a beam loading generally consistent with MASK calculations and measurements. The narrow output pulse and increased reflection were obtained by arbitrarily increasing beam loading by a factor of 2.5.

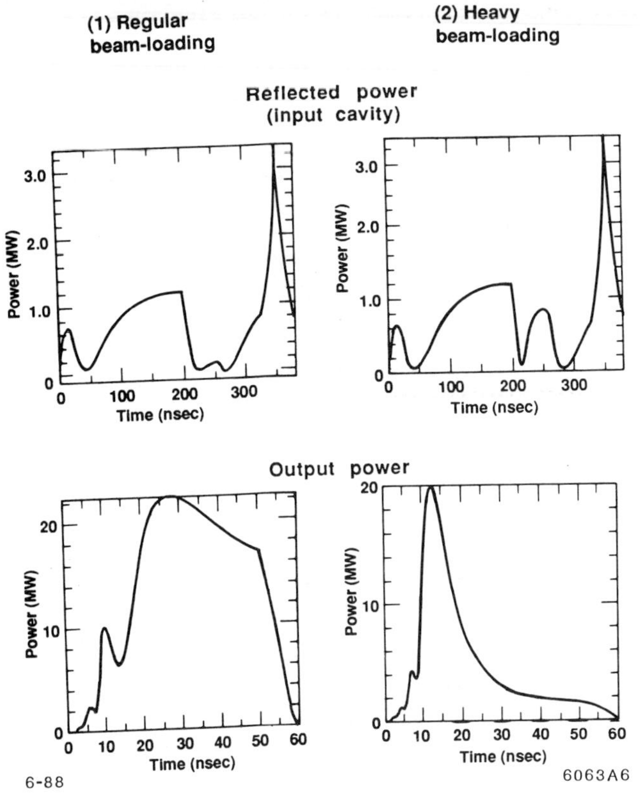

Fig. 7. Calculations of the effect of anomalous beam loading on SHARK reflected drive and output power.

The anomalous heavy beam loading which we believe causes pulse shortening appears to be a phenomenon similar to multipactor to the extent that it is sensitive to changes in focusing magnetic field and RF power level. The anomalous beam loading in fact may be multipactor.

Further studies of the SL4 relativistic klystron have demonstrated anomalous input cavity loading by charged particles when the RF drive level exceeds 40 W, even under cold cathode, beam-off conditions. Sometime after the onset of the RF drive pulse, the input cavity absorbs all of the incident RF drive that otherwise would be reflected from the beam-off, unloaded cavity. This power absorption is a function of drive level, and of axial magnetic field surrounding the cavity. No power absorption is present when the axial magnetic field is zero. This cavity loading phenomenon, which we have observed in the SL4 input cavity, both with and without beam, may occur in any of the cavities of a multicavity klystron and may be associated with pulse shortening.

The power threshold for RF pulse shortening was observed to increase with decreasing magnetic focusing field. Consequently, we tested a SHARK (5.7 GHz subharmonic) input cavity surrounded by iron which shunts the focusing field away from the beam axis in the region of the cavity gap. Figure 8 shows the effect of the iron on the axial magnetic field profile near the SHARK input cavity. The iron input cavity does not exhibit anomalous loading, with or without beam. Output pulse shortening continues to occur above a threshold in output power. However, the output pulse shortening in the configuration with the iron input cavity occurs at a higher power threshold than in the original SHARK tests with a copper input cavity, and the output pulse shortening apparently starts in the output cavity. The evidence that pulse shortening occurs in the output cavity is that no correlation is observed between the output pulse shortening and either the input cavity reflected power or the RF current as monitored by an RF probe in the drift downstream from the input cavity. Figure 9 shows multiple exposure photos of the output power pulses simultaneous with the input power reflected from the input cavity. The pulses in the figure were recorded during operation at the pulse shortening threshold. Both long and short output pulses are observed, while all reflected input pulses exhibit normal beam loading.

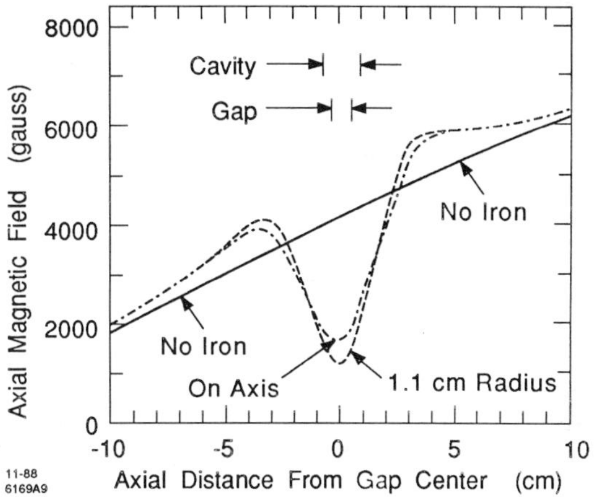

Fig. 8. Calculation of the axial magnetic field near the SHARK input cavity with and without iron. The cavity noses are at 1.1 cm radius.

4.4 SHARK-2

The observation that RF pulse shortening did not occur in the iron SHARK input cavity but did occur in the copper SHARK output cavity, as discussed in the previous section, prompted us to test a three cavity subharmonic drive klystron configuration consisting of the iron SHARK input cavity and copper penultimate and output cavities. (The penultimate and output cavities are the output section of the modularly designed SL4 relativistic klystron.) In this configuration we expect lower electric fields in the output cavity because the external Q of this output cavity is half that of the original SHARK output cavity. Results with this three cavity configuration, named SHARK-2, are summarized in Table 1.

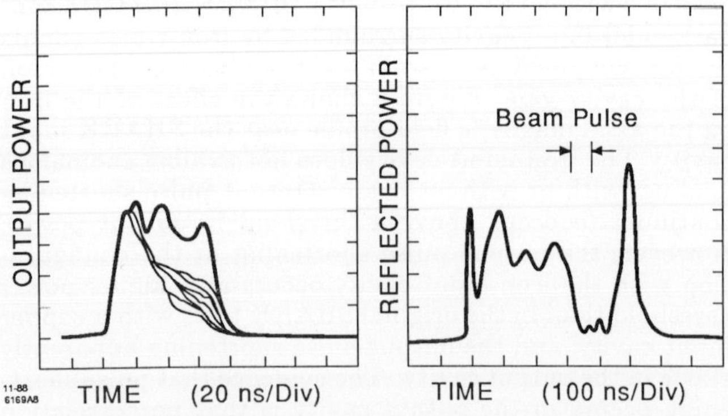

Fig. 9. Multiple exposure photos of long and short output power pulses simultaneous with "normal" reflected input power pulses in the SHARK with the magnetically shielded input cavity operating at the pulse shortening threshold.

Output pulse shortening is observed in the three cavity SHARK-2 at some point downstream from the iron input cavity. However, the power threshold is higher than in the two cavity iron input SHARK, presumably due to the lowered external Q of the output cavity. Flat output pulses of 80 MW amplitude and 45 nsec duration (FWHM) have been obtained from the three cavity subharmonic SHARK-2 before the onset of pulse shortening. This is 70% more flat-top RF power than we obtained from the two cavity SHARK.

Some improvement in the pulse shortening threshold has been observed with high power RF conditioning at 10 pulses/sec.

5. FUTURE PLANS

We are incorporating RF field probes in our designs of new intermediate cavities in order to ascertain where pulse shortening first occurs in multicavity relativistic klystrons, such as SL4.

In light of the apparent success of the iron magnetic shield at suppressing pulse shortening in the SHARK and SHARK-2 input cavities we plan to apply several techniques commonly used for multipactor suppression in order to raise the threshold for pulse shortening. We plan to add iron magnetic shields to downstream cavities. Consequently, we are studying the effect of the iron on the relativistic beam dynamics. We are fabricating and plan to test a cavity with slotted noses (as an alternative to iron). We plan to employ RF conditioning at higher pulse repetition rates with improved radiation shielding.

A traveling wave output structure which has been fabricated will be used in place of an output cavity in a relativistic klystron. This structure will have lower electric fields than a single output cavity and consequently the output power threshold for pulse shortening may be higher.

As soon as rectangular high power RF output pulses of sufficient duration are obtained, further studies of the high gradient accelerator structure described in Section 3.5 are planned. These tests will include studies of breakdown at 11.4 GHz and measurement of the accelerating gradient.

6. SUMMARY

We have been working to develop a high power (500 MW) short wavelength (2.6 cm) relativistic klystron with beam kinetic energy greater than 1 MeV. Four different klystrons have been tested. Peak RF power of 200 MW has been achieved, but only with an RF flat top much shorter than the beam current pulse. This pulse shortening phenomenon is by far the most serious problem encountered. It is clearly not beam breakup since it does not correlate with shortening of the DC current pulse. Experimental evidence from one of the klystrons (SHARK) indicates that pulse shortening is caused by loading of the input cavity by anomalous charged particle currents. The loading is believed to be due to either secondary electrons or to photoelectrons produced by the copious supply of x-rays caused by beam interception. The power threshold for pulse shortening is sensitive to magnetic focusing field and to RF field strength. Magnetically shielding a klystron input cavity raised the power threshold for pulse shortening. A second and perhaps related problem is rather poor beam transmission through the klystrons, which has not exceeded 65%. The shortened 200 MW peak RF pulses have been transmitted into a 26 cm long high gradient accelerator structure. This power corresponds to an accelerating gradient of 140 MV/m.

ACKNOWLEDGEMENTS

We thank R. J. Briggs, G. J. Caporaso, M. Chodorow, H. Gasquet, Y. Goren, J. Haimson, E. W. Hoyt, V. K. Neil, A. C. Paul, C. Pearson, D. S. Prono, L. L. Reginato, A. C. Smith, K. Whitham, D. Yu, and L. Zitelli for their assistance. The authors from SLAC thank LLNL for its continuing hospitality.

REFERENCES

1. L. L. Reginato and D. L. Birx, "Pulsed High Power Beams" and "Recent Advances in Magnetically Driven Induction Linacs" in Proceedings of the European Particle Accelerator Conference, Rome, Italy, June 7–11, 1988.
2. T. J. Orzechowski *et al.*, "High Efficiency Extraction of Microwave Radiation from a Tapered Wiggler Free Electron Laser," *Phys. Rev. Lett. 54*, 889 (1985).
3. A. M. Sessler and S. S. Yu, "Relativistic Klystron Two-Beam Accelerator," *Phys. Rev. Lett. 58*, 2439 (1987).
4. J. K. Boyd, "Snowtron Numerical Calculations," LLNL-RM-87-48 (November 1987).
5. K. Eppley, "Algorithms for the Self-Consistent Simulation of High Power Klystrons," SLAC-PUB-4622 (May 1988), to be published in the proceedings of the Linear Accelerator and Beam Optics Codes Workshop, San Diego, California, January 19–21, 1988.

RF PULSE COMPRESSION AND ALTERNATIVE RF SOURCES FOR LINEAR COLLIDERS*

P. B. Wilson

Stanford Linear Accelerator Center, Stanford University, Stanford, California 94309

ABSTRACT

Future linear colliders will require a very high peak power per meter of accelerating structure at a relatively high frequency—greater than 10 GHz—but at a relatively short pulse length—less than 100 ns. One technique for generating the required peak power is to use a more or less conventional microwave power source, which produces power at a pulse length typically on the order of 1 μs, together with RF pulse compression. Some parameters are given for a Binary Power Multiplier (BPM) pulse compression system operating at 17.1 GHz with an output pulse length of 60 ns. The peak power gain for a three stage system is estimated to be 6.6 (82% compression efficiency). Some possible long-pulse microwave sources which—when coupled with such a pulse compression system—would be suitable for driving a linear collider are briefly discussed.

1. INTRODUCTION

The peak power requirement for a recent TLC design[1] is 586 MW/m at a frequency of 17.1 GHz and a pulse length (equal to the structure filling time) of 60 ns. The relativistic klystron[2] is capable of producing sufficient peak power at this pulse length to power several meters of structure. More conventional microwave power sources operating at beam voltages of 500 kV or less could be expected to produce at most about 100 MW at this frequency. However, the natural pulse length for such sources, using conventional or near-conventional modulator technology, is on the order of 1–2μs. If a one microsecond pulse could be compressed to 62 ns by some technique, the gain factor in peak power would be $G = 16\ \eta_c$, where η_c is the compression efficiency. Thus, a 100 MW source coupled with pulse compression at 80% efficiency would produce 1.3 GW, which is comparable to the power to be expected from a relativistic klystron. In the next section we take a brief look at the capabilities and limitations of RF pulse compression.

2. PULSE COMPRESSION SYSTEMS

A number of pulse compression methods have been proposed and a few have been experimentally tested. The SLED scheme, for example, is in routine operation at SLAC. However, the peak power gain for SLED is limited to about a factor of three for a reasonable compression efficiency, and the system delivers a strongly time-varying output waveform. Energy storage cavities which are pumped up slowly by a relatively low power source with a long pulse length, and then dumped rapidly using a high power plasma switch, have also been proposed. However, the technological problems of a suitable switch have not been solved, and again the output waveform is highly nonuniform.

A pulse compression system with a potentially high peak-power gain, which gives a flat output waveform, is the Binary Power Multiplier (BPM) suggested by Z. D. Farkas[3]. The system uses a series of over-moded waveguide delay lines as low-loss energy storage elements, and 3 dB directional couplers—triggered by a pattern of phase reversals—as switching elements. It is important to note that in this pulse compression scheme two high power amplifiers and one set of delay lines are required for each BPM system. The phase reversal pattern is imposed at a low RF power level on the input drive to the amplifiers. The RF components used in a BPM system (3 dB directional couplers, transitions from rectangular to circular guide, 180° turn-arounds and right angle bends) must be carefully designed to avoid mode conversion. The theory of the BPM method is given in Ref. 3, and some initial experimental results on a prototype system (which did not, however, use all low-loss components) are reported in Ref. 4.

Let us look at a 17 GHz n–stage BPM system designed to compress an input pulse of $2^n \times 60$ ns to an output of 60 ns. Each stage compresses the pulse length by a factor of two and increases the peak power by a factor of $2\eta_n$, where η_n is the compression efficiency of the nth stage. To calculate the compression efficiency, we assume a loss of 0.18 dB (transmission efficiency $\eta = 0.96$) per stage for a coupler and turn-around. The delay-line loss at 17 GHz for standard 2.81 inch I.D. copper pipe is .06 dB/100 ns (transmission efficiency of 0.99/60 ns). The length of line per 60 ns is 17.1 m ($v_g/c = 0.95$). The delay line lengths and efficiencies are given in Table I. Two efficiency figures are given for each stage. The top figures give the component and total efficiencies for that particular stage. The bottom figures give the cumulative efficiencies for all stages up to and including the nth stage. Thus, a complete system of either one, two, three or four stages would have a net efficiency of 0.95, 0.89, 0.82 and 0.73, respectively.

*Work supported by the Department of Energy, contract DE-AC03-76SF00515.

Table I. BPM Pulse Compression System at 17.14 GHz with 60 ns output pulse length

Stage Number	0	1	2	3	4
Delay time (ns)	0	60	120	240	480
Line length (m)	0	17	34	68	137
Efficiencies:					
η_{coup}	—	0.96 0.96	0.96 0.92	0.96 0.89	0.96 0.85
η_{line}	—	0.99 0.99	0.98 0.97	0.96 0.93	0.92 0.86
η_n	1.00	0.95	0.94	0.92	0.88
η_{cum}	1.00	0.95	0.89	0.82	0.73
RF Source:					
T_k (ns)	60	120	240	480	960
$P_K/feed$ (MW)	938	493	263	143	80

Note that the coupler and delay line losses are equal in the third stage. The cumulative losses due to the couplers and delay lines are approximately equal in a four-stage system. The pulse length required from the RF source is given by T_k. The peak source power per feed, P_k, is given for Palmer's TLC design parameters:[1)] peak accelerating gradient, 186 MV/m; RF power per meter, 586 MW; distance between feeds, 1.60 m.

It is useful to note that a BPM system can be programmed to deliver more than one output pulse per RF source pulse. For example, suppose a three stage system is desired, which would nominally require a 480 ns source pulse. It may be more convenient and more efficient to design a modulator-source combination which delivers 3×480 ns $\approx 1.5\ \mu$s. The BPM phasing pattern is then simply repeated three times in succession, giving three 60 ns output pulses spaced 480 ns apart. The disadvantage is that the effective repetition rate of the collider is reduced by a factor of three; for example, from 360 Hz to 120 Hz in the TLC design in question[1)]. This, in turn, reduces the effectiveness of feedback in controlling emittance growth due to the high-frequency component of magnet jitter from ground motion. However, the 120 Hz rate may still be acceptable. This multipulse scheme also has implications for the damping rings, but the necessary design modifications are acceptable[5)] if the number of output pulses per klystron pulse is not too large.

Assembly of components for a low-loss 11.4 GHz BPM prototype system, which will compress 320 ns to 40 ns in three stages, is now underway at SLAC. The experiment will test the insertion loss of components (couplers, delay lines, 180° turn-arounds and transitions), the overall power gain of the system, and effects due to mode conversion and dispersion. Initial measurements at low power will be reported[6)] early next year. The expected compression efficiency is 80% to 85%, giving a peak power gain of 6.5 to 7.0. The next step will be to make the system vacuum tight, so that it can be tested at high power.

In order to test a BPM compression system, accelerating structures and other RF components at a frequency and power level of interest for a linear collider, an appropriate near-term high-power RF source is needed. A 100 MW, 11.4 GHz klystron is under design[7)] at SLAC which meets this need. The tube operates at 440 kV, and uses a combination of electrostatic and magnetic beam compression to achieve a high compression ratio and a reasonable cathode loading. K. Eppley[8)] has simulated the dynamics of this tube, and finds an efficiency of 43% using a double-gap output cavity. Diode tests on the tube will begin early in 1989.

Some other possible microwave sources which can be used together with RF pulse compression to power a linear collider are discussed in the next section.

3. ALTERNATIVE RF SOURCES

By "alternative" we mean RF power sources which produce moderate peak power levels at a pulse length on the order of one microsecond, as opposed to very high peak power devices such as relativistic klystrons which operate at a much shorter pulse length equal to the structure filling time. Alternative sources can be driven by modulators using presently existing technology, and must be followed by several stages of RF pulse compression.

An example of such an alternative RF source, mentioned in the previous section, is the 100 MW, 11.4 GHz klystron now under design at SLAC. This power level is probably close to the limit that can be produced at this frequency using conventional klystron technology and a beam voltage less than 500 kV. It is well known that the efficiency of a klystron depends on the perveance, defined as $K = I_0/V_0^{3/2}$, where I_0 is the beam current and V_0 the beam voltage. A typical plot of efficiency (defined as the RF output power divided by the beam power I_0V_0) as a function of perveance is given in Ref. 9. The efficiency typically falls below 50% at a perveance of about $2 \times 10^{-6} A/V^{3/2}$. This fact alone limits the output power of a klystron to about 180 MW, independent of frequency, if it is to have reasonable efficiency at a beam voltage less than 500 kV. However, at shorter wavelengths cathode loading imposes an additional constraint. The cathode area is related to the beam cross-section area by the area compression ratio R_A. In turn, the beam area cannot exceed about $\pi(\lambda/8)^2$ if the coupling to the rf gaps is to be adequate. The output power for an efficiency η greater than about 50% is therefore

$$P \approx (\lambda/8)^2 R_A I_A V_0 \eta \ , \tag{1}$$

where I_A is the cathode loading in A/cm^2. Choosing $I_A = 10A/cm^2$ (reasonable but not conservative), $R = 200$ (probably near the upper limit), $V_0 = 400$ kV and $\eta = 50\%$, we have

$$P(MW) \approx 20\,[\lambda(cm)]^2 \ . \tag{2}$$

Thus at 17 GHz we can expect at most about 60 MW from such a "pushed technology" klystron.

A number of devices have been suggested which get around the λ^2 scaling in Eq. (2). The gyroklystron, for example, employs an annular beam with a thickness which scales with wavelength, but a circumference which to first order can be independent of wavelength. The area of the beam, and hence the power output, then scales roughly in proportion to λ. Thus, the gyroklystron is more suited to the production of high power at high frequencies than the klystron. As an example, a gyroklystron is being developed at the University of Maryland which will produce 40 MW at 10 GHz. It is estimated that the device could be scaled to produce 110 MW at 17.1 GHz.[10)]

Strip beam, or sheet beam, tubes are another class of device which can get around the λ^2 scaling law that round beam tubes must follow. Consider, for example, a sheet beam of width w and thickness t. To couple to the RF circuit, the thickness must be a fraction of the wavelength, say $t \approx \lambda/5$. Thus the power output of a sheet beam device is roughly

$$P \approx 0.2\lambda w R_t I_A V_0 \eta \ . \tag{3}$$

Here, R_t is the transverse beam compression ratio (ratio of cathode width to beam thickness). We take R_t to be 10, although detailed gain design calculations are needed to determine a realistic limit. Again, take $I_A = 10A/cm^2, V_0 =$ 400 kV and $\eta = 0.5$. Equation (3) then gives

$$P(MW) \approx 4\,[\lambda(cm)]\,[w(cm)] \ . \tag{4}$$

At 17 GHz such a sheet beam device with a width of 25 cm would produce 175 MW.

The perveance of sheet beam devices is also an important consideration. Longitudinal and transverse space charge effects in a sheet beam depend on the perveance per square K_s, defined terms of the current in the beam per square, I_s, as

$$K_s = I_s/V_0^{3/2} \approx I_A R_t (\lambda/5)^2 V_0^{-3/2} \ . \tag{5}$$

For $I_A = 10A/cm^2, R_t = 10, V_0 = 400$ kV and $\lambda = 1.76$ cm (17 GHz), $K_s = 0.05$. Space charge effects are relatively small at this low a perveance. In fact, a sheet beam device with the same output power as a round beam device has a perveance per square which is lower than the round beam perveance by a ratio on the order of $\lambda/5w$. These concepts also apply to annular beams, where w is the circumference of the beam.

The design of a sheet beam X-band (11.4 GHz) klystron has been considered by Eppley, Herrmannsfeldt and Miller.[11] The device, shown in Fig. 1, uses permanent magnets with alternating polarity to focus the beam (wiggler focusing). The tube was not actually built, but in simulations it produced an output power of 2 MW per centimeter of beam width at a beam voltage of 200 kV with an efficiency of about 50%. The perveance per square was about 0.1 $\times 10^{-6}$. The calculations indicated that the output power could be raised above 10 MW/cm at somewhat higher voltage and perveance with no loss in efficiency, giving an output power of 250 MW for a 25 cm wide beam. A number of problems with the sheet beam klystron concept need more detailed investigation, in particular the design of suitable RF cavities and potential feedback between cavities which can result from construction asymmetries.

It is possible to envision a variety of other sheet beam devices, for example sheet beam traveling wave tubes. The concept of a deflection modulated sheet beam device also looks interesting. Such a device is based on the interaction of the beam with the transverse beam-breakup mode. The tube can use discrete cavities to amplify the deflection (the Deflectron), or a continuous periodic structure (the Deflection Wave Amplifier). All sheet beam devices could in principle also be constructed as annular beam devices.

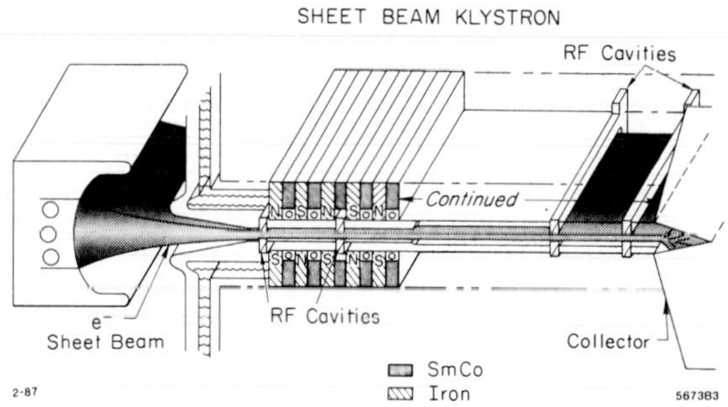

Fig. 1. Sheet beam X-band klystron.[11)]

To obtain high output power at a relatively low voltage from a linear beam device, it is clear that the beam cross section area must be increased so that in one dimension it becomes large compared to the RF wavelength. Only a few relatively simple possibilities for such extended beam area devices have been considered in this brief survey. More exotic concepts using annular or sheet beams have been proposed, such as the Gigatron[12]. Crossed-field amplifiers, which inherently operate at a high perveance (microperveance on the order of ten), have not been discussed. It has recently been suggested[13] that such devices are capable of being scaled to high power levels. In conclusion, there are a number of potential RF sources which can be used in conjunction with RF pulse compression to power a high gradient linear collider. The difficult problems are not so much in concept as in the detailed engineering and development needed to advance the technology on a reasonable time scale.

REFERENCES

1. R. B. Palmer, TLC Parameter List (1988). See also R. B. Palmer, "Energy Scaling of Linear Colliders," these Proceedings.
2. M. A. Allen *et al.*, these Proceedings; see also SLAC–PUB–4733.
3. Z. D. Farkas, IEEE Trans. MTT–34, 1036 (1986); also SLAC–PUB–3694.
4. Z. D. Farkas, and J. N. Weaver, SLAC/AP–59 (1987).
5. R. D. Ruth, private communication.
6. Z. D. Farkas, G. Spalek and P. B. Wilson, "RF Pulse Compression Experiments at SLAC," to be submitted to the 1989 Particle Accelerator Conference.
7. T. G. Lee, private communication.
8. K. Eppley, "Design of a 100 MW X-Band Klystron," to be submitted to the 1989 Particle Accelerator Conference.
9. Jean V. Lebacqz, Proceedings of the 1979 Linear Accelerator Conference, BNL Report 51134 (1979); pp. 327–335. Also SLAC–PUB–2380 (August 1979).
10. V. L. Granatstein, "Gyroklystrons for Linear Supercolliders: An assessment of Feasibility and Cost." Invited lecture at the International Workshop on Next Generation Linear Colliders, Stanford Linear Accelerator Center, Nov. 28–Dec. 9, 1988.
11. K. R. Eppley, W. B. Herrmannsfeldt and R. H. Miller, Proceedings of the 1987 Particle Accelerator Conference, IEEE Cat. No. 87CH2387-9 (IEEE, New York, 1987). pp. 1809–1811.
12. H. M. Bizek, P. M. McIntyre, D. Raparia and C. A. Swenson, IEEE Trans. Plasma Sci. 16, No. 2, 258 (1988).
13. M. A. Allen and J. Feinstein, private communication.

RF POWER AMPLIFICATION BY ENERGY STORAGE AND SWITCHING

WAYNE VERNON
University of California, San Diego, La Jolla, California 92093

During the last decade there have been several suggestions for RF storage and switching schemes.[1–6] The principle behind these schemes is simply that energy from a source which is on for a long time at moderate power can be stored in a resonant cavity and dumped ("switched") in a short time to yield higher power. This is also the basis of SLED[7] which is driving the SLC, but the major difference is in the switching and the proposed power gains. In the case of SLED there is no switch, only a "phase agile" RF source, and the maximum power gain is about a factor of 3. Proposed storage and switching schemes are often based on larger ratios of charge to discharge times, say 5 μsec/50 nsec = 100 which could be the power amplification ratio.

An early demonstration of the switching of a superconducting cavity was reported by Birx, et al.,[1] in 1977. They observed a peak power gain of 9 at low power levels with a cold cavity and a room-temperature switch. The switch was a He gas filled tube positioned in the leg of a waveguide T so that a $\lambda/2$ stub turned into a $\lambda/4$ stub when the gas broke down and became a good conductor. All switches encountered to date are some variant of this technique; the stub reflects back an out-of-phase signal which cancels the one from the cavity so that no power escapes while the low-loss dielectric tube is non-conducting.

Simple discharge of a cavity results in an exponential decay of the power, and it may be more useful to have a length of waveguide be the resonant element so that a constant power pulse is generated with a width equal to twice the waveguide's propogation time. Such a test was reported from the USSR in 1979;[5] power levels of 70 MW with a pulse width of 15 nsec were obtained for the S-band waveguide (1.2 m length) being driven by a 1.6 MW source. This was a room temperature system with a storage time constant of about 1 μsec. During the workshop Roger Miller suggested that a circular waveguide operating in low loss mode (TE_{01}) might allow the charging time constant to increase to 5 μsec which could result in large power amplification with 50 nsec output pulses.

A related development involves RF storage in superconducting cavities filled with sapphire.[8,9] The single-crystal pieces of sapphire fit inside the cavity with a small (1 mm) gap between the crystal and the superconducting wall and the electric and magnetic fields at the wall are quite a bit lower than in the empty cavity (for the same stored energy). If cavity modes can be found which keep the surface electric fields low, then energy densities of 25 mJ/cm^3 for X-band RF should be possible (the dielectric constant is of order 10 for these frequencies). If a dielectric loaded waveguide with a 6 cm^2 cross section and a length of 1.6 m is used as the resonant cavity, then that one liter crystal could store 25 Joules of RF and return it as a 50 nsec long pulse of 0.5 GW accelerator drive power. Single-crystal sapphire costs about \$ 10,000/liter, which is modest in this context. Power amplifications of 10^4 at GW output power levels might be possible.

It would appear that further evaluation of these schemes is warranted, especially if the switch part of the problem can be shown to work. The next generation of "SLED" will probably use "phase switching" with many stages of delay lines;[10] this rather complicated approach comes about because of the absence of trustworthy switch technology. One switching technique which seems to not have been tried with RF is the use of high resistivity silicon which is made to conduct with a laser pulse. It may be that the silicon can survive the high electric fields associated with the large stored energy in the resonant structure. If one of these energy storage approaches turned out to be moderately priced and reliable, then the high cost of peak RF power would cease to be the dominant problem in linear collider design.

REFERENCES

1. D. Birx, G. J. Dick, W. A. Little, J. E. Mercereau and D. J. Scalapino, *Microwave Power Gain Utilizing Superconducting Resonant Energy Storage*, Appl. Phys. Lett. **32**, 68 (1978).

2. Kazuo Minami and Kenji Hosoyama, *Proposal of a High-power Microwave Pulse Source Utilizing a Superconducting Cavity for Energy Storage*, Japanese Jour. of Appl. Phys. **18**, 85 (1979).

3. D. L. Birx and D. J. Scalapino, *A Cryogenic Microwave Switch*, IEEE Trans. on Magnetics, **MAG–15**, 33 (1979).

4. D. L. Birx and D. L. Scalapino, *Microwave Energy Compression Using a High-Intensity Electron Beam Switch*, J. Appl. Phys., **51**, 3629 (1979).

5. N. D. Devyatkov, A. N. Didenko, L. Ya. Zamyatina, S. V. Razin and Yu. G. Yushkov, *Formation of Powerful Pulses With Accumulation of UHF Energy in a Resonator*, Radio Eng. and Electron Phys., **25(6)**,87 (1980).

6. K. Saeki, T. Kimura, H. Kubo, K. Minami, M. Ohtsuka and M. Awano, *Microwave Pulse Source Utilizing a Superconducting Cavity*, Oyo Buturi, **54(8)**, 812 (1985); in Japanese.

7. Z. D. Farkas et al., *SLED: A Method of Doubling SLAC's Energy*, Proceedings of the 9th International Conference on High Energy Accelerators, p. 576, May 1976.

8. V. B. Braginsky and V. I. Panov, *Superconducting Resonators on Sapphire*, IEEE Trans. on Magnetics, **MAG–15**, 30 (1979).

9. D. G. Blair and S. K. Jones, *A High-Q Sapphire Loaded Superconducting Cavity Resonator*, J. Phys. D: Applied Physics **20**, 1559 (1987); see also IEEE Trans. on Magnetics, **MAG–21**, 142 (1985).

10. Z. D. Farkas, *Binary Peak Power Multiplier and its Application to Linear Accelerator Design*, IEEE Trans. on Microwave Theory and Tech., **MTT–34**, 1036 (1986).

A CLUSTER KLYSTRON*

ROBERT B. PALMER and ROGER MILLER

Stanford Linear Accelerator Center, Stanford University, Stanford, California 94309

ABSTRACT

We consider a honeycomb-like array of 126 individual klystrinos all fed from a single high-voltage source and all focused by a single axial magnetic field. It is argued that such a device may be a practical and economic power source for a TeV Linear Collider. It could be used either as a source of $\frac{1}{2}$ μsec rf pulses to be binary pulse compressed or, with a grid cathode and oil line energy storage, as a direct source of power.

1. INTRODUCTION

For a .5 on .5 TeV Linear Collider we need (for example)[1]

wavelength $\lambda = 17.5$ mm
power/length $P \approx 600$ MW/m
length/feed $\ell_{\text{feed}} = 1.6$ m
power/feed $P_{\text{feed}} = 930$ MW
no feeds $n_{\text{feed}} = 3.6$ K
pulse length $\tau = 60$ nsec
energy/length $J = 36$ Joules/m
total length $\ell_{\text{cavity}} \approx 6$ Km
total energy $J_{\text{TOTAL}} \approx 200{,}000$ Joules
total power $P_{\text{TOTAL}} \approx 3.4$ TWatts

If we consider the use of "conventional" klystrons operating at a voltage of the order of 400 KV then space-charge considerations (perveance ≤ 2) limit the current to 500 amps and the power output (at 50% beam-to-rf efficiency) per tube to 100 MW. And even this would be hard to achieve at 17 GHz. With 100 MW/tube, 34,000 tubes would be required. To avoid this two approaches are possible: either raise the voltage thus allowing higher currents; or use sheet or multiple beams so as to provide the very large number of "tubes" required without excessive cost.

The objections to raising the voltage are now becoming clear:

A) Higher voltage implies intense X-ray production, requiring significant shielding, and probably forcing the location of such klystrons in the accelerator tunnel rather than in a gallery.

B) Very high voltages are best obtained with Induction Linacs rather than a simple gun. Such Linacs are excessively expensive unless the drive current is relatively high (e.g., $\sim$ 10K amps). But such high currents require even higher beam energies with corresponding problems and development time.

C) The power supplies (e.g., magnetic compressors) to drive the required Induction Linacs with short pulses are also expensive.

For these reasons it is worth considering the other alternatives.

A) The Sheet-beam klystron has been studied by R. Miller et al.[2] Several problems were revealed: The efficiency was relatively low; there is a problem with maintaining the phase stability in the cavities, and there is a danger of feedback from the output to the input cavity of transversely polarized rf.

B) These problems are, of course, absent in a multi-klystrino approach as proposed by G. Loew.[3]

The present note represents an alternative packaging of Greg's basic idea.

2. PARAMETERS OF AN INDIVIDUAL KLYSTRON

We will consider two cases: 1) with a cathode of 8 amps/cm^2 and 2) with 32 amps/cm^2 and compare these

*Work supported by the Department of Energy, contract numbers DE-AC02-76CH00016 and DE-AC03-76SF00515.

with a conceptual design[4] of a 100 MW 12 GHz tube (refer to the table on the following page).

It is seen that the parameters of examples (1) are universally conservative, and of example (2) conservative in everything but the cathode current. Let me assume that we adopt example (1) for phase I of our TLC (i.e., 250 + 250 GeV) and hope that the higher current will become available for the full TLC (500 + 500 GeV). Then the number of tubes required is:

	Ref Design	Klystrino Expl(1)	Expl(2)
f gH_z	12	17	
j_{cathode} A/cm^2 [1]	8	8	(32)
diam (cathod) mm	90	18	
I amps	500	20	(80)
V kVolts	400	400	
Perveance [2]	2 10^{-6}	.08 10^{-6}	(.32 10^{-6})
efficiency [3]	50	75	(70)
P_{out} MWatts	100	6	(24)
B Tesla [4]	0.5	0.15	(0.3)
diam (drift tubes) mm	9	6	
R = $\frac{\text{cathode diam}}{\text{drift tube diam}}$ [5]	10	3	
J_{in} for 62 nsec		5	(20)
J_{in} for 75 nsec		6	(24)
J_{in} for 520 nsec		42	(166)

Notes:

[1] 8 amps/cm² is standard SLAC 5045 performance and thus very conservative. 32 amps/cm² is claimed to be now available from Varian, although its vacuum requirements and lifetime seem uncertain; it represents a less conservative but not unreasonable goal.

[2] Perveance is defined by

$$Pv = \frac{I(\text{amps})}{V^{\frac{3}{2}}(\text{Volts}^{\frac{3}{2}})}$$

and is a measure of the severity of space-charge effects.

[3] The efficiency of converting beam power to rf power could, if there were no space-charge effects, approach 100%. In practice the efficiency obtainable falls as the perveance rises (fig. 1),[6] and the assumed efficiencies used have reflected this empirical observation.

[4] The field required to balance the space-charge forces in the beam is given by:

$$B \propto \frac{1}{\lambda}\left(\frac{I}{\beta\gamma}\right)^{\frac{1}{2}} \quad .$$

In these examples I have scaled from the 17 GHz 100 MW design.[4]

[5] The ratio of cathode to drift tube diameters is a measure of the precision with which the gun must be designed and built. There may be no theoretical limit to this ratio but in practice ratios larger than about 7 are hard to achieve without beam loss.

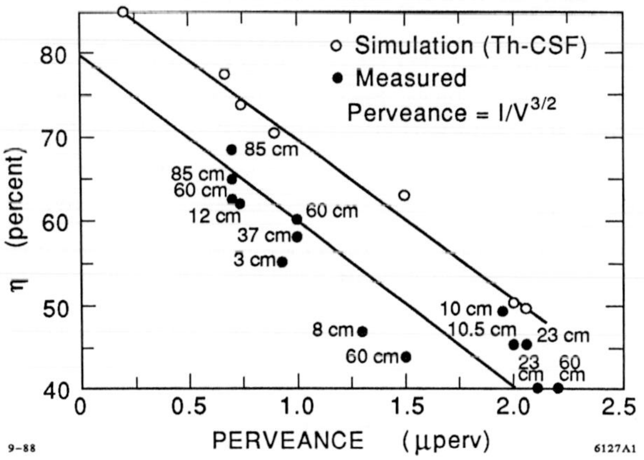

Fig. 1. Klystron efficiences plotted as a function of perveance. The wavelengths of examples are given by each measured point. The lines represent an approximate fit to the simulated and measured values.

a) With 8× pulse compression at 85% efficiency[5]

$$n = \frac{3.4\ 10\ 12\ \text{Watts}}{.85 \times 24\ \ 10\ 6\ \text{Watts}} \times \frac{1}{8} = 29,000 \quad ;$$

b) with no pulse compression

$$n = \frac{3.4\ \ 10\ 12}{24\ \ 10\ 6} = 140,000 \quad .$$

These are formidable numbers. How can one build them at reasonable cost?

3. THE HONEYCOMB CONCEPT

It is proposed here to build the klystrons into honeycomb-like structures. Each structure containing 126 individual klystrons. The number of such structures is now

a) with 8× pulse compression at 85% efficiency

$$w_s = \frac{18,000}{.85 \times 126} \approx 170 \quad (1 \text{ per } 35 \text{ m}) \quad ,$$

b) with no pulse compression

$$n_s = \frac{140,000}{126} = 1140 \quad (1 \text{ per } 6 \text{ m}) \quad .$$

Both of which seem reasonable. Indeed, if such clusters can be built for a reasonable cost, and if a method to power the tube for ≈ 60 nsec can be provided, then pulse compression would seem unnecessary.

It is proposed here to group the klystrons into clusters of 18 (see fig. 2a). Each cluster would have its own input and output wave guide. Within a cluster the cavities of all 18 tubes would be strongly coupled and thus locked in phase and amplitude. Each cluster of tubes would act as a single klystron, with the only difference that the current is broken up into sub beams, and, as a result the space charge forces greatly reduce.[7] Seven of these clusters would form a single structure with 126 individual tubes (see fig. 3a).

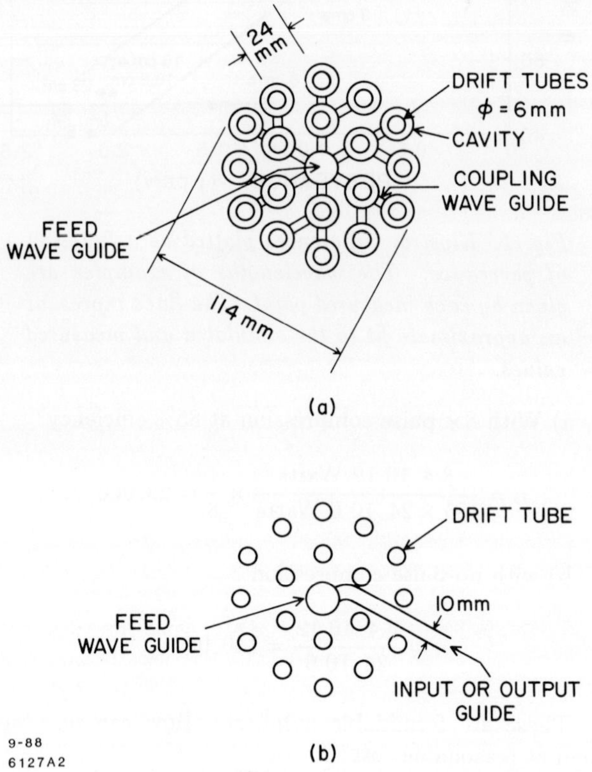

Fig. 2. A single cluster of 18 klystrinos; a) showing the cavities with rf feed at center; b) showing drift tubes between cavities with input or output guide to the feed.

The waveguides to feed, damp or extract power from a cluster of cavities could consist of short longitudinal mode guides at the center of each cluster. In the case of the output cavities these longitudinal guides could continue out through the beam dump. For the input guides they would extend just into the space before the next cavity, and there convert to conventional guides passing between the drift tubes (fig 3b). They could then couple again into longitudinal guides outside the honeycomb and pass thus out through the beam dump.

The object in this design is to minimize the number of parts. The input, output, or intermediate cavities, of all 126 tubes would be machined into discs of Copper, as would the feed waveguides (fig. 4). Though each disc would involve complex machining, the number of such discs would be no more than the number of parts in a single klystron.

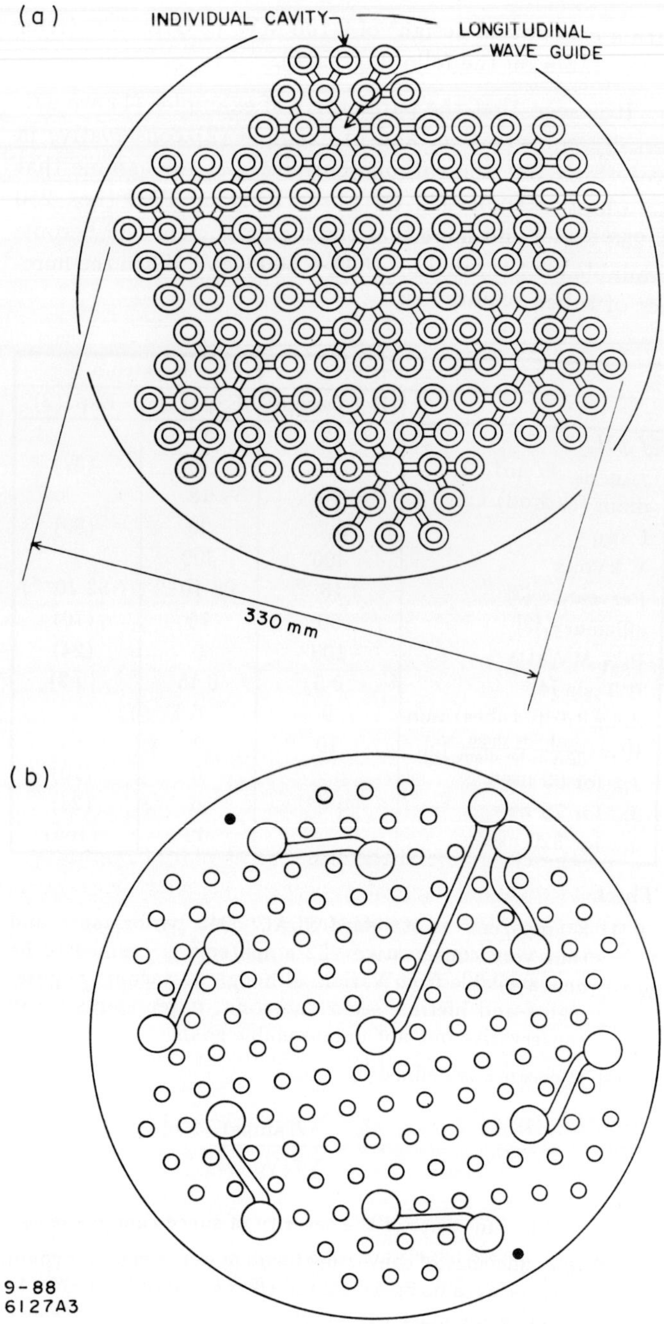

Fig. 3. Seven clusters in a single copper block; a) showing cavities; b) showing guides joining cluster feeds to external axial wave guides.

4. GUN DESIGN

Focusing within the honeycomb of klystrons can be provided by a uniform axial magnetic field. In order to keep the beams within the drift tubes the transverse fields must be small and the field quality requirement for the 275 mm diameter area should be of the order of:

$$\frac{\Delta B}{B} \le 2 \times 10^{-3} \quad .$$

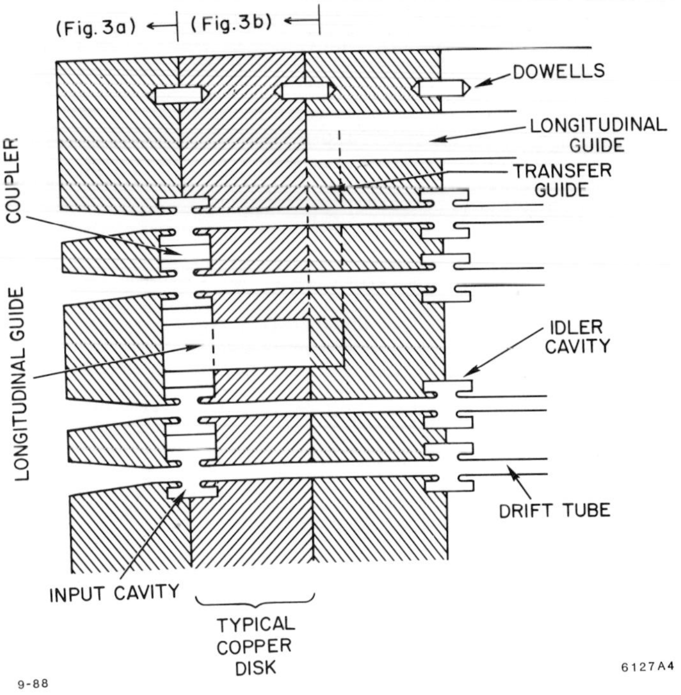

Fig. 4. Section showing construction.

At the cathode end the situation is more complicated. The field must be reduced in order to allow the current to converge from the larger cathode area into the smaller drift tubes. The method proposed is to use an iron end plate with holes in it to effectively end the axial fields (fig. 5).

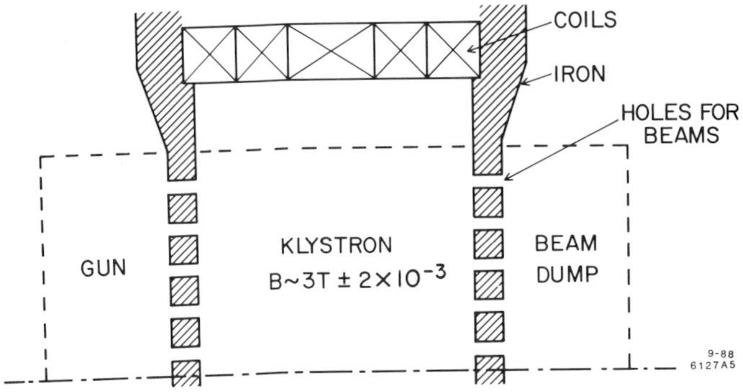

Fig. 5. Magnet with perforated iron end plates to restrict field to the klystron region.

The required thickness t of the iron plate depends on the focusing field B_k, the allowed field in the iron B_{Fe}, the hole radii r, the hole spacing s and the total honeycomb radius R

$$t = \frac{R}{2}\frac{B_k}{B_{Fe}}\frac{s}{s-r} \quad .$$

For the high j_c example: $B_k = 0.3$ Tesla, $B_{Fe} = 1$ Tesla R = 140 mm, r = 10 mm and s = 24 mm. Then, $t = 43$ mm (or 22 m for the lower current density).

Outside the iron plate the gun geometry would be relatively conventional (fig. 6). The design is complicated by the relatively late application of the magnetic field, but is eased by the low reduction ratio from the cathode to drift tube (3×).

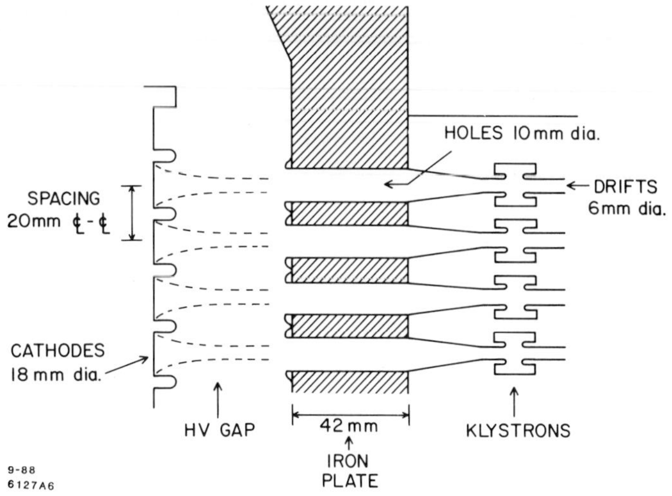

Fig. 6. Conceptual view of gun.

5. POWER SUPPLIES AND EFFICIENCY

A. Binary Pulse Compression

If used with 8× pulse compression then an initial pulse of 520 nsec would be required (60 nsec output pulses with 6 nsec between pulses to allow time for the phase to change). A 400 KV 520 nsec modulator could presumably be designed along more or less conventional lines. If the rise and fall times of such a modulator were ≈ 100 nsec, and the power supply efficiency were 90% than the overall efficiency would be:

$$\eta_a = \cdot 9 \times \cdot 8 \times \cdot 7 \times \cdot 9 \times \cdot 85 = \cdot 38$$

$\cdot 9$	$\cdot 8$	$\cdot 7$	$\cdot 9$	$\cdot 85$
↑	↑	↑	↑	↑
Power Supply	Rise & Fall	Klystron	Time to Change Phase	Pulse Compression

It would be reasonable to put a separate pulse compressor on each pair of clusters. For the 32 amp per square cm case, the output power is 432 MW per cluster of 18 tubes (18 * 24). Using two such clusters as input to an eight-fold compressor, the output (at 85% efficiency) would be 2.8 G watts emerging from each of two waveguides. These would have to be divided into three to give the required .9 G watt per feed. There would then be:

1 klystron/21 feeds, (1 klystron/35 m)

1 compressor/6 feeds, (1 compressor/10 m)

Total: 3.6 K feeds

600 compressors

171 klystrons

B. Magnetic-Pulse Compressor

If no pulse compression were used then a power supply is needed that can give 400 KV for a 60 nsec pulse. Magnetic compressors could do this. The rise and fall time would be of the order of 15 nsec leading to an overall efficiency:

$\eta_b =$	·9	×	·9	×	·75	×	·7	= ·42
	↑		↑		↑		↑	
	Power Supply		Magnetic Compressor		Rise & Fall		Klystron	

C. Grid-Controlled Cathode

Another way to obtain the required 60 nsec pulse of 10,000 Amps would be to leave essentially full voltage on the gap but switch the current with a grid placed close to the cathode. The gun gap would be connected to a suitable oil delay line (fig. 7a), recharged by a power supply between pulses. The voltage wave from the gap would be as shown in fig. 7b; and the maximum gap voltage V_0 will be given by:

$$V_0 = V_k \frac{1 + Z_0}{R_k} \quad ,$$

where V_k = klystron voltage = 400 KV, R_K = klystron impedance = $\frac{400\text{KV}}{10{,}000\text{A}}$ = 40 Ω 's (160 Ω 's for the lower current density) and Z_0 = impedance of the line.

The energy stored in the line U_0 is given by:

$$U_0 = U_K \frac{R_K}{4Z_0} \left(1 + \frac{Z_0}{R_K}\right)^2 \quad ,$$

where U_K is the energy required by the klystron.

$$Z_0 = 10\Omega (40\Omega \text{ for the lower current density})$$

and the energy stored in the line is

$$U_0 = 1.56\ U_K \quad .$$

It is not unreasonable to expect a rise time for the grid voltage, and thus currents, of the order of 2 nsec—significantly faster than that for the magnetic compressor. We will assume a 6 nsec rise time for the klystron. The efficiency of the oil line will be near 100%. Assuming a 90% efficiency for the power supply then the overall efficiency would be:

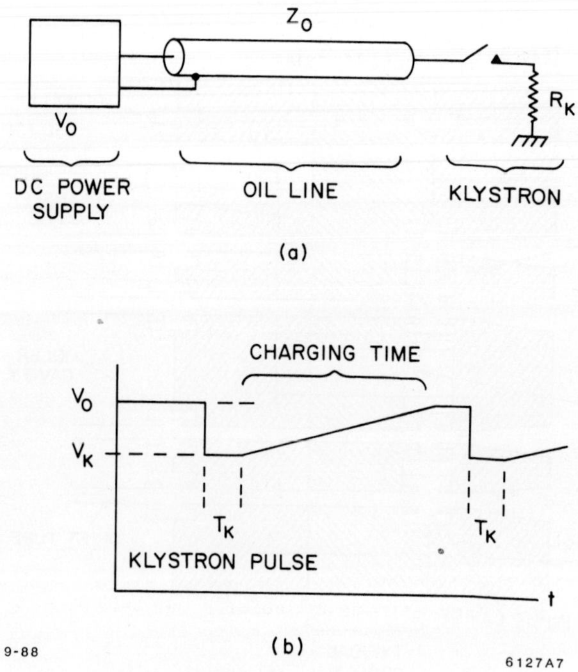

Fig. 7. Power supply for a grid controlled klystron a) schematic; b) voltage at end of oil line as a function of time.

$\eta_c =$	·9	×	·9	×	·7	= ·57
	↑		↑		↑	
	Power Supply		Rise & Fall		Klystron	

Although the grid-controlled cathode and oil lines require significant R&D the higher efficiency that could be obtained would be well worthwhile.

6. COST ESTIMATES

It is clearly very hard to estimate costs at this time. We do it only to see if there is any hope of the costs being significantly lower than other methods.

In the case using binary pulse compression the modulator can be scaled from SLAC at about .5 $/Joule input. For the pulse compressor we use 100K$ per unit which is twice P. Wilson's estimate (physicists are usually a factor of 2 low).

For the magnetic compressor we have used prices from W. Barletta at Livermore of .7 $/Joule input including the power supply. Bayless' estimates are similar.

For the oil line and power supply we took .4 $/Joule, a little less than the SLAC Modulator cost. This is a very unreliable guess.

For the klystron itself we have guessed a figure of 200K$ per unit. This is four times the SLAC 5045 cost, and is based on the observation that the cluster klystron is likely

Table 1.

A. With Binary Pulse Compression and Conventional Modulators			($\epsilon \approx 38\%$)
	Units	K$/Unit	M
Modulators	170	1000 (250)	180 (50)
Klystrons	170	200	30
Binary Compressors	600	100	60
Total rf Source			270 (140)

B. With Magnetic Compression and no BPC			($\epsilon \approx 42\%$)
	Units	K$/Unit	M
Magnetic Compressor + PS	1100	200 (50)	200 (60)
Klystrons	1100	200	220
Total rf Source			440 (280)

C. With Grid Cathode and no BPC ($\epsilon \approx 57\%$)			
	Units	K$/Unit	M
PS + Oil Line	1100	100 (25)	110 (30)
Klystrons	1100	200	220
Total rf Source			330 (250)

to weigh about four times more!

Using these assumptions, we estimate costs in Table 1. The parenthesized numbers in the table are for the lower current Phase I.

From Table 1 we conclude:

1) These costs are low and efficiencies higher, compared with those obtained for current moderate energy ($\lesssim$ 2 MeV) relativistic klystron designs.

2) There seems no advantage in using the magnetic compression example (B).

3) If 32 amps/cm^2 cathodes are okay then there is little significant cost difference between the approaches (A) and (C). The grid cathode and oil line (example C) is then preferred for its higher efficiency and pulse flatness.

4) If the cathodes only operate at 8 amps/cm^2 then there is probably a cost advantage in using binary pulse compression (example A).

But these conclusions may change when more is known. Note that other combinations such as grid control plus some Binary pulse compressions were not considered but may be a good compromise.

7. WORK TO BE DONE

1. See if the gun design is possible. Note that it may *not* be possible. Multiple Iron plates might be needed to raise the field more gently. Intermediate potential electrodes may be needed between cathode and anode to shape the fields and or shield space-charge effects.
2. Make a better cost estimate of the klystron. What will a 1.5 or 3 Kg magnet with 10^{-3} field quality cost? What are the machining costs? Can the iron be a brazed part of the klystron?
3. Design the Klystrino itself. Emphasis should be given to achieving the high efficiency that the low perveance should make possible. Double output cavities are essential and R. Ruth's phase-space swirling should be studied.
4. Get information on performance and cost of oil lines and 500KV DC supplies. It is important to gain order of magnitude cost estimates early to avoid wasted effort if they are too high.
5. Get better engineering, including installation cost estimates for binary pulse compression as a function of the number of stages.
6. Get more information on 32 amp/cm^2 cathodes, including information from LBL on LaB_6 thermionic cathodes.
7. Build a single cluster gun with a *single* klystrino on the central cathode. Tuning and variable damping should be available for all cavities to fix these parameters prior to #8.
8. Build a full single-cluster klystron with 18 tubes giving 108 MWatts.
9. Start R&D Program on high current cathodes.
10. Start R&D Program on grid-controlled cathodes (conditional on #4).

REFERENCES

1. Palmer, R.B., *The Interdependence of Parameters for TeV Linear Colliders*, SLAC–PUB–4295 (1987) and *Proceedings of the Workshop on New Developments in Particle Acceleration Techniques*, June 29–July 1987, Orsay, France; CERN 87–11; ECFA 87–110. R. B. Palmer, *Linear Collider Energy Scaling*, SLAC–PUB–4707 (1988) and *Proceedings of the DPF Workshop*, June 27, 1988, Snowmass, Colorado.
2. Eppley, K. R., W. B. Hermansfeldt and R. H. Miller, *Proceedings of the 1987 IEEE Particle Physics Conference*, March 16–19 1987, Washington D.C.
3. Loew, G., *Considerations of r.f. Frequency and Power Generation in Linear Colliders Using Conventional Technology*, SLAC Internal Report AAS #22.
4. Eppley, K. R., private communication.
5. Farkas, Z. D., *Binary Peak Power Multiplier and Its Application to Linear Accelerator Design*, SLAC–PUB–3694 and IEEE Trans. on Microwave Theory and Techniques, **MTT–34**, No. 10, p. 1036 (1986).
6. Simulation data is from Thompson–CSF, private communication.
7. The use of multiple channels to suppress space charge effects has been proposed for light ion acceleration by A. W. Maschke, Brookhaven National Laboratory Report, BNL 51209 (1979).

A MICROWAVE POWER DRIVER FOR LINAC COLLIDERS
GIGATRON

Hana M. Bizek*, Stephen M. Elliott, Peter M. McIntyre, Alireza Nassiri, Milorad B. Popovic, and Deepak Raparia
Department of Physics, Texas A&M University, College Station, Texas 77843-4242

and

Henry F. Gray
Code 6844, Naval Research Lab., Washington, D.C. 20375

*Present address: Advanced Photon Source, Argonne Nat'l Lab., 9700 S. Cass Ave., Argonne, IL 60439

ABSTRACT

The gigatron is a new rf amplifier tube designed for linac collider applications. Three design features permit extension of the lasertron concept to very high frequencies. First, a gated field-emitter array is employed for the modulated cathode. Second, a ribbon beam geometry mitigates space charge depression and facilitates efficient output coupling. Third, a traveling wave output coupler is used to obtain optimum coupling to the ribbon beam. This paper describes recent developments in the gigatron design, and progress towards experimental tests.

I. INTRODUCTION

This paper describes recent progress in the development of gigatron as a high-power microwave driver for future e^+e^- linac colliders.[1] The gigatron design concept has been described previously.[2] We have evaluated its design for the parameters required for most TeV linac collider designs:

rf frequency 10-30 GHz
rf power >100 MW
efficiency >50%.

The only rf devices with comparable capability are the relativistic klystron[3] and the gyroklystron[4]. By comparison to either of these the gigatron offers a simple, compact structure and low capital cost.

In the gigatron a bunched electron beam is extracted from a field emitter array (FEA) cathode, and accelerated through a pulse-modulated high-voltage diode as shown in Figure 1. The resulting beam is fully modulated at high energy without the requirements of modulator and drift region that characterize all conventional amplifier tubes. RF energy is then extracted from the beam in an output coupler. Since the bunching process is a dominant limitation to high power-high frequency operation in conventional tubes, this approach results in high performance for the above design parameters.

Three features of the gigatron design extend its performance to very high frequency and peak power. First, a ribbon beam geometry is adopted rather than the conventional round beam. Second, a traveling wave output coupler is used to achieve optimum output coupling across a wide beam. Third, a gated field emitter array (FEA) is employed for the cathode; this appears to offer a simple, reliable modulated cathode for microwave applications.

By the beginning of the past year we had calculated the beam dynamics in gigatron and verified the impressive performance estimated earlier. During the past year we have addressed three design issues raised in reviews of the gigatron technology. First, it was suggested that the ribbon beam might be unstable, since many previous ribbon beam devices have proven unstable. Detailed calculations have demonstrated that the gigatron beam is in fact extremely stable to a variety of perturbations over the full range of parameters contemplated. The reason for this is simply that the gigatron approaches a switch tube geometry: the beam has no time in which for instabilities to grow. From cathode to collector is ~6 rf cycles.

Second, it was suggested that the traveling wave coupler might have parasitic modes that could compromise the coupling to the desired loop-resonant traveling wave. We have shown that, by distributing the beam-wave coupling along ~ half the circumference of a loop-resonant circuit, higher-order modes (corresponding to different harmonics of the loop-resonant frequency) are in fact naturally suppressed. Only transverse harmonics remain; these occur at multiples of the operating frequency and are weakly coupled to the beam. The waveguide coupler thus appears to be a particularly benign structure for this traditional tube designer's headache.

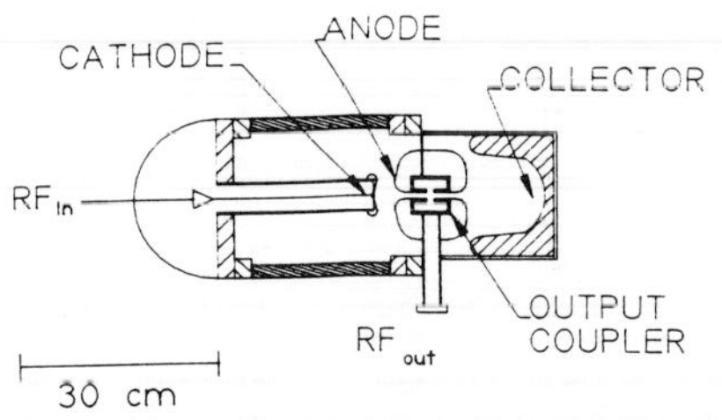

Figure 1. Cross section of gigatron.

The third problem is more substantial: microwave modulation of the FEA cathode. FEA cathodes have been built by several groups,[5,6,7] and have achieved impressive performance. Currents of >100A/cm² have been achieved. Extended life-cycle tests show no deterioration during a year of continuous operation. But these results were obtained in DC operation. The FEA gate structure presents a highly capacitive load to a modulation driver. The charging resistance R through the silicon substrate appears in series with the gate capacitance C and renders the FEA cathode configurations built by previous authors unworkable at microwave frequencies.

During the past year we have developed several specific improvements to FEA cathode design that together overcome this problem. First, a base layer metalization is incorporated so that charging currents flow in metal rather than silicon. Second, we configure the cathode as a lumped-constant resonant circuit (Q ~100), which is matched to a 50 Ω modulator input. With these improvements the system gain of gigatron is ~25-30 dB.

II. GIGATRON DESIGN CONCEPT

The gigatron is a design for a compact, efficient microwave power amplifier tube. It employs the lasertron concept,[8,9] in which a bunched electron beam is extracted from a modulated cathode and accelerated through a high-voltage diode structure. Several significant innovations have been introduced to achieve high efficiency, high frequency, and high power. The principle of operation is illustrated in Figure 2. A fully modulated ribbon electron beam is produced from a strip cathode of gated field-emitter array (FEA). The ribbon beam is accelerated through a high-voltage diode structure, and enters a drift slot leading to the output coupler. The output coupler consists of a slotted waveguide connected to form a resonant loop. The waveguide slot is configured to permit passage of the ribbon beam through the waveguide, parallel to the electric field of a traveling wave. The electron beam is decelerated by the wave, and thereby drives the wave amplitude. Optimum phase match to the electron beam is maintained over an arbitrary beam width by phasing the electron beam emission from the cathode to tilt the ribbons with respect to the direction of acceleration. By a

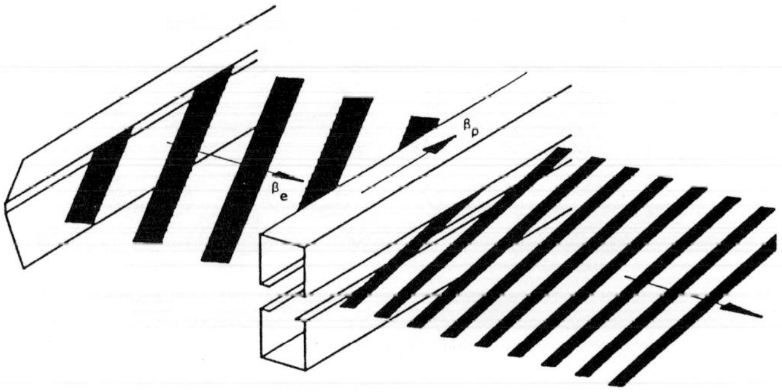

Figure 2. Ribbon beam geometry and traveling wave coupler.

suitable matching condition, the transverse position of the electron packet traversing the coupler slot can be made to move sideways at the same rate as the phase velocity of the traveling wave. The beam "surfs" with the wave, and optimum coupling can be preserved over an arbitrary beam width.

The advantages of the gigatron include a) low-loss coupling of input power to the FEA structure; b) e-beam fully modulated before acceleration; c) elimination of space-charge limits to high power/high frequency by using ribbon beam geometry; d) close coupling to the ribbon beam in the output coupler reduces phase dispersion at high frequency; e) provision of optimum output coupling over a wide beam front, using the traveling wave coupler; e) elimination of the requirement for static guide fields that characterize nearly all high frequency power tubes; f) simple, compact structure. Table I summarizes the important parameters of a particular design which has been evaluated recently.[2] The following text describes each aspect of the gigatron design. Numerical calculations are specific to the parameters of Table I.

1. Gated field-emitter cathode. The gated field-emitter (FEA) cathode appears to offer an attractive new technology for achieving a high current, fully modulated electron beam directly from a cathode structure.[6] C.S. Spindt et al.[5] and H. Gray et al.[6] have developed microfabrication techniques by which they can prepare planar arrays of field-emitting points. Gray uses directional etching techniques to produce atomically sharp needle and knife-edge arrays directly on silicon. A silicon-insulator-metal gate structure is then deposited on the cathode substrate, and plasma-etched to form a planar array of micro triodes. Application of a modest (25 V rms) gate-cathode voltage results in full modulation of emission current. Currents of >100 A/cm² have been routinely achieved. There is no evidence of in-service deterioration during extended life tests.[5]

We have improved upon the fabrication procedure of Gray in two respects that are critical to achieve high-frequency modulation.

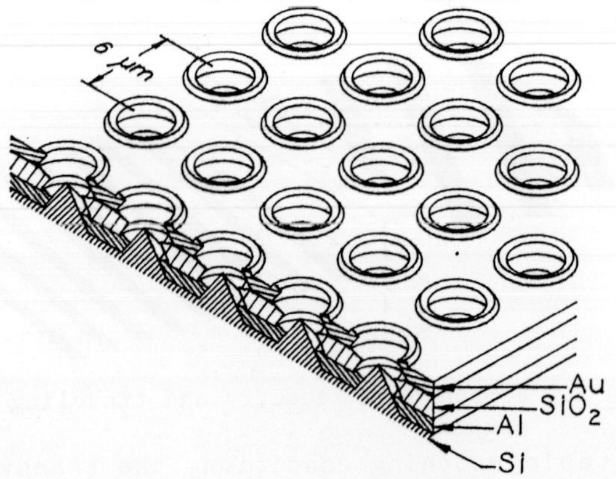

Figure 3. Gated field emitter array cathode.

First, the silicon surface is metalized after formation of the field-emitting tips, as shown in Fig. 3. This provides a metal base surface for low-impedance charging of the FEA tips. In the devices fabricated by Gray[6)] and Spindt[5)], charging current must flow through the bulk of the silicon substrate. Even with heavy doping, the silicon substrate presents an unacceptable charging resistance ($\rho > .01\ \Omega$ cm) at high frequency.

Second, a fabrication process has been devised whereby an optimum field geometry can be produced for quiesent electron optics. The optics of the electron beam emission from the FEA cathode determines the initial emittance of the beam in the gigatron. We have calculated the beam dynamics during field emission using the computer code MASK). Beam is assumed to be emitted from the tip into a cone of half-angle $\pi/4$. Figure 4 shows the transport at several time intervals after emission. The tip/gate geometry of the cathode forms an approximate point-to-parallel optical lens. As a result the emitted beam will have a much smaller emittance than the previous estimates based simply on assuming a transverse temperature $T_\perp \sim eV_m$. The beam divergence produces negligible transverse growth at the anode aperture, and hence does not pose a limitation to gigatron performance. If in fact the effective emittance is very small, the improved FEA cathode may be an excellent candidate for electron sources for linacs, FEL's etc.

2. Resonant input coupler. The field-emitter array presents a highly reactive load to a modulation driver. Matching can be optimized by configuring the cathode within a sequence of coupled lumped-constant resonant circuits, as shown in Figure 5. The gate/base junction of the FEA constitutes a capacitance C_1. A metalized quartz fiber is mounted adjacent to the FEA, and the metal-insulator-metal layers (without tips) are continued laterally to form a tuning capacitance C_2. The gate layer metalization and quartz fiber metalization are interrupted along the region where the fiber contacts the insulating layer, so that rf currents to/from the gate layer must flow around

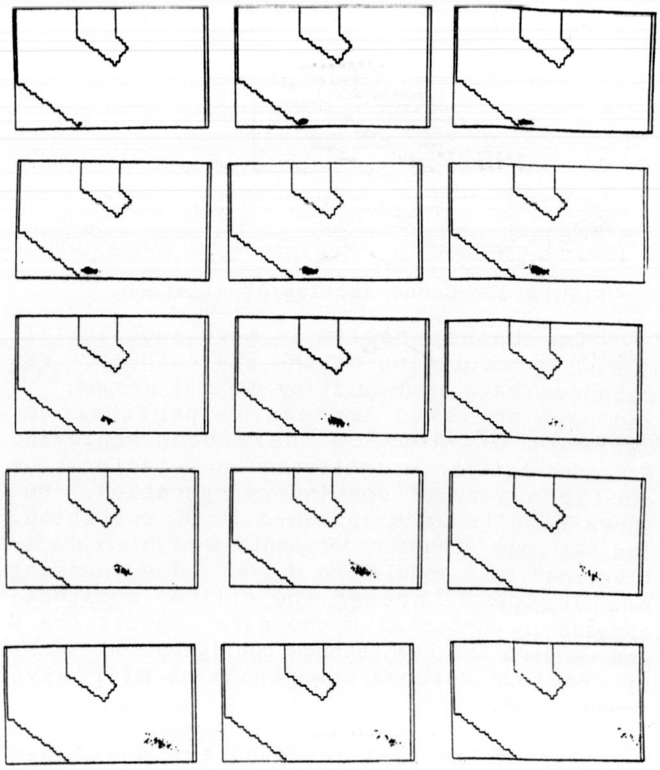

Figure 4. Acceleration of an electron bunch from an FEA cathode.

the fiber - it is an inductor L. The system $C_1 - L - C_2$ then forms a resonant circuit whose frequency can be controlled by choice of the radius r of the inductive fiber:

$$L = \mu_o \pi r^2 / x \qquad (1)$$

The cathode and resonant coupler are divided lengthwise into segments of equal length x = 5 mm. The metalized inductor is fabricated separately and bonded to the gate layers of C_1 and C_2 by conventional ultrasonic bonding.

Figure 6 shows an equivalent circuit of the cathode system, including dc bias network, resonant network, and impedance matching network. Each resonant circuit is capacitively coupled (C_c) to an impedance matching network which matches rf power from a waveguide into the sequence of input couplers. Resonance occurs at an angular frequency

$$\omega = \frac{1}{\sqrt{LC_o}} \qquad (2)$$

where C_o is the equivalent capacitance of C_1, C_2, C_c. The resistance R in the equivalent circuit is the effective series resistance to current flow in the resonant circuit:

$$R = \left(\tfrac{2}{3} w + 2\pi r\right) R_s / x \qquad (3)$$

where $R_s = .04\ \Omega$ is the surface resistance of the metalizations.

The Q of the resonant circuit is

$$Q = \frac{1}{\omega C_o R} \simeq \frac{2}{\omega C_1 R} \qquad (4)$$

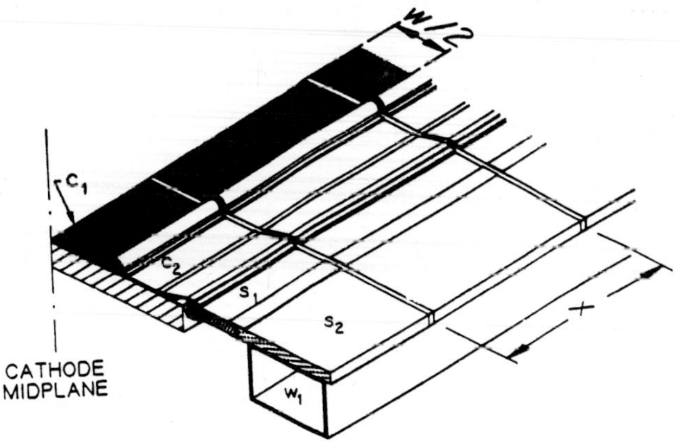

Figure 5. FEA cathode and resonant coupler.

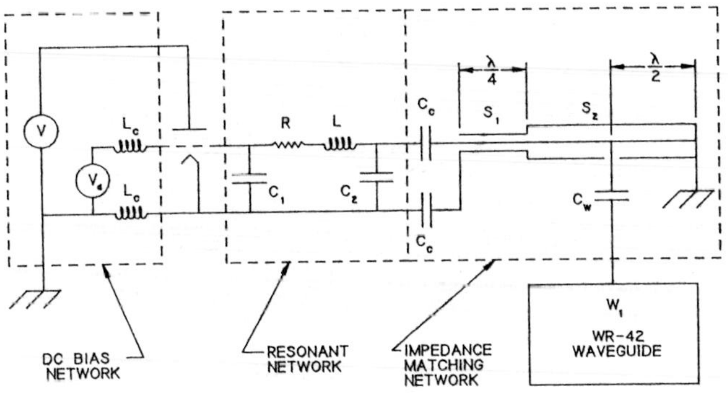

Figure 6. Equivalent circuit of FEA cathode and resonant coupler.

The resulting load impedance Z_L for critical coupling is

$$Z_L = \left(\frac{1}{1+i}\right) Q^2R/4 \tag{5}$$

Using parameters of Table I, each resonant network has

$$\begin{aligned} C_1 &= C_2 = 72 \text{ pF} \\ C_c &= 5 \text{ pF} \\ L_c &= 2.2 \text{ pH} \\ r &= 50 \text{ um} \\ R &= 9 \times 10^{-3}\ \Omega \\ Q &= 28 \\ Z_L &= 2\ \Omega \end{aligned} \tag{6}$$

The emission current corresponding to the above modulation characteristic is $I_m = 390$ A peak, $I_o = 8$ A from a (14 cm x 1 mm) cathode. The power supplied from the input modulator that is required to supply this emission is

$$P_e = I_m V_m \frac{\sin(\Delta\phi/2)}{\pi} = 2.2 \text{ kW} \tag{7}$$

The input power P_m required to drive each resonant network is

$$P_m = R(\omega C_1 V_m)^2/2 = 350 \text{ W} \tag{8}$$

In the overall cathode there are $N = 2\ell/x = 56$ networks. The total input rf power requirement is

$$P_{in} = P_e + NP_m = 22 \text{ kW} \tag{9}$$

The rf gain of gigatron is therefore

$$G = \log(P_{out}/P_{in}) = 27 \text{ dB.} \tag{10}$$

3. Traveling Wave Coupler. The traveling wave coupler naturally suppresses all parasitic modes. The resonant loop of waveguide has a total length chosen to be an integral number M of phase wavelengths:

$$2\ell = 2\pi M \frac{\beta_p c}{\omega} \tag{11}$$

For the parameters of Table I, an appropriate choice would be $M \simeq 14$. In principle, parasitic modes could be excited at frequencies corresponding to neighboring integral harmonics $M' \neq M$. The coupling to each mode M' must, however, be averaged over the length ℓ of the ribbon beam. The ribbon beam tilt is adjusted to achieve the phase-match condition.
The phase shift between the electron bunch and a traveling wave of harmonic M' varies linearly along the ribbon beam:

$$\Delta\phi(x) = 2\pi(M' - M)\left(\frac{x}{2\ell}\right) \tag{12}$$

Averaging over the width ℓ of the ribbon gives a coupling for each parasitic mode:

$$\frac{E}{Eo} = \int_0^{\ell} \cos\Delta\phi(x)\ \frac{dx}{\ell} = \begin{cases} 0 & M' \neq M \\ 1 & M' = M \end{cases} \tag{13}$$

All parasitic modes $M' \neq M$ thus receive no net coupling. This feature is a result of the distributed beam waveguide coupling, and is unique among electron tube designs.

III. DEVICE PERFORMANCE

We have calculated beam transport through the gigatron. Calculations were performed using the MASK computer code. Fig. 7 shows the trajectories of successive bunches through the diode region. Fig. 8 shows the trajectories of two successive ribbons through the coupler structure. Trajectories are shown on five points across each ribbon. Both transverse position and energy are shown for each trajectory. Slot width and height, coupler peak field, RF phase, and beam phase width were varied to optimize RF efficiency while transporting residual beam to the collector. For the optimized parameters, RF conversion efficiency of 74 % is obtained. No beam is intercepted before entering the collector structure.

We now have operational FEA cathodes at Texas A&M. The IV curves of these cathodes are presented below. The cathodes were made for us by C.S. Spindt, and employ the traditional fabrication approaches. We have built a vacuum assembly for testing the IV characteristics of cathodes in pulse mode, including provision for rf modulation. We have also built the complete vacuum structure for a round-beam gigatron prototype, with design parameters 12 GHz, 40 kW.

REFERENCES

1) P.B. Wilson, "Design Concepts for TeV Linac Colliders," SLAC-PUB-3674 (1985.
W. Schnell, "CERN Linear Collider Design Study," CERN-LEP-RF/86-06 (1986).

2) Hana M. Bizek, Peter M. McIntyre, Deepak Raparia and Charles A. Swenson, "Gigatron," IEEE Trans. on Plasma Sci., 16, No. 2, (1988).

3) S. Yu, "Physics of Relativistic Klystrons," Workshop on new developments in particle accelerator technology. Orsay, France (1987).

4) W. Lawson et al., "A High Peak Power, X-Band Gyroklystron for Linear Accelerators," Proc. 1986 Linac Conference.

5) C.A. Spindt et al., "Field Emission Array Development," Proc. 33rd Int. Field Emission Symp. (1986).

6) D.J Campisi and H. Gray, "Microfabrication of Field Emission Devices for Vacuum Integrated Circuits using Orientation Dependent Etching," Proc. Mat. Res. Soc. Meeting. 1986.

7) J. Fisher and T. Srinivasan-Rao, "Fabrication and Performance of Knife-Edge Field-Emitter Structures on Silicon," 1987 Workshop on New Technology for Future Accelerators. Orsay. 1987.

8) M.T. Wilson and P.J. Tallerico, U.S. Patent #4,313,072 (1982).

9) C.K. Sinclair, "The SLAC Lasertron Project," SLAC-PUB-4111 (1986).
Y. Fukushima et al., "Lasertron," Nucl. Instr. Math. A238, 215 (1985).

TABLE I
Parameters of Example Gigatron

$\omega/2\pi$	rf frequency	18 GHz
P	rf peak power	10 MW
G^{o}	power gain	27 dB
	rf efficiency	74%
V	beam voltage	200 kVDC
	peak beam current	390 A
ϕ	rf phase	230°
$\Delta\phi$	beam phase width	60°
$\ell \cdot w$	cathode size	14 x .1 cm²
$a \cdot b$	waveguide coupler (WR42)	1.1 x .4 cm²
g,h	slot width, height	.2 cm, .3 cm
β	electron velocity/c	.70
β^{ε}	phase velocity/c	1.37
θ^{p}	ribbon tilt angle	27°
Ψ	magnetic bend angle	22°
E_o	peak rf field	125 MV/m

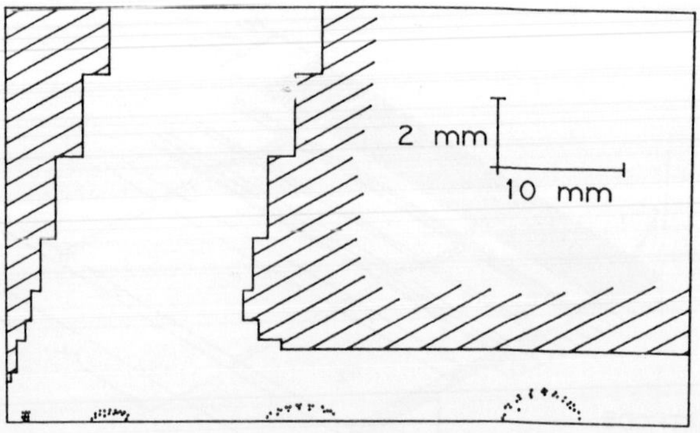

Figure 7. Calculated ribbon beam trajectories through diode region.

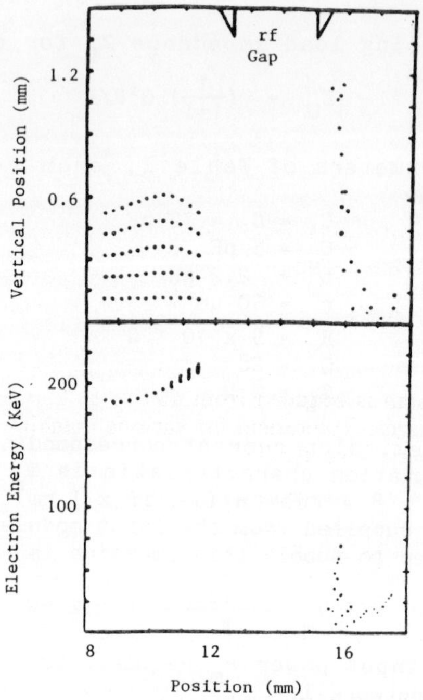

Figure 8. Electron energy and bunch motion through the traveling wave coupler

ACKNOWLEDGEMENTS

It is a pleasure to acknowledge stimulating discussions with W. B. Herrmannsfeldt, K. R. Epley, R. Miller, R. Taylor, J. Welch of SLAC, C. A. Spindt of SRI, K. Chang, D. L. Parker and R. Rogers of Texas A & M. This work is supported by U. S. Department of Energy contract number DE-AC02-85ER40236.

FINAL FOCUS SYSTEM FOR TLC*

Katsunobu Oide†

Stanford Linear Accelerator Center, Stanford University, Stanford, California 94309

ABSTRACT

A limit of the chromaticity correction for the final focus system of a TeV Linear Collider (TLC) is investigated. As the result, it becomes possible to increase the aperture of the final doublet with a small increase of the horizontal β function. The new optics design uses a final doublet of 0.5 mm half-aperture and 1.4 T pole-tip field. The length of the system is reduced from 400 m to 200 m by several optics changes. Tolerances for various machine errors with this optics are also studied.

1. INTRODUCTION

One of the merits of the flat-beam scheme for a future linear collider is that it makes the final focus system very simple, because one can mainly concentrate on the chromaticity correction of one dimension. Actually, a consistent design of the final focus system for TLC with R. B. Palmer's parameters[1] has been given,[2] and the number of optical elements is much fewer than that for a round beam final focus system like SLC[3] owing to the single-family sextupole chromaticity-correction scheme. Table 1 and Fig. 1 show the main parameters and these optics.

Although this design proved the existence of a final focus system for the TLC, it still needs many improvements and further optimization. One major problem is the aperture of the final quadrupole. It requires 100 μm half-aperture and 1.4 T pole-tip field, but until today only 5 mm with 1.25 T has been built as a high-gradient quadrupole.[4] The 100 μm aperture corresponds to 6 σ of the vertical beam size of this design, however, it is proved in this workshop that it must be at least 10 σ to prevent the synchrotron radiation in the final doublet from hitting the surface of itself.[5] A large aperture quadrupole can only be achieved by extending its length, unless one can get higher surface field. This makes the chromatic effect of the quadrupole larger in general and, thus, we need to know the limit of the chromaticity correction. This will be discussed in the next chapter and a design with larger aperture quads will be given in Chap. 3.

Another problem with the final focus system is the tolerances to various machine errors. In Chap. 4 we will briefly estimate several of them.

Table 1. Parameters for the final focus system of Fig. 1.

Center-of-mass energy	$E_{\rm CM} = 1$ TeV
Beta function at IP	$\beta_x^* = 14$ mm
	$\beta_y^* = 43\ \mu$m
Normalized emittance	$\varepsilon_{\rm Nx} = 2.5 \times 10^{-6}$ m
	$\varepsilon_{\rm Ny} = 2.5 \times 10^{-8}$ m
Momentum bandwidth	$\lvert\Delta p/p\rvert \leq 0.30$ %
Pole-tip field	$B_0 = 1.4$ T
Length of the experimental area	$\ell^* = 0.4$ m
Quadrupole half-aperture	$a = 100\ \mu$m

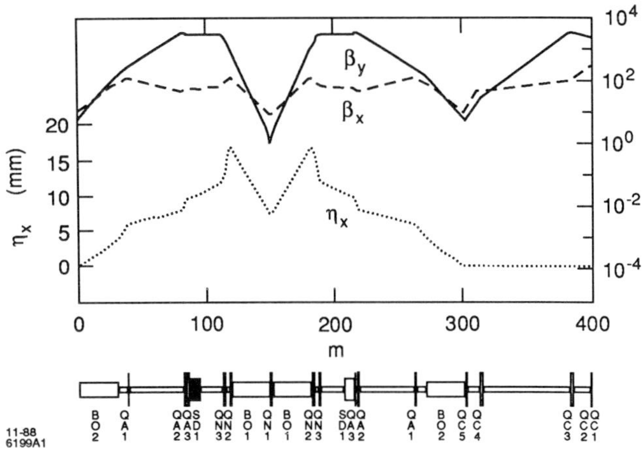

Fig. 1. A final focus optics for TLC with the single-family sextupole chromaticity correction scheme. The first character of each element specifies the type of the element as B:bend, Q:quadrupole, and S:sextupole.

*Work supported by the Department of Energy, contract DE-AC03-76SF00515.

†Visiting from KEK, National Laboratory for High Energy Physics, Oho, Tsukuba, Ibaraki 305, JAPAN.

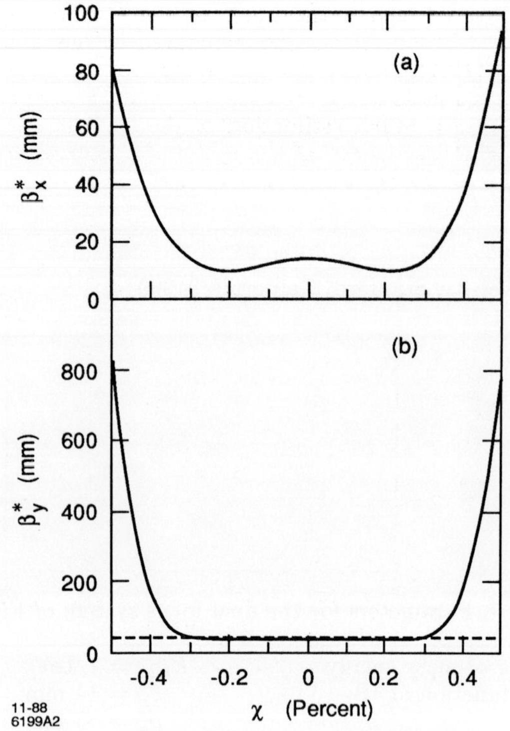

Fig. 2. The momentum dependence of the β functions at IP with the optics of Table 2. The horizontal axis is the parameter $\chi \equiv (p - p_0)/p$. The dashed line in β^*_y is the chromatic effect from the elements except the bad phase quadrupoles QC4 and QC5.

2. LIMIT OF THE CHROMATICITY CORRECTION

Figure 2 shows the momentum dependence of β functions at the interaction point with the previous optics. The behavior of the horizontal β function has been fully analyzed in Ref. 2 and the bandwidth for horizontal δ_x is determined by the horizontal chromaticity ξ_x

$$\delta_x = \frac{1}{\xi_x} \quad , \qquad \xi_x \equiv \int K\, \beta_x\, ds \quad , \tag{1}$$

where K is the focusing gradient of the final quadrupoles. When we use large aperture quadrupoles, we have to increase β^*_x to keep the desired bandwidth, as long as we use the single-family sextupole-correction scheme. Figure 3 shows how β^*_x depends on the aperture while keeping the bandwidth constant. We see in this figure that β^*_x is roughly proportional to $a^{\frac{1}{2}}$.

On the other hand, the behavior of the vertical β function has not been completely studied. We know that it is possible to cancel the momentum dependence up to fourth order by placing a sextupole at the phase difference $N\pi$ (N: integer) from the final doublet and setting the strength

$$k'\eta\, \beta_{Sy} = \xi_y \quad , \tag{2}$$

where η, β_{Sy} and ξ_y are the dispersion, vertical β function at the sextupole, and the vertical chromaticity of the final lenses. We see in Fig. 2 a large sixth-order term appears in β^*_y which gives the same bandwidth as β^*_x. We need to investigate this sixth-order term and determine the limit of the vertical chromaticity correction.

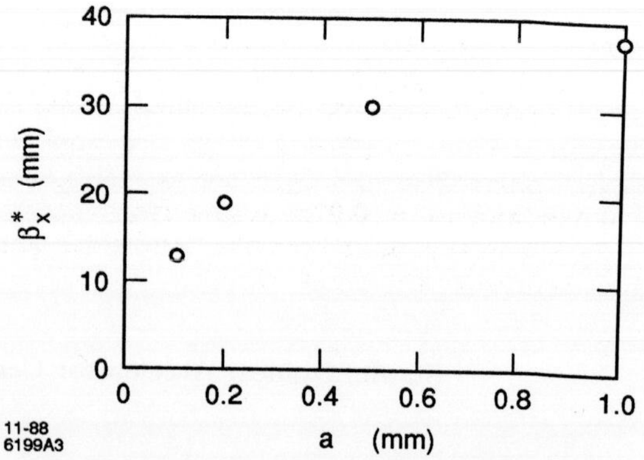

Fig. 3. The dependence of β^*_x on the aperture of the final doublet. The bandwidth is kept at $\pm\, 0.3$ %. Here we fixed ℓ^*, β^*_y, and B_0 to 40 cm, 40 μm, and 1.4 T, respectively.

There are a lot of candidates for the sixth-order term. The doublet scheme, the finite thickness of quadrupoles and a higher order term in the dispersion at the sextupole give this effect, but estimations on these effects give a few percent bandwidth in this optics, far smaller than the actual strength of the sixth-order term. An analysis below will reveal the real source: the answer is the chromatic effect of quadrupoles between the sextupole and the final quadrupoles.

Consider a quadrupole with a strength k_1 which is placed at a position with a phase difference μ_1 from the collision point. The Twiss parameters are β_1 and α_1 at the quadrupole. We consider the system by a thin-lens approximation as shown in Fig. 4. Here we omit the suffix y, but the arguments below are only applied to the vertical dimension. We represent the effect of the final doublet by one thin lens as Ref. 2. Therefore, the chromaticity of the final doublet is written as $\xi_0 = k_0\ell^2/\beta^*$. After a simple thin-lens calculation, we find that it is not possible to cancel the chromatic terms completely even below the fourth order, but the main aberration appears in the sixth-order term

$$\frac{\Delta\beta^*}{\beta^*} = \xi_0^4\xi_1^2 \left(\cos\mu_1 - \frac{\beta^*}{\ell}\sin\mu_1\right)^4 \chi^6 \quad , \tag{3}$$

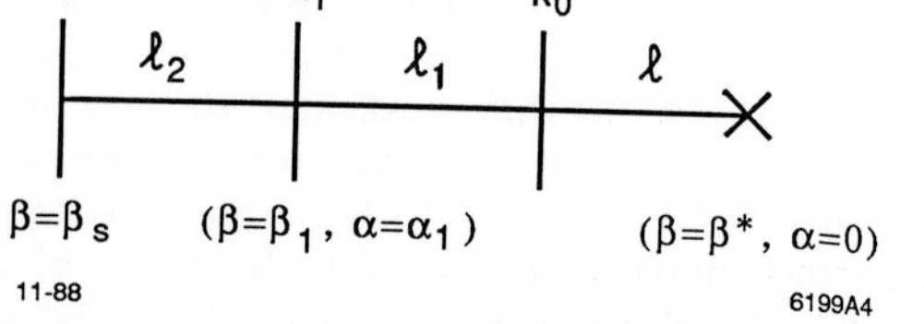

Fig. 4. A thin lens model of the final focus system. Here, ℓ is the length from the center of the final quadrupole to the collision point.

where $\xi_1 \equiv k_1\beta_1$ is the chromaticity of the intermediate lens and $\chi \equiv (p - p_0)/p$. Equation (2.3) tells us that a quadrupole gives a large sixth-order effect, unless the phase advance μ_1 is close to $(N + \frac{1}{2})\pi$. In our previous optics the quadrupoles QC4 and QC5 are at the bad positions $\mu_1 = 0.58\pi$ and $\mu_1 = 0.97\pi$, respectively. According to Eq. (2.3), the quadrupoles limit the bandwidth to ±0.27%, which is defined at the point where

$$\frac{\Delta\beta^*}{\beta^*} = 1 \quad . \tag{4}$$

This result agrees well with the behavior of β_y^* of Fig. 4. If both QC4 and QC5 were achromatic lenses, the behavior of the β function of this optics would become the dashed line of Fig. 4. It is a remarkable effect that despite the fact that the chromaticities of these lenses ($\xi_1 = 2.1$ and $\xi_1 = 0.4$ for QC4 and QC5, respectively) are much smaller than that of the final doublet ($\xi_0 = 14800$), they give the main contribution in the sixth-order term. For a lens which is at the phase advance $\mu_1 \approx (N + \frac{1}{2})\pi$, the sixth-order term given by Eq. (2.3) is smaller by the factor $(\beta^*/\ell)^4$ than those which are at the bad phase position.

What we have learned here is: "do not place a quadrupole at a position where the phase is not $(N + \frac{1}{2})\pi$ from the collision point." It is possible to remove these bad-phase quadrupoles, but at least one quadrupole is necessary to make the π phase advance between the sextupole and the final doublet.

We now consider the simplest system, which consists of one final lens, one sextupole and one intermediate lens which sets the phase advance to π. Let us denote the distances from the intermediate lens to the final lens and the sextupole as ℓ_1 and ℓ_2, respectively, as shown in Fig. 4. The π phase advance is achieved by setting the strength of the intermediate lens as

$$k_1 = \frac{\ell_1 + \ell_2}{\ell_1\ell_2} \quad ; \tag{5}$$

therefore the residual sixth-order given by Eq. (3) becomes

$$\frac{\Delta\beta^*}{\beta^*} = \xi_0^4 \, \frac{\beta^* L^2}{\beta_S \ell^2} \, \chi^6 \quad , \tag{6}$$

where L is the distance between the sextupole and the final doublet. Equation (6) gives the final limit of chromaticity correction. The bandwidth for the vertical correction is written

$$\delta_y = \left(\xi_{y0}^2 \sqrt{\frac{\beta_y^*}{\beta_{yS}}} \, \frac{L}{\ell}\right)^{-\frac{1}{3}} \quad , \tag{7}$$

which corresponds to the condition of Eq. (4), and we have restored the y suffix. This bandwidth is proportional to $\beta_y^{*(1/2)}$ if the other parameters are fixed. The strength of the sextupole is still expressed as Eq. (2), but ξ_y at the right hand side of Eq. (2) is given not only by the final lenses but also by all quadrupoles of the system.

Figure 5 shows the band widths for both x and y achieved by actual designs, compared to the calculations from Eqs. (2.1) and (2.7). The agreement in vertical bandwidth is a little worse than the horizontal one, but Eq. (2.) still gives a rough estimation of the vertical bandwidth. The total bandwidth of the system is determined by the smaller of the horizontal and vertical band widths.

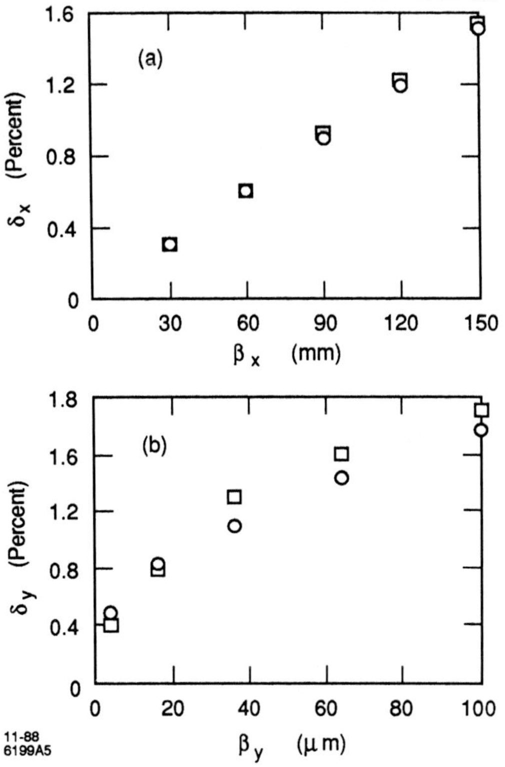

Fig. 5. The bandwidth δ_x and δ_y as functions of β_x^* and β_y^*. Here a final doublet with 0.5 mm half-aperture, 1.4 T pole-tip field, and $\ell^* = 40$ cm is used. The circles show the band widths achieved by actual designs, and squares the calculated values by Eqs. (2.1) and (2.7).

3. A LARGE APERTURE FINAL FOCUS SYSTEM

Although it is still not clear for us how small a quadrupole aperture is really available, here we use 0.5 mm half-aperture with 1.4 T pole-tip field for final doublet as an example. It is about one tenth of the aperture of quadrupole actually constructed.[4)] According to the results of the last chapter, it is possible to design an optics with this aperture and a bandwidth of 0.3%. In this case, the bandwidth is only determined by the horizontal chromaticity, and we have to increase β_x^* from 14 mm to 30 mm to achieve this bandwidth, as indicated in Fig. 3. The vertical β function is 40 μm and almost the same as before. Although we have enough acceptance in β_y^*, it is worthless to decrease β_y^* any more because it has already hit the synchrotron-radiation limit of the final quadrupole.[6)]

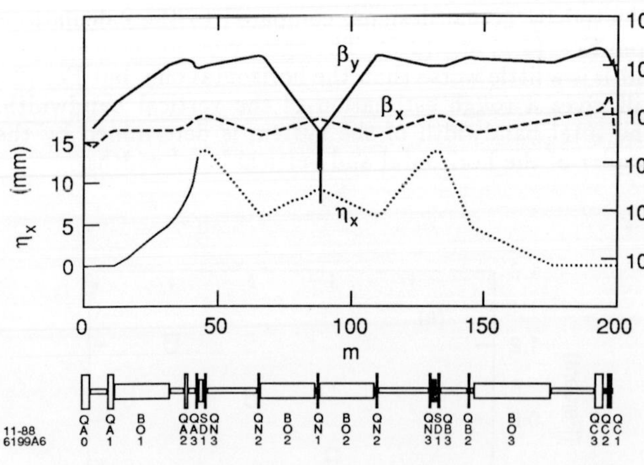

Fig. 6. A new design of the final focus optics with the 0.5 mm half-aperture final doublet.

Figure 6 shows a design of the final focus system with 0.5 mm half-aperture final doublet. This aperture is made by stretching the length of the final quadrupole QC1 from 40 cm to 120 cm and also QC2 from 40 cm to 68 cm. The pole-tip field and the length of the experimental area are the same as for the previous design. The system length is decreased to 200 m, which is just half of the previous one. In the previous design the mirror symmetry of the chromaticity correction section is kept from the first bend to the last bend, which is followed by a focusing section of 100 m length. The new design has kept the symmetry only between two sextupoles, and the dispersion suppressor and the telescope are merged into one complex. Another improvement is done to obtain a bigger dispersion/bending angle ratio. These two changes contribute to reducing the total length.

The vertical chromaticity is increased as the increase of the aperture and it needs strong sextupoles, which make the geometric aberrations large. One cure is to increase the dispersion at the sextupole, as in the paragraph above. Another cure is to shorten the length of the sextupoles. Although the shorter sextupole is the better in the geometric effects, owing to the $-I$ transformation between the two sextupoles, it is unclear how small an aperture sextupoles can be reduced to and still remain actually possible. Here we set the half-aperture to 0.4 mm, with a pole-tip field of 1.0 T, which is a modest value compared to those of the final quadrupoles.

Now the 200 m final focus system is short enough for the 500 GeV accelerator. This system has been checked by a multiparticle tracking simulation, and the total vertical aberration caused by synchrotron radiation and geometric nonlinearity is about 10% of the nominal beam size. The luminosity is decreased from the previous design due to the increase of β_x^*, but it can be recovered by changing other parameters of the collider.

Although a design with 1 mm half-aperture quadrupoles is also made, and it will be possible to increase the aperture more, a large aperture increases the chromaticity of the system and makes the tolerance of the system tighter, as shown in the next chapter.

Table 2. The optics parameters for the new design of the final focus system.

MARK

INPUT = (ALPHAX = 0 ALPHAY = 0
BETAX = 10.0 m BETAY = 10.0 m
EMIX = 2.5D–12 m EMIY = 2.5D–14 m
DP = 0.003)

DRIFT

LA0 = (L =	6.9938532)	LX0 = (L =	0.1000000)
LA1 = (L =	0.1000000)	LB2 = (L =	10.6758039)
LA2 = (L =	5.2890761)	LB1 = (L =	1.5095633)
LA3 = (L =	3.1396575)	LC4 = (L =	16.4694946)
LS2 = (L =	0.1000000)	LC3 = (L =	1.8851826)
LN2 = (L =	0.1000000)	LX2 = (L =	0.0400000)
LN1 = (L =	18.9086844)	LX1 = (L =	0.4000000)

BEND

B01 = (L = 21.0000000 ,	ANGLE = 0.000499346458)
B02 = (L = 21.0000000 ,	ANGLE = 0.000300000000)
B03 = (L = 29.0000000 ,	ANGLE = 0.000300000000)

QUAD

QA0 = (L = 3.0000000 ,	K1 = 0.2847124)
QA1 = (L = 2.0000000 ,	K1 = 0.2160445)
QA2 = (L = 1.0000000 ,	K1 = −0.1020459)
QA3 = (L = 1.0000000 ,	K1 = 0.0933806)
QN3 = (L = 1.0000000 ,	K1 = 0.0349031)
QN2 = (L = 1.0000000 ,	K1 = −0.0664416)
QN1 = (L = 1.0000000 ,	K1 = 0.0651954)
QB3 = (L = 1.0000000 ,	K1 = 0.0489054)
QB2 = (L = 1.0000000 ,	K1 = −0.0892923)
QC3 = (L = 3.0000000 ,	K1 = −0.2265658)
QC2 = (L = 0.6800000 ,	K1 = 1.1431554)
QC1 = (L = 1.2000000 ,	K1 = −2.0005762)

SEXT

SD1 = (L = 2.0000000 ,	K2 = −132.0000000)

LINE

FFS =	(INPUT	QA0	LA0	QA1	LA1
B01	LA2	QA2	LA3	QA3	LS2
SD1	LN2	QN3	LN1	QN2	LX0
B02	LX0	QN1	LX0	B02	LX0
QN2	LN1	QN3	LN2	SD1	LS2
QB3	LB2	QB2	LB1	B03	LC4
QC3	LC3	QC2	LX2	QC1	LX1)

Table 2 lists the parameters of these optics in the same format as Ref. 2.

4. TOLERANCES FOR MACHINE ERRORS

What we discuss in this chapter are the tolerances for various machine errors which reduce luminosity. In this section we discuss tolerances to abrupt changes of components from pulse to pulse. This is usually called jitter. For the flat beam collider, the tolerances are much tighter in the vertical dimension than the horizontal, because the final spot size is smaller and the chromaticity is higher in the vertical. We will show some simple expressions to estimate the tolerances for only the vertical, but these are easily applicable to the horizontal dimension. Figures in

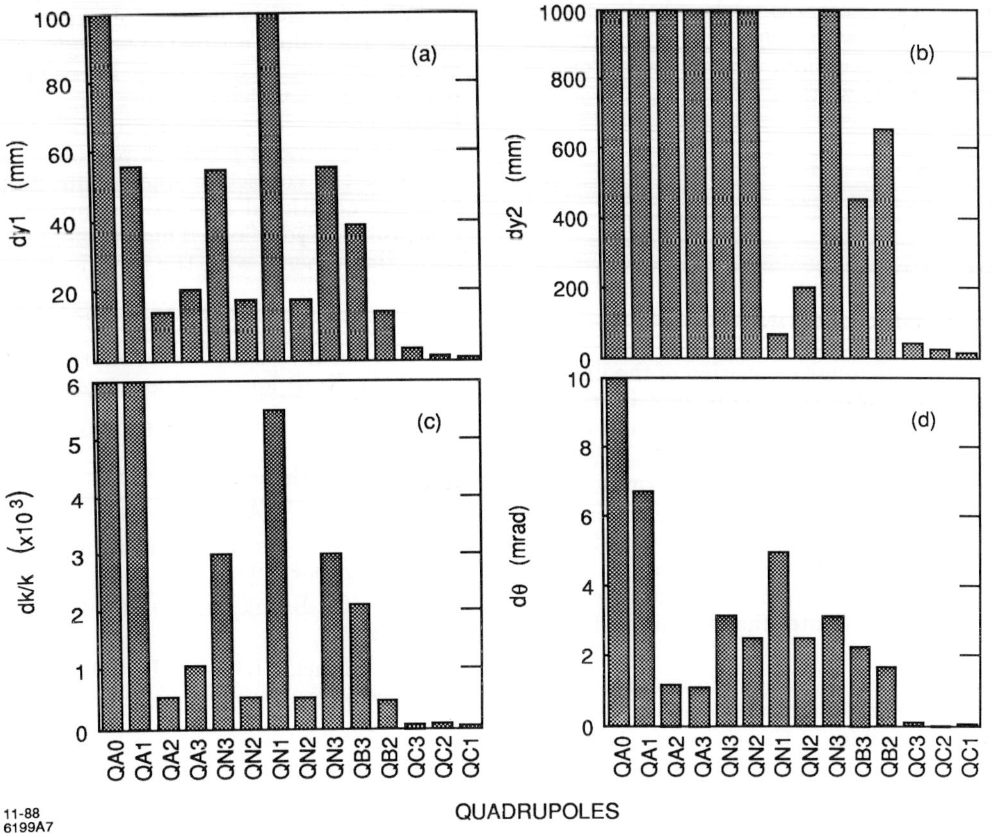

Fig. 7. Tolerances for vertical misalignments dy_1 and dy_2, strength error dk/k, and roll ang

this chapter show the values of machine errors for each quadrupole which reduce the luminosity to $1/\sqrt{2}$ of the nominal value. The new optics described in the last chapter is used in this calculation. We take into account only the first order optics and its first order momentum dependence.

The first kind of the error is the simplest one. The misalignment dy_1 of each quadrupole vertically shifts the beam spot and prevents the head-on collision. The amount which reduces the luminosity by $1/\sqrt{2}$ is

$$dy_{1i} = \frac{\sqrt{2\log 2}}{k_i \sin\mu_i}\sqrt{\frac{\varepsilon_y}{\beta_{yi}}} \quad , \tag{8}$$

where μ_i, β_{yi}, and k_i are the phase difference from the IP, the vertical β function, and the strength of the i-th quadrupole, respectively.

There is another effect of a misalignment of a quadrupole. A vertical displacement makes the vertical spot size larger due to the vertical dispersion created by the vertical displacement in the final doublet. The tolerance to this effect is written as

$$dy_{2i} = \frac{\sqrt{3}}{k_i\xi_y \cos\mu_i\delta}\sqrt{\frac{\varepsilon_y}{\beta_{yi}}} \quad , \tag{9}$$

where ξ_y is the chromaticity of the system, and we have assumed the beam has a uniform energy distribution of the full width 2δ. This effect becomes weaker for quadrupoles between two sextupoles, because about the half of the chromaticity is corrected, and ignorable for the quadrupoles before the chromaticity correction section.

Quadrupole strength error dk/k and roll $d\theta$ along its longitudinal axis also make the vertical spot size larger. The tolerances for them are written

$$\frac{dk_i}{k_i} = \frac{1}{k_i\beta_{yi}\sqrt{\sin^2\mu_i + \xi_y^2\delta^2\left[\frac{\sin^2 2\mu_i}{2} + \left(\frac{1}{3} + \frac{\xi_y^2\delta^2}{5}\right)\cos^4\mu_i\right]}} \tag{10}$$

and

$$d\theta_i = \frac{\sqrt{\frac{\varepsilon_y}{\varepsilon_x}}}{k_i\sqrt{\beta_{xi}\beta_{yi}\left(\sin^2\mu_i + \frac{\xi_y^2\delta^2\cos^2\mu_i}{3}\right)}} \quad . \tag{11}$$

Figure 7 shows above tolerances for each quadrupole used in this system. We see in this figure all tolerances are about one order more severe for the final three quadrupoles than other quadrupoles.

Tolerances for sextupoles can be estimated similarly. The tolerances for the horizontal displacement is calculated in the same expression as Eq. (4.3) by replacing dk

with $k'dx_S$. The vertical tolerances is also obtained from Eq. (11) by replacing $k_i d\theta$ with $k'dy_S$. These values are $dx_S = 0.82$ μm and $dy_S = 0.85$ μm. The tolerance for the sextupole strength dk'/k' is written

$$\frac{dk'}{k'} = \frac{\sqrt{3}}{k'\eta\beta_{yS}\delta} \quad , \tag{12}$$

which gives $dk'/k' = 3.4 \times 10^{-2}$.

5. DISCUSSIONS

In this paper we have a rough evaluation of the limit of the vertical chromaticity correction. This limit is still far from the present requirement of β_y^* and the actual bandwidth is determined only by β_x^*, due to the single-sextupole scheme. We have to abandon this scheme when we need smaller β_x^*, and use a two-family sextupole scheme instead. What limits the chromaticity correction with two-family sextupole scheme has not been studied, but another sixth-order terms from the coupling of these two families will appear, and the limit will be tighter than the single-family scheme.

We have achieved improvements of the final focus optics in the aperture of the final doublet and the total length. These will give more reality to TLC.

We have estimated the tolerances for several machine errors. The values estimated here give the maximum jitters of the components discussed. There are also other kinds of machine errors which have dc or very low frequency components. These are correctable by monitoring beam position and size in various points in principle, but there are some limits to the correction due to the finite errors of monitors and additional aberrations induced by correctors like synchrotron radiation. This problem also depends on the correction scheme with the specific optics.

ACKNOWLEDGMENT

The author would like to express his sincere thanks to Dr. R. D. Ruth for valuable suggestions.

REFERENCES

1. R. B. Palmer, SLAC–PUB–4295 (1987).
2. K. Oide, SLAC–PUB–4660 (1988).
3. Stanford Linear Accelerator Center, *SLC Design Handbook,* 1984.
4. W. O. Brunk and D. R. Walz, SLAC–PUB–3672 (1985).
5. T. Himel, SLAC–PUB–4804 (1988) and these Proceedings.

FINAL FOCUS SUPPORTS FOR A TEV LINEAR COLLIDER*

W. W. Ash

Stanford Linear Accelerator Center, Stanford University, Stanford, California 94309

ABSTRACT

Final focus quadrupoles supported from structures in the endcap region of a physics experiment appear to meet the high-frequency vibration and stability criteria for a TeV Linear Collider (TLC). The support stays within a ten-degree cone, minimizing interference with the experimental apparatus.

SUPPORTING BEAM

Final focus optics and a small-bore quadrupole have been discussed in early studies of a TLC.[1)] The quadrupoles would be about 2 centimeters tall and 40 centimeters long and would weigh about 1.5 kg. The support must keep the vertical position of each doublet stable to a few percent of the vertical collision-point height of about 10^{-9} m.

Consider supporting the quadrupole doublet on a mechanical beam that fills the ten-degree region and anchors to a firm, seismically damped base in the endcap region as shown in Fig. 1. (A model for this is the SLD detector at the SLC. The central detector and two endcaps are roughly comparable in size. Supporting machine components from the endcaps minimizes interference with physics components.)

The tapered, cantilevered mechanical supporting beam[2)] has a static deflection of about 0.6 μm and a simple vibration frequency, $\nu = \sqrt{g/y}$, of about 600 Hz.

SEISMIC ISOLATION

The seismic isolation is based on the system used in the Caltech gravity wave detector,[3)] illustrated in Fig. 2. Although their detector is elegant, the basic idea in the passive part of the support is not high-tech — it is comparable to the "granite table on rubber tires" mentioned in R. Palmer's talk at this conference.

With this arrangement the Caltech group measured a seismic vibration spectrum as shown in Fig. 3, which is

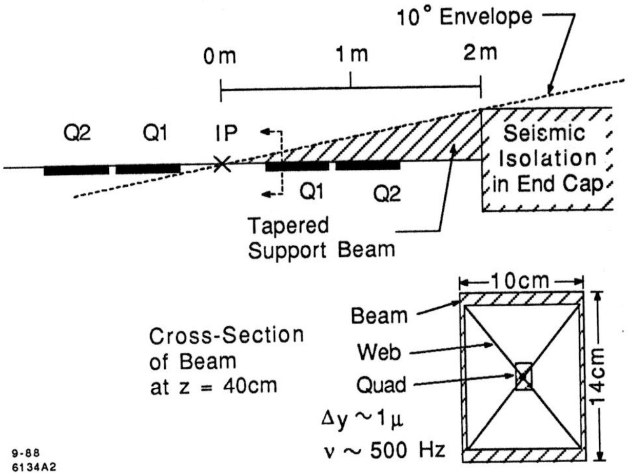

Fig. 1. Conceptual layout of a mechanical support beam for the small quadrupole doublet. The insert shows the cross section of the beam and the drawing shows the beam tapered to fill the available dead region. The beam is cantilevered from a heavy, seismically damped mass in the endcap region of a detector.

adapted from their report. (The isolation is too efficient to measure the effect directly so they used a mechanical thumper very close to their support and then scaled the vibrations down.)

The low-frequency curve is the normal Pasadena ground motion plotted directly on the curve in the region where the damping can do no good. The extrapolation is a guess at what to expect in between.

The adaptation here would exploit the mass of the endcap of the physics experiment, possible isolation in the floor under the endcap, and a suspension system inside the ten-degree cone. Seismic vibrations will drive the cantilevered beam with the amplitude enhanced at the beam's resonance. The Caltech data shows a similar resonance from the wire-support system around 800 Hz. Given the very rapid falloff of the motion, this effect is not serious.

* Work supported by the Department of Energy, contract number DE-AC03-76SF00515.

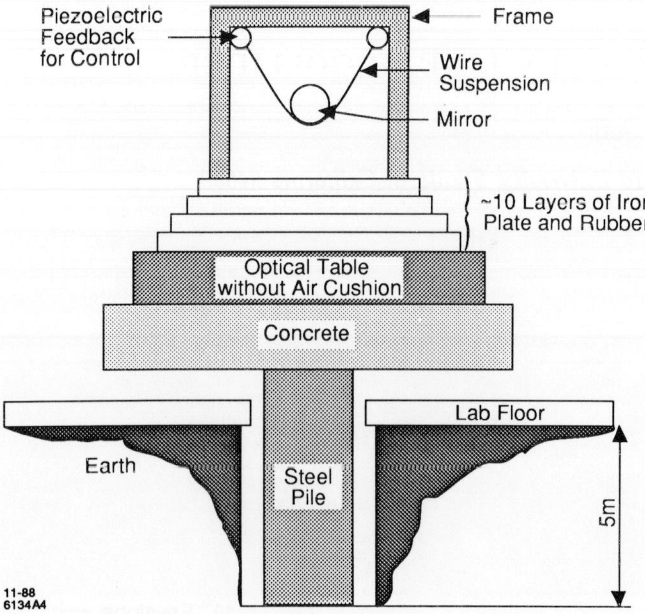

Fig. 2. Schematic layout of the passive seismic isolation for the Caltech interferometric gravity wave detector. In the application to a collider final focus, the heavy mass is the experiment's endcap, the floor might be appropriately modified, the optical table fits into the ten-degree dead region, and the final focus support beam hangs on the suspended "mirror".

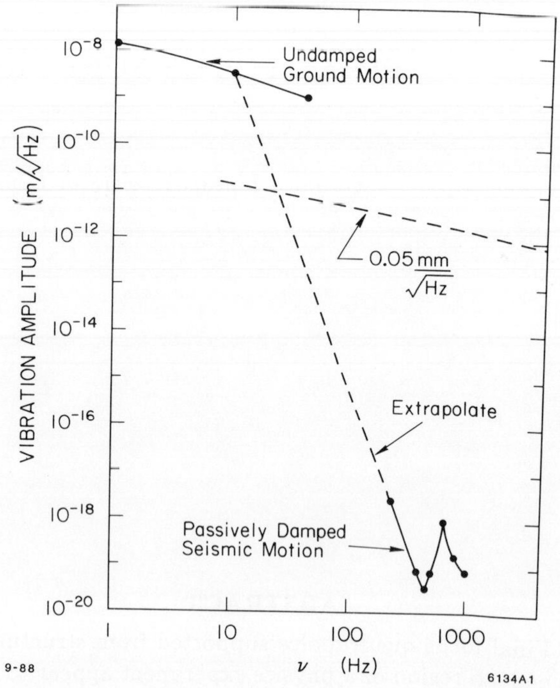

Fig. 3. This graph, adapted from the Caltech work, shows the noise measured on the kind of suspension sketched in Fig. 2. The extrapolation joins the indirect high-frequency data with undamped low-frequency seismic vibrations. The curve marked 0.05 $mm/\sqrt{Hz}$ *is an estimate of the collider requirement.*

The dashed curve falling off as $1/\sqrt{\nu}$ is the assumed criterion for supporting the linear collider final focus to give acceptably small vibrations over the frequency band.*

SLOW FEEDBACK

Above 100 Hz the isolation looks very good. Below around 10 Hz, however, the seismic noise becomes too large as the isolation becomes ineffective. Low frequency drifts, including rubber creep and thermal effects, will overpower the isolation of the doublets so something else is required to maintain alignment.

Steering coils would normally provide for this correction of the spot position. This is limited, however, to a few μm or the synchrotron radiation from the bending of the beam in the final quadrupoles may degrade performance.[1)] Something else is needed to keep the system mechanically stable to around 1 μm.

The Caltech experiment has a similar problem in maintaining the orientation of their detector components. A system of low-power lasers, reflecting mirrors and photodiode detectors operate below 30 Hz on piezoelectric transducers at the final suspension. (Note: this is a simple system and not the elegant high-power optical-cavity laser system used as the primary detector.) This achieved control in angles at the μrad level adequate to keep static tolerances of μm.

A combination of this active support control and steering correctors must be used to manage the low-frequency end. These problems are probably much harder to solve than the high frequency seismic case addressed in this note.

REFERENCES

1. Palmer, R. B., SLAC–AAS–36 (1988) and Oide, K., SLAC–PUB–4660 (June 1988).
2. Roark, R. J. and Young, W. C., *Formulas for Stress and Strain*, Fifth Edition, McGraw-Hill (1982). The standard formulas for simple bending are tabulated on page 98 and the conversion factors for tapered sections are given on page 174. The stiffening factor of about 50 requires some extrapolation.
3. Suspension systems and seismic background are discussed in the NSF proposals for a Laser Interferometer Gravitational Wave Observatory by Caltech and MIT. The relevance was noted by W. Atwood (SLAC) during visits to the site and in discussions there with R. W. P. Driver. For a more thorough discussion, see Fischer, G. E. and Mayoud, M., *Some Thoughts on Linear Collider Alignment and Mechnical Stability*, CERN–LEP–RF–SU–88–07 (March 1988).

* The curve comes by integrating a $1/\nu$ function for the square of the amplitude from frequencies too low to isolate up to those too high to affect the beam, and then requiring the overall rms amplitude to be of order 10% or less of the final spot size.

DISRUPTION, BEAMSTRAHLUNG, AND BEAMSTRAHLUNG PAIR CREATION*

PISIN CHEN

Stanford Linear Accelerator Center, Stanford, California 94309

ABSTRACT

The two major effects from the interaction of e^-e^+ beams — beamstrahlung and disruption — are reviewed, with emphasis on flat beam collisions. For the disruption effects we discuss the luminosity enhancement factor, the maximum and *rms* disruption angles, and the "kink instability". All the results are obtained from computer simulations, and scaling laws based on these are deduced whenever possible. For the beamstrahlung effects, we concentrate only on the final electron energy spectrum and the deflection angle associated with low energy particles. In addition to the generic studies on the beam-beam effects, we also list the relevant beam-beam parameters obtained from simulations on two sample designs: the TLC and the ILC. As an addendum, the newly discovered phenomenon of coherent beamstrahlung pair creation, together with the incoherent process, are discussed.

INTRODUCTION

There are two major phenomena induced by the beam-beam interaction which are important to the design of high energy linear colliders. Namely, the *disruption* process where particle trajectories are bent by the field provided by the oncoming beam, and the *beamstrahlung* process where particles radiate due to the bending of the trajectories. The most important impact of disruption is the deformation of the effective beam sizes during collision, which causes an enhancement on the luminosity. In addition, the disruption angle affects the constraints on the final quadrupole aperture. When the two beams are colliding with certain initial offset, the disruption effect between the two beams would induce a kink instability, which imposes a constraint on beam stability. Ironically, this instability helps to relax the offset tolerance for flat beams, because the offset beams tend to find each other during the initial stage of the instability. Under a multi-bunch collision mode, however, the kink instability will largely degrade the luminosity through the relatively long growth time. On the other hand, the direct impact of beamstrahlung is the loss of the available energy for high energy events, and the degradation of beam energy resolution because of the stochastic nature of the radiation.

* Work supported by the Department of Energy, contract number DE-AC03-76SF00515.

Most of the issues raised above can be studied by decoupling the disruption and the beamstrahlung effects. The energy loss due to beamstrahlung may modify the luminosity enhancement but this effect can be ignored since we are only interested in the case where the average energy loss is small. Conversely, the average energy loss and the final energy spectrum can be studied by assuming no disruption without compromising too much on accuracy. There is, however, one issue where the two effects are strongly coupled. This is the maximum disruption angle associated with the large deflections from particles that have suffered severe energy loss.

In this report we summarize what has been studied on these issues with emphasis on flat beam collisions. The computer simulations are performed using the code ABEL (Analysis of Beam-beam Effects in Linear colliders) described in Ref. 1, but improved considerably since it was first written. Some results given here are still preliminary and will be refined in later papers, but their qualitative features will not be changed.

LUMINOSITY ENHANCEMENT

Our primary interest is the enhancement of luminosity due to the mutual pinching of the two colliding beams. The details have been discussed in Ref. 2 for round beams and will be given in Ref. 3 for flat beams. As was pointed out in Ref. 2, the luminosity is infinite if the initial beam is paraxial and the computation is perfectly accurate. This is because a paraxial beam can be focused to a singular point. In reality, however, a beam will always have certain inherent divergence, and the singularity is only approached asymptotically. To account for this effect, a parameter $A_{x,y} = \sigma_z/\beta^*_{x,y}$ is introduced,[2)] which is proportional to the emittance for a given beam size $\sigma_{x,y}$. The computed enhancement factor $H_D = L/L_0$, where L_0 is the geometrical luminosity without the effect of the depth of focus related to $A_{x,y}$ taken into account, is plotted in Fig. 1 as a function of D_y and A_y for flat beams.

The data in Fig. 1 are obtained by using a distribution function which is uniform in x and Gaussian in y and z (UGG), instead of a three-dimensional Gaussian distribution (GGG), for easiness of computation. The enhancement factor of GGG distributions for a given D_y can be

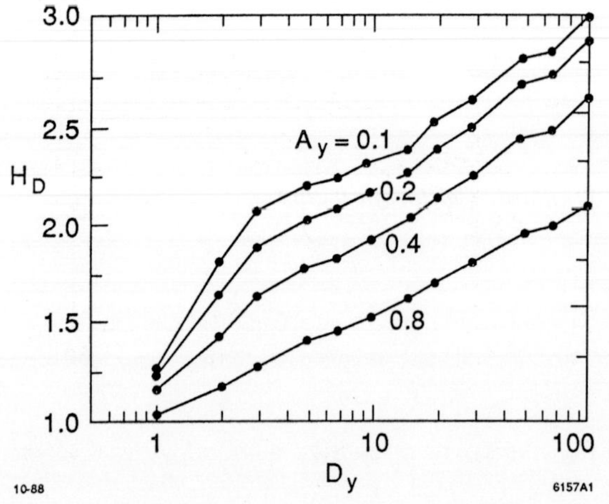

Fig. 1. Luminosity enhancement factor for flat beams.

deduced from a superposition of UGG results with disruption parameters ranging from 0 near the horizontal edge to $\sqrt{6/\pi}D_y$ at the beam center. The enhancement factor for round beams is shown in Fig. 2.

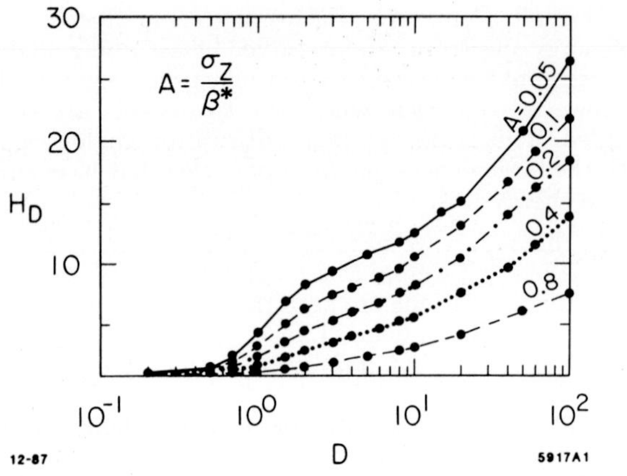

Fig. 2. Luminosity enhancement factor, round beams.

By comparing Figs. 1 and 2, one finds that the enhancement factor for flat beams scales roughly as the cube-root of the corresponding value for round beams; which obeys the following empirical scaling law that fits all data points in Fig. 2 to within 10% accuracy:

$$H_D = 1+D^{1/4}\left(\frac{D^3}{1+D^3}\right)\left[ln(\sqrt{D}+1)+2ln\left(\frac{0.8}{A}\right)\right] \quad . (1)$$

The reason for the flat beam enhancement not being scaled as a square root of the corresponding value for the round beam is because the horizontal focusing can enhance the vertical pinch effect (and vice versa) in the round beam case, whereas for flat beams the pinch in the major (horizontal) dimension can hardly affect the disruption in the minor dimension.

In both cases, our results indicate a logrithmic divergence of H_D as a function of A, or A_y. In addition, H_D is monotonically increasing as a function of D, or D_y, at least up to D =100. This second finding of ours is qualitatively the same as that found by Fawley and Lee[4] but in contradiction to Holebeek[5] and Solyak,[6] where the enhancement factor first saturates before eventually decreases at large D's.

The difference appears to be due to the different ways of handling stochastic errors. In a Monte Carlo simulation such as ours, the initial condition is generated by random numbers, which introduces a statistical fluctuation, and therefore an asymmetry, of the order $1/\sqrt{N_p}$, N_p being the number of macro particles. This asymmetry will be amplified during collision (i.e., kink instability) due to the beam-beam force, especially when the disruption parameter is large. The fact that the number of macro particles in a simulation is typically much smaller than the actual particle number, this fluctuation is artificially enhanced if no proper action is taken. To minimize this computation error, the particle distribution function is symmetrized at every time step in our calculation, so that the beam-beam force has the up-down symmetry at all times for the flat beam case. Similarly, in the round beam case only the radial force is computed. This process eliminates the possible instability triggered by computation errors.

The actual collisions are expected to have some unavoidable initial offset in alignment and skewness in distributions. Since the asymmetry in distributions tend to shift the center-of-gravity of the beams, it gives rise to the same effect as the initial alignment offsets. For this reason our study on the effect of imperfections is concentrated on initial offsets only.

As will be discussed in the next section, an initial offset triggers a kink instability, especially when the disruption parameter becomes large. As it occurs, this instability is not always harmful because, in the initial phase of the instability, the beams always tend to find each other, which prevents the otherwise rapid degradation of the luminosity for large initial offsets. Figure 3 shows the luminosity enhancement factor as a function of offset Δ_y (in units of σ_y) for various values of D_y. The dotted curve is the geometrical enhancement factor without beam-beam force, which is equal to $\exp(-\Delta_y^2/4)$. UGG distribution is used and A_y =0.2 for all curves. The up-down symmetry is not enforced except for the cases at Δ_y =0.

From Fig. 3 one finds that the tolerance on alignment offset reaches an optimum for values of D_y between 5 and 10. Within this range of D_y, H_D is still above unity even at $\Delta_y \sim 3$. Beyond this region of D_y the beam-breakup becomes severe while below which the beam-beam attraction is not yet strong enough.

The same data as in Fig. 3 is replotted in Fig. 4 as a function of D_y and each curve corresponds to a fixed value of Δ_y. (The region of large D_y and small Δ_y is not very accurate because of its sensitivity to computing errors.) One sees a saturation and decrease of H_D as a function of D_y unless $\Delta_y = 0$. One also notices that the curves with small offsets, e.g., $\Delta_y = 0.2$, resemble the results in Refs. 5 and 6, except that our offset was explicitly

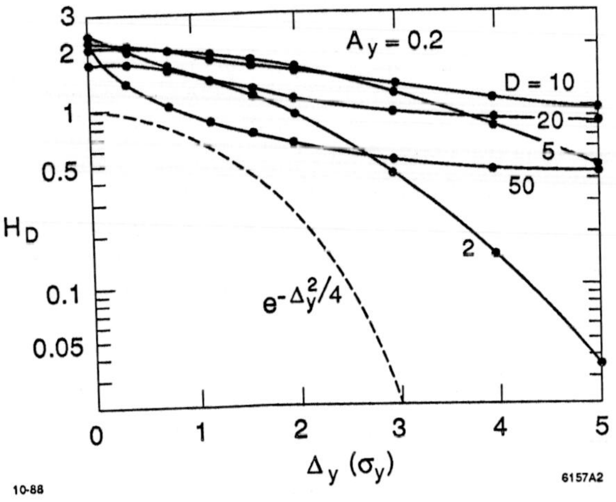

Fig. 3. H_D as a function of offset; Flat beams.

introduced. In designing a linear collider, one needs to estimate H_D for the chosen D and A. This depends on the assesment of potential imperfections of beam-beam collision. Though arbitrary, it may be safer to adopt the curve for $\Delta_y = 0.2$ or 0.4, instead of $\Delta_y = 0$, as the effective enhancement factor.

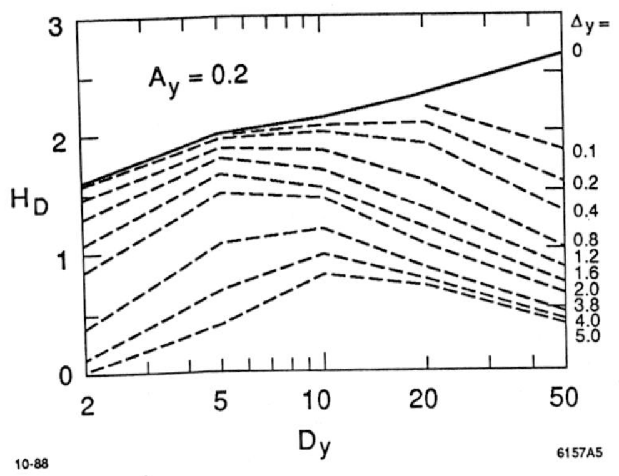

Fig. 4. H_D as a function of D_y; Flat beams.

Similar exercise for round beams are shown in Fig. 5 for D up to 50. Here we find the generic behavior as in the case for flat beams.

DISRUPTION ANGLE

Information on the final direction of the electron trajectory after collision is necessary in designing the interaction region, especially for the aperture of the final quadrupole magnets. If the disruption parameter is very small, the transverse location of a particle during collision is nearly constant. Then we can estimate the disruption angle θ_x

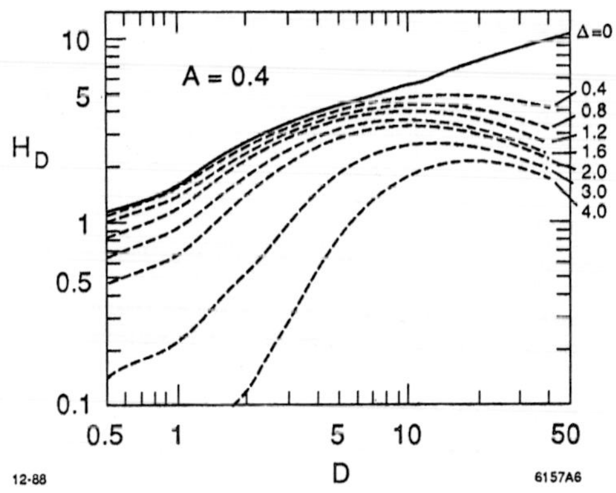

Fig. 5. H_D as a function of offset; Round beams.

and θ_y as functions of the initial transverse coordinates x_0 and y_0. For very flat Gaussian beams we have

$$\theta_x = -\sqrt{\frac{\pi}{2}} D_x \frac{\sigma_x}{\sigma_z} \mathcal{I}m \Big[\frac{i}{\pi} \int_{-\infty}^{+\infty} \frac{\exp(-t^2)dt}{(x_0/\sqrt{2}\sigma_x) - t - i0} \Big] \quad , \quad (2)$$

$$\theta_y = -\sqrt{\frac{\pi}{2}} D_y \frac{\sigma_y}{\sigma_z} \Big[\frac{2}{\sqrt{\pi}} \int_0^{y_0/\sqrt{2}\sigma_y} e^{-t^2} dt \Big] e^{-x_0^2/2\sigma_x^2} \quad , \quad (3)$$

where the quantities in the square brackets can be expressed by the complex error function $w(x_0/\sqrt{2}\sigma_x)$ and the real error function $\mathrm{erf}(y_0/\sqrt{2}\sigma_y)$. Here the emittance is ignored. One finds that the maximum and *r.m.s.* disruption angles to be

$$\theta_{x,max} = 0.765 D_x \frac{\sigma_x}{\sigma_z} \quad , \quad (x_0 = 1.31\sigma_x) \quad (4)$$

$$\theta_{y,max} = \sqrt{\pi/2} D_y \frac{\sigma_y}{\sigma_z} \quad , \quad (x_0 = 0,\ y_0 = \infty) \quad (5)$$

$$\theta_{x,rms} = \sqrt{\pi/(6\sqrt{3})} D_x \frac{\sigma_x}{\sigma_z} \quad ,$$

$$\theta_{y,rms} = \sqrt{\pi/(6\sqrt{3})} D_y \frac{\sigma_y}{\sigma_z} \quad . \quad (6)$$

(Rigorously speaking, for flat beams with large but finite aspect ratio, θ_y reaches a maximum near $y_0 \sim \sigma_x$ and then decreases; but this is not important.)

The distribution functions of θ_x and θ_y are shown in Fig. 6. The actual singularities at $\theta_x = \theta_{x,max}$ and $\theta_y = 0$ are not supposed to be as sharp as those in Fig. 6 because of finite emittance, various errors, and the disruption effect. However, we found from simulations that the qualitative difference between the horizontal and vertical angles still holds even for $D_{x,y}$ not much less than unity.

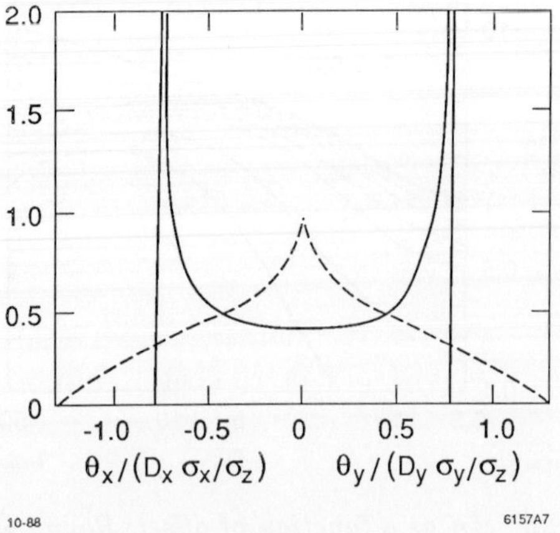

Fig. 6. Distribution of $\theta_{x,y}$ for small $D_{x,y}$.

Figure 7 shows the maximum and *r.m.s.* vertical disruption angle, in units of $D_y\sigma_y/\sigma_z$, as a function of D_y. Here we consider the case for small D_x only. The four curves corrrspond to $A_y = 0.1, 0.2, 0.4, 0.8$, respectively. The dependence on A_y is not as significant as in the case of H_D except for the small D_y region, where the beam divergence is emittance dominated. (The distribution of initial $\sigma_{x'}$ is truncated at 2.5 standard deviations in the simulation.)

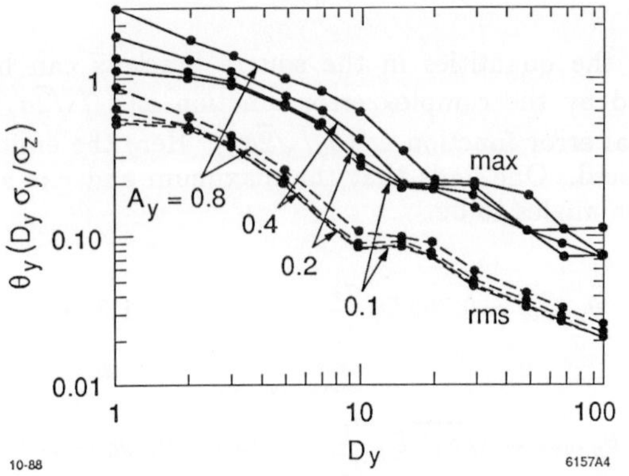

Fig. 7. Maximum and r.m.s. vertical disruption angle.

The simulation results can roughly be fitted by

$$\theta_{y,rms} \sim \sqrt{\frac{\pi}{6\sqrt{3}}}\frac{\sigma_y}{\sigma_z}\frac{D_y}{[1+(0.5D_y)^5]^{1/6}} \tag{7}$$

and $\theta_{y,max} \sim 2.5\theta_{y,rms}$. Here the contribution of the initial emittance ($= A_y\sigma_y/\sigma_z$ for $\theta_{y,rms}$) has not been included. The reason that the angle does not increase linearly in D_y is that the particle trajectories are bent backwards and oscillate when D_y is large.

So far, the collision is assumed to be head-on. For flat beams the disruption angle in the presense of vertical offset is also important in determining the aperture of the final quads. The mean deflection angle of the entire bunch can be written in the form[7]

$$\Theta_y = \frac{1}{2}\frac{\sigma_y}{\sigma_z}D_yH_c(D_y,\Delta_y) \quad , \tag{8}$$

where Δ_y is the vertical offset in units of σ_y and the weak dependence on A_y is ignored. For small disruptions, the function H_c approaches the following analytic form

$$H_c(D_y,\Delta_y) = \int_0^{\Delta_y} e^{-y^2/4}dy \quad . \tag{9}$$

Figure 8 shows H_c as a function of Δ_y computed by simulations, where UGG distribution is assumed.

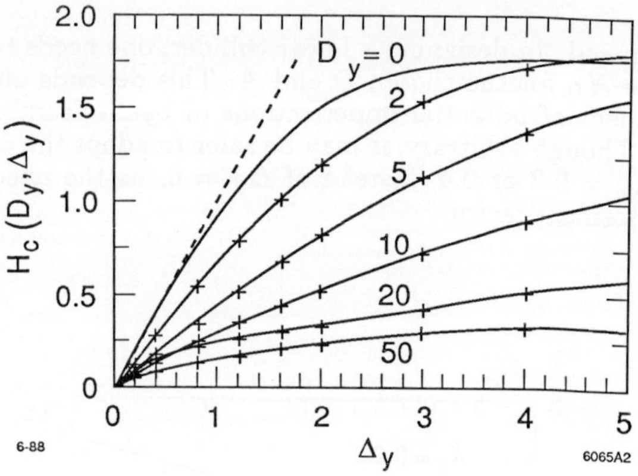

Fig. 8. Effective center-of-mass deflection.

Roughly speaking, the maximum disruption angle in the presense of offsets is the sum of the center-of-mass deflection angle Θ_y and the maximum angle in the absense of offsets, $\theta_{y,max}$.

KINK INSTABILITY

If one of the beams is displaced vertically for some reason, this offset triggers a vertical oscillation and, when D is large, the oscillation is enhanced by the beam-beam force. This phenomena is known as the *kink instability.* Figure 9 shows a specific example.

In the above figure the bunch is sliced longitudinally and the vertical coordinate y of the center-of-mass of each slice (in units of σ_y) is plotted against the longitudinal coordinate s (in units of σ_z). Each graph corresponds to a snapshot of the beam vertical position at a particular time t (in units of σ_z/c). The development of the instability can be seen in time sequence. The initial offset in this example is chosen to be $0.2\sigma_y$ (full) and the disruption parameter is $D_y = 20$.

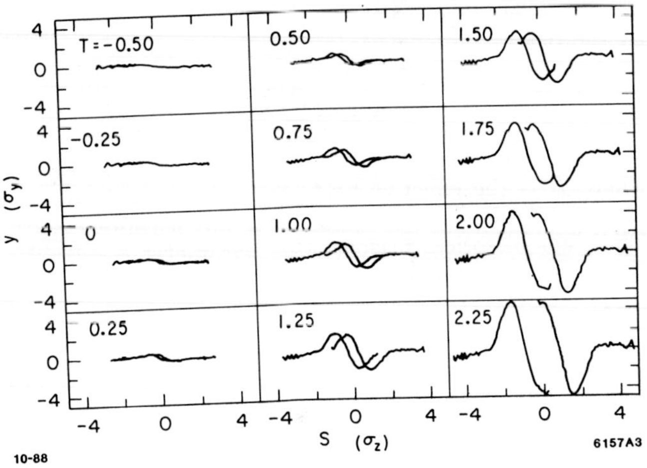

Fig. 9. An example of kink instability.

e⁻ and e⁺ beams is varified to be $\pi/3$, and the growth rate is as predicted. Furthermore, Fig. 9 clearly demonstrates the standing-wave nature of the kink instability, which agrees with the description of Eq. (11).

So far our discussion on the kink instability deals with collisions of two bunches. Another type of kink instability occurs during the collision of two bunch trains, each consists of N_B bunches. One of the major problems of such a multibunch operation is the interactions between bunches before and after their collisions at the central collision point. The ith bunch in the electron bunch train will collide not only with the ith bunch in the positron train, but also with the $j(<i)^{th}$ positron bunch before its coming to the central collision point. Colliding two flat beams at a relatively large crossing angle can help to avoid unwanted direct encounters between the outgoing bunch debris and the incoming fresh bunches. However, due to the long-

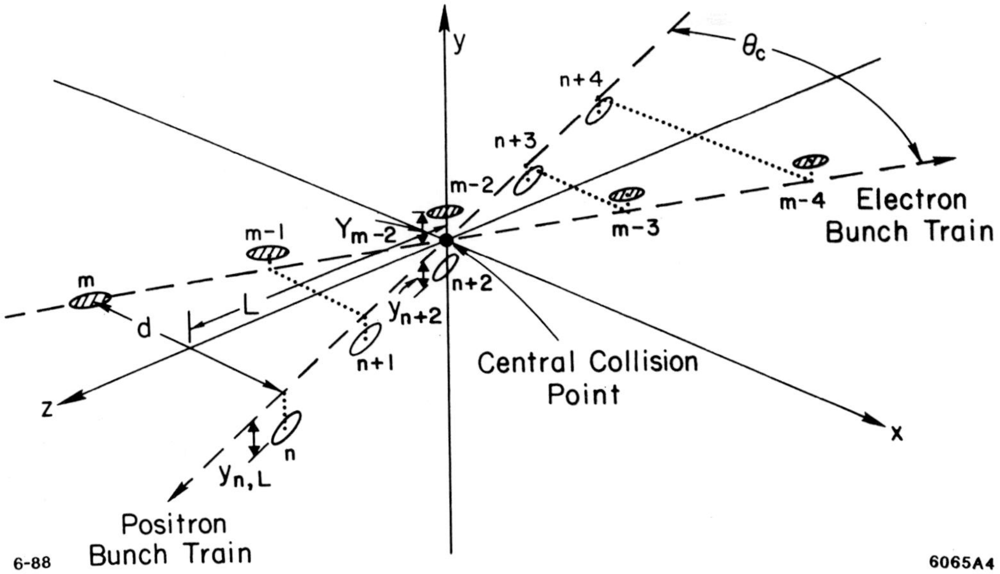

Fig. 10. Schematic diagram for collisions of bunch trains.

For uniform beams and small amplitude oscillations, the equation of motion for the beam particles can be obtained from fluid dynamics (the flat beam version of the equation is given in Ref. 8),

$$\left[\frac{\partial}{\partial t} \pm \frac{\partial}{\partial s}\right]^2 y_{\pm} = -\omega_0^2 (y_{\pm} - y_{\mp}), \quad \omega_0^2 = \frac{\sqrt{2\pi}}{6}\frac{D_y}{\sigma_z^2} \quad , \quad (10)$$

where $y_{\pm}$ is the y coordinate of e⁺and e⁻ beams. The most unstable solution is found to be

$$y_{\pm} = \text{const.} \times \exp\left[\pm i(\frac{\sqrt{3}}{2}\omega_0 s - \frac{\pi}{6}) + \frac{1}{2}\omega_0 t\right] \quad . \quad (11)$$

This solution is in reasonable agreement with the simulation shown in Fig. 9. Namely, the phase difference between range nature of the Coulomb interaction, there still exists undesirable interference between two separated bunches at a distance. Since the crossing angle cannot be made arbitrarily large due to the luminosity consideration, this long range interaction cannot be entirely suppressed. In fact, it imposes a severe restriction on the stability of the beams.

Consider the encounter between the nth positron bunch after collision and the mth $(m > n)$ electron bunch before collision at a distance L from the collision point. A schematic diagram of the system is shown in Fig. 10. We assume that all the bunch encounters occur within the drift space around the central collision point.

According to Eq. (8), the center-of-mass deflection angle for the nth positron bunch is

$$\Theta_{yn} = \frac{1}{2}\frac{\sigma_y}{\sigma_z} D_y H_c(D_y, \Delta_{yn}) \quad , \quad (12)$$

where Δ_{yn} is the relative offset between the mth electron and the nth positron bunch, in units of σ_y, at their closest encounter. The cumulative offset for the mth bunch before arriving at the central collision point is therefore

$$\Delta_m = C \sum_{n<m} H_c(D_y, \Delta_n) + \delta_m \quad , \tag{13}$$

where δ_m is the initial offset of the mth beam, and the coefficient C is

$$C = D_x D_y \left(\frac{\theta_d}{\theta_c}\right)^2 \quad , \tag{14}$$

and $\theta_d = \sigma_x/\sigma_z$ is the diagonal angle of the bunch.

The cumulative offset Δ_m (in units of $\delta(1+C)^{m-1}$) is plotted as a function of the number of bunches in Fig. 11. Since the factor θ_d/θ_c must be larger than unity in order that the crossing angle does not reduce the luminosity significantly, the condition for negligible growth of the instability, i.e., $\Delta_{yn} \lesssim \delta$, according to Fig. 11, is roughly

$$(N_B - 1)\ D_x D_y \lesssim 2 \quad . \tag{15}$$

This imposes a constraint on the allowable number of bunches per train.

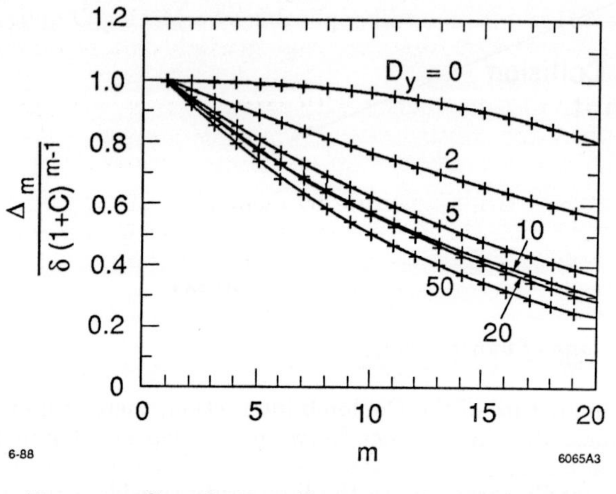

Fig. 11. Cumulative offset as a function of the number of bunches.

ENERGY SPECTRUM OF FINAL ELECTRONS

The energy spectrum of the electrons is important for two reasons: The *tip* of the spectrum, i.e., the distribution near the initial beam energy, provides information on the energy resolution for high energy physics events. On the other hand, the *tail* of the spectrum, i.e., the distribution of the low energy electrons, which had suffered severe energy loss through hard beamstrahlung, reveals the likelyhood of finding large disruption angles. This second issue will be addressed in the next section.

The energy spectrum of radiation can be characterized by the beamstrahlung parameter Υ, defined as

$$\Upsilon = \gamma \frac{B}{B_c} \quad , \tag{16}$$

where B is the effective field strength of the beam, and $B_c = m^2c^3/e\hbar \sim 4.4 \times 10^{13}$ Gauss is the Schwinger critical field. For historical reasons, this parameter is related to the parameter ξ introduced by Sokolov and Ternov, by a simple factor

$$\xi = \frac{\text{(critical energy)}}{\text{(initial energy)}} = \frac{3}{2}\frac{r_e\gamma^2}{\alpha\rho} = \frac{3}{2}\Upsilon \quad , \tag{17}$$

where ρ is the instantaneous radius of curvature. Since the two parameters are trivially related, we shall employ either of them depending on the convenience of the situation. The typical value of ξ during collision is

$$\xi_1 = \frac{r_e^2\gamma N}{\alpha\sigma_z\sigma_y}\frac{2}{1+R} \quad . \tag{18}$$

The average value of ξ is a bit smaller than Eq. (18) (by about a factor 2/3) but we adopt Eq. (18) for the better description of the spectrum tail which is contributed more effectively from beamstrahlung with larger ξ.

The number of emitted photons per electron is

$$N_\gamma = N_{cl}U_0(\xi_1), \text{ with } N_{cl} = 2.12\frac{\alpha r_e N}{\sigma_x + \sigma_y} \quad , \tag{19}$$

where N_{cl} is the number of photons computed by the classical formula and $U_0(\xi)$ is the ratio of the quantum-theoretical number of photons to that from the classical theory, and is found to be[9]

$$U_0(\xi) = \frac{1 - 0.598\xi + 1.061\xi^{5/3}}{1 + 0.922\xi^2} \quad , \tag{20}$$

where the relative error is within 0.7%.

An approximate formula for the energy spectrum of electrons after collision has been derived recently. The details will be given in Ref. 10. Here we only quote the results. The distribution function $\psi(\varepsilon)$ ($\varepsilon = E/E_0$), normalized as $\int \psi(\varepsilon)d\varepsilon = 1$, can be written as

$$\psi(\varepsilon) \simeq e^{-N_\gamma}\left[\delta(\varepsilon - 1) + \frac{e^{-y}}{1-\varepsilon}h(N_1 y^{1/3})\right] \quad , \tag{21}$$

with

$$\begin{aligned} h(x) &= \frac{1}{2\pi i}\int_{\lambda-i\infty}^{\lambda+i\infty} \exp(xp^{-1/3} + p)dp \quad (\lambda > 0) \\ &= \sum_{n=1}^{\infty} \frac{x^n}{n!\Gamma(n/3)} \end{aligned} \tag{22}$$

and

$$y = \frac{1}{\xi_1}(\frac{1}{\varepsilon} - 1) \quad ,$$

$$N_1 = \frac{1}{1+\xi_1 y} N_{cl} + \frac{\xi_1 y}{1+\xi_1 y} N_\gamma \quad . \tag{23}$$

(This formula does not exactly satisfy the normalization condition except for $\xi_1 \to 0$ which leads to $N_1 = N_\gamma = N_{cl}$.) The function $h(x)$ can be estimated very accurately by with relative error less than 2%. Figure 12 compares Eq. (21) with the simulation results using the parameters for the TLC and the ILC.[11] The design parameters of the two colliders are summarized in Table 1. The histograms in Fig. 12 are from simulations and the dotted data are computed from Eq. (21). The agreement is excellent.

Table 1. Parameters for TLC and ILC ($\lambda_{rf} = 17mm$)

	TLC	ILC
E_0[TeV]	0.5	0.25
N	8×10^9	7×10^9
σ_x[nm]	190	440
σ_y[nm]	1	3
σ_z[μm]	26	65
R	190	147
ϵ_x [mrad]	2.58×10^{-12}	5.2×10^{-12}
ϵ_y [mrad]	2.33×10^{-14}	5.2×10^{-14}
D_x	0.033	0.027
D_y	6.27	3.9
A_x	0.0002	0.0017
A_y	0.60	0.37
* L/L_0	1.61	1.71
* δ	0.15	0.01
* N_γ	1.33	0.38
ξ_1	3.43	0.19

* Quantities computed by simulations.

MAXIMUM DEFLECTION ANGLE UNDER BEAMSTRAHLUNG

The particle which once lost a large fraction of its initial energy through beamstrahlung would in principle be severely deflected by the beam–beam field and cause background problems for high energy experiments. Consider an electron which emits a hard photon at a particular time during the collision and results in an energy εE_0,

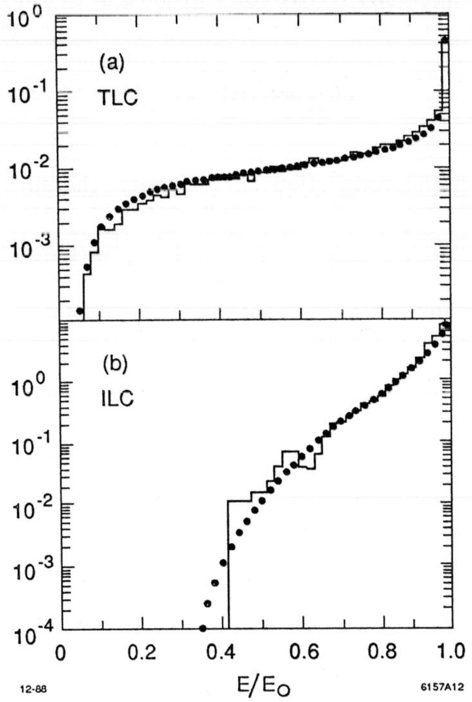

Fig. 12. Electron energy spectrum for TLC and ILC parameters.

with $\varepsilon \ll 1$. The effective disruption parameter for this particle becomes D_x/ε and D_y/ε. One might think that Eqs. (4) and (5) are still applicable by replacing D by D/ε. However, the collision of a single particle on a beam with the disruption parameter D/ε is different from the collision between two beams with D/ε, although the qualitative feature is the same; i.e., the disruption angle increases linearly in D for $D \lesssim 1$ and more slowly for $D \gtrsim 1$.

A simulation was done by monitoring low energy test particles through the collision process. The maximum deflection angle for a given ε is found to be roughly

$$\theta_{max} \sim \frac{\sigma}{\sigma_z} \frac{D/\varepsilon}{\sqrt{1+(0.75D/\varepsilon)^{4/3}}} \quad , \quad (\varepsilon \ll 1) \tag{24}$$

where $D = D_x(D_y)$ and $\sigma = \sigma_x(\sigma_y)$ for the horizontal (vertical) angle.

The minimum value of ε can in principle be as small as $1/\gamma$. But the real problem is about how small a ε should one care. Since the number of photons N_γ per beam particle for linear colliders in the near future is of order unity, the spectral function $\psi(\varepsilon)$ given in Eq. (21) is always dominated by the factor e^{-y} in the spectrum tail, where $y \gg 1$ (in logarithmic sense). Therefore, if the acceptable background counts is n out of N electrons, then the minimum ε of concern is approximately determined by $y = \log(N/n)$, or

$$\varepsilon_{min} = \frac{1}{1+\xi_1 \log(N/n)} \quad . \tag{25}$$

With this value of ε, one can directly estimate the maximum deflection angle using Eq. (24). Since the dependence on n is only logarithmic, one can set $n = 1$. Thus, for example, $\varepsilon_{min} = 0.013(0.188)$, $\theta_{x,max} = 10$ (0.95) mrad and $\theta_{y,max} = 0.4$ (0.15) mrad for TLC (ILC) parameters.

ADDENDUM: BEAMSTRAHLUNG PAIR CREATION

After the completion of this paper, the author identified a new phenomenon called "coherent beamstrahlung pair creation",[12] which, together with the incoherent process studied earlier,[13,14] would have impacts on linear collider designs. Recall that in the case of radiation by $e^-(e^+)$ during beam-beam collision, there are essentially two mechanisms that induce the radiation. Namely, there is an "incoherent" process, or Bremsstrahlung, associated with the individual e^-e^+ scatterings, and there is also a "coherent" process due to the interaction between the radiating charged particle and the macroscopic beam–beam EM field. At high energies and strong fields, the coherent process tends to dominate over the incoherent one. This is actually why our discussion on beam energy loss has been focused only on the beamstrahlung process.

The beamstrahlung photons once emmited would have to travel through the remainder of the oncoming beam before entering into free space, and would therefore turn themselves into e^-e^+ pairs. Analogous to the case of radiation, photon pair creation also involves coherent and incoherent processes. Here again, at high energies and strong fields the coherent process will dominate over the incoherent one. Once the e^-e^+ pairs are created with lower energies in general, one of the two particles in each pair will have the same sign of charge as the oncoming beam (For the sake of argument, consider a low energy e^+ moving against the positron beam). Unlike the case of low energy e^- moving against a positron beam, where the potential tends to confine the particle in the beam profile, in the case of a positron the potential is unconfining, and the particle can in principle be deflected by a large angle and thus create severe background problems. This effect would therefore impose a contraint on the final focus design.

It is well known that the cross section for incoherent pair creation is

$$\sigma(\gamma e \to ee^+e^-) \sim \frac{28}{9}\alpha r_e^2 \log\Big(\frac{2\omega E}{m^2}\Big)\mathrm{cm}^2 \quad , \qquad (26)$$

which is a very slowly varying function of the photon energy ω. For TLC, $\gamma = 1 \times 10^6$, the cross section is $\sim 5\times10^{-26}$ cm^2 for photons at full energy. The beam parameters for TLC listed in the above Table gives the average number of the beamstrahlung photon per beam particle as $N_\gamma \sim 1.3$. On the other hand, it can be shown[13] that the effective luminosity for such a cascading process is 1/2 of the original. Thus, the number of e^-e^+ pairs created per bunch crossing can be easily evaluated to be

$$N^i_{e^+e^-} = \frac{1}{2}\sigma(\gamma e \to ee^+e^-)\frac{N_\gamma L_{ee}}{f_{\mathrm{rep}}} \sim 2 \times 10^5 \quad , \qquad (27)$$

where $L_{ee} = 1.3 \times 10^{33}\mathrm{cm}^{-2}\mathrm{sec}^{-1}$, and $f_{\mathrm{rep}} = 220\ \mathrm{sec}^{-1}$ in this design. To be sure, this process provides a non-negligible amount of e^-e^+ pairs.

The rate of photon pair creation in a homogeneous magnetic field has been studied by many people,[15] and has been generalized to inhomogeneous fields by Baier and Katkov.[16] In the asymptotic limits the rate can be expressed as

$$\frac{dI}{dt} = \begin{cases} \frac{3\sqrt{3}}{16\sqrt{2}}\frac{\alpha\Upsilon}{\lambda_c\gamma}e^{-8/3\chi} & , \quad \chi \ll 1 \ ; \\ \frac{15}{7}\left(\frac{2}{3}\right)^{1/3}\frac{\Gamma(5/6)}{\Gamma(1/6)}\frac{\alpha\Upsilon}{\lambda_c\gamma}\chi^{-1/3}] & , \quad \chi \gg 1 \ . \end{cases} \qquad (28)$$

Here $\chi = \Upsilon\omega/E$ plays the similar role as Υ in the case of beamstrahlung. Notice that χ is independent of the initial particle energy γ, as the process does not care where the photon was originated from. Let

$$\frac{\mathrm{d}I}{\mathrm{d}t} \equiv \frac{\alpha\Upsilon}{\lambda_c\gamma}T(\chi) \quad . \qquad (29)$$

To a very good approximation,[17]

$$T(\chi) = 0.16\chi^{-1}K^2_{1/3}\Big(\frac{4}{3\chi}\Big) \quad , \qquad (30)$$

for all values of χ.

Integrating over the collision time — again, only half of the e^-e^+ collision time — we have

$$\begin{aligned} I &= \frac{\sqrt{3}}{2}\frac{\sigma_z\alpha\Upsilon}{\lambda_c\gamma}T(\chi) \\ &= \frac{1}{2}n_{cl}T(\chi) \quad . \end{aligned} \qquad (31)$$

Next we evaluate the mean value of $T(\chi)$ by weighting over the beamstrahlung spectral function,

$$\frac{\mathrm{d}n_b}{\mathrm{d}\omega} = \frac{1}{\pi}\frac{\alpha\sigma_z}{\gamma^2}\left\{\int_y^\infty K_{5/3}(x)\mathrm{d}x + \frac{\xi^2 y^2}{1+\xi y}K_{2/3}(y)\right\} \quad , \qquad (32)$$

and

$$\langle T(\Upsilon)\rangle = \int_0^E T(\chi)\frac{\mathrm{d}n_b}{\mathrm{d}\omega}\mathrm{d}\omega \Big/ \int_0^E \frac{\mathrm{d}n_b}{\mathrm{d}\omega}\mathrm{d}\omega \quad . \qquad (33)$$

The total number of e^-e^+ pairs created through this coherent process is therefore

$$N^c_{e^+e^-} = \frac{1}{2}N_\gamma n_{cl}\langle T(\Upsilon)\rangle \quad . \qquad (34)$$

A plot of $\langle T(\Upsilon)\rangle$ is shown in Fig. 13, where the solid curve is from the exact form of $dn_b/d\omega$ in Eq. (32), and the dashed curve corresponds to an asymptotic expression for $dn_b/d\omega$ at large y. The closeness between the two curves

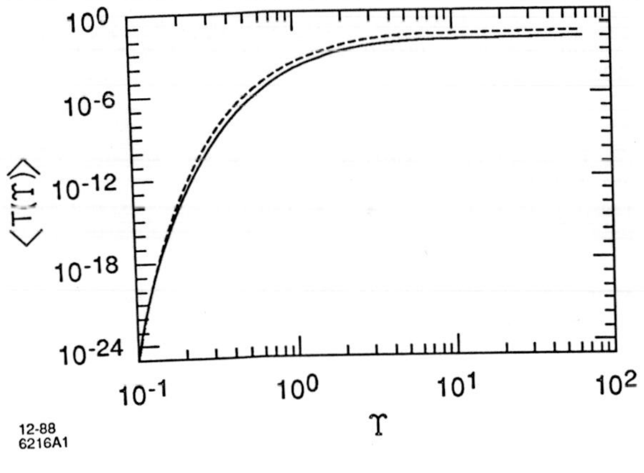

Fig. 13. The function $\langle T(\Upsilon)\rangle$ vs. Υ.

suggests that only the spectrum tip contribute effectively to the coherent pair creation process. From the TLC parameters, $n_{cl} \sim 1.9$, so we find that

$$N^c_{e^+e^-} \sim 5 \times 10^7 \quad , \tag{35}$$

which is much larger than the incoherent process.

It should be noticed, however, that $\langle T\rangle$ drops exponentially for $\Upsilon \lesssim 1$. Therefore, for next generation linear colliders at the range of 1 TeV, which would typically have $\Upsilon \sim 1$, it is not at all difficult to redesign the machine such that the coherent process can be entirely suppressed. For the above-mentioned TLC parameters the condition is $\Upsilon \leq 0.3$. This, ironically, is an over-kill of the issue since the incoherent process corresponds to $\Upsilon \sim 0.6$, as can be read from Fig. 13.

Since to a large extent N_γ is of the order unity and quite insensitive to other parameters, and since we usually choose to fix the luminosity in a design, the incoherent e^-e^+ pairs can not be easily suppressed. It is thus important to evaluate the energy spectrum of the pair created e^+. Assuming constant probability in finding the e^+ at energy $\varepsilon E \leq \omega$, the spectrum can be derived to be

$$N_{e^+}(\varepsilon) = \frac{7(3/2)^{2/3}}{18\pi^2\Gamma(1/3)}\frac{\alpha^3}{\gamma m} N D_y \Upsilon^{2/3} F(\varepsilon,\Upsilon) \quad . \tag{36}$$

The spectral function $F(\varepsilon,\Upsilon)$ is plotted in Fig. 14 for $\Upsilon = 0.2$. At the small ε limit, $F(\varepsilon,\Upsilon) \propto 1/\varepsilon$.

Finally, we evaluate the deflection angle of these low energy positrons by the beam–beam field. As a rough estimation, we assume that the vertical field beyond the beam height extents constantly to a distance equal to the beam width σ_x. It can be shown[18] that the deflection angle for the e^+ with energy ε is

$$\theta_\varepsilon = \begin{cases} \frac{2}{3^{1/4}}\frac{\sigma_x}{\sigma_z}\sqrt{\frac{D_x}{\varepsilon}} \quad , & \theta_\varepsilon \geq \frac{2}{\sqrt{3}}\theta_d \quad ; \\ 2\frac{\sigma_x}{\sigma_z}\frac{D_x}{\varepsilon} \quad , & \theta_\varepsilon \leq \frac{2}{\sqrt{3}}\theta_d \quad . \end{cases} \tag{37}$$

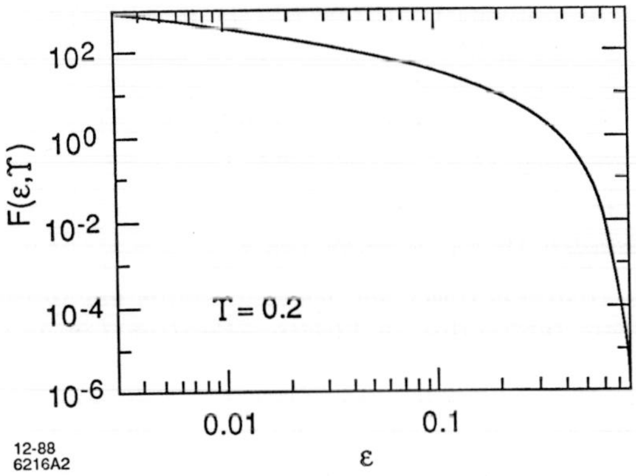

Fig. 14. The spectral function for incoherent pair created e^+.

The deflection angle in the above expression is plotted in Fig. 15. For a 1 GeV e^+, $\theta_\varepsilon \sim 45$ mrad. The information on the transverse monemtum can be easily deduced from the above expressions via $p_\perp = \varepsilon\theta_\varepsilon$.

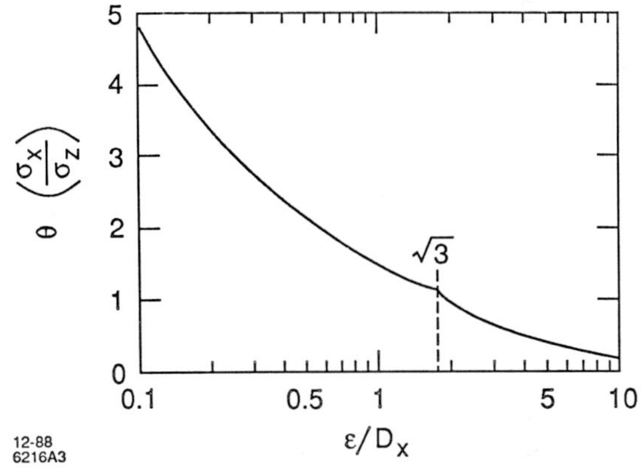

Fig. 15. The deflection angle as a function of e^+ *energy.*

REFERENCES

1. K. Yokoya, *A Computer Simulation Code for the Beam–Beam Interaction in Linear Colliders,* KEK Report 85–9, Oct 1985.
2. P. Chen and K. Yokoya, *Disruption Effects from the Interaction of Round* e^+e^- *Beams,* Phys. Rev. D **38**, 987 (1988).
3. P. Chen and K. Yokoya, *Disruption Effects from the Interaction of flat* e^+e^- *Beams,* in preparation.
4. W. M. Fawley and E. P. Lee, *Particle in Cell Simulations of Disruption,* in "New Developments in Particle Acceleration Techniques", Orsay 1987, CERN 87–11, ECFA 87/110.
5. R. Holebeek, *Disruption Limits for Linear Colliders,* Nucl. Ins. Meth. 184 (1981).

6. N. A. Solyak, *Collision Effects in Compensated Bunches of Linear Colliders,* Novosibirsk Preprint IYF–88–44, 1988.

7. K. Yokoya and P. Chen, *Multiple Bunch Crossing Instability,* SLAC–PUB–4653, June 1988.

8. Y. H. Chin, *Stability of a Colliding Beam in a Linear Collider,* DESY January 1987.

9. K. Yokoya and P. Chen, *Depolarization due to Beam-Beam Interaction in Electron–Positron Linear Colliders,* SLAC–PUB–4692, September 1988; and in the *Proceedings for the Eighth International Symposium on High Energy Spin Physics,* 1988.

10. P. Chen and K. Yokoya, *Final Electron Energy Spectrum and Maximum Deflection Angle under Beamstrahlung,* in preparation.

11. R. B. Palmer, *Interdependence of Parameters for TeV Linear Colliders,* SLAC–PUB–4295, April 1987; and SLAC–AAS–39, November 1988.

12. P. Chen, *Review of Beam-Beam Interaction,* summary talk given at the International Workshop on the Next Generation Linear Colliders, November 28–December 9, 1988, SLAC, Stanford.

13. M. S. Zolotarev, E. A. Kuraev and V. G. Serbo, *Estimates of Electromagnetic Background Processes for the VLEPP Project,* Inst. Yadernoi Fiziki, Preprint 81–63, 1981; English translation SLAC TRANS–0227, 1987.

14. I. F. Ginzburg, G. L. Kotkin, V. G. Serbo and V. I. Telnov, *Colliding γe and $\gamma\gamma$ Beams Based on the Single-Pass Accelerators,* Inst. Nucl. Phys. (Novosibirsk) Preprint 81–102, 1981.

15. See, for example, W. Y. Tsai and T. Erber, Phys. Rev. D **10**, 492, 1974; and references therein.

16. V. N. Baier and V. M. Katkov, Soviet Phys. JETP **26**, 854, 1968.

17. T. Erber, Rev. Mod. Phys. **38**, 626, 1966.

18. V. I. Telnov, private communications.

QUANTUM BEAMSTRAHLUNG: PULSE SHAPING PROSPECTS FOR A PHOTON-PHOTON COLLIDER*

RICHARD BLANKENBECLER AND SIDNEY D. DRELL

Stanford Linear Accelerator Center
Stanford University, Stanford, California 94309

Beamstrahlung is the fractional energy loss to radiation as one beam pulse is accelerated by the electromagnetic field of the other pulse through which it passes in a collider. It is very important for high energy electron-positron linear colliders ($\sqrt{s} \sim \frac{1}{2} - 1\, TeV$) that are being studied.[1)] In linear colliders the individual pulses must be formed with much larger charge densities to compensate for their much lower rate of interactions ($10^2 - 10^3$ per second) in comparison to storage rings (typically 10^5 per second). Therefore the electromagnetic fields generated by the charges in each pulse will be correspondingly larger, as will the acceleration they induce in the charges of the intersecting pulses; this in turn leads to a larger loss of energy to radiation. In the classical approximation the beamstrahlung is given by

$$\delta_{\text{classical}} = \frac{8\pi}{3} N^2 \gamma \frac{r_0^3}{\ell\{\pi B^2\}} \tag{1}$$

where N is the number of electrons (positrons) per pulse, for a uniformly charged cylindrical pulse of length ℓ and cross section πB^2; $\gamma = E/M$ is the beam energy and $r_0 \equiv 2 \cdot 8 \times 10^{-13}\, cm$ is the classical electron radius.

In this report we consider three questions:

1. What is the quantum result for beamstrahlung and when is the classical result valid?
2. How does the transverse beam geometry—*i.e.*, ribbon versus cylindrical pulse—affect the beamstrahlung?
3. Extending the quantum calculation to multiple photon emission, what is the photon flux and what are the prospects for a photon-photon collider?

* Work supported by the Department of Energy, contract DE-AC03-76SF00515.

1) Quantum beamstrahlung

Synchrotron radiation is a well-studied problem. Its application to beamstrahlung in the very high energy quantum limit was made in a seminal paper by Himel and Siegrist. The calculation can be done analytically[2)] for the physical case of small disruptions—*i.e.*, for parameters such that the impact parameter of a beam particle changes fractionally only by a small amount as it traverses an intersecting beam pulse. Making this approximation which is an appropriate one for envisioned TeV colliders the quantum result for beamstrahlung for uniformly charged colliding pulses with cylindrical geometry is

$$\delta = \delta_{\text{classical}}\, F(C) \tag{2}$$

where the form factor $F(C) \to 1$ in the classical limit ($\hbar \to 0$, $C \to \infty$) and $F(C) \to C^{4/3}(0.83)$ for $C \to 0$ in the extreme quantum limit. The physical interpretation of C is the ratio of a transverse coherence length, $\ell_\perp$, to the radiation length, ℓ_{rad}:

$$C \propto \frac{\ell_\perp}{\ell_{\text{rad}}}\,. \tag{3}$$

$\ell_\perp$ is defined as the length along the trajectory of a beam particle as it traverses a pulse for it to acquire a transverse momentum $p_\perp \sim m$ from the field produced by the charges of the pulse; after travelling a distance $\ell_\perp$ the trajectory turns through an angle m/E. This angle is the same as the width of the radiation pattern from a relativistic charged particle; $\theta_{\text{tr}} \sim 1/\gamma = m/E$. Thus the radiation ceases to be coherent after a particle moves a length $\ell_\perp$ along the curving path. The second length, ℓ_{rad}, is defined as the length of the target that the electron coherently scatters from during the radiation process. According to the uncertainty principle, it is the reciprocal of the minimum longitudinal momentum transfer to the target. As shown in ref. 2

$$C = \frac{1}{4\gamma y}\left\{\frac{\ell}{\hbar/mc}\right\} \tag{4}$$

where

$$y \equiv \frac{Nr_0}{B} = r_0\sqrt{\pi\mathcal{L}} \tag{5}$$

is a dimensionless classical variable proportional to the square root of the luminosity per pulse, $\mathcal{L}$, and we inserted $\hbar$ and c as appropriate to show that $C \propto 1/\hbar$. In the classical limit the radiation occurs locally and by (3) and (4), $C \propto 1/\hbar \to \infty$ as claimed above. In all cases currently under study as practical in the TeV energy range both lengths $\ell_{\rm rad}$, $\ell_\perp \ll \ell$ and end effects of the pulse are small.[2] Using (5) we can rewrite (1) and (2) in terms of scaling variable C which, for fixed luminosity per pulse, $\mathcal{L}$, and for pulses of circular cross-section is the only parameter appearing:

$$\delta = \frac{2}{3}\,\alpha y\,\frac{F(C)}{C}\ ; \quad \alpha \equiv \frac{1}{137}\,. \tag{6}$$

The classical approximation is valid for $C \gg 1$, $F(C) \to 1$, in which case the beamstrahlung is small since for practical values of $\mathcal{L} \lesssim 10^{30} cm^{-2}$, $y \lesssim 400$.

2) Ribbon Beams

The field strengths generated by the pulses are reduced when the circular cross section of a cylindrical pulse is distorted to an ellipse of the same area. Therefore the resulting acceleration of a beam particle is also reduced, as is the energy loss to radiation. As defined in ref. 2 this reduction can be summarized in the aspect ratio $1/G$ defined as the ratio of the mean value of the semi-major and minor axes to their average:

$$\frac{1}{G} = \frac{2\sqrt{a_x a_y}}{a_x + a_y} \le 1\,. \tag{7}$$

The result is[2]

$$\delta^{\rm ellipse} = \frac{2}{3}\,\frac{\alpha y}{G}\,\frac{F(CG)}{(CG)}\,. \tag{8}$$

The fact that $\delta^{\rm ellipse}_{\rm classical} \propto 1/G^2$ follows from the fact that $(\text{acceleration})^2 \propto (\text{field strength})^2 \propto (1/G)^2$. Since $G \ge 1$, the argument CG of the form factor increases with increasing values of the aspect ratio and we move toward the classical region for fixed C.

3) Multiple photon emission and prospects for a $\gamma - \gamma$ collider

Interest in this possibility is kindled by the fact that the expected gamma flux due to beamstrahlung for TeV linear colliders now under study is one to two orders of magnitude larger than that calculated in the Weiszacker-Williams equivalent-photon approximation, and the resultant photon spectrum is hard.[3]

This result introduces a new dimension into the design and application of $e^\pm$ linear colliders. For constant luminosity, the beam pulses can be shaped as desired in order to either maximize or minimize the fields and the resulting beamstrahlung. The beamstrahlung can be reduced by forming beams with very thin ribbon-like cross sections, *i.e.*, $G \sim 5$, permitting study of e^+e^- interactions at maximum energy and with narrow energy spread. Oppositely, in order to study high energy photon-photon processes, the collider can be tuned by shaping the beam pulses with smaller aspect ratio, *i.e.*, $G \sim 2$, in order to increase the flux of hard photons. We note that the multiple photon contributions substantially alter the spectral distribution of radiated photons but have little effect on the fractional energy loss δ.

The basic calculation proceeds from the rate equation for $\boldsymbol{P}_e(x,t)$, defined as the probability of finding an electron with a fraction x of its incident momentum at a fractional depth t within the pulse:

$$\frac{d\boldsymbol{P}_e(x,t)}{dt} = \int_x^1 dz\, T(x,z)\,\boldsymbol{P}_e(z,t) - p(x)\,\boldsymbol{P}_e(x,t)\,, \tag{9}$$

where the two terms on the right hand side are the standard "source" and "sink" contributions that represent falling into and out of the momentum fraction x as a result of radiating a photon. $T(x,z)\,dt\,dx$ is the probability that an electron with momentum fraction z will radiate a photon with momentum fraction $(z-x)$ and end up within dx of x after traversing a fraction dt of the pulse length $(0 < t < 1)$.

$$p(z)\,dt \equiv \int_0^z dx\, T(x,z)\,dt \tag{10}$$

is the probability that the electron with z radiates one photon while traversing dt.

Analogously, the rate equation for the probability of finding a photon with momentum fraction v at depth t is

$$\frac{d\boldsymbol{P}_{ph}(v,t)}{dt} = \int_z^1 dx\, T(x-v,x)\boldsymbol{P}_e(x,t)\,. \tag{11}$$

Equations (9) and (11) include only the first generation of shower development; neglected is subsequent pair conversion of the high energy radiation formed in the pulses. We have also proceeded classically above in calculating the probabilities for multiple photon emission since for $y \gg 1$ (see Eq. 5 and the subsequent discussion) we can chop the pulse into thin slices of thickness ℓ/y such that radiation from successive slices is incoherent.

Equations (9) and (11) can be integrated[3] numerically in terms of known single photon radiation probabilities, $T(x,z)$. We define the photon flux through a pulse by

$$F(v) \equiv \int_0^1 dl\, \boldsymbol{P}_{\text{ph}}(v,t) \ .$$

The equivalent luminosity for beamstrahlung photon-photon collisions at total mass W is given by

$$\frac{d\mathcal{L}_{\gamma\gamma}}{dW^2} = \int_0^1 dv_1 \int_0^1 dv_2\, F(v_1)F(v_2)\, \delta(sv_1v_2 - W^2) \ ,$$

or

$$\frac{d\mathcal{L}_{\gamma\gamma}}{dz} = 2z \int_0^1 \frac{dv}{v}\, F(v)\, F\left(\frac{z^2}{v}\right) \ , \qquad (12)$$

where $z^2 = W^2/s = W^2/4E^2$.

Using this definition of $\mathcal{L}_{\gamma\gamma}$, the cross section to form a final state X at mass W is given by

$$\frac{d}{dz}\sigma_{\text{beamstrahlung}}(e^+e^- \to e^+e^- X) = \left[\frac{d}{dz}\mathcal{L}_{\gamma\gamma}\right] \sigma_{\gamma\gamma\to X}(W) \ , \qquad (13)$$

In the narrow resonance approximation for a "particle" X of mass m_X (13) reduces to

$$\sigma_{\text{beamstrahlung}}(X) = \left[z\frac{d}{dz}\mathcal{L}_{\gamma\gamma}\right]_{z=m_X/\sqrt{s}} \left\{\frac{4\pi^2(m_X \to 2\gamma)}{m_X^3}\right\} \ . \qquad (14)$$

The equivalent luminosity is shown in Figs. 1 and 2 for two choices of luminosity

$$C = 1.5 \ \&\ y = 400, \text{ or } \mathcal{L} \sim 0.7\ 10^{30} cm^{-2} \ \&\ \ell \sim 0.9mm$$

$$C = 0.5 \ \&\ y = 800, \text{ or } \mathcal{L} \sim 2.8\ 10^{30} cm^{-2} \ \&\ \ell \sim 0.3mm \ . \qquad (15)$$

The sensitivity to the beam geometry is illustrated by the three curves plotted in each figure corresponding to different values of the pulse aspect ratio, $G = 1$, 2 and 5 respectively. The latter values correspond to a flat beam with aspect ratios $a_x/a_y = 15$ and and 100, respectively. The virtual photon curves are drawn as dotted lines for a 1 TeV collider. For a 600 GeV collider with the same values of C and y, the results differ only slightly since the beamstrahlung photon-photon flux does not depend explicitly on the beam energy, and the virtual flux decreases only slightly due to its logarithmic energy dependence.

This calculation shows that two photon reactions via beamstrahlung can be significantly larger than previously appreciated on the basis of virtual photon processes. Note also that the beamstrahlung source works with electron-electron as well as positron-electron colliders. These hard photons may lead to troublesome backgrounds or, conversely, raise the attractive possibility of doing interesting

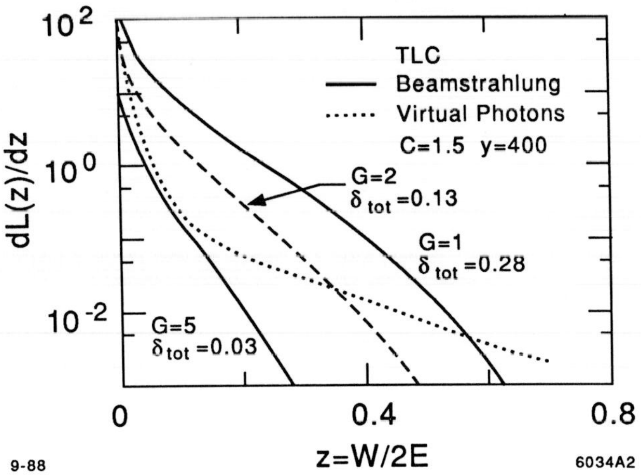

Figure 1. The photon-photon luminosity relative to the incident electron-positron flux is plotted for the case $C = 1.5$ and $y = 400$. The effect of pulse shaping is illustrated by the three values of G. The virtual two photon flux is also plotted as the dotted line for a collider energy of 1 TeV.

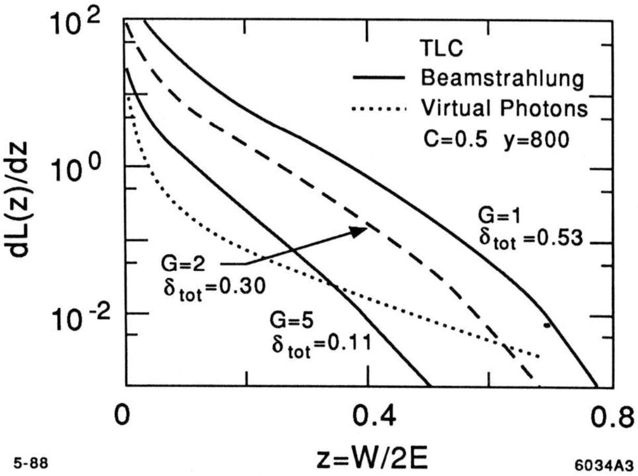

Figure 2. The same as Fig. 1 but for machine parameters of $C = 0.5$ and $y = 800$. The increase in luminosity is evident.

new physics with linear colliders in the 500 $GeV - 1\ TeV$ energy range that is now being studied. The colliders can be tuned by beam shaping to minimize or to maximize the beamstrahlung contribution as demanded by experiment.

REFERENCES

1. Palmer, R. B., SLAC-AAS-35 (1988), Stanford University preprint SLAC-PUB-4295 (1987), SLAC-AAS-31 (1987).
2. Blankenbecler, R. and Drell, S. D., Phys. Rev. D36, 277 (1987); D37, 3308 (1988).
3. Blankenbecler, R. and Drell, S. D., Phys. Rev. Lett. 61, 2324 (1988).

CALCULATION OF DETECTOR BACKGROUNDS AT TEV LINEAR COLLIDERS*

T. Himel

Stanford Linear Accelerator Center, Stanford University, Stanford, California 94309
and
Brookhaven National Laboratory, Upton, New York 11973

ABSTRACT

It is necessary to carefully design masks and beam lines to prevent the high energy physics detector from being inundated with background particles from a high energy linear collider. Presented here are preliminary calculations on two of the three expected backgrounds: (1) photons from synchrotron radiation produced in the final focus quadrupoles, and (2) electrons which lose energy due to beamstrahlung and are then bent into a mask or quadrupole by the field of the opposite beam. The former can be controlled with proper masking. The later may pose a problem so further calculations are needed. Work was also done on the third expected source of background: electrons in the tail of the beam which hit masks where showers are made whose products enter the detector. This work was very preliminary and is not included in this write-up. All the calculations here are based on the 1 TeV center-of-mass linear collider design of R. Palmer and the final focus design of K. Oide which can be found in these proceedings. Extrapolations to other accelerator designs should be straightforward.

BACKGROUNDS FROM SYNCHROTRON RADIATION

As an electron goes through the bend and quadrupole fields of the final focus, it emits synchrotron radiation photons. These photons may enter the detector either directly or by penetrating or being reradiated from a mask. The goal of the mask design is to minimize the number of photons entering the detector. From SLC design studies, this number must be kept below about 1000 photons or the occupancy of the detector is high enough to degrade its pattern recognition capabilities. This is the first of many times the existing 100 GeV SLC design will be used as a benchmark to which our new 1 TeV Linear Collider (TLC) design will be compared.

The first step in the calculation is to make a scale drawing of the quadrupoles near the Interaction Point (IP). For the TLC the final quadrupoles have an inner bore 100 μm in radius and cover a distance of 0.4 to 1.2 meters from the IP. Using the β functions from Oide's design, the trajectory of a particle at five standard deviations is superimposed on the drawing. Since synchrotron radiation (SR) is emitted tangentially to the electron's trajectory (the opening angle of $1/\gamma$ is negligible) it is easy to estimate where the radiation will hit. One can also estimate the range of standard deviations of electrons which will generate photons to hit a certain mask or quadrupole. The number of photons radiated by an electron is given by

$$N_\gamma = 2.06 \left(\frac{E_{\rm beam}}{100 \text{ GeV}}\right) \cdot \theta \ (\text{m}R) ,$$

where θ is the bend angle of the electron. Putting this all together with the fact that there are a total of 2×10^{11} incident electrons per pulse we can now calculate how many photons will hit the quadrupoles. Electrons between 2.8 and 3.6 σ will produce 1.1×10^9 photons which will hit the quadrupoles on the far side. Electrons outside 5σ will produce 1.4×10^5 photons which will hit the quadrupoles on the side of the incoming beam. The critical energy of these photons will be about 50 MeV.

This is a lot of photons especially when you consider that they are hitting a quadrupole which is inside the detector. Let's put in the best mask allowed by the constraints of detector acceptance (down to 10°) and see what happens. A 3.5 cm radius tungsten (because it is dense and has a large Z) rod which completely surrounds the quadrupoles and goes to within 20 cm of the IP just stays within the allowed 10°. In the center of this rod is drilled a 120 μm radius hole to a depth of 15 cm. For the next 5 cm the hole narrows to a 100 μm radius. For the next 80 cm we have the quadrupole with its 100 μm bore which is contained inside the tungsten mask. Making and aligning a mask with such a small hole in it may seem difficult, but compared to making the quadrupole it really isn't.

The purpose of this mask is to make it so the shower products from photons which hit the quadrupoles have a very small solid angle to enter the detector. The angles of the primary SR photons are so small that no photon can hit the 120 μm part of the hole since it is shadowed by the 100 μm part of the hole. So, to get to the detector a photon must either penetrate through the tungsten or make its way through a 15 cm long pin hole. The solid angle to get through the pin hole is 1000 times smaller than the equivalent in the SLC design. Photons penetrating the mask dominate.

*Work supported by Department of Energy contract DE–AC03–76SF00515.

Calculations of the penetration probability were done by H. DeStaebler using the electromagnetic shower simulation (EGS). Photons with a 50 MeV critical energy synchrotron radiation spectrum were simulated hitting the tungsten 15 cm from the end of the rod. For each incident photon only 0.004 came out the end. However, 0.17 came out the side of a 3.5 cm radius rod. Similarly, 0.02 and 0.002 came out of 6 and 10 cm rods, respectively. Given the 10^9 photons hitting the quadrupoles, this reduction factor is insufficient. Two improvements are needed to get an acceptable background.

The first improvement of 10^4 comes easily. We need to account for the 4 mR crossing angle of the beams, which is specified in the TLC design for other reasons. Because of this crossing angle, the outgoing beam doesn't have to go through the 100 μm bore of the quadrupole. Instead it goes 1.6 mm off to the side where there can be a larger hole. The crossing angle was originally put in the design so the disrupted electron beam would not have to pass through the center of the quadrupole. It has the same benefit for the outgoing SR photons. As long as the hole in the side of the quadrupole is at least 200 μm in radius no SR photons will hit it.

This still leaves us with the $0.17 \times 1.4 \times 10^5 = 2.4 \times 10^4$ photons which hit the quadrupole on the side of the incoming beam and penetrate the 3.5 cm thick mask around the quadrupole. This is a factor of 20 more than the detector can handle. The mask cannot be made thick enough to reduce this without encroaching on the solid angle reserved for the detector. The solution we arrived at was to increase the size of the quadrupole. Palmer redesigned the TLC with a quadrupole aperture of $10\sigma_{\text{beam}}$ instead of $6\sigma_{\text{beam}}$. This resulted in only a 10–20% reduction in the design luminosity while giving a very large reduction in the calculated synchrotron radiation background (there are many fewer particles outside 8σ than outside 5σ).

Having solved the toughest problem, let's look farther upstream to see if there are other magnets which will cause SR problems. The next quadrupole is 13 m from the IP. It is much weaker ($k = -.0249$ compared to the -3.416 of the quadrupole closest to the IP). It is also five times longer, so its field is very weak. The critical energy of the SR should be about 0.1 keV. This is so low that there is no problem stopping them from reaching the detector.

Finally, consider the last bend magnet which is 100 m from the IP. Only the last part of the bend which sweeps the beam past the 100 μm radius hole in the quadrupoles at the IP matters. This bend angle is $100\,\mu\text{m}/100\,\text{m} = 1\ \mu$ R. Hence, we expect 2×10^6 photons heading towards that hole. Note that only 1% of these photons will actually hit the inside bore of the quadrupole. The rest will pass right through and hit someplace well downstream of the detector. This still leaves 10^4 photons hitting the inside bore of the quadrupole inside the detector. Since the bend has a 0.144 kG field they will have a critical energy of 2.4 MeV. This is an uncomfortably large number of high energy photons being deposited inside the quadrupole. A significant number will penetrate the 3.5 cm thick mask. To fix this the last couple of microradians of the bend should be made with a lower field (a soft bend) to reduce the critical energy by about a factor of 25.

In summary, with moderate redesign we can avoid having synchrotron radiation contribute significantly to the detector's background. Note that the 100 μm radius of the quadrupoles sets the radius scale. Hence the minimum radius of a vertex detector will be determined by detector technology rather than the few hundred micron limit that background considerations set.

BACKGROUND DUE TO BEAMSTRAHLUNG FOLLOWED BY DISRUPTION

In the design of the TLC a maximum disruption angle is calculated. This is the maximum angle an outgoing electron has due to the kick it receives from the electromagnetic fields of the opposing positron bunch. The pinching of one beam by the other is included in the calculation. This maximum disruption angle is used to determine the crossing angle of the two beams to ensure there is a hole in the side of the quadrupole large enough to accept the outgoing beam. Not included in the calculation is the fact that an electron may lose energy due to beamstrahlung (synchrotron radiation in the field of the opposing beam) and thus be disrupted more. In some designs with a large beamstrahlung parameter it is possible for an electron to lose a large fraction of its energy in a single radiation event.

To see if beamstrahlung followed by disruption could be a source of background we made a simple computer model. The oncoming beam was considered to be Gaussian in z but very thin in y, and much wider in x than in y. Note that the TLC design has a 100:1 $x : y$ aspect ratio. Disruption of the oncoming beam was ignored. An electron was allowed to radiate up to 99.9% of its energy at any z from -0.5 to $3.5\ \sigma_z$. Electrons starting at 2, 3, 4, 6 and 8 σ_y were considered. The path of each electron was calculated by integrating the equations of motion out to $z = 5\sigma_z$. The outgoing angle was then calculated. For the case of no radiation the answer obtained agreed with previous calculations (0.25 mR). The largest disruption angle seen for an incident electron which radiated less than 99% of its energy at y less than $4\sigma_y$ was 1.1 mR. This should just manage to make it through the hole in the side of the quadrupole. However, the largest disruption angle seen when an electron was allowed to radiate up to 99.9% of its energy at a y up to $8\sigma_y$ was 2.7 mR. Such an electron would certainly hit the face of the quadrupole in the present design.

Time did not allow the calculation of the probability of such an event. The fact that there are 2×10^{11} electrons per pulse and that just a few electrons hitting the quadrupole will cause serious background problems for the detector certainly gives cause for worry. There are also arguments saying that the problem should be worse in the x direction than in the y direction. A more complete calculation is needed to see if we have a serious background here.

MULTIBUNCH INSTABILITIES IN SUBSYSTEMS OF 0.5 AND 1.0 TEV LINEAR COLLIDERS*

K. A. THOMPSON AND R. D. RUTH
Stanford Linear Accelerator Center, Stanford University, Stanford, California 94309

ABSTRACT

The acceleration of multiple bunches per RF fill of the linac accelerating structures is an important feature of the SLAC design for a next-generation linear collider, in order to use the RF as efficiently as possible and to obtain a higher luminosity. In this paper, we give simulation results on the control of multibunch instabilities in the subsystems of a 1.0 Tev linear collider (TLC) and an "intermediate" linear collider (ILC) of about 0.5 TeV center of mass energy.

INTRODUCTION

The acceleration of multiple bunches per RF fill of the linac accelerating structures is an important feature of the SLAC design for a very high energy linear collider (0.5 to 1.0 Tev). The reasons for multibunching are to use the RF as efficiently as possible and to obtain a higher luminosity. However, the problems of cumulative beam breakup in the linacs and coupled bunch instabilities in the damping rings must be addressed. In both cases, damped acceleration cavities may be a useful cure.[1] In such cavities, the transverse dipole wake is strongly damped by the use of axial slots through the irises of the RF structure coupled to radial waveguides. The Q of the fundamental transverse mode can be made as low as about 10 in this way, without significant adverse effect on the longitudinal accelerating mode, and the $Q's$ of higher order modes are expected to be at least as low as the Q of the fundamental.

Tuning the frequency of the fundamental transverse dipole mode to allow placement of the bunches near zero crossings of the wake fields will probably also be necessary in the main linacs. The transverse dipole wake for the accelerating structure considered here is strongly dominated by its fundamental mode and has zero crossings that are approximately equally spaced. Therefore it is possible to place all the bunches in a train near zero crossings of the wake field, if the ratio of the frequency of the fundamental dipole mode to the frequency of the accelerating RF is tuned to satisfy:

$$\frac{1}{2}\, n\lambda_{W_\perp} = m\lambda_{rf} = \ell \quad , \tag{1}$$

where ℓ is the bunch spacing, m and n are integers, and λ_{rf} and $\lambda_{W_\perp}$ are the wavelengths of the RF and the fundamental dipole wake mode.

The theoretical models and details of the computer simulations used in our study of multi-bunch instabilities have been discussed in more detail elsewhere.[2,3] The main emphasis of this paper will be on results for TLC and ILC parameter sets at the present stage of design.

MULTIBUNCH BEAM DYNAMICS

The bunches are taken to be point macroparticles, with an equal charge of N electrons in each bunch. The spacing ℓ between adjacent bunches is uniform and is an integral number of RF wavelengths in each subsystem.

The transverse dipole wake function is a sum of modes of the following form:

$$W_\perp(z) = \sum_m W_m \sin(K_m z)\exp\left(-\frac{K_m z}{2Q_m}\right) \quad , \tag{2}$$

where z is the distance behind the exciting bunch. K_m is the wavenumber and Q_m is the quality factor of mode m, and the $W_m's$ are constant coefficients. The units of $W_\perp(z)$ are V/Coul/m^2.

2.1 Transverse Dynamics in Linacs

The acceleration is assumed to be linear: $\gamma = \gamma_0 + Gs$, where γ is the energy divided by the rest energy and G is a constant. We use the smooth-focusing approximation $k(s) = 1/\beta(s)$ for the focusing function, where $\beta(s)$ is the "average" betatron function at longitudinal position s. The focusing is assumed to vary as

* Work supported by the Department of Energy, contract DE–AC03–76SF00515.

$$k(s) = \left(\frac{\gamma_0}{\gamma(s)}\right)^{1/2} k_0 \quad , \tag{3}$$

unless stated otherwise. Taking acceleration into account, the equation of motion for the transverse offset x_n of bunch n is:

$$\gamma(s)x_n'' + \gamma'(s)x_n' + \gamma(s)k_n^2(s)x_n = F_n(s) \quad . \tag{4}$$

where the driving term due to the wake is:

$$F_n(s) \equiv \frac{Ne^2}{mc^2}\sum_{j=1}^{n-1} W_\perp[(n-j)\ell]\; x_j(s) \quad . \tag{5}$$

Here primes denote derivatives with respect to s, and m is the rest mass of the electron. The transverse offsets are taken to be all in one plane. The WKB solutions of the homogeneous equation are

$$\begin{aligned} x_n^{\pm}(s) &= x_n^{\pm}(0)\left[\frac{\gamma_0 k_n(0)}{\gamma(s)k_n(s)}\right]^{1/2} \exp\left[\pm i\int_0^s k_n(s')ds'\right] \\ &= x_n^{\pm}(0)\left[\frac{\gamma_0}{\gamma(s)}\right]^{1/4} \exp\left[\pm i\int_0^s k_n(s')ds'\right] \quad , \end{aligned} \tag{6}$$

where we have used Eq. (3) for the variation of k_n with γ. Assuming the acceleration is adiabatic and the blowup of the transverse offset is slow compared to a betatron oscillation, it is straightforward to show (see Ref. 2 or 3 for details) that the general solution to the inhomogeneous equation for bunch n is

$$x_n(s) = a_n^+ x_n^+(s) + a_n^- x_n^-(s) + \int_0^s G_n(s,s')F_n(s')ds' \quad , \tag{7}$$

where a_n^+ and a_n^- are arbitrary constants, the Green function is:

$$G_n(s,s') = [\gamma(s)\gamma(s')k_n(s)k_n(s')]^{-1/2}\sin\psi_n(s,s') \quad , \tag{8}$$

and

$$\psi_n(s,s') \equiv \int_{s'}^{s} k_n(s'')ds'' \tag{9}$$

is the phase advance for bunch n. If we take the "positive phase" WKB solution

$$x_1(s) = x_1(0)\left(\frac{\gamma_0}{\gamma(s)}\right)^{1/4} \exp\left[\psi_1(s,0)\right] \quad , \tag{10}$$

as the motion for the first bunch, assume $a_n^- = 0$ for all $n > 1$, and drop a term with rapidly oscillating integrand, the solution for the transverse motion of bunch n may be written:

$$\begin{aligned} x_n(s) = \Bigg[x_n(0) + &\frac{Ne^2}{2i\gamma_0 mc^2 k_n(0)}\int_0^s \left(\frac{\gamma_0}{\gamma(s')}\right)^{1/4}\exp[-i\psi_n(s',0)] \\ &\times \sum_{j=1}^{n-1} W_\perp[(n-j)\ell]x_j(s')ds'\Bigg] \\ &\times \left(\frac{\gamma_0}{\gamma(s)}\right)^{1/4}\exp[+i\psi_n(s,0)] \quad . \end{aligned} \tag{11}$$

We note that this result is equivalent (see Refs. 2 and 3) to that obtained ignoring acceleration, from the equations of motion

$$x_n'' + k_n^2 x_n = \frac{Ne^2}{E}\sum_{j=1}^{n-1} W_\perp[(n-j)\ell]x_j(s) \quad , \tag{12}$$

with solution

$$\begin{aligned} x_n(s) = \Bigg[x_n(0) + \frac{Ne^2}{2iEk_n}\int_0^s e^{-ik_n s'} \times \\ \sum_{j=1}^{n-1} W_\perp[(n-j)\ell]x_j(s')ds'\Bigg] e^{ik_n s} \quad , \end{aligned} \tag{13}$$

provided that we interpret E and the k_n to be the energy and the focusing functions at the beginning of the linac, and s to be not the true distance along the accelerator, but rather an "effective distance" $\psi(s)/k_0$, where $\psi(s)$ is the phase advance in the actual distance s along the linac and k_0 is the focusing function at the beginning of the linac:

$$\begin{aligned} s_{eff} &= \frac{\psi(s)}{k_0} = \frac{1}{k_0}\int_0^s k(s)ds \\ &= \frac{2\gamma_0^{1/2}}{G}\left[\gamma^{1/2} - \gamma_0^{1/2}\right] \quad . \end{aligned} \tag{14}$$

Note that if $\gamma(L) \gg \gamma_0$ at $s = L$, the end of the linac, the effective length of the linac is approximately

$$L_{eff} = 2\left(\frac{\gamma_0}{\gamma}\right)^{1/2} L \quad . \tag{15}$$

Solutions for the transverse blowup of each bunch as a function of s may be obtained by numerical integration of the above equations for $x_n(s)$, and we shall present results in later sections for parameter sets of interest.

Finally, we note that in cases where the wake field is strongly damped, a bunch will only see a significant wake from the immediately preceding bunch, and a simple "daisy chain" model gives excellent agreement with the simulation results. Using the effective length approximation the equations of motion become:

$$x_1'' + k^2 x_1 = 0$$
$$x_n'' + k^2 x_n = \frac{Ne^2 W_\perp(l)}{E} x_{n-1} \qquad (n > 1) \quad . \tag{16}$$

Assuming $x_n(s) = a_n(s)e^{iks}$, the solution for the envelope function $a_n(s)$ for initial conditions $a_n(0) = 1$ is

$$a_n(s) = \sum_{j=0}^{n-1} \frac{(-i\sigma s)^j}{j!} \quad . \tag{17}$$

Since this is the first n terms of the Taylor series expansion of $\exp(-i\sigma_0 s)$, it is apparent that the criterion for little or no blow-up in the linac under these conditions is $|\sigma L| < 1$, that is:

$$\frac{Ne^2 |W_\perp(l)| L}{2kE} < 1 \quad , \tag{18}$$

where L is the effective length of the linac, and k and E are the values of focusing function and energy at the beginning of the linac.

In the main linacs of a TeV collider with damped acceleration cavities, we can have $|\sigma L| \sim 1$. Thus for n sufficiently large, $a_n(s)$ is approximately $\exp(-i\sigma s)$, and there is almost no blowup of the nth bunch.

2.2 Transverse Dynamics in Damping Rings

We assume the damping ring impedance is dominated by the contribution of the RF cavities and that damped acceleration cavities are used. If the Q of the transverse modes is sufficiently low that turn-to-turn wakes are negligible, then the equations of transverse motion are similar to those for a linac, except for the presence of a coherent damping term due to tune spread and/or head-tail damping, which we parameterize in a constant α_{cd} (units m^{-1}):

$$x_n'' + 2\alpha_{cd} x_n' + k^2 x_n(s) = \frac{Ne^2}{E} \sum_{j=1}^{n-1} W[(n-j)\ell] x_j(s) \quad . \tag{19}$$

We divide the motion on one turn through the ring into two parts: (1) the motion through the RF cavities, where the effect of the wake fields is modelled by a localized kick, and (2) a mapping around the rest of the ring, with specified average beta function and phase advance. Let us define $x_n^-(x_n^+)$ and $x_n'^-(x_n'^+)$ to be the transverse offset and angular divergence of bunch n just before(after) the RF cavities (assumed to be clumped together). Then, assuming that $\beta' = 0$ at the location of the cavities,

$$\begin{pmatrix} x_n^- \\ x_n'^- \end{pmatrix} = e^{-\alpha_{cd} C} \begin{pmatrix} \cos\mu & \beta \sin\mu \\ -\frac{1}{\beta}\sin\mu & \cos\mu \end{pmatrix} \begin{pmatrix} x_n^{old\,+} \\ x_n'^{old\,+} \end{pmatrix} \quad . \tag{20}$$

Here $x_n^{old\,+}$ and $x_n'^{old\,+}$ are the transverse offset and angular divergence just after passing through the RF cavities on the preceding turn, C is the circumference of the ring, and μ is the coherent phase advance around the ring. One may also show (see Ref. 3):

$$\begin{pmatrix} x_n^+ \\ x_n'^+ \end{pmatrix} = \begin{pmatrix} x_n^- \\ x_n'^- + \frac{Ne^2}{E} L_{rf} \sum_{j=1}^{n-1} W[(n-j)\ell]\; x_j^- \end{pmatrix} \quad , \tag{21}$$

where L_{rf} is the total length of the RF cavities and $W(z)$ is the wake per unit length in the cavities. The results presented later on damping ring transverse instabilities are obtained using a tracking program based on these equations.

PARAMETERS OF COLLIDER SUB-SYSTEMS

We assume a train of ten bunches is to be accelerated on each RF fill, and that the interbunch spacing within the train is 21.0 cm (i.e., two wavelengths at S-band or twelve RF wavelengths at 17.1 GHz).

Table 1. Parameters for TLC.

Subsystem	Energy (GeV)	Particles per bunch
Injection accelerators	$0.18 \rightarrow 1.8$	2×10^{10}
Damping rings	1.8	2×10^{10}
Preaccelerators	$1.8 \rightarrow 18.0$	1.8×10^{10}
Main linacs	$18.0 \rightarrow 500.$	1.67×10^{10}

Table 2. Parameters for ILC.

Subsystem	Energy (GeV)	Particles per bunch
Injection accelerators	$0.18 \rightarrow 1.8$	1×10^{10}
Damping rings	1.8	1×10^{10}
Preaccelerators	$1.8 \rightarrow 18.0$	9×10^{9}
Main linacs	$18.0 \rightarrow 250.$	8.33×10^{9}

Other parameters of individual collider subsystems are given in Table 1 for the TLC design, and in Table 2 for the ILC design.

We shall discuss results for each of these subsystems in the TLC design, and for the main linacs of an ILC design. Since the parameters for other subsystems of the ILC will be assumed the same as those for the TLC except for a lower charge per bunch, it is clear that instabilities can be controlled in these subsystems of the ILC if they can be controlled in the corresponding subsystems of the TLC.

RESULTS FOR MAIN LINACS IN THE TLC

We assume a main linac accelerating frequency of 17.1 GHz, and a disk-loaded accelerating structure with

Table 3. Parameters for main linacs at 17.1 GHz.

Number of bunches	10
Number of particles per bunch	1.67×10^{10}
Bunch spacing ℓ	21.0 cm
Initial energy of linac	18 GeV
Final energy of linac	500 GeV
Linac length	3000 m
Initial beta function	3.2 m ($k_0 = 0.3125 m^{-1}$)

cell length 5.83 mm, internal cell radius 7.47 mm, and relatively large iris radius of 3.47 mm.

The parameter set used is shown in Table 3. Each linac accelerates 10 bunches per RF fill, to an energy of 0.5 TeV. The spacing of 12 RF wavelengths between bunches (at the assumed RF frequency) is chosen in order to match the energy extracted by the bunch train to the energy input from the RF. This gives very nearly the same acceleration for every bunch in the bunch train.[4)]

The RF wavelength at 17.1 GHz is 1.75 cm, and the wavelength of the fundamental mode of the unmodified transverse dipole wake for our cavity design is 1.36 cm. If the frequency of the fundamental mode is shifted slightly, so that its wavelength is 1.31 cm, then Eq. (1) is satisfied with n=32, and we have

$$\frac{\lambda_{rf}}{\lambda_{W_\perp}} = \frac{4}{3} \quad . \tag{22}$$

When this relation is satisfied, the frequency of the fundamental transverse mode is 477.85 m^{-1}, which we shall denote by K_{W_0}.

In Fig. 1, we show "tuning curves" of the maximum transverse amplitude x_{max} in the bunch train as a function of the frequency of the fundamental transverse dipole mode, for values of Q = 30, 40, 50, and 60. The value of x_{max} is the maximum of the amplitudes reached by all bunches as they travel down the linac, normalized by dividing out the adiabatic damping factor $(\frac{\gamma_0}{\gamma})^{1/4}$. The central frequency, at which $\frac{\lambda_{rf}}{\lambda_{W_\perp}} = \frac{4}{3}$, is 477.85 m^{-1}. The range about the central frequency shown in the figure is only $\pm 0.1\%$, so the tolerance on tuning the frequency of the low-Q dipole mode with respect to the accelerating mode is rather tight compared to its bandwidth.

Note that at M bunch spacings behind a bunch, the wake field has damped by a factor of about

$$\exp\left(-\frac{K_{W_0}\ell M}{2Q}\right) \approx \exp\left(-\frac{50M}{Q}\right) \quad , \tag{23}$$

and so a given bunch can feel a significant wake field from several bunches ahead of it, unless it is near a node of these wakes and/or the value of Q is well below 50. Figure 2 shows tuning curves for the smaller values Q = 15, 20, and

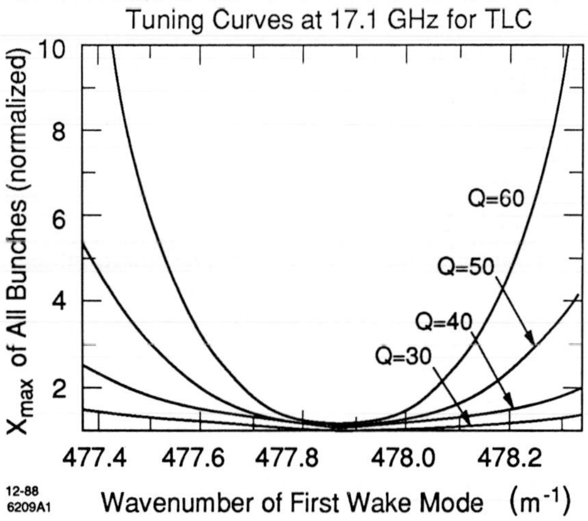

Fig. 1. Maximum transverse amplitude x_{max} (normalized) of all bunches as a function of the frequency of the fundamental transverse dipole mode, for values of Q = 30, 40, 50, and 60, at 17.1 GHz accelerating frequency. The central frequency, where $\lambda_{rf}/\lambda_{W_\perp}$ = 4/3, is 477.85 m^{-1}. The spread shown about K_{W_0} is $\pm$ 0.1%.

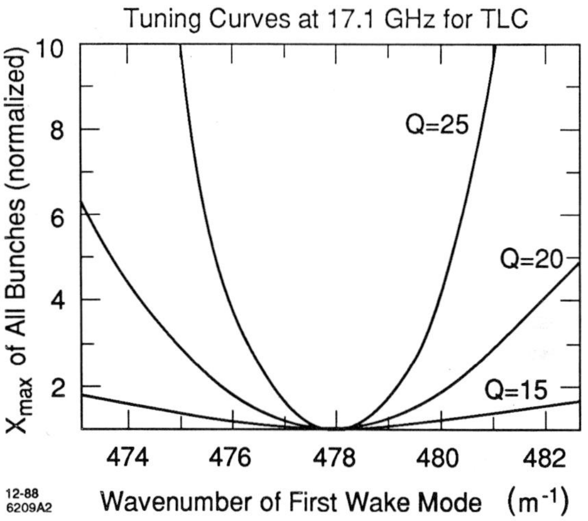

Fig. 2. Maximum transverse amplitude x_{max} (normalized) of all bunches as a function of the frequency of the fundamental transverse dipole mode, for values of Q = 15, 20, and 25, at 17.1 GHz accelerating frequency. The central value of the frequency, where $\lambda_{rf}/\lambda_{W_\perp}$ = 4/3, is 477.85 m^{-1}. The spread shown about K_{W_0} is $\pm$ 1%.

25. The spread around K_{W_0} in this figure is $\pm 1\%$, and as can be seen from the figure, for $Q = 15$, we have $x_{max} < 2$ for frequencies within this range.

For low values of Q, the frequency of the fundamental transverse mode becomes less sharply defined; the full width at half-maximum of the resonance around the central frequency K_{W_0} is $\Gamma \equiv K_{W_0}/Q$ (and the central frequency is shifted slightly from that of the undamped mode). So it is also of interest to compare the ratio R of the tuning tolerance for a given Q to the full width Γ of the resonance at that Q:

$$R \equiv \frac{\Delta K_{W_0}}{\Gamma} \quad . \tag{24}$$

Table 4. Tuning parameters for the fundamental transverse dipole mode for TLC main linacs.

Q	$\Delta K_{W_0}(\mathrm{m}^{-1})$	$\Gamma = \frac{K_{W_0}}{Q}(\mathrm{m}^{-1})$	$\frac{\Delta K_{W_0}}{K_{W_0}}$	$\frac{\Delta K_{W_0}}{\Gamma}$
15	>11.0	31.9	>2.3%	>34%
20	4.37	23.9	0.91%	18%
25	2.56	19.1	0.54%	13%
30	1.60	15.9	0.33%	10%
40	0.87	11.9	0.18%	7.3%
50	0.57	9.56	0.12%	6.0%
60	0.38	7.96	0.08%	4.8%

In Table 4, we show the full-width tuning tolerance ΔK_{W_0} for the criterion $x_{max} \leq 2$, the full-width Γ of the resonance peak, the tuning tolerance expressed as a percentage of the undamped central frequency, and the ratio R. The parameters used and the values of Q tabulated are those used in Figs. 1 and 2. The lower values of Q, say up to 30 or so, seem to be the most desirable in that the tolerance on tuning is at least 10% of the bandwidth of the resonance; this should be straightforward to do. Such $Q's$ have already been achieved in models (Ref. 1).

RESULTS FOR MAIN LINACS IN THE ILC

It is also of interest to consider a design for an "intermediate" energy linear collider with half the center of mass energy. The parameters for the main linacs are as shown in Table 3 for the TLC, except that the final energy of each linac is 250 GeV and the number of particles per bunch is 8.33×10^9. The lower charge per bunch more than compensates for the fact that since the energy is lower the beam is less "stiff" (also, the initial focusing function is the same as in the TLC, but weakens less quickly with distance, according to Eq. (3)). The tuning curves for the ILC case are shown in Fig. 3.

The tuning parameters corresponding to Fig. 3 are shown in Table 5.

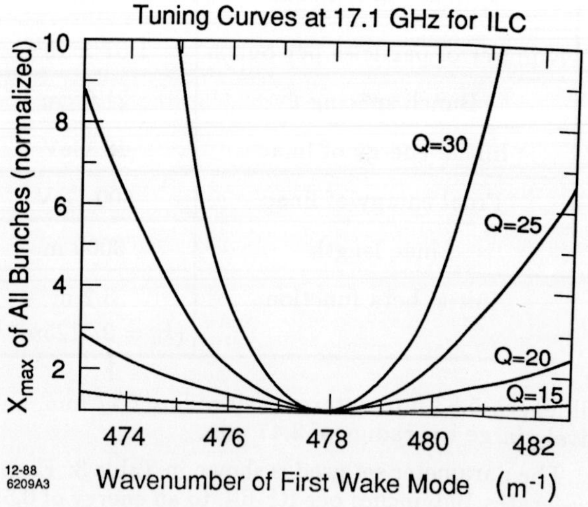

Fig. 3. Maximum transverse amplitude x_{max} (normalized) of all bunches as a function of the frequency of the fundamental transverse dipole mode, for values of Q = 15, 20, 25, and 30, using the ILC design parameters. The central frequency, where $\lambda_{rf}/\lambda_{W_\perp} = 4/3$, is 477.85 m^{-1}. The spread shown about K_{W_0} is ± 1%.

Table 5. Tuning parameters for the fundamental transverse dipole mode for ILC main linacs.

Q	$\Delta K_{W_0}(\mathrm{m}^{-1})$	$\Gamma = \frac{K_{W_0}}{Q}(\mathrm{m}^{-1})$	$\frac{\Delta K_{W_0}}{K_{W_0}}$	$\frac{\Delta K_{W_0}}{\Gamma}$
15	>15.0	31.9	>3%	>50%
20	7.20	23.9	1.5%	30%
25	3.73	19.1	0.78%	20%
30	2.51	15.9	0.53%	16%

RESULTS FOR THE PREACCELERATORS (TLC)

We turn now to the question of beam breakup in the preaccelerators of the linear collider. We assume the same type of large-irised, and possibly damped, accelerating structure as in the main linacs, but scaled to S-band.

The bunch spacing chosen for the main linacs, about 21.0 cm, is two S-band wavelengths, and other parameters are as shown in Table 6. In the preaccelerator the problem of beam breakup is much less acute than in the main linacs. It is not necessary to tune the frequency of the fundamental transverse dipole mode relative to the accelerating mode, so as to place all bunches near wake zero crossings. In our example the placement of the bunches is neither particularly favorable nor unfavorable.

Table 6. Parameters for preaccelerator.

RF accelerating frequency	2.856 GHz
Number of bunches	10
Number of particles per bunch	1.8×10^{10}
Bunch spacing ℓ	21.0 cm
Initial energy of preaccelerator	1.8 GeV
Final energy of preaccelerator	18 GeV
Acceleration gradient	50 MeV/m

Table 7. Transverse beam blowup at end of preaccelerators.

Initial β (meters)	Q	Max transverse blow-up factor
4.0	∞	1.22
2.0	∞	1.05
1.0	∞	1.01
4.0	40.0	1.05
2.0	40.0	1.01

Results for several choices of Q and focusing function are given in Table 7. Note that we are assuming that the transverse focusing weakens with distance along the linac according to Eq. (3), and that one might choose to keep it more nearly constant.

RESULTS FOR THE INJECTION ACCELERATORS (TLC)

Accelerators are also needed to take the bunches up to the damping ring energy of 1.8 GeV.

Table 8. Parameters for injection accelerator.

RF accelerating frequency	2.856 GHz
Number of bunches	10
Number of particles per bunch	2×10^{10}
Bunch spacing ℓ	21.0 cm
Initial energy	0.18 GeV
Final energy	1.8 GeV
Acceleration gradient	25 or 50 MeV/m

A representative parameter set is given in Table 8. We again assume an accelerating structure of the same type as in the main linacs, including damping if necessary, but scaled to S-band. Due to the lower energy, somewhat stronger focusing is needed here than is needed in the case of the preaccelerators.

Results are shown in Table 9 for two values of acceleration gradient, again assuming that the focusing scales according to Eq. (3).

In Table 10, we show the results obtained assuming the focusing function does not vary with s.

RESULTS FOR DAMPING RINGS (TLC)

We have carried out simulations for the damping ring parameters shown in Table 11, using two examples of RF cavity designs. One is the TLC main linac cavity described earlier, scaled to the damping ring frequency. The other is a scaled PEP cavity, which has nose cones and a relatively smaller iris. In both cases we assume the $Q's$ of the transverse modes may be lowered by modifying the cavities to include axial slots coupled to radial waveguides, as in the main linacs.

Table 9. Transverse beam blowup at end of injection accelerators.

Accel. gradient MeV/m	Initial β (meters)	Q	Max transverse blowup factor
50.0	4.0	∞	1.25
"	2.0	∞	1.05
"	4.0	40	1.05
25.0	4.0	∞	2.40
"	2.0	∞	1.20
"	1.0	∞	1.06
"	4.0	40	1.18
"	2.0	40	1.07

Table 10. Transverse beam blowup at end of Injection Accelerators with focusing assumed constant.

Accel. gradient MeV/m	β (meters)	Q	Max transverse blowup factor
50.0	4.0	∞	1.06
25.0	4.0	∞	1.32
"	2.0	∞	1.08
"	1.0	∞	1.02
"	4.0	40	1.08

Table 11. Damping ring parameters.

Number of bunch trains	10
Number of bunches per train	10
Number of particles per bunch	2×10^{10}
Particle energy	1.8 GeV
Ring circumference	154 m
RF frequency	1.428 GHz
Bunch spacing within a train	21.0 cm

For both cavities, the wavenumber K_1 of the fundamental transverse mode is approximately 40 m^{-1}. The number of e-foldings of this mode per turn around the circumference C of the ring is

$$\frac{K_1 C}{2Q} \approx \frac{3000}{Q} \quad , \tag{25}$$

and the number of e-foldings between bunch trains (assuming ten trains symmetrically placed in the ring) is about

300/Q. Thus we expect turn-to-turn and train-to-train wake effects to be negligible for sufficiently low Q's.

Simulations were done with the tracking program discussed earlier. For $Q = 30$ and a coherent damping parameter α_{cd} of at least 10^{-6}m^{-1}, none of the bunches ever get an amplitude larger than their initial injected amplitude, and the centroids of all the bunches have damped to small amplitudes after 50000 turns, due to the effect of α_{cd}. Note that this value of α_{cd} gives a coherent damping time of 3.33 msec, which corresponds to about 6500 turns. The amplitudes of the second and tenth bunches as a function of turn number are shown in Fig. 4, for the two cavities, for $Q = 30$ and $\alpha_{cd} = 10^{-6}\text{m}^{-1}$.

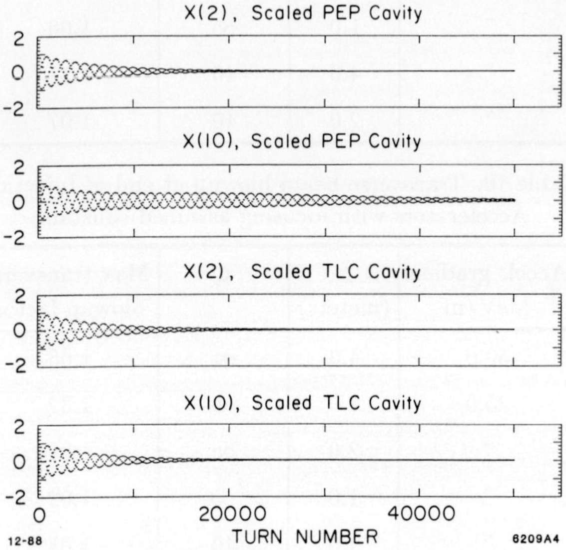

Fig. 4. Offset of second and tenth bunch as a function of turn number, for the scaled PEP and scaled TLC damping ring cavities. The value of Q for the cavity is 30 and the coherent damping parameter α_{cd} *is* $10^{-6}m^{-1}$.

CONCLUSIONS AND ACKNOWLEDGMENTS

Control of transverse multibunch instabilities appears to be feasible in all the subsystems of a linear collider studied here, namely the main linacs, injector linacs, preaccelerators, and damping rings. In all these subsystems, damped acceleration cavities may be a useful tool, although in the preaccelerators and injection accelerators one can probably get by without them. In the main linacs, both damped cavities and placing the bunches near wake zero crossings are essential. In the damping rings, it is highly desirable to use damped cavities to decouple the bunch trains from each other as well as to reduce the transverse wakes within each train.

We thank the members of the linear collider working group at Snowmass for useful discussion and comments.

REFERENCES

1. R. B. Palmer, SLAC–PUB–4542 (1988).
2. K. A. Thompson and R. D. Ruth, SLAC–PUB–4537, in *Proceedings of International Workshop on Linear Colliders*, Capri, Italy (1988).
3. K. A. Thompson and R. D. Ruth, SLAC–PUB–4801, to be submitted for publication (1988).
4. R. D. Ruth, SLAC–PUB–4541 (1988).

COMPUTING

Stewart C. Loken

Working Group Leader

COMPUTING FOR HIGH ENERGY PHYSICS IN THE 1990 s

Stewart C. Loken
for the
Computing Working Group†
Lawrence Berkeley Laboratory
1 Cyclotron Road, Berkeley CA 94720‡

Abstract

The working group on computing for high energy physics in the 1990 s has reviewed a wide range of issues related to computing and software development. Many of these are specific to the problems of the SSC. Many, however are more general. Foremost among these is the need to improve the quality of the software used in high energy physics and to improve the productivity of physicists and programmers working in the field.

1. Introduction

In this report we consider the computing problems of high energy physics in the 1990 s. It is clear that the computing needs of the field are increasing significantly and will continue to do so. We are building larger and more complex experiments. The detectors consist of more individual elements and the experiments wil collect more events.

The size and composition of experimental collaborations also present new problems. The collaborations are large and include members from many countries. There are more programmers involved in the computing effort and they are distributed through all institutions in the collaboration.

The Superconducting Super Collider (SSC) is just one example of the computing problem for the next decade. The reserch program of the 1990's will present similar challenges in many other areas.

In this report, we discuss the specific problems of the SSC computing including those of detector simulation. We also review the software development issues that will affect computing in the 1990 s.

2. Computer Planning for the SSC ERA

In the near future, the effort to design detectors for the SSC will increase significantly. This will result in a need to provide computer time and programs for detector simulation. Once the site is selected and the laboratory is formed we will need to establish a computing department. In this section, we consider these issues.

2.1 Computing for Detector Design

As the deadlines for letters of intent and for proposals approach it will be increasingly important to have sufficient computing capacity to satisfy the needs of collaborations to simulate SSC detectors and to optimize their design. We have made an estimate of the needed capacity based on recent simulation work. In doing this we make the following assumptions:

- Typical runs of 20,000 events
- Using fast shower Monte Carlo
- Time/event of 15 minutes (VAX-11/780)
- Total of 5000 VAX CPU hours per run

There is a great deal of effort going into improving the performance of the detector simulation program and the time to simulate a single event is expected to improve. The assumptions above are, however, useful to set the scale of the problem. For the next three fiscal years, we get the following:

- FY1989: We assume that there are ten simulation efforts. Each does ten runs. This gives a total requirement of 70 VAX-780 equivalents.
- FY1990: We assume that there are four collaborations with fifteen people in each working on simulation. Each does some full simulation in addition to the fas Monte Carlo. This doubles the time. The total requirement is 840 VAX equivalents.
- FY1991: We assume four collaborations with thirty people working on simulation. A large part of their effort involves full simulation. The Total requirement is expected to be 5000 VAX equivalents.

The total computing requirement is large even in the next year. Even with progress in speeding up the simulation or in making the code run efficiently on vector supercomputers, additional facilities must be provided to satisfy the needs of detector designers. We recommend that a center for detector simulation be established as soon as possible to provide the computing needed in FY1989. In addition to providing the needed computing, this center should manage program libraries and databases and save simulated data so that it can be shared by other groups. To

† R. Bock, F. Borcherding, D. Cutts, C. Day, D. Garelick, N. DiGiacomo, G. Hanson, S.-O. Holmgren, J. Linnemann, S. Loken, U. Mallik, K. McFarlane, G. Morgan, T. Nash, N. Oshima, F. Paige, D. Pellet, A. Palounek, L. Price, B. Roe, K. Roe, A. Sill, K. Sugano, T. Trippe, W. Vernon, Y. Watase, A. White, H. White, J. White, R. Wilson, E. Wolin

‡ This work was supported by the Director, Office of Energy Research, Office of High Energy and Nuclear Physics, Division of High Energy Physics, of the U.S. Department of Energy under Contract No. DE-AC03-76SF00098.

be useful to the community, the center must be available on HEPNET and other networks. As the use increases, higher speed networks will also be necessary.

As soon as the SSC laboratory is established, the computing facilities and support group should be moved there. We discuss below the need for an SSC Computing Division which will manage SSC computing facilities and act as a resource to the national high energy physics program.

2.2 Programs for Detector Simulation

There has been considerable progress in developing software for detector simulation. This was the focus of a workshop in August, 1987 [1] and was reviewed at Snowmass by Larry Price. Some of the important areas of progress are the following:

- Agreement to focus development on GEANT3 and ZEBRA
- Make Oak Ridge Calorimeter code CALOR available in GEANT3
- Outline of standard interface for event generators
- Plans for utilization of vector and parallel computers
- Standard particle numbering system coordinated with PDG
- Ideas for improving the ease of use of GEANT3

While there has been significant progress, much remains to be done to develop a simulation that is easy to use and which runs fast enough to generate the large numbers of events needed to study subtle backgrounds in SSC detectors. In addition, there needs to be a coordinated effort to compare simulations with test beam data to ensure that the simulations are correct.

3. SSC Computing Division

Our working group has concluded that a strong SSC Computing Division is necessary and that it should be formed as soon as possible, hopefully within the next year or two. Although there is a serious immediate need for this division it is also extremely important that it be formed with the right charter.

This division would have charge of computing, networking, software, and related facilities for SSC users. Because of the size and centrality of the SSC program this division would have a position of leadership for computing activities for the entire U.S. high energy physics community.

It is **essential** that this division be responsive to the needs and desires of the community. This must be reinforced structurally, perhaps by a policy committee of users, both SSC employees and outside users. **This division should lead the community by persuasion, not arbitrary fiat.**

The duties of the SSC Computing Division are described in the following sections.

3.1 Obtain and Operate SSC Computing Facilities

1. **Mainframe.** Although the computing environment is continually changing we expect that a mainframe computer to serve as a center around which the other activities are grouped will still be essential. It will need very large mass storage capabilities, and fast dataways for routing information internally and externally. The division may well decide that a whole set of multi-user computers is needed.

2. **Workstations.** The laboratory will need many workstations of widely varying power. Some of these will be very small computers, others will have signifigant computing power or be very fast graphics engines in their own right. The division will have many choices here. The division will have to decide quite early what level of commonality to establish between the various stations. They could be all members of a single computing family, or they could be entirely disparate, connected only by a data link to the central mainframe(s) or to other workstations. X-windows may be quite advantageous here since it is non-proprietary.

3. **Terminals.** The laboratory will have to provide terminals. It may well be that terminals will be largely replaced by various levels of workstations.

4. **Specialized off-line computing facilities.** At present it appears that many specialized arrays of processors or other facilities will provide a significant fraction of the computing power needed for SSC data analysis. Some of these will be designed and built by the computing division and some will be obtained from outside the division. A present day example of such facilities is the set of very successful Fermilab ACP systems, some of which are now operated as part of the Fermilab computing center.

5. **On-line facilities.** The division will have the very difficult problem of having to draw the line between the responsibilities of the division and that of individual groups. At the outer trigger levels the hardware is probably rather standard, but it gets more and more specialized for the inner trigger levels. Policy decisions will have to made about the level of laboratory equipment support for these facilities.

6. **Operating systems.** This is another gray area. The level of support that must go into non-standard (or standard) operating systems for on-line and off-line systems will depend on the state of computing at the time.

7. **Programming assistance.** It is clear that assistance must be available for specialized operating systems and specialized programming packages maintained by the division. However, programming assistance can also mean helping students with trivial errors, suggesting programming strategies to groups, or even developing and coding various packages for groups. We will discuss generic packages later, but it is clear that the activities listed here could take a great deal of time, and limits will have to be set.

8. **Evaluation of various commercial computing systems and options.** In the course of making evaluations for its own purchasing the division will have to develop good batch and interactive benchmarks and apply them to many systems. It will have to examine the differing advantages of scalar, vector, and parallel processing. The division should make available to the high energy physics community results from good benchmarks involving high energy physics type calculations, and should serve as a resource for groups trying to make decisions for their own computing.

Items 2, 4 and 8 will help standardize HEP facilities nationwide, improving computing efficiency, cost, and protability.

3.2 Provide and maintain networking facilities.

1. **Install and maintain networking within the laboratory including to the experiments.** Networks will have to be physically installed throughout the laboratory and switches and gateways maintained for efficient transfer of data between the various sites.

2. **Install and maintain networking for the community.** The SSC is the natural center of the network operations for high energy physics. This may mean nothing more than connecting to NSFNET or it may mean much more depending on the way networking evolves. Effective networking within the lab, between the lab and users, and between users is absolutely crucial for physics during the next decade. Networking outside of the U.S. will be very important for the large multi-national collaborations.

3. **Maintain and develop network software.** Good networking protocols, translation between various lower level protocols, and higher level software (e.g. coloured books or X.400) are all vitally necessary. They may all be available as a single commercial package or they may require considerable effort. The division should strive for uniform procedures for the high energy community. Special problems for users outside the U.S. will have to be considered.

3.3 Provide software utilities for the community.

For almost two decades the U.S. high energy community has depended to a large extent on CERN for development of software utilities. The working group feels it is necessary for the U.S. community to do a greater share of development than in the past. Utilities should be developed as needed. Sometimes this may involve collaboration with industry or with other labs. When feasible, utilities should be purchased from vendors. If utilities are to be purchased the division should attempt to arrange for discounts for the high energy physics community and should take into account the total cost to the community of decisions made by the division. Any general utilities developed or purchased must be available for a range of hardware, not just the machines at the SSC. They should work on the machines used by the bulk of the users. Some examples of possible utility packages follow:

1. Program development facility. This would perform the functions of PATCHY, HISTORIAN, etc. It should include a number of modern facilities developed to assist and organize large tasks being worked on by a number of people.

2. Graphics standard system.

3. Data base management system.

4. On-line software aids. This is a gray area. The old programs such as the Fermilab MULTI package were extremely useful. However, the modern trigger systems involve much more complicated structures. Some of it can be standardized. Much of it is still under development, but the line between standard and specialized utilities will be a constantly moving one as techniques develop.

5. Off-line workstation and graphics aids. There are a number of utilities developed or being developed at CERN in this area (PAW, ...). This area is in intense development now and packages will be changing quickly.

6. Standard large programs. An example of this is GEANT. Coordination with CERN will be very useful on some of these programs.

7. Program library. CERN has built a large library of standard subroutines. The division should coordinate with CERN on further developments.

8. I/O utilities. I/O utilities are needed to aid the portability of programs and data within and between experimental groups when the individual members have diverse facilities.

3.4 Suggest community programming standards

This has a strong educational aspect and also a strong relation to the preceeding section. New standards are needed in a number of areas as indicated below. These standards should be developed in conjunction with the community. Seminars, working groups, colloquia, schools, meetings, etc. are needed to define community desires and to obtain community agreement with and adherence to standards suggested. Some of the areas needing standards are:

1. Portability standards. These are standards of programming styles, of I/O, etc. which will make it easier to port programs from one hardware system to another.

2. Standards for designing new programs. Structured analysis, structured design (SASD) is an example. An educational effort here could result in programs which are better documented and easier to maintain then is the case for most of the present generation of high energy physics programs.

3. Programming styles, languages. It is time to consider the possibility of doing a part of particle physics programming in languages other than FORTRAN. Such a major

shift in style would require a considerable effort and investment by the community. The SSC division would be an appropriate place to have the community consider the problem and to lead an educational effort after a decision is reached.

4. Graphics. Graphics standards are in flux, but graphics is an increasingly large area of our programming.

5. Networking. It is clearly in the community interest to standardize networking protocols, gateways, higher level protocols, etc. as much as possible. Lines of the highest speed obtainable ($\geq$ T1) should be provided for the communication needs of the community.

6. Operating systems. This is at present a gray area. Is it possible to arrange to interface all systems with UNIX?

3.5 Investigate and develop advanced HEP facilities

The division will have to examine possible new hardware and software facilities. This may involve collaboration with industry, with other laboratories, or with Universities. Some examples of possible projects are given below. Even more than with the other subsections this is an area of rapid development and the items listed are only a few of the many projects that will undoubtedly develop.

1. Large processing arrays for on-line, off-line use. The Fermilab ACP is an example of a successful present day project in this area.

2. Specialized processors for on-line and off-line work. For trigger processors neural network or systolic processors or other special purpose processors might prove extremely useful.

3. New programming strategies. The computing community has developed a number of new programming strategies such as object oriented programming which may prove to be useful for some high energy physics programming. It would be worthwhile to explore their usefulness.

4. File servers, mass storage devices. This field is in great flux. The division should examine the field, attempt to select some appropriate technologies for high energy physics use and should attempt to set standards for these devices. (For example the VHS based tape cartridge systems do not have inter-vendor standards for how the data is stored. Portability requires standardization.)

4. Software

It is traditional that software problems receive less attention than hardware, especially in the early phases of an experiment. This has been a serious problem for the current generation of high energy physics experiments and the problem will be even worse in the future.

Each of the current generation of colliding beam experiments is developing a software base of $3 - 5 \times 10^5$ lines of code. This development will take an effort of 300–500 man-years. The developers are typically not professional programmers and they have little, if any, training in computer science. They are not located in one place and, for the most part, do not devote full time to their software responsibilities. Physicists working on the design of a hardware component often turn to an engineer or a technician for help. It is extremely rare that physicists will enlist professional help during any phase of the software development effort.

This situation is now changing. Recognizing the effort that software development will require, many groups are beginning to use software development methods that have been standard in the computing industry for years. Many are using specialized systems to maintain data structures within FORTRAN programs and database management tools to maintain constants. While it is still too early to assess the success of these efforts, the indications are that the software will be better and that it will be easier to maintain.

4.1 Traditional Development Methodology

Up to now, most experiments have followed a "bottom-up" strategy for software development; the software is put together from components which were written for isolated studies during the design phase of the experiment. The result is a system with little design documentation. There is agreement on a general outline but individuals work on components with little intercommunication. Little attention is paid to integration of modules until after they are complete.

4.2 The Software Life Cycle

A modern approach to the problem of software development for experiments begins by recognizing that the software project will have a long and complex life and that each phase must be planned in detail. In fact, the approach is very similar to the engineering of any hardware component of an experiment. The phases of the software life cycle are the following:

- Requirements analysis

Define what software is to do with out consideration for how it is to do it or what the hardware configuration will be.

Build a logical model of the software system

- Design

Define details of the software

Create a physical model which describes the implementation

Subdivide the problem into processing units, then into tasks, and finally into modules with in tasks

- Coding

Write code for modules

Test and document modules

- System integration

  Assemble modules and test
- Acceptance testing

  Demonstrate system performance
- Operation and maintenance

  Install in sites

  Update as necessary

The formal structure emphasizes the early phases of the development cycle. Many studies have showed that the cost to fix an error increases significantly in later phases of the project [2]. In the early phases, the group working on the project is smaller. The effort to create a new model is much less than the effort to recode, compile and retest a module.

It is important to recognize that there are real costs associated with software errors. These are not measured by the same criteria as in the commercial world. The cost, however, can be measured in terms of lost beam-time, the expense of rerunning data tapes, missed discoveries, or incorrect results. It is very much worth the effort to get it right.

4.3 Structured Analysis/Structured Design

One of the formal methodologies for the first two phases of the project is Structured Analysis/Structured Design or SA/SD[3]. This methodology has been adopted by the ALEPH collaboration at CERN [4] and the D0 experiment at Fermilab [5]. While it is too early to determine whether the software for these experiments will be significantly better than that developed using traditional methods, both groups agree that they are developing programs that are different from what they would have done without SA/SD.

In the analysis phase, SA/SD uses three graphical tools to model the software system. These tools describe the software in much the same way that engineering drawings model a hardware object. A Data Flow Diagram models the flow and transformation of data in the system. A State Transition Diagram describes the time dependence and is especially useful for online programs or control systems. The Entity Relationship Diagram defines the data elements of the system and the relationships between them. Together, these three tools describe the character of what is to be built. The graphical tools are augmented with textual specifications such as the Data Dictionary which specifies the characteristics of the stored data and Minispecs which describe the transformations in the diagrams.

The diagrams developed in the analysis phase are useful for describing the software system to people outside the development team. They should be the subject of a formal review, often called a Walkthrough. An error found by a review at this stage is usually fixed much more easily than later in the development cycle. There may be many ways to describe a system with the graphical tools. To choose the best model, or to refine a model, there are a number of criteria that may be applied. The most important of these is correctness. In addition, it is useful to simplify the interconnections in the system. The result of the analysis phase is a logical model of the system.

In the design phase, the constraints on the software are added. Pieces of the logical model will be allocated to various computers. The details of the human interaction will be incorporated. The elements of the logical model are assigned to modules and a physical model for the software is developed. The primary tool for the design phase is the Structure Chart. This is again a graphical tool and models the hierarchy, the partitioning and the interfaces of modules within a single program. The Structure Chart should also be reviewed in a Walkthrough. There are also techniques to evaluate and refine this physical model of the software.

4.4 Computer-Aided Software Engineering

The commercial interest in software development methodologies has spurred the development of new tools to support what is now called Computer-Aided Software Engineering (CASE). Many of these products support SA/SD. The tools include programs to draw the diagrams of SA/SD. There are routines to check the consistency of various diagrams and to ensure that there is a complete specification of each data element or transformation. The tools store the diagrams and specifications in databases which are available over a network from remote computers.

These products have improved substantially in the last year. The two experiments which have committed to SA/SD were not able to find products that would satisfy their requirements even two years ago and have developed some tools themselves. D0 has recently undertaken an evaluation of available products and will try to put existing documentation into a new system. ALEPH has developed a set of tools which use the MacIntosh to manipulate Entity Relationship Diagrams in their ALEPH DAta MOdel (ADAMO) [6]. The ADAMO package goes beyond many commercial products and provides FORTRAN tools to manipulate data within the analysis code.

4.5 Code Management

During the development cycle, the software library must be carefully managed. Stable versions must be distributed to all developers so that new software can be tested with the rest of the system. There must be a full record of changes and there must be set of rules to ensure that new versions in the library are fully tested and documented. In fact, the problems of management begin earlier in the project. The same considerations apply to the requirements documents and the design documents.

The development of software for an experiment presents some challenges which are often not found in a commercial organization. The software continues to evolve over the history of the experiment as the apparatus is upgraded, knowledge of the detector improves, or physics interests change. There are many people involved in the project and they are at many sites. There are often many different computers.

The problem of code management is a difficult one and there is no simple solution. The PATCHY system developed at CERN can run on many computer systems but is not adapted to interactive access and is not widely used in the United States. Its replacement at CERN is a commercial product, HISTORIAN, but it is expensive for groups outside CERN and has not been widely adopted, even by LEP experiments. Many groups in the United States have decided to manage their software only on the VAX computer systems. The VAX products, CMS (Code Management System) and MMS (Module Management System) provide most of the functionality needed for experiments and support library management over a network with DECnet. There is a significant problem, however, for groups that have other computer systems at their home institutions.

4.6 Software Testing

No one developing software would dispute the need to test new software. The problem comes as the program is changed. Each change is tested to ensure that it does what it is expected to do. Seldom is the program subjected to all the tests that were run at earlier stages to ensure that old problems have not been reintroduced or that the change did not have some unexpected impact. Doing this requires maintaining a series of test scripts that are run on each new version before it is distributed for general use. A set of standard output files must be maintained with the scripts so that a new file can be verified. Such a product is available as part of the VAX software system, DEC Test Manager (DTM) but again it works only for programs which run on the VAX computers.

Another aspect of testing that seldom gets the attention it deserves is the need to follow all possible paths through the program. A single program may have many thousands of possible paths depending, for example, on the nature of a complicated event. Often, these paths are not tried until a real event provides the set of parameters. If the program does not handle it properly, it is necessary to rerun the tape and chase the problem. It is more efficient to minimize the number of possible paths by the use of structured programming techniques, and then design the testing to ensure that all paths have been tried.

Various studies in the computer industry have indicated that up to half of the effort in a large project may be spent in testing programs. This statistic by itself should may it a high priority to improve the efficiency of software testing. The book by Myers [7] provides useful guidance for all aspects of software testing.

4.7 The Future

Software development for large projects will require a very significant effort. A example of how software is developed in industry was presented in this workshop [8].

Most physicists and programmers working in High Energy Physics are not familiar with the methods used in industry. We need to educate the physicists and programmers in these techniques. This can be done with sessions at major conferences comparable to the sessions on detector technology or accelerators. We should also encourage schools on computing techniques and workshops on special topics such as SA/SD. We should recognize computing as an important topic for seminars. These steps will help prepare us for the challenges of future experiments.

6. Conclusions

It is clear that computing will be increasingly important for the High Energy Physics program in the 1990 s. It will be necessary to provide facilities for computing, storage and networking to support the present and next generation experiments.

The software development effort for future experiments will require the use of new methodologies. We need to begin now to learn how to use these techniques effectively.

Acknowledgments

I want to express my appreciation to the members of the working group for their participation and for their many useful comments. I want to thank Byron Roe for preparing the summary of the discussion on the SSC Computing Division which is included in this report.

References

1. L.E. Price, editor, Proceedings of the Workshop on Detector Simulation for the SSC, ANL-HEP-CP-88-51
2. B.W. Boehm, Software Engineering, IEEE Transactions on Computing, C-25, No 12, 1976.
3. S.J. Mellor and P.T. Ward, Structured Development for Real-Time Systems. Yourdon Press, New York, 1986.
4. G. Kellner, Develpoment of software for ALEPH using structured techniques. In P. Kunz and T. Schalk, editors, Computing in High Energy Physics: Proceedings of the International Conference on Computing in High Energy Physics, Asilomar, 2-6 February 1987, page 229, 1987.
5. J.J. Linnemann et al., The use of SA/SD methods in D0 software development. In P. Kunz and T. Schalk, editors, Computing in High Energy Physics: Proceedings of the International Conference on Computing in High Energy Physics, Asilomar, 2-6 February 1987, page 245, 1987.
6. Z. Qian et al., Use of the ADAMO data management system within ALEPH. In P. Kunz and T. Schalk, editors, Computing in High Energy Physics: Proceedings of the International Conference on Computing in High Energy Physics, Asilomar, 2-6 February 1987, page 283, 1987.
7. Glenford J. Myers, The Art of Software Testing. Wiley-Interscience, New York, 1979.
8. J.S. Jenkins et al., Software Development in Large Engineering Projects: Lessons Learned, in these proceedings

AN INTEGRATED DESIGN, MODELING AND ANALYSIS RESOURCE FOR LARGE DETECTOR SYSTEM DEVELOPMENT

N.J. DiGiacomo, T. Adams, M. K. Anderson, M. Davis, J. Gliozzi, W. M. Hale, M. Kennedy, K. Killian, M. Krohn, R. Leitch, L. Mason, J. Mitchell, J. Pohlen and T. Wright

Martin Marietta Astronautics, Denver, Colorado 80201

ABSTRACT

We discuss the role that integrated computer-aided-engineering and computer-aided-design (CAE/CAD) can play to improve the design, modeling and analysis cycle in the development of "one-of-a-kind" high technology systems. We consider the specific requirements of SSC detector systems, and propose a candidate integrated CAE/CAD resource architecture for the SSC Laboratory. We report our progress in the development of such a resource, and demonstrate its capabilities on a model detector.

I "ONE-OF-A-KIND" HIGH TECHNOLOGY

The fundamental issue in understanding the difference between "one-of-a-kind" and "production" high technology is not the absolute number of units to be produced, but the extent to which economies of scale can be used to offset the impact of a given expenditure on the final cost. Of particular importance are expenditures in the early stages of development, such as those for extensive modeling, analysis and prototyping. Although both types of high technology can require early exploration of a broad range of design possibilities, prototyping and careful performance simulation, it is often judged that prototyping or extensive modeling and analysis is impractical (i.e. too expensive) for "one-of-a-kind" devices or systems.

Another difference between "one-of-a-kind" and "production" high technology involves the way in which one satisfies performance requirements. Scientific experiments such as modern collider detectors start with state-of-the-art sensor packages designed to make specific measurements. The component sensors are then "crafted" together into a system. The "production" philosophy, evident for example in modern aircraft development, emphasizes more the overall system design in meeting performance requirements. Components are only as complex or sophisticated as necessary to fulfill their functions within the system.

As "one-of-a-kind" high technology becomes more complex, more expensive and subject to more stringent performance requirements (e.g. the Space Station or an SSC detector), one must look for ways to do "production" quality modeling, analysis, prototyping and system design at reduced cost. One step in the right direction is to replace expensive prototyping with computer simulation. As we argue below, this is but one benefit of the improved design, modeling and analysis cycle that results from the implementation of integrated CAE/CAD.

II THE DESIGN, MODELING AND ANALYSIS CYCLE

The development of a complex device or system proceeds through a number of distinct, well-defined phases. Figure 1 depicts this evolution as a flow, starting with concept exploration and ending with system commissioning. The success of such a project can be viewed, in the context of Figure 1, as the convergence of the flow; careful, thorough concept exploration and requirements definition assure the expeditious assembly, test and commissioning of the system. Conversely, the process can well diverge if the early phases of the project are passed over too quickly.

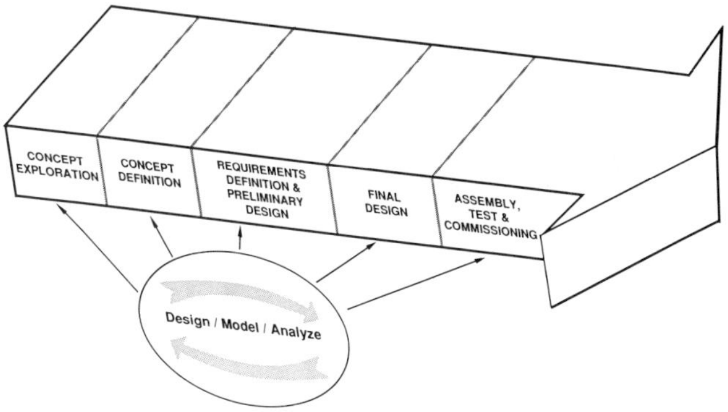

Figure 1 The project flow, with phases indicated. The design, modeling and analysis cycle reoccurs throughout the project lifetime, with requirements and goals that depend on the phase.

A constant feature of each project phase is the design, modeling and analysis cycle. The cycle is repeated continuously as the project proceeds, albeit with requirements and goals that are functions of the phase. Concept exploration, for example, requires fast turnaround analysis in order to explore a broad range of design possibilities, but can tolerate rather lower fidelity. The final design, however, demands thorough, high fidelity modeling and analysis.

The success of a project, then, can depend strongly on the availability of the design, modeling and analysis tools appropriate to the task. It is important to stress the words AVAILABLE and APPROPRIATE. Sophisticated analysis packages arc of little use in concept exploration if they are too cumbersome to use quickly. Likewise, simple, easy to use modeling packages may be inadequate to an engineer performing a detailed part design.

Islands of Automation

Individual design, modeling or analysis packages can be viewed as "Islands of Automation". Sophisticated analyses can be done by computer, but data transfer, manipulation and translation between programs must be done by hand. The fidelity of modeling and analysis is generally limited by the need for extensive manual setup, gridding and meshing. The analyses performed are quite independent of each other, and turnaround times between iterations are long compared to other timescales in the project, such as drawing preparation and technical or management milestones. In essence, the ability to exercise design and engineering judgement and to explore the space of design possibilities is limited by the time-consuming tedium of handling the data flow between the "Islands".

Integrated CAE/CAD

Integrated CAE/CAD provides designers and engineers in all phases of the project with linked design, modeling and analysis tools that can be tailored to the task at hand. The essential emphasis is on "value-added" contact with data. One only intervenes in the flow of data between packages to contribute engineering or design judgement. The user does not participate in routine data transfer, translation and manipulation. Interactive graphics are used extensively to input design, define modeling and analysis parameters and conditions and to evaluate results. The system is optimized to provide fast turnaround and to allow parallel analyses. The fidelity of modeling and analysis is determined by the task requirements and not by the tedium of setting up new grids or meshes.

Implementation of Integrated CAE/CAD

There are a number of "lessons learned" from various attempts at designing and implementing integrated CAE/CAD systems within scientific or engineering organizations, including Martin Marietta Astronautics [1]. These include the following:

- Integration of individual design, modeling and analysis packages into a useful, coherent tool requires a significant amount of work by systems professionals. The integration cannot be done solely by the users or by those responsible for the component packages.

- The design, modeling and analysis skills of the project team become less separable when integrated CAE/CAD is employed. An interdisciplinary approach seems to work best.

- Project management must adapt, both programmatically and organizationally, to the interdisciplinary environment. A major contributor to successful implementation of CAE/CAD in technical organizations is management commitment. In addition, management must believe that the up-front investment in CAE/CAD tools and an emphasis on more modeling and analysis in the early stages of a project will lead to a better design and faster commissioning.

- Careful hardware and software evaluation is critical, because one is building a SYSTEM, not a series of stand-alone codes. Again, systems professionals must assist users or specialists in making these choices.

The often conflicting requirements of resource distribution and centralization must be satisfied. The resource capabilities must be distributed to users wherever they are, but the main database must be controlled and code modifications, revisions and updates implemented by a central authority. One approach that seems to work is to have a fraction of the central facility staff rotate in from the distributed user facilities. This allows for good user familiarity with, and direct and timely user input to the central facility.

III AN INTEGRATED CAE/CAD RESOURCE FOR SSC DETECTOR SYSTEMS

We now consider integrated CAE/CAD in the context of SSC (or other large collider) detector system development. Figure 2 shows our project flow arrow, now tailored to the high energy physics world of accelerator laboratories and large detector collaborations. Figure 2 reflects the fact that much of the detector component development occurs away from the accelerator laboratory, at collaborating institutions. Recalling our earlier discussion on the need to bring more system design as well as component design, modeling and analysis capability to "one-of-a-kind" projects like collider detectors, we propose an integrated CAE/CAD resource centered at the SSC Laboratory, with distributed access to collaborating institutions. As such, the resource is available to the collaborations that will develop the detector system concepts and components, and to the Laboratory administration and staff that must evaluate, select and then integrate the detectors.

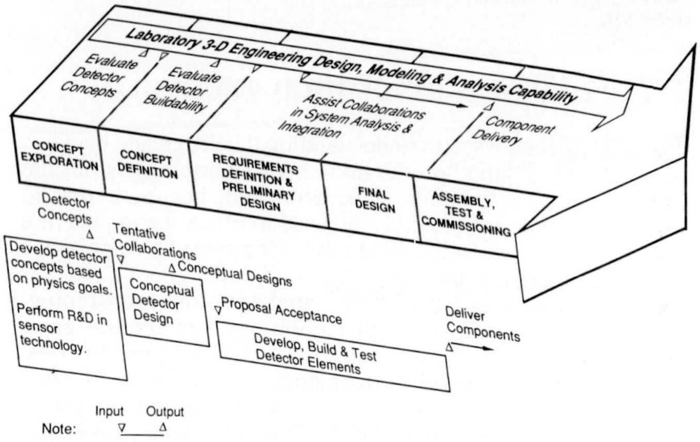

Figure 2. The project flow, adapted to detector system development in the world of accelerator laboratories and distributed collaborations.

In this section we examine the requirements for design, modeling and analysis during each phase of detector system development. We then propose a candidate integrated CAE/CAD resource architecture.

Concept Exploration and Definition

Groups (i.e. potential collaborations) develop a conceptual detector system design for a given set of physics requirements. For each design, one must:

- Evaluate the viability of different sensor/component detector technologies within the context of the particular system design.
- Simulate the overall response of the detector system
- Determine the engineering feasibility and implications of the design.

Groups should be able to perform the above quickly and easily enough to allow the exploration of the maximum fraction of "detector possibility space". In addition, the Laboratory must have the means to compare and evaluate competing designs.

Requirements Development and Preliminary Design

At this stage, collaborations have rather well-defined physics goals and preliminary detector designs. One must:

- Carefully determine the performance (i.e. physics response) and "buildability" of each detector system.
- Define the interfaces among components and sub-systems, and establish an interface control methodology.
- Generate preliminary drawing packages of the system and its components for distribution to collaborating institutions to allow evaluation of task magnitude and to constrain prototype development to overall system requirements.
- Feed back progress in prototype R&D into overall system design and analyze impacts on system performance, other detector components and cost.

Final Design

For each approved and funded detector system, one must:

- Efficiently evolve the detailed design, in an iterative manner, based on component development, interface control, fit checks and engineering modeling and analysis.
- Generate component drawings and manufacturing sequences from a central design database.
- Control quality and uniformity of components and parts through approved parts lists and a uniform cataloging system.

Assembly, Test and Commissioning

In this phase, components built at collaborating institutions, industry and the Laboratory are assembled, integrated and tested at the Laboratory. One must:

- Control changes and interfaces from a central authority.
- Redesign, reanalyze and recalculate detector response on a short timescale to adapt the detector system to changes in component design or performance, or to changes in physics requirements.
- Perform simulations of detector performance for data analysis purposes, from the highest fidelity model of the detector system.
- Generate assembly drawings and sequences.

Resource Architecture

Figure 3 is a schematic representation of an integrated CAE/CAD resource or "tool kit" for SSC detector development. The resource architecture, based on existing systems at Martin Marietta, is designed to satisfy the requirements discussed above. The schematic nature of Figure 3 must be emphasized; as noted earlier, integrating the design, modeling and analysis packages is a complex task for systems professionals. A graphic example of the actual complexity can be seen in Figure 4, where the thermophysics part of the "demonstration" integrated CAE/CAD resource (discussed in Section IV) is shown.

The resource has five basic functional areas: 3D modeler and database, CAD/CAM, CAE interface, analysis codes, and physics simulation. We discuss below the capabilities of each functional area, and the flow of data and information between them.

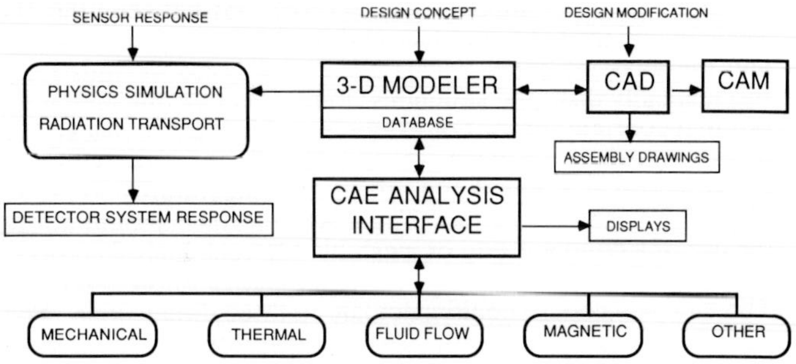

Figure 3. Candidate architecture of an integrated CAE/CAD resource for detector system development. The elements of the resource and the data flow between packages are described in Section III of the text.

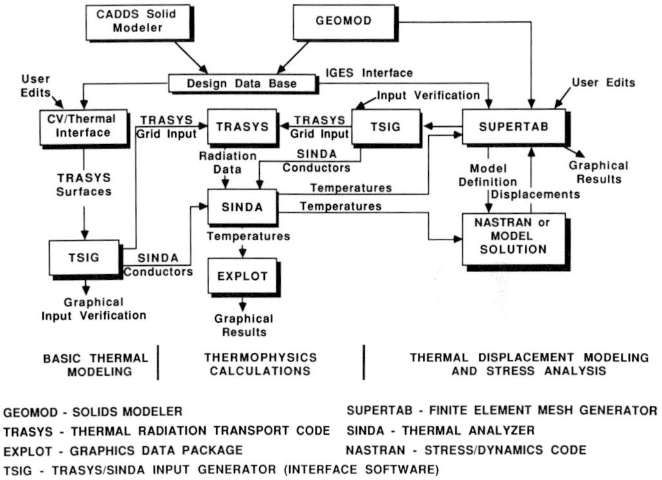

Figure 4. The thermophysics section of the integrated CAE/CAD resource. See Section IV of the text for details.

The 3D modeler and.database should have:

- Full interactive 3D modeling with view manipulation
- Easy entry of designs at all levels (piece part through system), with instancing
- 3D interference and fit checking at any stage of motion
- Material and mass properties
- A user-friendly operating environment
- Structured, two-way links to major CAD packages and to an analysis interface

The CAD/CAM packages allow generation of detailed dimensioned drawings and CAM sequences from the 3D design database. The 3D modeler should be capable of two way communication with a number of CAD packages, since collaborating institutions will not all have access to the same CAD systems. The path from CAD back to the 3D modeler is desireable so that design and engineering implications of changes in 2D drawings can be easily examined. The CAD package should also have cable design and management capability.

The CAE interface allows the engineer to mesh or grid designs for analysis, to view the results and to make 3D drawings and representations of designs both before and after analysis. This package is central to the efficient operation of the resource, since it is here one can minimize tedium and maximize "value-added" interaction with the data. Note that in Figure 3 communication is implied in both directions between the 3D modeler and the CAE interface and between the CAE interface and the analysis codes. This allows analyzed designs (e.g. mechanically distorted structures) to be directly compared with original configurations. For example, readout cable placement in a cryogenic calorimeter would be facilitated if drawings of the structure at ambient and operating temperatures could be viewed by cable designers. The CAE interface should allow the use of a common grid or mesh by different analysis tasks.

The analysis codes should calculate in three dimensions. For SSC detector development, analysis capability should include, but not necessarily be limited to:

- Mechanical: Stress, strain, loading, contacts and alignment
- Thermal and fluid flow: Forces, radiation, convection, conduction, mass transfer and distortions
- Magnetic: Field shape, quality and resultant forces

One should, of course, have access to a variety of codes for each analysis function, given variations in algorithms, methods and capabilities (e.g. one would have NASTRAN [2]), ABAQUS [3]) and ANSYS [4]) for structural and mechanical analysis).

The detector response section of the resource performs radiation transport, physics simulation, calculates coverage, shadowing and hermeticity and calculates electromagnetic signal response. It uses as input the 3D modeler database, individual sensor responses and event physics simulators. The result is the overall detector system response.

IV RESOURCE DEMONSTRATION ON A MODEL DETECTOR

As stated previously, the proposed resource architecture is based on software systems presently in existence at Martin Marietta. We are finding, in fact, that much of what is designed for aerospace high technology is appropriate for SSC detector development. Other capabilities, such as detector response and magnetic analysis, must of course be developed for the specific application. In this section, we report on the elements of the proposed architecture as they exist today and present some results that are indicative of the system's performance.

The 3D modeler and CAE interface in use at Martin Marietta are the IDEAS [5]) packages GEOMOD and SUPERTAB. Analysis codes attached to SUPERTAB include NASTRAN [2]), ABAQUS [3]) and ANSYS [4]) for mechanical analysis, POISSON [6]) and ANSYS [4]) for magnetic analysis and the SINDA/TRASYS packages [7]) for thermal and fluid flow analysis. CAD packages that communicate with GEOMOD include CADDS 4X [8]).

An example of code integration is shown in Figure 4, where the thermophysics capability of the resource is presented. A design created in GEOMOD is gridded in SUPERTAB for mechanical analysis by NASTRAN. SUPERTAB's gridding is translated for both TRASYS and SINDA by TSIG [7]) and the thermal analysis is performed. The temperature profiles are then passed back to NASTRAN via SUPERTAB for the mechanical analysis, with iteration occurring as necessary.

Figures 5-8 are representative results from our design, modeling and analysis of a model detector. Figure 5 shows a cut through our model detector. The model is based on the DØ liquid argon calorimeter [9]) and a cryostable superconducting coil similar to one proposed for the CDF detector [10]). Figure 6 depicts the GEOMOD-resident design calorimeter modules supported by a stainless steel strap and beam structure (the cradle) that attaches to the external coil support (not shown). Figure 7 shows the gridding selected in SUPERTAB for analysis of the strap/beam junction. Figure 8 shows the axial (i.e. along the beam axis) stresses on the junction under gravity, cryogenically loaded and in a 2 Tesla magnetic field. The indicated stresses are due to torque on the strap generated by the calorimeter modules. The analysis employed NASTRAN, TRASYS/SINDA and POISSON. Note that figures 5-8 are black and white prints of the original color pictures, so the true utility of the resource's graphic capability is not apparent; color is particularly important in making full use of contour plots such as Figure 8.

Our belief that existing Martin Marietta CAE/CAD tools and techniques are relevant to SSC detector development is supported by the breakdown of effort expended in the demonstration. Approximately 25% of our effort went to design and modeling, 62% to analysis and only 13% to code integration. The implication is that most of the necessary code integration had already been accomplished for other applications.

Figure 5. A GEOMOD section of our model detector, showing the calorimeter cryostat, calorimeter modules, and the superconducting coil.

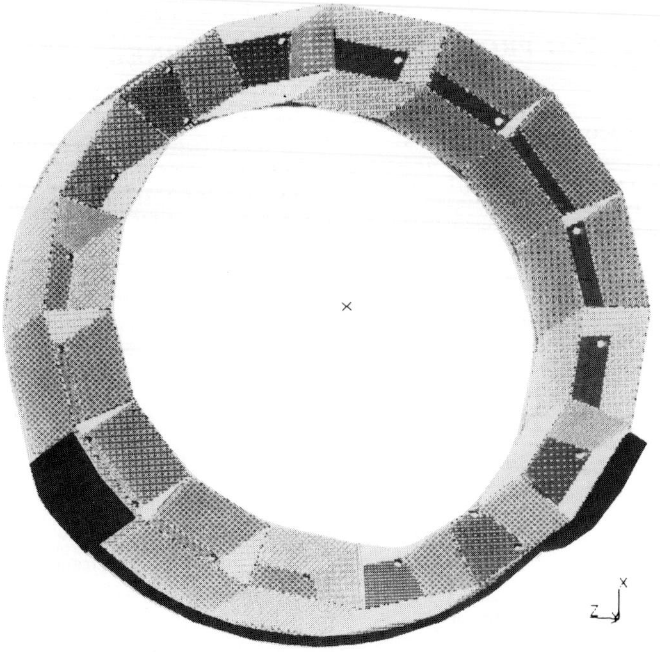

Figure 6. A GEOMOD detail of the calorimeter modules supported by a stainless steel cradle.

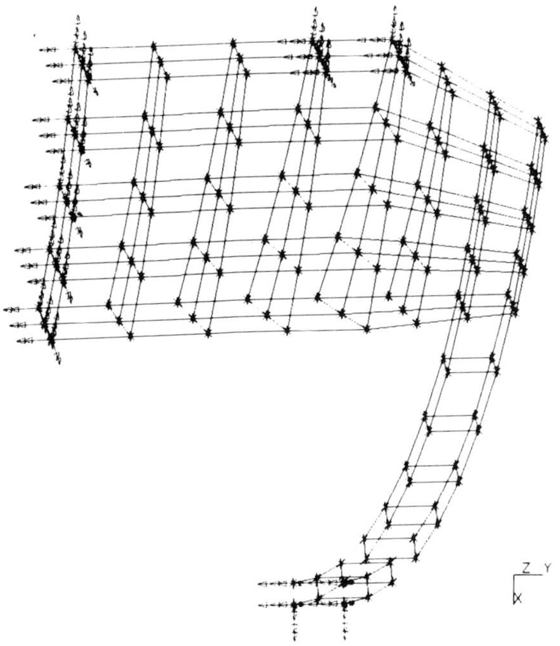

Figure 7. The FEM mesh generated in SUPERTAB for a detail of the cradle. The detail shows the intersection of the support strap and beam.

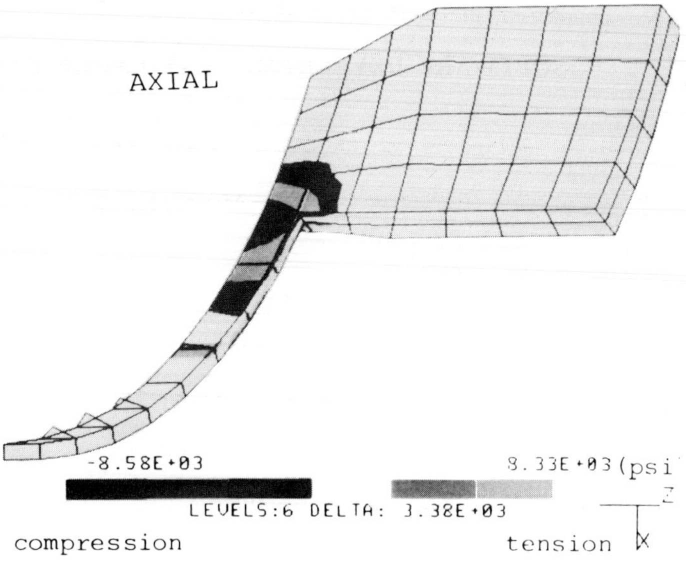

Figure 8. A SUPERTAB stress contour plot of the cradle detail, showing the axial forces of the junction. The compression/tension scale is shown at the bottom of the picture. The analysis was performed with gravity, under cryogenic load and in a magnetic field of 2 Tesla.

CONCLUSIONS

The complexity, cost and performance requirements of SSC detectors make it imperative that the design, modeling and analysis cycle be made as efficient, and cost effective as possible. To this end, and based on similar challenges with "one-of-a-kind" high technology in the aerospace industry, we propose an integrated CAE/CAD resource for SSC detector system development.

ACKNOWLEDGEMENTS

We wish to thank Gil Gilchriese for bringing this subject to our attention, Howard Gordon for DØ liquid argon calorimeter documentation and Ron Fast for the design study on a cryostable superconducting coil for CDF.

REFERENCES

(1) Project Challenge CAE/CAD/CAM Initiative, Martin Marietta Astronautics Group, Denver CO. (1987)
(2) NASTRAN, MacNeal-Schwender Corp., Los Angeles, CA, (1987).
(3) ABAQUS, Hibbit, Karlsson and Sorenson Co., Prividence, RI, (1987).
(4) ANSYS, Swanson Analysis Systems, Inc. Houston, PA, (1987).
(5) I-DEAS, Structural Dynamics Research Corp., Milford. OH, (1987).
(6) POISSON, Los Alamos National Laboratory, Group AT-6, Los Alamos, NM, (1987).
(7) SINDA and TRASYS, Martin Marietta Astronautics Group, Denver CO., (1987).
(8) CADDS 4X, Computervision Corp., Bedford, MA, (1987).
(9) DØ Revised Design Report, DØ Note 512, Fermilab, (1985).
(10) R. Fast et al., "Design Report for a Cryostable 3m Diameter Superconducting Solenoid for the Fermilab CDF Facility", TM-1075, Fermilab, (1981).

SOFTWARE DEVELOPMENT IN LARGE ENGINEERING PROJECTS: LESSONS LEARNED

J. S. Jenkins*, M. Davis, N.J. DiGiacomo and K. Killian

Martin Marietta Astronautics, Denver, CO 80201

ABSTRACT

We present here some observations on software development in large, complex space and ground-based high technology projects. We discuss various aspects of the the software life cycle, including requirements definintion, sizing and estimating, tool development, code implementation, performance measurement and testing. Particular emphasis is given to the role of project management in sucessful software development.

I INTRODUCTION

Software is rapidly becoming one of the most critical and expensive components in large engineering projects. For example, of a Department of Defense budget estimated to be \$120 to \$150 billion in 1990, 49% will go to software systems, and only 1-2% to computer hardware. There is still, however, a tendency in some circles to treat software development as a sub-element of the hardware, with the result that software does not receive adequate management visibility.

The difficulties with software development can be grouped into four categories: late delivery, cost overruns, reliability and maintainability. In this paper we offer some mitigation methodologies that impact management and engineering, controls, standards and documentation.

II THE SOFTWARE LIFECYCLE

Figure 1 portrays the evolution of a configuration item (CI), with particular emphasis on the parallel development of the hardware (HWCI) and software (SWCI). Hardware and software development are connected by the series of reviews that start with system requirements and end with test readiness. After the functional configuration audit the hardware and software is brought together, integrated, tested and commissioned. One important result of the review process is the maintenance of the functional, allocated and product baselines.

III REQUIREMENTS DEFINITION

A great many software development problems can be traced to poor requirements analysis. Specific actions that improve the effectiveness of the requirements phase include:

- Making software personnel part of the system analysis, operations analysis and system architecture teams.
- Maintaining detailed requirements traceability.
- Establishing user requirements during the system requirements definition.
- Developing as complete a requirements baseline as possible BEFORE design commences.
- Planning for implementation of changes due to evolving requirements.

* Director, Software Center of Excellence, Martin Marietta Astronautics, Denver CO 80201

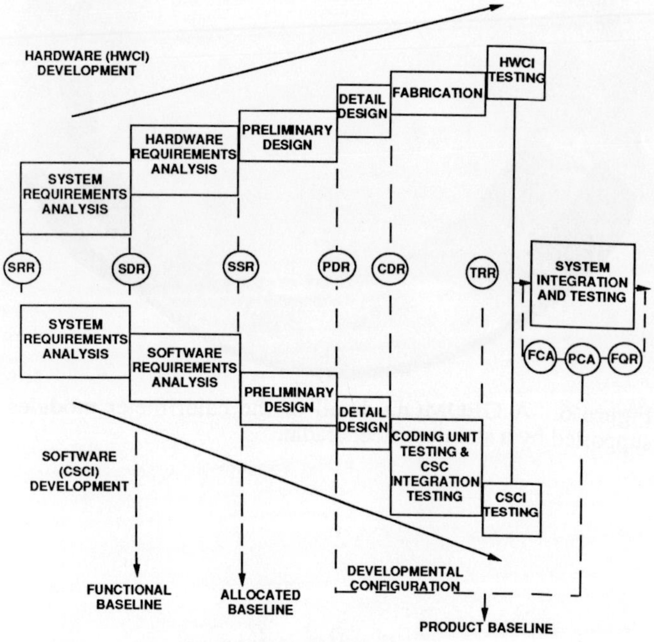

Figure 1. The software lifecycle that occurs between the system concept and the operational phases. Reviews are indicated by the circles, and are as follows:

- SRR System Requirements Review
- SDR System Design Review
- SSR Software Specification Review
- PDR Preliminary Design Review
- CDR Critical Design Review
- TRR Test Readiness Review
- FCA Functional Configuration Audit
- PCA Physical Configuration Audit
- FQR Formal Qualificaction Review

IV SOFTWARE SIZING AND ESTIMATING

Cost and schedule performance begins with credible estimating. Figure 2 depicts the process by which one estimates the scope and scale of a software project, develops a credible bid package and reliably determines risk. In the bottoms-up method of estimation and sizing, one integrates the test requirements, systems analysis and operations concepts into a software requirements baseline. The baseline is then subject to a functional anayisis and decomposition, yielding the schedule, code size, manpower and development resource requirements. Alternatively, one can use data-base derived parametric models to estimate cost, schedule and size. Two examples of parametric models, SASET and PCEM, are outlined in Figures 3 and 4, respectively.

One cannot overemphasize the importance of thorough requirements analysis, test and evaluation in software development. Indeed, as shown in Figure 5, actual coding constitutes less that 20% of the effort in a typical large software development program. Figure 5, in addition to breaking down the distribution of effort in software development, identifies the reviews associated with each phase of the program.

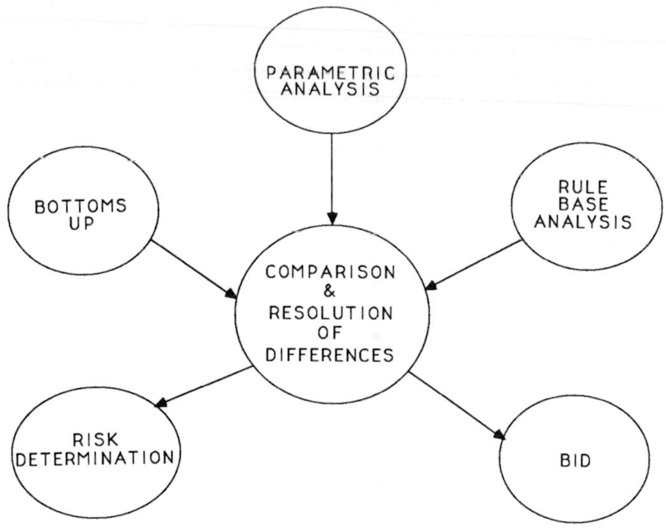

Figure 2. Approaches to software sizing and estimating.

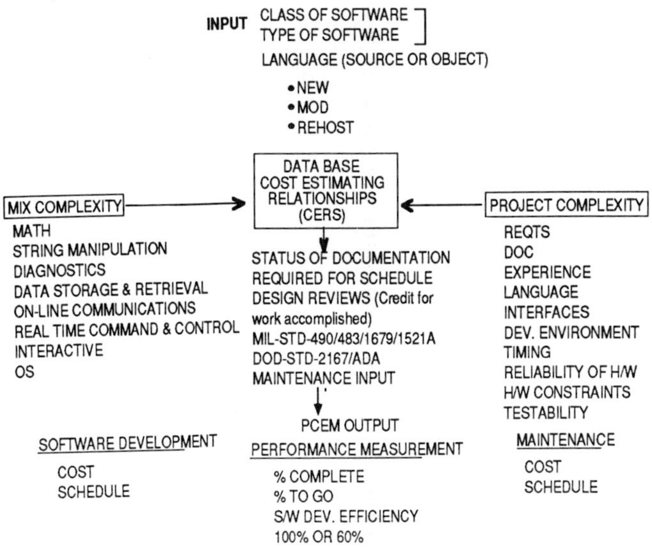

Figure 3. SASET, or Software Architecture Sizing and Estimating Tool. This is a data base derived expert/parametric cost estimating, scheduling and functional sizing model.

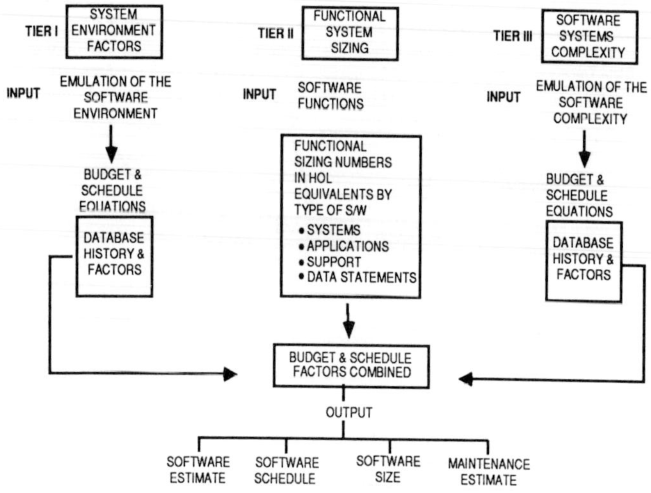

Figure 4. PCEM, or Parametric Cost Estimating Model. This is a Martin Marietta data base derived parametric cost estimating, schedule, maintenance and performance measurement model.

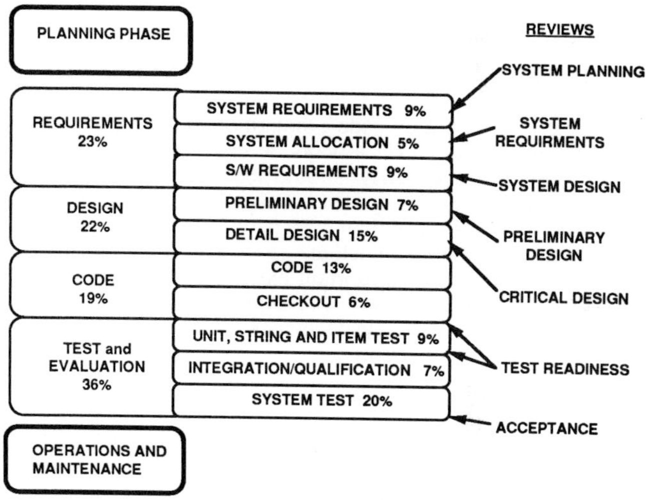

Figure 5. Decomposition of a typical large software project into effort as a function of project phase. Also indicated are major project reviews.

V TOOLS FOR SOFTWARE DEVELOPMENT

As in hardware development, successful software depends on having the right tools for the job. Figure 6 contains a generic "shopping list" of tools for each phase, along with a recommended minimum tool set. A more general way to view tool sets is as a software development environment. Figure 7 lays out a Comprehensive Software Development Environment (CSDE) Architecture that integrates many commercial tools (COTS).

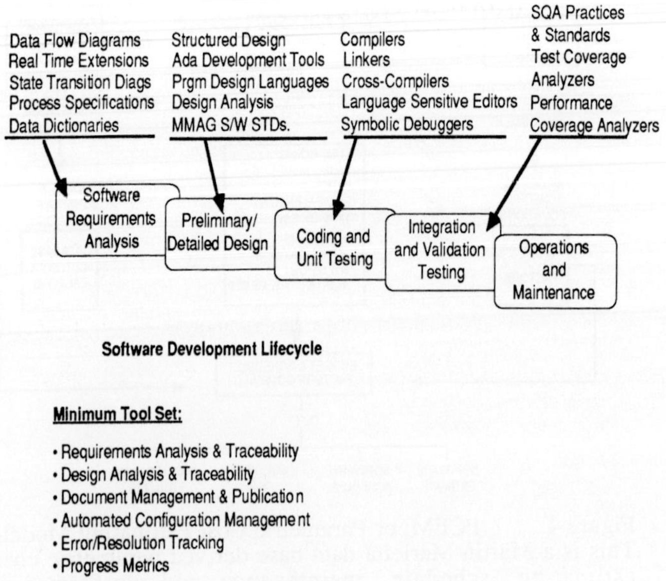

Figure 6. Tools necessary for efficient software development as a function of the development lifecycle.

VI IMPLEMENTATION (DESIGN AND CODE)

The use of "common code", though often the centerpiece of the proposal, is NOT always a cost effective option. Requirements traceability should be given priority over commonality. The oft-quoted "20% rule" - if modifying old software involves more than 20% new code, start over - is in fact quite a reasonable guideline.

Perhaps the most important exhortation is: "if it works, don't fix it or make it better". An important corollary is that management monitor code developers to ensure that they only implement the requirements.

VII PERFORMANCE MEASUREMENT

Software development must be carefully scheduled and closely tracked to insure a successful project. Figures 8,9 and 10 depict some important performance indicators. In Figure 8, resource utilization and requirements stability (i.e. requirements that are not testable or traceable) are shown versus time. Figure 9 shows the relationship between scheduled tests, passed tests and unresolved problem reports as a funtion of time in the test and evaluation phase of a program. A final example of performance measurement, one based on the variance in cost and schedule versus time, is given in Figure 10.

It is an unfortunate rule of software development that the software is always 90% complete. To avoid this syndrome, one must establish and utilize a process measurement methodology. Detailed schedules are required down to the code module level, and management must actively participate in, and contribute to, reviews.

VIII SOFTWARE TESTING

A sucessful software test program begins in the requirements phase. One must establish a baseline test plan early in the program, and build in the "testability" of code from the module to the system level. One must also determine, based on system requirements, the:

- level of testing required
- need for independent verification and validation
- error tracking and correction methodology

VIIII MANAGEMENT CONSIDERATIONS

Given that software and hardware development are tasks of similar magnitude, concomitant visibility must be given to software development within the project organization. Figure 11 shows a project organization that accomplishes this, with a software manager who has direct access to the project director. In addition, the software manager is organizationally in a position to educate other managers as to the requirements of software development.

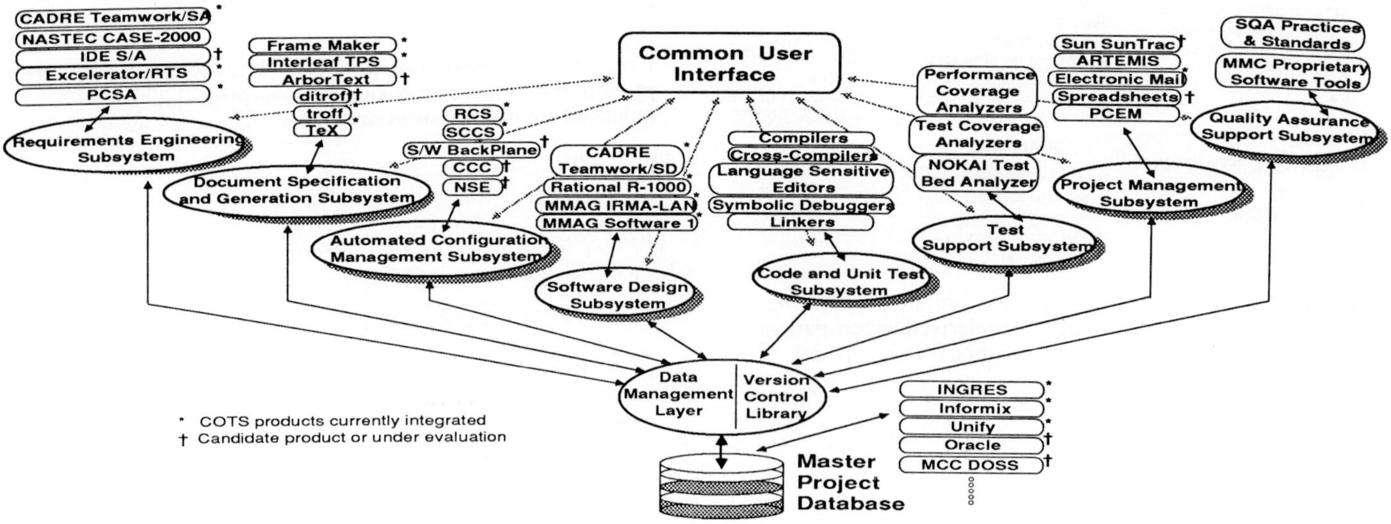

Figure 7. The CSDE (Comprehensive Software Development Environment) Architecture.

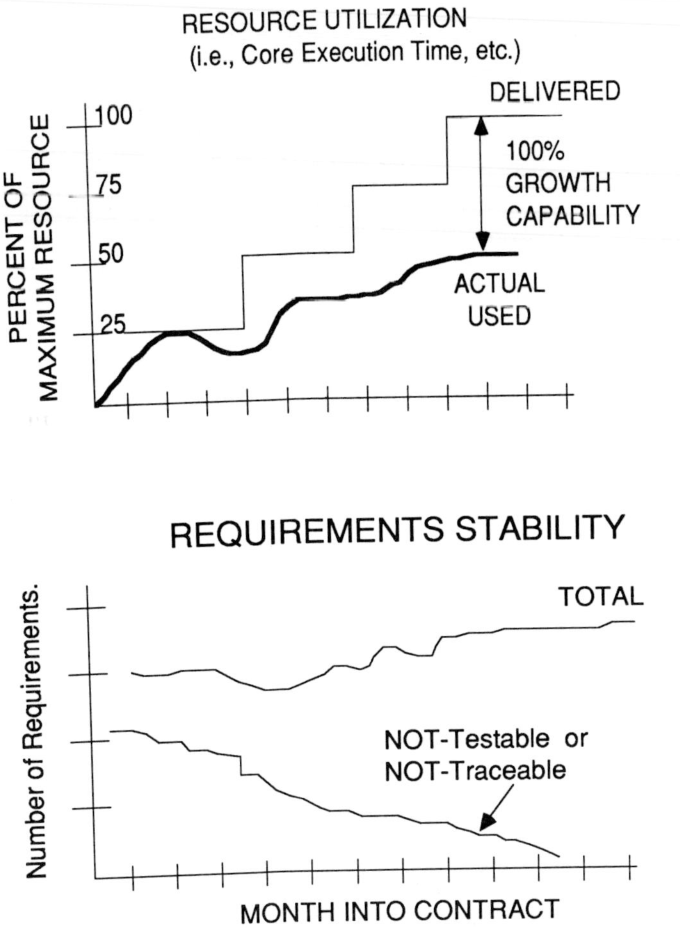

Figure 8. Two performance indicators for software development. The upper figure shows resource utilization versus time, while the lower tracks requirement stability versus time.

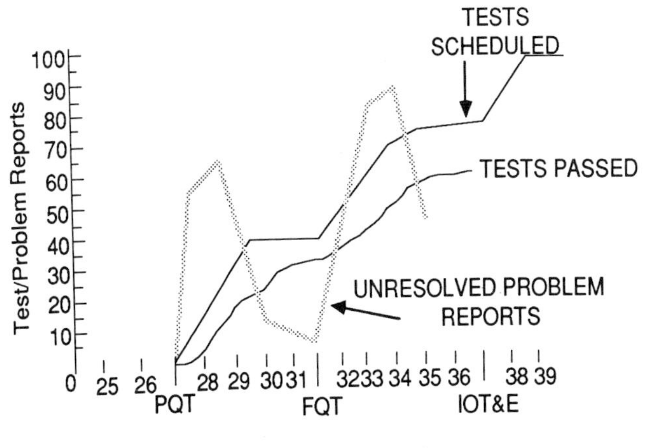

Figure 9. Test performance versus time. PQT, FQT and IOT&E are, respectively, the Preliminary Qualification Test, Final Qualification Test and Initial Operations Test and Evaluation milestones.

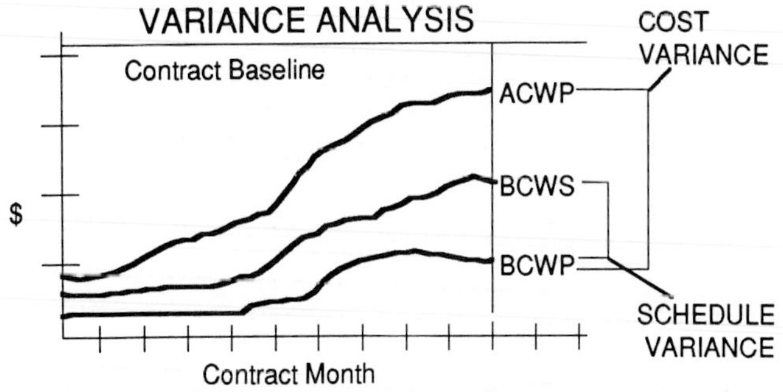

Figure 10. Cost and schedule performance versus time. ACWP is Actual Cost of Work Performed, BCWS is Budgeted Cost of Work Scheduled, and BCWP is Budgeted Cost of Work Performed.

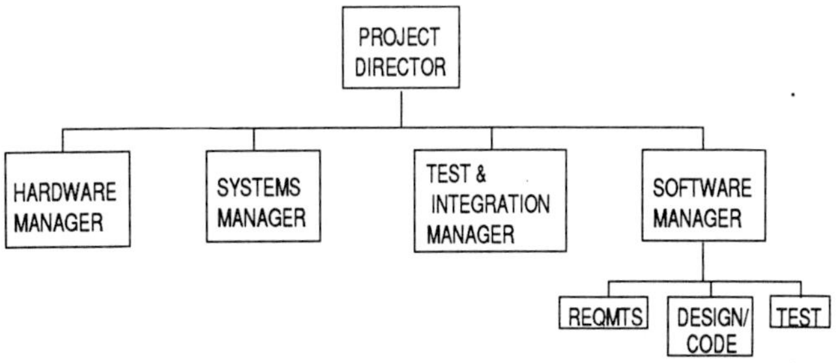

Figure 11. Organizational structure that gives appropriate visibility to software development.

A well thought-out and documented software development plan is crucial to project success. This plan must include a management methodology with clear roles and responsibilities, a subcontract management plan, a resource utilization plan and clear schedule definition. The software development plan is, of course, a living document, that will grow and evolve with the project. This is NOT, however, a good reason to discard the plan after its first modification. Adherence to the plan must be enforced and audited by management.

It is essential to involve the user (or customer) in software development from the beginning. Early agreement on data submittal format and content, and frequent progress versus schedule review are but two ways to smooth the communication between the vendor and customer. In addition, careful pre-review coordination can help minimize suprises at formal reviews.

Another important, yet often overlooked, way to improve efficiency and cost effectiveness is to insist that all subcontracted software be developed in a shared facility. This provides a transparency to management, eases the flow down of standards and data, and simplifies the implementation of the software development plan.

X SUMMARY

At the risk of oversimplifying the difficult and complex challenge presented by software development for large, high technology projects, we offer the following observation as a summary:
SOFTWARE DEVELOPMENT PROBLEMS ARE MORE OFTEN CAUSED BY POOR MANAGEMENT AND LACK OF DISCIPLINE IN THE DEVELOPMENT PROCESS THAN BY TECHNICAL DIFFICULTIES.

FUTURE DIRECTIONS IN DETECTOR R&D

M.D.G. Gilchriese

Conference Organizer

RADIATION LEVELS IN SSC DETECTORS*

Donald E. Groom
for the
Task Force on Radiation Levels in the SSC Interaction Regions†
SSC Central Design Group,‡
Lawrence Berkeley Laboratory 90-4040, Berkeley CA 94720

Abstract

Estimates of ionizing dose and neutron fluence have been made for typical SSC detector configurations exposed to radiation from p-p collisions. Using a description of "average events" in conjunction with simulations of secondary processes, it is found that in most cases the radiation exposure can be parameterized by functions of the form $A\,\langle p_\perp \rangle^\alpha \cosh^{\alpha'} \eta$, where η $(= -\ln\tan\theta/2)$ is the pseudorapidity, the exponent α lies between 0.5 and 1.0, and $\alpha' = \alpha$ or $\alpha + 2$, depending upon the problem. A calorimeter element 2 m from interaction point and 6° from the beam line is subjected to an annual dose is 0.1 MGy at shower maximum. The annual neutron fluence inside a central cavity of this radius exceeds 10^{12} cm^{-2}. The annual dose from the direct flux of charged particles scales inversely as the square of the distance from the beam line and is 0.4 MGy at 1 cm. A detector at an 8 TeV on 8 TeV p-p collider with the same "physics reach" would experience about 27 times as much radiation.

1. Introduction

High energy and high luminosity combine to create a radiation environment for experiments the Superconducting Supercollider (SSC) which is orders of magnitude more severe than at existing accelerators. Radiation problems must be addressed in any serious detector design. Accordingly, an SSC Central Design Group Task Force was formed to make as quantitative an assessment of the radiation level as is possible in the absence of a specific detector design. In the following we briefly summarize the approach and main results given in the Task Force Report[1].

In contrast to experience at present colliders, most of the radiation in an SSC interaction region (IR) comes from the 20 TeV on 20 TeV p-p collisions. At the design luminosity, about 10^8 collisions per second occur, and we assume that this luminosity is maintained for 10^7 s out of each year. At this luminosity the beam-beam collisions at each IR contribute (300 hr)$^{-1}$ to the reciprocal of the current lifetime, so collision products strike the detector at a rate which is equivalent to the accidental loss of one beam into the apparatus every 6 days. By comparison, beam-gas collisions and expected particle loss contribute very little. Accidental processes such as beam loss during injection are more problematical, but if they are significant severe machine damage will also result.

For purposes of this study, a detector has been modeled as a calorimeter surrounding an empty volume, with collisions occurring at the center. The calorimeter is a spherical shell extending down to some cutoff angle from the beam line, typically 6°. More distant endcap calorimeters provide coverage from this angle to a smaller angle. Since results may be scaled with distance from the interaction point in a fairly simple way, the exact geometry is not critical. It is also not difficult to apply the results to specialized detectors currently being discussed, such as a forward B spectrometer in an intermediate-luminosity IR.

An incident hadron produces both an ionizing dose and a neutron fluence in a calorimeter. Backscattered neutrons enter the central cavity, and there might be substantial rear leakage from the calorimeter. About a third of the particles from the primary collision are π^0's, so as many photons as hadrons strike the calorimeter. The resulting showers produce an ionizing dose which peaks more sharply and therefore more intensely than does the dose from an electromagnetic cascade. Backscattered photons re-enter the cavity.

A very simple approach has been used to describe all of these processes. In the first place, particle production is described—not particle production in the rare interactions of physics interest, but in the "minimum bias" events which occur 10^{15} times each year. Secondly, the interactions of single hadrons or photons with the detector are described and parameterized as a function of incident energy or momentum. The results are then combined to obtain the desired dose rate or flux. In a few cases "global" calculations have also been made; these corroborate the general approach but raise a few caveats. There are uncertainties in both particle production and in the effects of individual particles, and some of the assumptions are approximate. For example, the transverse size of a cascade is assumed to be small compared with the distances over which the incident flux changes appreciably, an assumption which is certainly not valid very close to the beam line. Uncertainties from all of these sources are such that the results are thought to be dependable to within factors of two to **four.**

* This paper closely follows SSC Central Design Group Report SSC-179, submitted for publication in Nucl. Instrum. and Meth. as part of the *Proceedings of the Inter. Conf. on Advanced Technology and Particle Physics* (Como, Italy, 13-17 June 1988). Corrections, minor revisions, and updates have been made.

† F. S. Alsmiller, R. G. Alsmiller, Jr., S. Ban, J. E. Brau, K. W. Edwards, A. Fassò, H. Fesefeldt, T. A. Gabriel , M. G. D. Gilchriese (Chairman), D. E. Groom, H. Hirayama, H. Kowalski, H.W. Kraner, N. V. Mokhov, D. R. Nygren, F. E. Paige, J. Ranft, J. S. Russ, H. Schönbacher, T. Stanev, G. R. Stevenson, A. Van Ginneken, E. M. Wang, R. Wigmans, and T. P. Wilcox, Jr.

‡ Operated by the Universities Research Association for the U. S. Department of Energy.

2. Particle production

Particle production at $\sqrt{s} = 40$ TeV has been studied with the aid of ISAJET[2], PYTHIA[3], and DTUJET[4]. In spite of some theoretical problems, a reasonable extrapolation from lower energies is possible because of the relatively slow variation of the data with $\ln s$. The increasingly important cross section for producing low-$p_\perp$ jets is included. For example, when ISAJET is used to generate a sample of "average" events, half MINBIAS and half TWOJET events are used. In all contexts, it is possible to write the charged multiplicity per average event as a factored function of pseudorapidity η (where $\eta \equiv -\ln\tan\theta/2$) and of the component of momentum perpendicular to the beam direction:

$$\frac{d^2N_{\text{charged}}}{d\eta dp_\perp} \approx H\, f(p_\perp)\ . \tag{1}$$

Equivalently,

$$\frac{d^2N_{\text{charged}}}{d\Omega dp_\perp} \approx \frac{H\, f(p_\perp)}{2\pi \sin^2\theta}\ . \tag{2}$$

Here $f(p_\perp)$ is a normalized function of $p_\perp$, and H, the height of the pseudorapidity plateau, is constant to within 10% or better for $|\eta| < 6$, corresponding to angles greater than 0.3°.

Several forms have been used for the function $f(p_\perp)$. A form proportional to $p_\perp \exp(-p_\perp/B)$ is usually sufficient, since the "skirt" appearing in the experimental distribution at high $p_\perp$ has little effect in radiation calculations. For most applications the distribution given by Eq. 1 is folded with a single-particle response of the form Np^α, where $0.5 < \alpha < 1.0$. Integration results are changed by less than 10% if realistic forms for $f(p_\perp)$ are replaced by

$$f(p_\perp)\ (= f(p/\sin\theta)) = \delta(p_\perp - \langle p_\perp\rangle)\ . \tag{3}$$

The best values for H and $\langle p_\perp\rangle$ are obtained by extrapolating experimental results. The central height of the rapidity plateau is shown in Fig. 1. With the exception of the circles indicating preliminary CDF data[5], the data points and fits are taken from Ref. 6. A reasonable extrapolation is to $H = 7.0$ at 40 TeV. Since the central value tends to be somewhat lower than the average, we have taken $H = 7.5$ as the average.

Similarly, $\langle p_\perp\rangle$ is shown in Fig. 2[5]. The extrapolation to 40 TeV is uncertain, but we have taken $\langle p_\perp\rangle = 0.60$ GeV/c for our present purposes.

About half as many π^0's as charged hadrons are produced in the primary collision, with a distribution given by Eq. 1 except with a value for H which is half as large. We therefore assume the same number of (decay) photons per rapidity interval as charged particles, with a momentum distribution which has half the mean value as that for charged hadrons.

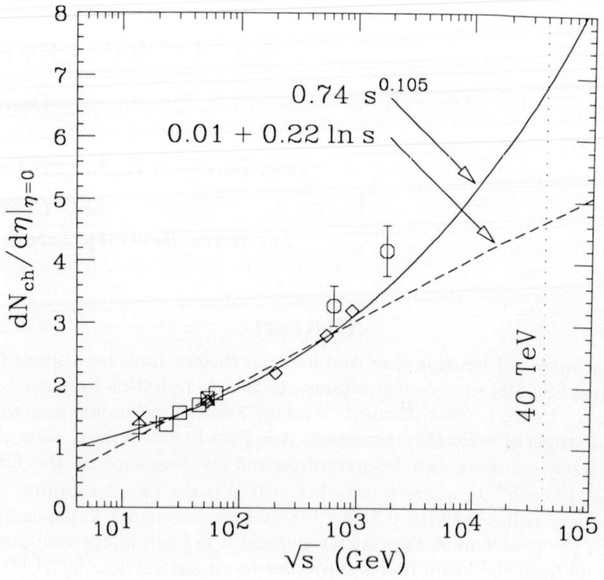

FIG. 1. $dN_{\text{charged}}/d\eta|_{\eta=0}$, based on FNAL fixed target (+ and ×), ISR (□), S$\bar{p}$pS (◇), and CDF (○) data. The preliminary CDF points[5] have been added to data taken from Fig. 2 in Alner *et al.*[6], and the two fits shown are to the data exclusive of CDF.

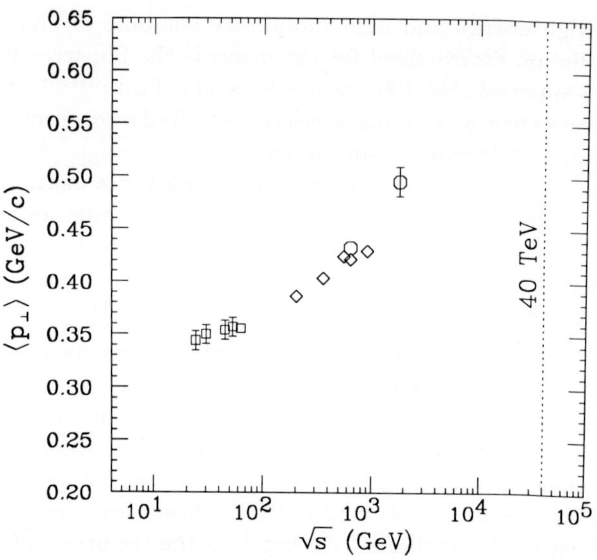

FIG. 2. The variation of the mean value of $p_\perp$ with center of mass energy for the distribution $d^2N_{\text{charged}}/dy\,dp_\perp$ for minimum bias events[5]. ISR (□), S$\bar{p}$pS (◇), and preliminary CDF (○) results are shown.

3. Effect of individual particles

Monte Carlo programs were used to simulate the results when a single photon or hadron strikes a calorimeter. EGS4[7] was used for electromagnetic simulations. For incident hadrons, the dose and neutron flux inside, the neutron leakage from the back, and neutron backscatter from the front were calculated, although not always at the same time. Members of the Task Force variously used CASIM[8], FLUKA86[9], GHEISHA[10], HETC[11], and MARS10[12] for hadronic cascade simulations. With the exception of GHEISHA, these codes do not transport neutrons with energies below some cutoff, typically chosen as 50 MeV. For several problems in which these neutrons were important the codes ANISN[13], MORSE[14], and COG[15] were used.

In a few cases experimental data were also available. The associated errors were quite large because of interpretative problems. For example, neutron flux might be measured using activation techniques, but usually by means of a reaction whose energy threshold is high compared with the neutron energies of interest, e.g. those which can produce dislocation damage in silicon. More and better experimental results are badly needed.

The calorimeters studied varied from uranium with silicon readout to iron with proportional wire chamber readout. As might be expected, neutron fluxes were highest for uranium with nonhydrogenous readout and lowest for iron with scintillator readout. A very rough comparison of neutron yield for several common configurations is shown in Table 1. The results should be used with caution at this stage, since (a) it was not possible to normalize all calculations to the same reference structure and (b) the scatter between different simulations was quite large. A systematic study of this problem would be quite useful.

Table 1

Relative maximum neutron flux and backscattered neutron flux for various calorimeter configurations.

Calorimeter	Relative flux
U-LiqAr or Si	1.0
Pb-LiqAr or Si	0.5
U-Scint.	0.3
Pb-Scint.	0.15

The most commonly modeled calorimeter was a "fine-sampling uranium/scintillator calorimeter," with alternating 3 mm uranium plates and 3 mm scintillator sheets. This was taken as a reference standard for the study.

A simple example of single-particle simulation results is shown in Fig. 3, in which the number of backscattered (albedo) photons above two energy thresholds is shown as a function of incident photon energy. There was no particular motivation for a power law fit in this and other cases, but it provided an adequate description. The yield cannot increase faster than linearly with E, and because higher energy showers peak deeper in the calorimeter a less than linear rise with energy is expected.

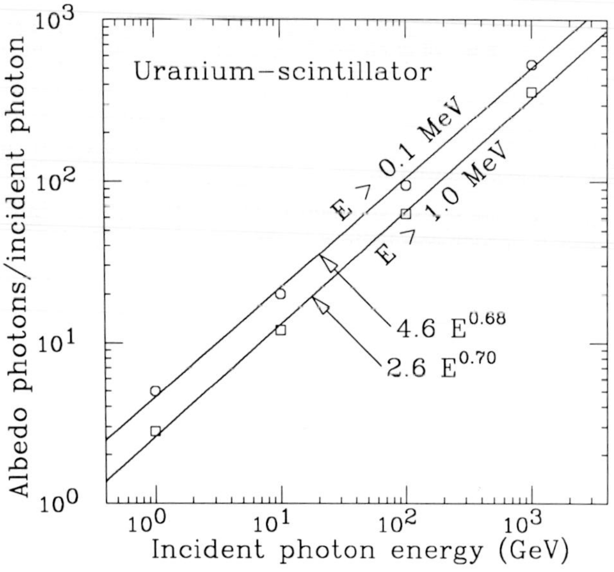

FIG. 3. The number of albedo photons for two different lower limits on the albedo photon energy vs. incident photon energy.

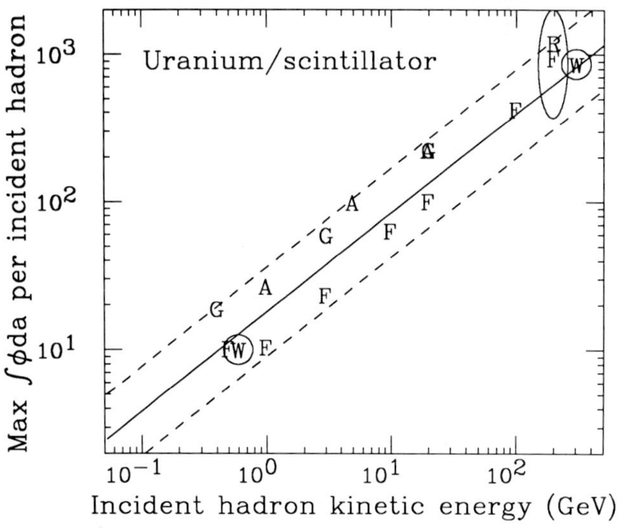

FIG. 4. Summary of selected maximum neutron flux data for a fine-sampling uranium/scintillator calorimeter. The solid curve represents $\int \varphi \, da = 18\,(E_k/(1\,\mathrm{GeV}))^{0.67}$, while the dashed curves are higher and lower by a factor of two. E_k may be replaced by p for practical calculations.

A more complicated case is shown in Fig. 4. The results from four simulations and two experiments (shown by the ellipses) are indicated by various letters. Hadronic cascades are more difficult to describe than electromagnetic showers, and the scatter indicates differences between codes. The solid line is drawn by eye, and the dotted lines are shifted by factors of two. It is from these and similar data that we have concluded that our results for the response to single incident particles are probably valid to within factors of two.

As a hadronic cascade progresses, charged secondaries

tend to "range out," so that below a few tens of MeV the only hadrons of any significance are neutrons. One may think of a "gas" of such neutrons, diffusing away from the shower core, out of the front face of the calorimeter, and out the back. A sampling of calculated spectra for neutrons from the front face is shown in Fig. 5. According to this "leaking gas" model, it is also representative of the neutron spectrum inside the calorimeter, although one might expect a harder spectrum near the shower core. The distribution in the logarithm of the kinetic energy is roughly gaussian, with a maximum at 1.3 MeV and a standard deviation corresponding to a factor of 3.7 from the mean, as shown by the smooth curve in Fig. 5. This "1 MeV" spectrum is characteristic of all processes we have examined. In some situations it is found to be slightly softer (e.g. the neutrons escaping from the Tevatron or SSC dipoles as a result of beam-gas losses[16]) and in some slightly harder, but in general most of the neutrons under the peak are above the effective threshold for damaging silicon devices (about 0.15 MeV) and have a silicon damage coefficient close to that for 1 MeV neutrons.

In a few cases neutron transport was extended to thermal energies. The thermal flux is very close to the "1 MeV" flux, and we believe it has little importance in the present context.

A final consideration concerns the enhancement of the neutron flux in the central cavity due to further backscattering of neutrons from the calorimeter. The calculations indicate that the flux in the cavity should be multiplied by about two to include this effect. The effect of holes in the detector near the beam line can be taken into account by multiplying the reflection probability by the fraction of the total solid angle subtended by the central calorimeter. In most practical cases this does not substantially change the answer.

4. Results

It follows from Eq. 2 that the flux of charged particles through a small counter with area da normal to the radius vector to the interaction point and at a distance $r_\perp$ from the beam line is given by

$$\frac{dN_{\text{charged}}}{da} = \frac{1.2 \times 10^8\,\text{s}^{-1}}{r_\perp^2} \qquad (4)$$

for r in cm and for the assumed interaction rate of 10^8 s^{-1}. In a light material $dE/dx \approx 1.8$ MeV g^{-1}cm^2, so 1 Gy corresponds to 3×10^9 particles/cm^2.* The annual dose in a thin detector $r_\perp$ from the beam line is then

$$D = \frac{0.4\,\text{MGy}}{r_\perp^2} \qquad (5)$$

for $r_\perp$ in cm and for assumed operation at design luminosity for 10^7 s per year.

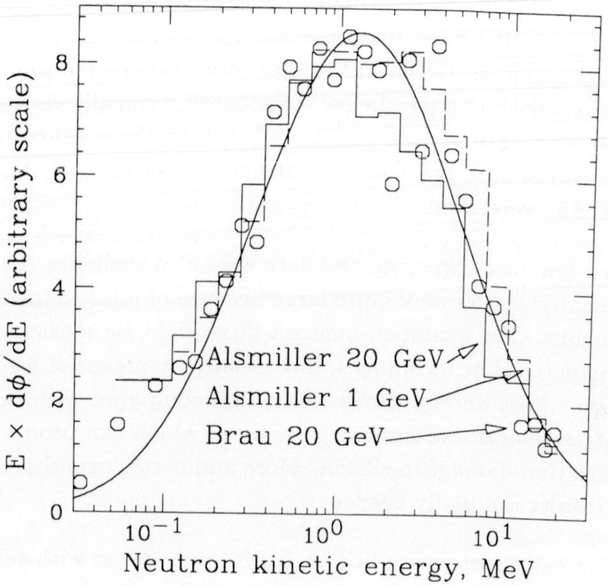

FIG. 5. The spectrum of neutrons backscattered from a uranium/scintillator calorimeter. The spectra at 1 GeV and 20 GeV were calculated by Alsmiller *et al.* (dashed and solid histograms), and a 20 GeV spectrum calculated by Brau and Gabriel is shown by the circles. Normalizations are chosen so that the peaks coincide. The solid curve is a gaussian in $\ln E_k$.

As discussed in the previous section, the effect (dose, fluence, number of backscattered particles, etc.) could usually be parameterized by a power law in the incident momentum. For example, the radially integrated neutron fluence for one incident hadron on a uranium/scintillator calorimeter can be written as

$$\int \varphi da = N_n p^\alpha \qquad (6)$$

with $N = 18$ and $\alpha = 0.67$, where p is measured in GeV/c ($p \approx E$) over most of the regions of interest). This integral is equal to the fluence at the same depth if one particle with momentum p is incident per unit area. The total response at that point may then be found by multiplying by the particle distribution (some variant of Eq. 1 or 2) and integrating over momentum. The integral may be carried out over all momenta, or over momenta above a cutoff imposed e.g. by a central solenoid. This has been done using a variety of forms for $f(p_\perp)$. As discussed in Section 2, in the absence of a momentum cutoff accuracy well within the uncertainties in single-particle response is obtained with $f(p_\perp) = \delta(p_\perp - \langle p_\perp \rangle)$. It follows immediately that the response *per rapidity interval* per average p-p collision can be written in the form

$$H\,N\,\langle p_\perp\rangle^\alpha \cosh^\alpha \eta\ , \qquad (7)$$

where the single-particle response is $N\,p^\alpha$ and where we

* One *gray* (1 Gy) is defined as 1 J/kg. It is equal to 100 rads.

have made use of the identity $\sin\theta \cosh\eta = 1$. Since $d\eta/d\Omega = 1/2\pi \sin^2\theta$, the effect per unit solid angle, per unit area, or per unit mass is of the form $HN \langle p_\perp \rangle^\alpha \cosh^{\alpha+2}\eta$. In general,

$$\text{Dose or fluence} = A \langle p_\perp \rangle^\alpha \cosh^{\alpha'} \eta \qquad (8)$$

where $\alpha' = \alpha$ or $\alpha + 2$, depending upon the problem. With few exceptions, all radiation fields in detectors may be written in this form.

The flux of particles incident on a detector at a given point scales inversely as the distance from the interaction point. Less obviously, the flux of backscattered neutrons and photons in the cavity defined by the calorimeter also scales inversely as the square of the linear dimensions. As a matter of convenience we have standardized either to an inside radius of 2 m or to a distance of 2 m to the interaction point, but the scaling of all radiation effects with the inverse square of typical detector dimensions must be kept in mind. The penalties for a compact detector design are obvious.

We illustrate the approach with two examples. Fig. 6 shows the maximum annual neutron fluence in a uranium/scintillator calorimeter under standard operating conditions. Fig. 7 shows the annual ionizing dose produced by hadrons (solid curve) and photons (dashed curve) under the same conditions. Although photons carry only about half the energy as do the charged particles, the showers peak more sharply and thus have higher maxima. As might be expected, the dose ratio is about the ratio of a nuclear interaction length to a radiation length. A calorimeter is exposed to the highest radiation levels near the electromagnetic shower maximum.

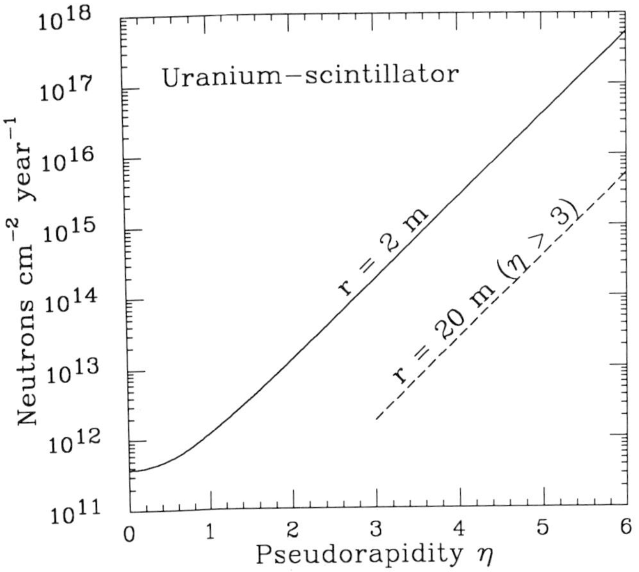

FIG. 6. The maximum neutron flux for a uranium/scintillator calorimeter. The solid curve shows the result assuming the maximum occurs at a radius of 200 cm. Also shown is the result for a radius of 20 m, typical of forward detectors, for rapidity > 3.

A similar distribution for the number of backscattered neutrons per pseudorapidity interval may be integrated to find the neutron flux inside the cavity. For a uranium/scintillator calorimeter with a 2 m inner radius extending to within $|\eta| = 3$ (or 6°) of the beam line, the annual neutron fluence is given by

$$\Phi = 1.2 \times 10^{12} (1 + \text{reflection})\,\text{cm}^{-2} , \qquad (9)$$

where "reflection" is mean number of subsequent reflections the neutron experiences before it loses substantial energy or is absorbed. Simulations of this problem indicate $(1 + \text{reflection}) \approx 2$ for most calorimeters.

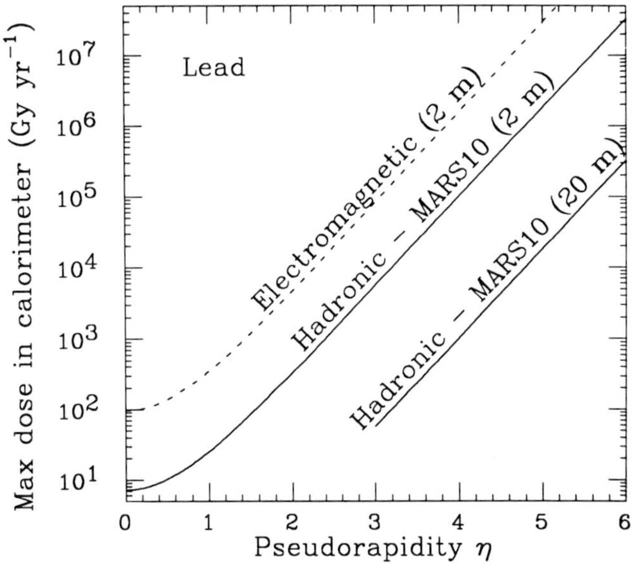

FIG. 7. The maximum hadronic dose as a function of pseudorapidity for a lead sphere, assuming that the maximum dose occurs at the indicated radius. The maximum electromagnetic dose in uranium/scintillator is shown by the dashed line. Since the radiation length, nuclear interaction length, and density are nearly identical for the two materials, dose (but not neutron flux) results may be compared directly.

5. Application to other hadron colliders

In the context of the approximations summarized in Eqs. 7 and 8, the dose or fluence at any point in a detector at a hadron collider scales as

$$\text{Dose or fluence} \propto \left(\int_T \mathcal{L}\, dt \right) \sigma_{\text{inelastic}} H \langle p_\perp \rangle^\alpha . \qquad (10)$$

For comparison purposes we assume the same duty cycle for all colliders and replace the luminosity integral by its nominal value. The factors appearing in Eq. 10 are evaluated in Table 2 for the Tevatron, the high-luminosity version of the proposed large hadron collider (LHC), and the SSC. The exponent α varies from 0.5 for neutron backscattering to 0.9 for the maximum electromagnetic and hadronic dose; for scaling purposes we take 0.7 as typical. An inelastic cross section of 100 mb has been adopted for 40 TeV[17]. This value has been crudely scaled with the aid of a fit to existing data to obtain the values given in the Table.* The values of H and $\langle p_\perp \rangle$ are obtained from Figs. 1 and 2. The final row gives radiation scale factors, normalized to 1.0 for the SSC. The nominal luminosity given for the Tevatron is fairly arbitrary, but in any case most of the radiation experienced by detectors at the Tevatron collider comes from other sources, such as the main ring.

Table 2

A rough comparison of beam-collision induced radiation levels at the Tevatron, SSC, and high-luminosity LHC.

	Tevatron	LHC	SSC
$\sqrt{s}$ (TeV)	1.8	16	40
$\mathcal{L}_{\rm nom}$ ($\rm cm^{-2}s^{-1}$)	2×10^{30}	4×10^{34}	1×10^{33}
$\sigma_{\rm inelastic}$	59 mb	86 mb	100 mb
H	4.1	6.3	7.5
$\langle p_\perp \rangle$ (GeV/c)	0.46	0.55	0.60
Scale factor	5×10^{-4}	27	1

6. Conclusions

A combination of uncertainties about the inelastic cross section, multiplicity, and mean transverse momentum lead to uncertainties of perhaps two in our description of particle production at the SSC. Similar uncertainties exist in parameterizing single-particle response. In most cases, estimates of radiation levels in SSC detectors are therefore thought to be valid to within factors ranging from two to four. In certain regions (e.g. very near the beam line) our approximations compromise the accuracy of the results.

Conclusions concerning the effect of this radiation on materials used in detectors are outside the scope of this study. However, it appears likely that with sufficient care in the choice of materials radiation damage will not be the limiting factor in the lifetime of most parts of a large general-purpose detector at the SSC[21, 22]. This qualified statement could not be made if the design luminosity were an order of magnitude higher.

* There is recent evidence from cosmic ray measurements at high energies[18] and coulomb-nuclear interference measurements at the $\rm S\bar{p}pS$[19] that the inelastic cross section at 40 TeV may be substantially larger than 100 mb, perhaps by as much as 50%[20]. We will await Tevatron results before revising any conclusions.

References

1. Report of the Task Force on Radiation Levels in the SSC Interaction Regions, SSC Central Design Group Report SSC-SR-1033 (June 1988).
2. F. E. Paige and S. D. Protopopescu, "ISAJET 5.35, A Monte Carlo Event Generator for p-p and p-$\bar{\rm p}$ Reactions," ISAJET PAM file; F. E. Paige and S. D. Protopopescu, in *Physics of the Superconducting Supercollider*, ed. by R. Donaldson and J. Marx, p. 320 (Snowmass CO, 1986).
3. H.-U. Bengtsson, Torbjörn Sjöstrand, "The Lund Monte Carlo for Hadronic Processes- PYTHIA version 4.8," LU TP 87-3 (Jan 1987); H. U. Bengtsson and T. Sjöstrand, in *Physics of the Superconducting Supercollider*, ed. by R. Donaldson and J. Marx, p. 311 (Snowmass CO, 1986).
4. J. Ranft, P. Aurenche, F. Bopp, A. Capella, K. Hahn, J. Kwiecinski, P. Maire and J. Tran Thanh Van, SSC Central Design Group Report SSC-149 (1987).
5. F. Abe *et al.*, Phys. Rev. Lett. **61**, 1819 (1988).
6. G. J. Alner *et al.* (UA5), Z. Phys. C **33**, 1-6 (1986).
7. W. R. Nelson, H. Hirayama and D. W. O. Rogers, "The EGS4 Code System," SLAC-265, Stanford Linear Accelerator Center (December 1985).
8. A. Van Ginneken, Fermilab Report FN-272, Fermilab, Batavia, IL, USA (1975).
9. P. A. Aarnio, A. Fassò, H-J. Möhring, J. Ranft and G. R. Stevenson, CERN TIS-RP/168 (1986).
10. H. Fesefeldt, "The Physics of Hadronic Showers—Physics and Applications," RWTH Aachen PITHA 85/02 (1985).
11. R. G. Alsmiller, Jr. *et al.*, "Modifications of the High Energy Transport Code (HETC) and Comparisons with Experimental Results," presented at the ANS Topical Conference on Theory and Practices in Radiation Protection and Shielding, 22-24 April 1987, Knoxville, Tennessee;
K. C. Chandler and T. W. Armstrong, "Operating Instructions for the High-Energy Nucleon-Meson Transport Code HETC," Oak Ridge National Laboratory Report ORNL-4744 (1972).
12. N. V. Mokhov, IHEP Preprint 82-168, Serpukov, USSR (1982); A. N. Kalinovsky, N. V. Mokhov and Yu. P. Nikitin, Penetration of High Energy Particles through Matter, (Energoatomizdat, Moscow, USSR) (1984).
13. W. W. Engle, Jr., "A User's Manual for ANISN, A One-Dimensional Discrete Ordinates Transport Code with Anisotropic Scattering," K-1693, Union Carbide Corporation, Oak Ridge, TN (1967).
14. M. B. Emmett, "The MORSE Monte Carlo Radiation Transport Code System," Oak Ridge National Laboratory Report ORNL-4972 (February 1975).
15. T. P. Wilcox, Jr., E. M. Lent, "COG: A Particle Transport Code Designed to Solve the Boltzman Equation for Deep-Penetration (Shielding) Problems—Volume I, Users Manual," Lawrence Livermore National Laboratory Report M-221-1.
16. T. A. Gabriel, F. S. Alsmiller, R. G. Alsmiller, Jr., B. L Bishop, O. W. Hermann, and D. E. Groom, "Preliminary Simulation of the Neutron Flux Levels in the Fermilab Tunnel and Proposed SSC Tunnel," SSC Central Design Group Report SSC-110 (1987).
17. D. E. Groom, "The Total p-p Cross Section at 40 TeV," SSC Central Design Group Note SSC-N-154 (1986).
18. T. K. Gaisser, U. P. Sukhatme, and G. B. Yodh, Phys. Rev. **D36**, 1350 (1987).
19. UA4 Collaboration, Phys. Lett. **B198**, 583 (1988).
20. M. Block, *Proc. of the 1988 Summer Study on High Energy Physics in the 1990's* (Snowmass CO, to be published).
21. Report of the Task Force on Radiation Effects at the SSC, SSC Central Design Group Report SSC-SR-1035 (1988).
22. H. Schönbacher and A. Stolarz-Iżycka, "Compilation of Radiation Test Data, Part I: Cable Insulating Materials," CERN Report CERN79-04 (1979);
H. Schönbacher and A. Stolarz-Iżycka, "Part II: Thermosetting and Thermoplastic Resins," CERN 79-08 (1979); P. Beynel, P. Meyer, and H. Schönbacher, CERN 82-10 (1982).

TRACKING REQUIREMENTS FOR THE SSC AND SILICON STRIP DEVELOPMENT

J. DEWITT, D. DORFAN, A. LITKE, H. SADROZINSKI, A. SEIDEN, A. WEINSTEIN

Santa Cruz Institute for Particle Physics
University of California, Santa Cruz, CA 95064

ABSTRACT

We quantify the design parameters for a central tracking system capable of good momentum measurements and high reconstruction efficiency for charged particles within 1 TeV jets. We express our results as a set of scaling rules. Based on these rules we outline a silicon strip tracking system which is well matched to the SSC high luminosity environment. Special challenges for the construction of this tracker are in the area of electronics readout. We present initial work we have done in this area.

GENERAL CONSIDERATIONS

In designing a tracking system it is useful to characterize the systems capabilities in terms of a few performance parameters which are determined by the geometry of the device and its detection method. Similarly, one can try to characterize the physics of particle production in terms of a few characteristic parameters, providing a heuristic, simplified picture of the events which have to be reconstructed. The physics and tracker parameters can be combined to yield scaling relations which can guide the choice of tracker technology and geometric layout. Given a design based on these heuristic ideas, a full Monte Carlo can then be performed to quantitatively evaluate the tracker performance and check that it is adequate

In the study presented below, we have chosen tracking within 1 TeV jets in the central rapidity range, $|Y| \leq 2.5$, as the goal for physics at the SSC. We assume that the detector is inside a solenoidal magnetic detector. Using the scaling rules which link the physics and the tracker parameters as a guide, we present a conceptual design of a tracking system made of thin silicon strip detectors. This detector provides the following features:

(1) High efficiency tracking within jets having transverse momentum ≥ 1 TeV in the central rapidity range,
(2) Allows a fast algorithm for usage of measurements and for track finding, which is important given the very large amount of information present,
(3) Allows precision momentum measurement,
(4) Has sufficient angular resolution to allow assignment of tracks to the correct vertex for multiple events in a time bucket and to match muons between the inner tracking detector and an outer muon identifier,
(5) Provides trajectory information for secondary vertex identification,
(6) Allows the measurement of charged energy flow per machine crossing to cross-check the calorimetry measurement which will probably sum over several crossings.

Tracking provides a much more fine grained picture of an event than does calorimetry and could be crucial in identifying new physics signatures. For example, calorimetry, because of tower sizes and shower spreading, will see the event energy in buckets which span a few degrees in angle. The tracking can provide a measurement of the event structure with roughly two orders of magnitude finer angular resolution and within each beam crossing.

Specific physics goals for the tracker are as follows:

(1) Leptons have often been indicative of new physics. The tracker, when combined with the rest of the detector, provides comparable μ and e detection efficiency and momentum resolution. Redundancy is provided for e detection by comparing the measured momentum from the tracker with the energy measured in the calorimeter, while for muons the tracker momentum can be compared to the momentum measured in an external muon identifier such as a toroid.
(2) Tracking provides the ability to search for long-lived particles or jets with unusual topologies.
(3) For the Standard Model Higgs, tracking is important for searches in all mass regions by providing:
    (a) Secondary vertex detection needed in the search for $H^0 \to b\bar{b}$ or $t\bar{t}$, for a Higgs mass less than about 150 GeV,
    (b) Good mass resolution for reconstruction of both the Z^0 and H^0 in the decays $H^0 \to Z^0Z^0 \to 4$

charged leptons. This provides enough rate for H^0 detection for Higgs masses up to about 600 GeV.

(c) Measurement of multiplicity and event shapes to be used to separate $H^0 \to W^+W^-$ or Z^0Z^0, where one gauge boson decays into hadrons, from the production of a single gauge boson accompanied by quarks and gluons. This is relevant for the Higgs search at masses larger than about 600 GeV.

SCALING RULES FOR TRACKER DESIGN

We take for our heuristic model of a jet a cone of half-angle $\theta_{1/2}$, with particles within the cone spaced by a mean angle $\langle\theta\rangle$, which is the angular spacing in the bend projection which for a solenoidal detector is the azimuthal angle. In addition, we assume the particles in the jet have momenta larger than a minimum, $P_{\min}$, which is determined by the typical transverse momentum relative to the jet axis and the jet angular size. We chose for $\theta_{1/2}$ the angle at which the energy flow drops by a factor of two for 1 TeV jets, according to the PYTHIA Monte Carlo.[1] This number is about 4 mrad for quark jets and 20 mrad for gluon jets. We therefore choose 10 mrad as a typical jet half-angle.

Since the jet core can be expected to contain 10 to 20 particles at these energies, we expect $\langle\theta\rangle$ to be about an order of magnitude smaller than $\theta_{1/2}$. To arrive at a number particularly well suited to tracking, we can use the extensive work done on B reconstruction at the Z^0.[2] For this case the tracker must have a two track resolution in the azimuthal angle of about 5 mrad for efficient B reconstruction. Scaling these jets up in energy to 1 TeV, we can estimate that $\langle\theta\rangle = .5$ mrad for the angular spacing of neighboring particles.

As the third and last parameter characterizing the jet core, we choose as the typical transverse momentum relative to the jet axis 500 MeV, giving $P_{\min} = 500\ \text{MeV}/\theta_{1/2} = 50$ GeV for a 1 TeV jet. Thus in this simple picture the jet core contains all particles with fractional momentum $\geq 5\%$ of the jet momentum.

Because of the very fine segmentation needed to resolve tracks, a tracker will probably have to use infrequent sampling to avoid ending up with an enormous number of channels. It will, in addition, have to go out to a moderately large radius, since it is critical not only to resolve tracks, but also to measure their momentum. Figure 1, below, shows our model of a section of the tracking detector, organized to provide local track samples which can be linked into tracks. The detectors are grouped to provide track segments called vector elements, allowing for efficient pattern recognition. In Fig. 1, the vector elements are two closely-spaced detectors; more generally, these can be several closely-spaced detectors. The parameters describing the detector are:

σ_m = position resolution for each measurement,
ϵ_m = two-track spatial resolving power,
δ = spacing of detector plane pairs forming vector elements,
$\sqrt{2}\sigma_m/\delta$ = error on track tangent,
Δ = spacing of vector elements,
r_m = outer radius,
N = number of layers,
B = magnetic field.

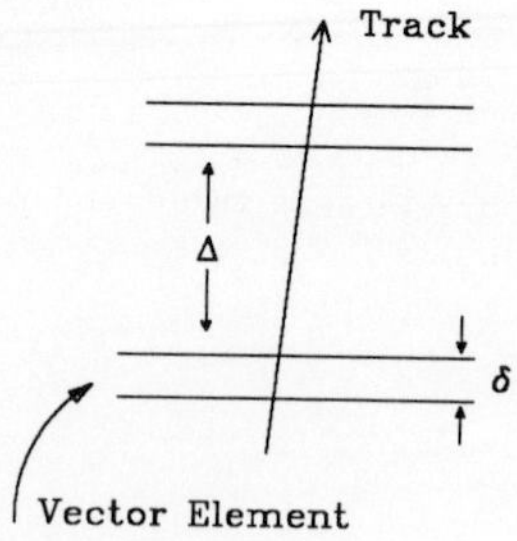

Fig. 1. Arrangement of detector made up of spaced local track-measuring vector elements.

We will assume below that measurements are of the azimuthal angle. It is necessary to also measure the z coordinate. We will assume this is done using stereo measurements and return to this question later. For the pattern recognition we require that:

- different tracks give distinct hits in each individual tracking detector of a vector element,
- hits in the vector elements can be locally associated into correct track segments,
- the local track segments have to be correctly linkable into tracks.

Using the three parameters characterizing the jet structure and the tracker characteristics, we quantify the three tracking requirements as follows:

(1) At a radius r, a track will typically have a nearest neighbor spaced by a distance $r\langle\theta\rangle$. The hits will be correctly resolved if the measuring device has a two track resolution $\epsilon_m < r\langle\theta\rangle$. This gives our first scaling rule for efficient tracking

$$\frac{\epsilon_m}{r\langle\theta\rangle} < 1,$$

where r = mean radius of the tracker.

(2) Figure 2 shows a track passing through a vector element. From the measurement in the inner layer, we can predict a hit on the second layer assuming infinite momentum, as shown by the dashed line in Fig. 2. The spatial mismatch between the prediction and the real trajectory is $r\delta/2\rho$, where r = radius of the layer and ρ = radius of curvature of the track. The largest mismatch for tracks in the jet will occur for tracks with the lowest momentum, $P_{\min}$. We will have confusion if a second track leaves a hit between the predicted location and the real hit location. Since a neighboring hit can be expected at a spacing $r\langle\theta\rangle$, this means that we require $r\delta/2\rho < r\langle\theta\rangle$, yielding the second scaling rule, $\delta/(2\rho\langle\theta\rangle) < 1$. Here ρ

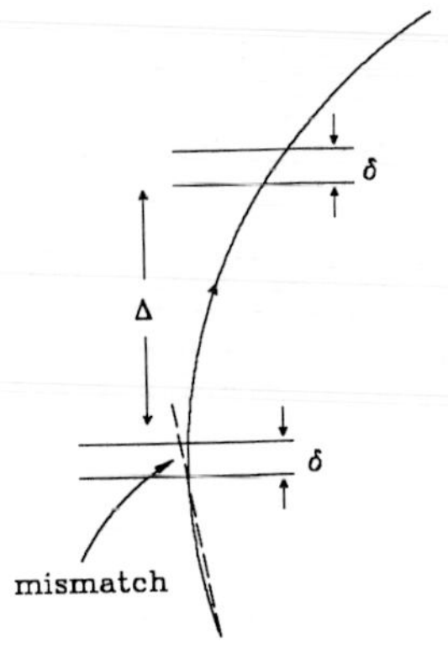

Fig. 2. A Track passing through two adjacent layer pairs.

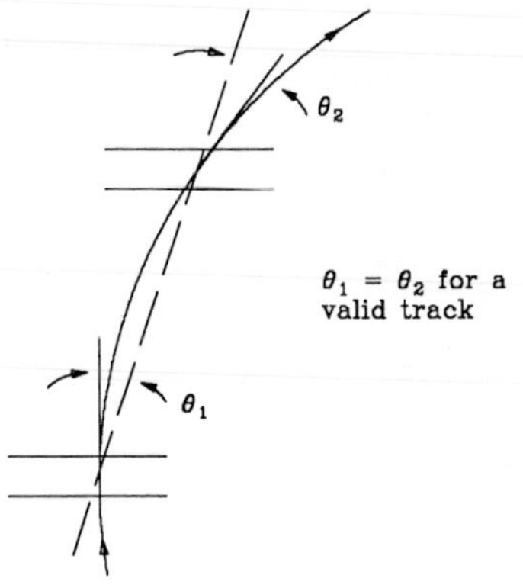

Fig. 3. Relation for vector segments lying on one track.

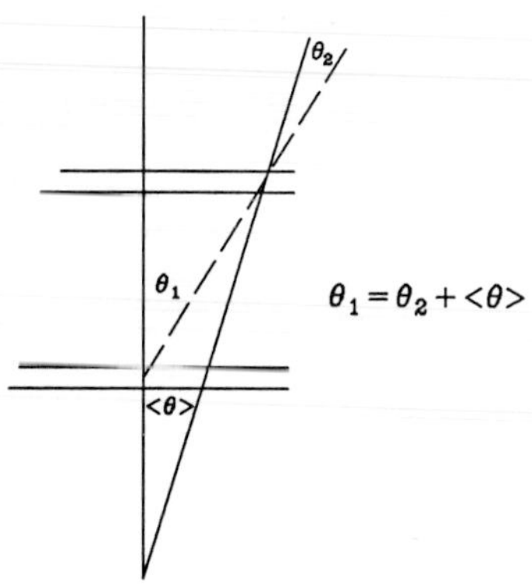

Fig. 4. Two tracks and spurious (dashed) segment linkage.

is calculated from $P_{\min}$ and the B field chosen for the detector.

(3) The basis for linking vector segments is illustrated in Fig. 3. Each angle θ_1 or θ_2 is measured with a precision given by $\sqrt{2}\sigma_m/\delta$. Linking can be done if this accuracy is adequate to exclude wrong combinations. We show in Fig. 4 the trajectories of two very high momentum tracks separated by a production angle $\langle\theta\rangle$. An incorrect link yields the dashed track whose angles don't match by an amount $\langle\theta\rangle$. Thus, a goal for the linking to work is $\sqrt{2}\sigma_m/\delta \lesssim \langle\theta\rangle$, giving the third scaling rule $\sqrt{2}\sigma_m/\delta\,\langle\theta\rangle < 1$.

Note that scaling rules (2) and (3) have an opposite dependence on δ. It has to be carefully chosen, since good matching for segments requires a small δ, whereas a well measured tangent vector in a segment requires a large δ.

The last scaling rule refers to the measurement of z coordinates using stereo tracking elements. We show an arrangement of vector elements in Fig. 5, which will allow full local association of axial and stereo measurements. The stereo layers can be characterized by a stereo angle $\alpha = \alpha_u = -\alpha_v$ and the stereo displacement $s = \ell\alpha$, where $\ell =$ detector length.

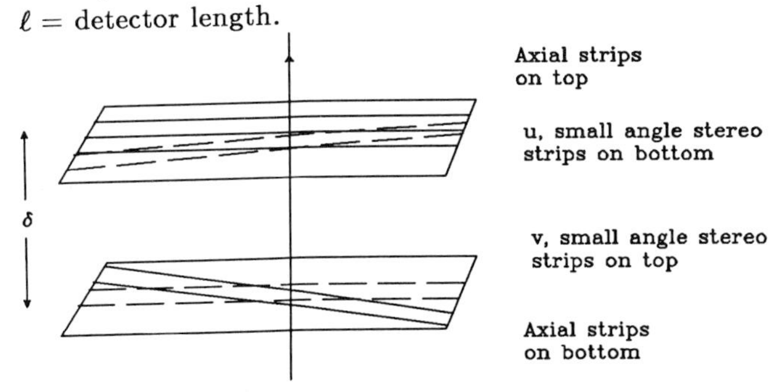

Fig. 5. Orientation of detector elements to allow a local reconstruction of the z coordinate using stereo measurements.

For the two vector elements we measure azimuthal angles ϕ_1, ϕ_2 and stereo measurements $\phi_v = \phi_1 + \alpha z_1/r_1$, $\phi_u = \phi_2 - \alpha z_2/r_2$. High momentum tracks have a trajectory $z = r\tan\lambda + z_0$, where z_0 is the vertex location in z_0 and $\tan\lambda = P_z/P_T$. For the case $z_0 = 0$, the correct stereo hits are given by the pair satisfying the matching equation $\phi_1 + \phi_2 = \phi_u + \phi_v$.

At the SSC, we expect the length of the interaction region $\sigma_{z_0} \simeq 5$ cm. Assuming complete ignorance of the vertex position in a given event (taking $z_0 = 0$), such a spread in vertex position will produce a mismatch in the angles in the above matching equation of $\sigma_{z_0}\alpha\delta/r^2$. This is smaller than the measurement error in the matching equation$=2\sigma_m/r$, except at small radii, even for a small σ_m. We include the finite beam size in the Monte Carlo studies discussed in the next section.

The confusion from many tracks can be quantified in terms of the detector and physics parameters. If a pair of axial strips have been successfully associated within the jet, all tracks within the stereo displacement region s (defined above) will leave stereo hits which can confuse the z matching. The number of tracks entering this region is $N_s = s/(r\,\langle\theta\rangle)$, since the mean displacement between tracks at a radius r is $r\,\langle\theta\rangle$. All of these tracks will leave stereo hits in the u plane which could accidentally associate with another hit in the v plane falling within a region of full width $2\sigma_m$, which is the resolution characterizing the correct choice. The probability of there being such a hit in the v plane is $P = N_s \cdot 2\sigma_m/s$, and the number of potential incorrect z pairings is $N_{inc-z} = N_s P$. Thus we arrive at the scaling rule for stereo association to work well

$$N_{inc-z} = N_s^2 \frac{2\sigma_m}{s} = \frac{2\sigma_m s}{(r\,\langle\theta\rangle)^2} < 1\ .$$

Note that if we increase s (by increasing the stereo angle α or the detector length l) N_{inc-z} will not necessarily continue to increase. Eventually s will exceed the size of the jet ($\simeq 2r\theta_{1/2}$) and any further increase in s will leave N_{inc-z} unchanged. In this limit, which is characteristic of long drift chamber detectors, the scaling rule is

$$N_{inc-z} = \frac{4\sigma_m r\theta_{1/2}}{(r\,\langle\theta\rangle)^2} = \frac{2\sigma_m}{(r\,\langle\theta\rangle)} \cdot \frac{2\theta_{1/2}}{\langle\theta\rangle} = \frac{4\theta_{1/2}\sigma_m}{r\,\langle\theta\rangle^2} < 1\ .$$

This completes the set of scaling rules, which we will use in the next section to evaluate our silicon strip tracker design.

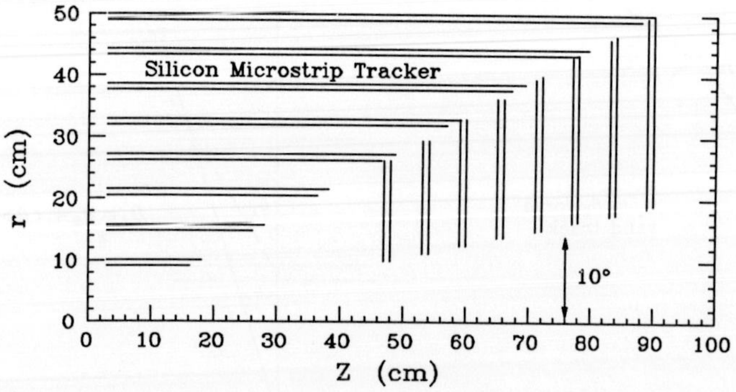

Fig. 6. Silicon tracking system.

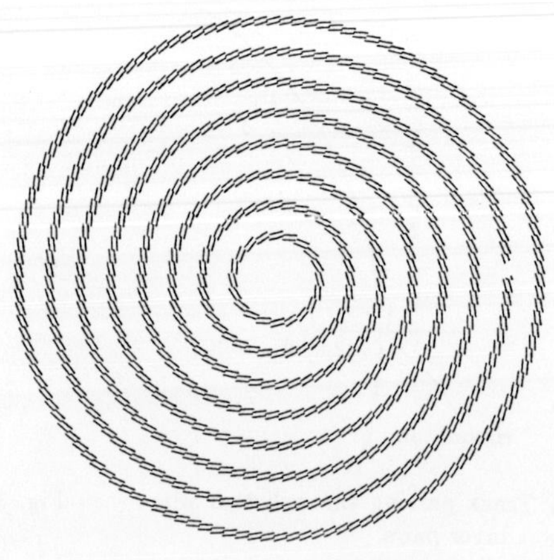

Fig. 7. Central silicon tracker.

SILICON STRIP TRACKING DETECTOR

To meet the very stringent tracking requirements for the SSC, we would like to use a technology offering an order of magnitude improvement in performance over the traditional drift chambers used in present experiments. We feel that silicon strip detectors with 25 μm pitch are a good choice for this problem since they offer:

(1) Intrinsic position resolution of 5 μm and two track resolution of 100 μm, well matched to the scaling rules derived above, thereby allowing efficient tracking,
(2) The time structure of the signal well matched to the SSC machine repetition rate,
(3) Reasonably radiation hard detectors,
(4) VLSI electronics and substantial multiplexing easily incorporated,
(5) A large community working with these types of detectors, yielding substantial technological improvements with time.

Figure 6 shows a proposed design made of 8 pairs of double sided-detectors, providing 32 measurements per track. The pairs yield 8 axial and 8 stereo vector segments for the track reconstruction. This compares, for example, to 6 axial and 6 stereo vector segments for the new MARK II drift chamber. The stereo arrangement is that shown previously in Fig. 5. Figure 7 shows a cross section of the central part of the detector. The individual silicon detectors are tilted to compensate for the Lorentz angle due to the magnetic field. The inner radius of the detector has been taken as 8 cm to limit the maximum radiation damage to less than 0.5 Mrad in an SSC year.[3] The outer radius has been chosen as 50 cm to provide adequate momentum resolution in a 2.5 Tesla field, as discussed below. The maximum occupancy, in the innermost layer, is 2×10^{-3} in a two-hit resolution element, $\epsilon_m = 100$ μm, assuming the detector is read out each beam crossing.

The stereo angle chosen for the detector is $\alpha = 5$ mrad. For a 10 cm long detector this gives a stereo displacement, s, of 500 μm. Tracks separated by a distance greater than s in the axial projection don't interfere with each other at all. The very small stereo angle still yields a $\sigma_z = 1.44$ mm per measurement and a resolution in z at the vertex of 835 μm.

Secondary vertex identification is available in the bend plane where the interaction region is very small. Note, for high momentum tracks, the impact parameter error is still $\sim 5\mu$m, even starting 8cm from the origin, since the long lever arm of the tracking devices provides a very well measured angle and curvature.

Table 1 presents a summary of the scaling rules for this detector. We have chosen $\delta = 1$ cm to provide a reasonably well measured tangent vector. The small stereo angle chosen is motivated by the desire to limit the confusion in the local stereo association, as quantified by the stereo scaling rule. The table indicates that the matching of vector segments into tracks has potentially the greatest confusion. For comparison, we also list the scaling parameters for the new MARK II drift chamber at the SLC, which has been carefully optimized by Monte Carlo studies for jets of 50 GeV.[4]

MONTE CARLO STUDIES

In this section we look in more detail, based on a Monte Carlo simulation, at the ability to do tracking in 1 TeV jets with a silicon tracker. The program ISAJET[5] was used to generate events with two quark or gluon jets, each with $P_T \geq 1$ TeV. The mean multiplicity within $|Y| < 2$ in these events is 330, of which 118 have $P_T \geq 1$ GeV. Figure 8 shows a high P_T event in a 3 Tesla magnetic field. A region near the core of the tight 1.5 TeV jet in the lower left quadrant of the event, at a radius of 26 cm, has been blown up as Fig. 9. The scale is matched to the size of a single strip detector containing 1024 25-μm elements. At this scale, which is the one relevant for the track finding, the stiff tracks are not confusing.

Vector Segment Finding

To quantify this, we have performed the first two steps of the pattern recognition for these Monte Carlo 1 TeV jets. As a first step, we keep only hits with no other hit within 100 μm in space. The detectors are assumed

TABLE 1: DETECTOR PARAMETERS AND SCALING RULES

Physics Parameters for:	1 TeV Jets	50 GeV Jets
Track angular separation $\langle\theta\rangle$	0.5 mrad	10 mrad
Jet angular width $\theta_{1/2}$	10 mrad	40 mrad
Track P_T in jet core	50 GeV	10 GeV
Detector Parameters	**Si Tracker**	**Mark II/SLC**
Magnetic field B	2.5 Tesla	0.5 Tesla
Radius of curvature $\rho = 3.3P_T/B$	66 m	66 m
Radial detector spacing δ	1 cm	0.8 cm
Superlayer thickness δ_{SL}	1 cm	4 cm
Maximum radius r_m	50 cm	150 cm
Mean radius r	29 cm	88 cm
Detector length l	10 cm	230 cm
Double track separation ε_m	100 μm	4 mm
Stereo angle α	5 mrad	74 mrad
Position resolution σ_m	5 μm	100 μm
z position resolution $\sigma_z = \sqrt{2}\sigma_m/\alpha$	1.4 mm	1.9 mm
Scaling Rule Requirements	**Si Tracker**	**Mark II/SLC**
Resolve hits: $\frac{\varepsilon_m}{r\langle\theta\rangle} < 1$	0.7	0.45
Find vector segments: $\frac{\delta}{2\rho\langle\theta\rangle} < 1$	0.15	0.006
Link vector segments: $\frac{\sqrt{2}\sigma_m}{\delta_{SL}\langle\theta\rangle} < 1$	1.4	0.4
Match z hits: $\frac{2\sigma_m l\alpha}{(r\langle\theta\rangle)^2} < 1$, or:	0.24	–
for $l\alpha > 2r\theta_{1/2}$: $\frac{4\sigma_m\theta_{1/2}}{r\langle\theta\rangle^2} < 1$	–	0.2

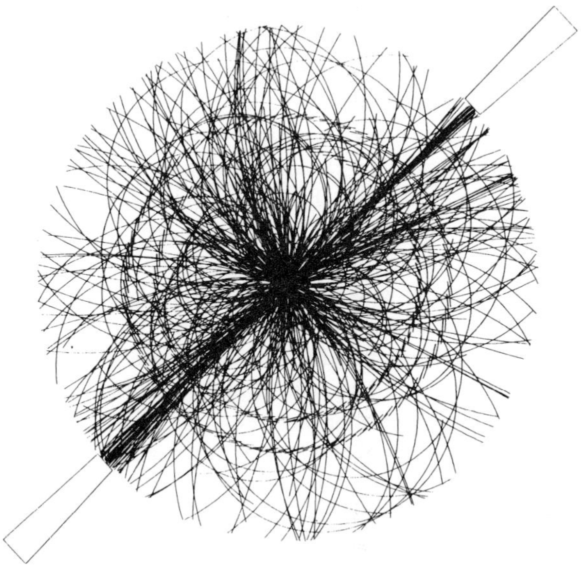

Fig. 8. Typical 1 TeV event from ISAJET.

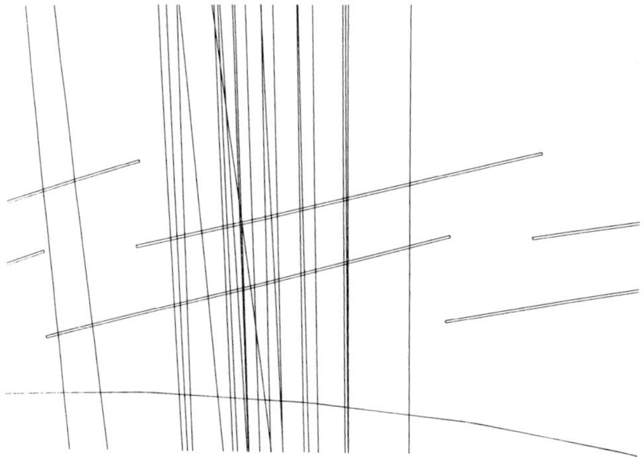

Fig. 9. Blowup of dense jet at mean detector radius.

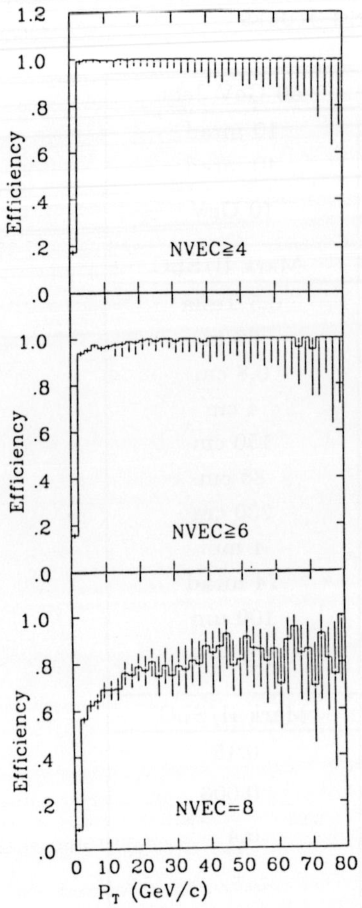

Fig. 10. Vector segment finding efficiency.

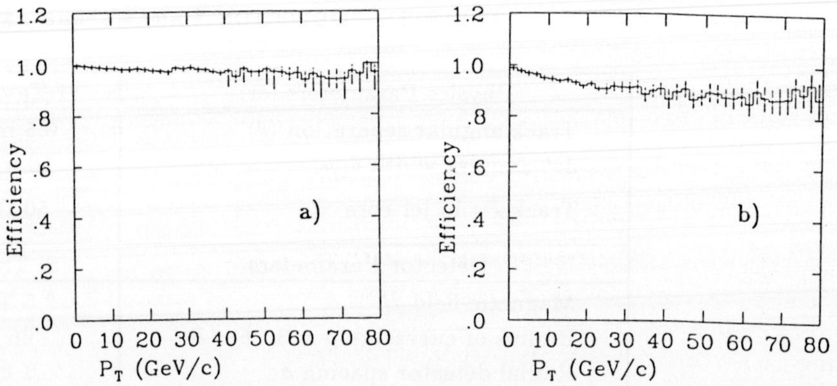

Fig. 11. Stereo association efficiency. a) α = 5 mrad, b) α = 20 mrad.

to be 10 cm in length, with the outer member of a pair staggered by one half of a detector length with respect to the inner layer. Using the simple algorithm presented above to link hits into vector segments, and considering only hits pairs which can come from tracks with $P_T > .5$ GeV, we evaluate the efficiency vs. P_T for finding tracks if we require at least NVEC $\geq 4, 6$, or 8 correctly found vector segments to call the track found. The results for δ chosen equal to 1 cm for a 2.5 Tesla B field are shown in Fig. 10, where the last bin includes overflows. The efficiency is close to 100% for high momentum tracks, if we require NVEC ≥ 6, which corresponds to ≥ 12 axial and 12 stereo measurements, assuming the stereo finding is fully efficient.

To test the scaling rule for the stereo z matching, we apply the stereo algorithm discussed above to the vector segments that were found correctly in the axial projection. The efficiency for making the correct local z match, as a function of P_T of the track is shown in Figs. 11a and b for the cases $\alpha = 5$ mrad and 20 mrad, respectively. The efficiency in the former case is $> 95\%$ for all tracks, falling at high P_T where tracks are in the center of jets. It is clear that the efficiency begins to degrade significantly when the stereo angle is increased to 20 mrad, as expected from the scaling rule. We have verified that the effect continues at larger stereo angles.

Dependence of Efficiency on the Magnetic Field

The presence of an axial magnetic field is required for the measurement of the track momentum. However, the magnetic field also has an impact on the pattern recognition efficiency. The field may improve the reconstruction efficiency by separating closely spaced tracks in the outer layers, or degrade the efficiency by causing tracks to cross in the outer layers, leading to confusion.

In order to determine which is the dominant effect for the tracking system discussed here, we evaluated the vector segment finding efficiency for various B fields. We find that this efficiency is degraded with larger magnetic fields, over the entire range of momenta. As an example, the efficiency for finding all 8 vector segments for tracks with $P_T > 15$ GeV falls from $\simeq 80\%$ to about 65% for $B = 2.5$ and 5.0 Tesla, respectively. For the case we have chosen, $B = 2.5$ Tesla, the inefficiency for vector segment finding in the inner layers due to the close spacing of tracks is approximately the same as that in the outer layers due to track crossings.

The magnetic field can also degrade the double track resolution by spreading the charge (electrons and holes) over several strips due to the Lorentz drift of the charges in the magnetic field. The Lorentz angle θ_L is defined by $\tan\theta_L = \mu B$, where μ is the Hall mobility of the charge carriers. However, the use of double-sided detectors means that we are collecting both electrons and holes, which have different Hall mobilities.[6] To minimize the problem, we can orient the detectors at an angle which is the average of the Lorentz angles for each charge, and choose a magnetic field which is low enough so that the residual Lorentz drift spreads the charge over a strip width or less. A detector of 150 μm thickness in a 2.5 Tesla field tilted by $\tan\theta_L = (\mu_e + \mu_h)B/2$ will have a charge spread of 22 μm on either side of the detector, just less than one strip width.

Vector Segment Linking

Once vector segments are found, they must be unambiguously linked together to form tracks. For tracks from the origin in the x, y plane, the track trajectory is given by

$$r = 2\rho \sin(\phi - \phi_0)$$
$$z = 2\rho(\phi - \phi_0)\tan\lambda + z_0$$

where the constants characterizing the track are ρ, $\tan\lambda$, ϕ_0 and z_0. A more convenient variable than the radius of curvature ρ is the curvature $\kappa = 1/2\rho$. For high momentum tracks, the above are approximately $\phi - \phi_0 = \kappa r$, $z = z_0 + r\tan\lambda$.

One way to link vector segments, which provides a visual picture of the information quality for a given event, is to use a technique employed successfully with the MARK II jet-cell central drift chamber at the SLC.[7] We describe the method for the barrel section of the silicon tracker. The idea is to calculate, for each vector measurement, the curvature and azimuth at the origin ϕ_0 of a track passing through the origin and tangent to the vector segment. A vector segment provides a measure, at average radius $\bar{r}$, of the track azimuth $\bar{\phi}$ and the rate of change of azimuth $d\phi/dr$. From these we calculate the track constants

$$\kappa = \frac{-d\phi/dr}{\sqrt{1 + (r\,\frac{d\phi}{dr})^2}}, \qquad \phi_0 = \bar{\phi} + \arcsin(\kappa\bar{r}) \ .$$

This will produce a mapping in which all vector segments found for a given track lie near a point in the $\kappa - \phi_0$ plane. The vector segments found for the event shown in Fig. 8 have been plotted in Fig. 12a. Up to 8 vector segments are plotted for each track. One can immediately see the two jets in the event. The effect of resolution in curvature is also seen from the spread in points for each track. For this plot we have used $\delta = 5$ mm. Each hit position is chosen as the center of the strip hit, so that the position resolution is $25\ \mu\text{m}/\sqrt{12} \simeq 7\mu\text{m}$.

Figure 12b is a blowup of the region around the jet at $\phi_0 \simeq 3.9$ radians. The effect of resolution is now more obvious, and it is also clear that the error in κ is strongly correlated with the error in ϕ_0. This can be parametrized by a 2×2 error matrix, which can be used to cluster the vector segments into tracks. Even by eye, it can be seen from Fig. 12b that vector segments from nearby tracks in the jet are becoming confused. This can be remedied by making the paired layer separation δ larger. Choosing $\delta = 1$ cm, the value used in Table 1 earlier, results in the measurements displayed in Fig. 12c. Here the individual tracks in the jet are generally separated. Note the spread in these plots of the measured ϕ_0 values for a given track is $\sqrt{2}\sigma_m/\delta$, the parameter introduced earlier to quantify the ability to link vector segments. A study of many such events indicates that $\delta = 1$ cm is a reasonably good choice.

The method outlined above can also be used to cluster vector segments and link them into tracks for tracks

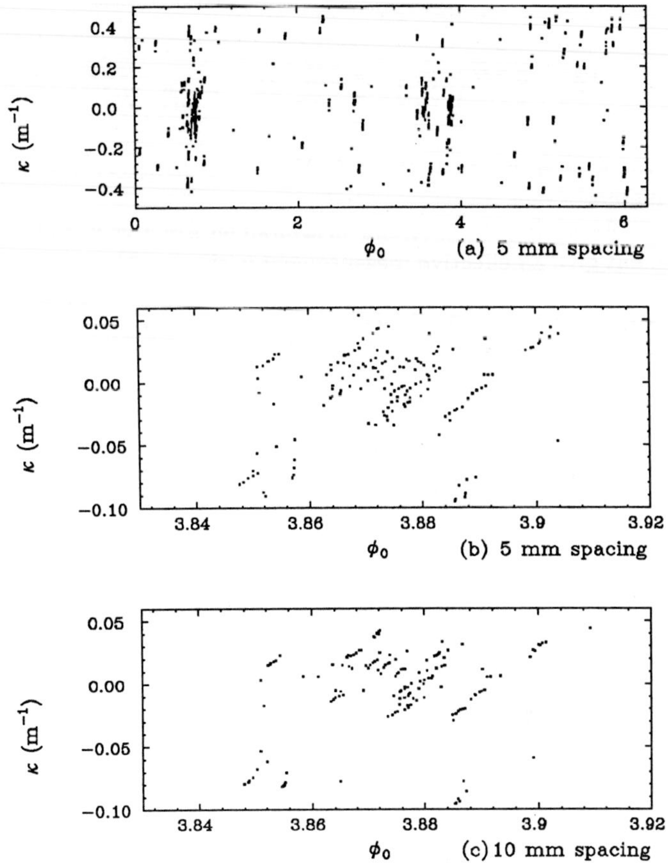

Fig. 12. Clustering for tracks in 1 TeV event.

passing into the forward detectors. Assuming that the strips are organized to measure ϕ at fixed z, the measurement $d\phi/dz$ and $\bar{\phi} - z\ d\phi/dz$ are constant along a track, being equal to $\kappa/\tan\lambda$ and $\phi_0 - (\kappa z_0)/\tan\lambda$, respectively. Thus the two dimensional clustering space is in these two parameters for these detectors.

MOMENTUM RESOLUTION

We look at the expected momentum resolution for the silicon tracker discussed above. We will assume a vertex constrained fit, as would be appropriate for the high momentum leptons from the heavy Higgs decay, which corresponds to an orbit equation at high momentum

$$\phi = \phi_0 + \frac{r}{2\rho}.$$

Measuring the track angle ϕ at m radii $r_1, r_2, \ldots r_m$, the value of ϕ_0 and ρ are extracted by minimizing

$$\chi^2 = \sum_{i=1}^{m} \frac{(r_i\phi_i - d_i)^2}{\sigma_i^2},$$

where d_i is the measured circumferential distance for the ith axial layer. The expected error on d_i is σ_i.

Relating ρ to P_T, and calculating the errors from the χ^2, gives a result:

$$\frac{\sigma_{P_T}}{P_T^2} = \frac{2}{0.3B}\sqrt{\frac{\sum_{i=1}^{m}\frac{r_i^2}{\sigma_i^2}}{\sum_{i=1}^{m}\frac{r_i^2}{\sigma_i^2}\sum_{i=1}^{m}\frac{r_i^4}{\sigma_i^2} - \left(\sum_{i=1}^{m}\frac{r_i^3}{\sigma_i^2}\right)^2}},$$

where B is in Tesla, P_T in GeV and all distances are in meters.

Assuming that the four measurements in a vector element give an effective measurement with an error σ, that the 8 pairs are approximately equally spaced out to a maximum radius r_m, gives after performing the sums above

$$\sigma_{P_T}/P_T^2 = 5.5\sigma/0.3B\ r_m^2\ .$$

Taking $\sigma = 3\ \mu$m, $r_m = 0.5$m, $B = 2.5$ Tesla, gives:

$$\sigma_{P_T}/P_T = 8.8\% P_T\ (\text{TeV})\ .$$

Figure 13 shows σ_P/P^2, where P is the full momentum, vs. rapidity for the silicon tracking detector, where the error at $Y = 0$ has been taken to be $\sigma_P/P^2 = 0.08$ TeV^{-1}.

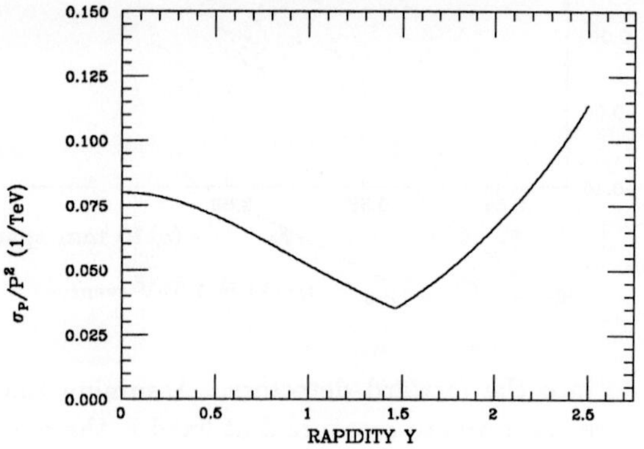

Fig. 13. Momentum resolution vs. rapidity.

MASS RESOLUTION FOR HIGGS DETECTION

The goals for momentum resolution are usually determined by the physics goals of an experiment. For the present case we choose the reconstruction of the Higgs via $H^0 \to Z^0Z^0 \to 4$ leptons, for a mass of the H^0 up to 1 TeV as a suitable goal for the SSC. Since the Z^0 has a finite width, arbitrarily good resolution does not improve background rejection. The width of the Z^0 therefore sets a goal for the resolution. The typical momentum for leptons from the 1 TeV H^0 is 250 GeV. The momentum resolution for these is $\sigma_P/P \simeq 2\%$, using Fig. 13 above.

Using a Monte Carlo for H^0 decay, we have calculated the expected mass resolution for the reconstructed Z^0, for decays into electrons and muons. The results are shown in Fig. 14. Also shown are the electron results for a calorimetric mass reconstruction and the muon results for an iron toroid. We see that the detector resolution is well matched to the limitations of the Z^0 width for Z^0 energies up to nearly 500 GeV. Fig. 15 shows similar quantities for the subsequently reconstructed H^0. The tracking system provides better resolution at all masses than the limitations imposed by the H^0 width. For the simplest Higgs scenario, the detector has about a factor 3 better resolution at about 500 GeV than is necessary. This should provide some safety margin in case the correct theory involves several Higgs particles. In this case each Higgs is narrower and has a smaller production cross section than the Higgs in the simplest model. With sufficiently good resolution, the signal to background will be equally good in this case since the narrower width makes up for the smaller rate in the mass plot.

ELECTRONICS ISSUES

Readout for the tracker must be via high-density (on the detector) fast amplification and discrimination, followed by digital event buffering to store information locally until a valid trigger is asserted.

Figure 16 shows a block diagram of the readout scheme we are working on. It uses two chips with VLSI circuits matched in pitch to the detector strips and wire bonded together. The first chip, to be implemented in bipolar technology, contains a fast amplifier and discriminator. The second chip, to be implemented in CMOS technology, contains level 1 and level 2 buffers and a shift register to readout the hit information. The level 1 buffer is a memory which is clocked at the machine frequency, providing a digital time slice of the detector output. The depth of the buffer, taken to be 64 clock cycles, is matched to the time necessary to make a first level trigger. If a trigger occurs, the information is stored in the second level buffer which can store up to 16 level one triggers. A successful second level trigger causes a readout of the appropriate time slice.

The use of a fast front-end amplifier is well matched to the signal collection time in the silicon as well as the beam crossing rate. It is also necessary since the strip detectors develop leakage current due to radiation damage. The signal expected is 12,000 e$^-$ for 150 μm thick detectors. The equivalent noise charge for the amplifier is expected to be between 700 to 1,000 e$^-$. For a 10 nsec shaping time, the noise from the detector leakage current is $\simeq 250$ e$^-$ $\sqrt{I(\mu\text{amp})}$. We therefore want to keep the leakage current $< 4\mu$amp.

The leakage current for the detectors decreases substantially over a period of several months if not irradiated, therefore an appropriate period for calculating dosage is ≤ 1 year of running. For strips with dimensions 25 μm $\times$ 10 cm $\times$ 150 μm, at the dosage expected,[8] the leakage current will be:

From Charged Particles: 2 μamp$/r_\perp^2$(cm)

From Neutrons: 2.5 $\times 10^{-14}\mu$amp $\times$ Neutron Fluence.[9]

Thus the nearest detector, located at $r_\perp = 8$ cm, for a neutron fluence $\leq 10^{13}$, will develop a leakage current $\ll 4\ \mu$amp, making the noise from the detector current a small contributor to the overall amplifier noise.

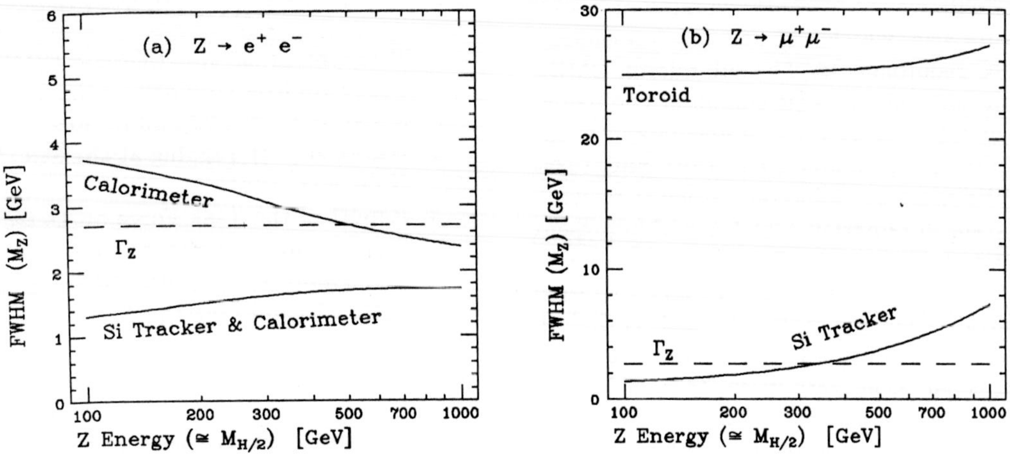

Fig. 14. FWHM resolution of the Z^0 mass as a function of Z^0 energy for a detector with calorimetry only, and for a detector including precision Si tracking. a) $Z^0 \to e^+e^-$. b) $Z^0 \to \mu^+\mu^-$.

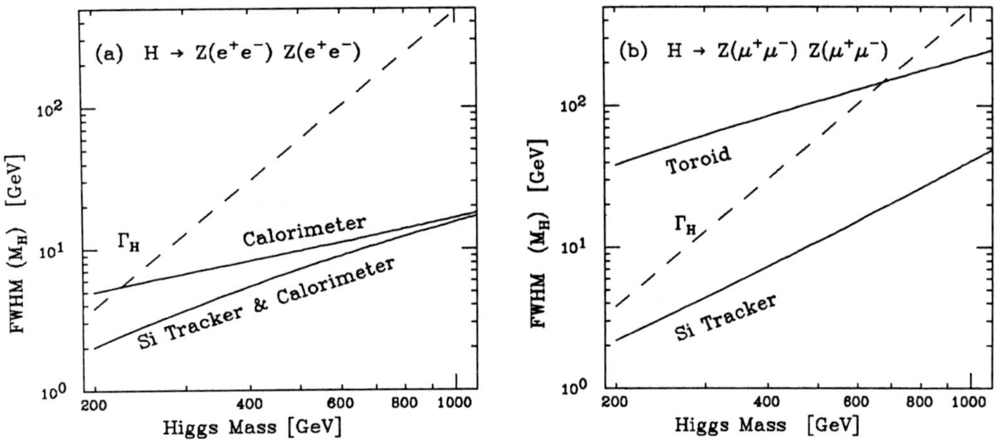

Fig. 15. Z^0Z^0 invariant mass resolution for different detector configurations, compared with the expected Higgs width Γ_H (dashed line). a) $Z^0 \to e^+e^-$. b) $Z^0 \to \mu^+\mu^-$.

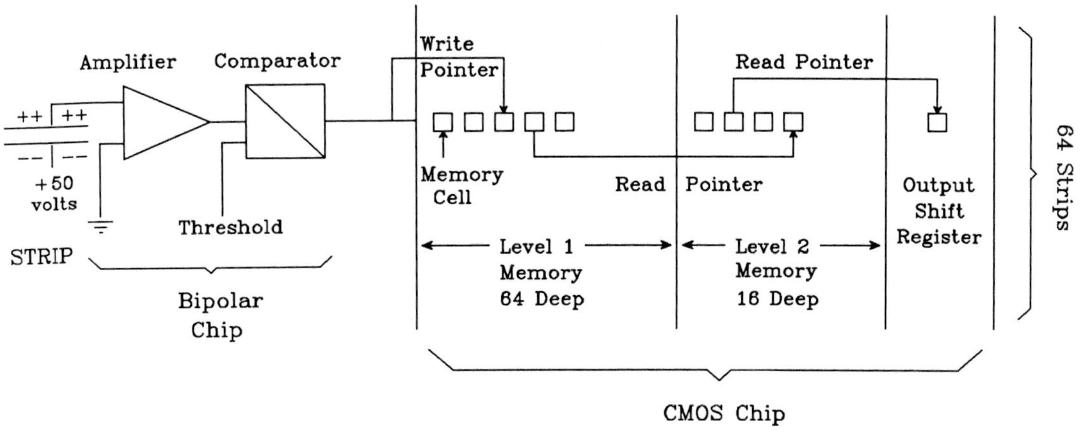

Fig. 16. Readout – VLSI chips.

The memory chip we are presently working on uses 2 μm CMOS feature size, which will not run fast enough to match the SSC requirements. The sub-micron CMOS presently used by some of the large computer companies should be well matched to SSC requirements and is expected to be commonly available well before SSC turn-on. These circuits must be radiation hardened to work after a few megarads of radiation, which is within present capabilities.[10)]

The analog circuitry must be carefully radiation tested and is more delicate than the memory which is digital. We are designing this chip using Dielectric Isolated Bipolar, which is expected to be very insensitive to damage from charged particles and neutrons. The latter sensitivity should improve as the speed of the available bipolar circuitry is increased[11)] as a result of decreased transistor geometry. It will be very important to minimize the power dissipation within the speed and noise constraints, since cooling of a large system will be a difficult problem.

CONCLUSIONS

A full scale silicon microstrip tracker, capable of resolving tracks in 1 TeV jets with high efficiency and good momentum resolution will provide invaluable information in the search for Higgs production and new physics signatures at the SSC. We present a design which we believe can be built and made to work in the SSC environment given a substantial amount of R&D effort.

REFERENCES

1. *Proceedings of the ECFA-CERN Workshop on a Large Hadron Collider in the LEP Tunnel*, CERN 84-10, 1984, p. 200 and p. 503.
2. Hayes, K., "B Tagging at the SLC," MARK II/SLC Note #73, 1984.
3. Report of the Task Force on Detector R&D for the Superconducting Super Collider, SSC-5R-1021, 1986.
4. Weinstein, A.J., "Performance of the New Central Drift Chamber for the MARK II at SLC," SCIPP Preprint 87/78, 1987.
5. Paige, F.E., and Protopopescu, S.D., ISAJET 5.34.
6. Belau, E., *et al.*, "Charge Collection in Silicon Strip Detectors," *Nucl. Instrum. & Methods* **A214**, 253 (1983).
7. Perl, J., *et al.*, "Track Finding with the MARK II/SLC Drift Chamber," *Nucl. Instrum. & Methods* **A252**, 616 (1986).
8. "Radiation Levels in the SSC Interaction Regions," D.E. Groom, editor, SSC-SR-1033, 1988.
9. Ohsugi, T., *et al., Nucl. Instrum. & Methods* **A265**, 105 (1988), and references cited therein.
10. Woodruff, R., United Technologies Microelectronics Center, private communication.
11. Gover, J.E. and Srour, J.R., "Basic Radiation Effects in Nuclear Power Electronics Technology," SAND 85-0766, Albuquerque, NM, 1986.

DEVELOPMENT AND PROSPECTS FOR CAPACITIVELY COUPLED SILICON DETECTORS

H. Dijkstra, R. Horisberger, L. Hubbeling, G. Maehlum, A. Peisert, P. Weilhammer and A. Zalewska
CERN, Geneva, Switzerland

T. Tuuva
Helsinki University, Helsinki, Finland

L. Evensen
SI, Oslo, Norway

Presented by A. Peisert

ABSTRACT

Capacitively coupled silicon μ-strip detectors were exposed to radiation from electrons of energy of up to 2.2 MeV and their parameters were measured. The increase in the leakage current per unit volume and radiation dose is 0.02 nA/cm^{3}rad for a total dose of 0.83 Mrad. An increase of the depletion layer capacitance and the interstrip capacitance was measured after a dose of about 190 krad. The polysilicon lines resistance decreased by about 6% at 120 krad. No change was measured for the coupling capacitance at the same dose.

DETECTOR DESIGN

The idea of the capacitively coupled detectors is to decouple the leakage current from the input of the amplifier by a capacitor integrated on the detector. This capacitor is created by a layer of silicon dioxide of a thickness of 200 nm between the diode and the metallization. The bias voltage is supplied to the diodes by individual polysilicon resistors connected on the other side to a common bias line (see Fig. 1) [1,2]. An aluminium pad with connection to the p$^+$-side of the diode and to the polysilicon resistor, allows us to access individual diodes.

The detectors used for the radiation hardness study have 257 diodes 39 mm long and a pitch of 25 μm. The thickness is about 300 μm. They are made of detector grade silicon of a resistance of about 4kΩ•cm.

For a later use of these counters in the DELPHI microvertex detector [3], we have chosen a readout pitch of 50 μm and only every second diode strip has metallization. For practical reasons diodes with metallization are connected to one side of the bias line on one end of the detector, and non-metallized interpolation strips on the other end. Every aluminium strip has bonding pads at both ends for connections either to electronics or another detector. The sensitive area of the detector is surrounded by a guard ring, which is a diode covered with aluminium on the longer sides and with a thin oxide layer on the shorter sides.

Another design of capacitively coupled silicon detectors was proposed by Kemmer and Lutz [4] who integrated the coupling capacitors on both sides of a detector with readout strips on both sides. On the side with p$^+$ diodes there is an orthogonal p$^+$-strip close to the readout strips for supplying the bias. Two nearby diodes will withstand a small voltage difference of a few volts before the current starts flowing between them through the bulk. The readout diodes will therefore be at a slightly more positive potential than the bias strip.

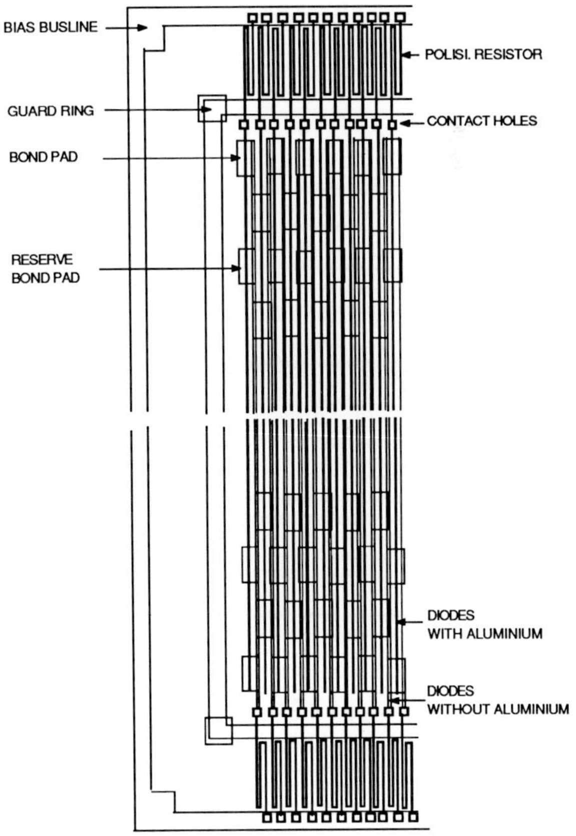

Fig. 1 Layout of Si detector.

The n$^+$-strips at the other side are surrounded by p-implants for insulation. A narrow gap towards the power supplying n$^+$-electrode is left and the electron accumulation layer builds the biasing resistors. First prototypes of these detectors have been produced and are being tested.

RADIATION STUDY

The radiation effects were studied using a ^{90}Sr source of an activity of about 10 mCi, giving electrons with an energy spectrum of up to about 2.2 MeV. The source was a disk of 12 mm diameter placed at a distance of 13 mm from the detector. For a uniform irradiation we placed the source at 2 positions above the detector at 6 mm from the edges. The average dose rate was about 1.3 krad/h for Si with a uniformity of ±10% over the full surface of the detector.

Measurement of The Detector Parameters

We have measured the coupling capacitance, polysilicon line resistance, and interstrip and depletion layer capacitance before and after radiation. For the boron-doped polysilicon lines we would expect a decrease in their resistance because a lattice damage created by the radiation acts as an acceptor, thus increasing the doping level. Figure 2 shows the polysilicon line resistance initially and after irradiation by a dose of 123 krad. A decrease of about 6% is observed.

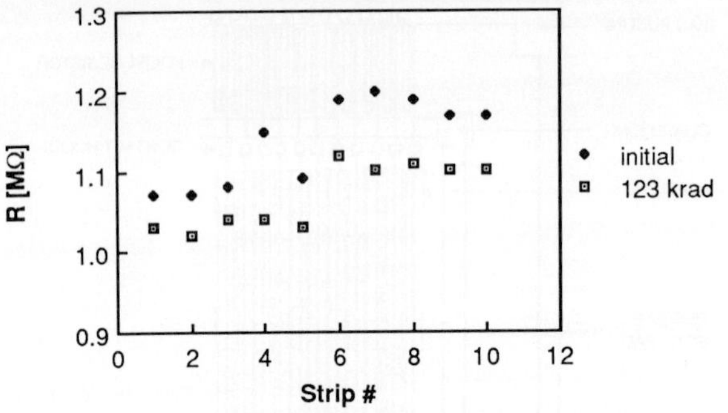

Fig. 2 Polisilicon resistance before and after irradiation for several strips.

The coupling capacitance C_C between aluminium strip and diode was measured by injecting a charge into a diode through the contact hole (see Fig. 1) and measuring the decay time of this charge. No systematic change is observed (Fig. 3).

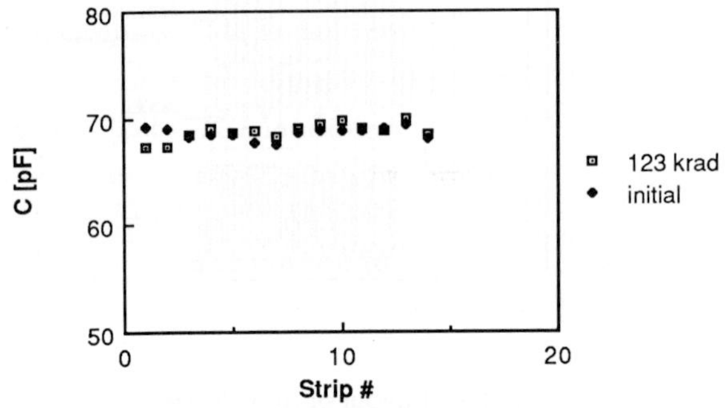

Fig. 3 Coupling capacitance.

The depletion layer capacitance was measured using a bridge with a generator frequency of 100 kHz. The capacitance of 39 aluminium strips against the backplane was measured as a function of bias voltage and is shown in Fig. 4. We calculate a capacitance of about 1 pF for 1 strip 39 mm long at the strip pitch of 50 μm. After a dose of 190 krad, the depletion layer capacitance increased by about 10%.

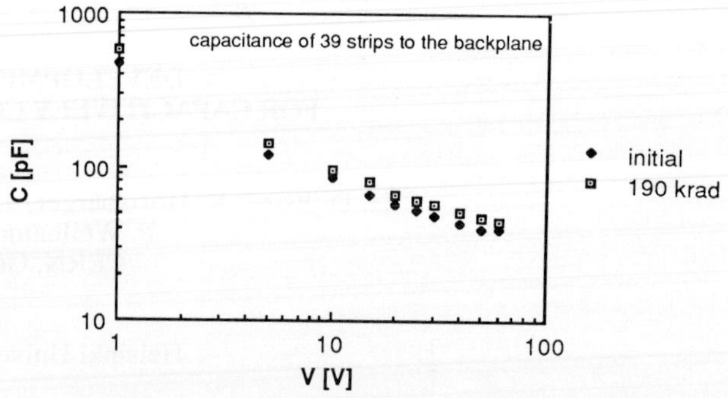

Fig. 4 Backplane capacitance as function of V_B.

We have measured the capacitance of 1 strip against its neighbours (about 40 on each side) using the same bridge. The bias voltage was provided to the diodes independently through the bias line and polysilicon resistors. Figure 5 shows the measured capacitance as a function of the bias voltage. At full depletion we measure about 2.5 pF before irradiation and about 6 pF after a dose of 190 krad. We can understand this increase assuming that the capacitance of 2 aluminium strips is a sum of 2 parallel capacitors : (see Fig. 6a) one with coupling through air and silicon dioxide, C_1, one through silicon,C_2. The main contribution comes from C_2 which is composed of 2 coupling capacitances, C_c, about 70 pF each in series, and the serial capacitance of the three diffusion lines separated by silicon. This picture is somewhat complicated through the presence of an electron accumulation layer generated by positive fixed charges in the $Si\text{-}SiO_2$ interface. The contribution from C_1 is negligible.

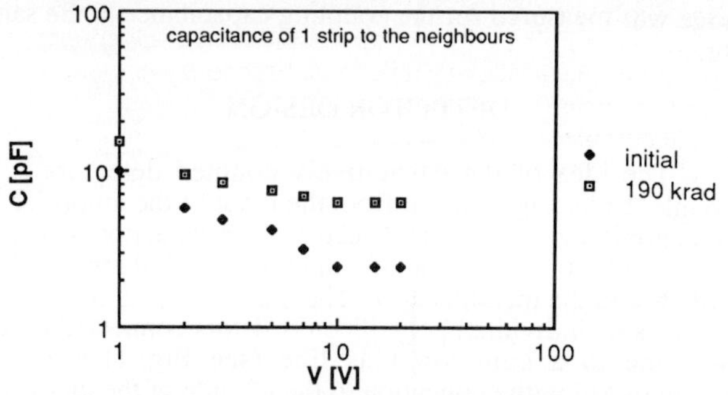

Fig. 5 Stray capacitance to neighbouring strips.

One of the effects of radiation on the oxide is the creation of additional positive fixed charges being trapped at the $Si\text{-}SiO_2$ interface which attract more electrons from the underlying silicon (see Fig. 6b).

Leakage Current

The influence from radiation damage on the leakage current of a detector with integrated coupling capacitance is expected to be more complicated than in the case of a conventional strip detector since radiation induced charges in the SiO_2 can cause a direct effect in the p^+ implant region. Trapping of radiation induced charges will particularly depend on the electric field between diode and metallization. Compared to standard detectors, most of the p^+-junction is covered by a thin oxide where positive charges can be trapped and cause direct effects on the p^+ implant region.

For a first measurement a good detector with a total leakage current of 20 nA was selected and put in the source with a bias voltage of 40 V applied between the diffusion lines and the backplane; the potential on the aluminium strips on top of the thin oxide was undefined.

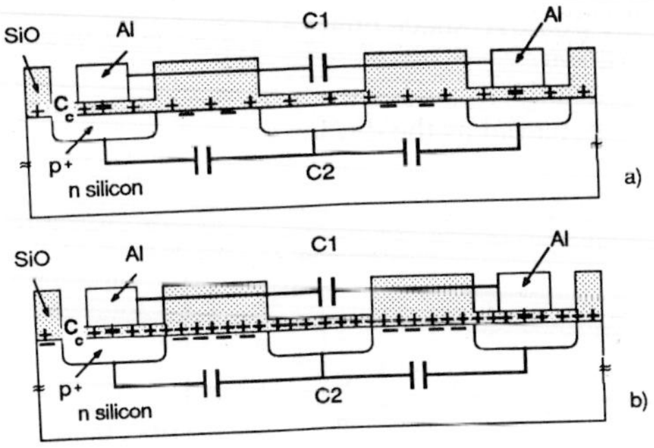

Fig. 6 Schematics of capacitive coupling to side strips.

After a first dose of about 200 rad the total current increased to 18 μA but the increase on single strips was very non-uniform, as can be seen in Fig. 7a for diodes with Aluminium on top. The same plot for diodes with no aluminization on top of the thin oxide is shown in Fig. 7b. The initial value of 0.2 nA per strip plotted in the figures is in fact the lower limit of the current due to limited range of the current-to-voltage converter we used. Comparison of Figs. 7a and b shows a big difference between the aluminized and non-aluminized strips. This rapid increase in leakage current can be qualitatively explained by the build-up of a strong positive space charge layer in the SiO_2 in a shallow region just above the Si-SiO_2 interface [5]. Under the influence of an electric field, caused by the charges accumulating on the floating surface of the detector, part of the hole electron pairs created in SiO_2 by radiation will separate before recombination is possible and drift in the electric field towards the Si-SiO_2 interface (holes) or the oxide surface (electrons). A certain fraction of these holes will be trapped at the Si-SiO_2 interface. This positive space charge will lead first to a field induced depletion layer in the p^+ region and with increasing positive charges accumulating an inversion layer of electrons (supplied from minority carriers) will be formed at the interface.

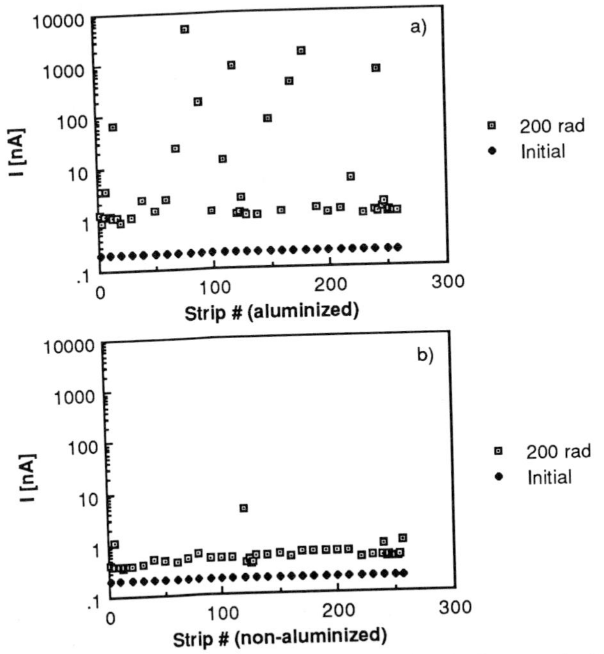

Fig. 7 Reverse current on several strips before and after irradiation.
a) with aluminium on top of diode
b) without aluminium on top of diode.

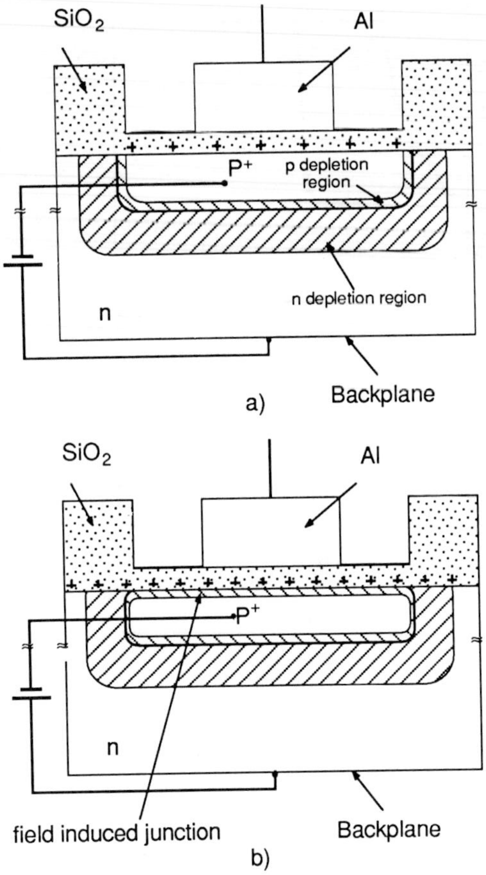

Fig. 8 a) Schematics of junction depletion layers in the p^+ junction
b) Development of additional field induced depletion zone with electron inversion layer after irradiation.

Figures 8a and 8b show schematically the cross-section of the p^+ implantation and silicon dioxide region before and after irradiation. When a high enough space charge has been built up, an avalanche breakdown will occur in the field induced junction upon applying a reverse bias voltage to the pn-junction and a channel current will flow from the electron inversion layer to the backplane. The difference between metallized and non-metallized strips may be explained by different field distributions in these regions. The electrons produced by radiation under the aluminium strips will accumulate on the surface of the aluminium strip, much further away from the Si-SiO_2 interface, than in the region where there is no aluminization (SiO_2 is 0.2 μm thick, aluminium strip is about 1 μm thick). A much higher dose of radiation is necessary to produce the field induced junction and the channel current on these strips.

To estimate the quantity of charge produced in the oxide and trapped in the Si-SiO_2 interface we have measured the flat band voltage shift of 3 MOS capacitors irradiated at 3 different gate voltages : +20 V, 0 V and -20 V on the gate. The capacitors consisted of 300 μm detector grade silicon, an oxide of 1 μm thickness, and an aluminium gate about 1 μm thick, 45 μm wide and 39 mm long. The dependence of the capacitance on the voltage is shown in Fig. 9 a, b and c for the three voltages. The total radiation dose was 30 krad, 45 krad and 289 krad for a, b and c respectively. The flat

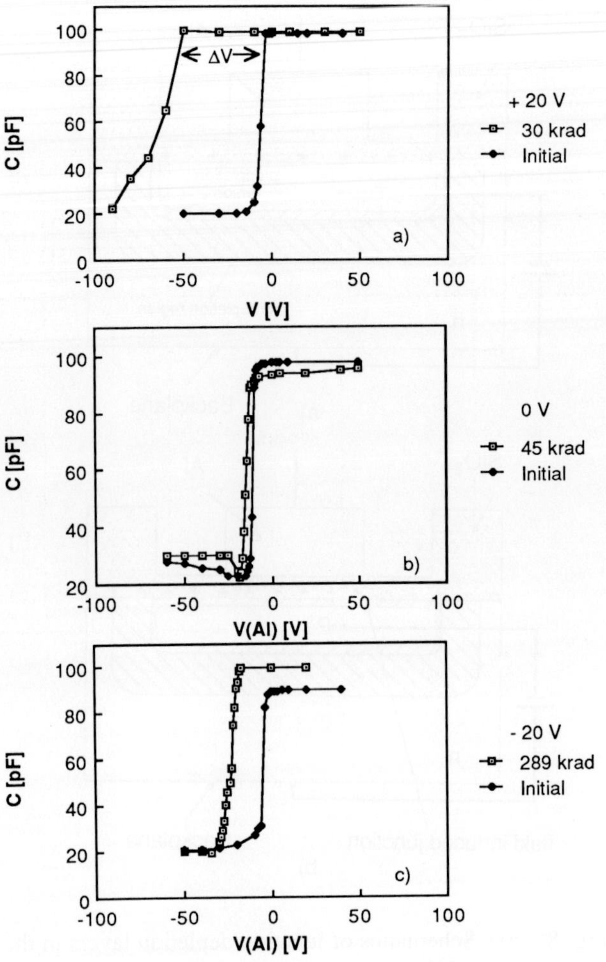

Fig. 9 High frequency capacitance voltage characteristics of MOS structures :
a) Positive gate voltage
b) Zero gate voltage
c) Negative gate voltage.

band voltage shift, ΔV, represents the additional potential necessary to neutralize the interface charge, before the silicon is beginning to deplete. We calculate this charge from $\Delta V \cdot C_{oxide}$ and we get 3.7 nC, 0.2 nC and 1.4 nC for +20 V, 0 V and -20 V voltages respectively and for the different doses. We calculate the charge generated in the oxide by the radiation and we can then deduce what fraction gets trapped in the interface. We find about 6% of trapped charge for a gate voltage of +20 V, and about 0.2% for 0 V and -20 V on the gate.

Attempting to neutralize the accumulated oxide charge on the detector we applied a variable voltage between aluminium and diode. All 257 aluminium strips on the detector were connected to each other by contacting the strips with a conductive rubber band of 5 mm width across the detector. At a constant biasing voltage of $V_B = 40$ V we have measured the current on individual strips as a function of the voltage between aluminium and diode (Fig. 10). At positive voltages on the aluminium the current increases slightly as expected, since the electron inversion layer is increased. For negative polarity the current decreases as one expects. Going from negative voltages on the aluminium back to 0 volts, the current is a factor 20 to 30 lower than at the beginning. Leaving the detector with 40 V of bias voltage and 0 V at aluminium the current decreased further by a factor of 5 after 12 hours. This indicates that the positive space charge is gradually neutralized during this operation.

Next we irradiated a detector with the aluminium strips kept at the same potential as the diodes.

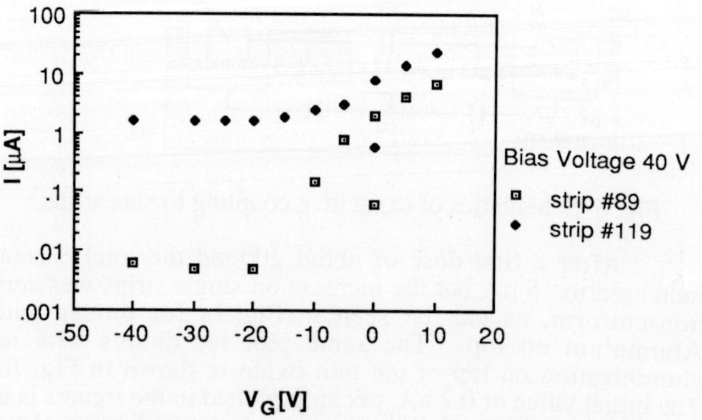

Fig. 10 Leakage current characteristics at fixed bias voltage as a function of gate voltage. There are 2 values at $V_G = 0$: The high one is obtained at the beginning of the measurement; the low one after applying first positive and then negative gate voltages.

Figure 11 shows the leakage current measured at 60 V bias voltage as a function of radiation dose for aluminized and non-aluminized strips and their sum. Calculating the increase of the total current for a unit volume and radiation dose we get $\Delta i/V \cdot D = 0.02$(nA/cm^3•rad), which is a very low value compared to other measurements (see Table 1). The absence of a strong increase in leakage current in contrast to the case of floating aluminium strips can be explained by the fact that in the absence of an electric field almost all electron hole pairs created by the irradiation recombine before the electron moves away far enough from the corresponding hole. We have not yet found an explanation for the visible difference in the leakage current increase on aluminized and non-aluminized strips.

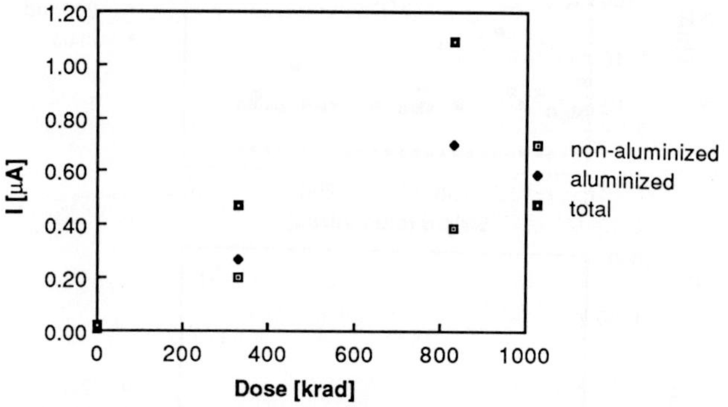

Fig 11 Leakage current as a function of dose with aluminium to diode voltage difference ΔV = 0.

These measurements show that it is important in an experimental environment with radiation to run these detectors with the diodes at the same potential as applied to the input stage of the LSI readout chip. At present we are studying radiation effects when a non-zero-field is applied between aluminium and diode during irradiation.

CONCLUSIONS

We have tested capacitively coupled silicon detectors for radiation resistance. When operated in conditions where there is no electric field across the silicon dioxide, we measure an increase of the leakage current per unit volume and unit dose to be 0.02(nA/cm^3•rad) at a total dose of 0.83 Mrad. Assuming we can operate the detector up to a leakage current from 1 strip of 20 nA, we get a tolerable radiation dose of about 5 Mrad.

The polysilicon lines resistance decreased by about 6% by a dose of 123 krad and no change in the coupling capacitance was observed.

A 10% increase of the depletion layer capacitance was measured after a dose of 190 krad. The capacitance of 1 strip towards its neighbours increased from about 2.5 pF to about 6 pF at the same dose.

ACKNOWLEDGEMENTS

I would like to thank Mr. R. Raffnsoe from the Radiation Protection group at CERN for allowing us to use the radioactive source. Without his help, the present study would have taken much longer.

REFERENCES

1) Caccia, M. et al., "A Si Strip Detector with Integrated Coupling Capacitors", NIM A260, 124 (1987).

2) Evensen, L. et al., "Recent Development of Detectors with Integrated Capacitors and Polysilicon Resistors", IEEE Trans. Nucl. Science Vol. 35, No. 1, 428 (1988).

3) DELPHI Technical Proposal, DELPHI 83-66/1, CERN/LEPC/83-3, LEPC/P2, 17 May 1983.

4) Kemmer, J. and Lutz, G., "New Structures for Position Sensitive Semiconductor Detectors", accepted for publication in Nucl. Instrum. and Methods, 1988.

5) Nicollian, E.H. and Brews, J.R., MOS Physics and Technology, (Wiley, 1982).

6) Dietl, H. et al., "Radiation Damage in Silicon Strip Detectors, NIM A253, 460 (1987).

7) Ohsugi, T. et al., "Radiation Damage in Silicon Microstrip Detectors", KEK preprint, 87-22 (1987).

8) Campanella, P. et al., "The Effect of Radiation on Ion-Implanted Silicon Detectors", NIM A243, 93 (1986).

9) Borgeaud, P. et al., "The Effect of Radiation on the Energy Resolution of Ion-Implanted Silicon Detectors", NIM 211, 363 (1983).

Table 1

Increase of the leakage current per unit volume and radiation dose

resisitivity [kΩ ·cm]	total dose	time	$\frac{\Delta i}{V \cdot dose}$ $\left[\frac{nA}{cm^3 \cdot rad}\right]$	ref.
3	~1 Mrad	~2 years	0.61	6)
4	5 Mrad	1 hour	1.4	7)
9	1.4 Mrad	several weeks	0.28	8)
3.7	70 krad	several weeks	0.063	8)
4	1.4 Mrad	30s	6.3	9)
4	0.83 Mrad	25 days	0.02	*)

*) Present work

RADIATION DAMAGE OF SILICON DEVICES BY NEUTRONS AND R&D FOR JCCD READOUT

Takahiko Kondo
Natinal Laboratory for High Energy Physics
1-1, Oho, Tsukuba-shi, Ibaraki-ken, 305, Japan

ABSTRACT

Two subjects are described in this article, (1) a development of a junction charge coupled device (JCCD) for readout of silicon microstrip detectors, and (2) a radiation damage study of high-resistivity silicon detectors by fast neutrons which were generated in a nuclear reactor or by spallation with 12 GeV protons.

1. JCCD READOUT FOR MICROSTRIPS

Fig.1 illustrates the principle of the CCD readout. Charge from silicon microstrip detector is injected into a CCD pixel element via a current injection circuit attached to each CCD array. Injected charge in each raw is transferred in parallel with high clock frequency. In a readout cycle, the parallel clock is slowed down, while the serial-out clock for the last column is activated to transfer charges stored in the last column. A floating diffusion amplifier transforms charge into voltage.

In comparison to an analog memory unit, CCD readout scheme is much simpler in its structure of charge transfer, and thus one can anticipate an uniform charge transfer as well as higher yield in mass production.

Junction CCD was first proposed by Schuermeyer et al.[1] and its gate structure is shown in Fig.2. In contrast to regular types of CCDs, there is no SiO_2 layer between gate and charge transfer area. This will enable a high transfer efficiency and a high speed charge transfer. The absence of SiO_2 layer makes the device radiation resistant for charged particles. Bipolar processing for the production of JCCD allows JFET and other bipolar circuits on the same chip.[2]

Five years ago, Ten Kate and Van der Klauw proposed an idea of application of JCCD to silicon microstrip readout.[3] Their idea is a use of vertical npn transistor as an injector as shown in Fig.2. The proposed injector plays an important role in allowing the charge transfer from a microstrip (C=1~5pF) to a CCD pixel (C~0.1pF). The speed of this charge transfer is approximately given by the time constant

$$\tau = C_d \cdot r_e = \frac{C_d\, k\, T}{q\, I_{bias}}$$

where I_{bias} is the collector-emitter bias current of the vertical transistor. I_{bias} must satisfy

$$I_{bias} < Q_{max} \cdot A \cdot f$$

in order to avoid charge saturation in CCD pixels, where A is the pixel area, Q_{max} the charge capacity per unit area and f is the clock frequency. Typically I_{bias} is at around 1μA and thus r_e~25kΩ resulting τ~125ns for C_d=5pF.

We have developed JCCDs with and without charge injectors in cooperation with Hamamatsu

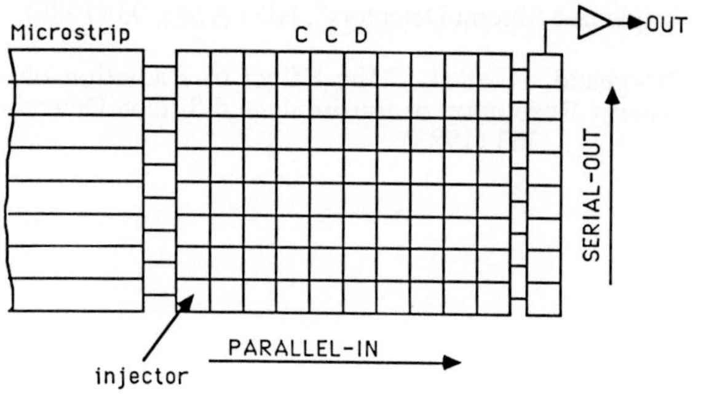

Fig.1. Principle of CCD readout for silicon microstrip detectors.

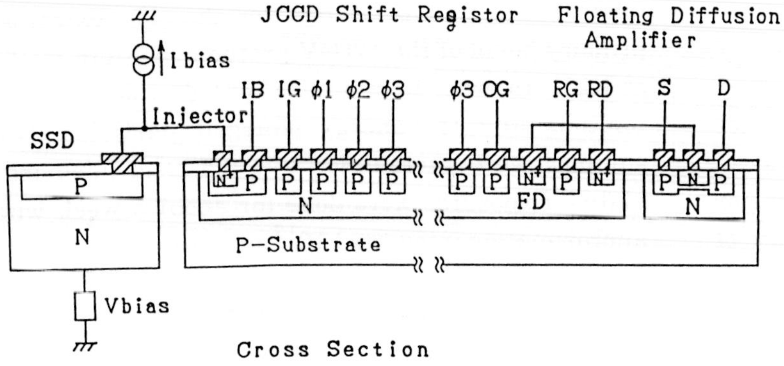

Fig.2. Cross section of JCCD readout circuit.

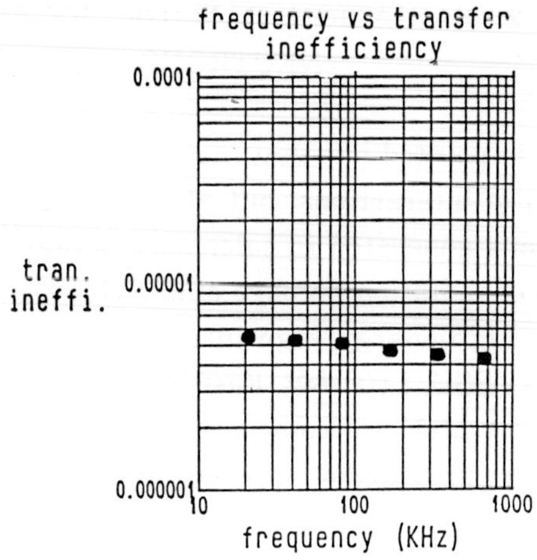

Fig.3a Transfer inefficiency of the JCCD as a function of clock frequency.

Photonics, Inc. starting from nothing. Up to now, several prototypes were fabricated in order to understand the basic properties such as transfer efficiency, linearity, dark current, frequency dependence etc. The circuit as shown in Fig.2 is fabricated on a N epilayer which was grown on a P substrate. Fig.3a shows the frequency dependence of the charge transfer inefficiency, indicating excellent charge handling capability up to 1 MHz. The linearity was demonstrated to be good as shown in Fig.3b for JCCDs without an injector, and in Fig.3c with an injector. These linearities were measured at I_{bias}=0.5μA. The charge injection is working in principle as shown in Fig.4, which is a picture of the pulse train for the case of 10fC charge injection with the clock frequency of 1MHz. The figure indicates that the noise level is about 1fC .

We summarize some of the characteristice of the JCCD accomplished by Hamamatsu Photonics Inc.

pixel size :	100 μm x 90 μm
transfer inefficiency :	5.10-6 at f=660kHz
charge handling capability :	2 pC
linearity :	good from 1000 e
CCD noise :	165 e at f = 660 kHz
clock frequency :	up to 8 MHz

We have a plan to improve the output floating diffusion amplifiers. Connection to silicon microstrip detectors will be done in near future.

This development has been performed by the collaboration of K.Niwa, M.Nakamura and Y.Tajima (Nagoya University) and K.Yamamoto and K.Miyaguchi (Hamamatsu Photonics).

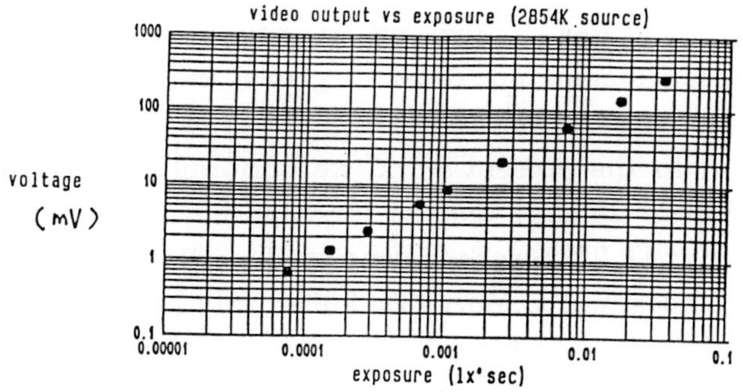

Fig.3b Linearity of the JCCD without injector at the clock frequency of 330 kHz.

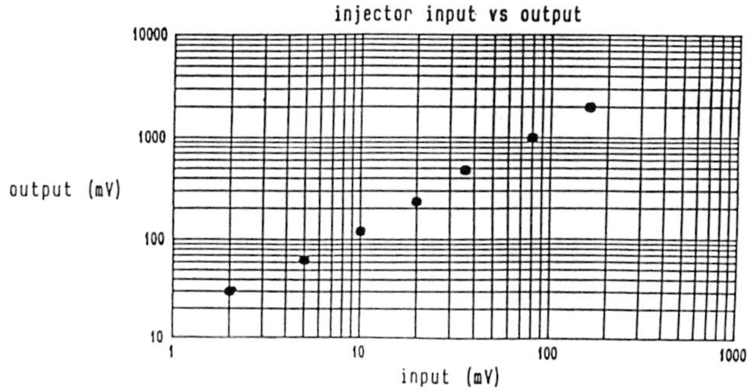

Fig.3c Linearity of the JCCD with charge injection circuit with I_{bias} = 0.5μA.

2. RADIATION DAMAGE BY FAST NEUTRONS

It is pointed out that, not only charged particles from p-p interactions, but also a quite high flux of neutrons generated in calorimeters may be a major cause of radiation damage of detectors in the case of super high-energy high-intensity hadron colliders such as SSC, LHC.[4,5] Fast neutrons with energies around 1 MeV are copiously generated through the nuclear spallation process in hadron cascade showers.[6] In an uranium hadron calorimeter, neutrons from fissions contribute significantly.[7] Silicon semiconductor devices could be sensitive to fast neutrons due to their high cross sections for vacancy creation in silicon crystals.[8] There have been numerous works of neutron damage effects on silicon devices and microelectronics.[9] However there seems a lacking of radiation damage measurement on detector-grade (namely high resistivity) silicon devices.

In the present work, we performed a simple neutron damage test on large pn-junction photodiodes[10]. This work is done by tha collaboration of M.Hasegawa, S.Mori (Tsukuba Univ.), T.Ohsugi, H.Kojima, A.Taketani (Hiroshima Univ.) and M. Noguchi (KEK), with kind cooperation by K. Yamamoto and H.Tanaka (Hamamatsu Photonics Inc.).

2-1. FAST NEUTRON SOURCES

Two neutrons sources were used in the test:

[1] JRR-4 : an experimental nuclear reactor at Tokai Research Establishment of the Japan Atomic Energy Research Institute. The reactor is a light water type with output power of 3.5 MW. Test samples were packed in a plastic capsule, which was sent to the edge of the reactor core through a gas pipe. Approximate rates of radiation at the test position were $4{\cdot}10^{13}$ thermal neutrons/sec/cm^2, $8{\cdot}10^{12}$ fast neutrons /sec/cm^2. Dose from γ rays was about 20 times of that from neutrons. Two tests were performed with exposure time of 1 and 10 seconds.

[2] 12 GeV Proton Synchrotron at KEK

Test samples were placed near (0.5-1.5m) but upstream of a copper target for the slow extracted primary beam of the 12GeV proton synchrotron (PS) at KEK. In order to minimize contribution from soft photons and low-energy charged particles, test samples were surrounded by 0.5~2cm-thick lead plates. Exposures were done for about a week with approximately integrated 10^{17} protons on the target.

2-2. FLUX MEASUREMENT

We employed a thin foil activation method for monitoring fast neutron flux. Thin Ti and Ni foils were used for the nuclear reactor test, while thin foils of Ti, Ni, and Fe were placed with test samples for 12GeV PS exposures. These elements were selected because they have fairly high cross sections to generate radioactive states with long decay time via the processes $Ti^{47}(n,p)Sc^{47}$, $Ni^{58}(n,p)Co^{58}$, $Fe^{54}(n,p)Mn^{54}$, $Ti^{48}(n,p)Sc^{48}$. In order to check the reproducibility, we placed two pieces of foils for each test. Each foil piece was weighed carefully before the exposure.

After the exposure, we have measured a gamma-ray spectrum from each foil using a Ge(Li) detector facility of the KEK radiation safety center. The absolute energy as well as the acceptance of this Ge(Li) detector was calibrated with the use of a standard radioactive source of ^{152}Eu. The energy-efficiency product thus obtained was well approximated by

$$EF = \text{energy} \cdot \text{efficiency} = e^{6.21-0.834\ln(E)}/100,$$

where E is the peak energy in keV. Then the number of an isotope N generated in a foil is calculated by

$$N = \frac{S_{peak}}{(e^{-t/\tau} - e^{-(t+\Delta t)/\tau}) \cdot Br \cdot EF}$$

where S_{peak} is the background subtracted peak area of corresponding γ-rays investigated, t the time elapsed from the exposure, Δt the real measurement time, τ the decay constant $(0.693/T_{1/2})$, and Br is the branching ratio. In the case of the exposure at the 12GeV PS, where the exposure time was over one week, we made a time dependent correction for isotope decays based on the recorded proton intensity on the target. Neutron flux is then obtained by

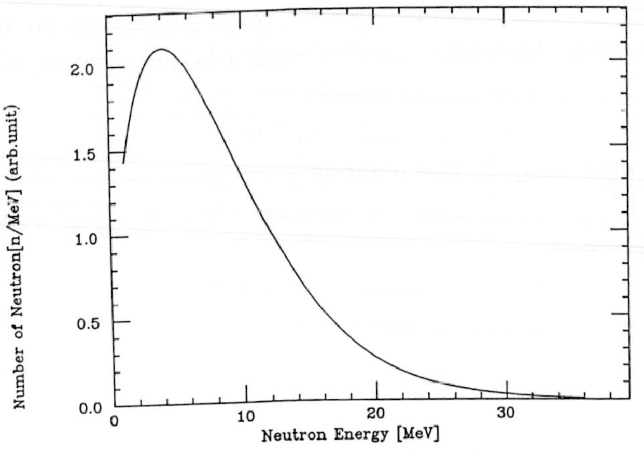

Fig.5 Energy spectrum of spallation/evaporation neutrons from 12GeV proton-nucleus interactions.

$$\text{flux} = \frac{N \cdot A}{M \cdot R \cdot N_A \cdot \int \sigma(E).f(E)\, dE}$$

where A is the atomic mass of the foil element, M the weight of the foil, R the fraction of abundance and N_A is the Avogadro number, $\sigma(E)$ is the production cross section of the isotope level by neutrons (and protons) of energy E,[11] and f(E) is the normalized energy spectrum of fast neutrons[12,13]. The energy spectrum of the spallation neutron is shown in Fig.5. The present method provided consistent results among different foils used for monitoring as summarized below.

exposure	expos. time	activity used	fast n flux
reactor #1	~ 1 sec	Sc^{47},Co^{58}	$2.5 \cdot 10^{13}$ n/cm^2
reactor #2	~10 sec	Sc^{47},Co^{58}	$9.8 \cdot 10^{13}$ n/cm^2
12GeVPS#1	~1week	Sc^{48},Co^{58},Mn^{54}	$1.1 \cdot 10^{13}$ n/cm^2
12GeVPS#2	~1week	Sc^{48},Co^{58},Mn^{54}	$7.0 \cdot 10^{12}$ n/cm^2
12GeVPS#3	~1week	Sc^{48},Co^{58},Mn^{54}	$9.1 \cdot 10^{12}$ n/cm^2
12GeVPS#4	~1week	Sc^{48},Co^{58},Mn^{54}	$2.7 \cdot 10^{12}$n/cm^2

The ratio of 1 to 3 for 1-sec and 10-sec reactor exposures is understandable since a certain amount of extra neutron dose is expected during capsule transportation. The flux rate $9.8 \cdot 10^{12}$cm^{-2}sec^{-1} obtained in our 10-sec exposure is not far from the standard value of $8 \cdot 10^{12}$ quoted for the present exposure spot of the reactor JRR-4.

For 12GeV PS exposure, we have used proton induced activations $Ti^{48}(p,n)V^{48}$ and $Fe^{56}(p,n)Co^{56}$ in order to estimate a possible slow (E~1-100MeV) proton contamination. The results indicated < 10^{11} protons/cm^2 for the 12GeV PS exposures #1 and #2 and < 10^{10} protons/cm^2 for #3 and #4.

2-3. TEST DEVICES

PIN photodiodes for this test were specially fabricated by Hamamatsu Photonics Inc. It has a sensitive area of 1 x 1 cm with a total thickness of 200μm and a full depletion depth of 170μm for a bias voltage of 30V. Its bulk silicon has a resistivity of 4KΩ·cm. The reverse-biased leakage current is less than 10nA at 30V. The packaging and performance as a photo-detector are identical to those reported elsewhere.[10] This detector works excellent not only for photon detection, but also for penetrating charged particle detection.

2-4. LEAKAGE CURRENT MEASUREMENT

A reverse-biased leakage current was used as an indicator of neutron damage effects. Fig.6 shows reverse-voltage versus leakage current relations before and after the irradiation, indicating no appreciable differences in the voltage dependence. This is in sharp contrast with the forward-bias V-I behaviour shown in Fig.7. We have observed large

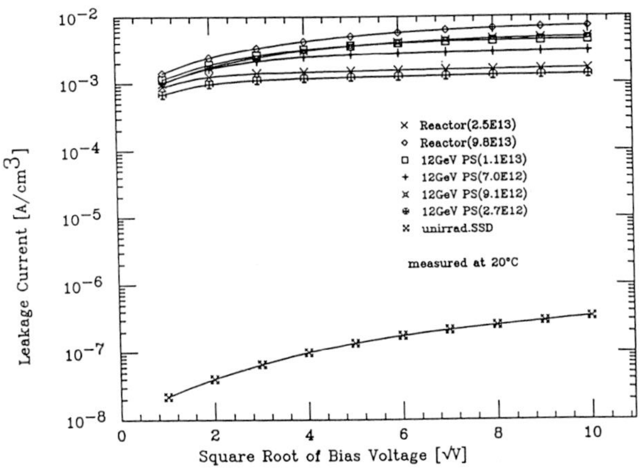

Fig.6 Reverse leakage current of silicon photodiodes before and after the radiation as a function of $\sqrt{V}$.

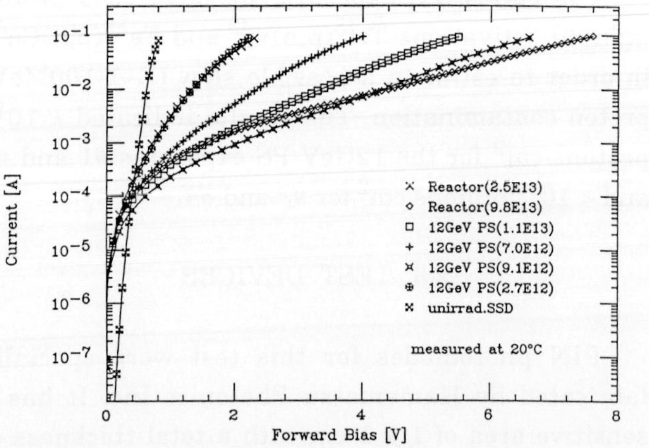

Fig.7 Forward current versus voltage for silicon photodiodes before and after the radiation.

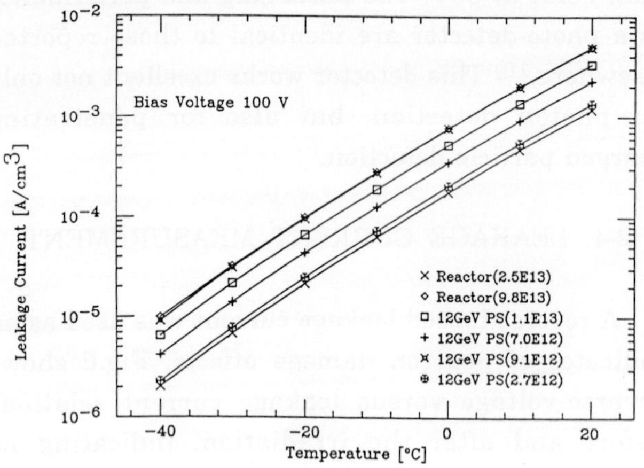

Fig.8 Temperature dependence the leakage current.

increase in low current region as well as apparent high resistance at higher forward current.

The temperature dependence of the leakage current is depicted in Fig.8 for the irradiated samples. The current varies approximately as exp($-E_g/kT$) with temperature, where E_g is close to the band gap width, k is the Boltzman constant, and T is the absolute temperature. The exponential behavior indicates that the leakage current is generated predominantly through generation-recombination processes by trap levels in the forbidden gap. Similar temperature dependence was observed for irradiated samples by charged particles.[14]

The leakage current density at the bias voltage of 100V is plotted as a function of neutron fluence in Fig.9. The leakage current is proportional to the fluence. However, we observed a large difference in the proportionality constants α_n defined by

$$I_{leak} = \alpha_n \cdot \Phi$$

where I_{leak} is the leakage current density (A/cm^3) and Φ is the neutron fluence (neutrons/cm^2). We obtained

$$\alpha_n = 7 \cdot 10^{-17}\ \text{A/cm} \quad \text{for reactor neutrons,}$$
$$= 5 \cdot 10^{-16}\ \text{A/cm} \quad \text{for spallation neutrons.}$$

at 20°C. These can be compared with the damage coefficient measured in the past,

$$\alpha_p = 3 \cdot 10^{-17}\text{A/cm} \quad \text{for high energy protons.}^{14)}$$

We have observed a factor of 7 larger leakage current coefficient in 12GeV PS exposure. The difference might be due to harsh radioactive environment of the 12GeV proton-extraction enclosure. However, slow protons and soft γ-rays were supposed to be attenuated well in our lead shield (0.5~2cm thick). In fact, as stated above, our foil activation indicated that slow protons were small in flux. Remaining possibilities are the differences of neutron energy spectrum and/or the exposure time. The mean energy of reactor neutrons is about 0.8 MeV while that of spallation/evaporation neutrons is around 8 MeV (see Fig.5).

The leakage current coefficient α can be converted to the damage constant K, defined by $1/\tau - 1/\tau_0 = \Phi/K$, where τ is the minority carrier lifetime. We get

$$K = 1.6 \cdot 10^7\ \text{neutron·s/cm}^2 \quad \text{for reactor,}$$
$$K = 2.2 \cdot 10^6\ \text{neutron·s/cm}^2 \quad \text{for 12GeV PS.}$$

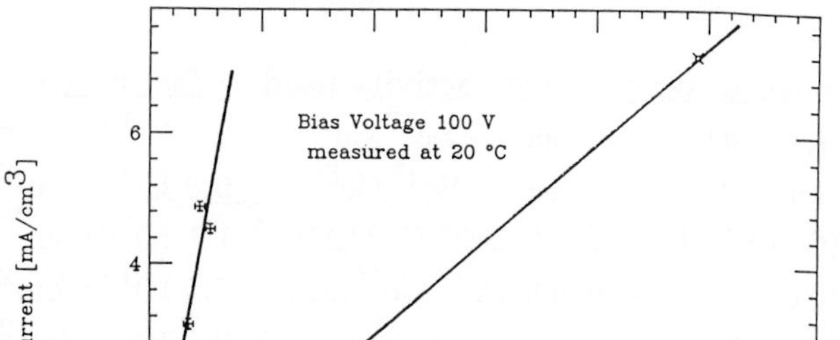

Fig.9 Leakage current density as a function of neutron fluence.

The latter value is close to the two-level model by Messenger for n-type silicon.[9)]

2-5. PENETRATING BETA-RAY SIGNALS

We have measured a pulse height spectrum using penetrating beta-rays from ^{106}Ru. Since the signal could not be separated from noise due to large leakage current, we cooled the samples to -30°C in order to reduce leakage current to be below 1μA. Fig.10 show the dependence on bias-voltage of the peak pulse height. One notices that the saturation voltage systematically shifts to higher voltage for increased neutron fluence. In order to verify that this saturation comes from the full depletion, we have plotted a C-V curve for each sample at 10kHz with a capacitance meter. The results are shown in Fig.11, indicating consistent correspondence of the pulse height saturation to the saturation in capacitance. Thus it is highly probable that the increase of depletion voltage for increased fluence suggests an apparent reduction of bulk resistivity by neutron damage. It should be reminded that an increase of bulk resistivity was observed for radiation damage by charged particles.[15,16)]

In addition, one notices in Fig.10 that the pulse height at the saturation decreases with increased neutron fluence.[14)] This could be explained by a decrease of minority carrier lifetime as for the case of radiation damage by charge particles.

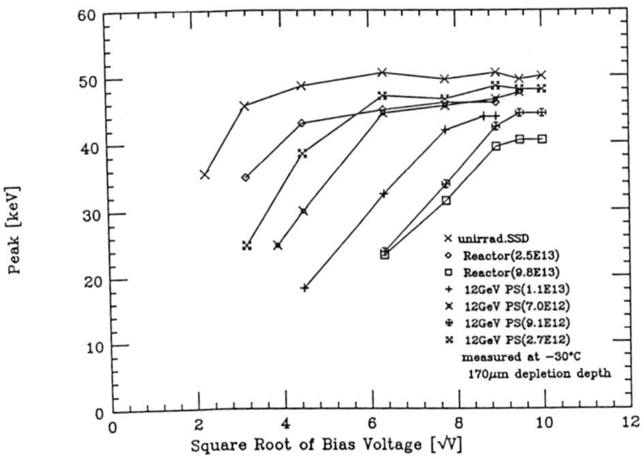

Fig.10 Peak pulse height of penetrating β-rays as a function of applied voltage measured at -30°C.

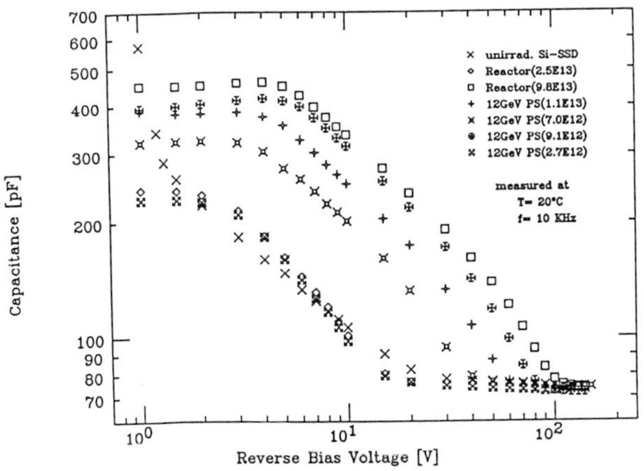

Fig.11 Capacitance-Voltage curve.

2-6. ROOM TEMPERATURE ANNEALING

A moderate effect of room temperature annealing was observed from our long-term measurement of leakage current. The devices irradiated by fast neutrons were kept at 20°C. No voltages were applied to the devices except for the time of leakage current measurement. Fig.12 shows the leakage current density as a function of passed time (days) from the irradiation. Unfortunately, data for the important transition period, namely from 80 to 200 days, are missing (we stopped our measurement), but it is clear that the device anneals itself with the rate of $e^{-t/160\text{days}}$ for 2 months and then it ceases self-annealing as far as the leakage current concerns.

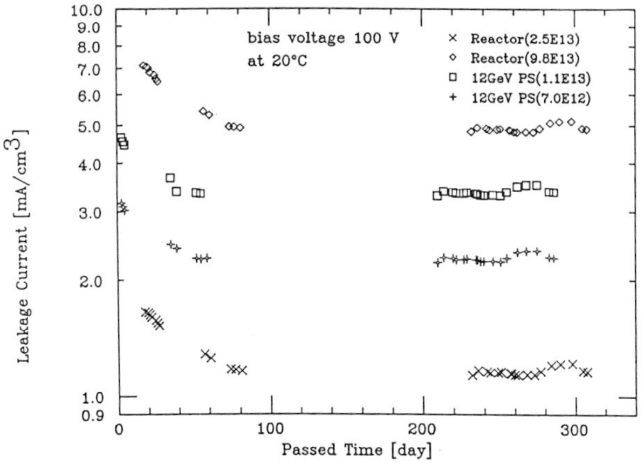

Fig.12 Leakage current as a function of days after the exposure showing the room temperature annealing effect for the initial 2 months.

3. SUMMARY

The current status was reported on the development of Junction CCD for the readout of silicon microstrip detectors. Basic properties of JCCD were mesured and the charge injection was successfully tested.

Radiation damage of high resistivity silicon by fast neutrons was studied using a nuclear reactor and the KEK 12GeV PS. Foils activation method was used to measure the fluxes of fast neutrons as well as slow protons. We have obtained the leakage current constants

$\alpha_n = 7 \cdot 10^{-17}$ A/cm for reactor neutrons,

$= 5 \cdot 10^{-16}$ A/cm for spallation neutrons.

The big difference can be attributed to the possible neutron energy dependence of radiation damage. The depletion voltage gets higher for higher neutron dose, indicating that the neutron damage would reduce the bulk resistivity. This is in contradiction to the observation of damage by charged particles as well as our forward V-I curves. A long-term room temperature annealing has been observed with time constant of 160days via leakage current measurement. Although the annealing ceases in 2 months after the neutron exposure.

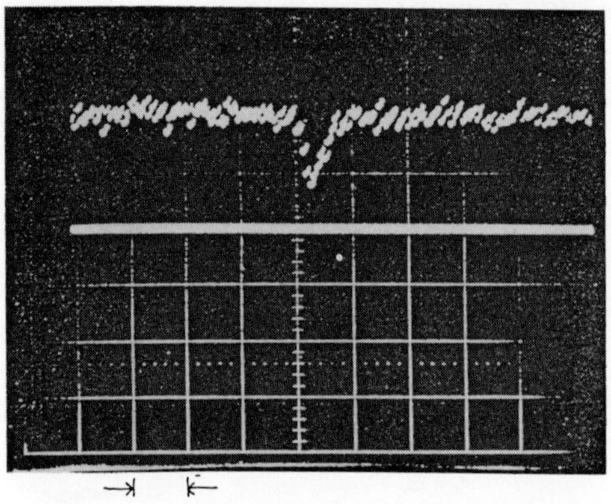

Fig.4 Output pulse train for the charge injection of 10 fC at the clock frequency of 1 MHz.

REFERENCES

1) F.L.Schuermeyer et al., IEEE(Letters)1972 p.1444.
2) E.A.Wolsheimer and M.Kleefstra,IEEE ED-29,1930 (1982).
3) W.R.Th.Ten Kate and C.L.M.Van der Klauw IEEE IEDM 1983 p.648, Nucl.Instr.and Meth. 28, 105(1984).
4) Report of the Task Force on Detector R&D for the Superconducting Super Collider, SSC-SR-1021 p.49 (1986).
5) D.Nygren, talk at the Workshop on Experiments, Detectors, and Experimental Areas for the Super- colliders, Berkeley, California, July 7-17, 1987.
6) R.Wigmans, Nucl. Instr. and Meth. A259, 389(1987). H.Bruckmann et al.,Nucl. Instr. and Meth. A263, 136(1988).
7) C.Leroy et al., Nucl. Instr. and Meth. A252, 4(1986).
8) J.R.Srour et al., Radiation Effects on and Dose Enhancement of Electronic Materials, Noyes Publications, N.J., 1984.
9) G.C.Messenger and M.S.Ash, The Effects of Radiation on Electronic Systems, Van Nostrand Reihold, New York, 1986.
10) S1723 by Hamamatsu Photonics,Inc. and K.Yamamoto et al.,Nucl. Instr. and Meth. A253, 542 (1987).
11) The energy dependent cross sections were taken from the Code ENDF/B-IV.
12) The neutron energy spectrum for nuclear reactor is given by $\Phi(E)=0.45e^{-E/0.265}\sqrt{\sinh(2.29E)}$ E in MeV, from "Radiation Dosemetry", ed. by F.Attix and E.Tochilin, Academic Press, Vol.3, 205 (1969).
13) For 12GeV PS exposure, we used a Monte Carlo simulation provided by S.Mori (Tsukuba Univ.).
14) T. Ohsugi et al., Nucl.Instr. and Meth.A265, 105 (1988).
15) M.Nakamura et al., Nucl.Instr. and Meth.A270, 42(1988).
16) H.Dietl et al., Nucl. Instr. and Meth. A235, 460(1987).

SILICON DETECTOR DEVELOPMENT IN EUROPE

Erik H.M. Heijne and Pierre Jarron

CERN, CH-1211 Geneva 23, Switzerland

ABSTRACT

Development and applications of segmented silicon particle detectors for high-energy physics are pursued in more than twenty places in Europe. Trends in development of technology are discussed. In high-energy physics experiments the silicon pixel detector will progressively replace simpler devices like the microstrip detector.

1. HISTORICAL OVERVIEW

Around 1979, several high-energy physics groups in the USA were discussing the use of silicon for high precision particle localization. The most consequent ideas were developed by a group of physicists preparing experiments for ISABELLE. Unfortunately, they were initially discouraged by the scepticism of one of the major detector manufacturing companies. As a result, it took some years before silicon detector development for elementary particle physics really started.

In the meantime, in Europe, a modest activity was going on. In Pisa some devices were produced in the INFN Laboratory and in the spring of 1980 some first results were published [1]. On the other hand, our group at CERN had been discussing a refinement of the old Philips checker board counter [2] since quite a while. In our opinion, the reason why this device had not been widely adopted was the unavailability of standard associated read-out electronics. Therefore, the design of a cheap, easy to use, low-noise amplifier chain was started at the same time as the detector development. In the first instance, a silicon surface barrier microstrip structure was produced in collaboration with Enertec-Schlumberger(*), Lingolsheim, France [3]. Very soon afterwards, the same company changed to the planar technology developed by Kemmer [4] and the planar silicon microstrip detectors were commercially available from the middle of 1981 [5]. This enabled several experiments in Europe and in the USA to use these new devices at modest investment cost.

Associated preamplifiers [6,7] and discriminators became available in the form of thick-film hybrid circuits, and made it relatively easy to equip detectors with a few thousand channels. The first monolithic multichannel preamplifier (Microplex) for silicon microstrip detectors was designed soon, in a collaboration between B. Hyams (CERN), S. Parker (Hawaï University) and J.T. Walker (Stanford Center for Integrated Systems) [8].

(*) Presently: Intertechnique

Now, nearly ten years later, silicon detectors are a feature in every new experiment, and there is a choice of commercial manufacturers. In many places development projects have been initiated, often in collaboration with these manufacturers, or with independent microelectronics research centres. Also the development of Application Specific Integrated Circuits (ASIC) for the read-out of silicon and other detectors has been stimulated by the synergy with the silicon detectors.

To promote some uniformity in terminology we propose to reserve the term <u>microstrip detector</u> for 1-D detectors with a pitch $<$ 1 mm. 1-D or true 2-D detectors with element size between 1 mm and 10 mm should be designated <u>pad detector</u>. If the smallest element dimension is $>$ 10 mm it is called <u>tile detector</u>. A <u>pixel detector</u> should include signal processing circuitry.

2. DETECTOR DEVELOPMENT ACTIVITY IN EUROPE

On the map in fig. 1 are indicated all places where a significant activity is deployed. The actual projects, as far as we know them, will be briefly reviewed. In general, it was not possible to obtain statements about future developments, and only some broad impressions will be given in sects 3-4, for the different directions that can be foreseen.

Fig. 1 Localities in Europe where significant silicon detector R & D or manufacturing exists.

2.1 Commercial manufacturers

In alphabetical order are given the European silicon detector manufacturers who have undertaken development of segmented silicon detectors and silicon drift detectors for use in high-energy particle physics. There exist many more suppliers in Europe and elsewhere, and it may be that some of these could be interested in this development as well.

AME,
N-3191 Horten (Norway)
AME has a close connection to the Senter for Industriforskning (SI), a non-profit research institute which has a department for semiconductor devices. In collaboration with the DELPHI Group at CERN, a new detector structure has been developed which includes the resistors and capacitors needed for AC coupling of individual elements in a microstrip detector [9]. This topic is dealt with in detail by A. Peisert, elsewhere in these proceedings.

ANSALDO, Semiconductor Dept.,
Via N. Lorenzi 8, I-16152 Genoa (Italy)
Within this large industry exists a production group for special semiconductor devices like thyristors, solar cells, etc., with significant capacity for volume production. Prototype development of silicon detectors for calorimeter modules has been going on since several years, in collaboration with the SICAPO Group. A large number of papers have been published concerning this development [10], and in this SSC Meeting a short report is given by J. Brau.

CANBERRA SEMICONDUCTOR NV,
Lammerdries 25, B-2430 Olen (Belgium)
This majority-owned daughter of Canberra Industries, Meriden, CT 06450, USA, manufactures various types of standard and custom planar silicon detectors. 3" and 4" high-resistivity wafers can be processed. Canberra supplied pad detectors for the UA2 inner silicon detector. In collaboration with IMEC (Leuven, Belgium) integrated amplifier circuits are under development.

INTERTECHNIQUE (formerly Enertec-Schlumberger),
1, ch. de la Roseraie, F-67300 Lingolsheim (France)
Various types of silicon and germanium detectors have been produced for particle physics experiments, in particular microstrip detectors and pad detectors. Detectors can be tailored to meet special requirements. Intertechnique supplied pad detectors for the UA2 inner silicon detector.

MBB (Messerschmitt-Bölkow-Blohm GmbH),
Postfach 801169, D-8000 München (West Germany)
This large industry has recently started an in-house silicon device manufacturing facility, where J. Kemmer among other developments, continues also his work on new types of particle detectors [11]. Some details will be mentioned later as these projects form, to a large degree, a collaboration with the MPI-München Group.

MICRON SEMICONDUCTOR Ltd.,
1, Royal Buildings, GB-Lancing BN15 8UN (Great Britain)
All types of silicon detectors are being fabricated to customer's specifications. Recently, drift detectors and double-sided structures were manufactured. Micropad detectors have been supplied to SLAC with 10 x 64 and 256 x 256 elements. These are used for constructing hybrid pixel detectors, as is described by S. Shapiro in this meeting. Micron supplied pad detectors for the inner and outer UA2 silicon detectors.

PHILIPS and SIEMENS
Both have in-house know-how in the field of silicon particle detectors, and they undertake small-scale production for application in some of their own products. At present, neither has ambition to enter this specialized market. However, under circumstances they may be willing to take on "custom" high volume production of sensors and dedicated electronics.

HAMAMATSU PHOTONICS KK,
1126-1 Ichino-Cho, JP-Hamamatsu City 435 (Japan)
Although not European, we should mention Hamamatsu, which to our knowledge, is at present the only commercial silicon manufacturer outside Europe who is actively pursuing development of microstrip detectors, pad detectors and other structures for particle position detection in high-energy physics and other applications. Some of the work at Hamamatsu is described by T. Kondo in this meeting.

2.2 Semiconductor R & D Institutes

CSEM (Centre Suisse d'Electron. et Microtechnique)
Maladière 71, CH-2000 Neuchâtel 7 (Switzerland)
The development of processing technology has resulted in a double-sided microstrip detector. CSEM has also extensive experience in low-power, low-noise IC design.

DELFT University of Technology,
P.O. Box 5031, NL-2600 GA Delft (Netherlands)
The study of silicon detector manufacturing technology and compatible electronics is pursued with the aim of applications in X-ray and electron detection. But results may be also applicable to tasks in high-energy particle physics. In particular, a capacitively coupled multiplexing scheme has recently been developed [12].

IMS (Fraunhofer Inst. für Mikroelektr. Schaltungen),
Finkenstr. 1, D-4100 Duisburg 1 (West Germany)
In collaboration with MPI München, work is in progress on the modelling of devices on high resistivity silicon [13]. Apart from this, CMOS front-end electronics for silicon detectors is developed.

IMEC (Interuniv. Micro-Electronic Center),
Kapeldreef 75, B-3030 Leuven (Belgium)
In collaborations with Canberra Semiconductor, with CERN and with others, several projects are under way concerning silicon detector technology, read-out electronics and integrated pixel detector design. Some publications will appear in the proceedings [14] of the recent "Workshop on silicon pixel detectors" held in IMEC. In sect. 3 a few more details will be shown.

NMRC (National Microelectronics Research Council),
IE-Cork (Eire)
A program for development of GaAs microstrip detectors is under way in collaboration with the CERN LAA Group.

RAL (Rutherford Appleton Laboratory),
GB-Didcot OX11 0QX (Great Britain)
The electronics group at RAL is engaged in scientific and practical support in microelectronic design for university groups in various disciplines. For

high-energy physics this has already resulted in the development of several detector read-out chips and, in addition, a pixel detector development program has started [15]. A HEP group is working on the installation of a large pixel detector at the SLC [16], using commercial CCD from GEC.

SI (Senter for Industriforskning),
P.O. Box 124, Blindern, N-0314 Oslo 3 (Norway)

SI has projects in detector technology development in collaboration with AME and others. There are also projects in the design of CMOS front-end chips for particle detectors, in particular the SAT detector in the DELPHI Experiment [17].

Southampton University,
GB-Southampton SO9 5NH (Great Britain)

Silicon detector process development is undertaken in collaboration with RAL. The Department of Electronics and Computer Science has a processing facility with an operational n-well CMOS process.

TUM, Technische Universität München
D-8046 Garching (West Germany)

The silicon detector activity in TUM is embodied by Dr. J. Kemmer who has (partly) moved his activity to MBB.

2.3 High-energy physics "user" groups

The development activities in the commercial or non-profit institutes mentioned before, have been initiated by the user groups. Very few user groups in Europe have themselves engaged in device manufacturing. This situation, in our opinion, favours technological transfer and early commercial availability of detectors and electronics. A "disadvantage" may be that an important fraction of the development cost is not hidden in the general budget but comes out explicitly.

Aarhus University, Institute of Physics,
DK-8000 Aarhus C (Denmark)

A continuous series of experiments with silicon and germanium detectors in high-energy beams has contributed to a better understanding of the energy loss and the energy deposit of minimum ionizing particles (m.i.p.) in these detectors [18]. Channeling and the emission of secondary X-rays have been studied. Aarhus has a close collaboration with the detector laboratory PHASE of the CRNS (Strasbourg), but this lab is presently not directly working on the development of segmented detectors for high-energy physics.

Institut für Hochenergiephysik der Akademie der Wissenschaften der DDR,
DDR-Berlin-Zeuthen (East Germany)

Prototype silicon microstrip detectors have been produced in collaboration with WF Berlin [19].

CERN (Organis. Europ. pour la Rech. Nucléaire),
CH-1211 Genève 23 (Switzerland)

Because most experiments take place at CERN, several groups use CERN as nominal base, while actually funding and technical support come from elsewhere. There are two indigenous groups. One is in the EP-Division and is responsible for the DELPHI silicon microvertex detector. It developed with AME-SI the integrated AC coupling [9]. New projects in this group concern a double-sided microstrip detector for a possible DELPHI upgrade and a true two-dimensional pixel detector development. The other group is our own group in the EF-Division, which is at present engaged in the UA2 silicon hodoscope (inner and outer layer) [20], provides support for continuing extension of the silicon detectors in the Omega spectrometer and proceeds with detector and electronics developments in the frame of the CERN LAA detector R & D project; leader professor A. Zichichi [21]. In this LAA effort the accent is on various microelectronic circuits for detector read-out, in particular for silicon pixel detectors. Also the radiation effects on detectors and read-out are studied.

Universität Hamburg, I. Institut für Exp. Physik,
D-2000 Hamburg 50 (West Germany)

Large area, surface barrier silicon detectors are developed for use in calorimeters. The detectors are built in the institute [22] which participates in the SICAPO Collaboration and in the H1 Experiment at HERA.

DESY (Hamburg),
Notkestr. 85, D-2000 Hamburg 52 (West Germany)

A large project is in preparation for the ZEUS calorimeter. A "pre-shower" detector with a total silicon detector area of 40 m² is to be constructed using the "tile detectors" shown in fig. 2. Prototype detectors from several commercial suppliers are presently tested. Preamplifiers will be mounted close to the detectors. In spite of the large number of channels, multiplexing is not applied at the front-end level because of the short time between beam crossings. A comprehensive overview of the ZEUS calorimetry is given by M. Derrick in this meeting.

Imperial College, Blackett Lab.,
GB-London SW7 2AZ (Great Britain)

The silicon detector group works in collaboration with RAL, Micron Semiconductor Ltd., and the processing group in Southampton University. Projects on microstrip detectors and silicon drift detectors are supported by an effort in device modelling.

Politecnico di Milano,
Piazza L. da Vinci 32, I-20133 Milano (Italy)

This group, which also includes people from Pavia University, engages mainly in special electronics for silicon detectors and in silicon drift chambers. The concept of the silicon drift chamber was proposed by Professor E. Gatti and P. Rehak [23].

INFN and Università di Milano,
Via Celoria 16, I-20133 Milano (Italy)

Several groups, participating in different experiments are engaged in silicon detector applications, both at CERN (Omega, DELPHI) and at Fermilab.

MPI, Max-Planck-Institut für Phys. and Astrophys.,
Föhringer Ring 6, D-8000 München 40 (W. Germany)

Following the successful application of the silicon microstrip detector in the ACCMOR Charm Physics Experiment (NA11-NA32) MPI has started several R & D projects [24] in collaboration with J. Kemmer (MBB and TUM), IMS Duisburg and others. A double-sided microstrip detector, incorporating AC coupling elements has been made and tested. A fully depletable CCD with junction capacitors has been tested [25]. The

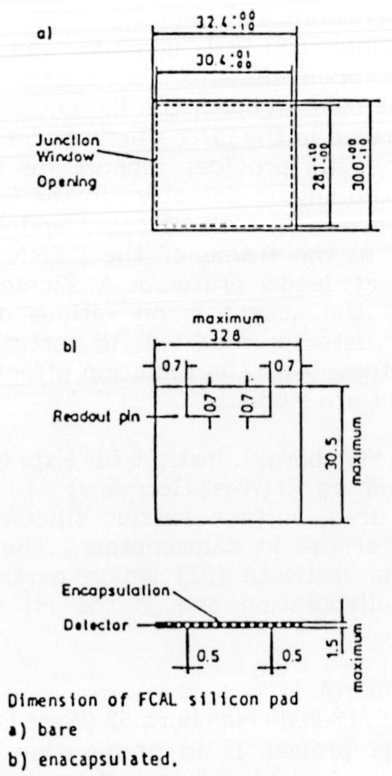

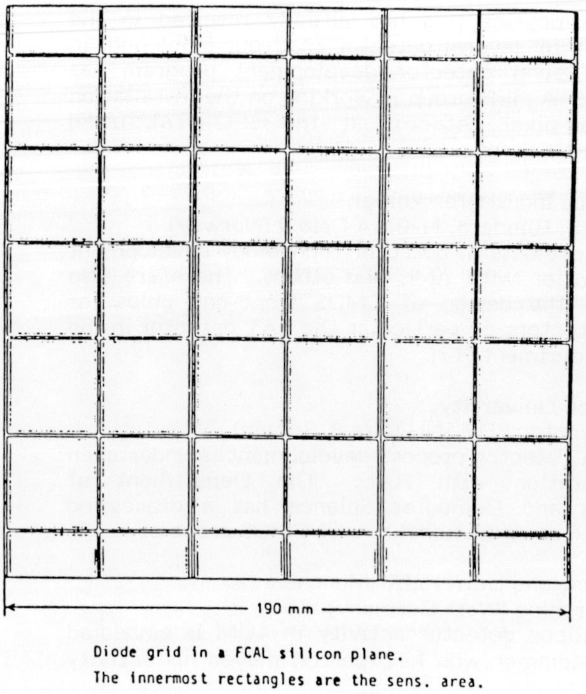

Fig. 2 Design drawing of the detector for the ZEUS calorimeter. The dimensions of each diode element at top left are 32.4 x 30.0 mm². The elements are mounted side to side in a "bathroom pattern".

basic configuration of such a device is illustrated in fig. 3. A potential is created at some depth in the bulk, under every third diode (φ_1 in fig. 3). By clocking the potentials φ, one can shift in a controlled way the signal charge packets towards the n^+ read-out contact. This structure can also be applied to build a true two-dimensional detector. Another example of such deep depletion devices is shown in fig. 4. The so-called DEPMOS and DEPFET devices use signal charges from the bulk instead of charges from the gate to steer the source-drain current. This is not a totally new idea [26] but will at present give better results due to the device miniaturization. The smaller capacitance should enable a faster response, in particular if the device is connected to a small pixel detector element, or to the output node of a silicon drift detector. At MPI, several projects concern the development of electronic components: JFET and MOSFET directly in the high resistivity silicon, JFET added to a standard CMOS process and various ASIC in CMOS. P. Sharp's talk about electronics will certainly mention the CAMEX-64 preamplifier circuit, developed by MPI and IMS Duisburg.

INFN, Sezione di Pisa,
S. Piero a Grado, I-56010 Pisa (Italy)

The Pisa Group participates in the ALEPH Experiment at CERN and the CDF Experiment at Fermilab. In both experiments there is a need for

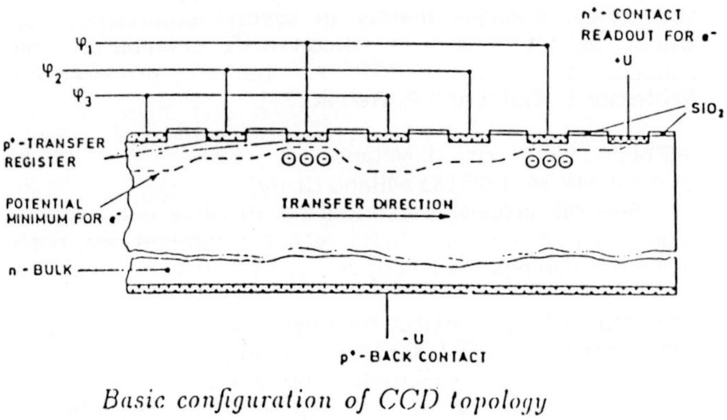

Basic configuration of CCD topology

Fig. 3 Fully depletable CCD structure, using p^+n junction capacitors with a buried potential well, created by the reverse biased p^+ back contact.

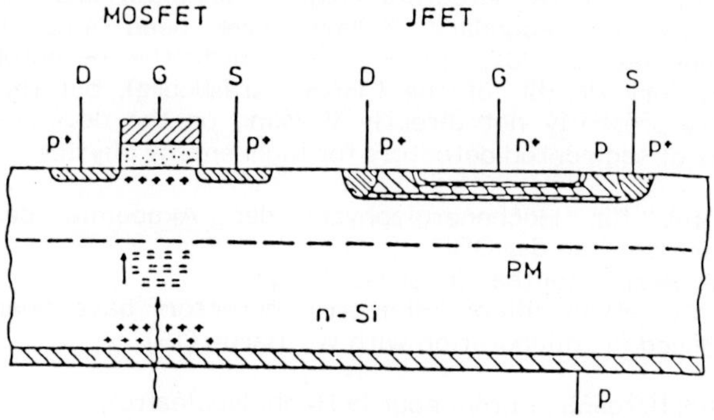

DEPMOS and DEPFET detector–amplification structures.

Fig. 4 Integrated transistor devices which give the possibility to amplify the charge signal generated in the bulk of the silicon wafer.

two-dimensional silicon vertex detectors. The development of double-sided surface barrier and planar detectors is pursued. A test set-up is shown in fig. 5, in which a surface barrier double-sided detector was operated directly inside the Tevatron vacuum. Good signals could be obtained only if some electrical shielding was surrounding the detector. Therefore, it has been decided for the future to mount the detector in a thin-walled metal box with a separated vacuum. This will hardly increase the scattering while providing safety against possible vacuum contamination.

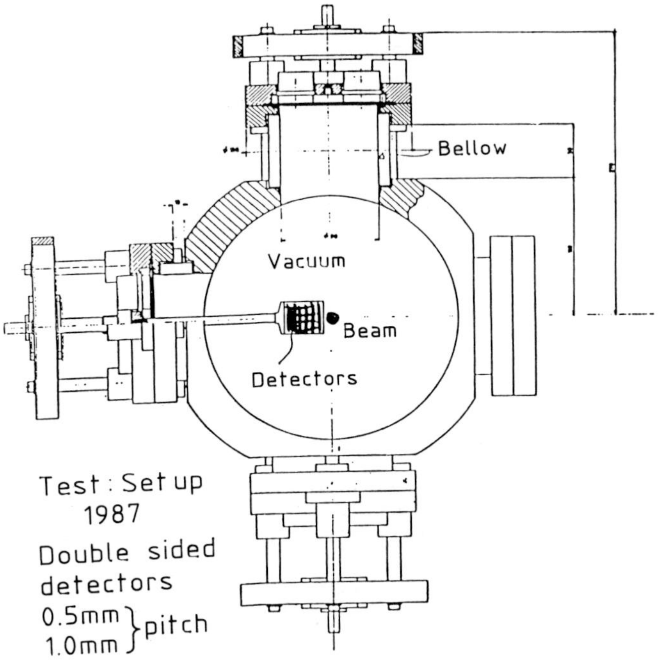

Fig. 5 Mounting of a double-sided surface barrier detector in the Tevatron beam pipe.

In an experiment with an electron source it is found that a double-sided detector does not really give unambiguous two-dimensional coordinates in case of multiple hits, but that there is some probability that similar pulse heights are correlated. This result is shown in fig. 6, where it should be noted that the charge on front (J-side) and on back (ohmic side) is the sum of signals on several adjacent strips [27]. It should be clear that double-sided detectors are not a definitive solution for two-dimensional position detection of relativistic particles, but they may be helpful as long as no better alternatives exist. For photon detection there is at present certainly no alternative.

2.4 Other activities

Silicon detector users are setting up small testing facilities in several places: Cambridge, Genoa, Perugia, etc. It seems that eventually silicon position detectors will be as widespread in high-energy particle physics as they are in nuclear physics. The likely difference is that HEP silicon detectors, with their segmentation, are more customized than the fairly standardized nuclear detectors.

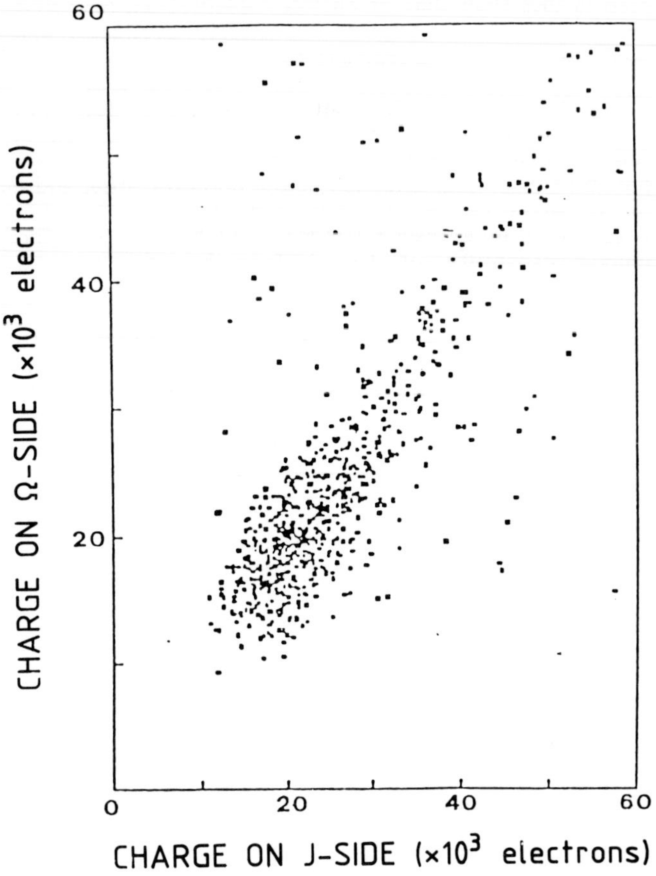

Fig. 6 Cluster charge on ohmic side of planar detector versus cluster charge on junction side. The signals were obtained by ^{90}Sr irradiation.

3. TRENDS IN THE DEVELOPMENT OF DETECTOR TECHNOLOGY

Detector development can be reviewed with the manufacturing technology as a guideline, or one may look at it from the viewpoint of applications (sect. 4).

3.1 Surface barrier technology

Very soon after purified high-resistivity monocrystalline silicon became available, it was found that a rectifying structure is obtained on an etched n-type surface after evaporation of a thin gold layer (~ 1955). A thin aluminium layer, on the contrary, produces an ohmic contact on high-resistivity n-type, but a rectifying contact on high-resistivity p-type. The characteristics of these surface barrier contacts depend very much on the chemical processing, but it turned out to be relatively easy to produce excellent particle detectors for energy spectrometry. The most difficult part is the edge protection, which is needed to avoid large surface leakage currents. One uses special expoxy resins, which can create n-type or p-type surfaces. Edge protection by thermally grown SiO_2, which is nowadays customary in silicon devices, has been used recently to produce large area silicon detectors for calorimetry [22]. The aim of this development is to produce low-cost detectors. However, it is questionable if the cost of a gold evaporation step is less than that of an ion implantation, now that the implantation equipment is widely available, and well adapted to mass production.

Although the first silicon microstrip detectors were made with the surface barrier technology [1,2,3] the planar technology offers more extensive possibilities of segmentation of the detector area, and a much better geometrical precision (typically ≦ 1 μm). Also the thin entrance window, which once was the unique capability of the surface barrier, can be obtained in ion-implanted devices [28].

The surface barrier technology is likely to remain attractive to experimental scientists who prefer to produce their own detectors in a simple in-house facility. Also, institutions in developing countries may prefer to use this type of detector, because of the modest investment needed for the production facility.

In Europe, at the University of Hamburg and in the INFN Pisa one continues to develop the fabrication of surface barrier detectors for use in HEP. For nuclear applications, such developments continue in many other places as well.

3.2 Diffused junction detector technology

From the earliest days of semiconductor technology the diffused junction detector has been manufactured, and its continuous study has contributed to the understanding of important issues, like surface conductivity (accumulation and inversion), reverse current components, crystal defects, etc.

Again, the edge protection of the diffused diode structure is the critical problem, and solutions have been found by making a mesa structure, or by ultrasonically cutting small diodes from a homogeneously-formed big diffused diode. Also guard ring structures and double-diffused diodes have been fabricated.

At present, dopant diffusion has been largely abandoned to the profit of dopant-ion-implantation, because the high temperature needed for diffusion together with the unavoidable impurity content of the dopant source lead to degradation of the original bulk silicon material. This degradation shows up as a high reverse diode current. Also, the diffusion profile cannot be controlled as closely as an implantation profile, and it is generally thicker: several hundreds of nanometres compared to ≦ 100 nm, using low-energy ion implantation.

Some manufacturers offer, nevertheless, competitive particle detectors or photodiodes made by diffusion technology, because they succeeded in finding an effective way of avoiding contamination by the use of gettering. One uses roughening of the wafer backside, heavy phosphorus doping, or other empirical methods.

To our knowledge, there are no European manufacturers of diffused detector structures, and certainly none who are making developments for elementary particle physics.

3.3 Planar, ion-implanted detector technology

In the middle of the sixties ion implantation was already used for making silicon diode particle detectors, as told by Mayer [29]. Kalbitzer et al. [30] in Heidelberg succeeded probably first in producing a detector with good energy resolution, and several other groups made research to obtain good implanted devices [31]. These efforts did not lead to commercially available detectors, except for the ion-implanted resistive surface position sensitive detector. However, this research supported the introduction of ion implantation in microelectronics and from 1970 this became a much more exciting field than that of particle detectors.

It took until 1980 for the silicon detector community to realize how useful the recent developments in silicon microelectronics could be for the detector manufacturing. Even today, there remains well-intented scepticism, probably based on the infructuous efforts in the early seventies. The key of the success is not just ion implantation, but the adoption of rigorously clean silicon processing, as developed for the manufacturing of MOS devices with stable characteristics. J. Kemmer published [1] detector performance with low noise at room temperature and very low leakage current, a few nA per cm^2 for fully depleted detectors of 300 μm thickness. Currents around 1 nA cm^{-2} are at present obtained in (selected) commercial devices. Surface barrier detectors usually have currents of at least an order of magnitude higher.

The real potential of the planar technology is in the possibility to produce a great variety of sophisticated devices, which marry the sensor function with the electronics read-out functions. This potential is worth developing in view of large-scale application of such devices in radiation imaging, in elementary particle physics, in synchrotron radiation experiments, in microbiology and in space astronomy.

The first steps towards sophistication were the microstrip detector [1,5] and the silicon drift detectors [23]. The microstrip detector with charge division, using floating elements [32] holds the palm of spatial precision in active particle detection, 2.5 μm, which approaches the precision of the passive nuclear emulsion. Besides the capacities and resistances inherent to a microstrip layout one may add capacitors or resistors, using polysilicon and metal structures [9] which enables to make capacitively-coupled (AC) read-out.

The next stage in detector development consists in adding active components into the high-resistivity detector. This is particularly useful for detectors with small segments, where external connections would add significant capacity to the amplifier input, and therefore degrade the noise performance. Several groups in Europe are working on the technology and characterization of transistors directly built in high-resistivity silicon, or built in a low-resistivity well: to our knowledge, such research is done by CERN/IMEC, RAL/IC/Southampton, Delft TU, and MPI/MBB in Munich. This last group has described a number of different structures and built and tested some of them [11, 25]. Using a circular transistor structure, 100 μm in diameter on backside depleted silicon, they showed also that direct signal amplification can lead to very low noise detection (60 e^- r.m.s. for ^{55}Fe). By creating a potential well deep under the transistor gate, the signal can be read out non-destructively, provided the carrier lifetime is sufficiently long (in present detector processing it is ~ 1 ms).

3.4 Silicon pixel detectors: microelectronic detectors

Once it is possible to include active elements in the detector chip, either directly or in a well, the way is open for multi-element two-dimensional arrays with microelectronic read-out. Such an array detector with monolithic (or hybrid) read-out is called "pixel detector". If the read-out architecture is intelligent and permits the recognition of selected patterns, we propose to speak of micropattern detector. All groups mentioned just above are engaged in designing detectors with integrated read-out.

A Workshop on the development of silicon pixel detectors has been held recently in Leuven, Belgium [14]. For illustration, one of the designs presently in processing is shown in fig. 7 [33].

Fig. 7 Part of the design of a silicon pixel detector read-out. This read-out could be coupled to a silicon detector in a hybrid form, or could be constructed in a well in the high restivity.

There are several approaches to the read-out architecture of a silicon pixel detector, and these have already been well developed for use in optical or infrared imagers. The Direct Read-Out (DRO) consists in addressing all pixel sites sequentially via read bus lines. The Charge Coupled Device (CCD) uses shifting of signal charge through the capacitor array towards one or several output nodes by a system of clock electrodes (fig. 3). With regard to the fundamental differences between lively images and elementary particle events, a new read-out with Pattern Recognition Architecture (PRA) will be needed to recognize in an economic way the useful information between the large number of empty pixel sites.

The CCD structure presents the advantage of simplicity as well as that of being commercially available for optical applications. For applications with low duty cycle, like SLC, existing devices can be used [16]. This helps to obtain physics results soon, and to avoid years of difficult R & D.

4. DEVELOPMENT WITH REGARD TO NEW DETECTOR APPLICATIONS

In the previous sections we sketched the development of ever more sophisticated detector array structures. These find application in position sensitive vertex detectors for fixed-target experiments or in colliders. The projected hadron colliders will have a high event rate and a fierce radiation environment. No presently available detector system is capable to handle properly vertex detection for these colliders. Silicon pixel detectors have some promise with respect to resolution and rate capability, but the reduction of the power dissipation of on-board electronics and the increase of radiation survivability create the need for extensive technological development. Even so, regular exchange of detector components is to be foreseen. Already in the mild environment of the SPS proton-antiproton collider at CERN partial exchange of detectors and electronics for the UA2 inner silicon array will be ineluctable after a dose of ~ 1 kGy (100 krad) which is accumulated after 5-10 months of running.

The second field of silicon detector application is in precision calorimetry. The possibility to make small, precise segmentation enables us to have a fine-grain shower sampling with significant redundancy. The silicon layers can be made very thin, even if the electronics is mounted on top of the sensitive silicon. Probably a thickness of 1-2 mm can be ultimately obtained, but at present already 3.5 m is possible, as in the UA2 inner silicon array [20]. The good linearity of charge collection in the silicon makes the task of calibration much easier. However, the cost of the silicon devices and associated electronics limit their application to a part of the calorimeter. A significant reduction of this cost will only be possible once a reliable design enables high yield production, and once the combined particle physics community guarantees a regular, high volume production for at least several years. Given that an average factory nowadays has a throughput of 500-1000 wafers/day, corresponding to ~ 10 m^2 of silicon devices, with maybe 50 personnel and 20 M$ investment, one could produce ~ 10^4 m^2 of sophisticated silicon detector in ~ 3 years for a total cost of ~ 80 M$, including an estimated 40 M$ for the silicon itself. Although this seems a large sum of money, it could be justified if the use of silicon in the calorimeter would provide significant advantages in compactness, read-out organization and calibration.

In between the extremes of high precision vertex detectors and very large area calorimeters, there are a number of applications which require pads of millimetre dimensions and moderate area. Examples are luminosity detectors, forward calorimeters, detectors for multiplicity measurement like the UA2 arrays, etc. At present, several projects in this intermediate range are under way, and these prepare the ground for a transition to large-scale applications, if these are judged to be needed.

5. CONCLUSION

Europe still seems to be leading in R & D for silicon devices in particle physics. The realization of new devices and new applications is accelerating, in spite of a relatively modest funding, which is estimated (for generic R & D) to be < 0.5 M$/year. This seems little if compared to expenditure for advanced detector development, e.g. by space research agencies.

In detector manufacturing there is a clear change from standard products towards "customized" devices. Using computer-aided design tools this change can be made without significantly higher production cost. If further development enables us to integrate electronics functions, the custom aspect will be even stronger. Present commercial detector manufacturers do not seem to be in a position to support the investments needed for compatible electronics production. The profit margin and the total turnover are both too small for this. The only way one could expect this small industry to engage in integrated electronics production is by placing development contracts which can partly cover investment cost. The alternative is to convince microelectronics manufacturers to produce high-resistivity detectors. This may turn out to be equally expensive.

In our opinion, it will not be a satisfactory solution to produce industrial quantities of devices in research laboratories. If one takes the hidden cost of investment and personnel into account, the expenses will turn out to be a multiple of those for industrial production. Also, in industry there is a better chance that technology transfer leads to applications outside the field of particle physics.

Acknowledgements

We are grateful for the discussions with many silicon enthusiasts, manufacturers and users alike, who helped us with providing data for this overview.

REFERENCES

[1] S.R. Amendolia, G. Batignani, F. Bedeschi, E. Bertolucci, L. Bosisio, C. Bradaschia, M. Budinich, F. Fidecaro, L. Foà, E. Focardi, A. Giazotto, M.A. Giorgi, M. Givoletti, P.S. Marrocchesi, A. Menzione, D. Passuello, M. Quaglia, L. Ristori, L. Rolanti, P. Salvatori, A. Scribano, R. Stanga, A. Stefanini and M.L. Vincelli, A multi-electrode silicon detector for high-energy physics experiments, Nucl. Instr. and Meth. 176 (1980) 449.

[2] W.K. Hofker, D.P. Oosthock, A.M.E. Hoeberechts, R. van Dantzig, K. Mulder, J.E.J. Oberski, L.A.Ch. Koerts, J.H. Dieperink, E. Kok and R.F. Rumphorst, The checker board counter: a semiconductor dE/dx detector with position indication, IEEE Trans. Nucl. Science NS-13, Nr 3 (1966) 208.

[3] E.H.M. Heijne, L. Hubbeling, B.D. Hyams, P. Jarron, P. Lazeyras, F. Piuz, J.C. Vermeulen and A. Wylie, A silicon surface barrier microstrip detector designed for high-energy physics, Nucl. Instr. and Meth. 178 (1980) 331.

[4] J. Kemmer, Fabrication of low-noise silicon radiation detectors by the planar process, Nucl. Instr. and Meth. 169 (1980) 449.

[5] J. Kemmer, P. Burger, R. Henck and E.H.M. Heijne, Performance and applications of passivated ion-implanted silicon detectors, IEEE Trans. Nucl. Science NS-29 (1982) 733.

[6] P. Jarron and M. Goyot, A fast current sensitive preamplifier (MSD2) for the silicon microstrip detector, Nucl. Instr. and Meth. 226 (1984) 156.

[7] J.P. Avondo, P. d'Angelo, P. Jarron, P.F. Manfredi, V. Speziali, Low-noise front-end electronics for microstrip position sensors and vertex detectors for experiments at TeV energies, Nucl. Instr. and Meth. A241 (1985) 107.

[8] J.T. Walker, S. Parker, B.D. Hyams and S.L. Shapiro, Development of high density read-out for silicon strip detectors, Nucl. Instr. and Meth. 226 (1984) 200.

[9] M. Caccia, L. Evensen, T.E. Hansen, R. Horisberger, L. Hubbeling, A. Peisert, T. Tuuva, P. Weilhammer and A. Zalewska, A Si strip detector with integrated coupling capacitors, Nucl. Instr. and Meth. A260 (1987) 124.

[10] P.G. Rancoita and A. Seidman, Silicon detectors in electromagnetic and hadronic calorimetry, Nucl. Instr. and Meth. A 263 (1988) 84.

[11] J. Kemmer and G. Lutz, New structures for position sensitive semiconductor detectors, Nucl. Instr. and Meth. A (in print).

[12] S.A. Audet and E.M. Schooneveld, High-purity silicon radiation-sensor array, Nucl. Instr. and Meth. A (in print).

[13] M. Berger and E. Stein, Two-dimensional simulation of integrated detectors and electronics devices, Nucl. Instr. and Meth. A 253 (1987) 382.

[14] Proceedings of the "Workshop on silicon pixel detectors", Leuven, Belgium, 1988, to be published in Nucl. Instr. and Meth. A.

[15] S.L. Thomas and P. Seller, A silicon pixel detector with routing for external VLSI read-out, to be published in Nucl. Instr. and Meth. A.

[16] C.J.S. Damerell, R.L. English, A.R. Gillman, A.L. Lintern, D. Phillips, Rong G., C. Sutton, F.J. Wickens, G. Agnew, P. Clarke, S. Hedges and S.J. Watts, A vertex detector for SLD, to be published in Nucl. Instr. and Meth. A.

[17] L. Bugge, T. Buran, F. Buschman, M. Dam, L. Evensen, B. Fjeld, E. Gjøtterud, T.E. Hansen, G. Maehlum, F. Nordby, E. Nygard, A. Olsen, B. Skaali, G. Skjevling and J. Wikne, Large area silicon detectors with front-end electronics in VLSI as a luminosity monitor at LEP, Institute of Physics, report series 86-10, Univ. Oslo (1986).

[18] J.F. Bak, A. Burenkov, J.B.B. Petersen, E. Uggerhøj, S.P. Møller, P. Siffert, Large departures from Landau distributions for high-energy particles traversing thin Si and Ge targets, Nucl. Phys. B288 (1987) 682.

[19] R. Leiste, Silicon semiconductor coordinate detectors, in Proceedings of Int. Symp. on Position Detectors in High-energy Physics, Dubna, September 1987, Joint Institute for Nuclear Research, Dubna (1988).

[20] R. Ansari, E. Beuville, K. Borer, P. Cenci, A.G. Clark, A. Federspiel, O. Gildemeister, C. Gößling, K. Hara, E.H.M. Heijne, P. Jarron, P. Lariccia, B. Lisowski, D.J. Munday, T. Pal, M.A. Parker, N. Redaelli, P. Scampoli, V. Simak, S.L. Singh, T. Vallon-Hulth and P.S. Wells, The silicon detectors in the UA2 Experiment, to be published in Nucl. Instr. and Meth. A.

[21] A. Zichichi, The LAA project, CERN/LAA 88-2 (1988) 176-184.

[22] E. Fretwurst, R. Grube, G. Lindström and J. Nagel, Development of large area silicon detectors - Special properties and radiation stability, Nucl. Instr. and Meth. A253 (1987) 467.

[23] Pavel Rehak, Emilio Gatti, Antonio Longoni, J. Kemmer, Peter Holl, Robert Klanner, Gerhard Lutz and Andrew Wylie, Semiconductor drift chambers for position and energy measurements, Nucl. Instr. and Meth. A235 (1985) 224.

[24] J. Kemmer and G. Lutz, New detector concepts, Nucl. Instr. and Meth. A253 (1987) 365.

[25] L. Strüder, G. Lutz, M. Sterzik, P. Holl, J. Kemmer, U. Prechtel, T. Ziemann and P. Rehak, First tests with fully depleted p-n CCDs, IEEE Trans. Nucl. Science NS-35 (1988) 372.

[26] T.C. Madden and G.L. Miller, A particle detector providing charge gain by means of a form of local transistor action, in: Semiconductor Nucl. Part. Detectors and Circuits, Ed. by W.L. Brown, W.A. Higinbotham, G.L. Miller and R.L. Chase, Publ. 1593, Nat. Ac. Sciences, Washington (1969) 314.

[27] G. Batignani, F. Bosi, L. Bosisio, A. Conti, E. Focardi, F. Forti, M.A. Giorgi, G. Parrini, E. Scarlini, P. Tempesta, G. Tonelli and G. Triggiani, Double-side read-out silicon strip detectors for the ALEPH minivertex, Report INFN-PI/AE 88-4, Pisa (1988).

[28] P. Burger, K. De Backker and W. Schoenmaekers, Performance of large area particle detectors, SPIE Proc. 591 (1985) 38.

[29] J.W. Mayer, Ion implantation in: Semiconductor nuclear particle detectors and circuits, Ed. by W.L. Brown, W.A. Higinbotham, G.L. Miller and R.L. Chase, Publ. 1593, Nat. Ac. Sciences, Washington (1969) 287.

[30] S. Kalbitzer, R. Bader and H. Herzer, The fabrication of high-quality silicon junction detectors by low-energy ion implantation, Zeitschr. für Phys. 203 (1967) 117.

[31] P. Zandveld, Some properties of ion-implanted p-n junctions in silicon, Solid St. Electron. 19 (1976) 659.

[32] U. Kötz, K.U. Pösnecker, E. Gatti, E. Belau, D. Buchholz, R. Hofmann, R. Klanner, G. Lutz, E. Neugebauer, H.J. Seebrunner, A. Wylie and J. Kemmer, Silicon strip detectors with capacitive charge division, Nucl. Instr. and Meth. A 235 (1985) 481.

[33] B. Dierickx, XYW detector: a smart two-dimensional particle sensor, to be published in Nucl. Instr. and Meth. A.

SILICON PIN DIODE ARRAY HYBRIDS FOR CHARGED PARTICLE DETECTION*

STEPHEN L. SHAPIRO, WILLIAM M. DUNWOODIE

Stanford Linear Accelerator Center, Stanford University, Stanford, California 94309

JOHN F. ARENS, J. GARRETT JERNIGAN

Space Sciences Laboratory, University of California, Berkeley, California 94720

STEPHEN GAALEMA

Hughes Aircraft Company, Carlsbad, California 92008

ABSTRACT

We report on the design of silicon PIN diode array hybrids for use as charged particle detectors. A brief summary of the need for vertex detectors is presented. Circuitry, block diagrams and device specifications are included.

INTRODUCTION

Hybrid silicon PIN diode arrays are promising devices for vertex and tracking detectors. Hybrid arrays having separate detector and readout structures have been developed for infrared radiation detection over the last 12 years. The technology can be extended readily to charged particle detection. The existing technology and the needs of detectors for use in high energy physics will be discussed.

THE SSC ENVIRONMENT

High resolution vertex and tracking devices at the SSC will be very important for revealing the detailed structure of complex events, and, in particular, for the reconstruction of secondary vertices close to the primary interaction point. The expected combination of high luminosity and large particle multiplicity will present a number of problems for these devices and their associated electronics because of the high radiation environment which will prevail in the vicinity of the collision point. Therefore, it is necessary to develop vertex and tracking devices which are radiation hard, have fast response times and have both fine spatial and multiple particle resolution.

The SSC beam crossing will occur approximately every 15 ns, and interesting events will occur every 1,000 to 10,000 crossings. The time of the hit will have to be recorded to an accuracy of one crossing time, and the readout time of these detectors should be less than the mean interval between interesting events.

For a proton–proton total cross section of approximately 100 mb[1)] the SSC will yield approximately 10^8 interactions per second when operating at its design luminosity of $10^{33}/\mathrm{cm}^2/\mathrm{s}$. At a distance of 5 cm from the beam, the resulting absorbed radiation dose will be approximately 1 Mrad per year.

DEVICE CANDIDATES

Candidates for vertex and tracking devices at the SSC include wire chambers, scintillating fiber detectors, silicon microstrip detectors, double-sided microstrip detectors, silicon drift chambers and pixel devices in the form of CCD arrays, monolithic silicon arrays and hybrid microdiode arrays. Silicon devices offer substantially higher space, momentum and multiple track resolution than wire chambers. It is also generally acknowledged that pixel devices are preferable to strip devices or drift devices for two track resolution.

Pixel devices, in particular silicon diode arrays, are a natural choice for vertex detectors. These devices provide three-dimensional coordinate information with spatial resolution of a few microns. As the distance from the collision point to the tracking detector increases, other devices mentioned above become attractive candidates for a variety of reasons.

A vertex detector based on the use of a pixel device would provide efficient trackfinding with a minimum number of layers in the high multiplicity environment of the SSC due to the three-dimensional nature of the coordinate information provided. The absence of ambiguities in coordinate matching which are present for non-pixel devices allows the number of detector layers to be minimized, thereby reducing the size and cost of the vertex detector.

The only pixel devices presently in use in high energy physics experiments are CCDs.[2)] These devices are inappropriate to the SSC environment for a variety of reasons. They have a small signal size of about 1,000 elec-

* Work supported by the Department of Energy, contract DE–AC03–76SF00515.

trons which necessitates their being cooled, renders them extremely susceptible to degradation caused by radiation, and requires low noise readout electronics. CCDs have varying efficiency across their face, no off gate during readout resulting in a time/space ambiguity and a serial readout resulting in a readout time of about 10 ms for a device with 10^5 pixels.

A number of researchers are presently working on the problem of fully integrated pixel detectors in monolithic arrays.[3,4)] These arrays have their readout electronics fabricated on the same high resistivity silicon as the detector diodes. Success in these efforts will result in detectors having the minimum thickness possible. However, there will be loss of flexibility. Each change in either the detector specification or the readout specification will result not only in a circuit change, but also in a re-analysis of the production process, since they are now intimately coupled. Monolithic arrays must be fabricated in high resistivity facilities that can handle both high resistivity silicon and also standard VLSI processing. There are far fewer of these facilities worldwide than those which can produce more conventional circuits.

MICRODIODES: SILICON PIN DIODE ARRAY HYBRIDS

An architecture which is appropriate for charged particle detection at the SSC is that of a hybrid.[5)] The charged particle detector, a silicon PIN diode array, and the readout electronics are constructed as two separate silicon chips, each optimized for its specific function. The two chips, indium bump bonded together, are the basic building blocks of a detector array.

The indium bump bonding process is one in which each diode of the detector array is bonded to an independent amplifier readout circuit on a mating VLSI chip via an array of aligned indium metal bumps that cold weld under pressure to form ohmic contact. This process allows flexibility in detector and readout electronics specifications.[6)] For instance, a change in the leakage current specification of the detector array will not affect the readout electronics, nor will a change in the VLSI chip oxide thickness to accommodate a radiation hardness specification affect the detector array. Figure 1 is a schematic representation of a Microdiode hybrid.

The present readout chip allows random access to any pixel, which operates as an independent detector. By virtue of its geometry alone, each pixel detector (PIN diode) provides about 3,000 times the radiation hardness of a microstrip detector. To complement this increase in the radiation hardness of the detector diodes, one of the present readout chips (the 10 × 64 array) has been fabricated in a technology which is radiation hard to 1 Mrad of Cobalt 60 gamma rays.

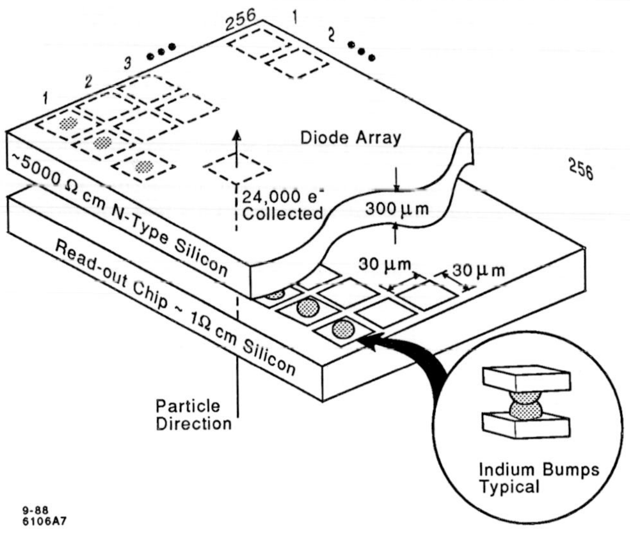

Fig. 1. Schematic representation of a hybrid detector.

The hybrid operates at room temperature, eliminating the need for a cryostat, and easing the problems of installation, alignment, and accessibility for power and control cabling.

The detector thickness itself can be optimized. The PIN diode array can be fabricated to give the requisite signal (80 electrons per μm of silicon traversed), while the VLSI chip can be thinned after bonding to less than 100 μm if necessary. This feature is important for minimizing multiple-scattering effects, and also for limiting the number of adjacent pixels turned on by tracks incident at large angles upon the active silicon volume.

It should be noted, that the diode anodes are indium bump bonded to the silicon readout chip, leaving the cathode available to be etched into a pattern of pads or strips. Signals picked up on the cathode pads would be available for fast triggering or for reducing the readout time by using the random access nature of the readout electronics. System considerations will determine if the cathode signals are useful.

In the present generation of readout chips, the power use during the write cycle is essentially zero. Power is used only when reading, and cooling needs are thus lessened.

Hardware Description

There are two chip geometries currently being built into PIN detector arrays; a 10 × 64 array with 120 μm square pixels, and a 256 × 256 array with 30 μm square pixels. However, 512 × 512 readout arrays have recently been fabricated by Hughes Aircraft Co. The diode arrays are identical as to process and differ only in that they must mate with their respective readout chips. The readout chips are similar in function, but differences between them exist. For instance, the 10× 64 array has been fabricated and optimized to be radiation hard to 1 Mrad at a temperature of 10K. It has a random access architecture in that a unique setting of its address lines will select one and only one row. The 256 × 256 array, on the other hand, is random access in that the pixels are addressed via row and column shift registers. This feature makes addressing a given pixel more complicated but allows an easy implementation of a sparse scan algorithm. This particular readout chip has been optimized to collect electrons but is bipolar at the signal levels we expect.

external circuit necessary for the correlated double sampling technique used for measuring the charge collected on the gate of the signal MOSFET. The voltage difference across the correlated double sampling capacitor C_{CDS} is determined by the difference in the output prior to the reset (time 1) and after the reset (time 5). S1 is held open during the reset, insulating C_{CDS} from the large reset pulse. The analysis time per pixel is 1 μs.

Figure 5 is a block diagram of the high energy physics data acquisition system. A Sun Microsystems Sun-3/110LC-4 workstation controls a VME system crate housing a Sun 3/E CPU, a Motorola 68020 bus converter board,

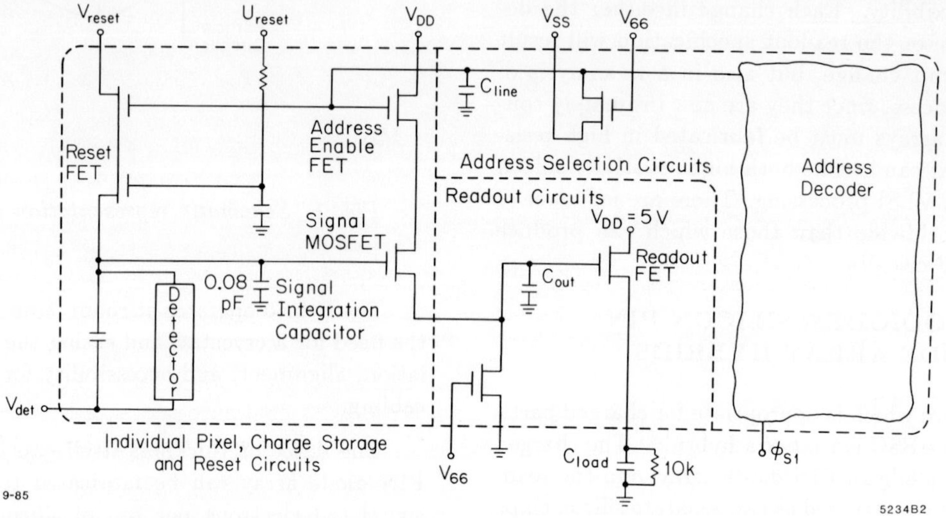

Fig. 2. Schematic drawing of the readout electronics of the 10 × 64 *readout array.*

Figure 2 is a schematic diagram of the MOSFET circuit of the 10 × 64 readout array. The diagram has been divided into its several functional portions. The section replicated for each pixel contains four MOSFETs. Signal charge is generated by the detector diode, and is fed to the gate of the signal MOSFET, where it stays until a readout is made. The pixel selection circuit indicates how a sequence of address lines can select an individual pixel by turning on the gates of the V_{DD} bias MOSFET and the enable gate of the reset MOSFET. The U_{reset} signal allows the gate of the signal MOSFET to be reset to the V_{reset} level for any pixel that is enabled by the reset MOSFET. All of the signal MOSFETs are connected to a readout MOSFET (one per column) in a source follower configuration which provides power for driving an external circuit.

Figure 3 is a simple schematic of the readout addressing architecture of the 256×256 array. This array contains only two output nodes for the entire chip. It can be seen that by turning on all pixels on a row or column, an OR circuit is created and one will observe a signal from that pixel which has the most charge.

Figure 4 indicates schematically the coordinated sequencing of the reset line and the switches S1 and S2 in the

amplifiers, ADCs, digital signal processors, and a clock generator. The digital signal processor is a Motorola DSP56001. This device acquires data at a rate of 10 MIPS, processes it, and passes it to the Sun 3/E. The bus converter board, a Parity systems AV20 dual port processor, interfaces the local analog bus (PECKBUS) to the VME standard.

The noise performance of the 10 × 64 readout chip at 10K has been measured to be 45 electrons rms referred back to the input. This is within a factor of two of the 1/f noise of the input signal MOSFET. At room temperature, the noise will probably be somewhat higher. However, with a moderate effort to maintain low system noise, a noise level of about 200 electrons is achievable. If the noise performance is as expected, detectors thinner than the traditional 300 μm thick PIN devices currently being fabricated for high energy physics would still offer excellent signal-to-noise performance.

RESEARCH PROGRESS

Dr. Arens, Dr. Jernigan, and their collaborators have designed and built cameras for 10 μm infrared wavelength.

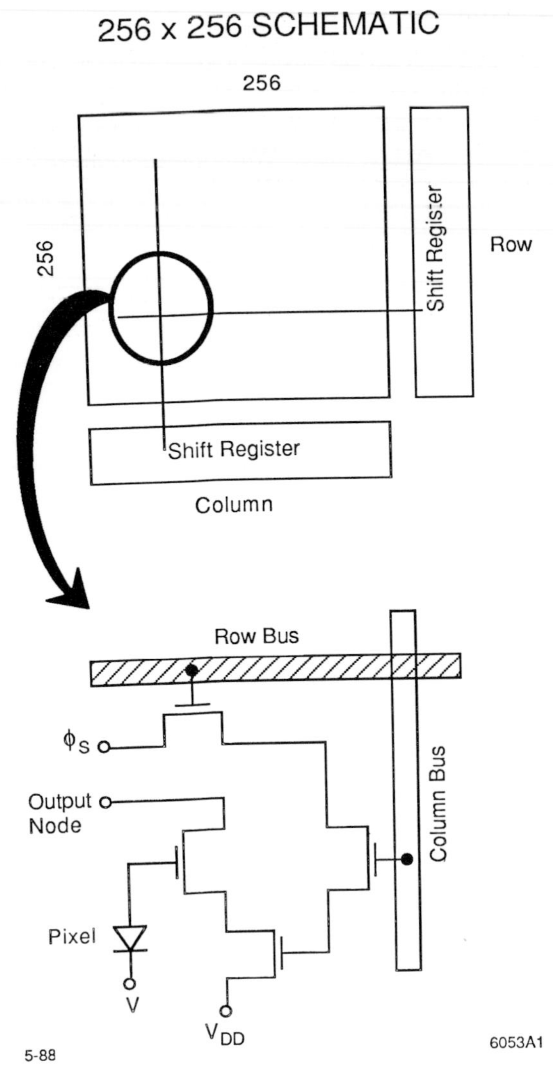

Fig. 3. Schematic representation of the readout architecture of the 256 × 256 *readout array.*

The details of one camera have been published in Applied Optics.[7] The camera, based on a 10 × 64 array of 120 μm square pixels, has been used at the University of Arizona 61″ and 90″ telescopes, the University of Wyoming 90″ telescope, and the NASA Infrared Telescope Facility on Mauna Kea, Hawaii. Two detector chips were used — an arsenic doped and a gallium doped silicon array — each bump bonded to a 10× 64 Hughes readout chip. The camera consisted of the detector, the Dewar, the readout electronics and computers, and the control computer. The computers were Sun Microsystems workstations with additional multibus cards for input and output.

In 1987, the infrared camera was reconfigured to include a germanium PIN diode array as the detector element. This detector was used at the National Laser Users Facility of the Laboratory for Laser Energetics at the University of Rochester to obtain x-ray images of laser targets.[8] Figure 6 is a photograph of one of the germanium detectors bump bonded to a 10 × 64 array.

Using Detector Development Funds from the Department of Energy, a contract was made with Micron Semiconductor to design and fabricate two silicon PIN diode arrays. The design and the masks necessary to fabricate the silicon chips have been completed. The PIN diodes are presently in production, with the first articles due this calendar year (1988). The array specifications were chosen to allow the detector chips to mate with the corresponding Hughes Aircraft Co. readout chips. The arrays being fabricated are a 10 × 64 array with 120 μm square pixels and a 256 × 256 array with 30 μm square pixels.

Additional Detector Development funds were used to begin the fabrication of the dedicated high energy physics data acquisition system described above. The amplifier/ADC/DSP56001 boards and the clock generator in the present system, however, are circuits remaining from the infrared data acquisition system. These need to be redesigned to high energy physics criteria.

A data acquisition and display software package based on the infrared system is being written. The operating system is UNIX, the DSP has been programmed in assembler language, and various control functions are written in Magic/L, an interactive language derived from Forth. The Parity AV20 dual port processor has been programmed in C, and the DSP will be reprogrammed in C.

At the present time, a system exists which can read either of our two detector geometries: 10 × 64 and 256 × 256. When the new detectors arrive, we will be ready to begin our program of testing and characterizing these devices. Figure 7 is a photograph of a 256 × 256 Hughes readout chip. Table I is a current summary of device characteristics.

FUTURE DIRECTION

Our efforts during the coming years will include testing and characterizing the new devices, designing and fabricating the remaining electronics to reflect the needs of these detectors, and exposing the new detectors to cosmic ray tests and beam tests.

To respond to data which may accumulate during the readout cycle, gating circuitry can be designed to inhibit the flow of charge between the detector and the readout chip after an event is tagged. Alternatively, an input buffer can be added to the design which leaves the detector alive but which does not allow alterations of the data being read out.

To avoid the confusion caused by data being recorded from various beam crossings, a design must be adopted which allows the data and the time of arrival to be correlated. The chips being built have been designed to provide a fast indication that a particle has been recorded, however, we have not yet explicitly studied this feature in the laboratory. This ability, interesting in its own right if confirmed, can be used in later design iterations to strobe shift registers which do, in fact, make use of the time information, while retaining the small pixel size of our present chips.

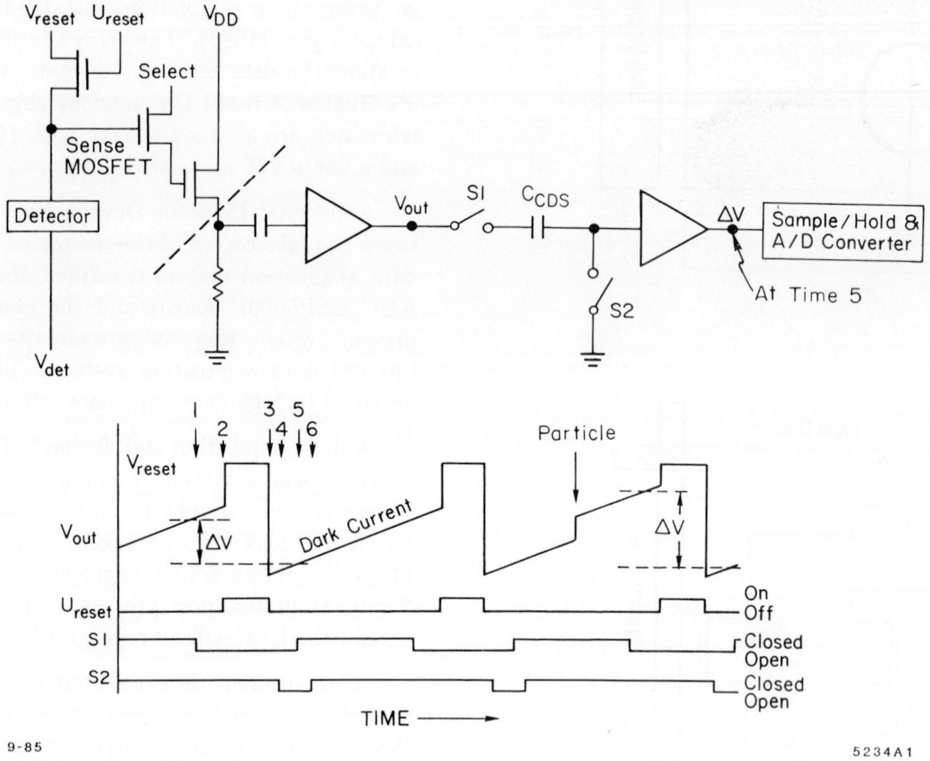

Fig. 4. Schematic drawing of the off-chip readout electronics highlighting the correlated double sampling technique, with a timing diagram included.

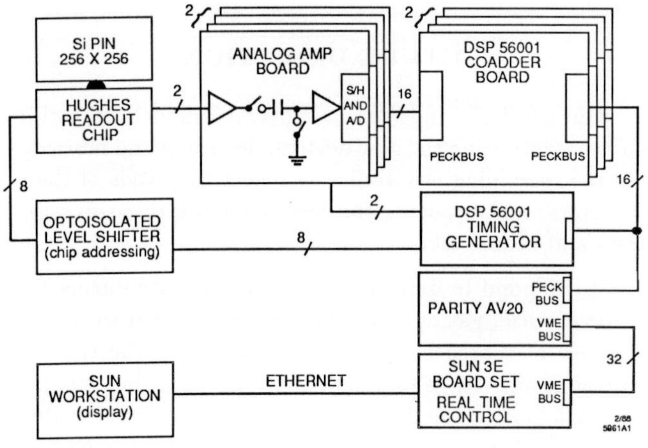

Fig. 5. A block diagram of the readout and display electronics.

Fig. 6. A photograph of a 10 × 64 readout chip bump bonded to a Germanium PIN diode array.

We also intend to propose a redesign of the Hughes readout chip to include some of these requirements. These include radiation hardness, increased readout speed, and data recording with time tagging.

We will address various data acquisition problems particular to the SSC. It is our intention to design a prototypical barrel vertex detector, including the specification of the global architecture. We hope to address questions relating to the usefulness of having signal processing local

Fig. 7. A photograph of a 256 × 256 readout chip.

Table 1. Summary of device parameters.

Array Dimension	10 × 64	256 × 256
Pixel Size	120 μm	30 μm
Detector Material	Germanium Silicon	Germanium Silicon
Number of Output Channels	10	2
Power During "Write" Cycle	0 mW	0 mW
Power During Read Cycle	10 mW	2 mW
Present Clock Speed	1 MHz	1 MHz
Theoretical Clock Speed	10 MHz	10 MHz
Readout Mode	Random Access	Random Access
Processing Power	10 MIPS/ channel	10 MIPS/ channel
Radiation Hardness	1 Mrad	?
Noise at 10K	45 electrons rms	?
System Noise at Room Temperature	200 electrons rms	?
Operating Temperature	Ge - Cool Si - Room Temp.	Ge - Cool Si - Room Temp.
Sparse Scan	No	Yes
Fast Signal	No	?
Input Gating	No	No
Smart Pixel	No	No

to the detectors, the degree of parallelism required of the readout structure, mounting and cooling schemes, and the design of the optimum host computer.

ACKNOWLEDGMENTS

The authors wish to thank Mike Peck, Ignas Ketvirtis and Will Coertnik for their assistance on this project, and Colin Wilburn of Micron Semiconductor for his continued support.

REFERENCES

1) *Proceedings of the Summer Study on the Design and Utilization of the Superconducting Super Collider*, edited by R. Donaldson and J. G. Morfin, Snowmass, Colorado (1984). *Report of the Task Force on Detector Research and Development for the Superconducting Super Collider*, SSC Central Design Group, SSC–SR–1021 (1986).

2) Experiment NA32 at the CERN SPS; Chris Damerell, Vertex Detectors, Rutherford Appleton Laboratory Report RAL 86–077, July 1986.

3) Sherwood Parker, *A Proposed VLSI Pixel Device for Particle Detection*, Paper submitted to the *International Workshop on Silicon Pixel Detectors for Particles and X-Rays*, Leuven, Belgium, June 1988.

4) G. Vanstraelen, I. Debusschere, C. Claeys and G. Declerck, *Fully Integrated CMOS Pixel Detector for High Energy Particles*, Paper submitted to the *International Workshop on Silicon Pixel Detectors for Particles and X-Rays*, Leuven, Belgium, June 1988.

5) S. Shapiro and T. Walker, *The Microdiode Array – A New Hybrid Detector*, SLD – New Detector Note, No. 122, October 1984.

6) Steve Gaalema, *Low Noise Random–Access Readout Technique for Large PIN Detector Arrays*, IEEE Trans. on Nucl. Sci., **NS-32**, No. 1, February 1985, 417.

7) John F. Arens, J. Garrett Jernigan, Michael Peck, Carl A. Dobson, Erik Kilk, John Lacy and Stephen Gaalema, *10 Micrometer Infrared Camera*, Applied Optics, **26**, 1987, 18.

8) J. Garrett Jernigan, *PIN Diode Array X-Ray Imaging*, a proposal to the National Laser Users Facility, Laboratory for Laser Energetics, University of Rochester, 1987.

SILICON TRACKING DEVICES FOR THE SSC

David R. Nygren
Lawrence Berkeley Laboratory
Berkeley, CA 94720

Abstract

The desirable properties of pixelized silicon charged particle detectors in the SSC context are discussed and compared with silicon strip detectors. A device architecture matched to the SSC environment, the "smart pixel array," is presented in some detail. Pattern recognition experience and Monte Carlo studies with QCD jet events are discussed. The radiation hardness advantage of pixelized detectors is developed quantitatively. A new high temperature process developed at LBL, permitting simultaneous fabrication of high quality transistors and detectors on high-resistivitiy material is presented. The current status of pixel and strip detector development at LBL is briefly reviewed.

1 INTRODUCTION

Tracking requirements at the SSC can be broadly characterized by the challenges of resolving many closely-spaced tracks within high multiplicity jet-like events, of measuring high momentum tracks accurately, of achieving efficient coverage of a large solid angle, of detecting secondary vertices, and of maintaining effective rejection of enormous instantaneous background rates. Together, these requirements drive the physical segmentation of tracking systems under consideration to an unprecedented level. In the case of silicon strips, tracking systems approaching a million addressable elements have been under study.[1–2] Systems with pixelized detectors involve at least an order of magnitude more.

Although the segmentation of a pixelized tracking system is very large, the advantages are compelling: track detection essentially free from overlap and complete absence of ambiguity in event reconstruction. Despite the expected complexity of SSC events, a non-projective tracking system with high resolution can reduce the necessary number of tracking layers to less than six. Additional advantages of pixelized detectors are a very large intrinsic signal/noise ratio and an effective radiation hardness that can easily tolerate decades of full luminosity SSC operations.

Despite large system segmentation, the quantity of primary tracking data that must be read out for even a rather complex event will be quite modest, on the order of 10 kilobytes. From the standpoint of system design, therefore, it is essential that the readout scheme be able to skip very efficiently over empty elements. In other words, **the system architecture must be matched to the ultra-sparse nature of the data field.**

With regard to random backgrounds, it is also clear that SSC detector systems must provide local time-slice buffering of information, at least until level I triggering decisions can be made. For example a detector without buffering will have contributions from about 100 events/microsecond at the design luminosity, adding perhaps 3600 extra background tracks (for ± 3 in pseudo-rapidity coverage) to any event of interest. Given the complexity anticipated for most SSC events of interest, any unnecessary addition of out-of-time tracks to the reconstruction task will be a heavy and perhaps overwhelming burden. Therefore, **the architectural concept must include time-slice buffering associated with pattern information.**

Strip detector readout systems can be equipped with time-slice buffering in a relatively straightforward way. The design of VLSI electronics for this purpose has been undertaken already by several groups.[2–5] For pixel devices, the situation is more complicated, and an architecture that capitalizes on the ultra-sparse character of the data seems appropriate. To provide some background for discussion, it is useful to review briefly the characteristics of some existing pixel devices. The suitability for the SSC of another interesting device, the silicon drift chamber, is treated elsewhere in these proceedings.[6]

The most familiar pixel device, and the only type to be used thus far as charged particle detectors for high energy physics research, is the optical CCD.[7] Optical CCD's are designed with the presumption that all pixels carry wanted information, and the devices are organized to read out every pixel. Readout times are in the millisecond range. In short, these devices have an architecture diametrically opposed to the needs of the SSC. In addition, the depletion depths are on the order of only 10 - 20 μm, resulting in relatively small signals for charged particles. Operation at cryogenic temperatures is required to eliminate substantial dark current levels associated with the long readout time. Finally, rather husky high-speed charge transport clock drivers must be located nearby, requiring careful system design to minimize induced noise. Nevertheless, impressive spatial resolutions have been obtained: ± 5 μm rms in x and y, and ~ 40 μm double track resolution.[8]

To ameliorate the serious limitations of the CCD in this context, various concepts have been considered, such as fast gating, fast clearing, and simultaneous sensing of an entire row for presence of hits. However, none of these provides a really fundamental solution to the limitations of the CCD concept for the SSC. It would appear that **conventional CCD's will play no significant role in tracking at the SSC.**

There also are less familiar pixel devices organized around a row-column intersect, random-access principle, such as those developed by Hughes Aircraft Company.[9] These elegant hybrid devices have addressable amplifiers on each pixel, and have also been used for scientific applications ranging from IR astronomy to x-ray detection. The amplifiers are unpowered until readout, and hence these devices lack altogether the capability to provide time-slice buffering with pattern information, an essential feature for the SSC. Because the pixels are normally quiescent and produce no prompt signals, there is no straightforward way to add the time-slice buffering capability. In addition, the random access feature by itself does not provide the means to search efficiently for struck pixels; a complete scan is still necessary. **A random access architecture by itself is not a solution.**

What is needed for the SSC is an architecture that provides unambiguous, direct registration and efficient readout of pixels which have been struck in the particular time slice of interest. The Smart Pixel array concept developed at LBL is intended to provide a solution matched to the challenge of the SSC environment.[10]

2 SMART PIXEL ARRAY

2.1 Concept

The basic concept of the Smart Pixel Array is that each pixel shall be equipped with continuously sensitive circuitry, so that prompt signals will be generated at the instant charged particles pass through a given pixel. These prompt signals are propagated directly to x-y pattern registers at the periphery of the chip. The pattern registers store the occurrence of these signals as fast time-sliced pattern information. The buffers within the pattern register are made deep enough to accommodate the time neccesary to generate a level I trigger. An event trigger will initiate pixel charge interrogation, directed efficiently by the stored pattern information for the time slice of interest. The direct interrogation of pixel charge content eliminates nearly all ambiguities at this stage, an especially important feature. The primary data are space points essentially free of pile-up and ambiguity. The device thus operates in two distinct modes, a pattern acquisition mode, and a random-access pattern-controlled readout mode. This functionality is illustrated schematically in figure 1.

The smart pixel array has been developed at LBL by D. Nygren and H. Spieler. In this concept, a smart pixel requires considerable circuitry to realize the needed functionality. A low-noise low-power amplifier, shaping element, and discriminator are needed to detect the occurrence of a hit. In addition, an analog memory element is required to provide hit verification during readout and interpolation information during reconstruction. Digital logic to propagate the "hit" signal to the pattern registers, to respond to interrogation, and to provide reset is also needed. Analog to digital conversion is done off-chip.

Nevertheless, for a standard 2 μm feature size CMOS foundry process, this circuitry fits into an area of about 100 $\times$ 100 μm^2. The circuitry can just as easily be packed into an equivalent rectangular area of about 40 $\times$ 250 μm^2, a more natural choice from the standpoint of circuit layout. The elongated rectangular shape can be used to provide higher resolution in phi using the narrow dimension, while the longer dimension along the z direction is a natural match of typical detector thickness ($\simeq$280μm) to the range of track angles associated with a long luminous region. This area requirement, though substantially larger than non-smart pixels, is still small enough to provide excellent pattern recognition, as discussed later. At a 10 cm radius the hit rate per pixel is just 100 Hz at full SSC luminosity. This allows plenty of time to retrieve the analog information during readout.

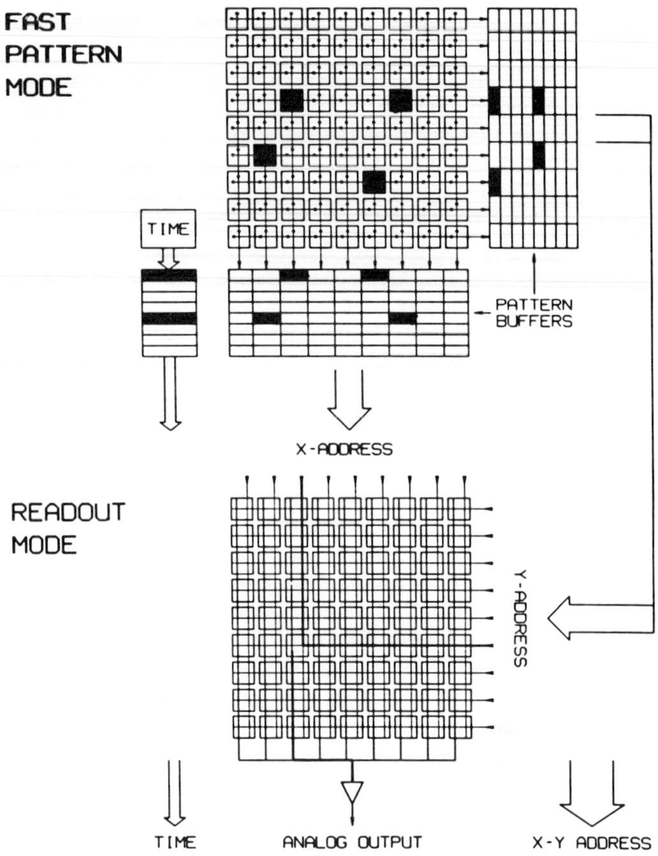

Figure 1: A schematic depiction of the smart pixel array concept modes of operation. During data acquisition, pattern and time information are stored in the buffers on the chip edge, while analog information is retained in pixels. For readout, row information for the time slice of interest efficiently directs interrogation of column information, including verification of hits by analog readout of struck pixels.

As is well-known, charge sharing among neighboring detector elements provides analog information that can be used to reduce the spatial resolution to a small fraction of the element size. It should be sufficient here simply to note that the pixel size in one dimension at least can be made as small as that of strip detector systems (with much poorer signal/noise) that have provided resolutions smaller than 5 μm.[11]

The implementation will be as a hybrid array, in which the detector wafer and readout integrated circuit (IC) are separately fabricated and are joined together using the well-established "indium bump bonding" technique. This allows separate optimization of the detector and readout IC, using standard technologies.

For the SSC in particular, this permits the use of available commercial rad-hard IC foundry processes for the readout IC. For the indium bump interconnection of detector and readout IC, a bonding pad of about 40 × 40 μm^2 will be needed; however, it is possible that some of the circuitry can be placed under the bonding pad aluminization to save some space. The trend toward smaller feature size, now at 1.0 μm for advanced commercial processes, offers the prospect both of improved performance and significantly smaller pixel area requirements.

Eventually, a monolithic implementation with the readout circuitry integrated onto the detector surface may be possible. As discussed briefly below, significant advances toward this goal have been made recently at LBL.

2.2 Pixel Performance

The principal issues here are signal/noise ratio, speed of response, power dissipation, and radiation hardness. In addition to this list of concerns for device design, there is a formidable set of problems associated with the implementation of a large system. A detailed discussion of these vital issues is outside of the scope of this paper. A more complete technical discussion can be found in our 1987 LBL DOE proposal.[10]

The signal/noise ratio for smart pixel arrays can be very high, primarily because the electronic noise in this case can be made very small. The capacitance presented to the amplifier by a pixel of 100 × 100 μm^2 is ≃100 fF, dominated by edge effects. Coupled to a well matched amplifier, this miniscule capacitance leads to extremely low predicted noise levels, *e.g.* ≤ 100 electrons rms for our designs. It is worth mentioning that similar noise levels have been regularly observed for various MOS circuits with similarly low input capacitance. Furthermore, it is important to note that an architecture that would gang n pixel amplifier outputs to individual row/column discriminators will increase the noise by $\sqrt{n}$, compromising the signal/noise ratio by at least a factor of 10.

With regard to signal, a fully depleted detector of 280 μm thickness will on average provide a signal corresponding to 22,500 electrons. The calculated signal/noise ratio thus exceeds 200. This favorable situation provides several circuit and system design advantages:

1. **Simplicity** The design of front-end circuitry can be significantly simpler than that for strip systems. This leads to smaller area requirements for the pixels. Alternatively, some signal/noise can be traded for a reduction in power dissipation.

2. **Threshold** The impact of unavoidable manufacturing variations on amplifier and discriminator threshold characteristics of a large-scale system should be negligible; in other words, it should be easy to find an operating point for a pixel array that rejects **all** electronic noise, while providing essentially 100% efficiency for minimum ionizing particle tracks. Recalling the ultra-sparse nature of the data, it is clear that the ability to reject all electronic noise is an important attribute for any system with a very large number of discrete detection elements such as the strip or pixel systems envisaged for the SSC. This may not be as easy to achieve for large-scale strip systems, which may be expected to have a signal/ noise ratio of ≃ 10-20. For example, setting the threshold at 3.5 σ on an otherwise ideal system with 10^6 strips, 10 nsec RC-CR shaping time, and gaussian noise will introduce 900 extra noise hits. To accommodate the inevitable circuit variations and noise conditions of a real system, the thresholds may have to be set considerably higher than this, with possible loss of efficiency for real tracks.

3. **Radiation Damage** A signal/noise ratio of several hundred provides considerable headroom for the accommodation of deterioration due to radiation damage. This deterioration can occur in both the detector and the readout circuitry. In detectors, the initial effect caused by radiation damage is an increase in shot noise due to increased dark current. A quantitative treatment of the noise problem due to radiation damage to the detector is given in a subsequent section. After irradiation, MOS transistors also display degraded noise characteristics although the sensitivity to damage depends upon the manufacturing process. There is some evidence that JFETs, which do not employ an oxide layer for their functionality, show very little deterioration in noise characteristics at the few Mrad level.[12] The relationships among noise characteristics, process, and radiation damage are not well-documented; more study is clearly warranted.

4. **Time Resolution** Another benefit of the very large intrinsic signal/noise ratio is that the risetimes of the pixel amplifier/discriminator can be long relative to the 16 nsec interbunch period of the SSC and still tag the correct time slice. The time resolution is determined by the amplifier integration time t_i and the signal/noise ratio (S/N = R):

$$\delta t = t_i/R$$

An integration time of 100 nsec is possible within the power dissipation constraint; the calculated time resolution is thus a small fraction of the SSC bunch period. Alternatively, longer peaking times can offer reduced power dissipation without loss of time resolution.

Speed of response, power dissipation, and electronic noise are strongly interrelated circuit design issues. A natural constraint is the maximum allowed power dissipation. Hefty power distribution busses and cooling mechanisms would add undesirable mass close to the vertex. A reasonable limit would appear to be 100 mW/cm^2, or 10 μW per pixel.

2.3 Discussion

Unlike other pixel concepts, the smart pixel array concept contains an inherent *economy of action* principle. During data acquisition, the smart pixel array is **quiet** until traversed by ionizing radiation. **There are no strobes or clock signals continuously scanning the amplifier array.** System power dissipation and induced noise are minimized. In the design we prefer, the SSC bunch crossing signal is continuously scaled by a single counter. Prompt pixel signals cause both the timing counter status and the pattern information to be strobed into augmented pattern+time registers, thereby linking the two kinds of information. The strobe pointer is then incremented to await the next hit. This signal-driven design avoids unnecessary activity, and very substantially reduces the depth requirement for the pattern+time registers. For example, a 4 cm^2 device (a size we hope can eventually be manufactured) at 10 cm radius will experience, on average, 4 hits in one microsecond, a likely level I trigger interval. Unless secondary backgrounds are dominant, 16 time slices should be more than sufficient toaccommodatenatural rate fluctuations. This may be compared with the depth of 64 time-slices needed to span one microsecond in a continuously clocked design.

3 PATTERN RECOGNITION

Pixel devices have the self-evident virtue of providing nonprojective spatial data. However, the actual level of performance will depend upon several factors such as pixel size, proximity to the beamline, the underlying physics of an event (*e.g.*, QCD jets vis-a-vis Higgs production), and ability to reject background.

Thus far the only example of the use of pixel devices in an experiment is with CCD's in CERN NA32.(13 NA32 also employed a silicon strip system, which permits a limited comparison of strip and pixel information quality. Monte Carlo studies at LBL have also provided useful information about pattern recognition.

3.1 NA32 Experience

NA32 is a fixed target experiment which had the goal, among others, to identify charm particles and measure their lifetimes. In essence, a short target was followed by an array of 7 planes of silicon strips positioned from 5 to 12 cm after the target. The experiment was operated at first without the CCD's. Under these conditions, two independent reconstruction efforts found charm events with low efficiency, and with an overlap of only 15%. The dominant problem was merged tracks in the early planes. Only about 60% of all clusters were distinct .

Two GEC P8600 CCD devices were subsequently added, at distances of 1 and 2 cm from the target. The pixel size of these 385 × 580 arrays was 22 × 22, μm^2, leading to ~5.0 μm rms spatial resolution and ~40 μm two-track resolution. With the information from the CCD's, efficiencies for identifying charm events increased very substantially, and the overlap rose immediately to more than 85%. The CCD's in combination with clean microstrip hits and the downstream drift chambers were sufficient to resolve a large sample of charm events without background. Both the spatial resolution and the two-track resolution contributed to the effectiveness of the CCD's. It should be noted that this experiment pioneered the utilization of both strip and pixel devices for physics; however, later experiments such as FNAL E691, using finer resolution microstrip planes, have also found large, clean charm event samples.(14

3.2 Monte Carlo Studies

Monte Carlo studies of some aspects of pixel performance have been made at LBL in collaboration with Ed Wang. The questions that these Monte Carlo studies addressed include the following:

- For QCD jets of 0.3 and 1.0 TeV/c, aimed normally at the center of a pixel array from a distance of 10 cm, what fraction F of events are registered with no more than 1 track per pixel?
- How does the fraction F depend on pixel area and shape?
- What is the relative performance of other kinds of pixel arrays that lack some capabilities of the smart pixel array?

The benchmark physics process chosen for these studies is the QCD jet, generated with the code Pythia 4.8. At the SSC high $P_\perp$ QCD jets will be tightly collimated, dense and frequent. Their study will contribute much to a more fundamental understanding of hadronic processes at high energy. These jets therefore would appear to provide a rather stringent yet straightforward test for pattern recognition efficiency. The detection of secondary vertices from heavy quark decays is another vitally important topic, but the software needed for a careful study of these processes was not yet available.

The geometry developed for this study typically consisted of a pixel array of area 1 x 1 cm^2, with options for variable pixel size and shape. The known axis of the QCD jet was aimed normally at the center of the array from a distance of 10 cm. Figure 2 shows a Lego plot of particle frequency for 1000 events; for clarity, large pixel sizes were chosen for this figure. The densely populated central peak riding on a broad background illustrates the inherent reconstruction difficulty. For the 1 TeV/c jets, the average charged particle multiplicity within this 1 cm^2 area is 12, with fluctuations up to 30.

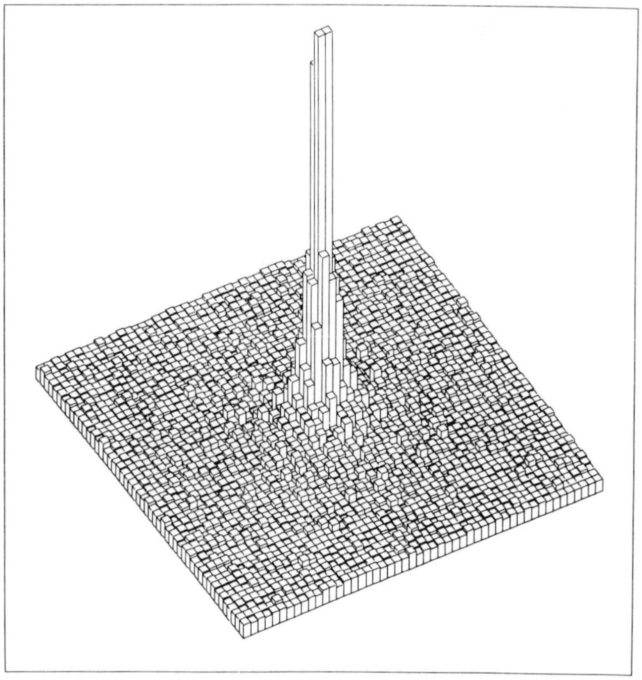

Figure 2: The density of tracks normally incident on a 1 cm^2 pixel array 10 cm from the point of production of 1 TeV/c QCD jets. A densely populated central region riding on a broader background makes this a particularly challenging class of events to reconstruct.

The fraction F of fully resolved events, *i.e.*, the fraction of events that are registered with no more than one track per pixel, versus pixel dimension (for square pixels) is given in Figure 3 for 1 TeV/c jets. It is noteworthy that even for rather large pixel dimensions such as 100 × 100 μm^2 the fraction F is still as high as 0.9. Most of the remaining events have just one pixel with two tracks. The lower energy 0.3 TeV/c jets gave somewhat higher

F values due to the lower multiplicities.

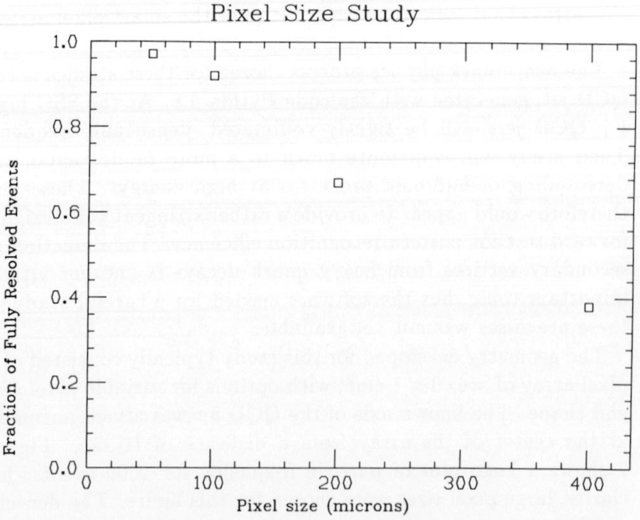

Figure 3: For 1 TeV/c QCD jets normally incident on a 1 cm^2 pixel array 10 cm from the production point, the fraction F of events with all tracks fully resolved is shown versus pixel dimension. Although this plot is for square pixels, equivalent results were found for rectangular pixels of equal area.

Rectangular shapes, such as 40 × 250 μm^2, were also studied with the conclusion that the F values depend only on the pixel area. While much work remains to be done, it seems clear from these studies that pixel devices with pixel areas on the order of 10^4 μm^2 can provide excellent pattern recognition capability even for very challenging processes such as QCD jets.

The smart pixel architecture retains analog information on the pixel for subsequent readout and pattern verification. Another architecture studied employed a three-axis pattern registration (x-y-u), but with no analog sample and hold. Pattern identification would be accomplished by an off-line 3-axis coincidence match. The motivation for this study was to see if the potential advantage of circuit simplification would be offset by difficulties in pattern identification due to chance coincidences. The results indicated, however, that the number of chance 3-fold coincidences was rather high: a 50% increase in apparent "hits" for a 50 × 50 μm^2, rising to 80% for 100 × 100 μm^2, for the 1 TeV/c QCD jets. For this reason and because this architecture has an intrinsic resemblance to a strip geometry, we chose not to study this particular architecture further.

4 RADIATION HARDNESS AND DETECTOR ELEMENT SIZE

4.1 Theoretical Background

In addition to basic material properties, the radiation hardness of semiconductor detectors depends on design parameters of the detector and the readout electronics system. Substantial gains in the ability of a silicon detector to tolerate radiation damage can be achieved through a high level of segmentation, such as is naturally achieved in a pixel array. This section summarizes the treatment presented by D. Nygren and H. Spieler in the *Report of the SSC Task Force on Radiation Effects at the SSC.*[(15] Two basic mechanisms are responsible for radiation damage in silicon detectors:

1. Displacement of atoms from their lattice sites (bulk damage).

2. Electron-hole pair generation in isolation layers, with trapping.

Displacement damage seems to be dominant in high energy physics applications of semiconductor detectors. These effects are easier to predict than the ionization mechanisms in dielectrics because they are much less sensitive to processing variations. Displacement damage manifests itself primarily through three phenomena:

1. Increased leakage current (shot noise).

2. Change in the resistivity of the material (depletion voltage effects).

3. Increased recombination or trapping of signal charge.

Increased leakage current is the most prominent initial effect. More work is needed for a quantitative assessment of the other two. The discussion here deals only with the impact of the leakage current.

The leakage current increase is due to the generation of carriers by radiation-induced lattice defects, which create intermediate energy levels in the band gap. This leakage current contributes shot noise Q_n

$$Q_n = (I_L \cdot t_i / q_e)^{1/2} \qquad (1)$$

where Q_n is expressed in electrons, q_e is the electron charge, I_L is the leakage current, and t_i is the effective integration time of the circuitry. The signal Q_s obtained from the detector is

$$Q_s = n_s \cdot W \qquad (2)$$

where n_s is the number of electrons per unit length created by the particle traversing the detector, and W is the thickness of the sensitive layer. It is useful to cast the discussion in terms of the signal-to-noise ratio $R = Q_s/Q_n$. From equations 1 and 2, the maximum current that can be tolerated is then

$$I_L = (n_s^2 \cdot W^2 \cdot q_e)/(R^2 \cdot t_i) \qquad (3)$$

For a silicon detector n_S = 80 electrons/μm for minimum ionizing particles; this yields the numerical equation

$$I_L = (100nA \cdot \sec \cdot \mathrm{cm}^{-2})W^2/(R^2 t_i) \qquad (4)$$

A typical case might be W = 300 μm and t_i = 100 ns, which allows a maximum current of 90 nA if we require a signal to noise ratio of R = 100, or a current of 9 μA for R = 10.

The detector leakage current is directly dependent on the generation lifetime t_g through the relation

$$I_L = q_e \cdot n_i A \cdot W/(2 \cdot t_g) \qquad (5)$$

where n_i is the intrinsic free carrier concentration (thereby introducing an implicit temperature dependence) and A is the area of the detector.[(16] The effect of displacement damage on t_{ig} is given by[(17]

$$1/t_g = 1/t_{go} + K \cdot F \qquad (6)$$

where t_{go} is the pre-irradiation generation lifetime, F is the particle fluence and K is the damage coefficient associated with the specific type of radiation and detector material. In a system designed to sustain substantial radiation damage $t_{go} \gg t_g$, and this expression may be simplified to

$$1/t_g = K \cdot F \tag{7}$$

Inserting (7) in (5) yields the detector leakage current as a function of radiation type and dose:

$$IL = q_e \cdot n_i \cdot A \cdot W \cdot K \cdot F/2 \tag{8}$$

Combining this expression with (3) defines the interdependence between signal-to-noise ratio R, radiation type and dose (K, F) and detector/readout parameters (n_s, A, W, t_i). The allowable fluence, for example, is

$$F = (2 \cdot n_s^2 \cdot W)/(K \cdot n_i \cdot A \cdot R^2 \cdot t_i) \tag{9}$$

For a silicon diode detecting minimum ionizing particles, n_s = 80 electrons/μm and $n_i = l.5 \cdot 10^{10}\text{cm}^{-3}$ at 300° K. Inserting these values provides a numerical relationship for the allowable fluence

$$F = (85\text{cm}) \cdot W/(K \cdot A \cdot R^2 \cdot t_i) \tag{10}$$

for a given detector/readout design. This equation provides a very instructive relationship for the evaluation of bulk effects in silicon detectors.

4.2 Discussion

Equation 9 shows that radiation hardness, in the practical sense of a system performance requirement R, is determined by several factors:

- Detector material (through n_s, n_i and the damage constant K)
- Detector geometry (through the thickness W and area A)
- Integration time of the electronics t_i

Radiation resistance can be enhanced by decreasing the area or increasing the thickness of the detector elements, or by reducing the integration time of the electronics. While all of these are design choices, detector area and integration time have the greatest flexibility; smaller elements of course bring greater electronic complexity, whereas reduced integration time leads to increased power dissipation.

For the evaluation of tracking devices at the SSC one can use equation 10 to calculate the maximum allowable area of a detecting element for a given lifetime under typical operating conditions:

$$A = (85\text{cm}) \cdot W/(K \cdot F \cdot R^2 \cdot t_i) \tag{11}$$

At distances greater than about 10 cm for the beam axis the background is expected to be dominated by neutrons from calorimeter albedo. At full luminosity the neutron fluence is roughly 10^{13} cm^{-2}yr^{-1} or 10^{14}cm^{-2} for a 10 year period. It is worth noting that the uncertainties on this value of the neutron fluence are probably quite substantial in either direction. For the following examples the values $K = 1.4 \cdot 10^{-7}\text{cm}^2 s^{-1}$ for the damage constant,[17] $W = 300$ μm for the detector thickness, and an integration time t_i of 100 nsec have been adopted.

1. **Silicon Strips** In this example, the system is assumed (idealistically) to display an initial signal-to-noise ratio $R \gg 20$, where the noise is determined by the strip capacitance at the preamplifier input. How large can the strip size be and still provide $R = 20$ after 10 years of operation? For a strip pitch of 50 μm, the calculated length using equation 11 is about 1 cm. This would be the case if there were no room temperature annealing; however, preliminary results presented at this conference by Kondo *et al.* indicate for their devices an anneal time constant of about 160 days.[3] This interesting result may, however, depend on processing choices. If this anneal time constant is a general result, then the corresponding length for $R = 20$ would be at least 10 cm. As noted above, faster electronic integration times can provide additional margin to preserve overall sytem performance. From the above discussion, and assuming both the generality of the annealing results and that no additional backgrounds exist, it appears possible that silicon strip detectors of $\simeq$10 cm length in a carefully designed system should survive the shot noise effect of radiation damage. Long strips near the beam pipe may not survive.

2. **Pixel Arrays** Clearly, a pixelized detector will be less prone to shot noise effects than a strip detector by the ratio of element area, all other factors assumed identical. The smart pixel element, with an area of *e.g.* 40 x 250 μm^2, is 500 times smaller than a 10 cm x 50 μm strip. The shot noise is essentially negligible after ten years, even without annealing.

5 A NEW ROBUST PROCESS FOR DETECTOR FABRICATION WITH INTEGRATED CIRCUITRY

5.1 Background

Silicon p-i-n detectors for high energy charged particles utilize high-resistivity silicon to obtain low leakage currents and large depletion depths at moderate applied voltage. Ideally, a modern detector process should offer minimum sensitivity to process-induced contaminants and would also be compatible with contemporary IC fabrication techniques. Such a process would allow the integration of front end circuitry with the detector, providing improved performance, reliability, and system design flexibility.

Until now, this has not been the case. Conventional detector processes have been found to be incompatible with the sequence of high temperature steps characteristic of IC fabrication. Very large leakage currents have been consistently observed if high temperature steps are included after the initial oxide growth. The reason is that the detector leakage current is extremely sensitive to impurities such as iron, nickel, copper, gold, etc., that introduce energy levels near the band gap center. For example, the presence of gold at a level of $1 \cdot 10^{-9}$ will increase leakage current density by three orders of magnitude. At typical IC fabrication temperatures of 1000° C, gold will migrate millimeters/hour through silicon, whereas at 600° C, the diffusion is very small.

5.2 The New High-Temperature Process

At LBL, Steve Holland has recently developed a new high-temperature process that turns the diffusion of impurities at high temperatures to advantage.[18] In his process, impurities are effectively removed from the active region by capture on the backside of the wafer, an electrically inert region. This technique is an example of "gettering," a concept which has a long history in vacuum tube and semiconductor electronics, but until now has not been applied with such success to the problem of detector fabrication. The LBL process appears to be fully compatible with high-quality transistor and IC fabrication steps.

Briefly, the LBL process utilizes a layer of polysilicon on the wafer backside, in-situ doped with phosphorus, and formed by chemical vapor deposition at 650° C. This 1 μm thick layer serves both as the gettering layer and backside contact. The gettering properties of this layer are due to the presence of both the polysilicon and the phosphorus. Detectors and transistors were simultaneously fabricated on the same wafer along with control devices that lacked the gettering layer.

5.3 Results

Figure 4 shows the leakage current versus bias voltage for both gettered and ungettered diodes; the current is reduced by a factor of 400 at low voltages and even more at higher voltages. The leakage current at 100V, which corresponds to a fully depleted detector, is on the order of 1-2 nA/cm^2. This compares very favorably with the results of Kemmer.[19] However, the Kemmer process limits processing temperatures to 600° C after the initial oxidation, whereas devices made by the LBL process have experienced approximately three hours at 900° C. The low-temperature Kemmer process leads to incomplete activation and high sheet resistance of the implanted layers; in turn this prevents monolithic integration of high performance circuitry.

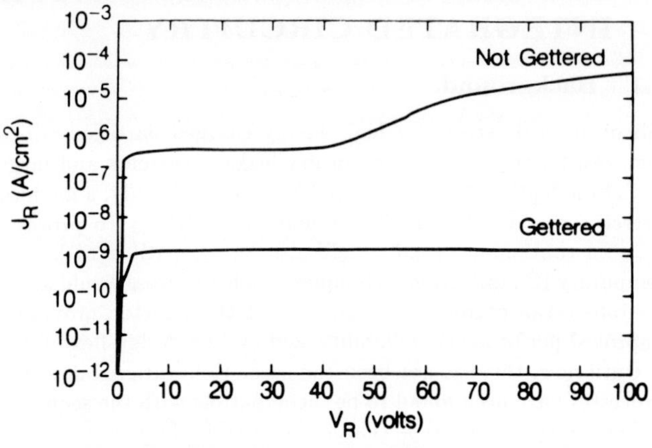

Figure 4: The detector diode reverse-leakage current for a device with backside gettering compared to one without. Both devices were fabricated on the same wafer.

The p-channel MOSFET transistors made by the LBL process were found to display good output current-voltage and frequency response characteristics. The application to the substrate of a 30V bias (sufficient for full depletion) causes only small deviations from the unbiased curves, as shown in figure 5. Diodes fabricated on the same wafer were tested up to 1000V bias, again with leakage currents on the order of 1-3 nA/cm^2.

An LBL-process detector was exposed to an Americium 241 source, demonstrating excellent energy resolution. The FWHM for the 59.4 keV gamma ray is approximately 1.8 keV, limited

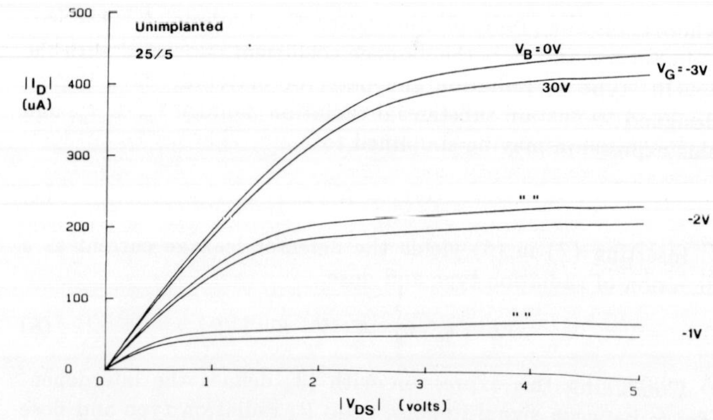

Figure 5: Current-voltage characteristics of a p-channel MOSFET fabricated on 10KΩ·cm silicon. The channel width to length ratio is $25\mu m/5\mu m$. The effect of the substrate bias is seen to be small; 30V is sufficient to deplete this material.

in this instance by the noise of the signal-processing electronics. A short detector with 128 strips has also been fabricated; preliminary results indicate low ($\leq$ 10 nA/cm^{-2}) and extremely uniform ($\leq$ 3% rms) strip leakage currents .

5.4 Conclusion

An extrinsic gettering technique has been developed that provides both a robust process for the fabrication of high-quality p-i-n diodes and the prospect of monolithic integration of these detectors with high performance CMOS and/or JFET circuitry.

6 SMART PIXEL ARRAY DEVELOPMENT STATUS

6.1 Device Analysis

CMOS technology offers a combination of speed, low power dissipation, and industrial maturity that makes it the clear choice for short term readout IC goals. However, nearly all commercial effort has been focussed on digital applications. The design of CMOS devices for unconventional and demanding applications such as detectors for high energy physics requires careful evaluation of device characteristics under a range of unusual operating conditions. Modeling of circuit response is quite likely to be seriously at variance with reality if standard parametrizations of device performance are employed.

Accordingly, significant effort has been spent in performing these measurements on a variety of custom devices. These include nmos and pmos transistors of various W/L ratios and elementary amplifiers, fabricated by the MOSIS organization. In addition to the DC characteristics for a family of drain – source curves and the gain-bandwidth response, a range of channel geometries has been evaluated to provide reliable scaling relationships.

6.2 SVX IC

In addition to its basic purpose as the readout device for the CDF silicon strip vertex detector, the SVX IC [20] project serves as an important stepping-stone in the realization of the smart pixel array. The SVX, designed mainly by Stuart Kleinfelder, offers similar noise performance to the pioneering Stanford μplex NMOS IC,[21] but dissipates a factor of 10 less power, and has a built-in sparse data readout feature that substantially reduces readout time by eliminating unhit channels. The sparse readout mode scans past empty channels at about 400 MHz. This same circuitry will be employed in the pattern readout of the smart pixel array. The input amplifier envisaged for the smart pixel array shares concepts with the SVX amplifiers.

6.3 Circuit Design Studies

A number of low-power low-noise amplifier designs have been modeled, with the aim of reducing power dissipation to the minimum compatible with response time and noise goals. Comparator designs, again building on the experience with the SVX IC have also been studied. The digital logic carried on each pixel has reached a fairly complete state. A one-dimensional array of pixel amplifiers and comparators has been submitted for fabrication through MOSIS. A two-dimensional version is planned for submission after a satisfactory evaluation of the one-dimensional chip has been completed.

6.4 Microsystems Laboratory

Considerable effort has been expended to bring into existence a new modern facility at LBL for the further development of detector process ideas and device fabrication. This new facility, the Microsystems Laboratory, provides four contiguous clean rooms offering $\simeq$700 ft^2 of Class 10 workspace. In addition to other novel detectors and devices, pixelized detector arrays will be fabricated here, possibly with integrated JFET's to provide a low-noise radiation-hard input stage for a pixel readout IC made by a standard rad-hard CMOS foundry process.

7 SUMMARY

Pixelized detectors offer several significant potential advantages relative to strip detectors in pattern recognition, systems engineering, and effective radiation hardness. A smart pixel architecture concept matched to the SSC environment has been developed, and pixel circuitry has been designed and implemented. A robust process for fabricating detectors with integrated circuitry has also been developed. With appropriate levels of effort devoted to device and system engineering, a powerful tracking system employing pixelized detectors can be realized.

8 ACKNOWLEDGEMENTS

This paper is largely concerned with the efforts and accomplishments of the LBL Advanced Detector Development Group. In addition to the explicit references to individuals within the text, I wish to acknowledge the seminal contributions made by Helmuth Spieler to the conceptual development and engineering advances of this program.

9 REFERENCES

1. *Report of the Task Force on Detector R&D for the SSC*, M. G. D. Gilchriese, ed., p. 74, June, 1986.

2. Seiden, A., "Tracking Requirements for the SSC and Silicon Strip Development," these proceedings.

3. Kondo, T., "Radiation Damage of Silicon Devices and R&D for JCCD Readout of Microstrips," these proceedings.

4. Heijne, E., "Future Silicon Detector Development in Europe," these proceedings.

5. Watase, Y., "Integrated Circuit Development in Japan for Future Experiments," these proceedings.

6. Rehak, P., "Silicon Detector and Readout Development at BNL," these proceedings.

7. Damerell, C. J., *et al.*, "Charge-Coupled Devices for Particle Detection with High Spatial Resolution," Nuc. Inst. and Meth. **185**, pp. 33-42 (1981).

8. Bailey, R., *et al.*, "First Measurement of Efficiency and Precision of CCD Detectors for High Energy Physics," Nuc. Inst. and Meth. **213**, pp. 201-215 (1983).

9. Gaalema, S., "Low-Noise Random-Access Readout Technique for Large PIN Detector Arrays," IEEE Trans. Nuclear Science **NS-32** pp. 417-418 (1985).

10. Nygren, D. R., and Spieler, H., "Semiconductor Detector and Integrated Electronics Development," LBL proposal to the DOE, July 1987.

11. England, J. B. A., *et al.*, "Capacitive Charge Division Readout with a Silicon Strip Detector," Nuc. Inst. and Meth. **185**, pp. 43-47 (1981).

12. Binkley, D. M., "A Radiation-Hardened Accelerometer Preamplifier for $2{\cdot}10^8$ rad(si) Total Dose," IEEE Trans. Nuclear Science **NS-29**, pp. 1500-1507 (1982).

13. Barlag, S., *et al.*, "Lifetimes of Charged and Neutral D Mesons," Z. Phys, C, **37-1**, 17-23 (1987).

14. Raab, J. R., *et al.*, "Measurement of the $D^0, D^+, and D_s^+$ Lifetimes," Phys. Rev. D, **37**, 2391-2402 (1988).

15. *Report of the Task Force on Radiation Effects at the SSC*, M. G. D. Gilchriese, ed., p. 74, June, 1988.

16. Sah, C-T., Noyce, R. N., and Shockley, W., "Carrier Generation and Recombination in P-N Junctions and P-N Junction Characteristics," Proc. IRE **45**, 1228 (1957).

17. van Lint, V. A. J., *et al.*, *Mechanisms of Radiation Damage on Electronic Materials*, Vol. 1, (Wiley, New York, 1980).

18. S. Holland, "Fabrication of Detectors and Transistors on High-Resistivity Silicon," presented at the Workshop on Pixel Detectors, Leuven, Belgium, May 31-June 2,1988; LBL preprint LBL-25520, June (1988).

19. Kemmer. J., "Improvement of Detector Fabrication by the Planar Process," Nucl. Inst. and Meth. **226**, pp. 89-93 (1984).

20. Kleinfelder, S., *et al.*, "A Flexible 128 Channel Silicon Strip Detector Instrumentation Integrated Circuit with Sparse Data Readout," IEEE Trans. Nuclear Science **35**, pp. 171-175 (1988).

21. Walker, J. T., *et al.*, "Development of High-Density Readout for Silicon Strip Detectors", Nucl. Inst. and Meth. **226**, pp. 200-203 (1984).

TRACKING SIMULATION AND WIRE CHAMBER REQUIREMENTS FOR THE SSC*

GAIL G. HANSON, BOGDAN B. NICZYPORUK† AND ANDREA P. T. PALOUNEK‡

Stanford Linear Accelerator Center, Stanford University, Stanford, California 94309

ABSTRACT

Limitations placed on wire chambers by radiation damage and rate requirements in the SSC environment are reviewed. Possible conceptual designs for wire chamber tracking systems which meet these requirements are discussed. Computer simulation studies of tracking in such systems are presented. Simulations of events from interesting physics at the SSC, including hits from minimum bias background events, are examined. Results of some preliminary pattern recognition studies are given.

1. INTRODUCTION

The primary motivation for the SSC is the expectation that it will lead to new discoveries, such as Higgs bosons, supersymmetric particles, heavy W's or Z's, new heavy fermions, or composite particles with masses in the TeV region. Such particles would be produced in the central rapidity region, that is, over $\pm$ 3 units of rapidity, and would decay to high-p_T electrons, muons, or jets, often with large missing transverse energy (E_T) due to undetectable neutrinos. In order to fully investigate the physics opportunities in this regime, a general-purpose detector which includes charged particle tracking is needed. Some of the most important functions of charged particle tracking include:

1. Identification of electrons.
2. Separation of multiple interactions within the same bunch crossing.
3. Matching electrons, muons, and jets to the correct vertex.
4. Electron charge sign determination.
5. Improving e/π separation.
6. Identification of secondary vertices.
7. Identification of τ leptons.
8. Invariant mass or momentum cuts.
9. Improving the missing E_T measurement and verifying calorimeter data.
10. Establishing the credibility of new physics and providing redundancy.

Many of these functions require momentum measurement in a magnetic field.

Tracking at the SSC at the full design luminosity of 10^{33} cm^{-2}s^{-1} is expected to be a difficult problem. The limitations imposed by rates and radiation damage are severe. However, the dominant constraint is the combination of occupancy and double-hit resolution. Single events from new physics at the SSC have many (several hundred) charged particle tracks and are further complicated by curling tracks in a magnetic field, photon conversions, hits from events from out-of-time bunch crossings, and multiple interactions within the same bunch crossing. These problems can probably be solved, but at the cost of mechanical complexity and many signal channels.[1)] It has not been established how well one can find tracks in complex SSC events. We report here on a computer simulation study which addresses these problems.

2. WIRE CHAMBER REQUIREMENTS

2.1 The SSC Environment

The design luminosity, $\mathcal{L}$, of the SSC is 10^{33} cm^{-2}s^{-1} with an energy of 40 TeV in the center of mass. The inelastic cross section, σ, at 40 TeV is expected to be about 100 mb, which gives 10^8 interactions per second at the design luminosity. The bunch separation is 4.8 m, so the time between bunch crossings, t_B, is 16 ns, which leads to an

* Work supported by the Department of Energy, contract DE–AC03–76SF00515.

† Permanent Address: CEBAF, Newport News, Virginia 23606.

‡ Present Address: Lawrence Berkeley Laboratory, Berkeley, California 94720.

average number of interactions per bunch crossing, n_I, of 1.6 at the design luminosity. Most of these interactions are minimum bias events or low-p_T hard scattering processes in which particle production is expected to be uniform in rapidity with an average charged particle multiplicity per unit of rapidity, n_c, of 7.5 over the rapidity range $|\eta| < 6$.[2] Figure 1 (from Ref. 2) shows the resulting charged particle flux and annual dose as a function of perpendicular distance from the beam for standard SSC operating conditions.

2.2 Rates and Radiation Damage

Radiation damage and rate limitations impose severe constraints on charged particle tracking detectors at the SSC, as described in several references.[1,3] These constraints are summarized here since they are necessary considerations for the design of any SSC tracking system.

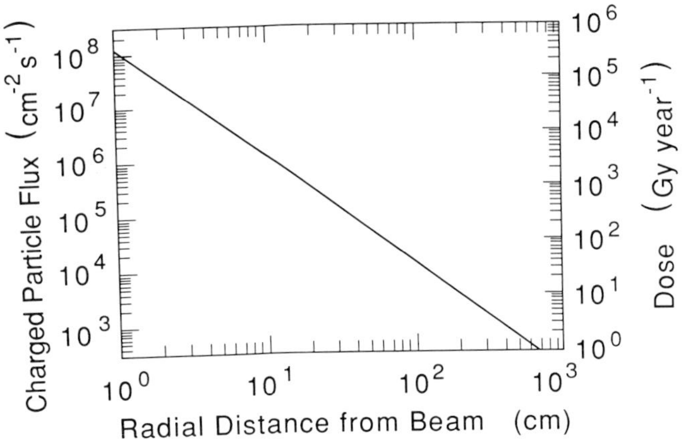

10-88 6164A1

Fig. 1. The charged particle flux and annual dose as a function of perpendicular distance from the beam under standard SSC operating conditions (from Ref. 2).

A tracking system for the SSC is assumed to be made up of wires running (nearly) parallel to the beam line. The width, w, of the cell is assumed to be equal to the height, h, and the drift distance, d, is half the cell width. The ionization rate, α, in the gas is assumed to be 100 electrons/cm. The gas gain, G, is assumed to be 2×10^4.

The flux of particles per unit length (ℓ) of wire in a cell at radius r is given by

$$\frac{d^2n}{d\ell\, dt} = \frac{n_c\, w\, \sigma\, \mathcal{L} \sin\theta}{2\,\pi\, r^2} , \quad (1)$$

where θ is the angle relative to the beam direction. The ionization produced by a charged particle at angle θ is $h\alpha/\sin\theta$, so the ionization per unit length of wire is independent of θ. Thus the current draw per wire, I, for a layer of wires of length L at radius r is given by

$$I = \frac{n_c\, w\, h\, \sigma\, \mathcal{L}\, G\, e\, \alpha\, L}{2\,\pi\, r^2} , \quad (2)$$

where e is the electron charge. A layer of 4 mm wide cells at a radius of 50 cm covering $|\eta| < 1.5$ ($L = 213$ cm) will draw 0.52 μA/wire. The limit of acceptable current draw before breakdown will occur is about 1 μA/wire.

Wire chamber lifetimes are measured in deposited charge per unit length of wire before a decrease in gain occurs. The decrease in gain is due to the buildup of material on the wires. For the above example of a 4 mm cell at 50 cm radius covering $|\eta| < 1.5$, the collected charge over a chamber lifetime of five years (5×10^7 s) would be 0.12 C/cm. Chamber lifetimes of 1.0 C/cm have been measured under very clean laboratory conditions.[4] For the purposes of a realistic experiment, it is probably best to assume a chamber lifetime about an order of magnitude below this.

Changes in gain for wire chambers have been observed at the level of 10^5 particles/cm-s at a gas gain of $\sim 4 \times 10^5$ due to space charge buildup.[5] The particle flux is given by Eq. (1). For the above example of 4 mm wide cells at a radius of 50 cm, the flux would be 1.9×10^4 particles/cm-s at $\theta = 90°$ where the flux is maximum. Since the gas gain must be much smaller than 4×10^5 because of current draw and lifetime considerations, space charge should not be an important limitation.

The hit rate per wire, R, for SSC central tracking chambers is quite large and is given by

$$R = \frac{n_c\, \eta_{max}\, \sigma\, \mathcal{L}\, w}{\pi\, r} \quad (3)$$

for chambers covering $|\eta| < \eta_{max}$. Thus a 4 mm cell at 50 cm radius covering $|\eta| < 1.5$ would have a hit rate per wire of 2.9 Mhz. Existing electronics can probably handle rates of $\sim$ 10 Mhz.

A very serious limitation for tracking systems at the SSC is occupancy. Since the time between bunch crossings at the SSC is shorter than the resolving time of a typical drift chamber cell, the cell is sensitive to several bunch crossings. The occupancy, O, is given by

$$O = \frac{2\, n_c\, \eta_{max}\, n_I\, n_B\, d}{\pi\, r} , \quad (4)$$

where n_B is the number of bunch crossings during the resolving time of the cell. n_B is given by

$$n_B = 1 + \text{int}\left(\frac{t_R}{t_B}\right)\left[2 - \frac{t_B}{t_R} - \left(\frac{t_B}{t_R}\right)\text{int}\left(\frac{t_R}{t_B}\right)\right] , \quad (5)$$

where t_R is the resolving time of the cell, d/v_D, for drift velocity v_D, and int(x) is the largest integer $\leq x$. Actually, n_B is very close to $t_R/t_B = d/(v_D t_B)$. A 4 mm wide cell (2 mm drift) has a resolving time of 40 ns for a typical drift velocity of 50 μm/ns and is therefore sensitive to 2.6 bunch crossings. A layer of such cells at a radius of 50 cm and covering a rapidity range $|\eta| < 1.5$ would have an occupancy of 12% per cell. It is guessed that an occupancy of $\sim$ 10% is reasonable, but a realistic answer depends on the effects on pattern recognition and track finding, which are discussed in more detail in Section 3. Occupancy limits the cell width at any radius in a tracking system since, even at larger radius, as one increases the cell width the cell becomes sensitive to more bunch crossings. Since the occupancy increases quadratically with d and decreases linearly with r, the maximum cell size can increase only as the square root of the radius. A faster gas,[6] such as mixtures of CF_4 with a saturated drift velocity of 125 μm/ns, would improve the situation considerably by giving lower occupancy for a given cell width.

One should keep in mind that these rates are based only on particles produced in an interaction and must be increased by a factor of 2–4 because of curling tracks in a magnetic field, converted photons, and albedo particles leaking out of the front face of the calorimeter. These limitations seriously affect the design of any SSC tracking device, regardless of pattern recognition considerations. For example, if one finds that an occupancy of 30% is reasonable for track finding, one cannot increase the cell width or move to a smaller radius without carefully considering the effects on current draw and chamber lifetime.

2.3 Tracking System Considerations

A large solenoid detector based on more-or-less "conventional" technology was discussed at the 1987 Berkeley Workshop.[7] An examination of the requirements for momentum resolution based on the physics led to the criterion that the sign of the charge for electrons should be measured for $p_T \leq 0.5$–1.0 TeV/c. The momentum resolution is given by[8]

$$\frac{\sigma_{p_T}}{{p_T}^2} = \sqrt{\frac{720}{1+5/N}} \left(\frac{\sigma_x}{0.3\, q\, B\, D^2 \sqrt{N}} \right) , \qquad (6)$$

where p_T is the transverse momentum of the particle in GeV/c, q is the charge in units of the electron charge, σ_x is the spatial resolution in m, B is the magnetic field in Tesla, D is the track length in m, and N is the number of measurements, assumed to be equally spaced. The outer radius for tracking was restricted to 1.6 m in order to place all of the calorimetry inside the magnet coil. The inner radius is constrained by the rate considerations discussed in the previous Section to 40–50 cm. A spatial resolution of 150 μm was assumed. The magnet was a large superconducting solenoid with 2 Tesla field. With $\sim$ 100 measurements the momentum resolution which can be obtained with such a system is $0.54 p_T$ (TeV/c) using only wires at radii larger than 50 cm. If the particles are constrained to come from the interaction region, the momentum resolution would improve to $0.26 p_T$.

The tracking detector design for the Large Solenoid Detector was divided into central tracking ($|\eta| \lesssim 1.2$) and intermediate tracking ($1.2 \lesssim |\eta| < 2.5$). The central tracking chambers were assumed to have an inner radius of 40 cm, but only the layers at radii greater than 50 cm were expected to be operable at the full design luminosity. For the reasons discussed in the previous Section, the cell widths are constrained to a few mm. Straw tube chambers are a natural candidate for a small cell design. Construction possibilities for a central tracking system made of straw tubes are discussed in Refs. 1 and 9. A central tracking system based on jet cells, such as shown in Fig. 2, might also be considered. However, since the jet cell would be only a few mm wide and would be tilted so that the electron drift trajectories in the 2 Tesla magnetic field would be perpendicular to the sense wire planes, the advantages of a jet cell, e.g., uniform drift electric field over most of the cell and long track segments in one cell, are lost. In addition, there would be large forces on the endplates due to the tension of the large number of field wires. Also, as in any open wire geometry, it would be difficult to devise a mechanism to support the long sense wires in order to maintain electrostatic stability.

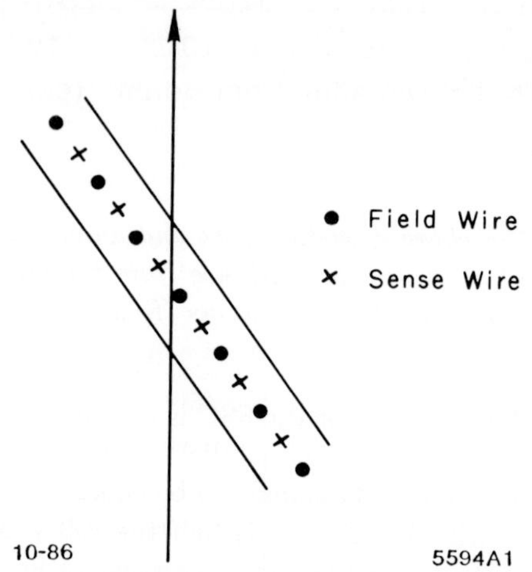

Fig. 2. Narrow tilted jet cell with radial track.

The central tracking system for the Large Solenoid Detector was assumed to be built of straw tubes of radii from 2 to 3.5 mm parallel or nearly parallel to the beam direction. The straws are made of aluminized polyester film (Mylar) or polycarbonate (Lexan) with wall thicknesses of about 30 μm. The straws were assumed to be at atmospheric pressure. Eight layers of straws are glued together to form superlayers. Within each superlayer the layers are staggered by half the cell width in order to allow hits from out-of-time bunch crossings to be rejected and resolve left-right ambiguities, as illustrated in Fig. 3. By dividing the chamber into eight-straw-thick superlayers locally identifiable track segments can be obtained with a high level of redundancy. Every other superlayer is small-angle stereo ($\sim 3°$) in order to measure the coordinate along the wire.

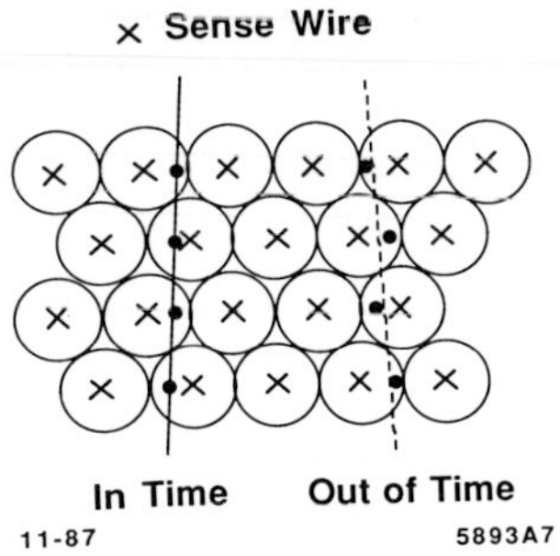

Fig. 3. Layers of straw tubes in a superlayer with every other layer staggered by the straw tube radius. A single in-time track will appear as a series of hits on the wires on alternate sides of the track. The left-right ambiguity is easily resolved locally. A track from an out-of-time bunch crossing will produce hits which are displaced from possible tracks by at least 16 ns in drift time.

Azimuthal cathode pad or strip readout, included on the outer surfaces of the superlayers, is needed for bunch assignment since the propagation time along the wires is 16 ns for the longer outer layers. Azimuthal cathode pads or strips will also be useful in reducing stereo ambiguities. There are 15 superlayers in all for a total of 120 measurements. The total number of cells is 122,368. The total number of radiation lengths is 8% for a particle traversing the central tracking chambers at 90°. The Large Solenoid Detector central tracking system geometry is summarized in Table I.[10] In order to provide momentum measurement for $1.2 \lesssim |\eta| < 2.5$, the Large Solenoid Detector included tracking in the intermediate region to take over where the central tracking ends. The intermediate tracking extends to $\pm$ 4 m along the beam line. Charged particles can be detected up to $|\eta| \leq 3.0$. The intermediate tracking consists of several superchambers, each with position measurements at several closely-spaced z values. Two options were considered: planes of parallel wires and radial chambers. Both options use cathode pads to read out the coordinate along the wires. The options for intermediate tracking have not yet been worked out in as much detail as the central tracking. Both require large numbers of readout channels.

The central and intermediate tracking systems for the Large Solenoid Detector are shown in Fig. 4. The entire Large Solenoid Detector is shown in Fig. 5. The momentum resolution for the Large Solenoid Detector as a function of polar angle and rapidity is shown in Fig. 6.

3. TRACKING SIMULATION

3.1 Simulation of a Central Tracking System for the SSC

The SSC central tracking system design used for this simulation was based on that in the Large Solenoid Detector[7], described in Section 2.3, although it is quite general and can be used for any system of cylindrically-oriented sensing elements. The central tracking system extended radially from 50 cm to 160 cm and covered $|\eta| <$ 1.2–1.5. The cylindrical layers of wires parallel to the beam direction were grouped into superlayers with eight layers each. Within each superlayer the layers were staggered by half the cell width, as shown in Fig. 3. All parameters of the detector, such as number of superlayers, number of layers in each superlayer, minimum and maximum radius for each superlayer, length

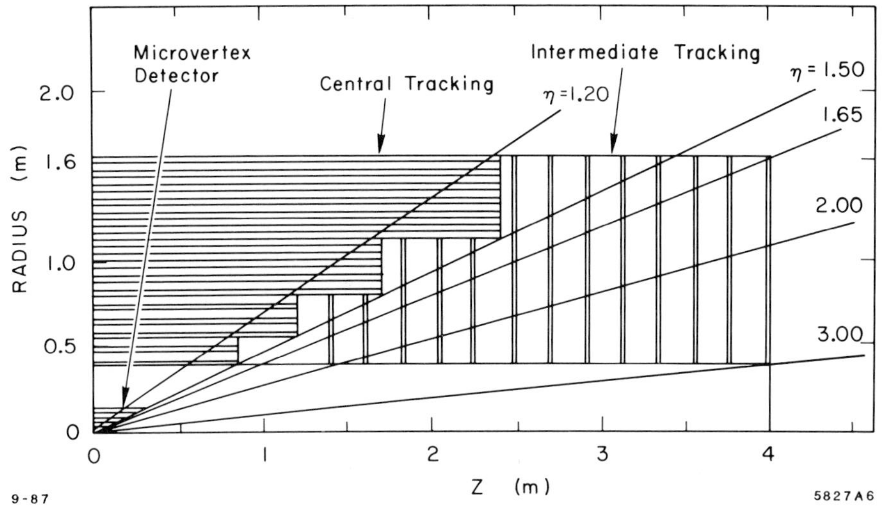

Fig. 4. Schematic view of central and intermediate tracking systems in the Large Solenoid Detector (from Ref. 7).

of each superlayer, and azimuthal spacing between sense wires can be specified independently. The parameters of the central tracking system used in this simulation are as shown in Table I, except that we included only the outer 13 superlayers. The design is based on a pattern recognition strategy of finding track segments in superlayers and then linking the segments to form tracks. Track segments in outer superlayers could be used in the trigger. We used a solenoidal magnetic field of 2 Tesla. The spatial resolution was taken to be 150 μm. So far, we have simulated only axial wires, that is, wires parallel to the cylinder axis.

Table I. Summary of Large Solenoid Detector Central Tracking System (from Ref. 7)

Super-layer Number	Inner Radius (cm)	Module Thickness (cm)	Half Length (cm)	Straw Diameter (mm)	Rapidity Range	Cell Occupancy (%)
1	40	2.7	85.2	3.92	1.50	12.1
2	48	2.7	85.2	3.92	1.34	9.1
3	56	2.7	119.0	3.92	1.50	8.8
4	64	2.7	119.0	3.92	1.38	7.0
5	72	4.1	119.0	5.89	1.28	13.0
6	80	4.2	170.0	6.04	1.50	14.5
7	88	4.2	170.0	6.17	1.41	12.9
8	96	4.3	170.0	6.28	1.34	11.6
9	104	4.4	170.0	6.38	1.27	10.5
10	112	4.5	238.5	6.47	1.50	11.9
11	120	4.5	238.5	6.55	1.44	10.9
12	128	4.6	238.5	6.61	1.38	10.0
13	136	4.6	238.5	6.68	1.33	9.3
14	144	4.6	238.5	6.73	1.28	8.5
15	152	4.7	238.5	6.78	1.23	7.9

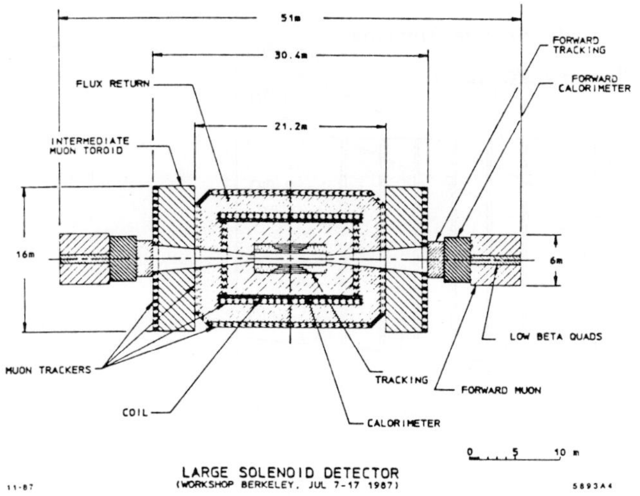

Fig. 5. Schematic view of the Large Solenoid Detector from the 1987 Berkeley Workshop.[7]

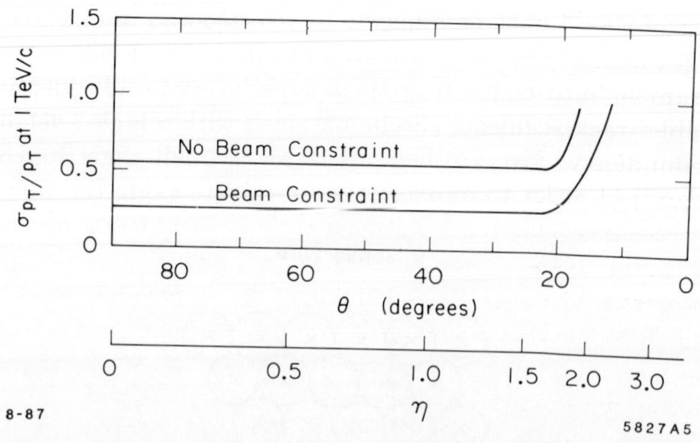

Fig. 6. Momentum resolution as a function of polar angle and rapidity in the Large Solenoid Detector for the 13 superlayers at radii > 50 cm in the central tracking system and including intermediate tracking (from Ref. 7).

We used ISAJET[11] to generate events from interesting physics processes, such as high-p_T two jet events or Higgs boson production, and from inelastic scattering background, for which we used minimum bias events. We used the GEANT3[12] general-purpose detector simulation package to simulate the interactions of the particles with the detector. One of the reasons for using GEANT3 was the hope that this tracking simulation software could be used with other GEANT-based subroutines to simulate SSC detectors. In fact, we have already realized some of the advantages of using a general-purpose detector simulation package in that we were able to use the GEANT3 environment, especially the graphics interface, set up by the SLD Collaboration on the SLAC IBM 3081 for their detector simulation.

Using GEANT, the particles interact in the 8% of a radiation length of material due to straw tube walls, wires, and gas[7] (the material was assumed to be distributed uniformly throughout the tracking volume), including photon conversion and multiple Coulomb scattering. Charged particles are bent in the 2 T magnetic field. Every time a charged particle crossed a layer boundary, a hit was produced representing the distance of closest approach to the nearest sense wire. Only the radial volumes were specified in GEANT, not the azimuthally-spaced straw tubes, in order to save computer time. The drift distances were smeared according to a Gaussian distribution with width equal to the spatial resolution. The drift distances were then converted to drift times using a drift velocity of 50 μm/ns. Digitizations were then produced representing a drift time and wire number for each track in each layer. If a track did not reach a layer because of bending in the magnetic field or leaving the tracking volume, no hit or digitization was produced. We included the background from inelastic scatterings in the same and out-of-time bunch crossings by superimposing the digitizations from minimum bias events. We included the number of bunch crossings indicated by the resolving time of the straw tube cell. At each bunch crossing the number of events to be included was determined from a Poisson distribution with a mean of 1.6 interactions per bunch crossing. Drift times from background events were then increased or decreased by the time difference between the bunch crossing of the background event and the bunch crossing of the event of interest. The double-hit resolution was equal to the cell width, that is, only one hit was allowed per wire. We discarded all but the earliest hit with positive time. In addition, we removed digitizations in the tails of those from previous bunch crossings, that is, we assumed that a wire was dead for a time equivalent to the cell width. The simulation program is described in more detail in Ref. 13.

3.2 Results of the Simulation

We used the simulation described above to study tracking in SSC events. The first type of events we examined were high-p_T ($p_T > 1$ TeV/c) two-jet events. Figure 7(a) shows such an event in the detector described in Section 2.3 with a 2 Tesla magnetic field. Figure 7(b) shows an enlargement of the same event in the outer two superlayers in the area of the dense jet. Figure 7(c) shows the earliest hits in the cells for the tracks shown in Fig. 7(b). When two tracks are close together, they produce hits only in every other layer because the cells in every other layer are staggered azimuthally so that one track passes closest to the sense wires in every other layer and the other track is closest in the alternate layers. Figures 7 and 8 do not show hits from background events or converted photons.

Some detector studies in the past have focused on nonmagnetic detectors.[14] Figure 8(a) shows the same event as in Fig. 7(a) with no magnetic field. Figure 8(b) shows the tracks and earliest hits in the cells in an enlarged region in the area of the dense jet in the outer two superlayers, as in Figs. 7(b) and (c).

Although we have not quantified these observations with high-statistics studies, we note:

1. Although these events have very dense jets which seem at first to be impossible to resolve, when viewed on the scale of the wire spacings most of the hits appear to lie on identifiable tracks with a 2 Tesla magnetic field, particularly in the outer superlayers.

2. Eight layers in a superlayer is probably close to the optimum number because two tracks which are as close as the wire spacings produce hits only on every other layer because of the staggering. Some of these hits may be lost due to nearby curling tracks or background hits. Three tracks within the wire spacing distance would not be resolvable.

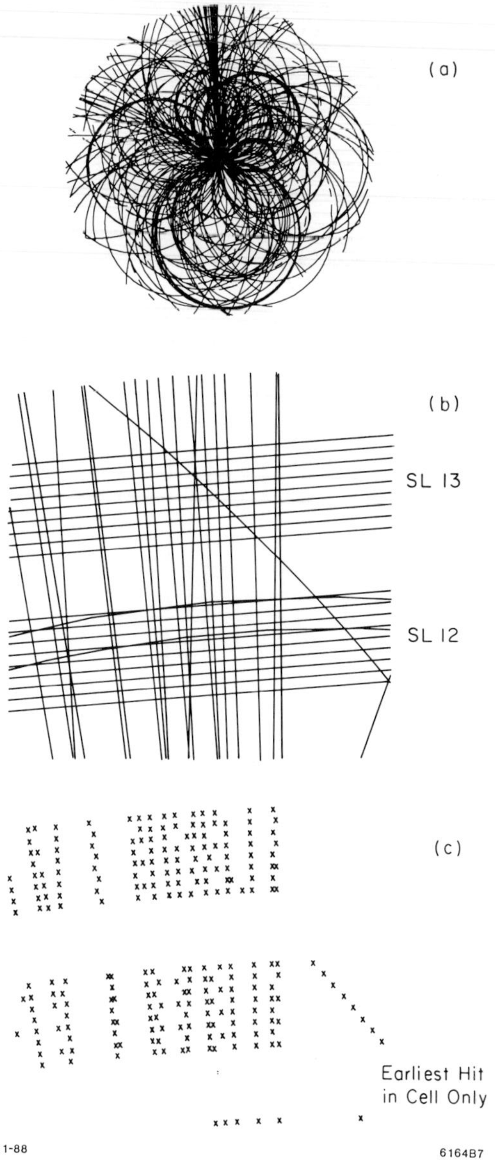

Fig. 7. (a) Two-jet event from ISAJET with $p_T > 1$ TeV/c in a 2 Tesla magnetic field in a detector of the geometry of the Large Solenoid Detector. There are 223 particles with $p_T > 200$ MeV/c and $|\eta| < 1.5$. Converted photons and background from minimum bias events are not shown. (b) Enlargement of the event in the outer two superlayers in the area of the dense jet at the top of the detector. (c) Earliest hit in each cell for the tracks shown in (b).

3. Although a 2 Tesla magnetic field produces curling tracks which obscure the high-p_T tracks to some extent, particularly in the inner superlayers, the effect in the outer superlayers is to spread out the tracks and, of course, remove the low-p_T tracks from consideration.

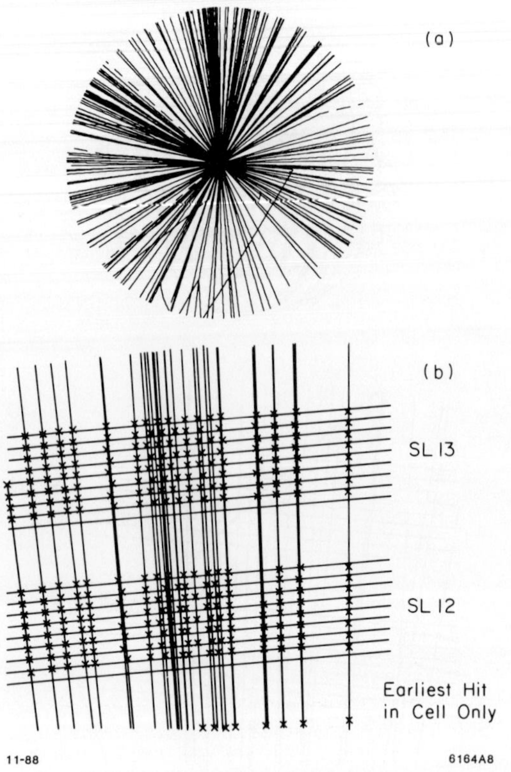

Fig. 8. (a) Same event as in Fig. 7(a) with no magnetic field. (b) Tracks and earliest hits in the cells in an enlarged region in the outer two superlayers in the area of the dense jet. Converted photons and background from minimum bias events are not shown.

We next turned our attention to events from Higgs boson production, $pp \to HX$, with the Higgs decaying to Z^0Z^0 and both Z^0's decaying to e^+e^- or $\mu^+\mu^-$. We used a Higgs mass of 400 GeV/c^2. Such events allowed us to focus on the measurement of the high-p_T particles from the Higgs decay. Any large solid angle SSC detector must be able to measure such events. Also, these events are not as trivial to deal with as might have been naïvely guessed. There are many tracks from the underlying event and from the particles recoiling against the Higgs boson, even before adding the hits from background interactions. For these events we used the full simulation as described in the previous Section. We generated $\sim$ 200 events. An example of a Higgs event in the simulated central tracking system is shown in Fig. 9. The leptons from the Higgs decay are indicated by the heavier lines.

The p_T distribution for all tracks in the Higgs events is shown in Fig. 10(a). Figure 10(b) shows the p_T distribution for leptons only. Nearly all of the tracks with $p_T > 50$ GeV/c are the leptons from the Higgs decay. The fully-simulated events, including adding digitizations from minimum bias background events and removing digitizations within the double-hit resolution, had 12,000–25,000 digitizations, as shown in Fig. 11(a). The fraction of digitizations from the minimum bias background events is shown in Fig. 11(b). On the average 57% of the digitizations were due to background events. The fraction of digitizations lost for all tracks in the Higgs events as a function of superlayer is shown in Fig. 12(a). On the average $\sim$ 12% of the digitizations were lost because of the double-hit resolution, and the loss was about the same in all superlayers. The fraction of digitizations lost for the leptons from the Higgs decay is shown in Fig. 12(b). An average of 7.3% of the digitizations were lost for the leptons from the Higgs decay with the worst losses being in the superlayers at 70–90 cm radius, presumably because of curling tracks.

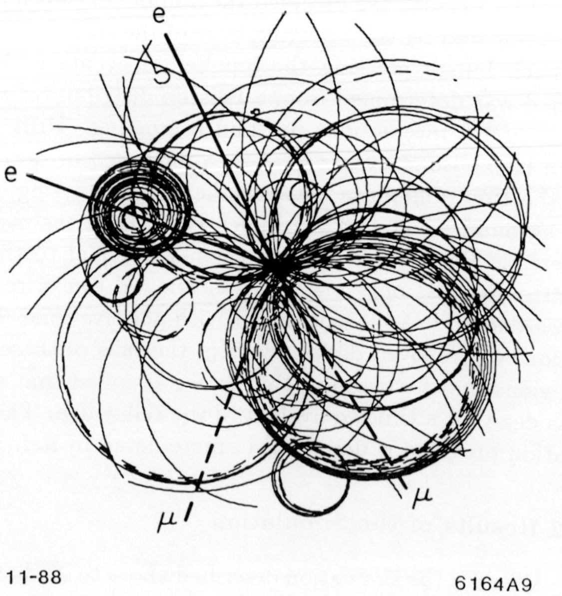

Fig. 9. Example of a Higgs event in the simulated central tracking system. The leptons from the Higgs decay are indicated by the heavier lines. Converted photons and other interactions with the material are included.

3.3 Pattern Recognition

We began working on pattern recognition algorithms in order to examine our original design goals of finding track segments in superlayers and removing hits from out-of-time bunch crossings. We also wanted to make the algorithm simple with the hope of using it in the trigger. The algorithm for finding track segments was the following:

1. In each superlayer we identified "roads" containing hits. There are two parameters which can be varied: the width of the road and the number of hits required on the road. We used a width of 5 wires (22 mrad in the outer superlayer) and required 3 or more hits out of 8 possible. For isolated tracks one could require more hits; however, if two tracks are close together, as in Fig. 7, they will produce hits only on alternate

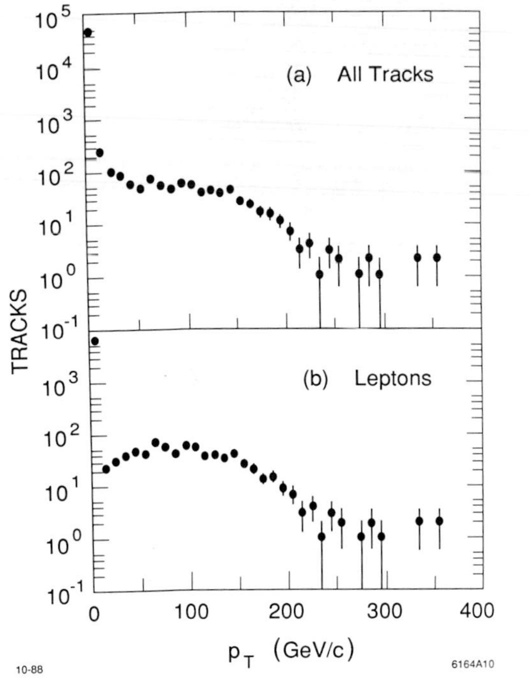

Fig. 10. (a) p_T distribution for all tracks in the Higgs events. (b) p_T distribution for the leptons in the Higgs events.

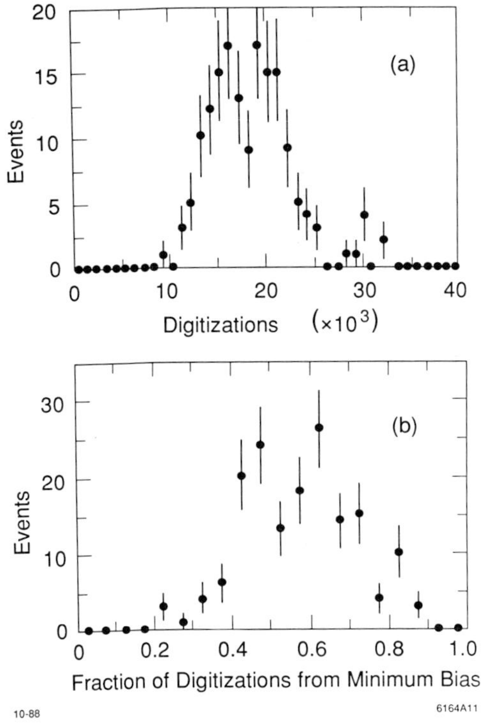

Fig. 11. (a) Total number of digitizations in Higgs events, including digitizations from minimum bias background events. (b) Fraction of digitizations from the minimum bias background events. The mean is 0.572 ± 0.011.

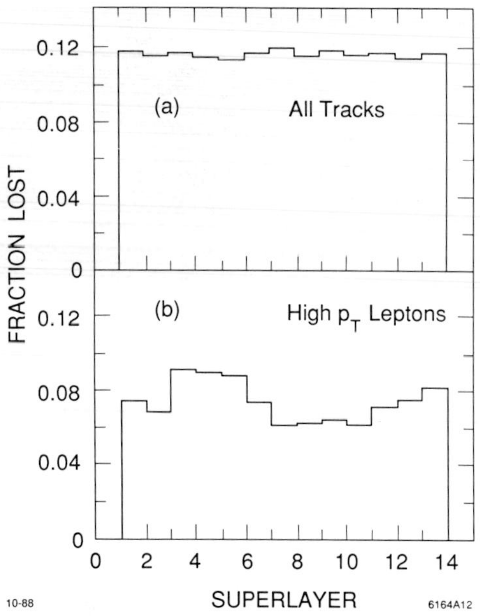

Fig. 12. (a) Fraction of digitizations lost for all tracks in the Higgs events as a function of superlayer. The mean is (11.58 ± 0.65)%. (b) Fraction of digitizations lost for the leptons from the Higgs decay as a function of superlayer. The mean is (7.33 ± 0.61)%.

layers and if one is lost due to the double-hit resolution there will be only 3 hits. The road requirement discriminates against low-p_T tracks.

2. We required that at least one of the hits be in a layer with the opposite wire stagger from the others so that the left-right ambiguities could be resolved and hits from out-of-time bunch crossings rejected.

3. We required that the hits be consistent with a straight line to within an error and in the process resolved the left-right ambiguities. Of course, the tracks approximate straight lines only locally within the superlayer, and the spatial resolution must also be taken into account.

Figure 13(a) shows all of the digitizations[15)] for the event shown in Fig. 9, including those from minimum bias background events. Figure 13(b) shows only those digitizations which are included in segments. Keeping only those digitizations which form segments cleans up the events considerably. Figure 13(c) shows the tracks from the original event in the outer five superlayers in the region around the muon at the lower right. Figure 13(d) shows all of the digitizations in the event in the enlarged region (the digitizations are displayed at the locations of the hit wires). Finally, Fig. 13(e) shows only those digitizations which form track segments; here, the left-right ambiguities have been resolved, the drift times have been converted to distances, and the digitizations are displayed at the positions of closest approach of the tracks to the wires. One can

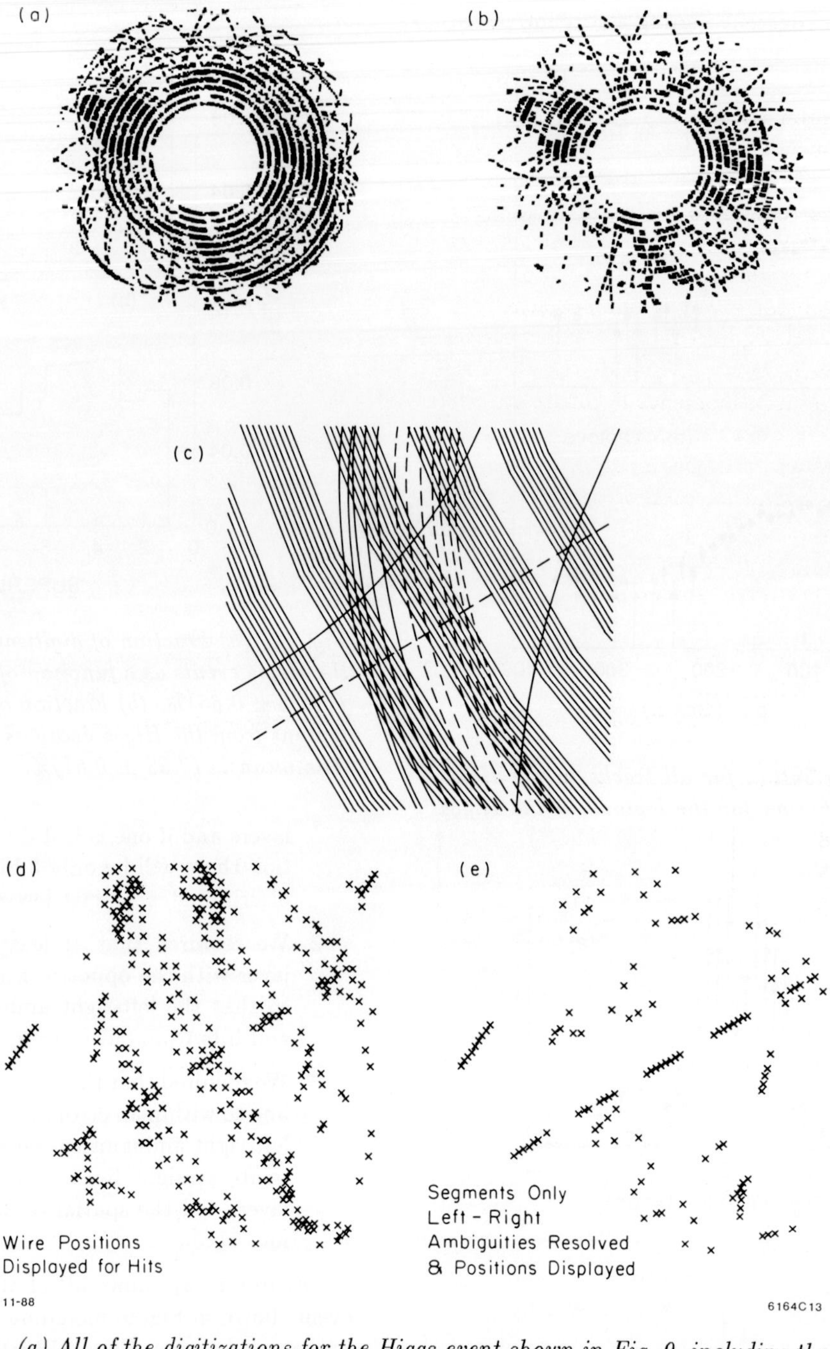

Fig. 13. (a) All of the digitizations for the Higgs event shown in Fig. 9, including those from minimum bias background events. (b) Digitizations for this event which are included in track segments, as defined in the text. (c) Tracks from the original event in an enlarged region in the outer five superlayers in the region around the muon at the lower right. (d) All of the digitizations in the event in the enlarged region of (c) (the digitizations are displayed at the locations of the hit wires). (e) Only those digitizations which form track segments in the enlarged region. Here, the left-right ambiguities have been resolved, the drift times have been converted to distances, and the digitizations are displayed at the positions of closest approach of the tracks to the wires.

clearly identify the muon track, and most of the extra hits have been removed.

Depending on the effects on the physics analyses, we might envision making this requirement at the processor level, reading out only the hits that form track segments.

Next, we applied our segment-finding algorithm to the e and μ tracks from Higgs boson decays. We defined two classes of segments: a "good" segment was one with at least five hits from a lepton track and no other hits, and an "OK" segment was one with at least five hits from the lepton track and one hit from another track. The effects of hits from other tracks remain to be studied; however, since all of the hits were consistent with a straight line we would expect the segment with one hit from another track to give nearly the same fitted straight line segment as it would without the hit. Besides, in reality, we will not know which hits are from which tracks. We will compare measured momenta with produced momenta in future studies. With these definitions, we counted the number of segments found for each lepton track. As an example, Table II gives the number of segments found for each of the four lepton tracks in the event shown in Figs. 9 and 13. The maximum number of segments is 13, the number of superlayers. As we shall see, this event is rather typical.

Table II. Segments Found for Four Lepton Tracks in Sample Event

e or μ Track	Good Segments	Total (Good or OK) Segments
1	4	9
2	8	11
3	5	8
4	9	11

The distribution of the number of good segments for the e's and μ's in the Higgs events is shown in Fig. 14(a). The corresponding distribution of total (good or OK) segments is shown in Fig. 14(b). We see that the lepton tracks from Higgs decay have an average of about 8 good segments and 10 total segments out of 13 possible. Figure 15(a) shows the fraction of segments which are good as a function of superlayer number, whereas Fig. 15(b) shows the same distribution for the fraction of segments which are either good or OK. Typically 30–50% of segments were good in the inner superlayers, increasing to almost 80% for the outer superlayers. When OK segments are counted as well, 50–60% of segments are accepted for inner superlayers and over 80% for outer superlayers. Segments were most likely to be lost in the superlayers at 70–90 cm, as noted in the previous Section for hits lost.

3.4 Future Work

We are planning to continue our tracking simulation studies using the software we have developed. Future work will include:

1. Continued studies of pattern recognition and track finding along the lines described in previous Sections, including linking the segments to form tracks.

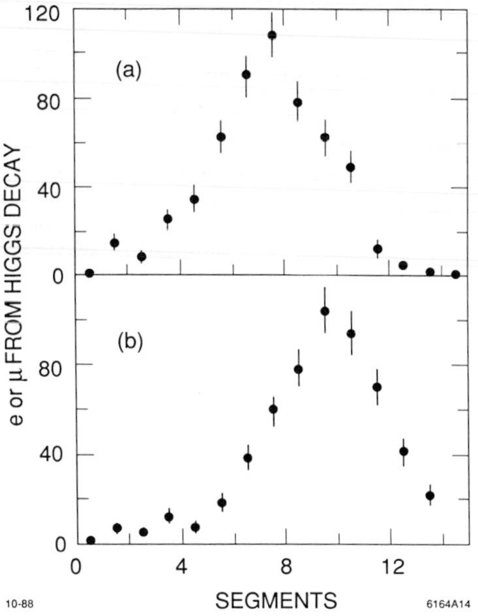

Fig. 14. (a) Distribution of the number of good segments out of 13 possible for the e's and μ's from the Higgs decays. (b) Distribution of the number of total segments (good or OK) for the leptons from the Higgs decay.

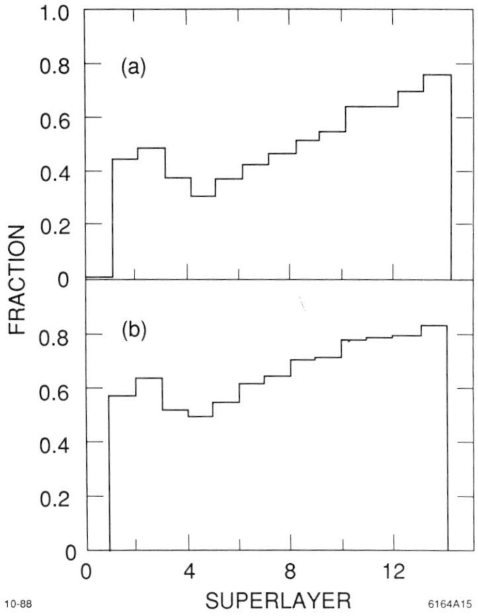

Fig. 15. (a) Fraction of segments which are good as a function of superlayer number for the lepton tracks from the Higgs decay. (b) Fraction of segments which are either good or OK as a function of superlayer.

2. Simulation of small-angle stereo wires and cathode strips for reconstruction of the direction along the wires. Stereo segments can then be found in the same manner as axial segments, and the stereo segments can be linked to axial segments to form tracks. At this point we will be able to study how much ad-

ditional information from cathode strips is needed in order to link the stereo hits.

3. A more realistic simulation of electron drift in small-cell or straw tube drift chambers is needed, including the effects of $\mathbf{E} \times \mathbf{B}$. For the study described here we have used only the geometric distance of closest approach.
4. Conceptual design and simulation of intermediate tracking,[7] ($1.2 < |\eta| < 2.5$) as described briefly in Section 2.3.
5. Tracking studies for different physics processes, such as new heavy fermions, supersymmetric particles, and high-p_T two-jet events.
6. Vary tracking system parameters such as cell radius to determine the effects on pattern recognition and tracking efficiency.
7. Begin to develop a realistic design for a tracking system for a complete SSC detector, including other detector components.

4. CONCLUSIONS

We have shown that an SSC tracking system design based on a pattern recognition strategy of finding track segments in superlayers appears to provide a powerful means of finding tracks in complex SSC events, even in an environment of multiple events from several bunch crossings. So far, detailed simulations have verified the concepts developed over several years for SSC tracking detectors. Although a great deal of work remains to be done, we are hopeful that an SSC tracking system based on conventional wire chambers will enable us to explore the new physics which awaits us in the SSC regime.

As an example of an existing tracking detector based on track segment finding, we show in Fig. 16(a) a Monte Carlo simulation of a multihadronic Z^0 decay in the Mark II central drift chamber.[16] Figure 16(b) shows a real SLC background event superimposed on the Z^0 event. The background event has an occupancy of $\sim 50\%$. Also shown in Figs. 16(a) and (b) are the tracks found by the standard Mark II pattern recognition and track fitting programs. Even in the presence of the very high background occupancy, the original tracks can be found with reasonably high precision. The Mark II detector would not take data under these circumstances because other parts of the detector, in particular, the calorimeters and time-of-flight scintillation counters, cannot operate with such high backgrounds. The SLC backgrounds originate from a much different source than SSC backgrounds, namely tails of the beam scraping various apertures. The Mark II central drift chamber, which is a wide cell jet chamber, would not be able to operate in the SSC environment. Nevertheless, this example shows the power of track segment finding in a high background situation. Another point which can be

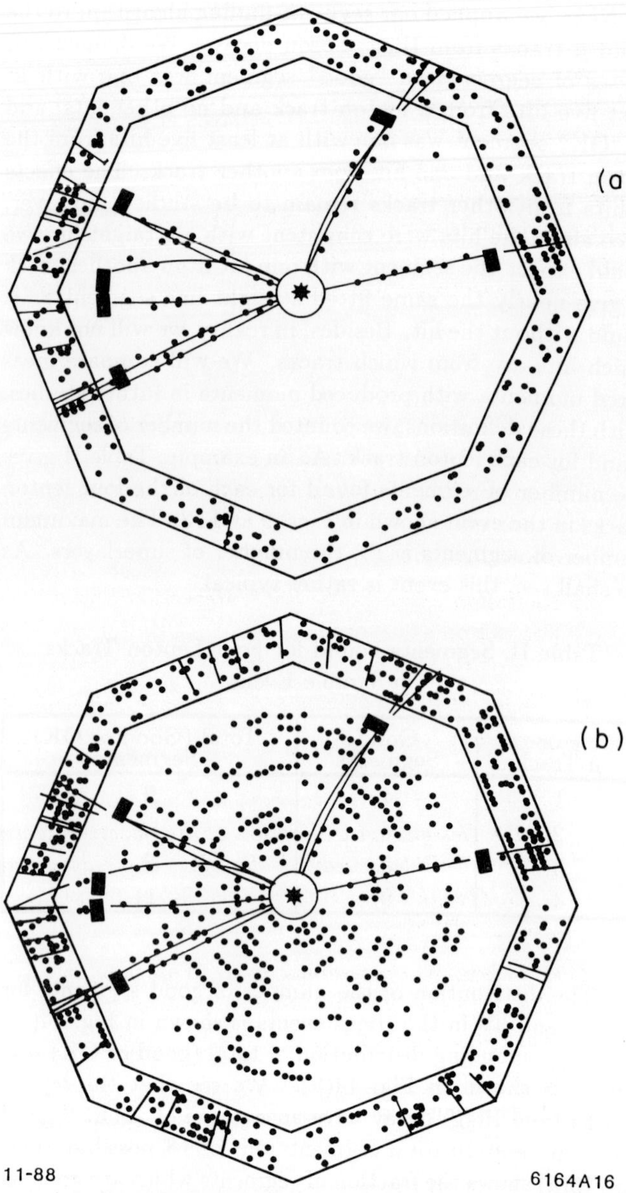

Fig. 16. (a) Monte Carlo simulation of a multihadronic Z^0 decay in the Mark II central drift chamber. (b) Real background event from the SLC superimposed on the Z^0 event. The background event has $\sim$ 50% occupancy. Also shown in (a) and (b) are the tracks found by the standard Mark II pattern recognition and track fitting programs.

made is that very high energy e^+e^- linear colliders which can probe physics in the same energy region as the SSC will be even more susceptible to backgrounds of the SLC type. Although the interactions themselves may be much cleaner than at the SSC, the backgrounds have the potential of being much worse. In either the SSC environment or at high-energy e^+e^- linear colliders a tracking system based on track segment finding has the best capability of surviving and yielding tracks in high background situations.

ACKNOWLEDGMENTS

We would like to thank members of the SLD Collaboration at SLAC, especially Dave Aston, for their help in getting GEANT running on the SLAC IBM computer and for providing the interface to GEANT graphics. We wish to thank Gil Gilchriese for organizing such a stimulating and informative conference. We gratefully acknowledge the support of the Department of Energy Program for Generic Detector Research and Development for the SSC.

REFERENCES

1) D. G. Cassel, G. G. Hanson *et al.*, "Report of the Central Tracking Group," in *Proceedings of the 1986 Summer Study on the Physics of the Superconducting Supercollider, edited by R. Donaldson and J. Marx, Snowmass, CO*, 1986, p. 377.

2) *Radiation Levels in the SSC Interaction Regions*, Task Force Report, edited by D. E. Groom, SSC–SR–1033, SSC Central Design Group, June, 1988.

3) *Report of the Task Force on Detector R&D for the Superconducting Super Collider*, SSC-SR-1021, SSC Central Design Group, June, 1986, pp. 44-60; M. G. D. Gilchriese, in *Proceedings of the 1984 Summer Study on the Design and Utilization of the Superconducting Super Collider, edited by R. Donaldson and J. G. Morfin, Snowmass, CO*, 1984, p. 607; G. Hanson and D. Meyer, *ibid.*, p. 585.

4) J. Va'vra, in *Proceedings of the Workshop on Radiation Damage to Wire Chambers, edited by J. Kadyk, Lawrence Berkeley Laboratory, Berkeley, CA*, 1986, p. 263.

5) A. H. Walenta, Nucl. Instr. and Meth. **217**, 65 (1983).

6) R. Thun, "Prospects for Wire Chambers at High Luminosity," in these Proceedings.

7) G. G. Hanson, S. Mori, L. G. Pondrom, H. H. Williams *et al.*, "Report of the Large Solenoid Detector Group," in *Proceedings of the Workshop on Experiments, Detectors, and Experimental Areas for the Supercollider, edited by R. Donaldson and M. G. D. Gilchriese, Berkeley, CA*, 1987, p. 340.

8) R. L. Gluckstern, Nucl. Instr. and Meth. **24**, 381 (1963).

9) R. DeSalvo, "A Proposal for an SSC Central Tracking Detector," CLNS 87/52.

10) The cell occupancies have been increased from those in Ref. 7 to reflect the increase of the number of charged particles per unit rapidity from 6 to 7.5.

11) F. E. Paige and S. D. Protopopescu, "ISAJET 5.30: A Monte Carlo Event Generator for pp and $\bar{p}p$ Interactions," in *Proceedings of the 1986 Summer Study on the Physics of the Superconducting Supercollider, edited by R. Donaldson and J. Marx, Snowmass, CO*, 1986, p. 320. (The current version of ISAJET is 6.12.)

12) R. Brun, F. Bruyant and A. C. McPherson, *GEANT3 User's Guide*, CERN DD/EE/84-1.

13) A. P. T. Palounek, "Simulating a Central Drift Chamber for a Large Solenoid Detector at the SSC," SLAC–PUB–4787.

14) T. Åkesson *et al.*, "Report of the Non-magnetic Detector Group," in *Proceedings of the Workshop on Experiments, Detectors, and Experimental Areas for the Supercollider, edited by R. Donaldson and M. G. D. Gilchriese, Berkeley, CA*, 1987, p. 472.

15) The GEANT3 graphics does not display digitizations, only hits. To display digitizations, they must be converted to spatial coordinates and stored in the hit bank.

16) G. Hanson, "The New Drift Chamber for the Mark II Detector at the SLAC Linear Collider," in *Proceedings of the Fourth International Wire Chamber Conference, Vienna, Austria, 1986*, Nucl. Instr. and Meth. **A252**, 343 (1986).

DEVELOPMENT OF MULTIDRIFT TUBES

Fabio Sauli

CERN, Geneva, Switzerland

ABSTRACT

The multidrift tube represents a modular approach to the problem of building a position-sensitive, high-accuracy gas detector that is capable of withstanding the very high fluxes foreseen at a short distance from the intersection region in future colliders. Several prototypes have been built and tested; these contain, within a hexagonal carbon-fibre envelope, around 70 independent drift cells (also of hexagonal geometry). Using dimethyl ether as the gas filling, we have obtained radial localization accuracies of ~ 50 μm r.m.s. per wire, and longitudinal accuracies (through charge division) of ~ 0.8% of the wire length. The radiation tolerance of the modules has been investigated; good operation up to about 1 C/cm of wire (at nominal gain, about 1 Mrad/cm) was demonstrated.

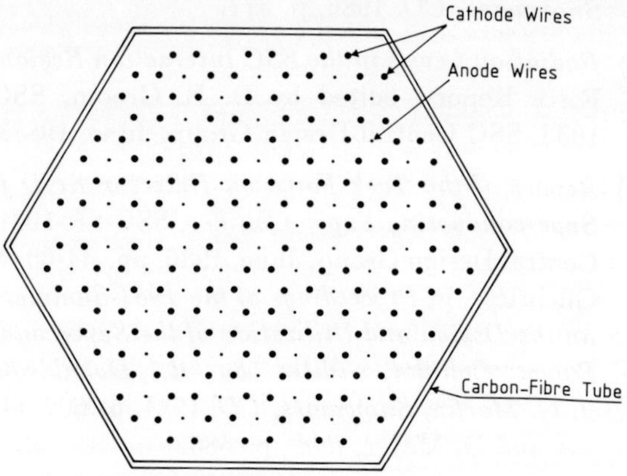

Fig. 1 Schematic showing the geometry of the MDM.

INTRODUCTION

The design of a central tracking device based on gaseous detectors for high-luminosity colliders has to satisfy two major imperatives: i) an excellent level of reliability and ii) endurance under very high radiation levels. The Multidrift Module detector (MDM) was proposed some time ago as a device capable of fulfilling those requirements.[1)]

The MDM consists of a thin, hexagonal envelope containing a large number of wires constituting individual drift cells, also of hexagonal geometry (Fig. 1). A measurement of the drift-time of electrons released in the gas by ionizing particles allows the radial coordinate of tracks to be deduced; the measurement of the charge ratio on the ends of each anode provides the longitudinal coordinate. The small radius of the cell (1.45 mm in the present design) makes it possible to obtain a high granularity with good multitrack resolution, and keeps the total drift-time at values (30–40 ns, depending on the gas filling) that are compatible with high-rate operation. Modules can be assembled around an intersect in a regular pattern, each element being easily replaceable in case of failure.

The present paper briefly describes some recent measurements of localization accuracy and rate tolerance realized with various MDM prototypes. This work is part of the LAA Project on Detector Research and Development at CERN (led by A. Zichichi).

CONSTRUCTION AND OPERATION OF THE MDM

As the method of constructing the MDM has been described in detail in previous references[2–4)], it will only be summarized here. Anode and cathode wires are stretched between two precision-drilled fibre-glass plates and soldered onto miniature brass pins having V-shaped grooves cut along their axes. We have verified an overall radial wire-positioning accuracy of better than 20 μm within each module. All cathode pins, and therefore the corresponding wires, are in electrical contact through a common thin copper layer pasted to the outside of the end-plates; anode wires are read out individually from both sides of the module. To avoid arcing, each end of the module is filled with epoxy, which is also used to fasten the carbon-fibre envelope to the end-plates; protruding

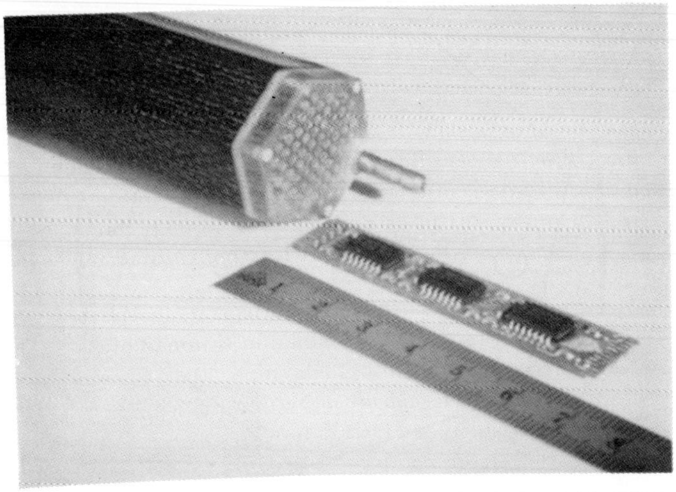

Fig. 2 One end of a completed module, showing the anode pins and the gas inlet. A 12-channel preamplifier card that plugs directly on the anode pins is also shown.

anode pins allow electrical contact for readout. Figure 2 is a close-up of one end of a completed MDM; the picture also shows a high-density preamplifier card that is designed to plug directly into the MDM anode pins (see below).

The operating characteristics of the MDM depend, of course, on the gas filling, which can be chosen in a wide range. For reasons that will be discussed below, we have preferred to use dimethyl ether [DME: $(CH_3)_2O$]. With a density of 2.4×10^{-3} g/cm^3 at atmospheric pressure, DME is a convenient medium for the detection of charged tracks in small cells [a minimum-ionizing particle releases, on the average, 160 ion pairs per centimetre].[5] Moreover, DME has very good quenching properties, allowing very high stable gains to be reached without secondary effects. We have built several prototype MDMs with anode wire diameters between 20 and 50 μm, and have obtained very comfortable, several hundred volts long, efficiency plateaux; the preferred choice at the present time is a 30 μm diameter stainless-steel anode wire, thick enough to have good mechanical properties and impervious to accidental sparking, but still with a resistivity high enough to permit localization through charge division (about 1 kΩ/m). A third, and even more important, reason for using DME is the expected high density of primary interactions as compared with other gases, given the total energy loss. Indeed, for ionizing trails close to the wires, one of the major sources of dispersion in the time measurement (and therefore in the localization accuracy) is the average distance between primary encounters. We have recently measured in DME a primary ionization density, produced by minimum-ionizing particles, of 55 ± 5 pairs per centimetre;[6] in argon, the same measurement has given the value of 25 pairs per centimetre, consistent with accepted values. Last, but not least, there is confirmed evidence that proportional counters with a DME gas filling are impervious to radiation damage due to polymerization;[7,8] this point will be discussed further.

In the course of development of the MDM, we have used various versions of home-made[9] or commercial amplifier–discriminators followed by time-to-digital and charge-to-digital converters. For example, the high-density preamplifier card, which is shown in Fig. 2, and which is based on a Fujitsu MB43458 quad-charge amplifier, is designed so as not to exceed the cross-section of the MDM itself, thus allowing the assembly of several modules in an array. Signals from each anode wire — around 70 per module — are then transported to a remote receiver and shaper, via miniature coaxial ribbon cables, and digitized.

Several fully instrumented MDM prototypes have been tested, both in the laboratory — with the help of radioactive sources — and in a high-energy charged particle beam. In the latter case, the trajectory of each track could be measured independently by a pair of high-precision silicon strip detectors.*) The gating for charge measurements and the reference for the drift-time measurement were provided by two small, fast scintillation counters in line. Data were collected and recorded on tape with the help of a small on-line computer.

EXPERIMENTAL RESULTS

Figure 3 shows a typical event for a single track crossing the detector. Circles represent the measured drift-time on each anode; the fitted track is also shown. With the present geometry, each track is sampled about ten times in each module. In order to deduce, from the drift-time

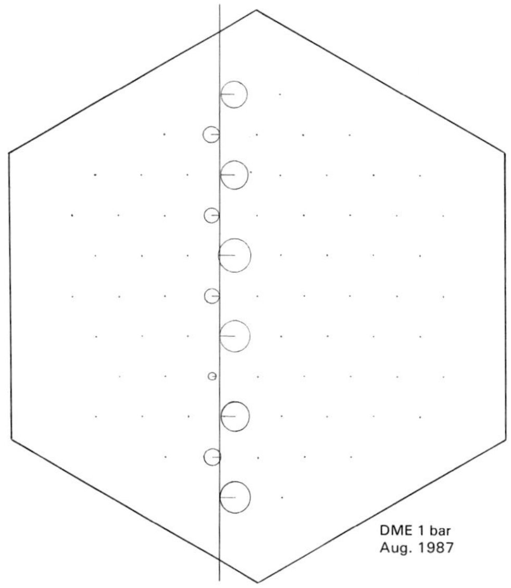

Fig. 3 A beam track recorded with a complete MDM prototype; circles represent the measured drift-time on the anodes.

*) Courtesy of the NA32 experiment at CERN.

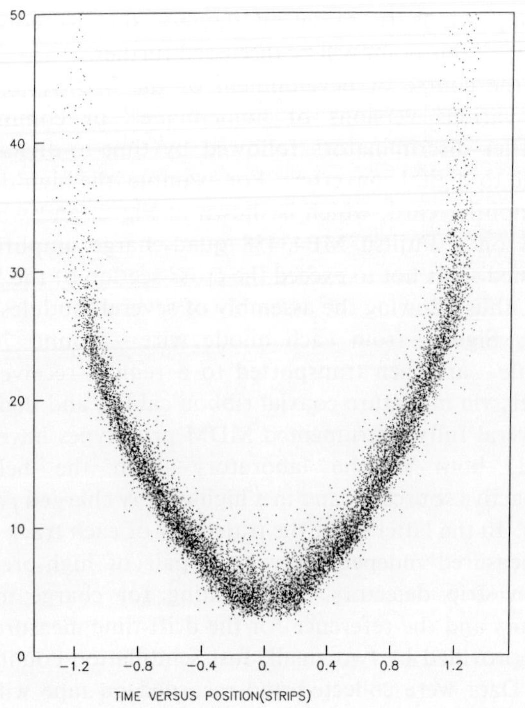

Fig. 4 Space–time correlation measured on a single wire in the MDM for minimum-ionizing particles. The true position (horizontal scale) is provided by a set of silicon microstrips.

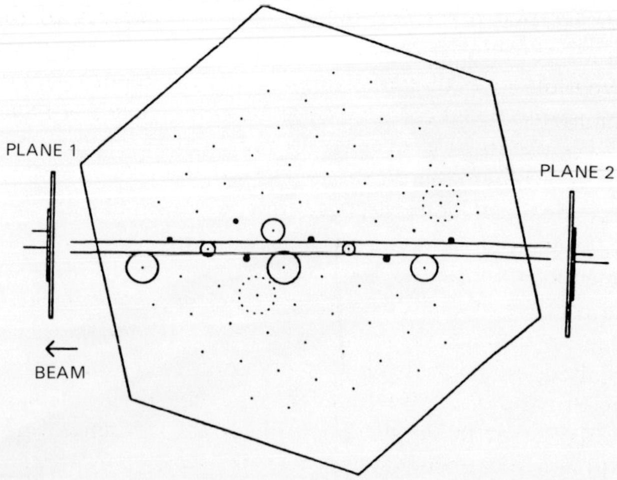

Fig. 5 Two tracks at about 1 mm, reconstructed using the MDM data only. The coordinates provided by the silicon strips are also shown; they do not line up with the MDM because of a compression of scale in the drawing.

measurement, the radial distance of the track from the anodes, one needs, of course, to know the space–time correlation. This could, in principle, be computed from the knowledge of the electric field geometry and of the drift velocity. However, the latter is not well known at the high field-values encountered around the anode wires in the MDM, and we have preferred to measure it directly by making a comparison with the coordinate provided by the silicon-strip detector. An example of a scatter plot correlating the real space coordinate—as provided by the silicon strips—and the measured drift-time for one wire is given in Fig. 4. Because of the large variation of field across the cell, the correlation is not linear, and moreover it depends slightly on the incidence angle of the track (with an hexagonal symmetry); the data analysis allows average correlations to be computed as a function of the angle, and these are parametrized for use by the fitting program. The width of the scatter plot provides information about the single-wire localization accuracy: it is around 50 μm r.m.s., including the accuracy of the silicon strips (which is probably not better than 20–30 μm r.m.s., owing to our coarse calibration procedure for the strips' electronics).

Because of the small cell size, it is not possible to resolve two tracks produced by the same event within one cell. However, in the case of very close tracks, the geometry of the MDM enables the coordinates to be measured alternatively by the rows of wires to the right and left of the tracks. The two-track resolution depends, then, essentially on the sophistication of the reconstruction program; we expect to be able to identify tracks that are only half a millimetre apart. Figure 5 shows an example of two tracks, about 1 mm apart, which have been recognized by the silicon strips and reconstructed in the MDM.

The longitudinal coordinate can be deduced from the ratio of charges measured on the two sides of the anodes. We have measured the localization accuracy in this direction in one of the MDM prototypes, which has 31 μm stainless-steel anode wires, 400 mm long, for both minimum-ionizing particles and soft X-rays. The single-wire accuracy distribution for fast particles in a MDM filled with DME has a width corresponding to 3.5 mm r.m.s. (or 0.9% of the wire length ℓ) (see Fig. 6). The correlation between the real source position and the measured longitudinal coordinate is linear within the experimental errors.

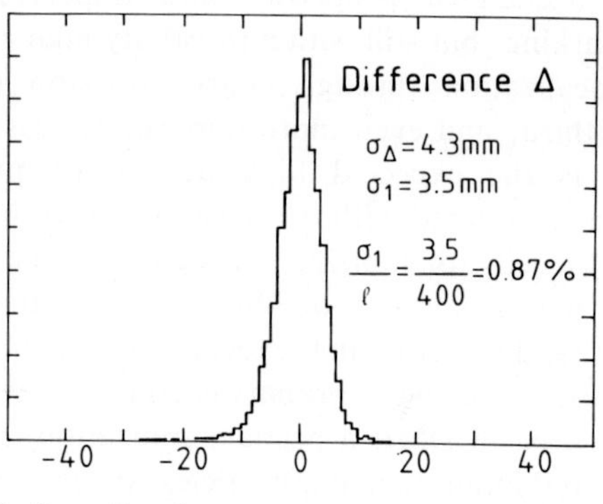

Fig. 6 Localization accuracy measured from charge division in a 40 cm long MDM for fast electrons; the width of the distribution, computed from a three-wire fit, allows us to deduce a single-wire accuracy of 3.5 mm r.m.s.

We have also verified the localization accuracy of the module for the detection of soft X-rays, using more conventional gas fillings (a noble gas and a quencher). For a collimated beam of 22 keV X-rays and in xenon–ethane, we have obtained a localization accuracy of 3 mm r.m.s. The MDM can therefore be a very powerful high-rate X-ray position detector, and this because of the time resolution (35 ns) and the density of the detection elements. In particular, we foresee applications as position detectors for crystal diffraction studies at the synchrotron radiation sources. Of course, the time information is not exploited, and localization is obtained from the current division and the wire number. The classic problem of parallax error in the detection of X-rays in conventional multiwire chambers having a thick conversion space is reduced by the high density of the wires. Another possible use of the device is for transition radiation detectors, because of the convenient ratio of energy loss on a given wire between X-rays (10 keV or more) and the charged component of energy loss for particles (1.5 keV per wire for a xenon filling).

OPERATION AT HIGH FLUXES, AND AGEING

Two problems arise when a gaseous proportional counter is exposed to high radiation fluxes. First, the large amount of slow positive ions released in the avalanches create a local space charge, modifying the electric field, and therefore the gain, above a certain flux. Secondly, the molecules of the gas (especially if it contains an organic quencher) break down and recombine, forming heavy polymers that deposit on electrodes and permanently modify the operating characteristics of the counter (ageing).

Because of the small drift region and the high values of the field in the cells of the MDM, we had reason to believe that the space-charge effects would be rather small compared with those of a conventional multiwire proportional chamber (MWPC). This is indeed the case with the MDM with a DME gas filling, as shown in Fig. 7. The curve provides the normalized charge gain as a function of charge rate per unit length of wire; a space-charge gain reduction intervenes at a charge rate of around 5×10^{-2} μA/cm. In a conventional MWPC with small wire spacing, an equivalent gain reduction is observed at a rate of 10^{-3} μA/cm, a factor of 20 lower.[10)]

The ageing of MWPCs is a subject of controversy between various groups who have obtained substantially different results under apparently similar experimental conditions. It appears that very small traces of unsuspected impurities—such as pollutants in the gas or outgassing of materials—can be dominant in the ageing process. Therefore we have tried to realize a measurement under well-controlled conditions, in collaboration with a

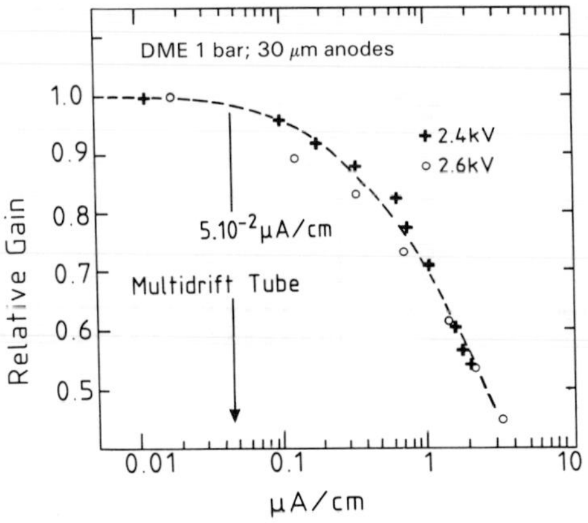

Fig. 7 Rate dependence of the gain measured in the MDM with DME gas filling.

group[*)] at the University of Florida, Gainesville. This group has set up a laboratory for ageing and radiation hardness measurements, which is complete with various instruments for analysing trace impurity measurements and for carrying out surface analysis of damaged wires. Detailed results of this investigation are described elsewhere.[8)] The major outcome is that with proper precautions in the choice of materials, particularly for the anode wires, and using high-purity DME, proportional counters survive without noticeable modification of gain up to integrated currents above 1 C/cm. Assuming an average charge per count of 10^6 electrons, this corresponds to 10^{13} minimum-ionizing particles, or about 1 Mrad.

In parallel with this systematic research, we have irradiated various prototype MDMs at CERN, using commercially available pure DME; the conditions of work were purposely kept 'normal', i.e. no special precautions were taken for ougassing, and we used a standard gas-distribution system. We have observed no detectable change in operating characteristics up to a total charge of around 0.5 C/cm of wire. However, at this point, a discharge tends to develop in the irradiated region (see Fig. 8), and we believe that this is due to the presence, in our DME, of various pollutants (e.g. Freons) that are known to accelerate the ageing process. The investigation will continue to try to identify the ageing agents and to eliminate them by means of a suitable purification technique. The same investigation is attempting to find other possible sources of pollution, such as the various epoxies used in the manufacturing process of the modules.

*) Led by S. Majewski.

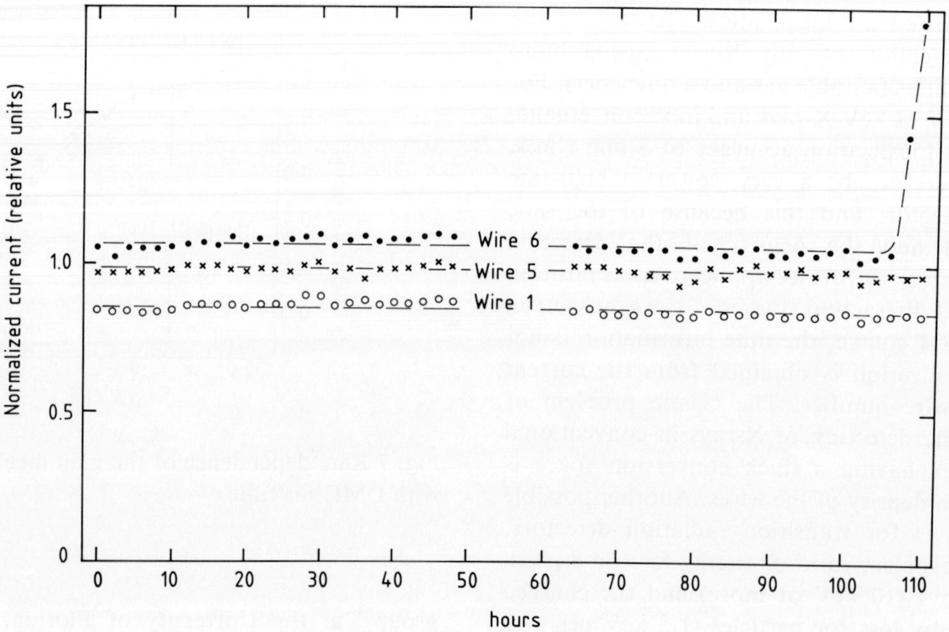

Fig. 8 Normalized current measured on several wires in a module as a function of exposure time to a strong radiation source, using standard-grade DME. Breakdown is observed after about 0.5 C/cm, probably due to the effect of polymerization of the pollutants in our gas.

REFERENCES

1) R. Bouclier, G. Charpak and F. Sauli, CERN EP Internal Report 84-03 (1984).
2) R. Bouclier et al., Nucl. Instrum. Methods **A252**, 373 (1986).
3) R. Bouclier et al., *in* Vertex Detectors, ed. F. Villa, Ettore Majorana Int. Science Series (Plenum Press, Inc., New York, 1988), Vol. 34, p. 101.
4) R. Bouclier et al., Nucl. Instrum. Methods **A265**, 78 (1988).
5) C. Del Papa, private communication.
6) P. Astier et al., Further developments of the multidrift module (in preparation).
7) S. Majewski, Proc. Workshop on Radiation Damage to Wire Chambers, Berkeley, 1986 (LBL-21170, Berkeley, 1986), p. 239.
8) M. Jibaly et al., Some results on the ageing of wire chambers with DME, to be published in Nucl. Instrum. Methods.
9) J.C. Santiard, CERN EP Internal Report 82-04 (1982).
10) H. Walenta, Phys. Scr. **23**, 354 (1980).

PROSPECTS FOR WIRE CHAMBERS AT HIGH LUMINOSITY

Rudolf P. Thun

University of Michigan, Ann Arbor, Michigan 48109

ABSTRACT

We summarize the radiation information relevant to operating wire chambers in an SSC experiment. By using small (< 4 mm wide) cells and fast gases based on CO_2 or CF_4 mixtures, wire chambers can be used for both tracking and calorimetry in the central ($\gtrsim 10°$) part of the detector. We report some results from measurements of operating characteristics of small proportional chambers using a variety of gas mixtures.

1.) INTRODUCTION

Wire chambers are the most commonly used detectors in high energy physics experiments with application in tracking, calorimetry and particle identification. The proposed Superconducting Super Collider (SSC) as well as other possible high-luminosity hadron colliders present a severe challenge to all detector technologies including those based on wire chambers. Event rates of 10^8 Hz, particle multiplicities around 100, and radiation exposures that can exceed the megarad/year level require a careful analysis of channel occupancies, wire aging and radiation damage to construction materials and electronics.

We present in this paper a summary of radiation levels expected in a prototype general purpose SSC detector which, for the sake of this study, is based entirely on wire chamber technology. This summary includes the expected yearly charge collected per cm of sense wire since this is a key parameter affecting wire aging. Such aging is typically caused by deposits on sense wires or cathode surfaces and leads to a loss of efficiency or to electrical discharges. The second part of the paper describes measurements performed at the University of Michigan with small drift cells and a variety of proportional chamber gases.[1] The goal of this research has been to find wire chamber configurations with maximum drift times approaching the inverse of the SSC event rate (~ 10 nsec) and with minimum sensitivity to low-energy (~ 1 MeV) neutrons. Such neutrons are produced copiously in hadronic showers and can lead to spurious energy measurements.

2.) SUMMARY OF RADIATION EFFECTS ON WIRE CHAMBERS

Beam-beam interactions are expected to be the primary source of radiation at the SSC. A recent study of this radiation [2] has yielded a very simple parametrization of the expected doses which can be summarized as follows (where a standard SSC year corresponds to an integrated luminosity of 10^4 pb^{-1}):

a.) Secondary charged particles are produced per collision with the following angular distribution:

$$dn/d\Omega \;=\; 1.25/\sin^2\theta. \qquad (1)$$

where θ is measured with respect to the beam direction.

b.) The yearly dose from electromagnetic showers is

$$D_{EM}(Gy/year) \;=\; 390(\cosh^{2.93}\eta)/r^2 \qquad (2)$$

where Gy=100 rad, η=-ln(tan(θ/2)), and r is the distance (in meters) of the calorimeter element from the interaction point.

c.) The corresponding quantity for hadronic showers is

$$D_H(Gy/year) \;=\; 29(\cosh^{2.89}\eta)/r^2 \qquad (3)$$

d.) The neutron fluence at hadronic shower maximum is approximately:

$$\phi_n(neutrons/cm^2 \text{ per year}) \;=\; 1.5\times10^{12}(\cosh^{2.67}\eta)/r^2 \qquad (4)$$

where the neutron energy spectrum (E dN/dE) peaks at about 1 MeV.

e.) Albedo neutrons and albedo photons inside the cavity enclosed by the calorimeter have a similar energy spectrum. The yearly albedo fluences are roughly independent of location and are given by:

$$\phi(\text{albedo neutrons}/cm^2 \text{ per year}) \sim 2.4\times10^{12} \qquad (5)$$

$$\phi(\text{albedo photons}/cm^2 \text{ per year}) \sim 10^{12} \qquad (6)$$

We apply this radiation information to a typical general detector geometry[3] as sketched in figure 1. We assume that the tracking and calorimeter active elements consist of proportional tubes with a cross section of 4x4 mm. As will be shown in the next section such tubes can have maximum drift times as low as 20 nsec with suitable gases. Furthermore, we assume that minimum-ionizing tracks generate a primary signal of 100 electrons/cm which is amplified by a chamber gain of 10^5 in the tracking and 10^4 in the calorimeter cells. In calculating cell occupancies we assume that signal gates are 20 nsec wide and that the event rate is 10^8 Hz.

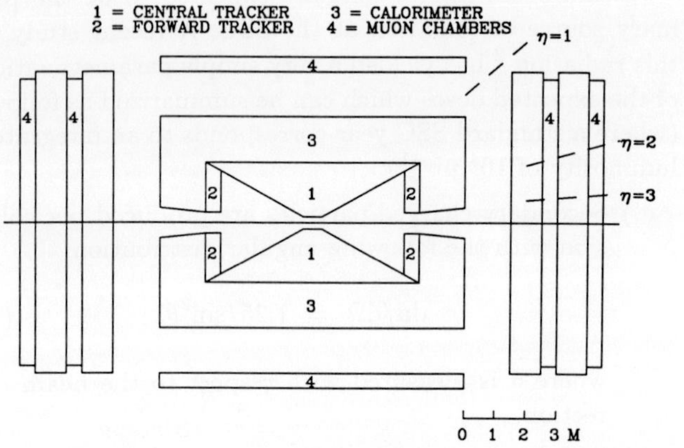

FIG. 1 SSC detector geometry used for the radiation study.

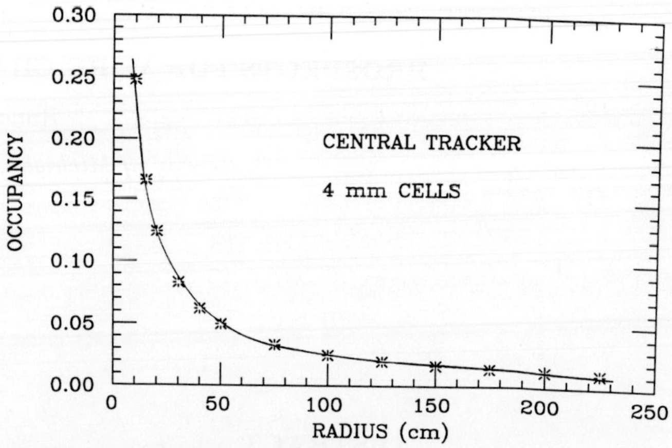

FIG. 2 Central tracker cell occupancies as a function of distance from the beam axis.

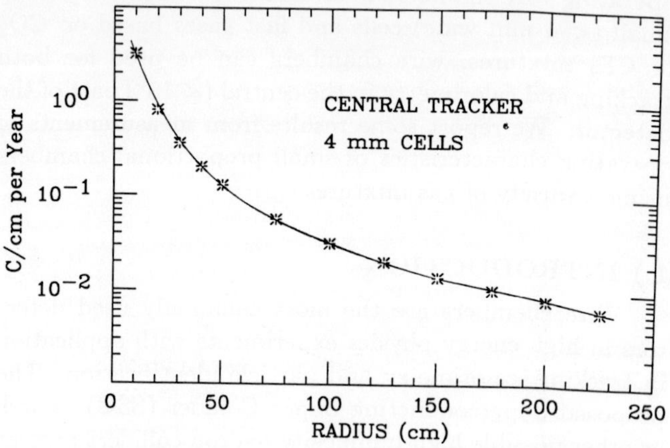

FIG. 3 Charge per cm of sense wire collected in the central tracker per year as a function of distance from the beam axis.

Results are shown in figures 2-8. Cell occupancies in the central tracking chamber are given in figure 2 as a function of radius from the beams. The effect of extra hits from looping tracks in a possible magnetic field are not included. Cell occupancies are at an acceptable level to within about 20 cm of the beams. The collected charge per cm of sense wire in the central tracker is independent of angle and is shown in figure 3 as a function of radius from the beams. Experience with wire chambers indicates that exposures of 0.1 C/cm can be tolerated without taking special precautions. Chambers can thus be operated at distances greater than about 50 cm from the beams. With special care exposures of 1.0 C/cm or more may perhaps be tolerated. This would allow operation of wire chambers to within 20 cm of the beams. The effect of albedo neutrons and photons is generally negligible compared to the flux of charged particles except for giving occasional large pulses from elastically scattered low energy protons or electrons. Such large pulses can lead to cross talk problems. The neutrons may also be a factor in damaging electronic components.

The calorimeter sketched in figure 1 has a cylindrical geometry with an interior radius of 2.0 m and endcaps starting at $z=\pm$ 3.5 m. Figures 4 and 5 show the yearly dose from electromagnetic and hadronic showers as a function of the pseudorapidity (angle) of the calorimeter element with respect to the beams. Figure 6 gives the yearly neutron fluence at hadronic shower maximum as a function of pseudorapidity. Converting the doses to a collected charge per cm of sense wire inside the calorimeter gives the results shown in figures 7 and 8. A conservative limit of 0.1 C/cm allows electromagnetic calorimetry at angles greater than 15° with respects to the beams. The corresponding angle for hadronic calorimetry is approximately 7°. Operation at smaller angles may be possible with chamber gains below the assumed 10^4 or with gases that exhibit good aging characteristics.

Muon tracking chambers located behind a reasonably thick absorber, calorimeter or solid iron magnet are not expected to suffer serious radiation effects. Effective shielding of the muon chambers at forward angles near the beam pipe may require special care.

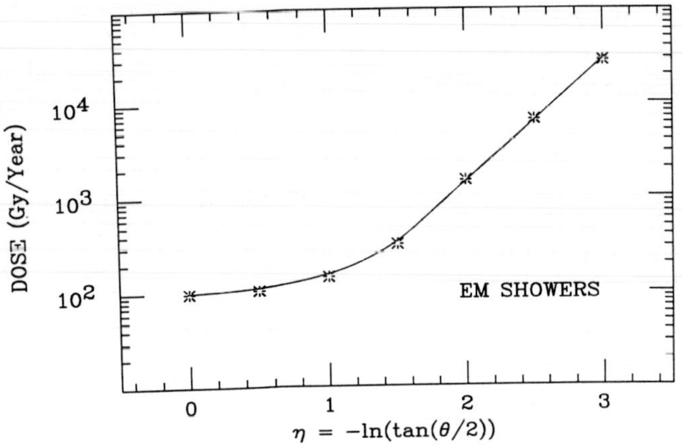

FIG. 4 Yearly maximum dose in the electromagnetic calorimeter as a function of pseudorapidity.

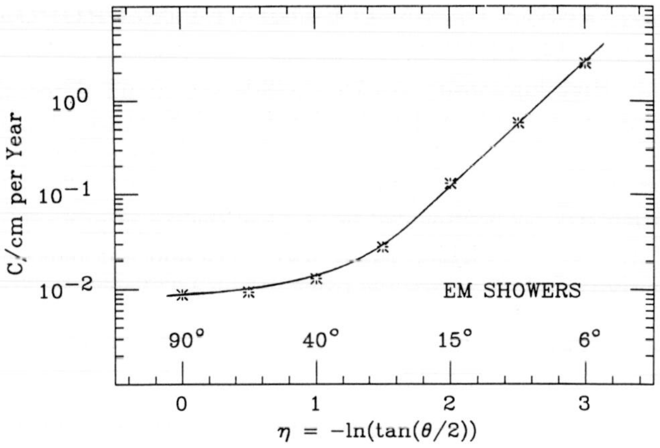

FIG. 7 Charge per cm of sense wire collected in the electromagnetic calorimeter per year as a function of pseudorapidity.

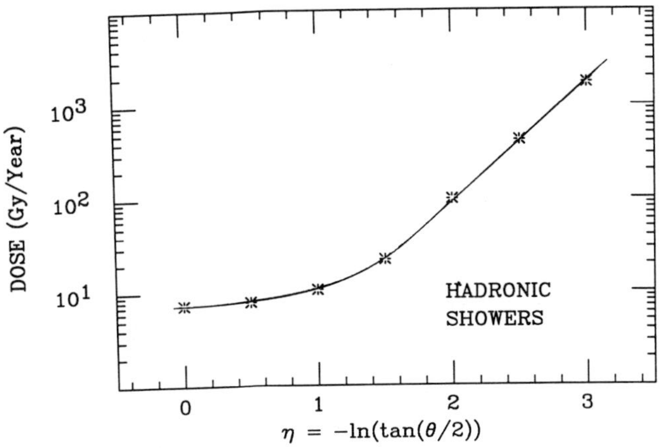

FIG. 5 Yearly maximum dose in the hadronic calorimeter as a function of pseudorapidity.

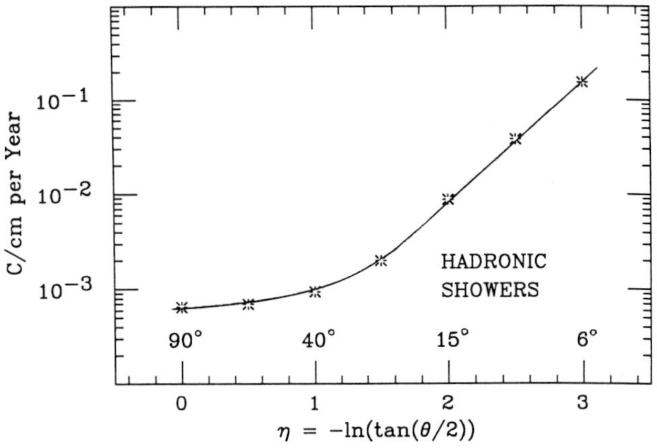

FIG. 8 Charge per cm of sense wire collected in the hadronic calorimeter per year as a function of pseudorapidity.

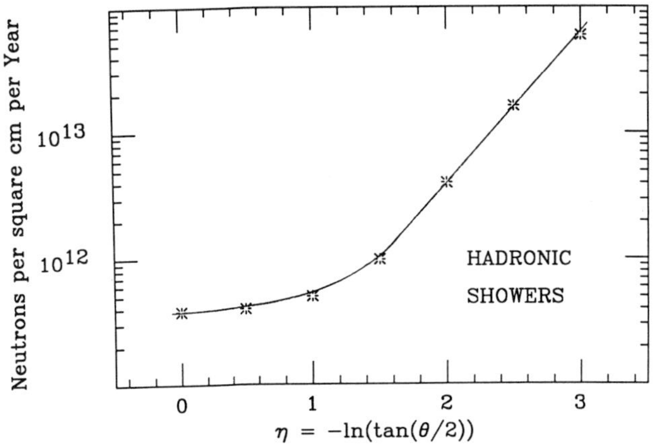

FIG. 6 Yearly neutron fluence at hadronic shower maximum as a function of pseudorapidity.

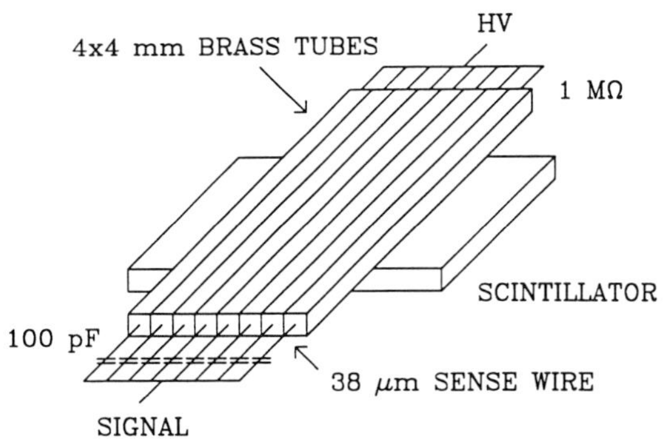

FIG. 9 Test chamber used for drift-time measurements.

3.) TESTS OF SMALL-DIAMETER PROPORTIONAL TUBES

High-luminosity hadron colliders demand detectors with fast signal collection times and resistance to high radiation levels. At the University of Michigan we (M. Bourns and R. Thun) have studied the operation of small-diameter proportional tubes with a variety of gases with the goal of achieving small drift times and minimal sensitivity to low-energy neutrons. Figure 9 shows the configuration of the test chamber. A cell width of 4 mm was chosen as a reasonable lower limit in terms of construction ease and signal size for operation at atmospheric pressure. Drift times were measured with cosmic rays for a variety of gas mixtures. Figure 10 gives drift-time distributions for HRS - CF_4 mixtures. HRS gas consists of 89% argon, 10% CO_2 and 1% CH_4 and has been used in a number of large detectors at SLAC because of its good operating characteristics.[4] It has been known for some time that CF_4 yields large drift velocities[5–7] and the results shown in figure 10 confirm this with pure CF_4 giving maximum drift times of around 16 nsec compared to 32 nsec for HRS gas.

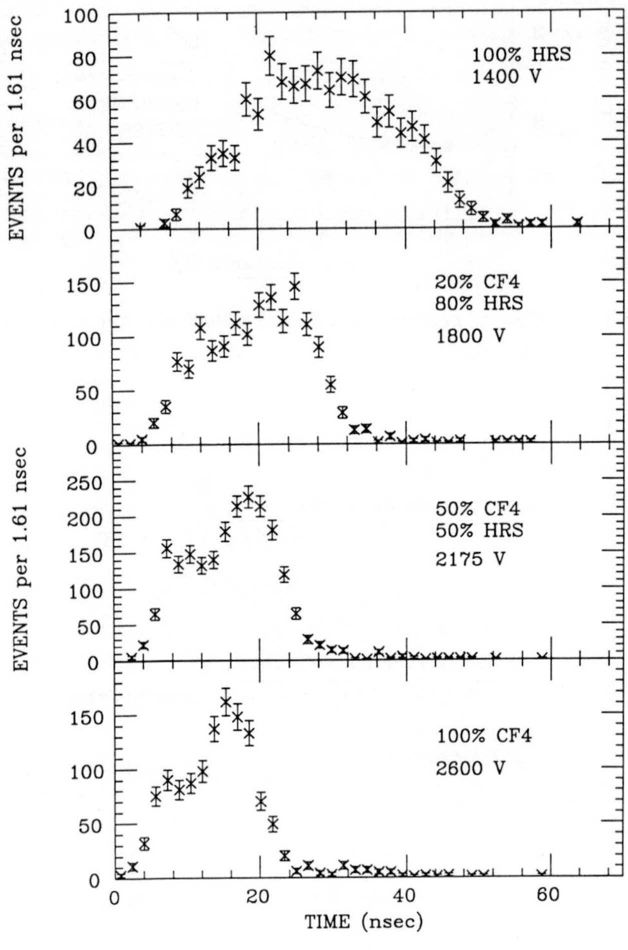

FIG. 10 Drift-time distributions for HRS-CF_4 mixtures.

We have also found that argon-CO_2 mixtures exhibit fast drift velocities in small-diameter proportional tubes. Figure 11 shows that maximum drift times for 50% argon-50% CO_2 are around 20 nsec, comparable to those found for 50% HRS-50% CF_4. The argon-CO_2 mixtures have some advantages over those containing CF_4 in terms of cost and energy resolution. Figures 12 and 13 show the signal from the 5.89 KeV x-rays of an Fe^{55} source. For CF_4 mixtures the ionization peak broadens with CF_4 content and, indeed, washes out at 100% CF_4. By comparison, the argon-CO_2 mixtures show a resolution essentially independent of CO_2 content. We note that argon-CO_2 gives short drift times only in small cells where the high electric fields (> 2 KV/cm) everywhere in the cell insure a large drift velocity.[8]

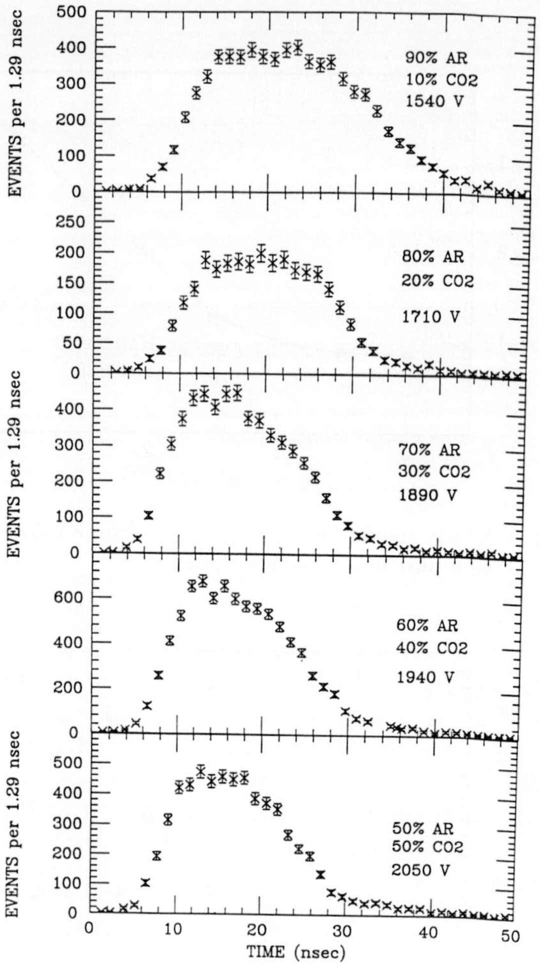

FIG. 11 Drift-time distributions for argon-CO_2 mixtures.

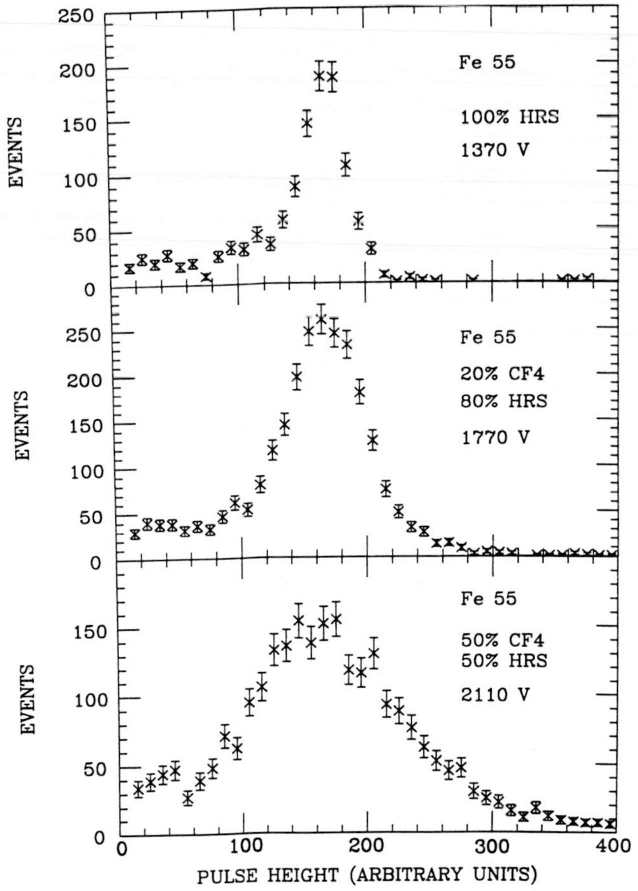

FIG. 12 Pulse-height distributions from an Fe^{55} source for HRS-CF_4 mixtures.

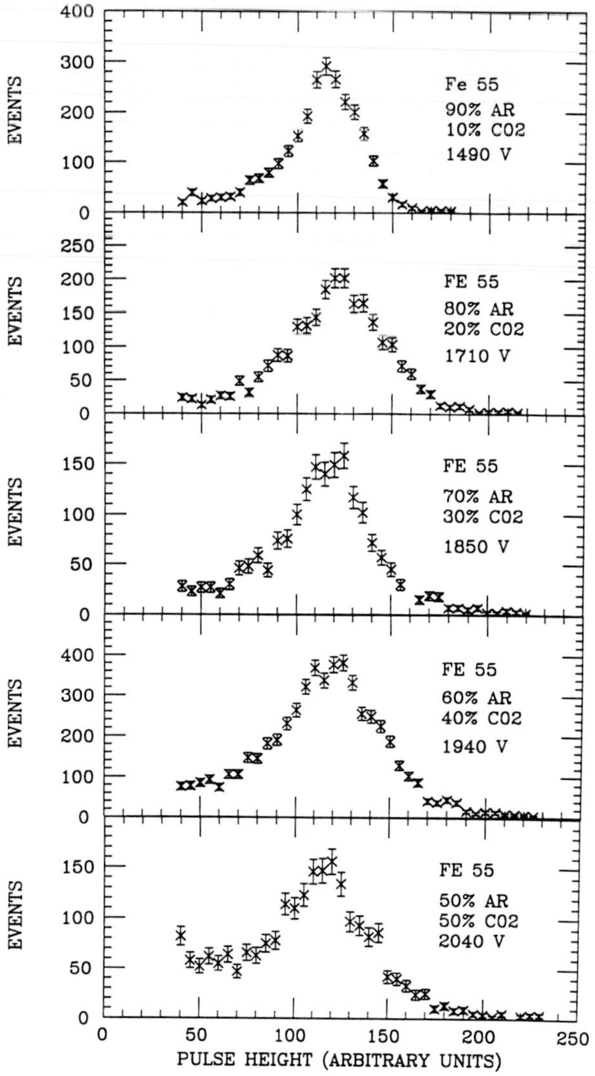

FIG. 13 Pulse-height distributions from an Fe^{55} source for argon-CO_2 mixtures.

We have also made a preliminary study of the effect of low-energy (~ 1 MeV) neutrons on the operation of wire chambers. Gases containing hydrogen can be expected to produce highly ionizing low-energy protons when neutron-proton elastic scattering occurs. Such protons can yield signals equivalent to that from several hundred minimum-ionizing particles and can therefore lead to spurious energy measurements in calorimetry. We exposed a test chamber to a 925 mCi Am-Be source which yielded approximately 2×10^6 neutrons per second with an energy spectrum from 1 to 5 MeV. The neutrons were moderated somewhat by several cm of paraffin. The test chamber was constructed with a minimum of scattering materials (thin windows, foil cathodes) to maximize gas related effects. Gamma rays from the $\alpha + \mathrm{Be} \rightarrow \mathrm{C}^* + \mathrm{n}$ reaction in the source were absorbed with 1.3 cm of lead placed between the chamber and source. The pulse-height spectrum of chamber signals obtained from exposure to the Am-Be source is shown in figure 14 for a number of gas mixtures. In each case, operating voltages were set to obtain equal chamber gain using the signal from an Fe^{55} source. Argon-CH_4 mixtures show a significant increase in the rate of large pulses as the CH_4 fraction is increased. A 50%-argon 50%-CO_2 mixture gives a rate of large pulses comparable to that from the argon-CH_4 mixture with the least amount of CH_4 (10%). A sense of scale is given to figure 14 by noting that the peak from the 5.89 KeV x-rays of an Fe^{55} source corresponds to a pulse height of 3.0 whereas the largest pulses from the Am-Be source gave counts above 100.

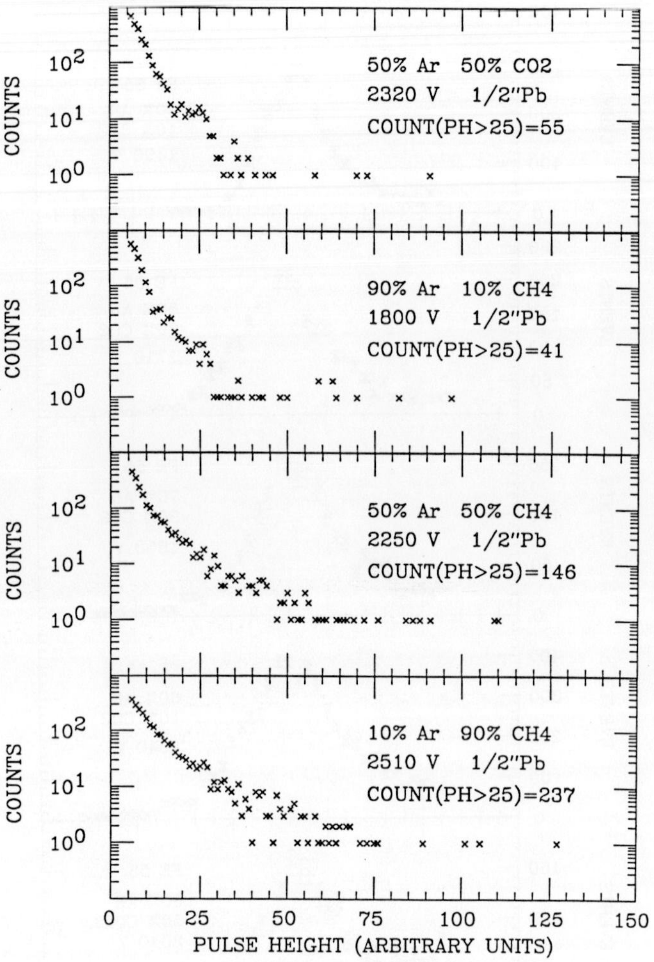

FIG. 14 Pulse-height distributions from an Am-Be neutron source for various gas mixtures.

4.) CONCLUSION

Wire chambers represent a very powerful technique for detecting particles in high-energy physics experiment. High-luminosity hadron colliders present a formidable challenge to these and all other detectors because of high rates and radiation levels. Nevertheless, the study summarized above indicates that wire chambers may continue to play an important role in experiments at the proposed large colliders such as the SSC. High rates can be handled by constructing small cells and using gas mixtures containing CO_2 or CF_4. Signal collection times of less than 20 nsec can be achieved, well matched to anticipated event rates. Wire aging effects are tolerable in tracking devices more than about 50 cm from the beams. Electromagnetic calorimeters based on wire chambers will survive radiation levels at angles greater than about 15° with respect to the beams, while hadronic calorimeters should be able to reach even smaller angles ($\sim 7°$). Recent wire aging studies[9] indicate that with some care wire chambers may be able to operate at even closer distances and angles with respect to the beams. Problems associated with large pulses from low-energy neutrons can be reduced by choosing gas mixtures that eliminate or minimize hydrogen content.

This study has not addressed the question of radiation damage to electronics placed near the wire chambers. As discussed elsewhere in these proceedings, we may expect that electronics of the required radiation hardness will be available for the experiments envisaged for the SSC. Of course, there is no substitute for building realistic prototypes of both the detectors and associated electronics and operating them in test beam environments that closely resemble those expected at the high-luminosity hadron colliders.

REFERENCES

1.) R. Thun, "Tests of Small Proportional Tubes with CF_4-HRS Gas Mixtures," to be published in Nucl. Inst. and Meth.

2.) "Radiation Levels in the SSC Interaction Regions," SSC-SR-1033 (1988).

3.) "Radiation Effects at the SSC," SSC-SR-1035 (1988).

4.) D. Koltick, "Gas Studies in the Prototype Outer Drift Chambers," HRS Internal Memo HRS-197 (1980).

5.) L.G. Christophorou, et al., "Fast Gas Mixtures for Gas-Filled Particle Detectors," Nucl. Inst. and Meth. 163 , 141 (1979).

6.) J. Fischer, et al., "Proportional Chambers for very High Counting Rates Based on Gas Mixtures of CF_4 with Hydrocarbons," Nucl. Inst. and Meth. A238 , 249 (1985).

7.) R. Henderson, et al., "A High Rate Proportional Chamber," IEEE Trans. on Nucl. Sci. NS-34 , 528 (1987).

8.) A. Peisert and F. Sauli, "Drift and Diffusion of Electrons in Gases: A Compilation," CERN 84-08 (1984).

9.) J. Kadyk and J. Va'vra, "Wire Chamber Aging," in SSC-SR-1035 (1988).

DEVELOPMENT OF PLASTIC SCINTILLATION FIBERS, PLASTIC SCINTILLATION FIBER CALORIMETER AND ITS READOUT SYSTEM

Fumihiko Takasaki
KEK, Tsukuba, Ibaraki-ken, Japan

Abstract

Development of plastic scintillation fibers(PSF), a calorimeter using PSF and its readout system with position sensitive photomultiplier tube(PSPMT) is reported.

Introduction

It is desirable that calorimeters to be used at SSC have following properties;

1) to be fast,
2) to be resistive against radiation,
3) to be hermetic,
4) to have an e/h ratio to be unity,
5) to be segmented finely,
6) to be immune to the magnetic field and
7) to be stable in short and long terms.

They have to be acceptable, of course, from the view point of construction and operation costs. It is not easy for any ideas of calorimeters to fulfill all requirements and therefore R & D is needed. A calorimeter using PSF has many good features and it deserves to be studied seriously. R & D efforts started a decade ago at KEK. This is a report of studies of some properties of plastic scintillation fibers, calorimeter with PSF and its readout system using a position sensitive photomultiplier tube undertaken at KEK last few years in collaboration with people of companies such as Kyowa Gas Chemical Ltd. and Hamamatsu K.K.

Plastic Scintillation Fibers

Plastic scintillator is a quite popular material which is quite stable in its performance and fast in its response to the charged particles. Although it has been already proposed some 30 years ago[1], its application to particle trackers and calorimeters has not been seriously considered because of the rapid development of a wire chamber technology. Therefore it was needed for us to start with the development of fiber itself. It has been made in collaboration with Kyowa Gas Chemical Ltd. in Japan. Since as given in reference 2, Kyowa people succeeded in fabricating fibers with reasonable light yield, 7 photoelectron for 1 mm fiber at 80 cm from one end of fiber, and long light attenuation length, about 2.8 m, emphasis of the recent development is placed on the resistivity of fibers against the radiation. There exists no solid material which are free from radiation effect. So the study is to find materials or combination of materials which are less affected by the radiation.

Development of the scintillator more resistive against radiation has been made by the following guiding principles;

1) to use polystyrene as the basic matrix,
2) to use stabler wave shifter,
3) to add oxygen quencher to minimize oxidation and
4) to shift wave length to longer one as possible.

It is known that polystyrene is more resistive than other popular materials such as PMMA and PVT. As for wave shifters, use of those organic materials is avoided which have oxazole or oxadiazole, for example, POP, POPOP and b-PBD. It is suspected that the oxidation of active radicals and ions generated by the irradiation form color centers and therefore materials get colored. So the quenching of oxygen may improve the radiation resistivity.

Kyowa people now provide us two kinds of plastic scintillators, SCSN38 and SCSN81. They are similar in light yield and light attenuation. Peak wave length of luminescence is 420 nm for SCSN38 and 430 nm for SCSN81. SCSN38 contains b-PBD and BDB and it has no oxygen quencher. We examined effects of gamma rays from Co^{60} to these scintillators. Exposure to the gamma ray was made in the air and at room temperature with the dose rate of 2×10^5 rad/hour. We also examined scintillators with additional wave shifter, Y7, to the SCSN81, SCSN81Y7. Peak wave length becomes 510 nm by adding Y7.

To see the effect of irradiation, we measured light yield when beta rays from Ru^{106} pass through scintillators. A photomultiplier tube, Hamamatsu R329, was used to measure the pulse height. It has the peak sensitivity to the light with the wave length of about 410 nm. In the measurement, scintillator bars with dimensions, 2.6 mm x 20 mm x 1000 mm, was wrapped with black paper and no reflector was used. Since it is observed that the darkening due to the radiation gets cured more or less in most plastic scintillators in short time, present measurement was made three weeks after the irradiation, leaving scintillators in air at room temperature. In figures 1, 2 and 3, we show variation of light yields for SCSN38, SCSN81 and SCSN81Y7 as a function of distance from one end of bar for different integrated doses. One observes that SCSN81 is more resistive against gamma rays than SCSN38 and that it is not much affected up to the integrated dose of 10^6 rad. The other observation is that SCSN81Y7 is least affected.

Transmittance of the scintillator was also examined for the irradiated samples. Results are shown for SCSN38, SCSN81 and SCSN81Y7 in figures 4, 5 and 6. The length of the samples was 40 cm. Clearly one observes that SCSN38 is most affected by the irradiation. It is noted that the photomultiplier tube used in the study, R329, has peak sensitivity around 420 nm. So it does not match the light spectrum of SCSN81Y7. Selection of the phototube with appropriate optical sensitivity or use of optical filter may reduce relatively effects of radiation to the fibers.

Similar study was made for PSF's made of SCSN38, SCSN81 and SCSN81Y7. Fibers have core diameter of 0.96 mm with clad thickness of 0.02 mm made of PMMA. Refractive index of polystyrene is 1.59 and that of PMMA is 1.49. Fibers were coated with white paint to enhance efficiency of light conversion to longer waves. In figures 7, 8 and 9, we give results of similar measurements of light yield as presented in figures 1 to 3. Similar trend was observed as for bars and again SCSN81Y7 is least affected by the radiation. However, we observed differences in performance between bars and fibers and they have be understood in the future study. It is noted that although PMMA is known to be more vulnerable to the radiation than polystyrene, no effect was observed in the performance of fibers. It is also noted that polystyrene used for fibers was not cross-linked.

Calorimeter with PSF

Using 1 mm ϕ PSF of SCSN38, we made a calorimeter module which is 2 m long with cross section of 8 cm square as shown in figure 10. PSF's are placed in lead with equal spacing between the neighboring fibers. The ratio of lead and PSF is 4 to 1 in volume. We have measured light yield for 4 GeV electrons to hit calorimeter module transversely and results is shown in figure 11. The position dependence resembles to that of fibers only. It demonstrates that no serious change in the performance of fibers occurred during the course of fabrication of calorimeter. The energy resolution is $\Delta E/E \sim 19\%/\sqrt{E}$ which does not depend so much on the position along the length of module. It hints that the photon statistics is not a dominating factor in determining the energy resolution of the present calorimeter module. It is planned to make 25 modules of this type to make performance test for high energy hadrons.

Readout System of PSF Calorimeter

The readout system of PSF calorimeter has to be fast and applicable in a magnetic field. It is also desirable to have capability to readout illumination pattern of fibers. This would provide us a detailed information on the development of shower and therefore it would be useful for the particle identification. UA2 group has used a readout system consisting of Image Intensifier with CCD readout[3] for its PSF calorimeter. Although it provides us a fine image pattern of light, it can not be operated fast due to the long decay time of the fluorescence and it will not be easy to use it in the magnetic field. So we propose here a readout system using a Position Sensitive Photomultiplier Tube(PSPMT).

We have developed a photomultiplier tube with transmissive dynodes in collaboration with Hamamatsu Photonics K.K.. It can be operated in a strong magnetic field when the line of field is parallel to the tube axis and it preserves a illumination pattern on the photocathode to the anode[4,5]. Its time response is as fast as or even faster than the conventional phototubes[6]. It was observed that the amplification gain dropped by about three order of magnitude at 10 kGauss as shown in figure 12. However, it was also observed that the gain drops could be recovered by applying higher electric voltage to the tube. Applicable maximum high voltage gets higher for higher magnetic field. Position resolution of the tube with 16 dynodes turned out to be about 2 mm in the recent study when a magnetic field of 450 Gauss was applied parallel to the tube axis. The dependence of position resolution on the magnetic field strength is shown in figure 13. Improvement of the resolution can be understood due to the fact that the magnetic field squeezes the spread of secondary electrons. From the figure one would expect further improvement of the resolution by applying the higher magnetic field. We try to understand how much position resolution can be improved.

It is noted that phototubes with quartz window are resistive against radiation. It is demonstrated in figure 14, in which we show response of tubes as a function of integrated gamma ray dose. Tubes with quartz window and borosilicate glass window were tested for the light generated from NaI or BaF_2 crystals when gamma rays from Cs^{137} were irradiated. One observes deterioration for a tube with borosilicate window but not for that with quartz window. Deterioration is due to the darkening of glass window. Similar results are reported for the irradiation of tubes with neutrons[7].

To readout signal pattern on the anode, we develop tubes with multi-anodes. To handle many anode signals, we develop a compact analog-multiplexer module. It accepts certain number of anode signals in parallel and reads them out serially with appropriate trigger signal. We start with a phototube with 64 anodes. It has a sensitive area of tube window of 40 mm x 40 mm. Since the anode is divided to 64 pieces, each anode occupies an area of about 5 mm x 5 mm. As for an analog memory uint, we use an Analytex- AN201[8]. It accepts 32 parallel input signals and each channel has 4 backets of time sampling. Its sampling speed is 20 ~ 30 MHz which is much slower than the speed of phototubes. To match the speed of the phototube and that of the AN201, we have a pulse shaper array to interface between the phototube and AN201. This inferface amplifies signals by about factor 10. Time sampling can be expanded by connecting many AN201's serially. For the moment, we plan to use 4 AN201's in series to have 16 time sampling. So we use 8 AN201's per tube. Fabrication of this multiplexer unit is currently under way by H. Ikeda et al. of KEK. They also consider to develop new analog memory unit[9] with CMOS technology which can be made more compact than the present design. It is noted that the segmentation of anode can be made finer and we also develop phototubes with 256 anodes with the same sensitive area of tube window.

This work was supported by the fund of US-Japan scientific cooperation and also by KEK, Japan. We are grateful to people of Kyowa Gas Chemical Ltd. and Hamamatsu K.K. for collaboration.

References

1) G.T.Reynolds and F. Reines, Methods of Experimental Physics, Vol. 5, ed. L.C.L. Yuan and C.S. Wu, Academic Press 1961,
2) F. Takasaki et al, Nucl. Instr. and Methods, A262, 224(1987),
3) UA2 collaboration, CERN-EP/85-203, December 1985,
4) F. Takasaki et al, Nucl. Instr. and Methods, 228, 369 (1985),
5) F. Takasaki et al, Nucl. Instr. and Methods, A260, 447(1987),
6) Hamamatsu Photonics K.K., private communication,
7) T. Iida et al, Journal of Nucl. Science and Tech., 24, 759(1987),
8) Analytex Ltd., 4123 Alpine Road, Portola Valley, CA94025,
9) F. Abe et al, KEK Internal Report 88-1, May 1988.

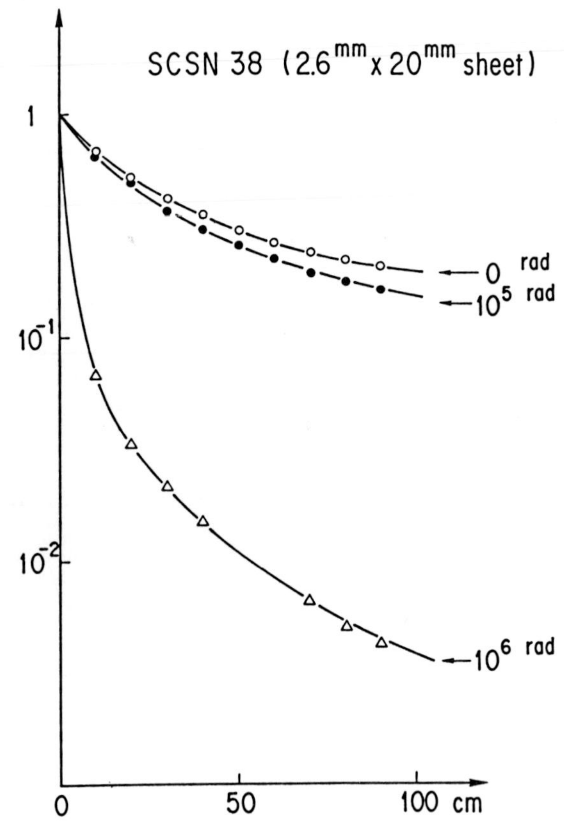

Fig. 1

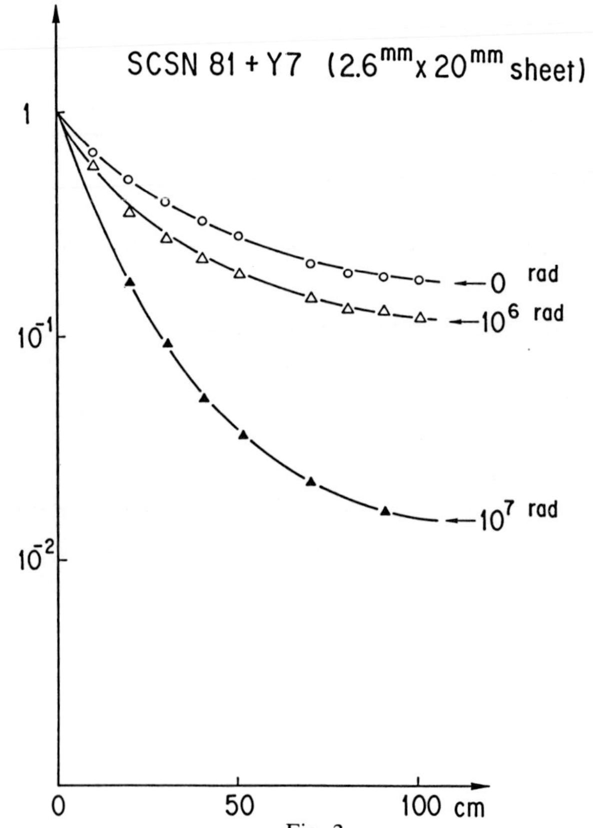

Fig. 3

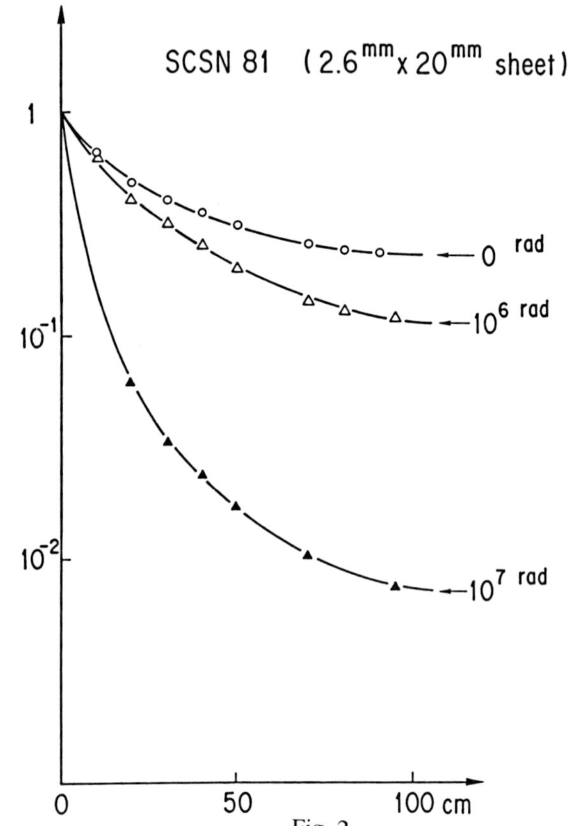

Fig. 2

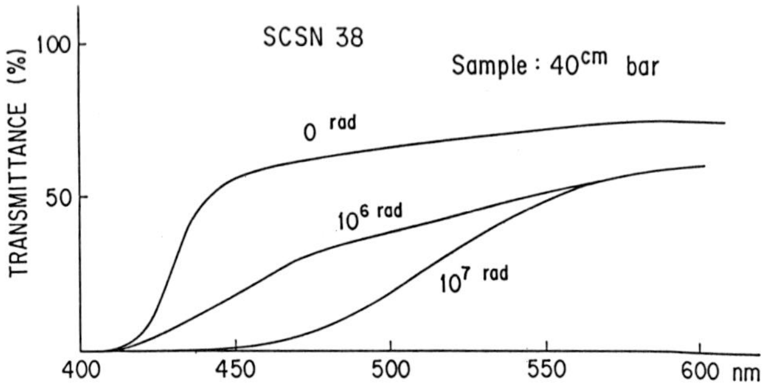

Fig. 4

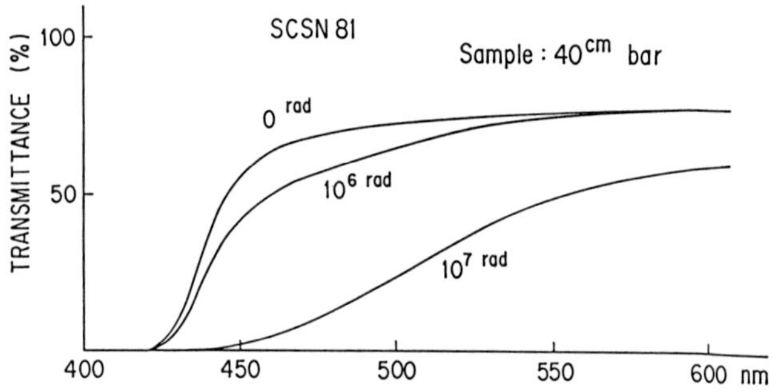

Fig. 5

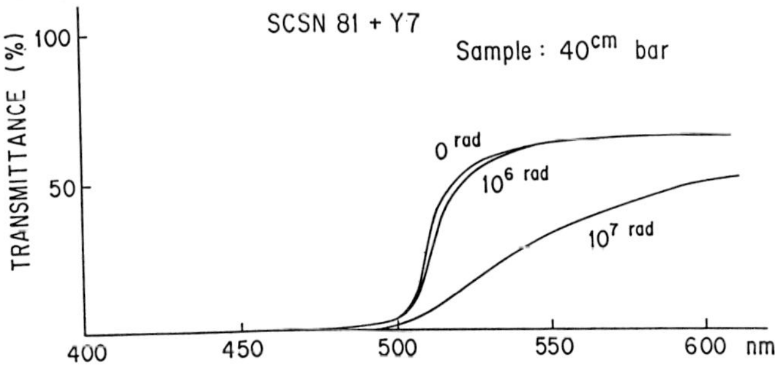

Fig. 6

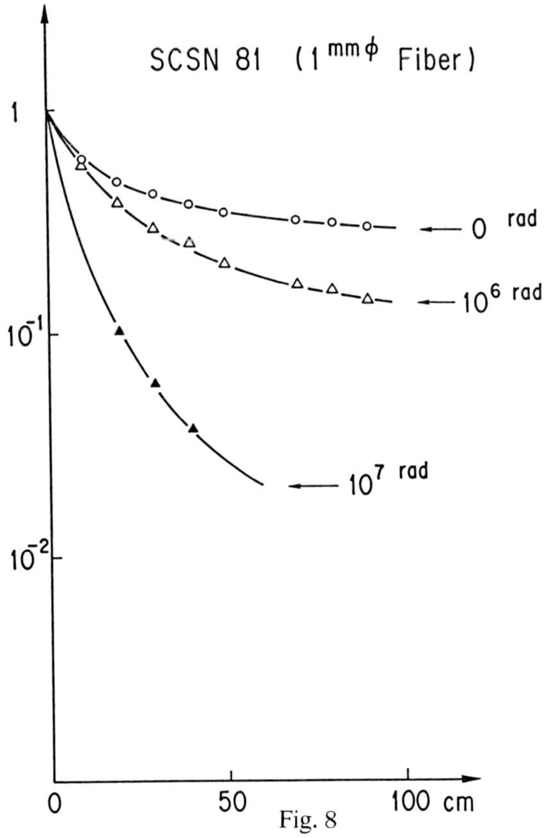

Fig. 8

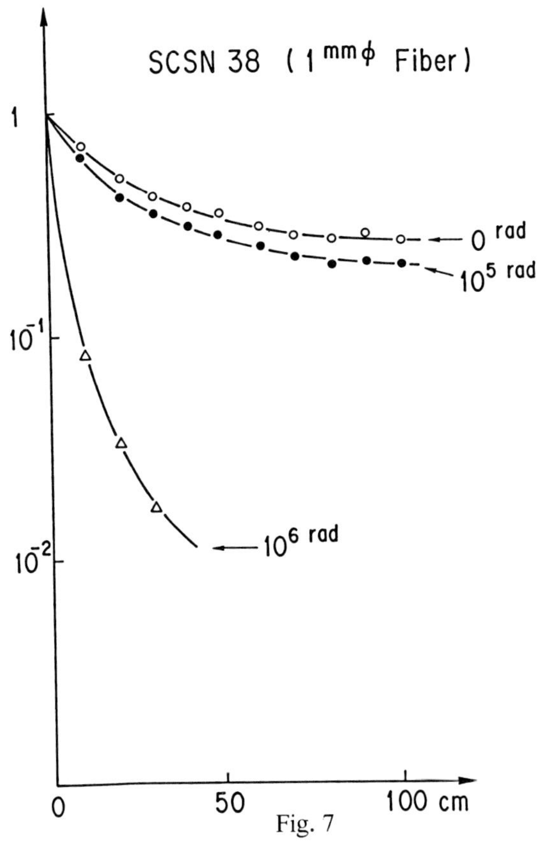

Fig. 7

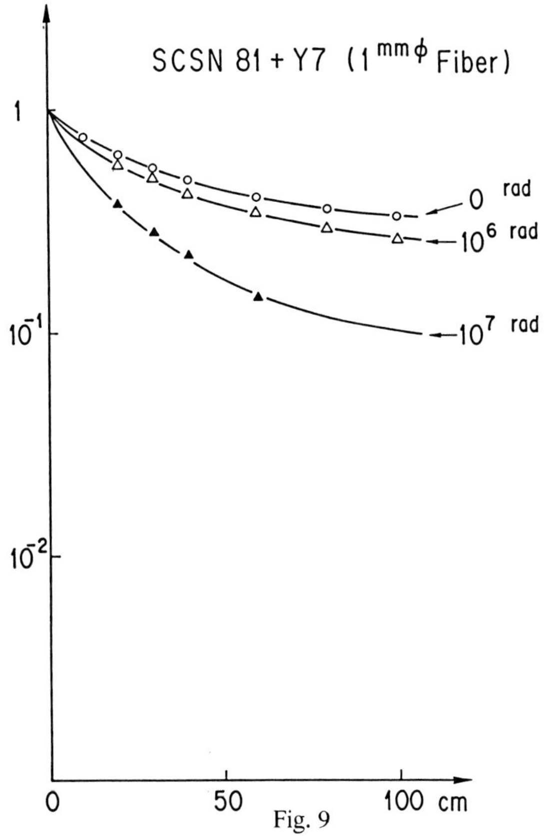

Fig. 9

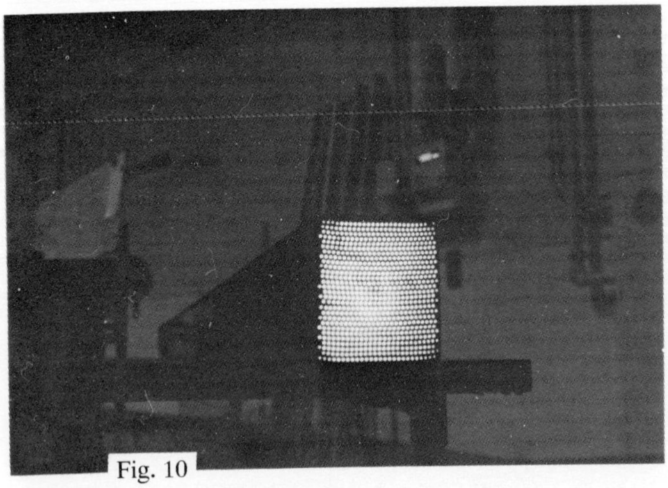

Fig. 10

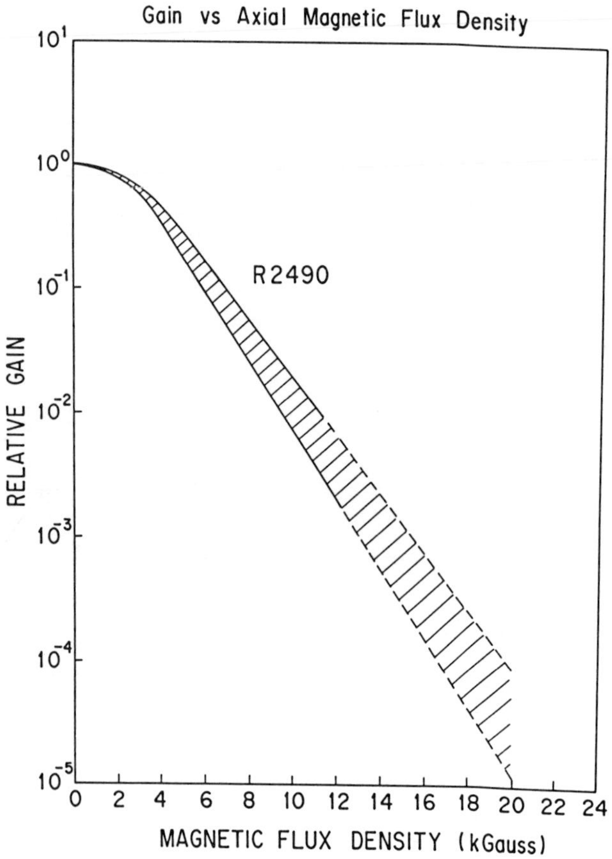

Fig. 12

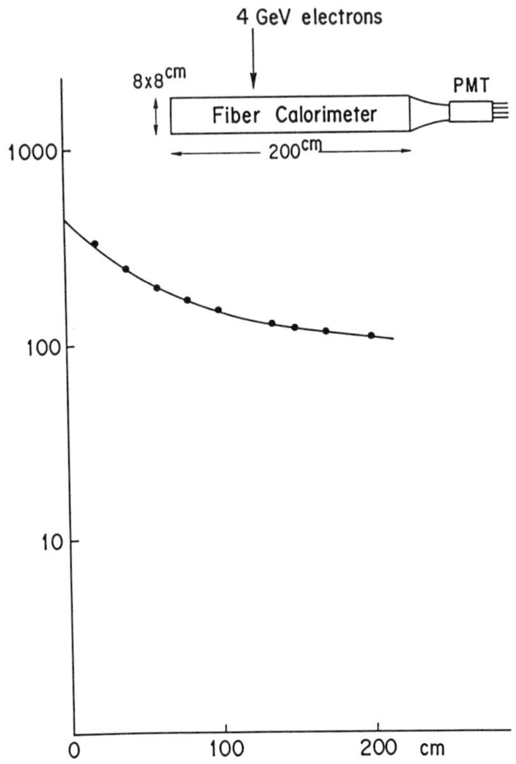

Fig. 11

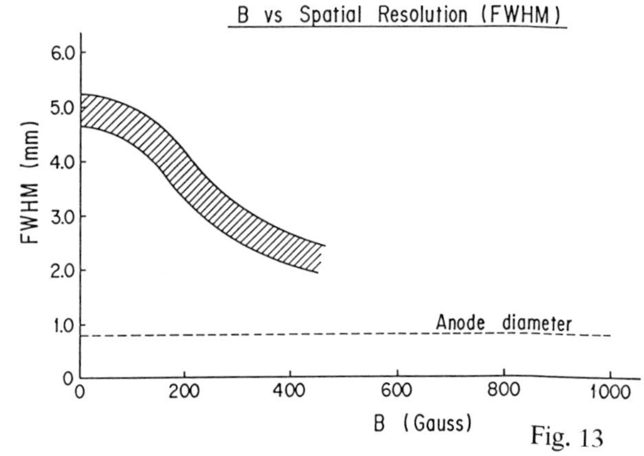

Fig. 13

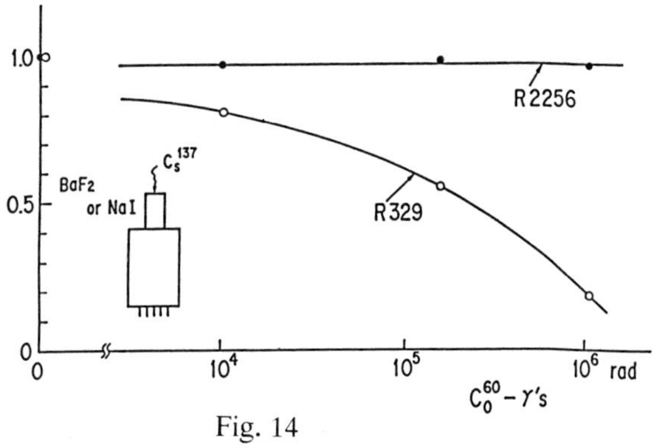

Fig. 14

CALORIMETRY IN ZEUS: LESSONS FOR THE FUTURE

M. DERRICK
ARGONNE NATIONAL LABORATORY, ARGONNE, IL 60439

INTRODUCTION

The storage ring facility HERA, under construction at the DESY Laboratory, will collide 30 GeV e^- with 820 GeV protons when the machine is first turned on in 1990. Four interaction regions are provided. Subsequently either e^- or e^+ beams can be used and a facility for producing longitudinal polarization of the electrons at the interaction points will be provided. Both the center-of-mass energy ($\sqrt{s}$ = 314 GeV) and the momentum transfer ($Q^2_{max} \sim 10^5$ GeV^2) are substantially higher than in any deep inelastic scattering studies to date. The HERA program will be the natural extension of both the neutrino and the muon scattering experiments done at CERN and Fermilab. With the design luminosity of 1.5 10^{31} cm^{-2} sec^{-1} experiments with 200 pb^{-1} of integrated luminosity will be possible.

HERA will be the first electron-quark collider. The energy is substantially higher than LEP, and the events will be cleaner than those resulting from $p\bar{p}$ collisions. It provides the third leg that will support the physics stool of the 1990 s. In neutral current events, the Z exchange graph is comparable in strength to the photon exchange and the charged current events will provide unique insights into the W coupling. There will be substantial rates of heavy quark production (10^8c, 10^6b, and t up to ~ 80 GeV for 200 pb^{-1}).

Two detectors are under construction: H1 uses a large solenoid with a liquid argon calorimeter inside the coil, ZEUS uses a smaller magnet with uranium-scintillator calorimeters outside of the coil.

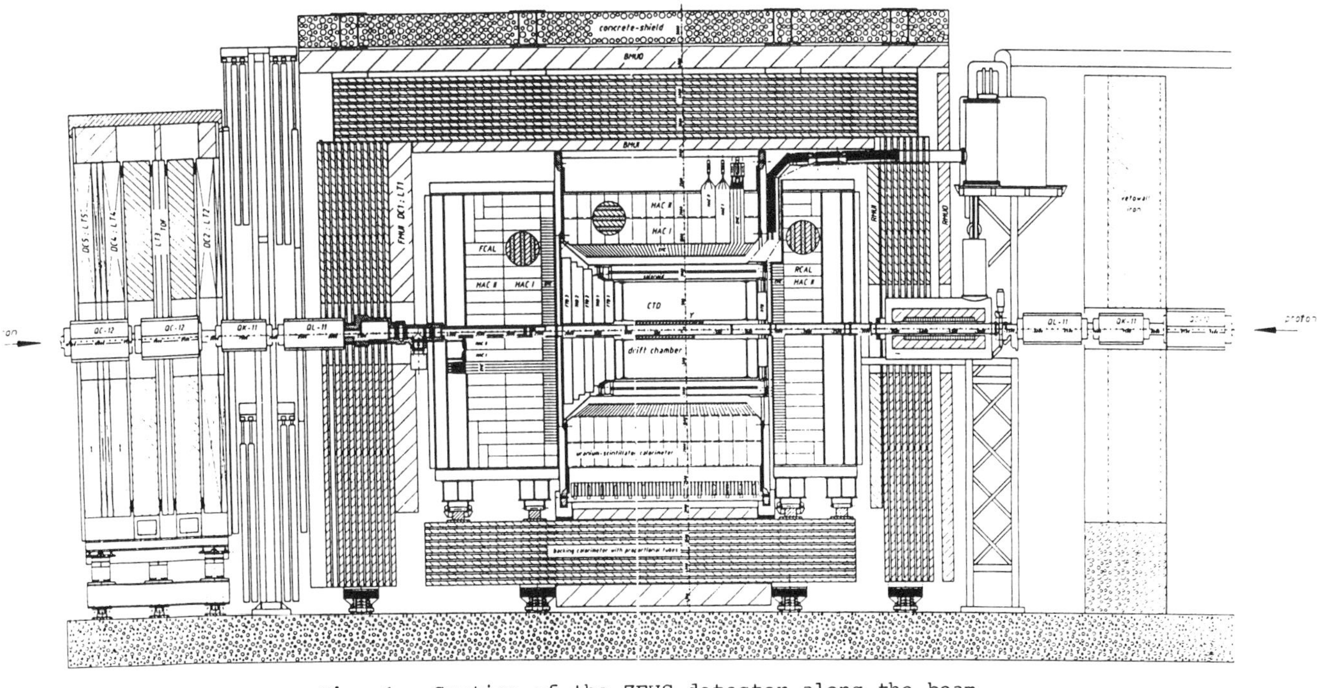

Fig. 1. Section of the ZEUS detector along the beam.

THE ZEUS DETECTOR

(a) Overview

Figure 1 shows a cross section of the ZEUS detector[1)] that is being built by a large International Collaboration: 10 countries, 43 institutions and about 300 people. The detector contains all the usual components: vertex chamber, central tracker, high resolution calorimeter, backing calorimeter, and muon detector. The high resolution calorimeters are the responsibility of groups from Canada, Germany, Japan, The Netherlands, and the United States.

As shown in Fig. 2, the maximum energies of the jets vary from 820 GeV in the proton beam direction to 20 GeV at backward angles. The kinematics of the charged current events $e^-p \to \nu X$ are measured from the quark jet so better calorimeter resolution translates into fuller coverage in the $x{:}Q^2$ plane and, in particular, to reconstruction of higher Q^2 events. A good knowledge (~ 2%) of the absolute energy scale is also required. A special challenge is presented by the beam crossing time of 96 nsec, which does not allow a simple implementation of the trigger. Solutions adopted here will have application to the SSC, where the crossing time will be even shorter.

New physics is often signed by combinations of leptons with leptons and/or jets, as well as missing E_T, so good electron identification within jets, full muon coverage, and an hermetic calorimeter are required.

The ZEUS detector is a simple and consistent design that uses existing technology as much as possible. The emphasis on achieving the best jet resolution requires an hadronic calorimeter giving the best possible energy measurements.

The calorimeter is divided into two sections: a front high resolution section varying in depth from 4λ in the rear to 7λ in the forward direction, as shown in Fig. 3. This depth is sufficient to contain 95% of the energy for 90% of the jets that have the maximum energy allowed by the kinematics. The backing calorimeter (BAC) consists of 10 layers of 7.5 cm steel plates interleaved with aluminum proportional tubes to give an energy resolution of $\sigma_E/E = 100\%/\sqrt{E}$. The BAC is used to classify events that have late developing showers and so a poorer energy measurement.

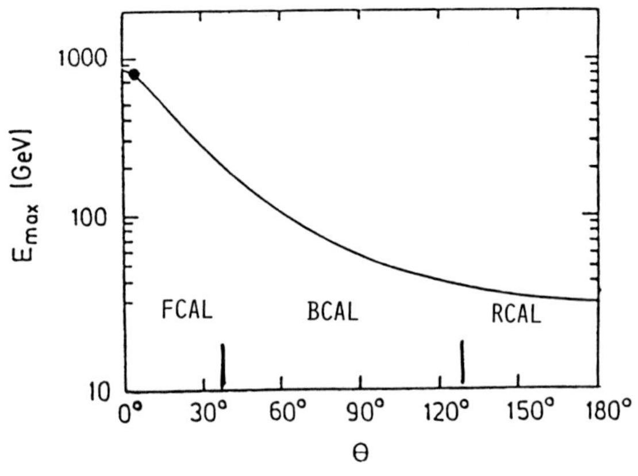

Fig. 2. Maximum jet energy as a function of angle.

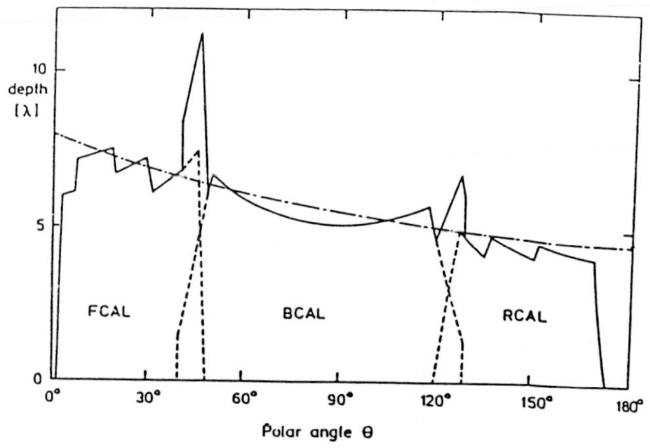

Fig. 3. Depth of high resolution calorimeter.

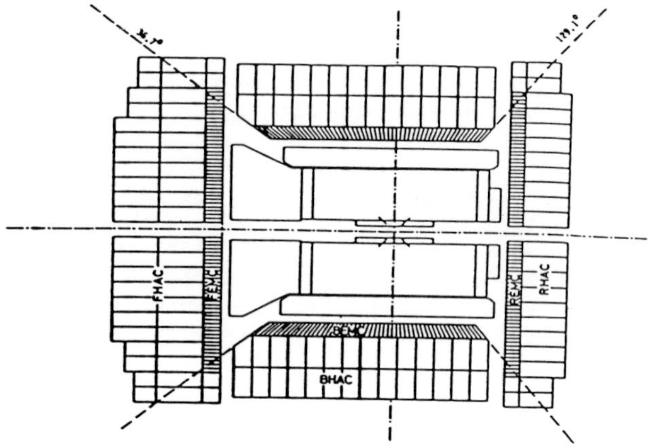

Fig. 4 High resolution calorimeter.

(b) Calorimeter Parameters

The High Resolution calorimeter is divided into three sections, as shown in Fig. 4. It is designed to cover the full solid angle (99% of 4π), have energy resolutions of $17\%/\sqrt{E}$ for incident electrons and photons, and $35\%/\sqrt{E}$ for incident hadrons. The latter requires that the systematic uncertainties be kept at the 1-2% level. With the jet angular resolution of 10 mr the energy uncertainty dominates the measurement of the jet kinematics. Electron/hadron separations in jets is an additional important design goal.

The EM resolutions require one radiation length sampling, and the necessity of having an equal response to incident photons, as well as photons from π^o's generated in the hadronic shower requires that the unit cell be the same throughout the calorimeter.

The jet angular resolution and the e/h rejection sets the tower size as typically 5 cm × 20 cm in the EMC section, and 20 cm × 20 cm in the hadronic section. One HAC tower covers four EMC towers. The readout in depth is typically 21 X_o or 1λ for the EMC section, and two hadronic sections (HAC1 and HAC2) that are each 2λ in BCAL and 3λ in FCAL.

(c) **Choice of Technology**

In order to obtain the best energy resolution, a compensating calorimeter is required, and we considered the options of lead or uranium as the radiator and liquid argon or scintillator as the readout medium. A comparison of detailed simulations[2)] with test beam measurements pointed to uranium/scintillator as a unique choice.

For a sampling calorimeter, the response to electrons is less than for muons in terms of the specific energy deposited in the sensitive medium. This reduction in the electromagnetic response comes about because low energy electrons and photons are preferentially absorbed in the high Z material. Typical values of e/μ are ~ 0.60 for u/scintillator or u/A combinations. The response to hadrons is even lower because of the invisible energy lost to nuclear excitation. The relative responses are illustrated in Fig. 5.

The compensation mechanism must enhance the hadronic component and a perfectly compensated calorimeter will have e/h = 1. Contributions to the compensation include:

(i) prompt photons from primary fission;
(ii) prompt photons from secondary fission;
(iii) delayed photons from (n, γ) capture;
(iv) recoil photons from (n,p) scatters;
(v) recoil (d, t, α) from (nd) reactions, etc.

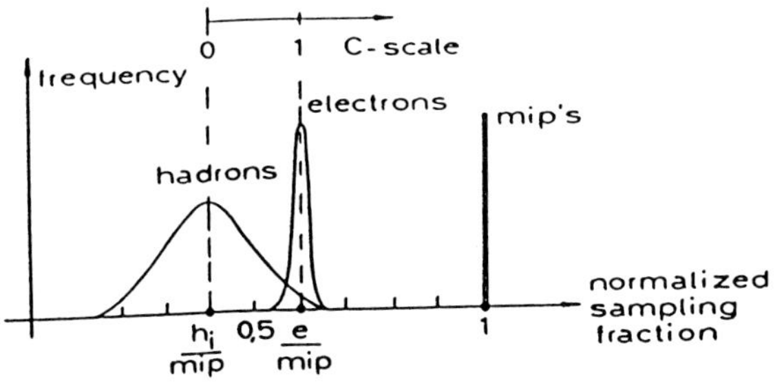

Fig. 5. Response of calorimeter to μ, e, and hadrons.

The relative contributions of these different components as a function of time are shown in Fig. 6, which comes from a detailed Monte Carlo simulation.[2)] It is clear that the compensation can be tuned somewhat by choice of the gate length.

A series of measurements in test beams at CERN,[3)] plus these simulations, clarified the compensation mechanism, confirmed the choice of depleted uranium plus scintillator, and set the unit cell of the calorimeter as 3.3 mm of uranium ($1X_o$) and 2.6 mm of scintillator (SCSN 38). Figure 7 shows the degree of compensation as a function of the ratio of the uranium-to-scintillator thickness. The lines show the results from the simulations and the points are from test beam measurements. The measured energy resolution and e/h ratios for a test calorimeter

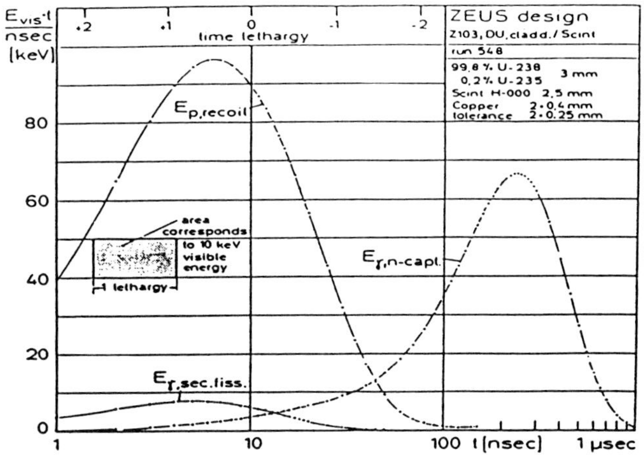

Fig. 6. Time evolution of compensation.

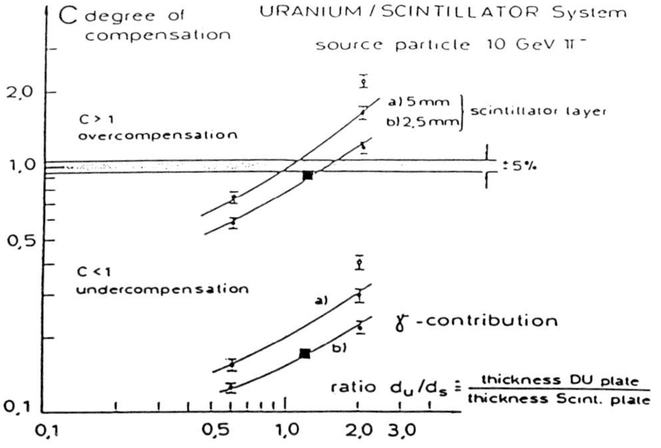

Fig. 7. Compensation as a function of uranium/ scintillator thickness.

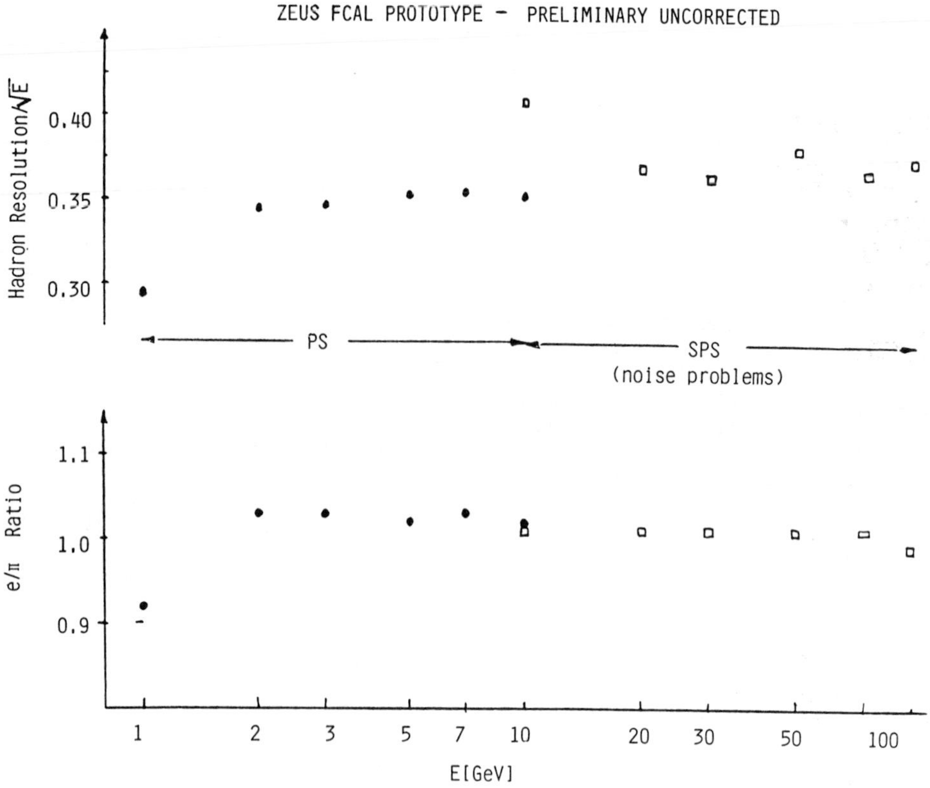

Fig. 8. Resolution and e/π ratio.

built to these specifications are shown in Fig. 8. It is clear that the design goals have been met.

(d) **Electron Identification**

Several techniques will be used to identify electrons. In the barrel region, the dE/dx in the central tracker and the E/P ratio, comparing the energy in the projective EMC towers with the track momentum, provide a good discrimination. In the forward direction, transition radiation detectors that alternate with the forward tracking detectors identify electrons. In both calorimeters, silicon pads 3 cm × 3 cm in area provide another independent e/π discriminator. In BCAL one layer will be used after $4X_o$, in FCAL two layers, one after $4X_o$ and one after $7X_o$. There are a total of about 50,000 such pads in the detector covering a total of 45 m^2. These pads provide a good discrimination against π:γ overlaps and also separate e from π by pulse height. Measurements in a 30-GeV test beam in a BCAL prototype show that for 90% electron acceptance, the hadron → electron confusion is at the 0.3% level when the data from two layers of silicon are used.

A final possibility is to use the time dependence of the energy deposition. Figure 9 shows the result of a Monte Carlo simulation[4], including the effects of the electronic pulse shaping. The fully compensated hadronic shower has more pulse height at late times coming from the compensation mechanism.

(e) **Readout**

The readout is done in the usual way by 2-mm thick Y7 wavelength shifter (WLS) plates on the two surfaces of the calorimeter modules. The uniformity, both in depth and laterally, is achieved by using a reflective mask. The response of every piece of scintillator (~ 80K in

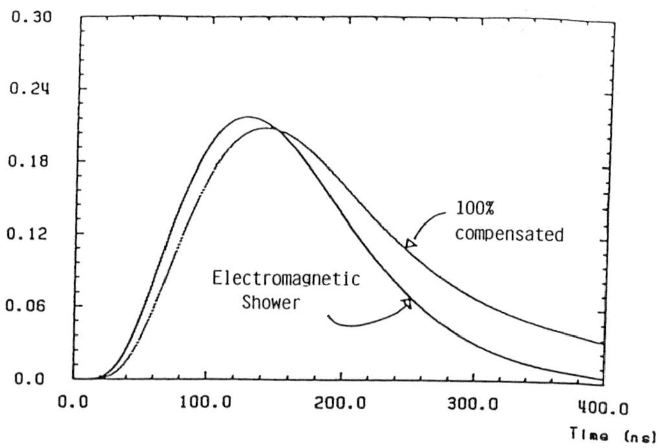

Fig. 9. Monte Carlo simulation of pulse shape.

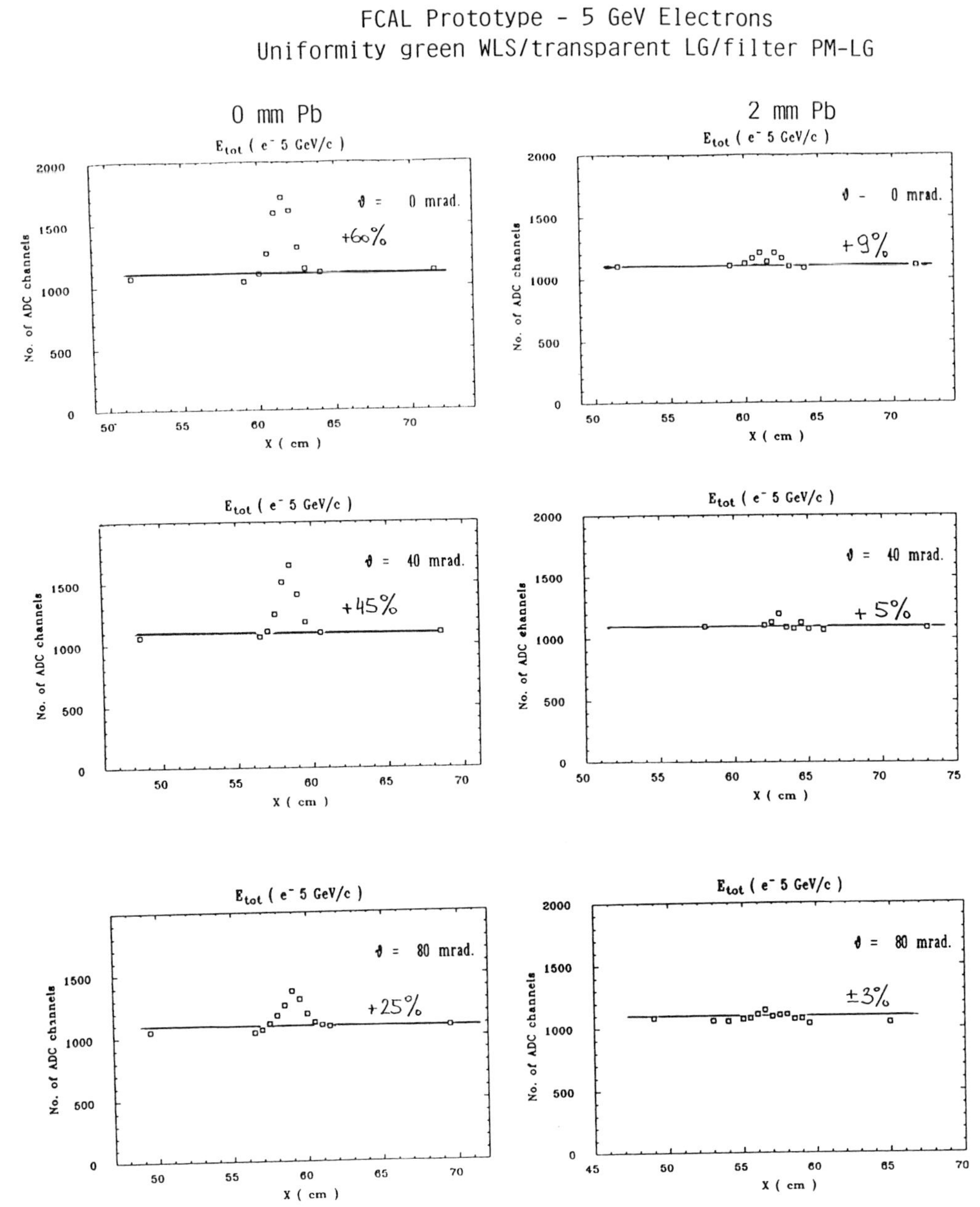

Fig. 10. Response of sweeping across crack between modules.

BCAL alone) are being individually measured. The thicknesses of all of the uranium plates at each tower will also be measured so that the masks can also compensate for the manufacturing variations.

A standard difficulty of such a readout scheme is that particles entering the calorimeter at and parallel to the WLS can give a large false signal in the photomultiplier tubes. This problem is worse in FCAL near the beam direction and between the 32 BCAL modules.

A series of test beam measurements made at CERN with 5 GeV electrons are shown in Fig. 10 for incident angles of 0°, 40 mr, and 80 mr. By placing a 2 mm thick lead sheet between the modules, the effect is reduced to the few percent level, except at normal incidence. One can think of this solution as utilizing an additional calorimeter placed between the modules but with only two samples and an unfavorable geometry.

The BCAL modules are rotated about a radius vector such that the angle of incidence at the front module boundaries is ~ 80 mr so that the effect at the BCAL boundaries will be small. This has been confirmed by EGS calculations and will be checked experimentally with a test beam at Fermilab.

(f) **Mechanical Structure**

The modules are built using small metal spacers between the uranium plates to provide a set of slots which contain the scintillator panels. The mechanical structure puts no pressure on the scintillator. The uranium plates are attached to a strong back-beam by a set of straps. The PM's are also located in this back beam. Figure 11 shows a cut-away of an FCAL module as an example.

(g) **Calibration and Stability**

Achieving 1-2% uniformity in space and time will be a formidable challenge. The first requirement is that every component be both individually measured and accurately located. Secondly, the uniformity of response of every tower should be measured using cosmic rays or accelerator test beams. Finally, a set of on-module calibrators will be provided. The uranium radioactivity provides a constant signal. To set this well above the pedestal, but below the MIP level, requires 0.4 mm (0.2 mm) of steel between the uranium and the scintillator for the HAC (EMC) towers. We therefore wrap each uranium plate in a thin stainless steel skin. Additional point radioactive sources are also provided that can be positioned anywhere in depth. Finally, a charge injector in the electronics and a laser flasher injecting light into the lightguide will be utilized to check the linearity of the system.

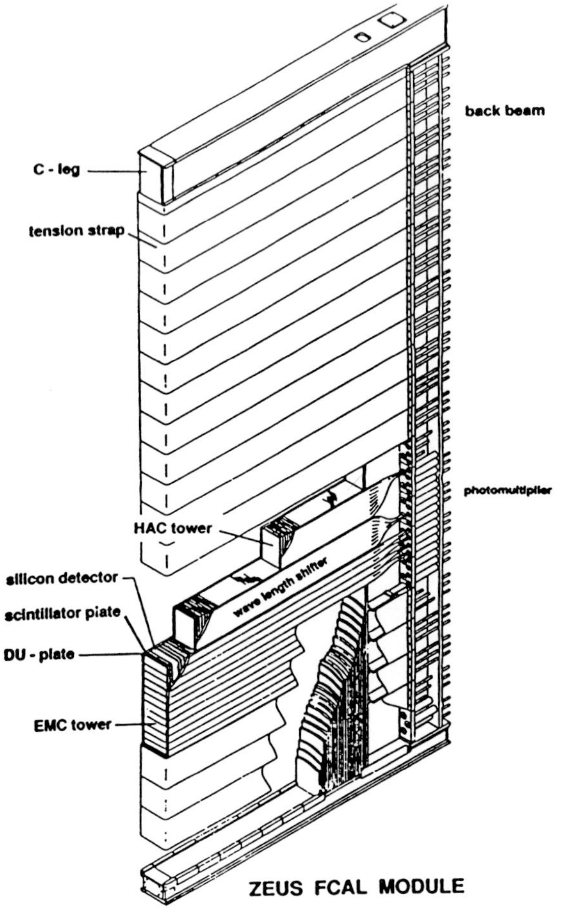

Fig. 11. FCAL module.

(h) **Readout**

The signal from each PM tube is fed into a pulse stretcher and shaper. The shaped pulse is sampled several times at 96 nsec intervals and the values put into a switched capacitor analogue pipeline with a 5μsec delay.

If the first level trigger requirement is satisfied, then the values are measured in an A to D converter and an on-board DSP calculates the time and amplitude of the pulse. These values are then fed to the second level trigger. The potential dynamic range is very large, ~ 10^5:1, since, in principle, all of the forward energy could be deposited in one EMC tower. Two ADC channels, high and low, will therefore be provided.

LESSONS FOR THE FUTURE

Since this conference is concerned with preparations for SSC detectors, I close with some subjective and personal remarks. We may hope that the current generation of detectors for LEP, SLC, the CERN collider, and TeV I, as well as HERA, are the last detectors to be built before the SSC program starts. Of these, ZEUS and H1 represent the latest models.

A perfect detector like a perfect spouse exists only in dreams; reality imposes some compromises. However, the detector design must be driven by the principal physics goals and not by technological hubris. To overstate the case, one should try to always use obsolete technology.

Some time pressure and the early availability of substantial equipment funds are essential so that the group does not spend several years in a "summer study" mode without coming to grips with the engineering compromises that are so important in a real detector.

The number of technologies utilized in the detector should be minimized in order to simplify both the operations and the data analysis: not to speak of the number of experts needed to operate the detector efficiently. All of the high resolution calorimetry in ZEUS uses the same uranium/scintillator system, and there is only one central tracker, plus vertex chamber, between the IP and the calorimeters. One may contrast this with the CDF detector.

The organization of such large collaborations provides special challenges. The financial, managerial, and oversight requirements seem, on the whole, to be in hand although there are difficulties for detectors funded largely by countries in a different continent. There is a special difficulty in balancing the need to have a critical mass of people in one place under proper leadership so that the problems can be addressed in a focussed way, with the equally essential need to have the full participation of the user and, particularly, the University community. This problem has not really been solved.

Finally, it is clear that any detector design and construction effort requires the upfront availability of proper tools. These include detailed detector simulation capabilities, constant access to a test beam, and the intimate full-time involvement of a critical group of physicists and engineers.

The ZEUS effort on the design of the calorimeters has been a well-focussed effort that has resulted in good design. What we have done less well is to understand in detail the effects of various structural elements on the physics and, in particular, the standard problem of the interfaces between say the forward and the barrel calorimeter. To address these problems in detail requires enormous computing resources, as well as a debugged Monte Carlo simulation program.

In the last several years, our field has invested in around a dozen very large and sophisticated colliding beam detectors. The next five years will show how well these detectors perform. A critical evaluation of the successes and shortcomings of the detectors resulting from the design decisions taken by these collaborations would be an appropriate topic for Snowmass '90. Let us hope that this meeting will provide the last chance for such discussions before the construction of the SSC detectors starts.

This work was supported by the U.S. Department of Energy, Division of High Energy Physics, under Contract W-31-109-ENG-38.

REFERENCES

1. The ZEUS Detector: Status Report 1987, Ed. J. Krueger. This report and the 1988 update contains a wealth of information on all aspects of the detector.

2. H. Brückmann, U. Behrens, and B. Anders, "Hadron Sampling Calorimetry: A Puzzle of Physics," DESY 86-155. Several groups contributed to the understanding of the compensation mechanism. J. Brau and T. A. Gabriel, Nucl. Instrum. and Meth. A238, 489 (1985); C. Leroy, Y. Sirois, and R. Wigmans, Nucl. Instrum. and Meth. A252, 4 (1986);

3. E. Ros and T. Tsurugai, ZEUS Note 88-086.

4. A. Caldwell and F. Sciulli, ZEUS Note 88-084.

The Spaghetti Calorimeter Project at CERN

Richard Wigmans

CERN, Genève, Switzerland

ABSTRACT

In the framework of the LAA project at CERN, an attempt is being made to develop a new type of calorimeter that could meet the Supercollider challenge. The detector is based on scintillating plastic fibres embedded in a lead matrix, and will be compensating. The goals and achievements of this project are discussed in some detail.

1. INTRODUCTION

In a previous contribution, M. Derrick discussed the lessons to be learned from the ZEUS experience on calorimeter development.[1] He did not mention one lesson that I consider particularly important for future applications, *i.c.* the experimental demonstration of compensating lead/plastic-scintillator calorimetry.

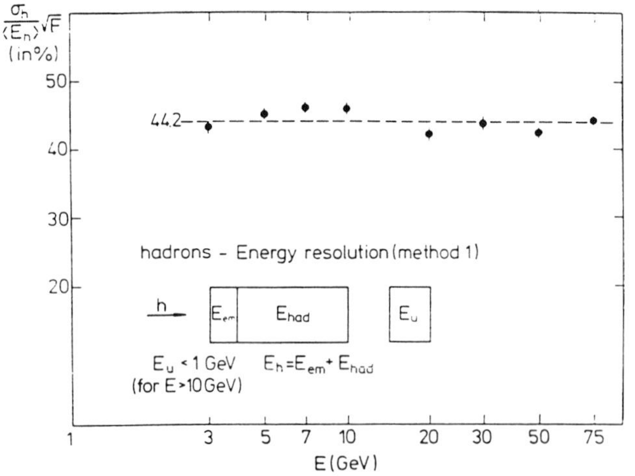

Figure 1. Hadronic energy resolution measured as a function of the pion energy with a 10/2.5 mm lead/plastic-scintillator calorimeter. Data from ref. 3.

In the framework of the ZEUS calorimeter *R&D* a detector was built consisting of 10 mm thick lead plates interleaved with 2.5 mm thick plastic scintillator sheets.[3] This detector showed all the characteristics of a compensating hadron calorimeter[2] : Signal linearity for pion detection, energy resolution σ/E scaling with $1/\sqrt{E}$ (fig. 1), and a gaussian response. The energy resolution was measured to be $43\%/\sqrt{E}$, and the e/h signal ratio 1.05 ± 0.04. These results confirmed the prediction[2] that inspired the construction of this unconventional instrument.

This breakthrough in the understanding of the working of hadron calorimeters made it possible to define a dedicated *R&D* effort to optimize the technique. The spaghetti calorimeter project, which is the subject of my contribution to this workshop, was proposed in this spirit. It is presently carried out in the framework of the LAA detector *R&D* project at CERN.

Although the idea of using scintillating fibres instead of scintillator sheets was primarily conceived because of the much more frequent shower sampling and, therefore, much better energy resolution achievable in this way, it turns out that the fringe benefits of using this technique are even more crucial in view of the Supercollider challenge: At present, I don't see which detector could be more hermetic, more compact, cheaper or faster than the one discussed here.

In section 2, the project is briefly described, the results obtained so far are discussed in section 3, and conclusions are given in section 4.

2. THE SPAGHETTI CALORIMETER PROJECT AT LAA

The fact that compensation can be achieved with lead absorber is in itself a major achievement. In contrast to the cumbersome uranium, lead is available in unlimited quantities, easy to handle and to machine, not radioactive and, last but not least, cheap. It is interesting to note that because of the fact that a much larger fraction of the detector has to consist of absorber in the case of lead, the *effective* nuclear interaction lengths of *compensating* lead/scintillator and uranium/scintillator calorimeters are about equal, ~ 20cm. Therefore, in spite of its much larger density uranium does not allow for constructing more compact compensating detectors than does lead.

Apart from the fact that it makes compensating calorimeters possible, plastic scintillator has one additional major advantage: it is very fast. Signal rise times of the order of 1 ns are feasible, and therefore it is one of the very few techniques with which the envisaged supercollider event rates can possibly be handled.

The large hadron calorimeters based on plastic-scintillator readout which are either operating or under construction today (HELIOS, CDF, UA2, ZEUS), all have the classical sandwich structure, *i.e.* alternating plates of absorber and plastic perpendicular to the direction of the incoming particles. In order to be able to read the signals, the light has to make a turn of 90 degrees, which is achieved with wave length shifting plastic plates that run perpendicular to the other ones. The scintillation light is absorbed in these plates, and re-emitted at a longer wave length. This inefficient process has the disadvantage of introducing regions without absorber, which usually span 5 – 10% of the total detector volume. Another source of inhomogeneity is caused by light attenuation in the plastic, which may be quite important since the attenuation lengths are typically considerably less than 1 meter.

The main disadvantage of the sandwich structure is, however, the impossibility to achieve a fine granularity. Although some experiments have demonstrated a lot of ingenuity to limit the effects of the mentioned disadvantages, the results are far from ideal.

In the LAA project, we are therefore using a different structure, based on scintillating plastic fibers. This eliminates in principle all the mentioned problems. The fibers are embedded in a lead matrix and are running (roughly) in the direction of the incoming particles. This technique was already successfully employed on a small scale, for electromagnetic shower detection in various experiments (Omega, NA38).[4)]

Apart from the ultrafast timing and the compensation capability common to all scintillator structures, the potential advantages of fiber calorimeters at a supercollider include:

a) The possibility of making a completely hermetic detector (no cracks).

b) An arbitrarily fine granularity.

c) An excellent energy resolution, due to the very frequent shower sampling. We expect $15\%/\sqrt{E}$ for electromagnetic showers and $30\%/\sqrt{E}$ for jets.

d) A large signal/noise ratio, due to the high light yield (no 90° wave length shifting).

e) The possibility of using very stable light detecting devices that can operate in a magnetic field, thanks to the same high light yield.

f) A very good homogeneity. Thanks to the long attenuation length in fibers (at least several meters), the signal practically does not depend on the position where the particles interacted.

g) Easy calibration and signal stability monitoring, a crucial point at very high energies.

A detailed discussion on all these points can be found in the proposal.[5)] In the first phase of the project, which is running since October 1987, a prototype calorimeter will be built, with a simple structure where all the fibers are running parallel to one another. This prototype will be used to extensively check the optimistic expectations concerning its performance. In order to be able to do proper tests, it will be made sufficiently large to contain the highest energy hadron showers available at CERN (450 GeV protons) at a level of 99%, such as to make the effects of shower leakage small compared to the expected energy resolution. This leads to a detector of about 15 tons.

If this phase turns out to be successful, a more sophisticated structure, optimized for operating in an experiment at a storage ring, is envisaged. The quasi-pointing geometry of the fibers will then allow obtaining additional shower information (see ref. 5), which will definitely lead to improved electron identification possibilities.

3. ACHIEVEMENTS SO FAR

Although the project was started less than a year ago, some major progress towards achieving its goals has already been made.

3.1 Compensation, energy resolution

In collaboration with some physicists from the ZEUS experiment, we have studied in more detail the properties of compensating lead/plastic-scintillator calorimetry, using the detector that was built to verify the original predictions[2,3)] This detector consists of 10 mm thick lead plates interleaved with 2.5 mm plastic-scintillator plates (the ratio 4 : 1 in these thicknesses is the secret of compensation).

The energy resolution of any hadron calorimeter consists of a component due to *sampling* fluctuations, arising from the fact that only a small fraction of the particle energy is deposited in the active layers, and an *intrinsic* component, due to the fact that the fraction of the particle energy transformed into ionization (the "visible" energy) fluctuates from event to event. The effects of these fluctuations on the energy resolution of the calorimeter add in quadrature:

$$\sigma_{tot}^2 = \sigma_{samp}^2 + \sigma_{intr}^2$$

The new measurements were intended to separate these

two contributions, to be able to make realistic predictions for the energy resolution of the fiber calorimeter. Since the showers are sampled much more frequently in the fiber case, σ_{samp} will be considerably smaller than in the rather crude-sampling sandwich detector, which will lead to a further improvement of the already impressive energy resolution obtained with the latter.

The contributions of sampling and intrinsic fluctuations were separated in the standard way, by blocking the light of half the scintillator plates. In doing so, the contribution of sampling fluctuations in the quadratic sum is doubled, and by comparing the total resolutions measured with and without blocking, one may disentangle σ_{samp} and σ_{intr}.

It turned out that the energy resolution of the sandwich calorimeter was completely dominated by sampling fluctuations. When half of the plates were read out, the resolution for hadron detection went up from $42.8\%/\sqrt{E}$ to $59.2\%/\sqrt{E}$, *i.e.* by almost a factor $\sqrt{2}$. The *intrinsic* energy resolution, *i.e.* the limit for an infinitely fine sampling device, was found to be $12.7^{+4.4}_{-7.5}\%/\sqrt{E}$, averaged over the energy range 10 - 75 GeV. This is significantly better than the intrinsic limit for a compensating uranium/scintillator detector, which was measured to be $22\%/\sqrt{E}$ with the same method.[6] This can be explained as follows (see ref. 2).

The intrinsic resolution is largely dominated by fluctuations in the amount of energy used to release nucleons from atomic nuclei in the nuclear spallation reactions (binding energy losses). Since most of these nucleons are neutrons in the case of high-Z target material, there is a correlation between these invisible nuclear binding energy losses and the kinetic energy carried away by neutrons. Efficient neutron detection, a crucial ingredient for compensation, therefore *reduces* the effect of fluctuations in the nuclear binding energy losses on the energy resolution. The extent to which this mechanism works depends then on the degree of correlation between the nuclear binding energy losses and the kinetic neutron energy. This correlation is considerably better in lead than in uranium, since in the latter case many of the neutrons come from fission processes. These fission neutrons are not correlated at all to the nuclear binding energy losses.

The results of the measurements are extremely encouraging, since they mean that the envisaged energy resolution for jet detection of $30\%/\sqrt{E}$, which in itself would already be a world record, is probably a conservative estimate (see fig. 2).

In a second series of measurements, half of the scintillator plates were physically *removed.* In doing so, the ratio lead/scintillator was changed by a factor 2 (8 : 1 instead of 4 : 1). It turned out that the e/h signal ratio was only marginally affected. This is also an important and encouraging result, since the detector envisaged for

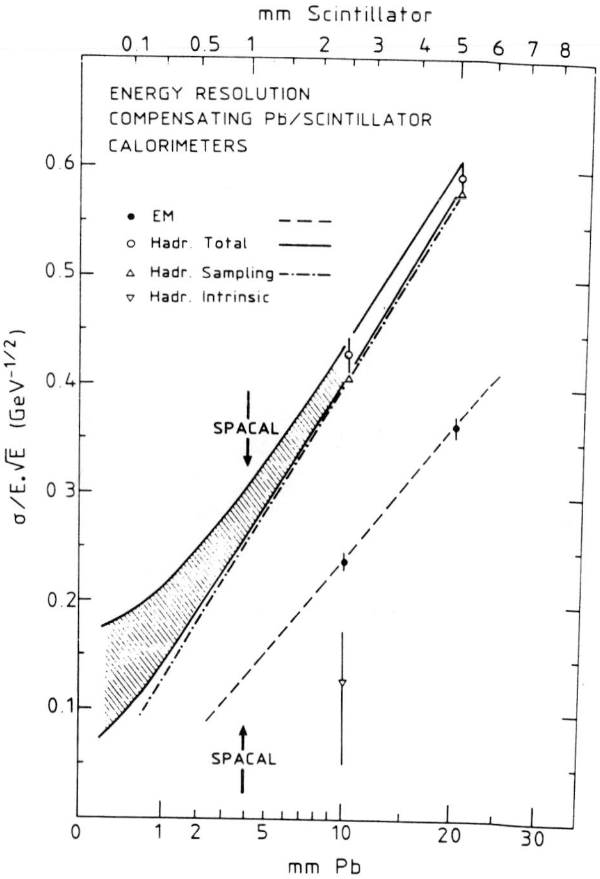

Figure 2. The energy resolution for electron and hadron detection with a compensating lead/plastic-scintillator calorimeter, as a function of the thickness of the readout layers. The experimental measurement results are included. The arrow denotes the working point of the fiber calorimeter under construction.

the second phase of the project will have a varying fraction of scintillator as a function of depth (pointing fibers). The measurements showed that this will not significantly affect the compensation.

3.2 Calorimeter design

Much thought has been given to the calorimeter design. We have chosen a modular structure, with each module having its own readout attached to it. The main advantage of this scheme is a very flexible setup, where modules can be rapidly exchanged, and many different things can be measured.

Previous fibre calorimeters were all built by machining grooves in lead sheets, and glueing fibres into these grooves. Especially in view of the future, more complicated projective structures, we investigated several alternative methods, among which a casting technique gave remarkably good results. In this method, a structure is set up consisting of very thin (50μm) stainless steel capillary tubes, held together by spacers. Liquid lead, or rather an alloy that contains 4% Sb for improving the mechanical properties, is poured around these tubes, yielding a Swiss cheese-like lead structure of any wanted shape. The fibres are slided into the holes afterwards. This method allows for a modular structure of the detector in a natural way.

The shape of the modules was chosen to be hexagonal. This shape matches the cylinder symmetric shower development better than rectangular shapes, it makes possible a very stable detector structure with minimal support, and avoids edge effects (fig. 3).

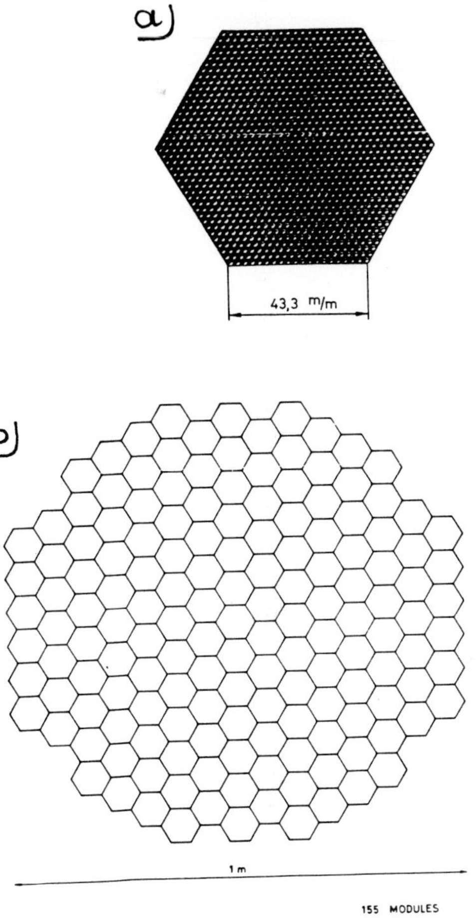

Figure 3. The structure of one calorimeter module (a), and the layout of the whole calorimeter (b).

The size of the modules, or the final granularity of the detector, was determined by many considerations, like the typical lateral dimensions of hadronic showers and the segmentation requirements for a calorimeter at a supercollider; moreover, the useful surface of standard photomultiplier tubes, the weight of the modules, and the possibility of calibrating them with a radioactive source were important criteria. We finally chose for a design that contains 1141 fibres, weighs 86 kg, and has a lateral cross section of 48.7cm^2 (about a factor 10 smaller than the cell size of the HELIOS and ZEUS hadron calorimeters). In total 155 modules are foreseen to be built (fig. 3b). The effective radiation and nuclear interaction lengths are 7.5 mm and 21.0 cm, respectively, almost equal to the figures of the HELIOS and ZEUS uranium detectors.

After many tests the casting technology, which is far from trivial, is now well under control. We have successfully built at CERN a first module with the nominal lateral size, and 20 cm (27 radiation lengths) deep. This module will primarily be used to study the optimization of the detector in terms of readout (what is the best way to couple the fibres to the PM?), light tightness, *etc.*

Apart from that, several other modules, among them a full size (2 m deep) one, are being constructed by industry. They are expected to arrive at CERN by the end of July and will be tested in particle beams during this summer.

3.3 Fiber studies

The quality of the scintillating fibres will be crucial for the success of this project. Therefore, we have put much effort in studying the relevant aspects of the fibre quality, which concern the signal uniformity, the rate capability and the radiation sensitivity of the detector.

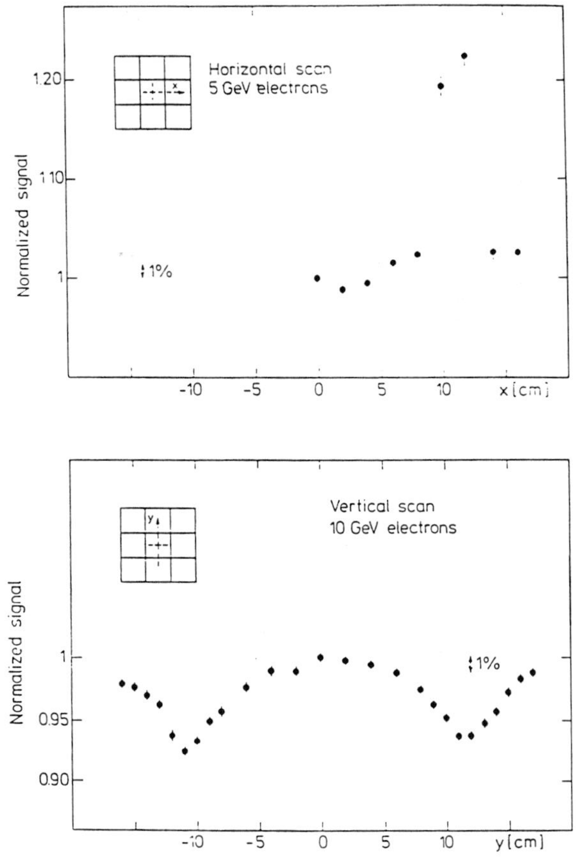

Figure 4. Results of a horizontal (a) or vertical (b) scan with an electron beam across a prototype module of the ZEUS hadron calorimeter. Data taken from ref. 3.

Signal uniformity implies that the calorimeter response should *not* depend on *which* fibres are hit by the shower particles, and *where* they are hit. Since we are aiming for calorimetry with 1% precision, the instrumental effects coming from non-uniformities should be kept below this level. That this is a very ambitious goal may be illustrated by the experience from ZEUS (and also HELIOS) where considerably larger non-uniformities have to be dealt with (fig. 4). Non-uniformities of the type shown in this figure will not only seriously affect the em performance, but will also make the energy resolution for jet detection considerably worse than for single hadrons.

In the fibre calorimeter, two types of non-uniformities have to be worried about:

a) Fibre-to-fibre response fluctuations. These will mainly affect the em energy resolution, since the average number of fibres contributing to the signal from a hadronic shower is so large (typically more than 1000) that the effects are completely washed out.

b) Light attenuation in the fibres. Because of the depth profile of the light production, this phenomenon will mainly affect the hadronic energy resolution.

Fibre-to-fibre response fluctuations may be caused by variations in the fibre diameter, by differences in the light yield or light transmission, and by non-uniformities in the quantum efficiency of the light detecting element. Simulations with the em shower programme EGS4 showed that in order to limit the systematic error on the energy measurement of em showers to 1%, fibre-to-fibre response fluctuations should be limited to 6%, which means that the average number of fibres effectively contributing to the em shower signal amounts to ~ 35 in our detector.

We have built a test setup that will allow us to study the response of individual fibres in our prototype detector. The variations in the fibre diameter have already been measured, and were found to be negligible ($\sigma_{diam} \sim 1\%$).

In order to limit the contribution of light attenuation to the energy resolution of hadron detection to the required level, we calculated that the attenuation length should be larger than 6 m for single hadrons and 3 m for jets. The fibres used by the UA2 Collaboration for their tracking detector did certainly not meet these requirements.[7)]

Measurements of the light spectra revealed that the light attenuation is strongly wave length dependent. Due to the Stokes shift of the dyes, light at short wave lengths (400 - 420 nm) is much more attenuated than the light beyond 450 nm. Therefore, one can considerably *increase* the light attenuation length by filtering the outcoming light, at the expense of a somewhat reduced signal (fig. 5). However, since photon statistics is by no means a limiting factor for the detector performance, this technique will be applied.

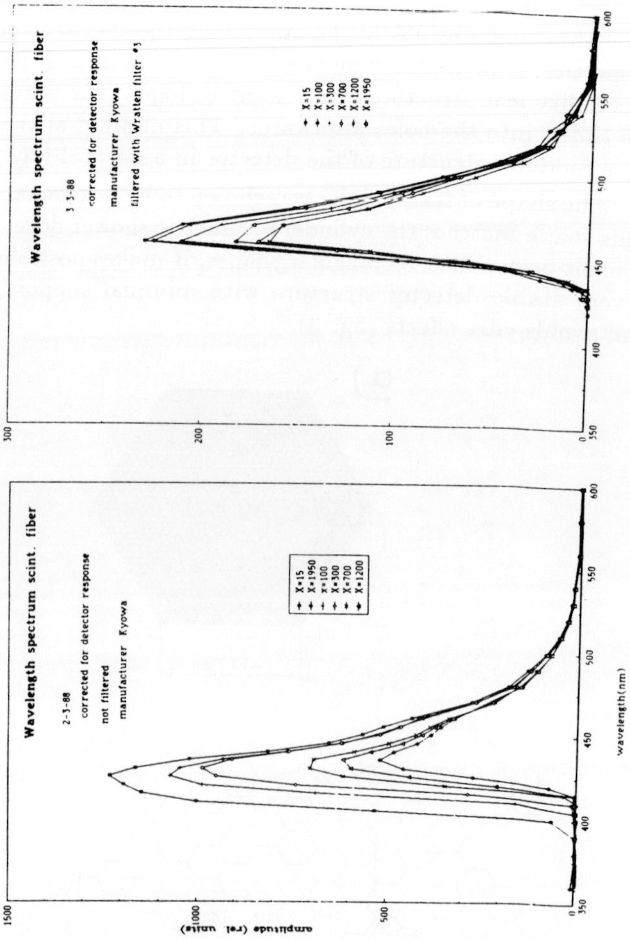

Figure 5. Wavelength spectra for SCSN-38 scintillating plastic fibres, measured at various distances from the point where the fibre was excited (a). Idem, using a yellow filter (b).

Another factor that may limit the attenuation length of the fibres, and determine the level of fibre-to-fibre fluctuations, concerns the optical quality of the core-cladding interface. This is apparently one of the most difficult aspects of fibre production, and not every company masters this technology equally well.

Apart from light filtering at the photomultiplier side, we also use a mirror at the open end of the fibres. This gives a further improvement of the uniformity as a function of depth. The fibre tests show that the calorimeter response can be made constant to within 6% over the full depth of 2 m (fig. 6), which is in fact better than needed and hoped for.

A very nice surprise is the extreme fastness of the fibre signals (fig. 7). The decay time of the light production by the dyes is only 2 ns, and the whole signal does not take more than ~ 7 ns. This is extremely important for experiments at high luminosity machines. The measurements were done with a 2 m long fibre, with a mirror at the open end. When the fibre was excited, the direct and reflected light signals were clearly recorded separately, even

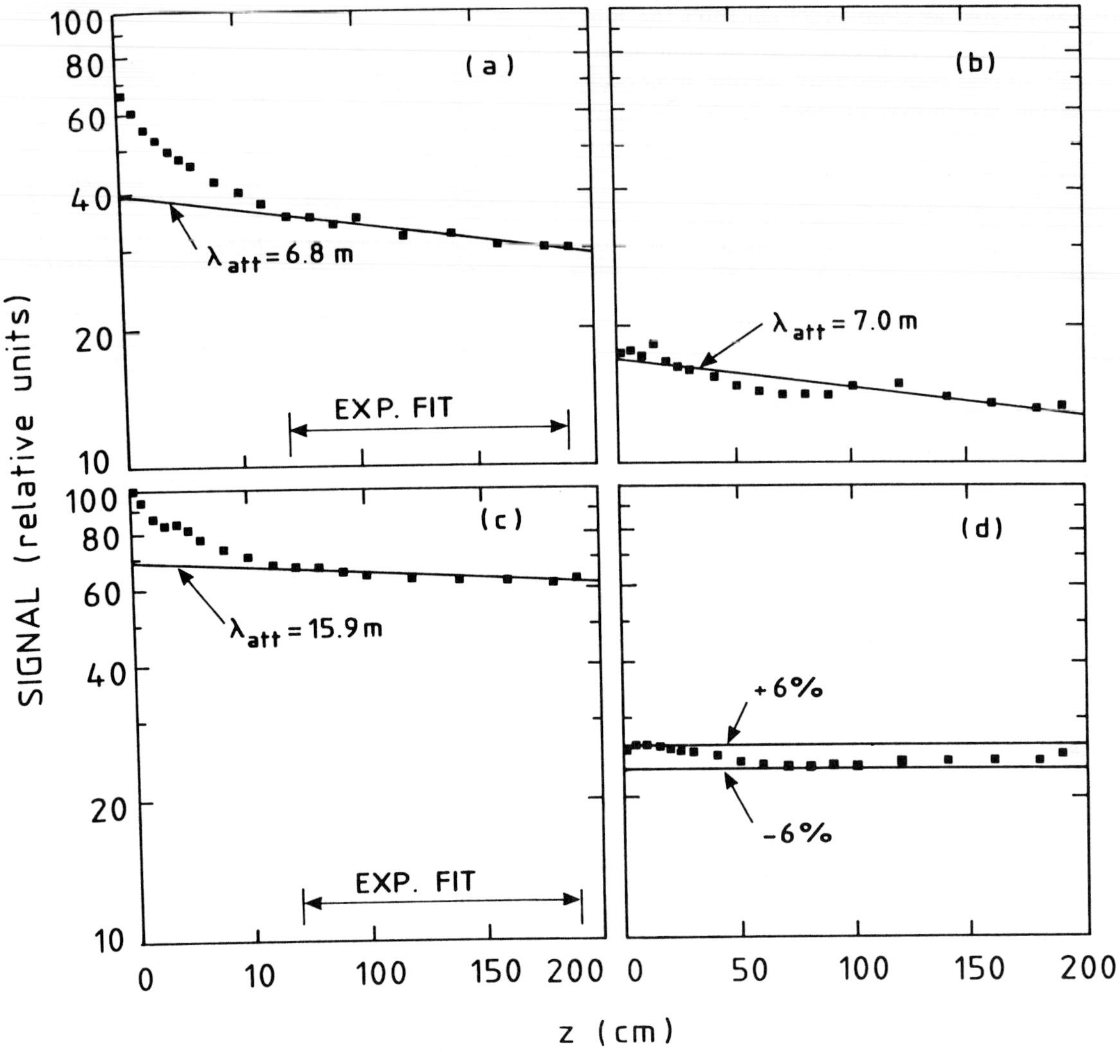

Figure 6. Effects of filter and mirror on the light attenuation curves of SCSN-38 scintillating plastic fibres. No filter/mirror (a). Effect of filter (b). Effect of mirror (c). Effect of filter plus mirror (d).

if the fibre was excited as close as 23 cm from the mirror. These measurements illustrate that the time structure of the calorimeter response may contain important depth information, which may be used *e.g.* for discriminating between em and hadronic showers without the need of splitting the detector into separate parts.

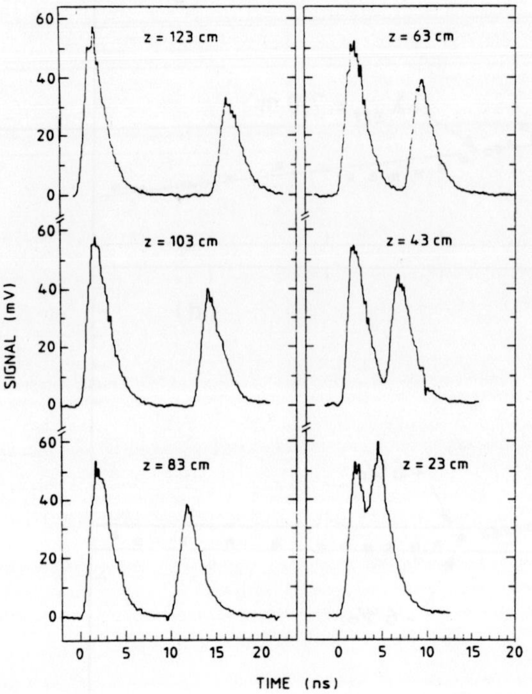

Figure 7. Time structure of the signal from SCSN-38 scintillating plastic fibres, for excitation at various distances z from the mirror at the open fibre end.

A crucial issue for experiments at a high-luminosity supercollider is the radiation sensitivity of the detector elements. Plastic scintillators do not have the best reputation in this respect. However, with the mechanisms of radiation damage becoming better understood, dedicated efforts to improve the radiation stability of scintillators are undertaken, sometimes with remarkably good results. Kyowa Gas Inc. have recently developed a fibre that is claimed to be an order of magnitude more resistant than the SCSN-38 ones that we are using in our prototype modules (fig. 8). Radiation stability up to a level of 1 MRad, which seems to be achievable for modern plastic scintillators, is considered acceptable for detectors that have to operate at the SSC.[8)]

Moreover, our modules are designed in such a way that replacement of damaged fibres is a relatively easy job, which could if necessary well be done for part of the detector in between runs.

And finally, the light filtering applied to improve the signal uniformity is also likely to improve the radiation resistance, since the effect of ionizing radiation on the light

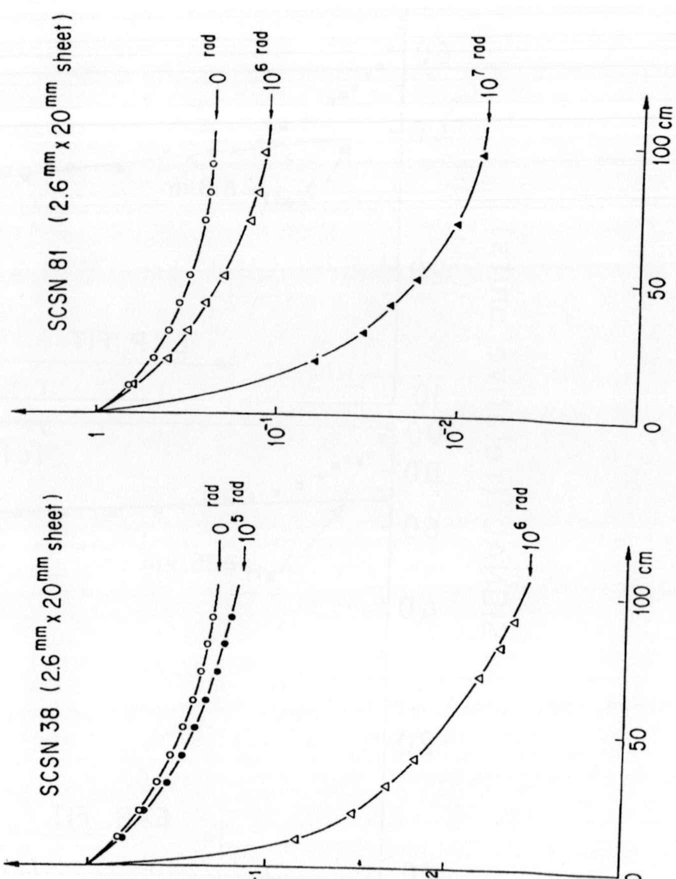

Figure 8. Light attenuation curves for two types of scintillating plastic fibres, before and after exposure to ionizing radiation at various levels.

transmission is wave length dependent (fig. 9). Summarizing, radiation damage seems to be a manageable problem for this type of detector.

4. CONCLUSIONS

The quality of the available calorimetry is of crucial importance for the success of experiments at future high-luminosity *pp* supercolliders. The spaghetti calorimeter investigated in the framework of the LAA project promises substantial improvements with respect to present-day technology in terms of energy resolution, hermeticity, and signal speed, which are all directly linked to the quality of the accessible physics.

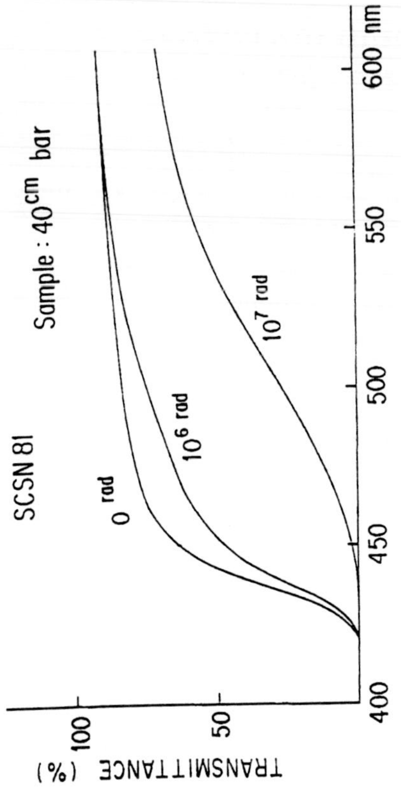

Figure 9. The light transmission as a function of wave length for two different types of scintillating plastic fibres, before and after exposure to ionizing radiation at various levels.

REFERENCES

1. M. Derrick, ZEUS Calorimetry - Lessons for the Future, contribution to this Workshop.
2. R. Wigmans, Nucl. Instr. and Meth. **A259** (1987) 389.
3. E. Bernardi *et al.*, Performance of a Compensating Lead-Scintillator Calori-meter, preprint DESY 87-041 (1987).
4. P. Sonderegger, Nucl. Instr. and Meth. **A257** (1987) 523.
5. P. Jenni *et al.*, The High Resolution Spaghetti Hadron Calorimeter, in Report on the LAA Project, presented by A. Zichichi, vol. 2, 25 June 1987, p. 603; preprint NIKHEF-H/87-7.
6. C.W. Fabjan and T. Ludlam, Ann. Rev. Nucl. Part. Sci. **32** (1982) 335.
7. J.M. Gaillard *et al.* (UA2 Collaboration), Proc. of the DPF Conf., Eugene, Oregon (1985) 912.
8. Radiation Levels in the SSC Interaction Regions, Task Force Report of the SSC Central Design Group, D.E. Groom, editor, Berkeley, June 1988.

SCINTILATTING NOBLE LIQUID DETECTORS
for Higgs -> $\gamma + \gamma$ and Higgs -> leptons
for Higgs mass from 80 GeV to 1 TeV

M. Chen, G. Herten
Laboratory of Nuclear Science and Phys. Dept., MIT, Cambridge, Mass. 02139, USA
C. Dionisi
Universita 'La Sapienza' and INFN, Rome, Italy
J. Seguinot, T. Ypsilantis
College de France, Paris, France

Abstract

Scintillating liquid xenon and argon detectors have been shown to be fast, radiation resistant and with good energy and spatial resolution. In addition, the large primary light can be converted into pulse height every 2 to 3 radiation lengths along the shower path using metal cathodes covered with a thin film of solid organic layers, such as neopentane + TMAE or TMPD. This measures the longitudinal shower profile of isolated electrons or γ's such as those from the Higgs decay and distinguishes them from neutral jet (neutral pions) background. The superior energy and spatial resolution of Xenon Detector also leads to excellent mass resolution for the Higgs and thus improve the signal to background ratio.

I. Introduction.

As we learned in this Workshop, the radiation damage to scintillators is very serious at the SSC, due to the very high luminosity and energy. It is clear that one needs high resolution calorimeter counters with fast response and radiation hardness. For example, the Xenon Olive Detector (Figs. 1 and 2) [1] is designed to be a compact precision electromagnetic detector for electrons and photons. Some of its physics goals are:

1. To identify high energy γ's from Higgs decay for Higgs masses below the two weak boson threshold and to reconstruct the Higgs mass precisely.

2. To reconstruct the invariant masses of di-leptons and di-jets from high p_t Z's to tag heavy new particles such as Higgs, new weak bosons, leptons and quarks, etc.

II. Noble Liquid Scintillation Detectors.

This type of detectors have the following unique properties necessary for selecting low rate electrons/γ's from heavy background at the SSC:

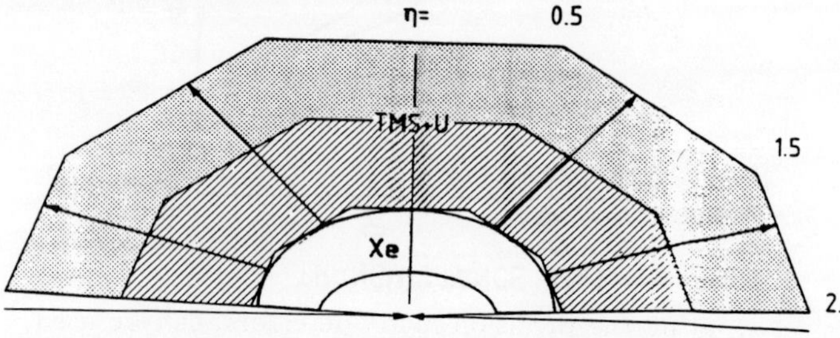

Fig.1 The side view of the Xenon Olive Detector, which is divided into 2 $x10^4$ cells. about 2.2 degrees in θ around 90° and 1 degree at θ around 10°, corresponding to a pseudo-rapidity interval of 0.03 at $\theta = 90^o$, 0.05 at $\theta = 45^o$ and 0.06 at $\theta = 15^o$. Each cell can be further divided into ten sub-cells along the radial direction to measure shower profiles.

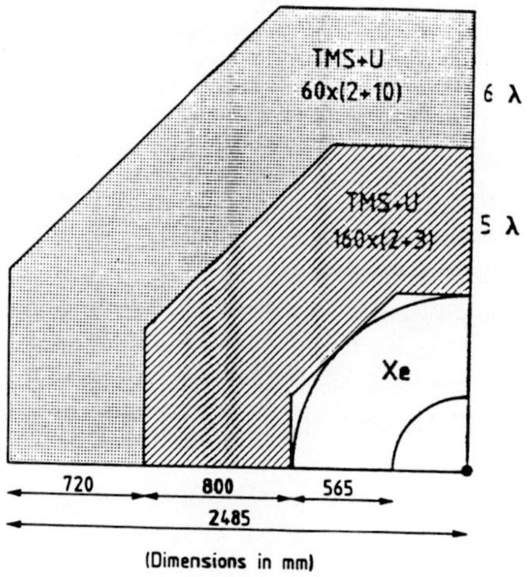

Fig.2 The end view of the Xenon Olive Detector, showing also the sampling thickness and the absorption lengths of a possible hadron calorimeter.

1. *Fast response time:* The decay time for the scintillation light of xenon is 3 ns for 96% of the light, which is much better than BaF_2 (see Table I).

2. *Radiation resistant:* Liquids such as argon and xenon are among the most stable and insensitive to radiation damage, in contrast to all other scintillation material. Xenon, argon and also in this respect, TMS, can be purified and recycled easily.

3. *Excellent linearity:* The scintillating light output from liquid xenon is much larger than that from BaF_2 and linear to 10^4 equivalent minimum ionization particles (mip), which is an order of magnitude better than both BaF_2 and NaI. This yields a large signal to noise ratio, which is important at the SSC environment. TMS is also known to be quite linear.

4. *Excellent energy resolution:* For electrons and photons, liquid xenon, with a large light output, can achieve an excellent energy resolution comparable to BGO, which yields δ E/E = 0.5% for E > 10 GeV. Unlike most sampling calorimeters, in xenon, the light output per unit energy loss, dL/dE, of an electron or photon is not reduced in comparison to a minimum-ionizing particle (mip). Thus cosmic ray muons, with their trajectories and momenta measured by the precision muon detectors, can be used to calibrate the xenon counters to better than 0.5% in situ. The ability to calibrate the detector frequently is crucial in obtaining a good energy resolution. Therefore a mass resolution $\delta M/M = 0.35\%$ is achievable for the decay of Higgs -> $\gamma + \gamma$.
   This superior energy resolution is important at the SSC to measure the electron pairs from new weak bosons and the two photons from the Higgs of intermediate masses.

   For jets, good angular and energy resolution are needed to measure the momenta, the masses of the jets and the jet-jet invariant masses. We consider a combined system consisting of a xenon shower counter followed by a compensating calorimeter such as uranium-TMS sandwich counter. Good energy resolution may be achieved through studies of the delicate compensation effects due to nuclear interactions in uranium and TMS. It also depends on the e/π ratio of the light output per unit energy loss, dL/dE. The ratio of dL/dE for alpha particles and electrons has been measured to be 1.1 ± 0.2 in xenon, in sharp contrast with BaF_2, NaI and BGO, which have ratios no larger than 0.5 (Table I).

   As seen in Table I, xenon saturates at much higher energy densities, and its α/e ratio is consistent with 1. Because these properties and the fact that the neutrino energy loss and the fluctuation of the nuclear binding energy loss in an EM shower counter are small (since the number of inelastic nuclear interactions is small), and the energy of neutrons will be efficiently measured by the uranium-TMS calorimeter, Monte Carlo results have shown that the e/π ratio is within 1± 0.1 for energy between 20 and 1000 GeV [2] for a combined system consisting of a xenon shower counter and a uranium-TMS hadron calorimeter. Thus an energy resolution of 1% at 500 GeV are achievable and is comparable to the resolution of a fully compensated TMS-U calorimeter alone.

5. *Good spatial resolution.* This is necessary for the identification of γ's and electrons with very low rate such as these from the Higgs decays. The Xenon Olive Detector, covering 4 π solid angle, is divided into about $2x10^4$ cells, about 2.2 degrees in both θ and ϕ angles at θ around 90° and the size in θ is reduced to 1 degree at θ around 10°, corresponding to a pseudo-rapidity interval of 0.03 at θ = 90°, 0.05 at θ = 45° and 0.06 at θ = 15°. For $\theta < 15°$, liquid xenon could be sandwiched with lead plates to reduce the equivalent radiation length to about 5 mm.

6. *Measurement of the 3-dimensional shower profile.* In order to obtain the longitudinal shower development profile, each xenon cell can be further segmented into ten separate readout units along the radial direction. Metal cathodes covered with a thin film of solid organic layers, such as neopentane + TMAE or TMPD [3], can be used to convert the UV light from xenon into pulse height with high quantum efficiency (> 10%) and high speed (a few ns). Both scintillation light (fast) and ionization (slow) signals of xenon give large outputs. It has been shown that the timing difference between the two signals from a single counter can yield precise space information with a resolution of about 30μ. These additional handles could be valuable for the identification of high energy electrons (photons), where one needs as much independent information as possible to reject hadrons and multi-photons from neutral pions.

III.Physics examples.

1. *FOR HIGGS -> γ + γ*

In this case, jets and neutral pions are rejected using the longitudinal and transverse shower profiles measured in the xenon detectors as well as by excluding charged particles in the tracking devices in front of the xenon detectors. We expect a mass resolution of δ M/M = 0.35%, or about 85% of the events are contained in a mass bin of 1%. The expected event rate from Higgs decay and background from quark and gluon productions [4,5] is shown in Fig. 3 as a function of the Higgs's mass. We note that the signal to background ratio is significantly improved in our case, (in comparison with the original calculations where typically a 2% mass resolution is assumed) due to better resolution of the xenon detector. We see that for top mass around 90 GeV, at 80 (160) GeV Higgs mass, we have 320 Higgs events with $1.6x10^4$ (4000) events of background for an integrated luminosity of $10^{40}cm^{-2}$. Thus it is possible to obtain a signal of 3 to 5 σ over the mass range between 80 and 160 GeV Higgs's mass. We note that LEP2 will be sensitive to Higgs mass below 80 GeV.

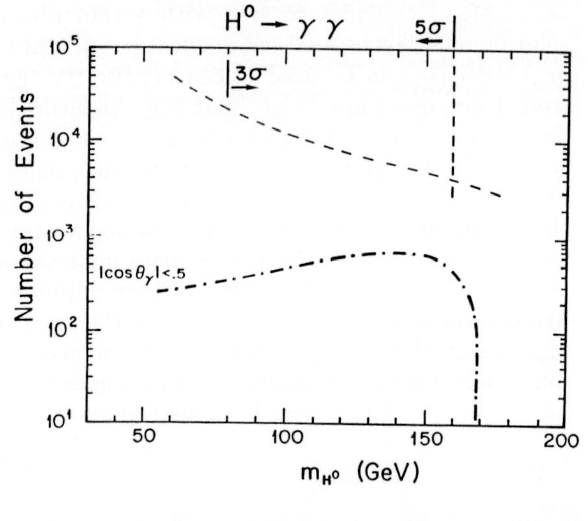

Fig.3 Expected number of events for Higgs $\rightarrow \gamma\gamma$ as function of the mass of the Higgs with an integrated luminosity of 10^4 pb^{-1} at $\sqrt{s}$ = 40 TeV. The dashed-dot curve is from Higgs for m$_t$=90 GeV [4]; the dashed curve is the background from QCD [5]. The momentum resolution of each lepton is assumed to be 0.5% which yields a Higgs mass resolution of 0.3%.

2. *Higgs -> 4 leptons*

In this case, electrons are measured in the xenon detectors while muons will have to be measured in a precision muon detector such as the L3+1 detector. We expect a mass resolution of δ M/M = 0.25% for Higgs mass below 500 GeV and 0.2% for Higgs mass below 200 GeV. The expected event rate from Higgs decay and from background is shown in Fig. 4 in the low mass region as a function of the Higgs's mass. We see that for top masses around 90 GeV, at 140 GeV Higgs mass, we have 100 events with 1250 events of background for an integrated luminosity of $10^{40}cm^{-2}$. At high Higgs mass (1 TeV), we expect about 20 events with negligible background after the constraints that the invariant mass of each lepton pairs must be within 5 GeV of the Z° mass. Thus it is possible to search for Higgs over the mass range between 130 and 1000 GeV Higgs's mass, with 3 σ at 130 GeV and 12 σ at 160 GeV.

3. *HIGGS -> 2 LEPTONS + JETS OR MISSING P_T*

In this case, electron pairs are measured in the xenon detectors, muon pairs in a precision muon detector such as the L3+1 detector (with a momentum resolution of < 2% at 500 GeV), while jets are measured in the TMS-U calorimeter. We expect a mass resolution of δ M/M better than 1.5%, for the lepton pairs. We expect more than 80 events for Higgs of mass up to 2 TeV with a comparable amount of QCD background. It seems possible to use the xenon detector to search for Higgs up to about 2 TeV even though a lot more work has to be done concerning the QCD background as well as the background from heavy quarks.

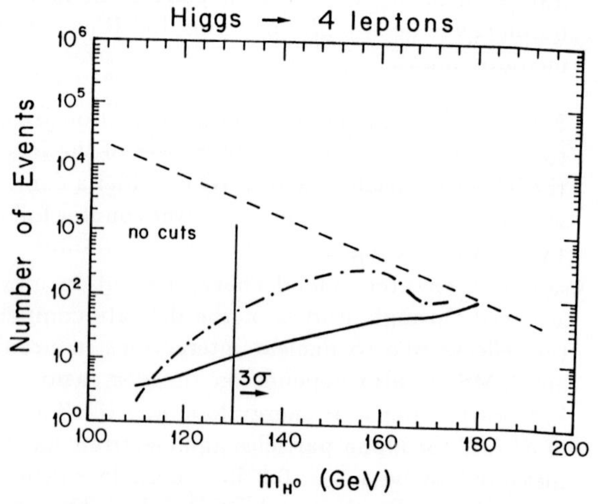

Fig.4 Expected number of events for Higgs $\rightarrow$ 4 leptons as function of the mass of the Higgs with an integrated luminosity of 10^4 pb^{-1} at $\sqrt{s}$ = 40 TeV. The dashed-dot curve is from Higgs for m$_t$=90 GeV [4]; the dashed curve is the background from QCD. The momentum resolution of each lepton is assumed to be 0.5% which yields a Higgs mass resolution of 0.2%.

IV. R&D Required

Compact, fast and radiation-hard calorimeters are obviously desirable in any detector at the SSC. The scintillating noble liquid detectors represent an idea in this direction. A large amount of information has recently become available from the data of nuclear physicists. It is clear that a lot more R&D work is needed before one can firmly establish such detectors for the SSC. A partial list covering several possibilities to detect the scintillating light in noble liquids is enclosed as follows:

1. To search for fast and radiation resistant wave length shifters which convert the scintillation light of xenon at 170 nm to visible light and match the sensitive region of the photo-detectors;

2. To test fast devices (phototubes or avalanche photo diodes etc.) to measure the light signals;

3. Since all the noble liquid detectors built so far are of small dimensions, one needs to build one large cell, say 2 x 2 x 60 cm^3 of liquid xenon, to measure the transmission efficiency of noble liquids.

4. To build one xenon cell with ten longitudial divisions, separated by metal cathodes covered with a thin film of solid organic layers, such as neopentane + TMAE or TMPD [3], to convert the UV light from xenon into pulse height with high quantum efficiency ($> 10\%$) and high speed (a few ns). Use both scintillation light (fast) and ionization (slow) signals of xenon to obtain precise space information where a resolution of about 30μ has been achieved for small devices.

5. To build and test an array of xenon shower counters, consisting of 3 x 3 cells at first and later 12 x 12 cells with a size of 2 x 2 x 60 cm^3 each. Find the optimum method to use both light and ionization signals for precise space and timing information for a large system.

6. For hadron calorimeter counters, one possibility would be to replace xenon with argon and to increase the thickness of each metal cathode in order to contain the hadronic showers.

Table I. The scintillation light decay times for both the fast and slow components and the ratio of their intensities; the saturation effect, K_B in Birk's law, for both scintillation and ionization outputs; and the light outputs of alpha and beta particles of several common media including BaF_2 in comparison with liquid xenon. T_1, T_2 are the decay time of the fast and slow component and A_1, A_2 their intensities respectively.

Scintillation	*Liquid Xe*	*BaF_2*	*NaI (T_1)*	*BGO*
fast decay time T_1 (ns)	3	0.6	250	60
slow decay time T_2(ns)	25-30	620	—	300
ratio A_1/A_2	25	0.09	—	0.05
rel. output from α	0.137	0.006	1.	0.08
ratio α/e	1.1 ± 0.2	0.34	0.5	0.23
K_B^{-1} in MeV/cm	30600	−−	3670	500
density (g/cm^3)	3.06	4.89	3.67	7.1
rad. length (cm)	2.8	2.05	2.59	1.12

Ionization	*Liquid Xe*	*Liq. Ar*	*TMS*	*TMP*
rel. output of a mip	12.6	4.6	1	1.1
K_B^{-1} in MeV/cm	300	200	-	70

References

[1] M. Chen et al., Nucl. Inst. and Meth. A267(1988)43 and MITLNS-163, 164, 165 and 166 and SSC Berkeley Workshop, 1987.

[2] H.S. Chen, MITLNS-164 and H. Fesefeldt, PITHA (Aachen) preprint, Nov. 1987.

[3] V. Peskov *et al.*, CERN-EP/87-218, Dec. 1987.

[4] J.F. Gunion *et al.*, UCD-87-28, Sept. 1987 and D.A. Dicus *et al.*, MAD/PH/389 Oct. 1987.

[5] H. Baer and J. F. Owens, FSU-HEP 88-02-02, Two Photon Background for the Intermediate Mass Higgs Boson.

LIQUID-ARGON CALORIMETRY — ITS POTENTIAL FOR SSC EXPERIMENTATION

C.W. Fabjan
CERN, Geneva, Switzerland

1. INTRODUCTION

The viability of liquid-argon calorimetric techniques in the experimental environment of the Superconducting Super Collider (SSC) is briefly analysed. We compare the required and achievable energy resolution with benchmark figures obtained using practical instruments. Comments on the desirable (i.e. required) temporal performance are made and compared with the state of the art. Some of the major engineering challenges are listed, for which solutions will have to be developed if such instruments are to find a place in an SSC experimental area.

2. ENERGY RESOLUTION: REQUIRED VERSUS PRACTICAL

The calorimetric energy measurement is thought to be the single, most important, spectroscopic tool for the SSC physics programme.[1] It requires instruments which must have excellent hermeticity over a large rapidity interval $|y| \leqslant 5.5$, and also 'sufficiently good' energy resolution. There are at least three 'fundamental' (i.e. detector-physics determined) contributions:

i) The intrinsic resolution σ(intrinsic), representing the accuracy on single, low-rate hadrons, for which in large, practical devices $\sigma(\text{intrinsic})/E \leqslant 0.3/\sqrt{E}$ has been achieved.[2] At the SSC, total or transverse energies will typically be in the few-hundred-GeV regime. A measurement with 10 GeV accuracy would imply an equivalent resolution of $\sigma(E)/E \approx 0.5/\sqrt{E}$(GeV). Similar resolution requirements are imposed for multijet spectroscopy, e.g. in the study of the reaction

$$\begin{array}{l} pp \rightarrow \text{Higgs} + \text{anything} \\ \quad \lfloor \rightarrow W \rightarrow \text{jet} + \text{jet} \\ \lfloor \rightarrow W \rightarrow 2\ell\,. \end{array}$$

ii) The reconstruction of the energy of a system based on signals in calorimeter cells implies the use of reconstruction algorithms and hence inherently some loss in information. The value of σ(reconstruction) has been studied,[3] and a performance of σ(reconstruction) $\approx$ 10 GeV for E $\approx$ 500 GeV has been estimated.

iii) The temporal signal response of the detector is at the origin of a further contribution to the energy resolution at very high particle rates since, during the 'occupation time' in the detector, on the average more than one particle is observed [σ(pile-up)]. This contribution will be further discussed in Section 3, where we show that this pile-up contribution can be limited to σ(pile-up) $\leqslant$ 5 GeV when operating calorimeters with today's technology at luminosities $L > 10^{33}\ \text{cm}^{-2}\,\text{s}^{-1}$.

Given the scale set by resolution contributions (ii) and (iii), it is acceptable to aim at values of the intrinsic calorimeter resolution $\sigma(\text{intrinsic})/E \leqslant 0.5/\sqrt{E}$(GeV), provided this $1/\sqrt{E}$ scaling law is applicable over the full energy range of interest at the SSC, i.e. E $\leqslant$ 0.5 TeV. The essential consequence of this requirement is summarized in Fig. 1, which displays σ(intrinsic) for various values of the relative electron-to-hadron response, e/π. For SSC instruments, it will be necessary to achieve *approximate compensation at the level* $e/\pi \leqslant 1.10$.

In the following we present the measured performance and e/π value for several of the present-generation LAr detectors. Some instrumental parameters are summarized in Table 1. Two of the three devices (D0, Ref. 4; HELIOS, Ref. 5) use uranium as the absorber medium, whilst the H1 detector[6] uses lead and copper absorber plates.

Preliminary results obtained in the D0 test are shown in Fig. 2.[4] The basic performance with respect to e/π is seen to be of 'SSC quality', $\langle e/\pi \rangle = 1.04 \pm 0.03$, whereas the signal response time of $\lambda \sim 3000$ ns is clearly not adequate at SSC luminosities. In view of the interest in

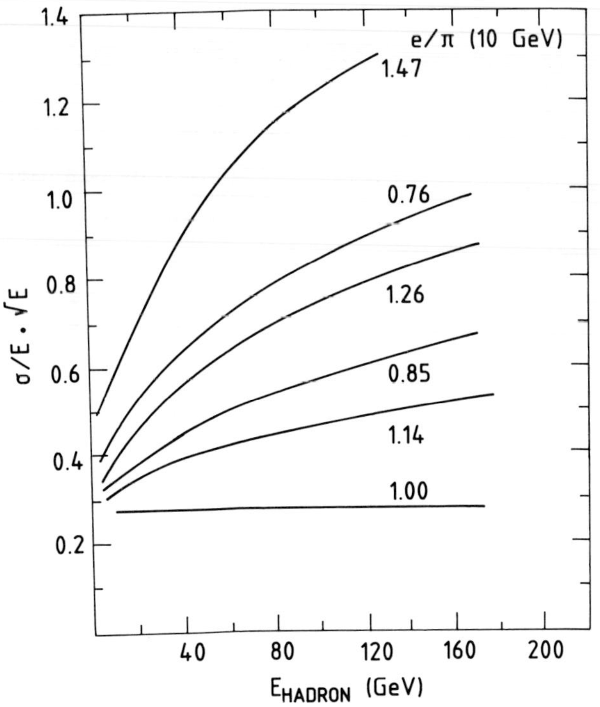

Fig. 1 The intrinsic energy resolution for hadrons as a function of particle energy and various values of e/π. Only for perfectly compensated instruments, $e/\pi = 1$, is the $1/\sqrt{E}$ scaling seen to be strictly true at all particle energies. A requirement for $(\sigma/E) \times \sqrt{E} \leqslant 0.5$ for energies up to a few hundred GeV imposes limits on e/π, so that $\sim 0.95 \leqslant e/\pi \leqslant 1.10$ at 70 GeV.

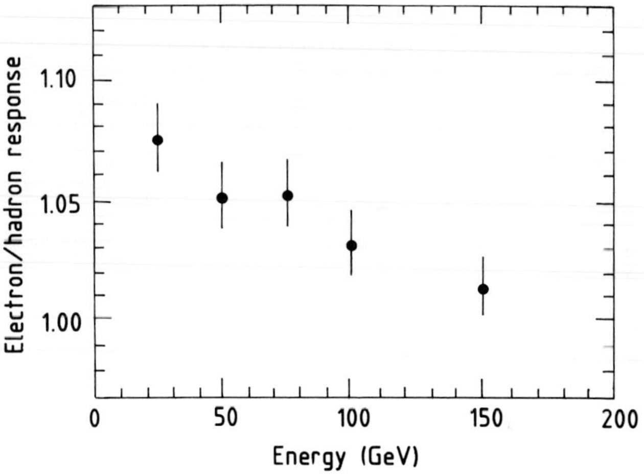

Fig. 2 The value e/π versus particle energy as observed in the D0 test (preliminary result). The signal-shaping time was $\lambda = 3\ \mu s$.[4]

'fast' liquids, the D0 team studied also the effect of using small admixtures of methane to liquid argon, which is known to reduce the drift-time by approximately a factor of 2 at concentration levels of $\leqslant 1\%\ CH_4$.[7]

These tests demonstrated that the increased recombination in the Ar/CH_4 system does not present any significant obstacle to using 'fast' LAr ($v_D \approx 1\ cm/\mu s$) in large calorimeter systems (an $\sim$ 30% loss in signal was observed at 0.5% CH_4 concentration).

Table 1

Instrumental parameters for three LAr calorimeters

Collaboration	Ref.	Mechanical parameters	Signal shaping time λ [ns]	Compensation method	'e/π' (10 GeV)
D0 at the Tevatron Collider (test device)	4	6 mm ^{238}U, 2 × 2.6 mm LAr	3000	Uranium	≈ 1.08 ± 0.02
HELIOS at the SPS (fixed target)	5	3.4 mm ^{238}U 2 × 2.5 mm LAr	400, 1000, 2000	Uranium	≈ 1.13 ± 0.02
H1 at HERA	6	Electromagnetic part: 2.4 mm Pb/2.8 mm LAr 5-fold longit. segmentation Hadronic part: 5 mm Cu/2 × 1.5 mm LAr 6-fold longit. segmentation		Local energy weighting	'e/π' [weighted] ≃ 1.03

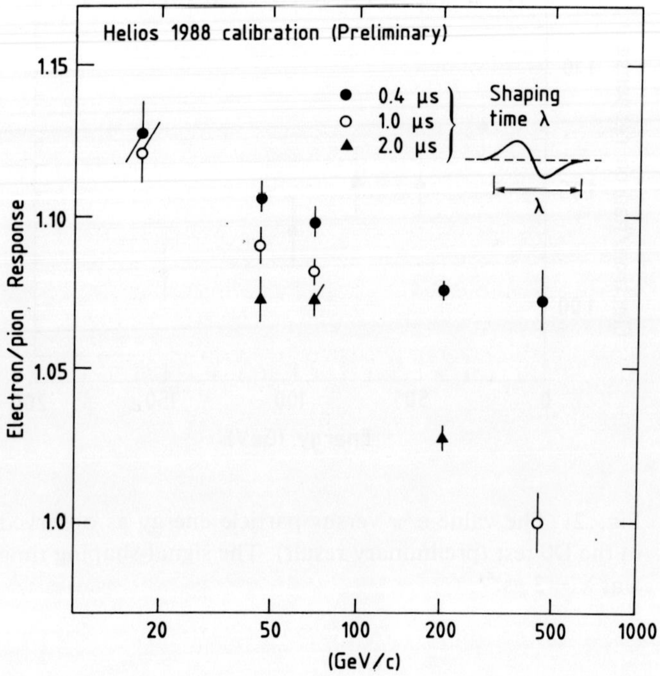

Fig. 3 Electron-to-pion response as a function of particle energy for three different values of the signal-shaping time; HELIOS test.[5)]

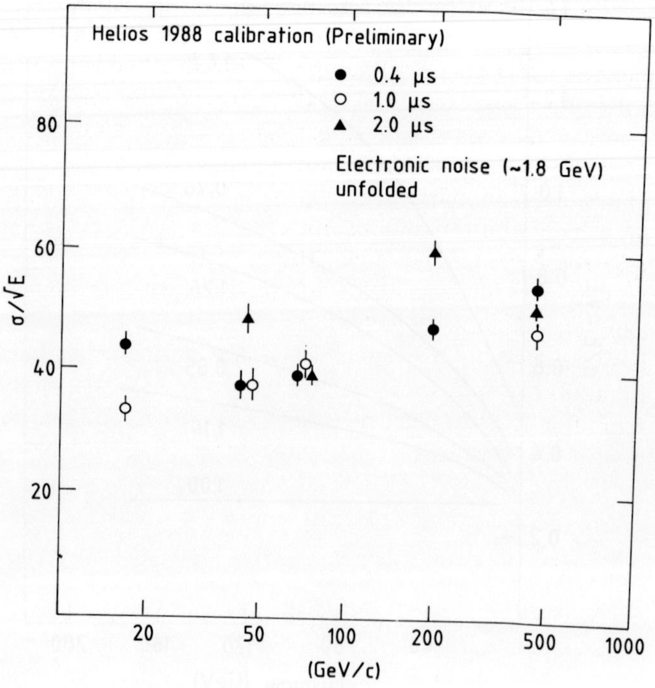

Fig. 4 Hadronic resolution versus particle energy for three different values of the signal-shaping time; HELIOS test.[5)]

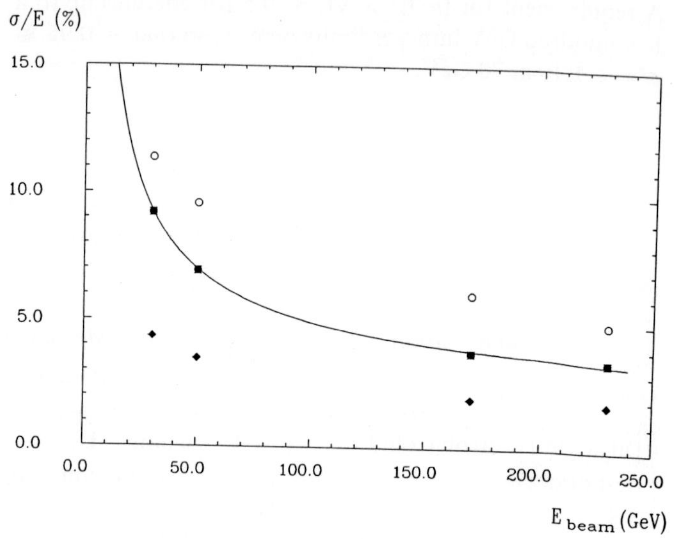

Fig. 5 Energy resolution with no π^0 weighting (open circles) and with π^0 weighting (full squares); sampling fluctuations (full rhombi); H1 test.[6)]

To date, the fastest LAr calorimeter is operated by the HELIOS Collaboration at CERN in a Super Proton Synchrotron (SPS) high-energy particle beam.[5)] It can be seen from Figs. 3 and 4 that even with the shortest shaping time of λ = 400 ns, a very good performance $[(\sigma/E) \times \sqrt{E} \leqslant 0.5$ up to E = 450 GeV] is obtained. In the next section we will discuss the technical implications of achieving these short (or even shorter) signal response times, and will argue that these values of λ are in a range that is acceptable for SSC operation.

The H1 Collaboration at HERA followed an entirely different tack. Their calorimeter uses lead and copper plates, instead of uranium, as absorber. This detector is characterized by an exceedingly high granularity (i.e. tower structure with sixfold longitudinal subdivision in the hadronic calorimeter) permitting the shower profile to be measured in detail. This capability suggests the use of an 'energy weighting' procedure[8)] that preferentially attenuates 'hot spots' of energy deposits (due to the electromagnetic component in the shower). This method has been tested for *single particles,* and is found to keep the linearity at the 1% level while considerably improving the energy resolution (Fig. 5). Preliminary Monte Carlo calculations indicate that the response to *particle jets* is also much improved, although this still has to be studied more thoroughly.

3. TEMPORAL RESPONSE: REQUIRED VERSUS ACHIEVABLE

The key to high-rate operation of calorimeters is the choice of bipolar pulse-shaping,[9)] so that the pile-up of several events within the sensitive time of the detector *does not, on the average,* produce a net shift in the measurement of a large signal. Pile-up will, however, contribute as noise σ(pile-up) to the overall energy resolution of the detector. The results of several quantitative studies[10] on σ(pile-up) as a function of interaction rate n and shaping time λ are summarized in Fig. 6. It is estimated that this noise contribution should not exceed a few GeV for jet spectroscopy at SSC luminosities ($n = 10^8\ s^{-1}$).

Apart from the value of the drift velocity and the charge collection, which determine the electronic noise level relative to the signal level, *the* principal technical limitation is imposed by the transfer time t_{tr} of the charge from the detector electrodes to the amplifier. The detailed analysis[9] indicates

$$t_{tr} \approx 3.5 \times (L_{cable} \times C_{detector})^{1/2} ,$$

$$t_{tr} \leqslant \lambda/5 .$$

These conditions can be satisfied provided the cable inductance and hence the total cable length L are kept very short: for acceptable values of $\lambda \leqslant 200$ ns, preamplifiers have to be placed in close proximity to the electrodes (< 1 m cable length). For SSC-type devices, amplifiers will have to be embedded inside the calorimeter and will have to operate at temperatures in the vicinity of ~ 100 K. The latter condition is not a particularly stringent requirement. The vicinity of the electrodes, however, renders the amplifiers almost inaccessible—a problem that will be common to all charge-collecting calorimeters.

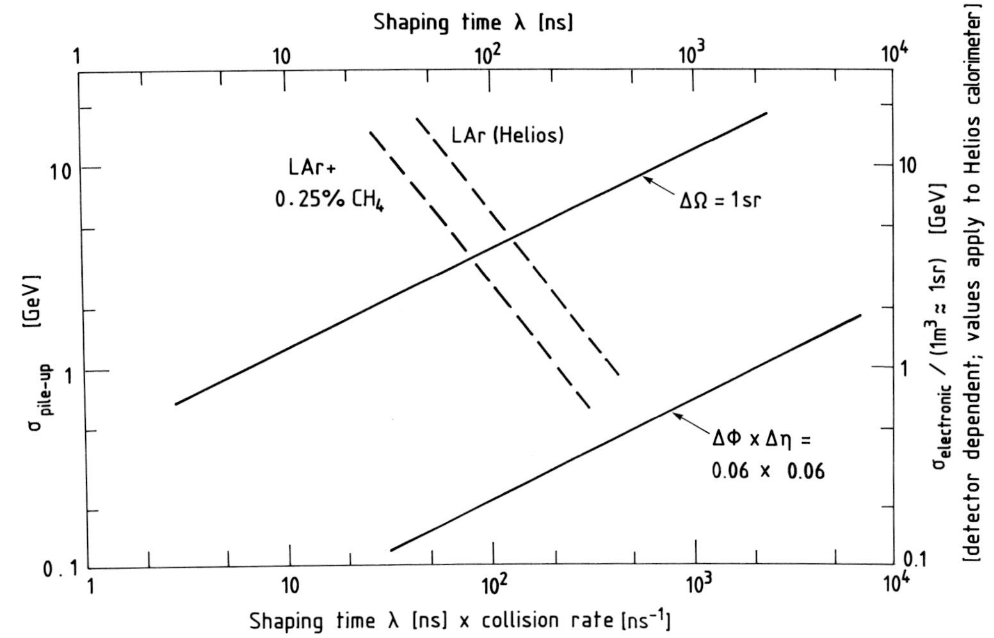

Fig. 6 Estimates (solid lines) of pile-up noise (left ordinate) as a function of the product of shaping time λ and the collision rate n (bottom abscissa), for two different sensitive detector areas. It is compared with the typical electronic noise (as realized in the HELIOS U/LAr calorimeter) (dashed lines) as a function of the signal-shaping time λ (right-hand ordinate and top abscissa).

4. ENGINEERING DEVELOPMENTS

In the previous sections we showed that the intrinsic performance of LAr calorimetry is commensurate with the required performance at the SSC. This is to be contrasted with several major, and as yet unresolved, engineering questions, which will need coordinated R&D to demonstrate the viability of LAr calorimetry. Here we list those of major concern:

1) Radiation resistance of active and passive components: this is, of course, the principal problem common to all SSC detectors. For charge-collecting devices, the electronics will have to be embedded in the detector and will therefore be exposed to the particle flux created in the absorber. Before the magnitude of this engineering issue can be adequately addressed, we need to develop a better understanding of all active components that are likely to be placed in this high-radiation environ-

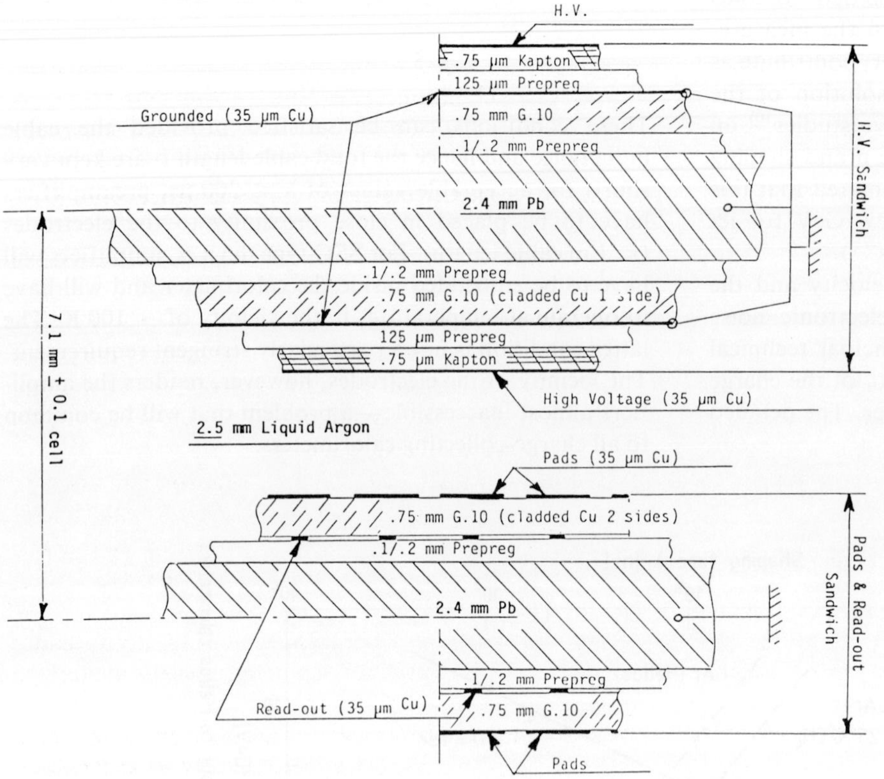

Fig. 7 Cross-section through the read-out cell of the electromagnetic part of the H1 calorimeter.

ment (preamplifiers, analog storage, trigger electronics, etc.). Radiation-hard electronics is *the* principal engineering challenge, which the high-energy physics community will have to tackle in collaboration with industry.

2) Readout electrodes: SSC calorimeters will be characterized by a very high granularity, typical of today's electromagnetic front-sections of advanced calorimeters.[4–6] The feasibility of such a design, satisfying the requirements of SSC calorimetry (very short cables, excellent hermeticity, ...) has yet to be demonstrated. As an example for the complexity of the readout structures, we show in Fig. 7 a cross-section through a readout gap of the electromagnetic section of the H1 calorimeter: it shows the complex multiboard structure of the electrodes; its fabrication is a major cost factor, taxing the capabilities of industry. The development of electrical prototypes of an SSC calorimeter segment should be envisaged with very high priority.

3) Cryostats: There is a wide-spread opinion that the cryogenic vessels required for LAr calorimetry preclude adequate hermeticity, acceptably thin entrance windows, and a minimum of dead space. In the absence of an adequate engineering evaluation for an SSC detector, only a few comments will be made:

i) Hermeticity: this is clearly an unresolved issue. In the present generation of detectors, it was not required to

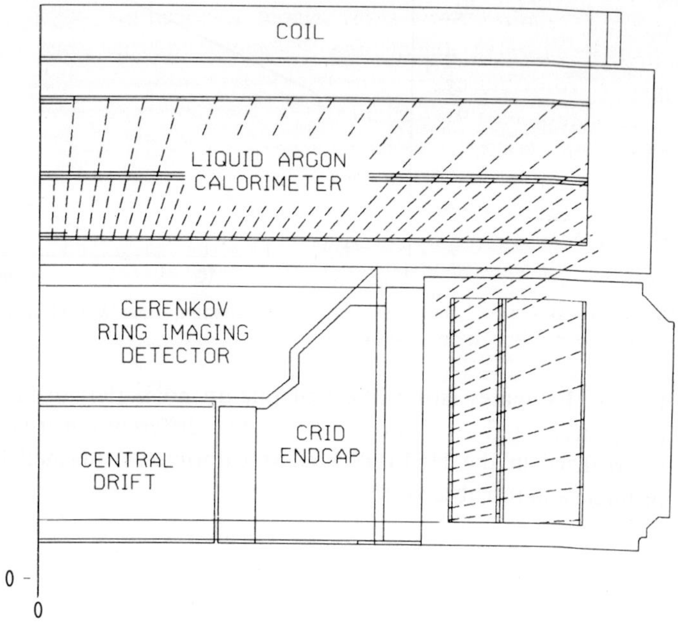

Fig. 8 Cross-section through the SLD detector, showing one possible approach towards hermeticity in LAr hadron calorimetry (Ref. 11).

the same extent.[4,11] In Fig. 8 we show, as an example, the SLD solution, which conceptually could possibly evolve into an SSC-hermetic design — a matter of creative engineering!

ii) Thickness of entrance window: I comment on this point by quoting the thickness (90° incidence) of the entrance window for several detectors: D0: $0.9X_0$; SLD: $0.9X_0$; UA1 $\sim 0.7X_0$ (TMP, room-temperature liquid!); H1 $\sim 0.25X_0$! It does not appear to be a problem.

iii) Dead space for cryogenic insulation: Typically $\sim$ 8 cm are required for purposes of insulation, which will increase the radial dimensions of the detector.

5. CONCLUSIONS

We have repeatedly stressed that the intrinsic performance of today's LAr calorimeters approaches SSC specifications.

The e/π ratio is adequately close to 1, if uranium is used. Alternatively, energy weighting in very granular devices may permit off-line compensations. The time response requires very short signal connections, with amplifiers located inside the detector. In this way, 'pile-up' contributions may be kept at an acceptable level.

It will be most important to face head-on some of the major engineering questions, such as radiation-hardness of the electronics, novel concepts of cryostats, and detailed work on readout planes.

I believe that LAr calorimetry may not be the only contender for an SSC interaction region. However, 15 years of experience with this technique has built up a solid base from which the challenge of the SSC can be faced.

REFERENCES

1) See, for instance, Cahn, R.N., et al., 'Detecting the heavy Higgs boson at the SSC', *Proc. Workshop on Experiments, Detectors, and Experimental Areas for the Supercollider,* Berkeley, 1987, eds. R. Donaldson and M.G.D. Gilchriese (World Scientific, Singapore, 1988), p. 20.
2) Åkesson, T., et al., *Nucl. Instrum. Methods* **A262** (1987) 243.
3) Åkesson, T., et al., 'Jets at the Large Hadron Collider', *Proc. ECFA-CERN Workshop on a Large Hadron Collider in the LEP Tunnel,* Lausanne and Geneva, 1984, ed. M. Jacob (ECFA 84/85, CERN 84-10, Geneva, 1984), p. 167.
4) Wimpenny, S.J., (D0 Collab.), 'Hadron and electron response of uranium/liquid argon calorimeter modules for the D0 detector', presented at the Int. Conf. on Advanced Technology and Particle Physics, Como, Italy (1988).
5) Gordon, H., et al. (HELIOS Collab.), 'Performance of the HELIOS uranium/liquid argon calorimeter', to be published in *IEEE Trans. Nucl. Sci.*
6) Braunschweig, W., et al., 'Results from a test of a Pb-Cu liquid argon calorimeter, DESY 87-098 (1987).
7) Shibamura, E., et al., *Nucl. Instrum. Methods* **131** (1975) 249.
8) Abramowicz, H., et al., *Nucl. Instrum. Methods* **180** (1981) 429.
9) Radeka, V., and Rescia, S., *Nucl. Instrum. Methods* **A256** (1988) 228.
10) Yamashita, A., and Kondo, K., *Proc. DPF Summer Study on the Physics of the Superconducting Super Collider,* Snowmass, 1986 (AIP, New York, 1987), p. 365.
Alversson, G.O., and Huston, J., ibid., p. 368.
11) SLD Design Report, Stanford report SLAC-273 (1984).

PRESENT STATUS OF WARM-LIQUID CALORIMETRY

James W. Rohlf

High Energy Physics Lab., Harvard University, 42 Oxford St., Cambridge, MA 02138

and

Physics Dept., Boston University, 590 Commonwealth Ave., Boston, MA 02215

ABSTRACT

The status of development of warm-liquid calorimetry for the UA1 experiment is described. The issues addressed are liquid performance, some technical aspects of the UA1 detector design, and recent test beam results.

INTRODUCTION

Research in warm-liquid calorimetry was motivated in UA1 mainly by the desire to construct an ionization chamber free of cryostats in order to have nearly 4π solid-angle coverage[1]. This is a very challenging problem on UA1 due in part to the geometrical constraints imposed by the existing central tracking chambers and the magnet.

Major considerations in the choice of liquids for an ionization calorimeter include free electron yield, electron lifetime and electron drift velocity. Some of the basic properties of tetramethylpentane (TMP), tetramethylsilane[2] (TMS), and argon (Ar) are shown in Table 1. TMP was chosen for UA1 because of its high boiling point. The free electron yield in TMP and TMS are about a factor of three smaller than in argon so that one has to work harder at the front end electronics to keep the noise down[3].

2,2,4,4 Tetramethylpentane

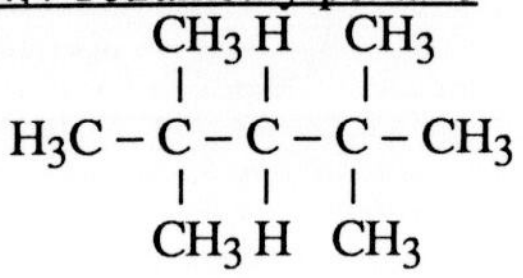

Tetramethylsilane

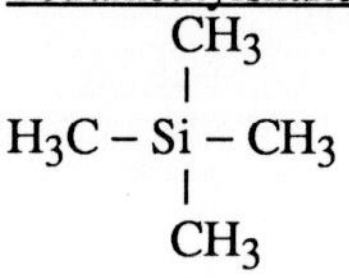

Table 1. Liquid Properties

	TMP	TMS	Ar
dE/dx (MeV/cm)	1.58	1.36	2.10
Free elec. yield (per 100 eV at 15KV/cm)	0.7	0.7	2.2
Drift Velocity (μ/ns at 15KV/cm)	4.5	15	5
Dielectric constant	2.0	1.8	1.5
boiling point	122	26.5	−186

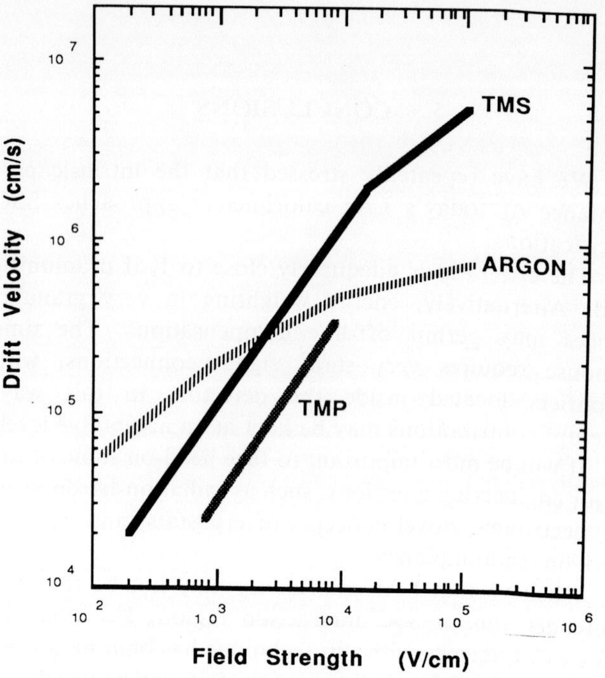

Figure 1. Electron drift velocities in TMS, TMP and argon as a function of electric field strength.

Figure 1 shows the dependence of electron drift velocities in TMS, TMP and argon on of electric field strength[4]. Both TMS and TMP are faster than argon at moderate values of electric field and TMS is an order of magnitude faster than argon at large field strengths.

WARM LIQUID PERFORMANCE

The first central issue to face in building a warm-liquid calorimeter is that of liquid purity. Extremely pure liquids are required in order to be able to drift electrons without significant signal attenuation through attachment to impurities. Certain contaminants such as chlorobenzene severely attenuate the electrons if present at the level of a few parts per billion (see figs. 2-4). It is often difficult to detect what the contaminant is. We are dealing with a tedious chemistry problem. We define the purity of a liquid by how fast drifting electrons are attenuated. The drift-time across a gap of thickness d, with applied voltage V is $t_d = d^2/\mu V$, where μ is the mobility (drift velocity divided by electric field) of the liquid. The current due to drifting electrons as a function of time is

$$i(t) = i_0 (1 - t / t_d) e^{-t/\tau}$$

where τ is defined to be the liquid lifetime which appears in an exponential attenuation due to attachment of the electrons to impurities[5-6]. After considerable development work, long electron lifetimes have been achieved in TMP. Lifetimes of greater than ten microseconds have been routinely achieved (hundreds of microseconds in the best cases) to be compared to an electron transit time across a liquid gap of a few hundred nanoseconds, which means that almost all the charge produced by the passage of an ionizing particle is collected[6].

Measurements have been made over the last couple of years[7] with two small prototype devices consisting respectively of 2 mm and 4 mm uranium plates interleaved with 10x10 cm^2 boxes containing 2.5 mm of tetramethylpentane (TMP), and more recently with full-sized calorimeter cells. The following basic results have been obtained:

1) The lifetime of electrons in the TMP has been found to be adequate for calorimetry and is stable over a period of about two years. These results show that adequate cleaning procedures have been achieved that are able to reduce the level of electronegative impurities in the TMP and its container to an acceptable level with no measurable long term poisoning of the liquid in the detector.

2) Amplifiers with sufficiently low noise level (~1000 electrons) have been constructed giving an signal/noise ratio of about 3:1 for a minimum ionizing particles passing through a single TMP box. (The noise is higher in the presence of uranium radioactivity.)

3) Variations of the pulse height across a single TMP box and from box to box are both at the percent level which is sufficient for high quality calorimetry. A single TMP box containing four electrodes shows uniformity of the total pulse height when an electron shower is scanned across an inter-electrode gap and less than 1% cross talk between electrodes when the beam is centered on one electrode.

4) The response to electrons (fig. 5) is linear to within 0.2% and the energy resolution is about $0.15/\sqrt{E}$ (E in GeV).

5) Measurements with incident hadrons (whose energy is not completely contained) are also in agreement with Monte Carlo predictions and have an e/π response ratio near unity.

DETECTOR DESIGN

In the upgraded UA1 detector the lead-scintillator calorimeters (gondola and bouchons) will be replaced by a uranium-TMP device. The calorimeter design is modular with the TMP contained in thousands of individual stainless steel boxes sealed by laser welding.

Uranium

The depleted-uranium raw material has been supplied by the United States government. A total of 300 tons has been manufactured into about 10K plates. Approximately 100 tons of this depleted uranium has been used to fabricate the plates of 2mm thickness at Rocky Flats, Colorado. Detailed thickness measurements have been made during stack

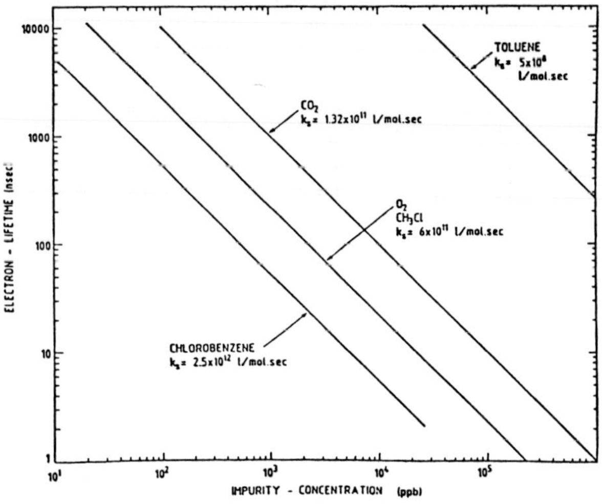

Figure 2. Electron lifetime limits in TMS as a for various impurities present.

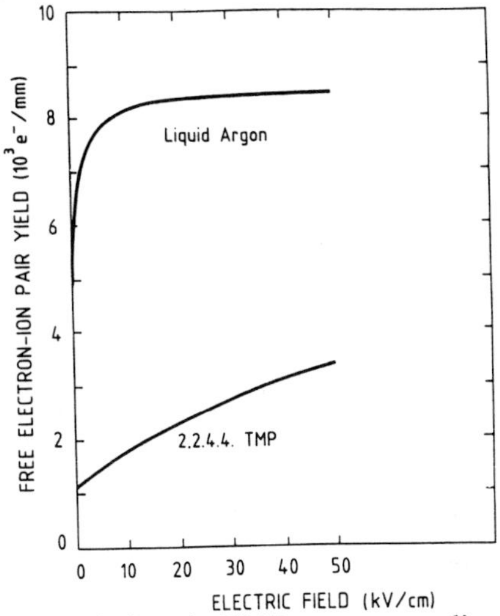

Figure 3. Signal sizes in TMP compare to liquid argon (see ref. 8).

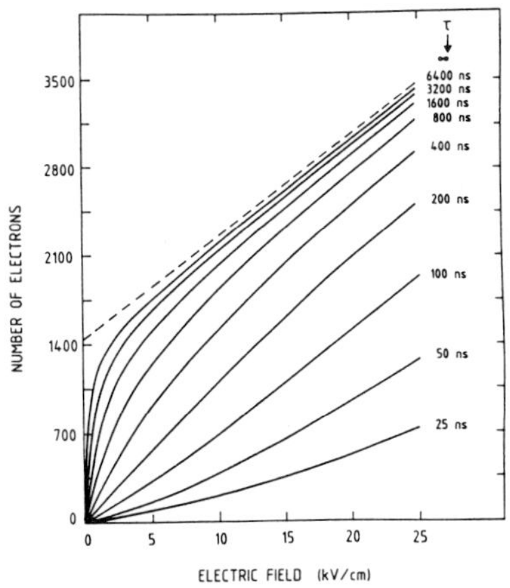

Figure 4. Expected signal sizes in TMP (1.25 mm gap) as a function of electric field for various liquid lifetimes.

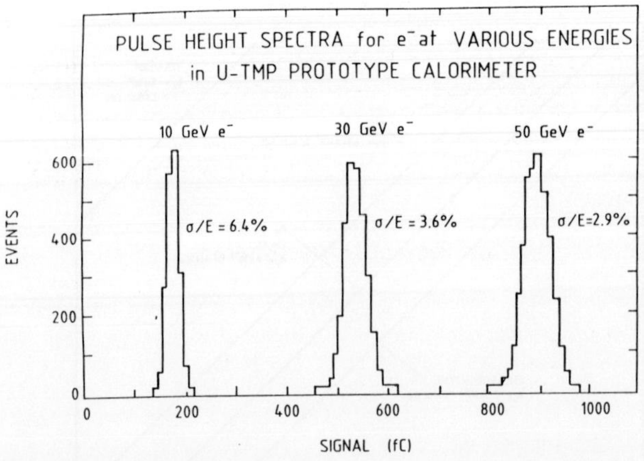

Figure 5. Electron signals in a uranium-TMP prototype calorimeter for three electron energies 10, 30 and 50 GeV.

assembly on all of the 2mm plates (10 micron accuracy) and stored in a computer data-base at CERN. For the 5-mm thickness (200 tons of depleted-uranium) two factories have been used to make plates: 1) Manufacturing Sciences Corporation in Oak Ridge, Tennessee and 2) C.E.R.C.A. in Romans, France.

Tetramethylpentane

The experiment needs 12,000 liters of pure TMP. The supplier (Wylie Organics) supplies TMP with a lifetime greater than 5 μs guaranteed, and has delivered TMP with lifetimes in excess of 15 μs. The lifetime is measured at the factory before shipment. The TMP is further purified at CERN by passing the vapor in vacuum through activated silica gel and 4 and 10 Angstrom molecular sieves. Eight purification lines operate in parallel and have a capacity of purifying 100 liters per week each. The best purity achieved has been 280 μs.

Boxes

In the manufacture of the boxes special care is taken in component cleanliness and box assembly. Components are cleaned in a ultra-sonic bath of hot detergent and deionized water and they are baked out under vacuum of better than 10^{-6} Torr (at low leak rate) at 900° C. Box welding is done by YAG laser under inert gas. Finished boxes are required to have a leak rate of less than 10^{-8} Torr L / s. The feed-throughs are required to have a surface leakage current of less than 1 nA. Boxes are manufactured at Hutchinson Technology (Minnesota) and at CERN. The box components and sizes are given in Table 2. Each box has a stainless steel frame with an additional center bar added to increase its rigidity and two thin covers (fig. 6). Inside the box there is a 0.5 mm thick stainless steel electrode, held in the middle of the gap by tiny insulating blocks made of ceramics. There are 8 such electrodes, roughly 10x10 cm^2, depending on box size. The high voltage, a maximum of 2.5 kV, is applied through a single feed-through which is welded to the electrode. After the assembly of the electrodes and insulators, the box is hermetically sealed by laser welding.

Table 2. Box Components

	Gondola	Bouchon
Frame (stainless)		
length (mm)	551	395 and 595
width (mm)	160 – 316	195
thickness (mm)	3.3	3.3
Covers (2, stainless)		
thickness (μ)	150	150
No. of electrodes (stainless)	8	8 and 12
thickness (μ)	500	500
Collection gap (2)		
thickness (mm)	1.25	1.25
No. of feed-throughs (ceramic)	8	8 and 12
No. of spacers (ceramic)	36	36 and 54
Total number of Boxes	22,000	12,000

Feed-throughs

The high-voltage feed-throughs have been made in France by Cice. The ceramic feed-throughs are individually produced with a brazed cover skirt and are subsequently laser welded to the stainless steel frames of the boxes. They need to have good mechanical properties (dimensional

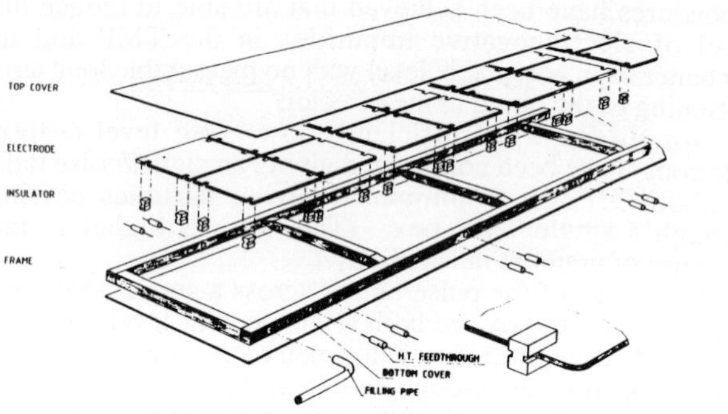

Figure 6. UA1 TMP calorimeter box

tolerances, robustness, etc...) and require two acceptance conditions to be satisfied: 1) a leak current of ≤ 1 nA at room temperature (~20° C) and humidity (60-70 %) at 3 kV; 2) a tightness, checked, with an helium leak detector, of < 10^{-8} Torr l/s for each bar. The H.V. test, in parallel to the measurement of the direct current, includes an observation of the noise at the output of the amplifier used in the final readout.

Cleaning and Filling

The final cleaning of assembled boxes is done at CERN. The main part of the cleaning is done with ultra-pure water of specific electrical resistance of 18M Ω-cm, which is also used for the cleaning of superconducting RF cavities for LEP. Ultra pure water is chemically almost inert, it is a neutral solvent with especially high dissolving capacity due to its high dielectric constant of $\mathcal{E} \approx 80$. It is used today in the electronics industry for cleaning of the crystalline surfaces of the circuits and in the pharmaceutical industry. The water purification system is mainly based on the use of an ultraviolet water sterilizer, an absorption filter and a mixed bed ion exchanger. Using this ultra pure water for stainless steel samples reduced the surface pollution, mainly Cl atoms, by a factor of five (fig. 7). The cleaning procedure for fully assembled boxes consists of:

- flushing the boxes with 60^0C acetone (3 hours)
- flushing with 60^0C ethylalcohol (3 hours)
- flushing with ultra-pure 80^0C water (36 hours)
- ultrasound during the whole operation
- drying with hot helium (3 days)

Filling is done in a large vacuum tank capable of holding four modules. The insides and outsides of the boxes are pumped to 10^{-7} Torr and heated to 300 degrees C. After cooling TMP is let into the boxes.

Position Detector

A position detector has been constructed for the purpose of measuring the precise location of electromagnetic showers. The position detector is a TMP liquid ionization chamber with a triple electrode construction. The parameters are given in Table 3. The center electrodes have a pad structure like the other calorimeter boxes with which they share a common readout. The other two electrodes have an orthogonal strip configuration. The position detector is the sixth box in the uranium-TMP stack, being placed after about three radiation lengths of absorber for particles at normal incidence in order to collect a large charge over a wide range of electron or photon shower energies. The position detector is compatible with the calorimeter design for many practical points such as mechanical placement, electronic readout, radiation resistance, high voltage distribution, hermeticity and energy resolution.

Position detectors have being constructed under our supervision at Hutchinson Technology. The position detector contains 24 horizontal strips, 56 vertical strips and 8 center electrodes (fig. 8). The electrodes have a width of 7 mm and a thickness of 0.5 mm. The space between electrodes is 2.1 mm. The thickness of the stainless steel covers is 0.150 mm. The electrodes are held apart by ceramic spacers and the high voltage is connected to a specially designed feed-through. The boxes are sealed to a leak rate of 10^{-9} torr-liter/s with high technology laser welding. The assembly is done in a clean room with special attention to the handling of all components during production.

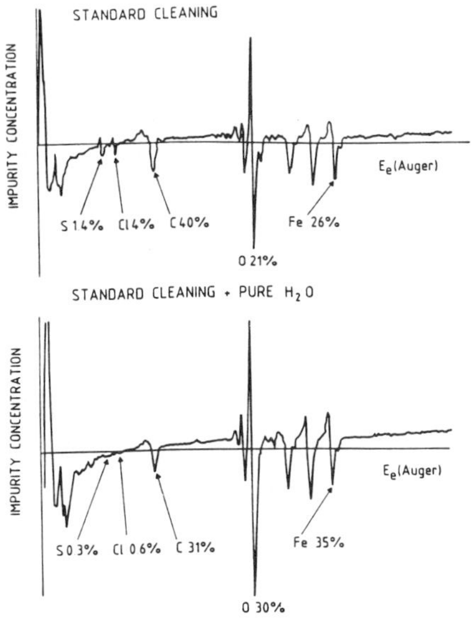

Figure 7. Auger spectra of steel surface before and after deionized water cleaning showing reduction of chlorine peak.

Table 3. Position Detector Parameters

Number of strips	24 x 58	19 x 42 and 19 x 63
Number of electrodes	8	8 and 12
Strip pitch (mm)	9.1	9.1
Strip width (mm)	7.1	7.2 and 6.9
TMP gap	2 x 1.25	2 x 1.25
Depth (rad. length)	3.1	3.1
Total number of boxes	320	168

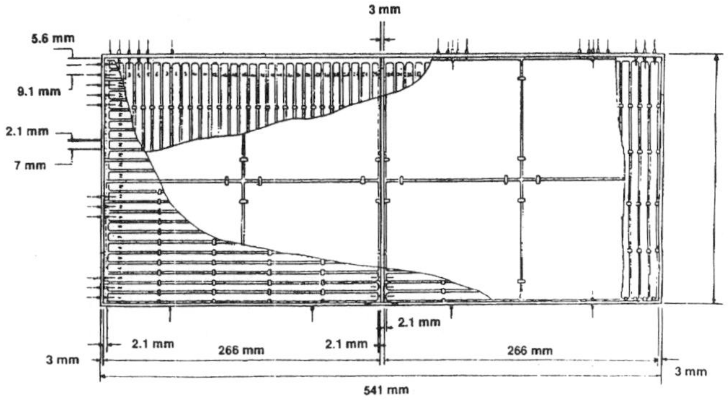

Figure 8. Position detector box.

Module Assembly

The number of boxes in each module is given in Table 4. Two modules make up a wedge, 16 wedges make up one super-gondola unit and 10 super gondolas make up the barrel region.

Table 4. Module Components

	Gondola	Bouchon
Boxes	62	62
U-plates (2 mm)	38	38
U-plates (5 mm)	24	24
Position detector	1	1

The module assembly goes as follows. First boxes are sandwiched with aluminum into substacks. Each substack has a common manifold TIG-welded onto the fill-tubes. Each substack is cleaned by rinsing with 80° C ultra-pure (18 MΩ-cm) water. It has been found that this washing reduces chlorine contamination by about a factor of ten as measured by Auger analysis. The substacks are then dried with pure Argon (O_2 content < 10 ppm) at 250° C. Then substacks are assembled into a wedge which undergoes a final vacuum bakeout at 300° C and a final leak test. The aluminum filler plates are then replaced with uranium and the wedge is filled with TMP. Each wedge is then cabled and has electronics mounted on the front face. Figures 9-11 show a uranium-TMP stack and wedge.

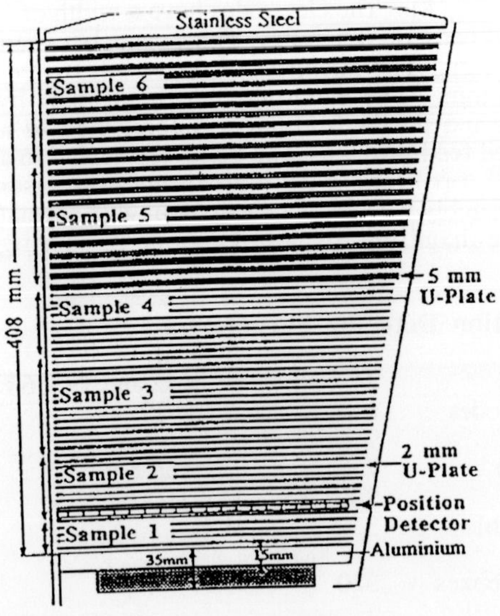

Figure 9. Uranium-TMP stack.

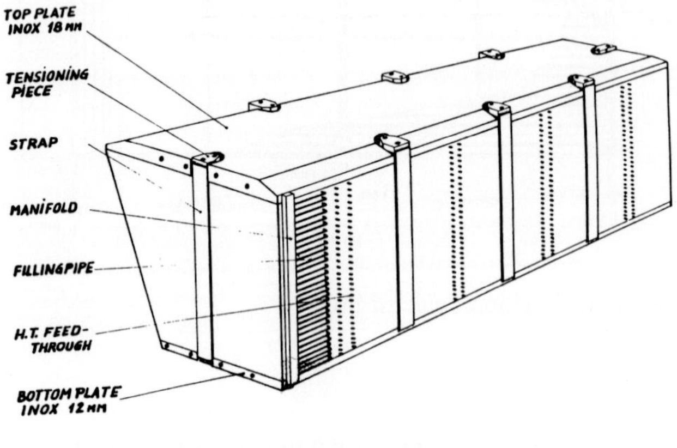

Figure 10. Uranium-TMP wedge.

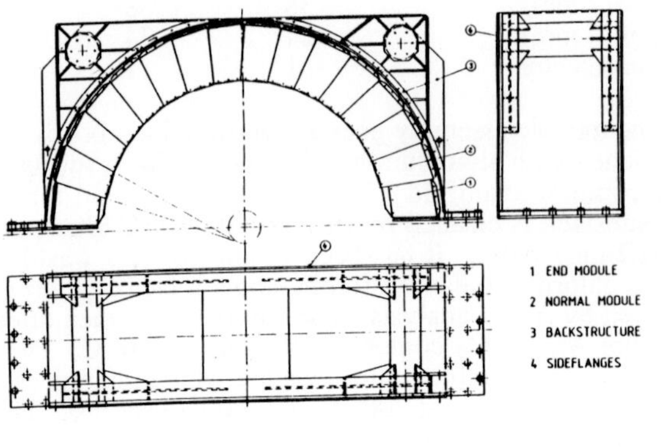

Figure 11. Uranium-TMP supergondola structure.

TEST BEAM RESULTS

Here I shall concentrate on results from the beam measurements with full-size position detector modules. Measurements were made in the SPS test beam at CERN during September 1987, and again in the PS test beam during November 1987. The position detector was mounted just upstream of the prototype uranium-TMP calorimeter. A target of 4 to 6 uranium plates of 2 mm thickness spaced by 3 mm was placed directly in front of the position detector. The incident angle of the beam was varied to simulate different parts of the barrel configuration of the final calorimeter. A multiwire proportional chamber was used to measure the position of the beam. Eight horizontal and 8 vertical strips of the position detector were connected to preamplifiers of the type designed for the DELPHI calorimeter[9], designed to have low noise at low detector capacitance. The signal from each preamplifier was then shaped using a UA1 calorimeter shaper (2 microsec shaping time constant) and the peak signal recorded with a LRS 2249 ADC. Electron trigger data were taken at beam energies of ranging from 3 to 50 GeV for a variety of detector voltages up to 2500V.

TMP Lifetime

Total charge was collected on a single strip (waveform data) averaged over many showers. The electronics used after the preamplifier (test pulse, filter, and flash-ADC) was the same as that used for measurements on the prototype calorimeter. Waveform data were taken at several voltages between 4 and 2500 volts. Charge was digitized in 50 ns bins except for the lowest voltage data which were digitized every 100 ns (see fig. 12). The purpose of the low voltage measurements is to measure the TMP lifetime. The high voltage measurements were taken in order to measure the drift time. The most sensitive measurement of the TMP lifetime comes from the data taken at a voltage of 4.4 volts. This measurement was found to be reproducible. The result of this fit to the TMP lifetime is

$$\tau = 57.7^{+31.2}_{-16.8}\ \mu s.$$

The errors on the fit are asymmetric; the result is consistent with a nearly infinite lifetime on the high-end and is inconsistent with about 40 μs on the low-end.

Drift Velocity

The electron drift time (t_d) across a gap of distance d, is given by the relationship $t_d = d^2/(\mu V)$, where μ is the TMP mobility of 30 cm^2/V-s and V is the voltage across the gap[10]. Figure 13 shows measured charge collection times. Figure 14 shows a plot of the measured inverse drift times t_d^{-1} vs. voltage on a log-log plot. The inverse drift times are seen to increase linearly with voltage without saturation over the measured range, showing that the TMP mobility is a constant. At 2.5 KV the charge is collected in about 200 ns in the 1.25 mm gap.

Response to Minimum Ionizing Particles

The position detector response to minimum ionizing particles was measured with a muon beam. Figure 15 shows the pedestal fluctuation together with the muon signal observed when the muon passes through the center of a strip (within ± 2 mm). The rms fluctuations in the noise are about 1500 electrons, and the mean value of the muon signal is about 3000 electrons. The noise has been further improved by nearly a factor of two by using better electronics.

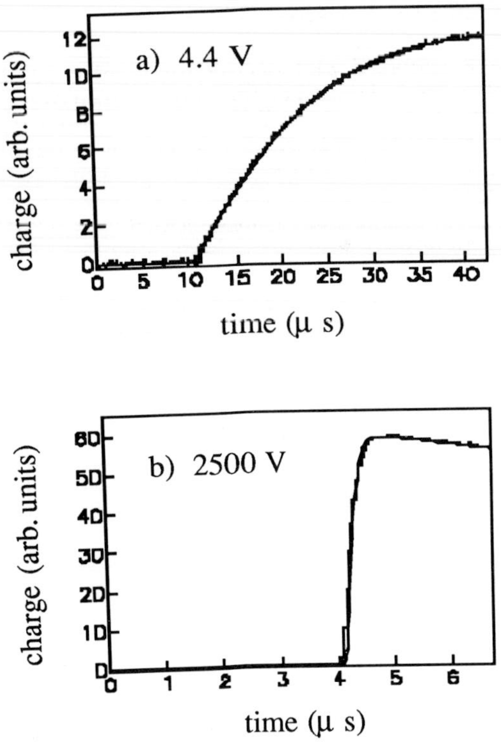

Figure 12. Integral charge collected vs. time on a single strip of position detector summed over 1000 showers for a) very low voltage to measure charge attenuation and b) high voltage.

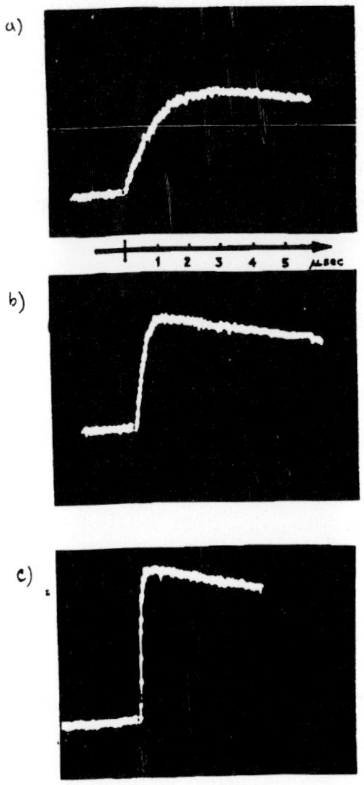

Figure 13. Charge on a single strip as a function of time for three different electric fields of a) 2 KV/cm, b) 8 KV/cm, and c) 20 KV/cm.

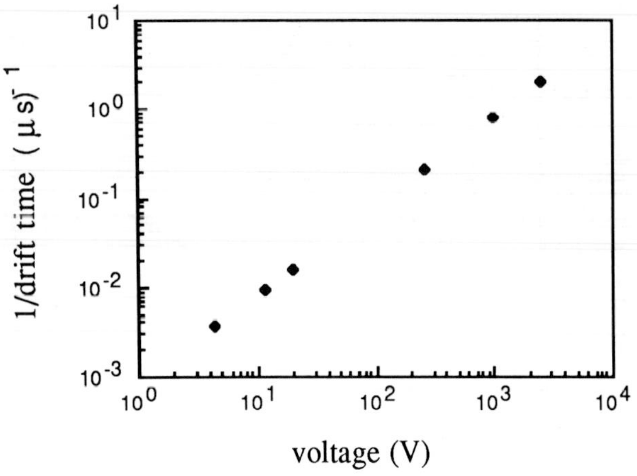

Figure 14. Measured inverse electron drift time across 1.25 mm gap vs. voltage applied across position detector gap.

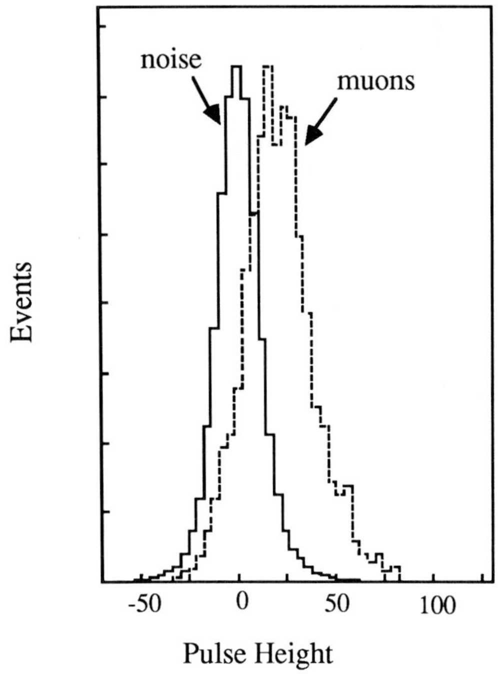

Figure 15. Response of minimum ionizing muons on single strip of position detector compared with noise measurements on the same strip.

Response to Different Beam Energies

The total position detector signal was measured (x and y strips separately) for a range of beam energies. The data taken at normal incidence with a target of 6 uranium plates is shown in fig. 16.

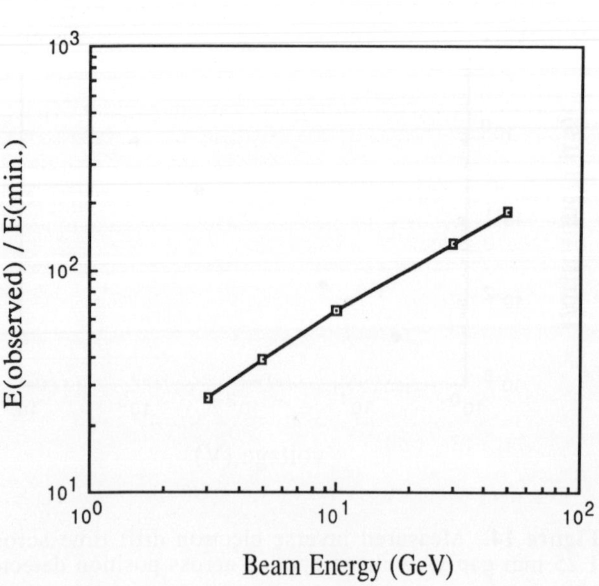

Figure 16. Observed average signals, normalized to single minimum ionizing, in position detector for different energy positron showers.

Strip Hit Distributions

Figure 17 shows histograms of the average pulse height fraction seen on each strip (% per strip) as a function of the beam position in 1 mm steps. The beam energy is 10 GeV.

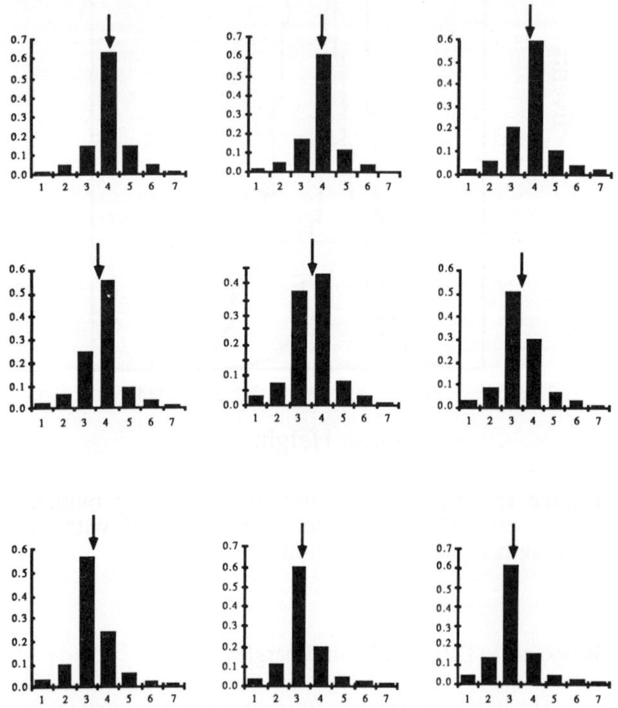

Figure 17. Average energy measured on each strip for 1 mm steps in beam position (indicated by arrows).

Uniformity of Detector Response

Figure 18 shows a plot of the total pulse height measured normalized to minimum ionizing summed over all y-strips vs. the beam position for a 10 GeV beam. A uniform total response is observed independent of shower location.

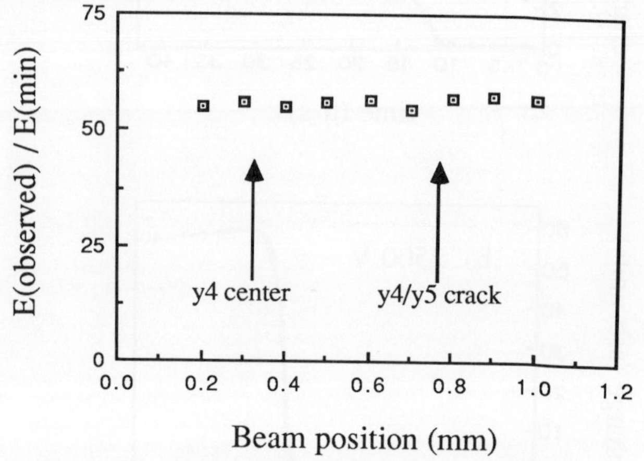

Figure 18. Observed signal, normalized to minimum ionizing, as a function of shower position; the arrow on the left indicates the position of a strip center and the arrow on the right indicates the position of a crack between two strips.

Position Resolution

The position resolution is a function of the amount of energy deposited in the position detector which depends not only on the shower energy (beam energy) but also on the local shower fluctuations. Figure 19 shows the resolution, calculated as full-width-half-maximum divided by 2.4, using the simple centroid calculation. The resolution is plotted vs. observed signal (normalized to minimum ionizing) for various beam energies. Figure 20 shows the resolution observed for the case of large pulse height deposited by a 50 GeV shower.

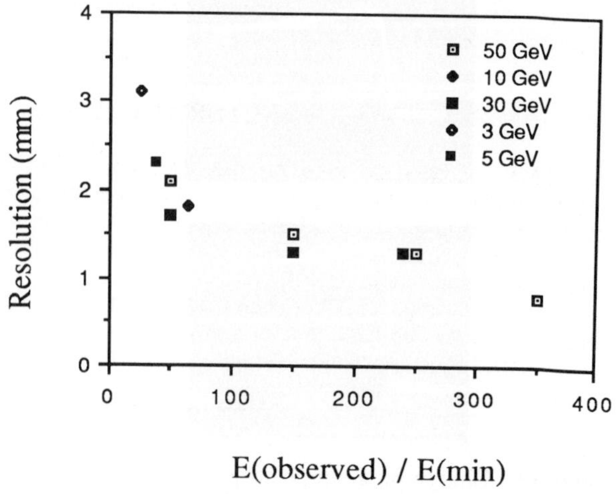

Figure 19. Position detector resolution (mm) measured for various observed signals from different shower energies.

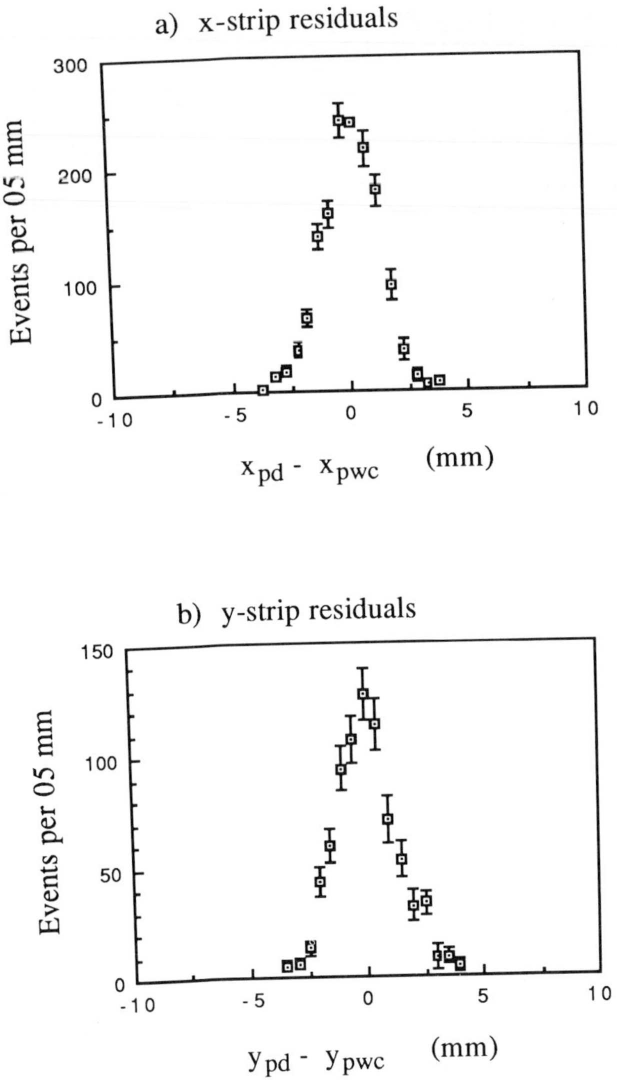

Figure 20. Residual of the shower position for 50 GeV showers as determined by center of gravity in the the position detector and the wire chamber for a) vertical strips and b) horizontal strips.

CONCLUSIONS AND FUTURE OUTLOOK

TMP has been demonstrated to work as a calorimeter liquid. Results from prototype devices show adequate signal to noise, good uniformity and linearity, and yield the expected energy resolution. The biggest practical problem in the construction of a large detector is that of maintaining the purity of the liquid. This has been mastered on a small scale but still must be demonstrated on a large scale.

Warm liquids such as TMP and TMS remain attractive candidates for use in calorimeters for SSC detectors[11]. However, many practical problems must still be addressed for an SSC detector[12]. One of these is just the geometry of the electrodes and the packaging of the liquid. Probably one would prefer a recirculating system. Another serious concern is that of radiation damage (radiation levels at the SSC are expected to be roughly 1000 times that of the SPS-Collider). Although TMS and TMP are expected to be radiation resistent[13], and in fact have been demonstrated to be so for the radiation doses expected at the SPS-Collider, more detailed measurements must be performed. The radiation damage may depend on how the liquid is packaged. Another issue is that of liquid speed due to the small time between collisions (16 ns) at the SSC. It would seem attractive to use TMS at very high voltages, although the ultimate limit may come from the time taken to get signals out of the device[14]. Here again, warm liquids because of the lack of cryostat may offer an advantage over argon.

REFERENCES

1. UA1 Collaboration, "Design Report of a U-TMP Calorimeter for the UA1 experiment with ACOL", UA1/86-112 (1986).

2. For early work on TMP, see J. Engler and H. Keim, Nucl. Instr. Meth. **223**, 47 (1984); J. Engler, H. Keim and B. Wild, Nucl. Instr. Meth. **A252**, 29 (1986).

3. C. Bacci, S. Centro and M. DeGiorgi, UA1-TN-86-116 (1986).

4. R. A. Holroyd and D. F. Anderson, Nucl. Instr. Meth. **A236**, 294 (1985).

4. A. Gonidec *et al.*, "Ionization Chambers with Room Temperature Liquids for Calorimetry", CERN-EP/88-36 (1988).

5. A. O. Allen and R. A. Holroyd, J. Phys. Chem. **78**, 796 (1974); A. O. Allen, T. E. Gangwer and R. A. Holroyd, J. Phys. Chem. **79**, 256 (1975); G. Bakale, U. Sowada and W. F. Schmidt, J. Phys. Chem. **79**, 3041 (1975); G. Bakale and W. F. Schmidt, Z. Naturforsch **36a**, 802 (1981).

6. For a review of mobility of electrons in hydrocarbons, see W. Schmidt, Can. J. Chem. **55**, 2197 (1977).

7. M. Albrow *et al.*, Nucl. Instr. Meth. **A265**, 303 (1988).

8. L. Onsager, Phys. Rev. **54**, 554 (1938).

9. G. Barichello *et al.*, " A Low-Noise Amplification Chain for the Vacuum Photo-Triode Read-Out of the Delphi Lead Glass Calorimeter, DFPD/3/86 (1986).

10. W. F. Schmidt and A. O. Allen, J. Chem. Phys. **52**, 4788 (1970); H. Jungblut and W. F. Schmidt, Nucl. Instr. Meth. **A241**, 616 (1985); T. G. Ryan and G. R. Freeman, J. Chem. Phys. **68**, 5144 (1978).

11. See various detector configurations considered in "Experiments, Detectors, and Experimental Areas for the Supercollider", World Scientific, ed. R. Donaldson and M. G. D. Gilchriese (1987).

12. M. Pripstein, "Requirements for the Development of Warm-Liquid Calorimetry" (these proceedings).

13. A. J. Swallow, "Radiation Chemistry of Organic Compounds", Permagon, Oxford (1960).

14. V. Radeka, "Electrode and Readout Considerations for Ionization Calorimetry" (these proceedings).

PROSPECTS FOR AND TESTS OF HADRON CALORIMETRY WITH SILICON*

James E. Brau
University of Oregon, Eugene, OR 97403

Tony A. Gabriel
Oak Ridge National Laboratory, Oak Ridge, TN 37831

Pier-Giorgio Rancoita
INFN-Milan, Italy

ABSTRACT

Hadron calorimetry with silicon may provide crucial capabilities in experiments at the high luminosity, high energy colliders of the future, particularly due to silicon's fast intrinsic speed and absolute calibration. The important underlying processes of our understanding of hadron calorimeters are reviewed to set the framework for the presentation of recent calculations of the expected performance of silicon detector based hadron calorimeters. Such devices employing uranium are expected to achieve the compensation condition (that is, $e/h \approx 1.0$) based on the understanding that has been derived from the uranium-liquid argon and uranium-plastic scintillator systems. In fact, even lead-silicon calorimeters are found to achieve the attractive value for the e/h ratio of 1.16 at 10 GeV. An experimental test of these predictions is underway at CERN by the SICAPO Collaboration.

INTRODUCTION

The physics results of future high luminosity, high energy colliders (eg. SSC or LHC) will depend on having very good calorimeters, which have been designed and constructed using the latest state of the art techniques. Many reviews have summarized the existing knowledge of calorimeter physics.[1–4] However, the understanding of the fundamental processes which limit calorimeter performance has undergone a significant advancement in recent years. It will be crucial to apply this understanding in the best ways. The present understanding has evolved through an interchange between theoretical calculations and experimental test programs, as the conclusions drawn from the theory in the calculations have been tested and found to be fundamentally sound. In this paper we will apply this understanding to assess the prospects for hadronic calorimetry with silicon and report on the experimental test underway at CERN.

Four years ago a new understanding of the underlying mechanisms of compensating calorimetry was introduced.[5–10] At that time the following critical points were realized:

1. prior to later experimental confirmation, it was argued that current designs of uranium liquid argon calorimeters were not fully compensating[5–9];
2. the importance of the hydrogen content in the active medium to couple the low energy neutrons to the output signal was stressed[5–9];
3. the significant role of "electromagnetic sampling inefficiencies" (which are the result of preferential photon absorption[11] and electron multiple scattering in the high-Z inactive material[6,9]) in reducing the ratio of electron to hadron response was explained[5–9];
4. the importance of the saturation of signal in the regions of high density energy deposition was emphasized[5–9]; and
5. these new understandings led us to "predict that a lead calorimeter may also give $EM/HAD \approx 1$"[12], where EM/HAD is the ratio of average electron to hadron response for the same incident kinetic energy, hereafter referred to as the e/h ratio. In other words, a compensating lead calorimeter was predicted.

At the time this new understanding was introduced in 1984, it was not generally accepted and met with much scepticism in the community. As a result of much experimental testing in the past four years, this scepticism has evolved to a general acceptance by the community. Shortly after the new understanding of compensation was announced, the experimental test programs of D0 at Fermilab and SLD at SLAC presented new data which supported the partial compensation expected for uranium liquid argon calorimeters. The predictions of the Monte Carlo calculations agreed very well with the new measurements.[9] Furthermore, the interpretation of the fundamental processes driving this behaviour had been determined and was presented at that time. The understanding of the role of hydrogen was expanded with the introduction of a prediction by Brückmann[13] of "tuning" of relative electron and hadron response by varying the ratio of uranium plate thickness to the scintillator thickness. Test beam studies of uranium scintillator calorimeters have confirmed the expected trends from the model of "tuning".[14] Considerations of the role of "tuning" in non-uranium scintillator calorimeters led to a specific design of a lead-scintillator compensating calorimeter with the ratio of lead thickness to scintillator thickness of approximately four; it was subsequently built and found to have $e/h = 1.05 \pm 0.04$.[15] Recently an activation analysis[16] of calorimeter-like stacks at CERN has confirmed the details of neutron-hydrogen coupling[7,8] which were key to the new understanding of calorimetry introduced in 1984. This new understanding has now been generally accepted and serves as the foundation of recent reviews of the phenomenology of compensating calorimeters.[17,18]

The challenge of doing physics at the next generation hadron colliders will be great. Event rates are being planned at the 100 MHz level, meaning that fast calorimeters must be built to minimize pileup. The experience at the Sp$\bar{p}$S of the importance of nearly full event containment (i.e. hermeticity) will be repeated. Many searches for new particles will demand good reconstruction of missing energy, for example the search for additional intermediate bosons such as $W' \to e\nu$. Excellent energy resolutions with good linearity and gaussian resolution functions are required for many experiments, such as the search for compositeness at the tens of TeV energy scale.[19] The search for such structure in the high-p_T jet signature will test the best calorimeters. While all these performance characteristics are necessary, they must exist in a very hostile radiation environment.[20] Neutron fluxes in excess of $10^{12} cm^{-2} yr^{-1}$ and doses of greater than $10^3 Grays/year$ in calorimeters situated at a radius of two meters from the interaction point are anticipated at the next generation of hadron colliders. These requirements of performance and survivability will test the present understanding of calorimeter physics and designers' ingenuity in building radiation hardened equipment. The understanding of all fundamental processes is essential to a successful program.

The option of employing silicon detector diodes as the sampling medium in hadron calorimeters has been suggested in a number of previous works.[21,22] Successful devices have been constructed and tested using silicon diodes in electromagnetic calorimeters.[23] The advantages of extending this success to hadron calorimetry are numerous.[24] As an ionization sampling calorimeter, a silicon readout calorimeter would be a system of absolute gain. Silicon is a nearly non-saturating medium so that sources can be used to get an absolute calibration of an energy scale. These devices can be operated at room temperature, eliminating the need for bulky and space wasteful cryostats required for other options. The signals have been shown to be relatively insensitive to magnetic fields. In principle the calorimeters can be made very compact due to the very thin sampling gaps required; since silicon diodes produce an electron-hole pair for every 3.7 eV deposited at room temperature, they require a much smaller sampling gap than most other systems. The technology of silicon diode fabrication makes fine lateral and longitudinal segmentation of silicon calorimeters easy and feasible. Finally, the property that provides the most promising feature of silicon diode readout is the speed of response. The intrinsic speed of normal diodes of thicknesses of 300 μm or so, is about 20 nanoseconds. Therefore, the speed of the front-end electronics is likely to limit the speed of response. However, achieving the optimal physics results at the future high rate colliders may well require calorimeters which exploit this intrinsic speed as fully as possible. It is for many of these reasons that the SICAPO Collaboration is working on the development of silicon hadron calorimeters.[25]

FUNDAMENTAL PHYSICS OF COMPENSATING CALORIMETERS

Basic Ideas

Before one is able to assess the prospects for hadron calorimetry with silicon, it is essential to understand the details of compensating calorimetry. This begins with an appreciation for the original motivation of the inventors of the technique.[26] It must be realized that hadronic cascades which contain large amounts of electromagnetic energy have less losses of energy due to nuclear break-up than those which contain little electromagnetic energy and therefore a large number of hadronic interactions. This comparison is illustrated in Figure 1 . Figure 1a shows the situation for non-compensating calorimeters. The overall distribution is created by summing a variety of different types of events. The two extremes, here called type A and type B, result in quite different responses. The type A events have a large fraction of the incident hadron energy converted to π^0s and other forms of electromagnetic energy, and therefore there are few hadronic nuclear interactions. Since the π^0s yield an observed signal which is comparable to the response of the calorimeter to incident electrons of the same energy and there is little energy lost to the breakup of nuclei, the e/h ratio for such events is nearly one, as shown in Figure 1a . Type B events in such non-compensating calorimeters, on the other hand, show large amounts of lost energy in the nuclear break up resulting from the many hadronic nuclear interactions. This includes nuclear binding energy losses, as well as losses from heavy fragment production, neutrino production, and low energy nucleon generation. All of these processes result in a reduction in the detectable energy in the calorimeter. These type B cascades have a minimal amount of generated electromagnetic energy and the resulting e/h ratio is much less than one. Combining all types of hadronic cascades, from type A to type B and everything in between, leads to a very broad distribution of responses, and a resulting poor resolution for the measured energy deposition.

Figure 1b illustrates the situation for compensating calorimeters where the response to events of type A and type B have been equalized by a cleverly designed calorimeter. This is the basis for the original ideas as presented in reference 26 . It can be seen from the figure that type A events respond with a signal nearly equal to an electron signal. However, the type B events also give an average response that is nearly comparable to the response of an electron. The nuclear breakup energy has been *compensated for*, by designing the calorimeter to preferentially respond to the low energy neutron component of the shower, which is correlated with the lost nuclear breakup energy. As will be shown later, it is not necessary to fully recover all of the nuclear breakup energy as the electromagnetic response is also suppressed in high Z materials due to processes occurring at the very low energy portion of the electromagetic shower. The role of these sampling inefficiencies which result from the preferential photon absorption[11] and the electron multiple scattering in the inactive material[6,9] has a very stong effect on the calorimeter response. A proper balance of the electromagnetic suppression and the preferential sampling of low energy neutrons leads to this equalization of the average response of type A and type B events. As a result, the sum of all the types of events, from A to B, yields a narrower distribution; that is, the energy resolution is greatly improved. This is the motivation for examining the ratio e/h as a measure of the degree of compensation.

Binding Energy Losses

In previous studies of the detailed processes occurring in hadronic cascades, extensive use has been made of the CALOR code system (HETC, EGS, MORSE, MICAP, SPECT, LIGHT, etc.- see the Appendix for details).[27] Figure 2 , for example, presents the calculated distribution from HETC of energy lost through the release of nuclear binding at the first interaction of a 5 GeV π^- in a uranium scintillator sampling calorimeter composed of alternating three millimeter uranium plates and scintillator sheets. An average energy loss of 380 MeV occurs at the first interaction sites. This represents 7.6% of the incident hadron's energy removed from the cascade at the first interaction. For lower energy interactions the loss is naturally going to be smaller, but only somewhat since for the same calorimeter HETC predicts an average loss of 200 MeV at the first interaction of 1 GeV incident neutrons. Notice that this represents twenty percent of the incident hadron kinetic energy. For lighter target materials, such as iron, the value will be somewhat less.[28]

As the shower progresses many additional interactions lead to an accumulated energy loss. Figure 3 shows the distribution of the number of nuclear interactions (excluding elastic collisions) in HETC for the incident 5 GeV π^- on this uranium scintillator stack. There are an average of 24 collisions,[29] while some showers terminate as the pion decays in the stack (notice three of these 500 showers had no nuclear collisions). Integrating the binding energy loss over these many interactions yields the distribution

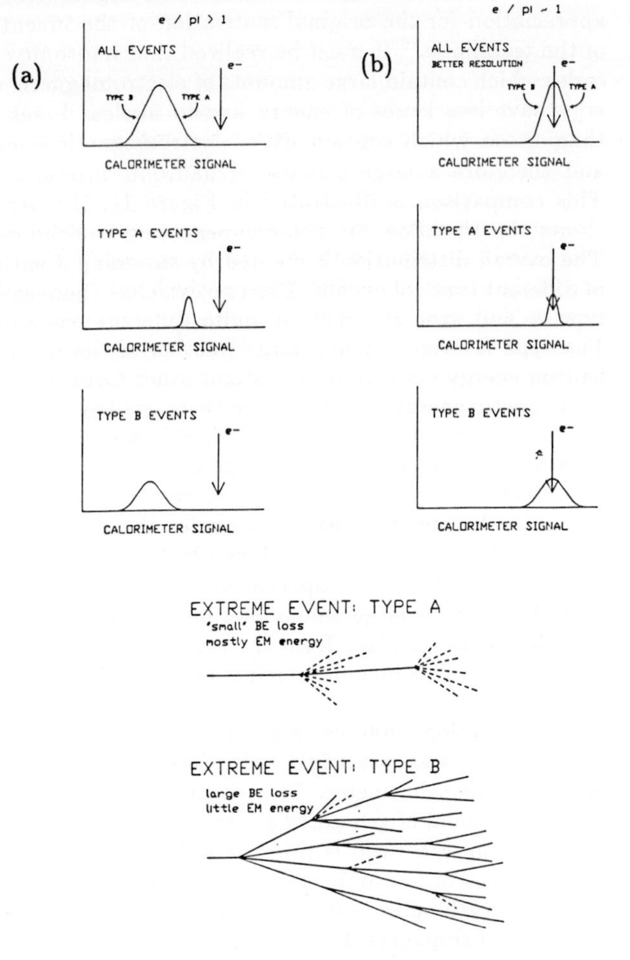

Figure 1. (a) An illustration of the origin of e/h > 1 in noncompensating calorimeters. (b) An illustration of the origin of e/h ≈ 1 in compensating calorimeters. This underlying role of electromagnetic and binding energy losses to the fluctuations in energy distributions was described in reference 26 to motivate the development of compensating calorimeters.

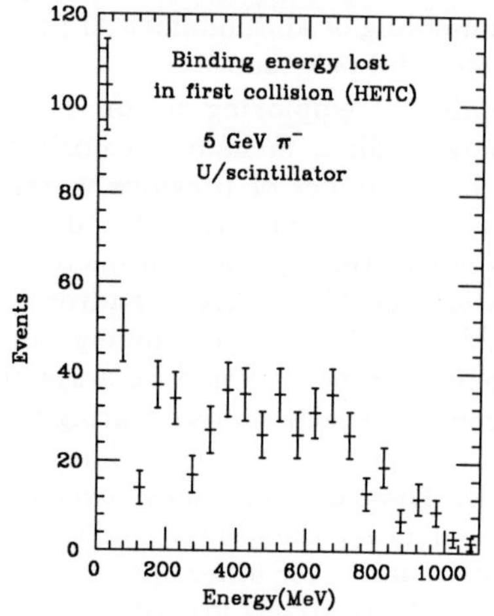

Figure 2. The distribution of binding energy lost in the first collision of a 5 GeV π^- on a uranium-scintillator calorimeter made of 3 mm uranium plates and 3 mm scintillator sheets as calculated by HETC. The average value is 380 MeV.

of binding energy lost for the entire showers as shown in Figure 4 . On average 32% of the incident kinetic energy is lost in binding energy associated with the ejection of nucleons (primarily low energy) from the nuclei. The 1600 MeV has been lost, but one remaining effect is the large number of low energy neutrons created in the nuclear disruption. Figure 5 shows the correlation of the energy content of the low energy spallation and evaporation neutrons (kinetic energy below 20 MeV) with the binding energy lost in each of the hadronic cascades. Events with small binding energy losses naturally have disrupted the nuclei to the smallest extent and have transferred the least energy to low energy nucleons. Events with large binding energy losses have resulted from a large number of very disruptive collisions, resulting in large amounts of energy being carried by the low energy nucleons. These trends are very evident in Figure 5 . It was this strong correlation which motivated the first attempts at the construction of compensating calorimeters.[26] It was reasoned that by increasing the neutron content through the fission process in the calorimeter, the binding energy losses could be overcome in a fashion which was proportional to the losses, thereby leading to an improved resolution. To achieve this end, uranium was employed as a radiator material.

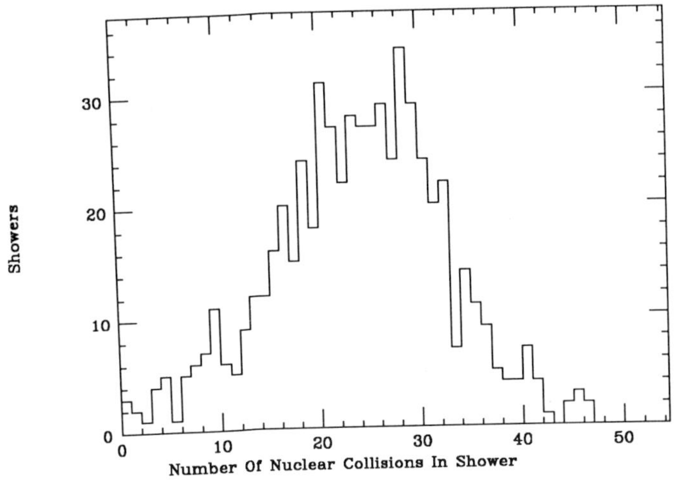

Figure 3. The distribution of the number of nuclear interactions by a 5 GeV π^- on a uranium-scintillator calorimeter made of 3 mm uranium plates and 3 mm scintillator sheets as calculated by HETC. The neutrons in the cascade are cutoff for this count at 20 MeV. The average number of collisions is 24.

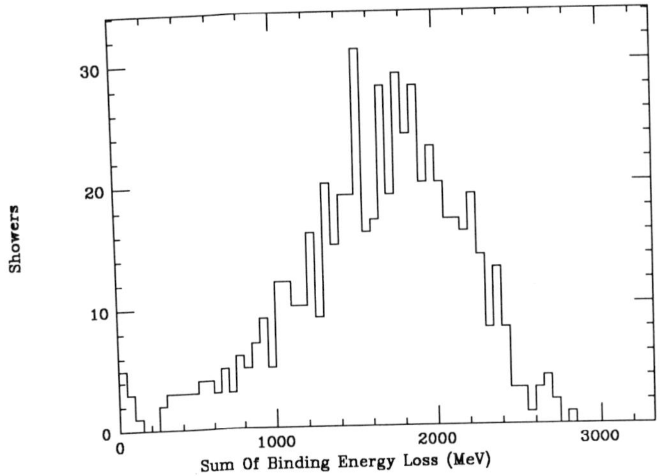

Figure 4. The distribution of total binding energy losses per shower from a 5 GeV π^- on a uranium-scintillator calorimeter made of 3 mm uranium plates and 3 mm scintillator sheets as calculated by HETC. The average energy lost is 1600 Mev, 32% of the incident particle energy.

Neutrons

The key to successfully capitalizing on this correlation is through the coupling of these low energy neutrons to the sampling medium. Eventually, most of these low energy neutrons are captured by U^{238}, yielding approximately eight MeV per neutron capture in prompt gammas. In addition, prompt fission gammas are produced in the fission process. Unfortunately, the sampling medium is very inefficient to sampling these soft gammas due to the effects mentioned earlier regarding preferential absorption in the high Z material. Also the time scale associated with thermalization of the neutrons is too long ($\sim \mu$sec). Therefore, the signal from the neutrons must come from their direct interaction in the active medium.[30] Figure 6 shows the observed neutron energy which our calculations predict will be sampled by the plastic scintillator in the three millimeter uranium-scintillator stack. This distribution results from detailed calculations which were per-

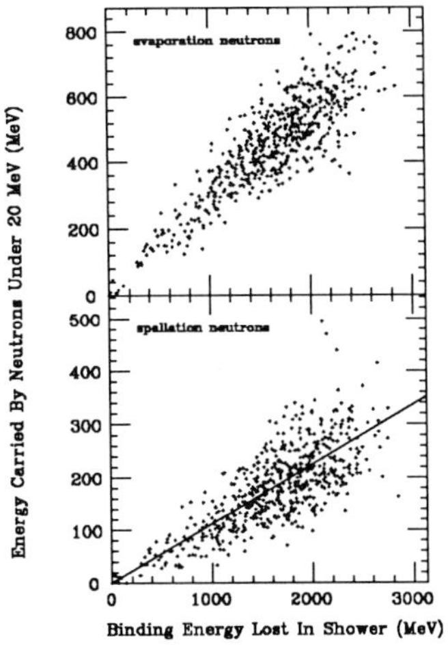

Figure 5. The correlation of energy carried by neutrons below 20 MeV to the binding energy lost in showers from 5 GeV π^-s on a uranium-scintillator calorimeter made of 3 mm uranium plates and 3 mm scintillator sheets as calculated by HETC. The distribution is presented both for the evaporation neutrons and the spallation neutrons. The solid line shows the average trend of the data and is plotted for comparison with Figure 6 .

formed with the Oak Ridge Monte Carlo Codes HETC and MORSE (see the Appendix for details.) The solid line drawn on this figure represents the average energy carried by the spallation neutrons before transport; this is the same line as is shown on Figure 5 . Notice that a sizeable fraction of the original neutron energy (even after the effects of saturation have been taken into account) has been observed (~ 12%). This is due to neutron amplification by fission of the uranium and the very efficient transfer of neutron energy to the hydrogen, which will be discussed later. For comparison, the fraction of electromagnetic energy observed is about 6%.

Since the transfer is so complete in the hydrogenous plastic scintillator, the relative contribution of the total signal can be varied by changing the relative thickness of the sampling layer and the uranium radiator, as was pointed out first in reference 13 . This is illustrated in our calculations shown in Figure 7 . A calorimeter composed of three millimeter uranium plates and three millimeter plastic scintillator plates gives approximately the correct fraction of neutron coupling to bring the e/h ratio to 1.0. When the radiator thickness is increased to six millimeters, the scintillator thickness must be considerably greater than six millimeters to achieve e/h = 1.0. The neutron energy has been significantly reduced by the first three millimeters of scintillator and therefore the doubling of thickness does not double the neutron coupled energy.

It is important to understand in detail the fission processes which take place in the calorimeter stack. Recent activation analysis measurements[16] have confirmed the basic level of fissions calculated with the CALOR code system.[31,32] Recently, for the Task Force on Radiation in the SSC Interaction Regions, calculations[33] were carried out to determine the numbers of fissions induced by neutrons of kinetic energies below 20 MeV for the two stacks studied experimentally by reference 16 . Reasonable agreement has been obtained for the massive uranium stack case (5.6 ± 0.3 /GeV calculated compared to the experimental 6.4 ± 0.7 /GeV), but the size of the measured number of fissions for the uranium-scintillator case remains a puzzle (2.4 ± 0.2 /GeV calculated compared to the experimental 5.1 ± 0.7 /GeV.) One would expect the number to drop by a larger fraction than measured as the coupling of the neutrons to the hydrogen in the plastic scintillator rapidly drops the neutron energies below the fast fission threshold of U^{238} at approximately 1 MeV.

Accurate predictions of performance also require a knowledge of the relative time of response for the various components of the calorimeter signal. The temporal response functions for the fission and neutron capture processes expected in an infinite natural uranium system were presented in reference 31 . Since the capture energy will be inefficiently transferred to the signal, the majority of the hadronic response of uranium calorimeters occurs in the first 50-100 nanoseconds.

Early Calculations

All of the effects mentioned above were included in our earlier studies of compensating calorimeters. A review of these earlier calculations is useful to understand the new results to be presented later. Figure 8 shows comparisons between calculations[7] and AFS measurements[34] for various ratios of iron (copper) and uranium in the calorimeter. Note that for these comparisons, the differences between iron and copper do not alter the basic results and conclusions. As the data and calculations show, the compensation resulting from increasing fractions of uranium radiator is well understood.

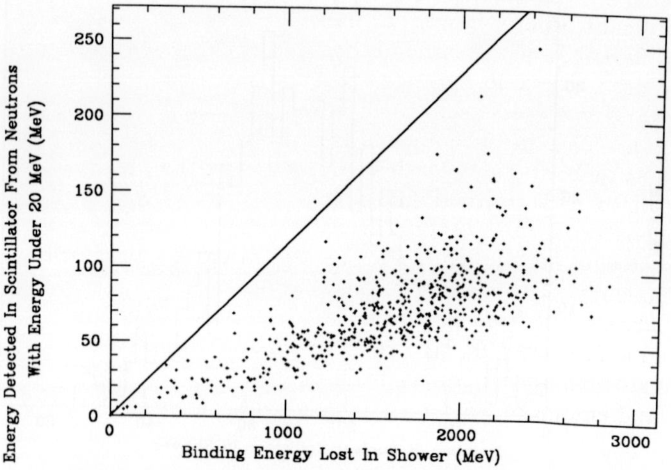

Figure 6. Calculations of the energy detected in a 3 mm uranium/ 3 mm scintillator calorimeter stack from the neutrons with energy below 20 MeV versus the binding energy lost in showers from 5 GeV π^-s. The solid line shows the average trend of the data in figure 5 . See text for discussion.

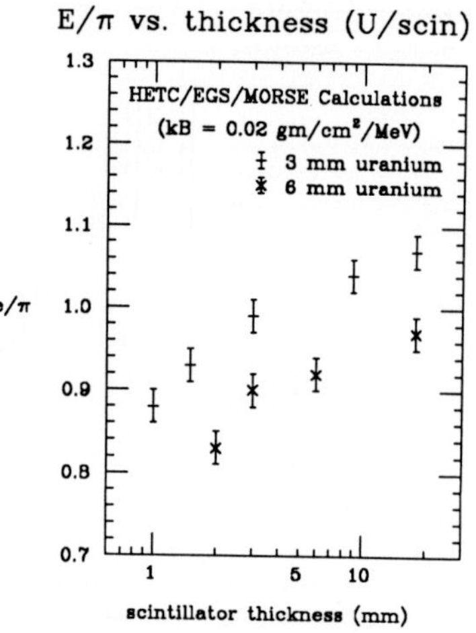

Figure 7. The variation of e/π with scintillator thickness for a set of uranium-scintillator calorimeter calculations for incident 5 GeV π^-s. This figure illustrates the "tuning" possible with scintillator, first noted in reference 13 .

In this same study, the response of a uranium-liquid argon calorimeter was found to be less compensating. The selection of events can alter the conclusion on the degree of compensation by reducing the e/h ratio and narrowing the energy distribution. References 7 and 8 pointed out this selection bias, and the importance of selection in the analysis was confirmed by the experimenters.[35]

Hydrogenous Media

This very different response of uranium calorimeters employing different readout media was interpreted[7,8] to yield the new understanding of compensating calorimetry. The underlying phenomena responsible for the significant difference were explained in terms of three basic differences of the sampling materials.[8] First, the neutron cross sections in hydrogen continue to rise below 1 MeV, while the cross sections in argon are falling (see Figure 9). This in itself means that a larger fraction of the soft neutron energy will be deposited in the plastic scintillator than in the liquid argon. Secondly, the kinematic constraint on the energy transferred in a neutron-nucleus collision highly favors the transfer of energy to a lighter nucleus. The maximum possible recoil energy ($E_R|_{max}$) transferred from an incoming nonrelativistic neutron of kinetic energy E_n to a recoil nucleus of atomic weight A can be shown[36] simply from conservation of momentum and energy to be

$$E_R|_{max} = \frac{4A}{(1+A)^2} E_n. \tag{1}$$

This means that in neutron-hydrogen collisions, the full energy of the neutron can be transferred, while in neutron-argon collisions the maximum energy which can be transferred is ten percent of the neutron energy. Thirdly, the energy coupling is even more suppressed in liquid argon due to the saturation which occurs in regions of dense energy loss. In scintillators this phenomenon has been studied for many years and was modelled by Birks[37] in a form that has come to be known as Birks' Law:

$$\frac{dL}{dx} = \frac{S\frac{dE}{dx}}{1 + kB\frac{dE}{dx}}, \tag{2}$$

where $\frac{dL}{dx}$ is the light yield per unit path length, $\frac{dE}{dx}$ is the specific energy loss for the charged particle, S is the normal scintillation efficiency, and the kB parameter describes the quenching which occurs for high density ionization. Typical kB factors for scintillators are 0.01-0.02 $gm\ cm^{-2}\ MeV^{-1}$. Following the suggestion of reference 7 , it has become customary to express the columnar recombination[38] which results in saturation in liquid argon ionization sampling calorimeters also in the form of the Birks' Law. In this form the equivalent kB factor for liquid argon is about 0.005 $gm\ cm^{-2}\ MeV^{-1}$, that is less saturating than scintillator. While the intrinsic saturation in liquid argon is less than in scintillator, the kinematic constraint on the energy transfers in neutron-nucleus collisions leads to a greater suppression of the transfer of neutron energy to observable signal. In liquid argon the highly ionizing recoil nuclei receive only one-tenth the relative energy of the less highly ionizing recoil protons in scintillator.

Qualitatively it is easy to see why the saturation has so much more of an effect by examining the calculations of references 7 and 8 . The 5 GeV protons studied in those calculations were found to deposit 402 MeV into the liquid argon gaps of 2.0 millimeters (the uranium thicknesses were 1.7 millimeters). However, due to saturation, only 308 MeV was detectable. The effect of this was to raise the e/h ratio to 1.2 from a potential value of 0.9 in the absence of saturation.[39] Of the 94 MeV lost to saturation, 72 MeV comes from collisions of the low energy (< 20 MeV) neutrons. Therefore, the saturation of the energy

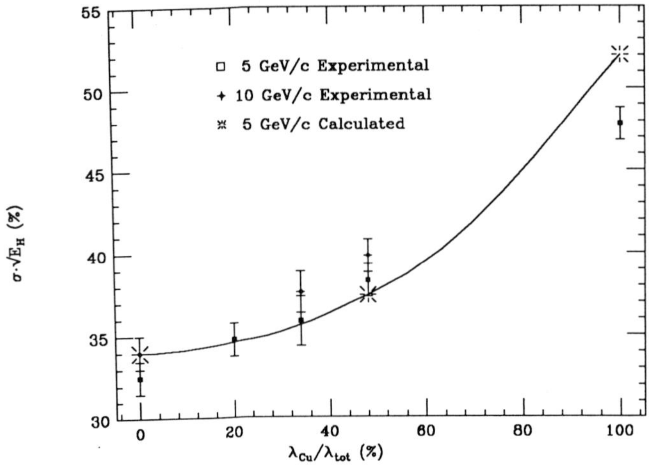

Figure 8. The comparison of the simulation[7] of the dependence of hadron energy resolution with uranium purity with the experimental measurements of the AFS Collaboration.[34]

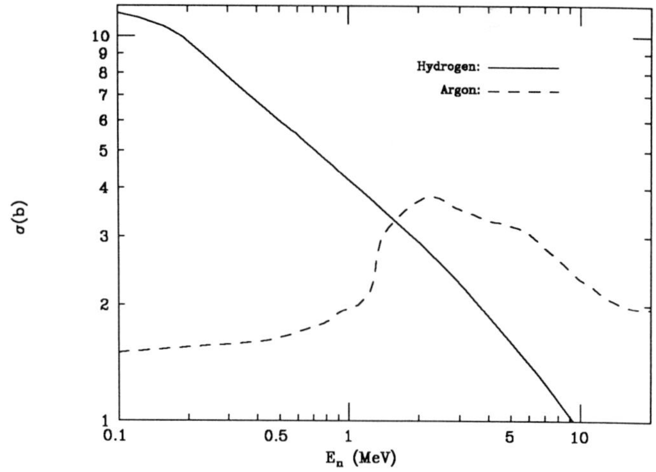

Figure 9. Comparison of the neutron total cross sections in hydrogen and argon. The argon cross section represents the average value in the regions of much structure.* (These are approximate curves presented only for comparison.)

delivered to the liquid argon has reduced the degree of compensation from an over-compensation condition to a value of 1.2. Note that this calorimeter was restricted in transverse dimension, particularly the uranium portion, and therefore a larger device would be expected to obtain smaller values of e/h. Furthermore, these calculations of e/h were subsequently confirmed by the experimenters.[35]

Electromagnetic Sampling

It is important to recognize the significant effects of reduced electromagnetic response in calorimeters composed of high Z radiator structures.[11] These "electromagnetic sampling inefficiencies" have been explained as resulting from the preferential absorption of low energy photons (< 1 MeV) in the high Z material and the effective electron path length stretching in the inactive media resulting from enhanced multiple scattering in the high Z material.[6,9,13] Figure 10 displays how the preferential absorption of the low energy photons in the high Z material comes about. Here the photon mass attenuation coefficient multiplied by the fraction of primary photon energy loss for argon and for lead can be seen.[40] Clearly in the high Z material, the relative absorption at low energy, say $\sim$ 100 keV, is much more effective in stopping the photons. The role of multiple scattering has also been clearly demonstrated by examining the effect of turning off multiple scattering in the electron-gamma shower program EGS.[41] Such calculations for 1 GeV electron showers in uranium liquid argon sampling calorimeters show an increase of 10% in the signal deposited in the liquid argon as the electron path length stretching disappears.[6,9] The magnitude of this "electromagnetic sampling inefficiency" is often measured by comparing the response of a calorimeter for incident electrons to that anticipated for minimum ionizing particles. The ficticious fully contained minimum ionizing particles can be measured by scaling up the response of the calorimeter to muons. Such measurements

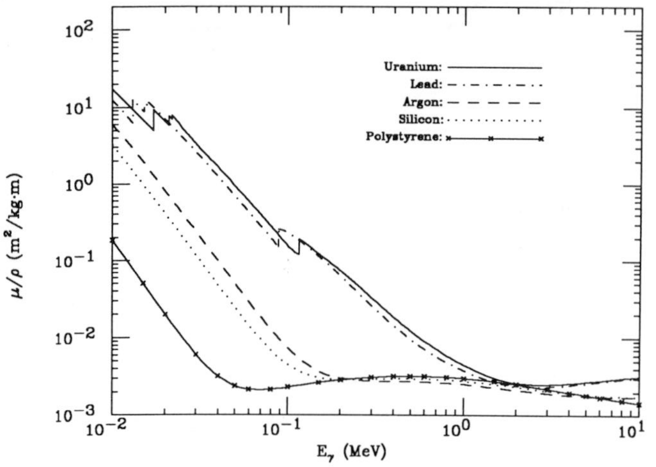

Figure 10. Comparison of the photon mass attentuation cofficients multiplied by the fraction of primary photon energy loss for polystyrene, silicon, argon, lead, and uranium.

find the ratio of average electron pulse height to the pulse height for the minimum ionizing particle to be roughly 0.6 to 0.7 for calorimeters constructed with high Z radiators. The magnitude of this effect led us to the prediction "that a lead calorimeter may also give EM/HAD $\approx$ 1."[12] This prediction has now become reality. A lead-scintillator compensating calorimeter has been built and tested with scintillator thicknesses of one-quarter the lead thicknesses, a ratio designed to give the optimal electron to hadron ratio.[15] The reported e/h ratio for this device is 1.05 ± 0.04 for energies over 10 GeV. Caution should be exercised, however, since this analysis is based on events selected by energy deposition. This selection could emphasize the electromagnetic parts of the hadronic shower and yield a biased e/h ratio.

The understanding of the underlying phenomenology of compensating calorimeters is on solid ground and calculations can be done reliably. The requirements of the future hadron colliders demand that wise use be made of this understanding.

CALCULATIONS OF SILICON HADRON CALORIMETERS

The understanding that now exists concerning compensating calorimeters can be applied to the silicon detector calorimeters. The fundamental question that needs to be addressed is what the e/h ratio for these silicon based calorimeters will be. One property of silicon that plays an important role in this study is the extreme linearity of silicon up to very large stopping power.[42] That is, silicon exhibits very little saturation. It has been demonstrated that saturation prevents full compensation in currently designed uranium-liquid argon calorimeters, so this could be a very important factor in silicon calorimeters. Naively it is expected that very good results can be obtained for silicon calorimeters.

The linearity of response of silicon to large energy deposition densities is summarized in Figure 11 . This figure shows the relative response of a recoil ion of maximal energy in neutron scattering.[43] One sees that even for silicon ions as low as 100 keV nearly one-half of the deposited energy is detected as observable output signal. Given these evidences of very limited saturation in silicon detectors a very good response to some components of the low energy development of hadronic showers can be expected.

It is pedagogically useful to consider the differences between the response of silicon and liquid argon, since except for the saturation in liquid argon, full compensation could possibly be achieved. The three critical factors which were reviewed earlier to compare the neutron coupling to scintillator and liquid argon are useful to compare again here. First the neutron cross sections[44] are compared in Figure 12 . It can be seen that the cross section in argon falls for energies below a few MeV, while the cross section in silicon remains large and even rises below 500 keV. This will result in a larger energy deposition by low energy neutrons in silicon than is observed in liquid argon. The typical neutron energy spectrum which has been calculated in these studies[33] is also shown in this figure and illustrates the importance of the cross sections at the lowest energies. Secondly, the effect of the kinematic limitation on neutron energy transfer can be considered. Equation 1 shows that while only one-tenth of the neutron energy can be transferred in liquid argon collisions, the restriction on energy transfer is somewhat less restricted in silicon, being limited to about 13 percent. This should lead to an additional increase in the energy transferred to the silicon. Thirdly and finally, the very limited saturation which occurs in silicon will enable these increased neutron depositions to be more observable.

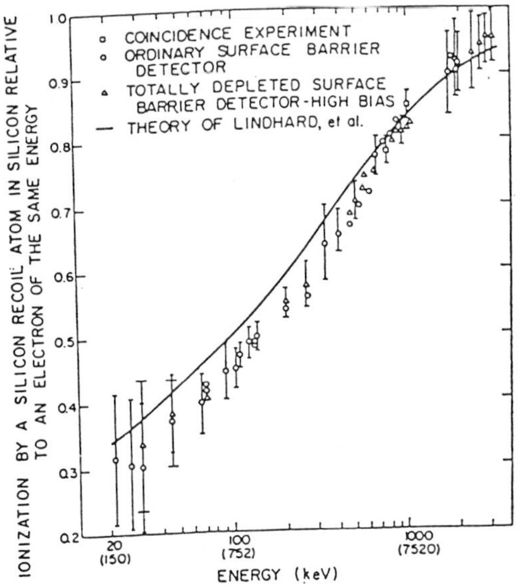

Figure 11. Pulse height produced by a Si recoil atom relative to that of an electron of the same energy in Si as a function of Si recoil energy. The incident monenergetic neutron energy necessary to produce the denoted recoil energy is shown in parenthesis. (From reference 43 .)

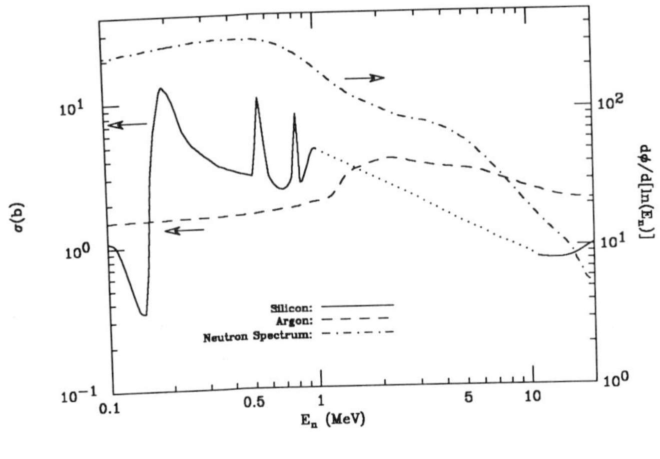

Figure 12. Comparison of the neutron total cross sections in silicon and argon. The silicon and argon cross sections represents the averages in regions of much structure. (These are approximate curves presented only for comparison.) Also shown on this figure is the typical calculated neutron energy spectrum from reference 33 .

The potential for compensation with silicon certainly seems better than for liquid argon. To make quantitative predictions it is necessary to make use of a Monte Carlo code[27] system which takes into account all of the basic processes and examines the details of the shower development. As in our earlier studies, the Oak Ridge system CALOR has been applied to this problem. The details of this system are described in the Appendix. It has been assumed that the effective kB parameter in Birks' law is equal to 0, an approximation which should be very good for these studies. The e/h ratio will be only slightly underestimated in these calculations.

Figure 13 shows the calculated response distribution for 10 GeV π^-s incident on a silicon calorimeter with 5 millimeter thick uranium radiators and 400 μm fully depleted silicon detectors sandwich between two layers of 5 millimeter thick G10. Therefore, the total layer thickness is 15.4 millimeters and the depth of the stack extends for 150 readout and radiator layers. The transverse dimensions of the stack have been taken to be 100×100 cm^2. The charge collection is cut off after 50 nanoseconds. The resulting energy resolution (σ/E) is 21.5% and the e/h ratio is 1.07, close to compensation.[45] In fact the two layers of G10 in this stack are increasing the value of e/h as the neutron energy from the uranium is being deposited in the G10 through the large energy transfers via hydrogen. Additionally, significant transfers occur to the carbon and oxygen in the G10.

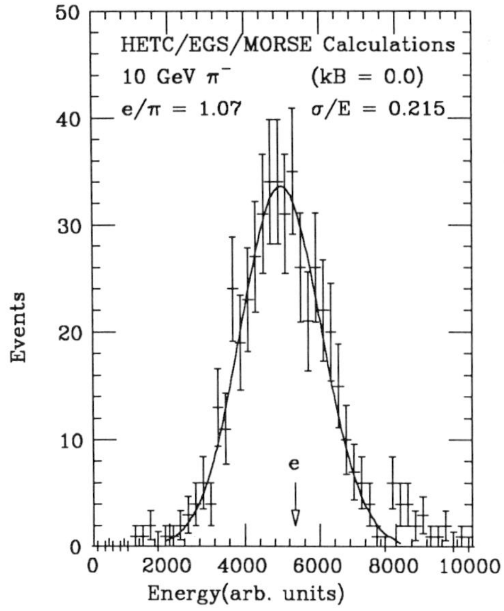

Figure 13. The calculation of the measured hadron energy distribution for 10 GeV π^-s incident on a uranium-silicon calorimeter with 5 mm thick uranium radiators and 400 μm fully depleted detectors imbedded in 5mm thick G10 supports.

Figure 14 presents the event by event correlation of the energy detected in the silicon from the low energy (<20 MeV) neutrons to the binding energy lost in the hadronic cascade. While the correlation is strong, this figure shows much larger fluctuations in detected energy than was seen in Figure 6 for the uranium-scintillator calorimeter. This fluctuation depends on the number of neutrons actually detected. Given the very thin detection layers, the number of neutrons actually detected is small. Unlike the normal shower fluctuations which vary as $\sqrt{t_{cell}}$ and are dominated by the thickness of the radiator, the fluctuation on neutron response is dominated by the thickness of the silicon (t_{Si}), and scales as $\sqrt{t_{Si}}$. Therefore, the sampling fluctuations for the charged hadrons in the hadron cascades will be similar to any other readout medium, but the additional larger fluctuations on the neutrons will add to the width of the energy distribution, preventing excellent low energy resolution even for perfectly compensating calorimeters.

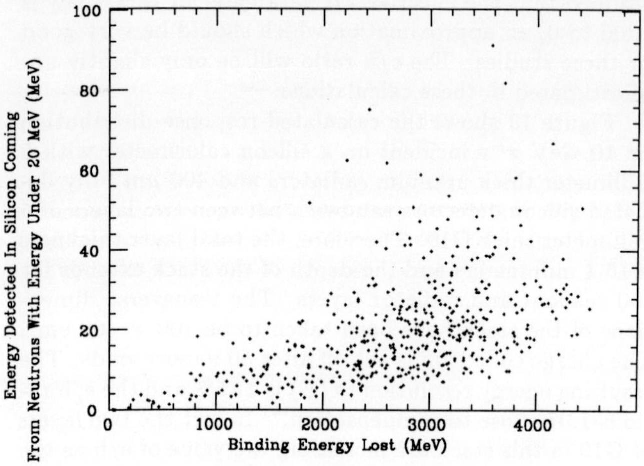

Figure 14. The correlation of detected low energy (E<20 MeV) neutron energy in the uranium-silicon calorimeter with the binding energy lost in the hadronic cascade.

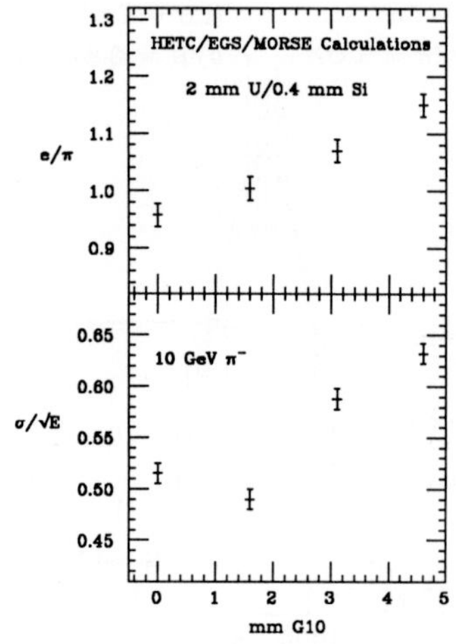

Figure 15. The calculated variation of e/π and $\sigma/\sqrt{E}$ with the thickness of a G10 sheet placed between the 2 mm uranium plates and the silicon detectors in a uranium-silicon calorimeter for incident 10 GeV π^-s. Without G10 a slight overcompensation condition exists.

Figure 15 presents a series of calculated results for calorimeters constructed with 2 millimeter uranium radiators, followed by a layer of G10 of varying thickness and a 400 μm fully depleted silicon detector. As before, kB = 0, the time cut is 50 nanoseconds, and the transverse size is taken to be 100 $\times$ 100 cm^2. The stacks contain 330 layers of uranium, G10, and silicon. It can be seen that with no G10 the condition of overcompensation is obtained due to the sensitivity of the silicon detectors to very low energy particles. As the G10 is added, the e/h ratio increases due to the removal of neutron energy by the G10, meaning less energy is available to be deposited in the silicon. Naively, one might expect that the neutrons interacting in the G10 would contribute to the response of the silicon by knocking protons into the detectors. It has been proposed[7] that coupling a hydrogenous material to liquid argon, for example, would be a possible method for restoring its potential for compensation. However, the efficiency and range of the protons produced in the G10 is not large enough to have much of an effect on the total response of a detector of 400 μm thickness.[46] It can also be seen in Figure 15 that the energy resolution reaches a minimum at approximately the point where the e/h ratio passes through one, as would be expected for a compensating calorimeter. The value of this resolution is not as small as the values achieved in scintillator because of the larger sampling fluctuations involved here, as described previously.

Several cases of silicon calorimeters have also been studied with different radiator materials, specifically one employing lead and one employing iron. For a lead calorimeter with 3 millimeter thick radiators and 400 μm fully depleted silicon detector with no air gaps and the same transverse geometrical constraints and time cuts as before, the calculated results yielded an expected e/h ratio of 1.16 with an energy resolution of $\frac{\sigma}{\sqrt{E}} = 47\%$ for 10 GeV π^-s. For an iron calorimeter with 7.5 millimeter radiators, followed by a 1.3 millimeter air gap, a 1 millimeter G10 layer, 400 μm of fully depleted silicon, a 1 millimeter G10 layer, and another 1.3 millimeter air gap, the e/h ratio increases to 1.29. For the 10 GeV π^-s the resolution is $\frac{\sigma}{\sqrt{E}} = 62\%$. Even for the iron, the role of low energy detection efficiency can be seen in the e/h ratio, and for the lead, the e/h ratio of 1.16 is within reach of the values which have been touted as acceptable for future "compensating" calorimeters.[47]

In connection with these calculations, the attractiveness of a low-Z radiator material such are iron should be pointed out. Iron has several advantages compared to uranium, for example. In iron, less binding energy loss in inelastic collisions is realized, and this factor therefore contributes less to the energy spread than in calorimeters employing higher Z radiators. Secondly, connected with this is the fact that fewer neutrons are produced in these collisions and the produced neutrons are a source of radiation damage at high luminosity colliders. Thirdly, uranium is more expensive than iron.[48] Finally, uranium is difficult to obtain and to work with. In spite of these advantages for low Z over high Z, it is difficult to anticipate a practical design of a compensating iron calorimeter. As we have seen, the electromagnetic sampling inefficiencies are an important factor in achieving compensation. These effects are small in iron due to the small value of the charge of the iron nucleus. As a result, the response of an iron calorimeter to electromagnetic and non-electromagnetic energy deposition is much different, yielding a large fluctuation in the pulse height distribution. The small binding energy losses are overcome by the fluctuations between electromagnetic and non-electromagnetic energy. It is conceivable to imagine a combination of low-Z and high-Z material which will combine the advantages of each material and approach a resolution determined by the small binding energy losses of iron.

EXPERIMENTAL TEST OF SI/U

The above Monte Carlo studies have predicted compensation in uranium silicon hadron calorimeters. If the experimental results should show that this cannot quite be achieved, further signal-equalization may be obtained with a silicon readout by tuning the electromagnetic component of the hadronic shower.[49–52] The experimental work of the SICAPO Collaboration at CERN is addressing this prospect. A reduction of the energy sensed in electromagnetic showers, caused by the fiberglass (G10) supports of the silicon mosaic, was observed by running the electromagnetic section of a full hadronic calorimeter using uranium absorbers and 0.25 m^2 of silicon active area during 1986.[51] To investigate this observation further, a systematic study of a silicon calorimeter response to electromagnetic shower was performed with small area Si/W and Si/U calorimeters. The measurements were performed at the CERN-PS, with incoming electrons of 2, 4 and 6 GeV, during 1987.[52] Figures 16 and 17 show the mean energy sensed by the calorimeter for Si/W and Si/U respectively, as a function of the incoming electron energy. From these figures it can be seen that in the case

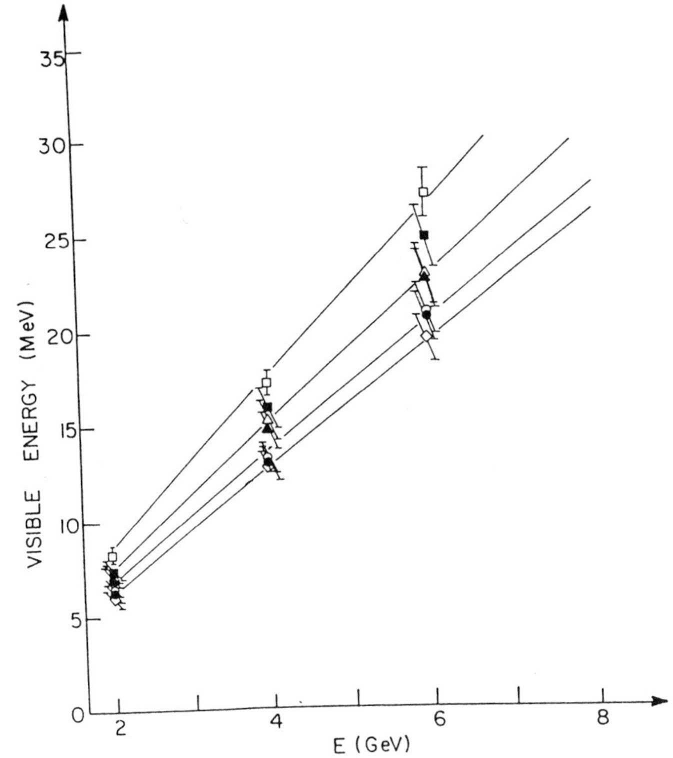

Figure 16. Mean energy deposited in Si/W calorimeter as a function of the incoming electron energy, from top to bottom:

no absorber in front of detector
0.5mm of G10 in front and rear of detectors
1.0mm of G10 in front and rear of detectors
1.5mm of G10 in front and rear of detectors
3.0mm of G10 in front and rear of detectors
4.0mm of G10 in front and rear of detectors
5.0mm of G10 in front and rear of detectors

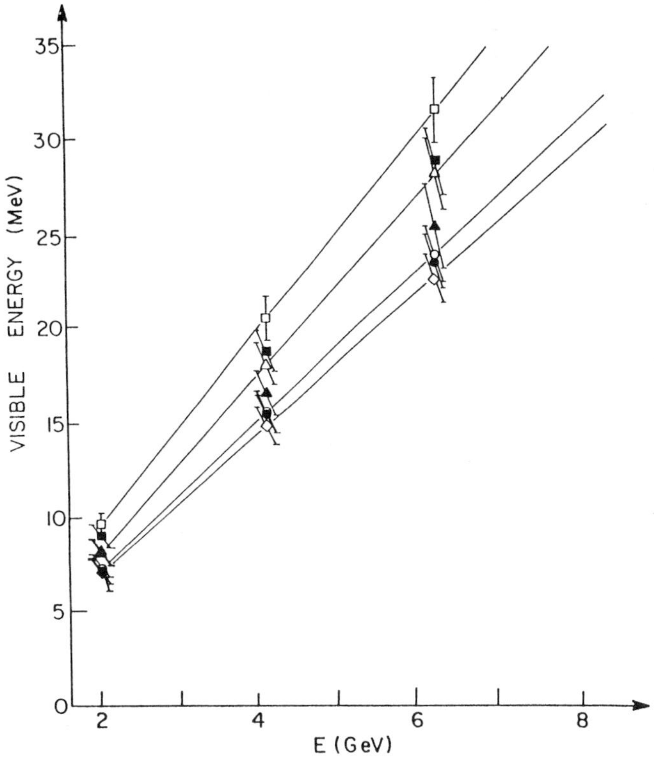

Figure 17. mean energy deposited in Si/U calorimeter as a function of incoming electron energy, from top to bottom:

no absorber in front of detector
0.5mm of G10 in front and rear of detectors
1.0mm of G10 in front and read of detectors
1.5mm of G10 in front and rear of detectors
3.0mm of G10 in front and rear of detectors
4.0mm of G10 in front and rear of detectors
5.0mm of G10 in front and rear of detectors

of the 5mm G10 absorber (located in the rear sides of the detector), the electromagnetic visible energy is reduced by about 29%. The reduction of visible energy (which is energy sensed by the silicon detectors) is greater than the 10% reduction one expected by the addition of the G10 plates.

This 29% reduction in the response of the calorimeter to electromagnetic showers (when 5 mm G10 plates are located on the rear side of the silicon detectors), can be compared to the 10% reduction expected for a purely hadronic shower in the same setup, resulting in a decrease in the value of e/h up to 18%. Thus the compensation condition for Si/U calorimeter is likely to be achieved. On this principle, a perfectly compensating calorimeter of the size needed for LHC/SSC projects is feasible using silicon detectors as active material. The SICAPO collaboration has built a prototype of a Si/U hadronic calorimeter which is 6 interaction lengths deep (Figure 18 .) The silicon mosaic active samplers (500 cm^2 active area, but with the possibility to double it as seen in Figure 19) are located next to 10 mm thick uranium plates. Data taking is expected to start during winter 1988/1989.

DISCUSSION OF POTENTIAL LIMITATIONS

It is important to keep in mind some of the other unsolved problems of silicon readout hadron calorimeters. One very important concern is the tolerance, or potential lack thereof, of silicon to the radiation environment that the future high luminosity, high rate colliders will present. Recent work in this area has illuminated the situation and shown that it is not as dismal as was earlier feared. The improvement in tolerance to neutrons which is expected to occur with higher resistivity materials[53] has been demonstrated.[54] Strong self-annealing effects have been observed following the exposure of such devices to intense neutron sources.[55,56] Finally, a study of the expected effects of the radiation environment of the SSC[20] on the operation of an imagined detector (the compact solenoidal detector) employing silicon calorimeters,[57] including the best understanding of these resistivity dependence effects and self annealing, has concluded that such an operation appears practical.[58] Preliminary results from the SICAPO Collaboration[55] using a DLTS analysis show that the concentration of trapping centers (generated by evaporative neutrons from Cf^{252}) is negligible with detectors annealing at about 150° C. Further radiation effects tests are called for, but the situation looks manageable.

The second major area of concern is in the cost of providing the detectors for such a project. The total area of silicon required for the calorimeter of the compact solenoidal detector is 20 million cm^2. To make such a project realizable means that devices must be available for about two dollars per cm^2. Presently, devices are being ordered for 6-12 dollars per cm^2 so that an improvement of a factor of 3-6 is needed. The prospects for such an improvement for a detector of the scale of the compact solenoidal detector look good, however.[59]

One final issue of concern must be mentioned and that is the problem of how to read out the detectors while preserving the intrinsic speed. Radeka and Rescia have noted the design constraints on speed for ionization sampling calorimeters.[60] Achieving calorimeter responses of ~20 nanoseconds requires that the product of detector capacitance and inductance of the connections be kept below 4 $nsec^2$. For a cell of one nanofarad this means the inductance of the connections must be kept below 4 nanoHenries, which requires less than a centimeter of

Figure 18. Silicon/U hadronic calorimeter of SICAPO collaboration at CERN PS

Figure 19. Silicon Mosaic active samplers

lead for practical means. The compact solenoidal detector design, in fact, proposes about 1 nanofarad per layer of silicon detectors in the hadron calorimeter. Therefore, preserving the intrinsic speed will probably require the mounting of active elements on every layer. While this involves millions of transistors, it appears achievable. The combining of pads in series is also a possible means to achieve reduced capacitance and is being pursued by the SICAPO Collaboration for the prototype silicon-uranium calorimeter. They will be controlling the noise in the experimental program of silicon-iron with detector pads of about 5.5 cm^2 and individual VLSI(BICMOS) preamplifier readouts. Some noise penalty must be accepted to preserve the speed, but it can be sufficiently limited by these procedures. This may prevent calibration of the calorimeter with muons, but shouldn't limit the energy resolution of real physics signatures.

CONCLUSIONS

The new understanding of compensating calorimeters which was introduced four years ago has now been experimentally confirmed in numerous tests. One important element of that understanding was the role of saturation in uranium-liquid argon calorimeters. High rate, high energy colliders are going to demand compensating, fast, finely segmented calorimeters. Our calculations indicate that the near absence of saturation in silicon should lead to compensation. Silicon's other attractive features (speed, ease of segmentation, and absolute gain calibration) make it an extremely attractive approach. An experimental program by the SICAPO Collaboration is underway at CERN to test these expectations.

APPENDIX

THE METHOD OF CALCULATION

The calculations reported here have been performed with the CALOR computer code system following approximately the procedures used in previous calculations.[61,31,7,8] The three-dimensional, multimedia, high-energy nucleon-meson transport code HETC[62] was used, with modifications to obtain a detailed description of the nucleon-meson cascade produced in the devices considered in this paper. The source distribution for the electromagnetic cascade calculation is provided by HETC. The transport of the electrons, positrons, and gammas from these sources is carried out using the EGS system.[41] Neutrons which are produced with energies below 20 MeV are transported using the MORSE[63,64] Monte Carlo transport code. Gamma rays (including those from capture, fission, etc.) produced during this phase of the calculations are stored for transport by the EGS code. The nonlinearity of the light pulse or charge collected (L) due to saturation effects is taken into account by the use of Birks' law:[37]

$$\frac{dL}{dx} = \frac{S\frac{dE}{dx}}{1 + kB\frac{dE}{dx}}, \qquad (3)$$

where kB is the saturation constant. Further details of these calculations can be found in the Appendix of Reference 32 .

* Supported in part by the U.S. Department of Energy.

REFERENCES

1 . S. Iwata, Nagoya University Report,DPNU-3-79 (1979).

2 . U. Amaldi, Phys. Scripta 23, 409 (1981).

3 . C.W. Fabjan and T. Ludlam, Ann. Rev. Nucl. Part. Sci. 32, 335 (1982).

4 . C.W. Fabjan, CERN-EP/85-54 (1985).

5 . J.E. Brau and T.A. Gabriel, SLD-New Detector Note no. 119, May 22, 1984.

6 . J.E. Brau, Proceedings of the September 1984 Seattle meeting of the SLD Collaboration.

7 . J. Brau and T.A. Gabriel, Nucl. Inst. and Methods A238, 489 (1985).

8 . T.A. Gabriel, B.L. Bishop, J. Brau, A. DiCiaccio, M. Goodman, and R. Wilson, IEEE Trans. Nucl. Sci. NS-32, 1 (1985).

9 . J.E. Brau, Proc. Workshop on Compensated Calorimetry, Pasedena (1985) CALT-68-1305.

10 . T.A. Gabriel, Proc. Workshop on Compensated Calorimetry, Pasedena (1985) CALT-68-1305.

11 . P.M. Mockett, Proceedings of the Eleventh SLAC Summer Institute on Particle Physics, July 1983, ed. Patricia M. McDonough, SLAC-267 (1984).

12 . H. Gordon and P. Grannis, Proceedings of the 1984 Summer Study on the Design and Utilization of the SSC, Snowmass, Colorado, p. 591 (1984), eds. R. Donaldson and J.G. Morfin.

13 . H. Brückmann, Proc. Workshop on Compensated Calorimetry, Pasedena (1985) CALT-68-1305.

14 . G. d'Agostini et al, NIKHEF-H/88-4 (April 1988).

15 . E. Bernardi et al, Nucl. Inst. and Methods A262, 229 (1987).

16 . C. Leroy, Y. Sirois, and R. Wigmans, Nucl. Inst. and Methods A252, 4 (1986).

17 . R. Wigmans, Nucl. Inst. and Methods A259, 389 (1987).

18 . H. Brückmann et al, Nucl. Inst. and Methods A263, 136 (1988).

19 . D. Soper, presentation at the Workshop on Experiments, Detectors, and Experimental Areas for the Supercollider, Berkeley, California, 1987.

20 . D. Groom (ed.), SSC-SR-1033 (1988).

21 . T. Kondo, Y. Watanabe, and S. Kim, Proceedings of the 1984 Summer Study on the Design and Utilization of the SSC, Snowmass, Colorado, p. 612 (1984), eds. R. Donaldson and J.G. Morfin.

22 . P.G. Rancoita and A. Seidman, Nucl. Inst. and Methods A263, 84(1988); S. Pensotti, P.G. Rancoita, A.Seidman, and L. Vismara, Nucl. Inst. and Methods A265, 266(1988).

23 . G. Barbiellini et al, Nucl. Inst. and Methods A235, 55(1985); G. Barbiellini et al, Nucl. Inst. and Methods A236, 316 (1985). M. Bormann et al, Nucl. Inst. and Methods A240, 63 (1985); and A. Nakamoto et al, Nucl. Inst. and Methods A251, 275 (1985).

24 . T. Akesson et al, CERN-EP/87-88.

25 . SICAPO Collaboration, presented by C. Leroy and F. Lemeilleur, International Conference on Advanced Technology and Particle Physics, Villa Olmo, Lake Como, Italy, June, 1988.
26 . C.W. Fabjan et al, Nucl. Inst. and Methods 141, 61 (1977);
27 . T.A. Gabriel, ORNL/TM-9727 (1985).
28 . G. Rudstam, Z. Naturforsch. 21a,1027 (1966); G. Rudstam, Nucl. Phys. A126, 401 (1969).
29 . This count does not include neutron nuclear collisions for neutrons below 20 MeV.
30 . Note that the prompt gammas mentioned above, although ineffective, are included in the calculations described here.
31 . T.A. Gabriel, Nucl. Inst. and Methods 150, 145 (1978).
32 . J.E. Brau and T.A. Gabriel, International Conference on Advanced Technology and Particle Physics, Villa Olmo, Lake Como, Italy, June, 1988 (to be published in Nuclear Instruments and Methods.)
33 . J.E. Brau and T.A. Gabriel, Appendix 15 of SSC-SR-1033, edited by D. Groom (1988).
34 . O. Botner et al, IEEE Trans. Nucl. Sci. NS-28, 510 (1981); O. Botner et al, Nucl Inst. and Methods 179, 145 (1981).
35 . P. Rehak, Proc. Workshop on Compensated Calorimetry, Pasedena (1985) CALT-68-1305.
36 . G. Knoll, Radiation Detection and Measurement, John Wiley and Sons, New York (1979), p. 570.
37 . J.Birks, Proc. Phys. Soc. A64, 874 (1951).
38 . W. Hofmann et al, Nucl. Inst. and Methods 135, 151 (1976).
39 . This value calculated for this calorimeter represents the lowest obtainable value for e/h. Note that liquid argon could be made to compensate if a method were available to retrieve all of the deposited energy.
40 . J.H. Hubbell, Int. J. Appl. Radiat. Isot. 33, 1269 (1982); J.H. Hubbell, H.A. Gimm, and I. Øverbø, J. Phys. Chem. Ref. Data 9, 1023 (1980); E. Storm and H.I. Israel, Nucl. Data Tables A7, 565 (1970).
41 . R.L. Ford and W.R. Nelson, SLAC-0210 (1978).
42 . B.D. Wilkins, M.J. Fluss, S.B. Kaufman, C.E. Gross, and E.P. Steinberg, Nucl. Inst. and Methods 92, 381 (1971).
43 . A.R. Sattler, Phys. Rev. 138, A1815 (1965).
44 . D. I. Garber and R.R. Kinsey, BNL 325 (1976).
45 . It is emphasized that the e/h ratio was calculated assuming no saturation in the silicon.
46 . T.A. Gabriel, J.O. Johnson, and J.E. Brau, Proceedings of the Workshop on Detector Simulation for the SSC, Argonne National Laboratory, August, 1987.
47 . M.G.D. Gilchriese, Proceedings of the 1987 SLAC Summer Institute on Particle Physics.
48 . C. Baltay, J. Huston, and B.G. Pope, Proceedings of the 1986 Summer Study on the Physics of the Superconducting Supercollider, Snowmass, Colorado, p. 355 (1986), eds. R. Donaldson and J. Marx.
49 . S. Pensotti, P.G. Rancoita, A. Seidman and L. Vismara, Nucl. Instr. and Methods, A265, 266 (1988)..
50 . C. Furetta, S. Pensotti, P.G. Rancoita, L. Vismara, G. Barbiellini and A. Seidman, IEEE Trans. Nucl. Sci., 35, 446 (1988).
51 . S. Pensotti, P.G. Rancoita, C. Simeone, L. Vismara, G. Barbiellini and A. Seidman, Nucl. Instr. and Methods, A270, 327 (1988).
52 . SICAPO Collaboration to be submitted for publication in Phys. Lett. B.
53 . G.C. Messenger and M.S. Ash, The Effects of Radiation on Electronic Systems, Van Nostrand Reinhold, New York, (1986), p. 157.
54 . H.W. Kraner, Z. Li, M. Momayezi, and K. Posnecker, International Conference on Advanced Technology and Particle Physics, Villa Olmo, Lake Como, Italy, June, 1988.
55 . SICAPO Collaboration, presented by L. Vismara, International Conference on Advanced Technology and Particle Physics, Villa Olmo, Lake Como, Italy, June, 1988.
56 . T. Kondo and T. Ohsugi, private communication.
57 . J. Kirkby, T. Kondo, S.L. Olsen et al, Proceedings of the Workshop on Experiments, Detectors, and Experimental Areas for the Supercollider, Berkeley, California, p. 388 (1988), eds. R. Donaldson and M.G.D. Gilchriese.
58 . T. Ohsugi, H.Ikeda, and T. Kondo, Appendix 4 of SSC-SR-1035, edited by M.G.D. Gilchriese (1988).
59 . private discussions with industry.
60 . V. Radeka and S. Rescia, Nucl. Inst. and Methods A265, 228 (1988).
61 . T. A. Gabriel and W. Schmidt, Nucl. Instr. and Meth. 134, 271(1976).
62 . K. C. Chandler and T. W. Armstrong, ORNL-4744, Oak Ridge National Laboratory (1972); H. W. Bertini, Phys. Rev. 188, 1711(1969).
63 . M. B. Emmett, ORNL-4972, Oak Ridge National Laboratory (1975).
64 . N. M. Greene et al., ORNL/TM-3706, Oak Ridge National Laboratory (1973).

PRELIMINARY STUDIES OF LARGE SOLENOIDS AND RELATED STRUCTURES

Shigeki Mori

Institute of Applied Physics
University of Tsukuba
Tsukuba, Ibaraki 305, Japan

ABSTRACT

Conceptual design studies of a large superconducting solenoid proposed by the large solenoid group at the 1987 Berkeley Workshop were carried out.[1] The present design allows the solenoid (8.7 m in diameter and 2 Tesla field) to quench safely without being damaged.

INTRODUCTION

Large solenoid detectors proposed for SSC experiments can be grouped into three categories according to the arrangement of a solenoid magnet and barrel calorimetry. Detectors in the first category are simple extension of the CDF detector in which the barrel calorimeter is placed outside a solenoid magnet. A typical example is a 4π solenoidal magnetic detector (Model B) presented in the CDG report of "Cost estimate of initial SSC experimental equipment".[2] The dimension of a solenoid is relatively small, typically 5m in diameter. Since both electromagnetic and hadron calorimeters are placed outside the solenoid, the material thickness of the solenoid is required to be as thin as possible. Considerable efforts to fabricate thin, large superconducting solenoids for collider detectors were made very successfully.[3]

In the second category the barrel sections of electromagnetic calorimeter and precision hadron calorimeter are placed inside a solenoid and the return yoke is used as a part of the absorber of the hadron calorimeter. Large solenoid detectors discussed in Snowmass-84 and 86 are typical examples of this category.[4,5]

A large solenoid detector proposed in Berkeley-87 is a detector of the third category in which the entire barrel section of the calorimetry is placed inside the solenoid. This arrangement is most likely to provide continuous and hermetic coverage by calorimeters in the barrel section and the end cap sections. This detector inevitably asks a larger solenoid magnet compared with solenoids of the other two categories. On the other hand there is essentially no limitation in material thickness of the solenoid. It can utilize fully stabilized, cryostable superconducting cables with a pool-boiling cryostat arrangement.

In the present paper preliminary results of feasibility studies of construction of the large solenoid magnet discussed in Berkeley-87 (2 Tesla, 8.7 m in diameter and 16 m in overall length) are described.

GENERAL CHARACTERISTICS OF THE MAGNET

Figure 1 shows the schematic diagram of the central section of the large solenoid detector proposed in Berkeley-87. The main parameters of the solenoid are given in Table I. The bore diameter is 8.2 m and the overall length is 16 m.

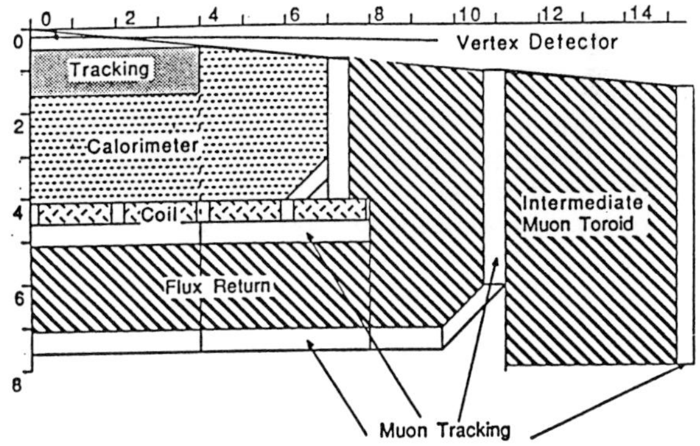

Fig.1. Schematic view of the central section of the large solenoid detector (Berkeley-87).

The total stored energy is 1 GJ with the central field of 2 Tesla. The magnet consists of 8 identical modules of 1.7 m in length each. The free space between two adjacent modules is 0.34 m. The parameters of the solenoid module are given in Table II. Each module weighs about 56 tons. Each coil is made of 6 layers of AMY-type conductor and it has in total 642 turns.[6] The parameters of the AMY conductor are given in Table III. With pool boiling the AMY conductor satisfies cryostable conditions for the degin current of 5000 A. The maximum field at the coil under the normal operating condition is estimated to be about 2.8 Tesla which corresponds to about 60% of the short sample value at 4.4K.

Table I. Main Parameters of the Solenoid.

Bore Diameter	8.2 meters
Total Length	16 meters
Central Field	2 Tesla
Number of Modules	8
Free Space Between Adjacent Modules	0.34 meters
Total Stored Energy	1 GigaJoule
Overall Inductance	80 Henries
Total Weight	450 × 10^3 Kilograms

Table II. Parameters of the Solenoid Module.

Length	1.7 m
Diameter (inner/outer)	8.2 m/9.2 m
Coil Length	1.2 m
Current	5000 A
Central Field	2.0 T
Winding	
Number of Layers	6
Total Turns of Winding	642
Stored Energy	125 MJ
Cryostat: He Vessel	
Material	SUS 304N
Wall Thickness (outer/inner)	3.5 cm/2.5 cm
Length	1.4 m
Conductor	AMY-type
Weight	56 × 10^3 Kg
Cold Mass	33 × 10^3 Kg
Temperature Rise for 125 MJ	75°K from 4.2°K

Table III. Parameters of the AMY Conductor.

Superconductor	Nb-46.6 wt% Ti
Cross Section	9.8 × 10.2 mm^2
Strand Diameter	1.35 mm
Number of Strands	7
Number of Filaments (30 μmϕ) per strand	1025
Critical Current:	
At 4.4 T (4.2°K)	11760 A
At 6.0 T (4.2°K)	8610 A

Figure 2 shows the schematic drawing of the arrangement of the vacuum chamber, an axial support, the LN_2 thermal radiation shield, the liquid helium cryostat, and the coil at one end of the solenoid module. The coil support system is a modification from that of the CDF solenoid.[7] An enlarged schematic drawing of the coil winding at the coil end is shown in Fig.3. The conductor will be wound helically on the inner shell of the helium cryostat with FRP insulator. The outer support shell and the other coil section will be assembled with a shrink-fit after the outer FRP layer is machined smoothly.[8] The schematic drawing of the cross section of the AMY conductor is shown in Fig.4.

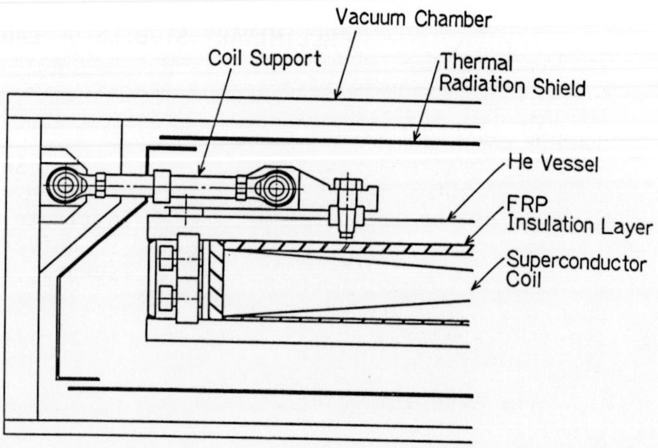

Fig. 2. Schematic drawing of the end section of the solenoid module.

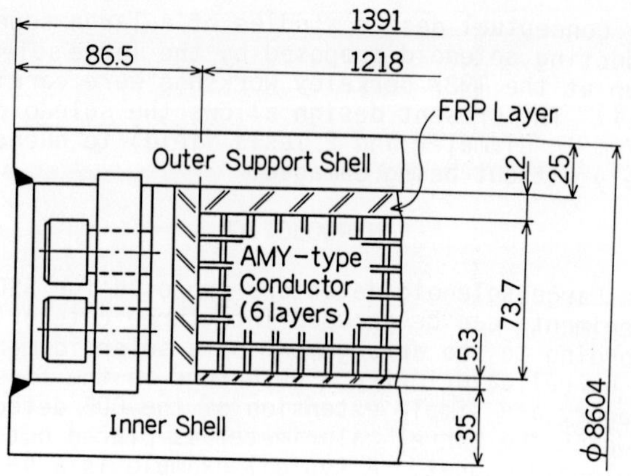

Fig. 3. Enlarged view of the liquid helium cryostat and superconducting coil at the coil end.

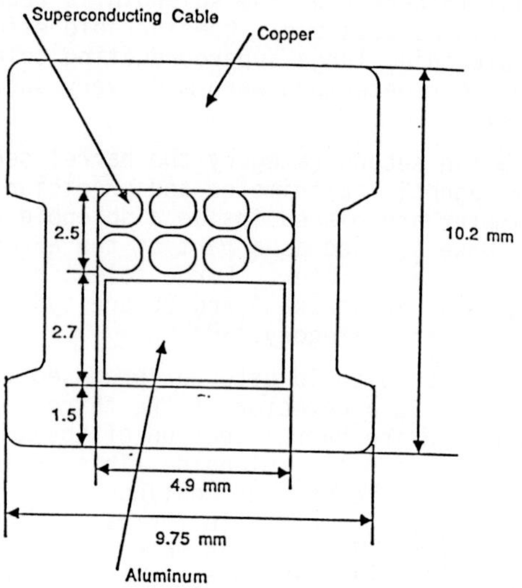

Fig. 4. Cross section of the AMY conductor.

QUENCH PROPERTIES

The magnet must be designed to be safe against quenches which can be caused by malfunctioning of the refrigeration system, for example. The safety requirements we impose for the coil during quenches are as follows:
(a) the local maximum temperature rise must be less than 100 K, and (b) the maximum voltage to ground must be less than 1000 V.

The first requirement is important to minimize inhomogeneous mechanical stresses due to sudden temperature rises during quenches. The thermal expansion coefficients of metals rise appreciably above 100 K.

The schematic diagram of the quench protection circuit is shown in Fig.5, where L is the coil inductance, R_d a protection resistor, SW a switch, and PS a power supply. We assume that the coil

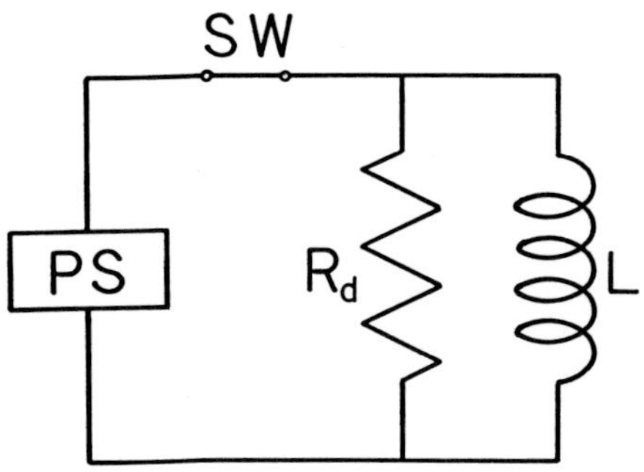

Fig. 5. Schematic drawing of the quench protection circuit.

resistance due to quenches is negligible. When the switch is opened, namely when the protection circuit is activated, the maximum voltage V_{max} is given by $V_{max} = R_d \times I_0$, where I_0 is the normal coil current. With the current decay constant $\tau = L/R_d$, we have a relation

$$\frac{I_0^2 \tau}{2} = \gamma \int^{T_{max}} \frac{C}{\rho}\, dT = \Gamma \qquad (1)$$

where γ, C, and ρ are respectively the density, specific heat, and resistivity of the conductor, the variable T is temperature and T_{max} is the maximum temperature rise. The parameter Γ depends upon T_{max} and the characteristics of the conductor. The above equation can be modified to

$$\Gamma = \frac{E_0 J_0^2}{V_{max} I_0} \qquad (2)$$

where J_0 is the current density, given by $J_0 = I_0/A$, A the conductor cross section area, and E_0 the coil stored energy, given by $E_0 = L \times I_0^2/2$. For T_{max} = 100 K the left-hand side of the equation is estimated to be approximately 1×10^{17}(Amp^2·s/m^4) for the AMY conductor.

The current density of the AMY conductor is 59 Amp/mm^2 for I_0 = 5000 Amp which corresponds to $E_0 J_0^2 = 40 \times 10^{23}$ J.Amp^2/m^4 for the whole solenoid Then, the maximum voltage becomes approximately 8000 V when the eight modules are connected in series to a single power supply. Therefore, this power supply arrangement does not satisfy the maximum voltage requirement of less than 1000 V. In the present studies we propose a power supply arrangement in which each module is connected to a separate power supply. Then, the maximum voltage becomes about 1000 V. The above calculation is consistent with the so-called hot spot model in which no thermal diffusion inside the conductor is assumed. Since Γ is a function of T_{max}, Eq. (2) gives the relation between V_{max} and T_{max} when E_0, J_0, and I_0 are fixed. Figure 6 shows this relation.

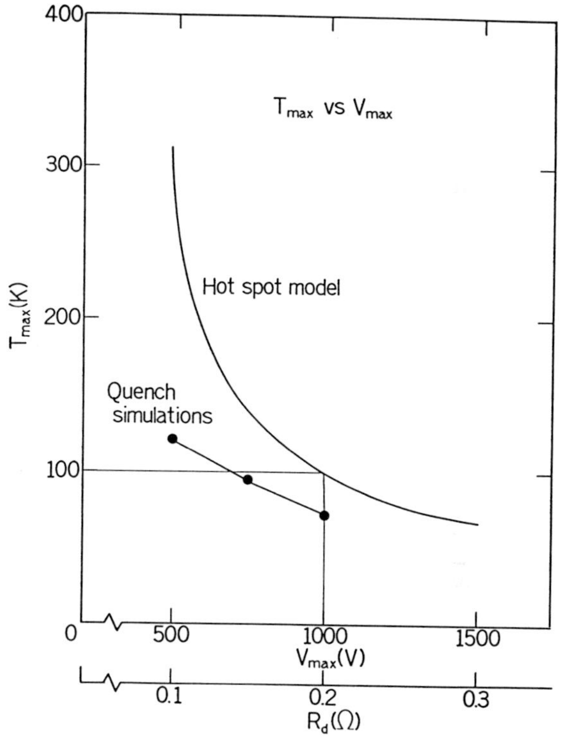

Fig. 6. T_{max} and V_{max} relation calculated by the hot spot model and quench simulations.

For T_{max} = 100 K and I_0 = 5000 Amp, Eq.(2) yields

$$V_{max} \propto J_0^2$$

Therefore, it is possible to have different power supply arrangements with fewer units.

In the calculations given above, the growth of the normal section of the coil due to a quench was not taken into account and the resistivity of the coil section was assumed to be negligible compared with R_d. In order to estimate this effect computer simulations of quenches were carried out. Figure 7 shows calculated current, internal voltage induced by the normal section, and maximum temperature rise as a function of time. We assumed that a quench started in one of the modules at t = 0 and a quench protection circuit was activated immediately. The dump resistor R_d was assumed to be 0.15 Ω. Since the current decayed faster than the rise of the internal voltage, the maximum voltage was 750 V which was determined by $I_0 \times R_d$ (= 5000 × 0.15). The maximum internal voltage reached to about 200 V and the maximum temperature rise was about 100 K. The current of one of the other seven modules which did not quench is also shown in Fig. 7. The current decay constant is slightly slower than that of the module with the quench due to absence of the normal section in the coil.

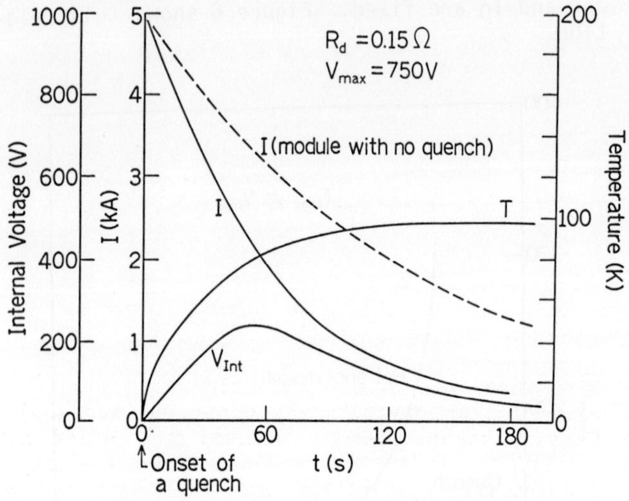

Fig. 7. Results of quench simulations for R_d = 0.15 Ω.

The maximum temperature rises obtained by quench simulations are shown in Fig. 6 for R_d=0.1, 0.15, and 0.2 Ω. The maximum temperature rise tends to be higher for a smaller dump resistor, whereas the maximum voltage is proportionally smaller. The results of the quench simulations give appreciably lower maximum temperature rises compared with those calculated by the hot spot model which tends to give conservative results. This is due to the fact that the thermal diffusion indeed lowers localized temperature rise.

REFRIGERATION SYSTEM

In the present design each module has a separate chimney, and eight chimneys are connected to a common control dewar with a liquid helium reservoir of about 1500 ℓ.

CONCLUSION

A large, 2 Tesla superconducting solenoid magnet of 8.7 m in diameter and 16 m in length with the stored energy of 1 GJ can be built with satisfactory results for safety requirements against quenches. The magnet consists of 8 modules. Fully stabilized, cryostable AMY-type conductor is used with a pool boiling cryostat arrangement. Each module is connected in parallel to a separate power supply through a common control dewar.

The author wish to thank Drs. H. Hirabayashi, R. Kephart, and D. Theriot for valuable discussions. He acknowledges cooperation of Hitachi Ltd. for the present work.

REFERENCES

1) Hanson, G.G., et al., "Report of the Large Solenoid Detector Group," Experiments, Detectors, and Experimental Areas for the Supercollider (World Scientific, Singapore, 1987), pp. 340-387.

2) Cost Estimate of Initial SSC Experimental Equipment, SSC-SR-1023, SSC Central Design Group (June 1986).

3) See, for example, H. Hirabayashi, "Detector Magnets in High Energy Physics," IEEE Trans. on Magnetics, 24 1256 (1988).

4) Feldman, G.J., et al., "4π Detectors," Proc. of the 1984 Summer Study on the Design and Utilization of the Superconducting Super Collider, edited by R. Donaldson and J.G. Morfin, Snowmass, CO (1984), p.623.

5) Theriot, D., et al., "4π Detectors," Proc. of the 1986 Summer Study on the Physics of the Superconducting Supercollider, edited by R. Donaldson and J. Marx, Snowmass, CO (1986), p.351.

6) Tsuchiya, K., et al., "A 3 Tesla Superconducting Solenoid for the AMY Particle Detector at TRISTAN," KEK Preprint 86-63, October 1986.

7) Minemura, H., et al., "Construction and Testing of a 3 m Diameter Superconducting Solenoid for the Fermilab Collider Detector Facility (CDF)," Nucl. Instr. and Meth. A238 18 (1985).

8) Minemura, H., et al., "Feasibility Test of a Shrink-Fit Assembly of a Large-Diameter Superconducting Solenoid for a Colliding Beam Detector," Nucl. Instr. and Meth. 219 472 (1984).

SSC DETECTOR SOLENOID

R.W. Fast, J.H. Grimson, R.D. Kephart, H.J. Krebs, M.E. Stone, E.D. Theriot, and R.H. Wands

Fermi National Accelerator Laboratory
Post Office Box 500
Batavia, Illinois 60510

ABSTRACT

A detector utilizing a superconducting solenoid is being discussed for the Superconducting Super Collider (SSC). A useful field volume of 8 m diameter x 16 m length at 1.5-2 T (~1 GJ at 2T) is required. It has been decided that all of the particle physics calorimetry will be inside the bore of the solenoid and that there is no need for the coil and cryostat to be "thin" in radiation lengths. An iron yoke will reduce the excitation required and will provide muon identification and a redundant momentum measurement of the muons. We have developed a conceptual design to meet these requirements. The magnet will use a copper-stabilized Nb-Ti conductor sized for a cryostable pool boiling heat flux ~0.025 W/cm^2. A thermosiphon from a storage vessel above the cryostat will be used to prevent bubble stagnation in the liquid helium bath. The operating current, current density, coil subdivision and dump resistor have been chosen to guarantee that the coil will be undamaged should a quench occur. The axial electromagnetic force will be reacted by metallic support links; the stainless steel coil case will support the radial force. The 5000 metric tons of calorimetry will be supported from the iron yoke through a trussed cylindrical shell structure separate from the cryostat. The coil and case, radiation shield and stainless vacuum vessel would be fabricated and cryogenically tested as two 8-m sections. These would be lowered into the underground experimental hall and installed into the iron flux return yoke to provide the required 16-m length.

INTRODUCTION

The SSC will have six interaction regions which, depending on the laboratory location, could be as deep as 120 meters. A large superconducting solenoid is being considered as a part of a detector facility for one of the interaction halls. Figure 1 is the SSC detector. The calorimetry and central tracking chambers, which will be located inside the solenoid, require a field volume 8 m in diameter by 16 m long. The field in the bore will be 2 T; the field in the iron flux return yoke will be ~1.5 T. Locating the calorimetry internal to the solenoid eliminates the need to have a "thin" coil and cryostat in terms of radiation lengths; we may therefore employ a cryostable pool boiling coil.

The required field volume will be provided by two solenoids, each 9.5 m in diameter and 8 m in length, connected in series electrically with a common power supply. Each 8-m unit will contain four identical 2-m long liquid helium/coil modules. The modules are mechanically connected and share a common vacuum vessel. Helium is supplied by thermosiphons to an 8-m assembly from a storage vessel located on top of the iron, with each of the four modules having its own helium supply and return piping.

The calorimetry and flux return iron will be split at the longitudinal center and the halves can be moved independently; therefore, we will use two 8-m support structures to support the 5000 metric tons of calorimetry and central tracking chamber. There are notches in the iron for hangers for the supports and for instrumentation cables. The support structures are immediately inside the solenoids but are completely independent of the cryostat and vacuum shell. The ends of each structure will be supported from the iron. The support structures, consisting of concentric stainless steel cylinders with a trussed web, having a radial thickness of 25 cm are expected to have a deflection of ~3 cm when loaded with the tracking chambers and calorimetry. The calorimeter will rest on rails and can be inserted from either end.

QUENCH PROTECTION

Four 2-m modules are connected in series with the superconducting bus between the modules being in liquid helium filled interconnecting pipes. It is essential that the coil survive quenches without damage and to this end a low current density and an external fast dump resistor were chosen. This resistor was chosen to be 0.1 Ω to limit the discharge voltage to 500 V across the terminals.

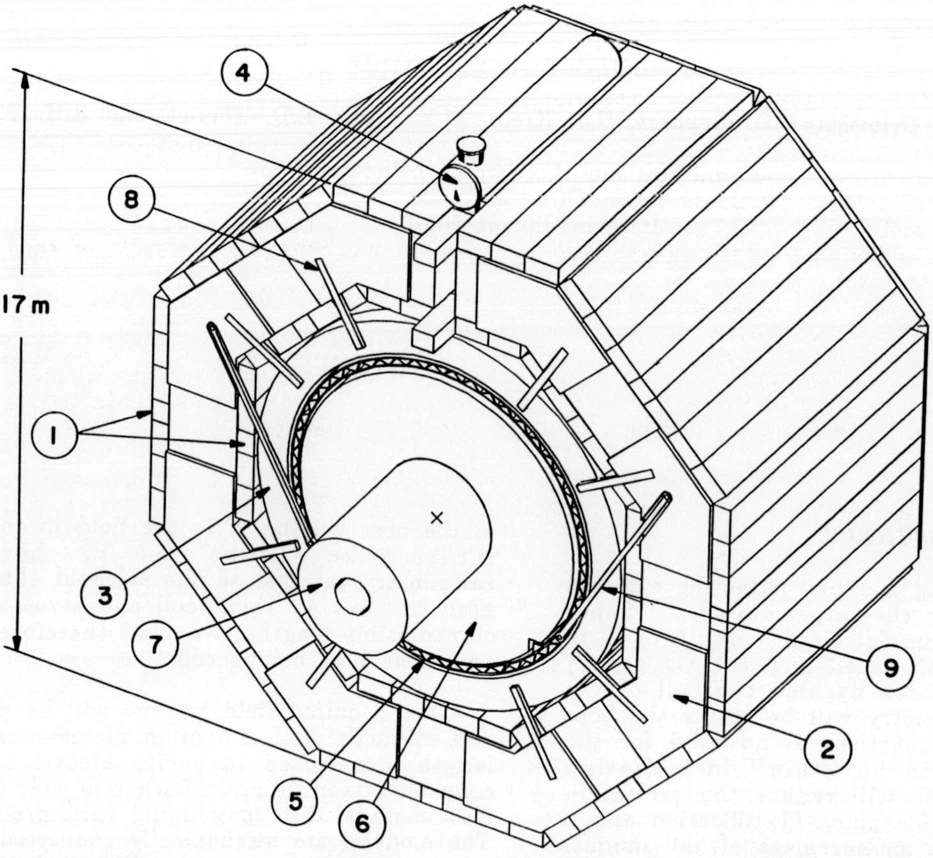

Figure 1. SSC Detector: 1, muon tracking; 2, flux return iron; 3, coil assembly; 4, 5000 L helium dewar; 5, calorimeter support structure; 6, calorimeter; 7, central tracking chambers; 8, coil supports; 9, support structure hangers.

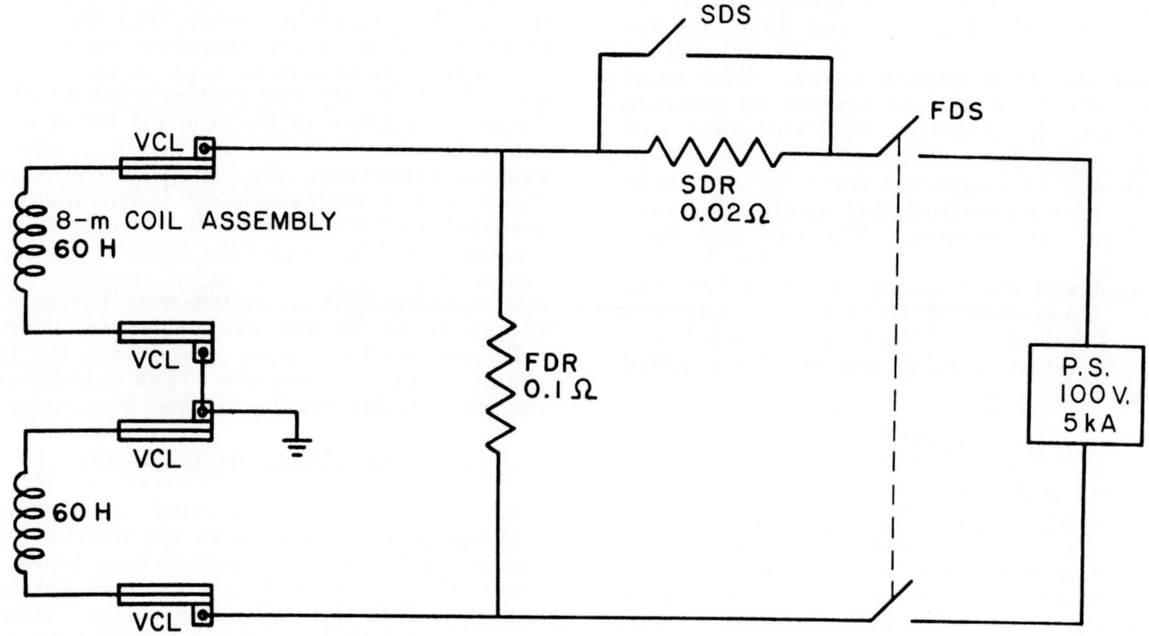

Figure 2. Electrical Schematic: FDR, fast dump resistor; FDS, fast dump switch (NC, open to initiate a fast dump); SDR, slow dump resistor; SDS, slow dump switch (NC, open to initiate a slow dump); VCL, vapor cooled leads; P.S., power supply.

The eddy current heating in both the coil and helium vessel during a fast dump is expected to be ~1300 W for an 8-m assembly. This might be sufficient to quench the coil and so the discharge will be through the fast dump resistor only if a normal zone is detected. The highest temperature reached in the coil due to a quench will be 100 K. A discharge for any other reason will be through a 0.02 Ω slow dump resistor with the power supply reversed to maintain a constant discharge voltage. The routine charge or discharge time, with 100 V across the terminals, would be 100 min and the eddy current heat load would be 26 W per 8-m assembly. Figure 2 is the electrical schematic.

CONDUCTOR AND COIL SPECIFICATIONS, WINDING AND ASSEMBLY SCHEME

The superconductor will be copper stabilized Nb-Ti wire which has a short sample current of 10 kA at 3 T and 4.5 K. This wire will be soldered into additional copper stabilizer. The final conductor will be 26 x 18 mm and will have a copper to superconductor area ratio of 146 to 1. The operating current will be 5000 A resulting in a current density of 1.56 x 10^5 A/cm^2 in the superconductor and 1068 A/cm^2 overall. The surface heat flux, with 25% wetting and a RRR of 100, is 0.025 W/cm^2.

There are seven conductor layers having a total of 651 turns in each 2-m coil module. Since the forces are radially outward, the coil will be wound starting with the outside layer, using the outer shell and flat annular heads as a coil form. Insulation will consist of G-10 buttons between turns, slotted G-10 sheets between layers and slotted G-10 and Kapton adjacent to the helium vessel. The packing factor will be 74% with 13% being helium space.

Each 2-m coil module will be designed in accordance with Section VIII of the ASME Boiler and Pressure Vessel Code for an internal pressure differential of 0.8 MPa (116 psia) and will be adequately relieved for a loss of insulating vacuum or a quench of the coil. The axial electromagnetic force will be transmitted through the shells to the supports, and the radial force will be resisted by both the conductor and outer shell.

Winding will be done with the coil axis vertical. The winding fixture will provide the radial preload and a series of compression bars at one end of each coil can be individually adjusted to provide a circumferentially uniform axial preload. These axial compression bars are adjustable only during assembly. After winding, the inner shell of the helium vessel will be welded to the coil form. After the four 2-m modules are bolted together and interconnections are made, the outer vacuum jacket and integral nitrogen shield are lowered over the 8-m cold mass. The coil supports will be attached, the 8-m assembly rotated to a horizontal position and the inner vacuum shell and nitrogen shield slid in. Figure 3 is a cross section of the coil, cryostat and vacuum vessel.

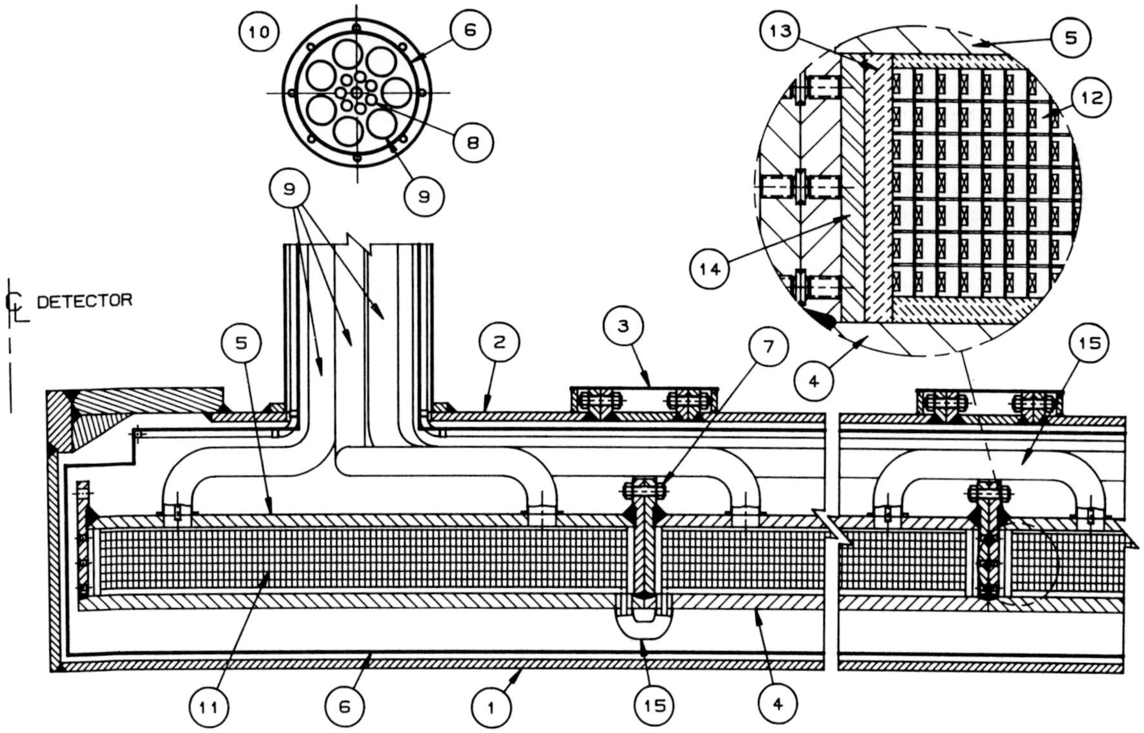

Figure 3. Axial cross section of coil, helium and vacuum vessels: 1, inner and 2, outer, vacuum shells; 3, assembly joint (if required); 4, inner and 5, outer helium vessel shell; 6, radiation shield; 7, coil module attachment; 8, liquid helium supply pipe; 9, helium return/vent pipe; 10, chimney to storage dewar; 11, coil winding; 12, conductor; 13, G-10 insulation; 14, axial preload bar; 15, electrical interconnect pipe.

SUPPORTS

The axial body force on the 8-m cold mass is dependent on the geometry of the end wall and studied thus far, calculations indicate that this force, which is towards the symmetry plane, cannot be eliminated but is a minimum of 7.3 MN (1.64 x 10^6 lb) when there is no re-entrant iron. The supports are designed for an axial or radial misalignment, or the equivalent due to iron non-homogeneity, of 2.5 cm. The force constants are 8.8 MN/m (5 x 10^4 lb/in) for radial and 149 MN/m (8.5 x 10^5 lb/in) for axial misalignment. The supports to the cold mass have been chosen to be metallic and to have both a forced flow liquid nitrogen and a thermosiphon liquid helium intercept.

The supports for the cold mass can be either combined function or separated function with different elements to react the axial and radial forces. We initially favored a combined function support because by properly adjusting the angle of the support with the axial direction, the force due to thermal contraction could be eliminated, but, since these supports are at an angle to the axial direction, they generate a large radial buckling force while reacting the axial body and decentering forces. This disadvantage does not occur with a separated function support, and therefore we now favor this style. An additional advantage is that both elements of the separated function supports can be much longer and therefore greatly reduce the heat load.

The warm ends of all supports will be attached near the ends of the vacuum vessels to avoid transmitting the loads and weight through the shells. Both the axial and radial supports are located on the outer diameter of the helium vessel. The axial supports will have their cold ends connected near the inside bolted flange on the outboard 2-m module; whereas the cold ends of the radial supports will be attached near either end of the 8-m assembly.

REFRIGERATION AND CRYOGENICS

The thermal shield cooling tubes and the support intercepts are fed with forced flow subcooled liquid nitrogen at an average temperature of 83 K. The inner and outer shields and each set of supports will have independent heat intercept circuits. The expected heat load to the nitrogen system could be as much as 5 kW for the complete 16-m magnet.

A refrigerator at the surface supplies helium to two 5000-L storage dewars on top of the iron. A cold compressor will be used if needed to maintain the storage dewar at about 30 kPa and 4.5 K. Each dewar has supply and return lines to the four modules in one 8-m assembly. These lines are sized such that the return helium is less than 1% gas by weight and the supply lines are carefully insulated to ensure the maximum liquid fraction entering the bottom of the magnet. The steady state thermosiphon flow will be about 25 g/s to each module. Separate helium circuits, which will have a return flow of less than 7% by weight, will intercept the supports. The heat load from the separated function support system might be as high as 100 W depending upon the design. The total steady state heat load for the magnet system could be 410 W (230 W plus 36 L/h for lead flows). The 4.5 K refrigerator will have a capacity of 1600 to 1800 W.

In order to facilitate cooldown, a separate refrigerator would probably be employed. It would consist of a GHe/LN_2 heat exchanger and a turbo-expander which could provide 400 g/s of 55 K helium gas. This would be adequate to cool down the total cold mass (both 8-m sections) of 1000 tons in approximately two and a half weeks.

CONCLUSIONS

Preliminary work has indicated that the size and field requirements of this magnet are a reasonable extrapolation of existing superconducting magnet technology. The magnet design must be very conservative to guarantee high reliability, since the time to fix even a small problem could have a major impact on the SSC physics program. More detailed study must be done to optimize parameters both in terms of reliability and cost-effectiveness. Fabrication techniques must be examined more closely and the necessity of on-site construction must be rigorously evaluated. However, at this point, we see nothing that would preclude the construction of such a magnet.

ACKNOWLEDGEMENTS

This work was sponsored by Universities Research Association, Inc. under contract with the U.S. Department of Energy.

A VERY HIGH FIELD MAGNET

S.Terada, K.Tsuchiya
H.Hirabayashi and A.Maki

KEK National Laboratory for High Energy Physics
Tsukuba, Ibaraki 305 Japan

T.Oba, T.Doi and H.Kakui

IHI Ishikawajima-Harima Heavy Industries Co., Ltd.
Chiyoda, Tokyo 100 Japan

Presented by A. Maki

ABSTRACT

The technical possibility of a very high field solenoidal magnet was investigated for the use of the experiment at the SSC. A magnet of 2-meter in diameter, 2.5 meter in length, and 6 tesla in the central field was designed. Its mechanical and thermal properties were studied.

INTRODUCTION

Superconducting solenoidal magnets are becoming essential components of modern experiments at particle colliders. As the collider machine becomes larger and larger, the detector which is used at the collider simultaneously becomes huge. This trend is, in some sense, natural because particles to be detected with the detector have higher momenta. On the other hand, the technologies of superconductivity have developed significantly in the last several years. By using a very high field solenoidal magnet, experimenters can enjoy several advantages over large scale detectors.

The only working example of the compact detector which utilizes a high field superconducting magnet is the AMY experiment[1)] at the KEK TRISTAN in Japan. The AMY magnet[2)] is about 2.5 meter in diameter, 1.5 meter in length and 3 tesla in the central field. As the result, the AMY weights only 800 tons compared to 2000 to 2500 tons of its competitors at TRISTAN. AMY has achieved very good momentum resolution, $\delta p_t/p_t \sim (0.7\%)p_t$ in spite of its small volume for tracking. Also due to the short flight path for hadrons ($\sim 80cm$), the muon background coming from the hadron decay-in-flight is as small as 0.025/p. In addition, it was easy to reduce the punch through background by increasing the thickness of the hadron absorber (up to 9 - 10.5 absorption lengths) without enormous additional cost for the iron. Two more advantages the AMY group are planning to extract from the high magnetic field and the small size of the detector are;
1) additional means to identify electrons from hadrons by using the synchrotron radiation from electrons in the strong magnetic field.
2) high luminosity configuration of the beam focusing magnets which could be placed as close as 1.7 meter from the interaction point.

DESIGN OF MAGNET

Requirements

Encouraged very much by the recent performance of the AMY experiment, people are investigating the possibility of a compact detector at the SSC based on the same philosophy as the AMY experiment[3)]. Considering the technological developments after the AMY magnet, the central field strength was chosen to be 6 tesla, twice as high as the AMY field. The physical demensions of the magnet were designed to be 2 meter in diameter and 2.5 meter in length. One more important parameter was the thickness of the magnet. Although we placed an electromagnetic calorimeter inside the magnet in the same way as the AMY experiment, we needed a hadron calorimeter outside the magnet for the SSC experiment. We assumed that the maximum allowable thickness of the magnet should be less than a half absorption length, $\sim$ 20 cm of aluminum. Other parameters for the design are given in table 1.

Coil Structure

From the simple formulas,

$$NI = \frac{B}{\mu_0} l \quad and \quad P = \frac{B^2}{2\mu_0}$$

we could easily find that this magnet needs about 4.8 x 10^6 ampere-turns and that it receives about 150 atmo-

Table 1: Requirements for the very high field magnet

Coil		
Inner Diameter	2.0	m
Length	2.5	m
Cryostat		
Inner Diameter	> 1.8	m
Outer Diameter	< 2.4	m
Total Thickness	< 20	cm (Al equiv.)
Central Field Strength	6.0	Tesla
Cool Down Time	< 1	week
Safety Margin	> 20	%
Rated Current	$\leq 10,000$	A
Induced Voltage a/c Coil	$< 2,000$	V
In Case of Quench		
Temp. Rise	< 100	K
Local Temp. Rise	< 200	K

spheres of pressure due to the magnetic field. This huge magnetic pressure has forced us to change the basic idea of designing the containing structure of the coil. The basic idea of this magnet is to adopt the same multilayer cylinder as that commonly used for high pressure gas vessels. As the rated current was set at about 10,000 amperes for the sake of practical operation, the coil is wound in 10 layers as 167 x 10 turns. Each coil layer is held by a single aluminum cylinder, the thickness of which varies from 13 to 1.5 milimeter depending upon the expansion force of the conductors. The conductor is assumed to be Ni/Ti superconductor, cross sectioal area of which is scaled from the conductor of the AMY magnet[2]. The wire has a copper stabilizer. The iron for the magnetic return yoke is assumed as shown in figure 1, although they give very little effect in terms of magnetic field and force.

Field Calculation

Using the POISSON[4] computer codes, the magnetic field distribution and the magnetic forces acting on the coil were calculated. The results are given in table 2. The radial and axial forces acting on each layer of the coil are shown in table 3. As expected by common sense, the radial forces decrease as the radius increases and the directions of the force are reversed at the last two layers. The axial compression forces are pretty much the same for different layers. Since the iron yokes are far away from the coil, the interaction between the iron and the coil is small. The decentering forces caused by possible misalignment of the coil relative to the iron were estimated with the POISSON and JMAG[5] codes. They are 1.5 and 0.75 tons respectively in axial and radial directions for 10 mm misalignment. These are very small compared with the weight of the coil itself ($\sim$ 8 tons). The stresses on the superconducting wires and the aluminum cylinders are shown in figure 2 together with the displacements of the superconducting wires due to the magnetic pressure.

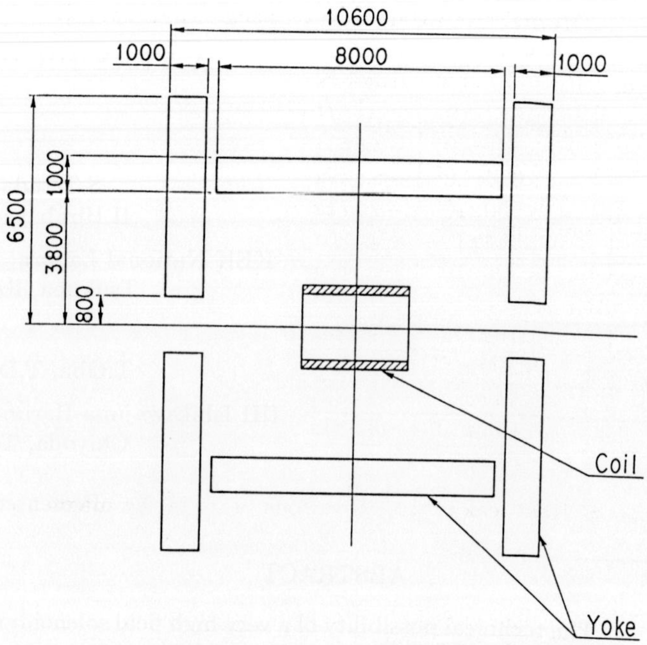

Figure 1; Arrangement of the coil and the return iron yoke. The space between them is filled with the hadron calorimeter. The electromagnetic part of the calorimeter is supposed to be placed inside the coil.

They are all well below the allowable stresses of the materials used.

Thermal Properties

Thermal properties of the magnet were studied with the computer code NASTRAN[6]. By assuming the coil structure shown in figure 3, the temperature differences in various parts of the magnet were estimated. As the input parameter, the temperature of the incomming coolant, two phase He, was taken to be 4.2 K. Then the highest temperature in the coil is 4.32 K which gives the superconducting wire the safety margin of the load line more than 20 %.

The initial cooling of the coil from the room temperature was simulated. A 10 x 18 inner dimension coolant pipe is soldered both on the inner and outer surface of the coil structure as shown in figure 4 a). The thermal con-

Table 2: Results of the field calculation with the POISSON code

Total Ampere Turn	$1.55\text{x}10^7$	AT
Central Field	6.0	Tesla
Max. Field	6.3	Tesla
Total Radial Expansion Force	12,700	Tonne
Total Axial Compression Force	5,850	Tonne
Inductance	3.4	H
Stored Magnetic Force	155.3	MJ

Table 3: Radial and axial forces on each individual layers of the magnet

Layer	Radial Force (kg/rad)	Axial Force (kg/rad)
1	$4.591x10^5$	$-8.643x10^4$
2	$4.086x10^5$	$-8.949x10^4$
3	$3.550x10^5$	$-9.171x10^4$
4	$2.984x10^5$	$-9.326x10^4$
5	$2.395x10^5$	$-9.431x10^4$
6	$1.785x10^5$	$-9.494x10^4$
7	$1.162x10^5$	$-9.525x10^4$
8	$5.290x10^4$	$-9.535x10^4$
9	$-1.117x10^4$	$-9.522x10^4$
10	$-7.622x10^4$	$-9.481x10^4$

Table 4: Heat leaks through various parts of the magnet

unit (W)

No.	Item	80 K LN_2 Shield	4.2 K L Temp	4.2 K H Temp
1	Rad fm Coil & Shild	61.68	7.76	
	Rad fm LH_e Trans Tube	7.55	10.0	
	Rad fm Current Leads	1.28	0.17	
2	Coil Support Rods	67.72	6.44	
	Shield Support Rods	12.22		
3	Current Leads			31.4
4	Others (plus 20 %)	29.55	9.22	
	Total	180	33.6	31.4
	Capacity of Refregirator	250 W	40 W	40 W

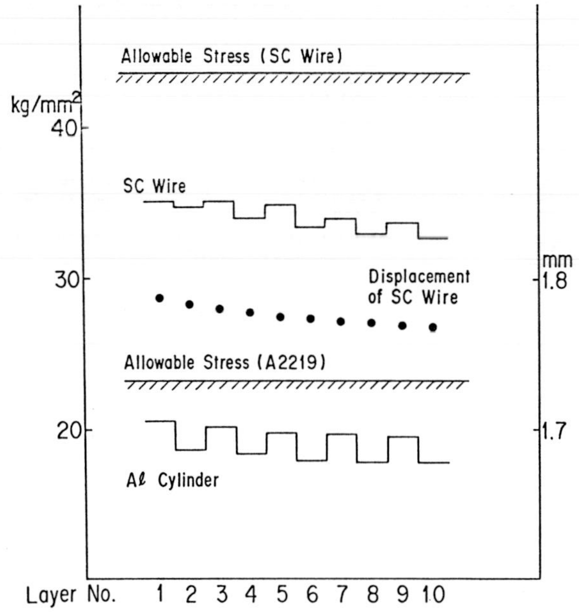

Figure 2; Stresses on the superconducting wires and the supporting aluminum cylinders as a function of the layer. The displacement of the wires are also given.

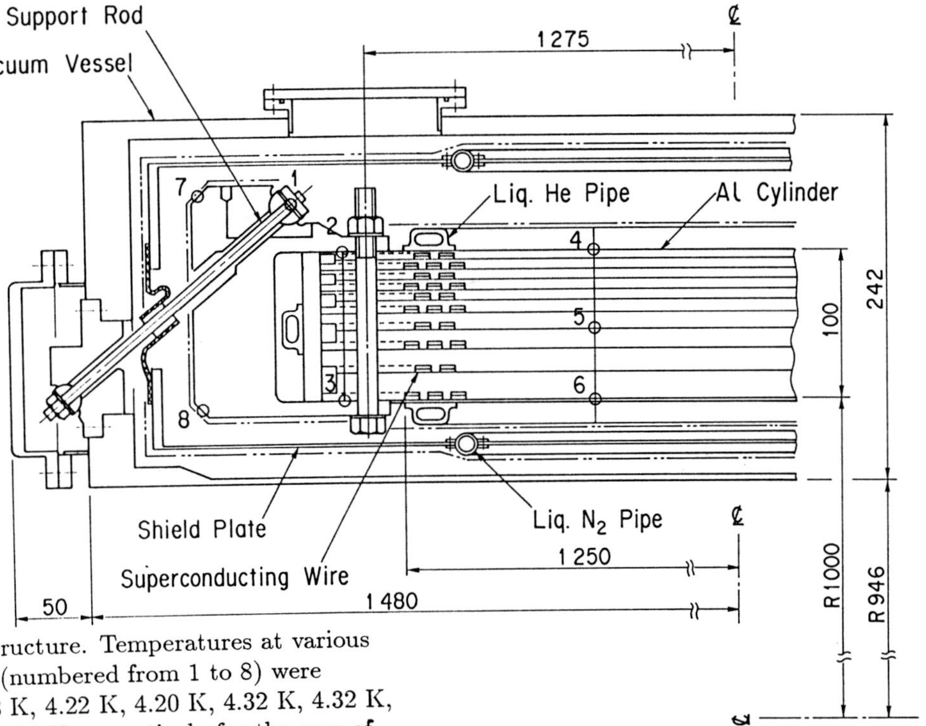

Figure 3; The coil structure. Temperatures at various locations of the coil (numbered from 1 to 8) were estimated to be 5.53 K, 4.22 K, 4.20 K, 4.32 K, 4.32 K, 4.32 K, 5.24 K and 4.24 K respectively for the case of the 4.2 K input coolant temperature flowing in the liquid helium pipe.

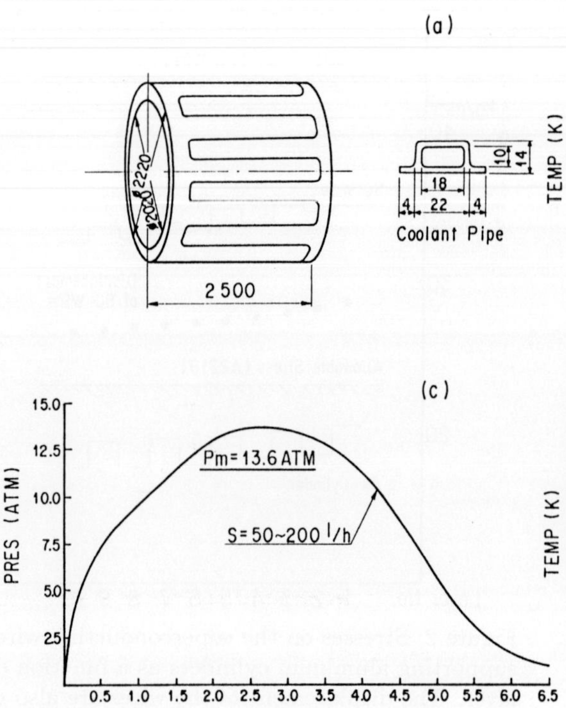

Figure 4; a) The coolant pipe arrangement. b) The coolant temperature and its mass flow as a function of time during the initial cooling. c) The pressure variation of the coolant pipe. and d) The temperature change of the coil.

tact between the pipe and the coil surface was assumed to be 20%. As shown in figure 4 b), the coolant temperature was lowered from 200 K to 4.2 K in four days. Accordingly the mass flow was increased from 25 to 200 l/h and was kept at this level afterwards. figure 4 c) and d) show the pressure variation of the coolant gas and the temperature decrease of the part of the coil where is the furthest from the coolant pipe. As one can see, the coil cools down to 4.3 K within a week.

Heat leaks through various paths were analysed and estimated. The results were given in table 4. The heat load through the current lead was estimated by minimizing the sum of the conductance and the Joule heating with the formula ;

$$Q = \frac{A}{l}\lambda\Delta T + \rho\frac{l}{A}I^2.$$

By considering the worst case like the accidental reduction of the coolant, we may better have a thicker current lead. In the table, the heat load through the transfer tubes were not considered. Therefore we may need a larger refregerator for actual operation.

Table 5: Materials in the radial path (aluminum equivalent)

Vacuum Chamber	:	Inner Wall	2.96 mm
		Outer Wall	11.82 mm
Shield Plate	:	Inner Plate	2.00 mm
		Outer Plate	2.00 mm
Al Cylinders	:	Average	70.13 mm
NbTi Filament	:	Average	24.00 mm
Copper	:	Average	82.31 mm
Total	:		195.20 mm

Thickness of Magnet

The total thickness of the coil was analyzed in table 5. The largest contribution comes from the copper stabilizer of the superconducting wires. That may be reduced by replacing the copper stabilizer with the aluminum one. One can see that it is possible to make the total thickness of the coil to be less than a half absorption length even with the copper stabilizer.

Summary of Magnet Parameters

The main parameters of the magnet are listed in table 6 together with those of the AMY magnet for comparison. Although the central field strength is twice as high as that of the AMY magnet, the maximum field at the coil is not quite so, 6.3 tesla vs. 4.0 tesla. Accordingly the magnetic pressure is just about 3 times larger than that of the AMY magnet.

CONCLUSION

A design study of a very high field solenoidal magnet was done for the possible use in the experiment at the SSC. It seems possible to build, without serious techinical problem, a solenoidal coil with 6 tesla central field, 2 meter in diameter and 2.5 meter in length, and less than a half absorption length in thickness. More studies are needed for the quench properties and some other detailed designs of current lead, transfer tube, power supply and others.

REFERENCES

1. Back, C. et al., "AMY-Chan, a High Resolution Lepton Detector for TRISTAN", TRISTAN-EXP-003 KEK (1984); Sagawa, H. et al. (AMY), Phys. Rev. Lett. 60 (1988) 93 - 96.

2. Tsuchiya, K. et al., Advances in Cryogenic Engineering 33 (1988) 33 - 40; Doi, Y. et al., Advances in Cryogenic Engineering 33 (1988) 551 - 557.

3. Kerkby, J. et al., "Report on the Compact Detector Subgroup" in Proceeding of the Workshop on Experiments, Detectors, and Experimental Areas for the Supercollider, July 7-17, 1987, Berkeley, California, pp. 388-471.

4. Cyr, M. and Iselin, C., POISSON, CERN (1976).

5. Developed by Japan Information Service Ltd., Tokyo, Japan.

6. NASTRAN, National Aeronautics Space Administration, USA.

Table 6: Comparison of the magnet parameters with those of the AMY magnet

		AMY	This Magnet
Current	(A)	5,000	9,558
Central Field	(T)	3.0	6.0
Maximum Field	(T)	4.0	6.3
Coil Inner Diameter	(m)	2.386	2.0
Coil Outer Diameter	(m)	2.584	2.2
Coil Length	(m)	1.54	2.50
Coil Winding			
# of layers		8	10
# of turns per layer		135	167
total # of turns		1080	1670
Total Ampere Turns	(AT)	5.4×10^6	1.55×10^7
Inductance	(H)	3.2	3.4
Stored Energy	(MJ)	40	155
Magnetic Pressure	(atm)	56.8	153.7
Total Axial Force	(ton)	1,734	5,850
Cryostat			
inner diameter	(m)	2.22	1.892
outer diameter	(m)	2.92	2.4
length	(m)	2.12	3.1
Heat Load (He)		33W +23l/h	34W +30l/h
Total Weight	(ton)	20	9

PROSPECTS OF TRANSITION RADIATION DETECTORS FOR ELECTRON IDENTIFICATION AND TRACKING

Y. Ducros

D.Ph.P.E. C.E.N. Saclay , 91191 Gif /Yvette Cedex, France

ABSTRACT

We present the performances of a Transition Radiation Detector for the electron identification in an experiment at the S.S.C. The X-ray detectors can be included in the tracking system.

1 - INTRODUCTION

Lepton identification provides a good signature of new phenomena in high energy proton-proton or proton-antiproton collisions. The decay of new particles such as heavy Quarks, Z', W', Higgs can be characterized by the emission of one or many leptons. Events with large missing transverse momentum without leptons signal supersymetric particles. In this case lepton identification is crucial in order to eliminate events with W production.

The detection of electrons rather than muons is advantageous because calorimetry can give a good measurement of the lepton energy. Their identification at high energy (E>20 GeV) depends crucially on the transversal and longitudinal granularity of the calorimeter. The pion rejection factor is currently of the order of one thousand, for a single track. This limitation is due to fluctuations of the energy deposition in the hadronic showers.

The isolation criteria, necessary for the identification of electrons, is difficult to satisfy in high energy proton-proton reactions at the S. S. C. where the mean number of charged tracks is more than 100 in a hard scattering. The superposition of energetic γ rays with low momentum tracks is the origin of one of the main backrounds. This pile up can be partly eliminated with a preshower detector (7) or by good matching of the position of the shower in the calorimeter with the track direction measured in the central tracking.

The Q.C.D. jet production in proton-proton collision can be the origin of an important hadronic background. As an example the Higgs particle of the standard model with a mass of 300 GeV decaying into two W's , one W going in electron-neutrino, has a cross section equal to 2 pb, 5 orders of magnitude below the two jet production with transverse momentum larger than 100 GeV.

The calorimeter and the tracking are not able to provide a rejection factor larger than few 10^4 against jets. In particular a fraction of jets fragments into one energetic neutral pion plus a few charged tracks. This process leads to the pile up problem mentioned above.

One solution consists of installing a Transition Radiation Detector to identify the particles before their destruction in the calorimeter. The T.R.D. can easily provide an additional pion rejection factor between one hundred and few thousand for low momentum charged tracks (P < 100 GeV/c)

2 - MAIN CHARACTERISTICS OF THE T. R. D.

Transition radiation is produced when a charged particle crosses the interface between 2 media. A good approximation to the rate of emitted X-rays produced by a stack of regularly spaced foils is given by the following equation (8)

$$\frac{d^2W}{d\Omega d\omega} = \frac{\alpha\,(\xi_1^2-\xi_2^2)^2\,\theta^2}{\pi\,(\gamma^{-2}+\xi_1^2+\theta^2)^2\,(\gamma^{-2}+\xi_2^2+\theta^2)^2} \times 4\sin^2\!\left(\frac{\Phi}{2}\right)\left(\frac{1-e^{-N\sigma}}{1-e^{-\sigma}}\right) \qquad (1)$$

where the two materials are noted 1 and 2 for the foils and the gas respectively. θ is the angle of the emitted X-rays, and is of the order of $1/\gamma$. $\xi=\omega_p/\omega$, with ω_p the plasma frequency of the material.

Φ is the phase difference between the 2 waves emitted at the two surfaces of the foil and has the following expression:

$$\Phi = (\gamma^{-2} + \theta^2 + \xi_1^2)\,\omega\, L_1/2$$

where L_1 is the thickness of the foil. L_1 must be sufficiently large to get a phase value of the order of π. (The "formation zone" is equal to 20 µm for the production of 6 keV X-rays in plastic foils with plasma frequency of 20 eV.)

N is the number of foils and σ the absorption in a single foil.

The spectrum of the radiation depends on the plasma frequency of the materials and on the Lorentz factor of the particle. The rate of the X-ray production is of the order of the fine structure constant α per interface, so hundreds foils or fibers of the chosen material are required for the radiator.

The above expression shows that $\xi_1 - \xi_2$ must be as large as possible, ξ_2 is very close to zero for a light gas such as Helium.

Since the photon can be reabsorbed in the succeeding foils with a cross section proportional to Z^3, the radiator material has to be chosen among the light elements.

Figure 1 shows the number of X-rays as a function of the Lorentz factor Gamma and produced with:

- An optimised radiator.
- A radiator made of thicker foils separated by larger gaps. In this second case the X-ray yield increases until a γ value of 10^4. This fact can be useful for pion identification above 150 GeV/c in proton-proton collisions at very high energy.

The angle between the direction of the emitted X-rays and the track is too small to separate the delta rays produced by the charged particle from the ionization due to the photo-absorption of the X-rays in the detector. Separation of the two phenomena using a magnetic field leads to large distances between the radiator and the detector incompatible with the construction of a central detector.

The optimisation of T.R.D.'s for high energy experiments has lead to detectors made of a few layers, each with a radiator followed by a proportional wire chamber. The rejection factor which can been obtained with these detectors is mainly a function of their total length. Fig 2 shows the variation of the pion rejection factor with the thickness of existing detectors using different radiator materials. The results can be fitted to the following expression : $R=1.2e^{0.12L}$, where L is the total length in cm.

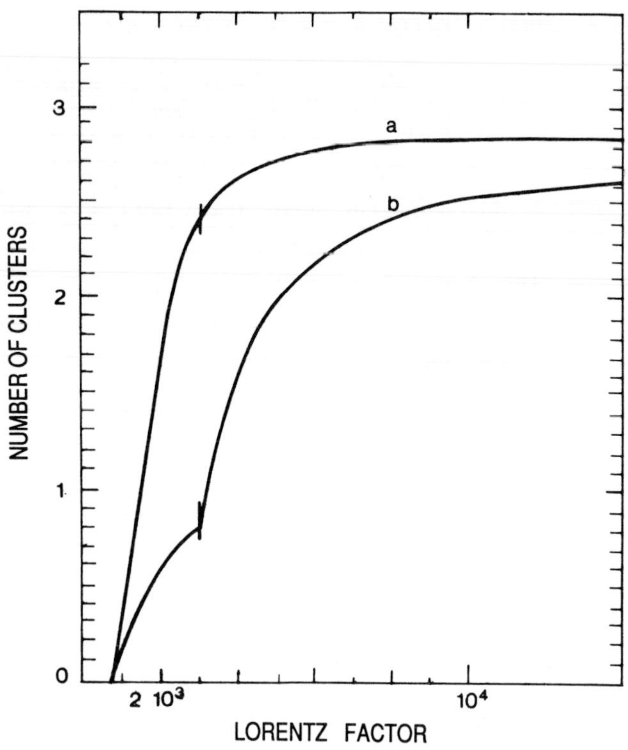

Fig. 1 - Yield of the X rays emitted from a radiator as a function of the Lorentz γ factor of the particle
a - Stack made of 353 foils 20 µm thick with a 150 µm gap between foils.
b - Stack of 156 foils 40 µm thick with a 600 µm gap between foils.

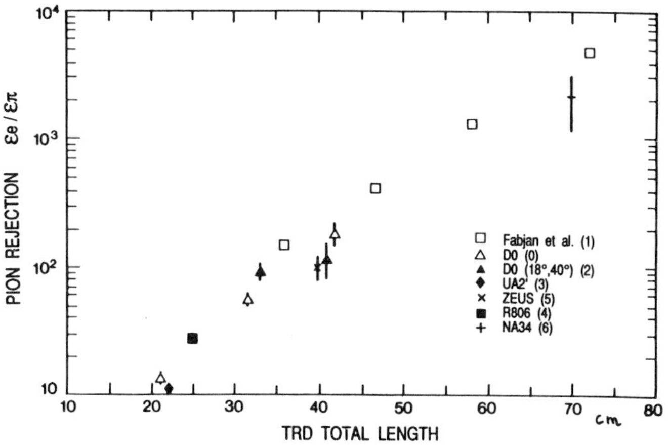

Fig. 2 - Pion rejection factor as a function of the total length of the detectors. ($\varepsilon_{el.} = 0.9$)

3 _ PROSPECTS FOR A T. R. D. AT THE S. S. C.

The construction of a T.R.D. in an experiment at the S.S.C. poses the following problems

- High rate : 1.5 interactions every 16 ns
- High multiplicity : The mean multiplicity is of the order of 70 charged particles in a proton-proton interaction,and more than 100 in a hard scattering.

The X-ray detector must have a very good granularity, and the information should be collected in a time interval shorter than a few bunch intervals.This last feature precludes the use of X-ray detectors with long drift distances. Radiation damage (aging) to the proportional counter leads to the division of the detector into small cells in azimuth and along the track. The gain of the proportional counter has to be as low as possible, compatible with the electronics (Gain $<10^4$). These few remarks show that the X-ray detectors have to satisfy the same kind of criteria as the tracking chambers.

A pion rejection factor larger than 100 requires more than 40 cm for an optimised T.R.D. Therefore the dimensions of this detector corresponds to a large fraction of the available space in front of the calorimeter. In addition the amount of material in the radiators has to be as low as possible in order to limit the γ conversions

All these remarks lead to a T.R.D. with fine segmentation of the radiators and to integration of the X-ray detectors in the central tracking. Radiator segmentation allows the construction of a T.R.D. with less material for the same number of X-rays, and helps to identify gamma conversions. The integration of the X-ray detectors in the central tracking reduces the total length required.

A solution for the construction of the radiators and their detectors is proposed in the following two paragraphs.

3 - 1 CHOICE AND CONSTRUCTION OF THE RADIATORS

Different types of radiators have been compared experimentally. Fig. 3 shows the result obtained with Lithium foils, and with CH2 foils and fibers, for the D0 T.R.D. prototype. The total length corresponds to 31.5 cm for the CH2 foils and fibers and 33.5 for the Lithium. The rejection obtained with 2 stacks of Lithium foils and with 3 stacks of polypropylene foils is very similar. On the contrary CH2 fibers appear to be less effective. These results have been obtained with the maximum-likelihood method using the total charge collected in the X-ray detectors. However other experiments (5) have published results showing equivalent rejection factors for fibers and foils.

CH2 foils or fibers have the advantage that they are easy to use, in contrast to the Lithium foils which have to be kept in a dry atmosphere. In addition a T.R.D. using polypropylene requires less space for the radiators than a T.R.D. with Lithium foils.

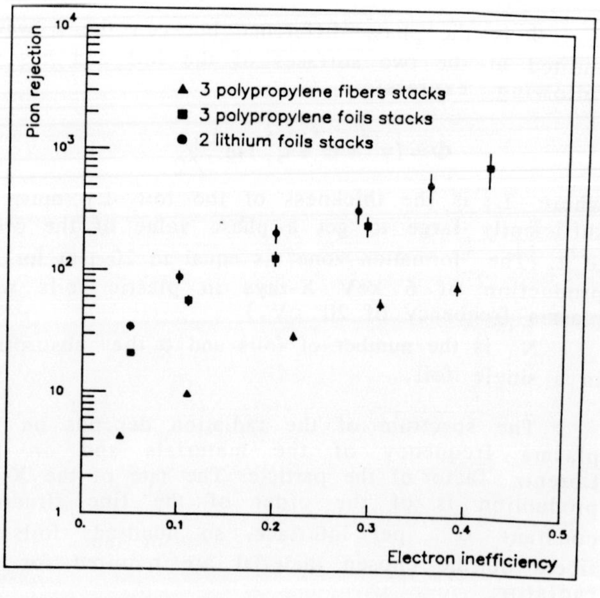

Fig. 3 - Pion rejection factor obtained with the D0 prototype for different radiator materials.

In the following we proposed using CH2 foils (Polypropylene) for the radiators of a T.R.D. at S.S.C. The distance between foils is obtained by embossing a pattern on the polypropylene (2). The gap varies from zero to a few times its mean value. The stacking of the foils is easy for plane radiators or for cylindrical shapes. In the latter case the mean value of the gap is adjusted by winding the indented foils under a constant load.

Figure 4 shows the total energy collected in an X-ray detector behind a stack of 380 polypropylene foils made according to the process described above (9). The experimental results are well reproduced by using an algorithm which takes into account the variation of the gaps and of the thickness of the foils (9). Simulation using equation 1 shows a deficit of 20% between the data and a perfect radiator with a constant gap.

3 - 2 X RAY DETECTOR

The experimental conditions at the S.S.C. constrain the X-ray detectors to have a very good granularity in the ϕ direction and along the tracks, similar to the central tracking. Measurement of the energy and of the interaction point of individual X-rays is also useful for maximum likelihood methods which can improve the pion rejection factor .

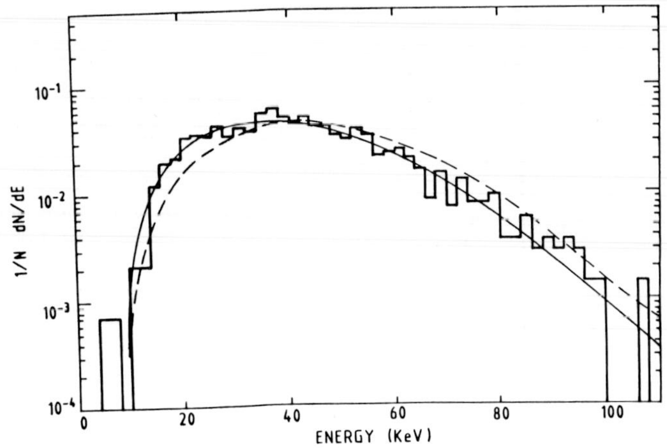

Fig. 4 - Distribution of the energy detected behind a stack 6.5 thick made of 380 polypropylene foils for a 5 Gev electron. (2) The direction of the track and the normal to the detector make an angle of 40 degrees. The distribution is the sum of the ionisation due to the charged track and the X rays detected in the chamber.

Solid line : Simulation of the X-ray production taking into account the distribution of the gap values;

Dashed line : Simulation using equation 1

Solid state detectors such as Silicon give good energy measurements but do not allow separation of the X-rays produced in one radiator. They give a measurement of the total energy emitted by the radiator for a given track. Their use in a T.R.D. requires important developments, in particular to reduce the noise and the ionisation of the charged track in the depleted zone. The amount of material in front of the sensitive volume has also to be transparent for few keV X-rays.

In the following we consider only proportional counters. These detectors need 2 to 3 cm of a heavy noble gas like Xenon. The drift time can reach 1μs for a detector adapted to the cluster counting method. This last value is clearly incompatible with the time interval between two bunch crossings.

One possible solution is to use the multidrift modules proposed by Sauli et al (10) for central tracking. The diameter of the cells can be as low as 2 mm and the radial segmentation is well adapted to the cluster counting method. (Straw tubes have the same granularity but are much less transparent to the photons). The position of the X-ray conversion point is given by the cell number. The total charge collected on the wire corresponds to the sum of the ionization energy of the charged particle in the cell and of the energy of the converted X ray in the same cell.

The detector gas has to present the following characteristics:

- Good photo electric absorption of the X ray.
- Short drift time.
- Energy loss of one track in the cell as low as possible.

Xenon is the gas which gives the best ratio between the photo-cross section and the ionizing energy deposition (including delta rays). A mixture of Xenon and CH4 (Or C2H6) 50%-50% can be used to get a drift time of 25ns per mm. CO2 can also be used as quencher, but the pulses have larger rise times and there is more loss from attachment due to Oxygen contamination.

The dilution of the Xenon leads to thicker detectors to collect the X-rays: 4 cm instead of 2 cm. The number of cells on one track varies between 10 and 20 according to their diameter. Thus the position of the clusters is known with a precision sufficient for maximum likelihood methods using the energy and the conversion point of the X-rays.

4_ RESULTS

The rejection factor has been calculated for a detector made of 4 layers. Each layer is 10 cm thick : 6 cm for the radiator and 4 cm for the detector.

The radiators are made of 350 polypropylene foils 20 microns thick. The mean value of the gaps is equal to 150 microns. The X-ray detectors are divided into 10 microdrift cells along the track. Each cell is 4 mm in diameter. This conservative value has been chosen to reduce the sharing of the X-ray energy between two cells.

The Z coordinate can be obtained using a stereo angle between different layers. For simplicity of construction we propose to divide the detector into modules of less than 1 meter length.

The position of the track is obtained by measuring the drift time in each cell. A good charge measurement with a shaper and an A.D.C. requires at least twice the rise time of the pulse. One alternative is to equip each channel with a discriminator and to register only cells which give a charge greater than a certain threshold. In this case the information is reduced to a single bit and can be collected in a shorter time. The electronics is also simpler, but we lose information.

Figure 5 shows a schematic of a the 4 cylindrical layers, extending from 30 cm to 70 cm in radius.

The rejection factor has been calculated using a simulation of the X-ray production in the radiator. The photo-absorption in the gas mixture and the energy deposition of the charged track are also simulated using the photo-absorption ionization model (11).

Figure 6 gives the distribution of the number of cells with an energy greater than 8 keV for pions and electrons of 5 GeV. The mean number of

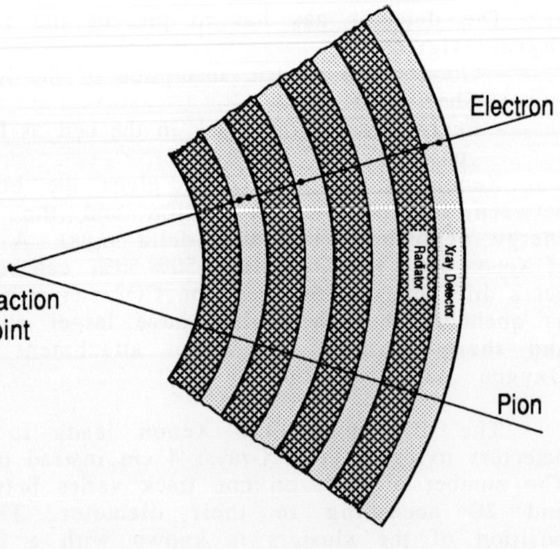

Fig. 5 - Schematic of a sector of the four layers of the T.R.D. The radiators are made of polypropylene foils wound on the inner mandrel. The X-ray detectors are micro drift cells 4 mm in diameter. The figure shows an electron track with 5 cells giving more than 8 keV and a pion track with no cell above this threshold.

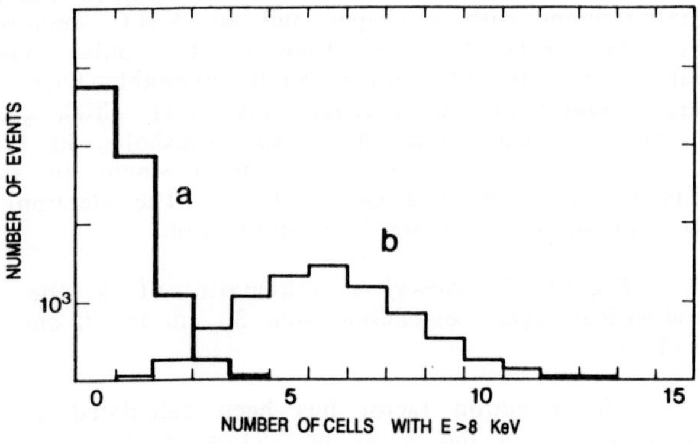

Fig. 6 - Distribution of the number of cells giving more than 8 keV for electrons (b) and pions (a). The calculation has been done with 4 T.R.D. layers for 5 Gev/c momentum particles.

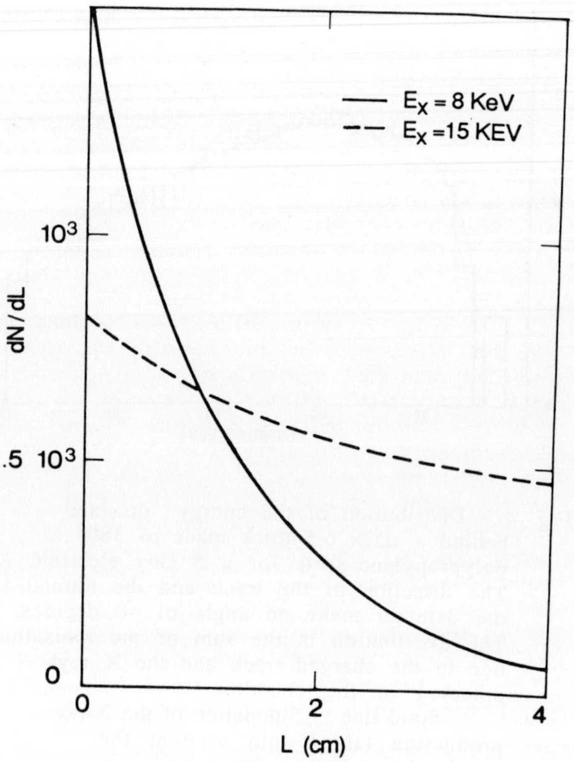

Fig. 7 - Position in depth of the X-rays converted in the chamber.
- Full line : 8 keV X rays.
- Dashed line : 15 keV Xrays.

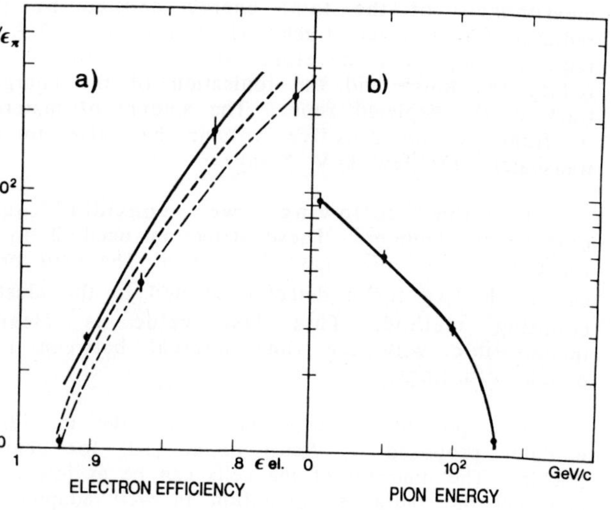

Fig. 8 - Pion rejection as a function of the electron efficiency.

Full line: 4 layers corresponding to 40 points per track

Dashed line and dotted-dashed line : 6 layers with thresholds of 6 keV and 8 keV, respectively

cells is 0.7 for pions and 6.0 for electrons. Figure 7 shows the distribution of the radial position of the X-ray clusters produced by an electron. The interaction point of the photon depends on its energy, X-ray of low energy being absorbed in the first cells of the detector. In contrast delta rays from ionization give a flat distribution. The correlation between the energy and the position of the cluster can also be used in this detector to improve the pion rejection factor.

Figure 8 shows the rejection using the number of cells only. A factor 90 can be obtained for 90% electron efficiency. In this case the position of the track is measured with 40 points. The rejection has also been calculated for a detector of the same thickness but divided into 6 layers instead of 4 ; the performance is slightly lower but the advantages are a better measurement of the tracks (60 points) and a better segmentation of the radiator with less material. (18 cm of radiator instead of 24 cm.). The same figure also shows the variation of the rejection factor with the pion energy. The slope between a few GeV and 100 Gev is due to the relastivistic rise of the ionization in the gas. Above 100 GeV the rejection drops to negligible values. Rejection for pions above 150 GeV requires longer radiators.

5 - CONCLUSION

The construction of a T.R.D. in an experiment at the S.S.C. leads to include the X-ray detector in the central tracking and to build radiators with good segmentation. Four or six layers gives a pion rejection of about 90 for a total length of 40 cm. The construction of such a detector is possible using rather conventional techniques. Pion rejection drops to zero above 100 GeV. Pion identification above 150 GeV is possible at the expense of a longer detector.

Drift time in the X-ray detectors is of the order of 40 ns for cells 4 mm in diameter. Cells 2 mm in diameter can be built, but their construction is more difficult, and the modules would be shorter.

Existing silicon detectors for H.E.P. are not adapted to the detector of X-rays of a few keV. The dE/dX energy deposition is 60 to 70 keV, one order of magnitude too large. The noise has to be reduced to less than1 keV. The use of silicon pads could be very appealing to adapt the spatial resolution of the T.R.D. to the granularity of the calorimeter and to get fast read-out,but more research is badly needed.

REFERENCES

1 - C. W. Fabjan et al. Nucl. Inst. and Meth. 185 (1981) 119

2 - J. F. Detoeuf et al. Nucl. Inst. and Meth. A265 (1988) 157
F. Feinstein (Thesis: Universite de Paris Sud 14 Dec. 1987)

3 - R. Ansari et al. Nucl. Inst. and Meth. A252 (1986) 357

4 - J. Cobb et al. Nucl. Inst. and Meth. A140 (1977) 413

5 - A. Schloesser : Talk given at Strasbourg Conf. (1985)
R.D. Appuhn et al. : T.R.D. for electron identification beyond 1 GeV/c
Bonn HE 87-13 July 1987

6 - B. Dolgosheim Nucl. Inst. and Meth. A252 (1986) 137

7 - Proceeding of the Workshop on Physics at Future Accelrators.
La Thuile 7-13 January 1987. (F. Palmonari page 233 Vol. 1)

8 - X. Artru et al. Phys. Rev. D12 (1975) 1289

9 - Y. Ducros et al. to be submitted to Nucl. Inst. and Meth. A

10 - R. Bouclier et al. " Fast tracking detector using the multidrift tubes."
C.E.R.N. E.P./ 87-89 12 May 1987
- This Conference

11 - V. C. Ermilova et al Nucl. Inst. and Meth. 145 (1977) 555

INTEGRATED CIRCUITS DEVELOPMENT IN JAPAN FOR FUTURE EXPERIMENTS

Y. Watase

National Laboratory for High Energy Physics(KEK)
1-1, Tsukuba, Ibaraki, 305 Japan

Abstracts

The recent projects of the LSI development in Japan are reviewed for the future experiments not only for the SSC but also for the KEK activities.

§ 1. Introduction

In the future experiments, the number of readout channels of the detector elements becomes extremely large and the packaging density of the electronics should be high due to the limited available space. Especially, in a typical detector at the SSC, the total number of the channels is almost one million. This large number of channels are resulting from the complex event topology with high particle multiplicity and small granularity required for each detector element to cope with the sharp jet events at the very high energy. In order to save the signal cables coming out of the detector, analog signals have to be processed into a digital data on the detector, which requires the high density implementation and radiation harden electronics.

A usage of large scale integrated circuits (LSI) in the frontend electronics is a key technology for these situations as well as in the data acquisition and processing. The application of the LSI to the detector electronics is getting easy in the past few years due to the industrial demands on application specific integrated circuits (ASIC) for various products. At the same time, tools for the LSI design are available on the workstation computers. But the production of LSI has to be placed with semiconductor companies, unless we build a semiconductor process facility keeping several excellent engineers trained for the update technology. The collaborations with companies are inevitable to use their technology properly.

In this report, I would like to review the status of LSI developments in Japan for the future experiments.

§ 2. Environment of LSI development in Japan

The semiconductor industry is one of the major parts of the recent economic growth. A number of companies in Japan compete with each other in various semiconductor products. Especially, memory chips are typical products to indicate the technical level of the individual company. Recent competition is on the development of 16-Mbit DRAM chip with 0.5 - 0.7 micro-rule CMOS.

Owing to the recent advancement of the semiconductor technology, ASIC (Application Specific Integrated Circuits) become popular in every field of the consumer products as well as in the industrial applications. More than 100 ASIC design centers are operational in Japan. But we have no organization similar with MOSIS of US which is very useful for our field and also for the universities in the educational aspects of the technology.

We had experiences on use of a semi-custom gate array for FASTBUS protocol chip (ADI) and an analog master slice for the drift chamber pre-amplifier. But we did not have a chance to develop a full custom LSI. The development of full custom LSI for readout electronics started in 1986. The main intention of the development is the introduction of the promising and advanced technology into the readout and data acquisition electronics for future experiments, especially for the SSC. The required conditions of our approach to the companies are as follows.

1. We want to work on the design stage of LSI's as much as possible for the levelup of our ability not only on the hardware, but also on the software tools. This reduces the turn around time and the development cost.
2. We do not need large quantity production in the application. It is order of 10,000 or less.
3. The company must be able to provide us an appropriate technology with a reasonable price and the reasonable conditions for a design by ourself.
4. We want to establish a way to develop various LSI's to support many experimental requirements for the electronics in the future experiments.

A number of big companies do not want to accept our approach for the LSI development due to busyness of their production lines. Other companies do not accept any circuits designed by people outside the company. On the other hand, smaller companies cannot provide advanced process technology. If we are satisfied with the current standard technology, a LSI development is available at reasonable cost and sufficient quality. As stated above, we want to apply current advanced technology for future applications. Since the technology of LSI production is changing rapidly, we have to keep our attention on the technological front. If companies find some interest in our project from their technical or economical points of view, they can cooperate on the developments with us.

The present activities on the LSI development are listed below. [1]

- Bipolar pre-Amplifier/Discriminator for silicon strip detectors
- Time memory cell for TDC
- Analog sampler (analog memory chip)
- 1 GHz GaAs shift register chip
- QAC/TAC chip for KAMIOKANDE II

[1]** Working people for individual project

1. Bipolar pre-Amp: (KEK) H.Ikeda, N.Ujiie, (NTT)Y.Akazawa
2. TMC: (KEK) Y.Arai,(Hirosima)T. Ohsugi, (NTT)T.Baba, T.Mano
3. Analog sampler: (KEK)H.Ikeda,S.Inaba
4. GaAs shift register:(KEK)T.Taniguchi,O.Sasaki,T.Ohska, (OKI)A.Tsukuda,H.Nishimurs,M.Takada H.Mori,K.Kaminishi,Y.Kawakami,Y.Sato
5. QAC/TAC: (KEK)H.Ikeda,T.Tanimori

The analog memory chip is now under consideration for the conceptual design by using CMOS process. That will start in a year for the circuit design.

The status of these projects are described in the next sections.

§ 3. Bipolar Pre-Amp/Discriminater Chip for SSD (PADC)

A bipolar pre-amplifier chip with full custom design is under development for readout of the silicon strip detector in a very high rate environment such as the SSC. In order to use the silicon strip detector in the environment, signals from the silicon detector have to be read out by a current sensitive mode in the pre-amplifier stage and followed with a digital readout scheme. This PADC consists of our stages of current sensitive pre-amplifier, filter amplifier, buffer amplifier, and a fast comparator. The scheme of the application is shown in Fig. 1. The position resolution of this method is determined by the spacing of the strip. The output pulse train from the comparator is fed to a memory chip. Digital Time Slice Chip (DTSC). The DTSC contains two stages of FIFO memories; the first stage FIFO with 32 bit in depth accepts the digital signal from each channel with the clock of beam crossing and the event data selected by the first level trigger condition are transferred to the second FIFO of 16 bit in depth. We are collaborating with a group from the University of California Santa Cruz to make the whole system. They share to develop the digital memory part DTSC. It is under development with CMOS process.

Major requirements on the silicon strip readout system are;

1. Low noise amplifier
2. Low power consumption
3. High density integration of microcircuit
4. Radiation hardness up to 10^6 - 10^7 rads/year
5. Wide frequency bandwidth to be compatible with high rate of particle flux.

The amplifier head is a current sensitive preamplifier, followed by a low pass filter that determines noise frequency band width. Finally, the signals are fed into a comparator through a buffer amplifier.

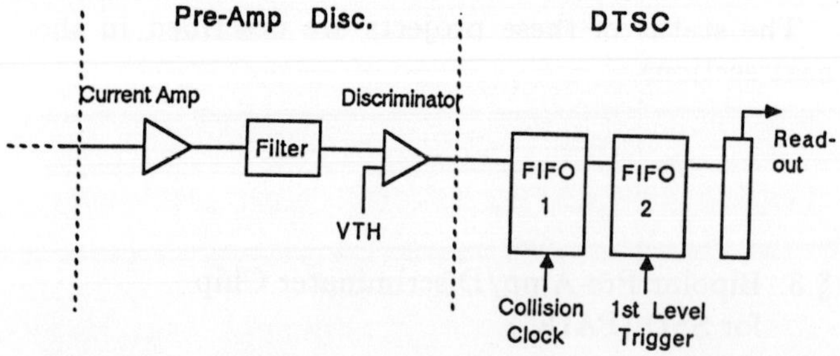

Figure 1: Digital readout scheme of silicon strip detector

The developing work has been proceeding under the collaborative relationship with the LSI Laboratory of NTT (Nippon Telegraph and Telephone Co.). They are very positive in cooperating with us on the pure scientific application of their advanced technologies. They provide an advanced bipolar process; SST (Super Self-aligned Technology) which technology is developed for GHz region applications.[1)] The SST process has particular features with a high current gain and a large frequency bandwidth by managing a submicron structure on the semiconductor process not on the mask design.[2)]

We designed the circuits and mask layouts of a test chip TEG(test element group) with their enough aids. This test chip does not contain the comparator part. The circuit schematics of the pre-amplifier and filter amplifier stages are shown in Fig. 2 and 3, respectively. The input stage of the current sensitive amplifier is a common emitter NPN transister whose quiscent current is 0.2 mA with h_{fe}=120, and low $r_{bb'}$ which is achieved with multi-rows of emitter. The collector signal is fed into cascode transister which in turn amplified at the differential amplifier stage. The open loop gain is about 500. Filter amplifier stage designed is a second order low pass filter which cuts off the high frequency components of the noise. The buffer stage with the gain of 5 follows the filter amplifier stage. The over all dynamic range of the whole stages is expressed by the equivalent input range which is about 12 μA. That is about 15 times larger than the response for a minimum ionizing particle penetrating through a silicon strip detector of 300-μm thick.

TEG chips have 5 channels of the whole stages described above and several test elements, such as the individual stages, bias voltage generators, typical transisters, and etc. The chip size is 2 x 2 mm^2. Fig. 4 shows a picture of the TEG chip.

In the TEG process, four types of circuits are designed and fabricated with the shaping time constants of 2.5 ns, 5 ns, 10 ns and 20 ns. Ten chips of each type were fabricated for test.

The frequency responses are measured. The frequency bandwidths defined by -3 dB are 80 MHz and 100 MHz for input pre-amplifier stage and buffer amplifier, respectively. The frequency responses of the filter amplifier with various shaping times were consistent with the SPICE simulation. The dynamic range measured is 20 dB relative to the standard signal that is equivalent to 25000 electrons. Typical noise characteristics are illustrated in Fig. 5 where equivalent input noise charges in the unit of electron charge are plotted as a function of the capacity of the input stage.

The characteristics of this pre-amplified chip are listed below.

Pre-amp	: current sensitive
Input impedance	: 50 Ω /150 Ω
Transconductance	: 250kΩ/ 750kΩ
Pulse rise time (typical)	: 5 nsec for the shaping time of 2.5 nsec
	: 20 nsec for the shaping time of 40 nsec
Total power consumption	: 7.0 mW/channel

We connected the pre-amplifier TEG chips with the silicon strip detector of 500 μm spacing with the depth of 200 μm. The typical pulse height distribution was measured as shown in Fig. 6 for β source ^{90}Sr in coincidence with 2 layers of silicon strip detectors. The noise level can be seen below the signal from the beta-ray.

Radiation damage test for TEG chips of bipolar pre-amplifier is underway by irradiation of thermal neutron and ^{60}Co source. The radiation hardness of the SST has been proved for ^{60}Co upto 10^6 rads [3)].

In the next step of the development, we will design the 64 channels per chip with the comparator stage for each channel.

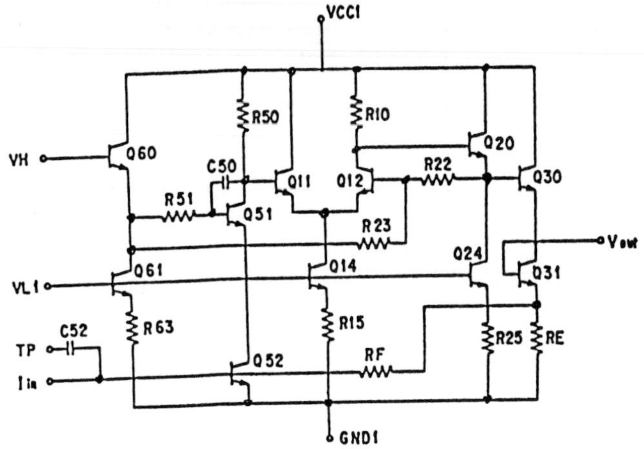

Figure 2: Circuit schematics of the pre-amplifier stage

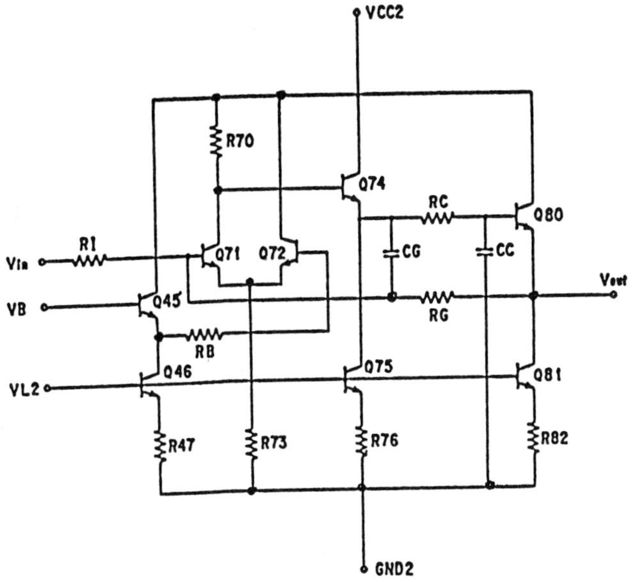

Figure 3: Circuit schematics of the filter amplifiler stage

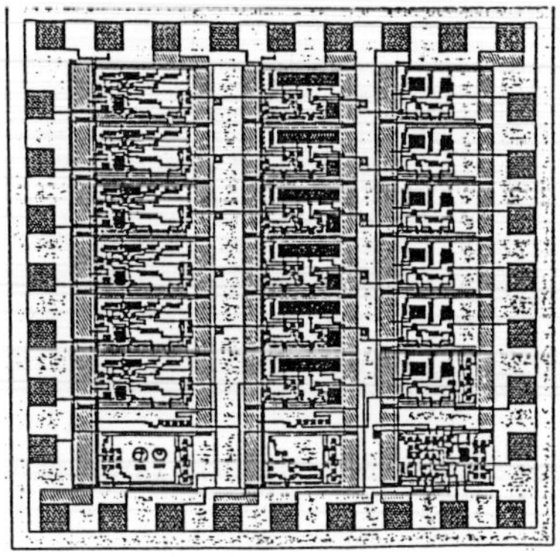

Figure 4: Photograph of the test chip

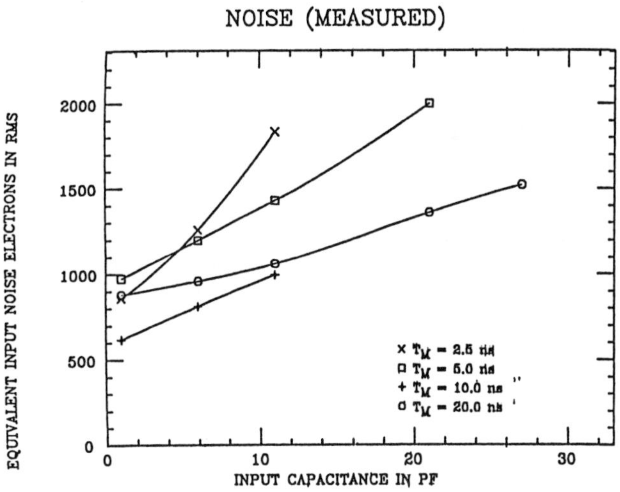

Figure 5: Noise characteristics vs. input chapacitance

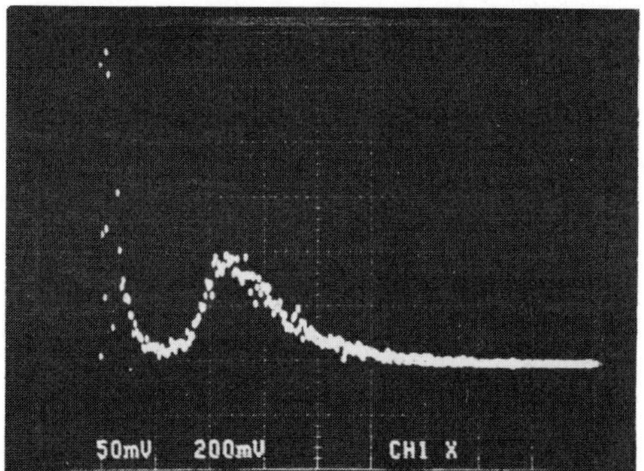

Figure 6: Typical pulse height disribution of 500-μm thick SSD for ^{90}Sr

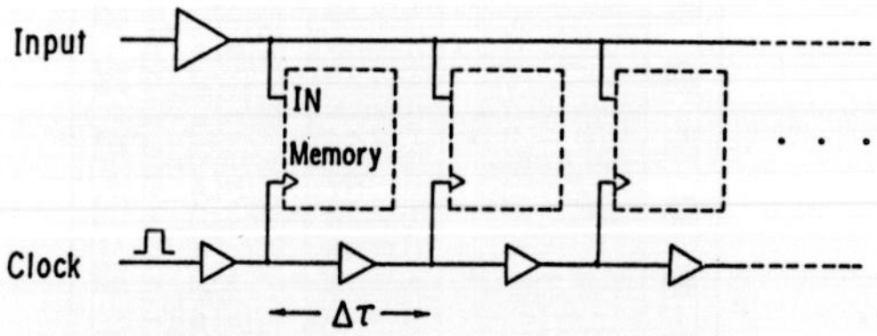

Figure 7: Basic function of Time Memory Cell

§ 4. CMOS Time Memory Cell (TMC)

A new LSI which can record a timing information with a 0.8 nsec resolution is under development using a submicron CMOS technology. The original design of TMC is for application at the SSC where capability for high repetition rate of the beam crossing and high channel density are required as well as low power consumption. The basic function of TMC is illustrated in Fig. 7. The basic element of the circuit is full-CMOS memory and delay lines composed of pairs of CMOS inverter. Signals from the detector electronics are fed to the data-INPUT pins of a line of memory cells through a buffer driver. The write-CLOCK are connected to a line of buffers (a pair of CMOS inverter) with a constant propagation delay of Δt. Then, a history of the input state can be recorded with the time division determined by the constant propargation delay of the CMOS buffer. Usually, the delay time varies more than 10 % due to the silicon process instabilities, power supply voltage variation, and temperature, although it is almost uniform within a chip. In order to achieve a good resolution, we implemented the feed back circuit in a chip by controlling the propagation time in the buffer.

The schematic diagram of the TMC chip is shown in Fig. 8. Since the beam crossing occurs every 16 nsec at the SSC, the each clock pulse of 62.5 MHz is fed to the different row of the 16 x 64 -memory cell array as shown in the figure. The array can store the time history of the input signal for the past 1 μsec. Within this time period, the first level trigger might be issued, which causes the read out cycle of the row of the memory array that contains an information of the specific beam crossing event.

The circuit of the basic cell is shown in Fig. 9. Transisters M1 - M4 compose a cross coupled inverter (full CMOS memory cell). A pair of CMOS inverter

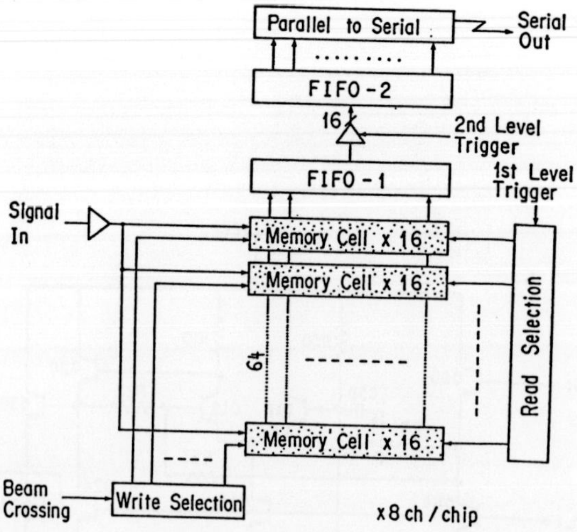

Figure 8: Schematic diagram of TMC chip

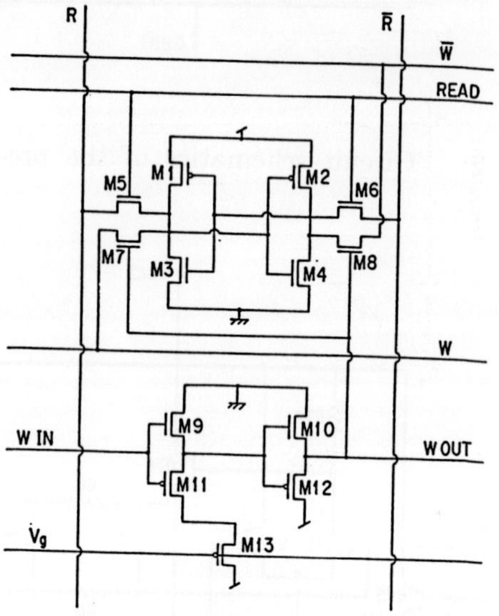

Figure 9: Circuit diagram of the basic cell

of M9 - M12 provides a delay element of the clock signal. M13 is used to control the propagation time by changing the rise time of the first inverter.

Test chips were designed and fabricated using the submicron CMOS process successfully. The test element consists of several key component circuits of TMC such as TMC array, memory, delay units, the feed back circuit, and the basic transisters. Fig. 10 shows a photograph of the fabricated chip. Fig. 11 shows a typical result of the linearity for the time digitization which gives linearity of less than 0.2 nsec. The differential linearity is shown in Fig. 12.

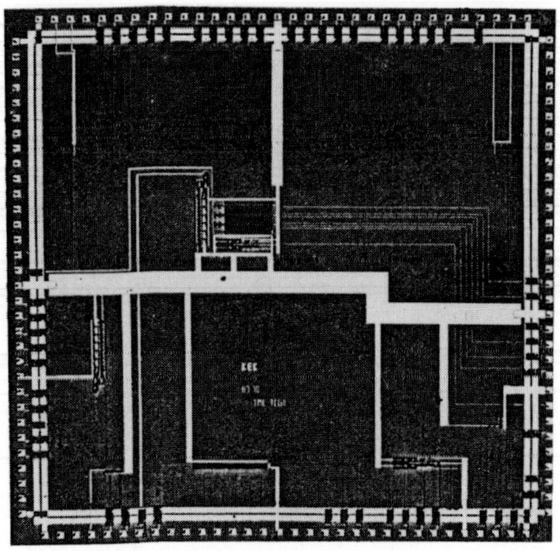

Figure 10: Photograph of the test chip of TMC

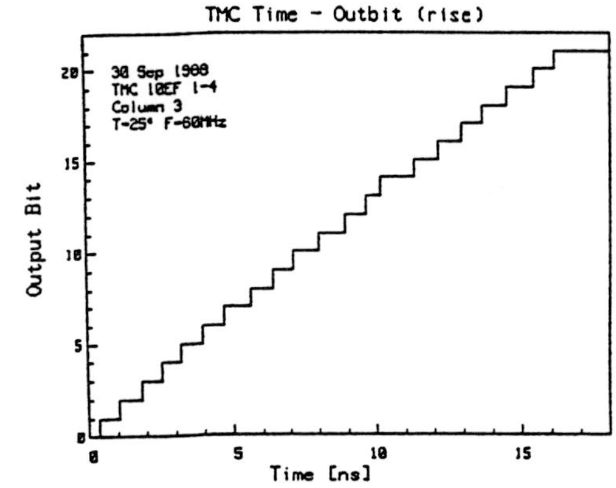

Figure 11: Linearity of the time digitizition of TMC

In the next step, we design a prototype chip with full function of TMC which can be used in a test experiment by connecting a bipolar preamplifier chip and a wire chamber.

§ 5. 1 GHz Shift Register for TDC

A dead time less TDC is required at fixed target experiments in the very high particle rate environment. One of such experiments is a search of a rare decay mode of K^0 using the high intensity neutral kaon beam of 12 GeV proton synchrotron at KEK. A drift chamber with the narrow spacing and fast gas has to be used as a tracking device. The experiment will run in 1989. In order to accommodate 1000 - 2000 channels in a reasonable number of TKO crate (KEK standard crate), the power consumption is strictly discussed on commercially available components and semi-custom gate arrays. Eventually, the

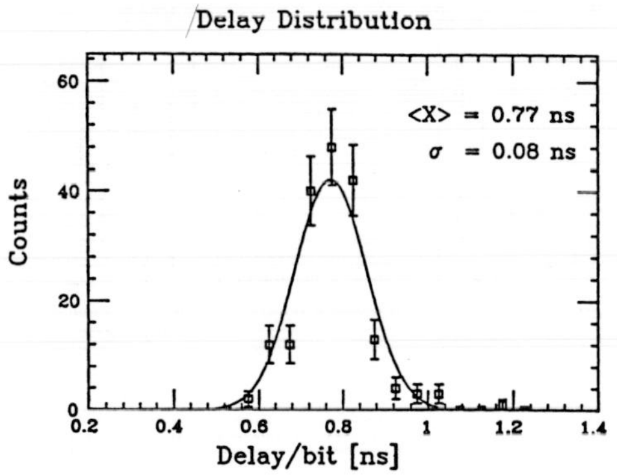

Figure 12: Differential linearity of TMC

development of 1-GHz shift register chip with a GaAs process has been contracted with OKI Electric Industry Co. Ltd. in 1987. OKI has long experiences on the GaAs semiconductor technology. They constructed GaAs device pilot plant in 1985 which produces GaAs-based products such as microwave amplifiers, multiplexer/demultiplexer for optical fiber communication. They had a potential plan to develop a large scale GaAs digital integrated circuit. They collaborate with us to make the shift register which has the size of 2000 gates equivalent. The development is at the final stage now and will be in the production phase by this summer (1988).

OKI has developed GaAs process with self-alignment, stable annealing and improvement of crystal growth. They used W-Al alloy Shottky gate with 1-micron length and the V_{th} control method by boron ion implantation and the following low temperature annealing. The basic circuit of the Flip-Flop consists of 6 NOR gates of DCFL (Direct Coupled FET Logic) as shown in Fig. 13. For the input and output stages which requires large driving capability, SBFL(Super-Buffered FET Logic) scheme was applied. From the TEG chips, the fabricated gate element has proved the propagation delay of 45 psec/gate and threshold voltage of - 650 mV and 70 mV for depletion mode FET and enhancement mode FET, respectively. Three types of the circuit scheme of the shift register were designed and processed for the test. These three types are shown in Fig. 14. In the straightforward drive of the clock signal along the train of the flip-flop circuits, a large deformation of the clock signal limited the frequency range of the clock. Type(3) utilizes the MUX-DEMUX circuits with four 32 bits shift registers in parallel which are clocked with the reduced clock frequency. In Fig. 15, the power consumption of several types of circuit configuration are shown as a function of the clock frequency. The most promising circuit type is Type (3). The power consumption of 128 bit shift register

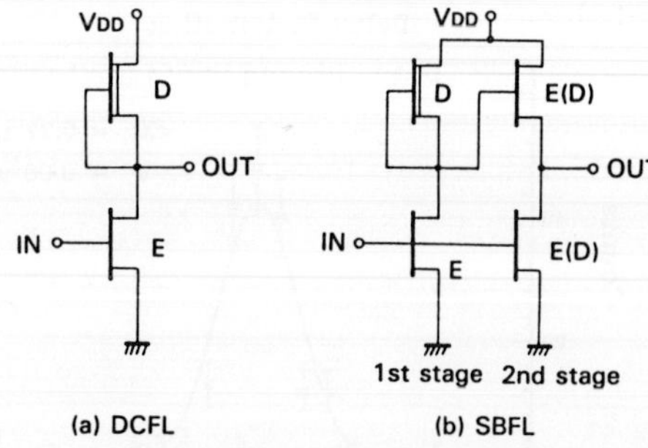

Figure 13: Circuit diagram of DCFL and SBFL of GaAs

is 0.47 W at 1.1 GHz clock with V_{dd}=1.7 V. Even for 256 bit shift register, the power consumption will be below 1 W/channel.

Now the final version of the 256 bit shift register is under process.

§ 6. QAC and TAC Chip

In the future experiment at KAMIOKANDE ; underground experiment for proton decay and cosmic/solar neutrino observation, more than ten thousand photomultiplier tubes will see the detail of the Cerenkov light emission in pure water of 40,000 tons. Signals from these PMT's have to be measured for its pulse height and the arrival time. This chip contains dual channels of a time-to-amplitude convertor and a charge-to-amplitude converter. The chip was designed by Toshiba Eletric Co Ltd. using FUJITSU analog master slice of bipolar process. The major specifications of the chip are sumarized below.

QAC/TAC specifications

QAC:

Convertor	: current integaration
Measurement range	: 0 - 1600 pC
Max output voltage	: + 5 V
Pedestal voltage	: < 0.3 V
Linearity	:±0.1%/full-scale
Clearing time	: 1.3 μsec

TAC:

Circuit	: current integration
Measurement range	: 100 nsec - 1 micro sec
Resolution	: 1 nsec

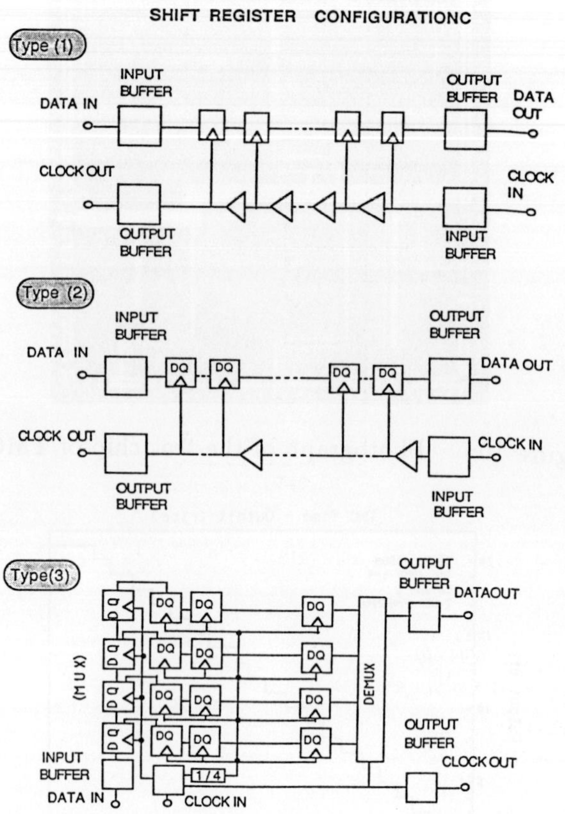

Figure 14: Three types of the test circuit diagram

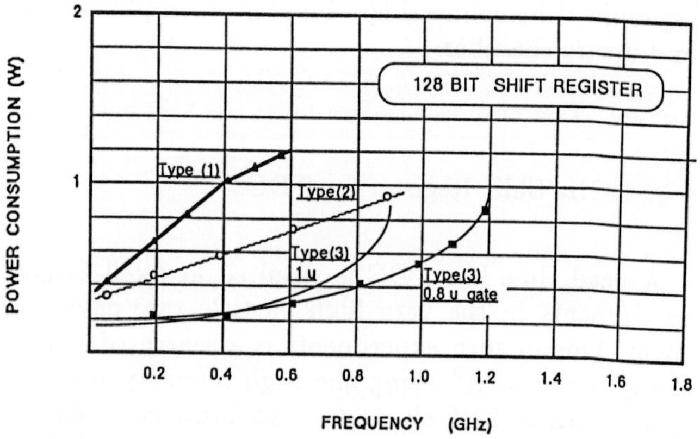

Figure 15: Measured power consumption of three different scheme of 1 GHz shift register

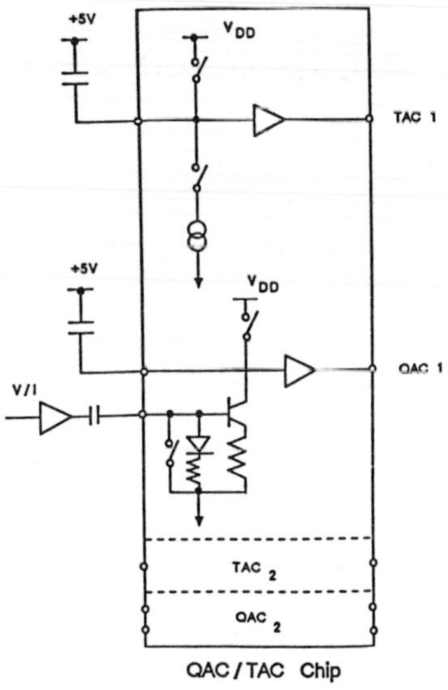

Figure 16: Schematic diagram of QAC/TAC chip

Output voltage	: + 5 V
Pedestal voltage	: < 0.3 V
Linearity	: ±0.1 %/full-scale
Clearing time	: 1.3 μsec

Two channels of both QAC and TAC are implemented in a chip as shown in the block diagram of Fig. 16. Twelve chips of this QAC/TAC will be mounted on a printed circuit board of TKO (KEK standard crate) with ADC chips to digitize the output voltages from the chips.

The first test samples were delivered and now under test. This chip will be tested on the module board within this year(1988).

§ 7. Conclusions

The application of various LSI's for signal processing and data acquisition is a key technology in the next generation experiments. The several projects of the LSI development are in progress under the cooperation with the semiconductor companies in Japan. Although the ways of the approach are different for individual projects, we would have valuable experiences for the new technical challenge.

Acknowledgements

The author wishes to express his thanks to Profs. S.Ozaki, K.Takahashi, H. Sugawara, and S. Mori for their continuous encouragement to these projects.

References

1) Y.Akazawa,N.Ishihara,T.Wakimoto,K.Kawarada, and S.Konaka, IEEE Jou. of Solid-State Circuits SC-21(1986)417
Y.Akazawa,A.Iwata,T.Wakimoto,T.Kamoto, H.Nakamura and H.Ikawa
ISSCC Dig. Tech. Paper (1987)98
2) S.Konaka,Y.Yamamoto, and T.Sakai
Proc. of 16th Conf. on Solid State Devices and Materials, Kobe Japan, (1984)209
N.Ohuchi,A.Kayanuma,A.Asano,H.Hayashi, and M. Noda
IEDM'84 Dig. Tech. Paper(1983)55
3) M.Umehira,H.Kikuchi,S.Konaka, and S.Kato
Proc. of 15th Int. Conf. on Space Technology and Science, Tokyo 1986 P.893

How can we get Intelligent Systems Close to Experiments ?

R.K.Bock, Y.Ermolin, W.Krischer, Ch.Ljuslin, S.Lone, A.Marchioro, K.Zografos

CERN, Geneva, Switzerland

Abstract

For the high data rates expected at future multi-TeV hadronic colliders like the SSC, it is of utmost importance to take decisions in real time on *partial data* and as fast as possible. At a first level and shortest timescale, some customized electronics will reduce the rates. In a second phase, decisions have to use concepts closer to physics and hence imply the presence of some *intelligence* in the trigger. We consider various parallel computer or computer-like systems for their possibilities to be embedded as critical active elements in future detectors. We also discuss the present activities and the pilot systems being built up as part of the LAA project at CERN. These activities aim at a better understanding of existing commercial systems, their design, and their limits of performance. The ultimate aim is to *integrate* suitable system designs in a flexible way into the data acquisition electronics of future detectors, most likely in VLSI technology.

The Data Acquisition Environment

The target detector we will have to consider is one which

– is *complex*, i.e. contains multiple detector elements including tracking and calorimetry for electromagnetic and hadronic particles, possibly track identification devices, at various production angles;

– is made of elements of *fine grain*, i.e. records interactions in considerable detail, producing a very large total amount of information;

– can operate in an environment of *high rates*, i.e. is capable of recording events at a hadronic collider with a luminosity of the order of 10^{33} $cm^{-2}sec^{-1}$.

With an assumed total inelastic cross section of around 100 mbarn, the assumed luminosity translates into a primary event rate of 10^8/sec. We will assume that custom-made fast (analogue) electronics will reduce this at least by a factor 1000. Further we ignore the event separation problems arising from the recording of multiple events. Event overlap in recording becomes likely as the average time between events shortens with increasing luminosity, and as event separation in time by bunching the colliding beams becomes ineffective.

We assume, then, that at a rate of 10^5/sec events are presented to some *second level trigger* processors, most likely in a pipeline just after analogue-to-digital conversion (see Fig.1). We postulate, that at a rate of one event every 10μsec or faster, these processors will have to decide with some degree of intelligence about the suitability of further processing.

Presently, no digital systems with computer-like characteristics exist to take on such a task. Existing trigger systems rely on full custom analogue or digital electronics to reduce event rates way below the 10^5/sec mark, and anything operating faster than, say, the msec level has little or no programmability. Computer-like systems used for triggering today are microprocessors, operating on full events and with several milliseconds of response time. Their parallelism is thus implemented at the event level, and we consider such system as *third level* triggers. We should note that substantial improvements in the computing capacity of the third level will also have to be an objective the SSC experiment planning (see also T.Nash, these proceedings).

It is our specific interest to find suitable ways to bridge the existing *serious time gap* between custom-made electronics, analogue or digital, and commercially available programmable systems.

Data Compression

The fine granularity of a detector translates into a large number of 'channels' and a large per-event data volume. Our assumption is that 10^6 bytes is a lower limit for the raw information content of an event, after removal of trivial zero channels. Both for making intelligent decisions and for economizing on transmission bandwidth, it is mandatory to *correlate* raw signals in order to extract *physics meaning* from them. Inasmuch as the extracted information like 'electromagnetic shower', 'local track segment', or 'parton jet', contains sufficient information for further processing,

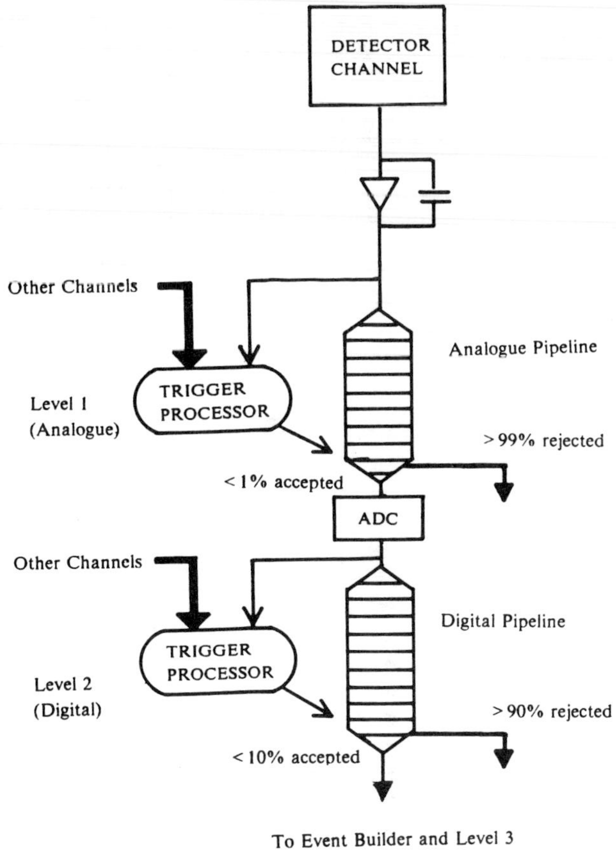

Figure 1: Schematic Data Flow for Data Acquisition

albeit in need of subsequent fine calibration, it may serve as a suitable local description of event features in the given detector element, and may thus replace the original raw information, much more voluminous (like amount of light recorded in calorimeter cells, drift times on wires, etc.). A *triggering* task executed in a processor may thus also become a *data compression* task, and this compression of data volume may even turn out to be a major asset of embedded intelligence. Today, the common use of digital signal processors (e.g. for extracting multi-hit information on single drift wires) could be viewed as a forerunner of the systematic use of image processors extracting features like track segments from two-dimensional information like a group of tracking signals.

Significant Trigger Tasks

Any novel technique to be introduced in a future detector will be expected to demonstrate its feasibility in a real-life test. Full scale trigger tests are of course detector dependent and thus not possible before accelerator and detector parts exist. We thus have to define for ourselves a number of *significant* triggering tasks that can be realistically simulated by existing hardware, in order to achieve the same purpose. A number of assumptions and educated guesses can not be avoided. Most prominent amongst the hypotheses we have to make (apart from the rate assumptions made above) are the *physics* objectives at a multi-TeV collider and the global *detector* characteristics we have to deal with.

As for physics, we can rely on a wealth of material collected at various workshops in the past. From the meetings at Snowmass (1986, proceedings edited by R.Donaldsen and J.Marx), La Thuile (1987, proceedings edited by J.Mulvey, CERN 87-07), and Erice (Eloisatron Project workshops 1988, proceedings to be edited by A.Ali and L.Cifarelli), we can extract as main physics objectives and their signatures the following.

— The Higgs boson. Depending on its mass which is subject to speculation, and possibly on the mass of the top quark, the Higgs will manifest itself in heavy quarks or W/Z. The most relevant signature in either case is a high-p_t lepton, possibly a high-p_t jet with lepton content.

— Heavy Sequential Quarks and Leptons. The expected signature consists of high-p_t jets, constrained by effective mass, and associated to high-p_t neutrinos.

— Additional Gauge Bosons. Their decay is expected to be analogous to the Z and W, i.e. into lepton pairs of high p_t (e/μ and ν for the W′, $e^+e^-/\mu^+\mu^-$ for the Z′). The dominant quark decay is considered as difficult to observe as for the known vector bosons.

— Supersymmetric Particles. The expected signatures from squark decay into photino and quarks consist of quark jets and missing p_t (photino). Electroweak sparticles leave a charged lepton and a photino signature.

— Leptoquarks. The expected signature would be jets associated to high-p_t leptons, charged or neutral.

These signatures can be abstracted to very few significant elements: The recognition of high-p_t charged leptons or jets, or of escaping transverse energy, possibly with constraints (mass) between them. Most high-p_t phenomena occur over a very central rapidity range, typically at angles larger than 40^0 with respect to the beam direction. 'High p_t' stands for at least 30 GeV/c. Charged leptons are single charged particles with a signature in an electromagnetic calorimeter or muon chamber (we ignore tau leptons), possibly supplemented by track identification information. Missing transverse energy is obtained by summing energy vectors over all cells of a hermetically closed hadronic calorimeter. Jets are defined by local clusters of energy (e.g. peaks), again in a hadronic calorimeter. Supporting information on jets from charged tracks is not considered relevant at this level.

Beyond these elements, the usual 'unexpected' physics must be catered for. At the trigger level, this can only mean that a maximum degree of flexibility of concepts and modularity of construction must be aimed at. Triggers must be expressed in terms of basic physics concepts, and operated by parameters and logical connections both easy to change.

As for data compression, the degree to which this must be a facility that can be 'turned off' to learn more about unforeseen details will have to be learned in the field. A detector which has the (idealized) facility to record physics-related quantities like complete shower information, jet and neutrino energy vectors, certified muons, etc., would seem an ideally flexible tool even for the unexpected physics, if the correlations and constraints on any of these quantities remain fully programmable.

Parallel Systems

No single computer system can be expected to handle the processing load arising from the task outlined above. This is only partly a problem of computing capacity, but one of making the large amounts of raw data available to the processor. Maximum parallelism of operation is therefore mandatory, and some structure has to connect the individual processing elements, providing the necessary synchronization.

In this paragraph, some general principles of parallel computing operations are recalled; the applications to high-energy physics problems and our specific triggering or data compression tasks will be discussed later.

Historical Note

The potential limits of the classical digital computer with one instruction and data memory and a single sequential processing unit were foreseen by J. von Neumann even before practical computers had been built, and many early publications dealt with the aspects of parallelism. S.H.Unger published a paper in 1958 (A Computer Oriented Towards Spatial Problems, Proceedings of the IRE, October 1958) which is the direct basis for a system available commercially today (the DAP from ICL, now from AMT). Different forms of parallelism have also been introduced early into the more complex and performing general computer systems: Concurrency between input/output and processing, separate units to deal simultaneously with mantissae and exponents in floating point operations, multiple registers to allow overlapping access to data, operation pipelines to achieve concurrency for different types of instructions, and, of course, multiple CPU-s run by a single operating system.

All designs of this type have as objective a simple *enhancement* of performance of a perfectly general system. No work is expected from the user of such a system to adapt his problem formulation to the architectural peculiarities, the parallelism remains *invisible*.

More recently, systems have appeared on the market that contain parallelism which *does* require architecture-specific code to make use of the system's potential. Vector processors belong into this category, and so do various other parallel architectures, some built for specific applications like the processing of two-dimensional images, others being, in a way solutions to problems yet to be defined.

In the following, the basic concepts which allow classifying parallel systems are introduced; the taxonomy of systems is, of course, not unique and simple, most systems being hybrids of one sort or another.

Pipelining Architectures

A frequent type of parallelism is to execute sequential functions concurrently on different data, like automobiles are assembled on an assembley line. A given data unit undergoes various successive processing steps; when processing step 1 on data unit 1 is terminated, the data unit 1 undergoes step 2, whilst unit 2 is subjected to step 1 (see Fig.2). A pipeline may result in remarkable throughput due to the possible high *functional* parallelism, but is also chracterized by its 'latency', the time that elapses between the beginning of the first and the end of the last processing step, for a given data unit.

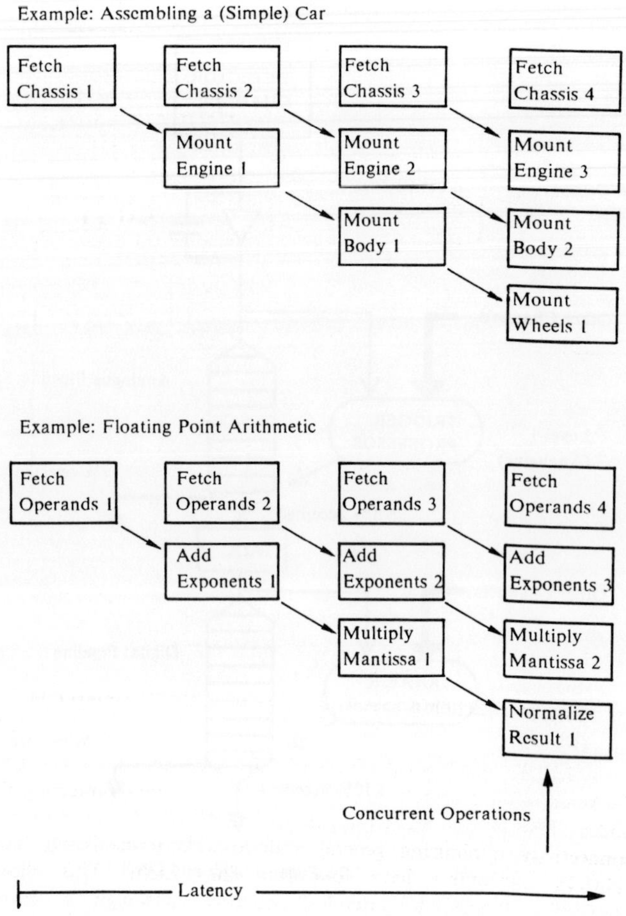

Figure 2: Principle of Pipelined Operations

SIMD (Single-Instruction-Multiple-Data) Architectures

Multiple processors can be built to operate in full lockstep on different data (Fig.3). Controlled from a single program, a SIMD architecture can achieve extraordinary peformance on problem formulations that keep all processors busy a good fraction of the available time. Depending on the available memory and the mechanism of flowing data into and from it, vectorial problems may be the most suitable for such architectures, and vector processors do, of course, exploit the SIMD principle. An example of a commercially available pure SIMD architecture is AMT's DAP 500: 1024 processing elements operate under a single master control unit, each unit performs single bit operations and has 32 Kbit of memory associated.

MIMD (Multiple-Instruction-Multiple-Data) Architectures

If multiple processors work independently on different data, general problems can take more easily advantage of the concurrency, at some level of 'data granularity'. Existing examples are multiple CPU-s in large systems (the data concurrency unit is the job, routing is done by the operating system), or high-energy physics event processing farms (emulators or the ACP, where the data concurrency unit is the event, and special dispatching software has been written into a host computer).

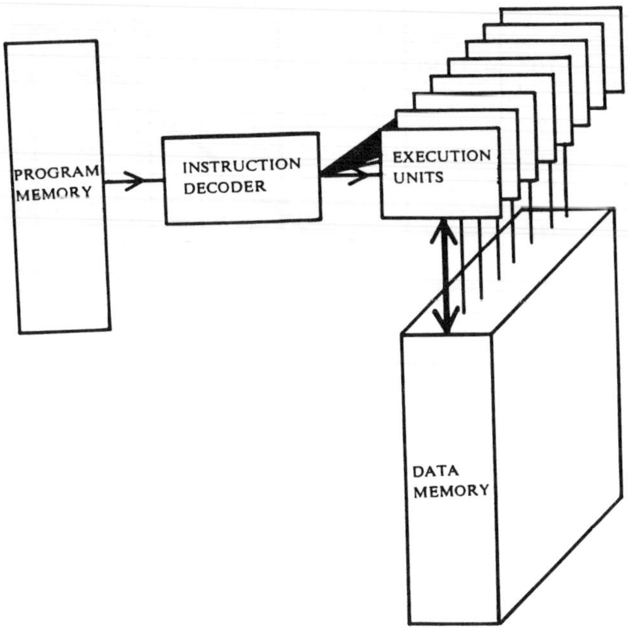

Figure 3: Schematic Architecture of a SIMD Computer

MIMD machines destined for finer data grain have also been marketed, with a variety of network structures interconnecting the nodes. The Intel iPSC/2 (Hypercube) systems have an 'n-space' connectivity minimizing general node-to-node transmission time, INMOS transputers have four fast links each. This allows connecting them into complex networks, although a simple two-dimensional grid layout is naturally favoured.

Examples of Implementing Parallelism

Image processors typically are hybrids between SIMD machines and pipelines, often custom-configured and thus allowing the application to determine the precise architecture. Operating on a single system clock and processing two-dimensional arrays of typically 8-bit information (pixels), the processing units have limited instruction sets and programmability taylored to image operations like convolutions, simple transforms, or thresholding. The processing proceeds at 'video rate', i.e. at least 30 images of 512x512 pixels per second, equivalent to 10 MHz.

Fig.4 shows an example of a simple image processing system which implements an edge finder. Two identical 3-by-3 convolvers and an arithmetic merger module are a realization of the 'Sobel operator', a transformation defining a zone boundary (edge) with its magnitude and direction. Modules of this type are available commercially, e.g. from Datacube Inc.

The rhythmic operation characteristic for image proceesing systems makes them part of a larger category of systems referred to as *systolic arrays* (systoles are the heart's rhythmic contractions). Such systems have been conceived for many other parallel tasks like matrix operations, and again are typically SIMD architectures with pipelining added at some higher level.

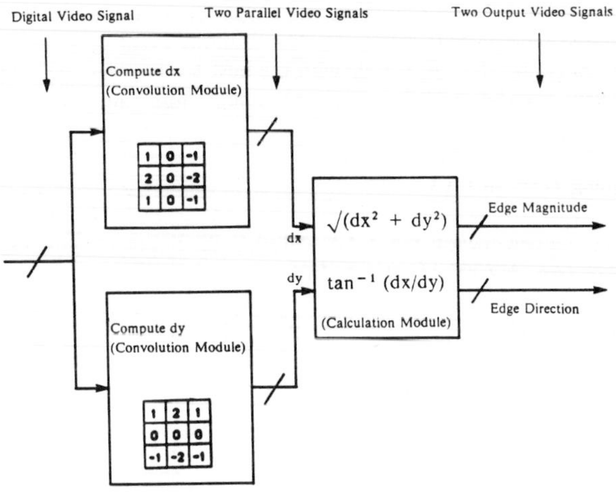

Figure 4: Implementation of an Edge Detection Algorithm (Sobel Operator)

Parallel Architectures in High-energy Physics

Physicists are near-insatiable in their requirements of computing power, where data analysis is concerned. More or less ingenious systems of the MIMD type have therefore been introduced for off-line data analysis in the past, the ACP system originated at FNAL and the emulator farms in the US and Europe are well known. Higher accelerator energies and increased event rates will undoubtedly accentuate the need for such systems in the future. It seems likely that the associated logistics for the expected large data volumes will push them in the direction of maximal use in real time.

The large number of microprocessors that are being used in the data acquisition systems of large experiments, do, of course, constitute another class of very successful MIMD systems operating in real time. More often than not their task is one of data flow control, many have data compaction tasks, in some cases a (third-level) trigger function is executed.

The Fast Triggering Problem

For genuine fast triggering, however, as was outlined above, fully programmable MIMD systems using the event as data concurrency unit are thinkable only at the last ('third') level. They require event builder functions to be performed before any computation can be launched, and the necessary compacting and pipelining or buffering operations have to be implemented further upstream. We believe that the second level triggering needs a different approach: the triggering functions need very deep embedding with those of detector readout, buffering, and data compaction. We think that a combination of special-purpose processors inspired by the image processing approach (and possibly following commercially available architectures very closely) with standard digital buffering techniques will have to be grafted locally onto groups of detector readout modules. Numerous such systems will have to operate independently on sub-event data units. The communication between systems will have to be as between nodes in MIMD architectures, to achieve the final trigger. Only the retained fraction of events will then require full event building, and may even consist largely of processed (compacted) data only. The resulting event and data rate must, of course, be compatible with the next-level processing, which is likely to be of pure MIMD architecture and with full programmability.

Ongoing Work in LAA

In the framework of the LAA detector development project at CERN (see A.Zichichi, ICFA Instrumentation Bulletin 3 (1987) p.17), we have started defining significant tasks and promising candidate architectures. Among the parallel systems available commercially, we concentrate on SIMD and image processing systems.

We have benchmarked, by simulation, a DAP 500 system with 32 by 32 processors, on various characteristic tasks (parameterized pixel transformations, histogramming, Hough transforms, peak finders for showers), and found, after some invested programming effort, a remarkable performance in execution (≥30 times Vax 8700 for the array). Also clearly apparent were difficulties to achieve the required rates in particular in data transmission and reordering into the DAP's fast memory.

We are also building up pilot systems based on three further manufacturer's hardware: a Datacube system for pure image processing, an NCR 45CG72 GAPP hybrid SIMD/image processing system (in collaboration with Laben, Milan), and an ASP (Associative String Processor) system (in collaboration with Brunel University, Uxbridge).

No practical or simulated experience exists for an application as we envisage. From specifications, the expected results should indicate that these systems are suitable from their architecture to take decisions on limited-size arrays of detector data, but not or only marginally within the necessary response time. We expect therefore, that both for data transfer logistics and performance, full integration of the systems or at least of systems' memory into the data flow of the detector readout will be necessary. The likely conclusion is that *existing* commercial parallel *systems* turn out *inadequate* without major modifications.

Technology Outlook

We therefore consider it our prime task to investigate in which direction improvements in the data flow can be expected or achieved, without deviating too far from marketed architectures.

As for performance, the increase in *component density* as observed over the last 20 years may be extrapolated towards ultra-large scale integration; the present yearly increase is about a factor 1.5 (see Fig.5). Manufacturers' chips announced today use 1μm geometry and contain $> 10^6$ transistors/chip; the next steps to sub-micron geometries are already under way (see also talks by H.H.Williams, Y.Watase, and P.Sharp, these proceedings). These numbers will directly translate into performance. Some manufacturers are also integrating more layers onto their chips; Mitsubishi has prototyped a three-layer chip with image processing functions, outperforming present-day components by orders of magnitude.

As for data *transmission speed*, the necessary steps into the future bandwidths of > Gbyte/sec are also being taken. Optical components like photodiode arrays may even become directly part

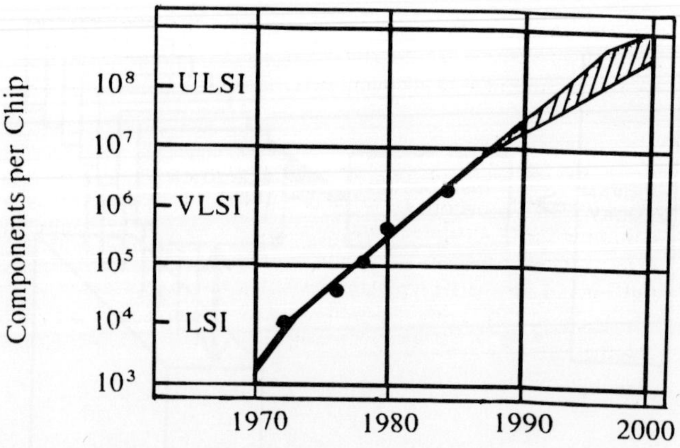

Figure 5: Evolution of Component Density on Chips since 1970

of some of our light-producing detectors (scintillators in calorimeters, fibres). Internal (on-chip) speed can take advantage of the increased component density and the absence of constraints imposed by interfacing standards. External 'buses' are likely to move increasingly towards high-performance point-to-point links (Meeting on 'Superbus' at CERN, 15 July 1988), with or without optical transmission.

Most importantly, perhaps, the possibilities of *integration* of components seem to be improving rapidly. Industry does talk about wafer-scale integration', implying fault-tolerant complex systems in which a faulty component does not cause the system to fail. In designing modern VLSI systems, increasing use is made of high-level ('behavioural') languages, of full simulation, and of automatic (computer-supported) layout. It is conceivable to accumulate libraries of modularly designed components, described in a standard language (e.g. VHDL), and translatable into hardware of a given technology. The integration of processing functions with digitization and buffering components, in a multi-chip system, may thus become a reality.

Summary

The *triggering problem* for future accelerators will be a *critical* one, particularly at the level where simple custom-made electronics becomes too complex and commercial computer systems do not offer enough performance. Unless decisions of some intelligence can be taken at a sustained rate of few microseconds per event, physics and/or accelerator time will be wasted.

Classical computer architectures and off-the-shelf solutions to computer parallelism will be inadequate to solve this problem. The *ingredients* for a future solution, however, do exist. Some of the existing parallel architectures, especially image processing devices, hold sufficient promise for being able to run the significant physics feature extraction tasks locally in a detector. Such tasks will be relevant both for triggering and data compaction.

With the foreseeable *advances of technology* and, in particular, with the trends towards standardizing the system design tools at the level of high level hardware description languages, we are hopeful that the *integration* of suitable processing architectures into future readout and buffering electronics can become reality. This technology transfer from industry, at the hardware design level, may be one of the most challenging tasks facing embedded systems of the future.

Several pilot projects are being built up as part of the LAA detector development project at CERN. It is their goal to evaluate commmercially available parallel architectures for their suitability to act as trigger processor elements, and to acquire the skills for their full integration as active elements in future acquisition electronics. To succeed on both accounts seems a necessary condition to channel the data rates at future colliders.

COMPUTING POSSIBILITIES IN THE MID-1990S*

Thomas Nash

Advanced Computer Program
Fermi National Accelerator Laboratory†
Batavia, IL 60510 USA

ABSTRACT

This paper describes the kind of computing resources it may be possible to make available for experiments in high energy physics in the mid and late 1990s. We outline some of the work going on today, particularly at Fermilab's Advanced Computer Program, that projects to the future. We attempt to define areas in which coordinated R&D efforts should prove fruitful to provide for on and off-line computing in the SSC era. Because of extraordinary components anticipated from industry, we can be optimistic even to the level of predicting million VAX equivalent on-line multiprocessor/data acquisition systems for SSC detectors. Managing this scale of computing will require a new approach to large hardware and software systems.

INTRODUCTION

In the mid-1990s, the complexity of high energy physics experiments will have reached the point that just dealing with the complexity itself will have become an important challenge.[1] Managing extremely complex systems will not just be an academic concern. The issue will be felt acutely by everyone involved with SSC detectors: physicists, lab and DOE managers, and even high tech aerospace corporations offering management solutions and tools.

The huge electronics, computer, and software requirements of SSC era experiments, both on and off-line, are what concerns us here. There are several areas: detectors and digitizing readout; trigger physics algorithms; on-line data acquisition and processing systems; off-line reconstruction software; and physics analysis. These are at the heart and soul of experimental activity. Yet, the scale of these systems, complex in the number, variety, and sophistication of their components, will not be amenable to conventional management techniques, certainly not the anarchistic, "we'll get to that later", approach traditional on HEP experiments.[2]

In the mid-90s we no longer will have the luxury to tolerate the extra complexity caused by the institutionally compartmentalized, bottom up, design which is now typical, encouraged, if not mandated, by the sociology, politics, and personality of our field. The fact that it may prove possible to operate large present-day experiments, handicapped as they are by excess complexity, does not mean that this will continue to be possible for SSC/LHC era detectors. It will be essential to streamline everything, without compromising necessary physics flexibility. In the jargon of modern software engineering, this means *structured* yet flexible systems.

The answers are far from all in. More R&D work aimed at the systems strategy for dealing with large HEP data acquisition and computing efforts is certainly in order. Nonetheless, in the spirit of this meeting, one can suggest how to move in the right direction. We will need to develop a standard processing environment suitable for most off-line and on-line computing. We need to steer away from hard-wired triggers and difficult to program, though fast, special devices often used for so called *second level* triggers. We need to use programmable and verifiable processors for all but the first level of triggers. This implies high-level lanquages and something like structured analysis, structured design (SASD), which we must make work for both hardware and software. We will need to make extensive use of expert systems for real time diagnosis of problems in big data acquisition systems, large group software packages, and all other large subsystems of the experiment. And we need excellent human interfaces on workstations to allow effective development of software, debugging of hardware, and monitoring of systems.

* Talk at *Future Directions in Detector R&D for Experiments at pp Colliders*, Snowmass, Colorado, July 5-7, 1988.

† Fermilab is operated by Universities Research Assoc., Inc. under contract with the U.S. Department of Energy.

Reasonable successes by HEP groups working on computing problems, now and in the past, encourages some confidence about attaining these future goals. High energy physicists, like most scientists, have always wanted more computer power than they could afford to buy.[3] In the commercial marketplace, the emphasis is on software backward compatibility, product differentiation, and isolation of a client "herd". On the scientist's computing agenda, raw processing power in a relatively easy to use form, not corporate profitability, is the dominant issue. Despite the fact that industry is not motivated by the rest of the market to provide the extremely cost-effective computer *systems* demanded by much of science, it does provide an extraordinary array of *components* (chips, modules, peripherals, workstations, software, etc.) that can be assembled into what science requires. For over 5 years the Advanced Computer Program (ACP) at Fermilab has drawn from industrial components to design and produce usable parallel computer systems of such cost-effectiveness that high energy physics (HEP) experiments are now being carried out that would otherwise be unthinkable.

The pioneering emulator work by Kunz *et al.* at SLAC demonstrated the feasibility of using multiprocessor systems to provide cost-effective computing for the reconstruction of experiment events.[4] Almost a decade after the first SLAC emulator, powerful 32-bit microprocessors allowed the ACP at Fermilab to develop even more convenient and cost-effective event-oriented parallel processing systems using many more CPUs.[5] Such systems are now an acknowledged important component of computing in high energy physics. First generation ACP systems are now at over 30 installations in universities and laboratories worldwide, primarily, though not exclusively, for HEP applications. The first of these systems which provide CPU power at a cost of less than $2500 per VAX equivalent, were brought on-line two years ago. New work at the ACP is now mainly in two areas[6]: a second generation multiprocessor targeted at experiments, and a multi-array processor "supercomputer" for the site-oriented problems of theoretical physics, principally lattice gauge calculations.

In this paper I will first briefly describe the new ACP systems for experimental HEP, in order to introduce the kind of powerful yet structured computing capabilities that should be attainable by the middle of the next decade. Early in the next year, these will provide well over an order of magnitude increase in cost-effectiveness over the original systems. This second generation project will also allow much higher bandwidth for both I/O and interprocessor communication, and will have software tools allowing almost any UNIX , VMS , or (potentially) VM based processor to be used equivalently as a node or "front end" in a multiprocessor ACP system. Perhaps most important is the way in which this new system allows for integration of powerful "back end" multiprocessors, now usable for both reconstruction and physics analysis, into a modern Ethernet-based workstation environment.

The new RISC processors that will be used by the ACP are on such a rapid growth curve, that it is not unreasonable to anticipate access to processing systems on-line of a million VAX 11/780 power in easy-to-program form. This makes it possible to avoid extensive use of difficult-to-program, elaborate special purpose processors. This is key, as suggested earlier, to streamlining on-line systems. We will discuss some of the main hardware and software development issues surrounding these huge increases in processing power as they affect system and module design and user applications. The ACP multi-array processor system for theory[7] will be briefly mentioned in the context of how its point-to-point switch architecture is relevant for future generations of experimental systems with extremely high performance processor nodes.

Industry is certainly providing us the opportunity, through the extraordinary array of new components being made available, to put together coherent systems that will meet the needs for SSC era computing *and* which will be manageable. At the end of this paper, there is a proposed list of projects that form a part of a coherent R&D program that should move us far in the direction we need to go for SSC experiment computing. The list is probably not complete, nor particularly well refined, but it is intended as a talking point so we can decide where to start, and get on with the job.

THE 2ND GENERATION ACP MULTIPROCESSOR

The continued saturation of computers (including ACP systems) by HEP experimenters motivates the development of a second generation of the ACP Multiprocessor system. This has been described in detail recently elsewhere.[6] A variety of new and increasingly powerful microprocessors are now available to incorporate into multiprocessors. Most of these are based on the Reduced Instruction Set Computer (RISC) philosophy. It was realized in the 1970s that many complex instructions in traditional machines with large instruction sets, like the IBM 360s and DEC VAXes, were rarely used. They effectively increased the cycle time for all instructions because they mandated extensive microcode. RISC machines are generally pipelined with no microcode and very fast instruction cycles. They defer complex instructions to software.

New processors will offer as much as a factor of twenty, more performance than the first generation ACP processors based on 68020s. Unlike earlier processors, some have usable FORTRAN and C compilers and UNIX operating systems even before the hardware appears. The broad availability of UNIX is particularly important. The new ACP architecture will support any processor running UNIX (or VMS) that can be connected via VME or Ethernet. It is very difficult to predict with any certainly which processor or commercial single board (or single slot) computer (SBC) will be most cost-effective in the future. The openness of the ACP system permits competitive purchase of processor nodes based on performance benchmarks and price.

The large increase in CPU power available for the Second Generation ACP System requires a redesign of the multiprocessor hardware and software system architecture to remove bottlenecks in current systems that would be felt at the higher performance levels. The bottlenecks are I/O and interprocessor communication bandwidth, and the CPU power available for the host process. Continuing the successful strategy of attacking computing limitations with parallelism, the new architecture solves I/O and host limitations by supporting parallel I/O and parallel host processing. Moreover, to avoid mini-computer bus bandwidth restrictions, any node may take on host functions including I/O, through controllers in its own local crate. Though significant, the changes in the new system are designed to be as transparent as possible to users of the original system.

As it has in the past, the ACP is encouraging competition in SBCs targeted at parallel processing by developing an extremely cost-effective VME SBC. The choice of the R3000 RISC processor from MIPS Computer Systems, Inc. for this design was based on performance evaluations. The ultimate standard is how real physics code runs in a high-level language, since it does not matter how many million instructions per second (MIPS) the CPU can execute if the instructions are not useful in FORTRAN or C and if the compilers fail to provide sufficient optimization. The ACP has performed benchmarks on several of the new chips using a suite of high energy physics FORTRAN programs. Based on these measurements, we expect HEP code to run at 12-15 VAXes on 25 MHz R3000 boards and 16-20 VAXes on the 33 MHz versions we plan to use.

The new CPU module will provide high level language processing power with a cost-effectiveness of well under $200/VAX 780 equivalent. The FORTRAN compiler is the best we have encountered for a microprocessor. It supports VMS extensions and compares favorably with the VMS compiler in convenience and sophistication. Since the MIPS CPU chip has on chip memory management, the board will be able to run the full UNIX operating system, booting either from a VME disk drive or using the Network File System (NFS) over the Branchbus. Full UNIX program development tools are available. This processor will form the cornerstone of the second generation ACP systems.

The original ACP multiprocessor used a single (MicroVAX) host which was the master of large numbers of microprocessor nodes.[5] The ACP Branchbus was developed to link several high performance commercial local bus crates (like VME) to a host and/or a data acquisition system. It is optimized for high speed (20 MBytes/sec) block transfers.[8] Improvements to the Branchbus system of interfaces allow higher performance and more complex interconnection schemes. Any VME master, in particular any node or smart I/O device controller in the system can now communicate with any other processor without host intervention, allowing the more elegant system architectures described later.

The new ACP Branchbus Switch allows full crossbar interconnection of up to 16 Branchbuses (or more using multiple switches). With this switch, any Branchbus master device can connect to any slave in the entire switch connected system. All channels of the switch can

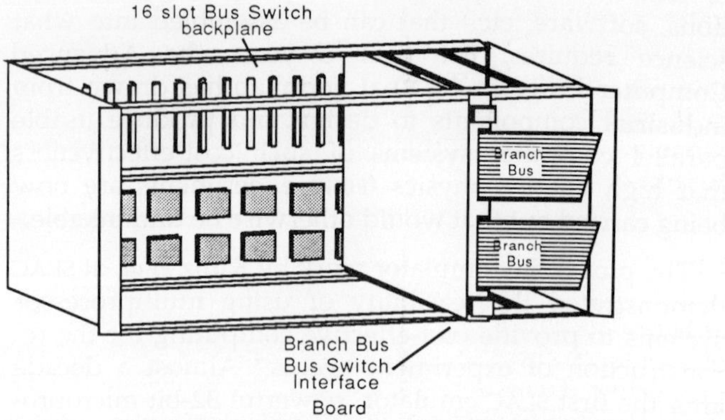

Figure 1. The ACP Branchbus Switch. The backplane uses single ended TTL Branchbus protocol.

be active simultaneously. For example, eight of the Branchbuses could be connected to the other eight, all transferring data simultaneously giving an aggregate bandwidth of 8 X 20 MBytes/sec or 160 MBytes/sec (in addition to any local bus activity on any of the VME crates in the system). Thirteen Texas Instrument 16 X 16 X 4 bit crossbar chips (TI 74AS8840) are used for the main switching elements. The Switch is a backplane in a 6U by 280 mm Eurocard crate. Modules may be plugged into the Switch Crate (see Figure 1) as with VME. However, instead of the signals being connected to a bus, each slot in the crate is a crossbar switch point. The first two Switch Crates are now in operation.

Two modules now exist that plug into the Switch Crate. One is the Branchbus Switch Interface Board (BSIB) which converts the differential RS485 signals on the standard Branchbus cables to the single ended TTL version of Branchbus used on the Switch backplane. The BSIB brings one standard Branchbus, with its VME crates, host, Fastbus, etc., into a port of the Switch, and allows them to be switched to whatever else is plugged into the Switch Crate, such as other Branchbus circuits. Use of the Switch in this way will allow multiprocessor systems to obtain extremely high bandwidth for interprocessor communication. The second existing module that plugs into the Switch is the Floating Point

Array Processor (FPAP), a 20 MFlop (peak) device that is used for theoretical physics, primarily lattice gauge, calculations. The FPAP is described elsewhere in detail, as is the innovative "better than a hypercube" architecture that the Switch allows.[7] Important future applications of the versatile Switch are described in the next section.

It will continue to be possible to read and write tapes through a VAX or MicroVAX into an ACP system. High performance operation, however, will take advantage of the potentially higher bandwidth of high capacity mass storage I/O devices that interface directly to the multiprocessor VME crate bus. The low cost of these devices particularly encourages parallel I/O. This is made possible by the availability of I/O directly in the node crates and by the new capability of a processor and/or its intelligent I/O controller to write from one crate to another. The multiprocessor system will have available to it many devices all reading and writing simultaneously, allowing the total I/O bandwidth to be increased to whatever level is required.

New I/O devices are replacing standard magnetic tape. These are very appealing to HEP experiments anticipating huge amounts of data, such as one approved at Fermilab which is planning to record tens of billions of events. At this time video tape cassettes appear most promising. The 8 mm format can pack well over 100 times the data in a given volume of shelf space as can conventional tapes, and the media is at least an order of magnitude more cost-effective. Current versions allow bandwidths comparable to those achieved with standard 6250 bpi tapes, with improvements expected. It is clear that with devices like these, one can count on far better I/O performance than has been available.

The redesign of the system software for the Second Generation ACP System will support the greatly improved performance and flexibility of the new processing, I/O, and communication hardware, while reducing the complexity encountered by both beginning and sophisticated users.[9] It will allow existing applications to run with minimal changes, yet will provide a variety of powerful new features so that users can realize the full potential of the new processors. Integration will be possible into a variety of computing situations, including large mainframe computer centers and the traditional VAX or MicroVAX host. Most important, in our view, is a distributed computing UNIX (or VMS) workstation environment in which the multiprocessor will function as a fully integrated back end engine directly controlled from the workstations. The general hardware environment supported is shown in Figure 2. Because of the emphasis on portable, nearly universal standards, it will be straightforward to incorporate any computer that runs UNIX and/or can communicate via TCP/IP over Ethernet. This makes the system open and receptive to future requirements and product options.

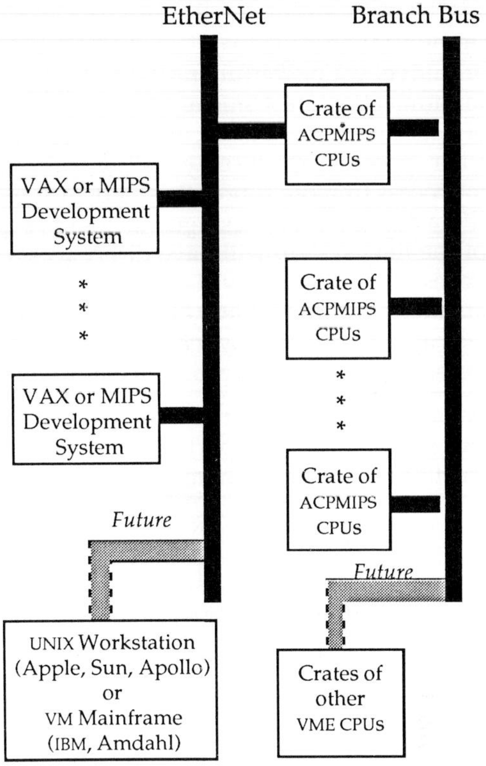

Figure 2. Second Generation ACP System: General hardware and network environment.

The ACP System Software is a tool to make it easy for physicists to bring programs to a high performance multiprocessor environment. An application is decomposed into a set of cooperating processes. Support for running these sets of processes as a job, including interprocess communication and synchronization, startup, etc., is provided by the ACP System Software. The processes run on an ACP Multiprocessor System interconnected by Branchbus and on any associated UNIX or VMS (later possibly also VM) computers connected via Ethernet and TCP/IP. They may be distributed as appropriate over the available computers. In this way a multi process job may be tested first in a single machine and then with increasing numbers of nodes. Program development is done using the full set of UNIX (or VMS) tools, including compilers, linkers and debuggers, of the computer on which the process will run.

Any node process can assume the functions previously exercised only by the VAX host processor, including reading or writing data tapes and accessing disk files. And any node process in the system can do `send` or `get` operations to or from an individual process (chosen by the system software from a class or rank of node processes) or set of processes in a given class or rank. As before, the system software will automatically find an available node process for the user. The ability to `send` and `get` to multiple nodes in a class allows `broadcast` and `accumulate` type operations.

Along with the traditional `send` and `get` type of ACP communication routines, there will be a variety of more primitive, yet powerful and easy to use, interprocess communication mechanisms. A process may send a block of data directly to a block of virtual address space in another process, or it may call a subroutine in another process (remote subroutine call), or it may send a small data packet (a message) to another process. Users of the new system will have direct access to process queues which they may define as they require or use in standard, traditional ACP defined ways (like node process *ready* or *complete*). Synch points provide a way for processes to synchronize program execution.

There are many possible process configurations that the new system can support. An example for a reconstruction problem with multiple input tapes is

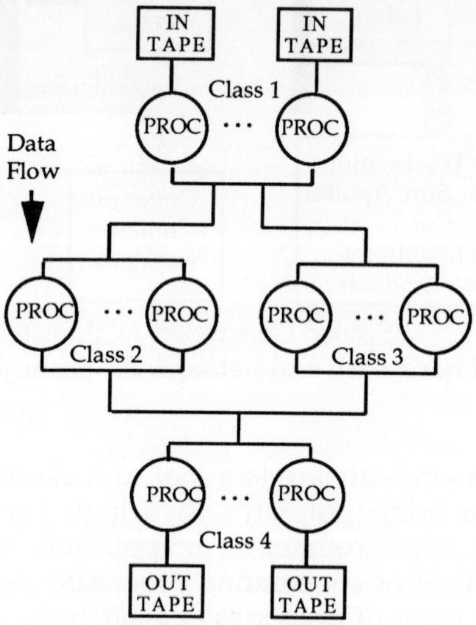

Figure 3. A multi rank configuration. Processes are indicated by circles.

shown in Figure 3. Note that this is a software configuration; the actual hardware connection of the nodes is over Branchbus including Bus Switches if necessary and is transparent to the programmer. Nodes in the top rank read events from data tapes and pass them along to either class 2 or class 3 nodes, which process events of different trigger types. Nodes in the bottom rank collect events from any nodes in the middle rank, either class 2 or class 3, for output to tape.

Another important configuration has enormous implications for physics data analysis (as opposed to reconstruction). Here, the top rank of nodes is the same as in Figure 3, and the second rank consists only of workstations and, perhaps, a single data recording process. With such a configuration, a whole experiment's data base of data summary tapes (DSTs) can be analyzed in parallel in much less than an hour. Traditional means of passing hundreds of DSTs through a computer center for a physics analysis pass often take weeks.

THE NEW WORLD OF RISC AND ITS IMPLICATIONS

The big jump in microprocessor power available in the last year or two has been a pleasant surprise, allowing a 20X increase in per node ACP multiprocessor performance. This has been the result of the new technology of reduced instruction set computers (RISC) we described earlier. Even this impressive *rate* of increase is now projected to continue well into the 1990s. For example, one of the three suppliers of MIPS RISC processor silicon, Integrated Device Technology, Inc. (IDT), is promising a 160 VAX Mips version in 1991. MIPS, Inc.'s own projections are indicated by the chart in Figure 4.

Although by industry standards these projections are conservative, high energy physics reconstruction code tends to run at about 25% below these numbers.

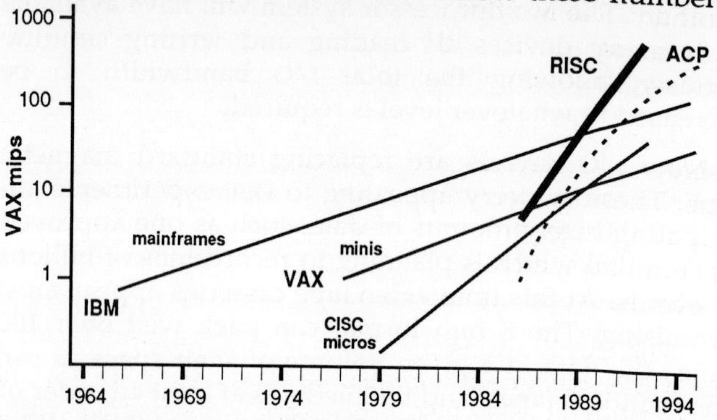

Figure 4. MIPS Inc.'s projections of future computer performance. Our projection of ACP style SBCs are added as a dashed line.

There is also an obvious delay from product announcement to when a group like the ACP can have usable processor modules available. Nonetheless, as indicated by the dashed curve we have added to Figure 4, ACP style SBC modules at a 500 VAX Mips level could be anticipated in 1995. It is important to emphasize that there are many companies playing in the RISC component field, including the semiconductor giant Motorola, and even IBM. The products from MIPS Inc. may well continue to be the most appropriate for our purposes, but this cannot be guaranteed. However, it seems likely, subject to the usual *caveat* about peering into crystal balls, that this technology, and the competition of industry, will produce RISC processors at and beyond 500 VAXes.

Where is all this performance coming from? We described earlier the philosophy behind RISC. The comparison with complex instruction set computers (CISC) is now well understood. Streamlining the instruction set can improve (reduce) the number of clocks per instruction by factors of 5 or more. Clock speeds can be maintained or even increased compared to CISC architectures. The cost of this simplification is an increase in the number of instructions to be executed, because complex instructions and functions are handled in software. RISC wins because this increase in instruction count is small.

Over the past few years dramatic increases in memory size have been more common than in processor speed. Although memory improvements are continuing, it is becoming clear that this is now happening at a slower rate than CPU performance -- and slower than the growing appetite for memory of high energy physicists, now acculturated to assuming endless increases in memory availability. Present generation ACP CPUs are based on 256K DRAMs. The second generation will use 1 Mbit. The conventional wisdom in the trade press is that 4 Mbit chips will make their appearance in 1989 and become a significant part of the market in 1991. 16 Mbit parts will appear in 1991 and we may guess that they will be readily available for projected 500 VAX performance ACP type module in 1995. This would correspond to 128 Mbytes per node. Not bad. But the ratio to processing power will be reduced somewhat below the first and second generation ACP nodes.

The processor and memory technology environment is changed. Experimentalists will no longer have their memory demands easily satiated while they hunger desperately for computer cycles. It will be important to take advantage of incredible micro-computer performance levels and cool it on memory profligacy.

The implications of this new environment for HEP computer designers will also be stimulating, to say the least. With 33 MHz RISC processors, we are already in a cache crisis. 20 nsec static memory is too slow and it is a struggle to find acceptably fast and large parts. Up to now lattice gauge processor[10,7] have avoided cache by matching DRAM speeds to the relatively slow floating point chip clocks. The next time around cache will be necessary, but theoretical problems make such regular accesses to memory that their cache miss rates are unacceptably high in standard cache.

Industry is sensitive to both aspects of the cache crisis and seems to be moving to multi-level cache. If this is not satisfactory for theorists, it may be necessary to use *anticipatory* cache which requires hardware and software to request memory in advance. In principle, this is not hard to accomplish for theoretical problems. In practice, supporting this at the compiler level would be a big effort.

The second technical crisis resulting from the new processor environment is the obsolescence of computer busses for multi-processor communication. Never mind the territorial religious wars about busses that have swept all HEP laboratories. No matter if we have crusaded for Fastbus, VME, Multibus, Nubus, or a homebrew, your or my favorite bus cannot handle the communication requirements of 100 VAX power processor nodes. Even for event-oriented reconstruction and trigger processing, just moving data in and out will saturate local crate busses.

The answer to the bus crisis is point-to-point communication between processors. Two variations on this solution are now being explored in the high energy physics community. There is considerable interest, strongest in Europe, in the INMOS Transputer architecture. These microprocessors incorporate several sophisticated communication channel ports on each chip that support direct links to neighboring Transputers. An example of how Transputers can be used in a point to point architecture is seen in the structure of the Global

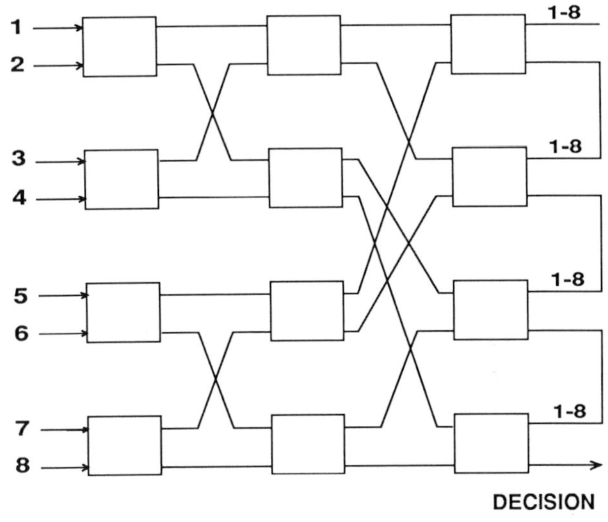

Figure 5. ZEUS 2nd level trigger box.

Second Level Trigger Box for ZEUS (Figure 5).[11] The architecture of the Transputer is certainly very appealing. However, no other semiconductor company besides INMOS has yet picked up on it. The processing performance of Transputers has lagged by over an order of magnitude behind the leading microprocessors at any given time, probably because design emphasis has been on the communication channels. Symptomatic is the fact that the FORTRAN compilers are still not very robust. This situation may change in the future as other manufacturers recognize the limitations of conventional communication mechanisms for their super fast processors and, possibly, pick up on the INMOS approach.

The ACP's Branchbus Switch Crate described earlier (Figure 1) represents an approach to point-to-point communication that does not depend on a particular processor's specialized communication ports. It thus allows a free choice of the highest performing processor at any given time. As already noted there are now two module types that plug into this crate, a Branchbus interface and the floating point array processor (FPAP) for theoretical calculations. It is very likely that the next version (1990) of an ACP processing node for experimentalists will also live in the Switch Crate so that there will be enough bandwidth to move data in and out.

The Switch Crate looks very much like a VME crate, using the same Eurocard hardware and having the same 6U height. It has all the advantages of a modular crate system and, with its cross bar switch back plane that allows up to 8 independent direct slot-to-slot connections, it does not have the bandwidth limitations of a bus. One essential characteristic of a commercial bus like VME is the wide spread availability in the standard of the latest in computer devices, such as I/O controllers and network interfaces. To make these available to the otherwise superior Branchbus Switch Crate environment, we expect to develop an interface to VME, using the extra depth of the Switch Crate. Any VME module could be plugged into the short interface card which in turn is plugged into the Switch Crate.

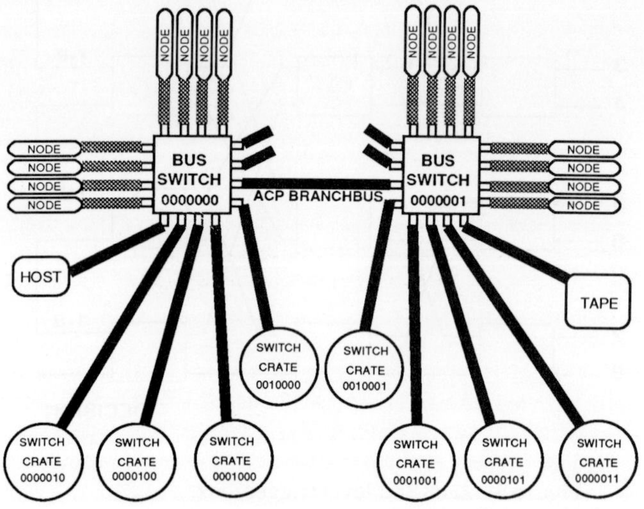

Figure 6. ACP Multi-Array Processor System: 256 node configuration.

In order to display the versatility of this approach, I can't resist describing briefly the ACP Multi-Array Processor architecture designed for theorists[7]. Figure 6 shows the 256 node configuration presently under construction. The individual single board FPAPs have peak performance of 20 MFlops. Performance of key kernels (SU(3) multiplies) have been measured on the prototypes to exceed 15 MFlops/node. The FPAPs are plugged into the Branchbus Switch Crates. The nodes can speak with each other in pairs at a full 20 MBytes/sec simultaneously. The architecture is a hypercube network of such crossbar Switch Crates each supporting 8-16 FPAPs. In a typical configuration 8 array processor nodes will be plugged into each switch crate along with up to 8 BSIB I/O modules (described earlier) that interconnect crates, via standard ACP Branchbus, in a hypercube (or better, if extra interconnects are desired).

Processing nodes do not participate in any communication activity other than their own. This is an important distinction from traditional hypercube implementations. The switches handle intra and intercrate routing automatically. The system therefore does not operate with all node programs (and/or communications) in lock step like an SIMD machine, as is the case in most of the other projects of this type[10]. It also does not strongly favor local communication (as existing hypercubes[12] do). It thus allows for any conceivable new lattice algorithm unconstrained by synchronous or local communication requirements. Despite its algorithmic flexibility the system ranks as the best (or nearly so, we won't argue) in terms of cost effectiveness of MFlops/$.

The first 16 node system is being assembled this summer. All components are working and tested. They have successfully run extensive physics code. Parts are being procured for a 256 node (5 GFlop for about $1 million) system which will be assembled at the end of the year. Maximum system size is 2048 nodes. The system is being designed in the ACP tradition to be commercialized and available to other institutions.

SSC TRIGGER AND DATA ACQUISITION

The key problems of SSC detector data acquisition and triggering are widely appreciated: the luminosity, the trigger reduction of 10^8, and the overlapping events in the detector. If the projections discussed earlier of 500 VAX equivalent processor nodes are realized in 1995, there is hope. A single module CPU, no matter what its capability, will always cost in the neighborhood of $2500. Given a willingness to spend $5 million (of the over $100 million cost of an SSC detector) for real time processing, a million VAX multiprocessor on-line is not only imaginable but likely.

The important thing about this megaVAX system is that it is all programmable in high level languages and, most important, it is accessible to modern software engineering tools. As was emphasized in the introduction, such tools, and a structured, homogeneous environment to work in, are nothing short of essential if we are to cope with the complex scale of SSC triggers and data acquisition. The same processors and programming environment would be available to off-line computing.

Given the opportunity and advantages of such a coherent and powerful system, how much of the data acquisition and trigger reduction task can one imagine handling in this way? This question requires study, but it is not unreasonable to hypothesize that a large fraction of the trigger reduction and essentially all of the standard data formatting could be accommodated.

The vision is of on-detector first level triggers based primarily on customized VLSI components and taking on a trigger reduction of 10 - 1000. The remaining 10^{5-7} would be written in structured high-level languages running on a multiprocessor, probably configured as a tree structure. Each branch of the tree would handle the data formatting and second level trigger tasks of a particular sub-detector. Unlike traditional second level triggers based on special purpose hardware and software, these would be easy to program and verify.

The opportunity to see structured, and manageable, first level triggers is also likely to be realized. This hope is based on the interesting development work now going on in several areas. At this workshop, Nygren and Shapiro described two different LBL projects that are moving rapidly toward pixel silicon detectors. These will ultimately incorporate on chip or hybrid (bump bonded) digital circuitry. Such circuits will take care of readout and data formatting, and will offer a fertile field for development of first level triggers.

There are two directions that such on-detector first level triggers may take (both for silicon and for other detectors). Perhaps futuristic, but certainly promising, is the hot new area of neural networks. Bruce Denby has described some early success in applying this to reconstruction problems.[13] Neural networks are patterned after the way the brain's system of neurons and synapses works by relaxing to a minimum energy state corresponding to recognition of a pattern. They are particularly appropriate for VLSI implementation and neural network chips are already being fabricated by Carver Mead at Cal Tech and L.D. Jackel at Bell Labs. Perhaps 1995 is a bit too optimistic, but one can easily imagine neural net readouts directly in terms of tracks and calorimeter clusters.

The other very strong candidate for implementation in VLSI and on-detector triggering is the data driven processing system approach, now available as PC board modules in two HEP variations.[14] At this workshop Rudi Bock described his interest at CERN in identifying components of a similar kind from the signal processing world. These systems are programmed by how modules, each carrying out a specific computing operation, are interconnected. When one module completes a task it puts its result, and a ready signal, onto its output cable. As soon as a following module has all its incoming data, indicated by all ready signals being on, it carries out its operation. The operations range from simple arithmetic to complex track finding and table lookups. It will be interesting to see how such a system when available in VLSI modules can be coupled to neural network chips directly on a detector.

TOWARD A STRUCTURED ON-LINE ENVIRONMENT

One need only look at the data acquisition system of a present-day large detector like CDF, and imagine it scaled up by an order of magnitude, to appreciate what we are getting into with SSC detectors. On-line systems coherence is far from being prevalent these days on experiments. Some of us have seen what happens when smaller experiments attempt to retrofit modern processors into existing electronics. Sergio Conetti at Trieste[15] described the extraordinary variety of ways in which ACP processors have been incorporated into experiments. Although widely used in standard configurations for off-line, there is no standard on-line implementation of these systems because there is nothing approaching a standard data acquisition system. Every experiment does its own thing.

In present-day large experiments, the sociology of multi-institution collaborations works against data acquisition system coherence. The typical scenario at design report time is to divide up responsibilities among the various groups allowing each to define the approach to be used in specific subsystems. A few years later the problem of interfacing these subsystems becomes a hot topic. The situation is, to some extent, due to the way approval committees operate. Expertise is compartmentalized by subdetector and institution. The overall system ends up being based on disparate subsystems, developed on the basis of correct, strongly held views, that don't mesh as a system.

Most experiments attempt to standardize at some level. Nonetheless, many still have systems that incorporate both Fastbus and VME, or otherwise mix standards. A similar situation exists with software standards. Much of this results from a plague of what might be called *standards evangelism* in our business. Individuals who have developed some degree of personal expertise in something like VME or Transputers or UNIX or VMS or Fastbus or ACP systems become strong advocates. At some level, this is an understandable tendency to protect an individual's intellectual investment in the expertise. However, the increasing complexity of detectors cries out for unbiased attitudes, a *secular humanist* approach, toward standards. We need to cool off the all too common electronics and computing standards religious wars which every laboratory has encountered.

The complexity of the behemoth detectors of the SSC/LHC era will require nothing short of a new and structured approach to on-line data acquisition and processor systems. Although there is much talk about code management for large off-line programs, it is pleasing that the first extensive application of serious software engineering tools, like structured analysis structured design (SASD), are happening in the on-line data acquisition world. The ALEPH on-line electronics system is a prime example of this. At Fermilab, DO is committed to SASD tools for on-line software. This is also the case at ZEUS. At OPAL and HERA's H1 on-line system management is being addressed through advanced human interfacing techniques taking advantage of Macintosh Hypercard stacks.

The present state of commercially available software engineering tools leaves much to be desired, as anyone who has seen them in use can tell you. There are a variety of packages available. Each of them has some desirable features and lacks others. The ability to automatically generate code from SASD bubbles, even just data defining statements and common blocks, is apparently not yet available, at least not to physicists. One suspects that certain large software development organizations (IBM, DEC, AT&T) are holding some pretty powerful tools close to their corporate chests. Physicists will have to pioneer in this area, encouraging vendors, acting as beta test sites, and being willing to spend money for software and workstations that may be discarded on relatively short time cycles.

Structured design tools and methodologies are advocated because they lead to software that has a dramatically reduced density of errors and which is easy to read and understand so that changes can be made with some hope they will work. Such structured code is also more readily testable, an issue we will return to later. It should be possible to apply the same philosophy to large hardware system design. Jon Thaler at the University of Illinois advocates extending PC board CAE/CAD tools to systems of modules.

At the very least, we need to structure and modularize our large systems in as homogeneous a way as possible. The ALEPH data acquisition system is a particularly good example of an existing well-structured system. LEP experiments do not require large amounts of processing power for triggers, so ALEPH has not had to deal with integrating large numbers of high-level language processors smoothly into a data acquisition system. For some reason, the traditional way to do this is through an event builder. Sub-events come from the many (usually different) data acquisition electronics sub-systems, each handling a sub-detector, to a single one of these very elaborate boxes. It assembles the event and passes it on to the high-level trigger.

The idea of assembling events directly into the memory of the processor seems so obvious. It has not been used because, in most cases, experiments treat the high level processor with trepidation and lack the confidence to incorporate it within the system, where it belongs. Trigger computers and data formatting and calibration processors should be integrated into a single system. Cutts and Zeller at DO have demonstrated that a system can be designed without an event builder. Eight data cables bring data from separate readout systems directly into special multi-port memory that sits on the Q Bus of a MicroVAX in the high level trigger farm. White, Barsotti, *et al.* in the FUSE project proposal at Fermilab have a similar two dimensional architecture approach. The system is extendible in one direction to allow more sub-detector read out controllers and in the other to allow a hierarchy of processing.

Using the Second Generation ACP software and hardware described earlier, one can picture a data acquisition system in which data is read out of VME digitizers and formatted by the same type processors as used in the high level trigger. Data is routed from the formatting processors to the trigger processors to data logging processors and even to monitoring work stations over Branchbus through one or more Branchbus Switch Crates at 20 MBytes/sec per channel. The software environment for all the computing activities is identical and modules are interchangeable throughout. The amount of intellectual effort required to understand and maintain such a homogeneous system is much less than has traditionally been required.

To us, this kind of system clearly points toward where we think we need to get for the far larger SSC data acquisition systems. What development areas should we be addressing? As noted earlier, future processors, and digitizers, should be in the Switch Crates themselves (or a successor technology) to handle the higher rates required by super fast processors. But how do we deal with the 2000 or more CPUs, and the associated digitizers, which probably will be arranged in a tree structure corresponding to the subdetector organization? How would one bring up such a monster when the different pieces of the hardware and software are in varying states of readiness, and robustness? We need a system skeleton that is easy to assemble in the sense that if a number of working subsystems are connected, the combination will work also, automatically. Or, equivalently, if a subsystem is removed, the remainder should continue to operate subject only to functional limitations caused by the missing piece.

The automatic routing ROM in the Branchbus Switch suggests a research direction that may well allow us to accomplish this very important requirement that a system skeleton be easy to assemble. One could imagine a smart switch router that looks at directives accompanying the data and replacing specific addresses. The directives would indicate the kind of processor to which the data should be routed (or from which data should be pulled). Such classifications are available now in software to users of ACP systems who need only request any free node, or any node that has completed a task, or all nodes in some class, or the same node as the last operation, etc. By handling this in hardware rather than software, it appears that a system would be transparently operable with or without all its subsystems. It would be easy to assemble in our way of speaking. This is not going to be an easy development, but it should not be thought impossible. Bruce Knapp has pointed out that this is similar to the data driven processor idea we described earlier for low-level triggers. Only here, the modules are high-powered, high-level language programmed, CPUs.

In addition to the structuring issues described above, one should take advantage of higher speed data transmission and switching technologies, such as serial optical fibers, that are likely to prove very attractive in the near future. Here, certainly, is an important and interesting development opportunity: developing an appropriate skeleton for SSC data acquisition digitizing and processing.

AI AND HIGH ENERGY PHYSICS

Very little has been done with "AI" in high energy physics. That's not because the issues that Artificial Intelligence researchers work on are not relevant to us. In fact, one can argue that almost all high energy physics computing (the exception is theory) is of the non-numerical kind that AI is really all about.[1] The difficulty is these problems are so hard that progress is slow and practical applications limited. In a number of critical areas, HEP's needs for practical AI type solutions are so acute that we will have to get involved in the effort.

I have argued that the way in which to deal with large detector electronics systems and big trigger reductions is to move as much specialized activity as possible into high-level language processors. This implies huge programs, and it is probably partially the case that people have avoided this approach because of fear of errors in the software. Rationally, for a given trigger and amount of effort there is no reason to expect less errors in a complex special trigger processor. Except, perhaps, because the expertise required limits access to a few, presumably very competent, individuals. This is in distinction to software where "everyone" can get their hands in. We can no longer afford such thinking. We will have to manage access along with everything else. But given the huge size that trigger programs will reach (CDF's off-line code is now a million-lines), we will have to learn how to establish their correctness.

There are two aspects of determining correctness. The first is, in principle, straight forward. It is the question of whether the program does what it has been specified to do. This is where computer-aided software engineering (CASE) tools, like SASD, become particularly important. It is likely they will form the platform for more sophisticated program verification tools. To take advantage of (and influence the development of) future CASE tools, we need to get heavily involved with what is presently available.

Though hardly solved, software verification is the easy part. The hard part is how to know whether the specification itself is correct. Everyone is familiar with how mistakes in understanding or definition or whatever, mistakes in the specification, cause major delays in developing off-line reconstruction and analysis code. This business of validating software specifications is a particularly advanced artificial intelligence issue because it involves the whole concept of a "common sense data base".

Both verification and validation are most acute in the real time situation (as many of us have pointed out in the SDI context). On experiments, it must be possible to change triggers as new information is learned. Even if such changes will no longer be tolerated on night shift, they will continue to happen on short time scales, and we have to know that they were both conceived and implemented correctly. While we wait for 21st century AI tools, we will have to discipline ourselves to maintain well-managed, well-structured programs and take advantage of whatever software engineering we can get.

AI is closer to providing practical solutions in some other areas we need, expert systems and human interfaces. Expert systems are already in use on CDF to help ordinary users maintain the Fastbus network during normal data acquisition. Carroll and Booth have provided a means for locating a faulty module without having to locate the university-based expert at home 2000 miles away in the middle of the night. Such systems will become widely used. They are very quick at disgorging expert information, but very slow at absorbing it. High energy physics has particularly dynamic requirements on expert data bases, as systems change rapidly. Because of this somewhat unique requirement, we may have to get involved in improving the information input aspects of expert systems.

Physicists have taken to the Macintosh with amazing speed and unanimity despite the fact that these personal workstations don't really do very much specialized for us yet. This is a clear example of the importance of good human interfaces. When analysis processors soon are available that can turn around a large experiment's DST data base in an hour or less, it will no longer be acceptable for physicists to spend days preparing their next batch of histogram subroutine call statements. At CERN, René Brun has developed the Physics Analysis Workstation (PAW) graphic front end to CERNLIB histogram and plotting software on Apollo and other workstations.

The ACP also has underway a project to develop particularly efficient work station tools for doing analysis. Macintosh like human interfaces, adjusted to physicist needs and abilities, will be used on Apple or, perhaps, Sun or other workstations. Dennis Hall and colleagues at LBL have proposed an Accelerator Designer's Workbench. A Mac would front end a Cray or other super processor. On the Mac screen users would move magnet components around to prepare for a new run of a transport calculation. They are also proposing development of a "software bus" that all such front ends, as well as CAD and project management tools, could plug into.

Most important in these projects is a top down design. "Designer screens" must be effectively prepared by specialists. Physicists are not too good to have information communicated to them efficiently.

A FIRST CUT AT A COHERENT R&D PROGRAM

To conclude, I list the key definable R&D projects suggested by the discussions in the preceding sections. The list is not necessarily complete or properly partitioned but is meant as a starting point in defining a program of R&D for SSC experiment computing needs. Included are guesses at appropriate manpower and annual nonsalary costs (averaged over 5 years).

- Low-level on-detector trigger system, based on VLSI, perhaps neural networks and/or data driven units. 10 persons, $300K.

- On-line data acquisition/multiprocessor system skeleton, pipelined, structured, and extensible (auto routing?). 7 persons, $200K.

- CPUs, multiprocessor software, and development environment commonly useful for on-line triggers and data formatting and off-line reconstruction and analysis. 10 persons, $200K.

- Computer-aided software engineering tools for high energy physics. Standard, usable, across the board program development, verification, and validation. 5 persons interacting closely with commercial CASE development, $500K.

- Expert systems for data acquisition, hardware and software testing, and diagnosis. 5 persons, $150K.

- Analysis workstation environment tools. 5 persons, $75K.

- Data management, code, and data file service tools. 5 persons interacting closely with industry, $250K.

These projects are likely to be very rewarding. At the same time as they represent tremendous leaps in technology, one can enter into them with confidence that there are excellent chances for success. And that success is essential to the physics of the SSC future.

ACKNOWLEDGEMENT

The ideas and speculations expressed in this paper have benefited from many useful discussions and a long collaboration with my colleagues on the Advanced Computer Program: H.Areti, R.Atac, J.Biel, J.Deppe, M.Edel, M.Fischler, I.Gaines, and D.Husby. They are only responsible for the good ideas.

REFERENCES

1. Nash, T., *Conference ... Summary* Proc.Comp. in HEP, Asilomar, CA, Feb. 2-6, 1987, Comp.Phys.Comm. 45, 9-14 (1987).
2. Nash, T., *Trieste Conference ..., Summary & Conluding Remarks*, Adriatico Conf...Processers in Part. Phys., Trieste, March 28-30 ,1988, to be published. FERMILAB-Conf-88/110.
3. Gaines, I., and Nash, T., *Use of New Computer Technologies in Elem. Part.Phys.*, Ann. Rev. Nucl. Part. Sci. 37, 177-212 (1987).
4. Kunz, P. F., Nucl. Instrum. Methods 135 435-440 (1976); Kunz, P.F., Gravina, M., Oxoby, G., Trang, Q., Fucci, A., Jacobs, D., Martin, B., and Storr, K., The 3081/E Processor in Proc. Three Day In-Depth Rev. on the Impact of Specialized Processors in Elementary Part. Phys., Padova, 1983, 83-100 (1983).
5. Gaines, I., Areti, H., Atac., R., Biel, J., Cook, A., Fischler, M., Hance, R., Husby, D., Nash, T., and Zmuda, T., *The ACP Multipr. Sys. at Fermilab*, Comp. Phys. Comm. 45 323-329 (1987); Biel, J., Areti, H., Atac., R., Cook, A., Fischler, M., Gaines, I.,Hance, R., Husby, D., Nash, T., and Zmuda, T., *Software for the ACP Multipr. Sys.*, Comp. Phys. Comm. 45 331-3379 (1987).
6. Nash, T. ,Areti, H., Atac, R., Biel, J., Cook, A., Deppe, J., Edel, M., Fischler, M., Gaines, I., Hance, R., Husby, D., Isely, M., Miranda, E., Miranda, M., Pham, T., Zmuda, T., Eichten, E., Hockney, G., Mackenzie, P., Thacker, H.B., and Toussaint, D., *High Performance Parallel Computers for Science: New Developments at the Fermilab ACP* in Proc. of Workshop on *Computational Atomic & Nucl. Phys. at One Gigaflop*, Oak Ridge, Apr.14-16, 1988, to be published. FERMILAB-Conf-88/97.
7. Mackenzie, P., Eichten, E., Hockney, G., Thacker, H.B., Atac, R., Cook, A., Fischler, M., Gaines, I., Husby, D., Nash, T., *ACPMAPS: The Fermilab Lattice Supercomputer Project*, Proc. Int. Symp. *Field Theory on the Lattice*, Seillac, France, Sept. 28- Oct. 2, 1987, Nucl. Phys.B (Proc. Suppl.)4, pp 580-584 (1988).; Nash, T., Areti, H., Atac, R., Biel, J., Cook, A., Deppe, J., Edel, M., Fischler, M., Gaines, I., Husby, D., Pham, T., Zmuda, T., Eichten, E., Hockney, G., Mackenzie, P., Thacker, H. B., and Toussaint, D., *The Fermilab ACP Multi-array Processor System (ACPMAPS) A Site Oriented Supercomputer For Theoretical Physics*, in Trieste Conf. *ibid*, to be published. FERMILAB-Conf-88/111.
8. Hance, R., Areti, H., Atac, R., Biel, J., Cook, A., Fischler, M., Gaines, I., Husby, D., Nash., T., and Zmuda, T., *The ACP Branchbus and Real Time Applications of the ACP Multiprocessor System*, IEEE Trans. Nucl. Sci, NS-34, 878-883 (1987).

9. *Second Generation ACP Multiprocessor System: System Specification Document*; July 25, 1988, and revisions;ACP unpubl. int. doc. Availability of ACP publications are listed at HEPNET location, `fnacp::acpdocs_root:[docs]doclist.doc`.

10. Christ, N.H., and Terrano, A. E., IEEE Trans. Comp. C-33(4) 344 (1984); Beteem, J., Denneau, M. and Weingarten, D., J. Stat. Phys., 43 1171 (1986); Albanese, M., *et al.*, The APE Computer, ROM 2F/87/005 (1987).

11. ZEUS collab., *ZEUS Status Report 1987*, DESY, Hamburg (1987).

12. Fox, G.C., Otto, S.W., *Phys. Today* 37(5): 50-59(1984);Fox, G., *The Performance of the Caltech Hypercube in Scientific Calculations*. In Supercomputers - Algorithms, Architectures, and Scientific Computation, ed. Masten, F.A., Tajima, T., Univ. Texas (1985); Seitz, C.L., Commun. ACM 28-22(1985).

13. Denby,B.,*Neural Networks & Cellular Automation in Exp. High Energy Physics*, Comp. Phys. Comm. 49, 429-448 (1988).

14. Barsotti, E., Appel, J.A., Bracker, S., *et al.*, IEEE Trans. Nucl. Sci. 26(1),686-96 (1979); Avilez, C., Borten, L., Christian, M., *et al*, Proc. Symp. on ... Computing ... for High Energy Physics, Guanauato, Mexico, 1984, Fermilab, Batavia, 45-54 (1984).

15. Conetti, S., in Trieste Conf. *ibid*, to be published.

REQUIREMENTS FOR CALORIMETRIC JET MASS RECONSTRUCTION FOR HIGH $p_\perp$ JETS AT THE SSC

A. Bay and E. M. Wang*
Lawrence Berkeley Laboratory, Berkeley CA 94720

D. Bintinger, M. G. D. Gilchriese, and D. E. Groom
SSC Central Design Group†
Lawrence Berkeley Laboratory, Berkeley CA 94720

Abstract

We have studied two aspects of calorimetry for high $p_\perp$ jets at the SSC. First, the impact of shower containment on very high $p_\perp$ jets and quark compositeness has been studied. Second, the effect of calorimeter granularity on the mass reconstruction of a high $p_\perp$ W ($p_\perp > 500$ GeV/c) producing jets has been examined in detail. We conclude, based on possible compositeness signatures only, that shower leakage has little effect on the compositeness signal for a calorimeter thickness greater than about eight absorption lengths. Mass resolution of very high $p_\perp$ W's reconstructed from hadronic decays of the W is strongly affected by intrinsic shower spreading as well as calorimeter granularity. These effects are quantified.

1. Introduction

Many previous SSC studies have used, as a simple calorimeter simulation, 4-vectors binned in rapidity and azimuth. At some level, energy smearing was applied to simulate effects of calorimeter resolution. In a real calorimeter, however, a given incoming hadron deposits its energy spatially as a shower of electromagnetic and hadronic particles. Some of this energy may even leak out of the calorimeter or be absorbed by inactive material (supports, cryostat walls, etc.) We have developed a more realistic calorimeter simulation program and have used it to revisit some aspects of high $p_\perp$ jets and calorimeter requirements for the SSC.

2. Parameterization and validation

The desire for a more realistic treatment of calorimeter response other than binned 4-vectors is satisfied, to some extent, by the use of parameterizations of the average behavior of electromagnetic (EM) and hadronic showers. Average EM and hadronic shower shapes are parameterized and factored into independent pieces, a longitudinal distribution along the particle direction and a transverse distribution perpendicular to the particle direction.

2.1 EM shower parameterization

The EM shower parameterization is given by [1]:

$$\frac{dE}{dz} = k_\parallel \left(\frac{z}{X_0}\right)^{a-1} e^{-bz/X_0}$$

where

$$b = 0.5$$
$$a = b\log\left(\frac{E}{E_{\text{critical}}}\right)$$
$$z = \text{longitudinal distance}$$
$$X_0 = \text{radiation length}$$
$$E = \text{incident particle energy}$$
$$E_{\text{critical}} = \frac{0.55}{Z}\ [\text{GeV}]$$
$$Z = \text{atomic number}$$
$$k_\parallel = \text{normalization constant}$$

A comparison of data and our longitudinal parameterization is shown in Fig. 1 [2].

The lateral EM shower shape (shown in Fig. 2) is parameterized by a sum of 2 gaussians:

$$\frac{dE}{dr} = k_\perp \left[e^{(-r/\sigma_1)^2} + e^{(-r/\sigma_2)^2}\right]$$

where

$$\sigma_1 = 0.12\ r_{\text{moliere}}$$
$$\sigma_2 = 0.80\ r_{\text{moliere}}$$
$$k_\perp = \text{normalization constant}$$

The Moliere radius, r_{moliere}, is the average transverse deflection of an electron with energy equal to the critical energy after travelling one radiation length longitudinally.

2.2 Hadronic shower parameterization

The longitudinal hadronic shower parameterization we have chosen [3] is:

$$\frac{dE}{dz} = c_\parallel \left[w\left(\frac{z}{X_0}\right)^{a-1} e^{-bz/X_0} + (1-w)\left(\frac{z}{\lambda}\right)^{c-1} e^{-dz/\lambda}\right]$$

where

$$z = \text{longitudinal distance}$$
$$X_0 = \text{radiation length}$$
$$\lambda = \text{absorption length}$$
$$a, b, c, d, w = \text{fitted constants}$$
$$c_\parallel = \text{normalization constant}$$

This parameterization has two terms, the first accounting for electromagnetic energy deposition and the second

* This work was supported by the Director, Office of Energy Research, Office of High Energy and Nuclear Physics, Division of High Energy Physics of the U.S. Department of Energy, Contract DE-AC03-76SF00098.

† Operated by the Universities Research Association for the U. S. Department of Energy.

for hadronic energy deposition. References 3 through 6 present studies of longitudinal hadronic shower development utilizing the above parameterization. We have chosen to use the fitted constants of Ref. 3 because they are based on data with the widest range of incident energy hadrons, including the highest energy of 400 GeV, and because the resultant parameterization behaves sensibly up to an incident energy of at least 1 TeV. Shower shapes using the fitted constants of Ref. 3 are shown in Fig. 3 for typical energies. Figure 4 presents the fraction of incident energy accounted for by the electromagnetic term versus incident energy for the four sets of fitted constants of Refs. 3 through 6. (An absorption length of 22 cm and a radiation length of 0.8 cm have been assumed.) It can be seen that results derived from Refs. 4 and 5 show unrealistic behavior beyond 100 GeV. Reference 6 gives reasonable behavior, but is based on incident energies up to only 150 GeV. At present, the model does not include longitudinal fluctuations of the showers except for fluctuations in the starting point of the shower. In particular, it neglects the fluctuations in a shower initiated by a hadronic particle which produces a large number of π^0's at an early stage of its development. These kinds of effects are only averaged when using parameterizations. Figure 5 presents a comparison at an incident energy of 50 GeV on solid lead of our parameterization to the longitudinal shower shape generated by the Monte Carlo programs FLUKA87 [7], CASIM [8], and MARS10 [9].

The transverse hadronic shape is a function of both incident energy and depth along the shower. Appropriate detailed data from measurements are scarce, particularly for large lateral distances. We have therefore used the shower program FLUKA87 to characterize the lateral shapes. For a given longitudinal slice at depth z, the transverse shape is parameterized as either a piecewise function of two exponentials ($z < 2\lambda$) or a single exponential ($z > 2\lambda$). A table of these fitted exponential slope constants based on calculations at 6 energies (1, 20, 50, 100, 500, and 1000 GeV) is used to characterize the entire energy and depth dependence of the lateral spread. Interpolation was not used for different energies. The material dependence of the shape (for reasonable choices of material)is weak when distances are expressed in absorption lengths. Figure 6 shows some typical profiles from FLUKA87. It is known that FLUKA87 agrees well with data in the lateral region for which data exist [10].

2.3 Implementation in simulation

These parameterizations are then implemented in a fast detector simulation program. The GEANT3 [11] package was used as the basic framework for description of the detector geometry. Detectors of arbitrary geometry and composition may then be studied. The typical shower simulation for a hadron incident upon a calorimeter is carried out using the following steps:

1. Choose shower start position from the distribution $e^{-z/\lambda}$.
2. Step through material with step size, $\Delta z \sim 0.1\lambda$, and at each longitudinal step position, z, divide the appropriate energy deposition ,

$$\Delta E = \int_{z-\Delta z/2}^{z+\Delta z/2} \frac{dE}{dz}\, dz \ ,$$

   into thirds and deposit each of the thirds at three different transverse and azimuthal positions according to the dE/dr distribution obtained from fits to FLUKA87 and a uniform azimuthal distribution. This choice gives a reasonable number of depositions and acceptable time for calculation. We have not looked at the sensitivity of all of the results to using more than three azimuthal and transverse points per step position.
3. The contribution to the overall resolution due to intrinsic effects such as the sampling or the e/π response are included by smearing the energy deposited by each incident particle, according to the parameterization

$$\frac{\sigma(E)}{E} = c_1 + \frac{c_2}{\sqrt{E}}$$

   where

$$E = \text{total energy deposited}$$
$$c_1, c_2 = \text{parameters characterizing response}$$

   An incident particle deposits energy E. A new total deposited energy, E_{new}, is chosen from a gaussian distribution of variance $\sigma^2(E)$ centered on E. The individual depositions which originally summed to E are rescaled to add to E_{new}.

3. Physics signals with high $p_\perp$ jets

3.1 Calorimeter thickness

The question of calorimeter thickness can be approached from the point of view of asking how many absorption lengths are needed to contain a given fraction of a single hadron's or a jet's incident energy or from the point of view of how many absorption lengths are needed to observe a physics signal. Given the parameterization of Section 2.2, the question of thickness to contain a fraction of a single hadron's or a jet's energy can be calculated. The point of shower initiation was chosen from the distribution $e^{-z/\lambda}$. The depth in absorption lengths versus incident particle energy to contain 95% or 99% of a single hadron's energy is presented in Fig. 7. These curves represent average shower containment. Figures 8 and 9 present histograms of calorimeter depth required to contain 95% and 99% respectively of detectable energy of average incident jet energies of 1.5, 3.5, and 5.5 TeV. Detectable energy means the jet energy less the energy of any muons or neutrinos. The histograms represent the effects of fluctuations in jet composition. The jets were generated by the program PYTHIA [12]. Figure 8, for example, shows that seven absorption lengths will contain 95% of the jet's detectable energy for 95% of generated 1.5 TeV jets. Similarly, Fig. 9 shows that 11.4 absorption lengths will contain 99% of the jet's detectable energy for 99% of generated 3.5 TeV jets. One should note that these conclusions depend on the longitudinal shower parameterization. For example, if the fitted constants of Ref. 6 are used, the calorimeter depths required for containment will increase by one to two absorption lengths. Of the fitted constants of Refs. 3 through 6, those of Ref. 6 yields the greatest containment depth. However they are based on data up to only 150 GeV. It should also be noted that a radiation length to absorption length ratio of 0.036 has been assumed for these and subsequent thickness calculations. The results presented are very insensitive to this ratio.

The observation of a quark compositeness signal via hadronic jets requires the reconstruction of TeV scale jets [13]. In order to study how calorimeter thickness affects the observation of this signal, $p_\perp$ spectra of jets were generated simulating QCD and QCD plus a compositeness signal. The compositeness scale (Λ) was 15 TeV and constructive interference was assumed. In each case jet masses were reconstructed, varying the calorimeter thickness. Figure 10 shows the QCD plus compositeness signal with an infinite thickness calorimeter and calorimeters of 8 and 11 absorption lengths. Included in this calculation were energy resolutions of $0.15/\sqrt{E}+0.01$ and $0.50/\sqrt{E}+0.02$ for the electromagnetic and hadronic portions respectively. Figure 10 shows little if any degradation of the signal even with a thickness of eight absorption lengths. This conclusion also holds for the QCD-only signal. This can be understood by the shallowness of the slopes of the $p_\perp$ distributions in comparison to the several percent of energy either lost or smeared. Indeed this same conclusion was reached in Ref. 13, when considering resolution only. Thus a calorimeter with a thickness of 8 absorption lengths seems sufficient to observe compositeness with hadronic jets. It should be noted that longitudinal shower fluctuations have not been included in these calculations and may change conclusions that do not depend only on average shower shape.

3.2 High $p_\perp$ W's and transverse segmentation and shower spreading

Reconstruction of a hadronically decaying $W^\pm$ or Z^0 may be important in some physics searches (such as a heavy Higgs search [14]). There have been a number of studies [15, 16] sometimes conflicting, of the effects of calorimeter granularity (size of $d\eta, d\phi$ bins) on W mass resolution. We have chosen to revisit this issue using as a "benchmark" the highest $p_\perp$ W's that might be produced at the SSC. The effects of the transverse segmentation of a calorimeter with tower geometry on the mass reconstruction of high $p_\perp$ $W's$ ($>$ 500GeV) have been studied. In addition, the effects of transverse shower spreading have also been examined.

Our procedure is as follows:

1. Centrally produced and hadronically decaying W's with $p_\perp > 500$ GeV and $\theta_W > 30°$ are generated with PYTHIA v4.9.
2. Two arrays binned in pseudorapidity versus azimuth weighted by $E_\perp$ are filled, one using 4-vectors and the other using the earlier described detector simulation with the full shower parameterizations. Neutrinos and muons appear as lost energy. Figure 11 displays the two methods. Only particles from the W decay are used in the reconstruction as these may be "tagged" in the Monte Carlo. The basic detector is a cylinder with $r_{\text{inner}} = 160$ cm and $r_{\text{outer}} = 400$ cm and a half-length = 400 cm. The absorption length is 22 cm and radiation length is 0.8 cm. These correspond roughly to the parameters of the Large Solenoid Detector (LSD) model from the 1987 Berkeley SSC Detector Workshop [17]. Figure 12 shows the detector geometry used.
3. The parameters that are varied in this study are the segmentation, $d\eta$ and $d\phi$, r_{inner} and r_{outer} and the absorption (λ) and radiation (X_0) lengths.
4. From an array of $E_\perp$ values, the mass m_W is calculated as

$$m_W = \sqrt{\left(\sum_i E_i\right)^2 - \left(\sum_i \vec{p_i}\right)^2}$$

where E_i and $\vec{p_i}$ are determined by the energy and position of the ith cell. The position is assigned to the center of the cell. Clustering algorithms are not employed since only particles from the W are used. This is done to clarify detector induced vs. clustering algorithm induced effects.

We have studied the m_W distribution versus granularity for a number of different cases which are described below. We will subsequently call the binned 4-vector study "ideal" and the more realistic detector with Large Solenoid Detector geometry and fast shower simulation "LSD".

In addition, we have studied the impact on W mass resolution from the pileup of "minimum bias" events for some of the cases to be described. Our "minimum bias" sample consisted of a 50-50 mixture of ISAJET [18] minbias and low $p_\perp$ QCD jet events ($>$ 5 GeV/c). The simplest case is for single bunch crossing pileup where the number of interactions is given by a poisson distribution with mean 1.5. This is ideal in the sense that it represents operating conditions for a calorimeter that has a resolving time less than the collider interbunch spacing time. In a more realistic case the total number of interactions is given by integrating over 7 bunch crossings corresponding roughly to a 100 ns integration time. If our simple mass reconstruction algorithm is used with no additional cuts, reconstruction of the W mass is severly degraded due to pileup. For this reason, two cuts were used in our study of pileup effects:

1. The contributions of cells with $E_\perp < (d\eta/0.05)(d\phi/0.05) \times 1$ GeV were eliminated.
2. Cells outside a cone around the W direction with radius $= 0.5$ in η and ϕ space were eliminated. NOTE: While this is not possible experimentally, the mean direction of hadron energy flow is usually not too far from this.

These two cuts were applied only for the cases concerning pileup effects.

The various cases studied were:

1. Ideal detector, perfect energy resolution.
2. (a) Ideal detector, smeared energies, $c_1 = 0.01$ (0.02), $c_2 = 0.15$ (0.50) for EM (hadronic) depositions.

   (b) $c_1 = 0.01$ (0.02), $c_2 = 0.30$ (1.00) for EM (hadronic) depositions.
3. LSD, perfect energy resolution, $X_0 = 0.8$ cm, $\lambda = 22$ cm, $r_{\text{inner}} = 160$ cm, $r_{\text{outer}} = 400$ cm, no magnetic field, no pileup.
4. (a) As in (3) but with smeared energies, $c_1 = 0.01$ (0.02), $c_2 = 0.15$ (0.50) for EM (hadronic) depositions.

   (b) As in (3) but with smeared energies, $c_1 = 0.01$ (0.02), $c_2 = 0.30$ (1.00) for EM (hadronic) depositions.
5. As in (3) but with $r_{\text{inner}} = 80$ cm.
6. As in (3) but with $X_0 = 0.37$ cm $\lambda = 12.3$ cm, $r_{\text{inner}} = 80$ cm. These parameters correspond roughly to that of a compact uranium-silicon calorimeter [19].
7. As in (3) but with a 2 Tesla magnetic field.
8. As in (3) but with single bunch pileup.
9. As in (3) but with 7 bunch pileup.

Cases 1, 2, 3 and 4, m_W and no pileup

Figures 13 and 14 show typical mass peaks and Fig. 15 shows the m_W rms resolution where resolution is defined as sigma for an m_W window between 60 and 120 GeV. This window was selected to reduce effects of low mass tails in the distribution due primarily to energy carried away by neutrinos. The resolution for the ideal calorimeter (binned 4-vectors) improves with finer segmentation until $d\eta = d\phi = 0.03$ and then flattens out with no further improvement, confirming earlier studies [15, 16]. The W mass resolution including the effects of shower spreading shows a weaker segmentation dependence. The effects of detector resolution have also been checked for reasonable values of energy resolution, $c_1 = 0.01$ (0.02), $c_2 = 0.15$ (0.50) for EM (hadronic). Energy resolution effects are a relatively small contribution to the W mass resolution. This is not unexpected for W's in this $p_\perp$ range, in which one anticipates that angular resolution will primarily determine mass resolution.

Table 1

W mass and rms resolution (using $60 < m_W < 120$ GeV window for calculation). Values in parentheses are the rms values. Typical errors for the rms values are about 1 GeV.

Case	Segmentation $d\eta$ $(= d\phi)$ 0.01	0.03	0.05	0.10
1	77.5 (5.5)	77.8 (5.6)	77.9 (6.4)	80.5 (9.5)
2(a)	77.1 (6.1)	77.9 (5.6)	77.9 (6.4)	80.4 (9.7)
2(b)	78.0 (5.8)	78.6 (6.2)	79.9 (6.8)	80.9 (10.1)
3	82.0 (9.0)	82.6 (9.4)	83.4 (9.2)	84.8 (11.1)
4(a)	82.3 (8.9)	82.6 (9.5)	83.2 (9.6)	84.8 (11.2)
4(b)	82.8 (8.9)	83.2 (9.0)	83.5 (9.8)	85.1 (11.5)
5	87.4 (11.8)	87.6 (11.7)	82.5 (13.1)	89.2 (12.6)
6	84.1 (10.0)	84.0 (10.0)	84.9 (10.7)	86.3 (11.8)
7	84.0 (9.9)	not run	84.7 (10.1)	not run
8	not run	not run	78.8 (8.3)	not run
9	81.9 (8.9)	79.6 (8.5)	79.0 (8.6)	not run

Cases 5 and 6 (m_W and detector size)

What happens if r_{inner} is changed to 80 cm and $\lambda = 12.3$ cm? This situation corresponds roughly to some of the most compact and dense of the calorimeters proposed thus far [19]. Figure 16 shows that the m_W resolution is still dominated (if not more so) by shower spreading.

Case 7, (m_W and a Magnetic Field) If a 2 Tesla magnetic field is used, the W mass resolution worsens slightly, as can be seen from the two "starburst" points in Fig. 16.

Cases 8 and 9 (m_W and event pileup)

We add in pileup events and apply the cell $E_\perp$ and cone cuts described earlier and try to reconstruct the W. As shown in Fig. 17, pile-up effects can be controlled using the above cuts in the sense that the resolution is not different to within statistics from the case where pileup is absent. For lower $p_\perp$ W's (> 150 GeV), however, the above algorithm is not useful since particles are dispersed over larger regions and the cone must be enlarged. Clustering algorithms are then necessary. This has not yet been pursued.

Table 1 summarizes the various scenarios described above. In addition, we tabulate reconstruction efficiencies in Table 2, where the efficiency is defined as the fraction of the events falling in a mass window ± 10 GeV around the peak.

Table 2

Fraction of events falling within ± 10 GeV of the peak.

Case	Segmentation $\delta\eta$ $(= \delta\phi)$ 0.01	0.03	0.05	0.10
1	0.63	0.63	0.64	0.55
2(a)	0.65	0.65	0.63	0.54
3	0.62	0.59	0.57	0.55
6	0.59	0.60	0.58	0.52
9	0.61	0.59	0.55	not run

4. Conclusions

We have studied two aspects of calorimeter requirements for detection of high $p_\perp$ jets at the SSC. Our studies of a possible quark compositeness signature, which will be manifested in the central rapidity region, indicate that a calorimeter depth of eight absorption lengths or more would be sufficient to detect a compositeness signature. Energy leakage from a calorimeter of this depth does not significantly perturb the shape of the jet $p_\perp$ spectrum. A similar conclusion [16] has been reached from studies of the mass resolution of high $p_\perp$ W decays to jets. Hence, in the central rapidity region at the SSC, calorimeter depths beyond eight absorption lengths must be justified on a basis other than energy containment, for example, hadronic punchthrough and muon detection.

We have studied the effects of calorimeter granularity and energy resolution on the mass resolution of W's with $p_\perp > 500$ GeV. We conclude that, for this $p_\perp$ range, intrinsic hadronic shower spreading has a significant impact on W mass resolution for a calorimeter inner radius/absorption length less than about eight. For a calorimeter with an absorption length of 20-25 cm (typical of uranium-liquid argon or compensating lead-scintillator) a calorimeter radius greater than 1.6 m is needed to take advantage of decreasing the calorimeter granularity ($\delta\eta = \delta\phi = 0.05$) to improve the W mass resolution. It may be possible to take advantage of finer calorimeter granularity at a smaller radius at the expense of reduced efficiency by employing clustering algorithms to select hadronic W decays with well separated jets [21]. We have not yet studied this possibility. For such high $p_\perp$ W's, the effect of pile-up of minimum bias events may be removed by requiring a minimum $E_\perp$ in the calorimeter cells used in the reconstruction of the W mass.

References

1. E. Longo *et al.*, Nucl. Instrum. Meth. **128** (1975) 283.
2. Bollini *et al.*, Nucl. Instrum. Meth. **171** (1980) 237.
3. R. K. Bock, T. Hansl-Kozanecka, and T. P. Shah, Nucl. Instrum. Meth. **186** (1981) 533.
4. S. Aronson *et al.*, Nucl. Instrum. Meth. **A269** (1988) 492-506.
5. E. Hughes, "Measurement of Hadronic and Electromagnetic Shower Development between 10 and 140 GeV by an Iron-Scintillator Calorimeter," unpublished CDHS Internal Report, (August 1986).
6. J. Freeman and A. Beretvas, *Physics of the Superconducting Supercollider*, ed. by R. Donaldson and J. Marx (Snowmass CO, 1986) 482.
7. P. A. Aarnio, A. Fassò, J. Lindgren, J. Ranft and G. R. Stevenson, "Enhancements to the FLUKA86 Program (FLUKA87)," CERN TIS-RP/190 (1987); see also P. A. Aarnio, A. Fassò, H-J. Möhring, J. Ranft and G. R. Stevenson, "FLUKA86 User's Guide," CERN TIS-RP/168 (1986).
8. A. Van Ginneken, Fermilab Report FN-272, Fermilab, Batavia, IL, USA (1975).
9. N. V. Mokhov, J. D. Cossairt. Nucl. Instrum. Methods, **A244**, 349 (1986); N. V. Mokhov, Soviet J. Particles and Nuclei, (Sept.-Oct. 1987), pp. 408-426.
10. N. V. Mokhov, Sov. J. Part. Nucl. **18** (Sept.–Oct. 1987) 5, and N. V. Mokhov, private communication.
11. R. Brun *et al.*, "GEANT3," CERN DD/EE/84-1 (Sept. 1987).
12. H. U. Bengtsson and T. Sjostrand, *Physics of the Superconducting Supercollider*, ed. by R. Donaldson and J. Marx (Snowmass CO, 1986) 311.
13. V. Barnes *et al.*, *Proc. of the Workshop on Experiments, Detectors, and Experimental Areas for the Supercollider*, ed. by R. Donaldson and M. G. D. Gilchriese, Berkeley CA July 7-17 (1987) 235.
14. R. N. Cahn *et al.*, *Proc. of the Workshop on Experiments, Detectors, and Experimental Areas for the Supercollider*, ed. by R. Donaldson and M. G. D. Gilchriese, Berkeley CA July 7-17 (1987) 20.
15. E. Fernandez *et al.*, *Proceedings of the 1984 Summer Study on the Design and Utilization of the Superconducting Super Collider*, Eds. R. Donaldson and J. Morfín, Snowmass, Colorado (1984) 107.
16. J. Freeman and C. Newman-Holmes, *Proc. of the Workshop on Experiments, Detectors, and Experimental Areas for the Supercollider*, ed. by R. Donaldson and M. G. D. Gilchriese, Berkeley CA July 7-17 (1987) 673.
17. G. G. Hanson, S. Mori, L. G. Pondrom, and H. H. Williams, *Proc. of the Workshop on Experiments, Detectors, and Experimental Areas for the Supercollider*, ed. by R. Donaldson and M. G. D. Gilchriese, Berkeley CA July 7-17 (1987) 340.
18. F. E. Paige and S. D. Protopopescu, *Physics of the Superconducting Supercollider*, ed. by R. Donaldson and J. Marx (Snowmass CO, 1986) 320.
19. J. Kirkby *et al.*, *Proc. of the Workshop on Experiments, Detectors, and Experimental Areas for the Supercollider*, ed. by R. Donaldson and M. G. D. Gilchriese, Berkeley CA July 7-17 (1987) 388.
20. J. Kirkby *et al.*, *Proc. of the Workshop on Experiments, Detectors, and Experimental Areas for the Supercollider*, ed. by R. Donaldson and M. G. D. Gilchriese, Berkeley CA July 7-17 (1987) 388.
21. C. Newman-Holmes and J. Freeman, Workshop on Detector Simulation for the SSC, Argonne National Laboratory Report ANL-HEP-CP-88-51 (1987) 285. See also Ref. 16.

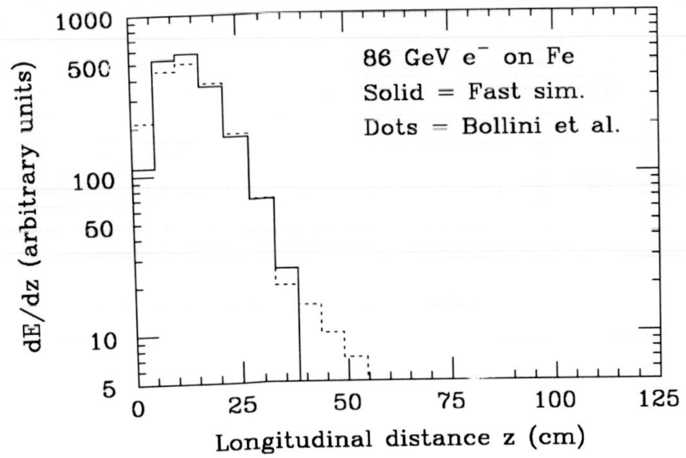

FIG. 1. A comparison of data from 86 GeV electrons incident on an iron calorimeter with our fast shower simulation. Energy deposition vs. longitudinal depth in shower.

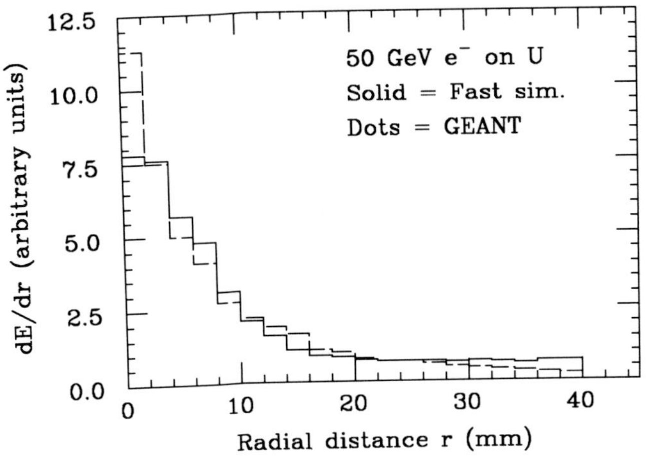

FIG. 2. A comparison of the radial electromagnetic shower shape for 50 GeV electrons on uranium from GEANT3 (solid histogram) and our parameterization (dashed histogram).

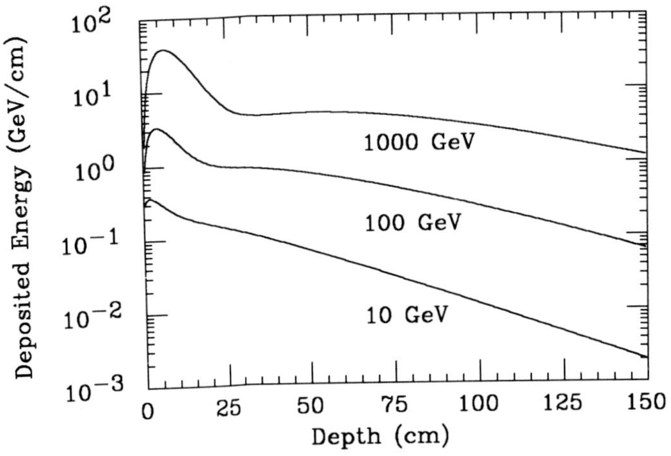

FIG. 3. The longitudinal hadronic shower shape for hadrons of several energies incident on a medium with an absorption length of 22 cm and a radiation length 0.8 cm, using the parameterization given in the text and the fitted constants of Ref. 3.

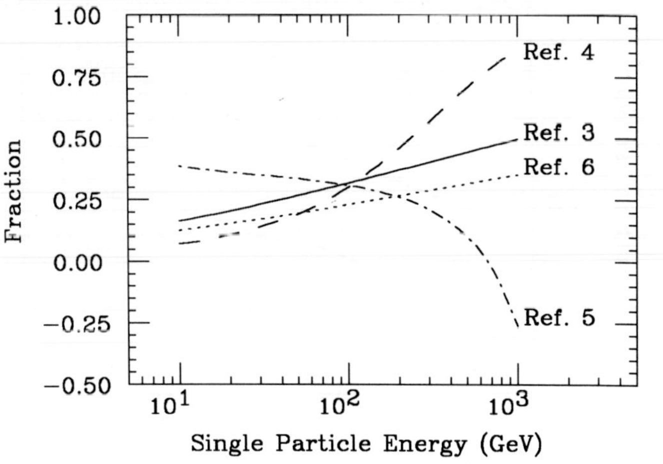

FIG. 4. Fraction of the incident energy accounted for by the electromagnetic term vs. incident energy using the parameterization given in the text and the fitted constants of Refs. 3 through 6.

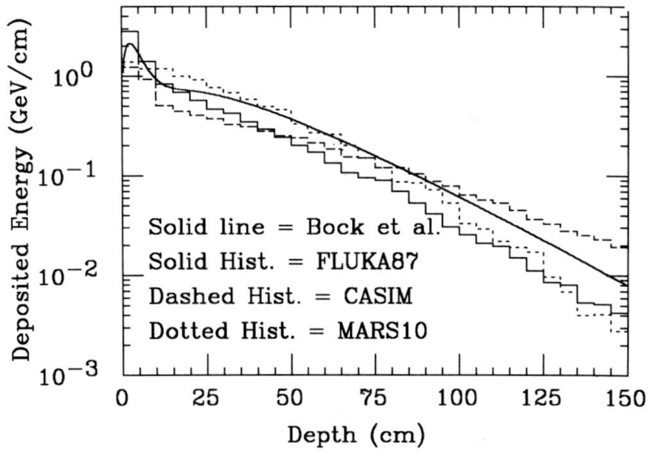

FIG. 5. A comparison of the longitudinal hadronic shower parameterization (Bock *et al.*) with three Monte Carlo programs (CASIM, FLUKA87, and MARS10) for 50 GeV hadrons incident on solid lead.

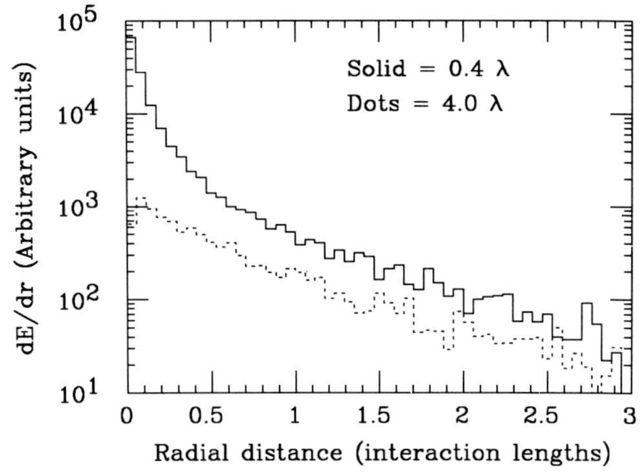

FIG. 6. Radial distributions as calculated by FLUKA87 (solid histogram) for 50 GeV incident hadrons at longitudinal depths of 0.4 λ and 4.0 λ.

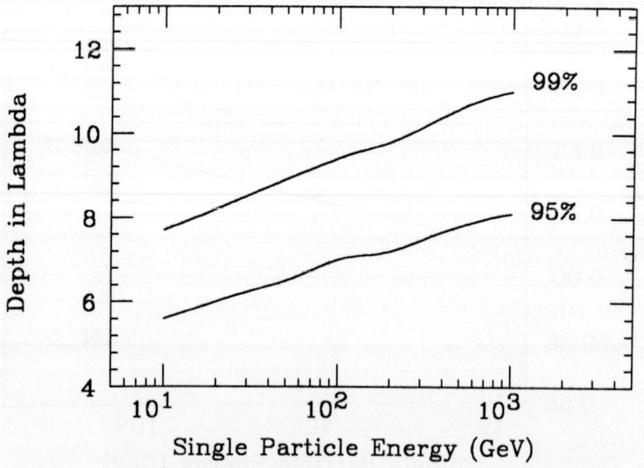

FIG. 7. The depth in absorption lengths vs. incident particle energy to contain 95% and 99% of a single hadron's energy, using the parameterization given in the text and the fitted constants of Ref. 4.

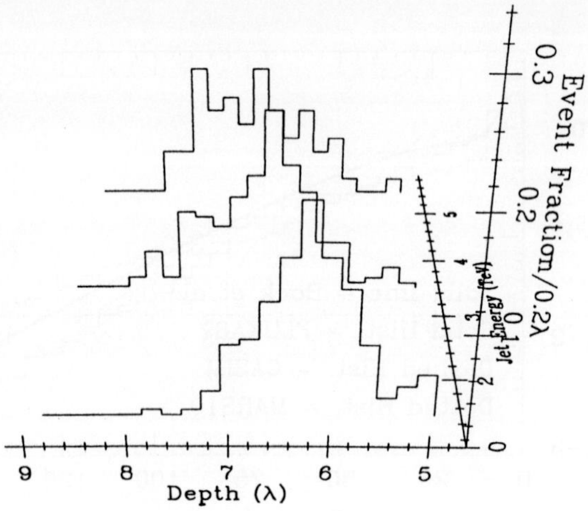

FIG. 8. Histograms of the fraction of jets vs. the number of absorption lengths to contain 95% of the jet's detectable energy for average jet energies of 1.5, 3.5, and 5.5 TeV. The histograms are normalized to unit area.

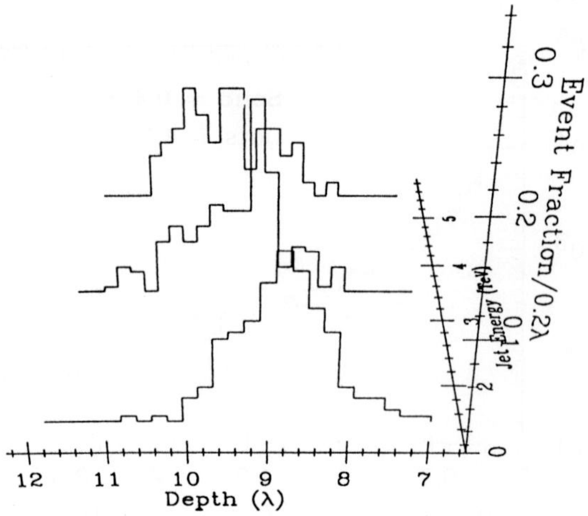

FIG. 9. Histograms of the fraction of jets vs. the number of absorption lengths to contain 99% of the jet's detectable energy for average jet energies of 1.5, 3.5, and 5.5 TeV. The histograms are normalized to unit area.

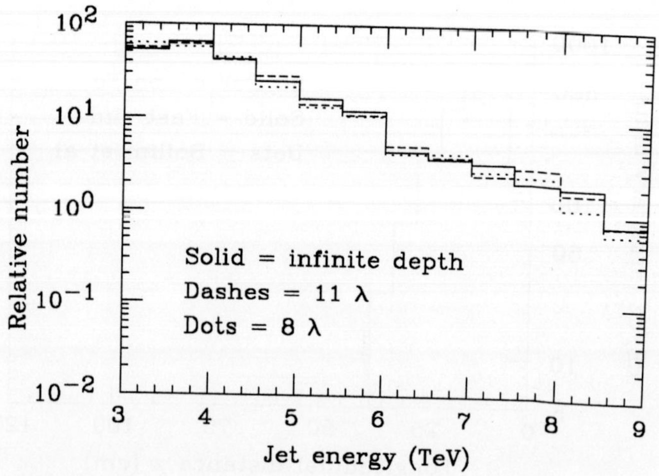

FIG. 10. The $E_\perp$ distribution of jets assuming a compositeness scale of 15 TeV with constructive interference for an infinite depth calorimeter and for calorimeter depths of 8 and 11 absorption lengths.

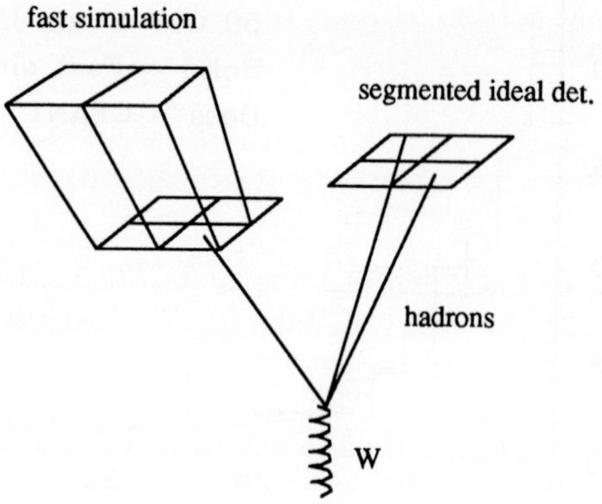

FIG. 11. A schematic representation showing the difference between an ideal detector and simulation with fast shower parameterization.

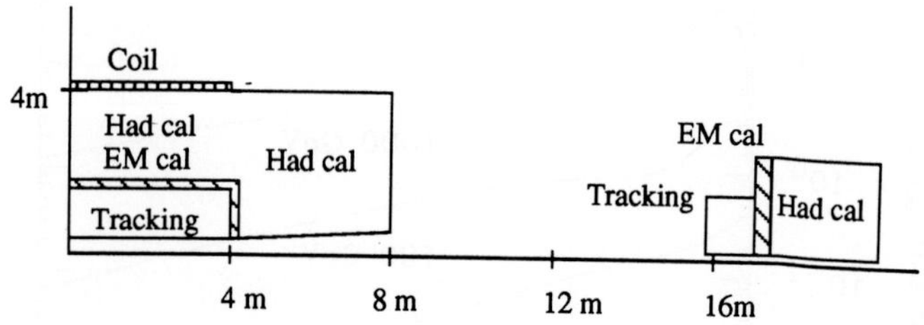

FIG. 12. Picture of the geometry assumed for simulations of a large solenoid detector.

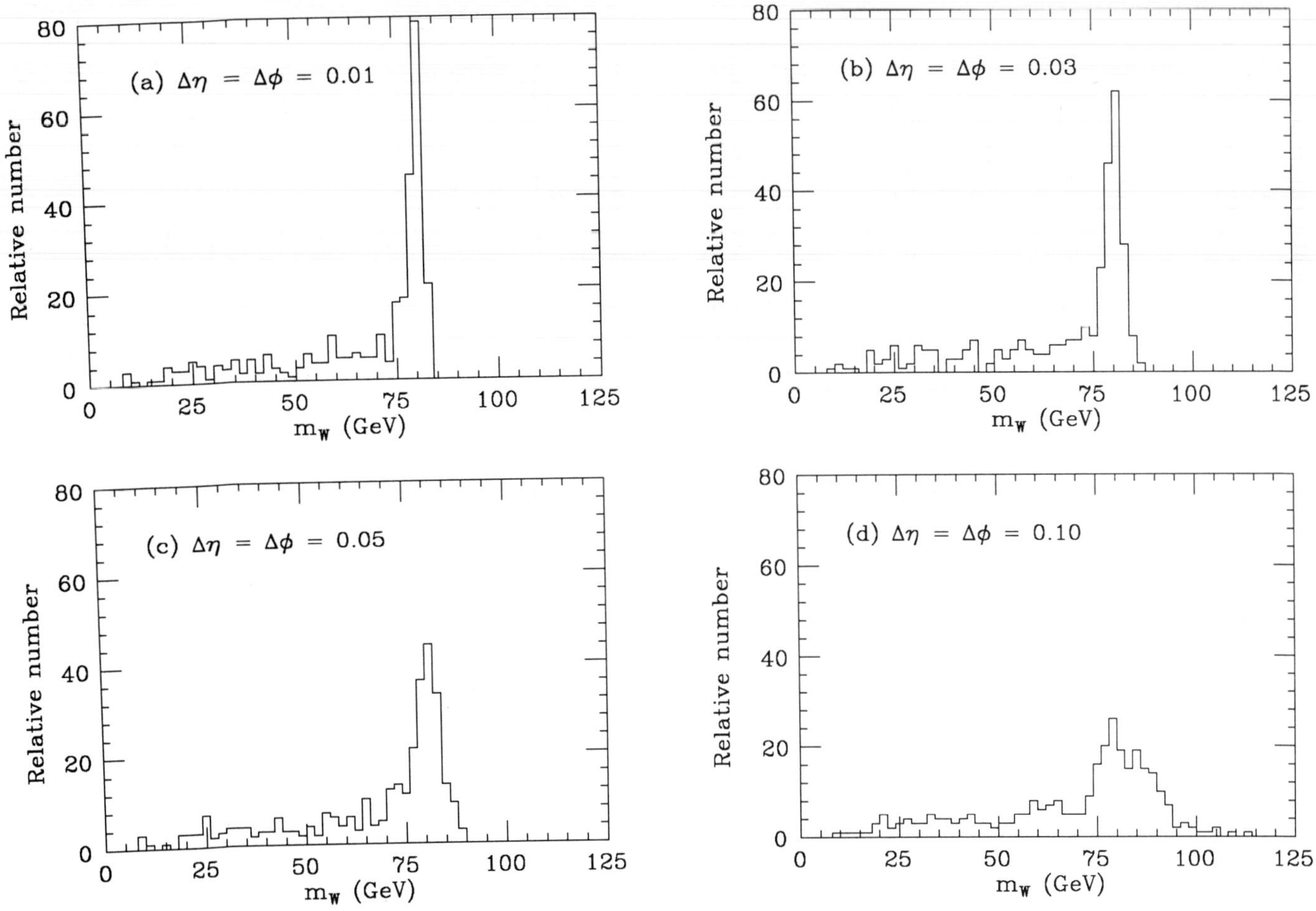

FIG. 13. Case 1: The W mass distribution assuming mass reconstruction using 4-vectors binned in rapidity and azimuth.

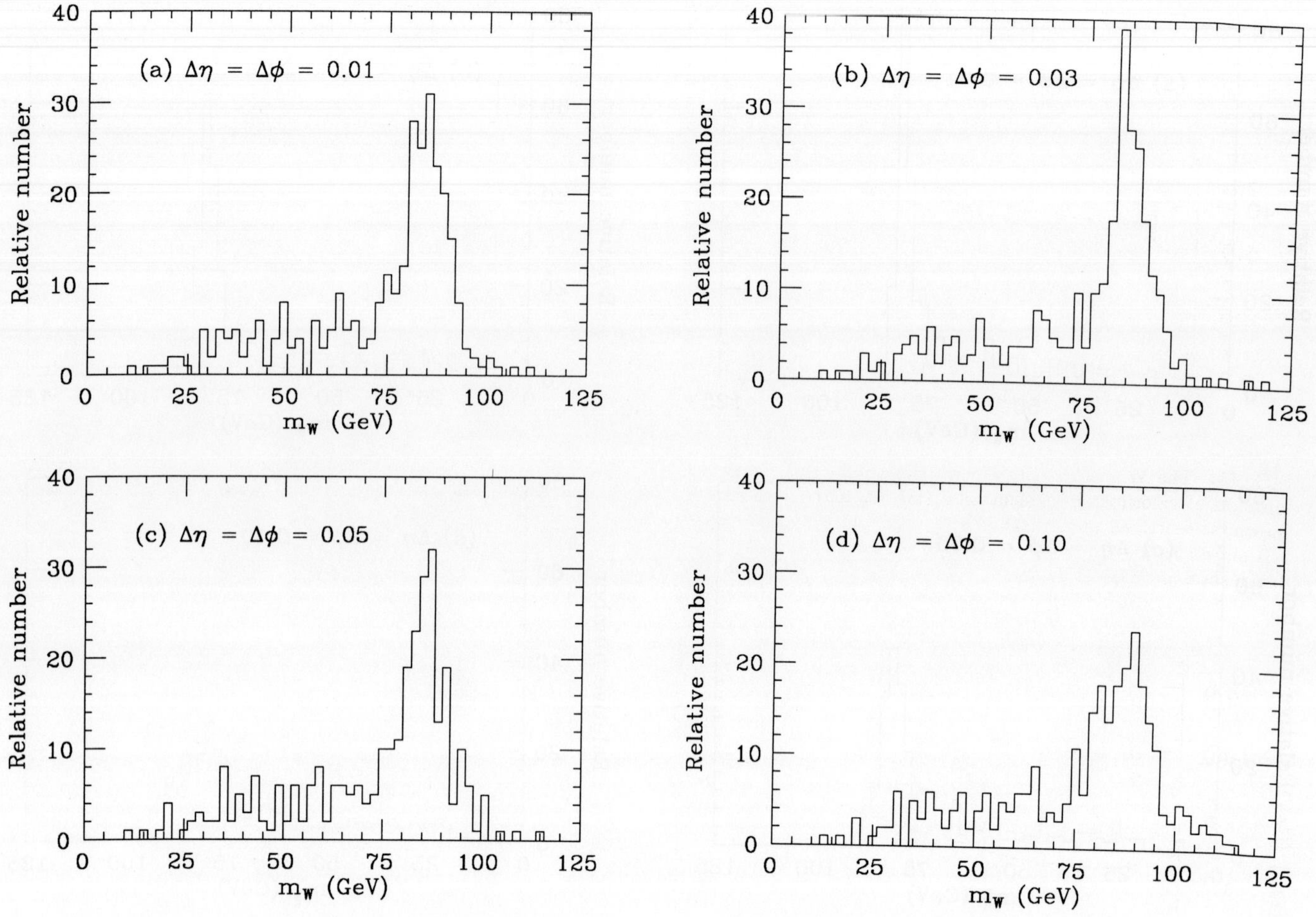

FIG. 14. Case 3: The W mass distribution with perfect energy resolution using the LSD detector simulation.

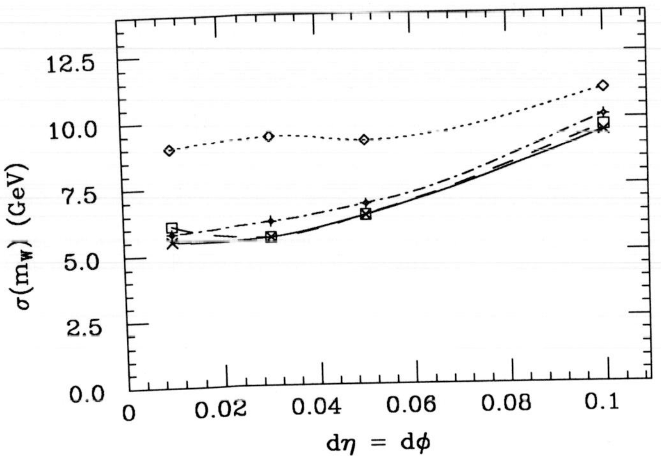

FIG. 15. The W mass resolution (as defined in the text) as a function of calorimeter segmentation ($\Delta\eta = \Delta\phi$) for the cases shown, as described in the text. No event pileup is present. Case 1 is shown as a solid curve, Case 2(a) as dashes, Case 2(b) as dot-dashes, and Case 3 as dots.

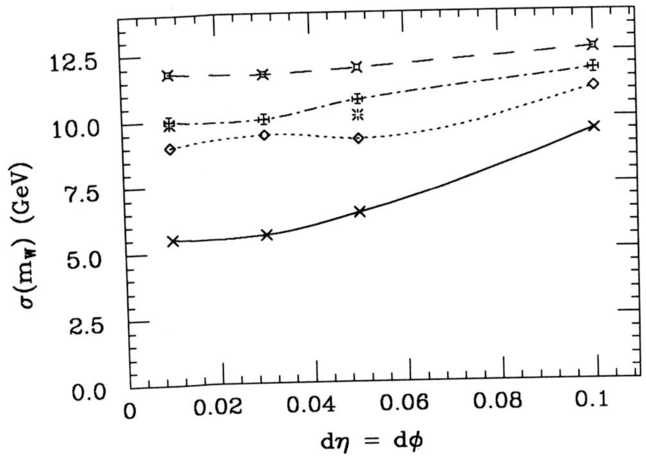

FIG. 16. The W mass resolution for different inner calorimeter radii and absorption lengths. For comparison, Case 1 is shown as solid, Case 3 as dots, Case 5 as dashes, Case 6 as dot-dashes and Case 7 as starburst points.

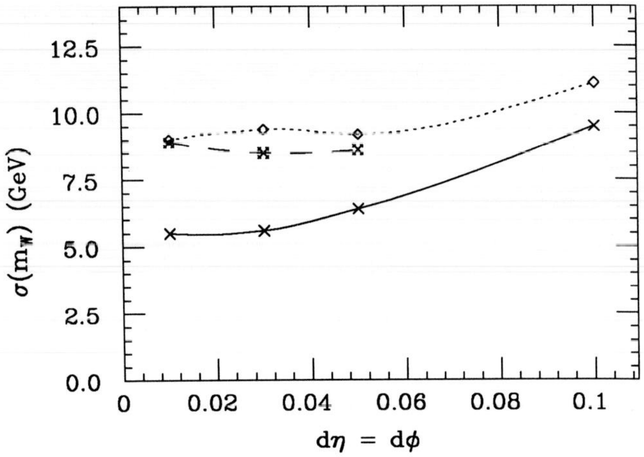

FIG. 17. The W mass resolution, including pileup as described in the text. Case 1 is shown as solid, Case 3 as dots and Case 9 as dashes.

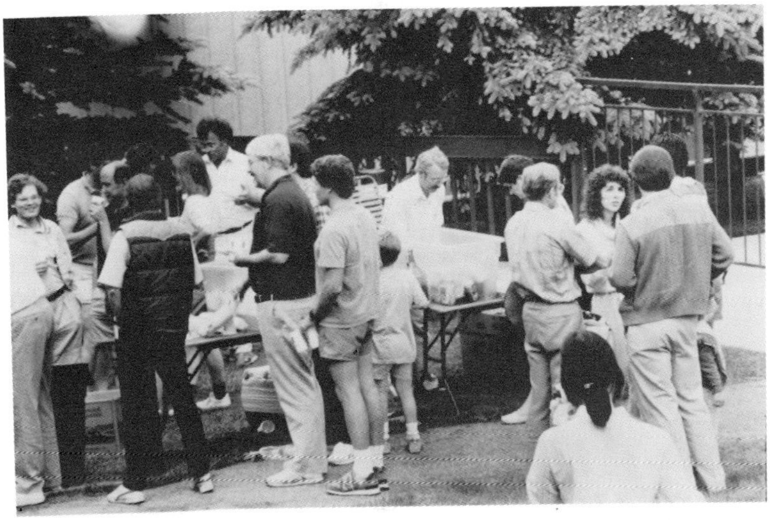

A - K

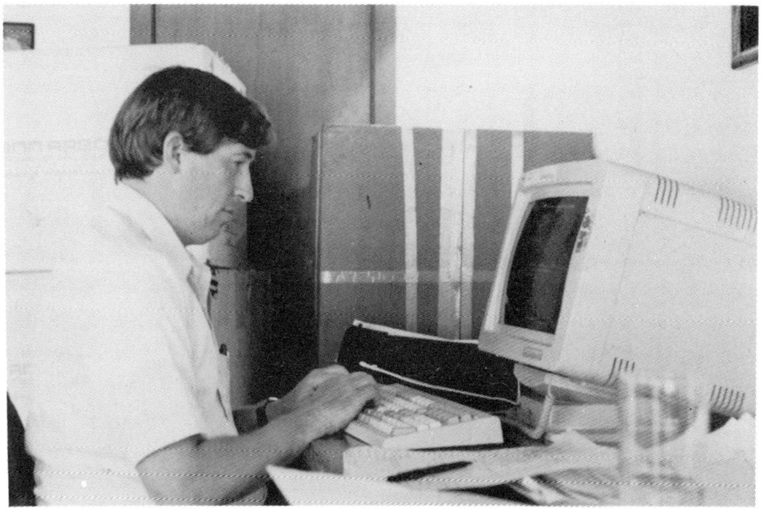

PARTICIPANTS

Snowmass Participants

1. Abashian, Al
   Virginia Polytechnic Institute
2. Abolins, Maris A.
   Michigan State University
3. Adair, Robert K.
   Brookhaven National Laboratory
4. Adams, Thomas S.
   Martin Marietta Astronautics
5. Akerlof, Carl W.
   University of Michigan
6. Alam, Mohammad S.
   SUNY, Albany
7. Aleksan, Roy
   Stanford Linear Accelerator Center
8. Alexander, Jim
   Stanford Linear Accelerator Center
9. Ankenbrandt, Charles M.
   Fermi National Accelerator Laboratory
10. Arens, John F.
    University of California
11. Ash, William
    Stanford Linear Accelerator Center
12. Atiya, Maged S.
    Brookhaven National Laboratory
13. Atwood, David
    McGill University
14. Atwood, William B.
    Stanford Linear Accelerator Center
15. Baartman, Rick
    TRIUMF
16. Baden, Drew
    Harvard University/FNAL
17. Baer, Howard A.
    Florida State University
18. Bailey, David
    University of Toronto
19. Baltay, Charles
    Columbia University
20. Barbaro-Galtieri, Angela
    Lawrence Berkeley Laboratory
21. Barbier, Louis M.
    Louisiana State University/NASA
22. Barger, Vernon
    University of Wisconsin
23. Barnett, Bruce
    Johns Hopkins University
24. Barnett, R. Michael
    Lawrence Berkeley Laboratory
25. Bartelt, John E.
    Stanford Linear Accelerator Center
26. Bay, Aurelio
    Lawrence Berkeley Laboratory
27. Belanger, Genevieve
    Université de Montréal
28. Berger, Edmond L.
    Argonne National Laboratory
29. Bergmann, Lewis
    Florida State University
30. Berkelman, Karl
    Cornell University
31. Berley, David
    National Science Foundation
32. Berz, Martin
    SSC Central Design Group
33. Bhadra, Sampa
    University of Illinois
34. Bhatia, T. S.
    Los Alamos National Laboratory
35. Bintinger, David L.
    SSC Central Design Group
36. Bisognano, Joseph J.
    CEBAF
37. Blackmore, Ewart
    TRIUMF
38. Blair, Robert
    Argonne National Laboratory
39. Blanar, George J.
    LeCroy Corporation
40. Blankenbecler, Richard
    Stanford Linear Accelerator Center
41. Block, Martin
    Northwestern University
42. Blondel, Alain
    CERN
43. Blüm, Peter
    Universität Karlsruhe
44. Blumenfeld, Barry
    Johns Hopkins University
45. Bock, Gregory
    Fermi National Accelerator Laboratory
46. Bock, Rudolf K.
    CERN
47. Borcherding, Fred
    Fermi National Accelerator Laboratory
48. Boyce, James R.
    CEBAF
49. Brashear, H. R.
    Oak Ridge National Laboratory
50. Brau, James E.
    University of Tennessee

51. Breidenbach, Martin
Stanford Linear Accelerator Center

52. Bross, Alan
Fermi National Accelerator Laboratory

53. Brown, Bruce C.
Fermi National Accelerator Laboratory

54. Bryman, Douglas
TRIUMF

55. Bugg, William M.
University of Tennessee

56. Bulos, Fatin
Stanford Linear Accelerator Center

57. Burchat, Patricia
Santa Cruz Institute for Particle Physics

58. Burge, Ray
TRIUMF

59. Burke, David L.
Stanford Linear Accelerator Center

60. Burnett, Thompson
University of Washington

61. Butler, Harold S.
Los Alamos National Laboratory

62. Caldwell, Allen
Columbia University

63. Campbell, William B.
University of Nebraska

64. Carey, Tom
Los Alamos National Laboratory

65. Cary, John R.
University of Colorado

66. Cassel, David G.
Cornell University

67. Cence, Robert J.
University of Hawaii

68. Chang, Chung Y.
University of Maryland

69. Chao, Alex
SSC Central Design Group

70. Charlton, Gordon
Department of Energy

71. Chen, Min
Massachusetts Institute of Technology

72. Chen, Pisin
Stanford Linear Accelerator Center

73. Chiba, Masami
Tokyo Metropolitan University

74. Childress, Sam
Fermi National Accelerator Laboratory

75. Chin, Yong Ho
Lawrence Berkeley Laboratory

76. Chivukula, R. Sekhar
Boston University

77. Cirio, Roberto
Università di Torino/INFN

78. Clark, Allan G.
Fermi National Accelerator Laboratory

79. Claus, Johannes
Brookhaven National Laboratory

80. Cline, David
University of California, Los Angeles

81. Cohn, Hans O.
University of Tennessee

82. Coignet, Guy
LAPP, Annecy

83. Cole, James B.
NASA/Goddard Space Flight Center

84. Collins, Tom
Fermi National Accelerator Laboratory

85. Colton, Eugene
Los Alamos National Laboratory

86. Coon, Darryl D.
University of Pittsburgh, Microtronics

87. Cooper, Martin
Los Alamos National Laboratory

88. Cornacchia, Max
Lawrence Berkeley Laboratory

89. Cornet, Fernando
Max-Planck Institut für Physik

90. Couture, Gilles
University of Guelph

91. Cox, Bradley
Fermi National Accelerator Laboratory

92. Craddock, Michael
TRIUMF

93. Craven, Robin
University of Wisconsin

94. Cresswell, John W.
TRIUMF

95. Cronin, James W.
University of Chicago

96. Cutts, David
Brown University

97. Davis, Michael
Martin Marietta Astronautics

98. Dawson, Bruce
University of Utah

99. Dawson, Sally
Brookhaven National Laboratory

100. Day, Christopher
Fermi National Accelerator Laboratory

101. Day, Joanne
Argonne National Laboratory

102. Dell, George S.
Brookhaven National Laboratory

103. Derrick, Malcolm
Argonne National Laboratory

104. Devlin, Thomas
Rutgers University

105. Dib, Claudio
Stanford Linear Accelerator Center

106. Diebold, Robert E.
Department of Energy

107. DiGiacomo, Nicholas
Martin Marietta Astronautics

108. Dornebosch, Jheroen
NIKHEF

109. Dragt, Alex
University of Maryland

110. Drees, Manual
University of Wisconsin

111. Drell, Persis
Cornell University

112. Ducros, Yves
CEN–Saclay

113. Edelstein, Richard M.
Carnegie-Mellon University

114. Edwards, Helen T.
Fermi National Accelerator Laboratory

115. Eichten, Estia
Fermi National Accelerator Laboratory

116. Eisner, Alan M.
University of California

117. Ekelof, Tord J.C.
University of Uppsala

118. Elbert, Jerome
University of Utah

119. Enegren, Terry
TRIUMF

120. Ernwein, Jean
CEN–Saclay

121. Errede, Debbie
University of Wisconsin

122. Evans, Lyndon
CERN

123. Fabjan, Chris
CERN

124. Feldman, Gary
Stanford Linear Accelerator Center

125. Field, Rick
University of Florida

126. Fields, Thomas H.
Argonne National Laboratory

127. Flynn, Jonathan M.
Rutherford Appleton Laboratory

128. Foley, Kenneth J.
Brookhaven National Laboratory

129. Fortner, Michael
Northern Illinois University

130. Frederiksen, Soren G.
Ohio State University

131. Freeman, James
Fermi National Accelerator Laboratory

132. Fricken, Raymond L.
Department of Energy

133. Fridman, Alfred
University of California, Los Angeles

134. Friedman, Jerome I.
Massachusetts Institute of Technology

135. Friedrichs, Carl
Los Alamos National Laboratory

136. Frisken, Bill
York University

137. Fukui, Yasuo
KEK–National Laboratory for HEP

138. Gabella, William E.
University of Colorado

139. Gabriel, Tony A.
Oak Ridge National Laboratory

140. Gabutti, Alberto
Argonne National Laboratory

141. Gaillard, Jean Marc
LAL–Orsay

142. Gaisser, Thomas K.
Bartol Research Institute

143. Gamberini, Giorgio
Lawrence Berkeley Laboratory

144. Gan, K. K.
Stanford Linear Accelerator Center

145. Ganguli, Som N.
Tata Inst. Fund. Research

146. Garbincius, Peter H.
Fermi National Accelerator Laboratory

147. Garelick, David
National Science Foundation

148. Garren, Al
SSC Central Design Group

149. Garvey, Gerald
Los Alamos National Laboratory

150. Gelmini, Graciela B.
SISSA/ICTP

151. Geng, Chaoqiang
TRIUMF

152. Georgi, Howard
Harvard University

153. Ghez, Philippe
LAPP, Annecy

154. Gilchriese, M.D.G.
SSC Central Design Group

155. Gilman, Fred
Stanford Linear Accelerator Center

156. Golden, Mitchell
Lawrence Berkeley Laboratory

157. Golutvin, I. A.
JINR, Dubna

158. Gomez-Cadenas, J. J.
Santa Cruz Institute for Particle Physics

159. Gonsalves, Richard J.
SUNY, Buffalo

160. Goodman, Jordan A.
University of Maryland

161. Gordon, Howard A.
Brookhaven National Laboratory

162. Goshaw, Alfred A.
Duke University

163. Griffin, James E.
Fermi National Accelerator Laboratory

164. Grifols, J. A.
Universidad Autónoma de Barcelona

165. Groom, Donald
SSC Central Design Group

166. Grosse-Wiesmann, Paul
Stanford Linear Accelerator Center

167. Gunion, John F.
University of California, Davis

168. Gurd, David
TRIUMF

169. Gustafson, Richard
University of Michigan

170. Haber, Howard
University of California, Santa Cruz

171. Haggerty, John S.
Brookhaven National Laboratory

172. Hagopian, Vasken
Florida State University

173. Haim, Dan
University of Illinois–Urbana

174. Hall, Dennis
Lawrence Berkeley Laboratory

175. Hall, G.
Imperial College–London

176. Han, Tao
University of Wisconsin

177. Hanson, Gail G.
Stanford Linear Accelerator Center

178. Harrison, Michael
Fermi National Accelerator Laboratory

179. Hartill, Don
Cornell University

180. Heijne, Erik
CERN

181. Heinz, Richard
Indiana University

182. Heitowit, Ezra
Universities Research Association

183. Herrera, John C.
Brookhaven National Laboratory

184. Hess, Wilmot N.
Department of Energy

185. Heusch, Clemens A.
University of California, Santa Cruz

186. Hewett, JoAnne L.
Iowa State University

187. Himel, Thomas M.
Stanford Linear Accelerator Center

188. Hinchliffe, Ian
Lawrence Berkeley Laboratory

189. Hitlin, David
California Institute of Technology

190. Hollebeek, Robert
University of Pennsylvania

191. Holmes, Stephen
Fermi National Accelerator Laboratory

192. Holmgren, Sven-Olof
LBL/University of Stockholm

193. Hoogland, W.
NIKHEF

194. Hoshi, Yoshimoto
Tohoku-Gakuin University

195. Hurlbut, Chuck
Bicron

196. Huson, F. R.
Texas A&M University/TAC

197. Huston, Joey
Michigan State University

198. Huth, John
Fermi National Accelerator Laboratory

199. Imlay, Richard
Louisiana State University

200. Irwin, Blake
McGill University

201. Ito, Masahito
University of Tsukuba

202. Iwata, Seigi
KEK–National Laboratory for HEP

203. James, Catherine
Fermi National Accelerator Laboratory

204. Jarlskog, Goran
Lund University

205. Jernigan, Garrett
University of California, Berkeley

206. Johnson, David E.
SSC Central Design Group

207. Jung, Chang Kee
Stanford Linear Accelerator Center

208. Kadyk, John A.
Lawrence Berkeley Laboratory

209. Kagan, Harris
Ohio State University

210. Kamon, Teruki
Texas A&M University

211. Kane, Gordon
University of Michigan

212. Kanofsky, Alvin S.
Lehigh University

213. Karchin, Paul
Yale University

214. Kass, Richard
Ohio State University

215. Katayama, Nobu
SUNY, Albany

216. Kayser, Boris
National Science Foundation

217. Kennedy, Michael
Martin Marietta Astronautics

218. Kenney, Robert W.
Lawrence Berkeley Laboratory

219. Kenney, V. Paul
Department of Energy

220. Kephart, Robert
Fermi National Accelerator Laboratory

221. Kephart, Tom
Vanderbilt University

222. Kheifets, S.
Stanford Linear Accelerator Center

223. Kienzle, W.
CERN

224. Killian, Ken
Martin Marietta Astronautics

225. Kim, Choong Sun
University of Wisconsin

226. Kim, Shinhong
University of Tsukuba

227. King, Stephen
Southampton University

228. Kirk, Thomas
SSC Central Design Group

229. Klanner, Robert
Deutsches Elektronen Synchrotron

230. Knapp, Edward
Universities Research Association

231. Kobrak, Hans
University of California, San Diego

232. Koiso, Haruyo
KEK–National Laboratory for HEP

233. Kolb, Edward W.
Fermi National Accelerator Laboratory

234. Komamiya, Sachio
Stanford Linear Accelerator Center

235. Kondo, Takahiko
KEK–National Laboratory for HEP

236. Koscielniak, Shane R.
TRIUMF

237. Kozanecki, Witold
Stanford Linear Accelerator Center

238. Kramer, Gordon
Hughes Aircraft

239. Kroll, Norman M.
University of California, San Diego

240. Kropp, William R. Jr.
University of California, Irvine

241. Kunori, Shuichi
University of Maryland

242. Kuroda, M.
Meiji-Gakuin University

243. Kuzminski, Joseph
McGill University

244. Lach, Joseph
Fermi National Accelerator Laboratory

245. Laintz, Dan
Martin Marietta Astronautics

246. Lamb, R. C.
Iowa State University

247. Lande, Kenneth
University of Pennsylvania

248. Lankford, Andrew J.
Stanford Linear Accelerator Center

249. Lanou, Robert E.
Brown University

250. Lanutti, Joseph E.
Florida State University

251. Lau, Kwong H.
University of Houston

252. Learned, John G.
University of Hawaii

253. LeBrun, Paul
Fermi National Accelerator Laboratory

254. Lederman, Leon
Fermi National Accelerator Laboratory

255. Leedom, Ian D.
Northeastern University

256. Leutz, Heinrich
CERN

257. Levi, Michael
Lawrence Berkeley Laboratory

258. Lillberg, John W.
Los Alamos National Laboratory

259. Linnemann, James T.
Michigan State University

260. Lipton, Ronald
Carnegie-Mellon University

261. Litke, Alan
University of California, Santa Cruz

262. Littenberg, Laurence S.
Brookhaven National Laboratory

263. Lockyer, Nigel S.
University of Pennsylvania

264. Loken, Stewart
Lawrence Berkeley Laboratory

265. London, David
Deutsches Elektronen Synchrotron

266. Louis, William C.
Los Alamos National Laboratory

267. Low, Francis E.
Massachusetts Institute of Technology

268. Lu, Al
University of California, Santa Barbara

269. Lubatti, Henry
University of Washington

270. Lundberg, Byron
Fermi National Accelerator Laboratory

271. Macek, Robert J.
Los Alamos National Laboratory

272. Machacek, Marie E.
Northeastern University

273. Majewski, Stan
University of Florida

274. Maki, Akihiro
KEK–National Laboratory for HEP

275. Malamud, Ernest
Fermi National Accelerator Laboratory

276. Malhotra, Prince K.
Tata Inst. Fund. Research

277. Mallik, Usha
University of Iowa

278. Mann, A. K.
University of Pennsylvania

279. Mantsch, Paul
Fermi National Accelerator Laboratory

280. Margulies, Seymour
University of Illinois/Chicago

281. Martin, John F.
University of Toronto

282. Marx, Michael
SUNY, Stony Brook

283. Mason, Lyle L.
Martin Marietta Astronautics

284. Masullo, Marie R.
INFN–Naples

285. Matsuda, Satoshi
Kyoto University

286. Matthews, James
University of Michigan

287. McBride, Patricia
Harvard University

288. McCarthy, Jack
Fermi National Accelerator Laboratory

289. McDaniel, B. D.
Universities Research Association

290. McDonald, Kirk
Princeton University

291. McFarlane, Kenneth W.
Temple University

292. McIntyre, Peter
Texas Accelerator Center

293. McLean, J. A.
SERC/Rutherford Appleton Laboratory

294. Mendez, Antoni
Universidad Autónoma de Barcelona

295. Mendiburu, Jean Pierre
Collège de France/CERN

296. Mes, Hans
National Research Council–Canada

297. Michelotti, Leo
Fermi National Accelerator Laboratory

298. Miller, Roger H.
Stanford Linear Accelerator Center

299. Mishina, Masa
KEK/Fermi National Accelerator Laboratory

300. Mistry, Nari B.
Cornell University

301. Mitchell, John
Martin Marietta Astronautics

302. Mokhov, Nikolai
Fermi National Accelerator Laboratory

303. Month, Melvin
Brookhaven National Laboratory

304. Morfin, Jorge G.
Fermi National Accelerator Laboratory

305. Morgan, Gerry
Brookhaven National Laboratory

306. Mori, Sigeki
University of Tsukuba

307. Morris, Duncan
California Institute of Technology

308. Morrison, Rollin J.
University of California, Santa Barbara

309. Mravca, Andrew
Department of Energy

310. Müller, Dietrich
University of Chicago

311. Murphy, C. Thornton
Fermi National Accelerator Laboratory

312. Nagae, Tomofumi
University of Tokyo

313. Nagashima, Yorikiyo
Osaka University

314. Nagle, Darragh
Los Alamos National Laboratory

315. Nakada, Tatsuya
Paul Scherrer Institute

316. Nash, E. Thomas
Fermi National Accelerator Laboratory

317. Nauenberg, Uriel
University of Colorado

318. Neuffer, David
Los Alamos National Laboratory

319. Newman-Holmes, Catherine
Fermi National Accelerator Laboratory

320. Ng, John N.
TRIUMF

321. Ng, King-Yuen
FNAL/SSC Central Design Group

322. Nicol, Tom
Fermi National Accelerator Laboratory

323. Niczyporuk, Bogdan
CEBAF

324. Nishikawa, Koichiro
University of Tokyo

325. Novaes, Sergio F.
Lawrence Berkeley Laboratory

326. Nygren, David R.
Lawrence Berkeley Laboratory

327. O'Fallon, John R.
Department of Energy

328. Oberg, Gary
Hutchinson Technology

329. Oberlack, Horst
Max-Planck Institute

330. Oddone, Pier
Lawrence Berkeley Laboratory

331. Odian, Allen
Stanford Linear Accelerator Center

332. Ogren, Harold
Indiana University

333. Oh, Seog
Duke University

334. Ohsugi, Takashi
Hiroshima University

335. Oide, Katsunobu
Stanford Linear Accelerator Center

336. Olness, Fredrick
Illinois Institute of Technology

337. Openshaw, Robert
TRIUMF

338. Oshima, Nobuaki
Fermi National Accelerator Laboratory

339. Owens, Jeff
Florida State University

340. Paar, Hans P.
University of California, San Diego

341. Paige, Frank E.
Brookhaven National Laboratory

342. Palmer, Robert
Brookhaven National Laboratory

343. Palounek, Andrea
Stanford Linear Accelerator Center

344. Panofsky, Wolfgang
Stanford Linear Accelerator Center

345. Panvini, Robert
Vanderbilt University

346. Parker, Sherwood
University of Hawaii

347. Partridge, Richard
Brown University

348. Paterson, James M.
Stanford Linear Accelerator Center

349. Peccei, Roberto
Deutsches Elektronen Synchrotron

350. Pederson, Flemming
CERN

351. Peisert, Anna
CERN

352. Pellett, David E.
University of California, Davis

353. Peoples, John
SSC Central Design Group

354. Perez, Patrice
University of Rochester

355. Peterson, Tom
Fermi National Accelerator Laboratory

356. Pevsner, Aihud
Johns Hopkins University

357. Phinney, Nan
Stanford Linear Accelerator Center

358. Pinsky, Stephen
Ohio State University

359. Pischalnikov, Yuri M.
IHEP, Serpukhov

360. Pisent, Andrea
Los Alamos National Laboratory

361. Pitman, Dale
Stanford Linear Accelerator Center

362. Planner, Charles
Rutherford Appleton Laboratory

363. Poirier, Roger
TRIUMF

364. Polychronakos, V. A.
Brookhaven National Laboratory

365. Pondrom, Lee
University of Wisconsin

366. Pope, Bernard G.
Michigan State University

367. Porter, Frank
California Institute of Technology

368. Prescott, Charles Y.
Stanford Linear Accelerator Center

369. Price, Lawrence
Argonne National Laboratory

370. Pripstein, Morris
Lawrence Berkeley Laboratory

371. Quigg, Chris
SSC Central Design Group

372. Radeka, Veljko
Brookhaven National Laboratory

373. Ragan, K. J.
University of Pennsylvania

374. Raja, Rajendran
Fermi National Accelerator Laboratory

375. Raka, Eugene
Brookhaven National Laboratory

376. Rameika, Regina
Fermi National Accelerator Laboratory

377. Rankin, Patricia
Stanford Linear Accelerator Center

378. Ratti, Sergio
Università di Pavia

379. Raubenheimer, Tor
Stanford Linear Accelerator Center

380. Raynor, Donna Gober
Martin Marietta Astronautics

381. Reardon, Paul
Science Applications International

382. Reay, Neville W.
Ohio State University

383. Reeder, Don D.
University of Wisconsin

384. Rees, Grahame H.
Rutherford Appleton Laboratory

385. Rees, John
Stanford Linear Accelerator Center

386. Rehak, Pavel
Brookhaven National Laboratory

387. Reidy, James
University of Mississippi

388. Rescia, Sergio
Brookhaven National Laboratory

389. Reucroft, Stephen
Northeastern University

390. Richter, Burton
Stanford Linear Accelerator Center

391. Riddiford, Alan
Fermi National Accelerator Laboratory

392. Ritchie, Jack
Stanford University

393. Rizzo, Thomas G.
Iowa State University

394. Roberts, B. Lee
Boston University

395. Roberts, Lee
Fermi National Accelerator Laboratory

396. Robinson, Jerry J.
McGill University

397. Roe, Byron P.
University of Michigan

398. Rohlf, James W.
Harvard University

399. Rosner, J. L.
University of Chicago

400. Rothberg, Joseph E.
University of Washington

401. Rubbia, Carlo
CERN

402. Rubin, David L.
Cornell University

403. Ruchti, Randy
University of Notre Dame

404. Russ, James S.
Carnegie-Mellon University

405. Ruth, Ronald
Stanford Linear Accelerator Center

406. Rutherford, John P.
University of Washington

407. Sadoulet, Bernard
University of California, Berkeley

408. Sadrozinski, Hartmut
Santa Cruz Institute for Particle Physics

409. Sanda, A.
Rockefeller University

410. Sauli, Fabio
CERN

411. Schachinger, Lindsay
SSC Central Design Group

412. Schellman, Heidi
University of Chicago

413. Schindler, Rafe H.
Stanford Linear Accelerator Center

414. Schmidt, Michael
Yale University

415. Seiden, Abraham
Santa Cruz Institute for Particle Physics

416. Selipsky, Stephen
Stanford University

417. Selove, Walter
University of Pennsylvania

418. Sessler, Andrew
Lawrence Berkeley Laboratory

419. Shaevitz, Mike
Columbia University

420. Shapiro, Marjorie
Harvard University

421. Shapiro, Stephen
Stanford Linear Accelerator Center

422. Sharp, Peter
Rutherford Appleton Laboratory

423. Sheaff, Marleigh
University of Wisconsin

424. Sheldon, Paul D.
University of Illinois

425. Sher, Marc
Washington University

426. Shibata, Edward I.
Purdue University

427. Shinkawa, Takao
KEK–National Laboratory for HEP

428. Sidorov, V.
INP, Novosibirsk

429. Sidwell, Ronald A.
Ohio State University

430. Siemann, Robert
Cornell University

431. Sill, Alan
University of Rochester

432. Simmons, Elizabeth H.
Harvard University

433. Sinervo, Pekka
University of Pennsylvania

434. Sinopoulos, Constantine
Stanford Linear Accelerator Center

435. Sivertz, Michael
Temple University

436. Skubic, Patrick
University of Oklahoma

437. Slaughter, A. Jean
Yale University

438. Smith, A.J.S.
Princeton University

439. Smith, Elton
Ohio State University

440. Smith, Wesley W.
University of Wisconsin

441. Smythe, Rodman
University of Colorado

442. Snoeys, Walter
Stanford University

443. Snow, George
University of Maryland

444. Sörgel, Volker
Deutsches Elektronen Synchrotron

445. Sondheim, Walter
Los Alamos National Laboratory

446. Spadafora, Anthony L.
Lawrence Berkeley Laboratory

447. Stairs, Douglas
McGill University

448. Stanfield, Ken
Fermi National Accelerator Laboratory

449. Stanton, Noel R.
Ohio State University

450. Stefanski, Raymond J.
Fermi National Accelerator Laboratory

451. Stenger, Victor J.
University of Hawaii

452. Strovink, Mark
Lawrence Berkeley Laboratory/FNAL

453. Stroynowski, R.
California Institute of Technology

454. Sugano, Katsu
Argonne National Laboratory

455. Sugimoto, Shojiro
Osaka University

456. Sulak, Lawrence
Boston University

457. Sutter, David F.
Department of Energy

458. Suzuki, Toshio
TRIUMF/KEK

459. Swain, George R.
Los Alamos National Laboratory

460. Takasaki, Fumihiko
KEK–National Laboratory for HEP

461. Talman, Richard
SSC Central Design Group

462. Tamura, Norio
Kyoto University

463. Tata, Xerxes
University of Wisconsin

464. Teng, Lee C.
Fermi National Accelerator Laboratory

465. Thaler, Jon J.
University of Illinois

466. Theilacker, Jay
Fermi National Accelerator Laboratory

467. Theriot, Dennis
Fermi National Accelerator Laboratory

468. Thiessen, Henry A.
Los Alamos National Laboratory

469. Thompson, Kathleen A.
Stanford Linear Accelerator Center

470. Thomson, Gordon
Rutgers University

471. Thun, Rudolf E.
University of Michigan

472. Tigner, Maury
SSC Central Design Group

473. Tonwar, Suresh C.
Tata Inst. Fund. Research

474. Toohig, Tim
SSC Central Design Group

475. Trilling, George
Lawrence Berkeley Laboratory

476. Trippe, Tom
Lawrence Berkeley Laboratory

477. Tschirhart, Robert
University of Michigan

478. Tung, Wu-Ki
Illinois Institute of Technology

479. Turner, Michael
University of Chicago

480. Van Asselt, Willem
Brookhaven National Laboratory

481. Van Kooten, Rick
Stanford Linear Accelerator Center

482. Vella, Eric
University of Washington

483. Vernon, Wayne
University of California, San Diego

484. Vidal, Jorge
University of Michigan

485. Vorobiev, A. P.
IHEP–Serpukhov

486. Voyvodic, Louis
Department of Energy

487. Wagoner, David E.
Prairie View A&M University

488. Wahl, Horst
Florida State University

489. Walenta, A. H.
Universität Gesamthoschschule Siegen

490. Walker, James K.
University of Florida

491. Wallenmeyer, William A.
Southeastern Univ. Research Asso.

492. Walsh, Thomas
University of Minnesota

493. Wang, Edward
Lawrence Berkeley Laboratory

494. Wang, Juwen
Stanford Linear Accelerator Center

495. Wang, Tai-Sen F.
Los Alamos National Laboratory

496. Ward, Bennie F.L.
University of Tennessee

497. Warnock, Robert
Stanford Linear Accelerator Center

498. Watase, Yoshiyuki
KEK–National Laboratory of HEP

499. Webber, Bryan
Fermi National Accelerator Laboratory

500. Weinstein, Alan
Santa Cruz Institute for Particle Physics

501. Weiss, Steve
University of California, Berkeley

502. Welch, James
Cornell University

503. Weng, W. T.
Brookhaven National Laboratory

504. Wenzel, William
Lawrence Berkeley Laboratory

505. Whitaker, Scott
Boston University

506. White, Andrew P.
University of Florida

507. White, D. Hywel
Los Alamos National Laboratory

508. White, James
Texas A&M University/TAC

509. Wienands, Ulrich
TRIUMF

510. Wigmans, Richard
CERN/NIKHEF

511. Wilburn, Colin
Micron Semiconductor

512. Willenbrock, Scott
University of Wisconsin

513. Williams, Hugh H.
University of Pennsylvania

514. Wilson, Perry B.
Stanford Linear Accelerator Center

515. Wilson, Robert J.
Boston University

516. Wimpenny, Steve
University of California, Riverside

517. Winn, David
Fairfield/Yale Universities

518. Witherell, Michael
University of California, Santa Barbara

519. Wojcicki, Stanley G.
SSC Central Design Group

520. Wolin, Elliott
Yale University

521. Woods, Robert
Department of Energy

522. Woodside, Jeffrey
Oklahoma State University

523. Wroton, Harrison C.
Martin Marietta Astronautics

524. Yager, Philip M.
University of California, Davis

525. Yamamoto, Hiroaki
University of California, Los Angeles

526. Yamanaka, Taku
Fermi National Accelerator Laboratory

527. Yehudai, Eran
Stanford Linear Accelerator Center

528. Yodh, Gaurang B.
University of California, Irvine

529. Yoh, John
Fermi National Accelerator Laboratory

530. Ypsilantis, Thomas
Collège de France/UCLA

531. Yuan, Chien-Peng
University of Michigan

532. Yuta, Harou
Tohoku University

533. Zeppenfeld, Dieter
University of Wisconsin

534. Zhao, Tianchi
University of Washington

535. Zhu, Ren-Yuan
California Institute of Technology

536. Zielinski, Marek
University of Rochester

AUTHOR INDEX

Author Index